10TH EDITION

& TOPLEY & WILSON'S

MICROBIOLOGY & MICROBIAL INFECTIONS

BACTERIOLOGY

VOLUME 1

TOPLEY & WILSON'S
MICROBIOLOGY & MICROBIAL INFECTIONS

10TH EDITION

Topley & Wilson's Microbiology and Microbial Infections has grown from one to eight volumes since first published in 1929, reflecting the ever-increasing breadth and depth of knowledge in each of the areas covered. This tenth edition continues the tradition of providing the most comprehensive reference to microorganisms and the resulting infectious diseases currently available. It forms a unique resource, with each volume including examples of the best writing and research in the fields of virology, bacteriology, medical mycology, parasitology, and immunology from around the globe.

www.topleyandwilson.com

VIROLOGY Volumes 1 and 2

Edited by Brian W.J. Mahy and Volker ter Meulen
Volume 1 ISBN 0 340 88561 0; Volume 2 ISBN 0 340 88562 9; 2 volume set ISBN 0 340 88563 7

BACTERIOLOGY Volumes 1 and 2

Edited by S. Peter Borriello, Patrick R. Murray, and Guido Funke
Volume 1 ISBN 0 340 88564 5; Volume 2 ISBN 0 340 88565 3; 2 volume set ISBN 0 340 88566 1

MEDICAL MYCOLOGY

Edited by William G. Merz and Roderick J. Hay
ISBN 0 340 88567 X

PARASITOLOGY

Edited by F.E.G. Cox, Derek Wakelin, Stephen H. Gillespie, and Dickson D. Despommier
ISBN 0 340 88568 8

IMMUNOLOGY

Edited by Stephan H.E. Kaufmann and Michael W. Steward
ISBN 0 340 88569 6

Cumulative index

ISBN 0 340 88570 X

8 volume set plus CD-ROM

ISBN 0 340 80912 4

CD-ROM only

ISBN 0 340 88560 2

For a full list of contents, please see the *Complete table of contents* on page 891

10TH EDITION

TOPLEY & WILSON'S
MICROBIOLOGY & MICROBIAL INFECTIONS

BACTERIOLOGY
VOLUME 1

EDITED BY

S. Peter Borriello PHD FRCPATH FFPH
Director, Specialist and Reference Microbiology Division; and
Interim Director, Health Protection Agency, Centre for Infections, London, UK

Patrick R. Murray PHD
Chief, Microbiology Service, Department of Laboratory Medicine
National Institutes of Health Clinical Center, Bethesda, MD, USA

Guido Funke MD
CEO, Gärtner & Colleagues Laboratories, Ravensburg, Germany

Hodder Arnold
A MEMBER OF THE HODDER HEADLINE GROUP

ASM
PRESS

First published in Great Britain in 1929
Second edition 1936; Third edition 1946
Fourth edition 1955; Fifth edition 1964; Sixth edition 1975
Seventh edition 1983 and 1984; Eighth edition 1990
Ninth edition 1998.
This tenth edition published in 2005 by
Hodder Arnold, an imprint of Hodder Education and a member of the Hodder Headline Group,
338 Euston Road, London NW1 3BH

http://www.hoddereducation.com

Distributed in the United States of America by ASM Press, the book publishing division of the American Society for Microbiology, 1752 N Street, N.W. Washington, D.C. 20036, USA

Hodder Headline's policy is to use papers that are natural, renewable and recyclable products and made from wood grown in sustainable forests. The logging and manufacturing processes are expected to conform to the environmental regulations of the country of origin.

Whilst the advice and information in this book are believed to be true and accurate at the date of going to press, neither the author[s] nor the publisher can accept any legal responsibility or liability for any errors or omissions that may be made. In particular (but without limiting the generality of the preceding disclaimer) every effort has been made to check drug dosages; however it is still possible that errors have been missed. Furthermore, dosage schedules are constantly being revised and new side-effects recognized. For these reasons the reader is strongly urged to consult the drug companies' printed instructions before administering any of the drugs recommended in this book.

British Library Cataloguing in Publication Data
A catalogue record for this book is available from the British Library

Library of Congress Cataloging-in-Publication Data
A catalog record for this book is available from the Library of Congress

Volume 1 ISBN-10   0 340 885 645   ISBN-13 978 0 340 88564 2
Volume 2 ISBN-10   0 340 885 653   ISBN-13 978 0 340 88565 9
Two volume set ISBN-10 0 340 885 661   ISBN-13 978 0 340 88566 6
Complete set and CD-ROM ISBN-10   0340 80912 4   ISBN-13   978 0 340 80912 9
Indian edition ISBN-10   0 340 88559 9   ISBN-13   978 0 340 88559 8

1 2 3 4 5 6 7 8 9 10

Commissioning Editor: Serena Bureau / Joanna Koster
Development Editor: Layla Vandenberg
Project Editor: Zelah Pengilley
Production Controller: Deborah Smith
Index: Merrall-Ross International Ltd.
Cover Designer: Sarah Rees
Cover image: *Helicobacter pylori* bacteria. A.B. Dowsett / Science Photo Library

Typeset in 9/11 Times New Roman by Lucid Digital, Salisbury, UK
Printed and bound in Italy

What do you think about this book? Or any other Hodder Arnold title? Please send your comments to www.hoddereducation.com

Contents

Please note: Chapter names shown in gray can be found in Bacteriology Volume 2

Contributors

Ben Adler BSc BA PhD MASM
Director, ARC Centre for Microbial Genomics
Department of Microbiology
Monash University
Victoria, Australia

Klaus Aktories MD PhD
Director, Institute of Experimental and
Clinical Pharmacology and Toxicology
University of Freiburg
Freiburg, Germany

Matthew B. Avison BSc PhD
Lecturer in Microbiology
Department of Pathology and Microbiology
University of Bristol
Bristol, UK

Hazel M. Aucken MA PhD
Formerly Clinical Microbiologist
Specialist and Reference Microbiology Division
Health Protection Agency, Centre for Infections
London, UK

Tom Baldwin
Institute of Infection, Immunity and Inflammation
Queen's Medical Centre
Nottingham, UK

Peter M. Bennett BSc PhD
Professor of Bacterial Genetics
Department of Pathology and Microbiology
School of Medical Sciences
University of Bristol
Bristol, UK

Anthony R. Berendt BM BCh FRCP
Consultant Physician-in-Charge, Bone Infection Unit
Nuffield Orthopaedic Centre
Headington, Oxford, UK

Ruth L. Berkelman MD
Department of Epidemiology
Rollins School of Public Health
Emory University
Atlanta, GA, USA

Norman T. Begg
Director of Vaccines
GlaxoSmithKline
Welwyn Garden City, UK

Frederick A. Bolton
Director, Health Protection Agency Laboratory
Manchester Royal Infirmary
Manchester, UK

S. Peter Borriello PhD FRCPath FFPH
Director, Specialist and Reference
Microbiology Division; and Interim Director
Health Protection Agency, Centre for Infections
London, UK

Jean-Paul Butzler MD PhD
Professor of Clinical Microbiology and Epidemiology
Department of Human Ecology
Faculty of Medicine, Vrije Universiteit Brussels
Brussels, Belgium

Jonathan S. Brazier MSc CBiol MIBiol PhD SRCS
Anaerobe Reference Laboratory
NPHS Microbiology Cardiff
University Hospital of Wales
Cardiff, UK

Keith A.V. Cartwright MA BM FRCPath FFPH
Head of Intervention Policy and
Research and Development
Local and Regional Services
Health Protection Agency South West
Stonehouse, UK

Barbara J. Chang BSc PhD FASM
Associate Professor
Microbiology, School of Biomedical and
Chemical Sciences
The University of Western Australia
Nedlands, Western Australia

Tom Cheasty BSc
Head, ESYV Reference Unit
Laboratory of Enteric Pathogens
Specialist and Reference Microbiology Division
Health Protection Agency, Centre for Infections
London, UK

Patricia S. Conville MS MT(ASCP)
Medical Technologist, Microbiology Service
Department of Laboratory Medicine
National Institutes of Health Clinical Center
Bethesda, MD, USA

Michael J. Corbel PhD DSc(Med) FIBiol FRCPath
Head, Division of Bacteriology
National Institute of Biological Standards and Control
Potters Bar, Hertfordshire, UK

David A. Dance MB ChB MSc FRCPath
Regional Microbiologist
Health Protection Agency (South West)
Plymouth, UK

Martin Day BSc, PhD
Reader in Microbial Genetics
School of Biosciences
Cardiff University
Cardiff, UK

Fiona E. Donald BMedSci BMBS FRCPath
Consultant Medical Microbiologist
Department of Microbiology
University Hospital
Queen's Medical Centre
Nottingham, UK

Julie F. Downes BSc
Research Assistant
Department of Microbiology
King's College London
London, UK

Bohumil S. Drasar PhD DSc FRCPath CBiol FBiol
DFC DHE DipHIC (Hon)
Emeritus Professor of Bacteriology
London School of Hygiene & Tropical Medicine
University of London
London, UK

Michael R. Driks MD
Baptist-Luthern Medical Center
Kansas City, MO, USA

J. Stephen Dumler MD
Division of Medical Microbiology
Department of Pathology
The Johns Hopkins University School of Medicine
Baltimore, MD, USA

Karen L. Elkins PhD
Senior Investigator, Laboratory of Mycobacteria
Division of Bacterial, Parasitic and Allergenic Products
CBER/US FDA
Bethesda, MD, USA

Meirion R. Evans BA MB BCh FRCP FFPH
Senior Lecturer
Department of Epidemiology, Statistics and
Public Health
College of Medicine, Cardiff Univeristy
Cardiff, UK

Richard R. Facklam PhD
Distinguished Consultant
Division of Bacterial and Mycotic Diseases
Centers for Disease Control and Prevention
Atlanta, GA, USA

Solly Faine
MediSci Consulting, Armadale; and Emeritus Professor
Department of Microbiology, Monash University
Melbourne, Australia

John J. Farmer III PhD
Scientist Director,
United States Public Health Service (Retired)
Foodborne and Diarrheal Diseases Branch
Division of Bacterial and Mycotic Diseases
Center for Infectious Diseases
Centers for Disease Control and Prevention
Atlanta, GA, USA

Mary Katherine Farmer MD
Palmetto Health Richland
University of South Carolina School of Medicine
Columbia, SC, USA

Roger Freeman†
Formerly Director, Public Health Laboratory
General Hospital
Newcastle Upon Tyne, UK

Guido Funke MD
Director, Department of Medical Microbiology
and Hygiene; and
CEO, Gärtner and Colleagues Laboratories
Ravensburg, Germany

Sören G. Gatermann
Head, Department of Medical Microbiology
Ruhr-Universität Bochum
Bochum, Germany

Nigel J. Gay
Modelling and Economics Unit
Communicable Diseases Surveillance Centre (CDSC)
Health Protection Agency, Centre for Infections
London, UK

Edwin E. Geldreich BS MS
Consulting Microbiologist, Retired Senior Microbiologist
US Environmental Protection Agency
Cincinnati, OH, USA

Saheer E. Gharbia BSC MSC PHD
Head, Genomics, Proteomics and
Bioinformatics Unit
Specialist and Reference Microbiology Division (SRMD)
Health Protection Agency, Centre for Infections
London, UK

Ian M. Gould BSc PHD MBCHB FRCPE FRCPATH
Consultant Microbiologist
Department of Medical Microbiology
Aberdeen Royal Infirmary
Aberdeen, UK; and
Honorary Full Professor of Microbiology
Epidemiology and Public Health
University of Trnava, Slovakia

Kim R. Hardie PGCAP PHD BSc(Hons)
Institute of Infection, Immunity and Inflammation
Centre for Biomolecular Sciences
Nottingham University
Nottingham, UK

Timothy G. Harrison BSc PHD
Deputy Director
Respiratory and Systemic Infection Laboratory
Specialist and Reference Microbiology Division
Health Protection Agency, Centre for Infections
London, UK

Tor Hofstad MD PHD
Emeritus Professor of Microbiology
Department of Microbiology and Immunology
The Gade Institute
University of Bergen
Bergen, Norway

Barry Holmes MSC PHD DSC FIBIOL
Head, National Collection of Type Cultures
Health Protection Agency, Centre for Infections
London, UK

Catherine Ison PHD FRCPATH
Director
Sexually Transmitted Bacteria Reference Laboratory
Specialist and Reference Microbiology Division

Health Protection Agency, Centre for Infections
London, UK

William M. Janda PHD
Department of Pathology
Division of Clinical Pathology
University of Illinois Medical Center at Chicago
Chicago, IL, USA

J. Michael Janda PHD
Chief, Microbial Diseases Laboratory
California Department of Health Services
Richmond, CA, USA

Eric A. Johnson ScD
Professor of Food Microbiology and Toxicology
Food Research Institute, University of Wisconsin
Madison, WI, USA

Judith A. Johnson PHD
Chief, Clinical Microbiology and Molecular Diagnostics
Veterans Affairs Maryland Health Care System; and
Associate Professor of Pathology
University of Maryland School of Medicine
Baltimore, MD, USA

Mogens Kilian DDS PHD
Professor of Bacteriology,
Department of Medical Microbiology and Immunology
University of Aarhus
Aarhus, Denmark

Keith P. Klugman MB BCH PHD
Professor of Global Health
The Rollins School of Public Health; and
Professor of Medicine
Division of Infectious Diseases, School of Medicine
Emory University
Atlanta, GA, USA

Guy R. Knudsen BS MS PHD
Soil and Land Resources Division
University of Idaho
Moscow, ID, USA

Nigel F. Lightfoot MBBS MRCPATH MSC
Director of Emergency Response
Centre for Emergency Preparedness and Response
Health Protection Agency, Porton Down
Salisbury, UK

Graham Lloyd PHD MS BS FBMS
Head of Special Pathogens Reference Unit
Novel and Dangerous Pathogens
Centre for Emergency Preparedness and Response
Health Protection Agency, Porton Dawn
Salisbury, UK

Niall A. Logan
Senior Lecturer and
University Biological Safety Adviser
Department of Biological and Biomedical Sciences
Glasgow Caledonian University
Glasgow, UK

John T. Magee PHD
Scientific Officer, Cardiff Public Health Laboratory
Cardiff, UK

Matthias Maiwald MD PHD FRCPA D(ABMM)
Associate Professor and Consultant Microbiologist
Department of Microbiology and Infectious Diseases
Flinders University and Medical Centre
Bedford Park, Australia; and
Department of Microbiology and Immunology
Stanford University School of Medicine
Stanford, California, USA

Scott A. Martin BS MS PHD
Professor, Department of Animal and Dairy Science
University of Georgia
Athens, GA, USA

Jim McLauchlin PHD
Food Safety Microbiology Laboratory
Specialist and Reference Microbiology Division
Health Protection Agency Center for Infections
London, UK

Timothy A. Mietzner PHD
Department of Molecular Genetics and Biochemistry
University of Pittsburgh School of Medicine
Pittsburgh, PA, USA

Per-Anders Mårdh PHD MD
Department of Obstetrics and Gynecology
Lund University
Lund, Sweden

Stephen A. Morse MSPH PHD
Associate Director for Science
Bioterrorism and Preparedness and Response Program
National Center for Infectious Diseases
Centers for Disease Control and Prevention
Atlanta, GA, USA

Reinier Mutters PHD
Professor, Institute of Medical Microbiology
and Hospital Hygiene
Philipps University
Marburg, Germany

Patrick R. Murray PHD
Chief, Microbiology Service
Department of Laboratory Medicine

National Institutes of Health Clinical Center
Bethesda, MD, USA

Francis E. Nano PHD
Professor, Department of Biochemistry
and Microbiology
University of Victoria
Victoria, BC, Canada

Angus Nicoll CBE
Director, Communicable Disease Surveillance
Centre (CDSC)
Health Protection Agency Centre for Infections
London, UK

Michael Noble MD FRCPC
Associate Professor
Department of Pathology and
Laboratory Medicine
University of British Columbia; and
Medical Microbiology and Infection Control
Vancouver Hospital
Vancouver, BC, Canada

Ingar Olsen DDS PHD
Professor, Department of Oral Biology; and
Dean of Research
Dental Faculty, University of Oslo
Oslo, Norway

Robert J. Owen PHD FRCPATH
Head, HPA Campylobacter and Heliobacter
Reference Unit
Specialist and Reference Microbiology Division
Health Protection Agency, Centre for Infections
London, UK

Norberto J. Palleroni PHD
Research Professor
Department of Biochemistry and Microbiology
Cook College, Rutgers University
New Brunswick, NJ, USA

Stephen R. Palmer MA MB BCHIR FRCP FFPHM
Mansel Talbot Professor of Epidemiology
and Public Health
Department of Epidemiology
Statistics and Public Health
College of Medicine, Cardiff University
Cardiff, UK

Roger Parton BSC PHD
Senior Lecturer,
Division of Infection and Immunity
Institute of Biomedical and Life Sciences
University of Glasgow
Glasgow, UK

Robin Patel MD
Associate Professor of Medicine; and
Associate Professor of Microbiology
Division of Infectious Diseases,
Division of Clinical Microbiology
Mayo Clinic College of Medicine
Rochester, MN, USA

Sharon J. Peacock PHD FRCP FRCPATH
Wellcome Trust Career Development Fellow in
Clinical Tropical Medicine
Faculty of Tropical Medicine
Mahidol University
Bangkok, Thailand

Gaby E. Pfyffer PHD
Professor of Medical Microbiology
Head, Department of Medical Microbiology
Center for Laboratory Medicine; and
Member of the Board of Hospital Directors
Kantonsspital Luzern
Luzern, Switzerland

Tyrone L. Pitt
Laboratory of Healthcare Acquired Infections
Specialist and Reference Microbiology Division
Health Protection Agency, Centre for Infections
London, UK

Victoria Pope PHD
Team Leader, Syphilis Serology Reference Laboratory
Laboratory Reference and Research Branch
Division of STD Prevention
National Center for HIV, STD, and TB Prevention
Centers for Disease Control and Prevention
Atlanta, GA, USA

Danièle Postic MD
Laboratoire des Spirochètes, Institut Pasteur
Paris, France

Didier Raoult MD PHD
Unité des Rickettsies CNRS UMR 6020
Faculté de Médecine
Marseille, France

Shmuel Razin
The Jacob Epstein Professor of Bacteriology
Department of Membrane and
Ultrastructure Research
The Hebrew-University Hadassah Medical School
Jerusalem, Israel

Robert C. Read BMEDSCI MBCHB MD FRCP
Professor in Infectious Diseases
Division of Genomic Medicine

University of Sheffield and Royal Hallamshire Hospital
Sheffield, UK

William Riley PHD D(ABMM)
Technical Director, Microbiology
Baptist Hospital of Miami
South Miami Hospital
Miami, FL, USA

Diane Roberts BSC PHD FIBIOL FIFST
Formerly Deputy Director of the PHLS
Food Safety Microbiology Laboratory

A. Denver Russell DSC PHD FRCPATH FRPHARMS†
Formerly Professor of Pharmaceutical Microbiology
Welsh School of Pharmacy, Cardiff University
Cardiff, UK

Anna Sander
Institut für Medizinische Mikrobiologie und Hygiene
Albert-Ludwigs Universität Freiburg
Freiburg, Germany

Phillippe J. Sansonetti MD
Professeur à l'Institut Pasteur
Unité de Pathogénie Microbienne Moléculaire
Unité INSERM 389
Institute Pasteur
Paris, France

Walter F. Schlech III MD FACP FRCPC
Head, Division of Infectious Diseases
Department of Medicine
Dalhousie University, QEII Health Sciences Center
Halifax, Nova Scotia, Canada

Harald Seifert MD
Professor of Medical Microbiology and Hygiene
Institute for Medical Microbiology,
Immunology and Hygiene
University of Cologne
Cologne, Germany

Bernard W. Senior BSC PHD FRCPATH
Senior Lecturer, Infection and Immunity Group
Department of Molecular and Cellular Pathology
University of Dundee Medical School
Ninewells Hospital
Dundee, UK

Haroun N. Shah BSC PHD FRCPATH
Head, Molecular Identification Services
Specialist and Reference Microbiology Division
Health Protection Agency, Centre for Infections
London, UK

James P. Shapleigh PHD
Department of Microbiology
Cornell University
Ithaca, NY, USA

Susan E. Sharp PHD
Assistant Professor, Department of Pathology
Oregon Health and Sciences University; and
Director of Microbiology, Kaiser Permanente-NW
Airport Regional Laboratory
Portland, OR, USA

Martin B. Skirrow MB PHD FRCPATH
Honorary Emeritus Consultant Microbiologist
Health Protection Agency Laboratory
Gloucestershire Royal Hospital
Gloucester, UK

Mary P.E. Slack MA MB BCHIR FRCPATH
Head, Haemophilus Reference Unit
Respiratory and Systemic Infection Laboratory
Specialist and Reference Microbiology Division
Health Protection Agency, London; and Head
WHO Collaborating Centre for Haemophilus influenzae

Henry R. Smith MA PHD
Director, Laboratory of Enteric Pathogens
Health Protection Agency, Centre for Infections
London, UK

Robert C. Spencer MBBS MSC FRCPATH FRCP(G)
HONDIPHIC
Laboratory Director
Health Protection Agency
South West Regional Laboratory
Bristol Royal Infirmary
Bristol, UK

Bret M. Steiner PHD
Laboratory Reference and Research Branch
Division of STD Prevention
National Center for HIV, STD, and TB Prevention
Centers for Disease Control and Prevention
Atlanta, GA, USA

Linda D. Stetzenbach PHD
Director, Microbiology Division
Harry Reid Center for Environmental Studies
University of Nevada, Las Vegas
Las Vegas, NV, USA

Lúcia M. Teixeira PHD
Associate Professor
Department of Medical Microbiology
Institute of Microbiology
Federal University of Rio de Janeiro
Rio de Janeiro, RJ, Brazil

Richard B. Thomson PHD
Professor of Pathology
Northwestern University Feinberg
School of Medicine; and
Director of Microbiology
Evanston Northwestern Healthcare
Evanston, IL, USA

E. John Threlfall PHD
Deputy Director, Laboratory of Enteric Pathogens
Specialist and Reference Microbiology Division
Health Protection Agency, Centre for Infections
London, UK

Olivier Vandenberg MD
Department of Microbiology
Saint-Pierre University Hospital
Brussels, Belgium

Véronique Vincent PHD
Institute Pasteur
Reference Library for Mycobacteria
Paris, France

William G. Wade BSC MSC DIPBM PHD
Professor of Oral Microbiology and
Clinical Consultant Scientist
Department of Microbiology,
King's College London
London, UK

Audrey Wanger PHD
Associate Professor
Department of Pathology
University of Texas-Houston, Medical School
Houston, TX, USA

Mark H. Wilcox BMEDSCI BM BS MD MRCPATH
Professor/Consultant
Professor of Medical Microbiology
Leeds General Infirmary and University of Leeds
Old Medical School
Leeds, UK

Paul Williams BPHARM PHD
Professor of Molecular Microbiology; and
Director, Institute of Infection, Immunity
and Inflammation, Centre for Biomolecular Sciences
University of Nottingham,
Nottingham UK

Hilmar Wisplinghoff MD
Institute of Medical Microbiology,
Immunology and Hygiene
University of Cologne
Cologne, Germany

Frank G. Witebsky MD
Assistant Chief, Microbiology Service
Department of Laboratory Medicine
National Institutes of Health Clinical Center
Bethesda, MD, USA

Michael W.D. Wren CSci FIBMS CBiol MIBiol
FRIPH
Clinical Microbiologist
UCL Hospitals
London, UK

Joseph D.C. Yao MD
Division of Clinical Microbiology
Mayo Clinic
Rochester, MN, USA

Preface

There has been a revolution in our understanding of infections and the infective process as well as in our ability to detect, identify, characterize, and understand commensal and pathogenic bacteria. Molecular methods for diagnosis and to identify and type microorganisms are becoming increasingly established in routine diagnostic laboratories, and in particular in specialist microbiology laboratories. The capability to detect and identify new pathogens is also becoming increasingly enabled. These new understandings are reflected in this 10th edition. As well as the opportunity and need to update the last edition, a complete revision was undertaken of the structure of the bacteriology volume.

The first key change was our decision to propose that the key immunology chapters, which have traditionally appeared in the volume on bacterial infections, should form the basis of a comprehensive volume on immunology. The second key change was to move from the separation of systematic bacteriology and bacterial infections into two separate volumes (118 chapters in total in the 9th edition). This approach had led to a degree of duplication and was not conducive to 'joined up' reading or integrated cohesive presentation and exposition. The approach adopted for the 10th edition is to have six major parts divided as general basic characteristics of bacteria (six chapters), general ecosystems such as soil, air, food (six chapters), general epidemiology (six chapters), organ and system infections such as bacterial infections of the respiratory tract (nine chapters), laboratory aspects (four chapters), and the organisms and their biology (fifty chapters). The only exception to this formula was for gastrointestinal infections. For some diseases, in particular those confined to the gastrointestinal tract, the coverage was restricted to the specific organism chapters to avoid redundancy.

The style and content are intended to make this volume of interest and value to students, research workers, and teachers of microbiology. As editors we hope that the excitement and wonder of bacteria is apparent to the reader. If it is, then it will match our wonder at the expertise and knowledge of the chapter authors and our excitement as editors in being involved in production of the bacteriology volume.

Peter Borriello
Patrick R. Murray
Guido Funke
London, Bethesda, and Ravensburg
May 2005

Abbreviations

Ab/HRP	antibody coupled to horseradish peroxidase
ABC	ATP-binding cassette
AC	adenylate cyclase
ACA	acrodermatitis chronica atrophicans
ACE	accessory cholera enterotoxin
ACES	*N*-2-acetamido-2-amino-ethanesulfonic acid
ACMSF	Advisory Committee on the Microbiological Safety of Food
ACP	acyl carrier protein
ACT	adenylate cyclase toxin
ADH	arginine dihydrolase
ADP	adenosine diphosphate
AE	attaching and effacing (lesions)
AFB	acid-fast bacilli
AFLP	amplified fragment length polymorphism
AGE	agarose gel electrophoresis
AGG	agglutinogen
AHL	acylhomoserine lactone
AIDS	acquired immune deficiency syndrome
AIP	autoinducing peptides
ALA	δ-aminolevulinic acid
ALAT	alanine aminotransferase
ALP	alkaline phosphatase
AMP	adenosine monophosphate
ANCA	antineutrophil cytoplasmic antibodies
APC	amino acid-polyamine choline; or antigen-presenting cell
APD	average pore diameter
AP-PCR	arbitrary primer PCR
APR	acute-phase response
APS	adenyl sulfate
ARD	antimicrobial removal device
ARDRA	amplified 16S ribosomal DNA restriction analysis
ASAT	aspartate aminotransferase
ASHP	accelerated and stabilized hydrogen peroxide
ASP	amnesic shellfish poisoning
AT	adenine–thymine
ATF	ambient temperature fimbriae
ATP	adenosine 5′ triphosphate
ATP/GTP	adenosine 5′-triphosphate/guanosine 5′-triphosphate
ATR	anthrax toxin receptor
ATS	American Thoracic Society

AUIC	area under the curve inhibitory concentration
AV	arterio-venous
BA	bacillary angiomatosis
BAP	bacillary angiomatosis peliosis
BB	mid-borderline (leprosy)
BC	blood culture
BCDMH	bromo-chloro-dimethyldantoin
BCESM	*B. cepacia* epidemic strain marker
BCG	bacille Calmette-Guérin
BCP	bromocresol-purple agar lactose
BCYE	buffered charcoal yeast extract
bDNA	branched DNA
BE	bile–esculin
BFP	biological false-positives; or bundle-forming pilus
BfPAI	*B. fragilis* pathogenicity island
BFT	*B. fragilis* toxin
BG	Bordet–Gengou
BGS	buffered glycerol saline
BHI	brain–heart infusion
BHIA	brain–heart infusion agar
BI	biological indicators; or bacterial index
BIG	botulism immune globulin
BLIS	bacteriocin-like inhibitory substances
BLNAR	β-lactamase-negative, ampicillin resistant
BLP	bacterial lipoprotein
BOD	biochemical oxygen demand
bp	base pairs
BPF	Brazilian purpuric fever
BPL	β-propiolactone
BPSU	British Paediatric Surveillance Unit
BR-PCR	broad-range polymerase chain reaction
BrkA	*Bordetella* resistance to killing protein A
BSA	bovine serum albumin
BSAC	British Society for Antimicrobial Chemotherapy
BSC	biological safety cabinets
BSE	bovine spongiform encephalopathy
BSI	bloodstream infections
BSL	biosafety level
BT	borderline tuberculoid (leprosy)
5-BU	5-bromouracil
BV	bacterial vaginosis
CAA	cold agglutinin antibodies

CABG	coronary artery bypass graft
CAMHB-LHB	cation-adjusted Mueller-Hinton broth with 2–5 percent lysed horse blood
CAMP	Christie–Atkins–Munch-Petersen (test)
cAMP	cyclic adenosine 5′-monophosphate
CAP	catabolite gene activator protein
CAPD	continuous ambulatory peritoneal dialysis
CAR	Cobas Amplicor, Roche
CAT	chloramphenicol acetyl transferase
CbpA	choline binding protein A
ccc	covalently closed circular
CCDC	consultant for communicable disease control
CCP	critical control point
CcpA	catabolite control protein
ccu	color-changing units
CDAD	*Clostridium difficile*-associated disease
CDC	Centers for Disease Control and Prevention
CDP	cytidine 5′-diphosphate
CDR	*Communicable Disease Report*
CDSC	Communicable Disease Surveillance Center
CEC	cation-exchange capacity
CF	complement fixation; or cystic fibrosis
CFA	cellular fatty acids; or colonization factor antigen
CFT	complement-fixation test
cfu	colony-forming units
CGD	chronic granulomatous disease
CGI	complicated gonococcal infection
CGRP	calcitonin gene-related peptide
CHEF	counter-clamped homogeneous electric-field electrophoresis
CIE	countercurrent immunoelectrophoresis
CIN	cefsulodin-Irgasan-novobiocin
CJD	Creutzfeldt–Jakob disease
CLO	*campylobacter*-like organism
cma	chromosome-mobilizing ability
CMBCS	continuously monitoring blood culture system
CMC	critical micelle concentration
CMGS	cooked meat medium containing glucose and starch
CMI	cell mediated immunity
CMR	chloroform–methanol residue
CMRNG	chromosomally-mediated resistant *Neisseria gonorrhoeae*
CMV	cytomegalovirus
CNA	colistin sulfate and nalidixic acid; or colistin–nalidixic acid
CNF1	cytotoxic necrotizing factor 1
CNS	central nervous system
CO_2	carbon dioxide
COL	chronic obstructive lung
CoNS	coagulase-negative staphylococci
COPD	chronic obstructive pulmonary disease
CP	capsular polysaccharide
CPE	cytopathic effect
CPS	carbohydrate phosphotransferase system
CPT	cycling probe technology

CRA	chlorine-releasing agent
CRBSI	catheter-related bloodstream infection
CRF	coagulase-reacting factor
CRMOX	Congo red magnesium oxalate
CRP	C-reactive protein; or cAMP receptor protein
CRS	congenital rubella syndrome
CS	coli surface
CSF	competence and sporulation factor
CSP	competence stimulating peptide
CSS	chlorhexidine silver sulfadiazine
CT	cholera antitoxin; or cholera toxin; or computed tomography
CTA	cystine trypticase agar
CTBA	cystine-tellurite blood agar
CVA	cefoperazone–vancomycin–amphotericin
CVC	central venous catheters
2D	two-dimensional
DAEC	diffusely adhering *Escherichia coli*
DAF	decay-accelerating factor
DAP	diaminopimelic acid
DAPI	4′, 6′-diamino-2-phenylindole hydrochloride
DBNPA	2,2-dibromo-3-nitropropionamide
DC	dendritic cell
ddNTP	dideoxynucleotide triphosphate
DEAC	di-ethyl aluminum chloride
DEFRA	Department of the Environment, Food and Rural Affairs (UK)
DFA	direct immunofluorescent antibody
DFA-TP	direct fluorescent antibody test for *Treponema pallidum*
DGI	disseminated gonococcal infection
DGlcDAG	diglucosyl diacylglycerol
DHFR	dihydrofolate reductase
DIC	disseminated intravascular coagulation
DNA	deoxyribonucleic acid
DNP	2,4-dinitrophenol
DNT	dermonecrotic toxin
DoH	Department of Health (UK)
DOTS	diseases other than syphilis; or directly observed therapy, short-cause
DPA	dipicolinic acid
DPG	diphosphatidylglycerol
DR	direct repeat
DRAG	dinitrogenase reductase-activating glycohydrolase
DRAT	dinitrogenase reductase ADP-ribosyltransferase
DS	double-stain
DSB	double-strand breaks
DSO	double-strand origin
DSP	diarrhetic shellfish poison
DT	definitive phage type
DTH	delayed-type hypersensitivity
dUMP	deoxyuridine 5′-monophosphatase
EAE	erythema arthriticum epidemicum
EAF	EPEC adherence factor

EAggEC	enteroaggregative *Escherichia coli*
EAST	enteroaggregative heat-stable enterotoxin
EB	elementary body; or ethidium bromide
EBSS	Earle's balanced salt solution
EC	*Escherichia coli*
ECM	extracellular matrix
ECP	extracellular products
ED	Entner–Doudoroff
EDTA	ethylene diamine tetraacetic acid
EF	edema factor
EF-2	elongation factor 2
EGF	epidermal growth factor
EGIA	evoked gamma interferon assays
***E*h**	redox potential
EHEC	enterohemorrhagic *Escherichia coli*
EIA	enzyme immunoassay
EIC	extracellular induction components
EIEC	entero-invasive *Escherichia coli*
ELISA	enzyme-linked immunosorbent assay
EM	erythema migrans
EMB	eosin methylene blue
EMP	Embden–Meyerhof–Parnas
ENL	erythema nodosum leprosum
ENT	ear, nose, and throat
EntFM	enterotoxin FM
EntK	enterotoxin K
EntT	enterotoxin T
EOP	efficiency of plating
EPEC	enteropathogenic *Escherichia coli*
EPP	exposure prone procedures
EPS	exopolysaccharide; or extracellular polysaccharide
ERIC	enterobacterial repetitive intergenic consensus
ERIC-PCR	enterobacterial repetitive intergenic consensus polymerase chain reaction typing
ESAT6	early secretory antigen type 6
ESBL	extended spectrum beta-lactamase
ESC	extracellular sensing components
ESM	extended-spectrum macrolide
Esp	enterococcal surface protein
ESR	erythrocyte sedimentation rate
ET	electrophoretic type; or epidermolytic toxin
ETA	exfoliatin A
ETB	exfoliatin B
ETBF	enterotoxigenic *B. fragilis*
ETEC	enterotoxigenic *Escherichia coli*
ETO	ethylene oxide
EU	European Union
EWGLI	European Working Group for *Legionella* Infections
FA	fluorescent antibody
FAD	flavin adenine dinucleotide
FAE	follicle-associated epithelium
FAFLP	fluorescence-based amplified fragment length polymorphism
FAME	fatty acid methyl ester
FBP	ferripyochelin-binding protein; or ferrous sulfate–sodium metabisulfite–sodium pyruvate; or fructose-1,6-biphosphate
FDA	Food and Drug Administration (USA)
FHA	filamentous hemagglutinin
FIGE	field-inversion gel electrophoresis
FISH	fluorescent in situ hybridization
FITC	fluorescein isothiocyanate
FIV	feline immunodeficiency virus
FMN	flavin adenine mononucleotide
FP	flavoprotein
FRET	fluorescence resonance energy transfer
FSA	Food Standards Agency (UK)
FTA	fluorescent treponemal antibody
FTA-ABS	fluorescent treponemal antibody-absorption
FUO	fever of unknown origin
FYSA	formalized yolk-sac antigen
3GC	third generation cephalosporins
G+C	guanine and cytosine
GABA	gamma-aminobutyric acid
GAP	GTPase-activating proteins
GBS	group B streptococci; or Guillain-Barré syndrome
GCAT	glycerophospholipid cholesterol acetyltransferase
GCFT	gonococcal complement fixation test
GDI	guanine nucleotide dissociation inhibitors
GEF	guanine nucleotide exchange factors
GERD	gastroesophageal disease; or gastroesophageal reflux disease
GFP	green fluorescent protein
GI	gastrointestinal
GLC	gas–liquid chromatography
GLP	glycolipoprotein
GM-CSF	granulocyte–monocyte colony-stimulating factors
GPAC	gram-positive anaerobic cocci
GSP	general secretory pathway
GTP	guanosine 5′-triphosphate
GUM	genitourinary medicine
GUS	glucuronidase
H$_2$	hydrogen
HAART	highly active antiretroviral therapy
HAI	hospital-acquired infection
Hbl	hemolysin BL
HBsAg	hepatitis B surface antigen
HBV	hepatitis B virus
HCFC	hydrochlorofluorocarbon
HCV	hepatitis C virus
HE	hektoen enteric; or hematoxylin–eosin
HEPA	high efficiency particulate air
Hfr	high frequency recombination
HFT	high-frequency transduction
HG	hybridization group
HGE	human granulocytic ehrlichiosis
hGH	human growth hormone
Hib	*Haemophilus influenzae* type b

HiPIP	high-potential iron protein
HIV-1	human immunodeficiency virus-1
HL	human lung
HLE	heat-labile enterotoxin
HLR	high-level resistance
HME	human monocytic (or monocytotropic) ehrlichiosis
HMP	hexose monophosphate pathway
HPA	Health Protection Agency (UK)
HPI	high-pathogenicity island
HPLC	high performance liquid chromatography
HPV	human papilloma virus
HSE	heat-stable enterotoxin
HSP	heat shock protein
HSPG	heparan sulfate proteoglycans
HSSA	heat-stable somatic antigens
HST	heat-stable toxin
HT	hemorrhagic toxin
HTE	hamster trachea epithelial
HTH	helix–turn–helix
HTIG	human tetanus immunoglobulin
HTM	*Haemophilus* test medium
HTST	high temperature, short time
HUS	hemolytic-uremic syndrome
HVAC	heating, ventilation, and air-conditioning
i.c.	intracerebral
IATS	International Antigenic Typing Scheme
ICAM-1	intercellular adhesion molecule-1
ICC	Infection Control Committee
ICD	infection control doctor
ICE	integrative conjugative elements
ICLN	infection control link nurses
ICN	infection control nurse
ICS	immunochromatographic strip
ICSB	International Committee for Systematic Bacteriology
ICSP	International Committee on Systematics of Prokaryotes
ICT	infection control team
ICU	intensive care unit
ID	immunodiffusion
IDSA	Infectious Diseases Society of America
idt	indeterminate (leprosy)
IEBC	International Entomopathogenic *Bacillus* Centre
IEOP	immunoelectro-osmophoresis
IF	immunofluorescence; or inactivation factor
IFA	indirect fluorescence antibody
IFAT	immunofluorescent antibody test; or indirect immunofluorescent antibody test
IFN-γ	interferon gamma
Ig	immunoglobulin
IHA	indirect hemagglutination
IHC	immunohistochemical
IHSS	idiopathic hypertophic subaortic stenosis
IID	infectious intestinal disease
IJSB	*International Journal of Systematic Bacteriology*

IJSEM	*International Journal of Systematic and Evolutionary Microbiology*
IL	interleukin
IL-1	interleukin-1
IL-10	interleukin-10
IMP	inosine-5′-phosphate
IMS	immunomagnetic separation
iNOS	inducible NO synthase
IP	intraperitoneally
IPTG	isopropyl-β-D-galactopyranoside
IR	inverted repeat; or intercept ratio
IRMA	immunoradiometric assay
IS	insertion sequence
ISF	immunosuppressive fraction
ITS	internal transcribed spacer
IU	international units
IUDR	iodoxyuridine
IUTLD	International Union Against Tuberculosis and Lung Disease
IV	intravenous
IVDU	intravenous drug users
IVET	in vivo expression technology
IVF	in vitro fertilization
kb	kilobases
KDO	keto-deoxy-octulonate; or 2-keto-3-deoxy-D-manno-oct-2-ulosonic acid; or 2-keto-3-deoxyoctonic acid
KDPG	2-keto-3-deoxy-6-phosphogluconate
KE	*Klebsiella/Enterobacter*
KIA	Kligler's iron agar
LAF	laminar air flow
LAM	lipoarabinomannan
LAMPf	*Mycoplasma fermentans* lipoproteins
LB	Lyme borreliosis
LBP	LPS-binding protein
LBSN	List of Bacterial names with Standing in Nomenclature
LCR	ligase chain reaction
LD	Legionnaires' disease
LDC	lysine decarboxylase
LDH	lactate dehydrogenase
LEM	leukocyte-endogenous mediator
LF	lethal factor
LGV	lymphogranuloma venereum
LH	light-harvesting
LJP	localized juvenile periodontitis
lktA	leukotoxin A
LL	borderline lepromatous (leprosy)
LLAP	*Legionella*-like amebal pathogens
LLO	listeriolysin O
LOS	lipo-oligosaccharide
LP	lipopolysaccharide; or lactoperoxidase
LPS	lipopolysaccheride (see LP)
LRN	laboratory Response Network
LRT	lower respiratory tract

LT	lethal toxin		**MPD**	maximum pore diameter
LTA	lipotechoic acid		**MPN**	most probable number
LTBI	latent TB infection		**MPTR**	major polymorphic tandem repeats
LTSF	low-temperature steam with formaldehyde		**MR**	mannose-resistant; or methyl red; or mino-cycline-rifampin; or multiresistant
LVS	live vaccine strain		**MREHA**	mannose-resistant and eluting hemagglutinins
mAb	monoclonally derived antibodies		**MRI**	magnetic resonance imaging
MAC	*Mycobacterium avium* complex; or membrane attack complex		**MR/K**	mannose-resistant *Klebsiella*-like
			MR/P	mannose-resistant *Proteus*-like
MALDI-	matrix-assisted laser desorption/ionization time of		**mRNA**	messenger RNA
TOF-MS	flight mass spectrometry		**MRSA**	methicillin resistant *Staphylococcus aureus*
MALP	mycoplasmal lipopeptides		**MRSE**	methicillin-resistant *Staphylococcus epidermidis*
MALT	mucosa-associated lymphoid tissue		**MRSP**	mapped restriction site polymorphisms
Map	mitochondrial associated protein		**MS**	macrolide, lincosamide, and streptogramin; or mannose-sensitive; or multiple sclerosis
MAT	microscopic agglutination test		**MSCRAMM**	microbial surface components recognizing adhesive matrix molecules
MATE	multidrug and toxic compound extrusion			
Mb	megabases		**MSHA**	mannose-sensitive hemagglutinin
MBC	minimal bactericidal concentrations		**MSM**	men who have sex with men
MCLO	*Mycobacterium chelonae*-like organism		**MSSA**	methicillin-sensitive *Staphylococcus aureus*
MCP	methyl-accepting chemotaxis proteins		**MW**	molecular weight
MCS	multiple cloning site			
MDH	malate dehydrogenase		N_0	initial number
MDP	muramyl dipeptide		**NAA**	nucleic acid amplification
***m*-dap**	*meso*-diaminopimelic acid		**NAAT**	nucleic acid amplification tests
MDR	multidrug resistance		**NAD**	nicotinamide adenine dinucleotide
MDT	multidrug therapy		**NADP+**	nicotinamide adenine dinucleotide phosphate
2ME	2-mercaptoethanol		**NAF**	non agglutinating fimbriae
MEE	multienzyme electrophoresis; or multilocus enzyme electrophoresis		**NAG**	*N*-acetylglucosamine; or nonagglutinable
			NAH	naphthalene
MEM	minimal Eagle medium; or minimal essential medium		**NALC-NaOH**	*N*-acetyl-L-cysteine-sodium hydroxide
methyl-CoM	methyl-coenzyme M		**NAP**	*p*-nitro-α-acetylamino-β-propiophenone
MF	membrane-filter		**NASBA**	nucleic acid sequence-based amplification
MFP	membrane-fusion protein		**NBT**	nitroblue tetrazolium
MFS	major facilitator superfamily		**NBTE**	nonbacterial thrombotic endocarditis
MGlcDAG	monoglucosyl diacylglycerol		**NCBI**	National Center for Biotechnology Information
MGP	methyl-α-D-glucopyranoside		**NCCLS**	National Committee for Clinical Laboratory Standards (USA)
MHA-TP	microhemagglutination assay for antibodies to *Treponema pallidum*			
			NCHI	noncapsulate *Hemophilus influenzae*
MHC	major histocompatibility complex		**NCV**	noncholera vibrios
MHRA	Medicines and Healthcare products Regulatory Agency (UK)		**NF**	nuclear factor
			NGU	non-gonoccocal urethritis
MI	morphological index		**Nhe**	non-hemolytic enterotoxin
MIC	minimal inhibitory concentration		**NHS**	National Health Service (UK)
MIF	microimmunofluorescence		**NIBSC**	National Institute of Biological Standards and Control (UK)
Mip	macrophage infectivity potentiator			
MIP-2	macrophage inflammatory protein-2		**NK**	natural killer
MIRU	mycobacterial interspersed repetitive units		**NNA**	neomycin nalidixic acid
MLEE	multilocus enzyme electrophoresis		**NNIS**	National Nosocomial Infection Surveillance (USA)
MLO	*Mycoplasma*-like organisms			
MLRT	multilocus restriction typing		**NO**	nitric oxide
MLST	multilocus sequence typing		**NPHS-ARL**	National Public Health Service Anaerobe Reference Laboratory for England and Wales
MMO	methane monooxygenase			
MMR	measles, mumps, and rubella		**NSAID**	nonsteroidal antiinflammatory drug
MOI	multiplicity of infection		**NSP**	neurotoxic shellfish poison
MOMP	major outer-membrane protein		**nt**	nucleotides
MOTT	mycobacteria other than tubercle			

NTM	nontuberculous mycobacteria		**PGU**	post-gonococcal urethritis
NTNH	nontoxic nonhemagglutinating protein		**PHB**	poly-β-hydroxybutyrate
NUD	nonulcer dyspepsia		**PHLS**	Public Health Laboratory Service, now HPA (UK)
nvCJD	new variant Creutzfeldt–Jakob disease			
NVFA	non-volatile fatty acid		**PHMB**	polyhexamethylene biguanide
NVS	nutritionally variant streptococci		**pI**	isoelectric point
			PI	gonococcal protein; or propamidine isethionate; or protein I
OA	oleic acid–albumin			
OC	Outbreak Committee		**PIA**	polysaccharide intercellular adhesin
ODC	ornithine decarboxylase		**PICC**	peripherally inserted central catheters
OE	outer envelope		**PID**	pelvic inflammatory disease
OF	opacity factor; or oxidation–fermentation		**PLET**	polymyxin-lysozyme EDTA-thallous acetate
OHPAT	outpatient and home parenteral antibiotic therapy		**PMC**	pseudomembraneous colitis
			PMF	*Proteus mirabilis* fimbriae
OL	ornithine amine lipid		**pmf**	proton motive force
OM	outer membrane		**PMMA**	polymethylmethacrylate
OMP	outer-membrane protein		**PMN**	polymorphonuclear cells; or polymorphonuclear neutrophils
ONPG	orthonitrophenol-β-D-galactopyranoside			
ONS	Office for National Statistics (UK)		**PMNL**	polymorphonuclear leukocytes
OOA	oxoline–esculin agar		**PMS**	pyrolysis mass spectrometry
OPAT	outpatient parenteral antibiotic therapy; or outpatient parenteral antibiotic treatment		**PNSG**	poly-N-succinyl beta-1-6 glucosamine
			pO₂	partial pressure of oxygen
			POGS	Parinaud's oculoglandular syndrome
OPCS	Office of Population Censuses and Surveys (UK)		**PPD**	purified protein derivative of tuberculin; or purified protein derivative
ORF	open reading frame		**PPE**	Pro-Pro-Glu
oriT	origin of transfer		**PPI**	proton pump inhibitors
oriV	origin of vegetative replication		**PPNG**	penicillinase-producing *Neisseria gonorrhoeae*
			PQS	pseudomonas quinolone signal
PA	protective antigen		**PRA**	polymerase chain reaction/restriction enzyme analysis
PAD	phage antibody display			
PAE	post antibiotic effect		**PRAS**	pre-reduced, anaerobically sterilized (media)
PAF	platelet-activating factor		**PRN**	pertactin
PAGE	polyacrylamide gel electrophoresis		**PROM**	premature rupture of the membranes
PAI	pathogenicity island		**PROS**	pathogen-related oral spirochetes
PAL	peptidoglycan-associated lipoprotein		**PRP**	penicillin resistant pneumonia; or polyribose-phosphate; or polyribosyl ribitol phosphate
Pap	pyelonephritis-associated pili			
PAS	periodic acid-Schiff		**PsaA**	pilus adhesin
PBL	peripheral blood lymphocytes		**PSP**	paralytic shellfish poison
PBP	penicillin-binding protein		**PspA**	pneumococcal surface protein
PCD	propamidine isothionate; or programmed cell death		**PT**	pertussis toxin; or phage type
			PTd	pertussis toxoid
PCR	polymerase chain reaction		**Ptl**	pertussis toxin liberation
PCT	procalcitonin		**PTS**	phosphoenolpyruvate phosphotransferase; or phosphoenolpyruvate-dependent phosphotransferase system; or phosphotransferase transport systems
PDH	pyruvate dehydrogenase complex			
PE	phosphatidylethanolamine; or Pro-Glu			
PEG	polyethylene glycol			
PEI	polyethyleneimine		**PVE**	prosthetic valve endocarditis
PEP	phosphoenolpyruvate; or post-exposure prophylaxis		**PVL**	Panton–Valentine leukocidin
			PYG	peptone yeast glucose; or peptone–yeast extract–glucose
PFGE	pulsed field gel electrophoresis			
Pfk	phosphofructokinase		**PYR**	pyrrolidonyl-β-naphthylamide
pfu	plaque-forming unit			
PGL-I	phenolic glycolipid I		**QAC**	quarternary ammonium compound
PGN	peptidoglycan		**QPCR**	quantitative polymerase chain reaction
PGPR	plant growth-promoting rhizobacteria		**QRDR**	quinolone resistance determining region
PGRS	polymorphic GC-rich repetitive sequences		**QRNG**	quinolone-resistant *Neisseria gonorrhoeae*

RA	rheumatoid arthritis		**SAI**	sexually acquired infection
RAPD	random amplification of polymorphic DNA		**SAK**	staphylokinase
RAPD-PCR	random amplified polymorphic DNA polymerase chain reaction		**SAL**	sterility assurance level
			SALT	skin-associated lymphoid tissue
RB	reticulate body		**SARA**	sexually acquired reactive arthritis
RBC	red blood cells		**SARS**	severe acute respiratory syndrome
RBS	ranitidine bismuth citrate; or ribosome binding sequence		**SASP**	small acid-soluble protein
			SAT	standard tube-agglutination test
RC	repeat clusters; or rolling circle		**SBA**	sheep blood agar
RCCS	random cloned chromosomal sequence		**sBCYE**	selective buffered charcoal yeast extract agar
RCI	reaction center I		**SBE**	subacute bacterial endocarditis
RCII	reaction center II		**SBT**	serum bactericidal titer
r-det	resistance determinant		**SCID**	severe combined immunodeficiency
rDNA	ribosomal DNA		**SCIEH**	Scottish Centre for Infection and Environmental Health
REA	restriction endonuclease analysis			
REAC	restriction endonuclease digestion of chromosomal DNA		**SCV**	small colony variants
			SD	smooth domed; or Shine–Dalgarno
REAP	restriction endonuclease digestion of plasmid DNA		**SDA**	strand-displacement amplification
			SDD	selective decontamination of the digestive tract
REP	repetitive element sequence; or repetitive extra-genic palindromic		**SDH**	succinate dehydrogenase
			SDS	sodium dodecyl sulfate
rep-PCR	repetitive DNA PCR; or repetitive extragenic palindromic element typing; or repetitive-element polymerase chain reaction		**SDS-PAGE**	sodium dodecyl sulfate polyacrylamide gel electrophoresis
			SEA	staphylococcal enterotoxin A
			SEM	scanning electron microscopy
RF	rheumatoid factor		**SFG**	spotted fever group
RFLP	restriction fragment length polymorphism		**SIDS**	sudden infant death syndrome
RGD	Arg-Gly-Asp		**sIgA**	secretory IgA
RGM	rapidly growing mycobacteria		**SIGN**	specific intracellular adhesion molecule-3 grabbing nonintegrin
RH	relative humidity			
RIA	radioimmunoassay		**SIRS**	systemic inflammatory response syndrome
RIL	rabbit ileal loop		**SLE**	systemic lupus erythematosus
Ris	regulator of intracellular stress		**SLT**	Shiga-like toxins
RIT	rabbit infectivity testing		**SMAC**	sorbitol MacConkey
RITARD	removable intestinal tie adult rabbit diarrhea		**SMD**	smooth domed (colony morphology)
RIVET	recombinase-based IVET		**SmD**	streptomycin dependent
RMP	reduction-modifiable protein		**SMEZ**	streptococcal mitogenic exotoxin Z
RNA	ribonucleic acid		**SMG**	*Streptococcus milleri* group
RNAP	RNA polymerase		**SMRF**	small multidrug resistance family
RND	resistance/nodulation/cell-division family		**SMT**	smooth transparent (colony morphology)
RODAC	replicate organism detection and counting		**SOD**	superoxide dismutase
ROS	reactive oxygen species		**SOFA**	sequential organ failure assessment
RPCFT	Reiter protein complement fixation test		**SOM**	soil organic matter
RPLA	reversed passive latex agglutination		**SPC**	standard plate count
RPR	rapid plasma reagin		**SPE**	streptococcal pyrogenic exotoxins
rpsU	ribosomal protein subunit		**SPG**	sucrose phosphate glutamate
rRNA	ribosomal RNA		**SPI**	*Salmonella* pathogenicity island
RSS	recurrent nontyphoidal salmonella septicemia		**SPS**	sodium polyanethol sulfonate
RT-PCR	reverse transcriptase polymerase chain reaction		**SRE**	small repetitive elements
			SRL	*Shigella* resistance locus
RTD	routine test dilution		**SRSV**	small round structured viruses
RTF	reduced transport fluid; or resistance transfer factor		**SS**	Stainer–Scholte
			SSA	streptococcal superantigen
RTX	repeats in toxin		**SSB**	single-strand break
RuMP	ribulose monophosphate		**SSI**	surgical site infection
			SSP	serotype-specific plasmid
SAFLP	single-enzyme amplified fragment length polymorphism fingerprinting		**SSp**	streptomycin/spectinomycin
SAg	superantigens			

SSSS	staphylococcal scalded skin syndrome		**TRNG**	tetracycline-resistant *Neisseria gonorrhoeae*
ST	Shiga toxin		**TRUST**	toluidine red unheated serum test
STD	sexually transmitted disease		**TSB**	tryptic-soy-broth; or trypticase–soy broth
STEC	Shiga toxin producing *E. coli*		**TSE**	transmissible spongiform encephalopathies
STI	sexually transmitted infection		**TSI**	triple sugar iron (agar)
STM	signature tagged mutagenesis		**TSS**	toxic shock syndrome
suPAR	soluble urokinase plasminogen activator receptor		**TSST-1**	toxic-shock syndrome toxin-1
			TST	tuberculin skin tests
TA	transaldolase		**TT**	tuberculoid
TAN	total adenine nucleotide		**TTE**	transthoracic echocardiography
TAT	twin arginine translocation		**TTGA**	taurocholate tellurite gelatin agar
Tb	Tbilisi		**TTP**	thrombotic thrombocytopenic purpura
TB	tuberculosis			
TBE	tick-borne encephalitis		**UBT**	urea breath test
TBSA	tuberculostearic acid		**UCA**	uroepithelial cell adhesin
TBW	tracheobronchial washings		**UCLA**	University of California, Los Angeles
TCA	tricarboxylic acid; or trichloracetic acid		**UGI**	uncomplicated gonococcal infection
TCBS	thiosulfate-citrate-bile salts-sucrose		**UHT**	ultra-heat-treated
TCF	tracheal colonization factor		**UMP**	uridine 5′-monophosphate
TCP	toxin coregulated pilus		**UNG**	uracil-*N*-glycosylase
TCSTS	two-component signal transduction systems		**uPAR**	urokinase-type plasminogen activator receptor
TCT	tracheal cytotoxin		**UPEC**	uropathogenic *E. coli*
TDE	transmissible degenerative encephalopathy		**USDA**	United States Department of Agriculture
TDHT	5-thyminyl-5,6-dihydrothymine		**USR**	unheated serum reagin
TDM	trehalose dimycolate		**UTI**	urinary tract infection
TDP	thermal death point		**UV**	ultraviolet
TDT	thermal death time			
TEE	transesophageal echocardiography		**VAP**	ventilator-associated pneumonia
TEM	transmission electron microscopy		**VBNC**	viable but nonculturable
TF	tissue factor; or transferrin		**VD**	venereal disease
TFSS	type IV secretion system		**VDRL**	Venereal Disease Research Laboratory
TG	typhus group		**VE**	vaccine efficacy
Th	T helper		**VIM**	Verona imipenemase
Ti	tumor-inducing		**VNTR**	variable number of tandem repeats
Tir	translocated intimin receptor		**VP**	Voges–Proskauer
TK	transketolase		**VPI**	vibrio pathogenicity island
TLC	thin-layer chromatography		**VRE**	vancomycin resistant enterococci
TLR	Toll-like receptor		**VRSA**	vancomycin-resistant *Staphylococcus aureus*
TLR2	Toll-like receptor 2		**VT**	verotoxin
Tm	DNA pseudo-melting point		**VTEC**	Vero cytotoxin-producing *Escherichia coli*
TMA	transcription-mediated amplification			
TNF	tumor necrosis factor		**WABO**	Whipple's-associated bacterial organism
TNF-α	tumor necrosis factor alpha		**Wb**	Weybridge
TOC	total organic carbon		**WBC**	white blood cell count; or white blood cell
TOL	toluene		**WCP**	whole-cell protein
topos	topoisomerases		**WHO**	World Health Organization
TPHA	*Treponema pallidum* haemagglutination assay		**WMD**	weapons of mass destruction
TPI	*Treponema pallidum* immobilization		**WR**	Wassermann reaction
TPP	thiamine pyrophosphate			
TPPA	*Treponema pallidum* particle assay; or *Treponema pallidum* particle agglutination		**XLD**	xylose-lysine-deoxycholate
tRNA	transfer RNA		**ZOT**	zonula occludens toxin

PART I

GENERAL BASIC CHARACTERISTICS

1

The bacteria: historical introduction

BOHUMIL S. DRASAR

HISTORY

The problem with writing about history is that developments appear inevitable and indeed when we present a chronological account this must be the case. However, it is as well to remember that at each point in the story the outcome could have been different. The advances in medicine in the twentieth and twenty-first centuries are the result of the demonstration of the validity of the germ theory in the nineteenth century and most significantly the establishment of its functional utility. It is easy to forget how recent these advances are. In 1922, Topley succeeded Sheridan Delépine in the Bacteriology Department of the University of Manchester. Delépine had worked with Pasteur. This was a world before molecular biology, before most immunology, and at the beginnings of antimicrobial chemotherapy. Here, some of the advances are documented, but the choice of what to include is determined by the concerns of the present and the recent past and may not reflect fully the concerns and priorities of the past.

MICROBIOLOGY

Microbiology is the study of living organisms ('microorganisms' or 'microbes'), simple in structure, and usually small in size, that are generally considered to be neither plants nor animals; they include bacteria, algae, fungi, protozoa, and viruses. 'Pure' microbiology concerns the organisms themselves and 'applied' microbiology their effects on other living beings, when they act as pathogens or commensals, or on their inanimate environment, when they bring about chemical changes in it. Thus, microbiology has applications in human and veterinary medicine, in agriculture and animal husbandry, and in industrial technology and even climatology.

Microorganisms were first seen and described by the Dutch lens-maker Antonie van Leeuwenhoek (1632–1723), who devised simple microscopes capable of giving magnifications of c. ×200. In a number of letters to the Royal Society of London between 1673 and his death, he gave clear and accurate descriptions and drawings of a variety of living things that undoubtedly included protozoa, yeasts, and bacteria (Dobell 1932). These striking observations did not lead immediately to great advances in the knowledge of microbes. These were delayed for nearly two centuries, until essential technical advances had been made by workers who nowadays would be described as industrial or medical microbiologists.

When bacteriology started in the nineteenth century the borders of the discipline were uncharted and much of what would now be called virology and immunology was included. More recently bacteriology has contributed to, or perhaps been subsumed by, the emergence of molecular biology. Such categories are largely arbitrary and the concern here is to present an outline that does justice to the roots of the subject.

COMMUNICABLE DISEASES

Long before microbes had been seen, observations on communicable diseases had given rise to the concept of contagion: the spread of disease by contact, direct or indirect. This idea was implicit in the laws enacted in early biblical times to prevent the spread of leprosy. It became less influential in the classical era, when supernatural and miasmatic causes were favored. In the later Middle Ages, there was renewed interest in contagion, reinforced at the end of the fifteenth century by the spread of syphilis in Europe, which was obviously associated with a specific form of contact. Fracastorius (Girolamo Fracastoro), a physician of Verona, published an influential analysis of contagion in 1546:

- by physical contact alone
- by formites, and
- at a distance.

He was led to conclude that communicable diseases were caused by living agents; these he spoke of as 'seminaria' or 'seeds,' but he was unable to give a more definite opinion about their nature.

In the subsequent 250 years, several authors speculated that the agents of contagious diseases were animate, but little evidence for this was produced. Even the recognition of parasitism of animals, e.g. scabies and some forms of helminthiasis (see Foster 1965), appears to have had little impact on thinking about the role of microorganisms as pathogens. Early in the nineteenth century, improvements had been made in the design of microscopes and between 1834 and 1850 numerous accounts were published of morphologically recognizable microorganisms in material from diseased animals or human subjects: of fungi subsequently called *Botrytis* in the silkworm disease 'calcino' by Bassi; of trichomonads in human vaginal discharges by Donné; of ringworm fungi by Schönlein and by Groby; of vibrios in cholera stools by Pouchet; and of large rod-shaped bacteria in anthrax blood by Rayer and Davaine (for references see Bulloch 1938).

THE 'GERM THEORY'

In 1840, Henle affirmed his belief in what came to be called the 'germ theory of disease,' which asserted that certain diseases were caused by the multiplication of microorganisms in the body, but he advanced little supporting evidence for this and his view was hotly disputed.

The lack of a germ theory probably led to the initial rejection by the General Board of Health in 1854 of Snow's explanation of the waterborne spread of cholera in London. In groundbreaking investigations, Snow established that cholera was spread by water contaminated with human feces; however, these findings conflicted with the miasmatic theory and their

significance was not fully realized until the germ theory was accepted. As late as 1894 Creighton, in his still unsurpassed 'History of Epidemics in Britain,' based his understanding on the miasmatic theory.

The intellectual and scientific case for the germ theory was established by the studies of Koch and Pasteur, which are set out below, but there was no immediate widespread acceptance. The Hamburg cholera outbreak and the dispute between Koch and Pettenkoffer both publicized and showed the strength of the theory. Drinking water from Hamburg was taken from the river Elbe above the town and sewage was discharged into the river. In Hamburg there were many cases of cholera. Altona, downstream from Hamburg, also drew drinking water form the Elbe; however, this water passed through slow sand filters before distribution. In Altona there were only a few cases of cholera. The demonstration that removal of contamination from the water supplies prevented infection convinced the public and the medical profession of the bacterial etiology of the disease. In spite of this elegant demonstration of the waterborne transmission of *Vibrio cholerae* and the mechanism for control, not all were convinced. Pettenkoffer and Emmerich both drank 1 ml of *V. cholerae* culture. Both survived (Pettenkoffer 1892). This publicity undoubtedly helped to popularize the theory.

FERMENTATION AND PUTREFACTION

In the first half of the nineteenth century, chemists became interested in fermentation: industrial processes in which organic substances underwent changes that yielded useful compounds, such as alcohol and acetic acid. A similar process, termed putrefaction, led to the decay of organic matter, usually with the production of an unpleasant odor and taste. In the 1830s, several observers, notably Cagniard-Latour and Schwann, saw yeasts in liquors undergoing alcoholic fermentation and concluded that these were living organisms and the cause of the process. This view was resisted by leading authorities of the day (Berzelius, Liebig, Wohler), who considered that fermentation was a purely chemical process and that the yeasts were a consequence rather than the cause of it. Between 1836 and 1860 controversy raged but without a clear outcome.

The work of Louis Pasteur

Louis Pasteur (1822–1895) (Figure 1.1) produced strong experimental evidence that microorganisms were the cause of fermentation and in so doing laid the foundations of microbiology as a science. He was a chemist whose early studies of fermentation aroused his interest in the molecular asymmetry of some of the compounds formed. He concluded that optically active chemical compounds, such as the stereoisomeric forms of tartaric

Figure 1.1 *Louis Pasteur (1822–1895).*

acid and amyl alcohol, never arose from the purely chemical decomposition of sugars but were formed from them by the action of microorganisms; these were always present in fermenting liquors and increased in number as the process continued. Different fermentation processes (e.g. alcoholic, acetic, butyric) were each associated with particular organisms, which were often recognizable by their morphology or requirements for growth.

To maintain that microorganisms caused fermentation, it was necessary to establish that they did not arise de novo. This was contrary to the widely held belief in the spontaneous generation of living things from dead animal or vegetable material ('heterogenesis'). Controversy about this in the eighteenth century had centered around the conditions under which putrefaction developed in organic matter that had been subjected to supposedly sterilizing temperatures in closed containers. This matter was unresolved in 1860. In a series of admirable experiments reported in the next 4 years (see Vallery-Radot 1922–33), Pasteur disposed of many purported instances of heterogenesis by showing that they could be attributed to failure of the initial sterilization or to subsequent recontamination. He emphasized the need for scrupulous sterilization of everything coming into contact with the material under examination

and demonstrated numerous sources of contamination from air, dust, and water. He showed that some organisms were not destroyed by boiling. For the sterilization of fluids, he advocated heating to 120°C under pressure, and for glassware, the use of dry heat at 170°C; he showed the value of the cottonwool plug for protecting material from aerial recontamination.

In the course of these experiments Pasteur used various forms of nutrient fluid to grow microorganisms and showed that a medium suitable for one might be unsuitable for another, so for successful cultivation it was necessary to discover a suitable growth medium and to establish optimal conditions of temperature, acidity or alkalinity, and oxygen tension.

The mass of experimental data produced by Pasteur carried general conviction, but a minority of adherents of heterogenesis continued to maintain their position, often supporting this by experiments in which inadequate heating had failed to destroy very heat resistant bacteria. The observations of Tyndall in the early 1870s, that all actively multiplying bacteria were easily destroyed by boiling, led to the introduction of a method of sterilization by repeated cycles of heating interspersed with periods of incubation. This method of 'tyndallization' served to eliminate many of the anomalies reported by the advocates of heterogenesis (see Bulloch 1938).

Pasteur devoted much effort to investigating the troubles of French winemakers, brewers, and vinegar makers. These studies often led him to perform experiments of fundamental scientific importance, as when his involvement in the problems of vinegar making led to valuable observations on the constancy of microbial characters in culture. His general conclusion was that fermentations owed their diversity to the characters of the several organisms responsible for them, but final proof of this was not obtainable until methods of obtaining pure cultures had been discovered.

PATHOGENIC MICROORGANISMS

In about 1865, Pasteur responded to an appeal to investigate a formidable disease of silkworms in southern France (pébrine); by 1869, his experiments had led him to the conclusion that this was a communicable disease transmitted by direct contact or fecal contamination. This, according to his biographer Vallery-Radot (1919), engendered in his mind the idea that communicable diseases of animals and man, like the 'diseases' of wine and beer, might be a consequence of microbial multiplication.

From 1857 onward there had been reports, notably by Brauell and Davaine, of the transmission of anthrax between animals by the injection of blood. At that time there was also much interest in the septic and pyemic diseases of man, including 'surgical fever' (see Bulloch 1938). In 1865, Coze and Felty began to publish a series of papers reporting the presence of bacteria in the blood of dogs and rabbits that had received injections of purulent material from human patients. In 1872, Davaine, starting with blood from patients suffering from 'putrid' infections, performed serial passage in experimental animals and demonstrated enhancement of virulence.

Joseph Lister (1827–1912) was aware of Pasteur's demonstration that both fermentation and putrefaction might be initiated by airborne organisms. On the assumption that 'putrefying' wounds might be similarly caused, he attempted to prevent surgical sepsis by denying access to wounds of microbes from the patient's surroundings, particularly from the air. His 'antiseptic' regimen, first described in 1867, was strikingly successful and transformed the prognosis of major surgical operations. Lister did not prove that this was due to the destruction of potentially pathogenic microbes, but this was rendered highly probable by the contemporary work of French and German bacteriologists.

In retrospect, the work of another surgeon, Alexander Ogston (1844–1929), who described staphylococci in pus, assumes perhaps equal importance and helped lay the foundations for the study of hospital infection. The history of hospital infection was reviewed by Selwyn (1991) and Ayliffe and English (2003).

The work of Robert Koch

In 1876, while a country physician at Wollstein in eastern Germany, Robert Koch (1843–1910) (Figure 1.2) published his first scientific work, a study of the anthrax

Figure 1.2 *Robert Koch (1843–1910).*

bacillus in experimental animals, of its growth in vitro, and of the formation and germination of its spores. This opened up a new era in bacteriology. In the following year, he described the fixing and staining of bacteria with the newly introduced aniline dyes. In 1878, his study of wound infections explored the role of animal experimentation in establishing the cause of bacterial infections. Then, in 1881, he described means of cultivating bacteria on solid media, thus making it possible to obtain pure cultures by transferring material from a single colony. First he used pieces of potato as his growth medium, then nutrient gelatin, and later agar gel media. In 1882 and 1884, he published classic papers on the tubercle bacillus, and in 1883, he described the cholera vibrio.

Koch had now assembled the techniques needed to investigate the bacterial causes of many communicable diseases. He had moved to Berlin, where Loeffler and Gaffky were already his assistants; later came Pfeiffer, Kitasato, Welch, and many others. Koch began to gather round him the group of followers who were destined to introduce his methods into many laboratories throughout the world.

We should remember that not all the techniques were devised by Koch; agar was first used by Hesse at the suggestion of his wife Fannie. In 1887, Petri described his culture dish, which remains a mainstay of bacteriological isolation.

The fruits of this technical revolution appeared with remarkable speed during the years 1876–90, the period described by Bulloch (1938) as 'the heyday of bacterial etiological discovery,' when most of the important groups of bacterial pathogens for man and animals were recognized.

Differential staining

In 1878, Paul Ehrlich had noted differences in the affinity for aniline dyes of various types of living cells, an observation that started him on his long search for chemotherapeutic agents. In 1882, he reported that tubercle bacilli stained with fuchsin retained the dye when subsequently treated with a mineral acid; this property of 'acid-fastness' formed the basis for methods later developed to detect mycobacteria in tissue sections, in sputum and other secretions, and in cultures.

A differential staining method of even wider applicability arose from the observation, reported in 1884 by the young Danish physician Christian Gram, that certain bacteria, when stained with methyl violet and treated with an iodine solution as a mordant, retained the violet dye when washed briefly with ethyl alcohol. The 'gram reaction' proved to be a useful means of dividing bacteria into two categories: 'gram-positive' organisms that were stained violet and 'gram-negative' organisms that lost the violet dye and were stained red by a counterstain applied after washing with alcohol. This property was later found to reflect differences in cell-wall composition and to be correlated with a number of other characters; organisms that retained the violet stain were in general less susceptible, than those that did not, to various chemical substances and to lysis by complement in the presence of specific antibody.

Establishing the pathogenicity of bacteria

As the number of different bacteria found in constant association with human and animal diseases grew, the question of how to establish their etiological role assumed importance. Already in the 1880s it was being recognized that, though the internal organs were normally sterile or nearly so, many surface sites and body cavities communicating with the outside had a rich bacterial flora, so the presence of an organism here was of little significance. When inflammatory lesions appeared in such places it was often difficult to decide which, if any, of the organisms present was responsible.

Koch's experience with anthrax, wound infection, and tuberculosis led him to place much reliance on the evidence of animal experimentation in establishing relationships between disease and isolate. A set of conditions, all of which must be fulfilled to justify such a conclusion, has been called Koch's postulates. They are as follows (see Topley and Wilson 1931), though Koch did not state them in precisely this form:

- the organism is regularly found in the lesions of the disease
- it can be grown in pure culture outside the body of the host for several generations, and
- such a culture will reproduce the disease in question when administered to a susceptible experimental animal.

It subsequently proved difficult or impossible to fulfill all these criteria in respect of many microbial diseases.

Immunity

Folk medicine had long established that exposure to certain infective agents might engender immunity to them (Parker 1998) and experience with Jennerian vaccination against smallpox had indicated the value of selecting a strain of the agent with low virulence for use as an inducer of immunity. While Koch and his pupils were continuing to characterize more and more pathogens, Pasteur turned his attention to the possibility of inducing prophylactic immunity by injections of live cultures of organisms, the virulence of which had been attenuated by prolonged culture or by growth under suboptimal conditions. Success was reported in 1877

with a live vaccine against the pasteurella of chicken cholera and in 1881 with one against anthrax in animals. In 1886, Pasteur reported on the use of an attenuated live vaccine against rabies. This consisted of a dried suspension of spinal cord from an infected rabbit. Its use was an extension of the original principle of vaccination in that the material was given after infection had taken place. It was used with apparent success to prevent disease in human subjects who had been bitten by a rabid animal. Most of the early attenuated living vaccines caused appreciable morbidity and even some deaths, but in 1886, Salmon and Smith showed that it was possible to protect pigeons against salmonella infection by the injection of heat-killed organisms. Pfeiffer demonstrated in 1889 that immunity conferred by vaccination was usually highly specific, but there were exceptions to this.

At a somewhat earlier date, Metchnikoff had observed the engulfment of bacteria and other microbes by phagocytes; in 1891, he expressed the view that immunity was primarily cellular. This conflicted with growing evidence for the importance of serum factors; the alternative humoral view of immunity was that 'antibodies' appeared in the serum of vaccinated or infected animals and that their specificity corresponded to that of the 'antigens' that elicited them.

Strong evidence for humoral immunity emerged after Roux and Yersin in Paris had demonstrated in 1888 the characteristic lethal effects of broth cultures of diphtheria bacilli on guinea-pigs and shown that these were caused by the liberation of a soluble toxin, an 'exotoxin.' In the following year, Behring in Koch's laboratory observed that chemically sterilized broth cultures of diphtheria bacilli retained their toxicity for guinea-pigs but animals given sublethal doses of them were subsequently immune to diphtheria. He also showed that the pleural fluid of animals dead of diphtheria was toxic but yielded no diphtheria bacilli on culture; however, injections of it rendered other guinea-pigs immune. By 1890, Behring had demonstrated that the blood of immunized guinea-pigs neutralized diphtheria toxin in vitro. Faber demonstrated tetanus toxin in 1889; the following year, Behring and Kitasato immunized rabbits against it and showed that their serum protected mice against tetanus.

By 1890, many of the basic areas of immunology had been outlined, though some concepts that are now considered as basic were discovered surprisingly late. The secondary response was described by Glenny and Sudmerson (1921) and the concept of herd immunity by Topley and Wilson (1923). These events set the study of immunity on a firm foundation and form the basis of the discipline of immunology; the further history can be found in the Immunology volume, chapter 2, History.

The chronology of these events is summarized in Tables 1.1 and 1.2. Figure 1.3 shows the Bacteriological

Table 1.1 *Bacteriology in the nineteenth century*

Year	Event
1834–50	Fungi, protozoa, and bacteria seen in diseased tissues or secretions (see text)
1836–37	Yeasts seen in liquors undergoing alcoholic fermentation
1840	Henle: 'germ theory of disease'
1844–57	Pasteur: studies optically active compounds from fermented fluids
1849–54	Snow: waterborne transmission of cholera
1857–63	Pasteur: reports that anthrax is transmitted by injections of blood from diseased animals
1860–64	Pasteur: experimental evidence that fermentation and putrefaction are effects of microbial growth
1865–67	Pasteur studies 'pébrine' of silkworms; concludes that it is caused by microbial action
1867	Lister: success of 'antiseptic surgery' supports view that microbes cause postoperative sepsis
1876	Koch: demonstrates pathogenicity and sporulation of anthrax bacilli
1877	Koch: staining of bacteria by aniline dyes
1877	Tyndall: heat-resistant bacteria destroyed by repeated cycles of moderate heating and incubation
1877	Pasteur: chicken cholera prevented by injections of live attenuated culture
1877	Soil nitrates replenished by microbial action
1878	Koch: studies of wound infection; use of experiments on animals to establish etiology
1879	Ehrlich: differences in affinity of chemical substances for various sorts of living cells
1880	Ogsten: staphylococci in pus
1881	Koch: use of solid media to obtain pure cultures of bacteria
1881	Pasteur: anthrax prevented by live attenuated vaccine
1882	Ehrlich: acid fastness of the tubercle bacillus
1882	Hesse: use of agar to solidify culture media
1882–84	Koch: etiological role of the tubercle bacillus; 'Koch's postulates'
1883	Koch describes the cholera vibrio
1883–91	Metchnikoff studies cellular defence mechanisms
1884	Chamberland filters
1884	Gram's stain
1886	Pasteur's rabies vaccine
1886	Salmon and Smith: killed bacterial vaccines effective
1887	Petri: double-sided culture dish
1888	Roux and Yersin: diphtheria bacillus forms exotoxin
1888	Nuttall: serum killing of bacteria
1889	Behring: antitoxic immunity to diphtheria
1889	Pfeiffer: specificity of immunity conferred by vaccines
1889	Faber: tetanus bacillus forms exotoxin
1889	Buchner: serum killing of bacteria inhibited by heating the serum
1890	Behring: diphtheria antitoxin neutralizes toxin in vitro
1890	Behring and Kitasato: antitoxic immunity to tetanus
1890	Winogradsky: nitrite- and nitrate-forming bacteria in soil
1892	Tobacco-mosaic disease transmitted by filtered material
1894	Pfeiffer's phenomenon: lysis of vibrios in peritoneal cavity
1895	Bordet: heat-stable and heat-labile factors (respectively, antibody and complement) in immune lysis
1897	Ehrlich: 'side-chain theory' of antibody production
1898	Foot-and-mouth disease transmitted by filtered material

Section of the Congress of Hygiene and Demography, London, 1891.

Bacterial filters and the origins of virology

An important technical advance in the latter part of the nineteenth century was the development of filters that held back bacteria but allowed the passage of smaller microorganisms and biologically important macromolecules: in 1884, Chamberland introduced filters made of unglazed porcelain and in 1891 Nordtmeyer introduced the Berkefeld-type of filter composed of kieselguhr. There were several important consequences of these innovations, as follows.

Filtration provided a convenient means of producing bacteria-free preparations of soluble toxins and thus greatly simplified the task of producing reagents for passive and active immunization against diphtheria and tetanus. It was also an essential preliminary to the

Table 1.2 *The twentieth century*

Year	Event
1900	Reed: yellow fever virus
1911	Rous: chicken sarcoma caused by a virus
1912	Ehrlich and Hata: Salvarsan for the treatment of syphilis
1915/17	Twort and d'Herrelle: 'bacteriophage'
1921	Glenny and Sudmerson: the secondary response to antigen
1923	Topley and Wilson: herd immunity
1928	Griffith: the transformation of pneumococci
1929	Fleming: penicillin – the first antibiotic
1935	Domagk: prontosil – the first suphonamide
1944	Avery, MacLeod, and McCarty: DNA as the agent of transformation
1944	Schartz, Bugie, and Waksman: streptomycin – the first antituberculosis treatment
1946	Lederberg and Tatum: conjugation in bacteria
1952	Lederberg and Zinder: phage transduction of bacteria
1953	Crick, Franklin, Watson, and Wilkins: structure of DNA
1960	Jacob, Perrin, Sanchez, and Monod: the operon concept
1961	Brenner, Jacob, and Meselson: ribosomes site of protein synthesis
1973	Cohen, Chan, Helling, and Boyer: plasmid vectors
1977	Fox, Pecham, and Woese: molecular systematics (16S RNA)
1977	Woese: Archaebacteria
1977	Gilbert and Sanger: DNA sequencing
1986	Mullis: the polymerase chain reaction (PCR)
1995	Venter, Smith, and Fraser: genome sequence of *Haemophilus influenzae*

purification of toxins and to chemical studies of their constitution.

It soon became apparent that some disease agents passed through bacteria-retaining filters; thus, the first viral pathogens were recognized. In 1892, Iwanowski described the transmission of mosaic disease to tobacco plants and, in 1898, Loeffler and Frosch described the transmission of foot-and-mouth disease in bovines by the injection of filtrates of infective material. In 1900, Walter Reed demonstrated that yellow fever is caused by a filterable agent (Reed 1902). The further history of virology is dealt with in the Virology volume, Chapter 1, A short history of research on viruses.

A later consequence of the use of bacterial filters was the discovery, independently by Twort in 1915 and by d'Herelle in 1917, of bacteriophages, subsequently shown to be viruses that multiply in bacterial cells. Intensive study of the interaction of bacteriophage and bacterium by Delbrück and Hershey in the early 1940s contributed much to the knowledge of viral infections and also led to important developments in bacterial genetics (see Chapter 4, Bacterial genetics).

NONMEDICAL APPLICATIONS OF BACTERIOLOGY

The discoveries of Pasteur and Koch had important applications for agriculture and industry. The replacement of nitrates lost from the soil by the washing action of rain had long been a mystery but it seemed to be connected in some way with the decomposition of organic matter. In 1877, Schloesing and Muntz, acting on a suggestion from Pasteur, showed by experiment that the formation of nitrates was due to the action of living organisms. Warington confirmed this in 1878 and 1879 and demonstrated that the process took place in two stages: first, the conversion of ammonia to nitrites, and then the oxidation of nitrites to nitrates. He believed that these two stages were performed by different organisms but failed to prove this. In 1890, Winogradsky isolated and described the nitrogen-fixing bacteria that caused the formation of nodules on the roots of leguminous plants. Later, Winogradsky described a free-living anaerobic organism that fixed atmospheric nitrogen and Beijerinck, some 10 years afterwards, described a large free-living nitrogen-fixing aerobe that he named *Azotobacter*.

The importance of bacteria in maintaining the fertility of the soil has thus been recognized for over a century. A more recent concept is that the chemical activities of primitive ancestral microbial forms may have created the atmospheric conditions essential for the appearance of plants and animals on earth (see Schlegel 1984).

Bacteria cause diseases of plants as well as animals. In 1878, Burrill described the organism responsible for pear blight and, in 1883, Wakker described the bacterial cause of 'yellows' of hyacinths. Recognition of the role of bacteria in the spoilage of foodstuffs and in the production of organic chemicals useful to man led to the entrance of the bacteriologist into numerous industrial fields.

The development of 'pure' bacteriology

Pasteur's studies of fermentation in the early 1860s revealed the physiological diversity of microbes and may be looked upon as the starting-point of 'pure' bacteriological studies. During the 1870s, he became more concerned with the role of microbes as pathogens, but he continued to be interested in their basic properties. For example, in 1878, he described under the name 'Vibrion septique' a pathogenic clostridium responsible for gangrenous conditions in animals and demonstrated that it was an obligate anaerobe. Within a few years it became possible to obtain pure cultures of many sorts of bacteria by colony selection on solid media and then to

	Bardach Odessa	Adami Cambridge	Nocard Paris	Watson Cheyne London	Cartwright Wood London		Frankland Dundee	Cunningham Calcutta	
Lehmann Würzburg	Buchner Munich	Gruber Vienna	Hankin Cambridge	Hueppe Prague	Metchnikoff Paris	Kitasato Tokyo	Fraenkel Koenigsberg	Ruffer London	Sherrington London
		Roux Paris	Burdon-Sanderson Oxford	Lister	Arloing Lyons	Fodor Budapest	Hunter London		

Figure 1.3 *Bacteriological Section, Congress of Hygiene and Demography, London, 1891.*

collect reliable data about their phenotypic characters. Accounts of their growth on various media under different physical conditions were soon supplemented by information about the range of their fermentative action on organic compounds and the products of fermentation and by the identification of chemical requirements for growth. Thus, the raw materials for systematic bacteriology began to be accumulated and basic studies of bacterial metabolism could begin.

For more detailed accounts of the early history of bacteriology, and for references, see Bulloch (1938), Clark (1961), Lechevalier and Slotorovsky (1965), and Foster (1970).

THE TWENTIETH CENTURY

The last years of the twentieth century were marked by a number of centenary retrospects; in the present context, the most important were those of the American Society for Microbiology founded in 1899 and the *Journal of Hygiene* (now *Epidemiology and Infection*) founded in 1901. Both these events resulted in the consideration of important events in microbiology, some of which are listed in Table 1.2 (Joklik et al. 1999; ASM 1999).

The twentieth century was the time when both pure and applied microbiology emerged as a science and produced radical developments across the whole field of biology. The most striking development has been the emergence of molecular biology and the way that this has altered our understanding of biology and medicine. This revolution is still at an early stage, but its impacts can be seen clearly in approaches to bacterial classification and typing.

SYSTEMATIC BACTERIOLOGY

Definition of the bacteria

The applied microbiologists of the time of Pasteur and Koch were not much interested in the classification of the microorganisms they considered responsible for fermentation or for communicable diseases. Contemporary biologists recognized two kingdoms of living things, plants and animals, but were uncertain in which to place the bacteria. In 1838, Ehrenberg had used the term 'bacteria' to describe rod-shaped organisms visible only with a microscope and considered them animals, but F. Cohn in 1854 claimed them for the botanists and Haeckel in 1866 thought that they should be placed, along with fungi, algae, and protozoa, in a third kingdom, distinct from plants and animals, the Monera or Protista. Haeckel's view did not receive wide acceptance and for the next 50 years and more the bacteria were in a taxonomic limbo. Then, technical advances, notably the introduction of the electron microscope in 1932 and of the phase-contrast microscope in 1935, led to the recognition, usually associated with the names of Stanier and van Niel (1941), that the bacteria and certain bacteria-like blue–green algae differed from most other microbes in that their genetic material was not separated from the cytoplasm by a nuclear membrane.

This view was later formalized into the concept that there were two sorts of living things differing fundamentally in cellular structure (Murray 1962; Gibbons and Murray 1978):

1 the Prokaryotae, comprising the bacteria (Eubacteria) and the blue–green algae, which were now recognized to be phototrophic bacteria and
2 the Eukaryotae, which included fungi, algae, protozoa, and all the metazoa of the plant and animal kingdoms.

The prokaryotes were characterized by the absence of a membrane-bounded nucleus and also of cellular organelles such as mitochondria and chloroplasts. As pointed out by Woese, the definition of bacteria as nonphototrophic prokaryotes is based entirely on negative characters and provides no grounds for distinguishing the conventional bacteria from the so-called Archaebacteria. These are organisms that inhabit environmentally 'hostile' habitats and include methanogens, extreme halophiles, and thermoacidophiles; they are said to form a coherent group of organisms with characteristic isoprenoid lipids and cell-wall components (see Woese and Wolfe 1985) and to be survivals from an earlier geological era. Woese (1994) considers that genetic evidence (see section on Bacterial genetics) should take precedence over cellular anatomy in defining the relationships of the prokaryotes and eukaryotes to each other and to these primitive forms.

Classification of bacteria

Linnaeus (1707–1778) classified plants and animals according to a hierarchical system based on Aristotle's theory of logical division (see Cain 1962) by placing individuals that were alike in 'essential' characters in the same species and then constructing genera and other higher taxa on the basis of progressively greater differences in characters. The selection of essential characters ('weighting') was at first made according to the intuition of the classifier, but post-Darwinian biologists used the fossil record, often supplemented by embryological evidence, to construct classifications of plants and animals that were wholly or in part phylogenetic. This

sort of evidence was not available to microbiologists and the apparent absence of sexual reproduction in bacteria meant that the biologists' favored criterion for the definition of the species, self-fertility, was denied them.

After 1880, the ability to study bacteria in pure culture led to the rapid accumulation of vast amounts of data about their phenotypic characters: colonial appearance, growth on various media, nutritional requirements, biochemical activities, serological relationships, pathogenicity for laboratory animals, and so on. Practical bacteriologists selected sets of key tests that seemed useful in identifying organisms of interest to them and in many cases attached Linnaean binomial epithets to species so characterized. What resulted was not a general classification of bacteria but a series of mini-classifications used by workers in laboratories studying medical, agricultural, or various sorts of industrial problems. There was a great deal of duplication in the naming of species, and the intuitional handling of complex collections of data led to some uncertainties in classification and identification.

NUMERICAL TAXONOMY

An alternative to seeking 'key' characters had been proposed in 1763 by Adanson, a contemporary of Linnaeus, who considered that biological classification should be based on general similarity in phenotypic characters. He rejected 'weighting' and determined, for each possible pair of individuals in a collection, the proportion of all ascertainable characters that were in accord: the so-called 'overall' similarity. Adanson found the manual computation of similarities between large numbers of pairs excessively laborious and his method could not be employed on a large scale until electronic computers became available. Then, the new discipline of numerical taxonomy was developed (Sneath 1957a, b; Sneath and Sokal 1974) and applied to collections of bacterial cultures that had been studied extensively. This made great contributions to bacterial classification at the levels of species and genus by providing an objective measure of the degree of similarity between large numbers of cultures. However, it is remarkable how often numerical-taxonomic studies supported earlier conclusions arrived at by the intuitional recognition of a 'good' classification as one that placed like organisms in the same taxon. Some taxonomists, for example Cowan (1962), expressed the view that the bacterial species was simply a man-made artifact, albeit a useful one, designed to put phenotypic data into manageable form. Numerical taxonomy provided a powerful impetus to the 'anti-essentialist' view of bacterial classification, but the practitioners of two other disciplines, chemotaxonomy and bacterial genetics, continue to search for 'key characters' that might form a basis for a broad classification of microorganisms (see Chapter 2, Taxonony and nomenclature of bacteria).

Antigenic specificity and chemotaxonomy

In the early years of the twentieth century the antibody response to bacterial antigens had been studied intensively in vitro (Parker 1998). In the 1920s, evidence began to appear of the chemical nature of some of these antigens. This was investigated eagerly by medical bacteriologists because the antigens in question appeared to have some association with pathogenicity. Thus, a great deal of information accumulated about certain classes of antigenic macromolecules and some of this was of significance for bacterial classification.

From 1923 onwards, Avery and Heidelberger, at the Rockefeller Institute in New York, studied the type-specific capsular polysaccharides of pneumococci and showed that antibodies to them conferred type-specific immunity on experimental animals. Rebecca Lancefield, in the same laboratory, described in 1933 the so-called group polysaccharides from the cell walls of hemolytic streptococci; some of these characterized streptococcal groups that caused disease only in certain species of mammals. These polysaccharides had several characters in common:

- Though determining the specificity of the antibody response, they were unable to elicit it when separated from the bacterial body and purified; in 1921 Landsteiner coined the term 'hapten' for such molecules.
- The specificity of the antibody response to them was relatively limited. Pneumococcal polysaccharides, for example, though defining clear-cut serotypes among pneumococci, cross-reacted widely with polysaccharides of otherwise dissimilar bacteria and even with nonbacterial polysaccharides, as shown by Heidelberger, Austrian, and colleagues. This was explained by the limited repertoire of specificities provided by the sequence of sugars in the terminal part of the polysaccharide chain. The role of the terminal sugars as antigenic determinants was further illuminated by the work of McCarty and Krause on the cross-reactions between streptococcal group antigens.
- Though sometimes clearly associated with virulence, the polysaccharides were not toxic for experimental animals.

In 1943, Rebecca Lancefield described a class of type-specific cell wall protein antigens in *Streptococcus pyogenes*. Like other proteins, they were fully antigenic when extracted and purified. Antibodies to them were highly specific and conferred type-specific immunity. Though nontoxic, these M proteins determined pathogenicity by interfering with phagocytosis.

Certain other cell-bound bacterial constituents proved to have toxic properties when injected into animals. These included the endotoxins of gram-negative bacteria, first described by Boivin and his associates at the Institut Pasteur in Paris in 1932–35. These were complex macromolecules in the cell envelope comprising:

- a polysaccharide responsible for antigenic specificity
- a lipid conferring toxicity, and
- a protein that, when linked to the polysaccharide, rendered it antigenic.

Studies of the amino-acid composition of the cell walls of gram-positive bacteria by Cummins and Harris in 1956 led to the recognition by Ghysen in 1965 of the structure of their main component, the peptidoglycan or mucopeptide. In 1972, Schleifer and Kandler showed that similarities in the cross-linking of the main components of the peptidoglycan molecule were of value in establishing relationships between bacterial genera that would have been difficult to ascertain by conventional serological means. It has since been noted that all eubacterial peptidoglycans contain *N*-acetyl muramic acid but that this is absent from the cell walls of archaebacteria.

Since 1970 the chemical study of bacterial macromolecules has advanced rapidly. New information about the distribution of particular classes of lipids and isoprenoid quinones has provided grounds for establishing relationships between higher taxa of gram-positive bacteria and for distinguishing eubacteria from archaebacteria (for references see Jones and Krieg 1984).

Protein antigens, when studied by conventional serological methods, showed such a narrow specificity as to limit their value to the identification of species or serotypes. Recent studies of the chemistry of some widely distributed classes of protein, e.g. the cytochromes (Jones 1980), have revealed differences relevant to the general classification of bacteria. The increasing ability to determine the sequence of amino acids in individual proteins will add information to that provided by antigenic analysis. If a constant rate of mutation is assumed, then this might be thought to provide evidence of the evolutionary 'distance' between taxa. Such considerations may be attractive to those who favor a phylogenetic approach to bacterial classification.

Bacterial genetics

At the beginning of the twentieth century it was generally recognized that the characters of bacterial strains in culture might vary, either temporarily, in response to changes in the environment ('adaptation'), or permanently, independent of environmental conditions ('mutation'). The latter phenomenon suggested the possession by bacteria of a genetic system analogous to that of larger organisms, but proof of this was lacking in the absence of a distinct nuclear apparatus.

In 1928, F. Griffith demonstrated 'transformation' among pneumococci, that is to say, transfer to a non-capsulate strain of the ability to form capsular poly-saccharide of a particular type by contact with a heat-killed culture of a capsulate strain of that type. This indicated a permanent modification of the genetic apparatus of one organism by inanimate material from another. The nature of the transforming material remained unknown until 1944, when Avery, Macleod, and McCarty showed that it was deoxyribonucleic acid (DNA).

In 1946, Lederberg and Tatum demonstrated 'recombination' between two biochemical mutants of a strain of *Escherichia coli*, each lacking different characters of the wild strain; when placed in cell-to-cell contact a strain with all the parental characters was formed. In this process, 'conjugation' had resulted in the transfer of genetic material between cells. In 1953, Hayes discovered that this was a one-way process in which a small amount of genetic material passed from a donor ('male') to a recipient ('female') cell and was thus different from sexual reproduction in plants and animals. The capacity to donate genetic material in this way depended on possession by the donor of an F (for fertility) factor.

Increasing interest in bacteriophages in the 1940s had revealed that their genetic material became incorporated into that of the bacteria they infected, and in 1952 Zinder and Lederberg showed that on occasion they might take a part of the bacterial genome with them when they left one organism and entered another, a process called transduction. Thus, by the early 1950s, it was possible to study the genetics of bacteria by exploiting the phenomena of transformation, recombination, and transduction.

Information gained from these genetic studies, together with physicochemical work on the constitution of nucleic acids, played a major role in Watson and Crick's formulation in 1953 of their hypothesis of the genetic code; according to this, the information that determines the phenotypic characters of an organism is embodied in paired linear strands of deoxyribonucleotides whose sequence in a particular region constitutes a gene. Most bacterial genes form part of a single, circular chromosome, but some are on smaller extrachromosomal closed circles of DNA called plasmids. Plasmid DNA may determine the inheritance of a single character or a few characters; examples are resistance to certain antimicrobial agents or heavy metals, individual biochemical properties, toxin production and virulence. Hayes's F factor, too, is a plasmid, one of a class of plasmids that mediate cell-to-cell transfer of genetic material, usually by means of conjugative pili.

The fact that the code is a sequence of four bases, guanine (G), cytosine (C), adenine (A), and thymine (T), had considerable significance for bacterial classification. In 1950, Chargaff had observed that, whatever the source of DNA, it contained equal amounts (in moles) of G and C, and of A and T. However, the ratio of G to A and C to T varied widely. Thus, the mol% of G + C came to be used as a measure of the genetic unrelatedness of groups of bacterial taxa. This was of most use in drawing attention to groups that were not easily distinguished by their phenotypic characters, e.g. the staphylococci and the micrococci, but had widely differing base compositions. Of course, the converse did not hold: similar base compositions did not necessarily indicate genetic 'nearness.'

Recognition that the base sequence of the DNA determines phenotypic characters suggested that it might provide the ultimate criterion for bacterial classification. At first, such sequences could be determined only in small lengths of genome. However, in 1961, Schildkraut and coworkers showed that it was possible to measure the total similarity in the sequences of the two organisms without knowing the actual sequence in either. This was done by denaturing their DNA into the single-stranded form, mixing preparations from the two organisms under appropriate conditions, and then measuring the amount of reassociation of the two strands. The amount of complementary base pairing is taken as a measure of the relatedness of the organisms. DNA–DNA pairing provided a great deal of information about the relations between organisms at or around the level of species but gave no grounds for relating the degree of homology to a definition of the species or genus. Indeed, the degree of homology within species previously defined on phenotypic grounds proved to be highly variable and proposals have been made to conflate a number of long-recognized species on the grounds of close genetic similarity.

Studies of DNA–DNA homology proved of little value in revealing broader groups among the bacteria. However, some parts of the genome, notably those coding for the production of ribosomal and transfer RNA, have come to be regarded as more 'conserved' than the rest, that is to say, to have evolved more slowly. Thus, taxonomists with a phylogenetic bent consider that studies of the homology of ribosomal RNA or of the oligonucleotide sequence on the RNA cistrons would be decisive evidence of the evolutionary 'distance' between the higher taxa of bacteria and bacteria-like organisms and thus provide a firm basis for classification (Johnson 1984).

In the eighth edition of this book it was thought that the debate between rival concepts of the phenospecies and genospecies (Hill 1990) is unresolved and perhaps unresolvable. The difference between the two schools of thought – the Lockean view that classification is a man-made device for handling information and the Aristotelian concept that the 'essence' of classification is the genetic code – was considered to be perhaps better looked upon as a philosophical rather than a scientific question.

It is a tribute to how knowledge has progressed that we now deal easily with use of molecular chronometers for the study of relatedness and how molecules other than 16S RNA are considered for this role.

The history of some other aspects of bacteriology – the mechanisms of pathogenicity and of immunity to infection, the antibacterial agents and resistance to them, and laboratory methods of investigating bacterial infections – needs to be reassessed in the light of recent understandings. The early history of bacterial genetics was reviewed in 1990 (Cliffs and Brock 1990), but since then change has accelerated.

Writing in 1938, Bulloch considered 1876–90 as the heyday of bacteriological discovery. It is clear that he underestimated the reach, both intellectually and practically, of our discipline. Pathogens have continued to be discovered. *Legionellae*, *Campylobacter*, and *Helicobacter* are prominent bacterial examples, but it is clear that without bacteriology we would lack much of modern science. The rest of these volumes will show why.

ACKNOWLEDGMENTS

This chapter is adapted and expanded from Parker, M.T. 1998. Bacteria as pathogens: historical introduction. In Hausler Jr, W.J. and Sussman, M. (eds) *Topley & Wilson's Microbiology and Microbial Infections*, Volume 3: *Bacterial Infections*, 9th edn. London: Edward Arnold, 1–14.

REFERENCES

ASM, 1999. Celebrating a century of leadership in microbiology. *ASM News* **65 (5)** and poster supplement.

Ayliffe, G.A. and English, M.P. 2003. *Hospital infection: from miasma to MRSA*. Cambridge: Cambridge University Press.

Bulloch, W. 1938. *The history of bacteriology*. London: Oxford University Press.

Cain, A.J. 1962. The evolution of taxonomic principles. In: Ainsworth, G.C. and Sneath, P.H.A. (eds), *Microbial classification*. Cambridge: Cambridge University Press, 1–13.

Clark, P.F. 1961. *Pioneer microbiologists of America*. Madison: University of Wisconsin Press.

Cliffs, N.J. and Brock, T.D. 1990. *The emergence of bacterial genetics*. New York: Cold Spring Harbor Press.

Cowan, S.T. 1962. The bacterial species – a macromyth? In: Ainsworth, G.C. and Sneath, P.H.A. (eds), *Microbial classification*. Cambridge: Cambridge University Press, 433–55.

Creighton, C. 1894. *A history of epidemics in Britain*. Cambridge: Cambridge University Press.

Dobell, C. 1932. *Antony van Leeuwenhoek and his 'Little Animals'*. London: John Bale, Sons and Danielson.

Foster, W.D. 1965. *A history of parasitology*. London: Livingstone.

Foster, W.D. 1970. *A history of medical microbiology and immunology*. London: Heinemann.

Gibbons, N.E. and Murray, R.G.E. 1978. Proposals concerning the higher taxa of bacteria. *Int J Syst Bacteriol*, **28**, 1–6.

Glenny, A.T. and Sudmerson, H.J. 1921. Notes on the production of immunity to diphtheria toxin. *J Hyg*, **20**, 176–220.

Hill, L.R. 1990. Classification and nomenclature of bacteria. In Parker, M.T., Duerden, BI. (eds.), *Topley and Wilson's principles of bacteriology, virology and immunity*, Vol. 2, 8th edn. London: Edward Arnold, 20–9.

Johnson, D.L. 1984. Nucleic acids in bacterial classification. In: Krieg, N.R. and Holt, J.G. (eds), *Bergey's Manual of Systematic Bacteriology*. Vol. 1. Baltimore: Williams & Wilkins, 8–11.

Joklik, W.K., Ljungdahl, L.G., et al. 1999. *Microbiology – a centenary perspective*. Washington, DC: American Society for Microbiology.

Jones, C.W. 1980. Cytochrome patterns in classification and identification, including their relevance in the oxidase test. In: Goodfellow, M. and Board, R.G. (eds), *Microbiological classification and identification*. London: Academic Press, 127–38.

Jones, D. and Krieg, N.R. 1984. Serology and chemotaxonomy. In: Krieg, N.R. and Holt, J.G. (eds), *Bergey's manual of systematic bacteriology*. Vol. 1. Baltimore: Williams & Wilkins, 15–18.

Lechevalier, H.A. and Slotorovsky, M. 1965. *Three centuries of microbiology*. New York: McGraw-Hill.

Murray, R.G.E. 1962. Fine structure and taxonomy of bacteria. In: Ainsworth, G.C. and Sneath, P.H.A. (eds), *Microbial classification*. Cambridge: Cambridge University Press, 119–44.

Parker, M.T. 1998. Bacteria as pathogens: historical introduction. In Hausler Jr, W.J. and Sussman, M. (eds), *Topley & Wilson's Microbiology and Microbial Infections*, Vol. 3: *Bacterial Infections*, 9th edn. London: Edward Arnold, 1–14.

von Pettenkofer, M. 1892. Ueber cholera, mit beruscksichtigung der jungsten Cholera-Epidemie in Hamburg. *Munchener Medizinischer Wochenschrift*, **46**, 807–17.

Reed, W. 1902. Recent researches concerning the etiology, propagation, and prevention of yellow fever, by the United States Army Commission. *J Hyg*, **2**, 101–19.

Schlegel, H.G. 1984. Global impacts of prokaryocytes and eukaryocytes. In: Kelly, D.P. and Carr, N.G. (eds), *The microbe*. Cambridge: Cambridge University Press, 1–32.

Selwyn, S. 1991. Hospital Infection: the first 2500 years. *J Hosp Infect*, **18**, Suppl. A, 5–64.

Sneath, P.H.A. 1957a. Some thoughts on bacterial classification. *J Gen Microbiol*, **17**, 184–200.

Sneath, P.H.A. 1957b. The application of computers to taxonomy. *J Gen Microbiol*, **17**, 201–26.

Sneath, P.H.A. and Sokal, R.R. 1974. *Numerical taxonomy: the principles and practice of numerical classification*. San Francisco: WH Freeman.

Stanier, R.Y. and van Niel, C.B. 1941. The main outlines of bacterial classification. *J Bacteriol*, **42**, 437–66.

Topley, W.W.C. and Wilson, G.S. 1923. The spread of bacterial infection, the problem of herd immunity. *J Hyg*, **21**, 249–53.

Topley, W.W.C., Wilson, G.S. 1931. *The principles of bacteriology and immunity*, Vol. 2, 1st edn. London: Edward Arnold, 591–4.

Vallery-Radot, R. 1919. *The life of Pasteur*, English translation by Mrs R.L. Devonshire. London: Constable.

Vallery-Radot, R. 1922-33. *Oeuvres de Pasteur*. Paris: Masson.

Woese, C.R. 1994. There must be a prokaryote somewhere: microbiology's search for itself. *Microbiol Rev*, **58**, 1–9.

Woese, C.R. and Wolfe, R.S. 1985. The bacteria, a treatise on structure and function. In: Gunsalus, I.C. and Stanier, R.Y. (eds), *Archaebacteria*. Vol. 8. London: Academic Press.

2

Taxonomy and nomenclature of bacteria

JOHN T. MAGEE

This chapter is intended as a brief introduction to the strange mixture of science, philosophy, and formal regulation that guides the division of bacteria into named groups. This is not an unbiased account. In the Alice in Wonderland of taxonomy, nothing is quite what it seems and, as in politics and theology, opinion is an indispensable ingredient. Anyone who reads this set of volumes has reached some viewpoint on bacterial taxonomy, consciously or subconsciously. There is considerable breadth of opinion in key arguments: utilitarian versus 'natural' classification, genomic versus phenotypic versus polyphasic classification, 'splitting' or 'lumping' of taxa (Cowan 1968), and so on. This range guarantees that each microbioilogist will have found some unique position in the spectrum of opinion, and so will disagree with at least some opinions expressed here. Constructive, logical criticism and discussion would be welcome.

AN INTRODUCTORY PERSPECTIVE

The practice of medical bacteriology is founded upon the ability to distinguish between the bacteria that are isolated and to predict their pathogenicity and epidemiology from in vitro characteristics via past clinical observations in infections with closely similar organisms. This predictive capability is essential and flows from the work of many microbiologists over two centuries. Their cumulative efforts have provided the current formal species nomenclature under the aegis of the International Committee for Systematic Bacteriology (ICSB), reviewed regularly in the editions of *Bergey's manual of systematic bacteriology* (currently 2nd edition in five volumes: Garrity 2001), and a wealth of vernacular schemes. Together, these schemes provide much of the technical vocabulary of microbiology and are the basis of interpretation and prediction from laboratory tests.

In a laboratory deprived of all knowledge of the classification of bacteria, their morphological and biochemical characteristics, and the relevance of individual characterization tests, each isolate would have to be regarded as unique. It would be impossible to distinguish between pathogens and commensals, or to assess the cross-infection potential of an isolate, from in vitro tests. All available antibiotics would have to be tested for each isolate. How long would it take, for example, to untangle the characteristics of the fecal pathogens from those of gut commensals, or to rediscover the link between diphtheria and toxin-producing corynebacteria? Clearly, it would be impossible to operate under such conditions. The laboratory would be forced to search for links between in vitro characteristics and effects on patients, to divide isolates into named groups, and so to reinvent the science of taxonomy.

Despite this key role, many diagnostic microbiologists profess a disregard for taxonomy (Magee 1993a). This attitude may derive from early specialist education, where the subject is often presented as an apparently endless list of genus and species names for rote learning, along with associated laboratory and medical characteristics. The underlying science, philosophy, and history of the subject are rarely taught; when they are, the students are often so inundated by the nomenclature and lacking in a context for their brief contact with taxonomy *per se* that these are soon forgotten. Further conscious contact with the subject is usually when rote learning is overturned by changes in the names of bacteria, for reasons that may appear obscure and irrelevant. It is therefore unsurprising that most medical microbiologists regard taxonomy as an unimportant, dull, and arcane science practiced by academics in isolation for their own edification.

This common view is grossly inaccurate but is also a self-fulfilling proposition. It builds barriers between routine microbiologists and taxonomists that limit communication and cooperation. This breaks the feedback process that is essential to the production of relevant, sensible classifications of the bacteria. Taxonomy and classifications are fundamental to all microbiological work, and an appreciation of the concepts involved is an absolute requirement for all practitioners of microbiology. Equally, taxonomists require the input of other microbiologists to direct them to areas of taxonomic confusion, to supply cultures, to moderate wilder taxonomic excursions, and, more importantly, to supply knowledge of the characteristics of the organisms in the broader context outside the laboratory.

ONE CLASSIFICATION OR MANY? THE IMPORTANCE OF CONTEXT

Cowan (1962) illustrates the range of possible classifications with the example of objects intended to join materials. He divides them first into animal (glues), vegetable (wooden pegs), and mineral (metal nails, bolts, or screws, or plastic adhesives, screws, and pins), then presents an alternative classification into solid devices and liquid glues. A dealer in scrap metals might classify these objects as ferrous, nonferrous, or nonmetal in the context of work, but quite differently in the context of repairing a broken china ornament or a broken chair. Humans have little difficulty in employing a multiplicity of overlapping, but distinct and context-specific, classifications for everyday objects. Microbiologists differ only in that most are trained to acknowledge a single formal scheme and so feel constrained to argue about its minutiae to the exclusion of other possibilities.

However, when presented with a set of clinical isolates, workers revert to practicality, happily employing classifications that have little in common with that embodied in the current edition of *Bergey's manual* (Garrity 2001). At the feces bench, isolates are divided into lactose and nonlactose fermenters, fecal pathogens and nonpathogens, aerobes, anaerobes, and microaerophiles. Staphylococci are divided in three distinct ways when dealing with blood cultures, skin swabs, or urines, reflecting experience of potential pathogenicity and the importance of further identification in the context of the specimen type, the sex of the patient, and catheterization status. Diverse gram-negative and gram-positive species are lumped together under the heading of respiratory pathogens, recognizing the need to extend this definition in the context of special cases: cystic fibrosis patients or patients on ventilators. These vernacular or 'trivial' classifications even intrude into reports in phrases such as 'coagulase-negative staphylococcus' or as multivalued context-specific phrases such as 'commensals only'.

Trivial classifications are as real and valid as those described in the successive editions of *Bergey's manual*. They exist because some worker perceived a new way of dividing bacteria that had practical use or significance, and many other microbiologists recognized the utilitarian value of the division. They reflect unavoidable deficiencies in the flexibility of nomenclature in the formal ICSB scheme, and the fact that a single classification cannot cater for all the myriad ways in which bacteria can be usefully grouped.

These trivial classifications have, by their nature, consensus acceptance and are strongly utilitarian. However, they lack mechanisms for definition and regulation of any nomenclature involved, and so are open to degradation by misuse. Terms such as 'fecal streptococcus' or 'enterobacteria' lack clear definition and are used in any of a range of senses from the all-inclusive to the restricted. Equally, these classifications tend to be restricted by their utilitarian nature to specific branches of applied microbiology.

By contrast, the formal scheme has a strictly defined, regulated nomenclature recognizable by the italicization of the formal species name. It is less robust on consensus acceptance, but it is broadly applicable in all branches of microbiology. The defined nomenclature makes this the vocabulary of choice for international and interdisciplinary communication. However, the rigidity of a defined nomenclature causes problems when changes become necessary (see below). In my view, a second context for the formal system is as a scheme supplying descriptions of the minimum-sized groups of bacteria about which one can make useful generalizations. These species definitions are the building blocks that can be assembled into larger groups in as many ways as utility demands. The ICSB provides an official system of genus and species nomenclature (Sneath 1992); by contrast the classifications into higher taxa outlined in successive editions of *Bergey's manual* are not official (Staley and Krieg 1984) and tend to change markedly between editions.

However, the formal scheme serves many interest groups. Those who wish the scheme to supply a phylogeny of the bacteria, a 'natural' classification, or to be bent to conform with the results of a specific characterization test, or to suit the requirements of some specialist branch of microbiology would all disagree.

STACK OR FLAT: THE PRACTICAL RELEVANCE OF HIERARCHICAL CLASSIFICATIONS

Classification embraces not only the division into named species but also the further grouping of species into genera, genera into families, and so on through orders, classes, and phyla to kingdoms. This hierarchical approach has much to commend it in the classification of higher organisms. The many morphological and developmental features allow division into higher groupings in agreement with fossil evidence of evolutionary development, and, for the most part, these groups can be neatly packaged into specialist areas relevant to the applied science. However, hierarchical classification has had a more chequered history in bacteriology. Successive editions of *Bergey's manual* have portrayed gross changes in the arrangement and membership of higher taxa. Equally, each of the major applied specialist areas of bacteriology – medical, veterinary, and industrial – deals with a host of diverse bacteria whose membership could not conceivably be grouped into any exclusive 'natural' or phylogenetic higher taxon, but rather forms a diverse portion of the all-inclusive group of 'the bacteria'.

One major difficulty is the lack of nonarbitrary definitions of the taxon concept at any division from species upwards. The definition of a species concept for the bacteria has been in dispute for more than a century, and definitions of the genus, order, etc. are successively more elusive. The value judgments that set these divisions are likely to be as controversial in the new era of phylogenetic classification as they were in the days of van Niel's physiological classifications (Van Niel 1946). However, despite these difficulties, the search for a phylogenetic classification of the bacteria and other prokaryotes is of enormous importance to biology. Results from this search are likely to shape our future views on the origins of life and of the eukaryote and prokaryote forms.

For medical microbiologists, much of the turmoil in the upper reaches of hierarchical classification has remained unnoticed. Changes at genus level have affected the naming of species in laboratory reports, causing considerable inconvenience. However, the higher taxonomic divisions form a province that

academics may remap with little fear of criticism from applied microbiologists, who are more interested in nomenclature and divisions at species level than in hierarchical classification.

This highlights an important, but rarely discussed contrast of attitude between academic taxonomy and applied bacteriology. The applied bacteriologist is interested almost solely in divisions at species level, and particularly in those species where identification yields an associated prediction of properties, such as pathogenicity or cross-infection potential, that cannot be examined readily or economically in the laboratory. Their requirements are for a division of the bacteria into readily identified species, each with clear and distinct properties significant to his or her application, and a stable, widely accepted species nomenclature, to allow concise communication of his or her findings. By contrast with the taxonomist, who attempts to group species in a hierarchy determined by phylogeny or similarity, the applied bacteriologist assembles species in a multitude of distinct context-specific classifications of convenience.

The single-tier classification of species nomenclature is common ground for the taxonomist and applied bacteriologist, and deserves a distinguishing name, such as the agnostic, agnarchic (by contrast to hierarchic, from the Greek root *hieros*, sacred, and *arche*, rule), or monadic (Gr. *monados*, unity) classification because of this unique unifying position. Adoption of one of these terms might eliminate the misunderstanding inherent in use of the term 'classification' to cover both the untiered division of bacteria into species, the sense in which many applied microbiologists instinctively interpret the term, and hierarchical schemes.

HISTORY AND OPERATION OF THE FORMAL CLASSIFICATION

In the early days of biology, Linnaeus understood that biologists required a concise, precise, and internationally recognized scheme of species nomenclature. He proposed a scheme of latinized binomials comprising a genus and species epithet, referring to a formal description of the species. This was universally accepted, and replaced the babel of trivial, poorly defined synonyms individual to each language with a defined vocabulary common to all biologists. This technical vocabulary now bridges language and speciality, providing an essential method of communication. Whatever other gulfs of language and misunderstanding may exist, biologists can guarantee that a species name will communicate a complex detailed description of an organism with precision, and in a remarkably concise form.

Microbes, observed earlier by van Leeuwenhoek, were assigned to six species in the class *Chaos* by Linnaeus. Work in the late nineteenth century led to the description of many bacterial pathogens and the first recognizable attempt to classify bacteria by the botanist Cohn (1872). The rapid growth in species descriptions required regulation, and individual international committees were eventually set up to govern the nomenclature of animals and plants, including bacteria. Important landmarks in the advance of bacterial taxonomy were the classifications of Chester (1901), Orla-Jensen (1919), and the work of Buchanan (1918, 1925). Committees for bacterial and viral taxonomy (Murphy et al. 1995) came later, with a breakaway group covering the blue-green algae, claimed originally by the botanists and now by the bacteriologists. Regulations covering validation of new names and changes in nomenclature are embodied in codes for each of these groups, and are supervised by the appropriate international committee.

The ICSB supervises the Bacteriological Code (Buchanan et al. 1948, 1958; Lapage et al. 1973; Sneath 1992), regularly providing lists of recent validly published species names and proposed changes in nomenclature, first in the *International Journal of Systematic Bacteriology* (IJSB) and later in the *International Journal of Systematic and Evolutionary Microbiology* (IJSEM). In an effort to rationalize nomenclature, remove the burden of early errors and synonyms, and simplify access to the nomenclature, a list of the approximately 2300 historical bacterial species names considered to be valid was published as the Approved Lists of Bacterial Names (Skerman et al. 1980, 1989). This, supplemented by the subsequent IJSB/IJSEM Validation Lists, defines the current, formal nomenclature of bacterial species.

The status of this scheme is reviewed about every 10 years in successive editions *of Bergey's manual*; the current classifications for each group of bacteria are summarized by experts, and the various genera and species are described. Although there is no formal link between the ICSB and *Bergey's manual*, both clearly draw upon the same pool of expertise, and the nomenclature of the Validation and Approved Lists tend to be closely conserved in *Bergey's manual*.

There are numerous subcommittees of the ICSB, most of which deal with specific groups of bacteria and give recommendations and advice on their area. Meetings of the ICSB and of the subcommittees occur regularly, usually associated with symposia organized by the International Union of Microbiological Societies. Most of the subcommittee meetings are open, allowing participation by any interested parties.

Those wishing to propose new species or changes in systematic nomenclature must follow the rules of the Bacteriological Code if their proposal is to be considered valid. Briefly, the proposal must be published in the IJSEM, or, if it is published elsewhere, a copy of the paper must be submitted to the IJSEM for inclusion of the proposal in the Validation Lists. If a new, culturable species is proposed, then a type culture should be designated and deposited with a recognized service culture collection. Further detailed rules exist, particularly on nomenclature, and it is advisable to read the most recent version of the code (Sneath 1992) in detail before submission of a proposal. In addition, minimal acceptable standards for species descriptions are being formulated (Lévy-Frébault and Portaels 1992; Ursing et al. 1994; ICSB Subcommittee on the Taxonomy of Mollicutes 1995), and the standards for the appropriate group should be consulted. Provisions for proposed species that are (currently) nonculturable but have been characterized by DNA sequencing have recently been clarified (Murray and Schleifer 1994; Murray and Stackebrandt 1995). Recent advice on characterization methodologies and the definition of the species level is presented by Stackebrandt et al. (2002). This paper reiterates the advice of Christensen et al. (2001), who argue that a species proposal should be based on the description of more than one and preferably at least five to ten strains.

THE PROCESS OF CLASSIFICATION

This begins with the selection of a limited series of isolates that appear to show restricted diversity (Figure 2.1). This is simplification by selection, and requires an intuitive recognition of probable groups. Inclusion of a range of reference and type cultures, and of many strains isolated from a broad range of sources, is advisable. The strains are then characterized in a large number of selected tests, producing descriptions for each. These descriptions are compared, either by calculation of a coefficient of similarity for each pair of isolates in numerical taxonomy (Sokal and Sneath 1963; Sneath and Sokal 1973), or intuitively. From these similarities, the existence of sets of isolates or taxa is postulated. It is desirable that the taxa should be cohesive and differentiable, i.e. that the differences between taxa should be greater than the variation within a taxon.

The taxa are then described and named, according to the rules of nomenclature (Bousfield 1993; MacAdoo 1993), allowing the cohesive (least variable) portions of the description to be communicated in the shorthand form of a name. There are some minor, but irritating problems with the current nomenclatural system. There is no maximum length in characters for a species binomial, but long species names can tax those installing computerized reporting systems. In the absence of an ICSB ruling on this, authors proposing new nomenclature should be mindful of the problem and avoid imposition of this inconvenience on their colleagues. Also, there is no nomenclatural guidance on standards for abbreviation of generic names, e.g. *Str.* for *Streptococcus* or *Streptomyces*, and this may be a problem in wide ranging papers, or general texts.

Finally, new isolates can be identified, i.e. named as being members of one of the postulated taxa, on the basis of their similarity to the taxon description, or to

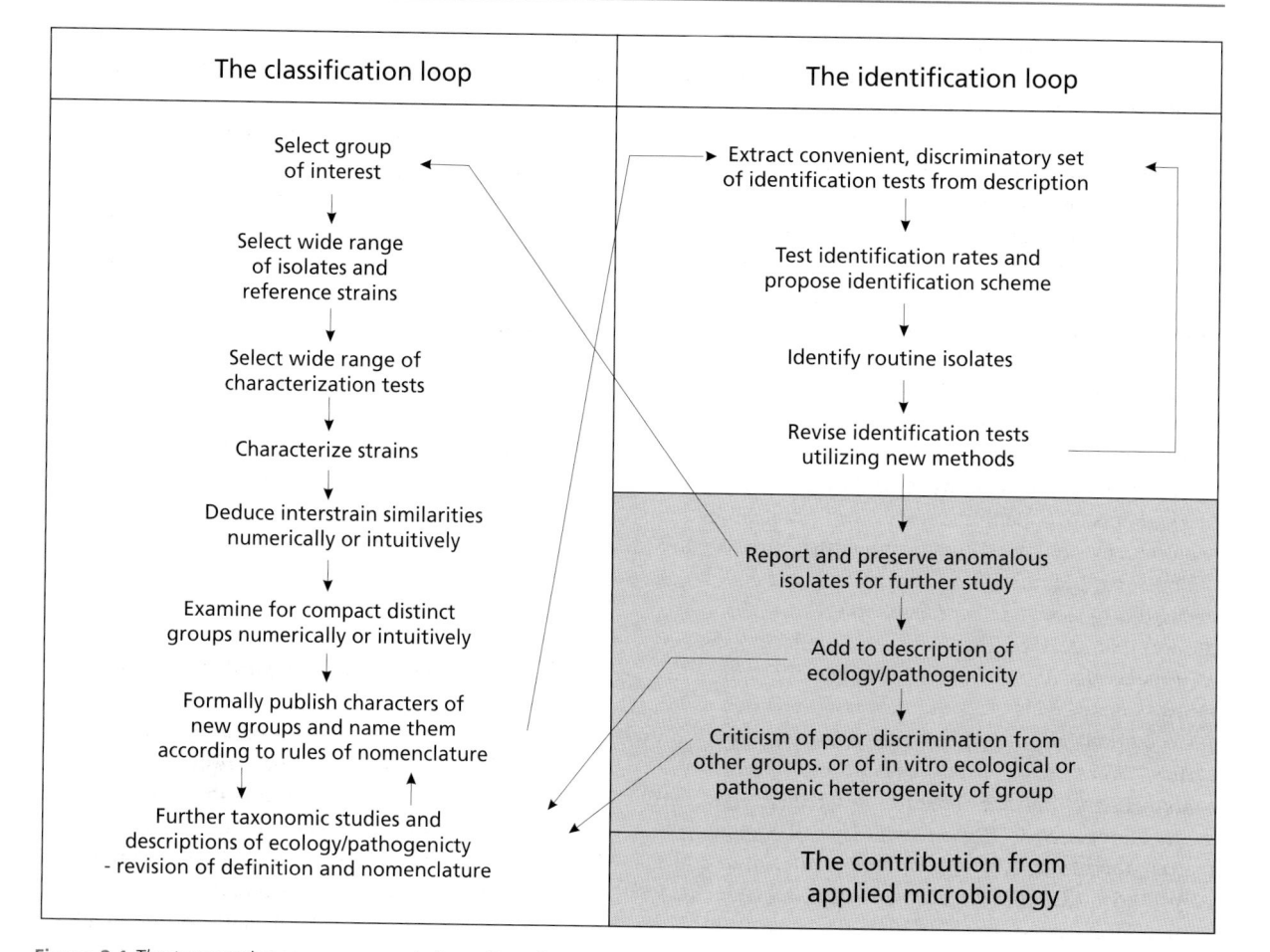

Figure 2.1 *The taxonomic process represented as a flow diagram.*

exemplar members of the taxon (type and reference strains). The taxon description will often have been enlarged by this stage. In particular, evidence on properties too difficult, time-consuming, or costly to assess in initial tests may have accumulated. In microbiology, this will often include a mass of anecdotal evidence ascribing particular pathogenic, ecological, or spoilage properties to a species. This evidence is a key ingredient in the prediction of significant pathogenic (and other) properties from identification, transforming a new taxon from the arena of solely academic interest to that of significance in applied bacteriology. New tests that are more convenient, reliable, or rapid may have been shown to differentiate the species. Usually, fewer tests are applied in identification than in classification; only those that have proved reproducible, discriminatory, convenient, and inexpensive are used. Note the crucial distinction between classification, which seeks to elucidate a group structure, and identification, which seeks to assign new isolates to an appropriate group within an existing classification.

IDENTIFICATION

There are three broad approaches to identification in general use. The first, and most important, is recognition

of colony morphology (and its variation over a range of isolation media) by an experienced microbiologist, sometimes with confirmation by a few rapid, simple tests. This is the basis on which most clinical isolates are divided into commensals or contaminants of no interest, possible pathogens, and probable pathogens. Without this rapid triage of isolates, most diagnostic laboratories would be faced with an overwhelming workload of formal identification. However, the apparent simplicity of the process conceals much complexity and has far-reaching consequences upon the practice of medical microbiology.

Intuitive human pattern recognition is notoriously difficult to document, is impossible to teach in a formal setting, and requires prolonged training, strong perceptive and intuitive talents, and considerable intelligence. This shapes a long training process of one-to-one teaching at the bench, with either tight selection or a high failure rate. Equally, the triage of isolates must be performed or checked by those with great expertise and experience. The screening process is an art, rather than a science; however, it is difficult to justify the expense of the costly but essential training in this art to paymasters. As in the production of hand-crafted furniture, the quality of the reports that a laboratory generates depends heavily on the expertise, skill, and time that staff can devote to each specimen. Surprisingly, there appears to be no study in the literature that documents the error rate of this process; whether this reflects an unquestionable accuracy, overweening ego, a profound fear of the consequences of documenting errors, or simple omission is difficult to ascertain.

After this triage, formal identification is performed upon potentially significant isolates. In earlier times this often used an 'identification key,' exemplified by the identification scheme for coagulase-negative staphylococci proposed by Kloos and Schleifer (1975). The key is a tree-like logic diagram, with a test name noted at each branch point, and two branches (one for isolates that are positive, and the other for isolates that are negative in that test) indicating the further logic pathway for identification. The investigator follows the logic tree to a terminal branch, where the designer of the key has decided that sufficient similarity has been established for identification and noted the appropriate species name. This was the favored approach in the early editions of *Bergey's manual*, and in the various earlier schemes (Chester 1901; Orla-Jensen 1919), reflecting the lasting botanical influence on early bacterial taxonomy. Keys depend heavily upon invariant characteristic reactions, although the later efforts, particularly the Kloos and Schleifer scheme, make some allowance for test variation. However, the work of numerical taxonomists brought general acknowledgment that bacterial species are polythetic, i.e. any individual property may vary within a species. Only the most sophisticated and complex keys could cope with this variation, and, with the introduction of numerical identification, formal keys have declined greatly in importance.

Numerical identification is based on surveys that define tables of the expected frequency of positivity in a series of tests for each species. The reaction pattern of an 'unknown' isolate is compared with this table, and its 'goodness of fit' with each species is determined. This comparison may be intuitive, as in the early schemes (Castellani and Chalmers 1919; Enterobacteriaceae Subcommittee 1958; Cowan 1956) and in the tables of the various editions of Cowan and Steel's classic identification manual (Cowan and Steel 1965), or by formal mathematical comparisons. The latter may be calculated concurrently by computer programs, or the calculations may be performed earlier, giving lists of common reaction patterns with the best-fit identification for each, as in the various identification listings for API identification kits. The underlying mathematical principles are described in the classic paper of Willcox et al. (1980) on numerical identification.

In contrast to classification, where a host of ill-defined mathematical assumptions must be made during data processing, the mathematical identification strategies described by Willcox et al. (1980) rely on only six clear prior assumptions. These are that:

1 the isolate under study cannot be a member of more than one taxon
2 the isolate is a member of a known taxon
3 each of the tests employed yields a result that varies independently of all others
4 the tests have roughly equal within-strain reproducibility
5 this within-strain reproducibility does not vary significantly between species or strains
6 the tests yield clear negative or positive results.

Two important parameters in assessing numerical identifications are the 'strangeness' or absolute likelihood of obtaining the observed pattern, often assessed as the estimated frequency of occurrence, logarithmic probability, or modal likelihood fraction, and the 'equivocality' or relative likelihood of the fit with known result patterns, often assessed as a percentage relative likelihood or 'identification score'. The 'strangeness' index is calculated for each known species (Table 2.1), and the species with the highest index is chosen as the best-fit identification. Caution is required if the strangeness index is low. This indicates that the observed pattern of reactions is most unusual, even for the best-fit species, and the isolate may well be a 'stranger' – a member of a new species or a species not listed in the table. The equivocality index indicates the level of uncertainty in ascribing the isolate to the best-fit species if it is not a 'stranger'. Again, if this index is low, then the identification is uncertain, split finely between two or more species, and it would be wise to proceed to further tests, particularly if the identification equivocates between species of significantly distinct pathogenic or cross-infection potential.

A common reason for unusual reaction patterns is that the culture is impure; close examination of prolonged subcultures on a wide variety of media, incubated under a range of conditions, will often reveal cryptic contaminants in these cases. Equally, isolates of *Bacillus* or *Clostridium* species that are asporogenous and show a negative gram-staining reaction present an insidious trap for the unwary, who may waste considerable effort under the false assumption that these organisms are true gram-negative bacilli. Conclusive proof of the true nature of such isolates may be difficult to obtain, although colony morphology often provides an indication of the problem.

Ascribing a species name to an isolate should be a carefully considered action. The name will usually communicate an implicit description of many properties that have not been, or cannot be, examined in the laboratory. Once identified, the isolate will be assumed to possess all these implied properties, whether these describe biochemistry, pathogenicity, capability of cross-infection, etc. A poor identification, arrived at with

Table 2.1 *Mathematical identification: a worked example*

a) Determine reaction pattern of unknown isolate in tests T1 to T7:

Test	T1	T2	T3	T4	T5	T6	T7
Result	+	+	−	+	−	+	+

b) Refer to a reputable identification table

Taxon	Known frequency of positive reactions in test:							Maximum likelihood
	T1	T2	T3	T4	T5	T6	T7	
A	.99	.01	.99	.01	.99	.99	.99	.932
B	.99	.75	.01	.90	.75	.90	.10	.442
C	.90	.85	.10	.95	.05	.99	.99	.609
D	.95	.99	.99	.99	.95	.01	.01	.858
E	.01	.01	.05	.01	.99	.01	.99	.894

The maximum likelihood is calculated for each taxon from the reaction pattern that yields a 'perfect' fit to that taxon (sequentially multiply 1−frequency for tests with frequency $\leqslant 0.5$, or frequency for tests with frequency >0.5; see below in c).

c) Calculate the likelihood of obtaining the isolate pattern for each taxon

For each taxon, calculate the likelihood by multiplying the frequencies for each test together, but substituting (1−frequency) for the frequency if the test result is negative

Taxon	Likelihood calculation	Likelihood
A	$.99\times.01\times(1-.99)\times.01\times(1-.99)\times.99\times.99$ =	9.70×10^{-9}
B	$.99\times.75\times(1-.01)\times.90\times(1-.75)\times.90\times.10$ =	$.147\times10^{-1}$
C	$.90\times.85\times(1-.10)\times.95\times(1-.05)\times.99\times.99$ =	$.609\times10^{-1}$
D	$.95\times.99\times(1-.99)\times.99\times(1-.95)\times.01\times.01$ =	4.66×10^{-9}
E	$.01\times.01\times(1-.05)\times.01\times(1-.99)\times.01\times.99$ =	9.41×10^{-11}
	Sum =	.756

d) Calculate the 'strangeness' indices, negative log probability, and modal likelihood score, which indicate how unlikely the test result pattern is for each taxon. Calculate the 'equivocality' indices, identification score, and relative likelihood, which indicate the equivocality of the best-fit result

Taxon	Negative log probability		Modal likelihood score	
	Calculation	Result	Calculation	Result
A	$-\log(9.70\times10^{-9})$ =	8.10	$9.70\times10^{-9}/.932$ =	1.04×10^{-8}
B	$-\log(1.47\times10^{-1})$ =	.83	$1.47\times10^{-1}/.442$ =	.33
C	$-\log(6.09\times10^{-1})$ =	.22	$6.09\times10^{-1}/.609$ =	1.00
D	$-\log(4.65\times10^{-9})$ =	8.33	$4.66\times10^{-9}/.858$ =	5.42×10^{-9}
E	$-\log(9.41\times10^{-11})$ =	10.03	$9.41\times10^{-11}/.894$ =	1.06×10^{-10}

Taxon	Identification score		Relative likelihood	
	Calculation	Result	Calculation	Result
A	$9.70\times10^{-9}/.756$ =	1.24×10^{-10}	$9.70\times10^{-9}/.609$ =	1.59×10^{-6}
B	$1.47\times10^{-1}/.756$ =	.194	$1.47\times10^{-1}/.609$ =	.24
C	$6.09\times10^{-1}/.756$ =	.805	$6.09\times10^{1}/.609$ =	1.00
D	$4.65\times10^{-9}/.756$ =	6.16×10^{-9}	$4.66\times10^{-9}/.609$ =	7.65×10^{-9}
E	$9.41\times10^{-11}/.756$ =	1.24×10^{-10}	$9.41\times10^{-11}/.609$ =	1.55×10^{-10}

e) Conclusion

All scores for taxa A, D, and E indicate extremely poor fit. The taxon most likely to yield this result pattern is C. The identification score is poor at .805 (values appraoching .99 are desirable), but this is expected with such a small number of tests that yield mostly variable results for this taxon. Correcting for taxon variability (division by the maximum likelihood for taxon C) yields a modal likelihood score of 1.00, indicating that no better fit can be obtained for this taxon with limited tests chosen. However, the 'equivocality' indices both indicate an equivocal identification, with the fit to taxon B being only one-quarter less likely than the fit to taxon C. The poor identification score and high equivocality indicate that this isolate requires further tests suited to discriminate between taxa B and C to produce a reliable identification result.

much equivocation, can acquire an undeserved aura of precision during the reporting process. In these cases it might well be more accurate to state that the isolate is unusual, and although most similar to the best-fit species, may well not share all the species properties.

Further, the perception of the properties implied in the species name will vary between individuals, according to their knowledge of systematics and the biases imbued in training and specialization. This is a problem of the 'shorthand' communication of data via a species name, which requires good to perfect agreement between the communicating parties on the definition implied in the species name. Clearly, the level of agreement on the implied description may well be unsatisfactory if one side is employing a current nomenclature and description, while the other has little contact or interest in microbiology and is struggling to remember information from a minor part of a syllabus learned decades before. The laboratory cannot assume that clinicians will understand the full implications of a species name for any but the most frequently encountered species and should take every opportunity to expand upon the properties that are intended to be communicated to the recipient. The possibilities of gross distortion of important clinical implications in this communication have become ever greater with reductions in resources, increasing time constraints, and the increasing speed of change in species nomenclature.

CHARACTERIZATION TESTS

The lack of a nonarbitrary, generally accepted definition of the species concept in bacteria (see below) inevitably, but wrongly in my view, directs bacteriologists to definitions based on some particular characterization technology. A broad description of the various technologies is given below and in Table 2.2. Note that none of these methods yields results that can be considered invariably to give the isologous comparisons of truly independent variables required in taxonomy.

Microscopic morphology

These characters comprise: cell shape, size, arrangement, staining, motility and capsule characteristics, spore position and morphology, and flagellar arrangement. Electron microscopy extends these somewhat to fimbriation, wall and cytoplasmic structures. Early microbiologists of the van Leeuwenhoek era were limited to describing the cell morphology and ecological occurrence of those few groups of bacteria distinguishable on the basis of light microscopy. Macroscopic organisms show a massive range of readily observed distinct morphological characters, sufficient to distinguish a huge number of groups. However, bacteria, at the limit of resolution in light microscopy, can be divided into few groups on the basis of cell morphology.

This may have been a major factor in forming a sharp difference of attitude between bacteriologists and macrobiologists. While eighteenth- and nineteenth-century botanists and zoologists sought to document the diversity of macroorganisms, ranging throughout the world in their search for specimens and describing hundreds of thousands of species, bacteriologists centered their work on groups that were involved in disease or had industrial importance. With few exceptions, bacteriologists did not participate in the eighteenth- and nineteenth-century 'collection fever' phase of biology because bacterial characterization was still primitive. By 1989, only about 2300 bacterial species were cited in the Approved Lists, indicating a combination of the low level of interest in discovery and documentation of new taxa during the previous century, and the difficulty of these tasks.

Table 2.2 *Characterization methods: features dictating their practical utility*

Method	High speed	Good throughput	Low skill	Automated	Low costs Capital	Low costs Running	Differentiation level < Genus Species Type >
Ecological	-	-	-	---	++	--	<<------------------------>
Morphological	++	++	-	---	+	++	<<----->
Colonial	+++	+++	+	---	+++	+	<<------------------->
'Biochemical'	+	++	++	p	++	++	<<----------------------->
Enzyme detection	+++	++	++	p	++	++	<<----------------------->
Serology	+++	++	++	p	++	-	<--------------->>
Chemotaxonomy	--	--	--	---	-	--	<<----------------------->
Fingerprinting	+++	+++	+	s	-	++	<<----------------------->>
DNA base ratios	-	-	-	--	+	++	<-------->
DNA Hybridization	--	--	--	---	+	+	<---------->
16S RNA sequencing	-	+	+	--	++	-	<<-------------->
Polyphasic studies	--	++	--	---	++	+	<<----------------------->>

Key: -, --, ---, increasingly less favorable; +, ++, +++, increasingly more favorable; p, possible; s, some.

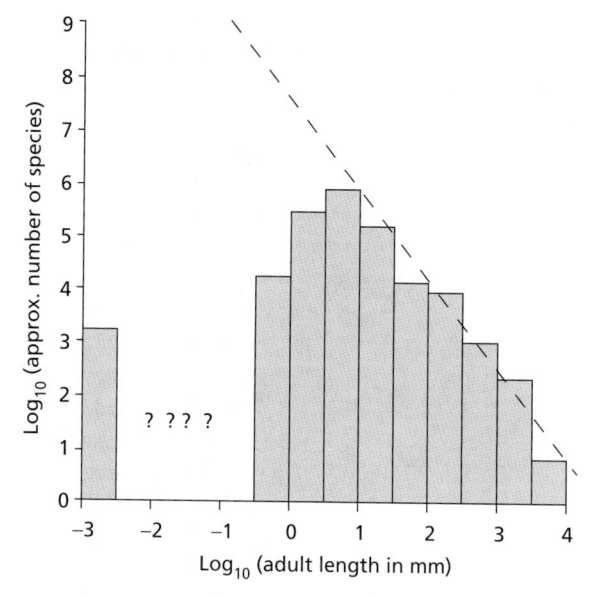

Figure 2.2 *Histogram representing the number of known species vs adult size. The dotted line indicates proportionality of the number of species with the inverse square of adult body length (modified from May 1988).*

This omission is reflected in the often-cited histogram (Figure 2.2) of the number of known biological species versus adult size. The diagram shows a maximum number of species at a size corresponding to the limit of resolution of the human eye. This may well be an artifact of limited endeavor. The approximately 5000 currently described species of bacteria probably represent a tiny fraction of bacterial diversity, a contention supported by recent studies of 16S rRNA diversity in natural environments (Stackebrandt et al. 1993; Liesack and Stackebrandt 1992).

Colonial morphology

With the invention of solid media and pure culture techniques came colonial morphological characters. For an experienced bacteriologist, the wealth of data from this source is often sufficient for identification at genus or finer taxonomic levels. Most isolates that are discarded in routine diagnostic work as not clinically significant are identified solely on this basis. However, although often highly discriminatory, particularly after prolonged incubation, the many subtle characters involved are often medium- and incubation-dependent and are notoriously difficult to document. Taxonomists should not ignore this elementary investigation, which will often delineate important divisions within a collection of strains and is crucial to exclusion of impure cultures.

Biochemical tests

Early experiments on optimization of medium composition and incubation conditions led naturally to

assessment of nutritional characteristics. The so-called 'biochemical tests' that evolved became an essential tool in bacterial taxonomy and identification. These include tests for fermentation or oxidation of sugars; waste products; metabolism of organic acids, proteins, amino acids, and lipids; temperature, pH, and redox range for growth or survival; and tolerance of various chemical agents. In short, they collectively define the nutritional and physiological interactions of the organism with its environment.

These tests are usually technically simple, are inexpensive, and have high throughput when adapted to multipoint inoculation, commercial test strip, or microtiter plate formats. Most yield bi-state, i.e. positive or negative, results that are convenient to read. However, it should be recognized that these bi-state results usually represent a simplification of a more complex situation.

As an example, the test for production of acid from sugars measures a fall in pH, separating as 'positive' those strains that produce sufficient excreted acidic metabolites to reduce the pH below a threshold where the indicator dye changes color. Measuring the pH of the culture (Figure 2.3) may reveal more complexity, with distinct groups that produce a large, moderate, or small fall in pH, no change, or even a rise in pH. On further examination, the results of the test depend upon many factors: the threshold at which the indicator changes color; the buffering capacity of the medium; the concentration of the sugar; the amount of acidic products excreted; the strength (pKa) and hence molecular structure of the acidic products; the parallel production of alkaline products (e.g. production of ammonia from amino acids or peptides in *Bacillus* and *Clostridium* spp.); the further metabolism of the acids; and the loss of volatile acids from the medium.

Clearly, the technical simplicity of this test conceals far-reaching complexity. Strains that produce small amounts of acetic acid, and others that produce large amounts of caproic acid, may both yield positive results,

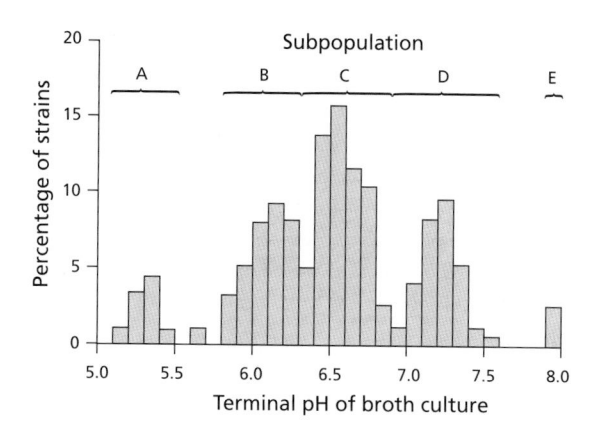

Figure 2.3 *Terminal pH of sugar-containing media: a histogram for 420 strains of anaerobic gram-negative bacilli showing multiple distinct populations (unpublished data provided by Prof. B.I. Duerden).*

because the quantitative difference is balanced by the qualitative difference in the nature and therefore strength of the acids. Equally, organisms with identical glycolytic pathways, but differing in the presence of a tricarboxylic acid cycle and an electron-transport chain capable of converting stronger acidic products to weakly acidic CO_2, would differ in their acidification of aerobic glucose-containing media. Strains with similar biochemistry may differ in their ability to acidify glucose-containing media, and strains with distinct biochemistry and end-products may yield identical results. In taxonomic studies, comparisons are assumed to be of isologous characters, yet acidification tests may reflect anything from a remote ecological similarity in acidification of sugar-containing environments, to truly isologous biochemistry.

A further problem results from this 'thresholding' of biochemical tests. Equivocal results and test irreproducibility are likely to be a problem if the test threshold does not coincide with a population minimum. In Figure 2.3, a detection threshold pH of 5.5 coincides with a minimum in the population distribution. A threshold at this level would give few equivocal results. However, a threshold at pH 6.2 would coincide with a subpopulation maximum, giving many equivocal results, and a division into positive and negative groups that would not reflect real underlying similarities in the biochemistry for subpopulation B. In practice, these thresholds are defined by the buffering capacity of the test medium, the pH range for color change in the indicator, and the concentration of the substrate.

Formal proof of adequate thresholding is rare in biochemical tests, but the range of genus-specific basal medium and indicator system formats of, for example, sugar acidification tests clearly indicates intuitive recognition of the problem. Finally, strains that give delayed metabolism of acid products or produce highly volatile acids may yield an early positive result but revert to negative on prolonged incubation.

Morphological, colonial, biochemical, and enzyme detection tests are the basis of current routine identification technology. For routine bacteriologists, it is essential that any species or genus name reaching the Validation Lists should have a description of these characters that allows clear identification and discrimination from similar taxa. Applied microbiologists have a clear interest in ensuring that this principle is a central feature of the minimal descriptions formulated by the ICSB committees to direct taxonomists on a line that is compatible with the current requirements of routine laboratories.

Tests for enzyme activity

Further characterization strategies were investigated. Research showed that some pathogenic effects of bacteria were due to toxins. These toxins included exoenzymes, and an offshoot of this work was a clear lineage of characterization tests based on detection of individual enzymes, from the early tests for proteolytic and lipolytic activity to the modern preformed enzyme tests (Manafi et al. 1991; James 1994). This line produced major advances in biology. Monod's work on the lactose fermentation system of *Escherichia coli* laid the foundations of our understanding of underlying control mechanisms of the cell biochemistry of all living things.

The trend in bacteriology from morphological to physiological and biochemical characterization contrasted with that in macrobiology. Here, morphology still retained practical precedence, but phylogeny, classification based on the postulated evolutionary history of organisms, was becoming a target for academic biologists. Bacteriologists, driven by a paucity of morphological characterization tests, and lacking the fossil material required for phylogeny, resorted to ever more biochemically oriented characterization technology. It is ironic that this technology ultimately led to the precedence of bacteriology in the discovery of DNA as the substance governing inheritance, and in much of the molecular biology that flowed from that event. Bacteriology, which had largely turned its back on phylogeny, provided the basis for the nucleic acid sequencing methods so essential to modern molecular phylogeny.

Chemotaxonomy

Early work on immunology produced serological characterization tests (Bowden 1993). Interest in the chemical basis of these tests, and work on the mode of action of beta-lactam antibiotics, gave a detailed understanding of the chemical structure of bacterial cell walls. The remarkably diverse structures that were found correlated with other properties. Crucially, this clarified grouping of genera into families and laid the foundations of chemotaxonomy, the examination and comparison of cell macromolecules. This approach to characterization has become steadily more important in taxonomy (Goodfellow and O'Donnell 1994).

Methods include comparison of cell-envelope macromolecular structure, particularly the murein polypeptide links, teichoic and teichuronic acids, and polar and nonpolar lipids (Hancock 1994; Embley and Wait 1994); electron-transport-chain components of oxidative and photosynthetic metabolism, the quinones, cytochromes, and bacteriochlorophylls (Collins 1994; Poole 1994; Richards 1994); and overall base ratios of DNA (Tamaoka 1994). For the most part these are costly, slow, low-throughput methods requiring considerable expertise, but they provide detailed chemical information relevant to classification at species, genus, and suprageneric levels.

Fingerprinting

Another characterization technology related to chemotaxonomy is termed 'fingerprinting'. The methods include sodium dodecyl sulfate polyacrylamide gel electrophoresis (SDS-PAGE) analysis of proteins (Vauterin et al. 1993; Pot et al. 1994), pyrolysis mass spectrometry (Hindmarch et al. 1990; Goodfellow et al. 1994), whole-cell matrix-assisted laser desorption ionization–time-of-flight (MALDI-TOF) mass spectrometry (Claydon et al. 1996; Lay 2001), and others of less familiar nature (Nelson 1991; Magee 1993b, 1994). These investigations yield results that reflect the chemical composition of the cell, but in comparative terms rather than by elucidation of detailed chemical structure. Fingerprinting methods utilize rapid physicochemical separation methods that produce profiles that can be compared to allow detection of similarities between strains. The profiles are often complex; the underlying nature of differences cannot be deduced from fingerprinting alone; and the methods are often based on costly instrumentation. However, these methods usually have a high throughput, automated processing, and low running costs, and produce many data relevant to cell composition, a facet that is otherwise difficult, slow, and costly to investigate. These data are pertinent at taxonomic levels from subspecies to supragenus.

Molecular genetics

The first inroads into DNA-based bacterial taxonomy came with assays for the ratio of guanine-cytosine (GC) to adenine-thymidine (AT) pairs in DNA (Tamaoka 1994), detecting the frequency of usage for GC-rich codons. The G+C content of bacterial DNA varies widely across the genera (see Tamaoka 1994), but less so within a genus. This is a useful tool to detect heterogeneity at genus level. Although significant differences clearly indicate dissimilarity, close similarity of G+C content does not necessarily indicate close similarity in other properties. Clearly, although the proportion of base pairs may be similar, the base sequence of the DNA may differ widely and code for a distinct metabolic pattern.

Later, DNA-DNA hybridization techniques allowed comparison of base sequence compatibility between strains (Kreig 1988; Johnson 1991; Stackebrandt and Goebel 1994). In this, DNA from a marker strain is annealed with DNA from the strain to be compared, and unbound DNA is removed. The extent of heterologous duplex formation indicates the extent of matching base sequences between the marker and test strain. As a control, 100 percent sequence compatibility is assessed by hybridization of marker and test DNA from the same strain. Annealing is performed under stringent conditions, at a high temperature, ensuring that only fragments with close sequence similarity remain bound, and at a lower, nonstringent temperature, where fragments may bind despite deletions and point-base differences.

Hybridization is currently recognized as the definitive test for similarity in bacteria. Strains showing more than 70 percent binding under stringent conditions with less than 5 percent difference between binding under stringent and nonstringent conditions are considered to belong to the same species (Wayne et al. 1987). Binding at this level indicates about 96 percent sequence identity (Stackebrandt and Goebel 1994). Anomalies between hybridization results and biochemical classifications occur in *Pseudomonas* (Roselló et al. 1991), the *Enterobacteriaceae* (Gavini et al. 1989), and *Bacillus* (Priest 1993).

However, hybridization is a prolonged, low-throughput approach requiring considerable expertise, and hampered by technical difficulties (Sneath 1983; Hartford and Sneath 1988; Johnson 1991) and potential distortion where repetitive sequences occur. Heteroduplex formation decreases markedly at supraspecies levels, and no inferences can be drawn on similarities at genus or further levels. Schleifer et al. (1993) have developed a rapid-lysis blotting hybridization system suited to identification.

Comparison of the base sequence for specific genes, usually that coding for the 16S rRNA of the small ribosome subunit, has become a standard approach. Various methods are available, including the original technique of direct sequencing of the 16S rRNA (Donis-Keller et al. 1977; Peattie 1979; Stackebrandt and Liesack 1993), but most groups currently utilize the polymerase chain reaction (PCR) to amplify a gene segment between two highly conserved sequences suited to PCR priming and DNA sequencing of the product.

The 16S rRNA gene is eminently suited to sequencing by this approach, as it contains 3′ and 5′ sequences that are virtually identical in all bacteria, are a suitable size for priming, and are separated by a sequence of a size (approximately 1.5 KB) that is readily copied by the PCR. Further, it is suggested that much of the sequence is highly conserved and so is unlikely to be subject to strain variation. Databases of known 16S rRNA sequences are available on the world wide web, notably on www.bdt.org.br/structure/molecular.html. Strains of the same species show sequence similarities greater than 97 percent. However, strains showing similarity at or above this level may belong to distinct species as defined by DNA hybridization (Ash et al. 1991; Fry et al. 1991; Fox et al. 1992; Martinez-Murcia et al. 1992; Stackebrandt and Goebel 1994). Advances in the technology have made 16S rRNA sequencing available as an identification method used in reference laboratories for particularly difficult strains. It is considerably less demanding in time, expertise, and cost than DNA hybridization.

Despite its current popularity, there may be several problems with 16S rRNA sequencing. Flawless transcription of a sequence of about 1500 base abbreviations to paper or computer file is no mean feat and the inevitable errors in proofreading these transcripts become embedded in the literature. This is no small source of inaccuracy, as noted elsewhere (Meissner 1995; Bernard et al. 1995). Sequences determined for the same species, or even the same strain, by different groups can differ significantly (Clayton et al. 1995). The source of these differences could be sequencing errors, intraspecies variation, or as many species contain multiple copies of the rRNA operon, interoperon differences. Several methods have been used in sequencing and the copy error level with different enzyme preparations is known to differ. Small differences in error rates or bias in the types of error from either source could lead to significant sequence differences. A level of about 3 percent sequence difference is suggested as a species cut-off, yet 18 percent of sequence pairs determined independently for the same strain differed by more than 1 percent in the study of Clayton et al. (1995), and 8 percent differed by more than 2 percent. Clearly, the overall error level in deposited sequences is close to the natural variation between closely similar species.

The sequence of individual product DNA molecules copied in PCR may differ slightly from that of the original template, or the operon copied may be atypical. With low-error-rate polymerase, these erroneous products are unlikely to affect direct sequencing of the pooled product. However, some groups select a single product molecule by cloning into a plasmid for later sequencing, and there is a possibility that an atypical product may be selected in cloning. A large blind-coded interlaboratory study to define the accuracy of sequence determination for the various methods and modifications and an effort to weed out or confirm any possibly inaccurate sequences in the databases would greatly enhance the utility of this approach.

The 'black-box' approach to numerical analysis that is evident in many papers on 16S rRNA sequencing has also been criticized. Goodfellow et al. (1997), reviewing a single issue of IJSB, found clear deficiencies in the numerical methods section of six papers on sequencing. Numerical analysis of similarity data is a complex subject unsuited to black-box processing. The deficiencies are highlighted by comparison with work on other characterization techniques, where data are often analyzed with four or more distinct numerical approaches and the homogeneity of the proposed consensus groups is then examined to verify the conclusions. Sequence data present greater difficulties in numerical analysis than most other characterization data. There are problems of alignment around deletions and insertions and in the interdependency of base sequences in distinct regions that hydrogen-bond together to form a functional tertiary structure. Sequence data require more effort in numerical analysis, and more caution in interpretation, rather than less. Unfortunately, the rise in availability of numerical taxonomic computer programs has encouraged cursory analysis, without the underlying understanding and caution required for interpretation.

Having said this, gene sequencing has opened the possibility that the evolutionary descent (phylogeny) of the current range of bacterial species may be determined. A phylogeny based on the 16S rRNA gene is now well-advanced (Woese 1987, 1992, 2000). This has had a major influence on the classification presented in the latest edition of *Bergey's manual* (Garrity 2001) and the few studies based on sequences of other conserved genes seem to verify this phylogeny.

Polyphasic taxonomy

Each of these individual characterization approaches has been hailed in its time as the last great advance in taxonomy, the discovery that would put every bacterium in to its proper place in a 'natural classification.' After experiencing four such 'revolutions,' I have come to a more moderate view. Each approach has its individual advantages and disadvantages. Each gives a different, and sometimes contrasting, insight into the differences between bacteria. Each has its place in taxonomic research and identification. Together, the results of all these approaches should give a robust general classification. The deficiencies of low-throughput, high-cost methods can be offset by testing the typical members of clusters detected in high-throughput, low-cost techniques. High-throughput techniques are suited to the search of large strain collections for potential new species.

Taxa that are distinct and cohesive in the large range of current techniques should stand the test of time. Colwell (1970) first suggested this polyphasic approach. However, despite its unarguable logic and repetition of the simple message by many experienced bacterial taxonomists, papers suggesting major taxonomic revisions based on a single technique continue to appear, only to be debunked later. The increasing instability of nomenclature at genus level in some areas of bacterial classification may well reflect the failure of workers to implement this polyphasic approach. A craftsperson uses many tools to fashion a useful object; taxonomists should recognize that each of the many approaches to characterization has an individual appropriate use in construction of classifications. Although a classification can be based on a single characterization method, it is unlikely to be of general use.

VISUALIZING THE RESULTS OF A CLASSIFICATION STUDY

Presentation of results is one of the greatest problems in taxonomy. Most authors choose diagrammatic representations – dendrograms, rooted or unrooted trees, or occasionally two- or pseudo-three-dimensional ordination diagrams (Figure 2.4). Each has advantages and disadvantages, but none can represent the full complexity of the data. The alternative tabular presentations (Figure 2.4) – full or shaded similarity matrices – show the full data but are unwieldy and difficult to interpret. Many applied microbiologists do not have the

knowledge to interpret the subtleties of these diagrams, and many would-be taxonomists use the programs that generate these diagrams as black boxes, in ignorance of the principles and shortcomings of the calculations involved (Goodfellow et al. 1997).

The dendrogram (Figure 2.4a) has a single axis indicating similarity between strains or between groups.

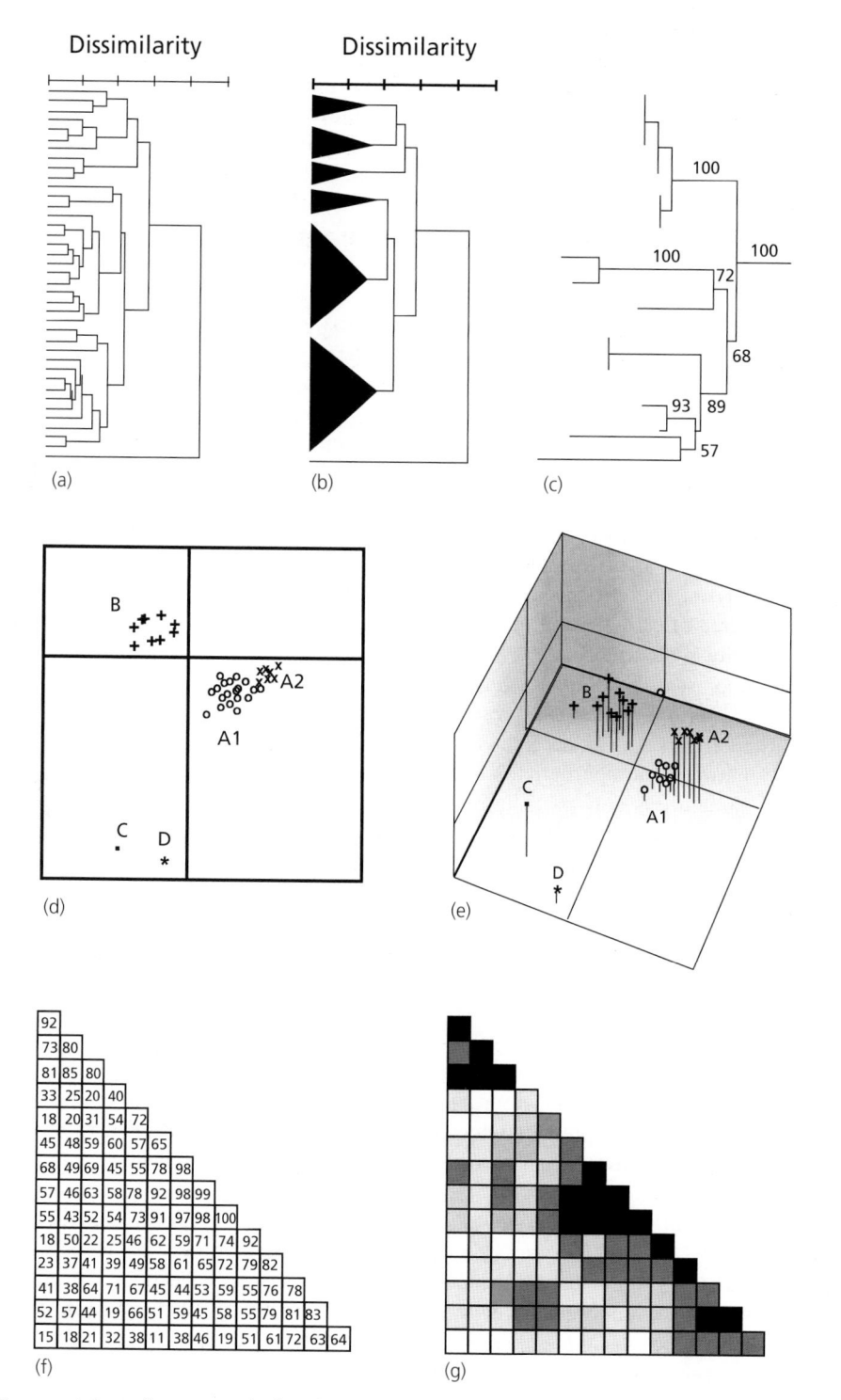

Figure 2.4 (a–g) *Representations of taxonomic data:* **(a, b)** *dendrograms;* **(c)** *tree (phenogram);* **(d)** *two-dimensional ordination;* **(e)** *pseudo-three-dimensional ordination;* **(f)** *similarity matrix;* **(g)** *shaded, ordered similarity matrix. (Data simplified from various biochemical and pyrolysis mass spectrometry studies by the author and colleagues.)*

The tree-like structure represents fusions of strains or groups, stretching from the multiple tips, each representing a single strain, to the root, a single line representing the fusion of all strains into a single group. Each branch point in the dendrogram represents the formation of a group with the level of intragroup similarity indicated by the position of the join relative to the axis. Where the join is between two single strains, the similarity of the strains is represented without distortion, but fusions between groups of strains become successively less representative of similarities between individual strains in the groups. In the unweighted pair-group method with arithmetic averages (commonly termed UPGMA) (Sneath and Sokal 1973), the most common method for constructing these figures, the mean group properties are compared at these group fusions. The groups formed at low levels of similarity may contain such diverse organisms that the mean group properties have little relevance to the properties of individual members.

Rooted and unrooted trees (Figure 2.4b) are essentially dendrograms that radiate from a central fusion out to radial tips representing individual strains. In this case the distance from the tip to a group fusion point represents the level of dissimilarity between the strain and the group mean. Often, each fusion point is annotated with a number showing the number (or proportion) of independent 'bootstrap' analyses that produced a similar fusion, indicating the degree of certainty that the fusion reflects a clear discontinuity in the original data.

Ordination diagrams represent interstrain dissimilarity as a distance. In Figure 2.4c, each strain is represented as a point, and the distance between pairs of points indicates the dissimilarity between the appropriate strains. However, the distance structure is only partially represented, and clusters of points may contain individual strains or small groups that are distant from the main cluster. This is illustrated in Figure 2.4d, where the data for Figure 2.4c are replotted on three axes, showing that group A1 is clearly resolved when a larger proportion of the distance structure is represented. The full distance structure from a large taxonomic study cannot usually be represented on fewer than eight axes.

Ordinations represent the distances between major groups well, but distances for some smaller groups and between individual strains are often represented poorly. This is the converse of tree representations, and comparison of the two types of diagram can highlight defects in the representations. The axes of ordination diagrams are usually complex statistically derived functions termed canonical discriminant variates or principal components. For the mathematically inclined, Manly (1994) gives an excellent explanation of these functions. For those averse to mathematics, each function is a composite measure of many properties that collectively distinguish between groups. These are similar to intuitively derived composite measures that are used in descriptions of

stature – ectomorphic, mesomorphic, or endomorphic – where the relative measurements of a complex set of features, rather than any single measurable feature, divide the groups. Composite measures are a common feature of languages; 'fair' complexion describes a composite of pigmentation features and 'gangly' describes a combined measure of limb length, posture, and 'slimness,' itself a composite measure.

Similarity matrices and shaded, ordered similarity matrices are shown in Figures 2.4e and f, respectively. These represent the interstrain similarities as numbers or graded shading in a tabular form, similar to the tables of intercity distances that can be found in road-map atlases. The shaded tables are particularly useful in highlighting intermingling at the edges of clusters and intergroup similarities that may not be apparent in the topology of a dendrogram. It should be noted, however, that the order of strains markedly influences the clarity of these matrices.

These diagrammatic representations all face the fundamental difficulty that it is usually impossible to fully represent the distance structure determined in a taxonomic study in a two-dimensional diagram. Mathematically, it can be shown that if n strains are examined in t discriminatory tests and $n<t$, then the distance structure may require up to $n-1$ dimensions for accurate representation and can be fully represented only rarely in $< (n-1)^{1/2}$ dimensions. For a taxonomic study involving about 100 strains, this would mean that the data would be fully representable only in ten or more dimensions. Clearly, pictorial two-dimensional representations of similarity data are likely to be deficient or misleading in some aspects. Failure to reveal differences may occur when clusters overlap in ordinations because the axes on which they separate are not plotted, or when isolates 'chain' into unsuitable groups in dendrograms.

THE 'TYPE' CONCEPT AND SERVICE CULTURE COLLECTIONS

The type concept originated from botany, where a single typical dead specimen can supply a wealth of morphological data sufficient to describe the species and to identify further examples. Sneath (1995) has suggested that, with modern characterization methods, preserved dead bacterial specimens might eventually prove equally useful. However, for more than a century bacteriologists have preserved cultures of bacteria as viable organisms. Private collections exist in most departments, but the national service collections, which supply preserved viable cultures, are the prime sources of reference and type cultures. Each valid culturable species has a type culture, which Buchanan (1955) considered to be the reference point for the species.

These service collections are of paramount importance to bacteriology. They provide reference cultures for comparative identification, quality control of characterization tests, teaching, and taxonomic studies. They often also provide focal expertise in taxonomy and identification and an essential service for identification of the more unusual organisms isolated in routine laboratories. Individual routine laboratories have insufficient time and see insufficient material to register rare, unrecognized pathogens as anything other than unusual. The long-standing high throughput of 'unusual' isolates for identification at national centers has repeatedly allowed identification of previously unnamed groups of bacterial pathogens as new species.

The Centers for Disease Control (CDC) laboratories in Atlanta have a notable record for the detection of such groups, recognized in the genus epithet *Cedecea*. It is particularly important that routine laboratories continue to send unusual isolates to such units, along with definitive information on their source and pathogenic effects, despite budgetary constraints. Equally, these units must continue to provide the identification service and maintain and review their collections of unnamed strains if we are to monitor new and rare pathogens adequately.

However, there are problems with these type and reference cultures. Live cultures maintained by serial subculture inevitably undergo selection and adaptation. Experimental evidence in chemostat (Helling et al. 1987) and batch culture (Atwood et al. 1951) strongly suggest that pure cultures undergo virtually complete clonal replacement with mutated variants after about every 100 divisions. Many older type strains were maintained by serial culture in early collections, and it is likely that most collection strains have undergone multiple predeposition culture cycles. Equally, the methods used to avoid continued subculture, freeze-drying and freezing in cryoprotectants, are highly selective. It is rare to recover more than 1 percent of the initial population from these preserved specimens, and such high kill rates must exert some selective pressure. This selection may be particularly severe for obligate anaerobes exposed to air during subculture. Early microbiologists recognized that biological variation was an inescapable consequence of the culture and storage of strains (Rahn 1929) and used the effect in attenuation of the pathogenicity of anthrax and tubercle bacilli. Modern microbiologists would do well to remember their conclusions.

Equally, reference cultures must be treated with the utmost care outside central collections. There have been several incidents of mislabeled reference strains being passed between research laboratories, with consequent errors in published work, notably with *Candida albicans* (Mackenzie and Odds 1991). All reference cultures should be obtained directly from a recognized culture collection and replaced regularly from the same source.

THE SPECIES CONCEPT

The definition of the species concept in bacteriology continues to be a favorite topic of debate, where those of a philosophical bent can argue free from the constraint of experimental data. A criticism is of the lack of enterprise in obtaining data that might elucidate a solution, rather than of the discussion of this central problem in classification.

In general biology, the best known species definition is that of a sexual interbreeding capability with production of offspring capable of further interbreeding. This definition clearly is not applicable to asexual organisms and is extremely difficult to test in the wild. Indeed, there are many examples where interspecies hybrids are viable, particularly in plants (Eckenwalder 1984) but also in animals (Hall 1978). Clearly, the capability of promiscuous genetic exchange in bacteria invalidates these definitions based on interfertility (Jones 1989). Interspecies exchange of chromosomal DNA has been demonstrated repeatedly (Maynard-Smith et al. 1991), and transposons and promiscuous plasmids can transfer large genetic elements across massive taxonomic gulfs (Natarajan and Oriel 1991; Kirby 1990; Arthur et al. 1987). Other species definitions of more general applicability exist (Simpson 1961; Hutchinson 1965; Van Valen 1976; Templeton 1989) but are not helpful in practical delineation of species.

In practice, identification and most classification work in general biology still relies almost exclusively on comparison of phenotypic characters. The various fine shadings of sexual species definitions are discussed with vigor, but their impact on practical aspects is small. Bacteriologists are probably further advanced on the species concept, with reasonable consensus agreement on a practically applicable, though arbitrary, definition based on DNA hybridization (Wayne et al. 1987). Cohan (2002) reviews species definitions and presents an interesting new definition for bacterial species based on the ecospecies concept.

However, even this consensus has been confused by the adoption of a series of terms that fragment the species concept into variants. Ravin (1963) proposed the terms 'genospecies,' encompassing mutually interfertile organisms, 'nomenspecies,' encompassing organisms 'similar' to the type strain, and 'taxospecies,' encompassing groups that share a high proportion of common properties. The term 'genomic species', readily confused with 'genospecies,' encompasses organisms showing high DNA sequence similarity. These concepts have pragmatic merit (Ursing et al. 1995), but they offer a dangerous attraction to an explosive diversification of concepts of convenience, each with its own classification and nomenclature. Medical microbiologists could easily find the concept of a medicospecies, encompassing a classification that they found particularly convenient, irresistible in the current milieu of rapid nomenclatural change.

The species concept that seems most universally accepted in practice, if not in theoretical discussion, is the taxospecies: a group of organisms that show high similarity in a broad range of characterization approaches and that are delimited from other organisms by an area of character combinations that are rarely found in nature. This definition is implicit in the numerical taxonomic approach and naturally entrains the concept of polyphasic taxonomy. It does not imply any arbitrary breakpoint of similarity in a specific test, but it requires proof of a discontinuity of variation at the species boundary.

Dividing organisms into species is an unavoidable necessity in biology; routine work invariably requires inference of more general properties from a limited number of observed characters. However, the organisms have no knowledge of, nor any obligation to obey, the divisions that we draw. The boundaries between species can often be diffuse, particularly in bacteriology. However, biologists often give the impression that the natural world can be segmented with absolute clarity into the divisions that we desire. Propagation of this erroneous impression in the general population may well encourage the view that humanity itself can be divided with equal clarity, with all the unsavory consequences that ensue. A definition of the species concept that denies the diffuse uncertainty of biological divisions is as undesirable politically as it is scientifically.

SPECIATION AND EVOLUTION

This is a much neglected area of research, but one in which bacteriology could make a considerable contribution to our understanding. In contrast to higher organisms, bacteria reproduce rapidly. It is possible to produce populations comparable with the entire worldwide population of man within a few days in a fermenter of modest size. Further, bacteriologists have evolved sophisticated techniques for detection and selection of variants forming a tiny proportion of a population. This offers the capability of observing evolution in action over short periods and in a level of detail that cannot be approached in macrobiology. It is unfortunate that the potential for useful work in this area has lain fallow for so long. The few studies to date (Atwood et al. 1951; Helling et al. 1987; Lenski and Travisano 1994) have been largely ignored, possibly because they emphasize a level of variability during successive culture cycles that many modern bacteriologists would prefer to ignore.

Equally, bacteriologists have observed evolution in the wild for many years but ignored its implications for evolutionary science. This is particularly true in medical microbiology. The introduction of antibiotics unintentionally produced the largest experiment in evolution ever undertaken by man, but its results have never been analyzed from an evolutionary viewpoint and the period in which a control population could be observed has long passed. What is obvious from this experiment is that bacteria are eminently suited to evolutionary studies. The high selective pressure of antibiotics and the rapid reproduction of bacteria has spawned, for example, at least five distinct successive races of *Staphylococcus aureus* suited to spread in the ecological niche offered by hospitals within 50 years. Fortunately, much material has been stored in culture collections, and this could provide an invaluable resource should sufficient funding and interest become available.

It is singularly unfortunate that the unique combination of the capabilities of bacteriological techniques and the highly selective environments found in hospitals, particularly intensive care units, has not been brought to the attention of evolutionary scientists, population geneticists, and microbial ecologists. The potential for mutual advantage from such a combination of expertise is enormous. However, the beginnings of interest are now obvious in studies of the genetics of evolutionary change in, for example, the acquisition of resistance to penicillin in commensal *Neisseria* spp. (Bowler et al. 1994) and in multilocus sequence typing of methicillin-resistant *S. aureus* (MRSA) (Oliveira et al. 2002) and glycopeptide-resistant enterococci (Woodford 2001). Population geneticists are beginning to study variation in bacteria (Baumberg et al. 1995; Feil and Spratt 2001; Vogel and Frosch 2002), and the impetus provided by the rapidly decreasing effectiveness of antibiotics against hospital-evolved pathogens may lead to funding of such projects.

ORIGINS, EVOLUTION, AND CONTROVERSY

The Earth is thought to have accreted as a planetary body in the solar disk about 5×10^9 years ago and to have cooled to form a solid surface at about 4.2×10^9 years ago. Prokaryotic life appeared less than 0.7×10^9 years later – a remarkably short time, considering that the subsequent appearance of eukaryotes was a further $1.5–2.3 \times 10^9$ years later. The earliest fossil evidence of life known at present occurs in the Apex chert in Australia (Schopf 1996). Among the diverse microfossils, some appear to be complex, highly evolved organisms that show close similarity in form and diversity to modern blue-green algae, a prokaryote lineage that appears to have originated midway in the bacterial line according to 16S rDNA phylogenies.

These phylogenies postulate three lineages of life (ignoring the viruses, which probably have multiple origins): the Bacteria (including the Cyanobacteria or blue-green algae), the Archaea, and the Eukarya (Woese et al. 1990). The time of origin of the Archaea is controversial (Cavalier-Smith 2002). These are prokaryotic organisms similar in morphology to bacteria but differing markedly in many other aspects. No archaeal microfossils have been found that can be unequivocally distinguished from bacteria by current techniques. The various 16S rDNA phyogenies often place the Archaea as diverging early from the bacterial lineage. However, from evidence of cell and biochemical organization, Cavalier-Smith (2002) argues that the Archaea originated later, possibly as late as the Eukarya.

The first unequivocal fossil evidence of eukaryotes occurs in rock about 1.2×10^9 years old, with less convincing evidence back to 1.8×10^9 years (Lipps 1993). The eukaryotic life forms diversified to give various unicellular lineages, the fungi, plants, and animals, with the first fossil evidence of multicellular eukaryotes occurring about 0.9×10^9 years ago. The Eukarya appear to have acquired prokaryotic symbionts early. There is strong evidence that the mitochondria present in most eukaryotes and the chloroplasts in photosynthetic eukaryotes are symbiotic prokaryotic organisms, descended from, respectively, gram-negative bacteria of the alpha-Proteobacteria line and Cyanobacteria (Margulis and Fester 1991). It is interesting to reflect that the dominant bacterium of the human flora is the mitochondrion, an obligate intracellular symbiont that is present in every cell, is essential to life, and represents a cooperative partnership 1 000 000 000 years old. Nature is not controlled solely by tooth and nail competition. Those who propagate the contrary view to politicians and the public should consider the longstanding symbiosis of mitochondria or chloroplasts, or study the wonderfully intricate symbiotic associations between prokaryotes and eukaryotes reviewed by Fenchel (1996) and Douglas (1996).

Beyond this broad picture of evolution of the three divisions of life, there is beginning to be controversy about the linear branching phylogenies that have dominated past thinking on evolution (Doolittle 1999; Jain et al. 2002; Gupta and Griffiths 2002). An increasing number of papers present evidence that widely divergent groups have exchanged genetic material coding for complex biochemical systems (Andersson and Roger 2002; Ragan and Charlebois 2002; Katz 2002). The evidence suggests that significant lateral genetic exchange between the most widely distinct evolutionary branches of life has occurred, possibly on many occasions. This implies that lateral exchange may be a common phenomenon in closely related lines, where similarities in exchange mechanisms and genome composition favor genetic exchange. There is strong evidence from the genetics of antibacterial resistance that sharing of genetic resources across widely divergent lines is a key option for evolutionary adaptation to adverse conditions. The current linear-branching model for evolution covertly assumes that competitive diversification is the sole mechanism involved. As cooperative redistribution of genes becomes recognized as an important additional mechanism, 'tree-of-life' representations

of evolution may be replaced by a reentrant ragged 'web-of-life' model. This type of model is likely to gain early favor among bacteriologists, where promiscuous genetic exchange mechanisms are already well documented, and rapid evolutionary processes, such as acquisition of antibacterial resistance, demonstrate the true complexity of evolutionary mechanisms.

Current views of evolutionary hierarchy for the bacterial lineage are based on 16S rDNA sequencing, which divides the following groups (only genera with members potentially pathogenic to man are named; 'others' indicates nonpathogenic genera not normally encountered in medical microbiology; the order of groups reflects order of branching where possible):

Thermotoga group – others
Bacteroides group – *Bacteroides, Prevotella, Flavobacterium, Cytophaga, Capnocytophaga*
High G+C gram +ves – *Mycobacterium, Nocardia, Streptomyces, Corynebacterium, Propionibacterium, Bifidobacterium, Gardnerella, Arthrobacter, Micrococcus*
Fusobacterium group – *Fusobacterium*
Low G+C gram +ves – *Mycoplasm, Ureaplasma, Acholeplasma, Streptococcus, Enterococcus, Lactobacillus, Listeria, Staphylococcus, Bacillus, Peptostreptococcus, Eubacterium, Clostridium*
Proteobacteria groups:
Gamma group – *Chromatium, Francisella, Xanthomonas, Coxiella, Legionella, Pseudomonas, Acinetobacter, Moraxella, Vibrio, Aeromonas, Haemophilus,* the Enterobacteriaceae + others
Beta group – *Burholderia, Alcaligenes, Bordetella, Neisseria, Eikenella* + others
Alpha group – *Wolbachia, Rickettsia, Brucella* + others + eukaryote mitochondria
Delta group – others
Eta group – *Campylobacter, Helicobacter, Wolinella* + others
Spirochaetes – *Treponema* + others
Fibrobacter – others
Green sulfur bacteria – others
Planctomyces – *Chlamydia* + others
Cyanobacteria – the blue-green algae + eukaryote chloroplasts

WHAT'S IN A NEW NAME?

Extrapolation of current trends is notoriously unreliable in science, but it is likely that bacterial taxonomy will have to cope with a major increase in the rate of description of new taxa. The background trend is already well established and may well be magnified by new influences. Automated acquisition of characterization data is becoming well established for an ever-wider range of characterization approaches, speeding experimental work, and changing the emphasis of problem-solving effort to data-handling. The international agreements on biodiversity may well increase funding for taxonomic studies, hitherto poorly funded by industry and public sources. The prospect that species may be patented will clearly lead to increased discovery efforts and provide a financial incentive to the 'splitter' faction. Pending patents have already caused difficulties in the deposition of type strains of commercial significance (Labeda et al. 1995). New technology may well redouble productivity from the molecular biology faction, and DNA sequencing studies of natural environments have already revealed an astounding diversity of previously undescribed and currently nonculturable bacteria, outnumbering described species (Derakshani et al. 2001; Eilers et al. 2000; Paster et al. 2001; Smit et al. 2001).

However, the speed of change in nomenclature has become a problem in applied bacteriology: those who deal with cystic fibrosis patients are, hopefully, aware of the multiple changes in the nomenclature of two major pseudomonad pathogens in the recent past (Hugh 1981; Swings et al. 1983; Yabuuchi et al. 1992; Palleroni and Bradbury 1993). Changes in nomenclature of pathogens generate considerable criticism among medical microbiologists. They are not merely inconvenient; they also increase the problems of communication and may well have unforseen implications in the complex legislation that touches on microbiology in each of the various countries of the world. A further increase in the rate of change could generate a major rift between those who use and those who produce classifications (Magee 1993a).

Key issues for clarification are the date from which proposed changes in nomenclature are to be regarded as active, in the sense that legal proceedings might judge a microbiologist to have erred by using an older nomenclature, and the process of consensus acceptance of change. Most taxonomists and many users regard the date of publication in the IJSB/IJSEM as the date of active change, but many may postpone judgment until the first date of inclusion in an edition of *Bergey's manual*. The latter includes sufficient time to judge the consensus acceptability of the change; the former clearly does not. However, neither solution seems completely acceptable, even at the present rate of change in nomenclature.

If changes in nomenclature are active from their appearance in the IJSEM, then each change is governed by a group comprising the authors of the paper, an editor, and a few referees. Is the judgment of so small a group invariably so reliable, and their insight into the ramifying consequences so broad, that the general community of microbiologists will find the change acceptable? Notwithstanding respect for fellow microbiologists, it is of concern that any group this small can institute changes that may have considerable legal and other consequences on the whole of bacteriology, and that these changes become active, without prior notice, on publication in a journal taken by a minority of applied bacteriologists.

Equally, if active change coincides with mention in *Bergey's manual*, then the interval between editions of about 10 years is too long and the number of concurrent changes to be made is too large. Even here, there is no defined mechanism for judging consensus acceptance of change. One or at most a few experts on each major group write on their subject. There is no guarantee that their interests are so all-inclusive that some change crucial to one branch of the applied science is missed or that their own biases reflect a consensus view of bacteriologists.

The ICSB has a pivotal role in guiding developments in bacterial systematics, including nomenclature, and its members face a new century that is likely to tax their powers of mediation. Negotiating satisfactory compromise between the wishes of applied microbiologists and taxonomists, two groups with widely differing views and requirements, could well require considerable diplomacy. However, the alternative to compromise and cooperation between taxonomists and users is to discard our only internationally accepted scheme of bacterial nomenclature, and this is clearly unacceptable. The solutions to these problems cannot be coercive; the nomenclatural code does not and cannot dictate the nomenclature used by individuals. The wisdom of preserving general acceptance of the formal nomenclature must be convincingly explained to a silent majority, who would much prefer to remain with the nomenclature they were taught years before.

Many of these communication problems could be solved by computerized communication of changes and opinions on the world wide web. One major advance is the availability of lists of all validated names on the world wide web (www.bdt.org.br/cgi-bin/bdtnet/bacterianame). A list of all validated bacterial names of human and veterinary importance is maintained by J.P. Euzéby at www.bacterio.cict.fr/. This site is particularly well maintained and user-friendly and is an essential reference for current names and other taxonomic information. In the interim, efforts are being made to disseminate notification of proposed changes more widely (Frederiksen and Ursing 1995a, b), and this should be supported and encouraged by the community of applied bacteriologists.

An associated problem is the growing gap between the characterization methods used by taxonomists and those available for identification in the routine laboratory. Base ratio determinations, DNA-DNA hybridization, 16S rRNA sequence determination, and chemotaxonomic methods are available in only a minority of routine laboratories and are, as yet, too specialized, tortuous, and slow to have major relevance to identification in routine bacteriology. However, an increasing number of taxonomic papers neglect any mention of characters that can be determined in routine laboratories. The bench microbiologist whose skills lie in identification from colony morphology, pure culture, and biochemical tests could be excused for regarding this change with horror. This generates a backlash, with regularly heard detrimental comments on the probable capabilities of taxonomists to perform conventional tests, or to examine plate cultures adequately. Clearly, this is an area where medical microbiological input to the formulation of minimal standards for description of new species is essential to our interests. Taxonomists and applied microbiologists must seek mixes of characterization methods that fulfill the requirements of both research and routine work, and revise these as technology advances.

However, medical microbiologists must recognize that they live in a time of great change in bacterial taxonomy. As in all science, taxonomic truth is approached by successive revision in the light of new facts, and the current turmoil in nomenclature is a reflection of growing interest in a subject that forms the basis of routine laboratory work. Nomenclatural change is inevitable and cannot be ignored. Many of the changes may increase the capability to predict important properties from identification results. However, applied microbiologists must also ensure that, when the changes cause problems, these are voiced to the ICSB with sufficiently convincing force, clarity, and representation. The widespread tradition of ignoring nomenclatural change, and restricting critical comment to colleagues in the same field is not helpful to either medical microbiology or taxonomy.

THE FUTURE

The future course of bacterial taxonomy is likely to be interesting and turbulent. We are beginning to discard attitudes and assumptions stemming from the late nineteenth century that have shackled progress for the greater part of the twentieth century. The introduction of numerical taxonomy marked the first major break with the early botanical influence, allowing recognition of the true nature of variation within bacterial species and ridding us of the baggage of monothetic groups and invariant diagnostic characters. Characterization tests now allow comparison of features other than the physiological and nutritional interactions with the environment. Together, these have encouraged the beginnings of exploration of the true diversity of bacteria, which may well produce many unforeseen discoveries of economic significance. Molecular methods are beginning to be used to explore diversification in bacteria, and there is now a possibility that the enormous potential of bacteria in the study of evolutionary processes may be recognized. The phylogeny of bacteria, or at least that of the 16S rRNA gene, is now becoming apparent.

However, these changes are bringing a host of problems. Nontaxonomists are increasingly bewildered by sudden changes in nomenclature and the alienation that this produces must be tackled. The tendency of taxonomists to adopt uniformly a single characterization technique, rather than a polyphasic approach, must be actively discouraged, for it will certainly lead to large, unsupportable swings in nomenclature as techniques fall in and out of style. A major problem of the polyphasic approach is that the full range of characterization techniques cannot be supported in a single laboratory. The consequences of this are that several groups must be involved in each study, and the paper produced must acknowledge a long list of workers in the authorship. The nonscientific problems of contact, cooperation, precedence of discovery, and adequate acknowledgment must be tackled if the polyphasic approach is to become the accepted method of taxonomic advance. Bacterial taxonomy is advancing into the next century in the midst of change. Hopefully, solutions to these problems will be found in a newly matured science that will continue to produce spin-off discoveries of enormous importance in general biology.

ACKNOWLEDGMENTS

My thanks to Professor P.H.A. Sneath and Professor M. Goodfellow for their moderating influence and their many helpful comments on the manuscript.

REFERENCES

Andersson, J.O. and Roger, A.J. 2002. Evolutionary analysis of the small subunit of glutamate synthetase: gene order conservation, gene fusions and prokaryote-to-eukaryote lateral gene transfers. *Eukaryot Cell*, **1**, 304–10.

Arthur, M., Brisson, N.A. and Courvalin, P. 1987. Origin and evolution of genes specifying resistance to macrolide, lincosamide and streptogramin antibiotics: data and hypotheses. *Antimicrob Chemother*, **20**, 783–802.

Ash, C., Farrow, J.A.E., et al. 1991. Phylogenetic heterogeneity of the genus *Bacillus* revealed by comparative analysis of small subunit-ribosomal RNA sequences. *Lett Appl Microbiol*, **13**, 202–6.

Atwood, K.C., Scheider, L.K. and Ryan, F.J. 1951. Periodic selection in *Escherichia coli*. *Genetics*, **37**, 146–55.

Baumberg, S., Young, J.P.W., et al. 1995. *Population genetics of bacteria*. Cambridge: Cambridge University Press.

Bernard, H.-U., Chan, S.-Y., et al. 1995. Reply to 'On the nature of papillomavirus hell'. *J Infect Dis*, **172**, 895–6.

Bousfield, I.J. 1993. Bacterial nomenclature and its role in systematics. In: Goodfellow, M. and O'Donnell, A.G. (eds), *Handbook of new bacterial systematics*. London: Academic Press, 318–38.

Bowden, G.H.W. 1993. Serological identification. In: Goodfellow, M. and O'Donnell, A.G. (eds), *Handbook of new bacterial systematics*. London: Academic Press, 29–62.

Bowler, L.D., Zhang, Q.Y., et al. 1994. Interspecies recombination between the PenA genes of *Neisseria meningitidis* and commensal *Neisseria* species during the emergence of penicillin resistance in *N. meningitidis*: natural events and laboratory simulation. *J Bacteriol*, **176**, 333–7.

Buchanan, R.E. 1918. Studies in the nomenclature and classification of the bacteria. V. Subgroups and genera of the *Bacteriaceae*. *J Bacteriol*, **3**, 27.

Buchanan, R.E. 1925. *General systematic bacteriology. History, nomenclature, groups of bacteria*. Baltimore: Williams & Wilkins.

Buchanan, R.E. 1955. Taxonomy. *Ann Rev Microbiol*, **9**, 20.

Buchanan, R.E., St John-Brooks, R., Breed, R.S. 1948. International bacteriological code of nomenclature. *J Bacteriol*, **55**, 287; *J Gen Microbiol*, **3**, 44.

Buchanan, R.E., Cowan, S.T., et al. 1958. *International Code of Nomenclature of Bacteria and Viruses*. Ames, IA: State College Press, reprinted with corrections 1959: Iowa State University Press.

Castellani, A. and Chalmers, A.J. 1919. *Manual of tropical medicine*, 3rd edn. London: Balliere, Tindall & Cox.

Cavalier-Smith, T. 2002. The neomuran origin of archaebacteria, the negibacterial root of the universal tree and bacterial megaclassification. *Int J Syst Evol Microbiol*, **52**, 7–76.

Chester, F.D. 1901. *A manual of determinative bacteriology*. New York: Macmillan.

Christensen, H., Bisgaard, M., et al. 2001. Is characterization of a single isolate sufficient for valid publication of a new species or group? Proposal to modify Recommendation 30b of the *Bacterological Code* (1990 Revision). *Int J Syst Evol Microbiol*, **51**, 2221–5.

Claydon, M.A., Davey, S.N., et al. 1996. The rapid identification of intact microorganisms using mass spectrometry. *Nat Biotechnol*, **14**, 11, 1584–6.

Clayton, R.A., Sutton, G., et al. 1995. Intraspecific variation in small-subunit rRNA sequences in GenBank: why single sequences may not adequately represent prokaryotic taxa. *Int J Syst Bacteriol*, **45**, 595–6.

Cohan, F.M. 2002. What are bacterial species? *Annu Rev Microbiol*, **56**, 457–87.

Cohn, F. 1872. Untersuchungen -ber Bacterien. *Bietr Biol Pfl. 1 Heft*, **2**, 127.

Collins, M.D. 1994. Isoprenoid quinines. In: Goodfellow, M. and O'Donnell, A.G. (eds), *Chemical methods in procaryotic systematics*. Chichester: John Wiley & Sons, 265–309.

Colwell, R.R. 1970. Polyphasic taxonomy of bacteria. In: Iizuka, H. and Hasegawa, T. (eds), *Culture collections of microorganisms*. Tokyo: University of Tokyo Press, 421–36.

Cowan, S.T. 1956. Taxonomic rank of the *Enterobacteriaceae* 'groups'. *J Gen Microbiol*, **15**, 345–9.

Cowan, S.T. 1962. The microbial species – a macromyth? In: Ainsworth, G.C. and Sneath, P.H.A. (eds), *Microbial classification*. Cambridge: Cambridge University Press, 433–55.

Cowan, S.T. 1968. *A dictionary of microbial taxonomic usage*. Edinburgh: Oliver & Boyd.

Cowan, S.T. and Steel, K.J. 1965. *Manual for the identification of medical bacteria*. Cambridge: Cambridge University Press.

Derakshani, M., Lukow, T. and Liesack, W. 2001. Novel bacterial lineages at the (sub)division level as detected by signature nucleotide-targeted recovery of 16S rRNA genes from bulk soil and rice roots of flooded rice microcosms. *Appl Environ Microbiol*, **67**, 623–31.

Donis-Keller, H., Maxam, A.M. and Gilbert, W. 1977. Mapping adenines, guanines, and pyrimidines in RNA. *Nuclic Acids Res*, **4**, 2527–37.

Doolittle, W.F. 1999. Phylogenetic classification and the universal tree. *Science*, **284**, 2124–9.

Douglas, A.E. 1996. Microorganisms in symbiosis: adaptation and specialization. In: Roberts, D.M., Sharp, P., et al. (eds), *Evolution of microbial life*. Cambridge University Press: Cambridge, 225–42.

Eckenwalder, J.E. 1984. Natural intersectional hybridisation between North American species of *Populus* (*Salicaceae*) in sections *Aigeiros* and *Tacamahaca*. III, Paleobotany and evolution. *Can J Bot*, **62**, 336–42.

Eilers, H., Pernthaler, J., et al. 2000. Culturability and *in situ* abundance of pelagic bacteria from the North Sea. *Appl Environ Microbiol*, **66**, 3044–51.

Embley, T.M. and Wait, R. 1994. Structural lipids of eubacteria. In: Goodfellow, M. and O'Donnell, A.G. (eds), *Chemical methods in procaryotic systematics*. Chichester: John Wiley & Sons, 121–61.

Enterobacteriaceae Subcommittee, 1958. Report of the Enterobacteriaceae Subcommittee of the Nomenclature Committee of the International Association of Microbiological Societies. *Int Bull Bacteriol Nomencl Taxon*, **8**, 25–33.

Feil, E.J. and Spratt, B.G. 2001. Recombination and the population structures of bacterial pathogens. *Annu Rev Microbiol*, **55**, 561–90.

Fenchel, T. 1996. Eukaryotic life: anaerobic physiology. In: Roberts, D.M., Sharp, P., et al. (eds), *Evolution of microbial life*. Cambridge: Cambridge University Press, 185–203.

Fox, G.E., Wisotskey, J.D. and Jurtshuk, P.J. Jr 1992. How close is close: 16S rRNA sequence identity may not be sufficient to guarantee species identity. *Int J Syst Bacteriol*, **42**, 166–70.

Frederiksen, W. and Ursing, J. 1995a. Proposed new bacterial taxa and proposed changes of bacterial names published during 1994 and considered to be of interest to medical or veterinary bacteriology. *J Med Microbiol*, **43**, 315–17.

Frederiksen, W. and Ursing, J. 1995b. Proposed new bacterial taxa and proposed changes of bacterial names published during 1994 and considered to be of interest to medical and veterinary bacteriology. *APMIS*, **103**, 651–4.

Fry, N.K., Saunders, N.A., et al. 1991. The use of 16S ribosomal RNA analyses to investigate the phylogeny of the family *Legionellaceae*. *J Gen Microbiol*, **16**, 1215–22.

Garrity, G.M. (ed.) 2001. *Bergey's manual of systematic bacteriology*, 2nd edn. New York: Springer Verlag.

Gavini, F., Mergaert, J., et al. 1989. Transfer of Enterobacter agglomerans (Beijerinck 1888) Ewing & Fife 1972 to Pantoea gen. nov. as Pantoea agglomerans comb. nov. and description of Pantoea dispersa. *Int J Syst Bacteriol*, **39**, 337–45.

Goodfellow, M. and O'Donnell, A.G. 1994. Chemosystematics: current state and future prospects. In: Goodfellow, M. and O'Donnell, A.G. (eds), *Chemical methods in procaryotic systematics*. Chichester: John Wiley & Sons, 1–20.

Goodfellow, M., Chun, J., et al. 1994. Curie-point pyrolysis mass spectrometry and its practical application to bacterial systematics. In: Priest, F.G., Ramos-Cormenzana, A. and Tindall, B.J. (eds), *Bacterial diversity and systematics*. New York: Plenum Press, 87–104.

Goodfellow, M., Manfio, G.P. and Chun, J. 1997. Towards a practical species concept for cultivable bacteria. In: Claridge, M.S., Dawah, H.A. and Wilson, M.R. (eds), *Species: the units of biodiversity*. London: Chapman & Hall.

Gupta, R.S. and Griffiths, E. 2002. Critical issues in bacterial phylogeny. *Theor Popul Biol*, **61**, 423–34.

Hall, R.L. 1978. Variability and speciation in canids and hominids. In: Hall, R.L. and Sharp, S.H. (eds), *Wolf and man, evolution in parallel*. New York: Academic Press, 153–77.

Hancock, I.C. 1994. Analysis of cell wall constituents of Gram-positive bacteria. In: Goodfellow, M. and O'Donnell, A.G. (eds), *Chemical methods in procaryotic systematics*. Chichester: John Wiley & Sons, 63–84.

Hartford, T. and Sneath, P.H.A. 1988. Distortion of taxonomic structure from DNA relationships due to different choice of reference strains. *System Appl Microbiol*, **10**, 241–50.

Helling, R.B., Vargas, C.N. and Adams, J. 1987. Evolution of *Escherichia coli* during growth in a constant environment. *Genetics*, **116**, 349–58.

Hindmarch, J.M., Magee, J.T., et al. 1990. A pyrolysis mass spectrometry study of *Corynebacterium* spp. *J Med Microbiol*, **30**, 137–49.

Hugh, R. 1981. *Pseudomonas maltophilia* sp. nov. nom. Rev. *Int J Syst Bacteriol*, **31**, 195.

Hutchinson, G.E. 1965. The niche: An abstractly inhabited hypervolume. In *The ecological theatre and evolutionary play*. New Haven: Yale University Press, 26–78.

ICSB Subcommittee on the taxonomy of Mollicutes, 1995. Revised minimal standards for description of new species of the class Mollicutes. *Int J Syst Bacteriol*, **45**, 605–12.

Jain, R., Rivera, M.C., et al. 2002. Horizontal gene transfer in microbial genome evolution. *Theor Popul Biol*, **61**, 489–95.

James, A.L. 1994. Enzymes in taxonomy and diagnostic bacteriology. In: Goodfellow, M. and O'Donnell, A.G. (eds), *Chemical methods in procaryotic systematics*. Chichester: John Wiley & Sons, 471–92.

Johnson, J.L. 1991. DNA reassociation experiments. In: Stackebrandt, E. and Goodfellow, M. (eds), *Nucleic acid techniques in bacterial systematics*. Chichester: John Wiley & Sons, 21–44.

Jones, D. 1989. Genetic methods. In: Williams, S.T., Sharp, M.E. and Holt, J.G. (eds), *Bergey's manual of systematic bacteriology*. Baltimore: Williams & Wilkins, 2310–12.

Katz, L.A. 2002. Lateral gene transfers and the evolution of eukaryotes: theories and data. *Int J Syst Evol Microbiol*, **52**, 1893–900.

Kirby, R. 1990. Evolutionary origin of aminoglycoside phosphotransferase resistance genes. *J Mol Evolution*, **30**, 489–92.

Kloos, W.E. and Schleifer, K.H. 1975. Simplified scheme for routine identification of human staphylococci. *J Clin Microbiol*, **1**, 82–91.

Kreig, N.R. 1988. Bacterial classification: an overview. *Can J Microbiol*, **34**, 536–40.

Labeda, D.P., Kurtzman, C.P. and Swezey, J.L. 1995. Taxonomic note: use of patent strains as type strains in the valid description of new bacterial taxa. *Int J Syst Bacteriol*, **45**, 868.

Lapage, S.P., Clark, W.A., et al. 1973. Proposed revision of the International Code of Nomenclature of Bacteria. *Int J Syst Bacteriol*, **23**, 83.

Lay, J.O. 2001. MALDI-TOF mass spectrmetry of bacteria. *Mass Spectrom Rev*, **20**, 172–94.

Lenski, R.E. and Travisano, M. 1994. Dynamics of adaptation and diversification: A 10,000-generation experiment with bacterial populations. *Proc Natl Acad Sci USA*, **91**, 6808–14.

Lévy-Frébault, V.V. and Portaels, F. 1992. Proposed minimal standards for the genus *Mycobacterium* and for description of new slowly growing *Mycobacterium* species. *Int J Syst Bacteriol*, **42**, 315–23.

Liesack, W. and Stackebrandt, E. 1992. Occurrence of novel groups of the domain Bacteria as revealed by analysis of genetic material isolated from an Australian terrestrial environment. *J Bacteriol*, **174**, 5072–8.

Lipps, J.H. 1993. *Fossil procaryotes and protests*. Boston: Blackwell Scientific Publications.

MacAdoo, T.O. 1993. Nomenclatural literacy. In: Goodfellow, M. and O'Donnell, A.G. (eds), *Handbook of new bacterial systematics*. London: Academic Press, 339–58.

Mackenzie, D.W.R. and Odds, F.C. 1991. Nonidentity and authentication of 2 major reference strains of Candida albicans. *J Med Vet Microbiol*, **29**, 225–61.

Magee, J.T. 1993a. Forsaking the tome: a worm's eye view of taxonomy. *J Med Microbiol*, **39**, 401–2.

Magee, J.T. 1993b. Whole organism fingerprinting. In: Goodfellow, M. and O'Donnell, A.G. (eds), *Handbook of new bacterial systematics*. London: Academic Press, 383–427.

Magee, J.T. 1994. Analytical fingerprinting methods. In: Goodfellow, M. and O'Donnell, A.G. (eds), *Chemical methods in prokaryotic systematics*. Chichester: John Wiley & Sons, 523–53.

Manafi, M., Kneifel, W. and Bascomb, S. 1991. Fluorogenic and chromogenic substrates used in bacterial diagnostics. *Microbiol Rev*, **55**, 335–48.

Manly, B.F.J. 1994. *Multivariate statistical methods: a primer*, 2nd edn. London: Chapman & Hall.

Margulis, L. and Fester, R. 1991. *Symbiosis as a source of evolutionary innovation speciation and morphogenesis*. Cambridge, MA: The MIT Press.

Martinez-Murcia, A.J., Benlock, S. and Collins, M.D. 1992. Phylogenetic interrelationships of members of the genera *Aeromonas* and *Plesiomonas* as determined by 16S ribosomal DNA sequencing: lack of congruence with results of DNA:DNA hybridization. *Int J Syst Bacteriol*, **42**, 412–21.

May, R.H. 1988. How many species are there on Earth? *Science*, **241**, 1441–9.

Maynard-Smith, J., Dowson, C.G. and Spratt, B.G. 1991. Localised sex in bacteria. *Nature*, **343**, 418–19.

Meissner, J.D. 1995. On the nature of papillomavirus hell. *J Infect Dis*, **72**, 95.

Murphy, F., Fauquet, C.M., et al. (eds) 1995. *Virus taxonomy. Sixth report of the International Committee on Taxonomy of Viruses*, supplement 10. Berlin: Springer-Verlag,

Murray, R.G.E. and Schleifer, K.H. 1994. Taxonomic notes: a proposal for recording the properties of putative taxa of prokaryotes. *Int J Syst Bacteriol*, **44**, 174–6.

Murray, R.G.E. and Stackebrandt, E. 1995. Taxonomic note: Implementation of the provisional status *Candidatus* for incompletely described prokaryotes. *Int J Syst Bacteriol*, **45**, 186.

Natarajan, M.R. and Oriel, P. 1991. Conjugal transfer of recombinant transposon Tn916 from *Escherichia coli* to *Bacillus stearothermophilus*. *Plasmid*, **26**, 67–73.

Nelson, W.H. (ed.) 1991. *Modern techniques for rapid microbiological analysis*. New York: VCH Publishers.

Oliveira, D.C., Tomasz, A. and de Lencastre, H. 2002. Secrets of success of a human pathogen: molecular evolution of pandemic clones of methicillin-resistant *Staphylococcus aureus*. *Lancet Infect Dis*, **2**, 315.

Orla-Jensen, S. 1919. *The lactic acid bacteria*. Copenhagen: Whøst & Son.

Palleroni, N.J. and Bradbury, J.F. 1993. Stenotrophomonas, a new bacterial genus for *Xanthomonas maltophilia* (Hugh0) Swings et al. 1983. *Int J Syst Bacteriol*, **43**, 606–9.

Paster, B.J., Boches, S.K., et al. 2001. Bacterial diversity in human subgingival plaque. *J Bacteriol*, **183**, 3770–83.

Peattie, D.A. 1979. Direct chemical method for sequencing RNA. *Proc Natl Acad Sci USA*, **76**, 1760–4.

Poole, R.K. 1994. Analysis of cytochromes. In: Goodfellow, M. and O'Donnell, A.G. (eds), *Chemical methods in prokaryotic systematics*. Chichester: John Wiley & Sons, 311–44.

Pot, B., Vandamme, P. and Kersters, K. 1994. Analysis of electrophoretic whole-organism protein fingerprints. In: Goodfellow, M. and O'Donnell, A.G. (eds), *Chemical methods in prokaryotic systematics*. Chichester: John Wiley & Sons, 493–521.

Priest, F.G. 1993. Systematics and ecology of *Bacillus*. In: Sonnenheim, A.L., Hoch, J.A. and Losick, R. (eds), *Bacillus subtilis and other gram-positive bacteria: biochemistry, physiology and molecular genetics*. Washington, DC: American Society for Microbiology, 3–16.

Ragan, M.A. and Charlebois, R.L. 2002. Distributional profiles of homologous open reading frames among bacterial phyla: implications for vertical and lateral transmission. *Int J Syst Evol Microbiol*, **52**, 777–87.

Rahn, O. 1929. Contributions to the classification of bacteria. *Zentbl Bakkt ParasitKde Abt II*, **78**, 8–27.

Ravin, A.W. 1963. Experimental approaches to the study of bacterial phylogeny. *American Naturalist*, **97**, 307–18.

Richards, W.R. 1994. Analysis of pigments: bacteriochlorophylls. In: Goodfellow, M. and O'Donnell, A.G. (eds), *Chemical methods in prokaryotic systematics*. Chichester: John Wiley & Sons, 345–401.

Rosselló, R., Garcia-Valdés, J., et al. 1991. Genotypic and phenotypic diversity of *Pseudomonas stutzeri*. *System Appl Microbiol*, **8**, 124–7.

Schleifer, K.H., Ludwig, W. and Amann, R. 1993. Nucleic acid probes. In: Goodfellow, M. and O'Donnell, A.G. (eds), *Handbook of new bacterial systematics*. London: Academic Press, 463–510.

Schopf, J.W. 1996. Are the oldest fossils cyanobacteria? In: Roberts, D.M., Sharp, P., et al. (eds), *Evolution of microbial life*. Cambridge: Cambridge University Press, 23–61.

Simpson, G.G. 1961. *Principles of animal taxonomy*. New York: Columbia University Press.

Skerman, V.B.D., McGowan, V. and Sneath, P.H.A. 1980. Approved lists of bacterial names. *Int J Syst Bacteriol*, **30**, 225–420.

Skerman, V.B.D., McGowan, V. and Sneath, P.H.A. 1989. *Approved lists of bacterial names*, (amended edition). Washington, DC: American Society for Microbiology.

Smit, Z., McCaig, A.E., et al. 2001. Species diversity of uncultured and cultured populations of soil and marine ammonia oxidising bacteria. *Microb Ecol*, **42**, 228–37.

Sneath, P.H.A. 1983. Distortions of taxonomic structure from incomplete data on a restricted set of reference strains. *J Gen Microbiol*, **129**, 1045–73.

Sneath, P.H.A. 1992. *International code of nomenclature of bacteria, 1990 revision*. Washington, DC: American Society for Microbiology.

Sneath, P.H.A. 1995. Taxonomic note: the potential of dead bacterial specimens for systematic studies. *Int J Syst Bacteriol*, **45**, 188.

Sneath, P.H.A. and Sokal, R.R. 1973. *Numerical taxonomy: the principles and practice of numerical classification*. San Francisco: Freeman.

Sokal, R.R. and Sneath, P.H.A. 1963. *Principles of numerical taxonomy*. San Francisco: Freeman.

Stackebrandt, E. and Goebel, B.M. 1994. Taxonomic note: a place for DNA:DNA reassociation and 16S rRNA sequence analysis in the present species definition in bacteriology. *Int J Syst Bacteriol*, **44**, 846–9.

Stackebrandt, E. and Liesack, W. 1993. Nucleic acids and classification. In: Goodfellow, M. and O'Donnell, A.G. (eds), *Handbook of new bacterial systematics*. London: Academic Press, 152–94.

Stackebrandt, E., Liesack, W. and Goebel, B.M. 1993. Bacterial diversity in a soil sample from a subtropical Australian environment as determined by 16S rDNA analysis. *FASEB J*, **7**, 232–6.

Stackebrandt, E., Frederiksen, W., et al. 2002. Report of the ad hoc committee for the re-evaluation of the species definition in bacteriology. *Int J Syst Evol Microbiol*, **52**, 1043–7.

Staley, J.T. and Krieg, N.R. 1984. Classification of prokaryotic organisms: an overview. In: Krieg, N.R., et al. (eds), *Bergey's manual of systematic bacteriology*. . Baltimore: Williams & Wilkins, 1602–3.

Swings, J., De Vos, P., et al. 1983. Transfer of *Pseudomonas maltophilia* Hugh 1981 to the genus *Xanthomonas* as *Xanthomonas maltophilia* (Hugh 1981) comb. nov. *Int J Syst Bacteriol*, **33**, 409–13.

Tamaoka, J. 1994. Determination of DNA base composition. In: Goodfellow, M. and O'Donnell, A.G. (eds), *Chemical methods in prokaryotic systematics*. Chichester: John Wiley & Sons, 463–9.

Templeton, A.R. 1989. The meaning of species and speciation: A genetic perspective. In: Otte, D. and Endler, J.A. (eds), *Speciation and its consequences*. Sunderland, MA: Sinauer Associates Inc., 3–27.

Ursing, J.B., Lior, H. and Owen, R.J. 1994. Proposal for minimal standards for describing new species of the family *Campylobacteriaceae*. *Int J Syst Bacteriol*, **44**, 842–5.

Ursing, J.B., Rosselló-Mora, R.A., et al. 1995. Taxonomic note: a pragmatic approach to the nomenclature of phenotypically similar genomic groups. *Int J Syst Bacteriol*, **45**, 604.

Van Niel, C.B. 1946. The classification and natural relationships of bacteria. *Cold Spring Harb Symp Quant Biol*, **11**, 285–301.

Van Valen, L. 1976. Ecological species, multispecies and oaks. *Taxon*, **25**, 233–9.

Vauterin, L., Swings, J. and Kersters, K. 1993. Protein electrophoresis and classification. In: Goodfellow, M. and O'Donnell, A.G. (eds), *Handbook of new bacterial systematics*. London: Academic Press, 251–80.

Vogel, U. and Frosch, M. 2002. The genus *Neisseria*: population structure, genome plasticity, and the evolution of pathogenicity. *Curr Top Microbiol Immunol*, **264**, 23–45.

Wayne, L.G., Brenner, D.J., et al. 1987. Report of the ad hoc committee on reconciliation of approaches to bacterial systematics. *Int J Syst Bacteriol*, **37**, 463-, 4.

Willcox, W.R., Lapage, S.P. and Holmes, B. 1980. A review of numerical methods in bacterial identification. *Antonie van Leeuwenhoek*, **46**, 233–99.

Woese, C.R. 1987. Bacterial evolution. *Microbiol Rev*, **51**, 221–71.

Woese, C.R. 1992. There must be a prokaryote somewhere: microbiology's search for itself. *Microbiol Rev*, **58**, 1–9.

Woese, C.R. 2000. Interpreting the universal genetic tree. *Proc Natl Acad Sci USA*, **97**, 8392–6.

Woese, C.R., Kandler, O. and Wheelis, M.L. 1990. Towards a natural system of organisms: proposal for the domains Archaea, Bacteria and Eukarya. *Proc Natl Acad Sci USA*, **87**, 4576–9.

Woodford, N. 2001. Epidemiology of the genetic elements responsible for acquired glycopeptide resistance in enterococci. *Microb Drug Resist*, **7**, 229–36.

Yabuuchi, E., Kosako, Y., et al. 1992. Proposal of *Burkholderia* gen. nov. and transfer of seven species of the genus *Pseudomonas* homology group II to the new genus, with the type species *Burkholderia cepacia* (Palleroni and Holmes 1981) comb. Nov. *Microbiol Immunol*, **36**, 1251–75.

3

Bacterial growth and metabolism

JAMES P. SHAPLEIGH

GROWTH OF BACTERIA

Bacterial nutrition and culture media

A minimal medium is one that contains the simplest chemical composition needed to support the growth of an organism. Some bacteria can grow quite well on a minimal medium consisting of distilled water, a carbon and energy source such as D-glucose, and some inorganic salts, for example sodium and potassium phosphates, ammonium sulfate, magnesium sulfate, and ferrous sulfate. Bacteria able to grow in such a medium must be able gain metabolic energy by the catabolism of glucose and synthesize all organic molecules needed for growth from carbon degradation products of the glucose and the inorganic salts. This carbohydrate–salts medium is chemically defined, in that the exact chemical composition is known and can be reproduced. In stating that such a medium is defined, we ignore the requirement for trace elements such as copper, zinc, cobalt, etc., which are required in only minute amounts and are provided as contaminants in the other ingredients.

If an organism lacks the enzymatic capability to synthesize any organic biosynthetic compounds from such a medium, then growth of that organism cannot occur using that medium unless such compounds are provided. If a mutant of *Escherichia coli* loses the ability to synthesize an amino acid, then that amino acid has become a required growth factor and must be added to the medium or growth of that mutant cannot occur. Many bacteria, especially those whose natural environment contains many complex organic compounds, such as pathogens, are unable to synthesize a wide range of biosynthetic intermediates, such as amino acids, nucleosides, vitamins, etc., and these must be supplied to permit growth. Such a growth medium may be complicated and difficult to assemble, but it is still chemically defined in that the exact chemical composition is known. Because a chemically defined growth medium is reproducible, it is often preferred for studies on the metabolism of bacteria.

Complex and rich bacterial growth media such as those made from nutrient broth (beef extract and peptone) or l-broth (peptone and yeast extract) will support the growth of many bacteria and are convenient to use, but such media are undefined in that the exact chemical composition is not known. Complex and undefined media of these types are often used for isolation and growth of bacteria but it should be realized that the exact composition of such media may vary from batch to batch.

Microbiologists have been ingenious in devising selective media that are designed for the isolation from natural sources of a particular type of microorganism. Such media are designed to facilitate growth of organisms targeted for isolation while discouraging, as far as possible, growth of contaminating organisms. Enrichment cultures are not only designed to possess the proper media for isolation of a desired organism, but growth conditions such as temperature, can be designed to aid in the isolation. A wide variety of media have also been designed to aid in the rapid isolation and identification of microorganisms, especially pathogens.

Growth in liquid culture

A growing bacterial cell increases in size until it separates into two daughter cells. If growth is balanced and the replication of the macromolecules has been carefully controlled, all of the macromolecules of the mother cell must have doubled in mass and each daughter cell is a duplicate of the mother cell. At that time, a cell generation or division cycle has taken place. The growth of bacterial cells may become temporarily unbalanced as a result of environmental stress, but regulatory systems will normally adjust growth rates so that synthesis of macromolecules once again becomes balanced. Continued unbalanced growth can lead to cell death.

Measurement of cell numbers is often performed by viable cell count. With the pour plate method, various dilutions of cells are mixed into tubes of a melted agar-growth medium that has been cooled to just above the temperature of solidification. The tubes are poured into Petri dishes and, after incubation, the number of colonies is taken as the number of viable cells in the original dilution. An alternate method is the spread plate method, where samples of 0.1 or 0.2 ml of the appropriate cell dilutions are spread over the surface of a solidified agar-growth medium plate. The spread plate method is reported to give slightly higher colony counts and has the advantage that individual colonies can be easily sampled for further testing.

Such information as the viable cell count or number is valuable to many microbiologists, especially those concerned with infectious diseases or food poisoning, because these techniques quantify cells capable of growth. However, such techniques really measure colony-forming units, not necessarily individual cells, as clusters of cells in packets or chains will produce a single colony. Also, it takes many generations of growth for a single cell to produce a colony visible to the unaided human eye. With this technique, a bacterial cell could produce hundreds of thousands of descendants and still be classified as nonviable. A direct cell count measures all of the cells present, including those not capable of growth. It is possible to count the total number of cells per unit volume, both living and dead, microscopically, using a special slide. Cell count as well as mass can also be determined by impedance counting, where cells passing through a narrow chamber disrupt the flow of an electrical current.

A problem with using viable or total cell counts to measure bacterial growth is the inability to detect increases in cell mass during the division cycle. Growing bacterial cells must increase their mass before division and an increase in cell mass must be considered as growth, even though the cell number may not have increased. Cell mass can be determined directly by measurement of cell weight after drying a sample of cells in an oven. Cell mass is often estimated by determining the light scattering of a suspension of bacterial cells with a standard curve to relate light scattering to mass.

The amount of bacterial growth that has taken place in a culture can be estimated by a variety of methods that measure a result of growth rather than cell number. The disappearance of a substrate such as glucose can be determined, or the utilization of oxygen during respiration. A product of growth such as cell deoxyribonucleic acid (DNA), ribonucleic acid (RNA), or protein can be determined, or the amount of acid produced during fermentation might be used as an estimate of cell growth.

The time for a division cycle to be completed is the time for a generation of growth to take place and is known as the generation time. If growth of a bacterial culture is synchronized, all cells are in exactly the same phase of the division cycle, and all will divide simultaneously. With synchronized cell division, the total cell number immediately increases from the initial number (N_0) to $2 \times N_0$. If the cells remain synchronized, another generation will simultaneously raise the cell number to $2 \times 2N_0$. Techniques that have been used for synchronizing cell cultures involve the germination of spores, filtration or centrifugation of cells to obtain a constant size cell, and prevention of initiation of DNA replication by amino-acid starvation. Special mutants have also been used, such as those with temperature-sensitive mutations in genes required for initiation of DNA replication, to prevent cell growth at the restrictive temperature. Upon change to permissive temperature, DNA replication in all cells in the culture will begin simultaneously and all cells will enter the same phase of the division cycle. Synchronization of cell growth can be useful in studying cell chemistry at different phases of the division cycle, but synchronization usually lasts for only a few generations.

CELL GENERATIONS AND GROWTH RATES

Because the total cell number doubles each division cycle or generation (g), the number of cells produced by growth (N) will be equal to the original number (N_0) times 2 to the power g. Therefore:

$$N = N_o \times 2^g. \tag{4.1}$$

Taking the $\log_{10}$ of both sides of the equation:

$$\mathrm{Log}_{10}N = \mathrm{Log}_{10}N_o + 0.301\, g. \tag{4.2}$$

Then to solve for g the equation becomes:

$$\frac{\text{Log}_{10}N - \text{Log}_{10}N_o}{0.301} = g. \qquad (4.3)$$

An important value that can be determined from these calculations is t, the generation time. Generation time is the time for each generation to occur so it equals the time elapsed (T) per generation (g), or $t = T/g$. The generation time is also the time required for the cell population to double.

The rate of increase of bacterial number (or mass) is exponential because the population doubles with each generation. If the bacteria are growing at a constant rate, then a plot of cell number or mass against time results in an exponential curve, as shown in Figure 3.1a. The plot of the $\log_{10}$ of bacterial number/mass against time will give a linear relationship, as shown in Figure 3.1b, provided the cells are growing at a constant rate. This is the conventional method of plotting bacterial growth.

The rate of increase in total cell number during growth is the initial cell number multiplied by the specific growth rate constant (μ):

$$\frac{dN}{dT} = \mu N. \qquad (4.4)$$

The value of the specific growth rate constant can be derived from the plot of log N versus t, where its value will be given by the slope of the linear plot. The specific growth rate constant is not the same as t but represents the growth rate at a specific instant.

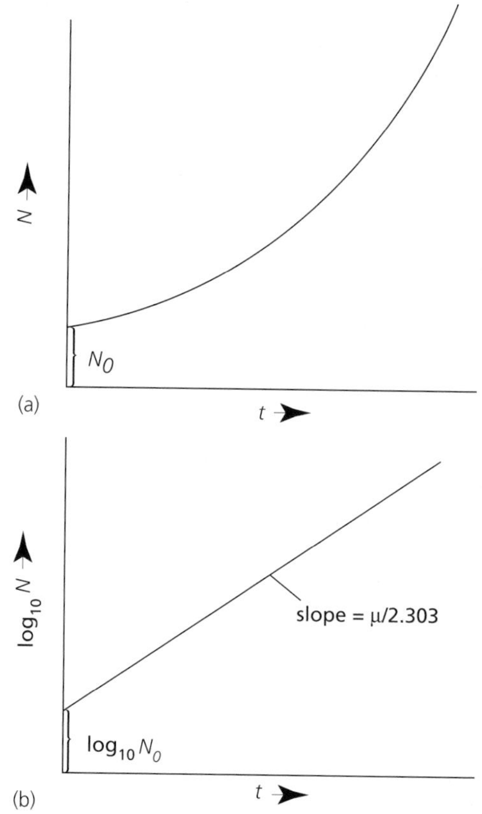

THE GROWTH PHASES AND THE GROWTH CURVE

The lag phase

When an inoculum of cells is transferred to a growth medium, a delay may take place before measurable growth occurs. This lag time is a period where cells are adjusting to new growth conditions, and growth may be temporarily unbalanced during this growth phase. The greater the change in growth conditions, experienced by the cells, the longer the predicted lag phase. Because cells must increase their size before cell division can take place, the exact length of the lag period may depend upon whether cell number or cell mass is being measured. A typical bacterial growth curve, with lag and exponential and stationary phases, is shown in Figure 3.2.

The exponential (logarithmic) growth phase

Once the cells have adjusted their regulatory systems to producing maximum growth under the new conditions, the culture enters the exponential or logarithmic growth phase where cells have approximately the same genera-

Figure 3.1 *Theoretical growth curves illustrating instantaneous exponential growth in a bacterial batch culture. Total cell number (N, i.e. number of bacterial cells per ml) plotted against time (t). N_o, inoculum cell concentration; μ, specific growth rate*

tion time. Growth is normally not synchronized and, at any moment, cells will be at different stages of the division cycle. In batch culture the growth rate eventually

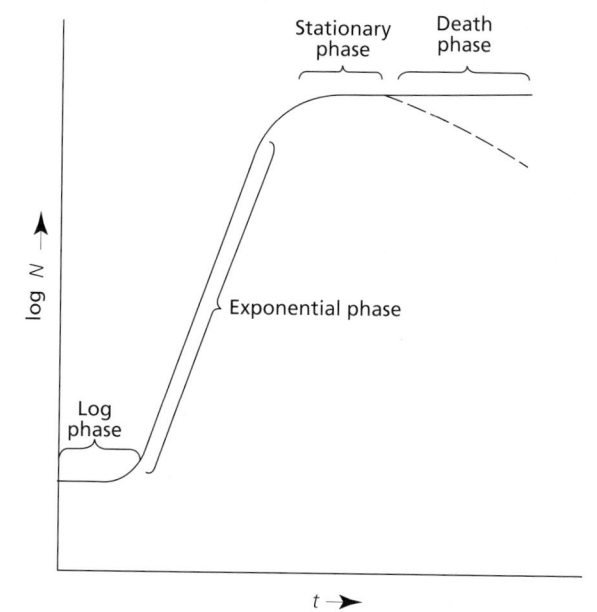

Figure 3.2 *Idealized growth curve of a bacterial batch culture. Unbroken line, total cell count; dashed line, viable cell count*

must begin to decrease as the availability of some essential nutrient becomes limiting or the accumulation of some product becomes toxic.

The stationary phase

Eventually the stationary phase is reached, a period where measurable growth ceases. Note that if a new viable cell was formed for each cell that lost viability, or if loss and increase in cell mass were balanced, then there would be no measurable increase in cell growth.

If *E. coli* cells are growing on a minimal salts medium containing a carbon and energy source and the energy source becomes exhausted, the pH will remain constant and growth and nitrogen utilization will both stop. However, under conditions where the nitrogen becomes depleted and the energy source is in excess, the cells will continue to use the energy source, even if the energy gained from catabolism is discarded. If possible, the bacteria will increase their internal supply of reserve food material under these conditions, while denying surrounding microorganisms a potential energy source. Traditionally, resting cells of bacteria are those incubated without any added nitrogen source. Such cells are still able to induce enzymes and catabolize carbohydrates, and these types of resting cells have been used for many metabolic studies.

The death phase

Some bacteria retain viability upon reaching the stationary phase, while others rapidly lose viability and enter a death phase. A variety of circumstances, including loss of energy and production of autolytic enzymes, may cause bacteria to lose viability. Reserve food sources that may be accumulated by bacteria, such as poly-β-hydroxybutyric acid and glycogen, are usually used quickly when the cells are starving but once such reserve food materials are exhausted, cellular RNA may serve as its major source of energy. Ribosomal RNA degradation may increase during starvation conditions. For example, *E. coli* can degrade 20–30 percent of its RNA within 4 h of starvation. The degradation of ribosomes will lead to the eventual death of the cell because cells lacking ribosomes are incapable of synthesizing protein. Starvation also increases the rate of protein degradation and peptidase-deficient mutants of *E. coli* have been shown to have poor survival during starvation for carbon and nitrogen. The amino acids produced by protein degradation can be used for the synthesis of new proteins to aid in survival under stress conditions.

A drop in internal energy levels may prevent transport of nutrients across the cytoplasmic membrane so that, upon transfer to recovery medium, the cell is unable to uptake nutrients. The energy charge of a cell has been reported as the total adenosine triphosphate (ATP) plus half of the adenosine diphosphate (ADP) over the totals of the cellular ATP, ADP, and adenosine monophosphate (AMP). Because two ADP are capable of forming one ATP and one AMP, the amount of ATP that can be obtained from ADP is equal to one-half of the ADP present. The equation

$$\text{energy charge} = \frac{\text{ATP} + \frac{1}{2}\,\text{ADP}}{\text{ATP} + \text{ADP} + \text{AMP}} \tag{4.5}$$

gives the ratio of available ATP to the total adenine nucleotides present. If all of these nucleotides were in the form of the triphosphate, the energy charge would be one. For the normal growth and metabolism of bacterial cells, the energy charge must be maintained in the range of 0.80–0.95; *E. coli* has been reported to grow only when the energy charge is over 0.8, it will maintain its viability when the energy charge is between 0.5 and 0.8, but it will die when the value is under 0.5. There is a certain amount of maintenance energy required just for cells to repair continuing damage but this may vary for different organisms. In experiments with *E. coli*, it was found that once the glucose was exhausted the organism maintained its energy charge at 0.8 by using acetate that had been excreted during growth on glucose. If no other substrates were present, then the capacity for protein synthesis and enzyme induction was lost along with cell viability.

Because a starving cell must maintain a certain level of metabolism and energy to remain viable, many bacteria activate special regulatory systems in response to starvation, regulatory systems that allow them to express 'starvation' genes (Matin 1992). These new proteins improve the cell's ability to use low levels of nutrients, regulate the degradation of their own polymers for carbon and energy, and aid the cells in surviving. Many of the genes encoding for these proteins require the nucleotide cyclic AMP for their induction. The starvation stress regulatory system may overlap with other stress regulatory systems because starved *E. coli* cells have been shown to acquire resistance to heat stress.

In *E. coli*, the nucleotide guanine triphosphate has an essential role in the regulatory adjustments made by cells starved for amino acids. This adjustment to amino-acid starvation is known as the stringent response. In cells with insufficient amino acids, the stringent response will cause a rapid reduction in the synthesis of ribosomal and transfer RNA.

CHANGING GROWTH RATES

The rate of growth of a bacterial strain may be shifted up or down with changes in growth conditions, such as medium composition or temperature. Shift-up experiments refer to a change in growth conditions with an increase in the growth rate of a bacterium. As an example, an improvement in nutrient conditions can lead to an increase in the internal amino acid concentration, thus resulting in an increase in rate of RNA synthesis, an increase in the rate of protein synthesis, and a more frequent initiation of DNA synthesis. Faster-growing cells tend to be larger, with an increase in the amount of the protein-synthesizing machinery. While the regulatory systems are adjusting to the new conditions, the cell length may continue to increase and the cell may become unusually long. Eventually, the cell length decreases to reach the proper size for the new growth rate (Cooper 1991).

By contrast, a transfer of cells from a complex undefined medium, such as a nutrient broth, to a minimal-salts medium can result in a shift-down or decrease in growth rate caused by a sudden decrease in internal concentration of nutrients, such as amino acids. A decrease in the internal concentration of amino acids leads to a stringent response with a decline in the rate of RNA synthesis, a decline in the rate of protein synthesis, and a less frequent initiation of DNA replication. In fact, DNA synthesis may halt altogether, until activation of biosynthetic pathways can produce the required building block compounds to permit cell growth.

Continuous culture of bacterial cells

The chemostat is a name given to devices that maintain the growth of bacteria in a constant exponential growth rate. A typical chemostat possesses a container containing sterile growth medium that is transferred into the growth chamber at a controlled rate. The growth chamber has an overflow device so that the entrance of new medium results in the overflow and exit of an equal volume of used medium, as shown in Figure 3.3.

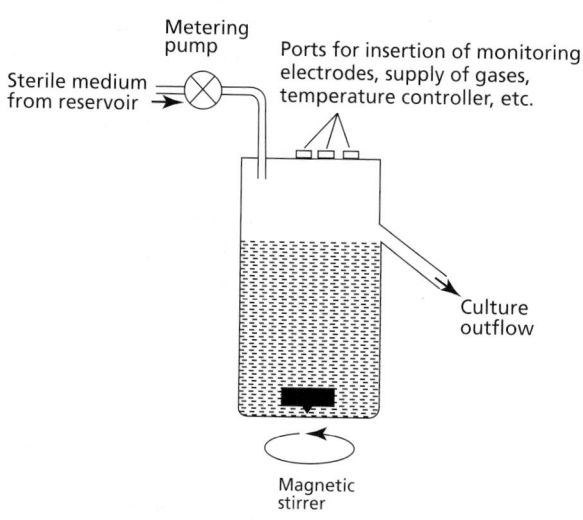

Figure 3.3 *Principal features of a continuous-flow culture apparatus*

CHEMOSTAT GROWTH RATES

The medium in the growth chamber of a chemostat can be inoculated with a bacterium to give an initial cell concentration of (N). If growth takes place, the rate of increase in cell mass against time can be represented by $dN/dt = \mu N$, where N is the cell number per unit volume and μ stands for the growth rate constant. The rate of increase in cell number can be represented as μN. If the growth rate constant is a positive number, the cell mass in the chemostat will be increasing. However, new medium is introduced into the chemostat at a constant flow rate and old medium is overflowing out of the chemostat at the same rate, removing cells. If the flow rate into the chamber is F volume per unit time period, and the volume of the growth chamber is V, then the fraction of the culture medium that overflows and is replaced each unit time period is represented by F/V (also known as the dilution rate). Therefore, the total amount of cells lost each time period by overflow would be $N \times F/V$. The real rate of change of cell number in the chemostat is the rate of increase resulting from growth (μN) minus the rate of decrease as a result of washout ($N \times F/V$). Therefore, the rate of change in the chemostat is $\mu N - N \times F/V$, or $N (\mu - F/V)$.

If F/V is greater than μ, the cells are being washed out of the chemostat faster than they can grow and eventually all of the cells will leave the chemostat. If F/V is smaller than μ, the number of cells will increase. However, as the cell mass increases, there are more cells competing for a constant food supply, because the flow rate into the growth chamber is constant and this competition causes a decrease in the growth rate constant. Eventually μ will decrease until it equals F/V and then a steady state will exist with μ held constant. In such a manner, by controlling the rate of flow of new medium into the chemostat, the experimenter can control the growth rate of the cells in the chemostat.

Chemostats are usually established with the growth rate limited by a single known factor, such as a limiting supply of a required growth factor or the carbon and energy source. Chemostats are very useful for obtaining mutants, because once a growth rate is established there is strong evolutionary pressure for selection of mutants that have increased their growth rates under the competitive chemostat conditions. However, chemostat experiments must often be terminated because of the selection of 'sticky' mutants that cling to the culture-vessel surface and are not washed out.

CHEMOSTAT CELL POPULATIONS

If the concentration of the limiting nutrient in the reserve chamber is designated as S_r and the unused amount of that nutrient that washes out of the chemostat is S, then the amount of nutrient used by the cells in the growth chamber is $S_r - S$. If Q represents the amount of that limiting nutrient needed to make one bacterial cell, then the total population of cells will be equal to $S_r/Q - S/Q$. However, S is usually close to zero and then the equation becomes S_r/Q, which represents the total population of cells. The growth rate in the chemostat is controlled by the rate of flow of new medium into the chemostat, but the total population of cells present is determined by the concentration of the limiting substrate.

Growth on solid surfaces

Good growth for a bacterium such as *E. coli*, on rich medium, under aerobic conditions, may range from 1×10^{10} to 5×10^{10} cells per milliliter. The final yield of cells per milliliter in liquid culture can be considerably increased if the cells are placed within a dialysis sack that is suspended in a large vessel of growth medium. The dialysis sack permits waste products to diffuse out of the sack, away from the cells, while fresh nutrients can diffuse into the sac. Cells growing on an agar surface can form colonies with high concentration of cells per unit volume for a similar reason. The agar permits nutrients to diffuse toward the colony, while waste products diffuse away.

The nutritional state of bacteria may also influence their adhesion to solid surfaces. Many bacteria have increased adhesion under starvation conditions but some bacteria actually display the opposite effect. Colonization of surfaces by bacteria may be influenced by motility and cell-surface hydrophobicity as well as by nutritional status (James et al. 1995). In nature, masses of cells commonly exist in films or mats and the cells in such masses may be able to regulate overall behavior by communication systems. Bacteria may grow in a manner that maximize cell-to-cell contact and, perhaps, cellular communication. For this reason, some microbiologists view the bacterial colony as a multicellular 'organism,' an organized population of communicating cells (Shapiro 1995).

STRUCTURE AND ORGANIZATION OF THE BACTERIAL CELL

General composition of bacteria

In order for bacterial cells to grow, they must be provided with all of the chemical elements needed for the synthesis of new cellular material. Once, organisms that could use simple molecules to satisfy their nutritional requirements, such as carbon dioxide (CO_2) as the sole carbon source, were thought of as 'simple' organisms. We now understand that the simpler the nutrients, the more complex the biochemical reactions required by the cell to convert those nutrients into complex cellular material.

The water content of bacterial cells can vary from 75 to 90 percent of the total weight. The dry weight, after removal of water, is composed of about 50 percent carbon, about 10 percent hydrogen, and about 20 percent oxygen. The nitrogen content of the dry weight can vary from 8 to 15 percent, sulfur from 0.1 to 1.0 percent, and phosphorus from 1 to 6 percent. The iron content is also variable and is highest in cells with heme proteins. Other elements found are potassium, calcium, magnesium, chlorine, sodium, zinc, cobalt, molybdenum, manganese, and copper. These and other trace elements total about 0.3 percent of the dry weight.

Cellular polymers

Most of the organic material in the bacterial cell is in the form of large polymers. However, the cell cytoplasm also contains a 'soluble pool' of smaller, unpolymerized organic molecules such as amino acids, nucleotides, carbohydrates, and fatty acids, as well as degradation products from the breakdown of any compounds providing carbon and energy sources for growth. Large organic polymers of the cell include those made of amino acids (proteins), those made of ribonucleotides (RNA), those made of DNA, and those composed of carbohydrates and lipids.

Detailed studies of polymer composition have been carried out in only a few bacteria. In *E. coli* protein represents about 50 percent of the dry weight of a typical bacterial cell. RNA represents about 20 percent of the dry weight of a bacterial cell but can vary depending upon growth conditions. About 90 percent of the RNA is ribosomal RNA (rRNA), which consists of three species, the 23S rRNA, the 16S rRNA, and the 5S rRNA. The transfer RNA (tRNA) represents about 9 percent of the total RNA, with 60 types of tRNA. The messenger RNA (mRNA) represents about 0.2–1 percent of the total RNA. There may be one to four molecules of DNA, depending upon the growth rate, representing about 3.3 percent of the dry weight (Neidhardt 1987). Carbohydrate polymers are present in the cell wall and may also be stored in the cytoplasm as food materials such as glycogen. Lipid material can vary from 1 to 8 percent of the cell dry weight, depending upon growth conditions, and can be found in the cell membrane and wall. Some bacteria, excluding *E. coli*, can store lipid-like substances as reserve food material, often in the form of polyhydroxybutyric acid granules in the cytoplasm. For the synthesis of this storage material, two acetyl-coenzyme A (acetyl-CoA) molecules are converted to acetoacetyl-CoA; this is changed to D-(-)-3-hydroxybutyryl-CoA, which is polymerized to form polyhydroxybutyric acid (Steinbuchel et al. 1991).

Eubacterial cell walls

PEPTIDOGLYCAN MATERIAL

The backbone structure of the eubacterial cell wall is the peptidoglycan material. Peptidoglycan is a polymer of N-acetyl glucosamine and N-acetyl muramic acid, linked together by β-1,4 or 1,6 alternating units and about

12 carbohydrates long. The disaccharide chains are linked together by polypeptide chains, usually from three to eight amino acids long and containing some D-amino acids such as D-alanine and D-glutamic acid. The polypeptide chain is attached by a peptide bond to the carboxyl group of the muramic acid. It normally contains one diamino acid, such as lysine or diaminopimelic acid but not both, and a short peptide bridge joining the free amino group of the diamino acid to the terminal carboxyl group of a similar polypeptide branch from a different disaccharide strand. In such a manner the backbone disaccharide strands are joined together. The cross-links in the peptidoglycan material form a 'web' capable of supporting stress in any direction. During synthesis of the peptidoglycan material, a disaccharide of N-acetylglucosamine and N-acetylmuramic acid is first formed and the amino-acid bridges are then attached with an extra D-alanine on the end of the peptide chain. Teichoic acid and peptide-bridge material may also be added to the unit. To build new wall material, this disaccharide–peptide unit is inserted into the existing cell wall in the space between the cytoplasmic membrane and the cell wall. A membrane-linked transpeptidase removes the final D-alanine unit from one peptide chain or peptide bridge and installs in its place the extra amino group on the diamino acid of another chain. The presence of D-amino acids in the cell wall is believed to give protection to external proteolytic enzymes. Enzymes that attack the bacterial cell wall are termed lysins, and autolysins are lysins, produced by the bacterium itself. Some lysins attack the peptidoglycan backbone, whereas others attack the peptide portion or the point where the peptide chain joins the glycan strands. Lysozyme is a lysin that cleaves at the β-1,4 linkage of N-acetylglucosamine.

GRAM-POSITIVE CELL WALLS

Gram-positive cell walls contain peptidoglycan and may also contain teichoic acid, polymers of glycerol or ribitol, usually about 10 to 50 units long, and linked together through phosphate bonds. Most teichoic acids are covalently linked to the peptidoglycan. Side groups on the teichoic acids can be carbohydrates such as glucose and N-acetylglucosamine and/or amino acids such as D-alanine. Teichuronic acid, a polymer consisting of alternating glucuronic acid and N-acetylgalactosamine, is sometimes found in the cell walls, especially when phosphate is limiting. Lipoteichoic acids are polymers of glycerophosphate covalently linked to a lipid molecule. The bound lipid allows the molecule to be anchored in the lipid bilayer of the cell membrane. However, the glycerophosphate polymer is not linked to the peptidoglycan.

GRAM-NEGATIVE CELL WALLS

The gram-negative cell wall is more complex than that of the gram-positive bacteria. One layer contains peptidoglycan material; the other layer is a convoluted outer membrane consisting of protein and at least two types of lipids: lipopolysaccharides (LPS) and phospholipids. The peptidoglycan material makes up only about 12 percent of the total cell wall material in E. coli and the four amino acids usually found in the peptidoglycan material of gram-negative cells are L-alanine, D-glutamic acid, meso-diaminopimelic acid, and D-alanine. LPS consists of three parts: an inner hydrophobic lipid A region, an outer hydrophilic O antigen polysaccharide region, and a core polysaccharide region that connects the A and O regions. The phospholipid and LPS form an asymmetric bilayer, with the phospholipid forming the inner layer and LPS forming the outer layer.

Proteins found in this outer wall include murine lipoproteins. The lipid end of this protein is embedded in the inner face of the bilayer formed by lipid and LPS. The protein end can be bound to the peptidoglycan. The main function of murine lipoprotein may be to help stabilize the outer membrane–peptidoglycan complex. Other important proteins found in the outer wall are porins, which form channels to allow the diffusion of hydrophilic molecules. Porins are sometimes grouped into three types (Nikaido 1992). Type 1 are nonspecific porins that distinguish substrates on the basis of size. A zone of constriction arising from the folding of a stretch of residues into the central pore allows porins to distinguish substrates on the basis of size (Cowan et al. 1992). Examples of this type are OmpF and OmpC of E. coli. Type 2 porins are related to type 1; both only permit diffusion of small substrates, but the type 2 porins preferentially transport certain substrates. The LamB porin of E. coli, which preferentially transports maltose and maltodextrins, is representative of this type of porin (Schirmer et al. 1995). LamB is also the binding site of the lambda phage. The third types of porins are TonB-dependent porins. TonB-dependent porins specifically transport certain substrates such as vitamin B_{12} or siderophores. These porins are unique in that transport is energy dependent. The energy is supplied through the TonB protein, which resides in both the cytoplasmic membrane and the outer cell wall (Moeck and Coulton 1998).

Between the outer and cytoplasmic membrane of gram-negative organisms is the periplasmic space, which contains many proteins, including transport-binding proteins, scavenging enzymes, detoxifying enzymes, and others. Periplasmic enzymes are synthesized in the cytoplasm and must be transported across the cytoplasmic membrane. Many periplasmic proteins contain a signal sequence of about 20–40 amino acid residues long at their N-terminus (Pugsley 1993). This sequence contains a positive region followed by a hydrophobic region and terminates in a peptidase cleavage site. The signal sequence targets the protein for transport via the Sec transport system (also known as the general secretory pathway). Once transport is completed, the signal sequence is cleaved from the transported protein.

A Sec-independent system for export of periplasmic proteins has been identified in *E. coli* (Weiner et al. 1998). This system recognizes proteins with a 'twin-arginine' motif at their N-termini. Proteins transported by this system appear to be mainly redox proteins containing cofactors. Related proteins are found in a wide range of bacteria, suggesting that this is an important transport system for periplasmic proteins.

Archaea cell walls

The third domain of life, the archaea, have distinctively different cell-wall polymers from the eubacteria (Kandler and Konig 1998). They do not possess the peptidoglycan polymer but may possess a slightly different polymer termed pseudopeptidoglycan. In the genus *Methanobacterium thermoautotrophicum*, this polymer contains N-acetyltalosaminuronic acid and N-acetylglucosamine and only L-amino acids, such as L-lysine, L-alanine, and L-glutamate. The cell wall of *M. ruminantium* has L-threonine in place of L-alanine, and N-acetylgalactosamine in place of N-acetylglucosamine. The synthesis of this polymer is unique to bacteria containing pseudopeptidoglycan. The cell walls in the genus *Methanosarcinaceae* are made of an acid heteropolysaccharide that contains galactosamine, neutral sugars, and uronic acids. No muramic acid, glucosamine, glutamic acid, or other amino acids typical of peptidoglycan are found. Most archaea also contain an S-layer (see below) as a cell envelope. Because the cell walls of archaea are structurally distinct from eubacterial walls, almost all of the antibiotics that target synthesis of components of the wall in eubacteria are ineffective against archaea; for example, β-lactam antibiotics and vancomycin do not inhibit growth of archaea.

Polymers external to the cell wall

Capsule or slime layers may exist as polymers outside the cell wall. They are usually made up of amino acids, carbohydrates, or both; for example, the polypeptide capsule of *Bacillus anthracis* is mostly D-glutamic acid, whereas the *E. coli* K-12 capsule consists of D-glucose, L-fucose, galactose, hexuronic acid, acetate, and pyruvate. Often the extracellular material is not tightly linked to the cell wall. Such loosely adhering material is referred to as a slime capsule. In pathogens surface polysaccharides are an important surface antigen. In *E. coli* the surface polysaccharide gives rise to the K antigens, of which there are more than 80 serotypes (Whitfield and Roberts 1999). Capsules are also important virulence factors. The capsule of *B. anthracis* is critical for cell survival during infection. Strong bonding forces rarely link capsule polymers to one another, although highly acidic types can be cross-linked by divergent metals such as Mg^{2+} or Ca^{2+}.

Another extracellular assemblage is known as the S-layer. The S-layers are made up of protein subunits that are sometimes glycosylated (Sleytr and Beveridge 1999). The unique feature of S-layers is that the subunits spontaneously form higher-order structures with crystalline lattices. S-layers are found in both eubacteria and archaea.

External appendages

Many bacteria contain appendages that extend beyond the outer cell wall. One of these is the bacterial flagellum. Flagella are used to propel bacteria in directions favorable for survival and growth. The flagellar structure can be broken down into three basic units, termed the basal body, the hook, and the filament (Manson et al. 1998). The basal body anchors the flagellum into the cytoplasmic membrane and contains numerous proteins required for assembly and function. The hook is a short curved region that links the basal body to the filament. Attached to the hook is a helical filament made up primarily of a single protein termed flagellin. The filament rotation which propels the cell forward, is driven by the ion gradient across the cytoplasmic membrane. Because the filament is helically shaped its rotation can propel the cell.

Flagella do not just randomly propel the cell through its environment. Cells have external receptors that detect environmental stimuli and then, using a phosphorelay system, this information modulates rotation of the flagellum. In *E. coli* counterclockwise rotation of the flagella produces forward movement termed 'runs,' while clockwise rotation causes cellular 'tumbling.' Stimuli that prolong runs are called attractants and stimuli that shorten the length of a swimming interval are called repellents. Changes in attractant or repellent levels cause changes in flagella rotation and result in movement in favorable directions. Other bacteria, for example *Rhodobacter sphaeroides*, can rotate their flagella in only one direction (Armitage and Macnab 1987). So, instead of runs and tumbles *R. sphaeroides* cells run and then stop and randomly re-orient. There are similarities in the signaling pathways in *E. coli* and *R. sphaeroides*, but the two pathways are not identical.

Many bacteria also possess external protein fibers known as pili or fimbriae composed of protein (pilin). The pili of pathogenic bacteria contain specific proteins that allow the bacteria to attach to and colonize host tissues. Binding of pili to external surfaces may be a key factor in activating expression of genes required for infection in pathogenic *E. coli* (Zhang and Normark 1996). There are other types of pili, referred to as sex pili, that are used by bacteria to attach to each other and exchange DNA.

Cytoplasmic membrane

The cytoplasmic membrane is made up of both lipid and protein and can constitute about 10 percent of the dry weight of the bacterial cell. In most cases the lipid is primarily phospholipid. The fatty-acid components of the phospholipid can be varied to maintain the appropriate fluidity of the membrane. Cyclopropane fatty acid is also a common modification of the lipids of the cytoplasmic membrane. It has been demonstrated that increases in content of cycloporane fatty acids render cells of *E. coli* more sensitive to acidic pH (Chang and Cronan 1999).

At least 200 different kinds of protein representing up to 6 to 9 percent of all the cellular proteins are located in the cytoplasmic membrane. Some of these proteins are cytochromes, dehydrogenases, NADH-oxidases, enzymes involved in cell-wall synthesis, enzymes involved in electron transport and oxidative phosphorylation, and proteins involved in the transport of specific molecules through the membrane. Typically, membrane-associated proteins are divided into two classes, integral and peripheral membrane proteins. Peripheral proteins associate with the surface of the membrane through ionic interactions. Therefore, they can be removed by treatment of the membrane with high concentrations of salt. Integral proteins are embedded in the lipid bilayer and can be dissociated from the bilayer only by the use of detergents. Integral proteins must contain surface-exposed hydrophobic regions to interact with the hydrophobic regions of the bilayer. In many cases these hydrophobic regions form α-helices and are termed membrane-spanning helices. In *E. coli* integral membrane proteins are targeted to the membrane by the signal recognition particle and the protein FtsY (Ulbrandt et al. 1997). Proteins to be translocated across the membrane by the Sec system are targeted by the SecB protein, differentiating the transport pathways of these two classes of proteins.

The cell membranes of archaea chemically differ from eubacterial membranes. The membrane lipids of archaea are isopranoid alcohols, typically 20 or 40 carbons long, and ether-linked to glycerol. The C20 alcohols can be linked together giving rise to the C40 molecules. The glycerol groups can be modified with a number of different polar head groups. It has been suggested that the use of the novel lipids by Archaea allows them to tolerate extreme environments better (van de Vossenberg et al. 1998).

The cytoplasm

A typical bacterial cell is able to synthesize over 1000 different enzymes to catalyze the many chemical reactions needed for cell growth. Catabolic pathways provide the cell with the energy for growth, while anabolic reactions are the biosynthetic reactions whose purpose is the formation of new cellular material. Although a bacterial cell may possess the ability to synthesize over 1000 different enzymes, its regulatory systems are

evolved to control the synthesis of such proteins, and the exact type and quantity of enzyme may be carefully regulated.

Enzymes function as catalysts and rapidly cause the equilibrium of thermodynamically feasible chemical reactions to be reached. Although an enzyme must participate in the reaction it catalyzes, binding substrates and releasing products, it is normally regenerated and can be reused when the reaction is completed. The activity of many enzymes is associated with a relatively low-molecular-weight compound termed a coenzyme (or prosthetic group or cofactor). The catalytic site of the enzyme binds the substrate and the enzyme–substrate complex changes to an enzyme–product complex with the release of the product. The specificity of enzyme action results from the specific recognition of the substrate by the catalytic site. Many enzymes will act upon compounds related in chemical structure to the natural substrate but usually with lower binding affinities. Sometimes such compounds are converted to products and released but some can bind to the enzyme and block activity. Enzyme activity can also be destroyed or inhibited by a variety of chemicals or conditions, including those that alter the structure of proteins in a nonspecific manner.

As increasing concentrations of a substrate are presented to an enzyme, the velocity of the enzyme-catalyzed reaction will increase until, at substrate concentrations sufficient to saturate the available catalytic sites, the rate of reaction will be sustained at some maximum value (V_{max}). Thus, a plot of reaction rate (v) versus substrate concentration [S] frequently approximates to a rectangular hyperbola. An enzyme displaying these kinetic properties is said to demonstrate Michaelis–Menten kinetics and the relationship between the reaction rate and substrate concentration can be described by the maximum velocity (V_{max}) constant and the Michaelis constant (K_m), which is the substrate concentration that limits the rate of the enzyme-catalyzed reaction to be one-half that of V_{max}. Studies of how V_{max} and K_m change under different chemical and physical conditions can provide information on the nature of the substrate–enzyme complex.

The regulatory mechanisms of bacterial metabolism are designed to permit a controlled synthesis of cellular polymers, usually as rapid as possible under the existing environmental conditions. Bacterial cells are not simply a 'bag of enzymes', as was once suggested. Indeed, there is careful organization of many proteins within the bacterial cell. Multienzyme complexes can be found within the cytoplasm and many enzymes and transport proteins are incorporated in the cell membrane or, in the case of gram-negative bacteria, within the periplasmic space. Some proteins involved in anabolic processes themselves are regulated. These allosteric proteins possess a site for binding a low-molecular-weight molecule termed an effector. The effector molecule may be quite different from the substrate molecule, but its binding to the enzyme results in a conformational change at the active site with a change in the affinity of the active site for the normal enzyme substrate. A positive effector will increase the binding capability of the enzyme for its substrate, while a negative effector will decrease the binding capability. With normal intercellular substrate concentrations, a positive effector will increase the rate of reaction, while a negative effector will decrease the rate of reaction. Proteins with nucleic acid binding sites can also have their binding ability increased or decreased by effector molecules.

BIOSYNTHESIS OF DNA, RNA, AND PROTEIN

The bacterial chromosome and DNA replication

The DNA of the chromosome in a typical bacterial cell can be 500–1 000 times as long as the cell itself. This molecule is compacted by folding and supercoiling. The initiation of chromosomal replication requires a certain cell mass before replication can begin. Replication of the chromosome proceeds bidirectionally from the origin of replication (in *E. coli* this is termed *oriC*). Replication requires that the DNA strands become unwound, forming an open complex. In *E. coli* formation of the open complex requires the protein DnaA and a few other proteins. DNA polymerases then synthesize DNA in the $5'–3'$ direction using a $3'$-hydroxyl primer. *E. coli* has three DNA polymerases, two of which are required for replication of the chromosome. Polymerase III replicates the chromosomal DNA. Because DNA polymerase III needs a primer to begin DNA polymerization on a single-stranded template, a polymerase is needed first to form an RNA primer. After DNA polymerase has used these primers for the elongation of the DNA, they are replaced by DNA. In the $3'–5'$ direction, RNA primers are formed and, after the formation of 1 000- to 2 000-long DNA strands termed Okazaki fragments, destroyed and replaced by DNA using DNA polymerase I. The nicks between the Okazaki fragments are repaired by DNA ligase (Bremer and Churchward 1991). Because replication is bidirectional, the two replication regions (termed replication forks) meet at a region in the chromosome called the termination region. This region blocks progression of the replication forks and allows termination to occur.

Some strains of bacteria have been reported to have more than one chromosome; for example, a strain of *R. sphaeroides* (strain 2.4.1) has been reported to contain two circular chromosomes as well as five plasmids (Suwanto and Kaplan 1989). The two chromosomes are 3.0 million base pairs (Mbp) and 0.9 Mbp in size. Another bacterium with multiple chromosomes is *Deinococcus radiodurans* (Lin et al. 1999). Interestingly this bacterium also contains multiple copies of each chromosome (Hansen 1978). This polyploidy has been suggested to play a role in the radiation-resistance of this bacterium. However, there are probably other factors involved because other bacteria, such as *Azotobacter* (Punita et al. 1989), are polyploid but are sensitive to ionizing radiation.

Protein synthesis

While the DNA contains the sequence information for the various proteins of the cell, the RNA is the machinery that translates that information into the polypeptide chains. The genetic code is triplet, meaning that three nucleotides determine the code 'word' for each amino acid. Because there are four nucleotide 'letters' (A, T, G, and C in the DNA alphabet and A, U, G, and C in the RNA alphabet), it is possible to have 64 different 'words' or codons for the 20 amino acids normally found in proteins. The genetic code is also degenerate in that there may be more than one codon for any one amino acid but no nucleotide triplet will code for more than one amino acid; for example, there are four mRNA codons for serine, UCU, UCC, UCA, and UCG. It has been speculated that the code was once a doublet rather than a triplet and, therefore, the original doublet code for serine might have been UC. Three codons called nonsense codons usually do not code for any amino acid and signal when to end synthesis of a polypeptide chain.

Amino acids are activated before being added to a polypeptide chain by being placed on specific tRNA molecules. An enzyme specific for each amino acid (aminoacyl tRNA synthetase) transfers the amino acid to the tRNA specific for the transfer of that amino acid. These tRNA molecules are about 80 nucleotides in length and each has a site recognized by its tRNA synthetase. The amino acid is first activated using ATP, resulting in the formation of an aminoacyl–AMP complex. This complex is transferred to the tRNA molecule and the tRNA molecule is said to be 'charged.' In such a manner, each tRNA can become 'charged' at its amino-acid attachment sites with the proper amino acid. The tRNA also possesses a ribosomal recognition site as well as a codon recognition site that will recognize the mRNA codon for that amino acid.

Polymerization of the amino acids takes place with the aid of the ribosome. A bacterial ribosome consists of two subunits, designated 30 and 50S, which combine to form the 70S ribosome. The complete ribosome is made up of three different ribosomal rRNAs and 52 different proteins (r-proteins). The smaller 30S subunit contains 21 proteins and a 16S rRNA, whereas the larger subunit has 21 proteins and two species of rRNA, 23S and 5S. The high-resolution structures of an intact 70S ribosome

(Cate et al. 1999) and the structure of the 50S (Ban et al. 1999) and 30S (Clemons et al. 1999) subunits of the ribosome have been reported. The functional elements, that is the tRNA and mRNA binding sites, of the ribosome are identified in these structures. The channel through which the extending peptide leaves the ribosome has also been visualized. Further analysis of these structures should permit detailed understanding of the chemistry taking place in this complex.

It is known that rRNA is initially transcribed as a single 30S precursor that is then processed into the three final rRNA products. The number of rRNA genes is variable; for example, in *E. coli* K-12 there are seven rRNA operons. Strains of *R. sphaeroides* have between three and five rRNA clusters (Nereng and Kaplan 1999). In *E. coli*, the ribosomal protein L7/L12 is present in four copies per ribosome, while other ribosomal proteins are present in one copy per ribosome. The genes responsible for the synthesis of the 52 r-proteins are organized into at least 20 translational units (operons) containing from 1 to 11 r-protein genes. At least six of the tRNA species are cotranscribed with rRNA species. The rates of synthesis of the 52 ribosomal proteins of *E. coli* and the three rRNA components are balanced so there is normally very little excess ribosomal protein or rRNA in the cells. The translation of ribosomal proteins from an operon can be inhibited by the proteins coded in that operon. These proteins normally bind to rRNAs but if their concentration is higher than required they will bind to their own mRNA, blocking translation. In addition, if sufficient ribosomal proteins are not present, rRNA will combine with other proteins but the complex formed will not function as ribosome (Nomura et al. 1984).

The mRNA represents only 2–3 percent of the total cellular RNA and its synthesis is controlled at the operon level. In translation of the mRNA into protein, a complex is formed between the 30S ribosomal subunit and certain protein initiation factors. The mRNA binds to the 30S subunit as does the first charged tRNA, the 'initiator' tRNA. The eubacterial initiator tRNA is N-formylmethionyl-tRNA, which binds to a site in the 30S subunit, termed the P site. The initiator tRNA appears to be methionine-tRNA for the archaebacteria. The ribosome also contains a binding site for a second charged tRNA (the A site). It is the codons found in the mRNA that determine the order of incorporation of the charged tRNAs. Once the charged tRNA binds at the A site, a peptide bond is formed between its amino group and the amino acid on the tRNA at the P site. The polypeptide chain is attached to the tRNA at the A site and the now uncharged tRNA moves to a third tRNA binding site in the ribosome referred to as the E site. The tRNA at the A site and its polypeptide chain move to the P site (translocation) as the ribosome moves to the next codon on the mRNA. The charged tRNA with the proper anticodon binds to the A site and the process of polypeptide chain elongation continues. The termination codons (UAA, UGA, and UAG) are not charged with any amino acid. When such codons appear on the mRNA, the polypeptide chain is released from the P site. Either the formyl group is removed from the first amino acid or the entire formylmethionine is removed.

PROTEIN FOLDING

The information required for the final structure of the synthesized protein resides in its primary sequence. While many proteins can spontaneously fold into their final tertiary structure, some are apparently assisted by accessory proteins referred to as chaperones. Several bacterial protein-folding chaperones have been studied in great detail. Perhaps the best studied chaperones are the GroEL/GroES chaperones, which have also been given the name chaperonins. The high-resolution structure of the GroEL/GroES complex has been determined. The complex is made up of two seven-membered rings of GroEL, each containing a central cavity (Xu et al. 1997). GroES forms a single seven-membered ring that caps one of the cavities of the GroEL complex. It is unclear exactly how targets of chaperones are recognized but exposed hydrophobic surfaces seem to be important. In many bacteria there are additional chaperones involved in protein folding. Two well-studied examples are DnaK (Hsp70) and DnaJ (Hsp40). These proteins help to limit aggregation of newly synthesized proteins (Fink 1999). They also probably function with the GroEL/GroES complex to carry out protein folding. These protein-folding chaperones also play a role in response to heat stress because they can assist in the refolding of incorrectly folded proteins.

Archaea also contain chaperones but so far have not been found to encode the GroEL/ES type. Archaeal chaperones are known to form stacked rings somewhat similar to the GroEL/ES ring. The high-resolution structure of the complex from the thermophile *Thermoplasma acidophilum* shows these ring structures (Ditzel et al. 1998). Both DnaK- and DnaJ-like chaperones have been found in archaea.

Proteins that are termed chaperones are involved in other cell functions besides protein folding; for example, chaperones are required for the secretion of proteins into or through the cell membrane. The protein SecB is required for recognition of proteins to be transported via the Sec system (Fekkes and Driessen 1998). Once SecB binds to targeted proteins it prevents them from folding and ferries them to the Sec translocation complex.

The eventual fate of almost all proteins is degradation. Interestingly, some proteins involved in the degradation of proteins form complexes similar in structure and function to GroEL/ES. ClpA and ClpP form rings that are probably required for the unfolding of proteins necessary for their degradation (Horwich et al. 1999). Another large complex involved in protein degradation has been termed the 20S proteasome (De Mot et al. 1999). This is a large structure consisting of four, stacked seven-member rings. This structure has been found in the archaea and eukaryotes but only among the gram-positive actinomycetes in the eubacteria.

PROTEIN SECRETION

It is obvious that bacteria must have systems for the transport of proteins across the cell membrane. The characteristics of some of these systems were discussed above. In addition to transport across the cytoplasmic membrane many proteins are secreted into the surrounding medium. The systems for protein export are referred to as the type I–IV secretion systems. Two of these systems, types II and IV, utilize the previously mentioned Sec machinery. The other two are Sec-independent. These systems are distinct from those that only transport proteins across the cytoplasmic membrane because they also must include pores for transport across the cell wall.

The *E. coli* hemeolysin is an example of a protein transported by a type I secretion system. A unique characteristic of these systems is that transport across the cytoplasmic membrane is driven by an ABC transporter. ABC transporters are a superfamily of proteins that contain an ATP-binding cassette (hence ABC transporter). The type II system is best represented by the pullunase secretion system of *Klebsiella oxytoca*. This system utilizes the Sec system to target proteins but includes additional proteins for transport across the outer membrane. The type III systems have been heavily studied because of their role in pathogenesis. The system used by *Yersinia* species to export a set of proteins known as Yop proteins is a good example of the type III system (Cornelis and Wolf-Watz 1997). Interestingly, the system used to export flagellar proteins is similar to the type III system. It is possible that the type III system may have evolved from the flagella export system. The type IV pathway is not as well studied but appears to include proteins that use the Sec system to cross the cytoplasmic membrane but do not rely on proteins of the type II system to cross the outer membrane.

The regulation of cell division

Cell division requires that cells reach a certain critical mass before division begins. To complete cell division the cell must replicate its chromosome, segregate the chromosome into opposite halves of the cell, and then divide. Replication of the *E. coli* chromosome was discussed above. The mechanisms used to segregate the replicated chromosomes are not yet understood. Once this has been achieved, however, the cell can form a septum at the site where division will occur. Formation of the septum involves the recruitment of certain proteins to the approximate middle of the cell to form a ring. In *E. coli* these ring-forming proteins include MinE and FtsZ. The protein MinC is also required to keep ring formation from occurring at inappropriate sites. The septal rings must then contract to allow two separate cells to form. The process of ring formation and septation is very complex but a number of the proteins responsible for this process have been identified and their roles elucidated (Rothfield et al. 1999).

TRANSPORT OF SOLUTES INTO THE CELL

Simple and facilitated diffusion

Most small solutes can diffuse through the peptidoglycan material and many hydrophilic compounds can pass through the porins of the outer membrane of gram-negative bacteria. However, the cytoplasmic membrane of both gram-negative and gram-positive bacteria is a barrier to hydrophilic compounds, and special transport machinery must be utilized to allow nutrients to pass through this membrane fast enough to allow competitive growth rates. If a compound moves through the cytoplasmic membrane and into the cytoplasm without utilizing the transport machinery this is termed simple or passive diffusion. For most compounds, with the exception of certain hydrophobic compounds, entry via diffusion will be very slow. The rate of entry of a diffusible compound into the cytoplasm by simple diffusion is typically dependent upon its concentration in the medium. Because most compounds are present at extremely low concentrations in the environment, diffusion cannot be used for the uptake of most nutrients required for growth.

One example of diffusion in bacteria is the uptake of glycerol by facilitated diffusion. Facilitated diffusion results when the cell possesses a protein (the facilitator protein) that aids or facilitates the transport of a compound through the membrane. This facilitator protein permits the compound to be transferred down a concentration gradient from one side of the membrane to the other. The process is not directly coupled to metabolic energy and the compound cannot be accumulated against a concentration gradient but it may be altered and trapped to prevent it from leaving once it enters the cell. Once inside the bacterial cell, glycerol is phosphorylated by glycerol kinase and glycerol phosphate can accumulate within the cells. The phosphorylation is essential in order to trap glycerol in the cells, and mutants lacking glycerol kinase are not able to accumulate glycerol. Another example of a facilitator protein is provided by AqpZ, the water channel of *E. coli* (Calamita et al. 1995). This protein is related to other water channels in plants and animals as well as to the glycerol facilitator. Inactivation of aqpZ is not lethal to cells but does limit cell growth under conditions of low osmolality (Calamita et al. 1998).

Active transport systems

Most solutes are brought into the cell by means of active transport systems that involve the use of metabolic energy. Because energy is used, these active transport systems can accumulate the compound within the cell to a concentration 1000 times greater than the concentration in the surrounding medium, greatly increasing the growth rate of the cell when nutrients are limited.

Substrates can be transported in various bacteria by several different active transport systems. Included among these are the ABC-type carriers. These are also known as periplasmic-binding protein-dependent transport systems or osmotic shock-sensitive transport systems. Examples of another type of active transport system are the chemiosmotic-driven, ion-gradient-linked permeases that use proton- or sodium-motive force to transport nutrients into the cell. The ATP and the proton-motive force systems can be differentiated by the use of inhibitors that prevent ATP from accumulating or by using proton ionophores, such as dinitrophenol (DNP), that collapse the proton gradient. A third type of transport, primarily associated with fermenting bacteria, is the phosphoenolpyruvate phosphotransferase system (PTS), where the compound is phosphorylated during transport through the cytoplasmic membrane.

ATP-BINDING CASSETTE, PERIPLASMIC-BINDING PROTEIN-DEPENDENT TRANSPORT SYSTEMS

These transport systems in gram-negative bacteria are composed of a substrate-binding receptor protein associated with the cell periplasm plus several membrane-bound components. This has led these types of transport systems to be referred to as osmotic shock-sensitive permeases, because a cold osmotic shock causes the loss of the substrate-binding protein, which is loosely attached in the periplasmic space. In gram-positive bacteria, the substrate-binding receptor protein is represented by a lipoprotein fastened to the membrane.

A likely scenario for transport is that substrate binds to the periplasmic-binding protein, which then interacts with the membrane-associated transport complex. This causes the membrane-associated subunits to undergo conformational changes, which, in turn, cause conformational changes in the ATP-binding subunit, which is located on the inner surface of the cytoplasmic membrane. These changes cause ATP hydrolysis, which is used to drive transport. Substrates reported to be transported by this system in *E. coli* include carbohydrates such as L-arabinose, maltose, D-ribose, D-xylose, succinate, and malate, and amino acids such as leucine, isoleucine, valine, histidine, lysine, and ornithine.

A good example of this type of transport system is provided by the histidine permease system of *Salmonella typhimurium*. This system consists of a periplasmic-binding protein, HisJ and a membrane-associated complex of HisQMP2 (Liu and Ames 1998). HisQ and HisM are the membrane-spanning subunits and HisP is the ATP-binding subunit. The high-resolution structure of the HisP subunit has been solved and has provided insight into the functional characteristics of this large family of proteins (Hung et al. 1998).

ION GRADIENT-LINKED OR CHEMIOSMOTIC-DRIVEN TRANSPORT SYSTEMS

Chemiosmotic-driven transport systems use an ion gradient, such as proton- or sodium-motive force, as the energy to drive the transport of the substrate through the membrane. The use of an ion gradient distinguishes them from the binding protein-dependent systems that use the energy of ATP directly. The binding protein-dependent systems are termed primary transport because they are driven by an energy-producing event. The ion-gradient transport systems are termed secondary transport systems because they utilize gradients that typically have been formed by an energy-producing event.

The three basic types of ion-gradient transport are symport, antiport, and uniport. With symport, a single carrier simultaneously transports two substances in the same direction. Antiport involves a common carrier transporting two substances at the same time but in opposite directions. Uniport is the situation where movement is independent of any coupled ion.

Electrochemical ion-gradient permeases are usually composed of a single, very hydrophobic membrane protein. Probably the best studied protein of this type is LacY, the lactose transporter of *E. coli*. Many of the residues in this protein have been mutagenized and only a few have been shown to be essential for transport (Kaback and Wu 1997). Cys-scanning mutagenesis has also been used to provide a better understanding of interactions among the various domains of the protein and to probe structural changes during transport. However, detailed insight into the catalytic mechanism is still lacking at this point.

In some of the thermophilic *Bacillus*, such as *B. sterothermophilus*, Na^+ is used for the coupling ion for amino-acid transport. The proton-motive force as well as an inward-directed Na^+ gradient drives the transport of glutamate and aspartate. Sodium/proton glutamate transporters have been found in a number of thermophilic *Bacillus* species. However, in *Bacillus subtilis*, glutamate uptake was coupled to proton-motive force (Tolner et al. 1995). In *E. coli* and *S. typhimurium*, carbohydrate–sodium cotransport systems have been reported for melibose and proline. Carbohydrate–sodium cotransport systems are common in halophilic bacteria.

THE PHOSPHOTRANSFERASE SYSTEM

PTS uses the phosphate from phosphoenolpyruvate (PEP) for the transport and phosphorylation of incoming sugars. The phosphate group is first transferred from PEP to a nonspecific enzyme component termed enzyme I or EI. The phosphorylated EI (P-EI) transfers the phosphate to histidine residue 15 on a small, heat-stable protein, designated HPr. The phosphate is then transferred from phospho-HPr to a sugar-specific, membrane-bound enzyme II (EII) complex, which catalyzes the transport of the carbohydrate through the membrane as well as its phosphorylation.

The EII complex consists of domains termed IIA, IIB, IIC (and sometimes IID) that may exist as separate proteins or components of the same protein. The IIA and IIB protein domains are associated with the cytoplasmic side of the membrane and are involved in phosphate transfer. If these domains are located in separate proteins, the IIA domain will have the first phosphorylation site and the IIB domain will have the second phosphorylation site. The IIC protein/domain is located in the cytoplasmic membrane, possesses the carbohydrate-binding site, and is responsible for the transport of the carbohydrate through the membrane. The overall effect of the PTS is to both phosphorylate the carbohydrate and transport it into the cytoplasm of the cell.

The EI and HPr components are the nonspecific portions of the PTS system. The other, carbohydrate-specific components of the PTS transport system can be indicated by placing the name of the transported sugar after the enzyme designation. The EII protein/domain that is specific for the transport of D-glucose into the cells as glucose-6-phosphate can be designated as enzyme $EII^{Glucose}$ or II^{Glc}. Some EII PTS transport systems may have activity upon more than one carbohydrate; for example, E. coli enzyme II^{Man} has activity for the transport and phosphorylation of glucose, mannose, 2-deoxyglucose, glucosamine, and fructose, in that order (Saier and Reizer 1994).

Because the PTS transport system requires phosphoenolpyruvate as the initial phosphate donor it tends to be found in facultative or anaerobic bacteria that utilize the hexose diphosphate pathway (see below) and produce two PEP molecules for each hexose catabolized. PTS is common in such bacteria as E. coli and other enteric bacteria, Bacillus, Clostridium, Bacteriodes, Staphylococcus, Lactobacillus, and Streptococcus. It is not common in aerobic bacteria employing other catabolic pathways, such as Arthrobacter, Azotobacter, Micrococcus, Pseudomonas, Nocardia, or Caulobacter. In E. coli, the PTS transport system is used for the transport of such carbohydrates as mannitol, sorbitol, galactitol, glucose, glucose-mannose, fructose, and N-acetylglucosamine. Many gram-positive bacteria transport lactose, sucrose, maltose, and/or pentitols via the PTS.

Alternative, non-PTS transport systems for glucose may also be present (Cvitkovitch et al. 1995).

The role of the PTS system in catabolite regulation

When presented with multiple carbon sources, most bacteria will preferentially utilize one of those carbon sources to the exclusion of other available substrates. In E. coli the preferred carbon source is glucose. This regulation of carbon metabolism is referred to as glucose repression or catabolite repression. One of the principal proteins involved in regulation of catabolite repression is IIA^{Glc}. When glucose and other sugars transported by the PTS are present, the HPr protein is rapidly being dephosphorylated as glucose is phosphorylated. Nonphosphorylated IIA^{Glc} will interact with, and inhibit the function of, enzymes required for metabolism of other compounds. This phenomenon is known as inducer exclusion. Examples of substrate whose metabolism is affected by inducer exclusion are maltose, lactose, melibiose, and glycerol. There are other causes of catabolite repression besides inducer exclusion. One of the main ways of achieving catabolite repression is via the small molecule, cyclic AMP (cAMP). This regulatory process is discussed below.

The HPr of Streptococcus pyogenes can also be phosphorylated on a serine residue (Ser-46) by a specific HPr kinase with ATP as the phosphate donor. This reaction has been found to be widespread among gram-positive bacteria but not gram-negative bacteria and even has been shown to occur in some species lacking a functional PTS. Phosphorylated HPr has some regulatory roles, regulating glucose and lactose permease activity in Lactobacillus brevis and inducer exclusion in Lactococcus lactis, and is involved in catabolite repression in B. subtilis (Thevenot et al. 1995).

BACTERIAL CATABOLIC PATHWAYS

The hexose diphosphate pathway

A very efficient pathway for the production of ATP during the catabolism of D-glucose and similar carbohydrates is often referred to as the EMP pathway to honor three scientists who contributed greatly to its elucidation, Embden, Meyerhof, and Parnas. One of the unique things about this pathway is the formation of a hexose diphosphate as an intermediate. Therefore, an alternative and descriptive name is the hexose diphosphate or HDP pathway (Figure 3.4).

THE PHOSPHORYLATION OF GLUCOSE

The catabolism of glucose by the HDP pathways begins by phosphorylation at the 6-carbon position. Various mechanisms have been reported for this reaction. If glucose is transported into the cells via the PTS, this

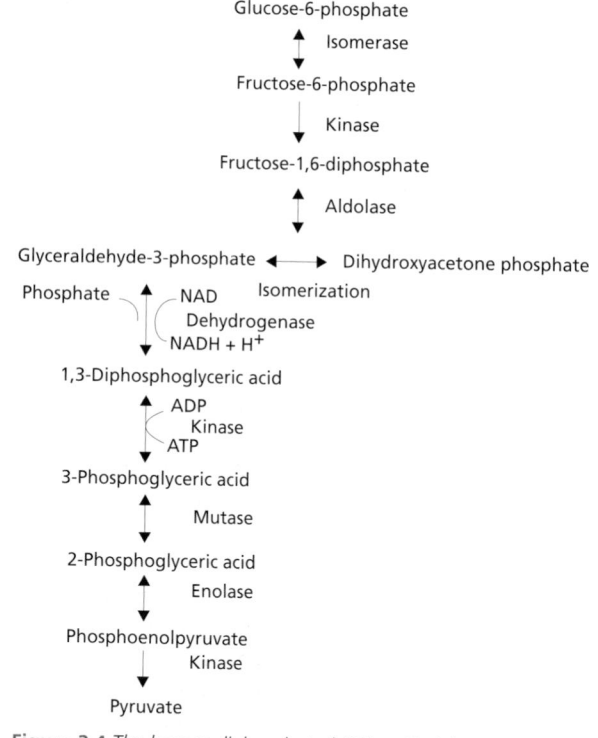

Figure 3.4 *The hexose diphosphate (HDP) or Embden, Meyerhof, and Parnas (EMP) pathway. Glucose-6-phosphate is converted to fructose-1,6-diphosphate, which is cleaved to produce the 3-carbon triose phosphates. Glyceraldehyde-3-phosphate is oxidized and the product of the oxidation converted to pyruvate.*

efficient transport system will convert glucose to the 6-phosphate upon passage through the cell membrane. If glucose enters the cell by a different transport system or is produced within the cell, it must be phosphorylated by other means. Hexokinase is a nonspecific kinase found in a variety of microorganisms. It can catalyze the phosphorylation of hexoses such as D-glucose, D-fructose, D-mannose, and D-glucosamine to the 6-phosphate, with ATP as the phosphate donor. Some bacteria, such as those in the genera *Staphylococcus*, *Streptococcus*, *Clostridium*, and *Lactobacillus*, possess a glucokinase that is specific for the phosphorylation of glucose to glucose-6-phosphate. At least some bacteria in the genus *Propionibacterium* can phosphorylate glucose using inorganic pyrophosphate as the phosphate donor. Transfer enzymes have been reported, which can transfer the phosphate group from a different hexose phosphate to glucose.

CONVERSION OF GLUCOSE-6-PHOSPHATE TO GLYCERALDEHYDE-3-PHOSPHATE

The reactions of the HDP pathway resulting in the conversion of glucose-6-phosphate to the two triose phosphates, glyceraldehyde-3-phosphate and dihydroxy-acetone phosphate, are illustrated in Figure 3.4. The first step is the isomerization of glucose-6-phosphate to the 2-ketosugar, fructose-6-phosphate. The enzyme catalyzing the reaction is glucose-6-phosphate (fructose-6-phos-

phate) isomerase. The equilibrium of this reaction is towards glucose-6-phosphate. However, the equilibrium of the next reaction, the conversion of fructose-6-phosphate to fructose-1,6-bisphosphate, is strongly towards product formation. The phosphorylating enzyme is phosphofructokinase (Pfk) and ATP normally is the phosphate donor. Two isozymes for Pfk have been found in *E. coli*. Most (90 percent) of the activity is provided by Pfk-1, an enzyme that is activated by nucleotide diphosphates such as ADP and inhibited by PEP. However, excess ATP and PEP are negative effectors of the enzyme and inhibit activity. Thus, the activity of the enzyme (and the HDP pathway) will be inhibited in cells with high energy charges (high levels of nucleotide triphosphates). The second enzyme, Pfk-2, is inhibited by ATP and the product of the reaction, fructose-1,6-bisphosphate.

Aldolase catalyzes the cleavage of fructose-1,6-bisphosphate to the two triose phosphates, glyceraldehyde-3-phosphate and dihydroxyacetone phosphate, as shown in Figure 3.4. The equilibrium of the reaction is back towards fructose-1,6 bisphosphate so that an accumulation of triose phosphates will lead to an accumulation of fructose-1,6 bisphosphate. The accumulation of this hexose diphosphate often serves as a regulatory signal that the oxidation of glyceraldehyde-3-phosphate is not functioning properly. Dihydroxyacetone phosphate is converted to glyceraldehyde-3-phosphate by the enzyme triosphosphate isomerase.

CONVERSION OF GLYCERALDEHYDE-3-PHOSPHATE TO PYRUVATE

The next reaction of the hexose diphosphate pathway involves the oxidation of glyceraldehyde-3-phosphate to 1,3-diphosphoglyceric acid. This oxidation is catalyzed by glyceraldehyde-3-phosphate dehydrogenase. The 1,3-diphosphoglyceric acid contains energy in an energy-rich bond between phosphoglyceric acid and phosphate. This reaction is reversible and nicotinamide adenine dinucleotide (NAD) is usually (not always) the electron acceptor. Inorganic phosphate is the phosphate donor.

ATP-phosphoglyceric transphosphorylase, also known as phosphoglycerate kinase, catalyzes the reversible transfer of the energy-rich phosphate at the one carbon of 1,3-diphosphoglyceric acid to ADP, to form ATP and 3-phosphoglyceric acid. Phosphate esterification in this manner, with the phosphate group being transferred to ADP to form ATP, is termed substrate-level phosphorylation. Another enzyme, phosphoglycerate mutase, catalyzes the transfer of phosphate from the 3 to the 2 position of glyceric acid, producing 2-phosphoglyceric acid. Then 2-phosphoglyceric acid dehydrase, also known as enolase, removes water from 2-phosphoglycerate to form PEP, which has the highest group-transfer potential of any intermediate produced by the HDP pathway. Pyruvate kinase transfers the phosphate to ADP to form ATP and pyruvic acid.

Two isozymes of pyruvate kinase have been reported in *E. coli*. They are Pyk-F, activated by fructose 1,6-bisphosphate, and Pyk-A, activated by AMP. Mutants lacking one of these two enzymes can still grow on glucose but mutants lacking both enzymes grow poorly on glucose, except aerobically on substrates transported into the cells by the PTS (Ponce et al. 1995). In the latter case, pyruvate is generated from PEP during substrate transport. The net gain of energy from the HDP pathway is two ATP for each glucose catabolized. This pathway is an efficient pathway for the fermentation of hexoses and is common in fermenting bacteria.

THE PYRUVATE DEHYDROGENASE COMPLEX

The pyruvate dehydrogenase complex (PDH) is a multienzyme complex that catalyzes the decarboxylation of pyruvic acid and an acetyl transfer from pyruvate to coenzyme A. The complex has a structural core of dihydrolipoate transacetylase (E2) components surrounded by independently bound pyruvate dehydrogenase (E1) and dihydrolipoate dehydrogenase (E3). The cofactor requirements of the complex include lipoic acid (dithiooctanoic acid), coenzyme A, NAD, flavine adenine dinucleotide (FAD), and thiamine pyrophosphate (TPP). The products of the reaction of the complex are CO_2, acetyl-CoA, and NADH.

The complex isolated from *E. coli* possesses a chain ratio of 1.0E1/1.0E2/0.5E3 with a core of 24 E2 chains. The genes encoding for the E1, E2, and E3 enzymes (*aceE*, *aceF*, and *ipdA*) are part of the same operon. The E3, the *ipd* gene product, is also a component of the α-ketoglutarate dehydrogenase complex of the tricarboxylic acid (TCA) cycle. The E1 and E3 subunits are assembled around an icosahedral core composed of E2 polypeptides and form the complex with its molecular size of about 10 MDa (Hiromasa et al. 1994).

This pyruvate dehydrogenase from *E. coli* is stimulated by HDP intermediates such as fructose-6-phosphate. It is inhibited by a high-adenylate energy charge but such inhibition is overcome by fructose-1,6-bisphosphate. The ratio of NADH/NAD seems to influence the flow of pyruvate between pyruvate–formate lyase and the PDH complex. The activity of the complex increases with increased NAD^+ concentration within the cells but is inhibited when intracellular NADH levels are high, such high levels indicating a problem in the disposal of electrons (Snoep et al. 1991). For these reasons, utilization of the system tends to increase under aerobic conditions or when the oxidation of NADH is not a problem for the cells.

The tricarboxylic acid cycle or citric acid cycle

The TCA cycle, also known as the citric acid cycle or the Krebs cycle, is a means of oxidizing acetate (as

acetyl-CoA produced by the HDP pathway) to CO_2 and electrons. The TCA cycle reactions have both anabolic and catabolic functions in many bacteria. The acetyl group from acetyl-CoA is transferred by citrate synthase to the α-keto, 4-carbon dicarboxylic acid, oxaloacetate, to form citric acid. The methyl carbon of the acetyl moiety is attached to the 2 or α-carbon of the oxaloacetate with the keto-group of oxaloacetate becoming a hydroxyl group. This hydroxyl group is transferred to a different carbon by isocitrate isomerase (aconitase) with the formation of isocitric acid. This hydroxyl group is then oxidized to an α-keto group by isocitrate dehydrogenase with the formation of oxalosuccinic acid. Oxalosuccinic acid is further oxidized with decarboxylation by isocitrate dehydrogenase to form α-ketoglutaric acid, a compound that is important in anabolic pathways. Reactions of the TCA cycle are shown in Figure 3.5.

In the catabolic TCA cycle, α-ketoglutaric acid is oxidized and decarboxylated by an α-ketoglutarate dehydrogenase enzyme complex, involving coenzyme A and lipoic acid. This complex is similar to the pyruvate dehydrogenase complex described previously, with CO_2 produced and a lipoic succinyl intermediate formed. The succinyl group is transferred to coenzyme A to give succinyl-CoA, which can be converted to succinic acid with the formation of a high-energy bond (ATP) in a manner similar to the formation of acetate from acetyl-CoA. Succinic acid is then oxidized to fumarate by succinate dehydrogenase with the reduction of a FAD^+ to $FADH_2$.

Succinate dehydrogenase (SDA) is a membrane-bound enzyme that is part of the respiratory chain and appears to be present in all aerobic cells. It is composed of two subunits, a flavoprotein (FP) and an iron sulfur

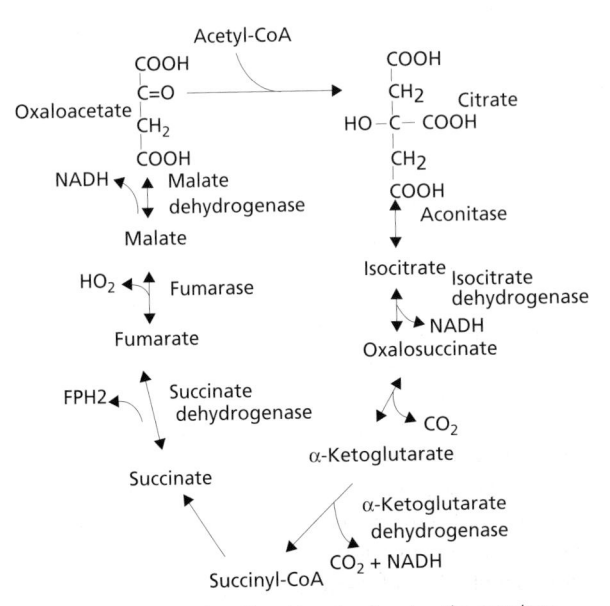

Figure 3.5 *The tricarboxylic acid cycle, showing the reactions resulting in the oxidation of acetate to carbon dioxide*

protein and, in prokaryotic cells, SDH is fastened to the inner side of the cytoplasmic membrane by a cytochrome. While the high-resolution structure of SDH has not been determined, the structure of fumarate reductase, which catalyzes the reverse reaction and has extensive structural and sequence similarity to SDH, has been determined (Iverson et al. 1999). Fumarate is hydrated by malate dehydrase (fumarase) to form malate, malate is oxidized by malate dehydrogenase to produce oxaloacetic acid, and the cycle is complete. For each acetate oxidized, provided as acetyl-CoA, two molecules of CO_2 and eight pairs of electrons are obtained. Two of these electrons are transported to FAD^+ and six to NAD^+. An ATP can be obtained from the succinyl-CoA.

Succinate and oxaloacetate, as well as α-ketoglutarate, are important compounds in biosynthetic reactions. Even anaerobic bacteria, which do not utilize the TCA cycle for complete oxidation of acetate, may need TCA-cycle enzymes for biosynthetic reactions. However, it would not be efficient for anaerobic bacteria to oxidize acetate completely because they would have difficulty in disposing of the electrons. Such bacteria often lack the α-ketoglutarate enzyme complex. This complex, usually missing in anaerobes, is known as the 'anaerobic lesion' of the TCA cycle.

Growth on acetate as a carbon source

E. coli has two pathways that can be used to convert acetate to acetyl-CoA. One pathway uses acetate kinase and ATP to produce acetyl phosphate, then phosphotransacetylase transfers the acetyl moiety from acetyl phosphate to CoA to form acetyl-CoA. However, this route is believed to function primarily in a catabolic role, generating ATP when acetate is being excreted. This system would be used to activate acetate only when acetate was present in large amounts outside the cells. The second pathway is believed to function by using extracellular acetate that is present in low concentrations. It uses the enzyme acetyl-CoA synthetase to convert acetate and ATP to acetyl-AMP and inorganic pyrophosphate. The acetyl-AMP reacts with CoA to release inorganic phosphate and form acetyl-CoA. Although wild-type E. coli cells grow well on acetate at concentrations ranging from 2.5 to 50 mM, mutants lacking acetyl-CoA synthetase grow poorly on low acetate concentrations. Mutants lacking acetate kinase and phosphotransacetylase grow poorly on high concentrations of acetate, and mutants lacking all three enzymes did not grow on acetate at any concentration tested (Kumari et al. 1995).

When growing on acetate as the sole carbon source, bacteria must have the enzyme capability to reverse the HDP pathway and form pentoses and hexoses from these simple 2- and 3-carbon compounds. While some strict anaerobes are able to reverse the pyruvate:ferredoxin oxidoreductase reaction to form pyruvate from

CO_2 and acetate, most bacteria that can grow with only acetate as the carbon source incorporate it into TCA-cycle intermediates (see Figure 3.5) by a series of reactions known as the glyoxylate bypass or glyoxylate cycle. During the glyoxylate cycle the acetyl-CoA produced by acetyl-CoA synthetase is combined with a molecule of existing oxaloacetic acid to form citrate and the citrate is isomerized to isocitrate by normal TCA cycle enzymes. A unique enzyme, isocitrate lyase, cleaves the isocitrate to produce succinate and glyoxylate. The glyoxylate is condensed with acetyl-CoA by the enzyme malate synthase to form malate, as shown in Figure 3.6. In this manner, one oxaloacetate and two acetyl groups have been converted to malate and succinate. These latter two compounds can be oxidized and converted to two oxaloacetates so the overall series of reactions form two oxaloacetates from one oxaloacetate and two acetates. Oxaloacetate can be used to form other TCA intermediates for biosynthetic purposes or can be decarboxylated to PEP by the enzyme phosphoenolpyruvate carboxykinase.

The genes coding for isocitrate lyase and malate synthase are located in the same operon in E. coli and the synthesis of the enzymes is regulated so they are formed when two carbon intermediates are the sole source of carbon. The enzymes of the bypass are induced by acetate and repressed by most other carbon compounds. In E. coli the flow of isocitrate through the glyoxylate bypass, rather than the TCA cycle, is controlled by activation and deactivation of isocitrate dehydrogenase. When bacteria are grown on glucose, isocitrate dehydrogenase is fully operational, but when cells are grown on acetate, the activity of the enzyme is decreased by 70 percent (Chung et al. 1988). This decrease is the result of the enzyme being phosphorylated at its active site by isocitrate dehydrogenase kinase/phosphatase (Hurley et al. 1990). The gene encoding the kinase/phosphatase is in the same operon

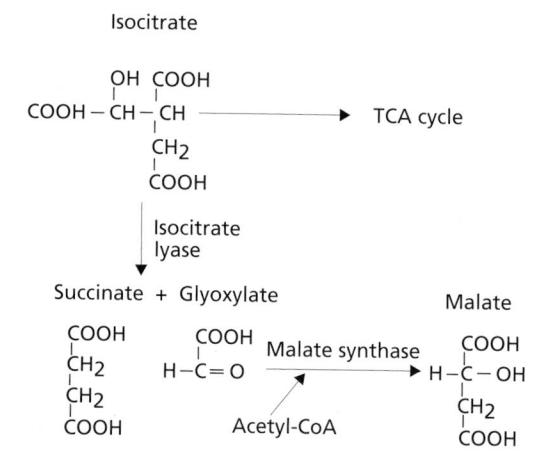

Figure 3.6 Reactions of the glyoxylate bypass. Oxaloacetate and acetate are converted to citrate and then isocitrate. The isocitrate is cleaved to succinate and glyoxylate and the glyoxylate is combined with another acetate to yield malate.

as the genes encoding isocitrate lyase and malate synthase. Some strains of *E. coli* cannot grow on acetate without the expression of this regulatory enzyme (Cortay et al. 1988).

If pyruvate or lactate are substrates, *E. coli* can use the enzyme phosphoenolpyruvate synthase and ATP to convert pyruvate to PEP, with AMP and inorganic phosphate as products. The HDP pathway can then be reversed (glucogenesis) to fructose-1,6-bisphosphate and a phosphatase can cleave phosphate from the 1-carbon position to yield fructose-6-phosphate. The PEP also can be combined with CO_2 (phosphoenolpyruvate carboxylase) to form oxaloacetate, which can be converted to other TCA-cycle intermediates.

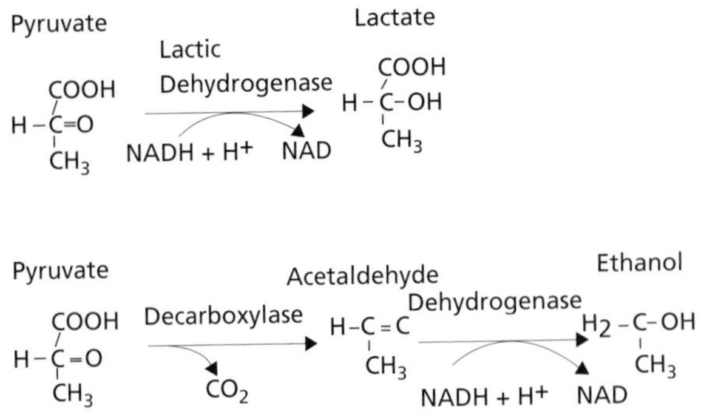

The glucose fermentations

In microbial physiology and metabolism the term 'fermentation' has a very specific meaning, i.e. the condition where the final electron acceptor of a catabolic pathway is an organic compound. An example of a process where this is important is in the oxidation of reduced NAD^+ produced by the HDP pathway. The NAD^+ reduced by the oxidation of glyceraldehyde-3-phosphate during the HDP pathway must be reoxidized if the pathway is to continue functioning. Therefore, fermentation is a process that can keep the HDP pathway functioning if no other electron acceptors are available. It should be pointed out that in industry the term 'fermentation' is used to denote many metabolic reactions caused by microorganisms, not just the specific definition described above.

THE HOMOLACTIC ACID FERMENTATION

Pyruvate, produced as a result of the HDP pathway, is an excellent acceptor for the electrons produced by the oxidation of glyceraldehyde-3-phosphate and most bacteria possess the metabolic capability of reducing pyruvate to lactic acid (see Figure 3.7). For some bacteria the reduction of pyruvate to lactate is the most important method of reoxidizing NADH and such bacteria produce large amounts of lactic acid as final products of glucose fermentation. For this reason, such bacteria are termed 'lactic-acid bacteria.' If glucose is the only carbon source provided to the cells, some of the products of glucose degradation by the HDP must be used for biosynthetic purposes to make new cell material. However, because the energy gained as a result of fermentation is relatively low, most of the carbon from the glucose must be converted to fermentation products. The term 'homolactic acid fermentation' means that significantly more than half of the carbon from the glucose fermented is in the form of lactic acid. It is apparent that any bacterium that carries out a homolactic acid fermentation must be able to survive in the presence of a large amount of lactic acid and at low pH.

Figure 3.7 *Formation of lactate and ethanol from pyruvate. Pyruvate can be directly reduced to lactate or decarboxylated to acetaldehyde, which can be reduced to ethanol.*

Lactic acid is the major product of the homolactic acid fermentation but other products may be formed in lesser amounts. Many homolactic-acid bacteria produce such fermentation products as acetate, ethanol, and formate, in addition to lactate. The conversion of pyruvate to formic acid, ethanol, and acetic acid rather than lactose actually increases the energy gained by the fermentation, but it also increases the amount of acid formed. Regulatory systems of homolactic-acid bacteria tend to shift the fermentation to larger amounts of these compounds at higher pH values, where acid production would be less likely to inhibit growth.

The purification of lactic acid dehydrogenase from a homolactic-acid bacterium led to the discovery that this enzyme was an allosteric enzyme with an absolute requirement for a positive effector. If the reduction of pyruvate to lactic acid does not proceed efficiently enough to reoxidize the NADH formed from the oxidation of glyceraldehyde-3-phosphate, the oxidation of the triose phosphate is inhibited. Because of the reversibility of the aldolase-catalyzed reaction, the internal pool of the positive effector of lactic acid dehydrogenase, fructose-1,6-bisphosphate, increases. This results in the activation of more molecules of lactic acid dehydrogenase and the more rapid reduction of pyruvate.

THE DIRECT ALCOHOL FERMENTATION

The direct alcohol fermentation or 'yeast type' of alcohol fermentation permits the oxidation of NADH from the HDP pathway without the formation of large amounts of lactic acid. Pyruvate is directly decarboxylated by pyruvate decarboxylase to yield acetaldehyde and CO_2. The enzyme alcohol dehydrogenase catalyzes the reduction of acetaldehyde to ethyl alcohol with electrons from NADH (see Figure 3.7). This mechanism of ethanol formation is common in yeast but, although many bacteria are capable of ethanol production, only a few do so by this pathway. Some of those bacteria using this yeast type of pathway for ethanol formation include those in the genera *Zymomonas*, *Acetomonas*, and *Sarcina ventriculi*.

THE FORMIC ACID FERMENTATION

Some bacteria have evolved the ability to obtain extra energy from pyruvate. In one mechanism for pyruvate fermentation, the enzyme pyruvate formate–lyase cleaves pyruvate to formic acid and a 2-carbon fragment that becomes acetyl-CoA. This enzyme is present in either an active or inactive state, with the active state having a free radical at its C-terminus (Sawers and Watson 1998). The cleavage of S-adenosylmethionine into methionine and 5'-deoxyadenosine is required for enzyme activation. The active form of the enzyme normally functions only under anaerobic conditions and is rapidly and irreversibly inactivated by oxygen (Sawer and Bock 1988). This enzyme is common in enteric bacteria and also accounts for the formic acid produced during some homolactic acid fermentations.

When pyruvate formate–lyase is active the NADH produced by the HDP pathway cannot be reoxidized by lactic-acid production. Instead, some of the acetyl group on the acetyl-CoA is reduced twice, first to acetaldehyde and then to ethanol. This reduction is catalyzed by a special alcohol dehydrogenase, a CoA-linked alcohol dehydrogenase. The energy of the acetyl-CoA bond is sacrificed in the reaction but two NADH can be oxidized for each ethanol produced. Facultative bacteria, such as *E. coli*, have a single large polypeptide responsible for both reductions. The *adhE* gene of *E. coli*, K-12, encodes a polypeptide of 891 amino acids, equivalent in size to a combined alcohol dehydrogenase and an acetaldehyde dehydrogenase, and may represent an evolutionary fusion product of separate genes. Some obligate anaerobes, such as *Clostridium*, have two proteins to catalyze the reduction of acetyl-CoA to ethanol, a CoA-linked acetaldehyde dehydrogenase and an alcohol dehydrogenase. These two separate polypeptides form a complex so that free acetaldehyde is not released.

The acetyl-CoA that is not needed for the oxidation of NADH can be used for the generation of additional ATP. The enzyme phosphotransacetylase catalyses the reversible transfer of acetate from acetyl-CoA to phosphate with the formation of acetyl phosphate. The enzyme acetate kinase, in a reaction that is also reversible, transfers the phosphate from acetyl phosphate to ADP, to form ATP and liberate acetate. Therefore, two pyruvate and two NADH can result in two formate, one ethanol and one acetate, with the gain of one ATP. This fermentation is also known as the mixed-acid fermentation and the metabolic reactions leading to these products are illustrated in Figure 3.8.

Mutants of *E. coli* lacking lactic dehydrogenase grow normally under anaerobic conditions, but if these mutants also lack pyruvate formate–lyase they cannot grow by fermenting glucose or other sugars (Mat-Jan et al. 1989). If mutants of *E. coli* lack the ability to produce ethanol from acetyl-CoA, they cannot grow anaerobically on

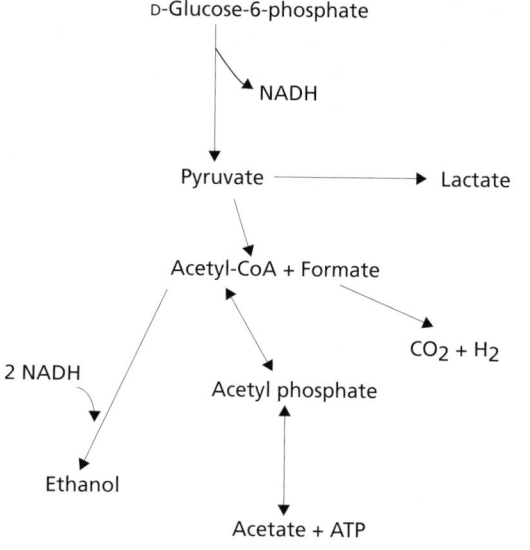

Figure 3.8 *Possible products of the formic-acid or mixed-acid fermentation. Pyruvate can be cleaved to yield acetyl-coenzyme A (acetyl-CoA) and formate. The acetyl-CoA can be converted to acetate, yielding a high-energy phosphate bond, or reduced to ethanol. If the proper enzymes are present, the formate can be oxidized to yield carbon dioxide and hydrogen gases.*

glucose unless provided with some alternative electron acceptor such as nitrate. However, an additional mutation also eliminating the pyruvate formate–lyase enzyme will restore the ability to grow on glucose anaerobically, with the cells now producing lactic acid.

BUTANEDIOL FORMATION

Some bacteria, for example some bacilli, have developed a pathway for the degradation of pyruvate that produces less acid. An α-acetolactate-synthesizing enzyme with a pH optimum at 6.0 converts two pyruvates to α-acetolactate and CO_2. This enzyme is sometimes named the pH 6.0 α-acetolactate-synthesizing enzyme to distinguish it from a biosynthetic enzyme that carries out the same reaction but possesses a pH optimum of 8.0. During the fermentative pathway the α-acetolactate is decarboxylated by α-acetolactate decarboxylase to produce another CO_2 and acetoin. The acetoin produced can be reduced to 2,3-butanediol by the action of butanediol dehydrogenase and the oxidation of NADH. This system of enzymes appears to be induced by acetate and at high pH values it is inhibited from operating (Zeng and Deckwer 1991). Because this pathway utilizes two pyruvate and results in the oxidation of only one NADH, it cannot be the sole fermentation route of glucose by a microorganism. The advantage of this pathway is the production of neutral products.

BUTYRATE AND BUTANOL FERMENTATION

In many strictly anaerobic bacteria and archaea, the oxidation and decarboxylation of pyruvate are catalyzed by pyruvate:ferredoxin oxidoreductase, an enzyme that catalyzes the oxidation of pyruvate to form acetyl-CoA and CO_2. The electrons are placed on the electron carrier ferredoxin, which may donate them to a hydrogenase with the eventual formation of hydrogen (H_2) gas. The class of electron carriers known as ferredoxins comprises small (6–13 kDa), iron–sulfur ([Fe-S]) proteins that function in a variety of reactions, including nitrogen fixation and photosynthesis.

The acetyl group from the acetyl-CoA obtained by the action of pyruvate:ferredoxin oxidoreductase can be used for the production of acetate and ATP, via phosphotransacetylase and acetate kinase, or utilized to generate electron acceptors for the NADH formed by the HDP pathway. In some bacteria, such as those of the genus *Clostridium*, an enzyme named acetyl coenzyme A acetyl transferase condenses two molecules of acetyl-CoA to produce one molecule of acetoacetyl-CoA. This latter compound can be reduced by β-hydroxybutyryl coenzyme A dehydrogenase and NADH to produce β-hydroxybutyryl-CoA. This latter compound can be dehydrated by 3-hydroxyacyl coenzyme A hydrolase (crotonase) to form crotonyl-CoA. The double bond in crotonyl-CoA is further reduced by butyl coenzyme A dehydrogenase to butyryl coenzyme A with electrons again coming from NADH. Thus, the overall series of reactions decarboxylates two pyruvates, with the acetyl groups becoming condensed into a four-carbon compound that undergoes two reduction steps to form butyryl-CoA.

The butyryl-CoA can have different fates. Acetyl transferase may transfer a free acetate on to the CoA group, producing acetyl-CoA and yielding free butyric acid. A second possibility is for phosphotransbutyrylase to transfer the butyryl group to phosphate to form butyryl phosphate. The phosphate can be donated to ADP to form ATP. Some clostridia produce free acetoacetate from acetoacetyl-CoA by replacing the acetoacetate with acetate. The free acetoacetate can be decarboxylated by acetoacetyl decarboxylase, producing acetone and CO_2, and the acetone may be reduced by isopropyl dehydrogenase yielding isopropyl alcohol.

The propionic acid fermentations

Some bacteria, for example those in the genera *Propionibacterium*, *Veillonella*, and *Bacteriodes*, produce propionic acid as a fermentation product. These bacteria can ferment lactic acid because reducing reactions involved in the production of propionic acid permit the oxidation of lactic acid to pyruvic acid. Some of the pyruvate formed as a result of fermentation is usually decarboxylated to yield acetate and CO_2. An additional product of this fermentation is succinic acid.

Oxaloacetate is an intermediate in the formation of propionate. To synthesize this four-carbon compound, a carboxyl group is recycled during the fermentation. The carboxyl group is transferred to the three-carbon compound pyruvate. The enzyme catalyzing this reaction is methylmalonyl-CoA-pyruvate transcarboxylase. The reduction of the oxaloacetate to malate, its conversion to fumarate, and the reduction of the fumarate to succinate can be used to dispose of electrons, such as those obtained from the HDP pathway or the oxidation of lactic acid. A CoA-transferase enzyme exchanges the succinate with the propionyl group on a molecule of propionyl-CoA to form free propionic acid and succinyl-CoA. An isomerase then converts the succinyl-CoA to methylmalonyl-CoA, which donates a carboxyl group to the enzyme–biotin complex to form a biotin–ADP–CO_2 complex and a molecule of propionyl-CoA. The activated carboxyl group is used to convert another pyruvate to oxaloacetate while propionic acid is liberated during the activation of another succinate. While this reaction is taking place, pyruvate is also being converted to acetate to balance the overall fermentation. By these reactions, a typical propionic acid bacterium such as *P. pentosoccium* will produce CO_2, acetic acid, propionic acid, and succinic acid as a result of glucose or lactic acid fermentation.

Clostridium propionicum also produces propionic acid as a fermentation product but the pathway differs from those above. Pyruvate is converted to propionic acid by a direct reduction route, involving reduction to lactate that is used to form lactyl-CoA. This latter compound is dehydrated, and reduced to propionyl-CoA. There is no formation of a four-carbon compound during this fermentation.

Another interesting use of oxaloacetate is as substrate for the Na^+-oxaloacetate decarboxylase, which is a membrane-bound Na^+ pump. The free energy of decarboxylation is sufficient to drive the development of a sodium gradient. This sodium gradient can be used to generate ATP by taking advantage of the electrochemical sodium gradient (see below). An example of a bacterium carrying out this reaction is *Propionibacterium modestum*, a strict anaerobe isolated from salt water.

The fermentation of ethanol

Some bacteria such as *Clostridum kluyveri* are able to gain energy by the fermentation of ethanol. Ethanol is oxidized and activated to form acetyl-CoA and the production of H_2 and the formation of fatty acids eliminate the electrons obtained by this oxidation. Two acetyl-CoA molecules are condensed to form acetoacetyl-CoA and this compound proceeds through the metabolic pathways described previously with two reduction steps giving butyryl-CoA. Additional acetyl molecules may be added with the production of larger fatty acids as final products. Pyruvate is synthesized by reversal of pyruvate:ferredoxin oxidoreductase.

The fermentation of pentoses

CONVERSION OF PENTOSES TO INTERMEDIATES OF THE HDP PATHWAY

Some bacteria are able to ferment certain 5-carbon sugars such as D-xylose and L-arabinose. The fermentation route of these pentoses by enteric bacteria generally involves isomerization of the aldopentose to the 2-keto-isomer or pentulose. A pentulose kinase then catalyzes phosphorylation to form the pentulose-5-phosphate. In such a manner, D-xylose is isomerized to D-xylulose and then phosphorylated to D-xylulose-5-phosphate, while L-arabinose is isomerized to L-ribulose and then phosphorylated to L-ribulose-5-phosphate. Epimerization reactions convert the pentulose-5-phosphates to D-xylulose-5-phosphate, which represents the common intermediate in the catabolism of these sugars. The exception is d-ribose, which is usually phosphorylated first and then isomerized to D-ribulose-5-phosphate.

The pentulose phosphates are converted to intermediates of the HDP pathway by a series of reactions known as the transketolase/transaldolase rearrangements. The enzyme transketolase (TK) cleaves 2-keto-sugars such as D-xylulose-5-phosphate between the 2- and 4-carbons and transfers the 2-carbon group to a proper aldehyde receptor, such as D-ribulose-5-phosphate, forming glyceraldehyde-3-phosphate and the seven-carbon carbohydrate sedoheptulose-7-phosphate. Transaldolase (TA) cleaves similar carbohydrates between the 3- and 4-carbons, transferring the 3-carbon moiety to an appropriate acceptor such as glyceraldehyde-3-phosphate. These rearrangements convert the 5-carbon sugar phosphate to intermediates of the HDP pathway such as glyceraldehyde-3-phosphate and fructose-6-phosphates with four- and seven-carbon carbohydrates as intermediates. The structures of some substrates and products of the TK/TA rearrangements are shown in Figure 3.9.

The conventional HDP pathway enzymes will convert glyceraldehyde-3-phosphate to pyruvate with the production of normal fermentation products. Because these enzyme-catalyzed reactions are reversible, pentose also can be synthesized from HDP pathway intermediates.

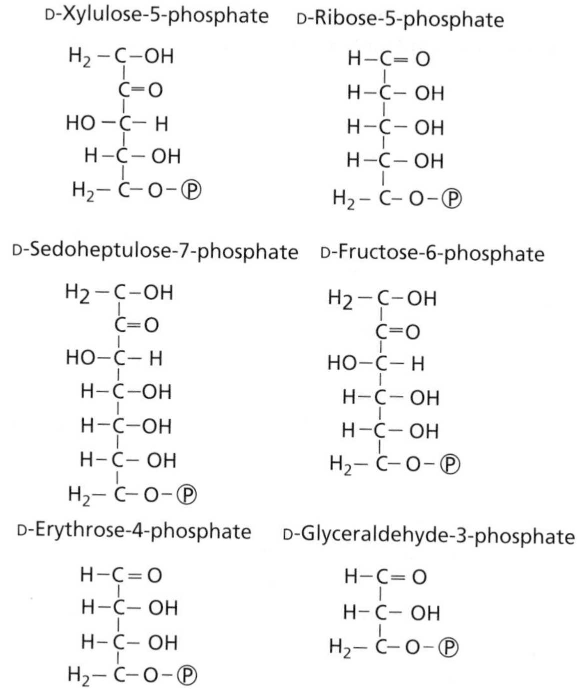

THE HEXOSE MONOPHOSPHATE OR PENTOSE PHOSPHATE PATHWAY

Many bacteria, including *E. coli*, can oxidize glucose-6-phosphate at the 1-carbon to form 6-phosphogluconic acid, which is then converted to a 5-carbon compound and metabolized using the TK and TA reactions. The enzyme that oxidizes glucose-6-phosphate, glucose-6-phosphate dehydrogenase, normally uses NADP, rather than NAD, as the electron acceptor. The 6-phosphogluconolactone formed by this reaction is converted to 6-phosphogluconate by a lactonase. The 6-phosphogluconate can be further oxidized and decarboxylated

Figure 3.9 *Carbohydrates involved in the transketolase/transaldolase reactions. D-Xylulose-5-phosphate, D-fructose-6-phosphate, and D-sedoheptulose-7-phosphate can serve as substrates to be cleaved by transketolase or transaldolase. D-Glyceraldehyde-3-phosphate, D-erythrose-4-phosphate, and D-ribose-5-phosphate can serve as acceptors for the 2- or 3-carbon moiety being transferred.*

by 6-phosphogluconate dehydrogenase to produce CO_2, D-ribulose-5-phosphate, and, usually, NADPH. In many bacteria, such as the enterics, this is primarily a biosynthetic route, providing the cells with pentose for RNA and DNA synthesis and NADPH for biosynthetic reducing power, and in aerobically growing *E. coli*, about 15 percent of the glucose is diverted to pentoses by this route. However, the pathway can be used for glucose catabolism with the pentose phosphates converted to HDP intermediates via the TK/TA rearrangements. This route of glucose catabolism does not involve a sugar diphosphate as an intermediate and is known as the hexose monophosphate pathway (HMP) or sometimes the pentose phosphate pathway. With an *E. coli* mutant unable to use the HDP, such as a glucose-6-phosphate isomerase-negative mutant, the cells can still grow on glucose using the HMP pathway, but at a slower rate than when the HDP pathway is employed. Some bacteria, such as *Thiobacillus novellus* and *Brucella abortus*, use only the HMP pathway for glucose metabolism.

Because fructose-6-phosphate can be one of the products produced from pentulose phosphate by the TK/TA reactions, the potential exists for some organisms to convert fructose-6-phosphate to glucose-6-phosphate and oxidize this compound to 6-phosphogluconate and then to ribulose-5-phosphate. In such a manner an HMP cycle may exist, with the oxidation of hexoses not requiring the oxidation of glyceraldehyde-3-phosphate.

THE HETEROLACTIC ACID FERMENTATION PATHWAY

A bacterium using the HDP pathway to ferment glucose to pyruvate can gain two ATP per glucose used by substrate-level phosphorylation, but the energy yield is less for the fermentation of a pentose. Theoretically, a homolactic acid fermenter using the HDP pathway could gain six ATP by the fermentation of three glucose to six lactate. However, it could only gain five ATP by the fermentation of three pentoses to five lactate using the pentose phosphate route and the TK/TA rearrangements.

Some lactic-acid bacteria have evolved a more efficient pathway for the fermentation of 5-carbon sugars. Instead of the D-xylulose-5-phosphate being cleaved by TK, it is cleaved by an enzyme termed phosphoketolase in a reaction that results in the esterification of inorganic phosphate and the formation of glyceraldehyde-3-phosphate and acetyl phosphate as products. The glyceraldehyde-3-phosphate can be converted to pyruvate by normal HDP pathway enzymes, yielding a gain of two ATP. The acetyl phosphate can be used to phosphorylate ADP to generate an ATP through the action of acetate kinase. Because one ATP is used to form the pentulose phosphate, the result is a net gain of two high-energy phosphate bonds for each pentose fermented. Bacteria that possess this fermentation route for pentoses normally possess high activity for lactic dehydrogenase and reduce most of the pyruvic acid to lactic acid. Therefore, the fermentation products from 5-carbon sugars are primarily lactate and acetate in equal amounts.

Bacteria possessing phosphoketolase are often lacking key enzymes of the hexose diphosphate pathway such as phosphofructokinase and aldolase and are unable to use the complete HDP pathway for the catabolism of glucose. If provided with D-glucose as a carbon and energy source, they use the enzymes glucose-6-phosphate dehydrogenase and 6-phosphogluconate dehydrogenase to form D-ribulose-5-phosphate and CO_2. The ribulose-5-phosphate can be epimerized to D-xylulose-5-phosphate, which is cleaved by phosphoketolase to produce glyceraldehyde-3-phosphate and acetyl phosphate. The glyceraldehyde-3-phosphate is converted to pyruvate and lactate in a normal manner for a homolactic acid fermentation. However, in order to dispose of all of the electrons from the oxidation of the six-carbon sugar to the pentulose, the acetyl phosphate must be reduced to ethanol in two reduction steps. Thus, the products of the fermentation of glucose by this route are CO_2, ethanol, and lactic acid and the energy yield is only one ATP per hexose fermented. This fermentation of glucose is termed a heterolactic acid fermentation because the three major products are produced in approximately equal amounts. The CO_2 is derived from the 1-carbon of

the hexose, the ethanol from the 2- and 3-carbons, and the lactic acid from the 4-, 5-, and 6-carbons.

Bacteria in the genus *Bifidobacterium* carry out a variation of this fermentation. These bacteria also lack the aldolase of the HDP pathway but possess a fructose-6-phosphate phosphoketolase that converts fructose-6-phosphate to acetyl phosphate and erythrose-4-phosphate. The acetyl phosphate can be used for the production of ATP with acetate being liberated. The transketolase and transaldolase enzymes can act upon the erythrose-4-phosphate and fructose-6-phosphate, producing 7- and 5-carbon carbohydrates and eventually forming some glyceraldehyde-3-phosphate that can be converted to pyruvate. Thus, these bacteria ferment glucose to large amounts of acetic acid with smaller amounts of lactic acid, ethanol, and formic acid.

Alternative routes of carbohydrate catabolism

THE KDPG PATHWAY

An alternative route for the conversion of glucose to pyruvate is known as the Entner/Douderoff pathway or the 2-keto-3-deoxy-6-phosphogluconate (KDPG) pathway, because this latter compound is an intermediate unique to this metabolic route. The two enzymes unique to the KDPG route are 6-phosphogluconate dehydratase and KDPG aldolase. The first reaction of the pathway is oxidation of glucose-6-phosphate to 6-phosphogluconate by the action of glucose-6-phosphate dehydrogenase. The 6-phosphogluconate is dehydrated by 6-phosphogluconate dehydratase, to form KDPG. This latter compound is cleaved by KDPG aldolase, producing pyruvate from the first three carbons and glyceraldehyde-3-phosphate from the 4-, 5-, and 6-carbons. These two reactions are shown in Figure 3.10. The pyruvate can be oxidized and decarboxylated to form acetyl-CoA and the acetyl group can be oxidized via the TCA cycle, while the glyceraldehyde-3-phosphate produced by the action of KDPG aldolase can be converted to pyruvate by the normal enzymes of the HDP pathway.

The KDPG route in enteric bacteria

The KDPG route is not normally used by *E. coli* cells catabolizing glucose but can be used when gluconic acid is the carbon source. *E. coli* catabolizes gluconate by first using gluconokinase to phosphorylate it to 6-phosphogluconate. Some of the 6-phosphogluconate may be oxidized to D-ribulose-5-phosphate and CO_2, and degraded by the HMP pentose route. Most, however, is dehydrated by 6-phosphogluconate dehydratase to KDPG, which is cleaved by KDPG aldolase. Because *E. coli* is unable to oxidize glucose to gluconate, the KDPG route is not induced when D-glucose is the substrate.

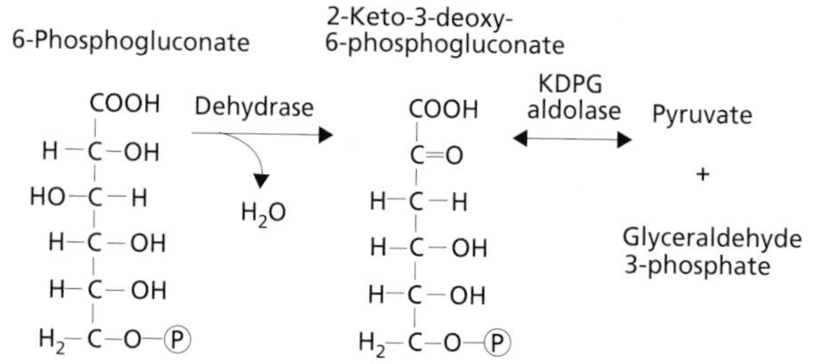

6-Phosphogluconate 2-Keto-3-deoxy-6-phosphogluconate

Figure 3.10 *Reactions of the 2-keto-3-deoxy-6-phosphogluconate (KDPG) route to pyruvate. Two reactions convert 6-phosphogluconic acid to pyruvate and glyceraldehyde-3-phosphate.*

With mutants of *E. coli* deficient in KDPG dehydratase, all gluconate must be metabolized through the pentose phosphate route. If another mutation occurs so that the organism also loses 6-phosphogluconate dehydrogenase activity, the cells cannot grow on gluconate at all.

Why does *E. coli* retain the capacity to use gluconate as a carbon source? One explanation is that it provides an advantage during growth in a specialized environment. Evidence has indicated that one such environment might be the large intestine. Mutants in which the gene encoding KDPG aldolase had been disrupted have been shown to be incapable of colonizing the large intestine of the mouse (Sweeney et al. 1996).

Analysis of the genome of *E. coli* has also revealed that there are two gluconate transporters. It has been shown, however, that one of these transporters is not used for gluconate but for the sugar acid L-idonic acid (Bausch et al. 1998). This compound is apparently transported without phosphorylation and then converted to 6-phosphogluconate. Enzymes of the KDPG pathway then metabolize the 6-phosphogluconate.

GLUCOSE CATABOLISM IN THE GENUS *PSEUDOMONAS*

The phosphorylated routes for glucose catabolism

One pathway of glucose catabolism in the pseudomonads involves phosphorylated intermediates. The phosphorylated route involves the phosphorylation of glucose to glucose-6-phosphate and its oxidation to 6-phosphogluconate. Much of the 6-phosphogluconate formed will be degraded through the KDPG route, although some may be diverted to the HMP route. Glucose-6-phosphate isomerase is usually present, and aldolase may also be present, but phosphofructokinase is often lacking or very weak, so the HDP route usually does not function

Aldolase can be present to form fructose-1,6-bisphosphate from triose phosphates; therefore, some of the pseudomonads have the ability to convert the glyceraldehyde-3-phosphate formed by KDPG aldolase back to fructose-1,6-bisphosphate. Phosphatase activity can

cleave phosphate to form fructose-6-phosphate and this can be isomerized to glucose-6-phosphate, which can be oxidized and converted to KDPG. This route, from hexose phosphate to KDPG, to triose phosphate, and back to hexose phosphate, can be considered as a KDPG–hexose cycle. Some mutants blocked in the conversion of glyceraldehyde-3-phosphate to pyruvate can still grow on glucose by using this cycle (Lessie and Phibbs 1984).

In many bacteria, the oxidation of glucose-6-phosphate to 6-phosphogluconate also serves an anabolic or biosynthetic function, producing pentoses and reducing power (NADPH) for biosynthetic reactions. In aerobes where glucose-6-phosphate dehydrogenase is also an important catabolic enzyme, the reaction may need to serve both catabolic and biosynthetic purposes. In some cases, either NAD or NADP may serve as the electron acceptor for glucose-6-phosphate dehydrogenase, depending upon the needs of the cell. In other cases, isozymes may be involved in the reaction, one isozyme reducing NAD and another reducing NADP.

The route of catabolism for 5-carbon carbohydrates by pseudomonads can be similar in many ways to the KDPG route of glucose catabolism. The aldopentose substrate can be oxidized directly to a pentonic acid and a dehydration reaction can follow to form the 2-keto-3-deoxy-pentonic acid. This may be cleaved by an aldolase to yield pyruvate and glycolaldehyde with the glycolaldehyde oxidized to glycolic acid. In this manner D-xylose can be oxidized to D-xylonic acid, which can be dehydrated to 2-keto-3-deoxy-xylonic acid. In a similar manner, D-ribose can be oxidized to ribonic acid, which can be dehydrated to 2-keto-3-deoxy-ribonic acid.

The oxidative pathway for hexoses

In addition to the phosphorylated pathways for the catabolism of sugars described above, many pseudomonads also possess oxidative pathways. In these cases, flavin or cytochrome-linked, membrane-bound oxidase enzymes, located in the periplasmic space, oxidize the sugars with electrons passing from the electron acceptor directly to O_2. Glucose oxidase oxidizes glucose to form gluconate, while other oxidases can convert the gluconate to keto-sugars such as 2-ketogluconate and 2,6-diketogluconate. The glucose oxidase (sometimes called dehydrogenase) of *Pseudomonas fluorescens* was found to be a quinoprotein, whereas the gluconate oxidase was found to be a flavoprotein. Such oxidation reactions, associated with the cell membrane, provide the cells with rapid energy. Pseudomonads lacking glucose oxidase are unable to produce gluconate and cannot induce the KDPG pathway enzymes.

The keto-sugars formed by the oxidative reactions may be later transported into the cells, phosphorylated, and further degraded by phosphorylated routes. Pseudomonads lacking gluconokinase cannot form glucose-6-phosphate from glucose but may utilize the oxidized products from the periplasm. Some pseudomonads can grow in the absence of oxygen by using nitrate as an electron acceptor. However, then they must use the direct phosphorylated pathways for catabolism because nitrate will not replace oxygen in the oxidative pathways.

If the oxygen concentration is low, the catabolism of glucose in *Pseudomonas aeruginosa* is decreased from the extracellular, direct oxidative route to the intracellular phosphorylative route. The change results in a decrease in the concentrations of glucose dehydrogenase and gluconate dehydrogenase activities and decreases the transport systems for gluconate and 2-ketogluconate, but there is increased activity for glucose transport (Mitchell and Dawes 1982). Global regulatory controls of these strict aerobes differ from those for the enteric bacteria. The major energy supply comes from oxidative phosphorylation reactions and substrates that can be quickly oxidized, such as acetate or TCA-cycle intermediates, cause catabolite repression of glucose catabolism enzymes (MacGregor et al. 1991).

GLUCOSE FERMENTATION BY *ZYMOMONAS*

Bacteria in the genus *Zymomonas* are obligate fermentative bacteria with substrate-level phosphorylation as the only source of energy. They are not, however, strict anaerobes and can grow in the presence of oxygen. The fermentation of glucose produces ethanol, CO_2, and lactic acid as primary products. The pathway of glucose catabolism is unusual in that it is the KDPG route, a pathway with a lower energy yield. The KDPG route is not an efficient route for the fermentation of glucose because only one ATP can be gained per glucose fermented. The pyruvate formed from glucose is oxidized to acetaldehyde and CO_2 by a pyruvate decarboxylase similar to the enzyme common in yeast and the acetaldehyde is directly reduced to ethanol by alcohol dehydrogenase. The pyruvate decarboxylase has an especially efficient promoter and represents about 4 percent of the total cellular protein (Neale et al. 1987). *Zymomonas mobilis* has no TCA cycle and only remnants of a respiratory chain. The KDPG pathway is constitutive and the glucose-6-phosphate dehydrogenase can use either NAD or NADP as the electron acceptor, so reducing power for biosynthetic reactions can be provided.

OTHER BACTERIA

A weak HDP pathway, an active KDPG route, and an active hexose monophosphate pathway characterize the metabolism of bacteria in the genus *Acetobacter*. These bacteria may utilize a hexose cycle with triose phosphate, originating either from the KDPG route or the pentose phosphate route, going to fructose-6-phosphate, glucose-6-phosphate, and 6-phosphogluconate. Bacteria classified as *Gluconobacter* do not possess a complete TCA cycle, a complete HDP pathway, or a KDPG pathway. They can use glucose either by a terminal oxidative route or a pentose phosphate cycle.

Other strictly aerobic bacteria tend to favor the degradation of glucose by the KDPG route and the HDP is rarely used, although some glucose carbon may pass through the pentose phosphate pathway. As examples, bacteria in the genus *Agrobacterium* have been reported to catabolize glucose 55 percent by the KDPG pathway and 44 percent by the pentose route. For bacteria in the genus *Azotobacter* there is no HDP pathway; the pentose pathway is of minor importance but the KDPG route is the major pathway for glucose degradation. Bacteria in the genus *Sphaerotilus* also have been reported to metabolize glucose via the KDPG and pentose phosphate pathways, as have bacteria in the genus *Rhizobium*. For bacteria in the genus *Arthrobacter*, however, an HDP pathway as well as a pentose pathway has been reported. For *Neisseria gonorrhoeae*, about 80–87 percent of glucose goes via the KDPG pathway, while 13–20 percent has been reported to pass through the HMP route.

BIOSYNTHETIC REACTIONS

Small-molecule biosynthesis

AMINO ACIDS

In general, the normal routes of synthesis of amino acids produce the α-oxoacid and a transamination reaction then adds the amino group. The amino acids may be grouped into six families based upon their routes of synthesis (Moat and Foster 1995). Aspartate, threonine, methionine, isoleucine, asparagines, and lysine make up the aspartic acid family, which originates from an intermediate in the TCA cycle, oxaloacetate. Another TCA-cycle intermediate, α-ketoglutarate (α-oxoglutarate), is converted to glutamic acid, glutamine, and arginine. Serine, glycine, and cysteine all require the triose 3-phosphoglycerate in their formation and thus are grouped together. The aromatic acid family comprises phenylalanine, tyrosine, and tryptophan. Erythrose-4-phosphate and phosphoenolpyruvate, both produced during carbohydrate metabolism, are required for the synthesis of the aromatic amino acids. Alanine, valine, and leucine make up the pyruvate family. Alanine is formed directly by transamination of pyruvate, whereas synthesis of valine and leucine requires condensation of two pyruvates to α-acetolactate, which is eventually converted to valine and leucine. Histidine synthesis is independent of the other families and requires ATP and phosphoribosylpyrophosphate.

PURINES AND PYRIMIDINES

The cells must have a supply of phosphorylated ribonucleotides and deoxyribonucleotides for the synthesis of

RNA and DNA. Purines and pyrimidines are synthesized by quite different routes. The purine ring has as its starting material ribose-5-phosphate while the pyrimidine ring originates in aspartate plus carbamoyl phosphate. The pathways of biosynthesis yielding inosine-5'-phosphate (IMP), a critical intermediate in the synthesis of purines, and uridylic acid (UMP), in the pyrimidine pathway, are complex. In bacteria such as *E. coli*, deoxyribonucleoside diphosphates are produced by the reduction of the corresponding ribonucleotide diphosphates in a reaction catalyzed by a ribonucleotide reductase. In a few species of bacteria, vitamin B_{12} is required for the reduction that occurs at the nucleoside triphosphate level. Methylation of the uracil ring to produce thymine occurs at the level of deoxyuridine monophosphatase (dUMP) with methylenetetrahydrofolate usually the donor of the C1 group and the reductant.

FATTY ACIDS AND LIPIDS

Lipids may serve as reserve food materials or as constituents of bacterial membranes. Lipids found in bacteria include phospholipids, lipoproteins, glycolipids, and the lipopolysaccharide of gram-negative bacteria. Most lipids are phospholipids that contain fatty acids ester-linked to glycerol phosphate. Bacteria synthesize fatty acids from acetyl-CoA using an acyl carrier protein (ACP) with a prosthetic group of 4'-phosphopantetheine. Interestingly, synthesis proceeds through the production of a 3-carbon intermediate, malonyl-CoA, produced by the carboxylation of acetyl-CoA. Growth of the fatty-acid chain is achieved by condensation of malonyl-ACP to the extending chain. The newly condensed molecule contains a β-keto group that is removed by reduction of the keto to a hydroxyl; then a dehydration removes the hydroxyl leaving a double bond followed by reduction of the double bond to leave a saturated chain. Synthesis of unsaturated fatty acids from the extended saturated chain can be accomplished in one of two ways. Under anaerobic conditions, the β-hydroxydecanoyl-ACP is acted upon by a special dehydrase to yield a *cis* double bond between carbon atoms 3 and 4. Two-carbon units are added until the proper length is obtained. Under aerobic conditions, molecular oxygen is used to introduce hydroxyl groups, which can be removed by dehydration reactions to produce double bonds.

Degradation of fatty acids is not the reverse of fatty-acid synthesis. Fatty acids are degraded using the β-oxidation pathway. This pathway involves introduction of a double bond into the fatty acid-CoA derivative. A keto group is the introduced β to the carboxyl group by hydration of the double bond and then oxidation of the hydroxyl. The keto group is then attacked by CoA, leading to the production of an acetyl-CoA and leaving the remaining fatty acid-CoA two carbons shorter.

TCA-cycle intermediates as biosynthetic compounds

Bacterial cells need many intermediates for biosynthetic purposes, including 6-carbon sugars for synthesis of cell walls, triose phosphates to make glycerol and lipids, and acetyl-CoA for the synthesis of fatty acids. In aerobic bacteria, the TCA-cycle functions to oxidize acetate and provide energy. However, another major function of the TCA-cycle enzymes is to produce intermediates such as α-ketoglutaric acid and oxaloacetic acid for biosynthetic reactions. During the TCA cycle, one oxaloacetate combines with the acetyl group from an acetyl-CoA to yield citric acid, which is eventually converted back to oxaloacetate and there is no net gain in di- or tricarboxylic acid intermediates. Because some of these intermediates are being used for biosynthetic reactions, mechanisms must exist for their net synthesis. A very important enzyme, especially in enteric bacteria, to replenish oxaloacetate loss as a result of anabolic reactions is phosphoenolpyruvate carboxylase, which forms oxaloacetate from phosphoenolpyruvate and CO_2. The enzyme phosphoenolpyruvate carboxylase is biosynthetic in its regulation and is inhibited by aspartate. Mutants of enteric bacteria that lack the enzyme are unable to grow on a glucose-salts minimal medium unless the medium is supplemented with a compound capable of providing a TCA-cycle intermediate. Pyruvate carboxylase also catalyzes the production of oxaloacetate but uses pyruvate as initial substrate. This enzyme is found in nonenterics such as *B. subtilis* and *Rhizobium etli*. It is also found in eukaryotic organisms.

If the TCA reactions are being used strictly for generation of biosynthetic precursors, oxaloacetate is a key intermediate that is metabolized via two separate pathways. In one pathway oxaloacetate is converted to succinyl-CoA, an essential biosynthetic precursor, by using a reductive TCA pathway. In this pathway malate dehydrogenase and malate dehydrase (fumarase) convert oxaloacetate to fumarate. Then fumarate reductase, instead of the normal TCA cycle enzyme succinate dehydrogenase, reduces the fumarate to form succinate. Succinyl-CoA is formed from succinate using succinate thiokinase. Oxaloacetate and acetyl-CoA can also yield citric acid, which can be converted to α-ketoglutaric acid, another essential precursor, by normal TCA-cycle enzymes. This splitting of the TCA cycle into a reductive and an oxidative branch is common among fermenting bacteria.

Pathways for incorporation of one-carbon compounds into cellular material

Bacteria that can grow on CO_2 as the sole source of carbon are termed autotrophs. However, bacteria that cannot grow on CO_2 as the sole carbon source also use CO_2 for biosynthesis of cell material; for example, phosphoenolpyruvate carboxylase, which was discussed above, incorporates CO_2 into cellular precursors. Because these reactions are essential for growth under certain conditions, CO_2 is an essential nutrient for most bacteria. However, enzymes like phosphoenolpyruvate carboxylase do not permit growth on CO_2 as the sole carbon source. Autotrophs rely on one of several unique pathways that allow CO_2 to be used as the sole carbon source.

THE CALVIN CYCLE FOR CO_2 FIXATION

The best-studied pathway of CO_2 incorporation for autotrophs is the Calvin cycle. The Calvin cycle, sometimes termed the reductive pentose phosphate pathway, is found in some photosynthetic bacteria and many chemoautotrophs and involves the conversion of a 5-carbon sugar and CO_2 to form two molecules of 3-phosphoglyceric acid. The unique enzymes of the Calvin cycle are phosphoribulokinase, converting ribulose-5-phosphate to ribulose-1,5-bisphosphate, and ribulose bisphosphate carboxylase (sometimes called ribulose-1,5-bisphosphate carboxylase/oxygenase). Ribulose bisphosphate carboxylase converts ribulose bisphosphate and CO_2 to two 3-phosphoglyceric acids. To incorporate the fixed CO_2, the 3-phosphoglyceric acid is phosphorylated, using ATP and phosphoglycerate kinase, to 1,3-diphosphoglyceric acid. This latter compound is reduced by glyceraldehyde-3-phosphate dehydrogenase to glyceraldehyde-3-phosphate, which can be converted to dihydroxyacetone phosphate by triosphosphate isomerase. Aldolase can catalyze the synthesis of fructose-1,6-bisphosphate from the triose phosphates and fructose-1,6-bisphosphate phosphatase can cleave the phosphate from the 1-carbon position to produce fructose-6-phosphate. Enzymatic reactions similar to those of the HMP pathway regenerate the pentose phosphates from hexose phosphates. The final product of these rearrangements is ribulose-5-phosphate, which is converted to ribulose-1,5-bisphosphate by phosphoribulokinase.

THE REDUCTIVE TCA CYCLE FOR CO_2 FIXATION

Bacteria in the genera *Cyanobacteria*, *Rhodospirillaceae*, *Chromatiaceae*, and *Chloroflexaceae* will use the Calvin cycle for CO_2 fixation when growing with CO_2 as the only carbon source. Some bacteria in the genus *Chlorobiaceae*, however, do not possess the enzymes of the Calvin cycle. This family of organisms uses the reductive TCA cycle or the reductive C_4 dicarboxylic acid CO_2 fixation pathway. In this pathway acetyl-CoA is carboxylated by a ferredoxin-dependent pyruvate synthase to form pyruvate. To convert the pyruvate to PEP the enzyme PEP synthetase and ATP are used to form PEP, AMP, and inorganic pyrophosphate. The PEP is carboxylated to oxaloacetate by PEP carboxylase and

the normal TCA cycle is reversed, with the oxaloacetate reduced to malate and the malate dehydrated to fumarate. The fumarate is reduced to succinate by fumarate reductase, and the succinate is phosphorylated and changed to succinyl coenzyme A by succinyl-CoA synthetase. The enzyme reaction producing α-ketoglutarate during the normal TCA cycle is irreversible, so a special enzyme is employed during the reductive TCA cycle, an α-ketoglutarate:ferredoxin oxidoreductase. The α-ketoglutarate is changed to citrate by normal enzymes of the TCA cycle, and an ATP-citrate lyase cleaves citrate to form oxaloacetate and acetyl-CoA. Through these reactions, four molecules of CO_2 have been converted to one molecule of oxaloacetate, which can be used for other biosynthetic purposes.

OTHER PATHWAYS FOR INCORPORATION OF ONE-CARBON COMPOUNDS INTO CELLULAR MATERIAL

Methylotrophs are bacteria that can use 1-carbon compounds other than CO_2 as their sole carbon source. Compounds of this type include methane, methanol, formate, and methylamine. When growing on methane it is first converted to methanol by methane monooxygenase (MMO). The methanol is then oxidized to formaldehyde, which can be further oxidized to CO_2 to provide energy or can be converted into cellular material. The MMO exists in either of two forms: a cytoplasmic (soluble) form and a membrane-bound (particulate) form. The soluble form is stable to purification and has been studied in detail. The particulate form is quite labile and little is known about its structure other than that it contains copper.

The incorporation of formaldehyde into cell carbon can proceed by either of two pathways. One pathway is referred to as the serine pathway. This pathway leads to the production of acetyl-CoA from formaldehyde and CO_2. The pathway is known as the serine pathway because the formaldehyde is added to glycine to form serine. This is eventually converted to PEP, which is carboxylated to give oxaloacetate using PEP carboxylase. Through a series of reactions this four-carbon compound is converted into glyoxylate, which is converted to glycine to keep the cycle going, and acetyl-CoA, which can be used for growth.

The other pathway for formaldehyde incorporation is the ribulose monophosphate (RuMP) pathway. In this pathway formaldehyde is added to D-ribulose monophosphate by the enzyme hexulose phosphate synthetase. The product is D-fructose-6-phosphate, which can be converted to other carbohydrates such as glucose-6-phosphate and glyceraldehyde-3-phosphate, or converted back to ribulose-5-phosphate by means of the TK and TA rearrangements. The hexulose phosphate synthetase is specific for D-ribulose-5-phosphate as a substrate but can use other aldehydes, such as glycolaldehyde, in place of formaldehyde (Kato et al. 1988;

Beisswenger and Kula 1991). Most methylotrophs use either the serine or RuMP pathway for incorporation of 1-carbon compounds into cell material.

Methanogens are bacteria capable of producing methane from mainly CO_2 and H_2 (see below). The methane-producing bacteria have the ability to synthesize acetate from 1-carbon compounds. The methyl group of acetate comes from methyltetrahydromethanopterin, which is a folate analog and an intermediate in the reduction of CO_2 to methane. Another molecule of CO_2 is reduced to a bound carbon monoxide by carbon monoxide dehydrogenase and the methane and carbon monoxide are condensed to form acetyl-CoA.

Nitrogen fixation

Many genera of bacteria, including both eubacteria and archaea, are capable of utilizing nitrogen gas (N_2) as the sole source of nitrogen for biosynthetic reactions. The reduction of N_2 to ammonia takes place using an enzyme complex termed nitrogenase. The major nitrogenase of most organisms is a molybdo-iron protein. The active complex has two component proteins, with one termed the iron protein and the other the molybdenum-iron protein (Georgiadis et al. 1992; Chan et al. 1993). The iron protein is a homodimer and the molybdenum-iron protein is a complex with an α-2β-2 structure. The intact complex contains several novel [Fe–S] centers and a novel iron-molybdenum cofactor.

The catalytic reaction involves the iron protein passing an electron to the molybdenum-iron protein, which is the site of nitrogen reduction. This step requires ATP hydrolysis, with a stoichiometry of two ATP per electron transferred. This electron transfer must be repeated six times to reduce the nitrogen to ammonia. The exact mechanism of nitrogen reduction at the iron-molybdenum cofactor remains unclear. In addition to the electrons required for nitrogen reduction there is also obligatory proton reduction during the catalytic cycle, leading to the net consumption of eight electrons per nitrogen reduced. This means that 16 ATP are hydrolyzed per nitrogen reduced.

One of the best studied model organisms for nitrogen fixation is *Azotobacter vinelandii*. This bacterium not only contains the iron-molybdenum nitrogenase but also can synthesize two other nitrogenases. If the growth medium contains molybdenum, then the molybdenum-containing nitrogenase I is synthesized. If the medium is lacking molybdenum but contains vanadium, then nitrogenase 2, which contains an iron-vanadium cofactor in place of the iron-molybdenum cofactor, is produced (Bishop et al. 1982). If the medium does not contain sufficient amounts of either of these two metals, then nitrogenase 3 is synthesized, which contains only iron (Pau et al. 1989). Alternative nitrogenases have also been found in other bacteria.

Nitrogenase activity is normally repressed when other nitrogen compounds are available, and its activity is inhibited by ADP, destroyed by O_2, and repressed by ammonia. In aerobic nitrogen fixers the enzyme must be protected from oxygen. A posttranslational regulation of nitrogenase activity has also been found in many different nitrogen-fixing bacteria (Ludden 1994). It involves the transfer of the ADP-ribose from NAD to the Arg-101 residue of one subunit of the dinitrogenase reductase by the enzyme dinitrogenase reductase ADP-ribosyltransferase (DRAT). This group can be removed by another enzyme, dinitrogenase reductase-activating glycohydrolase (DRAG), which restores nitrogenase activity.

RESPIRATION

Aerobic respiration

Aerobic respiration is the condition when oxygen is the final electron acceptor of catabolism. In bacteria that can utilize aerobic respiration, the presence of oxygen will result in the repression of any fermentation pathways or anaerobic respiration pathways. A number of different electron carriers are involved in bacterial respiration, including NAD and NADP, flavoproteins, quinones, [Fe-S] proteins, and cytochromes. Flavoproteins are oxidation/reduction enzymes that contain flavin adenine mononucleotide (FMN) or flavin adenine dinucleotide (FAD). Flavins are synthesized from riboflavin and, upon reduction, transport two protons and two electrons. The actual redox potential of the oxidation/reduction reaction of flavoproteins can vary depending upon the protein group, although a range of 0.0–0.2 V is normal. Quinones are hydrophobic electron carriers that can shuttle electrons and protons through the lipid bilayer. The two most common quinones are ubiquinone (coenzyme Q) and menaquinone (vitamin K or MK). Proteins containing non-heme iron- and sulfur-containing clusters, referred to as iron-sulfur proteins, are components of almost all respiratory chains. Some well-studied iron-sulfur proteins are ferredoxin, rubredoxin, and a high-potential iron protein (HiPIP). These [Fe-S] centers carry only one electron and with midpoint potentials ranging from -0.40 to $+0.35$ V can be involved throughout the respiratory chain. Cytochromes are redox proteins that contain heme and transport a single electron. Cytochromes have characteristic light-absorption bands in the visible range and are categorized on the basis of their absorption characteristics. One type of cytochrome, the *c*-type, is additionally distinguished because the heme is covalently attached to the protein by thioether linkages.

The standard mitochondrial respiratory chain serves as a useful model because it was originally derived from the bacterial respiratory chain. The chain is divided into four protein complexes, complexes I–IV, and additional components required to shuttle electrons. All of the components discussed below are found in bacteria. The major difference between bacterial and mitochondrial electron transport chains is that the bacterial respiratory complexes have a simpler subunit structure.

Complex I is the NADH-ubiquinone oxidoreductase, which oxidizes reduced NADH and transfers electrons to ubiquinone. This complex is made up of at least 20 different polypeptides and contains iron-sulfur proteins and flavoproteins. Electrons passed to the quinone along with two protons convert the quinone to quinol. This can then diffuse to other electron acceptors such as complex III. Complex III is also called the bc_1 complex or the ubiquinol-cytochrome c oxidoreductase. In bacteria this complex is made of three major subunits. Electron flow through the bc_1 complex is complex, with the two electrons from quinol eventually flowing to two different acceptors. This flow has been termed the Q cycle. Data from studies of the high-resolution structure of the bc_1 complex have suggested movement of the polypeptides during electron flow is critical for determining the eventual oxidant (Zhang et al. 1998). One of the terminal oxidants in the Q-cycle is a soluble cytochrome c. This cytochrome c can readily diffuse between different donors and acceptors to facilitate electron flow during respiration. Cytochrome c serves as a reductant for complex IV, also termed the cytochrome c oxidase, which reduces oxygen to water. This protein is a heme-copper protein whose structure has been solved (Iwata et al. 1995; Yoshikawa et al. 1998). A hallmark of this type of oxidase is the presence of a heme and copper in close proximity resulting in a structure termed a binuclear center. This is the site of oxygen reduction. Cytochrome c oxidase is also a vectorial proton pump with residues in and around the binuclear center critical for gating proton flow during turnover. Complex II is succinate dehydrogenase, also termed the succinate-ubiquinone oxidoreductase. This complex serves as an alternative site for entry of electrons into the respiratory chain.

Unlike mitochondria, bacteria can differ greatly in the composition of their respiratory chains. Even in individual bacteria, respiratory chains vary as a function of changes in growth conditions; for example, E. coli modulates the beginning and end of its aerobic respiratory pathway depending on available oxygen levels (Gennis and Stewart 1996). E. coli has two NADH dehydrogenases. The type I NADH dehydrogenase is the bacterial equivalent of the mitochondrial Complex I. The type II NADH dehydrogenase is a single subunit enzyme. It is a flavoprotein that contains one noncovalently bound FAD per polypeptide chain and also contains [Fe-S] centers. The type II oxidase is primarily expressed when oxygen is not limiting. There are also two terminal oxidases in E. coli. With higher levels of oxygen present, a cytochrome o oxidase is expressed.

This oxidase, which uses quinol as reductant, is a member of the same family of oxidases as the mitochondrial oxidase and contains a heme-copper binuclear center. Under microaerophilic conditions, however, the cytochrome d oxidase is expressed. This terminal oxidase is only found in bacteria. This protein contains the novel d heme and has an extremely high affinity for oxygen.

Other bacteria have respiratory chains almost exactly like that found in mitochondria. Probably the best studied example is found in bacterium *Paracoccus denitrificans* (Baker et al. 1998). Bacteria also have the capacity to use many compounds as reductants. Compounds ranging from methanol to iron can be used as electron donors and thus would require their own associated oxidoreductase.

Oxidative phosphorylation

It has been known for many years that aerobes gain ATP from the transport of electrons in the cell membrane, a phenomenon called oxidative phosphorylation. During respiration, oxidation/reduction reactions take place in the membrane of the cells so that an electrochemical potential (δp) or proton-motive force (pmf) is created with protons on the exterior side of the membrane. Because the respiratory chain consists of hydrogen carriers and electron carriers in alternating sequences across the cytoplasmic membrane and because the cell membrane was impermeable to OH^- and H^+, as oxidation reactions transferred protons from one side of the membrane to the other, a pH gradient and an electrical potential was established across the membrane (Mitchell 1961). The electrochemical gradient that occurs as a consequence of electron flow through the respiratory chain can be used to make ATP, drive solute transport, drive the rotation of the flagellum, and drive unfavorable electron-transport reactions.

Respiration-dependent ATP synthesis is carried out by the ATP synthase, also termed the F_1F_0 type ATPase. The ATPase is a complex multiprotein system and has been compared to a small organelle. When protons pass from the outside to the inside of the membrane, their passage through the ATP synthase causes inorganic phosphate to be esterified and ADP is converted to ATP.

ATP synthase can be dissociated into two complexes termed the F_0 and F_1. The structure named F_1 contains five different subunits with a ratio of $\alpha 3$, $\beta 3$, γ, δ, and ϵ. The F_1 contains six ATP binding sites, one each on the α and β subunits. However, only those on the β subunits are capable of ATP synthesis. By itself the F_1 complex functions as an ATPase, hydrolyzing ATP to ADP and Pi. The F_0 complex usually consists of three different subunits in bacteria with a stoichiometry of a, b2, and c10–12. The F_0 is an integral membrane complex that acts as an H^+ channel to translocate H^+ across the cytoplasmic membrane. Using purified protein it has been demonstrated that the complex functions as a rotary motor (Sambongi et al. 1999). The rotor is made up of the c subunits of the F_0 along with the γ and ϵ subunits of the F_1 complex. The remaining proteins are proposed to act as a stator. Models of ATP synthesis require that proton flow through the F_0 causes the rotor to turn. This drives rotation of the γ subunit and as it turns it causes structural changes in the ATP binding sites on the β subunits. The structural changes in the β subunits cause them to switch through low-, medium-, and high-affinity nucleotide binding sites. These shifts lead to the synthesis of ATP. High-resolution structures of both the F_1 and F_0 complexes have been determined and are generally consistent with models of ATPase function (Abrahams and Leslie 1994; Stock et al. 1999).

All bacteria, even strict anaerobes, must maintain an electrochemical gradient of protons across the membrane of the cell. This is needed to support the integrity of the cell membrane, to provide energy for motility, and to transport nutrients into the cell. Under anaerobic conditions, the F_1F_0 complex can function in the reverse direction, hydrolyzing ATP by means of an ATP hydrolase associated with the cell membrane to generate an electrochemical gradient. Some fermenting bacteria are also able to use the excretion of fermentation products through the cell membrane to generate a proton gradient.

In some bacteria, Na^+ substitutes for H^+ as the coupling ion and ATP-driven Na^+ pumps have been described in several genera. All cells appear to transfer energy obtained from external sources to energy stored as either ATP, proton-, or sodium-motive force. These energy sources may be interconvertible and it appears that any one can satisfy all of the energy sources of the cell (Skulachev 1994). An example of a Na^+-dependent respiratory chain was mentioned above in the discussion of oxaloacetate oxidation by *P. modestum*.

Oxygen toxicity

Cells growing in the presence of molecular oxygen produce some very reactive and toxic oxygen species, such as the superoxide radical, the hydroxyl radical, and hydrogen peroxide. These compounds are produced by autooxidation of components of the electron transport chain. In particular, flavoproteins have been shown to produce reactive oxygen species (Messner and Imlay 1999). In order to protect themselves against such compounds, bacteria have evolved special enzymes. Superoxide dismutase (SOD) converts the toxic superoxide anion (O_2^-) to hydrogen peroxide. *E. coli* is known to possess three species of SOD: Mn-SOD, Fe-SOD, and a periplasmic Cu, Zn-SOD (Benov and Fridovich 1994). Catalase and a class of enzymes named peroxidases

destroy hydrogen peroxide. Catalase converts hydrogen peroxide to O_2 and water. The synthesis of catalase is normally regulated and the level of activity kept low until needed. *E. coli* encodes two different catalases, hydroperoxidase I and hydroperoxidase II.

E. coli has been an informative model system for the study of how cells respond to oxidative stress. There are several regulons that control the synthesis of proteins responsible for mitigating toxic effects of oxygen radicals in *E. coli*. The two best-studied regulators are the SoxR and OxyR proteins. SoxR responds to increases in superoxide. When activated the SoxR protein activates transcription of SoxS, which activates transcription of target genes in the regulon. SoxR is activated by oxidation of a reduced [2Fe-2S] center in the protein (Hidalgo et al. 1997). It is unclear what molecule oxidizes SoxR but superoxide is a likely candidate. OxyR responds to increases in hydrogen peroxide. It is activated by the formation of a disulfide bond upon exposure to hydrogen peroxide (Zheng et al. 1998). Another important regulator of expression of oxygen-radical tolerance factors is the sigma factor (σ^s), which is synthesized at high levels during stationary phase growth and may control as many as 50 genes.

While the reactivity of the oxygen radicals allows them to react with many different cellular components, there have been a few especially sensitive targets identified. In *E. coli* lacking superoxide dismutase, cellular injury has been determined mainly to occur to a set of dehydrase enzymes (Storz and Imlay 1999). The common factor among this class of enzymes is that they contain superoxide-sensitive [Fe-S] centers. The precise targets of hydrogen peroxide accumulation have not been identified.

Anaerobic respiration

NITRATE RESPIRATION

As mentioned previously, fermentation is the condition in which the final electron is an organic compound. In aerobic respiration, oxygen is the final electron acceptor. Anaerobic respiration is the situation where the final electron acceptor is a compound other than oxygen. There are many compounds that can serve as alternative electron acceptors. One of the most heavily studied is nitrate. Nitrate reduction can proceed via two pathways (Berks et al. 1995). One pathway is the reduction of nitrate to ammonia, termed 'ammonification.' The other pathway is the reduction of nitrate to nitrogen gas, termed 'denitrification.' Both of these pathways allow growth in the absence of oxygen and are termed 'dissimilatory pathways.' Nitrate can also be reduced to ammonia for biosynthetic purposes in an assimilatory pathway. However, the biosynthetic enzyme pathway is not coupled to ATP formation.

Pathways of nitrate reduction

Bacteria that can use nitrate as a terminal electron acceptor for energy production normally will not do so in the presence of oxygen, because aerobic respiration can yield more energy. Possible electron donors for nitrate reduction include NADH formed from a variety of catabolic pathways, formate, H_2, glycerol-3-phosphate, succinate, and lactate. The electron flow is from NAD, through flavoproteins, to cytochromes, and finally to nitrate.

The first step in nitrate reduction is the reduction of nitrate to nitrite. There are two types of dissimilatory nitrate reductases. The enzyme typically associated with nitrate respiration is a membrane-bound three-subunit complex, although the membrane-anchoring subunit (γ or NarI) is sometimes lost during purification (Moreno-Vivian et al. 1999). The largest of the three subunits (α or NarG) contains molybdenum bound by the cofactor molybdopterin guanine dinucleotide and a [4Fe-4S] center. The remaining subunit (β or NarH) contains several [4Fe-4S] centers and a [3Fe-4S] center. The membrane-anchoring subunit typically contains a b-type heme. The other type of nitrate reductase is a periplasmically located enzyme. The periplasmic nitrate reductase is a heterodimer with prosthetic groups similar to those found in the membrane-bound nitrate reductase.

The ammonification and denitrification pathways diverge at the level of nitrite. During ammonification a single enzyme catalyzes the six-electron reduction of nitrite to ammonia. In *E. coli* and other enteric bacteria this enzyme is a periplasmically located *c*-type cytochrome (Gennis and Stewart 1996). Ammonia is the terminal product of ammonification so no other enzymes are involved. In denitrifiers, nitrite is reduced to nitric oxide, a gaseous nitrogen oxide (Zumft 1997). There are two types of nitric-oxide-producing nitrite reductases. One contains copper as its redox-active metal and the other contains heme. Both are periplasmically located. Nitric oxide is reduced to nitrous oxide by a membrane-bound nitric oxide reductase containing heme and non-heme iron. Interestingly, this protein is related to the family of proteins that includes the heme-copper binuclear center containing oxygen-reducing oxidases. Nitrous oxide is reduced to nitrogen gas by a copper-containing, periplasmically located enzyme. This enzyme is also related to the terminal oxygen-reducing oxidases.

Dissimilatory nitrate reduction is widespread amongst the bacteria. Ammonification is best studied in the enteric bacteria but is found in other bacteria, for example *B. subtilis* (Nakano and Zuber 1998). Denitrification occurs in both eubacteria and archaea. Ongoing sequencing projects have found portions of the denitrification electron transport chain in several pathogens including *Neisseria meningitidis*, *N. gonorrhea*, and *Corynebacterium diphtheriae*. The role of dissimilatory nitrate reduction in these pathogens has not been determined.

REGULATION OF NITRATE REDUCTION IN *E. COLI*

In *E. coli* several signals are required for the cells to activate expression of enzymes required for the reduction of nitrate. First, oxygen concentrations must be limiting because oxygen is always a preferred oxidant. Low oxygen activates the transcriptional regulatory protein Fnr. Fnr contains an oxygen sensitive [Fe-S] center that undergoes changes in structure in the presence of oxygen (Khoroshilova et al. 1997). Nitrate and nitrite are sensed by 2, 2-component regulatory systems. The interaction of these two systems and their target proteins is complex and is used to activate expression of some genes, such as those encoding nitrate and nitrite reductase, while at the same time repressing expression of other genes (Darwin et al. 1997).

Sulfate reduction

Another commonly used terminal oxidant is sulfate. Dissimilatory sulfate reducers are heterotrophs that are almost always strict anaerobes. Sulfate-reducing bacteria are found in both the eubacteria and archaea. The best studied sulfate reducers belong to the genus *Desulfovibrio*. The first step in the use of sulfur as electron acceptor is its activation on an AMP molecule. The enzyme adenyltransferase (ATP-sulfurylase) uses ATP and sulfate as substrates to form adenyl sulfate (APS) and inorganic pyrophosphate (Hansen 1994). The pyrophosphate is hydrolyzed by inorganic pyrophosphatase with the energy in the pyrophosphate bond lost to the cells. The enzyme APS reductase adds two electrons to APS to reduce the sulfate to sulfite with the liberation of AMP. Next, sulfite reductase reduces sulfite to form sulfide (S_2). A proton-motive force is generated by the appropriate topological arrangement of proton-producing and proton-consuming reactions on either side of the cytoplasmic membrane.

Desulfovibrio desulfuricans is a strict anaerobe that can ferment glucose and pyruvate and can also grow using lactate if sulfate is present to serve as an electron acceptor. Pyruvate, formed by either the HDP pathway or lactate oxidation, is cleaved by pyruvate/ferredoxin oxidoreductase, yielding CO_2, H_2, and acetyl-CoA. The acetyl-CoA can be converted to acetyl-phosphate and then acetate with ATP formation. *Desulfovibrio desulfuricans* can also grow using CO_2, H_2, and acetate if sulfate is present. In this case, electrons are transferred from H_2 to sulfate to gain energy and the pyruvate/ferredoxin oxidoreductase reaction is reversed to make pyruvate for biosynthetic reactions.

Additional terminal electron acceptors

Alternative terminal electron acceptors are not limited to inorganic oxides. Oxidized metals such as iron or manganese can be used as terminal oxidants; for example, members of the genus *Shewanella* can utilize both iron and manganese as terminal electron acceptors (Myers and Nealson 1990). Many carbon compounds can serve as terminal oxidants; examples include dimethyl sulfoxide, trimethylamine oxide, and fumarate. The crystal structure of the membrane-bound fumarate reductase from the microaerophile *Wolinella succinogenes* has been determined (Lancaster et al. 1999). It has also been demonstrated that halocarbons can serve as electron acceptors. Molecular studies of the halorespiring bacterium *Desulfitobacterium dehalogenans* are starting to identify components required for this form of anaerobic growth (Smidt et al. 1999). Some bacteria appear only to be capable of halorespiration (Maymo-Gatell et al. 1997). Another notable carbon compound that appears to serve as a terminal electron acceptor is humic acid (Lovley et al. 1996).

Additional electron donors

In addition to the capacity to utilize a wide variety of compounds as terminal electron acceptors, bacteria also have the capacity to utilize many compounds as electron donors for respiration. In addition to the wide variety of organic compounds that can be used, many inorganic compounds can be used as electron donors. Ammonia and nitrite can be oxidized by a group of bacteria collectively known as nitrifiers. Ammonia oxidizers, such as members of the genus *Nitrosomonas*, can oxidize ammonia to nitrite via the intermediate hydroxylamine (Hooper et al. 1997). Ammonia oxidizers cannot further oxidize nitrite. Nitrite oxidizers, such as members of the genus *Nitrobacter*, oxidize nitrite to nitrate but cannot oxidize ammonia. Reduced sulfur compounds can also be oxidized to provide electrons for respiration. Sulfate oxidation has been heavily studied in the genus *Thiobacillus* (Kelly et al. 1997). Some thiobacilli can couple sulfur oxidation to nitrate reduction, completely eliminating involvement of carbon oxidation or reduction in respiration. This type of metabolism is referred to as 'lithotrophy' and can be found in many other bacterial genera. Other potential reductants include iron, manganese, H_2, and carbon monoxide.

BACTERIAL PHOTOSYNTHESIS

Photosynthesis is the conversion of the energy available in light to chemical energy, which is available for cell growth. Many bacteria are capable of photosynthesis and there are several different bacterial photosynthetic electron-transport systems. In all of these systems light is absorbed (harvested) by accessory pigments and the energy transferred to a reaction center. The reaction center carries out the charge separation that permits light energy to be converted to chemical energy. The different electron-transport systems can be differentiated into three groups based on the types of bacteria having a particular system. The three types of bacteria are:

1   purple photosynthetic bacteria
2   green sulfur photosynthetic bacteria
3   cyanobacteria.

A brief description of the photosynthetic electron-transport chain of each group is presented below.

Purple photosynthetic bacteria

In the purple photosynthetic bacteria both organic and inorganic compounds, such as reduced-sulfur compounds, can be used as electron donors. However, some bacteria are limited in their choice of electron donor. Photosynthesis in the purple bacteria, as is true for most bacteria, is an anaerobic process. Therefore, expression of the components required for photosynthesis occurs when oxygen is limiting. Oxygen limitation causes the most significant increase in photosynthetic gene expression with light having a much smaller effect (Buggy et al. 1994).

Light absorption is carried out by a group of protein-associated pigments organized into light-harvesting (LH) complexes. In most bacteria there are two complexes, referred to as LHI and LHII. High-resolution and modeling studies of LHI and LHII have shown that both are roughly circular and contain both bacteriochlorophyll and carotenoids (Hu et al. 1998). Carotenoids absorb at lower wavelengths, in the 500-nm range, while the bacteriochlorophylls absorb at higher wavelengths (800–900 nm). Current models of the organization of LHII and LHI have many LHII complexes surrounding the LHI complex, which has the reaction center at its center. Therefore, the probable direction of excitation transfer about light absorption is LHII to LHII and LHI and then LHI to the reaction center.

Transfer of excitation energy to the reaction center causes the potential of an electron located at a dimer of bacteriochlorophylls, known as the special pair, to become excited to a lower reduction potential. This electron is then transferred through the electron-transfer components of the reaction center and eventually resides on a quinone molecule designated UQ_A. This quinone is tightly bound to the reaction center complex. UQ_A then transfers the electron to a second quinone referred to as UQ_B. This quinone is less tightly bound and, when fully reduced, can diffuse away from the reaction center complex. These steps are then repeated to reduce the quinone fully, which, after binding two protons, diffuses into the lipid bilayer. Because the special pair is located on the periplasmic surface and the ubiquinones are located on the cytoplasmic surface, electron transfer is vectorial. This vectorial transfer of electrons is referred to as charge separation because a negative charge is moved to the cytoplasmic side and a positive charge is left on the periplasmic side of the membrane. The details of electron flow through the reaction center are well defined because the high-resolution structure of the reaction center from purple photosynthetic bacteria has been solved (Deisenhofer et al. 1995).

Once UQ_B diffuses into the lipid bilayer it can pass its electrons to the bc_1 complex. Electron transfer through the bc_1 complex leads to the generation of a proton-motive force that can be used for the synthesis of ATP. The electrons are then passed to a periplasmic cytochrome c that then passes the electrons back to the reaction center. This type of electron flow during photosynthesis is termed 'cyclic photosynthesis.' Some photosynthetic bacteria use electron donors with reduction potentials higher than the $NAD^+/NADH$ couple so these cells use the proton-motive force to drive reverse electron transport to generate NADH.

Green sulfur bacteria

The basic mechanisms of light harvesting and electron flow through the reaction center are the same in both purple photosynthetic and green sulfur bacteria. The reaction center of the green sulfur also contains a special pair whose reduction potential is decreased upon transfer of excitation energy. The reaction center differs in the nature of its associated cofactors. The green sulfur bacteria reaction center includes iron-sulfur centers, which are not present in the reaction center of the purple photosynthetic bacteria (Oh-oka et al. 1995). In addition, the potential of the excited electron in the reaction center of the green sulfur bacteria is low enough to reduce NAD^+ to NADH. This allows electron transfer to be noncyclic. However, cyclic electron transfer can occur, generating a proton-motive force and ATP production. Sulfide is a common electron donor for the green sulfur bacteria. Its electrons are fed into the electron-transport chain via periplasmic cytochromes.

Cyanobacteria

Cyanobacteria differ from the other photosynthetic bacteria by using water as the source of electrons for photosynthesis. Another difference is that the photosynthetic electron-transport chain of the cyanobacteria is essentially a combination of the purple and green photosynthetic electron-transport chains. The electron-transport chain in cyanobacteria is differentiated into reaction center II (RCII), which is related to the reaction center from purple photosynthetic bacteria, and reaction center I (RCI), which is related to the green sulfur

bacteria reaction center. Electrons from water first flow through RCII, allowing formation of a proton-motive force. Electrons then flow to RCI via a copper protein called plastocyanin. The electron is then transferred to $NADP^+$ for biosynthetic purposes or back to RCI generating a proton-motive force. A continual supply of electrons is required because flow through RCII is not cyclic. This is generated by the manganese-containing water-splitting apparatus. The photosynthetic electron-transport pathway of plant chloroplasts is very similar to that of the cyanobacteria.

METABOLISM AMONG THE ARCHAEBACTERIA

Archaebacteria (domain archaea) encompass a diverse array of physiological processes. Probably the most prominent archaea are the methanogens. These bacteria grow by converting a limited set of substrates to methane. In addition to methanogens other archaea carry out aerobic or anaerobic respiration. No photosynthetic archaea have been isolated. However, some halophilic archaea can use light energy to generate a proton-motive force but this reaction is distinct from photosynthesis.

Methanogenesis

All known methanogens are members of the domain archaea and represent the largest group in that domain, with thermophiles, mesophiles, and halophiles included in the group. Methanogens are obligate anaerobes that reduce CO_2 and a few other substrates, such as formate, methanol, methylamines, or acetate, to produce the most reduced organic compound, methane. Most methanogens are autotrophs and the reduction of CO_2 is their only energy-yielding pathway.

The first step of methanogenesis is catalyzed by either a tungsten- or molybdenum-containing formyl-methanofuran dehydrogenase (Thauer 1998). CO_2 is activated and reduced to the level of formyl on methanofuran. The carbon is then transferred to the cofactor tetrahydromethanopterin. The carbon is reduced from the formyl to the methyl state while attached to this cofactor. At this stage the methyl is transferred to the cofactor coenzyme M. Methyl-coenzyme M (methyl-CoM) is reduced by methyl-CoM reductase to methane. The methyl-CoM reductase contains the unusual nickel tetrapyrrole, F_{430}. Methyl reduction also requires cofactor B (7-mercaptoheptanoylthreonine phosphate). During the reaction the cofactors coenzyme M and coenzyme B become linked via a disulfide bond. This bond is cleaved using the membrane-bound enzyme heterodisulfide reductase.

During methanogenesis from methanol about one out of four methanols are oxidized, by reversing part of the methanogenesis pathway, to provide electrons for the reduction of methanol to methane. Methanogenesis from acetate is more complex, requiring that acetate be first converted to acetyl-CoA using either the combination of acetate kinase and phosphotransacetylase or acetyl-CoA synthetase (Ferry 1999). Carbon monoxide dehydrogenase then breaks the carbon–carbon bond forming a carbonyl and a methyl group. The carbonyl is oxidized to CO_2 and the electrons used to reduce the methyl group to methane.

The energetics of methanogenesis are not completely understood. It seems likely that the heterodisulfide reductase step permits formation of a proton-motive force (Deppenmeier et al. 1991). The coenzyme M–coenzyme B disulfide serves as the terminal electron acceptor, with H_2 being a likely electron donor. It is not clear how the proton-motive force is generated during this reduction but it is possible that scalar production of protons by the hydrogenase is the source. Methanogens do contain ATPases similar in structure and organization to that of the F_1F_0-type ATPases, which are designated A_1A_0-type ATPases (A meaning 'archaeal' here). While there is no direct evidence indicating that the ATPases are ATP synthases, it seems likely they are used for that purpose. The transfer of carbon from tetrahydromethanopterin to coenzyme M may also be an energy-coupling step.

Methanogens must synthesize all cellular organic compounds from simple starting material and some methanogens can grow on a completely minimal medium and convert CO_2 to all required organic material. During growth with CO_2 as the carbon source it is incorporated into acetyl-CoA. This reaction utilizes the carbon monoxide dehydrogenase discussed above. One unit of CO_2 is bound to the carbon monoxide dehydrogenase and then oxidized to carbon monoxide. A second CO_2 unit is then reduced to a methyl group using the same reactions occurring during methanogenesis. The methyl group is then passed to carbon monoxide dehydrogenase where it is combined with the bound carbon monoxide. The dehydrogenase then catalyzes the production of acetyl-CoA. The acetyl-CoA is next converted to phosphoenolpyruvate, which is probably converted to further primary precursors by a modified reductive TCA cycle.

Methanogens such as *Methanobacterium* and *Methanosarcina* assimilate CO_2 but have an incomplete TCA cycle. In *Methanobacterium*, succinate is apparently converted to succinyl coenzyme A and then to α-ketoglutarate. In *Methanosarcina*, oxaloacetate appears to be converted to citrate, which is converted to isocitrate and then to α-ketoglutarate. Although glucose is not a carbon or energy source for growth, there are some reports of methanogens taking up and metabolizing glucose and the distribution of radioactive label that suggested the presence of the HDP pathway. However, there is no report of the presence of phospho-

fructokinase, a key enzyme of the catabolic HDP pathway.

Nonmethanogenic archaea

The nonmethanogenic archaea are a diverse mix of organisms that include thermophiles, acidophiles, strict aerobes, strict anaerobes, and facultative bacteria. Representatives of many of these groups have been studied in some detail and genome-sequencing efforts are also contributing to our understanding of these bacteria. Strict aerobes are somewhat rare, reflecting the preference of archaea for anaerobic growth. One obligately aerobic genus studied in some detail is the genus *Sulfolobus*. These bacteria are thermophilic acidophiles, some of which can use organic compounds as electron donors, while others use reduced sulfur compounds. The respiratory components have been characterized from some *Sulfolobus* species. These characterizations have shown that the respiratory complexes are orthologs of the eubacterial complexes; however, in many cases they are quite divergent. A good example is provided by the SoxABCD quinol oxidase from the bacterium *Sulfolobus acidocaldarius* (Lubben et al. 1992). Sequence analysis indicated that this complex contains a subunit with a heme-copper binuclear center and other components of binuclear center-containing quinol oxidases. However, in addition to those proteins it has a b-heme-containing protein that is similar to the b-heme-containing protein of the bc_1 complex. This suggests that the complex III and IV components have been modified into a single active complex. The observation that respiratory proteins in eubacteria and archaea are orthologs but that there are no photosynthetic archaea has led some to suggest that respiration arose before photosynthesis and that the archaea and eubacteria diverged before photosynthesis.

The aerobic respiratory chain of halophilic archaea has not been studied as extensively as that of the *Sulfolobus* strains. However, data suggest that the proteins are orthologs of other known respiratory proteins. Some halophilic archaea have been shown to use alternative electron acceptors such as nitrate and fumarate. Several halophilic archaea also have the capacity to use light energy for energy conservation. A single protein termed bacteriorhodopsin carries out this process. Bacteriorhodopsin occurs in high concentration in the membrane and the occurrence of these purple membranes in certain halophiles led to the interest in this protein. The structure and catalytic mechanism of bacteriorhodopsin has been studied in great detail. It is a 26-kDa integral membrane protein with seven transmembrane helices (Pebay-Peyroula et al. 1997). It contains the chromophore retinal covalently attached to the protein at a lysine in one of the transmembrane helices. The absorption of light shifts the *trans*-retinal to *cis*-retinal. This structural change begins a reaction scheme that ejects a proton into the periplasmic surface and causes the uptake of a proton from the cytoplasmic side of the membrane. Mutagenesis of bacteriorhodopsin has identified several acidic residues that make up the proton channel. In addition to bacteriorhodopsin these halophiles have several other retinal-containing proteins. One of these, halorhodopsin, is an internally directed chloride pump. The other two are photoreceptors involved in phototaxis.

Many of the nonmethanogenic archaea use proteins and amino acids as carbon and energy sources but a few have been shown to use carbohydrates. *Halobacterium saccharovorum* has been shown to possess a modified KDPG (Entner/Douderoff) pathway (compare with Figure 3.10, p. 59) where unphosphorylated glucose is oxidized to gluconate by a NAD-dependent dehydrogenase and a gluconate dehydrogenase converts the gluconate into 2-keto-3-deoxygluconate. Next, a kinase, using ATP as the phosphate source, phosphorylates the 2-keto-3-deoxygluconate to form KDPG. KDPG aldolase cleaves this to form glyceraldehyde-3-phosphate and pyruvate as in the normal KDPG pathway, and glyceraldehyde-3-phosphate is converted to pyruvate by the usual HDP pathway enzymes. Pyruvate is converted to acetyl-CoA using pyruvate/ferredoxin oxidoreductase.

Thermoacidophilic archaebacteria such as *Sulfolobus solfataricus* and *Thermoplasma acidophilum* have no phosphofructokinase and, therefore, no HDP pathway. Instead, they have been shown to use another variation of the KDPG pathway, a completely non-phosphorylated pathway. Glucose is oxidized to gluconic acid, which is dehydrated to 2-keto-3-deoxygluconate. This is cleaved by an aldolase to yield pyruvate and glyceraldehyde. The glyceraldehyde is oxidized to glyceric acid, which is phosphorylated to 2-phosphoglyceric acid and converted to phosphoenolpyruvate. The latter compound is cleaved by pyruvate/ferredoxin oxidoreductase. There is no evidence for acetyl phosphate formation as a precursor to acetyl-CoA formation (Danson 1988).

REGULATION OF PHYSIOLOGICAL PROCESSES

Operon organization and expression

Perhaps the best example of operon organization and expression is provided by the set of genes required for lactose metabolism. In order to grow using the disaccharide lactose as a carbon and energy source, *E. coli* requires the presence of the enzyme β-galactosidase (once termed lactase) to cleave lactose into the monosaccharides, D-glucose and D-galactose. This enzyme is inducible, meaning that without the presence of lactose in the growth medium, there is only a very low basal or endogenous level of the enzyme in the cells. However, within minutes of addition of lactose to the medium the differential rate of synthesis of the enzyme, which is the rate of enzyme synthesis compared with the rate of total cell protein synthesis, increases until the amount of enzyme in the cells is 10 000 times higher than previously. If the lactose is removed from the medium, the rate of β-galactosidase synthesis decreases to the previous basal level while at the same time the concentration of enzyme in the cell is diluted by the synthesis of new proteins during cell growth. When cells are provided with lactose, several proteins are induced at the same time, a phenomenon known as coordinate induction. These proteins include β-galactosidase, lactose permease (the protein responsible for the transport of lactose through the cell membrane), and a protein catalyzing the acetylation of certain disaccharides.

During studies on lactose catabolism by *E. coli*, mutants were obtained that changed the regulation of these three enzymes such that they were produced at high levels in the absence of lactose. These mutants where termed constitutive mutants and, because a single mutation affected the regulation of all three proteins at once, the expression of the genes coding for those proteins had to be under the control of a common regulatory gene. Such enzymes are said to be coordinately controlled. The analysis of structural gene and regulatory gene mutations that effected lactose catabolism in *E. coli* led Jacob and Monod (1961) to postulate the operon model of gene regulation. This model proposed that a single regulator controlled expression of all three genes because the genes are cotranscribed on a single transcript. An operon is a region of DNA containing cotranscribed genes and control sequences.

Further studies provided evidence that the regulatory gene produced a product that normally blocked the formation of the enzymes involved in lactose catabolism. This product was termed the repressor. When the lactose operon was induced, an RNA polymerase would bind to the DNA at a site named the promoter region. In the absence of the inducer, however, the repressor protein bound to a DNA site named the operator region and prevented transcription by the RNA polymerase. Thus, the regulation was a negative control system in that the normal function of the regulator gene was to prevent the formation of mRNA for the operon. Any mutation resulting in the loss of function of the repressor (the product of the *lacI* gene) would result in the loss of repression and the constitutive expression of the enzymes of the lactose operon (Jacob and Monod 1961).

THE LACTOSE OPERON OF *E. COLI*

The arrangement of the genes of the lactose operon of *E. coli* is shown in Figure 3.11. The order of the three

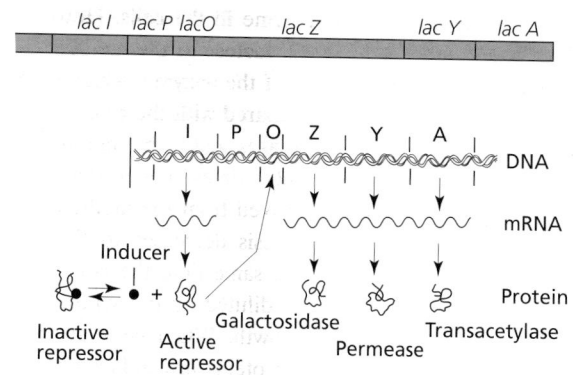

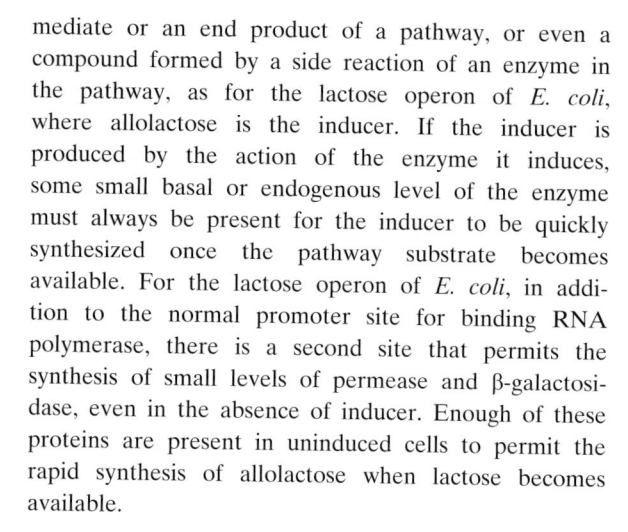

Figure 3.11 *The lactose operon of* Escherichia coli

structural genes of the *lac* operon is *lacZ*, *lacY*, and *lacA*, encoding β-galactosidase, lactose permease, and thiogalactoside transacetylase, respectively. The promoter (*lacP*) and operator (*lacO*) regions are adjacent to the *lacZ* gene and actually overlap. The regulatory gene (*lacI*) is situated next to the operon. The regulatory gene (*lacI*) is the structural gene for a protein that is termed the lactose repressor. The lactose repressor is an allosteric protein that is normally expressed constitutively and is a DNA-binding protein. When bound to the operator region the repressor blocks binding of RNA polymerase and blocks transcription of the operon. The repressor contains a binding site for the inducer of the lactose operon. The inducer serves as a negative effector and, when bound to the repressor, it decreases the affinity of the repressor for DNA. With the removal of the repressor from the operator region, RNA polymerase can transcribe the operon. The actual inducer molecule is not lactose but an isomer of lactose, allolactose, which is a disaccharide formed by a side reaction of β-galactosidase. Lactose is a D-glucose-D-galactose dimer with a β-1,4 linkage, whereas allolactose differs from this in that it possesses a β-1,6 linkage.

The purpose of the thiogalactoside transacetylase, coded by the *lacA* gene, remained a mystery for a number of years. This enzyme is now believed to be associated with the lactose operon for purposes of detoxification. The enzyme possesses the ability to acetylate a variety of different disaccharides, but lactose is one of the poorest substrates. The lactose permease is able to transport a number of disaccharides other than lactose into the cell and if these disaccharides cannot be further metabolized they accumulate and may be toxic for cell growth. Upon acetylation of these disaccharides by the thiogalactoside transacetylase, they are rapidly excreted by the cells and not transported back into the organism.

THE NATURE OF INDUCERS

The inducer of a catabolic enzyme pathway need not be the initial substrate of the pathway. It may be an inter-

mediate or an end product of a pathway, or even a compound formed by a side reaction of an enzyme in the pathway, as for the lactose operon of *E. coli*, where allolactose is the inducer. If the inducer is produced by the action of the enzyme it induces, some small basal or endogenous level of the enzyme must always be present for the inducer to be quickly synthesized once the pathway substrate becomes available. For the lactose operon of *E. coli*, in addition to the normal promoter site for binding RNA polymerase, there is a second site that permits the synthesis of small levels of permease and β-galactosidase, even in the absence of inducer. Enough of these proteins are present in uninduced cells to permit the rapid synthesis of allolactose when lactose becomes available.

THE LACTOSE OPERON OF *STAPHYLOCOCCUS AUREUS*

Staphylococcus aureus possesses a different pathway of catabolism for lactose, using enzyme *IILac* and enzyme *IIILac* of the phosphoenolpyruvate phosphotransferase (PTS) system to transport lactose through the cell membrane. During transport, lactose is phosphorylated to phospholactose, which is cleaved by a phospho-galactosidase enzyme to form D-glucose and D-galactose-6-phosphate. The genes coding for these three enzymes are designated *lacF*, *lacE*, and *lacG*, respectively, which form an operon that appears to be under the negative control of a repressor gene (*lacR*). Interestingly, these genes are in a larger operon including *lacABCD* that encodes genes whose products are involved in the conversion of galactose-6-phosphate to tagatose-6-phosphate and the further metabolism of tagatose-6-phosphate (Rosey et al. 1991). Galactose-6-phosphate is the inducer of the *lac* operon in *S. aureus*.

THE L-ARABINOSE OPERON OF *E. COLI*

While studying the catabolism of the pentose L-arabinose by *E. coli* strain B/r, Englesberg and coworkers found differences from the regulation of the lactose operon (Sheppard and Englesberg 1967). After the transport of L-arabinose into the cells it is isomerized by the enzyme L-arabinose isomerase to the 2-ketopentose, L-ribulose. The enzyme L-ribulokinase phosphorylates L-ribulose to form L-ribulose-5-phosphate, which is epimerized to D-xylulose-5-phosphate by L-ribulose-5-phosphate. The latter compound is further degraded by other cellular enzyme systems. These three enzymes are all coordinately induced when L-arabinose is present and the genes encoding for the three enzymes are termed *araA*, *araB*, and *araD*, respectively. These genes are organized into an operon with a regulatory gene, *araC*, located next to the operon. In contrast to the *lac* operon, deletions of the *araC* gene do not lead to constitutive expression of the enzymes. Instead, inactivation of *araC* causes the L-arabinose catabolic enzymes to become uninducible. Because the function of the regulator gene is essential for gene expression, this is termed a positive control system.

The regulatory protein, encoded by the *araC* gene, can interact with four different operator regions, O_1, O_2, I_1, and I_2, all connected to the L-arabinose operon. When *araC* binds to both the O_2 and I_1 sites, a loop is formed in the DNA preventing transcription of the operon (Huo et al. 1988). When the inducer, L-arabinose, binds to the allosteric repressor protein, the protein's DNA binding affinity changes and it now binds to operator sites I_1 and I_2, permitting transcription of the operon into mRNA.

REGULATION BY ATTENUATION

Another means of regulating the expression of biosynthetic pathways is by attenuation. The classic example of regulation by attenuation is the L-tryptophan operon in *E. coli* (Yanofsky 1981). Between the location where transcription of the operon initiates and the beginning of the first structural gene of the operon, there is a stretch of DNA termed the leader region. This region contains four stretches of sequence that, when transcribed, have the potential to form three RNA structures termed hairpins. Also located within the region are two L-tryptophan codons and an RNA polymerase pause site. As RNA polymerase proceeds through the leader region it stalls at the pause site. This delay permits a ribosome to bind to the mRNA and proceed towards the RNA polymerase. If internal L-tryptophan levels are high, then the level of tRNA charged with L-tryptophan will also be high and the ribosome can proceed to the stalled polymerase. This allows the formation of a hairpin from the downstream hairpin sites. This hairpin causes transcriptional termination. If L-tryptophan concentrations are low this causes the ribosome to stall at the L-tryptophan codons. This allows the formation of a hairpin from the middle two hairpin regions and prevents formation of the hairpin that terminates transcription. This form of regulation is different from regulation by repressor or activator proteins because transcription is allowed to proceed but the concentration of substrate determines whether the transcription will continue into the region containing the structural genes.

Reports of bacterial operons regulated by transcriptional attenuation mechanisms have been increasing (Yanofsky 2000). The *ilv* and *leu* genes in enteric bacteria are regulated mostly by a translational-dependent attenuation but the *ilv–leu* operon of *B. subtilis* is reported to be regulated, at least in part, by transcriptional attenuation (Lu et al. 1995). Regulation by attenuation has also been reported for the *E. coli* threonine and histidine operons. The *B. subtilis* aminoacyl-tRNA synthetase genes are also regulated by attenuation.

CELL-DENSITY-DEPENDENT GENE EXPRESSION

Certain genes are only expressed when the concentration of bacterial cells reaches a critical level. This cell density-dependent gene expression is termed 'quorum sensing' (Fuqua et al. 1996). The prototype quorum-sensing system is the *lux* system in the bacterium *Vibrio fischeri* (Engebrecht et al. 1983). *Vibrio fischeri* can exist as either a free-living bacterium or in the light organ of certain fishes or squid. When present in the light organs it activates the expression of the *lux* genes, which produce proteins required for bioluminescence. The expression of the *lux* genes only occurs in the light organ because cell density reaches much higher concentrations there than when the cells are present in the open ocean. Cells can sense cell density because they release a compound (the autoinducer) into the surrounding environment and the external concentration of autoinducer increases as the cell population increases. The autoinducer is freely diffusible across cell membranes, so when the concentration of autoinducer in the surrounding medium reaches the proper level, it can interact with a positive regulator in the cell that activates *lux* gene expression. Quorum-sensing systems have been found to be widespread in bacteria.

The autoinducer in *V. fischeri* and in many bacteria is a hydrophobic derivative of n-acyl homoserine lactone (Eberhard et al. 1981). In gram-positive bacteria quorum sensing takes place using a posttranslationally processed peptide (Dunny and Leonard 1997). Typically this peptide functions as a signal for the sensor component of a two-component sensor-regulator regulatory system. Examples of processes regulated by peptide-dependent quorum sensing are competence, the ability to take up DNA in *B. subtilis* and *Streptococcus pneumoniae*, the expression of virulence factors in *S. aureus*, and the production of antimicrobial peptides.

Global control networks

If more than one operon is under the control of a regulatory factor, those operons are said to constitute a regulon. Cells typically contain many overlapping regulons. Another term for a large collection of genes regulated by a common mechanism is global regulatory network. By having such global control networks, cells can quickly respond to sudden changes in the environment by increasing or decreasing the rate of synthesis of entire systems of proteins.

CATABOLITE REPRESSION

A good example of a global regulatory system is the catabolite repression system. It is not uncommon for an organism preferentially to utilize those catabolites most common in its natural environment. Therefore, bacteria given two possible carbon and energy sources simultaneously may not use them at the same time but may use one before beginning catabolism of the other. An example is the 'glucose effect' first reported in 1942 by Gale and Epps. They found that the synthesis of certain enzymes by *E. coli* was inhibited by the presence of D-glucose in the medium. In 1947, Monod reported that the presence of glucose prevented the induction of many other sugar-catabolic pathways in *E. coli*, especially those sugars catabolized by inducible pathways. If two carbohydrates, such as glucose and lactose, were both present in the growth medium, the growth curve showed two successive growth curves separated by a lag period (Monod 1947). This was termed 'a diauxic growth curve' and it is possible to classify carbohydrates as to whether or not a diauxic growth curve results when they are incubated with glucose.

In 1961, the expression 'catabolite repression' was used to describe this type of repression of enzyme pathways. It was later shown that catabolite repression by glucose of the synthesis of β-galactosidase involves a block in the synthesis of mRNA for the enzyme. Mutants with a *lacI* deletion and, therefore, constitutive for β-galactosidase production, are still subject to catabolite repression by glucose (Magasanik 1961).

Catabolite repression in *E. coli* and other enteric bacteria

In enteric bacteria, the rate of transcription of genes that are subject to regulation by catabolite repression was found to be controlled by the intracellular concentration of a nucleotide, cAMP. The internal level of cAMP in *E. coli* is very low in cultures grown on glucose and addition of cAMP to the medium results in partially overcoming the glucose catabolite repression. Mutants deficient in adenylate cyclase, the enzyme responsible for the synthesis of cAMP from ATP, can synthesize β-galactosidase only when cAMP is added to the medium (Pastan and Adhya 1976).

Mutants were obtained that could not make β-galactosidase or similar catabolite-repressible enzymes even when cAMP was added to the medium. These mutants were found to be deficient in the synthesis of a protein named the catabolite gene activator protein (CAP) or sometimes the cAMP receptor protein (CRP). This protein was found to be a homodimer, with each monomer capable of binding one molecule of cAMP. The cAMP–CRP complex binds to DNA at the operator-distal part of the promoter region of the lactose operon and increases RNA polymerase binding efficiency. Thus, the rate of transcription of the operon is dependent upon the presence of both the CAP protein and the intracellular level of cAMP. In the presence of glucose the internal concentration of cAMP is low and the transcription of those operons requiring the cAMP–CAP complex is repressed.

The CAP protein is a global regulator of gene expression and has been found to activate transcription at more than 50 promoters. The intracellular level of cAMP appears to be regulated at least partially in growing cells by the rate of activity of adenylate cyclase. Under conditions of catabolite repression, adenylate cyclase activity is inhibited and the cellular level of cAMP decreases and, as a result, the activity of the cAMP–CAP-dependent promoters also decreases. When adenylate cyclase is active the cAMP–CAP complex forms and activates transcription through direct contact between cAMP–CAP and RNA polymerase (Busby and Ebright 1999). The cAMP–CAP complex has different affinity for different promoters so the response of separate operons may differ at any concentration of cAMP. The genes controlled by CAP are sometimes referred to as members of the carbon/energy regulon.

The activity of adenylate cyclase in *E. coli* appears to be regulated by interactions with components of the cytoplasmic membrane transport system. In the presence of carbohydrates being brought through the membrane by the PTS, adenylate cyclase is almost inactive and the level of cAMP in the cells is very low. It has been suggested that when glucose becomes limiting the phosphorylated form of enzyme IIA^{Glc} activates adenylate cyclase, resulting in the activation of adenylate cyclase and the synthesis of cAMP (Hogema et al. 1999). Phosphorylated IIA^{Glc} accumulates because no glucose is present to allow dephosphorylation. Conversely, when cells are growing upon glucose, IIA^{Glc} will transfer its phosphate group to the incoming sugar and it will mostly be present in the cell in the unphosphorylated form that does not activate adenylate cyclase.

Other mechanisms of catabolite repression

Catabolite repression also exists in bacteria such as those in the genus *Pseudomonas*. However, several studies have shown that in *P. aeruginosa* and *P. putida*, cAMP is not involved with catabolite repression. Because these bacteria receive most of their energy by oxidative phosphorylation with electrons obtained from the oxidation of acetate by the TCA cycle, it is catabolites such as acetate and TCA-cycle intermediates that cause repression of other catabolic pathways including those for glucose (MacGregor et al. 1991).

The control of catabolite repression in gram-positive bacteria may be controlled by a number of different mechanisms. No cAMP has been detected in *B. subtilis* under aerobic conditions. The addition of cAMP does not effect catabolite repression in that organism and no CAP-like protein has been identified in any gram-positive bacteria (Saier et al. 1996). *Cis*-acting sequences named catabolite responsive elements or CRE seem to be responsible for catabolite repression in *B. subtilis* and these CREs elements also have been found to function in *B. megaterium* and *Staphylococcus xylosus*. A catabolite control protein (CcpA) has been identified and mutations in this protein have been found to prevent

CRE-dependent catabolite repression in both *B. subtilis* and *B. megaterium*. CcpA has been shown either to repress or activate expression of its targets.

The exact mechanism of action of CcpA is still not understood. Repression of gene activity apparently involves the PTS system, but in a different manner than in gram-negative bacteria. During the normal function of the PTS system, the HPr protein is phosphorylated at His-15. However, in a reaction that might be unique to gram-positive bacteria, the HPr protein is phosphorylated at Seryl residue 46 by an ATP-dependent kinase, a reaction that is required for catabolite repression (Deutscher et al. 1994). HPr(Ser-P) acts as a corepressor and may directly interact with CcpA in vivo (Kraus et al. 1998).

SIGMA FACTORS

Bacterial RNA polymerases consist of at least four polypeptides termed α, β, β', and sigma (σ). The core RNA polymerase consists of two α subunits and the β and β' subunits, and this core unit will bind randomly to DNA. If the σ subunit is also present, the complex is termed a holoenzyme. The presence of the σ subunit directs the holoenzyme to bind at specific promoter regions and initiate transcription. After the first few nucleotides of the RNA transcript have been synthesized, the σ factor is released and the core enzyme continues the polymerization of the ribonucleotides. By the use of multiple σ factors, the cell can initiate the transcription of different regulons, thereby synthesizing new families of enzymes in response to changing environmental conditions. The RNA polymerase holoenzymes have been designated as E-σx, with E representing the core polymerase and σx the particular σ factor attached.

Sigma factors in *E. coli*

Lonetto et al. (1992) studied the amino-acid sequence of bacterial σ factors and classified them into three broad groups. Group I proteins are the primary σ factors. Primary σ factors from different organisms are very similar and the regions involved in promoter recognition are especially highly conserved. These primary σ factors are required for survival of bacterial cells and are responsible for most RNA synthesis in exponentially growing cells. The primary σ factor in *E. coli* is σ^{70}, which recognizes the promoters of most of the genes expressed during exponential growth. A second group of σ factors are very similar to group I but are not essential for cell growth. The third group consists of the alternative σ factors that are responsible for the transcription of specific regulons. Apparently these σ factors all compete for RNA polymerase.

Sigma factors provide another mechanism for cells to establish global regulatory networks. The ability of the cell to produce a variety of alternative σ factors with different promoter specificity allows σ factors to regulate a wide variety of physiological responses. A good example of the role of σ factors in regulating cellular responses to environmental changes is provided by the heat shock response of *E. coli*. Bacteria have evolved mechanisms that permit them to resist a variety of different environmental stresses, such as heat or cold shock, acid pH, lack of oxygen, and starvation. Bacterial response to stress is to change the expression of groups of genes known under the general term of 'stress genes.' The response of microorganisms to elevated temperatures has been termed heat-shock response. Heat-shock response is a rapid, transient increase in the rate of synthesis of certain proteins (heat-shock proteins (HSP)) within seconds of a shift of the culture to higher temperatures. If the growth temperature of *E. coli* is shifted up, for example from 37 to 46°C, the rate of synthesis of proteins concerned with transcription and translation is decreased. However, the synthesis of HSPs is increased, some by as much as 100 times. If the growth temperature is returned to normal, the rate of synthesis of these proteins returns to normal. HSPs provide the cells with some resistance to high temperatures.

Heat-shock-induced transcription in *E. coli* is associated with the heat-shock-specific σ factor, σ^{32}. The heat-shock regulon consists of over 20 genes and at least 13 promoters are known to be specifically transcribed by the holoenzyme containing σ^{32} (Eσ^{32}). The actual amount of gene transcription depends upon the cellular concentration of σ^{32} and, under heat-shock conditions, σ^{32} levels increase by enhanced synthesis, elevated stability, and increased activity of the σ factor. During normal growth conditions, the σ^{32} polypeptide is very unstable in vivo, with a half-life of about 1 min. When the temperature is shifted from 30 to 42°C, the polypeptide rapidly becomes stabilized and has an eight times longer half-life (Mager and De Kruijff 1995). This change in stability is probably a result of changes in σ^{32} binding to the chaperone DnaK (Bukau 1993). When σ^{32} binds to DnaK it becomes more susceptible to protease degradation. However, when heat shock occurs, resulting in an increased association of DnaK with misfolded proteins, σ^{32} does not bind to DnaK and its turnover time of DnaK decreases. This permits it to bind to core polymerase and activate expression of target genes.

The gene encoding another sigma factor, σ^E, also appears to be essential for growth of *E. coli* at temperatures above 43.5°C. The activity of this σ factor appears to be directly modulated by two proteins, RseA and RseB. RseA is a membrane-bound protein that binds σ^E, preventing it from interacting with core polymerase. Upon heat stress RseA is probably degraded by a periplasmic protease (Ades et al. 1999). This frees up σ^E for interaction with core polymerase and gene activation. This regulatory mechanism suggests that genes activated by σ^E are involved in response to heat-shock response in the periplasm. Proteins that inactivate σ factors by directly binding them are termed antisigma factors.

The starvation of bacteria can also result in increased resistance to a variety of different stresses. This starvation effect depends to some extent upon alternative σ factors and involves the expression of starvation genes that code for special resistance factors (Nystrom 1999). This starvation-induced resistance depends upon expression of the regulons controlled by the alternative σ factors, σ^{32} and, a major regulator of the general starvation response in *E. coli*, σ^{38} (σ^{S}). Starvation causes a general response with the synthesis of 30–50 proteins. If cells of *E. coli* in the stationary growth phase are starved for glucose, they also gain an increased resistance to heat, to oxidizing agents such as hydrogen peroxide, and to sodium chloride. The 'starvation proteins' of *E. coli* are also induced by exposure to heat shock or oxidative stresses.

Sigma factors in *B. subtilis*

Another bacterium in which the role of sigma factors has been extensively investigated is *B. subtilis*. The *B. subtilis* σ factor termed σ^{A} (formally σ^{55}) is the primary σ factor present in vegetatively growing cells (Haldenwang 1995). The alternative sigma factor σ^{B} (σ^{37}) controls general stress response but is also active in cells in the stationary phase. Under nonstress conditions σ^{B} is expressed but inactivated by complexing with an anti-sigma factor. Interestingly, under stress conditions the antisigma factor is inactivated by a separate protein, which acts as an anti-antisigma factor (Vijay et al. 2000). A phosphorelay is involved in activating and inactivating the various factors during σ^{B} activation. Another well-studied σ factor is σ^{D}, which controls the chemotaxis-motility regulon.

Under conditions of starvation, *B. subtilis*, begins the differentiation process that results in the formation of an endospore. After the initiation of sporulation, there is an asymmetrical cell division forming two cells of different sizes. Sigma factors play an essential role in the developmental processes leading to sporulation (Haldenwang 1995). Available evidence indicates there is a cascade of interactions among the various σ factors required for sporulation. During activation of sporulation σ^{F} is the first prespore-specific σ factor, while in the mother cell, a different σ factor, σ^{E}, is the first σ factor activated. It has been suggested that σ^{F} is activated first; it somehow activates σ^{E}, and then sigma factors σ^{F}, σ^{G}, and σ^{K} are activated. The σ^{G} and σ^{K} proteins are required for activation of genes expressed late in the sporulation process.

Two-component signal-transduction systems

External environmental changes are often detected by bacteria using a system referred to as a two-component sensor regulator. This name refers to the fact that in many bacteria this sensing system is made up of a sensor kinase and a response regulator. The protein that detects the external signal is usually a transmembrane protein kinase, which transmits a signal by means of its cytoplasmic domain. Signal transfer involves transfer of a phosphate group from a histidine on the sensor kinase to the response regulator. Binding the phosphate alters the regulator's activity allowing it to bind to its target genes. A phosphatase may later remove the phosphate, restoring the system to its original state. There are many examples of two-component sensor regulator components in bacteria, although not all contain only two components. Some examples have been mentioned previously, including the nitrate-sensing system in *E. coli* and proteins sensing attractants or repellants during chemotaxis.

Altering the catalytic activity of existing enzymes

FEEDBACK INHIBITION

The regulatory processes described previously affect the expression of genes encoding proteins carrying out a particular physiological response. The process named feedback inhibition acts to decrease the activity of previously synthesized enzymatic pathways. Feedback inhibition occurs most commonly where the end product of a pathway inhibits the activity of the first enzyme in the pathway that is specific for the synthesis of that particular end product. The regulated enzyme is usually allosteric, with separate binding sites for substrate and effector molecules. In most cases the product of the pathway acts as a negative effector and, when bound to the enzyme, decreases the affinity of the enzyme for its substrate. To limit substrate flow through a pathway, it is only necessary to inhibit the activity of the first enzyme in the pathway. With the inhibition of that enzyme, the latter enzymes in the pathway will have no substrates and production of the end product must cease.

Feedback inhibition in branching biosynthetic pathways

In many biosynthetic pathways a single reaction can produce a product of which further metabolism leads to the synthesis of two or more different end products. Bacteria have developed different strategies to modulate the activity of these key reactions using feedback inhibition. One mechanism is termed cumulative inhibition. This is where the initial enzyme in the pathway has allosteric sites for each of the products eventually produced by metabolism of the product of the regulated enzyme. However, the enzyme can only be completely inhibited by the presence of multiple effectors. Because multiple pathways are dependent on the activity of the regulated enzyme, no single product should inactivate the enzyme. A variation on this regulation is concerted inhibition, which is when a single end product will not affect enzyme activity, but when multiple products are present the enzyme is inhibited.

Another type of inhibition is termed sequential inhibition. This occurs when two or more pathways use the same substrate produced by a shared enzyme, and the pathways are subjected to feedback inhibition, resulting in accumulation of the common substrate. If that substrate then causes feedback inhibition of the enzyme producing it, feedback inhibition of the terminal pathways will result in feedback inhibition of the initial common enzyme. An alternative, but very common, method for a bacterium to regulate the activity of an enzymatic reaction shared by several pathways, is for the organism to possess different enzymes to catalyze the same reaction. Different genes encode these isozymes (or isofunctional enzymes) and, therefore, can be both repressed and inhibited separately. Regulation by means of isozymes permits each structural gene coding for an isozyme to be part of a different regulon controlled by a separate end product. The possible disadvantage of the organism having to possess genes to code for two different enzymes catalyzing similar reactions may be offset by the efficiency of regulation.

Enzyme inhibition and activation in catabolic pathways

The regulation of the activity of allosteric enzymes by effectors also functions in catabolic pathways. The effectors may be negative, inhibiting enzyme activity, or positive, activating enzyme activity. An example is the regulation of the activity of lactate dehydrogenase in homolactic acid bacteria (see above). In these bacteria, the reduction of pyruvate to lactate by lactic dehydrogenase permits the rapid oxidation of NADH that has been reduced by the hexosediphosphate pathway. If the rate of NADH oxidation is insufficient to maintain a pool of NAD^+, there will be an increase in the internal pool of fructose-1,6-bisphosphate. Lactate dehydrogenase from these organisms is an allosteric enzyme that has an absolute requirement for fructose-1,6-bisphosphate as a positive effector. The increase in the internal pool of this diphosphate results in the activation of more lactate dehydrogenase enzymes and an increase in the rate of NADH oxidation.

COVALENT MODIFICATION

Another strategy used by cells to modulate the activity of active enzymes is covalent modification. Two examples of this were discussed previously. Enzymes in *E. coli* add a phosphate to the active site of isocitrate dehydrogenase to limit its activity during growth on acetate or other compounds metabolized through acetyl-CoA (Hurley et al. 1990). The other example discussed above was covalent modification in ADP-ribosylation of nitrogenase (Ludden 1994). Modification reduces activity of the nitrogenase.

One of the best-studied examples of covalent modification is the adenylylation of glutamine synthetase in enteric bacteria. Glutamine synthetase is a critical enzyme for the incorporation of ammonia into cellular constituents. This enzyme is used for nitrogen incorporation when nitrogen concentrations are low and uses the energy of ATP to drive the reaction. Because this ATP consumption can put a significant stress on energy levels in the cell, some bacteria, such as *E. coli* and *Salmonella typhimurium*, use covalent modification to fine-tune the activity of glutamine synthetase (Merrick and Edwards 1995). Glutamine synthetase is modified by addition of AMP to the enzyme; because the active enzyme is a dodecamer, as many as 12 AMPs can be added to the active complex (Yamashita et al. 1989). Addition of the AMP reduces enzyme activity, but the active site is not directly modified. A single enzyme, adenylyl transferase/adenylyl-removing enzyme, is responsible for modification of glutamine synthetase. The activity of this enzyme is controlled by another enzyme, designated PII, which is itself covalently modified by addition of UMP. The level of uridylylation of PII is dependent on the level of nitrogen in the cell. The net result is that when fixed nitrogen is high, the activity of glutamine synthetase is reduced to minimize ATP consumption. However, when nitrogen availability is low, glutamine synthetase is deadenylylated so it can be used for the incorporation of nitrogen.

ACKNOWLEDGMENTS

This chapter is adapted and expanded from Mortlock, R.P. 1998. Bacterial growth and metabolism. In: Balows, A. and Duerdon, B.I. (eds), *Topley & Wilson's Microbiology and Microbial Infections*. Vol. 2, *Systematic bacteriology* 9th edn. London: Edward Arnold, 85–124.

REFERENCES

Abrahams, J.P. and Leslie, A.G. 1994. Structure at 2.8 Å resolution of F1-ATPase from bovine heart mitochondria. *Nature*, **370**, 621–8.

Ades, S.E., Connolly, L.E., et al. 1999. The *Escherichia coli* σ(E)-dependent extracytoplasmic stress response is controlled by the regulated proteolysis of an anti-sigma factor. *Genes Dev*, **13**, 2449–61.

Armitage, J.P. and Macnab, R.M. 1987. Unidirectional, intermittent rotation of the flagellum of *Rhodobacter sphaeroides*. *J Bacteriol*, **169**, 514–18.

Baker, S.C., Ferguson, S.J., et al. 1998. Molecular genetics of the genus *ParaCoccus*: metabolically versatile bacteria with bioenergetic flexibility. *Microbiol Mol Biol Rev*, **62**, 1046–78.

Ban, N., Nissen, P., et al. 1999. Placement of protein and RNA structures into a 5 å-resolution map of the 50S ribosomal subunit. *Nature*, **400**, 841–7.

Bausch, C., Peekhaus, N., et al. 1998. Sequence analysis of the GntII (subsidiary) system for gluconate metabolism reveals a novel pathway for L-idonic acid catabolism in *Escherichia coli*. *J Bacteriol*, **180**, 3704–10.

Beisswenger, R. and Kula, M.R. 1991. Catalytic properties and substrate specificity of 3-hexulose phosphate synthase from *Methylomonas* M15. *Appl Microbiol Biotechnol*, **34**, 604–7.

Benov, L.T. and Fridovich, I. 1994. Escherichia coli expresses a copper- and zinc-containing superoxide dismutase. *J Biol Chem*, **269**, 25310–14.

Berks, B.C., Ferguson, S.J., et al. 1995. Enzymes and associated electron transport systems that catalyze the respiratory reduction of nitrogen oxides and oxyanions. *Biochim Biophys Acta*, **1232**, 97–173.

Bishop, P.E., Jarlenski, D.M. and Hetherington, D.R. 1982. Expression of an alternative nitrogen fixation system in *Azotobacter vinelandii*. *J Bacteriol*, **150**, 1244–51.

Bremer, H. and Churchward, G. 1991. Control of cyclic chromosomal replication in *Escherichia coli*. *Microbiol Rev*, **55**, 459–75.

Buggy, J.J., Sganga, M.W. and Bauer, C.E. 1994. CharaCterization of a light-responding trans-activator responsible for differentially controlling reaction center and light-harvesting-I gene expression in *Rhodobacter capsulatus*. *J Bacteriol*, **176**, 6936–43.

Bukau, B. 1993. Regulation of the *Escherichia coli* heat-shock response. *Mol Microbiol*, **9**, 671–80.

Busby, S. and Ebright, R.H. 1999. Transcription activation by catabolite activator protein. *J Mol Biol*, **293**, 199–213.

Calamita, G., Bishai, W.R., et al. 1995. Molecular cloning and charaCterization of AqpZ, a water channel from *Escherichia coli*. *J Biol Chem*, **270**, 29063–6.

Calamita, G., Kempf, B., et al. 1998. Regulation of the *Escherichia coli* water channel gene *aqpZ*. *Proc Natl Acad Sci USA*, **95**, 3627–31.

Cate, J.H., Yusupov, M.M., et al. 1999. X-ray crystal structures of 70S ribosome functional complexes. *Science*, **285**, 2095–104.

Chan, M.K., Kim, J. and Rees, D.C. 1993. The nitrogenase FeMo-cofactor and P-cluster pair: 2.2 Å resolution structures. *Science*, **260**, 792–4.

Chang, M.K. and Cronan, J.E. Jr 1999. Membrane cyclopropane fatty acid content is a major factor in acid resistance of *Escherichia coli*. *Mol Microbiol*, **33**, 249–59.

Chung, T., Klumpp, D.J. and LaPorte, D.C. 1988. Glyoxylate bypass operon of *Escherichia coli*: cloning and determination of the functional map. *J Bacteriol*, **170**, 386–92.

Clemons, W.M.J., May, J.L., et al. 1999. Structure of a bacterial 30S ribosomal subunit at 5.5 Å resolution. *Nature*, **400**, 833–40.

Cooper, S. 1991. Synthesis of the cell surface during the division cycle of rod-shaped, gram-negative bacteria. *Microbiol Rev*, **55**, 649–674.

Cornelis, G.R. and Wolf-Watz, H. 1997. The *Yersinia* Yop virulon: a bacterial system for subverting eukaryotic cells. *Mol Microbiol*, **23**, 861–7.

Cortay, J.C., Bleicher, F., et al. 1988. Nucleotide sequence and expression of the *aceK* gene coding for isocitrate dehydrogenase kinase/phosphatase in *Escherichia coli*. *J Bacteriol*, **170**, 89–97.

Cowan, S.W., Schirmer, T., et al. 1992. Crystal structures explain functional properties of 2 *E. coli* porins. *Nature*, **358**, 727–33.

Cvitkovitch, D.G., Boyd, D.A. and Hamilton, I.R. 1995. Glucose transport by a mutant of *Streptococcus mutans* unable to accumulate sugars via the phosphoenolpyruvate phosphotransferase system. *J Bacteriol*, **177**, 2251–8.

Danson, M.J. 1988. Archaebacteria. *Adv Micro Physiol*, **29**, 166–231.

Darwin, A.J., Tyson, K.L., et al. 1997. Differential regulation by the homologous response regulators NarL and NarP of *Escherichia coli* K-12 depends on DNA binding site arrangement. *Mol Microbiol*, **25**, 583–95.

Deisenhofer, J., Epp, O., et al. 1995. Crystallographic refinement at 2.3 Å resolution and refined model of the photosynthetic reaction center from *Rhodopseudomonas viridis*. *J Mol Biol*, **246**, 429–57.

De Mot, R., Nagy, I., et al. 1999. Proteasomes and other self-compartmentalizing proteases in prokaryotes. *Trends Microbiol*, **7**, 88–92.

Deppenmeier, U., Blaut, M. and Gottschalk, G. 1991. H_2:heterodisulfide oxidoreductase, a second energy-conserving system in the methanogenic strain Go1. *Arch Microbiol*, **155**, 272–7.

Deutscher, J., Reizer, J., et al. 1994. Loss of protein kinase-catalyzed phosphorylation of HPr, a phosphocarrier protein of the phosphotransferase system, by mutation of the *ptsH* gene confers resistance to several catabolite genes of *Bacillus subtilis*. *J Bacteriol*, **176**, 3336–44.

Ditzel, L., Lowe, J., et al. 1998. Crystal structure of the thermosome, the archaeal chaperonin and homolog of CCT. *Cell*, **93**, 125–38.

Dunny, G.M. and Leonard, B.A.B. 1997. Cell-cell communication in gram-positive bacteria. *Annu Rev Microbiol*, **51**, 527–64.

Eberhard, A., Burlingame, A.L., et al. 1981. Structural identification of autoinducer of *Photobacterium fischeri* luciferase. *Biochemistry*, **20**, 2444–9.

Engebrecht, J., Nealson, K. and Silverman, M. 1983. Bacterial bioluminescence: isolation and genetic analysis of functions from *Vibrio fischeri*. *Cell*, **32**, 773–81.

Fekkes, P. and Driessen, A.J. 1998. Protein targeting to the bacterial cytoplasmic membrane. *Microbiol Mol Biol Rev*, **63**, 161–73.

Ferry, J.G. 1999. Enzymology of one-carbon metabolism in methanogenic pathway. *FEMS Microbiol Rev*, **23**, 13–38.

Fink, A.L. 1999. Chaperone-mediated protein folding. *Physiol Rev*, **79**, 425–49.

Fuqua, C., Winans, S.C. and Greenberg, E.P. 1996. Census and consensus in bacterial ecosystems: the LuxR-LuxI family of quorum-sensing transcriptional regulators. *Annu Rev Microbiol*, **50**, 727–51.

Gennis, R.B. and Stewart, V. 1996. Respiration, *Escherichia coli* and Salmonella. In: Neidhart, F.C. (ed.), *Cellular and molecular biology*. Washington, DC: ASM Press, 217–61.

Georgiadis, M.M., Komiya, H., et al. 1992. Crystallographic structure of the nitrogenase iron protein from *Azotobacter vinelandii*. *Science*, **257**, 1653–9.

Haldenwang, W.G. 1995. The sigma factors of *Bacillus subtilis*. *Microbiol Rev*, **59**, 1–30.

Hansen, T.A. 1978. Multiplicity of genome equivalents in the radiation resistant bacterium *Micrococcus radiodurans*. *J Bacteriol*, **134**, 71–5.

Hansen, T.A. 1994. Metabolism of sulfate-reducing prokaryotes. *Antonie Van Leeuwenhoek*, **66**, 165–85.

Hidalgo, E., Ding, H. and Demple, B. 1997. Redox signal transduction: mutations shifting [2Fe-2S] centers of the SoxR sensor-regulator to the oxidized form. *Cell*, **88**, 121–9.

Hiromasa, Y., Aso, Y. and Yamashita, Y. 1994. Thermal disassembly of pyruvate dehydrogenase multienzyme complex from *Bacillus stearothermophilus*. *Biosci Biotech Biochem*, **58**, 1904–5.

Hogema, B.M., Arents, J.C., et al. 1999. Autoregulation of lactose uptake through the LacY permease by enzyme IIAGlc of the PTS in *Escherichia coli* K-12. *Mol Microbiol*, **31**, 1825–33.

Hooper, A.B., Vannelli, T., et al. 1997. Enzymology of the oxidation of ammonia to nitrite by bacteria. *Antonie Van Leeuwenhoek*, **71**, 59–67.

Horwich, A.L., Weber-Ban, E.U. and Finley, D. 1999. Chaperone rings in protein folding and degradation. *Proc Natl Acad Sci USA*, **96**, 11033–40.

Hu, X., Damjanovic, A., et al. 1998. Architecture and mechanism of the light-harvesting apparatus of purple bacteria. *Proc Natl Acad Sci USA*, **95**, 5935–41.

Hung, L.W., Wang, I.X., et al. 1998. Crystal structure of the ATP-binding subunit of an ABC transporter. *Nature*, **396**, 703–7.

Huo, L., Martin, K. and Schleif, R.F. 1988. Alternate DNA loops regulate the arabinose operon in *Escherichia coli*. *Proc Natl Acad Sci USA*, **85**, 5444–8.

Hurley, J.H., Dean, A.M., et al. 1990. Regulation of an enzyme by phosphorylation at the active site. *Science*, **249**, 1012–16.

Iverson, T.M., Luna-Chavez, C., et al. 1999. Structure of the *Escherichia coli* fumarate reductase respiratory complex. *Science*, **284**, 1961–6.

Iwata, S., Ostermeier, C., et al. 1995. Structure at 2.8 Å resolution of cytochrome *c* oxidase from *ParaCoccus denitrificans*. *Nature*, **376**, 660–9.

Jacob, F. and Monod, J. 1961. *On the regulation of gene activity*. Baltimore, MD: Waverly Press, 193–211.

James, G.A., Korber, D.R., et al. 1995. Digital image analysis of growth and starvation responses of a surface-colonizing *Acinetobacter* sp. *J Bacteriol*, **177**, 907–15.

Kaback, H.R. and Wu, J. 1997. From membrane to molecule to the third amino acid from the left with a membrane transport protein. *Q Rev Biophys*, **30**, 333–64.

Kandler, O. and Konig, H. 1998. Cell wall polymers in Archaea (Archaebacteria). *Cell Mol Life Sci*, **54**, 305–8.

Kato, N., Miyamoto, N., et al. 1988. Hexulose phosphate synthase from a new facultative methylotroph *Mycobacterium gastri* MB19. *Agric Biol Chem*, **52**, 2659–62.

Kelly, D.P., Shergill, J., et al. 1997. Oxidative metabolism of inorganic sulfur compounds by bacteria. *Antonie Van Leeuwenhoek*, **71**, 95–107.

Khoroshilova, N., Popescu, C., et al. 1997. Iron-sulfur cluster disassembly in the FNR protein of *Escherichia coli* by O_2: [4Fe-4S] to [2Fe-2S] conversion with loss of biological activity. *Proc Natl Acad Sci USA*, **94**, 6087–92.

Kraus, A., Kuster, E., et al. 1998. Identification of a co-repressor binding site in catabolite control protein CcpA. *Mol Microbiol*, **30**, 955–63.

Kumari, S., Tishel, R., et al. 1995. Cloning, characterization and functional expression of *acs*, the gene which encodes acetyl coenzyme A synthetase in *Escherichia coli*. *J Bacteriol*, **177**, 2878–86.

Lancaster, C.R., Kroger, A., et al. 1999. Structure of fumarate reductase from *Wolinella succinogenes* at 2.2 Å resolution. *Nature*, **402**, 377–85.

Lessie, T.G. and Phibbs, P.V. Jr. 1984. Alternate pathways of carbohydrate utilization in pseudomonads. *Annu Rev Microbiol*, **38**, 359–87.

Lin, J., Qi, R., et al. 1999. Whole-genome shotgun optical mapping of *Deinococcus radiodurans*. *Science*, **285**, 1558–62.

Liu, P.Q. and Ames, G.F. 1998. In vitro disassembly and reassembly of an ABC transporter, the histidine permease. *Proc Natl Acad Sci USA*, **95**, 3495–500.

Lonetto, M., Gribskov, M. and Gross, C.A. 1992. The σ^{70} family: sequence conservation and evolutionary relationships. *J Bacteriol*, **174**, 3843–9.

Lovley, D.R., Coates, J.D., et al. 1996. Humic substances as electron acceptors for microbial respiration. *Nature*, **382**, 445–8.

Lu, Y., Turner, R.T. and Switzer, R.L. 1995. Roles of the 3 transcriptional attenuators of the *Bacillus subtilis* pyrimidine biosynthetic operon in the regulation of its expression. *J Bacteriol*, **177**, 1315–25.

Lubben, M., Kolmerer, B. and Saraste, M. 1992. An archaebacterial terminal oxidase combines core structures of two mitochondrial respiratory complexes. *EMBO J*, **11**, 805–12.

Ludden, P.W. 1994. Reversible ADP-ribosylation as a mechanism of enzyme regulation in procaryotes. *Mol Cell Biochem*, **138**, 123–9.

MacGregor, C.H., Wolff, J.A., et al. 1991. Cloning of a catabolite repressor control (*crc*) gene from *Pseudomonas aeruginosa*, expression of the gene in *Escherichia coli* and identification of the gene product in *Pseudomonas aeruginosa*. *J Bacteriol*, **173**, 7204–12.

Magasanik, B. 1961. Catabolic repression. *Cold Spring Harbor Symp Quant Biol*, **26**, 249–59.

Mager, W.H. and De Kruijff, A.J.J. 1995. Stress induced transcriptional activation. *Microbiol Rev*, **59**, 506–31.

Manson, M.D., Armitage, J.P., et al. 1998. Bacterial locomotion and signal transduction. *J Bacteriol*, **180**, 1009–22.

Matin, A. 1992. Physiology, molecular biology and applications of the bacterial starvation response. *J Appl Bacteriol Symp Suppl*, **73**, 49S–57S.

Mat-Jan, F., Alam, K.Y. and Clark, D.P. 1989. Mutants of *Escherichia coli* deficient in the fermentative lactic dehydrogenase. *J Bacteriol*, **171**, 342–8.

Maymo-Gatell, X., Chien, Y., et al. 1997. Isolation of a bacterium that reductively dechlorinates tetrachloroethene to ethene. *Science*, **276**, 1568–71.

Merrick, M.J. and Edwards, R.A. 1995. Nitrogen control in bacteria. *Microbiol Rev*, **59**, 604–22.

Messner, K.R. and Imlay, J.A. 1999. The identification of primary sites of superoxide and hydrogen peroxide formation in the aerobic respiratory chain and sulfite reductase complex of *Escherichia coli*. *J Biol Chem*, **274**, 10119–28.

Mitchell, C.G. and Dawes, E.A. 1982. The role of oxygen in the regulation of glucose metabolism, transport and the tricarboxylic acid cycle in *Pseudomonas aeruginosa*. *J Gen Microbiol*, **128**, 49–59.

Mitchell, P. 1961. Coupling of phosphorylation to electron and hydrogen transfer by a chemi-osmotic type of mechanism. *Nature (Lond)*, **191**, 144–8.

Moat, A.G. and Foster, J.W. 1995. *Microbial physiology*, 3rd edn. New York: Wiley-Liss.

Moeck, G.S. and Coulton, J.W. 1998. TonB-dependent iron acquisition: mechanisms of siderophore-mediated active transport. *Mol Microbiol*, **28**, 675–81.

Monod, J. 1947. The phenomenon of enzymatic adaptation and its bearing on problems of genetics and cellular differentiation. *Growth Symp*, **11**, 192–289.

Moreno-Vivian, C., Cabello, P., et al. 1999. Prokaryotic nitrate reduction: molecular properties and functional distinction among bacterial nitrate reductases. *J Bacteriol*, **181**, 6573–84.

Myers, C.R. and Nealson, K.H. 1990. Respiration-linked proton translocation coupled to anaerobic reduction of manganese(IV) and iron(III) in *Shewanella putrefaciens* MR-1. *J Bacteriol*, **172**, 6232–8.

Nakano, M.M. and Zuber, P. 1998. Anaerobic growth of a 'strict aerobe' (*Bacillus subtilis*). *Annu Rev Microbiol*, **52**, 165–90.

Neale, A.D., Scopes, R.K., et al. 1987. Pyruvate decarboxylase of *Zymomonas mobilis*: isolation, properties and genetic expression in *Escherichia coli*. *J Bacteriol*, **169**, 1024–8.

Neidhardt, F.J. (ed.) 1987. *Escherichia coli and Salmonella typhimurium: cellular and molecular biology*. Washington, DC: ASM Press, 3–6, 56–69.

Nereng, S. and Kaplan, S. 1999. Genomic complexity among strains of the facultative photoheterotrophic bacterium *Rhodobacter sphaeroides*. *J Bacteriol*, **181**, 1684–8.

Nikaido, H. 1992. Porins and specific channels of bacterial outer membranes. *Mol Microbiol*, **6**, 435–42.

Nomura, M., Gourse, R. and Baughman, G. 1984. Regulation of the synthesis of ribosomes and ribosomal components. *Annu Rev Biochem*, **53**, 75–117.

Nystrom, T. 1999. Starvation, cessation of growth and bacterial aging. *Curr Opin Microbiol*, **2**, 214–19.

Oh-oka, H., Kakutani, S., et al. 1995. Highly purified photosynthetic reaction center (PscA/cytochrome *c*551)2 complex of the green sulfur bacterium *Chlorobium limicola*. *Biochemistry*, **34**, 13091–7.

Pastan, I. and Adhya, S. 1976. Cyclic adenosine 5-monophosphate in *Escherichia coli*. *Bacteriol Rev*, **40**, 527–51.

Pau, R.N., Mitchenall, L.A. and Robson, R.L. 1989. Genetic evidence for an *Azotobacter vinelandii* nitrogenase lacking molybdenum and vanadium. *J Bacteriol*, **171**, 124–9.

Pebay-Peyroula, E., Rummel, G., et al. 1997. X-ray structure of bacteriorhodopsin at 2.5 angstroms from microcrystals grown in lipidic cubic phases. *Science*, **277**, 1676–8.

Ponce, E., Flores, N., et al. 1995. Cloning of the 2 pyruvate kinase isoenzyme structural genes from *Escherichia coli*: the relative roles of these enzymes in pyruvate biosynthesis. *J Bacteriol*, **177**, 5719–22.

Pugsley, A.P. 1993. The complete general secretory pathway in Gram-negative bacteria. *Microbiol Rev*, **57**, 50–108.

Punita, S.J., Reddy, M.A. and Das, K.A. 1989. Multiple chromosome of *Azotobacter vinelandii*. *J Bacteriol*, **171**, 3133–8.

Rosey, E.L., Oskouian, B. and Stewart, G.C. 1991. Lactose metabolism by *Staphylococcus aureus*: characterization of *lacABCD*, the structural genes of the tagatose 6-phosphate pathway. *J Bacteriol*, **173**, 5992–8.

Rothfield, L., Justice, S. and Garcia-Lara, J. 1999. Bacterial cell division. *Annu Rev Genet*, **33**, 423–48.

Saier, M.H. Jr and Reizer, J. 1994. The bacterial phosphotransfease system: new frontiers 30 years later. *Mol Microbiol*, **13**, 755–64.

Saier, M.H. Jr, Chauvaux, S., et al. 1996. Catabolite repression and inducer control in gram-positive bacteria. *Microbiology*, **142**, 217–30.

Sambongi, Y., Iko, Y., et al. 1999. Mechanical rotation of the c subunit oligomer in ATP synthase (F0F1): direct observation. *Science*, **286**, 1722–4.

Sawer, G. and Bock, A. 1988. Anaerobic regulation of pyruvate formate-lyase from *Escherichia coli* K-12. *J Bacteriol*, **170**, 5330–6.

Sawers, G. and Watson, G. 1998. A glycyl radical solution: oxygen-dependent interconversion of pyruvate formate-lyase. *Mol Microbiol*, **29**, 945–54.

Schirmer, T., Keller, T.A., et al. 1995. Structural basis for sugar translocation through maltoporin channels at 3.1 å resolution. *Science*, **267**, 512–14.

Shapiro, J.A. 1995. The significance of bacterial colony patterns. *BioAssays*, **17**, 597–607.

Sheppard, D.E. and Englesberg, E. 1967. Further evidence for positive control of the l-*araB*inose system by gene *araC*. *J Mol Biol*, **25**, 443–54.

Skulachev, V.P. 1994. Chemiosmotic concept of the membrane bioenergetics: what is already clear and what is still waiting for elucidation? *J Bioenerg Biomembr*, **26**, 589–98.

Sleytr, U.B. and Beveridge, T.J. 1999. Bacterial S-layers. *Trends Microbiol*, **7**, 253–60.

Smidt, H., Song, D., et al. 1999. Random transposition by Tn916 in *Desulfitobacterium dehalogenans* allows for isolation and characterization of halorespiration-deficient mutants. *J Bacteriol*, **181**, 6882–8.

Snoep, J.L., Joost, M., et al. 1991. Effect of the energy source on the NADH/NAD ratio and on pyruvate catabolism in anaerobic chemostat cultures of *Enterococcus faecalis* NCTC 775. *FEMS Microbiol Lett*, **81**, 63–6.

Steinbuchel, A., Schlegel, H.G., et al. 1991. Physiology and molecular genetics of poly(beta-hydroxyalkanoic acid) synthesis in *Alcaligenes eutrophus*. *Mol Microbiol*, **5**, 535–42.

Stock, D., Leslie, A.G. and Walker, J.E. 1999. Molecular architecture of the rotary motor in ATP synthase. *Science*, **286**, 1700–5.

Storz, G. and Imlay, J.A. 1999. Oxidative stress. *Curr Opin Microbiol*, **2**, 188–94.

Suwanto, A. and Kaplan, S. 1989. Physical and genetic mapping of the *Rhodobacter sphaeroides* 2.4.1 genome: presence of 2 unique chromosomes. *J Bacteriol*, **171**, 5850–9.

Sweeney, N.J., Laux, D.C. and Cohen, P.S. 1996. *Escherichia coli* F-18 and *E. coli* K-12 edamutants do not colonize the streptomycin-treated mouse large intestine. *Infect Immun*, **64**, 3504–11.

Thauer, R.K. 1998. Biochemistry of methanogenesis: a tribute to Marjory Stephenson 1998 Marjory Stephenson Prize Lecture. *Microbiology*, **144**, 2377–406.

Thevenot, T., Brochu, D., et al. 1995. Regulation of ATP-dependent P-(Ser)-Hpr formation in *Streptococcus mutans* and *Streptococcus salivarius*. *J Bacteriol*, **177**, 2863–9.

Tolner, B., Ubbink-Kok, T., et al. 1995. CharaCterization of the proton/glutamate symport protein of *Bacillus subtilis* and its functional expression in *Escherichia coli*. *J Bacteriol*, **177**, 2863–9.

Ulbrandt, N.D., Newitt, J.A. and Bernstein, H.D. 1997. The *E. coli* signal recognition particle is required for the insertion of a subset of inner membrane proteins. *Cell*, **88**, 187–96.

Van de Vossenberg, J.L., Driessen, A.J. and Konings, W.N. 1998. The essence of being extremophilic: the role of the unique archaeal membrane lipids. *Extremeophiles*, **2**, 163–70.

Vijay, K., Brody, M.S., et al. 2000. A PP2C phosphatase containing a PAS domain is required to convey signals of energy stress to the sigmaB transcription factor of *Bacillus subtilis*. *Mol Microbiol*, **35**, 180–8.

Weiner, J.H., Bilous, P.T., et al. 1998. A novel and ubiquitous system for membrane targeting and secretion of cofactor-containing proteins. *Cell*, **93**, 93–101.

Whitfield, C. and Roberts, I.S. 1999. Structure, assembly and regulation of expression of capsules in *E. coli*. *Mol Microbiol*, **31**, 1307–19.

Xu, Z., Horwich, A.L. and Sigler, P.B. 1997. The crystal structure of the asymmetric GroEL-GroES-(ADP)7 chaperonin complex. *Nature*, **388**, 741–50.

Yamashita, M., Almassy, R., et al. 1989. Refined atomic model of glutamine synthetase at 3.5 å resolution. *J Biol Chem*, **264**, 17681–90.

Yanofsky, C. 1981. Attenuation in the control of expression of bacterial operons. *Nature*, **289**, 751–8.

Yanofsky, C. 2000. Transcription attenuation: once viewed as a novel regulatory strategy. *J Bacteriol*, **182**, 1–8.

Yoshikawa, S., Shinzawa-Itoh, K., et al. 1998. Redox-coupled crystal structural changes in bovine heart cytochrome c oxidase. *Science*, **280**, 1723–9.

Zeng, A. and Deckwer, W. 1991. A model for multiproduct-inhibited growth of *Enterobacter aerogenes* in 2,3-butanol fermentation. *Appl Microbiol Biotechnol*, **35**, 1–3.

Zhang, J.P. and Normark, S. 1996. Induction of gene expression in *Escherichia coli* after pilus-mediated adherence. *Science*, **273**, 1234–6.

Zhang, Z., Huang, L., et al. 1998. Electron transfer by domain movement in cytochrome bc_1. *Nature*, **392**, 677–84.

Zheng, M., Aslund, F. and Storz, G. 1998. Activation of the OxyR transcription factor by reversible disulfide bond formation. *Science*, **279**, 1718–21.

Zumft, W.G. 1997. Cell biology and molecular basis of denitrification. *Microbiol Mol Biol Rev*, **61**, 533–616.

4

Bacterial genetics

MATTHEW B. AVISON AND PETER M. BENNETT

INTRODUCTION

Comparison of the current bacterial genetic literature with that of a generation ago reveal a 'sea change' in technical approach. Where once the study of bacterial genetics largely involved isolating mutants that occurred by chance in cell populations, a procedure that often required considerable ingenuity in the design of mutant selections or screens, and investigating the bases of mutation by reversion and complementation analyses, current procedure brings to bear a panoply of powerful molecular genetic techniques that enable an investigator to determine precisely a gene's composition and its mode of expression. These techniques include genomic and proteomic analyses supported by the computational power of bio-informatic databases. This chapter will discuss some of these technologies, and will attempt to indicate the value of both old and new approaches.

The similarities displayed among bacteria and the differences that distinguish one from another reflect the content and expression of the genes possessed by each individual cell. The genes control every property of a cell, including morphology, physiology, biochemistry, and, in general, an exact copy of each one is inherited by the cell's progeny. Genetics is the study of genes, their structures, organizations, and expression.

With the exception of some bacteriophages (see Chapter 5, Bacteriocins and bacteriophages), the genetic information of bacterial systems is encoded in DNA. The characteristics expressed by a cell are referred to as its phenotype. The set of genes found in any particular cell constitutes its genome, which generally consists of a large circular DNA molecule, the bacterial chromosome, and possibly one or more smaller circular DNA molecules called plasmids (Charlebois 1999).

Bacteria are, in general, very adaptable and may alter their phenotype in response to environmental change while the genotype remains unchanged. The ability to adapt the phenotype to the prevailing environmental conditions is determined genetically, i.e. not all the genes of a bacterial cell are expressed all the time. For example, β-galactosidase, an enzyme that hydrolyzes lactose to its constituent sugars, glucose, and galactose, is required by *Escherichia coli* only when lactose is its main source of carbon and energy and normally it is produced only under such circumstances (Jacob and Monod 1961; Prescott et al. 2002). The gene (*lacI*) and associated sequences that control the expression of β-galactosidase are, together with the gene for β-galactosidase itself, *lacZ*, and two other associated genes, *lacY* and *lacA* encoded at essentially the same location on the *E. coli* chromosome. The three genes *lacZ*, *lacY*, and *lacA* together form the *lac* operon, a set of genes that is expressed from a single promoter to produce, in the first instance, a single transcript (polycistronic mRNA) (Chapter 3, Bacterial growth and metabolism). This, in turn, is translated to produce the three gene products of the *lac* operon. This is just one example of the many complex systems that can be switched on or off to allow bacteria to make best use of the nutrients available to them and to adjust their biochemical makeup rapidly in response to changes in their environments. Such switches in gene expression generate phenotypic variation; the phenotype of the cell changes while the genotype remains unaltered. By contrast, genotypic variation involves a change in the genetic information that determines the phenotype. This type of change can occur by mutation, i.e. an alteration of the nucleotide sequence of DNA, that can occur in one of several ways. Alternatively, gene content can be changed by acquisition of new genes from another bacterial cell, by one of three DNA transfer processes, transformation, transduction, or conjugation.

In general, all the cells in a bacterial culture respond phenotypically in the same way to a change (often nutritional) in their environment. The change in phenotype is normally rapid and reversible, e.g. synthesis of the enzyme β-galactosidase by *E. coli* starts within seconds of the addition of lactose to a suitable culture medium and ceases just as rapidly when lactose is withdrawn (or exhausted). In contrast, genotypic change (mutation) affects the individual cell, rather than the population, and is essentially permanent. There are, however, exceptions to these generalizations and it is sometimes difficult to determine into which category a particular variation fits. For example, lysozyme treatment of a culture of *Bacillus subtilis* converts the cells to protoplasts by removing the cell wall. Removal of lysozyme, however, does not always result in a return to normal cell physiology; the cells may continue to grow as protoplasts or may show mass reversion to the bacillary form, depending upon the conditions of culture. This is an example of a phenotypic change that, in some circumstances, can appear as heritable and permanent. By contrast, physiological changes resulting from mutation may give the impression of being phenotypic because bacteria can multiply so rapidly; a mutant cell, with a particular selective advantage, can overgrow a population of unaltered cells within a few hours. When the culture conditions are reversed, the mutant may be at a selective disadvantage and, in turn, may be overgrown by cells of the original type. If cells are examined on time scales that permit many generations, what is in reality genotypic variation may appear to be phenotypic variation.

Bacteria are haploid organisms with no true nucleus or sexual reproductive cycle. The first great expansion of knowledge of bacterial genetics followed the discovery that genes (DNA) can be passed from one bacterial cell to another by several mechanisms. The second major expansion of activity is now in progress and reflects the introduction of a raft of molecular genetic techniques such as restriction enzyme mapping, gene cloning, DNA sequencing, and site-directed mutation. The bacterium about which most is known is *E. coli* K12, whose single chromosome is a circular DNA molecule of approximately 2 mm, about 1 000 times longer than the bacterial cell in which it is contained. This DNA molecule contains c. 4 700 000 base pairs (bp) or 4 700 kilobases (kb), and accommodates approximately 4 300 genes. Other bacteria also have single circular chromosomes; this is true both of bacteria related to *E. coli*, such as *Shigella* and *Salmonella* spp. (Sanderson 1976),

and of unrelated ones such as *Pseudomonas aeruginosa* (Holloway et al. 1979) and *B. subtilis* (Henner and Hoch 1980). Chromosome size varies considerably from one species to another. The smallest recorded to date is that of the intracellular pathogen *Mycoplasma genitalium* at 600 kb, while the largest found so far is the 9 455 kb circular chromosome of *Myxococcus xanthus*, closely followed by that of *Stigmatella aurantiaca* (9 350 kb) (Cole and Saint-Girons 1999). However, circular chromosomes are not a universal feature of prokaryotes. Linear chromosomes have been discovered in several bacteria, including *Borrelia burgdorferi*, the Lyme disease agent (which has a single linear chromosome of approximately 950 kb), *Agrobacterium tumefaciens*, which induces the formation of galls on a number of plants and has one circular chromosome (3 000 kb) and one linear one (2 100 kb), and *Streptomyces* species, sources of many antibiotics, which have linear chromosomes of approximately 8 000 kb (Cole and Saint-Girons 1999; Leblond and Decaris 1999). Circularity of the *E. coli* chromosome was first established by gene linkage experiments, and this provided the paradigm for several other bacteria. In the case of *E. coli* K12, its chromosome was also extracted and visualized (by autoradiography) and shown unequivocally to be circular (Cairns 1963), so confirming the genetic data. In more recent times, the form of a bacterial chromosome has more commonly been established by physical techniques such as restriction mapping. Restriction endonucleases are used to fragment the DNA molecule and the fragments are separated and their sizes determined by pulsed-field gel electrophoresis (PFGE). Fragment contiguity (fragment order) is determined in a number of ways, including cross-hybridization, which involves isolating and labeling a genomic fragment and using it as a probe to identify neighboring sequences. A summary of the different approaches to genome analysis and results obtained have been reviewed by Cole and Saint-Girons (1999). Now, chromosome circularity is much more readily established by complete genome sequencing programs. In addition, many bacteria carry relatively small DNA molecules called plasmids. These may also be circular or linear. Many are itinerant molecules that can be passed from one bacterial cell to another. Examples par excellence of this type are the resistance (R) plasmids responsible for much of the acquired antibiotic resistance displayed by many human bacterial pathogens. However, carriage of some plasmids appears to be a universal feature of a particular bacterial species (Cole and Saint-Girons 1999), in which case they appear to be indispensable and can be thought of as mini-chromosomes. The results of all of these studies have established that circularity is the norm for bacterial DNA, but there are exceptions, as noted.

Bacterial chromosomes must replicate before cell division and be distributed (partitioned) so that each daughter cell has its own copy. A complex organization exists to ensure that this happens. Plasmids also encode replication functions that allow them to replicate independently of the bacterial chromosome of the host cell and many also encode partition functions, as well as other stabilization mechanisms. DNA molecules that replicate are called replicons (Jacob et al. 1963). The chromosome is the replicon(s) in which are encoded those genes that are essential for the cell's survival. It is also the largest replicon in the cell. However, as indicated in the previous paragraph, not all bacteria carry their essential genes on a single DNA molecule. In a few instances, such as *Brucella melitensis*, *Burkholderia cepacia*, and *Rhodobacter sphaeroides*, the genome comprises two or more large DNA molecules, which are essential to the cell (Cole and Saint-Girons 1999).

MUTATION

Mutation can be defined as any permanent change in the sequence of bases of DNA, irrespective of a detectable change in the cell phenotype. In practice, most mutations are detected by virtue of the change in phenotype generated, although now, with many DNA sequences being determined, it is found that silent mutation is very common, i.e. a change in the nucleotide sequence that has no detectable phenotypic effect. Indeed, silent mutation is a major feature of genetic drift within a bacterial species, and can be used for typing purposes (van Leeuwen et al. 2003). A given gene can exist in a variety of different forms – some functional, some not – as a result of mutation. These different forms are called alleles. The form in which a gene exists in a bacterium as it is first isolated from nature is defined as the wild-type

allele of the gene; all other forms resulting from mutation of that gene are, by definition, mutant alleles.

Mutations may occur spontaneously, i.e. in the absence of a specific treatment to generate them, or they may be induced deliberately (the cells of the bacterial culture may be treated in one of several ways so as to provoke changes in the base sequence of the DNA). Classically, induction of mutations has been achieved by chemical treatment with known mutagens such as ethylmethane sulfonate (EMS, $CH_3CH_2.O.SO_2.CH_3$) and *N*-methyl-*N'*-nitro-*N*-nitrosoguanidine (MNNG, NTG) or by irradiation with ultraviolet or X-rays. These treatments generate DNA lesions at random and the particular mutation desired is selected by judicious manipulation of the subsequent growth conditions of the culture. In recent years it has become possible to target mutations, i.e. to introduce specific changes into a particular nucleotide sequence, using recombinant DNA technology, a procedure termed site-specific mutagenesis.

The spontaneous nature of mutation was demonstrated by Luria and Delbruck (1943) by an experiment known as the fluctuation test. The experiment exploited the observation that in a large enough population of cells sensitive to a particular bacteriophage (phage) there will be a few cells resistant to the phage. When equal volumes of a broth culture, each containing approximately 10^9 *E. coli*, were spread on a nutrient agar seeded with the phage, the number of colonies arising from resistant cells showed a small fluctuation from plate to plate (Figure 4.1). In a parallel experiment, small inocula (c. 10^3 cells) of the same culture were used to produce a series of 50 separate broth cultures. When the same volumes of these (i.e. containing approximately 10^9 bacteria) were spread on agar seeded

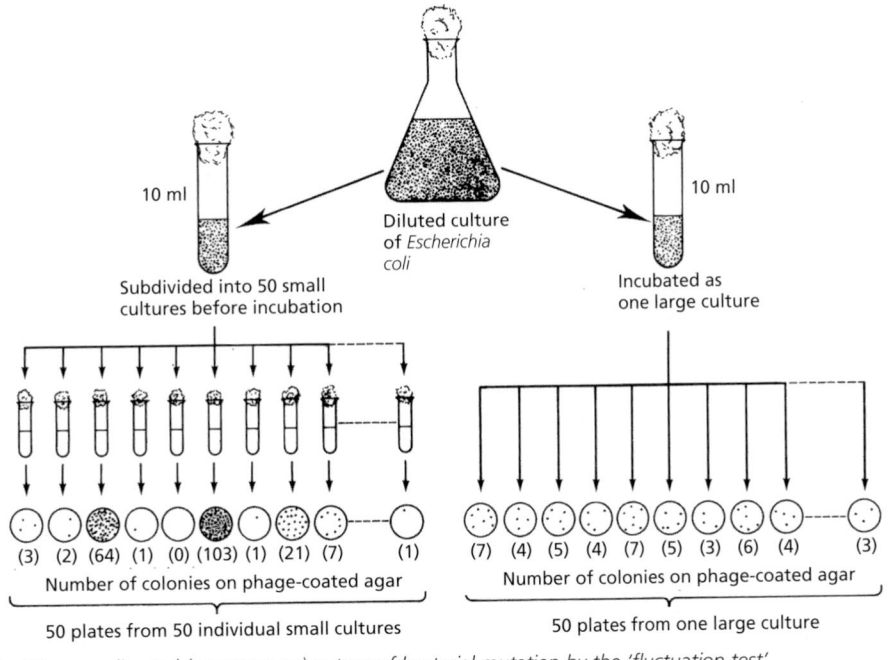

Figure 4.1 *Proof of the non-directed (spontaneous) nature of bacterial mutation by the 'fluctuation test'*

with the phage, much larger fluctuations were seen in the number of colonies obtained per plate (Figure 4.1). The results reflected the random nature of the mutations that had occurred in each culture, i.e. mutations arose at different times during the growth of the different cultures (the earlier the mutant appears the greater is the proportion of resistant cells in that population).

That mutations can arise spontaneously in a population of cells, independently of selective forces, was demonstrated directly and elegantly by Ester and Joshua Lederberg (1952). Approximately 10^8 phage-sensitive *E. coli* cells were inoculated onto nutrient agar and the plate was incubated. The resulting colonial growth was replica-plated to nutrient agar and nutrient agar seeded with the phage, where only phage-resistant cells grew to form colonies. Each colony that developed came from an inoculum of a phage-resistant cell (or cells) on the master plate. The locations of these were determined, and approximately 10^5 cells from the identical position on the phage-free nutrient agar were transferred into tubes of broth. From each of the resulting cultures 10^5 cells were spread on individual nutrient agar plates. The colonies that developed were again replica-plated to nutrient agar and nutrient agar seeded with phage and the entire process was repeated several times. By picking only phage-resistant colonies (identified by growth on the replica plate containing nutrient agar seeded with phage) from the phage-free nutrient agar to generate cultures for further testing, it was possible, eventually, to obtain a pure culture of a phage-resistant mutant. During the entire procedure the populations of cells used in the purification steps were never exposed to the phage, demonstrating unequivocally that resistant variants (i.e. mutants) emerged in the complete absence of the selective agent.

Mutagenesis

Much of our understanding of the mechanisms of mutation derives from experiments in which mutations have been induced by chemical agents that react with DNA in a known manner. Experiments designed to study the chemical basis of mutation have been carried out extensively with *E. coli* and several bacteriophages for which *E. coli* is a normal host. The protocols of mutation experiments vary according to the particular mutagen to be tested. For those reagents that react chemically with the constituent bases of DNA, bacteria are mixed with the mutagen and the suspension is incubated. The concentration of mutagen used and the time of incubation are determined pragmatically as those conditions under which approximately 5 percent of the input cells survive the treatment. The survivors are then permitted to grow in normal culture. Among the population that results, the proportion of mutants is high; in practice, however, the abundance of a particular mutation,

although considerably enhanced with respect to the untreated starting culture, may still be below the level at which a screen of individual cells in the population would readily detect it. Hence, populations of cells obtained after mutagenic treatment may be subject to culture on selective growth medium, constituted to favor the growth of the desired mutant.

Some chemicals cause mutation because they mimic one of the constituent bases of DNA and may be incorporated into DNA during replication when present in the growth medium. 5-Bromouracil (5-BU), an analog of thymine with a bromine atom at the C-5 position instead of a methyl (CH_3) group, is one such compound. Once part of DNA, such substitutions cause misincorporation of normal bases in subsequent rounds of replication, e.g. 5-BU causes misincorporation of guanine. For these mutagens the cells must necessarily grow during treatment because DNA replication is essential for the effect. The concentration of the reagent has to be chosen so that the level of incorporation is high enough to raise the frequency of mutation but not so high as to kill the majority of cells in the treated population. Again, the conditions of treatment are determined pragmatically.

One easy way to decide if the mutagenic treatment has been successful is to examine the culture for an increase in the proportion of a particular class of mutants, for example auxotrophic mutants or antibiotic resistant mutants. Experimentally, the percentage of auxotrophs generated, i.e. mutants that have acquired a nutritional requirement, or the percentage of antibiotic resistant derivatives generated, is determined. For example, samples of a suitable dilution of a mutagenized culture of *E. coli* are spread onto nutrient agar, so as to give 100–150 colonies per plate after a suitable period of incubation. These colonies are then tested individually, by replica plating, to determine if they can grow on a minimal-salts agar that supports growth of the parent bacterium. A satisfactory mutagenic treatment will generate 5–10 percent auxotrophs that will be unable to grow on the minimal agar. Alternatively, the increase in the proportion of cells resistant to certain antibiotics, e.g. rifampicin, streptomycin, or nalidixic acid, to which the cells in the treated culture were susceptible before treatment, can be monitored.

Various experimental 'tricks' can be used to enhance the chances of success in finding a particular mutant. Two are worthy of mention. The abundance of particular auxotrophic mutations in a population can be increased by penicillin enrichment to the point where they form the majority. Penicillin kills only growing bacterial cells. Non-growing cells survive the treatment, provided the medium is osmotically buffered. Hence, a whole mutagen-treated culture is inoculated into liquid medium lacking the particular growth factor (e.g. amino acid, vitamin, purine, or pyrimidine) needed by the desired auxotroph. After a short incubation (15–20 min), during which time the growth of the mutant type required is arrested, penicillin is added to the culture. The cells that do not need an exogenous supply of the growth factor grow and most are killed by the penicillin. The auxotrophs, because their growth has been arrested, survive. On removal of the penicillin and inoculation into a growth medium supplemented with the growth requirement, these cell types proliferate, together with any other survivors, to generate a new population of cells, among which the desired auxotrophs constitute an increased proportion. In practice, the procedure is not an all-or-none effect and a single round of treatment rarely increases the proportion of mutants sought by more than 100-fold. However, four or five cycles of treatment can usually enhance the proportion of the auxotroph required to greater than 10 percent of the surviving population of cells (Miller 1972), at which point mutants are easily identified. The procedure can be adapted to recover any mutant type where the mutation creates a conditional inhibition of growth that can be experimentally exploited.

The second technique that has been indispensable in the search for bacterial mutants is replica-plating (Lederberg and Lederberg 1952). Velvet is pressed gently onto the surface of agar on which bacterial colonies of interest have grown. A sample of each colony is imprinted on the velvet, which is then used to inoculate plates of sterile agar differing in composition from that on which the colonies originally appeared. The procedure preserves the pattern of colonies on the original plate. After incubation, the ability of individual colonies to grow under the new conditions can be assessed; e.g. a population of cells can be screened for auxotrophs of a particular type by replica-plating from a medium supplemented with the growth factor in question to a medium that lacks the growth factor but which is otherwise the same. The desired auxotrophs will fail to grow. An experienced person can easily screen 200–300 colonies per plate and, if necessary, several hundred plates per day.

Mutations can be divided into two types, micro- and macrolesions, depending on the extent of the alteration to the base pair sequence of DNA. Microlesions are also known as point mutations.

Microlesions

Point mutations are of two classes: base pair substitutions and frame-shift mutations. The first class comprises those mutants in which a single base pair has been altered, and can be subdivided into transitions and transversions. Such changes may be expressed (i.e. are detected as an altered phenotype) or silent. Frame-shift mutations are those in which one or a few base pairs have been added to or removed from the DNA with the consequence that the translational reading frame of the transcript, i.e. of messenger RNA (mRNA), of the

mutated gene is altered. Point mutations are, in general, revertible, i.e. a second mutational event at the same site can precisely reverse the first mutation and restore the original base sequence.

A transition occurs when one purine is replaced by another purine and a pyrimidine is replaced by a different pyrimidine, e.g. a change from a guanine–cytosine (GC) base pair to an adenine–thymine (AT) base pair, or vice versa; in this situation the purine–pyrimidine axis of the double helix at the point of mutation is preserved. Most chemically induced base substitutions are transitions. Transversions are those base pair substitutions in which the purine–pyrimidine axis is changed as a result of the mutation, e.g. a change from GC to CG or to TA. Such changes are much less common than transitions and probably arise as a consequence of occasional mistakes made during DNA replication and by one or more of the cell's DNA repair systems.

All four of the bases in DNA can exist in different tautomeric forms which are related to each other by single proton shifts. The tautomeric forms arise when the keto (C=O) group of a base is changed to the enol (C–OH) form or when the amino (–NH$_2$) group is changed to an imino (=NH) group. The enol and imino groups exist in equilibrium with the keto and amino forms, respectively, but the latter predominate under normal conditions, i.e. at ambient temperatures and at pH values near neutrality. Such tautomeric shifts are important because they can relax the normally strict Watson and Crick complementary base pairing conventions, permitting, for example, G to pair with T instead of with C, while retaining purine–pyrimidine base pairing. Such mistakes do occur during replication but they are kept to a minimum by the intrinsic accuracy of most DNA polymerases together with associated proofreading functions (Kunkel and Bebenek 2000) and by DNA mismatch repair systems (Harfe and Jinks-Robertson 2000; Marti et al. 2002). If these occasional mispairings are not corrected, on replication one daughter chromosome will have, for example, an AT base pair at a particular location instead of a GC base pair, and mutation will have occurred. These different forms are illustrated in Figure 4.2.

As well as occurring naturally, transitions may be formed by the incorporation of base analogs into the DNA. Among the most important analogs are the halogenated pyrimidines, particularly 5-BU. This compound can be incorporated into DNA in place of thymine. The halogen atom is strongly electronegative and can pull electrons from the oxygen of the keto group of 5-BU, converting it to an enol group. 5-BU and similar analogs, therefore, tautomerize much more readily than the natural base under normal physiological conditions. In its keto form, 5-BU can pair with A and be incorporated into DNA to form a pseudo-AT base pair. If 5-BU then undergoes a tautomeric shift to the enol form, at the next round of replication it will direct the incorpora-

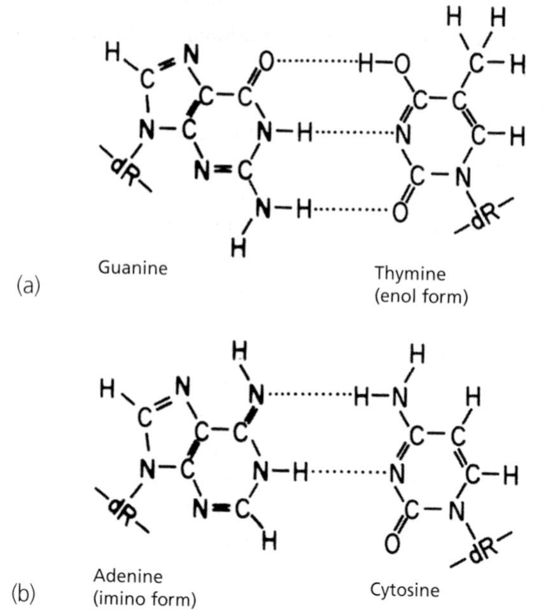

(a) Guanine   Thymine (enol form)

(b) Adenine (imino form)   Cytosine

Figure 4.2 *Changes in base pairing as the result of tautomeric shifts. In the enol form (a), thymine forms hydrogen bonds with guanine, instead of with adenine. In the imino form (b), adenine forms hydrogen bonds with cytosine, instead of with thymine. Similar shifts in guanine and cytosine will also cause changes in base pairing.*

tion of G rather than A into the newly synthesized strand of DNA. A further round of replication will establish an AT to GC transition. Because they are incorporated into DNA by replication, base analogs are effective as mutagens only in growing cells. By contrast, other chemicals such as nitrous acid and various alkylating agents interact with non-replicating and replicating DNA.

Transitions can be generated by the action of nitrous acid. This chemical reacts with amine groups, converting them to hydroxyl groups (oxidative deamination). Deamination of adenine converts it to hypoxanthine, which can base pair with cytosine. Therefore, upon replication, hypoxanthine may cause misincorporation of C. A further round of replication will then establish the AT to GC transition. In similar reactions, thymine is converted to uracil and guanine is converted to xanthine. The first of these changes can be detected by a dedicated repair system and reversed (Pearl 2000), whereas the latter appears to be lethal rather than mutagenic.

Alkylating agents are powerful mutagens that predominantly generate transition mutations, although some have been reported to generate transversions and frameshifts as well; the mechanisms by which the latter changes are achieved are not fully understood. Alkylating agents all carry one, two, or more reactive alkyl groups. An example, widely used to generate bacterial mutants, is EMS. The action of alkylating agents on DNA is complex. They are known to react with purine

bases, predominantly guanine, and the alkylated bases can cause base misincorporation if the damage is not corrected before replication. Bifunctional alkylating agents, i.e. those with two or more reactive alkyl groups, bring about cross-linking of the two strands of the double helix. This reaction is not, of itself, mutagenic but is lethal because cross-linking the strands of the double helix inhibits replication. However, attempts by the cell to repair the damage may result in mutation.

Transversions may arise spontaneously or may be induced at low frequency by some mutagenic treatments. However, no known chemical or irradiation treatment generates transversion exclusively, or even predominantly. Less is known about the mechanisms that generate transversions than about those that generate transitions, but many are probably due to replication or repair errors, rather than to the direct consequence of mutagenic treatment.

Acridine dyes induce frameshift mutations. The properties of these mutations are accounted for by the insertion (+) or deletion (−) of one, or a few, base pair(s) in DNA. Because the information encoded in mRNA is read as a sequence of triplets, insertion, or deletion of bases (other than multiples of three) causes a shift in the reading frame with the result that all codons (triplets) beyond the site of mutation are changed; a second frameshift, of similar magnitude but opposite sign, will correct the reading frame beyond the second mutation. If the amino acid sequence encoded by the base sequence between the two sites of mutation is not critical to the function of the gene product, the two mutations may cancel each other out and an altered but functional peptide may be produced. This is an example of pseudoreversion; the lost phenotype is restored by the second mutation but the original mutation remains. When this occurs, the second mutation is said to be a suppressor of the first and, because it is within the same gene, the phenomenon is termed intragenic suppression. When separated from the primary mutation by recombination, an intragenic frameshift suppressor mutation acts exactly like a primary frameshift mutation. Some frameshift mutations, particularly those that result from nucleotide insertions, will also revert precisely. Although frameshift mutants revert to the wild-type phenotype spontaneously and can be induced to revert at a higher rate by acridine dyes, frameshift mutations are never induced to revert by base analogs, nitrous acid, or hydroxylamine.

One of the best known acridine dyes is proflavine. Because it is a planar (flat) molecule with dimensions similar to that of a purine–pyrimidine base pair, it can intercalate between the stacked bases of DNA. It has been proposed that proflavine exerts its mutagenic effect on DNA undergoing recombination. This idea arose from the observation that while proflavine is a relatively poor mutagen for normal cells of *E. coli*, its mutagenic effect is considerably increased, (1) in the case of *E. coli* bacteriophage T4 which indulges in a considerable degree of recombination during the course of its development; and (2) with partial diploids of *E. coli* formed during bacterial conjugation as the result of incomplete transfer of the chromosome of one bacterium to another. In the latter system, frameshifts are generated at a higher frequency in that region of the genome that is temporarily duplicated than in the remainder. Recombination of DNA involves breakage and reformation of phosphodiester bonds and it has been proposed that frameshift mutagens act by temporarily stabilizing mispaired sequences that may form following strand cleavage and transient strand separation. All frameshift mutagens are intercalating agents, but not all intercalating agents are frameshift mutagens. Another common intercalating agent, widely used in molecular genetics, which is a frameshift mutagen is ethidium bromide.

In practice, when studying a mutagen or suspected mutagen, it is easier to gauge its effects on the frequency of reversion of a set of known point mutants than to analyze its ability to generate 'knock out' mutations in a particular gene(s). Although reversion frequencies are generally somewhat lower, analyzing reversion offers considerably more sensitivity and convenience. This is the basis of the well-known Ames test, used to test chemicals to determine if they are mutagenic (and so, potentially carcinogenic) (Ames et al. 1973; Ames 1979).

Transition, transversion, and frameshift mutations generate alterations in the sense of the mRNA transcript. A mis-sense mutation gives rise to a codon that specifies an amino acid different from that originally encoded. The effect on the polypeptide product depends on the position of the altered amino acid and the precise nature of the change. In some cases, the polypeptide may show no obvious change in its properties; in others, all protein function may be lost. The change may generate a protein with altered thermal stability such that at a relatively low (permissive) temperature the protein functions more or less normally, whereas at higher (nonpermissive) temperatures the protein is nonfunctional; if this results in cell death the mutation is said to be a conditional-lethal mutation, i.e. the mutation is lethal, but only at the elevated temperature. In the case of most enteric bacteria, such as *E. coli*, the permissive temperature would normally be 30°C and the nonpermissive 42°C. There are many other examples of conditional-lethal mutants in which a potentially lethal mutation can be tolerated under particular physiological conditions. Such mutations can be invaluable when studying a vital component of the cell.

Point mutations may change a codon from one specifying an amino acid to one specifying termination of protein synthesis, namely UAG, UAA, or UGA, when a truncated polypeptide is produced instead of the full-length protein. Such mutant codons, when they appear within a gene, are called nonsense codons because they do not code for an amino acid. The introduction of a nonsense codon into an early gene of an operon may produce an effect known as polarity (Zipser et al. 1970). An operon consists of a set of genes that are transcribed as a single mRNA. A polar mutation is one that results not only in the loss of expression of the mutated gene but also in loss or significantly lower expression of those genes in the operon located distal to (or downstream from) the nonsense mutation, i.e. located on the opposite side of the mutation to that of the operon promoter. Translation of mRNA ceases at a nonsense mutation, the nascent peptide is released and ribosomes are discharged. Some may reattach to the mRNA downstream of the mutation at an appropriate ribosome binding site preceding the next translation start signal. When fewer than normal reattach, polarity results. In some instances the polar effect is so strong that genes distal to the nonsense mutation are barely expressed, if at all. Transcription of most genes is coupled to translation of the transcript. If these processes are uncoupled, transcription may also be terminated prematurely as a consequence of what is termed rho-dependent transcription termination (Landick et al. 1996). Severe polar effects are often related to the location of the mutation close to the start of the mutated gene, when translation is uncoupled from transcription for a prolonged distance. In such instances, mRNA distal to the point mutation is either not synthesized, due to premature transcription termination (attenuation), or is synthesized in considerably reduced quantities (Adhya and Gottesman 1978). This coupling of mRNA synthesis to its translation has evolved in some systems to provide a sensitive mechanism of control of enzyme synthesis termed attenuation (for further details, see Yanofsky 1987; Landick et al. 1996).

Suppression of base substitution mutations

As in the example of frameshift mutations, it is sometimes possible to have a second mutation at a site different from the first, the result of which is restoration of the lost phenotype. Such second site mutations are examples of suppressors, and the double mutants are again referred to as pseudorevertants. Mutations due to base substitutions may be suppressed by either an intragenic mutation or by an intergenic mutation, in which case the second mutation is in an entirely different gene from the first. In the former case, the suppressor mutation alters a second amino acid in the polypeptide; the second substitution neutralizes the damaging effect of the first and restores protein activity. Again, if the second mutation is separated from the first by recombination, it may also impose a mutant phenotype. Such suppressors are quite specific for the original mutation.

Intergenic suppression operates primarily at the level of protein synthesis. Mutations of this type mostly give rise to altered transfer RNA (tRNA) molecules, although alterations in ribosome components, the so-called *ram* alleles of ribosomal protein genes, have been reported (Andersson and Kurland 1983). The altered tRNA molecules are able to read one, or sometimes two, of the three nonsense codons (UAG, UAA, and UGA) as an amino acid codon (Engelberg-Kulka and Schoulaker-Schwarz 1996). The change in specificity is, in many cases, the result of a base pair substitution in the DNA sequence that determines the anticodon of the tRNA species generating an anticodon to one of the terminating codons. Such suppressors will recognize the cognate codon, i.e. a particular nonsense mutation, wherever it occurs and will translate it. The efficiency with which a nonsense codon is translated by an appropriate suppressor tRNA will be determined by how effectively the mutant tRNA competes with the normal translation termination factors, which will also react to the nonsense codon, and the influence of the nucleotide context, i.e. the surrounding sequences within which the terminating codon appears (Bossi 1983). The affinities of both factors for the nonsense codon are known to alter with the context of the codon.

It was thought initially that suppression must be highly inefficient if the cell is to survive. This is now known not to be true and individual suppressors can translate nonsense codons with efficiencies of 60–75 percent. By contrast, some suppressors are relatively inefficient but, for many enzymes, even a small level of production, say 5 percent of the wild-type level, is sufficient to permit growth of the cell, albeit often with a lower growth rate than the original parent strain.

The generation of a suppressor, i.e. mutation of a tRNA gene, can be tolerated only if the cell produces an alternative tRNA with which it can translate the codons that the mutant tRNA would have recognized before its mutation, otherwise protein synthesis will be fatally compromised (Murgola 1985). In many cases this is possible because there is a redundancy of tRNA species for a number of codons. Hence, one of them can be mutated to recognize a different, but related, codon with little or no damage to the protein synthetic apparatus of the cell.

Since tRNA suppressors may be generated by mutation of the anticodon sequence, it should, in principle, be possible to isolate mis-sense suppressors as well. Several of these have indeed been isolated. They tend to involve minor species of tRNA and their efficiency of suppression is low but effective for the reasons given above. In addition, tRNA frameshift suppressors have also been isolated where the mutations are insertions in the anticodon sequences that increase their sizes from three to four bases (Engelberg-Kulka and Schoulaker-Schwarz 1996).

The importance of tRNA suppressors rests in the fact that they have been used extensively, particularly before gene cloning and sequencing was more or less routine,

to demonstrate that mutations were located within nucleotide sequences that encode peptide products, since the effectiveness of such suppressors depends on their abilities to decode mutant codons and restore function to the protein affected by the mutation. Their existence also helped to establish the current model of protein synthesis.

Macrolesions

Alterations of DNA involving large numbers of base pairs are of four types: deletions, duplications, inversions, and additions. Deletion mutations are recognized by their inability to revert and by their failure to recombine with two or more distinct point mutations. A number of mechanisms can be proposed to explain the generation of deletions, including errors in replication, in recombination and in DNA repair. In addition, the activities of transposable elements can and do generate deletions.

Errors in DNA replication (copy errors) involve sequence duplications and slippage of one strand of the replicating DNA in relation to the other. A disruption in the pairing between the parental template strand, DNA polymerase and the newly synthesized daughter strand could lead to either of the two results outlined in Figure 4.3. After having separated transiently, the two strands re-anneal, but misalign because of the sequence duplication. The misalignment is accommodated by one of the strands looping out part of its sequence. A 'jump ahead' in copying produces a deletion in the daughter strand; a 'jump back' produces a sequence duplication.

Deletions and inversions may result from intra- and inter-replicon recombination, as outlined in Figure 4.4. These events also require sequence duplications. Recombination between homologous sequences carried in the same orientation on the same DNA molecule results in the loss of one copy of the duplicated sequence and of the sequence between the duplication. Recombination between homologous but inverted sequences inverts the intervening sequence. Inter-replicon recombination between regions that have sequence duplications may result in sequence duplication in one of the DNA molecules involved and deletion in the other (what is lost by one molecule is gained by the other) (Figure 4.4).

Mutations induced by ultraviolet light

Mutations may be induced by ultraviolet (UV) light. DNA strongly absorbs UV light at an absorption maximum of c. 260 nm. Cells are killed by UV light and there is a high proportion of mutants among the survivors. The most frequent lesion introduced into DNA by UV irradiation is the formation of pyrimidine dimers; covalent bonds are formed between adjacent pyrimidines on the

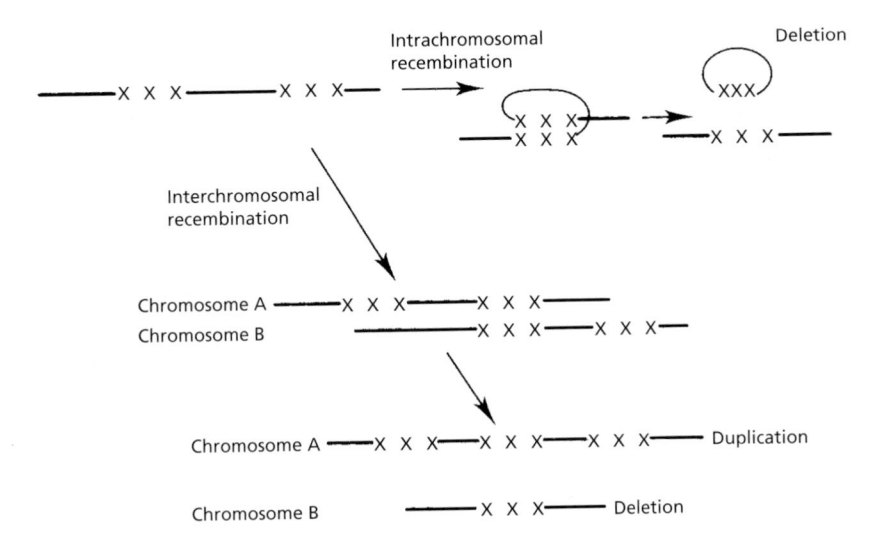

Figure 4.3 *Errors in DNA replication*

same DNA strand so that the two pyrimidine residues are joined by a cyclobutane ring. The most common are thymine dimers, but cytosine dimers and thymine–cytosine dimers are also generated. The appearance of these dimers distorts the shape of the DNA molecule (i.e. distorts the double helix) by interfering with normal base pairing. Because of the distortion, DNA replication stalls at thymine dimers. Replication may be reinitiated some distance away from the lesion, so generating a gap in the daughter strand. Unless repaired, the generation of a pyrimidine dimer is a lethal event. Not surprisingly, cells, from bacteria to higher eukaryotes, have evolved a number of mechanisms to repair pyrimidine dimers and restore the normal base sequence. If UV-treated bacterial cells are immediately exposed to visible light of wavelength 300–400 nm, both mutation frequency and lethality are greatly reduced, a phenomenon called photoreactivation. The visible light treatment activates an enzyme, photolyase, that hydrolyzes the cyclobutane ring, precisely reversing the reaction mediated by exposure to UV. The mechanism is accurate and error free. In addition, the cell has various repair systems that do not need light stimulation: the dark repair processes. One or more of these systems is error prone and it is the activity of this system (or systems) that can result in mutation (Altshuler 1993; Sinha and Hader 2002). DNA replication is involved and is mediated by DNA polymerases, which are intrinsically error-prone in terms of accommodating non-

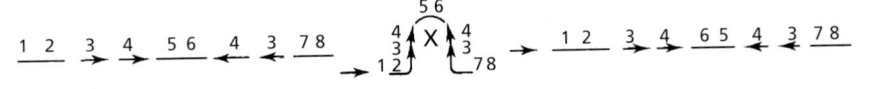

Figure 4.4 *Inter- and intrachromosomal recombination*

Crick–Watson base pairing during daughter-strand synthesis. This gives them the ability to replicate through damage to DNA, such as the presence of pyrimidine dimers, albeit with a high probability of introducing mistakes at that point (Goodman 2002; Woodgate 2001).

Other types of mutation

Spontaneous mutations may occur in the absence of known mutagenic treatment. There are probably many reasons for this. Various products or intermediates of the cell's own metabolism are demonstrably mutagenic; these include peroxides, nitrous acid, formaldehyde, and purine analogs. Some spontaneous mutations may, therefore, be induced by endogenously generated mutagens. Alkaline pH and elevated temperature have also been implicated in mutation.

Finally, it has been discovered that certain mutations of bacteria themselves predispose the cell to further mutation. Several loci are now known to be involved. They include functions connected with DNA replication and, particularly, with a process termed mismatch repair (Chopra et al. 2003), that operates to maintain the integrity of the genetic message (Schofield and Hsieh 2003).

In these cases the initial mutation diminishes the reliability of either those functions that ensure the fidelity of replication (i.e. the error rate is increased) or one of the DNA repair mechanisms, the purpose of which is constantly to scan the cell's DNA for damage and correct it when found. These mutations generate alleles called mutator (*mut*) genes because they enhance the frequency of mutation (Hutchinson 1996). There is a similar mutator gene in coliphage T4, whose product has been identified as a DNA polymerase. Mutator genes have been reported to induce transversions at relatively high frequencies, consistent with a partial breakdown in the mechanisms that maintain DNA fidelity. Recently, it has been discovered that the mutation frequency may rise significantly when the microbe is subjected to conditions of stress, particularly nutritional stress, which can occur on nutrient exhaustion. The phenomenon has been termed adaptive mutation and can, in part, be attributed to limitation of DNA mismatch repair in stationary phase culture (Notley-McRobb et al. 2003; Rosenberg 1997).

Site-directed mutation

Until the advent of gene cloning, the introduction of mutations into a particular gene was a strictly random process; mutant alleles could be selected or identified, but the location of the mutation could not be controlled. Now, by exploiting DNA synthesis in vitro, a technique known as polymerase chain reaction (PCR) (Mullis et al. 1986; Palmer 2004, see McPherson and Møller 2000 for a more detailed exposition), mutations can be targeted precisely to any base pair that is desired (Primrose et al.

2001), so as to introduce a missense, nonsense, or frame-shift mutation. Introduction of a specific macrolesion is also possible. DNA synthesis requires DNA polymerase, a template, deoxyribonucleotide triphosphates, and a priming sequence. In the cell this is usually a short RNA sequence but, in vitro, synthetic oligodeoxyribonucleotide primers of approximately 20 nucleotides are used instead. The gene to be mutated is often cloned into a suitable cloning vector. The construct constitutes the template DNA for the in vitro synthesis. Vectors based on the genome of bacteriophage M13, which is a circular, single-stranded DNA molecule, were initially favored, because the recovery of single-stranded DNA for the reaction is greatly facilitated, in that phage particles contain only one DNA strand (Zoller and Smith 1982). Now, however, the technology is much improved and site-directed mutation is routinely achieved using genes cloned in standard plasmid cloning vectors, or using PCR products.

With M13 based cloning vectors, the primer used is complementary to the sequence at the prospective site of mutation. By incorporating a mismatch into the primer, a targeted mutation can be engineered into the gene of interest. The primer is extended by the DNA polymerase, using the circular single-stranded DNA molecule recovered from phage particles as template until the 5′ end of the primer is encountered, when the newly synthesized DNA circle is closed by DNA ligase, generating a double-stranded DNA replicon with a mismatched nucleotide base pair. When introduced into a suitable bacterial host by transformation, this hybrid molecule is replicated to yield one daughter molecule carrying the original allele of the gene of interest and one carrying the mutant allele. Further replication amplifies both versions of the gene. The plasmids can then be isolated as a mixture and used to transform the same or another suitable host to yield clones carrying one or other of the two recombinant replicons. Clones carrying the mutant gene are identified and separated from transformants with the original allele of the gene of interest. The former clones can then be cultured to provide the mutant gene in quantity. The gene is then resequenced to ensure that the desired mutation is the only one that has been introduced. By using a set of primers, each with a different, but deliberately designed, mismatch in the reaction, a set of mutants with mutations directed to particular nucleotides can be generated. This allows a region of a gene, or indeed the entire gene if so desired, to be subjected to systematic mutation for structure–function studies of the gene product or for analysis of the sequences controlling gene expression. Various modifications of this procedure, designed to make it more efficient and to optimize recovery of the desired mutant, have been developed (see Watson et al. 1992, for simple descriptions)

A multistage PCR method for site-directed mutagenesis now more commonly employed uses double-stranded DNA as template (Figure 4.5) (see also

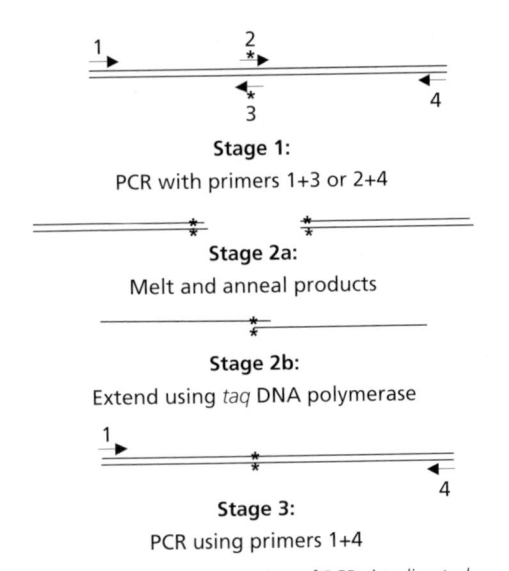

Stage 1:

PCR with primers 1+3 or 2+4

Stage 2a:

Melt and anneal products

Stage 2b:

Extend using *taq* DNA polymerase

Stage 3:

PCR using primers 1+4

Figure 4.5 *Schematic representation of PCR site-directed mutagenesis. In stage 1, two separate PCR reactions are used to generate, from the template sequence, two linear, overlapping fragments having a mutation in the overlapping region – with the mutation being the result of primers partially mismatched with their target sequence (noted as stars). In stage 2, the two fragments are melted, mixed, and allowed to anneal to each other. This product is then used as a template for further PCR. The resulting product will comprise copies of the gene of interest with the desired mutation (shown on the diagram).*

Primrose et al. 2001). To begin with, the gene of interest is amplified as two, partly overlapping fragments (stage 1). One fragment is generated using a primer pair with the forward primer a perfect match and targeted to the start of the gene (primer 1) and the reverse primer targeted to the point of mutation and incorporating the desired mutation (primer 3). PCR amplifies a fragment with the mutation. A second fragment is generated in a similar fashion, but with the forward primer (primer 2, the complement of the reverse primer in the first reaction and so incorporating the mutation required) targeted to the point of mutation while the reverse primer (primer 4), perfectly matched to the gene sequence, is targeted to the end of the gene. PCR amplifies the second fragment, with the same mutation within the same sequence as on the first fragment. The two fragments are then mixed and used as a template in a third PCR reaction together with the forward primer from the first PCR reaction and the reverse primer from the second, i.e. primers 1 and 4 (stage 2). When the template is melted and then allowed to reanneal, a proportion of strands from the different fragments anneal across a short common region generating a DNA molecule with a short central double-stranded section (corresponding to the mutational primers in the two initial PCR reactions) and two long single-stranded tails. These then become templates to extend the short double stranded section of DNA in both directions to encompass the full gene. A PCR reaction then amplifies this

product (stage 3), namely a full-length version of the mutant gene of interest. This is then recovered by ligation into a suitable cloning vector, where the mutant gene can either be examined directly, or used to mutate the original allele in the strain of origin by allelic replacement. For an up-to-date discussion of in vitro mutagenesis see Braman (2002).

Flip-flop mechanisms

Some functions expressed by bacteria and bacteriophages undergo changes that at first sight appear to be mutations. However, the basis of the changes is more like the operation of a switch than a random change in the DNA. Such changes occur as a result of site-specific recombination mechanisms that invert specific DNA sequences, the consequence of which is to turn off expression of one gene(s) and activate that of an alternative gene(s) (Craig 1988; Nash 1996). Two particularly well characterized systems, encoded by bacteriophages Mu and P1, operate to control the bacterial host ranges of the phages by controlling the production of alternative tail fibers for the phages (Giphart-Gassler et al. 1982). In *Salmonella* Typhimurium flagellar antigen phase variation, which allows the bacterium to evade the immune response of an infected host, is controlled by a very similar DNA inversion mechanism. Two genes, *H1* and *H2*, code for different forms of flagellin. Both proteins can polymerize to produce the cell's flagella. The cell can alternate expression of these two genes by inverting a short segment of the chromosome (Zieg et al. 1977). In phase I, flagellin encoded by *H1* is produced. In phase II, *H2*-encoded flagellin is synthesized and a gene called *rhI* is activated. The product of *rh1* is a transcriptional repressor of *H1*, so that production of flagellin encoded by *H1* is switched off. Hence, expression of *H1* and *H2* is mutually exclusive. The recombination inverts an approximate 1 kb sequence. In phase II orientation, the invertible sequence provides a promoter for *H2* and *rhI*. The sequence also encodes a gene called *hin*, the product of which mediates the inversion. The *hin* gene shows considerable homology with both *gin* and *cin*, the site-specific recombinase genes of bacteriophages Mu and P1, respectively, that mediate host specificity. The products of these three genes are functionally interchangeable.

Each of these three systems operates infrequently, the probability of inversion being about 10^{-8} per cell per generation. However, this low frequency is sufficient, within a population of cells or bacteriophages, to significantly extend the versatility of the organism. The ability of S. Typhimurium to switch flagellin production confers some protection against the immune response, whereas the ability to produce two sets of tail fibers broadens the host ranges of both Mu and P1, each set of tail fibers permitting infection of particular bacterial strains.

Other examples of phase variation controlled by invertible DNA segments are *E. coli* type I fimbriae (pili) (Abraham et al. 1985) and the fimbriae of *Moraxella bovis* (Marrs et al. 1988).

MOBILE GENETIC ELEMENTS

Insertion sequences and transposons

In the 1960s, evidence accumulated for the existence of a hitherto undescribed form of mutation. The first examples recognized in bacteria were found by Lederberg, who in 1960 described an unusual class of Gal⁻ mutants in *E. coli*. They reverted to Gal⁺ at frequencies of 10^{-6}–10^{-8}, but the reversion frequency was not enhanced by mutagenic treatments. These results indicated that the mutations were neither deletions nor classical point mutations (base substitutions or frameshifts). Further, when the mutation was in one of the first genes of a set it exerted an unusually strong polarity on expression of subsequent genes in the operon, virtually eliminating expression from them (unlike polar nonsense mutations that merely reduce expression of the distal genes). The new type of mutation was described as a strong polar mutation (Jordan et al. 1968). Molecular genetic analysis of these mutations, using λgal transducing phages, demonstrated that the *gal* operons of the mutants contained extra DNA sequences; revertants to Gal⁺ had lost the additional sequences. The conclusion from these studies was that the mutations were generated by the insertion of extra pieces of DNA, called insertion sequence (IS) elements (Jordan et al. 1968), at the point of mutation. IS elements are now known to be but one type of a class of genetic elements called transposable elements. The one characteristic that all such elements possess is the ability to transpose, i.e. they can insert into different sites, usually unrelated, on the same or on different DNA molecules. Unlike host-mediated homologous recombination, transposition requires no homology between the transposable element and its sites of insertion, and the mechanisms are independent of the *recA*+ gene product of the bacterial host.

Since the initial discovery of IS*1*, IS*2*, and IS*3* (Fiant et al. 1972) in *E. coli*, many IS elements have been described from a wide variety of bacteria (Galas and Chandler 1989; Chandler and Mahillon 2002). They are common in bacterial plasmids and are normal components of many, if not all, bacterial chromosomes. They are also cryptic, in the sense that they do not confer on the host bacterium a predictable phenotype. In the past, their presence was usually indicated by mutation, i.e. the genetic damage they cause. More recently, however, many elements have been discovered as a result of bacterial genome sequencing (Chandler and Mahillon 2002).

In 1974, the discovery of a second type of transposable element, one that encoded a recognizable gene product,

the TEM-2 β-lactamase determining resistance to several β-lactam antibiotics, including ampicillin (see Chapter 18, Antibacterial therapy), was reported by Hedges and Jacob (1974). It was termed a transposon, and many others carrying various resistance and other genes have since been identified and analyzed (see Berg and Howe 1989; Craig et al. 2002). Because of the ability of transposable elements to insert into a large number of sites, both on the bacterial chromosome and on plasmids, they assumed the popular name of 'jumping genes.' A few of these elements are described in Table 4.1.

It is now clear that there are several types of transposable element (see Craig et al. 2002). IS elements are generally small (1–2 kb) and encode only those functions needed for their own transposition, whereas transposons are larger, usually 4–25 kb (although considerably larger elements have been described), and they encode at least one function unconnected to transposition that alters the cell phenotype predictably (e.g. by conferring a particular antibiotic resistance). In addition, there are some bacteriophages, such as Mu and D108, that exploit transposition in their life cycles (Toussaint and Résibois 1983).

Bacteriophage Mu, mentioned earlier in relation to its site-specific recombination system, resembles a transposable element in that it inserts more or less at random into many different sites in the *E. coli* chromosome, as a consequence of its mode of replication; hence its name, Mu (for mutator phage). It is now known that phage Mu is a large transposable element that uses one form of transposition for replication.

Because transposition involves the movement of DNA sequences from one site to another, it is clear that whatever the mechanism, it is, by definition, a recombination event. Such events may be referred to as illegitimate recombination, to indicate that they occur independently of a classical recombination system and do not require extensive sequence homology between the sequences involved in the recombination.

The majority of known IS elements have a similar structure, namely a unique central sequence, comprising most of the element, flanked by short perfect or near-perfect inverted repeat (IR) sequences. Terminal IR sequences differ, both in sequence and size, from one element to another (c. 10–40 bp) (Table 4.1). For those elements that have been analyzed in detail, the unique central section encodes an element-specific transposition function(s). The IR sequences delineate the element, serving as specific sites for recombination mediated by the transposition enzyme(s). The inverted character of the terminal repeats and the asymmetric nature of the sequence ensures that both are retained in the transposed structure, i.e. the same sequence is recognized at both ends of the element and the DNA is cleaved at the same end of the IR sequence in both cases, i.e. precisely at the IS element–host replicon junctions.

Most transposons can be assigned to one of two broad groups, called composite transposons and complex trans-

Table 4.1 *Some transposable elements in bacteria*

Transposable element	Size (kb)	Terminal[a] repeats (bp)	Target size[b] (bp)	Phenotype[c] conferred	Origin
Insertion sequences					
IS1	0.768	18/23	9	None	Gram −ve
IS2	1.331	32/41	5	None	Gram −ve
IS3	1.258	32/38	3/4	None	Gram −ve
IS4	1.426	16/18	11/12	None	Gram −ve
IS10	1.329	17/22	9	None	Gram −ve
IS50	1.534	12/18	9	None	Gram −ve
IS903	1.05	18/18	9	None	Gram −ve
IS256	1.35	nd	nd	None	Gram +ve
Composite transposons					
Tn5	5.7	IS50(IR)	9	Km	Gram −ve
Tn9	2.5	IS1(DR)	9	Cm	Gram −ve
Tn10	9.3	IS10(IR)	9	Tc	Gram −ve
Tn903	3.1	IS903(IR)	9	Km	Gram −ve
Tn1681	2.1	IS1(IR)	9	Heat-stable toxin[d]	Gram −ve
Tn4001	4.7	IS256(IR)	nd	Gm Tb Km	Gram +ve
Complex transposons					
Tn1 family					
Tn1, Tn3	4.957	38/38	5	Ap	Gram −ve
Tn21	19.5	35/38	5	Hg Sm/Sp Su	Gram −ve
Tn501	8.2	35/38	5	Hg	Gram −ve
Tn1721	11.4	35/38	5	Tc	Gram −ve
Tn951	16.5		5	Lac	Gram −ve
Tn551	5.3	35/35	5	Ery	Gram +ve
γδ(Tn1000)	5.8	36/37	5	None	Gram −ve
Others					
Tn7	14	22/28	5	Tp Sm/Sp	Gram −ve
Tn554	6.2	None	0	Sp Ery	Gram +ve
Tn916	15	20/26	10	Tc, conjugal transfer[e]	Gram −ve
Tn4291	nd	nd	nd	Mec	Gram +ve

a) The number of identical base pairs in both terminal repeats/size of terminal repeats. DR, direct repeat; inverted repeat.

b) The size of the direct duplications which are found to flank the transposon in situ.

c) Entries which relate to an antibiotic(s) refer to resistance to that drug(s). Ap, ampicillin; Cm, chloramphenicol; Ery, erythromycin; Gm, gentamicin; Hg, mercuric ions; Km, kanamycin; Mec, methicillin; Sm, streptomycin; Sp, spectinomycin; Su, sulfonamide; Lac, lactose metabolism; Tb, tobramycin; Tc, tetracycline; Tp, trimethoprim. Sm/Sp indicates resistance to both antibiotics conferred by the same modifying enzyme.

d) Heat-stable toxin produced by some enterotoxigenic strains of *Escherichia coli*.

e) See text. nd, not determined.

posons. Composite transposons, as their name implies, are constructed from recognizable simpler units. Their overall form comprises a unique central sequence that encodes the function(s) by which the element is recognized and which lacks transposition functions, flanked by long sequence repeats (1–2 kb) which are arranged as direct repeats or, more commonly, as IRs. These terminal repeats provide transposition functions and, in all cases examined so far, comprise a pair of IS elements. The more common inverted arrangement possibly reflects the fact that it confers greater stability (coherence) on the composite structure than direct repeats where deletion of the transposon's distinguishing feature(s) as the result of recombination between the terminal IS elements, mediated by the host's homo-logous recombination machinery, is an ever-present possibility (Figure 4.4).

Complex transposons do not have modular structures. Rather, both transposition and non-transposition functions are located in the bulk of the sequence which is flanked by short (c. 40 bp) terminal IRs. Indeed, complex transposons resemble enlarged IS elements rather than composite transposons. The first transposon identified, Tn1 (formerly TnA; Hedges and Jacob 1974), typifies this type of element. Its structure is illustrated in Figure 4.7, p. 94. Tn1 is one of a family of transposons that are ancestrally related in terms of their transposition functions. This was first recognized because of striking similarities between their IRs. Subsequently, more extensive sequence analysis of several of these

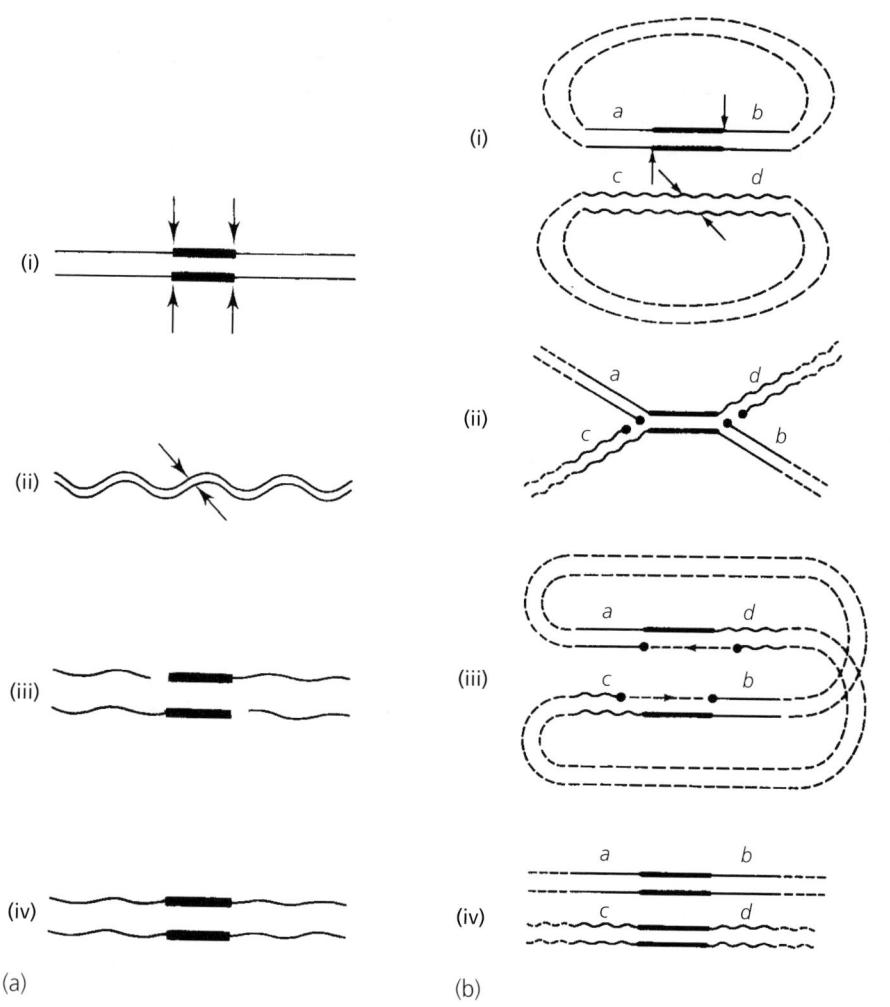

Figure 4.6 *Models for transposition.* **(a)** *Conservative transposition. (i) Flush double-stranded cuts (vertical arrows) are made at the ends of the transposable element (bold lines) and a staggered cut (diagonal arrows) is made at the target site. (ii) One end of the transposable element is attached by a single strand to each protruding strand at the staggered cut to insert the transposable element into the target site. (iii) Two small gaps, generated as the consequence of the staggered cut, are left on opposite strands, one at each end of the element. (iv) The gaps are filled in by DNA repair synthesis to produce short direct repeats that flank the transposable element.* **(b)** *Replicative transposition. (i) Single-stranded cuts are made at the ends of the transposable element, and a staggered cut of opposite polarity is made in the target DNA. (ii) One end of the transposable element is attached by a single strand to each protruding end of the staggered cut. Two replication forks are thus created, and replication may proceed to copy the transposable element. (iii) Semi-conservative replication has generated two new transposable elements from the original element and a short direct repeat of the target site. If ab and cd were circles, a, b, c, and d would now be covalently connected and transposable elements would form the joint regions of a fused replicon. (iv) A site-specific crossover between the transposable elements would resolve the cointegrate into the starting replicon ab and the target replicon cd, which now has a copy of the transposable element flanked by short direct repeats of the site of insertion (Shapiro 1979).*

elements demonstrated that the entire transposition mechanism of one is related to that of the others. In some cases the evolutionary divergence is small, so that the transposition functions are interchangeable. In other instances, elements are more distant in terms of evolution and analogous functions cannot be interchanged (Sherratt 1989; Grinsted et al. 1990; see also Craig et al. 2002)

Transposable elements have been grouped according to their similarities. Accordingly, IS elements and composite transposons are called class 1 transposable elements, whereas Tn*1* and related transposons are called class 2 transposable elements. Class 3 accommodates the trans-

posing bacteriophages such as Mu. Finally, class 4 at present accommodates those elements that do not naturally fall within classes 1–3. Perhaps two of the best known are Tn7, a compound transposon encoding resistance to trimethoprim and streptomycin prevalent in gram-negative bacteria, particularly the Enterobacteriaceae, and Tn*916*, an unusual transposon from *Enterococcus faecalis* that encodes resistance to tetracycline and which is conjugative, i.e. can mediate its transfer not only from one replicon to another by transposition, but also from one bacterial cell to another by a form of conjugation (Salyers et al. 1995). Tn*916* (Franke and Clewell

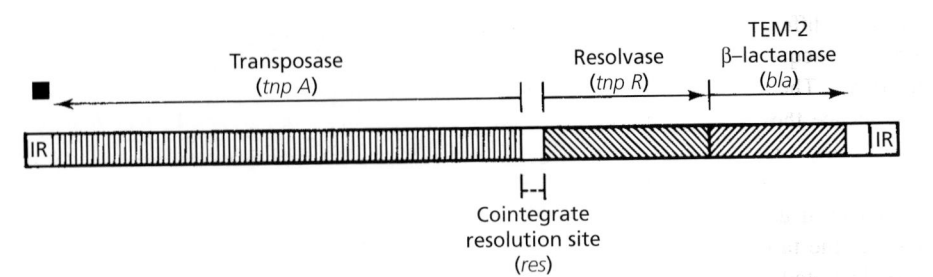

Figure 4.7 *Map of Tn1 showing the location of its three genes. The arrows indicate the direction of transcription. The cointegrate resolution site (res) is necessary for resolution of cointegrates, by resolvase, during transposition (see Figure 4.6). IR, perfect 38 bp inverted repeats*

1981) serves as the archetype for this kind of transposable element. Transposons with conjugative properties have also been found in *Bacteroides* spp. (Salyers and Shoemaker 1995), and, more recently, elements originally thought to be plasmids because they mediate transfer of antibiotic resistance genes between members of the Enterobacteriaceae are now considered to be conjugative transposons (Böltner and Osborn 2004).

Transposable elements not only mediate their own genetic mobility, but can be instrumental in bringing about important rearrangements in DNA molecules. Not only do they provide useful genes, but their transposition activities can generate macrolesions such as deletions, duplications, and inversions. Because many of them insert into DNA molecules more or less at random, they can generate mutation by disrupting the sequence continuity of genes; indeed, it was this consequence that first brought IS elements to the attention of bacterial geneticists. That the insertions are often polar has been noted for IS elements; this is also true for transposons. In addition, they can, in the cases of particular elements, activate previously silent genes by inserting upstream and adjacent to the gene to provide a functional promoter (Jaurin and Normark 1983; Pilacinski et al. 1977; Wood and Konisky 1985; Rasmussen and Kovacs 1991; Podglajen et al. 1994; Chandler and Mahillon 2002). The polar effects of transposable element insertion arise because the elements are, in general, self-contained and encode translational and transcriptional terminators that prevent transcription proceeding far into the element across the junction of the element and the adjacent host DNA. Thus the genes of an operon which are promoter distal to an insertion are separated from their promoter and normally become silent. Only when there is a promoter at the end of the element nearest to the orphan genes that can direct transcription towards that end of the element and beyond, are the remaining genes of the operon (downstream of the insertion) expressed.

IS2 exerts strong polar effects when in one of its two possible orientations (this was how it was discovered). In this orientation (designated orientation I), a rho-dependent transcription termination site prevents transcription proceeding through the element into adjacent sequences.

In the opposite orientation (orientation II), IS2 can provide a promoter sequence that can switch on downstream genes (Saedler and Reif 1974). Mutations in wild-type IS2 that create promoters have also been described (Sommer and Saedler 1979). It is of interest to note that cloned genes, either not expressed or expressed poorly from an original construct, may be activated by insertion of a transposable element upstream from them. However, it should be noted that this aspect of transposable element activity appears to be restricted to a minority of specific elements, of which IS2 is one.

Specificity and mechanism of insertion

When transposable elements insert into new sites, short duplications of recipient DNA sequence are usually generated. These direct repeats, commonly 5 or 9 bp (Table 4.1), flank the transposable sequence. The size of the duplication is specific to the element. The duplications arise because the recipient DNA at the target site is cleaved across both its strands as a short staggered cut. The transposon is inserted by joining it to the short single-strand extensions created by the cleavage. This generates small single-stranded gaps, on opposite strands, in the recipient DNA at the ends of the element. These gaps are filled in, probably by DNA gap repair synthesis (Sancar 1996) to generate the short flanking duplication. So typical is this arrangement that if, when analyzing a new bacterial nucleotide sequence, a section of DNA is found to be delimited by short IRs and flanked by short direct repeats (DR) then it is immediately considered likely to be a transposable element.

A wide range of insertion specificities has been found for different transposons and some display different characteristics with different DNA molecules. Some (e.g. Tn1) insert at many places on different plasmids and into the bacterial chromosome with little recognizable specificity save, perhaps, a preference for AT-rich regions; others appear to be rather more specific, e.g. IS4 has been found to insert into only one position in the *gal* region of the *E. coli* chromosome. Although

Tn*10* inserts at many different sites, they are all related to the 9 bp consensus sequence, NGCTNAGCN (where N denotes any base). The more closely a site approximates to this sequence the more readily it is used as a site of Tn*10* insertion. The preference for AT-rich regions shown by some transposons suggests that the transposition mechanism used may involve a degree of strand separation at the target, since AT bps are inherently more easily disrupted than GC bps.

Initially, research workers sought to describe transposition in terms of a single mechanism. It is now clear that this notion was misconceived and that there are several different mechanisms. Three fundamentally different strategies have been reported. One, used by several IS elements and their composite transposons, is called conservative, or cut-and-paste, transposition (Figure 4.6a). In this, the transposable element is cut from the donor DNA as a discrete double-stranded section of DNA with flush ends. The free ends are ligated to the target site using the single strand extensions generated by the staggered cut at the target site. The small gaps that remain are filled in to generate the direct flanking repeats typical of IS element insertions. In this mechanism the entire transposable sequence on one DNA molecule is moved to another DNA molecule (Bolland and Kleckner 1996). It is believed that the donor molecule minus the transposon is degraded (Berg 1989; Kleckner 1989). Some transposons, e.g. Tn7, use a variation of the cut-and-paste transposition system (Sarnovsky et al. 1996). Others, e.g. conjugative transposons, are excised from the donor site to generate a circular form of the element. This entity then either inserts into another site on the same or on another replicon in the same cell, or it transfers, by conjugation, to another cell where it is inserted into one of the resident replicons (Salyers et al. 1995). Although these elements display features of a bacterial plasmid, they differ from a plasmid in that they lack their own replication system. They are replicated passively as components of the replicons into which they insert.

A second mechanism of transposition involves semiconservative DNA replication. This was realized when it was found that certain mutants of Tn*1* generated end products that differed from those of the parental transposon. With these mutants, the end product of a transposition from one replicon to another was a cointegrate – a recombinant DNA molecule comprising both the donor and the recipient replicons with directly repeated copies of Tn*1* at the plasmid junctions, i.e. the transposon had been replicated (Figure 4.6b). These recombinants could then break down in a *rec*+ host, via normal recombination, to yield a replicon indistinguishable from the original plasmid donor and a transposon-carrying derivative of the recipient plasmid, i.e. a product indistinguishable from that generated by the parental transposon. Shapiro (1979) and Arthur and Sherratt (1979) synthesized these observations, and others involving bacteriophage Mu, into the now generally accepted model of replicative transposition (Figure 4.6b).

For some IS elements a small proportion of transposition events appear to occur via replicative rather than conservative transposition. This observation has been taken to imply that the two forms represent a split pathway proceeding from a common start (Ohtsubo et al. 1981). Consistent with this interpretation is the observation that c. 5 percent of transpositions of Tn*1* appear to be by a direct pathway rather than by replicative transposition (Bennett et al. 1983).

More recently, it has been discovered that members of a novel class of IS elements, of which IS*91* is the type element (Mendiola et al. 1992), and that lack terminal IR sequences, transpose by a third mechanism, RC transposition, that involves a form of rolling circle replication. Insertion is also site-specific, although the target site is small, 4 bp. The transposases of these elements are related to the initiator proteins of certain plasmids of *Staphylococcus aureus* (although the IS elements were isolated from gram-negative bacteria) (Mendiola and de la Cruz 1992). The plasmids replicate via a rolling circle mechanism that generates a single-strand version of the plasmid which is then converted to the double-stranded form. It is likely that this family of IS elements transposes by a mechanism that involves a single-stranded, possibly circular intermediate (Tavakoli et al. 2000; Garcillán-Barcia et al. 2002).

Deletions, duplications, and insertions

Deletion induced by a transposable element was first seen for IS*1*. In the *E. coli* chromosome, the frequency of deletions near a copy of IS*1* was found to be 100- to 1 000-fold higher than that observed elsewhere (Reif and Saedler 1975). Subsequently, other elements and transposons were also shown to stimulate the frequency of deletions in their vicinity. Several studies have demonstrated that the deletions always extend precisely from one end of the particular transposable element, leaving the element itself intact but removing one copy of the direct repeats that originally flanked it. Deletions such as these can, in principle, be formed in one of two ways. First, a second copy of the transposable element can insert into a different position on the same replicon, but in the same orientation as the first. Host-mediated recombination can then occur between the copies, effectively deleting one copy and the intervening DNA sequence (see Figure 4.4). Second, an element may transpose intramolecularly to a nearby site on the same DNA molecule by replicative transposition. If the insertion at the new site assumes the same orientation as that at the first site, the result is an automatic deletion of one of the copies of the element and the DNA sequences between the original site and the new site of insertion (Arthur and Sherratt 1979; Craig 1996).

DNA inversions may also arise by one of two mechanisms. Following insertion of a second copy of a transposable element into a DNA sequence to generate an inverted repeat, a single host-mediated recombination event across the repeated sequences will invert the intervening sequence (see Figure 4.4). Alternatively, when an element transposes intramolecularly by replicative transposition to a new site on the same DNA molecule in such a manner as to generate inverted repeats of the element, the intervening sequence between the old and new sites of insertion is automatically inverted, again as a consequence of the mechanics of the transposition (Arthur and Sherratt 1979). In practice, after the event, it is impossible to determine which mechanism generated a particular inversion because extended IRs act as recombination loci periodically to invert and reinvert the intervening sequence. As noted earlier, a similar phenomenon is exploited in the site-specific recombination systems that control P1 and Mu phage host ranges and *S.* Typhimurium flagellar type.

Transposition functions

Several transposons have now been studied in detail and, with very few exceptions, each element encodes its own transposition functions (see Craig et al. 2002). Perhaps the best studied to date is Tn*3* (which is virtually identical to Tn*1*), the class II element that confers resistance to a range of penicillins, including ampicillin and carbenicillin. The element is 5 kb and encodes three proteins: a TEM β-lactamase and two proteins that are necessary for transposition (see Figure 4.7). The larger of these, called transposase, is encoded by *tnpA* and is indispensable; it mediates cointegrate formation (Figure 4.6b). The second transposition enzyme, resolvase (encoded by *tnpR*), is normally responsible for resolution of the cointegrate to release the final transposition product. The promoters for both genes are located in the length of c. 100 bp that separates them. The site at which resolvase acts is called the *res* site, also located in the intergenic region between *tnpA* and *tnpR* (Sherratt 1989). A consequence of mutations in *tnpR* is that the frequency of transposition increases by approximately two orders of magnitude, i.e. resolvase functions not only as a site-specific recombination enzyme but also as a transcriptional repressor. This duality indicates a more general feature of prokaryotes that seems to be emerging, namely optimization of the genome. It is not uncommon to find the same DNA sequence performing more than one function.

In a *rec*[+] host (Watson et al. 1987), cointegrate resolution can be effected by the host's own recombination system, since the two copies of the element in the cointegrate are direct repeats and provide extensive sequence homology. Hence, *tnpR* mutants of Tn*3* generate cointegrates that are relatively stable in a recombination-deficient host, e.g. one which is *recA*, whereas normal transposition products are generated in a recombination-proficient host. Analyses of other class II transposons, e.g. γδ (Tn*1000*), Tn*21*, Tn*501*, and Tn*1721* (see Table 4.1), have shown that each encodes a transposase and a resolvase of approximately the same sizes as those of Tn*3*. The transposases are phylogenetically related, as are the resolvases. Some of these analogous enzymes are functionally interchangeable but others are not, indicating different degrees of evolutionary proximity (Grinsted et al. 1990).

Well-studied class I elements are IS*1*, IS*10* (Tn*10*), IS*50* (Tn*5*), and IS*903* (Tn*903*) (see Table 4.1). Each is genetically distinct and encodes its own unique transposition functions. The transposition of each is also regulated, but the strategies of regulation differ from one element to another. Thus IS*10* regulates both transcription and translation of its single transposition gene, whereas IS*50* regulates transcription of its transposition gene and modulates the activity of the enzyme with another IS*50*-encoded protein (Berg 1989; Kleckner 1989). Bacteriophage Mu, as might be expected of a lysogenic phage, has a variety of control systems to regulate its transposition (i.e. replication) activity (Toussaint and Résibois 1983), which is essentially dependent on a single phage gene designated *A*.

Not all transposition systems have so few element-encoded transposition functions. Tn*7*, for example, has five transposition genes, three of which are needed for all transposition events. Tn*7* displays two types of transpositional behavior. It can insert, more or less at random, into many different sites on plasmids at a relatively low frequency but it can also insert into one specific site on the *E. coli* chromosome at high frequency. Which transposition event occurs is determined by which one of the other two functions is used in the reaction (Craig 1989; Sarnovsky et al. 1996).

In addition to element-encoded functions and irrespective of the transposition mechanism, host functions are also required, such as DNA polymerase and DNA ligase activities. Several others have also been implicated.

Ubiquity of insertion sequences

The importance of insertion sequences and transposons in mediating bacterial variation cannot be overstated. Not only can they add useful genes to an organism directly (e.g. antibiotic resistance) but also, because of their ability to operate in pairs to transpose otherwise non-transposable genes (e.g. the unique central sequences of class I transposons), they can, in principle, effect the transposition of any bacterial DNA sequence from one DNA molecule to another. Indeed, IS elements have been instrumental in mobilizing chromosomal β-lactamase genes, such as the *blaSHV* of *Klebsiella pneumoniae* (Ford and Avison 2004) and the *blaCTX-M* genes of *Kluyvera* spp. (Walther-Rasmussen and Hoiby 2002) on to plasmids which have then carried these genes into *E. coli* and other Enterobacteria. These types of elements are not restricted to prokaryotes; indeed, the first reports of transposable elements in higher organisms, specifically maize, preceded description of those in *E. coli* by 10–15 years (McClintock 1953). Transposable elements are now known to exist in a number of eukaryotes, including yeast and *Drosophila* (see Shapiro 1983; Berg and Howe 1989; Craig et al. 2002).

Mutation by transposon insertion

Because of their ability to insert into replicons almost at random, transposons have been exploited as mutagens. They possess the advantage over other random forms of mutation that the site of mutation is grossly altered by the addition of new DNA sequences, which can be physically mapped by restriction enzyme analysis or by sequencing. Because of this, artificial transposons with a range of properties have been created from natural ones to provide tools to facilitate molecular genetic investigations (Berg et al. 1989; de Lorenzo et al. 1990).

Transposon mutagenesis has been adapted in an ingenious way to generate mutations in bacterial virulence genes (Hensel et al. 1995) and to allow negative selection. This technique, signature tagged mutagenesis (STM), is a negative selection method for bacterial virulence gene identification based on the ability to follow, simultaneously, the fates of many different mutants within a single animal. Individual copies of a transposon (mini-Tn5, de Lorenzo et al. 1990) were tagged with small, unique nucleotide sequences. The identity of any particular copy of the transposon could then be determined readily by hybridization techniques. These tagged transposons were then used to generate a bank of 1 510 random transposon mutants in a strain of *S.* Typhimurium. Each isolate in the bank was checked to confirm that it had acquired a uniquely tagged copy of the mini-Tn5, yielding a slightly smaller bank of 1 152 mutants. Cultures of each isolate were grown in the wells of microtitration trays and 12 mixtures, each with 96 strains, were prepared and used to inoculate BALB/c mice by intraperitoneal injection. Three days later, the mice were killed and the bacteria were recovered from each mouse by plating spleen homogenates onto a suitable medium. Approximately 10 000 colonies recovered from each mouse were pooled and the total DNA was extracted. The transposon tags were amplified and labeled by PCR and colony blots of the original 96 mutants in the subset were probed. The results were compared with the hybridization pattern obtained with a 'tag probe' obtained with DNA from the mixed inoculum of 96 mutants. Colonies which probed positive in the control, but not with the probe prepared from DNA from cells post-passage through mice, were shown to have had a virulence gene inactivated by the mini-Tn5. Because these strains were less virulent they survived less well in the mouse and so were poorly recovered from the spleen of the infected animals, in turn giving a much reduced yield of that particular transposon tag in the DNA used to prepare the 'tag probe.' Forty virulence mutants were identified in the bank of 1 152. Among these mutants, approximately half were in genes known to be involved in virulence. The remainder constitute a set of new virulence or colonization survival genes. These results accord well with an estimate of approximately 150 virulence genes in *S.* Typhimurium (Groisman and Saier 1990), of which more than half have been identified.

Integrons and gene cassettes

In addition to plasmids and transposable elements, bacteria have yet other systems for gene dissemination, namely, integrons (Stokes and Hall 1989; Bennett 1999). Integrons are essentially site-specific recombination systems that mediate the movement of small DNA elements called gene cassettes. Most of the gene cassettes described carry a single resistance gene. About 60 antibiotic resistance genes, including those for resistance to aminoglycosides, β-lactams, chloramphenicol, erythromycin, sulfonamides, tetracycline, and trimethoprim have been identified as being on gene cassettes (Recchia and Hall 1995). The most recently identified antibiotic resistance gene cassettes encode the metallo-β-lactamases IMP and VIM (Laraki et al. 1999; Lauretti et al. 1999; Poirel et al. 2000; Riccio et al. 2000; Nordmann and Poirel 2002), encoding resistance to the potent carbapenem β-lactams imipenem and meropenem.

Integrons were first identified on Tn*21* and closely related transposons (Grinsted et al. 1990), but are now known to be more widely distributed (Hall and Collis 1995). Molecular analysis of the resistance genes of Tn*21* and related elements revealed that although different transposons carried different complements of resistance genes, most appeared to be accommodated at a single point on the basic structure, i.e. irrespective of the identity of the resistance gene(s) carried on a particular transposon, the sequences flanking the gene or gene cluster were the same, called the 3′- and 5′-conserved regions. Analysis of the 5′-conserved region revealed a gene, *int*, encoding a protein of the 'integrase' family. This enzyme, called integrase, is a site-specific recombinase. Beside *int* is *attI*, a site that is recognized and utilized by the integrase to capture gene cassettes. The essential components of an integron, located in the 5′-conserved region, are an *int* gene, an *attI* site, and a promoter in the correct orientation to express the genes on captured cassettes, because most cassette genes identified to date lack their own promoters. Within the

3′-conserved region is often found the *sul1* gene, encoding resistance to sulfonamides. These integrons are called type 1 integrons (Hall and Collis 1995). Integrons are classified according to their integrase genes (Hansson et al. 2002; Collis et al. 2002). Members of the same class have closely related integrase genes.

A gene cassette can exist in two forms:

1 as a linear insert within an integron, or
2 transiently as a small double-stranded circle of DNA carrying one (or occasionally two) resistance genes and a recombination site called a 59-base element (Hall et al. 1991).

Integrase-mediated recombination between a 59-base element on a cassette and the *attI* site inserts the cassette gene(s) into the integron and orients it so that it is expressed from the integron promoter (Collis et al. 1993). Several cassettes may be captured in this way to form a resistance gene array in which each gene is separated from the next by a 59-base element. Arrays of five and six resistance genes have been found; 59-base pair elements do not have a unique sequence, but all conform to a consensus, which may be longer than 59 bases (Hall et al. 1991; Stokes et al. 1997).

Movement of gene cassettes is possible because the integration of a cassette into an integron is a reversible process, so what has been captured by one integron can subsequently be released and captured by another (Collis and Hall 1992). In this way the complement of resistance genes can be constantly rearranged and moved from one DNA molecule to another, providing it carries an integron or a suitable integration site.

In principle, there is no limit to the number of gene cassettes that can be incorporated into an integron, but functional constraints, such as reduced expression of genes located some distance from the integron promoter, may limit the number in practice. Nonetheless, class 1 integrons with eight gene cassettes have been reported (Naas et al. 2001), and super-integrons, with dozens of gene cassettes, reported first in *Vibrio cholerae* (Rowe-Magnus et al. 1999; Mazel et al. 1998) as a result of genome sequencing, have since been identified in a variety of bacteria (Nield et al. 2001; Rowe-Magnus et al. 2001; Vaisvila et al. 2001).

ACQUISITION OF NEW GENES

Genetic diversity in bacteria growing in monoculture in the laboratory can occur only by mutation, as described above. In mixed cultures, as are found in nature, diversity can be provided by the transfer of DNA from one cell to another, provided that the new genes can be stably inherited by the progeny of the transcipient cell. Four mechanisms of gene transfer have been described for bacteria, although not all necessarily apply to a particular bacterial species or strain; these are transformation, transduction, conjugation, and cell–cell fusion.

Genetic information (i.e. DNA) acquired by one of these four mechanisms is inherited, provided that one of two situations pertains; either the newly acquired DNA is incorporated into one of the replicons in the recipient cell by some form of recombination, or the newly introduced DNA is self-maintaining, i.e. is a replicon in its own right. In the former case, newly incorporated sequences are substituted for existing ones. The 'invading' DNA is usually linear and homologous to only part of the rescue replicon and is itself not a replicon. Homologous sequences on the acquired DNA and on the appropriate resident replicon (chromosome or plasmid) are efficiently paired and exchanged by recombination (Figure 4.8). DNA sequences not rescued by recombination are ultimately degraded. The mechanism is one of general recombination and requires that the host be recombination-proficient (Rec⁺) (Smith 1988, 1989). In particular, the host must have a *recA+* gene (Smith 1988), the product of which is essential for homologous recombination. In laboratory studies of this type, gene exchange often involves wild-type genes and their mutant alleles. For example, a length of DNA that includes the lactose fermentation genes (*lac+*) from *E. coli* K12 is introduced into a strain that cannot ferment lactose because of mutation in its *lac* operon. Recombination between the newly acquired chromosomal segment of DNA and the equivalent region on the chromosome of the recipient cell results in the substitution of the wild-type allele for the mutant allele and restoration of the recipient cell's ability to ferment lactose. Allelic variation is seen in naturally occurring bacteria, e.g. the genes that determine the antigens of *Salmonella*.

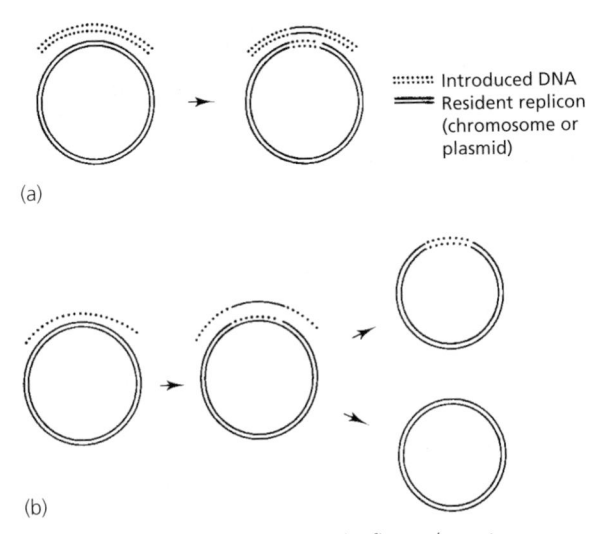

Figure 4.8 *General recombination. The figure shows two crossovers in homologous DNA having occurred between* **(a)** *two duplex molecules and* **(b)** *a single-stranded and a duplex molecule. Multiple crossovers (not illustrated) usually occur between homologous regions of DNA. The figure does not illustrate the mechanism of crossover. The RecA protein is required and some understanding of the mechanism is emerging (see text).*

When that part of the chromosome of *Salmonella* Para-typhi B that encodes antigen b is transferred into *S.* Typhimurium, recombination can lead to the substitution of the sequence encoding antigen b for the chromosomal sequence that encodes antigen i. In this way, new *Salmonella* serotypes are created. Phylogenetic relationships among Enterobacteriaceae have been reviewed by Sanderson (1976) (see also Chapter 51, The Enterobacteriacea: general characteristics).

DNA may also be added to the genome of the recipient cell. Plasmids (see next section, Plasmids) are autonomous DNA molecules and, to survive transfer, do not need to be rescued by recombination. Transposable elements, as discussed above, although not replicons, may insert into other DNA sequences with no requirement for homology or any loss of recipient DNA (although the insertion may result in loss of a particular function). Conjugative transposons, like plasmids, can transfer by conjugation from one bacterial cell to another, but unlike plasmids must integrate into one of the existing replicons in the new host to survive. Conjugative transposons belong to an extended group of mobile DNA elements called integrative conjugative elements (ICE) (Burrus et al. 2002) that includes elements originally thought to be plasmids (Osborn and Böltner 2002; Böltner and Osborn 2004) and the SXT resistance element of *Vibrio cholerae* (Hochhut et al. 2001; Beaber et al. 2002). In general, these elements integrate into particular sites on the host chromosome mediated by site-specific recombination mechanisms.

It is now known that all four mechanisms of gene transfer apply to both gram-positive and gram-negative bacteria, although not necessarily with any one particular bacterial strain. While some transfer systems that are widely used in the laboratory (e.g. transformation of *E. coli*) are likely to be laboratory artefacts rather than naturally occurring transfer systems, others (e.g. natural transformation of *Neisseria* spp. and *Streptococcus pneumoniae*) are known to have played important roles in gene modification and dissemination in nature (Dowson et al. 1994). For more information, see Grinsted and Bennett (1988).

Transformation

The first observation of bacterial transformation was by Griffith (1928) when studying pneumococcal infection in mice. The virulence of pneumococci depends on the capsular polysaccharide that gives the colonies a smooth appearance (S phenotype). Colonies of nonvirulent strains, without capsules, have a rough appearance (R phenotype). Griffith observed that when a mixture of living avirulent R cells and heat-killed virulent S cells was injected into mice, some of them died, whereas injection of either live R cells or heat-killed S cells was not lethal. From the dead animals, Griffith isolated virulent S cells whose capsular type was the same as that of the injected dead S cells. Griffith concluded that the heat-killed S cells had released something that had been taken up by the live R cells and had transformed them to the virulent S cell type. It was not until 1944 that Avery et al., using fractionated cell components, demonstrated that the transforming factor was DNA.

Cells that are able to take up DNA from their environment and incorporate it into the genome are said to be competent. In the case of some bacteria (e.g. *E. coli*), a state of competence can be developed by washing the cells with certain salt solutions (Saunders and Saunders 1988) but is otherwise absent. In others, competence is a natural feature of the life cycle as, for example, in the pneumococcus. Furthermore, some bacteria are competent at all times, whereas others display competence only at a particular phase of the growth cycle (Saunders and Saunders 1988). In the latter case, competence commonly develops towards the end of logarithmic growth in batch culture. Species that display natural, rather than laboratory-induced competence include *Acinetobacter calcoaceticus*, *Azotobacter vinelandii*, *Bacillus subtilis*, *Haemophilus influenzae*, *Moraxella* spp., *Neisseria* spp., *Pasteurella novicida*, some *Pseudomonas* spp., *Rhizobium* spp., *S. pneumoniae*, and *Xanthomonas phaseoi* (Lorenz and Wackernagel 1994). Species that can be made competent artificially, by one means or another, include *E. coli*, *Pseudomonas* spp., *S.* Typhimurium, and *S. aureus*.

Natural competence

Competence involves the ability to bind DNA to the cell surface and then transport it through the cell envelope into the cytoplasm prior to incorporation into the genome. The first of these activities, particularly in cases of natural competence, may involve specific cell receptors; this idea is supported by studies of interaction of competent pneumococci with DNA, in which the kinetics have been found to follow those of a classic enzyme–substrate reaction, and by the finding that new envelope proteins are synthesized as competence is acquired (Dubnau 1999).

The best characterized naturally occurring transformation systems are those of *S. pneumoniae*, *B. subtilis*, and *H. influenzae*. In all three systems, double-stranded donor DNA binds to the cell surface and is cut into discrete lengths giving fragments of 15–30 kb in the case of *B. subtilis*, 8–9 kb for *S. pneumoniae* and c. 18 kb for *H. influenzae*. The transforming activity of chromosomal DNA is better with large DNA fragments than with small, and there is a minimum size requirement for the donor DNA below which no transformation activity is seen. With *B. subtilis* and *S. pneumoniae* there is a period after DNA uptake during which the acquired DNA is incorporated into the genome. This period has been termed the eclipse phase.

B. subtilis and *S. pneumoniae* take up DNA from unrelated (heterospecific) bacteria as well as from closely related (homospecific) bacteria, whereas *H. influenzae* and *Neisseria gonorrhoeae* efficiently take up only homospecific DNA (Dubnau 1999). A membrane protein detectable only in competent *H. influenzae* and capable of binding DNA has been reported, and Sisco and Smith (1979) found that a specific nucleotide sequence of 8–12 bp mediates binding to the protein. It has been estimated that there are approximately 600 copies of this sequence on the *H. influenzae* chromosome, but it is rarely encountered on heterologous DNA. Such a specific mechanism is unlikely to have been preserved by chance, which implies that transformation plays an important role in the life-style of *H. influenzae*.

The fate of transformed donor DNA depends largely on its source. If the DNA is linear and homospecific, it can be incorporated into the chromosome of the recipient cell by recombination. If the linear DNA is heterospecific, its rescue will depend on whether there is sufficient homology with a resident replicon to permit recombination. Recombination with a resident replicon is not necessary when the donor DNA is itself a replicon, i.e. bacteriophage DNA (in which case the phenomenon is called transfection) or plasmid DNA. In these cases, however, successful transformation appears to require the uptake of more than one copy of the replicon, although complete duplication of sequence is not necessary. This finding indicates that replicon re-establishment also involves recombination (Saunders and Saunders 1988). In systems that rely on natural competence, transformation with homospecific chromosomal DNA is usually much more efficient than with intact heterospecific replicons, probably because of the requirement for sequence duplication to effect recombinational rescue, i.e. in the case of transformation with heterospecific DNA such as a plasmid, uptake of two plasmid molecules is necessary to provide the homology needed for recombinational rescue (unless the cell already contains a replicon with partial homology). Consistent with this interpretation is the finding that transformation with plasmid dimers (i.e. duplicate copies of the same plasmid joined head to tail in the same DNA molecule) is as efficient as transformation with homospecific chromosomal DNA (Saunders and Saunders 1988).

Artificial competence

The ability to transform bacteria is a valuable experimental tool but several commonly used laboratory bacteria (e.g. *E. coli*, *Pseudomonas putida*, *S.* Typhimurium, and *S. aureus*) do not display transformation competence at any stage of their life cycles in laboratory culture. Therefore, methods have been devised that can induce artificial competence. The common feature of these methods is the treatment of bacteria with salt solutions at 0°C. The most commonly used salt is calcium chloride (Saunders and Saunders 1988). The method was first developed to enable *E. coli* to take up noninfective phage DNA, and has since been used extensively to introduce plasmid DNA.

These transformation systems appear to work best with plasmid or bacteriophage DNA, rather than with chromosomal DNA. Indeed, *E. coli* does not transform efficiently with chromosomal DNA unless it is deficient in exonuclease V (encoded by the *recBCD* genes), since this enzyme degrades invading linear DNA; in addition, the cell must also be mutated at another locus, *sbcB*. This mutation restores recombination proficiency which is lost by mutations at *recB* and *recC*. Recombination is necessary to rescue linear homospecific DNA sequences.

Transformation with artificial competence works most efficiently with covalently closed circular (ccc) DNA, i.e. intact replicons such as bacteriophage genomes and plasmids (Cohen et al. 1972). These double-stranded DNA molecules are taken up intact and incorporated efficiently into the genome as autonomous replicating units. Any damage to the ccc form reduces the efficiency of transformation, as does increasing the size of the replicon; small plasmids transform the cell at much higher frequencies than large plasmids. It should be noted that induction of artificial competence is bacterial species-specific. Although very efficient systems have been developed for particular laboratory strains of a particular organism, these systems often do not work efficiently with other types of bacteria, or even with natural isolates of the same bacterium.

DNA may also be induced to enter bacterial cells when cell suspensions are mixed with free DNA and are subject to electrical discharge through the suspension, a procedure termed electroporation (or electrotransformation) (Miller 1994). This is now a commonly used method to introduce DNA molecules into many different bacterial species and electroporation apparatus is readily available commercially.

Conjugation

'Conjugation' is the term used to describe the transfer of DNA directly from one bacterial cell to another by a mechanism that requires cell-to-cell contact. DNA is transferred in a nuclease-resistant form that distinguishes the mechanism from transformation and without the aid of a bacteriophage, which distinguishes the mechanism from transduction.

Bacterial conjugation was discovered by Lederberg and Tatum (1946) when testing pairs of mutant strains of *E. coli* K12 for evidence for DNA transfer and recombination. From a mixture of a strain that required phenylalanine, cysteine, and biotin (a triple auxotroph), and another that required threonine, leucine, and thiamine, they obtained some clones that were prototrophs, i.e. they required none of the nutrients required by the two parental strains, and other clones with various combinations of these requirements. These recombinant clones could only have arisen as the result of acquisition of chromosomal DNA from one bacterial cell by the other. Further studies, over several years, showed that the recombination was not the result of sexual reproduction in the normal sense of equal participation of both parents in zygote formation. Rather, it appeared that only a small amount of DNA from the 'male' (donor) parent was incorporated into the offspring (recombinants) which otherwise showed all the characteristics of the 'female' (recipient) parent. The capacity to donate genetic information (DNA) in these experiments was shown to depend upon possession of the F (fertility) factor. Strains were found that had become infertile, having lost their F factor, but they could regain fertility by contact with an F-carrying (F⁺) strain (Hayes 1953a). At the time of these experiments, the role of DNA and the nature of the F factor were unknown; the F factor is now known to be a conjugative plasmid (see below).

In the first conjugation experiments the bacterial characters transferred were encoded by chromosomal genes. The F factor mediated transfer of part of the donor (F⁺) cell's chromosome and this was rescued by homologous recombination into the chromosome of the recipient (F⁻) cell, resulting in recombinant clones of bacteria with new assortments of characters. The recombinants remained F⁻, i.e. they did not also acquire the F factor.

Conjugation is not an intrinsic function of bacteria but is determined by a variety of plasmids of which the F factor is just one. Conjugation normally functions to transfer plasmids from one bacterial cell to another. Conjugal transfer of chromosomal DNA is a relatively infrequent event, but many different plasmids are able to mobilize chromosomal genes with various degrees of efficiency. Chromosome-mobilizing ability (cma) is a byproduct that arises from certain interactions of the plasmids concerned with the bacterial chromosome, just as transduction of chromosomal genes is a byproduct of phage particle production.

For many years, conjugation was thought to be the prerogative of plasmids found in gram-negative cells, and most of our knowledge derives from these systems (Wilkins 1995). However, conjugation systems have been reported in gram-positive bacteria, notably strains of *E. faecalis*. The mechanics of these systems appear to differ somewhat from those found in gram-negative bacteria, primarily in the way donor and recipient cells are paired (Clewell et al. 1986; Grohmann et al. 2003).

Conjugation determined by F

F was the first conjugative plasmid to be discovered and the functions that mediate conjugation have been analyzed in detail. The genes that determine these functions are arranged as a continuous array, mostly in a single operon on the F factor in a region c. 30 kb in length, called the transfer (*tra*) region (Firth et al. 1996). This cluster, comprising approximately 24 genes, is found not only on the F factor but also on several resistance (R) plasmids, found in members of the Enterobacteriaceae, e.g. Rl and R100. Other plasmids encode unrelated but analogous Tra functions. Several different conjugation systems have now been described (Wilkins 1995; Zechner et al. 2000).

Conjugative pili

For conjugation to occur, donor and recipient cells must be in physical contact. Establishing contact is part random and part directed, in the sense that although encounters between donor and recipient cells happen by chance, the cells are joined by design. In gram-negative systems, the first link is made by thin protein filaments on the donor cell, called conjugative or 'sex' pili. These can be visualized by electron microscopy (Figure 4.9). Cells that lack conjugative pill cannot transfer DNA. Sex pili are essentially fiber polymers of a protein subunit called pilin. The current view is that a sex pilus effects the initial contact between the donor and recipient cells, via an interaction between a binding ligand at the tip of the pilus and its cognate receptor on the

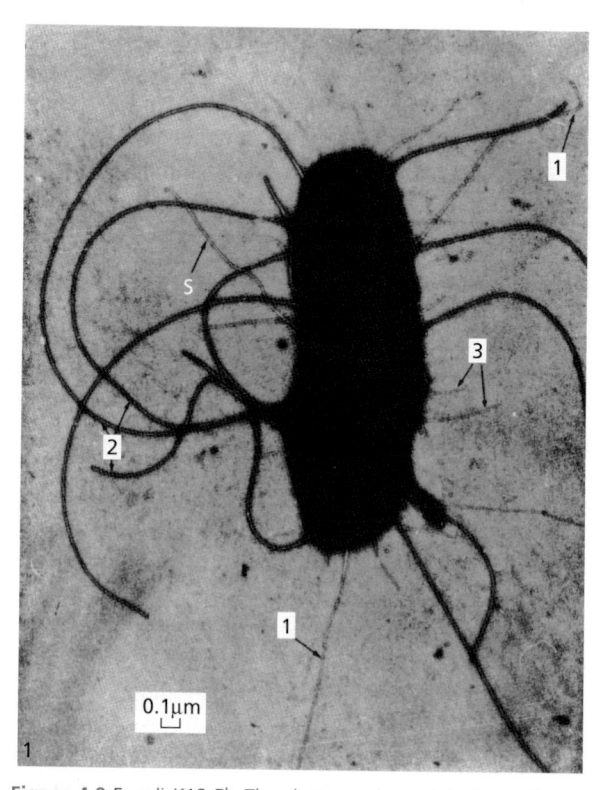

Figure 4.9 E. coli *K12 F⁺. The electron micrograph shows three kinds of appendage to the cell: 1, F pili; 2, flagella; and 3, common pili or fimbriae (from Meynell et al. 1968)*

surface of the recipient cell. The pilus is then retracted into the donor, by depolymerization, in the process drawing the two cells together until direct contact is made. Pilin protein has been found as a pool of unpolymerized molecules in the outer membrane of the cell.

Conjugative pili may act as receptors for bacteriophages. F⁺ cells are therefore susceptible to certain phages that cannot infect F⁻ bacteria whereas some other phages will infect only cells that carry one of another class of conjugative plasmid. Because these phages are specific for particular pili, they can be used as indicators for the presence of particular types of plasmids (Willetts 1988).

Mating pair formation

Random collision of bacteria is the first stage in conjugation or mating; plasmid transfer is rarely detected at low cell densities (Broda 1979). The initial link between donor and recipient cells, made by the sex pilus, is fragile and easily disrupted (e.g. by the shear force generated by pipetting), so not all donor–recipient encounters are productive. Before plasmid DNA is transferred, cell–cell attachment is strengthened, in the sense that much greater force than that generated by pipetting is needed to separate them. This state reflects the direct cell-to-cell attachment made after pilus retraction, and which is needed for formation of the DNA transfer channel between the participating cells. Such cell pairs can be forcibly split by vigorous treatment, e.g. by vortex mixing or in an electric blender. Once a stable mating pair is formed, initiation of F transfer is rapid and plasmid transfer takes about 2 min. When transfer has been completed, both cells contain F. Thereafter, the cells separate and both can initiate further rounds of F transfer.

High frequency of recombination (Hfr) and F' formation

The first bacterial conjugation experiments involved transfer of chromosomal genes mediated by F. For this to occur, F must first fuse with the bacterial chromosome to form one large circular molecule. F, like many other plasmids, carries several transposable elements, specifically IS2, IS3, and γδ (Tn1000). Copies of the IS sequences are also found on the chromosome of E. coli and so can serve as loci for recombination between F and the chromosome (Brooks Low 1996; Deonier 1996). A single crossover between an IS element on F and a copy of the same element on the chromosome inserts one replicon into the other to form one large conjugative DNA molecule, transfer of which can proceed from the origin of transfer (oriT) of the integrated plasmid. Plasmid integration into the chromosome can also result from transposition (by, for example, γδ) with the forma-

tion of a transposition cointegrate. Either mechanism can generate integration events that may be transient or relatively stable. E. coli clones with the F factor stably integrated into the chromosome are called high frequency of recombination (Hfr) strains because, in matings in which these are donors, chromosomal genes are transferred to recipient cells at high frequency and large numbers of recombinants can be recovered (Hayes 1953b). Although initiation of transfer of DNA from oriT is rapid once a stable mating pair is formed, because the chromosome is about 50 times as long as F, complete transfer of the plasmid-chromosome cointegrate takes about 100 min (cf. approximately 2 min for F itself). In practice, mating pairs usually separate before this, so the chromosome is rarely transferred in its entirety.

F can be inserted into the bacterial chromosome at numerous loci and in either orientation. For each Hfr strain the chromosome is transferred to recipients unidirectionally from the oriT of the integrated F factor. Genes nearest the transferring end of F enter the recipient first (Figure 4.10). As a consequence, these donor alleles are recovered in greatest numbers in recombinants. The further a gene is from the leading end of F, the lower the probability that it will be transferred because, as noted above, it is rare that the entire chromosome is transferred; i.e. the longer a mating proceeds, the greater the probability that the participating cells will separate. One consequence is that chromosomal genes are transferred as a gradient from the point in the chromosome where F is inserted. Accordingly, these gradients can be plotted to show the positions of the genes in relation to the oriT of the integrated F. The same gradient, more directly related

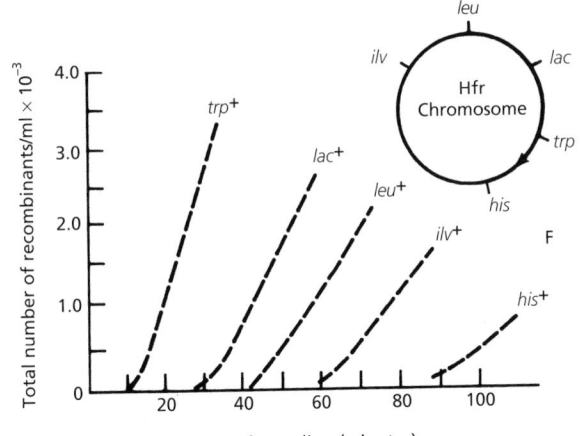

Figure 4.10 *Interrupted mating experiment. An Hfr strain of* E. coli, *with F inserted at the position shown, was mixed with an* F⁻ *strain that was* lac *(lactose nonfermenting),* trp, leu, ilv, his *(required tryptophan, leucine, isoleucine–valine and histidine for growth),* str *(streptomycin-resistant). Samples removed from the mixture at times shown were 'whirl-mixed' and plated on medium selective for the respective recombinants and containing streptomycin (to inhibit the Hfr donor).*

to the time of entry of each gene into recipient cells, can be determined from the results of interrupted matings. In these experiments, samples of a mating mixture are removed at intervals and diluted so as to prevent further mating pair formation. Mating pairs are separated by shearing in a blender or mixer and the cell suspension is then plated onto selective media to detect recombinants (Figure 4.10). When such experiments are performed with several different Hfr strains, the results indicate that the chromosome of *E. coli* is a circular molecule, because linear permutations of the same circular gene array are always obtained. As a consequence, the map of the *E. coli* chromosome is depicted as a circle with 100 min as reference points (Berlyn et al. 1996). Similar experiments with other bacteria, notably *S.* Typhimurium, *Proteus mirabilis*, *P. aeruginosa*, and *P. putida*, using F or a suitable analog, have shown that the chromosomes of these are also circular and similar in size (see O'Brien 1987).

When a plasmid integrates into the chromosome, either by transposition or by recombination between copies of a common sequence (e.g. an IS element), the overall result is the same, namely, formation of a cointegrate in which the plasmid is flanked by direct repeats of the transposable element or of the common sequence. Recombination across these direct repeats will restore the plasmid to its autonomous state. The integrated F of an Hfr strain can be excised from the chromosome in this way. Sometimes, however, excision is not precise and chromosomal DNA adjacent to one side or to both sides of the integrated plasmid is also removed and becomes part of the plasmid; sometimes a remnant of F is left in the chromosome. The derivative plasmids generated by the imprecise excision events are called plasmid primes; e.g. F'*lac* is F with the *lac* operon of *E. coli* incorporated into it. In this case the operon is functional so that when an F'*lac is* transferred to a lactose nonfermenting strain the transconjugants (i.e. the progeny of the cross) are *lac*[+] (i.e. are able to ferment lactose).

E. coli K12 strains that carry an F' plasmid are normally diploid for the chromosomal sequence carried on the plasmid. Recombination between these homologous sequences may be frequent and results in re-entry of the F sequence into the chromosome to generate an Hfr strain with the same origin and direction of chromosome transfer as the Hfr from which the F' originated. In practice, F' cultures transfer chromosomal DNA at frequencies between that of a normal F[+] culture (with autonomous F) and that of the originating Hfr strain, because the cells in the culture are a mixture of those with the F' integrated into the chromosome and others in which the plasmid is autonomous, i.e. the integrated and free plasmid forms in these cell lines are in dynamic equilibrium. This phenomenon is not, in principle, restricted to F. Any plasmid can be integrated into the bacterial chromosome by a suitable transposition with subsequent imprecise excision generating plasmid primes

that, in turn, can be used in mapping studies, provided that the original plasmid mediates its own transfer.

F or F' plasmids can be transferred to species other than *E. coli*, including ones belonging to different genera, such as *Yersinia* and *Vibrio* and the *E. coli* genes are expressed in these organisms. In the main, incorporation of chromosomal DNA from *E. coli* into the chromosomes of bacteria of other genera is not observed, unless there is a close phylogenetic relationship as, for example, between *E. coli* and *Shigella* spp., because there is insufficient homology to permit recombination. Even between *E. coli* and *Salmonella*, chromosomal recombination is rare. As might be expected, however, F can insert into the chromosomes of those bacteria in which it can be established, probably via the activities of the transposable elements it carries, to generate Hfr strains, so furnishing a tool for genetic analysis.

Conjugation in other genera

Conjugative plasmids have been identified in many different bacteria including all gram-negative enterobacteria, in species of *Pasteurella*, *Yersinia*, and many nonfermenting gram-negative aerobes such as *Pseudomonas* and *Acinetobacter*, as well as in *Haemophilus*, *Neisseria*, *Campylobacter*, and *Bacteroides* spp., and gram-positive organisms such as *Streptomyces*, *Streptococcus*, and *Clostridium*. Demonstration of conjugation in all these species does not necessarily imply that the mechanisms are the same; for example, conjugative pili have not been identified in all the genera listed above. However, conjugative plasmids in *Pseudomonas* (Holloway et al. 1979), *Streptomyces* (Hopwood et al. 1973), and *Acinetobacter* (Towner and Vivian 1976) spp. mediate chromosomal DNA transfer in ways that appear to be analogous to that of F in *E. coli*.

In some genera, conjugation and transfer of antibiotic resistance determinants have been observed although no plasmid DNA could be detected, before or after transfer (Stuy 1980). In these systems the mobile DNA is normally part of the bacterial chromosome, but the strains do not behave as Hfr strains. Rather, transfer involves a discrete DNA sequence. During transfer the element behaves like a plasmid, i.e. it separates from the chromosome of the donor cell, but it cannot be isolated by conventional plasmid isolation procedures. These transfer systems are now known to reflect the activity of conjugative transposons and other ICEs (see above). One example is typified by the transposon Tn*916* which was first identified in the chromosome of a strain of *E. faecalis*. Tn*916* encodes resistance to tetracycline and accommodates conjugation functions that mediate transfer of the transposon alone, even between Rec[−] hosts (Franke and Clewell 1981; Clewell et al. 1988). No plasmid is involved and the element transposes from the chromosome in the donor cell into the chromosome in the recipient (Salyers et al. 1995).

The conjugative functions of F are naturally derepressed, so that F transfers at high frequency. By contrast, most natural plasmids transfer at rather low frequencies. In some cases, mutant plasmids can be isolated that transfer at greatly increased frequencies, indicating that expression of the *tra* genes on these plasmids is normally genetically repressed. In other cases, genetic derepression has not been found, although the transfer frequency is relatively low; an example is the R plasmid RP4, which originated in a strain of *P. aeruginosa* but which can be transferred to virtually all gram-negative bacteria (Thomas and Smith 1987).

One consequence of regulating the expression of a set of transfer genes negatively, i.e. with a repressor, is that immediately after plasmid transfer expression of the transfer system is transiently derepressed in the recipient and remains so until sufficient repressor has been synthesized to re-establish control. During this time the recipient temporarily becomes capable of high-frequency plasmid transfer. If there are sufficient potential recipient cells available, the plasmid will spread rapidly from one cell to another in a chain reaction (epidemic spread) until the system is saturated, as was seen in early studies on the transfer kinetics of R plasmids such as R1 and R100. This finding has practical significance in the design of experiments to test for plasmid transfer (Broda 1979; Willetts 1988).

Conjugative plasmids in genera other than those of the Enterobacteriaceae may also repress expression of their transfer genes. In *E. coli* the only mechanism known to lift repression, other than mutation, is transfer to a new host. In other genera, however, induction of the conjugation functions has been reported, e.g. conjugation mediated by a tetracycline resistance plasmid in *Bacteroides fragilis* is induced by low concentrations of tetracycline together with induction of resistance to tetracycline (Privitera et al. 1981). This phenomenon does not appear to be widespread.

Conjugation in gram-positive bacteria appears to differ somewhat from that in gram-negative bacteria in the manner in which the donor and recipient cells are brought together. In *E. faecalis*, conjugation is mediated via sex pheromones excreted by potential recipient cells (Clewell 1993). These peptides cause donor and recipient cells to clump to form mating aggregates, a necessary prerequisite to plasmid transfer. Upon acquisition of the plasmid, the recipient promptly ceases to produce the particular pheromone that triggered plasmid transfer. Different plasmids respond to different pheromones and potential recipient cells produce several.

An unusual form of plasmid transfer, called phage-mediated conjugation, has been reported to occur with some strains of *S. aureus*. Lacey (1980) described the carriage of penicillin and cadmium resistance genes by elements with some of the characters of defective phages. Not only were these genetic elements themselves transmissible between non-lysogenic staphylococci, but they could also mediate transfer of unlinked resistance plasmids in mixed culture (but not in conventional transduction experiments).

Transduction

Gene transfer between bacteria can be mediated by some bacteriophages in a process known as transduction. Transduction is the transmission of a piece of host DNA from one cell (the donor) to another (the recipient) by a phage particle. Two forms have been described: generalized transduction (Masters 1996) and specialized transduction (Weisberg 1996). In the first, any part of the donor genome can be transferred to a suitable recipient; in the second, specific sequences are transferred.

Generalized transduction

Phage P1 replicates during infection of its *E. coli* host cell to generate long concatemers (head-to-tail multimers) of DNA that are subsequently cut into new phage chromosomes by the assembling phage particles. Transducing particles of P1 contain a length of the host chromosome, equivalent in size to the phage genome, in place of the phage DNA, indicating that some heads erroneously take in a section of bacterial rather than phage DNA during phage particle development. Such particles are morphologically normal and are released in the lysate along with ordinary phage particles; they can then infect fresh cells, injecting the contents of the head into the cell in the usual manner. Although the length of bacterial DNA transduced is similar to that of a normal phage chromosome, there is conflicting evidence as to whether particular regions of the host chromosome are preferentially transferred. Such transducing particles comprise approximately 1 in 100 000 of the phage particles in a phage lysate. A phage-infected cell has a roughly similar amount of phage DNA at the end of the replication period as there was bacterial DNA at the outset, yet only a small minority of phage particles contain bacterial DNA, indicating that the head-filling process preferentially packages phage rather than host DNA. Transduced genes are recovered by homologous recombination between the transferred DNA and the recipient cell's chromosome, although it is not uncommon for the transduced DNA to persist extrachromosomally but unreplicated in the cell. In this case it is inherited by only one daughter cell at cell division, a phenomenon called abortive transduction. The frequency of transduction of any one bacterial genetic marker is around 10^{-6} per surviving recipient cell.

Insofar as generalized (and specialized) transduction depends on infection by a phage, it is usual to find that bacterial genes can be transduced only to a recipient of the same species, or perhaps even strain, as the donor; the host range of a phage is generally delimited by the presence of specific receptors on the surface of the cell. Although best studied with phages P1 and P22, two temperate phages, generalized transduction is not a property of temperate phages alone. The lysates of even highly virulent phages, such as those in the T series in *E. coli*, are found to include some particles containing host DNA (Borchert and Drexler 1980; Young and Edlin 1983), but if such particles infect a recipient and establish a transductant, the latter is usually killed by superinfection with other lytic particles. In these cases, transduction is only seen at low multiplicities of infection, i.e. when bacteria outnumber bacteriophage.

Generalized transducing phages can also transduce plasmids, as well as fragments of the donor chromosome, provided these are sufficiently small to be contained within the phage head; where plasmids are too large to be accommodated in the phage head, variants with partial deletions may be transduced, a useful technique for the isolation of plasmid deletion mutants (transductional shortening) before the advent of genetic engineering. The transduction of antibiotic resistance plasmids in staphylococci is of considerable clinical significance (Lacey 1973, 1984). Unexpectedly, the frequency of cotransduction of two plasmids in *S. aureus* is often much higher than would be predicted from the transduction frequencies of each plasmid separately, a result that may reflect cointegrate formation between the two plasmids (Novick et al. 1981, 1986).

Specialized transduction

Phage λ lysogenizes *E. coli* when the phage genome integrates into the *E. coli* chromosome at a specific site by means of a recombination event mediated by the phage function Int and host function IHF (Arber 1983). On induction of the lytic cycle, the integration event is reversed and the phage genome is excised in the presence of an additional phage function, Xis. Both integration and excision are highly specific with respect to the sites in the DNA at which recombination occurs. Occasionally, however, recombination during the excision process occurs between two sites other than the normal attachment regions of phage and host DNA, usually one within the prophage and one just outside (see Figure 4.11). The consequence is that a ring of DNA is excised that consists primarily of λ genes plus a few host genes from one side of the attachment region. In the presence of other λ DNA molecules undergoing normal lytic replication in the cell, this hybrid DNA can be packaged in a λ head and released in the phage lysate where it is available to infect a further cell and inject into it DNA that contains some host genes.

The gene clusters on the bacterial chromosome that lie closest to the λ integration site and are the most

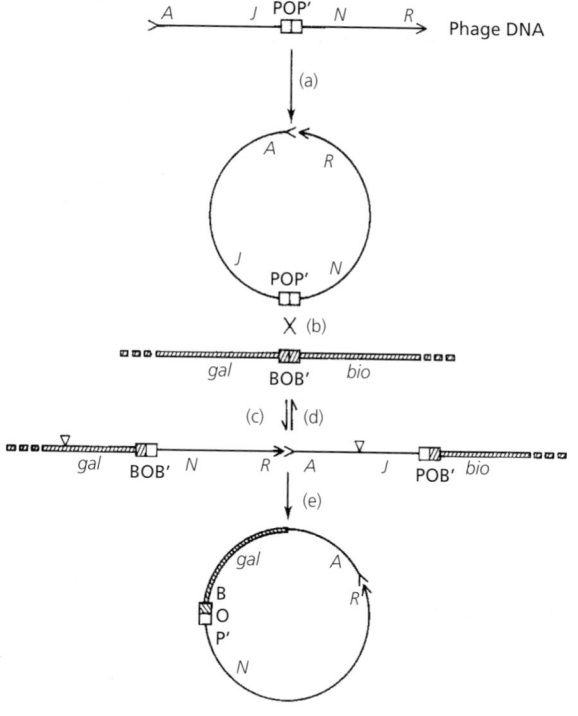

Figure 4.11 *The Campbell model for prophage λ integration into the bacterial chromosome.* **(a)** *Circularization of λ DNA by attachment of cohesive ends (arrows) followed by ligation.* **(b)** *and* **(c)** *Integration by site-specific recombination between phage (POP′) and host (BOB′) attachment regions; four phage genes (A, J, N, R) and two host gene clusters (gal, bio) are shown for reference, and phage integrase (int⁺ product) and integration host factor (IHF) are required. Note that the prophage genes are permuted relative to those in the phage DNA.* **(d)** *Normal excision: integrase and excisase (xis⁺ product) are required.* **(e)** *Abnormal excision, by recombination between two points (▽) other than the attachment regions, generating a defective λ particle carrying host gal genes a particle carrying bio genes can be formed similarly.*

easily selected in genetic crosses, i.e. *gal* and *bio* (Figure 4.11), and hence it is these genes that are readily found in specialized transducing particles represented as λ*gal* and λ*bio*. Lambdoid phages (Campbell 1994) are closely related to λ, but many have a different host integration site and are therefore able to transduce different genes; thus, φ80 acts as a specialized transducing phage for the tryptophan genes in *E. coli*. Less closely related phages are exemplified by P22 in S. Typhimurium, the prophage of which is integrated within the proline gene cluster; hence P22, which is a generalized transducing phage in *Salmonella*, is a specialized transducing phage for the *pro* region.

There are two constraints on the ability of a chromosomally integrated λ phage to form a specialized transducing particle: the recombination event must not exclude the cohesive end (*cos* site) of the phage genome and the total length of DNA in the particle must not significantly exceed that of wild-type λ itself, i.e. 5 percent increase or less, to allow for packaging in the head. A consequence of the second point is that, if the phage acquires some host genes, it will normally have to lose some phage DNA to accommodate the acquisition; hence such phage derivatives are often defective in their abilities to propagate themselves and so require a helper phage to supply the missing function(s). With the exception of the limitation on genome size, specialized transducing phage formation is closely similar to the process by which hybrid plasmids such as F-primes are formed (Hu et al. 1975).

Specialized transduction differs from generalized transduction in many respects. Two are significant. The transducible genes are limited to those near the prophage integration site and the transducing particle contains a hybrid of phage and bacterial DNA, rather than consisting entirely of the latter. As a consequence, the transduced genes can be added to the genome of the recipient cell rather than replacing recipient genes, as in generalized transduction of chromosomal markers, because the modified phage genome can integrate into the prophage attachment site of the recipient's DNA. When such transductants are induced to enter a lytic cycle, the integrated hybrid DNA is excised and replicated to generate a large number of copies (a process which may require the presence of superinfecting normal particles to supply missing phage functions), each of which carries bacterial genes; these copies are packaged to form phage particles and if this lysate is then allowed to infect a further population of sensitive cells, the transduction frequency of the phage-associated genes increases about 1 000-fold, giving rise to high-frequency transduction (HFT).

Protoplast fusion

When the cell wall of a bacterial cell is removed, an osmotically fragile form called a protoplast is generated. This remains intact as long as the growth medium is osmotically buffered. Bacterial protoplasts can fuse with one another, a process called protoplast fusion. This generates cytoplasmic bridges that allow mixing of the cytoplasms of the fused cells and exchange of genetic material. Fusions can be obtained between unrelated cells, even between members of different kingdoms, and gene transfer by this mechanism has been termed genetic transfusion.

As in the other forms of gene transfer, plasmid or phage DNA can be transferred interspecifically or intergenerically, but, for recombination between chromosomal DNAs to occur, a close relationship between the participating cells is necessary in practice. In the case of protoplast fusions, two complete chromosomes are brought together to generate an artificial diploid and from such fusions haploid recombinants with many permutations of the traits of the parent cells can be recovered. Protoplast fusion is an artificial system of gene transfer that has little or no demonstrated rele-

vance to gene transfer in nature. However, protoplast (or spheroplast) fusion has proved to be a valuable experimental tool that is likely to continue to prove useful in the advance of molecular genetics (Gokhale et al. 1993).

PLASMIDS

Plasmids are extrachromosomal, autonomous DNA molecules (i.e. replicons) found in many bacterial species (Stanisich 1988; Thomas 2000). Plasmids have also been found in lower eukaryotes, e.g. the 2 μm plasmid of yeast. The term 'plasmid' was proposed by Lederberg (1952) for all extrachromosomal genetic structures that can replicate autonomously. Here, the adjective 'autonomous' means self-governing and separate from the chromosome of the host, rather than an entity capable of separate existence, in that plasmids are obligate endosymbionts of bacteria, using the replication enzymes of their host to ensure their own propagation, but retaining control over replication. They differ from bacteriophages in having no independent, extracellular form. Most plasmids, whether from gram-negative or gram-positive bacteria, are circular elements, but linear plasmids have been reported in *Streptomyces* spp. and in *Rhodococcus* spp. (Hinnebusch and Tilly 1993). As more bacterial genomes have been sequenced, it has become clear that a number of species have multiple chromosomes rather than one (Cole and Saint-Girons 1999), some of which have sizes similar to a large plasmid. This raises the interesting question as to when such molecules should be considered as plasmids and when as minichromosomes. To date, a definition is lacking.

The term 'episome' has, in the past, been used as a synonym for plasmid. Episomes were defined as genetic elements that could exist and be replicated in either of two modes; autonomously or as part of a cointegrated structure with the bacterial chromosome. As it came to be realized that almost any plasmid can be integrated into the bacterial chromosome, the term lost its usefulness as the definition of a particular class of replicon (Novick et al. 1976). However, the adjective 'episomal' may be useful when applied in particular circumstances, e.g. to describe the state of F in an Hfr strain.

Plasmids are widespread (Stanisich 1988), but the practical importance of bacterial plasmids was first recognized in the early 1960s when transferable drug resistance was discovered (Watanabe 1963). Although plasmids represent only a fraction of the prokaryotic gene pool, they have attracted, and continue to attract, much research interest because the genes they carry determine many of the more interesting features displayed by bacteria, e.g. resistance to antibiotics and heavy metals, virulence, symbiosis and nitrogen fixation, exotic metabolic capacity, and many more (Stanisich 1988). They also serve as convenient models, particularly of replication and partition mechanisms (see Thomas 2000).

The extrachromosomal natures of F and of the first R plasmids discovered were deduced from their genetic behaviors. One feature is the tendency for clones in the culture to lose the plasmid, the set of plasmid-determined characters thus being lost simultaneously and irreversibly. The stability of inheritance of plasmids varies from one to another; some are lost very easily, at rates much higher than the rate of mutation, e.g. 1 percent or more of cell progeny may not inherit the plasmid in any one generation, compared with normal mutation frequencies in the range 1 in 10^6–10^8 per generation. Others are very stable indeed and are rarely, if ever, lost. Some plasmids can be eliminated from the host artificially (cured) by treatment with one of a variety of chemicals (curing agents) (Stanisich 1988). The quinolone drugs seem to be particularly effective (Weisser and Wiedemann 1985). However, no single curing agent or treatment can be relied upon; finding an appropriate agent or treatment to cure a specific plasmid is an empirical exercise that may well fail (Caro et al. 1984). The F plasmid is readily eliminated by treatment with acridine orange (Hirota 1960), but few others succumb to this agent. Some plasmids have temperature-sensitive replication systems and so are eliminated by growth of the cell at the nonpermissive temperature (May et al. 1964; Yokota et al. 1969), but this feature is not widespread. It has also been reported that the conversion of some bacteria to protoplasts (spheroplasts), e.g. *B. subtilis* and *S. aureus*, can result in plasmid loss. Again this is not a universal finding. Irreversible loss of a particular character, or set of characters, either spontaneously or after treatment, is still powerful genetic evidence for the existence of a plasmid but the inability to cure a specific feature, for the reasons given above, does not necessarily rule out plasmid involvement.

With the development of methods for isolating plasmid DNA, the presence of a plasmid can be determined directly. Modern small-scale preparative procedures (Sambrook et al. 1989) and the use of agarose gel electrophoresis allow plasmids to be detected in very small culture volumes (0.5–1.0 ml of a stationary phase culture). Plasmid DNA can be purified and reintroduced into bacterial cells by transformation (or electroporation) to establish definitively the connection between a phenotype of interest and the plasmid. However, this is still not always possible, particularly with large plasmids ($\geqslant$100 kb), or when investigating a poorly characterized bacterium, in which case it may be necessary to fall back on classical genetic techniques (conjugation, transduction, curing).

Plasmid genes

The only genes that are formally necessary to a plasmid are those needed for its replication. The amount of DNA devoted to this task is surprisingly small (Scott 1984; Thomas and Smith 1987) and rarely, if ever, exceeds a few kilobases. The minimum requirement is an origin sequence, origin of vegetative replication (*oriV*), at which replication can start, all enzyme functions being provided by the host. Several small plasmids have this arrangement, e.g. ColEI from *E. coli* and the related plasmid, CloDF13 from *Enterobacter cloacae*. In addition to an *oriV*, such plasmids also encode functions that control the extent of plasmid replication, i.e. ensure that the plasmid copy number per cell is maintained within fairly narrow limits specific to the particular plasmid (Sherratt 1986). In these plasmids no more than 1–2 kb is devoted to plasmid replication and its control. In others, more complex arrangements have evolved both to execute replication and to regulate it (Espinosa et al. 2000).

There are basically two modes of replication for circular plasmids: θ (theta) replication and RC (rolling circle) replication (Helinski et al. 1996). In the former, replication proceeds from a fixed origin in a unidirectional or bidirectional manner round the plasmid circle and both strands of the plasmid are replicated simultaneously, in a manner precisely equivalent to *E. coli* chromosome replication. Both strands of the parent plasmid remain intact during the replication process. The type of replication gets its name from the fact that a part replicated plasmid/chromosome resembles the Greek letter theta, θ, if opened out. For most of the plasmids in gram-negative organisms studied, this is the mode of replication utilized.

A few plasmids from gram-negative bacteria and the single-stranded coliphages (Baas and Jansz 1988) have been shown to use the alternative, RC replication mode, as does a family of highly related plasmids in gram-positive bacteria (del Solar et al. 1993). These plasmids always encode an initiator protein which introduces a site-specific nick at the double-strand origin (DSO) to generate a 3′-OH group, which serves as a priming terminus. In some cases, it has been shown that the initiator is covalently attached to the 5′-phosphate exposed by the cut. Replication proceeds by extension from the 3′-OH primer site, using the complementary strand (designated the plus-strand) as template. The old complementary strand (called the minus-strand) is displaced as replication proceeds. Once replication has traveled right round the circle (hence the designation RC) the minus-strand is circularized. Consequently, one double-stranded DNA molecule and one single-stranded molecule result. The circular single-stranded minus-strand is then converted to a double-stranded molecule using a second origin of replication that is specific to the minus-strand. Replication of plasmids, including linear plasmids, has been reviewed by Helinski et al. (1996) and by Espinosa et al. (2000).

Plasmids range in size from c. 1 kb to >400 kb (equivalent to c. 10 percent of the *E. coli* chromosome).

The smallest have only enough DNA to accommodate two or three small genes (e.g. Pl5A; Chang and Cohen 1978). Nevertheless, small plasmids are relatively abundant (Figure 4.12), though many confer no obvious phenotype on their host. Such plasmids (large or small) are said to be cryptic, a description that is largely an acknowledgment of ignorance of function.

Plasmids capable of mediating conjugation have a minimum size of c. 30 kb. Several have now been analyzed extensively with respect to the plasmid-encoded information needed for replication. Here, again, it is found that relatively little genetic content is devoted to replication. In addition to *oriV*, these plasmids encode one or two specific proteins necessary for their own replication, together with one or more replication-control functions (Scott 1984; Thomas and Smith 1987; see also Thomas 2000). These functions define the essential replicon of the plasmid and can be encoded in 4–5 kb (see Thomas 2000).

The relatively small amount of plasmid-encoded information needed for plasmid replication is a reflection of the fact that most of the replication apparatus is provided by the host. Most of the panoply of functions used to replicate the bacterial chromosome is borrowed to replicate the plasmid. Although not all plasmids have exactly the same requirements, they all use a common core of

enzymes, e.g. either all or the bulk of the DNA synthesis is carried out by the main DNA polymerase of the cell, e.g. in *E. coli* DNA polymerase III is used. In *E. coli*, a number of small plasmids (e.g. ColE1 and related plasmids) require DNA polymerase I (product of the *polA* gene) in the early stages of replication, before switching to DNA polymerase III for the bulk of the synthesis. Consequently, such plasmids cannot be established in a *polA* mutant strain of *E. coli*. Most plasmids described to date, however, do not show this requirement.

F and many other plasmids maintain themselves at a low copy number, i.e. one to four copies of the plasmid per bacterial chromosome. These plasmids have been described as being under stringent replication control. By contrast, other plasmids, usually small ones such as ColEI, maintain themselves at a relatively high plasmid–chromosome ratio (15–20). Plasmids of this type are said to be subject to relaxed replication control because it was initially thought that replication control of these latter elements was less rigorous than for low copy number plasmids. This is now known not to be true; rather, the threshold value at which inhibition of replication operates is set higher. Once activated it is just as efficient.

For each plasmid there is a characteristic copy number, although this can be altered by mutation. Some plasmids also show changes in copy number in response to the prevailing culture conditions; for example, some plasmids continue to replicate when the cells enter stationary phase (when chromosomal replication stops) with the accumulation of large numbers of plasmid molecules per cell. Treatment of a culture of *E. coli* carrying ColE1, or a related plasmid, with chloramphenicol (at a concentration sufficient to inhibit protein synthesis) has the same effect; whereas the initiation of new rounds of replication of the bacterial chromosome (and many plasmids) requires de novo protein synthesis, that of ColE1 does not (see Thomas 2000).

The genes that control plasmid replication include functions that act repressively, i.e. inhibit replication. One consequence is plasmid incompatibility. It was discovered early that two variants of the same plasmid cannot be maintained stably in the same bacterial cell, i.e. they are incompatible. This arises because each directs the production of replication inhibitors that act on the other replicon, as well as on themselves. The consequence is that the content of plasmid molecules in the cell is set to a lower level than the sum of the two individual copy numbers. If the plasmids involved are unit, or low copy number plasmids, plasmid segregation at cell division is a likely result, i.e. clones emerge in which one or other of the plasmids has not been inherited (Novick 1987). Plasmids with unrelated replication control systems do not, in general, interact in this way and so do not destabilize each other; they are said to be compatible. Plasmids that are incompatible are said to belong to the same incompatibility (Inc) group; plasmids that are compatible necessarily belong to different incompatibility groups.

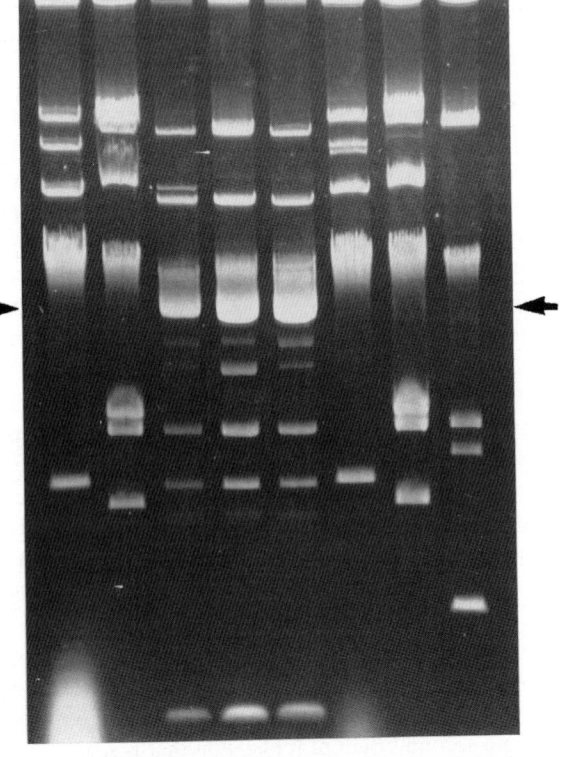

Figure 4.12 *Plasmid profiles from isolates of* E. coli *obtained from slurry waste from a calf-rearing unit. The diffuse band (arrows) in each track represents DNA fragments (chromosome and plasmid) that arose during the course of the preparation. Plasmid size decreases from top to bottom. The smallest plasmids seen in these preparations have sizes of 1–2 kb.*

Plasmid partition systems

To be inherited stably, plasmid DNA not only has to be replicated but at each cell division both progeny cells must obtain at least one copy of the plasmid (which will be replicated until the appropriate copy number has been achieved). For low copy number plasmids such as F and some common R plasmids (Rl, R100), plasmid distribution at cell division – namely, plasmid partition – is an active process involving both plasmid- and host-specified products (Hiraga 1992; Draper and Gober 2002).

The former are designated *par* functions. The components comprise a small sequence of a few hundred base pairs, carried on the plasmid, that is believed to function like a primitive centromere to pair plasmids prior to cell division and one or more proteins that interact with it. These may be plasmid- or host chromosome-encoded. The system may direct plasmid DNA membrane association at or near the point of septum formation. If two different plasmids have related partition systems, to the extent that each is unable to distinguish the two plasmids, incompatibility will result. This will reinforce any incompatibility between the two plasmids that arises as a consequence of 'common' replication systems. However, the replication and partition functions on a plasmid act independently, as can be demonstrated by creating synthetic replicons with the replication system of one plasmid and the partition system of an unrelated plasmid. These hybrids are stably maintained, i.e. they are efficiently replicated and partitioned (Austin and Abeles 1983).

In contrast to the active partition described above, small plasmids with high copy number do not encode specific *par* functions. They rely on the random distribution of units to daughter cells. When the copy number is 15–20, the plasmid content of a cell immediately before cell division will be at least 30–40. Assuming random distribution of plasmids throughout the cell, the probability of one daughter cell inheriting the entire complement of plasmid molecules while the other inherits none is negligible (this would not be true for low copy number plasmids). Nevertheless, and somewhat paradoxically, some small, very high copy number cloning vectors generated in the laboratory are not efficiently inherited, in contrast to the plasmid from which they were derived.

If low copy number plasmids, in particular, are to be partitioned accurately, it is desirable, if not essential in all cases, that the molecules be present in the cell as monomers. Cells in a natural environment are recombination proficient. Hence they will actively pair and recombine homologous sequences, including plasmid DNA. A single crossover between two plasmid molecules will fuse them to form a dimer. If this were to happen with a unit copy number plasmid immediately after it had replicated and no steps were taken to reverse the event, at cell division one daughter cell would inherit both copies of the plasmid. The accumulation of plasmid dimers (and higher multimers) is prevented by plasmid-encoded, site-specific recombination systems (Austin et al. 1981; Lane et al. 1986; Dodd and Bennett 1987; Garnier et al. 1987; Krause and Guiney 1991; Eberl et al. 1994; Guhathakurta et al. 1996). These are specialized recombination systems that comprise a specific DNA sequence and a recombination enzyme that acts at the sequence (Smith 1988). The site must be part of the plasmid, but the enzyme can be either a plasmid-encoded function or a host function (Sherratt 1986; Dodd and Bennett 1987). These recombination systems rapidly convert plasmid multimers to their momomeric units, so reducing the chance of plasmid loss at cell division.

Post-segregational killing

Some plasmids have a third type of mechanism that acts to maintain their stability. These systems act after segregational loss of the plasmid to kill the plasmidless cell (Jensen and Gerdes 1995; Yarmolinsky 1995; Gerdes et al. 1997, 2000; Rawlings 1999). Each comprises two genes; one encodes a toxin, the other an antidote. The toxin is more stable than the antidote, which means that to neutralize the toxin the antidote must be produced continuously. This is possible as long as the cell carries a copy of the plasmid. However, in those unfortunate progeny that fail to acquire a copy of the plasmid at cell division, the toxin activity is duly unmasked as the antidote decays and the cell is killed.

A variation on the theme is encoded by the resistance plasmid R1. In this instance, the control exerted over the toxin is at the point of production, rather than inhibition of activity. The control locus, *sok*, overlaps the toxin gene, *hok*, which is encoded on the opposite strand of the DNA duplex. Both sequences are transcribed, but translation of the *hok* mRNA is prevented as a consequence of the two RNA transcripts annealing to one another over the region of overlap, where they are complementary. This prevents ribosome binding to the Hok mRNA. However, the *sok* RNA decays faster than the *hok* mRNA, so if the R1 plasmid is lost then the *hok* mRNA is translated. The toxin produced causes the collapse of the transmembrane potential and the cell dies (Gerdes et al. 1986, 1988).

Plasmid replication functions together with stability functions (*par* and site-specific recombination functions) comprise the minimum requirements of a stable, low copy number replicon, i.e. one that will be inherited efficiently despite there being very few copies of the plasmid. Partition and post-segregational killing mechanisms act in concert to provide stability (Brendler et al. 2004). The whole can be accommodated in a sequence of <10 kb (a stable miniplasmid derived from F, which is 94 kb in size, carries just 9.2 kb of F, in addition to a selectable marker).

Genes determining conjugation

Conjugation determined by F and by many other plasmids is a complex process requiring many genes. Accordingly, conjugative plasmids cannot be very small. The smallest conjugative plasmids in the Enterobacteriaceae are about 30 kb (of which 15–20 kb are devoted to transfer compared with F which devotes approximately 30 kb to transfer functions) (Avila and de la Cruz 1988; Thomas 1989; Zechner et al. 2000). Conjugation can be viewed as a two step process involving synthesis and transfer of DNA, the latter process mediated by a type IV secretion system (Llosa et al. 2002).

Most natural plasmids that are unable to mediate their own transfer can, nevertheless, be transferred conjugally by co-opting the transfer functions of a conjugative plasmid. Such plasmids, e.g. ColEI, have their own *oriT* and encode functions specific to it (mob genes). To transfer by conjugation, these plasmids require a helper plasmid to provide the missing transfer functions; for example, F can serve as a helper plasmid for ColE1 (Dougan and Sherratt 1977). Such plasmids are said to be mobilizable and mainly lack the ability to direct attachment of donor and recipient cells. A few plasmids possess only an origin of transfer, and transfer only when the other necessary transfer functions are provided by helper plasmids; for example, the small plasmid Pl5A, the progenitor of the cloning vector pACYC184 and its derivatives, has its own *oriT*, but to transfer it requires both Mob and Tra functions provided by ColEI and F (or equivalents), respectively.

Other plasmid genes

Replication and its control, maintenance, and transfer functions may be considered as inherent plasmid characteristics. Together, they direct both vertical (inheritance) and horizontal (transfer) transmission in bacterial populations. However, these features alone, no matter how intrinsically interesting, would have been insufficient to attract the degree of interest that bacterial plasmids have engendered. Indeed, until the discovery of transmissible multiple drug resistance (Watanabe 1963), the study of plasmids and their activities was a minority interest. The veritable explosion of interest shown since then reflects the variety of other genes, expression of which can affect the human condition, carried on plasmids.

Antibiotic resistance determined by plasmids was first discovered in Japan in the late 1950s. Until then it had been assumed that antibiotic resistance in bacteria, which had already become a clinical issue, resulted from the selection of chromosomal mutants, multiple resistance resulting from the sequential accumulation of mutations. Then, unexpectedly, epidemiological analysis of drug resistance in *Shigella flexneri* indicated that resistance could be transmitted between bacteria. Bacillary dysentery was common in Japan after the Second World War. By the mid-1950s, most Japanese infections with *S. flexneri* involved bacteria that were resistant to sulfonamides, so antibiotic treatment (now considered inappropriate) was switched to streptomycin, tetracycline, or chloramphenicol (much of it self-prescribed). In due course, a few strains resistant to the four unrelated drugs were isolated; thereafter such strains became progressively more common (Table 4.2). Isolates resistant to two drugs, which one would expect to see if sulfonamide-resistant clones mutated to resistance to another drug, were rarely isolated. Instead, in outbreaks of infection with a single *S. flexneri* serotype, strains from some patients had the multiple-resistance pattern, whereas others did not. Furthermore, *E. coli* strains with the same pattern of resistance were found in the feces of patients with dysentery. Given these observations, Akiba and, independently, Ochiai in 1959 (cited by Watanabe 1963) tested the hypothesis that multiple antibiotic resistance might be determined by an infective agent analogous to the F factor of *E. coli*. Mixed culture experiments proved this idea correct: multiple drug resistance was transferred from *S. flexneri* to *E. coli*. The infective agents are now called resistance (R) plasmids.

The increasing frequency of drug resistance in *S. flexneri* in Japan was not an isolated phenomenon. The emergence of resistance to antibiotics was recorded in diverse strains of bacteria worldwide: in *Shigella sonnei* in the London area, in *Salmonella* strains in the Netherlands, and in *S.* Typhimurium in England (see Datta and Nugent 1983). These cases differed to some extent from the Japanese experience in that there was a stepwise accumulation of resistance, first to sulfonamides and streptomycin, then to tetracycline, and subsequently (and infrequently) to chloramphenicol (Anderson 1968). Nevertheless, the resistance was primarily plasmid mediated and transmissible; ampicillin resistance then emerged, and within a decade had become common in *Salmonella* and *Shigella* strains throughout Europe (although it remained uncommon in Japan).

Table 4.2 *Multiple drug resistance in* Shigella flexneri *isolated in Japan*

Year	No. of isolates tested	Percentage multiresistant
1956	4399	0.02
1958	6563	2.9
1960	497	15
1962	6853	23
1964	5388	45
1966	4292	75
1968	1237	64
1970	562	74
1972	824	76

Multiresistant strains were resistant to chloramphenicol, tetracycline, streptomycin, and sulfonamide. Figures from Mitsuhashi (1977).

At the time of the discovery of R plasmids in the Enterobacteriaceae, multiresistant strains of *S. aureus* were a cause of concern in hospital infection. Penicillinase-producing strains of *S. aureus*, first recognized in the 1940s, had become common and had acquired resistance to new antibiotics as each was introduced into clinical practice. In 1963, it was shown that penicillinase synthesis by *S. aureus* was a plasmid-determined characteristic (Novick 1963), as are many other antibiotic resistances in that species (Lacey 1975). Penicillin-resistance plasmids in *S. aureus* have been studied extensively. They encode not only the penicillinase, but also the means to regulate its production which is induced when the cells are exposed to penicillin and other β-lactam agents. The mechanism involves a form of signal transduction, because the inducers (β-lactams) do not penetrate the cytoplasmic membrane (Everett et al. 1990; Bennett and Chopra 1993; Zhang et al. 2001). These plasmids often also encode resistance to mercury or cadmium salts, or both, and sometimes to erythromycin or fusidic acid. Resistance to tetracycline, chloramphenicol, and neomycin is also extrachromosomal in *S. aureus*, but these markers are usually found on separate plasmids.

Not only are many resistance determinants in *S. aureus* encoded by plasmids, but several are also carried on transposons (Murphy 1989). Tn*551*, encoding resistance to erythromycin, was the first to be discovered in *S. aureus* (Novick et al. 1979). Later, a transposon encoding resistance to gentamicin, kanamycin, tobramycin, trimethoprim, ethidium bromide, and quaternary ammonium compounds, designated Tn*4001*, was discovered (Lyon et al. 1984). This element was found in a multiresistant methicillin-resistant *S. aureus* (MRSA) (see Chapter 32, *Staphylococcus*) strain isolated in Melbourne. Strains of this type have caused concern worldwide because they are resistant to all β-lactam agents and many other antibiotics that might be used against them. While many antibiotic resistance genes in *S. aureus* are found on plasmids and transposons, the methicillin-resistance gene, *mecA*, is found on acquired sequences of 21–67 kb (Hiramatsu et al. 2001; Ma et al. 2002), termed SCC*mec* (staphylococcal cassette chromosome *mec*) elements (Katayama et al. 2000), which are integrated into the chromosome of *S. aureus* at a unique site, *attBscc*, found in an open reading frame of unknown function near the origin of replication (Hiramatsu et al. 2001). Integration is mediated by a site-specific recombination mechanism comprising two SCC*mec* genes, *ccrA* and *ccrB*, closely linked to *mecA*. A possible precursor of SCC*mec* elements, SCC$_{12263}$, that lacks *mecA* has been identified in the methicillin-susceptible strain *Staphylococcus hominis* GIFU12263 (Katayama et al. 2003).

The *mecA* gene encodes a penicillin-binding protein (PBP-2a or PBP-2′) (Utsui and Yokota 1985; Reynolds and Fuller 1986) that is not inhibited by β-lactam agents at clinically achievable concentrations and this new PBP can substitute for the cell's normal PBPs at concentrations of β-lactam agents that would normally be lethal

to the cell (Archer and Niemeyer 1994; Berger-Bächi 1994). The drug of choice for these strains is often the glycopeptide antibiotic vancomycin. This inhibits peptidoglycan synthesis by binding to the terminal pair of D-Ala residues in the pentapeptide chain of the sugar–peptide cell wall precursor, sterically impeding its interaction with the transglycosylases which normally incorporate it into the peptidoglycan network. Unfortunately, resistance to vancomycin has emerged in the enterococci. The resistance mechanism is somewhat unusual in that it involves remodeling the pentapeptide of the cell wall precursor. The terminal D-Ala residue of the pentapeptide is changed to D-lactate. The loss of the amino group significantly reduces the affinity of the precursor for vancomycin. Because the D-lactate is cleaved from the precursor in the terminal stage of peptidoglycan synthesis, as the short peptide chains are joined together, the finished product is largely unaffected, although synthesis is slower. Remarkably, expression of vancomycin resistance can be inducible and the mechanism also involves signal transduction, mediated by what is known as a two-component regulatory system (see Hoch and Silhavy 1995). Two genes are involved, *vanS* and *vanR*. The former encodes a membrane protein, VanS, that senses the effect of the drug (Ulijasz et al. 1996) and records the fact by phosphorylating one of its histidine residues. The phosphate group is then transferred to an aspartate residue on the second component, VanR. This converts VanR to an activator specific for expression of the four genes needed for vancomycin resistance (Arthur and Courvalin 1993). A similar mechanism is used to regulate expression of virulence genes in *S.* Typhimurium and several other types of virulence gene (see Salyers and Whitt 2002). As if this degree of complexity were not sufficient, the whole system, vancomycin resistance and regulatory genes, is carried on a bacterial transposon, Tn*1546* (Arthur and Courvalin 1993). Recently, this form of vancomycin resistance has been detected in clinical strains of MRSA isolated in North America (Appelbaum and Bozdogan 2004), a worrying but not unexpected development because it has been shown in the laboratory that plasmids encoding vancomycin resistance transfer readily between enteroccocal strains and *S. aureus*. In addition, vancomycin resistance is evolving in *S. aureus* independently of acquisition of Tn*1546* or similar systems (Walsh and Howe 2002).

Many new antibacterial drugs have been introduced into clinical medicine in the last 25 years and plasmids conferring resistance to most of them have appeared within a relatively short time. Not only has the range of plasmid-determined resistance broadened but also more and more bacterial genera that carry resistance plasmids are being found (Tables 4.3 and 4.4). The appearance in the mid-1970s of plasmid-determined β-lactamase production, first in *H. influenzae* and then in *N. gonorrhoeae*, was of prime importance, not only to clinical

Table 4.3 *Some antibacterial drugs to which plasmids determine resistance*

Antibacterial drugs	
Penicillins, cephalosporins, carbapenems	
Erythromycin, lincomycin, and streptogramin B	
Streptomycin	Tetracyclines
Neomycin	Chloramphenicol
Kanamycin	Fusidic acid
Gentamicin	Sulfonamides
Tobramycin	Trimethoprim
Amikacin	

medicine where the strains presented a significant public health problem, but also to the study of plasmid evolution (Saunders et al. 1986; see also Chapter 25, Bacterial infections of the genital tract and Chapter 48, *Neisseria*), because some of the plasmids in both species were found to be closely related (Laufs et al. 1979). The TEM β-lactamase produced in both these organisms is encoded by a gene, *blaTEM*, carried on various transposons, among them Tn*1* and Tn*3* (Figure 4.7), which are widely distributed on plasmids. The production of a TEM β-lactamase by clinical isolates of a wide variety of gram-negative bacteria, particularly among the Enterobacteriaceae, of both human and veterinary origin, is common. Undoubtedly, the presence of this gene on a transposon, and its ability to transpose readily to many different plasmids, facilitated its rapid intergeneric spread. The route taken by such a resistance determinant, before it enters a particular bacterial species, may be circuitous, but if acquisition is desirable (i.e. of advantage to the survival of the organism), it is almost certainly inevitable. Indeed, reports indicate that transfer of resistance genes is not restricted to gram-positive–gram-positive and gram-negative–gram-negative exchanges, but can take place between gram-positive and gram-negative bacteria (Mazodier and Davies 1991), and even between bacteria and mammalian cells (Courvalin et al. 1995), although such events seem to be rare and any that do occur are likely to be nonproductive. Trieu-Cuot and Courvalin (1986) reported a kanamycin-resistance determinant encoding an APH(3′)III aminoglycoside phosphotransferase in a strain of *Campylobacter coli* identical to that previously found in streptococcal strains. For an account of antibiotic resistance in clinical practice, see Chapter 18, Antibacterial therapy.

Origin of resistance genes

The origin of resistance genes has been a puzzle for many years. Before the introduction of antibiotics these genes were rare, although plasmids similar to those now carrying resistance determinants were relatively common

(Hughes and Datta 1983). Perhaps the most persistent and logical suggestion is that many of them originated in the antibiotic-producing organisms themselves, since these organisms, mostly streptomycetes, are sensitive to the antibiotics and so must necessarily protect themselves. This suggestion has been given credence by the finding that the amino acid sequences of several classes of the APH(3′) aminoglycoside phosphotransferases, which confer resistance to kanamycin and neomycin, from both gram-positive and gram-negative bacteria, are evolutionarily related to another from the neomycin-producing *Streptomyces fradiae* (Trieu-Cuot and Courvalin 1986). An interesting variation on this theme is that in the production of antibiotics, small amounts of DNA from the producing organism remain as a contaminant and this may be the source of some resistance genes seen in clinical strains (Webb and Davies 1993). Clearly, if true, this has considerable significance for future antibiotic production, particularly in terms of quality control measures.

Some, if not all, β-lactamases have probably evolved from the same ancestral gene(s) that produced the β-lactam targets, the PBPs. Recent work has clearly shown the evolutionary relationship between the class A β-lactamase of *Bacillus cereus* and a PBP from *Streptomyces* (Samraoui et al. 1986; Joris et al. 1988, 1991). Furthermore, Ghuysen and his colleagues have shown that the protein structures of several β-lactamases and some PBPs are highly conserved, strongly indicating an evolutionary link (Ghuysen 1991, 1994; Frère 1995). The branches from the ancestral gene that yielded present-day β-lactamases, on the one hand, and PBPs, on the other, almost certainly diverged a long time before the discovery of penicillin, let alone its use in clinical medicine (Massova and Mobashery 1999). Nonetheless, the evolution of β-lactamase genes continues to the present day (Galleni et al. 1995; Bush and Jacoby 1997; Koch 2000; Bradford 2001; Gniadkowski 2001; Ford and Avison 2004).

Although most clinically significant β-lactamases are plasmid-encoded and constitutively expressed (particularly in gram-negative bacteria), β-lactamases are also relatively common chromosome-encoded enzymes in many types of bacteria, not only those causing disease, and their expression can be elaborately controlled; some species even have multiple, coordinately controlled β-lactamases (Imsande 1978; Normark 1995; Rossolini et al. 1996; Hanson and Sanders 1999; Philippon et al. 1998). These findings suggest that protection against β-lactam antibiotics is a common and desirable property in nature, although high concentrations of β-lactam agents in the soil, or in other natural habitats, have not been detected. So, it is not at all clear from where the selection pressure for the evolution, particularly of the inducible β-lactamases, comes. An alternative suggestion is that, although β-lactamases hydrolyze β-lactams effectively, this activity is not the normal function of the

Table 4.4 *Some genera in which R plasmids have been found*

Genera
Enterobacteriaceae, i.e.
Escherichia, Salmonella,
Shigella, Proteus,
Providencia, Klebsiella,
Serratia, etc.
Pseudomonas

Acinetobacter	*Staphylococcus*
Vibrio	*Streptococcus*
Yersinia	*Bacillus*
Pasteurella	*Clostridium*
Campylobacter	*Corynebacterium*
Haemophilus	
Neisseria	
Bacteroides	

chromosomally encoded enzymes, some of which, if not all, may be involved in peptidoglycan metabolism (Morosini et al. 2000).

An equally intriguing case is that of trimethoprim. This is a purely synthetic compound that has no known natural analog. Its mode of action is to inhibit dihydrofolate reductase, which catalyzes one of the steps in folic acid synthesis (see Chapter 18, Antibacterial therapy). The enzymes in many bacteria are readily compromised by this antibiotic. Resistance to trimethoprim, e.g. as encoded by Tn7 (Craig 1989), arises because of the acquisition of a gene encoding a dihydrofolate reductase that is much less sensitive to trimethoprim. The source of the gene is unknown. It may be one that naturally encodes a resistant enzyme; alternatively, it may be a mutant form of a naturally inhibited one. Several trimethoprim-resistance alleles, each encoding a dihydrofolate reductase, are now known (Huovinen 2001; Skold 2001). The speed with which resistance emerged, approximately 10 years after the introduction of the drug into clinical practice, is impressive and gives considerable cause for thought, because on an evolutionary time scale 10 years is but a blink.

The case of trimethoprim is not an isolated example. The emergence of resistance somewhere in the bacterial kingdom, following widespread use of a new antibiotic, and its subsequent dissemination among many bacterial species appears to be the norm rather than the exception, attesting to highly efficient gene transfer processes within populations of bacterial cells (Bennett 1995). The frequency of transfer does not need to be high to achieve efficient spread of a particular resistance gene. Even rare transfer events can be effective if environmental conditions impose a sufficiently strong selection pressure on those few progeny cells that do acquire the gene in question. The presence of the antibiotic in the cells' environment constitutes a powerful selective force. The few resistant cells will then rapidly reproduce, while the majority of their bacterial relatives are unable to do so, to become the dominant bacterial species. There is no reason to believe that the extensive interspecific gene transfer highlighted by the spread of resistance genes is confined to them. Rather, the view that all bacterial genes are potentially available to all bacteria is gaining more general acceptance, epitomized by the movement of some chromosomal β-lactamase genes on to plasmids and then into unrelated bacteria (Philippon et al. 2002; Walther-Rasmussen and Hoiby 2002; Bonnet 2004; Ford and Avison 2004), mediated by IS elements.

Resistance gene evolution by mutation

Resistance to an antibiotic may be the result of acquiring new genetic information in the form of a plasmid or transposon, or it may be due to a mutation in the gene for the antibiotic target that renders it less susceptible to the drug. For some drugs, e.g. streptomycin and rifampicin, this form of resistance is often readily selected in the laboratory; however, it is much less common in clinical experience, probably due in part to the success of plasmid–transposon systems and in part to the fact that alterations to key enzymes often come at a price (Andersson 2003). One clear exception to this is resistance to the fluoroquinolones, antibiotics that interfere with DNA synthesis through inhibition of enzymes called topoisomerases (Drlica and Malik 2003; Ruiz 2003; Sissi and Palumbo 2003). Here, resistance is acquired primarily by mutation of the GyrA component of DNA gyrase or of the ParC component of topoisomerase IV, or on occasion by a mutation that activates an efflux pump that removes the fluoroquinolones from the cell (Elliott et al. 2003; Eaves et al. 2004; Ricci et al. 2004).

There is one bacterial species for which acquisition of resistance by mutation is both the sole option and clinically significant, namely *Mycobacterium tuberculosis* in which plasmids appear to have little or no role to play (Cole and Telenti 1995). Multiple antibiotic resistance in *M. tuberculosis* is acquired by sequential accumulation of mutations in the genes encoding the antibiotic targets (Gillespie 2002; Wade and Zhang 2004). The apparent failure of *M. tuberculosis* to acquire R plasmids may reflect its unusually waxy outer surface that is markedly different from the outer surfaces of most bacteria (see Chapter 46, *Mycobacterium tuberculosis* complex, *Mycobacterium leprae*, and other slow growing mycobacteria).

Antibiotic synthesis

Plasmids determine the production of bacteriocins and microcins (Pugsley and Oudega 1987; see also Chapter 5, Bacteriocins and bacteriophages). They also play a part in the production of a variety of antibiotics, including some of those used in medicine (Hopwood 1978; Chater and Hopwood 1989). Plasmid SCP1, in *Streptomyces coelicolor* determines production of the antibiotic methylenomycin. The same plasmid confers resistance to the antibiotic. Plasmids can carry structural or control genes determining production by *Streptomyces* of oxytetracycline, chloramphenicol, and perhaps other antibiotics. The genetics of antibiotic-producing microorganisms is of great interest and of enormous potential importance in the pharmaceutical industry, and rapid progress is now being made in this area (see, for example, McDaniel et al. 1995; Paradkar et al. 2003; Weber et al. 2003).

Resistance to other agents

Plasmids may also confer resistance to antibacterial compounds other than clinically useful antibiotics, such as a variety of metal ions and organometallic compounds. Plasmid-determined resistance to mercuric ions (Hg^{2+}) is very common in bacteria found in soil and water (Bale et al. 1987) such as *Pseudomonas* spp., in members of the Enterobacteriaceae and in *Staphylococcus* spp.; this is frequently linked to antibiotic resistance. In different surveys, 25–60 percent of R plasmids determined Hg^{2+} resistance. In general, high frequencies of antibiotic resistance are found in environments where antibiotic concentrations are high, e.g. hospitals and farms on which livestock are reared intensively. The apparent linkage of Hg^{2+} resistance stems, in part at least, from the finding that a family of large transposable elements encoding clinically significant antibiotic resistance, mediated via integrons, appears to have evolved from an ancestral element that encoded Hg^{2+} resistance. These transposable sequences seem now to be widespread, particularly among the Enterobacteriaceae (Grinsted et al. 1990). Mercury resistance linked to antibiotic resistance may also be perpetuated by the use of dental amalgams that incorporate mercury (Summers et al. 1993).

Mercury resistance is effected by reducing ionic mercury to the pure metal and releasing it in gaseous form (volatilization). The mechanism also involves active concentration of mercuric ions prior to reduction (Barkay et al. 2003). The complexity of the system suggests one of some antiquity rather than one of recent evolution.

Resistance to cadmium, lead, antimony, arsenic, tellurium, and silver compounds are plasmid-determined in staphylococci and gram-negative bacteria (Summers and Jacoby 1977; Summers and Silver 1978; Foster 1983; Silver and Phung 1996; Nies 2003; Silver 2003). Resistance to silver salts was first reported after the use in burns units of silver sulfonamide compounds (McHugh et al. 1975). Resistance to copper salts displayed by bacteria isolated on pig farms where copper is used as a feed additive has also been reported (Tetaz and Luke 1983; Brown et al. 1995). In these cases, the immediate selective conditions are obvious; however, when and where the resistance mechanisms evolved is another question. Resistance to heavy metal ions may be mediated by innate or acquired efflux pumps.

Plasmids may also change the sensitivity of their bacterial host to UV light and other mutagens (Mortelmans and Stocker 1976) and to bacteriocins (Pugsley and Oudega 1987).

The degradation of several toxic organic compounds, such as camphor, naphthalene, octane, and toluene, may be determined by plasmids, particularly in soil bacteria. These bacteria, often *Pseudomonas* or *Flavobacterium* spp., have evolved the ability to use these toxic compounds as nutrients. The use of these and similar bacteria to effect biological detoxification and reverse environmental pollution has been attempted, with some success, but this area of bacterial exploitation is still very much in its infancy (Shannon and Unterman 1993; Janssen et al. 1994; Spain 1995).

Virulence factors

Virulence determinants are genes, the expression of which is necessary for bacteria to establish and maintain an infection (see Chapter 6, Molecular basis of bacterial adaptation to a pathogenic lifestyle and Roth 1988; Dorman 1994; Salyers and Whitt 2002). They include colonization mechanisms, mechanisms to overcome host defense systems, and the production of toxins. All these functions may be encoded by plasmids (Brubaker 1985). For example, enterotoxin production in *E. coli* and *S. aureus* (Betley et al. 1986) and production of surface antigens by *E. coli* that allow colonization of the small intestine in specific mammals, including humans, can both be plasmid-determined (Brubaker 1985). At least one enterotoxin gene found in *E. coli* has been found on a transposon (Tn*1681*) (So and McCarthy 1980).

Strains of *E. coli* isolated from systemic infections in humans and pigs are more frequently hemolytic and more frequently produce colicin V than comparable fecal strains; both characteristics are plasmid-determined. Carriage of a ColV plasmid increases the virulence of *E. coli* strains for calves, chickens, and mice (Smith and Huggins 1976); however, the effect is not due to the colicin itself, but to other determinants encoded by the plasmid. The ColV plasmid encodes functions that facilitate the uptake of iron by the bacterial host (Williams 1979) and others that increase bacterial resistance to serum (Binns et al. 1979). Plasmids other than ColV also confer serum resistance on *E. coli* (Taylor 1983). In one instance this seems to involve the acquisition of an additional outer-membrane protein, which is a component of the plasmid transfer system (Moll et al. 1980). A plasmid that confers serum resistance on *Salmonella* Dublin has also been described (Terakado et al. 1988; Foster and Spector 1995).

The smooth to rough (S→R) variation in *S. sonnei* is accompanied by loss of the form I (S) antigen and loss of virulence. Kopecko et al. (1980) demonstrated that this conversion is determined by a large plasmid, loss of which results in the S→R change. Transfer of the plasmid to *S. flexneri* 2a or *Salmonella* Typhi generated derivatives that were agglutinable both by their own specific antisera and by antiserum specific to *S. sonnei* form I antigen. In *S. sonnei* the form I antigen, and hence the plasmid that encodes it, is one of the requirements for virulence (Chapter 54, *Salmonella*), as demonstrated in the Serény test.

Strains of *S. aureus* responsible for scalded-skin syndrome produce exfoliating toxins that, in some cases, are encoded by plasmids (Wiley and Rogolsky 1977). Some *Staphylococcus* spp. produce enterotoxins that cause the syndrome known as intoxicating staphylococcal food poisoning. Some of the toxin genes responsible for this syndrome are encoded by plasmids (Couch et al. 1988). Other examples of pathogenic properties determined by or associated with plasmids have been described and reviewed elsewhere (Elwell and Shipley 1980; Brubaker 1985; see also Dorman 1994; Salyers and Whitt 2002).

The role of R plasmids as possible virulence factors has been discussed by Elwell and Shipley (1980). Resistant bacteria, such as chloramphenicol-resistant *S.* Typhi, penicillin-resistant *N. gonorrhoeae*, or multiresistant *S. aureus* may sometimes appear to be of unusual virulence simply because the infections they cause do not respond to the first line of antibacterial therapy. This property certainly may facilitate dissemination, but does not necessarily imply increased virulence. Particular R plasmids may well have specific effects upon virulence, either increasing or decreasing it, depending on how expression of plasmid-encoded functions affects other virulence determinants. Indeed Lacey and Kruczenyk (1986) have argued that the plasmid-carrying MRSA, although posing an undeniable clinical problem, are not as virulent as their predecessors. However, generalizations in this area without rigorous experimental evidence are inappropriate and the spread of MRSA into the community would argue differently (Palavecino 2004).

In plants, *Agrobacterium tumefaciens* has long been known as the causative agent of crown gall, a tumorous disease that affects many plants and causes serious losses in fruit growing. The transformation to a tumor-like growth requires transfer from the bacterium to the plant cell of genetic information, part of which is incorporated permanently into the genome of the plant cell. The requisite information is encoded on large tumor-inducing (Ti) plasmids which mediate the transfer. This process is an elegant example of natural genetic engineering since the transformed plant cells produce compounds that can be used as nutrients by the *Agrobacterium* responsible for the transformation (Hooykaas 1989; Hooykaas and Beijersbergen 1994).

Metabolic characters

The metabolic diversity of the pseudomonads is due, in part, to the carriage of plasmids that specify degradation of unusual, often toxic, organic compounds, as already noted (see Harayama and Timmis 1989). Often, the metabolic genes are associated with, or are parts of, transposable elements (Tan 1999). 'Metabolic' plasmids have also been reported in other genera. When they are present in clinical isolates, such plasmids can initially complicate strain identification in the diagnostic laboratory. Lactose fermentation by strains of *Salmonella* (including Typhi) is usually plasmid-determined. Plasmids encoding lactose catabolism have also been found in isolates of *Proteus* and *Yersinia enterocolitica* (Falkow et al. 1964; Cornelis et al. 1976), and *Klebsiella* (Reeve and Braithwaite 1973), although in *Klebsiella* they are not prominent because the chromosome has a *lac* operon. In the study involving *Y. enterocolitica*, the particular genes were not only carried on a plasmid but were also transposable (Cornelis et al. 1978).

Genes for the utilization of other sugars, especially sucrose (*suc*) and raffinose (*raf*) may be plasmid-borne. Plasmids with both *lac* and *suc* genes have been found in *Salmonella* (Johnson et al. 1976), and *raf* genes are associated with those determining the K88 antigen adhesion factor which has been shown to be important in porcine *E. coli* enteritis (Shipley et al. 1978; see also Chapter 52, *Escherichia*). Plasmids determining H_2S production (Ørskov and Ørskov 1973), urease production (Farmer et al. 1977; Wachsmuth et al. 1979), or citrate utilization (Smith et al. 1978) have also been found in clinical isolates. Their preponderance is difficult to assess from phenotypic observation. However, modern molecular genetic techniques such as DNA probing, or PCR analysis, could possibly supply an answer, if needed. Such plasmids are not restricted to clinically important bacteria. Lac plasmids in *Lactobacillus lactis* cheese starter strains are of major importance to the dairy industry (Farrow 1980; see also Chapter 10, The microbiology of milk and milk products).

Phage inhibition

Carriage of certain plasmids can alter the susceptibility of the host to certain bacteriophages, thus changing the phage type of that particular isolate, which may be important in diagnostic and epidemiological investigations (Stanisich 1988). Usually the molecular basis for the change is unknown; however, some plasmids encode one of a class of DNA-degrading enzymes known as restriction endonucleases that fragment foreign DNA as it enters the bacterial cell, rendering it less susceptible to attack by some, if not all, phages. Such plasmids also encode the means to protect the host cell DNA, including the plasmid itself, from similar attack. This usually involves methylation of particular bases within specific short sequences which are also the sites recognized by the restriction endonuclease (Primrose et al. 2001).

Mapping of plasmids

Because plasmids are small, relative to bacterial chromosomes, they are popular molecules with which to investigate numerous aspects of DNA metabolism, including replication mechanisms (Scott 1984), recombination (Dressler and Potter 1982; Hsu and Landy 1984) and gene expression (Primrose et al. 2001; Thomas 2000). This interest has led to the development of an impressive range of techniques designed to study DNA in general and plasmid DNA in particular (see Grinsted and Bennett 1988; Sambrook et al. 1989). In the past, electron microscopy has been employed extensively to analyze DNA molecules, both as a simple tool to estimate plasmid size and, by means of heteroduplex analysis (Burkardt and Puhler 1988), to examine DNA rearrangements such as insertions, deletions, replacements, and inversions. Electron microscopy has, however, largely been replaced by restriction enzyme analysis and, more recently, by DNA sequencing (Grinsted and Bennett 1988; Brown 2001; Primrose et al. 2001). This last technology has now evolved to the point where DNA sequencing of very large DNA molecules (e.g. human chromosomes) has been achieved. The complete sequences of many bacterial chromosomes, including several human and animal pathogens (e.g. *E. coli*, *H. influenzae*, *Mycoplasma genitalium*, and *Helicobacter pylori*) are now available, as are the complete sequences of several plasmids. These data allow individual genes to be identified and ordered on the plasmid and then to be cloned for further study. By means of in vitro mutagenesis, such genes can be systematically and deliberately altered (Braman 2002; Primrose et al. 2001; see also Watson et al. 1992). These mutations can then be analyzed for their effects in the same way as if they had occurred naturally (Grinsted and Bennett 1988; Brown 2001).

Evolution and classification of plasmids

Plasmids are found in most, if not all, bacterial genera. Their origins are a matter for speculation. Each bacterial genus or group of genera appears to have its set of plasmids. Host specificities of plasmids have not been much studied, with one or two notable exceptions such as RK2 (RP4) (Thomas and Smith 1987) and RSF1010 (Barth et al. 1981; Buchanan-Wollaston et al. 1987). However, R plasmids of *S. aureus* can replicate in *B. subtilis* (Ehrlich 1977) and some R plasmids in *E. faecalis* can transfer by conjugation to *S. aureus* (Engel et al. 1980) and to lactobacilli (Gibson et al. 1979). Plasmids found in the Enterobacteriaceae can usually be transferred between many, if not all, the genera within the family. Some transfer more widely. In general, plasmids resident in gram-positive bacteria appear not to be able to establish themselves in gram-negative bacteria and vice versa.

In classifying plasmids, host range might be expected to be an important criterion, but in practice it is not a very convenient one. When a newly discovered plasmid is not transferred to a new host in laboratory experiments it may be for any one of a variety of reasons, of which an inability to replicate in the new host is one. In practice, a much more convenient criterion is available. As already stated, certain pairs of plasmids cannot coexist stably in the same bacterial cell, a phenomenon termed incompatibility (Novick and Richmond 1965), which is expressed when two plasmids have the same, or very closely related, replication systems or partition systems, or both (Novick 1987). A strong incompatibility reaction generally indicates similarity in the former.

Replication systems and partition systems are not mutually dependent, and perfectly stable replicons with the replication system of one plasmid and the partition system of another can be constructed in the laboratory. In nature, however, it is likely that a particular replication system and a particular partition system will be found in association. Thus, two plasmids with essentially the same replication system are also likely to have related partition systems, reinforcing any incompatibility reaction.

When each plasmid in a collection is tested, in turn, against all others in the set, those which are incompatible can be identified and so the collection can be ordered into incompatibility (Inc) groups. This means of classification has been applied to plasmids in the Enterobacteriaceae, particularly *E. coli* (Jacob et al. 1977), in *Pseudomonas* spp. (Jacoby 1977), and in *S. aureus* (Novick et al. 1977). Plasmids that can replicate in both *E. coli* and *Pseudomonas* spp. are included in both classifications (Hedges and Jacoby 1980), for example, RP4 is IncP when in the Enterobacteriaceae but IncPI when in *Pseudomonas* spp.

Because it principally reflects the primary plasmid maintenance systems (replication and partition), incompatibility is a sound basis for grouping plasmids, and there is good evidence that it reflects evolutionary relationships. Within any one Inc group, plasmids, in general, show extensive DNA homology, i.e. common sequences are not restricted simply to replication and partition functions. By contrast, plasmids of different Inc groups rarely show extensive homology. Exceptions are plasmids of the IncF complex, which can be grouped into at least five Inc groups but which nevertheless display extensive homology because their transfer systems are closely related.

Plasmids that belong to the same incompatibility group in general encode the same or similar sex pili. By contrast, plasmids belonging to different Inc groups are more likely to encode sex pili that differ morphologically and serologically. Sex pili can act as receptors for bacteriophages. One practical consequence of this is that pili can often be detected because they permit specific bacteriophage adsorption (Figure 4.13). However, because of the normally repressed nature of a number of plasmid transfer systems it may be necessary transiently to derepress a specific system before its identity can be established by the use of specific bacteriophages (Willetts 1988). In electron microscopy studies of their morphology and serology, Bradley (1980) identified pili determined by conjugative plasmids of many known Inc groups of *E. coli* and several of *P. aeruginosa*. His work shows that, with very few exceptions, all conjugative plasmids within an incompatibility group determine the same type of pilus, a finding consistent with conclusions from DNA homology studies.

Plasmids of the same incompatibility group often share a similar host range, show considerable DNA homology, are of similar molecular sizes, and, with conjugative plasmids, possess related transfer systems. By contrast, other plasmid-encoded properties, in particular antibiotic resistance, are not indicative of a particular incompatibility group. Most antibiotic resistance genes, which can be identified either as DNA sequences or as their protein products, are found on a wide variety of otherwise unrelated plasmids. It seems that present-day plasmids have evolved, and continue to evolve, by the accretion of DNA

sequences of transient or lasting value to bacterial cells. These sequences are often components of bacterial transposons and integrons (Wiedemann et al. 1986; Hall and Collis 1995; see also Levy and Novick 1986; Berg and Howe 1989; Thomas 2000).

It is often stated that R plasmids consist of a resistance transfer factor (RTF) (a conjugative plasmid component) linked to a resistance determinant (r-det). The reason is historical (see Novick 1969; Helinski 1973) because the earliest studies of R plasmids involved members of one incompatibility group, IncFII, which display precisely this form. In the light of more diverse studies, it can be seen that the description is specific rather than general and many R plasmids have structures that do not conform to this simple compartmentalized scheme, e.g. RP1 (Figure 4.14) (Gorai et al. 1979; Villarroel et al. 1983; Thomas 1989).

BACTERIAL VARIATION AND EPIDEMIOLOGY

In epidemiological studies, the aim is to identify a particular clone within a bacterial species in order to trace its spread and, if possible, to determine its origin. In the past, biochemical and serological characters, phage sensitivities and lysogeny, production of or sensitivity to bacteriocins, sensitivity or resistance to antibiotics and other antibacterial agents, and plasmid profiles have all been exploited to this end. These characters are usually sufficiently stable to be useful when tracing bacterial strains over relatively short periods, but it should be remembered that none of them is immutable and several are encoded by extrachromosomal elements which may be lost over a period of time. Furthermore, new characters can be acquired by acquisition of a plasmid or a transposon. The isolation of variants indicates either genetic change in an epidemic strain or the discovery of a different strain of the same species. It is not always possible to distinguish between these alternatives (Seal et al. 1981), but the more characters used to identify the strain, the greater the chance that these alternatives can be distinguished. Today, molecular genetic techniques such as pulse-field gel electrophoresis (Gardiner et al. 1986; Morrison et al. 1999; Arakawa et al. 2000; Cohen et al. 2001) and multilocus sequence typing (Enright and Spratt 1999; Clarke 2002; Urwin and Maiden 2003; Cooper and Feil 2004) are assuming greater prominence in strain typing. However, these techniques are still technically demanding and relatively expensive, so serological, biochemical, and phage typing systems are still commonly used in the diagnostic laboratory.

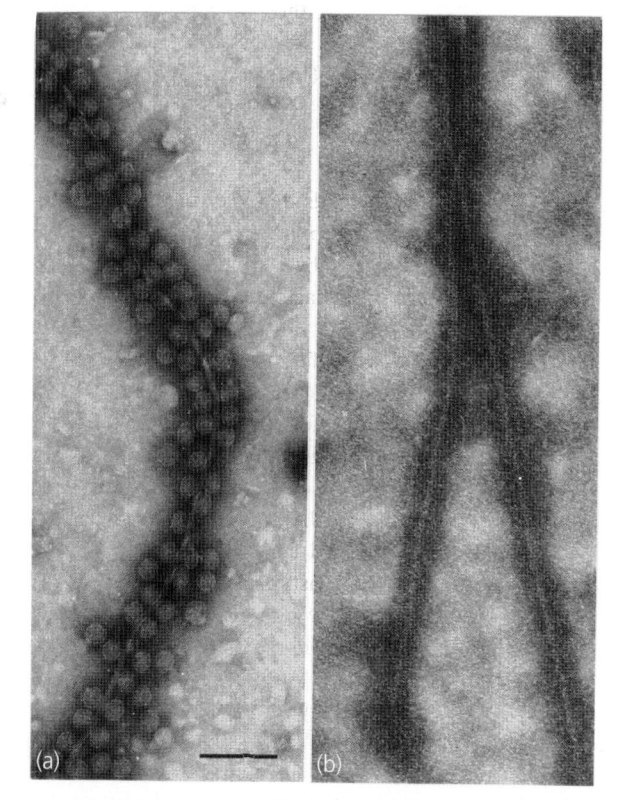

Figure 4.13 *Conjugative pili labelled with phage or antibody.* **(a)** *F pilus labelled with particles of phage R17. (Reproduced with permission from Bradley 1977).* **(b)** *D pilus labeled with specific antiserum (D.E. Bradley, unpublished). Bar = 100 μm*

Chromosome analysis

With the development of automated high throughput DNA sequencing systems, it has become possible to

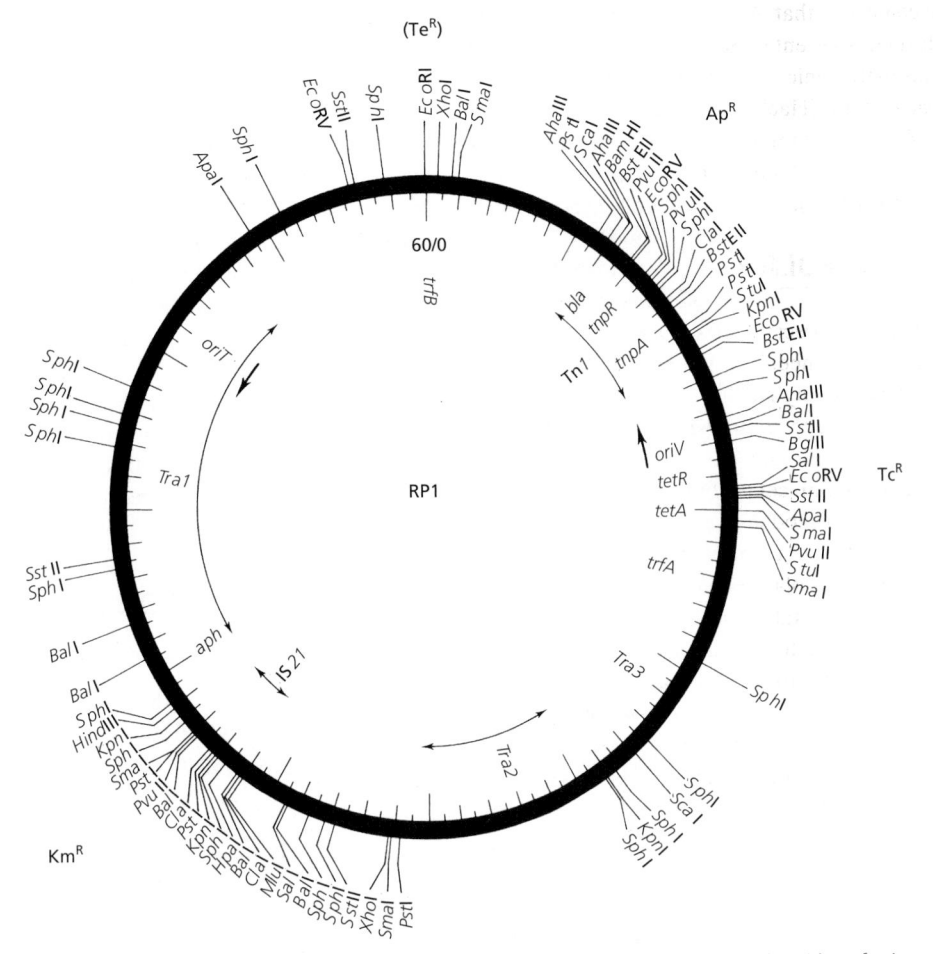

Figure 4.14 *Restriction and genetic maps of RP1. Plasmid RP1 is a 60 kb IncP conjugative resistance plasmid conferring resistance to ampicillin (Ap), kanamycin (Km), and tetracycline (Tc) that was isolated from a strain of* P. aeruginosa. *Genetic loci: aph, aminoglycoside phosphotransferase (resistance to kanamycin and neomycin); bla, TEM-2 β-lactamase (resistance to β-lactams including ampicillin and carbenicillin); tet, tetracycline resistance genes; oriT, origin of conjugative transfer; Tra, genes required for conjugative transfer; oriV, origin of replication; trf, plasmid replication genes; tnp, transposition genes of Tn1 (Figure 4.5). Te^R indicates the position of a cryptic gene that confers resistance to tellurite when active. The scale indicates plasmid coordinates in kb. The map is redrawn, with permission, from Thomas and Smith (1987).*

sequence entire bacterial chromosomes in a matter of months. Activity in this field has been intense in the last few years and now more than 200 have been completed, with more than 500 in progress (www.genomesonli-ne.org). Analyses of these sequences has led to the realization that most, if not all, bacterial chromosomes are mosaic structures (Lawrence and Roth 1999), comprising a backbone sequence, in which all the genes appear to derive from a common phylogenetic origin, and multiple inserted sequences from diverse sources, such as IS elements, transposons, ICEs, bacteriophage genomes, and genomic islands acquired by horizontal gene transfer.

A subset of genomic islands, detected first in human pathogens and then subsequently in animal and plant pathogens, contains genes for bacterial virulence. These genomic islands are called pathogenicity islands (PAI) (Hacker and Kaper 2000). The origins of some of these are more evident than others in that they are clearly

lysogenic phages or their remnants. Examples of virulence genes that are phage components are the cholera toxin genes of *V. cholerae* and the shiga toxin genes of *S. dysenteriae* (Waldor and Mekalanos 1996; Waldor 1998). PAIs encode a diversity of virulence determinants, such as adherence and invasion factors, iron uptake systems, protein-secretion systems, as well as bacterial toxins. Such genomic islands tend to be relatively large sequences (10–200 kb) and are often flanked by short direct repeats reminiscent of the result of site-specific recombination or transposition (Davis and Waldor 2002). Indeed, among the gene complement of a PAI are often found genes encoding putative transposases and integrases. A few PAIs have been shown to be mobile, and in some cases movement has undoubtedly been phage mediated, e.g. PAIs found in *S. aureus* and *V. cholerae* (Lindsay et al. 1998; Karaolis et al. 1999). In the case of the former organism, capture of the PAIs called SCCmecs generates MRSA strains (see above).

It is worth comment that it is possession of different PAIs that distinguishes enterohemorrhagic, enteropathogenic, and uropathogenic *E. coli* from avirulent strains and from each other (Hacker and Kaper 2000). For a more detailed and elegant discourse on the molecular genetics of a variety of human bacterial pathogens, see Salyers and Whitt (2002).

GENETIC MANIPULATION IN VITRO

Molecular genetics, especially recombinant DNA technology (or genetic engineering), provides powerful tools with which to analyze any genetic system. These techniques make it possible, in principle, to isolate any gene from any cell type and manipulate it so that it can be expressed in a bacterial cell such as *E. coli* and then return it, perhaps in a modified form, to the cell line of origin. Gene cloning is undertaken for two primary purposes: to allow the isolation of a desired fragment of DNA in quantities suitable for sequencing, and to enable the experimenter to insert a gene of interest into a suitable cell line where its expression can be studied or in which the gene product is produced in large amounts, so facilitating its isolation and purification.

The bacterial chromosomes of *E. coli* and similar bacteria contain almost 5×10^6 bp. In a human cell the DNA is packed into 46 chromosomes, each containing about $3\,000 \times 10^6$ bp. The problem is, therefore, how one gene a few thousand bp in length can be recovered from such a mass of DNA. Until the mid-1970s such a feat was not possible. This situation was changed by the discovery of the class of DNA endonucleases (type II restriction enzymes) (Primrose et al. 2001) that cleave double-stranded DNA at specific nucleotide sequences, cutting it into much smaller linear pieces that can then be recovered and identified. The target sites for these enzymes normally comprise 4, 5, or 6 bp of specific sequence and are usually rotationally symmetrical, often termed palindromic because the sequence of the site read in the 5′ to 3′ direction on one strand is the same as that read in the 5′ to 3′ direction on the opposite strand (Figure 4.15). There are two ways in which these enzymes cut double-stranded DNA. Some cut at positions on the DNA strands opposite each other, giving 'blunt' or flush ends. Others give a staggered cut, producing short single-stranded extensions (Figure 4.15) that, because they are generated from palindromic sites, are the same. Fragments generated by type II restriction enzymes (restriction fragments) can be joined together in vitro by DNA ligase. Ligation of fragments with single-stranded extensions generated by the same restriction enzyme proceeds more efficiently than ligation of fragments with flush ends because the complementary nature of single-strand extensions, sometimes called 'sticky ends,' encourages fragments to align end to end, so facilitating the ligation reaction. The most commonly used ligase commercially available is that encoded by

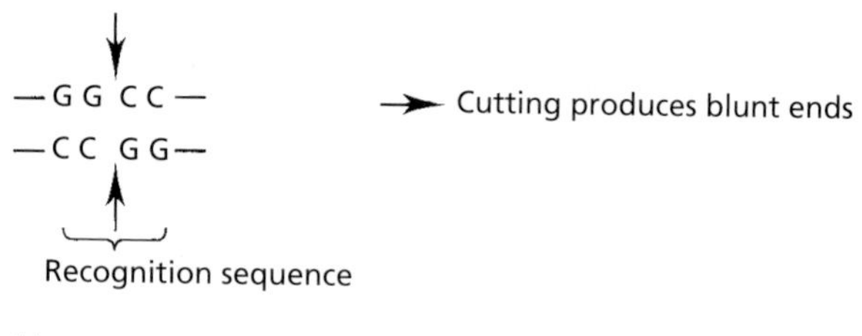

(a)

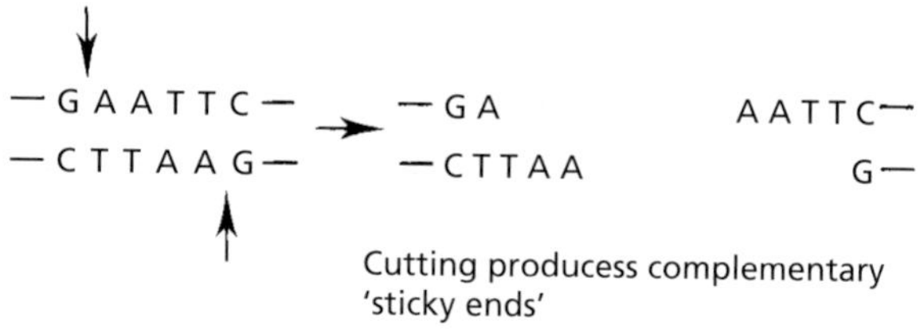

(b)

Figure 4.15 *Type II restriction enzymes.* **(a)** Hae*III from Haemophilus aegyptius.* **(b)** Eco*RI from E. coli RY13.*

phage T4. After manipulation in the test tube, desired recombinant DNA molecules are recovered by transformation of a suitable bacterial host. The formation of novel hybrid DNA molecules in vitro, coupled with their recovery and amplification in bacteria, is termed cloning (for a more detailed exposition, see Brown 2001; Primrose et al. 2001).

Cloning vectors

To be perpetuated, cloned genes must become part of a DNA molecule that is replicated in each division cycle. There are two convenient types of genome that can be used for cloning – plasmids and bacteriophages. In bacterial genetics the most commonly used plasmid cloning vectors are small, multicopy plasmids. Many are derivatives of two early vectors, pBR322 and pACYC184 (see, for example, Thompson 1988), constructed in the laboratory from small natural plasmids, pMBI and pl5, respectively. They have been manipulated to possess several desirable features not present on the original elements, such as a choice of suitable cloning sites (preferably associated with insertional inactivation) and a convenient and reliable marker for selection (usually a drug resistance gene); for example, one of the first cloning vectors, pBR322, has approximately 4 200 bp and single sites for a number of restriction enzymes, including *Bam*HI, *Hin*dIII, *Pst*I, and *Sal*I. The single *Pst*I site is within the ampicillin resistance gene which encodes a β-lactamase. Cloning into this site results in the loss of ability to confer resistance to several β-lactam antibiotics, including ampicillin, i.e. cloning into this site causes insertional inactivation of the Amp resistance determinant. Hence, clones containing recombinant plasmids are recognized because they acquire resistance to tetracycline but not to ampicillin. Similarly, cloning into the single *Bam*HI, *Hin*dIII, or *Sal*I sites creates recombinant plasmids that confer resistance to ampicillin but not to tetracycline.

The next incarnation of pMB1-derived cloning vectors, and one that has stood the test of time, up to the present day, was the pUC series of vectors, whose qualities are best displayed by pUC18, a vector commonly chosen for shotgun cloning for genomic sequencing projects. The size of the vector was reduced by removing the tetracycline resistance gene from pBR322, together with the *rop* region, the function of the latter being to control plasmid copy number. This single change increased the copy number more than 10-fold, greatly facilitating preparation of large amounts of the vector, and improved the transformation efficiency. Then a copy of the sequence encoding the N-terminal 145 amino acid peptide of LacZ (β-galactosidase), the alpha- (α-)peptide, that had previously been subject to in vitro mutagenesis to introduce a set of unique restriction endonuclease sites into the α-peptide coding region, the multiple cloning site (MCS), while retaining α-peptide function was inserted. In a host that produces the remainder of the β-galactosidase protein, the omega- (ω-) peptide, α-complementation occurs to give a functional β-galactosidase. The basis for the screen is the ability of β-galactosidase to hydrolyze the chromogenic substrate 5-bromo-4-chloro-3-indolyl-β-D-galactopyranoside, known as Xgal, to produce a deep blue, poorly diffusible product. When α-complementation occurs, i.e. the cell produces both the α- and ω-peptides, which combine to create functional β-galactosidase, the colonies are dark blue on agar containing Xgal. In contrast, cells in which there is no α-complementation fail to produce β-galactosidase and the colonies are white on Xgal-containing media. This allows ready identification of vectors containing inserts, because cells carrying such constructs fail to produce a functional β-galactosidase and the colonies are white. The process is referred to as blue-white selection, but it should be noted that the selection refers to experimenter choice, not use of a selective medium. A major strength of the system is that the distinction between blue and white can easily be made on the same plate, so no replica plating is required. Indeed, during construction of a genomic library using pUC18, blue/white screening is used simply to confirm that a high proportion of vectors in the library have inserts. One further development was the replacement of the ampicillin resistance marker of pUC18 with a kanamycin resistance gene to give pK18 (Pridmore 1987) for use when ampicillin selection is inappropriate.

Phages and genetically modified variants are also used extensively as vectors in genetic modification (cloning) experiments in *E. coli* and other organisms (Primrose et al. 2001). Two that have proved useful in this respect are the single-stranded DNA phage M13 (Messing 1991) and the double-stranded DNA phage λ (Brammar 1982), both parasites of *E. coli*.

Phage M13 has a 6.4 kb genome in which a 507 base intergenic region can be used for the introduction of up to 1.5 kb of cloned material. The great value of this particular phage is that, although it replicates as a double-stranded element that can be introduced into a host cell in the same way as plasmids are introduced by transformation (a process termed 'transfection'), its DNA, when packaged, is single-stranded – an ideal element for DNA sequencing and some mutagenesis procedures in vitro (Messing 1991).

Phage λ has a much larger genome (52 kb) than M13 (Arber 1983; Brammar 1982) and, by deleting the inessential integration functions and also sequences in the b2 region, it is possible to accommodate up to 25 kb of cloned DNA into the residue of the phage genome. One consequence of the large genome size is that there are several copies of all the common restriction endonuclease cut sites, so a further step in the design of λ cloning vectors was to remove all but one or two of each; this was accomplished by in vitro mutagenesis and also by selection of variants grown in a host that forms the endonuclease concerned. Two types of λ vector are in common use. Insertion vectors have a unique site (usually *Eco*RI) into which the DNA is cloned, thus inactivating the gene in which the site lies and allowing detection of the vector whose gene is inactivated; λ NM607 has such a site within the *cI*+ gene which, when inactivated, eliminates the immunity function and gives clear plaques (Brammar 1979), and members of the λgt11 series (λgt11–λgt23) and λZAP have that part of the *lacZ* gene which encodes the alpha fragment of β-galactosidase and can thus be used in blue-white plaque selection using X-gal (Chauthaiwale et al. 1992). In replacement vectors there are two endonuclease-sensitive sites and the region lying between them is replaced by the material to be cloned (e.g. the λcharon series 4A, 32, 35, 40, and λEMBL3, λEMBL4) (Chauthaiwale et al. 1992). The efficiency of the subsequent transfection step can be improved by utilizing a packaging procedure in vitro rather than introducing the entire vector molecule into its host (Primrose et al. 2001). A further development of λ as a cloning agent is the cosmid (Collins and Bruning 1978). This is a hybrid between λ and a plasmid, in which λ contributes the region containing its cohesive ends and the plasmid supplies an origin of DNA replication together with a gene that allows selection of the desired clone (Wahl et al. 1987). Cosmids do not contain the λ sequences needed for DNA replication and particle maturation and hence there is room for far more DNA to be cloned (up to about 40 kb). The recombinant molecules are packaged in vitro to produce transducing particles which are used to deliver the vector and its insert into a suitable strain of *E. coli*. The cosmid system has been used to construct genomic libraries of several organisms, a set of recombinant clones that between them contain all the DNA present in an individual organism (see, for example, Goodman et al. 1995; Jacobs et al. 1991; Little 1992). Among the phages specific to other organisms, the λ-like φC31 of *Streptomyces* has been used as the progenitor of several cloning vectors (see Pritchard and Holland 1985; Rodicio et al. 1985; see also Primrose et al. 2001).

Specialist plasmid cloning vectors

If a vector has been engineered to carry two origins of replication to allow it to be maintained in at least two different organisms, then it is called a shuttle vector, i.e. one origin supports replication in one cell type while the second supports replication in the other. Such vectors usually carry an origin of replication appropriate to *E. coli* (e.g. from pMB1), a selectable marker such as ampicillin resistance, also appropriate to *E. coli*, together with the *lacZ* α-sequence and multiple cloning site from pUC18, and a selectable marker for the second cell type. The point of incorporating these features is that recombinant plasmids, produced in vitro, can first be recovered by transforming *E. coli*, so maximizing transformation efficiency. Then, large amounts of the correct recombinant plasmid can be purified and used to transform the other target host. Using shuttle vectors, genes can easily be recovered in *E. coli* and then introduced into, for example, a human cell line (where the second origin of replication on the cloning vector often comes from the monkey virus, SV40, which can replicate in human cells), or plant or yeast cells. The purpose of shuttle vectors is to exploit the plethora of molecular genetic techniques that can be used with *E. coli* systems to recover and manipulate DNA sequences before reintroducing the particular sequence into the original cell line or an appropriate surrogate, where the effects of mutation or overexpression can be determined. When coupled with the use of expression systems (below), shuttle vectors can be used to overexpress a protein in a different background, when overexpression in *E. coli* is problematic.

Fulfilling a similar role to shuttle vectors, but generally limited to bacterial systems, broad host range bacterial vectors have a single origin of replication that can be used in many phylogenetically diverse bacteria. Broad host range vectors were originally derived from naturally occurring broad host range bacterial plasmids, which encode their own replication proteins that recognize their own origins of replication (see, for example, Bagdasarian et al. 1981; Morales et al. 1991). As long as the genes encoding the vector-specific replication proteins can be expressed in the target host, the vector will replicate. Broad host range vectors are, by necessity, larger than pUC18, but they have been modified over the years to reduce their size and in some cases, the origin of replication, *lacZ* α-peptide sequence and multiple cloning site from pUC18 have been added to facilitate initial manipulation of the vector in *E. coli*, so optimizing recovery of the sequence of interest, since many potential target bacteria are difficult to transform. Then, sufficient quantities of the correct recombinant plasmid can be produced to facilitate transformation of the final host. In some instances the recombinant plasmid can be mobilized into the final recipient by conjugation from the *E. coli* donor because the vector has also been supplied with an origin of transfer and the functions specific to it. The bulk of the transfer system is usually provided by a helper conjugative plasmid also present in the transformant.

Plasmid vectors for recovery of PCR amplicons

PCR is now the method of choice if what is needed is simple amplification of a DNA fragment. However, PCR cannot easily be used to amplify a section of DNA if the sequences at the ends of the fragment to be amplified have not been determined. Furthermore, to sequence a PCR amplicon is more difficult than to sequence the insert of a recombinant plasmid, particularly if it is more than 1 kb in length. Thus, PCR is generally used to amplify specific, small, known sequences of DNA, for surveys of sequence heterogeneity, and for more complex in vivo techniques, such as allelic replacement by homologous recombination (see below). As a preliminary to these more complex techniques, it is firstly necessary to insert the PCR amplicon into a cloning vector. PCR amplicons are linear, double-stranded molecules, and those produced using the most common form of *taq* DNA polymerase carry single adenine overhangs at the 3′ end of each DNA strand, generated by the inherent terminal transferase activity of DNA polymerase. These adenine residues complicate the task of inserting the PCR product (amplicon) into a cloning vector because they first have to be removed before the fragment can be blunt-end ligated into the vector. However, cloning vectors based on pUC18 are now available where the multiple cloning site contains a restriction site that, when cut, generates single T overhangs at the 3′ end of each strand (for example, see Marchuk et al. 1991). Thus, the PCR amplicon can be ligated directly into the vector, without the need to remove the A overhangs from the amplicon. Once ligated into the 'TA cloning vector,' for example pCR2.1, and replicated in *E. coli*, the insert can subsequently be removed by cleavage at other restriction sites within the multiple cloning site and subcloned into a more specialist vector.

Plasmid vectors for overproduction of proteins in vivo

Modern biochemical studies and biotechnological processes often require large amounts of a particular protein, and the most efficient way to produce proteins in large quantities is to overexpress recombinant genes in vivo. As with so many other molecular biology processes, *E. coli* was the first to be investigated as a surrogate host for protein production. Gene expression in *E. coli* requires an appropriate promoter sequence to allow transcription of the gene into mRNA, a ribosome binding sequence (RBS) to allow translation of the gene transcript, and an ATG start codon to initiate translation. It is sometimes possible to express a recombinant gene in *E. coli* simply by inserting it, together with its natural promoter, RBS and ATG start codon into a standard cloning vector such as pUC18 and transforming the construct into *E. coli*. In this case, the fact that pUC18 has a copy number of 50–100 relative to that of the *E. coli* chromosome means that the gene of interest can be expressed at levels well above that of a single copy gene. However, this approach is somewhat hit and miss. For many reasons, the expression of a cloned gene may not be optimal in the surrogate host. Because of phylogenetic drift, promoters, RBS sequences, and initiation codons associated with genes not native to *E. coli*, particularly those from eukaryotes and very different bacteria, may not be recognized by the *E. coli* RNA polymerase and ribosomes.

If the promoter or RBS associated with a cloned gene is not active in *E. coli* then, irrespective of the copy number of the recombinant vector, the gene will not be expressed. In such cases, a specialist expression vector can be used, for example one of the pET series of vectors, to provide an optimal *E. coli* promoter, an RBS, and, when necessary, an initiation codon upstream of the inserted gene to promote high-level expression (Studier et al. 1990).

Perpetual high-level expression of a recombinant gene in *E. coli* is rarely successful. For the bacterium to divert large amounts of energy into the production of a single protein tends to be debilitating to the organism as a whole. Furthermore, the presence of large quantities of 'foreign' protein often results in it being sequestered in insoluble aggregates or degraded, thus reducing overall yield. In some cases, the foreign protein directly disrupts the cell's normal physiology, and is said to be toxic. If toxicity is extreme, it may not be possible to transform the surrogate cell with a recombinant plasmid from which the gene encoding the toxic protein is constantly expressed. To help overcome the problems noted above, and to maximize protein yield, it is necessary to control gene expression. This is usually achieved by using an inducible promoter to drive transcription of the cloned gene. This strategy ensures a minimal level of transcription, and hence protein production, during normal cell growth, and allows the cell line to be cultured, but provides the means rapidly to increase production of the protein when the culture has attained a suitable cell density, thus generating large amounts of the protein. That this will also kill the cells is not important, because they are necessarily killed when the protein is harvested.

Many different transcriptional regulatory systems have been characterized, which it would be possible to adapt for incorporation into an expression vector. However, two inducible promoters dominate the field and their regulatory systems were among the first to be discovered. The most common is a derivative of the *E. coli lac* operon promoter, called *ptac*. This is a high efficiency, naturally generated hybrid promoter which is controlled by the *lac* operon repressor, LacI (de Boer et al. 1983). In the absence of an inducer (usually the gratuitous inducer isopropyl-β-D-galactopyranoside (IPTG)) promoter activity is shut off. However, repression is not complete and some escape synthesis of mRNA does occur, resulting in low-level expression of whatever gene is coupled to the promoter. For many cloned genes this level of control is adequate, but for a few even a low level of expression is potentially damaging. In these cases an alternative expression system is necessary.

Some expression vectors (e.g. the pET series) include a stylized promoter, which is recognized by the RNA polymerase of the *E. coli* bacteriophage T7 and not by the *E. coli* RNA polymerase. Regulated expression from this promoter is achieved by regulating production of the T7 RNA polymerase itself. To this end, an *E. coli* cell line carrying the T7 RNA polymerase gene, expression of which is driven by the *ptac* promoter, is used, e.g. BL21 cells (Dumon-Seignovert et al. 2004). Growth of BL21:pET in nutrient broth without IPTG allows little or no expression of the gene cloned into pET, because, whilst repression of transcription at the *ptac* promoter controlling production of the T7 RNA polymerase is leaky, the level of T7 RNA polymerase in the cell is never sufficient to activate transcription of the cloned gene from the T7 promoter. T7 RNA polymerase production is switched on when desired by addition of IPTG to the growth medium, when expression of the cloned gene is also triggered. To enhance production of the desired protein, once the cell has produced T7 RNA polymerase, the antibiotic rifampicin can be added to the medium. This inhibits the *E. coli* RNA polymerase, but not T7 RNA polymerase. So, whilst addition of rifampicin stops any further production of T7 RNA polymerase, because it blocks transcription of all *E. coli* genes by inhibiting the cell's own RNA polymerase, it stimulates transcription by T7 RNA polymerase by eliminating a major competitor for the substrates of RNA synthesis. With this strategy the cell is converted to one that synthesizes only the protein encoded by the cloned gene driven by the T7 promoter.

The second major regulatory system to be exploited for cloning vector development is the Ara system of *E. coli*, which is involved in metabolism of the sugar arabinose (Lee et al. 1974). In this system, a transcriptional activator, AraC, acts on the promoter controlling expression of the *araBAD* operon to enhance RNA polymerase binding and so promote transcription of the operon. Binding of AraC close to the promoter is absolutely dependent upon arabinose. When there is no arabinose in the cell, the *araBAD* promoter is silent. Genes cloned downstream of the *araBAD* promoter in expression vectors incorporating the *araC* gene, e.g. pBAD (Guzman et al. 1995) are only expressed in the presence of arabinose, and expression is completely shut off in the absence of the sugar, thus minimizing problems associated with toxic protein production.

High-level expression of some proteins in *E. coli* can be difficult to achieve, particularly if the proteins are normally functional in a eukaryotic cell. One reason for this is that eukaryotic proteins are often modified post-translationally; for example, they are processed to remove signal tags associated with protein targeting, or they have molecules added to their backbones, e.g. glycosylation. Without the correct modifications, the protein may lack activity and may even be degraded because it cannot fold properly. The solution to this problem is to express the protein in a eukaryotic host that can perform the required modifications. For this, a shuttle vector is used, which has the necessary promoter and translation signals for expression in the surrogate host, such as insect cells or *Xenopus* oocytes.

Plasmid vectors that allow targeted modification of chromosomal genes

Once a DNA sequence encoding a particular protein has been identified, it is possible to make targeted changes to the DNA, and then determine the effects of such changes on the activity of the protein product. Targeted changes include deletions, truncations, or point mutations in the gene, or in its promoter/operator sequences. Mutations can easily be introduced using PCR technology, as described earlier in the chapter. However, in order to determine the biological significance of a particular mutation, it is necessary to put the mutant gene into the original organism. If the mutant gene phenotype is dominant over that of the chromosomal wild-type allele, this experiment can be performed simply by transforming the original cell line with an appropriate construct carrying the mutant gene. However, in most cases, mutant alleles are recessive, even when present in multiple copies, and their presence is masked by the dominant wild-type chromosomal copy. In this case, the wild-type allele must be replaced with the mutant allele, a procedure termed allelic replacement, using homologous recombination.

Gene replacement by homologous recombination

Homologous recombination involves exchange between two very similar sequences of DNA (Eggleston and West 1996). If a gene knockout is needed, a modification of site-directed mutagenesis can be used to avoid the

need to clone the gene of interest when preparing DNA fragments for allelic gene replacement. Multistage PCR is used to create a DNA fragment carrying a version of the gene with a suitably located restriction site (see above). The fragment is cleaved with the restriction endonuclease and the two pieces are ligated to a third fragment encoding an antibiotic resistance gene. This ensemble can then be amplified by PCR to produce ample DNA to use directly to transform the cell line of interest, selecting for antibiotic resistant transformants. To maximize the frequency of transformation, electroporation is used to introduce the mutant allele of the gene of interest into the cell. The desired transformants are created by a double crossover between homologous sequences on the introduced DNA fragment and on the bacterial chromosome, one within each gene segment flanking the antibiotic resistance gene (Datsenko and Wanner 2000). This recombination exchanges a section of the chromosomal gene for its equivalent on the fragment including the interpolated antibiotic resistance gene. The drawback to this technique is that most bacteria are not easily transformed by linear DNA fragments. The fragments tend to be degraded rapidly once inside the cell. To circumvent this problem, the DNA fragment can be circularized before it is used to transform the cell. One way to do this is to clone the mutant allele and then use the recombinant plasmid to effect the transformation. When such a construct is put into a cell, then a single crossover within the sequence common to the recombinant plasmid and the chromosome will fuse the recombinant plasmid with the chromosome, generating in the process a direct duplication of the homologous sequence. A second crossover within the duplicated region, but at another point, will reverse the integration, leaving behind in the chromosome part of the cloned fragment while the equivalent part of the chromosomal copy becomes part of the excised plasmid, thus transferring the mutant allele to the chromosome. However, the system only works efficiently in practice if the recombinant plasmid is unable to replicate in the cell to be mutated (otherwise, excisions that essentially regenerate the mutant allele on the free recombinant plasmid and reconstitute the original chromosomal copy of the gene cannot be distinguished phenotypically from the desired mutants). This problem has been solved by the development of suicide vectors, i.e. cloning vectors that can be propagated only in special cell lines (permissive) and which otherwise are not maintained as extra-chromosomal elements (Datsenko and Wanner 2000).

Suicide vectors

Because suicide vectors are designed not to be maintained in most cells, when the DNA is electroporated into cells it will persist simply as nonreplicating circles of DNA, unless rescued by integration into one of the replicons in the cell. There are many such vectors, which all work in a similar manner, for example pKNG101 (Kaniga et al. 1991) and the pEX18 series of suicide vectors (Hoang et al. 1998). Each of these vectors carries a multiple cloning site, which allows insertion of a mutant allele of the gene of interest. They also possess the R6K origin of replication, ori_{R6K}, which will only function in cells that possess the R6K *pir* gene (Kolter et al. 1978), usually provided in *E. coli* by lysogenization with a specialized λ*pir* transducing phage. In addition, each vector is provided with a selectable marker gene, usually an antibiotic resistance gene, and a copy of the *B. subtilis sacB* gene. This last feature allows selection of derivatives that have lost the suicide vector, because expression of *sacB* is lethal when cells are exposed to sucrose. Finally, each also carries the transfer origin of the conjugative resistance plasmid RK2, which allows the construct to be mobilized easily by conjugation into the target cell line.

When the target host (nonpermissive) is mated with a *pir+ tra+ E. coli* (permissive host) carrying the pEX18 recombinant and a conjugative helper plasmid, the suicide vector and its passenger gene is transferred to the recipient. However, the antibiotic resistance encoded by the plasmid is only acquired if the plasmid is maintained, which is not possible if the plasmid remains autonomous. Accordingly, in order to be replicated, the cloning vector must be rescued by integration into the genome of the recipient cell. This normally occurs by homologous recombination between the cloned gene on the pEX18 recombinant and its homolog on the chromosome. This integrates the plasmid into the chromosome generating an arrangement where the suicide vector is sandwiched between two copies of the target gene, one of which is disrupted by the antibiotic resistance gene cassette. This first recombination event is rare, but because of the plasmid marker, the small number of bacteria carrying the integrated plasmid can easily be selected.

Derivatives with the chromosomally integrated pEX18 recombinant plasmid are then cultured in antibiotic-free medium, if necessary for several passages, to allow a second crossover between the two copies of the target gene. This excises the vector and one copy of the target gene from the chromosome. The excised sequence is lost as the cell multiplies, leaving a single chromosomal copy of the target gene. In a proportion of the derivatives the remaining copy of the target gene will be disrupted by the antibiotic resistance gene cassette. These are easily selected on medium containing sucrose, which kills all those cells that still carry the suicide vector, and the appropriate antibiotic, which kills all those that no longer possess the antibiotic resistance cassette used to disrupt the target gene.

One problem with this technique is that the mutation marker, i.e. the antibiotic resistance gene, remains in the chromosome, so the same antibiotic resistance gene cassette cannot be used to disrupt another gene in the same bacterium. To counter this problem, resistance cassettes have been developed that are flanked by specific recombination sites, so that the antibiotic resistance gene cassette can be excised by the appropriate site-specific recombinase. An inducible form of the gene for this enzyme is introduced into the cell on a helper plasmid and once the allelic replacement has been effected, expression of the recombinase is induced. Removal of the cassette is achieved in such a way that a frameshift mutation is generated in the gene of interest (Datsenko and Wanner 2000).

The technique can also be used to introduce mutations other than antibiotic resistance gene cassettes. Mutant genes are created by site-directed mutation as normal and the fragment inserted into an appropriate suicide vector. Then, the only modification to the final stages is that selection is for sucrose-resistant derivatives only. These are then screened for those carrying the mutant allele of the gene, because only approximately one in two will carry the mutant allele. Screening by change of phenotype may be possible; alternatively, the target gene is recovered by PCR amplification from the chosen clones, the fragments are sequenced and the mutants identified. The major advantage of this approach is that it can be used repetitively to introduce targeted point mutations into different genes of the same cell line or multiple mutations may be introduced sequentially into the same gene.

Transposon-delivery vectors for random insertional inactivation of chromosomal genes

Throughout the history of bacterial molecular genetics, the isolation of loss of function (null) mutants with easily detected phenotypes has been used to study the genetic basis of the particular phenotype. The main reason for selecting null mutants is to locate the mutated gene, and to confirm a role for that gene in a particular phenotype. A classic example is the selection of *E. coli lacI* null mutants, in which production of β-galactosidase is derepressed, i.e. lacZ is expressed at high level irrespective of the provision of an inducer such as IPTG. Although recovery of such mutants is relatively straightforward, locating the site of the mutation, and therefore the *lacI* gene, is somewhat more complicated. Originally this was achieved by classical bacterial genetics using Hfr crosses and recombination analysis, a tedious procedure. Hence, an easier process was developed that combined mutation with acquisition of a selectable marker. One way to achieve this is by transposon insertional inactivation, when a gene of interest is disrupted by an antibiotic resistance transposon. To facilitate this approach, suicide vectors carrying antibiotic resistance transposons, natural and artificial, have been constructed. The transposons chosen are those which insert more or less at random into any DNA sequence in the cell. Popular elements are Tn5 and Tn10 and their derivatives. The suicide vector is used to deliver the transposon into the target cell where it survives only by transposing from its carrier replicon into the host cell chromosome (or other viable replicon in the cell).

Transposition events are selected by plating on medium containing the appropriate antibiotic for the resistance gene on the transposon (e.g. kanamycin for Tn5 and tetracycline for Tn10). The survivors are screened for the knockout mutants of interest. Particular mutants are then chosen and genomic DNA from each of these is isolated, digested with a restriction enzyme that does not cut within the marker gene and no more than once within the transposon as a whole, and the fragments are ligated into a suitable cloning vector, e.g. pUC18, to create a gene bank, which is transformed into E. coli. Selection of particular transformants can then be for those that have recombinant plasmids carrying the transposon antibiotic resistance gene. This will have been recovered together with some sequences from the gene disrupted. Analysis of several such transposon mutants, where the transposon is inserted at different locations and in different orientations in the gene of interest, will allow the sequence of the intact gene to be determined.

'One of the most popular transposon delivery vectors is the pUT series (de Lorenzo et al. 1990), which have the same pir dependent origin of replication and mobilization functions as pEX18 suicide vectors, but carry the Tn5 transposase gene, and Tn5 repeat sequences flanking a selectable marker. These vectors have been designed to include many different antibiotic resistance genes, for use in different bacteria, since some are inherently resistant to certain antibiotics.

Fusion vectors

Derivatives of pUT have also been developed that allow the formation of protein fusions, i.e. hybrid proteins. These are constructed for a variety of reasons, including investigation of the cellular location of a particular protein peptide domain, study of transcription and translation control systems, and to facilitate purification of proteins of interest (de Lorenzo et al. 1990). These systems are typified by the phoA fusion system. The E. coli phoA gene encodes alkaline phosphatase, an enzyme that is easily assayed. Random insertions of the phoA gene into a gene of interest, achieved using a modified Tn5, can create hybrid genes, where the reading frames of the disrupted gene and that of phoA are in register. These hybrid genes produce proteins that have different lengths of the protein of interest at their N-termini fused to PhoA. If the fusion protein is expressed using the transcription and translation signals specific to the protein of interest, then changes in

level of expression can be monitored simply by determining alkaline phosphatase activity. This is a powerful tool with which to investigate production of proteins that lack an easily assayable activity. The phoA system is used particularly to study production of membrane proteins and their topologies. Alkaline phosphatase is normally secreted into the periplasmic space of E. coli, where it is activated. Enzyme protein that remains in the cytoplasm is generally inactive. Accordingly, alkaline phosphatase activity indicates that the PhoA domain of the fusion protein has been transported to the periplasm. In the study of membrane proteins, this technique is useful for mapping the topology of membrane proteins because, when PhoA is fused to a peptide domain that is normally in the cytoplasm then alkaline phosphatase activity remains dormant. In contrast, when it is fused to a peptide domain that is located in the periplasm, alkaline phosphatase activity is revealed (Hoffman and Wright 1985). The activation of alkaline phosphatase reflects the need for protein chaperones, found in the periplasm, to assist the enzyme to achieve active configuration.

One limitation of pUT vectors is that they can only be used in organisms where the transposase gene can be expressed. However, it is now possible to purchase purified transposon delivery vectors, comprising the Tn5 terminal repeat sequences flanking an antibiotic resistance gene, complexed with transposase enzyme. In the absence of magnesium ions (i.e. in the buffer supplied) the transposase is not active, but upon transformation of the vector/transposase complex into a bacterial cell, where magnesium ions are present, the transposase is activated and transposition occurs (Goryshin et al. 2000). Another potential use of this system is to introduce a gene of interest into the chromosome of a bacterium. This can be particularly important if the gene under investigation is normally located on a chromosome, when gene copy number is essentially one. The suicide vector pMOD carries Tn5 terminal repeats flanking a multiple cloning site, allowing the insertion any DNA sequence of interest between the repeats and has been provided with the pir-dependent replication origin and so replicates only in pir+ E. coli strains. The recombinant plasmid is isolated and purified and the plasmid DNA is mixed with purified Tn5 transposase enzyme, and the mixture is electroporated into the recipient bacterium (pir−), where the novel Tn5 construct is transposed into the chromosome of the new host. The suicide vector is lost because it fails to replicate.

One problem arising from the use of transposons to mark genes by insertional inactivation is that it is not always straightforward to clone out the transposon marker together with flanking sequences from the new site of insertion, and hence to identify the gene disrupted. To overcome this limitation, the pMOD vector has had the pir replication origin relocated and placed between the Tn5 repeat sequences, beside the MCS (York et al. 1998). Accordingly, not only is an antibiotic resistance gene transposed but, also, so is a tightly linked silent plasmid origin of replication. Once appropriate Tn insertion mutants have been identified, the chromosomal DNA of these can be purified, digested with an appropriate restriction enzyme, i.e. one that does not cut the transposed sequence, and the fragments can be end-ligated to circularize them. The circles generated are narrow host-range plasmids, that can be recovered on transformation of a *pir+ E. coli* strain. This strategy eliminates the need to construct a gene bank in a cloning vector such as pUC18 and provides a ready source of DNA from which to sequence the gene disrupted by the Tn insertion. This is because gene sequences, both promoter proximal and promoter distal to the point of insertion, are recovered simultaneously. If cleavage with one restriction endonuclease fails to recover the gene of interest in its entirety, then another can be tried.

Gene cloning

Foreign DNA is inserted into a plasmid vector by linearizing the circular plasmid molecule with a particular restriction enzyme. The DNA fragment to be cloned – usually generated by the same restriction enzyme used to cut the cloning vector – is joined (ligated) at both ends to the linearized vector to form a circular molecule. The hybrid DNA is then introduced into a suitable bacterial cell by transformation (see Brown 2001; Primrose et al. 2001). The desired recombinant clones then have to be identified. This may be done by selection if the gene of interest confers a selectable phenotype; otherwise, a set of recombinants called a gene bank is collected and the individual members of the set are screened to identify those with the gene of interest.

In contrast to plasmid vectors, which are circular, λ based vectors are linear DNA molecules. In these cases the vector is cut with a restriction enzyme to generate two (insertion vectors) or three (replacement vectors) fragments. If two, the sequence to be cloned (generated with the same restriction enzyme or one that generates compatible single-strand extensions) is inserted between them. This requires the three fragments to be ligated together in the correct order. The desired recombinants can then be recovered by transfecting a suitable cell line of *E. coli* with the recombinant DNA or by in vitro packaging of the recombinant DNA (i.e. generation of reconstituted λ phage particles in the 'test tube') and subsequent infection of a suitable *E. coli* cell line. The latter process is the more efficient and λ packaging kits are commercially available. If a replacement vector is used, restriction enzyme digestion of the vector produces three fragments, of which the two outside fragments are ligated to the fragment to be cloned and the central fragment, often a stuffer sequence (Chauthaiwale et al. 1992), is discarded. Again, the desired recombinants can

be recovered by transfection or by in vitro packaging and cell infection (see Primrose et al. (2001) for a more detailed presentation of gene cloning strategies).

Eukaryotic genes are not transcribed in bacteria from their own promoters because these promoters are not recognized by bacterial RNA polymerases. Hence, to express a eukaryotic gene in a bacterial cell it is necessary to join the gene to a suitable bacterial promoter sequence. A number of cloning vectors with cloning sites close to and downstream from active promoters have been developed for this purpose (see Thompson 1988). Many cloning vectors with a bewildering array of properties are now readily available commercially.

A fundamental problem that arises when genes are cloned on a fragment generated by a restriction endonuclease is that what is recovered may not simply be the gene(s) of interest, but also unwanted sequences flanking it. Expression of genes encoded on these additional sequences may complicate subsequent analysis, so it is frequently desirable that these be removed prior to analysis. This can be achieved by subcloning, i.e. the fragment recovered in the primary cloning is cut into smaller pieces using a different restriction enzyme and these, in turn, are recovered in the same or in a different cloning vector. The desired transconjugant is again detected by an appropriate screening process. If the fragment is still considered to be too large, the reducing process is repeated on the smaller fragment with another restriction enzyme until the smallest fragment that still encodes the intact gene(s) is recovered.

An alternative approach that bypasses the problem completely, and which is suitable for small peptides, is to synthesize the gene chemically and then insert it into a suitable cloning vector. This was done in the early days of cloning for the gene for somatostatin – a pituitary hormone of 14 amino acids, the sequence of which was known (Itakura et al. 1977). The genes for human growth hormone, thymosin and insulin have also been synthesized and cloned. A more recent solution to the problem of eliminating unwanted sequences that accompany a gene of interest recovered from a gene bank, because sequencing is now both straightforward and economical, is to sequence the insert and then amplify the gene of interest by PCR. The synthetic fragment can then be cloned into a specialized cloning vector, e.g. pTOPO, and the required recombinant can be recovered in the normal way by transformation of an appropriate cell line (Ramalingam et al. 2004). Such technology allows, in principle, examination of any system, gene by gene.

A second problem, that does not arise with bacterial genes, is encountered with many eukaryotic genes. Not only are they too large to synthesize with existing techniques but, unlike the genes of bacteria, they may also be discontinuous ('split' genes), the continuity of the coding sequence being disrupted by one or more unrelated sequences (introns) (see, for example, Lewin

2000). Such sequences are removed after transcript synthesis but before translation, a process referred to as RNA splicing, i.e. a functional mRNA is produced by splicing together those sequences necessary for the production of the particular peptide. Another approach must therefore be used if the gene is to be recovered and expressed in a bacterial cell. This involves isolating functional mRNA molecules (already processed), which are used as template to synthesize complementary DNA (cDNA) sequences using reverse transcriptase (an enzyme that synthesizes single-stranded DNA from an RNA template). The single-stranded cDNA is then converted to double-stranded DNA and inserted into a suitable cloning vector. The reconstructed, uninterrupted gene is then recovered by transformation of an appropriate bacterial cell line (Primrose et al. 2001). The necessary mRNA can be prepared by one of a number of techniques; which one is chosen depends largely on the task in hand (e.g. White and Radke 1997; Rosenow et al. 2001).

Several mammalian proteins are now being made in bacteria as the result of gene cloning. They include somatostatin, human growth hormone, insulin, several interferons, and tissue plasminogen activator. Many more can be expected in the future.

Genetic engineering is also likely to make a major impact in the field of vaccines. By making only the antigen against which antibody response is required, the risk associated with the intact pathogen is minimized. This technique is being pursued with respect to many pathogens, notably rabies and hepatitis B viruses, malaria, and *Bordetella pertussis* to name but four. The techniques of molecular biology that have been developed are so powerful, both with respect to unlocking the molecular biology of genetic systems and in furnishing the potential to allow specific manipulation of genes, that they offer the promise of major advances in our understanding of many diseases and of better informed positive intervention.

The power of bacterial genetics has also been harnessed to engineer plants to create novel cultivars with desirable characteristics. These genetic manipulations have resulted in the transfer of bacterial antibiotic resistance genes into these plants. Concern has been raised that large-scale cultivation of these new cultivars might exacerbate the spread of antibiotic resistance genes among bacterial pathogens and so hasten the time when antibiotic therapy is no longer effective. However, although it has been shown that bacteria can transfer DNA into plant cells, the reverse process has never been reported, except when artificial transformation systems have been used. Accordingly, while it is impossible to rule out gene transfer from plant to bacteria, it is exceedingly unlikely that it will have a significant impact on the further development of antibiotic resistance in bacteria (Bennett et al. 2004).

REFERENCES

Abraham, J.M., Freitag, C.S., et al. 1985. An invertible element of DNA controls phase variation of type 1 fimbriae of *Escherichia coli*. *Proc Natl Acad Sci USA*, **82**, 5724–7.

Adhya, S. and Gottesman, M. 1978. Control of transcription termination. *Ann Rev Biochem*, **47**, 967–96.

Altshuler, M. 1993. Recovery of DNA replication in UV-damaged *Escherichia coli*. *Mutat Res*, **294**, 91–100.

Ames, B.W. 1979. Identifying environmental chemicals causing mutations and cancer. *Science*, **204**, 587–93.

Ames, B.W., Durston, W.E., et al. 1973. Carcinogens are mutagens: a simple test system combining liver homogenates for inactivation and bacteria for detection. *Proc Natl Acad Sci USA*, **70**, 2281–5.

Anderson, E.S. 1968. The ecology of transferable drug resistance in the Enterobacteria. *Annu Rev Microbiol*, **22**, 131–80.

Andersson, D.I. 2003. Persistence of antibiotic resistant bacteria. *Curr Opin Microbiol*, **6**, 452–6.

Andersson, D.I. and Kurland, C.G. 1983. Ram ribosomes are defective proofreaders. *Mol Gen Genet*, **191**, 378–81.

Appelbaum, P.C. and Bozdogan, B. 2004. Vancomycin resistance in *Staphylococcus aureus*. *Clin Lab Med*, **24**, 381–402.

Arakawa, E., Murase, T., et al. 2000. Pulsed-field gel electrophoresis-based molecular comparison of *Vibrio cholerae* O1 isolates from domestic and imported cases of cholera in Japan. *J Clin Microbiol*, **38**, 424–6.

Arber, W. 1983. A beginner's guide to lambda biology. In: Roberts, J.W., Stahl, F.W. and Weisberg, R.A. (eds), *Lambda II*. New York: Cold Spring Harbor Laboratory, 381–94.

Archer, G.L. and Niemeyer, D.M. 1994. Origin and evolution of DNA associated with resistance to methicillin in staphylococci. *Trends Microbiol*, **2**, 343–7.

Arthur, A. and Sherratt, D. 1979. Dissection of the transposition process: a transposon-encoded site-specific recombination system. *Mol Gen Genet*, **175**, 267–74.

Arthur, M. and Courvalin, P. 1993. Genetics and mechanisms of glycopeptide resistance in enterococci. *Antimicrob Agents Chemother*, **37**, 1563–71.

Austin, S. and Abeles, A. 1983. Partion of unit-copy miniplasmids to daughter cells 1. P1 and F miniplasmids contain discrete, interchangeable sequences sufficient to promote equipartition. *J Mol Biol*, **169**, 353–72.

Austin, S., Ziese, M. and Sternberg, N. 1981. A novel role for site-specific recombination in maintenance of bacterial replicons. *Cell*, **25**, 729–36.

Avery, O.T., MacLeod, C.M. and McCarty, M. 1944. Studies on the chemical nature of the substance inducing transformation of pneumococcal types. *J Exp Med*, **79**, 137–57.

Avila, P. and de la Cruz, F. 1988. Physical and genetic map of the IncW plasmid R388. *Plasmid*, **20**, 155–7.

Baas, P. and Jansz, H. 1988. Single-stranded DNA phage origins. *Curr Topics Microbiol Immunol*, **136**, 31–70.

Bagdasarian, M., Lurz, R., et al. 1981. Specific-purpose plasmid cloning vectors. II. Broad host range, high copy number, RSF1010-derived vectors, and a host-vector system for gene cloning in *Pseudomonas*. *Gene*, **16**, 237–47.

Bale, M.J., Fry, J.C. and Day, M.J. 1987. Plasmid transfer between strains of *Pseudomonas aeruginosa* on membrane filters attached to river stones. *J Gen Microbiol*, **133**, 3099–107.

Barkay, T., Miller, S.M. and Summers, A.O. 2003. Bacterial mercury resistance from atoms to ecosystems. *FEMS Microbiol Rev*, **27**, 355–84.

Barth, P.T., Tobin, L. and Sharpe, G.S. 1981. Development of broad host-range vectors. In: Levy, S.B., Clowes, R.C. and Koenig, E.L. (eds), *Molecular biology, pathogenicity, and ecology of bacterial plasmids*. New York and London: Plenum Press, 439–48.

Beaber, J.W., Burrus, V., et al. 2002. Comparison of SXT and R391, two conjugative integrating elements: definition of a genetic backbone for the mobilization of resistance determinants. *Cell Mol Life Sci*, **59**, 2065–70.

Bennett, P.M. 1995. The spread of drug resistance. In: Baumberg, S., Young, J.P.W., et al. (eds), *Population genetics of bacteria*. Cambridge: Cambridge University Press, 317–44.

Bennett, P.M. 1999. Integrons and gene cassettes: a genetic construction kit for bacteria. *J Antimicrob Chemother*, **43**, 1–4.

Bennett, P.M. and Chopra, I. 1993. Molecular basis of beta-lactamase induction in bacteria. *Antimicrob Agents Chemother*, **37**, 153–8.

Bennett, P.M., de la Cruz, F. and Grinsted, I. 1983. Cointegrates are not obligatory intermediates in transposition of Tn*3* and Tn*21*. *Nature (London)*, **305**, 743–4.

Bennett, P.M., Livesey, C.T., et al. 2004. An assessment of the risks associated with the use of antibiotic resistance genes in genetically modified plants: report of the working party of the British Society for Antimicrobial Chemotherapy. *J Antimicrob Chemother*, **53**, 418–31.

Berg, C.M., Berg, D.E. and Groisman, E.A. 1989. Transposable elements and the genetic engineering of bacteria. In: Berg, D.E. and Howe, M.M. (eds), *Mobile DNA*. Washington, DC: ASM Press, 879–925.

Berg, D.E. 1989. Transposon Tn*5*. In: Berg, D.E. and Howe, M.M. (eds), *Mobile DNA*. Washington, DC: ASM Press, 185–210.

Berg, D.E. and Howe, M.M. 1989. *Mobile DNA*. Washington, DC: American Society for Microbiology.

Berger-Bächi, B. 1994. Expression of resistance to methicillin. *Trends Microbiol*, **2**, 389–93.

Berlyn, M.K.B., Brooks Low, K. and Rudd, K.E. 1996. Linkage map of *Escherichia coli* K-12, 9th edn, Escherichia coli *and* Salmonella: *Cellular and molecular biology*, 2nd edn, F.C. Neidhardt (ed.). Washington, DC: American Society for Microbiology, 1715-1902.

Betley, M.J., Miller, V.L. and Mekalanos, J.J. 1986. Genetics of bacterial enterotoxins. *Annu Rev Microbiol*, **40**, 577–605.

Binns, M.M., Davies, D.L. and Hardy, K.G. 1979. Cloned fragments of the plasmid ColV,I-K94 specifying virulence and serum resistance. *Nature (London)*, **279**, 778–81.

Bolland, S. and Kleckner, N. 1996. The chemical steps of Tn10/IS10 transposition involve repeated utilization of a single active-site. *Cell*, **84**, 223–33.

Böltner, D. and Osborn, A.M. 2004. Structural comparison of the integrative and conjugative elements R391, pMERPH, R997 and SXT. *Plasmid*, **51**, 12–23.

Bonnet, R. 2004. Growing group of extended-spectrum beta-lactamases: the CTX-M enzymes. *Antimicrob Agents Chemother*, **48**, 1–14.

Borchert, L.D. and Drexler, H. 1980. T1 genes which affect transduction. *J Virol*, **33**, 1122–8.

Bossi, L. 1983. Context effects: translation of UAG codon by suppressor transfer-RNA is affected by the sequence following UAG in the message. *J Mol Biol*, **164**, 73–87.

Bradford, P.A. 2001. Extended-spectrum beta-lactamases in the 21st century: characterization, epidemiology, and detection of this important resistance threat. *Clin Microbiol Rev*, **14**, 933–51.

Bradley, D.E. 1977. Characterization of pili determined by drug resistance plasmids R711b and R778b. *J Gen Microbiol*, **102**, 349–63.

Bradley, D.E. 1980. Determination of pili by conjugative bacterial drug resistance plasmids of incompatibility groups B, C, H, J, K, M, V and X. *J Bacteriol*, **141**, 828–37.

Braman, J. (ed.). 2002. Methods in molecular biology. In: *In vitro mutagenesis protocols*, Vol. 182, 2nd edn. Totowa, NJ: Humana Press.

Brammar, W.J. 1979. Safe and useful vector systems. *Biochem Soc Symp*, **44**, 13–27.

Brammar, W.J. 1982. Vectors based on bacteriophage lambda. In: Williamson, R. (ed.), *Genetic engineering*. London: Academic Press.

Brendler, T., Reaves, L. and Austin, S. 2004. Interplay between plasmid partition and postsegregational killing systems. *J Bacteriol*, **186**, 2504–7.

Broda, P. 1979. *Plasmids*. San Francisco: WH Freeman.

Brooks Low, K. 1996. Hfr strains of *Escherichia coli* K-12. In: Neidhardt, F.C. (ed.), *Escherichia coli and Salmonella: cellular and molecular biology*. Washington, DC: ASM Press, 2402–5.

Brown, T.A. 2001. *Gene cloning and DNA analysis: an introduction*, 4th edn. Oxford: Blackwell Science Ltd.

Brown, N.L., Barrett, S.R., et al. 1995. Molecular genetics and transport analysis of the copper-resistance determinant (PCO) from *Escherichia coli* plasmid pRJ1004. *Mol Microbiol*, **17**, 1153–66.

Brubaker, R.R. 1985. Mechanisms of bacterial virulence. *Annu Rev Microbiol*, **39**, 21–50.

Buchanan-Wollaston, V., Passiator, J.E. and Cannon, F. 1987. The mob and oriT mobilization functions of a bacterial plasmid promote its transfer to plants. *Nature (London)*, **328**, 172–5.

Burkardt, H.J. and Puhler, A. 1988. Electron microscopy of plasmid DNA. In: Grinsted, J. and Bennett, P.M. (eds), *Methods in microbiology*, Vol. 21. . London: Academic Press, 155–77.

Burrus, V., Pavlovic, G., et al. 2002. Conjugative transposons: the tip of the iceberg. *Mol Microbiol*, **46**, 601–10.

Bush, K. and Jacoby, G. 1997. Nomenclature of TEM β-lactamases. *J Antimicrob Chemother*, **39**, 1–3.

Cairns, J. 1963. The bacterial chromosome and its manner of replication as seen by autoradiography. *J Mol Biol*, **6**, 208–13.

Campbell, A. 1994. Comparative molecular biology of lambdoid phages. *Annu Rev Microbiol*, **48**, 193–222.

Caro, L., Churchward, G. and Chandler, M. 1984. Study of plasmid replication *in vivo*. In: Bennett, P.M. and Grinsted, J. (eds), *Methods in microbiology*, Vol. 17. . London: Academic Press, 97–122.

Chandler, M. and Mahillon, J. 2002. Insertion sequences revisited. In: Craig, N.L. and Craigie, R. (eds), *Mobile DNA II*. Washington, DC: ASM Press, 305–66.

Chang, A.C.Y. and Cohen, S.N. 1978. Construction and characterization of amplifiable multicopy DNA cloning vehicles derived from the P15A cryptic miniplasmid. *J Bacteriol*, **134**, 1141–56.

Charlebois, R.L. 1999. *Organization of the prokaryotic genome*. Washington, DC: ASM Press.

Chater, K.F. and Hopwood, D.A. 1989. Antibiotic synthesis in *Streptomyces*. In: Hopwood, D.A. and Chater, K.F. (eds), *Genetics of bacterial diversity*. New York: Academic Press, 129–50.

Chopra, I., O'Neill, A.J. and Miller, K. 2003. The role of mutators in the emergence of antibiotic-resistant bacteria. *Drug Res Updates*, **6**, 137–45.

Clarke, S.C. 2002. Nucleotide sequence-based typing of bacteria and the impact of automation. *Bioessays*, **24**, 858–62.

Chauthaiwale, V.M., Therwath, A. and Deshpande, V.V. 1992. Bacteriophage lambda as a cloning vector. *Microbiol Rev*, **56**, 577–91.

Clewell, D.B. 1993. Bacterial sex pheromone-induced plasmid transfer. *Cell*, **73**, 9–12.

Clewell, D.B., Ehrenfeld, E.E., et al. 1986. Sex-pheromone systems in *Streptococcus faecalis*. In: Levy, S.B. and Novick, R.P. (eds), *Antibiotic resistance genes: ecology, transfer and expression*. New York: Cold Spring Harbor Laboratory, 131–42.

Clewell, D.B., Senghas, E., et al. 1988. Transposition in *Streptococcus*: structural and genetic properties of the conjugative transposon Tn*916*. In: Kingsman, A.J., Chater, K.F. and Kingsman, S.M. (eds), *Transposition*. Symposium 43 of the Society for General Microbiology. Cambridge: Cambridge University Press, 43–58.

Cohen, S.N., Chang, A.C.Y. and Hsu, L. 1972. Nonchromosomal antibiotic resistance in bacteria: genetic transformation of *Escherichia coli* by R-facter DNA. *Proc Natl Acad Sci USA*, **69**, 2110–14.

Cohen, S.H., Tang, Y.J. and Silva, J. Jr 2001. Molecular typing methods for the epidemiological identification of *Clostridium difficile* strains. *Exp Rev Mol Diagn*, **1**, 61–70.

Cole, S.T. and Telenti, A. 1995. Drug-resistance in *Mycobacterium tuberculosis*. *Eur Respir J Suppl*, **20**, 701s–13s.

Cole, S.T. and Saint-Girons, I. 1999. Bacterial genomes-all shapes and sizes. In: Charlebois, R.L. (ed.), *Organization of the prokaryotic genome*. Washington, DC: ASM Press, 35–62.

Collins, J. and Bruning, H.J. 1978. Plasmids useable as gene-cloning vectors in an in vitro packaging by coliphage lambda: cosmids. *Gene*, **4**, 85–107.

Collis, C.M. and Hall, R.M. 1992. Gene cassettes from the insert region of integrons are excised as covalently closed circles. *Mol Microbiol*, **6**, 2875–85.

Collis, C.M., Grammaticopoulos, G., et al. 1993. Site-specific insertion of gene cassettes into integrons. *Mol Microbiol*, **9**, 41–52.

Collis, C.M., Kim, M.J., et al. 2002. Characterization of the class 3 integron and the site-specific recombination system it determines. *J Bacteriol*, **184**, 3017–26.

Cooper, J.E. and Feil, E.J. 2004. Multilocus sequence typing – what is resolved? *Trends Microbiol*, **12**, 373–7.

Cornelis, G., Bennett, P.M. and Grinsted, J. 1976. Properties of pGC1, a *lac* plasmid originating in *Yersinia enterocolitica* 842. *J Bacteriol*, **127**, 1058–62.

Cornelis, G., Ghosal, D. and Saedler, H. 1978. Tn*951*: a new transposon carrying a lactose operon. *Mol Gen Genet*, **160**, 215–24.

Couch, J.L., Soltis, M.T. and Betley, M.J. 1988. Cloning and nucleotide sequence of the type E staphylococcal enterotoxin gene. *J Bacteriol*, **170**, 2954–60.

Courvalin, P., Goussard, S. and Grillot-Courvalin, C. 1995. Gene transfer from bacteria to mammalian cells. *CR Acad Sci III*, **318**, 1207–12.

Craig, N.L. 1988. The mechanism of conservative site-specific recombination. *Annu Rev Genet*, **22**, 77–105.

Craig, N.L. 1989. Transposon Tn7. In: Berg, D.E. and Howe, M.M. (eds), *Mobile DNA*. Washington, DC: ASM Press, 211–25.

Craig, N.L. 1996. Transposition. In: Neidhardt, F.C. (ed.), *Escherichia coli and Salmonella: cellular and molecular biology*, 2nd edn. Washington, DC: ASM Press, 2339–62.

Craig, N.L., Craigie, R., et al. (eds). 2002. *Mobile DNA II*. Washington, DC: ASM Press.

Datsenko, K.A. and Wanner, B.L. 2000. One step inactivation of chromosomal genes in *E. coli* using PCR products. *Proc Natl Acad Sci USA*, **97**, 6640–5.

Datta, N. and Nugent, M.E. 1983. Bacterial variation. In: Wilson, G. and Dick, H.M. (eds), *Topley and Wilson's principles of bacteriology, virology and immunity*, Vol. 1, 7th edn. London: Edward Arnold, 145–76.

Davis, B.M. and Waldor, M.K. 2002. Mobile genetic elements and bacterial pathogenesis. In: Craig, N.L., et al. (eds), *Mobile DNA II*. Washington, DC: ASM Press, 1040–59.

de Boer, H.A., Comstock, L.J. and Vasser, M. 1983. The tac promoter: a functional hybrid derived from the trp and lac promoters. *Proc Natl Acad Sci USA*, **80**, 21–5.

de Lorenzo, V., Herrero, M., et al. 1990. Mini-Tn5 transposon derivatives for insertion mutagenesis, promoter probing, and chromosomal insertion of cloned DNA in gram-negative eubacteria. *J Bacteriol*, **172**, 6568–72.

del Solar, G., Moscoso, M. and Espinosa, M. 1993. Rolling circle-replicating plasmids from gram-positive and gram-negative bacteria: a wall falls. *Mol Microbiol*, **8**, 789–96.

Deonier, R.C. 1996. Native insertion sequence elements: locations, distributions, and sequence relationships. In: Neidhardt, F.C. (ed.), *Escherichia coli and Salmonella: cellular and molecular biology*. Washington, DC: ASM Press, 2000–11.

Dodd, H.M. and Bennett, P.M. 1987. The R46 site-specific recombination system is a homologue of the Tn3 and γδ(Tn*1000*) cointegrate resolution system. *J Gen Microbiol*, **133**, 2031–9.

Dorman, C.J. 1994. *Genetics of bacterial virulence*. Oxford: Blackwell Scientific Publications.

Dougan, G. and Sherratt, D.S. 1977. The transposon Tn*1* as a probe for studying ColE1 structure and function. *Mol Gen Genet*, **151**, 151–60.

Dowson, C.G., Coffey, T.J. and Spratt, B.G. 1994. Origin and molecular epidemiology of penicillin-binding-protein-mediated resistance to β-lactam antibiotics. *Trends Microbiol*, **2**, 361–6.

Draper, G.C. and Gober, J.W. 2002. Bacterial chromosome segregation. *Annu Rev Microbiol*, **56**, 567–97.

Dressler, D. and Potter, H. 1982. Molecular mechanisms of genetic recombination. *Annu Rev Biochem*, **51**, 727–61.

Drlica, K. and Malik, M. 2003. Fluoroquinolones: action and resistance. *Curr Top Med Chem*, **3**, 249–82.

Dubnau, D. 1999. DNA uptake in bacteria. *Ann Rev Microbiol*, **53**, 217–44.

Dumon-Seignovert, L., Cariot, G. and Vuillard, L. 2004. The toxicity of recombinant proteins in *Escherichia coli*: a comparison of overexpression in BL21(DE3), C41(DE3) and C43(DE3). *Prot Exp Purif*, **37**, 203–6.

Eaves, D.J., Randall, L., et al. 2004. Prevalence of mutations within the quinolone resistance-determining region of *gyrA*, *gyrB*, *parC*, and *parE* and association with antibiotic resistance in quinolone-resistant *Salmonella enterica*. *Antimicrob Agents Chemother*, **48**, 4012–15.

Eberl, L., Kristensen, C.S., et al. 1994. Analysis of the multimer resolution system encoded by the parCBA operon of broad-host-range plasmid RP4. *Mol Microbiol*, **12**, 131–41.

Eggleston, A.K. and West, S.C. 1996. Exchanging partners: recombination in *E. coli*. *Trends Genet*, **12**, 20–6.

Ehrlich, S.D. 1977. Replication and expression of plasmids from *Staphylococcus aureus* in *Bacillus subtilis*. *Proc Natl Acad Sci USA*, **74**, 1680–2.

Elliott, E., Oosthuizen, D., et al. 2003. Fluoroquinolone resistance in *Haemophilus influenzae*. *J Antimicrob Chemother*, **52**, 734–5.

Elwell, L.P. and Shipley, P.L. 1980. Plasmid-mediated factors associated with virulence of bacteria to animals. *Annu Rev Microbiol*, **34**, 465–96.

Engel, H., Soedirman, N. and Rost, J. 1980. Transferability of macrolide, lincomycin, and streptogramin resistances between Group A, B, and D Streptococci, *Streptococcus pneumoniae*, and *Staphylococcus aureus*. *J Bacteriol*, **142**, 407–13.

Engelberg-Kulka, H. and Schoulaker-Schwarz, R. 1996. Suppression of termination codons. In: Neidhardt, F.C. (ed.), *Escherichia coli and Salmonella: cellular and molecular biology*. Washington, DC: ASM Press, 909–21.

Enright, M.C. and Spratt, B.G. 1999. Multilocus sequence typing. *Trends Microbiol*, **7**, 482–7.

Espinosa, M., Cohen, S., et al. 2000. Plasmid replication and copy number control. In: Thomas, C.M. (ed.), *The horizontal gene pool: bacterial plasmids and gene spread*. Amsterdam: Harwood Academic Publishers, 1–47.

Everett, M.J., Chopra, I. and Bennett, P.M. 1990. Induction of the *Citrobacter freundii* group I β-lactamase in *Escherichia coli* is not dependent on entry of β-lactam into the cytoplasm. *Antimicrob Agents Chemother*, **34**, 2429–30.

Falkow, S., Wohlhieter, J.A., et al. 1964. Transfer of episomic elements to *Proteus*. II Nature of *lac*⁺ *Proteus* strains isolated from clinical specimens. *J Bacteriol*, **88**, 1598–601.

Farmer, J.J., Hickman, F.W., et al. 1977. Unusual Enterobacteriaceae: 'Proteus rettgeri' that 'change' into *Providencia stuartii*. *J Clin Microbiol*, **6**, 373–8.

Farrow, J.A.E. 1980. Lactose hydrolysing enzymes in *Streptococcus lactis* and *Streptococcus cremoris* and also in some other species of streptococci. *J Appl Bacteriol*, **49**, 493–503.

Fiant, M., Szybalski, W. and Malamy, M. 1972. 'Polar mutations in *lac*', *gal* and phage λ consist of a few IS-DNA sequences inserted with either orientation. *Mol Gen Genet*, **119**, 223–31.

Firth, N., Ippen-Ihler, K. and Skurray, R.A. 1996. Structure and function of the F factor and mechanism of conjugation. In: Neidhardt, F.C. (ed.), *Escherichia coli and Salmonella: cellular and molecular biology*. Washington, DC: ASM Press, 2377–401.

Ford, P.J. and Avison, M.B. 2004. Evolutionary mapping of the SHV beta-lactamase and evidence for two separate IS26-dependent blaSHV mobilization events from the *Klebsiella pneumoniae* chromosome. *J Antimicrob Chemother*, **54**, 69–75.

Foster, J.W. and Spector, M.P. 1995. How *Salmonella* survive against the odds. *Annu Rev Microbiol*, **49**, 145–74.

Foster, T.J. 1983. Plasmid determined resistance to antimicrobial drugs and toxic metal ions in bacteria. *Microbiol Rev*, **47**, 361–409.

Franke, A.K. and Clewell, D.B. 1981. Evidence for a chromosomal-borne resistance transposon (Tn*916*) in *Streptococcus faecalis* that is capable of 'conjugal' transfer in the absence of a conjugative plasmid. *J Bacteriol*, **145**, 494–502.

Frère, J.-M. 1995. β-Lactamases and bacterial resistance to antibiotics. *Mol Microbiol*, **16**, 385–95.

Galas, D.J. and Chandler, M. 1989. Bacterial insertion sequences. In: Berg, D.E. and Howe, M.M. (eds), *Mobile DNA*. Washington, DC: ASM Press, 109–62.

Galleni, M., Raquet, X., et al. 1995. DD-peptidases and β-lactamases: catalytic mechanisms and specificities. *J Chemother*, **7**, 3–7.

Garcillán-Barcia, M.P., Bernales, I., et al. 2002. IS*91* rolling-circle transposition. In: Craig, N.L., Craigie, R., et al. (eds), *Mobile DNA II*. Washington, DC: ASM Press, 891–904.

Gardiner, K., Laas, W. and Patterson, D. 1986. Fractionation of large mammalian DNA restriction fragments using vertical pulsed-field gradient gel electrophoresis. *Somat Cell Mol Genet*, **12**, 185–95.

Garnier, T., Saurin, W. and Cole, S.T. 1987. Molecular characterization of the resolvase gene, *res*, carried by a multicopy plasmid from *Clostridium perfringens*: common evolutionary origin for prokaryotic site-specific recombinases. *Mol Microbiol*, **1**, 371–6.

Gerdes, K., Rasmussen, P.B. and Molin, S. 1986. Unique type of plasmid maintenance function: postsegregational killing of plasmid-free cells. *Proc Natl Acad Sci USA*, **83**, 3116–20.

Gerdes, K., Helin, K., et al. 1988. Translational control and differential RNA decay are key elements regulating postsegregational expression of the killer protein encoded by the parB locus of plasmid R1. *J Mol Biol*, **203**, 119–29.

Gerdes, K., Jacobsen, J.S. and Franch, T. 1997. Plasmid stabilization by post-segregational killing. *Genet Eng (NY)*, **19**, 49–61.

Gerdes, K., Ayora, S., et al. 2000. Plasmid maintenance systems. In: Thomas, C.M. (ed.), *The horizontal gene pool: bacterial plasmids and gene spread*. Amsterdam, The Netherlands: Harwood Academic Publishers, 49–85.

Ghuysen, J.-M. 1991. Serine β-lactamases and penicillin-binding proteins. *Annu Rev Microbiol*, **45**, 37–67.

Ghuysen, J.-M. 1994. Molecular structures of penicillin-binding proteins and β-lactamases. *Trends Microbiol*, **2**, 373–80.

Gibson, E.M., Chace, N.M., et al. 1979. Transfer of plasmid-mediated antibiotic resistance from streptococci to lactobacilli. *J Bacteriol*, **137**, 614–19.

Gillespie, S.H. 2002. Evolution of drug resistance in *Mycobacterium tuberculosis*: clinical and molecular perspective. *Antimicrob Agents Chemother*, **46**, 267–74.

Giphart-Gassler, M., Plasterk, R.H.A. and van de Putte, P. 1982. G inversion in bacteriophage Mu: a novel way of gene splicing. *Nature (London)*, **297**, 339–42.

Gniadkowski, M. 2001. Evolution and epidemiology of extended-spectrum beta-lactamases (ESBLs) and ESBL-producing microorganisms. *Clin Microbiol Infect*, **7**, 597–608.

Gokhale, D.V., Puntambekar, U.S. and Deobagkar, D.N. 1993. Protoplast fusion: a tool for intergeneric gene transfer in bacteria. *Biotechnol Adv*, **11**, 199–217.

Goodman, M.F. 2002. Error-prone repair DNA polymerases in prokaryotes and eukaryotes. *Ann Rev Biochem*, **71**, 17–50.

Goodman, H.M., Ecker, J.R. and Dean, C. 1995. The genome of *Arabidopsis thaliana*. *Proc Natl Acad Sci USA*, **92**, 10831–5.

Gorai, A.P., Heffron, F., et al. 1979. Electron microscope heteroduplex studies of sequence relationships among plasmids of the W incompatibility group. *Plasmid*, **2**, 485–92.

Goryshin, I.Y., Jendrisak, J., et al. 2000. Insertional transposon mutagenesis by electroporation of released Tn5 transposition complexes. *Nature Biotechnol*, **18**, 97–100.

Griffith, F. 1928. The significance of pneumococcal types. *J Hyg*, **27**, 113–59.

Grinsted, J. and Bennett, P.M. 1988. *Methods in microbiology*, Vol. 21. London: Academic Press.

Grinsted, J., de la Cruz, F. and Schmitt, R. 1990. The Tn*21* subgroup of bacterial transposable elements. *Plasmid*, **24**, 163–89.

Grohmann, E., Muth, G. and Espinosa, M. 2003. Conjugative plasmid transfer in gram-positive bacteria. *Microbiol Mol Biol Rev*, **67**, 277–301.

Groisman, E.A. and Saier, M.H. Jr 1990. *Salmonella* virulence – new clues to intramacrophage survival. *Trends Biochem Sci*, **15**, 30–3.

Guhathakurta, A., Viney, I. and Summers, D. 1996. Accessory proteins impose site selectivity during ColE1 dimer resolution. *Mol Microbiol*, **20**, 613–20.

Guzman, L.M., Belin, D., et al. 1995. Tight regulation, modulation, and high-level expression by vectors containing the arabinose pBAD promoter. *J Bacteriol*, **177**, 4121–30.

Hacker, J. and Kaper, J.B. 2000. Pathogenicity islands and the evolution of microbes. *Ann Rev Microbiol*, **54**, 641–79.

Hall, R.M., Brookes, D.E. and Stokes, H.W. 1991. Site-specific insertion of genes into integrons: role of the 59-base element and determination of the recombination cross-over point. *Mol Microbiol*, **5**, 1941–59.

Hall, R.M. and Collis, C.M. 1995. Mobile gene cassettes and integrons: capture and spread of genes by site-specific recombination. *Mol Microbiol*, **15**, 593–600.

Hanson, N.D. and Sanders, C.C. 1999. Regulation of inducible AmpC beta-lactamase expression among Enterobacteriaceae. *Curr Pharm Des*, **5**, 881–94.

Hansson, K., Sundstrom, L., et al. 2002. IntI2 integron integrase in Tn7. *J Bacteriol*, **184**, 1712–21.

Harayama, S. and Timmis, K.N. 1989. Catabolism of aromatic hydrocarbons by *Pseudomonas*. In: Hopwood, D.A. and Chater, K.F. (eds), *Genetics of bacterial diversity*. New York: Academic Press, 151–74.

Harfe, B.D. and Jinks-Robertson, S. 2000. DNA mismatch repair and genetic instability. *Ann Rev Genet*, **34**, 359–99.

Hayes, W. 1953a. Observations on a transmissible agent determining sexual differentiation in *Bacterium coli*. *J Gen Microbiol*, **8**, 72–88.

Hayes, W. 1953b. The mechanism of genetic recombination in *Escherichia coli*. *Cold Spring Harb Symp Quant Biol*, **18**, 75–93.

Hedges, R.W. and Jacob, A.E. 1974. Transposition of ampicillin resistance from RP4 to other replicons. *Mol Gen Genet*, **132**, ?31–40.

Hedges, R.W. and Jacoby, G.A. 1980. Compatibility and molecular properties of plasmid Rms 149 in *Pseudomonas aeruginosa* and *Escherichia coli*. *Plasmid*, **3**, 1–6.

Helinski, D.R. 1973. Plasmid determined resistance to antibiotics: molecular properties of R factors. *Annu Rev Microbiol*, **27**, 437–70.

Helinski, D.R., Toukdarian, A.E. and Novick, R.P. 1996. Replication control and other stable maintenance mechanisms of plasmids. In: Neidhardt, F.C. (ed.), *Escherichia coli and Salmonella: cellular and molecular biology*. Washington, DC: ASM Press, 2295–324.

Henner, D.J. and Hoch, J.A. 1980. The *Bacillus subtilis* chromosome. *Microbiol Rev*, **44**, 57–82.

Hensel, M., Shea, J.E., et al. 1995. Simultaneous identification of bacterial virulence genes by negative selection. *Science*, **269**, 400–3.

Hinnebusch, J. and Tilly, K. 1993. Linear plasmids and chromosomes in bacteria. *Mol Microbiol*, **10**, 917–22.

Hiramatsu, K., Cui, L., et al. 2001. The emergence and evolution of methicillin-resistant *Staphylococcus aureus*. *Trends Microbiol*, **9**, 486–93.

Hiraga, S. 1992. Chromosome and plasmid partition in *Escherichia coli*. *Annu Rev Biochem*, **61**, 283–306.

Hirota, Y. 1960. The effect of acridine dyes on mating type factors in *Escherichia coli*. *Proc Natl Acad Sci USA*, **46**, 57–64.

Hoang, T.T., Karkhoff-Schweizer, R.R., et al. 1998. A broad-host-range Flp-*FRT* recombination system for site-specific excision of chromosomally-located DNA sequences: application for isolation of unmarked *Pseudomonas aeruginosa* mutants. *Gene*, **212**, 77–86.

Hoch, J.A. and Silhavy, T.J. (eds). 1995. *Two-component signal transduction*. Washington, DC: ASM Press.

Hochhut, B., Iotfi, Y., et al. 2001. Molecular analysis of antibiotic resistance gene clusters in *Vibrio cholerae* 0139 and 01 SXT constins. *Antimicrob Agents Chemother*, **45**, 2991–3000.

Hoffman, C.S. and Wright, A. 1985. Fusions of secreted proteins to alkaline phosphatase: an approach for studying protein secretion. *Proc Natl Acad Sci USA*, **82**, 5107–11.

Holloway, B.W., Krishnapillai, V. and Morgan, A.F. 1979. Chromosomal genetics of *Pseudomonas*. *Microbiol Rev*, **43**, 73–102.

Hooykaas, P.J.J. 1989. Tumorigenicity of *Agrobacterium* on plants. In: Hopwood, D.A. and Chater, K.F. (eds), *Genetics of bacterial diversity*. New York: Academic Press, 373–91.

Hooykaas, P.J.J. and Beijersbergen, A.G.M. 1994. The virulence system of *Agrobacterium tumefaciens*. *Annu Rev Phytopathol*, **32**, 157–79.

Hopwood, D.A. 1978. Extrachromosomally determined antibiotic production. *Annu Rev Microbiol*, **32**, 373–405.

Hopwood, D.A., Chater, K.F., et al. 1973. Advances in *Streptomyces coelicolor* genetics. *Bacteriol Rev*, **37**, 371–92.

Hsu, P.L. and Landy, A. 1984. Resolution of synthetic att-site Holliday structures by the integrase protein of bacteriophage λ. *Nature (London)*, **311**, 721–6.

Hu, S., Ohtsubo, E. and Davidson, N. 1975. Electron microscopic heteroduplex studies of sequence relations among plasmids of *Escherichia coli*: structure of F13 and related F-primes. *J Bacteriol*, **122**, 749–63.

Hughes, V.M. and Datta, N. 1983. Conjugative plasmids in bacteria of the 'pre-antibiotic' era. *Nature (London)*, **302**, 725–6.

Huovinen, P. 2001. Resistance to trimethoprim-sulfamethoxazole. *Clin Infect Dis*, **32**, 1608–14.

Hutchinson, F. 1996. Mutagenesis. In: Neidhardt, F.C. (ed.), Escherichia coli *and* Salmonella: *cellular and molecular biology*. Washington, DC: ASM Press, 2218–35.

Imsande, J. 1978. Genetic regulation of penicillinase synthesis in gram-positive bacteria. *Microbiol Rev*, **42**, 67–83.

Itakura, K., Hirose, T., et al. 1977. Expression in *Escherichia coli* of a chemically synthesized gene for the hormone somatostatin. *Science*, **198**, 1056–63.

Jacob, A.E., Shapiro, J.A., et al. 1977. Plasmids studied in *Escherichia coli* and other enteric bacteria. In: Bukhari, A.I., Shapiro, J.A. and Adhya, S.L. (eds), *DNA insertion elements, plasmids and episomes*. New York: Cold Spring Harbor Laboratory, 607–38.

Jacob, F. and Monod, J. 1961. Genetic regulation mechanisms in the synthesis of proteins. *J Mol Biol*, **3**, 318–56.

Jacob, F., Brenner, S. and Cuzin, F. 1963. On the regulation of DNA replication in bacteria. *Cold Spring Harb Symp Quant Biol*, **28**, 329–48.

Jacobs, W.R. Jr, Kalpana, G.V., et al. 1991. Genetic systems for mycobacteria. *Methods Enzymol*, **204**, 537–55.

Jacoby, G.A. 1977. Plasmids studied in *Pseudomonas aeruginosa* and other pseudomonads. In: Bukhari, A.I., Shapiro, J.A. and Adhya, S.L. (eds), *DNA insertion elements, plasmids and episomes*. New York: Cold Spring Harbor Laboratory, 639–56.

Janssen, D.B., Pries, F. and van der Ploeg, J.R. 1994. Genetics and biochemistry of dehalogenating enzymes. *Annu Rev Microbiol*, **48**, 163–91.

Jaurin, B. and Normark, S. 1983. Insertion of IS2 creates a novel *ampC* promoter in *Escherichia coli*. *Cell*, **32**, 809–16.

Jensen, R.B. and Gerdes, K. 1995. Programmed cell death in bacteria: proteic plasmid stabilization systems. *Mol Microbiol*, **17**, 205–10.

Johnson, E.M., Wohlhieter, J.A., et al. 1976. Plasmid-determined ability of a *Salmonella tennessee* strain to ferment lactose and sucrose. *J Bacteriol*, **125**, 385–6.

Jordan, E., Saedler, H. and Starlinger, P. 1968. 0° and strong-polar mutations in the *gal* operon are insertions. *Mol Gen Genet*, **102**, 353–63.

Joris, B., Ghuysen, J.-M., et al. 1988. The active-site-serine penicillin-recognising enzymes as members of the *Streptomyces* R61 DD-peptidase family. *Biochem J*, **250**, 313–24.

Joris, B., Ledent, P., et al. 1991. Comparison of the sequences of class A β-lactamases and of the secondary structure elements of penicillin-recognising proteins. *Antimicrob Agents Chemother*, **35**, 2294–301.

Kaniga, K., Delor, I. and Cornelis, G.R. 1991. A wide-host-range suicide vector for improving reverse genetics in gram-negative bacteria: inactivation of the blaA gene of *Yersinia enterocolitica*. *Gene*, **109**, 137–41.

Karaolis, D.K.R., Somara, S., et al. 1999. A bacteriophage encoding a pathogenicity island, a type IV pilus and a phage receptor in cholera bacteria. *Nature*, **399**, 375–9.

Katayama, Y., Ito, T. and Hiramatsu, K. 2000. A new class of genetic element, staphylococcus cassette chromosome *mec*, encodes methicillin resistance in *Staphylococcus aureus*. *Antimicrob Agents Chemother*, **44**, 1549–55.

Katayama, Y., Takeuchi, F., et al. 2003. Identification in methicillin-susceptible *Staphylococcus hominis* of an active primordial mobile genetic element for the staphylococcal cassette chromosome *mec* of methicillin-resistant *Staphylococcus aureus*. *J Bacteriol*, **185**, 2711–22.

Kleckner, N. 1989. Transposon Tn*10*. In: Berg, D.E. and Howe, M.M. (eds), *Mobile DNA*. Washington, DC: ASM Press, 227–68.

Koch, A.L. 2000. Penicillin binding proteins, beta-lactams, and lactamases: offensives, attacks, and defensive countermeasures. *Crit Rev Microbiol*, **26**, 205–20.

Kolter, R., Inuzuka, M. and Helinski, D.R. 1978. Transcomplementation-dependent replication of a low molecular weight origin fragment from plasmid RK6. *Cell*, **15**, 1199–208.

Kopecko, D.J., Washington, O. and Formal, S.B. 1980. Genetic and physical evidence for plasmid control of *Shigella sonnei* form I cell surface antigen. *Infect Immun*, **29**, 207–14.

Krause, M. and Guiney, D.G. 1991. Identification of a multimer resolution system involved in stabilization of the *Salmonella dublin* virulence plasmid pSDL2. *J Bacteriol*, **173**, 5754–62.

Kunkel, T.A. and Bebenek, K. 2000. DNA replication fidelity. *Ann Rev Biochem*, **69**, 497–529.

Lacey, R.W. 1973. Genetic basis, epidemiology, and future significance of antibiotic resistance in *Staphylococcus aureus*: a review. *J Clin Path*, **26**, 899–913.

Lacey, R.W. 1975. Antibiotic resistance plasmids in *Staphylococcus aureus* and their clinical importance. *Bacteriol Rev*, **39**, 1–32.

Lacey, R.W. 1980. Evidence for two mechanisms of plasmid transfer in mixed cultures of *Staphylococcus aureus*. *J Gen Microbiol*, **119**, 423–35.

Lacey, R.W. 1984. Antibiotic resistance in *Staphylococcus aureus* and streptococci. *Br Med Bull*, **40**, 77–83.

Lacey, R.W. and Kruczenyk, S.C. 1986. Epidemiology of antibiotic resistance in *Staphylococcus aureus*. *J Antimicrob Chemother*, **18**, Suppl C, 207–14.

Landick, R., Turnbough, C.L. Jr and Yanofsky, C. 1996. Transcription attenuation. In: Neidhardt, F.C. (ed.), Escherichia coli *and* Salmonella: *cellular and molecular biology*. Washington, DC: ASM Press, 1263–86.

Lane, D., de Feyter, R., et al. 1986. D protein of mini F plasmid acts as a repressor of transcription and as a site-specific resolvase. *Nucleic Acids Res*, **14**, 9713–28.

Laraki, N., Galleni, M., et al. 1999. Structure of In*31*, a *bla*IMP-containing *Pseudomonas aeruginosa* integron phyletically related to In*5*, which carries an unusual array of gene cassettes. *Antimicrob Agents Chemother*, **43**, 890–901.

Laufs, R., Kaulfers, P.-M., et al. 1979. Molecular characterization of a small *Haemophilus influenzae* plasmid specifying β-lactamase and its relationship to R factors from *Neisseria gonorrhoeae*. *J Gen Microbiol*, **111**, 223–31.

Lauretti, L., Riccio, M.L., et al. 1999. Cloning and characterization of *bla*VIM, a new integron-borne metallo-beta-lactamase gene from a

Pseudomonas aeruginosa clinical isolate. *Antimicrob Agents Chemother*, **43**, 1584–90.

Lawrence, J.G. and Roth, J.R. 1999. Genomic flux: genome evolution by gene loss and acquisition. In: Charlebois, R.L. (ed.), *Organization of the prokaryotic genome*. Washington, DC: ASM Press, 263–89.

Leblond, P. and Decaris, B. 1999. Unstable linear chromosomes: the case of Streptomyces. In: Charlebois, R.L. (ed.), *Organization of the prokaryotic genome*. Washington, DC: ASM Press, 235–61.

Lederberg, J. 1952. Cell genetics and hereditary symbiosis. *Physiol Rev*, **32**, 403–30.

Lederberg, E.M. 1960. Genetic and functional aspects of galactose metabolism, *10th Symposium on General Microbiology*. Cambridge, UK: Cambridge University Press, Cambridge, UK.

Lederberg, J. and Lederberg, E.M. 1952. Replica plating and indirect selection of bacterial mutants. *J Bacteriol*, **63**, 399–406.

Lederberg, J. and Tatum, E.L. 1946. Genetic recombination in *Escherichia coli*. *Nature (London)*, **158**, 558.

Lee, N., Wilcox, G., et al. 1974. In vitro activation of the transcription of *araBAD* operon by *araC* activator. *Proc Nat Acad Sci USA*, **71**, 634–8.

Levy, S.B. and Novick, R.P. 1986. *Banbury Report 24: antibiotic resistance genes: ecology, transfer and expression*. New York: Cold Spring Harbor Laboratory.

Lewin, B. 2000. *Genes VII*. Oxford, UK: Oxford University Press.

Lindsay, J.A., Ruzin, A., et al. 1998. The gene for toxic shock toxin is carried by a family of mobile pathogenicity islands in *Staphylococcus aureus*. *Mol Microbiol*, **29**, 527–43.

Little, P.F.R. 1992. Generating 'cloned DNA maps'. *Trends Biotechnol*, **10**, 33–5.

Llosa, M., Gomis-Ruth, F.X., et al. 2002. Bacterial conjugation: a two-step mechanism for DNA transport. *Mol Microbiol*, **45**, 1–8.

Lorenz, M.G. and Wackernagel, W. 1994. Bacterial gene transfer by natural genetic transformation in the environment. *Microbiol Rev*, **58**, 563–602.

Luria, S.E. and Delbruck, M. 1943. Mutations of bacteria from virus sensitivity to virus resistance. *Genetics*, **28**, 491–511.

Lyon, B.R., May, J.W. and Skurray, R.A. 1984. Tn*4001*: a gentamicin and kanamycin resistance transposon in *Staphylococcus aureus*. *Mol Gen Genet*, **193**, 554–6.

Ma, X.X., Ito, T., et al. 2002. Novel type of staphylococcal cassette chromosome *mec* identified in community-acquired methicillin-resistant *Staphylococcus aureus* strains. *Antimicrob Agents Chemother*, **46**, 1147–52.

Marchuk, D., Drumm, M., et al. 1991. Construction of T-vectors, a rapid and general system for direct cloning of unmodified PCR products. *Nucleic Acids Res*, **19**, 1154, .

Marrs, C.F., Ruehl, W.W., et al. 1988. Pilin gene phase variation of *Moraxella bovis* is caused by an inversion of the pilin genes. *J Bacteriol*, **170**, 3032–9.

Marti, T.M., Kunz, C. and Fleck, O. 2002. DNA mismatch repair and mutation avoidance pathways. *J Cell Physiol*, **191**, 28–41.

Massova, I. and Mobashery, S. 1999. Structural and mechanistic aspects of evolution of beta-lactamases and penicillin-binding proteins. *Curr Pharm Des*, **5**, 929–37.

Masters, M. 1996. Generalized transduction. In: Neidhardt, F.C. (ed.), Escherichia coli *and* Salmonella: *cellular and molecular biology*. Washington, DC: ASM Press, 2421–41.

May, J.M., Houghton, R.H. and Perret, C.J. 1964. The effect of growth at elevated temperatures on some heritable properties of *Staphylococcus aureus*. *J Gen Microbiol*, **37**, 157–69.

Mazel, D., Dychinco, B., et al. 1998. A distinctive class of integron in the *Vibrio cholerae* genome. *Science*, **280**, 605–8.

Mazodier, P. and Davies, J. 1991. Gene transfer between distantly related bacteria. *Annu Rev Genet*, **25**, 147–71.

McClintock, B. 1953. Induction of instability at selected loci in maize. *Genetics*, **38**, 579–99.

McDaniel, R., Ebertkhosla, S., et al. 1995. Rational design of aromatic polyketide natural-products by recombinant assembly of enzymatic subunits. *Nature (London)*, **375**, 549–54.

McHugh, G.L., Moellering, R.C., et al. 1975. *Salmonella* Typhimurium resistant to silver nitrate, chloramphenicol and ampicillin. *Lancet*, **1**, 235–40.

McPherson, M.J. and Møller, S.G. 2000. *PCR*. Abingdon, Oxford, UK: Bios Scientific Publishers.

Mendiola, M.V. and de la Cruz, F. 1992. IS*91* transposase is related to the rolling-circle-type replication proteins of the pUB110 family of plasmids. *Nucleic Acids Res*, **20**, 3521.

Mendiola, M.V., Jubete, Y. and de la Cruz, F. 1992. DNA sequence of IS*91* and identification of the transposase gene. *J Bacteriol*, **174**, 1345–51.

Messing, J. 1991. Cloning in M13 phage or how to use biology at its best. *Gene*, **100**, 3–12.

Meynell, E., Meynell, G. and Datta, N. 1968. Phylogenetic relationships of drug-resistance factors and other transmissible bacterial plasmids. *Bacteriol Rev*, **32**, 55–83.

Miller, J.H. 1972. *Experiments in molecular genetics*. New York: Cold Spring Harbor Laboratory.

Miller, J.F. 1994. Bacterial transformation by electroporation. *Methods Enzymol*, **235**, 375–85.

Mitsuhashi, S. 1977. Epidemiology of bacterial drug resistance. In: Mitsuhashi, S. (ed.), *R factor: drug resistance plasmid*. Baltimore: University Park Press, 1–24.

Moll, A., Manning, P.A. and Timmis, K.N. 1980. Plasmid-determined resistance to serum bactericidal activity: a major outer membrane protein, the *traT* gene product, is responsible for plasmid-specified serum resistance in *Escherichia coli*. *Infect Immun*, **28**, 359–67.

Morales, V.M., Bäckman, A. and Bagdasarian, M. 1991. A series of wide-host-range low-copy-number vectors that allow direct screening for recombinants. *Gene*, **97**, 39–47.

Morosini, M.I., Ayala, J.A., et al. 2000. Biological cost of AmpC production for *Salmonella enterica* serotype Typhimurium. *Antimicrob Agents Chemother*, **44**, 3137–43.

Morrison, D., Woodford, N., et al. 1999. DNA banding pattern polymorphism in vancomycin-resistant *Enterococcus faecium* and criteria for defining strains. *J Clin Microbiol*, **37**, 1084–91.

Mortelmans, K.E. and Stocker, B.A.D. 1976. Ultraviolet light protection, enhancement of ultraviolet light mutagenesis, and mutator effect of plasmid R46 in *Salmonella typhimurium*. *J Bacteriol*, **128**, 271–82.

Murgola, E.J. 1985. tRNA, suppression, and the code. *Ann Rev Genet*, **19**, 57–80.

Murphy, E. 1989. Transposable elements in gram-positive bacteria. In: Berg, D.E. and Howe, M.M. (eds), *Mobile DNA*. Washington, DC: ASM Press, 269–88.

Mullis, K., Faloona, F., et al. 1986. Specific enzymatic amplification of DNA in vitro: the polymerase chain reaction. *Cold Spring Harb Symp Quant Biol*, **51**, 263–73.

Naas, T., Mikami, Y., et al. 2001. Characterization of In53, a class 1 plasmid- and composite-transposon-located integron of *Escherichia coli* which carries an unusual array of gene cassettes. *J Bacteriol*, **183**, 235–49.

Nash, H.A. 1996. Site-specific recombination: integration, excision, resolution, and inversion of defined DNA segments. In: Neidhardt, F.C. (ed.), Escherichia coli *and* Salmonella: *cellular and molecular biology*. Washington, DC: ASM Press, 2363–76.

Nield, B.S., Holmes, A.J., et al. 2001. Recovery of new integron classes from environmental DNA. *FEMS Microbiol Lett*, **195**, 59–65.

Nies, D.H. 2003. Efflux-mediated heavy metal resistance in prokaryotes. *FEMS Microbiol Rev*, **27**, 313–39.

Nordmann, P. and Poirel, L. 2002. Emerging carbapenemases in gram-negative aerobes. *Clin Microbiol Infect*, **8**, 321–31.

Normark, S. 1995. beta-Lactamase induction in gram-negative bacteria is intimately linked to peptidoglycan recycling. *Microb Drug Resist*, **1**, 111–14.

Notley-McRobb, L., Seeto, S. and Ferenci, T. 2003. The influence of cellular physiology on the initiation of mutational pathways in *Escherichia coli* populations. *Proc R Soc Lond B*, **270**, 843–8.

Novick, R.P. 1963. Analysis by transduction of mutations affecting penicillinase formation in *Staphylococcus aureus*. *J Gen Microbiol*, **33**, 121–36.

Novick, R.P. 1969. Extrachromosomal inheritance in bacteria. *Bacteriol Rev*, **33**, 210–35.

Novick, R.P. 1987. Plasmid incompatibility. *Microbiol Rev*, **51**, 381–95.

Novick, R.P., Edelman, I. and Lofdahl, S. 1986. Small *Staphylococcus aureus* plasmids are transduced as linear multimers that are formed and resolved by replicative processes. *J Mol Biol*, **192**, 209–20.

Novick, R.P. and Richmond, M.H. 1965. Nature and interactions of the genetic elements governing penicillinase synthesis in *Staphylococcus aureus*. *J Bacteriol*, **90**, 467–80.

Novick, R.P., Clowes, R.C., et al. 1976. Uniform nomenclature for bacterial plasmids: a proposal. *Bacteriol Rev*, **40**, 168–89.

Novick, R.P., Cohen, S., et al. 1977. Plasmids of *Staphylococcus aureus*. In: Bukhari, A.I., Shapiro, J.A. and Adhya, S.L. (eds), *DNA insertion elements, plasmids and episomes*. New York: Cold Spring Harbor Laboratory, 657–62.

Novick, R.P., Edelman, I., et al. 1979. Genetic translocation in *Staphylococcus aureus*. *Proc Natl Acad Sci USA*, **76**, 400–4.

Novick, R.P., Iordanescu, S., et al. 1981. Transduction-related cointegrate formation between staphylococcal plasmids: a new type of site-specific recombination. *Plasmid*, **6**, 159–72.

O'Brien, S.J. *Genetic maps 1987*, Vol. 4. New York: Cold Spring Harbor Laboratory.

Ohtsubo, E., Zenilman, M., et al. 1981. Mechanisms of insertion and cointegration mediated by IS1 and Tn3. *Cold Spring Harb Symp Quant Biol*, **45**, 283–95.

Ørskov, I. and Ørskov, F. 1973. Plasmid-determined H_2S character in *Escherichia coli* and its relation to plasmid-carried raffinose fermentation and tetracycline resistance characters. *J Gen Microbiol*, **77**, 487–99.

Osborn, A.M. and Böltner, D. 2002. When phage, plasmids, and transposons collide: genomic islands, and conjugative- and mobilizable-transposons as a mosaic continuum. *Plasmid*, **48**, 202–12.

Palavecino, E. 2004. Community-acquired methicillin-resistant *Staphylococcus aureus* infections. *Clin Lab Med*, **24**, 403–18.

Palmer, C.J. 2004. Polymerase chain reaction (PCR). In: Schaechter, M. (ed.), *The desk encyclopaedia of microbiology*. Amsterdam: Elsevier Academic Press, 824–8.

Paradkar, A., Trefzer, A., et al. 2003. Streptomyces genetics: a genomic perspective. *Crit Rev Biotechnol*, **23**, 1–27.

Pearl, L.H. 2000. Structure and function in the uracil-DNA-glycosylase superfamily. *Mutat Res*, **460**, 165–81.

Philippon, A., Arlet, G. and Jacoby, G.A. 2002. Plasmid-determined AmpC-type beta-lactamases. *Antimicrob Agents Chemother*, **46**, 1–11.

Philippon, A., Dusart, J., et al. 1998. The diversity, structure and regulation of β-lactamases. *Cell Mol Life Sci*, **54**, 341–6.

Pilacinski, W., Mosharrafa, E., et al. 1977. Insertion sequence IS2 associated with int-constitutive mutants of bacteriophage lambda. *Gene*, **2**, 61–74.

Podglajen, I., Breuil, J. and Collatz, E. 1994. Insertion of a novel DNA sequence, 1S1186, upstream of the silent carbapenemase gene *cfiA*, promotes expression of carbapenem resistance in clinical isolates of *Bacteroides fragilis*. *Mol Microbiol*, **12**, 105–14.

Poirel, L., Naas, T., et al. 2000. Characterization of VIM-2, a carbapenem-hydrolyzing metallo-beta-lactamase and its plasmid- and integron-borne gene from a *Pseudomonas aeruginosa* clinical isolate in France. *Antimicrob Agents Chemother*, **44**, 891–7.

Prescott, L.M., Harley, J.P. and Klein, D.A. 2002. *Microbiology*, 5th edn. New York: McGraw-Hill Higher Education.

Pridmore, R.D. 1987. New and versatile cloning vectors with kanamycin-resistance marker. *Gene*, **56**, 309–12.

Primrose, S.B., Twyman, R.M. and Old, R.W. 2001. *Principles of gene manipulation*, 6th edn. Oxford: Blackwell Science Ltd.

Pritchard, R.H. and Holland, I.B. 1985. *Basic cloning techniques: a manual of experimental procedures*. Oxford: Blackwell Scientific.

Privitera, G., Sebald, M. and Fayolle, F. 1981. Common regulatory mechanism of expression and conjugative ability of a tetracycline resistance plasmid in *Bacteroides fragilis*. *Nature (London)*, **278**, 657–8.

Pugsley, A.P. and Oudega, B. 1987. Methods for studying colicins and their plasmids. In: Hardy, K. (ed.), *Plasmids, a practical approach*. Oxford and Washington, DC: IRL Press, 105–61.

Ramalingam, B., Baulard, A.R., et al. 2004. Cloning, expression, and purification of the 27 kDa (MPT51, Rv3803c) protein of *Mycobacterium tuberculosis*. *Protein Expr Purif*, **36**, 53–60.

Rasmussen, B.A. and Kovacs, E. 1991. Identification and DNA sequence of a new *Bacteroides fragilis* insertion sequence-like element. *Plasmid*, **25**, 141–4.

Rawlings, D.E. 1999. Proteic toxin-antitoxin, bacterial plasmid addiction systems and their evolution with special reference to the pas system of pTF-FC2. *FEMS Microbiol Lett*, **176**, 269–77.

Recchia, G.D. and Hall, R.M. 1995. Gene cassettes: a new class of mobile element. *Microbiology UK*, **141**, 3015–27.

Reif, H.J. and Saedler, H. 1975. IS1 is involved in deletion formation in the *gal* region of *E. coli* K12. *Mol Gen Genet*, **137**, 17–28.

Reeve, E.C.R. and Braithwaite, J.A. 1973. Lac-plus plasmids are responsible for the strong lactose-positive phenotype found in many strains of *Klebsiella* species. *Genet Res*, **22**, 329–33.

Reynolds, P.E. and Fuller, C. 1986. Methicillin-resistant strains of *Staphylococcus aureus*: presence of identical additional penicillin-binding protein in all strains examined. *FEMS Microbiol Lett*, **33**, 251–4.

Ricci, V., Peterson, M.L., et al. 2004. Role of topoisomerase mutations and efflux in fluoroquinolone resistance of *Bacteroides fragilis* clinical isolates and laboratory mutants. *Antimicrob Agents Chemother*, **48**, 1344–6.

Riccio, M.L., Franceschini, N., et al. 2000. Characterization of the metallo-beta-lactamase determinant of *Acinetobacter baumannii* AC-54/97 reveals the existence of *bla*(IMP) allelic variants carried by gene cassettes of different phylogeny. *Antimicrob Agents Chemother*, **44**, 1229–35.

Rodicio, M.R., Bruton, C.J. and Chater, K.F. 1985. New derivatives of the Streptomyces temperate phage phi C31 useful for the cloning and functional analysis of Streptomyces DNA. *Gene*, **34**, 283–92.

Rosenberg, S.M. 1997. Mutation for survival. *Curr Opin Genet Dev*, **7**, 829–34.

Rosenow, C., Saxena, R.M., et al. 2001. Prokaryotic RNA preparation methods useful for high density array analysis: comparison of two approaches. *Nucleic Acids Res*, **29**, E112.

Rossolini, G.M., Walsh, T. and Amicosante, G. 1996. The *Aeromonas* metallo-beta-lactamases: genetics, enzymology, and contribution to drug resistance. *Microb Drug Resist*, **2**, 245–52.

Roth, J.A. 1988. *Virulence mechanisms of bacterial pathogens*. Washington, DC: American Society for Microbiology.

Rowe-Magnus, D.A., Guerout, A.M. and Mazel, D. 1999. Super-integrons. *Res Microbiol*, **150**, 641–51.

Rowe-Magnus, D.A., Guerout, A.M., et al. 2001. The evolutionary history of chromosomal super-integrons provides an ancestry for multiresistant integrons. *Proc Nat Acad Sci USA*, **98**, 652–7.

Ruiz, J. 2003. Mechanisms of resistance to quinolones: target alterations, decreased accumulation and DNA gyrase protection. *J Antimicrob Chemother*, **51**, 1109–17.

Saedler, H. and Reif, H.J. 1974. IS2, a genetic element for turn-off and turn-on of gene activity in *E. coli*. *Mol Gen Genet*, **132**, 265–89.

Salyers, A.A. and Shoemaker, N.B. 1995. Conjugative transposons: the force behind the spread of antibiotic resistance genes among *Bacteroides* clinical isolates. *Anaerobe*, **1**, 143–50.

Salyers, A.A. and Whitt, D.D. 2002. *Bacterial pathogenesis: a molecular approach*, 2nd edn. Washington, DC: ASM Press.

Salyers, A.A., Shoemaker, N.B., et al. 1995. Conjugative transposons: an unusual and diverse set of integrated gene transfer elements. *Microbiol Rev*, **59**, 579–90.

Sambrook, J., Fritsch, E.F. and Maniatis, T. 1989. *Molecular cloning: a laboratory manual*, 2nd edn. Cold Spring Harbor: Cold Spring Harbor Press.

Samraoui, B., Sutton, B.J., et al. 1986. Tertiary structural similarity between a class A β-lactamase and a penicillin-sensitive d-alanyl carboxypeptidase-transpeptidase. *Nature (London)*, **320**, 378–80.

Sancar, A. 1996. DNA excision repair. *Ann Rev Biochem*, **65**, 43–81.

Sanderson, K.E. 1976. Genetic relatedness in the family Enterobacteriaceae. *Annu Rev Microbiol*, **30**, 327–49.

Sarnovsky, R.J., May, E.W. and Craig, N.L. 1996. The Tn7 transposase is a heteromeric complex in which DNA breakage and joining activities are distributed between different gene products. *EMBO J*, **15**, 6348–61.

Saunders, J.R. and Saunders, V.A. 1988. Bacterial transformation with plasmid DNA. In Grinstead, J. and Bennett, P.M. (eds), *Methods in microbiology*, Vol 21, *Plasmid technology*, 2nd edn. London: Academic Press, 79–128.

Saunders, J.R., Hart, C.A. and Saunders, V.A. 1986. Plasmid-mediated resistance to β-lactam antibiotics in gram-negative bacteria: the role of in-vivo recyclization reactions in plasmid evolution. *J Antimicrob Chemother*, **18**, Suppl C, 57–66.

Schofield, M.J. and Hsieh, P. 2003. DNA mismatch repair: molecular mechanisms and biological function. *Ann Rev Microbiol*, **57**, 579–608.

Scott, J.R. 1984. Regulation of plasmid replication. *Microbiol Rev*, **48**, 1–23.

Seal, D.V., McSwiggan, D.A., et al. 1981. Characterisation of an epidemic strain of *Klebsiella* and its variants by computer analysis. *J Med Microbiol*, **14**, 295–305.

Shannon, M.J.R. and Unterman, R. 1993. Evaluating bioremediation: distinguishing fact from fiction. *Annu Rev Microbiol*, **47**, 715–38.

Shapiro, J.A. 1979. Molecular model for the transposition and replication of bacteriophage Mu and other transposable elements. *Proc Natl Acad Sci USA*, **76**, 1933–7.

Shapiro, J.A. 1983. *Mobile genetic elements*. Orlando, FL: Academic Press.

Sherratt, D.S. 1986. Control of plasmid maintenance. In Booth, I.R. and Higgins, C.F. (eds), *Regulation of gene expression, 25 years on*. Symposium 39 of the Society for General Microbiology. Cambridge: Cambridge University Press, 239–50.

Sherratt, D.S. 1989. Tn3 and related transposable elements: site-specific recombination and transposition. In: Berg, D.E. and Howe, M.M. (eds), *Mobile DNA*. Washington, DC: ASM Press, 163–84.

Shipley, P.L., Gyles, C.L. and Falkow, S. 1978. Characterization of plasmids that encode for the K88 colonization antigen. *Infect Immun*, **20**, 559–66.

Silver, S. 2003. Bacterial silver resistance: molecular biology and uses and misuses of silver compounds. *FEMS Microbiol Rev*, **27**, 341–53.

Silver, S. and Phung, L.T. 1996. Bacterial heavy metal resistance: new surprises. *Annu Rev Microbiol*, **49**, 145–74.

Sinha, R.P. and Hader, D.P. 2002. UV-induced DNA damage and repair: a review. *Photochem Photobiol Sci*, **1**, 225–63.

Sisco, K.L. and Smith, H.O. 1979. Sequence-specific DNA uptake in *Haemophilus* transformation. *Proc Natl Acad Sci USA*, **76**, 972–6.

Sissi, C. and Palumbo, M. 2003. The quinolone family: from antibacterial to anticancer agents. *Curr Med Chem Anti-Canc Agents*, **3**, 439–50.

Skold, O. 2001. Resistance to trimethoprim and sulfonamides. *Vet Res*, **32**, 261–73.

Smith, G.R. 1988. Homologous recombination in procaryotes. *Microbiol Rev*, **52**, 1–28.

Smith, G.R. 1989. Homologous recombination in *E. coli*: multiple pathways for multiple reasons. *Cell*, **58**, 807–9.

Smith, H.W. and Huggins, M.B. 1976. Further observations on the association of the colicine V plasmid of *Escherichia coli* with pathogenicity and with survival in the alimentary tract. *J Gen Microbiol*, **92**, 335–50.

Smith, H.W., Parsell, Z. and Green, P. 1978. Thermosensitive H1 plasmids determining citrate utilization. *J Gen Microbiol*, **109**, 305–11.

So, M. and McCarthy, B.J. 1980. Nucleotide sequence of the bacterial transposon Tn1681 encoding a heat-stable (ST) toxin and its identification in enterotoxigenic *Escherichia coli* strains. *Proc Natl Acad Sci USA*, **77**, 4011–15.

Sommer, H. and Saedler, H. 1979. IS2-43 and IS2-44: new alleles of the insertion sequence IS2 which have promoter activity. *Mol Gen Genet*, **175**, 53–6.

Spain, J.C. 1995. Biodegradation of nitroaromatic compounds. *Annu Rev Microbiol*, **49**, 523–55.

Stanisich, V.A. 1988. Identification and analysis of plasmids at the genetic level. In: Grinsted, J. and Bennett, P.M. (eds), *Methods in microbiology*, Vol. 21. . Orlando FL: Academic Press, 11–47.

Stokes, H.W. and Hall, R.M. 1989. A novel family of potentially mobile DNA elements encoding site-specific gene-integration functions: integrons. *Mol Microbiol*, **3**, 1669–83.

Stokes, H.W., O'Gorman, D.B., et al. 1997. Structure and function of 59-base element recombination sites associated with mobile gene cassettes. *Mol Microbiol*, **26**, 731–45.

Studier, F.W., Rosenberg, A.H., et al. 1990. Use of T7 RNA polymerase to direct expression of cloned genes. *Methods Enzymol*, **185**, 60–89.

Stuy, J.H. 1980. Chromosomally integrated conjugative plasmids are common in antibiotic-resistant *Haemophilus influenzae*. *J Bacteriol*, **142**, 925–30.

Summers, A.O. and Jacoby, G.A. 1977. Plasmid-determined resistance to tellurium compounds. *J Bacteriol*, **129**, 276–81.

Summers, A.O. and Silver, S. 1978. Microbial transformations of metals. *Annu Rev Microbiol*, **32**, 637–72.

Summers, A.O., Wireman, J., et al. 1993. Mercury released from dental 'silver' fillings provokes an increase in mercury- and antibiotic-resistant bacteria in oral and intestinal floras of primates. *Antimicrob Agents Chemother*, **37**, 825–34.

Tan, H.M. 1999. Bacterial catabolic transposons. *Appl Microbiol Biotechnol*, **51**, 1–12.

Tavakoli, N., Comanducci, A., et al. 2000. IS1294, a DNA element that transposes by RC transposition. *Plasmid*, **44**, 66–84.

Taylor, P.W. 1983. Bactericidal and bacteriolytic activity of serum against gram-negative bacteria. *Microbiol Rev*, **47**, 46–83.

Terakado, N., Hamaoka, T. and Danbara, H. 1988. Plasmid-mediated serum resistance and alterations in the composition of lipopolysaccharides in *Salmonella dublin*. *J Gen Microbiol*, **134**, 2089–93.

Tetaz, T.J. and Luke, R.K.J. 1983. Plasmid-controlled resistance to copper in *Escherichia coli*. *J Bacteriol*, **154**, 1263–8.

Thomas, C.M. 1989. *Promiscuous plasmids of gram-negative bacteria*. London: Academic Press.

Thomas, C.M. (ed.) 2000. *The horizontal gene pool*. Amsterdam: Harwood Academic Publishers.

Thomas, C.M. and Smith, C.A. 1987. Incompatibility group P plasmids: genetics, evolution, and use in genetic manipulation. *Annu Rev Microbiol*, **41**, 77–101.

Thompson, R. 1988. Plasmid cloning vectors. In: Grinsted, J. and Bennett, P.M. (eds), *Methods in microbiology*, Vol. 21. Orlando, FL: Academic Press, 179–204.

Toussaint, A. and Résibois, A. 1983. Phage Mu: transposition as a life-style. In: Shapiro, J.A. (ed.), *Mobile genetic elements*. Orlando, FL: Academic Press, 105–58.

Towner, K.J. and Vivian, A. 1976. RP4-mediated conjugation in *Acinetobacter calcoaceticus*. *J Gen Microbiol*, **93**, 355–60.

Trieu-Cuot, P. and Courvalin, P. 1986. Evolution and transfer of aminoglycoside resistance genes under natural conditions. *J Antimicrob Chemother*, **18**, Suppl C, 93–102.

Ulijasz, A.T., Grenader, A. and Weisblum, B. 1996. A vancomycin-inducible LacZ reporter system in *Bacillus subtilis*: induction by antibiotics that inhibit cell wall synthesis and by lysozyme. *J Bacteriol*, **178**, 6305–9.

Urwin, R. and Maiden, M.C. 2003. Multi-locus sequence typing: a tool for global epidemiology. *Trends Microbiol*, **11**, 479–87.

Utsui, Y. and Yokota, T. 1985. Role of an altered penicillin-binding protein in methicillin-resistant and cephem-resistant *Staphylococcus aureus*. *Antimicrob Agents Chemother*, **28**, 397–403.

Vaisvila, R., Morgan, R.D., et al. 2001. Discovery and distribution of super-integrons among pseudomonads. *Mol Microbiol*, **42**, 587–601.

van Leeuwen, W.B., Jay, C., et al. 2003. Multilocus sequence typing of *Staphylococcus aureus* with DNA array technology. *J Clin Microbiol*, **41**, 3323–6.

Villarroel, R., Hedges, R.W., et al. 1983. Heteroduplex analysis of P-plasmid evolution: the role of insertion and deletion of transposable elements. *Mol Gen Genet*, **189**, 390–9.

Wachsmuth, I.K., Davis, B.R. and Allen, S.D. 1979. Ureolytic *Escherichia coli* of human origin: serological, epidemiological, and genetic analysis. *J Clin Microbiol*, **10**, 897–902.

Wade, M.M. and Zhang, Y. 2004. Mechanisms of drug resistance in *Mycobacterium tuberculosis*. *Front Biosci*, **9**, 975–94.

Wahl, G.M., Lewis, K.A., et al. 1987. Cosmid vectors for rapid genome walking, restriction mapping, and gene transfer. *Proc Nat Acad Sci USA*, **84**, 2160–4.

Waldor, M.K. 1998. Bacteriophage biology and bacterial virulence. *Trend Microbiol*, **6**, 295–7.

Waldor, M.K. and Mekalanos, J.J. 1996. Lysogenic conversion by a filamentous phage encoding cholera toxin. *Science*, **272**, 1910–14.

Walsh, T.R. and Howe, R.A. 2002. The prevalence and mechanisms of vancomycin resistance in *Staphylococcus aureus*. *Ann Rev Microbiol*, **56**, 657–75.

Walther-Rasmussen, J. and Hoiby, N. 2002. Plasmid-borne AmpC beta-lactamases. *Can J Microbiol*, **48**, 479–93.

Watanabe, T. 1963. Infective heredity of multiple drug resistance in bacteria. *Bacteriol Rev*, **27**, 87–115.

Watson, J.D., Gilman, M., et al. 1992. *Recombinant DNA*, 2nd edn. Scientific American Books, Scientific American, Inc, 191–211.

Watson, J.D., Hopkins, N.H., et al. 1987. *Molecular biology of the gene*, Vol. 1, 4th edn. Menlo Park, CA: Benjamin/Cummings.

Webb, V. and Davies, J. 1993. Antibiotic preparations contain DNA – a source of drug resistance genes? *Antimicrob Agents Chemother*, **37**, 2379–84.

Weber, T., Welzel, K., et al. 2003. Exploiting the genetic potential of polyketide producing Streptomycetes. *J Biotechnol*, **106**, 221–32.

Weisberg, R.A. 1996. Specialized transduction. In: Neidhardt, F.C. (ed.), Escherichia coli *and* Salmonella: *cellular and molecular biology*. Washington, DC: ASM Press, 2442–8.

Weisser, J. and Wiedemann, B. 1985. Elimination of plasmids by new 4-quinolones. *Antimicrob Agents Chemother*, **28**, 700–2.

White, M.W. and Radke, J.R. 1997. Methods to prepare RNA and to isolate developmentally regulated genes from Eimeria. *Methods*, **13**, 158–70.

Wiedemann, B., Meyer, J.F. and Zuhlsdorf, M.T. 1986. Insertions of resistance genes into Tn*21*-like transposons. *J Antimicrob Chemother*, **18**, Suppl C, 85–92.

Wiley, B.B. and Rogolsky, M. 1977. Molecular and serological differentiation of staphylococcal exfoliative toxin synthesized under chromosomal and plasmid control. *Infect Immun*, **18**, 487–94.

Wilkins, B.M. 1995. Gene transfer by bacterial conjugation: diversity of systems and functional specializations. In: Baumberg, S. and Young, J.P.W. (eds), *Society for General Microbiology Symposium 52, Population genetics of bacteria*. Cambridge: Cambridge University Press, 59–88.

Willetts, N. 1988. Conjugation. In: Grinsted, J. and Bennett, P.M. (eds), *Methods in microbiology*, Vol. 21. . Orlando, FL: Academic Press, 49–77.

Williams, P.H. 1979. Novel iron uptake system specified by ColV plasmids: an important component in the virulence of invasive strains of *Escherichia coli*. *Infect Immun*, **26**, 925–32.

Wood, A.G. and Konisky, J. 1985. Activation of expression of a cloned archaebacterial gene in *Escherichia coli* by IS*2*, IS*5*, or deletions. *Mol Gen Genet*, **198**, 309–14.

Woodgate, R. 2001. Evolution of the two-step model for UV-mutagenesis. *Mutat Res*, **485**, 83–92.

Yarmolinsky, M.B. 1995. Programmed cell death in bacterial populations. *Science*, **267**, 836–7.

Yanofsky, C. 1987. Operon-specific control by transcription attenuation. *Trends Genet*, **3**, 356–60.

Yokota, T., Kanamaru, Y., et al. 1969. Recombination between a thermosensitive kanamycin resistance factor and a nonthermosensitive multiple-drug resistance factor. *J Bacteriol*, **98**, 863–73.

York, D., Welch, K., et al. 1998. Simple and efficient generation *in vitro* of nested deletions and inversions: Tn*5* intramolecular transposition. *Nucleic Acids Res*, **26**, 1927–33.

Young, K.K.Y. and Edlin, G. 1983. Physical and genetical analysis of bacteriophage T4 generalized transduction. *Mol Gen Genet*, **192**, 241–6.

Zechner, E.L., de la Cruz, F., et al. 2000. Conjugative-DNA transfer processes. In: Thomas, C.M. (ed.), *The horizontal gene pool: bacterial plasmids and gene spread*. Amsterdam: Harwood Academic Publishers, 87–174.

Zhang, H.Z., Hackbarth, C.J., et al. 2001. A proteolytic transmembrane signaling pathway and resistance to beta-lactams in staphylococci. *Science*, **291**, 1962–5.

Zieg, J., Silverman, M., et al. 1977. Recombinational switch for gene expression. *Science*, **196**, 170–2.

Zipser, D., Zabell, S., et al. 1970. Fine structure of the gradient of polarity in the Z gene of the *lac* operon of *Escherichia coli*. *J Mol Biol*, **49**, 251–4.

Zoller, M.J. and Smith, M. 1982. Oligonucleotide-directed mutagenesis using M13-derived vectors: an efficient and general procedure for production of point mutations in any fragment of DNA. *Nucleic Acids Res*, **10**, 6487–500.

5

Bacteriocins and bacteriophages

MARTIN DAY

INTRODUCTION

Bacteriocins are a group of proteins secreted by bacteria that kill or inhibit competing strains. Bacteriophages are viruses that infect bacteria. Both agents have antimicrobial activity and they share some features, such as the uptake site on their host bacterium. However, examination of their modes of action, reproduction, and genetics reveals differences. The major difference is that only bacteriophages carry a genome (DNA or RNA) enabling them to reproduce in the cells they infect; bacteriocins are only proteins and cannot reproduce. Irradiating a phage will damage its genome irreparably; thus, an irradiated lysate will have no killing activity. Bacteriocins have no genome and thus are insensitive to this level of irradiation and retain their activity fully. Both bacteriocins and phages can kill susceptible cells.

Bacteriocins have for decades been utilized as food additives to limit spoilage. The continued demand for a longer shelf-life of food and other biologically degradable products has meant that these desirable properties of these bioactive agents have been more seriously investigated in recent years when research has shown phages of lactobacilli are a serious problem for the dairy industry. They have been used in the medical arena for phage typing of pathogens such as *Staphylococcus* spp. and as antibacterials. Some are also widely used as recombinant DNA vectors as cloning and expression vectors.

BACTERIOCINS

Microorganisms naturally produce a range of protein components from simple polypeptides to very complex macromolecules, such as antibiotics, toxins, pili, adhesins, siderophores, flagella, etc. Many of these form part of an array that forms a defense system. Bacteriocins are grouped under the term 'toxins' and provide a means of defense against, or a growth advantage over, other microorganisms in the same ecological niche. The term 'colicin' is used to describe those antagonistic compounds produced by *Escherichia coli* (Frederique 1957). The term 'colicin' was coined by Gratia and Frederique (1946) to describe these substances, the effects of which had been observed by Gratia (1925). Jacob and Woolman (1953) introduced the general term 'bacteriocin' to describe substances with similar activity produced by a wide range of bacteria. Tagg et al. (1976) defined bacteriocins, a subgroup of bacterial toxins, as proteinaceous compounds that kill closely related bacteria. Although this is true for most bacteriocins, it is evident that these molecules take many forms and may have bactericidal actions beyond closely related species. Individual bacteriocins are named by attaching 'cin' to the root of the genus or species name. For example, *Pseudomonas aeruginosa* produces aeruginocins and *Bacillus megaterium* produces megacins. The synthesis of most colicins (and this will probably be true of bacteriocins in general) are SOS-inducible (Pugsley 1984a,b). Frequently bacterial species carry genes that encode both the production of one or more bacteriocins and immunity to them on the chromosome or on plasmids (Frank 1994). The significance of the latter location is that the distribution of individual plasmid types in bacterial communities fluctuates under selection, for and against the plasmid and host cell, which provides opportunities for closely related and competing populations to gain and lose advantage. Loss and acquisition of bacteriocins is a consequence of plasmid stability, incompatibility and transfer. Thus, a great diversity of bacteriocins can be produced within a single species. Bacteriocins have a spectrum of sizes; generally proteinaceous agents, they are sometimes complexed with lipids, carbohydrates, or other distinctive proteins (Lewus et al. 1991; Nissen-Meyer et al. 1992; Jiminez-Diaz and Rios-Sanchez 1993). There are three general classes:

1 microcins, small molecules produced in stationary phase by gram-negative bacteria
2 lantibiotics, small molecules produced by gram-positive bacteria
3 bacteriocins, a group encompassing medium to large phage-like structures, whose synthesis generally appears to be induced by the SOS response.

It has been shown that bacteriocin synthesis provides the strain with a growth advantage over isogenic nonproducing strains (Ruiz-Barba et al. 1994). Since producing these antimicrobials is an energy-costly process, the benefits to the producer must be due to selective pressure. The energy cost must be less than the energy benefit resulting from their synthesis. Viewed in this light their synthesis should be well regulated and expression directed to times when it will be of benefit. Thus, bacteriocin synthesis reflects this by being sensitive to pH, temperature, medium composition, SOS response induction, etc.

These proteins share many characteristics. Many have a low molecular weight, are cationic, are amphiphilic, tend to aggregate, and are benign to the producing organism. They kill sensitive cells in distinct ways, targeting the cell membrane (e.g. phospholipases), affecting DNA tertiary structure (DNA gyrase), or affecting DNA/RNA integrity (ribonuclease or deoxyribonuclease activity). Since their sites of synthesis and modes of action are frequently intracellular, they must also be able to enter susceptible cells to exert their effects. Many are capable of killing activity at extremely low concentrations (one to a few molecules per susceptible cell) and target specific cellular pathways. Death of the cell is a consequence of transport collapse at membrane level, depleting proton-motive force (pmf) and resulting in a rapid reduction in indispensable metabolites or ions, or blocking of macromolecule synthesis (protein or DNA). Thus, due to the specificity of the target, the effect on cell metabolism is catastrophic. There is a further and historical complication in this analysis, as bacteriocins are effectively antibiotics, so there is an activity of structural and biochemical overlap between these molecules. Historically, some were initially regarded as antibiotics and have now been recognized as bacteriocins, and vice versa for others. Thus, their diversity in types and modes of action provides an indication that these molecules have great significance for microbial populations and the dynamics of interactions of these within communities. Bacteriocins are probably present in all bacterial species, and within each species there are probably hundreds of different kinds (Riley and Gordon 1992). For example eubacteria (*E. coli*), the archae (*Halobacteria* Torreblanca et al. 1994), and *Streptomycetes* all commonly produce broad-spectrum antimicrobials (Saadoun et al. 1999). Microbes invest considerable energy in their synthesis. What is less clear is how this diversity arose and what roles these agents serve in microbial communities.

Assay

A simple method for assaying bacteriocin activity is to take the supernate of an overnight culture of the putative bacteriocinogenic strain grown in broth culture, dilute this as a series of doubling dilutions, and place 10-μl drops on the surface of a dry nutrient agar plate pre-spread with a lawn of the indicator bacterium. After overnight incubation the lawn will have grown and the area containing bacteriocin will show a circular zone of clearing. The further down the dilution series the inhibition zones are found, the more bacteriocin is produced by the strain (Benkerroum et al. 1993). This demonstration of the inhibitory effect is also the basis of bacteriocin typing methods.

Modes of action

A neat and satisfying taxonomy of bacteriocins remains to evolve but may come through an interrogation of bacteriocin sequences. Bacteriocins have been described variously in terms of size, types of activity, and chemical nature. As there is no consistent theme it is difficult to obtain a proper grasp of their relationships. It seems sensible to categorize them through their modes of action, but the problem with this approach is that for most, the mechanisms of activity are unknown. However, the general strategies employed by these antimicrobial agents are illustrated by the examples that follow.

DESTRUCTION OF MEMBRANE POTENTIAL

The membrane is composed of many different molecules in complex functional associations. Bacteriocins that target individual proteins, in associations concerned with membrane potential, lead directly to a loss in pmf. Bacteriocins produced by lactic acid bacteria commonly act by this mechanism, but the role of receptor proteins and the mechanism by which pmf is depleted remain unresolved. Many that act in this way are lantibiotics and are produced only by gram-positive bacteria. The most striking feature of lantibiotics is the occurrence of intramolecular rings, introduced by the thioether amino acids lanthionine and 3-methyllanthionine, and unusual amino acids such as didehydroalanine and didehydrobutyrine (Bierbaum and Sahl 1993). These smaller bioactive molecules, which in gram-positive strains are 30–60 amino acids in size (Jack et al. 1995), as a group have both broad and narrow spectra of activity. Some are phospholipases that cause susceptible cells to leak intracellular contents through the damaged cell membranes (Kim 1993). In gram-negative bacteria some of the colicins destroy the membrane potential by forming pores in the membrane (Lau et al. 1992). Colicin A is an example of a water-soluble protein that inserts into lipid bilayers and produces this effect (Lakey et al. 1992, 1994).

NUCLEOLYTIC AND RELATED ACTIVITIES

Sano et al. (1993) examined pyocins S1 and S2, S-type bacteriocins of *P. aeruginosa* that have different receptor recognition specificities. The sequence homology suggests that pyocins S1 and S2 originated from a common ancestor of the E2 group colicins. Purified pyocins S1 and S2 make up a complex of the two proteins. Both pyocins cause breakdown of chromosomal DNA, as does colicin E2, as it has endonucleolytic activity (Lau et al. 1992). Colicin E3 has RNAse activity, cleaving the 16S RNA moiety of the 30S ribosome subunit. Microcin B17 inhibits DNA gyrase (Vizan et al. 1991).

INHIBITION OF PROTEIN SYNTHESIS

The large particulate bacteriocins, e.g. the pyocin-R (aeruginocins) produced by *P. aeruginosa* (Hayashi et al. 1994), are derived from defective phage particles; they contain the phage component capable of attaching to the cell surface. This class of agent generally shows a narrow spectrum of specificity, inhibits protein synthesis, but does not lyse the cell. The genes specifying this type of bacteriocin are encoded chromosomally.

Synthesis

The physiological state of the cell affects all cellular metabolism. The biosynthesis of bacteriocins is no exception and thus the yields of bacteriocin depend on local environmental influences on the producer cell physiology and the regulatory mechanisms governing the expression of the bacteriocin. For example, Malkhosyan et al. (1991) showed that levels of DNA supercoiling are influenced by anaerobic metabolism and that this regulates the level of expression of the colicin genes. The expression of the *cea* gene, which encodes colicin El, on the ColE1 plasmid was greatly increased when the cells were grown anaerobically (Eraso and Weinstock 1992). By using *cea–lacZ* fusions to quantitate expression, levels of β-galactosidase produced under aerobic conditions from the fusion were found to be only a few percent of the anaerobic levels. The gene is also

induced by the SOS response and subject to catabolite repression.

The majority of bacteriocins, such as colicins, aeruginocins, megacins, etc., are produced as the result of normal gene expression. Their synthesis starts with gene transcription, producing an mRNA, which is then translated into an active protein. This synthetic sequence is mediated by ribosomes (Pugsley 1984a, b). The lantibiotics class of bacteriocins (Hansen 1993) frequently contains unusual amino acids that contribute to their properties and functions. Most are synthesized as peptides by nonribosomal mechanisms (Klienkauf and von Dohren 1990), but some are synthesized by pathways that involve post-translational modification of ribosomally synthesized precursor peptides. Examples of the latter are nicin and subtilisin (Nishio et al. 1983).

RELEASE FROM CELLS

The microcins, e.g. colV, are small (6 kDa) and are not induced by the SOS system; their release coincides with the stationary phase. They do not utilize a lysis protein for release but have a dedicated pathway (Gilson et al. 1990; Van der Wal et al. 1995). The lantibiotics generally use a dedicated secretion pathway (Fath and Kolter 1993).

UPTAKE SITES

It has long been recognized that only a few molecules (perhaps only one) are necessary to kill a susceptible cell, and the receptor targets are known for several bacteriocins, in particular some of the colicins. Susceptible cells bear specific protein receptors in their outer membranes, which explains why colicins attach only to certain strains of bacteria.

'Tolerance' and 'resistance' are two terms that are often incorrectly and synonymously used in medical microbiology. The term 'tolerance' describes a state in which the bacteriocin fails to gain access to the sensitive target site within the cell because the receptor is absent or altered so the colicin remains unattached to the cell; thus, it cannot exert a lethal effect. Thus, tolerance may be produced by mutations (e.g. of *tolA*, *tolB*, or *tolC* genes) that affect the outermembrane proteins and the interactions between the cytoplasmic and outer membranes, which prevent the uptake of the bacteriocin into the cell. Resistance occurs when the bacteriocin can be taken up but, because the site is lost or altered by mutation, the cell remains unaffected. There is often a commonality in receptors, which also have a cellular functional role, utilized by phage and bacteriocins. For example, the site used by colicin κ and bacteriophage T6 is implicated in nucleoside uptake. The colicins E1, E2, and E3, and phage BF2 (phage T5-like) use a glycoprotein receptor involved in vitamin B12 uptake. TonB is a membrane-transport protein utilized by many substances and *tonB-* membrane mutants are tolerant to colV

(Pugsley and Schwartz 1985). This commonality signifies that overlap between resistance to inhibitory molecules can be shown.

Host range

Although the general statement that most bacteriocins have a narrow host range (Jack et al. 1995) is true for many, work on bacteriocins produced by *B. megaterium* (megacins), for example, shows that some have a broad spectrum and attack most strains of this species. These megacins are phospholipases and cause susceptible cells to leak intracellular contents by attacking the membrane phospholipids. The bacteriocins produced by lactic acid bacteria exhibit a relatively broad antimicrobial spectrum and are active against several food-spoilage and health-threatening microorganisms (Kim 1993).

Immunity

The production of bacteriocins is the result of deregulation of one or more genes and, in many instances, the release is by a dedicated export mechanism and not cell lysis (Lakey et al. 1994). Strains harboring plasmid or chromosomally encoded colicins, e.g. colicin E3, have immunity to the lethal activity of the colicin (Riley 1993; Frank 1994).

Clinical, technological, and ecological importance

Applications of bacteriocinogeny have been used as epidemiological tools through bacteriocin typing of clinical strains to aid in the identification or discrimination of opportunist and pathogenic bacterial strains. Bacteriocins that show narrow specificity within individual species or genera, such as the bacteriocins of gram-negative bacilli, have been particularly useful in bacteriocin typing. A strain may be characterized actively by the range of activity of its bacteriocins against a set of indicator strains of the same or closely related species, or passively in which it is characterized by the pattern of its susceptibility to the bacteriocins of a set of indicator strains. The former approach has proved particularly useful in typing isolates of *Shigella sonnei* (see chapter 53, *Shigella*) and *P. aeruginosa* (see chapters 62, Pseudomonas; and 63, Burkholderia spp. and related genera), and a combination of the two approaches has been used for *Proteus* spp. (see chapter 55, *Proteus, Morganella* and *Providencia*). In colicin and pyocin typing, strains to be typed are streaked across an agar plate and grown overnight so that their bacteriocins diffuse into the agar. Growth is removed from the surface and remaining viable bacteria are killed by exposure to chloroform, and then the indicator strains are streaked across the original line of growth and again incubated overnight. The

bacteriocin type is indicated by the pattern of inhibition of the indicator strains (Gillis 1964).

Bacteriocins have been used in the food industry for a long time. There is a potential for this to increase as more food is required to have a longer shelf-life. Applications stem from their lack of mammalian toxicity and the spectrum of activities available. For example, bacteriocins produced by lactic acid bacteria exhibit a relatively broad antimicrobial spectrum and are active against several food-spoilage and health-threatening microorganisms (Kim 1993) and some produce bacteriocins active against *Listeria monocytogenes*, a foodborne pathogen (Hechard et al. 1993). Nicin, produced by some strains of *Lactococcus lactis subsp. lactis*, was originally described in 1928 and is the most highly characterized bacteriocin produced by lactic acid bacteria. Nicin has been permitted as a food additive in the UK since the early 1960s and is currently an accepted food additive in at least 45 other countries (Harris et al. 1992). One example of its use is in the preservation of liquid egg (Delvesbroughton et al. 1992).

The activity or role of colicins in vivo is not clear. Pugsley (1984a) suggested that high proteolytic activity in the intestinal tract (an anaerobic environment) would degrade colicins and thus lead to an effective loss of their activity. Later Luria and Suit (1987) stated that 'colicinogeny appears an unnecessary complication in the life of coliform bacteria.' It seems a little premature to write off a group of proteins that are expressed from highly organized genetic systems and that selectively target specific pathways in selected species. It appears more likely that they do have an ecological role to promote the growth of the producer against its immediate competitors. These are likely to be strains with a similar metabolism (biochemistry and physiology), and will be related strains and species. As the particular and immediate requirement for bacteriocin activity is to inhibit cells adjacent to the one producing the bacteriocin, it is local spatial activity that is required. An agent that diffuses into larger volumes will help other competitors. This argument is supported by Frank (1994), whose hypothesis was that bacteriocin diversity in species could not occur if the organisms mixed freely. Spatial variation and poor habitats favored both susceptibility and bacteriocin diversity in strains. Bacteriocin producers are favored in good habitats where competition is likely to be high.

BACTERIOPHAGES

Bacteriophages (phages) are viruses that grow in bacterial cells utilizing their biosynthetic systems for reproduction. Their effects were first observed by Twort (1915), who described an infectious agent that distorted the morphology of staphylococcal colonies. D'Herelle (1917) found a filterable agent that sterilized broth cultures of *Shigella* spp., and he believed such agents might be useful in combating bacterial diseases, as they had no infectivity towards eukaryotic cells. More information about the history and current status of phage therapy can be obtained from a review by Summers (2001).

Phages are extremely common in the environment and can be found wherever their host bacteria are present. They are readily detected for most bacterial species. It is probable that all bacteria are sensitive to one or more phages which thus prey on their specific host strain or strains; it is rare to find one that infects a large number of species. The population density of their host cells is a major component determining the density and distribution of phage (Ogunseitan et al. 1990; Proctor and Fuhrman 1990). For example, high counts of phage occur in biofilms ($>10^8$/cm) (Ewert and Paynter 1980) where bacterial density is high, but are far lower in bulk waters (10^{3-4} cfu/ml) (Saye and Miller 1989) where bacterial density is also low. In sewage, enumeration by electron microscopy (Ewert and Paynter 1980) revealed 10^{8-10} phages/ml, compared with about 10^8/ml in freshwater and the open ocean (Bergh et al. 1989). Direct counts of bacteriophage in soil are also high, some 4 percent of the total bacterial counts (Ashelford et al. 2003).

Phage morphology

Examined as a group, phages show considerable morphological diversity (Ackermann 1983; www.virology.net/Big_Virology/BVHomePage.html). Some are filamentous, isometric, and superficially resemble animal viruses, whereas others show complex morphologies. The capsid or the phage head is the structure containing the genome; this can be single- or double-stranded DNA or RNA. Some phages are virulent, killing each cell they infect, whereas others are temperate. In the latter case, the phage resides, as a prophage, in a cell termed a lysogen and replicates synchronously with the cell. When phage replication is completed the particles are released from the cell. They usually contain a phage genome. The phage capsid recognizes an adsorption site on the outside of a susceptible cell, binds to it, and injects the genome, held within the capsid, into the cell. The genome then redirects the cellular metabolism to synthesize more phage. In the lysogenic state, the phage genome does not disrupt bacterial metabolism and replication but may contribute specific characteristics to its host, e.g. toxin production by *Corynebacterium diphtheriae* (see chapter 39, Corynebacteria and rare coryneforms) and *Streptococcus pyogenes* (see chapter 33, Streptococcus and lactobacillus).

Single-stranded RNA phages

These are the simplest of phages in terms of their genomic size and morphology. This is not to imply that the organization of their replicative cycle or their economy in genome size, relative to other phages, makes their biological activities simple, but it reflects an evolutionary adaptation. They rely on the host cell for all metabolic activities except for certain replicative functions to complete their life cycle. The RNA coliphage MS2 (*Leviviridae*), termed male-specific, utilizes the pilus of an F^+ *E. coli* cell as an infection site. The MS2 capsid is icosohedral, largely composed of a single protein species, which encloses a phage genome of 3 569 nucleotides (nt) of linear plus strand (sense) RNA. The capsid binds to the side of the F pilus and the phage genome is transferred into the cell. Once internalized, the phage RNA genome acts as messenger RNA and is translated by the ribosomes. There is no transcriptional control, as promoters and RNA polymerase are not involved, and yet there is temporal control over the expression of proteins. This is achieved by changes in conformation of the RNA genome. Relatively large numbers of coat protein molecules (van Duin 1988) are needed per virion. The RNA-directed RNA polymerase, a lysin gene, and an absorption protein are synthesized in smaller amounts and their expression is regulated by secondary structures formed in the phage genome (Kastelein et al. 1982; Berkhout et al. 1987).

On average, 200 phage particles are produced from one cell infected by one phage and this can be achieved in as little as 22 minutes. To achieve this efficiency, replication proceeds alongside expression and the phage polymerase acts in concert with some host accessory functions. The polymerase initiates repeatedly, at the $3'$ end of the RNA (Haruna and Spiegelman 1965), enabling multiple copies (of the minus strand) to be made. Concurrently these full-length minus strands are then copied to yield the plus strands, which are used initially for the expression of phage proteins. Later in the infection cycle the plus strands are individually packaged into a capsid. Cell lysis is induced by the final product, the lysis protein.

Single-stranded DNA phages

The tailless icosoheral phage ϕX174 (*Microviridae*) was the first single-stranded DNA phage to be identified (Hayashi et al. 1988). It has a circular genome (5.4 kb) that codes for ten proteins. Four different monomers are present in the capsid, one of which is an internal protein and is surrounded by a second (60 copies), which forms the icosahedral capsid. The other two monomers form 'spike' proteins, which are essential for infectivity. The phage receptor is a lipopolysaccharide that allows entry of the phage genome. This DNA is used as a template for single minus-strand synthesis and is entirely dependent upon the host cell to create a biologically active phage genome (Goulian et al. 1967). Once established as a duplex, the phage genome reproduces by rolling circle replication (Lewin 1987). This is initiated by a phage replication enzyme expressed from the newly formed duplex. Transcription proceeds in the same direction as replication. As the pools of capsid precursors and replicated genomes increase, the plus strands become encapsulated (Aoyama and Hayashi 1986). The reproductive cycle can be complete within 13 minutes, resulting in cell lysis yielding up to 180 phage particles

Although the replication cycles of the filamentous phages are similar to that of φX174, their structure, attachment, and the absence of cell lysis on phage release are different (Lindquist et al. 1993). The genome of phage M13 (*Inoviridae*) is circular (6407 nt) and encapsulated as a single strand by a flexible helical capsid composed largely of a single protein species (Model and Russell 1988). At the ends of this tube-like filamentous phage capsid are the 'minor' proteins. Those at one end are responsible for phage attachment to the F plasmid pilus via which the phage genome becomes internalized. Replication proceeds in the same manner as for φX174. The host cell continues to grow and divide and as the phage particles become 'mature' they are released through an intact cell envelope. Thus, phages are continually released after a latent period of 30 minutes, resulting in very high phage yields. The filamentous phages are good cloning vectors because foreign DNA can be spliced into the genome at appropriate sites and transformed into a host bacterial cell to give high yields of cloned material. The splicing of novel genes with the coat protein gives protein chimeras in which the novel peptide component may be exposed on the capsid surface (Barbas et al. 1991). As methods for purifying phages are well established, this provides an easy method for screening for protein activities.

Double-stranded DNA phages

The double-stranded DNA phages are divided into two groups, based on their virulent or temperate nature. Phage T4 (*Myoviridae*) is a virulent and morphologically complex phage; it has an elongated icosahedral capsid composed of one major and several minor (in terms of percentage) proteins (Mathews et al. 1983). The capsid is placed upon a tail comprising a core and contractile sheath on a base plate with six kinked tail fibres. In the capsid the 170-kb DNA genome is linear and double-stranded, with about a 5-kb terminal repeat. Free phage particles recognize a lipopolysaccharide receptor on the cell surface and become anchored to it. The proteins in the baseplate then reorientate to allow the core to be driven onto and through the cell surface as the sheath contracts. The DNA is then extruded from the head into the cell's cytoplasm. Phage replication uses the host polymerase and normal vegetative σ factor (σ_{70}). The phage reproductive cycle, from infection to lysis, is highly organized and complex. For example, transcription is temporally organized such that some genes are expressed early and in the middle of the replicative cycle (Mosig and Eiserling 1988). The late genes are transcribed by a phage-encoded σ factor and also govern DNA synthesis. Immediately after infection this phage degrades the host genome and inhibits host transcription. After replication, the concatomeric DNA is injected into the preformed heads and cut, and then the tail structures are added (Bhattacharyya and Rao 1993). Expression of the late phage genes results in cell lysis and release of ⩾200 mature phage particles 23 minutes after infection. The phage λ (*Siphoviridae*; 48.5 kb) has an icosohedral capsid and is termed a temperate phage. It interacts differently to phage T4 with its host. When carried by the host (λ replicates in-step with the host) it is termed a 'prophage' and the cell is termed a 'lysogen'. Phage replication and consequent cell lysis require phage replication to be induced. This is generally promoted by the SOS response (Ptashne and Gann 2001).

Experimental analysis

Phages may be investigated by a variety of molecular and standard laboratory techniques. Three terms describe key points in the analysis of phages. A plaque-forming unit (pfu) is a phage particle that is capable of forming a plaque. The efficiency of plating (EOP) is the proportion of pfu to total phage particles (enumerated by electron microscopy) in a lysate (Adams 1959). Finally, the multiplicity of infection (MOI) is the ratio of host cells to phage in a mixture.

ASSAY

It is a simple procedure to estimate the number of phage particles in a lysate. As titers are frequently 10^6–10^{11} pfu/ml, 10-fold dilution series of phage lysate is prepared to obtain 10–100 pfu/ml. A small volume (e.g. 100 μl) of diluted lysate is mixed with an equal volume of host bacterial cells in about 3.0 ml of 0.6 percent molten nutrient agar, held at 46°C. This is vortex-mixed and poured on to the surface of a standard nutrient agar plate. After overnight incubation, to allow the plaques to develop in the growing bacterial lawn, the number of plaques formed will reflect the original phage concentration and the dilution of the lysate. Thus, at high dilutions, no phage will be present and no plaques will develop. At a lower dilution, a few plaques (zones of lysis a few millimeter in diameter) will be present in the lawn, and at the next dilution, the plaque count will be high and the plaques may even merge in places. At the next dilution, the MOI will be about 1:1 and confluent lysis will occur, although it is likely that a few bacterial colonies, resistant to the phage, will grow (at a frequency of 10^{-8}). This may be due to phenotypic variation (the lack of expression of a phage uptake site) or to a resistant mutation (leading to a loss of the target site), or the cell may become infected by a normally lytic phage and yet become lysogenized. In each case, the cell remains unaffected by the phage and continues to grow. The titer of phage is calculated by dividing the dilution assayed into the plaque count from the plate by the highest number of countable plaques.

PLAQUE MORPHOLOGY

Phages can replicate only in an actively metabolizing host cell. A poor nutritional environment will reduce the plaque size and probably also reduce the phage yield. The morphology of the plaque is due to repeated cycles of infection, replication, and lysis of the phage; from the point of infection of a single cell, phages released by lysis infect adjacent cells. This process repeats itself while the bacteria in the lawn are growing, causing a circular plaque to form. Thus, the titer and plaque morphology remain consistent provided the media, incubation conditions, and host are constant. Changes in environmental conditions and the genetic status of the host bacterium or the phage can all influence both plaque morphology and the reproductive success of the phage. For example, virulent phage can display 'a temperate' nature when their host is growing slowly (Ripp and Miller 1998).

TRANSDUCTION

Transduction is achievable by only some phages. For gene transfer to occur, a proportion of the phage particles generated in a lysate have to package some host chromosomal DNA instead of a phage genome. The phage particle acts as a vector for host chromosomal genes, transferring the packaged sequence into the cytoplasm of a recipient cell. Providing the DNA sequence has homology with a recipient sequence (chromosomal or plasmid), recombination can then occur. This is mediated by the host-specified RecA protein, enabling replacement of the host sequence by integration of the transferred sequence. The process is termed transduction. If any gene on the bacterial chromosome may be transferred by the phage, this is termed generalized transduction (Margolin 1987), but if the phage will transfer only a few genes from one site on the genome, this is termed specialized or restricted transduction (Weisberg 1987).

About 2 percent of the phage particles produced by the *Salmonella typhimurium* phage P22 contain host DNA (Ebel-Tsipis et al. 1972), as do about 0.3 percent of *E. coli* phage P1 in a lysate (Lennox 1955). The frequencies of transfer of auxotrophic markers varies between 10^{-5} and 10^{-7} per pfu (Harriman 1971) for both of these phages. However, Schmeiger (1972) has shown that with different genes and hosts the frequencies can vary over 1000-fold.

There are various fates for DNA transduced into a recipient cell. The DNA that is recombined is stably inherited, giving rise to a stable transductant. The displaced strand is degraded and the nucleotides are reutilized. If the DNA sequence fails to become integrated, then it may persist for several generations, producing what is termed 'abortive transductants' (Ozeki 1959). These transductants show unilinear inheritance of the character transferred. This is explained by using a recipient cell that is an auxotroph (carries a mutation in the gene coding for an enzyme required for the synthesis of an amino acid), so it cannot synthesize a particular amino acid. This is transduced to prototrophy by an 'abortive fragment' that contains the prototrophic gene so the cell is able to synthesize the amino acid. When the cell divides, one daughter inherits the wild-type gene and continues to express the enzyme and consequently synthesizes the amino acid and grows. The other cell now has the mutant genotype but is phenotypically wild-type; it retains an active biosynthetic pathway while the enzyme retains its activity. This is lost gradually as it becomes diluted out through growth and cell division; the cell grows, but increasingly slowly. Hence, abortive transductants generate very small colonies, which appear some time after the wild-type has fully grown. The condition is relatively stable as less than 1 percent of cells go on to become complete transductants (Ozeki and Ikeda 1968). Schmeiger (1982) has shown that abortive transduction by P22 is a common phenomenon and varies over a range from 50 percent to less than 1 percent for different loci.

Specialized transduction occurs with *E. coli* phage λ. In this type of transduction the phage enters the chromosome by site-specific recombination at *attB*, using a homologous sequence on the phage termed *attP* (Weisberg and Landy 1983), to form a lysogen. Transducing particles are formed by the illegitimate excision of the phage genome, in which part of the phage sequence remains in the chromosome to be replaced by an adjacent host chromosomal sequence. Thus, genes to one side or the other of the *attB* site may be packaged in the capsid (hence the term specialized). The transducing phage genome is defective and cannot reproduce in the recipient cell. These transducing phages occur at a frequency of about 10^{-6}. If these defective phages co-infect a new recipient with a wild-type phage, 50 percent of the phage in the lysate produced from that cell will be capable of transducing the host gene. The frequency of transduction now is far greater; it can be 100 percent (Lewin 1987). Transducing phages of both types are common in other genera.

ENVIRONMENTAL ROLE

The role of phages in natural bacterial populations remains largely a point of conjecture. They seem to have a dramatic influence on population densities of bacteria in natural habitats (Ogunseitan et al. 1990; Proctor and Fuhrman 1990). Interactions between bacteriophages and planktonic bacteria in aquatic environments have shown that the abundance of free viruses can exceed that of planktonic bacteria by over an order of magnitude (Bergh et al. 1989). Short-term dynamics imply that bacteriophages can contribute to the control bacterial growth in situ (Bratbak et al. 1990). Transmission electron microscopic (TEM) studies indicate that 1 to 5 percent of planktonic bacteria were visibly infected by phage (Proctor and Fuhrman 1990). Bacterial mortality due to phage reproduction suggests phages can, at times, be entirely responsible for bacterial mortality (Bratbak et al. 1990; Proctor and Fuhrman 1990). Thus, it seems that phages can be significant moderators of microbial populations in a variety of natural environments (Ashelford et al. 2003). These include those in animal and plant as well as populations in water bodies and soil, and between bacterial growth, phage production, and chlorophyll (Cochlan et al. 1993).

Host range has been established by the observation of plaques on putative host strains and on this basis most phages have a narrow host range. There is a widespread assumption that the plating range (i.e. the ability of the phage to form plaques) is a direct reflection of the transduction range. The ambiguity of this has been illustrated with *E. coli* phage P1. Productive infections with phage P1 can be made from *Citrobacter freundii*, *Shigella* spp., *Salmonella* spp., Serratia spp., *Enterobacter liquefaciens*, *Erwinia* spp., *Proteus* spp., *Pseudomonas* spp., and *Klebsiella* spp. (Yarmolinsky and Sternberg 1988). In addition, phage P1 can naturally transduce DNA into *Yersina* spp., *Flavobacterium* sp. M46, *Agrobacterium tumefaciens*, *Alcaligenes faecalis*, and *Myxococcus xanthus*. Goldberg et al. (1974) have reported transduction by phage P1 into *S. typhimurium* even though they were unable to isolate stable lysogens and *Enterobacter amylovora* lysogens did not produce P1 phage. This provides evidence to support the proposal that the transduction and titration ranges for a phage are not unambiguously equivalent and the associations between phage and their host bacteria are not clear-cut. Thus, the contribution of phages to gene exchange remains an open and unanswered question. In addition, the host range can be influenced by the presence or absence of adsorption sites and the mechanism of phage replication. The restriction enzymes possessed by the new host target the phage genome (Roberts 1985). If the phage fortuitously evades the restriction enzymes and becomes modified, to shield the restriction sites, or if these sites are absent, then the phage may be able to replicate.

PHAGE IDENTIFICATION AND TYPING

The specificity and variety of phages that can infect bacterial species have enabled phages to be used for the identification of a species. Strain typing by determining phage sensitivity patterns within a species can be achieved when various phages will attack only a proportion of strains. Phages that lyse all members of a particular species have been used in the identification of *Brucella abortus* and *Bacillus anthracis*. The phages active on *Vibrio cholerae* are specific in their action on serotypes or biotypes. Phage typing within species has been most widely developed and most effectively used in the epidemiological investigation of *Staphylococcus aureus* infection and for strain identification within several *Salmonella* serotypes, including typhi. Typing phages are propagated on a susceptible host strain of the test species to give a high-titer phage suspension, which is then diluted to a concentration (the routine test dilution (RTD)), which gives barely confluent lysis of the propagating culture growing on a solid medium. In phage typing, a set of phages active on only some strains within a species is applied to a lawn of the isolate in question on an agar medium. After incubation, susceptibility is seen as lysis (i.e. plaques) produced by particular phages. The pattern of activity gives the phage type of the organism (Parker 1972).

CONCLUSIONS

Both bacteriocins and bacteriophages are universally present in microbial populations, affecting growth and biochemical relationships within and between bacterial populations (Ashelford et al. 2000). This relationship now clearly extends to interactions, like the pathogenicity endowing genotypes, with higher organisms. Frank (1994) has suggested that the great diversity of bacteriocins produced by a single species cannot occur if the organisms mix freely. His analysis suggests that spatial variation and poor habitats favor susceptible and polymorphic strains, populations with a variety of bacteriocin types. Toxin producers are favored in good habitats since the rate of competition for resources is lower. Although most bacteriocins inhibit the growth of closely related bacteria, some, typified by those produced by lactic acid bacteria, exhibit a relatively broad antimicrobial spectrum and are active against several food-spoilage and health-threatening microorganisms. Thus, there is circumstantial evidence for a role in the regulation of competing bacteria.

The most general use of bacteriophages in the medical sphere has been to phage type bacterial strains of clinical importance. They have been studied since the 1940s and have provided basic information on how DNA and RNA genomes replicate and express genes (Boyd and Brüssow 2002). Because of the relative ease with which they can be grown, purified, and now manipulated, they have become of primary importance (with plasmids) in the development of the molecular biological techniques used to clone and analyze genes and their products. Thus, they have provided a basis for the development of the biotechnological era.

Those debating the risks of recombinant genes being transferred from genetically manipulated bacteria into indigenous bacterial populations now accept that transfer will occur. Although bacteriophages may be one class of the mediators of such gene exchange, it is still an open question as to how efficient they might be as mechanisms for exchange in situ or in vivo. It will be an interesting step in our ecological appreciation of these genetic determinants when their true ecological value to microbial populations can be demonstrated.

REFERENCES

Ackermann, H.W. 1983. Current problems in bacterial virus taxonomy. In: Mathews, R.E.F. (ed.), *A critical appraisal of viral taxonomy*. Boca Raton, FL: CRC Press, 105–22.

Adams, M.H. 1959. *The bacteriophages*. New York: Interscience Publishers Inc.

Ashelford, K.E., Norris, S.J., et al. 2000. Seasonal population dynamics and interactions of competing bacteriophages and their host in the rhizosphere. *Appl Environ Microbiol*, **66**, 4193–9.

Ashelford, K.E., Day, M.J. and Fry, F.C. 2003. Elevated abundance of bacteriophage infecting bacteria in soil. *Appl Environ Microbiol*, **69**, 285–9.

Aoyama, A. and Hayashi, M. 1986. Synthesis of bacteriophage φX174 *in vitro* mechanism of switch from DNA replication to DNA packaging. *Cell*, **47**, 99–106.

Barbas, C.F.K., Kang, A.S., et al. 1991. Assembly of combinatorial antibody libraries on phage surfaces. The gene IV site. *Proc Natl Acad Sci USA*, **88**, 7978–82.

Benkerroum, N., Ghouati, Y., et al. 1993. Methods to demonstrate the bacteriocidal activity of bacteriocins. *Lett Appl Microbiol*, **17**, 78–81.

Bergh, O., Borsheim, K.Y., et al. 1989. High abundance of viruses found in aquatic environments. *Nature (London)*, **340**, 467–8.

Berkhout, B., Schmidt, B.F., et al. 1987. Lysis gene of bacteriophage MS2 is activated by translation termination at the overlapping coat gene. *J Mol Biol*, **195**, 517–24.

Bhattacharyya, S.P. and Rao, V.B. 1993. A novel terminase activity associate with the DNA packaging protein gp17 of bacteriophage T4. *Virology*, **196**, 34–44.

Bierbaum, G. and Sahl, H.G. 1993. Lantibiotics – unusually modified bacteriocin-like peptides from gram-positive bacteria. *Int J Med Microbiol Virol Parasitol Infect Dis*, **278**, 1–22.

Boyd, E.F. and Brüssow, H. 2002. Common themes among bacteriophage-encoded virulence factors and diversity among the bacteriophages involved. *Trends Microbiol*, **10**, 521.

Bratbak, G., Hedal, M., et al. 1990. Viruses as partners in spring bloom microbial trophodynamics. *Appl Environ Microbiol*, **56**, 1400–5.

Cochlan, W.P., Wikner, J., et al. 1993. Spatial distribution of viruses, bacteria and chlorophyll-A in neritic, oeanic and estuarine environments. *Marine Ecology Progress Series*, **92**, 77–87.

Delvesbroughton, J., Williams, G.C. and Williamson, S. 1992. The use of bacteriocin, nisin, as a preservative in pasteurised liquid whole egg. *Lett Appl Microbiol*, **15**, 133–6.

D'Herelle, R. 1917. Sur un microbe invisible antagoniste des bacilles dysentériques. *CR Acad Sci*, **165**, 373–5.

van Duin, J. 1988. The single-stranded RNA bacteriophages. In: Calendar, R. (ed.), *The bacteriophages*. vol. 1. New York: Plenum Publishing Corp, 117–68.

Ebel-Tsipis, J., Botsyein, D. and Fox, M.S. 1972. Generalised transduction by phage P22 in *Salmonella typhimurium*. 1 Molecular origin of transducing DNA. *J Mol Biol*, **71**, 433–48.

Eraso, J.M. and Weinstock, G.M. 1992. Anaerobic control of colicin E1 production. *J Bacteriol*, **174**, 5101–9.

Ewert, D.L. and Paynter, M.J.B. 1980. Enumeration of bacteriophage and host bacteria in sewage and activated sludge treatment processes. *Appl Environ Microbiol*, **39**, 576–83.

Fath, M.J. and Kolter, R. 1993. ABC transporters – bacterial exporters. *Microbiol Rev*, **57**, 995–1017.

Frank, S.A. 1994. Spatial polymorphism of bacteriocins and other allelopathic traits. *Evol Ecol*, **8**, 369–86.

Frederique, P. 1957. Colicins. *Annu Rev Microbiol*, **11**, 7–22.

Gillis, R.R. 1964. Colicin production as an epidemiological marker of Shigella sonnei. *J Hyg*, **62**, 1–9.

Gilson, L., Mahanty, H.R. and Kotter, R. 1990. Genetic analysis of an MDR-like export system: the secretion of ColV. *EMBO J*, **9**, 3875–84.

Goldberg, R.B., Bender, R.A. and Streicher, S.L. 1974. Direct selection for P1-sensitive mutants of enteric bacteria. *J Bacteriol*, **118**, 810–14.

Goulian, M., Kornberg, A. and Sinsheimer, R.L. 1967. Enzymatic synthesis of DNA XXIV. Synthesis of infectious phage φX174 DNA. *Proc Natl Acad Sci USA*, **58**, 2321–8.

Gratia, A. 1925. Sur un remarquable example d'antagonisme entre deux souches de colibacille. *C R Soc Biol*, **93**, 1040.

Gratia, A. and Frederique, P. 1946. Diversité des souches antibiotiques de B coli et étendue variable de leur champ d'action. *C R Soc Biol*, **140**, 1032–3.

Hansen, J.N. 1993. Antibiotics synthesised by posttranslational modification. *Annu Rev Microbiol*, **47**, 535–64.

Harriman, P. 1971. Appearance of transducing activity in P1 infected Escherichia coli. *Virology*, **45**, 324–5.

Harris, L.J., Fleming, H.P. and Klaenhammer, T.R. 1992. Developments in nisin research. *Food Res Int*, **25**, 57–66.

Haruna, I. and Spiegelman, S. 1965. Specific template requirements of RNA replicases. *Proc Natl Acad Sci USA*, **54**, 579–87.

Hayashi, M., Aoyama, A., et al. 1988. Biology of the bacteriophage φX174. In: Calendar, R. (ed.), *The bacteriophages*. vol. 2. New York: Plenum Publishing Corp, 1–72.

Hayashi, M., Matsumoto, H., et al. 1994. Cytotoxin converting phage, Phi-ctx and PS21 are R-pyocin related phages. *FEMS Microbiol Lett*, **122**, 239–44.

Hechard, Y., Renault, D., et al. 1993. Anti-listeria bacteriocins – a new family of proteins. *Lait*, **73**, 207–13.

Jack, R.W., Tagg, J.R. and Ray, M.R. 1995. Bacteriocins of Gram-positive bacteria. *Microbiol Rev*, **59**, 171-20, 0.

Jacob, F. and Woolman, E.L. 1953. Induction of phage development in lysogenic bacteria. *Cold Spring Harb Symp Quant Biol*, **18**, 101–21.

Jiminez-Diaz, R. and Rios-Sanchez, R.M. 1993. Plantaricins S and T, two new bacteriocins produced by *Lactobacillus plantarum* LPCO10 isolated from green olive fermentation. *Appl Environ Microbiol*, **59**, 1416–24.

Kastelein, R.A., Remaut, E., et al. 1982. Lysis gene expression of RNA phage MS2 depends on a frameshift during translation of the overlapping coat protein gene. *Nature (London)*, **285**, 35–41.

Kim, W.J. 1993. Bacteriocins of lactic-acid bacteria – their potentials as food. *Food Rev Int*, **9**, 299–313.

Klienkauf, H. and von Dohren, H. 1990. Nonribosomal biosynthesis of peptide antibiotics. *Eur J Biochem*, **192**, 1–15.

Lakey, J.H., Gonzalezmanas, J.M., et al. 1992. The membrane insertion of colicins. *FEBS Lett*, **307**, 26–9.

Lakey, J.H., Vandergoot, F.G. and Pattus, F. 1994. All in the family – the toxic activity of pore-forming colicins. *Toxicology*, **87**, 85–108.

Lau, P.C.K., Parsons, M. and Uchimura, T. 1992. Molecular evolution of colicin plasmids with emphasis on the endonuclease types. In: James, R. (ed.), *Bacteriocins and lanbiotics*. New York: Springer-Verlag, 353–78, NATO series.

Lewin, B.J. 1987. *Genes III*. New York: John Wiley & Sons.

Lewus, C.B., Kiaser, A. and Montville, T.J. 1991. Inhibition of food-borne bacteriocins by bacteriocins from lactic acid bacteria isolated from meat. *Appl Environ Microbiol*, **5**, 1683–8.

Lindquist, B.H., Dehó, G. and Calendar, R. 1993. Mechanisms of genome propagation and helper exploitation by satellite phage P4. *Microbiol Rev*, **57**, 683–702.

Luria, S.E. and Suit, J.L. 1987. Colicins and col plasmids. In: Neidhardt, F.C. (ed.), Escherichia coli *and* Salmonella typhimurium: *cellular and molecular biology*. Washington, DC: American Society for Microbiology, 1615–24.

Malkhosyan, R., Panchenko, Y.A. and Relesh, A.N. 1991. A physiological role for the DNA supercoiling in the anaerobic regulation of colicin gene expression. *Mol Gen Genet*, **225**, 342–5.

Margolin, P. 1987. Generalized transduction. In: Neidhardt, F.C. (ed.), Escherichia coli *and* Salmonella typhimurium: *cellular and molecular biology*. Washington, DC: American Society for Microbiology, 1154–68.

Mathews, C.K., Kutter, E.M., et al. 1983. *Bacteriophage T4*. Washington, DC: American Society for Microbiology.

Model, P. and Russell, M. 1988. Filamentous bacteriophage. In: Calendar, R. (ed.), *The bacteriophages*. vol. 2. New York: Plenum Publishing Corp, 375–456.

Mosig, G. and Eiserling, F. 1988. Phage T4 structure and metabolism. In: Calendar, R. (ed.), *The bacteriophages*. vol. 2. New York: Plenum Publishing Corp, 521–606.

Nishio, C., Komura, S. and Kurahashi, K. 1983. Peptide antibiotic subtilisin is synthesised via precursor proteins. *Biochim Biophys Res Commun*, **116**, 751–8.

Nissen-Meyer, J., Holo, H., et al. 1992. A novel lactococcal bacteriocin whose activity depends on the complimentary action of two peptides. *J Bacteriol*, **174**, 5686–92.

Ogunseitan, O.A., Sayler, G.S. and Miller, R.V. 1990. Dynamic interactions of *Pseudomonas aeruginosa* and bacteriophages in lake water. *Microb Ecol*, **19**, 171–85.

Ozeki, H. 1959. Chromosome fragments participating in transduction in Salmonella typhimurium. *Genetics*, **44**, 457-7, 0.

Ozeki, H. and Ikeda, H. 1968. Transduction mechanisms. *Annu Rev Genet*, **135**, 175-8, 4.

Parker, M.T. 1972. Phage typing of *Staphyloccus aureus*. In: Norris, J.R. and Ribbons, D.W. (eds), *Methods in microbiology*. vol. 7b. London: Academic Press, 1–28.

Proctor, L.M. and Fuhrman, J.A. 1990. Viral mortality of marine bacteria and cyanobacteria. *Nature (London)*, **343**, 60–2.

Ptashne, M. and Gann, A. 2001. *Genes and signals*. Cold Spring Harbor, New York: Cold Spring Harbor Laboratory.

Pugsley, A.P. 1984a. The ins and outs of colicins. Part 1: production and translocation across membranes. *Microbiol Sci*, **1**, 168–75.

Pugsley, A.P. 1984b. The ins and outs of colicins. Part 2: lethal action, immunity and ecological implications. *Microbiol Sci*, **1**, 203–5.

Pugsley, A.P. and Schwartz, M. 1985. Export and secretion of proteins by bacteria. *FEMS Microbiol Rev*, **32**, 3–38.

Riley, M.A. 1993. Positive selection for colicin diversity in bacteria. *Mol Biol Evol*, **10**, 1048–59.

Riley, M.A. and Gordon, D.M. 1992. A survey of col plasmids in natural isolates of *Escherichia coli* and an investigation into the stability of col-plasmid lineages. *J Microbiol*, **138**, 1345–52.

Ripp, S. and Miller, R.V. 1998. Dynamics of the pseudolysogenic response in slowly growing cells of *Pseudomonas aeruginosa*. *Microbiology*, **144**, 2225–32.

Roberts, R.J. 1985. Restriction and modification enzymes and their recognition sequences. *Nucleic Acids Res*, **13**, Suppl., r165–200.

Ruiz-Barba, J.L., Cathcart, D.P., et al. 1994. Use of *Lactobacillus plantarum* LPCO10, a bacteriocin producer, as a starter culture in Spanish-style green olive fermentations. *Appl Environ Microbiol*, **60**, 2059–64.

Saadoun, I., al-Momani, F., et al. 1999. Isolation, identification and analysis of antibacterial activity of soil streptomycetes isolates from north Jordan. *Microbios*, **100**, 41–6.

Sano, Y., Matsui, H., et al. 1993. Molecular-structures and functions of pyocin-s1 and pyocin-s2 in *Pseudomonas aeruginosa*. *J Bacteriol*, **175**, 2907–16.

Saye, D.J. and Miller, R.V. 1989. The aquatic environment: consideration of horizontal gene transmission in a diversified environment. In: Levy, S.B. and Miller, R.V. (eds), *Gene transfer in the environment*. New York: McGraw-Hill, 223–59.

Schmeiger, H. 1972. Phage P22 mutants with increased or decreased transduction abilities. *Mol Gen Genet*, **119**, 75–88.

Schmeiger, H. 1982. Packaging signals for phage P22 on the chromosome of *Salmonella typhimurium*. *Mol Gen Genet*, **187**, 516–18.

Summers, W.C. 2001. Bacteriophage therapy. *Annu Rev Microbiol*, **55**, 437-5, 1.

Tagg, J.R., Dajani, A.S. and Wannamaker, L.W. 1976. Bacteriocins of Gram-positive bacteria. *Bacteriol Rev*, **40**, 722-5, 6.

Torreblanca, M., Meseguer, I. and Ventosa, A. 1994. Production of halocin is a practically universal feature of archael halophilic rods. *Lett Appl Microbiol*, **19**, 201–5.

Twort, F.W. 1915. An investigation on the nature of ultramicroscopic viruses. *Lancet*, **189**, 1241–3.

Van der Wal, F.J., Luirick, J. and Oudega, B. 1995. Bacteriocin release proteins: mode of action, structure and biotechnological application. *FEMS Microbiol Rev*, **17**, 381–99.

Vizan, J.L., Hernadez-Chico, C., et al. 1991. The peptide antibiotic microcin B17 induces double stranded cleavage of DNA mediated by DNA gyrase. *EMBO J*, **10**, 467–76.

Weisberg, R.A. 1987. Specialized transduction. In: Neidhardt, F.C., Ingraham, J.L., et al. (eds), Escherichia coli *and* Samonella typhimurium. *cellular and molecular biology*, vol. 2. Washington, DC: American Society for Microbiology.

Yarmolinsky, M.B. and Sternberg, N. 1988. Bacteriophage P1. In: Calendar, R. (ed.), *The bacteriophages*. vol. 1. New York: Plenum Publishing Corp, 291–438.

Molecular basis of bacterial adaptation to a pathogenic lifestyle

KIM R. HARDIE, TOM BALDWIN AND PAUL WILLIAMS

INTRODUCTION

Infection is a complex and dynamic interaction, the outcome of which determines the onset or avoidance of disease. The term 'pathogen' denotes a microorganism that is able to cause disease in an animal, fish, plant, or insect host, while 'pathogenicity' or 'virulence' is the ability of that microorganism to induce disease in one or more target hosts. Virulence can be defined as the capacity of bacteria to cause disease; virulence determinants are those bacterial products or strategies that contribute to virulence, i.e. those cell-surface-associated or extracellular factors that promote tissue colonization or invasion, protect the bacterial cells from innate host immune defenses, and damage the host. While the expression of such virulence determinants is tightly regulated, they are not usually essential for bacterial viability or growth in vitro. The interplay between host- and pathogen-derived factors will determine the final outcome of an infection, which will depend on the relative degree of susceptibility or resistance of the host as well as the intrinsic virulence of the pathogen.

All living organisms are programmed with a biological capability to survive, replicate, and spread. From an evolutionary perspective, genetic modifications that improve any of these attributes will be selected for and these ultimately lead to the enrichment of such organisms within a given environmental niche. For pathogenic microbes, many of which have become highly adapted to

a specific host, the death of the host is not necessarily an indicator of pathogen success. This is because it signals the destruction of an environment that has provided nutrition and protection. Pathogenic microorganisms cannot be considered as 'malicious' since they do not 'deliberately' cause disease or the subsequent death of a host. Pathogenic bacteria are unicellular microorganisms that can ensure the survival of some of their number by infecting other hosts. If, during the process of infection, the host dies, pathogen survival will depend on transmission to a new host, a process that may depend on its capacity to survive in diverse environments. The manifestation of disease is very much symptomatic of the exploitation of host resources by the pathogen for promoting bacterial survival, replication, and dissemination. Thus, the lifestyle of pathogenic bacteria revolves around: (1) locating a host; (2) finding a replication niche; (3) initiating an infection; (4) progressing the infection; and (5) dispersal to a new host. For a bacterium to progress from one stage to the next, variations in sensory inputs that signal environmental change must be perceived and acted upon, for example by the induction of gene expression. New environmental signals may derive from movement from one environment to another, may be due to the actions of bacteria within a given environment, or may be a consequence of host responses to the activity of the bacteria. Thus, from a prokaryotic perspective, the successful interaction of bacterial cells with mammalian host tissues depends on a coordinated response not only to environmental cues such as nutrient availability, bacterial cell population density, temperature, osmolarity, and pH, but also to diverse host cell effector molecules. For example, bacterial pathogens secrete toxins that either release the nutrients from host tissues or alter the host internal environment to optimize bacterial growth and dissemination. The host responds to these changes by attempting to rectify the physiological and biochemical imbalances induced by the pathogen. Together, the damage caused by the pathogen and the processes elicited by the host during infection are manifested as disease.

MECHANISMS OF BACTERIAL PATHOGENICITY

The main characteristics of pathogenic bacteria that underlie the mechanisms by which they cause disease are:

- Colonization and infiltration of host tissues: the colonization of a specific host niche usually represents the first stage of infection. This may result in the carriage state or may lead to infiltration of the host tissues. The latter involves growth on or within host tissues, the ability to bypass or overcome host defense mechanisms, and the production of extracellular tissue-degrading exoenzymes.
- Toxin production: bacteria produce two categories of toxins, termed exotoxins and endotoxins. Exotoxins are released or secreted from bacterial cells and may act at tissue sites distant from the sites of bacterial infection, adhesion, and growth. Endotoxins are cell-associated structural components of the gram-negative bacterial envelope, which may be released from growing bacterial cells as blebs or from lysed cells (see Endotoxins). Hence, bacterial toxins, both soluble and cell-associated, may be transported by blood and lymph and induce cytotoxic effects at tissue sites remote from the original site of infection. Some bacterial toxins may also act at the site of colonization and play a role in promoting tissue invasion.

Colonization

The first stage of bacterial infection is colonization, which is essentially the establishment of the pathogen at a specific body site frequently followed by entry into host tissues. Pathogens usually colonize host tissues that are in contact with the external environment, for example the urogenital tract, the digestive tract, the respiratory tract, and the conjunctiva. Organisms that infect these sites have developed specific tissue-adherence mechanisms alongside the ability to overcome the local innate host defenses (see Host resistance to infection). This stage may, however, be bypassed as a consequence of tissue damage or introduction of medical devices, e.g. catheters.

ADHERENCE TO SURFACES

Bacterial adherence to host tissues involves at least two stages:

1 Nonspecific adherence. This physicochemical process can be considered as reversible attachment of the bacterium to the eukaryotic cell or tissue surface and involves nonspecific attractive forces that allow the bacterium to approach the eukaryotic cell/tissue surfaces. The interactions and forces involved include hydrophobic and electrostatic attractions, atomic and molecular vibrations resulting from fluctuating dipoles of similar frequencies between components of the bacterial and the host cell, Brownian movement, and bacterial cell recruitment or trapping by biofilm polymers (see Biofilms).

2 Specific adherence. This can be a reversible or permanent attachment of the microorganism to the cell or tissue surface. Specific adherence involves irreversible formation of many specific, noncovalent bonds between complementary adhesin and receptor molecules on each interacting surface. The latter must be accessible and arranged in such a way that many bonds form over the area of contact between the two cells. Once the bonds are formed, attachment under physiological conditions becomes virtually irreversible. Direct evidence that receptor and/or adhesin molecules mediate the specificity of adherence of bacteria to host cells or tissues has been derived from the competitive binding of bacteria to isolated receptors or receptor analogs in vitro. Purified bacterial adhesins and adhesin analogs that bind to eukaryotic cell/tissue surfaces have been shown to inhibit the interaction of bacteria with host cells. Such inhibition can also be achieved with purified receptors or antibodies to either the adhesins or to the eukaryotic receptor.

The eukaryotic receptors that have so far been identified are specific surface peptides or carbohydrates, which have functions in the host cell distinct from their functions as receptors for bacterial adhesins. Bacterial adhesins are macromolecular components of the bacterial cell envelope, and adhesins and receptors interact in a complementary and specific fashion.

HOST- AND SPECIES-SPECIFIC TROPISMS

It has long been recognized that bacteria exhibit different degrees of colonization or adherence tropisms. These can be broadly subdivided into (1) tissue- and host-specific tropisms and (2) species-specific tropisms.

1 Tissue and host tropisms. Some bacteria show a preference for certain tissues and specific hosts. For example, uropathogenic *Escherichia coli* adherence

patterns are determined by the binding specificity of the PapG adhesin (a component of the pyelone-phritis-associated pili (Pap)) and the differential distribution of the receptor isotypes in different hosts and tissues. Three different *papG* alleles exist in uropathogenic *E. coli* (class I, II, III). Each binds with a different specificity to distinct receptor isotypes, with class II being the allele predominantly associated with pyelonephritis, and class III being correlated with human cystitis. PapG binds to the Gal-α-1-4Gal-containing glycolipids present on the surface of human kidney epithelial cells, allowing the bacteria to gain a foot-hold in the tissue and resist displacement by the mechanical and physical forces operating within the kidney. The glycolipid receptor for PapG consists of a digalactoside core linked via a β-glucose residue to a ceramide group that anchors the receptor within the membrane. This minimum receptor isotype is called globotriasylceramide (GbO3). The various members of this receptor family differ by the addition of sugar residues distal to the Gal-α-1-4Gal core of GbO3. The addition of a single GalNAc sugar to GbO3 creates GbO4 (globoside), whereas the addition of two GalNAc sugars to GbO3 creates the Forssman antigen (GbO5) (Dodson et al. 2001).

2 Species-specific tropisms. Some pathogenic bacteria infect only certain animal species, e.g. *Neisseria gonorrhoeae*, *Neisseria meningitidis*, *E. coli* (expressing CFA I and CFA II), and group A streptococci are limited to infections of humans, whereas enteropathogenic *E. coli* K-88 infections are limited to pigs and *E. coli* K-99 strains predominantly infect calves. Although adhesin type and receptor specificity play an important role, it should be noted that other factors, such as nutritional requirements, may contribute to species-specific tropisms (see Iron scavenging). It should also be noted that certain strains or races within a species are genetically immune to specific pathogens. A good example is the lack of susceptibility of certain pig breeds to *E. coli* K-88 infections (Jann and Jann 1992). Although the differing nutritional or physiological conditions encountered in different animals is likely to influence the course of an infection, the outcome is related much more closely to the structural differences between, or absence of, a given eukaryotic receptor for a specific bacterial adhesin.

EXAMPLES OF BACTERIAL ADHESINS AND THEIR RECEPTORS

The adhesins of *E. coli* are varied and include pili, fimbriae, fibrillar, and afimbrial adhesins. A single strain of *E. coli* may be able to express several distinct adhesins encoded by distinct regions of the chromosome or of plasmids. This genetic diversity permits this organism

to adapt to its changing environment and exploit new opportunities presented by different host surfaces. Many of the adhesive fimbriae of *E. coli* have probably evolved from fimbrial ancestors resembling type 1 and type 4 fimbriae. Type 1 fimbriae enable *E. coli* to bind to D-mannose residues on eukaryotic cell surfaces. Although the primary fimbrial subunit (FimA) is the major protein component of type 1 fimbriae, the mannose-binding site is located in a minor protein (FimH) located at the tips or inserted along the length of the fimbriae. By genetically varying the minor 'tip protein' adhesin, the organisms gain the ability to adhere to different receptors. For example, tip proteins on Pap pili recognize a digalactose residue, while tip proteins on S-fimbriae recognize sialic acid. Differential binding can also be achieved by allelic variation of the *fimH* gene, for example, in *Salmonella enterica* serovar Typhimurium (Boddicker et al. 2002). The fibronectin-binding *E. coli* fimbriae known as curli (Olsén et al. 1989) are unusual since they are assembled via an extra-cellular nucleation precipitation pathway (Smyth et al. 1996). Production of curli increases internalization of bacterial cells by eukaryotic cells and triggers events leading to septic shock. Since curli also promote clumping of bacterial cells in culture and binding to abiotic surfaces, they are important for biofilm forma-tion (Chirwa and Herrington 2003) (see Biofilms). Although *E. coli*, *S. enterica* serovars Typhimurium and Enteriditis, and *Shigella* spp. have the genes required for curli synthesis, the expression of curli is not common to all strains due to insertion sequence disruptions, dele-tions, or other mutations. Unlike natural isolates of enterohemorrhagic, enterotoxigenic, and sepsis *E. coli* strains, enteroinvasive and enteropathogenic (EPEC) strains do not produce curli. EPEC combine the adhe-sive properties of bundle-forming pili (a type IV fimbria) (Tobe and Sasakawa 2001) with the intimate adherence mediated by effectors secreted by the type III secretion pathway (see Protein secretion) (Delahay et al. 2001). In addition, the flagella of EPEC can function as an adhesin for binding to mammalian cells, although the nature of the eukaryotic receptor(s) is not known (Girón et al. 2002). A fimbrial adhesins include Afa-I and Afa-III, which are made along with the Dr hemagglutinin and fimbrial F1845 adhesin by Afa/Dr diffusely adhering *Escherichia coli* (DAEC). The mechanism by which Afa/Dr DAEC infection is linked to inflammatory bowel diseases is currently unknown, however utilization of a glycosylphophatidylinositol-anchored protein (the complement-regulating decay-accelerating factor (DAF) or CD55) as a receptor for Afa/Dr adhesions offers one possible explanation. Afa/Dr DAEC strains recognize the short consensus repeat 3 (SCR3) domain of DAF, which plays a pivotal role in the regulatory function of DAF. In so doing, they induce a cascade of responses in T84 cells, including interleukin 8 (IL-8) production and polymorphonuclear leukocyte transepithelial migration.

Production of tumor necrosis factor-α (TNF-α) and IL1β follows, which upregulates DAF expression, thus increasing the brush border adhesion of Afa/Dr leading to structural and functional lesions (Bétis et al. 2003).

Pseudomonas, *Vibrio*, and *Neisseria* possess a fimbrial protein subunit that contains methylated phenylalanine at its amino terminus. These type IV pili have been established as virulence determinants of *Pseudomonas aeruginosa* lung infections in cystic fibrosis patients, and the receptor for such fimbriae produced by *N. gonorrhoeae* is the surface glycoprotein CD46 on urogenital cells (Tobiason and Seifert 2001). The major virulence factors of toxigenic *Vibrio cholerae* are cholera toxin (CT), which is encoded by a lysogenic bacteriophage (CTXφ), and toxin-coregulated pilus (TCP), which was also reported to be encoded on a filamentous phage genome. TCP is a type IV bundle-forming pilus essential for colonization and also serves as the receptor for CTXφ. Somewhat controversially, TCP has also been proposed to be a coat protein for a novel filamentous phage (Davis and Waldor 2003).

Gram-positive bacteria, such as *Staphylococcus aureus,* possess multiple cell wall proteins (microbial surface components recognizing adhesive matrix molecules (MSCRAMM)) which facilitate binding to host cells and tissues, thus promoting attachment of bacteria to surfaces. These include the fibronectin-binding proteins FnbA and FnbB, which promote adherence to host tissues and are essential for inducing the uptake of staphylococci into epithelial and endothelial cells. Other MSCRAMM can override the loss of fibronectin binding and allow adherence to host cells but not lead to efficient internalization into eukaryotic cells (Bayles and Bohach 2001).

Other gram-postive pathogens, such as Group A streptococci and *Streptococcus pyogenes*, also display cell-surface fibronectin-binding proteins (Bisno et al. 2003). The fibronectin binding proteins (e.g. F, M1 and M3) of *Str. pyogenes* in conjunction with lipotechoic acid (LTA) are involved in promoting attachment to the oral mucosa and buccal epithelial cells, where LTA loosely tethers streptococci to epithelial cells, whereas M proteins secure a much stronger, less reversible association.

Treponema pallidum, the causative agent of syphilis, has three related surface adhesins (P1, P2, and P3), which bind to a four-amino acid sequence (Arg-Gly-Asp-Ser) of the cell-binding domain of fibronectin. It is not clear whether *T. pallidum* uses fibronectin to attach to host surfaces or coats itself with fibronectin to avoid host defenses, such as phagocytic cells (Baughn 1987).

BIOFILMS

In most environmental niches, bacteria survive and multiply not as planktonic cells suspended in liquids but as surface-attached biofilms. These are complex communities of single or multiple species of microorganisms that develop on abiotic (e.g. rocks) and biotic (host mucosal tissues) surfaces. Since biofilms contaminate industrial pipelines, dental-unit water lines, catheters, ventilators, and medical implants, they act as a source of disease for humans, animals, and plants. They also represent a barrier to eradication as both the nature of the biofilm and the physiological state of bacterial cells within the biofilm confers high levels of resistance to antimicrobial agents.

Microbial development on surfaces is a dynamic stepwise process involving adhesion, growth, motility, and extracellular polysaccharide (EPS) production (Figure 6.1). Five distinct stages of biofilm development (namely (1) reversible attachment; (2) irreversible attachment; (3) maturation 1; (4) maturation 2; and (5) dispersion) (Sauer et al. 2002) culminate in a structured community sometimes characterized by mushroom and mound structures. Underlying this process is a genetically programmed series of events; however, bacteria appear to have evolved more than one pathway, probably due to particular environmental stimuli (growth media or substratum, etc.) triggering different developmental pathways. Despite this, all culminate with the same end point, i.e. a mature biofilm community (O'Toole 2003). For *P. aeruginosa*, biofilm formation involves the production of flagella and pili followed by the production of EPS. To become a member of the biofilm community, the bacterium must then differentiate into a biofilm-associated cell by repressing synthesis of the flagellum (O'Toole 2003; Watnick and Kolter 2000). Ultimately, however, flagella are produced once more to facilitate dispersal (Sauer et al. 2002).

Although redistribution of attached cells to form a structured community can be achieved by motility via flagella, pili, or fimbriae, this is not a prerequisite, since nonmotile organisms readily form biofilms, e.g. staphylococci, streptococci, and mycobacteria. As cells divide, daughter cells spread outward and upward from the attachment point to form cell clusters offering an alternative mechanism of biofilm formation. Surface-associated aggregation can also occur via the recruitment of single cells or cell flocs from the bulk fluid. Bacterial cells in biofilm microcolonies are held together by a slime-like matrix (EPS). This matrix is chemically complex, including polysaccharides, nucleic acids, and proteins. Intercellular adhesion within biofilms of *Staphylococcus epidermidis*, a major cause of medical device-related infections, is mediated by the polysaccharide intercellular adhesin (PIA). Synthesis of this linear glucosaminoglycan composed of *N*-acetylglucosamine in β-1,6-glycosidic linkages containing deacetylated amino groups and succinate and phosphate substituents requires expression of the *icaADBC* operon. These genes are subject to a complex regulatory control mechanism, which involves the alternative sigma factor σ^B, *icaR*, two other regulatory loci, plus phase variation due to IS element insertion. Cessation of PIA produc-

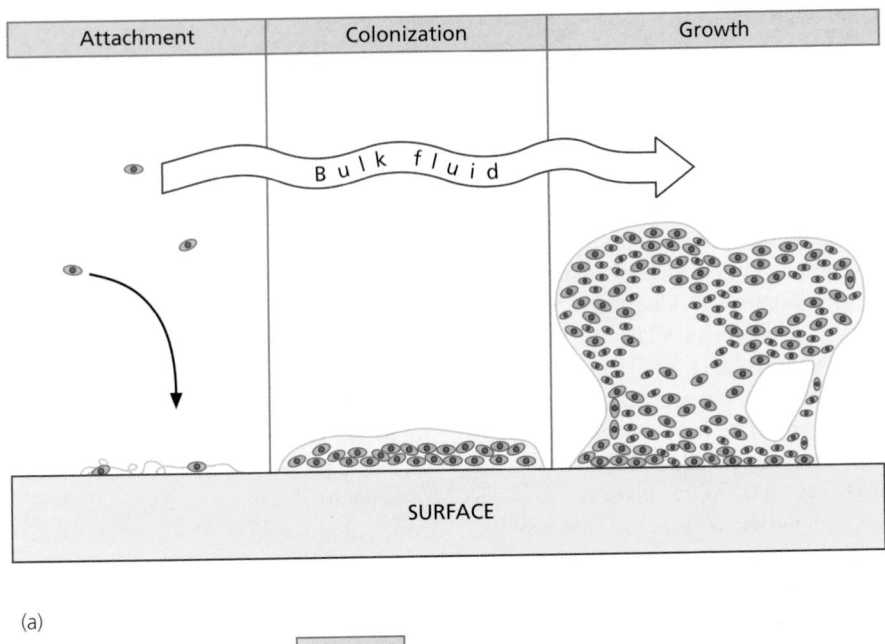

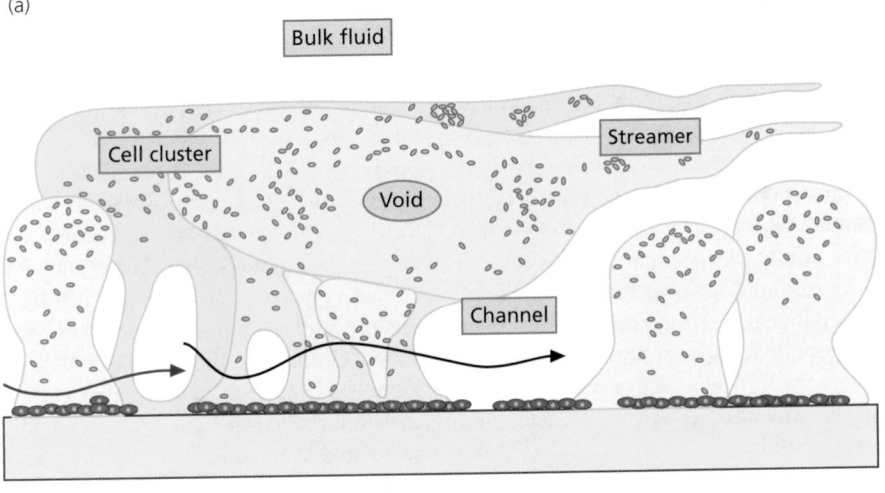

Figure 6.1 *Biofilm formation:* **(a)** *Bacteria attach to a surface and multiply. An extracellular polysaccharide is produced and the combined influence of this and the physical properties of the environment favor the development of complex structures resembling mounds and mushrooms. The bulk fluid flows between these structures through channels, and, if strong enough, streamer structures develop downstream.* **(b)** *Dispersal of detached bacterial cells finally occurs into the bulk fluid. Redrawn from originals supplied by Dr J.W. Costerton, Montana State University, USA/Artwork PEG DIRCKX from www.erc.montana.edu/Res-Lib99-SW/Image_Library/ Structure-Function/default.htm, copyright held by Center for Biofilm Engineering at Montana State University-Bozeman, Bozeman, MT 59717-3980, USA; Tel.: +1-406-994-4770; fax: +1-406-994-6098.*

tion by inactivational IS element insertion allows individual cells to leave the biofilm and colonize new surfaces (Dobinsky et al. 2003). Physical forces acting on the biofilm can also influence its structure, e.g. a faster flow can favor biofilm cell clusters, which tend to elongate in the downstream direction, forming filamentous 'streamers' (Hall-Stoodley and Stoodley 2002).

Proliferation

Bacterial proliferation within a colonization niche and also during the development of infection will depend on an ability not only to resist host defenses but also to acquire essential nutrients. Bacteria have therefore evolved a full complement of mechanisms for nutrient acquisition (e.g. to scavenge for carbon and nitrogen sources, phosphate, sulfur, and metal ions) and for the synthesis of complex organic molecules that cannot be scavenged from the local environment. Specific transport systems or cell-wall components capable of binding limiting substrates and transporting them into the bacterial cell are employed by both pathogenic and nonpathogenic bacteria. Pathogens that are highly host-adapted often dispense with biosynthetic pathways for complex organic molecules that can be obtained directly from the host. For example, *H. influenzae* requires heme (X factor) and nicotinamide adenine dinucleotide (NAD^+; V-factor) for growth.

For almost all bacterial pathogens, iron is one indispensable nutrient, the lack of which exerts considerable influence on growth in vivo. Iron is a redox reactive metal and is required as a cofactor for many important metabolic enzymes and cytochromes. Although there is an abundance of iron in the extracellular tissue fluids of man and animals, the amount of free ionic iron is of the order of 10^{-9} M, a level that is far too low to support the growth of most microorganisms. The injection of animals with iron compounds results in an increased susceptibility to experimental infection when compared with untreated controls, so that the lethal dose of infecting microorganism required is greatly reduced. Iron is also the only metal cation known to be capable of abolishing the antibacterial activity of body fluids in vitro. The ability of the host to withhold iron thus contributes to host defense against infecting microorganisms and in extracellular tissue fluids depends on the host iron-binding glycoproteins transferrin and lactoferrin (see Iron scavenging). However, most pathogens multiply successfully in body fluids and cause infection in the absence of exogenously supplied iron either because they possess efficient iron scavenging mechanisms capable of removing transferrin-bound iron or by invading host cells where intracellular iron is more freely available.

IRON SCAVENGING

The most intensively investigated bacterial high-affinity iron-transport systems are based on siderophores. These are low-molecular-mass iron chelators that bind iron with high affinity and specificity. They are employed by both gram-negative bacteria such as *E. coli*, *Shigella* spp., *Salmonella* spp., and *P. aeruginosa* and gram-positive pathogens such as the staphylococci, mycobacteria, and corynebacteria. *Shigella* species employ at least three distinct iron-transport systems, two of which involve siderophores (aerobactin and enterobactin) and their associated outer-membrane receptor proteins. A third pathway required for the transport and utilization of iron in the form of heme is employed by shigellae growing intracellularly within epithelial cells. While the enterobactin iron-transport system is found in all *E. coli* strains so far examined, aerobactin production and heme transport are more often associated with pathogenic *E. coli* strains and the corresponding genetic loci are encoded on virulence plasmids, within pathogenicity islands, or on chromosomal regions with the potential for horizontal transmission.

Certain bacterial pathogens and notably *H. influenzae* and *N. meningitidis* use a siderophore-independent receptor-mediated iron-uptake system that involves a direct interaction between a bacterial cell-surface receptor (consisting of two transferrin-binding proteins, TbpA and TbpB) and transferrin. These receptor-based mechanisms are further distinguished from siderophore-mediated mechanisms by their high degree of specificity for the iron-binding glycoprotein of the natural host; humans are the only natural reservoir for *N. meningitidis*, which will only bind human transferrin, while the pig pathogen *Actinobacillus pleuropneumoniae* can use porcine but not human transferrin as an iron source.

Apart from sequestering transferrin-bound iron, many pathogens have also evolved mechanisms for acquiring iron from heme-containing proteins (such as hemoglobin) or from heme- and hemoglobin-binding proteins (hemopexin and haptoglobin, respectively). These involve either specific surface receptors or secreted proteins such as HasA of *P. aeruginosa* and *Serratia marcescens*. The latter consists of an outer-membrane receptor (HasR) and a type I secretion apparatus (see below under Subversion of host cell signal transduction and on protein secretion) in addition to the extracellular heme-binding protein. HasR can bind and transport free heme or heme bound to hemoglobin, in a manner enhanced by the presence of HasA. Thus, it appears that the HasA–HasR interaction, rather than recognition of heme, is important for ligand uptake (Clarke et al. 2001).

For both high-affinity iron- and heme-scavenging systems in gram-negative bacteria that involve outer membrane receptor proteins, a common feature is the requirement for the energy-transducing protein TonB to promote the uptake of iron across the outer membrane together with an ABC transporter for transporting the iron–ligand complex across the periplasm and cytoplasmic membrane. Gram-positives need only the ABC transporter and a binding protein, which is usually a cytoplasmic membrane anchored lipoprotein

Motility

To establish infection in or on a susceptible host, a pathogen must reach a suitable infection site or breach in host tissues. For example, at mucus membranes a pathogen may need to penetrate the mucus blanket and subsequently adhere to the epithelium. Movement through mucus to the epithelial cells may be dependent on chemoattractants present in host tissues, the motility organelles on the bacterium (see below), and possibly mucus-degrading enzymes (see Infiltration of host tissues).

Motility can be achieved through the retraction of surface appendages (type IV pili-mediated twitching motility) (Mattick 2002) or rotation of specialized motors (flagella-mediated swimming) (Josenhans and Suerbaum 2002). Flagella are assembled by a machinery with significant similarity to the type III secretion pathway (see below under Subversion of host cell signal transduction and on protein secretion). The control of this machinery and the structural assembly of the flagella are subject to a complex hierarchical regulation. The flagellum comprises structurally of a hook filament protruding from the basal body, which anchors it to the cell membranes. Extending from this is the long rigid filament (Aldridge and Hughes 2002). Bacteria may have single or multiple flagella, which can be arranged in a polar fashion or distributed over the entire cell surface. Flagella rotation propels the bacterium, and this forward motion is punctuated with tumbling, during which reorientation may occur to enable bacteria to preferentially spread in certain directions (chemotaxis). The flagella of *E. coli* rotate in a counterclockwise direction to achieve a forward motion, and switch to a clockwise direction to effect stopping and tumbling (Berg 2003). Other bacteria, e.g. *Rhodobacter spirillum*, achieve this control by periodically ceasing flagella rotation (Armitage 1999).

Swarming is a specialized form of bacterial motility that is characteristic of species belonging to genera such as *Proteus*, *Vibrio*, *Bacillus*, and *Clostridium* but is also exhibited by *Serratia*, *Escherichia*, *Salmonella*, and *Yersinia* spp. Depending on the organism, swarming depends on flagella and pili, as well as surface components such as polysaccharides and surfactants, and involves differentiation of vegetative bacterial cells into hyperflagellate swarm cells that undergo rapid and coordinated population migration across solid surfaces. For pathogens such as *Proteus mirabilis*, swarming facilitates ascending colonization of the urinary tract and entry into host cells (Fraser and Hughes 1999).

Infiltration of host tissues

Motility is not the sole requirement for successful relocation. There are numerous factors that mediate the penetration of bacteria deep into host tissues. These factors function by altering the integrity of the tissues by degrading the extracellular matrix or tissue cells or by travelling directly through cell layers after uptake directly into mammalian cells, a process that leads to either tissue necrosis or programmed cell death (apoptosis). Such determinants are often termed 'spreading factors' and are usually bacterial enzymes that affect the physical properties of tissue matrices and intercellular spaces, thereby promoting the spread of the pathogen (see Table 6.1).

Once in contact with eukaryotic cells, bacteria may be taken up by the host cell or may mediate their own internalization to proceed further. Bacteria-mediated uptake into nonphagocytic host cells may be achieved by either bacterial surface proteins (invasins), which interact with specific receptors on the host cell (e.g. internalin A and B binding to E-cadherin during uptake of *Listeria monocytogenes*), or bacterial proteins injected into the host cell cytoplasm, which trigger membrane ruffling and concomitant bacterial uptake by macro-pinocytosis (e.g. *Salmonella* or *Shigella*) (see Subversion of host cell signal transduction). The invasin-mediated 'zipper' mode of entry is achieved by unrelated proteins in different bacteria (e.g. *Listeria* internalins and *Yersinia* invasin), whereas the mediators of the 'trigger' mechanism of uptake used by *Shigella* (Ipas) and *Salmonella* (Sips) are similar. All mechanisms proceed to localization of the bacterium within a membrane-bound vacuole. This vacuolar compartment is acidic, is poor in nutrients, and undergoes progressive vesicle fusion with early endosomes, late endosomes, and lysosomes that contain potent lytic enzymes. Some pathogens have evolved mechanisms to modify the vacuole or its process of maturation and remain localized within the vacuole, e.g. *Mycobacterium tuberculosis* and *Salmonella* spp. *M. tuberculosis* recruits tryptophan-aspartate-containing coat protein (TACO) to the membrane of the phagosome, thus mimicking the plasma membrane and thereby preventing maturation into, or fusion with, lysosomes (Pieters and Gatfield 2002). In contrast, *S. enterica* serovar Typhimurium appears to deliver effector molecules via the type III secretion machinery (*Salmonella* pathogenicity island; SPI-2) (see Subversion of host cell signal transduction) into the cytosol, which disrupt recruitment of the components of the noxious host enzyme NADPH oxidase to the phagosome (Amer and Swanson 2002).

Other pathogens escape from the vacuole into the cytosol, e.g. *Listeria monocytogenes*, *Shigella flexneri*, and rickettsiae (O'Riordan and Portnoy 2002). To escape from phagosomes, *L. monocytogenes* produces the pore-forming cytolysin listeriolysin O (LLO). The combined action of LLO with phosphatidylinositol-specific phospholipase C (PI-PLC) and phophatidylcholine-specific phospholipase C (PC-PLC) disrupts the phagosomal membrane leaflet, although the underlying mechanism employed is uncertain. *Sh. flexneri* appear to use the hemolytic activity of IpaB and possibly also the *icsB* gene product to achieve liberation from the phagosome (Goebel and Kuhn 2000).

Many cytosolically replicating intracellular bacteria possess the ability to spread from the primary infected cell into its neighbors. This ability depends on the polymerization of cellular actin at one bacterial cell pole mediated by a specific bacterial surface protein (ActA for *L. monocytogenes*; IcsA, IcsB, and OpaB for *Sh. flexneri*). *L. monocytogenes*, for example, move ahead of an actin comet tail at a speed of approximately 0.3 μm/s, upon contact with the host cell plasma membrane, a bacterium-containing protrusion is formed, which is internalized by the adjacent cell. The bacterium then lyses the two plasma membranes surrounding it in the newly infected cell to continue the cycle (see Figure 6.2) (Cossart and Lecuit 1998). Although *Sh. flexneri* spreads in a similar fashion, the proteins that mediate this phenomenon are unrelated to those of listeria (Bourdet-Sicard et al. 2000). This form of dissemination circum-

Table 6.1 *Spreading factors that aid bacterial invasion of tissues*

Spreading factor	Producing bacteria	Comments
Hyaluronidase	Streptococci, staphylococci, clostridia	Attacks the interstitial cement of connective tissue by depolymerizing hyaluronic acid
Collagenase	*Clostridium histolyticum, Clostridium perfringens*	Breaks down collagen, the framework of muscles, facilitating the development of gas gangrene
Neuraminidase	Intestinal pathogens: *Vibrio cholerae, Shigella dysenteriae*	Degrades neuraminic acid (sialic acid), an intercellular cement of the epithelial cells of the intestinal mucosa
Streptokinase and staphylokinase	Streptococci and staphylococci	Kinase enzymes that convert inactive plasminogen to plasmin, which digests fibrin and prevents clotting of the blood. The relative absence of fibrin in a spreading bacterial lesion allows more rapid diffusion of the infectious bacteria throughout the tissues of the host
Enzymes that cause hemolysis and/or leukolysis	Staphylococci (leukocidins), Streptococci (streptolysin)	Insert into membranes forming a pore that results in cell lysis, or enzymatically attack phospholipids, which destabilizes the membrane. They may be referred to as lecithinases or phospholipases, hemolysins (if they are able to lyse red blood cells), or leukocidins and streptolysin (lyse phagocytes and their granules)
Phospholipases	*Clostridium perfringens* (α toxin)	Hydrolyze phospholipids in cell membranes by removal of polar head groups
Lecithinases	*Clostridium perfringens*	Destroy lecithin (phosphatidylcholine) in cell membranes, leading to cell lysis
Hemolysins	Staphylococci (α toxin), Streptococci (streptolysin), Clostridia spp.	Channel-forming proteins or phospholipases or lecithinases that destroy red blood cells and other cells, such as phagocytes through cell lysis
Staphylococcal coagulase	*Staphylococcus aureus*	Cell-associated and diffusible enzyme that converts fibrinogen to fibrin, which causes clotting. Cell-associated coagulase may also provide an antigenic disguise by causing clotted fibrin to deposit on the cell surface. Also in the case of a staphylococcal lesion encased in fibrin, e.g. a boil or pimple, bacterial cells may be resistant to phagocytes or tissue bactericides or even antibiotics, which might be unable to reach their bacterial target
Extracellular digestive enzymes	Heterotrophic bacteria	Including proteases, lipases, glycohydrolases, and nucleases. May aid in invasion either directly or indirectly by damaging the cells of the tissue, or contribute to bacterial nutrition or metabolism
Toxins with short-range effects related to invasion	*Bordetella pertussis* (adenylate cyclase), *Bacillus anthracis*, (edema factor component of anthrax toxin)	Immediate effects on host cells that promote bacterial invasion of host tissues by causing increased levels of cAMP and disruption of cell permeability. These toxins may contribute to invasion through their effects on macrophages or lymphocytes in the vicinity, which are playing an essential role to contain the infection or on the cells of the tissue themselves

cAMP, cyclic adenosine monophosphate.

vents both the humoral and complement-mediated host defenses (see Host resistance to infection).

Obligate intracellular bacteria have taken these measures to the extreme. *Rickettsia prowazekii* (the causative agent of epidemic typhus), for example, is maintained in a cycle involving human and body lice (*Pediculus humanus*) cells, and is transmitted through direct contact of humans with contaminated body louse feces. Following entry into cells via induced phagocytosis, *R. prowazekii* escapes from the phagosome into the cytoplasm. Upon cell lysis, spread can continue to neighboring cells in a similar fashion; however, the related *Rickettsia rickettsii* (causative agent of Rocky Mountain spotted fever) achieves intracellular cell–cell spread powered by actin-based motility similar to that of *Listeria* and *Shigella* (Hackstadt 1998).

Avoidance of host immune defenses

During the processes of colonization and tissue invasion, pathogens must avoid host defenses (see Host resistance to infection). While some bacteria adopt an intracellular lifestyle (see Infiltration of host tissues) to evade defense mechanisms by their internal location, others employ

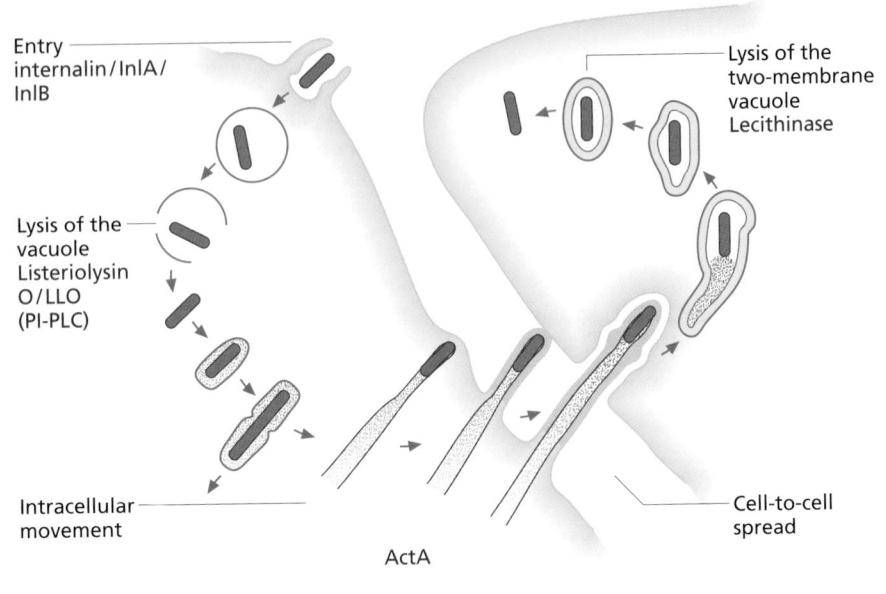

Figure 6.2 *Schematic representation of the cell infectious process of* Listeria monocytogenes. *Bacterial factors are indicated in blue. (Redrawn from Cossart and Lecuit 1998)*

protective surface coatings, e.g. exopolysaccharides and lipopolysaccharides (LPS) (see Toxins), which can be antiphagocytic and, in the case of surface macro-molecules, which mimic those of the host, function as 'camouflage.'

Exopolysaccharides include discrete capsules (e.g. the capsular polysaccharides (sometimes called K-antigens) of *E. coli, Klebsiella pneumoniae, Streptococcus pneumoniae,* and *S. aureus*) and the loosely associated slime polysaccharides produced by mucoid bacteria. Exopolysaccharides form highly hydrated water-insoluble gels composed of either homo- or heteropolysaccharides. The former consists of a single sugar monomer (e.g. the levans and dextrans of many oral streptococci), and the latter of repeating oligosaccharide units of more than one monomer. Many exopolysaccharides are acidic due to the possession of carboxyl groups, either from acidic sugars, such as uronic acids or neuraminic acid, or from nonsugar substituents, such as pyruvyl, acetyl, and formyl groups. Their acidic nature is reflected in many of their properties. A few of the bacilli (e.g. *Bacillus anthracis*) have a capsule made up of a single amino acid, poly-D-glutamic acid. In physical chemistry and function, however, it is similar to the polysaccharide capsules.

In the environment, the highly hydrated nature of exopolysaccharides offer bacteria protection from desiccation and a potential mechanism for trapping nutrients, although they are not employed as a reserve carbohydrate food source for use during starvation. Exopolysaccharides do, however, contribute to virulence in two different ways. First, they may contribute to bacterial adhesion. This is particularly obvious in the cariogenic organism, *Streptococcus mutans*. This bacterium can synthesize a dextran-like branched water-insoluble homopolymer of glucose from dietary sucrose. This forms a glutinous layer on the surface of teeth and contributes to the matrix of dental plaque. Second, capsular polysaccharides protect bacteria from phagocytic cells (including amoebae in the environment). This is because capsulate or slime-producing bacteria growing as aggregated microcolonies or biofilms cannot be engulfed for steric reasons, i.e. they are too large to be internalized. Mucoid *P. aeruginosa* strains, which produce the exopolysaccharide alginate in the lungs of infected individuals with cystic fibrosis (CF), provide a good example of this. In CF, lung damage arises in part from the tissue-damaging agents, e.g. elastase released by neutrophils unable to engulf the bacteria. Exopolysaccharides also render bacterial cells more resistant to phagocytic engulfment by increasing surface hydrophilicity, and blocking the action of complement (see LPS and virulence).

Several bacterial pathogens produce exopolysaccharide capsules, the chemical structure of which mimics a host tissue surface antigen. This protects the organism from the immune system by appearing to the host as 'self.' Perhaps the best known examples are the K1 capsule of *E. coli* and the immunologically and structurally identical capsule of *N. meningitidis* group B. These organisms have capsules of α-2,8-linked *N*-acetylneuraminic acid, which is partially *O*-acetylated and immunologically crossreacts with neonatal neural cell adhesion molecules.

Many pathogens have a highly organized proteinaceous surface layer, the S layer. The role of this in pathogenesis has not been fully elucidated, but the S-layer of *Aeromonas salmonicida* can provide resistance to the bactericidal activity of complement, that of *Rickettsia prowazekii* is responsible for humoral and cell-mediated immunity, and that of *Lactobacillus acidophilus* is required for adhesion to the intestinal epithelium (Sara and Sleytr 2000).

Toxins

Toxins are frequently responsible for the tissue damage characteristic of bacterial infections and hence make a significant contribution to pathogenicity. Indeed, for some pathogens, the adminstration of the toxin alone is sufficient to reproduce the same disease as that resulting from infection by the producer organism, e.g. botulism.

PROTEIN EXOTOXINS

Exotoxins are typically proteins secreted or released by bacteria during growth. They are usually produced only by a single bacterial species within a genus. For example, *Clostridium tetani* produces tetanus toxin, whereas *Clostridium botulinum* produces botulinum toxin. Usually, virulent strains of the bacterium produce the toxin while avirulent strains do not, indicating that in many cases the exotoxin is a major determinant of virulence. Bacterial exotoxins are produced by both gram-positive and gram-negative bacteria and have highly specific targets and modes of action. The target may be a component of specific cells, tissues, or organs and usually the site of toxin damage indicates the location of the substrate for that toxin. Terms such as enterotoxin, neurotoxin, leukocidin, and hemolysin are often used to indicate the target sites for protein toxins. Many protein toxins exhibit very specific cytotoxic activities, e.g. tetanus or botulinum toxins attack specific cell types (Table 6.2). However, other toxins, such as those produced by staphylococci, streptococci, clostridia, etc., have a broad cytotoxicity causing necrotic tissue cell death. Toxins that are phospholipases are particularly nonspecific in their cytotoxicity because they cleave phospholipids (which are components of all host cell membranes), resulting in the death of the cell by leakage of cellular contents. This is also true of pore-forming hemolysins and leukocidins.

A few highly potent protein toxins can kill the host in the absence of the producing organism and are known as lethal toxins. Even though the tissues affected, and the target sites for such toxins (e.g. anthrax toxin) may be known, the precise mechanism by which death occurs is not always understood.

As 'foreign' substances to the host, most protein toxins are strongly antigenic. In vivo, specific antibodies (antitoxins) can neutralize the actions of these bacterial proteins. Also, protein toxins are inherently unstable and can lose their toxic properties while retaining their antigenicity. This was first discovered by Paul Ehrlich who coined the term 'toxoid'. The formation of toxoids can be accelerated by treatment with a variety of chemical reagents, including formaldehyde. The resulting toxoids can be used for immunization against diseases caused by the corresponding pathogen where the primary determinant of bacterial virulence is the toxin.

A + B subunit toxins

Many protein toxins, notably those that act intracellularly, consist of two components termed A and B. Subunit A is responsible for the enzymatic activity of the toxin at the target site. Subunit B facilitates the binding of the toxin to a specific receptor on the host cell membrane and subsequent internalization of the A subunit. The enzymatic component is not active until it is released from the native toxin. Purified A subunits are enzymatically active but lack the ability to bind to host cells. Isolated B subunits may bind to target cells, but they are nontoxic. Toxin subunits may be synthesized and arranged in a variety of ways, for example A-B or A-5B indicates that subunits are synthesized separately and associated by noncovalent bonds, whereas A/B denotes that the subunit domains are formed from a single protein that may be separated only by proteolytic cleavage. The notation A + B indicates that the separate protein subunits interact at the target cell surface to form the toxin. More than one binding subunit is indicated as, 5B for example, which shows that the binding domain is composed of five subunits.

Attachment and entry of toxins

There are at least two mechanisms for toxin entry into target cells. In both cases, a large protein molecule must insert into and cross a membrane lipid bilayer. This activity is reflected by the ability of most A/B native toxins, or their B components, to insert into artificial lipid bilayers, creating ion-permeable channels. For direct entry, the B subunit of the native toxin binds to a specific receptor on the target cell and induces the formation of a pore in the membrane, through which the A subunit is transferred into the cell cytoplasm. The second mechanism relies on the native toxin binding to the target cell and the A + B structure is taken into the cell by the process of receptor-mediated endocytosis. The toxin is internalized in a membrane-enclosed vesicle called an endosome. H^+ ions entering the endosome lowers the internal pH, which causes the A + B subunits to separate. The B subunit effects the release of the A subunit from the endosome by forming a pore, facilitating association with its cellular target in the

Table 6.2 *Examples of two-component (A-B) exotoxins with intracellular targets*

Exotoxin	Comments
Adenylate cyclase toxin of *Bordetella* spp. Chromosomally encoded	Activated by eukaryotic intracellular calmodulin. Catalyzes conversion of ATP to cAMP, which perturbs cellular activity, inhibiting leukocyte chemotaxis and activity
Anthrax toxin of *Bacillus anthracis* Plasmid-encoded	Comprises three separate proteins, protective antigen (PA), edema factor (EF), and lethal factor (LF) EF + PA causes an increase in intracellular cAMP level, resulting in edema (fluid accumulation) LF + PA is a potent cytolethal toxin causing death of host cells and ultimately death of host
Botulinum toxins of *Clostridium botulinum* Phage-encoded	Comprises seven antigenically distinct toxins designated A to G. Among the most potent of all biological toxins of bacterial origin. The B-subunit binds to neuroreceptor gangliosides on cholinergic neurons, while the A-subunit irreversibly inhibits release of the stimulatory neurotransmitter acetylcholine, at muscle–nerve junctions, resulting in a paralysis and death
Cholera toxin from *Vibrio cholerae* Chromosomally encoded, A-5B structure	The B-subunit binds to GM_1 ganglioside receptors on the surface of small intestinal enterocytes. Reduction of the disulfide bond in the A-subunit activates A_1 fragment that ADP-ribosylates guanosine triphosphate (GTP)-binding protein (G_s) by transferring ADP-ribose from nicotinamide adenine dinucleotide (NAD). The ADP-ribosylated GTP-binding protein activates adenyl cyclase, resulting in an increase in intracellular cAMP, which triggers a profound movement of electrolytes (sodium, potassium, bicarbonate) and water into the lumen of the gut, leading to a profuse life-threatening diarrhea. In addition, the elevated cAMP levels in the gut cells also block the uptake of any further sodium and chloride from the lumen of the small intestine, ultimately resulting in hypovolemic shock and death in the absence of fluid and electrolyte replacement therapy
Diphtheria toxin from *Corynebacterium diphtheriae* Phage-encoded A-B structure	The toxin acts by ADP-ribosylation of intracellular targets, thereby inhibiting cell protein synthesis by catalyzing transfer of ADP-ribose from NAD to elongation factor-2 (EF-2)
Exotoxin A from *Pseudomonas aeruginosa* Chromosomally encoded	Very similar to diphtheria toxin
Heat-labile enterotoxins (LT) from enterotoxigenic *Escherichia coli* (ETEC) Plasmid-encoded A-5B structure	LT comprises two types (LT-I and LT-II). LT-1 is associated predominantly with human disease, while LT-II is produced only by strains isolated from animals. These toxins are very similar to cholera toxin
Pertussis toxin from *Bordetella pertussis* Chromosomally encoded A-5B structure	Different B-subunits where the S2 (B) subunit binds a glycolipid receptor on ciliated respiratory cells while the S3 (B) subunit binds to glycolipids on phagocytes. The S1 or A subunit inhibits cellular signal transduction via ADP-ribosylation of GTP-hydrolyzing protein (G_i), causing unregulated adenylate cyclase and large increases in cellular cAMP levels, resulting in hypersecretion of respiratory secretions, mucus, and paroxysmal cough. It is also known to inhibit leukocyte chemotaxis and activity
Shiga toxin of *Shigella dysenteriae* Chromosomally encoded A-5B structure	This very potent toxin has a similar mode of action to ricin, a toxin derived from castor beans. The B-subunit binds to Gb_3 glycolipid receptor on the surface of target cells. The A-subunit blocks protein synthesis by preventing binding of aminoacyl-transfer RNA through cleavage of the 28S rRNA from 60S ribosomal subunit
Shiga-like toxins of enterohemorrhagic *Escherichia coli* (EHEC) and *Shigella* spp. Phage-encoded A-5B structure	Two forms (SLT-I and SLT-II) in EHEC. SLT-I is identical to *S. dysenteriae* shiga toxin, with the exception of a single amino acid, while SLT-II has approximately 60% homology with shiga toxin. The B-subunit binds to target cell glycolipid globotriaosylceramide. The A-subunit is cleaved and the resulting A_1 fragment binds to 28S rRNA of 60S ribodomal subunit, thereby blocking protein synthesis in the same way as shiga toxin

(Continued over)

Table 6.2 *Examples of two-component (A-B) exotoxins with intracellular targets (Continued)*

Exotoxin	Comments
Tetanus toxin of *Clostridium tetani* Plasmid-encoded	This potent toxin is not secreted but is released upon lysis of the bacterial cell. The binding domain (B) binds to neuroreceptor gangliosides (GD_{1b}). The A-subunit is a zinc-dependent endopeptidase, which is internalized and migrates from peripheral nerves to central nervous system, across the synapses to presynaptic nerve endings this constitutes what is known as retrograde transport, i.e. against the normal direction of nerve impulses. Once at its target location, it accumulates in vesicles and irreversibly blocks the release of inhibitory transmitters. This results in continuous stimulation of muscles by excitatory neurotransmitters, resulting in spastic paralysis (spasms of bulbar and paraspinal muscles) with trismus (lockjaw; spasms of the masticatory muscles), risus sardonicus (spasms of the masseter muscles), and opisthotonos (spasms of back and neck muscles)

ADP, adenosine diphosphate; ATP, adenosine triphosphate; AMP, cyclic adenosine monophosphate; NAD, nicotinamide adenine dinucleotide.

cytoplasm. The B subunit remains in the endosome and is recycled to the cell surface.

Many of the genes encoding toxin production are located on plasmids or on lysogenic bacteriophages. Thus, genetic exchange in bacteria by conjugation and transduction can mobilize these genetic elements between different strains of bacteria and, therefore, may play a role in spreading pathogenic potential to other bacteria; see Table 6.2 for further examples of two-component exotoxins (Lahiri 2000).

Superantigens (type I toxins)

Conventional antigens are engulfed by antigen-presenting cells (APC), degraded into epitopes, e.g small peptides, and presented to the peptide groove of MHC-II molecules, where they are recognized only by specific T4-helper cells having a receptor (TCR) with a complementary shape. Superantigens (see Table 6.3), however, bind directly to the outside of the MHC-II molecules and are recognized by T4-lymphocytes. This activation of very large numbers of T4-lymphocytes results in the secretion of excessive amounts of interleukin-2 (IL-2), as well as the activation of self-reactive T-lymphocytes. High levels of IL-2 in the systemic circulation result in

fever, nausea, vomiting, diarrhea, and malaise. Excess stimulation of IL-2 secretion can also result in the production of other cytokines such as TNF-α, interleukin-1 (IL-1), IL-8, and platelet-activating factor (PAF) and can lead to the same endothelial damage, acute respiratory distress syndrome, disseminated intravascular coagulation, shock, and multiple organ system failure observed with LPS and other bacterial cell-wall components (see Endotoxins). Activation of self-reactive T-lymphocytes can also lead to autoimmune disease (Llewelyn and Cohen 2002).

ENDOTOXINS

Endotoxins are bacterial cell-envelope components that are not actively secreted by the cell but can be shed as membrane blebs or vesicles into the bloodstream during the course of infection. They are exemplified by the lipooligosaccharides (LOS) and LPS found in the outer membrane of gram-negative bacteria. They are toxic to most mammals and, regardless of the bacterial source, all endotoxins produce the same range of biological effects. The injection of living or killed gram-negative cells, or purified LPS, into experimental animals causes a wide spectrum of nonspecific pathophysiological reactions,

Table 6.3 *Examples of superantigens*

Superantigen	Associated pathogen	Effect on host
Toxic shock syndrome toxin-1 (TSST-1)	Some strains of *Staphylococcus aureus*	Toxic shock syndrome (TSS): excessive cytokine production leads to fever, rash, and shock
Streptococcal pyrogenic exotoxin (Spe)	Rare invasive strains and scarlet fever strains of *Streptococcus pyogenes* (group A β streptococci)	Toxic shock-like syndrome (TSLS): excessive cytokine production leads to fever, rash, and shock
Staphylococcal enterotoxins (SE)	Many strains of *Staphylococcus aureus*	Food poisoning: excessive IL-2 production results in fever, nausea, vomiting, and diarrhea (vomiting may also be due to these toxins stimulating the vagus nerve in the stomach lining)
Hormone analog (Sta; heat-stable toxin)	Enterotoxigenic *Escherichia coli* (ETEC)	Binds to guanylate cyclase on intestinal epithelial cells, stimulating the overproduction of cGMP and resulting in diarrhea

cGMP, cyclic guanosine monophosphate; IL-2, interleukin-2.

such as fever, changes in white blood cell counts, disseminated intravascular coagulation, hypotension, and shock. Injection of small doses of endotoxin results in the death of most mammals. The sequence of events follows a regular pattern: (1) latent period; (2) physiological distress (diarrhea, shock); and (3) death. How soon death occurs varies according to the dose of the endotoxin, route of administration, and species of animal. In humans, LPS/LOS levels in plasma and cerebrospinal fluid are associated quantitatively with the production of inflammatory mediators, clinical symptoms, and outcome of disease. Patients with persistent septic shock, multiple organ failure, and severe coagulopathy reveal extraordinarily high levels of LPS in plasma. Mortality related to shock increases from zero to >80 percent with an increase of plasma LPS from 10 to 100 endotoxin units/ml. Although gram-positive bacteria do not produce endotoxins of this type, they can stimulate cytokine release and elicit an inflammatory response identical to that triggered by LPS. Such a response involves cell-wall polymers, such as peptidoglycan and teichoic acid.

Structure and function of LPS/LOS

The gram-negative outer membrane is a permeability barrier that excludes large molecules and hydrophobic compounds and protects the bacterial cell from detergents, such as bile salts. LPS/LOS is located on the outer surface of this membrane and presents an effective barrier to peptidoglycan-degrading enzymes, such as lysozyme, and can confer resistance to bile salts, complement, and phagocytes. As a consequence, LPS plays an important role in host–pathogen interactions. LPS consists of three components or regions, lipid A (region I), an R (rough or core) polysaccharide (region II), and an O polysaccharide (or O-antigen, region III) (Figure 6.3). LOS characteristically lack the O-antigen.

Region I
The lipid component of LPS/LOS, lipid A, contains the hydrophobic membrane-anchoring region. Lipid A consists of a phosphorylated *N*-acetylglucosamine (NAG) dimer with six or seven fatty acids attached. All fatty acids in lipid A are saturated. Some are attached directly to the NAG dimer, while others are esterified to 3-hydroxy fatty acids. The structure of lipid A is highly conserved among gram-negative bacteria.

Region II
The core (R) antigen or R polysaccharide is attached to the 6 position of NAG. The R antigen consists of a short chain of sugars. For example, KDO–Hep–Hep–Glu–Gal–Glu–GluNAc. Two unusual sugars are usually present in the core region of LPS, heptose and 2-keto-3-deoxy-D-manno-oct-2-ulosonic acid (KDO). KDO is unique and invariably present in LPS and so has been an indicator in assays for endotoxins. With minor variations, the core polysaccharide is common to all members of the same bacterial genus, but it is structurally distinct from other genera. *Salmonella*, *Shigella*, and *Escherichia*, for example, have similar but not identical cores. Loss of the more proximal parts of the core, as in 'deep rough' mutants, makes the strains sensitive to a range of hydrophobic compounds, including antibiotics, detergents, bile salts, and mutagens. This region contains a large number of charged groups and is considered to be important in maintaining the permeability properties of the outer membrane. The LPS of bacteria such as *N. meningitidis*, *Neisseria gonorrhoeae*, and *H. influenzae* consists only of lipid A and a core polysaccharide and is often referred to as an LOS. The *N. meningitidis* LOS, for example, consists of a symmetrical hexa-acyl lipid A and a short oligosaccharide chain, which can be classified into 11 different 'immunotypes.'

Region III
The O- (or somatic) antigen or O-polysaccharide is attached to the core polysaccharide. It consists of repeating oligosaccharide subunits made up of between three and five sugars. The individual chains vary in length up to 40 repeat units. The O-antigen is much longer than the core polysaccharide, and it maintains the hydrophilic domain of the LPS molecule. Substantial variation occurs in the oligosaccharide composition between species and even between strains. At least 20 different sugars are known to be present, and many are unique dideoxyhexoses, which are found naturally only in gram-negative bacterial cell walls. Variations in the sugar content of the O-antigen contribute to the wide variety of antigenic types found in pathogens, such as *Salmonella* spp. and *E. coli*. Specific sugars, confer the immunological specificity of the O antigen and contribute to the 'smoothness' (colony morphology) of the strain. Loss of the O-specific region by mutation results in the strain acquiring a 'rough' colony morphology. Elucidation of the structural organization of LPS has relied heavily on the availability of mutants blocked in specific stages of LPS biosynthesis. Core sugars are added sequentially to lipid A followed by the O side chain, which is linked to the lipid A core polysaccharide as a preassembled unit. Mutants impaired in the assembly of lipid A usually cannot be isolated except as conditional lethal mutants, so this region is essential for cell viability and outer-membrane assembly. For example, *N. meningitidis* mutants deficient in lipid A due to mutation of *lpxA* (which encodes UDP-GlcNAc acyl transferase) require possession of a polysialic acid capsule for growth, since it can apparently substitute for the absent lipid A (Raetz and Whitfield 2002).

LPS and virulence

Lipid A (the toxic component) and the polysaccharide side chains (the nontoxic but immunogenic components) of LPS both contribute to the virulence of gram-negative

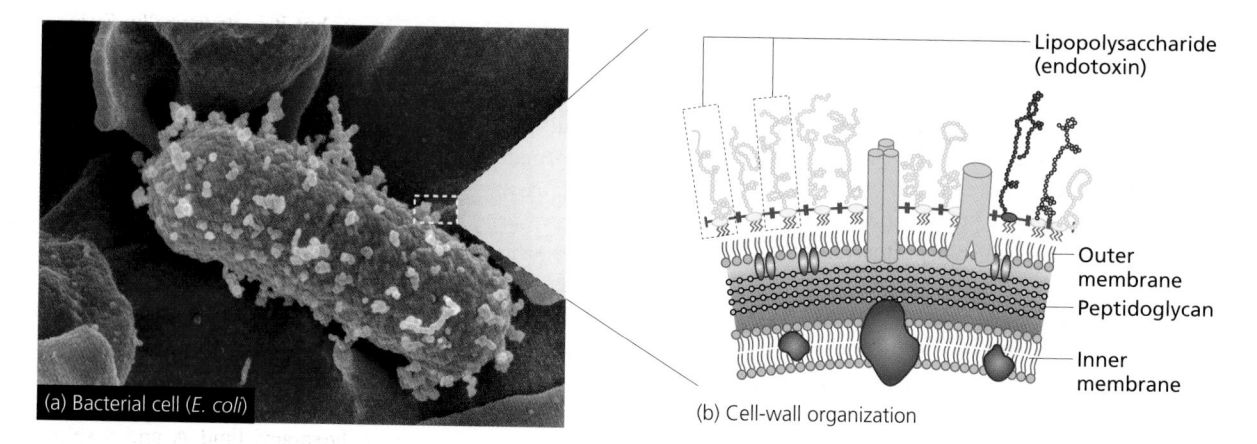

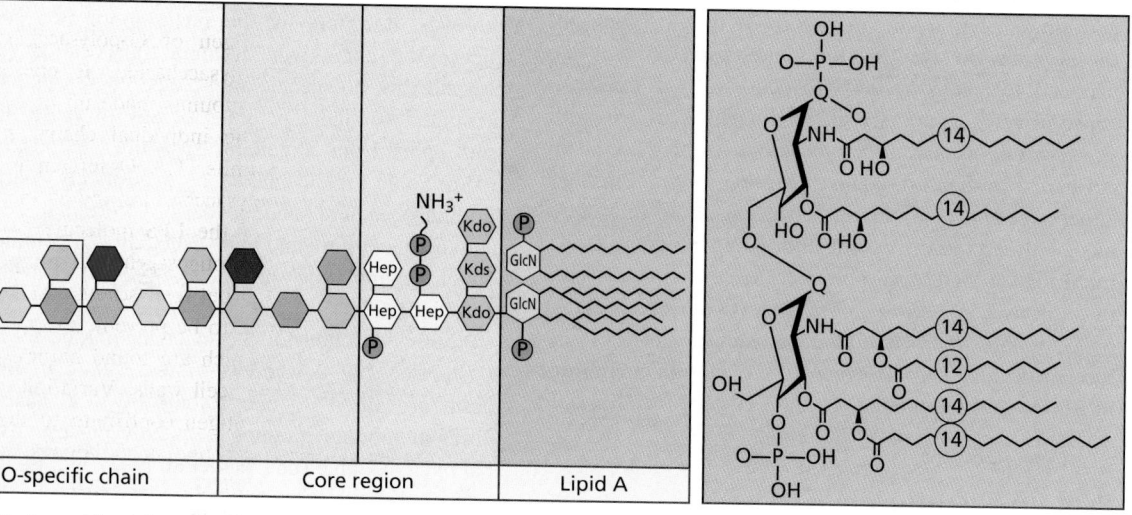

Figure 6.3 *The structure and localization of lipopolysaccharide (LPS). Electron micrograph of* Escherichia coli **(a)**, *together with a schematic representation of the location of the LPS in the bacterial cell wall* **(b)**, *and the architecture of LPS* **(c)**. *Also shown is the primary structure of the toxic center of LPS, the lipid A component* **(d)**. *GlcN,* D-*glucosamine; Hep,* L-*glycero-*D-*manno-heptose; KDO, 2-keto-3-deoxy-*D-*manno-oct-2-ulosonic acid; P, phosphate. Redrawn from Beutler and Rietschel 2003.*

bacteria. Loss of the O-antigen increases susceptibility to complement-mediated serum killing. This is because the C5b-9 membrane attack complex (MAC) can much more readily access the bacterial cell membrane of rough rather than smooth bacteria, since in the latter the MAC forms too far from the membrane to damage the cell. Rough strains are also generally more readily phagocytosed than smooth strains. Furthermore, the O-polysaccharide is also the basis of antigenic variation among many gram-negative pathogens, including *E. coli*, *Salmonella*, *Pseudomonas*, and *Vibrio*. The existence of multiple serotypes of the same organism offers opportunities to bypass the acquired immune response to a specific serotype. However, although the O-polysaccharides are highly antigenic, they seldom elicit immune responses that confer full protection to the host against secondary challenge.

The pathophysiological effects of endotoxin are mediated via lipid A. LPS, released into the bloodstream by gram-negative bacterial cells undergoing lysis, is first bound by LPS-binding plasma proteins. These interact with CD14 receptors on monocytes and macrophages and subsequently lead to the induction of inflammation, intravascular coagulation, hemorrhage, and shock through the stimulation of cytokines including IL-1, IL-6, IL-8, TNFα, and platelet-activating factor. These in turn stimulate production of prostaglandins and leukotrienes, which are powerful mediators of inflammation and the septic shock that accompanies endotoxinemia and activate both the complement and coagulation cascades. LPS can also function as a mitogen, stimulating the polyclonal differentiation and multiplication of B cells and the secretion of immunoglobulins, such as IgG and IgM.

Genetic loci involved in LPS biosynthesis

In *E. coli*, where lipid A biosynthesis is best characterized, the genes encoding the enzymes required are scattered around the genome and on the whole are not incorporated into coregulated operons. The genes involved include *lpxABCDHKLMP*. Subsequently, an inner-membrane ABC transporter (encoded by *msbA*) functions to achieve the initial stages of lipid A export. In *Enterobacteriaceae*, the *waa* locus encodes the enzymes required for the sequential assembly of the core oligosaccharide on to the lipid A acceptor. This locus is composed of three operons: (1) the *gmhD* operon directs inner core biosynthesis; (2) the central *waaQ* operon is responsible for the outer core; and (3) the *waaA* operon contains the structural gene for a bifunctional KDO transferase, which is required for the addtion of KDO to the inner core. With few exceptions, the enzymes involved in O-polysaccharide assembly are encoded by genes at the *rfb* locus and are expressed constitutively. There are three currently known pathways for O-polysaccharide biosynthesis, which are distinguished by their respective export mechanisms: (1) dependent on the Wzy-flippase to translocate the assembled polysaccharide across the inner membrane; (2) requiring an ABC transporter to achieve this transport (a pathway confined to unbranched O-polysaccharides); or (3) reliant upon simultaneous extension of the polysaccharide and extrusion across the plasma membrane by an integral membrane synthase (Raetz and Whitfield 2002).

Protein secretion

Many virulence factors are either located on the cell surface or are secreted into the extracellular medium, e.g. adhesins, sensory receptors, and exotoxins. Bacteria have evolved a wide range of mechanisms to translocate proteins across their membranes. It has been estimated that approximately 20 percent of the polypeptides synthesized by a bacterium are destined for a location external to the cytoplasmic membrane, and those in the outer membrane or beyond will be exposed to the host environment.

In gram-negative bacteria, there are three mechanisms for inner membrane translocation: the Sec machinery, the SRP-dependent machinery, and the Tat machinery. The Sec machinery is the most intensively studied, and polypeptides destined for it bear an N-teminal targeting signal, which is cleaved by a peptidase on the periplasmic face of the inner membrane. During Sec-dependent translocation the unfolded polypeptide chain is threaded through a channel (translocon) composed of SecYEG and a number of associated proteins (SecD, SecF, YajC, YidC) powered via SecA by the hydrolysis of ATP (Economou 2002; van Wely et al. 2001). Signal peptide binding to SecA is followed by ATP binding, which induces a profound conformational change in SecA, enabling it to interact with the translocon. ATP is hydrolyzed, reversing the conformational change and directing the release of the polypeptide. Binding of another ATP molecule enables SecA to bind to the polypeptide again, but this time 2.5 kDa nearer its C-terminus. This cycle is repeated many times, shunting the polypeptide through the channel. Some Sec-dependent proteins interact with the cytoplasmic chaperone SecB prior to translocation; others (particularly multiple-membrane-spanning membrane proteins) are recognized by SRP (Economou 1999). The Tat mechanism differs from those described above, since it affects the translocation of folded polypeptides that contain bound cofactors (Palmer and Berks 2003). In common with the Sec and SRP pathways, the Tat system requires an N-terminal targeting signal peptide; however, it differs by the inclusion of two arginines (hence the name twin arginine translocation (TAT)). Beyond this, no specific amino acid consensus has been attributed to signal peptides of any of these mechanisms apart from a general structure composing of a hydrophobic region flanked by a positively charged N-terminal segment and a basic C-terminal sequence contained within an approximately 20 amino acid stretch culminating in the peptidase cleavage site (Pugsley 1993).

Proteins destined for the inner membrane fold within it (Pugsley 1993), whereas those residing in the periplasm are aided in their folding by local chaperones. Outer membrane proteins frequently possess a C-terminal phenylalanine (de Cock et al. 1997), and their localization is usually assisted by periplasmic proteins, e.g. Skp (Buileris et al. 2003; Missiakas et al. 1996). Recently, a highly conserved protein required for outer-membrane protein assembly was shown to be essential in *N. meningitidis* and is likely to play a central role in this process (Voulhoux et al. 2003). In the case of lipoproteins, localization can be achieved by dedicated piloting proteins. Attachment of a lipid is signalled by an LXXC motif contained within the signal peptide adjacent to the peptidase cleavage site. Cleavage results in an N-terminal cysteine residue that is acylated. Inner and outer membrane lipoproteins are then distinguished by the second amino acid of the mature protein. If the second amino acid is not an aspartate, the lipoprotein is recognized by LolCDE and localized on the periplasmic face of the outer membrane (Masuda et al. 2002). LolCDE is an ABC transporter that releases the lipoprotein from the inner membrane, whereupon it interacts with LolA. The LolA–lipoprotein complex crosses the periplasm and interacts with the outer-membrane receptor LolB, which is essential for outer-membrane anchorage (Fukuda et al. 2002).

To secrete proteins across both the inner and outer membrane, gram-negative bacteria utilize one of five main pathways (Figure 6.4). The type I mechanism (see Table 6.4) is well known for its direction of the secretion of the hemolytic toxin HlyA from *E. coli*. This process proceeds in a single step from the cytoplasm using just three proteins. In the inner membrane are the HlyB and HlyD proteins. HlyB contains an ATPase activity and resembles the multidrug resistance protein of eukaryotes. In the outer membrane, and spanning the periplasm, is TolC, which docks into the membrane fusion protein (HlyB) forming a proteinaceous channel through which the unfolded HlyA polypeptide can move directed by its C-terminal targeting signal, which

remains uncleaved throughout (Thanassi and Hultgren 2000; Andersen et al. 2001).

Type II secretion (general secretory pathway (GSP)) utilizes two steps. First, proteins use either Sec or Tat to enter the periplasm; then, approximately 14 dedicated GSP proteins form the machinery required to secrete the fully folded protein across the outer membrane. This pathway is used by many bacteria to secrete virulence factors (see Table 6.5), such as the multisubunit cholera toxin. The targeting signal for the GSP is currently not known, and the machinery is composed of a number of

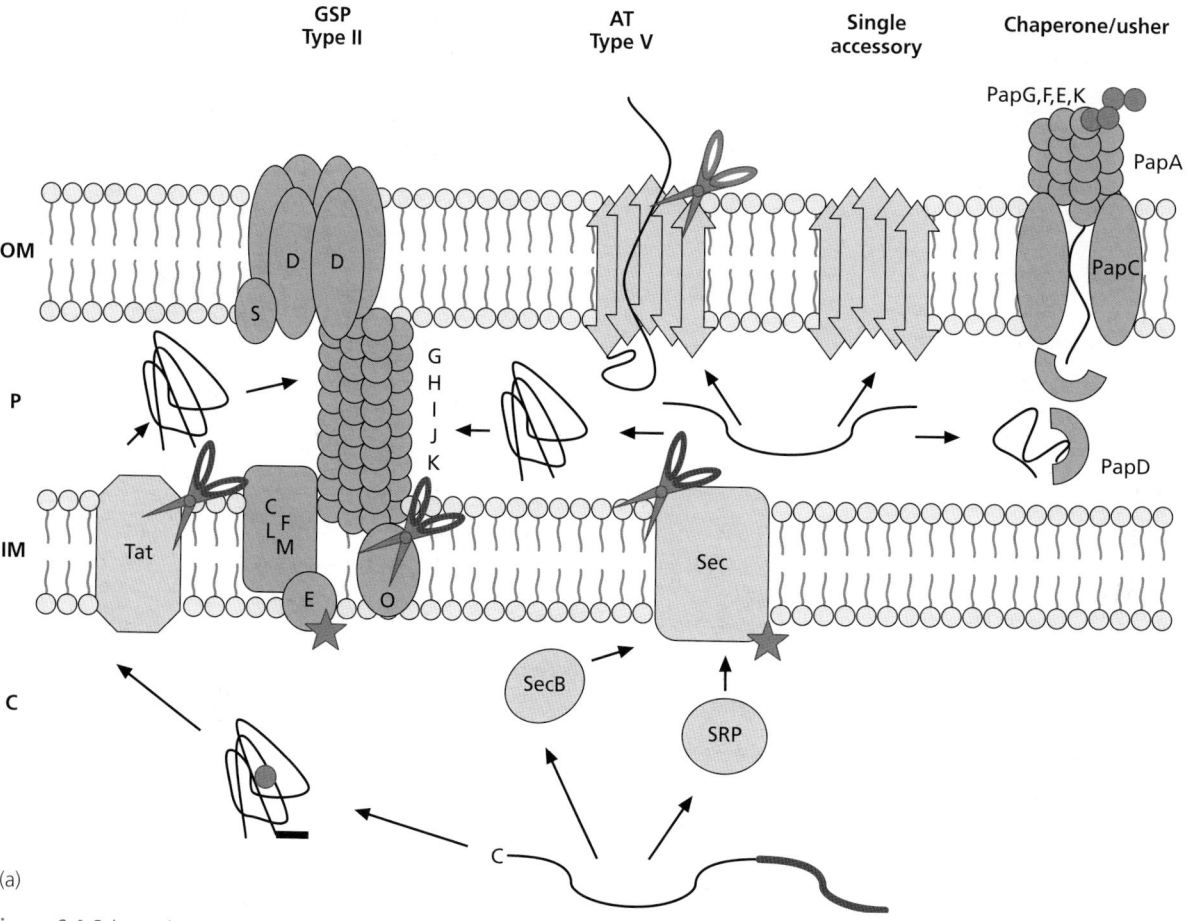

(a)

Figure 6.4 *Schematic representation of bacterial protein secretion machineries.* **(a)** *Depicts those pathways with a distinct step for translocation across the inner membrane. Polypeptides synthesized in the cytoplasm with an N-terminal signal peptide interact with either SecB or SRP and are targeted to the Sec machinery, which translocates them across the cytoplasmic membrane and concomitantly cleaves their signal peptide (as indicated by the scissors). This translocation occurs by threading the unfolded polypeptide across via the action of SecA and is driven by ATP hydrolysis (as indicated by the star). A subset of proteins with an N-terminal signal peptide frequently bearing twin arginines fold around their cofactors and are translocated through the Tat machinery. From the periplasm, a variety of pathways exist to achieve outer-membrane secretion. (1) Autotransported proteins (AT) insert their C-terminal β-barrel domain into the outer membrane to create a channel through which the rest of the polypeptide is threaded. This channel may be formed from multimers or involve Omp85. Cleavage at the cell surface liberates the mature protein, leaving the β-barrel domain attached to the cell. This pathway is also refered to as type V. (2) A similar mechanism is employed by ShlA. However, rather than being contained within a single polypeptide, the transporter domain is a separate protein and thus cleavage is not required (single accessory pathway). (3) Subunits of type IV pili interact with the PapD chaperone, which delivers them to the PapC usher, which in turn directs assembly of the pilus (chaperone/usher pathway). (4) Following folding, some proteins are exported by the general secretory pathway (GSP or type II). This requires 12 or more proteins; the system employed by pullulanase, which is secreted by Klebsiella oxytoca, is shown here. The single integral outer-membrane component (GspD) is piloted to its position by the lipoprotein (GspS), together these proteins form a gated channel through which the folded proteins are secreted. This complex is proposed to be linked to the inner-membrane components (GspC, E, F, L, M) by a pilus-like structure composed of GspG, H, I, J, K. The latter are processed by the methylating protease, GspO. GspE is likely to hydrolyze adenosine triphosphate (ATP) and contribute to energizing the system.*

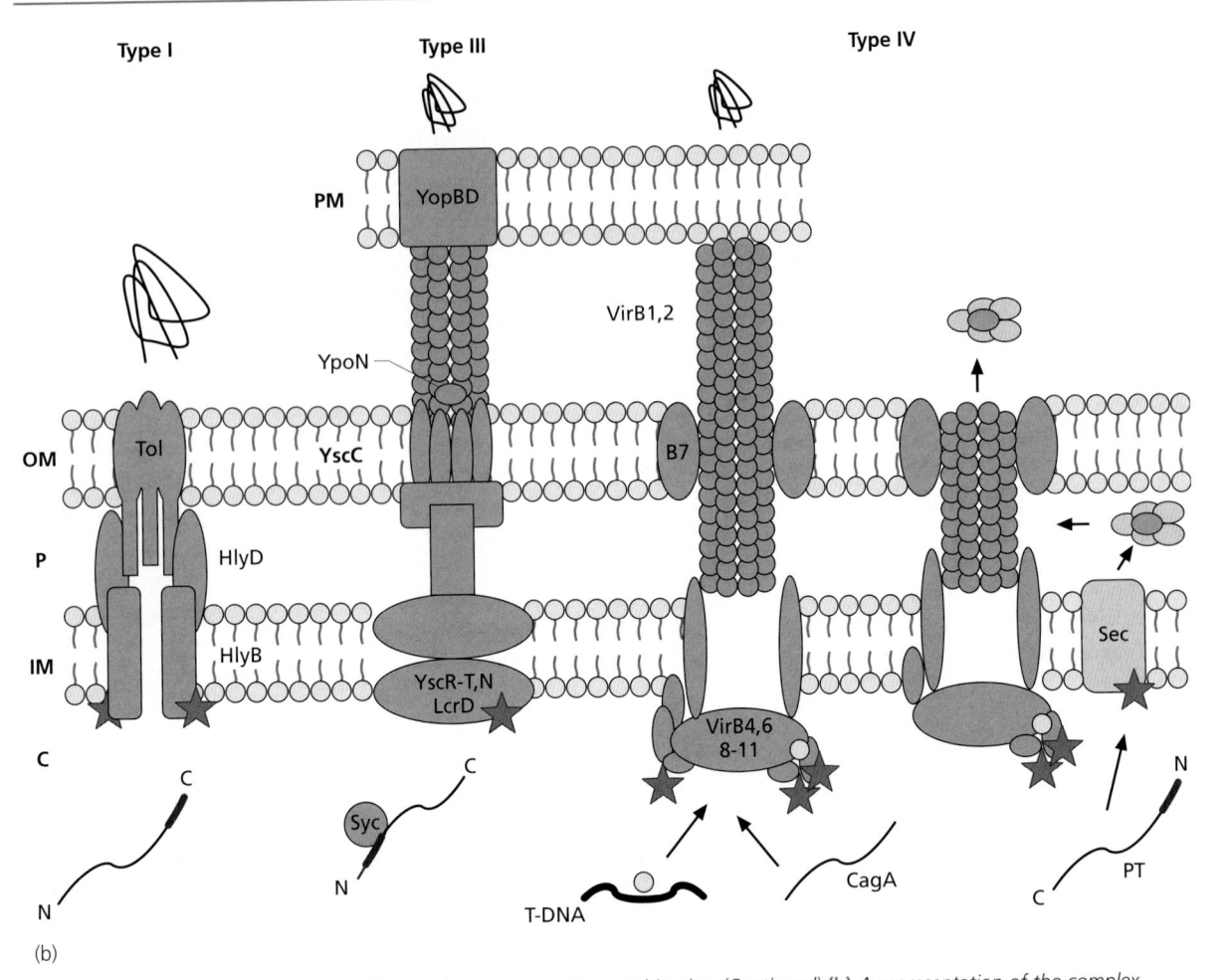

(b)

Figure 6.4 *Schematic representation of bacterial protein secretion machineries (Continued)* **(b)** *A representation of the complex machineries that comprise the type I, III, and IV pathways. Type I-dependent proteins (see Table 6.5) bear a C-terminal targeting peptide that is not cleaved during translocation of the unfolded polypeptide out of the cell in a single step, which avoids the periplasm. Secretion of the archetypical type I-dependent protein (HlyA from Escherichia coli) is achieved through a channel composed of the ATP binding dimer of HlyB, the membrane fusion protein (HlyB), and the outer-membrane protein TolC. The type III machinery uses a needle-like structure that resembles a flagellum to inject proteins directly into eukaryotic cells, and as shown here is used by Yops, which are secreted by Y. enterocolitica. Some type III substrates contain two N-terminal signals for targeting, one encoded by the mRNA, and the second by amino acids to which the cytoplasmic Syc chaperones bind. YscR-U and LcrD may form a central secretion apparatus energized by YscN-dependent ATP hydrolysis. The YsdC outer-membrane component resembles the GspD secretin, and the surface-located YopN is thought to act as a channel gate. Translocation of Yops into the target eukaryotic cell may occur via a channel formed in the plasma membrane by YopD and YopD. The type IV secretion pathway can function in alternative ways, depending on the bacterial host. Like the type III pathway, it can achieve direct injection of proteins (e.g. CagA, VirE2) or protein-bound DNA (T-DNA) into eukaryotic cells. Alternatively, proteins that have reached the periplasm via the Sec machinery can assemble and enter the type IV pathway to achieve extracellular localization (pertussis toxin (PT)). The machinery comprises multiple ATPases (VirB4, VirB11) and other inner-membrane structural components (VirB6, VirB8, VirB9, VirB10) plus an outer-membrane component (VirB7) that are linked together by a pilus-like structure composed of VirB1 and VirB2. The relative positions of the bacterial cytoplasm (C), inner membrane (IM), periplasm (P), outer membrane (OM), and the eukaryotic plasma membrane (PM) are indicated.*

inner membrane components, including an ATPase and a set of proteins that resemble pilin subunits (the pseudopilins) and that are proposed to form a retractable pilus, which pushes proteins through the outer membrane pore. The outer-membrane channel is formed by a multimer of a single protein (the GspD secretin) and is likely to be tightly gated (Thanassi and Hultgren 2000).

The type III secretion machinery achieves injection of proteins (see Table 6.6) with the aid of upwards of 14 components directly into eukaryotic cells (see below), as

does the type IV machinery (TFSS). A TFSS is possessed by several pathogens of plants and animals (see Table 6.7), where the secretion machinery transfers protein substrates intercellularly and appears to be related ancestrally to conjugation systems that have evolved the additional capacity to translocate DNA–protein complexes.

The TFSS machinery comprises 12 components that span the periplasm and membranes through the formation of a pilus, which takes differing forms in different organisms and can directly attach to the eukaryotic

Table 6.4 *Examples of proteins secreted by type I machinery*

Bacterium	Type I secreted proteins
Escherichia coli	Hemolysin (HlyA)
Serratia marcescens	Metalloprotease
Erwinia chrysanthemi	Metalloprotease
Pasteurella haemolytica	Leukotoxin
Bordetella pertussis	Cyclolysin
Pseudomonas aeruginosa	Metalloprotease
Rhizobium leguminosarum	NodO

Table 6.6 *Examples of proteins secreted by type III machinery*

Bacterium	Type III secreted proteins
Yersinia spp.	Yops
Salmonella typhimurium	Sips, Sops, SptP
Shigella flexneri	Ipas
Pseudomonas aeruginosa	ExoS, ExoT
Enteropathogenic/ Enterohemorrhagic *Escherichia coli*	EspA, EspB (EaeB), EspD, EspF, EspG, Map, EspE (Tir)
Phytopathogens (*Erwinia* spp., *Pseudomonas syringae*, *Xanthomonas campestris*)	Harpins, PopA, AvrB

target cell (Figure 6.4) (Christie 2001). The components can be divided into distinct activities: (1) substrate processing (the relaxosome); (2) substrate docking (coupling proteins/chaperones); (3) translocation (channel); and (4) cell–cell attachment (pilus/adhesions). The relaxosome function is not required for those TFSSs that do not transfer DNA since it functions to cleave the DNA strand destined for transfer at the plasmid origin of transfer. The relaxase remains attached to the DNA and may serve to pilot it for translocation. Coupling proteins and chaperones appear necessary for presentation of DNA and also proteins for translocation, and most contain Walker A and B nucleotide-binding motifs suggestive of ATPase activity (the TraG homologues). Other putative ATPases contributing to the function of TFSSs are (1) those showing homology to VirB4, which are proposed to transduce information in the form of ATP-induced conformational changes, across the cytoplasmic membrane to the extracytoplasmic subunits (Dang et al. 1999); and (2) those that form oligomeric structures in the cytoplasm, which are reminiscent of the GroEL family (homologs of VirB11), although it is unclear whether they function for substrate translocation or biogenesis of the translocase. It should be noted that the multisubunit pertussis toxin of *Bordetella pertussis* uses the GSP (Figure 6.4) to enter the periplasm and, following assembly, is secreted across the outer membrane via the TFSS in contrast to other effectors released via a TFSS, which enter the machinery from the bacterial cytoplasm (Thanassi and Hultgren 2000; Christie 2001).

In common with the type I secretion pathway, the type III secretion process is Sec-independent; proteins secreted by it do not accumulate in the periplasm, and nor are they subjected to N-terminal processing. However, the protein signal sequence for type III secretion is considered to reside within the N-terminal of the secreted proteins, since this region appears necessary for secretion and can direct the secretion of hybrid fusion proteins. Exhaustive mutational analysis has revealed a high degree of tolerance for sequence changes within the N-terminus of some secreted proteins, without loss of secretion. Moreover, the N-terminal sequences of proteins secreted via the type III pathway share no recognizable structural similarities that may constitute a common secretion signal. Therefore, it has recently been proposed that the secretion signal may actually reside in the $5'$ region of the mRNA that encodes the secreted proteins (Anderson and Schneewind 1997).

The process of protein secretion via this system requires the involvement of small cytoplasmic proteins termed chaperones, which protect the secreted proteins from premature interaction with components of the secretion machinery. Apart from a cytoplasmic, membrane-associated ATPase, and an outer-membrane

Table 6.5 *Examples of proteins secreted by type II machinery*

Bacterium	Type II secreted proteins
Klebsiella oxytoca	Pullulanase
Vibrio cholerae	Cholera toxin, protease
Aeromonas hydrophila	Aerolysin toxin, protease, acyltransferase
Erwinia chrysanthemi	Pectate lyases, cellulase
Erwinia carotovora	Pectate lyases, cellulase
Pseudomonas aeruginosa	Exotoxin A, lipases, proteases
Xanthomonus campestris	Cellulase, protease

Table 6.7 *Examples of type four secretion system (TFSS) and the effector proteins secreted by them*

Bacterium/plasmid	TFSS effector
Bordetella pertusis	Pertussis toxin
Helicobacter pylori	CagA
Brucella suis	Unknown, probably protein
Brucella abortus	Unknown, probably protein
Bartonella henselae	Unknown, probably protein
Campylobacter jejuni	Unknown, probably protein
Rhizobium etli	Unknown, probably protein
Rickettsia prowazekii	Unknown, probably protein
Wolbachia	Unknown, probably protein
Agrobacterium tumefaciens	T-DNA
pKM101	DNA (conjugal transfer)
R388	DNA (conjugal transfer)
RP4	DNA (conjugal transfer)
F	DNA (conjugal transfer)
Legionella pneumophila	DNA (conjugal transfer)

component that is homologous to the secretin of the type II secretion pathway, most of the components are located in the inner membrane. Interestingly, many of the components of the type III secretion apparatus show homology with those of flagella. In accordance with this, some type III secretion systems assemble super-molecular structures on the bacterial surface (Ginocchio et al. 1994). In contrast to type I secretion, where the secreted enzymes are active in the extracellular space, type III secretion systems appear to be dedicated to the translocation of pathogenicity proteins directly into the cytosol of eukaryotic cells and in some cases appear to be regulated by contact with the surface of the target eukaryotic cell (Charkowski et al. 1997). Type III secre-tion systems contribute to the virulence of diverse gram-negative pathogens (Table 6.6). Many of the type III secreted proteins interact directly with host cell compo-nents to alter host cell signal transduction, once inside the eukaryotic cytosol (see Subversion of host cell signal transduction).

The type V machinery is perhaps the simplest, since secretion appears to be determined within the primary sequence of the secreted protein, and hence it is often described as autotransportation. This is described later in Evolution of pathogens within the host; however, variations exist whereby the outer-membrane trans-porter domain is separated from the transported protein (e.g. the hemolysin secreted by *Serratia marcescens*) (Yang and Braun 2000). Finally, another terminal branch of the GSP exists, which is utilized for the forma-tion of Pap pili. This pathway is referred to as the chaperone/usher pathway and involves the interaction of pilus subunits with the PapD periplasmic chaperone, which facilitates their proper folding whilst preventing premature aggregation and guiding them to the PapC usher in the outer membrane, whereupon they assemble into a pilus (Stathopoulos et al. 2000; Thanassi et al. 1998).

Subversion of host cell signal transduction

Many of the proteins injected directly into eukaryotic cells by the type III secretion pathway interfere with signal transduction. For example, through YopH-medi-ated dephosphorylation of several macrophage proteins whose transient tyrosine phophorylation is required for normal phagocytosis, pathogenic *Yersinia* spp. resist the host's primary immune defenses by preventing their uptake into professional phagocytes. Other effectors are also injected into eukaryotic cells by *Yersinia* spp. and contribute to their cytotoxicity, e.g. by altering the cytos-keleton. In contrast to pathogenic *Yersinia* spp., *Shigella* spp. occupy predominantly intracellular locations, and although the pathogenic strategies and disease caused by these organisms differ entirely, both pathogens use type

III secretion systems as a key virulence mechanism. In vivo invasiveness, as well as the induction of membrane ruffles on cultured epithelial cells, depends on the *Shigella* type III secretion system, which delivers four proteins (IpaA, IpaB, IpaC, and IpaD) into eukaryotic cells. Together, these cause protein tyrosine phosphor-ylation of a number of cytoskeletal proteins, leading to the localized accumulation of a variety of cytoskeletal (F-actin, -vinculin, and -actinin) and signal transduction molecules (the protein tyrosine kinase $pp60^{c-src}$ and the GTP-binding protein rho) at the site of bacterial attach-ment. This results in tyrosine phosphorylation of paxillin, cortactin, and the focal adhesion kinase $pp125^{FAK}$. Shortly after internalization, the bacteria lyse the phagocytic membrane and gain access to the cyto-plasm and, in macrophages, induce apoptosis. In addi-tion, attraction of neutrophils to the site of bacterial infection occurs (Hueck 1998). Interestingly, salmonellae contain two type III secretion systems, which are encoded by two distinct gene clusters termed SPI-1 and SPI-2 (for *Salmonella* pathogenicity island). These two type III secretion systems appear to play different roles during pathogenesis, with SPI-1 being required for initial penetration of the intestinal mucosa and SPI-2 being necessary for subsequent systemic stages of infection. The effectors delivered by SPI-1 (SopE, SopE2, SopB, SipA, SipC, SptP) induce reversible actin cytoskeletal rearrangements by stimulating the host signal transduc-tion and directing modulation of actin dynamics (Zhou and Galán 2001). The effectors translocated by SPI-2 interfere with vesicular trafficking allowing survival within the eukaryotic cell (see Infiltration of host tissues) (Fang and Vazquez-Torres 2002). The opportu-nistic pathogen *P. aeruginosa* also employs type III secreted proteins to inflict epithelial cell damage and affect dissemination within infected hosts. Although their precise contribution to pathogenesis is unclear, two exotoxins with ADP-ribosyltransferase activity (exoen-zyme S, ExoS; exoenzyme T, ExoT) secreted via a type III secretion pathway by *P. aeruginosa* target members of the H-Ras and K-Ras family of GTP-binding proteins and vimentin (an intermediate filament protein) (Hueck 1998).

Another example of a bacterium that effectively subverts eukaryotic cell signaling is enteropathogenic *E. coli* (EPEC). EPEC belong to a growing family of diar-rheagenic *E. coli* strains that cause a characteristic attaching and effacing lesion on infected small-intestine enterocytes and cells grown in culture. This lesion consists of gross cytoskeletal damage to the brush-border microvilli, the deposition of filamentous actin beneath attached EPEC, and the formation of pedestal-like structures. In addition, many classical *EPEC* strains are characterized by their clustered or localized pattern of adherence to epithelial cells, which results in the formation of microcolonies on the cell surface. Intimate attachment, effacing of microvilli, and formation of pedestals require a bacterial adhesin (called intimin) and EPEC type III secretion. Although intimin is not secreted by the type III secretion pathway, the encoding gene (*eaeA*) is located within the gene cluster that encodes the EPEC type III secretion system, and intimin functions synergistically with type III secretion in pedestal and A/E lesion formation. Intimin specifically binds to translocated intimin receptor (Tir), which is secreted by the type III secretion apparatus. Three other type III secreted proteins, EspA, EspB, and EspD, are required for membrane insertion of Tir. Concomitant with pedestal formation, adherent EPEC strains induce tyrosine phosphorylation of several proteins in the eukaryotic cell, including Tir and phospholipase C-1. While Tir phosphorylation has been thought to facilitate intimin binding, activation of phospholipase C-1 induces inositol triphosphate and Ca^{2+} fluxes, which might be involved in the cytoskeletal rearrangements involved in pedestal formation. Another protein secreted by the type III machinery of EPEC is the mitochondrial associated protein (Map), which targets the mitochondria and disrupts their membrane potential. In addition, via activation of the host GTPase (Cdc42), Map stimulates nonpedestral-related cytoskeletal rearrangement, e.g. filopodia formation (Hueck 1998; Kenny 2002; Vallance and Finlay 2000; Campellone and Leong 2003) (Figure 6.5).

BACTERIAL SENSORY NETWORKS

Microorganisms inhabit diverse ecological niches where survival will depend on their capacity to sense the local environmental conditions and adapt by regulating the expression of specific genes and operons. Consequently, bacteria have developed multiple, highly sophisticated mechanisms to gather, process, and transduce environmental information, such as pH, temperature, nutrient availability, osmolarity, and cell population density. Pathogenic bacteria are therefore likely to encounter countless different environmental situations as they move through the host or between the host and external environment, and need to continually monitor diverse physical and chemical signals and integrate their responses. Such sensory transduction mechanisms operate at the transcriptional, post-transcriptional, and translational levels and bacteria employ multiple interconnected regulatory circuits to coordinate their responses. The 6.3-MB genome of *P. aeruginosa*, for example, contains around 5 500 genes, some 10 percent of which encode regulatory proteins. These enable the bacterium to express or repress genes as the conditions dictate. For example, if iron is plentiful, the energetically expensive synthesis of siderophores and their associated export and import pathways can be switched off. For bacteria such as *V. cholerae*, switching between an aquatic lifestyle and that required for survival within a human host has high-

lighted an inverse correlation between motility and virulence gene expression.

Pathogenic bacteria will therefore employ many different signal transduction mechanisms, some of which are required for 'house-keeping,' since they are required for replication, cell division, and the maintenance of cell viability. Others control the synthesis of the colonization factors, toxins, and tissue-degrading enzymes that characterize the organism as a pathogen and that may be required only in certain host compartments.

Given the multistage adaptive process leading to the establishment of infection, bacteria require multiple virulence factors that can be deployed at the appropriate time and place. For example, during the early stages of infection, bacteria express cell-surface adhesins to facilitate attachment to eukaryotic cell surfaces and extracellular matrices. If the infection becomes systemic, the expression of exotoxins and host-avoidance factors is upregulated, with the concomitant downregulation of colonization factors. This is important because virulence factors that contribute to infection in one host compartment may render the pathogen susceptible to host defenses in another. In this section, three main sensory pathways will be considered: iron-dependent repression, two-component signal transduction, and quorum sensing. Although treated as separate environmental parameters, it is important to understand that each is merely one parameter amongst many that a bacterial cell must integrate in order to determine subsequent behavior (Figure 6.6) (Withers et al. 2001).

Iron-dependent repression

To grow in vivo, bacterial pathogens deploy high-affinity iron-scavenging systems, which depend on either siderophores or surface receptors for host iron and heme-containing proteins (see Iron scavenging). The genes involved in the biosynthesis of siderophores and their transport in and out of the bacterial cell, as well as those encoding for surface receptors, are coordinately negatively regulated in response to the cellular iron status. This is because they are energetically expensive systems that confer an advantage for the bacterium only in environments where iron is not readily available. In *E. coli*, the 17-kDa Fur repressor protein controls transcription from iron-responsive promoters in an iron-dependent manner, i.e. to be active, Fur requires ferrous iron as a corepressor. When iron is plentiful, the Fur–Fe^{2+} complex interacts with an operator sequence called the *fur* box and prevents gene expression. During iron starvation, Fur is stripped of iron by the cell and is no longer able to function as a repressor, thus permitting expression of the iron-repressed structural genes. Homologs of Fur have been described in many different gram-negative pathogens including *Yersinia pestis*, *V. cholerae*,

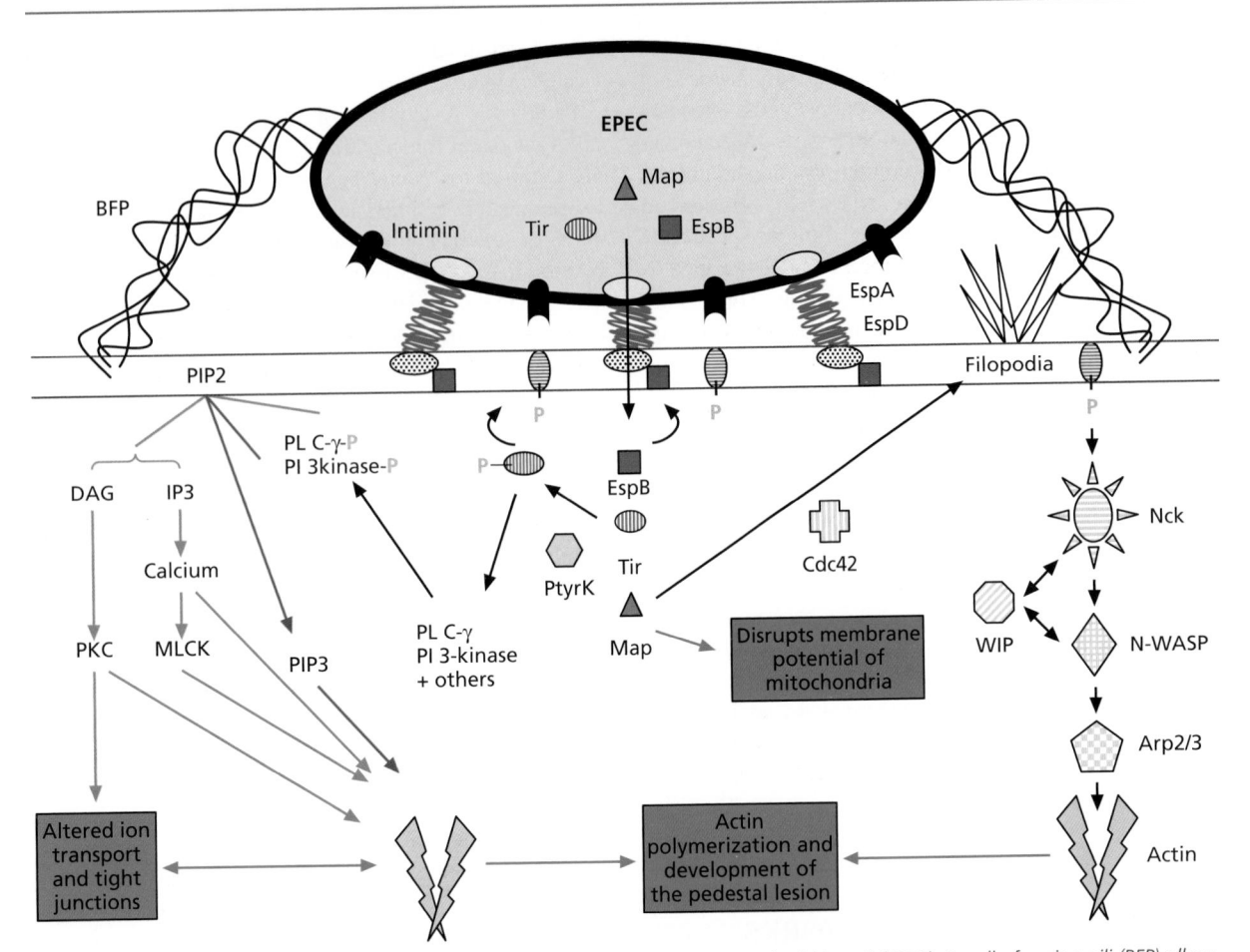

Figure 6.5 *Subversion of eukaryotic cell signaling cascades by enteropathogenic* Escherichia coli *(EPEC). Bundle-forming pili (BFP) allow initial aderence of EPEC to the eukaryotic cell surface. The type III secretion syringe composed of EspA (green wavy lines) allows injection of EspB (blue square) and EspD (black chequered oval), which form the translocon in the plasma membrane. Other injected proteins include Map (pink triangle) and Tir (red and white striped oval). Map, by direct or indirect mechanisms, activates the small GTPase protein Cdc42, converting it to its GTP-bound membrane-associating form that drives actin polymerization and filopodia formation. It is postulated that Cdc42 activation also leads to the modulation of other undefined cellular processes, again by direct and/ or indirect mechanisms. Map is also targeted to mitochondria, where it disrupts their membrane potential. Concurrently, Tir is proposed to undergo host- (PtyrK) and bacteria-dependent modification, leading to its insertion into the plasma membrane. The fully modified membrane-inserted Tir molecule interacts with intimin to trigger signaling events, leading to pedestal formation and other signaling events, such as phosphorylation of phospholipase C (PLC-1) and phosphoinositol 3 kinase (PI3-kinase). Changes in inositol phosphate fluxes lead to calcium release, which may in turn affect actin polymerization. Membrane-embedded tyrosine-phosphorylated Tir binds the adaptor protein Nick, which recruits N-WASP or WIP–N-WASP, complex to trigger activation of Arp2/3 complexes. This leads to actin assembly and pedestal formation with Nick, WIP, N-WASP, and Aro2/3 in its tip, myosin II and tropomyosin in the base, and F-actin, α-actinin, Talin, Exrin, and Villin distributed throughout. DAG, diacylglycerol; MLCK, myosin-like chain kinase; PKC, protein kinase C. See also Kenny (2002) Campellone and Leong (2003), Vallance and Finlay (2000), and www.biotech.ubc.ca/faculty/finlay/ EPEC.htm for animations and further illustrations.*

P. aeruginosa, N. meningitidis, and *H. influenzae.* In addition to regulating genes associated with iron acquisition, Fur also regulates expression of superoxide dismutases and exotoxins, such as shiga toxin, hemolysin, and exotoxin A. Gram-positive pathogens such as *Corynebacterium diphtheriae, M. tuberculosis,* and the staphylococci also possess iron-responsive regulatory proteins related to Fur. In *C. diphtheriae,* for example, DtxR regulates production of siderophores, a heme oxygenase (which enables the organism to strip the ferric iron from heme) as well as diphtheria toxin synthesis. Thus, in addition to regulating high-affinity iron-transport mechanisms, iron-dependent repressors

also regulate the expression of virulence genes unrelated to iron metabolism. However, defining their contribution to pathogenicity is not straightforward, since the constitutive expression of target structural genes may either enhance or reduce the fitness of the pathogen. For example, the *S. enterica* serovar Typhimurium *fur* mutant remains virulent in an experimental animal infection model, while many *E. coli* enteropathogens exhibit attenuated virulence when *fur* is inactivated. In other pathogens, e.g. *P. aeruginosa,* attempts to mutate the *fur* gene have not been successful, indicating that in these bacteria, Fur fulfils an essential function (Hantke 2001).

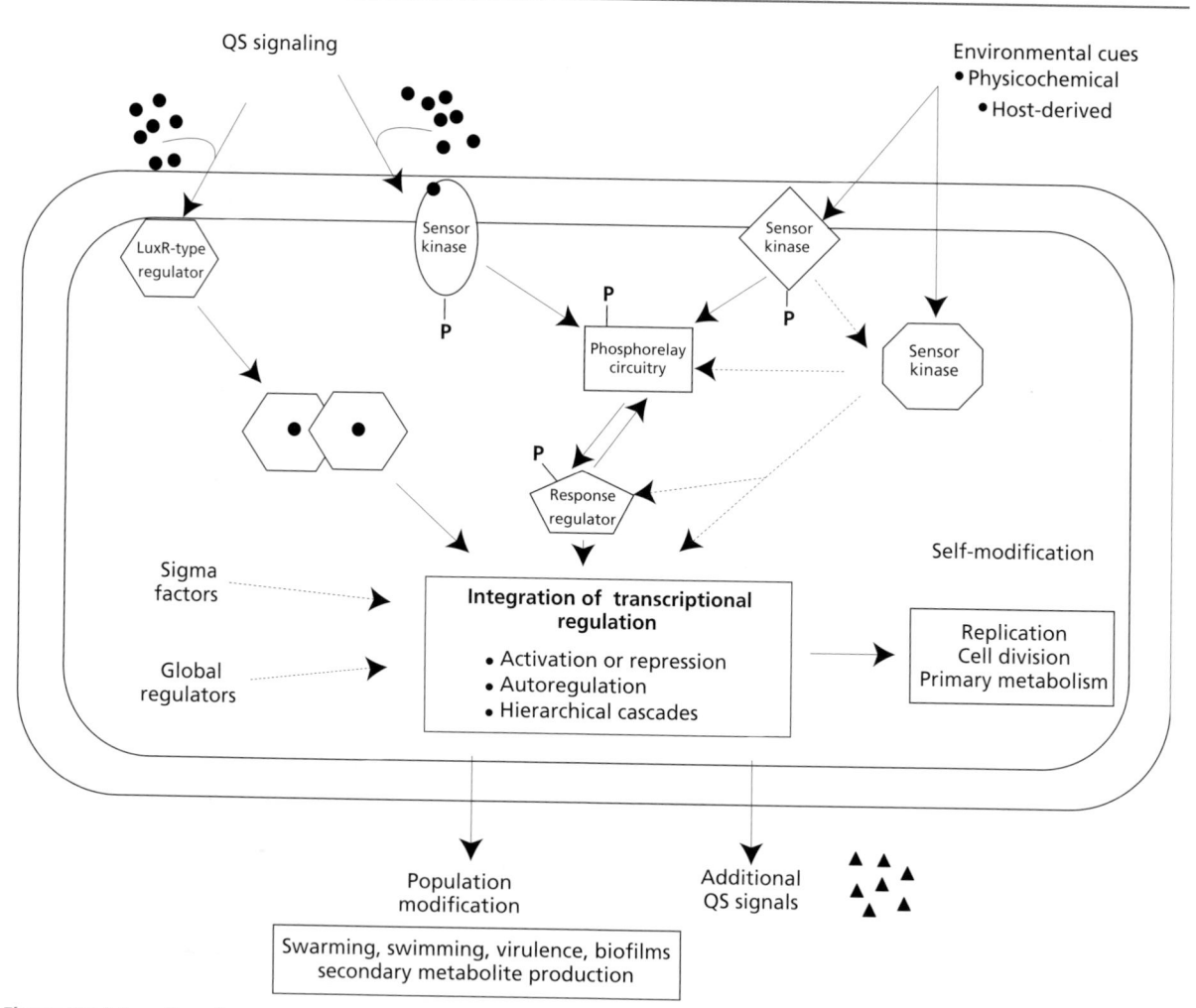

Figure 6.6 *Integration of quorum-sensing and environmental signals. Quorum sensing is one parameter amongst many that the cell integrates in order to determine subsequent behavior. Quorum-sensing signal molecules and environmental signals are internalized or detected at the cell envelope and initiate signal transduction, leading to gene expression changes. This diverse array of signaling cascades may act in parallel or be organized in a hierarchical manner to create a 'neural' regulatory network. Amended from Withers et al. (2001).*

Two-component signal transduction systems

Bacterial two-component signal transduction systems (TCSTS) are ubiquitous throughout the bacterial kingdom, are abundant within individual bacterial genomes, and have been adapted to regulate an enormous number of genes, including many virulence determinants (Table 6.8) (Stephenson and Hoch 2002; Dziejman and Mekalanos 1994; Hoch and Silhavy 1995). They consist of two regulatory proteins, a cytoplasmic membrane-associated sensor that samples the environment for a specific signal and a response regulator that mediates the internal response of the bacterial cell. Sensor and response regulator proteins exchange information by transferring a phosphate group from the sensor to the response regulator through conserved domains within the two proteins (Figure 6.7). In classical TCSTS, perception of an external signal results in the

phosphorylation of a histidine in the sensor (a histidine protein kinase). The phosphate group is subsequently transferred to a conserved aspartate within the receiver domain of the response regulator. Most response regulators are transcriptional activators or repressors (or both) that consist of two domains, an N-terminal receiver domain and a C-terminal DNA-binding domain.

TCSTS sense a variety of different factors including oxygen (Aer, *E. coli*), hydrogen (Hox, *Ralstonia eutropa*), quorum-sensing signal molecules (e.g. LuxQO, *V. cholerae*), Mg^{2+} (e.g. PhoPQ, *Salmonella*), and phosphate limitation (e.g. PhoBR, *E. coli*; PhoPR, *Bacillus subtilis*). The number of TCSTS produced by a particular bacterium appears to reflect the range of different environments it encounters during its lifecycle. For example, *H. pylori*, which uniquely inhabits the human stomach, has only three TCSTS pairs, whilst *P. aeruginosa*, which is found in a multitude of different environments, has an extraordinary number of putative two-

Table 6.8 *Examples of two-component signal transduction systems and their functions in selected pathogens. For more extensive information, see www.genome.ad.jp/kegg/ortholog/tab02020.html and www.uni-kl.de/FB-Biologie/AG-Hakenbeck/Tgrebe/HPK/Table3.html*

Process	Kinase	Regulator	Organism
Virulence/pathogenicity			
Virulence	BvgS	BvgA	*Bordetella*
Virulence	PhoQ	PhoP	*Salmonella*
Extracellular δ-toxin	AgrC	AgrA	*Staphylococcus*
Extracellular toxins	VirS	VirR	*Clostridium*
Alginate	AlgZ	AlgR	*Pseudomonas*
	kinB	AlgB	*Pseudomonas*
Autoinduction			
Competence	ComD	ComE	*Streptococcus*
	ComP	ComA	*Bacillus*
Quorum sensing	LuxQ	LuxO	*Vibrio cholerae*
Vancomycin resistance	VanS	VanR	*Enterococcus*
Catabolism	DegS	DegU	*Bacillus*
Chemotaxis	CheA	CheY	*Enterobacteria, Escherichia coli, Helicobacter pylori, Listeria, Pseudomonas, Salmonella, Treponema*
Capsule synthesis	RcsC	RcsB	*Enterobacteria*
Sporulation	Kin A,B,C	SpoOF	*Bacillus*
Nitrogen reduction	NarX	NarL	*Enterobacteria*
Nitrogen assimilation	NtrB	NtrC	*Salmonella, Vibrio*
Phosphate metabolism	CreC	CreB	*Enterobacteria*
	PhoR	PhoB	*Enterobacteria*
Citrate fermentation	CitA	CitB	*Enterobacteria*
	DpiA	DpiB	
Adaptation			
Anaerobic repression	ArcB	ArcB_RR	*E. coli*
Osmosensing	EnvZ	OmpR	*E. coli, Salmonella*
Turgor sensing (K+-transport)	KdpD	KdpE	*E. coli*
Cell-wall integrity	CiaH	CiaR	*Streptococcus*
	PmrB	PmrA	*Streptococcus*
Transport			
Hexose-phosphate	UhpB	UhpA	*E. coli*
Phosphoglycerate	PgtB	PgtA	*Salmonella*

component regulators, with 55 sensor proteins, 89 response regulators, and 14 sensor/response regulator hybrids (www.pseudomonas.com/suppdata.asp). The role performed by TCSTS also varies; for example, only three of the 16 TCSTS pairs contained within the genome of *S. aureus* appear to be essential (Throup et al. 2001). TCSTS can perform defined roles, e.g. LuxQO regulates the production of bioluminescence by *Vibrio harveyi*, or they can have a wide-ranging effect, e.g. the PhoPQ system of *Salmonella* governs the expression of at least 40 proteins including: (1) those mediating adaptation to Mg^{2+}-limiting environments; (2) virulence factors; and (3) components of the bacterial cell envelope. Indeed, *Salmonella* strains harboring null alleles of the *phoP* or *phoQ* gene are highly attenuated for virulence. This may be due to an inability to determine that cells have attained a subcellular location through the monitoring of the environmental Mg^{2+} concentration, although PhoPQ can control virulence

even in pathogens that do not have an intracellular lifestyle (Groisman 2001).

The BvgAS TCSTS is another example of one that has a major influence on virulence since it controls the expression of nearly all the known virulence genes in bordetella that cause respiratory infections in mammals. BvgAS functions like a molecular switch, mediating a biphasic transition between the Bvg+ and Bvg− phases in response to environmental cues, which are currently unknown. The Bvg+ phase is necessary and sufficient for respiratory infection by *Bordetella bronchiseptica*, whilst the Bvg− phase is adapted for survival under conditions of extreme nutrient deprivation (Stockbauer et al. 2001).

E. coli contains 36 TCSTS and among these are those that mediate chemotaxis, which are typically transmembrane homodimers composed of an external ligand-binding domain to monitor attractant compounds (such as serine or aspartate), four transmembrane helices, and internal methyl-accepting and signaling domains. These

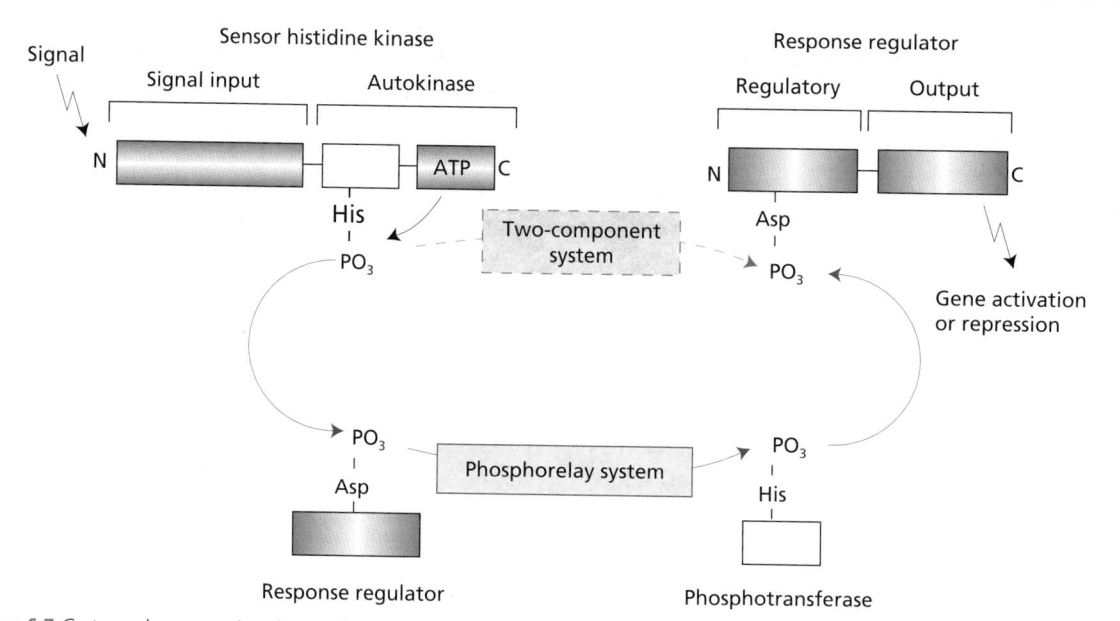

Figure 6.7 *Cartoon demonstrating the mechanism of two-component signal and phosphorelay signal-transduction systems. Signal ligands stimulate autophosphorylation of the sensor kinase on a histidine residue. In a two-component system (red), the phosphoryl group is subsequently transferred to an aspartate residue of a two-domain response regulator to activate its transcriptional properties. In a phosphorelay (blue), the phosphoryl group is first transferred to an aspartate of a single-domain response regulator that lacks a DNA-binding domain, and then to a histidine residue of a phosphotransferase that subsequently serves as the phosphoryl donor to activate the transcriptional properties of the response regulator. Redrawn from Stephenson and Hoch 2002.*

chemoreceptors (methyl-accepting chemotaxis proteins (MCP)) form higher order aggregates with other signaling proteins at the cell poles and control the phosphorylation state of the histidine protein kinase, CheA, via the linker, CheW, and the two response regulators, CheY and CheB. Phospho-CheY binds to FliM in the flagellar motor to augment clockwise rotation (Armitage et al. 2003; Parkinson 2003).

A family of proteins distinct from TCSTS, but also involved in signal transduction, are the hybrid sensor-regulators related to ToxR and TcpP, which regulate the genes encoding CT and TCP (see Examples of bacterial adhesins and their receptors) of *V. cholerae*. These unusual transcription regulators are localized to the cytoplasmic membrane and also regulate gene expression by binding to DNA and activating transcription. They share a bitopic arrangement with a cytoplasmic amino terminus and a periplasmic carboxy-terminus separated by a short transmembrane stretch of hydrophobic amino acids, and often work in conjunction with an effector protein that is localized on the periplasmic face of the inner membrane, e.g. ToxR interacts with ToxS and TcpP with TcpH. The cytoplasmic DNA-binding domains of ToxR and TcpP bear homology with the transcriptional regulator OmpR and consist of a helix–turn–helix (HTH) domain and two β-strands separated by a loop. Interaction of the loop between the two helices of the HTH domain with the carboxy-terminal portion of the α-subunit of RNA polymerase activates transcription. ToxR and TcpP act together to regulate CT and TCP expression via activation of the *toxT* gene.

It should be noted that other factors also influence the production of CT and TCP and that ToxR and TcpP can also regulate other genes in *V. cholerae*, e.g. biofilm formation (Crawford et al. 2003).

Quorum sensing: bacterial cell-to-cell communication

A crucial feature of many bacterial infections is a requirement for the pathogen to attain a cell population density sufficient to overcome host defenses and establish an infection. A mechanism for coordinating the control of multiple virulence genes as a function of population density offers a survival advantage to the pathogen, such that the host is overwhelmed before a defense response can be initiated fully. In this context, the ability of bacterial cells to communicate with each other offers the pathogen a survival advantage by orchestrating collective attack against the host immune system.

Many different gram-negative and gram-positive pathogens are now known to communicate using specific extracellular signal molecules, which facilitate the coordination of virulence and survival gene expression. The term 'quorum sensing' is commonly used to describe the phenomenon whereby the accumulation of a diffusible, low-molecular-weight signal molecule (sometimes called an 'autoinducer') enables individual bacterial cells to sense when the minimal number, or 'quorum,' of bacteria has been achieved for a concerted response to be initiated (for reviews see Swift et al. 2001; Camara et al. 2002; Williams et al. 2000). The accumulation of a diffusible signal molecule also indicates the presence of a diffusion barrier, which ensures that more molecules are produced than lost from a microhabitat (Winzer et al. 2002b; Redfield 2002). This could be regarded as a type of 'compartment sensing,' where signal molecule accumulation is both the measure for the degree of compartmentalization and the means to distribute this information among the entire population. Similarly, diffusion of quorum-sensing signal molecules between spatially separated bacterial subpopulations may convey information about their physiological state, their numbers, and the individual environmental conditions encountered.

Bacteria employ a number of different quorum sensing 'languages,' and several families of signal molecules have been characterized. These range from post-translationally modified peptides to quinolones, lactones, and furanones (Williams 2002). In addition, siderophores (see Iron-dependent repression), which previously were considered only in the context of iron transport, may also function in the producer organism as signal molecules capable of controlling genes unrelated to iron acquisition (Lamont et al. 2002). Thus, many bacteria produce multiple signal molecules (Fuqua et al. 2001; Withers et al. 2001; Camara et al. 2002) that are components of complex regulatory hierarchies responsible for controlling signal molecule production, detection, and response (Atkinson et al. 1999; Winzer and Williams 2001). At the molecular level, quorum sensing requires a synthase plus a signal transduction system for producing and responding to the signal molecule. The latter usually involves either a response regulator or sensor kinase protein or both (see Two-component signal transduction systems) (Swift et al. 2001; Williams 2002). While there is, as yet, no clear-cut evidence for a molecularly conserved quorum-sensing system throughout the bacterial kingdom, the luxS gene and a furanone-related molecule termed AI-2 (for autoinducer-2) have been suggested to fulfill such a role (Bassler 2002). However, many bacteria (e.g. P. aeruginosa) do not possess luxS and hence do not produce AI-2 (Winzer et al. 2002a). Furthermore, since LuxS is a key metabolic enzyme in the activated methyl cycle responsible for recycling S-adenosylmethionine (SAM) (Winzer et al. 2002a, b), phenotypes associated with mutation of luxS are often not a consequence of a defect in cell-to-cell communication but the result of the failure to recycle SAM metabolites (Winzer et al. 2002a, 2004).

The most intensively studied quorum-sensing systems in gram-negative bacteria utilize N-acylhomoserine lactone (AHL) quorum-sensing signal molecules (Swift et al. 2001; Whitehead et al. 2001; Fuqua and Greenberg 2002). AHLs consist of a homoserine lactone ring attached, via an amide bond, to an acyl side chain. Naturally occurring AHLs containing an acyl side chain ranging from four to 18 carbons in length have been identified, and this chain may be saturated or unsaturated and bear either a hydroxy-substituent or an oxo-substituent, or no substituent on the carbon at the 3 position (see Figure 6.8a). AHLs are produced by a number of human pathogens, including Aeromonas, Burkholderia, Brucella, Serratia, Pseudomonas, and Yersinia species (Table 6.9), but not by E. coli or Salmonella.

The two central players of AHL-dependent quorum-sensing systems typically belong to the LuxI and LuxR protein families, which function as AHL synthases and transcriptional regulators, respectively (Withers et al. 2001). AHLs with a short acyl side chain passively diffuse out of the cell, whilst those with a long acyl side chain are pumped out. Once in the extracellular milieu, AHLs accumulate until a sufficient concentration is reached, enabling the binding to and activation of the transcriptional regulator LuxR to form a complex responsible for activating or repressing specific sets of target genes. The gene encoding the AHL synthase may itself be activated resulting in a positive autoinduction circuit through which the AHL signal molecule controls its own synthesis (Figure 6.8b). AHL synthases distinct from the LuxI family have also been identified, e.g. the LuxM family of Vibrio species (Milton et al. 2001), and alternative mechanisms for responding to AHLs have also been described (Miller et al. 2002).

MULTISIGNAL QUORUM SENSING IN P. AERUGINOSA

In the opportunistic human pathogen P. aeruginosa, a complex quorum-sensing hierarchy plays a central role in the regulation of virulence and contributes to the late stages of biofilm maturation (Sauer et al. 2002; Kjelleberg and Molin 2002). P. aeruginosa possess two AHL-dependent quorum-sensing systems, termed LasRI and RhlRI. LasI and RhlI are LuxI homologs that direct the synthesis of N-(3-oxododecanoyl) homoserine lactone (3-oxo-C12-HSL) and N-butanoylhomoserine lactone (C4-HSL), respectively (Pearson et al. 1994; Winson et al. 1995). 3-Oxo-C12-HSL and C4-HSL are responsible for activating the LuxR homologs LasR and RhlR, respectively. Together, las and rhl form a regulatory hierarchy in which the LasR/3-oxo-C12-HSL system regulates the rhlRI genes. Moreover, the target genes regulated via las and rhl overlap considerably (Whiteley et al. 1999), and recently a third LuxR homolog (QscR) has been identified that further modulates their expression (Chugani et al. 2001). The important contribution that quorum sensing makes to the pathogenicity of P. aeruginosa has been clearly demonstrated in animal, plant, and nematode experimental infection models. Furthermore, the isolation of AHLs from infected body fluids (e.g. sputa) and tissues has confirmed that AHLs are produced by P. aeruginosa in vivo during human infections (Camara et al. 2002). Interestingly, AHLs

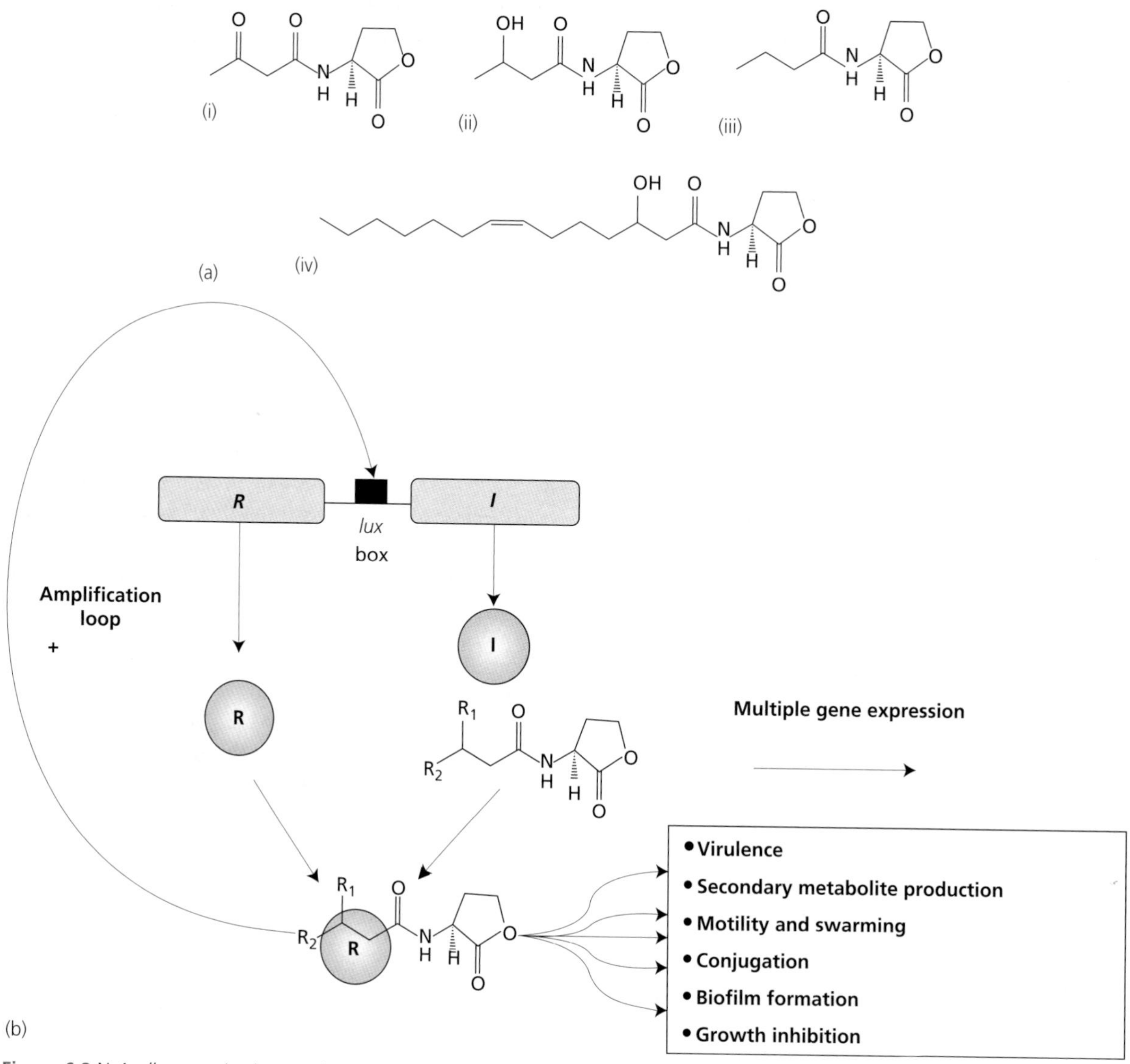

Figure 6.8 N-Acylhomoserine lactone (AHL)-mediated quorum-sensing. (a) Example structures showing diversity of AHLs with either (i) -oxo (3-oxo-C₄HSL), (ii) -hydroxy (3-hydroxy-C₄HSL), or (iii) no substitutent (BHL) at the 3-carbon position of the acyl side chain. An AHL with an unsaturated acyl side chain (3-hydroxy-C₁₄:₁HSL) is also shown (iv). (b) LuxRI/AHL-dependent signal circuitry. The 'R' and 'I' genes are homologs of the Vibrio fisheri luxR and luxI genes, which are arranged differently depending on bacterial species (they can be convergent, divergent, in tandem, or unlinked). The I protein directs the synthesis of AHLs, which exit the cell and are taken up into neighboring bacterial cells. Here, AHLs bind to and activate the response regulator (R) and this complex activates or represses the expression of multiple target genes by binding to conserved DNA sequences (lux box). In many cases the R protein/AHL complex also activates the I gene, creating an amplification loop resulting in a rapid increase in I gene expression and AHL production. Reproduced from Camara et al. (2002)

such as 3-oxo-C12-HSL may also function as virulence determinants in their own right since they exhibit proinflammatory, immune modulatory, and vasorelaxant properties (Telford et al. 1998; Lawrence et al. 1999; Smith et al. 2001). Since T-cell responses constitute an important component of the host immune defense against *P. aeruginosa*, the 3-oxo-C12-HSL-mediated suppression of T-cell activity is likely to be advantageous to this pathogen.

Apart from AHLs, *P. aeuginosa* also produces a third quorum-sensing signal molecule, which is essential for the expression of many *rhl*-dependent phenotypes as well as biofilm development (Diggle et al. 2003). This is the pseudomonas quinolone signal (PQS) (2-heptyl-3-hydroxy-4-quinolone), which structurally resembles the 4-quinolone antibiotics and several antimalarial drugs (Pesci et al. 1999). PQS, like the AHLs, has been detected in the sputum of CF patients infected with *P. aeruginosa*, indicating that this signal molecule is also produced in vivo during infection. *P. aeruginosa* also produces a number of cyclic dipeptides (diketopiperazines) that are capable of cross-activating or antagonizing AHL-regulated processes, although their physiological function remains unclear (Holden et al. 1999).

Table 6.9 N-*Acylhomoserine lactone (AHL)-mediated gene regulation in different organisms*

Bacterium	LuxR/I	Major AHL	Phenotypes
Pseudomonas aeruginosa	LasR, LasI	3-Oxo-C12-HSL	Exoenzymes, Xcp secretion apparatus, biofilm maturation, RlhR
	RhlR, RhlI	C4-HSL	Exoenzymes, cyanide, RpoS, lectins, pyocyanin, rhamnolipd
Burkholderia cepacia	CepR, CepI	C8-HSL	Protease, siderophore
Chromobacterium violaceum	CviR, CviI	C6-HSL	Antibiotics, violacein pigment, exoenzymes, cyanide
Yersina pestis	YpeR, YpeI	Unknown	Unknown
Yersina enterocolitica	YenR, YenI	C6-HSL, 3-Oxo-C6-HSL	Motility and swamming
Yersina pseudotuberculosis	YpsR, YpsI	3-Oxo-C6-HSL	Bacterial motility and clumping
	YtbR, YtbI	C8-HSL	Unknown
Aeromonas hydrophila	AhyR, AhyI	C4-HSL	Extracellular protease, biofilm formation
Brucella melitensis	Unknown	C12-HSL	Inhibition of *virR/virS*

The potential contribution of quorum-sensing to the development and maturation of *P. aeruginosa* biofilms has been the subject of much debate and controversy. *P. aeruginosa lasI* mutants, for example, were reported to be unable to form a highly structured biofilm community (Davies et al. 1998), which could be more readily dispersed by detergents than wild-type biofilms. The in vivo relevance of this finding has also been indicated by evaluation of quorum-sensing-dependent biofilm development in the murine model of corneal infection (Zhu et al. 2002). However, it is unlikely that quorum-sensing is the only important effector, since biofilm-associated communities will exploit all available adaptive mechanisms and the corresponding network of regulatory activities (Kjelleberg and Molin 2002). Indeed, mathematical models (Wimpenny and Colasanti 1997; Van Loosdrecht et al. 1997) have illustrated that biofilm formation is a predictable consequence of the physicochemical conditions in the biofilm environment, indicating that the bacterial cells do not perceive that they are within a biofilm but simply respond to local environmental conditions (such as stress and nutrient gradients). Nevertheless, it is clear that many genes important for biofilm development and maturation are subject to AHL-dependent quorum-sensing control in *P. aeruginosa* and in other gram-negative bacteria.

THE STAPHYLOCOCCAL *AGR* SYSTEM

In gram-positive bacteria, quorum-sensing signal molecules are generally small, modified peptides, and the transduction of such signals usually involves a TCST phosphorylation cascade that ultimately triggers target gene expression (see Two-component signal transduction systems). The opportunistic human pathogens *S. aureus*, *Enterococcus faecalis*, and *Str. pneumoniae*, as well as *Bacillus subtilis*, regulate many cellular processes in this way (Novick and Muir 1999; Dunny and Leonard 1997).

The staphylococci, for example, produce quorum-sensing signal molecules termed autoinducing peptides

(AIP). These are cyclic peptide thiolactones containing between seven and nine amino acid residues. In *S. aureus* and *Str. epidermidis*, these are produced via the *agr* regulatory locus, which also controls the production of multiple cell-wall colonization factors and exotoxins (Winzer and Williams 2001; Novick and Muir 1999; Otto 2001). The gene *agr* consists of two divergent transcription units, RNAII and RNAIII, controlled via the P2 and P3 promoters, respectively (Figure 6.9). While RNAIII is the Agr effector molecule, the P2 operon consists of four genes, *agrABCD*. AgrA and AgrC constitute a response regulator and sensor kinase, respectively, while the AIP is derived from an internal fragment of the AgrD protein. The transmembrane protein AgrB is responsible for processing AgrD and secreting the AIP signal molecule.

AIPs interact with the membrane-bound sensor kinase AgrC to activate the response regulator AgrA, located in the cytoplasm. Activation of *agr* occurs through the initiation of a phosphorylation cascade, resulting in the production of more AIP (via the P2 operon) and the effector RNA molecule, RNAIII via P3. At high *S. aureus* cell densities, *agr* represses the expression of genes coding for cell-surface proteins involved in colonization, while activating the expression of genes coding for exotoxins and tissue-degrading exoenzymes, Several experimental models of infection have highlighted a key role for the *agr* locus in the virulence of *S. aureus* (Novick and Muir 1999).

S. aureus strains can be subdivided into four AIP groups (I–IV), in which the AgrD-derived AIP of one group can cross-activate the *agr* response (and thus virulence) in other strains of that group but inhibit the *agr* response of members of other groups (Ji et al. 1997). While the primary amino acid sequence is different for each AIP, they all share a common central cysteine that is located five amino acid residues from the C-terminal amino acid. The AIPs are not linear peptides but contain an intramolecular thiol ester linkage between the sulfydryl group of the

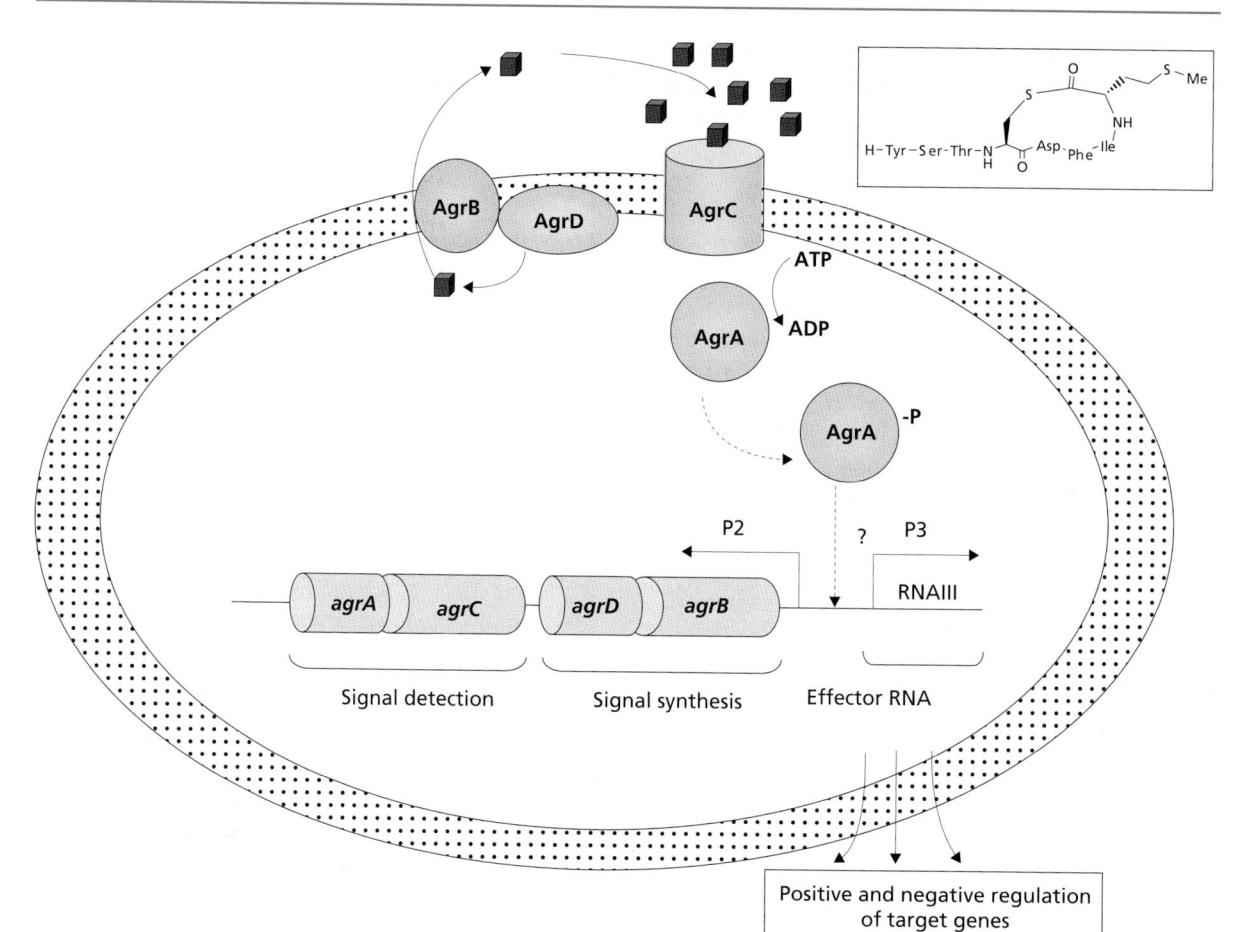

Figure 6.9 Agr-mediated quorum-sensing circuitry. The agr locus consists of two divergent transcriptional units. One is expressed from the P2 promoter and consists of agrABCD, which encodes for the proteins responsible for generating and sensing the peptide signal molecule. AgrB acts on the peptide derived from agrD to yield the signaling peptide and also to effect its export out of the cell. Once a critical extracellular concentration of the peptide is reached, it interacts with the AgrC membrane-associated sensor protein, causing it to autophosphorylate. This phophate is then transferred to the response regulator, AgrA, which then activates the expression of RNAIII from the P3 promoter. The effector molecule, RNAIII, mediates the changes in expression of multiple target genes, resulting in the agr response. An autoinduction feedback loop results in more peptide signal, since agrBD are amongst the activated genes. The insert in the top right-hand corner shows the structure of the quorum-sensing peptides produced by group I Staphylococcus aureus. Redrawn from Camara et al. (2002).

conserved cysteine and the α-carboxyl group of the C-terminal amino acid. This thiolactone structure is essential for full biological activity and, consistent with the concept of cell population density sensing, a critical threshold concentration (in the nanomolar range) of the AIP is required to trigger an *agr*-dependent response (Ji et al. 1997; McDowell et al. 2001). AIP antagonists therefore offer a novel means of controlling staphylococcal infections, since they inhibit exotoxin (e.g. toxin shock syndrome toxin, TSST-1) and exoenzyme (e.g. lipase) production. In a mouse subcutaneous abscess infection model, coinoculation of a virulent group I strain with the group II AIP resulted in reduction in the virulence of the group I strain (Mayville et al. 1999). Other staphylococcal species such as *S. epidermidis* and *S. lugdunensis* also produce AIPs that are cross-inhibitory with those of *S. aureus*. Thus, it seems that staphylococci have evolved an innate mechanism for coupling their

pathogenic potential with a means to ward off potential competitors.

EVOLUTION OF PATHOGENS WITHIN THEIR HOST

Although pathogens are capable of integrating the diverse environmental signals encountered during the progress of infection to control virulence factor deployment, other strategies are also employed. For example, the capsules of *N. meningitidis* undergo phase variation such that they are expressed at late stages of infection to provide protection against the immune system, and not when their presence would hinder intimate adhesion to host cells, which is required for invasion during the early stages (Deghmane et al. 2002). Phase variation in *N. meningitidis* can occur by several different mechanisms: capsule biosynthesis genes can be acquired or lost allowing antigenic variation, plus their expression can be regulated by slip-strand mispairing. This mechanism regulates a range of virulence factors of *N. meningitidis* and other bacteria such as *H. influenzae*, and relies upon the presence of an extended nucleotide repeat, which can loop out, resulting in a deletion that results in disruption of the reading frame and loss in gene expression (Moxon et al. 1994). The existence of such mechanisms suggests that the host exerts a certain selective pressure upon the pathogens it harbors. This is perhaps well illustrated by the apparent continuous microevolution of *H. pylori* within its cognate host (Israel et al. 2001). The highly diverse organism *H. pylori* persists in the human stomach and induces chronic gastritis, which may progress to peptic ulceration, noncardia gastric adenocarcinoma, or gastric lymphoma during the life of the host. The differential development of these clinical sequelae probably depends on differentially represented bacterial determinants, host characteristics, and interactions between the host and the bacterium throughout their time of coexistence.

Although *H. pylori* appear to exhibit a clonal population structure over short periods of time, isolates obtained from different individuals exhibit substantial genetic diversity, consistent with extensive recombination and a panmictic (a population derived from a single organism) population structure (Suerbaum et al. 1998). Strains may differ in their possession of point mutations in highly conserved genes (Achtman et al. 1999), the presence of nonconserved and/or mosaic forms of genes (Akopyants et al. 1998; Censini et al. 1996; Peek et al. 1998; Pride et al. 2001), and chromosomal organization (Taylor et al. 1992; Jiang et al. 1996; Alm et al. 1999). These variations may be acquired through horizontal genetic exchange or spontaneous mutation (Suerbaum et al. 1998; Achtman et al. 1999). Comparison of isolates of the sequenced *H. pylori* strain J99, after a 6-year interval from its human source patient, revealed genetic divergence amongst the isolates and between them and the original stain (Israel et al. 2001). Thus, it appears that within an apparently homogeneous population as determined by macroscale comparison (randomly amplified polymorphic DNA, polymerase chain reaction (PCR), and DNA sequencing of four unlinked genetic loci), genetic differences emerged among single-colony isolates of *H. pylori*. The acquisition of genes originating from distinct strains suggests that there may have been an additional subpopulation of *H. pylori* strain J99 possessing these open reading frames (ORF), or that another strain of *H. pylori* or closely related organism was transiently present in the stomach. The implied continuous state of genetic flux may allow *H. pylori* to adapt rapidly to changing conditions in its current host throughout the extended period of coexistence, as well as priming it for colonization of a new host. *H. pylori* strain variation may hinge upon gene mosaicism (see below) or the presence of segments of nonconserved DNA. This is perhaps best illustrated by the gene encoding the vacuolating cytotoxin (VacA) and the cytotoxin associated antigen (CagA), which is encoded on a pathogenicity island (PAI).

VacA is a member of the family of autotransporter proteins (Cover 1996; Telford et al. 1994) (see Table 6.10 for other members), and as such is produced as a precursor polypeptide in the cytoplasm. An N-terminal targeting domain delivers the autotransporter to the inner membrane and is cleaved during translocation to

Table 6.10 *Examples of members of the autotransporter family, amended from Henderson and Nataro (2001)*

Bacterium	Autotransporter
Neisseria meningitidis	IgA protease
Neisseria gonorrhoeae	IgA protease
Haemophilus influenzae	IgA protease, adhesin (Hap, Hia, Hsf)
Helicobacter pylori	VacA cytotoxin, BabA adhesin
Shigella flexneri	IcsA
Bordetella spp.	Adhesins (Pertactin, TcfA, Vag8), serum resistant factor (BrkA)
Dichelobacter nodusus	Putative elastases (BprV, BprB, AprV2, BprX)
Moracella catarrhalis	Adhesin (UspA1, UspA2h), serum resistance protein (UspA2)
Pasteurella haemolytica	Ssa1 protease
Pseudomonas aeruginosa	EstA esterase
Pseudomonas fluorescens	Proteases (PspA, PspB)
Ricettsiales	Adhesin (rOmpA, rOMPB)
Salmonella typhiumurium	Esterase ApeE
Serratia marcesens	Proteases (PrtS, PrtT, Ssp-H1, Ssp-H2)
Shigella flexneri	Proteolytic (SepA, SigA), mucinase (Pic), mediator of intracellular motility (IcsA)
Xenorhabdus luminescens	Lipase (PlaA)
Pathogenic *Escherichia coli*	AIDA-I adhesins (AIDA-I, TibA), proteolytic toxin (EspP, Pet, Sat), hemagglutinin (Tsh), mucinase (Pic), biofilm formation (Ag43), EspC.

the periplasm. This translocation may be achieved via the SRP/Sec pathway (Henderson et al. 1998; Sijbrandi et al. 2003). The C-terminal domain is proposed to insert into the outer membrane forming a β-barrel through which the mature protein (or passenger domain) is translocated to the extracellular medium. Cleavage of this C-terminal domain, which may be mediated by the passenger protein or surface proteases (Henderson et al. 1998; Henderson et al. 2000), then releases the mature protein (Figure 6.4). VacA shows homology to this family both by possession of an extended N-terminal signal peptide and by the presence of a C-terminal domain, which shows homology to other auto-transporters and is cleaved during secretion (Cover 1996; Schmitt and Haas 1994). Once secreted, the approximately 90-kDa passenger domain of VacA is processed further, yielding subunits of approximately 58 and 37 kDa. These monomers combine to form dodecamers or tetradecamers (Cover et al. 1997) with a flower-like rosette structure (Lupetti et al. 1996). In acidic conditions, these oligomers dissociate back into monomers, exposing hydrophobic regions and allowing efficient insertion into target membranes (Molinari et al. 1998b).

Once released from the bacterial cell, VacA damages eukaryotic cells, causing mitochondrial damage (Kimura et al. 1999; Galmiche et al. 2000), inducing apoptosis (Galmiche et al. 2000; Kuck et al. 2001), increasing permeability of polarized monolayers (Papini et al. 1998; Pelicic et al. 1999), disrupting antigen presentation (Molinari et al. 1998a), and interfering with vacuolar transport in the late endosomal pathway, resulting in the formation of characteristic vacuoles (Cover and Blaser 1992; Molinari et al. 1997). Although all *H. pylori* strains analyzed to date contain *vacA*, its sequence varies, a phenomenon known as mosaicism. The nucleotide sequence encoding the N-terminal signal region may be one of four variants (s1a, s1b, s1c, or s2), whilst the mature passenger domain may contain a segment of an m1, m2a, or m2b (Atherton et al. 1995; van Doorn et al. 1998) subtype. The s1 and m1 types have been associated with increased virulence of *H. pylori*, since strains bearing this combination are most frequently associated with more serious disease states in humans (Atherton et al. 1995). The more pathogenic strains of *H. pylori* (type I strains) also possess the *cag* PAI, an element lacking in strains associated with milder clinical outcomes (type II strains).

The *cag* PAI encodes a type IV secretion system (TFSS) (see Protein secretion and Figure 6.4) that facilitates the injection of bacterial proteins directly into eukaryotic cells. Following attachment to the surface of eukaryotic cells, CagA is delivered into the host cell by proteins encoded by other genes on the *cag* PAI (Asahi et al. 2000; Backert et al. 2000; Odenbreit et al. 2000; Segal et al. 1999; Stein et al. 2000). Intracellular CagA is tyrosine phosphorylated by a host cell kinase of the Src family (Selbach et al. 2002; Stein et al. 2002), whereupon

it interacts with the SHP-2 phosphatase initiating signaling events that lead to dramatic cellular elongation (the 'humming-bird' phenotype) (Higashi et al. 2002). The precise molecular events occurring throughout this alteration of the actin cytoskeleton remain unclear, although the involvement of small GTPases is possible (Covacci and Rappuoli 2003).

Independent of *cagA*, the *cag* PAI initiates the activation of NF-κB and the expression of proinflammatory cytokines, such as IL-8 (Crabtree 1998). The latter requires that the *cag* PAI encodes a functional TFSS.

HOST RESISTANCE TO INFECTION

Many of the bacterial virulence determinants described above contribute to pathogen evasion of host defenses. Resistance/immunity in higher animals can be divided into innate and acquired (adaptive) immunity. In an evolutionary context, innate immunity is considered to be the most ancient form of defense against invading microorganisms. It is present from birth, functions nonspecifically against diverse microorganisms, and, in contrast to acquired immunity, does not become more efficient on subsequent exposure to the same organisms or antigens. Acquired immunity, on the other hand, is based on the generation of an antibody response towards specific antigens associated with infecting pathogens. It is extremely flexible and is able to mount a response to virtually any antigen presented. A unique feature of the acquired immune response is the memory of previous encounters with an antigen and the capacity to mount an improved response during subsequent exposures.

Innate immunity

Innate immunity refers to the nonspecific defense mechanisms that a host uses immediately or within several hours following exposure to an antigen. There are numerous factors that contribute to the innate immunity of a higher organism and these can be subdivided into: (1) anatomical barriers; (2) mechanical removal; (3) nutritional restriction; (4) bacterial antagonism; and (5) common pattern recognition.

ANATOMICAL BARRIERS

Anatomical barriers are tough intact layers that prevent the entry and colonization by microorganisms. Examples include the skin and the mucous membranes.

The skin, consisting of the epidermis and the dermis, is dry and acidic and has a temperature less than 37°C. These conditions are not favorable for bacterial growth. The dead keratinized cells that make up the surface of the skin are continuously being sloughed off, so that microbes that do colonize these cells are constantly being removed. The hair follicles and sweat glands, which provide potential entry sites, produce lysozyme and toxic lipids that can kill bacteria. Also, beneath the surface, is the skin-associated lymphoid tissue (SALT), which contains cells that kill incoming microorganisms and initiate acquired immune responses by sampling antigens on the skin.

Mucous membranes line body cavities that are open to the exterior environment, such as the respiratory tract, the gastrointestinal tract, and the genitourinary tract. Mucous membranes are probably the most important site of entry of pathogenic bacteria into the host body. Mucosal surfaces are composed of an epithelial layer that secretes mucus, which acts as a physical barrier, trapping microorganisms and preventing deeper penetration into the tissues. Mucus contains the bactericidal protein lysozyme, which degrades bacterial peptidoglycan (leading to the lysis of bacterial cells), as well as secretory IgA, which prevents bacteria from attaching to the epithelial layer of cells. In addition, secretions from mucus membranes contain the iron-transporting protein lactoferrin, which plays a major role in mucosal defence. Lactoferrin not only withholds iron from infecting bacteria (see Motility), but also is capable of disrupting bacterial biofilms (Singh et al. 2002) and degrading bacterial virulence factors (such as the Hap adhesins and IgA proteases of *H. influenzae*) via its serine protease activity (Hendrixson et al. 2003). Lactoferrin also possesses immunoregulatory functions and bactericidal activity and can neutralize the toxic effects of LPS (Caccavo et al. 2002). Mucus also contains lactoperoxidase, which generates toxic superoxide radicals that rapidly kill bacteria. It has also been suggested that the resident normal microbial flora of the mucosa exclude colonization by potentially harmful bacteria. Beneath the mucosal surface is mucosa-associated lymphoid tissue (MALT), which contains cells that destroy invading bacteria and subsequently initiate an acquired immune response.

MECHANICAL REMOVAL OR THE PHYSICAL FLUSHING OF MICROORGANISMS FROM THE BODY

Cilia on the surface of the epithelial cells that line mucous membranes, such as those in the respiratory tract, propel mucus and trapped bacteria upwards towards the throat, where it is swallowed. The acid pH of the stomach breaks mucus down and kills any bacteria present. Coughing and sneezing, in a similar way, removes bacteria from the respiratory tract that have become trapped in the mucus. Although vomiting and diarrhea are considered characteristic symptoms of particular diseases, they also can be considered part of the host defense, since they effectively remove pathogens and toxins from the gastrointestinal tract. It should be noted that the flow of fluids, such as urine, tears, saliva, perspiration, and blood from injured blood vessels, is also important for the mechanical removal of pathogens from the host body.

NUTRITIONAL RESTRICTION

Nutrient-depriving host defense molecules include transcobalamins (which bind vitamin B12), calprotectin (which binds zinc), and lactoferrin (which sequesters iron). Iron is one essential nutrient for both host and pathogen (see Iron scavenging). During the course of infection, the human body makes considerable efforts to make iron unavailable to microorganisms for growth; this is called the hypoferremic response. Much of this is due to production of IL-1α (also known as leukocyte-endogenous mediator (LEM)) which acts to lower free iron in tissues and body fluids. However, there are other factors that contribute to the lowering of available free iron during infection, including: (1) decreased intestinal dietary iron uptake; (2) a reduction in plasma iron levels and increased synthesis of the iron storage protein ferritin; (3) increased synthesis of lactoferrin and transferrin; and (4) release of lactoferrin at sites of microbial infection, such as the mucous membranes.

BACTERIAL ANTAGONISM

Approximately 100 trillion bacteria and other microorganisms reside in or on the human body as the normal body flora. These are considered to exclude potentially harmful opportunistic pathogens by the production of, for example, fatty acids and antibacterial peptides, such as bacteriocins. The normal flora may also influence the growth of invading pathogens by depletion of, and competition for, the available nutrients. An example of the prevention of disease by bacterial antagonism is that of the severe antibiotic-associated colitis caused by *Clostridium difficile*. This anerobe, which cannot grow within a normal gut flora, will infect and colonize when the gut flora microecology is altered by, for example, antibiotics.

COMMON PATTERN RECOGNITION

Common pattern recognition has evolved to identify a few highly conserved molecules present in or on many different microorganisms. The structures recognized are termed pathogen-associated molecular patterns and include LPS, peptidoglycan, LTA, the sugar mannose (which is common in microbial polysaccharides but rare in those derived from higher organisms), bacterial DNA, double-stranded RNA (from viruses), and glucans (from fungal cell walls). Most body defense cells have pattern-recognition receptors (Toll-like receptors (TLR)) for these common pathogen-associated molecular patterns and so, upon infection, there is an immediate response to invading microorganisms. Pathogen-associated molecular patterns can also be recognized by the blood-borne proteins that initiate complement.

The lipid A component of LPS/LOS was the first agonist shown to bind and activate TLRs. These constitute a family of distinct proteins characterized by an intracellular domain with homology to that of the IL-1 receptor and containing extracellular leucine-rich repeats. TLRs activate the immune system to recognize conserved macromolecular components of pathogens and, through this interaction, activate the immune system. Pathogen recognition leads to the initiation of defensive protective actions, which include the production of reactive oxygen intermediates and the secretion of inflammatory cytokines, such as IL-1, interleukin-12 (IL-12), interferon-γ (IFN-γ), and TNF-α (Raetz and Whitfield 2002). These cytokines activate natural killer cells and also initiate a cascade of signals to cells of the adaptive arm of the immune response, preparing them for the development of antigen-specific immune responses. Activation of TLRs by lipid A also results in the production of defensins, which consist of several distinct families of antibacterial, antifungal, and antiviral peptides (Heine and Lien 2003).

TLRs direct these effects by activating host signal transduction processes, including the transcription factor NF-κB and the MAP kinases, p38, and c-jun. Differences between signals generated by alternative TLRs are emerging. Each TLR is activated by a different microbial surface component, e.g. TLR4 is activated primarily by bacterial LPS and LTA. Both LPS and LTA first bind to the CD14 receptor and are subsequently transferred to the TLR4 complex. LTA can bind directly to CD14, but LPS must be delivered to CD14 by LPS-binding protein (LBP). TLR4 signals through an accessory protein termed MyD88 to activate the transcription factors, NF-κB, Elk-1, and AP-1. TLR6 is also activated by LPS and shares the same signaling pathway as TLR4, as does TLR2 (activated by bacterial lipoarabinomannan (LAM) and peptidoglycans (PGN)). When activated by bacterial lipoprotein (BLP), TLR2 mediates both apoptosis and NF-κB activation. If NF-κB is inhibited downstream of MyD88, then apoptosis occurs through the activation of caspase 8 (Aliprantis et al. 2000). TLR2 is also responsible for the recognition of the yeast cell-wall particle zymosan. It is believed that zymosan acts through the CD14 receptor to influence TLR2, shown by high levels of downstream NF-κB in CD14$^+$/TLR2$^+$ macrophage exposed to zymosan, as compared with NF-κB levels observed in CD14$^-$/TLR2$^+$ macrophage exposed to zymosan. TLR2 is internalized after activation by zymosan; the yeast cell is also internalized with TLR2. TLR2 signals the production of TNFα, through NF-κB pathway, from the phagocytosed vesicle (Underhill et al. 1999).

TLR9 is responsible for the recognition of extracellular CpG islands of bacterial DNA. Two mechanisms for TLR9 activation are considered to exist. TLR9 bound to extracellular CpG fragments may be endocytosed, or alternatively phagosome located TLR may interact with phagosome internalized bacteria. The result of either of these is the activation of the NF-κB pathway from the endocytosed vesicle (Hemmi et al. 2000). TLR5 recognizes bacterial flagella and activates NF-κB pathway. The functions of TLR1, 3, 7, 8 and the recently cloned TLR10 (Chuang and Ulevitch 2001) are not yet known.

Acquired immunity

The lymphocyte-generated antibodies comprising the acquired immune system act in different ways. Some antibodies, such as IgG and IgE, function as opsonins, promoting phagocytic engulfment of bacteria. Antibodies, such as IgG, IgA, and IgM, bind to bacterial adhesins, pili, and capsules and block attachment to host cells. IgG and IgM can also activate the classical complement pathway, whilst IgA and IgM crosslink bacterial cells, enabling more efficient phagocytic clearance.

Bacteria evade antibody-mediated clearance in a number of ways (see Avoidance of host immune defenses), including antigenic variation of surface macromolecules, by the synthesis of surface components that mimic host tissue antigens, and by bacterial surface coating with host proteins, such as fibronectin and fibrinogen. By producing immunoglobulin-binding cell-wall proteins that bind the Fc portion of antibodies, bacteria such as *S. aureus* and *Str. pyogenes* can evade opsonization. Some bacteria produce immunoglobulin-degrading proteases, e.g. the IgA proteases of *H. influenzae*, *Str. pneumoniae*, *N. meningitidis*, and *N. gonorrhoeae*.

GENOMICS AND BACTERIAL PATHOGENICITY

Advances in molecular biology and, in particular, the ability to clone, express, and mutate specific genes have proved invaluable for defining the repertoire of virulence determinants employed by a pathogen. This in turn has resulted in the definition of Koch's molecular postulates (Falkow 1988; Joyce et al. 2002; Relman and Falkow 2001) for determining the relationship between a given virulence determinant and disease. The development of specific genetic strategies for identifying structural and regulatory genes that are essential for bacterial life in vivo, but not in vitro (in vivo expression technology (IVET) and signature-tagged mutagenesis (STM)) has provided many novel insights into the lifestyle of diverse pathogens. The information obtained from such experimentation can now be greatly enhanced by the availability of pathogen genome sequence information and the potential offered by DNA microarrays and proteomics for defining the global response of the pathogen to the host environment, and vice versa.

Several approaches can be used to analyze whole-genome sequences for candidate virulence factors, potential vaccine antigens, and novel antimicrobial targets. By comparing predicted coding sequences with sequences in databases such as GenBank through use of the BLAST program (www.ncbi.nlm.nih.gov:80/BLAST) identifies matches to known genes involved in virulence. Approximately 20 percent of the predicted ORFs in a genome do not match anything in GenBank, while another 10–20 percent match genes of unknown function. Therefore, the comparative approach has been useful for recognizing good candidates for virulence genes among genes whose functions have been previously defined. While this in silico approach has not proved particularly useful for the discovery of novel virulence genes, the availability of such information on easily accessible electronic databases has made it a routine tool for studies of pathogenic microbes and prompted the development of experimental techniques, such as IVET and STM.

STM is a powerful tool for the identification of genes that are required for virulence and in vivo survival. It is a method for identifying mutants that are unable to replicate in vivo and therefore display an attenuated phenotype. A transposon is used for random mutagenesis of the bacterial genome. Each transposon has been prepared to carry a unique index region or signature tag. The unique tag can be amplified by PCR using standard regions flanking the tag and the resulting products used as a hybridization probe to identify the unique transposon that encodes it. The initial set of random insertion mutants are arrayed as a master, pooled, and then grown under either in vitro or in vivo conditions in an experimental animal model of infection. All the mutants that emerge from the in vitro and in vivo growth regimens are collected, their index regions amplified by PCR, and used to hybridize against the master array of original mutants. Mutants in the master array that do not hybridize with the tag sequences amplified from bacteria recovered from in vivo grown bacteria represent mutants that are attenuated and are unable to survive in vivo. Regions flanking the transposon insertions in mutants of interest are sequenced and compared with the genomic sequence databases to identify the inactivated gene(s). This powerful technique has been used to identify novel genes involved in pathogenesis for numerous pathogens, including *Salmonella*, *Neisseria*, and *E. coli* (Tang et al. 2001).

IVET is a method based on trapping promoters. Random genomic fragments are ligated in front of a promoterless reporter gene, the activity of which is used as an indication of transcriptional activity of the fused gene. Reporter genes that have been used include those that can allow auoxotrophy complementation IVET selection (e.g. *purA*), antibiotic IVET selection (e.g. *cat*), and recombinase-based IVET (RIVET), with which screening is achieved using a promoterless *tnpR* allele, as this cleaves a tetracycline resistance gene from elsewhere in the bacterial genome when produced. The suicide plasmids recombine into the chromosome by insertion–duplication, and fusion strains are passaged through an appropriate animal model before being collected from infected tissues and/or fluids. This strategy has been successfully used for *S. entirica* serovar Typhimurium, *Yersinia entercolitica*, *Streptococcus gorndonii*, *V. cholerae*, and *S. aureus* (Merrell and Camilli 2000).

CONCLUSIONS

Pathogenic bacteria exhibit coordinated responses to stimuli received from the external environment and during life on or within the host. These stimuli in turn impose differential selective pressures, which direct pathogen evolution. Certain bacteria have evolved strategies for avoiding the host immune system by invading host cells, while others have adopted measures for immune evasion while remaining adherent to host cell surfaces. Many pathogens produce tissue-damaging exotoxins and exoenzymes, which aid bacterial survival by increasing the availability of nutrients for growth, while simultaneously destroying tissue and host defenses. The coordinated deployment of these multiple virulence determinants is subject to tight regulatory controls through a variety of sensory pathways, some of which involve bacterial cell-to-cell communication. The ongoing unveiling of genomic, proteomic, and metabolomic information is likely to improve substantially our understanding of pathogen adaptive behavior and the host response, which in turn should provide new opportunities for the development of novel anti-infective measures.

REFERENCES

Achtman, M., Azuma, T., et al. 1999. Recombination and clonal groupings within Helicobacter pylori from different geographical regions. *Mol Microbiol*, **32**, 459–70.

Akopyants, N.S., Clifton, S.W., et al. 1998. Analyses of the cag pathogenicity island of *Helicobacter pylori*. *Mol Microbiol*, **28**, 37–53.

Aldridge, P. and Hughes, K.T. 2002. Regulation of flagellar assembly. *Curr Opin Microbiol*, **5**, 160–5.

Aliprantis, A.O., Yang, R.-B., et al. 2000. The apoptotic signalling pathway activated by toll-like receptor 2. *EMBO J*, **19**, 13, 3325–36.

Alm, R.A., Ling, L.S., et al. 1999. Genomic-sequence comparison of two unrelated isolates of the human gastric pathogen *Helicobacter pylori*. *Nature (Lond)*, **397**, 176–80.

Amer, A.O. and Swanson, M.S. 2002. A phagosome of one's own: a microbial guide to life in the macrophage. *Curr Opin Microbiol*, **5**, 56–61.

Andersen, C., Hughes, C. and Koronakis, V. 2001. Protein export and drug efflux through bacterial channel-tunnels. *Curr Opin Cell Biol*, **13**, 412–16.

Anderson, D.M. and Schneewind, O. 1997. mRNA signal for the type III secretion of Yop proteins by *Yersinia enterocolitica*. *Science*, **278**, 1140–3.

Armitage, J.P. 1999. Bacterial tactic responses. *Adv Microb Physiol*, **41**, 229–89.

Armitage, J.P., Dorman, C.J., et al. 2003. Thinking and decision making, bacterial style: Bacterial neural networks, Obernai, France, 7–12 June 2002. *Molec Microbiol*, **47**, 583–93.

Asahi, M., Azuma, T., et al. 2000. *Helicobacter pylori* CagA protein can be tyrosine phosphorylated in gastric epithelial cells. *J Exp Med*, **191**, 593–602.

Atherton, J.C., Cao, P., et al. 1995. Mosaicism in vacuolating cytotoxin alleles of *Helicobacter pylori* – association of specific VacA types with cytotoxin production and peptic ulceration. *J Biol Chem*, **270**, 17771–7.

Atkinson, S., Throup, J.P., et al. 1999. A hierarchical quorum-sensing system in *Yersinia pseudotuberculosis* is involved in the regulation of motility and clumping. *Mol Microbiol*, **33**, 1267–77.

Backert, S., Ziska, E., et al. 2000. Translocation of the *Helicobacter pylori* CagA protein in gastric epithelial cells by a type IV secretion apparatus. *Cell Microbiol*, **2**, 155–64.

Bassler, B.L. 2002. Small talk: cell-cell communication in bacteria. *Cell*, **109**, 421–4.

Baughn, R.E. 1987. Role of fibronectin in the pathogenesis of syphilis. *Rev Infect Dis*, **9**, S372–85.

Bayles, K.W. and Bohach, G.A. 2001. Internalization of *Staphylococcus aureus* by nonprofessional phagocytes. In: Honeyman, A.L., Friedman, H. and Bendinelli, M. (eds), Staphylococcus aureus *infection and disease*. New York: Kluwer Academic/Plenum Publishers, 255–6.

Berg, H.C. 2003. The rotary motor of bacterial flagella. *Annu Rev Biochem*, **72**, 19–54.

Bétis, F., Brest, P., et al. 2003. Afa/Dr diffusely adhering *Escherichia coli* infection in T84 cell monolayers induces increased neutrophil transepithelial migration, which in turn promotes cytokine-dependent upregulation of decay-accelerating factor (CD55), the receptor for Afa/Dr adhesins. *Infect Immun*, **71**, 1774–83.

Beutler, B. and Rietschel, E.T. 2003. Innate immune sensing and its roots: The story of endotoxin. *Nat Rev Immunol*, **3**, 169–76.

Bisno, A.L., Brito, M.O. and Collins, C.M. 2003. Molecular basis of group A streptococcal virulence. *Lancet Infect Dis*, **3**, 191–200.

Boddicker, J.D., Ledeboer, N.A., et al. 2002. Differential binding to and biofilm formation on Hep-2 cells by *Salmonella enterica* serovar Typhimurium is dependent upon allelic variation in the *fimH* gene of the *fim* gene cluster. *Mol Microbiol*, **54**, 1255–65.

Bourdet-Sicard, P., Egile, C., et al. 2000. Diversion of cytoskeletal processes by *Shigella* during invasion of epithelial cells. *Microb Infect*, **2**, 813–19.

Buileris, P.V., Behrens, S., et al. 2003. Folding and insertion of the outer membrane protein OmpA is assisted by the chaperone Skp and by lipopolysaccharide. *J Biol Chem*, **278**, 9092–9.

Caccavo, D., Pellegrino, N.M., et al. 2002. Antimicrobial and immunoregulatory functions of lactoferrin and its potential therapeutic application. *J Endotoxin Res*, **8**, 403–17.

Camara, M., Williams, P. and Hardman, A. 2002. Controlling infection by tuning in and turning down the volume of bacterial small-talk. *Lancet Infect Dis*, **2**, 667–76.

Campellone, K.G. and Leong, J.M. 2003. Tails of two Tirs: actin pedestal formation by enteropathogenic *E. coli* and enterohemorrhagic *E. coli* O157:H7. *Curr Opin Microbiol*, **6**, 82–90.

Censini, S., Lange, C., et al. 1996. cag, a pathogenicity island of *Helicobacter pylori*, encodes type I-specific and disease-associated virulence factors. *Proc Natl Acad Sci USA*, **93**, 14648–53.

Charkowski, A.O., Huang, H.C. and Collmer, A. 1997. Altered localization of HrpZ in *Pseudomonas syringae* pv. syringae *hrp* mutants suggests that different components of the type III secretion pathway control protein translocation across the inner and outer membranes of gram-negative bacteria. *J Bacteriol*, **179**, 3866–74.

Chirwa, N.T. and Herrington, M.B. 2003. CsgD, a regulator of curli and cellulose synthesis, also regulates serine hydroxymethyltransferase synthesis in *Escherichia coli* K-12. *Microbiology*, **149**, 525–35.

Christie, P.J. 2001. Type IV secretion: intercellular transfer of macromolecules by systems ancestrally related to conjugation machines. *Mol Microbiol*, **40**, 294–305.

Chuang, T. and Ulevitch, R.J. 2001. Identification of hTLR10: a novel human toll-like receptor preferentially expressed in immune cells. *Biochim Biophys Acta*, **19**, 1–2, 157–61.

Chugani, S.A., Whiteley, M., et al. 2001. QscR, a modulator of quorum sensing signal synthesis and virulence in *Pseudomonas aeruginosa*. *Proc Natl Acad Sci USA*, **98**, 2752–7.

Clarke, T.E., Tari, L.W. and Vogel, H.J. 2001. Structural biology of bacterial iron uptake systems. *Curr Top Med Chem*, **1**, 7–30.

Cossart, P. and Lecuit, M. 1998. Interactions of *Listeria monocytogenes* with mammalian cells during entry and actin-based movement: Bacterial factors, cellular ligands and signalling. *EMBO J*, **17**, 3797–806.

Covacci, A. and Rappuoli, R. 2003. *Helicobacter pylori:* after the genomes, back to biology. *J Exp Med*, **197**, 807–11.

Cover, T.L. 1996. The vacuolating cytotoxin of *Helicobacter pylori*. *Mol Microbiol*, **20**, 241–6.

Cover, T.L. and Blaser, M.J. 1992. Purification and characterization of the vacuolating toxin from *Helicobacter pylori*. *J Biol Chem*, **267**, 10570–5.

Cover, T.L., Hanson, P.I. and Heuser, J.E. 1997. Acid induced dissociation of VacA, the *Heliocbacter pylori* vacuolating cytotoxin, reveals its pattern of assembly. *J Cell Biol*, **138**, 759–69.

Crabtree, J.E. 1998. Role of cytokines in pathogenesis in *Helicobacter pylori*-induced mucosal damage. *Dig Dis Sci*, **43**, 46S–55S.

Crawford, J.A., Krukonis, E.S. and DiRita, V.J. 2003. Membrane localization of the ToxR winged-helix domain is required for TcpP-mediated virulence gene activation in *Vibrio cholerae*. *Mol Microbiol*, **47**, 1459–73.

Dang, T.A., Zhou, X.R., et al. 1999. Dimerization of the *Agrobacterium tumefaciens* VirB4 ATP ase and the effect of ATP-binding cassette mutations on the assembly and function of the T-DNA transporter. *Mol Microbiol*, **32**, 1239–53.

Davies, D.G., Parsek, M.R., et al. 1998. The involvement of cell-to-cell signals in the development of a bacterial biofilm. *Science*, **280**, 295–8.

Davis, B.M. and Waldor, M.K. 2003. Filamentous phages linked to virulence of *Vibrio cholerae*. *Curr Opin Microbiol*, **6**, 35–42.

De Cock, H., Struyve, M., et al. 1997. Role of the carboxy-terminal phenylalanine in the biogenesis of outer membrane protein PhoE of *Escherichia coli* K-12. *J Mol Biol*, **269**, 473–8.

Deghmane, A.E., Giorgini, D., et al. 2002. Down-regulation of pili and capsule of *Neisseria meningitidis* upon contact with epithelial cells is mediated by CrgA regulatory protein. *Mol Microbiol*, **43**, 1555–64.

Delahay, R.M., Frankel, G. and Knutton, S. 2001. Intimate interactions of enteropathogenic *Escherichia coli* at the host cell surface. *Curr Opin Infect Diseases*, **14**, 559–65.

Diggle, S.P., Winzer, K., et al. 2003. The *Pseudomonas aeruginosa* quinolone signal molecule moderates production of rhl-dependent quorum sensing phenotypes and promotes biofilm development. *Mol Microbiol*, **50**, 1, 29–43.

Dobinsky, S., Kiel, K., et al. 2003. Glucose-related dissociation between *icaADBC* transcription and biofilm expression by *Staphylococcus epidermidis*: evidence for an additional factor required for polysaccharide intercellular adhesin synthesis. *J Bacteriol*, **185**, 2879–86.

Dodson, K.W., Pinkner, J.S., et al. 2001. Structural basis of the interaction of the pyelonephritic *E. coli* adhesin to its human kidney receptor. *Cell*, **105**, 733–43.

Dunny, G.M. and Leonard, B.A.A. 1997. Cell–cell communication in gram-positive bacteria. *Annu Rev Microbiol*, **51**, 527–64.

Dziejman, M. and Mekalanos, J.J. 1994. Analysis of membrane protein interaction: ToxR can dimerize the amino terminus of phage lambda repressor. *Mol Microbiol*, **13**, 485–94.

Economou, A. 1999. Following the leader: bacterial protein export through the Sec pathway. *Trends Microbiol*, **7**, 315–20.

Economou, A. 2002. Bacterial secretome: the assembly manual and operating instructions. *Mol Membr Biol*, **19**, 159–69.

Falkow, S. 1988. Molecular Koch's postulates applied to microbial pathogenicity. *Rev Infect Dis*, **10**, S274–6.

Fang, F. and Vazquez-Torres, A. 2002. *Salmonella* selectively stops traffic. *Trends Microbiol*, **10**, 391–2.

Fraser, G.M. and Hughes, C. 1999. Swarming motility. *Curr Opin Microbiol*, **2**, 630–5.

Fukuda, A., Matsuyama, S., et al. 2002. Aminoacylation of the N-terminal cysteine is essential for Lol-dependent release of lipoproteins from membranes but does not depend on lipoprotein sorting signals. *J Biol Chem*, **277**, 43512–18.

Fuqua, C. and Greenberg, E.P. 2002. Listening in on bacteria: acyl-homoserine lactone signalling. *Nat Rev Mol Cell Biol*, **3**, 685–95.

Fuqua, C., Parsek, M. and Greenberg, E.P. 2001. Regulation of gene expression by cell-to-cell communication: acylhomoserine lactone quorum sensing. *Annu Rev Genet*, **35**, 439–68.

Galmiche, A., Rassow, J., et al. 2000. The N-terminal 34 kDa fragment of *Helicobacter pylori* vacuolating cytotoxin targets mitochondria and induces cytochrome *c* release. *EMBO J*, **19**, 6361–70.

Ginocchio, C.C., Olmsted, S.B., et al. 1994. Contact with epithelial cells induces the formation of surface appendages on *Salmonella typhimurium*. *Cell*, **76**, 717–24.

Girón, J.A., Torres, A.G., et al. 2002. The flagella of enteropathogenic *Escherichia coli* mediate adherence to epithelial cells. *Mol Microbiol*, **44**, 361–79.

Goebel, W. and Kuhn, M. 2000. Bacterial replication in the host cell cytosol. *Curr Opin Microbiol*, **3**, 49–53.

Groisman, E.A. 2001. The pleiotropic two-component regulatory system PhoP-PhoQ. *J Bacteriol*, **183**, 1835–42.

Hackstadt, T. 1998. The diverse habitats of obligate intracellular parasites. *Curr Opin Microbiol*, **1**, 82–7.

Hall-Stoodley, L. and Stoodley, P. 2002. Developmental regulation of microbial biofilms. *Curr Opin Biotechnol*, **13**, 228–33.

Hantke, K. 2001. Iron and metal regulation in bacteria. *Curr Opin Microbiol*, **4**, 172–7.

Heine, H. and Lien, E. 2003. Toll-like receptors and their function in innate and adaptive immunity. *Int Arch Allergy Immunol*, **130**, 180–92.

Hendrixson, D.R., Qiu, J., et al. 2003. Human milk lactoferrin is a serine protease that cleaves *Haemophilus* surface proteins at arginine-rich sites. *Mol Microbiol*, **47**, 607–17.

Hemmi, H., Takeuchi, O., et al. 2000. A toll-like receptor recognizes bacterial DNA. *Nature*, **408**, 740–5.

Henderson, I.R. and Nataro, J.P. 2001. Virulence functions of autotransporter proteins. *Infect Immun*, **69**, 1231–43.

Henderson, I.R., Navarro-Garcia, F. and Nataro, J.P. 1998. The great escape: structure and function of the autotransporter proteins. *Trends Microbiol*, **6**, 370–8.

Henderson, I.R., Cappello, R. and Nataro, J.P. 2000. Autotransporter proteins, evolution and redefining protein secretion. *Trends Microbiol*, **8**, 529–32.

Higashi, H., Tsutsumi, R., et al. 2002. SHP-2 tyrosine phosphatase as an intracellular target of *Helicobacter pylori* CagA protein. *Science*, **295**, 683–6.

Hoch, J.A. and Silhavy, T.J. 1995. *Two-component signal transduction*. Washington, DC: American Society for Microbiology.

Holden, M.T.G., Chhabra, S.R., et al. 1999. Quorum-sensing cross talk: isolation and chemical characterization of cyclic dipeptides from *Pseudomonas aeruginosa* and other gram-negative bacteria. *Mol Microbiol*, **33**, 1254–66.

Hueck, C.J. 1998. Type III protein secretion systems in bacterial pathogens of animals and plants. *Microbiol Mol Biol Rev*, **62**, 379–433.

Israel, D.A., Salama, N., et al. 2001. *Helicobacter pylori* genetic diversity within the gastric niche of a single human host. *Proc Natl Acad Sci USA*, **98**, 14625–2630.

Jann, K. and Jann, B. 1992. Capsules of *Escherichia coli*, expression and biological significance. *Can J Microbiol*, **38**, 705–10.

Ji, G., Beavis, R. and Novick, R.P. 1997. Bacterial interference caused by autoinducing peptide variants. *Science*, **276**, 2027–30.

Jiang, Q., Hiratsuka, K. and Taylor, D.E. 1996. Variability of gene order in different *Helicobacter pylori* strains contributes to genome diversity. *Mol Microbiol*, **20**, 833–42.

Josenhans, C. and Suerbaum, S. 2002. The role of motility as a virulence factor in bacteria. *Int J Med Microbiol*, **291**, 605–14.

Joyce, E.A., Chan, K., et al. 2002. Redefining bacterial populations: A post-genomic reformation. *Nat Rev Genet*, **3**, 462–73.

Kenny, B. 2002. Enteropathogenic *Escherichia coli* (EPEC) – a crafty subversive little bug. *Microbiology*, **148**, 1967–78.

Kimura, M., Goto, S., et al. 1999. Vacuolating cytotoxin purified from *Helicobacter pylori* causes mitochondrial damage in human gastric cells. *Microb Pathog*, **26**, 45–52.

Kjelleberg, S. and Molin, S. 2002. Is there a role for quorum sensing signals in bacterial biofilms? *Curr Opin Microbiol*, **5**, 254–8.

Kuck, D., Kolmerer, B., et al. 2001. Vacuolating cytotoxin of *Helicobacter pylori* induces apoptosis in the human gastric epithelial cell line AGS. *Infect Immun*, **69**, 5080–7.

Lamont, I.L., Beare, P.A., et al. 2002. Siderophore-mediated signaling mediates virulence factor production in *Pseudomonas aeruginosa*. *Proc Natl Acad Sci USA*, **99**, 7072–7.

Lawrence, R.N., Dunn, W.R., et al. 1999. The *Pseudomonas aeruginosa* quorum sensing signal molecule, N-(3-oxododecanoyl)-homoserine lactone inhibits porcine arterial smooth muscle contraction. *Br J Pharmacol*, **128**, 845–8.

Lahiri, S.S. 2000. Bacterial toxins – an overview. *J Nat Toxins*, **9**, 381–408.

Llewelyn, M. and Cohen, J. 2002. Superantigens: microbial agents that corrupt immunity. *Lancet Infect Dis*, **2**, 156–62.

Lupetti, P., Heuser, J.E., et al. 1996. Oligomeric and subunit structure of the *Helicobacter pylori* vacuolating cytotoxin. *J Cell Biol*, **133**, 801–7.

Masuda, K., Matsuyama, S. and Tokuda, H. 2002. Elucidation of the function of lipoprotein-sorting signals that determine membrane localization. *Proc Natl Acad Sci USA*, **99**, 7390–5.

Mattick, J.S. 2002. Type IV pili and twitching motility. *Annu Rev Microbiol*, **56**, 289–314.

Mayville, P., Ji, G., et al. 1999. Structure-activity analysis of synthetic autoinducing thiolactone peptides from *Staphylococcus aureus* responsible for virulence. *Proc Natl Acad Sci USA*, **96**, 1218–23.

McDowell, P., Affas, Z., et al. 2001. Structure, activity and evolution of the group I thiolactone peptide quorum-sensing system of *Staphylococcus aureus*. *Mol Microbiol*, **41**, 503–12.

Merrell, D.S. and Camilli, A. 2000. Detection and analysis of gene expression during infection by in vivo expression technology. *Phil Trans R Soc Lond B*, **355**, 587–99.

Miller, M.B., Skorupski, K., et al. 2002. Parallel quorum sensing systems converge to regulate virulence in *Vibrio cholerae*. *Cell*, **110**, 303–14.

Milton, D., Chalker, V., et al. 2001. The LuxM homologue, VanM from *Vibrio anguillarum* directs the synthesis of N-(3-hydroxyhexanoyl)homoserine lactone and N-hexanoylhomoserine lactone. *J Bacteriol*, **183**, 3537–47.

Missiakas, D., Betton, J.M. and Raina, S. 1996. New components of protein folding in extracytoplasmic compartments of *Escherichia coli* SurA, FkpA and Skp/OmpH. *Mol Microbiol*, **21**, 871–84.

Molinari, M., Galli, C., et al. 1997. Vacuoles induced by *Helicobacter pylori* toxin contain both late endosomal and lysosomal markers. *J Biol Chem*, **272**, 25339–44.

Molinari, M., Salio, M., et al. 1998a. Selective inhibition of Li-dependent antigen presentation by *Helicobacter pylori* toxin VacA. *J Exp Med*, **187**, 135–40.

Molinari, M., Galli, C., et al. 1998b. The acid activation of *Helicobacter pylori* toxin VacA: Structural and membrane binding studies. *Biochem Biophys Res Comm*, **248**, 334–40.

Moxon, E.R., Rainey, P.B., et al. 1994. Adaptive evolution of highly mutable loci in pathogenic bacteria. *Curr Biol*, **4**, 24–33.

Novick, R.P. and Muir, T.W. 1999. Virulence gene regulation by peptides in staphylococci and other gram-positive bacteria. *Curr Opin Microbiol*, **2**, 40–5.

Odenbreit, S., Puls, J., et al. 2000. Translocation of *Helicobacter pylori* CagA into gastric epithelial cells by type IV secretion. *Science*, **287**, 1497–500.

Olsén, A., Jonsson, A. and Normark, S. 1989. Fibronection binding mediated by a novel class of surface organelles on *Escherichia coli*. *Nature*, **338**, 652–5.

O'Riordan, M. and Portnoy, D.A. 2002. The host cytosol: front line or home front? *Trends Microbiol*, **10**, 361–4.

O'Toole, G.A. 2003. To build a biofilm. *J Bacteriol*, **185**, 2687–9.

Otto, M. 2001. *Staphylococcus aureus* and *Staphylococcus epidermidis* peptide pheromones produced by the accessory gene regulator *agr* system. *Peptides*, **22**, 1603–8.

Palmer, T. and Berks, B.C. 2003. Moving folded proteins across the bacterial cell membrane. *Microbiology*, **149**, 547–56.

Papini, E., Satin, B., et al. 1998. *Helicobacter pylori* vacuolating cytotoxin increases the permeability of polarized epithelial cell monolayers. *J Clin Invest*, **102**, 813–20.

Parkinson, J.S. 2003. Bacterial chemotaxis: a new player in response regulator dephosphorylation. *J Bacteriol*, **185**, 1492–4.

Pearson, J.P., Gray, J.M., et al. 1994. Structure of the autoinducer required for expression of *Pseudomonas aeruginosa* virulence genes. *Proc Natl Acad Sci USA*, **91**, 197–201.

Peek, R.M. Jr., Thompson, S.A., et al. 1998. Adherence to gastric epithelial cells induces expression of a *Helicobacter pylori* gene, *iceA*, that is associated with clinical outcome. *Proc Assoc Am Phys*, **110**, 531–44.

Pelicic, V., Reyrat, J.-M., et al. 1999. *Helicobacter pylori* VacA cytotoxin associated with the bacteria increases epithelial permeability independently of its vacuolating activity. *Microbiology*, **145**, 2043–50.

Pesci, E.C., Milbank, J.B., et al. 1999. Quinolone signaling in the cell-to-cell communication system of *Pseudomonas aeruginosa*. *Proc Natl Acad Sci USA*, **96**, 11229–34.

Pieters, J. and Gatfield, J. 2002. Hijacking the host: survival of pathogenic mycobacteria inside macrophages. *Trends Microbiol*, **10**, 142–6.

Pride, D.T., Meinersmann, R.J. and Blaser, M.J. 2001. Allelic variation within *Helicobacter pylori babA* and *babB*. *Infect Immun*, **69**, 1160–71.

Pugsley, A.P. 1993. The complete general secretory pathway in gram-negative bacteria. *Microbiol Rev*, **57**, 50–108.

Raetz, C.R.H. and Whitfield, C. 2002. Lipopolysaccharide endotoxins. *Annu Rev Biochem*, **71**, 635–700.

Redfield, R.J. 2002. Is quorum sensing a side effect of diffusion sensing? *Trends Microbiol*, **10**, 365–70.

Relman, D.A. and Falkow, S. 2001. The meaning and impact of the human genome sequence for microbiology. *Trends Microbiol*, **9**, 206–8.

Sara, M. and Sleytr, U. 2000. S-layer proteins. *J Bacteriol*, **182**, 859–68.

Sauer, K., Camper, A.K., et al. 2002. *Pseudomonas aeruginosa* displays multiple phenotypes during development as a biofilm. *J Bacteriol*, **184**, 1140–54.

Schmitt, W. and Haas, R. 1994. Genetic analysis of the *Helicobacter pylori* vacuolating cytotoxin: structural similarities with the IgA protease type of exported protein. *Mol Microbiol*, **12**, 307–19.

Segal, E.D., Cha, J., et al. 1999. Altered states: involvement of phosphorylated CagA in the induction of host cellular growth changes by *Helicobacter pylori*. *Proc Natl Acad Sci USA*, **96**, 14559–64.

Selbach, M., Moese, S., et al. 2002. Src is the kinase of the *Helicobacter pylori* CagA protein *in vitro* and *in vivo*. *J Biol Chem*, **277**, 6775–8.

Sijbrandi, R., Urbanus, M.L., et al. 2003. Signal recognition particle (SRP)-mediated targeting and Sec-dependent translocation of an extracellular *Escherichia coli* protein. *J Biol Chem*, **278**, 7, 4654–9.

Singh, P.K., Parsek, M.R., et al. 2002. A component of innate immunity prevents bacterial biofilm development. *Nature*, **417**, 552–5.

Smith, R.S., Fedyk, E.R., et al. 2001. IL-8 production in human lung fibroblasts and epithelial cells activated by the Pseudomonas autoinducer N-3-oxododecanoyl homoserine lactone is transcriptionally regulated by NF-kB and activator protein-2. *J Immunol*, **167**, 366–74.

Smyth, C.J., Marron, M.B., et al. 1996. Fimbrial adhesisns: similarities and variations in structure and biogenesis. *FEMS Immunol Med Microbiol*, **16**, 127–39.

Stathopoulos, C., Hendrixson, D.R., et al. 2000. Secretion of virulence determinants by the general secretory pathway in gram-negative pathogens: an evolving story. *Microb Infect*, **2**, 1061–72.

Stein, M., Rappuoli, R. and Covacci, A. 2000. Tyrosine phosphorylation of the *Helicobacter pylori* CagA antigen after cag-driven host cell translocation. *Proc Natl Acad Sci USA*, **97**, 1263–8.

Stein, M., Bagnoli, F., et al. 2002. c-Src/Lyn kinases activate *Helicobacter pylori* CagA through tyrosine phosphorylation of the EPIYA motifs. *Mol Microbiol*, **43**, 971–80.

Stephenson, K. and Hoch, J.A. 2002. Two-component and phosphorelay signal-transduction systems as therapeutic targets. *Curr Opin Pharmacol*, **2**, 507–12.

Stockbauer, K.E., Fuchslocker, B., et al. 2001. Identification and characterization of BipA, a *Bordetella* Bvg-intermediate phase protein. *Mol Microbiol*, **39**, 65–78.

Suerbaum, S., Smith, J.M., et al. 1998. Free recombination within *Helicobacter pylori*. *Proc Natl Acad Sci USA*, **95**, 12619–24.

Swift, S., Downie, J.A., et al. 2001. Quorum sensing as a population-density-dependent determinant of bacterial physiology. *Adv Microb Pysiol*, **45**, 199–270.

Tang, C.M., Bakshi, S. and Sun, Y.H. 2001. Identification of bacterial genes required for in vivo survival. *J Pharm Pharmacol*, **53**, 1575–9.

Taylor, D.E., Eaton, M., et al. 1992. Construction of a *Helicobacter pylori* genome map and demonstration of diversity at the genome level. *J Bacteriol*, **174**, 6800–6.

Telford, J.L., Ghiara, P., et al. 1994. Gene structure of the *Helicobacter pylori* cytotoxin and evidence of its key role in gastric disease. *J Exp Med*, **179**, 1653–8.

Telford, G., Wheeler, D., et al. 1998. The *Pseudomonas aeruginosa* quorum sensing signal molecule, N-(3-oxododecanoyl)-homoserine lactone has immunomodulatory activity. *Infect Immun*, **66**, 36–42.

Thanassi, D.G. and Hultgren, S.J. 2000. Multiple pathways allow protein secretion across the bacterial outer membrane. *Curr Opin Cell Biol*, **12**, 420–30.

Thanassi, D.G., Saulino, E.T. and Hultgren, S.J. 1998. The chaperone/usher pathway: a major terminal branch of the general secretory pathway. *Curr Opin Microbiol*, **1**, 223–31.

Throup, J.P., Zappacosta, F., et al. 2001. The *srhSR* gene pair from *Staphylococcus aureus*: genomic and proteomic approaches to the identification and characterization of gene function. *Biochemistry*, **40**, 10392–401.

Tobiason, D.M. and Seifert, H.S. 2001. Inverse relationship between pilus-mediated gonococcal adherence and surface expression of the pilus receptor, CD46. *Microbiology*, **147**, 2333–40.

Tobe, T. and Sasakawa, C. 2001. Role of bundle-forming pilus of enteropathogenic *Escherichia coli* in host cell adherence and in microcolony development. *Cell Microbiol*, **3**, 579–85.

Underhill, D.M., Ozinsky, A., et al. 1999. The toll-like receptor 2 is recruited to macrophage phagosomes and discriminates between pathogens. *Nature*, **402**, 39–43.

Vallance, B.A. and Finlay, B.B. 2000. Exploitation of host cells by enteropathogenic *Escherichia coli*. *Proc Natl Acad Sci USA*, **97**, 8799–806.

Van Doorn, L.-J., Figueiredo, C., et al. 1998. Expanding allelic diversity of *Helicobacter pylori vacA*. *J Clin Microbiol*, **36**, 2597–603.

Van Loosdrecht, M.C.M., Picioreanu, C. and Heinen, J.J. 1997. A more unifying hypothesis for biofilm structures. *FEMS Microbiol Ecol*, **24**, 181–3.

Van Wely, K.H., Swaving, J., et al. 2001. Translocation of proteins across the cell envelope of gram-positive bacteria. *FEMS Microbiol Rev*, **25**, 437–54.

Voulhoux, R., Bos, M.P., et al. 2003. Role of a highly conserved bacterial protein in outer membrane protein assembly. *Science*, **299**, 262–5.

Watnick, P. and Kolter, R. 2000. Biofilm, city of microbes. *J Bacteriol*, **182**, 2675–9.

Whitehead, N.A., Barnard, A.M., et al. 2001. Quorum-sensing in gram-negative bacteria. *FEMS Microbiol Rev*, **25**, 365–404.

Whiteley, M., Lee, K.M. and Greenberg, E.P. 1999. Identification of genes controlled by quorum sensing in *Pseudomonas aeruginosa*. *Proc Natl Acad Sci USA*, **96**, 13904–9.

Wimpenny, J.W.T. and Colasanti, R. 1997. A unifying hypothesis for the structure of microbial biofilms based on cellular automation models. *FEMS Microbiol Ecol*, **22**, 1–16.

Winson, M.K., Cámara, M., et al. 1995. Multiple N-acyl-homoserine lactone signal molecules regulate production of virulence determinants and secondary metabolites in *Pseudomonas aeruginosa*. *Proc Natl Acad Sci USA*, **92**, 9427–31.

Winzer, K. and Williams, P. 2001. Quorum sensing and the regulation of virulence gene expression in pathogenic bacteria. *Int J Med Microbiol*, **291**, 131–43.

Winzer, K., Hardie, K.R., et al. 2002a. LuxS: Its role in central metabolism and the *in vitro* synthesis of 4-Hydroxy-5-methyl-3(2H)-furanone. *Microbiology*, **148**, 909–22.

Winzer, K., Hardie, K.R. and Williams, P. 2002b. Bacterial cell-to-cell communication: sorry, can't talk now – gone to lunch! *Curr Opin Microbiol*, **5**, 216–22.

Williams, P. 2002. Quorum sensing: and emerging target for antimicrobial chemotherapy? *Exp Opin Ther Targ*, **6**, 257–74.

Williams, P., Camara, M., et al. 2000. Quorum sensing and the population-dependent control of virulence. *Philos Trans R Soc Lond B Biol Sci*, **355**, 667–80.

Withers, H., Swift, S. and Williams, P. 2001. Quorum sensing as an integral component of gene regulatory networks in gram-negative bacteria. *Curr Opin Microbiol*, **4**, 186–93.

Yang, F.L. and Braun, V. 2000. ShlB mutants of *Serratia marcescens* allow uncoupling of activation and secretion of the ShlA hemolysin. *Int J Med Microbiol*, **290**, 529–38.

Zhou, D. and Galán, J. 2001. Salmonella entry into host cells: the work in concert of type III secreted effector proteins. *Microb Infect*, **3**, 1293–8.

Zhu, H., Thuruthyil, S.J., et al. 2002. Contribution of quorum sensing systems to the virulence of *Pseudomonas aeruginosa* during corneal infections. *Invest Ophthalmol Vis Sci*, **42**, S514.

PART II

GENERAL ECOSYSTEMS

7

Airborne bacteria

LINDA D. STETZENBACH

INTRODUCTION

Bacteria are ubiquitous in the environment, capable of growing on virtually any surface and in most liquids when nutritional and environmental factors are favorable. Their presence in the air is the result of dispersal from a site of colonization and growth. This chapter will review the study of bacteria in the air that impact human health and the environment.

BIOAEROSOLS

A bioaerosol is a collection of airborne biological material consisting of cells, cellular fragments, and by-products of metabolism. This material can be present as particulate, liquid, and volatile organic compounds, and it may have biological activity on other organisms (Cox and Wathes 1995). Bacterial aerosols are generated in a variety of indoor environments and facilities and as a result of a variety of activities (Table 7.1), including dispersal from heating, ventilation, and air-conditioning (HVAC) systems, water-spray devices (e.g. shower-heads, humidifiers), and flooring (Lighthart and Stetzenbach 1994). Bacterial aerosols are also generated by medical (Demers 2001; Nottrebart 1980) and dental (Depaola et al. 2002; Leggat and Kedjarune 2001) procedures, and patient (Shiomori et al. 2001, 2002) and animal care (Kaliste et al. 2002; Kowalski et al. 2002) practices, prompting cautionary measures to protect patients and staff from dispersed airborne bacteria. Talking and coughing generate bioaerosols from building occupants with droplet concentrations of 10^4–10^5 droplets/m^3 of air (Papineni and Rosenthal 1997). Manufacturing practices and biofermentation procedures can also generate bacterial aerosols (Abrams et al. 2000). Outdoors, bacterial aerosols are generated as a result of spray irrigation, wastewater treatment activity, wave splash, cooling towers, air-handling water-spray systems, and agricultural processes (Brandi et al. 2000; Burkhart et al. 1993; Hameed and Khodr 2001; Lighthart and Stetzenbach 1994; Lighthart and Mohr 1997; Pillai and Ricke 2002).

The particulate in a bioaerosol is generally 0.3–100 μm in diameter (Cox and Wathes 1995). Single bacterial cells range in size from 0.5 to 2.0 μm and are commonly spheres (cocci), rods (bacilli), or spirals. However, airborne bacterial cells are often present as aggregate formations of larger particles (Lighthart 1994). Aggregates of cells have different aerodynamic properties compared with single cells and they afford protection of individual cells from environmental stresses such as dessication, ultraviolet irradiation, and exposure to ozone in the atmosphere. Cell aggregates surrounded by a thin layer of water are commonly the result of dispersal from water sources by splash, rainfall, or mechanical disturbance (e.g. cooling towers, fountains). This often aids in the survivability of the cells while airborne. Airborne bacterial cells may also be associated with skin cells, dust, and other organic or inorganic material in a relationship that is commonly referred to as 'rafting.' Rafting cells are often found in bioaerosols generated from agricultural settings (e.g. during harvesting, tilling) and indoor environments. This formation affects the aerodynamic characteristics and the

Table 7.1 *Published concentrations and populations of airborne bacteria and bacterial components outdoors and in indoor environments*

Facility/activity	Organism or agent	Concentration	Reference(s)
Animal facilities	*Bacillus* species	10^1–10^3 CFU/m3	Wilson et al. 2002
	Corynebacterium species	10^0–10^3 CFU/m3	Wilson et al. 2002
	Culturable bacteria	10^4 CFU/m^3	Blom et al. 1984
		10^5–10^6 CFU/m^3	Chang et al. 2001
		10^3–10^5 CFU/m^3	Curtis et al. 1978
	Endotoxin		Duchaine et al. 2000
			Thorne et al. 1992
		100–500 pg/m^3	Milton et al. 1990
		0.3–41 ng/mg	Andersson et al. 1999
		10^1–10^4 EU/m^3	Duchaine et al. 2000
			Kullman et al. 1999
	Micrococcus luteus	10^1–10^3 CFU/m^3	Wilson et al. 2002
Clean rooms	Viable particles	10^0–10^2 CFU/m^3	Favero et al. 1966
Composting – agricultural or residential	Culturable bacteria	10^3–10^4 CFU/m^3	Hryhorczuk et al. 2001
		10^4–10^6 CFU/m3	Durand et al. 2002
	Endotoxin	0.12–0.41 ng/m^3	Hryhorczuk et al. 2001
	Gram-negative bacilli	10^3 CFU/m^3	Clark et al. 1983
Composting – sludge	Coliforms	<dl	Jones and Cookson 1983
	Culturable bacteria	10^0–10^3 CFU/m^3	Jones and Cookson 1983
Daycare center	Endotoxin	0.43 ng/m^3	Rylander et al. 1992
		5.7 ng/m^3	Wan and Li 1999
Effluent irrigation	Culturable bacteria	10^3 CFU/m^3	Applebaum et al. 1984
	Escherichia coli	10^1–10^3 CFU/m^3	Applebaum et al. 1984
	Fecal coliforms	10^2 CFU/m^3	Teltsch and Katzenelson 1978
	Fecal streptococci	10^2 CFU/m^3	Applebaum et al. 1984
	Total coliforms	10^2 CFU/m^3	Applebaum et al. 1984
Farming, harvesting, bailing, grain storage	Culturable bacteria	10^1–10^4 CFU/m^3	Lighthart 1994
			Swan and Crook 1998
		10^4–10^6 CFU/m^3	Hameed and Khodr 2001
	Endotoxin	10^2–10^3 EU/m^3	Buchan et al. 2002
	Gram-negative bacteria	10^3–10^5 CFU/m^3	Hameed and Khodr 2001
Food processing/ food storage	Endotoxin	0.0125–54.9 Φg/m^3	Dutkiewicz et al. 2000
		0.2–2681.0 Φg/m^3	Dutkiewicz et al. 2001b
		0.011–1893.9 Φg/m^3	Dutkiewicz et al. 2002
Heating, ventilation, and air-conditioning (HVAC) systems	Culturable bacteria	10^2–10^3 CFU/m^3	Hugenholtz and Fuerst 1992
		10^2–10^7 CFU/cm^2	Flaherty et al. 1984
	Endotoxin	NQ	Hugenholtz and Fuerst 1992
Hot-water heaters, hot water systems	*Legionella* sp.	10^0–10^1 CFU/m^3	Bollin et al. 1985
	L. pneumophila	NQ	Helms et al. 1983
Humans (skin fragments)	Culturable bacteria	10^2–10^3 CFU/m^3	Lundholm 1982
Humidifiers (portable)	*L. pneumophila*	10^2–10^4 CFU/ml	Zuravleff et al. 1983
	Pseudomonas	10^2–10^4 CFU/h	Couvelli et al. 1973
Manufacturing	Culturable bacteria	10^6 CFU/m^3	Travers Glass et al. 1991
	Gram-negative bacilli	10^3 CFU/m^3	Lundholm 1982
		10^2–10^4 CFU/m^3	Prazmo et al. 2000
	Endotoxin	10^1–10^3 EU/m^3	Dutkiewicz et al. 2001a
		16.4–234 EU/m^3	Abrams et al. 2000
	Total bacteria	0.02–3.6 ug/m^3	Walters et al. 1994
		10^1–10^3 CFU/m^3	Walters et al. 1994

(Continued over)

Table 7.1 *Published concentrations and populations of airborne bacteria and bacterial components outdoors and in indoor environments (Continued)*

Facility/activity	Organism or agent	Concentration	Reference(s)
Mortuary, necropsies	Coliforms	10^0–10^2 CFU/m^3	Newson et al. 1983
	Gram-positive cocci	10^1–10^2 CFU/m^3	Newson et al. 1983
Office buildings	Culturable bacteria	>10^5 CFU/m^3	Reynolds et al. 2001
	Endotoxin	3.7 ng/m^3	Wan and Li 1999
		0.5–3.0 EU/m^3	Reynolds et al. 2001
Operating room	Culturable bacteria	10^2 CFU/m^3	Tjade and Gabor 1980
	Particles	10^2–10^5 particles/m^3	Seal and Clark 1990
Post office	Endotoxin	0.19 ng/m^3	Rylander et al. 1992
Recycling plant	Culturable bacteria	10^5 CFU/m^3	Reinthaler et al. 1999
Residential	Endotoxin	0.03–5.4 ng/m^3	Gorny and Dutkiewicz 2002
	Gram-negative	<dl–10^2 CFU/m3	Gorny and Dutkiewicz 2002
	Gram-positive	10^2–10^3 CFU/m3	Gorny and Dutkiewicz 2002
Sanitary landfill	Culturable	10^1–10^4 CFU/m^3	Rahkonen et al. 1987
	Total coliforms	<dl–10^3 CFU/m^3	Rahkonen et al. 1987
	Fecal streptococci	<dl–10^4 CFU/m^3	Rahkonen et al. 1987
Schools	Culturable bacteria	7–10^4 CFU/m3	Daisey et al. 2003
	Endotoxin	0.21–0.26 ng/m^3	Rylander et al. 1992
Waste handling	Culturable bacteria	10^4 CFU/m^3	Kiviranta et al. 1999
		10^4–10^5 CFU/m^3	Neumann et al. 2002
	Endotoxin	<10–50 EU/m^3	Neumann et al. 2002
	Gram-negative bacilli	10^3 CFU/m^3	Kiviranta et al. 1999
Wastewater treatment – activated sludge	Coliforms	0.27–5.17 CFU/m^3	Fannin et al. 1985
	Culturable bacteria	10^2 CFU/m^2/sec	Kenline and Scarpino 1972
		10^2 CFU/m^3	Haas et al. 2002
	Endotoxin	0.1–350 ng/m^3	Laitinen et al. 1994
	Escherichia coli	10^2 CFU/m^3	Ranalli et al. 2000
	Gram-negative bacilli	10^1–10^5 CFU/m^3	Laitinen et al. 1994
Wastewater treatment – aeration tanks	Culturable bacteria	10^1–10^3 CFU/m^3	Brenner et al. 1988
			Brandi et al. 2000
			Crawford and Jones 1979
			Keline and Scarpino 1972
	Fecal streptococci	10^2 CFU/m^3	Crawford and Jones 1979
	Streptococcus faecalis	<dl–10^2 CFU/m^3	Crawford and Jones 1979
Wastewater treatment – trickling filter	Coliforms	10^2–10^4 CFU/m^3	Adams and Spendlove 1970
	Endotoxin	<dl–185 ng/m^3	Thorn et al. 2002
	Total bacteria	10^1–10^3 CFU/m^3	Adams and Spendlove 1970

dl = less than detection limit; NQ = not quantitated.

survival of the cells in the bioaerosol (Lighthart and Stetzenbach 1994).

Airborne particulate will remain airborne until settling on to surfaces occurs. Bioaerosols ranging in size from 1.0 to 5.0 μm generally follow the air stream while larger particles are more readily deposited on surfaces (Mohr 2002). The settling of the particles is affected by physical and environmental factors. Air currents, relative humidity, and temperature are the most important environmental parameters while the size, density, and shape of the cell or bacteria-containing particle are the most significant physical parameters (Mohr 2002;

Pedgley 1991). These parameters also affect the survivability of the cells and their ability to colonize surfaces or liquid when deposited.

Relative humidity is the most important factor in the survival of bacterial cells with low humidity resulting in desiccation. The bacterial cell membrane is adversely affected by loss of water, resulting in loss of viability (Cox 1987; Israeli et al. 1994), and low relative humidity has been shown to increase the resistance of some bacteria to ultraviolet irradiation (Ko et al. 2000; Mohr 2002). Increased temperatures also decrease the viability of airborne cells (Ko et al. 2000; Mohr 2002).

Vegetative bacterial cells are more susceptible to environmental stresses than bacterial endospores and vegetative gram-negative bacilli are more susceptible than gram-positive cocci. The greater resistance of vegetative gram-positive cocci is due in part to the presence of pigment and the composition of their cell wall. Carotenoid pigments and photoreactivation mechanisms in gram-positive cocci afford protection from sunlight (Henis 1987). The cell wall of gram-negative bacteria is actually an outer membrane composed of mucopeptides linked to lipopolysaccharide with a thin outermost layer of peptidoglycan that is only 5–10 percent of the structure. When exposed to heat, the lipopolysaccharide may be released resulting in injury and/or death to the cell. Gram-positive cells have a peptidoglycan layer that comprises up to 80 percent of the cell wall, affording more protection from drying and heat stress (Marthi 1994). The integrity of the ribosomes in various bacterial genera also contributes to recovery of heat-injured cells (Henis 1987). Several models have been presented to explain the complex interactions of physical, environmental, and physiological factors of bacterial survival in aerosols (Mohr 2002).

Bacterial aerosols can elicit adverse human health effects, including infection, allergic reaction, inflammation, and respiratory disease. The primary route of exposure to aerosols is inhalation. Approximately 10 m^3 of air per day are inhaled by the average human (Lynch and Poole 1979), but some ingestion and dermal exposure may also occur. Large airborne particles are lodged in the upper respiratory tract (nose and nasopharynx) and particles <6 μm in diameter are transported to the lung. One- and two-micrometer-sized particles have the greatest retention in the alveoli of the lungs.

Mycobacterium tuberculosis and *Legionella pneumophila* are recognized as respiratory pathogens that are capable of causing severe disease, but aerosol transmission of other bacteria can also result in illness. Allergic reactions and irritant responses to the foreign proteins of bacterial cells and fragments, and reactions to by-products of bacterial metabolism can result in adverse health. Inflammatory responses (Liu et al. 2000) caused by bacterial fragments present in airborne dust and allergic reactivity to *Pseudomonas* in wood dust (Skorska et al. 2002) have been reported. Disease transmission between farm animals via aerosols is also documented (Brockmeier and Lager 2002).

AIRBORNE BACTERIA ASSOCIATED WITH ADVERSE HUMAN HEALTH EFFECTS

Mycobacterium species

The dispersal of airborne infectious *Mycobacterium tuberculosis* (Kaufmann and van Embden 1993),

nontuberculosis mycobacteria (Contreras et al. 1988), and *M. leprae* (Bryceson and Pflatzgraft 1990) are associated with severe illness. The transmission of *M. tuberculosis* via bioaerosols from infected people occurs during coughing, talking, speaking, laughing, and sneezing (Rubin 1991), and exposure to infectious aerosols has been reported during air travel (Driver et al. 1994). Nosocomial *M. tuberculosis* infections have been reported resulting from aerosols generated by the operation of suctioning instruments during treatment, manipulation of tuberculosis ulcers, drainage of necrotic tissue, and exposure during autopsy (Nottrebart 1980). Installation of ultraviolet lights in areas of increased risk of exposure to *M. tuberculosis* and other infectious bioaerosols has been shown to be effective, but efficacy was affected by the number of lights and the air exchange rate and presence of air currents in the area being treated (Ko et al. 2002).

Laboratory experiments with *Mycobacterium terrae* isolated from a water damaged building demonstrated inflammatory responses in mouse lung (Jussila et al. 2002) and mycobacteria of the *M. avium* complex commonly isolated from soil, water, and the air are associated with respiratory infection in immunocompromised patients (Wolinsky 1979). Discharge of 10^8 *M. leprae* bacilli has been reported from the nose blow of an untreated person with leptomatous Hansen's disease (leprosy) (Stewart-Tull 1982).

Legionella

Legionella pneumophila is a vegetative gram-negative bacterium that is ubiquitous in freshwater environments worldwide (Fliermans et al. 1981; Steinert et al. 2002). Exposure to *L. pneumophila* via inhalation of the bacterium in contaminated aerosols and the replication of *L. pneumophila* in macrophages results in severe pneumonia termed legionnaires' disease. The first reporting of this disease followed the outbreak of respiratory illness among American Legion conventioneers in Philadelphia in 1976 (Fraser et al. 1977). Since that time an estimated 10 000–15 000 cases per year are reported in the USA (Marston et al., 1993). Nineteen other *Legionella* species are also human pathogens, often causing disease in immunosuppressed patients (Muder and Yu 2002). Pontiac fever is a mild flu-like syndrome attributed to the inhalation of *Legionella* that cannot replicate in host tissue (Fields 2002; Kaufman et al. 1981).

Fraser (1984) developed a framework for causation of a legionnaires' outbreak that detailed six factors that must occur. These factors are: (1) a reservoir of legionella must be present; (2) amplifying factors must occur to permit the concentration of legionella to increase; (3) the legionella must be dispersed resulting in human exposure; (4) the legionella must be virulent to humans;

(5) the legionella must be delivered to the appropriate site in the human host; and (6) the host must be susceptible to infection by the legionella. Although the organism is present in most freshwaters, it survives as an intercellular parasite of protozoa, thereby resulting in protection for the bacterium.

Exposure to *Legionella*-contaminated aerosols occurs as the result of mechanical air-conditioning systems with water-spray components, cooling towers, and aerosol-generating water devices (e.g. showerheads, faucets, whirlpool tubs, decorative fountains, humidifiers), but no human-to-human transmission has been documented (Mohr 2002).

Other gram-negative bacteria

Airborne gram-negative bacteria are associated with animal facilities with *Escherichia coli*, *Pantoea* (*Enterobacter*) *agglomerans*, *Pseudomonas* spp., and *Acinetobacter* spp. commonly isolated in cow barns, pig houses, and poultry barns (Zucker et al. 2000). Wastewater treatment plants (Brandi et al. 2000; Brenner et al. 1988; Crawford and Jones 1979; Kenline and Scarpino 1972) and recycling facilities (Reinthaler et al. 1999) also disperse gram-negative bacteria into the air. Dutkiewicz et al. (2001b) reported skin-test reactivity of herb-processing workers to extracts of *Alcaligenes faecalis*. These bacteria can cause a variety of health problems, especially to the young, elderly, and immunocompromised.

Endotoxin

Endotoxins are biologically active materials that affect humoral and cellular host mediation systems (Bradley 1979) and affect many organ systems (Hewett and Roth 1993). Endotoxin is an integral component of the gram-negative bacterial cell wall that is released during active cellular growth and after cell lysis (Bradley 1979). It has also been reported that endotoxin is shed in membrane vesicles when the organisms are associated with organic dust (Dutkiewicz et al. 1992).

Exposure to airborne endotoxin is a concern in occupational settings with high concentrations of organic dusts (e.g. cotton-processing facilities, agricultural settings) and industrial environments with water-spray components (e.g. machining fluids) (Gordon 1992). Airborne endotoxin is also a concern in indoor environments with humidified mechanical air-conditioning systems (Dutkiewicz et al. 1988; Flaherty et al. 1984), and the presence of dogs, moisture, and settled dust increases the likelihood of airborne endotoxin in the home (Park et al. 2001).

Symptoms of chest tightness, cough, shortness of breath, fever, and wheezing have been reported by individuals exposed to airborne endotoxin in industrial and agricultural settings (Donham et al. 1984; Lundholm and Rylander 1980; Thelin et al. 1984). Kline et al. (1999)

report airflow obstruction and exacerbation of asthma related to airborne endotoxin concentration. Atopic and nonatopic asthma have been associated with endotoxin exposure because of its ability to induce inflammatory reactions (Michel 1999), and repeated inhalation exposure has been shown to induce allergen-specific airway inflammation (Wan et al. 2000). However, childhood exposure to endotoxin in a rural farm setting has been shown to provide some protection from the development of atopic childhood diseases (von Mutius et al. 2000)

Gram-positive nonspore-forming bacteria

Airborne *Staphylococcus* spp. and *Micrococcus* spp. are nonendospore-forming bacteria that are commonly disseminated from the skin, oral and nasal surfaces, and hair of humans (Favero et al. 1966). The presence of these cells in the air increases the likelihood of nosocomial infections in healthcare settings (Schall 1991). Antibiotic-resistant gram-positive bacteria in hospital environments are transmitted among patients through aerosol dispersal from inanimate surfaces, patients, and staff members (Shiomori et al. 2001). Airborne dispersal of pathogens is also a concern in home-care and assisted-living environments, where infection-control practices may not be strictly followed. Actinomycetes are filamentous nonendospore-forming gram-positive bacteria that have been associated with illness in occupants of water-damaged buildings (Hyvarinen et al. 2002). Their presence is reported in association with numerous fungi that colonize following water intrusion and colonization on building materials and furnishings, and concentrations of 10^4 CFU/m^3 have been reported during repair of damaged structures (Rautiala et al. 1996).

Bioterrorism

The threat of purposeful aerosol release of pathogenic bacteria and bacterial toxins as weapons of mass destruction (WMD) has increased the awareness of governmental agencies, the medical community, and the public of the importance of bioaerosols (Sheeran 2002). Several pathogenic bacteria have been listed as possible bioaerosol WMDs, including *Francisella tularensis*, *Yersinia pestis*, and *Bacillus anthracis*, the causative agents of tularemia, plague, and anthrax, respectively. An alert public-health system is needed to recognize these diseases and minimize exposure of the public (Dennis et al. 2001), and additional research is needed to evaluate the risk of exposed populations (Dull et al. 2002).

MONITORING FOR AIRBORNE BACTERIA

Routine monitoring for airborne bacteria is performed in pharmaceutical and food-industry settings, but monitoring surveys in indoor environments are generally conducted in response to: (1) physician documented illness linked to occupancy of building; (2) visual evidence of biological contamination; (3) occupant complaints after a water damage incident; or (4) occupant complaints that are suggestive of biological contamination. However, documentation of bacteria in the air or on surfaces does not prove that they are responsible for disease in building occupants. Serologic tests or other diagnostic methods may be needed to directly correlate exposure with disease.

Prior to conducting the actual collection of samples, a strategy for testing should be formulated (Burge 1995) and an understanding of the water intrusion/water damage history of the building is needed (Macher 1999). Air sampling is used to document concentrations of airborne contaminants and assess exposure (Macher 1999), but air sampling alone does not provide assurance that a suspect area is free of biological contamination (Higgins et al. 2003). Cells growing on surfaces are aerosolized and settled cells become reaerosolized during routine activity (Buttner and Stetzenbach 1993; Weis et al. 2002). Therefore, surface sampling should be performed in addition to air sampling to assist in determining areas of contamination and in identifying the source of biocontamination. Surface sampling is also used in determining the effectiveness of remediation (Andersson et al. 1997; Buttner et al. 2002; Higgins et al. 2003).

Collection of airborne bacteria involves forced airflow sampling (Buttner et al. 2002). This type of sampling provides a means to determine the concentration of organisms in the bioaerosol per unit volume of air, but the specific collection method and the analysis method(s) used will influence the results. Commercially available forced airflow impaction samplers for use in the collection of bacteria deposit airborne particulate on to a semisolid agar surface. Impingement samplers collect airborne cells and other particulate into a liquid. Both impaction on to agar plates and impingement have been widely used for the collection of airborne bacteria (Buttner et al. 2002). Single-stage impactor samplers operated at 28.3 l/min for 2 min provide a lower detection limit of 18 colony forming units per cubic meter of air sampled (CFU/m^3) and an upper limit of 10^4 CFU/m^3 although at the higher concentrations considerable error is introduced as multiple cells are deposited on the same location of the agar surface. Increased sampling time beyond 3–5 min with agar-based impactor samplers results in drying of the agar surface decreasing the physical collection of airborne particles and decreasing the viability of the bacterial cells (Buttner et al. 2002). Impingement samplers collect cells in a buffered liquid. Commonly used impingement samplers designed for the collection of bacteria are operated at 10–12 l/min, but high-velocity samplers have been developed that collected larger volumes of air over extended sampling times (Radosevich et al. 2002). Collection of airborne cells into buffer permits manipulation of the sample by dilution or concentration to maximize accuracy in quantitation and affords the application of a variety of analytical methods, including culture, microscopy, immunoassay, flow cytometry, and molecular methods (Buttner et al. 2002). However, as observed with the agar-based impactors, extended sampling times may result in increased sampling stress, thereby decreasing the viability of the collected bacteria (Alvarez et al. 1994). The use of filtration sampling for monitoring of bacterial aerosols also results in dessication of bacterial cells, so this technique is generally used for sampling of airborne dust, fungal spores, and pollen (Crook 1995b). The use of settling plates or gravity sampling is not recommended for determination of airborne bacteria as it does not result in a representative sampling of airborne cell due to differential settling of particles from the air (Crook 1995a).

A variety of sampling methods have been used for the collection of surface-associated bacteria. Agar-filled replicate organism detection and counting (RODAC) with a convex sampling surface are used in hospital infection control to validate the efficacy of disinfection of surfaces. RODAC plates and agar-filled surface sampling swatches or strips are commercially available for sampling of smooth, hard surfaces in indoor environments. Sterile swabs are used to sample irregular, nonporous surfaces and they have been used to sample hard surfaces (Buttner et al. 2001; Higgins et al. 2003). Collection of settled dust in porous (e.g. carpeting, upholstery, clothing) and nonporous materials (e.g. flooring, horizontal surfaces, and furnishings) is a means to sample large surface areas and provides material for a variety of analysis methods (Macher 2001). Bulk sampling is a destructive sampling method that necessitates the removal of material from the sample site and, therefore, is used when materials are to be replaced regardless of the outcome of the monitoring (e.g. water-damaged carpet and pad, water-damaged or stained wallboard and wallpaper, and stained ceiling tile). Material collected with vacuum or bulk sampling is processed with a buffer solution containing a surfactant to assist in the separation of microorganisms from the dust and other particulates in the sample.

The agar-based impaction sampling method is limited to culture analysis. The other sampling methods can be analyzed using culture, chemical assay, and/or molecular biology. Limitations of culture-based assay include the requirement that the bacterial cells be viable/culturable at the incubation conditions and on the growth medium selected. For example, a series of media amended with antibiotics and growth factors are needed for the isolation and identification of *Legionella pneumophila* (Fields 2002), and the isolation of *Mycobacterium* species requires incubation for extended periods of time.

Stressed organisms in a bioaerosol may also require resuscitation (Crozier-Dodson and Fung 2002).

Limulus amebocyte lysate assays are used for the analysis of endotoxin (Olenchock 1990), but variations in extraction efficiency and the protocols for analysis result in differences in the results reported (Reynolds et al. 2002). A variety of analytical chemistry methods and flow cytometry can be used to identify and enumerate airborne cells (Spurny 1995). Assays for muramic acid, peptidoglycan, or fatty acids are used to detect markers of bacterial cells and endotoxin (Laitinen et al. 2001; Liu et al. 2000). Polymerase chain reaction (PCR) amplification and quantitative polymerase chain reaction (QPCR) have been shown to provide enhanced detection and rapid identification of specific airborne and surface-associated bacteria (Alvarez et al. 1994; Buttner et al. 2001; Higgins et al. 2003; Pascual et al. 2001). The use of molecular techniques requires the development of primers and probes to unique gene sequences and optimization of protocols, but the increased sensitivity and specificity of these methods, and the rapid reporting of data, provide a means to enhance the capability to monitor for contaminant airborne and surface-associated bacteria in indoor environments.

SUMMARY

Human-source and environmental bacteria can colonize and grow on a variety of indoor and outdoor surfaces when conditions are appropriate. Dispersal of bioaerosols from these surfaces and resulting exposure to bacteria and bacterial components may cause adverse human health effects. The dispersal, transport, and settling of bioaerosols is determined by physical and environmental factors, but health effects depend on the bacterial agent(s) within the aerosol and characteristics of the exposed population.

Monitoring for bioaerosols requires selection of appropriate sampling and analysis methods to determine the concentration and population of airborne bacteria and their components. Monitoring in indoor environments should also include surface sampling to determine the source(s) of bacterial contaminants. New information on human exposure to bioaerosols is being gathered with the assistance of enhanced technologies for collection and analysis, and this knowledge should provide assistance in determining risk of exposure and in understanding the fate and transport of airborne bacteria.

REFERENCES

Abrams, L., Seixas, N., et al. 2000. Characterization of metalworking fluid exposure indices for a study of acute respiratory effects. *Appl Occup Environ Hyg*, **15**, 492–502.

Adams, A.P. and Spendlove, J.C. 1970. Coliform aerosols emitted by sewage treatment plants. *Science*, **169**, 1218–20.

Alvarez, A.J., Buttner, M.P., et al. 1994. The use of solid-phase polymerase chain reaction for the enhanced detection of airborne microorganisms. *Appl Environ Microbiol*, **60**, 374–6.

Andersson, M.A., Nikulin, M., et al. 1997. Bacteria, molds, and toxins in water-damaged building materials. *Appl Environ Microbiol*, **63**, 387–93.

Andersson, A.M., Weiss, N., et al. 1999. Dust-borne bacteria in animal sheds, schools and children's day care centers. *J Appl Microbiol*, **86**, 622–34.

Applebaum, J., Guttman-Bass, N., et al. 1984. Dispersion of aerosolized enteric viruses and bacteria by sprinkler irrigation with wastewater. *Monogr Virol*, **15**, 193–201.

Blom, J.Y., Madsen, E.B., et al. 1984. Numbers of airborne bacteria and fungi in calf houses. *Nordic Vet Med*, **36**, 215–20.

Bollin, G.E., Plouffe, J.F., et al. 1985. Aerosols containing *Legionella pneumophila* generated by shower heads and hot-water faucets. *Appl Environ Microbiol*, **50**, 1128-113, 1.

Bradley, S.G. 1979. Cellular and molecular mechanisms of action of bacterial endotoxins. *Annu Rev Microbiol*, **33**, 67–04.

Brandi, G., Sisti, M. and Amagliani, G. 2000. Evaluation of the environmental impact of microbial aerosols generated by wastewater treatment plants utilizing different aeration systems. *J Appl Microbiol*, **88**, 845–52.

Brenner, K.P., Scarpino, P.V. and Clark, C.S. 1988. Animal viruses, coliphage, and bacteria in aerosols and wastewater at a spray irrigation site. *Appl Environ Microbiol*, **54**, 409-41, 5.

Brockmeier, S.L. and Lager, K.M. 2002. Experimental airborne transmission of porcine reproductive and respiratory syndrome virus and *Bordetella bronchiseptica*. *Vet Microbiol*, **6**, 267–75.

Bryceson, A. and Pflatzgraft, R.E. 1990. *Leprosy*. Churchill Livingstone: London.

Buchan, R.M., Rijal, P., et al. 2002. Evaluation of airborne dust and endotoxin in corn storage and processing facilities in Colorado. *Int J Occup Med Environ Health*, **15**, 57–64.

Burge, H.A. 1995. Bioaerosol investigations. In: Burge, H.A. (ed.), *Bioaerosols*. Boca Raton, FL: Lewis Publishers, 1–23.

Burkhart, J.E., Stanevich, R. and Kovak, B. 1993. Microorganism contamination of HVAC humidification systems: case study. *Appl Occup Environ Hyg*, **8**, 1010–14.

Buttner, M.P. and Stetzenbach, L.D. 1993. Monitoring of fungal spores in an experimental indoor environment to evaluate sampling methods and the effects of human activity on air sampling. *Appl Environ Microbiol*, **59**, 219-22, 6.

Buttner, M.P., Cruz-Perez, P. and Stetzenbach, L.D. 2001. Enhanced detection of surface-associated bacteria in indoor environments using quantitative PCR. *Appl Environ Microbiol*, **67**, 2564–70.

Buttner, M.P., Willeke, K. and Grinshpun, S.A. 2002. Sampling and analysis of airborne microorganisms. In: Hurst, C.J., Crawford, R.L., et al. (eds), *Manual of environmental microbiology*, 2nd edn. Washington, DC: ASM Press, 814–26.

Chang, J.C., Chung, H., et al. 2001. Exposure of workers to airborne microorganisms in open-air houses. *Appl Environ Microbiol*, **67**, 155–61.

Clark, C.S., Rylander, R. and Larsson, L. 1983. Levels of gram-negative bacteria, *Aspergillus fumigatus*, dust and endotoxin at compost plants. *Appl Environ Microbiol*, **45**, 1501–5.

Contreras, M.A., Cheung, O.T., et al. 1988. Pulmonary infection with nontuberculosis mycobacteria. *Am Rev Resp Dis*, **137**, 149–52.

Couvelli, H.D., Kleeman, J., et al. 1973. Bacterial emission from both vapor and aerosol humidifiers. *Am Rev Resp Dis*, **108**, 698–701.

Cox, C.S. 1987. *The aerobiological pathway of microorganisms*. Chichester, UK: John Wiley and Sons.

Cox, C.S. and Wathes, C.M. 1995. Bioaerosols in the environment. In: Cox, C.S. and Wathes, C.M. (eds), *Bioaerosols handbook*. Boca Raton, FL: Lewis Publishers, 11–14.

Crawford, G.V. and Jones, P.H. 1979. Sampling and differentiation techniques for airborne organisms emitted from wastewater. *Water Res*, **13**, 393–9.

Crook, B. 1995a. Inertial samplers: biological perspectives. In: Cox, C.S. and Wathes, C.M. (eds), *Bioaerosols handbook*. Boca Raton, FL: Lewis Publishers, 247–67.

Crook, B. 1995b. Non-inertial samplers: biological perspectives. In: Cox, C.S. and Wathes, C.M. (eds), *Bioaerosols handbook*. Boca Raton, FL: Lewis Publishers, 269–83.

Crozier-Dodson, B.A. and Fung, D.Y. 2002. Comparison of recovery of airborne microorganisms in a dairy cattle facility using selective agar and thin agar layer resuscitation media. *J Food Prot*, **65**, 1488–92.

Curtis, S.E., Balsbaugh, R.K. and Drummond, J.G. 1978. Comparison of Andersen eight-stage and two-stage viable air samplers. *Appl Environ Microbiol*, **35**, 208–9.

Daisey, J.M., Angell, W.J. and Apte, M.G. 2003. Indoor air quality, ventilation and health symptoms in schools: an analysis of existing information. *Indoor Air*, **13**, 53–64.

Demers, R.R. 2001. Bacterial/viral filtration: let the breather beware. *Chest*, **120**, 1377–89.

Dennis, D.T., Inglesby, T.V., et al. 2001. Working Group on Civilian Biodefense. Tularemia as a biological weapon: medical and pubic health management. *J Am Med Assoc*, **285**, 2763–73.

Depaola, L.G., Mangan, D., et al. 2002. A review of the science regarding dental unit waterlines. *J Am Dent Assoc*, **133**, 1199–206.

Donham, K.J., Zavala, D.C. and Merchant, J.A. 1984. Respiratory symptoms and lung function among workers in swine confinement buildings: a cross-sectional epidemiology study. *Arch Environ Health*, **39**, 96–101.

Driver, C.R., Valway, S.E., et al. 1994. Transmission of *Mycobacterium tuberculosis* associated with air travel. *J Am Med Assoc*, **272**, 1031–5.

Duchaine, C., Grimard, Y. and Cormier, Y. 2000. Influence of building maintenance, environmental factors, and seasons on airborne contaminants of swine confinement buildings. *AIHAJ*, **61**, 56–63.

Dull, P.M., Wilson, K.E., et al. 2002. *Bacillus anthracis* aerosolization associated with a contaminated mail sorting machine. *Emerg Infect Dis*, **8**, 1044–81047.

Durand, K.T., Muilenberg, M.L., et al. 2002. Effect of sampling time on the culturability of airborne fungi and bacteria sampled by filtration. *Ann Occup Hyg*, **46**, 113–18.

Dutkiewicz, J., Jablonski, L. and Olenchock, S.A. 1988. Occupational biohazards: a review. *Am J Ind Med*, **14**, 605–23.

Dutkiewicz, J., Tucker, J., et al. 1992. Ultrastructure of the endotoxin produced by gram-negative bacteria associated with organic dusts. *Syst Appl Microbiol*, **15**, 272–85.

Dutkiewicz, J., Krysinska-Traczyk, E., et al. 2000. Exposure of agricultural workers to airborne microorganisms and endotoxin during handling of various vegetable products. *Aerobiologia*, **16**, 193–8.

Dutkiewicz, J., Olenchock, S., et al. 2001a. Exposure to airborne microorganisms in fiberboard and chipboard factories. *Ann Agric Environ Med*, **8**, 191–9.

Dutkiewicz, J., Skorska, C., et al. 2001b. Response of herb processing workers to work-related airborne allergens. *Ann Agric Environ Med*, **8**, 275–83.

Dutkiewicz, J., Krysinska-Traczyk, E., et al. 2002. Exposure to airborne microorganisms and endotoxin in potato processing plant. *Ann Agric Environ Med*, **9**, 225–35.

Fannin, K.F., Vana, S.C. and Jakubowski, W. 1985. Effect of an activated sludge wastewater treatment plant on ambient air densities of aerosols containing bacteria and viruses. *Appl Environ Microbiol*, **49**, 1191-119, 6.

Favero, M.S., Puleo, J.R., et al. 1966. Comparative levels and types of microbial contamination detected in industrial clean rooms. *Appl Microbiol*, **14**, 539–51.

Fields, B.S. 2002. *Legionella* and legionnaires' disease. In: Hurst, C.J., Crawford, R.L., et al. (eds), *Manual of environmental microbiology*, 2nd edn. Washington, DC: ASM Press, 860–70.

Flaherty, D.K., Deck, F.H., et al. 1984. Bacterial endotoxin isolated from a water spray air humidification system as a putative agent of occupational-related lung disease. *Infect Immun*, **43**, 206–12.

Fliermans, C.B., Cherry, W.B., et al. 1981. Ecological distribution of *Legionella pneumophila*. *Appl Environ Microbiol*, **41**, 9–16.

Fraser, D.W. 1984. Sources of legionellosis. In: Thornsberry, C., Balows, A., et al. (eds), Proceedings of the Second International Symposium on *Legionella*. Washington, DC: American Society for Microbiology, 277–80.

Fraser, D.W., Rsai, T.F., et al. 1977. Legionnaires' disease: description of an epidemic of pneumonia. *New Engl J Med*, **297**, 1189–97.

Gordon, T. 1992. Acute respiratory effects of endotoxin-contaminated machining fluid aerosols in guinea pigs. *Fundam Am Appl Toxicol*, **19**, 117–23.

Gorny, R.L. and Dutkiewicz, J. 2002. Bacterial and fungal aerosols in indoor environment in central and eastern European countries. *Ann Agric Environ Med*, **9**, 17–23.

Haas, D.U., Reinthaler, F.F., et al. 2002. Comparative investigation of airborne culturable microorganisms in sewage treatment plants. *Cent Eur J Public Health*, **10**, 6–10.

Hameed, A.A. and Khodr, M.I. 2001. Suspended particulates and bioaerosols emitted from an agricultural non-point source. *J Environ Monit*, **3**, 206–9.

Helms, C.M., Massanari, R.M., et al. 1983. Legionnaires' disease associated with a hospital water system: a cluster of 24 nosocomial cases. *Ann Intern Med*, **99**, 172–8.

Henis, Y. 1987. Survival and dormancy of bacteria. In: Henis, Y. (ed.), *Survival and dormancy of microorganisms*. New York: John Wiley & Sons, 1–108.

Hewett, J.A. and Roth, R.A. 1993. Hepatic and extrahepatic pathobiology of bacterial lipopolysaccharides. *Pharmacol Rev*, **45**, 381–411.

Higgins, J.A. and Cooper, M. 2003. Field investigation of *Bacillus anthracis* contamination of U.S. Department of Agriculture and other Washington, D.C. buildings during the anthrax attack of October 2001. *Appl Environ Microbiol*, **69**, 593–9.

Hryhorczuk, D., Curtis, L., et al. 2001. Bioaerosol emissions from a suburban yard waste composting facility. *Ann Agric Environ Med*, **8**, 177–85.

Hugenholtz, P. and Fuerst, J.A. 1992. Heterotrophic bacteria in an air-handling system. *Appl Environ Microbiol*, **58**, 3914–20.

Hyvarinen, A., Meklin, T., et al. 2002. Fungi and actinobacteria in moisture-damaged building materials – concentrations and diversity. *Int Biodeter Biodegradation*, **49**, 27–37.

Israeli, E., Gitelman, J. and Lighthart, B. 1994. Death Mechanisms in bioaerosols. In: Lighthart, B. and Mohr, J. (eds), *Atmospheric microbial aerosols*. New York: Chapman & Hall, 166–91.

Jones, B.L. and Cookson, J.T. 1983. Natural atmospheric microbial conditions in a typical suburban area. *Appl Environ Microbiol*, **45**, 919-93, 4.

Jussila, J., Komulainen, H., et al. 2002. *Mycobacterium terrae* isolated from indoor air of a moisture-damaged building induces sustained biphasis inflammatory response in mouse lungs. *Environ Health Perspect*, **110**, 1119–25.

Kaliste, E., Linnainmaa, M., et al. 2002. Airborne contaminants in conventional laboratory rabbit rooms. *Lab Anim*, **36**, 43–50.

Kaufman, A.F., McDade, J.E., et al. 1981. Pontiac fever: isolation of the etiologic agent (*Legionella pneumophila*) and demonstration of its mode of transmission. *Am J Epidemiol*, **111**, 337–9.

Kaufmann, S.H.E. and van Embden, J.D.A. 1993. Tuberculosis: a neglected disease strikes back. *Trends Microbiol*, **1**, 2–5.

Kenline, P.A. and Scarpino, P.V. 1972. Bacterial air pollution from sewage treatment plants. *AIHAJ*, **May**, 346–52.

Kiviranta, H., Tuomainen, A.R., et al. 1999. Exposure to airborne microorganisms and volatile organic compounds in different types of waste handling. *Ann Agric Environ Med*, **6**, 39–44.

Kline, J.N., Cowden, J.D., et al. 1999. Variable airway responsiveness to inhaled lipopolysaccharide. *Am J Respir Crit Care Med*, **160**, 297–303.

Ko, G., First, M.W. and Burge, H.A. 2000. Influence of relative humidity on particle size and UV sensitivity of Serratia marcescens and Mycobacterium bovis BCG aerosols. *Tuberculosis Lung Dis*, **80**, 217–28.

Ko, G., First, M.W. and Burge, H.A. 2002. The characterization of upper-room ultraviolet germicidal irradiation in inactivating airborne microorganisms. *Environ Health Perspect*, **110**, 95–101.

Kowalski, W.J., Bahnfleth, W.P. and Carey, D.D. 2002. Engineering control of airborne disease transmission in animal laboratories. *Contemp Top Lab Anim Sci*, **41**, 9–17.

Kullman, G.J., Thorne, P.S., et al. 1999. Organic dust exposures from work in dairy barns. *AIHAJ*, **59**, 403–13.

Laitinen, S., Kangas, J.K., et al. 1994. Workers' exposure to airborne bacteria and endotoxins at industrial wastewater treatment plants. *AIHAJ*, **55**, 1055–60.

Laitinen, S., Kangas, J., et al. 2001. Evaluation of exposure to airborne bacterial endotoxins and peptidoglycans in selected work environments. *Ann Agric Environ Med*, **8**, 213–19.

Leggat, P.A. and Kedjarune, U. 2001. Bacterial aerosols in the dental clinic: a review. *Int Dent J*, **51**, 39–44.

Lighthart, B. 1994. Physics of bioaerosols. In: Lighthart, B. and Mohr, J. (eds), *Atmospheric microbial aerosols: Theory and applications*. New York: Chapman & Hall, 5–27.

Lighthart, B. and Mohr, J.A. 1997. Estimating downwind concentrations of viable airborne microorganisms in dynamic atmospheric conditions. *Appl Environ Microbiol*, **53**, 1580–3.

Lighthart, B. and Stetzenbach, L.D. 1994. Distribution of bioaerosols. In: Lighthart, B. and Mohr, J.A. (eds), *Atmospheric microbial aerosols: Theory and applications*. New York: Chapman & Hall, 68–98.

Liu, L.J., Krahmer, M.F., et al. 2000. Investigation of the concentration of bacteria and their cell envelope components in indoor air in two elementary schools. *J Air Waste Manag*, **50**, 1957–67.

Lundholm, I.M. 1982. Comparison of methods for quantitative determinations of airborne bacteria and evaluation of total viable counts. *Appl Environ Microbiol*, **44**, 179-18, 3.

Lundholm, M. and Rylander, R. 1980. Occupational symptoms among compost workers. *J Occup Med*, **22**, 256–7.

Lynch, J.M. and Poole, N.J. 1979. Aerosol dispersal and the development of microbial communities. In: Lynch, J.M. and Poole, N.J. (eds), *Microbial ecology: a conceptual approach*. New York: John Wiley & Sons, 140–70.

Macher, J. (ed.) 1999. *Bioaerosols: assessment and control*. Cincinnati, OH: American Conference of Governmental Hygienists.

Macher, J.M. 2001. Evaluation of a procedure to isolate culturable microorganisms from carpet dust. *Indoor Air*, **11**, 134–40.

Marston, B.J., Plouffe, J.F., et al. 1993. Preliminary findings of a community-based pneumonia incidence study. In: Barbaree, J.M., Breiman, R.F. and Dufour, A.P. (eds), *Legionella: Current status and emerging perspectives*. Washington, DC: American Society for Microbiology.

Marthi, B. 1994. Resuscitation of microbial bioaerosols. In: Lighthart, B. and Mohr, J.A. (eds), *Atmospheric microbial aerosols: Theory and applications*. New York: Chapman & Hall, 192–225.

Michel, O. 1999. Indoor endotoxin and asthma. *Allergy Clin Immunol Int*, **11**, 109–11.

Milton, D.K., Gere, R.J., et al. 1990. Endotoxin measurement: aerosol sampling and application of a new limulus method. *AIHAJ*, **51**, 331–7.

Mohr, A.J. 2002. Microorganisms fate and transport. In: Hurst, C.J., Crawford, R.L., et al. (eds), *Manual of environmental microbiology*, 2nd edn. Washington, DC: ASM Press, 827–38.

Muder, R.R. and Yu, V.L. 2002. Infection due to Legionella species other than *L. pneumophila*. *Clin Infect Dis*, **35**, 990–8.

Neumann, H.D., Balfanz, J., et al. 2002. Bioaerosol exposure during refuse collection: result of field studies in the real-life situation. *Sci Total Environ*, **3**, 219–31.

Newson, S.W.B., Rowlands, C., et al. 1983. Aerosols in the mortuary. *J Clin Pathol*, **36**, 127–32.

Nottrebart, H.S. 1980. Nosocomial infections acquired by hospital employees. *Infect Control*, **1**, 257–9.

Olenchock, S.A. 1990. Airborne endotoxin. In: Hurst, C.J., Crawford, R.L., et al. (eds), *Manual of environmental microbiology*, 2nd edn. Washington, DC: ASM Press, 853–9.

Papineni, R.S. and Rosenthal, F.S. 1997. The size distribution of droplets in the exhaled breath of health human subjects. *J Aerosol Med*, **10**, 105-11, 6.

Park, J.H., Spiegelman, D.L., et al. 2001. Predictors of airborne endotoxin in the home. *Environ Health Perspect*, **109**, 859–64.

Pascual, L., Perez-Luz, S., et al. 2001. Detection of *Legionella pneumophila* in bioaerosols by polymerase chain reaction. *Can J Microbiol*, **47**, 41–346.

Pedgley, D.E. 1991. Aerobiology: the atmosphere as a source and sink for microbes. In: Andrews, J.H. and Hirano, J.J. (eds), *Microbial ecology of leaves*. New York: Springer-Verlag, 43–59.

Pillai, S.D. and Ricke, S.C. 2002. Bioaerosols from municipal and animal wastes: background and contemporary issues. *Can J Microbiol*, **48**, 681–96.

Prazmo, Z., Dutkiewicz, J. and Cholewa, G. 2000. Gram-negative bacteria associated with timber as a potential respiratory hazard for woodworkers. *Aerobiologia*, **16**, 275–9.

Radosevich, J.L., Wilson, W.J., et al. 2002. Development of a high-volume aerosol collection system for the identification of air-borne micro-organisms. *Lett Appl Microbiol*, **34**, 162–7.

Rahkonen, P., Ettala, M. and Loikkanen, I. 1987. Working conditions and hygiene at sanitary landfills in Finland. *Ann Occup Hyg*, **31**, 505–13.

Ranalli, G., Principi, P. and Sorlini, C. 2000. Bacterial aerosol emission from wastewater treatment plants: culture methods and bio-molecular tools. *Aerobiologia*, **16**, 39–46.

Rautiala, S., Reponen, T., et al. 1996. Exposure to airborne microbes during repair of moldy buildings. *AIHAJ*, **57**, 279–84.

Reinthaler, F.F., Haas, D., et al. 1999. Comparative investigations of airborne culturable microorganisms in selected waste treatment facilities and in neighbouring residential areas. *Zentralbl Hyg Umweltmed*, **202**, 1–17.

Reynolds, S.J., Black, D.W., et al. 2001. Indoor environmental quality in six commercial office buildings in the midwest United States. *Appl Occup Environ Hyg*, **16**, 1065–77.

Reynolds, S.J., Thorne, P.S., et al. 2002. Comparison of endotoxin assays using agriculture dusts. *AIHAJ*, **63**, 430–.

Rubin, J. 1991. Mycobacterial disinfection and control. In: Block, S.S. (ed.), *Disinfection, sterilization, and preservation*, 4th edn. Philadelphia, PA: Lea & Febiger, 377–84.

Rylander, R., Persson, K., et al. 1992. Airborne Beta-1,3-glucan may be related to symptoms in sick buildings. *Indoor Environ*, **1**, 263–7.

Schall, K.P. 1991. Medical and microbiological problems arising from airborne infection in hospitals. *J Hosp Infect*, **18**, 451–9.

Seal, D.V. and Clark, R.P. 1990. Electronic particle counting for evaluating the quality of air in operating theatres: a potential basis for standards? *J Appl Bacteriol*, **68**, 225–30.

Sheeran, T.J. 2002. Bioterrorism. In: Bitton, G. (ed.), *Encyclopedia of environmental microbiology*. New York: John Wiley & Sons, Inc, 771–82.

Shiomori, T., Miyamoto, H. and Makishima, K. 2001. Significance of airborne transmission of methicillin-resistant *Staphylococcus aureus* in an otolaryngology-head and neck surgery unit. *Arch Otolaryngol Head Neck Surg*, **127**, 644–8.

Shiomori, T., Miyamoto, H., et al. 2002. Evaluation of bedmaking-related airborne and surface methicillin-resistant *Staphylococcus aureus* contamination. *J Hosp Infect*, **50**, 30–5.

Skorska, C., Krysinska-Traczyk, E., et al. 2002. Response of furniture factory workers to work-related airborne allergens. *Ann Agric Environ Med*, **9**, 91–7.

Spurny, K.R. 1995. Chemical analysis of bioaerosols. In: Cox, C.S. and Wathes, C.M. (eds), *Bioaerosols handbook*. Boca Raton, FL: Lewis Publishers, 317–34.

Steinert, M., Hentschel, U. and Hacker, J. 2002. *Legionella pneumophila*: an aquatic microbe goes astray. *FEMS Microbiol Rev*, **26**, 149–62.

Stewart-Tull, D.E.S. 1982. *Mycobacterium leprae* – the bacteriologist's enigma. In: Ratledge, C. and Stanford, J. (eds), *The biology of the mycobacteria*. New York: Academic Press, 273–307.

Swan, J.R.M. and Crook, B. 1998. Airborne microorganisms associated with grain handling. *Ann Agric Environ Med*, **30**, 7–15.

Teltsch, B. and Katzenelson, E. 1978. Airborne enteric bacteria and viruses from spray irrigation with wastewater. *Appl Environ Microbiol*, **35**, 290–6.

Thelin, A., Tegler, O. and Rylander, R. 1984. Lung reactions during poultry handling related to dust and bacterial endotoxin. *Eur J Resp Dis*, **65**, 266–71.

Thorn, J., Beiher, L., et al. 2002. Measurement strategies for the determination of airborne bacterial endotoxin in sewage treatment plants. *Ann Occup Hyg*, **46**, 549–59.

Thorne, P.S., Kiekhaefer, M.S., et al. 1992. Comparison of bioaerosol sampling methods in barns housing swine. *Appl Environ Microbiol*, **58**, 2543–51.

Tjade, O.H. and Gabor, I. 1980. Evaluation of airborne operating room bacteria with a Biap slit sampler. *J Hyg (Lond)*, **84**, 37–40.

Travers Glass, S.A., Griffin, P. and Crook, B. 1991. Bacterial contaminated oil mists in engineering woris: a possible respiratory hazard. *Grana*, **30**, 404–6.

von Mutius, E., Braun-Fahrlander, C., et al. 2000. Exposure to endotoxin or other bacterial components might protect against the development of atopy. *Clin Exp Allergy*, **30**, 1230–4.

Walters, M., Milton, D., et al. 1994. Airborne environmental endotoxin: a cross validation of sampling and analysis techniques. *Appl Environ Microbiol*, **60**, 996–1005.

Wan, G.H. and Li, C.S. 1999. Indoor endotoxin and glucan in association with airway inflammation and systemic symptoms. *Arch Environ Health*, **54**, 172–9.

Wan, G.H., Li, C.S. and Lin, R.H. 2000. Airborne endotoxin exposure and the development of airway antigen-specific allergic responses. *Clin Exp Allergy*, **30**, 426–32.

Weis, C.P., Intrepido, A.J., et al. 2002. Secondary aerosolization of viable *Bacillus anthracis* spores in contaminated US Senate office. *J Am Med Assoc*, **288**, 2853–8.

Wilson, S.C., Morrow-Tesch, J., et al. 2002. Airborne microbial flora in a cattle feedlot. *Appl Environ Microbiol*, **68**, 3238–42.

Wolinsky, E. 1979. Nontuberculosis mycobacteria and associated diseases. *Am Rev Resp Dis*, **119**, 107–59.

Zucker, B.A., Trojan, S. and Muller, W. 2000. Airborne gram-negative bacterial flora in animal houses. *J Vet Med B Infect Dis Vet Public Health*, **47**, 37–46.

Zuravleff, J.J., Yu, V.L., et al. 1983. *Legionella pneumophila* contamination of a hospital humidifier. *Am Rev Resp Dis*, **128**, 657–61.

8

Bacteriology of soils and plants

GUY R. KNUDSEN

INTRODUCTION

Prokaryotic organisms profoundly affect virtually every facet of life on our planet. While soil- and plant-associated microbes remain unseen and underappreciated by most people, all life on Earth depends on their activities. Indeed, literally billions of years before the arrival of our first animal ancestors, prokaryotes were busy transforming Earth into the habitable planet we now enjoy. Prokaryotic autotrophs were largely responsible for the cooling of our planet by removal of CO_2 from Earth's atmosphere. The evolution of oxygenic photosynthesis in the Cyanobacteria, long before the appearance of plants, created the aerobic conditions that enabled oxygen-using higher life forms to arise. Indeed, the widely accepted theory of serial symbiosis postulates that the first eukaryotic organelles (mitochondria, chloroplasts) arose from prokaryotic symbionts (Margulis 1971). In recent years our comprehension of the abundance and diversity of the soil- and plant-associated microbiota has increased dramatically, but it is clear that good stewardship of Earth's resources will require a still deeper understanding of the microbial communities around and within us.

Microbial ecology, the study of how prokaryotes and other microorganisms interact with their physical and biotic environment, spans a number of component disciplines. Historically, methodologies for the study of soil prokaryotes have evolved in a different framework from the study of plant-associated microbes, often deriving from a different scientific perspective and with different research goals. As Paul and Clark (1989) noted, the primary thrust of soil microbiology has been on the metabolic activities of soil-inhabiting organisms, especially their roles in the biogeochemical energy flow and nutrient cycling, e.g. primary production and carbon mineralization, nitrogen fixation, nitrification and denitrification, sulfur transformations, etc. Less attention has been focused on soils as reservoirs for some animal pathogens. In contrast, approaches to the study of plant-associated microbes have tended to focus on the interspecific interactions between plants and microbes, especially the mutualistic relationships (e.g. mycorrhizae, rhizobia–legume symbioses) and the vast array of plant diseases that are of microbial origin. By definition, plant-associated prokaryotes do not lend themselves to pure culture studies, and there has been relatively less attention paid to the measurement of metabolic activity of plant-associated microbes, compared with those in soil.

SOIL AS A HABITAT FOR PROKARYOTES

Soil quality is the primary determinant of plant productivity in all terrestrial ecosystems, whether they are natural or agricultural. Soils are as much biological entities as they are physical ones, formed by a variety of interactions that include the weathering of rock parent material, microbial autotrophic primary production, heterotrophic decomposition of plant and animal materials, and myriad mineral transformations and redistribution of materials. Viewed in cross-section, soils typically are described as series of horizons or layers (soil profile), with each layer being approximately parallel to the surface and with a unique set of characteristics. If a layer of organic litter is present it is designated the 'O' horizon; next below that is the 'A' horizon, which represents a zone where plant residues accumulate, and from which materials leach downwards. Below that is the 'B' horizon, which is a zone of accumulation of minerals and humic materials, but typically containing less total organic matter than the A horizon. Finally, the lower 'C' horizon consists of unconsolidated parent material and geologic deposits. There are a number of variations and subdivisions of the above classification scheme, which taken together (and with additional characteristics) provide a useful taxonomy of soils. Bacteria are usually the most numerous organisms in soils, with commonly as many as 10^8–10^9 cfu/g of soil (Metting 1993). In terms of biomass, prokaryote biomass typically is similar to that of fungi in the upper horizons of aerobic soils and may exceed the biomass of plant parts in soil. Typically, prokaryote soil biomass ranges from about 300 to 3 000 kg (wet weight) per hectare (Metting 1993). Microbial biomass in general is very much lower in the lower soil horizons but is dominated by prokaryotic organisms.

A variety of properties determine the suitability of a soil as a microbial habitat. These include, but are not limited to, texture, structure, water content, aeration, pH, mineral nutrient content, and soil organic matter (SOM) content. It is important to realize that each of these properties typically is discontinuous in space, often at a scale of only millimeters or even micrometers. The term 'soil texture' refers to the relative proportions of the three inorganic particulate constituents of soil: clay, silt, and sand. The sand and silt components of soil are relatively large and chemically inert, and by themselves are not sites of significant microbial activity. The nonliving colloidal materials in soil (particles less than 2 μm in size) consist of inorganic clay minerals, composed of aluminosilicate layers, as well as organic humic materials. These colloidal materials have a profound effect on microbial activity. Clay particles may have as much as 8×10^6 cm^2 of surface area per gram, compared with approximately 450 cm^2/g for silt particles and 20 cm^2/g for coarse sand. While the layers of so-called 1:1 clays (consisting of silica + alumina layers) are held together by hydrogen bonds, the layers of 2:1 clays (silica + alumina + silica) are not tightly held together, and thus these clays are highly expandable when water is present and water molecules are interspersed between the layers. Thus, the 2:1 clays have potentially much

greater reactive surface area than the 1:1 clays. This is critical for chemical activity at clay surfaces, including microbial biochemical activity, because the negatively charged surfaces of clay minerals are sites of active cation exchange and repulsion of anionic molecules. Humus represents the organic colloidal component of soils; it consists of organic molecules that originated from living organisms but that have been biochemically modified and are no longer recognizable. Humic materials, like clays, have a net negative charge and high cation-exchange capacity (CEC), which enhances their interactions with synthetic organic compounds, metals, nutrients, as well as bacteria.

Soil structure, which is critical to below-ground air and water movement, water-holding capacity, and soil stability, occurs through the formation and spatial arrangement of aggregates. Soil bacteria contribute greatly to soil structure through their production of extracellular polysaccharides and other cementing compounds that help hold aggregates together, although these same compounds may also damage structure by blocking pore spaces. The majority of active bacteria are adsorbed on to particle surfaces in soil or are located at gas–liquid interfaces. Adsorption on to particles provides some degree of protection from predation by protozoa and soil arthropods, and it also places the bacterial cells in close proximity to adsorbed mineral nutrients. Additionally, extracellular enzymes produced by the bacteria also may be adsorbed nearby, so that the products of those enzymes will be more readily available for absorption, rather than being lost by diffusion into the soil solution. It should be remembered that prokaryotes are primarily aquatic organisms, and that at the prokaryote scale, soils often are functionally aquatic habitats. About 30 percent of the volume of a typical well-drained soil is in the liquid phase with another 20 percent in the gaseous phase and the remaining 50 percent as solids. Water in a well-drained soil typically is discontinuous, and its distribution is a function of the soil pore size distribution as determined by soil texture and structure. Through their effects on soil structure, bacteria effectively determine much of their surrounding physical environment. Aerobic and facultative aerobic bacteria also play a role in the spatial distribution of oxygen within and around soil aggregates. Oxygen diffuses only slowly through water, so that water-filled pores form effective seals around portions of aggregates, and oxygen within those zones may be depleted by aerobic respiratory activity and then maintained at low levels by the scavenging of oxygen by facultative anaerobes. In this way, anaerobic microsites are formed and maintained, even in otherwise well-aerated soils. Facultative anaerobes and even strict anaerobes can thus be active within aerated soils, and important anaerobic processes (e.g. denitrification) can take place in these microsites. Even under conditions where microsites become oxic, strict anaerobes may be able to persist although active growth will not occur. There are two basic ways that strict anaerobes may survive oxic conditions: first, a number of strict anaerobes, for example *Clostridium* spp., form dormant spores. Also, vegetative dormancy may allow cells of a strictly anaerobic species to survive oxic periods. The compounds hydrogen peroxide and superoxide are byproducts of certain oxygen-utilizing reactions, and these compounds are destructive to cells. Aerobic organisms have enzymes that destroy these agents, for example the enzymes superoxide dismutase (which converts superoxide to the products O_2 and H_2O_2) and catalase (which then converts H_2O_2 to water and oxygen). Strict anaerobes lack these enzyme systems; however, because the toxic byproducts are only formed during active metabolism, physiologically dormant anaerobes may escape their effects.

PLANT-ASSOCIATED PROKARYOTES

The metabolic activity of soil bacteria contributes greatly to a soil's fertility and its ability to support plant growth. In turn, plant residues and root exudates are the most important food source for the free-living heterotrophic microbial populations in soils. Also, plant surfaces and interiors are themselves important habitats for microorganisms, and nutrients leaking from aerial plant parts support considerable microbial growth on plant surfaces. Indeed, some microorganisms are able to grow only in association with plants. I will use the terms 'rhizosphere' and 'phyllosphere' as nouns to describe plant-associated habitats and as adjectives to describe the microbial components of those habitats.

The rhizosphere

The term 'rhizosphere' was first used by Hiltner (1904) to describe the interaction between bacteria and the roots of legumes. A somewhat analogous term, 'spermosphere,' refers to the area of increased microbial activity around a seed. Organic and inorganic materials released from roots, leaves, and seeds are important determinants of microbial activity, numbers, and interactions on plant surfaces and in the adjacent soil. In turn, the plant-associated microbial community strongly influences plant health and productivity. Typically, the rhizosphere may extend up to 5 mm or slightly more away from the root surface. As Lynch (1987) noted, the rhizosphere environment actually is comprised of the interacting plant, adjacent soil, and associated microorganisms. Microbial habitats within the rhizosphere are sometimes further classified as endorhizosphere (within the root tissue itself) or rhizoplane (on the root surface). Populations of microbes in the rhizosphere differ quantitatively and qualitatively from those in the bulk soil; their numbers are generally higher, and different populations

commonly are represented. Plant roots release compounds, including simple sugars and amino acids, that encourage the growth of specific microbial communities. Because rhizosphere inhabitants rely heavily on organic exudates for their carbon and energy requirements, their metabolism may be closely tied to that of the plant (Bowen and Rovira 1976; Lynch 1987).

Although many rhizosphere microbes form apparently commensal relationships with plant roots, there are numerous well-known relationships involving mutualism and/or parasitism. Roots of many plants form an intimate relationship with certain fungi; the resulting mycorrhizal association results in enhanced uptake of certain mineral nutrients including phosphorus. Legumes and some other plants form symbioses with nitrogen-fixing bacteria. The mutualistic relationship between legumes and *Rhizobium* bacteria allows the plants to grow in otherwise nitrogen-deficient soil and enhances the soil's nitrogen status for other plants. Some microbes in the rhizosphere or phyllosphere are plant pathogens. With the exception of some vectored or seed-borne organisms, most plant pathogens first contact the plant at the phylloplane or rhizoplane. While some fungal plant pathogens are well equipped to breach these surfaces mechanically or enzymatically, many others spend much of their life cycle as nonparasitic rhizosphere or phyllosphere inhabitants and enter a parasitic phase only when conditions are conducive to passive entry and subsequent colonization of plant interiors. In contrast, some microbes in the rhizosphere or phyllosphere are antagonistic to plant pathogens or insect pests, and thus serve as potential biological control agents for agriculture and forestry.

Plant roots modify the surrounding soil environment in various ways. Almost all chemical components of the plant have been found to be lost from roots. Root deposition includes exudates that contain water-soluble compounds such as sugars, amino acids, organic acids, hormones, and vitamins, as well as sloughed whole cells and volatile compounds. It is generally believed that the zone just behind the root tip is the site of maximum root exudation. As roots elongate through soil, cells are sloughed off from the root cap, and root hairs and cortical cells are sloughed from the surface. Deposition from roots provides a readily used energy source for microbes, and spatial variation in the components of rhizodeposition over the root surface greatly influences variability in the composition of the rhizosphere community (Whipps 1990). Other plant influences on the rhizosphere community include the removal of soil water by plant uptake, release of extracellular enzymes, and reduced oxygen tension and the release of CO_2 due to aerobic respiration, with an accompanying localized increase in soil acidity. In turn, microbes associated with plant roots modify the rhizosphere environment by producing extracellular enzymes and plant growth factors and by forming parasitic or mutualistic relationships with the plant. As a consequence, the quantities and types of available substrates in the rhizosphere differ from those in the bulk soil, resulting in different compositions of the respective microbial communities. This so-called 'rhizosphere effect' reflects the influence of plant roots on the size and composition of the surrounding soil microbial community. One numerical measure of rhizosphere effect is the R/S (rhizosphere/soil) ratio, which compares the total number of microbes in a rhizosphere habitat with the number in the root-free soil. R/S ratios range from as low as 5 to 100 or more but most frequently are from 10 to 20 (Curl and Truelove 1986; Gray and Parkinson 1968; Katznelson 1965). Absolute size estimates of the rhizosphere vary greatly, depending in part on the plant species and spatial location on the root, which is reflective of physiological maturity (Darbyshire and Greaves 1970) as well as soil temperature and moisture content.

The phyllosphere

Aerial plant parts, both surfaces and interiors, are important habitats for bacteria and support both autochthonous and allocthonous microbial populations. The term 'phyllosphere' has a similar etymology to the term 'rhizosphere' and has been applied to the habitats including plant interiors (also called the endophytic habitat) as well as plant surfaces (sometimes referred to as the phylloplane). Microorganisms colonizing the aerial parts of plants are sometimes called epiphytic microbes (from the Greek, *epi* [on top of] and *phytos* [plant]), but unlike the epiphytic plants that derive no nutrition from the plant surfaces they inhabit, epiphytic microbes actively utilize compounds exuded from leaves and other aerial plant parts. In the tropics, species of nitrogen-fixing bacteria are important inhabitants of foliar surfaces. Although noncolonizing (allocthonous) microbes frequently can be found in the phyllosphere, there is substantial evidence of defined microbial succession and the development of specific phyllosphere communities (Blakeman 1981). On new leaves, bacteria predominate in the microbial community, while yeasts and filamentous fungi succeed them and become dominant later in the growing season (Kinkel 1991; Kinkel et al. 1987). The important environmental factors impacting phyllosphere populations include ambient and leaf surface temperature, relative humidity and leaf wetness, and solar radiation. Leaf surfaces can be harsh habitats, since microbes on them are exposed to radiation and the dessicating effects of wind and sunlight, yet populations of 10^6 or more bacteria per square centimeter of leaf surface may be supported. Generally, environmental factors fluctuate more rapidly in the phyllosphere than in soil, so that microbial numbers can differ greatly even between samples taken over a short time period.

With the exception of Cyanobacteria (which are able to fix both C and N) on the foliar surfaces of some tropical plants, plant foliar leachates serve as the primary source of carbon and mineral nutrients for epiphytic bacteria. These leachates include a variety of sugars, organic acids, and amino acids, and their transfer from the leaf interior to the leaf surface takes place via aqueous pores (Blakeman and Atkinson 1981; Tukey 1971). Some epiphytic bacteria may themselves increase leaf wettability by the production of biosurfactant compounds (Bunster et al. 1989). Many epiphytic bacteria appear to be commensals and do not harm the plant upon which they reside. However, aerial plant surfaces may be successfully colonized by virulent strains of phytopathogenic bacteria, as well as by nonpathogenic isolates of the same or related species (Hirano and Upper 1983).

Bacteria in several genera, whether naturally occurring or modified by recombinant DNA techniques, offer promise for foliar application to crops for control of insect pests, plant pathogens, and weeds, or to protect plants from frost injury (e.g. Armstrong et al. 1987; Blakeman and Atkinson 1981; Knudsen and Hudler 1987; Lindow et al. 1988). Research on microbial pest-control agents for foliar application has progressed to where field tests are needed to evaluate their performance. The first genetically engineered microorganisms to be released legally in the USA were strains of *Pseudomonas syringae* and *P. fluorescens* from which the ice nucleation gene was deleted; these were intended to competitively exclude indigenous ice nucleation-active bacteria involved in frost damage. Small-scale field releases of aerosolized ice-nucleation-minus bacteria were undertaken by the University of California at Berkeley, and there was a large investment in time, money, and effort in attempts to determine dispersal patterns and survival of the applied bacteria (Knudsen 1988; Lindow et al. 1988). Concerns about possible environmental effects of genetically engineered microorganisms applied to crop plants include uncertainty about their dispersal, survival, and interactions with indigenous organisms, and this interest has fueled much of the recent research on epiphytic bacteria.

Soil, the rhizosphere, and the phyllosphere are populated by complex communities of organisms, including microbes (bacteria, Archaea, fungi, Protists) but also typically including plant and animal components. The rhizosphere and phyllosphere are defined by the presence of a plant, so it is inherently impossible to characterize microbial activity in these habitats through the study of genetically uniform microbial strains in isolation. Determining microbial contributions to these communities may require a variety of techniques with different areas of focus. For example, it has long been known that microbes may be metabolically active even though they are not culturable (Colwell et al. 1985). Furthermore, quantification of the viable microbial biomass may help elucidate the potential for specific microbial metabolic activities in a particular habitat, but the nonviable biomass may also be important as a component of the food web, despite its lack of a direct metabolic contribution. Although it is possible and common to make inferences about microbial biomass based on microbial community metabolic activity, it is important to note that these may not be strongly correlated, i.e. high metabolic activity does not necessarily mean an actively growing, dividing microbial community. Also, it is important to remember that microbial physiological activities are often strongly affected by the physical arrangement of the members of a population, especially aggregations such as those that occur in biofilms, on root surfaces, and in soil microsites. Physiological and spatial heterogeneity have important implications for the ecology of microbial populations and communities, and the ability of recently developed molecular and microscopic methods to address heterogeneity in environmental samples is especially exciting.

There has been an explosion of interest in the subject of microbial biofilms in recent years. The majority of microbes in natural ecosystems are not free-floating (planktonic) but instead typically grow attached to surfaces and encased in an adhesive, usually a polysaccharide material, forming complex communities called biofilms (Costerton et al. 1986). Bacteria on plant surfaces such as leaves and roots may form biofilms, and development of plant-surface biofilm communities and regulation of biofilm polysaccharide production has been studied for several root-colonizing bacteria (e.g. Hax and Golladay, 1993; O'Toole and Kolter 1998; Sarand et al. 1998). There is great potential for the use of biofilm bacterial communities on plant roots for bioremediation of soil pollutants (Yee et al. 1998). Techniques used to investigate root biofilm communities have included confocal microscopy with or without targeted ribosomal DNA (rDNA) probing, scanning electron microscopy (SEM) and transmission electron microscopy (TEM), and polymerase chain reaction (PCR)-assisted rDNA restriction fragment length polymorphism (RFLP)/sequence analyses.

BIOGEOCHEMICAL TRANSFORMATIONS BY SOIL- AND PLANT-ASSOCIATED PROKARYOTES

Largely unheralded by humans, the Bacteria, Archaea, and Fungi drive the biogeochemical cycles that make this planet habitable for plants and animals. Entire volumes detail the roles of the prokaryotes in carbon fixation and decomposition or organic compounds, in the cycling of nitrogen, sulfur, phosphorus, and other elements essential to life, and in transformations of metals. We have only started to understand and appreciate the astounding degradative capacity of the prokaryotes and to apply it to remediation of natural and man-made environmental pollutants. Below is a short summary of a few of the major biogeochemical roles that prokaryotes play.

Prokaryote contributions to carbon cycling

Availability of mineral nutrients is an important limitation for most bacteria in soil and on plant surfaces. Available carbon typically is less limiting. Heterotrophy, which is the utilization of fixed organic substrates as a carbon source, is the dominant lifestyle for most soil- and plant-associated prokaryotes. The autotrophic representatives of these habitats (i.e. those organisms that are able to fix CO_2 or carbonates for their carbon needs) include relatively small numbers of the photoautotrophic Cyanobacteria, which inhabit some surfaces that are exposed to light (e.g. some soil and plant surfaces, swampy soils, and exposed surfaces as cosymbionts with ascomycetous fungi in lichens). Several chemoautotrophs are present in soils, most significantly the nitrate- and nitrite-oxidizing nitrifiers, as well as those bacteria that aerobically oxidize reduced forms of sulfur, iron, or manganese for their energy needs. Methanogens, which are members of the Archaea, are able to fix CO_2 and/or carbonates and are important in anaerobic sediments. However, the overall contribution of microbial autotrophy to the carbon balance of soil or plant habitats is relatively minor compared with the importance of heterotrophic decomposition of plant (and animal) materials. Within the constraints of mineral nutrient availability, especially nitrogen, heterotrophic bacteria in soil quickly colonize any organic matter that is added. Because bacteria are generally limited to the surfaces of organic particles, the size of particles (i.e. surface to volume ratio) is critical in determining the extent that this resource can be exploited. Soil arthropods can play an important role here, as their feeding activity shreds and mixes the plant litter, thus exposing it to bacterial activity. Arthropods may also increase available nitrogen by excreting uric acid.

Many bacteria escape limitations imposed by lack of nutrients by dormancy, either in the vegetative state or as spores. When a nutritional source becomes available in soil, typically there is a temporal succession of microbes that are able to utilize the resource, as the quality of the available substrates changes during decomposition. Those first able to colonize the resource usually are r-selected (Andrews and Harris 1986) bacteria such as *Bacillus* and *Pseudomonas*, as well as fungi such as *Penicillium* spp., which are able to rapidly utilize available sugars and amino acids. They are followed by a variety of K-selected (Andrews and Harris 1986) organisms that include prokaryotes such as Cyanobacteria but also a number of fungi. Many of these late-succession organisms either are stress-tolerant or possess highly competitive mechanisms such as cellulose- or lignin-degrading enzyme systems. Production of antibiotic compounds may also play a role in competition among soil microorganisms, although production of antibiotics in situ, as well as its ecological significance, have been difficult to demonstrate (Williams 1986). Only relatively recently has antibiotic production in soil been definitively shown to occur (Thomashow and Weller 1991), and it is possible that the level of antibiotics is still very low because of adsorption to clay minerals and humic materials. Many of the antibiotics that are so important in the treatment of infectious diseases in medical and veterinary practice, derive from soil microbes, especially the Actinomycetes.

Prokaryote contributions to nitrogen cycling

Nitrogen is a critical element for all forms of life. It is a building block of nucleic acids as well as amino acids, and thus is a major component of the genetics, physiological activity, and structural integrity of all organisms. Although the cellulose and lignin molecules that provide structural rigidity to plants are free of nitrogen, it still is the fourth most abundant element in plant composition, after C, H, and O. It is the mineral nutrient that most often is limiting for plant nutrition. Nitrogen is a cell-wall component of the majority of microbes, part of the chitin molecule found in the higher fungi, and the peptidoglycan molecule found in bacterial cell walls. The ratio of carbon to nitrogen varies among microorganisms, but generally it is significantly lower than in plants or in soil. Typical C:N ratios of fungi range from about 4.5:1 to 15:1, and for bacteria it usually is even lower, in the range of 3:1 to 5:1. By way of comparison, the C:N ratio in a typical arable soil is about 10:1, about 25:1 in green-plant material, and as much as 100:1 in dried plant residues such as straw. For this reason, nitrogen availability is usually limiting when plant materials are being decomposed in soil. The microbiota quickly mineralizes the nitrogen that is present in plant residues and immobilizes much of it in the form of microbial biomass.

Along with its occurrence as a component of organic molecules, the N atom can be found in a number of different inorganic molecules of differing oxidation states. Microbes mediate most of the transitions between these states and thus play essential roles in nitrogen cycling. The NH_4^+ (ammonium) form is the product of microbial mineralization of organic nitrogen compounds in soil and is readily assimilated by both plants and microorganisms. Because it is a cation, the ammonium ion may attach to negatively charged clay minerals and become entrapped or 'fixed' between the layers. This process, ammonium fixation, should not be confused with dinitrogen fixation, which is the transition (most commonly microbially mediated) of atmospheric N_2 into ammonia and subsequently to organic nitrogen in

biological systems. Under basic soil conditions, NH_4^+ ions will be deprotonated to NH_3; this form may volatilize and be lost from soil, especially under warm and dry conditions.

The aforementioned nitrifying bacteria are aerobic autotrophs that oxidize ammonium to nitrite (NO_2^-) and then oxidize nitrite to nitrate (NO_3^-), as a source of energy. The latter step occurs readily and nitrite rarely accumulates in soil. Different autotrophic nitrifying bacteria are responsible for each of these steps, e.g. *Nitrosomonas* oxidizes ammonium to nitrite, while *Nitrobacter* oxidizes nitrite to nitrate. Heterotrophic nitrification also occurs and may be especially significant in forest soils. Although heterotrophic nitrification primarily is done by fungi, a number of aerobic bacteria have been shown to nitrify heterotrophically, including species of *Arthrobacter*, *Aerobacter*, *Mycobacterium*, *Streptomyces*, *Thosphaera*, and *Pseudomonas*. Nitrate is a preferred nitrogen source for many plants and thus nitrification may increase soil fertility. However, because NO_3^- is an anion, it is repelled by the negatively charged surfaces of clay minerals and humic materials, and thus it is highly mobile in soil solution. Nitrate leaching and movement into groundwater is a significant health and environmental problem. Excessive nitrate levels in drinking water are the cause of infant methemoglobnemia, or 'blue-baby syndrome', as the infant gut favors formation of nitrite, which in turn binds to hemoglobin and prevents it from accepting and transferring oxygen. Nitrates in drinking water are also linked to formation of carcinogenic nitrosamines. Environmentally, nitrates in groundwater provide a nitrogen input to lakes and ponds and encourage growth of algae and subsequent eutrophication of the water body.

Nitrate is also susceptible to loss from soil through the process of denitrification, in which various facultatively anaerobic bacteria (e.g. *Pseudomonas denitrificans*, *Thiobacillus denitrificans*) utilize NO_3^- as a terminal electron acceptor in anaerobic respiration. The pathway of denitrification is: $NO_3^- \rightarrow NO_2^- \rightarrow NO \rightarrow N_2O \rightarrow N_2$. The gaseous intermediates NO (nitric oxide) and N_2O (nitrous oxide) are always released during denitrification and are atmospheric pollutants that are involved in catalytic destruction of atmospheric ozone.

DINITROGEN FIXATION

For most plants, the direct acquisition of nitrogen is primarily from uptake of the mineral forms NH_4^+ and NO_3^-, which are derived from organic matter decomposition in soil. However, a significant amount of the nitrogen used by some plants comes directly from the biological fixation of atmospheric nitrogen (N_2). Biological nitrogen fixation, which is the conversion of atmospheric dinitrogen to ammonia, is mediated by the nitrogenase enzyme system, which is unique to prokaryotes. A number of free-living and plant-associated bacteria fix nitrogen in terrestrial ecosystems. Approximately two-thirds of all the nitrogen fixed globally is through biological dinitrogen fixation.

Nitrogen fixation requires relatively large amounts of energy, either in the form of light energy for the phototrophic Cyanobacteria or as energy stored in the chemical bonds of organic compounds for the heterotrophic nitrogen-fixing bacteria. The symbiotic association with plants is valuable for associative nitrogen-fixers for this reason, and the symbiotic legume–rhizobia and alder–*Frankia* associations are thus especially effective and important in the amount of nitrogen they fix on a global basis. The legume–rhizobia and alder–*Frankia* associations are examples of mutualism, and both partners clearly benefit from the association. In each case, root nodules are formed containing the bacteria, which are only able to fix nitrogen in association with the plant host. Nodule formation is a complex process involving a series of physical and chemical interactions beween the symbiotic partners.

Certain Cyanobacteria also partner with ascomycetous fungi to form lichens, which are common in terrestrial habitats. The prokaryote partner in this mutualistic association is able to fix not only nitrogen but carbon as well, through photosynthesis, while the eukaryote partner obtains mineral nutrients from the substratum. The association allows lichens to thrive in austere habitats such as the surface of rocks, where both organic carbon and mineral nitrogen are extremely limited. Significant amounts of nitrogen are also fixed by free-living bacteria in a variety of habitats, including associated with plant roots (e.g. *Azospirillum* and roots of grasses) or foliage, the latter especially in the tropics. Tropical epiphytic microbial communities are made up of a number of taxonomic groups, including several genera of nitrogen-fixing bacteria, such as *Azotobacter*, *Beijerinckii*, *Klebsiella*, and Cyanobacteria such as *Nostoc* and *Scytonema*.

Prokaryote transformations of sulfur

A variety of primarily chemoautrophic and chemoheterotrophic prokaryotes are responsible for oxidizing sulfur in soils, under aerobic conditions. The most well known genus is *Thiobacillus*, which is found in a variety of habitats (one species, *T. denitrificans*, is also capable of anaerobic growth, using nitrate as a terminal electron acceptor). The phototrophic purple sulfur and green sulfur bacteria, which are dominant sulfur oxidizers in aquatic systems, are mostly absent from soil habitats due to their simultaneous requirements of light and anoxic conditions. However, they may be found in aquatic–soil interfaces, such as rice paddies and wetlands. *Sulfolobus*, a member of the Archaea, is an important sulfur oxidizer, especially in thermal habitats, but its importance in most soils is largely unknown. Soil microbes reduce sulfur compounds in an assimilatory process to

meet their sulfur needs. Dissimilatory reduction of sulfate to hydrogen sulfide, by organisms using sulfate as a terminal electron acceptor, occurs primarily in aquatic sediments but also may be significant in anaerobic, waterlogged soils and possibly in anaerobic soil microsites. Release of H_2S by sulfate reducers such as *Desulfovibrio* spp., *Desulfomonas* spp., and *Desulfomaculatum* spp., provides an energy source for nearby sulfur oxidizers in tightly linked microbial communities.

BIOTRANSFORMATION AND BIOREMEDIATION

In the very early twentieth century, the great microbial ecologist Martinus Beijerinck made his famous statement, 'Everything is everywhere, the environment selects.' Beijerinck recognized that microbes and the vast diversity of their metabolic capabilities are almost ubiquitous in the environment and that environmental influences select for certain microbial types. A variety of environmental pollutants are naturally occurring, nonsynthetic materials, such as petroleum products, nitrates, heavy metals, and acid mine drainage. In the latter part of the twentieth century, scientists became more aware of the prescience of Beijerinck's comments, as microbes in natural habitats were observed to successfully degrade a wide range of pollutants. The list of compounds that bacteria may successfully degrade in the environment includes petroleum or hydrocarbon products such as gasoline, diesel, and fuel oil, hazardous crude oil compounds such as benzene, toluene, xylene, and naphthalene, certain polynuclear aromatic compounds, certain pesticides (e.g. the insecticide malathion), coal compounds, some industrial solvents such as acetone, as well as ethers, methanol, methylethylketone, and ethylene glycol.

However, it was soon realized that a large number of organic chemicals synthesized by humans, which we now call 'xenobiotics,' have no close natural chemical analogs and often are able to persist in the environment because appropriate transport mechanisms and catabolic pathways have not evolved in natural microbial communities. Xenobiotic compounds include a number of pesticides and products generated from them, many plastics, industrial chemicals, and manufacturing byproducts, and compounds newly formed in nature from synthetic components. Typically, xenobiotic compounds are not found individually, but rather in simple or complex mixtures. Many of these compounds are toxic at low concentrations, and some are subject to biomagnification through the food chain. The potential for biodegradation of xenobiotic compounds is a subject of intense research. Fortunately, while a number of xenobiotic compounds may not support the growth of any single microorganism, they may nonetheless be degraded in the presence of other natural substrates, a process that is known as cometabolism. Also, a number of xenobiotics may be degraded through the cooperative efforts of several microbial strains comprising a microbial consortium.

The term 'bioremediation' refers to a variety of technologies that rely on the biodegradative capabilities of microorganisms for the mineralization or inactivation of chemical pollutants in contaminated sites. Complete mineralization, i.e. decomposition of a pollutant to inorganic ions and CO_2, is the most desirable situation because the end products are usually nontoxic. Although soil may be removed and transported from a site for ex situ bioremediation, this usually is prohibitively expensive, and in situ bioremediation (ISB) is the norm. There are two general approaches to bioremediation. The first, sometimes called 'bioaugmentation,' is intended to enhance the degradative potential of microbes already present at the site, by environmental modifications including nutrient application, aeration, mixing, etc. A second approach involves addition of appropriate xenobiotic degraders by seeding. Ideally, bacteria chosen as agents for bioremediation of organic pollutants will have a number of characteristics beyond the ability to digest the particular waste material rapidly and completely; these include the nonproduction of odors or noxious gases, nonpathogenicity to humans or other animals, and competitive ability in the natural environment. For some pollutants, commercial strains of bacteria and fungi have become available. These strains have been selected or in some cases genetically engineered for specific degradative capabilities.

BACTERIA AS PATHOGENS OF PLANTS AND ANIMALS

Despite the fact that prokaryotes are absolutely essential in making our planet a habitable place for plants and animals, most people still are most familiar with the small group of usually specialized bacteria that are pathogenic to humans, other animals, and the crops we grow for food and fiber.

Phytopathogenic bacteria

While the number of plant diseases caused by bacteria is smaller than those caused by fungal pathogens, nonetheless soil-borne and foliar bacterial plant pathogens are responsible for a large number of plant diseases, including various wilts, blights, galls, leafspots, and rots. The taxonomy of phytobacteriology is evolving continually, but pathogens are found in the genera *Pseudomonas*, *Burkholderia*, *Xanthomonas*, *Streptomyces*, *Agrobacterium*, *Erwinia*, *Corynebacterium*, and *Spiroplasma*. Some are obligate pathogens and many are seed-borne. A large number are saprophytic and capable of surviving and reproducing in soil. Pseudomonads especially are common and abundant in soil and the rhizosphere. A number of fluorescent pseudomonads have a permanent soil phase. Xanthomonads, some *Erwinia* spp., and a number of pseudomonads are associated with above-ground plant surfaces and can cause a number of serious foliar diseases. *Pseudomonas syringae*, for example, is responsible for a number of economically significant diseases of vegetables, fruits, and ornamentals.

Phytopathogenic bacteria gain entrance into plants through wounds or natural openings, which include stomata, lenticels, hydathodes, and nectarthodes. Once bacterial cells gain entrance into the plant, they may biochemically elicit plant-defense responses. The recognition phenomena that govern these responses typically are quite specific and are determinants of whether a particular bacterial strain is pathogenic. A number of virulence determinants are encoded by the pathogen genome, and govern the pathogen's 'weapons,' which may include toxins, hormones, extracellular polysaccharides, and enzymes such as cellulases and pectinases that are involved in degradation of plant cell walls.

One unique soil-borne plant pathogen that has proven to be of immense utility is the bacterium *Agrobacterium tumefaciens*, the causal agent of crown gall disease. *Agrobacterium* is a relatively common genus in soils, and the pathogen *A. tumefaciens* is easily disseminated by farm machinery, tools, irrigation, or rain-splashed soil. The pathogen gains entry through wounds and colonizes the wounded plant tissues. Then, a portion of the pathogen's own DNA in the form of a plasmid (Ti plasmid, for 'tumor-inducing') enters into a living host cell and somehow gains entry into the host nucleus. The newly introduced pathogen genes code for production of growth hormones, and large woody galls are formed on the plant roots (or, in some cases, plant stems). The galls form a highly nutritive environment for the pathogen, but the hypertrophic growth may result in collapse of plant vascular tissue and death of the host. Eventually, bacteria are released from decaying galls into surrounding soil or water. The ability of a prokaryotic organism to transfer some of its own DNA into the genome of a eukaryotic host is a unique characteristic. Plant molecular biologists have taken advantage of this ability, by generating variants of *A. tumefaciens* that are able to transfer DNA without the genes required to incite galls in the host plant. A wide variety of genes have been inserted into the bacterial plasmid, allowing plants to be transformed with any number of desirable traits.

Bacteria as biological control agents against plant pathogens

Many or most bacteria associated with the exterior surfaces of plants are nonpathogenic. Among those are a number of organisms that have been found to be antagonistic to bacterial and fungal plant pathogens, thus potentially acting as biological control (biocontrol) agents. Biological control may be defined as the use of one organism to suppress another organism that is an agronomic pest or parasite, or that is environmentally injurious. A number of mechanisms have been implicated in biological control of plant pathogens by beneficial bacteria, including production of antibiotic compounds, nutrient competition, hyperparasitism, and stimulation of host defenses. A variety of rhizosphere bacteria, which are effective colonizers of plant roots, have been designated plant growth-promoting rhizobacteria (PGPR). A number of PGPR have been identified, and several of these are being promoted as plant-disease-control products (Bahme and Schroth 1987; Kloepper and Beauchamp 1992; Suslow and Schroth 1982). Organisms that have received a great deal of attention include strains of *Pseudomonas fluorescens*, *Bacillus subtilis*, and *Streptomyces griseoviridis*, and they are being used especially for protection of high-value crops, including bedding plants, foliage plants, ornamentals, and some vegetable crops.

Soil- and plant-associated animal pathogens

Humans and other animals generally play a minor role as habitats for bacteria, but the effects on animal populations nonetheless can be devastating. A number of pathogenic prokaryotes are normally soil inhabitants but are facultative parasites of animals. Thus, soil (and to a lesser extent, plants) can serve as a reservoir for these bacterial pathogens.

Pathogens of medical and veterinary significance

Several important animal diseases are caused by pathogens that are indigenous to soil or plant surfaces. Some of the most important causal agents are spore-forming bacteria in the genera *Bacillus* and *Clostridium*, in part because of their ability to survive adverse environmental conditions. Anthrax is an acute infectious disease caused by the aerobic spore-forming bacterium *Bacillus anthracis*. Anthrax most commonly occurs in wild and domestic vertebrates (cattle, sheep, goats, camels, antelopes, and other herbivores), but it can also occur in humans when they are exposed to infected animals or tissue from infected animals. The anthrax organism is endemic in many agricultural regions throughout the world, including South and Central America, Southern and Eastern Europe, Asia, Africa, the Caribbean, and the Middle East. Horses, swine, deer, and humans are less susceptible than are cattle or sheep. Wild ruminants such as deer also may become infected. Infection in ruminants or horses is usually the result of grazing on infected pasture land. The organisms typically enter through the mouth, and less commonly via skin injury. After ingestion or inhalation, the bacteria spread rapidly throughout the body. Anthrax disease in humans typically results from occupational exposure to infected animals or by inhaling anthrax spores from contaminated animal products. Anthrax has become a major public-health issue because *B. anthracis* is considered to be a potential agent for use in biological warfare.

Members of the genus *Clostridium* are also spore-forming bacteria, but they are strict anaerobes. Clostridium is responsible for a number of important animal diseases, including blackleg of cattle and sheep (*C. chauvoei*), red-water disease of cattle (*C. haemolyticum*), enterotoxemias of various mammals caused by *C. perfringens*, botulism of cattle, horses, sheep, and occasionally humans (*C. botulinum*), and tetanus (all mammals) caused by *C. tetani*.

Clostridia are found naturally in soil, aquatic sediments, and animal gastrointestinal tracts. The majority are not pathogenic to animals, and species are significant for their ability to fix nitrogen. *Clostridium perfringens* causes gas gangrene disease, which results from contamination of wounds with soil containing spores of the bacterium. Clostridial food poisoning (botulism) results from eating food that is contaminated with soil or feces containing *C. botulinum*, and which then is stored under anaerobic conditions that permit reproduction of the organism. All forms of botulism can be fatal, although the disease is relatively rare. A number of other genera of soil bacteria, especially gram-negative bacteria in the genera *Pseudomonas*, *Enterobacter*, and *Acinetobacter*, may be of medical importance.

A number of plant- and soil-associated bacteria are pathogens of arthropods, including a number of insect pests. Many, like the gram-negative bacterium *Serratia marcescens*, are facultative parasites that, once ingested, may be able to invade the hemocoel if the integrity of the gut lining is compromised. Two members of the genus *Bacillus* have been widely used as microbial insecticides against a number of important insect pests. *Bacillus popilliae* causes 'milky-spore disease' of beetle larvae in soil. Larvae that ingest spores of this organism cease feeding, turn milky white in color from the profuse sporulation of the bacterium in the hemocoel, and then die. *Bacillus popilliae* has been commercialized and is widely used against Japanese beetle in lawns and turfgrass. More well known is the related organism, *B. thuringiensis*, which is arguably the most successful biological control agent in history. Along with spore formation, *B. thuringiensis* produces a parasporal crystalline protein. The organism is primarily an inhabitant of soil, but it is used primarily for foliar application as a biocontrol agent. When the crystalline protein, also known as the δ-endotoxin, is ingested by the herbivorous larvae of lepidopteran pests, it dissolves in the alkaline gut of the insect. The toxin disrupts the membranes lining the gut, and septicemia may follow as bacteria proliferate in the hemocoel. The larvae soon stop feeding and die soon thereafter. The organism is easily grown in mass culture, dried, and culture residues containing spores and crystals formulated for spraying on crops and forests. There are several commercial Bt formulations currently on the market, and certain subspecies of the organism have demonstrated efficacy against coleopteran (beetle) and dipteran (flies, including mosquitoes) pests. More recently, the toxigenic gene of Bt has been placed into strains of *Pseudomonas fluorescens* for use as seed or root inoculants, and into the maize endophyte *Clavibacter pyli* for protection against the European corn borer. Genes for toxin production have been inserted directly into crop plants via genetic engineering, and crops with Bt genes are now widely used in commercial agriculture. This approach is controversial, however, as it exerts a strong selection pressure on the pest population to develop Bt resistance. Resistance has already been observed in some important pest species, such as Colorado potato beetle.

Recombinant bacteria

Along with examples noted above, a number of soil- and plant-associated bacteria are candidates for genetic engineering, in order to enhance their environmental fitness or biological control efficacy. Environmental release of engineered bacteria has raised concerns about the possible spread of recombinant DNA sequences to other organisms present in the same environment. Bacterial plasmids and transposons occur in most prokaryotes; they carry the genetic determinants of many important phenotypic properties including antibiotic and heavy metal resistance, production of antibiotics, bacteriocins, and toxins, catabolism of xenobiotic compounds, nodulation and nitrogen fixation, plant tumorigenesis, and insecticidal toxin production. Knowledge about the potential of introduced organisms to transfer genetic information to other bacteria present in a given habitat is essential for assessment of possible environmental risk.

SAMPLING AND DETECTION METHODS FOR SOIL- AND PLANT-ASSOCIATED PROKARYOTES

A wide array of traditional and novel methods is now available to investigate the activities of microbes in soil and plant habitats. Particularly exciting advances have been made in our ability to observe, in situ, many aspects of microbial physiology that previously could only be demonstrated with pure cultures under highly controlled conditions. Various new strategies are available for detection of bacteria, their genes, and their gene products in natural environments. These strategies are equally useful for tracking the fate of introduced organisms and their genetic information in soil and the rhizosphere.

There are two general approaches to the quantitative estimation of microbial populations in the rhizosphere or phyllosphere: direct methods such as visualization techniques, and indirect methods such as dilution plating of root or leaf washings, or most probable number (MPN) estimates (Cochran 1950). Indirect methods have been reported to recover only about 0.01–10 percent of the numbers of bacteria enumerated by direct counts (Campbell and Porter 1982; Colwell et al. 1985; Richaume et al. 1993). However, direct counts are difficult to obtain from an opaque medium such as soil. Also, they may overestimate the number of bacteria in soil because it is difficult to differentiate between living and dead cells unless using a viability stain. In addition, a high level of concentration is needed by the observer; direct counts are labor-intensive if large numbers of samples need to be processed. Another disadvantage with direct counts is that different groups of organisms are difficult to identify unless their morphologies differ. Since precise identification of what constitutes rhizosphere soil is difficult or impossible, researchers are often forced to rely on an operational definition of the rhizosphere. In many studies, the rhizosphere has been defined as the thin layer of soil adhering to the root system after loose soil has been removed by shaking. However, depending on soil texture and moisture content, the amount of adhering soil may vary considerably. Parke et al. (1990) reported that the amount of adhering soil (from 0 to 1200 mg/cm of root) on roots and adhering soil mass (from 25 to 1000 mg) did not significantly affect estimates of root populations of a rhizobacterial strain of *Pseudomonas fluorescens*. If microbial numbers are to be expressed as colony forming units (CFU) per gram of rhizosphere soil, a subsample of the rhizosphere soil should be weighed, oven-dried, and weighed again to determine soil moisture content. For phyllosphere samples there are fewer difficulties, as generally the phyllosphere will be more or less synonymous with the phylloplane (plant surface).

A variety of methods have been used to separate rhizosphere microbes from root surfaces, and phyllosphere microbes from leaf surfaces. These methods include washing techniques (in flasks or tubes on a platform or wrist action shaker), vortexing (with or without additional agitation, for example with added glass beads), sonication, and blending or maceration. The method used for removal of microbes from either the rhizosphere or the phyllosphere affects the recovery of populations (Donegan et al. 1991; Kloepper et al. 1991). Washing or homogenizing diluents commonly include sterile tap or distilled water, saline solution, or buffer solutions. Regardless of the method, consideration must be taken of whether rhizosphere or phyllosphere microbes might be killed by the process, resulting in artificially low counts, and/or whether growth in the diluent may occur, possibly resulting in inflated counts. Freezing leaf samples for more than 3 days prior to processing is not recommended as recovery of bacteria has been shown to be adversely affected (Donegan et al. 1991).

The dilution plate count is one of the most widely used methods for enumeration of microbes. However, it is well established that because indirect counts take into account only culturable microbes, they generally underestimate actual microbial populations. Typically, only a small fraction (0.01–10 percent) of the number of cells observable by direct microscopy is recovered by the plate-count procedure. Skinner et al. (1952) listed several reasons for inadequacies of the plate count procedure for bacterial isolation; these include (1) clumps of bacterial cells may give rise to single colonies on a plate; (2) antibiosis and competition on plates cause some colonies to fail to develop; and (3) even so-called 'general' growth media are usually selective in some respect. Many cells that are metabolically active do not produce colonies on solid media. Plate counts estimate only those cells that are tolerant of the medium and incubation conditions to which they are subjected (Zuberer 1994). In contrast, the plate-count procedure may provide an apparent overestimation of fungal or actinomycete biomass if clusters of spores are present, since each spore may give rise to a colony on the plate. To accommodate the enormous size of most microbial populations in environmental samples, as well as the microscopic size of individuals, the main principle of enumeration is dilution of the population to countable numbers. Usually, an investigator will add components to the isolation medium which inhibit either fungi or bacteria.

From leaf or root surfaces, it is possible to directly isolate fungal and bacterial propagules by pressing the surface of the leaf or root segment on to an agar surface, then removing it. This method has the advantage of providing some information about the microbial spatial distribution on the leaf surface. It is also possible to pour a small volume of molten agar on to the leaf surface, allow it to solidify, and then to transfer it to fresh medium (Langvad 1980). Testing surfaces for fungal spores is done similarly, but with single-sided tape, and the spores are then identified microscopically (Knudsen and Hudler 1987). Additional information on sampling, including preparation of serial dilutions and plating techniques, is provided in Dandurand and Knudsen (2001) and Zuberer (1994).

QUANTIFYING PROKARYOTE METABOLIC ACTIVITY IN SOILS

Determining the metabolic status of microbial cells in natural habitats is significant, because metabolic status largely determines the effects that the microbes have on their environment. In contrast, determining the identity and distribution of microorganisms can only provide information on their potential environmental activity. Historically, a great deal of research in soil and plant microbiology has focused on biochemical processes, including microbial roles in the processes of biogeochemical cycles, but necessarily many studies were performed in vitro or by using artificially constructed systems. Frequently, the potential metabolic capabilities of microbial populations in natural systems have had to be extrapolated from capabilities observed in the laboratory setting.

Traditional methods, with respirometry being the most notable example, have provided a wealth of information on total microbial metabolic activity in soils but have not always provided a deeper understanding of soil microbial community structure and function. It has long been known that the majority of soil microbes are not amenable to recovery by standard cultural techniques, and often the environmental role of these organisms, if they were detected at all, could only be guessed at. However, recent advances in molecular biology, fluorescent stain technology, and microscopy have provided new and/or improved tools to detect and characterize soil microbes in situ, including their abundance, genetic makeup, and environmental activities. Particularly exciting advances have been made in our ability to observe, under natural conditions and with minimum perturbation, many aspects of microbial physiology that previously could only be demonstrated with pure cultures under highly controlled conditions. Also, by helping to escape the limitations of pure culture methods, these new techniques have provided insight into the roles that individual microbial populations play as contributing members of microbial consortia.

Microbial metabolism can be defined as the sum total of all the chemical reactions that occur in microbial cells, and the seemingly infinite variety of metabolic processes and products suggests potential for a correspondingly vast array of detection methodologies. The scope of available technology for monitoring microbial metabolic activity is increasing so rapidly that it is difficult to catalog the available methods. However, methods can be broadly categorized as those that detect end products or intermediate products of catabolic or anabolic activity (e.g. detection of carbon dioxide produced by aerobic respiration, via respirometry), those that attempt to correlate fluxes in specific structural components of the microbial biomass (e.g. phospholipid ester-linked fatty acids) with metabolic activity levels, and those that are designed to detect the biochemical indicators of metabolism (e.g. RNA, specific enzymes) or the specific products of gene expression.

Soil, the rhizosphere, and the phyllosphere are populated by complex communities of organisms, including microbes (bacteria, Archaea, fungi, Protists) but also typically including plant and animal components. The rhizosphere and phyllosphere are defined by the presence of a plant, so it is inherently impossible to characterize microbial activity in these habitats through the study of genetically uniform microbial strains in isolation. Determining microbial contributions to these communities may require a variety of techniques with different areas of focus. For example, it has long been known that microbes may be metabolically active even though they are not culturable. Furthermore, quantification of the viable microbial biomass may help elucidate the potential for specific microbial metabolic activities in a particular habitat, but the nonviable biomass may also be important as a component of the food web, despite its lack of a direct metabolic contribution. Although it is possible and common to make inferences about microbial biomass based on microbial community metabolic activity, it is important to note that these may not be strongly correlated; i.e. high metabolic activity does not necessarily mean an actively growing, dividing microbial community. Also, it is important to remember that microbial physiological activities are often strongly affected by the physical arrangement of the members of a population, especially aggregations such as those that occur in biofilms, on root surfaces (Dandurand et al. 1995, 1997), and in soil microsites. Physiological and spatial heterogeneity have important implications for the ecology of microbial populations and communities, and the ability of recently developed molecular and microscopic methods to address heterogeneity in environmental samples is especially exciting. Below, several of the most popular methodologies for quantifying microbial metabolic activity are briefly discussed. In each case, references are provided to original reports and more extensive, step-by-step protocols. The related subject area of bioreporter gene technology, which allows the analysis of in situ gene expression by supplying assayable gene products, and in which the gene product of interest is difficult or impossible to assay, is briefly discussed below. Similarly, biosensor technology is very promising for environmental microbiology; recent reviews on applications of biosensors for environmental microbiology are available (Burlage 1997).

Several methods for estimating metabolic activity levels of microbial populations involve quantifying cellular pools of specific biochemical components, including RNA, DNA, ATP, and total adenine nucleotide (TAN). Because the specific growth rate of bacteria is dependent on cellular amounts and synthesis rates of ribosomal RNA (rRNA), there is a strong correlation between cellular RNA content, or RND:DNA ratio, and growth rate (Kemp et al. 1993). ATP and TAN pool turnover rates can provide estimates of energy flux and specific growth rate in natural microbial populations. Kemp et al. (1993) described procedures to measure cell-specific quantities of rRNA and DNA in order to quantify the frequency distribution of activity among cells. In this procedure, fluorescently labeled oligonucleotide probes are hybridized to complementary 16S rRNA sequences in intact cells. Cell fluorescence, which is proportional to cellular rRNA content, is then measured, and compared with measurements of DNA content obtained by fluorescence of 4′, 6′-diamino-2-phenylindole hydrochloride (DAPI)-stained cells, generating a rRNA:DNA ratio for individual cells. Procedures for measuring rates of stable RNA synthesis have been applied to bacterial cultures as well as aquatic bacteria, based on the uptake and incorporation of radiolabelled adenine into cellular RNA (Karl 1979). Methods for simultaneous quantification of RNA and DNA synthesis rates also have been developed and used for estimating rates of growth, cell division, and biomass in microbial communities (Karl 1981, 1993). Recently, Lee et al. (1999) used a combination of fluorescent in situ hybridization (FISH) performed with rRNA-targeted oligonucleotide probes and microautoradiography, to quantify specific substrate-uptake profiles of individual bacterial cells within microbial communities.

Genetic methods

Recent advances in molecular biology have made it easier to analyze the genomes and quantify the genetic diversity of soil- and plant-associated microbes, as well as to track them in the environment (Bramwell et al. 1995; Louws et al. 1999). Much of this research has been motivated by the need for rapid and accurate detection of plant and animal pathogens, but the methods are applicable to many different aspects of environmental microbiology. Some of the DNA-based microbial characterization methods are dependent on PCR amplification, while others are not. Examples of the latter include DNA–DNA reassociation kinetics, characterization of plasmid profiles, and restriction enzyme analysis of total genomic or plasmid DNA. PCR-based approaches are often more useful because they require smaller amounts of DNA; also, they tend to be faster, more sensitive, and more specific. The main methods for analyzing and quantifying microbial genomes by PCR-based methodology include: repetitive DNA PCR (rep-PCR), rDNA-based PCR, amplified fragment length polymorphism (AFLP) analysis, and arbitrarily primed PCR (AP-PCR) randomly amplified polymorphic DNA (RAPD) (Louws et al. 1999).

A number of technical difficulties are associated with extraction and analysis of nucleic acids from soil; for example, DNA may be tightly bound to humic materials and clays. A number of protocols have been published for direct extraction of nucleic acids from soil, and the technology is rapidly evolving in this area (Levin et al. 1992; Pepper and Dowd 2002). Because PCR-based protocols are so highly sensitive, certain problems may arise. For example, only a very small quantity (e.g. a few molecules) of PCR-generated fragments can be sufficient to cross-contaminate samples of subsequent PCR runs, resulting in the detection of false positives. A number of recommendations have been suggested to avoid such contamination (Carrino and Lee 1995; Kwok 1990). Because PCR can amplify DNA from dead cells as well as live ones, it may produce false positives. Additionally, PCR cannot directly distinguish viable but not culturable (VBNC) cells from culturable cells. One approach, BIO-PCR, was developed to circumvent this problem when assaying for plant pathogenic bacteria; with BIO-PCR, samples are first plated on to growth medium to culture any viable cells, before PCR analysis (Schaad et al. 1995). This approach has been used successfully for bacteria in soil as well as for plant- and seed-associated bacteria. When PCR is used to detect plant-associated microbes, there is the possibility of the presence of sample-related compounds that can inhibit the polymerase reaction. Wilson (1997) reviewed methods to avoid potential problems with inhibitory compounds.

Use of marker and reporter genes

Recently, genetic engineering of biocontrol agents with reporter or marker genes has provided useful tools for detection and monitoring of introduced biocontrol agents in natural environments (Green and Jensen 1995; Lo et al. 1998; Bae and Knudsen 2000). For example, the selectable hygromycin B phosphotransferase (*hygB*) gene, coding for resistance to this antibiotic, has been used to detect fungal biocontrol agents in the rhizosphere and phyllosphere (Lo et al. 1998). The β-glucuronidase (GUS) marker gene is also a promising tool for ecological studies of biocontrol agents, because of the low background activity of GUS in fungi and plants, the relative ease and sensitivity of detection (Roberts et al. 1989), and the apparent lack of influence of GUS expression on biocontrol efficacy. However, some background GUS activity may be present in unsterile systems or natural soils. The green fluorescent protein (GFP) of the jellyfish *Aequorea victoria* has also been developed as a reporter for gene expression (Chalfie et al. 1994). It has been successfully cloned and expressed in a variety of different organisms including plants, animals, fungi, and bacteria. GFP requires only ultraviolet (UV) or blue light and oxygen to induce green-fluorescence. An exogenous substrate, such as GUS requires, is not needed for the detection of GFP, thus avoiding problems related to cell permeability and substrate uptake.

Advances in microscopy

A number of methods to quantify metabolic activity and/or biomass of individual microbes, populations, or microbial communities involve direct microscopic observation of cells. These include simple measurement of cell size and size distributions, formation of microscopically observable metabolic products, fluorochromes used as 'physiological stains,' and genetically based marker and reporter systems. Techniques used to investigate root biofilm communities have included confocal microscopy with or without targeted rDNA probing, electron microscopy (SEM and TEM), and PCR-assisted rDNA RFLP/sequence analyses. Most of the wide range of potential microscopic techniques have been attempted for visualization of microbes on root and foliar surfaces, including bright-field, phase-contrast, and differential interference, epifluorescence, scanning and transmission electron microscopy, and atomic-force microscopy. In general, direct visualization for quantitative purposes can be difficult because of opacity of the substrate and the small size of microorganisms; spatial associations are difficult to describe because of the short working distance and the high magnification needed to see them. Phase-contrast or interference microscopy of fresh material is possible, but associated soil and root material can give confusing background for observation of microorganisms. Fluorochromes and other stains are commonly used to observe microbes on plant surfaces. However, autofluorescence of plant tissue can be a problem when using fluorochromes to detect microbes on plant surfaces.

Sampling considerations

A number of special considerations arise when sampling microbes from the rhizosphere and phyllosphere. Parberry et al. (1981), Hirano and Upper (1983, 1986), Kinkel et al. (1995), and Kloepper and Beauchamp (1992) have reviewed some of the difficulties inherent in sampling the phyllosphere and the rhizosphere. Problems include the non-normal distribution of microorganisms, selection of sampling strategy, scale of sample, and sample unit. The choice of sampling strategy, scale of sample, and sample unit can all have a significant effect on the results of an experiment. The type of sampling strategy used in sampling the rhizosphere and the phyllosphere is often dependent on the question that is being asked. It is generally recommended that samples be taken at random so that the assumption of independence of samples for standard statistical tests (e.g. analysis of variance) is met. However, sampling of microbes from the rhizosphere, in many cases, cannot be done randomly. For example, colonization is often determined by taking root segments at a prespecified distance (e.g. 1-cm segments 2 cm below the soil line). Another example, which can result in autocorrelated data, is when roots are sequentially sampled in order to assess the root-colonizing ability of a microbe. In these situations, one must verify that the data meet the assumption of the statistical analysis prior to conducting the tests.

The selection of sample scale and sample unit is based on biological features of plants, reduction of sample variance, ease of processing, and other methodological constraints. When epiphytic bacterial populations were sampled, a high level of variability was reported among sample units at every scale investigated (leaf segments, leaflets, or whole leaves) (Kinkel et al. 1995). In contrast, variability of rhizobacteria was significantly less when whole roots were sampled compared with root segments (Kloepper et al. 1991). Thus, for phyllosphere sampling, Kinkel et al. (1995) suggested that selection of sample scale and unit should not be based only on the assumption that the variance is scale dependent but may be more appropriately based on methodological and biological considerations. Although sample variance may be reduced when sampling whole-root systems rather than root segments, this sample scale may not always be feasible (e.g. when sampling mature plants).

Natural populations and habitats are spatially heterogeneous, i.e. organisms and their resources are usually not uniformly distributed over space or time but are found in different degrees of aggregation. Several theoretical studies have demonstrated the importance of spatial heterogeneity for population and metapopulation persistence and abundance (Hanski 1991). Hirano and Upper (1993) demonstrated changes in frequency distributions of epiphytic bacterial populations on populations of habitats (individual leaflets) within a plant canopy. However, despite continuing demonstration of the theoretical and practical importance of spatial processes to ecology, relatively few mechanistic studies of the factors that create heterogeneous spatial distributions and the processes by which they occur in nature are available. As pointed out by Hirano and Upper (1993), an understanding of factors that regulate epiphytic bacterial population dynamics must address mechanisms that underlie variability.

Most natural environments are spatially structured by various energy inputs that result in patchy structures or gradients (Legendre and Fortin 1989). Thus, biological organisms are rarely distributed in a random or uniform manner. The rhizosphere is a good example of this, since energy input is largely due to root deposition. Spatial variability of nutrient deposition from seeds, as well as the spatial presence of roots, may influence sites of colonization by rhizosphere microbes and the quantity and composition of root deposition varies along the root. Sites may be preferentially colonized by some microbes, such as bacteria in cell junctions (Rovira 1956) or zoospore encystment in the zone of root-cell elongation (Hickman and Ho 1966; Mitchell and Deacon 1986; Zentmeyer 1961). Although the tendency for both phyllosphere and rhizosphere microbial populations to conform to lognormal or similar frequency distributions has been noted (Hirano et al. 1982; Loper et al. 1984; Newman and Bowen 1974), the actual spatial variability of these populations is less well documented.

Appropriate characterization of the spatial variability inherent in soil- and plant-associated microbial communities will allow us to improve quantification of the organisms and processes under study.

REFERENCES

Andrews, J.H. and Harris, R.F. 1986. R- and K-selection and microbial ecology. In: Marshall, K.C. (ed.), *Advances in microbial ecology*. vol. 9. New York: Plenum Publishing Corporation, 99–147.

Armstrong, J.L., Knudsen, G.R. and Seidler, R.J. 1987. Survival of recombinant bacteria associated with plants and herbivorous insects in a microcosm. *Curr Microbiol*, **15**, 229–32.

Bae, Y.S. and Knudsen, G.R. 2000. Cotransformation of *Trichoderma harzianum* with b-glucuronidase and green fluorescent protein genes provides a useful tool for monitoring fungal growth and activity in natural soils. *Appl Environ Microbiol*, **66**, 810–15.

Bahme, J.B. and Schroth, M.N. 1987. Spatial–temporal colonization patterns of a rhizobacterium on underground organs of potato. *Phytopathology*, **77**, 1093–100.

Blakeman, J.P. 1981. *Microbial ecology of the phylloplane*. London: Academic Press.

Blakeman, J.P. and Atkinson, P. 1981. Antimicrobial substances associated with the aerial surfaces of plants. In: Blakeman, J.P. (ed.), *Microbial ecology of the phylloplane*. London: Academic Press, 245–63.

Bowen, G.D. and Rovira, A.D. 1976. Microbial colonization of plant roots. *Ann Rev Phytopathol*, **14**, 121–44.

Bramwell, P.A., Barallon, R.V., et al. 1995. *Molecular Microbial Ecology Manual*. Dordrecht: Kluwer Academic.

Bunster, L., Fokkema, N.J. and Schippers, B. 1989. Effects of surface-active *Pseudomonas* spp. on leaf wettability. *Appl Environ Microbiol*, **55**, 1340–5.

Burlage, R.S. 1997. Emerging technologies: Bioreporters, biosensors, and microprobes. In: Hurst, C.J., Knudsen, G.R., et al. (eds), *Manual of environmental microbiology*. Washington, DC: American Society for Microbiology, 115–23.

Campbell, R. and Porter, R. 1982. Low-temperature scanning electron microscopy of microorganisms in soil. *Soil Biol Biochem*, **14**, 241–5.

Carrino, J.J. and Lee, H.H. 1995. Nucleic-acid amplification methods. *J Microbiol Meth*, **23**, 3–20.

Chalfie, M., Tu, Y., et al. 1994. Green fluorescent protein as a marker for gene expression. *Science*, **263**, 802–5.

Cochran, W.G. 1950. Estimation of bacterial densities by means of the most probable number. *Biometrics*, **March**, 105–16.

Colwell, R.R., Brayton, P.R., et al. 1985. Viable but non-culturable *Vibrio cholerae* and related pathogens in the environment: implications for release of genetically engineered microorganisms. *Biol Tech*, **3**, 817–20.

Costerton, J.W., Nickel, J.C. and Ladd, T.I. 1986. Suitable methods for the comparative study of free-living and surface-associated bacterial populations. In: Poindexter, J.S. and Leadbetter, E.R. (eds), *Methods and special applications in bacterial ecology. Bacteria in nature*. vol. 2. New York: Plenum Press, 49–84.

Curl, E.A. and Truelove, B. 1986. *The rhizosphere*. Berlin: Springer-Verlag, 288.

Dandurand, L.M. and Knudsen, G.R. 2001. Sampling microbes from the rhizosphere and phyllosphere. In: Hurst, C.J., Knudsen, G.R., et al. (eds), *Manual of environmental microbiology*, 2nd edn. Washington, DC: American Society for Microbiology.

Dandurand, L.M., Knudsen, G.R. and Schotzko, D.J. 1995. Quantification of *Pythium ultimum* var. *sporangiiferum* zoospore encystment patterns using geostatistics. *Phytopathology*, **85**, 186–90.

Dandurand, L.M., Schotzko, D.J. and Knudsen, G.R. 1997. Spatial patterns of rhizoplane populations of *Pseudomonas fluorescens*. *Appl Environ Microbiol*, **63**, 3211–17.

Darbyshire, J.F. and Greaves, M.R. 1970. An improved method for the study of the interrelationships of soil microorganisms and plant roots. *Soil Biol Biochem*, **2**, 63–71.

Donegan, K., Maytac, C., et al. 1991. Evaluation of methods for sampling, recovery and enumeration of bacteria applied to the phylloplane. *Appl Environ Microbiol*, **57**, 51–6.

Gray, T.R.G. and Parkinson, D. 1968. *The ecology of soil bacteria*. Toronto: University of Toronto Press, 681.

Green, H. and Jensen, D.F. 1995. A tool for monitoring *Trichoderma harzianum*. II. The use of a GUS transformant for ecological studies in the rhizosphere. *Phytopathology*, **85**, 1436–40.

Hanski, I. 1991. Single-species metapopulation dynamics: Concepts, models and observations. *Biol J Linn Soc*, **42**, 73–88.

Hax, C.L. and Golladay, S.W. 1993. Biofilm development and macroinvertebrate colonization of leaves and wood in a boreal river. *Freshw Bio*, **29**, 79–87.

Hickman, C.J. and Ho, H.H. 1966. Behavior of zoospores in plant-pathogenic phycomycetes. *Annu Rev Phytopathol*, **4**, 195–220.

Hiltner, L. 1904. Uber neuere Erfahrungen und Problem auf dem Gebiet der Bodenbakteriologie und unter besonderer Berucksichtigung der Grundungung und Brache. *Arb Dtsch Landwirt Ges*, **98**, 59–78.

Hirano, S.S. and Upper, C.D. 1983. Ecology and epidemiology of foliar bacterial plant pathogens. *Ann Rev Plant Pathol*, **21**, 243–69.

Hirano, S.S. and Upper, C.D. 1986. Temporal, spatial and genetic variability of leaf-assoceated bacterial populations. In: Fokkema, N.J. and Van Den Heuvel, J. (eds), *Microbiology of the phyllosphere*. New York: Cambridge University Press, 235–41.

Hirano, S.S. and Upper, C.D. 1993. Dynamics, spread, and persistence of a single genotype of *Pseudomonas syringae* relative to those of its conspecifics on populations of snap bean leaflets. *Appl Environ Microbiol*, **59**, 1082–91.

Hirano, S.S., Nordheim, E.V., Arny, D.C. and Upper, C.D. 1982. Lognormal distribution of epiphytic bacterial populations on leaf surfaces. *Appl Environ Microbiol*, **44**, 695–700.

Karl, D.M. 1979. Measurement of microbial activity and growth in the ocean by rates of stable ribonucleic acid synthesis. *Appl Environ Microbiol*, **38**, 850–4.

Karl, D.M. 1981. Simultaneous rates of ribonucleic acid and deoxyribonucleic acid syntheses for estmiating growth and cell division of aquatic microbial communities. *Appl Environ Microbiol*, **42**, 802–10.

Karl, D.M. 1993. Adenosine triphosphate (ATP) and total adenine nucleotide (TAN) pool turnover rates as measures of energy flux and specific growth rate in natural populations of microorganisms. In: Kemp, P.F., Sherr, B.F., et al. (eds), *Handbook of methods in aquatic microbial ecology*. Boca Raton, FL: Lewis Publishers, 483–94.

Katznelson, H. 1965. Nature and importance of the rhizosphere. In: Baker, K.J. and Snyder, W.C. (eds), *Ecology of soil borne plant pathogens – prelude to biological control*. Berkeley: University of California Press, 187–209.

Kemp, P.F., Lee, S. and LaRoche, J. 1993. Evaluating bacterial activity from cell-specific ribosomal RNA content measured with oligonucleotide probes. In: Kemp, P.F., Scherr, B.F., et al. (eds), *Handbook of methods in aquatic microbial ecology*. Boca Raton, FL: Lewis Publishers, 415–22.

Kinkel, L.L. 1991. Fungal community dynamics. In: Andrews, J.H. and Hirano, S.S. (eds), *Microbial ecology of leaves*. New York: Springer-Verlag, 253–70.

Kinkel, L.L., Andrews, J.H., et al. 1987. Leaves as islands for microbes. *Oecologia*, **71**, 405–8.

Kinkel, L.L., Wilson, M. and Lindow, S.E. 1995. Effect of sampling scale on the assessment of epiphytic bacterial populations. *Micro Ecol*, **29**, 283–97.

Kloepper, J.W. and Beauchamp, C.J. 1992. A review of issues related to measuring colonization of plant roots by bacteria. *Can J Microbiol*, **29**, 1219–32.

Kloepper, J.W., Mahafee, W.F., et al. 1991. Comparative analysis of five methods for recovering rhizobacteria form cotton roots. *Can J Microbiol*, **37**, 953–7.

Knudsen, G.R. 1988. Model for the dispersal of aerosolized bacteria in field trials. *Appl Environ Microbiol*.

Knudsen, G.R. and Hudler, G.W. 1987. Use of a computer simulation model to evaluate a plant disease biocontrol agent. *Ecol Modell*, **35**, 45–62.

Kwok, S. 1990. Procedures to minimize PCR-product carry-over. In: Innes, M.A., Gelfand, D.H., et al. (eds), *PCR protocols: a guide to methods and applications*. San Diego, CA: Academic Press, 142–5.

Langvad, F. 1980. A simple and rapid method for qualitative and quantitative study of the fungal flora of leaves. *Can J Microbiol*, **26**, 666–70.

Lee, N., Nielsen, P.H., et al. 1999. Combination of fluorescent in situ hybridization and microauto radiography: a new tool for structure-function analyses in microbial ecology. *Appl Environ Microbiol*, **65**, 1289–97.

Legendre, P. and Fortin, M.J. 1989. Spatial pattern and ecological analysis. *Vegetatio*, **80**, 107–38.

Levin, M.A., Seidler, R.J. and Rogul, M. 1992. *Microbial ecology: principles, methods and applications*. New York: McGraw-Hill.

Lindow, S.E., Knudsen, G.R., et al. 1988. Aerial dispersal and epiphytic survival of *Pseudomonas syringae* during a pretest for the release of genetically engineered strains into the environment. *Appl Environ Microbiol*, **54**, 1557–63.

Lo, C.-T., Nelson, E.B., et al. 1998. Ecological studies of transformed *Trichoderma harzianum* strain 1295-22 in the rhizosphere and on the phylloplane of creeping bentgrass. *Phytopathology*, **88**, 129–36.

Loper, J.E., Suslow, T.V. and Schroth, M.N. 1984. Lognormal distribution of bacterial populations in the rhizosphere. *Phytopathology*, **74**, 1454–60.

Louws, F.J., Rademaker, J.L.W. and de Bruijn, F.J. 1999. The three Ds of PCR-based genomic analysis of phytobacteria: diversity, detection, and disease diagnosis. *Annu Rev Phytopathol*, **37**, 81–125.

Lynch, J.M. 1987. Soil biology – accomplishements and potential. *Soil Sci Soc Am J*, **51**, 1409–12.

Margulis, L. 1971. The origin of plant and animal cells. *Am Sci*, **59**, 230–5.

Metting, F.B. (ed.) 1993. *Soil microbial ecology*. New York: Marcel Dekker.

Mitchell, R.T. and Deacon, J.W. 1986. Differential (host specific) accumulation of zoospores of Pythium on roots of graminaceous and non-graminaceous plants. *New Phytol*, **102**, 113–22.

Newman, E.I. and Bowen, H.J. 1974. Patterns of distribution of bacterial on root surfaces. *Soil Biol Biochem*, **6**, 205–9.

O'Toole, G.A. and Kolter, R. 1998. Initiation of biofilm formation in *Pseudomonas fluorescens* WCS365 proceeds via multiple, convergent signaling pathways: a genetic analysis. *Mol Microbiol*, **28**, 449–61.

Parberry, I.H., Brown, J.F. and Bofinger, V.J. 1981. Statistical methods in the analysis of phylloplane populations. In: Blakeman, J.P. (ed.), *Microbial ecology of the phylloplane*. London: Academic Press, 47–65.

Parke, J.L., Liddell, C.M. and Clayton, M.K. 1990. Relationship between soil mass adhering to pea taproots and recovery of *Pseudomonas fluorescens* from the rhizosphere. *Soil Biol Biochem*, **22**, 495–9.

Paul, E.A. and Clark, F.E. 1989. *Soil microbiology and biochemistry*. New York: Academic Press.

Pepper, I.L. and Dowd, S.E. 2002. PCR applications for plant and soil microbes. In: Hurst, C.J., Crawford, R.L., et al. (eds), *Manual of environmental microbiology*, 2nd edn. Washington, DC: American Society for Microbiology, 573–82.

Richaume, A., Steinberg, D., et al. 1993. Differences between direct and indirect enumeration of soil bacteria: the influence of soil structure and cell location. *Soil Biol Biochem*, **25**, 641–3.

Roberts, I.N., Oliver, R.P., et al. 1989. Expression of *Escherichia coli.* β-glucuronidase gene in industrial and phytopathogenic filamentous fungi. *Curr Genet*, **15**, 177–80.

Rovira, A.D. 1956. A study of the development of the root surface microflora during the initial stages of plant growth. *J Appl Bacteriol*, **19**, 72–9.

Sarand, I., Timonen, S., et al. 1998. Microbial biofilms and catabolic plasmid harbouring degradative fluorescent pseudomonads in Scots pine mycorrhizospheres developed on petroleum contaminated soil. *FEMS Microbiol Ecol*, **27**, 115–26.

Schaad, N.W., Cheong, S.S., et al. 1995. A combined biological and enzymatic amplification (BIO-PCR) technique to detect *Pseudomonas syringae* pv. *phaseolicola* in bean seed extracts. *Phytopathology*, **85**, 243–8.

Skinner, F.A., Jones, P.C.T. and Mollison, J.E. 1952. A comparison of a direct- and a plate-counting technique for the quantitative estimation of soil micro-organisms. *J Gen Microbiol*, **6**, 261–71.

Suslow, T.V. and Schroth, M.N. 1982. Rhizobacteria of sugar beets: effects of seed application and root colonization on yield. *Phytopathology*, **72**, 199–206.

Thomashow, L. and Weller, D. 1991. Role of antibiotics and siderophores in biocontrol of take-all disease of wheat. In: Keister, D.I. and Cregan, P.B. (eds), *The rhizosphere and plant growth.* Dordrecht: Kluwer, 245–51.

Tukey, H.B. 1971. Leaching of substances from plants. In: Preece, T.F. and Dickinson, C.H. (eds), *Ecology of leaf surface micro-organisms.* London: Academic Press, 67–80.

Whipps, J.M. 1990. Carbon economy. In: Lynch, J.M. (ed.), *The rhizosphere.* Chichester: J. Wiley & Sons, 59–97.

Williams, S.T. 1986. The ecology of antibiotic production. *Microb Ecol*, **12**, 43–52.

Wilson, I.G. 1997. Inhibition and facilitation of nucleic acid amplification. *Appl Environ Microbiol*, **63**, 3741–51.

Yee, D.C., Maynard, J.A. and Wood, T.K. 1998. Rhizoremediation of trichloroethylene by a recombinant, root-colonizing *Pseudomonas fluorescens* strain expressing toluene ortho-monooxygenase constitutively. *Appl Environ Microbiol*, **64**, 112–18.

Zentmeyer, G.A. 1961. Chemotaxis of zoospores for root exudates. *Science*, **133**, 1595–6.

Zuberer, D.A. 1994. Recovery and enumeration of viable bacteria. In: Weaver, R.W., Angle, J.S. and Bottomley, P.S. (eds), *Methods of soil analysis. Part 2. Microbiological and biochemical properties.* Madison, WI: Soil Science Society of America, Inc, 119–58.

Bacteriology of water

EDWIN E. GELDREICH

Bacteriology of water continues to be an important component in defining the quality of public health. In recent years, expanding methodology has led to the discovery of additional pathogens as the cause of water-borne outbreaks. As a consequence of these new discoveries, the focus on water quality is being redirected towards better characterization of the microbial risks associated with this vital resource. It is important to note that although water is essential to sustain life, it can be a hostile environment to humans when grossly polluted.

CHARACTERIZING THE BACTERIAL FLORA IN THE AQUATIC ENVIRONMENT

The bacterial flora of water is a composite of organisms indigenous to high-quality water or pristine waters plus all of the contributions from life's many controlled and uncontrolled uses or consumption of water.

Indigenous bacteria

These are naturally occurring bacteria found in waters that are remote from the activities of modern civilization. They are largely saprophytic organisms, some of which belong to the genera *Micrococcus*, *Pseudomonas*, *Serratia*, *Flavobacterium*, *Chromobacterium*, *Acinetobacter*, and *Alcaligenes*. There are indigenous bacteria that are difficult to detect in routine laboratory techniques because of their slow growth or fastidiousness. In the past, these organisms received little attention, but this attitude is changing because of problems with biofilm development in streams, ground-water extraction, water pipe networks, and industrial processes. An additional threat is that some opportunistic pathogens, such as *Legionella*, may be ubiquitous in the water environment.

Contributed bacteria

Rainfall runoff is a major contributor to changes in the bacterial flora found in water. Whereas rain or snow falling over the watershed contains only occasional bacteria acquired from air-borne particles (Table 9.1), the microbial and chemical qualities of rainwater drastically degrade upon contact with the soil (Geldreich et al. 1968). Various soil and fecal organisms are immediately accumulated into the drainage and merge as the bacterial flora. These organisms, though not normal inhabitants of high-quality water, either adapt to the aquatic environment or die out over time because of various environmental adversities. Predominant among these soil and vegetation organisms are the aerobic spore-forming bacilli, such as *Bacillus subtilis*, *Bacillus megaterium*, and *Bacillus mycoides*. Others, such as *Klebsiella pneumoniae*, *Enterobacter aerogenes*, and *Enterobacter cloacae*, may be found growing in the water- and food-conducting tissue of trees (Knittel et al. 1977), on crops, and on other vegetation. By use of special media, a variety of other specialized bacterial groups can be isolated. In the bottom sediments of

Table 9.1 *Seasonal variations (median values) for bacterial discharges in rainwater and stormwater from suburban areas, Cincinnati, Ohio*

| Source | Period | Total samples | Season | Organisms per 100 ml | | | Ratios FC:FS | FC/TC ×100 |
				Total coliforms (TC)	Fecal coliforms (FC)	Fecal streptococci (FS)		
Rain water	June 1965–February 1967	49	Spring	<1.0	<0.3	<1.0	–	–
			Summer	<1.0	<0.7	<1.0	–	–
			Autumn	<0.4	<0.4	<0.4	–	–
			Winter	<0.8	<0.5	<0.5	–	–
Wooded hillside	February 1962–December 1964	278	Spring	2 400	190	940	0.20	7.9
			Summer	79 000	1900	27 000	0.70	2.4
			Autumn	180 000	430	13 000	0.30	0.2
			Winter	260	20	950	0.02	7.7
Street gutters	January 1962–January 1964	177	Spring	1 400	230	3 100	0.07	16.4
			Summer	90 000	6 000	150 000	0.04	7.1
			Autumn	290 000	47 000	140 000	0.34	16.2
			Winter	1 600	50	2 200	0.02	3.1
Business district	April 1962–July 1966	294	Spring	22 000	2 500	13 000	0.19	11.4
			Summer	172 000	13 000	51 000	0.26	7.6
			Autumn	190 000	40 000	56 000	0.71	21.1
			Winter	46 000	4 300	28 000	0.15	9.4

Data from Geldreich et al. 1968.

streams and lakes where oxygen is restricted, nitrate, sulfate-reducing, and methanogenic bacteria can be detected and impart disagreeable earthy odors and tastes during summer periods of low stream flow.

The microbial flora of sewage is predominantly from fecal wastes including pathogens shed from individuals in the community. These would include species of *Salmonella, Shigella, Campylobacter, Yersinia, Leptospira, Streptococcus, Clostridium,* and *Vibrio cholerae* and pathogenic types of *Escherichia coli.* Other significant contributors to the flora of municipal sewage are introduced in combining stormwater runoff, laundry wash water, automatic car-washing water drainage, processing food wastes, and industrial discharges.

Within the complex flora of sewage-treatment sludge are organisms that are responsible for much of the biological breakdown of degraded waste products. Some of these beneficial organisms include the filamentous bacteria *Sphaerotilus natans, Haliscomenobacter hydrossis, Nostocoida limicola, Microthrix parvicella, Flexibacter, Microscilla,* and *Nocardia* (Eikelboom 1975) and the important floc-forming bacteria (*Zoogloea ramigera*), along with *Proteus* and the anaerobe *Clostridium sporogenes.*

Animal feedlot operations that require the confinement of cattle in small areas create a fecal-waste-removal problem equal to the domestic waste discharges of small cities (Geldreich 1972b). If the animal waste is not discharged to a lagoon or landfill, stormwater runoff over the animal feedlots will bring massive loads of fecal

pollution to the drainage basin. While cattle feedlot wastes may be a major source of bacterial pathogens (particularly, *E. coli* O157:H7), poultry-farm fecal wastes can contribute a variety of salmonellae into the drainage basin that could also be a resource for water supply downstream.

Food processing, beet-sugar extraction, and cane-sugar production have a drastic impact on the bacterial flora and the self-purification capacity of receiving waters. Elevated concentrations of nutrient wastes lead to oxygen depletion of the receiving waters plus an increase in environmental coliforms and fecal streptococci, particularly streptococci variants on biotypes (Mundt 1963; Mundt et al. 1966b). Most troubling is the extended persistence of *Salmonella* spp. under these conditions (Geldreich 1972a).

Discharges of poorly treated paper-mill waste to receiving streams or lakes can also have a severe impact on the microbial quality of surface waters. The association of environmental *Klebsiella* with paper manufacturing and the excessive nutrients in paper-mill waste cause a tremendous regrowth of this coliform in downstream waters. Environmental *Klebsiella* spp. colonize plants at flower pollination. As the seed germinates, *Klebsiella* spp. establish a coexistence within the plant by metabolizing nutrients from wood sugars (cyclitols) in the nutrient- and water-conducting tissues during the life of the tree (Seidler et al. 1977). Another byproduct of paper processing is wastewater sulfites, which encourage the growth of biofilm mats of the nuisance organism

Table 9.2 *Microbial characterization of solid wastes from various cities*

Waste collection site[a]	Number of samples	Organisms per g wet wt[b]			
		Total viable bacteria[c]	Spores	Total coliforms	Fecal coliforms
A	4	110 000 000	270 000	3 000 000	260 000
B	3	450 000 000	110 000	6 700 000	510 000
C	6	78 000 000	38 000	1 600 000	1 200 000
D	3	480 000 000	31 000	1 100 000	630 000
E	4	680 000 000	1 900 000	51 000 000	8 100 000
F	2	54 000 000	35 000	1 3000 000	5 600 000
G	2	4 000 000	25 000	340 000	15 000
H	3	300 000 000	160 000	8 600 000	3 000 000

Data from Peterson (1971).
a) Waste collection sites in Cincinnati, Chicago, Memphis, Atlanta, and New Orleans.
b) Average moisture content was 41 percent.
c) Spread plates (trypticase soy agar + blood) at 35°C for 48 h.

Sphaerotilus natans along the shoreline of receiving streams.

Solid waste, generally referred to as garbage or rubbish, contains a multitude of items such as food discards, garden rubbish, manufactured products (paper, plastic, rubber, leather, textile, wool, metals, glass, ceramics), rock, dirt and ash residues, and also fecal material (Table 9.2). Much of the fecal material in urban areas is derived from disposable diapers, pet-litter material, and feces of rodents foraging for food in these waste collections (Geldreich 1978). Therefore, it is not surprising that solid wastes contain a large and varied microbial population that includes a wide spectrum of bacteria (aerobes, anaerobes, thermophiles), actinomycetes, and fungi (Cook et al. 1967). If properly disposed of through sanitary landfills, many of the problems (foraging wildlife, odors, unsightliness) common to open dumping are avoided. However, poor placement of landfill sites may result in the migration of leachates into nearby surface waters and groundwater resources.

NATURAL WATERS

Natural waters occur either at or near the surface of the earth. Surface waters are frequently exposed to contamination from soil, stormwater runoff, domestic sewage, agricultural and industrial wastes, and decomposing vegetation. On the other hand, ground waters can be expected to be of excellent bacteriological quality due to the beneficial percolation into underground strata (aquifers). During downward passage to an aquifer, perhaps several hundred meters below the surface, most of the surface contaminants are entrapped in the soil or porous rock layers. One exception is a rock stratum composed of limestone, which is very porous and often results in sink holes and caverns through which surface water passes without effective entrapment of contaminants. In other situations, excessive application of minimally treated wastewaters to land surfaces may overwhelm the natural soil barrier (Kowal 1982). Once the aquifer becomes contaminated, restoration of water purity is very slow, even with accelerated intervention such as pumping the water to a treatment site and then returning it to the aquifer.

Surface waters

Natural water resources are replenished through rain, snow, and hail. This atmospheric moisture comes in contact with particles of suspended dust and smog that immediately contaminate the rain as it falls to the earth. The more dust encountered, the greater is the bacterial contamination. In remote areas away from the plumes of air pollutants, the total number of organisms may not exceed 10–20/l. Springtime dust storms originating in large farming areas may mix with atmospheric vapor to transport one to two total coliforms per liter of rain over great distances. Snow tends to be less pure than rain, probably because the snowflakes have greater surface area on which to collect suspended particles in the atmosphere, and also because their low temperature is conducive to the survival of bacteria. However, in snow on the tops of remote mountains, where the air is clean, very few organisms are present. Apparently hail contains more bacteria than either rain or snow. Hail examined from a storm over Padua during July 1901 was found to contain 140 000 microorganisms belonging to nine different types (Belli 1902). This surprising finding was probably caused by air currents that cycle the raindrops through periods of freezing and thawing as they repeatedly traverse plumes of dust in the air before falling to the ground. The number of organisms in ice depends on the nature of the water from which the ice was formed. With the exception of ice from glaciers, it is generally impure. Various heterotrophic bacteria, including coliforms and pathogens, have been found in ice made with contaminated water.

In remote areas, where human and farm animal populations are sparse, most organisms in water originate from soil with little evidence of contamination except

Table 9.3 *Indicator correlations with* Salmonella *occurrences in ambient waters (densities per 100 ml)*

Salmonella (%)	Clostridium[a]	Fecal streptococcus[a]	Fecal coliform[a]	Total coliform[a]
0	0	0	0	0
0	13	5	10	100
11	25	50	50	500
21	50	100	100	1 000
33	125	300	1 000	10 000
66	200	1 500	5 000	50 000
99	250	3 000	10 000	100 000
100	2 500	30 000	100 000	1 million
100	6 250	250 000	850 000	2.5 million

Data modified from Nemedi et al. (1984).
a) Densities per 100 ml.

for a few fecal coliforms (<100 organisms per 100 ml) originating from occasional wildlife inhabiting the immediate vicinity. Beaver and muskrats on the river banks, and significant concentrations of deer, elk, and other game animals in grazing plains and forest reserves, are responsible for a residual level of fecal microorganisms in water.

As these waters travel down a watershed, contact with agricultural and industrial activities increases and the river becomes laden with a variety of domestic and industrial wastes. As a consequence, rivers in most countries are heavily contaminated. Some measure of the potential occurrence of salmonellae in Hungarian ambient waters grouped at various average densities for several indicator groups is presented in Table 9.3. These data support the position that the greater the magnitude of fecal pollution, the greater the chance that some bacterial pathogens may also be present. Absence of any detectable fecal coliforms does not mean that there is no risk from *Giardia*, *Cryptosporidium*, or a virus. These pathogenic agents frequently occur in small densities requiring the testing of larger-volume (>10 liter) samples. For this reason, a surrogate indicator examination at only one sample volume of 100 ml is not going to provide sufficient precision to ascertain the potential for extremely low-density pathogen occurrences.

In a study of surface-water supplies used by 20 cities in the USA serving a total population of seven million people, it was estimated that the minimum wastewater component of the raw source water ranged from 2.3 to 16 percent and increased to predominantly wastewater for several cities during low-flow periods (Swayne et al. 1980). This public-health risk was verified by the finding of various serotypes of *Salmonella* at each water-plant intake. Bacteriological examination of raw water quality at the water treatment plant intakes of Omaha, Nebraska, St Joseph, Missouri, and Kansas City, Missouri, on the Mississippi River (Table 9.4) frequently revealed fecal coliform densities in excess of 2 000 organisms per 100 ml (Report 1971). This fecal pollution load resulted from inputs of raw sewage, effluents from

primary and secondary wastewater treatment plants of differing efficiencies, cattle feedlot runoff, and discharges from meat- and poultry-processing plants. Fecal discharges entrapped on soil also entered the drainage basin by the flushing action of storm events. The major raw-water quality concern is that human pathogens may also be present in surface waters used for water-supply sources or body-contact recreational pursuits.

Similarly high-elevation lakes in remote parts of a watershed will contain high-quality water (unless inhabited by flocks of aquatic birds), whereas those in the lower part of the watershed are fed with surface drainage from metropolitan areas and intense agricultural activity. For large lakes such as the North American Great Lakes, long retention time and vast volumes of water serve to buffer these magnificent water resources from the impact of stormwater runoff. Unfortunately, pollution plumes around waste-water discharges may spread to new areas by wind-driven currents. At greatest risk are the many small lakes surrounded by residential development that ultimately become polluted from sporadic drainage of septic systems and runoff from neighborhood lawns.

Ground waters

Groundwater resources are the major source of water supply for many communities, farms, and individual families world wide. In rural areas, many water supplies consist of dug wells less than 10 m deep. In these wells, source water is influenced by surface-water runoff that percolates through the soil. Since there is no protective bedrock perched on top to seal off surface contaminants, water quality is erratic. Many private wells are in this category, causing the untreated water supply to be an unsafe source of drinking water. Bore holes are often of better quality because casings are driven to depths of at least 30 m and soil depth does provide a barrier to much of the surface contamination. In general, public water supplies that use ground water attempt to reach high-

Table 9.4 *Fecal coliform densities and pathogen occurrence at Missouri River public water supply intakes*

Raw water intake	River mile	Date	Fecal coliforms (per 100 ml)[a]	Pathogen occurrence
Omaha, NE	626.2	7–18 October 1968	8 300	NT
		20 January–2 February 1969	4 900	NT
		8–12 September 1969	2 000	*Salmonella enteritidis*
		9–14 October 1969	3 500	*Salmonella anatum*
		3–7 November 1969	1 950	NT
St Joseph, MO	452.3	7–18 October 1968	6 500	NT
		20 January–2 February 1969	2 800	NT
		18–22 September 1969	4 300	NT
		9–14 October 1969	NT	*Salmonella montevideo*
		22 January 1970	NT	19 virus PFU
				Polio types 2, 3
				Eco types 7, 33
				3 virus PFU, not typed
Kansas City, MO	370.5	28 October–8 November 1968	6 500	NT
		20 January–2 February 1969	8 300	NT
		18–22 September 1969	3 800	*Salmonella newport*
				Salmonella give
				Salmonella infantis
				Salmonella poona

NT, no test PFU, plaque-forming units.
Data from Report (1971).
a) Geometric mean.

yield aquifers at greater depths, often 300 m or more, to satisfy a greater water demand and reliable source of bacteriologically safe drinking water.

The quality of water flowing from springs depends mainly on their source and surroundings. Unfortunately, in many instances the source is unknown and could be under the influence of surface water, so quality may vary as a result of rainfall events and other activities on the watershed.

Cisterns

This source of water is generally rainfall from some catchment surface, which often is a residential roof or paved hillside for water to drain down into a storage tank. Bacteriological quality is a reflection not only of rainwater and dust particles but also of fecal contamination from birds perched on the catchment surface. Table 9.5 illustrates the varying densities of several bacterial indicator groups, the standard plate count, and *Pseudomonas aeruginosa* that may be encountered in tropical cistern waters. In some areas of the world, cisterns have become a blend of rainwater augmented by purchased water during the dry season.

Swimming pools and therapeutic baths

Swimming pools most often use treated municipal supplies for filling, but the water may also be derived from thermal springs, mineral springs, or the ocean. Pool operation and construction materials (cement, plastic, and redwood) have a definite effect on microbial growth (Castle 1985; Davis 1985) as does variability in water temperature in both indoor and outdoor pools (Schiemann 1985). Outdoor pools can be contaminated through stormwater drainage, soil, vegetation debris, and pets and wildlife. Microorganisms of major concern are those from the bather's body and mucosa and include those causing infections of the ear, eyes, upper respiratory tract, skin, and intestinal tract (Favero and Drake 1966). Obviously the number of bathers is another serious contributor to the deterioration of water quality. Major microbial contributions are from saliva and sinus drainage, as well as fecal contamination from the defecation of infant bathers. Pool-water disinfection can suppress many microorganisms, but effectiveness is limited by short contact time, disinfectant demand created by bather body oils, particulates, and sunlight exposure.

Bottled water

This water is generally an acceptable alternative drinking water supply and the primary alternative for emergency use (public supply contamination, foreign travel in developing nations), recreational/sports, and use in reconstituting baby formulae. Bottled water

Table 9.5 *Microbial quality of cistern water, public housing, Virgin Islands*

| Cistern site | No. samples | Range of counts per 100 ml | | | Range of counts per 1 ml | |
		Total coliforms	Fecal coliform	Fecal streptococcus[a]	Standard plate count[b]	*Pseudomonas aeruginosa*
1	9	<1–50	<1–3	<1–2	11–24 00	<1–151
2	8	<1–340	<1–8	<1–160	10–5 300	<1–72
3	8	<1–4	<1	<1–14	<1–14 000	<1–112
4	9	<1–11	<1–8	<1–640	2–12 000	<1–799
5	8	<1–100	<1–9	<1	6–2 500	<1–1
6	6	<1–1	<1–1	<1	1–650	<1
7	10	<1–220	<1–27	<1–412	763–30 000	<1–91
8	5	<1–103	<1–20	<1–40	420–3 000	<1–617
9	9	3–1700	3–570	<1–475	400–30 000	1–207
10	9	<1–29	<1–8	<1–972	210–4 400	1–22

Data adapted from Ruskin et al. (1989).
a) Including Enterococcus.
b) SPC agar spread plates, 35°C for 48 h.

sources are most often ground water or springs, but in recent years reprocessed public water supply (distilled, carbon-filtered, deionized, ultraviolet-radiated, or reverse osmosis desalinated) has also become available. Quality of the source water is dependent upon watershed protection, limited access to the area of extraction, and protected housing around well heads and spring outflows. Bottling operations follow prescribed sanitary codes for dispensing a food product and disposable plastic containers are commonly used to avoid possible contamination introduced from returnable glass bottles. As might be anticipated, microbial quality of freshly bottled water (within 48 h of bottling) is usually excellent. The rate of microbial change is related to a variety of factors including assimilable organic nutrients in the water, organisms competing for dominance in the microbial flora, water pH, and storage temperature. High-quality bottled waters undergo a slow rate of change because growth of the indigenous organisms may take hours, not minutes, and the available nutrients are in trace amounts. Waters bottled from sources that have fluctuating microbial populations that exceed 1 000 organisms per milliliter may contain coliforms and some opportunistic pathogens. Such bottled waters are of concern because the higher densities of heterotrophic bacteria may mask the laboratory's ability to detect fecal pollution and pathogens that might be present. Table 9.6 illustrates the variability in bacterial density and frequency of coliform-positive samples among bottled waters of unknown age purchased from retail outlets.

Although reported disease outbreaks due to contaminated bottled water are rare, the lack of documentation does not lessen the concern posed by use of this alternative. Any contaminated bottled water supply presents a unique hazard because of the widespread distribution of the product, which is not necessarily confined to a single community. In April to November 1974, Portugal had a cholera epidemic that caused 2 467 confirmed cases and 48 deaths (Blake et al. 1977). Most of the country was affected and the source of the infections was traced to the consumption of a brand of commercially bottled water as well as to poorly cooked shellfish. The source water may have become contaminated from cholera-infected individuals living on the watershed or involved in the bottling operation. This outbreak clearly illustrates the importance of watershed protection for ground waters and springs plus the need for effective sanitation practices during bottling operations.

Estuarine areas

These are the areas where fresh water mixes with the salt water environment either through direct discharge to the sea or by tidal flooding of freshwater pools near the ocean. Water-quality protection and enhancement of estuarine waters are particularly important for shellfish cultivation. Release of wastes to shellfish-growing waters brings a variety of organisms to the water, some of which may remain in suspension for varying periods before becoming entrapped in water turbidity. Most of the organisms reaching the area are either in aggregates or adsorbed to fecal cell debris and quickly accumulate in the bottom sediments, aided by the settling action of recirculating bottom silts. This natural deposition process concentrates much of the contaminants at the water–sediment interface, where hard- and soft-shell clams browse for food. For these reasons, it is essential that shellfish-harvesting areas be given a degree of protection from pollution that almost parallels the requirements for drinking water supplies. A zero tolerance for fecal pollution is not achievable in most shellfish-growing waters, so there needs to be a maximum permissible contaminant level recognized for authorized harvesting to minimize the public-health risks from consumption of raw shellfish.

Table 9.6 *Comparison of standard plate-count variability among brands of bottled water*

Brand	Sample type[a]	No. of samples	Count range[b] (SPC/ml)	Coliform positive samples[c]
A	Fresh	91	<10–25 000	3/91
	Retail	16	<10–28 000	0/16
B	Retail	6	<10–260 000	0/4
C	Fresh	1	<10	0/1
	Retail	11	1 200–160 000	0/11
D	Retail	5	31–650	0/5
E	Fresh	1	<10	0/1
	Retail	2	10 000–390 000	2/2
F	Retail	2	<10–12 000	0/2
G	Fresh	2	<10	0/2
	Retail	1	<10	0/1
H	Retail	9	<10–390	1/9
I	Retail	2	<10	0/2
J	Retail	2	<10–2 000	1/2
K	Retail	2	<10–12	1/2
L	Retail	4	13–4 300	0/4
M	Retail	1	<10	0/1
N	Retail	1	<10	0/1
O	Retail	1	1×10^6	1/1
P	Retail	1	8 600	0/1
Q	Retail	1	<10	0/1
R[d]	Retail	1	<10	0/1
R[d]	Retail	1	12	0/1

Data from Geldreich et al. (1975).

a) Fresh, samples direct from bottler, examined within 24 h; retail, samples of unknown age purchased from retail outlets.
b) Standard plate count (SPC) range values represent average count of bacteria per milliliter calculated from five replicate plates incubated for 72 h at 35°C using plate-count agar.
c) One or more coliforms per 100 ml.
d) Imported bottled water, carbonated.

Coastal waters

Coastal marine waters deservedly receive much attention because of their use as recreational bathing waters and fishery resources. Major sources of pollution are stormwater runoff along the beach areas, offshore dispersion of sewage through outfalls, sanitary wastes from ships in harbor, oil spills, and disposal of municipal garbage outside designated ocean dumping areas. In the sea, off heavily polluted bathing beaches, the coliform content may rise to 200 000 or more per 100 ml and salmonellae are frequently present (Report 1959). Organic enrichment of the sea by pollution enhances the activities of sulfate-reducing bacteria in sediments near the water interface and methanogenesis in the sediments below. After discharge from a sewage outfall, fecal and enteric bacteria rapidly disperse if there is good diffuser design and high tidal energy. At Sidmouth, Devon, which has an outfall discharging 460 m from the high-water mark, the median coliform count at the shore near the outfall was 73 organisms per 100 ml in 1965, equivalent to dilution of the sewage by a factor of 400 000 (Gameson et al. 1967).

PATHOGENIC BACTERIA IN AQUATIC ENVIRONMENTS

Numerous pathogenic agents have been isolated from ambient waters used for water supply, recreational bathing, and irrigation of garden salad crops (Geldreich and Bordner 1971; Geldreich 1972a; Rosenberg et al. 1976; Cordano and Vergilio 1990; Rose 1990). The list of water-borne pathogens in polluted temperate and tropical drinking waters (Table 9.7) includes many of the same bacterial and viral agents but differs by the additional hazards of a variety of water-borne parasites with alternating life cycles between man and aquatic life forms. The list of water-borne agents will increase as new methodologies evolve to detect the more elusive organisms that cause gastroenteritis or other human illnesses.

The magnitude of pathogen reservoirs in the environment is exacerbated by expanding human populations worldwide. Population crowding and the unregulated development of satellite communities place undue pressures on sanitation infrastructural barriers (sewage collection and treatment, solid waste disposal, potable

Table 9.7 *Major infectious agents found in contaminated drinking waters world wide*

Bacteria	Viruses	Protozoa	Helminths
Campylobacter jejuni	Adenovirus (31 types)	*Balantidium coli*	*Ancylostoma duodenale*
Enteropathogenic *E. coli*	Enteroviruses (71 types)	*Entamoeba histolytica*	*Ascaris lumbricoides*
Salmonella (1 700 spp.)	Hepatitis A	*Giardia lamblia*	*Echinococcus granulosus*
Shigella (4 spp.)	Norwalk agent	*Cryptosporidium*	*Necator americanus*
Vibrio cholerae	Reovirus		*Strongyloides stercoralis*
Yersinia enterocolitica	Rotavirus		*Taenia solium*
	Coxsackie virus		*Trichuris trichiura*

Data from Geldreich (1990).

water supply processing and its distribution). Add to these problems the mobility of people throughout the world and it becomes apparent that an outbreak can quickly spread by person-to-person contact and through secondary water contamination, reaching epidemic proportions.

Pathogen pathways

Although sewage collection systems have decreased public-health risk in urban centers, this practice only serves to transport the collected wastes to some selected destination, hopefully where treatment is applied prior to release into a watercourse. Raw sewage discharges to receiving waters have often been shown to contain a variety of pathogens. The density and variety of human pathogens released are related to the population served by the sewage-collection system, seasonal patterns for certain diseases, and the extent of community infections at a given time. Some indication of the relative occurrences of various pathogens in raw sewage is given in Table 9.8 for two cities in South Africa (Grabow and Nupen 1972).

In major river systems receiving discharges of meat-processing wastes, raw sewage, and effluents from ineffective sewage treatment plants, the densities of *Salmonella* spp. may be substantial. It has been calculated that the Rhine and Meuse rivers carry approximately 50 million and 7 million *Salmonella* bacilli per second, respectively (Kampelmacher and Van Noorle Jansen 1973). The Missouri River represents another example of a pollution conduit, transporting a fecal pollution load from raw sewage, effluents from primary and secondary treatment plants of differing efficiencies, runoff from numerous cattle feedlots, and waste discharges from meat- and poultry-processing plants. As a consequence, it is not surprising that various *Salmonella* serotypes and viruses have been detected at the public water supply treatment plant intakes (see Table 9.4, p. 215).

Cattle feedlots and poultry operations result in an intense concentration of farm animals and their fecal wastes in a confined space. In cattle feedlot operations, the density of beef cattle per square mile may approach 10 000 animals. Under such restrictions, removal of fecal

wastes is a major disposal operation. The closeness of farm animals in confined feeding operations invites the spread of disease in a healthy herd or poultry flock. Some farm animal pathogens (*Salmonella*, *E. coli* O157:H7, *Giardia*, and *Cryptosporidium*) are also major human pathogens. Unless contour cultivation and storm-water retention ponds are undertaken to reduce runoff, fecal material transported in stormwater runoff from cattle feedlots and poultry farms becomes a major source of contamination in rural watersheds, polluting streams and lakes in its drainage path. Direct recycling of untreated farm animal wastes and sewage by application to fields as fertilizer can become a serious contributor of pathogens to salad crops and ground water.

Wildlife refuges may also be a significant source of fecal contamination, particularly on a seasonal basis, since many wild animals are migratory by instinct in their search for food. The largest reservoir of wildlife pathogens will be found in warm-blooded animals such as beaver, deer, coyote, ducks, and gulls that are permanent residents of watersheds (forests, prairies, lakes, reservoir impoundments). These animals and others serve as reservoirs for *Salmonella*, *Campylobacter*, *Yersinia*, *Giardia*, and *Cryptosporidium*. Wildlife is also attracted to protected watershed areas (forest reserves, private lands) where human activities over the water-shed are more restricted (Walter and Bottman 1967). Protected near-shore water environments are often the location of large beaver colonies, including individual animals infected with *Giardia*. Infected coyotes, musk-rats, and voles are other wild animals that may be involved in the shedding of *Giardia* cysts and other pathogens into the aquatic environment. Terrestrial birds and waterfowl can be sources of bacterial pathogens. Songbird populations include individuals that may be infected with *Salmonella*. Seagulls are scavengers that frequent open rubbish dumps, eat contaminated food wastes, and contribute *Salmonella* in their fecal droppings to coastal lakes (Alter 1954; Fennel et al. 1974). In one instance, seagulls were the contributors of *Salmonella* to an untreated surface supply in an Alaskan community, causing several cases of salmonellosis (Anonymous 1954). Pigeons colonizing two water storage towers in Gideon, Missouri, were considered the

Table 9.8 *Microbial densities in municipal raw sewage from two cities in South Africa*

Organism or microbial group	Average count per 100 ml	
	Worcester sewage	Pietermaritzburg sewage
Aerobic plate count (37°C, 48 h)	1 110 000 000	1 370 000 000
Total coliforms	10 000 000	–
E. coli, type 1	930 000	1 470 000
Fecal streptococci	2 080 000	–
C. perfringens	89 000	–
Staphylococci (coagulase-positive)	41 400	28 100
P. aeruginosa	800 000	400 000
Salmonella	31	32
Acid-fast bacteria	410	530
Ascaris ova	16	12
Taenia ova	2	9
Trichuris ova	2	1
Enteroviruses and reoviruses (TCID50)	2 890	9 500

TCID, tissue culture infective dose.
Source: Grabow and Nupen (1972).

source of *Salmonella typhimurium* that contaminated the water supply, causing 600 illnesses and four deaths (Clark et al. 1996). Wildlife is also believed to be the source of *Campylobacter* that contaminated streams and reservoirs of low turbidity in Vermont (Vogt et al. 1982) and British Columbia (Report 1981). Vacationers to national parks in Wyoming became ill after drinking water from mountain streams, resulting in a 25 percent increase in outbreaks state-wide involving *Campylobacter enteritis* (Taylor et al. 1983).

Pathogen persistence

Upon discharge into the aquatic environment, the persistence of pathogens becomes a variable determined by many factors. For example, *Salmonella* strains were detected with regularity in surface water up to 250 m downstream from a wastewater treatment plant, but never at sample sites 1.5–4 km upstream (Kampelmacher and Van Noorle Jansen 1976). Various *Salmonella* serotypes transported by stormwater through a residential storm sewer and a wash-water drain at the University of Wisconsin experiment farm were isolated with regularity at a swimming beach approximately 800 m downstream (Claudon et al. 1971). Salmonellae were also detected in coastal waters from several Staten Island beaches and from shellfish harvested in New York harbor (Brezenski and Russomanno 1969).

Excessive biochemical oxygen demand (BOD) or total organic carbon (TOC) in a poor-quality wastewater effluent combined with low stream temperatures can also affect pathogen persistence. For example, *Salmonella* were isolated in the Red River of the North (North Dakota, Minnesota), 35 km downstream of sewage discharges from Fargo, North Dakota, and Moorhead, Minnesota, during September (Report 1965; Spino

1966). By November, *Salmonella* strains were found 99 km downstream of these two sites. In January, with the beginning of the sugar-beet-processing season, wastes reaching the stream under cover of ice brought high levels of bacterial nutrients. *Salmonella* were then detected 117 km downstream – a flow time of 4 days from the nearest point-source discharges of warm-blooded animal pollution.

NATURAL SELF-PURIFICATION FACTORS

Natural waters provide a fragile purification buffer to limited amounts of raw wastes and stormwater runoff entering the drainage basin. Every stream, lake, estuary, and water aquifer has some limited capacity to self-purify; surface waters generally have greater capacity than ground waters. Stream self-purification is a complex and ill-defined process that involves bacterial adsorption with sedimentation, nutrients, predation, competitive microbial populations, dilution, aeration, water temperature, water pH, and solar radiation (Geldreich 1986).

Retention of surface water in lakes, impoundments, and streams is an important contribution to water-quality enhancement. Storage and sedimentation of surface water often improves the microbial quality. Many organisms are inactivated by natural self-purification processes, whereas others are removed from the water through natural siltation (Romaninko 1971; Dzyuban 1975; Geldreich et al. 1980). These water-quality improvements are variable and are directly proportional to retention time, dilution, and the number of contributing sources (Geldreich 1991). In temperate zones with pronounced seasonal temperature changes, many lakes and impoundments are subject to periods of thermal stratification in the summer and winter and destratification (water turnover) periods in spring and autumn. During stratification, water movement in the deeper portion of raw-water reservoirs becomes restricted, generally creating a zone of maximum bacterial contamination in the water layer above the thermocline (Weiss and Oglesby 1960; Collins 1963; Niewolak 1974; Drury and Gearheart 1975). With destratification, water from the bottom and top layers mixes and develops a more uniform water quality. This mixing process causes decaying vegetation and entrapped organisms in settled particulates to re-enter the water column (Weiss and Oglesby 1960; Niewolak 1987). Heterotrophic bacterial populations (including coliform bacteria), turbidity, and humics from partially decomposed vegetation will temporarily increase in the water.

Fresh seawater brings about a fairly rapid decline in fecal bacteria, salmonellae, and viruses. The main inactivating agent is solar radiation at wavelengths less than 400 nm (Gameson and Saxon 1967; Gameson and Gould 1975; Chamberlin and Mitchell 1978; Fujioka et al. 1981). Tidal currents, sedimentation, and sunlight intensity are variable factors that dictate the effectiveness of solar radiation as a force in natural self-purification of coastal marine water contamination.

Biofilms in water

A biofilm is a colonization of various organisms in a slimy matrix of extracellular organic polymers (glycocalyx) at sites that offer surfaces for attachment and protection from the shearing action of water turbulence (Costerton et al. 1978; Characklis and Marshall 1990). Patches of biofilm may be found on submerged rocks in slow-moving streams, within water-supply and wastewater treatment basin structures and their connecting flumes, in the water-pipe environment of sediments and tubercles, and in water-supply-attachment devices used in hospitals and clinics.

The microbial community in a water pipe begins with pioneering organisms in the passing flow of water becoming lodged in sediment, pipe corrosion, and at water sumps created in the design of various water delivery devices (Geldreich and Reasoner 1989). Colonization follows once a critical mass of assimilable organic carbon complexes, nitrogen occurring compounds, and traces of phosphorus become absorbed in associated sediments and tubercles. This growth is further accelerated as water temperatures rise above 15°C (Geldreich and LeChevallier 1999). Over time the consortium grows in complexity, the more fastidious organisms with unique nutrient requirements (*Legionella*) being supported by metabolic byproducts released from other bacteria in the microbial community.

As a biofilm expands in a section of water-distribution pipe, fragments of this bacterial growth are dislodged under certain situations into the bulk movement of water supply: shearing action of water velocity; flow reversals; abrasive action of pipe-cleaning procedures on tubercles; or abrupt shifts in water pH that softens sediment deposits. Among the variety of bacteria sloughed off the pipe surface are coliforms bacteria, particularly *Klebsiella pneumoniae*, *K. oxytoca*, *Pantoea agglomerans* (*Enterobacter agglomerans*), *E. cloacae*, or *Citrobacter freundii*. Coliform releases in biofilm fragments may not be of public health significance per se but should not be ignored because of perpetual concern for a hidden fecal contaminating event (Percival et al. 2000). Biofilms also induce corrosion and increase disinfectant demands and customer complaints of taste and odors. Other bacteria of concern in biofilm include a variety of heterotrophic organisms (*Pseudomonas*, *Aeromonas*, *Flavobacterium*, *Proteus*, *Bacillus*, *Actinomycetes*, and fungi), some of which may be opportunistic pathogens causing nosocomial infections in the hospital environment (Costerton et al. 1981). The bacterial pathogen *Salmonella* has been demonstrated in laboratory research on biofilm development to have a very tenuous persistence because of competition from other organisms in the biological consortium (Camper et al. 1998).

WATER-QUALITY CRITERIA

Bacterial criteria for determining water quality have been directed primarily towards concern for microbial hazards to human health, although some thought has been given to health hazards to farm animals and aqua cultures of fish stocks, shellfish, and turtles (Report 1968, 1985). Indicators of fecal pollution have long been the predominant microbiological tool used to define the microbial quality of a water supply. However, there is growing evidence that water may transport not only intestinal pathogens via ingestion but also respiratory agents by inhalation and skin diseases through body contact. Thus, the microbial risks in water quality may require not only traditional criteria (heterotrophic bacteria, total coliforms, fecal coliforms, *E. coli*, fecal streptococcus, enterococcus) but also special surrogates such as *Clostridium*, aerobic spore formers, bacteriophages, and species-specific pathogens to evaluate health-risk status related to water use.

BACTERIOLOGICAL ANALYSIS

Standard methods for the analysis of a variety of waters are described in specific detail by the World Health Organization (Report 1984, 1992), the American Public Health Association (Report 1998), and the United Kingdom Report on Public Health and Medical Subjects (Report 1994), among other national documents worldwide. These methods relate to drinking water, recreational waters, swimming-pool waters, shellfish-growing waters, wastewater discharges, recharge waters to aquifers, and other re-use applications.

For monitoring purposes, these indicators and various bacterial pathogens are detected by methods based on four principal procedures: pour or spread plates, selective cultivation in broth (multiple tube tests), membrane filter techniques, and enzyme–substrate reaction. More specific identification of selected heterotrophic organisms may be done by speciation of culture isolates.

Stressed organisms

Microbial detection must be based on methods that yield a high recovery of selected indicator groups or individual species. Recoveries are often 90 percent or better but are subject to some losses, which can be associated with injured bacteria in the water (McFeters et al. 1986; McFeters 1990). In various aquatic environments, competing organisms in the microbial community, toxic metal ions, toxic wastes, and inadequate disinfection are major causes of cell injuries. Many stressed organisms may thereupon become incapable of survival in standard laboratory procedures, a situation that can have a great impact on data interpretation. The public-health concern lies with bathing water, potable supplies, and shellfish beds that have indicator densities ranging below 200 organisms per 100 ml. It is important to note that while some enteropathogenic bacteria may also be susceptible to similar stresses, the potential for virulence remains and the injured cells recover after being injested by exposed indiciduals (LeChevallier et al., 1985; Singh et al. 1986; Singh et al., 1986).

Efforts to reverse these impacts on indicator organisms may include the use of a chelating agent and application of a dechlorinating compound in the sterile sample bottle, initially resuscitating the stressed cells by culturing in an enriched, nonselective medium with incubation for 2 h at room temperature prior to selective incubation on T7 agar at 35°C for the total coliform pupulation (LeChevallier et al. 1985). For recovery of injured fecal coliform cells, a two-layered agar using an overlay of tryptic soy agar on M-FC medium with incubation at 44.5°C provides a gradual sift to more selective nutrients (Rose et al. 1975). A similar principle applies to fecal streptococcus-enterococcus indicator selection (Report 1998). Initial verification of presumptive results for any of these special procedures is essential, using other procedures to establish test validity and to eliminate false-positive results that may break through the recovery protocol (Lin 1974).

Sample integrity is essential. Care must be exercised to collect samples representative of water use and to ensure that the sample does not become contaminated at the time of collection or before examination. Samples should be refrigerated or placed in an insulated cooler and promptly transported to the laboratory. Sample transit time limits for total coliform analysis of drinking water have been established at 24 h to limit any drastic changes in indicator occurrence impacted by antagonistic factors found in polluted natural waters. Sample processing within 8 h of collection should always be a goal for optimum recovery of organisms of significance. A dechlorinating agent (sodium thiosulfate) should be in the sterile sample bottle used to collect treated water supply or wastewater samples that often contain varying concentrations of disinfectant residues.

Heterotrophic bacteria

Heterotrophic bacteria are a diverse group of organisms. They encompass differences in morphology, gram-staining response, and biochemical reactions, but most of these organisms utilize carbon, nitrogen, phosphorus, and various trace elements necessary for the synthesis of living cell materials. The metabolic pathways to achieve these conversions are remarkably diverse, from those that use only inorganic sources of carbon, nitrogen and inorganic salts, to others that grow only when there are complex organics available. Such is the case for *Legionella* spp., which have fastidious nutrient requirements that are only made available in the metabolic biproducts from other bacteria in the aquatic community. Different groupings can be formed from this population of heterotrophs such as those that are pigmented (Reasoner et al. 1989), those that are antagonistic to coliform bacteria (Geldreich et al. 1978), and others that cause taste and odors (Burman 1965). Some of these bacteria are also known to be involved in plasmid transfers of R factors for antibiotics (Armstrong et al. 1982) or disinfection resistance (Ridgway and Olson 1982).

Predominate heterotrophic bacterial genera include *Acinetobacter*, *Aeromonas*, *Alcaligenes*, *Bacillus*, environmental coliforms (*Citrobacter*, *Enterobacter*, *Klebsiella*, *Escherichia*), *Flavobacterium*, *Pseudomonas*, *Micrococcus*, and *Moraxella*. Densities of heterotrophic bacteria in drinking water may be in a range from <10/ml to as high as 10 000/ml, while in raw sewage and food-processing wastes discharged to surface waters, densities may be in the millions per milliliter. In high-quality natural waters (protected ground water and springs) most of the organisms are indigenous and of no sanitary significance. In surface water (lakes and rivers), the heterotrophic composition will have varying inputs of contamination from stormwater runoff, domestic and industrial wastes, agricultural activities, wildlife habitation, soil contact, vegetation, and airborne pollutants.

Since the beginnings of bacteriology, the heterotrophic plate count has been used in an attempt to characterize water quality. Obviously, an excellent water was thought to have few bacteria per milliliter, whereas water with more than 1 000 bacteria per milliliter was considered of poor quality. In time, it was concluded that the general density of bacteria in water often had

Table 9.9 *Effect of incubation temperature and time to recover heterotrophic bacteria from water supply samples by different methods*

| Temperature (°C) | Medium | Method | Incubation time (days) | | | |
			2	4	6	7
20	SPC	Pour plate	22	130	570	900
	R2A	Spread plate	90	1 100	4 700	6 100
	R2A	Membrane filter	75	650	3 000	4 900
	M-HPC	Membrane filter	48	400	1 600	2 000
28	SPC	Pour plate	90	640	950	1 000
	R2A	Spread plate	360	2 800	6 700	7 200
	R2A	Membrane filter	160	2 200	3 500	4 000
	M-HPC	Membrane filter	140	1 000	1 700	1 900
35	SPC	Pour plate	22	100	110	115
	R2A	Spread plate	200	340	500	510
	R2A	Membrane filter	41	200	270	280
	M-HPC	Membrane filter	32	140	150	150

Data revised from Reasoner (1990).

little specific relationship to increased health risks and should be replaced by a search for organisms more specific to fecal contamination (Bartram et al. 2003). Characterization and interpretation of heterotrophic bacterial densities now places the focus on using the information to identify changes in the microbial quality of distribution water caused by accumulating pipeline sediments and subsequent loss of a protective disinfectant residue. Most often these organisms are not of immediate public-health significance but upon amplification in a protected pipe habitat become the source of customer complaints of offensive taste/odors or emerge as an opportunistic pathogenic threat to some segment of the population.

Traditionally, most water-plant laboratories have used a nutritive standard plate count (SPC) agar or equivalent formulations for years from which they can provide an extensive database on heterotrophic bacterial densities in various water treatment processes. More recently, the trend is to use a dilute but diverse nutrient formulation such as R-2A agar, which contains a diverse variety of biodegradable materials (peptone, casamino acids, glucose, soluble starch, and sodium pyruvate). Colony densities are often many times greater and consist of a more diverse variety of organisms, including pigmented bacteria (Reasoner and Geldreich 1985). To achieve this improvement, incubation for 5–7 days at a temperature of 28°C is preferred rather than the more conventional time of 48 h at 35°C. This change in cultivation provides more opportunities for the growth of slow-generating bacteria, stressed organisms, and those species that have unique nutrient requirements. Since pour plates limit the growth of obligate aerobes in agar and introduce the risk of heat shock from melted agar (Klein and Wu 1974), attention has turned to surface cultivation of bacteria by spread plate to optimize recovery of these sensitive bacteria.

The membrane-filter procedure provides another approach to surface cultivation of organisms. The unique advantage of this technique is that it permits the analysis of larger sample volumes of high-quality water containing too few organisms to be detected by a 1-ml sample. The only restrictions to the size of sample analyzed are turbidity and colony density on the filter surface.

Some indication of the impact of membrane filter surface cultivation in comparison with pour plates and spread plates can be seen in Table 9.9. Recovery densities are significantly better on spread plates than by the traditional pour plate method because of agar temperature (Stapert et al. 1962). Limited surface area on the membrane filter for discrete colony formation without developing confluence with the other colonies thereby restricts extended incubation time and is the major drawback for achieving results equivalent to the spread plate technique.

Total coliform bacteria

In an effort to obtain a better focus on the sanitary quality of water, coliform bacteria soon supplanted the heterotrophic plate count as the most widely used indicator for routine analysis. Coliform bacteria consist of an artificial grouping of organisms believed to be associated with fecal pollution, but in reality they may also include environmental organisms from soil, vegetation, and decaying organic matter (Randall 1956; Geldreich et al. 1962; Schubert and Mann 1968). In the aquatic environment, environmental coliforms persist longer than do *E. coli*, the coliform constantly present in large numbers in feces from humans and other warm-blooded animals (wildlife, farm animals, cats, and dogs). Major members of these environmental coliforms include *Klebsiella*, *Enterobacter*, and *Citrobacter*. Although not perfect, the total coliform concept continues to be a practical tool for providing basic microbial information on drinking-water treatment effectiveness and distribution system integrity.

Fecal coliforms/*E. coli*

Interest in further refining the total coliform group to exclude environmental coliforms, which are of no sanitary significance, led to the development of the fecal coliform (thermotolerant coliform) test based on lactose fermentation within 24 h at 44–44.5°C. The fecal coliform test proved to be a breakthrough with positive correlations of 93–99 percent to coliforms found in the feces of warm-blooded animals (Geldreich 1966).

In polluted waters, fecal coliform measurements relate more precisely to fecal contamination than to total coliforms and are significantly less susceptible to distortions caused by the regrowth characteristics of environmental strains in receiving waters. Essential factors that stimulate bacterial regrowth in polluted waters are nitrogen, carbon, and a warm water temperature (above 15°C). The regrowth phenomenon for fecal coliforms requires excessive nutrient discharges generally associated with poor treatment practices, particularly those used on some food-processing and paper-mill wastes.

Debate on the status of fecal *Klebsiella* and growth of the environmental component in the aquatic environment has led to attempts to search only for *E. coli*. Commercial biochemical kits for the identification of *E. coli* by detection of 4-methylumbelliferyl-β-D-glucuronidase (MUG) in either an amended medium or a chemically defined medium are now being used to further narrow the coliform group to one specific fecal organism – *E. coli*.

VERIFICATION OF COLIFORM BACTERIA

Verification of every positive coliform result is good laboratory practice that eliminates all doubts that the test results are perhaps caused by some analogous false reaction. Coliform analysis of drinking waters has revealed that as much as 30 percent of the samples having positive presumptive test results may also contain *Aeromonas* (Ptak et al. 1974; Leclerc et al. 1977). A few *Aeromonas* strains have been the cause of water-borne gastroenteritis (Moyer 1989), but most are saprophytic organisms in the aquatic environment. This situation illustrates the reason why test results must be confirmed to eliminate false-positive results in all methods even though this delay partly nullifies the rapidity of the results.

Rapid verification of coliform bacteria can be achieved in the demonstration of cytochrome oxidase and β-galactosidase activity. All coliform bacteria are negative for cytochrome oxidase while the enzyme β-glucuronidase appears to be highly specific for *E. coli*.

This latter metabolic response has formed the basis of a rapid confirmatory test using the 4-nitrophenyl-β-D-glucuronide (Perez et al. 1986) or MUG (Report 1998).

Coliform speciation is a more in-depth verification process that has been used to demonstrate the equivalence of methods to detect all coliforms. Speciation as a verification tool has its greatest use in the identification of a coliform biofilm event and in further tracing for the existence of a microbial pathway in the water supply distribution system. In either case, further exploration of the original sample may provide information on fecal contamination that might not be present in a repeat sample collected the next day. Most often, repeat samples are negative because the contaminating event is of short duration (Geldreich 1996).

Fecal streptococci

Fecal streptococci are distinguished by their ability to grow at 45°C in the presence of 40 percent bile, and in the concentrations of sodium azide or potassium tellurite that are inhibitory to most coliforms organisms. Methods are available using agar pour plates (KF or PSE agars), multiple tube (azide dextrose broth), and membrane filter (KF agar or broth). The occurrence of fecal streptococci in water generally indicates fecal pollution (Geldreich and Kenner 1969). Although fecal streptococci rarely multiply in polluted waters, they may persist for extended periods in irrigation waters with high electrolyte contents and favorable temperatures (Geldreich 1973). There is also some evidence that *Enterococcus fecalis* (*Streptococcus fecalis*) multiplies in water from vegetable-processing plants (Mundt et al. 1966a). The reason for such contrasting responses is that the fecal streptococcus group includes a wide spectrum of strains that have specific fecal origins and diverse survival rates and includes several environmental biotypes (Mundt et al. 1962; Mundt and Graham 1968). Within the fecal streptococcus group, *Streptococcus bovis* and *Streptococcus equinus* are specific indicators of farm animal pollution. This differential characteristic is particularly useful in pollution investigations involving cattle feedlot runoff, farmland drainage, discharge from meat- and poultry-processing operations, and dairy-plant wastes (Geldreich 1972b). In addition, *S. bovis* and *S. equinus* are the fecal streptococci that die off most rapidly outside the animal intestinal tract. Therefore, the detection of these strains in water indicates very recent farm animal contamination. By contrast, the ubiquitous *E. fecalis* var. *liquifaciens* may affect precision of this indicator system at counts below 100 fecal streptococci per 100 ml, because at these low population levels this biotype generally predominates in high-quality natural waters. Fecal streptococcus counts greater than 100 per 100 ml, however, indicate significant fecal pollution derived from some warm-blooded animal source. The densities for this indicator group in polluted waters approach the magnitude observed for coliforms, or at times exceed it by a factor of ten, depending upon the source of fecal pollution.

The density difference between fecal coliforms and fecal streptococci in fecal material is a unique relationship that can be useful in defining sources of pollution. The ratio of fecal coliform to fecal streptococcus in human feces and domestic wastes is greater than 4.0. The ratio of fecal coliforms to fecal streptococci in the feces of farm animals, cats, dogs, and rodents is less than 0.7. Such relationships between fecal coliform and fecal streptococci can only be developed from a fecal coliform medium (not an *E. coli* test) and from a fecal streptococcus medium such as KF or PSE agars (not from a medium dedicated to enterococcus). If cultivation of the more restrictive enterococcus subgroup is desired, mE agar or m-enterococcus agar may be considered.

Clostridium perfringens

The genus *Clostridium*, in particular *C. perfringens*, merits some attention as an indicator of fecal pollution because of its significant occurrence in feces (Bishop and Allcock 1960). *Clostridium* spp. spores are resistant to wastewater treatment practices and tolerant to extremes in temperature and environmental stress. *C. perfringens* is an indicator of present fecal contamination as well as a conservative tracer of past fecal pollution (Wilson and Blair 1925). *C. perfringens* has been used in the UK to monitor the quality of ground-water supplies that are examined infrequently, i.e. at intervals of 4–6 months (Wilson 1931). When *C. perfringens* is found in these well waters, sampling is repeated, possibly weekly, using total coliform or fecal coli measurements to search for intermittent fecal contamination and the need for a sanitary inspection to verify adequate well-head protection. The use of *C. perfringens* as a supplemental indicator to measure the efficiency of various water treatment processes has been proposed, but the greater abundance of total aerobic spore formers in raw water may be a more effective system in providing evidence of treatment barrier performance, particularly in reference to pathogenic protozoans(Rice et al. 1996).

The longer persistence of *C. perfringens* in the water environment creates a residual density of this organism, which can obscure the detection of low densities of recent polluting discharges to receiving waters. However, Bonde reported that *C. perfringens* densities in the sediment of marine waters were proportional to pollution discharges and the ratios of spores to vegetative cells of this bacterial species increased with distance from the pollution source (Bonde 1962, 1967, 1968). Thus, vegetative cells of *C. perfringens* may be expected to predominate in raw sewage.

Although *C. perfringens* is an anaerobic organism, it can tolerate up to 5 percent oxygen without significant loss in quantitative recovery (Fulton and Richardson 1971). Therefore, methodology has not been complicated by the restrictions of complete anaerobiosis. On sulfite–alum agar (to reduce available oxygen) incubated at 48°C, 79–100 percent of the black sulfite-reducing colonies recovered from a variety of waters were verified as *C. perfringens* (Bonde 1962). The other black colonies on this medium may be *Salmonella*, *Proteus*, *Bacteroides*, and sometimes *E. coli* (Johnston et al. 1964). This practical procedure has frequently been used to obtain a presumptive quantitation of *C. perfringens* in various fresh and marine waters and sediments. Another approach is to filter an appropriate volume of water sample through a membrane filter, place the membrane filter on a modified m-CP agar (Armon and Payment 1988), and incubate anaerobically for 24 h at 44.5°C. Upon exposure to ammonium hydroxide, the yellow to straw-colored *C. perfringens* colonies turn dark pink to magenta. Confirmation may be done by anaerobic growth in thioglycollate, a positive Gram stain reaction, and stormy fermentation of iron milk.

Pseudomonas species

Pseudomonads are ubiquitous bacteria that are able to flourish in a wide variety of habitats (surface waters, aquifers, seawater, soil, and vegetation). Some pseudomonads are among the prominent denitrifiers while others grow prodigiously in and on tertiary treatment devices such as reverse osmosis and electrodialysis membranes and in sand or carbon filtration beds. Pseudomonads reported in some drinking-water supplies include *Pseudomonas aeruginosa*, *Burkholderia cepacia* (*Pseudomonas cepacia*), *Pseudomonas fluorescens*, *Burkholderia mallei* (*Pseudomonas mallei*), *Stenotrophomonas maltophilia* (*Pseudomonas maltophila*), *Pseudomonas putida*, and *Comamonas testosteroni* (*Pseudomonas testosteroni*) (Geldreich 1990, 1996; Gambassini et al. 1990). To this list can be added *Pseudomonas stutzeri*, *Brevundimonas diminuta* (*Pseudomonas diminuta*), and *Delftia acidovorans* (*Pseudomonas acidovorans*), which have been found in bottled waters at densities ranging from 10^3 to 10^5 organisms per milliliter (Gavin and LeClerc 1974; Hernandez Duquino and Rosenberg 1987). These organisms metabolically adapt to survival on minimal nutrient concentrations typical of protected aquifers and treated drinking water. Routine enumeration of pseudomonads in water supply is not recommended but may be of value in certain industries, such as those manufacturing foods, drinks, and pharmaceutical products, where high-quality water is desirable. Monitoring the hospital building water-distribution system for *P. aeruginosa* is important because this organism is a major opportunistic pathogen and its presence in this environment is a matter for concern.

PRINCIPLES OF PATHOGEN DETECTION

Detection of pathogens in the aquatic environment requires techniques to concentrate a few organisms from large samples and reduce the interfering flora of many heterotrophic bacteria that are certain to be a factor in contaminated water. For bacterial pathogens, this is generally accomplished through filtration of large volumes of water (1–2 l) followed by cultivation in selective media for suppression of coliforms and other heterotrophic organisms. Another approach is to place gauze pads in the polluted stream for 3–5 days to entrap the pathogenic agent, then express the water from the pad to selective enrichment media (Moore 1948, 1950). A third approach is to filter samples of 1–2 l through diatomaceous earth held in a membrane-filter (MF) funnel by an absorbent pad, and then place sections of the plug in appropriate media for cultivation and species identification.

STRATEGIES FOR RAPID METHODS

Rapid laboratory methods to better characterize water-supply quality are an urgent research priority. The inability of microbiology to provide data within a few hours of sample processing has been a major deterrent to greater utilization of the science in water-plant operations and in bathing-water monitoring. With the discovery that disinfection byproducts may contribute carcinogens and possibly other toxins to the water supply, treatment schemes are being modified to minimize these undesirable formations. In so doing the effectiveness of treatment barriers required to avoid microbial contaminant passages becomes a matter of greater concern. Consequently, the need for rapid assessment of water quality is more urgent than ever before. Such information would also be of singular value as an aid in restoring the quality of drinking water after contamination caused by sudden source-water-quality deterioration, treatment failures, distribution line breaks, poor practices in disinfection line repair, and sudden occurrence of cross-connections.

Any breakthrough in the development of rapid tests must involve specificity, sensitivity, and precision with achievement of a test result within a few hours. Searching for rapid methods in environmental microbiology has revealed a variety of candidate methods that have potential in real-time monitoring (Geldreich and Reasoner 1985). As might be anticipated, rapid tests that can be performed in less than 1 h have little specificity and may possibly include nonviable cells. These limitations are being solved, but the tradeoff is in the need for more costly materials and instrumentation sensitive to trace concentrations of specific metabolic products produced by organisms of interest. One exception to this trend may be seen in a direct membrane-filter method incorporating MUG in a modified medium (M-7h agar), which provides detection of as few as one fecal coliform per 100 ml in 6 h (Berg and Fiksdal 1988).

Efforts to shorten the time required to detect bacterial indicator systems or individual pathogen species in a water sample flora to less than 5 h will require instrumentation capable of detecting viable organisms directly or after a brief period of amplification in a selective enrichment medium. Instrumentation and test materials for these methods will become more readily available to the laboratory in the near future.

Of the rapid test candidates currently in development, gene probes appear to be the most promising (Richardson et al. 1991). Research into microbial genetics has led to techniques that can detect minute quantities of nucleotide sequences unique to a single species of organism, whether bacteria, fungi, virus, or invertebrates. For example, ten enterotoxigenic *E. coli* per milliliter of canal water were detected in the grossly polluted canals in Bangkok, Thailand, by using gene-probe technology (Mosely et al. 1982).

The use of gene probes and automated microbiological techniques will become a major activity in the water plant laboratories of the future (Geldreich 1991). Perhaps development of multiplex probes to detect bacterial pathogens or enteroviruses will become sensitive enough to detect one organism per liter of sample within an hour. At the present time, some probe tests can be completed in 0.5–2 h, but they often require prior culture amplification for 7–18 h to achieve organism densities at levels where acceptable probe detection is possible (Tenover 1988; Report 1998).

REFERENCES

Alter, A.J. 1954. Appearance of intestinal wastes in surface water supplies at Ketchikan, Alaska. Proceedings of the Fifth Alaska Science Conference. Anchorage, AK: American Association for the Advancement of Science.

Anonymous. 1954. Ketchikan laboratory studies disclose gulls are implicated in spread of disease. *Alaska's Health*, **11**, 1–2.

Armon, R. and Payment, P. 1988. A modified M-CP medium for the enumeration of *Clostridium perfringens* from water samples. *Can J Microbiol*, **34**, 78–9.

Armstrong, J.L., Calomiris, J.J. and Seidler, R.J. 1982. Selection of antibiotic-resistant standard plate count bacteria during water treatment. *Appl Environ Microbiol*, **44**, 308–16.

Bartram, J., Cotruvo, J., et al. 2003. *Heterotrophic plate counts and drinking water safety: the significance of HPC's for water quality and human health*. London: IWA Pub, 256 pp.

Belli, C.M. 1902. Zentralbl Bakteriol. **11**, 8, 445.

Berg, D.J. and Fiksdal, L. 1988. Rapid detection of total and fecal coliforms in water by enzymatic hydrolysis of 4-methylumbelliferone-β-D-galactoside. *Appl Environ Microbiol*, **54**, 2112–18.

Bishop, R.F. and Allcock, E.A. 1960. Bacterial flora of the small intestine in acute intestinal obstruction. *Br Med J*, **5175**, 766–70.

Blake, P.A., Rosenberg, M.L., et al. 1977. Cholera in Portugal. *Am J Epidemiol*, **105**, 337–43.

Bonde, G.J. 1962. *Bacterial indicators of water pollution*. Copenhagen: Teknisk Forlag.

Bonde, G.J. 1967. Pollution of a marine environment. *J Water Poll Control Fed*, **39**, R45–63.

Bonde, G.J. 1968. Studies on the dispersion and disappearance phenomena of enteric bacteria in the marine environment. *Rev Int Oceanogr Med*, **9**, 17–44.

Brezenski, F.T. and Russomanno, R. 1969. The detection and use of *Salmonella* in studying polluted tidal estuaries. *J Water Poll Control Fed*, **40**, 725–37.

Burman, N.P. 1965. Taste and odour due to stagnation and local warming in long lengths of piping. *Soc Water Treat Exam*, **14**, 125–31.

Camper, A.K., Warnecke, M. et al. 1998. Pathogens in model distribution system biofilm. *Am Water Works Assoc Res Found Rep*, 1–68.

Castle, S. 1985. Public health implications regarding the epidemiology and microbiology of public whirlpools. *Infect Control*, **6**, 418–19.

Chamberlin, E. and Mitchell, R. 1978. A decay model for enteric bacteria in natural waters. In: Mitchell, R. (ed.), *Water pollution microbiology*. New York: John Wiley & Sons, 325–48.

Characklis, W.G. and Marshall, K.C. 1990. *Biofilms*. New York: John Wiley & Sons-Interscience.

Clark, R.M., Geldreich, E.E., et al. 1996. Tracking a *Salmonella* serovar typhimurium outbreak in Gideon, Missouri: role of contaminant propagation modelling. *J Water SRT-Aqua*, **45**, 171–83.

Claudon, D.G., Thompson, D.I., et al. 1971. Prolonged *Salmonella* contamination of a recreational lake by runoff waters. *Appl Microbiol*, **21**, 875–7.

Collins, V.G. 1963. The distribution and ecology of bacteria. *Proc Soc Water Treat Exam*, **12**, 40–73.

Cook, H.A., Cromwell, D.L. and Wilson, H.A. 1967. Microorganisms in household refuse and seepage water from sanitary landfills. *West Virginia Acad Sci*, **39**, 107–12.

Cordano, A.M. and Vergilio, R. 1990. Salmonella contamination on surface waters. Proceedings of the Second Biennial Water Quality Symposium Microbiological Aspects. Santiago: University of Chile.

Costerton, J.W., Geesey, G.G., et al. 1978. How bacteria stick. *Sci Am*, **238**, 86.

Costerton, J.W., Irvin, R.T., et al. 1981. The bacterial glycocalyx in nature and disease. *Ann Rev Microbiol*, **99**, 316.

Davis, B.J. 1985. Whirlpool operation and the prevention of infection. *Infect Control*, **6**, 394–7.

Drury, D.D. and Gearheart, R.A. 1975. Bacterial population dynamics and dissolved oxygen minimum. *J Am Water Works Assoc*, **67**, 154–8.

Dzyuban, A.N. 1975. The number and generation time of bacteria and production of bacterial biomass in water of the Saratov reservoir. *Gidrobiolog-icheskii Zh*, **11**, 14–19.

Eikelboom, D.H. 1975. Filamentous organisms observed in activated sludge. *Water Res*, **9**, 365–88.

Favero, M.S. and Drake, C.H. 1966. Factors influencing the occurrence of high numbers of iodine-resistant bacteria in iodinated swimming pools. *Appl Microbiol*, **14**, 627–35.

Fennel, H., James, D.B. and Morris, J. 1974. Pollution of a storage reservoir by roosting gulls. *J Soc Water Treat Exam*, **23**, 5–24.

Fujioka, R.S., Hashimoto, H.H., et al. 1981. Effect of sunlight on survival of indicator bacteria in seawater. *Appl Environ Microbiol*, **41**, 690–6.

Fulton, B.V. and Richardson, G. 1971. *Isolation of anaerobes*. London: Academic Press.

Gambassini, L., Sacco, C., et al. 1990. Microbial quality of the water in the distribution system of Florence. *Aqua*, **39**, 258–64.

Gameson, A.L.H. and Gould, D.J. 1975. Effects of solar radiation on the mortality of some terrestrial bacteria in sea water. In: Gameson, A.H.L. (ed.), *Discharge of sewage from sea outfalls*. Oxford: Pergamon Press, 209–19.

Gameson, A.L.H. and Saxon, J.R. 1967. Field studies on effect of daylight on mortality of coliform bacteria. *Water Res*, **1**, 279–95.

Gameson, A.L.H., Bufton, A.W.J. and Gould, D.J. 1967. Studies of the coastal distribution of coliform bacteria in the vicinity of a sea outfall. *Water Poll Control*, **66**, 501–24.

Gavin, F. and LeClerc, H. 1974. Etude des bacilles gram-pigmentés en joune isolés de l'eau. *Int Oceanogr Med*, **37**, 17–68.

Geldreich E.E. 1966. *Sanitary significance of fecal coliforms in the environment*. Cincinnati: Federal Water Pollution Control Administration, WP-20-3, 122.

Geldreich, E.E. 1972a. Water-borne pathogens. In: Mitchell, R. (ed.), *Water pollution microbiology*. New York: John Wiley & Sons, 207–41.

Geldreich, E.E. 1972b. Buffalo Lake recreational water quality: a study of bacteriological data interpretation. *Water Res*, **6**, 913–24.

Geldreich, E.E. 1973. The use and abuse of fecal streptococci in water quality measurements. Proceedings of the First Microbiological Seminar on Standardization of Methods. Washington, DC: Environmental Protection Agency, EPA-R4-73-022.

Geldreich, E.E. 1978. Bacterial populations and indicator concepts in feces, sewage, stormwater and solid wastes. In: Berg, G. (ed.), *Indicators of viruses in water and food*. Ann Arbor, MI: Ann Arbor Science Publishers, 51–97.

Geldreich, E.E. 1986. Control of microorganisms of public health concern in water. *J Environ Sci*, **29**, 34–7.

Geldreich, E.E. 1990. Microbiological quality of source waters for water supply. In: McFeters, G.A. (ed.), *Drinking water microbiology*. New York: Springer-Verlag, 3–31.

Geldreich, E.E. 1991. Visions of the future in drinking water microbiology. *J New Engl Water Works Assoc*, **106**, 1–8.

Geldreich, E.E. 1996. *Microbial quality of water supply in distribution systems*. Boca Raton, FL: CRC/Lewis Publishers, 504.

Geldreich, E.E. and Bordner, R.H. 1971. Fecal contamination of fruits and vegetables during cultivation and processing for market: a review. *J Milk Food*, **34**, 184–95.

Geldreich, E.E. and Kenner, B.A. 1969. Concepts of fecal streptococci in stream pollution. *J Water Poll Control Fed, Part II*, **41**, R336–52.

Geldreich, E.E. and LeChevallier, M. 1999. Microbiological quality control in distribution systems. In: Letterman, R.D. (ed.), *Water quality and treatment*. New York: McGraw-Hill, 18.1–18.49.

Geldreich, E.E. and Reasoner, D.J. 1985. Searching for rapid methods in environmental bacteriology. In: Meybeck, M., Chapman, D.V. and Helmer, R. (eds), *Rapid methods and automation and immunology*. Berlin: Springer-Verlag, 696–707.

Geldreich, E.E. and Reasoner, D.J. 1989. Home water treatment devices and water quality. In: McFeters, G.A. (ed.), *Drinking water microbiology*. New York: Springer-Verlag, 147–67.

Geldreich, E.E., Bordner, R.H., et al. 1962. Type distribution of coliform bacteria in the feces of warm-blooded animals. *J Water Poll Control Fed*, **34**, 295–301.

Geldreich, E.E., Best, L.C., et al. 1968. The bacteriological aspects of stormwater pollution. *J Water Poll Control Fed, Part I*, **40**, 1861–72.

Geldreich, E.E., Nash, H.D., et al. 1975. The necessity of controlling bacterial populations in potable waters – bottled water and emergency water supplies. *J Am Water Works Assoc*, **67**, 117–24.

Geldreich, E.E., Allen, M.J. and Taylor, R.H. 1978. Interferences to coliform detection in potable water supplies. In: Hendricks, C.W. (ed.), *Evaluation of the microbiology standards for drinking water*. Washington, DC: US Environmental Protection, 13–20, EPA-570/9-78-002.

Geldreich, E.E., Nash, H.D., et al. 1980. Bacterial dynamics in a water supply reservoir: case study. *J Am Water Works Assoc*, **72**, 31–40.

Grabow, W.O.K. and Nupen, E.M. 1972. The load of infectious microorganisms in the waste water of two South African hospitals. *Water Res*, **6**, 1557–63.

Hernandez Duquino, H. and Rosenberg, F.A. 1987. Antibiotic-resistant *Pseudomonas* in bottled drinking water. *Can J Microbiol*, **33**, 286–9.

Johnston, R., Harmon, S., et al. 1964. Method to facilitate the isolation of *Clostridium botulinum* Type E. *J Bacteriol*, **88**, 1521–2.

Kampelmacher, E.H. and Van Noorle Jansen, L.M. 1973. *Legionella* and thermotolerant *E. coli* in the Rhine and Meuse at their point of entry into the Netherlands. *H₂O*, **6**, 199–200.

Kampelmacher, E.H. and Van Noorle Jansen, L.M. 1976. *Salmonella* effluent from sewage treatment plants, wastepipes of butchers' shops and surface water in Walcheren. *Zentralbl Bakteriol Hyg Abt 1 Orig B*, **162**, 307–19.

Klein, D.A. and Wu, S. 1974. A factor to be considered in heterotrophic microorganisms enumeration from aquatic environments. *Appl Microbiol*, **27**, 429–31.

Knittel, M.D., Seidler, R.J., et al. 1977. Colonization of the botanical environment by *Klebsiella* isolates of pathogenic origin. *Appl Environ Microbiol*, **34**, 557–63.

Kowal, N.E. 1982. *Health effects of land treatment: Microbiological.* Cincinnati, OH: US Environmental Protection Agency, EPA-600/1-82-007, 58.

LeChevallier, M.W., Cameron, S.C., et al. 1985. New medium for the improved recovery of coliform bacteria from drinking water. *Appl Environ Microbiol*, **45**, 484–92.

Leclerc, H., Buttiaux, R., et al. 1977. *Microbiologie appliquée.* Paris: Doin Editeurs.

Lin, S.D. 1974. Evaluation of fecal streptococci tests for chlorinated secondary effluents. *J Environ Eng Div Proc Am Soc Civil Engr*, **100**, 253–67.

McFeters, G.A. 1990. Enumeration, occurrence, and significance of injured indicator bacteria in drinking water. In: McFeters, G.A. (ed.), *Drinking water microbiology.* New York: Springer-Verlag, 478–92.

McFeters, G.A., Kippin, M.W., et al. 1986. Injured coliforms in drinking water. *Appl Environ Microbiol*, **51**, 1–5.

Moore, B. 1948. The detection of paratyphoid carriers in towns by means of sewage examination. *Monthly Bull Ministry Health, Public Health Lab Serv*, **7**, 241–8.

Moore, B. 1950. The detection of typhoid carriers in towns by means of sewage examination. *Monthly Bull Ministry Health, Public Health Lab Serv*, **9**, 72–8.

Mosely, S.L., Echevaria, P., et al. 1982. Identification of enterotoxigenic *Escherichia coli* by colony hybridization using three enterotoxin gene probes. *J Infect Dis*, **145**, 863–9.

Moyer, N.P. 1989. *Aeromonas* gastroenteritis: another waterborne disease? Water Quality Technology Conference. Denver: American Water Works Association, 239–61.

Mundt, J.O. 1963. Occurrence of enterococci on plants in a wild environment. *Appl Microbiol*, **11**, 141–4.

Mundt, J.O. and Graham, W.F. 1968. *Streptococcus faecium* var. *casseliflavus*, nov. var. *J Bacteriol*, **95**, 2005–9.

Mundt, J.O., Coggin, J.H. and Johnson, L.F. 1962. Growth of *Streptococcus faecalis* var. *liquefaciens* on plants. *Appl Microbiol*, **10**, 552–5.

Mundt, J.O., Anandam, E.J. and McCarty, I.E. 1966a. Streptococcease in the atmosphere of plants processing vegetables for freezing. *Health Lab Sci*, **3**, 207–13.

Mundt, J.O., Larsen, S.A. and McCarty, I.E. 1966b. Growth of lactic acid bacteria in waste waters of vegetable processing plants. *Appl Microbiol*, **14**, 115–18.

Nemedi, L., Borbala, T., et al. 1984. Hydrobiological evaluation of water samples from 'standing waters' (lakes, reservoirs and ponds) near Budapest. *Budapest Kozegeszsegugy*, **2**, 40–8.

Niewolak, S. 1974. The occurrence of microorganisms in the waters of the Kortowskie Lake. *Pol Arch Hydrobiol*, **21**, 315–33.

Niewolak, S. 1987. Bacteriological water quality of an artificially destratified lake. *Rocz Nauk Rol Hig*, **101**, 115–54.

Percival, S.L., Walker, J.T. and Hunter, P.R. 2000. *Microbiological aspects of biofilms and drinking water.* Boca Raton, FL: CRC Press, 229.

Perez, J.L., Berrocal, C.I. and Berrocal, L. 1986. Evaluation of a commercial β-glucuronidase test for the rapid and economical identification of Escherichia coli. *J Appl Bacteriol*, **61**, 541–5.

Peterson, M.L. 1971. *Pathogens associated with solid waste processing: a progress report, SW-49r.* Cincinnati, OH: US Environmental Protection Agency.

Ptak, D.J., Ginsburg, W. and Willey, B.F. 1974. *Aeromonas*, the great masquerader. Water Quality Technology Conference, December 2–3. Denver, CO: American Water Works Association.

Randall, J.S. 1956. The sanitary significance of coliform bacteria in soil. *J Hyg*, **54**, 365–77.

Reasoner, D.J. 1990. Monitoring heterotrophic bacteria in potable water. In: McFeters, G.A. (ed.), *Drinking water microbiology.* Berlin: Springer-Verlag, 452–77.

Reasoner, D.J. and Geldreich, E.E. 1985. A new medium for the enumeration and subculturing of bacteria from potable water. *Appl Environ Microbiol*, **49**, 1–7.

Reasoner, D.J., Blannon, J.C., et al. 1989. Non-photosynthetic pigmented bacteria in a potable water treatment and distribution system. *Appl Envion Microbiol*, **55**, 912–21.

Report. 1959. *Sewage contamination of bathing beaches in England and Wales.* Medical Research Council, memorandum no. 37. London: HMSO.

Report. 1965. *Pollution of interstate waters of the Red River of the north (Minnesota, North Dakota).* Cincinnati, OH: Public Health Service.

Report. 1968. *Water quality criteria, National Technical Advisory Committee to the Secretary of the Interior.* Washington, DC: Federal Water Pollution Control Administration.

Report. 1971. *Report on the Missouri River water quality studies.* Kansas City, MO: US Environmental Protection Agency, Regional Office.

Report. 1981. Possible waterborne *Campylobacter* outbreak – British Columbia. *Can Dis Wkly Rep*, **7**, 223.

Report. 1984. *Guidelines for drinking water quality.* Geneva: World Health Organization.

Report. 1985. *Microbiological water quality criteria: a review for Australia.* Canberra: Australian Water Resources Council.

Report. 1992. *Revision of the WHO guidelines for drinking water quality.* Geneva: World Health Organization.

Report. 1994. Drinking water methods for the examination of waters and associated materials. *Rep Public Health Md Subj*, **1**.

Report. 1998. *Standard methods for the examination of water and wastewater.* Washington, DC: American Public Health Association.

Rice, E.W., Fox, K.R., et al. 1996. Evaluating plant performance with endospores. *J Am Water Works Assoc*, **88**, 122–30.

Richardson, K.J., Stewart, M.H. and Wolfe, R.L. 1991. Application of gene probe technology for the water industry. *J Am Water Works Assoc*, **83**, 71–81.

Ridgway, H.F. and Olson, B.H. 1982. Chlorine resistance patterns of bacteria from two drinking water distribution systems. *Appl Environ Microbiol*, **44**, 972–87.

Romaninko, V.I. 1971. Total bacterial numbers in Rybinsk reservoir. *Mikrobiologiya*, **40**, 707–13.

Rose, J.B. 1990. Emerging issues for the microbiology of drinking water. *Water/Eng Mgt*, **137**, 23–9.

Rose, R.E., Geldreich, E. and Litsky, W. 1975. Improved membrane filter method for fecal coliform analysis. *Appl Microbiol*, **29**, 532–6.

Rosenberg, M.L., Hazelt, K.K., et al. 1976. Shigellosis from swimming. *J Am Med Assoc*, **236**, 1849–52.

Ruskin, R.H., Knudsen, A.N. and Rinehart, F.P. 1989. *Water quality of public housing cisterns in the US Virgin Islands, technical report.* St Thomas, VI: Caribbean Research Institute, University of the Virgin Islands.

Schiemann, D.A. 1985. Experiences with bacteriological monitoring of pool water. *Infect Control*, **6**, 413–17.

Schubert, R.H.W. and Mann, S.W. 1968. Zum derzeitigen stand der Identifierungsmoglichkeiten von *Escherichia coli* als faekal Indikator bei der Trinkwasseruntersuchung. *Zentralbl Bakteriol*, **208**, 498–506.

Seidler, R.J., Morrow, J.E. and Bagley, S.T. 1977. *Klebsiella* in drinking water emanating from redwood tanks. *Appl Environ Microbiol*, **33**, 893–900.

Singh, A., Yeager, R. and McFeters, G.A. 1986. Assessment of in vivo revival, growth and pathogenicity of *Escherichia coli* strains after copper- and chlorine-induced injury. *Appl Environ Microbiol*, **52**, 832–7.

Spino, D.F. 1966. Elevated temperature technique for the isolation of *Salmonella* from streams. *Appl Microbiol*, **14**, 591–6.

Stapert, E.M., Sokolski, W.T. and Northam, J.I. 1962. The factor of temperature in the better recovery of bacteria from water by filtration. *Can J Microbiol*, **8**, 809–10.

Swayne, M.D., Boone, G.M. and Bauer, D. 1980. *Wastewater in receiving waters at water supply abstraction points*. Cincinnati: US Environmental Protection Agency, EPA-600/2-80-044, 189.

Taylor, D.N., McDermott, K.T. and Little, J.R. 1983. *Campylobacter enteritis* associated with drinking water in back country areas of the Rocky Mountain. *Ann Intern Med*, **99**, 38–40.

Tenover, F.C. 1988. Diagnostic deoxyribonucleic acid probes for infectious disease. *Clin Microbiol Rev*, **1**, 82–101.

Vogt, R.L., Sours, H.E., et al. 1982. *Campylobacter enteritis* associated with contaminated water. *Ann Intern Med*, **96**, 292–6.

Walter, W.G. and Bottman, R.P. 1967. Microbiological and chemical studies of an open and closed watershed. *J Environ Health*, **30**, 157–63.

Weiss, C.M. and Oglesby, R.T. 1960. Limnology and water quality of raw water in impoundments. *Public Works*, **91**, 97–101.

Wilson W.J. 1931. Official circular 96. British Waterworks Association.

Wilson, W.J. and Blair, E.McV. 1925. Correlation of the sulphite reduction test with other tests in the bacteriological examination of water. *J Hyg*, **24**, 111–19.

10

Bacteriology of milk and milk products

FREDERICK J. BOLTON

The milk of cows, buffaloes, sheep, goats, donkeys, horses, and camels has been an important item of human food for centuries. Pathogenic microorganisms often contaminate milk, which has been recognized as a vehicle for disease since the late nineteenth century (Department of Public Health 1886). Strategies such as improved hygiene, eradication of tuberculosis and brucellosis from cattle, and heat treatment are now employed in most developed countries to reduce the risk to human health. Liquid milk and milk products such as cheese, yogurt, and fermented milks are often stored for days, weeks, or months and finally consumed without cooking. It is not surprising, therefore, that these items are still implicated in major outbreaks of food poisoning and other communicable diseases (Djuretic et al. 1997; De Buyser et al. 2001; Gillespie et al. 2003a).

Because milk and milk products are such universal items of food, sporadic infections caused by them may go unnoticed. Wilson (1942) discussed the difficulties of ascertaining the frequency of milkborne disease. At that time, pasteurization was not widely practiced, and both milk-associated tuberculosis and undulant fever were major public health problems. Both diseases had long incubation periods, gradual onset, and variable infectivity. It was therefore difficult to prove their relationship with the consumption of milk. Moreover, if, in large explosive outbreaks of disease such as typhoid fever, scarlet fever, or food poisoning, cases were scattered over a wide area, other sources of infection were often suspected. Thus, unless careful epidemiological investigations were made, the association with contaminated

milk might be overlooked. Even nowadays, in developing countries, Wilson's observations remain true and where hygiene is poor and animal disease uncontrolled, milk is an important vehicle for various human diseases.

Most developed countries have sophisticated disease surveillance systems in place and resources to investigate outbreaks. In addition, the pathogenic microorganisms likely to contaminate milk differ from those prevalent in 1942 (Bell and Palmer 1983; Milner 1995; Djuretic et al. 1997; De Buyser et al. 2001; Gillespie et al. 2003a); some have decreased as a result of animal disease eradication programs and changes in patterns of human disease, but others have taken their place. It is clear that, although rare, public health problems with milk and milk products remain and where milk products are produced by large commercial establishments there is potential for a major outbreak should anything go wrong.

MILK

Microbiology of cows' milk

The major milk-producing animal is the cow and most of the available information and research into microbiology of milk relates to cows' milk. Microorganisms in raw milk are derived from various sources, including the commensal or pathogenic flora of the udder, teat canals, and skin, which varies depending on whether cattle are housed under cover or in yards or are out in pasture; fecal contamination; and environmental sources such as milking equipment, storage vessels or the water supply. Commensal or pathogenic organisms from the milkers may also contribute.

The importance of particular organisms depends on whether the milk is to be consumed raw or after heat processing, and on the temperature and duration of storage. In developed countries the commercial production of milk is subject to strict regulations regarding hygiene and refrigeration during transport and storage. The keeping quality and flavor of the heat-treated product is adversely affected by psychrotrophic bacteria (capable of growing at between 2 and 7°C), thermoduric organisms, and post-pasteurization contamination. Some of the enzymes produced by psychrotrophs are heat-tolerant and cause spoilage both before and after pasteurization. Post-pasteurization contamination is the most important factor influencing the quality of pasteurized milk but the quality and hygiene of raw milk also have important effects on the quality of the heat-treated product.

No matter how good the health of the dairy herd, the milking hygiene, and the storage conditions, the safety of raw milk cannot be guaranteed (Sharp et al. 1985; Maguire 1993; de Louvois and Rampling 1998; Jayarao and Henning 2001; Gillespie et al. 2003a). Detailed reviews of the microbiology of raw milk are to be found in specialist publications (Bramley and McKinnon 1990; Gilmour and Rowe 1990). Most of the microorganisms in hygienically produced raw milk with low bacterial counts are derived from the normal flora of the cow's teat canals and skin, and comprise coagulase-negative staphylococci, micrococci, and streptococci. Streptococci belonging to Lancefield group N cause natural souring of raw milk. Commercially cultured strains of these organisms are used as cheese starters in the dairy industry.

The skin of the teats and udder inevitably becomes contaminated by fecal and environmental bacteria, viruses, yeasts, and molds. Cows housed inside during winter have very heavily soiled udders and the bacterial counts in milk are high even after washing the teats. The most effective way of removing contamination is careful washing followed by drying with paper towels or with a disinfectant impregnated cloth. When the cattle are turned out to pasture there is a pronounced decline in the total bacterial count in bulk milk, even compared with milk from carefully washed cows (Bramley and McKinnon 1990). Inadequately disinfected milking machines, pipes, and tanks are also important sources of post-milking contamination with bacteria, which are mainly derived from milk previously in the system. In the UK, the water supply to all areas concerned with milking and milk collection must be potable, but in many countries farms obtain water from boreholes or springs which may at times be contaminated with pathogenic organisms such as *Campylobacter jejuni*, verocytotoxin producing *Escherichia coli* or *Salmonella* serotypes. Contaminated water can also be a source of

psychrotrophic organisms such as *Pseudomonas* spp., which cause spoilage of refrigerated milk. Under modern systems of dairy management and marketing, milk is stored in refrigerated farm tanks for up to 48 h prior to collection. Once at the creamery, it may again be stored overnight in large insulated or refrigerated silos. During this time there is considerable scope for multiplication of psychrotrophic organisms.

The level of bacteria in raw milk increases significantly if the cow has mastitis and many of the organisms associated with this condition are potential causes of human infection or intoxication. The most common causes of mastitis are *Staphylococcus aureus* (including enterotoxigenic strains); *Streptococcus agalactiae* (Lancefield group B); *Streptococcus uberis*; *Streptococcus dysgalactiae* (Lancefield group C); and *E. coli*. Less common causes are *Leptospira interrogans* serovar. *hardjo*, *Streptococcus zooepidemicus* (Lancefield group C), *Listeria monocytogenes*, *Bacillus cereus*, *Pasteurella multocida*, *Clostridium perfringens*, *Nocardia* spp., *Cryptococcus neoformans*, *Actinomyces* spp., and *Corynebacterium ulcerans*.

At one time, *Mycobacterium* spp. were an important cause of bovine mastitis in the UK and both *Mycobacterium bovis* and *Brucella abortus* were commonly present in milk even in the absence of mastitis (Galbraith and Pusey 1984). Both these pathogens are now virtually eradicated from cattle in most developed countries. Milkborne scarlet fever due to *Streptococcus pyogenes* (Lancefield group A) was also common at one time (Eyler 1986). Group A streptococci are not natural pathogens of cattle but cows can develop teat lesions and mastitis if infected by a human carrier. Other pathogens such as *L. monocytogenes*, *Coxiella burnetii*, and *S. zooepidemicus* may on occasions be excreted in milk in the absence of udder disease and *Salmonella* serotypes (Marth 1969) and *C. jejuni* occasionally cause clinical or subclinical mastitis (Morgan et al. 1985; Orr et al. 1995). Under natural conditions, fecal contamination is probably the most common source of salmonellae, *L. monocytogenes*, *C. jejuni*, and verocytotoxin producing *E. coli* O157.

A study conducted by the Public Health Laboratory Service (PHLS) in 1996–1997 showed that 41 of 1 097 samples of bottled raw milk were contaminated with potentially pathogenic organisms (de Louvois and Rampling 1998). Of these 41 samples, five contained *Salmonella* serotypes, three *E. coli* O157, 19 *C. jejuni*, 12 *S. aureus*, and two hemolytic streptococci. This study also confirmed that pathogens may be present in raw milk that satisfies other statutory criteria. In a study of bulk tank milk sampled from 131 dairy herds in South Dakota and western Minnesota, USA, 26 percent of samples contained one or more foodborne pathogens (Jayarao and Henning 2001). In this latter study *C. jejuni*, verocytotoxin producing *E. coli*, *L. monocytogenes*, *Salmonella* spp., and *Yersinia enterocolitica* were detected in 9.2, 3.8, 4.6, 6.1, and 6.1 percent of bulk tank milk samples, respectively. *E. coli* O157 was not detected in bulk tank milk samples in this study. These studies confirm that fecal contamination of raw milk and presence of potential pathogens in raw milk is a frequent event.

Microbiology of ewes' and goats' milk

Milk, and in particular milk products, made from ewes' and goats' milk have become very popular in recent years and the commercial market for cheese and yogurt is expanding. Surveys of the microbiological quality of goats' milk have been performed in Scotland (Hunter and Cruickshank 1984) and England (Roberts 1985; Little and de Louvois 1999). The 1984 and 1985 studies both showed that high standards of hygiene were possible but that problems arose from inadequate disinfection of milking equipment or from inadequate control of storage temperatures at the point of sale.

A study of goats' milk on six farms in Quebec over a one year period showed that the levels of aerobes, psychrotrophs, coliforms, yeasts, and molds increased during the summer months but that this could be controlled by good hygiene, rapid cooling, refrigeration, and frequent collections by refrigerated vehicles (Tirard-Collet et al. 1991). Somatic cell numbers were also measured and these confirmed that, in contrast to the finding with cows, variation with the season and the state of lactation of the animals, high levels of somatic cells were not necessarily an indication of mastitis (Hinckley 1991).

A more recent study undertaken by the Public Health Laboratory Service assessed the microbiological quality of unpasteurised goats' milk sold at retail (Little and de Louvois 1999). Although this was a pilot study, the results indicate that 47 percent (47/100) of goats' milk samples failed to meet the standards required in the Dairy Products (Hygiene) Regulations (HMSO 1995). *C. jejuni*, *E. coli* O157, and *Salmonella* spp. were not detected but *S. aureus*, hemolytic streptococci or enterococci were present in excess of 100 cfu/ml in 6, 2, and 12 percent of samples, respectively.

There are very few publications on the microbiology of sheep milk. Studies in Spain, Bulgaria, and Norway have indicated that significant contamination with Enterobacteriaceae and coliforms may occur during milking and as a result of poor cleaning practices on the farm (Gaya et al. 1987). A very limited study of ewes' milk was reported by Little and de Louvois (1999), and demonstrated that 50 percent (13/26) of ewes' milk failed to meet the standards required in the Dairy Products (Hygiene) Regulations (HMSO 1995). *C. jejuni*, *E. coli* O157, and *Salmonella* spp. were not detected, but *S. aureus*, or enterococci were present in excess of 100 cfu/ml in 11.5 and 27 percent of samples, respectively.

Even when hygiene is good, the risk of milkborne infection from unpasteurized goats' or ewes' milk is similar to that for cows' milk. *Brucella melitensis* and *Brucella ovis* are still common in goats and sheep in the Mediterranean countries, Asia, and Latin America and human infections are associated with the consumption of raw milk or cheese made from unpasteurized milk (Wallach et al. 1994; Report 1995). Other pathogenic microorganisms such as *Salmonella* Typhimurium, *S. aureus*, *Y. enterocolitica*, *C. perfringens*, *B. cereus*, and *C. jejuni* can survive in goats' milk under experimental conditions, and some of these pathogens multiply at temperatures of 22°C or more (Roberts 1985). Outbreaks and incidents of brucellosis, campylobacter infection, salmonellosis, viral tick-borne encephalitis, Q fever, staphylococcal food poisoning, and *C. ulcerans* sore throat have been associated with consumption of untreated goats' milk or goat cheese in Canada, Britain, and the USA (Hutchinson et al. 1985a; Sharp 1987; Fishbein and Raoult 1992; Report 1994a, 1995). There is also a high incidence of toxoplasma infection in goats (Skinner et al. 1987) and there is convincing evidence that consumption of unpasteurized goats' milk can transmit infection to man (Sacks et al. 1982). *E. coli* O157 has emerged as a significant pathogen in milk and dairy products and outbreaks associated with cheese made from unpasteurized goats' milk have been reported from the Czech Republic (Bielaszewska et al. 1997) and from Scotland (Curnow 1999). Both of these outbreaks were associated with infection in children. The outbreak in the Czech Republic was recognized because of a cluster of four cases of hemolytic uremic syndrome. The outbreak in Scotland resulted in 22 confirmed cases of infection in children attending a primary school. The most recent outbreak of *E. coli* O157 associated with unpasteurized goats' milk was reported in British Columbia, Canada (McIntyre et al. 2002). Following a review of infection with *E. coli* O157 in England and Wales and Scotland, it has been recommended that raw ewes' and goats' milk for sale for drinking should be heat treated (Task Force on *E. coli* O157 2001).

Sheep can also carry many microorganisms pathogenic for man, such as *Salmonella* spp., *C. burnetii*, *Campylobacter* spp., *E. coli* O157, *S. aureus*, *Toxoplasma gondii*, and *L. monocytogenes*. *S. aureus* seems to be a common contaminant of both ewes' and goats' milk and more prevalent than in cows' milk (Valle et al. 1990). A high proportion of isolates produce enterotoxin. An outbreak of staphylococcal food poisoning in Scotland was traced to cheese made from unpasteurized sheep milk (Sharp 1987). Although *L. monocytogenes* is a cause of meningoencephalitis in sheep and may also be excreted in the milk and feces of healthy sheep, the risk of contamination of ewes' milk on the farm does not seem to be any greater than for cows' milk. Rodriguez et al. (1994) sampled milk from 1 052 farm bulk tanks on 283

Spanish farms throughout a 1-year period and found that *Listeria* spp. could be detected in 4.56 percent and *L. monocytogenes* in 2.19 percent. This low prevalence was confirmed in the study described by Little and de Louvois (1999). *Listeria* species were not detected by enrichment culture in all 26 samples of ewes' milk and 100 samples of goats' milk. Therefore, the risk of acquiring *L. monocytogenes* infection from these products appears to be low.

Heat treatment of milk

PASTEURIZED MILK

The international definition of pasteurization (Report 1994b) is: 'A heat treatment process applied to a product with the aim of avoiding public health hazards arising from pathogenic microorganisms associated with milk. Pasteurization as a heat treatment process is intended to result in only minimal chemical, physical, and organoleptic changes.' The process was originally introduced by dairies to improve the keeping quality of milk (Wilson 1942). It soon became clear that heat treatment would also destroy pathogenic organisms such as tubercle bacilli and hemolytic streptococci. The public health aspects of the process then assumed the greater importance.

Several time–temperature combinations are effective. The Dairy Products (Hygiene) Regulations (HSMO 1995) for England and Wales state that milk shall be pasteurized by means of a heat treatment involving a high temperature for a short time (at least 71.7°C for 15 s, or any equivalent combination) or a pasteurization process using different time and temperature combinations to obtain an equivalent effect. The product must give a negative alkaline phosphatase test immediately after heat treatment and be cooled as soon as practicable to a temperature of 6°C or lower and stored at or below that temperature until it leaves the treatment establishment.

The high temperature short time (HTST) method is most commonly used both in large commercial dairies and on farms. The HTST plant consists of a plate heat exchanger and a holding tube (Figure 10.1). Raw milk

enters the heat exchanger and is warmed by the hot milk leaving the hot section along the other side of the plates. The warmed milk is then heated to the statutory temperature before it runs through the holding tube for at least 15 s. The hot milk then re-enters the heat exchanger, being cooled initially by cold raw milk, then by tap water and finally by chilled water. The law requires that any HTST system has an automatic temperature control and a safety device that automatically diverts milk back for reheating if it has not been raised to the statutory temperature. In addition, there must be indicating and recording thermometers on the plant and an automatic recording device for the safety system.

An alternative method for small quantities of milk is the holder or batch-heating process in which milk is heated to between 62.8 and 65.6°C and held at that temperature for 30 min, followed by rapid cooling to below 10°C. The heating vessel incorporates an insulated outer jacket through which the heating and cooling media can flow, an agitator to ensure uniform heating, thermometers, including a recording thermometer and a temperature control device that automatically regulates the heating of the milk to pasteurization temperature and the subsequent retention at this temperature for the holding period (Report 1994b). Nowadays, the holder or batch method is seldom practiced for the pasteurization of milk but is still used in some small on-farm dairies for the pasteurization of cream.

The combinations of time and temperature used for the heat treatment of milk are designed to kill all pathogenic vegetative cells including *C. burnetii* which is relatively heat-resistant (Zall 1990). Bacterial spores and some preformed bacterial or fungal toxins, however, are not destroyed by pasteurization temperatures. The keeping quality of pasteurized milk depends both on the quality before heat treatment and on the amount of post-pasteurization contamination. Thermoduric bacteria that survive pasteurization belong to the genera *Bacillus*, *Clostridium*, *Corynebacterium*, *Micrococcus*, *Streptococcus*, and *Lactobacillus*. The importance of thermoduric organisms depends on the storage temperature. Some thermoduric fecal streptococci, *Bacillus* spp. and *Clostridium* spp. can grow below 7°C and are therefore also psychrotrophic. Most thermoduric organisms are mesophilic with optimum growth between 20 and 32°C. If pasteurized milk is allowed to remain at ambient temperature for several hours, as commonly happens with doorstep deliveries, growth of mesophilic bacteria such as *B. cereus* will soon cause bitterness, sweet curdling, and 'bitty' cream. Contamination with *B. cereus* is very common and yet reports of bacillus food poisoning due to the consumption of milk or cream are rare (Gilbert 1979; Christiansson 1992). This is probably because enterotoxin is not usually produced by *B. cereus* until the concentration of organisms is so high and consequent spoilage so great that the milk is unusable.

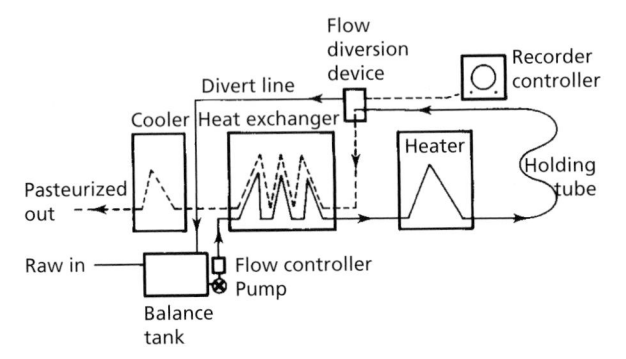

Figure 10.1 *Diagrammatic representation of milk flow through HTST pasteurizer*

The lactic acid-producing streptococci do not survive pasteurization and 'souring' does not therefore take place in pasteurized milk. Repasteurization of returned milk results in enrichment of the thermoduric organisms in the final product and so is bad practice.

The keeping quality and flavor of properly refrigerated pasteurized milk depends on the types and numbers of psychrotrophic organisms. The most important of these are species of *Pseudomonas*, *Flavobacterium*, *Alcaligenes*, *Acinetobacter*, yeasts, and molds. These organisms are not thermoduric but are derived from the post-pasteurization environment. They outgrow any thermoduric psychrotrophs that may be present, and for this reason, hygiene of the pipes, tanks, and bottle fillers is the most important factor influencing the keeping quality of milk stored below 10°C (Schröder 1984). Psychrotrophic organisms are responsible for various 'off' flavors, ropiness, and changes in color.

ULTRA-HEAT-TREATED MILK

According to the Regulations, the description ultra-heat-treated (UHT) may be applied to a continuous flow of milk that has been heated to a temperature of not less than 135°C and retained at that temperature for at least 1 s. The method of heating may be indirect or by direct injection of steam. After treatment, the milk must be packaged in sterile airtight containers that are filled and sealed with aseptic precautions. In practice, the milk is usually heated to more than 140°C for 2–4 s to produce 'commercial sterility' – which means that it has a shelf-life of several weeks because, although small numbers of spores may survive, they are usually unable to grow under the conditions of storage (Lewis 1994).

STERILIZED MILK

The term 'sterilized' is applied to milk that has been filtered, clarified, homogenized, and then heated to 100°C for long enough to denature all casein and whey proteins. Traditionally, sterilized milk is bottled and sealed before heating to 110–116°C for 20–30 min. The process inevitably results in a caramelized flavor and brownish color but this is popular with some consumers.

THERMIZED MILK

In situations where raw milk has to be held under chilled conditions for some time, its keeping quality can be improved by a low temperature heat treatment called thermization. Thermized milk must conform to strict definitions according to European law. It must be obtained from raw cows' milk which, if it is not treated within 36 h of acceptance by the establishment, has a plate count at 30°C prior to thermization which indicates a concentration of 3×10^5/ml or less. The thermization treatment consists of heating raw milk for at least 15 s at between 57 and 68°C so that, after such treatment, the milk still shows a positive reaction in the phosphatase test. This treatment is not sufficient to eliminate pathogenic microorganisms.

Legislation

In most developed countries, the hygiene and composition of milk and milk products are subject to statutory controls, which specify general standards for production, transport, and processing, and specific standards, which set microbiological and chemical limits for each product.

Council Directive 92/46/EEC (as amended) applies to milk from sheep, goats, buffaloes, and cows and covers milk and milk products produced within the European Union. Detailed microbiological standards including absence of pathogenic microorganisms are specified for raw milk, heat-treated milk, and milk-based products, which must be sampled at the production or processing establishments. Nonmicrobiological tests are also specified, including somatic cell count, which is an indicator of mastitis in raw milk from cows and buffaloes (no such standard is specified for milk from sheep and goats because the levels are subject to wide natural fluctuations in these animals), and the phosphatase and peroxidase test for pasteurized milk. Greenwood and Rampling (1995) list the microbiological standards and guideline criteria specified in the Dairy Products (Hygiene) Regulations (HMSO 1995), for England and Wales based on the European Directive (92/46/EEC). Greenwood (1995) has produced details of appropriate methods for the microbiological tests and the phosphatase and peroxidase tests.

Similar standards apply in the USA for milk and milk products from sheep, goats, and cows. The US Food and Drug Administration produces a model code for grade A milk which is specified in the Pasteurized Milk Ordinance and the Dried Milk Ordinance. These cover production, transport, and processing, as well as microbiological and chemical tests. Grade A standards are applicable to all products that are produced in plants that produce milk for shipment between states (more than 95 percent of fluid milk products). Milk is produced on the farm as either 'grade A' or 'manufacturing grade' but most milk for manufacture of dairy products is surplus grade A that has not been sold as liquid milk for drinking. The production of 'manufacturing grade' milk and of products such as cheese, butter, and ice cream is subject to regulatory oversight by the US Department of Agriculture.

QUALITY CONTROL TESTS

The plate count at 30°C and the coliform count form the basis of microbiological quality control tests for milk and milk products in most countries, including the USA. In Europe the plate count at 21°C after pre-incubation at 6°C for 5 days has been adopted as the statutory test for pasteurized milk. This test was designed to assess the level of psychrotrophic bacteria and thus keeping quality, but it has been subject to much criticism because it is time-consuming, expensive to perform, and has poor reproducibility (Scotter et al. 1993).

The coliform count is universally adopted as an indicator of poor hygiene and of post-pasteurization contamination. High counts in pasteurized milk may also be due to faults in the pasteurizer or contamination of pasteurized milk with raw milk. Pasteurization plants that produce milk which fails this test repeatedly, in spite of attention to cleaning and disinfection routines, should be examined carefully for leaks in the pasteurizer or faulty operation of the plant.

The absence of detectable alkaline phosphatase is the most important test to demonstrate adequate heat treatment of pasteurized milk or cream. Alkaline phosphatase is normally present in raw milk and is destroyed by heat treatment at a temperature slightly above that required to destroy *Mycobacterium tuberculosis* and slightly lower than those employed in the pasteurization of milk. Detection of phosphatase in pasteurized milk may therefore indicate inadequate heat treatment or contamination with raw milk.

There are several internationally recognized methods of testing that may be used for enforcement purposes (Rocco 1990). The statutory test in the Dairy Products (Hygiene) Regulations (HSMO 1995) is an analytical method based on the use of phenyl phosphate as the substrate. Milk that is satisfactorily pasteurized should liberate no more than 4 μg of phenol during a 1 h period. This is equivalent to the presence of 0.1 percent raw milk in pasteurized milk (Greenwood and Rampling 1997). A fluorimetric method is now available commercially which reports results in mU/l of fluorescent substrate released. A reading of 500 mU/l is equivalent to the presence of 0.1 percent raw milk, the statutory test limit of the previously mentioned statutory analytical method. The fluorimetric method is able to detect the equivalent of 0.006 percent raw milk in pasteurized milk and because of this increased sensitivity, it is recommended that a value of less than 100 mU/l should be used as an indicator of effective pasteurization and hence of microbiological safety (Greenwood and Rampling 1997).

It should be noted that not all of the alkaline phosphatase tests have the sensitivity to detect the presence of low levels of raw milk contamination due to small leaks in valves or heat exchange plates. To monitor the process efficiently, the test should be applied to samples taken from the cold milk exit side of the pasteurizer and immediately following an episode of flow diversion to check the integrity of the flow diversion valve.

Alkaline phosphatase in pasteurized milk may be due to any or a combination of residual, microbial, or reactivated alkaline phosphatase. The presence of residual phosphatase indicates inadequate pasteurization or post-pasteurization contamination with raw milk. Microbial phosphatase can occur when the milk has been stored for a long period of time before pasteurization and is the result of psychrotrophic bacteria growing and producing a heat stable alkaline phosphatase that survives the pasteurization process. Reactivated phosphatase may occasionally occur, particularly with cream. Reactivation is mediated by a heat-stable activator in the presence of magnesium and β-lactoglobulin (Lyster and Aschaffenburg 1962). The degree of reactivation depends on the initial concentration of phosphatase and, since the enzyme is adsorbed to the phospholipid of the fat globule membrane, it varies with the amount of fat and is greatest in cream. Statutory phosphatase tests for cream include a test which distinguishes between native (residual) and reactivated phosphatase.

The peroxidase test is a test for excessive heat treatment. If milk is subjected to significantly higher temperatures than those used for standard pasteurization, inactivation of the enzyme will occur. Thus a positive peroxidase test indicates that milk has not been overheated.

RESTRICTIONS ON THE SALE OF UNPASTEURIZED MILK

Most milk for drinking and for the manufacture of milk products is pasteurized but few countries have formal restrictions on the sale of raw milk. In the USA, however, very few milk products can be sold in the unpasteurized state. Some states allow limited sale of raw fluid milk and some hard cheeses may be manufactured from raw milk (Childer R.N., personal communication 1996). In England and Wales, legislation enacted in 1985 and 1986 limited sales of unpasteurized cows' milk to sales by farmhouse caterers, from farm gates, and from local retail deliveries of milk bottled on the farm. In Scotland, legislation in 1983 and 1986 prohibited all sale of unpasteurized cows' milk and cream. There are no restrictions on the sale of raw liquid milk from goats, sheep, or buffaloes or on any milk products made from raw milk anywhere in the UK and no restrictions on the sale of any unpasteurized milk or milk products elsewhere in Europe.

Outbreaks and incidents of infection associated with milk

A recent review of bacterial foodborne disease in France and six other countries estimated that outbreaks associated with milk and dairy products accounted for 1–5 percent of all foodborne outbreaks (De Buyser et al. 2001). A recent review of 69 milk and dairy product associated outbreaks in France during the period 1992–1997 confirmed that 87 percent were linked with cheese, 10 percent with milk, and 3 percent with other products. *S. aureus* was the major problem accounting for 85.5 percent of the outbreaks. Raw milk and raw milk products accounted for 48 percent of the 69 outbreaks. The situation is different in England and Wales where most of the outbreaks are associated with liquid milk and fewer with dairy products. Milk and dairy products account for 2.5 percent (46/1774) of all foodborne outbreaks in England and Wales (Gillespie, I. CDSC, personal communication). Milkborne outbreaks in England and Wales account for 1.5 percent of all foodborne outbreaks (Gillespie et al. 2003a).

In the UK most outbreaks of infection due to the consumption of liquid milk continue to be caused by pathogens associated with the gastrointestinal tract of animals. *E. coli* O157, *Salmonella* spp. or *C. jejuni* are most prevalent. There have also been incidents due to less common animal-associated organisms such as *Y. enterocolitica*, *L. monocytogenes*, *S. zooepidemicus*, *Streptobacillus moniliformis*, *C. ulcerans*, and *Cryptosporidium parvum*. Table 10.1 lists outbreaks of infection associated with milk and milk products in England and Wales from 1992 to 2000. In comparison to the data

summarizing outbreaks in England and Wales from 1989 to 1994 (Rampling 1998) there has been a significant change in that verocytotoxin producing *E. coli* O157 has emerged as the major cause of outbreaks associated with milk and dairy products. In spite of legislation limiting sales, and even though less than 3 percent of the total milk production is unpasteurized in England and Wales, 53 percent of outbreaks continue to be caused by the consumption of raw milk.

On the other hand, the situation is different in Scotland where, in the face of the epidemiological, microbiological, and research evidence that heat treatment of all cows' milk would be cost-effective, the retail sale of untreated cows' milk was prohibited from August 1983. Thereafter, incidents of infection related to milk were controlled in the general community but a small number persisted in the farming community where farm workers and their families received untreated milk as

Table 10.1 *Milk- and milk product-associated outbreaks and incidents of infection; reports to the Health Protection Agency Communicable Disease Surveillance Centre, England and Wales 1992–2000*

Year	Organism	Vehicle of infection	Number of persons infected	Evidence
1992	*Salmonella* Enteritidis PT4	Ice cream	7	S
	Salmonella Enteritidis PT4	School milk[a]	44	D
	Campylobacter jejuni	Sold as pasteurized milk	110	S
	Campylobacter jejuni	Raw milk	72	S
1993	*Salmonella* Typhimurium	Raw milk	13	M, D
	Escherichia coli O157	Raw milk	11	M
	Campylobacter jejuni	Raw milk	22	D
1994	*Salmonella* Typhimurium	Raw milk	11	M
	Salmonella Typhimurium	Raw milk	4	M, S
	Salmonella Typhimurium	Sold as pasteurized milk	26	M
	Campylobacter jejuni	Raw milk	23	S
	Streptococcus zooepidemicus	Raw milk	4	M
1995	*Salmonella* Typhimurium	Raw milk	26	M, D
	Campylobacter jejuni	Milk[a,b]	12	S
	Campylobacter jejuni	Raw milk[b]	35	S
	Cryptosporidium spp.	Sold as pasteurized milk	52	S
1996	*Salmonella* Typhimurium	Raw milk	5	M
	Escherichia coli O157	Sold as pasteurized milk	12	D
	Escherichia coli O157	Sold as pasteurized milk and raw milk	6	M
	Campylobacter jejuni	Raw milk	5	S
1997	*Salmonella* Anatum	Milk (infant formula)	17	M, S
	Salmonella Java	Milk (toddler)	14	S, D
	Escherichia coli O157	Raw milk	8	D
1998	*Salmonella* Typhimurium	Sold as pasteurized milk	121	M, D
	Escherichia coli O157	Raw milk	3	M
1999	*Escherichia coli* O157	Sold as pasteurized milk	114	S, D
	Escherichia coli O157	Sold as pasteurized milk	11	S,D
2000	*Escherichia coli* O157	Raw milk	4	M
	Escherichia coli O157	Raw milk	2	M

a) Not known whether raw or pasteurized.

b) Campylobacters: a number of incidents attributed to bird-pecked bottled milk were recorded. D, descriptive; M, Microbiological; S, statistical. Data supplied by the Health Protection Agency, Centre for Infections, London.

wage benefit. Between 1992 and 2000, only four milk-associated outbreaks occurred in Scotland. All four were associated with drinking raw milk and were restricted to farms and did not involve the general community.

In a recent study of milkborne outbreaks in England and Wales, milk sold as pasteurized was associated with 37 percent of all milkborne outbreaks (Gillespie et al. 2003a). Despite the introduction of a British Standards code of practice for pasteurization of milk on farms and in small dairies (BS7771:1994) and guidelines for 'on farm' milk processors (ADAS 1996) there is a concerning trend of outbreaks associated with milk sold as pasteurized. Most of the outbreaks linked to milk sold as pasteurized in Table 10.1 were associated with farms and in particular those with on farm dairies. Of the nine outbreaks linked to consumption of milk sold as pasteurized, six were associated with failures of the pasteurization equipment and three with post-pasteurization contamination. In a study in the northwest of England, in which farm-produced pasteurized milk was monitored during a $2\frac{1}{2}$ year period from 1999–2001, 18 percent (24/130) of on-farm dairies produced milk that failed the alkaline phosphatase test (Allen et al. 2004). In the absence of timely control checks and regular maintenance of equipment, farm-bottled milk will continue to be a microbiological safety issue (Gillespie et al. 2003a). The study by Gillespie et al. (2003a) also showed a bimodal seasonality with milkborne outbreaks occurring in spring and autumn which the authors highlight may be due to changing animal husbandry practices.

Evidence from a number of outbreak investigations has shown that it is not uncommon for operators to switch HTST pasteurizers to manual/clean mode during running, thereby removing the safety system and preventing flow diversion. This practice occurs most often with small producers who may be unaware of the exact sequence of events during flow diversion. In dairies that experience pasteurization failures, examination of thermograph records may show that soon after the milk temperature drops, initiating flow diversion, the plant has been switched to manual/clean mode, thus preventing the inevitable delay while the correct temperature is re-established. At the same time this has allowed inadequately heated milk to flow forward (Figures 10.1 and 10.2). The Dairy Products (Hygiene) Regulations (HMSO 1995) for England and Wales state that HTST pasteurizers must have an automatic safety device to prevent insufficient heating and an automatic recording device that records operation of the safety system. The regulations for Scotland also state that the record must include a record of when the cleaning mode is in operation. Some of the older designs of farm pasteurizers do not always incorporate chart recorders, which include a record of the operation of the safety system.

Information from some of the recent outbreaks associated with milk sold as pasteurized confirms that there

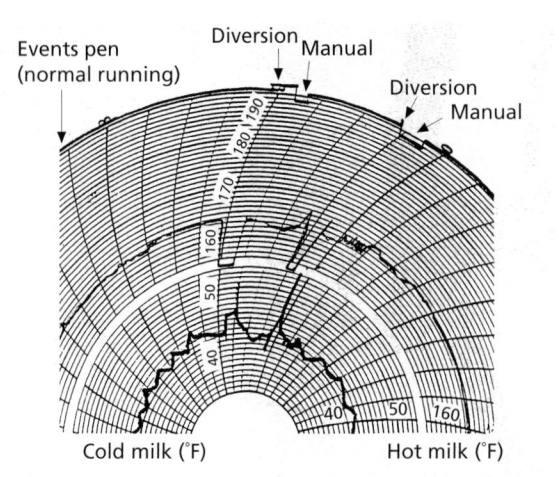

Figure 10.2 *Part of a pasteurizer thermograph chart demonstrating inappropriate use of manual/clean mode following flow diversion*

are common faults. Integrity of heat exchange plates, problems with flow restriction valves, and incorrect repair of equipment by untrained operatives have contributed to these phosphatase failures. Poor design of dairy plants and addition of other equipment such as homogenizers and cream separators can also alter the efficacy of pasteurization units.

Process control in both large and small plants must include the whole plant. The hygiene of pipes, tanks, bottle or carton fillers, and bottle washers should be monitored and the water supply must be of potable quality. Most dairies use mains tap water but occasionally the supply comes from another source. Whatever the source, water should be sampled regularly to check the microbiological quality. Where bottles are used, the final rinse consists of cold tap water and the bottles are not dried before filling with milk. This could potentially cause recontamination of milk and such a problem occurred many years ago when an outbreak of *Salmonella* Paratyphi B infections that continued for 2 years was eventually traced to river water supplying a pasteurization plant in south Wales (Thomas et al. 1948).

MICROBIOLOGICAL INVESTIGATION OF OUTBREAKS ASSOCIATED WITH ON-FARM DAIRIES

As explained above, most of the outbreaks in England and Wales associated with milk sold as pasteurized are linked to on-farm dairies. Microbiological investigation of these farms is essential to establish the etiology and source of the infecting organism. Microbiological examination should be carried out in conjunction with an evaluation of the process, equipment, and structure of the dairy. Each on-farm producer will have slightly different systems and equipment and therefore each investigation will be different. The microbiological investigations require a team approach and must involve environmental health officers (responsible for licensing

Figure 10.3 *In-line milk filters sampled during milkborne outbreaks.*

the premises), public health microbiologists, and veterinary microbiologists. The key areas to address are:

- Ensure that the pasteurization and other equipment is functioning correctly and take samples for statutory tests to confirm this.
- Take milk samples for pathogen examination from all pasteurization units, from the bulk milk tank.
- Collect filters or washings from the in-line filter between the milking parlor and the bulk tank (Figure 10.3). The latter is very important because it is a natural concentration step and will frequently result in isolation of pathogens when other samples are negative.
- Take swabs of any equipment and valve joints that appear to be faulty.
- Environmental samples or swabs from the milking parlor, the processing area and milk bottle/container storage area. Bottle rinses may also be valuable in determining post-pasteurization contamination. Specific sample points can only be decided when visiting the premises.
- Other samples that should be taken from around the farm may include the water supply, water from drinking troughs, samples of soiled bedding, and samples from the midden.
- It may also be necessary to ask a veterinary microbiologist to collect fecal or rectal swab specimens from individual cows. It is important not to limit this to only those cows that are being milked. For example, calves may be the source of the pathogen on an incriminated farm.

SALMONELLA OUTBREAKS

Cattle are the main reservoir of *Salmonella* Typhimurium and it is not surprising that this and other salmonella serotypes are frequently present in raw milk

(Threlfall et al. 1980). In the UK study of bottled raw milk, 0.5 percent of samples were salmonella positive and isolates included *Salmonella* Typhimurium phage type DT104, *Salmonella* Dublin, *Salmonella* Virchow PT26, *Salmonella* Anatum, *Salmonella* Havana, and *Salmonella* Ruiru (de Louvois and Rampling 1998). This diversity of serotypes is not unexpected and in a study of 131 dairy herds in the USA Salmonella of O serogroups B, C, D, and E were detected in 6.1 percent of bulk tank milk samples (Jayarao and Henning 2001). These studies confirm that raw milk will from time to time be contaminated with *Salmonella* spp. and will result in outbreaks of infection. *Salmonella* Typhimurium DT104 is now established as a multiresistant clone, although there are subtle differences amongst strains (Threlfall et al. 1997).

During the period from 1992 to 2000, *Salmonella* Typhimurium has been the predominant cause of salmonella associated milkborne outbreaks (Table 10.1). The growing trend is for these outbreaks to be associated with multiresistant strains of *Salmonella* Typhimurium DT104. Of the nine salmonella outbreaks associated with liquid milk, six were associated with multiresistant *Salmonella* Typhimurium DT104. A large outbreak in 1998 that affected 121 cases was caused by a ciprofloxacin-resistant multiresistant strain of *Salmonella* Typhimurium phage type DT104. There was evidence from molecular typing that this strain developed ciprofloxacin resistance following treatment of cattle in the herd with enrofloxin a few months prior to the outbreak (Walker et al. 2000).

However, large processing dairies produce thousands of gallons of pasteurized milk daily and have the potential for causing major outbreaks if there is an unrecognized fault in the structure or operation of the plant. In 1985 such an outbreak occurred in Illinois, resulting in more than 16 000 confirmed cases of *Salmonella* Typhimurium infection (Ryan et al. 1987; Report 1985).

Salmonella Typhimurium has recently caused a multistate outbreak in the USA (MMWR 2003). Sixty-two cases were linked to consumption of raw milk from a combination dairy–restaurant in Ohio during November 2002 and January 2003. Cases were identified in four states by searching for isolates with a specific pulsed field gel electrophoresis (PFGE) profile. The dairy was the only place in Ohio that sold raw milk in jugs and served raw milk and milk shakes made from raw milk to customers. *Salmonella* Typhimurium with the same PFGE type was also isolated from raw skim milk, butter made with raw milk, and cream. The sale of raw milk is permitted in 27 of the US states and this outbreak has prompted a recommendation that the retail milk regulations should be reviewed and strengthened to minimize the risk from consumption of raw milk.

During the investigation of the multistate outbreak described above, 16/211 workers at the dairy were found to be positive for the outbreak strain. This illustrates the difficulty in determining the role of food handlers in such outbreaks. There is still no evidence that any milk-associated outbreak was due to asymptomatic carriage of a salmonella by a person. Both the dairy industry and those responsible for enforcing standards should concentrate attention on process control, particularly of pasteurization, instead of screening staff to detect harmless asymptomatic excreters of salmonellae (Editorial 1987).

For further information on salmonella infections, see Chapter 54, Salmonella.

CAMPYLOBACTER

Fecal carriage of campylobacters is common in cattle and frequently results in contamination of milk. Typing of isolates from cows within a herd usually reveals that several strains are endemic on the farm and individual cows may excrete one or more strains intermittently. Thus, contamination of milk will reflect the strains circulating within each dairy herd. Several surveys have investigated the prevalence of campylobacters in milk from individual cows, farm bulk tanks, and bottled unpasteurized milk. Between 4.5 and 12.3 percent of untreated milk samples from either bulk tanks or from bottled milk have been shown to be contaminated (Beumer et al. 1988; Humphrey and Hart 1988; Rohrbach et al. 1992). This range of results reflects differences in study design, the volume of milk examined, the sensitivity of the method employed, and on seasonal variations. The most recent microbiological studies of the prevalence of C. jejuni in untreated milk report contamination of 2 percent (19/1097) of bottled milk in England and Wales (de Louvois and Rampling 1998) and 9.2 percent (12/131) of bulk milk samples in an American study (Jayarao and Henning 2001). The risk of infection following consumption of unpasteurized milk is substantial and milk has been responsible for large numbers of outbreaks of campylobacter infection in many parts of the world (Skirrow 1991). Robinson (1981) demonstrated on himself that as few as 500 organisms mixed with 180 ml of pasteurized milk were sufficient to induce abdominal cramps, diarrhea, and an antibody response. Thus, a low degree of contamination in a batch of milk could cause a large outbreak. Although C. jejuni does not multiply in milk, many strains survive for long periods and can be recovered from milk that has been stored for 14–21 days at 4°C. Pasteurization destroys even large numbers of organisms (Waterman 1982; D'Aoust et al. 1988) and properly controlled heat treatment is the most satisfactory way of ensuring that milk is free from campylobacters.

Most milkborne outbreaks of campylobacter infection reported in the literature have been due to unpasteurized milk (Wood et al. 1992; Morgan et al. 1994b; Orr et al. 1995; Hutchinson et al. 1985b) and the outbreaks listed in Table 10.1 confirm that in England and Wales from 1992–2000 this is still the case. The outbreaks shown in Table 10.1 from 1992–1995 have been reviewed by Djuretic et al. (1997) and Peabody et al. (1997) and those from 1996–2000 by Gillespie et al. (2003a). Although the last C. jejuni outbreak in England and Wales was reported in 1996, there is evidence from other countries of the continuing risk associated with drinking unpasteurised milk (Kalman et al. 2000; MMWR 2003). Although most milkborne outbreaks are usually caused by a single strain, more than one serotype may be incriminated in a single outbreak. This should not be unexpected if fecal contamination is the source of the organism in the milk.

Another important risk factor for campylobacter infection is consumption of unpasteurized milk during visits to farms for educational, agricultural, or recreational purposes. Wood et al. (1992) identified 20 outbreaks of campylobacter infection amongst children who had consumed raw milk during school field trips and organized youth activities. This mode of transmission is still a problem as highlighted by a recent outbreak in the Netherlands (van den Brandhof et al. 2003). Fifty-seven of 92 children and three teachers drank raw milk provided during the visit to the farm. Forty-seven percent of the 57 children that drank milk developed diarrhea or vomiting and other symptoms. There was a clear dose–response relationship; 17 percent of children that took a mouthful of milk became ill whereas 100 percent of children that drank two cups of milk were ill. In this outbreak, isolates were typed by amplified fragment length polymorphism (AFLP) and they all showed the same pattern. The authors recommend heat treatment of all milk provided during farm visits. Kalman et al. (2000) reported an outbreak in 1998 in Hungary during which they identified 34 confirmed cases of campylobacter infection amongst 500–600 visitors to a farm sale. Similarly, Morgan et al. (1994b) reported the investigation of an outbreak of campylobacter infections in 1992. At a large music festival in southwest England, two local farms had been issued with Temporary Producers Licenses to enable them to sell raw milk at the festival and pasteurized milk was also available for sale. A case–control study based on 72 people with microbiologically confirmed C. jejuni infection found that only the consumption of unpasteurized milk was shown to be associated with campylobacter infection. Despite recommendations about the sale of unpasteurized milk, there was another outbreak at the festival the following year that resulted in 22 members of staff being infected (Peabody et al. 1997). Fortunately, on this occasion, unpasteurized milk was not sold to the general public. Campylobacter isolates from cases and the farms collected during both outbreaks were typed by several methods and shown to be indistinguishable. This confirms the endemic nature of pathogenic C. jejuni strains within herds for long periods and the potential problem of fecal contamination of milk. In these outbreaks, the supply of unpasteurized milk resulted in significant but preventable public health problems. It is strongly recommended that visitors to such events and to open farms do not consume raw milk whilst on the premises.

There is also good evidence that *C. jejuni* may cause clinical or subclinical mastitis resulting in the shedding of large numbers of organisms into the milk (Hudson et al. 1984; Hutchinson et al. 1985b; Orr et al. 1995). Outbreaks of infection have also been associated with the consumption of milk from herds in which one or more animals have asymptomatic campylobacter mastitis. In this situation an outbreak could occur even though dairy hygiene is faultless and the milk achieves the microbiological standards for total plate count and coliforms. It is also more likely that milk will be contaminated with greater numbers of campylobacters and hence that the outbreaks will affect more people.

Outbreaks due to improperly pasteurized milk have also been documented on many occasions. In these circumstances, large numbers of people may be put at risk. In 1979, 2 500 children became ill after drinking free school milk supplied from a single milk processing plant in Luton. The campylobacter could not be isolated from the dairy or from raw milk but epidemiological studies strongly indicated milk as a source (Jones et al. 1981). An outbreak of *C. jejuni* enteritis affecting schoolchildren and staff at a boarding school in Vermont in 1986 was traced to inadequate batch pasteurization of milk. Two serotypes of *C. jejuni* were isolated, from both the patients and the dairy cows (Birkhead and Vogt 1988). During 1987, there were two more outbreaks of campylobacter infection in England and Wales, due to improperly pasteurized milk. More than 100 cases were confirmed by culture and serotyping following consumption of milk from a large dairy in East Anglia (Rampling et al. 1987). Another outbreak in 1987 affected 346 students and staff at a field study center in North Wales. Campylobacters were isolated from raw milk prior to pasteurization (Sockett 1991). In 1992 an outbreak in Northamptonshire affected at least 110 people of whom 41 had confirmed infection with *C. jejuni*. Consumption of pasteurized milk from a single farm dairy was implicated by case–control study (Fahey et al. 1995). In all of these outbreaks there was evidence of faulty processing resulting in phosphatase failures. The factors contributing to these outbreaks were: incorrect use of the bypass system (Jones et al. 1981), a leaking valve and inappropriate use of the plant with the safeguards switched off (Rampling et al. 1987), and an incorrectly wired element resulting in plant that had been malfunctioning for some time (Fahey et al. 1995).

Post-pasteurization contamination with campylobacters can also occur as a result of birds pecking milk delivered to the doorstep. This route of contamination was first observed in the late 1980s when magpies and jackdaws were incriminated. This route of transmission was confirmed microbiologically by Hudson et al. (1990,

1991) and epidemiologically by a case control study in south Wales in 1990 (Southern et al. 1990). A further point source outbreak in Exeter in May 1991 affected 11 children at a day nursery and was attributed to consumption of milk that had been pecked by magpies (Riordan et al. 1993). Evidence was obtained from a case–control study and by isolation of campylobacters from two bottles of milk that had been pecked. The most recent outbreak related to bird-pecked milk occurred in 1995 in a residential school in Gloucestershire, England. A retrospective cohort study showed that drinking pasteurized milk from bottles with damaged tops was associated with illness (Stuart et al. 1997). Bird-pecked milk has been suggested as the cause of many of the sporadic cases of campylobacter infection recorded during the annual spring, early summer rise in infections. The demise of doorstep delivery of milk and the growing trend to purchase milk from supermarkets should have an impact on this route of infection.

Peabody et al. (1997) stated that outbreaks of infection are rare events for such a common pathogen and this may in part be due to our inability to recognize that sporadic cases are linked. Two recent epidemiological studies have addressed this issue. A case–control study in Sweden with 101 cases and 198 controls, found that drinking unpasteurized milk was the major risk factor (OR 3.56, 95 percent CI 1.46–8.94) (Studahl and Andersson 2000). Enhanced surveillance of campylobacter cases in England and Wales was launched in May 2000 and analysis of these data also identified that illness in the community was associated with drinking unpasteurized milk (OR 1.55, 95 percent CI 1.06–2.27) (Gillespie et al. 2003b). These studies confirm that the risk associated with drinking unpasteurized milk is still prevalent. It may also be possible that some of the cases currently defined as sporadic cases are in fact part of local unrecognized outbreaks, a question that has been raised by Gillespie et al. (2003b).

For further information on campylobacter infections, see Chapter 60, Campylobacter and arcobacter.

VERO CYTOTOXIN-PRODUCING *ESCHERICHIA COLI*

There are many serogroups of *E. coli* that produce verocytotoxins (also called shiga-like toxins) and the bovine is a major reservoir for these organisms (Clarke et al. 1989; Chapman et al. 1993). Not all serogroups of verocytotoxin producing *E. coli* (VTEC) are pathogenic for man, and hence presence of a VTEC in milk or diary product does not necessarily imply a risk to human health. In the UK and North America, VTEC O157:H7 has emerged as the significant public health problem but on mainland Europe and in Australia, non-O157 serogroups are more problematic.

Raw milk contaminated with VTEC is a potential source of human infection. A study in Germany found that 3.9 percent of 127 samples of raw milk and 2.1

percent of 146 samples of certified milk were positive for VTEC (Klie et al. 1997). *E. coli* O157:H was isolated from one raw milk sample, the other isolates comprised five different serogroups with different combinations of virulence markers. The authors concluded that drinking raw milk would continue to be a potential public health problem. Detection of non-O157 VTEC is problematic due to the lack of specific and sensitive culture methods. However, a French study using a PCR-ELISA assay determined that 21.5 percent of 205 raw milk samples contained non-O157 VTEC (Fach et al. 2001).

Most other studies have concentrated on determining the prevalence of *E. coli* O157 in milk. In England and Wales, *E. coli* O157 was detected in 3/1 097 samples of bottled unpasteurized milk (de Louvois and Rampling 1998). However, *E. coli* O157 was not detected in a study in Holland of 1 011 bulk milk samples from different dairy herds (Heuvelink et al. 1998) and in a more recent study in Scotland from 500 samples of raw milk (Coia et al. 2001). Contamination of raw milk is therefore a relatively rare event. Verocytotoxin producing *E. coli* have also been associated with coli mastitis in cows (Stephan and Kuhn 1999). These workers found that 2.8 percent (4/145) of cows with coli mastitis were infected with VTEC but none of the isolates were *E. coli* O157. Although many different VTEC are associated with clinical infection in farm animals, especially in calves and pigs, *E. coli* O157:H7 does not seem to cause clinical infection of animals but has been isolated from healthy cattle and dairy calves (Wells et al. 1991; Chapman et al. 1993). A longitudinal study of infection in the dairy commenced in 1993 and showed that although 29 of 130 cows excreted *E. coli* O157 on one or more occasions during the following year, there was no evidence of clinical disease (Report of the Advisory Committee on the Microbiological Safety of Food 1995).

Growth and survival of *E. coli* O157 in unpasteurized and pasteurized milk has been investigated by several groups of workers. Wang et al. (1997) found that *E. coli* O157 did not grow at 5°C in raw milk and that there was a gradual decrease in numbers after storage for 28 days. In this study, growth was observed at 8°C with an increase of about 1–2 $\log_{10}$ CFU/ml after 4 days and at 15°C about 3–5 $\log_{10}$ CFU/ml was noted after 3 days. In pasteurized milk, *E. coli* O157 continued to grow to populations of greater than 8 $\log_{10}$ CFU/ml at day 7. A report from Holland produced similar findings in that *E. coli* O157 was shown to grow in raw milk at 7 and 15°C (Heuvelink et al. 1998). In a later study, Massa et al. (1999) confirmed that growth at 8°C was strain dependent, some strains producing a 2–3 $\log_{10}$ CFU/ml increase during storage for 9–17 days. The recommendations from all these studies are to store milk at (5°C but, more importantly, to ensure that there is effective pasteurization without the risk of post-pasteurization contamination.

Outbreaks of *E. coli* O157 infection associated with raw milk have been reported from several countries but particularly from the USA, Canada, and the UK. In April 1986, there was a large outbreak of colitis in 60 kindergarten children who had visited a dairy farm in southwest Ontario, where they drank unpasteurized milk (Borczyk et al. 1987). Evidence of infection with *E. coli* serotype O157:H7 was demonstrated by isolating the organism or by detecting verocytotoxin in the feces of 43 patients and by isolation of the organisms from the feces of two cows. Martin et al. (1986) investigated two cases of hemolytic uremic syndrome (HUS) in young children. *E. coli* serotype O157:H7 was isolated from a fecal specimen from one child and also from cattle in both dairy herds which supplied the milk. In December 1992, there was a community outbreak in Oregon, USA which was associated with the consumption of raw milk. Eight people were affected and two were hospitalized (Report of the Advisory Committee on the Microbiological Safety of Food 1995). Keene et al. (1997) described a protracted outbreak, affecting 14 people that occurred over a 2-year period in the Portland area of Oregon, USA. These infections may have been reported as sporadic cases but molecular typing using PFGE demonstrated that isolates from cases had the same PFGE profile. The public health message is clear; raw milk can be contaminated intermittently and unpredictably and this can result in infections in the community from a common source that may go unrecognized as part of an outbreak. Enhanced epidemiological and microbiological surveillance are necessary if such cases are to be recognized.

Table 10.1 illustrates that *E. coli* O157 has emerged as the most common cause of milkborne outbreaks in England and Wales. Outbreaks are, in the main, associated with drinking unpasteurized milk but have also been linked to milk sold as pasteurized (Gillespie et al. 2003a). The first reported outbreak associated with raw milk in the UK occurred in Sheffield in 1993. Eleven cases, mostly children, were affected with *E. coli* O157:H7, phage type (PT)2, VT2, and three developed HUS (Table 10.1). The organism was also isolated from farmyard slurry, ten of 105 rectal swabs from dairy cattle, and milk from one rectal culture-positive cow at the farm that supplied the milk (Chapman and Wright 1993). In 1996, cases of infection with *E. coli* O157, PT1 in West Yorkshire were associated with the consumption of unpasteurized milk and the organism was isolated from a bottle of milk (Report 1996). Outbreaks associated with raw milk have continued to occur; one in 1997 resulted in eight cases, another in 1998 in three cases, and of two further outbreaks in 2000, one resulted in four cases and the other two cases (Table 10.1). The outbreak with four cases occurred in the Oldham area of Greater Manchester and was related to consumption of milk supplied by a single farm. Investigations on the farm recovered *E. coli* O157 from the in-line milk filter and from 67/127 bovines sampled. All strains from human cases, the milk filter, and the cattle were confirmed as *E. coli* O157 PT 21/28 VT 2 with indistinguishable PFGE profiles. Similarly outbreaks associated with raw milk have been reported in Scotland. One outbreak involving three cases occurred in 1999 and another with two cases in 2000 (Task Force on *E.coli* O157 2001).

Of greater concern are the outbreaks in the UK associated with milk sold as pasteurized that have resulted in large numbers of cases. In May 1994, an outbreak of *E. coli* infection involved a pasteurized milk supply in West Lothian, Scotland (Upton and Coia 1994). One hundred people were affected, 69 had infection confirmed by culture, and one was confirmed by serology. Forty-six were under 15 years of age, and of these, 32 were under 5 years. Almost one-third of patients required hospital admission and HUS developed in nine children. Six required dialysis of whom two now have chronic renal failure. One child who did not have HUS died due to perforation of the large bowel. One elderly woman developed thrombotic thrombocytopenic purpura. More than 90 percent of patients had consumed pasteurized milk from the local dairy and at least seven cases resulted from secondary spread within households. Environmental investigation at the dairy and at the farms that supplied the milk yielded isolates of *E. coli* O157 from fecal specimens from cattle, raw milk from the bulk tank at one of the farms, from a pipe that carried milk from the pasteurizer to the bottling machine, and from a discarded bottling machine rubber at the dairy. The isolates from the patients, the farm, and the dairy were indistinguishable from one another by typing. All were *E. coli* O157:H7, PT2, VT2, and displayed identical PFGE profiles.

There have been four outbreaks in England and Wales associated with drinking milk sold as pasteurized, two in 1996 and two in 1999 (Table 10.1). One of the outbreaks in 1996 occurred in a rural community in the northwest of England (Clark et al. 1997). Of the 12 cases associated with the outbreak, nine were primary cases that had consumed milk from the farm, and three others were symptomless excreters. Three children <5 years old and one adult were admitted to hospital and one child developed HUS. Investigation at the farm highlighted that a flow restriction valve was faulty on one of the pasteurizers and allowed milk to pass through the heating coil too quickly resulting in inadequate pasteurization of milk. *E coli* O157 was isolated from the in-line milk filter and from two of the cows in the herd. All of the isolates were the same phage type but the isolates from the filter and the cattle had a slightly different PFGE profile than the isolates from the human cases. Distribution of milk from the farm was discontinued and the faulty pasteurization unit repaired. This action prevented further cases. Although no conclusive evidence of the origin of the outbreak was found, the farm was the most likely source. This outbreak highlights the importance of taking preventative action quickly on the basis of good descriptive epidemiological evidence.

The largest milkborne outbreak of O157 infection in the UK occurred in north Cumbria in March 1999 and was associated with drinking milk sold as pasteurized from a local farm dairy (Goh et al. 2002). A total of 114 cases were reported to the outbreak control team and 88 of these were confirmed microbiologically. Twenty-eight (32 percent) of the confirmed cases were admitted to hospital including three children with HUS. A case control study found that illness was strongly associated with drinking pasteurized milk from a local farm dairy ($p = <0.0001$). Microbiological investigations on the farm yielded isolates from calves and the farm environment of *E. coli* O157 PT 21/28 VT2 with a PFGE profile that was indistinguishable from the human isolates. Investigation of the pasteurization unit revealed that at the time of production of the implicated batch, there had been a fault with the heat exchange plates on the pasteurizer and the flow diversion pen on the chart recorder was also broken and hence during routine operation it would have been difficult to ensure that milk was being pasteurized effectively. These outbreaks associated with pasteurized milk raise issues about the effectiveness of current legislation for the prevention and control of milkborne infections.

These milkborne outbreaks, in addition to outbreaks incriminating yogurt and cheese (see later sections), confirm that raw milk, or inadequately pasteurized milk, is an important source of human infection with *E. coli* O157 in the UK. In the light of this evidence, the Advisory Committee on the Microbiological Safety of Food (Report of the Advisory Committee on the Microbiological Safety of Food, 1995) and the Task Force on *E. coli* O157 (2001) have recommended that the government should reconsider banning all sales of raw cows' milk in England and Wales and that raw cows' cream should be subject to the same regulations as raw cows' milk. They also recommend that the industry ensures that the pasteurization of milk and milk products is carefully controlled and that post-pasteurization contamination is avoided.

There is evidence that other serotypes of VTEC that occasionally cause outbreaks of human illness may be derived from milk. In 1994, four people in Helena, MT, USA developed bloody diarrhea due to *E. coli* O104:H21. Further case finding investigations revealed 11 confirmed and seven suspected patients and a case–control study implicated one specific brand of pasteurized milk as the source of infection (Moore et al. 1995). Recently, Allerberger et al. (2003) reported two cases of HUS in Austrian children caused by verocytotoxin producing *E. coli* O26. Both cases, one an 11-month-old boy and the other a 28-month-old girl, had consumed raw milk from a breakfast buffet at the same hotel.

Investigations at the farm supplying the milk resulted in isolation of *E. coli* O26:H- from a cow that had an indistinguishable PFGE profile from the isolates from the two cases.

Non-O157 VTEC characteristically ferment sorbitol and hence culture methods used to detect *E coli* O157 are not suitable for the detection of other VTEC. It is therefore possible that sporadic cases or even outbreaks due to these organisms may be associated with milk, yet go undetected.

For further information on VTEC, see Chapter 52, Escherichia.

BACILLUS CEREUS AND BACILLUS ANTHRACIS

Spores of *B. cereus* are common in the farm environment and inevitably contaminate raw milk. The extent of contamination can be controlled by careful washing of the udder and teats prior to milking. Post-pasteurization contamination may also contribute to the numbers of spores, and the organism can grow readily and produce toxin in milk. *B. cereus* is a common cause of sweet curdling and bitty cream but documented cases of food poisoning due to *B. cereus* in milk are rare. However, in 1972, an outbreak of abdominal pain and diarrhea of short duration affected 221 schoolchildren in Romania after drinking contaminated milk. Two incidents documented in 1981 in Denmark involved a boy aged 12 months who drank curdled milk and a baby aged 6 months who developed diarrhea and vomiting after drinking contaminated expressed breast milk (Christiansson 1992).

Milk-transmitted anthrax is more of a theoretical than a real danger but difficulties may arise regarding the action to be taken with milk that has been derived from a dairy herd in which anthrax infection is subsequently diagnosed. The chance that such milk could be contaminated is very remote because anthrax bacilli are not excreted in milk until bacteremia develops. This occurs just prior to death at which stage it is likely that milk secretion will have ceased and, furthermore, the animal would be obviously ill and unsuitable for milking. Experiments have also shown that both spores and vegetative cells of *B. anthracis* (unlike those of *B. cereus*) disappear rapidly in milk (Bowen and Turnbull 1992) and infectivity of *B. anthracis* is low by the oral route.

Advice from the Department of Health (London, UK) (PL/CO(95)3 plus appendix and addendum) is that milk from animals known or suspected of being infected with anthrax at the time of milking must be excluded from the rest of the milk supply and sterilized (Anthrax Order 1991). The rest of the milk from the herd or flock must be pasteurized before being offered for sale. Where milk has already entered the public supply or where there is a suspicion that bulk milk may contain milk from an animal that could be suffering from anthrax or has died from the disease, pasteurization should suffice to protect the consumer against the risk of anthrax after consuming the milk.

CORYNEBACTERIUM ULCERANS

Although *C. ulcerans* usually causes only a mild sore throat, it can cause an illness indistinguishable from classic diphtheria. The organism is culturally similar to *Corynebacterium diphtheriae* var. *gravis* but is identified by the production of urease and of the exotoxins of both *C. diphtheriae* and *Corynebacterium ovis*. Cattle seem to be the main reservoir of infection and occasionally excrete the organism in milk. Isolated cases of human infection, including one of classic diphtheria, were identified by Meers (1979) in Devon and Cornwall. He postulated that some of the outbreaks of milkborne diphtheria, which at one time were common in Britain, may have been caused by *C. ulcerans*. Hart (1984) isolated *C. ulcerans* from a patient with a sore throat and from one symptomless carrier, both of whom lived in a farming community. Of the 30 cattle on the farm, eight were excreting toxigenic *C. ulcerans*. Hart also surveyed 52 other dairy herds in north Devon and found toxigenic *C. ulcerans* in milk from four of the herds. The organism does not seem to be prevalent in cattle in the UK nowadays and there have been no new reports of infection associated with milk.

CRYPTOSPORIDIOSIS

An outbreak of acute gastroenteritis due to cryptosporidiosis occurred amongst children at a primary school in Yorkshire in September 1995. There were 52 cases of whom 50 were pupils at the school and two were secondary cases. Initially, water was suspected as the likely source of infection but a case–control study implicated school milk supplied by a small producer with an on-farm pasteurizer. Subsequent investigation at the farm revealed that the pasteurizer had been failing intermittently and had not been working properly at the time of the outbreak (Gelletlie et al. 1997). Cryptosporidia are killed by effective pasteurization and hence this is an important critical control point.

LISTERIA MONOCYTOGENES

The first conclusive evidence that listeriosis was acquired from food was reported by Schlech et al. (1983). Since then it has become clear that many foods, including milk and, in particular, soft cheese, may be contaminated with *L. monocytogenes* and may occasionally cause outbreaks or sporadic cases of human listeriosis (McLauchlin 1993).

L. monocytogenes is prevalent in farm environments, in the feces of cattle, and in raw milk (Slade et al. 1989; Rampling et al. 1992; Sanaa et al. 1993; Fenlon et al. 1995). In addition, the organism survives well in moist environments and multiplies slowly at refrigerator temperatures. Although it is killed by pasteurization, there may be opportunity for post-process contamination from plant or machinery. Food manufacturers, including manufacturers of dairy products, are very much aware of the problems of *L. monocytogenes* and are now very active in carrying out surveillance for this organism in environments and products.

Samples of milk from farm bulk tanks have shown that a high proportion (up to 16 percent) may be contaminated with *L. monocytogenes* at any one time. However, contamination is usually sporadic and persistent contamination seems to be rare. It can result from fecal contamination or from direct shedding by a cow with subclinical mastitis (Slade et al. 1989; Fenlon et al. 1995). Sanaa et al. (1993) carried out a detailed study of 128 selected dairy farms to assess the risk factors for contamination of milk with *L. monocytogenes*. They confirmed the findings of previous studies that poor quality silage (pH >4.0) was a major risk factor. Other predisposing conditions were infrequent cleaning of the cattle exercise area, obvious fecal soiling of udders, thighs, and anal regions, insufficient lighting in milking barns or parlors, which presumably made it difficult to ensure optimal hygiene during milking, and inadequate disinfection of udder cloths between milkings.

Most of the isolates found in surveys of raw milk belong to serovar 1. More detailed subtyping by multienzyme electrophoresis (MEE) has indicated that many isolates belong to subtypes that are unique to milk or to the environment and are not associated with human or animal disease. However, Fenlon et al. (1995) showed that whereas isolates from 25 of 160 milk producers could be subgrouped into nine different electrophoresis types, most of which were unique to milk, some isolates belonged to a type that had been associated with human infection.

In most instances, liquid milk is probably the original source of *L. monocytogenes* in a dairy product but the environment of dairy plants may also provide a reservoir of contamination from milk and other sources (Cotton and White 1992; Jacquet et al. 1993; Peeler and Bunning 1994). In particular, processing plants that are adjacent to dairy farms are more likely to be contaminated with *Listeria* spp. than processing plants without on-site dairy farms (Pritchard et al. 1994).

The first evidence to suggest that raw milk could be a source of human infection came from Halle in Germany where Reiss et al. (1951) described a pathological condition of stillborn infants which they called *granulomatosis infantiseptica* on account of the distinctive focal necrosis found throughout the bodies. A series of outbreaks of this condition occurred over a long period from 1949 to 1957. Initially, the organism responsible was identified as a new species of *Corynebacterium* but was later recognized as *L. monocytogenes*. These outbreaks were considered to be due to the ingestion of unpasteurized milk (Seeliger 1961).

In 1983, pasteurized milk was implicated in an outbreak in the USA: 49 patients were admitted to hospital in Massachusetts with meningitis, septicemia, or abortion due to *L. monocytogenes* serotype 4b (Fleming et al. 1985). Case–control studies demonstrated that infection was associated with consumption of a specific brand of whole or semi-skimmed pasteurized milk; there had been an outbreak of encephalitis due to *L. monocytogenes* in one of the herds that supplied milk to the dairy. There was, however, no evidence of failure of pasteurization.

Luchansky (1995) reported an investigation in Wisconsin in 1994 which followed recall of four dairy products due to detection of contamination with *L. monocytogenes*. One of these products was a 1 percent low fat chocolate milk which had been stored at inappropriate temperatures and contained 10^9 cfu/ml of serovar 1/2b. It had been implicated in a small outbreak of fever and gastroenteritis in a group of picnickers. The Wisconsin Division of Health then identified four patients retrospectively who had suffered typical invasive listeriosis and had also drunk the recalled chocolate milk.

MYCOBACTERIUM PARATUBERCULOSIS

Mycobacterium paratuberculosis commonly infects dairy cattle, leading to Johne's disease, also known as paratuberculosis, a chronic enteritis in cattle (Collins 1997). *M. paratuberculosis* has been suggested as a possible cause of Crohn's disease in humans but the etiology of Crohn's disease has not yet been established. The association with *M. paratuberculosis* has created significant debate in the last 5–10 years. There is evidence both for and against this theory (Chiodini 1989; Thompson 1994). One of the more convincing cases was reported by Hermon-Taylor et al. (1998). A young boy diagnosed with *M. paratuberculosis* cervical lymphadenitis developed terminal ileitis similar to Crohn's disease 5 years later. The organism was detected by polymerase chain reaction (PCR) in biopsies from both the lymph gland and the inflamed intestine. The infection was treated with a combination of rifabutin and clarithromycin. The debate continues and currently there is no substantial evidence to prove or disprove that infection with *M. paratuberculosis* is a major cause of Crohn's disease (European Commission 2000; Rubery 2001). The possible association with Crohn's disease has also raised the issue that this may be a foodborne disease transmitted in the milk of infected cows. This hypothesis has generated great interest in the organism, its prevalence in dairy herds and milk, and the organism's susceptibility to pasteurization.

Prevalence of *M. paratuberculosis* in dairy herds in Alberta, Canada has been investigated by Sorensen et al. (2003). These workers tested 50 herds for the presence of the organisms by fecal culture and a serum enzyme immunoassay (EIA). A total of 1 500 cattle from the 50 herds was studied, individual sera were tested by the EIA but fecal samples were pooled so that 500 pooled samples were examined. The true herd prevalence, determined by EIA, was 26.8 percent and that determined by fecal culture ranged from 27.6 to 57.1 percent depending on the number of individual samples positive in each pooled sample.

Grant et al. (2002a) reported the incidence of *M. paratuberculosis* in bulk raw milk and commercially pasteurized cows' milk in the UK. Samples were collected over a 17-month period from March 1999 to July 2000 from 241 approved dairy processing establishments. *M. paratuberculosis* was detected by an immunomagnetic PCR method and a culture method. *M. paratuberculosis* was detected by the PCR assay in 19/244 (7.8 percent) raw milk samples from bulk tanks and in 67/567 (11.8 percent) of pasteurized milk samples. The culture method detected confirmed *M. paratuberculosis* isolates from 4/244 (1.6 percent) of the raw milk samples and 10/567 (1.8 percent) of the pasteurized milk samples. Finding *M. paratuberculosis* in pasteurized milk raises the issue of heat resistance of the organism and its survival after pasteurization.

The effect of pasteurization temperatures and holding times has been the subject of several studies (Grant et al. 1999; Grant et al. 2002b; Lund et al. 2002). Grant et al. 1999 conducted experiments with raw cows' milk spiked with 10^6 CFU/ml of *M. paratuberculosis*. These workers investigated heat treatments ranging from 72 to 90°C for 15 s and also 72°C for 20 and 25 s. Although all treatments achieved a 5–6 log_{10} reduction small numbers of the organisms (4–16 CFU/10 ml) survived even when heated at 90°C for 15 s. A longer holding time of 25 s at 72°C was more effective and it was concluded that this is more likely to achieve complete inactivation of *M. paratuberculosis* than a higher temperature. In a follow-up study using a commercial-scale HTST pasteurization unit, Grant et al. (2002b) concluded that *M. paratuberculosis* in naturally contaminated milk are capable of surviving commercial HTST pasteurization. Lund et al. (2002) reviewed data from eight studies on the effect of heat treatment of *M. paratuberculosis*. There was significant variation in the strains, culture medium, and methods used which probably explains the differences in the results obtained. These authors highlighted the need for a set of defined performance criteria for pasteurization of milk in relation to *M. paratuberculosis*. Furthermore, information about the heat resistance of this bacterium will enable a decision to be made on the process criterion that is necessary to eliminate the organism in commercially pasteurized milk.

STREPTOBACILLUS MONILIFORMIS

This organism causes a febrile septicemic illness in man, characterized by an erythematous rash, most prominent on the hands and feet, arthralgia, and sore throat. The disease was first described in Haverhill, Massachusetts, where an explosive outbreak affected 86 people and was termed erythema arthriticum epidemicum (EAE). The outbreak was due to consumption of contaminated raw milk (Place and Sutton 1934). In February 1983 an outbreak of EAE affected 130 children attending a boarding school in Chelmsford, Essex. *S. moniliformis* was isolated from the blood of four of the children and unpasteurized milk from a local farm was again the suspected source of infection (Shanson et al. 1983).

STREPTOCOCCUS ZOOEPIDEMICUS

S. zooepidemicus is an animal pathogen commonly associated with infection in horses but occasionally causing clinical or inapparent mastitis in cows. It has been associated with severe milkborne illness in man. In 1968, there was an outbreak of infection among 74 adults and 11 children in a town in Romania, causing sore throat, fever, and lymphadenitis, complicated by post-streptococcal glomerulonephritis in one-third of patients. Consumption of improperly pasteurized milk was implicated and *S. zooepidemicus* was isolated from cows with mastitis (Duca et al. 1969). In one week, three unrelated patients presented to the General Infirmary, Leeds, with *S. zooepidemicus* bacteremia. One patient had meningitis, two had endocarditis, and one subsequently died (Ghoneim and Cooke 1980). Investigations eventually confirmed that unpasteurized milk had been the source of infection in all three cases. In 1983, *S. zooepidemicus* infection was again associated with acute glomerulonephritis in three members of a North Yorkshire family who drank unpasteurized milk from their own cows (Barnham et al. 1983).

The most severe milk-associated outbreak of streptococcal infection in Britain was in West Yorkshire in 1984 (Edwards et al. 1988); 12 people developed *S. zooepidemicus* septicemia or meningitis, and eight died following consumption of raw milk from a single supplier. Three cows in the herd were excreting the organism in the milk. In Australia, a man developed septicemia and glomerulonephritis due to *S. zooepidemicus* and two other members of the family were found to be asymptomatic throat carriers. The source of infection was unpasteurized milk from the house cow and typing demonstrated that the isolates from the three individuals and the cow's milk were indistinguishable from one another (Francis et al. 1993).

In England in 1994, an elderly couple who owned a dairy farm became ill and were admitted to hospital within days of one another (Report 1994c). The man developed fatal meningitis and *S. zooepidemicus* was isolated from his cerebrospinal fluid (CSF). His wife had septic arthritis and the organism was isolated from pus from a knee joint aspirate. Another man who drank milk purchased at the farm gate became ill with empyema and subsequently died, and a man who had eaten a chocolate cake made with cream from the farm developed endocarditis. *S. zooepidemicus* was also isolated from milk from the bulk tank and from one cow that appeared to have subclinical mastitis. Ribotyping of the isolates from the patients and the milk demonstrated that they were indistinguishable from one another (Burden, personal communication 1995). There have been no further reports in the literature of infections associated with milk in the UK.

It is clear that *S. zooepidemicus* is a rare but extremely serious cause of human illness. When it is diagnosed, milk should be suspected as the source of infection.

For further information on streptococci, see Chapter 33, Streptococcus and lactobacillus.

YERSINIA ENTEROCOLITICA

The reported isolation rates of *Y. enterocolitica* from samples of raw milk vary from 14.3 to 48.1 percent. Most strains have the biochemical characteristics of environmental, nonpathogenic strains and are nontypable or belong to serotypes not associated with human illness. A few isolates from milk are pathogenic, however, and yersinia infection has been associated with consumption of raw milk in England and Wales and in the USA.

Y. enterocolitica can survive and replicate at refrigeration temperatures but although one strain of yersinia able to survive laboratory pasteurization has been reported, there is good evidence that HTST pasteurization will destroy all viable organisms (Francis et al. 1980). The presence of pathogenic yersiniae in pasteurized milk implies a fault during pasteurization or contamination afterwards (D'Aoust et al. 1988; Greenwood et al. 1990).

An outbreak in 1976 (Black et al. 1978) resulted in admission to hospital of 36 schoolchildren, 16 of whom had appendectomies. Evidence from a case–control study implicated chocolate milk but chocolate syrup was added to the milk after pasteurization and this may also have been a source of the organism. Case–control studies implicated pasteurized milk in a large multistate outbreak of yersiniosis in the USA in 1982 (Tacket et al. 1984); 172 infections were identified by culture. Enteritis occurred in 148 patients, mostly children, and extraintestinal infection in 24 adults. Appendectomy was performed on 17 patients with intestinal infection. Investigations did not reveal any fault in the pasteurization process but indirect evidence that milk was the source was obtained by isolating the same unusual serotype (O:13,18) from the outside of a milk crate on a pig farm where outdated milk was used as pig feed.

The most recent outbreak associated with *Y. enterocolitica* was reported in the USA (Ackers et al. 2000). Ten patients were infected with *Y. enteroocolitica* O:8,

three patients were admitted to hospital and one underwent an appendicectomy. Consumption of pasteurized milk from a local dairy was associated with the illness. Post-pasteurization contamination was the most likely source of the problem originating from either the farm environment, one sample from a pig on the farm was positive, or from contaminated water used to rinse the milk bottles.

MILK POWDER

Commercial drying of milk was introduced at the beginning of the twentieth century. The roller method was developed first followed by the spray method in the 1930s. Both were in operation until the 1950s when the spray method assumed popularity because it gave a more palatable product. For a detailed account of the various processes and designs of plant see Knipschildt and Anderson (1994).

The process is in two stages: preliminary concentration during which 90 percent of the water is removed by evaporation, followed by rapid drying. Before concentration, the milk must be preheated to pasteurization temperatures to kill vegetative pathogenic bacteria. For the spray-drying process, concentrated milk is fed from vacuum evaporators to a balance tank and then through fine jets or rotary atomizers into the top of a large chamber supplied with filtered hot air. Dry powder is collected from the bottom of the chamber.

The development of processing conditions that minimize nutrient change and improve physical and organoleptic properties has increased the microbiological hazards (Lovell 1990). Spray-dried milk has been responsible for many outbreaks of food poisoning due to enterotoxigenic *S. aureus* and to salmonellae. The safety of the process relies on adequate preheating to destroy vegetative pathogenic organisms; if, however, they survive this stage, there is opportunity for them to multiply in the evaporators and the balance tank and thus for heavily contaminated milk to enter the spray-drier. Although hot air enters the spray-drier at over 160°C, cooling is rapid. There is good evidence that salmonellae can survive the drying process (McDonough and Hargrove 1968) and that the heat treatment is insufficient to inactivate *S. aureus* or its enterotoxin. Because of the complexity of the plant, it is very difficult, if not impossible, to eradicate contamination once established in the plant or its environment.

In 1953, several outbreaks of staphylococcal food poisoning in schools near London were traced to powdered milk from the same source. Later that year, another outbreak in Yorkshire was traced to milk powder from a different source (Anderson and Stone 1955). Tests on batches of milk powder from the plant involved in the first outbreak demonstrated that on two days during May the total bacterial concentration was considerably higher than on other days and that there were approximately 10^4 *S. aureus* per gram. The organism was not detected in batches with low bacterial concentrations. Other studies indicated that toxin was probably produced at some stage before the drying process. Investigations at another plant not implicated in the outbreaks showed that bacteria multiplied in the balance tank throughout the 12 h of continuous processing. There was also evidence that small amounts of contamination remained after cleaning and recontaminated the pasteurized concentrated milk as it passed through the tank (Hobbs 1955). Other studies have indicated that *S. aureus* contamination may occur during manufacture and that isolates from milk powder may not necessarily be the same as those in the raw milk (Lovell 1990). Harvey and Gilmour (1990) examined milk powders from five different processors in Northern Ireland for both *S. aureus* and coagulase-negative staphylococci. They found that low levels of total viable organisms were associated with low total staphylococcal levels but that five of 37 samples had high total levels and high concentrations of coagulase-negative staphylococci. The powders with high levels contained added fat which was mixed into the milk after pasteurization and concentration, but prior to drying. No *S. aureus* or staphylococcal enterotoxins were detected but this study highlighted a potential source of contamination with microorganisms which are able to survive the drying process. Even coagulase-negative staphylococci are undesirable as some strains can produce enterotoxins.

However, staphylococcal intoxication from dried milk does not seem to have caused many problems since the general introduction of refrigerated farm bulk milk tanks, which maintain raw milk at a temperature that inhibits multiplication and toxin production by staphylococci (Morgan-Jones 1987). However, the problem has not totally disappeared. One report from El-Dairouty (1989) describes a series of food poisoning incidents involving 21 patients who consumed non-fat dried milk in Egypt in 1986. The largest outbreak associated with a milk product has recently been reported in Japan (Asao et al. 2003). Over 13 000 cases are likely to have occurred as a result of contaminated powdered milk used as an ingredient in a range of dairy products. Staphylococcal enterotoxin A (SEA) was detected in low fat milk and also in powdered skim milk. The total intake of SEA per capita was estimated at about 20–100 ng. SEA exposed at least twice to pasteurization at 130°C for 2–4 s retained both immunological and biological activity. This outbreak is a reminder that viable *S. aureus* can be eliminated by pasteurization but that the enterotoxin can retain sufficient activity to cause intoxication.

Salmonella infection has been associated with the consumption of powdered milk on several occasions and its occurrence in dried milk is well documented. In 1966, an outbreak of *Salmonella* Newbrunswick infection in the USA was associated with spray-dried milk (Collins et al. 1968). Following this episode, the US Food and Drug Administration instituted a large-scale testing program (Marth 1969) and 156 milk plants in 56 states were examined. Of 2 741 samples tested, 24 samples of non-fat dried milk and 27 environmental samples contained salmonellae. Later, in 1967, 200 factories in 19 states were surveyed: 1 percent of 3 315 product samples and 8.2 percent of 1 475 environmental samples were positive for salmonellae. In total, 13 percent of the plants tested were manufacturing milk powder contaminated with salmonellae. These surveys also demonstrated the value of taking environmental rather than product samples to monitor for pathogenic bacteria.

Outbreaks have continued to occur in the USA. Pickett and Agate (1967) isolated a lactose-positive strain of *Salmonella* Newington from instant non-fat milk powder which had been responsible for nine cases of salmonellosis. *Salmonella* Newport was isolated from dried skimmed milk and from 20 patients with diarrhea and vomiting in Newfoundland (Marth 1969) and an outbreak of *Salmonella* Agona and *Salmonella* Typhimurium infection was traced to powdered milk in Oregon (Furlong et al. 1979).

These episodes in North America are not unique. In England and Wales, *Salmonella* Typhimurium infection has been traced to spray-dried milk on several occasions (Galbraith et al. 1982). In Australia, *Salmonella* Bredeney, harbored in the insulating material of a spray-drier, provided a continuing source of contamination for the milk powder via cracks in the stainless steel lining of the drier (Craven 1978). In Trinidad, approximately 3 000 people, predominantly infants and small children, were infected with *Salmonella* Derby which was traced to contaminated powdered milk produced on the island (Weissman et al. 1977).

In November and December 1985, there was a cluster of *Salmonella* Ealing infections in infants in England and Wales (Rowe et al. 1987). Investigations showed that 76 people, including 48 babies, had been infected and that a smaller cluster of cases had occurred in May, June, and July 1985. Seven of the babies were admitted to hospital and one died. A case–control study implicated dried milk from one manufacturer. *Salmonella* Ealing was isolated from an unopened packet of baby food from the home of a patient and from four other packets from the same batch. Investigations at the factory revealed that the inside of the spray-drier had several minor pinholes and weld cracks and that there was an irregular hole 1 × 3 cm in the metal lining. Removal of the metal outer casing at the site of the hole showed stained insulation material and a large collection of discolored milk powder. *Salmonella* Ealing was isolated from the insulation material and the milk powder in continuity with the hole, but not from the dry insulation material on the opposite side of the drier. It was thus likely that infected milk had gained access to the insulation material from the inside of the drier through the hole in the metal lining and had then remained within its wall as a continual source of contamination. *Salmonella* Ealing was also found in the factory vacuum system and in a silo containing waste milk powder and sweepings. Examination of the pasteurizer revealed pinhole defects in the plate heat exchanger which could have been sufficient to allow contamination of the pasteurized milk with raw milk. *Salmonella* Ealing was known to be present in one of the herds that supplied milk to the dairy in April 1985 and the organism may have gained access to the factory at that time.

The many incidents of infection associated with spray-dried milk have caused much concern among manufacturers and public health microbiologists. There has also been controversy about standards for total bacterial levels and their relevance as a method of monitoring quality and the frequency of sampling required. A standard for total concentrations of no more than 5×10^4/g after incubation for three days at 30°C is generally accepted (Lovell 1990) but in practice concentrations of less than 1×10^3/g can be achieved. It is customary to perform quality control tests on each batch of powder before releasing it for sale. As was clearly demonstrated by the surveys in the USA in 1966 and 1967, however, this practice is by no means as sensitive for detecting pathogens as testing the dust and waste powder in the factory environment. An obviously dangerous practice in some milk powder factories is the custom of blending powder with high bacterial levels into powder with low bacterial levels in order to give an apparently satisfactory final product (Rowe et al. 1987).

Mettler (1989) has reviewed the issues raised by the *Salmonella* Ealing and other salmonella incidents. He points out that production of pathogen-free milk powder requires detailed committed attention to good manufacturing practice involving every aspect of the process, including design of equipment, management, staff training and discipline, and control of the environment, including air quality. In addition, sampling programs should be detailed and targeted and include the environment as well as the product; the test methodology should be assessed critically in relation to the detection of pathogens. Regular and thorough crack testing of the spray dryer should also be an essential part of the quality control program. In the long term, it should be possible to design spray-drying chambers that do not have this intrinsic fault and it should be possible to improve the processing equipment at all stages so that a build up of microorganisms does not occur.

An outbreak of *Salmonella* Anatum infection in babies in late 1996 and early 1997 associated with infant dried formula, is a stark reminder of the potential of dried milk products to cause widely dispersed outbreaks. This outbreak was only recognized because of the joint national referral of salmonella isolates to the reference laboratory for typing and the national surveillance system in England and Wales. A case–control study on the first 12 cases strongly associated consumption of infant dried formula. Subsequently, the isolates from the cases were shown to have indistinguishable PFGE profiles. A wider investigation revealed that 22 cases had occurred from September 1996 to January 1997 in four different countries: 13 in England, four in Scotland, four in France, and one in Belgium (Anon 1997). The infant formula was produced at a factory in France and upon further Europe-wide investigation it became apparent that there were also cases in France. Investigation revealed that the spray-dried milk used to make the infant formula may have originated from either a plant in France or the Netherlands. This outbreak illustrates the complexity that has developed in the modern food chain and dairy industry and the essential need for Europe-wide microbiological and epidemiological surveillance.

Although guidelines for the hygienic manufacture of spray-dried milk powders have been issued by the International Dairy Federation (Report 1991a) incidents continue to occur. In 1993, Louie et al. reported that three cases of *Salmonella* Tennessee infection in infants had been traced to the consumption of one particular brand of powdered infant formula and that this organism had also been isolated from production equipment at the spray-drying plant and from cans of powdered milk formula.

Salmonella spp. and *S. aureus* are the pathogens most often associated with dried milk, but other organisms have also caused public health problems. Three cases of neonatal meningitis due to *Enterobacter sakazakii* from infant milk formula were diagnosed in 1986 and 1987 at the National University Hospital, Reykjavik, Iceland (Biering et al. 1989). Two neonates who were normal at birth survived with severe brain damage and a third, who had Down's syndrome, died. *E. sakazakii* was isolated from cerebrospinal fluid from the three babies with meningitis and from the urine of a fourth infant who had no signs of infection. It was also isolated by enrichment culture from five unopened packages of milk powder. All isolates from the babies, and all but one of 23 isolates from the infant formula, had identical biotypes, antibiograms, and plasmid profiles. Other workers have isolated *E. sakazakii* from powdered milk (Muytjens et al. 1988) and powdered infant milk has been suspected in the past as a source of neonatal infection with this organism (Muytjens et al. 1983). These incidents confirm the importance of implementing the strictest possible measures for the control of microbiological quality during the manufacture of milk powder and, in particular, of infant milk. The Reykjavik outbreak demonstrated that even low numbers of *E. sakazakii* were significant perhaps because of the extreme susceptibility of the patients but also possibly because there was opportunity for multiplication in the reconstituted formula.

B. cereus is another contaminant of raw milk that is commonly present in milk powder (Becker et al. 1994; Crielly et al. 1994). The endospores are heat-resistant and able to survive the pasteurization and drying process. Indeed, replication may be encouraged at various stages of the production process and, in the case of infant formula, further contamination may be introduced with added ingredients. Once reconstituted, there is opportunity for germination of spores, growth, and enterotoxin production depending on the temperature of storage. In spite of this, *B. cereus* has rarely been documented as a cause of food poisoning associated with dried milk (Christiansson 1992; Becker et al. 1994). Finally, *C. perfringens* is another sporing organism which is a theoretical risk but has rarely been implicated in food poisoning due to powdered milk (Lovell 1990).

CREAM

Unless hygiene is strict, environmental contamination of cream may be considerable because of the many stages of handling during processing. When cream is separated by centrifugation, most of the bacteria from the whole milk are separated with the cream. In addition, separation is often performed at 40–50°C which is a temperature that encourages the growth of many microorganisms (Milner 1995). However, cream is normally subjected to higher pasteurization temperatures (up to 80°C) than milk and is expected to have a longer shelf-life (Davis and Wilbey 1990).

In 1970, a PHLS working party examined 4 385 heat-treated, 282 clotted, and 517 raw samples of cream (Working Party to the Director of the Public Health Laboratory Service 1971). Bacterial levels were frequently high and, although the unheated creams were the most unsatisfactory, many heat-treated creams also contained large numbers of bacteria including coliforms, thereby indicating either post-pasteurization contamination or inadequate heat treatment. From 3 417 samples examined, *S. aureus* was isolated from 59, 54 of which were unpasteurized. Other pathogens found in untreated samples were one isolation each of *Salmonella* Typhimurium, *B. abortus*, and *E. coli* serotype O126. *C. perfringens* was isolated from one heat-treated sample.

Infections and intoxications associated with cream

Raw cream is as likely as raw milk to contain pathogenic bacteria (Rothwell 1979). It is, however, surprising that few outbreaks have been attributed either to raw or to heat-treated cream. This may be because relatively small volumes are produced and sporadic infections are seldom reported. Furthermore, cream is usually incorporated into desserts, cakes, sauces, etc., and it may be difficult to incriminate the cream alone in the event of an outbreak. Food poisoning due to enterotoxigenic *S. aureus* was a well-recognized hazard associated with cream before the introduction of farm refrigeration tanks and refrigerated storage in shops and homes. When detailed investigations have been made, the source of the organism has usually proved to be a cow or cows with symptomatic or asymptomatic *S. aureus* mastitis. Isolation of the same phage type of *S. aureus* from individual cows, from the cream and from patients with food poisoning has been reported (Steede and Smith 1954).

Raw and pasteurized cream have also been associated with many small incidents of salmonella food poisoning, often related to overt infections in cows or calves on the farms supplying the milk (Vernon and Tillett 1974; Sharp et al. 1985). There have also been incidents of *B. cereus* food poisoning associated with cream (Galbraith et al. 1982; Sockett 1991).

The potential risk associated with the consumption of raw cream was demonstrated in the outbreak of infections with *S. zooepidemicus* which occurred in the home counties of England in 1994. An elderly patient who developed endocarditis due to this organism acquired his infection by eating a chocolate cream cake made with contaminated raw cream (Burden, personal communication 1995). It is probable that he was unaware that the cream was unpasteurized.

Raw cream has also been incriminated as a source of *E. coli* O157. An outbreak in England in 1997 was linked to consumption of raw cream purchased from a farm shop. Descriptive epidemiological evidence indicated that consumption of raw cream from the one outlet was associated with infection in the seven confirmed cases (Task Force on *E. coli* O157 2001).

YOGURT

A wide variety of drinks and semi-solid foods are prepared by the fermentation of milk and yogurt is by far the most popular. Nearly all yogurt is made in large dairies under carefully controlled conditions but there is increasing interest in small-scale farm production, particularly from sheep and goat milks.

There are two systems for commercial manufacture (Robinson and Tamime 1993). In one, incubation and fermentation are performed in the retail container, resulting in *set yogurt*. In the other, fermentation takes place in large tanks and the coagulum is stirred to produce a thick viscous fluid (*stirred yogurt*) which is cooled and packaged by automatic fillers. Fruits or flavorings are added to stirred yogurt after fermentation is completed and to set yogurt before fermentation. Stirred yogurt is sometimes pasteurized to give it an extended shelf-life.

The first stage of manufacture for both products is preparation and standardization of the milk mix. Fresh milk is preheated and separated through a centrifugal separator. The concentration of total solids in the skimmed milk is increased, either by evaporation to 15–16 percent or by adding milk powder. If a medium or full fat yogurt is required, a monitored amount of cream and sugar is added. Stabilizers, such as pregelatinized starch or plant gums, may also be included at this stage. The mix must be homogenized if cream is added, otherwise separation will occur during incubation. Homogenization also increases the viscosity.

The mix is heat treated to kill pathogenic and spoilage organisms and to expel air in order to produce microaerobic conditions for the fermentation. The optimum heat treatment for improving the physical state of the coagulum is a high temperature, prolonged treatment (85°C for 30 min). This is done in large tanks containing mechanical stirrers and heated by water jackets. The tanks are then cooled to 40–45°C, inoculated with the starter cultures and held at constant temperature until fermentation is complete after $3^1/_2$–$4^1/_2$ h. Many dairies find this system too slow and prefer to treat the mix in a UHT pasteurizer (85°C for 2 s) before filling the fermentation tanks or the small volume containers used for set yogurt.

Almost without exception, commercial producers use *Streptococcus thermophilus* and *Lactobacillus bulgaricus* in a ratio of 1:1 as a starter culture for fermenting yogurt. The cultures are grown in milk in bulk starter vats and added to the basic mix to give a final concentration of 1–2 percent (v/v). An incubation temperature of 40–42°C is usually selected for the fermentation. This temperature lies between the optima for acid production by the two species (39°C for *S. thermophilus* and 45°C for *L. bulgaricus*). The two organisms grow symbiotically; *S. thermophilus* produces metabolites that encourage the growth of *L. bulgaricus*. *S. thermophilus* produces most of the lactic acid early in the fermentation, but later, both organisms produce aldehydes and other fermentation products important for flavor. The final acidity is greater if the two cultures are grown together than if either is grown alone. A final concentration of 1 percent lactic acid (pH 4–4.2) should be achieved. Strict quality control is essential to ensure that starters are active and free from bacterial and bacteriophage contamination, and that the milk is free from antibiotics which would inhibit growth of the starter.

Microbiological quality of yogurt

Yogurt mix is heated to very high temperatures before fermentation. It should therefore be free from pathogenic and spoilage organisms but opportunities exist for contamination during fermentation and packaging. Additions such as fruit or flavorings may also carry contaminants. However, it is generally a very safe product because most potential pathogens are inactivated by the high level of acidity and antibiotic substances produced by the starter cultures (Robinson and Tamime 1990). *Campylobacter* spp. and *Salmonella* spp. are inactivated rapidly in yogurt or fermented milks, but under experimental conditions both *Y. enterocolitica* and *L. monocytogenes* can survive for several weeks in the refrigerated product (Schaack and Marth 1988; Ashenafi 1994; Aytac and Ozbas 1994). Between 1988 and 1991, a survey of the occurrence of *Listeria* spp. in milk and dairy products in England and Wales included 180 samples of yogurt. *L. monocytogenes* was isolated from four yogurt samples (one made from cows' milk, two from goats' milk, and one from ewes' milk) (Greenwood et al. 1991). However, yogurt does not seem to have been associated with any cases of human clinical infection due to *L. monocytogenes* or *Y. enterocolitica*.

Provided that yogurt is produced under carefully controlled conditions with the correct starter cultures, pathogenic bacteria, but not their toxins or spores, should die out rapidly. The quality of added ingredients is equally important and these should be subject to the same quality control. An outbreak of staphylococcal food poisoning involving at least 67 people was traced to *dahi* (Indian yogurt) made in a shop in Calcutta (Saha and Ganguli 1957). The suspected source was contaminated milk powder added to the mix just before incubation. It was clear that there was a fault in the fermentation process because it took 36 h to complete. The staphylococcal enterotoxin was presumably produced during this time.

Until 1989 there were no records of outbreaks of human infection associated with yogurt in England or Wales or Europe (Galbraith et al. 1982; Sharp 1987). In June 1989, however, yogurt was associated with the biggest outbreak of botulism in Britain during this century (O'Mahony et al. 1990). Between 30 May and 13 June 1989, 27 cases of botulism, with one death, were traced to the consumption of low fat hazelnut yogurt. *Clostridium botulinum* type B toxin was detected in the contents of an unopened can of the hazelnut purée used to flavor the yogurts, and from unopened cartons of yogurt. The yogurt was made in the northwest of England, and the hazelnut purée was supplied in 6 lb cans by a factory in Folkestone. The hazelnut purée was sweetened with aspartamine, rather than sugar, which might have prevented bacterial growth, and insufficient heat had been used in the manufacturing process to destroy spores of *C. botulinum*.

In the autumn of 1991, an unusual outbreak of *E. coli* O157:H7 PT 49 infections was associated with consumption of a brand of live yogurt made specially for children by a farm producer in the northwest of England (Morgan et al. 1993). Sixteen cases of culture confirmed infection were identified from 1 September to 1 November 1991. Eleven of the children were age 10 years or less and five developed HUS. A formal case–control study confirmed a close association between consumption of the yogurt and development of *E. coli* O157 infection. However, the organism was not isolated from raw or pasteurized milk on the farm nor from another batch of the implicated brand of yogurt sampled at a later date.

BUTTER

Butter contains approximately 81 percent fat and, under commercial conditions, is prepared from pasteurized cream with or without the addition of starter cultures which increase the acidity and add flavor. Dairy spreads may be defined as products with more than half their ingredients derived from milk. They have a higher water content than butter and low fat spreads contain additional milk proteins which stabilize the product (Wilbey 1994).

Most of the bacteria that are present in raw milk separate with cream and potentially would contaminate butter, so pasteurization is a very important part of the manufacturing process. The physicochemical characteristics of butter appear to inhibit the growth of many microorganisms but it has been shown that under experimental conditions *Salmonella* spp. can grow in butter at room temperature and may survive for long periods at refrigeration temperatures (El-Gazzar and Marth 1992). It has also been demonstrated that *L. monocytogenes* can increase in numbers during refrigerated storage of butter made from contaminated cream and that both *L. monocytogenes* and *Y. enterocolitica* can grow in some refrigerated dairy spreads (Milner 1995).

Spoilage of butter and dairy spreads may occur due to surface growth of *Pseudomonas* spp. or yeasts and molds (Milner 1995). There have been only a few incidents of food poisoning associated with butter and most of these have been due to *S. aureus* enterotoxin. In 1970, 24 customers and staff at a department store in Alabama became ill with acute gastroenteritis soon after consuming whipped butter in the restaurant. The butter had been prepared at the restaurant by whipping milk with softened butter but microbiological tests showed that the butter, not the milk, contained *S. aureus*. The same brand of butter was also implicated in a single case of typical staphylococcal food poisoning in Tennessee a few days later; tests on the butter demonstrated the presence of staphylococcal enterotoxin A (Wolf et al. 1970). A second outbreak of staphylococcal food poisoning, also associated with commercially produced whipped butter, occurred in Kentucky in 1977. More than 100 cases were diagnosed some of whom were admitted to hospital (Francis et al. 1977).

In 1991, an outbreak of food poisoning in California and Nevada involved more than 265 people and was traced to butter blend and margarine that was contaminated with enterotoxin-producing *Staphylococcus intermedius*. Isolates from patients and product produced staphylococcal enterotoxin A and PFGE analysis of restriction enzyme digests of DNA from five clinical and ten food isolates provided evidence that all the isolates were derived from a single strain (Bennett et al. 1994).

Another very serious outbreak of gastroenteritis followed by HUS affected nine children age 1–3 years at a nursery school in Germany (Tschäpe et al. 1995). All nine children required admission to hospital and dialysis and one boy died shortly after the onset of illness. In addition, six children age 4–6 years developed gastroenteritis without HUS. The children had eaten sandwiches prepared with 'green' butter that was made by mixing parsley with the butter. The parsley originated from a garden enriched with pig manure. The organism responsible for the infection was a verocytotoxin producing isolate of *Citrobacter freundii* which was isolated from patients, some symptomless contacts, and from the parsley. The isolates from the patients and the parsley were shown to be toxigenic by specific DNA probes and PCR and to be genetically indistinguishable from one another. This outbreak again demonstrates that the quality and hygiene of added ingredients are as important as those of the dairy product.

ICE CREAM

Ice cream usually contains 8–18 percent fat, 9–12 percent milk solids-not-fat, sugar, small amounts of stabilizer, and emulsifier. If the fat is entirely milk fat the product is known as dairy ice cream. When vegetable oils or other animal fats are used in manufacture the product cannot be described as 'dairy' ice cream.

Ice cream is a complex product and manufacture consists of several stages. Detailed descriptions of the process and plant are given by Rothwell (1990) and Mitten and Neirinckz (1993). Initially, the mix is blended to give a homogenous product. Compositional and heat treatment legislation varies considerably around the world but most countries require pasteurization (Rothwell 1990). Following pasteurization, the mix is homogenized and then cooled to 4°C or below and held in a tank for a variable time ranging from 4 to 12 or more hours to undergo aging. During the aging process, protein is adsorbed to fat globules, fat crystallization occurs, and stabilizers increase in hydration. In large, highly mechanized, and sophisticated ice cream plants, the aging process is carefully controlled and relatively short. However, some processors use a longer aging time of 24 h or more and at this stage there is opportunity for psychrotrophic microorganisms to multiply unless careful attention is paid to hygiene and prevention of post-pasteurization contamination.

Following aging, the mixture is frozen rapidly with agitation and air is incorporated into the mix. It is then extruded and may be dispensed as soft-serve ice cream or packaged into bricks or large containers which are hardened at between −20 and −40°C and sold as hard ice cream. Flavorings are usually added to the mixing or aging tanks, but pieces of fruit, nuts, syrups, or purées are added to the ice cream after extrusion from the freezer.

Microbiological quality of ice cream

There were many small and large outbreaks of food poisoning due to ice cream in the UK prior to the introduction of heat treatment legislation for commercial ice cream (Rothwell 1990; Nichols and de Louvois 1995). Outbreaks of both *S. aureus* and salmonella food poisoning are recorded. In 1946, a large outbreak of typhoid fever in Aberystwyth in Wales affected more than 100 people with four deaths and was due to contamination of the ice cream by the manufacturer who was a typhoid carrier (Evans 1947). Following this incident, heat treatment regulations were introduced but did not become effective immediately and between 1950 and 1955 there were 19 recorded outbreaks associated with consumption of ice cream and ice lollies (11 due to *Salmonella* spp., two due to *S. aureus*, and six of unknown cause) (Rothwell 1990). Between 1955 and 1984, there were no outbreaks reported in the UK but since 1984 there have been a number due to home-made ice cream. In 1984, an outbreak of *Salmonella* Enteritidis PT4 infection involving 12 people was attributed to home-made ice cream (Barrett 1986). In 1987, six members of a family were ill with *B. cereus* poisoning after eating home-made ice cream containing egg white and fresh cream. The product had >10^{10} *B. cereus* per ml (Sockett 1991).

Several outbreaks of *Salmonella* Enteritidis PT4 infection have occurred due to the use of raw shell eggs in home-made ice cream. In 1988, an outbreak affected 18 of 75 people at a bridge party (Cowden et al. 1989). In 1991 there were three family outbreaks (Report 1991b; Morgan et al. 1994a) and in 1993 another outbreak affected seven staff at a hotel (CDSC 1993, unpublished data). In the UK, it is illegal to use unpasteurized egg in commercial ice cream but it is clear that the continued use of raw egg in domestic homes and by commercial caterers poses a serious threat of food poisoning which may sometimes involve large numbers of people. A large number of outbreaks of salmonellosis in other countries have also been associated with unpasteurized ice cream and the use of contaminated eggs (Nichols and de Louvois 1995). A more recent very large outbreak involving more than 2 000 people in the USA was attributed to the use of eggs in nationally distributed ice cream products (Report 1994d).

Ice cream may also become contaminated with *L. monocytogenes*. In 1986, contamination of ice cream bars in the USA was thought to be responsible for a flu-like illness affecting at least 40 people in four states. This resulted in a recall of the product. Listeria contamination requiring recall of product has also been reported on other occasions in ice cream in the USA and France (Rothwell 1990).

Most countries have strict standards for hygiene and heat treatment requirements for commercial ice cream and it is clear that careful control of both ingredients and the production process is needed to produce a hygienic product and prevent post-pasteurization contamination. In 1992, in the UK, the Ice Cream Federation and the Ice Cream Alliance in collaboration with the Milk Marketing Board produced guidelines for the hygienic manufacture of ice cream and for the safe handling of products by retailers (Report of the Ice Cream Alliance and the Ice Cream Federation 1992a, b). These outline the main principles of quality assurance and attention to premises, raw materials, equipment, and finished products.

A survey of the microbiological quality of 2 612 samples of ice cream, ice lollies, and similar frozen products, was performed in England and Wales in 1993 as part of the European Community Coordinated Food Control Programme (Nichols and de Louvois 1995). The results of this survey demonstrated wide variations in microbiological profile but concluded that, generally, ices were of good quality and unlikely to cause food poisoning. However, nonbranded hard ice creams and soft ice creams were more likely to contain high levels of aerobic bacteria, indicator organisms, and potential pathogens than branded ice cream bars. *L. monocytogenes* was found in four of 1 964 samples of nonbranded hard ice cream and soft ice cream and *Salmonella* Enteritidis PT4 was isolated from a sample of 'real fruit' ice cream.

CHEESE

Cheeses are a heterogeneous group of foods made by coagulating the solids from high or low fat milk by the action of lactic acid and rennet. The curds so formed are treated with varying degrees of heating ('scalding'), cutting, salting, and pressing to produce greater or less acidity and moisture content. The resulting cheese can vary in size, softness, consistency, and degree of ripening. Further modifications depend on the type of lactic acid bacteria used to initiate souring and on whether other bacteria or molds are encouraged to grow within the cheese or on the surface. Survival of pathogens or development of spoilage organisms depends on the physical and chemical characteristics of the cheese and on the degree of maturation. Cheeses may be classified broadly into three main types: hard and semi-hard pressed cheeses, semi-soft cheeses which are usually not pressed, and soft cheeses. Hard cheeses include Parmesan, Gruyère, Cheddar, Derby, Leicester, Caerphilly, Edam, Gouda, and Fynbo. Soft cheeses include Camembert, Brie, Livarot, Cambazola, Cambridge, and Coulommiers, and semi-soft include Munster, Stilton, Roquefort, Gorgonzola, and Mozzarella. There are also a few unshaped curd cheeses such as cottage and lactic cheese which have high moisture content and acidity, and a very limited shelf-life (Shaw 1993; Tamine 1993).

Hard cheeses such as Cheddar or Swiss are typically large, with low moisture content (c. 39 percent) and a clear rind. The curds are scalded and then matted together while in the vats, 'milled', or chopped, salted, and pressed into molds under heavy weights. Ripening takes 3–9 months and proceeds evenly throughout the cheese. Flavor depends on enzymes from the starter culture and from rennet. The starter is usually a mixture of subspecies of *Lactococcus lactis*. Swiss cheeses such as Emmenthal and Gruyère are manufactured at high temperatures. Thermoduric starters such as *Streptococcus salivarius* subsp. *thermophilus* or *Lactobacillus* spp. are used. In addition, cultures of propionibacteria are added for development of the essential and characteristic flavors.

Semi-hard cheeses such as Caerphilly, Lancashire, and the Dutch cheeses, are lightly pressed cheeses with a moisture content of around 45–50 percent. The textures vary from smooth to crumbly. Most semi-hard cheeses are matured for 1–3 months, but Caerphilly is usually eaten while quite fresh (around 2 weeks old) and Lancashire may be matured for as long as 12 months.

Soft cheeses have a high moisture content of 55–60 percent. The curds are not cut or scalded but are drained into small molds without pressing. Cheeses such as Coulommiers or Cambridge are eaten fresh within a few days but surface-ripened cheeses such as Camembert or Port Salut are surface-ripened for 1–6 weeks by molds (*Penicillium camemberti*) or bacteria (*Brevibacterium linens*).

Internally mold-ripened cheeses such as Gorgonzola or Stilton are semi-soft cheeses with a moisture content of 45–55 percent. They are made with high acid curd to which either *Penicillium glaucum* or *Penicillium roqueforti* is added. The curd is filled into tall molds, which are turned frequently so that it is pressed down by its own weight. After a few weeks, the cheeses are pierced with skewers to admit air and encourage growth of molds. Ripening is complete after 3–6 months, during which the pH rises to 6.0 or more, and lipolysis and proteolysis contribute to the flavor.

Defects and spoilage of cheese

Cheeses made from unpasteurized milk may develop numerous defects depending on the microbial quality of the raw milk and on subsequent contamination from vats and other equipment, and from the curing rooms in which they are matured. The quality of cheese made with pasteurized milk depends mainly on the extent of contamination during manufacture and storage (Chapman and Sharpe 1990). Off flavors and gas production early in manufacture are usually due to lactose-fermenting Enterobacteriaceae or, sometimes, to lactose-fermenting yeasts. Swiss-type cheeses are liable to develop gas some weeks later due to clostridial contamination. Surface growth of yeasts, molds, or proteolytic bacteria can cause softening or discoloration of the rind and spoilage of the cheese. Blue vein cheeses are particularly susceptible to this defect. Growth of yeasts and molds or slime production by psychrotrophic organisms is common in cottage cheese. Production requires a very high standard of plant hygiene.

FUNGAL TOXINS

Cheese-making experiments show that aflatoxin M_1 in milk will survive pasteurization and the cheese-making process, and will not deteriorate on storage. Growth of toxin-producing molds in cheese is also possible. *Penicillium* spp. can produce toxins such as penicillic acid, patulin, cyclopiazonic acid, and roquefortins both at room temperature and at temperatures below $10°C$ (Chapman and Sharpe 1990). Toxin-producing strains of *Penicillium* have been isolated from both Cheddar and Swiss cheese. There is concern that some of the strains of *P. camemberti* and *P. roqueforti* used for making blue cheese may be toxigenic, but conditions in the cheese tend to be unfavorable for toxin production because of the low content of carbohydrate. Many countries, however, now screen for and select nontoxigenic strains for cheese production. The development of aflatoxin in cheese is probably of less concern because *Aspergillus flavus* and *Aspergillus parasiticus* do not produce toxins at temperatures below $10°C$, at which most hard and semi-hard cheeses are ripened.

BACTERIAL TOXINS

Histamine poisoning associated with the consumption of cheese was first reported by Doeglas et al. in 1967. In 1980, a small outbreak occurred in New Hampshire, USA. Specimens of suspect cheese contained 187 mg of histamine per 100 g and a strain of *Lactobacillus buchneri* isolated from the cheese was found to have very high histidine carboxylase activity (Sumner et al. 1985). Staphylococcal enterotoxin has also been the cause of many outbreaks of food poisoning associated with cheese.

Pathogenic bacteria in cheese

The efficient pasteurization of milk should eliminate the risk from all viable pathogenic organisms but many cheese makers still use raw milk or add raw milk to the cheese milk, believing it to be essential for good flavor. Survival of pathogenic organisms in cheese depends on the method of manufacture, the competitive effect of the lactic acid-producing starter bacteria and on conditions of storage and length of maturation of the finished product. Safety cannot be guaranteed if cheese is made from raw milk.

There have been many outbreaks and sporadic cases of infection or food poisoning, including brucellosis and botulism, associated with the consumption of cheese. Most detailed investigations have demonstrated that the source of contamination was raw milk, inadequately pasteurized milk, or post-pasteurization contamination with organisms originally derived from raw milk. The most recent botulism outbreak occurred in Italy and was linked to eating a dessert made with mascarpone cream cheese (Aureli et al. 2000). The outbreak was caused by *Clostridium botulinum* type A and affected eight young people, one of whom died, having eaten Tiramisu. Type toxin A was detected in five sera and six stools from the patients and in samples of Tiramisu. The authors postulate that a break in the cold storage of the dessert probably lead to germination of the *Clostridium botulinum* spores and toxin production.

Tuberculosis in man has not been traced directly to the consumption of cheese but experimental work has shown that *M. tuberculosis* can survive in many varieties of cheese.

BRUCELLOSIS

Brucellosis is still an important cause of human and animal infection; 500 000 new cases of human infection are reported annually throughout the world. Many cases are associated with consumption of cheese made with raw milk. It has been shown that *B. abortus* and *B. melitensis* can survive and remain viable for many months even in hard cheese such as Cheddar (Chapman and Sharpe 1990).

Wallach et al. (1994) described an outbreak affecting an Argentine family after eating unpasteurized goats' milk cheese. Nine of 14 became ill with *B. melitensis* and blood cultures were positive from seven patients.

Human brucella infection is rare in England and Wales and, with very few exceptions, is acquired abroad after consumption of raw milk and cheese. Two cases of *B. melitensis* were reported in a family who stayed in Malta in April 1995 and ate a locally produced cheese pastry. They were among those affected in an outbreak involving at least 135 people (one death) which was linked to consumption of a soft cheese made with unpasteurized goats' and ewes' milk (Report 1995).

STREPTOCOCCUS ZOOEPIDEMICUS

In 1983, 16 cases of *S. zooepidemicus* occurred in New Mexico and were traced to a home-made cheese made with raw cows' milk which was sold at several stores. A case–control study confirmed the association between *S. zooepidemicus* infection and eating the suspect cheese. In addition, the organism was isolated from multiple samples of cheese and from milk from the cattle at the farm (Espinosa et al. 1983).

STAPHYLOCOCCUS AUREUS

Cheese is the milk product most often associated with food poisoning due to *S. aureus* enterotoxin. This organism is a common cause of mastitis in cattle and common in raw milk. Toxin may be present before pasteurization if the milk has been stored at ambient temperature. If sufficient numbers of *S. aureus* are present at the start of cheese-making, either because raw milk is used or because of post-pasteurization contamination, they may multiply and produce toxin during the cheese-making process (Minor and Marth 1972). Experiments with cheddar cheese-making have shown that even heavily contaminated milk is unlikely to become toxic if the starter culture is working normally. If, however, development of acidity is slow because of the presence of bacteriophage, the numbers of *S. aureus* may attain values of 10^6–10^8/g and produce toxin in the curd. During ripening, the numbers of *S. aureus* decline over the course of a few days or weeks in normally acidic cheese but may continue to multiply in cheese of low acidity during the first few weeks of ripening and survive for months thereafter.

Between 1951 and 1980, cheese was implicated in 16 recorded outbreaks of *S. aureus* food poisoning involving 507 people in England and Wales (Galbraith et al. 1982). Cheddar, Stilton, home-made soft cheese, and Romanian hard cheese were included. In ten outbreaks it was suspected that contamination occurred during manufacture. In the 1950s and 1960s, *S. aureus* intoxication was the most common cause of food poisoning in the USA, and numerous cases were traced to cheese (Hendricks et al. 1959; Walker et al. 1961). Since the almost universal introduction of refrigerated storage for milk, there have been very few incidents. One small outbreak involving only two people occurred in 1983 (Barrett 1986) and in 1987 cheese was suspected as a source of *S. aureus* intoxication in an outbreak involving 28 of 350 people who attended a buffet lunch of a variety of food items (Communicable Disease Report 1987, unpublished). Sharp (1987), in his survey of infections associated with dairy products in Europe and North America from 1980 to 1985, found no other incidents associated with cheese made from cows' milk. Cheese made from contaminated sheep milk was, however, responsible for outbreaks in France and Scotland.

A series of 36 outbreaks of food poisoning in England from November 1988 to January 1989, affecting 155 people, were linked to consumption of Stilton cheese made from raw cows' milk. Although the microbiology and tests for toxin were negative, the clinical illness was typical of staphylococcal intoxication. Control measures included withdrawal of the implicated cheese from sale and subsequent change to use of pasteurized milk for cheese production (Maguire et al. 1991).

For further information on *S. aureus* food poisoning, see Chapters 12, Bacteriology of foods, excluding dairy products and 32, Staphylococcus.

SALMONELLA

Salmonellae can multiply rapidly during the cheese-making process and will survive in the curd unless a high degree of acidity is reached. During ripening, an initial rapid decline in numbers is followed by a relatively stationary phase. Small numbers have been detected even after storage for 6–10 months at refrigeration temperatures (Medina et al. 1982). It may be difficult to isolate the organisms because small numbers are slow to revive and are not distributed evenly through the cheese. Survival tends to be greater in the middle of the cheese and poor near the surface. The infective dose seems to be low. About 10^4 organisms can cause illness, presumably because the high lipid content of the cheese protects the bacteria from the acidity of the stomach (Ratnam and March 1986). Many outbreaks of salmonellosis have been traced to cheeses of all types. It is clear that effective pasteurization of the milk is the only dependable way of ensuring that salmonellae do not contaminate cheese (Fontaine et al. 1980). From 1980 to 1985, outbreaks were recorded in Italy, Canada, Finland, and Switzerland (Sharp 1987).

In 1984, there was a huge outbreak of infection with *Salmonella* Typhimurium PT10 in the four Atlantic provinces of Canada and in Ontario (Ratnam and March 1986). The outbreak was the largest common source outbreak ever to occur in Canada and involved an estimated 10 000 persons. *Salmonella* Typhimurium PT10 was isolated from the factory-packed contaminated lots of Cheddar cheese as well as leftover cheese recovered from homes of human cases. Estimated numbers of *Salmonella* Typhimurium in the cheese were low but the lots of cheese that were positive also contained high levels of coliforms and *E. coli*. Defective pasteurization was shown to be the underlying cause of contamination and the outbreak strain was isolated from milk from a cow on one of the farms supplying the factory. A careful investigation of the pasteurization process revealed that an employee manually overrode the controller and allowed unpasteurized milk to go through to the vats. The pasteurizer was shut down after filling three vats and later restarted to fill the next three vats so that the first and third vats contained *Salmonella* Typhimurium when contaminated raw milk was being pasteurized (Johnson et al. 1990).

In 1989, an outbreak of *Salmonella* Dublin affected 42 people in England and Wales and was traced to imported Irish soft cheese made with unpasteurized cows' milk. *Salmonella* Dublin was isolated from samples of cheese and from the milk of four cows in the milking herd. The cheese was made on the dairy farm (Maguire et al. 1992).

In 1993, *Salmonella* Paratyphi B infection in southwest France was traced to unpasteurized cheese made from goats' milk. *Salmonella* Paratyphi B was isolated from a batch of cheese and from milk from a goat in a herd that supplied milk to the cheese makers (Desenclos et al. 1996).

The most recent outbreak in the UK occurred in the north of England in late 1996. Nineteen cases of *Salmo-nella* Goldcoast infection were linked by microbiological and epidemiological investigations to consumption of cheddar cheese (Goh, S.K. personal communication). Although the initial cases were in the north of England, additional cases were later identified in other parts of England. The cheddar cheese, made with pasteurized milk, originated from a large commercial cheese manufacturer in Somerset. Microbiological investigations confirmed that batches of the cheese were contaminated with *Salmonella* Goldcoast. Investigations at the factory highlighted process failures during the production of the infected cheese. Veterinary investigation of the farm supplying milk to the cheese manufacturer detected *Salmonella* Goldcoast in cattle on the farm.

Cheese continues to be a possible source of infection with *Salmonella* as demonstrated by an outbreak associated with the consumption of a Tyrolean cheese in Austria. Sixteen cases of *Salmonella* Oranienburg were linked to consumption of the cheese made on a local Alpine farm (Eurosurveillance 2000).

LISTERIA MONOCYTOGENES

The fairly recent association of *L. monocytogenes* infections with soft cheese may be due to changes in marketing practices for dairy products. In recent years, a wider range of dairy products, including many varieties of continental soft cheese, have become popular in Britain and North America and are available in most supermarkets. Many of these products are imported from the country of origin and all have been stored at refrigeration temperatures for days or weeks. *L. monocytogenes* multiplies at low temperatures and minor contamination may increase significantly during storage. *L. monocytogenes* appears to be common in raw milk and many soft cheeses are made from raw milk or have small amounts of raw milk added to them during manufacture. Post-process contamination with *Listeria* spp. may also occur because of the wide environmental distribution of the organism; it can be prevented by strict attention to hygiene and the prevention of cross-contamination during handling and storage (Marfleet and Blood 1987).

In 1985, 86 cases of *L. monocytogenes* infection were identified in Los Angeles and Orange Counties, California (James et al. 1985); 58 cases were in mother–infant pairs. There were eight neonatal deaths and 13 stillbirths out of a total of 29 deaths. Case–control studies implicated Mexican-style soft cheeses from one manufacturer. *L. monocytogenes* serotype 4b was isolated from patients and from packets of cheese. Although most of the cheese milk was pasteurized, about 10 percent was not.

Surveys in the USA and Europe (including the UK) have shown that *L. monocytogenes* may be found in a wide variety of soft cheeses (Gilbert 1987; Pini and Gilbert 1988; Greenwood et al. 1991). Experimental production of cottage cheese with artificially contaminated milk confirmed that a proportion of the inoculum survived manufacture (which included cooking at 57.2°C for 30 min). Organisms could be recovered from more than one-third of samples by direct plating after storage at 3°C (Ryser et al. 1985). Ryser et al. (1985) also showed that listeria survive during the manufacture of Cheddar cheese and that numbers increase during the first 14 days of ripening. Thereafter, numbers decrease but may persist for months.

Sporadic cases of listeria infection have been associated with the consumption of soft cheeses (Bannister 1987; Azadian et al. 1989; McLauchlin et al. 1990). In 1987, cases of listeriosis in Switzerland were associated with Vacherins Mont d'Or soft cheese and investigations revealed that an outbreak lasting from 1983 to 1987 had involved 122 persons (with 34 deaths) (Bille 1990).

In an outbreak of listerosis reported in France (Goulet et al. 1995), 20 cases were identified between 2 April and 16 May 1995, including 11 in pregnant women, resulting in two spontaneous abortions, four premature births, and two stillbirths. The source of infection was one specific chain of Brie de Meaux (a raw milk soft cheese) production. This is the first outbreak of listeriosis, documented in France, which has been linked to the consumption of a raw milk cheese.

A study to identify the main hazards associated with the spread of *L. monocytogenes* in dairy products in Switzerland reviewed data from 1990 to 1999. Overall, 3 722/76 271 (4.9 percent) of samples were positive (Pak et al. 2002). Cheese-ripening facilities had the highest proportion of positive samples (7.6 percent) followed by small-scale dairies (4.4 percent). Contamination of these products is frequently associated with environmental contamination and in this study 9.5 percent of the samples of water, used for cheese washing, were positive. This may indicate that a preventative step can be introduced. 1 328/3 722 (35.7 percent) of isolates were serotyped with serotypes 1/2a, 1/2b, and 4b, accounting for over 90 percent of isolates. Hard and semi-hard cheeses were more likely to be contaminated with serotype 1/2b and soft cheeses with serotype 1/2a.

An outbreak of listeriosis associated with homemade Mexican-style cheese occurred during October to January 2000 in North Carolina, USA (MMWR 2001). Noncommercial, homemade, Mexican style fresh soft cheese produced from contaminated raw milk from a local dairy was implicated as the source. Once again, this outbreak highlights the risk of using raw milk in the preparation of dairy products.

An outbreak of febrile gastroenteritis following consumption of on-farm manufactured fresh cheese has been reported from Sweden (Carrique-Mas et al. 2003). Thirty-three consumers showed an attack rate of 52 percent. Illness was characterized by presentation of diarrhea (88 percent), fever (60 percent), stomach cramps (54 percent), and vomiting (21 percent) in the

cases. Although gene sequences of verocytotoxin producing *E. coli* were detected in six subjects, 27/32 fecal specimens were positive for *L. monocytogenes*. Molecular profiles of *L. monocytogenes* from dairy, fecal, and clinical isolates were indistinguishable. Both microbiological and epidemiological evidence pointed to *L. monocytogenes* as the most likely source of this unusual outbreak. It is a timely reminder that *L. monocytogenes* can cause gastrointestinal symptoms and should be considered as a possible etiology in outbreaks of gastroenteritis associated with dairy products.

ESCHERICHIA COLI

Outbreaks of illness due to pathogenic *E. coli* have been recorded in association with cheese and it has become clear that serious infection with both enteroinvasive and VTEC may be acquired from contaminated cheese.

In 1971, there were 107 episodes of enteroinvasive *E. coli* illness involving 387 people in the USA (Marier et al. 1973). Imported French Brie, Camembert, and Coulommiers cheeses made at a single factory over a period of 2 days were implicated. A slowly lactose-fermenting strain of *E. coli* serotype O124:B17 was isolated from samples of cheese and from the feces of patients. Laboratory tests confirmed that the strain was invasive but not enterotoxigenic. There is no mention of whether the cheese milk had been pasteurized, but river water, used to wash the vats, was the suspected source of infection.

Epidemiological data incriminated French Brie and Camembert cheeses in a large outbreak of enterotoxigenic *E. coli* infection in 1983. This affected more than 3 000 people in the USA (Washington, Illinois, Wisconsin, and Georgia), as well as in Denmark, Holland, and Sweden (MacDonald et al. 1985; Nooitgedagt and Hartog 1988). *E. coli* serotype O25:H20 was isolated from fecal specimens and heat-stable (ST) toxin production was demonstrated. The plasmid profile of isolates from patients in Washington, Illinois, Georgia, and Wisconsin were similar. The illness was associated with cheese from two separate batches made at the same factory 46 days apart. There was no information about pasteurization and the source of contamination was not discovered. Cultures of cheese samples did not grow *E. coli* serotype O27:H20 but organisms binding antiserum to *E. coli* O27 were detected in one cheese specimen by solid-phase radioimmunoassay at a concentration of 10^4 organisms per gram.

There have been several reports of outbreaks of *E. coli* O157:H7 infection due to the consumption of cheese. One cluster of four cases occurred in a rural community of France and was linked to 'fromage frais' made with unpasteurized goats' and cows' milk on a farm. Four children aged from 9 to 15 months were affected, three during the spring of 1992 and one during the spring of 1993. All children had diarrhea, three developed renal failure, and one child died. Samples taken from the farm yielded isolates of VTEC but not of *E. coli* O157 (Report 1994e). The first outbreak in the UK occurred in Scotland in 1994. Twenty cases of *E. coli* O157:H7 phage type 28 infection, in the Grampian region, were associated with a farm-produced cheese made with unpasteurized milk. Most of the patients had bloody diarrhea and one had HUS. *E. coli* O157 PT28 was also isolated from a sample of the cheese and typing of the isolates from the patients and the cheese by PFGE demonstrated that they were indistinguishable from one another (Curnow 1994). An outbreak associated with a hard cheese made from unpasteurized milk was identified in Lancashire, England in 1997. Five cases were associated with a strain of *E. coli* O157 PT 8 (Aird et al. 2000). Two cases had consumed the cheese, two cases had consumed different cheese from retail outlets that sold the contaminated batch, and there was no apparent connection in the fifth case. All five isolates from the human cases, from the cheese, and from cattle on the dairy farm producing the milk, had PFGE profiles that were indistinguishable. Enumeration of *E. coli* O157 in contaminated cheese revealed that the cheese was contaminated with between 5 and 11 CFU/g (Bolton, F. personal communication). This confirms the low infectious dose for this organism. This outbreak also highlights the potential for cross-contamination of other cheese products at retail outlets.

An outbreak involving four cases in a private house in Scotland in 1998 was also reported to be associated with the consumption of cheese (Task Force on *E. coli* O157 2001). Another outbreak was reported in Wisconsin, USA in 1998, associated with the consumption of fresh cheese curds. A total of 55 cases were linked to the outbreak of which 25 were admitted to hospital (MMWR 2000). Three outbreaks of *E. coli* O157 infection associated with cheese were reported in the UK in 1999. One of these was the outbreak linked to cheese made from goats' milk (see section on Goats' milk). The other two occurred in the north of England. One of these was associated with Cotherstone cheese, a specialist cheese made in the northeast of England from unpasteurized milk (CDR 1999). There were three cases of infection with *E. coli* O157 PT 21/28. Isolates were also obtained from several cheese samples but not from the farm producing the milk. All of the isolates had an indistinguishable PFGE profile. The remaining outbreak in 1999 occurred in the northwest of England. Four cases, one of which was a food handler, were linked to the consumption of a homemade cheese sold at a local restaurant (Aird et al. 2000).

The outbreaks in Scotland prompted a survey of *E. coli* O157 in high-risk foods such as raw meats, raw milk, and raw milk cheeses. The survey examined 739 samples of raw milk cheeses sampled at the point of sale by an enrichment culture-immunomagnetic capture method and did not find any positive samples (Coia et al.

2001). In contrast, a French survey that examined 180 cheeses made from unpasteurized milk, found that for all VTEC 30.5 percent were positive by PCR. None of these were *E. coli* O157 (Fach et al. 2001).

MICROBIOLOGICAL QUALITY CONTROL DURING MANUFACTURE OF CHEESE

In recent years, there has been increasing concern about pathogens in cheese and particularly of the risk of contamination with *L. monocytogenes* and *E. coli* O157 (Report of the Committee on the Microbiological Safety of Food 1990; Report of the Advisory Committee on the Microbiological Safety of Food 1995). It is clear that soft and fresh cheeses that develop pH values in excess of 5.0 should be considered as high risk for *L. monocytogenes* whereas soft cheese which maintains a pH of less than 5.0 is of lower risk. Hard-pressed cheese is also of low risk particularly if there is a long maturation time of 60 days or more, but no cheese is entirely without risk for contamination with *L. monocytogenes* or other pathogens if the hygiene during manufacture is poor or if the cheese has been manufactured from unpasteurized milk.

The industry has taken the problem of hygiene very seriously in the UK. The Creamery Proprietors' Association has published 'Guidelines for good hygienic practice in the manufacture of soft and fresh cheeses' (Report 1988) and the Milk Marketing Board has produced similar guidelines for the manufacture of soft and fresh cheeses in small and farm-based units (Report 1989). In addition, the joint FAO/WHO Codex Alimentarius Commission is at present drafting a code of practice for hygienic practice for the manufacture of uncured, unripened, and ripened soft cheeses (Alinorm 95/13). European Standards (Council Directive 92/46/EEC) and Dairy Products (Hygiene) Regulations (HMSO 1995) reflect the current anxiety about *L. monocytogenes* and are very stringent for soft cheese whether it is made from raw or pasteurized milk. *L. monocytogenes* must be absent from five samples of 25 g taken at the manufacturing premises and each 25-g sample must consist of five pieces of 5 g from different parts of the same product. In the USA, *L. monocytogenes* must be absent from cheese and other milk products, either home-produced or imported, and this is monitored by testing performed by the state and federal authorities.

In early 1995, 13 public health laboratories in England and Wales carried out a survey of the quality of a variety of imported and local soft and semi-soft cheeses purchased from retail outlets such as supermarkets, shops and delicatessens (Nichols et al. 1996). The cheeses were examined by internationally accepted methods according to a standard protocol to see how they compared when tested for *Salmonella* spp., *L. monocytogenes*, *S. aureus*, coliforms, and *E. coli*. Data obtained for 1 437 samples showed that generally the standard was good and that most cheeses would satisfy the legal requirements.

Cheeses made from unpasteurized milk were of a lower microbiological quality than those prepared from pasteurized milk and it was of concern that for 67 percent of the cheeses sampled, there was no information to enable the customer to determine whether the cheese was made from pasteurized milk or not. However, the isolation rate for *L. monocytogenes* was low for all cheeses (1.1 percent overall and 1.4 percent for cheese made from raw milk). This was significantly lower than that found during a previous survey in 1988–89 (Greenwood et al. 1991).

ACKNOWLEDGMENTS

This chapter is adapted and expanded from Rampling, A. 1998. The microbiology of milk and milk products. In: Balows, A. and Duerden, B.I. (eds) *Topley & Wilson's Microbiology and Microbial Infections* Vol. 2: *Systematic Bacteriology*. 8th edn. London: Edward Arnold, p. 395–416.

REFERENCES

Ackers, M.L., Schoenfeld, S., et al. 2000. An outbreak of Yersinia enterocolitical O:8 infections associated with pasteurized milk. *J Infect Dis*, **181**, 1834–7.

ADAS, 1996. *Pasteurised milk. Assured hygienic quality*, Issue 3, ADAS, Oxford, 1–21.

Aird, H.C., Bolton, F.J. and Wright, P.A. 2000. *Escherichia coli* O157:H7 and dairy products in the north-west of England. *Supplement to SCIEH Weekly Report*, **34**, 23.

Allen, G., Bolton, F.J., et al. 2004. Assessment of pasteurisation of milk and cream produced by on-farm dairies using a fluorimetric method for alkaline phosphatase activity. *Commun Dis Publ Hlth*, **7**, 96–101.

Allerberger, F., Friedrich, A.W., et al. 2003. Hemolytic uremic syndrome associated with enterohemorragic *Escherichia coli* O26:H infection and consumption of unpasteurised cow's milk. *Int J Infect Dis*, **7**, 42–5.

Anderson, P.H.R. and Stone, D.M. 1955. Staphylococcal food poisoning associated with spray-dried milk. *J Hyg Camb*, **53**, 387–97.

Anon, 1997. Preliminary report of an international outbreak of Salmonella anatum infection linked to an infant formula milk. *Eurosurveillance Monthly Archives*, **2**, 22–4.

Asao, T., Kumeda, Y., et al. 2003. An extensive outbreak of staphylococcal food poisoning due to low-fat milk in Japan: estimation of enterotoxin A in the incriminated milk and powdered skim milk. *Epidemiol Infect*, **130**, 33–40.

Ashenafi, M. 1994. Fate of *Listeria monocytogenes* during the souring of Esgo, a traditional Ethiopian fermented milk. *J Dairy Sci*, **77**, 696–702.

Aureli, P., Di Cunto, M., et al. 2000. An outbreak in Italy of botulism associated with a desert made with mascarpone cream cheese. *Eur J Epidiol*, **16**, 913–18.

Aytac, S.A. and Ozbas, Z.Y. 1994. Survey of the growth and survival of *Yersinia enterocolitica* and *Aeromonas hydrophila* in yoghurt. *Milchwissenschaft*, **49**, 322–5.

Azadian, B.S., Finnerty, G.T. and Pearson, A.D. 1989. Cheese-borne listeria meningitis in immunocompetent patient. *Lancet*, **1**, 322–3.

Bannister, B.A. 1987. *Listeria monocytogenes* meningitis associated with eating soft cheese. *J Infect*, **15**, 165–8.

Barnham, M., Thornton, T.J. and Lange, K. 1983. Nephritis caused by *Streptococcus zooepidemicus* (Lancefield group C). *Lancet*, **1**, 945–7.

Barrett, N.J. 1986. Communicable disease associated with milk and dairy products in England and Wales: 1983–1984. *J Infect*, **12**, 265–72.

Becker, H., Schaller, G., et al. 1994. *Bacillus cereus* in infant foods and dried milk products. *Int J Food Microbiol*, **23**, 1–15.

Bell, J.C. and Palmer, S.R. 1983. Control of zoonoses in Britain: past, present and future. *Br Med J*, **287**, 591–3.

Bennett, R., Khambaty, F.M. and Shah, D.B. 1994. *Staphylococcus intermedius*: etiologic association with foodborne intoxication from butter blend and margarine. *Dairy Food Environ Sanit*, **14**, 604.

Beumer, R.R., Cruysen, J.J.M. and Birtantie, I.R.K. 1988. The occurrence of *Campylobacter jejuni* in raw cows' milk. *J Appl Bacteriol*, **65**, 93–6.

Bielaszewska, M., Janda, J., et al. 1997. Human *Escherichia coli* O157:H7 infection associated with the consumption of unpasteurised goat's milk. *Epidemiol Infect*, **119**, 299–305.

Biering, G., Karlsson, S., et al. 1989. Three cases of neonatal meningitis caused by *Enterobacter sakazakii* in powdered milk. *J Clin Microbiol*, **27**, 2054–6.

Bille, J. 1990. Epidemiology of human listeriosis in Europe with special reference to the Swiss outbreak. In: Miller, A.J., Smith, J.L. and Sumkuti, G.A. (eds), *Foodborne listeriosis*. Amsterdam: Elsevier, 71–4.

Birkhead, G. and Vogt, R.L. 1988. A multiple-strain outbreak of *Campylobacter enteritis* due to consumption of inadequately pasteurized milk. *J Infect Dis*, **157**, 1095–7.

Black, R.E., Jackson, R.J., et al. 1978. Epidemic *Yersinia enterocolitica* infection due to contaminated chocolate milk. *N Engl J Med*, **298**, 76–9.

Borczyk, A.A., Karmali, M.A., et al. 1987. Bovine reservoir for verotoxin-producing *Escherichia coli* O157:H7. *Lancet*, **1**, 98.

Bowen, J.E. and Turnbull, P.C.B. 1992. The fate of *Bacillus anthracis* in unpasteurised and pasteurised milk. *Lett Appl Microbiol*, **15**, 224–7.

Bramley, A.J. and McKinnon, C.H. 1990. The microbiology of raw milk. In: Robinson, R.K. (ed.), *Dairy microbiology*, Vol. 1. 2nd edn. London and New York: Elsevier Applied Science, 163–208.

Carrique-Mas, J.J., Hokeberg, I., et al. 2003. Febrile gastroenteritis after eating on-farm manufactured fresh cheese – an outbreak of listeriosis? *Epidemiol Infect*, **130**, 79–86.

CDR, 1999. *Escherichia coli* O157 associated with eating unpasteurised cheese. *CDR Weekly*, **9**, 113–16.

Chapman, H.R. and Sharpe, M.E. 1990. Microbiology of cheese. In: Robinson, R.K. (ed.), *Dairy microbiology*, Vol. 2. 2nd edn. London and New York: Elsevier Applied Science, 203–89.

Chapman, P.A. and Wright, D.J. 1993. Untreated milk as a source of verotoxigenic *E. coli* O157. *Vet Rec*, **133**, 171–2.

Chapman, P.A., Siddons, C.A., et al. 1993. Cattle as a possible source of verocytotoxin-producing *Escherichia coli* O157 infections in man. *Epidemiol Infect*, **111**, 439–47.

Chiodini, R.J. 1989. Crohn's disease and the mycobacteriosis: a review and comparison of two disease entities. *Clin Micribiol Rev*, **2**, 90–117.

Christiansson, A. 1992. The toxicology of *Bacillus cereus*. In: Bacillus cereus *in milk and milk products*. Internation Dairy Federation Bulletin 275. Brussels: IDF, 30–5.

Clark, A., Morton, S., et al. 1997. A community outbreak of Vero cytotoxin producing *Escherichia coli* O157 infection linked to a small farm dairy. *Commun Dis Rep CDR Rev*, **7**, R206–11.

Clarke, R.C., McEwen, S.A., et al. 1989. Isolation of verocytotoxin-producing *Escherichia coli* from milk filters in south-western Ontario. *Epidemiol Infect*, **102**, 253–60.

Coia, J.E., Johnston, Y., et al. 2001. A survey of the prevalence of *Escherichia coli* O157 in raw meats, raw cow's milk and raw milk cheeses in south-east Scotland. *Int J Food Microbiol*, **66**, 63–9.

Collins, M.T. 1997. Mycobacterium paratuberculosis: a potential food-borne pathogen? *J Dairy Sci*, **80**, 3445–8.

Collins, R.N., Treger, M.D., et al. 1968. Interstate outbreak of *Salmonella newbrunswick* infection traced to powdered milk. *J Am Med Assoc*, **203**, 838–44.

Cotton, L.N. and White, C.H. 1992. *Listeria monocytogenes, Yersinia enterocolitica* and *Salmonella* in dairy plant environments. *J Dairy Sci,* **75**, 51–7.

Cowden, J.M., Chisholm, D., et al. 1989. Two outbreaks of *Salmonella enteritidis* phage type 4 infection associated with consumption of fresh shell-egg products. *Epidemiol Infect,* **103**, 47–52.

Craven, J.A. 1978. Salmonella contamination of dried milk products. *Victorian Vet Proc,* **9**, 56–7.

Crielly, E.M., Logan, N.A. and Anderton, A. 1994. Studies on the bacillus flora of milk and milk products. *J Appl Bacteriol,* **77**, 256–63.

Curnow, J. 1994. *E. coli* O157 phage type 28 infections in Grampian. *Commun Dis Environ Health Scot,* **28**, 94/96, 1.

Curnow, J. 1999. *E. coli* O157 Outbreak in Grampian. *SCIEH Weekly Report,* **33**, 156.

D'Aoust, J.Y., Park, C.E., et al. 1988. Thermal inactivation of *Campylobacter* species, *Yersinia enterocolitica* and haemorrhagic *Escherichia coli* O157/H7 in fluid milk. *J Dairy Sci,* **71**, 3230–6.

Davis, J.G. and Wilbey, R.A. 1990. Microbiology of cream and dairy desserts. In: Robinson, R.K. (ed.), *Dairy microbiology*, Vol. 2, 2nd edn. London and New York: Elsevier Applied Science, 41–108.

De Buyser, M.L., Dufour, B., et al. 2001. Implications of milk and milk products in food-borne disease in France and in different industrialized countries. *Int J Food Microbiol,* **67**, 1–17.

de Louvois, J. and Rampling, A. 1998. One fifth of unpasterised milk are are contaminated with bacteria. *Br Med J,* **316**, 625, letter.

Department of Public Health, 1886. On the relation between milk-scarlatina in the human subject and disease in the cow. *Practitioner,* **37**, 61–80.

Desenclos, J.C., Bouvet, P., et al. 1996. *Salmonella enterica* serotype paratyphi B in goat milk cheese, France 1993: a case finding and epidemiological study. *Br Med J,* **312**, 91–4.

Djuretic, T., Wall, P.G. and Nichols, G. 1997. General outbreaks of infectious intestinal disease associated with milk and dairy products in England and Wales:1992 to 1996. *Commun Dis Rep CDR Rev,* **7**, R41–54.

Doeglas, H.M.G., Huisman, J. and Nater, J.P. 1967. Histamine intoxication after cheese. *Lancet,* **2**, 1361–2.

Duca, E., Teodorovici, G., et al. 1969. A new nephritogenic streptococcus. *J Hyg Camb,* **67**, 691–8.

Editorial, 1987. Food handlers and salmonella food poisoning. *Lancet,* **2**, 606–7.

Edwards, A.T., Roulson, M. and Ironside, M.J. 1988. A milk-borne outbreak of serious infection due to *Streptococcus zooepidemicus* (Lancefield group C). *Epidemiol Infect,* **101**, 43–51.

El-Dairouty, K.R. 1989. Staphylococcal intoxication traced to non-fat dried milk. *J Food Prot,* **52**, 901–2.

El-Gazzar, F.E. and Marth, E.H. 1992. Salmonellae, salmonellosis and dairy foods: a review. *J Dairy Sci,* **75**, 2327–43.

Espinosa, F.M., Ryan, W.M., et al., 1983. Group C streptococcal infections associated with eating home-made cheese – New Mexico, *MMWR,* **32**, 510, 515–16.

European Commission, 2000. Possible links between Crohn's disease and paratuberculosis. SANCO/B3/R16 2000.

Eurosurveillance, 2000. Salmonella enterica serotype Oranienburg infections associated with consumption of locally produced Tyrolean cheese, **5**, 4–5

Evans, D.I. 1947. An account of an outbreak of typhoid fever due to infected ice-cream in Aberystwyth Borough in 1946. *Med Officer,* **77**, 39–44.

Eyler, J.M. 1986. The epidemiology of milk-borne scarlet fever: the case of Edwardian Brighton. *Am J Public Health,* **76**, 573–84.

Fach, P., Perelle, S., et al. 2001. Comparison between a PCR-ELISA test and the vero cell assay for detecting Shiga toxin-producing *Escherichia coli* in dairy products and charcaterisation of virulence traits of the isolated strains. *J Appl Microbiol,* **90**, 809–18.

Fahey, T., Morgan, D., et al. 1995. An outbreak of *Campylobacter jejuni* enteritis associated with failed milk pasteurisation. *J Infect,* **31**, 137–43.

Fenlon, D.R., Stewart, T. and Donachie, W. 1995. The incidence, numbers and types of *Listeria monocytogenes* isolated from farm bulk tank milks. *Lett Appl Microbiol,* **20**, 57–60.

Fishbein, D.B. and Raoult, D. 1992. A cluster of *Coxiella burnetii* infections associated with exposure to vaccinated goats and their unpasteurised dairy products. *Am J Trop Med Hyg,* **47**, 35–40.

Fleming, D.W., Cochi, S.L., et al. 1985. Pasteurised milk as a vehicle of infection in an outbreak of listeriosis. *N Engl J Med,* **312**, 404–7.

Fontaine, R.E., Cohen, M.L., et al. 1980. Epidemic salmonellosis from cheddar cheese: surveillance and prevention. *Am J Epidemiol,* **111**, 247–53.

Francis, A.J., Nimmo, G.R., et al. 1993. Investigation of milk-borne *Streptococcus zooepidemicus* infection associated with glomerulonephritis in Australia. *J Infect,* **27**, 317–23.

Francis, B.J., Langkop, C., et al. 1977. Presumed staphylococcal food poisoning associated with whipped butter. *MMWR,* **26**, 268.

Francis, D.W., Spaulding, P.L. and Lovett, J. 1980. Enterotoxin production and thermal resistance of *Yersinia enterocolitica* in milk. *Appl Environ Microbiol,* **40**, 174–6.

Furlong, J.D., Lee, W., et al. 1979. Salmonellosis associated with consumption of nonfat powdered milk, Oregon. *MMWR,* **18**, 129–30.

Galbraith, N.S. and Pusey, J.J. 1984. Milkborne infectious disease in England and Wales 1938–1982. In: Freed, D.L.G. (ed.), *Health hazards of milk*. Eastbourne: Baillière Tindall, 27–59.

Galbraith, N.S., Forbes, P. and Clifford, C. 1982. Communicable disease associated with milk and dairy products in England and Wales 1951–80. *Br Med J,* **284**, 1761–5.

Gaya, P., Medina, M. and Nuñez, M. 1987. Enterobacteriaceae, coliforms, faecal coliforms and salmonellas in raw ewes' milk. *J Appl Bacteriol,* **62**, 321–6.

Gelletlie, R., Stuart, J., et al. 1997. Cryptosporidiosis associated with school milk. *Lancet,* **350**, 1005–6.

Ghoneim, A.T.M. and Cooke, M. 1980. Serious infection caused by group C streptococci. *J Clin Pathol,* **33**, 188–90.

Gilbert, R. 1979. *Bacillus cereus* gastroenteritis. In: Riemann, H. and Bryan, F.L. (eds), *Foodborne infections and intoxications*. New York: Academic Press, 495–518.

Gilbert, R. 1987. Foodborne infections and intoxications – recent problems and new organisms. *Microbiological and environmental health problems relevant to the food and catering industries.* Proceedings of a Joint Symposium (Campden Food Preservation Research Association and the Public Health Laboratory Service) 19–21 January, Stratford-upon-Avon. Chipping Campden: Campden Food Preservation Research Association, 1–22.

Gillespie, I.A., Adak, G.K., et al. 2003a. Milkborne general outbreaks of infectious intestinal disease, England and Wales, 1992–2000. *Epidemiol Infect,* **130**, 461–8.

Gillespie, I.A., O'Brien, S.J., et al. 2003b. Point source outbreaks of *Campylobacter jejuni* infection – are they more common than we think and what might cause them? *Epidemiol Infect,* **130**, 367–75.

Gilmour, A. and Rowe, M.T. 1990. Micro-organisms associated with milk. In: Robinson, R.K. (ed.), *Dairy microbiology*, Vol. 1, 2nd edn. London and New York: Elsevier Applied Science, 37–75.

Goh, S., Newman, C., et al. 2002. *E. coli* O157 phage type 21/28 outbreak in North Cumbria associated with pasteurized milk. *Epidemiol Infect,* **129**, 451–7.

Goulet, V., Jacquet, C., et al. 1995. Listeriosis from consumption of raw-milk cheese. *Lancet,* **345**, 1581–2.

Grant, I.R., Ball, H.J. and Rowe, M.T. 1999. Effect of higher pasteurisation temperatures, and longer holding times at 72 degrees C, on the inactivation of *Mycobacterium paratuberculosis* in milk. *Lett Appl Microbiol,* **28**, 461–5.

Grant, I.R., Ball, H.J., et al. 2002a. Incidence of *Mycobacterium paratuberculosis* in bulk raw and commercially pasteurised cows' milk from approved dairy processing establishments in the United Kingdom. *Appl Environ Microbiol,* **68**, 2428–35.

Grant, I.R., Hitchings, E.I., et al. 2002b. Effect of commercial-scale high-temperature, short-time pasteurisation on the viability of

Mycobacterium paratuberculosis in naturally infected cows' milk. *Appl Environ Microbiol*, **68**, 602–7.

Greenwood, M. 1995. Microbiological methods for examination of milk and dairy products in accordance with the Dairy Products (Hygiene) Regulations 1995. *PHLS Microbiol Dig*, **12**, 74–82.

Greenwood, M. and Rampling, A. 1995. A guide to the Dairy Products (Hygiene) Regulations 199[5] for public health laboratory microbiologists. *PHLS Microbiol Dig*, **12**, 7–10.

Greenwood, M. and Rampling, A. 1997. Evaluation of a fluorimetric method for detection of phosphatase in milk. *PHLS Microbiol Dig*, **14**, 216–17.

Greenwood, M.H., Hooper, W.L. and Rodhouse, J.C. 1990. The source of *Yersinia* spp. in pasteurized milk: an investigation at a dairy. *Epidemiol Infect*, **104**, 351–60.

Greenwood, M.H., Roberts, D. and Burden, P. 1991. The occurrence of *Listeria* species in milk and dairy products: a national survey in England and Wales. *Int J Food Microbiol*, **12**, 197–206.

Hart, R.J.C. 1984. *Corynebacterium ulcerans* in humans and cattle in north Devon. *J Hyg Camb*, **92**, 161–4.

Harvey, J. and Gilmour, A. 1990. Isolation and identification of staphylococci from milk powders produced in Northern Ireland. *J Appl Bacteriol*, **68**, 433–8.

Hendricks, S.L., Belknap, R.A. and Hausler, W.J. 1959. Staphylococcal food intoxication due to cheddar cheese. 1. Epidemiology. *J Milk Food Technol*, **22**, 313–17.

HMSO. 1995. *The Dairy Products (Hygiene) Regulations* 1995. Statutory Instrument No. 1086. London: HMSO.

Hermon-Taylor, J., Barnes, N., et al. 1998. Mycobacterium paratuberculosis cervical lymphodenitis followed five years later by terminal ileitis similar to Crohn's disease. *Br Med J*, **316**, 449–53.

Heuvelink, A.E., Bleumink, B., et al. 1998. Occurrence and survival of verotoxin-producing *Escherichia coli* O157 in raw cow's milk in the Netherlands. *J Food Protect*, **61**, 1597–601.

Hinckley, L.S. 1991. Quality standards for goat milk. *Dairy Food Environ Sanit*, **11**, 511–12.

Hobbs, B.C. 1955. Public health problems associated with the manufacture of dried milk. 1. Staphylococcal food poisoning. *J Appl Bacteriol*, **18**, 484–92.

Hudson, P.J., Vogt, R.L., et al. 1984. Isolation of *Campylobacter jejuni* from milk during an outbreak of campylobacteriosis. *J Infect Dis*, **150**, 789.

Hudson, S.J., Sobo, A.O., et al. 1990. Jackdaws as potential source of milk-borne *Campylobacter jejuni* infection. *Lancet*, **335**, 1160.

Hudson, S.J., Lightfoot, N.F., et al. 1991. Jackdaws and magpies as vectors of milkborne human campylobacter infection. *Epidemiol Infect*, **107**, 363–72.

Humphrey, T.J. and Hart, R.J.C. 1988. *Campylobacter* and *Salmonella* contamination of unpasteurised cows' milk on sale to the public. *J Appl Bacteriol*, **65**, 463–7.

Hunter, A.C. and Cruickshank, E.G. 1984. Hygienic aspects of goat milk production in Scotland. *Dairy Food Sanit*, **4**, 212–15.

Hutchinson, D.N., Bolton, F.J., et al. 1985a. *Campylobacter enteritis* associated with consumption of raw goats' milk. *Lancet*, **1**, 1037–8.

Hutchinson, D.N., Bolton, F.J., et al. 1985b. Evidence of udder excretion of *Campylobacter jejuni* as the cause of a milk-borne campylobacter outbreak. *J Hyg Camb*, **94**, 205–15.

Jacquet, C.L., Rocourt, J. and Reynaud, A. 1993. Study of *Listeria monocytogenes* contamination in a dairy plant and characterisation of the strains isolated. *Int J Food Microbiol*, **21**, 253–61.

James, S.M., Fannin, S.L., et al. 1985. Listeriosis outbreak associated with Mexican-style cheese – California. *MMWR*, **34**, 357–9.

Jayarao, B.M. and Henning, D.R. 2001. Prevalence of foodborne pathogens in bulk tank milk. *J Dairy Sci*, **10**, 2157–62.

Johnson, E.A., Nelson, J.H. and Johnson, M. 1990. Microbiological safety of cheese made from heat-treated milk. Part II Microbiology. *J Food Prot*, **53**, 519–40.

Jones, P.H., Willis, A.T., et al. 1981. Campylobacter enteritis associated with the consumption of free school milk. *J Hyg Camb*, **87**, 155–62.

Kalman, M., Szollosi, E., et al. 2000. Milkborne Campylobacter Infection in Hungary. *J Food Protect*, **63**, 1426–9.

Keene, W.E., Hedberg, K., et al. 1997. A prolonged outbreak of *Escherichia coli* O157:H7 infections caused by commercially distributed raw milk. *J Infect Dis*, **176**, 815–18.

Klie, H., Timm, M., et al. 1997. Detection and occurrence of verotoxin-forming and/or shigatoxin producing *Escherichia coli* (VTEC and/or STEC) in milk. *Berl Munch Tierarztl Wochenschr*, **110**, 337–41.

Knipschildt, M.E. and Anderson, G.G. 1994. Drying of milk and milk products. In: Robinson, R.K. (ed.) *Modern dairy technology*, Vol. 1, 2nd edn. London: Chapman and Hall, 159–254.

Lewis, M.J. 1994. Heat treatment of milk. In: Robinson, R.K. (ed.), *Modern dairy technology*, Vol. 1, 2nd edn. London: Chapman and Hall, 1–60.

Little, C.L. and de Louvois, J. 1999. Health risks associated with unpasteurised goats' and ewes, milk on retail sale in England and Wales. A PHLS Dairy Products Working Group Study. *Epidemiol Infect*, **122**, 403–8.

Louie, K.K., Paccagnella, A.M., et al. 1993. *Salmonella* serotype Tennessee in powdered milk products and infant formula – Canada and United States 1993. *MMWR*, **42**, 516–17.

Lovell, H.R. 1990. The microbiology of dried milk powders. In: Robinson, R.K. (ed.), *Dairy microbiology*, Vol. 1, 2nd edn. London and New York: Elsevier Applied Science, 245–69.

Luchansky, J.B. 1995. Use of pulsed-field gel electrophoresis to link sporadic cases of invasive listeriosis with recalled chocolate milk. *FRI Newslett*, **7**, 2, 1–2.

Lund, B.M., Gould, G.W. and Rampling, A.M. 2002. Pasteurisation of milk and the heat resistance of *Mycobacterium avium* subsp. *paratuberculosis*: a critical review of the data. *Int J Food Microbiol*, **77**, 135–45.

Lyster, R.L.J. and Aschaffenburg, R. 1962. The reactivation of milk alkaline phosphatase after heat treatment. *J Dairy Res*, **29**, 21–35.

MacDonald, K.L., Eidson, M., et al. 1985. A multistate outbreak of gastrointestinal illness caused by enterotoxigenic *Escherichia coli* in imported semisoft cheese. *J Infect Dis*, **151**, 716–20.

McDonough, F.E. and Hargrove, R.E. 1968. Heat resistance of salmonella in dried milk. *J Dairy Sci*, **51**, 1587–91.

McIntyre, L., Fung, J., et al. 2002. *Eschericia coli* O157 outbreak associated with the ingestion of unpasteurised goat's milk in British Columbia, 2001. *Can Commun Dis Rep*, **28**, 6–8.

McLauchlin, J. 1993. Listeriosis and *Listeria monocytogenes*. *Environ Policy Pract*, **3**, 201–14.

McLauchlin, J., Greenwood, M.H. and Pini, P.N. 1990. The occurrence of *Listeria monocytogenes* in cheese from a manufacturer associated with a case of listeriosis. *Int J Food Microbiol*, **10**, 255–62.

Maguire, H. 1993. Continuing hazards from unpasteurised milk products in England and Wales. *Int Food Safety News*, **2**, 21–2.

Maguire, H., Boyle, M., et al. 1991. A large outbreak of food poisoning of unknown aetiology associated with stilton cheese. *Epidemiol Infect*, **106**, 497–505.

Maguire, H., Cowden, J., et al. 1992. An outbreak of *Salmonella dublin* infection in England and Wales associated with a soft unpasteurized cows' milk cheese. *Epidemiol Infect*, **109**, 389–96.

Marfleet, J.E., Blood, R.M. 1987. Listeria monocytogenes *as a foodborne pathogen*. British Food Manufacturing Industries Research Association Scientific and Technical Surveys no. 157. Leatherhead, Surrey: BFMIRA.

Marier, R., Wells, J.G., et al. 1973. An outbreak of enteropathogenic *Escherichia coli* foodborne disease traced to imported French cheese. *Lancet*, **2**, 1376–8.

Marth, E.H. 1969. Salmonellae and salmonellosis associated with milk and milk products. A review. *J Dairy Sci*, **52**, 283–315.

Martin, M.L., Shipman, L.D., et al. 1986. Isolation of *Escherichia coli* O157:H7 from dairy cattle associated with two cases of haemolytic uraemic syndrome. *Lancet*, **2**, 1043.

Massa, S., Goffredo, E., et al. 1999. Fate of *Escherichia coli* O157 in unpasteurised milk stored at 8°C. *Lett Applied Microbiol*, **28**, 89–92.

Medina, M., Gaya, P. and Nuñez, M. 1982. Behaviour of salmonellae during manufacture and ripening of manchego cheese. *J Food Prot*, **45**, 1091–5.

Meers, P.D. 1979. A case of classical diphtheria, and other infections due to *Corynebacterium ulcerans*. *J Infect*, **1**, 139–42.

Mettler, A.E. 1989. Pathogens in milk powders – have we learned the lessons? *J Soc Dairy Technol*, **42**, 48–55.

Milner, J. 1995. *LFRA microbiology handbook, 1, dairy products*. Leatherhead, Surrey: Leatherhead Food RA.

Minor, T.E. and Marth, E.H. 1972. *Staphylococcus aureus* and staphylococcal food intoxications. III Staphylococci in dairy foods. *J Milk Food Technol*, **35**, 77–82.

Mitten, H.L. and Neirinckz, J.M. 1993. Developments in frozen-products manufacture. In: Robinson, R.K. (ed.), *Modern dairy technology*, Vol. 2, 2nd edn. London: Elsevier Applied Science, 281–329.

MMWR, 2000. Outbreak of *Escherichia coli* O157:H7 infection associated with eating fresh cheese curds – Wisconsin, June 1998. *MMWR*, **49**, 911–13.

MMWR, 2001. Outbreak of listeriosis associated with homemade Mexican-style cheese – North Carolina, October 2000–January 2001. *MMWR*, **50**, 560–2.

MMWR, 2003. Multistate outbreak of Salmonella serotype Typhimurium infections associated with drinking unpasteurised milk – Illinois, Indiana, Ohio, and Tennessee, 2002–2003. *MMWR*, **52**, 613–16.

Moore, K., Damrow, T., et al. 1995. Outbreak of acute gastroenteritis attributable to Escherichia coli serotype O104:H21 – Helena, Montana, 1994. *MMWR*, **44**, 501–3.

Morgan, D., Newman, C.P., et al. 1993. Verotoxin producing *Escherichia coli* O157 infections associated with consumption of yoghurt. *Epidemiol Infect*, **111**, 181–7.

Morgan, D., Mawer, S.L. and Harman, P.L. 1994a. The role of home-made ice cream as a vehicle of *Salmonella enteritidis* type 4 infection from fresh shell eggs. *Epidemiol Infect*, **113**, 21–9.

Morgan, D., Gunneberg, C., et al. 1994b. An outbreak of *Campylobacter* infection associated with the consumption of unpasteurised milk at a large festival in England. *Eur J Epidemiol*, **10**, 581–5.

Morgan, G., Chadwick, P., et al. 1985. *Campylobacter jejuni* mastitis in a cow: a zoonosis-related incident. *Vet Rec*, **116**, 111.

Morgan-Jones, S. 1987. *Staphylococcus aureus* in milk and milk products. *Commun Dis Environ Health Scot*, **21**, 87/43, 7–9.

Muytjens, H.L., Zanen, H.C., et al. 1983. Analysis of eight cases of neonatal meningitis and sepsis due to *Enterobacter sakazakii*. *J Clin Microbiol*, **18**, 115–20.

Muytjens, H.L., Roelofs-Willemse, H. and Jaspar, G.H.J. 1988. Quality of powdered substitutes for breast milk with regard to members of the family Enterobacteriaceae. *J Clin Microbiol*, **26**, 743–6.

Nichols, G. and de Louvois, J. 1995. The microbiological quality of ice-cream and other edible ices. *PHLS Microbiol Dig*, **12**, 11–15.

Nichols, G., Greenwood, M. and de Louvois, J. 1996. The microbiological quality of soft cheese. *PHLS Microbiol Dig*, **13**, 68–75.

Nooitgedagt, A.J. and Hartog, B.J. 1988. A survey of the microbiological quality of Brie and Camembert cheese. *Neth Milk Dairy J*, **42**, 57–72.

O'Mahony, M., Mitchell, E., et al. 1990. An outbreak of foodborne botulism associated with contaminated hazelnut yoghurt. *Epidemiol Infect*, **104**, 389–95.

Orr, K.E., Lightfoot, N.F., et al. 1995. Direct milk excretion of *Campylobacter jejuni* in a dairy cow causing cases of human enteritis. *Epidemiol Infect*, **114**, 15–24.

Pak, S.I., Spahr, U., et al. 2002. Risk factors for *L. monocytogenes* contamination of dairy products in Switzerland, 1990–1999. *Prev Vet Med*, **53**, 55–65.

Peabody, R.G., Ryan, M.J. and Wall, P.G. 1997. Outbreaks of campylobacter infection: rare event for a common pathogen. *CDR Rev*, **7**, R33–7.

Peeler, J.T. and Bunning, V.K. 1994. Hazard assessment of *Listeria monocytogenes* in the processing of bovine milk. *J Food Prot*, **57**, 689–97.

Pickett, G. and Agate, G.H. 1967. Lactose-fermenting salmonella infection. *MMWR*, **16**, 18.

Pini, P.N. and Gilbert, R.J. 1988. The occurrence in the UK of *Listeria* species in raw chickens and soft cheeses. *Int J Food Microbiol*, **6**, 317–26.

Place, E.H. and Sutton, L.E. 1934. Erythema arthriticum epidemicum (Haverhill fever). *Arch Intern Med*, **54**, 659–84.

Pritchard, T.J., Beliveau, C.M., et al. 1994. Increased incidence of *Listeria* species in dairy processing plants having adjacent farm facilities. *J Food Prot*, **57**, 770–5.

Rampling, A.M. 1998. The microbiology of milk and milk products. In: Collier, L., Balows, A. and Sussman, M. (eds), *Topley & Wilson's microbiology and microbial infections*, Vol. 2, 9th edn. London: Arnold, 367–93.

Rampling, A., Taylor, C.E.D. and Warren, R.E. 1987. Safety of pasteurised milk. *Lancet*, **2**, 1209.

Rampling, A., Mackintosh, M., et al. 1992. A microbiological hazard assessment of milk production on three dairy farms. *PHLS Microbiol Dig*, **9**, 116–19.

Ratnam, S. and March, S.B. 1986. Laboratory studies on salmonella-contaminated cheese involved in a major outbreak of gastroenteritis. *J Appl Bacteriol*, **61**, 51–6.

Reiss, H.J., Potel, J. and Krebs, A. 1951. Granulomatosis infantiseptica. *Z Gesamte Inn Med*, **6**, 451–7.

Report, 1985. *Salmonellosis outbreak, Hillfarm Dairy, Melrose Park, Illinois*, Final Task Force report. Washington, DC: US Food and Drugs Administration.

Report, 1988. *Guidelines for good hygienic practice in the manufacture of soft and fresh cheeses – 1988*. 19 Cornwall Terrace, London NW1 4QP: Creamery Proprietors Association.

Report, 1989. *The Dairy Farmers Co-operative. Guidelines for good hygienic practice for the manufacture of soft and fresh cheeses in small and farm based production units*. Thames Ditton, Surrey: Milk Marketing Board.

Report, 1991a. *IDF recommendations for the hygienic manufacture of spray dried milk powders*. Brussels: International Dairy Federation Bulletin no. 267, IDF.

Report, 1991b. *Salmonella enteritidis* associated with home-made ice cream. *Commun Dis Rep Weekly*, **1**, 175.

Report, 1994a. Outbreak of the tick-borne encephalitis; presumably milk borne. *Weekly Epidemiol Rec*, **69**, 140–1.

Report, 1994b. *Code of practice for pasteurisation of milk on farms and in small dairies*. BS 7771.

Report, 1994c. Unpasteurised milk and *Streptococcus zooepidemicus*. *Commun Dis Rep Weekly*, **4**, 241.

Report, 1994d. Outbreak of *Salmonella enteritidis* associated with nationally distributed ice cream products – Minnesota, South Dakota, and Wisconsin 1994. *MMWR*, **43**, 740–1.

Report, 1994e. Two clusters of haemolytic uraemic syndrome in France. *Commun Dis Rep Weekly*, **4**, 29.

Report, 1995. Brucellosis associated with unpasteurised milk products abroad. *Commun Dis Rep Weekly*, **5**, 151.

Report, 1996. VTEC O157 infection in West Yorkshire associated with the consumption of raw milk. *Commun Dis Rep Weekly*, **6**, 181.

Report of the Advisory Committee on the Microbiological Safety of Food, 1995. *Report on Verocytotoxin-producing* Escherichia coli. London: HMSO.

Report of the Committee on the Microbiological Safety of Food, 1990. *The microbiological safety of food, part 1*. London: HMSO.

Report of the Ice Cream Alliance and the Ice Cream Federation, 1992a. *Code of practice for the hygienic manufacture of ice cream*. Thames Ditton, Surrey: Milk Marketing Board.

Report of the Ice Cream Alliance and the Ice Cream Federation, 1992b. *Code of practice for the safe handling and service of scoop and soft serve ice cream*. Thames Ditton, Surrey: Milk Marketing Board.

Riordan, T., Humphrey, T.J. and Fowles, A. 1993. A point source outbreak of campylobacter infection related to bird-pecked milk. *Epidemiol Infect*, **10**, 261–5.

Roberts, D. 1985. Microbiological aspects of goat's milk. A Public Health Laboratory Service survey. *J Hyg Camb*, **94**, 31–44.

Robinson, D.A. 1981. Infective dose of *Campylobacter jejuni* in milk. *Br Med J*, **282**, 1584.

Robinson, R.K. and Tamime, A.Y. 1990. Microbiology of fermented milks. In: Robinson, R.K. (ed.), *Dairy microbiology*, Vol. 2, 2nd edn. London and New York: Elsevier Applied Science, 291–343.

Robinson, R.K. and Tamime, A.Y. 1993. Manufacture of yoghurt and other fermented milks. In: Robinson, R.K. (ed.), *Modern dairy technology*, Vol. 2. 2nd edn. London and New York: Elsevier Applied Science, 1–48.

Rocco, R.M. 1990. Fluorometric analysis of alkaline phosphatase in fluid dairy products. *J Food Prot*, **53**, 588–91.

Rodriguez, J.L., Gaya, P., et al. 1994. Incidence of *Listeria monocytogenes* and other listeria spp. in ewes' raw milk. *J Food Prot*, **57**, 571–5.

Rohrbach, B., Draughon, A., et al. 1992. Prevalence of *Listeria monocytogenes*, *Campylobacter jejuni*, *Yersinia enterocolitica* and *Salmonella* in bulk tank milk: risk factors and risk of human exposure. *J Food Prot*, **55**, 93–7.

Rothwell, J. 1979. Food poisoning risks associated with foods other than meat and poultry. *Health Hyg*, **3**, 3–10.

Rothwell, J. 1990. Microbiology of ice cream and related products. In: Robinson, R.K. (ed.), *Dairy microbiology*, Vol. 2, 2nd edn. London and New York: Elsevier Applied Science, 1–39.

Rowe, B., Begg, N.T., et al. 1987. *Salmonella* Ealing infections associated with consumption of infant dried milk. *Lancet*, **2**, 900–3.

Rubery, E. 2001. A review of the evidence for a link between exposure to Mycobacterium paratuberculosis (MAP) and Crohn's disease (CD) in humans. A report for the Food Standards Agency. http:// www.jims.cam.ac.uk/people/faculty/erubery.htm

Ryan, C.A., Nickels, M.K., et al. 1987. Massive outbreak of antimicrobial-resistant salmonellosis traced to pasteurized milk. *J Am Med Assoc*, **258**, 3269–74.

Ryser, E.T., Marth, E.H. and Doyle, M.P. 1985. Survival of *Listeria monocytogenes* during manufacture and storage of cottage cheese. *J Food Prot*, **48**, 746–50.

Sacks, J.J., Roberto, R.R. and Brooks, N.F. 1982. Toxoplasmosis infection associated with raw goat's milk. *J Am Med Assoc*, **248**, 1728–32.

Saha, A.L. and Ganguli, N.C. 1957. An outbreak of staphylococcal food poisoning from consumption of dahi. *Ind J Public Health*, **1**, 22–6.

Sanaa, M., Poutrel, B., et al. 1993. Risk factors associated with contamination of raw milk by *Listeria monocytogenes* in dairy farms. *J Dairy Sci*, **76**, 2891–8.

Schaack, M.M. and Marth, E.H. 1988. Survival of *Listeria monocytogenes* in refrigerated cultured milks and yoghurt. *J Food Prot*, **51**, 848–52.

Schlech III, W.F., Lavigne, P.M., et al. 1983. Epidemic listeriosis – evidence for transmission by food. *N Engl J Med*, **308**, 203–6.

Schröder, M.J.A. 1984. Origins and levels of post pasteurisation contamination of milk in the dairy and their effects on keeping quality. *J Dairy Res*, **51**, 59–67.

Scotter, S., Aldridge, M., et al. 1993. Validation of European Community methods for microbiological and chemical analysis of raw and heat treated milk. *J Assoc Public Analysts*, **29**, 1–32.

Seeliger, H.P.R. 1961. *Listeriosis*, 2nd edn. New York: Hafner Publishing.

Shanson, D.C., Gazzard, B.G., et al. 1983. *Streptobacillus moniliformis* isolated from blood in four cases of Haverhill fever. *Lancet*, **2**, 92–4.

Sharp, J.C.M. 1987. Infections associated with milk and dairy products in Europe and North America, 1980–85. *Bull WHO*, **65**, 397–406.

Sharp, J.C.M., Paterson, G.M. and Barrett, N.J. 1985. Pasteurisation and the control of milkborne infection in Britain. *Br Med J*, **291**, 463–4.

Shaw, M.B. 1993. Modern cheesemaking: soft cheeses. In: Robinson, R.K. (ed.), *Modern dairy technology*, Vol. 2, 2nd edn. London and New York: Elsevier Applied Science, 221–80.

Skinner, L.J., Chatterton, J.M.W., et al. 1987. Toxoplasmosis in the home and on the farm. *Commun Dis Environ Health Scot*, **21**, 87/41, 5–7.

Skirrow, M.B. 1991. Epidemiology of *Campylobacter enteritis*. *Int J Food Microbiol*, **12**, 9–16.

Slade, P.J., Fistrovici, E.C. and Collins-Thompson, D.L. 1989. Persistence at source of *Listeria* spp. in raw milk. *Int J Food Microbiol*, **9**, 197–203.

Sockett, P.N. 1991. Communicable disease associated with milk and dairy products: England and Wales 1987–89. *Commun Dis Rep Rev*, **1**, R9–12.

Sorensen, O., Rawlula, S., et al. 2003. Mycobacterium paratuberculosis in dairy herds in Alberta. *Can Vet J*, **44**, 221–6.

Southern, J.P., Smith, R.M.M. and Palmer, S.R. 1990. Bird attack on milk bottles: possible mode of transmission of *Campylobacter jejuni* to man. *Lancet*, **336**, 1425–7.

Steede, F.D.W. and Smith, H.W. 1954. Staphylococcal food-poisoning due to infected cow's milk. *Br Med J*, **2**, 576–8.

Stephan, R. and Kuhn, K. 1999. Prevalence of verotoxin-producing *Escherichia coli* (VTEC) in bovine coli mastitis and their antibiotic resistance patterns. *Zentalbl Veterinarmed B*, **46**, 423–7.

Stuart, J., Sufi, F., et al. 1997. Outbreak of campylobacter enteritis in a residentail school associated with bird pecked bottle tops. *CDR Rev*, **7**, R38–40.

Studahl, A. and Andersson, Y. 2000. Risk factors fro indigenous campylobacter infection: a Swedish study. *Epidemiol Infect*, **125**, 269–75.

Sumner, S.S., Speckhard, M.W., et al. 1985. Isolation of histamine-producing *Lactobacillus bucheri* from Swiss cheese implicated in a food poisoning outbreak. *Appl Environ Microbiol*, **50**, 1094–6.

Tacket, C.O., Narain, J.P., et al. 1984. A multistate outbreak of infections caused by *Yersinia enterocolitica* transmitted by pasteurised milk. *J Am Med Assoc*, **251**, 483–6.

Tamine, A.Y. 1993. Modern cheesemaking: hard cheeses. In: Robinson, R.K. (ed.), *Modern dairy technology*, Vol. 2, 2nd edn. London and New York: Elsevier Applied Science, 49–220.

Task Force on *E. coli* O157, 2001. Final report. Edinburgh: Scottish Executive Health Department, Food Standards Agency Scotland.

Thomas, W.E., Stephens, T.H., et al. 1948. Enteric fever (paratyphoid B) apparently spread by pasteurised milk. *Lancet*, **2**, 270–1.

Thompson, D.E. 1994. The role of mycobacteria in Crohn's disease. *J Med Microbiol*, **41**, 74–94.

Threlfall, E.J., Ward, L.R., et al. 1980. Plasmid-encoded trimethoprim resistance in multiresistant epidemic *Salmonella typhimurium* phage types 204 and 193 in Britain. *Br Med J*, **280**, 1210–11.

Threlfall, E.J., Ward, L.R. and Rowe, B. 1997. Increasing incidence of resistance to trimethoprim and ciprofloxacin in epidemic *Salmomella typhimurium* DT104 in England and Wales. *Eurosurveillance*, **2**, 81–4.

Tirard-Collet, P., Zee, J.A., et al. 1991. A study of the microbiological quality of goat milk in Quebec. *J Food Prot*, **54**, 263–6.

Tschäpe, H., Prager, R., et al. 1995. Verotoxinogenic *Citrobacter freundii* associated with severe gastroenteritis and cases of haemolytic uraemic syndrome in a nursery school: green butter as the infection source. *Epidemiol Infect*, **114**, 441–50.

Upton, P. and Coia, J.E. 1994. Outbreak of *Escherichia coli* O157 infection associated with pasteurised milk supply. *Lancet*, **344**, 1015.

Valle, J., Gomez-Lucia, E., et al. 1990. Enterotoxin production by staphylococci isolated from healthy goats. *Appl Environ Microbiol*, **56**, 1323–6.

van den Brandhof, W., Wagenaar, J., et al. 2003. An outbreak of campylobacteriosis after drinking unpasteurised milk, 2002, Netherlands. *Int J Med Microbiol*, **293**, 142.

Vernon, E. and Tillett, H.E. 1974. Food poisoning and salmonella infections in England and Wales 1969–1972. *Public Health*, **88**, 225–35.

Walker, G.C., Harmon, L.G. and Stine, C.M. 1961. Staphylococci in Colby cheese. *J Dairy Sci*, **44**, 1272–82.

Walker, R.A., Lawson, A.J., et al. 2000. Decreased susceptibility to ciprofloxacin in outbreak associated multiresistant *Salmonella typhimurium* DT104. *Vet Rec*, **147**, 395–6.

Wallach, J.C., Miguel, S.E., et al. 1994. Urban outbreak of *Brucella melitensis* infection in an Argentine family: clinical and diagnostic aspects. *FEMS Immunol Med Microbiol*, **8**, 49–56.

Waterman, S.C. 1982. The heat-sensitivity of *Campylobacter jejuni* in milk. *J Hyg Camb*, **88**, 529–33.

Weissman, J.B., Deen, R.M.A.D., et al. 1977. An island-wide epidemic of salmonellosis in Trinidad traced to contaminated powdered milk. *West Ind Med J*, **26**, 135–43.

Wells, J.G., Shipman, L.D., et al. 1991. Isolation of *Escherichia coli* serotype O157:H7 and other shiga-like toxin producing *Escherichia coli* from dairy cattle. *J Clin Microbiol*, **29**, 985–9.

Wilbey, R.A. 1994. Production of butter and dairy based spreads. In: Robinson, R.K. (ed.), *Modern dairy technology*, Vol. 1, 2nd edn. London: Chapman and Hall, 107–58.

Wang, G., Zhao, T., et al. 1997. Survival and growth of *Escherichia coli* O157:H7 in unpasteurised milk. *J Food Prot*, **60**, 610–13.

Wilson, G.S. 1942. *The pasteurization of milk*. London: Edward Arnold.

Wolf, F.S., Floyd, C., et al. 1970. Staphylococcal food poisoning traced to butter – Alabama. *MMWR*, **19**, 271.

Wood, R.C., MacDonald, K.L. and Osterholm, M.T. 1992. *Campylobacter* enteritis outbreaks associated with drinking raw milk during youth activities. A 10-year review of outbreaks in the United States. *J Am Med Assoc*, **268**, 3228–30.

Working Party to the Director of the Public Health Laboratory Service, 1971. The hygiene and marketing of fresh cream as assessed by the methylene blue test. *J Hyg Camb*, **69**, 155–68.

Zall, R.R. 1990. Control and destruction of micro-organisms. In: Robinson, R.K. (ed.), *Dairy microbiology*, Vol. 1, 2nd edn. London and New York: Elsevier Applied Science, 115–61.

11

Bacteriology of foods, excluding dairy products

DIANE ROBERTS

GENERAL PRINCIPLES OF FOOD MICROBIOLOGY

The field of food microbiology is very diverse, but it can be subdivided into three broad areas of study:

1 the utilization of microorganisms in the production of foods in order to enhance their nutritive value, organoleptic properties (color, odor, flavor, texture, etc.) and shelf-life
2 the control of spoilage of foods by reducing or eliminating their microbial content or by making the food environment unsuitable for the growth of spoilage microorganisms
3 the microbiological safety of foods, protecting of the food supply from those pathogenic bacteria that may be naturally present in foods or that may be introduced during processing and service.

The area of food safety is covered in other chapters of this treatise so will not be discussed in detail here. Local, national, and international food legislation is in place to protect the consumer from food-borne infections and intoxications of both microbial and nonmicrobial origin. These are aimed mainly at protecting the consumer from pathogenic microorganisms, so they will not be discussed here.

Not all microorganisms in foods are harmful and without some of them there would not be the great variety of foods that is enjoyed today. Some of the foods for which microorganisms (bacteria, yeasts, and molds)

are essential to their production include cheese, yogurt, fermented sausages, sauerkraut, and a whole array of fermented delicacies (tempeh, tofu, soy sauce). In addition, the production of beer, wines, and other alcoholic beverages is dependent on the fermentation of appropriate substrates by yeasts. Production of a fermented product with the desirable characteristics already mentioned, but without the problems associated with the growth of spoilage organisms and pathogens, requires an intimate knowledge of the microorganisms involved in the three areas. This is particularly true with respect to their behavior in various food systems and their susceptibility to environmental conditions and chemicals. The branch of microbiology that deals with this specific area is called microbial ecology and forms the central focus of this chapter. This topic is addressed in a concise manner with important references for additional information. The two volumes on *Microbial ecology of foods* developed by the International Commission on Microbiological Specifications for Foods (ICMSF) are highly recommended as authoritative sources for additional information on this subject (ICMSF 1980, 1998).

FACTORS INFLUENCING MICROBIAL GROWTH IN FOODS

Foods are very complex biological materials that span a tremendous range in texture, fluidity, and nutritional content. Nevertheless, there are several universal characteristics that can exert an influence on the growth of microorganisms. The major factors that must be considered are temperature, pH, and water activity (a_w). In addition, redox potential (Eh) and the presence of organic acids, salts, and other chemicals can all influence the growth of microorganisms in foods. These factors can be considered as two groups, the intrinsic and the extrinsic factors (Mossel et al. 1995; Montville 1997a). Intrinsic factors are those inherent to the food itself such as naturally occurring compounds that stimulate or retard growth, compounds that are added as preservatives, redox potential, water activity, and pH. The primary extrinsic factors, those imposed from the outside, are temperature, humidity, and gas composition (controlled and modified atmosphere packaging).

Extrinsic factors

TEMPERATURE

Although microbial growth can occur over a temperature range of −8 to 90°C, in food microbiology the range of greatest importance is 1–50°C. Within this range the temperature can affect the duration of the lag phase, the rate of growth, the final population of cells, and the nutritional requirements. The temperatures immediately above and below the growth ranges of organisms may just retard growth, but if the temperature exceeds the upper limit, cell injury and cell death occur. The response of microorganisms to freezing temperatures is more complex and will depend on the composition of the microbial cell wall, the final temperature, and, more importantly, the rate of freezing.

Bacteria can be grouped according to their temperature ranges for growth: thermophiles (optimum 55–75°C), mesophiles (30–45°C), psychrophiles (5–15°C), and psychrotrophs (25–30°C). Many food-spoilage and most pathogenic bacteria are mesophiles. Psychrophiles are sensitive to temperatures higher than 20°C; therefore, they are not as important in foods as psychrotrophs, which have temperature maxima in the range of 30–35°C. The ability of psychrotrophs to grow at refrigeration temperatures make them important contributors to food spoilage. Bacteria in the genera *Acinetobacter*, *Alcaligenes*, *Bacillus*, *Chromobacterium*, *Enterobacter*, *Flavobacterium*, *Proteus*, *Pseudomonas*, and *Serratia* are important causes of the spoilage of foods, particularly those stored at refrigeration temperatures in order to increase their shelf-life. However, some food-borne pathogenic bacteria (*Listeria monocytogenes*, *Yersinia enterocolitica*) are psychrotrophic and disease outbreaks have occurred from contaminated food products that have been stored at refrigeration temperatures.

Effect of chilling

As the temperature is lowered to freezing point, the lag phase of bacterial growth increases greatly. All except the psychrotrophic group stop growing, thus giving the group a competitive advantage in foods held at low temperatures. Although the generation times of psychrotrophic organisms increase significantly at temperatures near freezing (for example, the generation time for *Pseudomonas fluorescens* at 0.5°C is 6.68 h), during refrigerated storage of foods for several days or weeks their populations can reach several millions. These bacteria grow using the food as their source of nutrients, with the result that the lipids, proteins, and carbohydrates in the food are broken down, producing degradation products of food components and the bacterial metabolic byproducts that are associated with food spoilage.

Effects of freezing

The response of bacteria to freezing ranges from little or no effect to varying degrees of injury and ultimately cell death (see Montville 1997a). Most spores and some vegetative cells can survive with almost no effect, but most non-spore-forming bacteria are affected to some extent. In general, gram-negative bacteria are more sensitive to freezing than gram-positive bacteria. Among those that are sensitive the consequences of freezing are sublethal cell injury and cell death. Cells that are sublethally injured are highly sensitive to the selective agents used in the selective enrichment and plating media used for the isolation and enumeration of food-borne bacteria. The use of these microbiological media to determine bacterial counts in frozen food products is most likely, therefore, to result in an underestimate of the microbial populations because of the failure of these sublethally injured bacteria to produce colonies. However, under appropriate cultural conditions, these sublethally injured bacteria may repair their damage and become fully viable and thus remain a threat to the integrity of the food.

The rate of freezing and the final storage temperature have an influence on bacterial injury and survival with the temperature range of −2 to −10°C being the most detrimental for bacteria. Slow freezing to approximately −10°C will cause maximal damage, whereas rapid freezing to −30°C or lower is likely to cause only minimal damage. Slow freezing also has a detrimental effect on the organoleptic characteristics of foods and is therefore not generally used as a means of bacterial inactivation.

Whether bacteria will grow in frozen foods after thawing will depend on the numbers and types that survived the freezing process and the conditions used for thawing. Uncontrolled thawing may result in a significant zincrease in the numbers of bacteria. When a large block of frozen food is thawed, the areas near the surface will thaw much more rapidly than the core. Thus, bacteria near the surface of the food may start to grow and go through several generations before the center of the food has completely thawed. This is why it is considered safer to thaw frozen foods, particularly large blocks such as frozen turkeys, either at refrigeration temperature or rapidly by microwave heating rather than at ambient temperature.

Effects of heating

As the temperature rises above the optimum growth range, growth of bacteria ceases and injury and, ultimately, cell death occur. As with the effects of freezing,

a mild exposure to heat can cause a sublethal injury in vegetative cells that is repairable. Spores are much more refractile to heat and may be activated to germinate. An exposure of 5 min at 80°C is typically used to inactivate vegetative cells and induce germination of spores. With further increase in temperature there is destruction of the bacterial population at an exponential (logarithmic) rate. Figure 11.1 shows a typical survivor curve for bacteria. By plotting the logarithm of the number of survivors against the time of heating at a specific temperature, a straight line is obtained because the rate of death is constant at a given temperature and is independent of the initial population of cells. The decimal reduction time (the time required to destroy 90 percent of the cells, D value) may be calculated from the

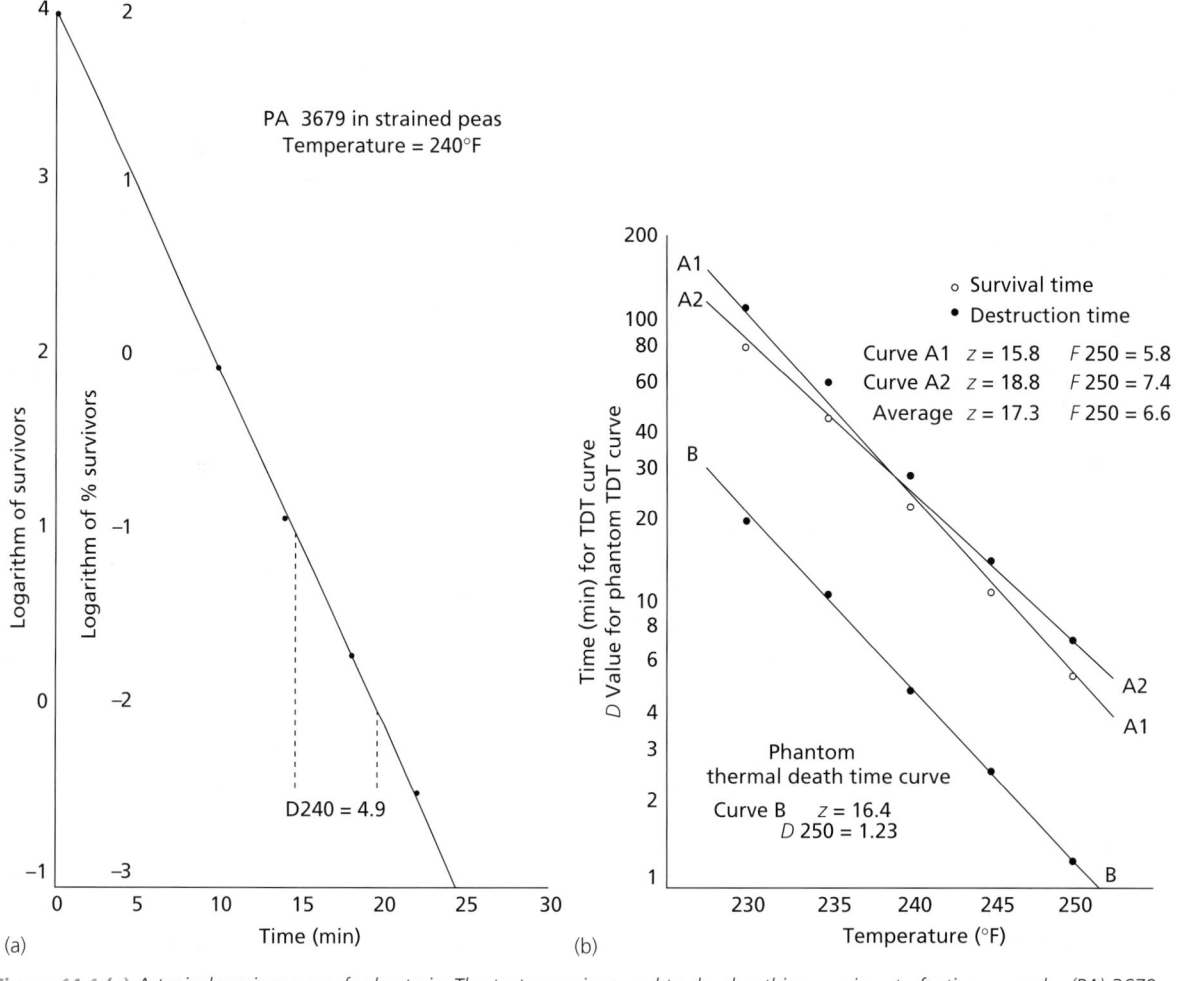

Figure 11.1 (a) A typical survivor curve for bacteria. The test organism used to develop this curve is putrefactive anaerobe (PA) 3679, which is a non-toxigenic Clostridium species that has heat-resistance characteristics similar to those of C. botulinum. Therefore, PA 3679 is often used to determine or verify the thermal process conditions required to inactivate C. botulinum in specific foods. In this illustration, the D value for PA 3679 at 115°C (240°F) is 4.9 min. (b) An example of a thermal death time (TDT) curve. The D values are plotted on a logarithmic scale against the corresponding temperature on a linear scale to obtain the phantom TDT curves. These lines are described by the D value at a reference temperature (usually 60 or 121°C) and the slope of the curve (denoted by z). z is defined as the temperature that is necessary to bring about a ten-fold change in the D value. F denotes the time of treatment necessary to destroy microorganisms at a reference temperature, usually 60°C for vegetative cells and 121°C for spores. (Parts a and b based on Figures 7-7 and 7-8 from Laboratory Manual for Food Canners and Processors, vol. 1, National Canners Association Research Laboratories, 1968, AVI Publishing Company, Westport, CT, with permission from National Food Processors Association, Washington, DC.)

survivor curve. The D value for the bacteria in Figure 11.1 is 4.9 min at 121°C. The ordinate of the plot is on a logarithmic scale, thus it is evident that the population of bacteria would never be reduced to zero. It follows, therefore, that if a known population of bacteria in each of several unit volumes or containers were exposed to lethal heat treatments, a probability will always remain of a survivor in any one container. Such calculations are used to determine that the heat treatment to be applied to a specific type of food in a specific type of container will result in a probability of survival so low that it is acceptable. For example, a $12D$ process is generally used in commercial canning operations to assure that canned low-acid foods are free from the most resistant spores of *Clostridium botulinum* and the hazard of botulinum toxin.

The survivor curve gives information on the time of heat treatment required to destroy a certain proportion of bacteria at a specific temperature. The thermal death time curve is used to determine the effect of heat treatment of bacteria suspended in a medium at several temperatures. A thermal death time curve is constructed by plotting the $\log_{10} D$ values against the treatment temperature. The slope of the curve is denoted by the term z, which is the temperature in degrees Celsius required for the thermal death curve to traverse one log cycle. D and z values are used extensively in determining the heating requirement for foods preserved by pasteurization and sterilization.

Many factors influence the heat resistance of bacteria. These may be divided into three broad types:

1 the differences in heat-resistance characteristics of different species within a genus, strains within the same species, or spores versus vegetative cells
2 the environmental conditions to which the bacteria were previously exposed during growth
3 the properties of the food to which they are exposed during heat treatment (pH, a_w, salts and other compounds, type of food).

If these factors are not considered during the development of a thermal treatment process for foods, the outcome may be food spoilage and food-borne infection or intoxication.

GASEOUS ATMOSPHERE

The growth-limiting value of lowered partial pressure of oxygen (pO_2) or redox potential (Eh) is often combined with and reinforced by another inhibitory parameter, such as raised pCO_2. The composition of the gaseous atmosphere in which foods are stored may affect the fate of their microbial association more than by the oxygen partial pressure alone. A number of gases are utilized in the commercial preservation of foods: carbon dioxide, ethylene oxide, propylene oxide, sulfur dioxide, and ozone.

Intrinsic factors

WATER ACTIVITY

Microorganisms, in order to grow and metabolize, require water in an available form. The most useful measurement of available water is water activity (a_w). The a_w of a food or solution is the ratio of the water vapor pressure of the food (p) to that of pure water (p_0) at the same temperature. The a_w of pure water is assumed to be 1. When a solution becomes more concentrated with the addition of more solute (for example, salt, sugar, or other chemical) or the removal of water by drying, the vapor pressure decreases and so the a_w becomes a value of less than 1. The water molecules are oriented about solute molecules or become adsorbed on to insoluble food constituents and are thus less available for reactions. Most bacteria and fungi are unable to grow at an a_w of less than 0.90. Thus the dehydration of foods and the addition of salt and sugar are long-established food-preservation methods particularly in relation to food safety. Freezing can also reduce the a_w. However, some bacteria and fungi can grow at a_w values below 0.90 and are thus the organisms to be considered in preventing food spoilage. They are variously referred to as halophiles, xerophiles, and osmophiles, but these terms strictly should apply only to those organisms that can grow well at reduced levels of a_w but fail to grow at high levels. Gram-negative bacteria generally do not grow below a_w values of 0.97. Between a_w values of 0.98 and 0.93, gram-positive bacteria – Lactobacillaceae, Bacillaceae, and Micrococcaceae – become important. Below 0.93, food spoilage is caused primarily by yeasts and fungi. The only bacterial pathogen growing within this range is *Staphylococcus aureus*. Xerophilic fungi and osmophilic yeasts cause spoilage at a_w values below 0.85. Below an a_w of 0.60, microorganisms are unable to grow but they can remain viable for long periods.

pH

Bacteria have a minimum, optimum, and maximum pH for growth. The cell membrane is only slightly permeable to hydrogen or hydroxyl ions, and the cytoplasmic contents have a buffering effect, thus the pH inside bacterial cells is usually close to 7.0. The minimum pH values at which bacteria can initiate growth do vary. Some bacteria, such as the lactic acid bacteria, create an acidic environment by producing acids as metabolic products; they therefore gain a selective advantage over bacteria that are less acid-tolerant. Most bacteria are unable to grow at pH levels of less than 3.5 and those of importance in food-borne illness are unable to grow below about pH 4.5. High pH values also inhibit microorganisms. Egg white develops a pH of 9 because of the loss of CO_2 after the egg is laid; this protects the egg from bacterial invasion. Proteolytic organisms such as

Pseudomonas spp. can grow in moderately alkaline substrates. The antimicrobial effect of reduced pH depends on interactions with other factors (temperature, a_w, salt concentration).

REDOX POTENTIAL

The redox potential (*E*h) of a biological system is an index of its degree of oxidation and is an important selective factor in the food environment. Redox potential can denote the oxygen relations of a living microbe and may be used to specify an environment in which the organism can produce energy and new cells without recourse to molecular oxygen. Bacteria are classified as aerobic, anaerobic, facultative, or micro-aerobic based on the *E*h (ranging from positive values of about +300 mV (aerobe) to negative values of −420 mV (anaerobes)) required for their growth and metabolism. The *E*h of a food is related to its chemical composition (pH, concentration of reducing substances such as ascorbic acid, mercapto groups in proteins, reducing sugars, etc.) and the oxygen partial pressures existing over the food during storage. Gradients of *E*h may occur from the surface to the depth of a food where gaseous diffusion is restricted.

NUTRIENTS

The nutritional composition of a food has an influence on the type of bacteria that are likely to be present. Also, the nutrient requirements of bacteria may change depending on the environmental conditions in which they are grown. For example, bacteria generally become more nutritionally exacting at higher or lower growth temperatures or at lower a_w values. This may be due to inability to utilize particular nutrients as a result of impaired transport processes or from markedly reduced activity of critical enzymes. In general, most foods provide sufficient nutrients for most microorganisms. However, there is a wide range in nutrient complexity and availability; meats are generally considered highly nutritious and vegetables are considered to provide minimal nutrients.

Spoilage of starchy foods, such as potatoes and cereal grains, is caused primarily by bacteria that are able to break down the complex carbohydrates and that thrive in a minimum nitrogen and salts environment. Fresh vegetables are spoiled primarily by cellulolytic organisms. Bacteria with high lipolytic activity are favored in lipid-containing foods. Highly proteinaceous foods favor the growth of bacteria with complex nutritional needs. Nutrients control the rate of growth only when present at very low levels.

OTHER FACTORS

In addition to the factors discussed above, other factors that influence the growth of bacteria in foods are those natural components of foods, which have antibacterial properties (essential oils, tannins, and lectins). Animal foods contain many inhibitors of bacteria (lactenin in milk, lysozyme, conalbumin, ovomucoid, and avidin in eggs) but they usually have a narrow spectrum of activity and are quite labile. Therefore, they do not play a major role in preventing food spoilage.

Inhibitory factors may also be produced in foods as a result of processing. Sugar syrups undergo a browning reaction during storage, leading to the formation of furfural and its derivatives, which have antimicrobial properties. Similarly, the oxidative chemical deterioration of foods during storage also leads to the formation of products with antimicrobial properties.

FOOD PROCESSING AND PRESERVATION

Most agricultural products used for human food cannot be used directly without some form of further processing. Even fresh produce such as apples, potatoes, tomatoes, and beans must be washed, trimmed, sorted, and packaged. Often, more extensive processing is needed to convert raw agricultural commodities such as beef cattle, wheat kernels, and sunflower seeds into meat, flour and bread, and cooking oil, respectively. Thus, the primary objective of food processing is to convert raw agricultural commodities into processed foods or food ingredients. The food ingredients (flour, sugar, oil, etc.) may then be used in a secondary food-processing operation to manufacture products that are ready for human consumption. Generally, food-processing operations, particularly the secondary operations, integrate food-processing with food preservation. The two objectives of preservation are firstly to prolong the shelf-life of the processed food or food ingredient in order that it may be stored for long periods of time and distributed efficiently over great distances, and secondly to assure the safety of processed foods. Therefore, the food-preservation aspect of food processing deals with the control of microorganisms in foods to enhance the quality of foods, retard their spoilage, and eliminate pathogens. Strategies to achieve these objectives are developed on the basis of our knowledge of microorganisms and their response to the various extrinsic and intrinsic environmental factors discussed earlier (see also Table 11.1).

Preservation methods can be subdivided into three broad groups:

1 Physical methods involving treatments that inhibit, destroy, or remove undesirable organisms without antimicrobial additives or products of microbial metabolism, e.g. drying, temperature, radiation, high hydrostatic pressure, etc. (Farkas 1997).

2 Addition of chemical preservatives and natural antimicrobial compounds or utilization of their natural presence (Davidson 1997).

Table 11.1 *Preservation methods used in food processing*

Process	How preservation is accomplished
Well-established processes	
Drying (reduced water activity)	Reducing water activity below the level required for microbial growth
Chilling	Lowering of temperature to a point at which microbial growth is retarded or inhibited
Freezing	Changing of water to a form that microbes are unable to utilize. Freezing temperatures also inhibit enzymatic activity and thus growth
Heating 1. Heat sterilization in hermetically sealed containers	Inactivating all viable forms of bacteria followed by storing under conditions that will prevent recontamination and chemical degradation
Heating 2. Pasteurization and other heat processes that do not completely eliminate microorganisms	Reducing spoilage bacteria to a significantly low level so that the shelf-life of the product is extended; vegetative cells of pathogenic bacteria will also be eliminated
Irradiation (ionizing)	a Inactivating all viable bacteria (sterilization) b Eliminating pathogens or prolonging storage time (pasteurization)
Pickling, salting, and sugaring	Reducing the acidity and/or water activity to levels that inhibit the growth of microorganisms
Fermentation	Controlling environmental conditions in order to favor the growth of beneficial microorganisms and retarding or inhibiting the growth of spoilage and harmful microorganisms
New/developing processes	
Gases	
Controlled atmosphere storage	Reducing oxygen content and increasing carbon dioxide content combined with chilled storage inhibits growth of spoilage organisms
Modified atmosphere packaging (MAP)/vacuum packing (VP)	Using combinations of CO_2, N_2, and O_2 to sustain visual appearance and extend shelf-life (MAP) Reducing the pressure of air and thus the O_2 levels to inhibit growth of some organisms (VP) Combined with the use of low-permeability packaging film
Heat treatments:	
Microwaves	Reducing process times and energy and water usage in some food processing areas. Microwaves create internal cellular friction that produces heat. Thermal nonuniformity is a problem
Ohmic heating	Using direct electrical heating by passing an electric current through the product. The heat produced results in microbial death
High hydrostatic pressure	Applying high or ultrahigh hydrostatic pressure that transfers through food instantly and uniformly. Probably alters permeability of cell membranes, leading to cell leakage
Electric-field effects	Applying external pulsed electric field (PEF) causing increase in membrane permeability and leakage of cell contents
Magnetic-field effects	Applying oscillating magnetic field impulses induces changes in ionic drift across the plasma membrane. May affect growth and reproduction of bacteria
Chemical preservatives (traditionally propionates, sorbates, benzoates, nitrites)	Adding to food retards growth of, or kills, microorganisms. Mainly bacterio- or fungistatic so will not preserve food indefinitely
Naturally occurring compounds, e.g. lactoperoxidase system, avidin, lysozyme	Formation of antimicrobial compounds naturally by enzyme activity or interference with transport or causes cell-wall degradation

3 Use of biologically based systems and probiotic bacteria, e.g. use of lactic acid bacteria (Montville 1997b).

Food dehydration

The basic principle of food preservation by dehydration is that microbial growth is inhibited if the water available for growth is removed, i.e. if the a_w is reduced. Among bacteria, gram-negative species have the highest a_w requirements, while the gram-positive, nonsporing bacteria are less sensitive to reduced a_w. Some yeasts and molds have lower a_w requirements than the bacteria.

Removal of moisture from a food by natural sun drying is the oldest food-preservation technique used by humans. The method is still used to preserve large quantities of food in many developing countries, particularly those with tropical climates. In modern food processing, however, dehydration is accomplished in more controlled environments such as by the use of hot air (spray drying, fluidized bed drying, etc.), by contact with a hot surface (roller drying), or by sublimation of ice from frozen food (freeze drying). In addition to the inhibition of growth of microorganisms, dehydration also reduces the bulkiness of foods, making them more economical to store and transport. However, drying techniques have disadvantages: the acceleration of undesirable chemical reactions that occur in foods during exposure to high temperatures. Some foods withstand dehydration better than others. Freeze drying has the least undesirable effects on the food quality, but it is the most expensive and energy-intensive drying process.

It is important to control carefully the handling of foods both before and after the drying operation in order to prevent microbial spoilage and/or the spread of organisms that would lead to food-borne illness. The ingredients for dried food products must be of good microbiological quality and they should be stored under conditions that minimize microbial growth before the drying process. Some foods, such as meats, require heat treatment prior to drying. Cooking will reduce the moisture content by approximately 20 percent and will also reduce the number of viable bacteria present. Similarly, the blanching of fruits and vegetables or the pasteurization of liquid egg before dehydration will also reduce the total number of viable bacteria in the final product.

The bacterial numbers in dehydrated foods are usually lower than those found in the food before drying due to the inactivation of some, but not all, of the organisms by the process itself. Of the surviving bacteria, a proportion may be sublethally injured. The extent of cell death and cell injury occurring during the process will depend on the type of food (fat content, pH, presence of inhibitors, previous treatment history), the organism (species, strain, physiological age, state of the cells, and the concentration of cells), and the type of drying process.

Freeze drying is the most gentle of freezing methods, so it is likely to result in the greatest survival of bacteria. Thus, for example, the total bacterial levels in freeze-dried shrimps may be reduced by only one or two $\log_{10}$ units. If the shrimp had 10^6 bacteria per gram at the start of the freeze-drying process, the level at the end is likely to be of the order of 10^4 organisms per gram. In general, gram-negative bacteria such as *Pseudomonas* spp., *Escherichia coli*, and *Vibrio* spp. do not survive freeze drying as well as gram-positive bacteria. Cells of *Lactobacillus acidophilus* that survived freeze drying revealed damage to both cell membrane and cell wall (Brennan et al. 1986); cellular DNA may also be damaged.

During spray drying, finely aerosolized liquid food is brought into contact with hot air to remove the moisture. Bacteria present are affected by the aero-solization, heating, and drying. There may be changes in the outer layer of the cell envelope, cytoplasmic constituents may leak out, and they may lack control of ion transport (Banwart 1987). Survival of bacteria in the spray-drying process is related to the outlet temperature (Miller et al. 1972; Thompson et al. 1978). During the spray drying of liquid egg, a lower temperature during drying together with rapid cooling of the product after drying enhance the qualities of the dried egg product. However, the same conditions are favorable for the survival of salmonellae. Over 99 percent of salmonellae cells are killed during spray drying if the inlet temperature is maintained at $121°C$ and the outlet temperature at $60°C$ (Banwart 1987). Although dried fish products can be produced with bacterial loads as low as 10^2 cfu/g, commercial dried fish may have bacterial loads as high as 10^7 cfu/g.

All the bacteria present in dried foods are not inactivated during the drying process so growth of the surviving viable and sublethally injured bacteria will be initiated when environmental conditions become appropriate. It is therefore critical to maintain product integrity by storing dried foods under the correct conditions. Dried foods must be stored in moisture-impermeable containers under low humidity to prevent unacceptable increases in the a_w. Salmonellae that survive the drying of eggs are inactivated during the storage of the dried product even at low temperatures. In dried albumen, salmonellae can be inactivated by storage of the product at $50–60°C$ for up to 7 days. Generally, bacteria survive better in dried foods under vacuum than in the presence of oxygen.

Chilling and freezing

Clarence Birdseye in the USA and Bill Heeney in Canada pioneered the frozen-food industry in North America. The rapid expansion in frozen-food technology paralleled the engineering developments in the refrigeration industry. Although frozen foods are expensive to process and store, food-quality deterioration is greatly diminished by freezing compared with dehydration and heat sterilization.

The obvious major difference between refrigerated food and frozen foods is the temperature at which the food is held. Another important difference is that the water in the refrigerated food is still available for growth of psychrotrophic spoilage bacteria and pathogens, i.e. it still has a high a_w.

The freezing of food makes water unavailable to microorganisms by immobilizing it in the form of ice. The low temperature at which frozen food must be held also deters the growth of microorganisms. Freezing is not effective in inactivating bacteria, although some reduction in bacterial content is achieved. Gram-negative bacteria are more susceptible than gram-positive bacteria to freezing-induced injury and death.

Several factors influence the effectiveness of freezing as a preservation technique. These include the initial microbial load, the physical dimensions of the product to be frozen, the packaging material, the freezing rate, the time and temperature of subsequent storage, the degree of fluctuation in the storage temperature, and the time–temperature conditions used for thawing the product. The freezing process can be divided into three stages:

1 cooling down from the initial temperature of the food product to the temperature at which freezing begins
2 chilling at the latent heat plateau when there is no temperature change but phase change of water to ice occurs
3 further cooling to the final storage temperature.

The initial cooling of the food may inactivate a proportion of the microorganisms present due to cold shock; the decrease in temperature also reduces the growth rate of psychrotrophic bacteria. During the phase transition from water to ice, bacteria may be killed as a result of mechanical injury to the cell wall and cell membrane through the formation of ice. The growth of most cold-tolerant microorganisms is inhibited during further cooling until it ceases at about $-8°C$. Slow freezing to a final temperature of $-10°C$ is more lethal to bacteria than rapid freezing to $-20°C$. During slow freezing, ice crystals form and soluble solids are concentrated that affect the stability of cellular proteins. However, there are also adverse effects on the texture and structural integrity of the food.

The freezing process may cause inactivation of 10–60 percent of the total bacterial population depending on the characteristics of the food and the organisms present and the conditions of freezing. Bacterial spores are unaffected by freezing; and neither are bacterial toxins inactivated. It is possible, therefore, for food intoxication to result from consumption of a frozen food in which the toxin had been produced by bacterial growth prior to the freezing process.

Food preservation by heat treatment

HEAT STERILIZATION IN HERMETICALLY SEALED CONTAINERS

The technology of the conventional canning process may be summarized as filling food into containers of metal, glass, or thermostable plastic, or into multilayered flexible pouches, sealing the containers, and then applying the heat treating process. This process is sometimes called 'terminal sterilization' and is designed to destroy large numbers of *Clostridium botulinum* spores should they be present. It also reduces the chances of survival of spores of spoilage organisms that are even more heat-resistant that *C. botulinum* spores. After heating, the

containers must be cooled and handled in a manner that will assure container integrity and avoid contamination resulting from leakage.

The sterilization parameters vary greatly and depend on the food composition. One of the most important factors in determining the correct time–temperature combinations to use for a particular food is the pH of the product. Low-acid foods (pH >4.5) and foods of a_w >0.85 are likely to support the germination and outgrowth of spores of *C. botulinum* and the subsequent production of botulinal toxin. Specific regulations exist, therefore, in most countries for the processing of low-acid foods in hermetically sealed containers. The minimum heat treatment process for these foods is a 12D reduction process based on the thermal death times determined for each specific food product. The 12D reduction process is sufficient to reduce the probability of survival of *C. botulinum* spores to no more than 1 in 10^{12} cans (containers) processed.

The manufacturing process must include appropriate good manufacturing practices before the heat-sterilization step. The heat-sterilization step will not inactivate preformed heat-resistant toxins. Therefore, the quality of the ingredients must be evaluated and their handling must be designed to prevent bacterial contamination and/or growth. If acidification of the product is possible, i.e. it does not adversely influence the organoleptic characteristics of the food, it should be included because acidified food products require much less heat treatment than low-acid foods to assure their safety. Examples of food products that are amenable to acidification include artichoke hearts, pimientos, and many pickles.

The D and z values were described earlier in this chapter. Another term that figures in the calculation of heat treatment parameters for foods is the F value. This is defined as the time required to inactivate a given number of bacteria at a specified temperature and is usually 121°C for spores and 60°C for vegetative cells. F60 is the thermal death time (TDT) at 60°C and F121 the TDT at 121°C. In 1922, Esty and Meyer indicated that the F121 for *C. botulinum* spores was 2.78 min. Thus, the generally accepted minimum treatment for achieving a 12D reduction of *C. botulinum* spores is 3 min at 121°C. The thermal treatments applied to commercially canned cured meat products is well below this because the heat treatment is supplemented by the inhibitory effect of curing salts on the germination of spores of *C. botulinum*. Low-acid canned uncured meats are treated at an F121 of 6 min. Using the D, F, and z values, the rate of heat penetration in a specific food, the pH of the food, and the temperature of the pressurized container in which the cans of food are heated, a thermal process for that specific food can be calculated. Experimentally inoculated packs are processed at the calculated thermal process conditions to evaluate the adequacy of the process. *Clostridium sporogenes* strain PA3679 is the usual test organism in inoculated pack studies because the spores of this non-toxigenic strain have a greater heat resistance than those of *C. botulinum*.

Container integrity is essential for maintaining the safety of commercially sterilized food products and for preventing spoilage. If the cans develop leaks, air or water may enter and introduce spoilage bacteria. In some instances the water may also carry pathogenic bacteria. Effective control measures to prevent post-processing contamination of cans include assuring the proper function of sealing equipment, disinfection of the water used to cool the containers after heat treatment, maintaining the containers in a dry condition as far as possible, and avoiding abuse of the containers.

Heat-sterilized foods in cans are subject to a number of different kinds of spoilage. Spoilage of the food may occur prior to processing or may be due to an inadequate heat treatment or to post-process contamination. Spoilage resulting from inadequate heat treatment is due to heat-resistant spores while inadequate cooling or a high product storage temperature leads to spoilage by thermophilic bacteria. Spoilage following water leakage into the can usually involves nonspore-forming bacteria, yeasts, and molds. The visual signs of spoilage may be bulging of the can at one or both ends, although the growth of the 'flat sour' organisms (*Bacillus coagulans* and *Bacillus stearothermophilus*) does not result in bulging of cans. Occasionally cans may swell due to the effect of acid foods on poorly coated internal can surfaces; the hydrogen produced causes the swelling.

ASEPTIC PACKAGING

In aseptic packaging, the food is sterilized, cooled, and transported under aseptic conditions to the presterilized container. The product is filled into the container and sealed under aseptic conditions. The end product is thus a hermetically sealed container holding a commercially sterile food that can be stored at ambient temperature for prolonged periods.

PASTEURIZATION

Heat treatment of foods below the temperatures needed for sterilization is termed pasteurization. Generally, this type of heat treatment is used to prolong the shelf-life of specific foods that do not withstand the sterilization process. Pasteurization is usually used to inactivate specific groups of bacteria, particularly pathogenic bacteria, and to reduce the numbers of spoilage bacteria. The process is most commonly applied to milk and dairy products, but in addition pasteurization is used to control microorganisms in liquid egg products, alcoholic beverages, smoked fish, and high-acid products (fruit juices, pickles, sauerkraut, and vinegar).

Food regulations in developed countries require that all liquid, frozen, and dried whole egg, yolk, and white be pasteurized to destroy viable salmonellae. However, the temperature and time requirements vary from country to country, ranging from 62°C for 150 s to 68°C for 180 s.

Irradiation

The process of food irradiation uses high-energy electromagnetic radiation (γ-rays) from radioactive isotopes such as caesium-137 or cobalt-60, or electrons from linear accelerators (Banwart 1987). These types of radiation are chosen because they produce the desired effects with respect to the food, they do not induce radioactivity in foods or packaging materials, and they are commercially economical (Farkas 1997). Radiation treatment of foods can have a range of effects depending on the level of the dose applied. Low doses of only 1–2 kGy can inhibit sprouting of vegetables (potatoes and onions) and delay ripening of fruits and vegetables. Higher doses (1–5 kGy) destroy parasites and pathogenic bacteria in meats, poultry, and fish, reduce viable bacteria in dried products such as spices and herbs, and destroy spoilage organisms to extend the shelf-life of perishable and semiperishable foods. The highest doses (20 kGy and above) can produce commercially sterilized foods (WHO 1988).

The ionizing radiation doses required for the destruction of microorganisms in foods will vary widely and are dependent on the type and numbers of cells, type of substrate or suspending medium, moisture content, oxygen tension, temperature, or the presence of protective compounds. Ionizing radiation causes cell death directly by attacking the DNA and causing single- or double-strand breaks or indirectly by the primary water radicals $^-$H, $^-$OH, and e_{aq}^-. The most important in DNA damage is the OH radical: those formed in the hydration layer around the DNA molecule are responsible for 90 percent of the DNA damage (Farkas 1997). The resistance of an organism to ionizing radiation is, in part, dependent on the capacity of the organism to repair the damage to its DNA. Most can repair single-stranded breaks, but the more sensitive organisms (e.g. *E. coli*) cannot repair double-stranded breaks. Bacterial spores are highly resistant to ionizing radiations due to the low water content of the spore protoplast; reported D values (decimal reduction dose) for spores of *C. botulinum* type A are 2.18–2.35 kGy in cured ham. Generally, gram-positive bacteria are more resistant than gram-negative bacteria: *Pseudomonas* spp., *E. coli* and *Proteus* spp. are very sensitive to ionizing radiation. Microbial radiation resistance in frozen foods is about two- to three-fold higher than at ambient temperature due to immobilization of the free radicals.

Although ionizing radiation has the potential to enhance the quality of our food supply and to increase its safety, it has not gained wide acceptance for use in food processing because of the public's fear of anything 'nuclear.' However, its use is permitted in a number of countries with doses up to 10 kGy for extension of refrigerated shelf-life and inactivation of nonsporing pathogens for poultry, pork, and sausages (WHO 1994). Ionizing radiation treatment has been approved in the USA for fruits, vegetables, spices, poultry, and pork, but it is not commercially used. In the UK, regulations (Statutory Instrument 1990a, b) came into force in 1991 introducing food labeling and a licensing system for food-irradiation plants including cover of all irradiated food from the processing plant to the supermarket shelf. The regulations state that food must be of good microbiological quality before it is irradiated.

Hurdle technology

Manipulation of the effect of a number of environmental factors together to inhibit microbial growth, for example pH, salt concentration, and temperature, is the basis of multiple 'hurdle technology.' Instead of setting one environmental factor to the extreme limit for growth, hurdle technology 'de-optimizes' a variety of factors. Whereas growth of food-borne pathogens is limited at a water activity of 0.085 or a pH of 4.6, similar inhibition may be obtained at pH 5.2 and a_w 0.92. In maintaining intracellular pH and accumulating compatible solutes in the low a_w, i.e. in maintaining homeostasis, the cells channel energy needed for biosynthesis. Thus their growth is inhibited. If this energy demand exceeds the energy producing capacity of the cells, they die (Montville 1997b).

BACTERIOLOGY OF SPECIFIC FOOD COMMODITIES

Meats

Reviews of the microbiology of meat and meat products are given by Davies and Board 1998 and ICMSF 1998.

Meat can be defined as the flesh of animals used as food and now tends to exclude poultry and fish. It is primarily the skeletal musculature tissue but commonly includes various organs that are regarded as edible (this may vary between countries). Thus meat comprises the musculature of 'red-meat' mammals, most commonly bovines, ovines, and porcines. Depending upon country and religion, the meat supply may be beef, pig, lamb, goat, camel, buffalo, deer, or horse. Within a species, the proportions of muscle, fat, and connective tissues can vary from one breed to another, and with age and rearing conditions, particularly the energy content and quantity of feed. Young animals deposit mainly protein; fat deposition increases with physiological age at the expense of protein. The amount of connective tissue varies in different muscles: those with little are tender to

eat even after mild cooking (e.g. grilling), whereas those containing larger amounts are considered tough unless subjected to more severe and prolonged cooking to gelatinize the collagen. With increasing age the collagen of the musculature becomes more heavily cross-linked. A proportion of cross-links are heat-stable, causing meat from older animals to be less tender unless cooked thoroughly (e.g. by cooking under pressure). Some fat is necessary for maximum eating quality, particularly for succulence, and for the full development of the cooked meat flavor, but there is sufficient intramuscular fat for these purposes in all but unusually lean animals.

The contractile elements of the muscle, the myofibrils, also contribute to meat texture. The ultimate pH of meat is proportional to the amount of lactic acid formed during postmortem glycolysis, which is dependent on the amount of glycogen in the muscles at death. In a muscle rested before slaughter almost all the glycogen is converted to lactic acid (a concentration of approximately 1 percent lactic acid corresponds to pH 5.5), giving a muscle of normal appearance and texture. Low glycogen levels result in muscle of high ultimate pH. Such muscles often appear dark and, in addition, are firm and dry (DFD). In stress-susceptible animals, especially certain cross-bred pigs, excitement or stress before slaughter causes rapid conversion of muscle glycogen to lactic acid, resulting in a low pH before the muscles have cooled. This denatures sarcoplasmic protein and reduces the water-holding capacity of the tissue, giving a pale, soft, exudative (PSE) condition.

Meat comprises contractile myofibrillar elements and soluble sarcoplasmic proteins with up to one-quarter by weight connective tissue and as much as one-third fat. The presence of connective tissue seems unimportant microbiologically, but the relevant properties of the fat differ considerably from those of the muscle.

Intensive rearing of cattle, sheep, and pigs has resulted in a large international trade in meat. This international trade developed initially by transporting carcasses but, in transporting prime cuts, has become predominantly an intermediate stage between the carcass and the retail-sized portion. Muscle contains approximately 75 percent water and a variety of substrates for microbial growth, including carbohydrates, amino acids, lactic acid (Dainty et al. 1983; Lawrie 1985; ICMSF 1998). The a_w of meat is high, approximately 0.99, which makes it a suitable growth medium for most microbes. Only if most of the water is removed (e.g. by drying) will the a_w be reduced sufficiently to affect microbial growth.

Within the normal pH range of meat (approximately 5.4–7.0), values approaching 5.4 are less favorable to the growth of many of the important bacteria; the lower pH values are particularly important in maintaining the microbiological stability and safety of cured meat products.

The redox potential of meat has been claimed to be important in determining the nature of microbial spoilage, but its exact role in permitting or preventing the growth of microbes has not been fully elucidated, probably due to unresolved experimental problems of controlling and measuring redox potential. Tissue respiration of oxygen continues after death; because the supply of oxygen via the blood has ceased, the oxygen content and redox potential of muscle fall, leading to anaerobic conditions even at only a few millimeters below the surface. However, microbial growth in the deep musculature is rare, except where refrigeration is inadequate (Gill 1979; Roberts and Mead 1986).

MICROBIOLOGY AT PRODUCTION

In industrially developed countries, the change in eating patterns and in technological developments in animal husbandry, meat production, food processing, and preservation have led to an enormous increase in the range of meat products available. Novel methods have also been developed to maximize the recovery of meat from the carcass. Many meat products spend weeks, or even months, in distribution and storage before being offered for sale alongside the fresh product. The lower the number of microbes present initially, the longer the shelf-life of the perishable product providing storage conditions are correctly controlled.

Bacterial contamination of carcass surfaces can occur during slaughter and dressing procedures from a variety of sources, such as the hide, intestinal contents, contact surfaces, and handling by workers. Fecal contamination, routinely experienced during the normal slaughtering process, contributes high levels of bacteria (typically >10 000 cfu/cm^2). Good sanitary practices during slaughter, trimming of visible fecal contamination, and washing of carcasses with or without sanitizers are the approaches usually employed to minimize the contamination of carcasses (Dickson and Anderson 1992). The addition of chlorine, acetic acid, or lactic acid to the wash water reduces the bacterial load on the surface but does not significantly increase the shelf-life or completely eliminate pathogenic bacteria (Prasai et al. 1995). Higher water pressures (>13.8 bar) during spray washing are more effective in reducing bacterial loads on the surfaces of carcasses. Also, a two-step washing procedure with a low-pressure, hot-water (72°C) wash followed by a high-pressure, low-temperature (30°C) wash has been reported to be effective in reducing surface bacterial contamination (Gorman et al. 1995; Dorsa et al. 1996).

The microbial flora of the surface of freshly slaughtered carcasses, usually in the range of 10^1–10^3 cfu/cm^2, is primarily mesophilic, having originated in the alimentary tract and on the external surfaces of the live animal. Contamination from the slaughtering environment is also predominantly by mesophiles; psychrotrophs originating from soil and water are also present, but at much lower levels. Usually carcasses are cut into smaller portions in refrigerated work rooms, where most bacteria on the surfaces of processing equipment are psychrotrophic in nature. During refrigerated holding, the microflora of meats begins to shift towards psychrotrophs of the Pseudomonas–Acinetobacter–Moraxella group. These organisms are ultimately responsible for the spoilage of refrigerated meats. The spoilage is primarily a surface phenomenon resulting in the formation of slime and off-odor. The shelf-life of refrigerated meats is extended by controlling the factors contributing to the growth of psychrotrophs, namely moisture levels at the surface of the meat, initial load of psychrotrophs, pH, oxygen tension, and temperature. Wrapping meat in oxygen-impermeable films retards surface growth and favors the growth of microaerophilic bacteria, such as lactobacilli and Brochothrix thermosphacta, over the Pseudomonas–Acinetobacter–Moraxella group. The bacterial load of chilled meats at the retail market may bear no relationship to the load at the processing plant because the psychrotrophs continue to grow during transportation and storage.

The microbiology of meat other than the musculature is relatively poorly researched, with few publications on the microbiology of edible offal such as heart, liver and kidney, despite their increasing use in meat products. The initial contamination of offal is often higher and there is more likely to be contamination with potential pathogens than in skeletal muscle. The chilling process is often done poorly and offal may be packed while still warm before transport to chillers or freezers. Slow cooling results in significant increase in microbial numbers. Contacting surfaces remain moist and may be an important factor in the lower incidence of Campylobacter jejuni (C. coli) on retail carcass meats compared with offal.

Coliforms, E. coli, enterococci, Campylobacter spp., Staphylococcus aureus, Clostridium perfringens, L. monocytogenes, Y. enterocolitica, and Salmonella spp. are often present on fresh tissues because the slaughtering process does not currently include a bactericidal step. The enterohemorrhagic or verotoxigenic strain (EHEC or VTEC) E. coli O157:H7 is a recently emerged pathogen that has caused major food-borne disease outbreaks, with undercooked ground beef, particularly as hamburgers, commonly implicated. Enterohemorrhagic E. coli, of which E. coli O157:H7 is the most frequently encountered serotype (although other serotypes including O26:H11, O103, O104, and O111 have been implicated in cases of bloody diarrhea), have the ability to adhere intimately to human intestinal cells by an attaching and effacing mechanism and produce one or more phage-encoded Shiga-like toxins. In addition to hemorrhagic colitis, E. coli O157:H7 may cause serious life-threatening clinical diseases such as hemolytic–uremic syndrome and thrombotic thrombocytopenic purpura in children and the elderly. *E. coli* O157:H7 has a low infectious dose (estimated to be of the order of 10–100 cfu) and can be spread by person-to-person contact in day-care settings (Griffin 1995).

Cattle are primary reservoirs of *E. coli* O157:H7. The carcass may become contaminated during slaughter and dressing operations. Meat from a single carcass may contaminate large batches of ground beef because of the tendency to mix and blend meat from several animals from different farms during production. Witholding feed, as is normally done before shipping or slaughter, to reduce fecal contamination, can lead to intestinal outgrowth and colonization by both *E. coli* O157:H7 and salmonellae due to the disruption of normal fermentation in the rumen (Rasmussen et al. 1993). Shedding of salmonellae and *E. coli* O157:H7 by feedlot cattle after shipping appears to be directly related to their length of stay in pens (National Animal Health Monitoring System 1995). *E. coli* O157:H7 have also been isolated from lamb products (Chapman et al. 2000).

Few studies have been conducted on the prevalence of pathogenic bacteria at different locations on pork carcasses. Epling et al. (1993) reported that the prevalence of *Salmonella* spp. and *Campylobacter* spp. on shoulder and ham surfaces was as great or greater after 24 h of chilling at 4°C when compared with the numbers at slaughter. Swine are the main reservoir of *Y. enterocolitica* where the organism appears to have a predilection for the tonsils and tongue. Careful removal is needed to prevent carcass contamination. Outbreaks of *Y. enterocolitica* infections have occurred in the USA caused by raw chitterlings (pork intestines); adults handle the chitterlings and then transmit the organisms to their infants and children (Lee et al. 1990). Consumption of undercooked pork in some European countries has also been implicated (Tauxe et al. 1987; Ostroff et al. 1994).

Salmonellae, campylobacters, motile aeromonads, and *Y. enterocolitica* have been isolated from lamb carcasses sampled at commercial abattoirs (Sierra et al. 1995; Chapman et al. 1997). Horizontal transmission of infection and persistent shedding (for 2 weeks) of *E. coli* O157:H7 were demonstrated after oral inoculations of lambs with 10^5–10^9 cfu of the organism. As observed in cattle, withholding of feed from lambs causes increased shedding of *E. coli* O157:H7 (Kudva et al. 1995). In nature, however, sheep appear to harbor *E. coli* O157:H7 and other enterohemorrhagic *E. coli*.

Irradiation treatment has the potential to reduce contamination of meat with pathogens signficantly, while the shelf-life of the meats is also prolonged. The effects will vary with animal species and type of packaging. Irradiation of ground beef in oxygen-permeable packaging with 1 kGy of γ-rays will reduce the numbers of *Salmonella* spp. *Y. enterocolitica*, and *Campylobacter* spp. but may cause objectionable odor, color, and flavor changes, but there are no odor problems with pork.

In zvacuum-packaged beef a 2-kGy treatment at 25°C eliminates pseudomonads, enteric bacteria, and enterococci and causes fewer odor and flavor problems (Lee et al. 1996).

SHELF-STABLE RAW SALTED AND SALT-CURED MEATS

Salting of foods is one of the oldest methods of food preservation. Salting of meats was extremely common practice before the introduction of home refrigeration units and raw salted products continue to be produced and sold, especially in predominantly agricultural areas. Salt pork, dry cured bacon, and country-cured hams are still produced in volume. In the process of dry curing, pieces of meat are coated with salt and stored in bins below 10°C. Special care must be taken to ensure that the salt is applied to all folds and crevices in the meat to prevent bacterial growth leading to spoilage. Water is released from the meat by the action of the salt. At intervals, the pieces of meat are recoated with salt. At the end of the salting period, the product contains high levels of salt and is racked and held at ambient temperature until the surface dries. Finally, the meat is rubbed with salt and spices, netted, and sold. With such a high salt content, the product does not require refrigeration. The main microbial flora of these products after curing and drying are members of the Micrococcaceae, with lactic acid bacteria and yeasts as minor components. *Staphylococcus aureus* can be found in low numbers. Products sold as country-cured hams or bacon are similarly processed, except that sodium nitrite is used in combination with a lower level of sodium chloride. The product is hung to dry at ambient temperature for 35–140 days before removal for sale as a shelf-stable product. During the curing process, the salts penetrate and equilibrate in the tissue. The combined action of reduced water activity and the preservative effect of nitrite inhibit or inactivate the bacteria that cause spoilage or human illness. During the subsequent drying period, salt-tolerant enterococci and micrococci begin to grow and predominate; they also suppress the growth of the undesirable bacteria.

PERISHABLE RAW SALTED AND SALT-CURED MEAT

Fresh sausages are the most commonly encountered perishable raw salted meat products, but the salt content used is not sufficient to make the product shelf-stable. The method of packaging is the major factor that determines the predominant spoilage flora. Fresh sausage sold in bulk on trays or stuffed in casings is highly perishable, with a shelf-life of just a few days; psychrotrophic pseudomonads are the primary spoilage organisms. The shelf-life of fresh sausage can be extended to 1 or more weeks by wrapping the sausage in oxygen-impermeable film. The restriction of oxygen favors the growth of lactic-acid-producing bacteria, which impart a tangy flavor to the product. Large quantities of perishable meat products are cured with salt, sodium nitrite and nitrate, ascorbic acid, and other flavoring materials. Most of these are cooked before shipping from the processing plants, although some are sold with little or no heat treatment and must be cooked before eating. Some raw cured meats are heated or smoked to different degrees to produce dry surfaces and a smoked flavor; these procedures may also extend their shelf-life by reducing the numbers of bacteria.

PERISHABLE COOKED UNCURED MEATS

Most pork products are given a heat treatment sufficient to destroy all nonspore-forming bacteria present, with only spores at a level of approximately 10^2/g surviving. Beef products are usually processed at a lower temperature sufficient to destroy nonspore-forming pathogens, but not necessarily some of the thermoduric bacteria such as enterococci. Microbial levels in the final product depend on the initial microbial load and types of bacteria, the cooking time and temperature, and the storage temperature and time. In meats that are only lightly cooked, i.e. where the center is still essentially raw ('rare'), the center temperature may not have been sufficient to eliminate even relatively heat-sensitive organisms, including those of concern such as *Salmonella* spp., *Campylobacter* spp., *E. coli*, and *Y. enterocolitica*. The color change in meat from red to gray associated with cooking occurs at temperatures approaching 60°C. Freshly cooked meat products may contain only about 10^2 cfu/g, but post-cooking handling and packaging may result in contamination with spoilage bacteria and pathogens. Cooked uncured meat products are ideal substrates for microbial growth. They are highly nutritious, they have favorable pH and a_w, and the cooking process reduces or destroys the competing microflora. Many of these foods are frozen for shipment and distribution. However, if they are held at above freezing point for 1 to several days, spoilage will occur due to the growth of enterococci, pseudomonads, lactic acid bacteria, and other psychrotrophs.

PERISHABLE COOKED CURED MEATS

Precooked cured meats include a variety of popular luncheon meats including frankfurters, bologna, and ham. The heating process applied destroys the normal meat microflora apart from spores and possibly thermoduric bacteria. During chilling, holding, and packaging, exposed surfaces may become contaminated. Salt and nitrite used in the curing process may inhibit the growth of any surviving or contaminating organisms. On prolonged storage at refrigeration temperatures, spoilage with slime formation occurs primarily due to enterococci, micrococci, lactic acid bacteria, and yeasts. If oxygen-impermeable film is used for wrapping the product, spoilage is primarily due to lactic acid bacteria.

Brochothrix thermosphacta can form a significant part of the spoilage flora of vacuum-packed products, but growth of this organism in relation to lactic acid bacteria is reduced by a lower nitrite concentration, lower pH, and lower packaging film permeability.

CANNED CURED MEATS

Canned cured meats may be shelf-stable or perishable. Shelf-stable products include canned wieners, corned beef, frankfurters, meat spreads, luncheon meat, small canned hams, canned sausages covered with oil, and vinegar-pickled meats. Stability and safety of these products depend on the combined effect of heating and other factors that will inhibit the growth of surviving spores (e.g. salt, nitrite, a_w) and a sealed container to prevent post-process contamination. Some of these products, e.g. canned wieners and frankfurters, are given a 'botulinum cook' under the regulations applicable for low-acid foods packed in hermetically sealed containers. Canned sausages covered in oil do not spoil unless the a_w is higher than recommended (ranging from 0.86 to 0.92, depending on the product) and the vacuum seal is broken. Pickled pigs' feet and pickled sausages are immersed in vinegar-brine and are preserved by low pH, acetic acid, little or no fermentable sugar remaining in the tissue, and/or an airtight package.

Perishable canned cured meats made from pork must be stored refrigerated. They contain nitrite and salt at levels used to prepare shelf-stable products, but their heat treatment is inadequate to inactivate spores. Perishable canned cured meats may be shelf-stable for up to 3 years if properly processed and refrigerated. Spoilage is primarily due to psychrotrophic, thermoduric nonspore-forming bacteria (e.g. enterococci, *Lactobacillus viridescens*) that were present at abnormally high levels or survived an inadequate processing.

FERMENTED AND ACIDULATED SAUSAGES

Fermented products such as Thuringer, summer sausage, pepperoni, Lebanon bologna, Genoa salami, and cervelat depend on lactic fermentation and low water activity for preservation. At the end of fermentation, lactic acid bacteria exceed 10^8 cfu/g. Previously, the primary safety concern for these products was the production of staphylococcal enterotoxins. Improved industrial practices have greatly reduced or eliminated this problem. However, these practices are not adequate to control acid-tolerant bacteria, such as verotoxigenic *E. coli*, which can survive the fermentation and drying process (Glass et al. 1992), as evidenced by outbreaks due both to serotypes O157:H7 (Anon 1995) and O111 (Paton et al. 1996; Desmarchelier 1997).

Use of starter organisms in the fermentation of these meat products allows a greater level of control of the final product. A naturally fermented unheated meat product was the vehicle of infection for an outbreak of

Salmonella serotype Typhimurium infection in the UK when the a_w and pH was insufficient to prevent survival of the organisms present in the raw meat (Cowden et al. 1989).

Spoilage of naturally fermented sausage is due primarily to fungi; *Penicillium* spp. are dominant but species belonging to *Aspergillus* and *Scopulariopsis* are also present (Anderson 1995). Acidulated products are produced by adding acidulants (citric acid, lactic acid, glucono-δ-lactone) to the meat. Because no fermentation is involved, the products require a higher heat treatment than fermented sausages.

DRIED MEATS

Commercial dried meat products, such as beef jerky, include a cooking step that destroys normal vegetative cells and a rapid drying step that reduces the water activity to a level at which microorganisms of concern cannot grow. Thus such products are microbiologically stable at ambient temperature. Some traditionally dried meats such as biltong may include a salting step and a lesser heat process during drying. In such products bacterial numbers, especially micrococci, yeasts, and molds, may be high.

Poultry and poultry products

Poultry production and processing range from small farm or yard operations in developing countries to almost exclusively large-scale, highly integrated, and automated operations in developed countries. Poultry meat is now one of the cheapest forms of meat and consumption has increased dramatically in the past 50 years. It is estimated that 38 million metric tons are produced annually (ICMSF 1998), of which 80 percent is chicken. To meet increased demand, production and processing operations have been integrated and centralized. Intensive production integrates breeding, hatching, rearing, feeding, and slaughtering. The large numbers and close proximity of birds make control of hygiene difficult, and introduction of pathogens into a flock has very serious consequences, although it also provides the opportunity for preventive medicine, efficiencies of sale, and genetic improvements.

In industrialized countries, the microbiological condition of raw poultry at retail sale is a reflection of hygienic measures taken at the breeder farm, at the hatchery, through the growing period, and throughout slaughtering and the subsequent handling and storage.

MICROBIOLOGY AT PRODUCTION

Although several pathogenic bacteria have been associated with outbreaks or sporadic disease caused by poultry products, *Salmonella* spp., *C. jejuni*, and *C. coli* are considered predominant poultry-associated pathogens (Bryan and Doyle 1995), although a thorough examination of raw poultry may yield many other pathogens, including *L. monocytogenes*, *S. aureus*, *E. coli*, and *C. perfringens*. Microbial contamination of the egg can occur during its development in the ovary or later by penetration of the shell after laying. *Salmonella* serotype Enteritidis can infect the ovaries or oviduct of apparently healthy birds, leading to internal contamination of eggs (Coyle and Palmer 1988; Humphrey 1994), but the frequency and extent of such infection are uncertain and the laying of contaminated eggs may be sporadic (Humphrey et al. 1989). Salmonellae may also be present on the egg shell due to fecal contamination and may penetrate the shell during cooling after laying. Chicks and poults may become contaminated from the shell during or after hatching. In contrast, eggs derived from hens that were fecal excreters of *C. jejuni* failed to yield this organism from either the homogenates of yolks and albumen or shell surfaces (Bryan and Doyle 1995). Thus, *C. jejuni* is unlikely to be transmitted by eggs.

Poultry-improvement programs for the genus *Salmonella* identify infected breeder flocks (USDA 1990; EC 1992; Statutory Instrument 1989a, b, 1993a). Infections may be controlled by treatment with antibiotics or by slaughter, particularly if serotype Typhimurium or Enteritidis is identified, and more recently by vaccination programs. Breeder stock should be given feed made free from salmonellae by heating (Williams 1981a, b) or by treatment with organic acids (Hinton and Linton 1988). Other sources of salmonellae, such as rodents and wild birds, should be excluded. New stock should be disease-free and should be quarantined before being mixed with other birds. Contamination of the shell has been reduced by the use of cages and automated feeding systems, belt collection of eggs, and manure removal equipment. Other improvements include in-line washers, mass egg candling, and automated packing equipment (Bell 1995).

The development of the microflora of the live bird is reviewed by Grau (1986) and the origins and composition of the contamination of the carcass by Mead (1980, 1982) and the ICMSF (1998).

Within 24 h of oral inoculation with *C. jejuni* a single chick can contaminate 70–100 percent of fellow chicks held in transport boxes. *C. jejuni* colonizes primarily the lower intestinal tract of chicks, principally the cecum, large intestine, and cloaca. *C. jejuni* are chemotactically attracted to mucin; highly active flagella assist them to move into mucus-filled crypts, where the organisms establish themselves. The colonization of poultry by *C. jejuni* and intervention strategies for prevention of the colonization have been described by Stern (1992).

Salmonellae have been isolated from the yolk sacs of chicks, setters, and chick belts sampled at hatcheries. The cecum is considered the principal site of colonization. Chicks and poults are readily infected with salmonellae from contaminated feeds or drinking water, or by pecking in contaminated soil or litter; infected chicks may shed as many as 10^8 cfu/g of feces. As they mature, the birds tend to shed fewer salmonellae and fewer birds remain infected unless salmonellae are continuously reintroduced through contaminated feed or other environmental sources (Bryan and Doyle 1995).

Live poultry showing no symptoms of illness may carry and excrete low numbers of salmonellae (symptomless excreters), thereby contaminating the environment. Litter becomes contaminated with droppings, feathers, and soil; with time and use, the water activity, ammonia content, and pH of the litter become unfavorable for the survival of salmonellae.

Transmission of microorganisms between birds continues during transportation from the farm to the slaughtering facilities. Intestinal colonization is the most important factor contributing to carcass contamination. Feed is usually withheld for several hours before transport to reduce the amount of fecal droppings. However, considerable defecation still occurs during transit. Birds stand, walk, and fall on the fecal material present on crates and truck beds, thus allowing fecal bacteria including pathogens to be transferred to the feathers and skin. Stress caused by gathering, transporting, crowding, and holding in crates greatly increases the bacterial load; the levels of *C. jejuni* on unprocessed chicken carcasses increase by 3–4 logs after transportation from the farm (Stern et al. 1995). Cleaning and disinfection of cages and truck beds after each use minimizes the transfer of contamination between flocks.

In countries where birds are killed after purchase by the consumer, the hygienic conditions of preparation may be primitive but the brief time between slaughter and consumption of the cooked product usually prevents most microbiological problems.

Most modern poultry-slaughter operations are largely automated with electrical stunning and mechanical severing of the blood vessels in the neck. Most dressing and processing steps are accomplished with minimal handling. Broilers entering the slaughtering plant are frequently contaminated with *Salmonella* spp., *C. jejuni*, *C. coli*, and *E. coli* (Kotula and Pandya 1995; ICMSF 1998). Dressing procedures are designed to reduce contamination, and controls such as chlorinated rinses of equipment are aimed at minimizing cross-contamination between carcasses.

Carcasses are scalded to facilitate the removal of feathers, the most common method being immersion scalding. Microbes from the exterior surface of the birds and from the intestinal and respiratory tracts are continually released into the scald water. Hot-water sprays, steam, and simultaneous hot-water spray and plucking are alternatives to immersion scalding but are rarely used. Build-up of contamination is prevented by controlled overflow and countercurrent replacement of scald water and by the lethal effects of the high temperature. High scalding temperatures (greater than 60°C) are more effective in eliminating vegetative microorganisms than are the lower temperatures commonly used (50°C), but the lower temperatures are sometimes preferred because they cause less damage to the appearance of the carcass (Slavik et al. 1995).

Plucking (picking) to remove feathers is another process that spreads bacterial contamination between carcasses. Effective cleaning of defeathering machines (pluckers) and the plucking fingers is an important step to prevent transmission of contaminants between carcasses. Spray washing of carcasses after plucking removes only loosely attached contaminating material together with a proportion of the microbial flora, many microbes remaining bound to the surface tissues. Removal of the viscera may lead to further spread of contamination over the carcasses, particularly with intestinal organisms, commonly as a consequence of gut breakage. With time, bacteria become attached more firmly to the skin, so post-evisceration washing should be undertaken as soon as possible and should include washing of both the body cavity and the outer surface of the carcass. Reduction of the levels of bacteria is usually limited to about 90 percent; the remaining bacteria are either entrapped within or adhere to skin and flesh surfaces.

Carcass chilling delays the growth of spoilage bacteria and prevents the growth of food-borne pathogens. Continuous countercurrent immersion chilling, with or without ice, generally decreases microbial numbers on carcasses and minimizes cross-contamination between the carcasses if it is correctly controlled and the water is properly chlorinated. Air chilling using various combinations of temperature, humidity, and time may also be employed. During such dry chilling contact between carcasses is less; however, there is also less dilution of pathogens. Campylobacters, however, are sensitive to drying; therefore, a reduction in the numbers of *C. jejuni* may be expected.

The number of microbes detected on the finished poultry carcass varies with both the site sampled and the method of sampling. Excised skin samples will yield a microflora of approximately 10^3–10^5 aerobic mesophiles/cm, 10^1–10^3 psychrotrophs/cm, and 10^3–10^4 Enterobacteriaceae/cm (Grau 1986; ICMSF 1998). The main causes of spoilage when carcasses are stored in air at approximately 1°C are pseudomonads, *Shewanella putrefaciens*, *Acinetobacter* spp., and *Moraxella* spp. At higher temperatures atypical lactobacilli and *Serratia liquefaciens* (formerly *Enterobacter liquefaciens*) may predominate. *S. putrefaciens* and acinetobacters grow better in leg muscle (pH 6–6.7) than in breast muscle (pH 5.7–5.9), whereas *Pseudomonas* spp. grow well in either. If carcasses are wrapped in oxygen-impermeable films, then *Alteromonas* spp., *B. thermosphacta*, and atypical lactobacilli are the principal causes of spoilage. Atmospheric concentrations of 10–25 percent carbon dioxide delay the growth of pseudomonads and other spoilage organisms when the product is held below approximately 4°C. The increase in shelf-life is approximately proportional to the concentration of carbon dioxide up to 25 percent, at which point discoloration occurs.

Irradiation of poultry has been extensively investigated with respect to its effects on spoilage and pathogenic bacteria. Treatment of whole eviscerated chickens with 2.5 kGy of γ-rays resulted in an increase in the shelf-life from 10 to 40 days. The same dosage reduces the numbers of salmonellae, campylobacters, and *Y. enterocolitica* by 2.5–4 $\log_{10}$ units. Irradiation is less effective for killing salmonellae on frozen carcasses than on chilled. The US Department of Agriculture issued regulations in 1992 permitting the irradiation of packaged, fresh or frozen poultry and poultry products, including ground and mechanically separated poultry products at a dosage level of 1.5–3.0 kGy (Lee et al. 1996). The UK regulations of 1990 (Statutory Instrument 1990b) permit the sale of specified types of irradiated food (seven types including poultry) where irradiation has taken place under licence. The regulations also specify the limits of overall dose of ionizing radiation which apply.

Eggs and egg products

Since the pioneering study of egg microbiology in 1873 by Gayon, an associate of Louis Pasteur, a number of investigators have shown that the hen's egg is endowed with many chemical and physical defences against microorganisms. The shell and its membranes provide a physical barrier to penetration in addition to possessing bactericidal activity (Board et al. 1994), while the antimicrobial components of the egg white make it an antagonistic medium for microbial growth. The yolk is an excellent medium for bacterial growth; either liquid whole egg or egg yolk permits rapid growth of bacteria if the temperature is appropriate.

Bacterial contamination of eggs can occur at different stages ranging from production to storage, processing, distribution, and preparation (Board and Fuller 1994; ICMSF 1998). Transovarian or 'vertical' transmission takes place when infection occurs during formation of the egg in the hen's ovaries. Horizontal transmission refers to contamination when eggs are exposed to microorganisms that subsequently penetrate the shell (Fajardo et al. 1995). *Salmonella* serotype Enteritidis (*S. enteritidis*) emerged as a transovarian pathogen in 1985 (St Louis et al. 1988) and has caused many egg-related outbreaks of salmonellosis in North America and Europe (Mishu et al. 1991, 1994; ACMSF 1993a, 2001; Humphrey 1994). Phage types (PT) 4, 8, and 13a are the three most frequently isolated phage types of *S. enteritidis*. PT4 causes more severe disease than the other two phage types and during the 1990s essentially displaced the other two phage types in Europe. In the USA the predominant types were PT8 and PT13a, and PT4 did not appear until 1993 (Mason 1994). *S. enteritidis* contamination of eggs may occur by both vertical and horizontal transmission. The frequency of salmonella contamination of shell eggs produced in the UK in 1991 was reported just under 1 percent (de Louvois 1993a, b). Despite extensive and costly measures adopted by the industry to address the problem, this had not improved by 1995 (ACMSF 2001).

The washing of shell eggs does not assure the total removal of bacteria. The temperature of the wash water is not high enough to effect any killing of bacteria, but its pH may affect bacterial growth. Gram-negative bacteria on the egg shell can be resistant to detergents in the wash water; even chlorine rinses of shell eggs may not prevent contamination by salmonellae.

Success in preserving safe and wholesome eggs depends on three steps:

1 preventing spoilage organisms from entering the egg
2 preventing pathogenic bacteria from entering the egg
3 maintaining egg quality by preventing the loss of CO_2 and water.

Refrigeration is the most popular method of preserving egg quality. The low storage temperature retards loss of CO_2 and slows the growth of pathogenic bacteria. Previously the recommended storage temperature for shell eggs in the USA was 13°C. However, *S. enteritidis* grows in eggs stored at 13°C but not at 7°C. Therefore, the 1995 Food Code currently requires the receipt of shell eggs at 5°C (Food Code 1995). The reported decline of infections in the USA (ACMSF 2001) may be explained by the interventions introduced in the egg and or meat/poultry industry (Department of Health and Human Services 1999). In Europe the marketing of eggs is controlled by EC Directives (EC 1989). These allow on-farm storage of eggs for a limited time at ambient temperature not exceeding 18°C before delivery. At retail the directives prohibit storage below 5°C to ensure quality standards are maintained. In the UK, it is recommended that eggs are stored in home refrigeration at less than 8°C (ACMSF 1993b).

Liquid whole egg is a blend of egg albumen and yolk and contains 23–25 percent solids. Liquid egg is used in bakeries, candy makers, and many catering establishments. Contamination of liquid egg products is generally due to contaminants within the shell egg, the cleanliness of the shells on breaking out, or the presence of small egg shell particles that often drop into the liquid. At the time of laying, the egg temperature is above 41°C. The cooling and drying of the moist egg surface, especially the cuticle, and the continued cooling of the contents result in a negative pressure inside the egg. Thus bacterial contaminants are pulled from the surface of the shell through the pores. Shell contaminants therefore make an important contribution to the microbial flora of the liquid egg; washing eggs in water is an effective method of removing shell contaminants. Spoilage bacteria occurring in liquid eggs are mainly gram-negative rods (*P. fluorescens*, *Alcaligenes bookeri*, *Paracolobacterium intermedium*, *Proteus melanovogenes*, *S. putrefaciens*, *Proteus vulgaris*, *Flavobacterium* spp.). Many *Salmonella* serotypes are associated with egg products. Both *L. monocytogenes* and *Y. enterocolitica* may also be found in liquid eggs (Stadelman 1994).

In the USA and in the European Union, all liquid, frozen, and dried whole egg, yolk, and albumen must be pasteurized or otherwise treated to reduce the number of viable salmonellae to very low levels. The US regulations require that liquid whole egg be heated to at least 60°C and held at that temperature for 3.5 min. The UK regulations specify a heat treatment of not less than 64.4°C for at least 2.5 min or another temperature/time combination that will give the same degree of destruction of vegetative pathogens (Statutory Instrument 1993b; EC 1989). Either temperature–time combination process significantly reduces the number of spoilage bacteria and pathogens without affecting the functional properties of liquid egg. As liquid egg white may have a pH of between 7.6 and 9.3, it should be stabilized by the addition of ammonium sulfate before applying the heating step. Processes using heat and hydrogen peroxide combinations are also acceptable (Baker 1994).

In 1993, a commercially produced mayonnaise was epidemiologically implicated as the source of *E. coli* O157:H7 in an outbreak of bloody diarrhea. However, the evidence suggested that contaminated meat was the more likely source of the organism. Subsequent studies have shown that *E. coli* O157:H7 is more acid-tolerant than other gram-negative enteric bacteria and can survive in refrigerated acidified foods, including mayonnaise. Interestingly, *E. coli* O157:H7 was rapidly inactivated in mayonnaise when it was stored at room temperature (Weagent et al. 1994; Zhao and Doyle 1994; Erickson et al. 1995).

Fish and fish products

The consumption of seafood products has increased dramatically in recent years. This increase has focused greater attention on both the quality and the safety of these products and has led to the development of hazard analysis critical control points (HACCP, see section on New approach to assuring the safety of foods) systems and inspections.

The term 'fish' is used commonly as both a specific term for 'finfish' and also as a generic term covering all edible aquatic and marine finfish, molluscan shellfish, and crustaceans. In this section the individual groups will be referred to by their specific names.

Finfish are generally regarded as more perishable than other high-protein-muscle foods. This high perishability is due primarily to the presence of high concentrations of free nonprotein nitrogenous compounds in fish muscle. These compounds, e.g. ammonia, urea, trimethylamine oxide, creatine, taurine, etc., are utilized actively by bacteria during spoilage, resulting in fishy, ammoniacal smells (Jay 2000). In addition, the perishability of cold-water fish is not effectively reduced by refrigeration because of the preponderance of psychrotrophs in their microflora.

The internal flesh of live, healthy fish is sterile; the natural bacterial flora reside primarily in the outer slime layer of the skin, on the gills, and in the intestinal tract of feeding fish. Bacterial numbers will depend on the temperature of the waters but will range from 10^2 to 10^6 cfu/cm on the skin, from 10^3 to 10^5 cfu/g on the gills, and from very few (nonfeeding fish) to 10^7 cfu/g or greater in the intestines. These initial microflora are directly related to the environment, whereas the total microbial load is subject to seasonal variation. Warmwater fish tend to have a more mesophilic, gram-positive microflora (micrococci, bacilli, coryneforms), whereas cold-water fish harbor predominantly gram-negative psychrotrophs (*Moraxella*, *Acinetobacter*, *Pseudomonas*, *Flavobacterium*, *Vibrio*). Regardless of the initial microflora, spoilage of finfish during iced storage is primarily caused by *Pseudomonas* spp. and *S. putrefaciens*.

The level of bacteria present on a fish is not a reliable indicator of its sensory quality or shelf-life. Gram-negative bacteria are the major causes of fish spoilage. Their metabolism produces the unpleasant odors and flavors associated with fish spoilage. *Shewanella putrefaciens* (previously classified in the genera *Alteromonas* and *Pseudomonas*) is the most important fish-spoilage bacterium in marine fish stored at 0°C; it produces H_2S and causes off-odors. Spoilage bacteria belonging to the family Vibrionaceae are important in spoilage of fish stored at temperatures higher than 0°C (Gram 1992).

The only bacteria that are both pathogens for humans and indigenous members of the normal microflora of the marine environment are *C. botulinum* and various *Vibrio* spp. Other bacterial species may contaminate fish taken from waters subject to human or animal pollution; most of the recognized food-borne pathogens have been isolated from finfish or shellfish.

CRUSTACEANS

Crustacean species of commercial importance include crabs, lobsters, shrimps, and prawns. Crabs and lobsters are generally trapped and transported live to the processing plant. Shrimps and prawns are usually captured by trawlers and iced for transport to the processing plant. As crustaceans may be caught close inshore, a proportion will be from polluted waters and will be contaminated with microorganisms derived from sewage.

Shrimps and prawns are common causes of food-borne illness involving *S. aureus*, *Salmonella* and *Shigella* species and *Vibrio parahaemolyticus*. They are harvested from coastal waters in many parts of the world and may be captured either by fishing from boats or by hand-netting from the shore. Increasingly they are grown in ponds (aquaculture). The microorganisms present will depend on where the crustacean shellfish are caught. There may be a variety of microbial pathogens originating from untreated sewage, such as *Salmonella* and *Shigella* species, or the organisms that are part of the normal flora of the animal, such as *V. parahaemolyticus* or *Vibrio vulnificus*. Crustaceans are extremely susceptible to microbial deterioration and ideally should be harvested from unpolluted water and cooled rapidly to a temperature of −1 to +2°C. The ice used for chilling should be of good microbiological quality; alternatively clean refrigerated seawater may be used. If water quality is suspect, it should be chlorinated or otherwise treated, but excessive use of chlorine or other biocidal agents should be avoided. On fishing vessels, holds and container boxes should be cleaned thoroughly between catches to prevent a build-up of contamination.

MOLLUSCS

Molluscan shellfish include bivalves, such as oysters, mussels, cockles, clams, and scallops, and gastropods such as whelks and periwinkles. With the exception of some oysters and scallops, which are harvested from deeper waters, most molluscs grow in estuarine and nearshore coastal waters where exposure to fecal contamination is both possible and likely. Clams and cockles are usually dug from the sand; oysters and scallops are either raked or trawled from the seabed; and mussels are hand picked. Oysters and mussels are frequently farmed. They are transported live in the shell, often with little refrigeration. Molluscan shellfish are filter feeders and can concentrate microorganisms from the water in their bodies. Hence the need for depuration in clean water, especially if the mollusc is to be consumed raw.

Vibrio spp. are important pathogens often associated with the consumption of clams and oysters. In 1989, a co-ordinated surveillance for human infections caused by *Vibrio* spp. in four US Gulf Coast states (Alabama, Florida, Louisiana, and Texas) revealed 121 cases of molluscan shellfish associated illness, of which 71 had gastroenteritis, 29 wound infections, and 14 primary septicemia. *Vibrio cholerae* non-O1, *V. hollisae*, *V. alginolyticus*, *V. fluvialis*, *V. parahaemolyticus*, and *V. vulnificus* were isolated from the patients (Levine and Griffin 1993). *V. vulnificus* causes wound infections and primary septicemia with high mortality in susceptible persons, particularly those with impaired liver function. Shellfish, particularly clams and oysters, taken from contaminated waters have also been incriminated as the source of outbreaks of infectious hepatitis A (Bryan 1980) and may serve to transmit other viruses, including caliciviruses, enteroviruses, noroviruses (Norwalk-type viruses), snow mountain agent, and non-A/non-B hepatitis viruses (Cliver 1971, 1988). Noroviruses may be the most frequent cause of shellfish-borne illness.

Toxins have been responsible for a number of seafood-associated outbreaks of illness in the USA, the UK, and elswhere. The principal intoxications having a microbiologial origin include paralytic shellfish poisoning (PSP), diarrhetic shellfish poisoning (DSP), neurotoxic shellfish poisoning (NSP), amnesic shellfish poisoning (ASP) (or domoic acid poisoning), ciguatera, and scombrotoxic (biogenic amine) fish poisoning (Scoging 1997). PSP, DSP, and NSP are caused by toxins produced by dinoflagellates, ASP by a diatom toxin, and all result from the consumption of bivalve molluscs that have been feeding on toxigenic algae, accumulating the toxin in their organs. Ciguatera toxin originates from toxic microalgae that are passed up the marine food chain from herbivorous to carnivorous fish, magnifying the toxin levels. The toxigenic organisms are not usually detectable in the seafood and identification depends on toxin analysis (Scoging 1997). Scombroid poisoning is caused by consumption of fish containing high levels of histamine (and other biogenic amines) resulting from the histidine decarboxylating activity of bacteria multiplying on fish after catching or during the food processing and preparation chain. The toxins are not denatured by cooking or eliminated by cleansing of shellfish in purification tanks. Hence the only available control measure is to prohibit collection of shellfish when, during periods of dinoflagellate blooms, the toxins in the shellfish approach dangerous concentrations.

Vegetables and fruits

Fresh vegetables and fruits are an essential part of the diet of people worldwide. If land is available, families grow their own fruits and vegetables. Otherwise, produce is purchased from local farmers or retail establishments. In developed countries, the production, processing, storage, and distribution of fruits and vegetables are highly integrated and are consolidated in a few commercial companies. This has resulted in year-round availability of such produce; efficient production and processing have also enabled the industry to meet the increased demand from a health- and nutrition-conscious public. However, these integrated practices have given rise to some new public-health problems.

VEGETABLES

Bacterial contamination of vegetables is a public-health concern because of soil microorganisms and the possibility of contamination from water used for irrigation and the various materials, including night soil, used as fertilizers. Several studies have shown that bacteria of public-health concern can be isolated from whole and prepared vegetables. Control of the quality of plant foods begins at production with suitable geographical location; choice of seed, fertilizers, and pesticides; effective use of irrigation or drainage systems; and crop rotation. In developed countries there is little obvious association between agricultural practices and food-borne illness. The relatively large scale of many agricultural operations results in minimal human or animal contact with crops and constitutes a useful safety factor. Practices such as prolonged cooking and rejection of bruised or rotted foods also minimize the consumption of potentially hazardous material. In developed countries the control measures for quality of plant foods and for minimizing product loss from bruising, browning, wilting, or rot result in foods with minimal microbiological hazards.

The nutrient content of vegetables is suitable for the growth of molds, yeasts, and bacteria (Jay 2000). The number of bacteria on vegetables upon arrival at a processing plant ranges from 10^4 to 10^7/g (ICMSF 1998); number of fungi on fresh vegetables at harvest ranges from 0 to 5×10^5, and at market from 5×10^3 to 2×10^6 (Brackett 1987). Fresh vegetables are inherently susceptible to spoilage because the a_w is above the minimum (0.90) needed to support the growth of most microorganisms, and the pH of most vegetables is within the range allowing multiplication. Washing can remove up to 90 percent of the surface microorganisms, depending on the smoothness of the surface, but those trapped on the vegetable remain and any residual water facilitates their rapid multiplication. Microorganisms multiply faster in cut produce because of the ready availability of nutrients and water. Further handling of the produce increases the likelihood of additional contamination from the handler, work surfaces, or utensils, which may have been contaminated previously by contact with other foods. Most fungal spoilage is caused by the genera *Penicillium*, *Sclerotinia*, *Botrytis*, and *Rhizopus*. A more complete listing is given by the ICMSF (1998). Fungal spoilage of fresh vegetables is considered more an economic than a health threat because few spoilage fungi produce toxic metabolites. Exceptions include aspergilli, penicillia, and *Alternaria*. Stress metabolites that are sometimes produced by plants in response to fungal infections can affect humans, e.g. furanocoumarin compounds, called psoralens, which are produced when *Sclerotinia sclerotiorum* produces 'pink rot' of celery. The effects of these compounds include blisters or lesions of the skin of consumers and those handling the diseased celery.

Bacteria cause a range of spoilage conditions known as rots, spots, blights, and wilts. Soft rots, which occur during transport and storage, are usually caused by coliforms, *Erwinia carotovora*, and certain pseudomonads. Bacteria causing spoilage other than soft rot include corynebacteria, xanthomonads, and pseudomonads (Lund 1983). Soft rot of potatoes by *Clostridium* spp. has also been reported (Lund 1972, 1986). These organisms break down pectins, causing a soft consistency, sometimes with an objectionable odor and a watery appearance.

Storage

Harvested vegetables and fruits continue to respire actively and produce heat. Handling and storage procedures therefore should minimize respiration and water loss and maintain an environment in which cells of the product remain healthy. Rapid chilling is important for leafy vegetables. Correct temperature and humidity control are essential in order to promote wound healing and to minimize spoilage of root crops. A weak link in the distribution chain for fruits and vegetables is the use of unsuitable containers or packaging material (e.g. large sacks, rough wooden boxes, bamboo baskets), which can result in crushing, bruising, and puncturing of the product. Long journeys, hot weather, unventilated trucks, heavy loads, and unpaved roads in developing countries are responsible for huge losses of fruit and vegetable harvest (Ryall and Lipton 1979; FAO 1981; Proctor et al. 1981).

The only means of control of spoilage microorganisms is by strict sanitation of equipment and control of the temperature, relative humidity, and gas composition of the atmosphere in which the raw vegetable is stored. Further examples of vegetable spoilage are given in Dennis (1987), ICMSF (1998), and Jay (2000).

Modified-atmosphere packaging or vacuum packaging is used to suppress the proliferation of aerobic spoilage microorganisms on vegetables. In addition, these processes reduce the rate of oxidative deterioration, enzymatic degradation, and water loss. Modified-atmosphere or vacuum packaging is used to market fresh-cut packaged vegetables including cabbage, lettuce, onions, green and red peppers, carrots, cauliflower, broccoli, and other leafy salad vegetables. Spoilage of ready-to-use salads is frequently caused by pectinolytic pseudomonads such as *Pseudomonas marginalis*.

Freezing is a preferred means of preserving vegetables, since both the organoleptic and the nutritional quality are better retained than by other processes. Before freezing vegetables it is normal practice to blanch, usually at 86–98°C for several minutes, to inactivate plant enzymes, thereby stabilizing the product during subsequent frozen storage. Blanching commonly reduces the microbial load 10^3–10^5 times. In most frozen vegetables, lactic acid bacteria are numerically dominant, although micrococci and gram-positive and gram-negative rods, including coliforms, comprise a considerable proportion of the total microflora of certain products. Microbial spoilage of frozen vegetables is rare, and they are rarely involved in food-borne illness because nonspore-forming pathogens do not survive blanching and most frozen vegetables are cooked before consumption.

Dried vegetables are shelf-stable and are rarely involved in food-borne illness although they can be a source of organisms, particularly spores of clostridia and bacilli, when rehydrated or when added in the dried form to other dishes. Modified atmospheres have been used to control fungal spoilage of vegetables, with mixed results. An increased carbon dioxide concentration is fungistatic for many fungi, including *Candida albicans*, *Botrytis cinerea*, *Botrytis parasiticus*, *Penicillium* spp., *Mucor*, and *Aureobasidium pullulans*, but there are examples of some vegetables being either more or less sensitive to increased concentrations of carbon dioxide, or off flavors developing due to the different gases used for preservation.

Vegetables can be contaminated with pathogenic microorganisms during growth, at harvesting, post-harvesting handling and storage, and distribution. Beuchat (1996) has published a comprehensive review on this subject. Food-borne disease outbreaks of bacterial origin are associated more frequently with vegetables than fruits. Among the bacterial pathogens of greatest concern in vegetables are *Shigella* spp., salmonellae, enterotoxigenic, enteropathogenic, enteroinvasive, and enterohemorrhagic *E. coli*, *Campylobacter* spp., *Y. enterocolitica*, *L. monocytogenes*, *Aeromonas* spp., *V. cholerae*, *S. aureus*, *Bacillus cereus*, *C. perfringens*, and *C. botulinum*.

Several large outbreaks of shigellosis have been traced to contaminated vegetables such as shredded or cut lettuce and fresh green onions. With the current popularity of fresh-cut packaged produce, there is a potential for causing large, geographically widespread outbreaks. *Shigella* spp. can survive on vegetables for several months at ambient and refrigerator temperatures.

Vegetables implicated as vehicles of *Salmonella* spp. in food-borne disease outbreaks include raw tomatoes and bean sprouts. *Salmonella* spp. can grow on raw tomatoes (pH 4.0) at 20–30°C. *L. monocytogenes* was established as a food-borne pathogen after a large outbreak of listeriosis in Nova Scotia in 1981 in which contaminated coleslaw was implicated (Schlech et al. 1983). The cabbage that was used to prepare the coleslaw had been grown on fields that were fertilized with contaminated sheep manure. In addition, the cabbage was stored at refrigeration temperatures over winter; this probably allowed the psychrotrophic pathogen to multiply. *L. monocytogenes* is a common natural contaminant of raw vegetables. The ability of the organism to survive and multiply at low temperatures makes it a difficult organism to control in the cool, moist environments of food-processing establishments. The organism can also exist in biofilms, which impede the process of cleaning and disinfection. The cut surfaces and exudate from some chopped vegetables, such as lettuce and cabbage, included in ready-prepared salads, both prepacked and from delicatessen counters, may provide an ideal substrate for the growth of *L. monocytogenes* (Beuchat and Brackett 1990a). The juice of other vegetables, for example carrots, may be inhibitory (Beuchat and Brackett 1990b). Many cases of travelers' diarrhea are attributed to enterotoxigenic *E. coli* acquired from contaminated salads eaten at restaurants. Salad-bar vegetables and lettuce have been implicated in instances of infection with *E.coli* O157:H7.

Vegetables sold in modified-atmosphere packages may pose a threat of botulism. *C. botulinum* type A spores have been isolated from shredded cabbage, chopped green pepper, and a salad mix. There is likely to be a hazard only if they have been prepared or stored under conditions that will lead to spore germination, cell multiplication, and toxin production. The presence of normal spoilage microflora in vegetables packaged under modified atmosphere may actually prevent the growth of *C. botulinum* and production of toxin. Packaging under modified atmosphere does not prevent the growth of *E. coli* O157:H7 on shredded lettuce or sliced cucumbers.

FRUITS

Losses in fresh-fruit harvests due to rots and other defects are usually the result of mold growth. Some molds are able to invade and infect the intact healthy tissue, whereas others can become established only after the fruit has been infected by a pathogen or has been damaged. Several diseases are named after the causative mould (e.g. *Alternaria*, *Botrytis*, and *Fusarium* rots) and others are derived from the appearance of the infected fruit (e.g. brown rot, stem-end rot). In some cases diseases of different fruits may have the same name despite the involvement of different molds (e.g. black rot is caused by *Physalospora obtusa* in apples and by *Alternaria citri* in oranges). The market diseases of fruits, and their causative fungi, are tabulated by the ICMSF (1998). Yeast spoilage of fresh fruits is usually from the fermentative action of, for example, *Hanseniaspora valbyensis*, *Candida* spp., *Candida (Torulopsis) stellata*, *Pichia kluyveri*, and *Kloeckera apiculata*.

At the orchard or vineyard pruning, certain cultural practices and application of fungicides are the main means by which infection and spoilage prior to harvest are minimized. After harvesting at optimum maturity, fruit should be handled gently to prevent physical injury and fruit-contact surfaces should be cleaned regularly and treated to destroy fungal spores. Moldy or bruised fruits are removed during sorting and grading. Fruit may be treated with hot water or with permitted fungicides to control, for example, *Phomopsis*, *Diplodia*, *Penicillium*, and *Botrytis* (Salunkhe and Desai 1984).

Storage

Fruits are stored cooled but at temperatures that do not cause chill injury. The concentration of carbon dioxide in the atmosphere may be increased to inhibit *Alternaria*, *Botrytis*, and *Rhizopus* spp. Fruit products can be preserved by heating to approximately 80–90°C because most aciduric microbes are relatively sensitive to heating. Spoilage of canned soft drinks has occurred because ascospores of *Saccharomyces cerevisiae* or spores of *Kluyveromyces marxianus* survived the heat process. Spoilage of canned-fruit products has occurred because ascospores survived the heating process, e.g. *Penicillium vermiculatum* and *Byssochlamys* (Splittstoesser 1987).

Freezing is often used to preserve fruits and fruit products and results in some reduction in microbial numbers. Growth of yeasts and molds does not occur if temperatures are maintained below −18°C, but exposure to automatic defrost cycles is undesirable in the long term because some multiplication may occur.

Reduction of the water activity is a common means of fruit preservation, either by natural exposure to the sun or by a controlled dehydration procedure. Fungal spoilage of dried fruit does not occur when the water content is below 25 percent. Some high-moisture dried fruits contain 30–35 percent moisture and are therefore heated or treated with potassium sorbate to prevent mold growth. Many fruits are treated with sulfite before drying to prevent browning and this destroys a high proportion of the contaminating microbes. Yeasts and molds able to grow on dried fruit include *Zygosaccharomyces bisporus*, *Zygosaccharomyces rouxii*, *Hanseniaspora*, *Candida*, *Penicillium* spp., *Aspergillus glaucus*, and *Aspergillus niger*. *Monascus bisporus* is one of the most xerotolerant molds and has caused spoilage of dried prunes (Splittstoesser 1987). The relationship between water activity and fungal growth is reviewed thoroughly by Corry (1987).

Pathogenic bacteria are not normally associated with fruit; however they may be present due to fecal contamination. Survival times of enteric pathogens on fruits will depend on pH and temperature. On high-acid fruits, e.g. lemons, it will be shorter than on low-acid types, such as melons. Growth is more likely to occur on cut or damaged surfaces.

Sliced fresh fruits support the growth of *Shigella* spp. and thus may serve as the vehicle of infection. Several outbreaks of salmonellosis have been associated with fresh fruits, particularly raw tomatoes and melons. *Salmonella* serotypes Oranienburg and Javiana have caused outbreaks associated with the consumption of contaminated watermelons; serotypes Chester and Poona have been linked with cantaloupes. Organisms can enter either through fissures or cracks in the fruit surface or via the blade or other implement used to slice the fruit. Outbreaks of *E.coli* O157:H7 were attributed to unfermented apple juice, orange juice, and cantaloupe (ICMSF 1998).

CANNED FRUIT AND VEGETABLE PRODUCTS

Preservation by canning is exploited to increase the shelf-life and availability of fruit and vegetables. The heat processing of these products and the importance of pH are discussed thoroughly elsewhere (Hersom and Hulland 1980; Lund 1986). Most vegetables intended for canning fall into the category of foods known as 'low-acid' because their pH is above 4.6. Low-acid foods must be heated sufficiently to destroy the spores of *C. botulinum*, which are inevitable contaminants of agricultural products, including vegetables. The canning process is a good example of where a high margin of safety is achieved by control of the process rather than by microbiological tests on the raw material and final product. The critical control points in the process, i.e. initial microbial contamination, time between filling and processing, cleanliness of equipment, heat process, cooling cycle and disinfection of cooling water, maintenance of container integrity by seam or seal control, and appropriate handling are outlined in ICMSF (1988). Leaker spoilage of canned vegetables is usually characterized by the presence of a mixture of microbial types, mainly bacteria, while with spoilage due to underprocessing there is often only one species present.

Canned vegetables that have been heat processed to commercial sterility may still contain viable spores that are unable to grow under the normal temperature of storage of the product, e.g. thermophilic spores of *B. stearothermophilus* and *B. coagulans*, which cause flat sour spoilage with production of acid but not gas. Thermophilic anaerobic spoilage may be caused by obligately thermophilic spore-forming anaerobes, such as *Clostridium thermosaccharolyticum*, which forms large quantities of hydrogen and carbon dioxide, or *Desulfotomaculum nigrificans*, which produces hydrogen sulfide, often with blackening of the food if iron is present. Thermophilic spoilage usually occurs after improper cooling of heat-processed vegetables (Hersom and Hulland 1980; Lund 1986; Dennis 1987; ICMSF 1988, 1998).

Cereals, nuts and seeds, and herbs and spices

Cereals are the main staple diet of a large proportion of the world's population, the main ones being wheat, maize, oats, rye, rice, barley, millet, and sorghum. They may be eaten directly after cooking, principally rice, and sometimes maize, or they may be converted into flours for further processing by baking into breads and biscuits or molding into doughs for pasta or noodles. They may be fermented as gruels, manufactured into western-style cereal products and eaten as breakfast foods, or fermented into beers and wines. The principal micro-organisms of concern are fungi and spore-forming bacteria.

The microbiology of cereals is summarized by Pitt and Hocking (1986) and considered in greater detail in ICMSF (1998). Cereals are of near-neutral pH and provide good nutrition for bacteria. Microbial activity is controlled by drying the harvested cereals to a sufficiently low a_w (below 0.70) at which bacteria are unable to multiply, and growth of molds is largely prevented in temperate zones. At the humidity levels common in tropical countries, mold growth is a serious problem. Historically, the molds able to spoil cereals have been grouped into 'field' and 'storage' fungi, but some (e.g. *Aspergillus flavus*) grow in both situations. Field fungi are plant pathogens that invade the crop as a consequence of a combination of favorable circumstances. Particular molds may cause problems in one geographical area, but not in another. For example in Japan an important disease of wheat and barley, (red mold disease) is caused by *Fusarium* spp., predominantly *Fusarium graminearum*, whereas in Australia *Fusarium* spp. appear to cause little problem in wheat, probably due to the drier climate.

Fungi occurring commonly on wheat, barley, and oats in Scotland and England include *Alternaria*, *Cladosporium*, *Epicoccum*, and *Penicillium* spp., while in Egypt wheat carries mainly *Aspergillus* and *Penicillium*, together with *Alternaria*, *Cladosporium*, and *Fusarium*, and barley mainly *Aspergillus* and *Penicillium*, together with *Rhizopus*, *Alternaria*, *Fusarium*, and *Drechslera* (Pitt and Hocking 1986).

Bacteria generally found on cereals are from the Pseudomonadaceae, Micrococcaceae, Lactobacillaceae, and Bacillaceae. The level of fecal contamination will be low in field grains unless there is considerable animal activity. Groups that are nearly always present include psychrotrophic bacteria (10^4 to $>10^7$ cfu/g), actinomycetes (up to $>10^6$ cfu/g), and aerobic sporing bacteria (10^0 to 10^5 cfu/g).

WHEAT

Wheat grows best in fairly dry and mild climates. After harvesting, it is dried and stored and may be transported long distances before final milling. Before consumption, milled wheat is cooked or baked. Spoilage by bacteria is not usually a problem because of the relatively low a_w, but wheat may act as a passive carrier for bacterial pathogens (ICMSF 1998). Some that are part of the normal flora such as *C. perfringens*, *B. cereus* and other *Bacillus* spp. can, under the right set of conditions, become harmful to man when consumed in a final cereal-based dish. Mold growth, spoilage, and the possible presence of mycotoxins are more serious hazards. Under particular weather conditions myco-toxins can be produced while the wheat is growing (Saito and Ohtsubo 1974; Ichinoe et al. 1983) or during storage if the a_w becomes sufficiently high.

Microbiology at production

The diverse surface microflora of wheat in the field does no damage. The grain may, however, be infected by *F. graminearum* to produce wheat scab, red mold disease, or fusarium head blight. Perithecia of *Gibberella zeae* (the perfect stage) maturing in the field liberate ascospores into the air, which infect wheat at the heading stage. Under high relative humidity and low temperatures *F. graminearum* grows and may produce trichothecenes (nivalenol or deoxynivalenol) and zearalenone (Saito and Ohtsubo 1974; Ichinoe et al. 1983; Hagler et al. 1984). There are unresolved geographical differences in the occurrence of trichothecenes, perhaps related to which strains of *Fusarium* occur (Osborne and Willis 1984). *A. flavus* and *A. parasiticus* are common contaminants of certain grains.

The crop is harvested as soon as possible after maturity, before rain can damage the grain (e.g. by causing sprouting). In a single operation, combine harvesters reap, thresh, clean, and load the wheat kernels into trucks for transport to storage. The rate and extent of subsequent drying of the crop depend on the moisture content at harvest (15–22 percent). It is important to harvest mature grain of known moisture content in dry weather without grain breakage and to separate kernels from hull and straw.

The moisture content of harvested wheat may be as high as 22 percent (a_w approximately 0.95), with the temperature reaching nearly 40°C. The grain must then be dried. Also it may need cooling in order to prevent spoilage and mold growth, to lower the respiration rate of the wheat, and to prevent overheating and damage of the grain. Wheat is dried initially to about 13 percent moisture (Zeleny 1971), to an a_w of less than 0.70, and later to 11–12 percent for long-term storage (ICMSF 1998). The higher the moisture content of the grain, the lower the air temperature must be, to prevent heat damage (Kent 1966).

Storage

After drying and before primary storage, wheat is cleaned by screening and aspiration to remove dust, broken and light grain, and foreign material. Broken grain is more susceptible to mold and insect attack (Williams 1983).

When grain is moved from primary storage to long-term storage in terminal elevators by truck or railcar, the vehicles are sources of insects or storage molds (Williams 1983). They may also be a source of contamination by salmonellae if the previous cargo was animals, meat scraps, or meat or fish meal (ICMSF 1998).

Cleaning and fumigation are necessary to keep transport vehicles clean and insect-free. Grain must be protected from rain. On grains at harvest time, mold propagules range from a few to 10^5/g and bacteria from 10^3 to 10^6/g. The bacteria are usually from the families Pseudomonadaceae, Micrococcaceae, Lactobacillaceae, and Bacillaceae. Indicators of fecal pollution are rare unless there has been animal access to fields. The presence of *Bacillus subtilis* and *B. cereus* has been demonstrated and other bacterial spores from the soil should be expected (e.g. *C. botulinum, C. perfringens*). Actinomycetes commonly exceed 10^6/g and psychrotrophs 10^4/g. Yeasts occur but do not cause problems during storage. During artificial drying, the temperatures achieved (40–80°C) destroy most mold spores. Sun drying is less effective in this respect.

Flour may be contaminated with spores of *B. subtilis* and *Bacillus licheniformis*, which can give rise to 'ropey' bread particularly if the preservatives commonly added in commercial bakeries are omitted. There is evidence associating large numbers of bacilli other than *B. cereus* with food-borne illness, especially in relation to foods with a flour-based sauce or encased in pastry (Kramer and Gilbert 1989).

RICE

Unlike wheat, rice (*Oryza sativa*) crops are largely grown in relatively small production units without great benefit of mechanization. Also, there is often much human and animal contact with the crop. However, salmonellae and other vegetative pathogens from such sources should be destroyed by the cooking process. Rice is often harvested and stored under hot, humid conditions that increase the risk of fungal attack and present a significant hazard from mycotoxins and spoilage losses. As the a_w is low, bacterial spoilage is rarely a problem. Spores of *B. cereus* survive cooking and have caused food-borne illness (Kramer and Gilbert 1989). The illness that has been recognized as usually associated with the consumption of rice contaminated with *B. cereus* has a rapid onset (1–5 h); the predominant symptoms are nausea, vomiting and malaise and are referred to as the 'emetic syndrome;' occasionally diarrhea may occur. Approximately 95 percent of 'emetic syndrome' episodes have been associated with the consumption of Cantonese-style cooked rice served by Chinese restaurants usually as take-away meals (Kramer and Gilbert 1989). The restaurants store cooked rice for reheating on demand at room temperature for long periods of time. There is a reluctance to store cooked rice in the refrigerator because the grains become sticky and clump together. Heat-resistant spores of *B. cereus* survive the initial cooking period and, during subsequent storage in the warm kitchen, germinate and multiply rapidly and produce toxin. The emetic toxin is extremely heat-resistant (Melling and Capel 1978) and will not be inactivated by the rapid cooking before consumption.

Fungi can invade rice in the field and spoil the developing grain (Fazli and Schroeder 1966; Bhat et al. 1982), while other fungi (storage fungi) attack the rice during drying and storage (Mallick and Nandi 1981; Mossman 1983). A variety of mycotoxins have been found in rice (Jarvis 1976; ICMSF 1998). It is crucial to reduce the a_w of rice quickly to approximately 0.65 to prevent mold growth and to maintain it at that level during storage even in unfavorable climates. Molds can penetrate the endosperm and discolor the kernels, but most mold growth and aflatoxin formation occurs in the bran layers that are removed during milling to produce white rice (Schroeder et al. 1968). Milling also tends to break moldy kernels, which can then be removed, thereby improving the microbial quality of the rice (Ilag and Juliano 1982).

Crop losses caused by plant diseases and insect infestations can be reduced by the planting of resistant varieties and by the application of insecticides (Mikkelsen and DeDatta 1980). Provision of adequate drying and storage facilities for crops harvested in the wet season is critically important (Greeley 1983; Mossman 1983).

Microbiology at production

Rice is harvested when the moisture content is 18–23 percent (DeDatta 1981). Delaying harvest increases field losses (shattering and lodging) and causes subsequent milling losses due to sun-cracking (Chandler 1979; Mikkelsen and DeDatta 1980). Early harvesting gives a more moist grain, which is difficult to hand-thresh and requires more drying.

In traditional tropical operations, rice stalks are pulled out by hand, bundled into sheaves for prethreshing drying, or threshed immediately (Mossman 1983). Manual harvesting causes less damage to the paddy, results in fewer weeds being incorporated into the crop, and allows the gathering of a higher percentage of mature grain. However, the longer the grain is left in the field for prethreshing drying, the greater the risk of cracking and attack by insect molds (Hall 1970). Drying is the most important step in production; if not undertaken or if delayed, the whole crop can spoil. Nevertheless, improper drying can cause grain breakage, partial spoilage, or spoilage losses in storage.

Most rice is sun-dried, and if done correctly this reduces the microbial population (Kuthubutheen 1984). The usual procedure in the tropics is to spread rice on a drying floor exposed to the sun for several days until the moisture level is 14–16 percent (a_w approximately 0.76–0.84). However, at this moisture content, rice cannot be stored safely so further drying during storage is needed. Traditional crops are harvested in the dry season at the time of maximum sun, but this process is often inadequate with increased yields and newer rice varieties that mature during the wet season. Artificial drying is then necessary, enabling harvests to be dried more uniformly with less loss from grain cracking, and to safe moisture levels (<13 percent moisture or a_w <0.70) regardless of the vagaries of the weather.

Storage

Significant losses can occur during storage if rice is not protected from rodents and insects, or if it is not dry enough to prevent mold growth. Mold attack is exacerbated by the high temperatures and humidities of tropical areas. At the start of storage, up to 85 percent of paddy rice kernels may be internally infected with field molds. During storage, field molds gradually disappear or are replaced by storage fungi (Mallick and Nandi 1981; Kuthubutheen 1984). The type and extent of growth of storage fungi depend on the initial load, temperature, a_w or relative humidity, and time (ICMSF 1998). Paddy rice can be safely stored at 30°C with a constant moisture content of 12.5 percent (a_w 0.67) for 12 months with no fungal growth (Mallick and Nandi 1981). A moisture content of 14.3–14.5 percent (a_w approximately 0.79) produces significant mold growth and deterioration within a few months.

Storage containers such as sacks or baskets give little protection from rodents or insects and may be a source of eggs, larvae, and mold spores (Williams 1983). Some modern rice varieties have thinner husks than traditional cultivars and are more susceptible to insect attack (Greeley 1983). Storage in bins or solid-walled containers provides protection from ingress of rodents, insects, and external moisture from rain or high humidity. However, failure to dry paddy rice to near 12 percent moisture (a_w 0.65) before, or soon after, bin storage causes considerable crop loss. Since rice insulates well, temperature gradients within the bin are readily established and lead to condensation and raised moisture contents in the cooler regions. The resultant increased metabolic activity and mold growth generate heat, carbon dioxide, and more water, resulting in 'hot spots,' which spread upwards and outwards. Bins made of metal are particularly susceptible to temperature gradients resulting from fluctuations in external temperatures.

Insect infestation is controlled by treating sacks, storage areas, and stocks of stored rice with insecticides or fumigants, and by keeping storage areas free of dust, fines, and spilt rice. Paddy rice is not generally transported long distances but transportation should be regarded as an extension of storage, with the same principles of preventing exposure to rain, dew, insects, and rodents.

Milling can improve the microbiological quality of rice. Much of the bacterial and mold contamination and aflatoxin, if present, is on the husk and in the surface layers of the kernel and will be removed by milling in the husk and bran fractions (Ilag and Juliano 1982). Grain heavily infested with mold will also break more readily during milling, yielding discolored, broken rice that is separated in the broken-grain fraction. Much dust is generated during milling; it may absorb enough moisture to allow microbial growth and can be a source of insects. It is essential that dust is controlled by regular sweeping, vacuuming, or aspiration.

The absence of husk makes milled rice more susceptible to insect attack. During storage it must be protected from insects, rodents, and increases in moisture content. Thus the same principles of control apply to milled rice as for paddy. Brown rice may also go rancid as a result of scratches in the bran, allowing lipases access to oils within it (Mossman 1983), possibly because of the presence and activity of a wide range of lipolytic bacteria and molds (DeLucca et al. 1978).

NUTS AND SEEDS

Nuts are dry, one-seeded fruits that are usually enclosed in a rigid outer casing or shell. They mostly grow on shrubs or trees (tree nuts) and include almonds, hazelnuts, pistachios, Brazil nuts, pecans, coconuts, and macadamia nuts. The peanut or groundnut is the only major nut that is a legume, but it tends to be grouped together with tree nuts.

Oilseeds from a range of botanical families are grown primarily for oil production and include palm nuts, rapeseed, sesame, sunflower, safflower, cottonseed, cacao seeds, and maize.

Many types of leguminous seeds are dried, such as a wide variety of beans and also soybeans.

The shells of nuts present an effective barrier to entry of bacteria during growth, and after natural drying the low a_w of most nuts restricts bacterial spoilage or toxin production. Bacterial contamination can occur after harvest, e.g. coconut flesh with salmonellae, with resultant concern over products with high a_w to which nuts are added. The high oil content of nuts and oilseeds results in high susceptibility to attack by lipolytic bacteria and spoilage fungi and possible mycotoxin production. Dried legumes, apart from soybeans, are low in oils so microbiologically are similar to cereals.

Microbiology at production

Tree nuts are usually dried in situ before harvesting. Coconuts are pierced or broken, drained of water, and the kernels cut into slices and sun dried. They may be further shredded as desiccated coconut. Peanuts are pulled from the ground and dried at least partially before threshing. There are few data on the microflora of nuts in the field. The shells provide a strong, protective barrier and tree nuts are essentially sterile before harvest apart from possible systemic fungal infection from the tree. Limited data on peanuts suggest that the flora is mainly field fungi, with *Fusarium* spp., *Lasio-*

diplodia theobromae, and *Macrophomina phaseolina* being dominant (MacDonald 1970).

Primary processing of nuts, seeds, and legumes usually involves natural drying. If there is good sun-drying, the initial microflora may be reduced, but if conditions are adverse, the mycoflora may increase in both kind and numbers and there may be mycotoxin formation. The bacterial flora of raw commodities, as would be expected, reflects the growing and harvesting environment and contamination from soil, equipment, etc. The fungal flora isolated from copra, cashew nuts, peanuts, and dried legumes are listed by the ICMSF (1998).

Storage

Stored dried nuts are very susceptible to spoilage because slight increases in moisture content lead to sharp rises in a_w due to the low levels of soluble carbohydrate and high lipid levels. These increases may be due to uneven storage temperatures; therefore, refrigerated storage is widely used to retard the development of rancidity. Increases in moisture may also lead to growth of spoilage fungi and mycotoxin production. The opening and cutting of coconuts allows contamination with bacteria and also many species of fungi.

A variety of bacteria have been isolated from nuts with damaged cells or contaminated with soil (King et al. 1970) including species of *Bacillus*, *Brevibacterium*, *Streptococcus*, and *Xanthomonas*, as well as *E. coli*. There may also be contamination with bacterial pathogens from animals. Contamination of hazelnuts with *C. botulinum* led to human cases of botulism from yogurt containing inadequately processed hazelnut puree (O'Mahoney et al. 1990). The principal problem with nuts and oilseeds lies with the potential for production of mycotoxins, notably aflatoxins. Peanuts, maize, and cottonseed are the three crops of greatest economic importance affected.

Salmonella has frequently been isolated from desiccated coconut (Gilbert 1982) and has been traced to unhygienic preparation at collection centers at the farms. The salmonella problem has largely been overcome as a result of the hygiene controls that have been implemented. However, a large outbreak of *Salmonella* serotype Java infection traced to a single batch of desiccated coconut was reported in the UK in 1998–99 (PHLS Communicable Disease Surveillance Centre 1999).

The most important step from the microbiological viewpoint in the further processing of nuts is roasting, which results in a large reduction in numbers and types of organisms including enteric pathogens. Control of mycotoxins in nut products consists essentially of control of the raw material or primary processing stages. Sorting and diverting discolored, rejected nuts is an important commonly used means to reduce mycotoxins to accep-table levels in finished products. End-product testing for mycotoxins is widely practiced.

HERBS AND SPICES

Spices are any of a variety of aromatic plant products used to season, flavor, or impart an aroma to foods. They may be fruits (peppers, allspice, coriander), arils (mace), flower buds (cloves), rhizomes (ginger), barks (cassia, cinnamon), or seeds (nutmeg, fenugreek, mustard, caraway, celery, and aniseed). Herbs are generally the leafy parts of soft-stemmed plants (oregano, marjoram, basil, curry leaves, mint, rosemary, and parsley) (ICMSF 1998).

Spices are of microbiological interest for a number of reasons:

- Antimicrobial activity and preservative properties. This activity, due to the essential oils, is of little significance in food preservation. Levels are generally too low to prevent microbial growth but can sometimes augment other microbial activities or act synergistically. Spices containing the most inhibitory of essential oils include cloves, thyme, oregano, cinnamon, allspice, cumin, and caraway. Yeasts are more readily inhibited than bacteria and the gram-positive bacteria are more sensitive than the gram-negative. The activity of a range of essential oils is listed by the ICMSF (1998).
- Stimulation of microbial metabolism.
- Tendency to go moldy at improper humidity and temperature.
- Contain large numbers of microorganisms that may cause spoilage or disease when introduced into foods.

Spices and herbs contain the organisms that are common to the soil and plants in which they are grown and that will survive the drying process. Other sources of contamination include dust, insects, fecal material from birds and rodents, and possibly water. Microbial counts will vary widely depending on original microbial load, opportunity for growth, subsequent die-off, stage of storage, origin of product, etc. Most of the microbial flora consists of aerobic mesophilic sporing bacilli originating from the soil as spores due to their ability to survive the drying process, most commonly *B. subtilis*, *B. licheniformis*, *B. megaterium*, *B. pumilus*, *B. brevis*, *B. polymyxa*, and *B. cereus*. A wide variety of nonsporing organisms may be found including coliforms, *E. coli*, fecal streptococci, and, rarely, staphylococci and lactic acid bacteria. The main components of the mold flora tend to be the *Aspergillus glaucus* group, *A. niger*, and *Penicillium* spp.

As the different spices vary widely in form they require diverse harvesting and post-harvesting methods; these will have a variable effect on their microbial content. The most important consideration is to dry the products rapidly to prevent spoilage while maintaining their desirable characteristics. ICMSF (1998) gives examples of some harvest and post-harvest treatments.

Spoilage of herbs and spices is mainly by molds, chiefly *Aspergillus* spp.; bacterial spoilage is rare. Insect infestation may aid the transfer of mold spores. However, other foods may be spoiled by the addition of herbs and spices bearing large numbers of bacterial spores; the degree of spoilage will depend on the amount of further processing or heat treatment that the final product receives.

Some of the spore-forming bacteria found in herbs and spices are capable of causing gastroenteritis (*B. cereus*, *B. subtilis*, *B. licheniformis*) (Kramer and Gilbert 1989). *C. perfringens* may also be present. As these spores will survive cooking temperatures, spices harboring these organisms must be considered as a potential health hazard if the foods to which the spices have been added are not properly prepared and handled. Salmonellae have also been found in a variety of spices and have been the source of a number of food-borne outbreaks, for example in black and white pepper and paprika. Aflatoxins have also been found.

NEW APPROACH TO ASSURING THE SAFETY OF FOODS

Food is essential to sustain life. People expect safe foods and expect their food supply to be protected and the safety and wholesomeness assured by governments. These expectations are the basis for food laws and food regulations, the inspection programs that monitor compliance with the laws and regulations, and the judicial system that punishes those who violate the laws. Until recently, governments took the responsibility for inspections of foods, food-processing establishments, and their environment. In several countries, government inspectors are assigned to meat, fish, and poultry processing plants. Inspectors of the US Food and Drug Administration (FDA) perform unannounced inspections of packaged or canned food-processing plants and examine finished product to monitor compliance; similar policies exist in other countries.

Currently the approach to assuring the safety of the food supply being implemented in the USA and several other developed countries is the HACCP system, a preventive approach designed in the 1950s to assure the safety of foods in manned space flights (Bernard 1997). The system goes beyond the limited effectiveness of sampling and examining finished goods.

HACCP is a systematic approach to the assurance of safety in food production. It consists of a number of sequential steps:

1 Hazard analysis.
2 Determination of critical control points (CCP).
3 Determination of the critical limits to be met at each identified CCP.

4 Establishment of procedures to monitor critical limits.
5 Establishment of corrective action to be taken when a deviation is identified by the monitoring.
6 Implementation of effective record-keeping-systems.

The concept adopted is based on understanding how a product becomes unsafe. Then control measures can be developed to prevent or detect such failures and keep them from reaching the consumer.

Hazard analysis is an assessment of risks associated with all aspects of food production from growing in the fields to consumption. The hazards may be physical, chemical, or biological. The biological hazard category includes bacterial, viral, fungal, and parasite contamination. Pathogenic bacteria are classified on the basis of the severity of risk: *C. botulinum*, *Salmonella* serotype Typhi, *E. coli* O157:H7, and *V. cholerae* are classified as severe hazards; *L. monocytogenes* and *Salmonella* serotypes other than Typhi are classified as moderate hazards with potential for extensive spread; *B. cereus* and *C. perfringens* are classified as moderate hazards with limited potential for spread. The hazard analysis also includes the ranking of food and its raw materials or ingredients according to six hazard characteristics followed by assigning a risk category to the food and its raw materials based on the hazard rankings.

Critical control point determination is required to control the previously identified hazards. A CCP is defined as any point or procedure in a specific food production system where a loss of control may result in an unacceptable risk.

Critical limits must be established that must be met at each identified CCP. For example, the thermal treatment in a retort in the production of low-acid canned foods is a CCP. The components of the CCP include pressure of the retort, retort temperature and time the food is held at the retort temperature.

Monitoring procedures must be established and also the corrective actions to be taken when a deviation is identified by monitoring critical components of a CCP. Implementation of *effective record-keeping systems* that document the HACCP plan is also an essential requirement of HACCP.

HACCP-based regulations were published first in the USA for low-acid foods in hermetically sealed containers and then later for smoked fish products. In February 1995, the Food Safety and Inspection Service of the United States Department of Agriculture (USDA) published proposed regulations (FSIS 1995). These essentially overhaul the meat and poultry inspection system and change it from a reactive, federal inspection-based approach to a proactive HACCP approach in which the manufacturer or processor assumes greater responsibility for food safety. Both the US FDA and the USDA have published final regulations mandating the development and implementation of

HACCP plans to cover a variety of domestic and imported foods (Bernard 1997).

The US initiatives have been paralleled in other countries. The European Union (EU) has adopted a number of product-specific directives that require specific foods introduced into commerce within EU countries to be produced in accordance with HACCP principles (e.g. fishery products, milk and milk-based products, meat products). In addition the EU has adopted a horizontal directive that requires consistency with HACCP principles for a range of food items (Roberts and Greenwood 2003). Countries outside the EU have also adopted HACCP-based food-safety control systems.

CONCLUSIONS

The processing and regulation of foods are, at the commencement of the twenty-first century, poised to undergo significant changes that will have profound effects on the quality and safety of the food supply. Changes in food-safety regulation are the driving force and changes in the microbiological quality of foods will occur as a result of the changes in the food-safety approach, as an extension of prevention-based HACCP implementation. For example, in July 1996, the US government published a sweeping reform of food-safety rules for meat and poultry (FSIS 1996). The four major elements of the new rules were:

- Every plant must develop and implement a system of preventive controls, known as HACCP, designed to improve the safety of their products. The plant must demonstrate the effectiveness of the HACCP plan, which federal inspectors will continually verify.
- Every plant must regularly test carcasses for *E. coli* of fecal origin to verify the effectiveness of the plant's procedures for preventing and reducing fecal contamination, which is the major source of pathogenic bacteria.
- All slaughter plants and plants producing raw ground meat and poultry products must ensure that the *Salmonella* contamination rate is below the current national baseline incidence.
- Every plant must develop and implement written sanitation standard operating procedures as the foundation for its specific HACCP program.

The time-frame for implementation by large processing plants was 18–42 months depending on the size of the plant.

As systematic programs to reduce and eliminate contamination of products of animal origin are implemented, food-processing plants will critically review their processing operations as part of the hazard analysis component of HACCP. Inevitably steps in their processing will be identified that will affect the keeping quality (related to microbial spoilage) of their products and appropriate control procedures will be implemented to enhance the shelf-life and microbiological quality of the foods.

Despite decades of effort by the food industry and food regulatory agencies, it has not been possible to raise food animals free of pathogenic bacteria. In addition, many more problems of food-borne disease, caused by contaminated fresh fruits and vegetables, are being encountered, possibly as a result of increased worldwide production and movement of these commodities around the globe to meet consumer demands for year-round supplies. The growing conditions for fruits and vegetables vary widely in different parts of the world, thus it may be very difficult to avoid contamination with pathogenic bacteria. At the same time, populations in many developed countries are becoming more susceptible to infectious diseases because of immune deficiencies caused by age, chronic diseases, and immune-depressing infectious diseases. How can consumers be assured of a safe food supply? At least for the foreseeable future, an effective approach seems to be the decontamination of the final raw product. The use of γ-irradiation or electron accelerators for terminal decontamination of food of plant and animal origin may be among the best available options (Corry et al. 1995). Unfortunately, the use of γ-irradiation has been very limited due to public misconceptions about the radiation treatment of foods and the resulting reluctance of the food industry to utilize the process. There are other logistical problems relating to the use of radiation treatment for foods, including the need for transport of foods from processing plants to a central radiation treatment facility, safety problems associated with the use of radioactive materials, and problems associated with the disposal of radioactive wastes. Electron accelerators are more convenient as they do not use radioisotopes. However, their primary limitation is the inability to penetrate more than 1–2 cm below the surface of the food. The ideal decontamination method for uncooked foods still remains to be developed.

ACKNOWLEDGMENTS

This chapter is an updated version of Swaminathan, B. and Sparling, P.H. 1998. The bacteriology of foods excluding dairy products. In Balows, A. and Duerden, B.I. (eds), *Topley & Wilson's Microbiology and Microbial infections*. Vol. 2, *Systematic Bacteriology*, 9th edn. London: Arnold, 395–416.

REFERENCES

ACMSF (Advisory Committee on the Microbiological Safety of Food). 1993a. Interim report on Campylobacter. London: Her Majesty's Stationery Office.

ACMSF (Advisory Committee on the Microbiological Safety of Food). 1993b. Report on *Salmonella* in eggs. London: Her Majesty's Stationery Office.

ACMSF (Advisory Committee on the Microbiological Safety of Food). 2001. Second report on *Salmonella* in eggs. London: Her Majesty's Stationery Office.

Anderson, S.J. 1995. Compositional changes in surface microflora during the ripening of naturally fermented sausages. *J Food Protect*, **58**, 426–9.

Anon, 1995. *Escherichia coli* O157:H7 outbreak linked to commercially distributed dry-cured salami – Washington and California, 1994. *MMWR*, **44**, 157–60.

Baker, R.C. 1994. Effect of processing on the microbiology of eggs. In: Board, R.G. and Fuller, R. (eds), *Microbiology of the avian egg*. London: Chapman & Hall, 153–73.

Banwart, G.J. 1987. *Basic food microbiology*, 2nd edn. Westport, CT: AVI Publishing Co.

Bell, D. 1995. Forces that have helped shape the US egg industry: the last 100 years. *Poult Trib*, **3**, 33–43.

Bernard, D.T. 1997. Hazard analysis and critical control point system. Use in controlling microbiological hazards. In: Doyle, M.P., Beuchat, L.R. and Montville, T.J. (eds), *Food microbiology. Fundamentals and frontiers*. Washington, DC: ASM Press, 740–51.

Beuchat, L.R. 1996. Pathogenic microorganisms associated with fresh produce. *J Food Protect*, **59**, 204–16.

Beuchat, L.R. and Brackett, R.E. 1990a. Survival and growth of *Listeria monocytogenes* on lettuce as influenced by shredding, chlorine treatment, modified atmosphere packaging and temperature. *J Food Sci*, **55**, 755–758, 870.

Beuchat, L.R. and Brackett, R.E. 1990b. Inhibitory effects of raw carrots on *Listeria monocytogenes*. *Appl Environ Microbiol*, **56**, 1734–42.

Bhat, R.V., Deosthale, Y.G., et al. 1982. Nutritional and toxicological evaluation of 'black tip' rice. *J Sci Food Agric*, **33**, 41–7.

Board, R.G. and Fuller, R. 1994. *Microbiology of the avian egg*. London: Chapman & Hall.

Board, R.G., Clay, C., et al. 1994. The egg: a compartmentalized aseptically packaged food. In: Board, R.G. and Fuller, R. (eds), *Microbiology of the avian egg*. London: Chapman & Hall, 43–61.

Brackett, R.E. 1987. Vegetables and related products. In: Beuchat, L.R. (ed.), *Food and beverage mycology*, 2nd edn. New York: Van Nostrand Reinhold, 129–54.

Brennan, M., Wanismail, B., et al. 1986. Cellular damage in dried *Lactobacillus acidophilus*. *J Food Protect*, **49**, 47–53.

Bryan, F.L. 1980. Epidemiology of foodborne diseases transmitted by fish, shellfish and marine crustaceans. *J Food Protect*, **43**, 859–868, 873.

Bryan, F.L. and Doyle, M.P. 1995. Health risks and consequences of *Salmonella* and *Campylobacter jejuni* in raw poultry. *J Food Protect*, **58**, 326–44.

Chandler, R.F. 1979. *Rice in the tropics*. Boulder, CO: Westview Press.

Chapman, P.A., Siddons, C.A., et al. 1997. A one-year study of *Escherichia coli* O157 in cattle, pigs, sheep and poultry. *Epidemiol Infect*, **119**, 245–50.

Chapman, P.A., Siddons, C.A., et al. 2000. A 1-year study of *Escherichia coli* O157 in raw beef and lamb products. *Epidemiol Infect*, **124**, 207–13.

Cliver, D.O. 1971. Transmission of viruses through foods. *Crit Rev Environ Control*, **1**, 551–79.

Cliver, D.O. 1988. Virus transmission via foods. *Food Technol*, **42**, 241–8.

Corry, J.E.L. 1987. Relationships of water activity to fungal growth. In: Beuchat, L.R. (ed.), *Food and beverage mycology*, 2nd edn. New York: Van Nostrand Reinhold, 51–100.

Corry, J.E.L., James, C., et al. 1995. *Salmonella*, *Campylobacter* and *Escherichia coli* O157:H7 decontamination techniques for the future. *Int J Food Microbiol*, **28**, 187–96.

Cowden, J.M., O'Mahoney, M., et al. 1989. A national outbreak of *Salmonella typhimurium* DT124 caused by contaminated salami sticks. *Epidemiol Infect*, **103**, 219–25.

Coyle, E.F. and Palmer, S.R. 1988. *Salmonella enteritidis* phage type 4 infection associated with hen's eggs. *Lancet*, **2**, 1295–6.

Dainty, R.H., Shaw, B.G. and Roberts, T.A. 1983. Microbial and chemical changes in chill-stored red meats. In: Roberts, T.A. and Skinner, F.A. (eds), *Food microbiology: advances and prospects*. London: Academic Press, 151–78.

Davidson, P.H. 1997. Chemical preservatives and natural antimicrobial compounds. In: Doyle, M.P., Beuchat, L.R. and Montville, T.J. (eds), *Food microbiology, fundamentals and frontiers*. Washington: ASM Press, 520–56.

Davies, A. and Board, R.G. 1998. *The microbiology of meat and poultry*. London: Blackie Academic & Professional.

DeDatta, S.K. 1981. *Principles and practice of rice production*. New York: John Wiley & Sons.

de Louvois, J. 1993a. *Salmonella* contamination of eggs, a potential source of human salmonellosis. *PHLS Microbiol Dig*, **10**, 158–62.

de Louvois, J. 1993b. *Salmonella* contamination of eggs. *Lancet*, **342**, 366–7.

DeLucca, A.J., Plating, S.J. and Ory, R.L. 1978. Isolation and identification of lipolytic microorganisms found on rough rice from two growing areas. *J Food Protect*, **41**, 28–30.

Dennis, C. 1987. Microbiology of fruits and vegetables. In: Norris, J.R. and Pettipher, G. (eds), *Essays in agricultural and food microbiology*. Chichester: John Wiley & Sons, 227–60.

Department of Health and Human Services, Food and Drug Administration. 1999. Food labeling: safe handling statements: labeling of shell eggs: refrigeration of shell eggs held for retail distribution; proposed rule. Preliminary regulatory impact analysis and initial regulatory flexibility analysis of the proposed rule to require refrigeration of shell eggs at retail and safe handling labels; proposed rule. *Federal Register*, **64**, 36491–516.

Desmarchelier, P.M. 1997. Enterohaemorrhagic *Escherichia coli* – the Australian perspective. *J Food Protect*, **60**, 1447–50.

Dickson, J.S. and Anderson, M.E. 1992. Microbiological decontamination of food animal carcasses by washing and sanitizing systems: a review. *J Food Protect*, **55**, 133–40.

Dorsa, W.J., Cutter, C.N., et al. 1996. Microbial decontamination of beef and sheep carcasses by steam, hot water-spray washes, and a steam-vacuum sanitizer. *J Food Protect*, **59**, 127–35.

EC (European Community). 1989. Council of the European Communities Directive No. 89/437/EEC. Hygiene and health problems affecting the production and placing on the market of egg products. *Offical Journal of the European Communities*, **L212**, 87–100.

EC (European Community). 1992. Council of European Communities Directive No. 92/117/EEC concerning measures for protection against specified zoonoses and specified zoonotic agents in animals and products of animal origin in order to prevent outbreaks of food-borne infections and intoxications. *Official Journal of the European Communities*, **L62**, 38–48.

Epling, L.K., Carpenter, J.A. and Blankenship, L.C. 1993. Prevalence of *Campylobacter* spp. and *Salmonella* spp. on pork carcasses and the reduction effected by spraying with lactic acid. *J Food Protect*, **56**, 537–40.

Erickson, J.P., Stamer, J.W., et al. 1995. An assessment of *Escherichia coli* O157:H7 contamination risks in commercial mayonnaise from pasteurized eggs and environmental sources, and behavior in low-pH dressings. *J Food Protect*, **58**, 1059–64.

Esty, J.R. and Meyer, K.F. 1922. The heat resistance of spores of *B. botulinus* and allied anaerobes XI. *J Infect Dis*, **31**, 650–63.

Fajardo, T.A., Anantheswaran, R.C., et al. 1995. Penetration of *Salmonella enteritidis* into eggs subjected to rapid cooling. *J Food Protect*, **58**, 473–7.

FAO (United Nations Food and Agricultural Organization). 1981. *Food loss prevention in perishable crops*. Rome: FAO Agricultural Services Bulletin no. 43.

Farkas, J. 1997. Physical methods of food preservation. In: Doyle, M.P., Beuchat, L.R. and Montville, T.J. (eds), *Food microbiology, fundamentals and frontiers*. Washington, DC: ASM Press, 449–519.

Fazli, S.F.I. and Schroeder, H.W. 1966. Effect of kernel infection of rice by *Helminthosporium oryzae* on yield and quality. *Phytopathology*, **56**, 1003–5.

Food Code. 1995. *1995 Recommendations of the United States Public Health Service, Food and Drug Administration*. Washington, DC: US Department of Health and Human Services.

FSIS (Food Safety and Inspection Service). 1995. Pathogen reduction; hazard analysis and critical control point (HACCP) systems. *Federal Register*, **60**, 6774.

FSIS (Food Safety and Inspection Service). 1996. Pathogen reduction; hazard analysis and critical control points (HACCP) systems. *Federal Register*, **61**, 38806.

Gilbert, R.J. 1982. The microbiology of some foods imported into England and Wales and through the port of London and London (Heathrow) Airport. In: Kurata, H. and Hesseltine, C.W. (eds), *Control of the microbiological contamination of foods and feeds in international trade: microbiological standards and specifications*. Tokyo: Saikon Publishing Company, 105–19.

Gill, C.O. 1979. Intrinsic bacteria in meat. *J Appl Bact*, **47**, 367–78.

Glass, K.A., Loeffelholz, J.M., et al. 1992. Fate of *Escherichia coli* O157:H7 as affected by pH or sodium chloride in fermented dry sausage. *Appl Environ Microbiol*, **58**, 2513–16.

Gorman, B.M., Morgan, J.B., et al. 1995. Microbiological and visual effects of trimming and/or spray washing for removal of fecal material from beef. *J Food Protect*, **58**, 984–9.

Gram, L. 1992. Evaluation of the bacteriological quality of seafood. *Int J Food Microbiol*, **16**, 25–39.

Grau, F.H. 1986. Microbial ecology of meat and poultry. In: Pearson, A.M. and Dutson, T.R. (eds), *Advances in meat research: meat and poultry microbiology*. Westport, CT/London: AVI Publishing/Macmillan, 1–47.

Greeley, M. 1983. Solving third world food problems: the role of post-harvest planning. In: Lieberman, M. (ed.), *Post-harvest physiology and crop protection*. New York: Plenum Press, 515–35.

Griffin, P.M. 1995. *Escherichia coli* O157:H7 and other enterohemorrhagic *Escherichia coli*. In: Blaser, M.J., Smith, P.D., et al. (eds), *Infections of the gastrointestinal tract*. New York: Raven Press, 739–61.

Hagler, W.M. Jr., Tyczkowska, K. and Hamilton, P.B. 1984. Simultaneous occurrence of deoxynivalenol, zearalenone, and aflatoxin in 1982 scabby wheat from the midwestern United States. *Appl Environ Microbiol*, **47**, 151–4.

Hall, D.W. 1970. *Handling and storage of food grains in tropical and sub-tropical areas*. FAO Agriculture Development Paper no. 90. Rome: FAO.

Hersom, A.C. and Hulland, E.D. 1980. *Canned foods, thermal processing and microbiology*, 7th edn. London: Churchill Livingstone.

Hinton, M. and Linton, A.H. 1988. Control of *Salmonella* infections in broiler chickens by the acid treatment of their feed. *Vet Rec*, **123**, 416–21.

Humphrey, T.J. 1994. Contamination of egg shell and contents with *Salmonella enteritidis*: a review. *Int J Food Microbiol*, **21**, 31–40.

Humphrey, T.J., Baskerville, A., et al. 1989. *Salmonella enteritidis* phage type 4 from the contents of intact eggs. A study involving naturally infected hens. *Epidemiol Infect*, **103**, 415–23.

Ichinoe, M., Kurata, H., et al. 1983. Chemotaxonomy of *Gibberella zeae* with special reference to production of trichothecenes and zearalenone. *Appl Environ Microbiol*, **46**, 1364–9.

ICMSF (International Commission on Microbiological Specifications for Foods). 1980. *Microbial ecology of foods*, vol. 1 *Factors affecting the life and death of microorganisms*. New York: Academic Press.

ICMSF (International Commission on Microbiological Specifications for Foods). 1988. *HACCP in microbiological safety and quality*. Oxford: Blackwell Scientific.

ICMSF (International Commission on Microbiological Specifications for Foods). 1998. *Microorganisms in foods 6. Microbial ecology of food commodities*. London: Blackie Academic & Professional.

Ilag, L.L. and Juliano, B.O. 1982. Colonization and aflatoxin formation by *Aspergillus* spp. on brown rices differing in endosperm properties. *J Sci Food Agric*, **33**, 97–102.

Jarvis, B. 1976. Mycotoxins in food. In: Skinner, F.A. and Carr, J.G. (eds), *Microbiology in agriculture, fisheries and food*. London: Academic Press, 251–67.

Jay, J.M. 2000. *Modern food microbiology*, 3rd edn. New York: Chapman & Hall.

Kent, N.L. 1966. *Technology of cereals with special reference to wheat*. London: Pergamon Press.

King, A.D., Miller, M.J. and Eldridge, L.C. 1970. Almond harvesting, processing and microbial flora. *Appl Microbiol*, **20**, 208–14.

Kotula, K.L. and Pandya, Y. 1995. Bacterial contamination of broiler chickens before scalding. *J Food Protect*, **58**, 1326–9.

Kramer, J.M. and Gilbert, R.J. 1989. *Bacillus cereus* and other *Bacillus* species. In: Doyle, M.P. (ed.), *Foodborne bacterial pathogens*. New York: Marcel Dekker, 21–70.

Kudva, I.T., Hatfield, P.G. and Hovde, C.J. 1995. Effect of diet on the shedding of *Escherichia coli* O157:H7 in a sheep model. *Appl Environ Microbiol*, **61**, 1363–70.

Kuthubutheen, A.J. 1984. Effect of pesticides on the seed-borne fungi and fungal succession on rice in Malaysia. *J Stored Prod Res*, **20**, 31–40.

Lawrie, R.A. 1985. *Meat science*, 4th edn. Oxford: Pergamon Press.

Lee, L.A., Gerber, A.R., et al. 1990. *Yersinia enterocolitica* O:3 infections in infants and children, associated with the household preparation of chitterlings. *New Engl J Med*, **322**, 984–7.

Lee, M., Sebranek, J.G., et al. 1996. Irradiation and packaging of fresh meat and poultry. *J Food Protect*, **59**, 62–72.

Levine, W.C. and Griffin, P.M. 1993. *Vibrio* infections on the Gulf Coast: results of first year of regional surveillance. *J Infect Dis*, **167**, 479–83.

Lund, B.M. 1972. Isolation of pectolytic clostridia from potatoes. *J Appl Bacteriol*, **35**, 609–14.

Lund, B.M. 1983. Post-harvest pathology of fruits and vegetables. In: Dennis, C. (ed.), *Post-harvest physiology of fruits and vegetables*. London: Academic Press, 219–57.

Lund, B.M. 1986. Anaerobes in relation to foods of plant origin. In: Barnes, E.M. and Mead, G.C. (eds), *Anaerobic bacteria in habitats other than man*. London: Academic Press, 351–72.

MacDonald, D. 1970. Fungal infection of groundnut fruit before harvest. *Trans Br Mycol Soc*, **54**, 453–60.

Mallick, A.K. and Nandi, B. 1981. Research: rice. *Rice J*, **84**, 8–13.

Mason, J. 1994. *Salmonella enteritidis* control programs in the United States. *Int J Food Microbiol*, **21**, 155–69.

Mead, G.C. 1980. Microbiological control in the processing of chickens and turkeys. In: Mead, G.C. and Freeman, B.M. (eds), *Meat quality in poultry and game birds*. Edinburgh: British Poultry Science Ltd, 99–104.

Mead, G.C. 1982. Microbiology of poultry and game birds (processed carcass). In: Brown, M.H. (ed.), *Meat microbiology*. London: Applied Science Publishers, 67–101.

Melling, J. and Capel, B.J. 1978. Characteristics of *Bacillus cereus* emetic toxin. *FEMS Microbiol Lett*, **4**, 133–5.

Mikkelsen, D.S. and DeDatta, S.K. 1980. Rice culture. In: Luh, B.S. (ed.), *Rice: production and utilization*. Westport, CT: AVI Publishing, 147–234.

Miller, D.L., Goepfert, J.M. and Amundson, C.H. 1972. Survival of salmonellae and *Escherichia coli* during the spray drying of various food products. *J Food Sci*, **37**, 828–31.

Mishu, B., Griffin, P.M., et al. 1991. *Salmonella enteritidis* gastroenteritis transmitted by intact chicken eggs. *Ann Intern Med*, **115**, 190–4.

Mishu, B., Kochler, J., et al. 1994. Outbreaks of *Salmonella enteritidis* infections in the United States 1985–1991. *J Infect Dis*, **169**, 547–52.

Montville, T.J. 1997a. Principles which influence microbial growth, survival and death in foods. In: Doyle, M.P., Beuchat, L.R. and Montville, T.J. (eds), *Food microbiology, fundamentals and frontiers*. Washington, DC: ASM Press, 13–29.

Montville, T.J. 1997b. Biologically based preservation systems and probiotic bacteria. In: Doyle, M.P., Beuchat, L.R. and Montville, T.J. (eds), *Food microbiology, fundamentals and frontiers*. Washington, DC: ASM Press, 557–77.

Mossel, D.A.A., Corry, J.E.L., et al. 1995. *Essentials of the microbiology of foods*. Chichester: John Wiley and Sons.

Mossman, A.P. 1983. In: Chan, H.T. (ed.), *Handbook of tropical foods*. New York: Marcel Dekker, 489.

National Animal Health Monitoring System, 1995. Salmonella *shedding by feedlot cattle*. Fort Collins, CO: US Department of Agriculture, Animal and Plant Health Inspection Service, Veterinary Services.

O'Mahoney, M., Mitchell, E., et al. 1990. An outbreak of foodborne botulism associated with contaminated hazelnut yoghurt. *Epidemiol Infect*, **104**, 389–95.

Osborne, B.G. and Willis, K.H. 1984. The occurrence of some tricothecene mycotoxins in UK home grown wheat and in imported wheat. *J Sci Food Agric*, **35**, 579–83.

Ostroff, S.M., Kapperud, G., et al. 1994. Sources of sporadic *Yersinia enterocolitica* infections in Norway: a prospective case–control study. *Epidemiol Infect*, **112**, 133–41.

Paton, A.W., Ratcliff, R.M., et al. 1996. Molecular microbiological investigation of an outbreak of hemolytic-uremic syndrome caused by dry fermented sausage contaminated with shiga-like toxin-producing *Escherichia coli*. *J Clin Microbiol*, **34**, 1622–7.

PHLS Communicable Disease Surveillance Centre. 1999. *Salmonella java phage type Dundee – rise in cases: update*. *Commun Dis Rep CDR Week*, **9**, 105 and 108.

Pitt, J.I. and Hocking, A.D. 1986. *Fungi and fungal spoilage*. London: Academic Press.

Prasai, R.K., Phebus, R.K., et al. 1995. Effectiveness of trimming and/or washing on microbiological quality of beef carcasses. *J Food Protect*, **58**, 1114–17.

Proctor, F.J., Goodliffe, J.P. and Coursey, D.G. 1981. In: Spedding, C.R.W. (ed.), *Vegetable productivity*. London: Macmillan, 139.

Rasmussen, M.A., Cray, W.C. Jr, et al. 1993. Rumen contents as a reservoir of enterohemorrhagic *Escherichia coli*. *FEMS Microbiol Lett*, **114**, 79–84.

Roberts, D. and Greenwood, M. 2003. *Practical food microbiology*, 3rd edn. Oxford: Blackwell Publishing.

Roberts, T.A. and Mead, G.C. 1986. Involvement of intestinal anaerobes in the spoilage of red meats, poultry and fish. In: Barnes, E.M. and Mead, G.C. (eds), *Anaerobic bacteria in habitats other than man*. London: Academic Press, 333–50.

Ryall, A.L. and Lipton, W.J. 1979. *Handling, transportation and storage of fruits and vegetables*, 2nd edn. Westport, CT: AVI Publishing Co.

Saito, M. and Ohtsubo, K. 1974. Tricothecene toxins of *Fusarium*. In: Purchase, I.F.H. (ed.), *Mycotoxins*. Amsterdam: Elsevier Scientific, 263–81.

Salunkhe, D.K. and Desai, B.B. 1984. *Postharvest biotechnology of vegetables*. Boca Raton, FL: CRC Press.

Schlech, W.F., Lavigne, P.M., et al. 1983. Epidemic listeriosis – evidence for transmission by food. *N Engl J Med*, **308**, 203–6.

Schroeder, H.W., Boller, R.A. and Hein, H. 1968. Reduction in aflatoxin contamination of rice by milling procedures. *Cereal Chem*, **45**, 574–80.

Scoging, A.C. 1997. Fish toxins. In: Elmadfa, I. and Walter, P. (eds), *Naturlich vorkommende toxische stoffe in lebensmitteln*. XII Symposium der Internationalen Stftung zur Forderung der Ernahrungsforschung und Ernahrungsaufklarung (ISFE) 26/27 September 1997, Vienna: AMC Verlags, 61–80.

Sierra, M.-L., Gonzales-Fandos, E., et al. 1995. Prevalence of *Salmonella*, *Yersinia*, *Aeromonas*, *Campylobacter*, and cold-growing *Escherichia coli* on freshly dressed lamb carcasses. *J Food Protect*, **58**, 1183–5.

Slavik, M.F., Kim, J.-W. and Walker, J.T. 1995. Reduction of *Salmonella* and *Campylobacter* on chicken carcasses by changing scalding temperature. *J Food Protect*, **58**, 689–91.

Splittstoesser, D.F. 1987. Fruits and fruit products. In: Beuchat, L.R. (ed.), *Food and beverage mycology*, 2nd edn. New York: Van Nostrand Reinhold, 101–28.

St Louis, M.E., Morse, D.L., et al. 1988. The emergence of grade A eggs as a major source of *Salmonella enteritidis* infections: new implication for control of salmonellosis. *J Am Med Assoc*, **259**, 2103–7.

Stadelman, W.J. 1994. Contaminants of liquid egg products. In: Board, R.G. and Fuller, R. (eds), *Microbiology of the avian egg*. London: Chapman & Hall, 139–51.

Statutory Instrument. 1989a. The Poultry Breeding Flocks and Hatcheries (Registrations and Testing) Order 1989. SI No. 1963. London: Her Majesty's Stationery Office.

Statutory Instrument. 1989b. The Poultry Laying Flocks (Testing and Registration etc.) Order 1989. SI No. 1964. London: Her Majesty's Stationery Office.

Statutory Instrument. 1990a. The Food Labelling (Amendment) (Irradiated Food) Regulations 1990. SI No. 2489. London: Her Majesty's Stationery Office.

Statutory Instrument. 1990b. The Food (Control of Irradiation) Regulations 1990. SI No. 2490. London: Her Majesty's Stationery Office.

Statutory Instrument 1993a. The Egg Products Regulations 1993. SI No. 1520. London: Her Majesty's Stationery Office.

Statutory Instrument. 1993b. The Poultry Breeding Flocks and Hatcheries Order 1993. SI No. 1898. London: Her Majesty's Stationery Office.

Stern, N.J. 1992. Reservoirs of *Campylobacter jejuni* and approaches for intervention in poultry. In: Nachamkin, I., Blaser, M.J. and Tompkins, L.S. (eds), *Campylobacter jejuni: current status and future trends*. Washington, DC: American Society for Microbiology, 49–60.

Stern, N.J., Clavero, M.R., et al. 1995. *Campylobacter* spp. in broilers on the farm and after transport. *Poult Sci*, **74**, 937–41.

Tauxe, R.V., Vandepitte, J., et al. 1987. *Yersinia enterocolitica* and pork: the missing link. *Lancet*, **i**, 1129–33.

Thompson, S.S., Harmon, L.G. and Stine, C.M. 1978. Survival of selected organisms during spray drying of skim milk and storage of non fat dry milk. *J Food Protect*, **41**, 16–19.

USDA (US Department of Agriculture), 1990. *National poultry improvement plans and auxilliary provisions*, Animal and Plant Health Inspection Service edn. Washington, DC: US Government Printing Office.

Weagent, S.D., Bryant, J.L. and Bark, D.H. 1994. Survival of *Escherichia coli* O157:H7 in mayonnaise and mayonnaise-based sauces at room and refrigerator temperatures. *J Food Protect*, **57**, 629–31.

WHO (World Health Organization) 1988. Food irradiation. A technique for preserving and improving the safety of food, Geneva: World Health Organization.

WHO (World Health Organization) 1994. Safety and nutritional adequacy of irradiated foods. Geneva : World Health Organization.

Williams, J.E. 1981a. Salmonella in poultry feeds – a worldwide review. *World Poult Sci J*, **37**, 6–25.

Williams, J.E. 1981b. Salmonella in poultry feeds – a worldwide review. Methods in control and elimination. *World Poult Sci J*, **37**, 97–105.

Williams, P.C. 1983. Maintaining nutritional and processing quality in grain crops during handling, storage and transportation. In: Liebermann, M. (ed.), *Post-harvest physiology and crop preservation*. New York: Plenum Press, 425–44.

Zeleny, L. 1971. Criteria of wheat-m quality. In: Pomeranz, Y. (ed.), *Wheat: chemistry and technology*. St Paul, MN: American Association of Cereal Chemists, 19–49.

Zhao, T. and Doyle, M.P. 1994. Fate of enterohemorrhagic *Escherichia coli* O157:H7 in commercial mayonnaise. *J Food Protect*, **57**, 780–3.

12

Human microbiota

PATRICK R. MURRAY

GENERAL COMMENTS

For the first 9 months of life, the human fetus lives in a sterile environment protected from microbes, except when pathogens such as cytomegalovirus, rubella virus, or *Toxoplasma* are able to infect the child transplacentally. This state of sterility comes to an abrupt end, however, at the time of birth when the newborn is confronted with the mother's vaginal microbes and environmental organisms. The infant's skin surface is initially colonized and then the oropharynx, gastrointestinal tract, and other mucosal surfaces rapidly become populated. Through the duration of the individual's life, the microbial population evolves, with many organisms, transient colonizers, and others becoming well-established, permanent residents. It is important to recognize that this is a normal phenomenon. Even those microbes with the well-recognized ability to cause serious disease are normally found in and on the human body. Indeed, one hallmark of pathogenesis is not the recovery of a specific organism, but rather the recovery of the organism in a normally sterile site and associated with a pathologic response (e.g. inflammatory cells). For example, *Escherichia coli* is a normal resident in the gastrointestinal tract. To find it there can be anticipated. However, *E. coli* should remain confined to the gastrointestinal tract. If it is found in the abdominal cavity or the patient's blood, this would be considered abnormal. Likewise, organisms such as *Streptococcus pneumoniae*,

Staphylococcus aureus, *Neisseria meningitidis*, and *Haemophilus influenzae* are all capable of causing lower respiratory tract infection. Recovery of these organisms in lower airway secretions would be consistent with their role in disease; however, isolation of the same organisms in a throat washing could be considered insignificant. Naturally, there are certain organisms whose recovery in humans is always associated with clinically significant disease (e.g. *Bacillus anthracis*, *Brucella* spp., *Francisella tularensis*, and *Salmonella typhi*, to name just a few). However, the majority of microbes responsible for human disease are commonly part of the normal microbiota that are able to invade normally sterile tissues and spaces. For that reason, it is important to know the normal habitats of the organisms associated with humans and to understand their capacity for producing disease.

It should also be appreciated that microbes serve a useful purpose in their human hosts. The normal microbiota maintain a protected environment that prevents colonization with potentially pathogenic organisms. For example, *Clostridium difficile* is able to produce gastrointestinal disease ranging from diarrhea to pseudomembranous colitis only when the normal intestinal flora has been reduced or eliminated by antibiotics. The production of proteolytic enzymes by microbes augments host factors in the digestion of foods. Likewise, the disruption of the normal microbial flora can interfere with this process. For example, when the bacteria in the small intestine are replaced by colonic bacteria following small-bowel stasis (blind loop syndrome), bile salts that are secreted into the intestines are metabolized by *Bacteroides* spp., producing a resultant malabsorption syndrome. Intestinal bacteria can also synthesize vitamins (e.g. biotin, pantothenic acid, pyridoxine, riboflavin, vitamin K), many of which are required for the growth of other bacteria. For example, strains of *E. coli* can produce vitamin K, which is utilized as a required growth factor by *Porphyromonas melaninogenica*, which in turn produces penicillinases that protect other bacteria from penicillin. Thus, a complex relationship among the microbial species can be established for the mutual benefit of each organism and the host.

It is important to define certain terms. The interaction between microbes and humans can result in three general outcomes: disease, transient colonization, and prolonged colonization. Another term for colonization is infection, which does not imply disease but rather the association of the microbe with the human host for a period of time. Disease results when the interaction between microbe (e.g. bacterium, fungus, virus, or parasite) and human host results in a pathological process. This process can be due to microbial factors (e.g. hydrolytic enzymes, toxins) or the host's immune response to the presence of the organism. A few organisms have a high virulence potential and are always associated with disease, whereas most organisms are opportunistic pathogens and cause disease only when the host's immunity is suppressed or when the organisms are introduced into tissues in which they are able to express their virulence potential. The other two outcomes of microbe and host interaction result in colonization, either transiently or prolonged. It should be noted that transient and prolonged colonization implies a distinction based on the duration of the interaction. Although no specific time limit can be used to define colonization, it may be as short as hours to a few days (e.g. organisms passing through the gastrointestinal tract) or extend to weeks, months, or even years. The complex population of microbes that become established in a host are called a variety of names, including natural, indigenous, resident, and normal flora. In previous editions of this text, they were referred to as 'normal microbiota,' a term that has been retained here.

An individual is exposed to numerous organisms that colonize other human hosts and animals or are present in the air that is breathed and food and liquids that are consumed (refer to other chapters in this text that discuss the microbiology of air, soil, water, milk, and food products). Factors that determine whether exposure to a microbe results in transient passage through a human host or prolonged colonization are complex, involving environmental factors, host characteristics, and microbial properties. Nutrients and environmental conditions must favor the survival of microbes. For example, vegetative forms of *Bacillus* and *Clostridium*

spp. cannot survive prolonged exposure to desiccation or heat. However, spores of these bacteria can exist in nature for months to years. Gram-positive bacteria have a thick peptidoglycan layer in their cell wall, which promotes osmotic stability and enables these organisms to exist on dry surfaces (e.g. household articles, hospital bed linens, skin surfaces). By contrast, gram-negative bacteria have a relatively thin peptidoglycan layer and a lipoprotein outer membrane. These structures reduce gram-negative bacteria to a more fragile state, restricting their survival to moist, protected areas (e.g. standing water, vegetation, sinks, and showers). Thus, exposure to gram-positive bacteria is frequently by person-to-person contact or exposure to contaminated fomites, whereas gram-negative bacteria are usually acquired following consumption of contaminated food or drink or exposure to a contaminated water source.

Various host factors determine the success of colonization with a microbe. Although the skin forms a barrier, preventing penetration of microbes into the deeper tissues, colonization of the surface is accomplished by organisms that can tolerate the dry surface and that are resistant to fatty acids produced by anaerobic bacteria and from the metabolism of sebum triglycerides. Some organisms are adapted to grow in skin areas with a high moisture content (e.g. axilla, skin folds, perianal area), whereas other organisms survive in hair follicles. The presence of lysozyme in tears and other secretions is highly toxic for many bacteria and restricts the organisms that can colonize these areas. Inhaled organisms can be rapidly trapped in the nasal passages, engulfed with secreted mucus, and either expelled or swallowed. Organisms ingested with food or drinks are frequently eliminated when exposed to the acidic environment of the gastric system or simply passed through the digestive tract. Other organisms fail to survive interaction with locally secreted antibodies, exposure to bacteriocins produced by the normal microbiota, or the localized environment (e.g. acidic pH of the vagina). The age of the host influences microbial colonization. The presence or absence of teeth, hormone secretions that are initiated at the time of puberty or altered in menopause, sexual activity, person-to-person interactions in day-care facilities, the military, or nursing homes, alteration of dietary habits, and many other age-related factors determine which organisms an individual is exposed to and which will become successfully established as a part of the normal microbiota.

The final factors that determine the success of an organism to colonize the human body are properties of the specific organism. For example, the terrain of the oropharynx is diverse, with opportunities for organisms to colonize saliva, the mucosal surface, the tongue, the gingiva above and below the tooth line, and teeth. Extremely oxygen-sensitive bacteria are able to proliferate in the gingival crevices where the Eh is appropriately low. Organisms such as *Streptococcus mutans* and *Streptococcus sanguis* adhere to the hard, smooth surface of teeth by producing extracellular polysaccharides (e.g. glucans, dextrans, fructans) from dietary carbohydrates. A pellicle consisting of proteins from saliva and crevicular fluids initially forms on the enamel surface, which forms the anchor for the streptococcal polysaccharides. Subsequent layering of proteins, bacterial polysaccharides, and mixed populations of bacteria leads to the formation of plaque on the tooth surface. Bacteria can also bind to cells lining the oropharynx, intestine, and vagina via specific receptors for the bacterial pili. This ability to adhere can prevent the mechanical elimination of organisms when saliva washes the oropharynx and food and drink pass through the intestines.

With the complex environmental, host, and microbial factors that shape the normal microbiota, it should be appreciated that this population of organisms is constantly changing in an individual. For example, performing tests to determine which microbes colonize an individual's oropharynx provides data for that individual at a discrete point in time. Surveys that estimate the prevalence of colonization with microbes are at best an approximation for the individuals sampled. The ability to apply microbial colonization data from the individual studied to another person is dependent on the similarities between the two individuals. The data collected become more useful if the number of individuals studied is increased and the host factors minimized (e.g. study individuals with a broad age range, include immunocompetent and immunosuppressed patients, etc.). Furthermore, the utility of the data is determined by the thoroughness of the microbiological procedures. Adequate types and volume of specimens must be collected and transported to the laboratory in a manner that preserves the viability of the microbial population; isolation, detection, and identification procedures that will permit recognition of all significant organisms must be used. Unfortunately, studies of the normal microbiota are limited by the numbers of individuals that can be studied using comprehensive microbiological techniques. In addition, refinements in microbiology procedures and changes in the taxonomic classification of microbes make it difficult to compare results of recent studies with those published in the older literature. Finally, molecular techniques have led to the recognition of many bacteria that cannot be isolated in culture. Despite these cautions, extensive literature exists that estimates the prevalence of organisms in different body sites. Table 12.1 is a summary of this literature for the four major body sites that are populated with the most common bacteria colonizing humans: respiratory tract, gastrointestinal tract, genitourinary tract, and the body surface. The table indicates whether the organisms are commonly present or not present in healthy individuals. The frequency with which these organisms are

recovered is discussed in the following text. The source of this information is from the accompanying chapters in this edition, as well as the review articles and reference works that are listed at the end of this chapter (Balows et al. 1988, 1992; Boone and Castenholz 2001; Clarke and Bauchop 1977; Drasar and Hill 1974; Holt et al. 1994; Jousimies-Somer et al. 2002; Maibach and Hildick-Smith 1965; Mandell et al. 2000; Murray 1998; Noble and Sommerville 1974; Rosebury 1961; Skinner and Carr 1974). The taxonomic classification of the organisms listed herein is consistent with the *Manual of clinical microbiology* (Murray et al. 2003) and websites listing currently accepted bacterial nomenclature: www.bacterio.cict.fr (Society for Systemic and Veterinary Bacteriology) and www.dsmz.de/bactnom/bactname.htm.

NORMAL MICROBIOTA OF THE RESPIRATORY TRACT

The respiratory tract can be divided anatomically into the upper airways, which includes the anterior and posterior nares and the nasopharynx, the middle airways, comprised of the oropharynx and tonsils, and the lower airways, with the larynx, trachea, bronchi, and lungs. This classification serves as a useful foundation for examining the dynamics of airway colonization. The structural and physiological differences at each site provide an environment compatible for some organisms and hostile for others.

Nares and nasopharynx

Relatively small numbers of organisms are present in the nares, with *Staphylococcus*, including *S. aureus* and coagulase-negative species, *Corynebacterium*, *Peptostreptococcus*, and *Fusobacterium* species being the most numerous. The carriage rate of *S. aureus* in the anterior nares is higher in preadolescent children than in adults. The recovery of other organisms in the nares is common, but they are usually present transiently and in small numbers. The microbial population in the nasopharynx is more complex, with a predominance of streptococci and *Neisseria* species. The viridans streptococci can be readily recovered in the nasopharynx, with *Streptococcus salivarius* and *Streptococcus parasanguis* most commonly isolated, as well as *Streptococcus pneumoniae*. Many species of *Neisseria*, including *N. meningitidis*, have been recovered in nasopharyngeal cultures. Colonization with *N. meningitidis* varies from less than 10 percent to as great as 95 percent, with the highest incidence in young adults confined to institutions and in the military, where focal epidemics of meningococcal disease are observed. The most commonly isolated species of *Neisseria* are *Neisseria subflava*, *Neisseria sicca*, *Neisseria cinerea*, *Neisseria mucosa*, and *Neisseria*

Table 12.1 *Human microbiota*

Organism	Site of carriage			
	Respiratory tract	GI tract	GU tract	Skin, ear, eye
Abiotrophia defectiva	+	0	0	0
Acholeplasma laidlawii	+	0	0	0
Acidaminococcus fermentans	+	+	0	0
Acinetobacter spp.	+	+	+	+
Actinobacillus spp.	+	0	0	0
Actinomyces spp.	+	+	+	0
Aerococcus christensenii	0	0	+	0
Aerococcus viridans	0	0	0	+
Aerococcus urinae	0	0	+	0
Aeromonas spp.	0	+	0	0
Alloiococcus otitis	0	0	0	+
Anaerorhabdus furcosus	0	+	0	0
Anaerococcus hydrogenalis	0	+	+	+
Anaerococcus lactolyticus	0	+	+	0
Anaerococcus prevotii	0	+	+	0
Arcanobacterium spp.	+	0	0	+
Bacillus spp.	0	+	0	+
Bacteroides caccae	0	+	0	0
Bacteroides distasonis	0	+	0	0
Bacteroides eggerthii	0	+	0	0
Bacteroides fragilis	0	+	+	0
Bacteroides merdae	0	+	0	0
Bacteroides ovatus	0	+	0	0
Bacteroides splanchnicus	0	+	0	0
Bacteroides thetaiotaomicron	0	+	0	0
Bacteroides vulgatus	0	+	0	0
Bifidobacterium adolescentis	0	+	0	0
Bifidobacterium bifidum	0	+	+	0
Bifidobacterium breve	0	+	+	0
Bifidobacterium catenulatum	0	+	+	0
Bifidobacterium dentium	+	+	+	0
Bifidobacterium longum	0	+	+	0
Bilophila wadsworthia	+	+	+	0
Brevibacterium casei	0	0	0	+
Brevibacterium epidermidis	0	0	0	+
Burkholderia cepacia complex	+	0	0	+
Butyrivibrio fibrisolvens	0	+	0	0
Campylobacter concisus	+	+	0	0
Campylobacter curvus	+	+	0	0
Campylobacter gracilis	+	+	0	0
Campylobacter jejuni	0	+	0	0
Campylobacter rectus	0	+	0	0
Campylobacter showae	+	+	0	0
Campylobacter sputorum	+	0	0	0
Capnocytophaga granulosum	+	0	0	0
Capnocytophaga gingivalis	+	0	0	0
Campylobacter haemolytica	+	0	0	0
Capnocytophaga ochracea	+	+	+	0
Capnocytophaga sputigena	+	0	0	0
Cardiobacterium hominis	+	0	0	0
Centipeda periodontii	+	0	0	0
Citrobacter freundii	0	+	0	0

(Continued over)

Table 12.1 *Human microbiota (Continued)*

Organism	Site of carriage			
	Respiratory tract	GI tract	GU tract	Skin, ear, eye
Citrobacter koseri	0	+	0	0
Clostridium spp.	0	+	0	0
Corynebacterium accolens	+	0	0	+
Corynebacterium afermentans	+	0	0	+
Corynebacterium amycolatum	0	0	0	+
Corynebacterium auris	0	0	0	+
Corynebacterium diphtheriae	+	0	0	+
Corynebacterium durum	+	0	0	0
Corynebacterium glucuronolyticum	0	0	+	0
Corynebacterium jeikeium	0	0	0	+
Corynebacterium macginleyi	0	0	0	+
Corynebacterium matruchotii	+	0	0	0
Corynebacterium minutissimum	0	0	0	+
Corynebacterium propinquum	+	0	0	0
Corynebacterium pseudodiphtheriticum	+	0	0	0
Corynebacterium riegelii	0	0	+	0
Corynebacterium simulans	0	0	0	+
Corynebacterium striatum	+	0	0	+
Corynebacterium ulcerans	+	0	0	0
Corynebacterium urealyticum	0	0	+	+
Dermabacter hominis	0	0	0	+
Dermacoccus nishinomiyaensis	0	0	0	+
Desulfomonas pigra	0	+	0	0
Dysgonomonas spp.	0	+	0	0
Eikenella corrodens	+	+	0	0
Enterobacter aerogenes	0	+	0	0
Enterobacter cloacae	0	+	0	0
Enterobacter gergoviae	0	+	0	0
Enterobacter sakazakii	0	+	0	0
Enterobacter taylorae	0	+	0	0
Enterococcus spp.	0	+	0	0
Escherichia coli	0	+	+	0
Escherichia fergusonii	0	+	0	0
Escherichia hermannii	0	+	0	0
Escherichia vulneris	0	+	0	0
Eubacterium spp.	+	+	0	0
Ewingella americana	+	0	0	0
Finegoldia magnus	0	+	+	+
Fusobacterium alocis	+	0	0	0
Fusobacterium gonidiaformans	0	+	+	0
Fusobacterium mortiferum	0	+	0	0
Fusobacterium naviforme	0	+	+	0
Fusobacterium necrophorum	+	+	0	0
Fusobacterium nucleatum	+	0	0	0
Fusobacterium sulci	+	0	0	0
Fusobacterium russii	0	+	0	0
Fusobacterium varium	0	+	0	0
Gardnerella vaginalis	0	+	+	0
Gemella haemolysans	+	0	0	0
Gemella morbillorum	+	+	0	0
Granulicatella spp.	+	0	0	0
Haemophilus spp.	+	0	0	0

(Continued over)

Table 12.1 *Human microbiota (Continued)*

Organism	Site of carriage			
	Respiratory tract	GI tract	GU tract	Skin, ear, eye
Hafnia alvei	0	+	0	0
Helcococcus kunzii	0	0	0	+
Helicobacter spp.	0	+	0	0
Kingella spp.	+	0	0	0
Klebsiella spp.	+	+	0	0
Kocuria spp.	0	0	0	+
Kytococcus sedentarius	0	0	0	+
Lactobacillus acidophilus	+	+	+	0
Lactobacillus breve	+	0	0	0
Lactobacillus casei	+	0	+	0
Lactobacillus cellobiosus	0	0	+	0
Lactobacillus fermentum	+	+	+	0
Lactobacillus reuteri	0	+	0	0
Lactobacillus salivarius	+	+	0	0
Lactococcus spp.	0	0	+	0
Leclercia adecarboxylata	0	+	0	0
Leminorella spp.	0	+	0	0
Leptotrichia buccalis	+	0	+	0
Listeria monocytogenes	0	+	0	0
Leuconostoc spp.	0	0	+	0
Megasphaera elsdenii	0	+	0	0
Micrococcus luteus	+	0	0	+
Micrococcus lylae	+	0	0	+
Micromonas micros	+	0	0	0
Mitsuokella multiacidus	0	+	0	0
Mobiluncus curisii	0	+	+	0
Mobiluncus mulieris	0	+	+	0
Moellerella wisconsensis	0	+	0	0
Moraxella catarrhalis	+	0	0	0
Morganella morganii	0	+	0	0
Mycoplasma buccale	+	0	0	0
Mycoplasma faucium	+	0	0	0
Mycoplasma fermentans	+	0	+	0
Mycoplasma genitalium	+	0	+	0
Mycoplasma hominis	+	0	+	0
Mycoplasma lipophilum	+	0	0	0
Mycoplasma orale	+	0	0	0
Mycoplasma penetrans	0	0	+	0
Mycoplasma pneumoniae	+	0	0	0
Mycoplasma primatum	0	0	+	0
Mycoplasma salivarium	+	0	0	0
Mycoplasma spermatophilum	0	0	+	0
Neisseria cinerea	+	0	0	0
Neisseria flavescens	+	0	0	0
Neisseria lactamica	+	0	0	0
Neisseria meningitidis	+	0	0	0
Neisseria mucosa	+	0	0	0
Neisseria polysaccharea	+	0	0	0
Neisseria sicca	+	0	0	0
Neisseria subflava	+	0	0	0
Oligella ureolytica	0	0	+	0
Oligella urethralis	0	0	+	0

(Continued over)

Table 12.1 *Human microbiota (Continued)*

Organism	Site of carriage			
	Respiratory tract	**GI tract**	**GU tract**	**Skin, ear, eye**
Pantoea agglomerans	0	+	0	0
Pastuerella bettyae	0	0	+	0
Pasteurella multocida	+	0	0	0
Peptococcus niger	0	0	+	+
Peptoniphilus asaccharolyticus	0	+	+	+
Peptoniphilus lacrimalis	+	0	0	0
Peptostreptococcus anaerobus	+	+	0	0
Peptostreptococcus productus	0	+	0	0
Peptostreptococcus vaginalis	0	0	+	+
Porphyromonas asaccharolytica	0	+	+	0
Porphyromonas catoniae	+	0	0	0
Porphyromonas endodontalis	+	0	0	0
Porphyromonas gingivalis	+	0	0	0
Prevotella bivia	0	0	+	0
Prevotella buccae	+	0	0	0
Prevotella buccalis	+	0	+	0
Prevotella corporis	+	0	0	0
Prevotella dentalis	+	0	0	0
Prevotella denticola	+	0	0	0
Prevotella disiens	0	0	+	0
Prevotella enoeca	+	0	0	0
Prevotella heparinolytica	+	0	0	0
Prevotella intermedia	+	0	0	0
Prevotella loescheii	+	0	+	0
Prevotella melaninogenica	+	0	+	0
Prevotella nigrescens	+	0	0	0
Prevotella oralis	+	0	+	0
Prevotella oris	+	0	0	0
Prevotella oulorum	+	0	0	0
Prevotella tannerae	+	0	0	0
Prevotella veroralis	+	0	+	0
Prevotella zoogleoformans	+	0	0	0
Propionibacterium acnes	0	0	0	+
Propionibacterium avidum	0	0	0	+
Propionibacterium granulosum	0	0	0	+
Propionibacterium propionicum	+	0	0	0
Propionferax innocuum	0	0	0	+
Proteus mirabilis	0	+	+	0
Proteus penneri	0	+	+	0
Proteus vulgaris	0	+	+	0
Providencia rettgeri	0	+	0	0
Providencia stuartii	0	+	0	0
Pseudomonas aeruginosa	0	+	0	0
Retortamonas intestinalis	0	+	0	0
Rothia dentocariosa	+	0	0	0
Rothia mucilaginosa	+	0	0	0
Ruminococcus productus	0	+	0	0
Selenomonas spp.	+	0	0	0
Serratia liquefaciens	0	+	0	0
Serratia marcescens	0	+	0	0
Serratia odorifera	0	+	0	0
Staphylococcus aureus	+	0	+	+

(Continued over)

Table 12.1 *Human microbiota (Continued)*

Organism	Respiratory tract	GI tract	GU tract	Skin, ear, eye
Staphylococcus auricularis	0	0	0	+
Staphylococcus capitis	0	0	0	+
Staphylococcus caprae	0	0	0	+
Staphylococcus cohnii	0	0	0	+
Staphylococcus epidermidis	+	0	+	+
Staphylococcus haemolyticus	0	0	0	+
Staphylococcus hominis	0	0	0	+
Staphylococcus lugdunensis	0	0	0	+
Staphylococcus pasteuri	0	0	0	+
Staphylococcus saccharolyticus	0	0	0	+
Staphylococcus saprophyticus	0	0	+	+
Staphylococcus simulans	0	0	0	+
Staphylococcus xylosus	0	0	0	+
Staphylococcus warneri	0	0	0	+
Streptococcus agalactiae	0	+	+	0
Streptococcus anginosus	+	+	+	0
Streptococcus bovis	0	+	0	0
Streptococcus constellatus	+	+	+	0
Streptococcus criceti	+	0	0	0
Streptococcus crista	+	0	0	0
Streptococcus equisimilis	+	0	0	0
Streptococcus gordonii	+	0	0	0
Streptococcus intermedius	+	+	+	0
Streptococcus mitis	+	0	0	0
Streptococcus mutans	+	0	0	0
Streptococcus oralis	+	0	0	0
Streptococcus parasanguis	+	0	0	0
Streptococcus pneumoniae	+	0	0	0
Streptococcus pyogenes	+	0	0	+
Streptococcus salivarius	+	0	0	0
Streptococcus sanguis	+	0	0	0
Streptococcus sobrinus	+	0	0	0
Streptococcus vestibularis	+	0	0	0
Succinivibrio dextrinosolvens	0	+	0	0
Tissierella praeacuta	0	+	0	0
Treponema denticola	+	0	0	0
Treponema maltophilum	+	0	0	0
Treponema minutum	0	0	+	0
Treponema phagedenis	0	0	+	0
Treponema refringens	0	0	+	0
Treponema socranskii	+	0	0	0
Treponema vincentii	+	0	0	0
Turicella otitidis	0	0	0	+
Ureaplasma urealyticum	+	0	+	0
Veillonella spp.	+	+	0	0
Weeksella virosa	0	0	+	0

GI, gastrointestinal tract; GU, genitourinary tract; skin, includes ears and eyes as well as skin.

lactamica. Multiple species coexist in more than half of all individuals. Related gram-negative coccobacilli that colonize the nasopharynx include *Moraxella catarrhalis*, now recognized as a common cause of respiratory tract infections (e.g. sinusitis, bronchitis), and *Kingella* spp. Unencapsulated strains, as well as occasional encapsulated strains, of *H. influenzae* are commonly found in the nasopharynx, as is *Cardiobacterium hominis*, an organism associated with infections of previously damaged heart valves.

Oropharynx and tonsils

The oropharynx is a complex mixture of ecosystems, each with a distinctive microbial population. Thus, predictable differences in organisms will be found in saliva, gingival crevices, surfaces of teeth, the tongue, and the mucosal lining. Gram-positive and gram-negative cocci predominate in the oropharynx. Overall, anaerobes outnumber aerobic bacteria 100 to 1. The most common anaerobic bacteria are *Peptostreptococcus*, related gram-positive cocci (i.e. *Micromonas, Peptoniphilus*), *Veillonella, Actinomyces,* and *Fusobacterium*; the most common aerobic bacteria are *Streptococcus* and *Neisseria*.

Relatively small numbers of staphylococci are present, although it has been estimated that as many as 20 percent of individuals are colonized with *S. aureus*. Streptococcal species are more numerous, particularly members of the viridans group. *S. salivarius* is present in high numbers in saliva (as the name implies) and on the surface of the tongue. Tooth surfaces are colonized with *S. sanguis* and *S. mutans*, whereas the oral mucosa is populated with *Streptococcus vestibularis* and *S. sanguis*. Other streptococci that colonize the oropharynx include *S. pneumoniae*, commonly present in children and adults with children, and β-hemolytic streptococci, including groups A, C, F, and G. Group A *Streptococcus* (*Streptococcus pyogenes*, the organism responsible for streptococcal pharyngitis) can transiently colonize healthy individuals or become a more permanent member of the oral microflora. Other gram-positive cocci that are present in the oropharynx include *Rothia mucilaginosa*, *Gemella* species, *Peptostreptococcus anaerobius*, *Micromonas micros*, and *Abiotrophia* species. Both *Rothia* and *Peptostreptococcus* are found in virtually all individuals.

As in the nasopharynx, gram-negative cocci and coccobacilli can colonize the oropharynx. *Veillonella* species are the most numerous gram-negative cocci found in the oropharynx, representing as much as 15 percent of the total bacterial population. *Veillonella atypica* and *Veillonella dispar* are present on the tongue, oral mucosa and in saliva. *Veillonella parvula* is present in subgingival spaces and in dental plaque, in higher numbers in dental caries. Other gram-negative cocci and coccobacilli include *Neisseria, Moraxella, Kingella, Cardiobacterium,* and *Eikenella*. *Eikenella corrodens* can establish residence in the oropharynx by adhering to buccal epithelial cells.

Gram-positive bacilli are also prominent members of the oropharyngeal flora. *Actinomyces* are present in large numbers, comprising 20 percent of the bacterial flora in saliva and on the tongue, 35 percent in gingival crevices, and 40 percent of the bacteria in dental plaque. The *Actinomyces* spp. present in the oropharynx are *Actinomyces israelii, Actinomyces naeslundii, Actinomyces odontolyticus, Actinomyces meyeri, Actinomyces*

georgiae, and *Actinomyces gerencseriae*. *A. israelii* and *A. naeslundii* also colonize the surface of the tonsils. The ability of *Actinomyces* spp. to colonize the various surfaces is mediated by the presence of fimbriae that can adhere to mucosal cells and the production of extracellular polysaccharides that form a slime over tooth surfaces, entrapping organisms and preventing their removal by the flow of saliva and food through the oropharynx. Other genera of gram-positive bacilli present in the oropharynx include *Bifidobacterium, Corynebacterium, Eubacterium, Lactobacillus, Propionibacterium,* and *Rothia*. *Bifidobacterium dentium* is isolated in dental plaque and species of *Lactobacillus* are prominently associated with carious teeth, thriving in the acid environment produced by the streptococci responsible for the development of caries. *Eubacterium* spp. are isolated from subgingival crevices, dental plaques, and calculus.

The predominant gram-negative bacilli in the oropharynx are anaerobes, including *Fusobacterium, Bacteroides, Porphyromonas, Prevotella,* and *Selenomonas*. *Fusobacterium nucleatum* is the most common fusobacterium in the mouth, *Fusobacterium alocis* and *Fusobacterium sulci* are isolated in the gingival crevices, and the other species are present in smaller numbers throughout the mouth. Many of the anaerobic gram-negative bacilli in the mouth that were formerly classified as *Bacteroides* are now *Prevotella* and *Porphyromonas* spp. However, at least four species found in the mouth have been retained in the genus *Bacteroides*, while multiple species of *Prevotella* and two *Porphyromonas* species colonize the mouth. At least six species of *Selenomonas* are present in the oropharynx, primarily in gingival crevices.

Haemophilus spp. are present in almost all individuals although in small numbers (<5 percent of the microbial population). *H. parainfluenzae* is the most common *Haemophilus* species present in the mouth. *H. influenzae* is less commonly present, with most strains nonencapsulated. Other species are found on the tongue, palate, cheeks, and teeth, associated with dental plaque and periodontal disease. *Actinobacillus actinomycetemcomitans* is commonly isolated in the oropharynx and is an important cause of juvenile periodontitis. Another group of bacteria commonly found in the oropharynx, particularly in the gingival crevices, and associated with periodontal disease are *Treponema* species.

Members of the Enterobacteriaceae and nonfermentative gram-negative bacilli, such as *Pseudomonas* spp. and *Acinetobacter* are present in the oropharynx of healthy individuals, but usually only in small numbers or transiently. This changes in debilitated or hospitalized patients in whom these organisms can become the predominant bacteria in the oropharynx and are frequently responsible for lower respiratory tract disease.

Trachea, larynx, bronchi, and lungs

Colonization of the lower airways is generally transient, with relatively few organisms present at any one time. The only time long-term colonization occurs is when the ciliated epithelial cells are damaged through infection (e.g. with influenza virus) or disease (e.g. chronic obstructive pulmonary disease). This permits drainage of respiratory secretions into the bronchials and lower airways, with subsequent proliferation of the microbes.

NORMAL MICROBIOTA OF THE GASTROINTESTINAL TRACT

As with the respiratory tract, the gastrointestinal tract can be subdivided into distinct anatomical areas, each harboring its own indigenous microbiota. These would include the esophagus, stomach, jejunum and upper ileum, distal small intestine, and large intestine. It should be obvious that within each of these areas, the microbial flora on the mucosa, within crypts, and in the lumen can also be different. Unfortunately, the ability to sample each area without contamination from other sites in healthy individuals is limited. The most complete information has been obtained for the stomach and the large intestine. The microbial flora at other sites is estimated by studying the microbial composition of feces or the specimens collected at the time of intraabdominal surgery.

Esophagus

Insufficient information is available about the microbial flora of the esophagus. However, oropharyngeal bacteria can be isolated from this site, as can the organisms that colonize the stomach. Transient colonization occurs with these organisms in healthy individuals. In the diseased state, bacteria are uncommon causes of esophagitis and esophageal infections, with yeasts (e.g. *Candida*) and viruses (e.g. herpes simplex, cytomegalovirus) playing a more prominent role.

Stomach

The stomach is an inhospitable organ, containing hydrochloric acid and pepsinogen (a precursor of pepsin) secreted by parietal and chief cells, respectively, which line the gastric mucosa. For this reason, the normal microbial microbiota are sparse, primarily associated with the surface epithelium, and protected by secretions of mucus and bicarbonate. The organisms present in the stomach are acid-tolerant *Lactobacillus* spp., *Streptococcus* spp., and *Helicobacter pylori*. Whereas the first two organisms are not associated with gastric disease, *H. pylori* causes gastritis and gastric (peptic) and duodenal ulcers, and is associated with gastric malignancies. Other organisms may be isolated in the stomach, particularly a few hours after a meal, but are believed to represent transient passage through the stomach.

Jejunum and upper ileum

The number of microbes in the upper portion of the small intestine is low (generally fewer than 10^5 organisms per milliliter of fluid) and predominantly anaerobic, consisting primarily of *Lactobacillus, Streptococcus, Peptostreptococcus, Anaerococcus, Finegoldia, Peptoniphilus, Porphyromonas,* and *Prevotella*. If upper-tract obstruction and stasis occur (e.g. blind loop syndrome), then the microbial flora can shift to resemble colonic bacteria (e.g. *Bifidobacterium, Bacteroides, Clostridium, Escherichia, Enterococcus*) and lead to a malabsorption syndrome.

Distal small intestine

This is the transition area between the relatively sparse numbers of acid-tolerant bacteria that occupy the upper portion of the intestinal tract and the plethora of microbes that exist in the large intestine. Although it is unclear which organisms colonize this portion of the gastrointestinal tract permanently, it is known that the microbial population is large (approximately 10^8–10^9 organisms per gram of feces) and diverse, with a distinct predominance of strict anaerobes.

Large intestine

This is the most densely populated organ in the human body, with more than 10^8 aerobic bacteria and 10^{11} anaerobic bacteria per gram of feces. The most numerous bacteria in the large intestine are *Bifidobacterium* spp., *Bacteroides* spp., *Eubacterium* spp., *Enterococcus* spp., and *E. coli*. It has been estimated that feces consist of 10^{11} bacteroides per gram. Although *B. fragilis* is the most virulent species, *Bacteroides thetaiotaomicron* is more numerous in the colon. Other *Bacteroides* spp. in the colon include *Bacteroides capillosus, Bacteroides coagulans, Bacteroides putredinis,* and *Bacteroides ureolyticus*. *Eubacterium* spp. are the second most commonly isolated bacteria in the intestine, with more than 10^{10} organisms per gram of feces and 16 distinct species described. The most commonly isolated species are *Eubacterium aerofaciens, Eubacterium contortum, Eubacterium cylindoides, Eubacterium lentum,* and *Eubacterium rectale*. Ten species of *Bifidobacterium* have been isolated in feces, some species isolated preferentially from infants and others from adults. The most commonly isolated species are *B. bifidum, B. longum* and *B. adolescentis*. Many species of *Enterococcus* have been described, most of which are present in the human intestines. *Enterococcus faecalis,*

Enterococcus faecium, *Enterococcus casseliflavus*, and *Enterococcus gallinarum* are most frequently isolated. *E. coli* inhabits the intestine of virtually all humans, establishing intestinal colonization soon after an infant is born. Although it represents a relatively minor position quantitatively (approximately 1 percent of the total bacterial population), it is the most common facultative organism responsible for intra-abdominal infections.

A large number of other organisms have been demonstrated to colonize the large intestine. *Actinomyces* are frequently isolated in fecal specimens, even though intestinal colonization has not been clearly demonstrated. *Streptococcus* spp., including *Streptococcus bovis*, which is associated with intestinal malignancies, have been isolated from fecal specimens.

Gemella morbillorum is a member of the normal intestinal flora, as are a variety of gram-positive anaerobic cocci (e.g. *Anaerococcus prevotii*, *Finegoldia magnus*, *Peptoniphilus asaccharolyticus*). The spore-forming *Bacillus* spp. and *Clostridium* spp. are isolated in fecal specimens. Although this may represent simple transit through the gastrointestinal tract following ingestion with food or drink, most authors believe that *Clostridium* spp. are part of the permanent intestinal population. *V. parvula* can colonize the intestinal tract of humans, although it is generally present in small numbers. Like *E. coli*, other members of the Enterobacteriaceae can establish residence in the intestines. *Citrobacter* spp., *Klebsiella* spp., *Enterobacter* spp., *Proteus* spp., and various other genera can be consistently isolated in fecal specimens. *Haemophilus* spp. can be recovered in fecal specimens if selective media are used. Other organisms commonly isolated, but in small numbers, include species of *Fusobacterium*, *Porphyromonas*, and *Prevotella*. Uncultivable species of *Treponema* have been observed in fecal specimens.

NORMAL MICROBIOTA OF THE GENITOURINARY TRACT

The genitourinary tract is relatively sterile, with the exception of the female urethra and vagina. Microbes can migrate up the urethra into the bladder, but these are rapidly cleared in healthy individuals by the action of localized antibodies, microbicidal activity of the epithelial cells lining the bladder, and the flushing action of voided urine. The ureters, kidneys, prostate, and cervix are normally sterile. The female urethra is colonized with large numbers of lactobacilli, streptococcal species, and coagulase-negative staphylococci. Fecal organisms such as *Escherichia coli* and *Enterococcus* spp. can also colonize the female urethra but are generally transient and present in small numbers. When these latter organisms migrate up into the bladder, they are able to proliferate in urine and can establish a urinary tract infection.

The microbial flora in the vagina are more numerous and diverse. Lactobacilli are the predominant organisms because they are able to proliferate in the acidic environment. The most commonly isolated species include *Lactobacillus acidophilus*, *Lactobacillus fermentum*, *Lactobacillus casei*, and *Lactobacillus cellobiosus*. Other anaerobes commonly isolated in vaginal secretions are *Bifidobacterium*, anaerobic cocci, *Porphyromonas*, and *Prevotella*. Six species of *Bifidobacterium* have been recovered in the vagina; *B. bifidum* and *B. longum* are the most numerous. Likewise, eight species of anaerobic cocci (i.e. *Peptococcus niger*, *Peptoniphilus asaccharolyticus*, *Anaerococcus hydrogenalis*, *A. lactolyticus*, *A. prevotii*, *A. tetradius*, *Finegoldia magna*, and *Peptostreptococcus vaginalis*) reside in the vagina. *Porphyromonas asaccharolytica*, *Prevotella bivia*, and *Prevotella disiens* are also important residents of the vagina. *Actinomyces* spp. are believed to be present in the vagina because they are associated with vaginal infections. However, demonstration of their presence is controversial. *A. israelii* is most commonly associated with genital actinomycotic infections. *Propionibacterium*, particularly *Propionibacterium propionicus*, is also present in the vagina. Finally, *Mobiluncus* spp. are relatively uncommon in healthy women but a significant cause of bacterial vaginosis. Thus, it is likely that this anaerobe is present in the vaginal microbial flora in small numbers.

Common aerobic bacteria present in the vagina include *Staphylococcus* (primarily coagulase-negative species), *Streptococcus*, and *Corynebacterium* spp. The viridans group and β-hemolytic strains (e.g. groups B, C, and G) of streptococci are present in the vagina. *Gardnerella vaginalis* can also colonize the human genital and urinary tract, although it is only present in high numbers in women with vaginosis and their male partners. *Neisseria* spp., including *N. meningitidis*, are recovered in vaginal secretions, as are species of *Haemophilus* (including *H. influenzae* and *H. parainfluenzae*). Although these organisms can be recovered from a large proportion of individuals, they are generally part of the minor bacterial population. Members of the Enterobacteriaceae, particularly *E. coli*, can be found in the vaginal flora, although generally in small numbers. *Weeksella virosa* is almost exclusively isolated in genital specimens, particularly from sexually active women. Three species of nonpathogenic *Treponema* (*Treponema phagedenis*, *Treponema refringens*, and *Treponema minutum*) are isolated in vaginal specimens.

Six species of *Mycoplasma* (*Mycoplasma hominis*, *Mycoplasma genitalium*, *Mycoplasma fermentans*, *Mycoplasma primatum*, *Mycoplasma spermatophilum*, and *Mycoplasma penetrans*) primarily colonize the genitourinary tract. In addition, the related organism *U. urealyticum* is a common inhabitant of this site. The role of these organisms in disease is controversial, but clearly

M. hominis and *U. urealyticum* have pathogenic potential.

NORMAL MICROBIOTA OF THE BODY SURFACE

The surface of the skin is relatively inhospitable in comparison with other body sites. It is exposed to extremes in temperature and moisture and to chemical disinfectants such as soaps and shampoos. It is not surprising, therefore, that the microbial population is less numerous and complex than at other body sites. Despite this observation, many organisms, particularly gram-positive bacteria, are able to establish permanent residence on the skin surface. Transient colonization with a diverse array of environmental and endogenous microbes can also occur.

As with other areas of the body, the skin should not be considered a homogeneous surface, but rather a landscape of mountains and valleys, each with a specific environment and microbial population. Thus, there are relatively dry, hairless areas such as palms and soles, areas with a proliferation of apocrine glands such as the axillae, inguinal, and perineal areas, and areas rich with sebaceous glands such as the forehead and nasolabial folds. Each area is associated with distinct microbes. Most microorganisms proliferate in a moist environment, hence higher densities of microbes are present in areas rich with sweat glands or on occluded surfaces. The skin proximal to the oral cavity (face), gastrointestinal tract (perirectal area), or genitourinary tract (groin) has a more complex microbial flora than at other surface sites, although most of these organisms result from surface contamination and only transiently colonize the skin surface.

Anaerobic bacteria are ten- to 100-fold more numerous on the skin surface compared with aerobic bacteria; gram-positive bacteria predominate over gram-negative bacteria. The bacteria most commonly recovered on the skin surface are members of the genera *Staphylococcus*, *Micrococcus*, *Corynebacterium*, *Peptococcus*, *Peptoniphilus*, *Finegoldia*, *Peptostreptococcus*, and *Propionibacterium*. *Staphylococcus epidermidis* is the most frequently isolated bacterium, present on surfaces where the moisture is highest. Other coagulase-negative staphylococci that are found on the skin include *Staphylococcus hominis*, *Staphylococcus haemolyticus*, *Staphylococcus warneri*, *Staphylococcus capitis* (particularly on the forehead and face after puberty), *Staphylococcus saprophyticus*, *Staphylococcus caprae*, *Staphylococcus saccharolyticus*, *Staphylococcus pasteuri*, *Staphylococcus lugdunensis*, *Staphylococcus simulans*, and *Staphylococcus xylosus*. *Staphylococcus auricularis* is the most common species found colonizing the exterior auditory canal. The coagulase-positive *S. aureus* can colonize the skin adjacent to the nares and in moist folds and, less frequently, at other sites.

Micrococcus luteus is the most common *Micrococcus* sp. present on the skin, found in virtually all adults and representing as much as 20 percent of the bacterial population on the head, legs, and arms. However, relatively few micrococci are present in areas with a high moisture content and other competitive bacteria (e.g. axillae, nares). Additional gram-positive cocci (formerly classified as *Micrococcus*) that colonize the skin surface include *Dermacoccus nishinomiyaensis*, *Kocuria kristinae*, *Kocuria rosea*, *Kocuria varians*, and *Kytococcus sedentarius*.

Aerococcus viridans, *S. pyogenes*, and various anaerobic *Peptostreptococcus* spp. can also establish residence on the skin surface. *Aerococcus* is an airborne bacterium, so the incidence of true colonization versus contamination and transient colonization is unknown. *S. pyogenes* differ from other streptococci in their tolerance to the dry surfaces of the skin and to the bactericidal fatty acids from sebum and also produced by the anaerobic cocci and bacilli. Thus, they are particularly well suited for dry skin surfaces. Anaerobic cocci survive in the anaerobic niches of hair follicles and skin glands.

Many species of *Corynebacterium* spp. are present on the skin surface, including *Corynebacterium striatum*, *Corynebacterium minutissimum*, *Corynebacterium pseudodiphtheriticum*, *Corynebacterium xerosis*, *Corynebacterium urealyticus* (especially in the groin area), and *Corynebacterium jeikeium* (on moist surfaces rich in apocrine glands in hospitalized patients). The anaerobic counterpart to these aerobic gram-positive bacilli are the *Propionibacterium*. *Propionibacterium acnes* is found in high concentrations in areas rich in sebaceous glands, as is *Propionibacterium granulosum*. *P. acnes* is the predominant species, found in high numbers on virtually all individuals, whereas *P. granulosum* is recovered in small numbers and on fewer than 20 percent of individuals sampled.

Propionibacterium avidum, by contrast, requires an area high in moisture for survival and is most commonly found in axilla and perineum. Other resident gram-positive bacilli include *Dermabacter hominis*, *Brevibacterium epidermidis*, *Brevibacterium casei*, and poorly defined *Brevibacterium* spp. *Turicella otitidis* colonizes the external ear. Some species of *Bacillus* and *Clostridium*, particularly *Clostridium perfringens*, can colonize the skin surface due to their ability to form spores and withstand desiccation and detergents. However, they are generally present in small numbers and only transiently.

Gram-negative bacteria are generally not recovered from the surface of the skin, except during transient colonization. The outer portion of the cell wall of gram-negative bacilli consists of a lipid membrane that is unable to survive exposure to detergents or the dry surfaces of the skin. However, *Acinetobacter* spp. have adapted to survive in moist areas such as toe webs, the groin, and axillae. *Burkholderia* species can also colonize the skin surface but generally is found only transiently.

NORMAL MICROBIOTA OF BLOOD, CEREBROSPINAL FLUID, AND OTHER BODY FLUIDS

The human body is bathed in a variety of fluids, including blood, cerebrospinal, synovial, pleural, pericardial, peritoneal, and other exudates and transudates. All of these fluids are normally sterile or only transiently infected. Whereas microbes from the mouth or gastrointestinal tract can invade the bloodstream in healthy individuals (e.g. during tooth brushing or a bowel movement), these organisms are rapidly removed and generally of little or no significance. Thus, the isolation of an organism from a body fluid should be considered significant unless the specimen is contaminated during the process of collection.

NORMAL MICROBIOTA OF BODY TISSUES

As with body fluids, organ tissues are generally sterile unless they are infected following the systemic spread of an organism in the bloodstream. Some organisms (e.g. *Mycobacterium tuberculosis*) may be disseminated to tissues such as the lungs, liver, or kidneys at the time of initial infection and remain dormant for many years. In this situation it is possible that the organism could be recovered from tissue samples, but even that would be unlikely unless the disease process was active.

CONCLUDING COMMENTS

From the moment of birth until the terminal breath, the human body serves as a home for qualitatively and quantitatively numerous microbes. They cover the skin and mucosal surfaces of the body, occasionally invade into sterile tissues and fluids to produce disease, but more typically coexist with each other and their human host. These microbiota function as a microbial barrier to more virulent microorganisms, supplement the human host's mechanical and enzymatic digestion of food, provide vitamins and other required growth factors for their host, or simply exist as a commensal inhabitant on their host. The complexity of the microbiota is shaped over time by environmental, host, and microbial factors, but it is remarkably predictable for a general population. Certainly the dominant microbial species at each of the major body sites (e.g. respiratory tract, gastrointestinal tract, genitourinary tract, and body surface) are now well known. Variations from the expected population can predictably lead to disease (e.g. replacement of the microbial population of the small intestine with bacteria from the large intestine, leading to a malabsorption syndrome). Likewise, spread of bacteria from their normal habitat into sterile tissues and fluids can initiate well-characterized disease (e.g. colonic perforation leads to a polymicrobic peritonitis and intraabdominal abscess formation by *B. fragilis*). Thus, knowledge of the human microbiota forms a fundamental building block for our knowledge of the normal physiological processes in the human body and our understanding of infectious diseases.

REFERENCES

Balows, A., Hausler, W.J., et al. 1988. *Laboratory diagnosis of infectious diseases: Principles and practice.* New York: Springer-Verlag.

Balows, A., Truper, H.G., et al. 1992. *The prokaryotes*, 2nd edn. New York: Springer-Verlag.

Boone, D.R. and Castenholz, R.W. 2001. *Bergey's manual of systematic bacteriology*, 2nd edn. New York: Springer.

Clarke, R.T.J. and Bauchop, T. 1977. *Microbial ecology of the gut.* London: Academic Press.

Drasar, B.S. and Hill, M.J. 1974. *Human intestinal flora.* London: Academic Press.

Holt, J.G., Krieg, N.R., et al. 1994. *Bergey's manual of determinative bacteriology*, 9th edn. Baltimore: Williams & Wilkins.

Jousimies-Somer, H.R., Summanen, P., et al. 2002. *Wadsworth anaerobic bacteriology manual*, 6th edn. Belmont, California: Star Publishing Co.

Maibach, H.I. and Hildick-Smith, G. 1965. *Skin bacteria and their role in infection.* New York: McGraw-Hill.

Mandell, G.L., Bennett, J.E. and Dolin, R. 2000. *Principles and practice of infectious diseases*, 5th edn. New York: Churchill Livingstone.

Murray, P.R. (ed.) 1998. *Topley and Wilson's Principles of bacteriology, virology and immunity.* 9th edn. London: Edward Arnold.

Murray, P.R., Baron, E.J., et al. 2003. *Manual of clinical microbiology*, 8th edn. Washington, DC: ASM Press.

Noble, W.C. and Sommerville, J.A. 1974. *Microbiology of human skin.* London: Lloyd Luke.

Rosebury, T. 1961. *Microorganisms indigenous to man.* New York: McGraw-Hill.

Skinner, R.A. and Carr, J.G. 1974. *The normal microflora of man.* London: Academic Press.

PART III

GENERAL EPIDEMIOLOGY, TRANSMISSION, AND THERAPY

Epidemiology of infectious diseases

STEPHEN R. PALMER AND MEIRION R. EVANS

Epidemiology is the study of the patterns of occurrence and causes of disease in populations. It does not stand alone, but is complementary to microbiology and environmental risk assessment methods of investigation. For general accounts of epidemiology see Hennekens and Buring (1987), Beaglehole et al. (1993), Detels et al. (2002), and Rothman (2002). The occurrence of an episode of microbial disease is a result of the interaction of the agent, host, and environmental factors that leads to the exposure of the host to sufficient numbers of the agent in the appropriate transmission mode. Successful investigation and control require that not only the microbiological, but also behavioral, genetic, environmental, and social factors which influence the occurrence and presentation of the disease be taken into account. Close collaboration between epidemiologists, microbiologists, and other public health workers is therefore essential. Accurate laboratory diagnosis is usually a major factor in successful control, though epidemiological methods alone may be sufficient to indicate appropriate interim control measures. For example, the modes of transmission and the risk behavior leading to the acquired immune deficiency syndrome (AIDS) were identified by epidemiological methods 2 years before the causative agent was discovered. Advice to the public about reducing the risk of infection did not need to be revised substantially when the human immunodeficiency virus (HIV) was identified. However, the discovery of HIV was a major advance and allowed the development of diagnostic tools that are now used to study epidemiological characteristics of the infection, e.g. incubation period, progression rate to AIDS, spectrum of clinical disease, period of infectivity. Increasingly, the epidemiological method is seen by microbiologists as a necessary tool in setting priorities for the use of scarce resources. Applications of epidemiology in microbiological disease include population surveillance to monitor trends in disease and infections and to detect case clusters, the investigation of sources and modes of transmission of microorganisms, applying public health measures to control outbreaks of disease, and devising and evaluating preventive and control programs.

CONCEPTS IN THE EPIDEMIOLOGY OF INFECTIOUS DISEASE

Reservoir of infection

This is where the infectious agent normally lives and where it may multiply or survive; it may be human (e.g. in chickenpox), animal (e.g. in brucellosis) or the inanimate environment (e.g. in tetanus).

Source of infection

Infection may be derived from the patient's own microflora (endogenous), or from another human being, or an animal (zoonosis) or an environmental source (exogenous). The source of an exogenous infection may sometimes be different from its reservoir. For example, in an outbreak of cryptosporidiosis in Maine, the reservoir of infection was a herd of cattle on a farm that supplied apples for making apple cider. The apple cider then became the source of infection for human subjects. *Cryptosporidium* oocysts were detected in the apple cider, on the cider press, and in a stool specimen of a calf on the farm which supplied the apples (Millard et al. 1994). When the source of infection is inanimate, e.g. food, water, or fomites, it is termed the vehicle of infection.

Mode of transmission

The mechanism by which an infectious agent passes from the reservoir or source of infection to the person can be classified as follows.

FOOD-, DRINK-, OR WATER-BORNE INFECTION

Examples of this type of infection are typhoid and cholera. The term 'food poisoning' has in the past sometimes been restricted to incidents of acute disease in which the agent has multiplied in the food vehicle before ingestion (e.g. food-borne salmonellosis), or where it may have formed toxins (e.g. botulism).

DIRECT OR INDIRECT CONTACT

This includes spread from cases or carriers, animals, or the environment to other persons who are 'contacts'. Within this category possible routes include: feces-to-hand-to-mouth spread (e.g. shigellosis); sexual transmission (e.g. gonorrhea); skin or mucous membrane contact (e.g. wound infection, cutaneous anthrax).

PERCUTANEOUS INFECTION

This includes: insect-borne transmission via the bite of an infected insect either directly from saliva (e.g. malaria); or indirectly from insect feces contaminating the bite wound (e.g. typhus); transfusion of contaminated blood and inoculation of blood or blood products from needle stick injuries (e.g. hepatitis B); direct transmission through intact skin (e.g. schistosomiasis), or broken skin (e.g. leptospirosis).

AIR-BORNE INFECTION

Infectious organisms may be inhaled as: droplets (e.g. streptococcal pharyngitis); droplet nuclei (e.g. tuberculosis); aerosols (e.g. Legionnaires' disease); dust (e.g. ornithosis); or spores (e.g. anthrax).

TRANSPLACENTAL INFECTION

Examples include listeriosis, rubella, cytomegalovirus (CMV), and toxoplasmosis.

Occurrence

An infection that is continuously in a population is said to be endemic, whereas an increase in incidence above the endemic level is described as an epidemic, or pandemic when the epidemic is worldwide. Cases may be sporadic, that is, not known to be related to other cases or infections, or clustered in outbreaks which may be defined as two or more related cases or infections, suggesting the possibility of a common source or transmission between cases. Three commonly used measures of occurrence of disease or infection are the incidence rate, the rate of occurrence of new cases in a defined population (e.g. 10 cases per 100 000 persons per year); cumulative incidence or risk, which is the proportion of people who become diseased during a specific period; and prevalence, the proportion of a defined population with the disease at a point in time (point prevalence) or during a defined period of time (period prevalence). The prevalence of a disease depends upon its incidence and duration. In chronic diseases, prevalence may be high although incidence is low, but in short-duration infectious diseases prevalence approximates to incidence.

The attack rate during an outbreak is a type of cumulative incidence, the proportion of the population at risk at the beginning of a time period who became ill during the period. The secondary attack rate is the attack rate in the contacts of primary cases due to person-to-person spread.

Incubation period

This is the time from infection to the onset of symptoms (Richardson et al. 2001). For each organism there is a characteristic range within which infecting dose and portal of entry, as well as host factors such as age (Glynn and Palmer 1992) and immunosuppression, give rise to individual variability.

Host response

This depends upon the dose of the infecting agent and the susceptibility of the host, perhaps influenced by genotype, age, sex, other concurrent disease, immunity, and 'risk factors' such as smoking for Legionnaires' disease. In an outbreak of infection there is often a spectrum of clinical response ranging from no symptoms to fulminant disease and death.

Communicability

The infectious agent may be passed to others over a variable time, the period of communicability (Richardson et al. 2001); in some infections even from symptomless temporary or chronic carriers (e.g. *Salmonella* Typhi).

EPIDEMIOLOGICAL METHODS

Collection of observations

Investigation of the occurrence and distribution of disease requires accurate definition of the disease and its possible determinants and the measurement of their frequency in the population. A clear case definition is essential in epidemiological studies. If serological or cultural confirmation of the diagnosis is not possible, because a specific laboratory test is lacking or appropriate samples were not collected, the case definition will depend upon the clinical features as, for example, with AIDS before the discovery of HIV. During the investigation of an outbreak, laboratory confirmation of the diagnosis may be possible only in a few cases; a clinical definition will then need to be used in further studies. The specificity of the diagnosis in some patients may be doubtful; in these circumstances it is helpful to classify cases as 'definite', 'probable', or 'possible'. When case definitions are based on the presence of symptoms these must be precisely defined, e.g. 'by diarrhea we mean at least $\geqslant 3$ loose or watery stools in a 24 h period', so that all respondents understand the questions in the same way and the comparability of different studies can be assessed.

Routine sources of data on infectious diseases are described in a later section. In detailed epidemiological enquiries, however, other sources must be used: medical records, interviews with patients, and questionnaires.

Medical records

Clinical records may contain data on symptoms, investigations performed and their results, and personal details of patients, but for epidemiological purposes clinical records are seldom sufficient. The data recorded are likely to be accurate but incomplete, since they are not normally collected in a standard way from each patient. Scrutiny of these records may be useful in confirming the reported diagnosis and assessing whether the patient meets the case definition, and laboratory records usually provide an invaluable source of microbiological data and data on age, sex, and the geographical distribution of cases. However, to obtain detailed, accurate clinical information, interviews with patients are usually required. For example, in a survey of hydatid disease, a review of clinical records yielded data on the patients' age, sex, and home address, and on the results of clinical investigations and pathological tests. However, data on past history of residence, occupation, and exposure to dogs were seldom recorded; this could be obtained only by questioning the patients (Palmer and Biffin 1987).

Epidemiological interviews and questionnaires

To ensure accurate and comparable records of all persons included in the enquiry and to facilitate analysis, the data should be collected on a carefully designed standard form or questionnaire. Whenever possible, the questionnaire should be tested on a few cases and changes made as necessary before use in the main study. Administration of the questionnaire will often be by direct face-to-face interview by a single investigator or group of investigators trained to administer the questionnaire. Interview by telephone may be useful in obtaining data quickly. When numbers are large and the enquiry is straightforward, a self-administered postal questionnaire is cheaper and quicker to administer, but the response rate and accuracy of the data may be less than those obtained by interview.

When designing questionnaires it is important to take into account the limitations of people's recall of events. For example, in one study, food consumption recalled by people 2–3 days after a luncheon was compared with that observed at the time and recorded on tape (Decker et al. 1986). Only four of 32 patients made no errors in reporting. The sensitivity of the food-history questionnaire was 87.6 percent and its specificity 96.1 percent. Thirteen percent of the respondents reported eating one or both of two food items that were on the questionnaire but not served at the luncheon. Such errors in recall can be reduced by providing background details of events, by design of good questionnaires, and by making use of other sources to check data, such as diaries, menus, discussion with relatives, etc. Random errors in recall are unlikely to give rise to false associations between illness and, for example, an item of food, but they will reduce the power of the study to identify the true vehicle of infection. For some types of data, such as history of immunization, recall by patients or parents is of very limited value; studies of vaccine efficacy usually require validation of vaccine history from medical records.

Descriptive analysis

Variables measured may be:

- fixed or discrete (e.g. sex, occupation, nationality) or
- continuous (e.g. age, white blood cell count).

Analysis of the distribution will usually be by calculation of proportions of people who fall within certain categories or rates of occurrence of disease within subgroups of the population. Analysis of continuous variables is more complex because values obtained from a population will form a continuous distribution, usually approximating to a normal or skewed normal distribution (Figure 13.1). This distribution may be summarized

Value	8	9	10	11	12	13	14	15	16	17	18	19	20	21
Number of observations	2	0	6	5	8	16	13	21	11	11	8	2	2	1

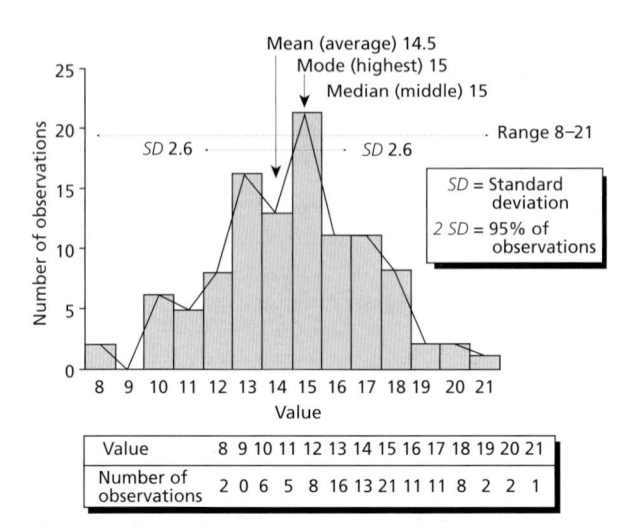

Figure 13.1 *Frequency distribution and measures of continuous variable*

by the mean, median, mode, and the standard deviation; these measures are commonly used to compare continuous variables in different populations.

The data should first be analyzed within the three classical epidemiological parameters of time, place, and person, taking into account interactions of these variables.

TIME

The epidemic curve is the most useful and immediate means of assessing the type of outbreak (Figure 13.2). In point-source outbreaks in which all cases are exposed at a given time, onset of symptoms of all primary cases will cluster within the range of the incubation period. For example, in the winter term of 1982, two campylobacter outbreaks were reported in boarding schools in the south of England. In the first of these (Figure 13.2a), 102 of 780 boys were admitted to the sanatorium with gastrointestinal illness, 46 of them on 1 day; this explosive outbreak was probably due to post-pasteurization contamination of the milk supply on one particular day. The other outbreak occurred in a school supplied with unpasteurized milk and was due to a continuing or recurring source of contamination. In this case the epidemic curve extended over several incubation periods (Figure 13.2b); 35 of 370 boys were admitted to the sanatorium over the first weeks of the term. In some outbreaks, a point-source of infection may be followed by person-to-person spread as in an outbreak of *Salmonella* Typhimurium, phage-type 10, infection affecting 66 students and one member of staff in a university hall of residence in Bristol in March 1980 (Palmer et al. 1981). The main wave of the outbreak was due to the

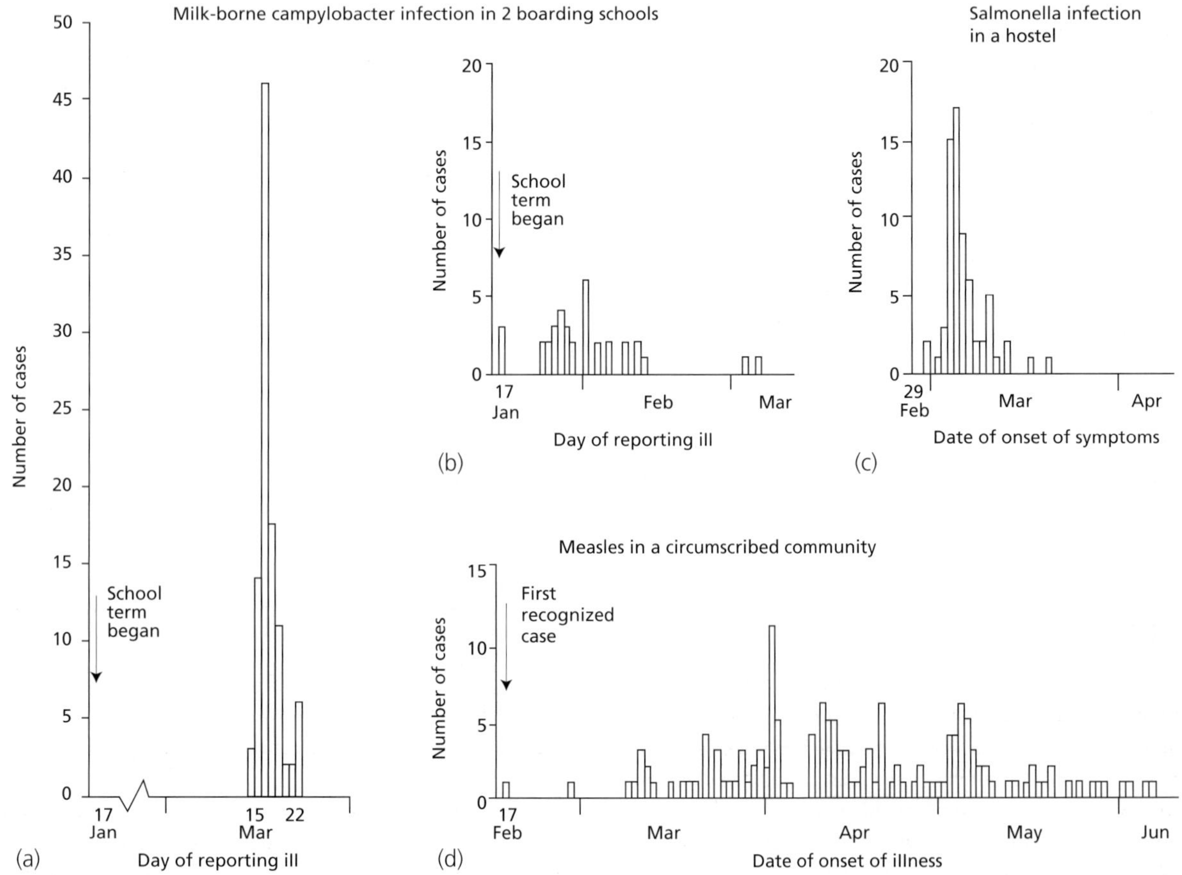

Figure 13.2 *The epidemic curve*

consumption of contaminated meat pie, but subsequent cases were due to person-to-person spread, prolonging the decline in the epidemic curve (Figure 13.2c). In outbreaks propagated from person to person the occurrence of cases will be spread over several incubation periods with peaks at intervals of the incubation period. For example, an outbreak of measles affecting 151 persons in a circumscribed rural community in Oxfordshire between February and June 1981 (Figure 13.2d) showed a smooth epidemic curve but with distinct peaks at 1, 2, 3, and 4 incubation-period intervals after the case (Knightley and Mayon-White 1982). In larger community outbreaks of diseases spread from person to person, the epidemic curve is usually smoother and the peaks at the generation time of the new cases less obvious.

The onset of disease and the epidemic curve should be studied in relation to other events in the environment of the patients; this may draw attention to possible sources of the infection. For example, investigation of a hospital outbreak of Legionnaires' disease in 1983 revealed that a few weeks earlier the domestic hot-water temperatures in the building had been reduced from approximately 55°C at outlets to 45°C, a temperature at which legionellae flourish. Subsequent investigations confirmed that the domestic water supply in the hospital was the source of infection (Palmer et al. 1986).

PLACE

The geographical distribution of disease may provide evidence of its source or method of spread. For example, the crucial factor in the recognition of Lyme disease and the role of tick bites in transmission of *Borrelia burgdorferi* was geographical clustering of presumed childhood rheumatoid arthritis in Old Lyme, Connecticut (Steere et al. 1978). Investigation revealed that the incidence of illness was higher in communities on the east than on the west side of the Connecticut River, and field studies showed that *Ixodes* ticks were particularly abundant in the former area. Case clustering in a particular place of work or neighborhood may indicate the existence of a point-source of infection or of person-to-person spread. In outbreaks of hospital-acquired infection, movement of patients between wards may hide clusters. It is therefore necessary to identify and plot the location of the patients at the time of likely exposure.

PERSON

This includes analysis by age, sex, occupation, and any other relevant characters which preliminary enquiries indicate may be relevant (e.g. food histories, history of travel, leisure activities, and medical or nursing care). For example, food-borne outbreaks of infection due to milk, ice cream, and confectionery characteristically affect children rather than adults. A sudden increase in isolations of *Salmonella* Ealing mainly affecting infants led to the recognition and early control of a nationwide outbreak of salmonellosis due to infant-formula dried milk (Rowe et al. 1987). In contrast, in an outbreak of *Salmonella* Oranienburg infection in Norway in 1981 and 1982, 83 percent of 121 cases were aged 25 years or more, suggesting a food vehicle restricted to adults. This proved to be home-cured meats, the organism originating from contaminated black pepper used as one of the ingredients (Gustavsen and Breen 1984). In another outbreak, of *Salmonella* Cubana infection in hospital patients, the predominance of patients with gastrointestinal dysfunction led to identification of carmine dye, used in investigations, as the vehicle of infection (Lang et al. 1967).

Epidemiological surveys and analytical studies

Descriptive analysis may suggest hypotheses about the source or mode of transmission of an infection, but is not always a sufficient base for introducing control measures. Analytical epidemiology refers to the use of epidemiological techniques to answer specific questions or to test specific hypotheses. The epidemiological approach is complementary to the microbiological. For example, when microbiological investigations reveal legionellae in a cooling tower this does not in itself identify the source of infection because legionellae commonly colonize water systems without causing disease. Epidemiological evidence is necessary to demonstrate an association between exposure to contaminated water and disease.

EXPERIMENTAL AND INTERVENTION STUDIES

Studies of the efficacy of treatments and vaccines are usually undertaken by randomized controlled trials. Random allocation of people to treatment and nontreatment groups is used to overcome bias that can arise if treated and untreated groups differ in underlying factors. For example, if the groups differed in susceptibility to disease, this might produce a favorable result wrongly attributed to the treatment. Treated and untreated groups are then followed to determine the outcome, and incidence rates are compared. With the exception of vaccine trials, most measures of infectious-disease control have not been subject to randomized controlled trials (US Preventive Services Task Force 1996).

Figure 13.3 shows an epidemiological study design for an experimental study.

PREVALENCE AND INCIDENCE STUDIES

In prevalence or cross-sectional studies (Figure 13.4) the aim is to measure the proportion of a population with disease or other variable at a point in time. Prevalence studies are often used in descriptive epidemiology, for

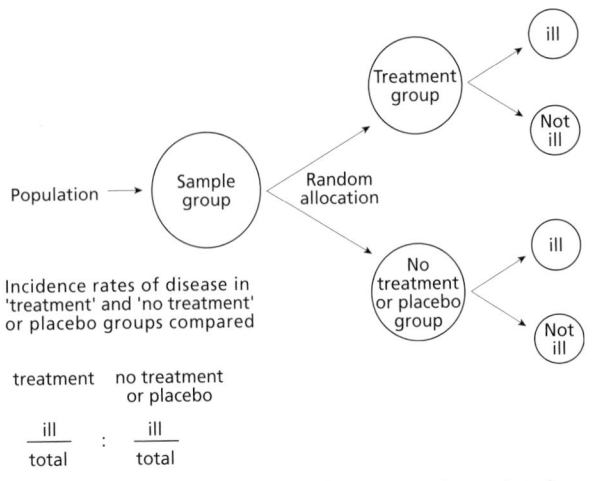

Figure 13.3 *Epidemiological study design – experimental study*

example, in the surveillance of HIV infection in which the proportions of different risk groups who are HIV-antibody positive are calculated at different times to monitor the spread of infection in the population. Crucial to the success of such studies is the representativeness of the sample of the population studied. For example, studies of AIDS and HIV infection are hampered by an inability to identify the population of homosexuals and drug abusers from which to select a representative sample. Patients attending clinics for sexually transmitted infections (STI) probably represent the extreme end of the spectrum of sexual activity and are likely to have the highest prevalence of HIV infection. On the other hand, high risk patients may attend private medical facilities, so that STI clinic populations cannot be considered necessarily representative even of highly sexually active homosexuals, although they may provide a suitable population for monitoring trends in infection.

The methodology of prevalence studies requires the definition of a population and the study sample. The list of names of the population (e.g. electoral register, school register, general practice age-and-sex register) is referred to as the 'sampling frame'; if a sample is to be taken this should be selected either by random, systematic, stratified or cluster sampling (Abramson 1999), so that findings are representative of the practice population. Information on the presence of disease or symptoms is collected by questionnaire or pro-forma, together with other data on personal characteristics or potential risk factors. The proportions of persons with various characteristics or exposures are then compared to identify high risk groups or exposures associated with high prevalence.

Incidence, longitudinal, or follow-up studies measure the rate of occurrence of disease. For these purposes, observations on the population must be made at more than one point in time. Incidence studies may be used descriptively, for example, in following the age incidence of measles to monitor the impact of measles vaccination in a community, in describing the natural history and fatality rates from HIV infection by following cohorts of infected people over several years, or in monitoring crossinfection rates in hospitals. However, they are often also used to test specific hypotheses.

Analytical cohort studies

The analytical cohort study is an application of the incidence study; it attempts to investigate causes of disease by using a natural experiment in which a proportion of a population is exposed and a proportion unexposed. It differs from the true experiment in that exposure is not controlled and may not be random; caution is therefore needed when interpreting the data. If the at-risk population is large, a random sample can be investigated and the results extrapolated to the total population.

Cohort studies may be prospective, when the disease occurs after the study has begun and the characteristics of the population have been identified. For example, in a study to identify risk factors for HIV infection, a serological survey of homosexual men was carried out and seronegative men were enrolled in the study. Baseline data on sexual activity were recorded, and 6 months later the sera of the same men were again tested and seroconversion rates were calculated for groups with a particular sexual behavior. Seroconversion rates were significantly higher in men practicing receptive anal intercourse (Kingsley et al. 1987) than in those who practiced other forms of sexual activity. In a study in the USA of the influence of socio-economic factors on the incidence and outcome of cytomegalovirus infection in pregnancy, two cohorts of women were selected by enrolling those attending a private clinic and those attending a state health department clinic. Seroconversion rates were significantly higher in the lower socio-economic group (Stagno et al. 1986).

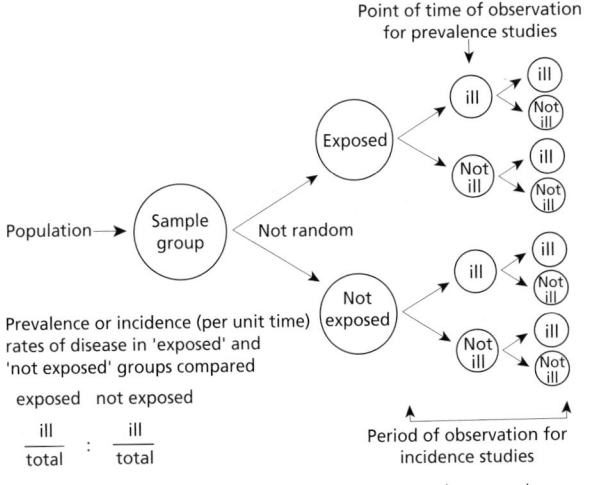

Figure 13.4 *Epidemiological study design – prevalence and incidence studies*

Retrospective or historical cohort studies are possible when the population has been defined and identified previously for other purposes. In investigations of food-poisoning outbreaks that have taken place in an institution or after attendance at receptions it is usually possible to identify retrospectively all those exposed and to relate the attack rates to food consumed (food-specific attack rates).

Several possible types of bias may lead to misinterpretation of results of analytical studies and should be considered at the design stage. One potential problem is that of misclassification of cases and non-cases or misclassification of exposure. For example, in the outbreak already described, of *Salmonella* Typhimurium in 1981 in a university hostel, it was not possible to detect the vehicle of infection by comparing food-specific attack rates if a case definition based upon the presence of gastrointestinal symptoms was used. Feces samples were obtained from the whole cohort and only when symptomless excreters had been excluded from the well group did a significant difference in food-specific attack rates emerge (Palmer et al. 1991). To help overcome this problem it is usual to ask all the individuals in the cohort about symptoms over the appropriate time period, so that those who may have been unrecognized cases can be excluded from the analysis or reclassified as cases. The feasibility of serotesting, swabbing, or otherwise testing non-cases to exclude symptomless infected persons should always be considered. To avoid misclassification of exposure, it is desirable to verify exposures by obtaining records from independent sources (such as medical records, employment records) on all or a sample of the individuals in the study. Even quite small errors of classification can lead to misinterpretation of results as in a cryptosporidiosis outbreak in Nevada in 1994 that may have been wrongly attributed to waterborne transmission (Craun and Frost 2002).

Another important possible bias may arise from 'loss to follow-up'. The loss of cases or non-cases from the study because of refusal to be interviewed or failure to trace patients can seriously bias results since exposures in nonresponders may differ from those of responders. A poor response rate may invalidate the results of a study.

CASE-CONTROL STUDIES

The essential difference between a case-control study (Figure 13.5) and a cohort study is that the former begins with the identification of people with and without the infection and then attempts retrospectively to identify factors associated with disease. In cohort studies, on the other hand, groups of people are identified other than by the presence of disease, and then information on disease occurrence in this group is sought. When the population affected cannot be accurately determined or cases are few, case-control studies are appropriate. Their

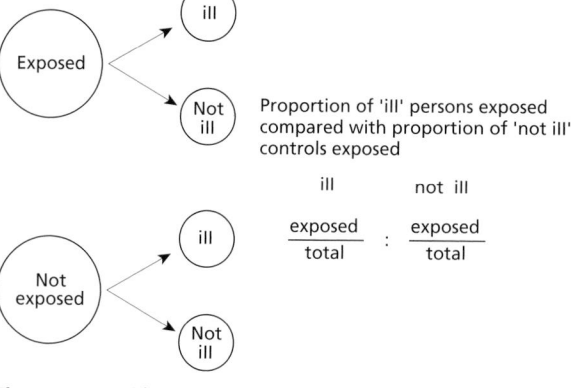

Figure 13.5 *Epidemiological study design – case-control study*

use should be confined to the testing of specific hypotheses. For example, in the outbreak of *Salmonella* Ealing infection in infants, the hypothesis that the vehicle of infection was infant-formula dried milk was tested by a case-control study that showed a strong association between illness and consumption of one particular brand. Subsequently, *Salmonella* Ealing was cultured from the milk powder and from the factory where it was produced (Rowe et al. 1987). Case-control studies are relatively quick and cheap to perform, but their design and analysis can be complex and special attention must be paid to potential sources of bias.

Cases and controls should be representative of the infected and uninfected population, respectively, from which they came and should have had equal opportunity for exposure to the suspected source. Possible bias in the detection of cases may occur if, for example, only patients admitted to hospital are studied; those who have died of fulminant disease or have only mild illnesses may as a result be excluded. Variables significantly associated with disease in a biased sample may merely reflect the factors that caused the bias, for example, admission to hospital. 'Sampling frames' commonly used to select controls include electoral registers, hospital admissions lists, general practitioner age-sex registers, hotel and reception guest lists, family members of cases, neighbors of cases, acquaintances nominated by cases, and persons investigated by the laboratory, but who were negative for the disease in question. When only a few cases are identified the statistical power of the study can be increased by increasing the number of controls per case up to five before the efficiency of the study falls.

Interpretation of differences in the proportions of cases and controls with a particular variable must take into account the possible effect of confounding factors. A confounding factor is one that is not the source of infection but is associated both with the cases and independently with the suspected source. The association with the occurrence of disease may lead to a misinterpretation of the source of the infection. When confounding factors can be reasonably predicted they may be excluded by selecting controls that are matched to cases for exposure to those factors. For example, in the first recognized outbreak of hemorrhagic colitis in the USA in 1982, interviews with cases suggested that food eaten at one fast-food restaurant chain were associated with illness. Since exposure to the particular restaurant chain depended upon the location of the restaurants, and probably also on age, controls were matched with cases for neighborhood of residence and age. A significant association between illness and the restaurant chain was found and subsequently frozen hamburger meat from one restaurant yielded the causative organism *Escherichia coli* O:157 H:7 (Riley et al. 1983).

One development of case-control methodology made possible by laboratory typing techniques is case–case comparison to study causation of common infectious diseases (McCarthy and Giesecke 1999). Cases with the same disease can be divided into etiologically meaningful subgroups by subtyping the pathogen. Cases with one organism subtype are then compared with cases with a different subtype of the same organism. This removes several biases since both groups are drawn from the same population and no differential recall of exposures should occur. For example, a comparison of cases of *Campylobacter coli* infection with cases of *C. jejuni* infection from standardized, population-based sentinel surveillance information in England and Wales found that persons with *C. coli* infection were more likely to have drunk bottled water and were more likely to have eaten paté (Gillespie et al. 2002). Similar studies have compared cases of antibiotic-resistant and antibiotic-sensitive infection in order to identify risk factors for antibiotic resistance (Smith et al. 1999).

The major drawback of case-control studies is that accurate and complete data may not be available retrospectively. Medical records are notoriously incomplete and the recall of patients may be faulty. The latter problem is lessened in acute incidents, because there is usually little delay between the event and the interview. A particular problem is that of 'rumination bias'; because of their illness, sufferers will have gone over in their minds possible exposures and recall may be biased by their own preconceptions or by speculation in the press or other 'media'. Cases may also have been interviewed on many occasions; as well as promoting a more detailed recall, this may introduce bias from suggestions made by interviewers.

STATISTICAL ANALYSIS

In both cohort and case-control studies the basic analysis is made by a comparison of proportions. The data can be presented in a contingency table (Table 13.1).

In cohort studies the ratio $a/a + b$ is the attack rate in the exposed. The ratio $[a/(a + b)]:[c/(c + d)]$ is the ratio

Table 13.1 *Contingency table for cohort and case-control studies*

	Case	Not case	Total
Exposed	a	b	a+b
Not exposed	c	d	c+d
Total	a+c	b+d	a+b+c+d

of the attack rates in the exposed and the unexposed and is called the relative risk. The size of the relative risk is an indication of the causative role of the factor concerned. In case-control studies there are usually no denominators from which risks can be estimated and it is necessary to work with the 'odds' of infection or exposure. The odds of a case having been exposed are a/c and the odds of a control having been exposed are b/d. The odds ratio, ad/bc, or crossproduct ratio, is a very useful measure of association. It approximates to the relative risk when the disease is rare. If there is no association between exposure and infection, the odds ratio will be unity.

In both cohort and case-control studies, dose response effects can be examined if data on different levels of exposure are collected. If the relative risk or odds ratio increases with increasing exposure, the strength of the evidence to determine causality is greatly enhanced. Dose response may be the only way of showing an association between exposure and disease if the exposure is universal (e.g. water consumption).

In analytical studies it is desirable that confidence intervals are presented together with the relative risk or odds ratio, rather than depending solely on *p* values. The *p* value is the probability of obtaining a difference between the proportion of cases and non-cases who are exposed, which is as large or larger than that observed in the study if there is no association between disease and exposure. The *p* value depends not only on the size of the effect but also on the sample size. A study may fail to show an odds ratio or relative risk significantly greater than unity even when a real difference exists because the study size is too small (type II error). Analysis of data sets usually begins by looking at exposure variables one at a time (univariate analysis). Associations between variables and the outcome measures may be causal or they may be due to shared associations with another variable, such as age or sex, known as a confounding variable. The latter can be investigated by stratified analysis in which associations between exposure and disease are examined within subcategories of the confounding variable. The Mantel–Haenszel method is frequently used and logistic regression modelling methods are increasingly used. When matched studies are performed the matching should be preserved in analysis and McNemar's test and the exact binominal probability should be calculated in place of χ^2 and Fisher's exact test. For further discussion of statistical methods see Breslow and Day (1980, 1987) and Altman (1991).

EPIDEMIOLOGICAL SURVEILLANCE

Surveillance needs to be distinguished from research. The former focuses on problem detection and characteristics, whereas research is mainly to do with hypothesis testing. Surveillance systems generate hypotheses and should not be expected to give detailed answers to research questions. Surveillance data provide information for action and as such should stimulate investigation. Emphasis has to be on speed of detection of a potential problem, rather than full and accurate documentation. Consequently, great care has to be exercised in interpreting surveillance statistics. Surveillance systems by definition require ongoing collection of data and consequently rely on minimal data which are often incomplete. National surveillance systems have often gone into decline because data collection has been driven by a desire for full documentation rather than the overriding requirements of timeliness and sensitivity to trends.

One of the earliest recorded surveillance programs was developed in the City of London in the sixteenth and seventeenth centuries to detect the appearance of plague, so that the City administration could decide when to close the theaters of the City to limit the assembly of large crowds of people, and the Royal Court could be advised if and when it was desirable to leave London to escape the disease. The parish clerks of the City were responsible for data collection; each parish appointed two lay searchers to ascertain burials of plague victims, which were then recorded by the clerks along with other burials in the parish burial registers. These data were summated each week by the parish clerks in returns to the Warden of the Hall of Parish Clerks, who then prepared a statistical tabulation of burials by parish. This was published in a weekly bulletin, the 'Bill of Mortality', together with information on the total plague burials compared with the previous week and the number of parishes affected by plague and the number free of the disease (Wilson 1927). The long-term trends in burials in London derived from these Bills of Mortality were studied in 1662 by John Graunt, a City draper, who assessed the validity of the data and demonstrated a higher mortality in towns than in the country by comparing London with Romsey in Hampshire, and drew attention to the high mortality in young children. Greenwood (1948) regarded John Graunt and William Petty, with whom Graunt worked, as the founders of present-day medical statistics.

Thacker and Berkelman (1988) wrote of the USA that: the 'basic elements of surveillance were present in Rhode Island in 1741 when the colony passed an act requiring tavern keepers to report contagious diseases among their patrons. Two years later, the colony passed a law requiring the reporting of small pox, yellow fever and cholera.' Systematic reporting of disease began in the USA in 1874 in Massachusetts, when the State Board of Health introduced voluntary weekly reporting of disease by physicians using a standard postcard.

In the UK, the need for more accurate and complete mortality data led to the introduction of medical certification of death and the civil registration of births, marriages, and deaths in 1836. The General Register Office for England and Wales was established in London, later known as the Office of Population Censuses and Surveys (OPCS) (Nissel 1987) and in 1996 it became part of the Office for National Statistics (ONS). Similar national register offices exist in the other countries of the UK. William Farr, the first Compiler of Abstracts (medical statistician) at the General Register Office, during his 41 years in office, initiated the present international classification of causes of death and developed further the surveillance of communicable disease, establishing a method of influenza surveillance that remains in use today. Dr John Simon, the first medical officer to the Local Government Board (Chief Medical Officer), said of William Farr: 'Eminently he was the man to bring into statistical relief, and to make intelligible to the common mind, whatever broad lessons were latent in the life-and-death registers of that great counting house . . .' (the General Register Office). This provision of readily comprehensible information to those who need it for prompt action remains a crucial component of surveillance.

Surveillance has now assumed even greater importance because of the increased threat of national and international spread of infections arising from the escalating speed, distance, and volume of human travel and the expanding national and international distribution of foodstuffs and other materials which may carry pathogenic organisms as well as from the threat of deliberate release (Lederberg et al. 1992). For example, surveillance may be the only means of detecting outbreaks when the victims have traveled during the incubation period from the place of exposure to many different destinations, or when the vehicle of infection is widely distributed geographically and sometimes also in time. In the face of this threat, national and international surveillance of communicable disease was revived and developed beginning in the 1950s (Langmuir 1963; Raska 1966). More recently, it has been appreciated that increasing human social, technical, environmental, and population change is likely to promote the evolution of new pathogens and facilitate the return of old diseases, now termed 'emerging and re-emerging infections', and that this requires increased surveillance on a global scale to ensure rapid detection, investigation, and control (Heymann and Rodier 1998).

The six main objectives of epidemiological surveillance for communicable disease are:

1 early detection of changes in disease pattern to enable rapid investigation and application of appropriate control measures

2 monitoring long-term trends in disease and infection, including serological surveillance, to assess the need for intervention and to predict future trends

3 determining the prevalent infections in a population so that clinicians may be alerted

4 collation of data about newly recognized or rare diseases at national or international level so that their epidemiology can be described and a basis for research is provided

5 evaluation of disease-control measures and preventive programs

6 planning and costing of health services for the prevention and control of communicable disease.

The method of surveillance remains similar to that of plague surveillance in the seventeenth century, namely:

- the systematic collection of data
- analysis of these data to produce statistics
- interpretation of the statistics to provide information
- timely distribution of this information in a readily assimilable form to all those who require it so that action can be taken
- continuing surveillance to evaluate the action.

The main principles of successful surveillance are simplicity, timeliness, accuracy, and regular analysis and reporting to those who provide data and to those who are responsible for control action (Teutsch and Elliott Churchill 2000). Surveillance is, by definition, an ongoing activity and can be sustained only when the burden placed on the data provider is light. Surveillance data should be limited to the minimum required to meet its specified objectives. Reporting methods should be simple and streamlined. Electronic data collection should be linked to electronic systems for dissemination of high quality surveillance information. For accuracy, surveillance data ideally require clear case definitions as have been developed by the Centers for Disease Control in the USA (Wharton et al. 1990).

Guidelines for the evaluation of surveillance systems have been proposed (Centers for Disease Control 2001). These include the following features:

- a description of the public health importance of the health event including incidence and prevalence, severity of disease as measured by mortality rates and case fatality rates, and preventability
- a description of the system including the objectives, the population under surveillance, case definitions, a flow chart of data collection, details of data transfer, data analysis, and dissemination of information
- a measure of the usefulness of the surveillance system including decisions and actions taken as a result of the information generated
- evaluation of key attributes of the system including: simplicity, flexibility, acceptability, sensitivity, predictive value positive, respresentativeness, and timeliness
- costs of the system.

Data-collection systems

Surveillance data may be sought actively or acquired passively by making use of routinely generated data.

Passive data-collection systems are based upon clinical or microbiological diagnoses which often do not have precise definitions and despite the absence of case definitions, these are invaluable for detecting episodes or cases for further study. For example, in all countries in the UK, cases of typhoid and paratyphoid fevers are detected nationally by statutory notifications, laboratory reports, and referral of cultures of the organisms for identification to the Health Protection Agency (HPA), formerly the Public Health Laboratory Service (PHLS). These three data sources are then linked (Wall et al. 1996). Further active enquiries are made by questionnaire to find out the country where the infection was acquired and possible sources of infection, so that preventive action may be taken. Similarly, cases of Legionnaires' disease are detected by laboratory and incident reports and subsequent active enquiries made, nationally and internationally, to discover cases associated with a common environmental source of infection, so that control measures can be quickly applied.

There is also a surveillance system for general outbreaks of infectious intestinal disease introduced in 1992 by the HPA Communicable Disease Surveillance Center (CDSC). This excludes family outbreaks affecting members of the same private residence only. The CDSC is made aware of outbreaks from laboratory reports, consultants in communicable disease control (CCDC), environmental health officers, and others. Outbreaks are then followed up using a standard request form which documents basic detail which is completed by the local investigator. Participation in the surveillance scheme is entirely voluntary (Wall et al. 1996).

Most active data-collecting systems are based on carefully designed standard case definitions. For example, the surveillance of certain rare childhood disorders, including Reye's syndrome, in the UK is maintained by the British Paediatric Surveillance Unit (BPSU), by monthly mailing of pediatricians to detect cases, followed by subsequent detailed clinical and epidemiological enquiry. These data are then assessed to ensure that all cases meet the standard case definitions before being analyzed to describe the epidemiology of the diseases (Verity and Preece 2002). A unit, similar to the BPSU, was created in 1994 for the active surveillance of neurological disease, the British Neurological Surveillance Unit. A clinical reporting system, set up in 1982 for surveillance of the AIDS epidemic, relies partly on the passive reporting of clinical cases, deaths, and laboratory data for the detection of cases and partly on the active collection of data when these were incomplete. The cases are then scrutinized to ensure that they meet the internationally agreed case definition so that the changing epidemiology of the syndrome can be accurately described (PHLS AIDS Centre 1991).

MORTALITY DATA

Mortality data on communicable diseases have limited use because they do not usually cause death. However, they can be made available quickly, are usually accurate, and probably nearly complete. They have been used in the surveillance of influenza since the epidemic of 1847, now in combination with morbidity data from general practitioners and laboratory reports (Tillett and Spencer 1982), and in the surveillance of AIDS (McCormick 1994). The death entry is a public document and this may sometimes deter the doctor from entering the correct diagnosis on the death certificate, for example in deaths due to syphilis or AIDS, although the doctor may subsequently provide further information about a death in confidence, after the death entry has been completed.

STATUTORY NOTIFICATION

Notification of infectious disease was first introduced in Huddersfield in 1876 by local act of Parliament; other local authorities followed and in 1899 it became mandatory throughout England and Wales. Weekly summaries of these data began in 1910 and were first published in 1922. The Public Health (Control of Disease) Act 1984, Section 11, states 'if a registered medical practitioner becomes aware, or suspects, that a patient whom he is attending . . . is suffering from a notifiable disease' he/she shall notify forthwith the proper officer of the local authority (usually the CCDC). The CCDC in turn sends each week a return of the number of notifications received in the preceding 7 days to the ONS (except leprosy which is reported in strict confidence to CDSC). Weekly summaries of these data are available on the HPA website and are published electronically by CDSC in the weekly *Communicable Disease Report* (CDR). The data are later corrected and quarterly and annual summaries created. Similar systems operate in Scotland and Northern Ireland. The chief advantages of these data are that they are available quickly, they relate to defined populations so that rates by age and sex can be calculated, and they provide an invaluable means of monitoring trends for diseases that are not often confirmed in the laboratory, for example whooping cough and mumps. The defects of the data are that the clinical diagnosis may not always be correct; the diagnosis may vary between clinicians especially because there are no case definitions (except for ophthalmia neonatorum and food poisoning); under-notification frequently occurs and may vary from place to place and at different times (Doyle et al. 2002). These deficiencies can be partially overcome by training notifying clinicians and by regular feedback of local and national information, but it appears that neither the legal obligation to notify nor the payment of a fee promotes more complete notification (McCormick 1993).

In the USA, systems for the notification of selected diseases such as cholera, smallpox, plague, and yellow fever began to be introduced in 1878. By 1903 notification to local authorities for selected diseases was required in all states. In 1925 all states joined a national morbidity reporting system which was taken on after 1948 by the National Office of Vital Statistics which continued to produce weekly morbidity statistics. These reports have been developed as the *Morbidity and Mortality Weekly Report,* which since 1961 has been the responsibility of the Centers for Disease Control (CDC) in Atlanta. Currently, notifiable diseases are reported by states to CDC through the National Notifiable Diseases Surveillance System using the National Electronic Telecommunications System for Surveillance (Koo and Wetterhall 1996).

LABORATORY REPORTING OF MICROBIOLOGICAL DATA

The routine voluntary reporting of laboratory-diagnosed infections forms the core of communicable disease surveillance in England and Wales and also in Scotland, where a somewhat similar reporting system operates. This reporting system was originally developed by the PHLS in the 1940s and 1950s (Grant and Eke 1993) and comprises confidential electronic reporting each week by medical microbiologists to the director of CDSC in England and Wales and the Scottish Centre for Infection and Environmental Health (SCIEH) in Scotland each week of specified infections diagnosed in their laboratories. At CDSC, data are analyzed within a week of receipt to produce tables and line lists and to compile narrative reports for publication in the CDR. Statistics derived from these data are available on-line to staff in CDSC, and via the HPA website to CCDCs and medical microbiologists. The main benefits of laboratory reports are that they are very precise, in that they are based on laboratory-diagnosed infections and the fine typing of the infecting organisms; they often include clinical and epidemiological details; and they allow for free-text comment. Furthermore, the reporting system is flexible, so that any important, unusual or new infections can be reported, even though they were not necessarily included in the original reporting instructions. However, the reports have some drawbacks: they are limited to infections in which there is a suitable laboratory test; infections which are easily diagnosed clinically tend to be poorly covered; and the reports are not population based, that is the data do not usually have a population denominator, so that incidence rates cannot be calculated. Moreover, as with all routine morbidity reporting systems, the data are incomplete, not all laboratories report, and the completeness of the reports received may vary between laboratories and over time, so that trends are sometimes difficult to interpret.

In order to gain a better appreciation of the true incidence of gastrointestinal infections, a national study, the study of Infectious Intestinal Disease (IID) in England was commissioned by the Department of Health in 1990 (Wheeler et al. 1999; Food Standards Agency 2000). This provides the best data to date of the true incidence of enteric pathogens in the country. The objectives of the study were: first, to estimate incidence and etiology of intestinal infectious disease in people presenting to their general practitioners and from whom stool specimens are routinely sent for laboratory examination; second, to compare these data with the data from the national reporting surveillance system; and third, to estimate the prevalence of asymptomatic infection with these agents. Seventy practices were selected to be representative of the socioeconomic characteristics of the area and cases of infectious intestinal disease were defined as persons with loose stools or significant vomiting lasting less than 2 weeks in the absence of a known noninfectious cause and preceded by a symptom-free period of 3 weeks. Vomiting was considered significant if it occurred more than once in the 24-h period and if it incapacitated the case or was accompanied by other symptoms such as cramps or fever. One component of the study was to draw at random a cohort of people registered with the 70 GPs, who were followed up for a period of 6 months. These volunteers agreed to fill out diary cards every week and return them to the GPs. These cards stated whether the person had suffered gastrointestinal (GI) illness or not. If someone developed an illness, a stool sample was submitted.

Of particular interest in this study was the estimate of the carriage of potential pathogens in otherwise healthy people. Ten out of the 2 264 controls (0.4 percent) were found to be excreting *Salmonella*, 16 out of the 2 264 (0.7 percent) were excreting *Campylobacter*, but none out of 2 264 were excreting *E. coli* 0157. These data confirm that verotoxin-producing *E. coli* are a rare infection in the UK at the moment.

The IID study was also able to quantify the reporting pyramid referred to above using the cohort component of the study. It was estimated that for every case the IID reported to the CDSC, six patients were investigated by routine laboratory tests, 23 presented with GI symptoms to their GP, and there were 136 actual cases in the community. For *Salmonella* cases, however, the ratios were much smaller. For every case reported to CDSC, it was estimated that there were 2.3 cases presenting to their GP and 3.2 actual cases in the community.

GENERAL PRACTICE REPORTING OF CLINICAL DATA

Morbidity data from general practice were studied first in 1955, following which several national morbidity studies have taken place, the last in 1991–1992 (McCormick et al. 1995). Continuously collected clinical data from general practice, which is more useful in communicable disease

surveillance, first became available in 1966 when the Royal College of General Practitioners (RCGP) set up a reporting system based on first consultations in a limited number of volunteer practices (Fleming 1999). In 2003 there were 73 participating general practices serving a population of about 615 000 people providing weekly data to the RCGP Research Unit in Birmingham for analysis. These weekly analyses are then sent to the ONS, the CDSC, the Department of Health (DoH), and other organizations concerned with national surveillance. They are published annually by the RCGP Research Unit in the annual reports of the unit. Analogous general practitioner reporting schemes operate in Wales (Palmer and Smith 1991) and some other parts of the UK. These general practitioner reporting systems act primarily as early warning systems providing data rapidly within 10 days of reporting and have the advantages that the data are related to defined practice populations and are unique for some common diseases which are not notifiable and for which laboratory tests are not usually performed, such as chickenpox, herpes zoster, and infectious mononucleosis (Fleming et al. 2002) (see Table 13.2). In the RCGP system, guidelines to diagnosis are provided and in the Welsh system a set of standard case definitions, but the precise definition of an infection is less important than the speedy recognition of an emerging outbreak which may require prompt investigation. However, there are some deficiencies: reporting may not always be complete; the population covered may not be representative geographically or demographically of the whole country and is too small for the surveillance of less common diseases (Harcourt et al. 2004).

REPORTS OF SEXUALLY TRANSMITTED DISEASES

Legislation in the nineteenth and early twentieth centuries required the registration, regular examination and, if necessary, detention of prostitutes, and compulsory admission to hospital for other sufferers from venereal disease (VD) was recommended. However, when a national VD service came into being in 1916, these legal powers were discontinued and a free confidential service for diagnosis and treatment was established at VD clinics, now genitourinary medicine (GUM) clinics. At the same time, a system of quarterly clinic returns began, providing the number of new episodes of specified infections by gender and for some diseases in age groups, which, with variations and additions, has continued ever since. Sexually transmitted infections, including AIDS, have not been made statutorily notifiable in the UK because of fears that a legal obligation to report individual patient data might lead to some individuals concealing their infections, consequently hindering the control of these infections. The clinic returns are analyzed by the UK departments of health and are collated by CDSC and SCIEH; annual and periodic

Table 13.2 *The main routine data-collecting systems*

Disease	Data collecting system				
	Death registration	Statutory notification	Laboratory reports	RCGP reports	GUM clinic reports
Anthrax		++	+		
Brucellosis			++		
Chickenpox	+		+	++	
Herpes zoster	+		+	++	
Cholera		+	++		
Diphtheria		++	++		
Food poisoning	+	++	++	+	
Gonorrhea			+		++
Hepatitis A	+	+v	++	+	
Hepatitis B	+	+	++		+
HIV infection (AIDS)	+		++		+
Hydatid disease	+		++		
Influenza	++		++	++	
Infectious mononucleosis			+	++	
Legionnaires' disease			++		
Leptospirosis	+	+	++		
Malaria	+	+	++		
Measles	+	++	++	++	
Meningitis	+	++	+v	+	
Mumps		++	+	++	
Ornithosis			++		
Pneumonia	+		+	+	
Poliomyelitis		++	++		
Rubella	+	++	++	++	
Shigellosis		++	++		
Syphilis	+		+		++
Tetanus	++	++	++		
Tuberculosis	+	++	++		
Typhoid and paratyphoid		+	++		

GUM, genitourinary medicine; RCGP, Royal College of General Practitioners.

reviews of trends are published, the most recent for the decade 1991–2001 (PHLS, DHSS & PS and Scottish ISD(D)5 Collaborative Group 2001). The clinic returns provide data that are probably accurate, being based on specialist clinical diagnosis, often supported by laboratory tests, are unlikely to vary between clinics and over time, and provide a unique set of data extending over nearly three-quarters of a century for some infections. However, the reports record episodes of infection rather than patients, are incomplete because they do not include patients treated outside GUM clinics, and the proportion of patients treated outside the clinics is unknown and may vary by diagnosis, between clinics and over time.

HOSPITAL DATA

Data from a 10 percent sample of hospital discharges and deaths were available from 1955 to 1985, but had limited use in communicable disease surveillance because the data did not become available until about 2 years after collection (Ashley et al. 1991). In 1995, these data were replaced by Hospital Episode Statistics which are currently being assessed for their possible value in communicable disease surveillance. Paget (1897) in his book *Wasted Records of Disease*, referring to hospital records, commented: 'At present these records of disease are rarely utilized for public purposes, and they represent in their present circumstances little more than so much waste of time, material, and intelligence.' Over a century later, a satisfactory method of capturing these data for the surveillance of communicable diseases and other acute diseases has still to be found.

ROUTINE SEROLOGICAL SURVEILLANCE

In 1990 a serological study to measure the spread of HIV infection in the population was begun; it has continued since and become a routine surveillance system. Samples from sera collected for clinical purposes are unlinked from personal identifiers but remain linked to epidemiological information; sera remaining unused are then tested for HIV infection. This Unlinked Anonymous Prevalence Monitoring Programme has since been extended to include viral hepatitis. The data have helped identify a continuing high rate of transmission among homosexual and bisexual men, and increased hepatitis C transmission in injecting drug users (Report 2002).

OTHER ROUTINE DATA-COLLECTING SYSTEMS

Work-absence data have been used in the surveillance of influenza, but surveillance is now based principally on mortality data, laboratory reports, and general practitioner reports. CCDCs report voluntarily to the CDSC outbreaks occurring in their districts and are required to report serious outbreaks; these reports are usually of food-borne diseases, notably salmonellosis, and provide supplementary data for surveillance of these diseases.

More recently, data from NHS Direct, a national telephone helpline service for health-related enquiries operated by nurses in the UK, have been explored for their potential use in communicable disease surveillance. NHS Direct data on calls for 'influenza-like illness' have been found to correlate well with other routinely available influenza surveillance data (Harcourt et al. 2001). Despite its limitations, the timeliness and population coverage of the service make it a valuable potential source of data for diseases such as influenza or gastro-intestinal illness that are mostly treated by primary care (Cooper et al. 2003).

Action consequent on surveillance

Changes in numbers of reports were hitherto usually discovered by examining the data in either numerical or graphical form, but statistical and computer techniques are now often used to assist in the detection of significant variations from previously recorded experience (Tillett and Spencer 1982; Farrington and Beale 1993). Such variations may indicate changes in the incidence of disease, but they may also result from changing interest of disease, altered diagnostic techniques or reporting methods, or fluctuations in the number of microbiologists, laboratories, clinicians, practices, or clinics participating in the reporting system. It is usual, therefore, to compare the data from all the relevant data-collecting systems to validate the observed trends. Often, especially if there is a rapid increase in reports, field investigation is necessary to substantiate an increase and determine its cause.

Routine, regular, and systematic dissemination of information to the providers of the data and to disease control authorities is essential for successful surveillance (Langmuir 1963). In the United Kingdom this is maintained through the weekly CDR and the monthly *Communicable Disease and Public Health*, which are available without charge to professional staff concerned with the control of communicable disease. These bulletins contain narrative reports of newly identified or suspected outbreaks of disease, reviews of prevalent communicable diseases and related topics as well as numerical data on certain organisms, notifiable diseases, and general practitioner reports. More extensive reviews are published in quarterly reviews, annual reports, and in the medical press. Sometimes when an acute episode of infection is identified, more rapid transmission of information electronically, by phone, or facsimile is required.

The uses of epidemiological surveillance may be illustrated by examples that refer to its main objectives.

EARLY DETECTION FOR RAPID CONTROL

In November 2000, an outbreak detection algorithm identified a cluster of 12 cases with *Salmonella* Enteritidis phage type 4b in data from the National Reference Center for Salmonella at the National Institute of Public Health and the Environment (RIVM) in The Netherlands. A rapid descriptive epidemiological study identified chicken, eggs, or bean sprouts as the possible vehicles of infection. These hypotheses were tested by a case-control study which showed a statistically significant association between infection and eating bean sprouts. Two days after the cluster was detected, the company that produced the sprouts voluntarily took several measures including intensified testing, removal of the raw materials from which the contaminated bean sprouts were grown, and disinfection of the factory premises. No further cases of *Salmonella* Enteritidis phage type 4b were reported (Van Duynhoven et al. 2002).

ASSESSING THE NEED FOR INTERVENTION

Notifications of meningococcal disease increased in UK in the early 1990s with a much higher proportion being notified as meningococcal septicemia. Laboratory surveillance also indicated that there was an increase in the proportion of cases due to *Neisseria meningitidis* serogroup C, particularly serotype C2a infections (Ramsay et al. 1997). In response to this, an enhanced surveillance system for meningococcal disease was begun in January 1998 that combined data from statutory notifications, routine laboratory reporting, and the PHLS Meningococcal Reference Unit (Davison et al. 2002a). The greatest burden of disease was in young children and teenagers, and there was some evidence that sequelae were more common than occurred following serogroup B meningococcal disease. This information was provided to the Joint Committee on Vaccination and Immunisation and provided the basis for a decision to introduce a national meningococcal serogroup C conjugate vaccination program in the UK in November 1999, the first country in the world to do so (Davison et al. 2002b).

The vaccine was incorporated into the routine infant immunization schedule and was offered to all under 18 year olds in a catch-up campaign. The vaccine was well accepted with coverage around 90 percent in infants and 85 percent in schoolchildren up to the age of 14 years. The incidence of serogroup C meningococcal disease in the targeted age groups fell by 80 percent in 2000/2001, and the number of deaths in laboratory-confirmed cases under 20 years of age fell from 78 to eight between 1998/1999 and 2000/2001 (Trotter et al. 2002). The incidence of serogroup B disease increased slightly over this same time period, but the reduction in serogroup C disease has been sustained.

DETERMINING PREVALENT INFECTIONS AS AN AID TO CLINICAL CARE

Following the introduction of methicillin-resistant *Staphylococcus aureus* (MRSA) into Wales in the early 1990s, continuous total population surveillance was introduced. The objectives were to gain an understanding of the extent of MRSA, to describe variations in incidence, to measure the burden of disease, and to identify possible risk factors for infection (Morgan et al. 1999). All first isolates of MRSA from both hospital and community settings and all isolates of methicillin-sensitive *Staphylococcus aureus* (MSSA) from blood or cerebrospinal fluid (CSF) were reported electronically by microbiology laboratories, and the data analyzed at CDSC (Wales). The incidence of MRSA was found to be highest in older patients and in men. Blood and CSF isolates were more likely to come from postsurgical patients. The majority of isolates were resistant to at least two antibiotics in addition to methicillin, most frequently erythromycin and the fluoroquinolones. The surveillance system has thus allowed a simple and intelligible picture of the MRSA problem to be determined, and this information has been fed back to hospitals to assist decisions on infection control.

Important developments in communicable disease may be brought to the attention of clinicians when required, by Chief Medical Officer letters, by reports of specialist associations, and by the medical press. The use of electronic communication is increasingly giving clinicians rapid access to communicable disease surveillance information via the Internet.

DETECTING NEW OR RARE DISEASES FOR STUDY

The surveillance of Reye's syndrome began in 1981. This showed that the annual incidence varied between 0.3 and 0.6 per 100 000 in the British Isles with the highest rate in Northern Ireland. No clear seasonal peaks were seen, but 59 percent of patients had an onset of disease in autumn or winter. The median age was 14 years and the sex distribution equal. The national collection of data on cases that met the standard case definition enabled an analytical study to be made of risk factors in

the disease. This suggested an association with the use of aspirin, similar to that seen previously in the USA (Starko et al. 1980), and led to the withdrawal of pediatric aspirin preparations and cessation of the general use of aspirin in childhood in 1986. The surveillance scheme continued and demonstrated a decline in the number of cases in subsequent years despite more active data collection (Newton and Hall 1993).

EVALUATION OF PREVENTIVE PROGRAMS

In October 1992, *Haemophilus influenzae* type b (Hib) conjugate vaccine was introduced to infants in the United Kingdom with a catch-up programme for those aged under 4 years. An enhanced prospective survey using data from laboratory surveillance showed that during the prevaccination period the majority of cases of invasive Hib occurred in children under 5 years of age and that it was the most common cause of meningitis in this age group. After the introduction of routine Hib immunization, there was a 16-fold reduction in the annual attack rate of invasive Hib disease recorded in children under 5 years of age, but no decrease in the number of infections caused by other serotypes (Slack et al. 1998).

However, although the rate of invasive Hib disease initially decreased dramatically, surveillance showed a resurgence from 1999 onwards. To identify possible reasons for this increase, the effectiveness of the vaccine was investigated. Vaccine efficacy was found to be lower in children vaccinated during infancy, compared with those who were vaccinated during the catch-up campaign, to decline with time since vaccination and to be lower in children born during 2000–2002 (Ramsay et al. 2003). Lower efficacy during 2000–2002 was subsequently found by means of a case-control study to be due to the introduction of combination vaccines that contained acellular pertussis (DTaP-Hib) (McVernon et al. 2003). In February 2003, the Department of Health announced a second catch-up campaign offering all children between 6 months and 4 years a further dose of Hib vaccine (Trotter et al. 2003).

PLANNING AND COSTING OF HEALTH SERVICES

Surveillance data on AIDS and HIV infection have been used to make future projections of the likely numbers of cases of AIDS and severe HIV disease for planning purposes (Report 1996). By employing surveillance data on salmonellosis Sockett (1995) has assessed the costs of the disease and estimated the cost–benefits of preventive measures.

Surveillance of immunization programs

After licencing and general release of a vaccine, continuing surveillance is necessary, not only to demonstrate the effect of the immunization programs on disease incidence, but also to monitor any subsequent changes in vaccine efficacy under field conditions, to identify groups of susceptible subjects, and to detect unsuspected, rare or new vaccine reactions (Begg and Miller 1990). The main objectives of surveillance of immunization programs are, therefore, the continual measurement of efficacy, safety, and uptake of vaccines.

VACCINE EFFICACY

Potency testing

In the UK, vaccine manufacturers are required by the Medicines Act 1976 to submit to the Medicines and Healthcare products Regulatory Agency (MHRA) a detailed application for licencing of all new vaccines. These applications include information about the quality assurance laboratory tests that they will subsequently carry out on all batches of vaccines that they produce before they are released, to ensure they meet agreed standards of efficacy and safety and conform with World Health Organization (WHO) and European Union standards. The National Institute of Biological Standards and Control (NIBSC), acting on behalf of MHRA, receives samples of all batches before release and verifies that they meet the required standards. After release, samples taken from the field are sometimes retested by NIBSC, for example, in the event of apparent vaccine failure, vaccine-associated disease or problems with the vaccine cold chain. In warm climates special measures are required to ensure the stability of vaccines, particularly live virus vaccines, because of their sensitivity to environmental conditions, and cold-chain monitors (temperature-sensitive color cards) are usually included in vaccine packs to monitor these circumstances. These are now also used in many temperate areas to ensure careful attention to vaccine handling and storage.

Causative organisms

Organisms that cause vaccine preventable diseases need to be continually monitored as they may undergo modifications in antigenic structure that may render vaccines less effective. For example, because of continual small antigenic changes in the influenza A virus (antigenic drift) and occasional larger changes (antigenic shift), viral isolates from all parts of the world are collected and studied in WHO collaborating centers, so that the most appropriate strains may be selected for vaccine production (Pereira 1979). Organisms may also undergo changes in the prevalence of different serotypes. For example, the efficacy of whooping cough vaccine appeared to decrease in the 1960s and was found to be due to a change in the prevalent serotypes of B. pertussis (Preston 1965); it was subsequently corrected by the inclusion of the new serotypes in the vaccine.

Serological surveillance

Serological studies provide valuable information about the duration of vaccine-induced immunity in individuals and the extent of immunity in the population. For example, serological surveillance in England and Wales between 1986 and 1991 showed that the proportion of school-aged children who were susceptible to measles was increasing following the introduction of measles, mumps, and rubella (MMR) vaccination in 1988 and, using mathematical models, a major epidemic was predicted in the mid-1990s (Ramsay et al. 1994). A national vaccination program for school-aged children ensued in November 1994 and an epidemic was averted.

Epidemiological studies

Vaccine efficacy (VE) can be calculated from the following equation (where ARu is attack rate in the unvaccinated and ARv is attack rate in the vaccinated):

$$VE = \frac{(ARu - ARv)}{ARu} \times 100$$

The epidemiological techniques for measuring vaccine efficacy in the field were reviewed by Orenstein et al. (1985). There are five main methods:

1 Screening. If a vaccine has a 90 percent efficacy, an attack rate of over 10 percent in vaccinated subjects indicates the need for investigation. Farrington (1993) discusses the method in more detail.
2 Outbreak investigation. The assumption which must be satisfied in the above equation is that both vaccinated and unvaccinated subjects have had equal exposure to infection, an assumption which is likely to be satisfied in outbreaks in confined populations when the attack rate is high (e.g. in institutional outbreaks of measles where case ascrtainment will also probably be complete and immunization records easily available) (Haber et al. 1995). The method can also be applied in community-wide outbreaks in a defined population or in samples of that population.
3 Secondary attack rates in households. This method has been used to determine the efficacy of varicella vaccine, because here again the contacts of the index cases in households are likely to have an equal exposure to infection. By adding together the data from several households, sufficient numbers become available for analysis (Vessey et al. 2001).
4 Routine data-collecting systems. These can be used to estimate vaccine efficacy if the data can be linked with immunization histories. The poportion of cases in the vaccinated (PCV) is related to VE and the proportion of the population vacinated (PPV). Using a variation of the formula:

$$PCV = \frac{PPV - (PPV \times VE)}{1 - (PPV \times VE)}$$

Orenstein et al. (1985) showed that by means of this formula it was possible to estimate vaccine efficacy when the attack rates in vaccinated and in unvaccinated individuals were not known (Figure 13.6). For example, analysis of routine surveillance data on hepatitis A infections linked to immunization histories demonstrated the efficacy of routine hepatitis A vaccination to control the disease in a community with recurrent epidemics (Averhoff et al. 2001).

5 Case-control and cohort studies (Smith et al. 1984). Case-control methods have been used to determine measles and pertussis vaccine efficacy by comparing the immunization status of notified cases with that of children on the child-health computer file, which is available in most parts of the UK (Clarkson and Fine 1987).

VACCINE SAFETY

Common, local or systemic vaccine reactions are likely to be identified in large-scale clinical trials before the introduction of vaccines into general use. Post-licensure data on vaccine reactions are collected both passively and actively in the UK. The principal passive data-collecting system relies on medical practitioners reporting suspected reactions following immunization to the Committee on the Safety of Medicines, using specially designed yellow cards which are available in everyedition of the *British National Formulary*. These reports are reviewed regularly by the Joint Committee on Vaccination and Immunization. Other sources of data are complaints to vaccine manufacturers and informal reports to NIBSC, CDSC, and the National Poisons Centres. Unfortunately, none of these passive data collecting systems normally involves the doctors reponsible for immunization, the immunization coordinators,

and consequently they are liable to remain unaware of adverse reactions taking place locally in their immunization programs (Miller et al. 1998). Active data collection has been used in the surveillance of smallpox vaccine adverse events in civilians in the United States (Centers for Disease Control 2002) and in special studies of bacille Calmette-Guerin (BCG), measles, and MMR vaccines (Begg and Miller 1990).

Investigating whether an adverse event is causally related to immunization poses considerable difficulties. The following four methods have been developed:

1 Case clusters. When clusters of disease or reactions are detected in apparent association with imunization, field investigation may discover the cause and lead to prevention of a recurrence. For example, a cluster of cases of aseptic meningitis associated with measles/mumps/rubella vaccine in Nottingham UK, triggered an investigation that led to the withdrawal of Urabe strain mumps vaccine (Miller et al. 1993). Similarly, after a cluster of infant deaths due to severe metabolic acidosis following immunization was reported in a farming village in Egypt, investigations found that the deaths, and other previously unrecognized illness following immunization, were associated with excessive topical application of methanol as an antipyretic and anti-inflammatory agent following injections (Darwish et al. 2002). The practice was subsequently stopped.

2 Etiological investigations. These may detect the agent of a live vaccine in the lesion which has followed immunization, for example, the CSF in meningoencephalitis after mumps vaccination may be shown to contain vaccine virus (Maguire et al. 1991).

3 Case-control studies. The comparison of the vaccine histories of cases experiencing vaccine reactions with those of a control group, matched for age and sex, may reveal a temporal association between vaccination and the reaction, but this association may not be causal. For example, a strict case-control study showed an association between whooping cough immunization and serious acute neurological illness (Miller et al. 1981), but after several years of further studies and continuing debate it was concluded that this association was not causal (Griffith 1989). Cohort studies, comparing disease in a vaccinated group with that in an unvaccinated group, are not suitable for studying rare vaccine reactions because the incidence in the study population is likely to be too low to find significant differences between the two groups.

4 Case series analysis. Farrington et al. (1996) have described a method for estimating the incidence of clinical events after vaccination compared with a control period, based upon data from the cases alone, which may be used for monitoring vaccine safety when a vaccine is in widespread use. This method has the great advantage of reducing the need to follow up

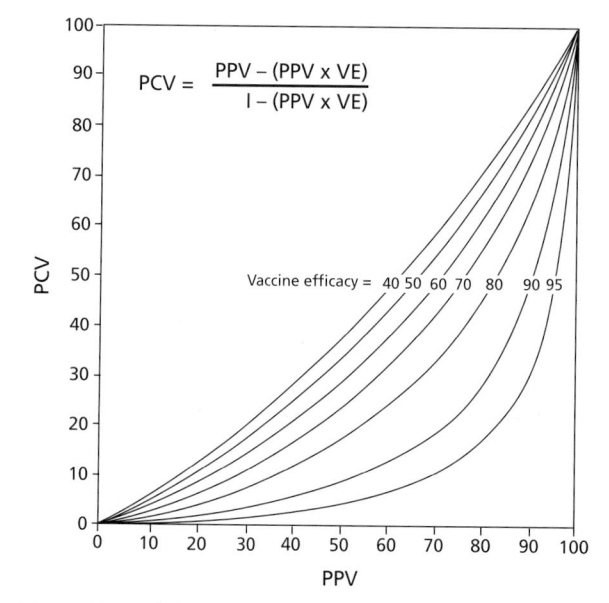

Figure 13.6 *Relationship between PCV and PPV*

large population cohorts or of selecting and investigating control groups. It has been used, for example, to investigate adverse events after pertussis and MMR vaccinations (Farrington and Pugh 1995) and a possible link between oral polio vaccine and intussusception (Andrews et al. 2001).

VACCINE UPTAKE

Vaccine uptake is measured by four main methods:

1 Vaccine usage gives a coarse measure of vaccine uptake. Despite being inaccurate because vaccine wastage and incomplete courses are not taken into account, the method enables major changes to be recognized quickly, which can then be investigated in the field. This method has been used to investigate variation in pneumococcal vaccine uptake between countries (Fedson 1998).
2 Repeated health surveys can also be used to estimate and monitor uptake. They are useful when data on individual immunization status is not readily available from medical data systems. An example is the Behavioral Risk Factor Surveillance System (BRFSS) used in the United States to estimate coverage of influenza and pneumococcal vaccines (Report 2003).
3 Calculated vaccine coverage gives a much more precise measure but requires accurate data on immunized individuals and up-to-date population denominators. An example is the Cover of Vaccination Evaluated Rapidly (COVER) program used in the UK to calculate uptake in the routine childhood immunization programme. Using a computerized national child-health register, quarterly cohorts of children are studied just after reaching the age of immunization to determine the numbers immunized. Quarterly uptake rates are calculated using the resident population denominators and the results made available quickly to immunization coordinators (Begg et al. 1989).
4 The WHO Expanded Programme of Immunization has developed a variety of techniques for rapid assessment of overall population vaccination coverage. Cluster sampling has been used to randomly select groups of children whose immunization histories are then recorded and used as an estimate of the vaccine uptake of the area sampled. Repeated sampling over time can reveal trends in uptake (Henderson and Sundaresan 1982). Lot quality assessment sampling is a method based on industrial techniques for quality control of manufactured products. It is a form of stratified sampling. Rather than check every item in a lot to determine the number of defective items, a sample of the lot is taken and a predetermined level of risk defined in order to decide whether to accept or reject the entire lot (Hoshaw-Woodward 2001). This method has been used to assess tetanus vaccine coverage in the WHO neonatal tetanus elimination campaign (Cotter et al. 2003).

THE INVESTIGATION OF OUTBREAKS

In the past, the detection of outbreaks of disease relied mainly on the appearance of groups of cases associated in time or place. With the advent of epidemiological surveillance it became possible to search actively for less apparent outbreaks and other variations in disease pattern, recently aided by the use of computer programs (Farrington and Beale 1993). The investigation of outbreaks, or indeed a single case, requires a systematic approach (Goodman et al. 1990) to achieve rapid and effective disease control. This can also be assisted by computer programs, for example, 'Epi-Info' produced by the Centers for Disease Control, Atlanta, now widely used throughout the world. The epidemiological methods used for this purpose were pioneered by John Snow in his famous study of cholera near Golden Square, London in 1854 (Snow 1855) and remain in use today. They may be considered under six main headings:

1 preliminary enquiry
2 management
3 identification of cases and collection and analysis of data
4 control
5 communication
6 further epidemiological and laboratory studies (Palmer and Swan 1991).

It is not always appropriate to follow this sequence of action; the order will depend upon the particular circumstances of an outbreak and often several of the steps are taken at the same time. There is a useful list of sources of information available in the UK for communicable disease control, designed particularly for CCDCs (Morgan et al. 1992). A detailed account of the methods of field investigation can be found in Gregg (2002).

Preliminary enquiry

The objects of the preliminary enquiry are:

- to confirm that there is in reality an outbreak
- to verify the provisional diagnosis of the disease
- to agree a case definition for epidemiological investigation
- to formulate tentative hypotheses of the source and spread of the infection and
- to initiate immediate control measures if required.

CONFIRMING THE OUTBREAK

An increase in the reported number of cases of a disease may not necessarily be caused by an outbreak, but could be due to changes in recognition or reporting. For example, the increase may be related to improved ascertainment of cases following the introduction of new or more sensitive diagnostic procedures, to the need to detect as many cases as possible because of the availability of a new specific treatment, or to more extensive investigation of a disease because of special interests of new clinicians or microbiologists. Reporting may increase because of changes in the population size or structure, or as a consequence of improved data handling procedures such as computerization, or as a result of false positive laboratory or other tests, or because of the misinterpretation of the original data (Shears 1996).

CONFIRMING THE DIAGNOSIS

The clinical diagnosis can usually be established by a study of the case histories of a few affected persons. Laboratory tests are essential to confirm this in most infections, but epidemiological investigations should begin as soon as possible and not normally be delayed until the laboratory results become available.

CASE DEFINITION

A clear case definition should be agreed at this stage and applied consistently throughout the investigation by all investigators. This is of particular importance in a previously unrecognized disease or one in which there are no satisfactory confirmatory laboratory tests.

TENTATIVE HYPOTHESIS

The preliminary enquiry should include detailed interviews with a few of the affected persons, so that obvious common features may be identified quickly; for example, the association with a specific food, or contact with a particular person or place. Hypotheses can then be developed of the sources and mode of spread of the infection and a questionnaire designed to test these hypotheses in subsequent analytical tudies.

IMMEDIATE CONTROL

It may be possible in some circumstances to take immediate control measures based on the tentative hypothesis before this can be confirmed, so that further cases may be prevented. For example, in serious infections that spread from person to person, such as diphtheria, hepatitis B or poliomyelitis, as soon as the diagnosis is suspected it is necessary to identify individuals who may have been the source of infection so that they are isolated if appropriate, and to identify those who may have been exposed to infection so that they can be traced and given protection by vaccines or chemotherapy. If a common vehicle or source of infection is suspected, appropriate action should be taken to interrupt the spread and control the source.

Management of an incident

If the preliminary enquiry confirms that the outbreak is indeed real, an early decision should be made on the management of the investigation. Small outbreaks will usually be managed by the CCDC with the appropriate and usually essential assistance of a consultant microbiologist and an environmental health officer. In hospitals, an outbreak will usually be managed by the infection control doctor assisted by the infection control nurse (Hospital Infection Working Group 1995). The CCDC is obliged by the Public Health (Infectious Diseases) Regulations 1988 to inform the Chief Medical Officer for England, or for Wales as appropriate, and the CDSC of 'any serious outbreak of any disease' and of cases of disease subject to the International Health Regulations (meaning here cholera, plague, smallpox, and yellow fever) (see also Table 13.3). He/she should also report directly to CDSC any case of leprosy, malaria, or rabies contracted in Great Britain, and viral hemorrhagic fever. In serious incidents, that is large outbreaks, severe diseases, geographically widespread outbreaks, and those of public interest, an outbreak control team should be formed. This should include, in addition to the CCDC, a consultant medical microbiologist, a CDSC consultant epidemiologist, and a consultant physician in infection. Other agencies may also be involved, for example in suspected water-borne disease these may include the local water companies, river authorities, veterinary services, and the Department of the Environment, Food and Rural Affairs (DEFRA) (Report of the Group of Experts 1990). The control team will require an administrator or nonmedical epidemiologist to manage an 'incident room' where information on the outbreak should be collated and made available to those who require it. The control team should define the responsibilities of its members and allocate to one person the task of spokesperson for the press and other news media, often through the local authority, health authority, or DoH press officers. The control team should meet frequently until the acute incident is over when its tasks may be devolved to the local authorities, health authorities, and the CDSC.

Identification of cases, collection, and analysis of data

The cases first reported in an outbreak usually comprise only a small proportion of the total and may not be a representative sample. Investigation of these cases alone may be misleading for three main reasons:

1 in diseases spread from person to person, such as diphtheria, missed cases or carriers may be responsible for spreading the infection

2 in point-source outbreaks, such as those of Legionnaires' disease or food-borne diseases, the presenting cases may have come to light because of a chance association with a place or potential vehicle of infection

3 without knowing the population from which the cases have come, the denominator of 'persons at risk' is not

Table 13.3 *Statutorily notifiable diseases*

Notifiable diseases	
Under the Public Health (Control of Disease) Act 1984	
Cholera	Relapsing fever
Food poisoning	Smallpox
Plague	Typhus
Under the Public Health (Infectious Diseases) Regulations 1988	
Acute encephalitis	Ophthalmia neonatorum
Acute poliomyelitis	Paratyphoid fever
Anthrax	Rabies
Diphtheria	Rubella
Dysentery (amoebic and bacillary)	Scarlet fever
Leprosy	Tetanus
Leptospirosis	Tuberculosis
Malaria	Typhoid fever
Measles	Viral hemorrhagic fever
Meningitis	Viral hepatitis
Meningococcal septicemia (without meningitis)	Whooping cough
Mumps	Yellow fever

Notes: 'Viral hemorrhagic fever' means Argentine hemorrhagic fever (Junin), Bolivian hemorrhagic fever (Machupo), Chikungunya fever, Congo/Crimean hemorrhagic fever, dengue fever, Ebola virus disease, hemorrhagic fever with renal syndrome (Hantaan), Kyasanur forest disease, Lassa fever, Marburg disease, Omsk hemorrhagic fever, and Rift Valley disease.
There are minor differences in notifiable diseases in Scotland and Northern Ireland (see Ashley et al. 1991).
Some diseases are notifiable locally, for example, psittacosis in Cambridge.
AIDS is not statutorily notifiable, but clinicians report cases voluntarily, in strict confidence, to the directors of the CDSC in England and Wales and of the SCIEH in Scoland. Advice about reporting is available from these centers from genitourinary medicine physicians.

available to calculate attack rates and so determine whether or not a particular exposure was associated with an increased incidence of the disease.

Identification of the exposed population will enable thorough case finding to be accomplished, for example, by scrutinizing school or hotel registers, lists of institutional residents, pay-rolls, and other occupational records, nominal rolls of travelers, and lists of persons attending functions associated with the disease. If such studies are not possible, as is often the case in community outbreaks, case finding may be undertaken by reviewing routine mortality and morbidity data used for communicable disease surveillance, by individual household enquiry, and by special appeal to medical and lay persons through the press and other news media.

The aim of the enquiry will be to collect data from those affected and those at risk but not affected, by careful questionnaire. The data routinely sought from cases include name, date of birth, gender, occupation, recent travel, immunization history, date of onset of symptoms, description of the illness, and the names and addresses of the medical attendants. Other details will depend on the nature of the infection and possible modes of spread. The case data will then be analyzed by time, place, and person to determine the mode of spread, source of infection, and persons who may have been exposed. The date or time of onset of symptoms of cases should be plotted on a graph so that the type of epidemic curve can be recognized, and on a map to reveal the geographical distribution of the cases. Cases that do not conform to the time or geographical distribution may sometimes provide valuable evidence of the source of infection. For example, in John Snow's classic study of the outbreak of cholera near Golden Square, Soho, in 1854, a widow died of the disease in West Hampstead, several miles from Soho, and provided the most conclusive evidence of the source of infection because she had previously resided in Broad Street and had delivered to her daily a large bottle of water from her favourite supply, the Broad Street pump (Snow 1855). Comparisons of attack rates by age, gender, location, food history, and other appropriate parameters may provide evidence of the source and spread of infection which may then be confirmed by analytical epidemiological studies already described.

Control

Control measures may be directed towards the source of infection, the mode of transmission, people at risk, or a combination of these.

CONTROL OF THE SOURCE

The source may be human, animal, or environmental.

Human source

Infections derived from a human source can be controlled by the physical isolation of cases and carriers and, if necessary, treatment until they are free from infection, provided that the cases and carriers are easily identified and carrier rates are low. For example, it is possible to control diphtheria and typhoid fever in this way because cases can be recognized clinically and confirmed by laboratory tests and because carrier rates are low and carriers can readily be detected microbiologically. In contrast, meningococcal infection is not susceptible to control by isolation because pharyngeal carrier rates in excess of 20 percent in the population are common during outbreaks and it is usually impractical to detect and isolate all the carriers. The method of physical isolation used depends upon the mode of spread and severity of the disease; for example, negative-pressure plastic isolators are used in the UK for the African viral hemorrhagic fevers, room isolation for diphtheria, and special precautions in the disposal of secretions and excretions (isolation by 'barrier nursing') in typhoid fever. Isolation of human sources of infection can also be accomplished in some diseases by 'ring immunization', that is, by encircling the case or carrier with a barrier of immune persons which blocks the spread of infection to susceptible people outside the 'ring'. This method was first used in Leicester in the nineteenth century and later throughout the UK for smallpox control; when modified and accompanied by meticulous surveillance worldwide it helped to achieve the eradication of the disease (Fenner et al. 1988). It has also been applied in the control of poliomyelitis by mass immunization and in measles by immunization of contacts in school and institutional outbreaks.

Animal source

When animals are the source of an infection, it is sometimes possible to control an outbreak by eradication of that source, for example, in ornithosis by slaughtering the birds and disinfecting the cages and premises. Rabies may be controlled by the destruction of rabid animals and of wild or stray animals that may harbor the virus, and by the muzzling of domestic dogs. Outbreaks of food-borne zoonoses are usually controlled by removing the vehicle of infection: eradication plays a major part in the long-term control of these zoonoses, for example, bovine tuberculosis and brucellosis.

Environmental source

The source of Legionnaires' disease and primary amoebic meningoencephalitis, both recognized in the 1970s, is water in the environment. Outbreaks of Legionnaires' disease are usually associated with water-cooling systems of air-conditioning plants, water-distribution systems in large buildings, or whirlpool spas. Primary amoebic meningoencephalitis is usually contracted from warm unchlorinated natural water sources, such as warm springs, sometimes used to supply

swimming pools. Both these infections may be controlled by appropriate cleansing and disinfection. Other pathogens, mainly gram-negative bacteria, may contaminate the inanimate environment of man; for example, salmonellae are found on work surfaces, utensils and equipment in kitchens, and coliform organisms and pseudomonads on surfaces and equipment in hospitals. Control measures include appropriate cleansing, disinfection or sterilization of these environmental sources.

CONTROL OF SPREAD

Food- and water-borne infections can be curtailed by withdrawal from sale of the contaminated product or treatment of the product to render it safe, for example, the recall of a contaminated meat product (Macdonald and Drew 2000), and the pasteurization of a contaminated milk supply (Proctor and Brosch 1995).

Direct contact infections may be reduced by avoiding contact, for example, in cutaneous anthrax by not handling potentially contaminated animal products such as unsterilized bone meal.

Indirect contact infections, for example, staphylococcal infections in hospitals, may be reduced by strict aseptic techniques and meticulous hand washing, accompanied in the case of wound infection by the use of 'no touch' dressing techniques.

Fecal–oral spread, such as in bacillary dysentery and hepatitis A infection, may also be restricted by scrupulous attention to hand washing and by the careful and frequent disinfection of surfaces in lavatories and toilet areas.

Percutaneous spread by insects, such as in malaria, typhus, and yellow fever, may be controlled by vector destruction, protective clothing, and insect repellents. Infections spread by percutaneous inoculation, such as hepatitis B and HIV, may be prevented by measures to avoid accidental inoculation and the contamination of broken skin by infected blood or tissue fluids.

Air-borne spread may be limited by ventilation in buildings and in special circumstances the physical isolation of highly susceptible subjects.

CONTROL OF PERSONS AT RISK

It is sometimes possible to control disease by active or passive immunization or by chemoprophylaxis in persons exposed to risk. For example, active immunization against measles within 72 h of exposure may prevent the disease. Human immunoglobulins are commonly given for passive protection, for example, rabies immunoglobulin in addition to active immunization as part of post-exposure prophylaxis given as soon as possible after exposure, hepatitis B immunoglobulin in post-exposure prophylaxis after accidental inoculation injury and for infants of carrier mothers, and zoster immunoglobulin in the newborn and immunodeficient patients after exposure to chickenpox. Chemoprophylaxis is offered to close contacts of cases of meningococcal disease, such as household contacts and children in the same class at school, and long-term chemoprophylaxis against pneumococcal infection is given to patients after splenectomy.

Communication

Accurate and timely information about an outbreak is of major importance, especially if the outbreak is of national interest and is receiving wide publicity. To achieve this, it is helpful for the CCDC to maintain a check list of authorities and individuals to be informed (Morgan et al. 1992). These would normally include the local consultant physician in infection, microbiologist, chief environmental health officer, and director of public health, but depending upon the type of outbreak might also include the local divisional veterinary officer, employment medical adviser, education officer, and clinicians in hospital and general practice, and similar persons in neighboring areas. In a major incident, national authorities including the HPA and its CDSC, the DoH, the Food Standards Agency (FSA), DEFRA, the Health and Safety Executive, and others may be involved. An initial written report should normally be completed within 24–48 h after the preliminary enquiry and circulated to all the authorities and individuals concerned. A final report suitable for publication should be produced at the end of the investigation and interim reports may be necessary if the investigation is complex or prolonged. In a major incident it may also be necessary for frequent bulletins of updated information to be issued by the spokesperson of the 'outbreak control team' from the 'incident room'.

Further epidemiological and microbiological studies

Analytical studies may be necessary to confirm the association of disease with a particular vehicle or source of infection. Often case-control studies are undertaken early in an enquiry because they can be accomplished quickly and cheaply, but more time-consuming cohort studies may be required to determine the relative risk so that the most cost-effective long-term control measures can be implemented.

An outbreak may suggest a previously unknown source or vehicle of infection and microbiological surveys may be needed to discover the extent, frequency, and mode of transmission so that long-term control measures can be designed and implemented. For example, the discovery that bulk egg products were a vehicle of infection of paratyphoid B and other salmonella infections led to extensive studies of the bacteriology of these products and to improved hygienic methods of production and pasteurization. The identification of Legionnaires' disease and its association with cooling towers of air-conditioning systems and with piped water systems in large buildings led to studies of the growth of legionellae in these systems and subsequently to improvements in design and maintenance to prevent colonization and growth of the organism.

MATHEMATICAL MODELS OF INFECTIOUS DISEASE

The idea of modelling epidemic diseases dates back to a seminal paper published at the beginning of the twentieth century by Hamer (1906), which suggested that cycles of measles infections and deaths were not due to changes in the pathogenicity of the measles virus, but to changes in the number of susceptible and immune individuals over time. Mathematical models have grown in sophistication over the last few decades particularly with increases in computing power; however, all are based on the logical deductions proposed by Hamer. Infectious disease models are used not only to predict epidemics of disease, but also to help understand disease transmission patterns, to measure the burden of infection, and to anticipate the impact of infection control measures and immunization policies.

The potential for an infectious disease to spread from person to person is known as the basic reproductive rate (R_0). It represents the average number of secondary infections produced in a totally susceptible population by a single case during the entire infectious period. The value of R_0 depends on:

- the probability, for a given disease, that an infected person will transmit infection to a susceptible person
- the frequency of contacts with susceptible persons
- the period of time for which an infected person is infectious.

The value of R_0 is normally signified by the equation:

$$R_0 = \beta \times \kappa \times D$$

where β is the risk of transmission per contact, κ is the number of contacts, and D is the duration of infectivity of the case.

When a new disease enters a virgin population, if $R_0 > 1$ there will be an epidemic, if $R_0 = 1$ the disease will become endemic, and if $R_0 < 1$ the disease will eventually disappear. Childhood diseases generally have a high R_0 and confer lifelong immunity, since nearly everyone is exposed to them early in life, but subsequently protected from disease. For example, classic studies by Hope Simpson (1952) on the spread of childhood infectious diseases among household contacts in a rural area of England derived values for attack rates (β) of 0.80 for measles, 0.72 for chickenpox, and 0.38 for mumps. A case of measles is thought to be capable of infecting 15 other people on average in a susceptible western population, and so has an $R_0 = 15$.

In practice, populations are seldom totally susceptible. Furthermore, if the disease is one that gives rise to immunity and no more susceptible people enter the population, then the number of susceptible people will decline with time and more and more of a person's contacts will be with people who are immune. The chance of spreading the disease gradually becomes less and less, and the actual reproductive rate (R) of the infection decreases (although the basic reproductive rate (R_0) remains the same) until the epidemic eventually dies out.

The steps that are involved in developing and using a model for an infectious disease have been described in detail (Habbema et al. 1996). They are illustrated in Figure 13.7.

The kind of model structure chosen will depend on the various transition states of the infection, for example, between the categories susceptible, infectious, and immune. A simple model could be represented as follows:

Susceptible → Infected → Infectious → Immune

This model would be appropriate for a disease such as measles. Estimates are required for the transition intervals between each category in order to begin to develop the model. If the model is intended to describe the long-term dynamics of an infection then other parameters such as births, migration patterns, and deaths may need to be incorporated. The model population may also need to be split into subgroups if, for example, the dynamics of infection in different age groups are to be studied.

The next step is to choose the model type, usually either a deterministic or stochastic model. Deterministic or compartmental models aim to describe broadly what happens in the population as a whole. They tend to categorize individuals into subgroups or compartments and describe the transition between categories using average transition rates. They require fewer data and are relatively easy to set up. For example, an age-structured deterministic model showed that routine varicella immunization of preschool children would probably result in a substantial reduction in uncomplicated chickenpox and in cases requiring hospitalization, but that the number and age distribution of cases would depend strongly on the characteristics of the vaccine (Halloran et al. 1994). Stochastic models (or Monte-Carlo simulation models) often try to describe the effect of disease on every individual in the population. They do this by incorporating chance variation into the disease transmission process in order to model a range of possible outcomes. Compared to deterministic models they are more intuitive, they can follow disease dynamics in all individuals in the population, and they can provide estimates of the range within which an outcome may lie, taking account of chance variation. They are, however, laborious to set up and

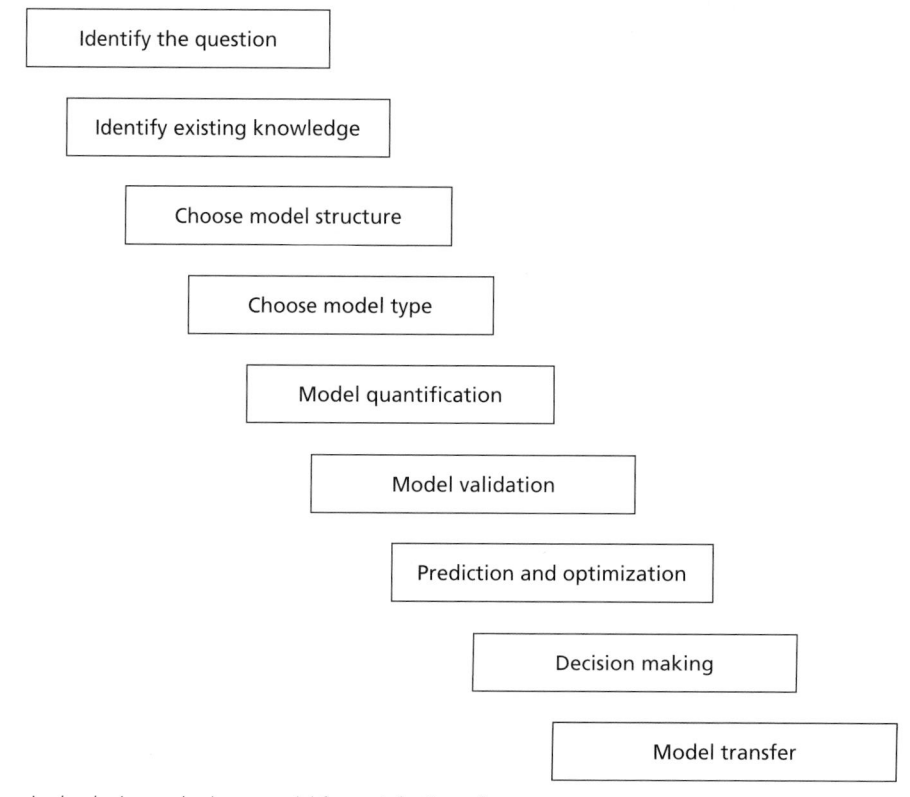

Figure 13.7 *Steps in developing and using a model for an infectious disease*

many simulations need to be run in order to obtain useful predictions. For example, this approach has been used to model the epidemic of hepatitis C in France based on the temporal pattern of deaths from hepatocellular cancer (Griffiths and Nix 2002).

Once the form of the model has been determined, the rates and distribution of flows between disease transition states need to be quantified. This stage is often problematic because of the lack of reliable data. Key data requirements include the probability that a specific infectious and susceptible individual comes into sufficient contact to transmit infection during a given time period (β), the risk of a specific individual becoming infected per unit time (related to the incidence rate in person years or force of infection), becoming infectious per unit time, and becoming immune per unit time. Sometimes primary data collection is required to make up for deficiencies in the data available. Models then need to be validated by checking outputs against known data, and optimized by testing various model assumptions in a process of sensitivity analysis, before the results of the model can be applied to decision-making and public health practice.

REFERENCES

Abramson, J.H. 1999. *Survey methods in community medicine*, 5th edn. Edinburgh: Churchill Livingstone.

Altman, D.C. 1991. *Practical statistics for medical research*. London: Chapman and Hall.

Andrews, N., Miller, E., et al. 2001. Does oral polio vaccine cause intussusception in infants? Evidence from a sequence of three self-controlled cases series studies in the United Kingdom. *Eur J Epidemiol*, **17**, 701–6.

Ashley, J.S.A., Cole, S.K. and Kilbane, M.P.J. 1991. Health information resources: United Kingdom – health and social factors. In: Holland, W.W., Detels, R. and Knox, G. (eds), *Oxford textbook of public health*, Vol. 2. . Oxford: Oxford University Press, 30–53.

Averhoff, F., Shapiro, C.N., et al. 2001. Control of hepatitis A through routine vaccination of children. *JAMA*, **286**, 2968–73.

Beaglehole, R., Bonita, R. and Kjellstrom, T. 1993. *Basic epidemiology*. Geneva: WHO.

Begg, N. and Miller, E. 1990. Role of epidemiology in vaccine policy. *Vaccine*, **8**, 180–9.

Begg, N.T., Gill, O.N. and White, J.M. 1989. COVER (Cover of Vaccination Evaluated Rapidly): description of the England Wales scheme. *Public Health*, **103**, 81–9.

Breslow, N.K. and Day, N.E. 1980. *The analysis of case-control studies. Statistical methods in cancer research*, Vol. 1. . Lyon: WHO International Agency for Research on Cancer.

Breslow, N.E. and Day, N.E. 1987. *The design and analysis of cohort studies. Statistical methods in cancer research*, Vol. 2. . Lyon: WHO International Agency for Research on Cancer.

Centers for Disease Control. 2001. Updated guidelines for evaluating public health surveillance systems: recommendations from the Guidelines Working Group. *MMWR*, **50**, (No. RR-13).

Centers for Disease Control. 2002. Smallpox vaccine adverse events monitoring and response system for the first stage of the smallpox vaccination program. *MMWR*, **52**, 88–9.

Clarkson, J.A. and Fine, P.E.M. 1987. An assessment of methods for routine local monitoring of vaccine efficacy, with particular reference to measles and pertussis. *Epidemiol Infect*, **99**, 485–99.

Cooper, D.L., Smith, G.E., et al. 2003. What can analysis of calls to NHS direct tell us about the epidemiology of gastrointestinal infections in the community? *J Infect*, **46**, 101–5.

Cotter, B., Bremer, V., et al. 2003. Assessment of neonatal tetanus elimination in an African setting by lot quality assurance cluster sampling (LQA-CS). *Epidemiol Infect*, **130**, 221–6.

Craun, G.F. and Frost, F.J. 2002. Possible information bias in a waterborne outbreak investigation. *Int J Environ Hlth Res*, **12**, 5–15.

Darwish, A., Roth, C.E., et al. 2002. Investigation into a cluster of infant deaths following immunization: evidence for methanol intoxication. *Vaccine*, **20**, 3585–9.

Davison, K.L., Crowcroft, N.S., et al. 2002a. Enhanced surveillance scheme for suspected meningococcal disease in five regional health authorities in England: 1998. *Commun Dis Pub Hlth*, **5**, 205–12.

Davison, K.L., Ramsay, M.E., et al. 2002b. Estimating the burden of serogroup C meningococcal disease in England and Wales. *Commun Dis Pub Hlth*, **5**, 213–19.

Decker, M.D., Booth, A.L., et al. 1986. Validity of food consumption histories in a foodborne outbreak investigation. *Am J Epidemiol*, **124**, 859–63.

Detels, R., McEwen, J., et al. 2002. *Oxford textbook of public health*, 4th edn. Vol. 2. *Methods of public health*. Oxford: Oxford Medical Publishers.

Doyle, T.J., Glynn, M.K. and Groseclose, S.L. 2002. Completeness of notifiable infectious disease reporting in the United States: an analytical literature review. *Am J Epidemiol*, **155**, 866–74.

Farrington, C.P. 1993. Estimation of vaccine effectiveness using the screening method. *Int J Epidemiol*, **22**, 742–6.

Farrington, C.P. and Beale, A.D. 1993. Computer-aided detection of temporal clusters of organisms reported to the Communicable Disease Surveillance Centre. *Commun Dis Rep*, **3**, R78–82.

Farrington, P. and Pugh, S. 1995. A new method for active surveillance of adverse events from diphtheria/tetanus/pertussis and measles/mumps/rubella vaccines. *Lancet*, **345**, 567–9.

Farrington, C.P., Nash, J. and Miller, E. 1996. Case series analysis of adverse reactions to vaccines; a comparative evaluation. *Am J Epidemiol*, **143**, 1165–73.

Fedson, D.S. 1998. Pneumococcal vaccination in the United States and 20 other developed countries. *Clin Infect Dis*, **26**, 1117-1123, 1981–96.

Fenner, F., Henderson, D.A., et al. 1988. *Smallpox and its eradication*. Geneva: WHO.

Fleming, D.M. 1999. Weekly returns service of the Royal College of General Practitioners. *Commun Dis Pub Hlth*, **2**, 96–100.

Fleming, D.M., Smith, G.E., et al. 2002. Impact of infections on primary care – greater than expected. *Commun Dis Pub Hlth*, **5**, 7–12.

Food Standards Agency. 2000. *A report of the study of infectious intestinal disease in England*. London: The Stationery Office.

Gillespie, I.A., O'Brien, S.J., et al. 2002. A case-case comparison of *Campylobacter coli* and *Campylobacter jejuni* infection: a tool for generating hypotheses. *Emerg Infect Dis*, **8**, 937–42.

Glynn, J.R. and Palmer, S.R. 1992. Incubation period, severity of disease, and infecting dose: evidence from a *Salmonella* outbreak. *Am J Epidemiol*, **136**, 1369–77.

Goodman, R.A., Buehler, J.W. and Koplan, J.P. 1990. The epidemiologic field investigation: science and judgment in public health practice. *Am J Epidemiol*, **132**, 9–16.

Grant, A.D. and Eke, B. 1993. Application of information technology to the laboratory reporting of communicable disease in England and Wales. *Commun Dis Rep*, **3**, R75–8.

Greenwood, M. 1948. *Medical statistics from Graunt to Farr*. Cambridge: Cambridge University Press.

Gregg, M.B. 2002. *Field epidemiology*, 2nd edn. Oxford: Oxford University Press.

Griffith, A.H. 1989. Permanent brain damage and pertussis vaccination: is the end of the saga in sight? *Vaccine*, **7**, 199–210.

Griffiths, J. and Nix, B. 2002. Modelling the hepatitis C virus epidemic in France using the temporal pattern of hepatocellular carcinoma deaths. *Hepatology*, **35**, 709–15.

Gustavsen, S. and Breen, O. 1984. Investigation of an outbreak of *Salmonella oranienburg* infections in Norway, caused by contaminated black pepper. *Am J Epidemiol*, **119**, 806–12.

Habbema, J.D., De Vlas, S.J., et al. 1996. The microsimulation approach to epidemiologic modeling of helminthic infections, with special reference to schistosomiasis. *Am J Trop Med Hyg*, **55**, Suppl. 5, 165–9.

Haber, M., Orenstein, W.A., et al. 1995. The effect of disease prior to an outbreak on estimates of vaccine efficacy following the outbreak. *Am J Epidemiol*, **141**, 980–90.

Halloran, M.E., Cochi, S.L., et al. 1994. Theoretical epidemiologic and morbidity effects of routine varicella immunization of preschool children in the United States. *Am J Epidemiol*, **140**, 81–104.

Hamer, W.H. 1906. Epidemic disease in England – the evidence of variability and persistence of type. *Lancet*, **March 17**, 733–9.

Harcourt, S.E., Edwards, D.E., et al. 2004. How representative is the population used by the RCGP spotter practice system. Using Geographical Information Systems to assess. *J Publ Hlth*, **26**, 88–94.

Harcourt, S.E., Smith, G.E., et al. 2001. Can calls to NHS Direct be used for syndromic surveillance? *Commun Dis Pub Hlth*, **4**, 178–82.

Henderson, R.H. and Sundaresan, T. 1982. Cluster sampling to assess immunization coverage: a review of experience with a simplified sampling method. *Bull WHO*, **60**, 253–60.

Hennekens, C.H. and Buring, J.E. 1987. *Epidemiology in medicine*. Boston/Toronto: Little Brown and Company.

Heymann, D.L. and Rodier, G.R. 1998. Global surveillance of communicable diseases. *Emerg Infect Dis*, **4**, 362–5.

Hope Simpson, R.E. 1952. Infectiousness of communicable diseases in the household (measles, chickenpox and mumps). *Lancet*, **ii**, 549–54.

Hoshaw-Woodward, S. 2001. *Description and comparison of the methods of cluster sampling and lot quality assessment sampling to assess immunization coverage*. Geneva: World Health Organization.

Hospital Infection Working Group. 1995. *Hospital infection control. Guidance on the control of infection in hospitals*. London: DOH and PHLS.

Kingsley, L.A., Detels, R., et al. 1987. Risk factors for seroconversion to human immunodeficiency virus among male homosexuals. Results from a Multi-center AIDS Cohort Study. *Lancet*, **1**, 345–9.

Knightley, M.J. and Mayon-White, R.T. 1982. Measles epidemic in a circumscribed community. *JR Coll Gen Prac*, **32**, 675–80.

Koo, D. and Wetterhall, S.F. 1996. History and current status of the National Notifiable Diseases Surveillance System. *J Pub Hlth Mgmt Pract*, **2**, 4–10.

Lang, D.J., Kunz, J., et al. 1967. Carmine as a source of nosocomial salmonellosis. *New Engl J Med*, **276**, 829–32.

Langmuir, A.D. 1963. The surveillance of communicable disease of national importance. *New Engl J Med*, **268**, 182–92.

Lederberg, J., Shope, R.E. and Oaks, S.C. 1992. *Emerging infections. Microbial threats to health in the United Slates*. Washington, DC: National Academic Press.

Macdonald, C. and Drew, J. 2000. Outbreak of *Escherichia coli* O157:H7 leading to the recall of retail ground beef – Winnipeg, Manitoba, May 1999. *Can Commun Dis Rep*, **26**, 109–11.

McCormick, A. 1993. The notification of infectious diseases in England and Wales. *Commun Dis Rep*, **3**, R19–25.

McCormick, A. 1994. The impact of human immunodeficiency virus on the population of England and Wales. *Pop Trend*, **76**, 1–7.

McCormick, A., Fleming, D. and Charlton, J. 1995. *Morbidity statistics from general practice: fourth national study 1991–1992, Series MBS, No.3*. London: HMSO.

McVernon, J., Andrews, N., et al. 2003. Risk of vaccine failure after *Haemophilus influenzae* type b (Hib) combination vaccines with acellular pertussis. *Lancet*, **361**, 1521–3.

Maguire, H.C., Begg, N.T. and Handford, S.G. 1991. Meningoencephalitis associated with MMR vaccine. *Commun Dis Rep*, **1**, R60–1.

McCarthy, N. and Giesecke, J. 1999. Case–case comparisons to study causation of common infectious disease. *Int J Epidemiol*, **28**, 764–8.

Millard, P.S., Gensheimer, K.F., et al. 1994. An outbreak of cryptosporidiosis from fresh-pressed apple cider. *JAMA*, **272**, 1592–6.

Miller, D.L., Ross, E.M., et al. 1981. Pertussis immunization and serious acute neurological illness in children. *Br Med J*, **282**, 1595–9.

Miller, E., Goldacre, M., et al. 1993. Risk of aseptic meningitis after measles, mumps, and rubella vaccine in UK children. *Lancet*, **341**, 979–82.

Miller, E., Waight, P. and Farrington, P. 1998. Safety assessment post-licensure. *Dev Biol Stand*, **95**, 235–43.

Morgan, D., O'Mahony, M. and Stanwell-Smith, R.E. 1992. From the briefcase to the bookshelf: information sources for communicable disease control. *Commun Dis Rep*, **2**, R91–5.

Morgan, M., Salmon, R., et al. 1999. All Wales surveillance of methicillin-resistant *Staphylococcus aureus* (MRSA): the first year's results. *J Hosp Infect*, **41**, 173–9.

Newton, L. and Hall, S.M. 1993. Reye's syndrome in the British Isles: report for 1990/91 and the first decade of surveillance. *Commun Dis Rep*, **199**, R11–16.

Nissel, M. 1987. *People count. A history of the General Register Office*. London: HMSO.

Orenstein, W.A., Bernier, R.H., et al. 1985. Field evaluation of vaccine efficacy. *Bull WHO*, **63**, 1055–68.

Paget, C.E. 1897. *Wasted records of disease*. London: Edward Arnold, pp. 78–9.

Palmer, S.R. and Biffin, A. 1987. The changing incidence of human hydatid disease in England and Wales. *Epidemiol Infect*, **99**, 693–700.

Palmer, S.R. and Smith, R.M.M. 1991. GP surveillance of infections in Wales. *Commun Dis Rep*, **3**, R25–8.

Palmer, S.R. and Swan, A.V. 1991. The epidemiological approach to infection control. *Rev Med Microbiol*, **2**, 187–93.

Palmer, S.R., Jephcott, A.E., et al. 1981. Person-to-person spread of *Salmonella typhimurium* phage type 10 after a common-source outbreak. *Lancet*, **1**, 881–4.

Palmer, S.R., Zamiri, I., et al. 1986. Legionnaire's disease cluster and reduction in hospital hot water temperatures. *Br Med J (Clin Res)*, **292**, 1494–5.

Pereira, M.S. 1979. Global surveillance of influenza. *Br Med Bull*, **35**, 9–14.

PHLS AIDS Centre. 1991. The surveillance of HIV-1 infection and AIDS in England and Wales. *Commun Dis Rep*, **1**, R51–R6.

PHLS, DHSS & PS, and Scottish ISD (D) 5 Collaborative Group. 2001. *Sexually transmitted infections in the UK: new episodes seen at genitourinary medicine clinics, 1991–2001*. London: Public Health Laboratory Service.

Preston, N.W. 1965. Effectiveness of pertussis vaccines. *Br Med J*, **2**, 11–13.

Proctor, M.E. and Brosch, R. 1995. Use of pulsed-field gel electrophoresis to link sporadic cases of invasive listeriosis with recalled chocolate milk. *Appl Environ Microbiol*, **61**, 3177–9.

Ramsay, M., Gay, N., et al. 1994. The epidemiology of measles in England and Wales: rationale for the 1994 national vaccination campaign. *Commun Dis Rep*, **199**, 141–6.

Ramsay, M., Kaczmarski, E., et al. 1997. Changing patterns of case ascertainment and trends in meningococcal disease in England and Wales. *Commun Dis Rep*, **7**, 49–54.

Ramsay, M.E., McVernon, J., et al. 2003. Estimating *Haemophilus influenzae* type b vaccine effectiveness in England and Wales by use of the screening method. *J Infect Dis*, **188**, 481–5.

Raska, K. 1966. National and international surveillance of communicable diseases. *WHO Chron*, **20**, 315–21.

Report. 1996. The incidence and prevalence of AIDS and prevalence of other severe HIV disease in England and Wales for 1995 to 1999: projections using data to the end of 1994. *Commun Dis Rep*, **6**, R1–R24.

Report. 2002. *Prevalence of HIV and hepatitis infections in the United Kingdom 2001*. Annual report of the Unlinked Anonymous Prevalence Monitoring Programme. London: Department of Health.

Report. 2003. Influenza vaccine coverage among adults aged ⩾50 years and pneumococcal vaccine coverage among adults aged ⩾65 years – United States, 2002. *MMWR*, **52**, 987–92.

Report of the Group of Experts. 1990. *Cryptosporidium in water supplies,* London: HMSO, 45–56.

Richardson, M., Elliman, D., et al. 2001. Evidence base of incubation periods, period of infectiousness and exclusion policies for the control of communicable diseases in schools and preschools. *Ped Infect Dis J*, **20**, 380–91.

Riley, W., Remis, R.S., et al. 1983. Hemorrhagic colitis associated with a rare *Escherichia coli* serotype. *New Engl J Med*, **308**, 681–5.

Rothman, K.J. 2002. *Epidemiology: an introduction.* New York/Oxford: Oxford University Press.

Rowe, B., Begg, N.T., et al. 1987. *Salmonella ealing* infections associated with consumption of infant dried milk. *Lancet*, **2**, 900–3.

Shears, P. 1996. Pseudo-outbreaks. *Lancet*, **347**, 138.

Slack, M.P., Azzopardi, H.J., et al. 1998. Enhanced surveillance of invasive *Haemophilus influenza* disease in England, 1990 to 1996: impact of conjugate vaccines. *Ped Infect Dis J*, **17**, Suppl, S204–7.

Smith, K.E., Besser, J.M., et al. 1999. Quinolone-resistant *Campylobacter jejuni* infections in Minnesota, 1992–1998. *New Engl J Med*, **340**, 1525–32.

Smith, P.G., Rodrigues, L.C. and Fine, P.E.M. 1984. Assessment of the protective efficacy of vaccines against common diseases using case-control and cohort studies. *Int J Epidemiol*, **13**, 87–93.

Snow, J. 1855. *On the mode of communication of cholera.* London: Churchill.

Sockett, P.N. 1995. The epidemiology and costs of diseases of public health significance, in relation to meat and meat products. *Food Safety*, **15**, 91–112.

Stagno, S., Pass, R.F., et al. 1986. Primary cytomegalovirus infection in pregnancy, incidence, transmission to fetus and clinical outcome. *JAMA*, **256**, 1904–8.

Starko, K.M., Ray, C.G., et al. 1980. Reye's syndrome and salicylate use. *Pediatrics*, **66**, 859–64.

Steere, A.C., Broderick, T.F. and Malawista, S.E. 1978. Erythema chronicum migrans and lyme arthritis: epidemiologic evidence for a tick vector. *Am J Epidemiol*, **1980**, 108, 312–21.

Teutsch, S.M. and Elliott Churchill, R. (eds) 2000. *Principles and practice of public health surveillance*, 2nd edn. Oxford/New York: Oxford University Press.

Thacker, S.B. and Berkelman, R.L. 1988. Public health surveillance in the United States. *Epidemiol Rev*, **10**, 164–90.

Tillett, H.E. and Spencer, I. 1982. Influenza surveillance in England and Wales using routine statistics. *J Hyg Camb*, **88**, 83–94.

Trotter, C.L., Ramsay, M.E. and Kaczmarski, E.B. 2002. Meningococcal serogroup C conjugate vaccination in England and Wales: coverage and initial impact of the campaign. *Commun Dis Pub Hlth*, **5**, 220–5.

Trotter, C.L., Ramsay, M.E. and Slack, M.P. 2003. Rising incidence of *Haemophilus influenzae* type b disease in England and Wales indicates a need for a second catch-up vaccination campaign. *Commun Dis Pub Hlth*, **6**, 55–8.

US Preventive Services Task Force. 1996. *Guide to clinical preventive services,* 2nd edn. Washington, DC: US Department of Health and Human Services.

Van Duynhoven, Y.T.H.P., Middowson, M.A., et al. 2002. *Salmonella enterica* serotype Enteritidis phage type 4b outbreak associated with bean sprouts. *Emerg Infect Dis*, **8**, 440–3.

Verity, C. and Preece, M. 2002. Surveillance for rare disorders by the BPSU. The British Paediatric Surveillance Unit. *Arch Dis Child*, **87**, 269–71.

Vessey, S.J., Chan, C.Y., et al. 2001. Childhood vaccination against varicella: persistence of antibody, duration of protection, and vaccine efficacy. *J Pediatr*, **139**, 297–304.

Wall, P.G., DeLouvois, J., et al. 1996. Food poisoning: notifications, laboratory reports, and outbreaks -- where do the statistics come from and what do they mean? *Commun Dis Rep*, **6**, R93–R100.

Wharton, M., Chorba, T.L., et al. 1990. Case definitions for public health surveillance. *MMWR*, **39**, RR–13.

Wheeler, J.G., Sethi, D., Infections Intestinal Disease Study Executive, et al. 1999. A study of infectious intestinal disease in England: rates in the community, presenting to general practice, and reported to national surveillance. *Br Med J*, **318**, 1046–50.

Wilson, F.P. 1927. *The plague in Shakespeare's London.* London: Clarendon Press.

14

Theory of infectious disease transmission and control

ANGUS NICOLL, NIGEL J. GAY, AND NORMAN T. BEGG

INTRODUCTION

Successful control of an infectious disease requires an understanding of the characteristics and behavior of the agent causing disease and its interaction with the hosts it infects and with the environment. This may seem an obvious statement, however history is littered with failed attempts at controlling infectious diseases through ignorance of basic issues such as the mode of transmission, natural reservoirs of infection, and the role of carriers (Table 14.1).

For example, in the early nineteenth century cholera was thought to be an airborne disease. Control efforts at that time were based on remedies to ward off contaminated atmosphere ('miasma'). It was not until 1854 that John Snow, a British epidemiologist, showed that cholera was a water-borne infection. Snow's discovery was based on the simple observation, during a cholera epidemic, that attack rates were highest in people who obtained their drinking water from one particular pump (Figure 14.1). His remedy – to remove the handle of the pump and thereby stop the outbreak – revolutionized the control of cholera and presaged a series of public health measures aimed at improving the safety of drinking water. Unfortunately, ignorance about cholera control persists to this day. When the seventh cholera pandemic reached Latin America in 1991, the reaction of many countries was to introduce a requirement for all visitors to be vaccinated. Cholera vaccine may provide some individual protection against disease, but does not prevent cases from excreting *Vibrio cholerae*. As a public health measure it therefore has no value in controlling the spread of infection (Finkelstein 1984).

A particular area of medicine in which knowledge of disease transmission and so how control can be effected, is immunization. There are several characteristics of an infection that determine how readily it may be controlled by vaccines. These characteristics include the infectivity of the disease, the length of the incubation period, the duration of naturally acquired and vaccine-induced immunity, and the presence or absence of subclinical infection and nonhuman hosts (Noah 1988). An especially important concept for immunization is herd immunity, the proportion of a population or group that are immune to a specific infection (Topley and Wilson 1923; Anderson and May 1985b; Last 2001). Mathematical models that incorporate these effects can be used to predict the outcome of alternative vaccine strategies, under varying assumptions about vaccine efficacy and vaccine coverage. This type of information has been pivotal in the design of successful immunization programs against diseases such as smallpox, poliomyelitis, and measles, and predicting the effectiveness or otherwise of proposed interventions.

In this chapter we will describe the basic theory of infectious disease transmission, and consider how the application of theory contributes to the control of communicable disease in man.

Table 14.1 *Factors affecting transmission and determining means of infectious disease control*

Factors affecting transmission	
Reservoirs of infection	Human
	Animal (Zoonoses)
	Environmental
Modes of transmission	Respiratory – droplets, aerosols
	Oral–gastrointestinal
	Parenteral
	Fomites
	Sexual
	Vectors
	Vertical (mother-to-child)
Period and duration of infectivity	
Subclinical infectious carriers	
Level of infectiousness (R_0 and R)	
Behavior and contact patterns of those infected	
High risk situations and core groups	
Effectiveness of treatments in reducing infectiousness	
Resulting immunity following infection or immunization	
Genetic susceptibility	

HISTORICAL PERSPECTIVE

The theory of infectious disease transmission has its origins in the first decade of the twentieth century. Following the discovery of germ theory in the nineteenth century, scientists tried to understand the cycles in the incidence of many infectious diseases. In particular, they sought an explanation for the regularly recurring epidemics of diseases, such as measles, and why these epidemics peaked and died out before all susceptible persons had been infected. In 1906, William Hamer proposed a model in which he considered the cases to occur in discrete, 2-week generations. The number of cases of disease in one generation was assumed to be a constant multiple of the number of cases in the previous generation and the number of susceptibles. On this simple assumption, Hamer was able to explain the epidemic nature of measles: the regular cycles were caused by the depletion of the number of susceptibles during an epidemic followed by a gradual reaccumulation as more children were born. His argument, however, was not widely accepted at the time. An alternative, empirically derived theory, due to Brownlee (1907, 1909), was based on Farr's (1840) observations of the shape of epidemic curves. According to this theory, the infectivity of the agent was reduced by a constant factor with each successive generation of cases in the course of an epidemic: the decay of an epidemic was due to this attenuation of the infective agent and not to a

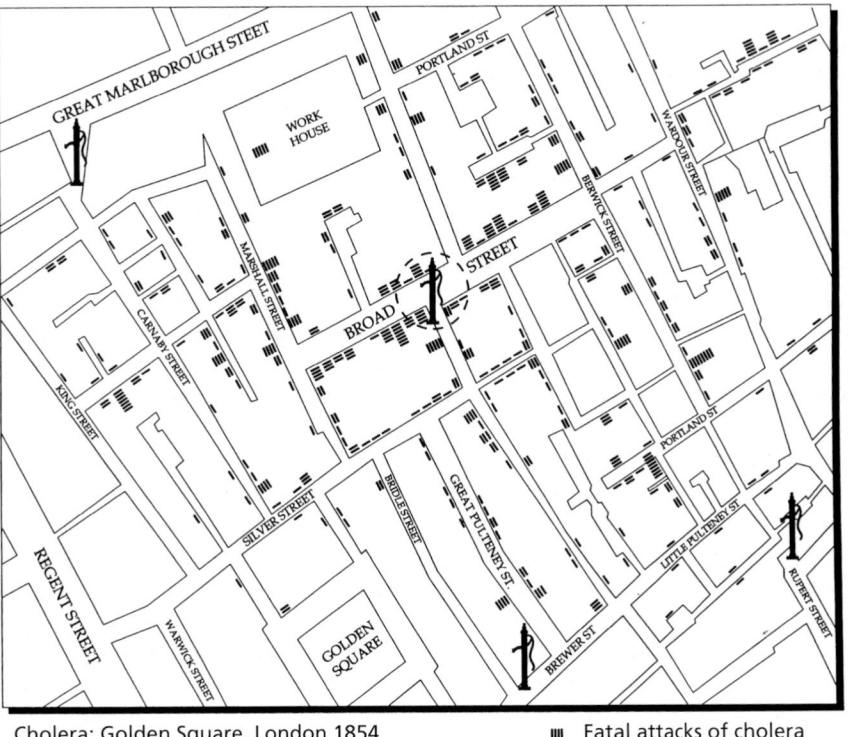

Cholera: Golden Square, London 1854   ▥ Fatal attacks of cholera

Figure 14.1 *Distribution of cholera deaths, Golden Square, 1854*

lack of susceptible persons. (An account of Brownlee's work is given by Fine (1979).) Contemporary with this debate, Ross (1909, 1911) proposed a model of malaria transmission along similar lines to Hamer's theory of measles. Both Ross and Hamer were well aware that their models were a great simplification of the heterogeneity of real populations, but realized that they captured the fundamental mechanisms underlying the observed patterns of disease incidence.

It was against this background that Topley and Wilson commenced their classic series of experimental studies of the spread of infection among mice. The carefully controlled conditions in the mouse populations were ideal for testing the competing theories. At the outset of this research Topley (1919a, b, c) set out the main problems to be investigated in his Goulstonian Lectures to the Royal College of Physicians of London. It is clear that he was influenced by Brownlee's ideas, but did not dismiss Hamer's theory: 'Though reasons have been given for believing that the outstanding feature in the subsidence of an epidemic is a loss of infectivity by the bacterial virus [sic], yet the resistance of the host cannot be a negligible factor. It will operate by decreasing the concentration of susceptible individuals, and hence the chances of successful transference.' After 5 years of experiments, Topley and Wilson (1923) were led to the conclusion that 'the question of immunity as an attribute of a herd should be studied as a separate problem, closely related to, but in many ways distinct from, the problem of the immunity of an individual host'. Theirs was the first published reference of the term 'herd immunity'. Its use indicates that the focus of the research was moving away from the properties of the infective agent alone and towards also considering properties of the herd or population. Their most seminal finding was that the addition of a number of susceptible mice to a population in which a bacterial parasite was at equilibrium produced an outbreak of disease (Topley 1923). Similar observations in a human population, on the incidence of diphtheria in a boarding school, were made by Dudley (1926): outbreaks recurred when the admission of sufficient susceptible new boys had removed the herd immunity. Two decades later experiments by Webster (1946) in the United States confirmed this. Gradually, Hamer's theory became accepted as the experimental evidence mounted, while no evidence to support Brownlee's theory of varying infectivity was obtained. A full account of the series of mouse experiments in Britain is given in the final Medical Research Council report (Greenwood et al. 1936).

During this period there was further theoretical progress. Kermack and McKendrick (1927) proved the existence of an epidemic threshold, i.e. that an epidemic would occur if, and only if, the proportion susceptible exceeded a given level. Soper (1929) provided a clear statement of Hamer's model and its continuous time equivalent. He showed that the continuous model

produced damped oscillations around the equilibrium solution and derived the period of small oscillations. He also noted the similarity with some laws for chemical reactions, hence the model is sometimes referred to as being of 'mass action'. The inability of this model to yield the undamped epidemic cycles observed for measles was seen as a major flaw, until Bartlett (1956) demonstrated that a stochastic formulation of the Hamer–Soper model overcame this problem. His numerical simulations generated recurrent epidemics with no tendency to the equilibrium. He noted that the size of the population influenced the pattern of epidemics: larger towns had more regular cycles, whereas 'fade out' of infection often occurred in simulations with a small population. This led him to derive a critical community size above which fade out was unlikely (Bartlett 1957), a result which accorded well with observations. Bailey provided the first comprehensive account of 'The mathematical theory of epidemics' in which these developments were discussed (Bailey 1957). The text also included methods for estimating parameters (such as latent and infectious periods) from epidemiological data.

Meanwhile, in the United States, Reed and Frost were developing a model which has become the basis for the study of outbreaks in small populations. This included a nonlinear term to discount the effect of two or more infectious individuals contacting the same susceptible person (Frost 1976). Fox et al. (1971) explored the effects of superimposing family and other structures on a Reed–Frost model, the first study to include heterogeneity in contacts between members of a population. Their stochastic simulations led them to emphasize the important role that small pockets of susceptible individuals could play in sustaining transmission.

The concept of the basic reproduction number (R_0) was introduced to infectious disease epidemiology by Macdonald (1957) in the context of his studies of malaria. R_0 has become a central feature of the theory of infectious diseases since Dietz (1975) and Hethcote (1983) first used it in the study of directly transmitted infections. It provides a simple method of comparing diseases and the difficulty of eliminating them. Dietz (1975) showed how, in a simple homogeneous model, R_0 could be estimated from the average age at infection. Diekmann et al. (1990) provided a rigorous mathematical framework for the definition of R_0 in models of disease transmission in heterogeneous populations. These include the models with age-dependent transmission rates developed by Anderson and May (1984, 1985a, b) and others (Schenzle 1984; Hethcote 1983) that have been used to investigate the dynamics of a wide range of infections and to explore the effects of different vaccination programs (Anderson and Grenfell 1986; Babad et al. 1995; Anderson and May 1985a, 1990; Anderson et al. 1987; Halloran et al. 1994). These considerations are now applied to many other infections apart from those for which effective vaccines are available and have become increasingly influential on policy decisions relating to control of infectious disease, including when traditional control measures are applied to new infections such as severe acute respiratory syndrome (SARS) and avian influenza (WHO 2003; Anderson et al. 2004). Mollison et al. (1994) has reviewed developments across the entire spectrum of modeling approaches and highlighted areas in which further research is needed. Particular developments took place in the 1980s and 1990s when attention was focused on the epidemiology of human immunodeficiency virus (HIV) and acquired immune deficiency syndrome (AIDS). Models for sexually transmitted diseases and other sexually transmitted infections (STI) often require many population subgroups with widely differing activity levels and the long incubation period for AIDS and its changes in infectivity present further complications (Johnson et al. 1992; Boily et al. 2000; Aral 2000). Most recently, field observations were combined with modeling approaches in real time to rapidly determine how best to control severe acute respiratory syndrome (SARS) (WHO 2003).

BASIC PRINCIPLES

Infection in the individual

There are several stages in an episode of infectious disease (Figure 14.2). The process begins when a susceptible individual is exposed to an infectious case. If transmission of the infectious organism occurs, and the organism begins to multiply in the tissues of the exposed individual, then the individual can be considered to have been infected. Not all exposures result in transmission, and of these only a proportion progress to clinical disease. At this early stage the individual is not yet infectious – this is known as the latent period of infection.

The next stage is when the infected individual begins to shed the organism from the respiratory or gastro-intestinal tract or in body fluids, such as blood or urine. At this point the individual is infectious, i.e. capable of transmitting the organism to others. The period of latency has now ended. This next stage marks the beginning of the period of infectivity.

Throughout both the latent period and the early stages of the period of infectivity, the individual has no symptoms. The onset of the first symptom heralds the end of the incubation period, which is defined as the interval between the initial exposure and the onset of symptoms. For some infections, symptoms appear at the same time as the patient becomes infectious. For others, infectiousness rises as symptoms worsen, while for a third group it is more usual for infectivity to precede symptoms by a few days. It is unusual for infectivity to be constant through the infectious period.

The next stage is reached when the individual is no longer infectious. Normally this occurs while symptoms are present. In many infections the organisms are cleared from the body and immunity develops. In some diseases, shedding of the organism can continue after symptoms have resolved. This tends to occur particularly in immunosuppressed individuals. When shedding of the organism becomes persistent, then the individual has become a carrier. Knowledge of carrier states is particularly important in understanding the dynamics of disease transmission. Some patients may be asymptomatic carriers for years, for example as occurs in typhoid fever, while for some persistent viral infections such as hepatitis B and C and HIV, individuals may remain infectious, but without symptoms for many years (Heyman 2004; Aral 2000).

In some infections the symptoms resolve and the patient is no longer infectious, but the organism remains dormant, i.e. the infection re-enters the latent period. The disease may then subsequently reactivate, some-

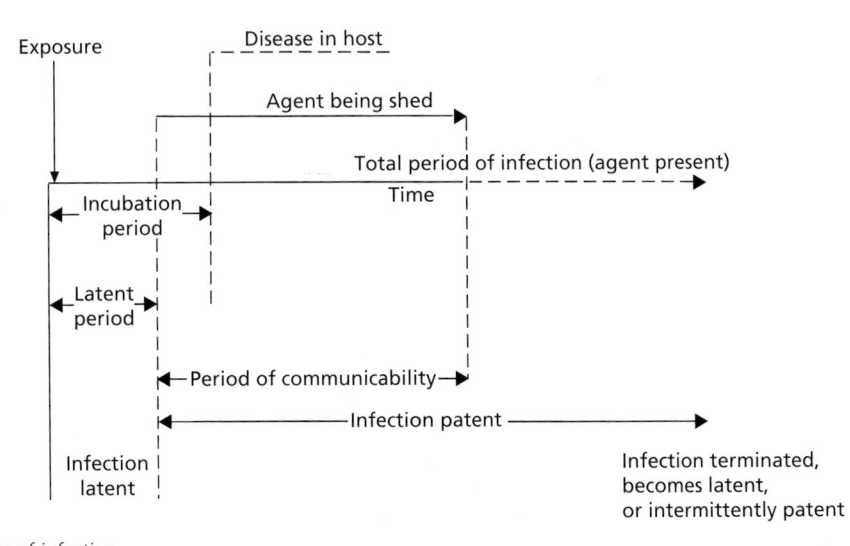

Figure 14.2 *Stages of infection*

times years later, with shedding of the organism. Varicella and malaria are examples of infections that commonly become latent. In other chronic infections, such as HIV, there may be an initial illness followed by a prolonged latent period with a more serious later illness years later (see Chapter 60, Human immunodeficiency virus, in the *Virology* volume in this series). For hepatitis B the infection is commonly cleared, but in some cases the infection and infectiousness becomes chronic perhaps leading to severe disease later (Chapter 55, Hepatitis B, in the *Virology* volume in this series).

The duration of the latent and infectious periods are important parameters of mathematical models. Classic epidemiological studies which observed the interval between primary and secondary cases within households enabled these parameters to be estimated for measles and other infections (Hope Simpson 1948; Bailey 1956a, b).

DEVELOPMENT OF IMMUNITY

The immunological basis for the development of immunity to bacterial infections is considered in the *Immunology* volume in this series. From the perspective of infectious disease transmission theory and control, there are a number of key components to be considered.

Some infections do not result in significant natural immunity and consequently can be acquired repeatedly. Gonorrhea and chlamydia are examples (Aral et al. 1999; Stoner et al. 2000). For infections that do stimulate an immune response, the duration of immunity influences the rate at which susceptibles accumulate in a population. Generally speaking, naturally acquired immunity is longer lasting than vaccine-induced immunity and in some infections immunity may be lifelong.

It is particularly important to distinguish between immunity against infection and immunity against disease. For example, live oral polio vaccine protects against both infection and disease, whereas inactivated polio vaccine protects predominantly against disease. This difference has significant implications for control

programs – control of a polio epidemic can be achieved much more rapidly, and with lower vaccine coverage, with live vaccine than with inactivated vaccine.

Cross-immunity may occur with one prior infection giving some immunity to another different infection. Breastfeeding reduces susceptibility to a number of infections that affect infants and young children. The influence of maternal antibody which has crossed the placenta in utero is a highly significant factor in infections where transmission is common among babies. For example, the incidence of meningococcal disease in infants bears an inverse relationship to the decline in maternal antibody (Figure 14.3). The peak incidence is at 6 months of age, coinciding with the nadir in maternal antibody. After 6 months, naturally acquired immunity begins to develop and the disease incidence declines. The significance for control measures is that any vaccine for use in a mass immunization program would need to be effective during the window of susceptibility in early life.

Infection in the population

The course of infection in the individual has a strong influence on the pattern of disease in a population. Diseases that confer permanent immunity after a short infection, such as measles and rubella, occur naturally in a cycle of regularly recurring epidemics. Diseases for which individuals become susceptible to reinfection on recovery (e.g. gonorrhea), or some infected individuals become long-term carriers (e.g. hepatitis B, tuberculosis), do not show such a regular pattern (Noah et al. 1995).

Factors specific to each population, including demographics, behavioral patterns, and environmental factors also affect the incidence of the disease. These determine whether, and at what level, an infection can remain endemic in a particular population. There are many infections that have a widely differing level of endemic infection in industrialized and resource-poor countries. Hepatitis B is commonly acquired in childhood in

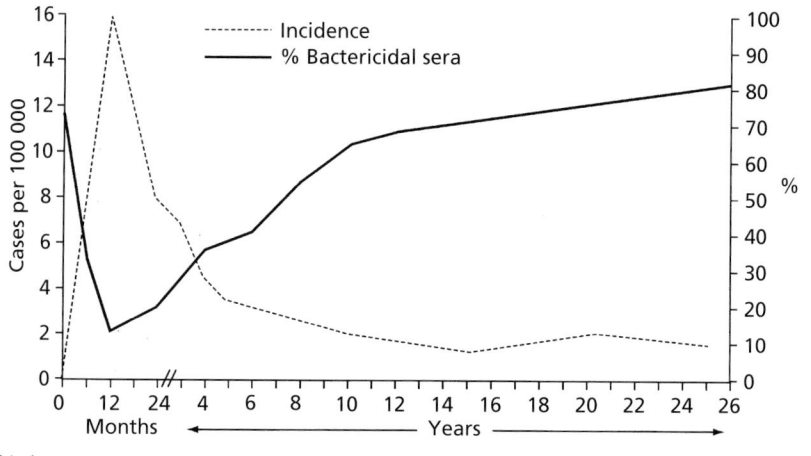

Figure 14.3 *Relationship between age-specific incidence of meningococcal infection and immunity (after Goldschneider et al. 1969).*

sub-Saharan Africa, but is largely confined to adults in particular risk groups in Northern Europe. Hepatitis A, and other infections transmitted by the fecal–oral route, are acquired at a younger age in countries where sanitation is poor (see Control of disease by hygiene). The zoonoses mostly occur where their specific animal hosts are to be found and humans are exposed, Legionnaires' disease is almost unheard of where its environmental reservoir of piped water is not found (Giesecke 2002). The differences are less striking for infections such as measles and rubella, but these too are acquired at a younger age in less developed countries. The contrasting epidemiology of an infection in different populations may necessitate entirely different approaches to disease control.

BASIC REPRODUCTION NUMBER

The basic reproduction number (R_0) for an infection within a population summarizes the effect of many of these factors in a single parameter. (It is variously termed the basic reproductive/reproduction rate/ratio/number by different authors, but almost always denoted R_0; and pronounced R nought, R zero or R subzero!) It is defined as the number of secondary infections that a typical infectious individual would produce in a completely susceptible population (Anderson and May 1991; Dietz 1993). Thus, R_0 depends both on the properties of the infective organism and on the social and demographic characteristics of the population. It may vary between infections, but also for the same infection in different populations. R_0 can be interpreted as the average number of contacts made by an individual during the infectious period; a contact being defined as any encounter in which an infectious individual would transmit infection to a susceptible individual (a definition that is clearly disease specific). This is a crucial variable for designing rational control programs. Some accepted values of R_0 for different common infections are shown in Table 14.2. Infections with high values of R_0 are regarded as highly infectious. Vaccine preventable diseases with a high R_0 are much more difficult to eliminate than those with low R_0. For example, a human infectious disease with an $R_0 = 20$ and without an environmental reservoir requires a population immunity of 95 percent or higher to achieve elimination, whereas one with $R_0 = 4$ requires a population level immunity of only around 75 percent.

Where infections have relatively low values for R_0 and are not infectious during the latent period, control may be effected by case finding and isolating those affected or influencing them to reduce infection risk (for example by condom use for sexually transmitted infection). This will be less successful or impossible if R_0 is high and transmission can take place before symptoms appear or immunity emerges. For example compare SARS, sexual acquisition of HIV, influenza, and measles (Table 14.2).

EFFECTIVE REPRODUCTION NUMBER

In most real situations, not all of the population are susceptible to infection: some are immune either as a result of previous infection or through vaccination. Some of the contacts of an infectious person will not result in new infections but be 'wasted' on immune individuals. The effective reproduction number R is defined as the number of secondary cases produced by a typical infectious individual (Anderson and May 1991). Clearly R depends on R_0 and the susceptibility of the population. For a simple homogeneous model (in which contact between any two individuals is assumed to be equally likely) $R = R_0 x$, where x denotes the proportion of the population who are susceptible.

Persistence thresholds

An infection can only establish itself in a population if $R_0 > 1$. Any such infection has an endemic equilibrium state at which $R = 1$, i.e. where a typical infectious individual produces on average one secondary infection. At this equilibrium state only one of the R_0 contacts made by an infective is with a susceptible individual. In the simple model, the proportion of the population who are susceptible is $1/R_0$. The equilibrium at $R = 1$ is a threshold: if $R > 1$ the number of infections increases; if $R < 1$ the number of infections decreases. Many infections demonstrate this threshold behavior in their natural epidemic cycles. During these cycles R oscillates around 1 as the proportion susceptible oscillates around the equilibrium value.

The $R = 1$ threshold is the key to designing vaccination and other control programs. If a vaccination program aims to eliminate a disease in a population, it must maintain $R < 1$, so that the number of infections decreases with each successive generation of cases. The necessary level of immunity could be achieved, in the long term, by immunizing a proportion p of the population at birth,

Table 14.2 *Selected infections – what stage infection can take place at and some estimated values for* R_0

Infection	At what stage infection generally takes place	Basic reproduction No. R_0
HIV (sexual transmission)	At any stage	<1 (per single sexual act)
Influenza	Prior to or during symptoms	3–4
Measles	Prior to or during symptoms	Around 10
Severe acute respiratory syndrome	Not prior to symptoms, infectiousness peaks in second week	2–3

where $p = 1 - 1/R_0$ (so that only a proportion $1 - p = 1/R_0$ of newborns remain susceptible).

HERD IMMUNITY

The above arguments suggest that it is possible to eliminate a vaccine preventable infection without immunizing every member of the population. It is this concept of indirect protection of susceptible individuals to which the term 'herd immunity' has been attached, although many authors have been deliberately vague in any definition. Fox et al. (1971) discussed a dictionary definition of herd immunity as 'the resistance of a group to attack by a disease to which a large proportion of the members are immune, thus lessening the likelihood of a patient with a disease coming into contact with a susceptible individual'. In a wide-ranging review, Fine (1993) points out the ambiguity of this definition – it fails to distinguish between total protection (when the immunity of the herd is above the threshold and the disease cannot persist) and partial protection (where the presence of some immune individuals lessens the risk to a susceptible). The threshold concept is the more appealing, and it leads to a definition of a population having herd immunity if $R < 1$. An equivalent statement of this definition is 'a population is said to have herd immunity to an infection if a typical primary infection produces less than one secondary infection'. Such a concise definition has the advantage of clarifying a sometimes vague concept; either a population has herd immunity, or it does not. The amount by which the population is above or below the herd immunity threshold is quantified by the effective reproduction number R.

Although the concept of herd immunity has been defined, there is still a need to clarify what is meant by a 'typical' infective. In a simple homogeneous model all individuals in the population are assumed to have the same number of contacts. Every infective is typical. In more complex models, which incorporate heterogeneity in the contact process, the typical infective is defined as some suitably weighted average of all infectives. (The typical infective is determined through a defined mathematical framework as the fastest growing stable distribution of infectives (Diekmann et al. 1990.) The definition of a population having herd immunity if $R < 1$ is applicable to these more complex models. The factors determining the effective reproduction number are the number of contacts made by infectives and the levels of susceptibility. The immunity of the herd does not depend necessarily on the immunity of individuals. For example, a population in which an infection has an $R_0 < 1$ has herd immunity to that infection even if no individuals are immune.

ESTIMATION OF R_0

Dietz (1993) has reviewed the variety of methods that have been used to estimate the basic reproduction number for infectious diseases. Many require data on the endemic state of the disease before the introduction of any control measures. Perhaps the most simple of these is (Anderson and May 1991):

$$R_0 = \frac{N}{B(A - m)},$$

where N is the population size, B is the annual number of births, A the average age at infection and m is theaverage duration of protection from infection provided by maternally derived antibodies. For stable populations this expression reduces to $R_0 = L/(A - m)$, where L denotes the life expectancy. (A brief heuristicjustification of this formula is possible for populations where virtually all individuals are infected during their lifetime: $A - m$ is the average duration of susceptibility so the proportion susceptible $Rx = (A - m)/L$, and at equilibrium $R_0 = 1/x$.) In deriving these simple formulae it is assumed that the risk of infection for a susceptible individual (often called the force of infection) is independent of age. The complications introduced by an age-dependent force of infection, which preclude the derivation of an explicit formula for R_0, are discussed later.

Mathematical modeling

The essential requirement of a mathematical model isthat it describes the mechanism underlying the processof interest. The key to models of infectious diseases is an understanding of the interactions between infectious and susceptible individuals that result in the transmission of infection. These should be described by a set of clearly stated assumptions, which can then be expressed as mathematical equations relating the variables and parameters of the model. The model is simply the means of evaluating the consequences of a set of assumptions.

When formulating a model, a compromise must be reached between simplicity and overcomplication. Simple models, which may have explicit solutions, clearly demonstrate the effect of each parameter, but may grossly oversimplify the complexities of the transmission process. More complicated models provide greater flexibility, but each additional refinement introduces more parameters thereby increasing the task of parameter estimation and making it more difficult to discern the effect of any individual parameter. In general, simple models are most useful for investigating qualitative effects, whereas more detailed models are needed for quantitative predictions.

When interpreting the results of any modeling exercise, it is essential to bear in mind the assumptions on which the model is based. No model can fully describe all the complexities of the transmission of a disease in a population. However, models can greatly enhance the understanding of the processes underlying disease transmission, and play an important role in the rapid design of control programs, for example, during the control of SARS, modeling, along with observational techniques, was used to quickly estimate the disease's R_0 and this, with other parameters, was used to inform the control strategy (WHO 2003; Anderson et al. 2004).

The examples in this part of the chapter will concentrate on infections for which humans are the only host, that have short latent and infectious periods and confer permanent immunity, as exemplified by measles. A flow diagram illustrates the basic structure of a model for such an infection showing the compartments that may be included (Figure 14.4). Models for other infections, such as those that confer no immunity or which have a carrier state, require this structure to be adapted accordingly. Anderson and May (1991) present and discuss models with different structures, which are used for a wider range of infectious diseases such as zoonoses, sexually transmitted infections, and infections with environmental reservoirs for which modeling approaches are also used (Anderson and May 1991). Some of these will be considered later in this chapter.

DYNAMIC BEHAVIOR

Numerical simulations of infectious disease transmission proceed in discrete timesteps to describe the changes over time in the number of individuals in each of the model's compartments. The values at each timestep are calculated from the values at the previous timestep, starting from some specified initial values. This procedure is essentially the same for all models, although complex models may have more compartments or further stratification by age, sex, etc.

The rates at which transitions between compartments occur are determined by the model's assumptions and parameters. For example, the rate at which individuals move from the latent to the infectious compartment is governed by the latent period, and the rate of movement from infectious to immune by the infectious period. The crux of all models of disease transmission is the rate at which susceptibles are infected, referred to as the force of infection, and how this is assumed to depend on the number of infectious individuals.

The mass action assumption

The force of infection, λ, is the rate at which susceptible persons acquire infection. The probability that a given susceptible individual will acquire infection in a short period δt is $\lambda \delta t$. For a directly transmitted infection, the force of infection at any time depends on the infectives present in the population at that time. The simplest assumption is that the force of infection is directly proportional to the total number of infectives. The rate at which new infections occur in the population is then proportional to the product of the number of susceptibles and the number of infectives. This model is known as the mass action model because of its similarity with the model for chemical reactions.

Hamer's discrete time model

The simplest version of a mass action model is that originally used by Hamer (1906) to describe measles epidemics in London. He considered infections as occurring in discrete generations, denoting the number of susceptibles and infectives in the nth generation by S_n and I_n, respectively. The number of susceptibles and cases in the following generation were calculated from the difference equations

$$I_{n+1} = \frac{1}{m} I_n \, ThickSpace S_n$$
$$S_{n+1} = S_n - I_{n+1} + ThickSpace b.$$

Thus the number of new infections is assumed to be proportional to the number of susceptibles and the number of infections (the 'mass action' assumption); all infections from the previous generation are removed (by recovery or death). The number of susceptibles is decreased by the number of new infections and increased by the influx of b susceptibles at each timestep; it is assumed that there is no mortality of susceptibles.

The only parameters of the model are the time between generations, the number of susceptibles added at each timestep, and the critical number of susceptibles, m. (The critical number of susceptibles m can be written in terms of the basic reproduction number R_0 and the population size N as $m = N/R_0$. The mass action assump-

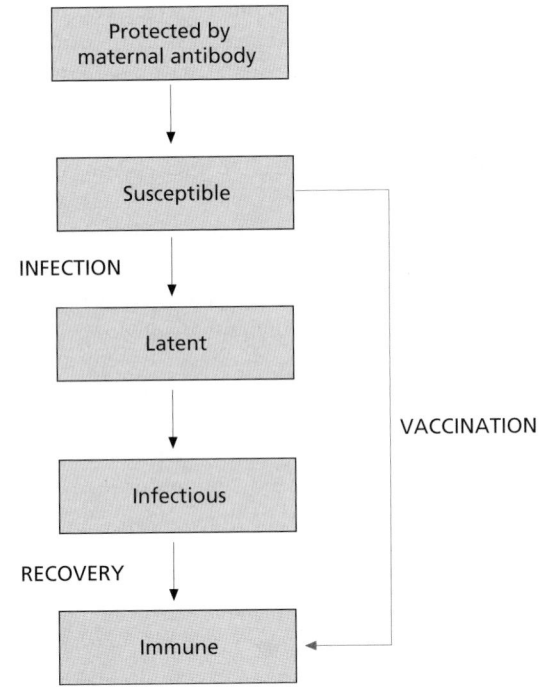

Figure 14.4 *Flow chart showing the compartmental structure of mathematical models*

tion then has the interpretation that each of the I_n cases makes R_0 contacts, of whom a proportion S_n/N are susceptible; leading to $I_{n+1} = R_0 I_n S_n/N$ cases in the next generation.) Hamer used a 2-week generation time (the sum of the latent and infectious periods for measles), and the values $b = 4\,400$ (there were approximately 2 500 births per week, but 300 of these babies died before losing protection from maternal antibodies at 6 months of age) and $m = 150\,000$ (determined empirically).

The model produces a series of undamped regularly recurring epidemics, whose magnitude and period depend on the parameters and initial conditions. Hamer used initial values of $I_0 = 12\,800$, $S_0 = 150\,000$ to produce a cycle with a period of 18 months (i.e. 39 2-week periods) (Figure 14.5). He noted that the number of susceptibles was fairly stable during the course of an epidemic cycle, oscillating by just 20 percent around the equilibrium number of 150 000. If the number of susceptibles exceeds this threshold there are more infections in the next generation, i.e. $I_{n+1} > I_n$ if $S_n > m$; if the number of susceptibles is below the threshold, the number of infections decreases. (The effective reproduction number at the nth generation, $R_n = S_n/m$, and oscillates around 1 during the epidemic cycle.)

Continuous time model

The continuous time equivalent of Hamer's model produces damped epidemic cycles (Soper 1929). The period of these cycles, T, can be calculated analytically (Soper 1929) and expressed in terms of the average age at infection, A, and the combined length of the latent and infectious periods D;

$$T = 2\pi\sqrt{AD}.$$

Predictions of the interepidemic period using this formula correspond well with observed values for many infections (Anderson and May 1991).

Reed–Frost model

The Reed–Frost model is preferred to the mass action model for outbreaks in small populations. Again, cases are considered as occurring in discrete generations, and let p be the probability of contact between any two individuals in each timestep. The probability of a susceptible being infected is modeled as the probability of meeting at least one infective, which is $1 - (1 - p)^{I_n}$. In this formulation the number of new infectives is not proportional to the number of infectives in the previous generation, as the model accounts for the possibility that a susceptible may be contacted by more than one infectious individual.

HETEROGENEITY

The models considered so far have been homogeneous models in which all susceptibles in the population experience the same force of infection. This is undoubtedly a gross simplification of most real situations. For infections such as measles, mumps, and rubella, the force of infection is highly age dependent due to heterogeneity in contact patterns (Farrington 1990; Grenfell and Anderson 1985; Anderson and May 1985a); in schools, children have most contact with other children of the same age. Changes in transmission rates as schools open and close can account for the seasonality of measles (Fine and Clarkson 1982a; Schenzle 1984). For sexually transmitted diseases, the degree of mixing between different subgroups of the population may be crucial in determining the spread of infection (Aral 1999; Aral et al. 1999; Gregson et al. 2002). For other infections, spatial factors such as population density may play a role, or genetic differences in susceptibility may exist (Anderson and May 1984). Control of infections spread by intimate contact, notably sexually transmitted infections, have to take into account factors such as sexual networks, partnership patterns, and diverse behaviors with varying risk of infection. Heterogeneity in the

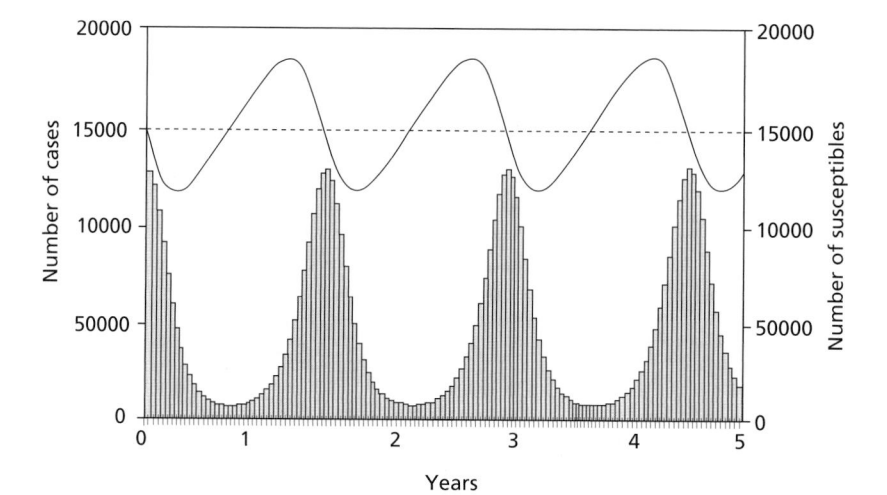

Figure 14.5 *Number of cases and susceptible individuals predicted by Hamer's discrete time model for measles in London.*

application of control measures such as vaccination may introduce further problems, e.g. pockets of susceptibility from unvaccinated communities or clustering of vaccination failures. The simple homogeneous model can be generalized in order to investigate these effects (Anderson and May 1991).

DETERMINISTIC AND STOCHASTIC MODELS

The transmission of an infection between individuals in a population depends on a number of chance events. In a large population, with many such events, the observed behavior will be close to that expected from averaging random effects. Deterministic models examine the expected behavior of a system; running a deterministic simulation under the same conditions gives the same result every time. This makes them easy to work with.

The major weakness of deterministic models is their behavior in a situation with a small number of susceptibles or infectives, where individual random events may influence the outcome. This may arise from an outbreak in a small population such as a school, or from a very low level of infection in a large population, perhaps as the result of a successful vaccination program. An example of infection outside vaccine preventable diseases is SARS where a small proportion of cases results in superspreading events (WHO 2003; Anon 2003). Stochastic models, which use randomly generated numbers to determine the outcomes of individual events, are better suited to such situations. Stochastic simulations are usually repeated many times to generate a range of results, so that the probability of an outbreak of a particular size or duration can be inferred.

The difference between deterministic and stochastic formulations is illustrated by the case of introducing a single infective into a population in which there is no infection. If $R > 1$, a deterministic model will always generate the same outbreak, whereas with a stochastic model one possibility is that no secondary transmission occurs (this may even be the most likely outcome). Alternatively, if $R < 1$ no significant spread occurs in a deterministic model, whereas a stochastic model has a finite probability of generating a large outbreak.

A combination of stochastic and deterministic approaches may be required at different stages of the same problem. The investigation of the effects of introducing a mass vaccination program in a population is an appropriate application for a deterministic model. If the program was implemented successfully to eliminate endemic infection, stochastic modeling could be used to yield the distribution of the size of outbreaks expected from any infections that come into the population, for example from abroad.

Strategies for disease control

The strategy adopted to control an infectious disease will depend on many factors including the aim of the program, the availability of resources, the acceptability of possible measures, and all the factors around the epidemiology of the disease (Table 14.1). This section outlines the possible aims, describes the strategies available, and discusses when they are appropriate.

AIM OF A CONTROL PROGRAM

Programs for disease control need to have clearly defined aims. The program can be designed to meet this aim and its success in achieving the aim can be assessed (Hawker et al. 2001). The three broad aims are containment, elimination, and eradication (Noah 1988) and a variety of control measures may be available (Table 14.3).

Containment

Containment is the most modest control option. Infection and infection risk remain endemic in the population or with an imported infection its spread is limited, but morbidity from the disease is reduced to an 'acceptable' level. An example of a containment aim was the WHO measles target for 1995: a 90 percent reduction in measles

Table **14.3** *Possible control strategies to interrupt or reduce transmission of infection*

Control strategy	Example
Removal of infections from distribution	Contaminated food source
Controlling infection in a nonhuman reservoir	Immunizing chicken flocks against *Salmonella* species
Isolation or cohorting	Isolating infectious patients in hospital – methicillin resistant *Staphylococcus aureus*
Quarantine	Quarantining smallpox contact
Increasing or inducing immunity	Routine immunization
Treatment of those infected	Whooping cough in early stages
Treatment of those potentially exposed	Immunoglobulin given to newborns born to hepatitis B infectious women
Influencing behavior through education and establishing a facultative environment	HIV – condom use by those infected
Reducing transmission in high risk situations	Protective clothing in operating theaters
Preventing spread between populations	Travel restrictions –SARS

cases and a 95 percent reduction in measles deaths from the levels before vaccine was introduced. With many infections, containment is the only possible aim.

Elimination

Elimination of an infectious disease from a population requires that chains of transmission either do not occur or are readily broken. This will mean that there is no endemic transmission of infection within the population, and that an infection imported into the population produces no more than an isolated outbreak. This may be achieved through the herd immunity of the population or by vigorous application of control measures. Either way the effective reproduction number must be maintained below one. Elimination is generally more difficult to achieve where there are reservoirs of infection outside humans, though it may be achieved if inter-human transmission is inefficient, i.e. R_0 is low.

Eradication

Eradication of an infectious disease requires the global destruction of the pathogen that causes it. There has to be no effective environmental or animal reservoir. It eventually allows the cessation of all control measures, potentially providing huge savings for future generations. Smallpox is the only disease to be eradicated so far. The last case of wild smallpox infection occurred in 1977. A WHO program to eradicate polio is in progress. Elimination from the Americas and the Western Pacific has been achieved and significant advances have been made in all other regions, though as of 2004 the infection is stubbornly persisting in six particular countries in sub-Saharan Africa and South Asia (WHO 2004a).

ROUTINE VACCINATION

Routine vaccination is the most common method for administering vaccine to a population. Individuals (usually children) are vaccinated when they reach a target age. Such a program requires maintenance of a suitable infrastructure (to ensure a regular supply of vaccine, have trained staff, informed families, etc.). If the vaccine provides protection against infection (and not just against the disease), the introduction of a vaccination program will affect the circulation of infection.

Age at infection

The dynamic interaction between susceptible and infectious individuals in the population can lead to counter-intuitive results. A vaccination program that leads to reduced circulation of infection, but not elimination, will cause the average age of infection to rise (Figure 14.6). Unprotected individuals experience a reduced force of infection and so remain susceptible longer, and thus are older when they eventually become infected. The change in the age distribution of cases is even more marked if transmission rates are higher in older age groups. This effect is beneficial if disease is most severe at a young age (e.g. pertussis, measles in developing countries), but it can lead to an increase in morbidity if disease severity increases with age or if it puts fetuses at risk, for example by shifting the age of infection in females into their child-bearing years as is the case with rubella (Anderson and Grenfell 1986). Vaccination programs against diseases such as polio and rubella (see later under Rubella: the prevention of congenital rubella syndrome) must achieve high coverage so that the reduction in circulation outweighs the problem of increasing age at infection.

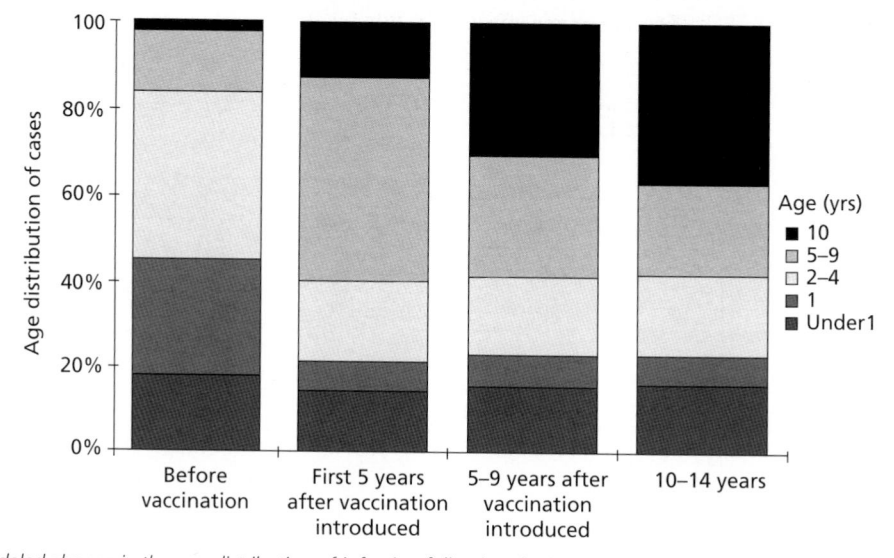

Figure 14.6 *Modeled change in the age distribution of infection following the introduction of vaccination. Simulation of measles in a developing country before and after the introduction of a routine vaccination program that immunizes 80 percent of children at age 9–12 months. Parameter values: basic reproduction number, R_0 = 20; life expectancy, L = 50 years; latent period, 1 week; infection period, 1 week; duraction of protection by maternal antibody, m = 6 months.*

Honeymoon periods

The 'honeymoon period' is the name sometimes given to the period of low incidence that is frequently observed following the introduction of a routine vaccination program, but which (if vaccination coverage or efficacy is not sufficiently high) is inevitably followed by a resurgence of infection (McLean and Anderson 1988; McLean 1995; Chen et al. 1994). When the vaccine is first introduced, many of those above the age of vaccination will be immune through natural infection. Vaccination reduces the proportion susceptible well below the threshold. However, the consequent low force of infection allows unprotected individuals (those who are not vaccinated or in whom vaccination fails to induce immunity) to accumulate. Once the threshold is exceeded, a post-honeymoon epidemic can occur (Figure 14.7). If the vaccination program is maintained, the infection will settle into a new epidemic cycle with a considerably longer epidemic period than before vaccination.

VACCINATION CAMPAIGNS

Vaccination campaigns aim to vaccinate all children in a target age range and geographical locality within a short period of time (often a day, a week or a month). Campaigns may be used to supplement routine programs (either as a one-off or as a method of giving a second dose) or as an alternative to them. They are often favored in countries that do not have the infrastructure to maintain a high coverage through routine vaccination program, but they are now also used in industrialized countries where it becomes necessary to quickly reduce susceptibility in a particular age group (Gay 1996).

The advantage of a suitably targeted campaign is that it removes most of the susceptible individuals from the population, thus reducing the reproduction number R well below the $R = 1$ threshold. This has a dramatic effect

on transmission which may be interrupted if sufficient coverage is achieved. The follow-up program must maintain herd immunity, either through routine immunization or through periodic campaigns. If the latter approach is chosen, the interval between follow-up campaigns should be determined so as to maintain $R < 1$. A campaign is needed whenever the herd immunity threshold is about to be exceeded. Serological surveillance data enable a campaign to be targeted to maximal effect.

This basic strategy is behind the WHO's attempt to eradicate polio, and the attempt of the Pan American Health Organization to eliminate measles from the Americas (de Quadros et al. 1996). The elimination of measles through mass campaigns was pioneered in the Gambia in 1968–1970, based on modeling simulations by Macdonald (Thacker and Millar 1991), and revived in Cuba (Sabin 1991). Elimination of measles from the Americas may pave the way for eventual eradication.

OUTBREAK CONTROL

Outbreak control is a key activity for many types of infections. The techniques and tools that can be used are various and are dependent on the infection and its characterization though it needs to be recognized that the evidence base for a number of the older and more established infections are not as well established as they are for more recently described infections (Tables 14.1–14.4). For vaccine-preventable diseases, it may be used to complement or even replace mass vaccination programs when disease incidence is low. However, it is only suitable for infections where outbreaks can be identified at an early stage and/or where the infection is not highly transmissible (a low R_0), so that control measures can be taken before widespread transmission occurs. This requires that cases can be diagnosed rapidly and reasonably accurately (preferably clinically, and there is likelihood that measures are likely to be successful, i.e.

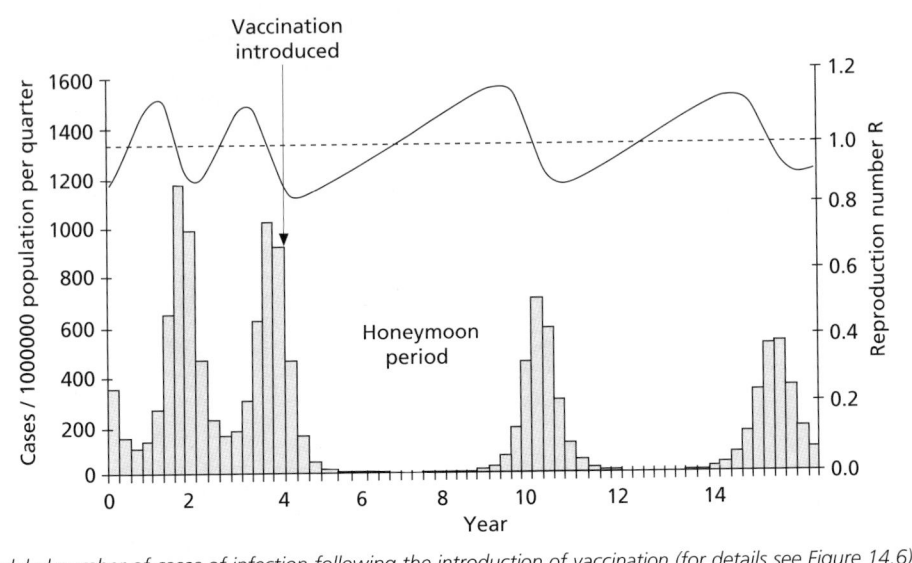

Figure 14.7 *Modeled number of cases of infection following the introduction of vaccination (for details see Figure 14.6)*

Table 14.4 *Techniques used for outbreak control*

Control techniques
Establishing an outbreak control team with defined authorities
Reporting the outbreak
Microbiological investigation of cause
Epidemiological investigation
Determining the appropriate control measures
Case-finding
Contact-tracing
Isolation of those infected and infectious
Quarantine of those exposed for the incubation period
Immunization of those at risk or to induce herd immunity
Source identification and control
Screening to detect those infected or at risk
Treatment of those infected
Protective infection control measures
Cohorting of hospital patients
Informing professionals and public
Reducing risk behaviors
Monitoring application of control measures
Monitoring effectiveness of control measures
Communicating with the authorities, media, and public

without needing laboratory confirmation) and that asymptomatic infections are uncommon. Low transmissibility of infection and a long latent period are also advantageous as they cause outbreaks to develop slowly, permitting time for interventions. As always, the control measures must be acceptable to the population and be within the law.

Aggressive outbreak control was an integral part of the smallpox eradication strategy and for the control of SARS; as both satisfy all the above criteria. However, attempts in the United States to control outbreaks of measles through targeted vaccination proved largely unsuccessful (Davis et al. 1987). Infections for which vaccine is currently used to control community outbreaks include hepatitis A, although it is not ideally suited to such an approach due to the high proportion of subclinical infections, but the low incidence of disease in the population does not justify the expense of mass vaccination.

QUARANTINE AND ISOLATION

Quarantine means to isolate or control the behavior of healthy people who are thought to have been exposed, are probably or possibly infected, and are in the incubation period (Heyman 2004). A quarantine strategy may be used to prevent the introduction of an infection into a population from which it is absent. It is only appropriate if the population is isolated from the potential source of infection. No attempt is made to create herd immunity in the population, so a failure of the quarantine measures could lead to a sizeable outbreak. Quarantine policies should therefore be backed up by outbreak control procedures. Quarantine measures were

used extensively in North America in the nineteenth century to prevent ships from Europe introducing infections, such as cholera, smallpox, and typhus. They are still used to exclude rabies from the UK and are part of plans to control viral hemorrhagic fevers (Department of Health 1996). They are also a key part of emergency plans if smallpox should re-emerge and during the 2003 SARS outbreaks were used in some affected areas (Centers for Disease Control 2002; WHO 2003).

Isolation is generally taken to mean separating symptomatic infectious persons from those who are susceptible, as a policy can only be successful if infectives are identifiable and identified before they transmit infection.

CASE-FINDING AND SCREENING

Case-finding is essential to isolation and caring for those infected. However, it is also used more widely in controlling infections where identification contributes to interrupting transmission. It requires that cases are readily identifiable – some diseases, like SARS, pose difficulties because of the problem in distinguishing cases from other diseases causing similar symptoms (WHO 2003).

Case-finding is considered an important part of the strategy to control HIV transmission in industrialized countries where it is used to effect behavior change amongst those infected so as to make them less likely to pass on infection. Another example is when there has been health care worker infections for hepatitis B or C whilst undertaking exposure-prone procedures.

Screening is a special form of case-finding where it is considered useful to test particular population groups for specific infections (Table 14.4). It may be systematic or opportunistic (testing patients who are attending medical care for another purpose) and usually has to conform to the guidelines applied to screening (Calman 1994). An example of systematic screening is the testing of all pregnant women for infections they may transmit to their babies and for rubella immunity. Donor screening is used to ensure infections are not transmitted to recipients of blood and tissues.

Herd immunity and vaccination programs

Most communicable diseases have a regular cyclical pattern (Noah 1988). The interval between epidemics is determined by the rate at which susceptibles accumulate in the population. If epidemics occur in regular cycles it is possible to predict in advance when the next epidemic will occur. For example, pertussis epidemics in the UK have occurred at 4-year intervals. Other infections are less predictable. Epidemics of meningococcal disease occurred during both world wars, in the early 1970s, and again since the mid-1980s. Influenza epidemics are even less predictable and are associated with major antigenic shift in the influenza virus (see the *Virology* volume in this series). In general, the epidemic cycles of viral infections are more regular than bacterial diseases.

Introducing a vaccine which protects against transmission of infection disrupts the natural epidemic cycle of a disease. Models which simulate the transmission of infection within a population may be used to predict the effects of vaccination programs on the incidence of disease. The predictability of epidemics has been greatly enhanced in recent years by serological surveillance, where the proportion susceptible to an infection by age is determined through large cross-sectional antibody prevalence studies. In the UK, routine serological surveillance for measles, mumps, and rubella antibody was established in 1987, prior to the introduction of combined measles, mumps, rubella (MMR) vaccine (Morgan-Capner et al. 1988) using the unlinked-anonymous technique. The recent application of mathematical models to serological surveillance data enabled the anticipation of a measles epidemic in the UK which was averted by a national immunization campaign (Gay et al. 1997) (see later under Measles: vaccination strategy in England and Wales). Data from other serological surveillance contribute to the control of many other infections (Nicoll et al. 2000; Osborne et al. 2000).

Diphtheria

Diphtheria is an example of an infection where the use of high herd immunity has changed over time. The symptoms of diphtheria are caused by the production of an exotoxin which is produced by the infecting organism, *Corynebacterium diphtheriae*. Naturally acquired antitoxic immunity provides protection which is usually lifelong. The traditional method for measuring diphtheria immunity – the Schick test – is no longer available and has been superseded by the assessment of neutralizing antitoxin in serum. A level of 0.1 units of antitoxin per ml or greater indicates long-term immunity. In the prevaccination era, most babies were born with maternally derived natural immunity, and infection was relatively uncommon in the first year of life. Attack rates were highest in the 1–5 year age group and by early adulthood most individuals were immune to diphtheria (Figure 14.8).

Close contact is required for transmission of diphtheria and therefore epidemics occurred particularly under conditions of crowding and poor hygiene. The importance of closeness and duration of contact in determining the spread of disease was first described by Dudley in 1926. Children sleeping in the same school dormitory were at greater risk than those in casual contact during working hours. In temperate countries, the disease was most common in the colder winter months, whereas in tropical countries transmission occurred all year round. In tropical countries, the principal source of infection is from cutaneous diphtheria lesions, whereas in temperate countries pharyngeal diphtheria was more common.

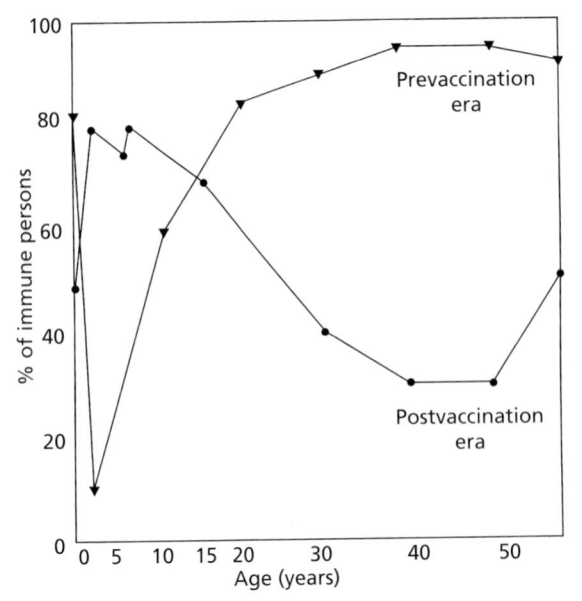

Figure 14.8 *Diptheria immunity by age in the pre- and post-vaccine eras*

Routine immunization with diphtheria toxoid (which protects against the toxic effects of the disease) was introduced in the 1940s. Immunization very rapidly reduced disease incidence in these countries and by the 1960s diphtheria had virtually been eliminated. As a result, the pattern of immunity in the population has changed completely. Because the organism no longer circulates in the community, adults born since the introduction of immunization rely solely on vaccine-induced immunity, which is generally lower than naturally acquired immunity. Immunization is commenced at 2 months of age so young infants very rapidly acquire vaccine-induced immunity, which is boosted by further immunizations given at school entry and (since 1993) on leaving school. However, older adults, in whom the last dose of vaccine was given many years ago, have waning immunity and up to a third of older adults may now be susceptible to diphtheria in Britain (Maple et al. 1995). Fortunately, this immunity gap does not appear to have led to significant outbreaks despite the fact that cases of toxigenic diphtheria continue to be imported from resource-poor countries (Begg and Balraj 1995). In the states of the former USSR, diphtheria re-emerged in the 1990s. This is only partly explained by waning immunity in adults. Other factors include mass population movements and lower vaccine coverage in young children.

Because vaccine-induced immunity protects against the toxic effects of the disease, one might expect that diphtheria immunization would not reduce circulation of the organism. Nevertheless, isolation of a toxigenic strain of *C. diphtheriae* is very rare nowadays, therefore it appears that vaccine does interrupt transmission as well as prevent disease. This is probably related to the fact that transmission of *C. diphtheriae* is much more efficient from symptomatic cases than from subclinical carriers. The level of diphtheria immunity in the population that is required to achieve herd immunity is not known precisely, but is generally considered to be between 70 and 90 percent (Simonsen et al. 1987).

Pertussis

Immunity to pertussis is complicated, and many components of *Bordetella pertussis* appear to have a role. At present the serological correlates of immunity are not known, thus population immunity data are of little value in mathematical modeling.

In Britain, pertussis epidemics occurred at 4-year intervals before the introduction of routine immunization in the 1950s (Figure 14.9). There were over 100 000 reported cases a year and about 1 per 1 000 patients died from the disease. The mortality rate and the risk of complications from pertussis is greatly dependent on age. Most deaths occur in young children under the age of 6 months. After this age the severity of the disease declines progressively and in adults it may be a milder, self-limiting though long-lasting illness. Because the risk of severe disease is greatest in the first few months of life it is important that pertussis vaccines are given as early as possible. For this reason, an accelerated schedule of immunization, at 2, 3, and 4 months of age, was introduced in the United Kingdom in 1992.

Whole call pertussis vaccines were introduced in the 1950s. Coverage was initially high (over 80 percent) and the magnitude of epidemics decreased considerably. In 1972 only 2 000 cases were recorded. However, in the mid-1970s doubts were cast over the safety of the vaccine and vaccine coverage dropped from 80 to 30 percent. This led to a resurgence in epidemics, the size of which had not been seen for over 20 years. Subsequently, vaccine coverage has improved and the epidemics have again decreased in size. In 2000, only 883 cases were notified in England and Wales, the second lowest annual figure ever recorded (the lowest was 712 in 2000).

The interesting feature throughout this period is that despite the introduction of vaccine with great

fluctuations in coverage there has been little change in the length of the interepidemic cycle. This led to the suggestion that pertussis vaccine may have no effect on carriage, but simply protects against disease (Fine and Clarkson 1982b). However, a decrease in the incidence of pertussis in children too young to be vaccinated points to some reduction in transmission (Miller et al. 1994). A recent study, which used an age-structured model to simulate the transmission of pertussis in England and Wales, suggested that both these observations are consistent with vaccine reducing transmission by approximately 80 percent (Miller and Gay 1997). In this scenario, the recent improvements in coverage to more than 90 percent would lead to a discernible lengthening of the epidemic period and an increasing proportion of infections would occur in adults. There is increasing evidence that adults with clinically inapparent or mild infection may act as important reservoirs of infection.

The question of vaccine providing protection against the transmission of infection has important implications for the control of pertussis. If the current subunit vaccines do not protect against transmission then elimination of the infection would be impossible. It remains to be seen whether the recently developed subunit vaccines affect transmission. Some preliminary evidence suggests that they may do so. Field trials of efficacy have recently established the efficacy of acellular pertussis vaccines in infants from 2 months of age (Olin et al. 1997).

Rubella: the prevention of congenital rubella syndrome

Most cases of rubella present as a mild illness with rash. Infection in pregnancy, however, can cause congenital rubella syndrome (CRS), resulting in severe malformation of the fetus, including cataracts, deafness, and heart defects, and an outcome of death or severe handicap (Gregg 1941; Miller et al. 1982). The purpose of a rubella vaccination program is to reduce the number of cases of CRS by preventing infections in pregnant women. A number of different strategies has been used to achieve this goal since the vaccine became available in 1970. The risks and benefits of each vaccination program can be evaluated using mathematical models (Anderson and Grenfell 1986), enabling informed policy decisions to be taken.

Before vaccination was introduced in developed countries, rubella epidemics occurred every 4–5 years and the average age of infection was approximately 10 years. About 85 percent of the population experienced the disease by age 16, leaving 15 percent who remained susceptible to infection as adults. A significant number of infections in pregnancy and consequent cases of CRS resulted. In developing countries, rubella is acquired at a much younger age – almost everyone is infected as a

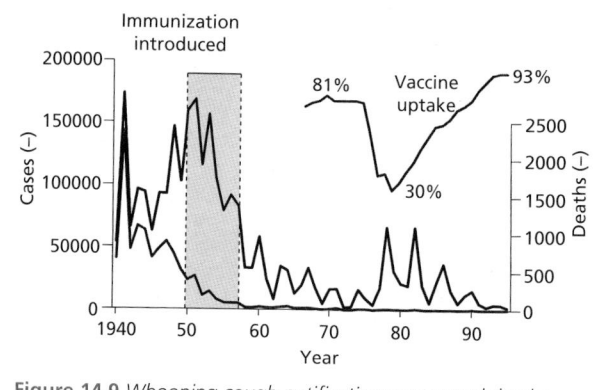

Figure 14.9 *Whooping cough notifications: cases and deaths, England and Wales 1940–1995 (prepared by CDSC; source OPCS)*

child and cases of CRS are uncommon. However, rubella vaccination may become necessary if social and environmental changes lead to a reduction in transmission rates.

POLICY OPTIONS

The number of rubella infections in pregnancy depends on the number of susceptible pregnant women and the risk of infection in pregnancy. Vaccination programs were designed to reduce the number of infections in pregnancy by reducing one of these contributing factors. The two main options of universal or selective vaccination are often termed the US and UK policies, respectively, after the countries in which they were originally used.

The universal policy

The universal policy was designed to eliminate CRS through the elimination of rubella in the population, achieved by a high vaccine coverage of all children at a young age. Susceptible pregnant women would be protected from infection by the herd immunity of the population.

The selective policy

The selective policy was designed to reduce the incidence of CRS by reducing the level of susceptibility among pregnant women, achieved by vaccinating girls only at age 10–14 and screening all pregnant women for rubella immunity. This would allow the virus to remain endemic, so that many girls would be infected and acquire immunity before they were vaccinated. Vaccination would serve to reduce the number of susceptibles. Further reductions could be achieved by screening

pregnant women, and vaccinating those found to be susceptible before their next pregnancy.

RISKS AND BENEFITS

A universal policy offers the greater benefit (the elimination of CRS), but also entails more risk. If vaccination coverage is low, the number of infections in pregnancy may actually be increased by the vaccination program. Vaccination reduces the force of infection acting on the unvaccinated population, causing them to remain susceptible longer until they acquire infection at an older age. The number of susceptible pregnant women may increase. If the increase in susceptibility among pregnant women outweighs the reduction in the force of infection, more cases of CRS will occur. This occurred, for example, in Greece in the 1990s when a universal policy was incompletely implemented (Panagiotopoulos et al. 1999).

In contrast, a selective policy can only eliminate CRS with 100 percent coverage. The risk to susceptible pregnant women is not reduced by the vaccination program. However, the policy does not carry the risk of increasing the incidence of CRS; the reduction in the incidence of CRS is the same as the coverage achieved.

Anderson and Grenfell (1986) used an age-structured model to calculate the number of cases of CRS expected at different levels of coverage for each of the two policies. At the long term steady state under the universal policy, there was an increase in the number of cases of CRS for coverages less than 30 percent, but rubella and CRS were eliminated for coverage greater than 80–85 percent (Figure 14.10). If coverage exceeded 75 percent, the universal policy prevented more cases of CRS than a selective policy with the same coverage. Short-term dynamic simulations produced similar results.

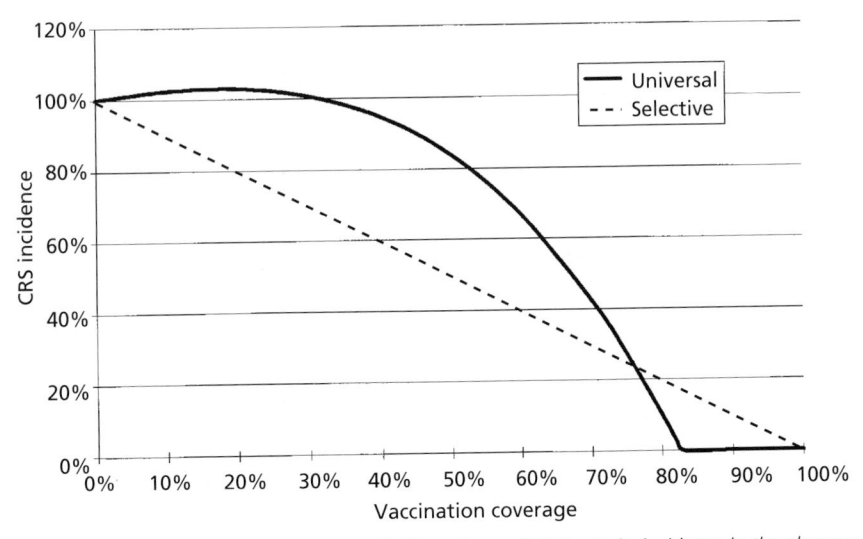

Figure 14.10 *Predicted equilibrium incidence of congenital rubella syndrome (relative to its incidence in the absence of vaccination) by vaccination coverage for selective and universal rubella vaccination programs (after Anderson and Grenfell 1986)*

Policy implications

The United States introduced a universal rubella vaccination program in 1969. Following a decline in the number of infections during the 1970s and 1980s to very low levels, a resurgence of rubella occurred in 1989–1991 (Lindegren et al. 1991; Centers for Disease Control 1994). Many cases were in unvaccinated young adults (including pregnant women; 56 cases of CRS were reported in 1990–1991) or in children from communities who refused vaccination. Since these outbreaks the incidence of rubella has returned to low levels. The US aimed to eliminate indigenous rubella and CRS by 1996, maintaining herd immunity through high vaccination coverage. To a large extent this has been successful with only 26 infants reported as born with CRS between 1997 and 1999 and 24 of these were in foreign-born mothers who are more likely to have missed out on immunization programs (Centers for Disease Control and Prevention 1994, 2001).

The UK introduced a selective vaccination program in 1970, and achieved 80–85 percent coverage at a time when the coverage of measles vaccination was less than 60 percent. Although this resulted in a considerable reduction in the number of infections in pregnancy, their continuing occurrence prompted a switch to a universal policy in 1988. A vaccination campaign amongst 5–16 year olds in 1994 removed many susceptibles from these cohorts, but in 1995 10–15 percent males aged 18–30 remained susceptible to rubella. This led to a resurgence of rubella in 1996, which included many outbreaks in universities and military barracks. The legacy of the selective vaccination program is that still in the mid-1990s less than 2 percent of women were susceptible to rubella, thus limiting the number of infections in pregnancy (Miller et al. 1996).

Sao Paulo State, Brazil, introduced a universal rubella vaccination program in 1992, following a modeling study to explore strategy options. This is an exemplary approach to vaccination program design. At the start of the program (in which vaccine is given to children at 15 months of age) a one-off vaccination campaign was carried out among older children to prevent the resurgent epidemics experienced in the US. The upper age limit of the campaign was set as 10 years as models suggested that little benefit was gained from including older children, despite the estimated seroprevalence in this age group being only 70–80 percent (Massad et al. 1994). The model's transmission rates were based on an age-specific force of infection estimated from a serological survey that included only 54 samples from persons aged 10 years or more (de Azevedo Neto et al. 1994). The experience of the next 10–15 years following the intervention will determine whether the decision not to include children aged 11–15 years in the campaign is an example of modeling saving unnecessary expenditure, or of too much faith being placed in the predictions of insufficiently robust models.

Measles: vaccination strategy in England and Wales

Measles is the most infectious of the vaccine-preventable diseases (Table 14.2) and causes significant morbidity and mortality if not controlled by vaccination. In this section we review the history of measles vaccination in England and Wales. In particular, we discuss the role of models in the planning of a national vaccination catch-up campaign conducted in 1994.

BACKGROUND

Measles vaccination was introduced in England and Wales in 1968. Before this time, measles epidemics occurred in alternate years causing an average of 100 deaths per year. Almost everyone experienced measles infection as a young child: 55 percent of notified cases were in those under 5 years, 42 percent in 5–9 year olds, and only 3 percent were in persons aged over 10 years. Vaccine uptake was initially low with only 50 percent of children being vaccinated up to 1980. Coverage then increased steadily reaching 80 percent by 1988. Over this period measles notifications and deaths showed a downward trend, but coverage was sufficiently low for the virus to remain endemic. Children who were not vaccinated became infected, resulting in continuing morbidity and mortality.

In 1988 combined measles, mumps, and rubella vaccine (MMR) replaced single antigen measles vaccine, following which coverage increased to 93 percent by the second birthday. This, together with an MMR catch-up program targeted at preschool children, resulted in a marked reduction in measles incidence in all age groups. Unvaccinated children, therefore, had little opportunity to acquire immunity through infection, and remained susceptible. The aging of cohorts with higher levels of susceptibility caused an increase in the proportion of school children who were susceptible to measles, detected through serological surveillance (Figure 14.11).

The shift of measles susceptibility was also reflected by a change in the age distribution of measles infections. A study using laboratory salivary diagnosis demonstrated that notifications (which are based on clinical diagnosis) were unreliable, especially in young children (Brown et al. 1994). Most infections occurred in persons aged 10 years or more and outbreaks in secondary schools were becoming more common (Ramsay et al. 1994; Calvert et al. 1994; Morse et al. 1994).

MODELING

Two separate approaches were used to predict the incidence of measles. Both models summarize their results using the reproduction number R. If $R < 1$, the population has herd immunity and no resurgence of disease can occur.

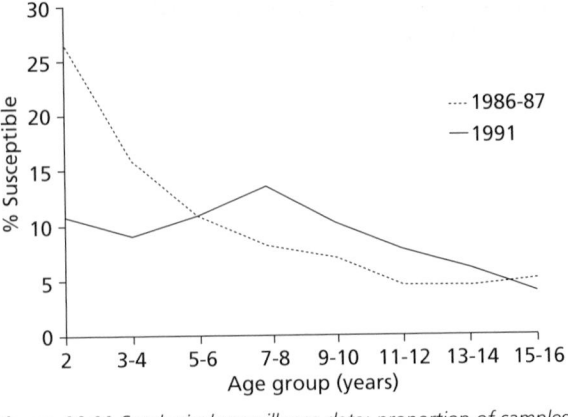

Figure 14.11 *Serological surveillance data: proportion of samples negative for measles IgG antibody by age, England 1986–1987 and 1991*

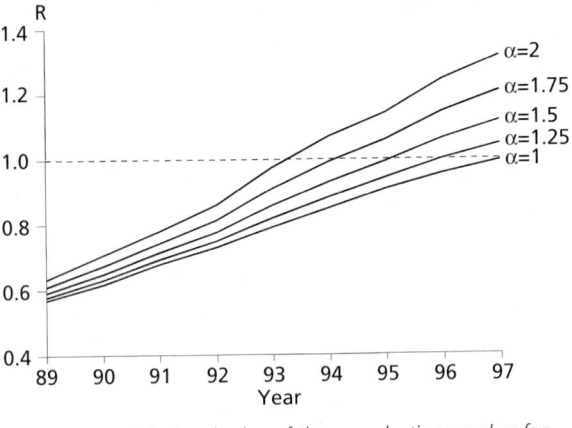

Figure 14.13 *Calculated value of the reproduction number for measles in England for a range of values of the age-specific transmission rates (Gay et al. 1995)*

The first study used models to interpret the serological surveillance data, to examine whether the increase in susceptibility would be sufficient to exceed the herd immunity threshold (Gay et al. 1995). The model divided the population into five age groups. The reproduction number is determined by the level of susceptibility in each age group and the transmission rates within and between age groups. (This calculation is the age-stratified equivalent of $R = R_0 x$.) The level of susceptibility expected each year in each age group was projected from the serological data assuming no infection occurred. Transmission rates were derived from age-specific notification data from the prevaccination period (Figure 14.12). A range of values for the transmission rate within the 10–14 year age group was used due to the difficulty in obtaining a precise estimate. The model predicted that as the most susceptible cohorts moved into the age group with the highest transmission rates (10–14 year olds), the reproduction number, R, would exceed one. This would occur at some point before 1998, depending on the transmission rate assumed in the 10–14 year age group (Figure 14.13). The scenario that best reflected the observed changes in the

age distribution of cases suggested that this would occur sooner rather than later. This provided the potential for a resurgence of measles in 1995/1996 involving more than 100 000 cases with most occurring in persons aged 10 years or more. Such a resurgence would have resulted in a high level of morbidity and mortality, because of the severity of measles in older age groups.

The second study simulated measles transmission in England and Wales using a dynamic model fed with vaccination coverage statistics (Babad et al. 1995). The model provided a good reflection of historic data. The study concluded that a single dose vaccination policy would not be sufficient to eliminate measles and investigated the options for supplementary vaccination measures. A vaccination campaign covering all children aged 5–16 years would reduce R well below the threshold. The introduction of a second routine vaccine dose at age 4 years, immediately after the campaign, would maintain the herd immunity of the population.

EFFECT OF THE CAMPAIGN

A national catch-up campaign to immunize children aged 5–16 with a combined measles and rubella vaccine was carried out in 1994 (Miller 1994). All children were offered vaccine, irrespective of a previous history of vaccination or disease; 92 percent coverage was achieved. The incidence of measles dropped dramatically following the campaign and the few cases that have occurred are in a pattern consistent with limited spread from imported infections (Gay et al. 1997). The introduction of a second dose into the routine vaccination schedule at school entry should sustain the herd immunity achieved. Continued surveillance is essential to monitor the effect of the program: salivary investigation of all suspected cases enables outbreaks to be identified; serological surveillance monitors the maintenance of herd immunity. Probably the most damaging impact on immunization for measles and mumps have been continuing controversies, much as two decades earlier similar

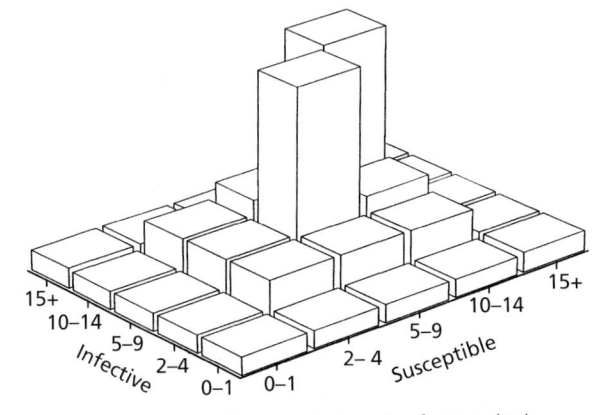

Figure 14.12 *Age-specific transmission rates for measles in England and Wales (with α = 1) (Gay et al. 1995)*

stories were for pertussis. Though there has been a decline in uptake of MMR in the UK, the fall is much less than that seen for pertussis (Gangarosa et al. 1998; Nicoll 2001; Miller 2003).

ENVIRONMENTAL AND SOCIAL EFFECTS AND POPULATION IMMUNITY

Changes in environmental and social conditions may affect the population in a number of ways. Poor hygiene facilitates spread of gastrointestinal infections, such as dysentery, by the fecal–oral route. Crowded accommodation facilities spread airborne infections, such as tuberculosis and meningococcal disease. Excessive throughput of patients in hospital makes it difficult to attend to infection control and control outbreaks. Temperature and other environmental conditions affect the survival of infectious disease agents. Social conditions, such as sexual behavior, and patterns of occupation influence the likelihood of being exposed to specific infectious agents (Hawker et al. 2001).

Control of disease by hygiene

In populations and settings where hygiene is poor, enteric infections spread easily. In settings such as refugee camps and following natural disasters, intense outbreaks may occur. For example, hepatitis A is endemic in many developing countries where basic hygiene measures such as hand washing and safe disposal of sewage are lacking. Most children become infected in the first few years of life. In this young age group, a large proportion of infections are subclinical and icteric cases are relatively uncommon. Where hygiene is good, infection in young children is relatively uncommon and most reach adulthood with no immunity to hepatitis A (Figure 14.14). In these countries, most infections occur in adults, at which age symptomatic infection is very common. Two main patterns of disease are observed – slowly evolving community outbreaks lasting for months or even years in which infection is transmitted from person-to-person and explosive point source outbreaks arising from a contaminated food or water source (Tang et al. 1991). Countries in the transition between the highly and lowly endemic states often experience a rapid reduction in the incidence of infection (Perez-Trallero et al. 1994; Lim and Yeoh 1992; Yap and Guan 1993). Models suggest that this may be followed by a substantial resurgence of endemic disease (Gay 1996).

Hepatitis A vaccines have recently become available and are of value in protecting those who are likely to be most at risk, such as travellers to highly endemic areas. No country has yet implemented hepatitis A for routine use (largely on the grounds of cost), however vaccines against hepatitis A, which could be incorporated into

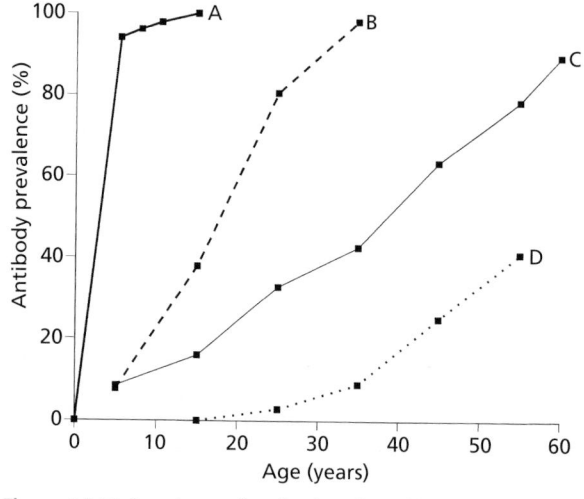

Figure 14.14 *Prevalence of antibody to hepatitis A in several countries: A, Cameroon, 1989 (Stroffolini et al. 1991); B, Spain, 1986 (Perez-Trallero et al. 1994); C, England, 1987 (Gay et al. 1994); D, Sweden, 1977 (Iwarson et al. 1978)*

other vaccines as combination products, are being developed. In countries with good hygiene, the basic reproduction number for hepatitis A is relatively low (approximately two in England and Wales; Gay et al. 1994) and herd immunity could be achieved with modest vaccination coverage. However, mass vaccination programs may be most appropriate for countries in the transitional phase.

Poliomyelitis is another example of a disease that is affected by changes in hygiene. During the early part of this century, when hygiene was poor in Britain, polio circulated widely among preschool children. In this age group paralysis was relatively uncommon and therefore the number of paralytic cases was small. During the 1940s and 1950s, as hygiene improved, and the circulation of wild virus diminished, the average age of infection rose. This led to epidemics of paralytic disease in older children and young adults (Figure 14.15). Polio vaccines became available in the 1950s, and rapidly reduced the incidence of the disease. Global eradication of polio is now planned by the World Health Organiza-

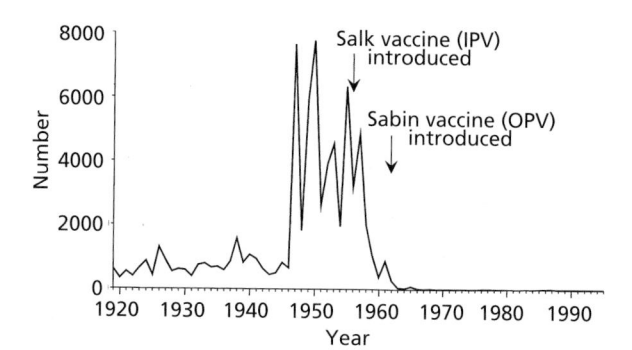

Figure 14.15 *Poliomyelitis notifications, England and Wales 1919–1995 (prepared by CDSC).*

tion. Eradication of polio is feasible, as there are no known nonhuman reservoirs of the disease, the virus does not survive long in environmental sources, such as sewage, live oral vaccine is able to both interrupt transmission as well as prevent disease, and is relatively cheap and easily delivered. However, at present (2004), it is proving difficult to achieve elimination in some countries experiencing civil strife and certain other populous countries, such as India and Nigeria (WHO 2004a).

Hospital hygiene is a major driver of health care-associated infections, though other factors come into play including surgical techniques undertaken as well as the level of dangerous pathogens in the surrounding population.

Sexually transmitted diseases and blood-borne viruses

Sexually transmitted diseases are an example of infections whose incidence is affected by behavior. The incidence of a disease is affected not only by the mean rate of partner acquisition, but especially by the degree of heterogeneity in behavior within a population. The importance of core groups (groups accounting for disproportionately large amounts of transmission) in the transmission of infection is well documented (Yorke et al. 1978; Jacques et al. 1988). Changes in the behavior of the core groups or their contact with noncore groups can have a dramatic effect on the incidence of infection and close attention to core groups is important for any STI control program.

Core groups are not the same in every country. In industrialized countries, men who have sex with men (MSM) are often important, while in resource-poor countries female sex workers often have a greater role though it may be that MSM are present, but not recognized because of severe stigma. Not all STIs conform to a pattern determined by core groups. Transmission of chlamydia and human papilloma virus (genital warts) is mostly accounted for by contact in the general sexually active population, especially among young adults. Control of STIs has to be through case-finding and treatment and effecting behavior change, including condom use (Mayaud et al. 1997; Samuel et al. 1991; Wawer et al. 1999; Cohen 1995). The blood-borne viruses, hepatitis B and C and HIV, follow a similar pattern of transmission to the STIs, but in addition are spread by equipment sharing among injecting drug users. Control is through prevention of drug injection, targeted promotion of safer injecting (needle exchanges, etc.), and treatment of those addicted.

The studies of gonorrhea by Yorke et al. (1978) and Hethcote and Yorke (1984) were the first detailed investigations of the transmission dynamics of a sexually transmitted disease. They described how changes in behavior affected the level of endemic disease. Since the mid-1980s the public health priority of controlling the HIV/AIDS epidemic has produced explosive growth in the literature. However, the complexity of the interactions between subgroups of the population and the problem of collecting suitable data from which to estimate parameters make modeling particularly difficult. Population-based surveys (Johnson et al. 1992) have provided a basis for initial parameter estimates. However, transmission models remain most useful for assessing the relative contribution of various factors to future trends in the incidence of AIDS, rather than for making quantitative predictions (Williams and Anderson 1994). STI control programs have shown how control of one bacterial infection can in some circumstances reduce transmission of a viral infection, HIV (Wasserheit 1994; Grosskurth et al. 1995; Cohen 1995).

Environmental infections

Some important infections persist in the environment and this is the key to their control. For example, Legionnaires' disease is a persistent problem in large buildings and requires scrupulous attention to water regulations (Giesecke 2002). For other infections, environmental spread via food, water, or the air means careful attention to water sources and food distribution chains are required (Hawker et al. 2001). In these infections par excellence contributions by social scientists and modelers have become crucial to understanding spread and devising control methods (Garnett et al. 1999; Hahn 1991; Klovhdah 1985; Robinson et al. 1995; Morris and Kretzschmar 1997; Schinazi 2001).

Vector control

A variety of infections are transmitted indirectly through vectors, such as insects. One of these, malaria, was the first infection in which population dynamics were studied extensively. Transmission of malaria depends on several factors, but the most important relate to the anopheline mosquito vector. The calculation of the basic reproduction number and eradication threshold are different with reference to vector-borne infections than to directly communicable infections. The contact parameter is a function of the density, survival rate, and feeding behaviors of the vector populations. The estimates of the basic reproduction number for malaria range between 5 and 100. These reflect considerable variations in the epidemiology of malaria, which is observed even within relatively small geographic areas. Macdonald developed a formula for the likelihood of infection based on the proportion of anopheline mosquitoes with sporozoites in their salivary glands. Malaria transmission in Macdonald's model was proportional to the density of the vector, the number of times each day that the mosquito bites man, and the probability of the mosquito surviving for 1 day. Macdonald's model has been refined in recent years, but it does illustrate certain important points relevant to the control of malaria. Vector longevity is the most important factor in determining transmission and therefore focuses control measures on the adult mosquito. Control programs are most likely to be effective where vector longevity is short and the basic reproduction number is low. This was the situation in Europe where the reproduction level was low in many areas and the vector was found mainly inside houses where it could be attacked with insecticides. There is also considerable variation between different species of anopheline mosquitoes with respect to their ability to transmit malaria (the vectorial capacity). Each vector has its own behavior patterns and breeding grounds. Malaria is often seasonal, coinciding with the rainy season which provides water for mosquito breeding and increased humidity favoring mosquito survival.

Vector-borne human infections are not confined to tropical countries and there are many other organisms that are spread by mosquitoes (Ludwig and Cooke 1975). West Nile virus has, since the late 1990s, appeared and spread extensively from east to west across the USA and is proving very difficult to control (Granwehr et al. 2004).

Current malaria control strategies focus on protecting individuals against bites (bed-nets, etc.) or disease (human chemoprophylaxis). These are not very successful at a population level. Several malaria vaccines are currently under development and these may have individual, as well as population, actions. They are being designed to act at various stages of the cycle of infection. Some of the candidate vaccines may only protect against transmission and not against disease.

GENETIC FACTORS IN POPULATION IMMUNITY

Studies of herd immunity must clearly take into account genetic differences in resistance. Much of the work in this area was done on mice in the 1920s developing so-called Webster mice. These workers were able to show genetic effects in resistance to infection which emerged as a result of inbreeding and other genetic manipulations. The extent to which this work can be extrapolated to man is not clear. Observations in human populations are beset by confounding factors. There is no doubt that some populations are more susceptible to certain infectious diseases than others. Of particular note is the fact that there are sex differences in disease attack rates for many childhood infectious disease, for example mumps is more common among boys than among girls. The incidence of meningococcal disease and other infections caused by bacteria with polysaccharide capsules varies enormously in different parts of the world. Certain

populations, for example Australian Aboriginal, have very high rates of disease and respond less well to polysaccharide vaccines than white children (McIntyre et al. 2000).

ZOONOSES AND EMERGING INFECTIONS

A number of the most important human infections have their origins in animals which either act as a continuing reservoir of infections (e.g. salmonellae, *Escherichia coli* O157), or were the original source, but now transmission is established between humans (HIV, measles). For the former, theories of transmission and control rely on animal husbandry, as well as on protecting humans. For example, some countries have been very successful in reducing levels of salmonellae pathogenic to man through vaccinating chicken flocks (Methner et al. 1999).

Zoonoses contribute disproportionately to the continuing stream of emerging infections that threaten human health. Many of these infections are of limited consequence, but a small number, such as HIV, the new viral hemorrhagic fevers, multidrug resistant organisms, and West Nile virus, has posed major threats. These infections require control measures which have to be based on good theoretical work tested against rapidly gathered data. The rapid institution of control measures against SARS shows how modern laboratory and epidemiological science can quickly combine even with an atypical and novel infection (WHO 2003; Anderson et al. 2004). Equally the persistent threat and near occurrence of pandemic influenza from avian influenza in domestic fowls as far away as China and as near as the Netherlands indicates no scope for complacency (WHO 2004b).

REFERENCES

Advisory Committee on Dangerous Pathogens. 1996. *Management and control of viral haemorrhagic fevers*. London: The Stationery Office.

Anderson, R.M. and Grenfell, B.T. 1986. Quantitative investigations of different rubella vaccination policies for the control of congenital rubella syndrome (CRS) in the United Kingdom. *J Hyg Camb*, **96**, 305–33.

Anderson, M. and May, R.M. 1984. Spatial, temporal and genetic heterogeneity in host populations and the design of vaccination programmes. *IMA J Math Appl Med Biol*, **1**, 233–66.

Anderson, R.M. and May, R.M. 1985a. Age-related changes in the rate of disease transmission: implications for the design of vaccination programmes. *J Hyg Camb*, **94**, 365–435.

Anderson, R.M. and May, R.M. 1985b. Vaccination and herd immunity to infectious diseases. *Nature*, **318**, 323–9.

Anderson, R.M. and May, R.M. 1990. Immunisation and herd immunity. *Lancet*, **335**, 641–5.

Anderson, R.M. and May, R.M. 1991. *Infectious diseases of humans: dynamics and control*, 2nd edn. Oxford: Oxford University Press.

Anderson, R.M., Crombie, J.A. and Grenfell, B.T. 1987. The epidemiology of mumps in the United Kingdom: a preliminary study of virus transmission, herd immunity and the potential impact of immunisation. *Epidemiol Infect*, **99**, 65–84.

Anderson, R.M., Fraser, C., et al. 2004. Epidemiology, transmission dynamics and control of SARS: the 2003–2004 epidemic. *Phil Trans R Soc Lond B*, **359**, 1091–105.

Anon. 2003. Severe acute respiratory syndrome. *Singapore Wkly Epi Rec*, **78**, 157–62.

Aral, S.O. 1999. Sexual network patterns as determinants of STD rates: Paradigm shift in the behavioural epidemiology of STDs made visible. *Sex Transm Dis*, **26**, 5, 262–4.

Aral, S.O. 2000. Patterns of sexual mixing: mechanisms for or limits to the spread of STIs? *Sex Transm Inf*, **76**, 415–16.

Aral, S.O., Hughes, J.P., et al. 1999. Sexual mixing patterns in the spread of gonococcal and chlamydial infections. *Am J Public Health*, **89**, 6, 825–33.

Babad, H.R., Nokes, D.J., et al. 1995. Predicting the impact of measles vaccination in England and Wales: model validation and analysis of policy options. *Epidemiol Infect*, **114**, 319–41.

Bailey, N.T.J. 1956a. On estimating the latent and infectious period of measles. I. Families with two susceptibles only. *Biometrika*, **43**, 15–22.

Bailey, N.T.J. 1956b. On estimating the latent and infectious periods of measles. II. Families with three or more susceptibles. *Biometrika*, **43**, 322–31.

Bailey, N.T.J. 1957. *The mathematical theory of epidemics*. London: Griffin.

Bartlett, M.S. 1956. Deterministic and stochastic models for recurrent epidemics. *Proceedings of the Third Berkeley Symposium on Mathematical Statistics and Probability*. Berkeley, CA: University of California Press, 81–109.

Bartlett, M.S. 1957. Measles periodicity and community size. *J R Stat Soc A*, **120**, 48–60.

Begg, N. and Balraj, V. 1995. Diphtheria: are we ready for it? *Arch Dis Child*, **73**, 568–72.

Boily, M.C., Poulin, R. and Masse, B. 2000. Some methodological issues in the study of sexual networks: from model to data to model. *Sex Transm Dis*, **27**, 10, 558–71.

Brown, D.W., Ramsay, M.E., et al. 1994. Salivary diagnosis of measles: a study of notified cases in the United Kingdom, 1991–3. *Br Med J*, **308**, 1015–17.

Brownlee, J. 1907. Statistical studies in immunity. The theory of an epidemic. *Proc Roy Soc Edinb*, **26**, 484–521.

Brownlee, J. 1909. Certain considerations of the causation and course of epidemics. *Proc Roy Soc Med (Epid Sec)*, **2**, 243–58.

Calman, K. 1994. Developing screening in the NHS. *J Med Screening*, **1**, 101–5.

Calvert, N., Cutts, F.T., et al. 1994. Measles among secondary school children in West Cumbria: implications for vaccine policy. *Commun Dis Rep*, **4**, R70–3.

Centers for Disease Control. 1994. Rubella and congenital rubella syndrome – United States, January 1, 1991–May 7, 1994. *MMWR*, **43**, 391–401.

Centers for Disease Control. 2001. *Control and prevention of rubella: evaluation and management of suspected outbreaks, rubella in pregnant women, and surveillance for congenital rubella syndrome 2001/50* (RR12), 1–23

Centers for Disease Control. 2002. *Smallpox response plan and guidelines*. MMWR (Version 3.0) www.bt.cdc.gov/agent/smallpox/response-plan/index.asp.

Chen, R.T., Weierbach, R., et al. 1994. A 'post-honeymoon period' measles outbreak in Muyinga sector, Burundi. *Int J Epidemiol*, **23**, 185–93.

Cohen, M.S. 1995. Sexually transmitted diseases enhance HIV transmission: no longer a hypothesis. *Lancet*, **351**, Suppl. iii, 5–7.

Davis, R.M., Whitman, E.D., et al. 1987. A persistent outbreak of measles despite appropriate prevention and control measures. *Am J Epidemiol*, **126**, 438–49.

de Azevedo Neto, R.S., Silveira, A.S., et al. 1994. Rubella seroepidemiology in a non-immunized population of Sao Paulo State, Brazil. *Epidemiol Infect*, **113**, 161–73.

de Quadros, C.A., Olivé, J.M., et al. 1996. Measles elimination in the Americas: evolving strategies. *J Am Med Assoc*, **275**, 224–9.

Diekmann, O., Heesterbeek, J.A.P. and Metz, J.A.J. 1990. On the definition and the computation of the basic reproduction ratio R_0 in models for infectious diseases in heterogeneous populations. *J Math Biol*, **28**, 365–82.

Dietz, K. 1975. Transmission and control of arboviruses. In: Ludwig, D. and Cooke, K.L. (eds), *Mathematical models for the spread of infectious diseases*. Philadelphia: Society for Industrial and Applied Mathematics, 104–21.

Dietz, K. 1993. The estimation of the basic reproduction number for infectious diseases. *Stat Meth Med Res*, **2**, 23–41.

Dudley, S.F. 1926. *The spread of droplet infection in semi-isolated communities*. Medical Research Council special report series 111. London: HMSO.

Farr, W. 1840. *Progress of epidemics*. Second report of the Registrar General of England and Wales, 16–20.

Farrington, C.P. 1990. Modelling forces of infection for measles, mumps and rubella. *Stat Med*, **9**, 953–67.

Fine, P.E.M. 1979. John Brownlee and the measurement of infectiousness: an historical study in epidemic theory. *J R Stat Soc A*, **142**, 347–62.

Fine, P.E.M. 1993. Herd immunity: history, theory, practice. *Epidemiol Rev*, **15**, 265–302.

Fine, P.E.M. and Clarkson, J.A. 1982a. Measles in England and Wales – I: An analysis of factors underlying seasonal patterns. *Int J Epidemiol*, **11**, 5–14.

Fine, P.E.M. and Clarkson, J.A. 1982b. The recurrence of whooping cough: possible implications for assessment of vaccine efficacy. *Lancet*, **i**, 666–9.

Finkelstein, R.A. 1984. In Germaner, R. (ed.), *Bacterial vaccines*. Orlando, FL: Academic Press, p. 107.

Fox, J.P., Elveback, L., et al. 1971. Herd immunity: basic concept and relevance to public health immunization practices. *Am J Epidemiol*, **94**, 179–89.

Frost, W.H. 1976. Some conceptions of epidemics in general. *Am J Epidemiol*, **103**, 141–51.

Gangarosa, E.J., Galazka, A.M., et al. 1998. Impact of anti-vaccine movements on pertussis control: the untold story. *Lancet*, **351**, 356–61.

Garnett, G.P., Mertz, K.J., et al. 1999. The transmission of dynamics of gonorrhoea: modelling the reported behaviour of infected patients from Newark, New Jersey. *Philos Trans R Soc Lond B Biol Sci*, **354**, 1384, 787–97.

Gay, N.J. 1996. A model of long term decline in disease transmissibility: implications for the incidence of hepatitis A. *Int J Epidemiol*, **25**, 854–61.

Gay, N.J., Morgan-Capner, P., et al. 1994. Age-specific antibody prevalence to hepatitis A: implications for disease control. *Epidemiol Infect*, **113**, 113–20.

Gay, N.J., Hesketh, L.M., et al. 1995. Interpretation of serological surveillance data for measles using mathematical models: implications for vaccine strategy. *Epidemiol Infect*, **115**, 139–56.

Gay, N., Ramsay, M., et al. 1997. The epidemiology of measles in England and Wales since the vaccination campaign. *Commun Dis Rep*, **7**, R17–21.

Giesecke, J. 2002. *Modern infectious disease epidemiology*, 2nd edn. London: Arnold.

Granwehr, B.P., Lillibridge, K.W., et al. 2004. West Nile Virus: where are we now? *Lancet Infect Dis*, **4**, 547–52.

Goldschneider, I., Gotschich, E.C. and Artenstein, M.S. 1969. Human immunity to the meningococcus. II. Development of natural immunity. *J Exp Med*, **129**, 1327–48.

Greenwood, M., Bradford Hill, A. et al. 1936: *Experimental epidemiology*. Medical Research Council special report series 209. London: HMSO.

Gregg, N.M. 1941. Congenital cataract following German measles in the mother. *Trans Opthalmol Soc Aust*, **3**, 35.

Gregson, S., Nyamukapa, C.A., et al. 2002. Sexual missing patterns and sex-differentials in teenage exposure to HIV infection in rural Zimbabwe. *Lancet*, **359**, 1896–902.

Grenfell, B.T. and Anderson, R.M. 1985. The estimation of age related rates of infection from case notifications an serological data. *J Hyg Camb*, **95**, 419–36.

Grosskurth, H., Mosha, F., et al. 1995. Impact of improved treatment of sexually transmitted diseases on HIV infection in rural Tanzania: randomised controlled trial. *Lancet*, **346**, 530–6.

Hahn, R.A. 1991. What should behavioural scientists be doing about AIDS? *Soc Sci Med*, **33**, 1, 1–3.

Halloran, M.E., Cochi, S.L., et al. 1994. Theoretical epidemiologic and morbidity effects of routine varicella immunisation of preschool children in the United States. *Am J Epidemiol*, **140**, 81–104.

Hamer, W.H. 1906. The Milroy Lectures on epidemic disease in England – the evidence of variability and of persistency of type. *Lancet*, **i**, 733–9.

Hawker, J., Begg, J., et al. 2001. *Communicable disease control handbook*. Oxford: Blackwell.

Hethcote, H.W. 1983. Measles and rubella in the United States. *Am J Epidemiol*, **117**, 2–13.

Hethcote, H.W. and Yorke, J.A. 1984. Gonorrhea: transmission dynamics and control. *Lect Notes Biomath*, **56**, 1–105.

Heyman, D. (ed.) 2004. *Control of communicable disease manual*, 18th edn. Washington: American Public Health Association.

Hope Simpson, R.E. 1948. The period of transmission in certain epidemic diseases. *Lancet*, **255**, 755–60.

Jacques, J.A., Simon, C.P., et al. 1988. Modelling and analysing HIV transmission: the effect of contact patterns. *Math Biosci*, **92**, 119–99.

Johnson, A.M., Wadsworth, J., et al. 1992. Sexual lifestyles and HIV risk. *Nature*, **360**, 410–12.

Kermack, W.O. and McKendrick, A.G. 1927. Contributions to the mathematical theory of epidemics, part 1. *Proc Roy Soc Edinb*, **115**, 700–21.

Klovdhah, A.S. 1985. Social networks and the spread of infectious diseases: the AIDS example. *Soc Sci Med*, **21**, 11, 1203–16.

Last, J. 2001. *Dictionary of epidemiology*. Oxford: International Epidemiology Association.

Lim, W.L. and Yeoh, E.K. 1992. Hepatitis A vaccination. *Lancet*, **339**, 304.

Lindegren, M.L., Fehrs, L.J., et al. 1991. Update: rubella and congenital rubella syndrome, 1980–1990. *Epidemiol Rev*, **13**, 341–8.

Ludwig, D. and Cooke, K.L. (eds) 1975. *Transmission and control of arboviruses*. Philadelphia: Society for Industrial and Applied Mathematics, pp. 104.

Macdonald, G. 1957. *The epidemiology and control of malaria*. London: Oxford University Press.

Maple, P.A., Efstration, A., et al. 1995. Diphtheria immunity in UK blood donors. *Lancet*, **345**, 963–5.

Massad, E., Burattini, M.N., et al. 1994. A model-based design of a vaccination strategy against rubella in a non-immunized community of Sao Paulo State, Brazil. *Epidemiol Infect*, **112**, 579–94.

Mayaud, P., Mosha, F., et al. 1997. Improved treatment services significantly reduce the prevalence of sexually transmitted diseases in rural Tanzania: results of a randomised controlled trial. *AIDS*, **11**, 1873–80.

McIntyre, P., Isaacs, D., et al., 2000. *Invasive haemophilus influenzae infection*. Australian, Paediatric Surveillance Unit. Final report. apsu.inopsu.com/hib2000.pdf

McLean, A.R. 1995. After the honeymoon in measles control. *Lancet*, **345**, 272.

McLean, A.R. and Anderson, R.M. 1988. Measles in developing countries. Part II. The predicted impact of mass vaccination. *Epidemiol Infect*, **100**, 419–22.

Methner, U., Barrow, P.A., et al. 1999. Combination of vaccination and competitive exclusion to prevent *Salmonella* colonization in chickens: experimental studies. *Int J Food Microbiol*, **49**, 35–42.

Miller, E. 1994. The new measles campaign. *Br Med J*, **309**, 1102–3.

Miller, E. 2003. Measles-mumps-rubella vaccine and the development of autism. *Semin Pediatr Infect Dis*, **14**, 199–206.

Miller, E. and Gay, N.J. 1997. Epidemiological determinants of pertussis. *Dev Biol Stand*, **89**, 15–23.

Miller, E., Cradock-Watson, J.E. and Pollock, T.M. 1982. Consequences of confirmed maternal rubella at successive stages of pregnancy. *Lancet*, **ii**, 781–4.

Miller, E., White, J.M. and Fairley, C.K. 1994. Pertussis vaccination. *Lancet*, **344**, 1575–6.

Miller, E., Waight, P., Gay, N. et al. 1996. The epidemiology of rubella in England and Wales before and after the vaccination campaign. *Commun Dis Rep*, **6**, R.

Mollison, D., Isham, V. and Grenfell, B.T. 1994. Epidemics: models and data. *JR Stat Soc A*, **157**, 115–49.

Morgan-Capner, P., Wright, J., et al. 1988. Surveillance of antibody to measles, mumps and rubella by age. *Br Med J*, **297**, 770–2.

Morris, M. and Kretzschmar, M. 1997. Concurrent partnerships and the spread of HIV. *AIDS*, **11**, 641–8.

Morse, D., O'Shea, M., et al. 1994. Outbreak of measles in a teenage school population: need to immunise susceptible adolescents. *Epidemiol Infect*, **113**, 355–65.

Nicoll, A. 2001. Benefits, safety and risks of immunisation programmes. *Interdisc Sci Rev*, **26**, 20–30.

Nicoll, A., Gill, O.N., et al. 2000. The public health applications of unlinked anonymous seroprevalence monitoring for HIV in the United Kingdom. *Int J Epidemiol*, **29**, 1–10.

Noah, N.D. 1988. Disease elimination or reduction? In: Silman, A.J. and Allwright, S.P.A. (eds), *Opportunities for health service action in Europe*. Oxford: Oxford Medical Publications.

Noah, N.D., Pearce, M.C., et al. 1995. The cyclical nature of communicable diseases. *Lancet*, **346**, 20–3.

Olin, P., Rasmussen, F., Ad Hoc Group for the Study of Pertussis Vaccines, et al. 1997. Randomised controlled trial of two-component, three-component, and five-component acellular pertussis vaccines compared with whole-cell pertussis vaccine. *Lancet*, **350**, 569–77.

Osborne, K., Gay, N., et al. 2000. Ten years of serological surveillance in England and Wales: methods, results, implications and action. *Int J Epidemiol*, **29**, 362–8.

Panagiotopoulos, T., Antoniadou, I. and Valassi-Adam, E. 1999. Increase in congenital rubella occurrence after immunisation in Greece: retrospective survey and systematic review. *Br Med J*, **319**, 1462–7.

Perez-Trallero, E., Cilla, G., et al. 1994. Falling incidence and prevalence of hepatitis A in northern Spain. *Scand J Infect Dis*, **26**, 133–6.

Ramsay, M.E., Gay, N.J., et al. 1994. The epidemiology of measles in England and Wales: rationale for the 1994 national vaccination campaign. *Commun Dis Rep*, **4**, R141–6.

Robinson, N.J., Mulder, D.W., et al. 1995. Modelling the impact of alternative HIV intervention strategies in rural Uganda. *AIDS*, **9**, 1263–70.

Ross, R. 1909. *Report on the prevention of malaria in Mauritius*. London: J and A Churchill.

Ross, R. 1911. *The prevention of malaria*, 2nd edn. London: Murray.

Sabin, A.B. 1991. Measles, killer of millions in developing countries: strategy for rapid elimination and continuing control. *Eur J Epidemiol*, **7**, 1–22.

Samuel, M.C., Guydish, J., et al. 1991. Changes in sexual practices over 5 years of follow-up among heterosexual men in San Francisco. *J Acquir Immun Def Synd*, **4**, 896–900.

Schenzle, D. 1984. An age-structured model of pre- and post-vaccination measles transmission. *IMA J Math Appl Med Biol*, **1**, 169–91.

Schinazi, R. 2001. On the importance of risky behaviour in the transmission of sexually transmitted diseases. *Math Biosci*, **173**, 25–33.

Simonsen, O., Kjeldsen, K., et al. 1987. Susceptibility to diphtheria in populations vaccinated before and after elimination of indigenous diphtheria in Denmark. A comparative study of antitoxic immunity. *Acta Pathol Microbiol Immunol Scand*, **95**, 225–31.

Soper, H.E. 1929. The interpretation of periodicity in disease prevalence. *JR Stat Soc*, **92**, 34–73.

Stoner, B.P., Whittington, W.L., et al. 2000. Comparative epidemiology of heterosexual gonococcal and chlamydial networks: implications for transmission patterns. *Sex Transm Dis*, **27**, 4, 215–23.

Tang, Y.W., Wang, J.X., et al. 1991. A serologically confirmed, case-control study of a large outbreak of hepatitis A in China associated with consumption of clams. *Epidemiol Infect*, **107**, 651–7.

Thacker, S.B. and Millar, J.D. 1991. Mathematical modeling and attempts to eliminate measles: a tribute to the late Professor George Macdonald. *Am J Epidemiol*, **133**, 517–25.

Topley, W.W.C. 1919a. The Goulstonian Lectures on the spread of bacterial infection. *Lancet*, **ii**, 1–5.

Topley, W.W.C. 1919b. The Goulstonian Lectures on the spread of bacterial infection. *Lancet*, **ii**, 45–9.

Topley, W.W.C. 1919c. The Goulstonian Lectures on the spread of bacterial infection. *Lancet*, **ii**, 91–6.

Topley, W.W.C. 1923. The spread of bacterial infection: some general considerations. *J Hyg Camb*, **21**, 226–36.

Topley, W.W.C. and Wilson, G.S. 1923. The spread of bacterial infection. The problem of herd immunity. *J Hyg Camb*, **21**, 243–9.

Wasserheit, J.N. 1994. Effect of changes in human ecology and behaviour on patterns of sexually transmitted diseases, including human immunodeficiency virus infection. *Proc Natl Acad Sci USA*, **91**, 2430–5.

Wawer, M.J., Sewankambo, N.K., The Rakai Project Study Group, et al. 1999. Control of sexually transmitted diseases for AIDS prevention in Uganda: a randomised community trial. *Lancet*, **353**, 525–35.

Williams, J.R. and Anderson, R.M. 1994. Mathematical models of the transmission dynamics of human immunodeficiency virus in England and Wales: mixing between different risk groups. *J R Stat Soc A*, **157**, 69–87.

World Health Organization. 2003. Consensus document on the epidemiology of severe acute respiratory syndrome. *WHO*. SARS WHO/cds/csr/gar/2003. (October 17th 2003) www.who.int/csr.sars.en/whoconsensus.pdf

World Health Organization, Rotary International et al. 2004. Global polio eradication initiative strategic plan, 2004–2008. www.polioeradication.org/content/publications/2004stratplan.pdf

World Health Organization, 2004. Avian influenza and human health. Report to Executive Board www.who.int/gb/ebwha/pdf_files

Yap, I. and Guan, R. 1993. Hepatitis A sero-epidemiology in Singapore: a changing pattern. *Trans Roy Soc Trop Med Hyg*, **87**, 1, 22–3.

Yorke, J.A., Hethcote, H.W. and Nold, A. 1978. Dynamics and control of the transmission of gonorrhea. *Sex Transm Dis*, **5**, 51–6.

15

Emergence and resurgence of bacterial infectious diseases

RUTH L. BERKELMAN AND KEITH P. KLUGMAN

In recent years, the world has witnessed an emergence and resurgence of infectious diseases. Despite predictions earlier in this century to the contrary, infectious diseases remain the leading cause of death in developing countries, and the potential threats posed by infectious diseases are increasing (Institute of Medicine (IOM) 1992, 2003). The factors contributing to the emergence of infectious diseases are multiple and complex. They include demographic changes, changing lifestyles, political instability, unprecedented global travel and commerce, development of new technologies, ecological changes, microbial adaptation and natural variation or mutation of microorganisms, deterioration of national and international infrastructures for the control of infectious agents, and intent to do harm. These factors may contribute to the emergence of a wide array of microorganisms, including bacteria, viruses, parasites, and fungi.

As broadly defined by the 1992 US Institute of Medicine report, *Emerging Infections: Microbial Threats to Health in the United States*, emerging infections are those whose incidence in humans has increased within the past few decades or whose incidence threatens to increase in the near future. These include the emergence of new agents, the re-emergence of agents that previously had declined in incidence, and the development of antimicrobial resistance. The term also includes the recognition that an established disease has a previously unknown infectious origin.

DEMOGRAPHIC CHANGES

Increasing population density and urban poverty in many areas of the world are major factors favoring the emergence and re-emergence of diseases such as tuberculosis, shigellosis, and cholera. The population of the world, less than two billion at the beginning of the twentieth century, exceeded six billion by the end of the century and by the year 2050 is expected to grow to between 7.3 and 10.5 billion (Figure 15.1) (Rousch 1994; Lutz and Qiang 2002).

Along with the population surge, the geographical distribution of the world's population is shifting. Since 1950, an increasing percentage of people are living in Asia and Africa. It is predicted that Africa's percentage of the world's population will continue to increase, from 12 percent in 1990 to 23 percent by the year 2050 (Rousch 1994). In turn, the percentage of the world's population living in Europe or North America is declining. There is also a dramatic trend towards urbanization, with most of the population increase expected between 2000 and 2030 to be absorbed by urban areas in developing countries. By the year 2015, the United Nations (2001) projects that six cities will have 20

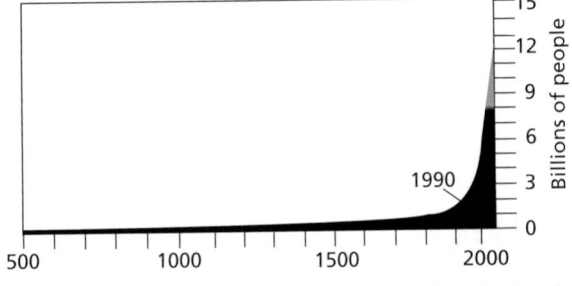

Figure 15.1 *World population through 1990, with projections by United Nations demographers of high and low scenarios. (Abstracted with permission from Roush, 1994. Copyright 1994 American Association for the Advancement of Science.)*

million population or more, including Tokyo, Dhaka, Mumbai, Sao Paulo, Delhi, and Mexico City. Overcrowding and development of slum areas are growing factors in a complex set of interactions that contribute to disease emergence.

INTERNATIONAL TRAVEL AND COMMERCE

The dramatic increases in international travel and commerce are also playing a significant role in the emergence of infectious diseases. The current volume of global traffic, together with the speed at which destinations can be reached, is unprecedented. In 2000, the number of international trips on commercial airplane flights was estimated to reach nearly 700 million (IOM 2003), and flights carried people with drug-resistant tuberculosis, cholera, and other diseases. Occasionally, the transport vehicles are themselves the sites of dissemination of disease, as with legionnaires' disease on cruise ships (Centers for Disease Control and Prevention (CDC) 1994c; Castellani et al. 1999) or food-borne illnesses or tuberculosis on airlines (Driver et al. 1994; Hedberg et al. 1994; CDC 1995c).

There are many examples of human movement resulting in geographical dispersion of a pathogen; recent advances in molecular epidemiology have been instrumental in documenting the course of a number of pathogens. In the early 1980s, a pneumococcus multiply resistant to penicillin, tetracycline, chloramphenicol, and trimethoprim-sulphamethoxazole was described in Spain, and its spread to the USA was documented in 1992 (Munoz et al. 1991), South Africa in 1994 (Klugman et al. 1994), and South Korea in 1997 (McGee et al. 1997). It is now recognized as a global clone (McGee et al. 2001) and the contribution to this single clone to the burden of penicillin-resistance is remarkable – in three studies, this clone alone comprised 39 percent (Corso et al. 1998), 16 percent (Gherardi et al. 2000), and 18 percent (Richter et al. 2002) of all penicillin-resistant strains from multiple geographic sites across the continental USA. Recently

this clone has acquired resistance to the fluoroquinolones. In Hong Kong, which has the highest rate of fluoroquinolone resistance in the world, all the resistant strains belonged to this single clone (Ho et al. 2001).

In the early 1990s, an estimated 20 million people were refugees, a group that may be at higher risk of many infectious diseases (Wilson 1995). More recent estimates are even higher (Summerfield 1997), and tuberculosis, *Shigella*, and cholera are among the diseases causing deaths among refugees. Drug-resistant tuberculosis has been common among refugees entering developed countries from developing countries and has contributed to the resurgence of tuberculosis witnessed in some developed countries.

Acute disasters such as earthquakes and displacement of people through civil strife, may have an impact on food and water supplies and often require temporary living quarters, which frequently resemble slums. Following civil strife in Rwanda in 1994, an estimated 50 000 Rwandan refugees died within the first month as cholera and *Shigella dysenteriae* type 1 swept through the camps in Zaire (Goma Epidemiology Group 1995). This strain spread widely down through Africa, associated with epidemic hemolytic-uremic syndrome (HUS) (Bloom et al. 1994).

Global commerce has also been an important factor in transporting pathogens. Importation of turtles, iguanas, and other reptiles carrying *Salmonella* (CDC 1995b), the carriage of *Vibrio cholerae* in the ballast water of ships, and shipment of fresh fruits and vegetables contaminated with *Shigella* or enterotoxigenic *Escherichia coli* from one geographical area to another (Hedberg et al. 1994) have resulted in outbreaks of disease geographically distant from the source of the pathogen (Wilson 1995).

Although most fruits and vegetables used to be grown and consumed locally, the increasing trend towards global commerce has led to expanded importation of fresh fruits and vegetables from developing countries into more developed countries, and workers within developed countries today are often low-paid migrant workers (Hedberg et al. 1994). These factors increase the potential for products to be contaminated in the field, during packing, and during distribution to retail markets and have contributed to the changes in the epidemiology of food-borne diseases, which has witnessed an increasingly wider array of pathogens including bacteria, parasites, and viruses.

Travel, migration, and commerce allow global mixing of microbial species. Disease emergence may occur if circumstances in the new environment are propitious, allowing survival and proliferation of an introduced microbial pathogen.

SOCIETAL CHANGES

Changes in individual and collective behavior can affect the risk of exposure to infectious agents. For example, as the association was made between cardiovascular disease risk and consumption of selected foods such as beef, people changed their dietary habits, including the consumption of different food items. In the USA and many other developed countries, the per capita consumption of whole milk and beef declined, and consumption of cheese, poultry, and fresh fruits and vegetables increased (Hedberg et al. 1994). Changes in dietary habits continue as new issues emerge. The concern following the recognition of variant Creutzfeld–Jakob disease related to bovine spongiform encephalopathy led to an abrupt decline in beef consumption in some countries; at the same time, an inceased focus on addressing obesity in the USA has led to increases in the consumption of high-protein foods, including beef. The variety of food has also changed; compared with 50 years ago, a markedly increased number of items are stocked in the average grocery store in developed countries. The place of consumption has also changed; the number of fast-food restaurants has increased, as have such features as salad bars to accompany the changing dietary preferences.

The proliferation of fast-food restaurants with uniform methods of preparation and frequent use of single sources of food has further expanded the opportunities for a large number of people to be exposed to contaminated food. In an outbreak of HUS and hemorrhagic colitis in western USA caused by hamburger contaminated with *E. coli* O157:H7 (Bell et al. 1994), scores of restaurants affiliated with a single fast-food chain were implicated.

In some economies, there has been an expanding participation of women in the workforce and a rising proportion of single-parent families, accompanied by growth in the use of child-care services. In the USA, 90 percent of families with preschool children use full- or part-time child daycare services (Thacker et al. 1992). Because children are frequently in close or direct physical contact and have poor personal hygiene, pathogens may be transferred more readily from one child to another in these facilities. Use of child-care centers has been associated with an increased risk of transmission of many enteric and respiratory pathogens

Outbreaks of diarrheal illness in child-care centers have resulted from many enteric pathogens including *Shigella, Salmonella, Campylobacter,* and *E. coli* O157:H7 (Thacker et al. 1992). Children attending daycare centers are at a 1.6–3.4 times greater risk of having a diarrheal illness compared with children who receive care at home. The risk is greatest for those not yet toilet-trained. Secondary spread to household and community contacts may also occur.

The incidence of respiratory illnesses, including those caused by *Streptococcus pneumoniae* (Levine et al. 1999) and group A *Streptococcus* (Espinosa de los Monteros et al. 2001), is also elevated for children attending child-care centers. Accumulating evidence suggests that daycare attendees are at greater risk for acute otitis media, which often follows upper respiratory infections, than children cared for at home (Rovers et al. 1999; Baraibar 1997). Increased use of child-care centers may account partially for the marked increase in visits to physicians for otitis media documented in the USA between 1975 and 1990, from approximately ten million visits to over 24 million visits per year (Schappert 1992). Outbreaks of tuberculosis in family daycare homes have also been reported.

Changes in sexual behavior have also affected changes in disease patterns (Wasserheit 1994; Donova 2002). The rates of sexually transmitted diseases are closely associated with changes in the number of sexual partners, use of drugs and adequacy of health-care services. Recent increases in reported rates of chlamydial infection are related not only to improvement in surveillance but also to increases in sexual activity of very young women. High rates of oral contraceptive use may foster rapid transmission of chlamydial infection.

CHANGES IN TECHNOLOGY

Changes in technology have also resulted in the emergence of newly recognized pathogens. New methods of food production have led to new niches for pathogens. Mass production, with its increased complexity of operations, can greatly magnify the public-health significance of microbial contamination. A pathogen present in some of the raw material may contaminate a large batch of final product. For example, there have been dramatic changes in how hamburger meat is produced in the USA. Today, a single hamburger patty typically includes meat from many cattle, and cattle from many different farms (Boyce et al. 1995). Over a quarter of a million hamburgers contaminated with *E. coli* O157:H7 were recalled in one single outbreak (Bell et al. 1994; Berkelman 1994). Contaminated cheese from one plant distributed to four other processors, which subsequently shredded it and thereby contaminated cheeses from other sources, led to a widespread outbreak of salmonellosis (Hedberg et al. 1992).

Medical and technological advances in the medical care of patients have resulted in benefits to many patients, but many of these advances have also been accompanied by an increased risk of infections. The escalation in the use of chemotherapy, radiation, and other immunosuppressive therapy has increased the frequency of opportunistic infections. Bacteremias and fungemias associated with intravascular devices have increased dramatically (Edgeworth et al. 1999). Renal dialysis units expose susceptible patients and personnel to complex equipment that frequently is difficult to decontaminate (Favero et al. 1992). Surgery is performed on patients who are already highly susceptible to infections.

Special attention must be given to the risk of transmission of infection through transfusion of blood or blood products (Kuehnert et al. 2001; Walsh et al. 2002). In the USA, there has been an increase in the number of episodes of sepsis caused by *Yersinia enterocolitica*; prolonged storage of the packed red blood cell units have resulted in high bacterial and endotoxin concentrations in the transfused unit (Tipple et al. 1990; CDC 1991). In addition, there have been more reports of infections with staphylococcal species resulting in sepsis or death associated with contaminated platelets; most of these units were pooled units with long storage times (Zaza et al. 1994).

The use of organ transplants has also grown dramatically in many countries in the past decade. Organ transplants pose special risks: the organ may harbor a pathogen, and the transplantation procedure is accompanied by immunosuppressive therapy (Hibberd and Rubin 1992). The number of opportunistic bacterial, fungal, and other infections has also risen as a result of the emergence of human immunodeficiency virus (HIV). Although the advent of highly active antiretroviral therapy (HAART) has had a major impact on the burden of opportunistic infections in developed countries, these infections threaten to overwhelm health services in sub-Saharan Africa.

Legionnaires' disease has emerged in developed countries as a result of technological change. Cooling towers, air conditioners, whirlpool spas, respiratory therapy equipment, ultrasonic mist machines, and industrial settings in which aerosols are produced have all provided new opportunities for the *Legionella* bacterium to become aerosolized and infect humans (CDC 1994b, Fry et al. 2003).

Technology related to tampon development and use was associated with the emergence of menstrual-associated toxic shock syndrome. This illness is due to in vivo production of a unique toxin, toxic shock syndrome toxin-1, by *Staphylococcus aureus*, and the epidemic of toxic shock syndrome witnessed in 1980 was associated with the introduction and marketing of hyperabsorbable tampons. The molecular basis for increased toxin production in the presence of tampon fibers has been debated (Kass 1987).

LAND-USE PATTERNS AND ECOLOGICAL CHANGE

Ecological changes have contributed to the emergence of tick-borne diseases in many areas of the world. Lyme disease, caused by transmission of *Borrelia burgdorferi* through tick bites, has become the most common vector-borne bacterial disease in the USA, and reported cases of disease continue to increase. The disease is also well recognized across Europe and northern Asia; cases have also been reported from other continents (Berglund et al. 1995). Sites of intense transmission of *B. burgdorferi*

often represent newly reforested areas that had been farmed recently. The suburban encroachment on the newly reforested areas, along with increased recreational use of forested areas, has also increased the exposure of humans to ticks.

Human ehrlichiosis was first described in the USA in the mid-1980s (Blanco and Oteo 2002; Dumler and Bakken 1995). *Ehrlichia chaffeensis* causes human monocytic ehrlichiosis, a syndrome characterized by fever, headache, and laboratory findings such as leukopenia and thrombocytopenia, and it may be life-threatening if untreated (Paddock and Childs 2003). It is primarily transmitted by the lone star tick, *Amblyoma americanum*. A closely related organism, *Anaplasma phagocytophilum* (formerly classified in the genus *Ehrlichia equi*), causes a similar syndrome known as human anaplasmosis (formerly human granulocytic ehrlichiosis). People affected in the USA have spent time in tick-infested areas where *Ixodes scapularis* and *Ixodes pacificus* are common. The European vector is the same as that of Lyme borreliosis, *Ixodes ricinus*.

Alterations in the aquatic environment may result in changes in the occurrence of waterborne diseases. The seasonality of cholera has been related to coastal algae blooms (Epstein et al. 1994). Other changes in human ecology, such as urbanization and the development of periurban slums, may also contribute to an increased risk of diseases such as cholera (Levine and Levine 1994).

Both long-term climate changes and weather have been shown to alter the risk of various infectious diseases. Heavy rainfall contributed to a large outbreak of leptospirosis in Nicaragua in 1995, with exposure of humans to flood waters contaminated by urine from infected animals, particularly dogs (Trevejo et al. 1998). El Niño provided unusual conditions that favored growth of *Vibrio parahaemolyticus*, which led to an outbreak of disease related to shellfish (CDC 1998).

The incidence of Buruli ulcer, caused by *Mycobacterium ulcerans*, has dramatically increased in West Africa in recent years (Dobos et al. 1999). The increase is attributed to environmental changes including regional flooding. The natural reservoir is unknown but suspected to be an environmental exposure, since many affected individuals live near slow-moving or stagnant water.

MICROBIAL ADAPTATION AND CHANGE

New pathogens may emerge as a result of a mutation causing enhanced virulence. Such a mutation may explain the emergence of Brazilian purpuric fever in 1984, caused by *Haemophilus aegyptius*; it was the first time that this agent had been shown to cause invasive disease (CDC 1985; Brazilian Purpuric Fever Study Group 1987). Since its recognition in Brazil, this virulent pathogen has been found in Australia; other parts of the world may be at risk for epidemics.

In 1993, the clonal spread of a novel non-O1 serotype of *V. cholerae* was documented (Ramamurthy et al. 1993); *V. cholerae* O139 was first detected in southern Asia and quickly replaced *V. cholerae* O1 strains in many affected areas. Previous natural infection or receipt of cholera vaccine was found to afford little or no protective benefit.

The development of resistance to antimicrobial agents is also a powerful example of bacteria's capacity to adapt. Almost all major bacterial pathogens acquire antibiotic-resistance genes (Tomasz 1994) and the problem has been amplified by the increased use of antibiotics worldwide.

The introduction of penicillin in the 1940s was soon followed by the detection of resistant bacteria. Bacterial resistance has grown with the addition of large numbers of antibiotics with distinct mechanisms of action. These antibiotics have increasingly had a wider antibacterial spectrum. Antibiotics have been introduced not only for therapeutic use in people but also on a large scale in animal feed. Endtz et al. (1991) demonstrated the rise in prevalence of *Campylobacter* strains that were resistant to ciprofloxacin in both poultry products and in human stools in the Netherlands. Extensive use of fluoroquinolones in the poultry industry between 1982 and 1989 is likely to have resulted in the resistance observed. The ban of antimicrobials for growth promotion in Denmark has led the way to reducing resistance amongst enterococci in that country (Aarestrup et al. 2001).

Once genes for resistance have been acquired, the progeny of these bacteria with their resistance genes tend to spread with great rapidity within the species. Production of β-lactamase by staphylococci was rare before the introduction of penicillin. More than 90 percent of all isolates of *S. aureus* now carry the β-lactamase gene, in both the USA and Europe, and methicillin resistance increased rapidly in the 1980s. Strains with and without genes for resistance demonstrate similar virulence in terms of toxin production. Methicillin resistance is mediated most frequently by the production of a novel penicillin-binding protein. Noble et al. (1992) demonstrated conjugative transfer of high-level vancomycin resistance from *Enterococcus faecalis* to *Staphylococcus aureus* in the laboratory and raised concern that resistance may be transferred to wild-type *S. aureus*. The vancomycin-resistance determinant *vanA* has now indeed spread from the enteroccocus to *Staphylococcus aureus* to create the first high-level vancomycin-resistant *Staphylococcus aureus* (VRSA) infections (CDC 2002).

The incidence of infections with vancomycin-resistant enterococci has also increased. These strains are now very common in intensive care units, and amongst dialysis units the proportion reporting a patient with vancomycin resistant enterococci (VRE) increased from 11.5 percent in 1995 to 32.7 percent in 2000 (Tokars et al. 2002). Spread of resistant strains within hospitals may be limited by aggressive infection control measures and prudent vancomycin use.

Drug-resistant *S. pneumoniae* was first documented in Australia in the 1960s and multiply resistant strains were reported in South Africa in the 1970s; reports of similar isolation were made with increasing frequency from other parts of the world in the 1980s. By 1998, 24 percent of pneumococcal isolates submitted to CDC were resistant to penicillin and 14 percent were resistant to at least three drug classes. The emergence of penicillin resistance in Europe has been correlated closely with antimicrobial use in those countries (Bronzwaer et al. 2002) (Figure 15.2). The rapid emergence of drug-resistant strains is complicating the management of pneumococcal infections and demonstrates the need for more judicious use of antimicrobial agents (Heffelfinger et al. 2000).

Penicillinase-producing *Neisseria gonorrhoeae* was first recognized in 1976. By the 1990s, 32 percent of *N. gonorrhoeae* isolates were resistant to penicillin or tetracycline, and resistance to fluoroquinolones was recognized (CDC 1993). Fluoroquinolone-resistant strains now have a global distribution. *N. gonorrhoeae* also demonstrates the increasingly diverse array of mechanisms of resistance, including both plasmid- and chromosomally mediated resistance to the penicillins and tetracyclines.

Antibacterial resistance is also a problem for the developing world. In central and southern Africa, isolates of *S. dysenteriae* 1 have been demonstrated to be resistant to all readily accessible oral antibiotics,

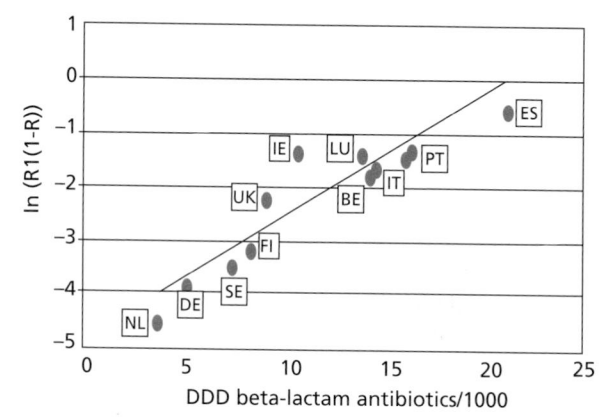

Figure 15.2 *Association of antibiotic use with resistance in pneumococci: The log odds of resistance to penicillin among invasive isolates of* Streptococcus pneumoniae *(ln RI/(1-R)) is regressed against outpatient sales of beta-lactam antibiotics in 11 European countries. DDD, defined daily dose; BE, Belgium; DE, Germany; FI, Finland; IE, Ireland; IT, Italy; LU, Luxembourg; NL, the Netherlands; PT, Portugal; ES, Spain; SE, Sweden; UK, United Kingdom. (Redrawn from Bronzwaer et al. 2002, www.cdc.gov/ncidod/EID/vol8no3/01-0192.htm, with permission.)*

including ampicillin, chloramphenicol, nalidixic acid, tetracycline, and trimethoprim-sulphamethoxazole (Ries et al. 1994; Tuttle et al. 1995). The need to monitor, regularly and frequently, the prevailing antimicrobial resistance patterns in different geographical areas has been repeatedly demonstrated. Preventive measures may be more successful than therapy for control of many epidemics.

BREAKDOWN OF PUBLIC-HEALTH INFRASTRUCTURE

Complacency has also resulted in the re-emergence of pathogens previously controlled. Basic tenets of public health, such as close monitoring of disease in a population accompanied by a rapid response for diseases such as shigellosis and salmonellosis, and control programs for diseases such as tuberculosis and diphtheria, have been neglected in many areas of the world as public-health attention towards the control of all but a few selected conditions declined in the 1970s and 1980s (Berkelman et al. 1994). In addition, the emergence of new pathogens such as *E. coli* O157:H7 has been addressed inadequately.

Cases of tuberculosis in the USA increased 18 percent from 1985 through 1991; yet, by 1992, most public-health laboratories had not incorporated the more rapid radiometric methods for routine culture or drug-susceptibility testing of mycobacteria. The erosion of public-health services together with the mistaken perception of tuberculosis as a disease of declining public-health significance resulted in decreased funding for many laboratories in the USA and resulted in the delay in implementation of newer, more rapid diagnostic technologies for tuberculosis (Bloom and Murray 1992; Dowdle 1993; Huebner et al. 1993) as well as a decline in resources for control programs.

In 1994, reports of widespread transmission of pneumonic plague within India resulted in estimated economic losses to India of over $1 billion dollars (Campbell and Hughes 1995). Intense investigation revealed that most reported cases of plague were incorrectly diagnosed and that no transmission of pneumonic plague was identified in any major city except Surat. Effective surveillance coupled with laboratory expertise in plague might have averted some of the economic loss.

The epidemic levels of cholera reached in South America in the early 1990s, after almost a century of absence of the disease from the continent, were attributed, in part, to breakdowns in public-health measures (IOM 1992). Contaminated municipal water supplies contributed to the spread of the epidemic after the pathogen was introduced along the coastal waters of Peru (Blake 1993). As both the population and the percentage of the population in poverty grow in South America, the potential for spread of diarrheal illnesses increases.

In the early 1990s, Russia and many of the newly independent states experienced a dramatic resurgence of diphtheria (CDC 1995a) (Figure 15.3). Inadequate immunization of the population, crowding and low socioeconomic conditions, and high mobility of the infected individuals may have contributed to this outbreak, which has resulted in thousands of deaths. Also, from 1990 to 1995, tuberculosis rates in Russia increased by 70 percent, with more than 25 000 people dying from the disease each year (Netesov and Conrad 2001). The increased incidence has been compounded by the spread of multidrug resistant tuberculosis.

INTENT TO DO HARM: BIOTERRORISM

The growing threat of intentional attacks using biological agents is a serious one and must be addressed (IOM 2003). CDC has developed three categories of biological agents, prioritized according to their likelihood of bioterrorist use and the severity of the diseases they produce. Three of the six CDC category A (high priority) agents are bacteria or their toxins: *Yersinia pestis*, *Bacillus anthracis*, and *Clostridium botulinum* toxin. Outbreaks may occur as a result of natural occurrence, unintentional release by those working on these agents in laboratories, or intentional use. All of these agents have been weaponized to cause disease when released in aerosol form. Pneumonic plague, inhalation anthrax, and botulism may all occur naturally but are quite rare in developed countries today.

In 1979, an unintentional release of anthrax spores from a Soviet bioweapons facility caused an outbreak of inhalational anthrax, the largest recorded outbreak of inhalation anthrax, with at least 77 cases diagnosed (Meselson et al. 1994). In 2002, 22 cases of anthrax (11 inhalational, 11 cutaneous) were identified as a result of bioterrorism; envelopes containing *Bacillus anthracis* spores were mailed to news media companies and government officials in the USA (Jernigan et al. 2002). No naturally occurring cases of inhalation anthrax had

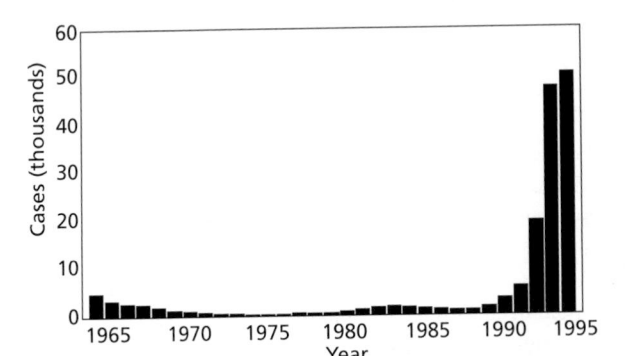

Figure 15.3 *Reported cases of diphtheria in New Independent States of the Former Soviet Union, 1965–1995. Redrawn from www.cdc.gov/mmwr/preview/mmwrhtml/00043378.htm, with permission.*

been diagnosed in the previous 25 years in the USA. In addition, Salmonella and Shigella have also been used to deliberately contaminate food and cause illness in the USA (Carus 2001; Torok et al. 1997).

INFECTIOUS ORIGINS OF CHRONIC DISEASES

A number of diseases or conditions, previously considered to be noninfectious, have recently been recognized to have an infectious etiology. *Helicobacter pylori* was first isolated from humans in 1982 and is now known to play a central role in the development of diffuse gastritis and duodenal ulceration (Marshall 1989). More than 90 percent of patients with duodenal ulceration are infected with *H. pylori*, and treatment with antimicrobial therapy heals duodenal ulcers; additionally, lower recurrence rates are associated with eradication of *H. pylori*. Presence of *H. pylori*, as determined by serological studies, is also associated with a three- to six-fold higher risk of gastric cancer (Fuchs and Mayer 1995); gastric cancer is the second most common cause of cancer-related deaths in the world. Evidence is still accumulating that *Chlamydia pneumoniae* may have an etiological role in the development of atherosclerosis (Campbell and Kuo 2003).

Guillain–Barré syndrome, a leading cause of acute paralysis, has been associated with infection with *Campylobacter jejuni*; in a study conducted in England and Wales, 26 percent of 96 patients with Guillain–Barré syndrome had evidence of infection with *C. jejuni* (Rees et al. 1995). Peripheral nerves may share epitopes with *C. jejuni*. Other syndromes that are ill-defined and often considered to be immune-mediated may soon also be discovered to have infectious microorganisms playing a central role in etiology, possibly through molecular mimicry (Bolton 1995).

Techniques to identify pathogens that rely on molecular techniques rather than on cultivation have been developed. To identify the bacillus causing Whipple's disease, the bacterial 16S ribosomal RNA sequence from infected tissue was amplified and its phylogenetic relations have been described (Relman et al. 1992; Relman 2002).

SUMMARY

Following the 1992 publication of the US Institute of Medicine report on emerging infections, CDC released a prevention strategy to address emerging infectious disease threats for the USA (CDC 1994a). In 1994, the World Health Organization defined goals to guide implementation of a global effort on emerging infectious diseases (World Health Organization 1994); in 1995, the World Health Assembly passed a resolution calling for attention to the problem of new and re-emerging infections. The original 1992 IOM report was revisited by a new committee in 2003 and its report (IOM 2003)

finds that the threat of emerging infections had increased over the past decade. The need to increase global capacity for response, to develop inexpensive and easy diagnostic tests for field use, and to have available the diagnostic reagents, drugs, and vaccines needed has been emphasized, both for naturally occurring diseases and for those diseases likely to result from bioterrorism. As the speed of change in the world continues to accelerate, the challenges confronting us by the re-emergence of old pathogens or the recognition of new ones will likewise be heightened.

REFERENCES

Aarestrup, F.M., Seyfarth, A.M., et al. 2001. Effect of abolishment of the use of antimicrobial agents for growth promotion on occurrence of antimicrobial resistance in fecal enterococci from food animals in Denmark. *Antimicrob Agents Chemother*, **45**, 2054–9.

Baraibar, R. 1997. Incidence and risk factors of acute otitis media in children. *Clin Microbiol Infect*, **3**, SS3, S13–22.

Bell, B.P., Goldoft, M., et al. 1994. A multistate outbreak of *Escherichia coli* O157:H7 associated bloody diarrhea and hemolytic uremic syndrome from hamburgers: the Washington experience. *J Am Med Assoc*, **272**, 1349–53.

Berglund, J., Eitrem, R., et al. 1995. An epidemiologic study of Lyme disease in southern Sweden. *N Engl J Med*, **333**, 1319–24.

Berkelman, R.L. 1994. Emerging infectious diseases in the United States, 1993. *J Infect Dis*, **170**, 272–7.

Berkelman, R.L., Bryan, R.T., et al. 1994. Infectious disease surveillance: a crumbling foundation. *Science*, **264**, 368–70.

Blake, P.A. 1993. Epidemiology of cholera in the Americas. *Gastroenterol Clin North Am*, **22**, 639–60.

Blanco, J.R. and Oteo, J.A. 2002. Human granulocytic ehrlichiosis in Europe. *Clin Microbiol Infect*, **8**, 763–72.

Bloom, B.R. and Murray, C.J.L. 1992. Tuberculosis: commentary on a reemergent killer. *Science*, **257**, 1055–64.

Bloom, P.D., MacPhail, A.P., et al. 1994. Haemolytic-uraemic syndrome in adults with resistant *Shigella dysenteriae* type I. *Lancet*, **344**, 206.

Bolton, C.F. 1995. The changing concepts of Guillain–Barré syndrome (editorial). *N Engl J Med*, **333**, 1415–17.

Boyce, T.G., Swerdlow, D.L. and Griffin, P.M. 1995. *Escherichia coli* O157:H7 and the hemolytic–uremic syndrome. *N Engl J Med*, **333**, 364–8.

Brazilian Purpuric Fever Study Group, 1987. *Haemophilus aegyptius* bacteraemia in Brazilian purpuric fever. *Lancet*, **2**, 761–3.

Bronzwaer Stef, L.A.M., Cars, O., et al. 2002. A European study on the relationship between antimicrobial use and antimicrobial resistance. *Emerg Infect Dis*, **8**, 278–82.

Campbell, G.L. and Hughes, J.M. 1995. Plague in India: a new warning from an old nemesis. *Ann Intern Med*, **122**, 151–3.

Campbell, L.A. and Kuo, C.C. 2003. Chlamydia pneumoniae and atherosclerosis. *Semin Respir Infect*, **18**, 48–54.

Carus, W.S. 2001. *Working Paper: Bioterrorism and biocrimes: The illicit use of biological agents since 1900*. Washington, DC: National Defense University.

Castellani, P.M., Lo Monaco, R., et al. 1999. Legionnaires' disease on a cruise ship linked to the water supply system: clinical and public health implications. *Clin Infect Dis*, **28**, 39–41.

Centers for Disease Control, 1985. Preliminary report: epidemic fatal purpuric fever among children – Brazil. *Morb Mortal Wkly Rep*, **34**, 217–19.

Centers for Disease Control, 1991. Update: *Yersinia enterocolitica* bacteremia and endotoxin shock associated with red blood cell transfusions – United States, 1991. *Morb Mortal Wkly Rep*, **40**, 176–8.

Centers for Disease Control and Prevention, 1993. Sentinel surveillance for antimicrobial resistance in *Neisseria gonorrhoeae* – United States, 1988–1991. *Morb Mortal Wkly Rep*, **42**, SS-3, 29–39.

Centers for Disease Control and Prevention. 1994a. Addressing emerging infectious disease threats: a prevention strategy for the United States. Atlanta, Georgia: US Department of Health and Human Services, Public Health Service, 1–46.

Centers for Disease Control and Prevention. 1994b. Legionnaires' disease associated with cooling towers – Massachusetts, Michigan, and Rhode Island, 1993. *Morb Mortal Wkly Rep*, **43**, 491–3.

Centers for Disease Control and Prevention, 1994c. Update: outbreak of Legionnaires' disease associated with a cruise ship, 1994. *Morb Mortal Wkly Rep*, **43**, 574–5.

Centers for Disease Control and Prevention, 1995a. Diphtheria epidemic – New Independent States of the former Soviet Union, 1990–1994. *Morb Mortal Wkly Rep*, **44**, 177–80.

Centers for Disease Control and Prevention, 1995b. Reptile-associated salmonellosis – selected states, 1994–1995. *Morb Mortal Wkly Rep*, **44**, 347–50.

Centers for Disease Control and Prevention, 1995c. Exposure of passengers and flight crew to *Mycobacterium tuberculosis* on commercial aircraft, 1992–1995. *Morb Mortal Wkly Rep*, **44**, 137–40.

Centers for Disease Control and Prevention, 1998. Outbreak of *Vibrio parahaemolyticus* infections associated with eating raw oysters – Pacific Northwest, 1997. *Morb Mortal Wkly Rep*, **47**, 457–62.

Centers for Disease Control and Prevention, 2002. *Staphylococcus aureus* resistant to vancomycin – United States, 2002. *Morb Mortal Wkly Rep*, **51**, 565–7.

Corso, A., Severina, E.P., et al. 1998. Molecular characterization of penicillin-resistant *Streptococcus pneumoniae* isolates causing respiratory disease in the United States. *Microb Drug Resist*, **4**, 325–37.

Dobos, K.M., Quinn, F.D., et al. 1999. Emergence of a unique group of necrotizing mycobacterial diseases. *Emerg Infect Dis*, **5**, 367–78.

Donova, B. 2002. Rising prevalence of genital *Chlamydia trachomatis* infection in heterosexual patients at the Sydney Sexual Health Centre, 1994 to 2000. *Commun Dis Intell*, **261**, 51–5.

Dowdle, W.R. 1993. The future of the public health laboratory. *Ann Rev Public Health*, **14**, 649–64.

Driver, C.R., Valway, S.E., et al. 1994. Transmission of *Mycobacterium tuberculosis* associated with air travel. *J Am Med Assoc*, **272**, 1031–5.

Dumler, J.S. and Bakken, J.S. 1995. Ehrlichial diseases of humans: emerging tick-borne infections. *Clin Infect Dis*, **20**, 1102–10.

Edgeworth, J.D., Treacher, D.F. and Eykyn, S.J. 1999. A 25-year study of nosocomial bacteremia in an adult intensive care unit. *Crit Care Med*, **27**, 1421–8.

Endtz, H.P., Ruijs, G.J., et al. 1991. Quinolone resistance in *Campylobacter* isolated from man and poultry following the introduction of fluoroquinolones in veterinary medicine. *J Antimicrob Chemother*, **27**, 199–208.

Epstein, P.R., Ford, T.E., et al. 1994. Marine ecosystem health: implications for public health. In: Wilson, M.E., Levins, R. and Spielman, A. (eds), *Disease in evolution*, 1st edn. New York: The New York Academy of Sciences, 13–23.

Espinosa de los Monteros, M., Bustos, I.M., et al. 2001. Outbreak of scarlet fever caused by an erythromycin-resistant *Streptococcus pyogenes* emm22 genotype strain in a day-care center. *Pediatr Infect Dis J*, **20**, 807–9.

Favero, M.S., Alter, M.J. and Bland, L.A. 1992. Dialysis-associated infections and their control. In: Bennett, J.V. and Brachman, P.S. (eds), *Hospital infections*, 3rd edn. Boston/Toronto/London: Little, Brown and Company, 375–403.

Fry, A.M., Rutman, M., et al. 2003. Legionnaires' disease outbreak in an automobile engine manufacturing plant. *J Infect Dis*, **187**, 1015–18.

Fuchs, C.S. and Mayer, R.J. 1995. Gastric carcinoma. *N Engl J Med*, **333**, 32–41.

Gherardi, G., Whitney, C.G., et al. 2000. Major related sets of antibiotic-resistant pneumococci in the United States as determined by pulsed-field gel electrophoresis and pbp1a-pbp2b-pbp2x-dhf restriction profiles. *J Infect Dis*, **181**, 216–29.

Goma Epidemiology Group, 1995. Public health impact of Rwandan refugee crisis: what happened in Goma, Zaire, in July, 1994. *Lancet*, **345**, 339–44.

Hedberg, C.W., Korlath, J.A., et al. 1992. A multistate outbreak of *Salmonella javianna* and *Salmonella oranienburg* infections due to consumption of contaminated cheese. *J Am Med Assoc*, **268**, 3203–7.

Hedberg, C.W., MacDonald, K.L. and Osterholm, M.T. 1994. Changing epidemiology of food-borne disease: a Minnesota perspective. *Clin Infect Dis*, **18**, 671–82.

Heffelfinger, J.D., Dowell, S.F., et al. 2000. Management of community-acquired pneumonia in the era of pneumococcal resistance: a report from the Drug-Resistant *Streptococcus pneumoniae* Therapeutic Working Group. *Arch Intern Med*, **160**, 1399–408.

Hibberd, P.L. and Rubin, R.H. 1992. Infection in transplant recipients. In: Bennett, J.V. and Brachman, J.S. (eds), *Hospital infections*, 3rd edn. Boston/Toronto/London: Little, Brown and Company, 899–921.

Ho, P.L., Yam, W.C., et al. 2001. Fluoroquinolone resistance among *Streptococcus pneumoniae* in Hong Kong linked to the Spanish 23F clone. *Emerg Infect Dis*, **7**, 906–8.

Huebner, R.E., Good, R.C. and Tokars, J.I. 1993. Current practices in mycobacteriology: results of a survey of state public health laboratories. *J Clin Microbiol*, **31**, 771–5.

Institute of Medicine, 1992. *Emerging infections: Microbial threats to health in the United States*. Washington, DC: National Academy Press.

Institute of Medicine, 2003. *Microbial threats to health*. Washington, DC: National Academies Press.

Jernigan, D.B., Raghunanthan, P.L., et al. 2002. Investigation of bioterrorism-related anthrax, United States, 2001: epidemiologic findings. *Emerg Inf Dis*, **8**, 1019–28.

Kass, E.H. 1987. On the pathogenesis of toxic shock syndrome. *Rev Infect Dis*, **9**, Suppl 5, S482–9.

Klugman, K.P., Coffey, T.J., et al. 1994. Cluster of erythromycin-resistant variant of the Spanish multiply resistant 23F clone of *Streptococcus pneumoniae* in South Africa. *Eur J Clin Microbiol Infect Dis*, **13**, 171–4.

Kuehnert, M.J., Roth, V.R., et al. 2001. Transfusion-transmitted bacterial infection in the United States, 1998 through 2000. *Transfusion*, **41**, 1493–9.

Levine, M.M. and Levine, O.S. 1994. Changes in human ecology and behavior in relation to the emergence of diarrheal diseases, including cholera. *Proc Natl Acad Sci USA*, **91**, 2390–4.

Levine, O.S., Farley, M., et al. 1999. Risk factors for invasive pneumococcal disease in children: a population-based case-control study in North America, Abstract. *Pediatrics*, **103**, e28.

Lutz, W. and Qiang, R. 2002. Determinants of human population growth. *Phil Trans R Soc Lond*, **357**, 1197–210.

McGee, L., Klugman, K.P., et al. 1997. Spread of the Spanish multi-resistant sterotype 23F clone of *Streptococcus pneumoniae* to Seoul, Korea. *Microb Drug Resist*, **3**, 253–7.

McGee, L., McDougal, L., et al. 2001. Nomenclature of major antimicrobial-resistant clones of *Streptococcus pneumoniae* defined by the Pneumococcal Molecular Epidemiology Network. *J Clin Microbiol*, **39**, 2565–7.

Marshall, B.J. 1989. History of the discovery of *C. pylori*. In: Blaser, M.J. (ed.), *Campylobacter pylori in gastritis and peptic ulcer disease*. New York: Igaku Shoin Medical Publishers, 7–23.

Meselson, M., Guillemin, J., et al. 1994. The Sverdlovsk anthrax outbreak of 1979. *Science*, **266**, 1202–8.

Munoz, R., Coffey, T.J., et al. 1991. Intercontinental spread of a multiresistant clone of serotype 23F *Streptococcus pneumoniae*. *J Infect Dis*, **164**, 302–6.

Netesov, S.V. and Conrad, J.L. 2001. Emerging infectious diseases in Russia, 1990–1999. *Emerg Infect Dis*, **7**, 1–5.

Noble, W.C., Virani, Z. and Cree, R.G.A. 1992. Co-transfer of vancomycin and other resistance genes from *Enterococcus faecalis* NCTC 12201 to *Staphylococcus aureus*. *FEMS Microbiol Lett*, **93**, 195–8.

Paddock, C.D. and Childs, J.E. 2003. *Ehrlichia chaffeenis*: a prototypical emerging pathogen. *Clin Microbiol Rev*, **16**, 37–64.

Ramamurthy, T., Garg, S., et al. 1993. Emergence of novel strain of *Vibrio cholerae* with epidemic potential in southern and eastern India (corresp.). *Lancet*, **341**, 703–4.

Rees, J.H., Soudain, S.E., et al. 1995. *Campylobacter jejuni* infection and Guillian–Barré syndrome. *N Engl J Med*, **333**, 1374–417.

Relman, D.A. 2002. New technologies, human–microbe interactions, and the search for previously unrecognized pathogens. *J Infect Dis*, **186**, suppl 2, S254–258.

Relman, D.A., Schmidt, T.M., et al. 1992. Identification of the uncultured bacillus of Whipple's disease. *N Engl J Med*, **327**, 293–301.

Richter, S.S., Heilmann, K.P., et al. 2002. The molecular epidemiology of penicillin-resistant *Streptococcus pneumoniae* in the United States, 1994–2000. *Clin Infect Dis*, **34**, 330–9.

Ries, A.A., Wells, J.G., et al. 1994. Epidemic *Shigella dysenteriae* type 1 in Burundi: panresistance and implications for prevention. *J Infect Dis*, **169**, 1035–41.

Rousch, W. 1994. Population: the view from Cairo. *Science*, **265**, 1164–7.

Rovers, M.M., Zielhuis, G.A., et al. 1999. Day-care and otitis media in young children: a critical overview. *Eur J Pediatr*, **158**, 1–6.

Schappert, S.M. 1992. *Office visits for otitis media: United States, 1975–90*. Advance data from vital and health statistics, no. 214. Hyattsville, MD: National Center for Health Statistics.

Summerfield, D. 1997. The social, cultural, and political dimensions of contemporary war. *Med Confl Surviv*, **13**, 3–25.

Thacker, S.B., Addiss, D.G., et al. 1992. Infectious diseases and injuries in child day care: opportunities for healthier children. *J Am Med Assoc*, **268**, 1720–6.

Tipple, M.A., Bland, L.A., et al. 1990. Sepsis associated with transfusion of red cells contaminated with *Yersinia enterocolitica*. *Transfusion*, **30**, 207–13.

Tokars, J.I., Frank, M., et al. 2002. National surveillance of dialysis-associated diseases in the United States, 2000. *Semin Dial*, **15**, 162–71.

Tomasz, A. 1994. Multiple-antibiotic-resistant pathogenic bacteria: a report on the Rockefeller University Workshop. *N Engl J Med*, **330**, 1247–51.

Torok, T.J., Tauxe, R.V., et al. 1997. A large community outbreak of salmonellosis caused by intentional contamination of restaurant salad bars. *J Am Med Assoc*, **278**, 389–95.

Trevejo, R.T., Rigau-Perez, J.G., et al. 1998. Epidemic leptospirosis associated with pulmonary hemorrhage – Nicaragua, 1995. *J Infect Dis*, **178**, 1457–63.

Tuttle, J., Ries, A.A., et al. 1995. Antimicrobial-resistant epidemic *Shigella dysenteriae* type 1 in Zambia: modes of transmission. *J Infect Dis*, **171**, 371–5.

United Nations, 2001. *World urbanization prospects: the 2001 revision*. New York: United Nations.

Walsh, A.L., Molyneux, E.M., et al. 2002. Bacteraemia following blood transfusion in Malawian children: predominance of Salmonella. *Trans R Soc Trop Med Hyg*, **96**, 276–7.

Wasserheit, J.N. 1994. Effect of changes in human ecology and behavior on patterns of sexually transmitted diseases, including human immunodeficiency virus infection. *Proc Natl Acad Sci USA*, **91**, 2430–5.

Wilson, M.E. 1995. Travel and the emergence of infectious diseases. *Emerg Infect Dis*, **1**, 39–46.

World Health Organization, 1994. Emerging infectious diseases. *Wkly Epidemiol Rec*, **69**, 234–6.

Zaza, S., Tokars, J.I., et al. 1994. Bacterial contamination of platelets at a university hospital: increased identification due to intensified surveillance. *Infect Control Hosp Epidemiol*, **15**, 82–7.

16

Healthcare-associated infections

MARK H. WILCOX AND R.C. SPENCER

INTRODUCTION

Of the infections dealt with in this volume, very many may be acquired in hospital, but these hospital infections have special features. Hospitals bring together uniquely vulnerable hosts and subject them to particular risks of infection from animate and inanimate sources. Apart from being the cause of morbidity and mortality, infection limits the effectiveness and adds greatly to the cost of medical treatment.

The term 'hospital-acquired infection' (syn. nosocomial infection) is applied to any infection causing illness that was not present or in its incubation period when the subject entered hospital or received treatment in an outpatient or accident and emergency department. The increasing blurring between primary and secondary health care systems, and thus often the lack of clarity about the origin of infection, means that the term 'healthcare-associated infection' is preferably used. It includes not only incidents in which a single microorganism spreads from person to person ('cross-infection') or from a common source in the hospital, but also single and apparently unconnected infections. Sporadic and endemic hospital-acquired infections may be less striking than epidemic ('outbreaks'), but they predominate numerically and their control or reduction present the greater challenge.

Some healthcare-associated infections do not differ from infections with the same microorganism in the general population, but many are profoundly affected by the patient's underlying illnesses or by medical or surgical treatments to which they are subjected while in hospital. The source of the infecting organism may be exogenous, from another patient or a member of the hospital staff, or from the inanimate environment in the hospital; or it may be endogenous, from the patient's own flora, which at the time of infection may include organisms brought into hospital at admission and others acquired subsequently. In either case, the infecting organisms may spontaneously invade the tissues of the patient or be introduced into them by surgical operation, instrumental manipulation, or nursing procedure. The concentration of patients with particular illnesses and undergoing similar treatments in specialist hospital units often creates unique niches for hospital pathogens. Healthcare-associated infection may also affect discharged inpatients, outpatients, and staff, and an episode of hospital infection may be initiated by the admission of an infected patient from the general population. Hospital infection may spill over into the community, necessitating investigation and control in both populations. In all considerations of healthcare-associated infection, the source, the mode of infection, and host susceptibility must all be kept in mind. It is clear from a review of the current literature base that much modern day infection control practice is unfortunately not evidence based. A step forward in this respect is the production of true evidence-based infection control guidelines, such as The epic project (developing national evidence-based guidelines for preventing healthcare-associated infections) (Pratt et al. 2001).

OCCURRENCE, CONSEQUENCES, AND COST OF HEALTHCARE-ASSOCIATED INFECTION

Information about the extent of hospital infection in general may be obtained from incidence or prevalence studies (see Chapter 13, Epidemiology of infectious diseases). The relationship between the results of these two types of survey is complex (Freeman and Hutchison 1980). Most investigations are limited to particular sites of infection, such as surgical wounds, or to individual hospital services, such as intensive care. However, in the USA it has been estimated that more than 2 million, and perhaps as many as 4 million, patients are infected in hospital each year. Study of a random sample of patients from 6 449 acute care hospitals gave an incidence of 5.7 infections per 100 admissions (Haley et al. 1985a, b). Ongoing nationwide surveillance of nosocomial infection in the USA is reported in terms of specific infection sites (National Nosocomial Infections Surveillance (NNIS) System 1995). Recent trends have been an increase in bloodstream infections and pneumonias, and an increase in the rate of infections per 1 000 patient-days because of shorter periods of hospitalization (Weinstein 1998). A large proportion of surgical site infections in particular, present in the community even though they were acquired in hospital. This percentage is likely to increase as hospital stays decrease, and therefore, unless an effective form of post-discharge surveillance is carried out then there will be marked under-ascertainment of some hospital-acquired infections (Weigelt et al. 1992; Mangram et al. 1999).

In Great Britain and Ireland, a cross-sectional prevalence survey (Emmerson et al. 1996) suggested an overall nosocomial infection prevalence rate of 9.0 percent, with significantly higher prevalence in teaching (11.2 percent) than in nonteaching (8.4 percent) hospitals. A previous survey (Meers et al. 1981) found a similar overall prevalence (9.2 percent), but Emmerson et al. (1996) point out differences in the study methods and medical practice in the two periods. For example, earlier patient discharge in the more recent survey implied that many surgical wound infections would be detected after discharge from hospital and were therefore not included (Bailey et al. 1992; Weigelt et al. 1992; Byrne et al. 1994; Emmerson 1995). Nationwide prevalence surveys in Spain (EPINE 1992, 1995) have shown an overall nosocomial infection rate of 8.5 percent in 1990, falling to 7.2 percent in 1994. A similar investigation in Norway found a prevalence of 6.3 percent, which was a fall in comparison with previous years (Aavitsland et al. 1992). A prevalence survey in Hong Kong (Kam and Mak 1993) found 8.6 percent of patients to have a hospital-acquired infection. Caution is needed when comparing these and similar figures, as there may be differences in the spectrum of patients, in medical practice, and in definitions of infection. Indeed, there were important differences between the individual hospitals in all these investigations. A similar body site distribution of infections was seen in the studies, with four sites predominating: urinary tract, respiratory tract, surgical wound, and skin (Figure 16.1). Improvements in prevalence figures are often due mainly to control of urinary tract infection.

A prevalence of hospital infection between 6 and 9 percent has frequently been observed in recent years and it has been questioned whether the 'irreducible minimum' of hospital-acquired infections has been reached (Ayliffe 1986). This stability may indicate some success, in that an increase in the infection rate might have been expected because modern methods of treatment create more opportunities for infection and highly susceptible patients are surviving in greater numbers. Nevertheless, an examination of many aspects of current hospital practice suggests that the incidence of infection could be reduced.

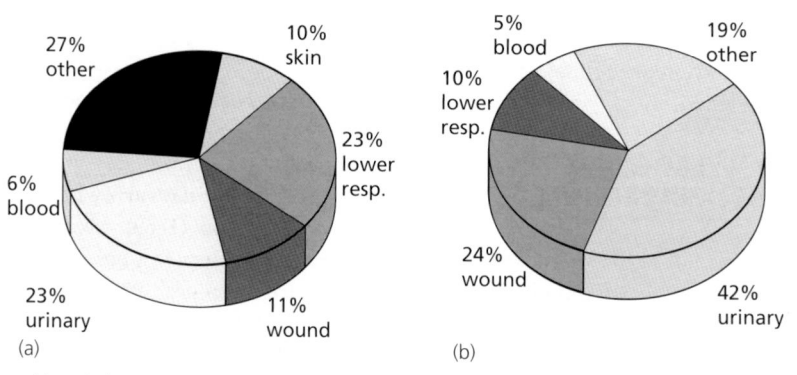

Figure 16.1 *Occurrence of hospital-acquired infection by site:* **(a)** *UK, prevalence survey (Emmerson et al. 1996);* **(b)** *USA, incidence survey (Haley et al. 1985b)*

Most types of hospital infection cause appreciable morbidity and may lead to residual disability and even death, although a significant worsening of outcome may be difficult to demonstrate statistically except in very large samples of patients (Freeman et al. 1979). Some impression of the impact of hospital infection can be gained from the large prevalence and incidence studies, as can an estimate of the increased cost of care. Martone et al. (1992) suggested that, in the USA, a hospital-acquired infection results in an average of 4.0 extra days of stay. The average extra charges (1992) were $2 100, giving an estimated annual national cost of more than $4.5 billion. It was estimated that more than 19 000 deaths were due directly to hospital infection and that this was contributory to a further 58 000 deaths. Similar figures are not available in Europe. In a 1988 case–control study in England it was established that hospital infection in surgical patients added an average of 8.2 days to hospital stay with added costs of £1 041 (Coello et al. 1993). Antibiotic resistance in the infecting organism adds to costs and reduces the effectiveness of treatment (Holmberg et al. 1987). A socio-economic study of hospital-acquired infection in the National Health Service (NHS) in England in 1994/1995 used modeling based on logistic regression to calculate confidence limits for the estimated increased lengths of stay (Plowman et al. 2001). Extrapolating from measurements in one district general hospital, the calculated annual cost to the NHS was £1 000 million (Plowman et al. 2001). The study examined approximately 4 000 inpatients and identified 7.8 percent as having acquired at least one hospital-acquired infection. Additionally, 30 percent of this hospital-acquired infection cohort, and 19 percent of the remainder reported post-discharge symptoms consistent with the diagnosis of urinary tract, chest, or surgical infections. On average, a patient with hospital-acquired infection spent 2.5 times longer in hospital, and cost £3 000 more to treat than a similar nonhospital-acquired infection patient. This sum is of a similar order of magnitude to other studies, and highlights both the socioeconomic burden of hospital-acquired infections and the potential for savings and/or health gains if even only a small proportion of these can be prevented. Indeed, the control of hospital-acquired infection has been characterized as one of the most cost-effective of health care interventions (Haley et al. 1985b; Chaudhuri 1993; Wenzel 1995).

HISTORY OF KNOWLEDGE OF HEALTHCARE-ASSOCIATED INFECTION

A chapter devoted to hospital-acquired infection appeared for the first time in the seventh edition of this book, reflecting the growth of interest in hospital epidemiology and infection control, but the subject is not new. The need to isolate patients with obviously infectious diseases has been recognized since ancient times and the spread of infection that might ensue from the introduction of such patients into hospitals has been known for centuries. However, segregation of fever hospitals from general hospitals dates only from the early nineteenth century and the usefulness of this measure was not demonstrated statistically until much later. Hospital-acquired infection and the consequent mortality probably reached their peak in the nineteenth century. The following brief account includes only the salient references, but the more recent papers referred to (e.g. Cruickshank 1944; Williams 1956; Ayliffe and English 2003) supply more detail.

The nineteenth century

Urban overpopulation and hospital overcrowding made hospitals places of dread for the poor. There was uncontrolled puerperal sepsis, and surgical sepsis caused death in most cases of compound fracture admitted to hospital. Crossinfection in children's hospitals was associated with mortality rates of 25–40 percent. The now well known work of Semmelweiss (1861) on puerperal sepsis was largely disregarded at the time. He observed its association with medical staff and students who attended patients and also performed autopsies. Semmelweiss deduced that the disease was spread by 'morbid matter' on their hands derived from cadavers or other affected patients. A dramatic reduction in infection rates was achieved by the introduction of hand-washing with chlorinated lime.

At the same time, and to greater immediate effect, Florence Nightingale, after her experience of hospital sepsis at Scutari and her reform of the army medical services, turned her attention to British hospitals. In a much quoted remark in her book *Notes on hospitals* (Nightingale 1863) she states:

> It may seem a strange principle to enunciate as the very first requirement in a Hospital that it should do the sick no harm ... the actual mortality in hospitals, especially in those of large crowded cities, is very much higher than any calculation founded on the mortality of the same class of diseases among patients treated *out of* hospital ...

Florence Nightingale established important principles of nursing, hospital design, and hygiene, while remaining sceptical of the germ theory of disease. Further evidence was provided by the survey of the sequelae of amputation by Simpson (1869) which established that sepsis, gangrene, and pyemia were very much more common in large urban hospitals than in rural practice.

At about this time Lister (1867a, b) introduced his 'antiseptic surgery', with extensive use of carbolic acid to pack wounds, especially of compound fractures, to sterilize instruments and sutures, to decontaminate his hands, as well as use as an air spray. He observed a considerable improvement in the results of treatment of compound fractures and of surgical operations. Lister came to appreciate that the air spray did not add greatly to the effectiveness of the other measures, and that what he was introducing was a barrier between the patients' tissues and infection. 'Antiseptic surgery' was later replaced by von Bergman's 'asepsis' (Schimmelbusch 1894) and by the end of the century, with the introduction of surgical gloves in the USA (see Halsted 1913), the measures in use in a modern operating theater had to a large extent been introduced.

During these years, when many fundamental discoveries in bacteriology were being made, other principles of hospital infection control were established. Flügge (1897, 1899) showed the importance of droplet and aerial spread of tuberculosis and, with Hutinel (1894) and others, established basic isolation systems for diphtheria and other infectious diseases in children's and fever hospitals.

The early twentieth century

The discovery of pathogenic bacteria provided a new basis for the study of hospital infection, and the importance of *Streptococcus pyogenes* was demonstrated during this period, in burn (Cruickshank 1935) and postoperative (Okell and Elliott 1936) infections. Cubicle and barrier nursing was introduced widely (Crookshank 1910) and was shown to be effective in preventing the spread of childhood fevers, except chicken pox and measles.

Aseptic surgery was for a time deemed adequate to keep wound infection to a low level, but this complacency was shattered by the experience of two world wars, when large open wounds readily became infected in the base hospitals (Cruickshank 1944). The local application of mild antiseptics to the wounds proved ineffective (Cruickshank 1944; Williams 1956). The arrival of penicillin in the later years of the Second World War gave considerable benefit (Fraser 1984).

Though Dukes (1929) recognized the importance of the indwelling catheter as a means of introducing infection into the bladder, this was forgotten. The idea was revived much later with the introduction of closed drainage systems (Gillespie 1956; Kunin and McCormack 1966).

The antibiotic era

The introduction of penicillin banished from hospitals the terrible cases of chronic sepsis, mainly caused by *Staphylococcus aureus* (Fletcher 1984). Nevertheless, the era of antibiotics ushered in for the first time a period in which staphylococcal, rather than streptococcal, infection dominated the scene (Williams 1956). Penicillin-

resistant, and later multiply-resistant, *S. aureus* strains (Clarke et al. 1952) which had additional properties of transmissibility and virulence, caused serious wound, burn, and other sepsis. Interest in air-borne and dust-borne spread, as well as transmission on the hands of attendants, revived. From this period date many of the established methods of infection control: the supply of clean air for operating theaters, procedures for wound dressings, and the provision of isolation units, for example (Williams et al. 1960). So does the present general system for hospital-infection control: the appointment in the UK of medical control of infection officers (hospital epidemiologists), infection control nurses (nurse practitioners), and control of infection committees (Colebrook 1955, Subcommittee of the Central Health Services Council 1959; Gardner et al. 1962). Similar arrangements were established in the USA, where the first nurse practitioner in the discipline was appointed in 1963 (Osterman 1981).

Antibiotic treatment and hospital infection control are intimately entwined. The use of antibiotics, although not the only influence, has altered the prevailing pathogens in hospital infection. *S. pyogenes* was almost banished and, later, resistant gram-negative bacilli replaced *S. aureus*. Then gram-positive bacteria, *S. aureus*, coagulase-negative staphylococci, enterococci, certain corynebacteria, and *Clostridium difficile*, re-emerged. The emergence of resistant gram-negative bacteria, such as *Klebsiella* spp. with extended spectrum β-lactamases, *Stenotrophomonas maltophilia* and *Acinetobacter baumannii*, has occupied much attention recently. Changes in hospital practice may favor particular pathogens irrespective of their susceptibility to antibiotics. An example is the association of coagulase-negative staphylococci of varied antibiotic sensitivity with prostheses and cannulae. The availability of antibiotics has allowed the development of modern medical treatment, for example major surgery and the use of immunosuppression, associated with new infection risks and a new range of pathogens.

Antibiotic resistance may increase the impact of hospital infection that would otherwise be of small importance and may necessitate a change to less satisfactory and more expensive antibiotics. It also appears to worsen the prognosis of infection (Hart 1982; Holmberg et al. 1987; French et al. 1990). On occasion it may exclude the possibility of using antibiotics to remove the source of infection in an outbreak. Hospital infection control measures, such as isolation, are extremely important in preventing the spread of resistant bacteria.

The development of multiple resistance in hospital pathogens, without the prospect of new antibiotics to deal with them, is alarming. There have been reports of infection by strains resistant to all established antibiotics (Murray 1991; Tomasz 1994), accounts of untoward failures of treatment (Armstrong et al. 1995), and trends towards resistance to agents favored for treatment and prophylaxis, such as the late generation cephalosporins and the fluoroquinolones (Burwen et al. 1994; Jones et al. 1994; Goldstein and Acar 1995). It has been suggested that the end of the antibiotic era may be near, or that the safety of some forms of hospital treatment may be severely compromised (Kunin 1993; Tomasz 1994). Much evidence suggests that particular resistance problems in hospital-acquired infection have resulted from the routine, unthoughtful, or 'blanket' use of antibiotics in courses that are too long. Restraint, enlightened prescribing, and control of antibiotic use in hospitals may prevent a worsening of the problem (Gould 1988; Report 1995; Editorial 1996; Tenover and McGowan 1996). This has been recognized by the medical and scientific community for some years and has also been acknowledged by public bodies. An important landmark was the inquiry into antimicrobial resistance by the House of Lords in the UK. This report called on governments to recognize the threat, educational authorities to move antibiotic resistance up the agenda, and food producers to curb use in animals (Select Committee on Science and Technology 1998).

More recent developments

For a period, the importance of multiply-resistant *S. aureus* appeared to fade (Ayliffe et al. 1979) and interest shifted to gram-negative bacilli: antibiotic-resistant enterobacteria, such as *Klebsiella* and later *Serratia* spp., which caused large outbreaks of colonization, with some clinical infections. Infection by *Pseudomonas aeruginosa* came to prominence with the increasing number of patients rendered susceptible by illness or treatment. The infecting bacteria appeared to be favored by the antibiotics in current use in the hospitals. More recently, possibly as a result of the introduction of new antibiotics, and the extensive use of indwelling medical devices, gram-positive cocci have again become the predominant causes of infection. *S. aureus* strains resistant to even more antibiotics ('methicillin-resistant *S. aureus*', see later under Staphylococcus aureus) have been as difficult to control as those encountered in the 1950s and multiply-resistant strains of *Staphylococcus epidermidis* have been the most common pathogens in some units. *Enterococcus* spp. resistant to all current antibiotics including vancomycin (see later under Enterococcus spp.) have presented problems in specialized units. Outbreaks of diarrhea and colitis caused by *C. difficile* (see section on Clostridium difficile below) have followed the use of broad spectrum antibiotics, such as cephalosporins. The increase in tuberculosis in immunosuppressed patients (see later under Enterobacteriaceae) and the occurrence of multiply-resistant strains have revived the need for isolation facilities and procedures to prevent air-borne spread of infection to patients and staff. On the other hand, the rediscovery of the principles of effective antibiotic prophylaxis in surgery has done much to prevent endogenous wound infection, for example, in colonic surgery.

The spread of viruses from patients admitted to hospital to other already sick patients remains an uncontrolled problem, particularly with respiratory syncytial virus (RSV) and other respiratory viruses in children's hospitals. Spread of viral gastrointestinal infection has been a major problem (see later under Viruses). The larger numbers of immunosuppressed patients has been associated with an increasing incidence and variety of fungal infection (see later under Fungi and protozoa).

Methods of controlling hospital infection established earlier have been questioned. The UK Medical Research Council (Medical Research Council, Subcommittee of the Committee on Hospital Infection 1968) conducted an inquiry into the measures used in operating theaters. In the USA, efforts to establish the cost-effectiveness of infection control measures culminated in the SENIC study (see later under Surveillance) which demonstrated cost savings by the employment of effective numbers of infection control staff and the institution of preventive measures. Methods of surveillance have been further refined and computerization of records has assisted in this process. It has been important to focus the system of surveillance and control of hospital infection on areas of most effective cost benefit. The need to integrate infection control in the hospital with the wider community has been re-emphasized.

Severe acute respiratory syndrome

Following reports in February 2003 of a mystery respiratory illness affecting some 300 people in Guangdong province, mainland China, a new disease entity was described – severe acute respiratory syndrome (SARS). This disease has subsequently been shown to be caused by a novel coronavirus (Ksiazek et al. 2003; Drosten et al. 2003; Peiris et al. 2003). As of 7 August 2003, some 8 422 cases had been notified to the World Health Organization (WHO) of whom 916 had died. The majority of cases occurred in mainland China (5 327 cases, 349 deaths), Hong Kong (1 755 cases, 300 deaths), Taiwan (665 cases, 180 deaths), Singapore (238 cases, 33 deaths) and Canada (251 cases 41 deaths) (Lee et al. 2003; Tsang et al. 2003; Poutanen et al., 2003; Booth et al. 2003; Ho et al. 2003). The deaths were age-stratified: age <24, <1 percent mortality; 25–44, 6 percent mortality, 45–64, 15 percent mortality; >65, 50 percent mortality. Underlying diseases, such as diabetes, were important cofactors in morbidity and mortality. Spread is thought to be predominantly by large droplets and therefore close contacts are most at risk. The possibility of fecal–oral transmission is still debated and for how long patients are infectious remains unknown. It is possible that some patients become 'super spreaders' and are more infectious than others. There is not much evidence concerning the role of fomites, though studies have shown that the virus can remain viable for over 24 h on inanimate surfaces. Survival in feces may be as long as 96 h, but the relevance of this fact is unknown.

Meticulous attention to infection control procedures were shown to be most effective in preventing the spread of SARS within a healthcare setting (Ho et al. 2003; Seto et al. 2003; Wenzel and Edmond 2003). Ideally, SARS patients should be nursed in an isolation room with negative pressure relative to the surrounding area or a single room with own bathroom and toilet facilities. The personal protective equipment for all healthcare workers includes for routine use, an N95 EN 149:2001 approved respirator/mask or equivalent (95 percent filter efficiency against particles 1 micron in size or smaller, a tight facial seal – less than 1 percent leak); an FFP2 mask with minimum filter efficiency of 92 percent is deemed equivalent. Masks should be fit tested according to the manufacturer's recommendations and fit checked each time the mask is put on. (To check test the mask, the wearer takes a quick, forceful inspiration to determine if the mask seals tightly to the face.) For high risk procedures which generate aerosols and require prolonged very close contact with affected patients (e.g. intubation prior to ventilation, taking respiratory samples, suctioning whilst the patient is being ventilated), a P-100 half face respirator mask (FFP3 is deemed equivalent, minimum filter efficiency, 98 percent). Eye protection (goggles or face shield) when providing direct patient care during cough producing and aerosol generating procedures, and where there is potential for spattering or spraying of body substances. Droplet and contact precautions (including use of gowns and gloves). Instructions are given for removing and discarding masks to avoid transmission of virus during this process. Hand hygiene is also emphasized. More stringent infection control recommendations may be required in specific situations, e.g. outbreak management in a healthcare facility.

MODE OF SPREAD OF INFECTION IN HOSPITALS

Air-borne spread

With the tradition of the spread of disease by effluvium or miasma, it is not surprising that air-borne spread of infection, for example of tuberculosis, was demonstrated early in the scientific era of bacteriology (Flügge 1897, 1899). Interest in this route went into eclipse in the early years of this century (see Williams 1956) to be revived in the later 1930s and the 1940s. The contribution of the air-borne route to common hospital infection remains the subject of controversy. Clearly, the effectiveness of this route depends on the source, on the number of

microorganisms present, and the degree of dispersal, whether in droplets, in droplet nuclei or on skin scales; on survival and retention of pathogenicity by the microorganisms in the air or environment (or their death, impairment, or dilution there); on the size of the infecting dose; and on the local or general susceptibility of the persons exposed to infection. Air-borne spread of infection has been reviewed in detail by Ayliffe and Lowbury (1982) (Chapter 7, Airborne bacteria).

Bacteria can be counted in air by the slit sampler, which draws in air and impacts the particles on to a moving culture plate. Settle plates, which are agar plates exposed on horizontal surfaces for measured periods of time, capture a limited range of the particles from the air and are not quantitative. Small hand-held samplers (e.g. centrifugal samplers) are now available. They are easier and less disturbing to use in a busy operating theater than the slit sampler, and are adapted to making counts during operations and in different areas near the operation itself. Some, however, extract only some of the particles and give results that are not as accurate as, or comparable with, those obtained with the slit sampler (Whyte 1981). Outside the operating theater, the provision of good quality air for severely neutropenic patients in the absence of high efficiency particulate air (HEPA) filters, especially in the context of preventing aspergillus infection, is the other major indication for occasional air sampling. The absence of agreed standards, however, makes interpretation of air sampling outside the operating theater difficult (Humphreys 1993).

TUBERCULOSIS

Clinical experience and experiments with guinea pigs indicate that air-borne spread of tuberculosis can occur by the transfer of very few microorganisms. Patients differ greatly in their ability to transmit tuberculosis (Riley et al. 1962) and the susceptibility of the recipient is also very significant. Only patients with smear-positive pulmonary tuberculosis are regarded as constituting an infection risk and requiring single room isolation. Infectivity declines rapidly after effective treatment is initiated so that only 2 weeks of isolation during treatment is recommended (but see later under *Mycobacterium* spp.). Multidrug-resistant (MDR) tuberculosis in HIV patients during recent years has been associated with nosocomial spread to patients and healthcare workers and emphasizes the need for early identification of positive patients and patient isolation (Beck-Sagué et al. 1992).

PNEUMOCOCCAL INFECTION

Much infection by *Streptococcus pneumoniae* is endogenous and it has not been customary to isolate patients with pneumococcal pneumonia. The appearance of penicillin-resistant and multi-resistant strains has been associated with significant mortality (Pallares et al. 1995). The ready transmission of these emerging organisms has demonstrated the need to contain such infections (Gould et al. 1987), but clusters of cases also occur with some more sensitive strains, which appear to have the ability to cause lobar pneumonia in the relatively healthy (Davies et al. 1984). When the disease has been convincingly diagnosed by a Gram stain of sputum, it may therefore be advisable to isolate patients with pneumococcal pneumonia for the first 24 h of treatment. Protection of vulnerable patients with the currently available polyvalent vaccine becomes more important as resistance increases (Austrian 1994).

MENINGOCOCCAL INFECTION

Meningococcal infection acquired in hospital is uncommon, but isolation for the first 48 h of treatment is advised. Among staff, only those who have had particularly close contact with the patient, as in mouth-to-mouth resuscitation, need be offered prophylactic antibiotics.

OTHER BACTERIAL INFECTIONS

The evidence for spread by air of *S. aureus* and *S. pyogenes* is contradictory. These bacteria are carried by many normal people (Williams et al. 1960); streptococci are readily shed from the upper respiratory tract, by coughing, sneezing, and singing (Lidwell 1974), as are staphylococci on skin squames during physical activity. There is considerable variation in the degree of shedding. *S. aureus* is carried particularly in the perineum and nares and males tend to shed more staphylococci than females. Infected lesions on patients provide rich sources of these bacteria and release into the air is brought about by movement of secondarily contaminated articles, such as bedclothes, protective clothing, bed curtains, dressings, and dust. The numbers in the air can be reduced by disinfection with ultraviolet light or chemicals, by such simple measures as damp, rather than dry, dusting or cleaning, and by control of air circulating between infected and susceptible persons. These measures are not, however, always effective in reducing infection rates. Striking results are most common when the prevailing infection rate is very high, as where dressings are handled, or where there is a large exposure of undefended tissue, as in burns. The important experiments in burned patients by Lowbury et al. (1970) suggested that infection by gram-negative bacilli, e.g. *P. aeruginosa*, was almost entirely by contact, infection by *S. pyogenes* by air and staphylococcal infection by both routes. Measures to control ventilation have on occasion resulted in a decline in the rate of nasal acquisition of organisms by patients. Several workers have found it impossible to control the spread of resistant strains of staphylococci by contact barriers alone, but have achieved success with the use of a separate isolation unit or decontamination of the environment (Pearman et al. 1985). It may be that for these strains the selection

pressure of antibiotics administered to the recipient is of importance in allowing colonization from very small inocula. These are situations in which air control is likely to yield good results.

Gram-negative bacilli tend to die when desiccated and infection by the aerial route is confined mainly to spread by nebulized spray.

VIRAL INFECTION

Smallpox was notorious as being readily transmitted by very few infective particles by the air-borne route. The most common viral infections transmitted by air in hospital are chicken pox, measles, influenza, and respiratory syncitial virus (RSV). Availability of immuno-fluorescence techniques for the rapid diagnosis of RSV renders isolation or 'cohort nursing' (placing patients with similar infections in the same ward or area) shortly after admission to hospital is an increasing possibility and it is recommended (Editorial 1992). It seems that transmission of other childhood exanthems requires closer contact. Small round structured viruses (SRSV) (now referred to as noroviruses) are an increasing cause of nosocomial diarrhea and vomiting. Large inocula are present in vomitus and the air-borne inhalation route as well as the fecal route is now recognized as an important means of transmission (Caul 1994).

FUNGAL INFECTION

Dispersal by spores is a feature of most filamentous fungi, but only *Aspergillus* spp. have been shown to be a significant cause of air-borne infection. This may occur after cardiac surgery or in immunosuppressed patients. Outbreaks have been associated with building work in hospitals (Opal et al. 1986) or inexpert filter changing in operating theater ventilation systems. Other fungi occasionally cause clusters of infections by aerial spread, e.g. phycomycetes (del Palacio Hernanz et al. 1983). Exceptionally, other fungi that cause cardiac valve infection are thought to have arrived via air. Effectively filtered air should be supplied to certain classes of susceptible patients (Rhame et al. 1984). *Cryptococcus neoformans* is frequently found in pigeon droppings around hospitals, but there is no convincing evidence of air-borne spread from this source to susceptible patients.

INFECTION IN THE OPERATING THEATER

The pathogenesis of postoperative infection is complex, but most infections arise from the patient's own flora and the remainder are acquired mainly from staff in the operating theater (Ayliffe 1991). At one time 'aseptic surgery' appeared to offer all that was needed to minimize postoperative infection. In the 1940s, partly stimulated by cases of clostridial infection thought to be due to dust, interest was revived in the possibility that improved ventilation of operating theaters and other measures designed to limit air-borne infection would reduce it further. This was reflected in the principles of modern systems of operating theater ventilation and the design of modern operating areas, including plenum ventilation to ensure flow of air from the theater to the outside rather than the reverse (Humphreys 1993). Satisfactory operating theater clothing, small operating teams with the minimum of movement and restriction of the patient's blankets and dressings to anterooms are other important factors in minimizing postoperative infection, because all these measures reduce the air levels of bacteria in the theater (Dharan and Pittet 2002). Even so, any human activity raises the levels in a conventionally ventilated theater, but these bursts should be quickly suppressed by effective ventilation. Recent publications have addressed the problems of behavior and rituals in operating theaters (Woodhead et al. 2002) and the microbiological commissioning and monitoring of operating theater suites (Hoffman et al. 2002).

The bacteria found in the air of a properly ventilated operating theater are rarely pathogens in the usual sense. For example, they rarely include *S. aureus* or *S. pyogenes*. Indeed, it is difficult to find proof that reducing microorganisms in the air by conventional ventilation systems is beneficial in preventing surgical wound infection, although there have been instances in which the introduction of the systems described above have apparently resulted in a fall from an undesirably high incidence of wound infection (Blowers et al. 1955; Shooter et al. 1956).

It was an early suggestion that in surgical operations on healthy tissues with good defenses, a small degree of contamination with air-borne bacteria is unlikely to result in infection, but that less healthy tissues and areas such as the meninges may not be able to deal with such contamination (Cairns 1939). In the very specialized field of orthopedic surgery, the incidence of infection has fallen with increased expertise, and it is now clear that the spectrum of infecting agents includes *S. epidermidis* and propionibacteria, such as are found in significant numbers in the air of a conventional operating theater. Charnley and Eftekhar (1969) and others found very low rates of clinical infection when ultraclean air was provided, but increased operative skill and the prophylactic use of antibiotics may have played a part in this. A multicenter trial (Lidwell et al. 1982) compared conventional theater conditions, provision of ultraclean air, exhaust-ventilated clothing and, incidentally, use of prophylactic antibiotics, in 8 055 operations (6 781 hip replacements; 1 274 knee replacements). The results (Figure 16.2) showed a cumulative benefit for the three measures and indicated that antibiotic prophylaxis alone would achieve more than the expensive provision of ultraclean air alone. A close relation was found between the levels of bacteria in the air and the risk of infection, and between staphylococci isolated in infected joints and those found in the patient or operating team at the time of operation. The benefit of very low levels of air-borne

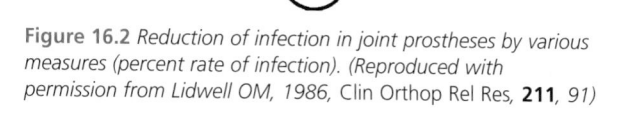

Figure 16.2 *Reduction of infection in joint prostheses by various measures (percent rate of infection). (Reproduced with permission from Lidwell OM, 1986,* Clin Orthop Rel Res, **211**, *91)*

bacteria has not yet been proven for other types of surgery, but it would probably apply to the insertion of other prostheses, such as artificial heart valves and cerebrospinal fluid shunts.

Infection associated with water

LEGIONNAIRES' DISEASE

The Legionellaceae are widespread in water (Chapter 68, Legionella), including potable water supplies, and many cases of Legionnaires' disease occur sporadically in the community. In one study in the UK they accounted for 15 percent of pneumonia cases (Macfarlane et al. 1982), but in other studies they were much less common. Hospitals are not spared the occurrence of legionellae in their water supplies (Tobin et al. 1981) and have suffered a series of outbreaks (Bartlett et al. 1986), mainly of infection with *Legionella pneumophila* (particularly serotype 1) but also with *Legionella micdadei*, *Legionella bozemanii*, and *Legionella longbeachae*.

The factors that favor hospital-acquired Legionnaires' disease are as follows. Hospital patients include a number of individuals with an impaired immune system who are more likely to become clinically infected with legionellae and who have a higher mortality than otherwise healthy subjects. Hospital-acquired infection occurs in patients rather than in staff who may nevertheless have circulating antibodies, which suggests exposure. Water in medical equipment that delivers a nebulized spray may become contaminated (Arnow et al. 1982). Hospitals are large buildings with extended hot-water systems. It has been customary to run these at lower than usual temperatures, for economy and for the safety of patients. Peak demand may lead to a further fall in temperature. Legionellae can tolerate temperatures to 55°C and can grow at 35–45°C and thus are adept at surviving and flourishing in the water systems of buildings (Latham et al. 1992). There have been many examples of the association of cases in hospital with hot-water systems, particularly with those supplying showers (Tobin et al. 1980), or run at a low temperature (Meenhorst et al. 1985). Ancillary systems used only intermittently may discharge water that has stagnated, allowing growth of legionellae in sludge or biofilm, perhaps assisted by substances from plumbing materials (Bartlett et al. 1986). Cases associated with release of stagnant water into the general system have been described (Fisher-Hoch et al. 1982). Hospital ventilation systems often have air-cooling towers in which potentially contaminated water flows over pipes through which air is circulated. Faults in such a system may allow direct access of the water to the air, or drift from the tower may contaminate the air at a later stage. Outbreaks caused by contaminated cooling towers can be explosive, occurring over a short period of time, whereas those due to domestic water systems may be more insidious and may only be revealed after active surveillance.

The source of infection of patients in hospital is thus environmental water; person-to-person spread is unknown. As legionellae are ubiquitous in water systems, regular monitoring is of little value in hospitals that have had no previous problems, but this is more problematical in hospitals where documented cases or outbreaks of nosocomial infection have occurred (Fallon 1994). Prevention depends on the good design, installation, and maintenance of hot-water and ventilation systems and of the use wherever possible of air-cooled condensers rather than water-cooling towers; on keeping domestic hot water at high temperatures (>60°C for storage and >50°C at the point of delivery) unless patients are thereby exposed to risk of scalding, as for example, in mental handicap departments; on flushing those parts of the hot-water system that are subject to stagnation; on chlorination and biocide treatment of water; and on the use of sterile water in nebulizers, etc. A recent outbreak was terminated by a series of measures including the use of copper and silver ions introduced into the water system as disinfectants (Colville et al. 1993). Preventive measures, which should involve clinicians, microbiologists, and engineers, must be accompanied by good surveillance for the occurrence of Legionnaires' disease with adequate diagnostic facilities, and by contingency plans to be brought into action when possible hospital-acquired infections occur. In the UK, the 1995 Approved Code of Practice (ACoP) (L8), The prevention or control of legionellosis (including legionnaires' disease), and the guidance booklet

HS(G)70 The control of legionellosis including legionnaire's disease, have both been recently revised and combined into a single document (Health and Safety Executive 2001). In its new form, the document is split into two parts. Part 1 is the ACoP itself, and part 2 of the document then provides guidance specific to particular systems, notably wet cooling systems (systems incorporating cooling towers and evaporative condensers), hot and cold water systems, spa/whirlpool baths, and humidifiers/air washers. In addition, a table in appendix 1 of this document provides a useful list of most other systems known to pose a risk, along with a summary of the main recommended control measures and the frequency of their application. For further information about Legionnaires' disease see Chapter 68, Legionella.

OTHER BACTERIA IN WATER AND IN DAMP AREAS IN THE HOSPITAL

Other gram-negative bacteria may be present in hospital water supplies, and their multiplication is encouraged by warmth and the other factors enumerated above. They may then act as opportunist pathogens in hospital patients. *Aeromonas hydrophila*, for example, causes pneumonia and sepsis in immunologically impaired patients and Picard and Goullet (1987) found *A. hydrophila* in maximal numbers in hospital water in hot weather. Risks of infection exist whenever aqueous fluids or damp objects, in or on which gram-negative bacteria have had an opportunity to multiply and have not subsequently been destroyed, come into intimate contact with susceptible patients. Examples include dental water-cooling systems, cooling cycles in autoclaves, pressure-monitoring systems on arterial lines, unsterilized lotions, and irrigation solutions.

Pseudomonas spp. are to be found in mains water supplies and may multiply in any moist area. They have simple growth requirements and often show considerable resistance to antibiotics and may even survive in disinfectants, such as those prepared for multiple use. *P. aeruginosa* is ubiquitous in wet hospital sites. It has been found in food and water, ice machines (also implicated recently with infections due to *Stenotrophomonas maltophilia* in neutropenic patients in the UK), pharmacy preparations, plaster of Paris, mouthwash, dental water units, nebulizers, whirlpools and other hot baths, mattresses, sinks, taps, potted plants, flowers, flower vases, and so on. In some instances, particularly when the organisms have been introduced into the gastrointestinal or respiratory tract or the patients were particularly susceptible, colonization and later infection have been shown to have arisen from such sources. In other instances, however, the strains carried by the patients were different from those found only at inanimate sites in the wards. The patients' strains contaminated the immediate surroundings of the carrier patients and the hands of members of the staff and were occasionally transferred to other patients (Levin et al. 1984; Allen et al. 1987). A more recent potential hazard that has been highlighted is pseudomonas infection following birth in water baths.

Burkholderia (Pseudomonas) cepacia, and other bacteria such as *Burkholderia pickettii*, possess the attributes that make them suitable for the role of hospital opportunists. Under certain conditions *B. cepacia* grows in distilled water and in the presence of aqueous chlorhexidine and quaternary ammonium compounds and may cause infections in neonates (Kahyaoglu et al. 1995). It has frequently been isolated from badly formulated, unsterilized supplies of these antiseptics, and even from povidone-iodine. Infection from a contaminated disinfectant or other fluid is generally associated with application to an open wound (Bassett et al. 1970), intravenous injection (Phillips et al. 1971), or urethral catheterization (Speller et al. 1971). *B. cepacia* is increasingly recognized as an important pathogen in patients with cystic fibrosis. In this group of patients the most important source is respiratory secretions with subsequent patient-to-patient spread. There is increasing evidence that some genomovars of *B. cepacia* complex are more virulent than others (e.g. genomovar III), and consequently patient segregation may be advisable (Govan et al. 1996; Mahenthiralingam et al. 2002).

Infection acquired from food

As in the general population, gastrointestinal pathogens may be transmitted by foods served to patients. Hospital food can also be a source of antibiotic-resistant bacteria, which may colonize the gut and later cause infection in susceptible patients. Of 1 280 outbreaks of infectious intestinal disease reported to the Communicable Disease Surveillance Centre in England during the period 1992–94 (Djuretic et al. 1996), 189 occurred. Salmonella infections were particularly associated with poultry or eggs, and *Clostridium perfringens* infections with meat. Outbreaks of salmonella infection in hospitals are important because they may cause serious effects in the very young, the elderly, and in patients with impaired immunity (Taylor et al. 1982) and may disrupt the working of the hospital (Kumarasinghe et al. 1982). The catering faults most often responsible are failures of catering staff to follow good practice: incomplete defrosting of frozen meats and poultry, insufficient cooking of large amounts of food, use of raw or insufficiently cooked egg products, inadequate chilling and storage, and contact between food to be consumed without further cooking and raw poultry. Effective processing of all foods, including poultry and eggs, and safe preparation, storage and distribution of food, are important in hospitals, where particularly susceptible patients are exposed to risk (Wilkinson 1988). The system of centralized preparation of food that is immediately chilled to safe storage

temperatures and later regenerated in a standard system after distribution ('cook-chill' or variations on this theme) allows, and demands, careful control and monitoring of bacterial content (Wilkinson 1988; Wilkinson et al. 1991; Shanaghy et al. 1993).

Most outbreaks of salmonellosis in hospital are not food-borne, but are caused by person-to-person spread or contaminated fomites. In the 3-year survey by Djuretic et al. (1996), only 19 of 189 hospital episodes were mainly food-borne, whereas 156 were mainly person to person; in seven, both routes were equally involved and in seven the mode of spread was unknown. Similar figures have been reported in the USA (Baine et al. 1973). In a prospective investigation during a period of 2 years, food-borne spread probably accounted for only 11 percent of hospital outbreaks, although these tended to be the larger incidents (Palmer and Rowe 1983). Careful epidemiological investigation is required to determine the mode of spread and to define the population for further study (Palmer and Rowe 1983). There was a recent report by a London hospital of an outbreak of *Salmonella* Enteritidis phage type (PT) 6a. There were 27 hospital-acquired cases over 1 month in 2002. The source was thought to be imported raw shell eggs. A second cluster of approximately 25 hospital-acquired cases predominantly due to S. Enteritidis PT 1 occurred within the following month. During the outbreaks there were six deaths among patients identified as having salmonella, although none of these was directly attributable to salmonella infection (Public Health Laboratory Service 2002).

Hospital food is well known to often be a source of antibiotic-resistant gram-negative bacilli that colonize patients: *P. aeruginosa*, *Escherichia coli*, *Klebsiella* spp., and others (Shooter et al. 1969; Cooke et al. 1970, 1980; Casewell and Phillips 1978). It is possible for dietary components, equipment, and preparation areas to become contaminated with nosocomial bacteria, particularly in facilities for the preparation of individualized diets, including enteral feeds (Casewell 1982). Colonization of the bowel is an important source of infection of other sites, for example of bacteremia in granulocytopenic patients, who may benefit from a diet of low microbial content, with avoidance of foods likely to contain pathogenic or antibiotic-resistant bacteria (Schimpff 1993).

Infection by contact

FROM STAFF

Microorganisms on the hands of staff may be resident (persistent over time and not readily removed by handwashing) or transient (recently acquired from another source). It is generally accepted that the hands of staff are an important vehicle and that hand-washing makes a significant contribution to the control of hospital-acquired infection (Reybrouck 1983; Larson 1988). The relevant microorganisms can readily be demonstrated on the hands of staff and may easily be transferred to the skin of others by brief contact (Marples and Towers 1979). Introduction of hand-washing has been accompanied by reduction of the infection rate in many studies, from that of Semmelweiss (1861) to more recent investigations of infection by gram-negative bacilli, e.g. Casewell and Phillips (1977). The literature of more than 100 years was reviewed by Larson (1988), who concluded that the evidence strongly supported a causal connection between hand-washing and control. The effect has been observed in a variety of settings under many different hygienic circumstances (Larson 1988).

The most important microorganisms spread by hand contact are *S. aureus* and gram-negative bacilli, such as *Klebsiella* and *Serratia* spp. Carriage on the hands has been suggested as important in the transmission of antibiotic-resistant enterococci (Rhinehart et al. 1990) and of *C. difficile* (Kim et al. 1981; McFarland et al. 1989), but this has not been a consistent finding. In addition, the route has been described for many other hospital-spread pathogens, including *Candida albicans* (Burnie 1986) and even respiratory pathogens: *Corynebacterium diphtheriae* (in the past) and viruses, such as RSV and rhinoviruses (Pancic et al. 1980). With *S. aureus*, attention has been focused on long-term carriage (Williams et al. 1960); gram-negative bacilli also have been demonstrated to persist for long periods on the hands of some persons (Casewell and Phillips 1977; Adams and Marrie 1982). Transient carriage resulting from recent contact with an infected or colonized patient may often be more important. Casewell and Desai (1983) demonstrated that strains of gram-negative bacilli, particularly of *Klebsiella* and *Serratia*, that had caused outbreaks in which infection was apparently spread by staff from patient to patient, survived longer on the hands of volunteers than did strains capable of causing outbreaks only from a point source in food or the environment. The ability of glycopeptide-resistant enterococci to survive for 1 h on fingers has also been shown (Noskin et al. 1995b).

It should be noted that healthcare workers sometimes erroneously wear gloves as a substitute for hand hygiene. Also, hand hygiene products may degrade gloves, and healthcare workers need to be aware of the need to clean hands after glove removal, because of the inevitable contamination that occurs during degloving. The preparations available for hand decontamination and the methods for assessing them have been reviewed (Ayliffe 1980; Reybrouck 1986). Hand hygiene may be designed for the removal of transient flora (Price 1938), which may be achieved by a simple wash with soap or detergent, or even better by an aqueous preparation of an antiseptic with a detergent. Sometimes a residual disinfectant action is needed, with or without reduction of the resident flora, as in the preoperative preparation of a surgeon's hands, or in high-dependency units in the hospital. Iodophors, the bisguanide chlorhexidine, and trichlorohydroxyphenol and related compounds may all be used and all have some residual activity (Ayliffe 1980). Alcoholic preparations, or alcohols alone, are more rapid in action than aqueous preparations and are more effective (Lowbury et al. 1974). Their drying action on the skin can be prevented by the addition of glycerol or other emollients. Alcohol-based hand rubs are quicker to use and aid compliance with hand hygiene policies, but they are unsuitable for use on soiled hands. A minimum of skin flora cannot be reduced further by these processes (Lilly et al. 1979). Methods should be adopted that are suitable for the particular purpose and that avoid damage to the skin and development of resistant flora on repeated use (Ayliffe 1980; Reybrouck 1986). One clinical trial (Doebbeling et al. 1992) compared different hand-washing methods in consecutive periods in intensive care units. A reduction in nosocomial infection was directly demonstrated when chlorhexidine (rather than soap or alcohol) was used; the effect may have been partly due to better compliance by staff. Poor compliance with the prescribed measures is frequently observed and may vary with the professional group, the hand-washing system prescribed, and its ready availability (Albert and Condie 1981; Doebbeling et al. 1992; Gould 1994; Wurtz et al. 1994; Pittet et al. 1999). Importantly, it has been shown that as the intensity of required patient care increases, the compliance with hand hygiene falls (Pittet et al. 1999). Hand cleaning should be carried out immediately before and after each contact with patient of their close surroundings. Efforts are increasingly focused on how to improve healthcare workers' compliance with hand hygiene. Pittet and colleagues recently examined >20 000 opportunities for hand hygiene and successfully improved compliance by healthcare workers from 48 percent in 1994 to 66 percent in 1997 ($p < 0.001$) (Pittet et al. 2000). This was attributed primarily to the promotion of bedside antiseptic handrubs. Hand hygiene improved significantly among nurses and nursing assistants, but remained poor among doctors. Importantly, the sustained improvement in compliance with hand hygiene coincided with a reduction of nosocomial infections and MRSA transmission.

The clothing of personnel can be shown to become contaminated with potential pathogens, such as *S. aureus* and, less frequently, gram-negative bacilli, particularly after the handling of heavily colonized patients (Babb et al. 1983). The significance of this in the spread of infection is unknown.

FROM THE PATIENT'S ENVIRONMENT

The immediate environment readily becomes contaminated with the bacteria carried by a patient. It has sometimes been difficult to prove that organisms from environmental surfaces cause infection (Maki et al. 1982), partly because the virulence of the bacteria may become impaired. Nevertheless, if heavily contaminated (Sanderson and Rawal 1987), such surfaces must be regarded as a potential source of air-borne organisms and also of infection by direct contact with another patient or via the hands of staff. Whilst it has been known for many years that staphylococci are desiccation resistant and can survive in the environment, thus prolonging outbreaks, emerging pathogens such as *Acinetobacter baumanni* may also persist. In a recent study the mean survival times of sporadic and outbreak strains on glass coverslips were 21–33 days (Jawad et al. 1998). Particularly in the case of bacteria carried in the bowel, which can survive in the environment for long periods, contamination of the environment or equipment may be responsible for the clustering of cases, as with *C. difficile* (Cartmill et al. 1994) and *Enterococcus* spp. (Karanfil et al. 1992). In a review of intensive care unit design and environmental factors in the acquisition of infection, O'Connell and Humphreys (2000), make the point that the inanimate environment of air, water, food, floors, walls, and ceilings can contribute to the risk of acquisition of infection in intensive care patients, although their actual role can be difficult to quantify in most instances. However the design of the intensive care unit (ICU) should be such as to minimize the entry and persistence of microorganisms into this environment and allow the thorough cleaning of any surfaces which may become contaminated (Dieckhaus and Cooper 1998).

Dancer (1999) found that cleaning in the hospital setting is often a neglected component of infection control and many outbreaks occur secondary to suboptimal cleaning. Hospitals provide a reservoir of microorganisms adapted to the hospital environment including the ability to survive harsh environmental conditions. The immediate environment of patients in sometimes heavily contaminated and may include vancomycin-resistant enterococci (VRE), MRSA, *A. baumanii*, *P. aeruginosa*, and *C. difficile*. The hardy spores of *C. difficile*, an organism that can cause serious outbreaks throughout the hospital, can withstand extremes of temperatures and some disinfectants. They also remain viable for a considerable length of time in the environment and are responsible for many outbreaks of infection in general wards and in the ICU (Spencer 1998). Other nonspore-forming microbes, although not as resilient as *C. difficile*, can survive in areas that are damp and poorly cleaned. Standards for environmental cleanliness in hospitals were published in the UK in 1999 with a view to raising the profile of hygiene in hospitals, developing an audit tool and producing prescriptive recommendations from which standards for cleaning in hospitals could be derived (ICNA/ADM 1999). Cleaning and disinfection of the rooms of infected patients are recommended when the patients leave the room or are discharged from hospital. There are ongoing studies as to whether materials with intrinsic antibacterial activity could potentially

be used as finishes for walls, floors, etc., and thus help to control environmental contamination (Talon 1999).

Rampling and coworkers (2001) again highlighted the importance of general hospital hygiene in bringing to an end a prolonged outbreak of MRSA. They showed that a dusty ward was an important source of MRSA for their surgical patients. During the last 15 years in the UK, hospital cleaning services have been targeted for cost-cutting in the NHS (Dancer 1999). However, it is difficult to demonstrate that a dusty or soiled environment can actually be hazardous to patients, with little evidence to support this (Ayliffe et al. 1967; Maki et al. 1982; Collins 1988). Nevertheless, a high standard of hygiene should be an absolute requirement in hospitals for aesthetic, as well as clinical reasons. In the long term, cost-cutting on cleaning service is neither cost-effective nor common sense (Blyth et al. 1998; Plowman et al. 1999; Wilcox and Dave 2000). All this emphasizes that basic hospital hygiene, such as cleaning, remains an important element in preventing infection and attempts at cost containment in this area are likely to be counterproductive.

FROM EQUIPMENT

Infection from surgical instruments is now extremely rare, but other items of equipment, even if they do not penetrate the tissue, may convey infection from one patient to another. Some of these escaped attention because the risks associated with them appeared to be low or had not been perceived. Others are pieces of equipment that are difficult to clean and disinfect adequately, or are expensive and in short supply.

Many items that come into contact with patients usually do not need to be supplied sterile, but patients with impaired immunity may be susceptible to opportunist organisms present on or in these. For example, patients with debilitating illnesses may develop superficial and subcutaneous phycomycosis in macerated areas under adhesive plaster contaminated with *Rhizopus* spp. (Gartenberg et al. 1978). Premature babies have become infected by *Rhizopus* spp. from unsterile wooden tongue depressors, used as limb splints (Mitchell et al. 1996).

Bedpans and urinals have been blamed for the spread of enteric bacteria, such as *C. difficile* and antibiotic-resistant gram-negative bacilli (Curie et al. 1978). The bacterial load may be high if the washing process is not efficient, and bedpan washer-disinfectors often do not maintain an adequate temperature for the intended time. The practice of decontaminating urinals in tanks of phenolic disinfectant is often ineffective because of inaccurate dilution of the disinfectant, or its inactivation by organic matter, or by the survival of bacteria in biofilm on surfaces (Curie et al. 1978). Special decontamination measures may be necessary for particularly resistant pathogens, e.g. *C. difficile* and *Cryptosporidium* spp. Rectal thermometers have been believed to be responsible for the spread of salmonellae (Im et al. 1981) and enterococci (Livornese et al. 1992); all thermometers should be decontaminated after use (Nystrom 1980).

Many fiberoptic endoscopes for gastrointestinal and bronchial use are difficult to disinfect because they do not withstand heating; also their complex channels are difficult to clean. Numerous species have been transmitted by such endoscopes (Spach et al. 1993). Among the most common have been *Salmonella* spp. and *P. aeruginosa* in the gastrointestinal tract and *Mycobacterium tuberculosis* and other mycobacteria in the respiratory system. Endoscopes are readily contaminated with *Helicobacter pylori* and transmission by endoscopy has been demonstrated (Langenberg et al. 1990). Recent updated guidelines on endoscopy disinfection have increased the disinfection contact times, e.g. from 4 to 10 min for 2 percent gutaraldehyde, to provide a greater margin of safety, and have emphasized the importance of water quality of the final rinse (Report of a Working Party of the British Society of Gastroenterology Endoscopy Committee 1998). In the impaired patient infection of other sites and bacteremia may result (Schimpff 1993); bacteremia is particularly likely to follow biliary endoscopy (Struelens et al. 1993). Hepatitis B has rarely been transmitted on endoscopes although they readily become contaminated with the virus (see Ridgway 1985). Spread of human immunodeficiency virus via endoscopes has not been reported. Such accidents can be avoided by a thorough regimen of cleaning and disinfection with a suitable chemical agent, which may be carried out manually or by special apparatus (Fraser et al. 1993; Bradley and Babb 1995) if sufficient endoscopes can be provided (Anon 1988; Fantry et al. 1995; Babb and Bradley 1995).

The ever expanding use of endoscopic procedures in diagnosis and surgical treatment of patients has reinforced the need for safe and effective methods of cleaning, disinfection, and sterilization of these instruments (Ruddy and Kibbler 2002). Reports on the decontamination of minimally invasive surgical endoscopes and accessories have been published (Ayliffe 2000). As the number of patients undergoing endoscopy has increased, the requirements for the safe use of glutaraldehyde and other potentially irritant or sensitizing instrument disinfectants has led to an increased use of automated washer disinfector systems which provide safe containment of hazardous vapor and can remove toxic residues effectively. These automated systems have been shown to be more effective in removing bacteria than manual methods but they are not without problems, and the following issues need to be recognized:

- Many systems are available on the market. Irrespective of whether or not the system has its own wash cycle prior to disinfection, it is recognized that manual precleaning of endoscopes is essential (Babb and Bradley 1995).

- There is a need to ensure endoscope compatibility with the autodisinfector fittings. Endoscopy staff must follow instructions strictly in order to ensure that all endoscope channels are adequately irrigated with disinfectants. There is a need to ensure endoscopes and washer disinfectors are compatible with the disinfectants used.

- Automated systems require regular and adequate maintenance which is time consuming and expensive. Furthermore, some parts of the decontamination process (e.g. brushing lumens and wiping insertion tubes) cannot be automated and thus manual cleaning is still an essential prerequisite.

The most significant problem which has not been addressed by the increased use of automated systems is that of contaminated rinse water derived from either the internal pipework of the washer disinfector or mains water, as well as intermediate tanks and poorly maintained filters. Recent reports have attempted to address this and give recommendations on possible ways of dealing with this problem (MacKay et al. 2002; Report from a Joint Working Group of the HIS and PHLS 2002).

Infection by inoculation

Since the universal introduction of single-use disposable needles and other devices, and satisfactory procedures for the sterilization of surgical instruments, with a wide safety margin, transmission of infection by this route in the developed world has been infrequent. Few agents can resist the standard sterilization processes. The prions responsible for transmissible spongiform encephalopathies (e.g. Creutzfeldt–Jakob disease (CJD)), previously associated with neurosurgery, are highly resistant to most sterilization procedures and therefore instruments used on patients with these conditions should either be disposed of or subjected to prolonged autoclaving at high temperature (Advisory Committee on Dangerous Pathogens 1994). However, there remain infections (1) transmitted by blood transfusion or tissue donation; (2) resulting from accidental injury from contaminated sharp instruments; (3) from contaminated blood; and (4) from other contaminated infusion fluids.

INFECTION TRANSMITTED BY BLOOD TRANSFUSION AND TISSUE DONATION

Infectious agents may from time to time be transmitted to patients by donors of blood, blood products, or tissue. The risk of transmission of the three most important agents, hepatitis B and C viruses and human immunodeficiency virus (HIV), has been reduced to a low level by a combination of donor selection, screening of donations for antigen or antibody, and heat treatment of blood products, but not before countless cases of 'serum hepatitis' had occurred and a considerable proportion of hemophilia sufferers received factor VIII infected with HIV (Editorial 1984b). Efforts should also be made to exclude other, less common, blood-transmitted infections such as other forms of hepatitis, including δ agent and possibly hepatitis G in the future, cytomegalovirus, Epstein–Barr virus, parvovirus, HTLV-1, brucellosis, syphilis, salmonellosis, malaria, trypanosomiasis, toxoplasmosis, babesiosis, and filariasis (see Mollison et al. 1987). Acquisition of transmissible spongiform encephalopathies is not considered likely to follow blood transfusion. An epidemiological survey of CJD in the UK revealed that only 16 of 202 cases had a history of blood transfusion and this was not significantly different from controls (Esmonde et al. 1993). However, cases of blood transfusion-acquired CJD have now been reported in England (Llewelyn et al. 2004). Further efforts to exclude infections less common than those already currently screened for in the blood supply must be balanced against the considerable cost of such screening and the impact this might have on blood donation.

Tissue donations have been responsible for the transmission of cytomegalovirus, HIV (L'Age-Stehr et al. 1985), and rabies (Helmick et al. 1987) and may also transmit hepatitis B and hepatitis C. Creutzfeldt–Jakob disease may be acquired during surgical procedures that involve tissue transfer and when this occurs, the incubation period is a matter of months rather than years (Hart 1995). Creutzfeldt–Jakob disease first described by Jakob in 1921, is one of four transmissible spongiform encephalopathies (TSE) found in man. The others being Gerstmann–Straussler–Scheinker syndrome, Kuru (associated with cannibalism in Papua New Guinea), and fatal familial insomnia. TSEs are rare, fatal neurogenerative diseases that cause degeneration of the brain resulting in loss of coordination and faculties. The infectious agents are thought to be unique proteins and are termed 'prions'. Prions replicate by transforming normal cellular prion proteins into abnormal isoform proteins. These abnormal prion proteins, PrP^{res}, then accumulate in the central nervous system causing spongiform changes in gray matter where they trigger neurological symptoms.

Sporadic CJD occurs worldwide. It is rapidly progressive and ultimately results in a fatal disorder of the central nervous system with a long incubation period of 15 months to over 30 years. There are approximately 0.5 to 1 new case per million of the population per year. The similarities between TSEs found in different animal species (e.g. bovine spongiform encephalopathy (BSE)) and those found in man suggest that the disease, in some form or another, may be able to cross the species barrier. Sporadic CJD usually presents in late middle age, mean age 66 years (range, 16–95) with progressive dementia, a duration of illness of <4 months (range, 1–74 months). Diagnosis is based on clinical signs and characteristic waveforms when recording electrical events in the brain (EEG), and as no treatment is available for CJD, medical management remains limited to supportive care.

The new variant of Creutfeldt-Jakob disease (vCJD) was described in the UK in 1996 after the occurrence of ten atypical cases of CJD in patients under 40 years of age, nine of them under 30; median age 28 years (1–74 years). All these patients had common features: young age, a special clinical presentation with psychiatric features at onset, an abnormal duration in the evolution of the disease (14 months on average compared with 2 to 6 months in classical cases of CJD), and a pathognomic neuropathology. Particular neuropathological features were amyloid plaques surrounded by vacuoles (florid plagues) whose distribution in the central nervous system was remarkably similar from one patient to another; a potent PrP immunostaining was identified in the injured areas, and the presence of PrP^{res} was found by Western blot, Moreover, all these patients were homozygous methionine/methionine to codon 129 of the human *PRNP* gene, and did not have a medical history to suggest an iatrogenic or familial origin or disease. In addition, unlike sporadic CJD, no periodical electrophysiological abnormalities were observed; further, the protein 14-3-3, a nonspecific marker usually found positive in sporadic forms of CJD, was not detected. Finally, hyperdense signals to cerebral magnetic resonance imaging (MRI) have been described in the post-hypothalamus area (Pulvinar sign, not seen in sporadic) and can be a helpful diagnostic feature in cases with suggestive clinical picture. This form of CJD was suspected to be the consequence of infection in man with the bovine spongiform encephalopathy (BSE) agent as no other etiological hypothesis could reasonably be considered. The evidence that this is indeed the case is compelling, with BSE manifesting itself as a novel human prion disease – variant Creutzfeldt-Jakob Disease (vCJD), with temporal association with consumption of bovine central nervous system material in food. First exposure was thought to have occurred in 1983, peak exposure 1989, last exposure in 1995. First cases of variant CJD, identified retrospectively, occurred in 1993. Incubation of vCJD is governed by infective dose and methionine/methionine at codon 129. Experimental transmission of BSE and vCJD in bovine transgenic mice have shown similar brain pathology characteristics (possibly satisfying Koch's postulates).

Despite many sensational reports in the media, TSEs although infectious are not easily transmissible as this requires specific material from the affected individual's tissue from or adjacent to their central nervous system. For this reason isolation of sufferers is not necessary. Thorough cleaning and sterilization of surgical instruments is strongly recommended for instruments used on known or suspected CJD cases. Cases of iatrogenic transmission of classical CJD have been reported (corneal grafts, pituitary growth hormone, dura mater grafts). As of 2002 there was no evidence implicating iatrogenic transmission of vCJD (Spencer and Ridgway 2002). This picture has now changed (see above).

INFECTION FROM ACCIDENTAL INOCULATION

In the past, tuberculosis of the skin ('prosector's wart') was a common consequence of inoculation and is once again a potential risk from HIV patients heavily infected with *M. tuberculosis* (Kramer et al. 1993). In the pre-antibiotic era, infection by a needlestick injury with *S. pyogenes* might result in septicemia and death, as observed by Semmelweiss in his colleague, Kolletschka. Such events remain a hazard, though a much reduced one (Hawkey et al. 1980). The main risks of such injuries, however, are infection with hepatitis B and C viruses and HIV.

HEPATITIS B AND C AND HIV INFECTION IN HOSPITALS

Hepatitis B and C viruses and HIV are the pathogens most likely to be transmitted from blood and tissue fluids (Zuckerman 2002). Bodily secretions from infected patients contain much lower concentrations of virus, or none at all. The risk of transmission to staff is greatest for those with the highest exposure to blood and to injuries with sharp instruments, but a history of a definite incident is not always given by hospital staff members who develop hepatitis B. The risk of acquiring hepatitis B from an infectious carrier in an inoculation incident is significant, especially if the patient is e-antigen positive (HBeAg). An investigation following the identification of an HBeAg-positive cardiac surgeon revealed that 13 percent of susceptible patients contracted hepatitis B despite good infection control practice (Harpaz et al. 1996). Until recently it was considered that the risk of nonsocial acquisition from HBeAg-negative staff was low, but a recent report describes spread from HbeAg-negative surgeons with precore mutations. These mutations prevented the expression of HBeAg, but not the assembly of infectious virus and quantitative measurements of HBV DNA may be necessary to fully assess the risk of transmission in surgical and other settings (Incident Investigation Teams and others 1997). Currently, in the UK and Ireland, all healthcare workers performing exposure prone procedures (EPP) are required to be able to demonstrate immunity to hepatitis B. Hepatitis B-infected healthcare workers who are e-antigen negative and who perform EPPs should have their viral loads measured, and those with viral loads exceeding 10^3 genome equivalents per ml should not perform EPPs in future (NHS Executive 2000).

It is clear that the risk of transmission of HIV to staff is very small; of 1 036 subjects included in several series, with adequate epidemiological and serological investigation of percutaneous or mucous-membrane exposure to blood or tissue fluid from known HIV-positive patients, only three (0.3 percent) showed seroconversion (Centers for Disease Control 1987). More recently, only four of 1 103 healthcare workers with percutaneous exposure to HIV-infected blood seroconverted and in one case this occurred following post-exposure zidovudine (Tokars

et al. 1993). The risk of a patient acquiring HIV from an infected healthcare worker has been the subject of much debate, some of it ill-informed, but has raised the issue of screening healthcare workers for HIV. A retrospective study of more than 6 400 patients of an infected Florida dentist revealed 28 with HIV infection, of whom 24 had behavioral risk factors. Analysis of the viral gene sequences from the dentist and his patients excluded a possible link with the 24 with behavioral risk factors, but not with the remaining four (Jaffe et al. 1994). A survey of 22 171 patients of 51 infected healthcare workers conducted by the Centers for Disease Control, Atlanta, USA failed to reveal evidence of transmission from healthcare worker to patient (Robert et al. 1995). A cost–benefit analysis of screening surgeons for HIV as a strategy to minimize transmission to patients did not support this approach as cost-effective (Owens et al. 1995). In the UK, recommendations on the investigation of patients who have been cared for by a member of staff who is subsequently identified as HIV-positive, are available (Expert Advisory Group on AIDS and the Advisory Group on Hepatitis 1998).

Post-exposure prophylaxis (PEP) against HIV should be considered if there has been possible exposure to HIV-infected blood or body fluid (Expert Advisory Group on AIDS 2000). The initial risk assessment is based on the potential for viral transmission, i.e. the type of body fluid involved (e.g. blood is high risk), the route (e.g. deep injury by a blood-containing or stained device) or/and severity of exposure (e.g. injury with a needle that has been in the source patient's artery or vein; and/or terminal HIV-related illness in the source patient). Some occupational exposures after careful assessment may not be considered to be significant, that is they do not have potential for HIV transmission, i.e. splashes on to intact skin. In such circumstances PEP is not indicated. If a significant exposure has occurred and the source patient is known to be HIV-positive, PEP should be commenced, the first dose should be given within 1 h of the incident. Based on the results of a HIV test on the donor (with informed consent) the healthcare worker, advised by a specialist clinician can make an informed decision on whether to continue with PEP. However, the exact mechanisms of HIV infection are not fully understood, therefore if more than 1 h has elapsed after exposure this does not automatically preclude starting PEP. It is worth considering starting PEP up to 2 weeks following exposure, however specialist advice should be sought.

Hepatitis C (see Chapter 54, Hepatitis C in the *Virology* volume in this series), recognized in the last decade as the major cause of non-A non-B hepatitis following blood transfusion, is encountered worldwide with especially high prevalence in Japan and the southern part of the USA, and may persist in 80 percent of infected patients (van der Poel et al. 1994). Some patient groups have a higher seroprevalence; 3.3 percent of Dutch hemophilia patients were positive compared with 0.03 percent of blood donors (Schneeberger et al. 1993). Acquisition associated with particular occupations is also described; oral surgeons had a higher incidence (9.3 percent) than dentists (0.97 percent) or unselected blood donors (0.14 percent) in one study (Klein et al. 1991). The risk of seroconversion following a needlestick incident is approximately 3 percent, which is intermediate between HIV (0.3 percent) and hepatitis B (up to 30 percent for hepatitis B eAg positive donors). At present, the risk of a hepatitis C-positive healthcare worker transmitting infection to patients is not certain. An investigation of an incident involving a hepatitis C-positive cardiac surgeon revealed that six of 222 patients followed up contracted postoperative hepatitis; in five patients the viral gene sequences were similar to those of the surgeon (Esteban et al. 1996). Several other reports of hepatitis C transmission from healthcare workers to patients have since occurred (Gunson et al. 2003). The lower risk of transmission of hepatitis C compared with hepatitis B is unfortunately counterbalanced by the greater risk of chronic hepatitis C infection (80 percent), which may in turn lead to cirrhosis or hepatocellular carcinoma. In the UK, healthcare workers entering training to specialities that require practice of exposure prone procedures (EPPs) are now required to be screened for hepatitis C. Healthcare workers that are found to have high levels of hepatitis C RNA in blood are required to cease performing EPPs until RNA levels decline, possibly following interferon/ribivirin treatment. This policy approach is, however, not universal, and in some other countries hepatitis C infected healthcare workers are being allowed to continue to practice after disclosure of their infection status (Barrigar et al. 2001).

Active hepatitis B immunization should be given to those at greatest risk (Mulley et al. 1982) and this and hepatitis B immune globulin (e.g. vaccine staus incomplete/unknown) may be used as post-exposure prophylaxis (Grady et al. 1978). The action needed in the immediate post-exposure period to protect a healthcare worker against hepatitis B is dependent on the vaccination status of the healthcare worker and (where known) the hepatitis B surface antigen status of the source patient. Other measures are clearly to be directed against the risk of inoculation of blood. With the recognition of new agents (e.g. hepatitis C) transmissible by blood and the likelihood of others to follow, control measures have more and more emphasized good practice when handling blood, blood products, and other fluids such as cerebrospinal fluid, irrespective of patient source (Expert Advisory Group on AIDS and the Advisory Group on Hepatitis 1998). These are referred to as universal precautions and include greater awareness of the risks by staff, staff covering cuts or abrasions, careful disposal of sharps and clinical waste, and the use of gloves and gowns when in contact with body fluids

(Board of Science and Education 1996; Expert Advisory Group on AIDS and the Advisory Group on Hepatitis 1998). Additional precautions may be taken with known infected patients, or those suspected of being infected (e.g. intravenous drug abusers), but the adoption of universal precautions at all times minimizes the risks to staff and patients when a patient is not known or considered to be in an at-risk group. Guidelines issued in the UK to minimize acquisition of hepatitis B and HIV during surgery, which would also largely apply to hepatitis C, emphasize good practice at all times, e.g. avoidance of passing sharps by hand, with additional precautions such as double gloving when operating on positive patients or those in at-risk groups (Joint Working Party of the Hospital Infection Society and the Surgical Infection Study Group 1992). Continuing education on basic techniques to avoid inoculation risk incidents is important; taking sharps boxes for the disposal of needles to the patient's bedside reduced the number of recapped needles (a practice likely to place the healthcare worker at risk of a needlestick incident) from 33 to 18 percent (Makofsky and Cone 1993).

Testing for hepatitis B surface antigen (HBsAg) has been used to define more closely the patients who present an 'inoculation risk' and identification of HBeAg and antibody (HBeAb) to define the degree of infectivity. Screening of patients for hepatitis C is not usually indicated as an infection control measure. Testing for HIV with consent is not recommended at present, partly because of the implications of detecting seropositivity in a symptomless patient and partly because of the lack of a sufficiently sensitive and specific antigen test to identify infected patients before antibody appears. Testing for hepatitis B, hepatitis C, and HIV is essential before tissue donation and recommended in certain other defined situations.

Containment measures for identified patients should concentrate on the risks associated with invasive procedures and bleeding, and especially with surgical operations and obstetric deliveries. Single-room isolation is required only for certain patients, e.g. those with mental disturbance, uncontrolled bleeding, or dual infection with HIV and another disease, such as tuberculosis. In deciding whether a patient requires single-room isolation, consideration should also be given to the psychological effects this may have. The standard decontamination measures for crockery, bedpans, etc., are adequate. For invasive procedures, limiting the number of staff present, providing adequate protective clothing, including precautions against splash on mucous membranes, limiting and containing blood loss as far as possible, care with sharp instruments, and adequate decontamination of instruments are the principles to apply. Detailed recommendations for the care of HBsAg or HIV risk patients have been formulated (Expert Advisory Group on AIDS and the Advisory Group on Hepatitis 1998; Expert Advisory Group on AIDS 2000).

INFECTION FROM CONTAMINATED BLOOD

Extrinsic contamination of donor blood with small numbers of bacteria is not uncommon and serious consequences of this are rare. Occasionally, stored blood may contain large numbers of bacteria (Braude et al. 1955), often psychrotrophs, and septicemia and shock may occur in the recipient (McEntegart 1956). The range of possible organisms and the risks associated with erythrocytes, whole blood, and platelets together with advice on how to investigate outbreaks and prevent such occurrences have been well reviewed (Wagner et al. 1994).

INFECTION FROM CONTAMINATED INFUSION FLUIDS

Major outbreaks of serious infection and endotoxic shock have been caused by heavily colonized infusion fluids other than blood. Some accidents were due to failure of sterilization during commercial manufacture (Meers et al. 1973), and others to the introduction of microorganisms from cooling water in sterilizers (Phillips et al. 1972) or from contaminated closures which release their bacteria into the infusion fluid during manipulation in the ward. Infusion fluid in reservoirs or administration sets can become colonized during use, from cracks in containers or during manipulations and additions (Duma et al. 1971). Indeed, a significant proportion of containers becomes contaminated with small numbers of bacteria (Maki et al. 1973). Fluids for total parenteral nutrition, such as protein hydrolysate with or without dextrose, readily support the growth of microorganisms, particularly of *Klebsiella* and *Enterobacter* spp., and of *Candida* spp. Disseminated fungal infection has been particularly associated with total parenteral nutrition, but not all episodes have resulted from contamination of fluid in the reservoir or administration set (Goldmann and Maki 1973). The risk can be reduced by care with all manipulations or by having all additions to intravenous fluids made by skilled staff working in controlled environmental conditions. Containers and administration sets should be changed every 48 h; more frequent changing is unnecessary (Buxton et al. 1979). The use of bacterial filters, i.e. in administration sets, has done little to reduce infection (Collin et al. 1973). This may indicate that infection related to the intravascular cannula is more common than colonization of fluid in the rest of the system.

HOSPITAL INFECTIONS AT VARIOUS BODY SITES

Surgical wounds and other soft tissue sites

Surgical site infections (SSI) are important numerically and as a cause of morbidity and prolonged hospital stay. In a prevalence survey (Emmerson et al. 1996) it accounted for 12.3 percent of hospital-acquired infections. In the USA incidence study, SSI accounted for 24 percent of nosocomial infection. In some surveys the definition of SSI was based on a simple, easily observed character such as the presence of pus in the wound and this has the advantages of minimizing observer variability. A more complex definition, making use of several other signs of inflammation (e.g. erythema), can be used with an elaborate scoring system (A.P. Wilson 1995). This may increase the likelihood that significant differences will be observed in comparative trials. The definition should be based on clinical observation. It may be useful to culture all infected wounds for microbiological surveillance purposes, but the presence of pathogenic bacteria in itself does not imply infection. The recorded incidence of infection will also depend on the length of postoperative stay and the degree of follow-up after discharge from hospital (Bailey et al. 1992; Weigelt et al. 1992; Byrne et al. 1994; Emmerson 1995).

Several investigations have been made, in sufficiently large numbers of patients, of the factors predisposing to SSI, to determine where improvements in practice should be directed (e.g. Cruse and Foord 1973). Regression analysis has been employed to separate variables. Surgical procedures may be crudely divided as follows: 'clean', when no inflammation is encountered and no colonized body system is entered; 'clean contaminated', when a colonized system, e.g. gastrointestinal or respiratory, is entered, but there is no significant spillage; 'contaminated', when inflammation, but not pus, is encountered or spillage from a viscus occurs; and 'dirty', when a perforated viscus or pus is encountered. The incidence of infection should be less than 5 percent for clean operations; some surgical teams may record a rate of less than 1 percent (Cruse and Foord 1973). On the other hand, operations on the left colon or rectum may be followed by wound infection in more than 30 percent of patients if antibiotic prophylaxis is not given. Refinements of this classification system have been suggested. It may be more useful to consider individual types of operation separately. For example, Cesarean section, though included in 'clean' surgery, may have a high infection rate (Beattie et al. 1994; Henderson and Love 1995).

In general investigations, the factors most consistently associated with an increased incidence of postoperative infection are age over 60 years, long preoperative stay in hospital, long duration of operation, pre-existing infection at the site of the wound, and bacteria in the wound at the end of the operation. Underlying disease, such as diabetes, immunosuppression or irradiation, and in some studies, malnutrition and administration of adrenocorticosteroids, are also important. In preoperative preparation, shaving of hair from the site, rather than treatment with depilatories or clipping, has been associated with a much greater frequency of infection (Seropian and Reynolds 1971), but other aspects of skin preparation appeared not to exert a consistent effect. Factors of significance in some studies but not in others are male sex, emergency operations, and the use of surgical drains. It is generally agreed that good surgical technique is most important, but it is difficult to single out particular surgical practices. The use of diathermy was associated with increased infection in one series (Cruse and Foord 1973).

Overall, *S. aureus* is the dominant species in SSI, followed by the enterobacteria (Meers et al. 1981). UK surveillance data show that approximately two-thirds of *S. aureus* SSIs are caused by MRSA. Even higher MRSA prevalence in SSIs is seen in subspecialities such as vascular surgery (Public Health Laboratory Service 2002). Sternal wound infection after cardiac surgery, though difficult to recognize clinically in the early stages, may later be associated with mediastinitis, osteomyelitis, septicemia, and colonization of prosthetic material, and thus with significant morbidity and mortality. Infection occurs at the time of operation from the many sources outlined above; it is probable that a relatively small inoculum is needed in the inevitably damaged tissue. Prevention may be a major role for the prophylactic antibiotics that are used almost universally in cardiac surgery (Farrington 1986). Among factors that have been cited as predisposing to this type of postoperative infection are long and double procedures, excessive trauma to the sternum, postoperative hematoma formation, and reopening of the wound (Sarr et al. 1984), as well as the features already described as predisposing to surgical wound infection in general. It has been reported that the incidence of infection is greater in coronary artery bypass surgery when leg veins are used, and that the infecting organisms, particularly gram-negative bacilli, may be transferred from the leg. Infection of the leg wound itself is also common (Wells et al. 1983; Farrington et al. 1985).

The microbiological findings vary with the site and type of operation and also with the adequacy of culture methods, which often underestimate the role of anaerobic bacteria, such as *Bacteroides* spp. These, with other gut bacteria, often in mixed growth, are typical of infections after a colonized viscus is entered (Dunn and Simmons 1984). *S. aureus* may occur in all types of wound and is the typical cause of the less frequent wound infection in 'clean' surgery. Most infection of surgical wounds occurs at the time of operation. Evidence from animal experiments (Burke 1961) and studies of the effectiveness of antibiotic prophylaxis (e.g. Stone et al. 1976) suggest that there is a short period after bacterial contamination during which infection is initiated. In the great majority of cases, the origin of the bacteria appears to be the patient's own body flora. Much less often it is from a member of the operating team, but in many instances the origin is obscure. Occasionally a cluster of infections can be related to an organism present in a septic lesion or at a carrier site in a member of the operating team, but this is uncommon. The route by which a patient becomes infected is often not clear. The air-borne route is important in the implantation of prostheses, and in rare episodes in general surgery, which were linked with an identifiable source among the theater staff, the circumstances strongly suggested the air as the vehicle. More usual routes are direct spread from incised organs, and intraoperative contamination of instruments and of surgeons' gloves and clothing. Contamination from apparatus such as that providing an extracorporeal circulation has occasionally been described.

BURNS

Burns provide a suitable site for bacterial multiplication. When this has taken place, the burn is a richer and more persistent source of infection than the surgical wound because a larger area of tissue is exposed for a longer time. The clinical consequences of infection in burns may be very serious; a large proportion of the mortality in burned patients who have survived the initial trauma and shock has been due to infection. Multiple-antibiotic resistance in the infecting agents has for many years presented difficulties in the treatment of burned patients.

S. aureus and *P. aeruginosa* are the most common isolates in most burns units, followed by various enterobacteria and other gram-negative bacilli, such as *Acinetobacter* species *S. pyogenes*, once a cause of serious trouble in burns units (Cruickshank 1935), now colonizes comparatively few patients (Lawrence 1985). Nevertheless, *S. pyogenes* is important both in episodes of acute infection and in causing the failure of skin grafts. As in other hospital areas, the emergence of resistant gram-positive bacteria, including multiresistant enterococci and *S. aureus*, has been seen in burns units (Phillips et al. 1992; Donati et al. 1993).

As in many other sites, colonization without invasion is far more common than invasive infection, but the frequency and clinical severity of septicemic infections caused by most of the potential pathogens, particularly by *P. aeruginosa*, encourage regular monitoring by the culture of swabs from burns. A greater understanding of the relationship between the bacteria present and the host tissues can be obtained by examining punch biopsies (Pruitt and McManus 1992). Septicemia may follow surgical intervention; this may be prevented by systemic antibiotic prophylaxis, as is applied in most burns units (Papini et al. 1995).

Bacteria reach burns mainly by indirect contact. Airborne infection has been demonstrated on occasions and is more important for *S. aureus* than for gram-negative bacilli. Occasionally mattresses, baths, and hydrotherapy pools have been incriminated (Ayliffe and Lilly 1985; Tredget et al. 1992). In some studies, standard infection control measures, including controlled ventilation, have not been shown to greatly influence the colonization of burns by bacteria, including antibiotic-resistant strains (Ayliffe and Lawrence 1985), although it has been suggested that single-room isolation reduces the problem of infection by gram-negative bacilli (Pruitt and McManus 1992; McManus et al. 1994). It is clear that improvements in the surgical management of burns, including judicious use of excision and timely closure, and the use of topical chemotherapy with agents such as silver sulfadiazine, have considerably reduced infection in many units.

Infection associated with extensive skin disease and major plastic surgery presents problems that are to some extent similar to those with burns.

THE EYE

The cornea is considered sterile, but the conjunctiva is colonized by a number of different bacterial genera, predominantly coagulase-negative staphylococci, *Corynebacterium* spp., and *Propionibacterium* spp. (Willcox and Stapleton 1996). Resident flora and other physiological factors, e.g. immunoglobulins in tears, prevent colonization by more pathogenic bacteria. The use of contact lenses is associated with colonization by *S. aureus* and gram-negative bacilli. Postoperative infections of the eye are uncommon, but may result in considerable disability and are difficult to treat. An etiological diagnosis is usually possible only if material for culture is obtained from within the eye. Preoperative monitoring of the eye flora is not usually indicated unless there are clinical signs of infection on the surface of the eye. It may, however, be useful to examine swabs from the skin round the eye and from the anterior nares for *S. aureus*, particularly if the patient has a skin disease, so that extra care can be taken with skin preparation if *S. aureus* are recovered.

Endophthalmitis is an infection of the intraocular structures and sclera and usually occurs following intraocular surgery or penetrating trauma, or as part of a generalized infection, such as systemic candidiasis. Infection after cataract surgery has declined in recent years and is now less than 0.1 percent (Das and Symonds 1996). Gram-positive bacteria predominate; a survey of 64 centers in France identified *S. epidermidis* as the most important pathogen (Fisch et al. 1991). A similar finding was reported in a UK study (Hassan et al. 1992). The somewhat surprising emergence of *S. epidermidis*, traditionally associated with the immunocompromised host and with foreign body-related infection, may represent improved diagnostic techniques. Various fungi cause occasional infections, presumably by the air-borne route. Many donor corneas yield microbial growth, but few patients develop infection with organisms similar to those found on the cornea. The importance of making a microbiological diagnosis should not be underestimated as management is difficult, partly because antimicrobial

agents penetrate poorly into the anterior and vitreous chambers of the eye.

Keratoconjunctivitis is usually community-acquired and is most commonly caused by *S. aureus*, but *P. aeruginosa* may be associated with corneal ulceration in hospital and *Haemophilus influenzae* with infection in children (Limberg 1991). Epidemics caused by adenovirus type 8 are essentially an infection of hospitals and clinics, possibly spread by instruments such as tonometers, but also by the hands of staff, and otherwise only by close, person-to-person, family, or other contact. Scrupulous hand-washing, decontamination of equipment – by chemical means if it is heat-sensitive – and use of individual containers for ophthalmic preparations, will terminate an outbreak (Barnard et al. 1973). Other adenoviruses may also be transmitted in this way (Tullo and Higgins 1980). Hospital-acquired bacterial conjunctivitis is common only in neonatal units, where minor conjuctival infection with *S. aureus* is more common than neonatal ophthalmitis due to *Chlamydia trachomatis* or *Neisseria gonorrhoeae* acquired from the birth canal.

THE PERITONEUM

Peritonitis is one of the classical associations of surgery: a breach in the bowel wall is followed by invasion of the peritoneum by a mixture of enteric bacteria, possibly acting synergistically. Advances in the classification and pathogenesis of sepsis, including peritonitis (Bone 1991), have raised the possibility of new therapeutic strategies. Patients who require intensive care have a mortality of 63 percent; amongst the poor prognostic features are advancing age, the presence of septic shock, failure to clear the source of sepsis and an upper gastrointestinal source (McLauchlan et al. 1995).

With the development of peritoneal dialysis for renal failure, and particularly of continuous ambulatory peritoneal dialysis (CAPD) for its chronic management, a very different form of peritonitis has become common. Short-term peritoneal dialysis, in patients unsuitable for hemodialysis, is often complicated by peritoneal infection caused by gram-negative bacilli. CAPD peritonitis has a different spectrum of etiological agents and coagulase-negative staphylococci, particularly *S. epidermidis*, account for about half the cases (Spencer 1988; von Graevenitz and Amsterdam 1992). Factors important in the pathogenesis of such infections include the size of the bacterial inoculum, adaptation of bacteria to the local microenvironment, especially adherence to plastics, and diminished phagocytic activity in peritoneal fluid (von Graevenitz and Amsterdam 1992). Many factors have been invoked to explain the success of *S. epidermidis* as a pathogen in CAPD and there is increasing interest in bacterial phenotypic variation in peritoneal dialysis fluid. Differences in iron-regulating proteins, cell wall and cytoplasmic membrane protein profiles have been demonstrated when strains are grown in peritoneal dialysis fluid rather than in nutrient broth (Wilcox et al. 1991). Other gram-positive cocci and other skin bacteria, enterobacteria, and the Pseudomonadaceae, account for most other cases. Anaerobic bacteria and fungi are less common. Special cultural methods may be needed to reveal the presence of intracellular organisms or organisms that for other reasons are difficult to grow. The major cause of infection is a lapse in technique by the patient or attendants in changing the containers of dialysis fluid. Patient motivation is therefore important in minimizing infection. A number of specific measures, including good surgical technique at the time of catheter insertion and the use of occlusive dressings (Ludlam et al. 1989), have been advocated to reduce infections, especially those due to S. aureus, which are difficult to treat. The infecting organisms may reach the peritoneum from the abdominal skin, with or without overt 'tunnel' infection. Other sources are the bowel – mixed infection with multiple enterobacteria or anaerobes, which suggests perforation (Editorial 1982) – the bloodstream, and the female genital tract. Episodes of infection may be treated with antibiotics without removal of the catheter (Working Party of the British Society for Antimicrobial Chemotherapy 1987), but repeated infection in association with tunnel infection and certain pathogens, such as S. aureus, P. aeruginosa, and fungi, make success less likely.

Prevention of postoperative wound infection

The measures taken in the operating suite to prevent postoperative wound infection have been classified into the following categories (Medical Research Council Subcommittee of the Committee on Hospital Infection 1968):

- established
- provisionally established
- rational methods
- rituals.

Since then, some time-honored practices have been discarded. Nevertheless, there is still a lack of firm evidence to justify a number of the 'rational' measures that are taken to protect the patient. This chapter gives only an outline that applies to operations generally and not to special situations, such as orthopedic implants. Many measures can be deduced from the risk factors already outlined. Thus, an operation carried out expeditiously with meticulous technique will have a reduced risk of infection.

In elective surgery, patients may be investigated preoperatively at assessment clinics, to avoid a long period in hospital before the operation. Underlying disease and existing infections should, as far as possible, be controlled. Nasal carriage of S. aureus is associated with postoperative infection and a review of recent studies found that the relative risk of infection in S. aureus carriers was 7.1 (95 percent, CI 4.6–11.0). von Eiff et al. (2001) recently showed that the great majority of cases of S. aureus bacteremia appear to be of endogenous origin. Of 14 patients who had nasal colonization with S. aureus and subsequent S. aureus bacteremia, 12 (86 percent) had nasal isolates indistinguishable from those recovered from blood cultures 1 day to 14 months later. Preoperative monitoring for pathogens, such as S. aureus or S. pyogenes, is commonly carried out before high-risk operations, such as those on the heart, but its usefulness has not been established (Ridgway et al. 1990). Recent studies have highlighted the potential use of nasal mupirocin prophylaxis to prevent SSIs, notably in patients undergoing cardiothoracic surgery (Kluytmans et al. 1997; Cimochowski et al. 2001), and possibly upper gastrointestinal surgery (Yano et al. 2000). However, a placebo-controlled study of nasal mupirocin prophylaxis in patients undergoing orthopedic prosthetic implant surgery found a reduction in S. aureus nasal carriage, but no effect on SSI incidence (Kalmeijer et al. 1999). Wilcox et al. (2003b) observed a marked decrease in the incidence of MRSA SSIs, from 23/1 000 operations to 3.3–4.0/1 000 operations, following the introduction of empirical perioperative prophylaxis based on nasal mupirocin and also including topical skin disinfection with triclosan. A recent placebo-controlled, randomized study of perioperative prophylaxis with nasal mupirocin failed to demonstrate a reduction in the overall rate of S. aureus SSIs, but a significant reduction in S. aureus nosocomial infections was seen in patients given mupirocin who were S. aureus nasal carriers (Perl et al. 2002).

The value of total-body antiseptic treatment, as with chlorhexidine detergent, is also controversial (Lynch et al. 1992). Antiseptic skin preparation in the operating theater, with alcoholic solutions of chlorhexidine or an iodophor, applied with some friction, is very effective (Davies et al. 1978). There is no evidence that longer applications are beneficial, except perhaps when clostridial spores must be removed (Lowbury et al. 1964). The use of specially impermeable material, whether woven or adhesive plastic, or of disinfectant-impregnated material, for the drapes around the operative field, is not supported by firm evidence (Lewis et al. 1984). Hair should, whenever possible, be clipped or removed by depilatory methods rather than by shaving.

A full discussion of the prophylactic use of antibiotics to prevent surgical wound infection is beyond the scope of this chapter (Leaper 1994; Sheridan et al. 1994). The usefulness of very short perioperative courses of aptly chosen antibiotics has been decisively established for operations with a high incidence of endogenous infection. Examples are colonic surgery, upper gastrointestinal surgery in conditions of bacterial overgrowth, some operations on the biliary system, and vaginal hysterectomy. Prophylactic antibiotics are also justified for operations with a lower incidence of infection when the consequences would be very serious as, for example, for the insertion of orthopedic and other implants, cardiac surgery, and dental procedures in those at risk of endocarditis. The value of such prophylaxis has been established only for joint replacement surgery (Lidwell et al. 1982). Prophylaxis with cephalosporins, which is associated with a very low complication rate, is beneficial for patients undergoing other categories of 'clean' operations. The degree to which this extension of prophylaxis should be encouraged has been debated (Page et al. 1993; Lewis et al. 1995).

The measures customarily taken to establish a safe environment in the operating theater have been reviewed by Hambraeus and Laurell (1980), and the principles of theater suite design by Humphreys (1993). Briefly, ventilation is provided with clean filtered air sufficient in volume to suppress as far as possible the airborne particles produced by human movement, in particular in the region of the operative field. This is far less important in other areas, such as near the floor. The effectiveness of conventional ventilation systems is readily upset by unsatisfactory building design or by undisciplined human activity. Transfer of bacteria from the operating team to the wound is somewhat reduced by the exclusion of staff with clinical infection and by the wearing of clean, frequently changed clothes and an operation gown. However, if the latter is made of conventional fabric, its effectiveness as a barrier to skin microorganisms is small. Masks to filter or deflect upper respiratory particles and caps to reduce contamination from the hair should be worn by the immediate surgical team (Dineen and Drusin 1973). Hand-washing with chlorhexidine or iodophor solutions before donning gloves should be careful and thorough, but not necessarily prolonged (Larson 1984). Gloves are very commonly perforated during operations, and this may be associated with an increased risk of infection for the patient (e.g. Cruse and Foord 1973) and of blood-borne infection for the surgeon (Palmer and Rickett 1992).

Effective sterilization of surgical instruments, with a large margin of safety (Russell et al. 1992), is now universal, but it was not always so (Howie 1988). Heat-sensitive equipment may give rise to difficulties and the need for a rapid turnover may force the use of less stringent decontamination methods, even for instruments that penetrate tissue.

THE URINARY TRACT

Most healthcare-associated infections of the urinary tract are associated with urethral catheterization. Even the single passage of a catheter is associated with a definite, though usually low, infection risk (Kunin 1987). The need for catheterizations should be carefully assessed in each patient and alternatives considered such as self in–out catheterization in females and external catheter in males (Pearman 1998). With an indwelling catheter eventual colonization of the bladder is almost inevitable, but it may be considerably delayed by the system of closed drainage, first introduced by Dukes (1929) and developed by Gillespie and others (Gillespie 1956; Miller et al. 1958). The closed system is not interrupted for the taking of urine specimens, and irrigation after operation can be provided without opening the system (Miller et al. 1958).

In spite of closed drainage, bladder colonization nevertheless occurs. It may reach about 10 percent of patients per day (Kunin and McCormack 1966). The route of infection is not via the lumen of the catheter, but between the catheter and the urethral wall. Early infection is by local commensals, such as *E. coli*, coagulase-negative staphylococci and enterococci. Later, more resistant hospital-associated gram-negative bacilli such as *Klebsiella*, *Proteus*, *Serratia*, *Pseudomonas*, and *Providencia* spp. may invade, particularly under the selective influence of antibiotics. Careful passage of the catheter by skilled staff under conditions of strict asepsis and scrupulous catheter care thereafter are also important (Falkiner 1993; Pearman 1998). Much attention has been given to the use of disinfectants in preventing infection (Stickler and Chawla 1987). Chlorhexidine may usefully be included in the local anesthetic applied to the urethra, or used as an initial 'flush' after the catheter has been passed. Chlorhexidine irrigations reduce the colonization rate (Kirk et al. 1979), but the prolonged application of chlorhexidine may damage the bladder mucosa. Little evidence supports the use of disinfectants in established infection (Davies et al. 1987) or in the long-term protection of indwelling catheters (Stickler and Chawla 1987). Periurethral disinfection, beyond simple toilet, has not been shown to be beneficial (Burke et al. 1983), nor has the inclusion of disinfectants in the catheter-drainage bag (Gillespie et al. 1983). They may, however, prevent the spread of resistant strains of bacteria (Noy et al. 1982), presumably via the hands of those who empty the drainage bags. Such spread is better avoided by an improved nursing regimen. The bacteria that cause urinary tract infection are variably sensitive to the antiseptics that can be used in contact with tissues (Hammond et al. 1987), especially when the bacteria are present in biofilm on the surface of the bladder (Stickler et al. 1987). It has been suggested, but not proved, that extensive use of chlorhexidine may account for the spread of relatively resistant gram-negative bacilli in certain hospital departments (Dance et al. 1987).

The presence of bladder bacteriuria is benign and symptomless in many patients, but its subtle long-term effects have not been studied, and upper tract infection may itself be silent. There is no doubt that urinary infection prolongs hospital stay and worsens the prognosis (Jepsen et al. 1982). The prevalence of hospital-acquired urinary infection was lower in a recent study in the UK and Ireland when compared with 1980 (Emmerson et al.

1996) and this may reflect improved practice and catheter care. A proportion of patients develop pyelonephritis and other complications, such as epididymoorchitis. Urinary tract infection is one of the most common causes of bacteremia in hospital. It frequently follows urethral procedures (catheter passing, catheter removal, cystoscopy, transurethral prostatectomy) in patients with colonized urine and may lead to septicemia and even death. Thus, perioperative bacteremia was demonstrated by Murphy et al. (1984) in 60 percent of patients with infected urine undergoing urethral operations and, in similar patients in the same hospital, Cafferkey et al. (1982) recorded 6 percent of clinical septicemia arising in these circumstances. Of 31 patients with postoperative septicemia investigated by Cafferkey et al. (1980), eight suffered severe shock and four died.

The risk of septicemia may be avoided by the preoperative or perioperative administration of suitable antibiotics to patients with colonized urine, especially those due to undergo urological procedures, such as prostatectomy and bladder resection (Amin 1992). There may also be a case for the administration of antibiotics to patients with sterile urine who have had indwelling catheters for some time, because damage to the urethra may encourage significant passage of bacteria into the bloodstream. In a study of prophylactic ciprofloxacin in carefully selected patients undergoing vaginal repairs, there was a reduction in symptomatic and asymptomatic bacteriuria compared with controls (van der Wall et al. 1992), but the long-term implications in terms of cost and the emergence of resistance should be carefully considered before this can be recommended for routine use. Antibiotic prophylaxis has no place in other situations and treatment of urinary infection in the catheterized should be limited to dealing with clinical illness (Gillespie 1986).

Silver impregnated antimicrobial urinary catheters have been proposed as a way of reducing associated infection. However, a recent review of the studies in this area concluded that general poor study design means that claims of efficacy cannot be accepted with confidence (Niel-Weise et al. 2002). The one high quality study identified by the reviewers did not demonstrate superior efficacy of silver coated versus standard catheters (Thibon et al. 2000). Alternatives to bladder catheterization that may be feasible are suprapubic catheterization, which has in some studies been accompanied by less bladder infection than urethral catheterization, and intermittent catheterization, including self-catheterization, which may be successful in patients with neurogenic bladder disorders (Pearman 1984). When catheterization appears necessary to achieve dryness in incontinent patients, external, condom drainage may be used, or effective incontinence pads may be substituted with success (Nordqvist et al. 1984), with other measures to encourage regular micturition.

THE RESPIRATORY TRACT

Infections of the respiratory tract represent a significant proportion of all healthcare-associated infections. In the recent prevalence study, respiratory tract infection accounted for approximately 25 percent of hospital infection and was as common as that of the urinary tract (Emmerson et al. 1996). Mortality is between 25 and 35 percent, but depends on the patient population, the diagnostic approach, and whether the mortality rate is crude or takes into account other factors that contribute to a fatal outcome.

A firm diagnosis of pneumonia is not easy to make and it is even more difficult to establish its microbial etiology. Sputum may be contaminated with bacteria from the upper respiratory tract and cultures may give misleading results. The use of bronchoscopic or alternative nonbronchoscopic approaches in the intensive care unit to obtain representative samples from the lower respiratory tract with quantitation of the organisms isolated is increasingly emphasized (Baselski and Wunderink 1994). The results from brush and lavage specimens correlate with histology even in patients on antimicrobial therapy (Chastre et al. 1995). Bronchoscopy may not, however, always be possible and it requires experience. Nonbronchoscopic lavage, with a catheter inserted through an endotracheal or tracheostomy tube, combined with semiquantitation, can be incorporated into routine practice relatively easily (Humphreys et al. 1996).

The etiology of pneumonia varies according to the patient population, diagnostic approaches used, local practices and procedures, and geographical location. Gram-negative bacilli, especially *P. aeruginosa* and *S. aureus*, account for the majority of cases (Dal Nogare 1994), but *S. pneumoniae* may be responsible for up to 20 percent of nosocomial pneumonia (Berk and Verghese 1989). Resistant enterobacteria, e.g. *Klebsiella*, *Serratia*, and *Enterobacter* spp., *P. aeruginosa*, *Acinetobacter* spp., and methicillin-resistant *S. aureus* (MRSA) pose particular difficulties, especially in high dependency areas of hospitals, because of fewer treatment options and the propensity of some of these organisms to spread. For a more general account of the etiology of pneumonia, see Chapter 24, Bacterial infections of the lower respiratory tract.

The general conditions that predispose to pneumonia are similar to those for wound infection: obesity, advanced age, and some underlying diseases. The most important factors, however, are respiratory intubation and tracheostomy. Tracheal intubation removes an important barrier to invasion of the lower respiratory tract, and even brief intubation allows oropharyngeal bacteria to pass into the trachea (Nair et al. 1986). High gastric pH, often induced deliberately to prevent stress ulcers in patients in intensive therapy, encourages overgrowth of potential pathogens that may then pass into the respiratory tract of the intubated patient. It has been suggested that fibronectin is a barrier to colonization by *P. aeruginosa* and that it is breached by the increased protease content of saliva during illness (Woods et al. 1981). Molecular approaches to understanding the pathogenesis of pneumonia reveal a complex interaction between host factors, such as cytokine release, phagocytosis and lymphocyte activation, and virulence determinants of the pathogen, including lipopolysaccharide and bacterial enzymes, such as elastase. For patients in the ICU this is further complicated by other conditions such as the adult respiratory distress syndrome (Meduri and Estes 1995; Harmanci et al. 2002).

Sporadic cases and outbreaks of pneumonia may be associated with contamination of respiratory equipment, such as anesthetic machines, ventilators, humidifiers, nebulizers, resuscitation equipment, and suction catheters (Hovig 1981). Better design of equipment, improved decontamination methods and the use of filters may lessen the importance of these sources. Breathing circuit change intervals of 7 and 30 days have been found to be associated with lower risks of ventilator-associated pneumonia than a 2-day interval, and the difference yielded marked reductions in morbidity and labor and acquisition costs (Fink et al. 1998). In general, there are two main targets within ventilators where filters may be used: firstly, to protect against contamination of the internal parts of the ventilator itself by microorganisms in respiratory secretions (internal filter), and second, to prevent dissemination of microorganisms in respiratory secretions into the external environment, i.e. filter situated on the exhalation limb of the tubing (external filter). To change filters between operations is expensive and there is no conclusive evidence that this will reduce crossinfection. Also, there are no data to prove whether contamination of ventilator circuits is implicated in the etiology of pneumonia. Both pleated hydrophobic and electrostatic filters can be shown to be permeable to bacteria and viruses in vitro. In short, there are no evidence-based guidelines that support the use of filters, and no recommendations about filter type can be made on infection control grounds alone (Das and Fraise 1997; Thomachot et al. 1999). The importance of maintaining respiratory equipment in good order and of cleaning/decontamination has become better recognized after outbreaks with highly resistant bacteria such as *Acinetobacter* spp. (Vandenbrouke-Grauls et al. 1988; Getchell-White et al. 1989; Hanberger et al. 2001). The causative bacteria may also be spread by indirect hand contact, and from there colonize the oropharynx before causing infection. Antibiotic administration encourages colonization by more resistant gram-negative bacilli.

The above account indicates some of the appropriate preventive measures. These include hand-washing, early removal of nasogastric and endotracheal tubes, maintenance of gastric acidity, and elevating the head of the bed to 30° (Dal Nogare 1994). This conventional approach to prevention, with an assessment of which practices have been validated, has been expanded upon elsewhere (Tablan et al. 1994). Nebulized antibiotics for prophylaxis may at first give satisfactory results, but later lead to problems with resistant bacteria (Feeley et al. 1975). Selective decontamination of the digestive tract (SDD), incorporating the use of topical antimicrobial agents to preserve colonization resistance and the protective effect of normal flora (van der Waaij et al. 1990), has been advocated to reduce ICU-acquired infection, including pneumonia (van Saene et al. 1992). The results have, however, been conflicting and have led some to advise against its routine introduction as a preventive measure particularly because of the potential for selection of multiresistant bacteria, such as MRSA (Hamer and Barza 1993; Hurley 1995). In a recent re-evaluation of SDD, de Jonge et al. (2003) carried out a prospective, controlled, randomized, unblinded study on 934 patients admitted to a surgical and medical ICU who were randomly assigned oral and enteral polymyxin E, tobramycin, and amphotericin B combined with an initial 4-day course of intravenous cefotaxime or standard treatment. They found that SDD use was associated with a significant reduction in ICU and hospital mortality and colonization with resistant gram-negative aerobic bacteria. Importantly, the authors note that their findings may not be applicable to units with endemic vancomycin-resistant enterococci or MRSA.

Though legionellosis (see section on Legionnaires' disease) is predominantly community-acquired, nosocomial cases accounted for 15 percent of cases in England and Wales between 1980 and 1992 (Joseph et al. 1994). Hospital-acquired infection should prompt a search for its source and for active preventive measures, including chemical disinfection, control of circulating water temperature, and surveillance (Hoge and Breiman 1991; Joseph et al. 1994). In young children, RSV and at times other viruses, such as enteroviruses, are significant causes of hospital-acquired respiratory infection. Without special precautions, nosocomial acquisition of RSV can be high, and a combination of cohort nursing and the use of gloves and plastic aprons significantly reduces spread (Madge et al. 1992).

Infection associated with indwelling medical devices

The increasing practice of inserting devices, usually made of various plastics, into the body of patients, and the selection pressure of antibiotics, have shifted the spectrum of significant microorganisms commonly encountered in diagnostic laboratories.

Almost every type of indwelling device may become colonized. The predominant bacteria are coagulase-negative staphylococci, particularly *S. epidermidis*. Othercolonizing organisms are *S. aureus* and other skin organisms such as corynebacteria, including

Corynebacterium jeikeium (see Chapter 12, Human microbiota; 32, *Staphylococcus*; and 39, Corynebacteria and rare coryneforms), propionibacteria, streptococci, many of low-grade pathogenicity, various gram-negative aerobic bacilli, yeasts, and filamentous fungi. *S. epidermidis* strains associated with such infections appear to produce greater quantities of extracellular slime than do strains from other sources. The slime consolidates bacteria adherent to the polymer surface and interferes with the body's response to infection (Gristina et al. 1993; Jansen and Peters 1993).

INTRAVASCULAR CANNULAE

Plastic cannulae allow infusions to be given via the same vessel for much longer than do steel needles, but they are also associated with bacterial colonization and resulting infection (Maki et al. 1973). At first it was recommended that the site of infusion should be changed frequently, but with better materials, better cannula design, and careful placement, cannulae such as the Hickman and Broviac catheters may now be left in situ for long periods, albeit with a low but significant infection rate (Darbyshire et al. 1985).

The colonization of intravascular cannulae may remain silent or be associated with a simple fever. However, if it is due to more pathogenic organisms, especially *S. aureus*, it may result in septicemia or disseminated infection. Even with *S. epidermidis*, metastatic infection, particularly of other prosthetic material, endocarditis or even immune-complex nephritis, such as that associated with cerebrospinal fluid shunts, may occur. The rate of colonization of intravascular cannulae is very variable, depending on the site of cannulation, the fluid infused, the plastic used in the cannula, the design of the cannula, and the care to achieve asepsis at insertion. Investigations up to 1973 have been reviewed (Maki et al. 1973). The use of lower limb veins is associated with an increased incidence of infection, as is umbilical cannulation in babies. Subclavian catheters are associated with lower infection rates than those placed in the internal jugular vein. Solutions for total parenteral nutrition readily support microbial growth and, alone or in admixture with circulating plasma near the cannula, they encourage colonization and a high risk of catheter-related bloodstream infection (CRBSI). The use of the same line for infusion of blood products and clear fluids, or its use for pressure monitoring and the taking of blood specimens, also increases the colonization rate.

The cuff of the implanted Hickman or Broviac catheter may prevent infection from tracking between catheter and tissues. The importance for cannula colonization of a side port closely related to the cannula is the subject of controversy. Reports of side port colonization are common, but there seems to be no significant difference in the rate of CRBSI between cannulae with and without side ports (Cheesbrough et al. 1984). Careful

adherence to instructions for the setting up and care of the infusion reduces the rate of colonization and subsequent infection. The source of colonizing organisms is usually the skin of the patient. Several studies have shown a correlation between microorganisms on the skin and those recovered from the catheter after removal (e.g. Fuchs 1971). For this reason, careful antiseptic preparation of the skin before cannula insertion, preferably with alcoholic chlorhexidene, toilet of the area, and careful choice of dressings to prevent skin maceration is advocated. The application of antibiotic or antiseptic preparations to the entry site once the catheter is in situ is not recommended and may be associated with candidal overgrowth (Norden 1969). In some circumstances, as with Hickman and Broviac catheters with their cuff and protective tunnel, and other central lines, but also at times with simple peripheral cannulae, contamination of the connections of the cannula with the administration set and other reservoirs appears to be of importance (Sitges-Serra et al. 1983; Cheesbrough et al. 1984; Weightman et al. 1988). Care with manipulation of the apparatus may be important in cannula colonization, as it is in contamination of fluid in the rest of the system. Occasionally, cannulae are infected by the hematogenous route (Maki et al. 1973). Preventive measures and their success, depends on the postulated entry route for microorganisms (Weightman et al. 1988; Parras et al. 1994). The route of infection may determine the preponderant site of colonization, whether the microorganisms are on the outside of the cannula or in the lumen, and thus the best means of culturing the cannula tip after its removal. The technique devised by Maki and colleagues (1977) for peripheral cannulae may not be as suitable for central lines. This approach requires line sacrifice and it is known that the majority of central venous catheters removed on suspicion of CRBSI are ultimately shown not to be infected. Catheter colonization may be detected in lines already being used for the taking of blood, by comparing the microbial concentration in a specimen of blood taken through the cannula with one from a distant vein; a simple pour-plate method will suffice (Weightman et al. 1988). An even simpler approach is to compare the time taken for central versus peripheral blood cultures to become positive with the same microorganism. If the time difference is at least 2 h (in favor of the central blood cultures) then this is highly predictive of CRBSI (Blot et al. 1999; Seifert et al. 2003). Other new approaches to the diagnosis of CRBSI which do not rely on catheter removal include the acridine orange leukocyte centrifugation test and the endoluminal brush (Kite et al. 1999). Luminal colonization of intravascular catheters is probably almost universal, particularly as dwell time increases, and thus it is CRBSI which is the clinically important endpoint and not catheter colonization.

Minocycline-rifampin (MR) and chlorhexidene silver sulfadiazine (CSS) containing catheters have both been shown to reduce the incidence of CRBSI compared with uncoated/impregnated central venous catheters (CVC). A study comparing MR and CSS catheters found significant differences in favor of the former in terms of both prevention of catheter colonization (22.8 versus 7.9 percent, $p < 0.001$) and CRBSI (3.4 versus 0.3 percent, $p < 0.002$) (Darouiche et al. 1999). The enhanced efficacy of the MR catheters is probably due to the fact that these have antimicrobial coatings on both the outer and inner surfaces, whereas only the outer surfaces of CVCs were coated with CSS. Meta-analyses of studies on the efficacy of antimicrobial impregnated CVCs have reached differing conclusions. A recent review of 11 randomized studies identified several methodological flaws, including inconsistent definitions of CRBSI, failure to account for confounding variables, suboptimal statistical and epidemiological methods, and rare use of clinically relevant endpoints (McConnell et al. 2003). This review also failed to demonstrate any significant clinical benefit associated with the use of antimicrobial-impregnated CVCs for the purpose of reducing CRBSI or improving patient outcomes.

INTRAVASCULAR GRAFTS

A similar spectrum of microorganisms is seen in infected cardiac valve prostheses and patches of polymers, and grafts in vessels (Liekweg and Greenfield 1977). This may be influenced by the antibiotic prophylaxis in use, because this may alter the flora of patients and hospital staff (Archer and Armstrong 1983). A multicenter randomized controlled trial of rifampicin-bonded Dacron grafts in aorto-femoral surgery demonstrated a reduction in total early wound and graft infection rates. However, results after 2 years showed a small, nonsignificant reduction in graft infection (1.7 percent in study group, 2.3 percent in control group) (Pratesi et al. 2001).

Endocarditis on prosthetic valves may be caused by microorganisms implanted at the time of operation or soon afterwards (Marples et al. 1978) when many intravascular lines are in situ and the graft has not yet become endothelialized. It may follow sternal-wound infection (Braimbridge and Eykyn 1987) or it may arise later from hematogenous seeding, as happens with diseased natural valves. This latter gives a microbial spectrum close to that seen in infections of natural valves (Heimberger and Duma 1989). Though the incidence of early prosthetic valve endocarditis is now less than 1 percent, the mortality is high. Perioperative infection may manifest after a delay and the various types of endocarditis cannot be separated by an arbitrary time after the operation (Braimbridge and Eykyn 1987; Freeman 1995). An incidence in the first year of 1.4–3.0 percent has been calculated (Karchmer 1991). Reliance should be placed on general measures to prevent perioperative infection and on antibiotic prophylaxis at critical times when bacteremia is likely (Karchmer 1991).

JOINT PROSTHESES

Infection of total joint replacements is a very serious and costly complication and a variety of preventative measures that contribute to minimizing this have recently been well reviewed (An and Friedman 1996; Gosden et al. 1998). As techniques have improved, the incidence has fallen from approximately 12 percent to less than 2 percent for hip replacements, and from approximately 15 percent to less than 4 percent for knee replacements. The effects of ultraclean air and of prophylactic antibiotics on the incidence of infection have been discussed above. In addition to the use of systemic antibiotics, incorporation of antibiotics in the bone cement is now standard practice for revision operations and is used by some in primary operations. Gentamicin may be used, or alternatively a heat-stable antibiotic can be chosen to correspond to known infecting strains (Strachan 1993). The spectrum of infecting bacteria (An and Friedman 1996) is dominated by *S. aureus* and coagulase-negative staphylococci, with smaller numbers of enterobacteria and other gram-negative bacilli. The prognosis of infections with gram-negative organisms is worse after reoperation than infections with gram-positive organisms (Buchholz et al. 1981). Fungal infection is rare. Some recent series have shown increased numbers of streptococci (especially of enterococci) and of anaerobic cocci. Infection of a joint prosthesis almost always requires its removal, with replacement as a one- or two-stage procedure, under local and general antibiotic cover. Occasionally, with a low-grade pathogen sensitive to oral antibiotics infecting a stable joint, medical treatment may succeed, but this form of management should rarely be undertaken as it is beset with complications.

CEREBROSPINAL FLUID SHUNTS

In ventriculoatrial or ventriculoperitoneal shunts, infection may give rise to systemic illness from bacteremia, to ventriculitis or to shunt blockage. *S. epidermidis* is by far the most common infecting agent (Ersahin et al. 1994). As in the case of cannula infections, there may be a long lag between colonization and clinical illness. Most infecting organisms are probably implanted in the tissues at the time of operation (Pople et al. 1992), but other routes may occasionally be followed (Price 1984; Ronan et al. 1995). These infections are managed by a combination of surgical management, including shunt removal and chemotherapy (Bayston 1985; Walters 1992). There is no consensus about the benefits of prophylactic antibiotic use in shunt insertion (Brown 1993), but a meta-analysis suggests a statistically significant advantage (Haines and Walters 1994).

Bacteremia

Hospital-acquired bloodstream infections constitute a serious health problem and are associated with high morbidity and mortality which results in increased healthcare costs. Each year, as many as 35 million patients are admitted to hospital in the United States. Of these, at least 2.5 million will develop a hospital-acquired infection with almost 250 000 attributable to bloodstream infections (Correa and Pittet 2000). Extensive studies in the USA have reported hospital-acquired bacteremia in 0.2–0.4 percent of hospital admissions. These account for about half the total bacteremias observed. Much higher rates have been reported for tertiary referral centers than for general hospitals (Bryan et al. 1986).

The microorganisms that cause hospital-acquired bacteremia and the circumstances in which they do this are fully discussed in Chapter 19, Bloodstream infection and endocarditis. Briefly, the infecting organism may arise from a focus of infection in another system (secondary bacteremia) or no source may be identified clinically or after additional investigations, such as ultrasound (primary bacteremia). The organisms isolated and their resistance to antibiotics to a large extent reflect those prevailing in the hospital. In a study conducted in a UK tertiary referral center, *E. coli* was the most common cause and accounted for 27 percent of cases, followed by other enterobacteria and *S. aureus* (11 percent); polymicrobial bacteremia was responsible for 7 percent of cases (Ispahani et al. 1987). Between 1980 and 1992, the crude bacteremia rate in a US center increased from 6.7 to 18.4 per 1 000 patient discharges and a considerable proportion of this increase was due to gram-positive cocci (Pittet and Wenzel 1995). An increasing proportion of bloodstream infections due to coagulase-negative staphylococci, enterococci, and *Candida* spp. has been documented in recent years. This may reflect the changing hospital patient population and the increasing use of invasive devices. Central venous lines and blood transfusion were independent risk factors for hospital-acquired bacteremia in Australia (Duggan et al. 1993). Coagulase-negative staphylococci are especially important in the pediatric age group and were responsible for 43 percent of bacteremia in Great Ormond Street Hospital, London, UK (Holzel and de Saxe 1992). They were the third most common bloodstream pathogen after group B streptococci and *S. aureus* in a Swedish neonatal unit (Faxelius and Ringertz 1987).

The increasing proportion of bacteremia due to gram-positive bacteria is seen in immunocompromised patients who may require an in situ intravascular device for a long period and for whom repeated courses of broad spectrum antimicrobial agents are often necessary. Viridans streptococcal bacteremia in neutropenic patients is associated with chemotherapy-induced mucositis as shown by the finding that blood and oral isolates were indistinguishable by ribotyping (Richard et al. 1995). Enterococcal bacteremia, especially that due to *Enterococcus faecium*, is closely linked with the presence of cancer, neutro-

penia, renal failure, steroids, and recent antibiotics (Noskin et al. 1995a). The risk factors for candidemia are similar, but include recent abdominal surgery (Nielsen et al. 1991). On the other hand, the incidence of bacteremia due to *Bacteroides* spp. has declined in many places since the introduction of effective prophylaxis for operations on the large gut (Young 1982).

Clusters of cases may reflect a hospital outbreak of infection. Thus, nine cases of gram-negative bacteremia involving different organisms and occurring after open heart surgery were traced to an aqueous disinfectant used to clean pressure-monitoring equipment that was assembled before surgery and left open and uncovered overnight in the operating theater (Rudnick et al. 1996). Care must be taken to exclude pseudobacteremia, due to the contamination of blood culture media or their additives, contaminated skin disinfectants, specimen containers or blood-gas analyzers or, in the laboratory, from contaminated apparatus or carriers among the staff (Maki 1980).

Mortality rates depend largely on the age and underlying illnesses of the patients, whether there is a removable or remediable focus of infection, and on the microbial cause. Thus, infection with enterobacteria has a worse prognosis than infection with coagulase-negative staphylococci. Polymicrobial bacteremia also has a poor prognosis. Candidemia is associated with a mortality of 76 percent (Nielsen et al. 1991), which reflects the severity of underlying disease in such patients. In the intensive care patient with severe sepsis and shock, bacteremia is associated with a poor outcome, but the category of organisms appears to be less important (Brun-Buisson et al. 1995). Recent efforts to develop and use monoclonal antibodies directed against the gram-negative bacterial cell wall or cytokine mediators, such as tumor necrosis factor (TNF) in these patients, have been disappointing. A multicenter study in 15 centers revealed that blocking TNF activity with high doses of a competitive inhibitor led to a higher mortality compared with placebo (Fisher et al. 1996). Recently, recombinant activated protein C has been made commercially available for use as adjunctive treatment of severe sepsis accompanied by multiple organ failure. Further work on the pathogenesis of severe sepsis and bacteremia will be necessary before useful and safe immunomodulatory agents can routinely be used in management.

For bacteremia resulting from the contamination of intravenous fluids or additives to these, see the section on Infection from contaminated infusion fluids.

Since April 2001, all English hospital trusts have been obliged to report the number of patients with *S. aureus* bacteremia, along with methicillin/oxacillin sensitivity of the isolate. The current 3-monthly report considers the MRSAs amongst these, and dividing the total by the average daily bed occupancy of each hospital trust produces an index of MRSA bacteremia rates per 1 000 bed-days. The 186 hospital trusts are then ranked according to their bacteremia rates. At the bottom are a few centers that can congratulate themselves on rates of zero, whilst at the other extreme, high ranking trusts will have to explain MRSA bacteremia rates of up to 0.69 per 1 000 bed-days (Barrett and Spencer 2002).

The wide release of these figures adds to the increasing profile of hospital infections in the English public's awareness, which in turn should be viewed in the context of longstanding media interest in health-service shortcomings. Although the exact relevance of MRSA is debated, collecting MRSA bacteremia data has a certain logic. As a mainly hospital-acquired organism, it can be inferred that MRSA in blood culture suggests that a serious hospital-acquired infection has occurred. However, this does not allow for the unknown but increasing reservoir of common MRSA entering hospitals. In one locality, 17 percent of nursing home residents carried MRSA (Fraise et al. 1997). The higher the rate, by inference, the worse must be the individual hospital trust's problems with hospital-acquired infection – or at least its prevalence of MRSA. Moreover, extracting the numbers from laboratory computers and workload databases is vastly cheaper and quicker than surveillance of infections by clinical documentation. Appropriate surveillance techniques constitute the cornerstone of infection control and make possible the implementation of specific preventive approaches. Furthermore, accurate surveillance may identify risk factors for infection and other adverse hospital-acquired events (Correa and Pittet 2000). Surveillance of hospital-acquired bacteremia may appear straightforward because of the simple definition of the infection. Review of the microbiology laboratory records certainly allows identification of most of the cases. However, clinical sepsis is a recognized entity that may not be associated with positive blood cultures. In 2003, mandatory reporting of glycopeptide-resistant enterococcal bacteremia also began.

HOSPITAL INFECTION AT THE EXTREMES OF LIFE

Geriatric and long-stay facilities

Hospital infection and infection control in aged patients have been somewhat neglected, but their importance is undeniable. Patients older than 65 years account for a disproportionate number of healthcare-associated infections (Smith 1988). The relative frequency of infection at different sites depends, for example, on whether simple colonization of the bladder or of an ulcer is included, but the three most frequently infected systems are the respiratory tract, the urinary tract, and the skin. The last is more common in long-stay units (Darnowski et al. 1991; Jackson et al. 1992) than in acute geriatric units (Hussain et al. 1996).

Particular difficulties may be the declining defenses and multiple underlying chronic diseases in the patients and, in some long-stay hospitals, a lack of facilities, including isolation facilities, low staffing levels, and a comparative lack of access to medical assessment. Patients may be ambulant and have poor hygienic practices, and there may be close contact among patients. This may result in the spread of infections, such as adenovirus conjunctivitis, ringworm, scabies, etc., not seen in acute units. Chronic infections, such as tuberculosis, may spread insidiously. Carriage of MRSA may cause more severe problems when patients are transferred to surgical units or other services with vulnerable patients than in the geriatric unit itself (Combined Working Party of the British Society for Antimicrobial Chemotherapy and the Hospital Infection Society 1995). Infections in nursing homes have been reviewed by Nicolle et al. (1996).

Long-term indwelling urethral catheters are common in the elderly. Most surveys have shown a high prevalence of bacteriuria due to a number of different species. The most common are of *Proteus* or *Providencia* spp. rather than *E. coli* (Standfast et al. 1984; Lee et al. 1992). The clinical significance of most of this bacteriuria may be doubted (Jewes et al. 1988).

Nosocomial pneumonia in the elderly resembles that in younger age groups, in terms of etiology and in risk factors such as tracheal intubation, but it is associated with malnutrition and neuromuscular disease and has a particularly poor prognosis (Craven et al. 1992; Hanson et al. 1992; Woodhead 1994). Infection by respiratory viruses, including influenza A, is common and may cause serious illness and death.

The grave consequences of salmonella infection in aged patients have been reported frequently (e.g. Committee of Inquiry 1986). Clusters of cases of diarrheal disease are common. When an etiological agent is identified it is frequently a Norwalk-like virus (Stevenson et al. 1994; Augustin et al. 1995), but rotavirus outbreaks have also been reported (Holzel and Cubitt 1982).

Over 80 percent of *C. difficile* diarrhea cases in the UK occur in hospitalized patients aged over 65 years. Notably, the frail elderly are more prone to severe *C. difficile* infection (Kyne et al. 1999). Antibiotic use is the prime risk factor, but others linked to severe *C. difficile* diarrhea include functional disability, cognitive impairment, recent endoscopy, and possibly enteral tube feeding. *C. difficile* infection appears to be a marker, but not a cause per se, of increased risk of dying (Wilcox et al. 1996; Kyne et al. 1999, 2002). Further details about *C. difficile* infection can be found below and in Chapter 44, *Clostridium perfringens*, *Clostridium difficile*, and other *Clostridium* species.

Neonatal units

Intrauterine and perinatal infections of maternal origin, such as rubella, listeriosis, group B streptococci infection, candidiasis, and infection by HIV will not be considered here. However, many of these agents may, occasionally, be spread among babies via staff or equipment.

Unlike well-baby nurseries, the prevalence of infection may exceed 20 percent in neonatal intensive therapy units, and on this there may be superimposed outbreaks of infection by particularly virulent and usually antibiotic-resistant strains. The babies in these units, particularly when premature, have undeveloped defenses and lack a normal flora. Ill babies require much handling by staff, and it is therefore difficult to prevent the spread of infection by contact from one to another.

Infection with *S. aureus* was the main concern in the 1950s. Babies in nurseries for the newborn were regularly colonized and some developed septic skin lesions of varying severity that sometimes progressed to fatal generalized infections. The causative staphylococci showed considerable differences in virulence. The problem was ameliorated by standard infection control measures, early discharge of healthy babies from hospital, and prophylactic use of topical hexachlorophane (Alder et al. 1980). A large neonatal outbreak of *S. aureus* identified by routine phage typing required multiple, staged, infection control measures including strict emphasis on hand hygiene, application of topical triclosan solution and hexachlorophane powder, aseptic handling of a skin protectant material, and use of topical mupirocin for staff nasal carriers of the endemic strain and for babies colonized or infected with *S. aureus* (Wilcox et al. 2000b). Withdrawal of hexachlorophane after overdosing had been shown to give rise to neurological damage and was followed by a recrudescence of staphylococcal infections (De Souza et al. 1975; Allen et al. 1994). Control of *S. aureus* in a neonatal unit was maintained when topical umbilical hexachlorophane powder was substituted by 1 percent chlorhexidine powder (Wilcox et al. 2003a). The neonatal unit has often been a focus of infection by MRSA (Reboli et al. 1989; Tam and Yeung 1988). In neonatal intensive therapy units there may occasionally be outbreaks of infection by enterobacteria such as *Klebsiella* spp., *Serratia* spp., and *P. aeruginosa*, sometimes resistant to many antimicrobial agents and with the ability to spread rapidly and cause serious illness (Davies and Bullock 1981; Lewis et al. 1983; Coovadia et al. 1992). Colonization by such bacteria without infection is very common. Routine monitoring of carriage sites contributes little to preventing illness (Jolley 1993), but may help in formulating a rational policy for antibiotic prescribing. Surveillance is better directed towards identifying strains with the potential to spread widely and cause serious illness (Goldmann 1988).

In recent years, *S. epidermidis* or other coagulase-negative staphylococci have been the most frequent cause of infection in many neonatal units (Thompson et al. 1992; Nataro and Corcoran 1994). Such infections are generally more benign than those due to gram-negative aerobic bacilli. The reason for their current prevalence is unknown, but may partly reside in their resistance to commonly used antibiotics and the increased use of intravascular cannulae and respiratory apparatus. Some strains spread within a unit, but there is no evidence of the emergence of particularly virulent or transmissible strains of coagulase-negative staphylococci. Indeed, several different strains may simultaneously infect a single patient (Simpson et al. 1986; Kacica et al. 1994).

Necrotizing enterocolitis has increased in incidence in recent years and has features, such as clustering of cases, that suggest an infective etiology. At times tightening of infection control procedures appeared to abort an outbreak (Willoughby and Pickering 1994). However, attempts to demonstrate a specific microbial cause have been unsuccessful (de Louvois 1986; Gupta et al. 1994).

Outbreaks of virus infection in neonatal units can be serious and difficult to control. Enteroviruses such as echovirus 11 (Nagington et al. 1978), adenovirus 7 (Finn et al. 1988), or adenovirus 8 (Piedra et al. 1992), rotaviruses (Holzel and Cubitt 1982), and RSV (Hall et al. 1979) may be responsible. *Candida* spp. may on occasion cause crossinfection in neonatal units (Finkelstein et al. 1993; Reagan et al. 1995).

THE ROLE OF INDIVIDUAL PATHOGENS

Staphylococcus aureus

This organism has probably always been an important cause of hospital-acquired infection, but it was first seriously investigated in the early years of the antibiotic era (see Williams et al. 1960). Certain strains that could be identified by phage typing (Chapter 32, *Staphylococcus*) and were resistant to penicillin, and often also to other antibiotics, became prevalent in hospitals. Some of these multiresistant strains were exceptionally virulent, able to cause severe infections in otherwise healthy persons, and colonized widely. These so-called 'epidemic strains' caused many outbreaks of skin sepsis, most notably in the postnatal period, and were the main cause of postoperative wound infection in the 1950s (Shanson 1981).

During the 1960s, hospital-acquired infection by multiresistant strains became progressively less common (Ayliffe et al. 1979). MRSA strains (Chapter 32, *Staphylococcus*) were recognized at this time, but they seldom spread widely. In the late 1970s, however, strains that were methicillin-resistant, and often also resistant to aminoglycosides, became prevalent in many parts of the world and have subsequently spread widely in hospitals. Patients particularly affected are those in hospital for prolonged periods, elderly or debilitated patients, those with wounds or burns, and those who have had recent treatment with antibiotics (Brumfitt and

Hamilton-Miller 1989). Clinical and laboratory evidence suggests that MRSA are not less virulent than sensitive isolates (French et al. 1990) and in severely ill patients under intensive care, mortality from pneumonia may be high (Rello et al. 1994). The financial impact of MRSA is increasingly recognized (Casewell 1995); the additional measures necessary to control an outbreak with a new epidemic strain that involved over 400 patients in England amounted to £400 000 (Cox et al. 1995).

Control of MRSA can be difficult and ward closures can result in the severe disruption of clinical services (Barrett et al. 1993). Nonetheless, active intervention combined with effective surveillance is successful (Hartstein et al. 1995). In the face of endemic infection, control measures are usually based upon a risk assessment of the likelihood of spread and invasive infection, and include selective screening of patients (less so, staff), isolation or cohort nursing, and eradication of carriage by the judicious use of topical antibiotics (Ayliffe 1996; Combined Working Party of the British Society for Antimicrobial Chemotherapy, the Hospital Infection Society and the Infection Control Nurses Association 1998). Recognition of epidemic strains may be assisted by noting the origin of the patient, the clinical illness, the type of infected lesion, the antibiotic-sensitivity pattern, and phage type. Collecting this information and the use of a simple computerized database can provide useful information on local epidemiology (Rossney et al. 1994). The complexity of strains isolated and the presence of phage nontypable strains means that a combination of phenotypic and genotypic typing methods (see later under Epidemiological typing) is now often required to determine source and spread (Tenover et al. 1994). The emergence of MRSA strains with reduced vancomycin susceptibility is of some concern (Hiramatsu et al. 1997). There have been recent reports of outbreaks of community-acquired MRSA infection, particularly in the US, which are caused by strains with a novel methicillin resistance gene and often also a hitherto rare exotoxin, Panton–Valentine leucocidin, that is associated with primary skin infection and pneumonia (Dufour et al. 2002; Mongkolrattanothai et al. 2003; Anonymous 2003). This development raises the specter of the transmission and acquisition of MRSA clones outside the traditional secondary healthcare setting. Basic infection control practice, including the rational use of antimicrobial agents, becomes even more important as we face the possible dissemination of strains for which there are few effective therapeutic agents.

Colonized or infected patients should be treated in isolation, preferably in a single room or designated isolation facility, with precautions against both contact and aerial spread. Contaminated environmental surfaces may be a source for continuation of an outbreak. When large numbers of patients are involved, cohort nursing in separate areas of the hospital may be necessary, but control may be impossible without a dedicated isolation unit

(Pearman et al. 1985). Mupirocin seems particularly effective in eliminating carriage (Casewell and Hill 1986), but resistance to this agent can emerge following prolonged or repeated use in MRSA patients and patients with chronic skin disease (Cookson 1990). MRSA carriage may be unresponsive to a variety of antiseptics and increasing numbers of carriers are found in nonacute units and in the community.

Enterococcus spp.

Recent interest in *Enterococcus* spp. as hospital pathogens has focused on unusually antibiotic-resistant strains of *Enterococcus faecium* and *Enterococcus faecalis*. Resistance may be multiple and include high-level resistance to penicillins and to aminoglycosides, such as gentamicin, which are important in providing synergistic bactericidal combinations with penicillins. They may also be resistant to glycopeptides such as vancomycin, which had been regarded as universally active against gram-positive bacteria that may infect humans, apart from intrinsically resistant genera of minor importance such as *Pediococcus*, *Leuconostoc*, and *Lactobacillus* (Woodford et al. 1995; Chavers et al. 2003).

Resistance to vancomycin in enterococci was first reported in a cluster of hospital-acquired infections in the UK (George et al. 1989) and such strains are increasingly found in oncology, renal, transplant, and intensive care units (Hospital Infection Control Practices Advisory Committee 1995). The epidemiology of resistant enterococci is not entirely clear. Treatment, particularly with cephalosporins and glycopeptides, may encourage colonization with enteroccci, which may then spread from patient to patient by indirect contact, via staff and environment, and thus cause persistent problems.

For epidemiological purposes, enterococci may be typed by ribotyping, but pulsed field gel electrophoresis is more discriminating. The strains are heterogeneous, but the evidence suggests that particular epidemiological types spread between hospitals (Chadwick et al. 1996; Morrison et al. 1996). Guidelines for the prevention of spread have been issued in the USA (Hospital Infection Control Practices Advisory Committee 1995) and discussed in the UK (Chadwick et al. 1996). These guidelines are similar to those for the control of MRSA (see above).

Clostridium difficile

C. difficile has emerged during the last decade as a significant hospital-acquired pathogen. It was recognized in the late 1970s as the main cause of antibiotic-associated diarrhea, which may be associated with pseudomembraneous colitis (PMC) (Chapter 44, *Clostridium perfringens*, *Clostridium difficile*, and other *Clostridium* species). The number of laboratory reports of *C. difficile* infection sent to the UK Communicable Disease Surveillance Centre increased ten-fold between 1982 and 1991 (Joint DH/ PHLS Working Group 1994). Laboratory reports for *C. difficile* positive samples for England and Wales currently (2002) stand at approximately 29 000 per year (provisional data). This is certainly an under-reporting of the true burden of this disease. *C. difficile* may be part of the normal bowel flora, especially in the elderly, and may also be recovered from the hospital environment. Antibiotics, especially broad spectrum agents, disrupt the normal bowel flora and render patients susceptible to colonization. *C. difficile* rarely, if ever, affects healthcare personnel and spread from patient to staff is unusual.

The increasing numbers of elderly and debilitated patients admitted to hospitals, the misuse of antimicrobial agents and changes in healthcare, including greater diagnostic and therapeutic interventions, have all probably contributed to the increased colonization with *C. difficile*. Whilst most cases are sporadic, clusters or outbreaks also occur. In the UK, a large outbreak involving 175 patients occurred in three hospitals in Manchester in 1992. Most of the patients were over 60 years old and infection was believed to have contributed to 17 deaths (Cartmill et al. 1994). In smaller outbreaks or clusters, epidemiological evidence usually suggests a traceable chain of person-to-person spread between wards or hospitals (Nolan et al. 1987). Apart from significant morbidity and mortality in an already vulnerable population, management and control of this condition have significant financial implications. In a case-control study it was calculated that 94 percent of the additional costs associated with *C. difficile* infection were due to increased duration of hospital stay and the total cost was in excess of £4 000 per case (Wilcox et al. 1996).

Even before the advent of newer typing techniques, clinical and epidemiological evidence suggested that patient-to-patient spread occurred by the fecal–oral route and that precautions were necessary to prevent spread. More than 20 percent of patients may carry *C. difficile* at some stage during hospitalization, especially if they share a room with a positive patient, and environmental and patient isolates are often indistinguishable (McFarland et al. 1989; Samore et al. 1996). Control measures should be aimed at minimizing the risk of antibiotic-associated diarrhea by good antimicrobial prescribing, scrupulous hand-washing, the isolation or cohorting of affected patients, treatment of persistently symptomatic patients, careful disposal of contaminated laundry and, finally, environmental disinfection/decontamination (Joint DH/PHLS Working Group 1994; Gerding et al. 1995; Wilcox 1998). The role of asymptomatic carriers in spread, especially amongst the elderly, remains unclear. Some *C. difficile* strains are more virulent than others. For example, ribotype 1 is epidemic in the UK, being responsible for about 60 percent of reported cases, and possesses several phenotypes that distinguish it from other strains; for example, ribotype 1 sporulates more than

comparators, a phenomenon that is enhanced when exposed to detergents (Wilcox and Fawley 2000), and is more resistant to fluorquinolones (Wilcox et al. 2000a). It is known that environmental contamination with *C. difficile* is widespread despite routine detergent-based cleaning (Fawley and Wilcox 2001; Verity et al. 2001). There is some evidence that environmental cleaning with hypochlorite may be more effective than detergent as a control measure for *C. difficile* diarrhea (Wilcox and Fawley 2000; Wilcox et al. 2003b).

Enterobacteriaceae

E. coli is one of the most frequently encountered bacteria in hospital-acquired infection (Jarvis and Martone 1992). It causes urinary tract infection (Chapter 26, Bacterial infections of the urinary tract), intraabdominal and gut-related wound infection and bacteraemia (Chapters 19, Bloodstream infection and endocarditis and 52, *Escherichia*), but infection is almost always endogenous and sporadic; even resistant strains seldom appear to spread between hospital patients (Hart 1982). However, cross-infection by strains with the potential to cause enteric infection, including verocytotoxin-producing *E. coli* strains, is common in hospitals and other healthcare facilities (Carter et al. 1987; Kohli et al. 1994).

Other Enterobacteriaceae are more important in the spread of nonenteric infection in hospitals. This is due to antibiotic resistance, transmissibility, and virulence, which interact in hospital units that accumulate patients with similar medical problems who undergo similar procedures and receive similar antibiotics. For example, *Klebsiella* spp. have come into prominence in outbreaks caused by multiresistant strains that produce extended spectrum β-lactamases (Johnson et al. 1992; Meyer et al. 1993; Quinn 1994). Ease of transmission from patient to patient, resulting in large-scale epidemics and inter-ward spread rather than small clusters of infections or common-source outbreaks, may be related to better survival on hands (Casewell and Desai 1983) or on inanimate surfaces (Hart et al. 1981), but other factors must be involved (Fryklund et al. 1995). The characteristics of hospital infection by members of the various enterobacterial genera are outlined in Table 16.1. Infection by *Salmonella* spp. has been mentioned under infection transmitted by food (Chapters 11, Bacteriology of foods excluding dairy products and 54, *Salmonella*).

Pseudomonas aeruginosa

P. aeruginosa has been recognized as a pathogen of hospital patients only in the modern era of intensive treatment and antibiotic administration. Its ability to grow in moist conditions with simple nutrients and its comparative resistance to antibiotics and disinfectants have allowed it to become established, often in very large numbers, in fluids and wet places in the hospital

Table 16.1 *Enterobacteriaceae of importance in hospital infection*

Genus[a]	Source and mode of infection	Circumstances of infection	Antibiotic characteristics of hospital strains	References
Escherichia (*E. coli*)	Endogenous; crossinfection uncommon (except enteric infection)	Common (especially urinary, bloodstream)	Ampicillin R common	Olesen et al., 1995
Citrobacter (*C. freundii*, *C. diversus*)	Infrequent, endogenous infection	Bloodstream, etc.	Often multi-R including cephalosporins	Thurm and Gericke, 1994
Klebsiella (*K. pneumoniae*, *K. oxytoca*)	Endogenous, but crossinfection common, with fecal carriage; transmission via hands	Bloodstream, urinary (especially catheterized), respiratory; ICUs especially neonatal, urological units	Ampicillin R; frequently aminoglycoside R; increasingly R to newer cephalosporins	Johnson et al. 1992; Meyer et al. 1993; Quinn 1994
Serratia (nonpigmented *S. marcescens*)	Crossinfection common, also environmental sources implicated; skin, throat carriage, but fecal carriage less common; hand spread	Urinary, respiratory; ICUs, especially neonatal	R to ampicillin, early cephalosporins; frequently R to aminoglycosides and newer cephalosporins; emergent R during treatment; characteristic polymyxin R	Lewis et al. 1983; Coria-Jimenez and Ortiz-Torres 1994; Luzzaro et al. 1995
Enterobacter (*E. cloacae*, *E. aerogenes*)	Endogenous and hospital infection from common source, e.g. medicaments, food, apparatus; hand spread less common; throat and fecal carriage	Respiratory, urinary, bloodstream; ICUs especially neonatal, burns units	Ampicillin R; frequently R to cephalosporins; emergent R during treatment	Gaston 1988; Weischer and Kolmos 1992
Proteus (*P. mirabilis*, *P. vulgaris*, also *Morganella*)	Endogenous and small clusters; fecal carriage	Urinary, bloodstream; ICUs, urological units	Variable (see Chapter 55, *Proteus*, *Morganella*, and *Providencia*)	Williams et al. 1983; Watanakunakorn and Perni 1994
Providencia (*P. stuartii*)	Endogenous; slow transmission among patients; fecal carriage	Urinary, occasionally other sites; geriatric, paraplegia units; mixed infection with *Proteus* spp. common	Aminoglycoside R common	Hawkey et al. 1984; Woods and Watanakunakorn 1996

ICUs, intensive care units; R, resistance or resistant.
a) Only more important species are listed. For *Salmonella*, see Chapter 54.

and to colonize the mucous membranes and skin of patients. Respiratory infection in intensive therapy units, infection of burns, in neonates, in abdominal surgery, and after trauma, are among the most characteristic presentations, but cases may be encountered in any acute hospital unit.

Strains of *P. aeruginosa* can be distinguished by serotyping, pyocine typing, phage typing (Chapter 62, *Pseudomonas*) and by genotypic methods. Strains may differ markedly in virulence and some epidemics are caused by particularly virulent strains or by exposure of a sensitive body site, such as the eye, to heavily contaminated fluids or apparatus. Mostly, however, infections are endemic or sporadic, and colonization is more common than clinical disease. Indeed, in some hospital units clinical infection may be rare, though colonization may be almost universal. It may, therefore, be difficult to interpret the significance of the isolation of *P. aeruginosa* from patients' specimens (Davies and Bullock 1981).

Normal subjects may carry *P. aeruginosa* in their bowel. The apparent prevalence depends on the sensitivity of isolation methods (Shooter et al. 1966) and antibiotic treatment encourages such carriage. The importance of strains at inanimate sites in the hospital varies, but their spread to patients has been reported (e.g. Doring et al. 1993). Strains found at such sites are, however, often different from those found in the patients (Kropec et al. 1993). Clearly, there is much greater likelihood of infection when the patient comes into close contact with the infected source or object, such as contaminated medicaments, inhalants, or splashing from sinks. It is, therefore, essential that medicaments and aqueous disinfectants intended for application to patients are supplied sterile and preferably in single-use portions. Patient-to-patient spread may occur when there is a breakdown in infection control measures, such as failure to disinfect urinals and bedpans and failure of hand hygiene (Lowbury et al. 1970). In these situations rectal and other colonization usually precedes infection.

P. aeruginosa is the most common persistent pathogen in chest infection in cystic fibrosis, but the degree to which crosscolonization occurs in hospital units appears to differ widely.

Mycobacterium spp.

In the past, tuberculosis was a serious occupational risk for healthcare staff. This is no longer the case in most clinical situations where tuberculin testing and immunization, supplemented by selective chest X-ray, are available for staff in contact with patients and those handling pathological material (Joint Tuberculosis Committee of the British Thoracic Society 1994). Nosocomial acquisition of tuberculosis by patients in the developed world is also uncommon. Risk factors for transmission to patients

and staff include delayed diagnosis, drug-resistant strains and lapses in administrative, engineering, and infection control practices (Menzies et al. 1995). Multidrug-resistant tuberculosis in HIV patients is of increasing concern as presenting a risk to staff and other patients. Previous admission to hospital, failure to isolate positive patients in a single room, and the absence of positive pressure ventilation are associated with spread (Pearson et al. 1992).

The risk of infection depends on the infectivity of the primary source, the duration and proximity of the exposure, and the susceptibility of the recipient (George et al. 1986). Infectivity depends on the amount of cough and the concentration of bacteria in the sputum. Thus, 'three-smear-negative' culture-positive patients with pulmonary tuberculosis and patients with nonpulmonary tuberculosis are usually regarded as noninfective. Sputum infectivity declines rapidly during chemotherapy, including that with rifampicin, with a 99 percent decline in viability after 2 weeks, but scanty bacteria may survive for much longer (Jindani et al. 1980). Isolation for 2 weeks in a well-ventilated side ward, without recirculation of the air to other parts of the hospital, is recommended while patients are infectious (Joint Tuberculosis Committee of the British Thoracic Society 1994). Susceptible individuals are the young and those whose immunity is depressed by disease or treatment. Care should be exercised in the handling of fomites from sputum-positive patients and in protecting highly susceptible patients from exposure (George 1988). This may entail prolonged isolation of smear-positive patients and the segregation of smear-negative or nonpulmonary cases of tuberculosis. Contact tracing should be undertaken after incidents in which staff or patients were exposed to infection (Joint Tuberculosis Committee of the British Thoracic Society 1994). Control measures can halt the transmission of multidrug-resistant strains (Wenger et al. 1995). The use by healthcare staff of HEPA respirators is expensive, inconvenient, and is not believed to be cost-effective (Adal et al. 1994).

Hospital infection, including keratitis and systemic disease, caused by opportunist mycobacteria, such as *Mycobacterium chelonei*, *Mycobacterium gordonae*, and *Mycobacterium fortuitum*, has been observed (Watt 1995). Contributory factors are the ability of these mycobacteria to survive and multiply in moist environments, inadequate decontamination procedures, and poor hygienic precautions in renal dialysis, bronchoscopy, hydrotherapy, and cardiac surgery units (George 1988). Contamination of specimens by the use of tap water to rinse disinfected bronchoscopes has led to 'pseudo-outbreaks' (Nye et al. 1990). The extent to which crossinfection is responsible for the *Mycobacterium avium* complex infections seen in patients with AIDS (Chapter 46, *Mycobacterium tuberculosis* complex, *Mycobacterium leprae*, and other slow growing mycobacteria) is unknown. The *M. avium* complex may be recovered from standing water and food sources (Horsburgh 1991) and clusters of cases may therefore reflect acquisition from a common source rather than person-to-person spread.

Other bacteria

The main features of hospital infection by other bacterial species are given in Table 16.2.

Viruses

The prevalence of nosocomial viral infection is probably underestimated. Most surveys of hospital-acquired infection concentrate on bacteria and the subclinical nature of many viral illnesses together with limited diagnostic facilities partly explains why viruses do not figure prominently. Viral infections are, however, important in neonatal and pediatric patients and in immunosuppressed patients. In pediatrics, rotavirus may be the most common individual pathogen encountered.

Community-acquired infection by respiratory syncytial virus is almost universal in children in the early years of life. In hospital this is a particular danger to premature neonates and children with cardiorespiratory abnormalities. Staff and adult patients may be reinfected, with minor respiratory symptoms, but severe disease may occur in the immunodeficient adult. The virus spreads by respiratory secretions entering the eye or nose, with or without an intervening period on surfaces or fomites. The risks of infection may be reduced by a combination of cohort nursing and the wearing of gowns and gloves (Snydman et al. 1988; Madge et al. 1992). Such a proactive approach is made possible by the availability of rapid antigen detection facilities (Editorial 1992). Similar principles apply to control of the childhood exanthemata, when infected patients are admitted to hospital (Breuer and Jeffries 1990). Staff working with pregnant women should be immune to rubella. Apart from classical influenza, influenza virus A can cause a variety of syndromes in hospital patients, including a particularly dangerous form of pneumonia in the elderly. Selective immunization of patients can be helpful and a case can be made for the routine immunization of some health workers (Heimberger et al. 1995).

Viral gastroenteritis is common in neonatal and pediatric units, in general wards, and in long-stay accommodation for the elderly (Holzel and Cubitt 1982). Rotavirus is the most frequent causative agent in children. In the elderly, outbreaks of diarrhea caused by SRSV, such as noroviruses, and other agents, can affect large numbers of patients. Spread occurs via the fecal–oral and probably also via the air-borne routes. Prevention of spread may be achieved by early diagnosis, for example by electron microscopy, and strict precautions that include

wearing of aprons and gloves, strict hand-washing routines, availability of separate toilet facilities, and restriction on the movement of symptomatic patients and staff caring for them, as well as good communications between hospitals to minimize inadvertent spread (Caul 1994; Rao 1995; Chadwick et al. 2000).

Enterovirus infections, though usually subclinical, are also common in hospitals (Dowsett 1988). Besides hepatitis A, coxsackie A and B and echoviruses may cause generalized febrile infections and meningitis, particularly in the newborn, generalized illness with bowel symptoms in immunodeficient persons, and minor infections in staff. Spread may occur by the oral and respiratory routes and by hands, fomites, and such shared facilities as hydrotherapy pools. The same groups of patients are particularly susceptible to the herpes group of viruses. Cytomegalovirus, as well as being transmitted in utero and at birth from mother to baby, may be conveyed in donations of blood or tissue, and infections in the immunosuppressed often result from reactivation. Varicellazoster may be conveyed by the vesicular exudate or respiratory secretions of patients with chicken pox and, less frequently, zoster. Both diseases necessitate barrier precautions and patients must be segregated from susceptible persons, who may, when appropriate, be protected with immune globulin (Burns et al. 1998). This will also protect healthcare staff, of whom approximately 15 percent may not be immune. A register of staff immunity may be useful, especially in clinical areas where varicella infection may be more common, such as pediatric wards (Breuer and Jeffries 1990). In the USA, vaccination of susceptible healthcare staff likely to be exposed to patients at high risk of complications of chicken pox is recommended (Bolyard et al. 1998). Herpetic whitlows occasionally develop in staff not protected from exposure to the respiratory secretions of patients.

Viral hemorrhagic fevers such as Lassa, Marburg, and Ebola, have been responsible for infections of hospital staff and patients in tropical countries where equipment and staffing are inadequate and medical practice poor (Fisher-Hoch et al. 1995). In the UK, patients are categorized as minimum, moderate, or high risk. Minimum and moderate risk patients may be cared for in an isolation unit, but those in the high risk group should be tranferred to a high security infectious disease unit (Advisory Committee on Dangerous Pathogens 1996). This may not always be feasible and is in contrast to US recommendations which emphasize isolation in a single room with negative pressure ventilation and the implementation of universal precautions, as spread by the air-borne route is considered relatively rare (MMWR 1995).

For other blood-borne viruses (see section Hepatitis B and C and HIV infection in hospitals).

The control and prevention of viral infection amongst hospital staff is best managed by occupational health departments, by staff screening, immunization, and

Table 16.2 *Other bacteria[a] associated with hospital infection*

Bacteria	Source and mode of spread	Circumstances of infection	Antibiotic characteristics of hospital strains	References
Gram-positive bacteria				
Coagulase-negative staphylococci	Endogenous; 'hospital' strains by contact, air	Implants; intravascular cannulae; sternal wounds; neonates	Variable; often multi-R especially to isoxazolylpenicillins, aminoglycosides; usually S to vancomycin, rifampicin	Stillman et al. 1987; Patrick et al.
Streptococcus				
S. pyogenes (streptococcus Lancefield group A)	Endogenous; uncommon hospital spread from cases and throat, nose, skin, rectal carriers; contact, air	Wounds, burns, skin lesions; parturient women	Invariably S to benzylpenicillin; variable to erythromycin, tetracycline	Garrod 1979; Easmon 1984
S. agalactiae (streptococcus Lancefield group B)	Endogenous (feces, vaginal); occasionally contact in neonatal units	Usually congenital in neonates; postpartum infection; skin and soft tissue lesions	Consistently benzylpenpenicillin S	Easmon 1984; Zangwill et al. 1992
S. pneumoniae	Endogenous; occasionally air spread of more virulent or antibiotic-resistant strains	Pneumonia, etc., in respiratory or general impairment; children	Hospital clusters showing penicillin R or intermediate R and multi-R	Millar et al. 1994
Other streptococci: viridans or indifferent; *S. anginosus* (syn. *milleri*); β-hemolytic streptococci groups C and G	Endogenous; rarely contact	Abdominal surgery; implants; neutropenic patients	May be relatively R to β-lactams	Editorial 1984a, 1985
Corynebacteria (especially *C. jeikeium*)	Endogenous; skin colonization may precede infection	Implants; cannulae; immune impairment; antibiotic treatment	Multiple R; vancomycin S	McGowan 1988
Clostridium				
C. perfringens, etc.	Usually endogenous (feces); indirect contact spread of enteritic strains	Abdominal surgery; amputations; diabetics; elderly (enteritic)	Benzylpenicillin S	Lowbury and Lilly, 1958; Borriello and Barclay, 1986
C. tetani	In the past, environmental and operative materials; now possibly endogenous	Very rare postoperative infection	Benzylpenicillin S	Parker and Mandal 1984
Bacillus spp.	Environmental; may resist sterilization processes	Opportunistic; soft tissue and occasional systemic infection	Often R to β-lactams	Weber and Rutala 1988
Rhodococcus equi	Unknown, possibly by air from soil, etc.	Pneumonia, etc., in immunocompromised, e.g. HIV infection		Verville et al. 1994

(Continued over)

Table 16.2 *Other bacteria[a] associated with hospital infection (Continued)*

Bacteria	Source and mode of spread	Circumstances of infection	Antibiotic characteristics of hospital strains	References
Nocardia asteroides	Inanimate environment; possibly air-borne spread from cases	Respiratory tract infection; immunosuppression		Houang et al. 1980; Javaly et al. 1992
Gram-negative bacteria				
Pseudomonas spp. and similar 'environmental' gram-negative bacilli	Water; damp areas; liquid medicaments; moist equipment; fecal carriage; direct and indirect contact (hands)	Colonization more frequent than infection. Chronic disease; renal failure; immune impairment; major surgery; burns; broad spectrum antibiotic treatment; neonates; implants; cannulae; catheters; respiratory equipment	Often multi-R	Chapter 62, *Pseudomonas*
Burkholderia cepacia	Unsterilized aqueous solutions, e.g. disinfectants; nebulisers; patient-to-patient infection, probably air-borne	Various from solutions; late infection in cystic fibrosis	Often R to all classes	Pankhurst and Philpott-Howard 1996
Stenotrophomonas (Xanthomonas) maltophilia	Multiple epidemiological types; source and spread not clear	Bacteremia; device-related infections; broad spectrum antibiotics, including penems	Multi-R; characteristic R to penems	Vartivarian et al. 1994; Laing et al. 1995; Gerner-Smidt et al. 1995; Spencer 1995
Acinetobacter spp. (*A. baumannii, A. calcoaceticus*)	Endogenous; also indirect contact staff and equipment	Bacteremia; respiratory infection; ICUs; broad spectrum antibiotics	Increasingly multi-R	Bergogne-Berezin 1995; Mulin et al. 1995; Seifert et al. 1995
Achromobacter xylosoxidans		Uncommon bacteremia, respiratory infection	Aminoglycoside R and trimethoprim S typical	Schoch and Cunha 1988; Legrand and Anaissie 1992
Flavobacterium spp.		*F. meningosepticum* associated with infantile meningitis; others as *Pseudomonas* spp.	Often S to some antibiotics usually thought of as 'anti-gram-positive'	Ratner 1984
Aeromonas spp.		Bacteremia; soft tissue infection; outbreaks and invasive infection uncommon. Toxigenic strains may be enteritic		Mellersh et al. 1984; Murphy et al. 1995
Capnocytophaga spp.	Endogenous; usually from oropharynx	Bacteremia; immunosuppression, neutropenia	Aminoglycoside and trimethoprim R, β-lactam S typical	Kristensen et al. 1995

R, resistance; S, sensitivity.

a) Other than anaerobic gram-negative bacilli, *Legionella* spp., *S. aureus, Enterococcus* spp., *C. difficile, Mycobacterium* spp. and *P. aeruginosa*.

education programs, preferably working in tandem with infection control personnel.

Fungi and protozoa

Fungal infection continues to increase in hospitals (Pfaller and Wenzel 1992), as more effective and broader spectrum antibacterial agents and immunosuppressive regimens are deployed and increase the number of patients at risk. However, it remains likely that many fungal infections go undiagnosed because of a lack of awareness of risk factors and shortcomings in conventional laboratory techniques, e.g. fungemia is difficult to confirm microbiologically. As has recently been described in the HIV population, where *Candida dubliniensis* is now recognized as an important oral pathogen (Sullivan and Coleman 1998), studies combining molecular techniques and good epidemiology are likely to reveal a higher incidence of nosocomial infection and both new pathogens and new strains. The most common significant fungal isolate is *Candida albicans*, a bowel commensal that in most cases gives rise to endogenous infection. Members of other species, such as *Candida tropicalis*, *Candida parapsilosis*, *Candida krusei*, and *Candida glabrata*, are becoming more common. Their source is particularly associated with the skin, and therefore with intravascular cannulae. Less common fungi, such as *Fusarium* and *Trichosporon beigelii* are emerging causes of infection in cancer and bone marrow transplant recipients (Pfaller and Wenzel 1992). Clusters of infections with the same strain of *Candida* with evidence of colonization of the hands of staff, may occur under selective pressure from broad spectrum antibiotics (Burnie et al. 1985). Typing of *C. albicans* for epidemiological purposes is difficult. Phenotypic typing, such as biotyping, may be useful but DNA-based approaches are likely to be increasingly used. *Aspergillus* infection is related to the air-borne spread of spores, particularly when these reach high concentrations, as during building repairs near patients with neutropenia (Rogers and Barnes 1988) and occasionally other vulnerable groups, such as ICU patients (Humphreys et al. 1991). Molecular typing techniques allow the comparison of isolates from patients and the environment (Leenders et al. 1996). Clusters of infections due to non-*A. fumigatus* may indicate an environmental source (Loudon et al. 1996). Strategies for the prevention of aspergillus infection include pre-emptive antifungal therapy. Isolation of especially vulnerable patients in laminar flow ventilated facilities has been recommended.

There is no good evidence for the hospital spread of *Cryptococcus neoformans*. Phycomycetes may be transmitted to susceptible patients by dressings (Gartenberg et al. 1978) or by air (del Palacio Hernanz et al. 1983). Dermatophytes occasionally spread in geriatric units by close contact, or on clothing or apparatus (Peachey and English 1974).

The epidemiology of *Pneumocystis carinii* is obscure. Dormant infection has been said to be ubiquitous and reactivated by immune suppression, as may be infection by *Toxoplasma gondii*. On the other hand, air-borne spread of *P. carinii* can be shown in animals and there have been suggestions of clusters of linked cases (Rhame et al. 1984; Haron et al. 1988).

Cryptosporidium, a coccidian protozoon, is an important cause of debilitating diarrhea in the immunosuppressed, such as AIDS patients (Goodgame 1996). The oocysts are comparatively resistant to disinfection and may survive well in the inanimate environment. There is evidence of case-to-case spread and of transmission to staff and to other patients in hospitals (Martino et al. 1988), presumably by the fecal–oral route. A nosocomial outbreak in Denmark was linked to a ward ice machine (Ravn et al. 1991).

Scabies can be very disruptive in chronic institutions or hospital wards for the elderly (Holness et al. 1992). Infestation in patients with underlying skin disease or Norwegian scabies (extensive hyperkeratotic lesions) are especially contagious (Moberg et al. 1984). Early and reliable diagnosis, effective treatment, for example with permethrin, applied extensively to the body, and decontamination of bed linen and fomites are important to limit spread.

INVESTIGATION AND SURVEILLANCE OF INFECTION

Investigation of outbreaks

An outbreak is defined as two or more cases of symptomatic infection or of symptomatic infection and colonization, suspected on clinical or laboratory evidence. An outbreak may be obvious, such as a number of cases of acute diarrhea occurring in a ward over a few hours, or it may not be apparent for some time, such as deep staphylococcal infection after hip replacement in several patients over a period of months.

Outbreaks may be detected by simple laboratory surveillance with informal sources of clinical intelligence, but more subtle occurrences may be detected only by a more formal investigation. This should, at the same time, follow the epidemiological and the microbiological approaches. A first essential is definition of a case in clinical or microbiological terms. Outbreaks defined by an increase in the infection rate at a single body site may have a multiple bacterial etiology and the question arises whether there is in fact an outbreak. There may simply have been heightened awareness of infective episodes, a series of unlinked but significant clinical events, unrelated cases of infection by similar microorganisms or a 'pseudo-outbreak' due to specimen contamination (McGowan and Metchock 1996).

The spread of infection may be:

- from a point-source at one time
- from a point-source with continued exposure
- by person-to-person spread
- from infected persons with secondary contamination of the environment, or
- by a mixture of these.

Outbreaks should be defined epidemiologically in terms of time, place, and persons involved, and an epidemic curve should be plotted where the outbreak is relatively large (Palmer 1989). The shape of this curve and the microbiological cause may yield information for follow-up.

Small outbreaks purely of hospital concern may be dealt with by the Infection Control Team or equivalent (see Organization of hospital infection control) in association with clinical staff and local managers. More serious outbreaks and those with community implications require the involvement of those responsible for infection control in the community, i.e. local public health authorities and public health laboratories. When a major incident is found or suspected that is of potential seriousness for patients, or involves many patients or more than one hospital, or is liable to cause public concern, a formal action group or outbreak committee should be established (Hospital Infection Working Group of the Department of Health and Public Health Laboratory Service 1995). The responsibilities of such a group are to investigate and control the outbreak; to ensure that affected patients are adequately cared for, and that there are sufficient staff and supplies in the wards and laboratory; to disseminate information to patients, staff, relatives, and the general public; to record the episode, prepare a report, and to consider future prevention (Hospital Infection Working Group of the Department of Health and Public Health Laboratory Service 1995).

Depending on the nature of the outbreak (e.g. nosocomial legionellosis) or its extent (e.g. a large number of cases of hospital-acquired food poisoning), national bodies with specialized epidemiological skills such as the Health Protection Agency's Centre for Infections, Communicable Disease Surveillance Centre (UK) or Centers for Disease Control (USA) may need to be involved. Larger outbreaks, particularly those arising from a single source, whether or not microbiological indicators are available, may require formal epidemiological studies, particularly case-control studies (Rothman 1986; see also Chapter 13, Epidemiology of infectious diseases). The descriptive epidemiology may have produced hypotheses to be tested. This approach may be more satisfactory than the use of extended case-control studies to investigate a variety of possible factors. However, case-control studies provide a powerful means to study some types of hospital outbreak and may give more rapid answers than microbiological methods, particularly if microbiological diagnosis of the particular infection is difficult or slow, e.g. some forms of viral gastroenteritis. Particularly in point-source outbreaks, 'cohort' studies (Rothman 1986) may be possible. In such studies, two or more groups that differ in exposure to a postulated source of infection may be compared for incidence of the infection. An example is the identification of the responsible food consumed in a food-borne outbreak. Such approaches are less useful in many hospital outbreaks that are due to person-to-person spread.

The increasing numbers of compromised hospital patients, the greater awareness of the hospital environment as a source and means of transmission of infection, and the emergence of more resistant microbes may require the use of considerable laboratory support to identify or confirm the extent of the outbreak and to confirm that preventive measures have been effective. The successful control of an outbreak requires good clinical investigation and in some instances up-to-date typing techniques such as ribotyping, polymerase chain reaction based techniques, etc. (Emori and Gaynes 1993; Wendt and Wenzel 1996).

Epidemiological typing

Epidemiological typing, characterization of isolates to intraspecies level, is useful (Parker 1978; Pitt 1994) to define the extent of an outbreak and to elucidate the sources and spread of infection and the effectiveness of prevention. Established typing methods that permit stable and discriminatory subdivisions are available for some common hospital pathogens, such as *S. aureus* (Chapter 32, *Staphylococcus*). However, the continued appearance of new strains that are not typable by the existing systems, and the progress of newly significant species, diminish the value of classically established methods. This is a stimulus for the development of new methods, preferably those of more general applicability. The need is for simple, reproducible and discriminatory methods; fine discrimination can sometimes be attained by applying two or more typing methods sequentially. The aim is to define, with reasonable confidence, the strain responsible for an incident of infection by distinguishing all epidemiologically irrelevant isolates (Anderson and Williams 1956).

Well-established phenotypic methods of typing include serotyping, phage typing (including 'reverse' phage typing; de Saxe and Notley 1978), and bacteriocin typing. Methods in common use for more local studies include biotyping and comparison of antibiotic-sensitivity patterns, which is often the first clue to a common-source outbreak, but it is an unreliable epidemiological tool. Other methods that have proved to be of value for typing particular organisms include morphotyping of *C. albicans* (Brown-Thomsen 1968), the Dienes phenomenon for swarming strains of *Proteus* spp. (Chapter 55,

Proteus, *Morganella*, and *Providencia*) and resistotyping (Elek and Higney 1970).

Analysis of DNA and other cellular constituents may provide a means of discriminating between strains that are similar by conventional typing methods, and makes possible the 'fingerprinting' of diverse species for which standard methods have not been established. The ease with which such methods can be applied has sometimes encouraged their use without proper critical evaluation of their suitability (Pitt 1994; Struelens et al. 1996). Electrophoretic analysis of cellular proteins (total cellular proteins, outer-membrane proteins, or essential enzymes) has been applied to several species, with discrimination by pattern recognition or automated mathematical analysis. Electrophoretic demonstration of plasmids, with or without treatment by restriction endonucleases, has been widely used (Mayer 1988). Chromosomal analysis is more difficult because of the larger size of the genome and the many fragments produced by endonuclease action. Pulsed field gel electrophoresis (PFGE) overcomes this problem by separating relatively few large-sized DNA fragments created by rare-cutting restriction endonucleases. PFGE has generally become the gold standard method for fingerprinting bacteria. Nucleic acid probes can also be applied to endonuclease digests (Bingen et al. 1994). The polymerase chain reaction can be used in a wide variety of ways to characterize the genome of microorganisms (van Belkum 1994). The various epidemiological typing and 'fingerprinting' methods have been recently reviewed elsewhere (Struelens 2002), and criteria for their use and evaluation were proposed by a Study Group of the European Society for Clinical Microbiology and Infectious Diseases (Struelens et al. 1996).

Surveillance

Surveillance is the systematic observation and recording of disease. It is an active process and implies the analysis and dissemination of data, so that direct or indirect action can subsequently be taken to improve patient care. Unlike the investigation of an outbreak or epidemic, surveillance is ongoing and the results can be used as an indicator of the quality of care, e.g. surgical wound infection rates (Glenister 1993; Smyth and Emmerson 2000). In his summary of the Proceedings of the First International Conference on Nosocomial Infections, Williams (1971) stated: 'Quite clearly, the first message from this conference is the need for surveillance. It is essential that hospital staff know what is going on in the hospital'. It is important to analyze surveillance activities critically to determine their usefulness in control, rather than to undertake them for their own sake, or with a false idea of their usefulness. The increasing availability of computer databases makes the analysis of data easier, but more traditional and time-consuming methods such as a card recording

system can be used equally well. In particular, the use of optical scanning technology has been shown to improve the speed and accuracy of data entry and can greatly facilitate analysis and the dissemination of results (Smyth et al. 1997). In Britain, laboratory-based surveillance, with a variable degree of clinical reporting of infection, especially of organisms such as MRSA, has been the norm. In the USA, partly because of accreditation requirements and the fear of litigation, much staff time is devoted to the continuous gathering of data (Casewell 1980; Wenzel 1986). Justification for the latter approach comes from the Study of the Efficacy of Nosocomial Infection Control (SENIC) (Haley et al. 1985b). This large and expensive investigation determined the infection rates in 1970 and 1976, by a validated retrospective case-paper review, in a large sample of patients from 338 US hospitals. The rates were compared with the very variable infection control activities of the hospitals, expressed in terms of two indices, surveillance and control. The conclusion was that optimal measures would prevent 32 percent of hospital-acquired infections of the urinary tract, the lower respiratory tract, surgical wounds, and the bloodstream.

It should be recognized that comparisons of data collected, for example, in different institutions, are of limited value (Haley 1988). Even the most basic of surveillance activities, the detection of cases of hospital-acquired infection, is difficult and expensive, and many systems give results far below the actual rate. Furthermore, differences in case definitions between centers make reliable comparisons difficult. A comparison of different approaches, with a 'standard' of prospective identification, by an infection control physician who examined patients and records is shown in Table 16.3 (Freeman and McGowan 1981). Physician self-reporting is the least reliable method of recording infections. In the UK, Glenister (1993) assessed a number of different methods and compared these with a reference method. A laboratory-based telephone method took 3.1 h for every 100 beds, but was only 51 percent sensitive in detecting hospital-acquired infection, whereas a laboratory-based ward liaison method was 76 percent sensitive but more time-consuming.

The dissemination of the results of surveillance is crucial if a beneficial effect on patient care in the form of improved practices and reduced hospital-acquired infection rates are to result. Wound infection rates should be reported to individual surgeons (Cruse et al. 1980). The prevalence of all types of nosocomial infection fell from 10.5 to 5.6 percent over a period of 3 years following surveillance in the form of repeat prevalence surveys accompanied by improvements in infection control policies (French et al. 1989).

It is not always feasible to conduct unit or hospital-wide surveillance. Consequently, targeted or selective surveillance is recommended (Hospital Infection Working Group of the Department of Health and Public

Table 16.3 *Hospital-acquired infection: sensitivity of methods of case finding*

Method	Sensitivity
Physician self-report forms	0.14–0.34
Fever	0.47
Antibiotic use	0.48
Fever plus antibiotic use	0.59
Microbiology reports	0.33–0.65
Selected chart review with 'Kardex' clues	0.85
Total chart review	0.90
SENIC pilot project	
Prospective data collection	0.52–0.90
Retrospective chart review	0.66–0.80
Standard method	1.00

SENIC, Study on the Efficiency of Nosocomial Infection Control. Standard method, clinical detection of infection by a trained physician, who examined all the patients and reviewed all the data in respect of them, is given a sensitivity of 1.00 by definition. Modified from Freeman J., McGowan J.E. 1981. *Rev Infect Dis*, **3**, 658 (© The University of Chicago) with permission.

Health Laboratory Service 1995; Glynn et al. 1997). Targeted surveillance concentrates on areas where infection is particularly important and where it has a major impact on patient morbidity and mortality, e.g. intensive care and oncology units (Hanberger et al. 2001). Selective surveillance addresses problem areas or units or specialties with potential problems, outside the occurrence of a clearly defined epidemic, where infection rates might be higher than expected, or following the opening of a new specialist unit or facility. As the duration of hospital stay becomes shorter, it is important not to underestimate hospital infection presenting in the community after discharge. With the availability of newer typing techniques (see above), it should become easier to trace the spread of organisms that cause nosocomial infection so that preventive strategies to reduce crossinfection may be based on clearer scientific evidence. The continuing challenge for infection surveillance schemes is to give ownership of infection prevention to clinicians. Surveillance systems need to be cognizant of the time restraints and consequent issues surrounding compliance with data collection. It is therefore important that minimum datasets are used wherever possible to facilitate the process of surveillance. The mandatory *S. aureus* bacteremia scheme in England and Wales referred to above (see Bacteremia) is an example of how a simple collection system can provide easily accessible data, which in turn should allow hospitals to focus resources on problem areas and then monitor the effects.

ORGANIZATION OF HOSPITAL INFECTION CONTROL

Systems for the management of infection control in hospitals developed in parallel in Great Britain and in the USA, and essentially similar systems have been adopted in other European countries and in other parts of the world (Hambraeus 1995; Astagneau and Brucher 2001; Bassetti et al. 2001; Bijl and Vos 2001; Frank et al. 2001; Hryniewicz et al. 2001; Jepsen 2001; Lim 2001; Reybrouck et al. 2001; Rodriguez-Bano and Pascual 2001; Leblebiciogh and Unal, 2002; Melo-Cristino et al. 2002; Reed et al. 2003).

In the UK, the staff who provide infection control expertise in hospitals have been reviewed and guidance has been issued (Hospital Infection Working Group of the Department of Health and Public Health Laboratory Service 1995). Executive action in infection control, though a responsibility of the chief executive (CE) of the unit, is the function of an infection control team (ICT), which must maintain close liaison with the CE, or appointed deputy, and other managers. In its simplest form the ICT consists of an infection control doctor (ICD) and an infection control nurse (ICN) (Gardner et al. 1962). The ICD is the control of infection officer (Colebrook 1955; Subcommittee of the Central Health Services Council 1959) and the leader of the team, and may be formally appointed with responsibility to the medical consultants' committee or to the CE, and may report directly to the Strategic Health Authority or Primary Care Trusts. The ICT may be augmented by the active participation of medical microbiologists, infectious disease physicians, scientists, medical laboratory scientific officers, and others, and should be supported by an Infection Control Committee (ICC) (Hospital Infection Working Group of the Department of Health and Public Health Laboratory Service 1995).

The ICD has a wide-ranging role in hospitals and is the specialist adviser on hospital infection to the executive officers of hospital trusts and to managers and planners. The ICD is expected to develop an infection control program, policies, and procedures and to ensure that all staff are educated in infection control and that suitable audit is carried out. The ICD has the primary responsibility for surveillance and the investigation of outbreaks and advises about infectious disease in hospital staff and its prevention, including immunization policy. The ICD is the leader of the ICT and works closely with and supports the ICN. In the USA, the equivalent is the infection control epidemiologist.

The ICN is responsible for infection control matters to the ICD and undertakes similar activities either with the ICD or independently under his or her general guidance. The ICN may provide clinical surveillance, day by day and in outbreaks, and has particular responsibility for the application of policies and procedures and for the training of staff. The ICN often has close links with various ancillary departments, such as the pharmacy, domestic services, laundry, etc. The counterpart of the ICN in the USA is the infection control practitioner (Goldmann 1986). On 5 December 2003, Sir Liam Donaldson the Chief Medical Officer of England announced wide ranging proposals in an attempt to reduce the burden of healthcare-associated infections. The plan, 'Winning Ways: Working together to reduce Healthcare Associated Infection in England' means that every NHS trust will get a designated director of infection control – a senior figure with experience in the discipline with the power to impose tough new rules on each hospital in order to reduce infection levels. He or she will lead dedicated infection control teams, charged with ensuring every possible step is taken to minimize the risk of infection. In addition, the NHS watchdog, the Healthcare Commission, will be asked to make infection control a key priority when assessing hospital performance.

The infection control link nurse

All healthcare workers need to ensure that effective infection control practices are implemented in the care of patients to achieve a reduction in healthcare-associated infection. It is important for wards and directorates to develop 'ownership' of infection control. One way of assisting this is by the use of infection control link nurses (ICLN) who have been implemented by infection control teams (ICT) as a method of improving practice at a clinical level (Cooper 2001; Dawson 2003).

ICLNs are a link or intermediary between the clinical areas/wards and ICTs. A key part of their role is to provide information to assist in the early detection of outbreaks of infection and to help increase awareness of infection control issues on their ward. Attention should be drawn to ICNs of changes of practice and equipment which could have implications for infection control (Hospital Infection Working Group, Department of Health and Public Health Laboratory Service 1995; Comptroller and Auditor General, National Audit Office 2000). In some trusts they have been trained to collect surveillance data on healthcare-associated infections for the ICT (Teare and Peacock 1996). It is essential that they have sufficient clinical experience and standing to have authority with managers and colleagues. ICLNs should not be seen as a substitute for the ICN, they are a ward-based resource who act under the supervision of the ICNs as a role model for colleagues. The ideal ICLN should be a keen, enthusiastic, motivated volunteer with a special interest in infection control (Charalambous 1995). The role of ICLNs is one that is still evolving. Many trusts have used them with success and they have been of particular value in the light of controls assurance and clinical governance (Hill et al. 2001). It is important for infection control to become the responsibility of directorates and the use of ICLNs at ward level can assist with this.

The ICC, which includes the augmented ICT, together with representatives of clinical departments and service departments, is not to be confused with the Outbreak Committee (OC), but is usually the body for discussing, monitoring, and approving the activities of the ICT listed above and lends it authority. It produces an action plan for infection control. It also produces general policies, encourages education, and reviews the implementation of the measures (Jenner and Wilson 2000; Naikoba and Hayward 2001).

It is important that the hospital ICT should have strong links with the community. This connection has been responsible for preventing many outbreaks of infection both in hospitals and outside. The consultant for communicable disease control (CCDC) has a general responsibility for infection control in the whole district and provides the main communication channel for the hospital ICT on such matters. In major outbreaks with implications wider than the hospital unit both the ICD and the CCDC will be actively involved. In outbreaks of typical nosocomial infection mainly restricted to hospital patients the ICD will lead the OC, with information to the CCDC.

The present arrangements for infection control in England and Wales were reviewed by Howard (1988) for the Hospital Infection Society and of the 207 Health Districts and London postgraduate hospitals approached, 93 percent responded. They were responsible for 95 percent of 'acute' hospital beds and 85 percent of other beds. Their replies identified 264 ICDs, of whom 82 percent were medical microbiologists; ICCs had been established in 92 percent of districts; 205 ICNs were in post and 93 percent of these worked closely with a medical microbiologist who was usually the ICD.

Thus the ICT, however constituted, has responsibility for the investigation of untoward occurrences and, even more importantly, for establishing microbiologically safe conditions in the hospital, for the provision of necessary supplies, and for ensuring safe practices in the day-to-day care of patients. Written statements of policies and procedures must be provided; the ICC and ICT must oversee the implementation of these and the education of staff in their use and interpretation. For detailed recommendations, the reader is referred to reviews (Bennett and Brachman 1992; Wenzel 1993; Philpott-Howard and Casewell 1994; J. Wilson 1995). Individual procedures may be required for special units, e.g. neonatal, pediatric, obstetric, intensive therapy, dialysis, transplantation, and oncology. In addition infection control is now an important part of risk management (Farrington and Pascoe 2001) and clinical governance (Masterton and Teare 2001).

Matters to be dealt with, in many cases in liaison with other groups of staff, are listed in Table 16.4.

In spite of the uniformity of approach to infection control in Europe, practices are diverse, as was revealed during a survey of infection in intensive care units (Vincent et al. 1995). In the USA and in Europe, there is an increasing movement towards providing and

Table **16.4** *General subjects for which infection control policies may be required*

Requirements
Provision of safe air supply (to operating theaters and accommodation for immunosuppressed patients)
Safe provision of water and food
Cleaning and general hygiene
Pest control
Waste disposal
Laundry
Decontamination (cleaning, disinfection, sterilization) of all equipment and materials
Ward procedures, including dressings and minor invasive procedures, urinary catheterization
Disposal of excreta
Venepuncture and vascular cannulation
Operating theater procedures
Antibiotic use, including prophylaxis
Isolation of patients and 'barrier nursing', to prevent spread of infection ('source isolation') or to protect susceptibles ('protective isolation')
Notification of infective disease and other liaison with community officers
Laboratory and pharmacy procedures
Staff health, immunization, and carriage of pathogenic microorganisms

unifying standards in hospital infection control against which practice can be judged and audit can be carried out (Infection Control Standards Working Party 1993; Cookson 1995; Jepsen 1995; McGowan 1995). It has been suggested that these should be 'outcome' rather than 'process' oriented (McGowan 1995). Although the methods available for measuring health benefit are limited, improving the health of the population by reducing the morbidity, mortality, and use of resources incurred by healthcare-associated infection must remain the aim of all the systems outlined above (Ward et al. 1997; Plowman et al. 1999; Comptroller and Auditor General, National Audit Office 2000). The National Audit Office has recently reviewed the extent of progress in the delivery of infection control in England since its original 2000 report (Comptroller and Auditor General, National Audit Office, 2004).

REFERENCES

Aavitsland, P., Stormark, M. and Lystad, A. 1992. Hospital-acquired infections in Norway. A national prevalence survey in 1991. *Scand J Infect Dis*, **24**, 477–83.

Adal, K.A., Anglim, A.M., et al. 1994. The use of high-efficiency particulate air-filter respirators to protect hospital workers from tuberculosis. *N Engl J Med*, **331**, 169–73.

Adams, B.G. and Marrie, T.J. 1982. Hand carriage of gram-negative rods may not be transient. *J Hyg*, **89**, 33–46.

Advisory Committee on Dangerous Pathogens. 1994. *Precautions for work with human and animal transmissible spongiform encephalopathy*. London: HMSO.

Advisory Committee on Dangerous Pathogens. 1996. *Management and control of viral haemorrhagic fevers*. London: HMSO.

Albert, R.K. and Condie, F. 1981. Hand-washing patterns in medical intensive-care units. *N Engl J Med*, **304**, 1465–6.

Alder, V.G., Burman, D., et al. 1980. Comparison of hexachlorophane and chlorhexidine powders in neonatal infection. *Arch Dis Child*, **55**, 277–80.

Allen, K.D., Bartzokas, C.A., et al. 1987. Acquisition of endemic Pseudomonas aeruginosa in an intensive therapy unit. *J Hosp Infect*, **10**, 156–64.

Allen, K.D., Ridgway, E.J. and Parsons, L.A. 1994. Hexachlorophane powder and neonatal staphylococcal infection. *J Hosp Infect*, **27**, 29–33.

Amin, M. 1992. Antibacterial prophylaxis in urology: a review. *Am J Med*, **92**, Suppl. 4A, 114S–7S.

An, Y.H. and Friedman, R.J. 1996. Prevention of sepsis in total joint arthroplasty. *J Hosp Infect*, **33**, 93–108.

Anderson, E.S. and Williams, R.E.O. 1956. Bacteriophage typing of enteric pathogens and staphylococci and its use in epidemiology. *J Clin Pathol*, **9**, 94–127.

Anon. 1988. Cleaning and disinfection of equipment for gastrointestinal flexible endoscopy: interim recommendations of a Working Party of the British Society of Gastroenterology. *Gut*, **29**, 1134–51.

Anonymous. Outbreaks of community-associated methicillin-resistant *Staphylococcus aureus* skin infections – Los Angeles County, California. *MMWR* 2002–2003 02/07/2003; 52: 88. Available at: http://www.cdc.gov/mmwr/preview/mmwrhtml/mm5205a4.htm. Last accessed 27 November 2003.

Archer, G.L. and Armstrong, B.C. 1983. Alteration of staphylococcal flora in cardiac surgery patients receiving antibiotic prophylaxis. *J Infect Dis*, **147**, 642–9.

Armstrong, D., Neu, H., et al. 1995. The prospects of treatment failure in the chemotherapy of infectious disease in the 1990s. *Microb Drug Resist*, **1**, 1–4.

Arnow, P.M., Chou, T., et al. 1982. Nosocomial Legionnaires' disease caused by aerosolized tap water from respiratory devices. *J Infect Dis*, **146**, 460–7.

Astagneau, P. and Brucher, G. 2001. Organisation of hospital-acquired infection control in France. *J Hosp Infect*, **47**, 84–7.

Augustin, A.K., Simor, A.E., et al. 1995. Outbreaks of gastroenteritis due to Norwalk-like virus in two long-term care facilities. *Can J Infect Control*, **10**, 111–13.

Austrian, R. 1994. Confronting drug-resistant pneumococci. *Ann Intern Med*, **121**, 807–9.

Ayliffe, G.A.J. 1980. The effect of antibacterial agents on the flora of the skin. *J Hosp Infect*, **1**, 111–24.

Ayliffe, G.A.J. 1986. Nosocomial infection – the irreducible minimum. *Infect Control*, **7**, Suppl., 92–5.

Ayliffe, G.A.J. 1991. Role of the environment of the operation suite in surgical wound infection. *Rev Infect Dis*, **10**, Suppl. 10, S800–4.

Ayliffe, G.A.J. 1996. *Recommendations for the control of methicillin-resistant* Staphylococcus aureus *(MRSA)*. Geneva: World Health Organization.

Ayliffe, G. 2000. Decontamiantion of minimally invasive surgical endoscopes and accessories. *J Hosp Infect*, **45**, 263–77.

Ayliffe, G.A.J. and English, M.P. 2003. *Hospital infection: from miasmas to MRSA*. Cambridge: Cambridge University Press.

Ayliffe, G.A.J. and Lawrence, J.C. 1985. Infection in burns. *J Hosp Infect*, **6**, Suppl. B, 1–66.

Ayliffe, G.A.J. and Lilly, H.A. 1985. Cross infection and its prevention. *J Hosp Infect*, **6**, Suppl. B, 47–57.

Ayliffe, G.A.J. and Lowbury, E.J.L. 1982. Airborne infection in hospital. *J Hosp Infect*, **3**, 217–40.

Ayliffe, G.A.J., Collins, B.J., et al. 1967. Ward floors and other surfaces as reservoirs of hospital infection. *J Hyg Camb*, **65**, 515–36.

Ayliffe, G.A.J., Lilly, H.A. and Lowbury, E.J.L. 1979. Decline of the hospital staphylococcus? Incidence of multiresistant *Staph. aureus* in three Birmingham hospitals. *Lancet*, **1**, 538–41.

Babb, J.R. and Bradley, C.R. 1995. Endoscope decontamination: where do we go from here? *J Hosp Infect*, **30**, Suppl., 543–51.

Babb, J.R., Davies, J.G. and Ayliffe, G.A.J. 1983. Contamination of protective clothing and nurses' uniforms in an isolation ward. *J Hosp Infect*, **4**, 149–57.

Bailey, I.S., Karran, S.E., et al. 1992. Community surveillance of complications after hernia surgery. *Br Med J*, **304**, 469–71.

Baine, W.B., Gangarosa, E.J., et al. 1973. Institutional salmonellosis. *J Infect Dis*, **128**, 357–60.

Barnard, D.L., Hart, J.C., et al. 1973. Outbreak in Britain of conjunctivitis xcaused by adenovirus type 8, and its epidemiology and control. *Br Med J*, **2**, 165–9.

Barrett, S.P. and Spencer, R.C. 2002. MRSA bacteraemia surveillance scheme in England. *J Hosp Infect*, **50**, 241–2.

Barrett, S.P., Teare, E.L. and Sage, R. 1993. Methicillin resistant *Staphylococcus aureus* in three adjacent Health Districts of south-east England 1986–91. *J Hosp Infect*, **24**, 313–25.

Barrigar, D.L., Flagel, D.C. and Upshur, R.E. 2001. Hepatitis B virus infected physicians and disclosure of transmission risks to patients: a critical analysis. *BMC Med Ethics*, **2**, 4 (Epub 2001, October 25).

Bartlett, C.L.R., Macrae, A.D. and Macfarlane, J.T. (eds) 1986. *Legionella infections*. London: Arnold, pp. 92, 100.

Baselski, V.S. and Wunderink, R.G. 1994. Bronchoscopic diagnosis of pneumonia. *Clin Microbiol Rev*, **7**, 533–58.

Bassett, D.C.J., Stokes, K.J. and Thomas, W.R.G. 1970. Wound infection with *Pseudomonas multivorans*. *Lancet*, **1**, 1188–91.

Bassetti, M., Topal, J., et al. 2001. The organisation of infection control in Italy. *J Hosp Infect*, **48**, 83–5.

Bayston, R. 1985. Hydrocephalus shunt infections and their treatment. *J Antimicrob Chemother*, **15**, 259–61.

Beattie, P.G., Rings, T.R., et al. 1994. Risk factors for wound infection following caesarean section. *Aust NZ J Obstet Gynaecol*, **34**, 398–402.

Beck-Sagué, C., Dooley, S.W., et al. 1992. Hospital outbreak of multidrug-resistant *Mycobacterium tuberculosis* infections. Factors in transmission to staff and HIV-infected patients. *JAMA*, **268**, 1280–6.

Bennett, J.V. and Brachman, P.S. 1992. *Hospital infections*, 3rd edn. Boston: Little Brown.

Bergogne-Berezin, E. 1995. The increasing significance of outbreaks of *Acinetobacter* spp: the need for control and new agents. *J Hosp Infect*, **30**, Suppl., 441–52.

Berk, S.L. and Verghese, A. 1989. Emerging pathogens in nosocomial pneumonia. *Eur J Clin Microbiol Infect Dis*, **8**, 11–14.

Bijl, D. and Vos, A. 2001. Infection control in the Netherlands. *J Hosp Infect*, **47**, 169–72.

Bingen, E., Denamur, E. and Elion, J. 1994. Use of ribotyping in epidemiological surveillance of nosocomial outbreaks. *Clin Microbiol Rev*, **7**, 311–27.

Blot, F., Nitenberg, G., et al. 1999. Diagnosis of catheter-related bacteraemia: a prospective comparison of the time to positivity of hub-blood versus peripheral-blood cultures. *Lancet*, **354**, 1071–7.

Blowers, R., Mason, G.A., et al. 1955. Control of wound infection in a thoracic surgery unit. *Lancet*, **2**, 786–94.

Blyth, D., Keenlyside, D., et al. 1998. Environmental contamination due to methicillin resistant *Staphylococcus aureus* (MRSA). *J Hosp Infect*, **38**, 67–70.

Board of Science and Education. 1996. *A guide to hepatitis C*. London: British Medical Association.

Bolyard, E.A., Tablan, O.C., et al. 1998. Guidelines for infection control in healthcare personnel. *Infect Control Hosp Epidemiol*, **19**, 407–63.

Bone, R.C. 1991. The pathogensis of sepsis. *Ann Intern Med*, **115**, 457–69.

Booth, C.M., Matukas, L.M., et al. 2003. Clinical features and short-term outcomes of 144 patients with SARS in the greater Toronto area. *JAMA*, **289**, 2801–9.

Borriello, S.P. and Barclay, F.E. 1986. An in-vitro model of colonisation resistance to *Clostridium difficile* infection. *J Med Microbiol*, **21**, 299–309.

Bradley, C.R. and Babb, J.R. 1995. Endoscope decontamination: automated vs manual. *J Hosp Infect*, **30**, Suppl., 537–42.

Braimbridge, M.W.V. and Eykyn, S.J. 1987. Prosthetic valve endocarditis. *J Antimicrob Chemother*, **20**, Suppl. A, 173–80.

Braude, A.I., Carey, F.J. and Siemienski, J. 1955. Studies of bacterial transfusion reactions from refrigerated blood: the properties of cold-growing bacteria. *J Clin Invest*, **34**, 311–25.

Breuer, J. and Jeffries, D.J. 1990. Control of viral infections in hospitals. *J Hosp Infect*, **16**, 191–221.

Brown, E.M. 1993. Antimicrobial prophylaxis in neurosurgery. *J Antimicrob Chemother*, **31**, Suppl. B, 49–63.

Brown-Thomsen, J. 1968. Variability in *Candida albicans* (Robin) Berkhout studies on morphology and biochemical activity. *Hereditas*, **60**, 355–98.

Brumfitt, W. and Hamilton-Miller, J. 1989. Methicillin-resistant *Staphylococcus aureus*. *N Engl J Med*, **320**, 1188–96.

Brun-Buisson, C., Doyon, F., et al. 1995. Incidence, risk factors, and outcome of severe sepsis and septic shock in adults. *JAMA*, **274**, 968–74.

Bryan, C.S., Hornung, C.A., et al. 1986. Endemic bacteremia in Columbia, South Carolina. *Am J Epidemiol*, **123**, 113–27.

Buchholz, H.W., Elson, R.A., et al. 1981. Management of deep infection of total hip replacement. *J Bone Joint Surg [Br]*, **63**, 342–53.

Burke, J.F. 1961. The effective period of preventive antibiotic action in experimental incisions and dermal lesions. *Surgery*, **50**, 161–8.

Burke, J.P., Jacobson, J.A., et al. 1983. Evaluation of daily neonatal care with poly-antibiotic ointment in prevention of urinary catheter-associated bacteriuria. *J Urol*, **129**, 331–4.

Burnie, J.P. 1986. *Candida* and hands. *J Hosp Infect*, **8**, 1–4.

Burnie, J.P., Odds, F.C., et al. 1985. Outbreak of systemic *Candida albicans* in intensive care unit caused by cross-infection. *Br Med J*, **290**, 746–8.

Burns, S.M., Mitchell-Heggs, N. and Carrington, D. 1998. Occupational and infectional control aspects of varicella. *J Infect*, **36**, Suppl. 1, 73–8.

Burwen, D.R., Banerjee, S.N. and Gaynes, R.P. 1994. Ceftazidime resistance among selected nosocomial gram-negative bacilli in the United States. National Nosocommial Infections Surveillance System. *J Infect Dis*, **170**, 1622–5.

Buxton, A.E., Highsmith, A.K., et al. 1979. Contamination of intravenous infusion fluid: effects of changing administration sets. *Ann Intern Med*, **90**, 764–8.

Byrne, D.J., Lynch, W., et al. 1994. Wound infection rates: the importance of definition and post-discharge wound surveillance. *J Hosp Infect*, **26**, 37–43.

Cafferkey, M.T., Conneely, B., et al. 1980. Post-operative urinary infection and septicaemia in urology. *J Hosp Infect*, **1**, 315–20.

Cafferkey, M.T., Falkiner, F.R., et al. 1982. Antibiotics for the prevention of septicaemia in urology. *J Antimicrob Chemother*, **9**, 471–7.

Cairns, H. 1939. Bacterial infection during intracranial operations. *Lancet*, **1**, 1193–8.

Carter, A.O., Borczyk, A.A. and Carlson, J.A.K. 1987. A severe outbreak of Escherichia coli O157:H7-associated hemorrhagic colitis in a nursing home. *N Engl J Med*, **317**, 1496–500.

Cartmill, T.D.I., Panigrahi, H., et al. 1994. Management and control of a large outbreak of diarrhoea due to *Clostridium difficile*. *J Hosp Infect*, **27**, 1–15.

Casewell, M.W. 1980. Surveillance of infection in hospitals. *J Hosp Infect*, **1**, 293–7.

Casewell, M.W. 1982. Bacteriological hazards of contaminated enteral feeds. *Hosp Infect*, **3**, 329–31.

Casewell, M.W. 1995. New threats to the control of methicillin-resistant *Staphylococcus aureus*. *J Hosp Infect*, **30**, Suppl., 465–71.

Casewell, M.W. and Desai, N. 1983. Survival of multiply-resistant *Klebsiella aerogenes* and other gram-negative bacilli on finger-tips. *J Hosp Infect*, **4**, 350–60.

Casewell, M.W. and Hill, R.L.R. 1986. The carrier state: methicillin-resistant *Staphylococcus aureus*. *J Antimicrob Chemother*, **18**, Suppl. A, 1–12.

Casewell, M. and Phillips, I. 1977. Hands as a route of transmission for *Klebsiella* species. *Br Med J*, **2**, 1315–17.

Casewell, M.W. and Phillips, I. 1978. Food as a source of *Klebsiella* species for colonisation and infection of intensive care patients. *J Clin Pathol*, **31**, 845–9.

Caul, E.O. 1994. Small round structured viruses: airborne transmission and hospital control. *Lancet*, **1**, 1240–2.

Centers for Disease Control. 1987. Recommendations for prevention of HIV in health care settings. *MMWR*, **36**, Suppl. 2S, 3S–18S.

Chadwick, P.R., Chadwick, C.D. and Oppenheim, B.A. 1996. Report of a meeting on the epidemiology and control of glycopeptide-resistance enterococci. *Hosp Infect*, **33**, 83–92.

Chadwick, P.R., Beards, G., et al. 2000. Management of hospital outbreaks of gastro-enteritis due to small round structural viruses. *J Hosp Infect*, **45**, 1–10.

Charalambous, L. 1995. Development of the link nurse role in clinical settings. *Nurs Times*, **91**, 36–7.

Charnley, J. and Eftekhar, N. 1969. Postoperative infection in total prosthetic hip replacement arthroplasty of the hip joint. With special reference to the bacterial content of the air of the operating room. *Br J Surg*, **56**, 641–9.

Chastre, J., Fagon, J.-Y., et al. 1995. Evaluation of bronchoscopic techniques for the diagnosis of nosocomial pneumonia. *Am J Respir Crit Care Med*, **152**, 231–40.

Chaudhuri, A.K. 1993. Infection control in hospitals: has its quality-enhancing and cost-effective role been appreciated? *J Hosp Infect*, **25**, 1–6.

Chavers, L.S., Moser, S.A., et al. 2003. Vancomycin-resistant enterococci: 15 years and counting. *J Hosp Infect*, **53**, 159–71.

Cheesbrough, J.S., Finch, R.G. and Macfarlane, J.T. 1984. The implications of intravenous cannulae incorporating a valved injection side port. *J Hyg*, **93**, 497–504.

Cimochowski, G.E., Harostock, M.D., et al. 2001. Intranasal mupirocin reduces sternal wound infection after open heart surgery in diabetics and nondiabetics. *Ann Thorac Surg*, **71**, 1572–8.

Clarke, S.K.R., Dalgleish, P.G. and Gillespie, W.A. 1952. Hospital cross-infections with staphylococci resistant to several antibiotics. *Lancet*, **1**, 1132–4.

Coello, R., Glenister, H., et al. 1993. The cost of infection in surgical patients: a case-control study. *J Hosp Infect*, **25**, 239–50.

Colebrook, L. 1955. Infection acquired in hospital. *Lancet*, **2**, 885–91.

Collin, J., Tweedle, D.E.F., et al. 1973. Effect of a Millipore filter on complications of intravenous infusions: a prospective clinical trial. *Br Med J*, **4**, 456–8.

Collins, B.J. 1988. The hospital environment: how clean should a hospital be? *J Hosp Infect*, **11**, Suppl. A, 53–6.

Colville, A., Crowley, J., et al. 1993. Outbreak of Legionnaires' disease at University Hospital, Nottingham. Epidemiology, microbiology and control. *Epidemiol Infect*, **110**, 105–16.

Combined Working Party of the British Society for Antimicrobial Chemotherapy and the Hospital Infection Society. 1995. Guidelines on the control of methicillin-resistant *Staphylococcus aureus* in the community. *J Hosp Infect*, **31**, 1–12.

Combined Working Party of the British Society for Antimicrobial Chemotherapy, the Hospital Infection Society and the Infection Control Nurses Association. 1998. Revised guidelines for the control of methicillin-resistant *Staphylococcus aureus* infection in hospitals. *J Hosp Infect*, **39**, 253–990.

Committee of Inquiry. 1986. *Report of the committee of inquiry into an outbreak of food poisoning at the Stanley Royd Hospital, Wakefield*. London: HMSO.

Comptroller and Auditor General. 2000. *The management and control of hospital acquired infection in acute NHS trusts in England*. London: National Audit Office.

Comptroller and Auditor General. 2004. *Improving patient care by reducing the risk of hospital acquired infection: a progress report*. London: National Audit Office.

Cooke, E.M., Kumar, P.J., et al. 1970. Hospital food as a possible source of *Escherichia coli* in patients. *Lancet*, **1**, 436–7.

Cooke, E.M., Sazegar, T., et al. 1980. *Klebsiella* species in hospital food and kitchens: a source of organisms in the bowel of patients. *J Hyg*, **84**, 97–101.

Cookson, B.D. 1990. Mupirocin resistance in staphylococci. *J Antimicrob Chemother*, **25**, 497–503.

Cookson, B.D. 1995. Progress with establishing and implementing standards for infection control in the UK. *J Hosp Infect*, **30**, Suppl., 69–75.

Cooper, T. 2001. Educational theory into practice: development of an infection control link nurse programme. *Nurs Educ Pract*, **1**, 35–41.

Coovadia, Y.M., Johnson, A.P., et al. 1992. Multiresistant *Klebsiella pneumoniae* in a neonatal nursery: the importance of maintenance of infection control policies and procedures in the prevention of outbreaks. *J Hosp Infect*, **22**, 197–205.

Coria-Jimenez, R. and Ortiz-Torres, C. 1994. Aminoglycoside resistance patterns of *Serratia marcescens* strains of clinical origin. *Epidemiol Infect*, **112**, 125–31.

Correa, L. and Pittet, D. 2000. Problems and solutions in hospital-acquired bacteraemia. *J Hosp Infect*, **46**, 89–95.

Cox, R.A., Conquest, C., et al. 1995. A major outbreak of methicillin-resistant *Staphylococcus aureus* caused by a new phage-type (EMRSA-16). *J Hosp Infect*, **29**, 87–106.

Craven, D.E., Steger, K.A., et al. 1992. Nosocomial pneumonia: epidemiology and infection control. *Intensive Care Med*, **18**, Suppl. 1, S3–9.

Crookshank, F.G. 1910. Control of scarlet fever. *Lancet*, **1**, 477–80.

Cruickshank, R. 1935. The bacterial infection of burns. *J Pathol Bacteriol*, **41**, 367–9.

Cruickshank, R. 1944. Hospital infection: a historical review. *Br Med Bull*, **2**, 272–6.

Cruse, P.J.E. and Foord, R. 1973. A five-year prospective study of 23,649 surgical wounds. *Arch Surg*, **107**, 206–10.

Cruse, P.J.E., Foord, R., et al. 1980. The epidemiology of wound infection. A 10-year prospective study of 62,939 wounds. *Surg Clin North Am*, **60**, 27–40.

Curie, K., Speller, D.C.E., et al. 1978. A hospital epidemic caused by gentamicin-resistant *Klebsiella aerogenes*. *J Hyg*, **80**, 115–23.

Dal Nogare, A.R. 1994. Nosocomial pneumonia in the medical and surgical patient. Risk factors and primary management. *Med Clin North Am*, **78**, 1081–90.

Dance, D.A.B., Pearson, A.D., et al. 1987. A hospital outbreak caused by a chlorhexidine and antibiotic-resistant *Proteus mirabilis*. *J Hosp Infect*, **10**, 10–16.

Dancer, S.J. 1999. Mopping up hospital infection. *J Hosp Infect*, **43**, 85–100.

Darbyshire, P.J., Weightman, N.C. and Speller, D.C.E. 1985. Problems associated with indwelling central venous catheters. *Arch Dis Child*, **60**, 129–34.

Darnowski, S., Gordon, M. and Simor, A. 1991. Two years of infection surveillance in a geriatric long-term care facility. *Am J Infect Control*, **19**, 185–90.

Darouiche, R.O., Raad, I.I., Catheter Study Group, et al. 1999. A comparison of two antimicrobial-impregnated central venous catheters. *N Engl J Med*, **340**, 1–8.

Das, I. and Fraise, A.P. 1997. How useful are microbial filters in respiratory apparatus? *J Hosp Infect*, **37**, 263–72.

Das, I. and Symonds, J.M. 1996. Endophthalmitis. *Rev Med Microbiol*, **7**, 133–42.

Davies, A.J. and Bullock, D.W. 1981. *Pseudomonas aeruginosa* in two special care baby units – patterns of colonization and infection. *J Hosp Infect*, **2**, 241–7.

Davies, A.J., Hawkey, P.M., et al. 1984. Pneumococcal cross-infection in hospital. *Br Med J (Clin Res Ed)*, **288**, 1195.

Davies, A.J., Desai, H.N., et al. 1987. Does instillation of chlorhexidine into the bladder of catheterized geriatric patients help reduce bacteriuria? *J Hosp Infect*, **9**, 72–5.

Davies, J., Babb, J.R., et al. 1978. Disinfection of the skin of the abdomen. *Br J Surg*, **65**, 855–8.

Dawson, S.J. 2003. The role of the infection control link nurse. *J Hosp Infect*, **54**, 251–7.

de Jonge, E., Schultz, M.J., et al. 2003. Effects of selective decontamination of digestive tract on mortality and acquisition of resistant bacteria in intensive care: a randomised controlled trial. *Lancet*, **362**, 1011–16.

de Louvois, J. 1986. Necrotising enterocolitis. *J Hosp Infect*, **7**, 4–12.

del Palacio Hernanz, A., Fereres, S., et al. 1983. Nosocomial infection by *Rhizomucor pusillus* in a clinical haematology unit. *J Hosp Infect*, **4**, 45–9.

de Saxe, M.J. and Notley, C.M. 1978. Experiences with the typing of coagulase-negative staphylococci and micrococci. *Zentralbl Bakteriol [Orig A]*, **241**, 46–59.

De Souza, S.W., Lewis, D.M., et al. 1975. Hexachlorophane dusting powder for newborn infants. *Lancet*, **1**, 860–1.

Dharan, S. and Pittet, D. 2002. Environmental controls in operating theatres. *J Hosp Infect*, **51**, 79–84.

Dieckhaus, K.D. and Cooper, B.W. 1998. Infection control concepts in critical care. *Crit Care Med*, **14**, 55–70.

Dineen, P. and Drusin, L. 1973. Epidemics of postoperative infection associated with hair carriers. *Lancet*, **2**, 1157–9.

Djuretic, T., Wall, P.G., et al. 1996. General outbreaks of infectious intestinal disease in England and Wales 1992 to 1994. *CDR Rev*, **6**, R57–63.

Doebbeling, B.N., Stanley, G.L., et al. 1992. Comparative efficacy of alternative hand-washing agents in reducing nosocomial infections in intensive care units. *N Engl J Med*, **327**, 88–93.

Donati, L., Scamazzo, F., et al. 1993. Infection and antibiotic therapy in 4000 burned patients treated in Milan, Italy, between 1976 and 1988. *Burns*, **19**, 345–8.

Doring, G., Horz, M., et al. 1993. Molecular epidemiology of *Pseudomonas aeruginosa* in an intensive care unit. *Epidemiol Infect*, **110**, 427–36.

Dowsett, E.G. 1988. Human enteroviral infections. *J Hosp Infect*, **11**, 103–15.

Drosten, C., Gunther, S., et al. 2003. Identification of a novel coronavirus in patient with severe acute respiratory syndrome. *N Engl J Med*, **348**, 1967–76.

Dufour, P., Gillet, Y., et al. 2002. Community acquired methicillin-resistant *Staphylococcus aureus* infections in France: emergence of a single clone that produces Panton–Valentine leukocidin. *Clin Infect Dis*, **35**, 819–24.

Duggan, J., O'Connell, D., et al. 1993. Causes of hospital-acquired septicaemia – a case control study. *Q J Med*, **86**, 479–83.

Dukes, C. 1929. Urinary infections after excision of the rectum: their cause and prevention. *Proc R Soc Med*, **22**, 259–69.

Duma, R.J., Warner, J.F. and Dalton, H.P. 1971. Septicemia from intravenous infusion. *N Engl J Med*, **284**, 257–60.

Dunn, D.L. and Simmons, R.L. 1984. The role of anaerobic bacteria in intra-abdominal infections. *Rev Infect Dis*, **6**, Suppl. 1, S139–46.

Easmon, C.S.F. 1984. What is the role of beta-haemolytic streptococcal infection in obstetrics? Discussion paper. *J R Soc Med*, **77**, 302–8.

Editorial. 1982. Ambulatory peritonitis. *Lancet*, **1**, 1104–105.

Editorial. 1984a. Group G streptococci. *Lancet*, **1**, 144.

Editorial. 1984b. Blood transfusion, haemophilia and AIDS. *Lancet*, **2**, 1433–35.

Editorial. 1985. *Streptococcus milleri*, pathogen in various guises. *Lancet*, **2**, 1403–404.

Editorial. 1992. Nosocomial infection with respiratory syncytial virus. *Lancet*, **340**, 1071–72.

Editorial. 1996. Thoughtful drug use can overcome antibiotic resistance. *ASM News*, **62**, 12–13.

Elek, S.D. and Higney, L. 1970. Resistogram typing, a new epidemiological tool: application to *Escherichia coli*. *J Med Microbiol*, **3**, 103–10.

Emmerson, A.M. 1995. The impact of surveys on hospital infection. *J Hosp Infect*, **30**, Suppl., 421–40.

Emmerson, A.M., Enstone, J.E., et al. 1996. The Second National Prevalence Survey of Infection in Hospitals – overview of the results. *J Hosp Infect*, **32**, 175–90.

Emori, T.G. and Gaynes, R.P. 1993. An overview of nosocomial infections, including the role of the microbiology laboratory. *Clin Microbiol Rev*, **6**, 428–42.

EPINE (Working Group). 1992. Prevalence of hospital-acquired infections in Spain. *J Hosp Infect*, **20**, 1–13.

EPINE (Groupo de Trabajo EPINE). 1995. *Prevalencia de las Infecciones Nosocomiales en los Hospitales Españoles*, Sociedad Española de Hygiene y Medicina Preventiva Hospitalanas y Grupo de Trabajo EPINCAT. Barcelona: Grafimed Publicidad.

Ersahin, Y., Mutluer, S. and Guzelbag, E. 1994. Cerebrospinal fluid shunt infections. *J Neurosurg Sci*, **38**, 161–5.

Esmonde, T.F.G., Will, R.G., et al. 1993. Creutzfeldt–Jakob disease and blood transfusion. *Lancet*, **341**, 205–7.

Esteban, J.I., Gomez, J., et al. 1996. Transmission of hepatitis C virus by a cardiac surgeon. *N Engl J Med*, **334**, 555–60.

Expert Advisory Group on AIDS and the Advisory Group on Hepatitis. 1998. *Guidance for clinical health care workers: protection against infection with blood-borne viruses. Recommendations of the Expert Advisory Group on AIDS and the Advisory Group on Hepatitis.* UK Health Departments.

Expert Advisory Group on AIDS. 2000. *HIV post exposure prophylaxis: Guidance from the UK Chief Medical Officers' Expert Advisory Group on AIDS.* UK Health Departments.

Falkiner, F.R. 1993. The insertion and management of indwelling urethral catheters – minimizing the risk of infection. *J Hosp Infect*, **25**, 79–90.

Fallon, R.J. 1994. How to prevent an outbreak of Legionnaires' disease. *J Hosp Infect*, **27**, 247–56.

Fantry, G.T., Zheng, Q.X. and James, S.P. 1995. Conventional cleaning and disinfection techniques eliminate the risk of endoscopic transmission of *Helicobacter pylori*. *Am J Gastroenterol*, **90**, 227–32.

Farrington, M. 1986. The prevention of wound infection after coronary artery bypass surgery. *J Antimicrob Chemother*, **18**, 656–9.

Farrington, M. and Pascoe, G. 2001. Risk management and infection control – time to get our priorities right in the United Kingdom. *J Hosp Infect*, **47**, 19–24.

Farrington, M., Webster, M., et al. 1985. Study of cardiothoracic wound infection at St Thomas's Hospital. *Br J Surg*, **72**, 759–62.

Fawley, W.N. and Wilcox, M.H. 2001. Molecular typing of endemic *Clostridium difficile* infection. *Epidemiol Infect*, **126**, 343–50.

Faxelius, G. and Ringertz, S. 1987. Neonatal septicemia in Stockholm. *Eur J Clin Microbiol*, **6**, 262–5.

Feeley, T.W., Du Moulin, G.C., et al. 1975. Aerosol polymyxin and pneumonia in seriously ill patients. *N Engl J Med*, **293**, 471–5.

Fink, J.B., Krause, S.A., et al. 1998. Extending ventilator circuit change interval beyond 2 days reduces the likelihood of ventilator-associated pneumonia. *Chest*, **113**, 405–11.

Finkelstein, R., Reinhertz, G., et al. 1993. Outbreak of *Candida tropicalis* fungemia in a neonatal intensive care unit. *Infect Control Hosp Epidemiol*, **14**, 587–90.

Finn, A., Anday, E. and Talbot, G.H. 1988. An epidemic of adenovirus 7a infection in a neonatal nursery: course, morbidity and management. *Infect Control Hosp Epidemiol*, **9**, 398–404.

Fisch, A., Salvanet, A., et al. 1991. Epidemiology of infective endophthalmitis in France. *Lancet*, **338**, 1373–6.

Fisher, C.J., Agosti, J.M., et al. 1996. Treatment of septic shock with the tumor necrosis factor receptor: Fc fusion protein. *N Engl J Med*, **334**, 1697–701.

Fisher-Hoch, S.P., Smith, M.G. and Colbourne, J.S. 1982. *Legionella pneumophila* in hospital hot water cylinders. *Lancet*, **1**, 1073.

Fisher-Hoch, S.P., Tomori, O., et al. 1995. Review of cases of nosocomial Lassa fever in Nigeria: the high price of poor medical practice. *Br Med J*, **311**, 857–9.

Fletcher, C. 1984. First clinical use of penicillin. *Br Med J*, **289**, 1721–3.

Flügge, C. 1897. Ueber Luftinfektion. *Z Hyg Infektionskr*, **25**, 179–224.

Flügge, C. 1899. Die Verbreitung der Phthise durch staubförmiges Sputum und durch beim Husten verspritzte Tröpfchen. *Z Hyg Infektionskr*, **30**, 107–24.

Fraise, A.P., Mitchell, K., et al. 1997. Methicillin-resistant *Staphylococcus aureus* (MRSA) in nursing homes in a major UK city: An anonymised point prevalence survey. *Epidemiol Infect*, **118**, 1–5.

Frank, U., Gastmeier, P., et al. 2001. The organisation of infection control in Germany. *J Hosp Infect*, **49**, 9–13.

Fraser, I. 1984. Penicillin: early clinical trials. *Br Med J*, **289**, 1723–5.

Fraser, V.J., Zuckerman, G., et al. 1993. A prospective randomized trial comparing manual and automated endoscope disinfection methods. *Infect Control Hosp Epidemiol*, **14**, 383–9.

Freeman, J. and Hutchison, G.B. 1980. Prevalence, incidence and duration. *Am J Epidemiol*, **112**, 707–23.

Freeman, J. and McGowan, J.E. 1981. Methodologic issues in hospital epidemiology. 1. Rates, case-finding and interpretation. *Rev Infect Dis*, **3**, 658–67.

Freeman, J., Rosner, B.A. and McGowan, J.E. 1979. Adverse effects of nosocomial infections. *J Infect Dis*, **140**, 732–40.

Freeman, R. 1995. Prevention of prosthetic valve endocarditis. *J Hosp Infect*, **30**, Suppl., 44–53.

French, G.L., Cheng, A.F.B., et al. 1989. Repeated prevalence surveys for monitoring effectiveness of hospital infection control. *Lancet*, **2**, 1021–3.

French, G.L., Cheng, A.F.B., et al. 1990. Hong Kong strains of methicillin-resistant and methicillin-sensitive *Staphylococcus aureus* have similar virulence. *J Hosp Infect*, **15**, 117–25.

Fryklund, B., Tullus, K. and Burman, L.G. 1995. Survival on skin and surfaces of epidemic and non-epidemic strains of enterobacteria from neonatal special care units. *J Hosp Infect*, **29**, 201–8.

Fuchs, P.C. 1971. Indwelling intravenous polyethylene catheters. Factors influencing the risk of microbial colonization and sepsis. *JAMA*, **216**, 1447–50.

Gardner, A.M.N., Stamp, M., et al. 1962. The Infection Control Sister. A new member of the Control of Infection Team in general hospitals. *Lancet*, **2**, 710–11.

Garrod, L.P. 1979. The eclipse of the haemolytic streptococcus. *Br Med J*, **1**, 1607–8.

Gartenberg, G., Bottone, E.J., et al. 1978. Hospital-acquired mucormycosis (*Rhizopus rhizopodiformis*) of skin and subcutaneous tissue: epidemiology, mycology and treatment. *N Engl J Med*, **299**, 1115–18.

Gaston, M.A. 1988. Enterobacter: an emerging nosocomial pathogen. *J Hosp Infect*, **11**, 197–208.

George, R.H. 1988. The prevention and control of mycobacterial infections in hospitals. *J Hosp Infect*, **11**, Suppl. A, 386–92.

George, R.H., Gully, P.R., et al. 1986. An outbreak of tuberculosis in a children's hospital. *J Hosp Infect*, **8**, 129–42.

George, R.C., Uttley, A.H.C., et al. 1989. High-level vancomycin-resistant enterococci causing hospital infection. *Epidemiol Infect*, **103**, 173–81.

Gerding, D.N., Johnson, S., et al. 1995. Clostridium difficile-associated diarrhea and colitis. *Infect Control Hosp Epidemiol*, **16**, 459–77.

Gerner-Smidt, P., Bruun, B., et al. 1995. Diversity of nosocomial *Xanthomonas maltophilia* (*Stenotrophomonas maltophilia*) as determined by ribotyping. *Eur J Clin Microbiol Infect Dis*, **14**, 137–40.

Getchell-White, S.I., Donowitz, L.G. and Gröschel, D.H.M. 1989. The inanimate environment of an intensive care unit as a potential source of nosocomial bacteria: evidence for long survival of *Acinetobacter calcoaceticus*. *Infect Control Hosp Epidemiol*, **10**, 402–7.

Gillespie, W.A. 1956. Infection in urological patients. *Proc R Soc Med*, **49**, 1045–7.

Gillespie, W.A. 1986. Antibiotics in catheterized patients. *J Antimicrob Chemother*, **18**, 149–51.

Gillespie, W.A., Simpson, R.A., et al. 1983. Does the addition of disinfectant to urinary drainage bags prevent infection in catheterized patients? *Lancet*, **1**, 1037–9.

Glenister, H.M. 1993. How do we collect data for surveillance of wound infection? *J Hosp Infect*, **24**, 283–9.

Glynn, A.A., Ward, V., et al. 1997. *Hospital acquired infection: surveillance, policies and practice*. London: PHLS.

Goldmann, D.A. 1986. Nosocomial infection control in the United States of America. *J Hosp Infect*, **8**, 116–28.

Goldmann, D.A. 1988. The bacterial flora of neonates in intensive care – monitoring and manipulation. *J Hosp Infect*, **11**, Suppl. A, 340–51.

Goldmann, D.A. and Maki, D.G. 1973. Infection control in total parenteral nutrition. *JAMA*, **223**, 1360–4.

Goldstein, F.W. and Acar, J.F. 1995. Epidemiology of quinolone resistance: Europe and North and South America. *Drugs*, **49**, Suppl. 2, 36–42.

Goodgame, R.W. 1996. Understanding intestinal spore-forming protozoa: cryptosporidia, microsporidia, *Isospora* and *Cyclospora*. *Ann Intern Med*, **124**, 429–41.

Gosden, P.E., MacGowan, A.P. and Bannister, G.C. 1998. Importance of air quality and related factors in the prevention of infection in orthopaedic implant surgery. *J Hosp Infect*, **39**, 173–80.

Govan, J.R., Hughes, J.E. and Vandamme, P. 1996. *Burkholderia cepacia*: medical, taxonomic and ecological issues. *J Med Microbiol*, **45**, 395–407.

Gould, D. 1994. Nurses' hand decontamination practice: results of a local study. *J Hosp Infect*, **28**, 15–30.

Gould, F.K., Magee, J.G. and Ingham, H.R. 1987. A hospital outbreak of antibiotic-resistant *Streptococcus pneumoniae*. *J Infect*, **15**, 77–9.

Gould, I.M. 1988. Control of antibiotic use in the United Kingdom. *J Antimicrob Chemother*, **22**, 395–7.

Grady, G.F., Lee, V.A., et al. 1978. Hepatitis B immune globulin for accidental exposures among medical personnel: final report of a multicenter controlled trial. *J Infect Dis*, **138**, 625–38.

Gristina, A.G., Giridhar, G., et al. 1993. Cell biology and molecular mechanisms of artificial device infections. *Int J Artif Organs*, **16**, 755–63.

Gunson, R.N., Shouval, D., European Consensus Group, et al. 2003. Hepatitis B virus (HBV) and hepatitis C virus (HCV) infections in health care workers (HCWs): guidelines for prevention of transmission of HBV and HCV from HCW to patients. *J Clin Virol*, **27**, 213–30.

Gupta, S., Morris, J.G., et al. 1994. Endemic necrotizing enterocolitis: lack of association with a specific infectious agent. *Pediatr Infect Dis J*, **13**, 728–34.

Haines, S.J. and Walters, B.C. 1994. Antibiotic prophylaxis for cerebrospinal fluid shunts: a metanalysis. *Neurosurgery*, **34**, 87–92.

Haley, R.W. 1988. The vicissitudes of prospective multihospital surveillance studies: the Israeli Study of Surgical Infections 1988. *Infect Control Hosp Epidemiol*, **9**, 228–31.

Haley, R.W., Culver, D.H., et al. 1985a. The nationwide infection route. A new need for vital statistics. *Am J Epidemiol*, **121**, 159–67.

Haley, R.W., Culver, D.H., et al. 1985b. The efficiency of infection surveillance and control programs in preventing nosocomial infections in US hospitals. *Am J Epidemiol*, **121**, 182–205.

Hall, C.B., Kopelman, A.E., et al. 1979. Neonatal respiratory syncytial virus infection. *N Engl J Med*, **300**, 393–6.

Halsted, W.S. 1913. Ligature and suture materials: the employment of fine silk in preference to catgut and the advantages of transfixion of tissues and vessels in control of hemorrhage, also an account of the introduction of gloves, gutta-percha tissue and silver foil. *JAMA*, **60**, 1119–26.

Hambraeus, A. 1995. Establishing an infection control structure. *J Hosp Infect*, **30**, Suppl., 232–40.

Hambraeus, A. and Laurell, G. 1980. Protection of the patient in the operating suite. *J Hosp Infect*, **1**, 15–30.

Hamer, D.H. and Barza, M. 1993. Prevention of hospital acquired pneumonia in critically ill patients. *Antimicrob Agents Chemother*, **37**, 931–8.

Hammond, S.A., Morgan, J.R. and Russell, A.D. 1987. Comparative susceptibility of hospital isolates of gram-negative bacteria to antiseptics and disinfectants. *J Hosp Infect*, **9**, 255–64.

Hanberger, H., Diekenia, D., et al. 2001. Surveillance of antibiotic resistance in European ICUs. *J Hosp Infect*, **48**, 161–76.

Hanson, L.C., Weber, D.J. and Rutala, W.A. 1992. Risk factors for nosocomial pneumonia in the elderly. *Am J Med*, **92**, 161–6.

Harmanci, A., Harmanci, O. and Akova, M. 2002. Hospital acquired pneumonia: challenges and options for diagnosis and treatment. *J Hosp Infect*, **S1**, 160–7.

Haron, E., Bodey, G.P., et al. 1988. Has the incidence of *Pneumocystis carinii* pneumonia in cancer patients increased with the AIDS epidemic? *Lancet*, **2**, 904–5.

Harpaz, R., von Seidlein, L., et al. 1996. Transmission of hepatitis B virus to multiple patients from a surgeon without evidence of inadequate infection control. *N Engl J Med*, **334**, 549–54.

Hart, C.A. 1982. Nosocomial gentamicin- and multiply-resistant enterobacteria at one hospital. Factors associated with carriage. *J Hosp Infect*, **3**, 165–72.

Hart, C.A. 1995. Transmissible spongiform encephalopathies. *J Med Microbiol*, **42**, 153–5.

Hart, C.A., Gibson, M.F. and Buckles, A.M. 1981. Variation in skin and environmental survival of hospital gentamicin-resistant enterobacteria. *J Hyg*, **87**, 277–85.

Hartstein, A.I., Denny, M.A., et al. 1995. Control of methicillin-resistant *Staphylococcus aureus* in a hospital intensive care unit. *Infect Control Hosp Epidemiol*, **16**, 405–11.

Hassan, I.J., MacGowan, A.P. and Cook, S.D. 1992. Endophthalmitis at the Bristol Eye Hospital: an 11-year review of 47 patients. *J Hosp Infect*, **22**, 271–8.

Hawkey, P.M., Pedler, S.J. and Southall, P.J. 1980. *Streptococcus pyogenes*: a forgotten occupational hazard in the mortuary. *Br Med J*, **281**, 1058.

Hawkey, P.M., Malnick, H., et al. 1984. *Capnocytophaga ochracea* infection: two cases and a review of the literature. *J Clin Pathol*, **37**, 1059–65.

Health and Safety Executive. 2001. *The control of legionella bacteria in water systems: Approved Code of Practice and Guidance*. UK Health and Safety Executive.

Heimberger, T.S. and Duma, R.J. 1989. Infection of prosthetic heart valves and cardiac pacemakers. *Infect Dis Clin North Am*, **3**, 221–45.

Heimberger, T., Chang, H.-G., et al. 1995. Knowledge and attitudes of healthcare workers about influenza: why are they not getting vaccinated? *Infect Control Hosp Epidemiol*, **16**, 412–14.

Helmick, C.G., Tauxe, R.V. and Vernon, A.A. 1987. Is there a risk to contacts of patients with rabies? *Rev Infect Dis*, **9**, 511–18.

Henderson, E. and Love, E.J. 1995. Incidence of hospital-acquired infections associated with caesarean section. *J Hosp Infect*, **29**, 245–55.

Hill, D.E., Martin, T.A. and Adams, D.H. 2001. Infection control link nurses: how a system can work. *Br J Infect Control*, **2**, 14–17.

Hiramatsu, K., Hanaki, H., et al. 1997. Methicillin-resistant *Staphylococcus aureus* clinical strain with reduced vancomycin susceptibility. *J Antimicrob Chemother*, **40**, 135–6.

Hoffman, P.N., Williams, J., et al. 2002. Microbiological commissioning and monitoring of operating theatre suites. *J Hosp Infect*, **52**, 1–28.

Ho, A.S., Sung, J.Y. and Chan-Yeung, M. 2003. An outbreak of severe acute respiratory syndrome among hospital workers in a community hospital in Hong Kong. *Ann Intern Med*, **139**, 564–7.

Hoge, C.W. and Breiman, R.F. 1991. Advances in the epidemiology and control of *Legionella* infections. *Epidemiol Rev*, **13**, 329–41.

Holmberg, S.D., Solomon, S.L. and Blake, P.A. 1987. Health and economic impacts of antimicrobial resistance. *Rev Infect Dis*, **9**, 1065–78.

Holness, D., DeKoven, J.G. and Nethercott, J.R. 1992. Scabies in chronic health care institutions. *Arch Dermatol*, **128**, 1257–60.

Holzel, H.S. and Cubitt, W.D. 1982. Enteric viruses in hospital-acquired infection. *J Hosp Infect*, **3**, 101–4.

Holzel, H. and de Saxe, M. 1992. Septicaemia in paediatric intensive-care patients at the Hospital for Sick Children, Great Ormond Street. *J Hosp Infect*, **22**, 185–95.

Horsburgh, C.R. 1991. *Mycobacterium avium* complex infection in the acquired immunodeficiency syndrome. *N Engl J Med*, **324**, 1332–8.

Hospital Infection Control Practices Advisory Committee (HICPAC). 1995. Recommendations for preventing the spread of vancomycin resistance. *Infect Control Hosp Epidemiol*, **16**, 105–13.

Hospital Infection Working Group of the Department of Health and Public Health Laboratory Service. 1995. *Hospital infection control. Guidance on the control of infection in hospitals*. London: Department of Health.

Houang, T., Lovett, I.S., et al. 1980. Nocardia asteroides infection – a transmissible disease. *J Hosp Infect*, **1**, 31–40.

Hovig, B. 1981. Lower respiratory tract infections associated with respiratory therapy and anaesthesia equipment. *J Hosp Infect*, **2**, 301–15.

Howard, A.J. 1988. Infection control organization in hospitals in England and Wales 1986. Report of a survey undertaken by a Hospital Infection Society Working Party. *J Hosp Infect*, **11**, 183–91.

Howie, J. 1988. Upgrading surgeons' autoclaves – a personal recollection. *J Infect*, **16**, 231–4.

Hryniewicz, W., Grzesiowski, P. and Ozorowski, T. 2001. Hospital infection control in Poland. *J Hosp Infect*, **49**, 94–8.

Humphreys, H. 1993. Infection control and the design of a new operating theatre suite. *J Hosp Infect*, **23**, 61–70.

Humphreys, H., Johnson, E.M., et al. 1991. An outbreak of aspergillosis in a general ITU. *J Hosp Infect*, **18**, 167–77.

Humphreys, H., Winter, R., et al. 1996. Comparison of bronchoalveolar lavage and catheter lavage to confirm ventilator-associated lower respiratory tract infection. *J Med Microbiol*, **45**, 226–31.

Hurley, J.C. 1995. Prophylaxis with enteral antibiotics in ventilated patients: selective decontamination or selective cross-infection? *Antimicrob Agents Chemother*, **39**, 941–7.

Hussain, M., Oppenheim, B.A., et al. 1996. Prospective survey of the incidence, risk factors and outcome of hospital-acquired infections in the elderly. *J Hosp Infect*, **32**, 117–26.

Hutinel, V. 1894. La dipthérie aux enfants-assistés de Paris; sa suppression; étude de prophylaxie. *Rev Mens Mal Enfants*, **12**, 515–30.

Im, S.W.K., Chow, K. and Chau, P.Y. 1981. Rectal thermometer mediated cross-infection with *Salmonella wandsworth* in a paediatric ward. *J Hosp Infect*, **2**, 171–4.

Incident Investigation Teams and others. 1997. Transmission of hepatitis B to patients from four infected surgeons without HBe antigen. *N Engl J Med*, **336**, 178–84.

Infection Control Nurses Association/ADM Standards Working Group. 1999. *Standards for environmental cleanliness in hospitals*.

Infection Control Standards Working Party. 1993. *Standards in infection control in hospitals*. London: Central Public Health Laboratory.

Ispahani, P., Pearson, N.J. and Greenwood, D. 1987. An analysis of community and hospital acquired bacteraemia in a large teaching hospital in the United Kingdom. *Q J Med*, **241**, 427–40.

Jackson, M., Flerer, J., et al. 1992. Intensive surveillance for infections in a three-year study of nursing home patients. *Am J Epidemiol*, **135**, 685–96.

Jaffe, H.W., McCurdy, J.M., et al. 1994. Lack of HIV transmission in the practice of a dentist with AIDS. *Ann Intern Med*, **121**, 855–9.

Jansen, B. and Peters, G. 1993. Foreign body associated infection. *J Antimicrob Chemother*, **32**, Suppl. A, 69–75.

Jarvis, W.R. and Martone, W.J. 1992. Predominant pathogens in hospital infections. *J Antimicrob Chemother*, **29**, Suppl. A, 19–24.

Javaly, K., Horowitz, H.W. and Wormser, G.P. 1992. Nocardiosis in patients with human immunodeficiency virus infection. Report of 2 cases and review of the literature. *Medicine (Balt)*, **71**, 128–38.

Jawad, A., Seifert, H., et al. 1998. Survival of *Actinetobacter baumannii* on dry surfaces: comparison of outbreak and sporadic isolates. *J Clin Microbiol*, **36**, 1938–41.

Jenner, E.A. and Wilson, J.A. 2000. Educating the infection control team – past, present and future. A British perspective. *J Hosp Infect*, **46**, 96–105.

Jepsen, O.B. 1995. Towards European Union standards in hospital infection control. *J Hosp Infect*, **30**, Suppl., 64–8.

Jepsen, O.B. 2001. Infection control in Danish healthcare: organisation and practice. *J Hosp Infect*, **47**, 262–5.

Jepsen, O.B., Larsen, S.O., et al. 1982. Urinary tract infection and bacteraemia in hospitalized medical patients – a European multicentre prevalence survey on nosocomial infection. *J Hosp Infect*, **3**, 241–52.

Jewes, L.A., Gillespie, W.A., et al. 1988. Bacteriuria and bacteraemia in patients with long-term indwelling catheters – a domiciliary study. *J Med Microbiol*, **26**, 61–5.

Jindani, A., Aber, V.R., et al. 1980. The early bactericidal activity of drugs in patients with pulmonary tuberculosis. *Am Rev Respir Dis*, **121**, 939–49.

Johnson, A.P., Weinbren, M.J., et al. 1992. Outbreak of infection in two UK hospitals caused by a strain of *Klebsiella pneumoniae* resistant to cefotaxime and ceftazidime. *J Hosp Infect*, **20**, 97–103.

Joint DH/PHLS Working Group, 1994. *The prevention and management of Clostridium difficile infection*. London: Department of Health and Public Health Laboratory Service.

Joint Tuberculosis Committee of the British Thoracic Society. 1994. Control and prevention of tuberculosis in the United Kingdom: Code of Practice 1994. *Thorax*, **49**, 1193–200.

Joint Working Party of the Hospital Infection Society and the Surgical Infection Study Group. 1992. Risks to surgeons and patients from HIV and hepatitis: guidelines on precautions and management of exposure to blood or body fluids. *Br Med J*, **305**, 1337–43.

Jolley, A.E. 1993. The value of surveillance cultures on neonatal intensive care units. *J Hosp Infect*, **25**, 153–9.

Jones, R.N., Kehrberg, E.N., et al. 1994. Prevalence of important pathogens and antimicrobiol activity of parenteral drugs at numerous medical centers in the United States, I. Study on the threat of emerging resistance: real or perceived. *Diagn Microbiol Infect Dis*, **19**, 203–15.

Joseph, C.A., Watson, J.M., et al. 1994. Nosocomial Legionnaires' disease in England and Wales, 1980–92. *Epidemiol Infect*, **112**, 329–45.

Kacica, M.A., Morgan, M.J., et al. 1994. Relatedness of coagulase-negative staphylococci causing bacteremia in low-birthweight infants. *Infect Control Hosp Epidemiol*, **15**, 658–62.

Kahyaoglu, O., Nolan, B. and Kumar, A. 1995. *Burkholderia cepacia* sepsis in neonates. *Pediatr Infect Dis J*, **14**, 815–16.

Kalmeijer, M.D., Coertjens, H. et al. Perioperative eradication of nasal carriage of *Staphylococcus aureus* by mupirocin nasal ointment as prevention of surgical site infections in orthopedic surgery. Abstracts of the 39th Interscience Conference on Antimicrobial Agents and Chemotherapy, September 1999, p. 591, Abstr. 514.

Kam, K.M. and Mak, W.P. 1993. Territory-wide survey of hospital infection in Hong-Kong. *J Hosp Infect*, **23**, 143–51.

Karanfil, L.V., Murphy, M. and Josephson, A. 1992. A cluster of vancomycin-resistant *Enterococcus faecium* in an intensive care unit. *Infect Control Hosp Epidemiol*, **13**, 195–200.

Karchmer, A.W. 1991. Prosthetic valve endocarditis: a continuing challenge for infection control. *J Hosp Infect*, **18**, Suppl. A, 355–6.

Kim, K.H., Fekety, R., et al. 1981. Isolation of *Clostridium difficile* from the environment and contacts of patients with antibiotic associated colitis. *J Infect Dis*, **143**, 42–50.

Kirk, D., Dunn, M., et al. 1979. Hibitane bladder irrigation in the prevention of catheter-associated urinary infection. *Br J Urol*, **51**, 528–31.

Kite, P., Dobbins, B.M., et al. 1999. Rapid diagnosis of central venous catheter related blood stream infection without catheter removal. *Lancet*, **354**, 1504–7.

Klein, R.S., Freeman, K., et al. 1991. Occupational risk for hepatitis C virus infection among New York City dentists. *Lancet*, **338**, 1539–42.

Kluytmans, J., van Belkum, A. and Verbrugh, H. 1997. Nasal carriage of *Staphylococcus aureus*: epidemiology, underlying mechanisms, and associated risks. *Clin Microbiol Rev*, **10**, 505–20.

Kohli, H.S., Chaudhuri, A.K.R., et al. 1994. A severe outbreak of *E. coli* O157 in two psychogeriatic wards. *J Public Health Med*, **16**, 11–15.

Kramer, F., Sasse, S.A., et al. 1993. Primary cutaneous tuberculosis after a needlestick injury from a patient with AIDS and undiagnosed tuberculosis. *Ann Intern Med*, **119**, 594–5.

Kristensen, B., Schonheyder, H.C., et al. 1995. *Capnocytophaga* (*Capnocytophaga ochracea* group) bacteremia in hematological patients with profound granulocytopenia. *Scand J Infect Dis*, **27**, 153–5.

Kropec, A., Huebner, J., et al. 1993. Exogenous or endogenous reservoirs of nosocomial *Pseudomonasaeruginosa* and *Staphylococcus aureus* infection in a surgical intensive care unit. *Intens Care Med*, **19**, 161–5.

Ksiazek, T.G., Erdman, D., et al. 2003. A novel coronavirus associated with severe acute respiratory syndrome. *N Engl J Med*, **348**, 1953–66.

Kumarasinghe, G., Hamilton, W.J., et al. 1982. An outbreak of *Salmonella muenchen* infection in a specialist paediatric hospital. *J Hosp Infect*, **3**, 341–4.

Kunin, C.M. 1987. *Detection, prevention and management of urinary tract infection*, 4th edn. Philadelphia: Lea & Febiger.

Kunin, C.M. 1993. Resistance to antimicrobial drugs – a worldwide calamity. *Ann Intern Med*, **118**, 557–61.

Kunin, C.M. and McCormack, R.C. 1966. Prevention of catheter-induced urinary-tract infection by sterile closed drainage. *N Engl J Med*, **274**, 1156–61.

Kyne, L., Merry, C., et al. 1999. Factors associated with prolonged symptoms and severe disease due to *Clostridium difficile*. *Age Ageing*, **28**, 107–13.

Kyne, L., Hamel, M.B., et al. 2002. Health care costs and mortality associated with nosocomial diarrhoea due to *Clostridium difficile*. *Clin Infect Dis*, **34**, 346–53.

L'Age-Stehr, J., Schwarz, A., et al. 1985. HTLV-III infection in kidney transplant recipients. *Lancet*, **2**, 1361–2.

Laing, F.P., Ramotar, K., et al. 1995. Molecular epidemiology of *Xanthomonas maltophilia* colonization and infection in the hospital environment. *J Clin Microbiol*, **33**, 513–18.

Langenberg, W., Ramos, E.A., et al. 1990. Patient-to-patient transmission of *Campylobacter pylori* infection by fiberoptic gastroduodenoscopy and biopsy. *J Infect Dis*, **161**, 507–11.

Larson, E. 1984. Current handwashing issues. *Infect Control*, **5**, 15–17.

Larson, E. 1988. A causal link between handwashing and mode of infection. Examination of the evidence. *Infect Control*, **9**, 28–36.

Latham, R.H., Schaffner, W., et al. 1992. Nosocomial Legionnaires' disease. *Curr Sci*, **62**, 512–17.

Lawrence, J.C. 1985. The bacteriology of burns. *J Hosp Infect*, **6**, Suppl. B, 3–17.

Leaper, D.J. 1994. Prophylactic and therapeutic role of antibiotics in wound care. *Am J Surg*, **167**, 15S–20S.

Leblebiciogh, H. and Unal, S. 2002. The organisation of hospital infection control in Turkey. *J Hosp Infect*, **51**, 1–6.

Lee, Y.L., Thrupp, L.D., et al. 1992. Nosocomial infection and antibiotic utilization in geriatric patients: a pilot prospective surveillance program in skilled nursing facilities. *Gerontology*, **38**, 223–32.

Lee, N., Hui, D., et al. 2003. A major oubreak of severe acute respiratory syndrome in Hong Kong. *N Engl J Med*, **348**, 1986–94.

Leenders, A., van Belkum, A., et al. 1996. Molecular epidemiology of apparent outbreak of invasive aspergillosis in a hematology ward. *J Clin Microbiol*, **34**, 345–51.

Legrand, C. and Anaissie, E. 1992. Bacteremia due to *Achromobacter xylosoxidans* in patients with cancer. *Clin Infect Dis*, **14**, 479–84.

Levin, M.H., Olson, B., et al. 1984. Pseudomonas in the sinks in an intensive care unit: relation to patients. *J Clin Pathol*, **37**, 424–7.

Lewis, D.A., Hawkey, P.M., et al. 1983. Infection with netilmicin resistant *Serratia marcescens* in a special care baby unit. *Br Med J*, **287**, 1701–5.

Lewis, D.A., Leaper, D.J. and Speller, D.C.E. 1984. Prevention of bacterial colonization of wounds at operation: comparison of iodine-impregnated ('Ioban') drapes with conventional methods. *J Hosp Infect*, **5**, 431–7.

Lewis, R.T., Weigand, F.M., et al. 1995. Should antibiotic prophylaxis be used routinely in clean surgical procedures: a tentative yes. *Surgery*, **118**, 742–6, discussion 746–47.

Lidwell, O.M. 1974. Aerial dispersal of micro-organisms from the human respiratory tract. In: Skinner, A. and Carr, J. (eds), *The normal microbial flora of man*. London: Academic Press, 135–54.

Lidwell, O.M., Lowbury, E.J.L., et al. 1982. Effect of ultraclean air in operating rooms on deep sepsis in the joint after total hips or knee replacement: a randomised study. *Br Med J*, **285**, 10–14.

Liekweg, W.G. and Greenfield, L.J. 1977. Vascular prosthetic infections: collected experience and results of treatment. *Surgery*, **81**, 335–42.

Lilly, H.A., Lowbury, E.J. and Wilkins, M.D. 1979. Limits to progressive reduction of resistant skin bacteria by disinfection. *J Clin Pathol*, **32**, 382–5.

Lim, V.K.E. 2001. Hospital infection control in Malaysia. *J Hosp Infect*, **48**, 177–9.

Limberg, M.B. 1991. A review of bacterial keratitis and bacterial conjunctivitis. *Am J Ophthalmol*, **112**, 2S–9S.

Lister, J. 1867a. On a new method of treating compound fracture, abscess, etc., with observations on the conditions of suppuration. *Lancet*, **1**, 326–29, 357–59, 387–89.

Lister, J. 1867b. On a new method of treating compound fracture, abscess, etc. *Lancet*, **2**, 95–6.

Livornese, L.L., Dial, S. and Samel, C. 1992. Hospital-acquired infection with vancomycin-resistant *Enterococcus faecium* transmitted by electronic thermometers. *Ann Intern Med*, **117**, 112–16.

Llewelyn, C.A., Hewitt, P.E., et al. 2004. Possible transmission of variant Creutzfeldt-Jacob disease by blood transfusion. *Lancet*, **363**, 417–21.

Loudon, K.W., Coke, A.P., et al. 1996. Kitchens as a source of *Aspergillus niger* infection. *J Hosp Infect*, **32**, 191–8.

Lowbury, E.J.L. and Lilly, H.A. 1958. The sources of hospital infection of wounds with *Clostridium welchii*. *J Hyg*, **56**, 169–82.

Lowbury, E.J.L., Lilly, H.A. and Bull, J.P. 1964. Methods for disinfection of hands and operation sites. *Br Med J*, **2**, 531–6.

Lowbury, E.J.L., Thom, B.T., et al. 1970. Sources of infection with *Pseudomonas aeruginosa* in patients with tracheostomy. *J Med Microbiol*, **3**, 39–56.

Lowbury, E.J.L., Lilly, H.A. and Ayliffe, G.A.J. 1974. Preoperative disinfection of surgeons' hands: use of alcoholic solutions and effects of gloves on skin. *Br Med J*, **4**, 369–72.

Ludlam, H.A., Young, A.E., et al. 1989. The prevention of infection with *Staphylococcus aureus* in continuous ambulatory peritoneal dialysis. *J Hosp Infect*, **14**, 293–301.

Luzzaro, F., Pagani, L., et al. 1995. Extended spectrum beta-lactamases conferring resistance to monobactams and oxyimino-cephalosporins in clinical isolates of *Serratia marcescens*. *J Chemother*, **7**, 175–8.

Lynch, W., Davey, P.G., et al. 1992. Cost-effectiveness analysis of the use of chlorhexidine detergent in preoperative whole-body disinfection in wound infection prophylaxis. *J Hosp Infect*, **21**, 179–91.

Macfarlane, J.T., Finch, R.G., et al. 1982. Hospital study of adult community acquired pneumonia. *Lancet*, **2**, 255–8.

MacKay, W.G., Leamond, A.T. and Williams, C.L. 2002. Water, water everywhere nor any a sterile drop to rinse your endoscope. *J Hosp Infect*, **51**, 256–61.

Madge, P., Paton, J.Y., et al. 1992. Prospective controlled study of four infection-control procedures to prevent nosocomial infection with respiratory syncytial virus. *Lancet*, **340**, 1079–83.

Mahenthiralingam, E., Baldwin, A. and Vandamme, P. 2002. *Burkholderia cepacia* complex infection in patients with cystic fibrosis. *J Med Microbiol*, **51**, 533–8.

Maki, D.G. 1980. Through a glass darkly. Nocosomial pseudo-epidemics and pseudobacteremia. *Arch Intern Med*, **140**, 26–8.

Maki, D.G., Goldman, D.A. and Rhame, F.S. 1973. Infection control in intravenous therapy. *Ann Intern Med*, **79**, 867–87.

Maki, D.G., Weise, C.E. and Sarafin, H.W. 1977. A semiquantitative culture method for identifying intravenous catheter-related infection. *N Engl J Med*, **296**, 1305–9.

Maki, D.G., Alvarado, C.J., et al. 1982. Relation of the inanimate hospital environment to endemic nosocomial infection. *N Engl J Med*, **307**, 1562–6.

Makofsky, D. and Cone, J.E. 1993. Installing needle disposal boxes closer to the bedside reduces needle-recapping rates in hospital units. *Infect Control Hosp Epidemiol*, **14**, 140–4.

Mangram, A.J., Horan, T.C., Centers for Disease Control and Prevention (CDC) Hospital Infection Control Practices Advisory Committee, et al. 1999. Guideline for prevention of surgical site infection. *Am J Infect Control*, **27**, 97–132.

Marples, R.R. and Towers, A.G. 1979. A laboratory method for the investigation of contact transfer of microorganisms. *J Hyg*, **82**, 237–48.

Marples, R.R., Hone, R., et al. 1978. Investigation of coagulase-negative staphylococci from infection in surgical patients. *Zentralbl Bakteriol Parasitenkd Infektionskr Hyg Abt Orig*, **241**, 140–56.

Martino, P., Gentile, G., et al. 1988. Hospital-acquired cryptosporidiosis in a bone marrow transplantation unit. *J Infect Dis*, **158**, 647–8.

Martone, W.J., Jarvis, W.R., et al. 1992. Incidence and nature of endemic and epidemic nosocomial infections. In: Bennett, J.V. and Brachman, P.S. (eds), *Hospital infections*. Boston: Little Brown, 577–96.

Masterton, R.G. and Teare, E.L. 2001. Clinical governance and infection control in the United Kingdom. *J Hosp Infect*, **47**, 25–31.

Mayer, L.W. 1988. Use of plasmid profiles in epidemiologic surveillance of disease outbreaks and in tracing the transmission of antibiotic resistance. *Clin Microbiol Rev*, **1**, 228–43.

McConnell, S.A., Gubbins, P.O. and Anaissie, E.J. 2003. Do antimicrobial-impregnated central venous catheters prevent catheter-related bloodstream infection? *Clin Infect Dis*, **37**, 65–72.

McEntegart, M.G. 1956. Dangerous contaminants in stored blood. *Lancet*, **2**, 909–11.

McFarland, L.V., Mulligan, M.E., et al. 1989. Nosocomial acquisition of *Clostridium difficile* infection. *N Engl J Med*, **320**, 204–9.

McGowan, J.E. 1988. JK coryneforms: a continuing problem for hospital infection control. *J Hosp Infect*, **11**, Suppl. A, 358–66.

McGowan, J.E. 1995. Success, failures and costs of implementing standards in the USA – lessons for infection control. *J Hosp Infect*, **30**, Suppl., 76–87.

McGowan, J.E. and Metchock, B.G. 1996. Basic microbiologic support for hospital epidemiology. *Infect Control Hosp Epidemiol*, **17**, 298–302.

McLauchlan, G.J., Anderson, I.D., et al. 1995. Outcome of patients with abdominal sepsis treated in an intensive care unit. *Br J Surg*, **82**, 524–9.

McManus, A.T., Mason, A.D., et al. 1994. A decade of reduced gram-negative infections and mortality associated with improved isolation of burned patients. *Arch Surg*, **129**, 1306–9.

Medical Research Council, Subcommittee of the Committee on Hospital Infection, 1968. Aseptic methods in the operating suite. *Lancet*, **1**, 705–9, 763–68, 831–39.

Meduri, G.U. and Estes, R.J. 1995. The pathogenesis of ventilator-associated pneumonia: II. The lower respiratory tract. *Intens Care Med*, **21**, 452–61.

Meenhorst, P.L., Reingold, A.L., et al. 1985. Water-related nosocomial pneumonia caused by *Legionella pneumophila* serogroups 1 and 10. *J Infect Dis*, **152**, 356–64.

Meers, P.D., Calder, M.W., et al. 1973. Intravenous infusion of contaminated dextrose solution: the Devonport incident. *Lancet*, **2**, 1189–92.

Meers, P.D., Ayliffe, G.A.J., et al. 1981. Report on the national survey of infection in hospitals, 1980. *J Hosp Infect*, **2**, Suppl. 1, 1–39.

Mellersh, A.R., Norman, P. and Smith, G.H. 1984. Aeromonas hydrophila: an outbreak of hospital infection. *J Hosp Infect*, **5**, 425–30.

Menzies, D., Fanning, A., et al. 1995. Tuberculosis among health care workers. *N Engl J Med*, **332**, 92–9.

Meyer, K.S., Urban, C., et al. 1993. Nosocomial outbreak of klebsiella infection resistant to late generation cephalosporins. *Ann Intern Med*, **119**, 353–8.

Millar, M.R., Brown, N.M., et al. 1994. Outbreak of infection with penicillin-resistant *Streptococcus pneumoniae* in a hospital for the elderly. *J Hosp Infect*, **27**, 99–104.

Miller, A., Gillespie, W.A., et al. 1958. Postoperative infection in urology. *Lancet*, **2**, 608–12.

Mitchell, S.J., Gray, J., et al. 1996. Nosocomial infection with *Rhizopus microsporus* in preterm infants: association with wooden tongue depressors. *Lancet*, **348**, 441–3.

Moberg, S.A., Löwhagen, G.E. and Hersle, K.S. 1984. An epidemic of scabies with unusual features and treatment resistance in a nursing home. *J Am Acad Dermatol*, **11**, 242–4.

Mollison, P.L., Engelfriet, C.P. and Contreras, M. 1987. *Blood transfusion in clinical medicine*, 8th edn. Oxford: Blackwell, pp. 764.

Mongkolrattanothai, K., Boyle, S., et al. 2003. Severe *Staphylococcus aureus* infections caused by clonally related community-acquired methicillin-susceptible and methicillin-resistant isolates. *Clin Infect Dis*, **37**, 1050–8.

Morrison, D., Woodford, N. and Cookson, B.D. 1996. Epidemic vancomycin-resistant *Enterococcus faecium* in the UK. *Clin Microbiol Infect*, **1**, 146–7.

Mulin, B., Talon, D., et al. 1995. Risks for nosocomial colonization with multiresistant *Acinetobacter baumanii*. *Eur J Clin Microbiol Infect Dis*, **14**, 569–76.

Mulley, A.G., Silverstein, M.D. and Dienstag, J.L. 1982. Indication for use of hepatitis B vaccine, based on cost effectiveness analysis. *N Engl J Med*, **307**, 644–52.

Murphy, D.M., Stassen, L., et al. 1984. Bacteraemia during prostatectomy and other transurethral operations: influence of timing of antibiotic administration. *J Clin Pathol*, **37**, 673–6.

Murphy, O.M., Gray, J. and Pedler, S.J. 1995. Non-enteric aeromonas infections in hospitalized patients. *J Hosp Infect*, **31**, 55–60.

Murray, B.E. 1991. New aspects of antimicrobial resistance and the resulting therapeutic dilemmas. *J Infect Dis*, **163**, 1185–94.

Nagington, J., Wreghitt, T.G., et al. 1978. Fatal echovirus 11 infections in outbreak in special-care baby unit. *Lancet*, **2**, 725–8.

Naikoba, S. and Hayward, A. 2001. The effectiveness of interventions aimed at increasing handwashing in health-care workers a systemic review. *J Hosp Infect*, **47**, 173–80.

Nair, P., Jani, K. and Sanderson, P.J. 1986. Transfer of oropharyngeal bacteria into the trachea during endotracheal intubation. *J Hosp Infect*, **8**, 96–103.

Nataro, J.P. and Corcoran, L. 1994. Prospective analysis of coagulase-negative staphylococcal infection in hospitalized infants. *J Pediatr*, **125**, 798–804.

NHS Executive. 2000. *Hepatitis B infected health care workers*. Health Service circular HSC 2000/020.

National Nosocomial Infections Surveillance (NNIS) System. 1995. National Nosocomial Infections Surveillance (NNIS) Semiannual report, May 1995. *Am J Infect Control*, **23**, 377–85.

Melo-Cristino, J., Marques-Lito, L. and Pina, E. 2002. The control of hospital infection in Portugal. *J Hosp Infect*, **51**, 85–8.

Nicolle, L.E., Strausbaugh, L.J. and Garibaldi, R.A. 1996. Infections and antibiotic resistance in nursing homes. *Clin Microbiol Rev*, **9**, 1–17.

Niel-Weise, B.S., Arend, S.M. and van den Broek, P.J. 2002. Is there evidence for recommending silver-coated urinary catheters in guidelines? *J Hosp Infect*, **52**, 81–7.

Nielsen, H., Stenderup, J. and Bruun, B. 1991. Fungemia in a university hospital 1984–1988. *Scand J Infect Dis*, **23**, 275–82.

Nightingale, F. 1863. *Notes on hospitals*. London: Longman.

Nolan, N.P.M., Kelly, C.P., et al. 1987. An epidemic of pseudomembranous colitis: importance of person to person spread. *Gut*, **28**, 1467–73.

Norden, C.W. 1969. Application of antibiotic ointment to the site of venous catheterization – a controlled trial. *J Hosp Infect*, **120**, 611–15.

Nordqvist, P., Ekelund, P., et al. 1984. Catheter-free geriatric care. Routines and consequences for clinical infection, care and economy. *J Hosp Infect*, **5**, 298–304.

Noskin, G.A., Peterson, L.R. and Warren, J.R. 1995a. *Enterococcus faecium* and *Enterococcus faecalis* bacteremia: acquisition and outcome. *Clin Infect Dis*, **20**, 296–301.

Noskin, G.A., Stosor, V., et al. 1995b. Recovery of vancomycin-resistant enterococci on fingertips and environmental surfaces. *Infect Control Hosp Epidemiol*, **16**, 577–81.

Noy, M.F., Smith, C.A. and Watterson, L.L. 1982. The use of chlorhexidine in catheter bags. *J Hosp Infect*, **1**, 365–7.

Nye, K., Chadha, D.K., et al. 1990. *Mycobacterium chelonei* isolation from broncho-alveolar lavage fluid and its practical implications. *J Hosp Infect*, **16**, 257–61.

Nystrom, B. 1980. The disinfection of thermometers in hospitals. *J Hosp Infect*, **1**, 345–8.

O'Connell, N.H. and Humphreys, H. 2000. Intensive care unit design and environmental factors in the acquisition of infection. *J Hosp Infect*, **45**, 255–62.

Okell, S.C. and Elliott, S.D. 1936. Cross-infection with haemolytic streptococci in otorhinological wards. *Lancet*, **2**, 836–42.

Olesen, B., Kolmos, H.J., et al. 1995. Bacteraemia due to *Escherichia coli* in a Danish university hospital, 1986–1990. *Scand J Infect Dis*, **27**, 253–7.

Opal, S.M., Asp, A.A., et al. 1986. Efficacy of infection control measures during a nosocomial outbreak of disseminated aspergillosis associated with hospital construction. *J Infect Dis*, **153**, 634–7.

Osterman, C.A. 1981. The infection control practitioner. In: Wenzel, R.P. (ed.), *CRC handbook of hospital acquired infections*. Boca Raton, FL: CRC Press, 19–32.

Owens, D.K., Harris, R.A., et al. 1995. Screening surgeons for HIV infection. *Ann Intern Med*, **122**, 641–52.

Page, C.P., Bohnen, J.H., et al. 1993. Antimicrobial prophylaxis for surgical wounds. Guidelines for clinical care. *Arch Surg*, **128**, 79–88.

Pallares, R., Linares, J., et al. 1995. Resistance to penicillin and cephalosporin and mortality from severe pneumococcal pneumonia in Barcelona, Spain. *N Engl J Med*, **333**, 474–80.

Palmer, J.D. and Rickett, I.W. 1992. The mechanisms and risks of surgical glove perforation. *J Hosp Infect*, **22**, 279–86.

Palmer, S.R. 1989. Epidemiology in search of infectious diseases: methods in outbreak investigation. *J Epidemiol Commun Health*, **43**, 311–14.

Palmer, S.R. and Rowe, B. 1983. Investigation of outbreaks of salmonella in hospitals. *Br Med J*, **287**, 891–3.

Pancic, F., Carpentier, D.C. and Came, P.E. 1980. Role of infectious secretions in the transmission of rhinovirus. *J Clin Microbiol*, **12**, 467–71.

Pankhurst, C.L. and Philpott-Howard, J. 1996. The environmental risk factors associated with medical and dental equipment in the transmission of *Burkholderia* (*Pseudomonas*) *cepacia* species in cystic fibrosis patients. *J Hosp Infect*, **32**, 249–55.

Papini, R.P., Wilson, A.P., et al. 1995. Wound management in burn centres in the United Kingdom. *Br J Surg*, **82**, 505–9.

Parker, L. and Mandal, B.K. 1984. Postoperative tetanus. *Lancet*, **2**, 407.

Parker, M.T. 1978. *Hospital-acquired infections: guidelines to laboratory methods*. Copenhagen: WHO, WHO Regional Publications European Series No. 4.

Parras, F., Ena, J., et al. 1994. Impact of an educational program for the prevention of colonization of intravascular catheters. *Infect Control Hosp Epidemiol*, **15**, 239–42.

Patrick, C.H., John, J.F., et al. 1992. Relatedness of strains of methicillin-resistant coagulase-negative staphylococcus colonizing hospital personnel and producing bacteremias in a neonatal intensive care unit. *Pediatr Infect Dis J*, **11**, 935–40.

Peachey, R.D.G. and English, M.P. 1974. Outbreak of *Trichophyton rubrum* infection in a geriatric hospital. *Br J Dermatol*, **91**, 389–97.

Pearman, J.W. 1984. Infection hazards in patients with neuropathic bladder dysfunction. *J Hosp Infect*, **5**, 355–8.

Pearman, J.W. 1998. Catheter care. In: Brumfitt, W., Hamilton-Miller, J.T. and Bailey, R.R. (eds), *Urinary tract infections*, 1st edn. London: Chapman and Hall Medical, 303–16.

Pearman, J.W., Christiansen, K.J., et al. 1985. Control of methicillin-resistant *Staphylococcus aureus* (MRSA) in an Australian metropolitan teaching hospital complex. *Med J Aust*, **142**, 103–8.

Pearson, M.L., Jereb, J.A., et al. 1992. Nosocomial transmission of multidrug-resistant *Mycobacterium tuberculosis*. *Ann Intern Med*, **117**, 191–6.

Peiris, J.S., Lai, S.T., et al. 2003. Coronavirus as a possible cause of severe acute respiratory syndrome. *Lancet*, **361**, 1319–25.

Perl, T.M., Cullen, J.J., et al. 2002. Intranasal mupirocin to prevent postoperative *Staphylococcus aureus* infections. *N Engl J Med*, **346**, 1871–7.

Pfaller, M. and Wenzel, R. 1992. Impact of the changing epidemiology of fungal infections in the 1990s. *Eur J Clin Microbiol Infect Dis*, **11**, 287–91.

Phillips, I., Eykyn, S., et al. 1971. *Pseudomonas cepacia* (*multivorans*) septicaemia in an intensive-care unit. *Lancet*, **1**, 375–7.

Phillips, I., Eykyn, S. and Laker, M. 1972. Outbreak of hospital infection caused by contaminated autoclaved fluids. *Lancet*, **1**, 1258–60.

Phillips, L.G., Heggers, J.P. and Robson, M.C. 1992. Burn and trauma units as sources of methicillin-resistant *Staphylococcus aureus*. *J Burn Care Rehabil*, **13**, 293–7.

Philpott-Howard, J. and Casewell, M. 1994. *Hospital infection control policies and practical procedures*. London: WB Saunders.

Picard, B. and Goullet, P. 1987. Seasonal prevalence of nosocomial *Aeromonas hydrophila* infection related to aeromonas in hospital water. *J Hosp Infect*, **10**, 152–5.

Piedra, P.A., Kasel, J.A., et al. 1992. Description of an adenovirus type 8 outbreak in hospitalized neonates born prematurely. *Pediatr Infect Dis J*, **11**, 460–5.

Pitt, T.L. 1994. Bacterial typing systems: the way ahead. *J Med Microbiol*, **40**, 1–2.

Pittet, D. and Wenzel, R.P. 1995. Nosocomial bloodstream infections. *Arch Intern Med*, **155**, 1177–84.

Pittet, D., Mourouga, P. and Perneger, T.V. 1999. Compliance with handwashing in a teaching hospital. *Ann Intern Med*, **130**, 126–30.

Pittet, D., Hugonnet, S., et al. 2000. Effectiveness of a hospital-wide programme to improve compliance with hand hygiene. Infection Control Programme. *Lancet*, **356**, 1307–12.

Plowman, R., Graves, N., et al. 1999. *The socio-economic burden of hospital acquired infection*. MWL Print Group Ltd: Public Health Laboratory Service.

Plowman, R., Graves, N., et al. 2001. The rate and cost of hospital-acquired infections occurring in patients admitted to selected specialties of a district general hospital in England and the national burden imposed. *J Hosp Infect*, **47**, 198–209.

Pople, I.K., Bayston, R. and Hayward, R.D. 1992. Infection of cerebrospinal fluid shunts in infants: a study of etiological factors. *J Neurosurg*, **77**, 29–36.

Poutanen, S.M., Low, D.E., et al. 2003. Identification of severe acute respiratory syndrome in Canada. *N Engl J Med*, **348**, 1995–2005.

Pratesi, C., Russo, D., et al. 2001. Antibiotic prophylaxis in clean surgery: vascular surgery. *J Chemother*, **13**, 1, 123–8.

Pratt, R.J., Pellowe, C.M., The Epic Guideline Development Team, et al. 2001. The epic project: developing national evidence-based guidelines for preventing healthcare associated infections. Phase 1: guidelines for preventing hospital-acquired infections. *J Hosp Infect*, **47**, Suppl, S1–S82.

Price, E.H. 1984. *Staphylococcus epidermidis* in cerebrospinal fluid shunts. *J Hosp Infect*, **5**, 7–17.

Price, P.B. 1938. The bacteriology of normal skin: a new quantitative test applied to a study of the bacterial flora and the disinfectant action of mechanical cleaning. *J Infect Dis*, **63**, 301–18.

Pruitt, B.A. and McManus, A.J. 1992. The changing epidemiology of infection in burn patients. *Can J Surg*, **16**, 56–67.

Public Health Laboratory Service. 2002. Nosocomial outbreak of *Salmonella* Enteritidis PT 6a (Nx, CpL). *Commun Dis Rep CDR Wkly*, [serial online]; **12** (43): news. Available at http://www.phls.co.uk/publications/cdr/archive02/News/news4302.html#salmupdate

Quinn, J.P. 1994. Clinical significance of extended-spectrum β-lactamases. *Eur Clin Microbiol Infect Dis*, **13**, Suppl. 1, S39–42.

Rampling, A., Wiseman, S., et al. 2001. Evidence that hospital hygiene is important in the control of methicillin-resistant *Staphylococcus aureus*. *J Hosp Infect*, **49**, 109–16.

Rao, G.G. 1995. Control of outbreaks of viral diarrhoea in hospitals – a practical approach. *J Hosp Infect*, **30**, 1–6.

Ratner, H. 1984. *Flavobacterium meningosepticum*. *Infect Control*, **5**, 237–9.

Ravn, P., Lundgren, J.D., et al. 1991. Nosocomial outbreak of cryptosporidiosis in AIDS patients. *Br Med J*, **302**, 277–80.

Reagan, D.R., Pfaller, M.A., et al. 1995. Evidence of nosocomial spread of *Candida albicans* causing bloodstream infection in a neonatal intensive care unit. *Diagn Microbiol Infect Dis*, **21**, 191–4.

Reboli, A.C., John, J.F. and Levkoff, A.H. 1989. Epidemic methicillin-gentamicin-resistant *Staphylococcus aureus* in a neonatal intensive care unit. *Am J Dis Child*, **143**, 34–9.

Reed, C.S., Goorie, G. and Spelman, D. 2003. Hospital infection control in Australia. *J Hosp Infect*, **54**, 267–71.

Rello, J., Torres, A., et al. 1994. Ventilator-associated pneumonia by *Staphylococcus aureus*: comparison of methicillin-resistant and methicillin-sensitive episodes. *Am J Respir Crit Care Med*, **150**, 1545–9.

Report. 1995. Report of the ASM Task Force on Antibiotic Resistance. *Antimicrob Agents Chemother*, Suppl., 1–23.

Report from a Joint Working Group of the Hospital Infection Society (HIS) and the Public Health Laboratory Service (PHLS). 2002. Rinse water for heat labile endoscopy equipment. *J Hosp Infect*, **51**, 7–16.

Report of a Working Party of the British Society of Gastroenterology Endoscopy Committee. 1998. Cleaning and disinfection of equipment for gastrointestinal endoscopy. *Gut*, **42**, 585–93.

Reybrouck, G. 1983. Role of the hands in the spread of nosocomial infection. *J Hosp Infect*, **4**, 103–10.

Reybrouck, G. 1986. Handwashing and hand disinfection. *J Hosp Infect*, **8**, 5–23.

Reybrouck, G., Vaude, N., et al. 2001. The organisation of infection control in Belgium. *J Hosp Infect*, **47**, 32–5.

Rhame, F.S., Streifel, A.J., et al. 1984. Extrinsic risk factors for pneumonia in the patient at high risk of infection. *Am J Med*, **76**, Suppl. 5A, 42–52.

Rhinehart, E., Smith, N., et al. 1990. Rapid dissemination of beta-lactamase-producing aminoglycoside-resistant *Enterococcus faecalis* among patients and staff on an infant-toddler surgical ward. *N Engl J Med*, **323**, 1814–18.

Richard, P., Del Valle, G.A., et al. 1995. Viridans streptococcal bacteraemia in patients with neutropenia. *Lancet*, **345**, 1607–9.

Ridgway, E.J., Wilson, A.P. and Kelsey, M.C. 1990. Preoperative screening cultures in the identification of staphylococci causing wound and valvular infections in cardiac surgery. *J Hosp Infect*, **15**, 55–63.

Ridgway, G.L. 1985. Decontamination of fibreoptic endoscopes. *J Hosp Infect*, **6**, 363–8.

Riley, R.L., Mills, C.C., et al. 1962. Infectiousness of air from a tuberculosis ward. *Am Rev Respir Dis*, **85**, 511–25.

Robert, L.M., Chamberland, M.E., et al. 1995. Investigations of patients of health care workers infected with HIV. *Ann Intern Med*, **122**, 653–7.

Rodriguez-Bano, J. and Pascual, A. 2001. Hospital infection control in Spain. *J Hosp Infect*, **48**, 258–60.

Rogers, T.R. and Barnes, R.A. 1988. Prevention of airborne fungal infection in immunocompromised patients. *J Hosp Infect*, **11**, Suppl. A, 15–20.

Ronan, A., Hogg, G.G. and Klug, G.L. 1995. Cerebrospinal fluid shunt infection in children. *Pediatr Infect Dis J*, **14**, 782–6.

Rossney, A.S., Pomeroy, H.M. and Keane, C.T. 1994. *Staphylococcus aureus* phage typing, antimicrobial susceptibilty patterns and patient data correlated using a personal computer: advantages for monitoring the epidemiology of MRSA. *J Hosp Infect*, **26**, 219–34.

Rothman, K.J. 1986. *Modern epidemiology*. Boston: Little Brown, pp. 57, 62 and 237.

Ruddy, M. and Kibbler, C.C. 2002. Endoscopic decontamination: an audit and practical review. *J Hosp Infect*, **50**, 261–8.

Rudnick, J.R., Beck-Sagué, C.M., et al. 1996. Gram-negative bacteremia in open-heart-surgery patients traced to probable tap-water contamination of pressure-monitoring equipment. *Infect Control Hosp Epidemiol*, **17**, 281–5.

Russell, A.D., Hugo, W.B. and Ayliffe, G.A.J. 1992. *Principles and practice of disinfection, preservation and sterilisation*. Oxford: Blackwell.

Samore, M.H., Venkataraman, L., et al. 1996. Clinical and molecular epidemiology of sporadic and clustered cases of nosocomial *Clostridium difficile* diarrhea. *Am J Med*, **100**, 32–40.

Sanderson, P.J. and Rawal, P. 1987. Contamination of the environment of spinal cord injured patients by organisms causing urinary-tract infection. *J Hosp Infect*, **10**, 173–8.

Sarr, M.G., Gott, V.L. and Townsend, T.R. 1984. Mediastinal infection after cardiac surgery. *Ann Thorac Surg*, **38**, 415–23.

Schimmelbusch, C. 1894. *The aseptic treatment of wounds*. London: Lewis, trans A.T. Rake.

Schimpff, S.C. 1993. Gram-negative bacteremia. *Support Care Cancer*, **1**, 5–18.

Schneeberger, P.M., Vos, J. and van Dijk, W.C. 1993. Prevalence of antibodies to hepatitis C virus in a Dutch group of haemodialysis patients related to risk factors. *J Hosp Infect*, **25**, 265–70.

Schoch, P.E. and Cunha, B.A. 1988. Nosocomial *Achromobacter xylosoxidans* infections. *Infect Control Hosp Epidemiol*, **9**, 84–7.

Seifert, H., Strate, A. and Pulverer, G. 1995. Nosocomial bacteremia due to *Acinetobacter baumannii*. Clinical features, epidemiology and predictors of mortality. *Medicine (Balt)*, **74**, 340–9.

Seifert, H., Cornely, O., et al. 2003. Bloodstream infection in neutropenic cancer patients related to short-term nontunnelled catheters determined by quantitative blood cultures, differential time to positivity, and molecular epidemiological typing with pulsed-field gel electrophoresis. *J Clin Microbiol*, **41**, 118–23.

Select Committee on Science and Technology. 1998. *Resistance to antibiotics and other antimicrobial agents*. London: House of Lords.

Semmelweiss, I.F. 1861. *The aetiology, the concept and the prophylaxis of childbed fever*. Birmingham: Classics of Medicine Library, trans F.P. Murphy.

Seropian, R. and Reynolds, B.M. 1971. Wound infections after preoperative depilatory versus razor preparation. *Am J Surg*, **121**, 251–4.

Seto, W.H., Tsang, D., et al. 2003. Effectiveness of precautions against droplets and contact in prevention of nosocomial transmission of severe acute respiratory syndrome (SARS). *Lancet*, **361**, 1519–20.

Shanaghy, N., Murphy, F. and Kennedy, K. 1993. Improvements in the microbiological quality of food samples from a hospital cook-chill system since the introduction of HACCP. *J Hosp Infect*, **23**, 305–14.

Shanson, D.C. 1981. Antibiotic-resistant *Staphylococcus aureus*. *J Hosp Infect*, **2**, 11–36.

Sheridan, R.L., Tompkins, R.G. and Burke, J.F. 1994. Prophylactic antibiotics and their role in the prevention of surgical wound infection. *Adv Surg*, **27**, 43–65.

Shooter, R.A., Taylor, G.W., et al. 1956. Post-operative wound infection. *Surg Gynecol Obstet*, **103**, 257–62.

Shooter, R.A., Walker, K.A., et al. 1966. Faecal carriage of *Pseudomonas aeruginosa* in hospital patients. Possible spread from patient to patient. *Lancet*, **2**, 1331–4.

Shooter, R.A., Gaya, H., et al. 1969. Food and medicaments as possible sources of hospital strains of *Pseudomonas aeruginosa*. *Lancet*, **1**, 1227–9.

Simpson, J.Y. 1869. Our existing system of hospitalization and its effects. Part III. Provincial hospitals of Great Britain, etc. Chapter XII Some comparisons etc., between the limb amputations in country practice and in the practice of large and metropolitan hospitals. *Edinburgh Med J*, **15**, 523.

Simpson, R.A., Spencer, A.F., et al. 1986. Colonization by gentamicin-resistant *Staphylococcus epidermidis* in a special care baby unit. *J Hosp Infect*, **7**, 108–20.

Sitges-Serra, A., Jaurrieta, E., et al. 1983. Bacteria in total parenteral nutrition catheters: where do they come from? *Lancet*, **1**, 531.

Smith, P.W. 1988. Nosocomial infections in the elderly. *Infect Dis Clin North Am*, **3**, 763–77.

Smyth, E.T.M. and Emmerson, A.M. 2000. Surgical site infection surveillance. *J Hosp infect*, **45**, 173–84.

Smyth, E.T.M., McIlvanny, G., et al. 1997. Automated entry of hospital infection surveillance data. *Infect Control Hosp Epidemiol*, **18**, 486–91.

Snydman, D.R., Greer, C., et al. 1988. Prevention of nosocomial transmission of respiratory syncytial virus in a newborn nursery. *Infect Control Hosp Epidemiol*, **9**, 105–8.

Spach, D.H., Silverstein, F.E. and Stamm, W.E. 1993. Transmission of infection by gastrointestinal endoscopy and bronchoscopy. *Ann Intern Med*, **118**, 117–28.

Speller, D.C.E., Stephens, M.E. and Viant, A.C. 1971. Hospital infection by *Pseudomonas cepacia*. *Lancet*, **1**, 798–9.

Spencer, R.C. 1988. Infection in continuous ambulatory peritoneal dialysis. *J Med Microbiol*, **27**, 1–9.

Spencer, R.C. 1995. The emergence of epidemic, multiple antibiotic-resistant *Stenotrophomonas* (*Xanthomonas*) *maltophilia* and *Burkholderia* (*Pseudomonas*) *cepacia*. *J Hosp Infect*, **30**, Suppl., 453–64.

Spencer, R.C. 1998. Clinical impact and associated costs of *Clostridium difficile*-associated disease. *J Antimicrob Chemother*, **41**, Suppl. C, 5–12.

Spencer, R.C. and Ridgway, G.L. 2002. Sterilisation issues in vCJD – towards a consensus. *J Hosp Infect*, **51**, 168–74.

Standfast, S.J., Michelsen, P.B., et al. 1984. A prevalence survey of infections in a combined acute and long-term care hospital. *Infect Control*, **5**, 177–84.

Stevenson, P., McCann, R., et al. 1994. A hospital outbreak due to Norwalk virus. *J Hosp Infect*, **26**, 261–72.

Stickler, D.J. and Chawla, J.C. 1987. The role of antibiotics in the management of patients with long-term indwelling bladder catheters. *J Hosp Infect*, **10**, 219–28.

Stickler, D.J., Clayton, C.L. and Chawla, J.C. 1987. The resistance of urinary tract pathogens to chlorhexidine bladder washouts. *J Hosp Infect*, **10**, 28–39.

Stillman, R.I., Wenzel, R.P. and Donowitz, L.C. 1987. Emergence of coagulase negative staphylococci as major nosocomial bloodstream pathogens. *Infect Control*, **8**, 108–12.

Stone, H.H., Hooper, C.A., et al. 1976. Antibiotic prophylaxis in gastric biliary and colonic surgery. *Ann Surg*, **184**, 443–52.

Strachan, C.J.L. 1993. Antibiotic prophylaxis in peripheral vascular and orthopaedic prosthetic surgery. *J Antimicrob Chemother*, **31**, Suppl. B, 65–78.

Struelens, M.J., Rost, F., et al. 1993. *Pseudomonas aeruginosa* and Enterobacteriaceae bacteremia after biliary endoscopy: an outbreak investigation using DNA macrorestriction analysis. *Am J Med*, **95**, 489–98.

Struelens, M.J., Study Group on Epidemiological Markers (ESGEM) of the European Society for Clinical Microbiology and Infectious Diseases (ESCMID), et al. 1996. Consensus guidelines for appropriate use and evaluation of microbial epidemiologic typing systems. *Clin Microbiol Infect*, **2**, 2–11.

Struelens, M. 2002. Molecular typing: a key tool for the surveillance and control of nosocomial infection. *Curr Opin Infect Dis*, **15**, 383–5.

Subcommittee of the Central Health Services Council, 1959. *Staphylococcal infection in hospital*. HMSO: London.

Sullivan, D. and Coleman, D. 1998. *Candida dubliniensis*: Characteristics and identification. *J Clin Microbiol*, **36**, 329–34.

Tablan, O.C., Anderson, L.J., et al. 1994. Guidelines for prevention of nosocomial pneumonia. Part I. Issues on prevention of nosocomial pneumonia – 1994. *Am J Infect Control*, **22**, 247–92.

Talon, D. 1999. The role of the hospital environment in the epidemiology of multiresistant bacteria. *J Hosp Infect*, **43**, 13–17.

Tam, A.Y. and Yeung, C.Y. 1988. The changing pattern of severe neonatal staphylococcal infection: a 10-year study. *Aust Paediatr J*, **24**, 275–9.

Taylor, D.N., Bied, J.M., et al. 1982. Salmonella dublin infections in the United States, 1979–1980. *J Infect Dis*, **146**, 322–7.

Teare, E.L. and Peacock, A. 1996. The development of an infection control link nurse programme in a distrct general hospital. *J Hosp Infect*, **34**, 267–78.

Tenover, F.C. and McGowan, J.E. 1996. Reasons for the emergence of antibiotic resistance. *Am J Med Sci*, **311**, 9–16.

Tenover, F.C., Arbeit, R., et al. 1994. Comparison of traditional and molecular methods of typing isolates of *Staphylococcus aureus*. *J Clin Microbiol*, **32**, 407–15.

Thomachot, L., Vialet, R., et al. 1999. Do the components of heat and moisture exchanger filters affect their humidifying efficacy and the incidence of nosocomial pneumonia? *Crit Care Med*, **27**, 923–8.

Thibon, P., Le Coutour, X., et al. 2000. Randomized multi-centre trial of the effects of a catheter coated with hydrogel and silver salts on the incidence of hospital-acquired urinary tract infections. *J Hosp Infect*, **45**, 117–24.

Thompson, P.J., Greenough, A., et al. 1992. Nosocomial bacterial infections in very low birth weight infants. *Eur J Pediatr*, **151**, 451–4.

Thurm, V. and Gericke, B. 1994. Identification of infant food as a vehicle in a nosocomial outbreak of *Citrobacter freundii*: epidemiological subtyping by allozyme, whole-cell protein and antibiotic resistance. *J Appl Bacteriol*, **76**, 553–8.

Tobin, J.O'H., Beare, J., et al. 1980. Legionnaires' disease in a transplant unit: isolation of the causative agent from shower baths. *Lancet*, **2**, 118–21.

Tobin, J.O'H., Swann, R.A. and Bartlett, C.L.R. 1981. Isolation of *Legionella pneumophila* from water systems: methods and preliminary results. *Br Med J*, **282**, 515–17.

Tokars, J.I., Marcus, R., et al. 1993. Surveillance of HIV infection and zidovudine use among health care workers after occupational exposure to HIV-infected blood. *Ann Intern Med*, **118**, 913–19.

Tomasz, A. 1994. Rockefeller University Workshop special report. Multiple antibiotic-resistant pathogenic bacteria. *N Engl J Med*, **330**, 1247–51.

Tredget, E.E., Shankowsky, H.A., et al. 1992. Epidemiology of infections with *Pseudomonas aeruginosa* in burn patients: the role of hydrotherapy. *Clin Infect Dis*, **15**, 941–9.

Tsang, K.W., Ho, P.L., et al. 2003. A cluster of cases of severe acute respiratory syndrome in Hong Kong. *N Engl J Med*, **348**, 1977–85.

Tullo, A.B. and Higgins, P.G. 1980. An outbreak of adenovirus type 4 conjunctivitis. *Br J Ophthalmol*, **64**, 489–93.

Update: Management of patients with suspected viral hemorrhagic fever – United States. 1995. *MMWR Morbid Mortal Wkly Rep*, **44**, 475–479.

van Belkum, A. 1994. DNA fingerprinting of medically important microorganisms by the use of PCR. *Clin Microbiol Rev*, **7**, 174–84.

Vandenbrouke-Grauls, C.M.J.E., Kerver, A.J.H., et al. 1988. Endemic *Acinetobacter anitratus* in a surgical intensive care unit: mechanical ventilators as reservoir. *Eur J Clin Microbiol Infect Dis*, **7**, 485–9.

van der Poel, C.L., Cuypers, H.T. and Reesink, H.W. 1994. Hepatitis C virus six years on. *Lancet*, **344**, 1475–9.

van der Waaij, D., Manson, W.L., et al. 1990. Clinical use of selective decontamination: the concept. *Intens Care Med*, **16**, S212–15.

van der Wall, E. and Verkooyen, R.P. 1992. Prophylactic ciprofloxacin for catheter-associated urinary-tract infection. *Lancet*, **339**, 946–51.

van Saene, H.K.F., Stoutenbeek, C.C. and Stoller, J.K. 1992. Selective decontamination of the digestive tract in the intensive care unit: current status and future prospects. *Crit Care Med*, **20**, 691–703.

Vartivarian, S.E., Papadakis, K.A., et al. 1994. Mucocutaneous and soft tissue infections caused by *Xanthomonas maltophilia*. A new spectrum. *Ann Intern Med*, **121**, 969–73.

von Eiff, C., Becker, K., et al. 2001. Nasal carriage as a source of *Staphylococcus aureus* bacteremia. Study Group. *N Engl J Med*, **344**, 11–16.

von Graevenitz, A. and Amsterdam, D. 1992. Microbiological aspects of peritonitis associated with continuous ambulatory peritoneal dialysis. *Clin Microbiol Rev*, **5**, 36–48.

Verity, P., Wilcox, M.H., et al. 2001. Prospective evaluation of environmental contamination by *Clostridium difficile* in isolation side rooms. *J Hosp Infect*, **49**, 204–9.

Verville, T.D., Huycke, M.M., et al. 1994. *Rhodococcus equi* infection of humans. 12 cases and a review of the literature. *Medicine (Balt)*, **73**, 119–32.

Vincent, J.L., Bihari, D.J., EPIC International Advisory Committee, et al. 1995. The prevalence of nosocomial infection in intensive care units in Europe. Results of the European Prevalence of Infection in Intensive Care (EPIC) study. *JAMA*, **274**, 639–44.

Wagner, S.J., Friedman, L.I. and Dodd, R.Y. 1994. Transfusion-associated bacterial sepsis. *Clin Microbiol Rev*, **7**, 290–302.

Walters, B.C. 1992. Cerebrospinal fluid shunt infection. *Neurosurg Clin North Am*, **3**, 387–401.

Ward, V., Wilson, J., et al. 1997. *Preventing hospital-acquired infection – clinical guidelines*. London: PHLS.

Watanakunakorn, C. and Perni, S.C. 1994. *Proteus mirabilis* bacteremia: a review of 176 cases during 1980–1992. *Scand J Infect Dis*, **26**, 361–7.

Watt, B. 1995. Lesser known mycobacteria. *J Clin Pathol*, **48**, 701–5.

Weber, D.J. and Rutala, W.A. 1988. Bacillus species. *Infect Control Hosp Epidemiol*, **9**, 368–73.

Weigelt, J.A., Dryer, D. and Haley, R.W. 1992. The necessity and efficiency of wound surveillance after discharge. *Arch Surg*, **127**, 77–81.

Weightman, N.C., Simpson, E.M., et al. 1988. Bacteraemia related and indwelling central venous catheters: prevention, diagnosis and treatment. *Eur J Clin Microbiol Infect Dis*, **7**, 125–9.

Weinstein, R.A. 1998. Nosocomial Infection update. *Emerg Infect Dis*, **4**, 416–20.

Weischer, M. and Kolmos, H.J. 1992. Retrospective 6-year study of enterobacter bacteraemia in a Danish university hospital. *J Hosp Infect*, **20**, 15–24.

Wells, F.C., Newsom, S.W.B. and Rowlands, C. 1983. Wound infection in cardiothoracic surgery. *Lancet*, **1**, 1209–10.

Wendt, C. and Wenzel, R.P. 1996. Value of the hospital epidemiologist. *Clin Microbiol Infect*, **1**, 154–9.

Wenger, P.N., Otten, J., et al. 1995. Control of nosocomial transmission of multidrug-resistant *Mycobacterium tuberculosis* among healthcare workers and HIV-infected patients. *Lancet*, **345**, 235–40.

Wenzel, R.P. 1986. Old wine in new bottles. *Infect Control*, **7**, 485–6.

Wenzel, R.P. 1993. *Prevention and control of nosocomial infections*. Baltimore: Williams & Wilkins.

Wenzel, R.P. 1995. The Lowbury Lecture. The economics of nosocomial infections. *J Hosp Infect*, **31**, 79–87.

Wenzel, R.P. and Edmond, M.B. 2003. Listening to SARS: lessons for infection control. *Ann Intern Med*, **139**, 592–3.

Whyte, W. 1981. The Casella slit sampler or the biotest centrifugal sampler – which is more efficient? *J Hosp Infect*, **2**, 297–9.

Wilcox, M.H. 1998. Antibiotic use and abuse. *Lancet*, **352**, 1152.

Wilcox, M.H. and Dave, J. 2000. The cost of hospital-acquired infection and the value of infection control. *J Hosp Infect*, **45**, 81–4.

Wilcox, M.H. and Fawley, W.N. 2000. Hospital disinfectants and spore formation by *Clostridium difficile*. *Lancet*, **356**, 1324.

Wilcox, M.H., Williams, P., et al. 1991. Variation in the expression of cell envelope proteins of coagulase-negative staphylococci cultered under iron-restricted conditions in human peritoneal dialysate. *J Gen Microbiol*, **137**, 2561–70.

Wilcox, M.H., Cunniffe, J.G., et al. 1996. Financial burden of hospital-acquired *Clostridium difficile* infection. *J Hosp Infect*, **34**, 23–30.

Wilcox, M.H., Fawley, W., et al. 2000a. In vitro activity of new generation fluoroquinolones against genotypically distinct and indistinguishable *Clostridium difficile* isolates. *J Antimicrob Chemother*, **46**, 551–6.

Wilcox, M.H., Fitzgerald, P., et al. 2000b. A five year outbreak of methicillin-susceptible *Staphylococcus aureus* phage type 53,85 in a regional neonatal unit. *Epidemiol Infect*, **124**, 37–45.

Wilcox, M.H., Fawley, W.N., et al. 2003a. Comparison of effect of detergent versus hypochlorite cleaning on environmental contamination and incidence of *Clostridium difficile* infection. *J Hosp Infect*, **54**, 109–11.

Wilcox, M.H., Hall, J., et al. 2003b. Use of perioperative mupirocin to prevent methicillin-resistant *Staphylococcus aureus* (MRSA) orthopaedic surgical site infections. *J Hosp Infect*, **54**, 196–201.

Wilkinson, P.J. 1988. Food hygiene in hospitals. *J Hosp Infect*, **11**, Suppl. A, 77–81.

Wilkinson, P.J., Dart, S.P. and Hadlington, C.J. 1991. Cook-chill, cook-freeze, cook-hold sous vide: risks for hospital patients. *J Hosp Infect*, **19**, 225–30.

Willcox, M.D.P. and Stapleton, F. 1996. Ocular bacteriology. *Rev Med Microbiol*, **7**, 123–31.

Williams, E.W., Hawkey, P.M., et al. 1983. Serious nosocomial infection caused by *Morganella morganii* and *Proteus mirabilis* in a cardiac surgery unit. *J Clin Microbiol*, **18**, 5–9.

Williams, R.E.O. 1956. The progress of ideas on hospital infection. *Bull Hyg*, **31**, 965–79.

Williams, R.E.O. 1971. Summary of conference. In: Brachman, P.S. and Eickhoff, T.C. (eds), *Proceedings of the International Conference on Nosocomial Infections*. American Hospital Association, Chicago, 318.

Williams, R.E.O., Blowers, R., et al. 1960. *Hospital infection. Causes and prevention*. London: Lloyd-Luke, p. 30.

Willoughby, R.E. and Pickering, L.K. 1994. Necrotizing enterocolitis and infection. *Clin Perinatol*, **21**, 307–15.

Wilson, A.P. 1995. Surveillance of wound infection. *J Hosp Infect*, **29**, 81–6.

Wilson, J. 1995. *Infection control in clinical practice*. London: Baillière Tindall.

Woodford, N., Johnson, A.P., et al. 1995. Current perspectives on glycopeptide resistance. *Clin Microbiol Rev*, **5**, 585–615.

Woodhead, M. 1994. Pneumonia in the elderly. *J Antimicrob Chemother*, **34**, Suppl. A, 85–92.

Woodhead, K., Taylor, E.W. and Baumsler, G. 2002. Behaviours and rituals in the operating theatre. *J Hosp Infect*, **51**, 241–55.

Woods, D.E., Straus, D.C., et al. 1981. Role of salivary protease activity in adherence of gram-negative bacilli to mammalian buccal epithelial cells in vivo. *J Clin Invest*, **68**, 1435–40.

Woods, T.D. and Watanakunakorn, C. 1996. Bacteremia due to *Providencia stuartii*: review of 49 episodes. *South Med J*, **89**, 221–4.

Working Party of the British Society for Antimicrobial Chemotherapy, 1987. Diagnosis and management of peritonitis in continuous ambulatory peritoneal dialysis. *Lancet*, **1**, 845–9.

Wurtz, R., Moye, G. and Jovanovic, B. 1994. Handwashing machines, handwashing compliance, and potential for cross-contamination. *Am J Infect Control*, **22**, 228–30.

Yano, M., Doki, Y., et al. 2000. Preoperative intranasal mupirocin ointment significantly reduces postoperative infection with *Staphylococcus aureus* in patients undergoing upper gastrointestinal surgery. *Surg Today*, **30**, 16–21.

Young, S.E.J. 1982. Bactaeraemia 1975–1980: a survey of cases reported to the PHLS Communicable Disease Surveillance Centre. *J Hosp Infect*, **5**, 19–26.

Zangwill, K.M., Schuchat, A., et al. 1992. Group B streptococcal disease in the United States, 1990: report from a multistate active surveillance system. *Morbid Mortal Weekly Rep CDC Surveill Summ*, **41**, 25–32.

Zuckerman, M. 2002. Surveillance and control of blood-borne virus infections in haemodialysis units. *J Hosp Infect*, **50**, 1–5.

17

Microbial susceptibility and resistance to chemical and physical agents

A. DENVER RUSSELL

INTRODUCTION

Chemical and physical agents have, in one form or another, been used for many centuries to destroy microorganisms. In ancient times, methods were empirical and it is only within the last 150 years or so that a scientific basis for them has been developed and their applications expanded. Early studies, considered in part by Hugo (1991a, 1999a) and Block (2001a), form the basis of the scientific consideration of antisepsis, disinfection, preservation, and sterilization.

This chapter will consider various types of chemical and physical agents, their medical and other uses, and the mechanisms of both microbial inactivation and microbial resistance. Because the subject area is so vast, appropriate review articles will be cited frequently together with salient research papers when necessary.

CHEMICAL AGENTS USED IN ANTISEPSIS, DISINFECTION, PRESERVATION, AND STERILIZATION

There are several types of chemical antimicrobial agents (for chemical structures, see Russell and Chopra 1996; Hugo and Russell 1999; Russell et al. 1999). They have a variety of uses, both medical and otherwise; their activity is influenced by many factors (physical, chemical, and biological); their efficacy may be evaluated by various techniques; and their mechanism of action varies from one type of chemical to another.

Terminology

Several important terms (see also BSI 1991; Block 2001b) are in regular use, including the following:

- *Sterilization*: the destruction or removal (by filtration) of all microorganisms.
- *Disinfection*: the destruction of most microorganisms, but not usually bacterial spores. Usually applied to the treatment of inanimate objects, it can also refer to the disinfection of the skin, e.g. before an operation. In some countries, a disinfectant is considered to be a substance that will also destroy bacterial spores.
- *Antisepsis*: the destruction or inhibition of microorganisms on living tissues, thereby limiting or preventing the harmful effects of infection.
- *Preservation*: the prevention of multiplication of microorganisms in formulated products (e.g. food, pharmaceuticals, cosmetics), thereby preventing spoilage or contamination, which could render the product a possible hazard to the consumer.

Other terms are more specific, e.g. those having the suffix 'stat' or 'cide.' A static agent, e.g. bacteriostat, fungistat, or sporistat, inhibits the growth of bacteria or fungi or germination of spores, respectively, whereas a cidal agent, e.g. bactericide, fungicide, sporicide, or virucide, kills bacteria, fungi, spores, or viruses, respectively. The general term 'biocide' refers to a chemical (nonantibiotic) agent that inactivates microbes. The term 'chemosterilant' (or sometimes 'chemosterilizer' or 'sterilant') is used to describe a chemical agent that is used to kill all microorganisms, including spores (Block 2001b).

Types and uses of chemical agents

Antimicrobial agents are widely used for many purposes (Table 17.1). Comprehensive data are provided by Gardner and Peel, (1991), Ayliffe et al. (1993), Gould (1995), Rutala (1990, 1995), McDonnell and Russell (1999), Russell (1999a, b, c, d, e), and Block (2001b). Fraise (1999) describes factors influencing the choice of hospital disinfectants, Coates and Hutchinson (1994) consider the relevant aspects of a suitable hospital disinfection policy and Rutala and Weber (1999a, b), the role of disinfection and sterilization in infection control, and the disinfection of endoscopes.

PHENOLS, BISPHENOLS, AND CRESOLS

Most phenols used as disinfectants are obtained from tar, a byproduct in the destructive distillation of coal. Fractionation of the tar yields a group of products including phenols (tar acids). A coal tar produced in a low-temperature carbonization process contains (in parentheses, boiling range in °C) phenol (182), cresols (189–205), xylenols (210–230), and high-boiling tar acids (230–310). The combined fraction of cresols plus xylenols is also available commercially as cresylic acid. Noncoal tar phenols are also available, e.g. 2-phenylphenol (o-phenylphenol). Phenol itself is now made in large quantities synthetically, as are some of its derivatives (Hugo and Russell 1999).

Para (4)-substitutions in phenols of an alkyl chain up to six carbon atoms in length increase antimicrobial activity. Activity is also improved by halogenation and by a combination of alkyl and halogen substitution; greatest activity is achieved when the alkyl group is in the *ortho*(2) and the halogen *para*(4) positions in relation to the phenolic group. Nitration also increases antimicrobial potency but unfortunately also enhances systemic toxicity; nitrophenols act as uncoupling agents and interfere with oxidative phosphorylation. Phenols are more active in the undissociated forms at acid than at alkaline pH (Russell and Chopra 1996; Hugo and Russell 1999).

Phenols and cresols are bactericidal to gram-positive and gram-negative bacteria, but bacterial spores are resistant at ambient temperatures. 2-Phenylphenol and the black fluids (see below) are particularly effective against mycobacteria, but bisphenols are ineffective. Phenols and cresols, especially when halogenated, possess antifungal activity. Lipid-enveloped viruses may be sensitive to phenolics whereas nonenveloped viruses are more resistant: 2-phenylphenol is active against both types. Triclosan (2,4,4'-trichlor-2'-hydroxydiphenylether) is a chlorinated phenylether (sometimes referred to as a

bisphenol) used in some medicated soaps and hand-cleansing gels and as a body wash; it is predominantly bacteriostatic rather than bactericidal, but has a broad spectrum of activity, except for *Pseudomonas aeruginosa* (Schweizer 2001). Hexylresorcinol (HCP) is active mainly against gram-positive bacteria (Russell and Chopra 1996).

Black fluids consist of a solubilized crude phenol fraction prepared from tar acids (boiling range 250–310°C). On dilution with water, black fluids give either clear solutions or emulsions. White fluids consist of emulsified phenolic compounds. On dilution with water, they form weaker emulsions and are more stable in the presence of electrolytes than are black fluids.

ACIDS, ACIDULANTS, AND ESTERS

Many aliphatic and aromatic acids are employed as preservatives, especially in the food industry, and to some extent in pharmaceutical and cosmetic products. They include acetic, benzoic, propionic, and sorbic acids and the methyl, ethyl, propyl, and butyl esters (the parabens) of *para*(4)-hydroxybenzoic acids. They are not sporicidal and the activity of the acids, but not the esters, is very pH-dependent, since activity is associated mainly with the undissociated form. Acids may be added to various foods as acidulants, notably the organic lipophilic acids (benzoic, sorbic, propionic) referred to above. Organic acids of chain length greater than C_{10} or C_{11} are highly effective against gram-positive but not against gram-negative bacteria (Russell and Gould 1988), but their low solubility restricts their use (Sofos and Busta 1999). Lactic acid, an acidulant exhibiting preservative activity, is the main product of many food fermentations.

Citric acid is an approved disinfectant against foot-and-mouth virus. Lactic acid, CH_3-$CH(OH)COOH$, has been employed as an aerial disinfectant against nonsporing bacteria. Formic acid, HCOOH, and propionic acid have been used for controlling salmonellae in feedstuffs. Two mineral acids employed in veterinary work are hydrochloric and sulfuric acids. The former, but not the latter, is sporicidal and has been used at a concentration of 2.5 percent for disinfecting hides and skin contaminated with anthrax spores. In some countries, a 5 percent solution of sulfuric acid has been used, usually in combination with phenol, for decontaminating floors, feed boxes, and troughs (Russell and Hugo 1987).

ALKALIS

The antimicrobial action of alkalis is related to hydroxyl ion concentration. Sodium hydroxide (caustic soda, lye, soda lye, NaOH) possesses strong alkali properties. It kills most common vegetative bacteria and high concentrations (5 percent and above) are lethal to anthrax spores. Calcium hydroxide (hydrated lime, air-slaked

Table 17.1 *Summary of uses of some antimicrobial agents*

Group	Use(s)	Example(s)
Acids and esters	Preservation	Organic acids, parabens
	Aerial disinfection	Lactic acid
	Veterinary disinfection	Hydrochloric, sulfuric, citric acids
	Salmonella control in feedstuffs	Formic acid, propionic acid
Alcohols	Working surfaces, equipment, gloved hands (rapid action)	Ethanol, isopropanol
	Preservation	Bronopol, ethanol
Aldehydes	Disinfection/sterilization of thermolabile medical equipment	Glutaraldehyde, succinaldehyde-based products, orthophthalaldehyde
	Virucide in preparation of some human and veterinary vaccines; removal of warts; antiseptic mouthwash	Formaldehyde solution
	Topically; irrigation solutions; treatment of peritonitis	Formaldehyde-releasing agents, e.g. noxythiolin, taurolin[a]
	Cosmetic preservatives	Formaldehyde-releasing agents, e.g. imidazole derivatives
Alkalis	Vehicle disinfection	Sodium carbonate, trisodium phosphate
	Whitewashing surfaces	Calcium hydroxide
Alkylating agents	See Vapor-phase disinfectants	Ethylene oxide, propylene oxide, β-propiolactone
Amphoteric surfactants	Skin 'disinfection;' disinfection of surgical instruments; sanitizers and disinfectants in food industry	Dodecyl-di(aminoethyl)-glycine derivatives
Antibiotic	Food preservative	Nisin
Biguanides	Antiseptic, disinfectant, preservative in some ophthalmic products; antiplaque agent; veterinary teat dip	Chlorhexidine, alexidine
	Swimming-pool disinfection; application to surfaces in food industry; preservation of leather	Polyhexamethylenebiguanide (PHMB), a polymeric biguanide
Bisphenols	Surgical scrubs; medicated soaps; limited uses as preservative in cosmetics	Hexachlorophane
	Surgical scrubs; soaps; deodorants; hand-cleansing gels	Triclosan
Diamidines	Topical application to wounds	Propamidine, dibromopropamidine, as isethionates
Halogen-releasing agents	Disinfection of blood spillages containing HIV or HBV; industrial sanitizing compounds (food, dairy, restaurant, swimming pool); veterinary disinfection	Hypochlorites, dichloro- and trichloro-isocyanuric acids
	Disinfection of hands preoperatively; antiseptics; cleansing of dairy plant; veterinary teat dip	Iodophors (including povidone–iodine)
Heavy-metal derivatives	Algicides; fungicides; wood, paint, cellulose and fabric preservation	Copper derivatives
	Fungicides; bactericides; textile and wood preservation	Organotin compounds
	Pharmaceutical preservation	Organomercurials
	Prevention of infection in burns	Silver nitrate, silver sulfadiazine
Isothiazolones	Preservatives for cosmetics, toiletries and pharmaceuticals, fabrics	Mixture of chloromethyl and methyl derivatives
Peroxygens		
	Oxidizing agent used as antiseptic and disinfectant	Hydrogen peroxide
	Disinfection/chemosterilization of endoscopes	Peracetic acid
Phenols and cresols	Preservation	Phenol, cresol, chlorocresol
	Disinfection	Black fluids, white fluids
	Aerial disinfection	Hexylresorcinol
	Food preservatives	Phenolic antioxidants, e.g. butylated hydroxyanisole, butylated hydroxytoluene

(Continued over)

Table 17.1 *Summary of uses of some antimicrobial agents (Continued)*

Group	Use(s)	Example(s)
Quaternary ammonium compounds	Preoperative disinfection, bladder and urethra irrigation, ophthalmic preservation, skin disinfection, oral and pharyngeal antisepsis, cosmetic preservation of emulsions	Cetrimide, benzalkonium chloride, cetylpyridinium chloride (as appropriate)
Sulfites and nitrites	Food preservation	Sodium metabisulfite, sodium nitrite (carcinogenic?)
Vapor-phase disinfectants	Medical and pharmaceutical sterilization	Ethylene oxide
	Disinfection/sterilization of heat-sensitive materials	Low-temperature steam with formaldehyde
	Veterinary fumigation	Formaldehyde
	Decontamination	β-Propiolactone (carcinogenic?)
	Surface sterilization	Hydrogen peroxide (vapor phase)
	Range of disinfection and sterilization purposes	Ozone

HBV, hepatitis B virus; HIV, human immunodeficiency virus.
a) There is some doubt as to whether taurolin acts as a formaldehyde releaser.

lime, $Ca(OH_2)$) is produced and heat is generated when calcium oxide (lime, quicklime, CaO) is moistened with water. As a 20 percent suspension, calcium hydroxide is effective for whitewashing surfaces, which kills most types of nonsporing bacteria. Sodium carbonate (Na_2CO_3) is used primarily as a cleaning agent, a 4 percent w/v solution being used for washing vehicles before disinfecting them after a shipment of animals. It has also been used extensively as a cleaning agent in outbreaks of foot-and-mouth disease. Trisodium phosphate (Na_3PO_4) has similar properties and uses. Comprehensive data on the veterinary uses of alkalis are provided by Huber (1982) and Linton et al. (1987).

CHLORINE-RELEASING AGENTS

The stability of free available chlorine in solution depends upon a number of factors, in particular chlorine concentration, pH, presence of organic matter, and light (Dychdala 2001).

Hypochlorites and organic *N*-chloro compounds are the two most widely used types of chlorine-releasing agents (CRAs). The hypochlorites have a wide antimicrobial spectrum and are among the most potent sporicidal agents. They are active against enveloped and nonenveloped viruses (Springthorpe and Sattar 1990; Sattar et al. 1994; Maillard 2001; Prince and Prince 2001) but at one time were considered to be rather ineffective against mycobacteria (Croshaw 1971). More recent studies demonstrate, however, that they are mycobactericidal (Favero and Bond 1993; Dychdala 2001).

Two factors with pronounced effects on their antimicrobial action are (a) the presence of organic matter, chlorine being highly reactive; and (b) pH, the hypochlorites being more active in acid than in alkaline conditions: the active factor is undissociated hypochlorous acid, HClO (Dychdala 2001). Hypochlorite

solutions gradually lose strength on storage, so fresh solutions must be prepared before use. Methanolic solutions buffered to pH 7.6–8.1 appear to show maximal stability and sporicidal activity and are worthy of further investigation. In addition to their medical uses, hypochlorites are used widely in the dairy industry and as disinfectants of farm buildings, e.g. for concrete floors, walls, and ceilings (Linton et al. 1987). Sodium hypochlorite is normally used for the disinfection of swimming pools.

N-chloro compounds, containing the =N–Cl group, possess microbicidal activity; they include chloramine T, dichloramine, halazone, and the sodium salts of dichloroisocyanuric (NaDCC) and trichloroisocyanuric acids. All appear to hydrolyze in water to produce an imino (=NH) group. Their action is slower than that of the hypochlorites, but this can be increased under acidic conditions. The disinfecting action of chloramine decreases less significantly than that of the other compounds in the presence of organic matter, and it has been used in veterinary practice for washing and spraying surfaces and for soaking items to be decontaminated. The uses of chlorine compounds are summarized in Table 17.1. NaDCC has been recommended as a disinfectant for use against various body spillages from acquired immune deficiency syndrome (AIDS) patients (Bloomfield et al. 1990; Bloomfield 1996; Dychdala 2001; Prince and Prince 2001).

Chlorine dioxide, a recently introduced instrument disinfectant (Knapp and Battisti 2001; McDonnell 2001), is more active than the hypochlorites at alkaline pH and in the presence of organic matter, although its activity may be reduced under dirty conditions. It is sporicidal, bactericidal, mycobactericidal (including activity against glutaraldehyde-resistant *Mycobacterium chelonae* strains), and fungicidal (Fraise 1999; Griffiths et al. 1999).

SUPEROXIDIZED WATER

Superoxidized water contains a mixture of oxidizing substances, the presence of hypochlorous acid probably contributing to a marked degree to its mechanism of action (Rutala and Weber 2001a, b). It is prepared by the electrolysis of saline solutions, the end product being water. It is rapidly microbicidal to a range of micro-organisms in the absence (Selkon et al. 1999) and presence (Shetty et al. 1999) of serum, although activity against *Clostridium difficile* spores is greatly reduced (Shetty et al. 1999).

IODINE AND IODOPHORS

Iodine was first employed for the treatment of wounds some 150 years ago. Normally it is used in aqueous or alcoholic solution but it is only sparingly soluble in cold water; solutions can be made with potassium iodide. Iodine is an efficient microbicidal agent rapidly lethal to bacteria and their spores, molds, yeasts, and viruses. Antiseptic-strength iodine solutions are not sporicidal (Russell 1990a, b; Favero and Bond 1993). Iodine is less reactive than chlorine but, whereas the activity of high iodine concentrations is little affected by the presence of organic material, that of low concentrations is significantly reduced. The activity of iodine is greater at acid than at alkaline pH; the most active form is diatomic iodine (I_2). At acid and neutral pH, hypoiodous acid (HIO) is less bactericidal. At alkaline pH, hypoiodite ion (HIO^-) is even less active and iodiate (IO_3^-), iodide (I^-), and triiodide (I_3^-) ions are all inactive (Trueman 1971; Bloomfield 1996).

Because iodine has certain limitations in use, namely toxicity and staining of fabrics, attention has been turned towards the iodophors (literally, 'iodine carriers'), solutions in which iodine is solubilized by surface-active agents and that retain the microbicidal, but not the undesirable, properties of iodine (Bloomfield 1996). In most iodophor preparations the carrier is usually a nonionic surfactant, whereas in poloxamer iodine formulations the carriers are poloxamers, a series of nonionic polyoxethylene–polyoxypropylene polymers. When an iodophor is diluted with water, dispersion of the micellar aggregates of iodine occurs and most of the iodine is liberated slowly (Gottardi 1985). Dilutions of commercial povidone–iodine solutions may be more bactericidal (Berkelman et al. 1982) or sporicidal (Williams and Russell 1993) than undiluted stock solutions. The reasons are complex but iodine complexation is involved since the concentration of free iodine (I_2) determines activity (Gottardi 1985).

The iodophors are microbicidal with activity over a wide pH range. Provided that the pH does not rise above about 4, iodophors retain their antimicrobial potency in the presence of organic matter. There is a pronounced decrease in activity if solutions are diluted excessively with water that has a high alkaline hardness.

The presence of a surface-active agent as carrier improves the wetting capacity. Iodophors are used in the dairy industry; when employed in the cleansing of dairy plant, it is important to keep the pH acidic with phosphoric acid, to ensure adequate removal of milkstone (dried residue of milk). They are also used for skin and wound disinfection. In some countries, alcoholic solutions of iodophors are widely used for disinfection of operation sites. In the veterinary context, iodophors formulated with phosphoric acid are employed as antiseptics, disinfectants, and teat dips (Huber 1982; Russell and Hugo 1987).

SURFACE-ACTIVE AGENTS

Surface-active agents (surfactants) have hydrophobic and hydrophilic regions in their molecular structure. On the basis of the charge or the absence of ionization of the hydrophilic group, these surfactants are classified into anionic, cationic, nonionic, and ampholytic (amphoteric) compounds.

Nonionic surfactants are not antimicrobial, but low concentrations of polysorbates (Tweens) are claimed to affect the permeability of the outer membrane of gram-negative bacteria (Brown 1975). High concentrations, however, neutralize the activity of some types of antimicrobial agents (Russell et al. 1979; Russell 1999a). *Anionic surfactants* usually have strong detergent but weak antimicrobial properties, except at high concentrations, which induce lysis of gram-negative bacteria (Salton 1968). Fatty acids are considerably more active against gram-positive than gram-negative organisms (Russell and Gould 1988).

Amphoteric agents combine the detergent properties of anionic with the antimicrobial properties of the cationic compounds. Activity remains virtually constant over a wide pH range and they are inactivated less readily than cationic surfactants by proteins. Examples of amphoteric surfactants are the Tego series of compounds (Hugo and Russell 1999).

For microbiological use, most important surface-active agents are the *cationic surface-active agents*, quaternary ammonium compounds (QAC) (Hugo and Russell 1999; Merianos 2001). They possess strong bactericidal but, at normal in-use concentrations, weak detergent properties. QACs may be considered as organically substituted ammonium compounds in which the nitrogen atom has a valency of five; four of the substituent radicals (R^1–R^4) are alkyl or heterocyclic and the fifth is a small anion. The sum of the carbon atoms in the four R groups is greater than ten. For a QAC to have high antimicrobial activity, at least one of the R groups must have a chain length in the range C_8–C_{18} (Hugo and Russell 1999). The QACs are primarily active against gram-positive, nonsporing bacteria; at high concentrations they are lethal to gram-negative organisms, although *P. aeruginosa* tends to be particularly resistant (Russell 1999a).

They are sporostatic but not sporicidal, and fungistatic rather than fungicidal. They are active against viruses with lipid envelopes, e.g. herpes and influenza, but are much less so against nonenveloped viruses, e.g. enteroviruses (Narang and Codd 1983; Resnick et al. 1986). A newer QAC product is claimed to have mycobactericidal activity (Nicholson et al. 1995) but no neutralizing agent was included in the test methodology.

The QACs are incompatible with a wide range of chemical agents, including nonionic and anionic surfactants, and phospholipids such as lecithin. Use is made of this property in evaluating the lethal effects of QACs by employing a combination of lecithin and a nonionic surfactant as a neutralizing agent (Russell et al. 1979). Antimicrobial activity is affected greatly by organic matter and by pH. Activity is greater in alkaline conditions because there is an increase in the degree of ionization of bacterial surface groups so that the cell surface becomes more negatively charged.

Organosilicon-substituted quaternary ammonium salts, organic amines, or amine salts with antimicrobial activity in solution are also highly effective on surfaces. One such, 3-(trimethoxysilyl)propyloctadecyldimethyl ammonium chloride, exhibits powerful antimicrobial activity while chemically bonded to a variety of surfaces (Malek and Speier 1982; Speier and Malek 1982). Uses of the QACs are summarized in Table 17.1.

BIGUANIDES AND POLYMERIC BIGUANIDES

The most important member of the family of N^1,N^5-substitued biguanides is chlorhexidine, which is available as dihydrochloride, diacetate, and gluconate, the last-named being the most water-soluble. Chlorhexidine has a wide spectrum of activity against gram-positive and gram-negative bacteria but is not sporicidal or mycobactericidal; it is also generally considered as having a low activity against fungal spores and viruses (Russell and Day 1993; Ranganathan 1996; Hugo and Russell 1999; Russell 1999a, c; Denton 2001). Potency is reduced in the presence of serum, blood, pus, and other organic matter. Because of its cationic nature, activity is also reduced in the presence of soaps and other anionic compounds. Chlorhexidine is more active at alkaline pH because an increase in the degree of ionization of bacterial surface groups renders the cell surface more negatively charged (Hugo 1999b). The main uses of chlorhexidine are as a medical, dental, and veterinary antiseptic, as a disinfectant, and as a preservative in some types of pharmaceutical products (Russell and Day 1993).

Alexidine differs from chlorhexidine in that it posesses ethylhexyl end groups (Russell and Chopra 1996); it is also more rapidly bactericidal and produces a significantly faster alteration in bacterial permeability (Chawner and Gilbert 1989a,b).

Vantocil is a heterodispersed mixture of polyhexamethylene biguanides (PHMB) with a molecular weight of approximately 3 kDa. It is active against gram-positive and negative bacteria (*P. aeruginosa* and *Proteus vulgaris* are less sensitive) but is not sporicidal. Because of the residual positive charges on the polymer, PHMB is precipitated from aqueous solutions by anionic compounds. The activity of PHMB increases with increasing polymer lengths but synergistic effects have been observed in a mixture of polymeric biguanides (Gilbert et al. 1990a, b).

DIAMIDINES

The aromatic diamidines are organic cationic agents that show antimicrobial activity (Hugo 1971). The two most important members are propamidine (4,4′-diamidinophenoxypropane) and the more active dibromopropamidine, both used as the soluble isethionate salts. Gram-positive bacteria are considerably more sensitive than gram-negative organisms and activity decreases at acid pH and in the presence of organic matter. Bacteria cultured in the presence of increasing doses of a given diamidine rapidly acquire resistance to it and to other diamidines (Hugo 1971). The past decade or so has seen the isolation of multiple antibiotic-resistant strains of *Staphylococcus aureus* (Townsend et al. 1984) and *S. epidermidis* (Leelaporn et al. 1994), which possess plasmid-mediated resistance to cationic agents, including propamidine.

ALDEHYDES

Three aldehydes are important, namely the monoaldehyde, formaldehyde (methanal), and two dialdehydes, glutaraldehyde (pentane-1,5-dial) and *ortho*-phthalaldehyde, although other alkdehydes, e.g. succinaldehyde, also possess antimicrobial activity.

Formaldehyde

Formaldehyde (CH_2O) is employed in the liquid and vapor states; use of the latter is described below (see section on Gaseous and vapor-phase disinfectants).

Formaldehyde solution (formalin) is an aqueous solution containing 34–38 percent w/w of formaldehyde (CH_2O). The presence of methyl alcohol (methanol, CH_3OH) delays polymerization. Formaldehyde is lethal to bacteria and their spores (but less so than glutaraldehyde), fungi, and viruses. It combines readily with proteins and is less effective in the presence of organic matter. Formaldehyde is employed as a virucidal agent in the preparation of many human and veterinary vaccines (Russell and Hugo 1987), as an antiseptic mouthwash, for the disinfection of membranes in dialysis equipment, and as a preservative in hair shampoos.

Formaldehyde is often employed in the form of formaldehyde-releasing agents. Examples are (a) noxythiolin (hydroxymethylenethiourea), a bactericidal agent widely used both topically and in accessible body cavities, e.g. an irrigation solution in the treatment of peritonitis (Browne and Stoller 1970); (b) taurolin, in which the amino acid taurine acts as a formaldehyde carrier (Browne et al. 1976, cf. Myers et al. 1980); (c) hexamine (methenamine), imidazole derivatives, triazines, and oxazolo-oxazoles. For further information, see Hugo and Russell (1999).

Glutaraldehyde

Glutaraldehyde (GTA) is a saturated five-carbon dialdehyde with an empirical formula of $C_5H_8O_2$ and a molecular weight of 100.12. It is highly active, reacting with enzymes and proteins but only slightly with nucleic acids; it prevents dissociation of free ribosomes. Probably because of increased interaction with $-NH_2$ groups, such activities increase with increasing pH, a factor of considerable importance to microbicidal activity (Russell 1994; Power 1997).

GTA possesses high microbicidal activity against bacteria and their spores, mycelial and spore forms of fungi, and various types of viruses, including human immunodeficiency virus (HIV) and enteroviruses (Hanson et al. 1994; Russell 1994). Despite earlier reports to the contrary, GTA is now considered to be mycobactericidal (Russell 1998a), although *Mycobacterium avium–intracellulare* might be of above-average resistance (Hanson 1988) and highly GTA-resistant strains of *Mycobacterium chelonae* have been isolated (van Klingeren and Pullen 1993; Walsh et al. 1999a, b, 2001; Fraud et al. 2001, 2003). In solution, GTA is more stable at acid than at alkaline pH, whereas the converse is true for its antimicrobial activity. In practice, 2 percent (or greater) solutions of GTA are alkalinated when required and are used within a stipulated period. Some formulations are available that appear to have overcome the problem of stability (Babb et al. 1980).

The dialdehyde is an important fixative in leather tanning, in electron microscopy and in biochemistry. It is employed widely for the disinfection or sterilization of medical equipment liable to damage by heat, in particular endoscopes. Nevertheless, there is the possibility of severe toxic reactions arising in personnel and its replacement by other equally microbiologically active agents is being actively pursued.

Ortho-phthalaldehyde

Ortho-phthaladehyde (OPA) (Alfa and Sitter 1994) is a recently inroduced (1998) cyclic dialdehyde. It is used, at a concentration of 0.55 percent and unadjusted pH, predominantly for the disinfection of endoscopes. OPA has potent lethal activity against gram-positive (including mycobacteria) and -negative bacteria (Alfa and Sitter 1994; Gregory et al. 1999; Walsh et al. 1999a, b, 2001; Fraud et al. 2001, 2003). At ambient temperatures, in-use concentrations and unadjusted pH, OPA has a low order of sporicidal activity (Walsh et al. 1999a, b),

but sporicidal activity is observed when the concentration is increased and solutions made more alkaline.

Significantly, OPA is lethal towards GTA-resistant *M. chelonae* strains (Walsh et al. 1999a, b, 2001; Gregory et al. 1999; Fraud et al. 2001, 2003).

ALCOHOLS

Generally, alcohols rapidly kill bacteria (Morton 1983), including acid-fast bacilli, but are not sporicidal and have poor activity against some viruses, although HIV type 1 is susceptible to ethanol and isopropanol in the absence of organic matter (van Bueren et al. 1994). The presence of water is essential to the antimicrobial action of ethanol, which is most effective at concentrations of 60–70 percent (Price 1950).

Isopropanol (propan-2-ol) is a more effective bactericide. Benzyl alcohol (phenylmethanol) is a weak local anesthetic and also possesses antimicrobial properties. Phenylethanol (phenylethyl alcohol) is selectively active against gram-negative bacteria in mixed flora and is sometimes used as a preservative in ophthalmic solutions. Phenoxyethanol (phenoxetol) has significant activity against *P. aeruginosa* but less against other bacteria (Hugo and Russell 1999). Bronopol (2-bromo-2-nitropropane-1,3-diol) has a broad spectrum of activity, which is reduced in the presence of serum and especially of sulfydryl compounds. The effect of pH on its activity is complex (Croshaw and Holland 1984). Bronopol is widely used as a cosmetic preservative. Chlorbutanol (chlorbutol, trichloro-*t*-butanol) has been used as a bactericide in solutions for injection, but its instability presents a problem.

ISOTHIAZOLONES

Three isothiazolones have been studied comprehensively: 1,2-benzisothiazol-3-one (BIT), 5-chloro-*N*-methylisothiazol-3-one (CMIT), and *N*-methylisothiazol-3-one (MIT). They are widely used as industrial preservatives. Their activity is rapidly quenched by thiol-containing compounds and by valine and histidine, nonthiol amino acids (Collier et al. 1990).

DYES

Three groups of dyes find application as antimicrobial compounds:

- The acridines, which are heterocyclic compounds and which have been studied extensively. Albert (1979) showed that small changes in their chemical structure cause significant changes in biocidal properties, the most important factor being ionization, which must be cationic in nature. Acridines compete with H^+ ions for anionic sites on the bacterial cell and are more effective at alkaline than at acid pH.
- The triphenylmethane dyes (e.g. crystal violet, brilliant green, and malachite green), which have been used as topical antiseptics. However, their uses were

limited because they are effective only against gram-positive bacteria (Hugo and Russell 1999). This property does, however, have a practical application in the formulation of selective media for diagnostic purposes. Like the acridines, these dyes are more active at alkaline pH.

- The quinones, which are natural dyes imparting color to many forms of plant and animal life. Some members are important agricultural fungicides, notably chloranil and dichlone (Owens 1969; D'Arcy 1971).

HEAVY-METAL DERIVATIVES

The salts of heavy metals are sometimes employed as antimicrobial agents. The main antimicrobial use of copper derivatives is as algicides and fungicides; some copper compounds are used as preservatives in wood, cellulosics, paints, and fabrics (Hilditch 1999; Springle 1999).

Mercury compounds, in the form of organic derivatives (e.g. phenylmercuric nitrate and acetate), are still, despite criticism, sometimes used as preservatives for parenteral and ophthalmic solutions. Thiomersal serves this purpose for various immunological products. They are active against both gram-positive and gram-negative bacteria but are sporistatic, not sporicidal, at ambient temperatures (Russell 1998b; Hugo and Russell 1999).

Organotin compounds are used as biocides (e.g. fungicides, bactericides) and textile and wood preservatives (Hilditch 1999; McCarthy 1999).

Silver and its salts have long been used as antimicrobials. In recent years, silver nitrate has been employed to prevent infection of burns. Lowbury (1992) reviewed the use of this compound, at a concentration of 0.5 percent, for topical antimicrobial prophylaxis, and as a more satisfactory topical prophylactic, silver sulfadiazine (which, however, has the disadvantage that many bacteria are sulfonamide-resistant).

Russell and Hugo (1994) have described the antimicrobial properties and actions of various silver compounds.

PERMEABILIZING AGENTS

A permeabilizing agent is a chemical that increases the permeability of the outer membrane of gram-negative bacteria.

Chelating agents are not usually considered as being antimicrobial agents, but ethylenediamine tetraacetic acid (EDTA) enhances the activity of many antiseptics and disinfectants. The potentiating effects by EDTA on antibacterial agents acting against staphylococci (Kraniak and Shelef 1988) are likely to be the result of sequestration of metal ions present in the growth media rather than an increase in cellular permeability. In medicine, EDTA is used to treat chronic lead poisoning and pharmaceutically as a stabilizer in certain parenteral and ophthalmic preparations (Russell 1999a). In the context of disinfection, EDTA potentiates the effects of many antibacterial agents against gram-negative, but not usually gram-positive, bacteria (Russell and Furr 1977; Ayres et al. 1993, 1998a, b, c, d).

Polyethyleneimine (PEI) is a newer type of permeabilizer (Helander et al. 1997, 1998) that increases the permeability of gram-negative bacteria to several biocides.

ANILIDES

Anilides have the general structure C_6H_5NHCOR. In salicylanilide, R is C_6H_4OH and in carbanilide (diphenylurea) it is C_6H_5NH. Salicylanilide was introduced in 1930 as a fungistat for use in textiles and has also been used in ointment form for treating ringworm. Of the many substituted salicylanilides tested, the tribromo- and tetrachlorosalicylanilides have been the most widely used as antimicrobial agents. Their photosensitizing properties have, however, restricted their use in situations in which they come into contact with human skin. Trichlorocarbanilide, although one of the most potent members of the substituted carbanilides, has the same disadvantage.

QUINOLINE AND ISOQUINOLINE DERIVATIVES

There are three main groups: 8-hydroxyquinoline, 4-aminoquinaldinium, and isoquinoline derivatives. 8-Hydroxyquinoline (oxine) is a chelating agent active only in the presence of certain metal ions (Albert 1979). The 4-aminoquinaldinium derivatives are QACs that contain one or more quinoline ring systems; examples are laurolinium acetate and dequalinium chloride (a bis-QAC), both of which are active against gram-positive bacteria and many species of yeast and fungi (D'Arcy 1971). The most important isoquinoline derivative is hedaquinium chloride, another bisquaternary salt, which possesses antibacterial and antifungal properties.

PEROXYGENS

A peroxygen is a compound containing an –O–O– group. Hydrogen peroxide and peracetic acid are important peroxygens but others also find application.

Hydrogen peroxide

Hydrogen peroxide, H_2O_2, is active against gram-positive and gram-negative bacteria. Because they are unable to produce catalase, anaerobes are particularly susceptible. At high concentrations, hydrogen peroxide is sporicidal; this activity increases as the temperature is raised (Bloomfield 1999; Block 2001b). Activity is less affected by pH than many other biocides and is enhanced in the presence of cupric (Cu^{2+}) of ferric (Fe^{3+}) ions, which promote free-radical oxidation of organic compounds with peroxide. A mixture of peroxide and Cu^{2+}, but neither alone, produces strand breaks in DNA (Sagripani and Kraemer 1989). It is environmentally friendly because its decomposition products are oxygen and water (Fraise 1999). Some uses are described in Table 17.1.

New hydrogen peroxide products have been described. A product based on accelerated and stabilized hydrogen peroxide (ASHP) had broad-spectrum activity, the undiluted form being sporicidal (>7/8 log reduction in 6 h), mycobactericidal (>6 log reduction in 20 min), and fungicidal (5 log reduction in 5 min) at 20°C (Sattar et al. 1999). AHSP was shown to be an excellent hard-surface cleaning agent with potent microbicidal activity (Rochon and Sullivan 1999; Alfa and Jackson 2001).

Peracetic acid

The best known example of a peracid is peracetic acid, CH_3COOOH, a strong oxidizing agent that is rapidly lethal for a wide range of microbes, including bacterial spores, viruses, and fungi (Baldry 1983; Broadley et al. 1995; Fraise 1999; Block 2001d). It has greater lipid solubility and more potent antimicrobial activity than hydrogen peroxide. It is sporicidal even at low temperatures and effective in the presence of organic matter (Block 2001c). The lethal species is the ●OH radical (Clapp et al. 1994). Peracetic acid is frequently used as a chemosterilant (see section on New low-temperature sterilization technologies).

Performic acid

Performic acid, HCOOOH (see section on New low-temperature sterilization technologies) is a new proprietary liquid sterilant that is rapidly sporicidal (Rutala and Weber 2001a).

Ozone

Ozone, O_3, is an allotropic form of oxygen. Because of its powerful oxidizing properties (Schneider 1998; Weavers and Wickramanayake 2001), ozone is bactericidal, virucidal and sporicidal, although spores are some 10–15 times more resistant than vegetative cells. Gaseous ozone reacts with amino acids, especially those containing sulfur, and with RNA and DNA. In water, ozone is unstable chemically but activity persists because of the production of free radicals, including ●OH. Its unstable nature means that ozone has to be produced on site (Schneider 1998).

SULFITES AND NITRITES

Traditional chemical food preservatives include common salt, sucrose, spices, and smoke and its components. Sulfur dioxide has for centuries been used as a fumigant and as a wine preservative. Metabisulfites have been widely used as antioxidants in foods (and in some pharmaceutical products), and sulfur dioxide and sulfites are used to preserve a variety of food products (Gould and Russell 1991; Sofos and Busta 1999).

Nitrite (NO_2^-) and nitrate (NO_3^-) have been used for centuries in meat processing. By reaction with heme proteins, nitrites are responsible for color formation in cured meat. They are antimicrobial, especially against outgrowing *Clostridium botulinum* spores. Nitrates function solely as a source of nitrite. The antimicrobial activity of nitrite is affected by many factors (Roberts et al. 1991). Carcinogenic nitrosamines are formed in some cured meat products cooked under certain conditions.

GASEOUS AND VAPOR-PHASE DISINFECTANTS

Gaseous and vapor-phase chemical agents have long been used to achieve disinfection and sterilization. Sulfur dioxide (obtained by burning sulfur) and chlorine found early application for fumigating sickrooms, but the scientific basis for their use was established only comparatively recently. A brief historical account is provided by Richards et al. (1984). Kaye and Phillips (1949) reviewed early studies on ethylene oxide (ETO).

The most important vapor-phase agents are ETO and formaldehyde.

Ethylene oxide

Ethylene oxide, $(CH_2)_2O$, is a colorless gas that is soluble in water, most organic solvents, and oils and diffuses into rubber in a manner similar to entering a solution (Joslyn 2001). It is inflammable when more than 3 percent is present in air, but this hazard can be overcome by mixing it with carbon dioxide or an appropriate hydrochlorofluorocarbon (HCFC) compound. Chlorofluorocarbons originally used for this purpose are no longer permitted (Alfa et al. 1998) and HCFCs are banned in some countries.

The antimicrobial activity of ETO depends upon several factors, notably relative humidity (RH), temperature, concentration, and time, and especially on the presence of water vapor (Ernst 1974; Joslyn 2001). Its high toxicity must be borne in mind when devising safe sterilization procedures. Furthermore, porous materials absorb the gas to various degrees during the sterilization cycle, so various periods of time after sterilization must be allowed for desorption of residual ETO. ETO is used as a decontaminating agent, for sterilizing ophthalmic and anesthetic equipment, crude drugs, and powders, and in veterinary practice for fumigating egg shells.

Formaldehyde

Formaldehyde gas can be generated by various means, namely evaporation of commercial formaldehyde solution (formalin), addition of formalin to potassium permanganate, or volatilization of paraformaldehyde,

$HO(CH_2)_nH$, where $n = 8$–100. Its antimicrobial activity depends upon several factors, including RH; it increases with RH up to a figure of 50 percent, but higher RH values confer little further advantage (Nordgren 1939). Formaldehyde is toxic, and inhalation of the vapor may pose a risk of carcinogenesis; adequate precautions should be taken to protect personnel. Low-temperature steam with formaldehyde (LTSF) has been used for disinfecting or sterilizing heat-sensitive materials (Russell 1999a). In the veterinary field, formaldehyde vapor is an important fumigant of animal buildings (Russell and Hugo 1987).

Other gases and vapors

These include

- β-Propiolactone (BPL). Its activity depends primarily on RH, concentration, and temperature. BPL may be carcinogenic, but it has been claimed to have a use in decontamination of animal premises. Liquid BPL is used widely in the preparation of many veterinary viral vaccines.
- Methyl bromide, CH_3Br. This is less active than ETO and is highly toxic but has been used as a fumigant.
- Propylene oxide, C_3H_6O. This is also less active than ETO. It has been used as a decontaminating agent, e.g. for animal feeds.
- Ozone. This is discussed above.
- Carbon dioxide, CO_2. This inhibits the growth of bacteria, including slime-producing bacteria, in soft drinks. Its activity depends upon low temperatures, its addition at an early stage, and its concentration.
- Hydrogen peroxide vapor. This has been used for the sterilization of surfaces in flexible and rigid-wall isolators and small rooms (Groschel 1995) and for endoscopes; it also has dental applications (Schneider 1998).
- Gaseous chlorine dioxide. This is unstable at high concentrations and must thus be generated on site (Schneider 1998).

Gas and vapor-phase plasmas are discussed in the section on New low-temperature sterilization technologies.

AERIAL DISINFECTANTS

An effective aerial disinfectant should be capable of being dispersed so as to ensure its complete and rapid mixing with infected air. An effective concentration should be maintained in the air and the disinfectant must be highly and rapidly active against air-borne microorganisms at different relative humidities. In addition, it should be nontoxic and nonirritant.

Early samples of aerial disinfectants were fumigants, such as sulfur dioxide and chlorine, employed in sickrooms. However, aerosols – which consist of a very fine dispersed liquid phase in a gaseous (air) disperse phase – are the most important form of aerial disinfectant.

Examples of aerial disinfectants are: (a) hexylresorcinol, which is vaporized from a thermostatically controlled hotplate; (b) lactic acid, which is effective but unfortunately an irritant at high concentrations; (c) propylene glycol, which may be used as a solvent for dissolving a solid disinfectant prior to atomization, but which is also a fairly effective and nonirritating antimicrobial agent in its own right; and (d) fumigants such as formaldehyde.

NATURAL ANTIMICROBIAL SYSTEMS

Natural antimicrobial systems (reviewed by Wilkins and Board (1989), Board (1995)) occur in animals, plants, and microorganisms. Major components of natural defense systems are classified broadly into inducible, e.g. complement, and constitutive. Principal representatives of the latter are enzymes, lysozyme (a cell-wall lytic agent, see section on Mechanisms of action of antimicrobial agents) and lactoperoxidase (LP), which mediates the production of compounds toxic to foreign, but not to host, cells. LP is most abundant in bovine cells. With a halide or thiocyanate and hydrogen peroxide, LP forms a potent antimicrobial lactoperoxidase system.

Additionally, transferrin in mammalian blood cells and milk, and lactoferrin in milk, may show antibacterial activity. They both act as iron-binding proteins, inducing lipopolysaccharide loss from the outer membranes of gram-negative bacteria and an increased cellular permeability (Russell and Chopra 1996).

Factors influencing activity

The activity of antimicrobial compounds may be influenced by various extraneous factors, including concentration, pH, time, temperature, type, number, growth phase, and location of microorganisms, and the presence of other agents that may reduce or increase their potency (Russell 1999a). A sound knowledge of these factors is essential for their most effective deployment, e.g. in the design of hospital disinfection policies (Coates and Hutchinson 1994) and in selecting disinfectant agents for particular purposes (Fraise 1999; Rutala and Weber 1999a, b).

CONCENTRATION OF ANTIMICROBIAL AGENT

The effect of concentration, or dilution, on the activity of an antimicrobial compound is not a simple arithmetic one. As was first pointed out by Krönig and Paul (1897), microbial death is not an all-or-nothing response but depends greatly on the period of contact and concentration (Figure 17.1).

In kinetic studies, the concentration exponent (dilution coefficient, η) is used to express the effect of

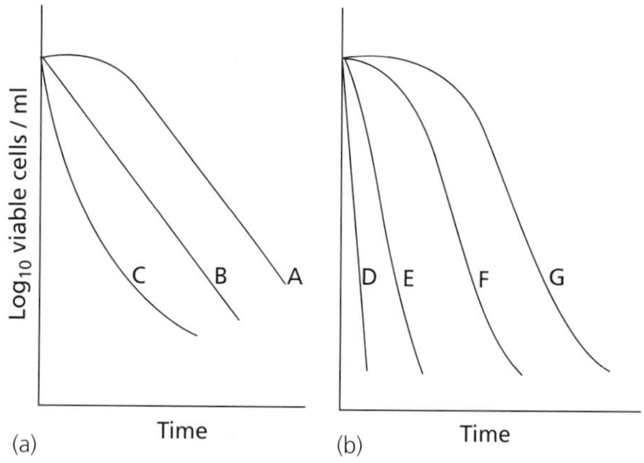

(a)   Time

(b)   Time

Figure 17.1 *Examples of time–survivor curves of bacterial suspensions exposed to disinfectants.* **(a)** *Bacterial responses: A, initial 'shoulder' followed by exponential death; B, exponential death; C, exponential death followed by 'tailing.'* **(b)** *Effects of different concentrations (D, highest; G, lowest) of phenol acting on Escherichia coli. At the lowest concentration, a sigmoidal curve is produced. This becomes less pronounced as the concentration increases, leading to an apparent straight-line response at the highest concentration.*

changes in concentration (or dilution) on cell-death rate. Its value may be determined by measuring the times necessary to kill the same number of bacteria in a suspension exposed to two concentrations of the antimicrobial agent. If C_1 and C_2 represent the two concentrations and t_1 and t_2 the respective times to reduce the viable population to the same end point, then

$$C_1 \eta t_1 = C_2 \eta t_2 \qquad (18.1)$$

from which

$$\eta = (\log t_2 - \log t_1)/(\log C_1 - \log C_2). \qquad (18.2)$$

Alternatively, η may be calculated from the slope of the straight line resulting when $\log_{10}$ death time ($\log t$) is plotted against $\log_{10}$ concentrations ($\log C$) (see Russell 1999a). Examples of η values are provided in Table 17.2.

A decrease in concentration of compounds with high η values results in a pronounced increase in the time necessary to achieve a comparable kill, other conditions

remaining constant. By contrast, agents with low η values are much less affected (Table 17.2). Concentration is the single most important factor affecting biocidal activity (McDonnell and Russell 1999; Russell and McDonnell 2000).

Knowledge of the effect of concentration on antimicrobial potency is essential:

- in evaluating activity
- in testing medical and pharmaceutical products for sterility
- in ensuring adequate concentrations of preservative in pharmaceutical, cosmetic and food products, and
- in providing adequate concentrations of antiseptics and disinfectants for practical purposes (Russell 1999a).

TEMPERATURE

The activity of a disinfectant or preservative is usually increased when the temperature at which it acts is raised, although compounds vary considerably in response to temperature changes (Table 17.2).

A useful formula for measuring the effect of temperature on activity is given by

$$\theta^{T2-T1} = k_2/k_1 = t_1/t_2 \qquad (18.3)$$

in which θ is the temperature coefficient, k_2 and k_1 the rate (velocity) constants at temperatures T_2 and T_1, respectively, and t_2 and t_1 the respective times to bring about a complete kill at T_2 and T_1.

The velocity constant (k) can itself be obtained from a knowledge of the time (t) taken to reduce the initial viable number of cells (N_0) to a value N_t at time t, thus:

$$k = \frac{1}{t}\log_e \frac{N_o}{N_t} = \frac{1}{t}2.303 \log_{10} \frac{N_o}{N_t}. \qquad (18.4)$$

The temperature coefficient, θ, refers to the effect on activity per 1°C rise and is nearly always between 1.0 and 1.5 (Bean 1967). It is more meaningful to specify the θ^{10} (also known as Q^{10}) value, which is the change in activity per 10°C rise in temperature (Table 17.2). At temperatures above about 40°C, there is little, if any,

Table 17.2 *Effects of concentration and temperature on activity of some antimicrobial agents*

Concentration exponents (η)[a]	Temperature coefficients (Q_{10} [or θ^{10}] values)[b]
Group A ($\eta < 2$)	θ^{10} 1.5–5
Hydrogen peroxide, mercurials, chlorhexidine, formaldehyde, QACs	Phenolics, ethylene oxide, β-propiolactone, phenols, cresols
Group B ($\eta = 2$–4)	θ^{10} 30–50
Parabens, sorbic acid	Aliphatic alcohols
Group C ($\eta > 4$)	
Aliphatic alcohols, phenolics, benzyl alcohol, phenethanol	

QAC, quaternary ammonium compound.
a) Based on Hugo and Denyer (1987).
b) Based on Bean (1967).

difference in activity between acid and alkaline glutaraldehyde, although the latter formulation is less stable (Gorman et al. 1980).

ENVIRONMENTAL PH

pH can influence the activity of antimicrobial agents in the following ways:

- Changes may occur in the molecule. Phenols and benzoic, sorbic, and dehydroacetic acids are effective predominantly in the unionized form and, as pH rises, their degree of dissociation increases. It is claimed that the dissociated form of sorbic acid may make a small contribution to activity.
- Changes may occur in the microbial cell surface. As pH increases, the number of negatively charged groups on the cell surface increases, causing enhanced binding of positively charged molecules, e.g. QACs (Hugo 1999b), dyes (Moats and Maddox 1978), acridines (Albert 1979), and chlorhexidine (Hugo 1999b). It must also be pointed out that the sporicidal activity of sodium hypochlorite is potentiated in the presence of methanol, although there is no simple relationship between activity, stability, and change of pH of the mixture (Coates and Death 1978). Maximal sporicidal activity and stability are achieved by buffering hypochlorite alone or a hypochlorite–methanol mixture to within a pH range of 7.6–8.1 (Death and Coates 1979).

INTERFERING SUBSTANCES

Organic matter

Organic soiling matter may be present as, for example, serum, blood, pus, soil, food residues, dried milk (milkstone), or fecal material; it can interfere with the action of an antimicrobial agent, usually because an interaction results in a reduced effective concentration of the latter (Russell 1999a).

Such reduction in activity is most noticeable with highly reactive compounds such as chlorine disinfectants. Iodine and iodophors, because of their lower chemical reactivity, are influenced to a rather lesser extent. Cationic agents such as QACs and chlorhexidine show a considerable reduction in activity, whereas the efficacy of phenols depends upon the actual phenolic compound; lysol, for example, retains much of its activity in the presence of feces and sputum (Russell 1999a).

Adequate precleaning before employment of a disinfectant, or a combination of disinfectant with a suitable detergent, may overcome the problem caused by the presence of organic soil. However, it should be noted that the activity of disinfectants may suffer considerably in the presence of certain detergents.

Surface-active agents

Cationic bactericides and anionic surfactants are incompatible. Low concentrations of nonionic surfactants such as polysorbates, tritons, and tergitols increase the activity of cationic agents and of parabens (Allwood 1973), whereas at higher concentrations of the nonionic detergents, significantly greater concentrations of the antimicrobial substances are necessary to inhibit or kill microbes. It is believed that below the critical micelle concentration (CMC) of the nonionic compound, potentiation occurs by action on the surface layers of the bacterial cell, resulting in enhanced permeability to the antimicrobial agent. Above the CMC, it either forms a complex with the antimicrobial compound or the latter is partitioned between the aqueous and micellar phases, only the concentration in the aqueous phase being available for microbial attack.

Because of their ability to inactivate various types of antimicrobial compounds, nonionic surfactants are frequently employed as neutralizing agents in sterility testing and in antimicrobial evaluation (Russell et al. 1979; Sutton 1996; Fassihi 2001).

Oils

A problem encountered in the formulation of pharmaceutical and cosmetic creams and emulsions is that the antimicrobial activity of a preservative may be high in aqueous conditions but much less so when an oil is present. The reason for this is that the preservative is partitioned between the oily and aqueous phases (Bean 1972; Fassihi 2001).

Other factors

The hardness of the water with which disinfectants are prepared or diluted is a contributory factor in reducing the effectiveness of QACs and iodophors; black (but not white) fluids are incompatible with hard waters. Partitioning into rubber may cause problems in the preservation of multiple dose parenteral and ophthalmic solutions. Relative humidity has a profound influence on the activity of gaseous disinfectants (see section on Gaseous and vapor-phase disinfectants).

CONDITION, GROWTH PHASE, AND SIZE OF MICROBIAL POPULATION

It is clearly easier for an antimicrobial agent to be effective when there are few microorganisms against which to act. Adequate and thorough precleaning is usually an important prerequisite to a disinfection process. Exponential phase cells of *S. aureus* (Luppens et al. 2002) and *Listeria monocytogenes* (Luppens et al. 2001) are more susceptible to cationic and oxidizing biocides than cells in the decline phase.

The association of bacteria (or other microorganisms) with solid surfaces leads to the production of a biofilm, a consortium of bacteria organized within an extensive exopolymer (glycocalyx). Within the depths of the biofilm, growth rates are likely to be reduced as a consequence of nutrient limitation (Poxton 1993), which can alter the bacterial cell surface and hence be responsible for modified sensitivity to antibacterial agents (Costerton et al. 1987; Carpentier and Cerf 1993; Stickler and King 1999). This is one reason for the reduced sensitivity of the sessile cells found in biofilms; other reasons include chemical reaction with glycocalyx and prevention by the glycocalyx of access of biocide to the underlying cells. This aspect is considered in more detail below (see Biofilms and resistance).

Types of microorganism

The activity of antimicrobial agents depends greatly upon the type of microorganism present. It is therefore pertinent to examine the sensitivity of various types of microbes which, for convenience, have been classified into the following groups.

GRAM-POSITIVE BACTERIA (COCCI)

Generally, cocci are more sensitive to antimicrobial agents than are gram-negative bacteria (Baird-Parker and Holbrook 1971; Russell and Gould 1988). Cocci are readily killed by halogens, phenols, especially bisphenols, and QACs. However, multiple antibiotic-resistant *S. aureus* and *S. epidermidis* may show reduced sensitivity to cationic-type agents (see Types and uses of chemical agents). Clinical isolates of enterococci (*Enterococcus faecium* and *E. faecalis*) are often antibiotic-resistant and include vancomycin-resistant enterococci (VRE), but do not appear to be more resistant to chlorhexidine than staphylococci (Baillie et al. 1992; Alqurashi et al. 1996). The effects of biocides on antibiotic-resistant cocci are discussed by Day and Russell (1999).

MYCOBACTERIA

The sensitivity of acid-fast bacteria to disinfectants is intermediate between that of other nonsporulating bacteria and bacterial spores (Ayliffe et al. 1993; Favero and Bond 1993; Russell 1998a; Lauzardo and Rubin 2001). QACs and dyes inhibit tubercle bacilli but do not kill them. *Mycobacterium tuberculosis* is also insensitive to chlorhexidine, acids, and alkalis, but moderately sensitive to amphotericsurface-active agents, including the Tego compounds. Of the phenols, 2-phenylphenol is particularly effective, but the bisphenols are inactive. Alcohols, liquid and vapor-phase formaldehyde, formaldehyde–alcohol, iodine–alcohol, and ethylene oxide are all tuberculocidal; despite earlier doubts to the contrary, so is GTA (Favero and Bond 1993). However, *M. avium–intracellulare* (Hanson 1988) and *M. chelonae* (van Klingeren and Pullen 1993) may be resistant to the dialdehyde (see also Terminology).

GRAM-NEGATIVE BACTERIA

Many types of gram-negative bacteria, especially *Escherichia coli*, *Klebsiella* spp., *Proteus* spp., *P. aeruginosa*, and *Serratia marcescens*, are increasingly implicated as hospital pathogens. In the veterinary context, bacteria causing zoonotic diseases include *Campylobacter* spp., salmonellae, and brucellae (Linton 1983; Russell et al. 1984). Gram-negative bacteria are usually less sensitive than gram-positive bacteria, and *P. aeruginosa* and *Proteus* spp. may contaminate solutions of QACs (Russell et al. 1986). These bacteria, and others such as *Providencia stuartii*, are also more resistant to chlorhexidine than is *E. coli* (Ismaeel et al. 1986).

Legionella pneumophila is frequently found in domestic water and cooling-water systems of large buildings. Although outbreaks of legionnaires' disease are uncommon, they are associated with high morbidity and some deaths when they occur. Adequate control measures must be taken. Neither raising the temperature of hot-water systems nor continuous chlorination alone is completely effective. Control of the organism in recirculating water systems has also been studied, but extrapolation from laboratory conditions to such systems is often unrealistic because of possible interaction between the biocide and other water-treatment chemicals as well as slime (Elsmore 1999). Wang et al. (1979) found hypochlorite to be effective against *L. pneumophila*, but Kurtz et al. (1982) showed a chlorinated phenol to be more effective than a QAC–tributyl tin oxide mixture or sodium dichloroisocyanurate. Bronopol is effective both under laboratory conditions and in cooling towers (Elsmore 1999).

BACTERIAL SPORES

Many antibacterial compounds have no or little sporicidal activity. Nevertheless, such chemicals are not entirely without effect on spores because they may prevent germination or outgrowth, or both. Chemicals that are sporistatic rather than sporicidal include phenols and cresols, parabens, QACs, biguanides, alcohols, dyes, and mercury compounds. Comparatively few substances are sporicidal and even then the process may be comparatively slow. Examples of sporicides are hydrogen peroxide, hypochlorites, glutaraldehyde, formaldehyde, iodine compounds, ethylene oxide, peracetic acid, and β-propiolactone (Russell 1990a, 1998a, b; Bloomfield and Arthur 1994; Setlow 1994; Bloomfield 1999). Vegetative, rod-shaped cells of *Bacillus* spp. are considerably more sensitive to biocides than the corresponding spore forms (Russell 2003b, c).

MOLDS AND YEASTS

Several species of molds and yeasts are pathogenic. Others are important in spoiling foods and pharmaceutical and cosmetic products.

Many compounds show both antibacterial and antifungal activity, although the latter may be fungistatic rather than fungicidal (Russell and Furr 1996; Russell 1999c). These include phenolics (notably the halogenated members and hexachlorophane), QACs, oxine (8-hydroxyquinoline), diamidines, organic mercury derivatives, and the parabens. Sorbic acid shows significant antifungal activity at low pH values when it occurs in solution mainly in the undissociated form. At higher pH values, activity is lost. Fungal spores are often resistant to chemical disinfectants (Russell 1999c).

PROTOZOA

Several distinctly different types of protozoa, including *Giardia*, *Cryptosporidium*, *Naegleria*, *Entamoeba*, and *Acanthamoeba*, are potentially pathogenic and may be acquired from water. Furthermore, their life cycles may contain a resistant cyst stage (Jarroll 1999). Agents, in ascending order of efficacy, that are cysticidal towards *Giardia muris* cysts are monochloramine, free chlorine, iodine, chlorine dioxide, and ozone (Jarroll 1988, 1999). Chlorine dioxide is more effective against *Cryptosporidium* oocysts than is either free chlorine or monochloramine; these cysts are also sensitive to ozone (Korich et al. 1990).

Acanthamoeba spp. may cause corneal keratitis, sometimes associated with the use of contaminated contact lenses or contact-lens solutions. Whereas trophozoites are readily inactivated by contact-lens disinfecting solutions, cysts are more refractory (Khunkitti et al. 1997, 1998a, b, 1999; Turner et al. 1999, 2000a, b) and the presence of proteinaceous matter is an additional hazard (Seal and Hay 1992).

VIRUSES

Several bactericidal agents are virucidal, but antibacterial activity does not necessarily imply antiviral potency. An excellent comprehensive treatise of virus disinfection is that presented by Grossgebauer (1970). Viruses are classified primarily as to whether they contain DNA or RNA, but the presence or absence of a lipid envelope is in general more important in relation to disinfection. An important hypothesis was put forward in 1963 and modified some 20 years later by Klein and Deforest (1983) (see also Prince and Prince 2001), who classified viruses into three groups:

1 lipid-enveloped viruses, e.g. herpes simplex virus (HSV), HIV, hepatitis B virus (HBV)
2 small nonenveloped viruses, e.g. picornaviruses, parvoviruses
3 larger nonenveloped viruses, e.g. rotaviruses.

The order of these groups, in terms of biocide sensitivity, is 1 > 3 > 2.

Detailed knowledge of the inactivation of viruses has, in the past, been important in the preparation of vaccines such as inactivated (Salk) poliomyelitis vaccine, in which formaldehyde was used as the virucidal agent. Nowadays, with the current importance of HIV, HBV, and hepatitis A virus (HAV), plus the need to prevent the transmission of viral infection, it is essential to have a sound knowledge of viral inactivation. Important virucidal agents include CRAs, formaldehyde (in inactivated vaccine production), glutaraldehyde at alkaline pH, and peracetic acid (Springthorpe and Sattar 1990; Sattar et al. 1994, 2001; Maillard and Russell 1997; Maillard 2001; Prince and Prince 2001). Bacteriophages have been considered as indicator 'organisms' for assessing the virucidal activity of biocides (Davies et al. 1993; Maillard et al. 1995; Maillard and Russell 1997) (see also Evaluation of antimicrobial activity).

Viruses also cause many diseases of animals. Lists of approved dilutions of approved disinfectants for veterinary use in the UK are frequently published by the Department for Environment, Food and Rural Affairs (DEFRA) (formerly the Ministry of Agriculture, Fisheries and Foods). These are for general use and in connection with certain Statutory Orders applying to viral infections (foot-and-mouth disease, swine vesicular disease) and bacterial disease (tuberculosis). Porcine parvovirus, which causes reproductive failure in swine, is quite resistant to disinfectants, only sodium hypochlorite and sodium hydroxide being rapidly virucidal; by contrast, two enveloped swine viruses (pseudorabies and transmissible gastroenteritis viruses) are highly sensitive to a wide range of disinfectants (Quinn and Markey 1999).

Russell (1990b) has provided a comprehensive list of the susceptibility of different types of viruses to heat and chemical biocides, as well as considering acid stability and sensitivity to the organic solvent ether, which inactivates most lipid-enveloped viruses.

PRIONS

Prions are responsible for a distinct group of unusual neurological diseases, the transmissible degenerative encephalopathies (TDE) or transmissible spongiform encephalopathies (TSE), and are sometimes referred to as unconventional transmissible agents. Examples of TDEs are Creutzfeldt–Jakob disease (CJD) and new variant Creutzfeldt–Jakob disease (nvCJD), scrapie in sheep, bovine spongiform encephalitis (BSE), and kuru.

Prions are composed of an altered protein and are inordinately resistant to most chemical and physical agents, including glutaraldehyde, formaldehyde, organic solvents, chlorine dioxide, iodine, strong acids, normal autoclaving conditions, dry heat, and ionizing and ultraviolet radiations (Taylor 1999; Baron et al. 2001). However, as prions have yet to be purified (Taylor 2004), it is difficult to state whether this is an intrinsic property or to what extent it is influenced by the protective effect of host tissue. They are inactivated by strong alkali, guanidinium hydrochloride (or isocyanate), and sodium hypochlorite (Baron et al. 2001).

Evaluation of antimicrobial activity

A detailed survey of the numerous published methods for evaluating antimicrobial activity is outside the scope of this chapter. A summary is provided in Table 17.3 and the following should be consulted for additional information: Quinn (1987), CEN (1996), Quinn and Carter (1999), Reybrouck (1999), and Cremieux et al. (2001).

The evaluation of virucidal activity is not an easy matter (Bellamy 1995). Viruses are unable to grow in artificial laboratory culture media. Another appropriate system, usually involving living cells, must thus be employed. In essence, the principles of testing follow methods of evaluating bactericidal or fungicidal activity in that the virus is exposed to appropriate concentrations of disinfectant in suspension, or on a carrier, e.g. glass coverslips, as used with herpes virus and poliovirus (Tyler and Ayliffe 1987). Inactivation is then tested in an appropriate manner, care being taken to ensure that residual disinfectant is neutralized and that the antimicrobial agent has no toxic effects on host cells.

Estimation of virucidal activity can be made using the following systems (see also Quinn and Carter 1999; Sattar and Springthorpe 2001):

- *Tissue culture* or fertile eggs, which after incubation are examined for signs of viral infection.
- *Plaque-counting* procedures, which have been used for various viruses, e.g. herpes and poliovirus (Tyler and Ayliffe 1987) and human rotaviruses (Lloyd-Evans et al. 1986; Springthorpe et al. 1986).
- Use of an *'acceptable' animal model*, e.g. the chimpanzee, for studying survival of HBV. The discovery of hepadnaviruses that infect ducks and small animals might provide a suitable alternative (Tsiquaye and Barnard 1993).
- *Immunological reactions*: tests for the inactivation of the hepatitis B surface antigen (HBsAg) were at one time used to assess the efficacy of disinfection procedures for this virus. However, since this component is much more stable to chemical and physical agents than those viral parts conferring infectivity, the method is no longer used.
- *Endogenous reverse transcriptase*: HIV is an enveloped RNA retrovirus. After HIV enters a cell, RNA is converted to DNA under the influence of an enzyme, reverse transcriptase. HIV replicates in certain T lymphocytes and in the assay reverse transcriptase activity is not a satisfactory alternative to tests in which infectious HIV can be detected in systems employing fresh human peripheral blood mononuclear cells (Resnick et al. 1986).

Table 17.3 *Examples of methods used for evaluating antimicrobial activity*

Test/evaluation	Examples of methods available[a,b]	Comments
Bacteriostatic/fungistatic potency (MIC determination)	Broth/agar serial dilution methods; agar diffusion methods	Provide preliminary information only
Bactericidal/fungicidal activity	(1) Suspension tests with determination of numbers of survivors. Include sporicidal and mycobactericidal estimations (AFNOR 1989)	(1) Neutralization of agent essential to prevent 'carry-over' into recovery medium;[a] rapid methods increasingly being studied
	(2) European Suspension Test	(2) 5 Log_{10} reduction in specified period
Phenol coefficient tests	RW, CM, AOAC methods	Compare activity of a test phenolic vs. a standard (phenol)[a]
Capacity-type tests	Kelsey–Sykes test	Refers to capacity of a disinfectant to retain its activity when repeatedly challenged by test organisms
Carrier tests	AOAC use-dilution confirmatory test; DGHM test	Organisms are dried on to appropriate carriers
Virucidal tests	Tissue culture, eggs, plaque counting; animal models; enzyme activity	See text
Skin disinfection	MIC and lethal effects (presence of serum or blood); skin test at 37°C	
Aerial disinfection	Closed-chamber evaluation of microbial population before and after exposure to aerial disinfectant	
Preservative activity	MIC evaluation, lethal activity; challenge testing of preserved pharmaceutical and cosmetic products	See British Pharmacopoeia (2000)

AFNOR (1989), Association Française de Normalisation; AOAC (1990), Association of Official Analytical Chemists; CM, Chick–Martin; DGHM, Deutsche Gesellschaft für Hygiene und Mikrobiologie; MIC, minimum inhibitory concentration; RW, Rideal–Walker.
a) See also text.
b) Suitable tests must be done where relevant to demonstrate adequate neutralization of antimicrobial agents.

- *Use of bacteriophages* as model systems: bacteriophages have been suggested as model systems for human viruses (Davies et al. 1993). Certainly, they have several advantages such as ease of handling and fairly rapid and reproducible results (Maillard et al. 1995). Their resistance to biocides should, of course, mimic a particular virus (or viruses), e.g. coliphage MS2 has the same order of response to some biocides as poliovirus.

Mechanisms of action of antimicrobial agents

Considerable progress has been made in understanding the mechanisms of action of many antibacterial agents (Denyer and Stewart 1998; Denyer 2002, Russell 1998a, b, 2002b, 2003b; Hugo 1999b). The same is not always true when the target organisms are yeasts, fungi, bacterial spores or viruses (Russell et al. 1997; Russell 2003b).

The initial reaction between an antimicrobial agent and a microbial cell takes place at the cell surface. This is followed by penetration of the chemical to reach its site(s) of action, usually within the cell itself, while some act by 'physical' puncture of cell walls. Unlike most antibiotics, however, many biocides have more than one site of action and it is thus often difficult to elucidate the exact mechanism whereby cellular inactivation is achieved. Furthermore, secondary effects may make a significant contribution to the overall process. Conversely, McMurray et al. (1998a) found that the phenylether, triclosan, acts on a specific bacterial target, inhibiting lipid synthesis in *E. coli*.

ACTION ON NONSPORULATING BACTERIA

Uptake of an antibacterial agent by bacteria represents an early manifestation of its effect, although there is little information available about uptake by mycobacteria (Russell 1996, 2003b).

Different agents have different sites and modes of action. Data are presented below, and summarized in Table 17.4, of the ways in which they act (Russell and Russell 1995; Russell and Chopra 1996). It must again be emphasized, however, that little is known about the mechanisms of inhibition and inactivation of mycobacteria (Russell 1996).

Cell-wall or cell-envelope effects

EDTA is not a powerful bactericide in its own right. Generally, it has little effect on gram-positive bacteria or fungi but is active against certain gram-negative bacteria, especially *P. aeruginosa* (Hugo and Russell 1999). Its usefulness resides in the fact that it is a chelating agent that combines with cations associated with the outer membrane (OM) of gram-negative

bacteria with the release of some 30–50 percent of OM lipopolysaccharide (LPS). These changes in the OM render the cells more sensitive to many chemically unrelated compounds such as lysozyme, QACs, dyes, and phenols (Hugo 1999b).

EDTA thus increases the permeability of gram-negative bacteria. Other chemicals that act in a similar manner include sodium hexametaphosphate, gluconic acid, citric acid, and malic acid (Vaara 1992; Ayres et al. 1993, 1998a, b) and PEI displaces Mg^{2+}, thereby opening up the OM of gram-negative bacteria (Helander et al. 1997, 1998).

GTA has several effects on nonsporulating bacteria (Russell 1994). It causes cross-linking of the peptidoglycan in gram-positive bacteria so that lysis of this component is reduced on subsequent treatment with lysozyme (*Bacillus subtilis*) or lysostaphin (*S. aureus*). GTA interacts with protein in the cell envelope of gram-negative bacteria, with consequent reduction of lysis by sodium lauryl sulfate or EDTA–lysozyme.

The dialdehyde increases resistance to lysis of protoplasts and spheroplasts, by stabilizing the cytoplasmic membrane. OPA is much less efficacious as a cross-linking agent (Walsh et al. 1999a, b, 2001).

Membrane-active agents

Various chemical agents effect perturbation of homoeostatic mechanisms in bacteria in one of two ways (Gould 1988; Russell and Hugo 1988). In the first, leakage of intracellular materials is promoted by their physical interaction with the cytoplasmic membrane. In the second, there is a specific attack on the insulatory function of the membrane, such that the proton-motive force (pmf) across it is either discharged or prevented from forming.

Leakage can be considered as being a measure of the generalized loss of function of the cytoplasmic membrane as a permeability barrier, the rate and extent of leakage usually depending on the concentration of the inhibitor and the time and temperature of exposure. Leakage may be related to bacteriostasis but not necessarily to cell death (Denyer and Stewart 1998). Examples of antibacterial agents that induce leakage are cationic, anionic, and polypeptide surface-active agents, phenol, chlorhexidine, parabens, hexachlorophane (hexachlorophene), phenoxyethanol salicylanilides, and triclosan. These effects are not, however, necessarily responsible for cell death. Triclosan, for example, has been shown to have a specific effect on enoyl-ACP reductase (FabI) in *E. coli* (McMurry et al. 1998a, b). Heath et al. (2001) proposed that this is the only target for triclosan in this organism, and that the phenylether does not possess membrane-disruptive properties (cf. McMurry et al. 1998a). FabK and FabL are the respective enoyl-ACP reductase isoforms from *Streptococcus pneumoniae* and *B. subtilis*. FabK is triclosan-resistant

Table 17.4 *Mechanisms of antibacterial action*[a]

Target site	Example(s)[b]	Mechanism(s)
Outer cell layers (cell wall/ outer membrane)	GTA, OPA	Interaction with $-NH_2$ groups, e.g. proteins, peptidoglycan (OPA less effective than GTA)
	EDTA	Chelation of divalent cations, especially Mg^{2+}
	PEI	Displaces Mg^{2+}
	Lysozyme	β,1–4 links in peptidoglycan
	Cationic biocides	
	CHA, QACs, DBPI, PHMB	Outer-membrane damage, thereby promoting own entry postulated
Cytoplasmic (inner) membrane	Phenols	Generalized membrane damage
	Triclosan	Membrane damage, but at low concentrations inhibits enoyl reductase
	QACs	Generalized membrane damage
	CHA	Low concentrations affect membrane integrity, high concentrations inhibit congeal protoplasm
	Alexidine, PHMB	Phase separation and domain formation of acidic phospholipids
	Hexachlorophane	Inhibits membrane-bound electron-transport chain
	Phenoxyethanol	Proton-conducting uncoupler
	Sorbic acid	Transport inhibitor (effect on pmf); another unidentified mechanism?
	Parabens	Low concentrations inhibit transport, high concentrations affect membrane integrity
	Metallic compounds[c]	
	mercury, silver, copper	Interaction with $-SH$ groups in proteins and enzymes
	Isothiazolones[c]	Interaction with $-SH$ groups in proteins and enzymes
	Bronopol[c]	Oxidizes thiol groups to disulfides
Cytoplasmic constituents	*Alkylating agents*	
	ETO, BPL, PO, formaldehyde	Combination with amino, carboxyl, sulfydryl, and hydroxyl groups in protein; ETO also interactions at N-7 guanine moieties in DNA
	Cross-linking agents	
	GTA, OPA, formaldehyde	Intermolecular protein cross-links; OPA less effective but enters mycobacteria more readily than GTA
	Intercalating agents, e.g. acridines	Acridines intercalate between two layers of base pairs in DNA
	Phenethyl alcohol	Inhibits DNA synthesis
	Oxidizing agents	
	1 Hypochlorites[c]	Progressively oxidize thiol groups to disulfides, sulfoxides, or disulfoxides
	2 Hydrogen peroxide[c]	Formation of free hydroxyl radicals ($\bullet OH$) causes oxidation of thiol groups in enzymes and proteins
	3 Peracetic acid[c]	Possibly disrupts thiol groups in proteins and enzymes
	Iodine	Interaction with cytoplasmic protein

BPL, β-propiolactone; CHA, chlorhexidine diacetate; DBPI, dibromopropamidine isethionate; EDTA, ethylenediamine tetraacetate; ETO, ethylene oxide; GTA, glutaraldehyde; OPA, *ortho*-phthalaldehyde; PEI, polyethyleneimine; PHMB, polyhexamethylene biguanide; pmf, proton-motive force; PO, propylene oxide; QACs, quaternary ammonium compounds.
a) For further information, see Russell and Chopra (1996).
b) Some agents affect more than one target site.
c) These compounds might affect enzymes and proteins in both the cytoplasmic (inner) membrane and the cytoplasm.

whereas FabL is reversibly inhibited by triclosan. The expression of either FabK or FabL in *E. coli* produces cells with increased triclosan resistance of three orders of magnitude. In *S. aureus*, only FabI is expressed and overexpression results in an increased MIC of triclosan (Slater-Radosti et al. 2001). Suller and Russell (1999, 2000) found that triclosan caused leakage from *S.*

aureus. Triclosan and isoniazid inhibit InhA (an FabI analogue) in *M. smegmatis* and *M. tuberculosis*, but unlike isoniazid, triclosan may have a second target in *M. smegmatis* (Heath et al. 2001). Two aspects of triclosan action are important, namely its growth-inhibitory and lethal effects. It is unlikely that inhibition of enoyl reductase is solely responsible for its lethal effect and membrane damage, including leakage, must play an important role at higher triclosan concentrations.

Chlorhexidine has a biphasic effect on membrane permeability in which an initial high rate of leakage occurs as the biguanide concentration increases, with a progressive decrease in leakage at higher chlorhexidine concentrations as a consequence of coagulation or precipitation of the cytosol (Hugo 1999b).

Both alexidine and the polymeric biguanide PHMB produce lipid phase separation and domain formation of the acidic phospholipids in the cytoplasmic membrane (Broxton et al. 1984, Chawner and Gilbert 1989a, b; Gilbert et al. 1990a, b). Chlorhexidine and QACs also combine with membrane phospholipids but do not bring about lipid phase separation and domain formation.

Ethanol and isopropanol are membrane disrupters, inducing a rapid leakage of intracellular constituents; disorganization of the membrane probably results from their penetration into the hydrocarbon core of the membrane (Seiler and Russell 1991).

Mitchell's (1961) chemiosmotic theory seeks to explain active transport, synthesis of adenosine triphosphate (ATP) and flagellar movement. During metabolism, protons are extruded to the exterior of the bacterial cell, resulting in acidification of the exterior and positivity in charge relative to the cell interior. It may be expressed mathematically as

$$\Delta p = \Delta\psi - Z\Delta pH \tag{18.5}$$

in which Δp is the pmf, ψ the membrane electric potential, ΔpH the transmembrane pH gradient, and Z a constant (2.303 RT/F) with a value of 61 at 37°C that is used to convert pH values to millivolts.

Several chemicals that are antiseptics, disinfectants, or preservatives behave in a manner similar to 2,4-dinitrophenol (DNP), i.e. they act as uncouplers of oxidative phosphorylation. Examples include 2-phenoxyethanol, bisphenols, and pentachlorophenol. Tetrachlorosalicylanilide discharges the membrane potential, $\delta\psi$, in E. faecalis. The thiobisphenol, fentichlor, behaves very much like DNP, having a direct action on the collapse of a proton potential (Bloomfield 1974). Phenoxyethanol and the alkyl phenols act as proton-conducting uncoupling agents. Pharmaceutical and food preservatives such as the parabens and lipophilic acids (propionic, sorbic, 4-hydroxybenzoic) inhibit the active uptake of some amino and oxo acids in E. coli and B. subtilis. Sorbic acid affects the pmf in E. coli and accelerates the movement of H^+ ions from low pH media into the cytoplasm. Sorbic acid appears to dissipate ΔpH, having a much smaller effect on $\Delta\psi$ (Eklund 1983, 1985a, b; Salmond et al. 1984).

Agents acting on nucleic acids and proteins

Intercalation: The antimicrobial activity of the acridines increases with degree of ionization (Albert 1979). Ionization is the most important factor governing their activity, but it must be cationic in nature, because acridine derivatives that are ionized to form anions or zwitterions are only poorly antibacterial by comparison. The acridines induce filamentous forms in gram-negative bacteria, inhibit DNA synthesis, and combine strongly with DNA, although binding to other sites such as RNA, cell envelopes, and ribosomes has also been reported. Binding to DNA has been studied extensively. The classic studies of Lerman (1961) suggested that the planar drug molecules become intercalated between adjacent base pairs in DNA. Binding to DNA occurs by two distinct mechanisms (Peacocke and Skerrett 1956). In the first, there is a first-order reaction with one proflavine molecule binding to every four or five nucleotides, whereas in the second, there is a slower higher order of reaction with one molecule per single nucleotide.

Alkylation: Alkylation is defined as the conversion of H-X→R-X, where R is an alkyl group. Biological activity of alkylating agents is indicated by reaction with nucleophilic groups.

Epoxides, of which ethylene oxide is an example, interact with amino acids and proteins. Ethylene oxide causes hydroxyethylation of amino acids and combines with the amino, carboxyl, sulfydryl, and hydroxyl groups of proteins (Russell 1976). Alkylation of phosphated guanine in nonsporing bacteria has been proposed as the primary reason for the lethal effect of ethylene oxide (Michael and Stumbo 1970; Russell 1976).

It has been suggested that formaldehyde also acts by its alkylating effect. Binding to RNA is reversible up to a point (Staehelin 1958). Interaction of formaldeyde with T2 bacteriophage DNA and with protein has been described (Grossman et al. 1961).

The reaction of GTA with nucleic acids follows pseudo-first-order kinetics at high temperatures but there is little evidence for the formation of intermolecular cross-links. GTA inhibits synthesis of protein, DNA and RNA in E. coli, but this is believed to arise from an inhibition of precursor uptake as a consequence of protein–dialdehyde interaction in the outer structures of the cell.

Interaction with enzymes: Metals such as mercury and silver interact with thiol (–SH) groups on enzymes to form mercaptides (Hugo 1999b). This reaction may be reversed by excess of an –SH compound such as sodiumthioglycollate or cysteine, an important finding since such agents are used to inactivate metals in microbicidal testing (Russell et al. 1979; Russell and Hugo 1994). Bronopol also interacts with –SH groups (Hugo 1999b).

Other effects: Silver salts produce structural changes in the cell envelope of P. aeruginosa and the Ag^+ ion reacts preferentially with the bases rather than the phosphate groups in DNA (Russell and Hugo 1994).

ACTION ON BACTERIAL SPORES

Bacterial spores are not inactivated rapidly or readily at ambient temperatures (Bloomfield and Arthur 1994;

Bloomfield 1999) and comparatively few chemicals are sporicidal (Russell 1990a, b, 1998b). The mechanism of action of those that are sporicidal is often poorly understood, although progress continues to be made.

Although it is possible to correlate changes in cell structure with the development of resistance (sporulation) or sensitivity (germination and outgrowth), it is rather more difficult to identify the mechanism whereby death is brought about (Knott et al. 1995). Superoxidized water does not cause DNA damage and dipicolinic acid (DPA) is not released; it is considered that *B. subtilis* spores are inactivated by a modification of the inner membrane, which becomes nonfunctional in germinated spores (Loshon et al. 2001). Hydrogen peroxide-treated spores maintain their permeability barrier, there being no DPA release even after subsequent heat treatment. It is unlikely that significant damage occurs to the spore inner membrane (Melly et al. 2002), although Shin et al. (1994) found that peroxide sensitizes spores to heat damage. Usual targets for peroxide attack on membranes are polyunsaturated acids, which are present in extremely low levels in spores (Melly et al. 2002). Prolonged exposure of spores to 15 percent peroxide at elevated but sublethal temperatures, or to high peroxide concentrations, causes spore lysis with major damage to coat, cortex and protoplast (Shin et al. 1994). Imlay and Linn (1988) are of the opinion that one mechanism of hydrogen peroxide action is the generation of •OH radicals that cleave the DNA backbone. Peracetic acid was found by Clapp et al. (1994) to be the most potent peroxygen in their test series, the lethal moiety being •OH radicals. Reduced transition-metal ions sensitize spores to peracetic acid (Marquis et al. 1995).

Low concentrations (0.01 percent) of alkaline GTA inhibit germination and outgrowth, whereas much higher concentrations must be used for long periods to kill spores (Russell 1990a, 1999c). In fact, as with formaldehyde (Spicher and Peters 1981), it has been shown (Power et al. 1988; Williams and Russell 1993) that it is possible to revive a very small proportion of GTA-treated spores by means of a sublethal heat shock or by sodium hydroxide or other treatment. It is thus possible that the dialdehyde is a less effective sporicide than orginally thought, but that its cross-linking effect on bacterial spores will, under ordinary circumstances, be sufficient to prevent spore germination and subsequent vegetative development. Cabrera-Martinez et al. (2002) provided evidence that GTA does not damage spore DNA but eliminates the spore's ability to germinate and that OPA-treated spores that cannot germinate are not recovered by artificial germinants or by treatment with sodium hydroxide or lysozyme.

The enzyme, lysozyme, hydrolyzes β,1–4-linkages between *N*-acetylmuramic acid and *N*-acetylglucosamine in the spore peptidoglycan in coatless spores. Sodium nitrite breaks the peptidoglycan chain at the muramic

lactam residues unique to spores. Hypochlorites interact strongly with spore coats, but their major site of sporicidal action is believed to be on the cortex (Bloomfield and Arthur 1994; Bloomfield 1999). The alkylating agents, ethylene and propylene oxides, are considered to inactivate bacteria and their spores by combining with amino, carboxyl, sulfydryl, and hydroxyl groups of proteins, as mentioned above.

The general antimicrobial activity of ethylene oxide (CH_2CH_2O) and related substances parallels their activity as alkylating agents (Phillips 1952). Thus, cyclopropane ($CH_2CH_2CH_2$), which is not an alkylating agent, is devoid of antimicrobial activity, whereas ethylene sulfide (CH_2CH_2S) and ethylene imine (CH_2CH_2NH) are potent alkylating compounds that demonstrate antimicrobial properties.

ACTION ON FUNGI

It is often assumed that the same effects as those in nonsporulating bacteria are responsible for fungal inactivation, but in view of the considerable structural and biochemical differences between these microbes this is undoubtedly an oversimplification. However, as with bacteria, it is likely that an initial interaction at the cell surface is followed by passage of a biocide across the fungal cell wall to reach its target site(s). Little information is available about the mechanisms of biocide uptake into fungal cells despite long-standing studies of biocide adsorption to yeasts and molds (Lyr 1987; Gadd and White 1989; Russell 1999c).

Although the fungal cell wall may be a prime target for developing new antifungal antibiotics (Hector 1993), few biocides are likely to have the wall as a sole target. Chitin has been suggested as a potentially reactive site for glutaraldehyde action in yeasts (Gorman and Scott 1977). Glutaraldehyde is also known to cause agglutination of yeast cells (Navarro and Monsan 1976).

Chlorhexidine induces K^+ release from bakers' yeast (Elferink and Booij 1974) and affects the ultrastructure of budding *Candida albicans* with the loss of cytoplasmic constituents (Bobichon and Bouchet 1987). Uptake of chlorhexidine by and membrane damage in *S. cerevisiae* have been described (Hiom et al. 1993, 1995a, b, 1996). The QACs also affect membrane integrity. Heavy metals probably bind to key functional groups of enzymes (Lyr 1987).

Very few relevant studies have been made on other target sites, but DNA and RNA would be expected to be targets for a number of biocides, such as cationic agents that interact strongly with nucleic acids (Hugo 1999b).

ACTION ON VIRUSES

The mechanisms whereby biocidal agents inactivate viruses are still poorly understood, although recent progress has been encouraging. In an excellent paper,

Grossgebauer (1970) discussed the possible effects of biocides on viruses and proposed that (a) water-saturated phenol caused separation of protein from infectious RNA in poliovirus; (b) low formaldehyde concentrations produced an antigenic but noninfectious particle (for use in a vaccine); and (c) higher aldehyde concentrations gave a noninfectious destroyed particle.

Thurmann and Gerba (1988, 1989) considered the possible interactions between viruses and biocides as being:

- adsorption to capsid receptors
- release of infectious nucleic acid
- destruction of capsid leading to release of infectious nucleic acid, or
- nucleic acid rendered noninfectious
- capsid remaining intact, but nucleic acid rendered noninfectious.

They pointed out that CRAs could inactivate viruses by attacking either capsid proteins or nucleic acid.

GTA reduces the activity of HBsAg and especially core antigen in HBV (Adler-Storthz et al. 1983) and interacts with lysine residues on the surface of hepatitis A virus (Passagot et al. 1987). Low concentrations (<0.1 percent) of alkaline GTA act against purified poliovirus, whereas poliovirus RNA is highly resistant to higher concentrations (Bailly et al. 1991). From this, it may be inferred that changes to the capsid are responsible for loss of infectivity. Support for this contention has been obtained by demonstrating that the capsid proteins of poliovirus and echovirus react with low concentrations (0.05 and 0.005 percent, respectively) of the dialdehyde, the ten-fold difference in aldehyde concentration probably reflecting major structural alterations in the two viruses (Chambon et al. 1992). Some biocides (hypochlorite, 70 percent ethanol and cetrimide) induce a rapid loss of the outer capsid layer, whereas chlorhexidine and phenol affect morphology only after extended periods of exposure (Rodgers et al. 1985).

Mechanistic studies of the effects of biocides on bacteriophages have been undertaken. These investigations (Maillard et al. 1995; Maillard and Russell 1997; Maillard 2001) have provided useful information about the interaction of biocides with phage protein and nucleic acid and on the inhibition of transduction, but much remains to be done to understand fully the mechanisms of inactivation.

ACTION ON PROTOZOA

Little is known about the mechanisms of antiprotozoal action of biocides, a disappointing gap in our knowledge and one that needs to be rectified. It has, however, been demonstrated (Khunkitti et al. 1997, 1998a, b, 1999) that chlorhexidine affects the membrane of trophozoites and cysts of *Acanthamoeba castellanii*, although cysts are less sensitive (Furr 1999).

Mechanisms of microbial resistance to biocides

Three levels, low, intermediate, and high, of biocidal activity have been recognized (Favero and Bond 1993). In high level, bacterial spores, fungi, mycobacteria, and viruses are inactivated; in intermediate level, mycobacteria, fungi, and viruses are inactivated; and in low level, nonsporulating bacteria (except mycobacteria) and lipid enveloped viruses are rendered nonviable.

Thus, different organisms vary considerably in their susceptibility to biocides (see Types of microorganism).

RESISTANCE OF NONSPORULATING BACTERIA

Vegetative bacterial cells respond to chemical and other stresses in a variety of ways that include the activation and expression of new groups of genes (Gould 1989). When *E. coli* is subjected to nutrient limitation or to antimicrobial agents, the growth rate decreases and gene expression is markedly altered (Gilbert et al. 1990c). This is essential for the long-term survival of cells and is partly mediated by alternative sigma factors such as σ^s encoded by the *rpoS* gene. Positive regulation of this expression results in σ^s synthesis being activated by guanosine 3'-diphosphate-5'-diphosphate (ppGpp). In an insertional mutant of *E. coli* that cannot produce σ^s, and in *E. coli* and other gram-negative organisms that produce ppGpp, the cells are more resistant to biocides and antibiotics than isogenic mutants (Greenaway and England 1999a, b). Tolerance to stress by hydrogen peroxide arises from intrinsic resistance mechanisms. The oxidative stress response involves the production of neutralizing enzymes to prevent cellular damage, with the SOS defense system response coming into operation when DNA damage takes place (Imlay 2002).

Drug efflux is a major mechanism involved in antibiotic resistance (Poole 2000, 2002; Levy 2002a, b). There are several types of efflux pumps: (1) P-glycoprotein, encoded by a human or rodent gene, mediated by ATP-dependent exporters, and a member of the ATP-binding cassette (ABC) superfamily of transporters, is responsible for resistance to many cytotoxic drugs; (2) small multidrug resistance family (SMR); (3) major facilitator superfamily (MFS); (4) resistance/nodulation/cell division family (RND), unique to gram-negatives and working in conjunction with a periplasmic membrane-fusion protein (MFP) and an outer membrane protein; and (5) multidrug and toxic compound extrusion (MATE) family. Families 2–5 are secondary transporters driven by the protonmotive force.

Enzymatic inactivation occurs at low biocide concentrations (Hugo 1991b) but is unlikely to be a problem with in-use concentrations, although Meade et al. (2001)

consider inactivation to be one mechanism of triclosan resistance in some bacteria.

Bacterial resistance mechanisms fall into two general categories. *Intrinsic resistance* (intrinsic susceptibility) is a natural propert of an organism, whereas *acquired resistance* results from genetic changes to a cell. This general concept (Table 17.5) has been of value but should perhaps be reconsidered because some biochemical mechanisms of resistance, e.g. efflux, can be associated with both types.

It has been known for many years that repeated exposure of bacterial cells to an antibiotic can result in stable bacterial resistance (Levy 2002a). It has also been demonstrated that bacteria also have this ability to acquire resistance to a variety of biocidal agents (Russell and Chopra 1996; Russell 1998a, 1999d, 2003a, b, c, 2004). The resistance thus developed may be low- to high-level, stable or unstable (in which there is reversion to susceptibility or the adaptive resistance is lost on removal of biocide). The clinical relevance of stable decreased susceptibility to biocides is a matter of conjecture. Cross-resistance to chemically related biocides occurs, but occasionally to chemically unrelated biocides nd to some antibiotics also (Tattawasart et al. 1999a, b, 2000a, b; Thomas et al. 2000; Lambert et al., 2001; Joynson et al. 2002; Lear et al. 2002; see below).

Biocides are generally considered to have concentration-dependent multiple actions on bacteria and other microbial cells. Thus, single mutations are unlikely to lead to large increases in biocide resistance to biocides. However, a mutation in the primary target site, enoyl reductase (FabI) in *E. coli* and *S. aureus*, might be the reason for the reduced susceptibility of these organisms to triclosan (McMurry et al. 1998b) (see also Mechanisms of action of antimicrobial agents).

Mycobacterial resistance

The mycobacterial cell wall is highly hydrophobic with a mycoylarabinogalactanpeptidoglycan skeleton. Thus, hydrophilic agents have difficulty in penetrating the wall in sufficiently high concentrations to achieve a mycobactericidal effect. Low concentrations must, however, traverse the wall because inhibitory concentrations of nonmycobactericidal agents such as chlorhexidine and QACs are generally of the same order as those against other, nonmycobacterial organisms (Russell 1996). The component(s) of the mycobacterial wall responsible for the high level of biocide resistance is (are) unknown, but both the mycolic acids and arabinogalactan may be involved (Broadley et al. 1995; Russell 1996). Acquired mycobacterial resistance by repeated exposure to a biocide is a possibility, but there appears to be no evidence that plasmids are involved.

The new aromatic cyclic dialdehyde, OPA (see Types and uses of chemical agents) is believed to readily traverse the cell wall of GTA-resistant *M. chelonae* strains, whereas the less hydrophobic GTA is unable to do so (Walsh et al. 1999a, b, 2001; Simons et al. 2000; Fraud et al. 2001, 2003).

Resistance of other nonsporulating gram-positive bacteria

The cell wall of gram-positive bacteria is a thick, fibrous layer pressed against the cytoplasmic membrane; the wall consists mainly of an inelastic peptidoglycan interspersed with which may be lipids and teichoic and teichuronic acids (Hammond et al. 1984). The peptidoglycan comprises at least 50 percent of the dry weight of the walls. Mechanisms whereby biocidal agents enter gram-positive bacteria have been little studied, but passive diffusion across the wall is probably a major factor. Thus, organisms such as staphylococci and streptococci are usually highly susceptible to biocides (Russell 1995, 1998b, 1999d, 2003a; Russell and Russell 1995; Russell and Chopra 1996). However, bacteria grown under different conditions may show a wide response to biocides. For instance, fattened cells of

Table **17.5** *Mechanisms of bacterial resistance to biocides*[a]

Bacteria	Mechanism(s) of Intrinsic resistance	Acquired resistance
Mycobacteria	Permeability barrier associated with hydrophobic cell wall	Not described, unlikely
Other nonsporulating gram-positive bacteria	Cell-wall modulation 1 Peptidoglycan changes 2 Lipid increase	Plasmid-mediated efflux of cationic biocides in multiple antibiotic-resistant *S. aureus*; enzymatic inactivation of Hg compounds
Bacterial spores	Permeability barrier associated with outer membrane	Not described, unlikely
	SASPs and DNA protection	Unlikely
Gram-negative bacteria	1 Reduced uptake associated with outer membrane permeability barrier and efflux	Enzymatic inactivation of Hg compounds; efflux of cationic biocides?
	2 Constitutive biocide-degrading enzymes?	?

a) For further information, see Russell and Russell (1995); Russell and Chopra (1996); Russell (1999a, 2003a) and following text.

S. aureus produced by repeated subculturing in glycerol-containing media are more resistant to benzylpenicillin and higher phenols (Hugo 1999a). Additionally, nutrient limitation and reduced growth rates (Gilbert et al. 1990a; Poxton 1993) can alter sensitivity to biocides by changes in peptidoglycan thickness and cross-linking. Such changes in these two examples can be regarded as the expression of intrinsic resistance brought about in response to a phenotypic (physiological) adaptation.

Acquired resistance is also known in some strains of *S. aureus*. Inorganic mercury (Hg^{2+}) resistance is a common property of clinical isolates of *S. aureus* containing penicillinase plasmids. These plasmids are either narrow-spectrum, specifying resistance to Hg^{2+} and some organomercurials, or broad-spectrum, encoding additional resistance to other organomercurials. The enzymes involved are mercuric reductase (Hg^{2+}) and lyase (hydrolase) and reductase (organocompounds) (Foster 1983). Resistance to other metals has been described (Silver et al. 1989) and often involves efflux mechanisms. Silver (Ag^+) reduction may not form the basis of silver or silver sulfadiazine (AgSD) resistance (Belly and Kydd 1982; Foster 1983; Silver et al. 1989), but may be associated with accumulation (Trevor 1987; Russell and Hugo 1994).

Some methicillin-resistant strains of *S. aureus* (MRSA) containing plasmids encoding gentamicin resistance (methicillin–gentamycin-resistant *S. aureus*, MGRSA) also have increased minimum inhibitory concentration (MIC) values towards such biocides as QACs, chlorhexidine, ethidium bromide (EB), acridines, and propamidine isethionate (PI) (Lyon and Skurray 1987; Cookson and Phillips 1990; Cookson et al. 1991; Day and Russell 1999). Several genetic determinants are responsible for cationic biocide resistance in clinical isolates of *S. aureus*. There is evidence that there is an efflux of cationic agents from such antibiotic-resistant strains (Midgley 1986, 1987). Several genetic determinants are responsible for cationic biocide resistance in clinical isolates of *S. aureus*. These are *qacA*, which specifies resistance to QACs, acridines, EB, propamidine isethionate, and low-level chlorhexidine resistance; *qacB*, which is similar; *smr* (formerly *qacC/qacD/ebr*) encoding QAC and low-level EB resistance; and *qacG* and *qacH*. The corresponding gene products belong to the MFS (qacA, qacB) and SMR (smr, qacG, qacH) efflux families.

Resistance of gram-negative bacteria

Gram-negative bacteria such as pseudomonads and Enterobacteriaceae are usually less sensitive to a variety of chemically unrelated inhibitors than are gram-positive cocci. One reason for this lies in the different structural and chemical composition of the outer layers of the organisms, thereby conferring an intrinsic resistance

mechanism on gram-negative bacteria. The cell envelope of gram-negative bacteria is more complex: the inner (cytoplasmic) membrane is adherent to the peptidoglycan which is covalently linked to Braun's elongated lipoprotein. The peptidoglycan–lipoprotein complex partly occupies the periplasmic space that exists between the outer and inner hydrophobic membranes.

The OM acts as a permeability barrier in limiting or preventing the entry of many chemically unrelated types of antibacterial compounds into gram-negative bacteria. Some antibacterial agents, e.g. polymyxins and possibly chlorhexidine and the QACs, damage the OM, thereby promoting their own entry into the cell (Hancock 1984).

Acquired chromosomal resistance by mutation has been described for chlorhexidine resistance with *P. stuartii* (Chopra et al. 1987) and *S. marcescens* (Lannigan and Bryan 1985), alcohols with *E. coli* (Fried and Novick 1973), and cationic bactericides with various organisms (Chopra 1987). Resistance is not always stable. Temporary resistance by phenotypic adaptation to alcohols is described by Chopra (1987), who also considers that, in general, a nongenetic adaptive type of resistance is unlikely to play an important part in determining the long-term survival of bacteria to antiseptics and disinfectants.

Strains of *P. stuartii* isolated from paraplegic patients often harbor plasmids conferring resistance to Hg^{2+} and several antibiotics and the organisms may also express resistance to cationic biocides. However, attempts to transfer chlorhexidine and QAC resistance to suitable recipients have failed and the occurrence of a plasmid-linked association between antibiotic and antiseptic resistance has not been substantiated.

High-level resistance to biocides has been observed in hospital isolates of other gram-negative bacteria but no clear role for plasmid-mediated biocide resistance (Russell 1985, 1997) has emerged in the following instances: (1) chlorhexidine- and antibiotic-resistant *P. mirabilis*; (2) chlorhexidine-resistant strains of *Burkholderia cepacia* and *Alcaligenes denitrificans*; and (3) chlorhexidine- and QAC-resistant *P. aeruginosa* and several members of the Enterobacteriaceae (Russell and Chopra 1996).

'Emerging' gram-negative bacteria that may prove to be of concern include *Acinetobacter* spp. Isolates of these organisms do not, however, appear to date to have a particularly high level of biocide resistance (Higgins et al. 2001; see also Weber and Rutala 1999; Rutala and Weber 2001a, b; Russell 2002a).

BACTERIAL SPORE RESISTANCE

Bacterial spores differ fundamentally in both structure and composition from vegetative cells. In essence, a spore usually consists of a central core (protoplast, germ cell) and germ cell wall surrounded by a cortex, external to which are an inner and an outer spore coat (Foster 1994; Setlow 1994). Degradative changes take place during spore germination and biosynthetic processes during outgrowth (Paidhungat and Setlow 2002). Vegetative cells thus produced have the typical form and composition of nonsporulating bacteria (Russell 1990a).

Bacterial spores are amongst the most resistant of all microbial forms to inactivation by chemical or physical agents. Several factors contribute to this high resistance (Russell 1990a; Bloomfield and Arthur 1994; Bloomfield 1999; Loshon et al. 2001). These are (a) the thick, proteinaceous coats that limit the uptake of some biocides; (b) the reduced permeability of the spore core/protoplast to hydrophilic agents; (c) the protection of spore DNA by saturation with α, β-type (but not γ-type) small, acid-soluble proteins (SASP); and (d) the repair of DNA damage during germination. SASPs are basic proteins present in the spore core, but are rapidly degraded during germination. They play an important role in determining the response of bacterial spores to biocides. Spores ($\alpha^-\beta^-$) lacking such SASPs are much more sensitive to hydrogen peroxide and hypochlorites than parent spores (Tennen et al. 2000; Loshon et al. 2001).

Several studies (Power et al. 1988; Shaker et al. 1988; Knott et al. 1995; Cabrera-Martinez et al. 2002) have shown that during sporulation, an early resistance develops to formaldehyde, whereas resistance to chlorhexidine and QACs is an intermediate, and to glutaraldehyde and OPA a late, event.

BIOFILMS AND RESISTANCE

A biofilm is a consortium of bacteria or other microbes organized within an extensive exopolysaccharide exopolymer (glycocalyx: Costerton et al. 1987, 1994; Carpentier and Cerf 1993; Denyer et al. 1993; Poxton 1993). Biofilms may consist of monocultures, diverse species, or mixed phenotypes of a given species. Originally envisaged as homogeneous structures, biofilms are considered to be highly structured habitats with spatial and physiological heterogeneity (Wimpenny et al. 2000). Organisms can deposit enzymes within the matrix (Gilbert and McBain 2001a, b; Gilbert et al. 2002a, b) and cell-to-cell signalling (Davies et al. 1998; Swift et al. 2001) is an important feature in biofilm regulation, with the quorum-sensing systems of *P. aeruginosa* providing an excellent example (Wimpenny et al. 2000).

Attachment to surfaces is a vital element in bacterial infections, including those pertaining to medical devices (McBain and Gilbert 2001; Denyer 2002). Sessile bacteria on surfaces or present within biofilms are much less readily inactivated than planktonic cells (Carpentier and Cerf 1993; Poxton 1993; Donlan and Costerton 2002; Dunn 2002). Several reasons have been put forward for this difference. Early stress responses may be involved (see above). Diffusion of biocides into

biofilms and interaction with biofilms are both important aspects (Gilbert et al. 2001; Stewart et al. 2001), but are not the only factors responsible. A biocide gradient is produced throughout the biofilm, so that in a thick biofilm there will be an 'in-use' concentration of biocide at the surface, but a greatly reduced concentration as the biocide penetrates into the community (Gilbert and McBain 2001b). Degradative enzymes might be more effective at these lower concentrations within a biofilm than against in-use concentrations acting on planktonic cells

A physiological gradient is also produced such that nutrients and oxygen will be consumed at the periphery of the biofilm with limits to both occurring within the deepest parts of a biofilm. Nutrient-depleted cells are less sensitive to biocides, so that greater numbers of biocide-resistant cells are present towards the interior of a biofilm where they are exposed to lower biocide concentrations. In addition, pockets of surviving organisms may occur as small clusters in biocide-treated biofilms, although neighboring cells have been inactivated (Huang et al. 1995). These might have arisen as a result of the occurrence of biocide and physiological gradients. The clusters might also contain mutants in addition to genotypes with modification in single gene products (Gilbert et al. 2002b).

Furthermore, sublethal treatment with a biocide could induce the expression of multidrug efflux pumps and efflux mutants. Thus, *mar* expression is greatest within the depths of a biofilm where growth rates are lowest (Maira-Litran et al. 2000).

Lewis (2000, 2001) has proposed that the inactivation of a cell does not occur as a result of the direct action of a biocide, but rather from programmed cell death (PCD). Persisters of such a programme are then regarded as cells deficient in PCD that can grow rapidly in the presence of exudate released from lysed community cells.

Thus, many complex factors contribute to the relative insusceptibility to biocides of cells within biofilm communities (Gilbert and McBain 2001a, b; Gilbert et al. 2001, 2002a). In addition to the factors outlined above, cellular impermeability of some types of bacteria will also play a role.

POSSIBLE LINKED BIOCIDE-ANTIBIOTIC RESISTANCE

Biocides and antibiotics share some common mechanisms or sites of action. Acridines, phenylethanol, phenoxyethanol, fluoroquinolones, and some β-lactams all induce filamentation in gram-negative bacteria (Ng et al. 2002), although not necessarily by the same means. However, no cross-resistance was found between the different chemical agents.

Another example occurs with triclosan and the antitubercular drug, isoniazid (isonicotinylhydrazine, INH) (McMurry et al. 1999). Triclosan inhibits enoyl reductase involved in fatty acid synthesis in gram-negative and gram-positive bacteria (including mycobacteria), whereas INH has a specific effect on *M. tuberculosis* and some other mycobacteria. INH is essentially a prodrug that is converted into an active form by a *katG*-encoded oxidase-peroxidase system in *M. tuberculosis*, but not in other bacteria. Absence of this enzyme is the major mechanism for INH resistance in this organism. In *M. smegmatis*, mutations of the *inhA* gene that encodes a protein target (InhA) of INH involved in mycolic acid biosynthesis confers resistance to both INH and triclosan. Thus, the potential exists for the development of resistance to triclosan in mycobacteria to also be translated into INH resistance. However, it has not been demonstrated that triclosan is responsible for selecting for INH-resistant strains of *M. tuberculosis*.

Possible shared mechanisms of resistance to biocides and antibiotics are impermeability, uptake into gram-negative bacteria of cationic agents, mutations at a specific target site, enzymatic degradation, and extrusion (efflux) (Russell 1995, 2000, 2001c, 2002b, c, 2003a, b, c, 2004; Paulsen et al. 1996; Levy 2000, 2001, 2002a, b; Chuanchen et al. 2001; Health Council of the Netherlands 2001; Sulavik et al. 2001; Schweizer 2001; White and McDermott 2001; Fraise 2002a, b; Koljalg et al. 2002; Loughlin et al. 2002; Walsh et al. 2003a, b). Three biocidal types of agent, namely chlorhexidine, QACs, and triclosan, have been suggested as being involved in antibiotic resistance (Heir et al. 1998, 1999; Sundheim et al. 1998). Lambert et al. 2001 found that biocide/antibiotic-resistance in *P. aeruginosa* occurred especially in hospital, but not in industrial, isolates. However, antibiotic resistance resulted from the selective pressure exerted from antibiotic usage with no evidence that disinfection procedures produced biocide-resistant organisms. In a subsequent study (Joynson et al. 2002), it was demonstrated that an organism, *P. aeruginosa*, that is resistant to a common biocide (the QAC, benzalkonium chloride) could become sensitive to an antibiotic, whereas the converse was not true.

Veterinary clinical isolates of *P. aeruginosa* are resistant to high levels of triclosan, carbenicillin, and tetracycline, all of which are good substrates for the MexAB-OprM efflux system (Beinlich et al. 2001). In an earlier study (Chuanchen et al. 2001), ciprofloxacin resistance had been shown to be possibly associated with triclosan resistance, the phenylether inducing an efflux system. However, this fluoroquine antibiotic is a marginal substrate for this efflux system but a better substrate for MexCD-OprJ and MexEF-OprN (Beinlich et al. 2001).

S. aureus strains are commonly highly sensitive to triclosan (MICs around 1 µg/ml). However, some (but not all) clinical isolates, including MRSA strains, have elevated MICs of 1–2 µg/ml but do not necessarily show antibiotic resistance and are not killed less rapidly than triclosan-sensitive strains by higher concentrations of this compound (Suller and Russell 1999, 2000). VRE (Hiramitsu 1998) are not more resistant to biocides than are vancomycin-sensitive strains (Alqurashi et al. 1996; Day and Russell 1999).

Paulsen et al. (1998) proposed that *qacA* had evolved from *qacB*, that benzalkonium chloride (a QAC) induced the expression of *qacA* and *qacB*, and that their chronological emergence in clinical isolates of *S. aureus* mirrored the introduction and usage of cationic biocides in hospitals. Akimitsu et al. (1999) have claimed that QAC-resistant mutants of MRSA strains show greatly increased resistance to some β-lactam antibiotics.

Ishikawa et al. (2002) suggested that after *E. coli* had mutated to resist a relatively low concentration of a QAC under laboratory conditions, it could be converted to a multidrug-resistant type. However, whilst several authors have shown an association between biocide and antibiotic resistance in laboratory experiments, the 'real-life' situation is more difficult to assess. Currently, it must be stated that the evidence of a linkage between biocide usage and clinical antibiotic resistance is far from proven (Russell 2000; Gerba and Rusin 2001; Gilbert and McBain 2001a; McBain and Gilbert 2001). Antibiotic resistance is far more likely to occur as a result of the inadequate and improper usage of these drugs in clinical practice and their incorporation into animal feed. Many different acquired genes responsible for resistance of gram-negative bacteria to many antibiotics are part of gene cassettes that can move into or out of companion elements (integrons) (Hall and Collis 1998).

The possible role of biocide rotation in preventing biocide resistance from developing in hospitals has been considered by Murtough et al. (2001).

FUNGAL RESISTANCE

Two basic mechanisms of fungal resistance to biocides can be envisaged: intrinsic and acquired (Dekker 1987; Russell 1999d). The fungal cell wall contains various types of polymers, including chitin and chitosan (Zygomycetes), chitin and glucan (mycelial forms of ascomycetes and deuteromycetes), and glucan and mannan (yeast forms of ascomocytes and deuteromycetes). Thus, there is ample opportunity for a cell to exclude biocide molecules. Studies on the sensitivity of *C. albicans* to the polyene antibiotic, amphotericin, suggests that glucan may have a role to play in limiting drug uptake (Gale 1986). Comparable studies with biocides are sparse, but there is tentative evidence that glucan, but not mannan, in yeast cell walls could have a role to play in limiting chlorhexidine uptake (Hiom et al. 1993, 1995a, b, 1996).

There is no evidence linking the presence of plasmids in fungal cells and the ability of the organisms to acquire resistance to fungistatic or fungicidal agents, although acquired resistance of yeasts to organic acids is known (Warth 1986, 1989).

VIRAL RESISTANCE

Conflicting results have been reported about the action of biocides on different types of viruses. In part, this can probably be explained on the basis of different methodologies. However, the penetration of biocides into viruses and phages of different types has not been examined in depth, nor has interaction with viral protein and nucleic acid. It is, therefore, difficult to provide adequate reasons to explain the relative response, or resistance, of different virus types to biocides. Nevertheless, some progress has been made on the basis of the scheme of Klein and Deforest (1983), as described in Types of microorganism. Thurmann and Gerba (1988) suggested that the structural integrity of a virus is altered by an agent that reacts with viral capsids to increase viral permeability. A 'two-stage disinfection' could offer an efficient means of viral inactivation whilst overcoming the possibility of multiplicity reactivation first put forward in 1947 to explain an initial reduction and then an increase in titer of biocide-treated bacteriophage.

RESISTANCE OF PROTOZOA

The cyst form represents the stage in the life cycle that is resistant to biocides. Only a few significant studies have been undertaken about biocide sensitivity or resistance during encystment and excystment and the relative uptake of biocides into cysts and trophozoites. It has been shown that resistance develops as encystment proceeds and is lost during excystment of *Acanthamoeba castellanii*. It may be concluded that the outer regions of cysts limit entry of biocides, thus providing one mechanism of intrinsic resistance (Khunkitti et al. 1997,

1998a, b, 1999; Turner et al. 1999, 2000a, b; Lloyd et al. 2001).

PHYSICAL AGENTS AND OTHER STERILIZATION METHODS

Physical agents (Table 17.6) are normally used in preference to chemical agents for sterilization, either by microbial destruction or by removal of microorganisms (Dewhurst and Hoxey 1990; Soper and Davies 1990; Kowalski 1993; Russell 1993, 1999d, 2001a, b; Denyer and Hodges 1999; Hoxey and Thomas 1999). It is also practicable, however, to employ cold, desiccation, and freeze-drying as methods of preservation. Traditionally, heat in one form or another has been employed as a sterilization procedure. It still occupies a key role and is the method of choice whenever possible. Filtration is also based on ancient knowledge but suffers several limitations. Ionizing radiation is widely used for the sterilization of single-use, disposable items. Newer procedures employ various types of gas plasmas.

Moist heat

Heating in the presence of water has long been used as a method of sterilization, or sometimes of disinfection. It can be employed in different ways, e.g. at temperatures below, at, or above 100°C.

TERMINOLOGY

Thermal death time (TDT) is the time (in minutes) required to kill all cells in a suspension at a given temperature. Thermal death point (TDP) is the temperature needed to kill all cells in suspension after a

Table 17.6 *Types and applications of sterilization processes*[a,b]

Process	Application(s)
Moist heat (autoclave)	Sterilization of many ophthalmic and parenteral products, surgical dressings, rubber gloves (high-vacuum autoclave)
Flash sterilization	Modification of conventional steam sterilization
Dry heat	Sterilization of glassware, glass syringes, oils and oily injections, metal instruments; depyrogenation at high temperatures
Infrared radiation	Sterilization of heat-resistant instruments (high temperature responsible)
Ionizing radiation	Sterilization of single-use disposable medical items
Ultraviolet radiation	Poor sterilizing agent; use restricted to air sterilization (in conjunction with air filtration) and water disinfection
	Surface sterilization
Pulsed light (non-ionizing radiation)	
Ethylene oxide	Sterilization of fragile, heat-sensitive equipment, powders, components of spacecraft
Low-temperature steam with formaldehyde (LTSF)	Disinfection/sterilization of some heat-sensitive materials
Filtration	Air sterilization; sterilization of thermolabile ophthalmic and parenteral solutions, and of sera
Gas plasmas	

a) See text for information about most important processes.
b) Chemical agents such as glutaraldehyde and peroxygens are sometimes employed as 'chemosterilizers.'

fixed exposure time, e.g. 10 min. There are many variables associated with these terms, notably inoculum size, and as such they are usually of little value.

The D value (decimal reduction time, DRT), is the time (in minutes) needed at a particular temperature to reduce the viable organisms by 90 percent, i.e. to 10 percent or by 1 $\log_{10}$ unit. D value is independent of inoculum size and is related inversely to temperature. Examples are given in Table 17.7.

Inactivation factor (IF) is the degree of reduction in the number of viable cells and is obtained by dividing the initial viable count (N_0) by the final viable count (N_u), i.e. IF = N_0/N_u. An alternative procedure is to use the D value approach; IF = $10^{t/D}$ (in which t represents the treatment dose, i.e. time). There are, however, pitfalls with this method (Russell 1982).

The z value, defined as the number of degrees (°C) to bring about a 10-fold reduction in TDT or D value, is obtained from the slope of the curve in which temperature is plotted against time (see Russell 1982). An alternative method is based on a knowledge of the temperature coefficient (Q_{10}) per 10-fold rise in temperature, i.e. $Q_{10} = 10^{10/z}$ from which z = 10/log Q_{10}.

The F value is the time in minutes to destroy an organism in a specified medium at 121°C (250°F). F_0 is the F value when $z = 10$°C (18°F). Calculations employing F_0 values are used in validating thermal sterilization processes (Soper and Davies 1990).

Exponential rate of inactivation is depicted in Figure 17.2(A). Deviations from this occur frequently (Figure 17.2(B, C, D): Cerf 1977; Smerage and Teixeira 1993; Stringer et al. 2000). With convex- or concave-type curves, extrapolation of the line to cut the Y-axis gives the Y-intercept value (Y_0). If N_0 represents the initial number of cells, then (1) when $Y_0 = N_0$, there is a straight line response; (2) when $Y_0 > N_0$, there is an initial shoulder; (3) when $Y_0 < N_0$, a decreasing death rate is obtained.

The intercept ratio (IR) is a ratio of the two values, i.e. IR = log Y_0/log N_0, and is thus useful in characterizing time–survivor curves.

MICROBIAL SENSITIVITY TO MOIST HEAT

Nonsporulating bacteria are heat-sensitive and are usually destroyed at temperatures of 50–60°C (Russell 2003b). Thermal death times of fungi and protozoa are similar and most viruses are inactivated at 60°C for 20 min (Russell 1999d). D and z values for some bacterial spores are listed in Table 17.7.

Bacterial spores are considerably more resistant to moist heat. Among *Bacillus* spp., *B. stearothermophilus* is the most heat-resistant. Considerable variation in sensitivity occurs among *Clostridium* spp., *C. botulinium* type E being far more sensitive than types A, B or C (Russell 1982, 2003d).

Sensitivity to moist heat depends on the conditions of exposure and recovery. In medical, surgical, and veterinary practice, microorganisms are often embedded in organic debris, which may to some extent insulate them from the effect of heat, and thorough cleaning prior to sterilization is essential (Roberts 2001).

TYPES OF MOIST HEAT PROCESSES

Temperatures below 100°C

During his studies on spoilage of French wines, Louis Pasteur employed a temperature of 60°C. Subsequently, this process (60°C for 30 min) was used for the pasteurization of milk. A 'flash' method (71.1°C for 15 s) was introduced later. A procedure similar to pasteurization has been employed in the production of certain bacterial vaccines, e.g. typhoid vaccine, the aim being to kill the organisms while preserving their immunogenicity.

Although some types of spore are killed by moist heat at about 80°C, temperatures below 100°C cannot be relied upon to achieve sterilization.

Temperatures around 100°C

These temperatures cannot be relied upon to kill bacterial spores, and steaming is thus of limited use. Two variations have been used. The first (tyndallization) involved steaming for 30 min (originally 80°C for

Table 17.7 Responses[a] of some bacterial spores to heat and to ionizing radiation

Bacterial spore	Moist heat		Dry heat		Ionizing radiation (D)	
	D (min) at 121°C	z	D (min) at 160°C	z	Mrad	kGy
Bacillus stearothermophilus	4–5	8–10	<1	15–25	...	...
B. subtilis	<1	5.5–9.5	1–5	20	0.15	1.5
B. cereus	<1	...	...	20	0.1	1
B. pumilus	...	...	...	25	0.2	2
Clostridium sporogenes	<1	10	2	20	0.15–0.2	1.5–2
C. botulinum type A	<1	9.5	...	...	0.1	1

a) Values are approximate responses to a particular process. The actual response will depend upon several parameters (see text and Russell 1982, where comprehensive details are provided). *D*, time (min) at a specified temperature or radiation dose needed to reduce the viable population by 1 $\log_{10}$ unit; *z*, temperature (°C) needed to reduce the *D* value by 90 percent.

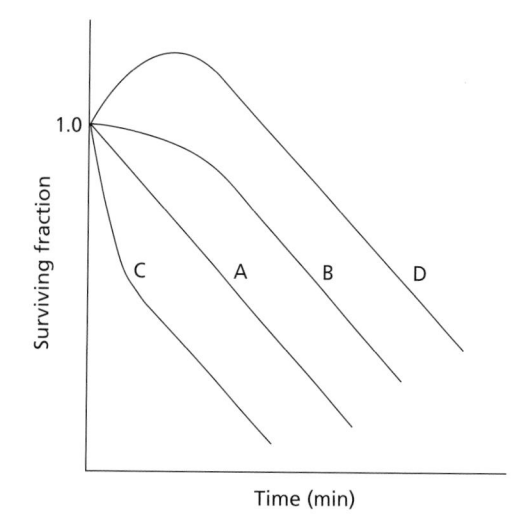

Figure 17.2 *Thermal inactivation curves of bacterial spores. A, exponential death; B, increasing death rate; C, decreasing death rate; D, heat activation.*

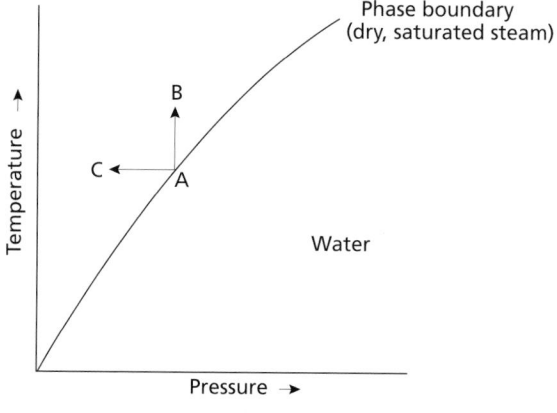

Figure 17.3 *Phase diagram for water vapor, depicting the phase boundary for dry saturated steam and the production of superheated steam (points B and C). A, any point on the phase boundary; B, pressure kept constant, temperature increased; C, temperature kept constant, pressure increased.*

60 min) on each of 3 consecutive days, the principle being that spores that survived a heating process would germinate before the next thermal exposure and would then be killed. Spores will not, however, germinate in a non-nutrient medium and, although this process was originally an official method for sterilizing injections, it is no longer used. The second variation involved heating to 98–100°C in the presence of a specified antimicrobial agent (the activity of which is considerably potentiated at the high temperature; see Factors influencing activity). This method is no longer permitted in the British Pharmacopoeia (1988, 1993, 1998, 2000) for sterilizing any injectable or eyedrop formulations.

Temperatures above 100°C (autoclaving)

Sterilization by steam under pressure depends upon four properties of dry saturated steam: high temperature, wealth of latent heat, ability to form water by condensation, and instantaneous contraction in volume that occurs during condensation. These properties are optimal only in steam on the phase boundary between itself and condensate at the same temperature (Figure 17.3). Steam formed at any point on the phase boundary has the same temperature as the boiling water from which it was derived. However, it holds an extra, and relatively heavy, load of latent heat, which, without a drop in temperature, is available instantly and entirely as soon as it wets a cooler surface.

Superheated steam is hotter than dry, saturated steam at the same pressure, the process becoming akin to dry heat, which is much less efficient. Superheated steam behaves as a gas and only slowly yields its heat to cooler objects. In Figure 17.3, steam at any point A on the phase boundary is saturated: if, however, the temperature is raised to B (keeping pressure constant), then superheated steam is generated. A small amount (5°C) of superheat may be tolerated in practice.

When air is present in a space with steam, the air will carry part of the load so that the pressure of the steam is reduced, creating a condition equivalent to that defined by the horizontal line AC in Figure 17.3. This can also be deduced from Dalton's law of partial pressures, which, applied to the present context, can be written as $P_{total} = P_{steam} + P_{air}$. The removal of air is, in fact, important in ensuring efficient autoclaving and can be achieved in various ways. In downward-displacement sterilizers, air is displaced downwards, being heavier than steam; this process may, however, be wasteful of steam and time-consuming. Its widest application is in the sterilization of bottled fluids (Owens 1993). Bulky packages of, for example, surgical dressings are, however, air-retentive and the downward-displacement method of removing air is not efficient. Removal of air before admission of steam is far more effective and can be achieved by the use of high prevacuum autoclaves or by means of a pulsed evacuation and steam admission. Thermal processing is also widely employed in the food industry (Brown 1994; Gould 1999). Table 17.8 lists the time–temperature relationships provided by the British Pharmacopoeia (2000). Sterilizers designed for bottled fluids generally employ temperatures of 121°C and those for porous loads 126–134°C.

Flash sterilization

Flash sterilization is a modification of conventional steam sterilization. In this process, a flashed item is placed unwrapped in an open mesh tray or in a specially designed rigid container to allow for rapid penetration of steam. Either gravity or prevacuum autoclaves can be used for this purpose (Barrett 2001).

Low-temperature steam with formaldehyde

Low-temperature steam (LTS) at subatmospheric pressure was developed originally for disinfecting heat-

Table 17.8 *Permitted time–temperature relationships in moist-heat and dry-heat sterilization processes*[a]

Process	Temperature (°C)	Holding period (min)
Moist heat (autoclave)	121	15
	126	10
	134	3
Dry heat	Minimum of	Not less than
	160	120
	170	60
	180	30

a) Based on recommendations in British Pharmacopoeia (2000).

sensitive materials. LTS at 80°C was found to be much more effective than water at the same temperature. The addition of formaldehyde to LTS to produce low-temperature steam with formaldehyde (LTSF) achieved a sporicidal effect (Dewhurst and Hoxey 1990; Soper and Davies 1990), and LTSF is thus suitable for sterilizing thermolabile equipment, although there are now doubts about the efficacy of this process.

MECHANISMS OF MICROBIAL INACTIVATION BY MOIST HEAT

Heat induces a multiplicity of injuries in sporing and nonsporing bacteria (Russell 2003b), and every cellular component (outer layers, membranes, enzymes, proteins, RNA, DNA) is likely to be affected to some degree. In thermophiles and hyperthermophiles, enzymes and proteins are much more heat-stable (Madigan et al. 1997) because of critical amino acid substitution in one or a few locations. It is thus unlikely that the lethal effect can be attributed to a single event in an organism.

In nonsporulating bacteria, mild heat treatment (approximately 45–50°C) damages the outer membrane of *E. coli*, rendering cells more sensitive to hydrophobic inhibitors (Hitchener and Egan 1977; Mackey 1983). Lethal damage occurs to the cytoplasmic membrane with RNA breakdown and protein coagulation (Tomlins and Ordal 1976) and damage to the bacterial chromosome (Pellon and Sinskey 1984). Virtually all structures and functions can be damaged by heat, but repair to non-DNA structures can occur only if DNA remains functional, thereby providing the necessary genetic information. DNA is implicated in heat damage, directly or as a result of enzymatic action after thermal injury (Pellon and Sinskey 1984). Using repair-deficient *E. coli* mutants, Mackey and Seymour (1987) proposed that heat-induced DNA damage occurred indirectly via an oxidation mediated by hydrogen peroxide (or free radicals derived from it) in peroxide-generating complex recovery media. Their results also provide a satisfactory explanation for 'minimal medium recovery,' which appears to involve protection from this peroxide-induced effect. By potentiating the effectiveness of DNA repair systems, *Deinococcus radiodurans* (see Physical agents and other sterilization methods) has an extraordinary ability to withstand the lethal and mutagenic effects of DNA-damaging treatments. A relationship exists between bacterial sensitivity to ionizing radiation and mild heat shock, and *D. radiodurans* is considered to be a heat-resistant organism (Battista et al. 2000).

The bacterial cell senses exposure to heat by means of:

- Extracellular alarmones by switching on inherent tolerant mechanisms. Extracellular sensing components (ESC) are directly converted into extracellular induction components (EIC), with very rapid responses occurring at increasing levels of stress (Rowbury 2001).
- Inducible intracellular heat-shock proteins (HSP) Hecker and Volker 2001). In *E. coli*, most genes require the sigma factor σ^{70} for transcription. Genes induced by heat shock have different sequences and RNA polymerase (RNAP) requires σ^{32} to recognize them. The amount of this alternative sigma factor regulates the heat-shock response. σ^{32} is encoded by *rpoH*, which specifically directs RNAP to transcribe from the heat-shock promoters (Polissi et al. 1995; Yura et al. 2000).

In bacterial spores (Figure 17.2), thermal injury has been attributed to a variety of reasons. These include denaturation of vital spore enzymes, impairment of germination or outgrowth or both, membrane damage (leading to leakage of calcium dipicolinate), increased sensitivity to inhibitory agents, structural damage (as observed by electron microscopy), and damage to the spore chromosome (as evinced by mutations or DNA strand breaks). Effects on DNA are much more pronounced during dry heating (see section on Dry heat); which generates a high level of mutants in spore populations (Gould 1989). DNA is, however, an important target, although the exact mechanism of spore inactivation is unknown. Deficiencies in DNA repair mechanisms are known to render spores more heat-sensitive (Hanlin et al. 1985). SASPs are of two types, the α,β-type associated with spore DNA, and the γ-type, not associated with any spore macromolecule. Spores lacking α,β-type SASPs are more susceptible than wild-type spores to moist heat (Setlow 1995). The DNA in these mutants is more sensitive, with a high frequency of single-strand breaks (Setlow and Setlow 1996). Spore proteins have also been implicated (Belliveau et al. 1992; Setlow 1994), although the nature of these is unclear.

MECHANISMS OF RESISTANCE OF BACTERIAL SPORES TO MOIST HEAT

The resistance of spores to heat can be manipulated over several orders of magnitude by exposure to extreme pH values and cationic exchange treatment (H-form and Ca-form, respectively) (Alderton and Snell 1963; Alderton et al. 1980). The content and location of water in spores play important roles in their heat resistance. Spores have a low water content (see below) and this is an essential factor in resistance. Spore coats do not contribute to thermal resistance since coatless spores are not more heat-sensitive than ordinary spores. Gould and his colleagues (reviewed by Gould 1989) found that resuspension of newly germinated spores in high concentrations of sucrose or sodium chloride (but not of glycerol) restored resistance to heat and ionizing radiation. In such an osmoregulatory mechanism the cortical structure would have to maintain a uniform pressure upon, and thereby control, the water content of the spore core.

Warth (1985) considered three types of mechanisms that might contribute to the stability of proteins in the spore. They could be intrinsically stable; substances might be present that help stabilize them (there is no strong evidence to support this notion but calcium dipicolinate may play a role); and the removal of water could alter their stability. A direct role for dipicolonic acid (DPA) seems unlikely, but it might help establish and maintain dormancy and heat resistance (Tovar-Rojo et al. 2002).

Several spore properties are, in fact, important for heat resistance, notably protein thermotolerance, dehydration, mineralization, thermal adaptation, and cortex function (Murrell 1981; Beaman and Gerhardt 1986; Beaman et al. 1988, 1989; Gerhardt and Marquis 1989; Marquis et al. 1994, 1995). In addition, SASPs found in the spore core have some role to play. Spores lacking α/β-type SASPs are more sensitive than wild-type spores and it is likely that heat-induced DNA damage is the mechanism involved (Setlow 1994). However, these SASPs are not a major determinant of spore resistance to moist heat because heat resistance during sporulation is attained well after their synthesis (Setlow 1994; Setlow et al. 2001).

The single most important factor contributing to spore resistance to moist heat is the reduced water content of the spore core, with other factors playing a supporting role (Setlow 2000).

Dry heat

Dry heat is a less efficient sterilization process than moist heat. The terminology used in describing moist heat sterilization is also employed here. Bacterial spores are the most resistant organisms to dry heat and D and z values are much higher than for moist heat (Table 17.7) (Wood 1993; Russell 1999e, 2003d).

TYPES OF DRY HEAT PROCESSES

Normally, dry-heat sterilization takes place in hot-air ovens, a number of combinations of temperature and time being permitted (Table 17.8). It is essential that the hot air can circulate between objects being sterilized, which must therefore be loosely packed with adequate air spaces to ensure optimum heat transfer. A fan is essential to prevent the wide variation in temperature that would otherwise occur. Industrially, sterilization equipment can also consist of forced-convection ovens, rapid heat-transfer sterilizers, and continuous-belt sterilizers (Wood 1993). Dry-heat sterilization (Table 17.6) is the method of choice in the pharmaceutical industry for many heat-stable objects and materials (Dewhurst and Hoxey 1990) and at higher temperatures it can also destroy pyrogens (depyrogenation; Wood 1993). The use of infrared ovens has been described (Mata-Portuguez et al. 2002).

MECHANISMS OF MICROBIAL INACTIVATION BY DRY HEAT

Microbial inactivation by dry heat has been considered primarily an oxidation process. Spores heated in oxygen would thus be expected to be more sensitive than when heated in the presence of other gases, but this is not necessarily true (Pheil et al. 1967). Thus, although oxidation may play an important part, other possibilities must be considered.

An effect on DNA is one possibility (Setlow 1994), sublethal temperatures inducing mutants in *B. subtilis* spores (Zamenhof 1960) as a consequence of depurination (Northrop and Slepecky 1967). The water content of spores is an important factor in determining inactivation by dry heat. Rowe and Silverman (1970) postulated that only a relatively small amount of water is needed to protect the heat-sensitive site in spores and that resistance to dry heat depends mainly on the location, rather than on the amount, of water in the spore, and on its association with other molecules.

Inactivation by cold

Growth of microorganisms is retarded and eventually ceases if they are held at reduced temperatures. Some can grow at temperatures approaching 0°C and these psychrophiles may be important food-spoilage organisms. Environmental factors such as nutrient status, pH, salt concentration and a_w (water activity or moisture content) can alter the minimum growth temperature (Herbert 1989).

Freeze-drying involves rapid freezing with subsequent drying in a vaccum and is used widely for preserving microbial cultures and certain foodstuffs, but its use subjects microorganisms to stresses such as freezing, drying, storage, and eventual rehydration (Mackey 1984). Single-strand breaks in DNA and an increase in the frequency of mutation may be caused by freeze-drying (Asada et al. 1979).

Freezing and thawing can inactivate bacteria. Damage to the outer and inner membranes envelope has been shown in *E. coli* (Mackey 1983, 1984). For comprehensive accounts, the papers by MacLeod and Calcott (1976) and Mackey (1984) should be consulted.

Cold shock is a process in which organisms are suddenly chilled without freezing. Gram-positive and gram-negative bacteria, but not yeasts, may be killed (MacLeod and Calcott 1976; Rose 1976). Several factors influence the response of cells, notably the age of the culture (exponential phase, but not stationary phase, cells being susceptible) and the composition of the medium. Divalent cations can protect cells against chilling. Low-molecular-weight materials are released from chilled cells as a consequence of increased membrane permeability caused by a phase transition in membrane lipids (Rose 1976). A modification of this treatment is 'cold osmotic shock,' in which bacteria are suspended in a hypertonic sucrose solution containing EDTA and are then suspended in ice-cold magnesium chloride solution. This treatment does not kill the cells but does induce the release of periplasmic enzymes, including β-lactamases, from gram-negative bacteria.

Hydrostatic pressure

Some organisms can exist at the bottom of deep oceans where they are subjected to high pressure (Dring 1976). Bacterial spores are more resistant to hydrostatic pressure than are germinated spores or nonsporulating bacteria. Temperature profoundly affects bacterial sensitivity to hydrostatic pressure, the rate of inactivation increasing as the temperature rises. This combined effect can be potentiated if an additional heat treatment is given after pressurization (Gould, 1970; Gould and Jones, 1989; see also New low-temperature sterilization technologies). The application of hydrostatic pressure is a useful technique because it enables investigations to be made of germination in the absence of physiological germinants. In view, however, of the occurrence of 'superdormant' spores (Gould et al. 1968), which are resistant to hydrostatic pressure, it is unlikely that this process will find use as a sterilization procedure.

Ionizing radiation

Ionizing radiations, e.g. X-rays, γ-rays, and high speed electrons (β-rays), strip electrons from the atoms of the material through which the radiations pass; essentially all the chemical changes produced in the microbial cell are due to these stripped-off electrons, which initiate a chain of chemical reactions. Ionizations occur principally in water, resulting in the formation of short-lived but highly reactive radicals (hydroxyl, •OH) and protons (H^+). Single-strand, and sometimes double-strand, breakage in DNA ensues. By contrast, ultraviolet (UV) radiation does not possess enough energy to eject an electron to produce an ion, although there is an alteration of electrons within their orbits. It is not therefore an ionizing radiation. Infrared radiations rapidly raise the temperature of objects that they strike and are thus employed for their heating effect. X-rays and γ-rays are of very short wavelengths and are generated by machines or radioactive sources (e.g. ^{60}Co). High-speed electrons were originally produced from radioactive isotopes but had little penetration; various machines have been developed that accelerate atomic particles to give them the energies for penetrating deeply. X-rays and γ-rays have considerable penetrating power. By contrast, α-particles, which consist of helium nuclei ($^4_2$He), have little penetration and thus are not used for sterilization. Industrially, radioisotopes in the form of ^{60}Co (Herring and Saylor 1993) and accelerated electrons (Cleland et al. 1993) are the main sources of ionizing radiation.

The unit of radiation most widely used in microbiology was previously the rad, a measurement of the energy absorbed from ionizing radiation by the matter through which the radiation passes. The newer, more modern (SI) unit is the gray (Gy; 1 Gy = 100 rad), which is defined as the deposition of 1 J/kg energy in tissue.

MICROBIAL SENSITIVITY TO IONIZING RADIATIONS

Bacterial spores are generally more resistant than nonsporulating bacteria, although *D. radiodurans* is the most resistant organism known (Moseley 1989; Russell 1999f). Among the clostridia, *C. botulinum* types A and B are the most resistant, type E being highly sensitive. Among *Bacillus* spp., *B. pumilus* E601 is probably the most resistant (see Table 17.7). During bacterial sporulation, resistance to γ-radiation develops about 2 h before the onset of heat resistance (Durban et al. 1970). Although spore-coat disulfide bonds were believed to be radioprotective, it is now clear that the –S–S-rich protein structure does not afford resistance, since coatless spores are as resistant as normal spores (Hitchins et al. 1966; Gould and Sale 1970; Farkas 1994).

MECHANISMS OF MICROBIAL INACTIVATION BY IONIZING RADIATIONS

Ionizing radiations induce structural defects in microbial DNA, which, unless repaired, are likely to inhibit DNA synthesis, leading to cell death (Hutchinson 1985). DNA is not the only target relevant to inactivation, although it is undoubtedly the principal one (Lindahl 1982), with single-strand breaks (SSB) and double-strand breaks (DSB) depending on the intensity of the radiation dose (Moseley 1989).

The state of DNA in the cell is important in relation to bacterial inactivation. Spores are generally more resistant to ionizing radiation than are nonsporing organisms. There are several possible reasons, notably that:

- spores contain a radioprotective substance (but there is no evidence to support this supposition)
- spore coats confer protection, but coatless spores are not less resistant
- DNA is present in a different state in spores.

DNA in bacterial spores exists in the A-form, associated with a low a_w value, and it is certainly true that DNA in the intact spore is more resistant to SSBs and DSBs than DNA in the intact vegetative cell. However, DNA extracted from spores shows the same response in vitro to ionizing radiation as DNA extracted from nonsporulating bacteria. Furthermore, newly germinated spores exposed to ionizing radiation in an environment of high osmotic pressure again become radiation-resistant (Gould 1983, 1984, 1985). Damage to DNA is less extensive under such circumstances.

α/β-type SASPs have some role to play in conferring heat resistance on spores, presumably by stabilizing DNA; however, neither they nor γ-type SASPs are involved in ionizing radiation resistance (Setlow 1994).

MECHANISMS OF MICROBIAL REPAIR FOLLOWING IONIZING RADIATION

Several microorganisms show above-average resistance to ionizing radiation (and to UV radiation, see section on Ultraviolet radiation) because they possess enzymes capable of repairing damage to DNA (Bridges 1976; Moseley 1984, 1989). Repair has been widely studied in *D. radiodurans* and in *E. coli* mutants in which sensitivity to ionizing radiations is conferred by genes such as *rec (A, B, C)*, *polA*, and *lon*.

SSBs in bacterial spores can be repaired during post-irradiation germination (Terano et al. 1969, 1971). In *C. botulinum* 33A, which is highly radiation-resistant, direct repair (rejoining) of SSBs takes place during or after radiation in spores existing under nonphysiological conditions at 0°C and in the absence of germination (Durban et al. 1974). Such repair of SSBs within these dormant, ungerminated spores may result from DNA ligase activity (Gould 1984).

USES OF IONIZING RADIATION

Ionizing radiation has found particular use as a means of sterilizing single-use medical devices (Russell 1999f). Food irradiation has long been considered as a means of sterilizing or preserving food (Russell 1999f) but has met with considerable consumer resistance.

Ultraviolet radiation

UV radiation has a wavelength of approximately 328–210 nm. Its maximum bactericidal effect is documented at 240–280 nm (Sykes 1965). Modern mercury-vapor lamps emit more than 95 percent of their radiation at 253.7 nm, which is at or near the maximum for microbicidal activity. The quantum of energy liberated is low, and consequently UV radiation has less penetrating ability and is less effective than other types of radiation. It is not, therefore, considered as being an effective method of sterilization (Russell 1993, 1999g).

MICROBIAL SENSITIVITY TO UV RADIATION

Bacterial spores are generally more resistant to UV light than are vegetative cells (an exception being *D. radiodurans*: Battista 1997), although mold spores may be even more resistant. Viruses are also inactivated; they tend to be more sensitive than bacterial spores, but are often more resistant than nonsporulating bacteria (Morris and Darlow 1971). HIV is not inactivated by UV radiation (Report 1986).

MECHANISMS OF MICROBIAL INACTIVATION BY UV RADIATION

Exposure of nonsporulating bacteria results in the formation of purine and pyrimidine dimers between adjacent molecules in the same strand of DNA (e.g. the forward reaction in Figure 17.4a). The photoproduct, 5,6-hydroxydihydrothymine, is also found in *D. radiodurans* exposed to ionizing and UV radiation (Figure 17.4b). In bacterial spores, another type of photoproduct, 5-thyminyl-5,6-dihydrothymine (TDHT) (Figure 17.4c) accumulates in DNA. A less well-known and less frequent photoproduct, pyrimidine (6,4) pyrimidone, is the relevant lesion involved in the toxic and mutagenic effects of UV on *E. coli* (Mitchell and Nairn 1989; Koehler et al. 1996). Unless removed, photoproducts form noncoding lesions in DNA and bacterial death results.

MECHANISMS OF MICROBIAL REPAIR IN NONSPORULATING BACTERIA

The SOS network, controlled by the RecA and LexA proteins, is responsible for the expression of genes whose products play roles in radiation-induced excision, daughter strand gap, and DSB repairs. There are three major mechanisms that are involved in repair of UV-induced DNA damage:

1 Photoreactivation (light repair), in which the exposure of UV-irradiated cells to light of a higher wavelength, but below 510 nm, results in the induction of a photoreactivating enzyme that monomerizes thymine

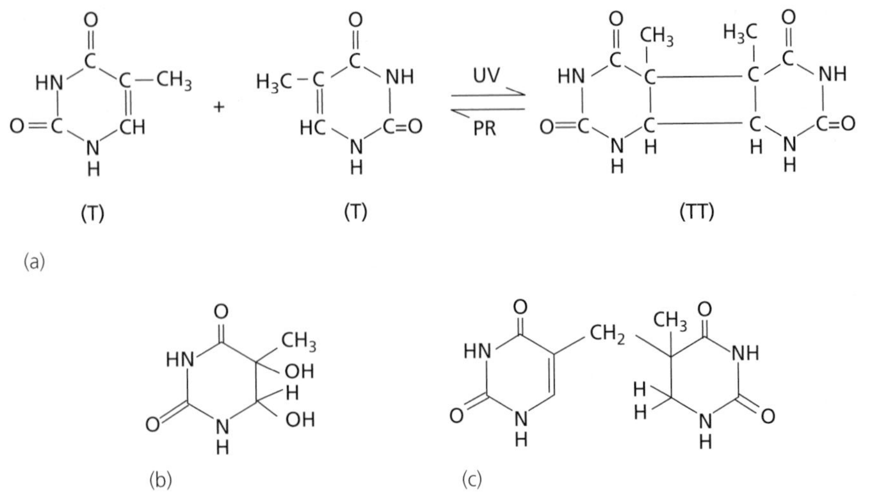

Figure 17.4 *Ultraviolet (UV)-induced changes in DNA:* **(a)** *pyrimidine dimer formation and reversal (photoreactivation, PR);* **(b)** *5,6-dihydroxydihydrothymine;* **(c)** *5-thyminyl-5,6-dihydrothymine (TDHT). T, thymine; TT, thymine dimer.*

dimers in situ (Figure 17.4a, reverse direction) and results in the recovery of a large proportion of cells that would otherwise have been inactivated (Moseley 1989; Russell 1999g).

2 Dark repair (excision), in which the stages are first, the action of a dimer-specific endonuclease, which 'nicks' the DNA; second, a dimer-specific exonuclease, which excises the damaged portion together with a number of nucleotides on each side, followed by the action of DNA polymerase I; and, finally, DNA ligase, which is responsible for the final joining together.

3 Dark repair (post-replication recombination), in which replication proceeds normally until it reaches an unexcised dimer, which is a noncoding lesion. A gap is left opposite the dimer and replication recommences at a new initiation site some 800–1000 bases downstream. Because dimers occur in both parental strands of DNA, both daughter strands contain gaps, which are repaired by recombinational insertion of parental DNA from the sister duplex. The process requires a functional recA gene product. Recombination-deficient mutants are abnormally sensitive to UV and ionizing radiations.

For additional information on repair, mutants and error-prone repair, see Bridges (1976), Moseley (1984, 1989), and Sancar and Sancar (1988).

MECHANISMS OF MICROBIAL REPAIR IN BACTERIAL SPORES

The spore photoproduct, TDHT, is identical to one that accumulates in hydrolysates of DNA exposed dry, or as a frozen solution, to UV radiation (Rahn and Hosszu 1968; Varghese 1970). With the exception of *D. radiodurans*, vegetative cells in the frozen state are supersensitive to UV and a photoproduct (presumably

TDHT) other than thymine dimers accumulates, which is less susceptible to repair. UV-sensitive mutants of UV-resistant *B. subtilis* spores form the same photoproduct (TDHT) and to the same extent for a given dose of radiation as do the resistant spores. UV resistance is thus linked to the ability to remove TDHT (Munakata and Rupert 1972), and two genetically controlled mechanisms have been described (Munakata and Rupert 1972, 1974):

1 'Spore repair' involving the elimination of TDHT during germination, although vegetative growth is not required

2 Excision repair, in which TDHT disappears slowly from the large-molecular-weight trichloracetic acid (TCA)-insoluble fraction and appears in the TCA-soluble fraction.

PRACTICAL USES OF UV RADIATION

UV radiation has little penetrative power through solids and is extensively absorbed by glass and plastics. Sterilization is achieved only by doses of radiation beyond the limits of practicability. It does, however, have some use as a disinfection procedure. It has been used to disinfect drinking water (Sykes 1965); as a possible means of obtaining pyrogen-free water; and especially for air disinfection, notably in hospital wards and operating theaters, in aseptic laboratories and in ventilated safety cabinets in which dangerous microorganisms are being handled (Morris and Darlow 1971).

Sunlight also has disinfecting properties, rays of wavelengths 254–257 nm possessing the greatest bactericidal activity. It has been claimed that sunlight can inactivate *B. anthracis* spores on wooden and metal surfaces, although only after prolonged exposure. Infectivity of viruses may be reduced on exposure to direct sunlight (Russell et al. 1984).

Sterilization by filtration

Filtration in various forms was used in ancient times, e.g. in attempts to purify water and sewage by allowing it to percolate through beds of sand, gravel, or cinders. These and other applications are considered by Denyer and Hodges (1999) and Jornitz et al. (2002), who also describe the development of modern filtration devices.

TYPES OF FILTER

Several types of filters are used, including (1) *unglazed ceramic filters*, e.g. the Chamberland and Doulton filters, which are manufactured in several different grades of porosity and have been used industrially for the large-scale clarification of water. After use, they can be cleaned with sodium hypochlorite solutions and will withstand scrubbing; (2) *compressed diatomaceous earth filters*, examples of which are the Berkefeld and Mandler filters, which are manufactured in different grades of porosity. They have been used industrially for the large-scale clarification of water. After use, they can be cleaned with sodium hypochlorite but without scrubbing; (3) *asbestos filters*, which include Seitz, Carlson, and Sterimat filters. They have high adsorbing capacities and tend to alkalinate solutions being filtered. There is also a possibility of toxic effects (including carcinogenesis) in handling these filters and they are now practically obsolete; (4) *sintered glass filters*, which are prepared by size-grading finely powdered glass followed by heating. The pore size can be controlled by the general particle size of the glass powder. The filters are easily cleaned, have low adsorption properties, and do not shed particles, but they are fragile and relatively expensive; (5) *membrane filters*, which consist of cellulose esters. They are used routinely in water analysis and purification, sterilization, and sterility testing and are the most suitable for preparing sterile solutions for parenteral use. Membrane filters are available in a range of pore sizes from approximately 12 μm down to 0.5 μm, sterilizing filters usually having an average pore diameter (APD) of 0.2–0.22 μm. The ratio of maximum pore diameter (MPD) to APD is in general 3.5–5. Membrane filters have several advantages over conventional depth filters (Denyer and Hodges 1999).

High-efficiency particulate air (HEPA) filters remove particles of 0.3 μm or larger (and probably particles <0.1 μm) (Denyer and Hodges 1999) and are widely used in air filtration, especially that type incorporating laminar air flow (LAF) (White 1990). LAF units are of two types, horizontal and vertical, depending upon the direction of the air flow.

MECHANISMS OF STERILE FILTRATION

Membrane filters are often described as 'screen' filters in contrast to media such as sintered glass, asbestos fiber, and ceramic, which are believed to retain organisms and particles by a 'depth' process in which particles are trapped or adsorbed within the interstices of the filter matrix. Screen filtration, on the other hand, has been considered to exclude (sieve out) all particles larger than the rated pore size. However, it is now apparent that the filtration characteristics of many membrane filters cannot be accounted for in terms of the sieve retention theory alone, but that other factors are involved, including van der Waals' forces and electrostatic interactions.

Additional information is provided by Lukaszewicz and Melzer (1979), Tanny et al. (1979), and Denyer and Hodges (1999).

APPLICATIONS OF FILTRATION

Filtration can be used for sterilization of thermolabile parenteral and ophthalmic solutions, sterility testing of pharmaceutical products, clarification of water supplies, microbiological evaluation of water purity, viable counting procedures (including estimation of survivors following exposure to antibiotics or disinfectants), and for determination of viral particle size (see also Table 17.6). For most of these applications, membrane filters are usually employed.

Additionally, filters are used in air sterilization, including LAF cabinets as described above.

New low-temperature sterilization technologies

Several new low-temperature procedures have been described in recent years that have proved to be useful sterilization procedures.

GAS PLASMAS

Low-temperature hydrogen peroxide gas plasmas are generated by applying radiofrequency waves to hydrogen peroxide vapor. The vapor is broken up into reactive free radicals that form the gas plasma (Groschel 1995; Jacobs and Lin 2001). This gas plasma is sporicidal and is eventually broken down to oxygen and water (Rutala and Weber 2001a).

Low-temperature plasma sterilizers that utilize ion plasmas formed from peracetic acid have also been described (Alfa et al. 1998).

LIQUID STERILIZATION SYSTEMS

New liquid sterilization systems have been described that are proving to be of benefit in sterilization of some medical devices. These include automated machines that utilize peracetic acid for the sterilization of lensed flexible and rigid endoscopes (Groschel 1995; Holton et al. 1995; Alfa et al. 1998).

Performic acid, a new proprietary liquid sterilant, is part of a new automated endoscope-reprocessing system

(Rutala and Weber 2001a). It is formed by mixing the two component solutions, hydrogen peroxide and formic acid.

Combined synergistic treatments

Enhanced antimicrobial efficacy is often achieved by combining two or more processes, chemical or physical. There are various types of combined treatment (Gould and Jones 1989).

THERMOCHEMICAL TREATMENT

The antibacterial activity of a compound usually increases as the temperature rises (see Factors influencing activity). This was exemplified by using certain preservatives at high temperatures in the sterilization of certain injectable and ophthalmic products (Russell 1999a), although this process is no longer approved. The gaseous sterilant ethylene oxide is also employed at higher temperatures (approximately 60°C) and LTSF has been claimed to be a useful sterilization procedure. In the food industry, acid–heat treatment is an important method of sterilization (Gould and Jones 1989).

CHEMICAL TREATMENT AND IRRADIATION

The presence of various chemicals during irradiation may either sensitize bacterial spores to, or protect them from, the consequences of ionizing radiation. Spores are more sensitive under oxic than under anoxic conditions. Ketonic agents of widely differing electron affinities sensitize spores suspended in anoxic buffers to subsequent radiation, although the maximum sensitization achieved is only about 40 percent of that in oxygen alone (Tallentire and Jacobs 1972). Sensitizing agents appear to act by removing electrons from the conversion of •OH to OH⁻, which is harmless to the spore.

THERMORADIATION

Thermoradiaton is the simultaneous use of heat and ionizing radiation. Careful selection of temperature is, however, important because a 'paradoxical inversion' or 'thermorestoration' occurs at certain temperatures; for example, with *C. botulinum 33A* spores, there is an increase in sensitivity above 80°C (Grecz 1965).

OTHER COMBINATIONS

Other combined processes that have been examined include hydrostatic pressure combined with heat or radiation (Wills 1974), ultrasonic waves with GTA (Boucher 1979), and combinations of chemicals (Russell 1982).

Process validation, monitoring, and parametric release

It is important that strict control measures are employed to validate and monitor the efficacy of sterilization processes. In addition, sterility tests on samples of a batch of treated product are necessary (Favero, 1998; British Pharmacopoeia 2000), although undue reliance should not be placed on such tests alone.

Process validation (Favero 1998) is a means of demonstrating that a process actually works, i.e. that it kills (or removes) microorganisms, thereby achieving what it purports to do (achieve sterilization). For this reason, the choice of a test organism is important: it must possess above-average resistance to the process in question, although it need not necessarily be the most resistant organism known. Biological indicators (BI) in the form (usually) of bacterial spores are used to validate those processes designed to inactivate microbes. *D. radiodurans* is even more resistant to ionizing radiation than spores (see Ionizing radiation). It is not, however, used as a BI for this process because (a) it is not a normal contaminant (bioburden) on pharmaceutical or medical products and (b) its inactivation would require unacceptably high radiation doses that could damage many such products. Examples of BIs are provided in Table 17.9.

The principle involved in validating sterile filtration is to use a smaller-than-average test organism. If this is retained by the filter, then larger organisms also will not pass through the filter.

Table 17.9 *Validation and monitoring of sterilization processes*

Process	Validation[a]	Monitoring[b,c]
Moist heat	B. stearothermophilus	Temperature recording charts; thermocouples; CI
Dry heat	B. subtilis var. niger	Temperature recording charts; thermocouples; CI
Ionizing radiation	B. pumilus	Dosimeters; CI
Ethylene oxide	B. subtilis var. niger	Temperature probes; relative humidity; B. subtilis var. niger
Filtration	S. marcescens; P. diminuta	Bubble point pressure test

a) In addition to the biological indicators listed, the desired physical and chemical conditions to achieve sterilization must also be validated, e.g. measuring devices for heat, dosimeters for radiation, physical methods for determining filter integrity and temperature, and relative humidity for gaseous sterilization.
b) CI, chemical indicator (used as a visual check that a sterilization process has been undertaken: a color change does not necessarily mean that sterilization has been achieved).
c) Samples taken at random from a batch may also be subjected to sterility testing.

In addition to validating a process with BIs, it is also necessary to measure and record the required physical or chemical conditions reached throughout the sterilization cycle within every part of the load (Habarer and Wallhaeusser 1990).

Routine monitoring of a process is carried out by appropriate physical methods, except for ethylene oxide where difficulties still occur (Table 17.9). Sterility can then be assessed, except with ethylene oxide, by monitoring only the physical conditions of the process. Such a system (parametric release) is defined as the release of sterile products based on process compliance to physical specification (Dewhurst and Hoxey 1990).

Sterility assurance level (SAL) is a term employed to provide an expected level of safety with a sterilization process. It is generally accepted that terminally sterilized products, i.e. those sterilized in their final containers, have a safety level (SAL) of not more than one unsterile article per 10^6 items processed (Habarer and Wallhaeusser 1990; Bruch 1993; Graham and Boris 1993). SAL is thus based upon a sound knowledge of microbial inactivation kinetics, rather than upon sterility testing, which can detect only gross contamination.

REFERENCES

Adler-Storthz, K., Schultster, L.M., et al. 1983. Effect of alkaline glutaraldehyde on hepatitis B virus antigens. *Eur J Clin Microbiol*, **2**, 316–20.

AFNOR. 1989. *Association Française de Normalisation – Official French Standards*, T72-301. Paris: AFNOR.

Akimitsu, N., Hamamoto, H., et al. 1999. Increase in resistance of methicillin-resistant *Staphylococcus aureus* to β-lactams caused by mutations conferring resistance to benzalkonium chloride, a disinfectant widely used in hospitals. *Antimicrob Agents Chemother*, **43**, 3042–3.

Albert, A. 1979. *Selective toxicity: The physiochemical basis of therapy*, 6th edn. London: Chapman & Hall.

Alderton, G. and Snell, N.S. 1963. Base exchange and heat resistance in bacterial spores. *Biochem Biophys Res Commun*, **10**, 139–43.

Alderton, G., Chen, J.K. and Ito, K.A. 1980. Heat resistance of the chemical resistance forms of Clostridium botulinum 62A spores over the water activity range 0 to 0.9. *Appl Environ Microbiol*, 511–15.

Alfa, M.J. and Jackson, M. 2001. A new hydrogen peroxide-based medical-device detergent with germicidal properties: comparison with enzymatic cleaners. *Am J Infect Contr*, **29**, 168–77.

Alfa, M.J. and Sitter, D.L. 1994. In-hospital evaluation of ortho-phthalaldehyde as a high level disinfectant for flexible endoscopes. *J Hosp Infect*, **26**, 15–26.

Alfa, M.J., Olson, N., et al. 1998. New low temperature sterilization technologies: microbicidal activity and clinical efficacy. In: Rutala, W.A. (ed.), *Disinfection, sterilization and antisepsis in health care*. Washington, DC: APIC, 67–78.

Allwood, M.C. 1973. Inhibition of *Staphylococcus aureus* by combinations of non-ionic surface-active agents and antibacterial substances. *Microbios*, **7**, 209–14.

Alqurashi, A.M., Day, M.J. and Russell, A.D. 1996. Susceptibility of some strains of enterococci and streptococci to antibiotics and biocides. *J Antimicrob Chemother*, **38**, 745.

AOAC. 1990. *Disinfectants – Official methods of analysis of the association of agricultural chemists*, 15th edn. Arlington, VA: AOAC.

Asada, S., Takano, M. and Shibasaki, I. 1979. Deoxyribonucleic acid strand breaks during drying of *Escherichia coli* on a hydrophobic filter membrane. *Appl Environ Microbiol*, **37**, 266–73.

Ayliffe, G.A.J., Coates, D. and Hoffman, P.N. 1993. *Chemical disinfection in hospitals*, 2nd edn. London: Public Health Laboratory Service.

Ayres, H., Furr, J.R. and Russell, A.D. 1993. A rapid method of evaluating permeabilizing activity against *Pseudomonas aeruginosa*. *Lett Appl Microbiol*, **17**, 149–51.

Ayres, H., Furr, J.R. and Russell, A.D. 1998a. Effect of permeabilizers on antibiotic sensitivity of *Pseudomonas aeruginosa*. *Lett Appl Microbiol*, **28**, 13–16.

Ayres, H.M., Furr, J.R. and Russell, A.D. 1998b. Effect of divalent cations on permeabilizer-induced lysozyme lysis of *Pseudomonas aeruginosa*. *Lett Appl Microbiol*, **27**, 372–4.

Ayres, H., Payne, D.N., et al. 1998c. Use of the Malthus-AT system to assess the efficacy of permeabilizing agents on the activity of antibacterial agents against *Pseudomonas aeruginosa*. *Lett Appl Microbiol*, **26**, 422–6.

Ayres, H., Payne, D.N., et al. 1998d. Effect of permeabilizing agents on antibacterial activity against a simple *Pseudomonas aeruginosa* biofilm. *Lett Appl Microbiol*, **27**, 79–82.

Babb, J.R., Bradley, C.R. and Ayliffe, G.A.J. 1980. Sporicidal activity of glutaraldehydes and hypochlorites and other factors influencing their selection for the treatment of medical equipment. *J Hosp Infect*, **1**, 63–75.

Baillie, L.W.J., Wade, J.J. and Casewell, M.W. 1992. Chlorhexidine sensitivity of *Enterococcus faecium* resistant to vancomycin, high levels of gentamicin, or both. *J Hosp Infect*, **20**, 127–8.

Bailly, J.-L., Chambon, M., et al. 1991. Activity of glutaraldehyde at low concentrations (<2%) against poliovirus and its relevance to gastrointestinal endoscope disinfection procedures. *Appl Environ Microbiol*, **57**, 1156–60.

Baird-Parker, A.C. and Holbrook, R. 1971. The inhibition and destruction of cocci. In: Hugo, W.B. (ed.), *Inhibition and destruction of the microbial cell*. London: Academic Press.

Baldry, M.G.C. 1983. The bactericidal, fungicidal and sporicidal properties of hydrogen peroxide and peracetic acid. *J Appl Bacteriol*, **54**, 417–23.

Baron, H., Safar, J., et al. 2001. Prions. In: Block, S.S. (ed.), *Disinfection, sterilization and preservation*, 5th edn. Philadelphia: Lippincott Wiliams & Wilkins, 659–74.

Barrett, T.L. 2001. Flash sterilization: What are the risks? In: Rutala, W.A. (ed.), *Sterilization, disinfection and antisepsis in health care*. Washington, DC: APIC, 70–6.

Battista, J.R. 1997. Against all odds: The survival strategies of *Deinococcus radiodurans*. *Ann Rev Microbiol*, **51**, 203–24.

Battista, J.R., Earl, A.M. and White, O. 2000. The stress responses of *Deinococcus radiodurans*. In: Storz, G. and Hengge-Aronis, R. (eds), *Bacterial stress responses*. Washington, DC: ASM Press.

Beaman, T.C. and Gerhardt, P. 1986. Heat resistance of bacterial spores correlated with protoplast dehydration, mineralization and thermal adaptation. *Appl Environ Microbiol*, **52**, 1242–6.

Beaman, T.C., Pankratz, H.S. and Gerhardt, P. 1988. Heat shock affects permeability and resistance of *Bacillus stearothermophilus* spores. *Appl Environ Microbiol*, **54**, 2515–20.

Beaman, T.C., Pankratz, H.S. and Gerhardt, P. 1989. Low heat resistance of *Bacilluss phaericus* spores correlated with high protoplast water content. *FEMS Microbiol Lett*, **58**, 1–4.

Bean, H.S. 1967. Types and characteristics of disinfectants. *J Appl Bacteriol*, **30**, 6–16.

Bean, H.S. 1972. Preservatives for pharmaceuticals. *J Soc Cosmet Chem*, **23**, 703–20.

Beinlich, K.I., Chuanchen, R. and Schweizer, H.P. 2001. Contribution of multidrug efflux pumps to multiple antibiotic resistance in veterinary isolates of *Pseudomobas aeruginosa*. *FEMS Microbiol Lett*, **198**, 129–34.

Bellamy, K. 1995. A review of the test methods used to establish virucidal activity. *J Hosp Infect*, **30**, Suppl, 389–96.

Belliveau, B.H., Beaman, T.C., et al. 1992. Heat killing of bacterial spores analyzed by differential scanning calorimetry. *J Bacteriol*, **174**, 4463–74.

Belly, R.T. and Kydd, G.C. 1982. Silver resistance in microorganisms. *Dev Ind Microbiol*, **34**, 751–6.

Berkelman, R.L., Holland, B.W. and Anderson, R.L. 1982. Increased bactericidal activity of dilute preparations of povidone-iodine solutions. *J Clin Microbiol*, **15**, 635–9.

Block, S.S. 2001a. Historical review. In: Block, S.S. (ed.), *Disinfection, sterilization and preservation*, 5th edn. Philadelphia: Lippincott Williams & Wilkins, 5–17.

Block, S.S. 2001b. Definition of terms. In: Block, S.S. (ed.), *Disinfection, sterilization and preservation*, 5th edn. Philadelphia: Lippincott Williams & Wilkins, 19–28.

Block, S.S. (ed.) 2001c. *Disinfection, sterilization and preservation*, 5th edn. Philadelphia: Lippincott Williams & Wilkins.

Block, S.S. 2001d. Peroxygen compounds. In: Block, S.S. (ed.), *Disinfection, sterilization and preservation*, 5th edn. Philadelphia: Lippincott Williams & Wilkins, 185–204.

Bloomfield, S.F. 1974. The effect of the phenolic antibacterial agent fentichlor on energy coupling in *Staphylococcus aureus*. *J Appl Bacteriol*, **37**, 117–34.

Bloomfield, S.F. 1996. Chlorine and iodine formulations. In: Ascenzi, J.M. (ed.), *Handbook of disinfectants and antiseptics*. New York: Marcel Dekker, 133–58.

Bloomfield, S.F. 1999. Bacterial sensitivity and resistance. C. Resistance of bacterial spores to chemical agents. In: Russell, A.D., Hugo, W.B. and Ayliffe, G.A.J. (eds), *Principles and practice of disinfection, preservation and sterilization*, 3rd edn. Oxford: Blackwell Science, 303–20.

Bloomfield, S.F. and Arthur, M. 1994. Mechanisms of inactivation and resistance of spores to chemical biocides. *J Appl Bacteriol*, **76**, 91S–104S.

Bloomfield, S.F., Smith-Burchnell, C.A. and Dalgleish, A.G. 1990. Evaluation of hypochlorite-releasing agents against the human immunodeficiency virus (HIV). *J Hosp Infect*, **15**, 273–8.

Board, R.G. 1995. Natural antimicrobials from animals. In: Gould, G.W. (ed.), *New methods of food preservation*. London: Blackie, 40–57.

Bobichon, H. and Bouchet, P. 1987. Action of chlorhexidine on budding *Candida albicans*: scanning and transmission electron microscopic study. *Mycopathologia*, **100**, 27–35.

Boucher, R.M.G. 1979. Ultrasonics. A tool to improve biocidal efficacy of sterilants or disinfectants in hospital and dental practice. *Can J Pharm Sci*, **14**, 1–12.

Bridges, B.A. 1976. Survival of bacteria following exposure to ultraviolet and ionizing radiations. In: Gray, T.G.R. and Postgate, J.R. (eds), *The survival of vegetative microbes. 26th Symposium of the Society for General Microbiology*. Cambridge: Cambridge University Press, 183–208.

British Pharmacopoeia. 1988. London: HMSO.

British Pharmacopoeia. 1993. London: HMSO.

British Pharmacopoeia. 1998. London: HMSO.

British Pharmacopoeia. 2000. London: Pharmaceutical Press.

British Standards Institute. 1991. BS5283, *Glossary of terms relating to disinfectants*. London: BSI.

Broadley, S.J., Jenkins, P.A., et al. 1995. Potentiation of the effects of chlorhexidine diacetate and cetylpyridinium chloride on mycobacteria by ethambutol. *J Med Microbiol*, **43**, 458–60.

Brown, M.R.W. 1975. The role of the cell envelope in resistance. In: Brown, M.R.W. (ed.), *Resistance of* Pseudomonas aeruginosa. London: John Wiley & Sons, 71–99.

Brown, K.L. 1994. Spore resistance and ultra heat treatment processes. *J Appl Bacteriol*, **76**, 67S–80S.

Browne, M.K. and Stoller, J.L. 1970. Intraperitoneal noxythiolin in faecal peritonitis. *Br J Surg*, **57**, 525–9.

Browne, M.K., Leslie, G.B. and Pfirrman, R.W. 1976. Taurolin, a new chemotherapeutic agent. *J Appl Bacteriol*, **41**, 363–8.

Broxton, P., Woodcock, P.M. and Gilbert, P. 1984. Interaction of some polyhexamethylene biguanides and membrane phospholipids in *Escherichia coli*. *J Appl Bacteriol*, **57**, 115–24.

Bruch, C.W. 1993. The philosophy of sterilization validation. In: Morrissey, R.F. and Phillips, G.B. (eds), *Sterilization technology. A practical guide for manufacturers and ssers of health care products*. New York: Van Nostrand Reinhold, 17–35.

Cabrera-Martinez, R.-M., Setlow, P. and Setlow, P.S. Studies on the mechanisms of the sporicidal action of *ortho*-phthalaldehyde. *J Appl Microbiol*, **92**, 675–81.

Carpentier, B. and Cerf, O. 1993. Biofilms and their consequences, with particular reference to hygiene in the food industry. *J Appl Bacteriol*, **75**, 499–511.

CEN. 1996. (European Committee for Standardization). *Chemical disinfectants and antiseptics*, EN 1040, 1275.

Cerf, O. 1977. Tailing of survival curves of bacterial spores. *J Appl Bacteriol*, **42**, 1–19.

Chambon, M., Bailly, J.-L. and Peigue-Lafeuille, H. 1992. Activity of glutaraldehyde at low concentrations against capsid proteins of poliovirus type 1 and echovirus type 25. *Appl Environ Microbiol*, **58**, 3517–21.

Chawner, J.A. and Gilbert, P. 1989a. A comparative study of the bactericidal and growth inhibitory activities of the bisbiguanides alexidine and chlorhexidine. *J Appl Bacteriol*, **66**, 243–52.

Chawner, J.A. and Gilbert, P. 1989b. Interaction of the bisbiguanides chlorhexidine and alexidine with phospholipid vesicles: evidence for separate modes of action. *J Appl Bacteriol*, **66**, 253–8.

Chopra, I. 1987. Microbial resistance to veterinary disinfectants and antiseptics. In: Linton, A.H., Hugo, W.B. and Russell, A.D. (eds), *Disinfection in veterinary and farm animal practice*. Oxford: Blackwell Science, 43–65.

Chopra, I., Johnson, S.C. and Bennett, P.M. 1987. Inhibition of *Providencia stuartii* cell envelope enzymes by chlorhexidine. *J Antimicrob Chemother*, **19**, 743–51.

Chuanchen, R., Beinlick, K., et al. 2001. Cross-resistance between triclosan and antibiotics in *Pseudomonas aeruginosa* is mediated by multidrug efflux pumps: exposure of a susceptible mutant strain to triclosan selects *nfxB* mutants overexpressing MexCD-OprJ. *Antimicrob Agents Chemother*, **45**, 428–32.

Clapp, P.A., Davies, M.J., et al. 1994. The bactericidal action of peroxides: an E.P.R. spin-trapping study. *Free Rad Res*, **21**, 147–67.

Cleland, M.R., O'Neill, M.T. and Thompson, C.C. 1993. Sterilization with accelerated electrons. In: Morrissey, R.F. and Phillips, G.B. (eds), *Sterilization technology. A practical guide for manufacturers and users of health care products*. New York: Van Nostrand Reinhold, 218–253.

Coates, D. and Death, D.E. 1978. Sporicidal activity of mixtures of alcohol and hypochlorite. *J Clin Pathol*, **31**, 148–52.

Coates, D. and Hutchinson, D.N. 1994. How to produce a hospital disinfection policy. *J Hosp Infect*, **26**, 57–68.

Collier, P.J., Ramsey, A.J., et al. 1990. Growth inhibitory and biocidal activity of some isothiazolone biocides. *J Appl Bacteriol*, **69**, 569–77.

Cookson, B.D. and Phillips, I. 1990. Methicillin-resistant staphylococci. *J Appl Bacteriol*, **69**, 55S–70S.

Cookson, B.D., Bolton, M.C. and Platt, J.H. 1991. Chlorhexidine resistance in *Staphyloccus aureus* or just an elevated MIC? An *in vitro* and *in vivo* assessment. *Antimicrob Agents Chemother*, **35**, 1997–2002.

Costerton, J.W., Chang, K.J., et al. 1987. Bacterial biofilms in nature and disease. *Ann Rev Microbiol*, **41**, 435–64.

Costerton, J.W., Lewandowsik, Z., et al. 1994. Biofilms, the customized niche. *J Bacteriol*, **176**, 2137–42.

Cremieux, A., Freney, J. and Davin-Regli, A. 2001. Methods of testing disinfectants. In: Block, S.S. (ed.), *Disinfection, sterilization and preservation*, 4th edn. Philadelphia: Lippincott Williams & Wilkins, 1305–27.

Croshaw, B. 1971. The destruction of mycobacteria. In: Hugo, W.B. (ed.), *Inhibition and destruction of the microbial cell.* London: Academic Press, 420–49.

Croshaw, B. and Holland, V.R. 1984. Chemical preservatives: use of bronopol as a cosmetic preservative. In: Kabara, J.J. (ed.), *Cosmetic and drug preservation. Principles and practice.* New York: Marcel Dekker, 31–62.

D'Arcy, P.F. 1971. Inhibition and destruction of moulds and yeasts. In: Hugo, W.B. (ed.), *Inhibition and destruction of the microbial cell.* London: Academic Press, 614–86.

Davies, D.G., Parsek, M.R., et al. 1998. The involvement of cell-to-cell signals in the development of a bacterial biofilm. *Science,* **280**, 295–8.

Davies, J.G., Babb, J.R., et al. 1993. Preliminary study of test methods to assess the virucidal activity of skin disinfectants using poliovirus and bacteriophages. *J Hosp Infect,* **25**, 125–31.

Day, M.J. and Russell, A.D. 1999. Bacterial sensitivity and resistance. F. Antibiotic-resistant cocci. In: Russell, A.D., Hugo, W.B. and Ayliffe, G.A.J. (eds), *Principles and practice of disinfection, preservation and sterilization,* 3rd edn. Oxford: Blackwell Science, 344–59.

Death, J.E. and Coates, D. 1979. Effect of pH on sporicidal and microbicidal activity of buffered mixtures of alcohol and sodium hypochlorite. *J Clin Pathol,* **32**, 148–52.

Dekker, J. 1987. Development of resistance to modern fungicides and strategies for its avoidance. In: Lyr, H. (ed.), *Modern selective fungicides.* Harlow: Longman, 39–52.

Denton, G.C. 2001. Chlorhexidine. In: Block, S.S. (ed.), *Disinfection, sterilization and preservation,* 5th edn. Philadelphia: Lippincott Williams & Wilkins, 321–36.

Denyer, S.P. 2002. Close encounters of the microbial kind. *Pharm J,* **269**, 451–4.

Denyer, S.P. and Hodges, N.A. 1999. Filtration sterilization. In: Russell, A.D., Hugo, W.B. and Ayliffe, G.A.J. (eds), *Principles and practice of disinfection, preservation and sterilization,* 3rd edn. Oxford: Blackwell Science, 733–66.

Denyer, S.P. and Stewart, G.S.A.B. 1998. Mechanisms of action of disinfectants. *Int Biodeter Biodegrad,* **41**, 261–8.

Denyer, S.P., Gorman, S.P. and Sussman, M. (eds) 1976. *Microbial biofilms: formation and control. Society for Applied Bacteriology Technical Series No. 30.* Oxford: Blackwell.

Dewhurst, E. and Hoxey, E.V. 1990. Sterilization methods. In: Denyer, S. and Baird, R. (eds), *Guide to microbiological control in pharmaceuticals.* Chichester: Ellis Horwood, 182–218.

Donlan, R.M. and Costerton, J.W. 2002. Biofilms: Survival mechanisms of clinically relevant microorganisms. *Clin Microbiol Rev,* **15**, 167–93.

Dring, G.J. 1976. Some aspects of the effects of hydrostatic pressure on micro-organisms. In: Skinner, F.A. and Hugo, W.B. (eds), *Inhibition and inactivation of vegetative microbes. Society for Applied Bacteriology Symposium Series No. 5.* London: Academic Press, 257–78.

Dunn, W.M. Jr. 2002. Bacterial adhesion: Seen any good biofilms recently? *Clin Microbiol Rev,* **15**, 155–66.

Durban, E., Goodnow, R. and Grecz, N. 1970. Changes in resistance to radiation and heat during sporulation and germination of *Clostridium botulinum* 33A. *J Bacteriol,* **102**, 590–2.

Durban, E., Grecz, N. and Farkas, J. 1974. Direct enzymatic repair of DNA single strand breaks in dormant spores. *J Bacteriol,* **118**, 129–38.

Dychdala, G.R. 2001. Chlorine and chlorine compounds. In: Block, S.S. (ed.), *Disinfection, sterilization and preservation,* 5th edn. Philadelphia: Lippincott Williams & Wilkins, 135–57.

Eklund, T. 1983. The antimicrobial effect of dissociated and undissociated sorbic acid at different pH levels. *J Appl Bacteriol,* **54**, 383–9.

Eklund, T. 1985a. The effect of sorbic acid and esters of *p*-hydroxybenzoic acid on the protonmotive force in *Escherichia coli* membrane vesicles. *J Gen Microbiol,* **131**, 73–6.

Eklund, T. 1985b. Inhibition of microbial growth at different pH levels by benzoic and propionic acids and esters of *p*-hydroxybenzoic acid. *Int J Food Microbiol,* **2**, 159–67.

Elferink, J.G.R. and Booij, H.L. 1974. Interaction of chlorhexidine with yeast cells. *Biochem Pharmacol,* **23**, 1413–19.

Elsmore, R. 1999. Bacterial sensitivity and resistance. E. *Legionella.* In: Russell, A.D., Hugo, W.B. and Ayliffe, G.A.J. (eds), *Principles and practice of disinfection, preservation and sterilization,* 3rd edn. Oxford: Blackwell Science, 333–43.

Ernst, R.R. 1974. Ethylene oxide sterilization kinetics. *Biotechnology and Bioengineering Symposium,* No. 44, Supplement. 865–78.

Farkas, J. 1994. Tolerance of spores to ionizing radiation: mechanisms of inactivation, injury and repair. *J Appl Bacteriol,* **76**, 81S–90S.

Fassihi, R.A. 2001. Preservation and microbiological attributes of nonsterile pharmaceutical products. In: Block, S.S. (ed.), *Disinfection, sterilization and preservation,* 5th edn. Philadelphia: Lippincott Williams & Wilkins, 1263–81.

Favero, M.S. 1998. Developing indicators for monitoring sterilization. In: Rutala, W.A. (ed.), *Disinfection, sterilization and antisepsis in health care.* Washington, DC: APIC, 119–32.

Favero, M.S. and Bond, W.W. 1993. The use of liquid chemical germicides. In: In Morrissey, R.F. and Phillips, G.B. (eds), *Sterilization technology. A practical guide for manufacturers and users of health care products.* New York: Van Nostrand Reinhold, 309–34.

Foster, T.J. 1983. Plasmid-determined resistance to antimicrobial drugs and toxic metal ions in bacteria. *Microbiol Rev,* **47**, 361–409.

Foster, S.J. 1994. The role and regulation of cell wall structural dynamics during differentiation of endospore-forming bacteria. *J Appl Bacteriol,* **76**, 25S–39S.

Fraise, A. 1999. Choosing disinfectants. *J Hosp Infect,* **43**, 255–64.

Fraise, A. 2002a. Susceptibility of antibiotic-resistant cocci to biocides. *J Appl Microbiol,* **92**, 158S–62S.

Fraise, A. 2002b. Biocide abuse and antimicrobial resistance – cause for concern? *J Antimicrob Chemother,* **49**, 11–12.

Fraud, S., Maillard, J.-Y. and Russell, A.D. 2001. Comparison of the mycobactericidal activity of *ortho*-phthalaldehyde, glutaraldehyde and other dialdehydes by a quantitative suspension test. *J Hosp Infect,* **48**, 214–21.

Fraud, S., Hann, A.C., et al. 2003. Effects of *ortho*-phthalaldehyde, glutaraldehyde and chlorhexidine diacetate on *Mycobacterium chelonae* and *Mycobacterium abscessus* strains with modifceied permeability. *J Antimicrob Chemother,* **51**, 575–84.

Fried, V.A. and Novick, A. 1973. Organic solvents as probes for the structure and function of the bacterial membrane; effects of ethanol on the wild type and on an ethanol-resistant mutant of *Escherichia coli* K-12. *J Bacteriol,* **114**, 239–48.

Furr, J.R. 1999. Sensitivity of protozoa to disinfectants. A. Acanthamoeba and contact lens solutions. In: Russell, A.D., Hugo, W.B. and Ayliffe, G.A.J. (eds), *Principles and practice of disinfection, preservation and sterilization,* 3rd edn. Oxford: Blackwell Science, 237–50.

Gadd, G.M. and White, C. 1989. Heavy metal and radionuclide accumulation and toxicity in fungi and yeasts. In: Poole, R.K. and Gadd, G.M. (eds), *Metal–microbe interactions. Special Publication of the Society for General Microbiology No. 26.* Oxford: Oxford University Press, 19–38.

Gale, E.F. 1986. Nature and development of phenotypic resistance to amphotericin B in *Candida albicans. Adv Microb Physiol,* **27**, 277–320.

Gardner, J.F. and Peel, M.M. 1991. *Introduction to sterilization and disinfection.* Edinburgh: Churchill Livingstone.

Gerba, C.P. and Rusin, P. 2001. Relationship between the use of antiseptics and disinfectantsand the development of antimicrobial resistance. In: Rutala, W.A. (ed.), *Disinfection, sterilization and antisepsis.* Washington, DC: APIC, 187–94.

Gerhardt, P. and Marquis, R.E. 1989. Spore thermoresistance mechanisms. In: Smith, I., Slepecky, R. and Setlow, P. (eds), *Regulation of procaryotic development.* Washington, DC: American Society for Microbiology, 17–63.

Gilbert, P. and McBain, A.J. 2001a. Biocide usage in the domestic setting and concern about antibacterial and antibiotic resistance. *J Infect*, **43**, 85–91.

Gilbert, P. and McBain, A.J. 2001b. Biofilms : their impact upon health and their recalcitrance towards biocides. *Am J Infect Cont*, **29**, 252–5.

Gilbert, P., Pemberton, D. and Wilkinson, D.E. 1990a. Barrier properties of the Gram-negative cell envelope towards high molecular weight polyhexamethylene biguanides. *J Appl Bacteriol*, **69**, 585–92.

Gilbert, P., Pemberton, D. and Wilkinson, D.E. 1990b. Synergism within polyhexamethylene biguanide biocide formulations. *J Appl Bacteriol*, **69**, 593–8.

Gilbert, P., Collier, P.J. and Brown, M.R.W. 1990c. Influence of growth rate on susceptibility to antimicrobial agents : biofilms, cell cycle, dormancy and stringent response. *Antimicrob Agents Chemother*, **34**, 1865–8.

Gilbert, P., Das, J.R., et al. 2001. Assessment of resistance towards biocides following the attachment of micro-organisms to, and growth on, surfaces. *J Appl Microbiol*, **91**, 248–54.

Gilbert, P., Allison, D.G. and McBain, A.J. 2002a. Biofilms *in vitro* and *in vivo*: Do singular mechanisms imply cross-resistance? *J Appl Microbiol*, **92**, S98–S110.

Gilbert, P., Maira-Litran, T., et al. 2002b. The physiology and collective recalcitrance of microbial biofilm communities. *Adv Microb Physiol*, **46**, 205–56.

Gorman, S.P. and Scott, E.M. 1977. Uptake and media reactivity of glutaraldehyde solutions related to structure and biocidal activity. *Microbios Lett*, **5**, 163–9.

Gorman, S.P., Scott, E.M. and Russell, A.D. 1980. Antimicrobial activity, uses and mechanism of action of glutaraldehyde. *J Appl Bacteriol*, **48**, 161–90.

Gottardi, W. 1985. The influence of the chemical behaviour of iodine on the germicidal action of disinfectant solutions containing iodine. *J Hosp Infect*, **6**, Supplement A, 1–11.

Gould, G.W. 1970. Potentiation by halogen compounds of the lethal action of γ-radiation on spores of *Bacillus cereus*. *J Gen Microbiol*, **64**, 289–300.

Gould, G.W. 1983. Mechanisms of resistance and dormancy. In: Hurst, A. and Gould, G.W. (eds), *The bacterial spore*, Vol. 2. . London: Academic Press, 173–209.

Gould, G.W. 1984. Injury and repair mechanisms in bacterial spores. In: In Andrew, M.H.E. and Russell, A.D. (eds), *The revival of injured microbes. Society for Applied Bacteriology Symposium Series No. 12.* London: Academic Press, 199–220.

Gould, G.W. 1985. Modifications of resistance and dormancy. In: Dring, G.J., Ellar, D.J. and Gould, G.W. (eds), *Fundamental and applied aspects of bacterial spores*. London: Academic Press, 371–82.

Gould, G.W. 1988. Interference with homeostasis: food. In: Whittenbury, R., Gould, G.W., et al. (eds), *Homeostatic mechanisms in microorganisms*. Bath: Bath University Press, 200–28.

Gould, G.W. 1989. Heat-induced injury and inactivation. In: Gould, G.W. (ed.), *Mechanisms of action of food preservation procedures*. London: Elsevier Applied Science, 11–42.

Gould, G.W. (ed.) 1995. *New methods of food preservation*. London: Blackie.

Gould, G.W. 1999. D. Applications of thermal processing in the food industry. In: Russell, A.D., Hugo, W.B. and Aycliffe, G.A.J. (eds), *Principles and practice of disinfection, preservation and sterilization*, 3rd edn. Oxford: Blackwell Science.

Gould, G.W. and Jones, M.V. 1989. Combination and synergisitic effects. In: Gould, G.W. (ed.), *Mechanisms of action of food preservation procedures*. London: Elsevier Applied Science, 401–22.

Gould, G.W. and Russell, N.J. 1991. Sulphite. In: Russell, N.J. and Gould, G.W. (eds), *Food preservatives*. Glasgow and London: Blackie, 72–88.

Gould, G.W. and Sale, A.J.H. 1970. Initiation of germination of bacterial spores by hydrostatic pressure. *J Gen Microbiol*, **60**, 335–46.

Gould, G.W., Jones, A. and Wrighton, C. 1968. Limitations of the initiation of germination of bacterial spores as a spore control procedure. *J Appl Bacteriol*, **33**, 357–66.

Graham, G.S. and Boris, C.A. 1993. Chemical and biological indicators. In: Morrissey, R.F. and Phillips, G.B. (eds), *Sterilization technology. A practical guide for manufacturers and users of health care products*. New York: Van Nostrand Reinhold, 36–69.

Grecz, N. 1965. Biophysical aspects of clostridia. *J Appl Bacteriol*, **28**, 7–35.

Greenaway, D.L.A. and England, R.R. 1999a. ppGpp accumulation in *Pseudomonas aeruginosa* and *Pseudomonas fluorescens* subjected to nutrient limitation and biocide exposure. *Lett Appl Microbiol*, **29**, 298–302.

Greenaway, D.L.A. and England, R.R. 1999b. The intrinsic resistance of *Escherichia coli* to various antimicrobial agents requires ppGpp and σ^s. *Lett Appl Microbiol*, **29**, 323–6.

Gregory, A.W., Schaalje, G.B., et al. 1999. The mycobactericidal efficacy of ortho-phthalaldehyde and the comparative resistances of *Mycobacterium bovis*, *Mycobacterium terrae* and *Mycobacterium chelonae*. *Infect Control Hosp Epidemiol*, **20**, 324–30.

Griffiths, P.A., Babb, J.R. and Fraise, A.P. 1999. Mycobactericidal activity of selected disinfectants using a quantitative suspension test. *J Hosp Infect*, **41**, 111–21.

Groschel, D. 1995. Emerging technologies for disinfection and sterilization. In: Rutala, W.A. (ed.), *Chemical germicides in health care*. Washington, DC: APIC, 73–81.

Grossgebauer, K. 1970. Virus disinfection. In: Benarde, M.A. (ed.), *Disinfection*. New York: Marcel Dekker, 103–48.

Grossman, L., Levine, S.S. and Allison, W.S. 1961. The reaction of formaldehyde with nulceotides and T2 bacteriophage DNA. *J Molec Biol*, **3**, 47–60.

Habarer, J. and Wallhaeusser, K.-H. 1990. Assurance of sterility by validation of the sterilization process. In: Denyer, S. and Baird, R. (eds), *Guide to microbiological control in pharmaceuticals*. Chichester: Ellis Horwood, 219–40.

Hall, R.M. and Collis, C.M. 1998. Antibiotic resistance in gram-negative bacteria: the role of gene cassettes and integrons. *Drug Resist Update*, **1**, 109–19.

Hammond, S.M., Lambert, P.A. and Rycroft, A.N. 1984. *The bacterial cell surface*. London: Croom Helm.

Hancock, R.E.W. 1984. Alterations in membrane permeability. *Ann Rev Microbiol*, **38**, 237–64.

Hanlin, J.H., Lombardi, S.J. and Slepecky, R.A. 1985. Heat and UV light resistance of vegetative cells and spores of *Bacillus subtilis* Rec⁻ mutants. *J Bacteriol*, **163**, 774–7.

Hanson, P.J.V. 1988. Mycobacteria and AIDS. *Br J Hosp Med*, **40**, 149.

Hanson, P.J.V., Bennett, J., et al. 1994. Enteroviruses, endoscopy and infection control: an applied study. *J Hosp Infect*, **27**, 61–7.

Health Council of the Netherlands. 2001. *Disinfectants in consumer products*. Publication No. 2001/05E, The Hague, Netherlands.

Heath, R.J., White, S.W. and Rock, C.O. 2001. Lipid biosynthesis as a target for antibacterial agents. *Prog Lipid Res*, **40**, 467–97.

Hecker, M. and Volker, U. 2001. General stress response of *Bacillus subtilis* and other bacteria. *Adv Microb Physiol*, **44**, 35–91.

Hector, R.F. 1993. Compounds active against cell walls of medically important fungi. *Clin Microbiol Rev*, **6**, 1–21.

Heir, E., Sundheim, G. and Holck, A.L. 1998. The *Staphylococcus qacH* gene product; a new member of the SMR family encoding multidrug resistance. *FEMS Microbiol Lett*, **163**, 49–56.

Heir, E., Sundheim, G. and Holck, A.L. 1999. The *qacG* gene on plasmid pST94 confers resistance to quaternany ammonium compounds in staphylococci isolated from the food industry. *J Appl Microbiol*, **86**, 378–88.

Helander, I.M., Alakomi, H.-L., et al. 1997. Polyethyleneimine is an effective permeabilizer of Gram-negative bacteria. *Microbiology*, **143**, 3193–9.

Helander, I.M., Latva-Kala, K. and Lounatmaa, K. 1998. Permeabilizing action of polyethyleneimine on *Salmonella typhimurium* involves disruption of the outer membrane and interactions with lipopolysaccharide. *Microbiology*, **144**, 385–90.

Herbert, R.A. 1989. Microbial growth at low temperature. In: Gould, G.W. (ed.), *Mechanisms of action of food preservation procedures.* London: Elsevier Applied Science, 71–96.

Herring, C.M. and Saylor, M.C. 1993. Sterilization with radioisotopes. In: Morrissey, R.F. and Phillips, G.B. (eds), *Sterilization technology.* New York: Van Nostrand Reinhold, 196–217.

Higgins, C.S., Murtough, S.M., et al. 2001. Resistance to antibiotics and biocides among non-fermenting Gram-negative bacteria. *Clin Microbiol Infect*, **7**, 308–15.

Hilditch, E.A. 1999. Preservation in specialized areas. E. Wood preservation. In: Russell, A.D., Hugo, W.B. and Ayliffe, G.A.J. (eds), *Principles and practice of disinfection, preservation and sterilization*, 3rd edn. Oxford: Blackwell Science, 604–20.

Hiom, S.J., Furr, J.R., et al. 1993. Effects of chlorhexidine diacetate and cetylpyridinium chloride on whole cells and protoplasts of *Saccharomyces cerevisiae. Microbios*, **74**, 111–20.

Hiom, S.J., Furr, J.R. and Russell, A.D. 1995a. Uptake of ^{14}C-chlorhexidine gluconate by *Saccharomyces cerevisiae, Candida albicans* and *Candida glabrata. Lett Appl Microbiol*, **21**, 20–2.

Hiom, S.J., Hann, A.C., et al. 1995b. X-ray microanalysis of chlorhexidine-treated cells of *Saccharomyces cerevisiae. Lett Appl Microbiol*, **20**, 353–6.

Hiom, S.J., Furr, J.R., et al. 1996. The possible role of yeast cell walls in modifying cellular response to chlorhexidine diacetate. *Cytobios*, **86**, 123–35.

Hiramitsu, K. 1998. Vancomycin resistance in staphylococci. *Drug Resist Update*, **1**, 135–50.

Hitchener, B.J. and Egan, A.F. 1977. Outer membrane damage to sublethally heated *Escherichia coli* K-12. *Can J Microbiol*, **23**, 311–18.

Hitchins, A.D., King, W.L. and Gould, G.W. 1966. Role of disulphide bonds in the resistance of *Bacillus cereus* spores to gamma irradiation and heat. *J Appl Bacteriol*, **29**, 505–11.

Holton, J., Shetty, N. and McDonald, V. 1995. Efficacy of 'Nu-Cidex' (0.35% peracetic acid) against mycobacteria and cryptosporidia. *J Hosp Infect*, **27**, 105–15.

Hoxey, E.V. and Thomas, N. 1999. Gaseous sterilization. In: Russell, A.D., Hugo, W.B. and Ayliffe, G.A.J. (eds), *Principles and practice of disinfection, preservation and sterilization*, 3rd edn. Oxford: Blackwell Science, 703–32.

Huang, C.T., Fu, Y.P., et al. 1995. Non-spatial patterns of respiratory activity within biofilms during disinfection. *Appl Environ Microbiol*, **61**, 2252–6.

Huber, W.B. 1982. Antiseptics and disinfectants. In: Booth, N.H. and McDonald, L.E. (eds), *Veterinary pharmacology and therapeutics.* Ames, IA: Iowa State University Press, 693–716.

Hugo, W.B. 1971. Amidines. In: Hugo, W.B. (ed.), *Inhibition and destruction of the microbial cell.* London: Academic Press, 121–36.

Hugo, W.B. 1991a. A brief history of heat and chemical preservation and disinfection. *J Appl Bacteriol*, **71**, 9–18.

Hugo, W.B. 1991b. The degradation of preservatives by microorganisms. *Int J Biodeteri*, **27**, 185–94.

Hugo, W.B. 1999a. Historical introduction. In: Russell, A.D., Hugo, W.B. and Ayliffe, G.A.J. (eds), *Principles and practice of disinfection, preservation and sterilization*, 3rd edn. Oxford: Blackwell Science, 1–4.

Hugo, W.B. 1999b. Disinfection mechanisms. In: Russell, A.D., Hugo, W.B. and Ayliffe, G.A.J. (eds), *Principles and practice of disinfection, preservation and sterilization*, 3rd edn. Oxford: Blackwell Science, 258–83.

Hugo, W.B. and Denyer, S.P. 1987. The concentration exponent of disinfectants and preservatives (biocides). In: Board, R.G., Allwood, M/C/ and Banks, J.G. (eds), *Preservation in the food, pharmaceutical and environmental industries. Society for Applied Bacteriology Technical Series No. 22.* Oxford: Blackwell Science, 281–91.

Hugo, W.B. and Russell, A.D. 1999. Types of antimicrobial agents. In: Russell, A.D., Hugo, W.B. and Ayliffe, G.A.J. (eds), *Principles and practice of disinfection, preservation and sterilization*, 3rd edn. Oxford: Blackwell Science, 5–94.

Hutchinson, F. 1985. Chemical changes induced in DNA by ionizing radiation. *Prog Nucl Acid Res Molec Biol*, **32**, 115–54.

Imlay, J.A. 2002. How oxygen damages microbes: Oxygen tolerance and obligate anaerobiosis. *Adv Microb Physiol*, **46**, 111–53.

Imlay, J.A. and Linn, S. 1988. DNA damage and oxygen radical toxicity. *Science*, **240**, 1302–9.

Ishikawa, S.Y., Matsumura, Y., et al. 2002. Characterization of a cationic surfactant-resistant mutant isolayted spontaneously from *Escherichia coli. J Appl Microbiol*, **92**, 261–8.

Ismaeel, N., El-Moug, T., et al. 1986. Resistance of *Providencia stuartii* to chlorhexidine: A consideration of the role of the inner membrane. *J Appl Bacteriol*, **60**, 361–7.

Jacobs, P.T. and Lin, S.M. 2001. Sterilization processes utilizing low-temperature plasma. In: Block, S.S. (ed.), *Disinfection, sterilization and preservation*, 5th edn. Philadelphia: Lippincott Williams & Wilkins, 747–63.

Jarroll, E.L. 1988. Effect of disinfectants on *Giardia* cysts. *CRC Rev Environ Cont*, **18**, 1–28.

Jarroll, E.L. 1999. Sensitivity of protozoa to disinfectants. In: Russell, A.D., Hugo, W.B. and Aycliffe, G.A.J. (eds), *Principles and practice of disinfection, preservation and sterilization*, 3rd edn. Oxford: Blackwell Science, 251–7.

Jornitz, M.W., Soelkner, P.G. and Meltzer, T.H. 2002. Sterile filtration – a review of the past and present technologies. *J Pharm Sci Technol*, **56**, 192–5.

Joslyn, L.J. 2001. Gaseous chemical sterilization. In: Block, S.S. (ed.), *Disinfection, sterilization and preservation*, 5th edn. Philadelphia: Lippincott Williams & Wilkins, 337–59.

Joynson, J.A., Forbes, B. and Lambert, R.J.W. 2002. Adaptive resistance to benzalkonium chloride, amikacin and tobramycin: the effect on susceptibility to other antimicrobials. *J Appl Microbiol*, **93**, 96–106.

Kaye, S. and Phillips, C.R. 1949. The sterilizing action of gaseous ethylene oxide. IV. The effect of moisture. *Am J Hyg*, **50**, 296–306.

Khunkitti, W., Avery, S.V., et al. 1997. Effects of biocides on *Acanthamoeba castellanii* as measured by flow cytometry and plaque assay. *J Antimicrob Chemother*, **40**, 227–33.

Khunkitti, W., Hann, A.C., et al. 1998a. Biguanide-induced changes in *Acanthamoeba castellanii*: An electronic study. *J Appl Microbiol*, **84**, 53–2.

Khunkitti, W., Lloyd, D., et al. 1998b. *Acanthamoeba castellanii*: Growth, encystment, excystment and biocide susceptibility. *J Infect*, **36**, 43–8.

Khunkitti, W., Hann, A.C., et al. 1999. X-ray nicroanalysis of chlorine and phosphorus content in biguanide-treated *Acanthamoeba castellanii. J Appl Microbiol*, **86**, 453–9.

Klein, M. and Deforest, A. 1983. Principles of viral inactivation. In: Block, S.S. (ed.), *Disinfection, sterilization and preservation*, 3rd edn. Philadelphia: Lea and Febiger, 422–34.

Knapp, J.E. and Battisti, D.L. 2001. Chlorine dioxide. In: Block, S.S. (ed.), *Disinfection, sterilization and preservation*, 5th edn. Philadelphia: Lippincott Williams & Wilkins, 215–27.

Knott, A.G., Russell, A.D. and Dancer, B.N. 1995. Development of resistance to biocides during sporulation of *Bacillus subtilis. J Appl Bacteriol*, **79**, 492–8.

Koehler, D.R., Courcelle, J. and Hanawalt, P.C. 1996. Kinetics of pyrimidine (6-4) pyrimidone photoproduct repair in *Escherichia coli. J Bacteriol*, **178**, 1347–50.

Koljalg, S., Naaber, P. and Mikelsaar, M. 2002. Antibiotic resistance as an indicator of bacterial chlorhexidine susceptibility. *J Hosp Infect*, **51**, 106–13.

Korich, D.G., Mead, J.R., et al. 1990. Effects of ozone, chlorine dioxide, chlorine and monochloramine on *Cryptosporidium parvum* oocyst viability. *Appl Environ Microbiol*, **56**, 1423–8.

Kowalski, J.B. 1993. Selecting a sterilization method. In: Morrissey, R.F. and Phillips, G.B. (eds), *Sterilization technology. A practical guide for manufacturers and users of health care products.* New York: Van Nostrand Reinhold, 70–8.

Kraniak, J.M. and Shelef, L.A. 1988. Effect of ethylenediaminetetraacetic acid (EDTA) and metal ions on growth of *Staphylococcus aureus* 196E in culture media. *J Food Sci*, **53**, 910–13.

Krönig, B. and Paul, T. 1897. Die chemischen Grundlagen der Lehre von der Giftwirkung und Desinfektion. *Zeitsch Hyg*, **25**, 1–112.

Kurtz, J.B., Bartlett, C.L.R., et al. 1982. *Legionella pneumophila* in cooling water systems. *J Hyg*, **88**, 369–81.

Lambert, R.J.W., Joynson, J. and Forbes, B. 2001. The relationships and susceptibilities of some industrial, laboratory and clinical isolates of *Pseudomonas aeruginosa* to some antibiotics and biocides. *J Appl Microbiol*, **91**, 972–84.

Lannigan, R. and Bryan, L.E. 1985. Decreased susceptibility of *Serratia marcescens* to chlorhexidine related to the inner membrane. *J Antimicrob Chemother*, **15**, 559–65.

Lauzardo, M. and Rubin, R. 2001. Mycobacterial disinfection. In: Block, S.S. (ed.), *Disinfection, sterilization and preservation*, 5th edn. Philadelphia: Lippincott Williams & Wilkins, 513–28.

Lear, J.C., Maillard, J.Y., et al. 2002. Chloroxylenol- and triclosan-tolerant bacteria from industrial sources. *J Ind Microbiol Biotechnol*, **29**, 238–42.

Leelaporn, A., Paulsen, I.T., et al. 1994. Multidrug resistance to antiseptics and disinfectants in coagulase-negative staphylococci. *J Med Microbiol*, **40**, 214–20.

Lerman, L.S. 1961. Structural considerations in the interaction of DNA and acridines. *J Molec Biol*, **3**, 18–30.

Levy, S.B. 2000. Antibiotic and antiseptic resistance. *Ped Infect Dis J*, **19**, S12–22.

Levy, S.B. 2001. Antibacterial household products: Cause for concern. *Emerg Infect Dis*, **7**, Suppl. 3, 512–15.

Levy, S.B. 2002a. Factors impacting on the problem of antibiotic resistance. *J Antimicrob Chemother*, 25–30.

Levy, S.B. 2002b. Active efflux: A common mechanism for biocide and antibiotic resistance. *J Appl Microbiol*, **92**, S65–71.

Lewis, K. 2000. Programmed cell death in bacteria. *Microbiol Molec Biol Rev*, **64**, 503–14.

Lewis, K. 2001. Riddle of biofilm resistance. *Antimicrob Agent Chemother*, **45**, 997–1007.

Lindahl, T. 1982. DNA repair enzymes. *Ann Rev Biochem*, **51**, 61–87.

Linton, A.H. 1983. *Guidelines on prevention and control of salmonellosis, VPH/83.42.* Geneva: World Health Organization.

Linton, A.H., Hugo, W.B. and Russell, A.D. (eds) 1987. *Disinfection in veterinary and farm animal practice.* Oxford: Blackwell Science.

Lloyd, D., Turner, N.A., et al. 2001. Encystation in *Acanthamoeba castellenii*: development of biocide resistance. *J Eukaryot Microbiol*, **48**, 11–16.

Lloyd-Evans, N., Springthorpe, V.S. and Sattar, S.A. 1986. Chemical disinfection of human rotavirus-contaminated inanimate surfaces. *J Hyg*, **97**, 163–73.

Loshon, C.A., Melly, E., et al. 2001. Analysis of the killing of spores of *Bacillus subtilis* by a new disinfectant, Sterilox[R]. *J Appl Microbiol*, **91**, 1051–8.

Loughlin, M.F., Jones, M.V. and Lambert, P.A. 2002. Pseudomonas aeruginosa adapted to benzalkonium chloride show resistance to other membrane-active agents but not to clinically relevant antibiotics. *J Antimicrob Chemother*, **44**, 631–9.

Lowbury, E.J.L. 1992. Special problems in hospital antisepsis. In: Russell, A.D., Hugo, W.B. and Ayliffe, G.A.J. (eds), *Principles and practice of disinfection, preservation and sterilization,* 3rd edn. Oxford: Blackwell Science, 310–29.

Lukaszewicz, R.C. and Melzer, T.H. 1979. Concerning filter validation. *J Parenteral Drug Assoc*, **33**, 187–94.

Luppens, S.B.I., Abee, T. and Oosterom, J. 2001. Effect of benzalkonium chloride on viability and energy metabolism in exponential- and stationary-growth-phase cultures of *Listeria monocytogenes*. *J Food Sci*, **64**, 476–82.

Luppens, S.B.I., Rombouts, F.M. and Abee, T. 2002. The effect of growth phase on *Staphylococcus aureus* on resistance to disinfectants in a suspension test. *J Food Sci*, **65**, 124–9.

Lyon, B.R. and Skurray, R.A. 1987. Antimicrobial resistance of *Staphylococcus aureus*: genetic basis. *Microbiol Rev*, **51**, 88–134.

Lyr, H. 1987. Selectivity in modern fungicides and its basis. In: Lyr, H. (ed.), *Modern selective fungicides*. Harlow: Longman, 31–58.

Mackey, B.M. 1983. Changes in antibiotic and cell surface hydrophobicity in *Escherichia coli* injured by heating, freezing, drying or gamma radiation. *FEMS Microbiol Lett*, **20**, 395–9.

Mackey, B.M. 1984. Lethal and sublethal effects of refrigeration, freezing and freeze-drying on micro-organisms. In: Andrew, M.H.E. and Russell, A.D. (eds), *The revival of injured microbes. Society for Applied Bacteriology Symposium Series No. 12.* London: Academic Press, 45–76.

Mackey, B.M. and Seymour, D.A. 1987. The effect of catalase on recovery of heat-injured DNA-repair mutants of *Escherichia coli*. *J Gen Microbiol*, **133**, 1601–10.

MacLeod, R.A. and Calcott, P.H. 1976. Cold shock and freezing damage to microbes. In: Gray, T.R.G. and Postgate, J.R. (eds), *The survival of vegetative microbes. 26th Symposium of the Society for General Microbiology.* Cambridge: Cambridge University Press, 81–109.

Madigan, M.T., Martinko, J.M. and Parker, J. 1997. *Brock's Biology of microorganisms*, 8th edn. New Jersey: Prentice Hall International.

Maillard, J.-Y. 2001. Virus susceptibility to biocides: an understanding. *Rev Med Microbiol*, **12**, 63–74.

Maillard, J.Y. and Russell, A.D. 1997. Viricidal activity and mechanisms of action of biocides. *Sci Prog (Oxford)*, **80**, 287–315.

Maillard, J.-Y., Beggs, T.S., et al. 1995. Electronmicroscopic investigation of the effects of biocides on *Pseudomonas aeruginosa* PAO bacteriophage F116. *J Med Microbiol*, **42**, 415–20.

Maira-Litran, T., Allison, D.G. and Gilbert, P. 2000. An evaluation of the potential role of the multiple antibiotic resistance operon (*mar*) and the multidrug efflux pump *acrAB* in the resistance of *Escherichia coli* biofilms towards ciprofloxacin. *J Antimicrob Chemother*, **45**, 789–95.

Malek, J.R. and Speier, J.L. 1982. Development of an organosilicone antimicrobial agent for the treatment of surfaces. *J Coat Fabric*, **12**, 38–45.

Marquis, R.E., Sim, J. and Shin, S.Y. 1994. Molecular mechanisms of resistance to heat and oxidative damage. *J Appl Bacteriol*, **76**, S40–8.

Marquis, R.E., Rutherford, G.C., et al. 1995. Sporicidal action of peracetic acid and protective effects of transition ions. *J Ind Microbiol*, **15**, 486–92.

Mata-Portuguez, V.H., Perez, L.S. and Acosta-Gio, E. 2002. Sterilization of heat-resistant instruments with infrared radiation. *Infect Cont Hosp Epidemiol*, **23**, 393–6.

McCarthy, B.J. 1999. Textile and leather preservation. In: Russell, A.D., Hugo, W.B. and Ayliffe, G.A.J. (eds), *Principles and practice of disinfection, preservation and sterilization*, 3rd edn. Oxford: Blackwell Science, 565–76.

McBain, A.J. and Gilbert, P. 2001. Biocide tolerance and the harbingers of doom. *Int Biodeter Biodegrad*, **47**, 55–61.

McDonnell, G. 2001. New and developing chemical antimicrobials. In: Block, S.S. (ed.), *Disinfection, sterilization and preservation*, 3rd edn. Philadelphia: Lippincott Williams & Wilkins, 431–43.

McDonnell, G. and Russell, A.D. 1999. Antiseptics and disinfectants: activity, action and resistance. *Clin Microbiol Rev*, **12**, 147–79.

McMurry, L.M., Oethinger, M. and Levy, S.B. 1998a. Triclosan targets lipid synthesis. *Nature (Lond)*, **394**, 531–2.

McMurry, L.M., Oethinger, M. and Levy, S.B. 1998b. *marA, soxS* or *acrAB* produces resistance to triclosan in laboratory and clinical strains of *Escherichia coli. FEMS Microbiol Lett*, **166**, 305–9.

McMurry, L.M., McDermott, P.F. and Levy, S.B. 1999. Genetic evidence that InhA of *Mycobacterium smegmatis* is target for triclosan. *Antimicrob Agent Chemother*, **43**, 711–13.

Meade, M.J., Waddell, R.L. and Callahan, T.M. 2001. Soil bacteria *Pseudomonas putida* and *Alcaligenes xyloxidans* subsp. *Denitrificans* inactivate triclosan in liquid and solid substrates. *FEMS Microbiol Lett*, **204**, 45–8.

Melly, E., Cowan, A.E. and Setlow, P. 2002. Studies on the mechanisms of killing of *Bacillus subtilis* spores by hydrogen peroxide. *J Appl Microbiol*, **93**, 316–25.

Merianos, J.J. 2001. Surface-active agents. In: Block, S.S. (ed.), *Disinfection, sterilization and preservation*, 5th edn. Philadelphia: Lippincott Williams & Wilkins, 283–320.

Michael, G.I. and Stumbo, C.R. 1970. Ethylene oxide sterilization of *Salmonella senftenberg* and *Escherichia coli*; death kinetics and mode of action. *J Food Sci*, **35**, 631–4.

Midgley, M. 1986. The phosphonium ion efflux system of *Escherichia coli*: relationship to the ethidium efflux system and energetic studies. *J Gen Microbiol*, **132**, 1387–93.

Midgley, M. 1987. An efflux system for cationic dyes and related compounds in *Escherichia coli. Microbiol Sci*, **14**, 125–7.

Mitchell, P. 1961. Coupling of phosphorylation to electron and hydrogen transfer by a chemiosmotic type of mechanism. *Nature (Lond)*, **191**, 144–8.

Mitchell, D.L. and Nairn, R.S. 1989. The biology of the (6-4) photoproduct. *Photochem Photobiol*, **49**, 805–19.

Moats, W.A. and Maddox, S.E. Jr. 1978. Effect of pH on the antimicrobial activity of some triphenylmethane dyes. *Can J Microbiol*, **24**, 658–61.

Morris, E.J. and Darlow, H.M. 1971. Inactivation of viruses. In: Hugo, W.B. (ed.), *Inhibition and destruction of the microbial cell*. London: Academic Press, 687–702.

Morton, H.E. 1983. Alcohols. In: Block, S.S. (ed.), *Disinfection, sterilization and preservation*, 3rd edn. Philadelphia: Lea and Febiger, 225–39.

Moseley, B.E.B. 1984. Radiation damage and its repair in non-sporulating bacteria. In: Andrew, M.H.E. and Russell, A.D. (eds), *The revival of rnjured microbes. Society for Applied Bacteriology Symposium Series No. 12*. London: Elsevier Applied Science, 43–70.

Moseley, B.E.B. 1989. Ionizing radiation: action and repair. In: Gould, G.W. (ed.), *Mechanisms of action of food preservation procedures*. London: Elsevier Applied Science, 43–70.

Munakata, N. and Rupert, C.S. 1972. Genetically controlled removal of 'spore photoproduct' from deoxyribonucleic acid of ultra-violet-irradiated *Bacillus subtilis* spores. *J Bacteriol*, **111**, 192–8.

Munakata, N. and Rupert, C.S. 1974. Dark repair of DNA containing 'spore photoproduct' in *Bacillus subtilis. Molec Gen Genet*, **130**, 239–50.

Murrell, W.G. 1981. Biophysical studies on the molecular mechanisms of spore heat resistance and dormancy. In: Levinson, H.S., Sonenshein, A.L. and Tipper, D.J. (eds), *Sporulation and germination*. Washington, DC: American Society for Microbiology, 64–77.

Murtough, S.M., Hiom, S.J., et al. 2001. Biocide rotation in the healthcare setting: is there a case for policy implementation? *J Hosp Infect*, **48**, 1–6.

Myers, J.A., Allwood, M.C., et al. 1980. The relationship between structure and activity of taurolin. *J Appl Bacteriol*, **48**, 89–96.

Narang, H.J. and Codd, A.A. 1983. Action of commonly used disinfectants against enteroviruses. *J Hosp Infect*, **4**, 209–12.

Navarro, J.M. and Monsan, P. 1976. Étude du mécanismes d'interaction du glutaraldehyde avec les micro-organismes. *Ann Microbiol (Paris)*, **127B**, 295–307.

Nicholson, G., Hudson, R.A., et al. 1995. The efficacy of the disinfection of bronchoscopes contaminated in vitro with *Mycobacterium tuberculosis* and *Mycobacterium avium-intracellulare* in sputum: a comparison of Sactimed-I-Sinald and glutaraldehyde. *J Hosp Infect*, **29**, 257–64.

Ng, E.G.L., Jones, S., et al. 2002. Biocides and antibiotics with apparently similar actions on bacteria: Is there the potential for cross-resistance? *J Hosp Infect*, **51**, 147–9.

Nordgren, C. 1939. Investigations on the sterilising efficacy of gaseous formaldehyde. *Acta Pathol Microbiol Scand*, **Suppl XL**, 1–165.

Northrop, J. and Slepecky, R.A. 1967. Sporulation mutations induced by heat in *Bacillus subtilis. Science*, **155**, 838–9.

Owens, R.G. 1969. Organic sulphur compounds. In: Torgeson, D.C. (ed.), *Fungicides*. Vol. 2. New York : Academic Press, 147–301.

Owens, J.E. 1993. Sterilization of LVPs and SVPs. In: Morrissey, R.F. and Phillips, G.B. (eds), *Sterilization technology. A practical guide for manufacturers and users of health care products*. New York: Van Nostrand Reinhold, 254–85.

Paidhungat, M. and Setlow, P. 2002. Spore germination and outgrowth. In: Sonenshin, A.I., Hoch, J. and Losick, R. (eds), *Bacillus subtilis and its closest relatives*. Washington, DC: ASM Press, 537–48.

Passagot, J., Crance, J.M., et al. 1987. Effect of glutaraldehyde on the antigenicity and infectivity of hepatitis A virus. *J Virol Meth*, **16**, 21–8.

Paulsen, I.T., Brown, M.H. and Skurray, R.A. 1996. Proton-dependent multidrug efflux systems. *Microbiol Rev*, **40**, 575–608.

Paulsen, I.T., Brown, M.H. and Skurray, R.A. 1998. Characteristics of the earliest known *Staphylococcus aureus* plasmid encoding a multidrug efflux system. *J Bacteriol*, **180**, 3477–9.

Peacocke, A.R. and Skerrett, J.N.H. 1956. The interaction of amino acridines with nucleic acids. *Trans Faraday Soc*, **52**, 261–79.

Pellon, J.R. and Sinskey, A.J. 1984. Heat-induced damage to the bacterial chromosome and its repair. In: Andrew, M.H.E. and Russell, A.D. (eds), *The revival of injured microbes. Society for Applied Bacteriology Symposium Series No. 12*. London: Academic Press, 105–26.

Pheil, C.G., Pflug, I.J., et al. 1967. Effect of various gas atmospheres on destruction of microorganisms in dry heat. *Appl Microbiol*, **15**, 120–4.

Phillips, C.R. 1952. Relative resistance of bacterial spores and vegetative bacteria to disinfectants. *Bacteriol Rev*, **16**, 135–8.

Polissi, A., Goffin, L. and Georgopoulos, C. 1995. The *Escherichia coli* heat shock response and bacteriophage lambda development. *FEMS Microbiol Rev*, **17**, 159–69.

Poole, K. 2000. Efflux-mediated resistance to fluoroquinolones in Gram-negative bacteria. *Antimicrob Agent Chemother*, **44**, 2233–41.

Poole, K. 2002. Mechanisms of bacterial biocide and antibiotic resistance. *J Appl Microbiol*, **92**, 55–64.

Power, E.G.M. 1997. Aldehydes as biocides. *Prog Med Chem*, **34**, 149–201.

Power, E.G.M., Dancer, B.N. and Russell, A.D. 1988. Emergence of resistance to glutaraldehyde in spores of *Bacillus subtilis* 168. *FEMS Microbiol Lett*, **50**, 223–6.

Poxton, I.R. 1993. Prokaryote envelope diversity. *J Appl Bacteriol*, **67**, S91–8.

Price, P.B. 1950. Re-evaluation of ethyl alcohol as a germicide. *Arch Surg*, 492–502.

Prince, H.N. and Prince, D.L. 2001. Principles of viral control and transmission. In: Block, S.S. (ed.), *Disinfection, sterilization and preservation*, 5th edn. Philadelphia: Lippincott Williams & Wilkins, 543–71.

Quinn, P.J. 1987. Evaluation of veterinary disinfectants and disinfection processes. In: Linton, A.H., Hugo, W.B. and Russell, A.D. (eds), *Disinfection in veterinary and farm animal practice*. Oxford: Blackwell Science, 66–116.

Quinn, P.J. and Carter, M.E. 1999. Virucidal activity of biocides. C. Evaluation of viricidal activity. In: Russell, A.D., Hugo, W.B. and Ayliffe, G.A.J. (eds), *Principles and practice of disinfection, preservation and sterilization*, 3rd edn. Oxford: Blackwell Science, 197–206.

Quinn, P.J. and Markey, B.K. 1999. Virucidal activity of biocides. C. Activity against veterinary viruses. In: Russell, A.D., Hugo, W.B. and Ayliffe, G.A.J. (eds), *Principles and practice of disinfection, preservation and sterilization*, 3rd edn. Oxford: Blackwell Science, 187–96.

Rahn, R.O. and Hosszu, J.L. 1968. Photoproduct formation in DNA at low temperatures. *Photochem Photobiol*, **8**, 53–63.

Ranganathan, N.S. 1996. Chlorhexidine. In: Ascenzi, J.M. (ed.), *Handbook of disinfectants and antiseptics*. New York: Marcel Dekker, 235–64.

Report. 1986. *LAV/HTLV III – The causative agent of AIDS and related conditions, revised outlines.* Advisory Committee of Dangerous Pathogens. London: DHSS.

Resnick, L., Veren, K., et al. 1986. Stability and inactivation of HTLV-III/LAV under clinical and laboratory environments. *J Am Med Assoc*, **255**, 1887–91.

Reybrouck, G. 1999. Evaluation of the antibacterial and antifungal activity of disinfectants. In: Russell, A.D., Hugo, W.B. and Ayliffe, G.A.J. (eds), *Principles and practice of disinfection, preservation and sterilization*, 3rd edn. Oxford: Blackwell Science, 124–44.

Richards, C., Furr, J.R. and Russell, A.D. 1984. Inactivation of microorganisms by lethal gases. In: Kabara, J.J. (ed.), *Cosmetic and drug preservation. Principles and practice.* New York: Marcel Dekker, 209–22.

Roberts, C.G. 2001. Studies on the bioburden of medical devices and the importance of cleaning. In: Rutala, W. (ed.), *Disinfection, sterilization and antisepsis*. Washington, DC: APIC, 63–9.

Roberts, T.A., Woods, L.F.J., et al. 1991. Nitrite. In: Russell, N.J. and Gould, G.W. (eds), *Food preservatives*. Glasgow and London: Blackie, 89–110.

Rochon, M., Sullivan, S. 1999. Products based on accelerated and stabilized hydrogen peroxide. *Can J Infection Control* 51–55.

Rodgers, F.G., Hufton, P., et al. 1985. Morphological response of human rotavirus to ultraviolet radiation, heat and disinfectants. *J Med Microbiol*, **20**, 123–30.

Rose, A.H. 1976. Osmotic stress and microbial survival. In: Gray, T.R.G. and Postgate, J.R. (eds), *The survival of vegetative microbes. 26th Symposium of the Society for General Microbiology*. Cambridge: Cambridge University Press, 155–82.

Rowbury, R.J. 2001. Extracellular sensing components and extracellular induction component alarmones give early warning against stress in *Escherichia coli. Adv Microb Physiol*, **44**, 215–57.

Rowe, A.J. and Silverman, G.J. 1970. The absorption-desorption of water by bacterial spores and its relation to dry heat resistance. *Dev Ind Microbiol*, **11**, 311–26.

Russell, A.D. 1976. Inactivation of non-sporing bacteria by gases. In: Skinner, F.A. and Hugo, W.B. (eds), *Inhibition and inactivation of vegetative microbes. Society for Applied Bacteriology Symposium Series No. 5*. London: AcademicPress, 61–88.

Russell, A.D. 1982. *The destruction of bacterial spores*. London: Academic Press.

Russell, A.D. 1985. The role of plasmids in bacterial resistance to antiseptics, disinfectants and preservatives. *J Hosp Infect*, **6**, 9–19.

Russell, A.D. 1990a. The bacterial spore and chemical sporicides. *Clin Microbiol Rev*, **3**, 99–119.

Russell, A.D. 1990b. The effects of chemical and physical agents on microbes: disinfection and sterilization. In: Parker, M.T. and Collier, L.H. (eds), *Topley & Wilson's Principles of bacteriology, virology and immunity*. Vol. 1. 8th edn. London: Edward Arnold, 71–103.

Russell, A.D. 1993. Theoretical aspects of microbial inactivation. In: Morrissey, R.F. and Phillips, G.B. (eds), *Sterilization technology. A practical guide for manufacturers and users of health care products*. New York: Van Nostrand Reinhold, 3–16.

Russell, A.D. 1994. Glutaraldehyde: Current status and uses. *Infect Control Hosp Epidemiol*, **15**, 724–33.

Russell, A.D. 1995. Mechanisms of bacterial resistance to biocides. *Int Biodeter Biodegrad*, **36**, 247–65.

Russell, A.D. 1996. Activity of biocides against mycobacteria. *J Appl Bacteriol*, **81**, S87–S101.

Russell, A.D. 1997. Plasmids and bacterial resistance to biocides. *J Appl Microbiol*, **82**, 155–65.

Russell, A.D. 1998a. Mechanisms of bacterial resistance to antibiotics and biocides. *Prog Med Chem*, **35**, 133–97.

Russell, A.D. 1998b. Assessment of sporicidal efficacy. *Int Biodeter Biodegrad*, **41**, 281–7.

Russell, A.D. 1999a. Factors influencing the efficacy of antimicrobial agents. In: Russell, A.D., Hugo, W.B. and Ayliffe, G.A.J. (eds), *Principles and practice of disinfection, preservation and sterilization*, 3rd edn. Oxford: Blackwell Science, 95–144.

Russell, A.D. 1999b. Bacterial sensitivity and resistance. D. Mycobactericidal agents. In: Russell, A.D., Hugo, W.B. and Ayliffe, G.A.J. (eds), *Principles and practice of disinfection, preservation and sterilization*, 3rd edn. Oxford: Blackwell Science, 321–32.

Russell, A.D. 1999c. Antifungal activity of biocides. In: Russell, A.D., Hugo, W.B. and Ayliffe, G.A.J. (eds), *Principles and practice of disinfection, sterilization and reservation*, 3rd edn. Oxford: Blackwell Science, 149–67.

Russell, A.D. 1999d. Bacterial resistance to disinfectants: Present knowledge and future problems. *J Hosp Infect*, **43**, Suppl, 121–35.

Russell, A.D. 1999e. Heat sterilization. A. Sterilization and disinfection by heat methods. In: Russell, A.D., Hugo, W.B. and Ayliffe, G.A.J. (eds), *Principles and practice of disinfection, preservation and sterilization*, 3rd edn. Oxford: Blackwell Science, 629–39.

Russell, A.D. 1999f. Radiation sterilization. A. Ionizing radiation. In: Russell, A.D., Hugo, W.B. and Ayliffe, G.A.J. (eds), *Principles and practice of disinfection, preservation and sterilization*, 3rd edn. Oxford: Blackwell Science, 675–87.

Russell, A.D. 1999g. Radiation sterilization. B. Ultraviolet radiation. In: Russell, A.D., Hugo, W.B. and Ayliffe, G.A.J. (eds), *Principles and practice of disinfection, sterilization and preservation*, 3rd edn. Oxford: Blackwell Science, 688–702.

Russell, A.D. 2000. Do biocides select for antibiotic resistance? *J Pharm Pharmacol*, **52**, 227–33.

Russell, A.D. 2001a. Principles of antimicrobial activity and resistance. In: Block, S.S. (ed.), *Disinfection, sterilization and preservation*, 5th edn. Philadelphia: Lippincott Williams and Wilkins, 31–55.

Russell, A.D. 2001b. Chemical sporicidal and sporostatic agents. In: Block, S.S. (ed.), *Disinfection, sterilization and preservation*, 5th edn. Philadelphia: Lippincott Williams and Wilkins, 529–42.

Russell, A.D. 2001c. Mechanisms of bacterial insusceptibility to biocides. *Am J Infect Cont*, **29**, 259–61.

Russell, A.D. 2002a. Emerging infectious organisms and their susceptibility to disinfectants. *Steriliz Aust*, **20**, 12–19.

Russell, A.D. 2002b. Mechanisms of antimicrobial action of antiseptics and disinfectants: An increasingly important area of investigation. *J Antimicrob Chemother*, **49**, 597–9.

Russell, A.D. 2002c. Introduction of biocides into clinical paractice and the impact on antibiotic-resistant bacteria. *J Appl Microbiol*, **92**, S121–35.

Russell, A.D. 2003a. Bacterial resistance to biocides: Current knowledge and future problems. In: Lens, P., Moran, A.P., et al. (eds), *Biofilms in medicine, industry and environmental biotechnology*. London: IWA Publishing, 512–33.

Russell, A.D. 2003b. Similarities and differences in the responses of microorganisms to biocides. *J Antimicrob Chemother*, **52**, 750–63.

Russell, A.D. 2003c. Biocide use and antibiotic resistance: The relevance of laboratory findings to clinical and environmental situations. *Lancet Infect Dis*, **3**, 794–803.

Russell, A.D. 2003d. Effects of heat on bacterial physiology and structure. *Sci Prog (Oxford)*, **86**, 115–37.

Russell, A.D. 2004. Bacterial adaptation and resistance to antiseptics, disinfectants and preservatives is not a new phenomenon. *J Hosp Infect*, in press.

Russell, A.D. and Chopra, I. 1996. *Understanding antibacterial action and resistance*, 2nd edn. London: Chapman and Hall.

Russell, A.D. and Day, M.J. 1993. Antibacterial activity of chlorhexidine. *J Hosp Infect*, **25**, 229–38.

Russell, A.D. and Furr, J.R. 1977. The antibacterial activity of a new chloroxylenol preparation containing ethylenediamine tetraacetic acid. *J Appl Bacteriol*, **43**, 253–60.

Russell, A.D. and Furr, J.R. 1996. Biocides: Mechanisms of antifungal action and fungal resistance. *Sci Prog (Oxford)*, **79**, 27–48.

Russell, A.D. and Gould, G.W. 1988. Resistance of Enterobacteriaceae to preservatives and disinfectants. *J Appl Bacteriol*, **65**, S167–95.

Russell, A.D. and Hugo, W.B. 1987. Chemical disinfectants. In: Linton, A.H., Hugo, W.B. and Russell, A.D. (eds), *Disinfectants in veterinary and farm animal practice*. Oxford: Blackwell Science, 12–42.

Russell, A.D. and Hugo, W.B. 1988. Perturbation of homeostatic mechanisms in bacteria by pharmaceuticals. In: Whittenbury, R., Gould, G.W., et al. (eds), *Homeostatic mechanisms in microorganisms*. Bath: Bath University Press, 206–19.

Russell, A.D. and Hugo, W.B. 1994. Antimicrobial activity and action of silver. *Prog Med Chem*, **31**, 351–71.

Russell, A.D. and McDonnell, G. 2000. Concentration: A major factor in studying biocidal action. *J Hosp Infect*, **44**, 1–3.

Russell, A.D. and Russell, N.J. 1995. Biocides: Activity, action and resistance. In: Hunter, P.A., Darby, G.K. and Russell, N.J. (eds), *Fifty years of antimicrobials: Past perspectives and future trends. 53rd Symposium of the Society for General Microbiology*. Cambridge: Cambridge University Press, 327–65.

Russell, A.D., Ahonkhai, I. and Rogers, D.T. 1979. Microbiological applications of the inactivation of antibiotics and other antimicrobial agents. *J Appl Bacteriol*, **46**, 207–45.

Russell, A.D., Yarnych, V.S. and Koulikovskii, A.V. 1984. *Guidelines on disinfection in animal husbandry for prevention and control of zoonotic diseases, VPH/84.4*. Geneva: World Health Organization.

Russell, A.D., Hammond, S.A. and Morgan, J.R. 1986. Bacterial resistance to antiseptics and disinfectants. *J Hosp Infect*, **7**, 213–25.

Russell, A.D., Furr, J.R. and Maillard, J.-Y. 1997. Microbial susceptibility and resistance to biocides. *ASM News*, **63**, 481–7.

Russell, A.D., Hugo, W.B. and Ayliffe, G.A.J. (eds) 1999. *Principles and practice of disinfection, preservation and sterilization*, 3rd edn. Oxford: Blackwell Science.

Rutala, W.A. 1990. APIC guidelines for selection and use of disinfectants. *Am J Infect Cont*, **18**, 99–117.

Rutala, W.A. (ed.) 1995. *Chemical germicides in health care*. Morin Heights, PQ, Canada: Polyscience.

Rutala, W.A. and Weber, D.J. 1999a. Infection control: the role of disinfection. *J Hosp Infect*, **43**, Suppl, S43–55.

Rutala, W.A. and Weber, D.J. 1999b. Disinfection of endoscopes: review of new chemical sterilants used for high-level disinfection. *Infect Cont Hosp Epidemiol*, **20**, 69–76.

Rutala, W.A. and Weber, D.J. 2001a. An overview of the chemical germicides in healthcare. In: Rutala, W.A. (ed.), *Disinfection, sterilization and antisepsis*. Washington, DC: APIC, 1–15.

Rutala, W.A. and Weber, D.J. 2001b. Management of equipment of equipment contaminated with the Creutzfeldt–Jakob disease agent. In: Rutala, W.A. (ed.), *Disinfection, sterilization and antisepsis*. Washington, DC: APIC, 167–72.

Sagripani, J.L. and Kraemer, K.H. 1989. Site-specific oxidative DNA damage at polyguanosines produced by copper plus hydrogen peroxide. *J Biol Chem*, **264**, 1729–34.

Salmond, C.V., Kroll, R.G. and Booth, I.R. 1984. The effect of food preservatives on pH homeostasis in *Escherichia coli*. *J Gen Microbiol*, **130**, 2845–50.

Salton, M.R.J. 1968. Lytic agents, cell permeability and monolayer penetrability. *J Gen Physiol*, **52**, 227–52.

Sancar, A. and Sancar, G.B. 1988. DNA repair enzymes. *Ann Rev Biochem*, **57**, 29–67.

Sattar, S.A. and Springthorpe, S. 2001. Methods of testing the virucidal activity of chemicals. In: Block, S.S. (ed.), *Disinfection, sterilization and preservation*, 5th edn. Philadelphia: Lippincott Williams & Wilkins, 1391–428.

Sattar, S.A., Springthorpe, V.S., et al. 1994. Inactivation of the human immunodeficiency virus: An update. *Rev Med Microbiol*, **5**, 139–50.

Sattar, S.A., Springthorpe, V.S. and Rochon, M. 1999. A product based on accelerated and stabilized hydrogen peroxide: evidence for broad-spectrum germicidal activity. *Can J Infect Cont*, **13**, 4, 123–30.

Sattar, S.A., Tetro, J., et al. 2001. Preventing the spread of hepatitis B and C viruses: where are germicides relevant? *Am J Infect Cont*, **29**, 187–97.

Schneider, P.M. 1998. Emerging low temperature sterilization technologies (non-FDA approved). In: Rutala, W.A. (ed.), *Disinfection, sterilization and antisepsis in health care*. Washington, DC: APIC, 79–92.

Schweizer, H.P. 2001. Triclosan: a widely used biocide and its link to antibiotics. *FEMS Microbiol Lett*, **202**, 1–7.

Seal, D.V. and Hay, J. 1992. Contact lens disinfection and Acanthamoeba: Problems and practicalities. *Pharm J*, **248**, 717–19.

Seiler, D.A.L. and Russell, N.J. 1991. Ethanol as a food preservative. In Russell. In: Russell, N.J. and Gould, G.W. (eds), *Food preservatives*. Glasgow: Blackie, 153–71.

Selkon, J.B., Babb, J.R. and Morris, R. 1999. Evaluation of the antimicrobial activity of a new super-oxidized water, Sterilox[R] for the disinfection of endoscopes. *J Hosp Infect*, **41**, 59–70.

Setlow, P. 1994. Mechanisms which contribute to the long-term survival of spores of *Bacillus* species. *J Appl Bacteriol*, **76**, S49–60.

Setlow, P. 1995. Mechanisms for the prevention of damage to the DNA in spores of *Bacillus* species. *Ann Rev Microbiol*, **49**, 29–54.

Setlow, P. 2000. Resistance of bacterial spores. In: Storz, G. and Hengge-Aronis, R. (eds), *Bacterial stress responses*. Washington, DC: ASM Press, 217–30.

Setlow, B. and Setlow, P. 1996. Role of DNA repair in *Bacillus subtilis* spore resistance. *J Bacteriol*, **178**, 3486–95.

Setlow, B., Loshon, C.A., et al. 2001. Mechanism of killing spores of *Bacillus subtilis* by acid, alcohol and ethanol. *J Appl Microbiol*, **92**, 36.

Shaker, L.A., Dancer, B.N., et al. 1988. Emergence and development of chlorhexidine resistance during sporulation of *Bacillus subtilis* 168. *FEMS Microbiol Lett*, **51**, 73–6.

Shetty, N., Srinivasan, S., et al. 1999. Evaluation of microbicidal activity of a new disinfectant: Sterilox[R] against *Clostridium difficile* spores, *Helicobacter pylori*, vancomycin resistant *Enterococcus* species, *Candida albicans* and several *Mycobacterium* species. *J Hosp Infect*, **41**, 101–5.

Shin, S.Y., Calvisi, E.G., et al. 1994. Microscopic and thermal characterization of hydrogen peroxide killing and lysis of spores and protectionby transition metal ions, chelators and antioxidants. *Appl Environ Microbiol*, **60**, 3192–7.

Silver, S., Nucifora, G., et al. 1989. Bacterial ATPases: Primary pumps for exporting toxic cations and anions. *Trends Biochem Sci*, **14**, 76–80.

Simons, C., Walsh, S.E., et al. 2000. A note: *ortho*-phthalaldehyde: Mechanism of action of a new antimicrobial agent. *Lett Appl Microbiol*, **31**, 299–302.

Slater-Radosti, C., Van Aller, G., et al. 2001. Biochemical and genetic characterization of the action of triclosan on *Staphylococcus aureus*. *J Antimicrob Chemother*, **48**, 1–6.

Smerage, G.H. and Teixeira, A.A. 1993. Dynamics of heat destruction of bacterial spores: a new view. *J Ind Microbiol*, **12**, 211–20.

Sofos, J.N. and Busta, F.F. 1999. Chemical food preservative. In: Russell, A.D., Hugo, W.B. and Ayliffe, G.A.J. (eds), *Principles and practice of disinfection, preservation and sterilization*, 3rd edn. Oxford: Blackwell Science, 485–541.

Soper, C.J. and Davies, D.J.G. 1990. Principles of sterilization. In: Denyer, S. and Baird, R. (eds), *Guide to microbiological control in pharmaceuticals*. Chichester: Ellis Horwood, 157–81.

Speier, J.L. and Malek, J.R. 1982. Destruction of micro-organisms by contact with solid surfaces. *J Coll Interfac Sci*, **89**, 68–76.

Spicher, G. and Peters, J. 1981. Heat activation of bacterial spores after inactivation by formaldehyde. Dependance of heat activation on temperature and duration of action. *Zentralbl Bakteriol Mikrobiol [B]*, **173**, 188–96.

Springle, W.R. 1999. Preservation in specialized areas. C. Paint and paint films. In: Russell, A.D., Hugo, W.B. and Ayliffe, G.A.J. (eds), *Principles and practice of disinfection, preservation and sterilization*, 3rd edn. Oxford: Blackwell Science, 577–82.

Springthorpe, V.S. and Sattar, S.A. 1990. Chemical disinfection of virus-contaminated surfaces. *Clin Rev Environ Cont*, **20**, 169–229.

Springthorpe, V.S., Grenier, J.L., et al. 1986. Chemical disinfection of human rotaviruses: efficacy of commercially available products in suspension tests. *J Hyg*, **97**, 139–61.

Staehelin, M. 1958. Reactions of tobacco mosaic virus nucleic acid with formaldehyde. *Biochim Biophys Acta*, **29**, 410.

Stewart, P.S., Rayner, J., et al. 2001. Biofilm penpenetration and disinfection efficacy of alkaline hypochlorite and chlorosulfamates. *J Appl Microbiol*, **91**, 525–32.

Stickler, D.J. and King, J.B. 1999. Bacterial sensitivity and resistance. A. Intrinsic resistance. In: Russell, A.D., Hugo, W.B. and Ayliffe, G.A.J. (eds), *Principles and practice of disinfection, preservation and sterilization*, 3rd edn. Oxford: Blackwell Science, 284–96.

Stringer, S.C., George, S.M. and Peck, M.W. 2000. Thermal inactivation of *Escherichia coli* O157:H7. *J Appl Microbiol*, **88**, S70–89.

Sulavik, M.C., Houseweart, C., et al. 2001. Antibiotic susceptibility profiles of *Escherichia coli* lacking multidrug efflux pumps. *Antimicrob Agent Chemother*, **45**, 1126–36.

Suller, M.T.E. and Russell, A.D. 1999. Antibiotic and biocide resistance in methicillin-resistant *Staphylococcus aureus* and vancomycin-resistant staphylococcus. *J Hosp Infect*, **43**, 281–91.

Suller, M.T.E. and Russell, A.D. 2000. Triclosan and antibiotic resistance in *Staphylococcus aureus*. *J Antimicrob Chemother*, **46**, 11–18.

Sundheim, G., Langrsud, S., et al. 1998. Bacterial resistance to disinfectants containing quaternary ammonium compounds. *Int Biodeter Biodegrad*, **41**, 235–9.

Sutton, S.V.W. 1996. Neutralozer evaluation in control experiments for antimicrobial efficacy tests. In: Ascenzi, J.M. (ed.), *Handbook of disinfectants and antiseptics*. New York: Marcel Dekker, 43–62.

Swift, S., Downie, J.A., et al. 2001. Quorum sensing as a population-density-determinant of bacterial physiology. *Adv Microb Physiol*, **45**, 199–270.

Sykes, G. 1965. *Disinfection and sterilization*, 2nd edn. London: E&FN Spon.

Tallentire, A. and Jacobs, G.P. 1972. Radiosensitization of bacterial spores by ketonic agents of differing electron affinities. *Int J Radiat Biol*, **21**, 205–13.

Tanny, G.B., Strong, D.K., et al. 1979. Adsorptive retention of *Pseudomonas diminuta* by membrane filters. *J Parenter Drug Assoc*, **33**, 40–51.

Tattawasart, U., Maillard, J.-Y., et al. 1999a. Development of resistance tochlorhexidine diacetate and cetylpyridinium chloride in *Pseudomonas stutzeri* and changes in antibiotic susceptibility. *J Hosp Infect*, **42**, 219–29.

Tattawasart, U., Maillard, J.-Y., et al. 1999b. Comparative responses of *Pseudomonas stutzeri* and *Pseudomonas aeruginosa* to antibacterial agents. *J Appl Microbiol*, **87**, 323–31.

Tattawasart, U., Maillard, J.-Y., et al. 2000a. Outer membrane changes in *Pseudomonas stutzeri* resistant to chlorhexidine diacetate and cetyhlpyridinium. *Int J Antimicrob Agent*, **16**, 233–8.

Tattawasart, U., Hann, A.C., et al. 2000b. Cytological changes in chlorhexidine-resistant isolates of *Pseudomonas stutzeri*. *J Antimicrob Chemother*, **45**, 145–52.

Taylor, D.M. 1999. Inactivation of unconventional agents of the transmissible degenerative encephalopathies. In: Russell, A.D., Hugo, W.B. and Ayliffe, G.A.J. (eds), *Principles and practice of disinfection, preservation and sterilization*, 3rd edn. Oxford: Blackwell Science, 222–36.

Taylor, D.M. 2004. Transmissible degenerative encephalopathies: inactivation of the unconventional causal agents. In: Fraise, A.P., Lambert, P.A. and Maillard, J.-Y. (eds), *Russell, Hugo and Ayliffe's Principles and practice of disinfection, preservation and sterilization*, 4th edn. Oxford: Blackwell Science, 325–41.

Tennen, R., Setlow, B., et al. 2000. Mechanism of killing of spores of *Bacillus subtilis* by iodine, glutaraldehyde and nitrous acid. *J Appl Microbiol*, **89**, 330–8.

Terano, H., Tanooka, H. and Kadota, H. 1969. Germination induced repair of single strand breaks of DNA in irradiated *Bacillus subtilis* spores. *Biochem Biophys Res Commun*, **37**, 66–71.

Terano, H., Tanooka, H. and Kadota, H. 1971. Repair of radiation damage to deoxyribonucleic acid in germinating spores of *Bacillus subtilis*. *J Bacteriol*, **106**, 925–30.

Thomas, L., Lambert, R., et al. 2000. Development of resistance to chlorhexidine diacetate in *Pseudomonas aeruginosa* and the effect of a 'residual' concentration. *J Hosp Infect*, **46**, 297–303.

Thurmann, R.B. and Gerba, C.P. 1988. Molecular mechanisms of viral inactivation by water disinfectants. *Adv Appl Microbiol*, **33**, 75–105.

Thurmann, R.B. and Gerba, C.P. 1989. The molecular mechanisms of copper and silver ion disinfection of bacteria and viruses. *CRC Crit Rev Environ Cont*, **18**, 295–315.

Tomlins, R.I. and Ordal, Z.J. 1976. Thermal injury and inactivation in vegetative bacteria. In: Skinner, F.A. and Hugo, W.B. (eds), *Inhibition and inactivation of vegetative microbes. Society for Applied Bacteriology Symposium Series No. 5.* London: Academic Press, 153–90.

Tovar-Rojo, F., Chander, M., et al. 2002. The products of the *spoVA* operon are involved in dipicolinic acid uptake into developing spores of *Bacillus subtilis*. *J Bacteriol*, **184**, 564–7.

Townsend, D.E., Ashdown, N., et al. 1984. Transposition of gentamicin resistance to staphylococcal plasmids encoding resistance to cationic agents. *J Antimicrob Chemother*, **14**, 115–34.

Trevor, J.T. 1987. Silver resistance and accumulation in bacteria. *Enzyme Microb Technol*, **9**, 331–3.

Trueman, J.R. 1971. The halogens. In: Hugo, W.B. (ed.), *Inhibition and destruction of the microbial cell*. London: Academic Press, 137–184.

Tsiquaye, K.N. and Barnard, J. 1993. Chemical disinfection of duck hepatitis B virus: a model for inactivation of infectivity of hepatitis B virus. *J Antimicrob Chemother*, **32**, 313–23.

Turner, N.A., Russell, A.D., et al. 1999. Acanthamoeba species, antimicrobial agents and contact lenses. *Sci Prog*, **82**, 1–8.

Turner, N.A., Russell, A.D., et al. 2000a. Emergence of resistance to biocides during differentiation of *Acanthamoeba castellanii*. *J Antimicrob Chemother*, **46**, 27–34.

Turner, N.A., Harris, J., et al. 2000b. Microbial cell differentiation and changes in susceptibility to antimicrobial agents. *J Appl Microbiol*, **89**, 751–9.

Tyler, R. and Ayliffe, G.A.J. 1987. A surface test for virucidal activity: preliminary study with herpes virus. *J Hosp Infect*, **9**, 22–9.

Vaara, M. 1992. Agents that increase the permeability of the outer membrane. *Microbiol Rev*, **56**, 395–411.

van Bueren, J., Larkin, D.P. and Simpson, R.A. 1994. Inactivation of human immunodeficiency virus type 1 by alcohols. *J Hosp Infect*, **28**, 137–48.

van Klingeren, B. and Pullen, W. 1993. Glutaraldehyde-resistant mycobacteria from endoscope washers. *J Hosp Infect*, **25**, 147–9.

Varghese, A.J. 1970. 5-Thyminyl-5,6-dihydrothymine from DNA irradiated with ultraviolet light. *Biochem Biophys Res Commun*, **38**, 484–90.

Walsh, S.E., Maillard, J.-Y. and Rusell, A.D. 1999a. *Ortho*-phthalaldehyde: A possible alternative to glutaraldehyde for high level disinfection. *J Appl Microbiol*, **86**, 1039–46.

Walsh, S.E., Maillard, J.-Y., et al. 1999b. Studies on the mechanisms of the antibacterial action of *ortho*-phthalaldehyde. *J Appl Microbiol*, **87**, 702–10.

Walsh, S.E., Maillard, J.-Y., Russell, A.D. and Hann, A.C. 2001. Possible mechanisms for the relative efficacies of *ortho*-phthalaldehyde and glutaraldehyde against glutaraldehyde-resistant *Mycobacterium chelonae*. *J Appl Microbiol*, **91**, 80–92.

Walsh, S.E., Maillard, J.-Y., et al. 2003a. Activity and mechanisms of action of selected biocidal agents on Gram-positive and -negative bacteria. *J Appl Microbiol*, **94**, 240–7.

Walsh, S.E., Maillard, J.-Y., et al. 2003b. Development of bacterial resistance to several biocides and effects on antibiotic susceptibility. *J Hosp Infect*, **55**, 98–107.

Wang, W.L.L., Blaser, M.J., et al. 1979. Growth, survival and resistance of the Legionnaires' disease bacterium. *Ann Intern Med*, **90**, 614–18.

Warth, A.D. 1985. Mechanisms of heat resistance. In: Dring, G.J., Gould, G.W. and Ellar, D.J. (eds), *Fundamental and applied aspects of bacterial spores*. London: Academic Press, 209–25.

Warth, A.D. 1986. Effect of benzoic acid on growth yield of yeasts differing in their resistance to preservatives. *Int J Food Microbiol*, **3**, 263–71.

Warth, A.D. 1989. Relationships among cell size, membrane permeability and preservative resistance in yeast species. *Appl Environ Microbiol*, **55**, 2995–9.

Weavers, L.K. and Wickramanayake, G.B. 2001. Disinfection and sterilization using ozone. In: Block, S.S. (ed.), *Disinfection, sterilization and preservation*, 5th edn. Philadelphia: Lippincott Williams & Wilkins, 205–14.

Weber, D.J. and Rutala, W.A. 1999. The emerging nosocomial pathogens *Cryptosporidium, Escherichia coli* O157:H7, *Helicobacter pylori* and hepatitis C: Epidemiology, environmental survival and control measures. *Infect Cont Hosp Epidemiol*, **22**, 306–15.

White, P.J.P. 1990. The design of controlled environments. In: Denyer, S. and Baird, R. (eds), *Guide to microbiological control in pharmaceuticals*. Chichester: Ellis Horwood, 87–124.

White, D.G. and McDermott, P.F. 2001. Biocides, drug resistance and microbial evolution. *Curr Opin Microbiol*, **4**, 313–17.

Wilkins, K.M. and Board, R.G. 1989. Natural antimicrobial systems. In: Gould, G.W. (ed.), *Mechanisms of action of food preservation procedures*. London: Elsevier Applied Science, 285–362.

Williams, N.D. and Russell, A.D. 1993. Injury and repair in biocide-treated spores of *Bacillus*. *FEMS Microbiol Lett*, **106**, 183–6.

Wills, P.A. 1974. Effects of hydrostatic pressure and ionising radiation on bacterial spores. *Atom Energy Aust*, **17**, 2–10.

Wimpenny, J., Manz, W. and Szewzyk, U. 2000. Heterogeneity in biofilms. *FEMS Microbiol Rev*, **24**, 661–71.

Wood, R.T. 1993. Sterilization with dry heat. In: Morrissey, R.F. and Phillips, G.B. (eds), *Sterilization technology. A practical guide for manufacturers and users of health care products*. New York: Van Nostrand Reinhold, 81–119.

Yura, T., van Schaik, M. and Morita, M.T. 2000. The heat shock response: Regulation and function. In: Stortz, G. and Hengge-Aronis, R. (eds), *Bacterial stress responses*. Washington, DC: ASM Press, 3–18.

Zamenhof, S. 1960. Effects of heating dry bacteria and spores on their phenotype and genotype. *Proc Natl Acad Sci USA*, **46**, 101–5.

18

Antibacterial therapy

IAN M. GOULD

PHILOSOPHY OF ANTIBACTERIAL USE

Prospects for the future

At the start of the twenty-first century, more than 250 antibacterial agents are in use throughout the world. With such an abundance at the prescribers' disposal it might be imagined that bacterial infection has become a minor problem and that bacterial drug resistance would be unable to keep pace with resources available to counteract it. Nothing could be further from the truth. Bacteria appear to possess a limitless ingenuity in avoiding the effects of antimicrobial agents, as well as in finding new ways to invade the compromised host. Moreover, mutations conferring resistance to one antibiotic can, at a stroke, render a whole drug family impotent. Bacteria resistant to one sulfonamide or one tetracycline are usually resistant to all sulfonamides, or all tetracyclines; 'methicillin-resistant' staphylococci are resistant to all β-lactam antibiotics. Even worse, bacteria can assemble resistance genes for unrelated classes of agents on integrons and plasmids that can be readily transmitted between bacterial species or, sometimes, genera.

For these reasons it is unwise to be complacent about our ability to control infection. Resistance is a serious global problem that must be countered by discriminating use of antimicrobial agents by all prescribers, by vigorous application of control of infection measures in health care institutions, and at government level by effective regulation of drug production, distribution, and use (European Union Conference 1998; Gould et al. 2000).

Ecological issues

Antibiotic use is huge with estimates as much as 100 000 tons/year worldwide, the majority of this in farming including veterinary use, growth promotion, crop spraying, and fish farming (Levy 2002; Harrison and Lederberg 1998). Potentially there is significant environmental pollution as much antibiotic is excreted in active form in urine and feces and is stable for prolonged periods (weeks to months) in the environment (Harrison and Lederberg 1998). No doubt this pollution revisits us through the food-chain in the form of antibiotic residues, antibiotic-resistant bacteria, and their resistant-determinate genes contaminating food. The scope of the problem has only been investigated on a very superficial basis.

It seems likely, nevertheless, that with a few notable exceptions, most of the antibiotic-resistant bacteria that cause problems in human medicine are generated by use of antibiotics in human medicine. Within human medicine, the intensity of antibiotic use is greatest in the hospital and this is where (magnified by crossinfection) the most significant problems are seen (Gould 1999, 2002a). Within intensive care units (ICU), these problems of crossinfection and intensity of use are further magnified and even added to, by increasingly invasive and immunosuppressive diagnostic and therapeutic procedures. It is no surprise then that ICUs are considered the genesis units for the creation, maintenance, and dissemination of multiply-resistant bacteria (Gould 2002b).

Community antibiotic use is much less intense, although greater in overall volume (Cars et al. 2001). While the limited data available suggest not much difference in usage patterns between different hospitals across Europe, when case mix is taken into account, community use between different countries differs up to four-fold, tending to be lowest in northern Europe (Harbarth et al. 2002). A European average is approximately one prescription per head of population per year. Some countries have managed to lower community prescribing in the past few years with the help of educational campaigns directed both at doctors and their patients and concentrating on reducing the demand for and use of antibiotics for self-limiting respiratory tract infection which are usually viral in origin (Department of Health 2000).

In most countries of the industrially developed world, antibiotics used in human and veterinary practice are presently available only on prescription. It is most important that this status is maintained if resistance problems are to be kept under control. In developing countries where regulation of pharmaceutical products is generally more lax, more responsible national drug policies need to be enforced. The World Health Organization is actively engaged in promoting rational policies of drug use, but much remains to be done. One important initiative has been the formulation of an essential drug list which is updated at regular intervals (World Health Organization 1995). Only a handful of antibacterial agents appear on this list (Table 18.1); significantly, there are no cephalosporins or quinolones on the main list, emphasizing the restricted need for these extensively promoted and commonly prescribed drugs.

How use relates to resistance

While it is often said that there is little absolute proof that antibiotic use causes resistance, the link has been well established in numerous studies at individual patient level, group level such as family or day care center, in hospitals, community medical practices, regions, and whole countries (Baquero et al. 2002). The association applies to all antibiotics and bacteria but not to all bug/drug combinations. *Streptococcus pyogenes*, for example, has famously failed to become resistant to penicillins. Even here, however, while strains still yield the same minimum inhibitory concentration (MIC) of penicillin as they did 60 years ago there is some evidence of tolerance developing to the bactericidal activity of penicillins (Slater and Greenwood 1983). The speed of development of resistance and its degree will depend on many things. Probably most important will be the mechanism of action of the antibiotic and the ease with which the bacteria can develop resistance, e.g. bactericidal antibiotics, by eradicating a pathogen, are probably less likely to select resistance than bacteriostatic agents which will merely suppress division (Stratton 2003). If the bacterial strain has a high yield of resistant mutants or resistance spreads easily on plasmids then resistance will, presumably, arise and even spread easily. Intensity of use, spread by crossinfection, high inoculum infections, and immunosuppression are other factors likely to lead to rapid selection and/or

Table **18.1** *Antibacterial agents (excluding topical agents) on the World Health Organization's list of essential drugs (WHO 1995)*

Main list			
Penicillins	**Other antibacterial agents**	**Antimycobacterial agents**	**Complementary list**[a]
Amoxycillin[b]	Chloramphenicol	Clofazimine	Chloramphenicol (oily suspension)
Ampicillin	Co-trimoxazole[b]	Dapsone	Ciprofloxacin
Benzathine penicillin	Doxycycline	Ethambutol	Clindamycin
Benzylpenicillin	Erythromycin[b]	Isoniazid	Nalidixic acid
Cloxacillin[b]	Gentamicin[b]	Pyrazinamide	Nitrofurantoin
Phenoxymethylpenicillin	Metronidazole[b]	Rifampin	Trimethoprim
Piperacillin[b]	Spectinomycin	Rifampin + isoniazid	Thioacetazone + isoniazid (antimycobacterial)
Procaine penicillin	Sulphadimidine[b]	Streptomycin	
	Tetracycline[b]		

a) For use when drugs on the main list are known to be ineffective or inappropriate for a given individual, or for use in exceptional circumstances (e.g. chloramphenicol oily suspension in epidemics of meningitis) when the health services are overwhelmed. The need for additional reserve agents, e.g. a cephalosporin or vancomycin, is acknowledged.

b) Example of a therapeutic group for which acceptable alternatives exist.

spread of resistance and these are all factors operating in hospitals (Gould 1999; Gould and MacKenzie 2002). Long term, low-dose penicillin therapy has been shown to select efficiently for penicillin resistance in pneumococci (Guillemot et al. 1998). Patients receiving prior trimethoprim or amoxicillin for urinary and chest infections are several times more likely to have antibiotic-resistant infections if they present again within a few months (Steinke and Davey 2001). Within hospitals, most organisms causing problems of hospital-acquired infection (HAI) are selected by intensive antibiotic use; extended spectrum beta-lactamase (ESBL) producing *Klebsiella* by third generation cephalosporins (3GC) (MacKenzie and Gould 1998), derepressed *Enterobacter* and *Citrobacter* by 3GC (Wagenlehner et al. 2002), quinolone-resistant coliforms by quinolones (Scheld 2003; Kern et al. 1994), methicillin resistant *Staphylococcus aureus* (MRSA) by 3GCs, quinolones and macrolides (Monnet 1998; Monnet et al. 2004). In most cases, the organisms are resistant to the antibiotics selecting for them. Things may become more complicated in future as integrons, coding for multiple antibiotic-resistant determinants, are well dispersed in European hospitals (Gruteke et al. 2003; Leverstein-van Hall et al. 2003).

Can resistance be controlled?

It has long been assumed that if the high use of antibiotics is reversed, bacteria will normally return to their susceptible state. There is, however, no clear answer to this as it is usually a multifactorial issue (Austin et al. 1999). There are many examples in the literature, both in communities and hospitals where resistance problems have been reversed (although not necessarily returning to zero) when antibiotic use is controlled (Austin et al. 1999). Generally, resistance problems fall more quickly in hospitals than in the community because of the dilutional effect of patient discharge. Nevertheless, there are examples in the literature where control of use fails to reverse resistance problems and there may be many other cases where such negative results have not been published (Bonhoeffer et al. 1997). Mathematical models and in vitro experiments suggest that bacteria can rapidly learn to carry resistance without too much cost so that in many cases the resistance may be stable, even in the absence of the selection pressure of antibiotic prescribing (Bowler et al. 1994). Similarly, if a virulent clone becomes resistant, it may spread without the need for antibiotic selection pressure. Other factors to consider have been discussed above. Nevertheless, the consensus is that if antibiotic use is inappropriate it is correct to reduce it. If the high use is associated with a resistant problem then it will probably recede along with the reduction in use, but the role of crossinfection should also be addressed, particularly in healthcare institutions.

Antibiotic stewardship

Every healthcare institute and primary care practice should have an antibiotic formulary (list of antibiotics easily available and which the doctors should be familiar with) and antibiotic guidelines (details of diagnosis and treatment for specific infections) (Gould et al. 1994; Gould 2001; Keuleyan and Gould 2001). Within any region it is helpful if the formulary and guidelines are drawn up in consultation between primary and secondary care. Their primary aim is to act as an educational tool for doctors to improve their prescribing and in the UK and USA, most hospitals and primary care centers will have them (Gould et al. 1994; Shlaes et al. 1997; Gould 2001; Keuleyan and Gould 2001). There is also helpful guidance available on design and implementation of guidelines (www.agreecollaboration.org).

The formulary is a good basis for pharmacists to work to, only stocking formulary agents and within this list, monitoring the use of key 'restricted' agents, often with the help of infection specialists with an interest in prescribing. Inappropriate prescriptions can then be targeted with an educational intervention to limit inappropriate use (Solomon et al. 2001). In the near future it is hoped that electronic prescribing will allow easier monitoring of prescriptions (Burke and Pestotnik 1996). Other measures of a restrictive or educational nature that have variable success include special antibiotic order forms, automatic stop orders, therapeutic substitution, antibiotic rotation (or cycling) (Gould et al. 1994; Gould 2001; Keuleyan and Gould 2001; Shlaes et al. 1997), although the evidence base for these measures is often questionable. Currently, there are two Cochrane groups assessing the literature in this area in hospitals and primary care (Arnold 2004; Brown 2004). Help on assessing the effectiveness of interventions is available at www.abdn.ac.uk/hsru/epoc/. There is an increasing tendency to appoint pharmacists and microbiologists with a special interest in antibiotic therapy to monitor quality of use and to create multidisciplinary teams with hospital executive authority (Saizy-Callaert et al. 2003; Fluckiger et al. 2000).

Particular problems of antibiotic use differ between primary and secondary care. In primary care it tends to be unnecessary use for minor, self-limiting bacterial or viral respiratory infections (Steinman et al. 2003a, b; McCaig et al. 2003). In secondary care it is often too much use of powerful broad spectrum, injectable antibiotics without any evidence of serious infections (Kumarasamy et al. 2003). These are often continued for too long and added to irrationally if the patient does not respond (spiralling therapeutic empiricism) (Kim and Gallis 1989).

Whereas use of the laboratory is relatively rare when treating infection in primary care it could (arguably) be increased (Shackley et al. 1997). Providing specimens reach the laboratory the same day, it might be possible to give useful preliminary results the next day for the great majority of specimens. In most cases, even where antibiotic is thought to be indicated by the primary care doctor, the patient will not come to any harm waiting overnight to see if culture indicates that one is really needed. With modern electronic links, results can easily be communicated to the doctor the following morning (MacKenzie et al. 2003). In secondary care the laboratory is used much more, but clinicians are often reluctant to act on the results, e.g. streamline successful therapy, even with susceptibility results from clinically significant bacterial growths to guide them (Kumarasamy et al. 2003). They are, generally, even more reluctant to stop therapy early in the presence of negative cultures which may be strong evidence of a nonbacterial etiology for a patient's illness (Lawrence et al. 1973; Bartlett et al. 1991). Much is written about the importance of rapid molecular methods to guide clinicians in the therapy of infectious diseases. Much more efficient use, however, could be made of conventional test results, the great majority of which can be made available in a useful clinical timeframe: antigen tests, Gram stain, microscopy all within an hour, 6-h urine culture, Gram stain of blood cultures – 90 percent positive within 24 h, other specimen preliminary cultures available overnight. Conventional susceptibility testing methods can often give good indications of the best agents to use within 6–12 h, although final results take longer to read (NCCLS, 2000, 2002).

Good clinical liaison by the laboratory is crucial and there is an increasing tendency for microbiologists to undertake clinical consultations and ward visits to report on positive blood cultures, make daily visits to ICU, other high dependency units, and hematology/cancer wards to discuss important results and confirm correct choice of therapy.

General principles of use of antimicrobial agents for treatment and prophylaxis

The general principles underlying the choice of antimicrobial agents in infection are discussed by Lambert and O'Grady (1997), Finch (1995), and Keuleyan and Gould (2001). In brief, antibiotics should be used therapeutically only after thorough clinical assessment of the need, whenever possible on the basis of laboratory evidence of infection. Factors to be considered should include: the type of infection; the age and condition of the patient; the local prevalence of resistance (in the absence of specific laboratory guidance); the pharmacological properties of the agent in its various formulations; the likelihood of adverse reactions; possible interactions with other medications; and cost. Antimicrobial agents should be used for chemoprophylaxis only in

individuals in whom the risk of infection is high. The agents chosen should be reliably active against the organisms likely to be encountered and use should be confined to the period of greatest risk (Slack 1995b).

A helpful definition of appropriate antibiotic therapy might include the following: (1) Appropriate antibiotic prescribing should potentially benefit the patient. (2) There should be clinical evidence, supported where possible by laboratory tests of bacterial infection. (3) Sepsis parameters should be documented and should support the need for antibiotic treatment. (4) Critical patients should receive appropriate treatment as quickly as possible. (5) Treatment should be limited to bacterial infections, using antibiotics directed against the causative agent, given in optimal dosage, interval, and length of treatment, with steps taken to ensure maximum patient compliance with the treatment regimen and only when the benefit of treatment outweighs the individual and global risks. (6) Antibiotic therapy should be streamlined at the earliest opportunity using the results of laboratory tests where possible (Gould 2001). There is an increasing number of evidence-based guidelines on the prophylaxis and treatment of infectious diseases now available (www.sign.ac.uk; www.cochcrane.org).

Pharmacodynamics of antibiotics (the effect of antibiotic concentration)

Over the past 20 years great strides have been made in understanding how to optimize antibiotic treatment by varying dose and dosing interval. This not only improves patient outcome, but also reduces the likelihood of selecting resistant strains during therapy. The study of pharmacodynamics (the effect of concentration) developed from a knowledge of the pharmacokinetics of different antibiotics (that is, the concentration achieved at different target sites) and the realization that antibiotics act in different ways. In particular, some antibiotics are primarily concentration dependent in their mode of action, most importantly the aminoglycosides and quinolones (Gould 1998; Drusano 2003). This has led us away from traditional three or four times daily administration of gentamicin to once daily administration to optimize its bactericidal effect. In addition, aminoglycosides and quinolones have a prolonged suppressive effect on regrowth of surviving bacteria (post antibiotic effect (PAE)) which allows a safe period of several hours of subinhibitory concentrations before it is necessary to give the next dose. Conveniently this also appears to ameliorate toxicity which is more dependent on high trough than high peak concentrations, due to the uptake of aminoglycosides by sensitive renal, vestibular, and cochlear cells being concentration limited (Mingeot-Leclecq and Tulkens 1999). Finally, high peak concentrations appear to be important in preventing selection of resistant mutants which otherwise might arise during therapy and cause treatment failure (Blondeau et al. 2001; Zhao and Drlica 2001; Scheld 2003).

A great deal of work has been carried out recently on the mutant prevention concentration of new quinolones against pneumococci and other bacteria. Although purely theoretical at the moment, it is probable that this work will lead to further reassessment of dosing schedules for quinolones, where mutants selected during therapy appear to be the only way of resistance arising. In the meantime, dosing schedules of quinolones are considered flexible and in general terms can either be given one or two times daily despite widely varying half-lives. The single most important pharmacodynamic predictor of both clinical and microbiological success is achieving an area under the curve inhibitory concentration (AUIC) >125 (AUIC = AUC_{0-24}/MIC) (Thomas et al. 1998).

Other antibiotics, most importantly the β-lactam drugs and macrolides are time-dependent drugs. Once the concentration at the site of infection is above the MIC there is usually little advantage in increasing concentration further. Most of the antibacterial effect will be determined by the duration of exposure. Animal experiments suggest optimal effects if the concentration is maintained above the MIC (T > MIC) for approximately 50 percent of the dosing interval, although this figure is probably lower for most gram-positive infections and higher for gram-negative infections because of the absence of a significant PAE in the latter. For life-threatening or high inoculum infection, such as meningitis, where the host immune response is limited and for more resistant organisms, it is possibly safer to try to ensure T > MIC for the whole dosing interval and to aim for concentrations around 8× MIC to ensure maximum bactericidal effect.

Antimicrobial drug combinations

The use of combinations of antimicrobial agents is increasing but makes more likely the risk of adverse reactions from the antibiotics themselves and from interactions with other drugs that the patient may be receiving. They include use of combination therapy in tuberculosis and leprosy; use of compounds that interact synergically to achieve improved bactericidal activity in certain serious infections, notably cystic fibrosis and infective endocarditis; use of combinations to achieve adequate cover of mixed infections or in the 'blind' therapy of serious undiagnosed sepsis and lastly, treatment of infections that are already multiresistant to prevent selection of further resistant mutants. Thus, combinations are frequently used for MRSA infections and particularly when rifampin or fucidin is used as resistant mutants are known to arise easily to these agents when used alone. Such a strategy has been very successful over the years with tuberculosis.

CLASSIFICATION OF ANTIBACTERIAL AGENTS IN CLINICAL USE

Since antibacterial agents exhibit a very wide variety of properties, the simplest way to categorize them is according to their site of action in the bacterial cell: agents acting on the bacterial cell wall; agents acting on bacterial protein synthesis; agents acting on nucleic acid synthesis; and agents acting on the bacterial cell membrane (Table 18.2).

AGENTS ACTING ON BACTERIAL CELL WALL SYNTHESIS

The bacterial cell wall is a structurally unique feature that is absent from mammalian cells and is thus a prime target for selectively toxic agents. The cell wall of gram-negative organisms is very different from that of gram-positive bacteria and the complex of lipopolysaccharide and lipoprotein that forms the gram-negative outer-membrane confers properties of differential permeability that have a profound effect on the susceptibility to antibacterial agents of all kinds (Nikaido and Vaara 1985;

Table 18.2 *Classification of antibacterial agents according to their site of action*

Cell wall synthesis	Protein synthesis	Nucleic acid synthesis	Cell membrane
Penicillins	Aminoglycosides	Sulphonamides[a]	Polymyxins
Cephalosporins	Chloramphenicol	Diaminopyrimidines[a]	Gramicidin
Carbapenems	Tetracyclines	Quinolones	Tyrocidine
Other β-lactams	Macrolides	Rifamycins	Valinomycin[b]
Glycopeptides	Lincosamides	Nitroimidazoles	Monensin[b]
Bacitracin	Fusidic acid	Nitrofurans	
Cycloserine	Streptogramins	Novobiocin	
Fosfomycin	Mupirocin		
Isoniazid	Ketolides		
	Oxazolidinones		
	Glycylcyclines		

a) Indirect action on nucleic acid synthesis.
b) Not used in human medicine.

Hancock and Bellido 1992). Within the gram-negative outer membrane are hydrophilic channels (porins) that allow the differential passage of many agents whose target is the underlying peptidoglycan. Permeation of these porin channels by β-lactam antibiotics depends on molecular size and ionic charge and these factors may strongly influence the spectrum of activity and potency (Nikaido 1985). Glycopeptides are too large to penetrate the gram-negative outer membrane and consequently their spectrum of activity is virtually restricted to gram-positive organisms.

β-Lactam antibiotics

STRUCTURE

Antibacterial agents that share the structural feature of a β-lactam ring are now known to be very diverse. The classic penicillins are penams, characterized by a fused heterocyclic structure composed of a β-lactam ring and a five-membered sulfur-containing thiazolidine ring. In the cephalosporins the fused dihydrothiazine ring has an extra carbon with an unsaturated bond between C-3 and C-4, giving a cephem structure. The cephamycins are similar, but the β-lactam ring is substituted with a methoxy group which confers stability to many β-lactamase enzymes. Other structural variants represented among clinically useful compounds are: carbapenems, carbacephems, oxacephems, clavams, sulfones, and monocyclic monobactams (Figure 18.1).

MODE OF ACTION

All β-lactam antibiotics interfere with bacterial cell wall synthesis, but the effect on gram-positive and gram-

negative bacteria is very different because of the differing nature of their cell walls. In the simplest terms, β-lactam antibiotics interfere with the final transpeptidation reaction that forms the crosslink between adjacent peptidoglycan strands and gives the cell wall its essential rigidity. Various forms of transpeptidase exist depending on whether cell wall structure is engaged in extension of the cylinder of rod-shaped cells, forming the poles of the cell, or separating the two daughter cells during the division process. Realization that β-lactam antibiotics affected all of these processes came from studies that showed concentration-dependent effects on the morphological response of gram-negative bacilli (Greenwood and O'Grady 1973a); by observation that certain β-lactam agents were anomalous in the morphological effects they produced (Greenwood and O'Grady 1973b); and by the demonstration of differential affinity of β-lactam agents for penicillin-binding proteins (PBP) in isolated cell membranes (Spratt 1975). The morphological consequences of binding to various PBPs of *Escherichia coli* are shown in Figure 18.2. PBPs have been investigated in many gram-positive and gram-negative bacteria. Although their number and molecular size vary considerably, they are always present in multiple forms.

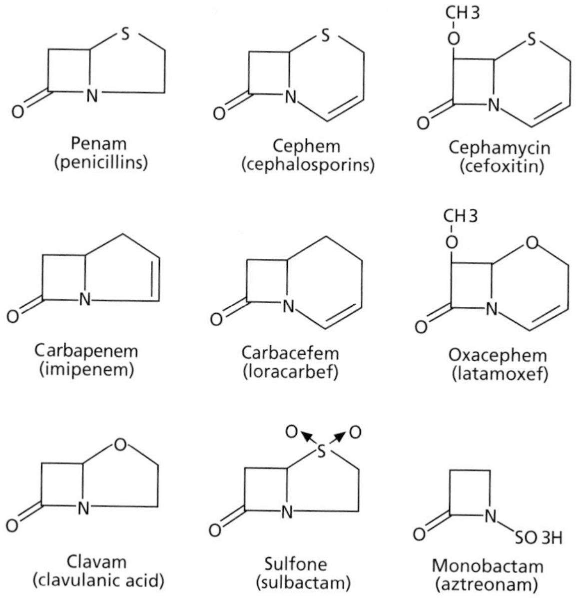

β-Lactam antibiotics are normally bactericidal drugs, but the mechanism of bactericidal activity seems to be different in gram-negative and gram-positive bacteria. In gram-negative rods, cell death can be quantitatively prevented by provision of adequate osmotic protection (Greenwood and O'Grady 1972), so that the mechanism appears to be simple osmotic lysis of bacteria deprived of their normal cell wall. In gram-positive cocci the situation is more complicated: exposure to β-lactam agents causes a loss of lipoteichoic acid from the wall and this seems to remove control from normal autolytic processes that dismantle the peptidoglycan (Tomasz 1979).

Certain strains of gram-positive cocci succumb to the bactericidal effects of penicillin and other β-lactam agents more slowly than usual. Such strains have been dubbed *tolerant* to penicillin (Sabath et al. 1977; Handwerger and Tomasz 1985). Penicillin tolerance may have some relevance in the treatment of bacterial endocarditis or in other situations in which a bactericidal effect is crucial to therapeutic success.

RESISTANCE TO β-LACTAM ANTIBIOTICS

The most common form of resistance to β-lactam agents is caused by enzymes that render the molecules inactive by opening the β-lactam ring. In staphylococci the enzymes involved are inducible exoenzymes that conform to a small number of related types. In contrast, the β-lactamases of gram-negative bacilli vary greatly in their physicochemical characteristics. Indeed, all gram-negative bacilli appear to exhibit some β-lactamase

Figure 18.1 *Core molecular structures of various kinds of β-lactam antibiotics (examples in parentheses). (Redrawn from Greenwood 1995)*

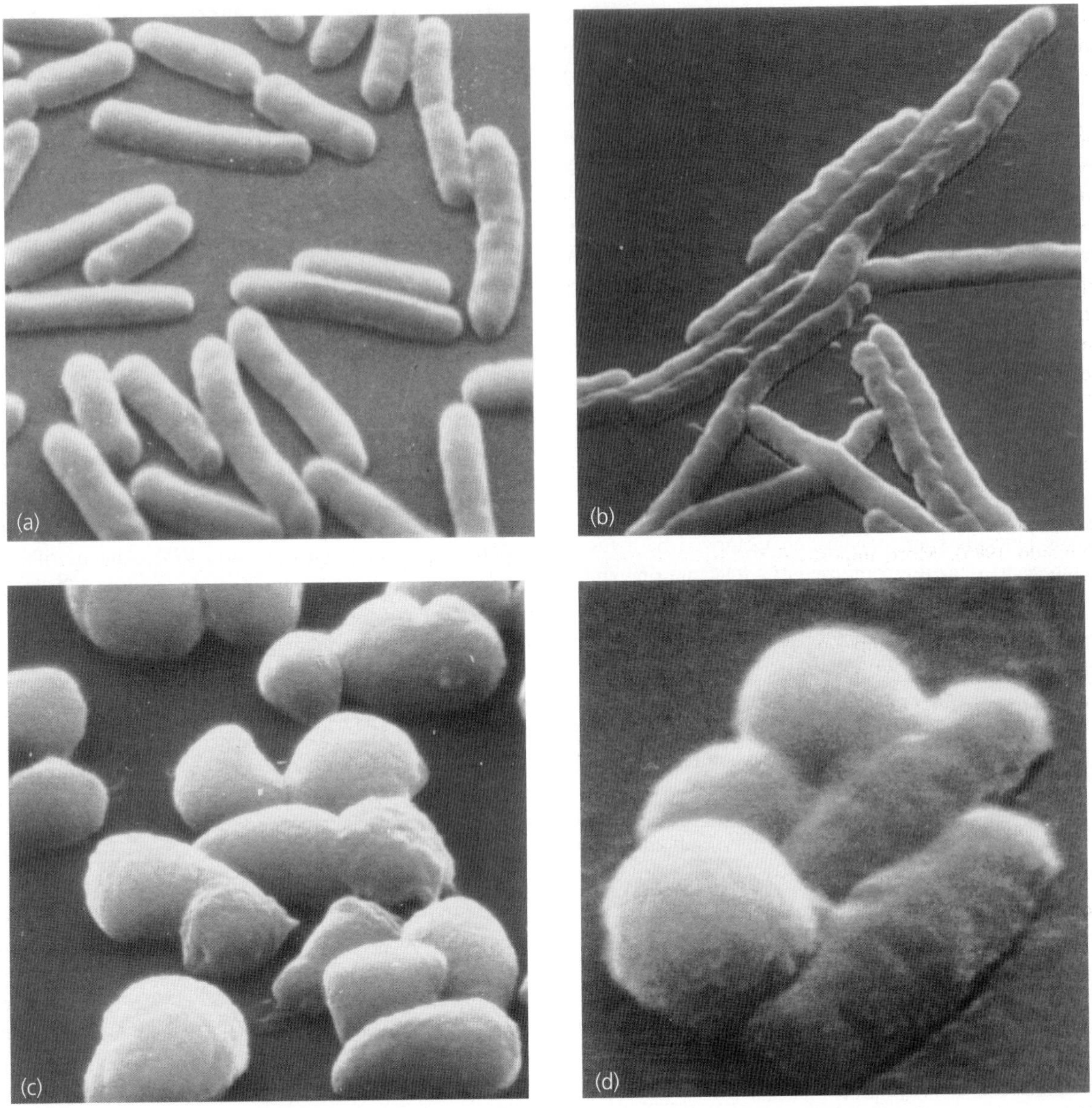

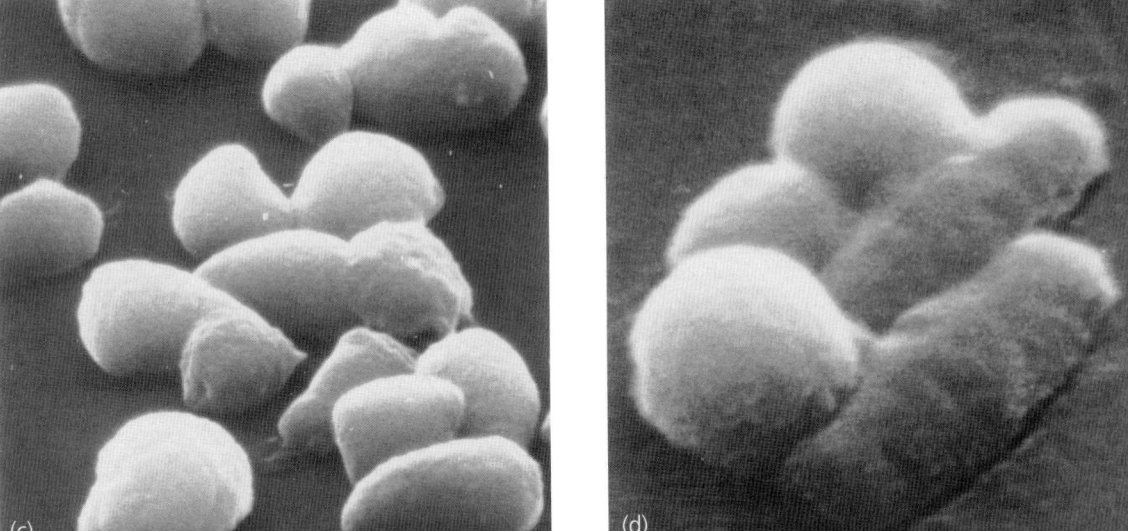

Figure 18.2 *Morphological effects of β-lactam antibiotics in* Escherichia coli: **(a)** *normal cells (no antibiotic);* **(b)** *filamentation caused by cephalexin (binding to PBP-3);* **(c)** *generalized effect on cell wall caused by mecillinam (binding to PBP-2);* **(d)** *formation of osmotically fragile spheroplasts caused by cephalexin and mecillinam in combination (binding to PBP-2 and PBP-3). Most β-lactam antibiotics, in sufficient concentration, also bind to the PBP-1 complex and cause rapid lysis of susceptible gram-negative bacilli.*

activity as a chromosomally encoded genetic feature. These inherent β-lactamases may be inducible and certain β-lactam antibiotics, notably cefoxitin, are particularly efficient inducers.

There have been various attempts to classify β-lactamases, particularly the many types found in gram-negative rods (Ambler 1980; Richmond and Sykes 1973; Bush 1989). The latest comprehensive classification scheme recognizes four separate molecular types and several functional groupings (Bush et al. 1995).

Although these classification schemes attempt to include all the known types of β-lactamases and are therefore quite complex, the majority of the enzymes that commonly cause problems in clinical isolates of gram-negative bacilli belong to a few related types that share the feature of having serine at the active site and are susceptible to inhibition by clavulanic acid and other β-lactamase inhibitors. Most common of all are the TEM-1 and TEM-2 enzymes found widely among enterobacteria and elsewhere, and SHV-1, found predominantly in *Klebsiella pneumoniae*. Many variants of the TEM enzymes have arisen, often by single amino acid substitutions and these may exhibit extended, or idiosyncratic, substrate ranges. Most problematic at present for clinicians are the clavulanic acid-sensitive ESBLs that hydrolyze third and to a lesser extent, fourth generation cephalosporins (Gould 1999) and the clavulanic acid resistant enzymes – the so-called inhibitor-resistant TEM enzymes as they are usually derived from TEM-1 (Chaibi et al. 1999). These enzymes give resistance to co-amoxiclav and other β-lactamase inhibitors. They may be widespread, but are difficult to test for on a routine basis. TEM-1 itself, and some other β-lactamases, have found their way on to transposons and this has aided their transmission and spread. Bacterial β-lactamases that are able to hydrolyze carbapenems, like imipenem, are still unusual in many countries. They are usually zinc-requiring metalloenzymes that are not susceptible to β-lactamase inhibitors (Payne 1993). They are presently found most commonly in *Acinetobacter* spp., *Aeromonas* spp., and in *Stenotrophomonas maltophilia* and are classified by Bush et al. (1995) as a separate molecular class B (group 3) (see section on Carbapenems).

Resistance to β-lactam antibiotics can arise by means other than β-lactamases. Alterations in porins in the outer membrane of gram-negative bacilli can affect transport of β-lactam antibiotics to their site of action (Nikaido 1985). More importantly, mutations affecting the structure of PBPs can alter their affinity to β-lactam compounds so that inhibitory effects become much less efficient. This, for example, is the mechanism of resistance in methicillin-resistant *Staphylococcus aureus* and in strains of *Streptococcus pneumoniae* that display reduced susceptibility to penicillin (Klugman 1990; Jacoby and Archer 1991).

TOXICITY AND SIDE EFFECTS

Since the β-lactam antibiotics act on a target (bacterial peptidoglycan) that is absent from mammalian cells, they are among the least toxic of all antimicrobial agents. The most troublesome side effect is hypersensitivity, which can range from relatively trivial rashes to life-threatening anaphylactic reactions. Less than 10 percent of patients hypersensitive to penicillins are cross-allergic to cephalosporins (Anne and Resman 1995). Ampicillin and amoxicillin commonly give rise to a maculopapular rash when given to patients with glandular fever; this is not a true hypersensitivity reaction and is not a contraindication to subsequent use of a penicillin. Various other side effects are occasionally encountered. A fuller discussion is provided in the standard texts (Kucers and Bennett 1987; Dollery 1991).

Penicillins

CLASSIFICATION AND SPECTRUM OF ACTIVITY

The various penicillins differ in the nature of the side chain at the C-6 position of the molecule. All except mecillinam (amdinocillin), in which the side chain is joined in an amidino linkage, are acyl derivatives of 6-aminopenicillanic acid (Figure 18.3).

The original penicillin, benzylpenicillin (penicillin G), has a phenylacetamido group at the 6-amino position. It exhibits exceptionally good activity against the classic pyogenic cocci: staphylococci, streptococci (including pneumococci and enterococci), meningococci, and gonococci. It is also very active against spirochetes, most anaerobes (but not *Bacteroides* spp.), and many gram-positive bacilli. Enterobacteria, *Pseudomonas aeruginosa*, and obligate intracellular bacteria are mostly insensitive.

The chief imperfections of penicillin are: lability to gastric acid, so that it cannot be given orally; an exceptionally short plasma half-life; susceptibility to staphylococcal β-lactamase; and a restricted spectrum of activity. Subsequent developments of penicillin have sought to overcome these shortcomings and presently available compounds can be categorized into several groups with distinctive properties (Table 18.3).

The long-acting salts of penicillin are poorly soluble depot preparations that are injected intramuscularly and release benzylpenicillin slowly from the injection site. In this way, procaine penicillin can maintain a concentration of penicillin inhibitory to many sensitive organisms for up to 24 h. Benzathine and benethamine penicillin are even less soluble and liberate small amounts of penicillin over several days. An alternative way of sustaining penicillin levels in plasma is by the concurrent administration of probenecid, which competes for the active tubular secretion sites in the kidneys.

Oral derivatives of penicillin, notably phenoxymethylpenicillin (penicillin V), have antibacterial properties very similar to those of benzylpenicillin.

The antistaphylococcal penicillins, of which the isoxazolyl derivatives, oxacillin, cloxacillin, dicloxacillin, and flucloxacillin are most widely used, were designed to overcome the problem of enzymic resistance in staphylococci. They are intrinsically less active than benzylpenicillin and although they retain adequate antistreptococcal activity, their use is largely restricted to infections with penicillinase-producing staphylococci, which now account for more than 90 percent of all isolates in many European countries.

The first successful extension of the spectrum of benzylpenicillin was achieved in 1961 with the β-amino derivative, ampicillin (Rolinson and Stevens 1961). Unlike the parent compound, ampicillin exhibited useful activity against some gram-negative bacilli, including *Haemophilus influenzae*, *Escherichia coli*, *Salmonella* spp., and *Shigella* spp., but not *Klebsiella* spp. or *P. aeruginosa*. Amoxicillin, which is better absorbed when administered orally, followed in 1970. In the meantime, the first penicillin to exhibit (albeit weak) antipseudomonal activity, carbenicillin, the β-carboxy derivative of benzylpenicillin, was described (Knudsen et al. 1967). The thienyl variant of carbenicillin, ticarcillin, and a series of acylureido derivatives of ampicillin (azlocillin, mezlocillin, piperacillin, and apalcillin), all of which display improved antipseudomonal activity, were developed later. None of these compounds is stable to staphylococcal β-lactamase.

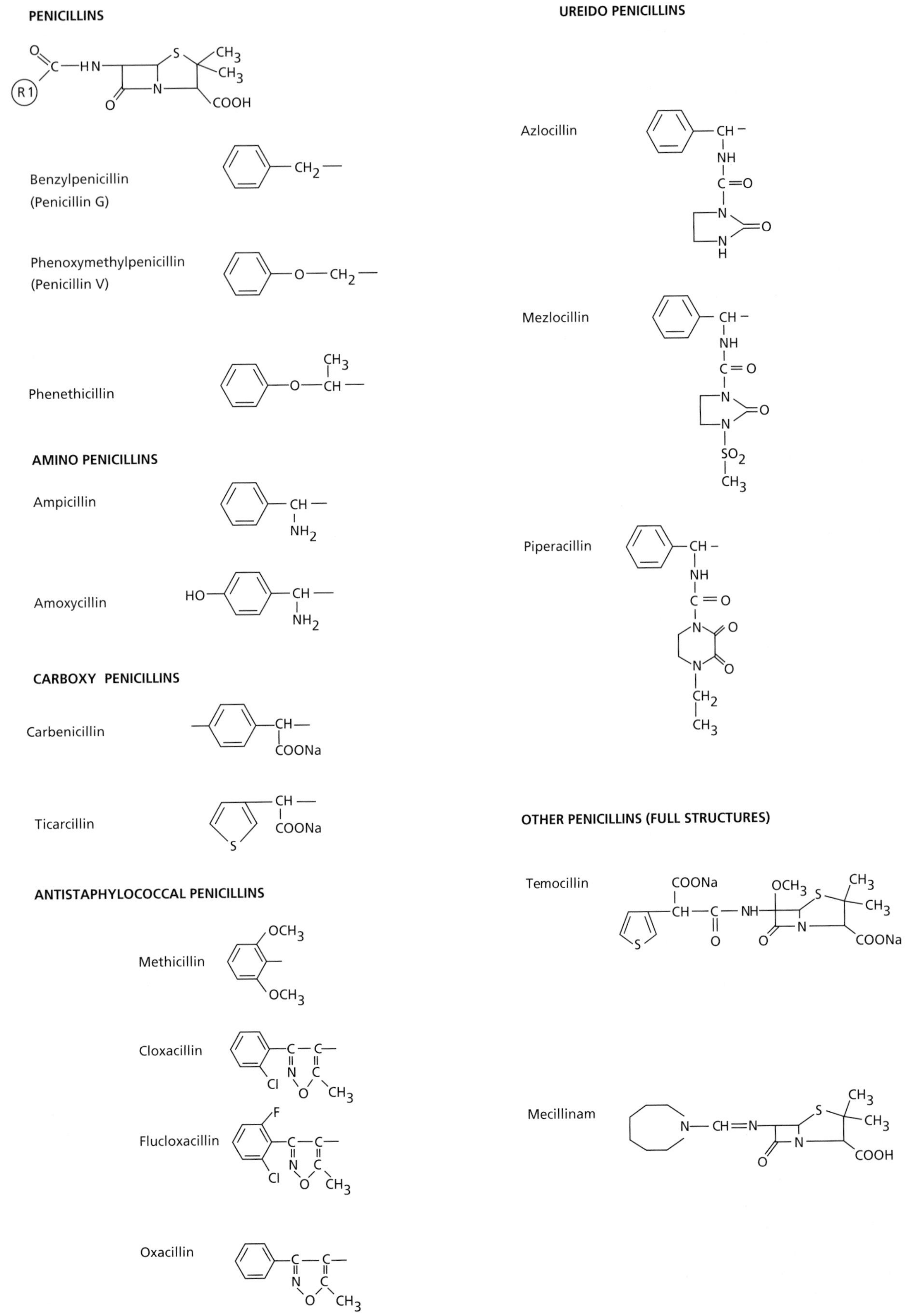

PENICILLINS

Benzylpenicillin
(Penicillin G)

Phenoxymethylpenicillin
(Penicillin V)

Phenethicillin

AMINO PENICILLINS

Ampicillin

Amoxycillin

CARBOXY PENICILLINS

Carbenicillin

Ticarcillin

ANTISTAPHYLOCOCCAL PENICILLINS

Methicillin

Cloxacillin

Flucloxacillin

Oxacillin

UREIDO PENICILLINS

Azlocillin

Mezlocillin

Piperacillin

OTHER PENICILLINS (FULL STRUCTURES)

Temocillin

Mecillinam

Figure 18.3 *Structures of the most important penicillins.*

Table 18.3 *Categorization of the most important pencillins in clinical use*

Most important penicillins in clinical use	Availability
Benzylpenicillin (penicillin G)	Available
Long-acting salts (depot preparations)	
Procaine penicillin	Available via import
Benzathine penicillin	Not available in UK
Benethamine penicillin	Not available in UK
Acid-stable (oral) derivatives	
Phenoxymethylpenicillin (penicillin V)	Available
Azidocillin	Not available in UK
Phenethicillin	Not available in UK
Propicillin	Not available in UK
Antistaphylococcal penicillins (stable to staphylococcal β-lactamase)	
Methicillin	Not available in UK
Cloxacillin	Not available in UK
Flucloxacillin	Available
Oxacillin	Not available in UK
Dicloxacillin	Not available in UK
Nafcillin	Not available in UK
Penicillins with activity against enterobacteria	
Ampicillin (and esters)	Available
Amoxycillin	Available
Mecillinam (and ester)	Available as pivmecillinam
Ciclacillin	Not available in UK
Epicillin	Not available in UK
Penicillins that are stable to enterobacterial β-lactamases[a]	
Temocillin	Not available in UK
Penicillins with activity against *Pseudomonas aeruginosa*	
Carbenicillin (and esters)	Not available in UK
Ticarcillin	Available only with clavulanic acid (Timentin)
Azlocillin	Not available in UK
Mezlocillin	Not available in UK
Piperacillin	Available only with tazobactam (Tazocin)
Apalcillin	Not available in UK

a) Ampicillin, amoxicillin, ticarcillin, and piperacillin are also formulated with β-lactamase inhibitors that confer the property of stability to enterobacterial β-lactamases.

Among penicillins that possess an idiosyncratic spectrum of activity that excludes gram-positive cocci and anaerobes are the amidinopenicillins, mecillinam and temocillin, a compound structurally similar to ticarcillin, but rendered stable to enterobacterial β-lactamases by incorporation of a methoxy group on to the β-lactam ring (compare with cefoxitin and other cephamycins). Neither mecillinam nor temocillin has any useful activity against *P. aeruginosa*.

PHARMACOLOGICAL PROPERTIES

Most penicillins are absorbed erratically when given orally, often because of hydrolysis by gastric acid. Phenoxymethylpenicillin and related compounds are acid stable and are used as oral substitutes for benzylpenicillin. Among isoxazolyl penicillins, flucloxacillin achieves the highest plasma concentrations after oral administration. About 30 percent of an oral dose of ampicillin is absorbed, but this can be much improved by esterification of the carboxyl group. These ampicillin esters, talampicillin, pivampicillin, bacampicillin, and lenampicillin, are inactive prodrugs that are de-esterified during passage through the intestinal mucosa with liberation of ampicillin into the bloodstream. Prodrug esters of carbenicillin and mecillinam have also been produced as oral formulations.

Benzylpenicillin and phenoxymethylpenicillin are extremely rapidly excreted, with a plasma half-life of about 30 min. Most other penicillins exhibit elimination half-lives of around 30 min to 1.5 h. Temocillin has an unusually extended half-life of approximately 4–5 h. Protein binding is generally less than 50 percent; the isoxazolylpenicillins are extensively (>90 percent) protein-bound in plasma, but this does not seem to adversely affect their therapeutic activity.

Penicillins do not penetrate the intact blood–brain barrier well, but in the presence of meningeal inflammation concentrations are often sufficient to treat pyogenic meningitis and benzylpenicillin (meningococcal and pneumococcal meningitis) and ampicillin (*Haemophilus meningitis*) have been widely used for this purpose in the past.

Cephalosporins

CLASSIFICATION AND SPECTRUM OF ACTIVITY

Since they first became available in the mid-1960s, the cephalosporins have been developed more diversely than any other group of antimicrobial agents. They have overlapping properties that defy any rigid classification, but can usefully be divided into six broad classes (Table 18.4). In the cephalosporins, the extra carbon of the dihydrothiazine ring carries an additional side chain, the nature of which often affects the pharmacokinetic behavior of the molecule and in some cases, the toxicity. Structures of some of the commonly used cephalosporins are shown in Figure 18.4. As a group, the cephalosporins have certain properties in common: relative stability to staphylococcal β-lactamase; broad spectrum activity that encompasses most enterobacteria, including *Klebsiella* spp.; and lack of activity against enterococci.

Like the penicillins, cephalosporins interfere with bacterial cell wall synthesis through binding to PBPs. Some cephalosporins, notably cephalexin and cephradine, bind almost exclusively to PBP-3, which has the effect of halting division, but not growth, of susceptible gram-negative bacilli. Consequently, these cephalo-sporins are much more slowly bactericidal than the others and the build up of endotoxin from filamentation may well have deleterious clinical effects (Gould and MacKenzie 1997).

PARENTERAL CEPHALOSPORINS

The first cephalosporins, cephalothin and cephaloridine, modestly expanded the spectrum of ampicillin (e.g. to include *Klebsiella* spp.) and possessed relative stability to staphylococcal β-lactamase. Some later derivatives offered little, if any, improvement, but cephazolin exhibited the unusual characteristic of achieving enhanced concentrations in bile and cephamandole offered partial resistance to some enterobacterial β-lactamases.

The first major improvement in cephalosporins came with the introduction of compounds that displayed stability to a wide range of enterobacterial β-lactamases. One such compound, cefoxitin, is a semisynthetic derivative of a naturally occurring compound in which the β-lactam ring is substituted with a methoxy grouping that confers stability to the structure. Such compounds are called cephamycins; various examples, including cefotetan, cefbuperazone, cefmetazole, and cefminox, are in use around the world. They are unusual among cephalo-sporins in their activity against *B. fragilis*.

β-Lactamase stability was achieved in a different way by pharmaceutical chemists who found that addition of a methoximino group to the side chain on 7-aminocepha-losporanic acid also protected the β-lactam ring from enzymic attack. The first cephalosporin of this type was

Table 18.4 *Categorization of cephalosporins in clinical use*

Cephalosporins				
Parenteral compounds	**Oral compounds**			
Cephalothin	Cephacetrile	Ceforanide	Cephalexin	Cephaloglycin
Cephaloridine	Cefapirin	Cefonicid	Cephradine	Cefatrizine
Cephazolin	Cefazedone		Cefaclor	Cefroxadine
Cephamandole	Ceftezole		Cefadroxil	Cefprozil
Compounds with improved β-lactamase stability				
Cefuroxime	Cefmetazole[a]	Cefotiam	**Non-esterified**	**Esterified**
Cefoxitin[a]	Cefbuperazone[a]		Cefixime	Cefuroxime axetil
Cefotetan[a]	Cefminox[a]		Ceftibuten	Cefpodoxime proxetil
			Cefdinir	Cefetamet pivoxil
				Cefteram pivoxil
Compounds with improved intrinsic activity and β-lactamase stability				
Cefotaxime	Cefmenoxime			
Ceftizoxime	Cefodizime			
Ceftriaxone	Latamoxef[b]			
Compounds distinguished by activity against *Pseudomonas aeruginosa*				
Broad spectrum	**Medium spectrum**	**Narrow spectrum**		
Ceftazidime	Cefoperazone	Cefsulodin		
Cefpirome	Cefpimazole			
Cefepime	Cefpiramide			

Adapted from Greenwood (1995).
a) Cephamycins
b) Oxacephem

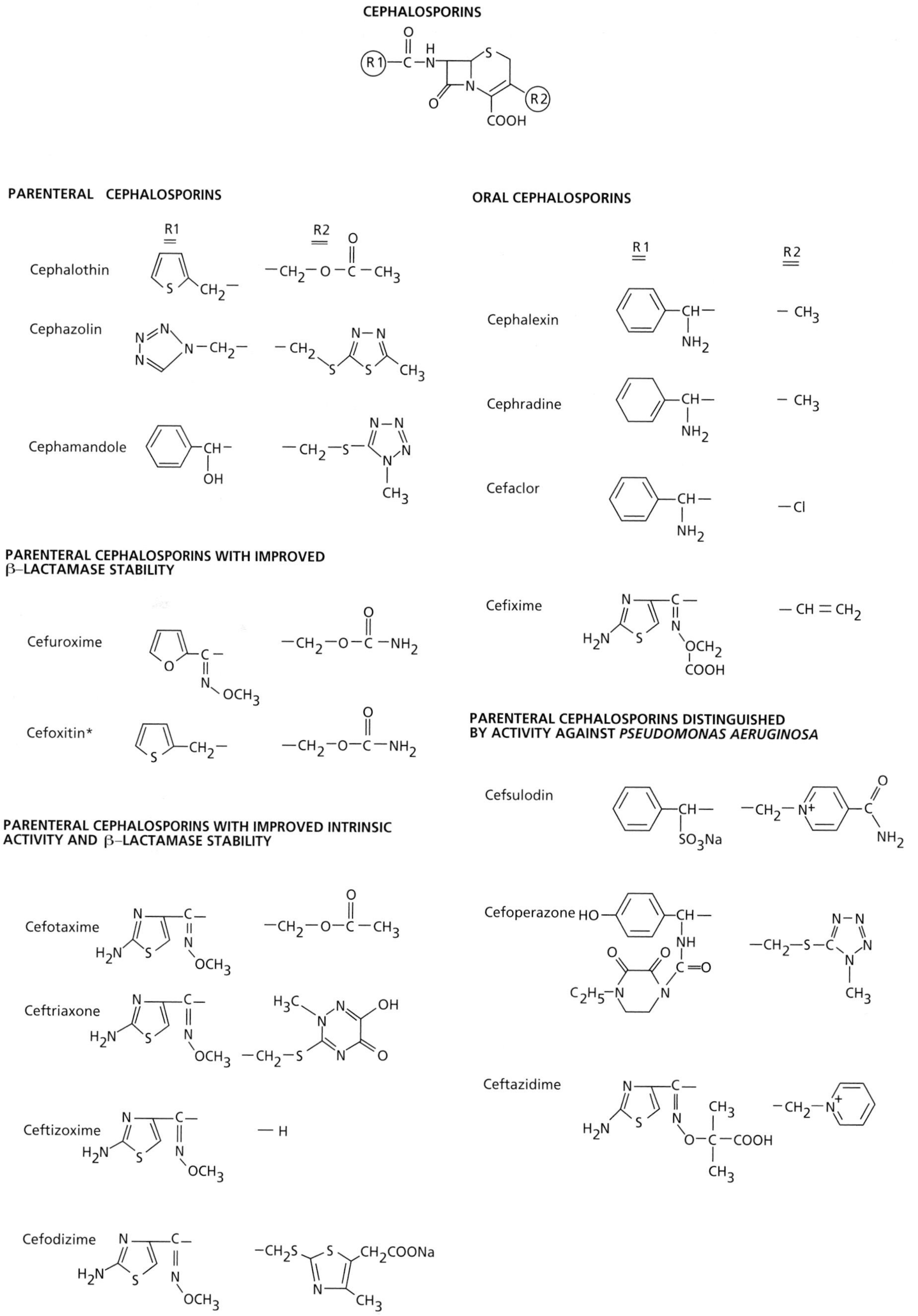

CEPHALOSPORINS

PARENTERAL CEPHALOSPORINS

Cephalothin

Cephazolin

Cephamandole

PARENTERAL CEPHALOSPORINS WITH IMPROVED β–LACTAMASE STABILITY

Cefuroxime

Cefoxitin*

PARENTERAL CEPHALOSPORINS WITH IMPROVED INTRINSIC ACTIVITY AND β–LACTAMASE STABILITY

Cefotaxime

Ceftriaxone

Ceftizoxime

Cefodizime

ORAL CEPHALOSPORINS

Cephalexin

Cephradine

Cefaclor

Cefixime

PARENTERAL CEPHALOSPORINS DISTINGUISHED BY ACTIVITY AGAINST *PSEUDOMONAS AERUGINOSA*

Cefsulodin

Cefoperazone

Ceftazidime

* Cefoxitin (a cephamycin) also has a methoxy group on the β-lactam ring.

Figure 18.4 *Structures of the most important cephalosporins.*

cefuroxime, but this was quickly followed by a family of compounds, of which cefotaxime was the forerunner, in which the side chain carried an aminothiazole group, as well as the methoximino substitution. The effect of this was not only to provide β-lactamase stability, but also to much improve the intrinsic antibacterial activity, particularly against enterobacteria. Activity of these compounds against *P. aeruginosa* is, however, poor, but some later derivatives, notably ceftazidime, include this organism in the spectrum. Cefsulodin is unique in exhibiting good antipseudomonal activity, but has little or no useful activity against other organisms. Later developments in the form of cefepime and cefpirome not only provide antipseudomonal activity but some resistance to most ESBLs.

ORAL CEPHALOSPORINS

Few cephalosporins are absorbed when administered by the oral route. Cephalexin was the first of those that are well absorbed; it exhibits modest antibacterial activity against a wide spectrum of gram-positive and gram-negative bacteria, but is slowly bactericidal to enterobacteria because of its preferential affinity for PBP-3 (see above). It has no useful activity against *P. aeruginosa*, *H. influenzae* or *B. fragilis*. Most other oral cephalosporins are structurally similar to cephalexin and not surprisingly, share its limitations. Cefaclor is unusual in displaying useful activity against *H. influenzae* and is especially well absorbed. Attempts have been made to improve the intrinsic activity of oral cephalosporins, or to esterify parenteral compounds to enhance their absorption. Cefixime and ceftibuten display much improved activity against enterobacteria, but have some important weaknesses in their gram-positive spectrum, in particular in their poor activity against staphylococci, enterococci, and some streptococci. Cefpodoxime retains better activity against staphylococci and streptococci. None have useful activity against *P. aeruginosa*. Oral cephalosporins have recently been reclassified (Williams et al. 2001).

PHARMACOLOGICAL PROPERTIES

Cephalosporins, like penicillins, are generally excreted rapidly by the kidneys with elimination half-lives that usually range from 1 to 2 h. The expanded spectrum compound, ceftriaxone, is unusual in that its plasma half-life is about 6–8 h. Cephazolin, cefoperazone, and ceftriaxone are notable in achieving significant concentrations in bile. Those cephalosporins that carry an acetoxymethyl group at C-3 (including cephalothin and cefotaxime) are susceptible to hepatic enzymes which deacetylate the molecule to the corresponding hydroxymethyl derivative. This reduces the inherent antibacterial activity, but there is little evidence that this affects the therapeutic potency. Some other cephalosporins, cephamandole, cefotetan, and cefoperazone among them, have a nitrogen-rich tetramethylthiomethyl

substituent at C-3. Compounds with this feature have been implicated in hypoprothrombinemia that is reversible by administration of vitamin K (Lipsky 1988).

Cephalosporins penetrate poorly into cerebrospinal fluid after intravenous administration, but some, notably cefotaxime and ceftriaxone achieve sufficient concentration in the cebrospinal fluid (CSF) in the presence of inflammation for them to be useful for intravenous treatment in bacterial meningitis.

Other β-lactam antibiotics

Several variants on the β-lactam theme other than penams (penicillins) and cephems (cephalosporins and cephamycins) are in therapeutic use (Figure 18.5).

β-LACTAMASE INHIBITORS

Certain molecules that possess a β-lactam ring, including the naturally occurring antibiotic, clavulanic acid and the penicillanic acid sulfones, sulbactam and tazobactam, are noteworthy not for their intrinsic antibacterial activity, but because they have a high affinity for and stability to, bacterial β-lactamases. These agents are formulated with β-lactamase labile partner compounds to secure their activity against otherwise resistant organisms: clavulanic acid with amoxicillin (co-amoxiclav) or ticarcillin; sulbactam with ampicillin (or with cefoperazone in some countries); tazobactam with piperacillin.

CARBAPENEMS, CARBACEPHEMS, OXACEPHEMS

In the mid-1970s, a potent, broad spectrum β-lactam compound, thienamycin, was found as a naturally occurring product of *Streptomyces cattleya* (Kahan and Kahan 1979). The substance, which was found to possess an unusual carbapenem ring structure, was, unfortunately, inherently unstable, but this was overcome by synthesis of the *N*-formimidoyl derivative, imipenem (Kahan et al. 1983) (see Figure 18.5). Imipenem has exceptionally good activity against most gram-positive and gram-negative bacteria, including organisms like enterococci, *P. aeruginosa*, and *B. fragilis* that are resistant to most cephalosporins. It is very stable in the presence of bacterial β-lactamases, but by a strange trick of nature it is hydrolyzed by a dehydropeptidase located in the brush border of the mammalian kidney and has to be given with a dehydropeptidase inhibitor, cilastatin (Kahan et al. 1983). A related carbapenem, meropenem, is not susceptible to dehydropeptidase and is therefore administered alone. These compounds are not absorbed when given by mouth and are administered parenterally. Resistance to carbapenems is still uncommon, but they are hydrolyzed by naturally occurring metallo-β-lactamases produced by *S. maltophilia*, *Flavobacterium* sp., *Aeromonas* sp., and *B. fragilis*. Methicillin-resistant staphylococci and penicillin-resistant enterococci are

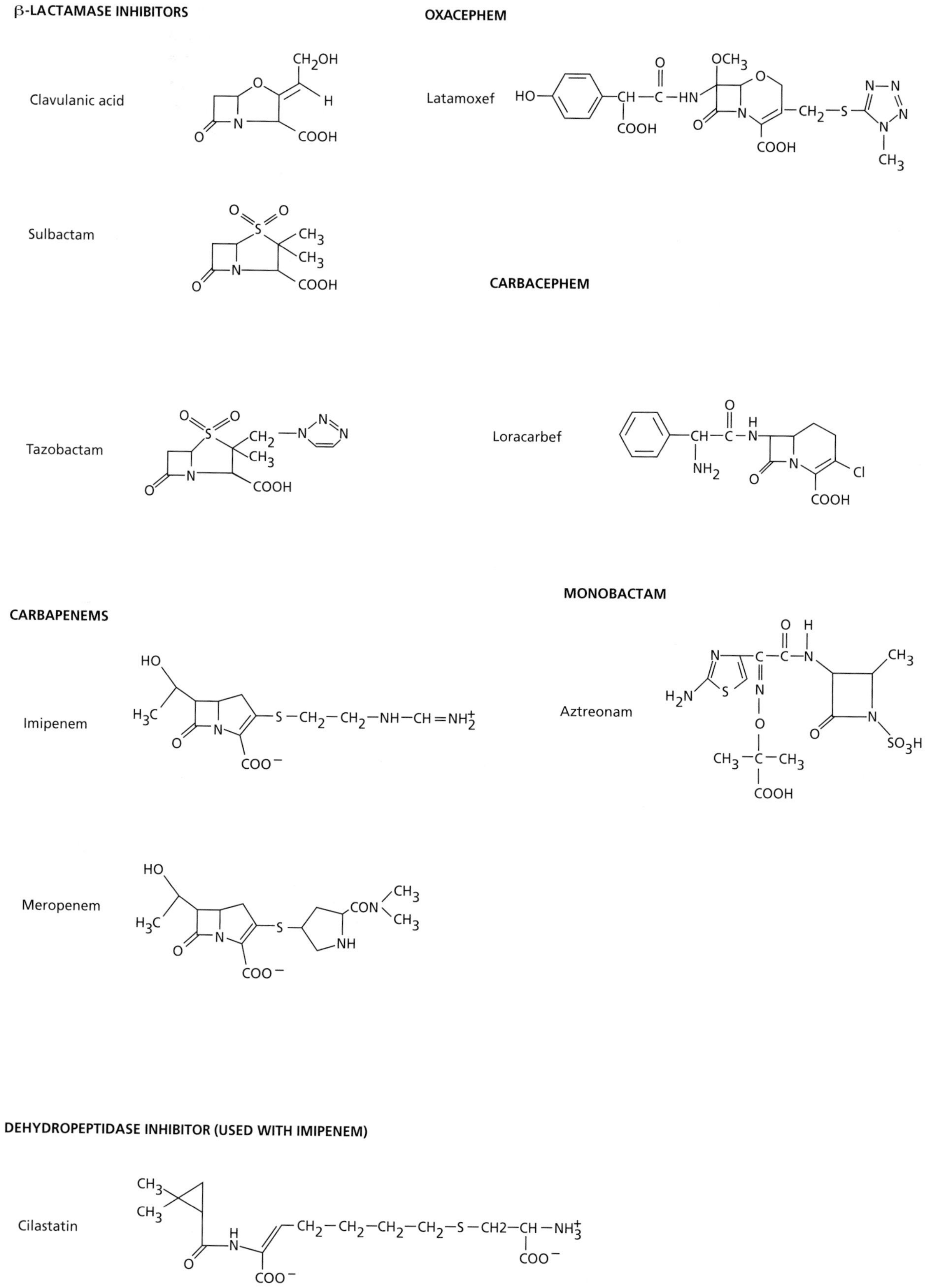

Figure 18.5 *Structures of important β-lactam compounds other than penicillins and cephalosporins, and of the dehydropeptidase inhibitor cilastatin.*

usually refractory to carbapenems. A new, once daily carbapenem, ertapenem has reduced antipseudomonal activity. Carbapenem resistance in Acinetobacter, Pseudomonas, and some Enterobacteriaceae is becoming an increasing clinical problem. Two types of class B metallo β-lactamase are prevalent: Verona imipenemase (VIM) and IMP. They are now becoming widespread and a VIM-producing *P. aeruginosa* recently spread among 200 patients in Greece (Giakkoupi et al. 2003; Docquier et al. 2003; Jeong et al. 2003). They are very worrying not only because they hydrolyze all β-lactams except the monobactams and are not susceptible to inhibitors, but they can be coded for on integrons.

The possibility of synthesizing carbacephems and oxacephems has also been explored. Two such compounds, loracarbef (a carbacephem) and latamoxef (an oxacephem), have been developed (Figure 18.5). Loracarbef can be given orally, but does not have the breadth of spectrum or the β-lactamase stability of the carbapenems; its activity and use are similar to those of cefaclor, to which it is related structurally. Latamoxef (also known as moxalactam) is a parenteral compound that exhibits properties similar to those of the expanded spectrum cephalosporins such as cefotaxime, but including *B. fragilis* within the spectrum. It has a methyl-tetrazole group at C-3 and the associated bleeding problems (see above) have restricted its popularity.

MONOBACTAMS

The idea of developing therapeutically useful compounds in which the β-lactam ring was not fused with another cyclic structure was once thought to be unlikely, if not impossible. Therefore, it came as a surprise when naturally occurring compounds with this feature were described. Only one such compound is in therapeutic use, the semi-synthetic monobactam, aztreonam (Figure 18.5). In marked contrast to the ultra-broad spectrum of the carbapenems, aztreonam is a narrow spectrum agent, the useful activity of which is restricted to enterobacteria and *P. aeruginosa*. It is administered parenterally. Aztreonam specifically inhibits PBP-3 in *E. coli* and causes filamentation of susceptible gram-negative bacilli, which are killed slowly.

Glycopeptide antibiotics

STRUCTURE AND SPECTRUM OF ACTIVITY

Two glycopeptides, vancomycin and teicoplanin, are in clinical use. A related compound, avoparcin, is used as a growth promoter in animal husbandry. They are complex heterocyclic compounds composed of a hepta-peptide substituted with certain sugars, one of which, in teicoplanin, carries a fatty acid chain (Figure 18.6). The molecule is too large to penetrate the outer membrane of gram-negative bacilli, although some gram-negative anaerobic bacilli are anomalously sensitive, particularly

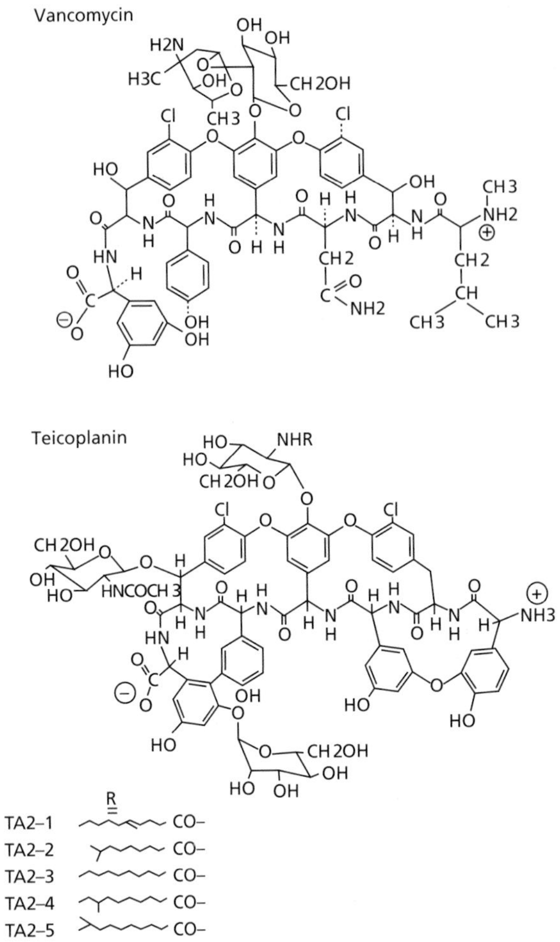

Figure 18.6 *Structures of the glycopeptide antibiotics, vancomycin and teicoplanin.*

to teicoplanin (Greenwood et al. 1988). Nearly all gram-positive organisms are susceptible, although some genera, including *Lactobacillus*, *Pediococcus*, and *Leuconostoc*, are intrinsically resistant (Ruoff et al. 1988). They are mainly used in serious infection with staphylococci and enterococci that are resistant to other drugs although bactericidal activity is slow. Oral vancomycin has been used in antibiotic-associated colitis in which *Clostridium difficile* is implicated and is still used for refractory cases. New derivatives such as daptomycin are being developed with improved bactericidal activity and daptomycin has already been licensed in some countries.

MODE OF ACTION

Glycopeptides prevent the transfer of peptidoglycan building blocks to the growing cell wall by binding to the acyl-D-alanyl-D-alanine terminus of the pentapeptide side chain (Reynolds 1989).

RESISTANCE

There are many reports of resistance in enterococci (Johnson et al. 1990), and coagulase-negative staphylococci (Sanyal et al. 1993). In enterococci, resistance is associated with enzymic alteration of the D-alanyl-D-alanine target to D-alanyl-D-lactate (Walsh 1993), but other mechanisms are found in coagulase-negative staphylococci. The most common type of resistance in enterococci (*vanA*) is inducible and transferable (Shlaes et al. 1989; Healy and Zervos 1995); high level resistance to both vancomycin and teicoplanin is conferred. Transfer to *S. aureus* has been reported (Noble et al. 1992) and at the time of writing three clinical isolates of this glycopeptide-resistant *S. aureus* have been described (Chang et al. 2003; Public Health Dispatch 2002; Hiramatsu 2004; Kacia, 2004) Heteroresistant strains are also increasingly described with thickened cell walls probably acting as a sump for both vancomycin and teicoplanin (Manuel et al. 2002; MacKenzie et al. 2003). Some vancomycin-resistant isolates of enterococci retain susceptibility to teicoplanin, but *Staphylococcus haemolyticus* isolates are often more resistant to teicoplanin than to vancomycin.

PHARMACOLOGICAL PROPERTIES AND TOXICITY

Vancomycin has to be administered by slow intravenous infusion, since bolus injection is apt to cause 'red-man syndrome' owing to the release of histamine. Teicoplanin is less likely to cause this complication (Sahai et al. 1990). In addition, teicoplanin does not cause tissue necrosis that is associated with intramuscular injection of vancomycin and it can be safely given by this route. The elimination half-life of teicoplanin (approximately 40 h) is much longer than that of vancomycin (approximately 7 h), and it is much more extensively protein bound (90 versus 50 percent). Renal and ototoxicity may be more common with vancomycin, but serum monitoring of both is recommended to ensure efficacious trough levels.

OTHER CELL WALL ACTIVE AGENTS

Bacitracin

This is a cyclic peptide that prevents the dephosphorylation of the lipid carrier molecule that transfers newly formed peptidoglycan across the cell membrane during cell wall synthesis. It is too toxic for systemic use, but is found in several topical preparations. It is mainly active against gram-positive cocci. The exquisite sensitivity of *S. pyogenes* is exploited in a laboratory screening test for that organism.

Cycloserine

This is an analog of D-alanine, which prevents the racemization of L-alanine and the ligation of D-alanyl-D-alanine. It is quite toxic and is now used only as a reserve agent for drug-resistant *M. tuberculosis*, although

it used to be prescribed for prophylaxis of recurrent urinary tract infections.

Fosfomycin

Fosfomycin (epoxypropylphosphonic acid) is a broad spectrum antibiotic that blocks the formation of *N*-acetylmuramic acid from *N*-acetylglucosamine by inhibition of pyruvyltransferase. In vitro activity against *E. coli* and some other gram-negative rods is potentiated by glucose-6-phosphate, which induces a hexose phosphate transport pathway. The sodium salt is used for parenteral administration; for oral use, the earlier calcium salt has been superseded by the much more soluble trometamol formulation, which is particularly suitable for the treatment of urinary tract infection (Reeves 1994). Resistance to fosfomycin emerges readily in vitro, but the drug has been extensively used in some countries without apparent problem. Plasmid-mediated resistance has been reported (Suárez and Mendoza 1991), but is as yet uncommon. A somewhat similar phosphonic acid derivative, fosmidomycin, is available in Japan.

Ethambutol, isoniazid, and pyrazinamide

These are agents used specifically in mycobacterial disease; they have no useful activity against other bacteria. Although the mode of action has not been definitely established, it is likely that they act on the mycobacterial cell wall, which is unusual in its composition. Isoniazid is most active against *M. tuberculosis* against which it exerts bactericidal activity. It is well absorbed when given orally and is eliminated in the urine, largely in an acetylated form. In persons with a genetically determined ability to acetylate the drug rapidly, excretion is hastened and plasma concentrations correspondingly low. Unfortunately, resistance is an increasing problem (Pablos-Mendez et al. 1998; Warburton et al. 1993). Pyrazinamide is also mycobactericidal, but only at an acid pH and after intracellular conversion by a bacterial amidase to pyrazinoic acid (Mitchison 1992). Ethambutol, by contrast, is predominantly mycobacteristatic, but it has a wider spectrum of activity within the mycobacteria, including activity against organisms of the *Mycobacterium avium* complex.

INHIBITORS OF BACTERIAL PROTEIN SYNTHESIS

Many of the naturally occurring antibiotic families that were discovered by mass screening of soil samples in the 1940s and 1950s turned out to achieve their antibacterial effect by interfering with various stages of the process of protein synthesis. Some, but not all, owe their selective toxicity to the difference in structure between bacterial and mammalian ribosomes.

Aminoglycosides

STRUCTURE AND SPECTRUM OF ACTIVITY

Aminoglycosides are complex heterocyclic compounds that usually possess an aminocyclitol group in addition to one or more amino sugars. Spectinomycin is unusual in being a pure aminocyclitol and has properties, including predominantly bacteristatic activity, that separate it from the aminoglycosides proper. It is used only as a reserve agent in gonorrhea. The remaining aminoglycosides can be divided into two main types depending on the structure of the aminocyclitol ring: those (of which streptomycin and dihydrostreptomycin are the only surviving examples) in which the aminocyclitol is streptidine; and those in which the aminocyclitol moiety is deoxystreptamine. The deoxystreptamine-containing aminoglycosides can, in their turn, be divided into neomycin derivatives, kanamycin derivatives, and gentamicin derivatives (Figure 18.7).

The neomycins (including framycetin; neomycin B), are mainly used in topical preparations because of their toxicity, but one member of the group, paromomycin, is more conspicuous for its activity against the protozoa *Entamoeba histolytica* and *Leishmania* spp. Most of the remaining aminoglycosides in common therapeutic use, including gentamicin, tobramycin (deoxykanamycin B), netilmicin and the semisynthetic antibiotic, amikacin, share a common spectrum of activity that encompasses staphylococci and enterobacteria, but excludes streptococci, enterococci, anaerobes, and intracellular bacteria. Certain aminoglycosides exhibit activity against *M. tuberculosis* (streptomycin, kanamycin, and amikacin) or against *P. aeruginosa* (gentamicin, tobramycin, netilmicin, amikacin). Members of this last group have been widely used in the management of serious infection, often in combination with a β-lactam agent with which they may interact synergically. Bactericidal synergy with penicillin is also exploited in the therapy of streptococcal and enterococcal endocarditis.

MODE OF ACTION

Aminoglycosides bind to bacterial ribosomes. A single amino acid change in a protein of the 30S ribosomal subunit renders the cells completely resistant to streptomycin, but not to deoxystreptamine-containing aminoglycosides, which bind to both subunits. The mechanism of action is uncertain. Binding induces misreading of messenger RNA so that defective proteins are produced, but it is likely that aminoglycosides also interfere with the formation of functional initiation complexes. Neither of these mechanisms satisfactorily explains the potent bactericidal activity of these compounds and this may follow from membrane-related effects (Davis 1988).

RESISTANCE

Aminoglycosides are taken up by bacterial cells by an active transport process that involves respiratory processes. The absence of such a transport mechanism in anaerobes, streptococci, and enterococci accounts for the relative resistance of these organisms.

Alterations in permeability characteristics of sensitive bacteria may lead to relative resistance to aminoglycosides, and such bacteria are usually resistant to all members of the group. However, acquired resistance in otherwise sensitive species is most commonly due to the production of enzymes that adenylate or phosphorylate hydroxyl groups on the aminoglycoside molecule, or acetylate exposed amino groups. Since these antibiotics vary in the availability of vulnerable groupings on the molecule, they exhibit variable susceptibility to the many aminoglycoside-modifying enzymes that have been described. Some examples are shown in Table 18.5.

PHARMACOLOGICAL PROPERTIES

Aminoglycosides are very poorly absorbed by the oral route and are administered by intramuscular injection or intravenous infusion. They do not penetrate into cells or cross the blood–brain barrier to enter the cerebrospinal fluid. They are excreted almost entirely in the urine via the glomerular filtrate with an elimination half-life of about 2–4 h. Protein binding is generally low (<25 percent).

TOXICITY AND SIDE EFFECTS

All members of this antibiotic family are nephrotoxic and ototoxic, but they vary in their propensity to cause these adverse effects. For example, gentamicin and tobramycin are less likely to cause vestibular toxicity than is streptomycin, but are somewhat more likely to cause renal damage. Because of the risk of toxicity, it is common practice to assay aminoglycoside levels after intramuscular injection or intravenous infusion ('peak' level) and immediately before the next dose ('trough' level). If single large daily doses of aminoglycosides are used, as is increasingly advocated (Levison 1992), peak levels become difficult to interpret, but trough levels still need to be monitored (Nicolau et al. 1995).

Chloramphenicol

STRUCTURE AND SPECTRUM OF ACTIVITY

Chloramphenicol is a relatively simple antibiotic (Figure 18.8) that is nowadays synthesized rather than being obtained by fermentation from the producer organism. Modification of the molecule has not been very productive, although thiamphenicol, in which a sulphomethyl substituent replaces the nitro group, is available in some countries. Fluorinated derivatives,

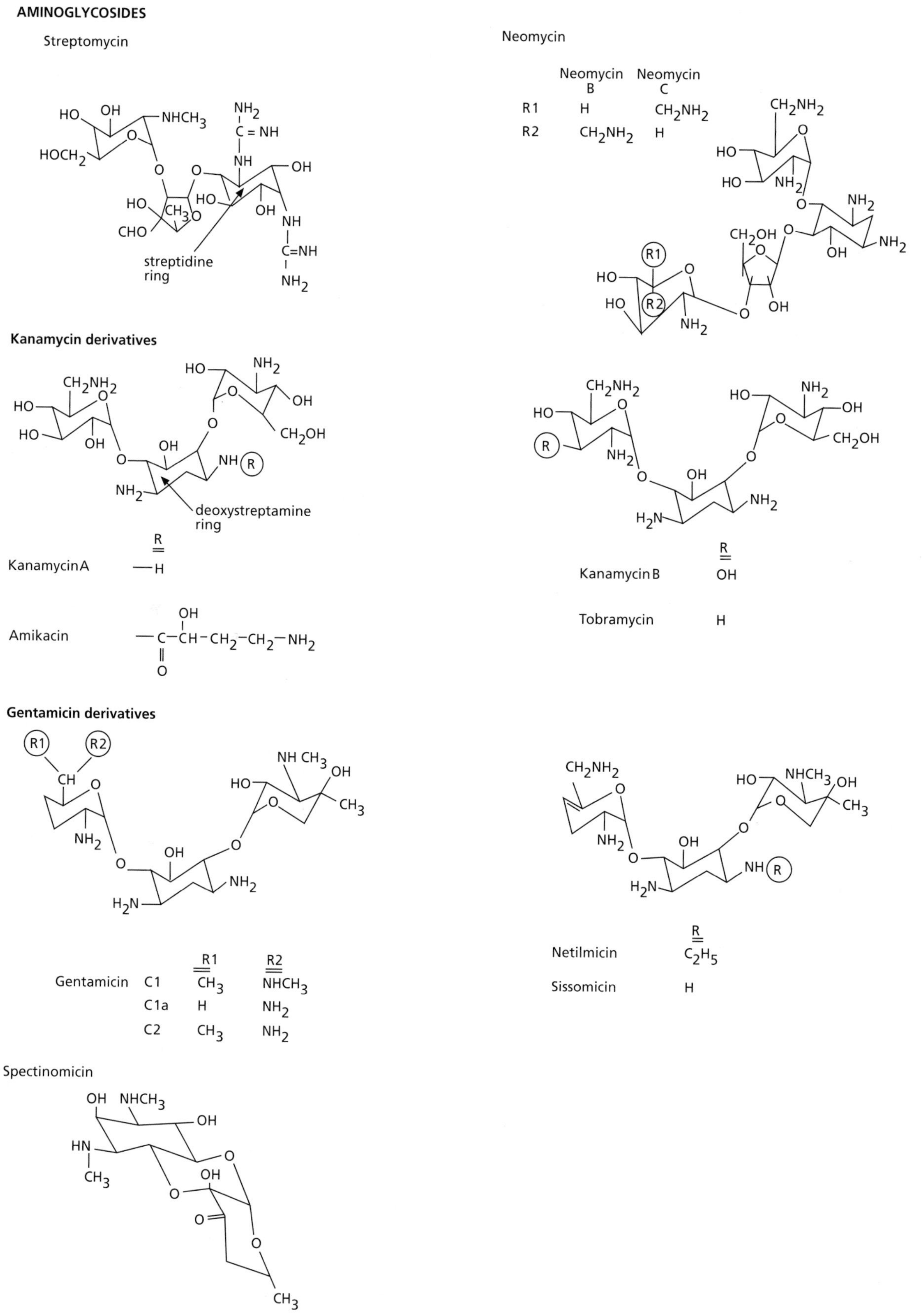

AMINOGLYCOSIDES

Streptomycin

Neomycin

	Neomycin B	Neomycin C
R1	H	CH$_2$NH$_2$
R2	CH$_2$NH$_2$	H

streptidine ring

Kanamycin derivatives

deoxystreptamine ring

	R
Kanamycin A	—H
Amikacin	—C(=O)—CH(OH)—CH$_2$—CH$_2$—NH$_2$

	R
Kanamycin B	OH
Tobramycin	H

Gentamicin derivatives

	R1	R2
Gentamicin C1	CH$_3$	NHCH$_3$
C1a	H	NH$_2$
C2	CH$_3$	NH$_2$

	R
Netilmicin	C$_2$H$_5$
Sissomicin	H

Spectinomicin

Figure 18.7 *Structures of the most important aminoglycosides.*

Table 18.5 *Some common aminoglycoside modifying enzymes*

Enzyme	Preferred substrates	Typical bacterial distribution	
		Gram-positive	Gram-negative
Acetyltransferases			
AAC(3)-I	Gen, Sis	−	+
AAC(3)-II	Gen, Kan, Net, Sis, Tob	−	+
AAC(2′)	Gen, Neo, Net, Sis, Tob	−	+
AAC(6′)-I	Amk, Net, Tob	+	+
AAC(6′)-II	Gen, Kan, Net, Sis, Tob	−	+
Nucleotidyltransferases (adenylyltransferases)			
AAD(6)	Str	+	−
AAD(4′)(4′′)	Amk, Kan, Neo, Tob	+	−
AAD(2′′)	Gen, Kan, Sis, Tob	−	+
AAD(3′′)(9)	Spc, Str	−	+
AAD(9)	Spc	+	−
Phosphotransferases			
APH(6)	Str	−	+
APH(3′)	Kan, Neo	+	+
APH(2′′)	Gen, Net, Sis, Tob	+	−
APH(3′′)	Str	+	+

Amk, amikacin; Gen, gentamicin; Kan, kanamycin; Neo, neomycin; Net, netilmicin; Sis, sissomicin; Spc, spectinomycin; Str, streptomycin.
The figures in parentheses indicate the sites of modification of exposed amino or hydroxyl groups according to the international numbering system for the heterocyclic ring structure of the aminoglycosides. Bifunctional enzymes that act at more than one site have two such numbers. Roman numerals indicate different forms of the enzyme acting at that site.

such as florphenicol, have also been made, but are not used in human medicine.

The spectrum is very broad and includes intracellular pathogens, such as chlamydiae and rickettsiae, as well as most conventional gram-positive and gram-negative bacteria. *P. aeruginosa* and *M. tuberculosis* are usually resistant.

MODE OF ACTION

Chloramphenicol inhibits the enzyme peptidyl-transferase which links new amino acids from aminoacyl transfer RNA to the growing peptide chain. It does not bind to mammalian ribosomes except, perhaps, those within mitochondria.

Chloramphenicol

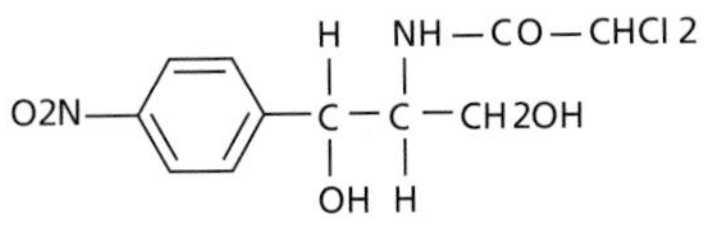

Thiamphenicol

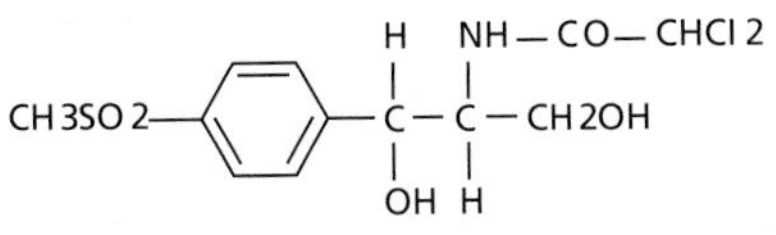

Figure 18.8 *Structures of chloramphenicol and thiamphenicol.*

RESISTANCE

Target site alterations and reduction in drug uptake have been described, but the most common form of resistance is due to the production of chloramphenicol acetyltransferases. These enzymes acetylate the C-3 hydroxyl group preferentially, but also attack the C-1 hydroxyl; interest in the fluorinated derivatives centers on their resistance to these enzymes. Inactivation of chloramphenicol by a nitroreductase has also been described in *B. fragilis* (Tally and Malamy 1984).

PHARMACOLOGICAL PROPERTIES

Because chloramphenicol is poorly soluble and tastes very bitter, prodrugs are used in therapy: water-soluble chloramphenicol succinate is used for injection and the more palatable chloramphenicol palmitate or stearate for oral administration. These salts release chloramphenicol in the body, but are themselves inactive and unsuitable for laboratory tests. The oral route of administration is very efficient and is generally preferred. The drug is well distributed, with excellent penetration into the cerebrospinal fluid. It is excreted into urine, largely as inactive glucuronide conjugates, with a half-life of about 2–5 h in adults. Plasma protein binding is approximately 50 percent.

TOXICITY AND SIDE EFFECTS

The popularity of this excellent antibiotic suffered a severe setback when it was realized that it occasionally caused an irreversible aplastic anemia. The overall incidence of this lethal side effect is about 1 in 40 000 courses of therapy, but there may be a genetic component that makes it more likely in some patients. There is concern that sufficient chloramphenicol may be absorbed systemically after topical use, e.g. in eye ointments, for aplastic anemia to be a hazard (Doona and Walsh 1995), but the degree of risk is disputed (Mulla et al. 1995). Thiamphenicol appears to be free of this side effect, but is more likely to cause reversible depression of the bone marrow.

Young infants have a limited capacity to conjugate and excrete chloramphenicol; accumulation of the drug may lead to 'gray baby' syndrome, with circulatory collapse. For this reason chloramphenicol levels should be assayed if the drug is used in neonatal meningitis or other life-threatening conditions in young infants.

Tetracyclines

STRUCTURE AND SPECTRUM OF ACTIVITY

The tetracyclines are a closely related group of naturally occurring and semisynthetic antibiotics that differ according to the nature of chemical substituents on the basic tetracyclic skeleton (Figure 18.9). They are distinguished more for differences in pharmacokinetic behavior than for variations in antimicrobial activity. The spectrum of activity is broad. They have been widely and successfully used in many types of infection including those in which chlamydiae, rickettsiae, and mycoplasmas are involved. They have some antiprotozoal activity and are sometimes used in drug-resistant malaria.

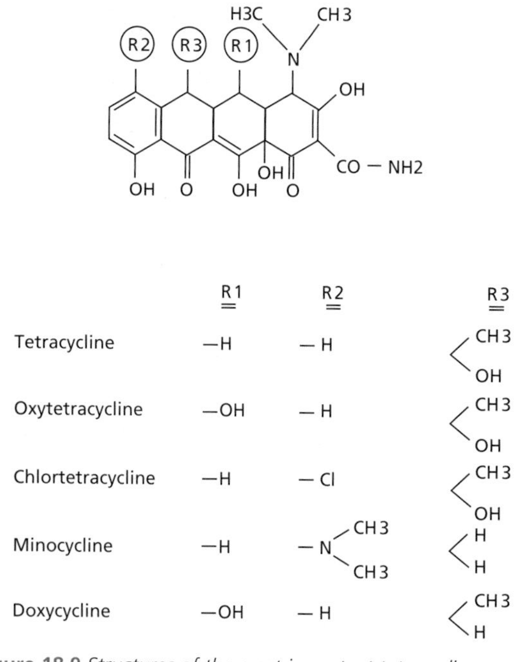

	R1	R2	R3
Tetracycline	—H	—H	CH3 / OH
Oxytetracycline	—OH	—H	CH3 / OH
Chlortetracycline	—H	—Cl	CH3 / OH
Minocycline	—H	—N(CH3)CH3	H / H
Doxycycline	—OH	—H	CH3 / H

Figure 18.9 *Structures of the most important tetracyclines.*

MODE OF ACTION

Tetracyclines enter bacterial cells by an active uptake process. They bind to the 30S ribosomal subunit and prevent access of aminoacyl transfer RNA to the acceptor site by a mechanism that has not been fully elucidated (Chopra et al. 1992). Mammalian cells do not concentrate tetracyclines in the way that bacteria do and the ribosomes (other than mitochondrial ribosomes) are relatively insusceptible.

RESISTANCE

The emergence of resistance in gram-positive and gram-negative bacteria has seriously undermined the value of tetracyclines although most epidemic strains of MRSA in the UK retain susceptibility. Resistance is generally plasmid mediated and is commonly found on transposons. Several mechanisms of resistance have been described, but the most prevalent appears to be due to production of a novel cytoplasmic membrane protein that mediates active efflux of the drug so that inhibitory levels are not maintained within the cell (Chopra et al. 1992; Speer et al. 1992).

Resistance usually affects all tetracyclines equally, although minocycline may retain activity against some strains. Efforts have been made to devise tetracyclines that overcome resistance mechanisms. Several candidate molecules have been described, including the glycylcyclines which are under investigation (Tally et al. 1995). One of these, tigecycline is in an advanced stage of development (Milatovic et al. 2003). It is a derivative of minocycline and offers useful activity for problem multiresistant gram-positive cocci such as MRSA, penicillin resistant pneumonia (PRP), and vancomycin-resistant enterococci (VRE).

PHARMACOLOGICAL PROPERTIES

The tetracyclines are usually given orally, but since they form nonabsorbable chelates with divalent cations, administration with milk or other food may interfere with absorption. Doxycycline and minocycline are among the best absorbed; these compounds also have a longer plasma half-life (approximately 16–18 h) than other congeners (generally approximately 6–12 h), but are more extensively (minocycline 75 percent; doxycycline 90 percent) protein bound. They penetrate well into tissues, but not into cerebrospinal fluid. They are excreted by glomerular filtration into the urine and into feces via the bile.

TOXICITY AND SIDE EFFECTS

Diarrhea and other forms of gastrointestinal intolerance are common. Renal failure may occur in patients who already have impaired renal function, but doxycycline, which is predominantly excreted by the hepatobiliary route, may be safely given to such patients. Since tetracyclines are yellow compounds that chelate calcium, the pigment is deposited in growing bone and teeth. For this reason, tetracyclines should not be given to young children when the dentition is being formed as permanent discoloration of teeth may occur.

Macrolides

STRUCTURE AND SPECTRUM OF ACTIVITY

The macrolides are a family of related compounds that feature a large macrocyclic lactone structure substituted with various unusual sugars (Figure 18.10). Erythromycin (the oldest and best known member of the group), oleandomycin, clarithromycin (6-O-methyl erythromycin), dirithromycin, and roxithromycin have a 14-membered lactone ring. Some others that are used in various parts of the world, including spiramycin, josamycin, midecamycin, kitasamycin, and rokitamycin, possess a 16-membered lactone structure. In azithromycin, the 14-membered structure of erythromycin has been expanded by insertion of a methyl-substituted nitrogen atom; the 15-membered ring thus produced is sometimes referred to as an azalide structure.

Erythromycin and other macrolides are unable to penetrate easily through the outer membrane of enteric gram-negative bacilli, although they exhibit a variable degree of activity against *H. influenzae*, *Legionella pneumophila*, and *Campylobacter jejuni*. They exhibit good activity against staphylococci and streptococci and useful activity against *Mycoplasma pneumoniae* (but not *M. hominis*) and some environmental mycobacteria. Chlamydiae are susceptible and azithromycin is effective in

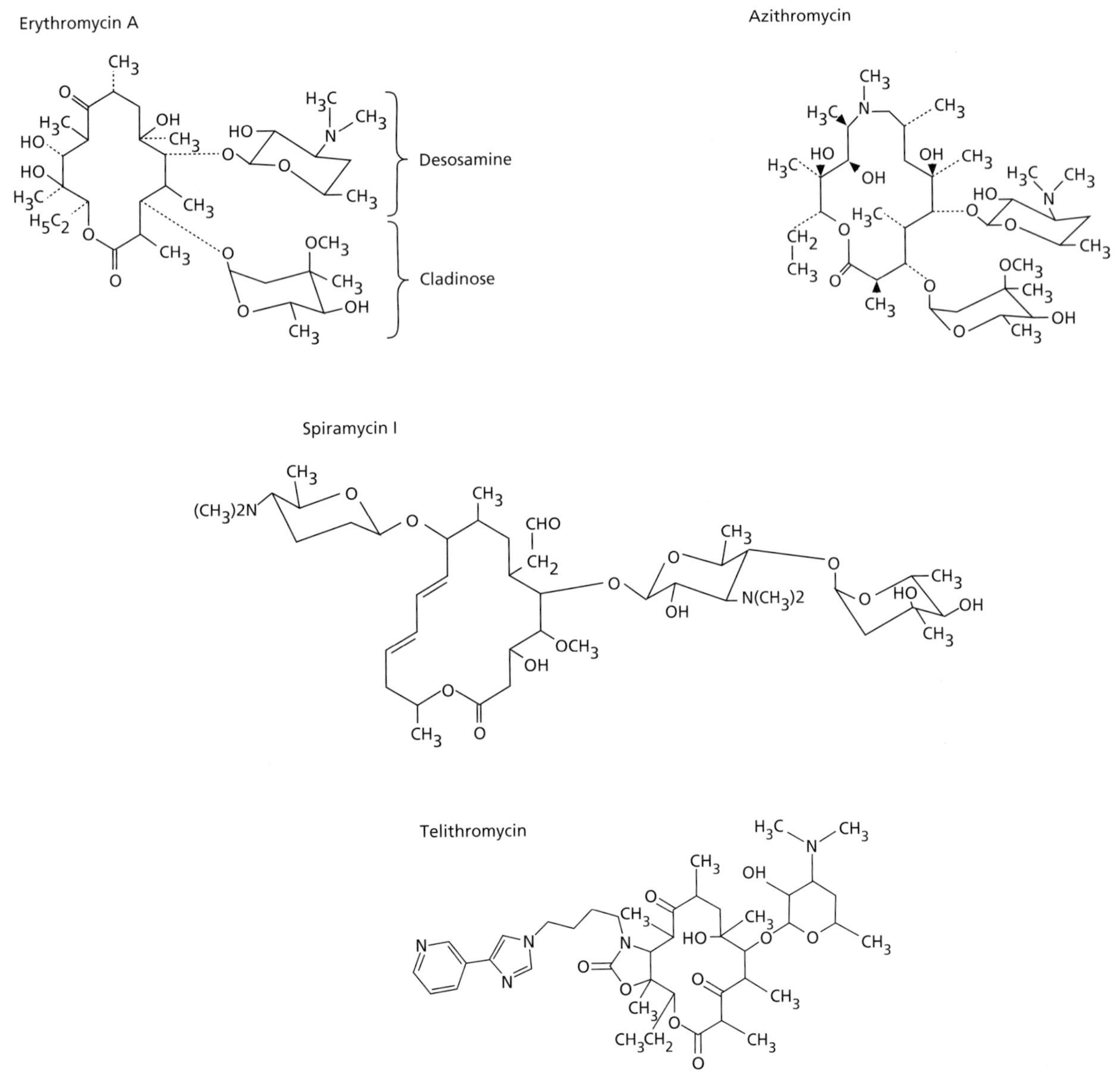

Figure 18.10 *Structures of erythromycin A (a 14-membered macrolide); azithromycin (a 15-membered macrolide and an azalide); spiramycin I (a 16-membered macrolide); and telithromycin.*

the single-dose treatment of genital infections (Martin et al. 1992; Stamm et al. 1995) and trachoma (Bailey et al. 1993). Spiramycin has been successfully used in toxoplasmosis and clarithromycin has a place in regimens for the eradication of *Helicobacter pylori*. The ketolide derivatives have undergone development recently and one of them, telithromycin is being used clinically. Its spectrum of activity and structure is very similar to clarithromycin, although it is bactericidal rather than static, like traditional macrolides (Fogarty et al. 2003).

MODE OF ACTION

Macrolides bind to 50S ribosomal subunits in bacteria and interfere with the translocation process during synthesis of polypeptides, probably by causing dissociation of peptidyl transfer RNA from the ribosome (Mazzei et al. 1993).

RESISTANCE

Resistance to macrolides is quite commonly encountered in staphylococci and streptococci but the prevalence varies considerably from country to country. Resistance may arise from alterations in ribosomal proteins. It is, though, more commonly due to an inducible, plasmid-mediated enzyme that methylates an adenine residue in ribosomal RNA (Weisblum 1995a) or alternatively, due to active efflux (Alós et al. 2003; Low 2002). The former is due to the *erm (B)* gene which exhibits the MLS_B phenotype, the latter to the *mef (A)* gene which exhibits the M phenotype. This latter mechanism tends to give lower levels of resistance although probably still clinically significant (Lonks et al. 2002). Both efflux and the methylase gene are widespread, but geographical distribution varies. Only 14- and 15-membered ring macrolides are susceptible to efflux but methylase also confers resistance to lincosamides and streptogramins when constituitively expressed although these antibiotics do not act as inducers, so the organisms appear sensitive in laboratory tests, unless erythromycin is also present to induce the enzyme. Variants readily emerge on exposure to lincosamides or streptogramins exhibiting constitutive resistance (Weisblum 1995b). It has been suggested that the newer long-acting macrolides are responsible for recent increases in resistance because of their lack of cidal activity and their prolonged but low tissue levels (Baquero et al. 2002). The ketolides, so far, retain activity against the great majority of erythromycin-resistant streptococci, although not MRSA (Fogarty et al. 2003).

PHARMACOLOGICAL PROPERTIES

Macrolides are mostly irregularly absorbed when given orally; in the case of erythromycin this is because it is unstable at gastric pH. Newer macrolides, such as azithromycin and clarithromycin, are more acid stable and their oral absorption is consequently much improved. Oral formulations of erythromycin are either coated to avoid destruction in the stomach, or are presented as prodrug salts or esters (estolate, stearate, or ethylsuccinate). The lactobionate or gluceptate salts are used for parenteral administration.

All these drugs are well distributed in the body, but azithromycin and clarithromycin are said to achieve particularly good intracellular levels, especially in the lung (Honeybourne and Baldwin 1992). Penetration into cerebrospinal fluid is poor. They are extensively metabolized (except azithromycin) and excreted largely by the hepatobiliary route. The plasma half-life is variable and may be dose-dependent. Azithromycin is unusual in that it has a much extended terminal half-life. The poor serum levels, particularly of azithromycin, make it less desirable for the treatment of bacteremic pneumococcal pnemonia (Lonks et al. 2002).

TOXICITY AND SIDE EFFECTS

Nausea and other gastrointestinal side effects are common, but newer derivatives, including azithromycin and clarithromycin, are better tolerated. Cholestatic jaundice is described as an uncommon complication, particularly with erythromycin formulations. The immunomodulatory and anti-inflammatory properties of macrolides are being investigated in cystic fibrosis and there is good evidence of their benefit in diffuse panbronchiolitis (Krishnan et al. 2002).

Lincosamides

STRUCTURE AND SPECTRUM OF ACTIVITY

The original lincosamide antibiotic, lincomycin, has been largely superseded by the 7-deoxy-7-chloro derivative, clindamycin (Figure 18.11), which is more active and better absorbed by the oral route. The most potent activity is against staphylococci, streptococci, and anaerobes, including *B. fragilis*. Enterobacteria and *P. aeruginosa* are resistant. Activity against *Propionibacterium acnes* and corynebacteria has led to the use of clindamycin in acne, but its success in this condition may be equally due to the anti-inflammatory and antiphagocytic activity that it is known to possess (Oleske and Phillips 1983) or to effects on skin lipids. Clindamycin possesses some antiprotozoal activity and has been used in malaria and toxoplasmosis.

MODE OF ACTION

The mechanism of action of lincosamides is not known for certain. Early work suggested that these drugs, like chloramphenicol, interfere with the peptidyltransferase reaction (Weisblum and Davies 1968), but later evidence pointed to effects on peptide chain initiation (Pestka 1971; Reusser 1975). Although the structure of lincosamides is completely different from that of macrolides,

Lincomycin

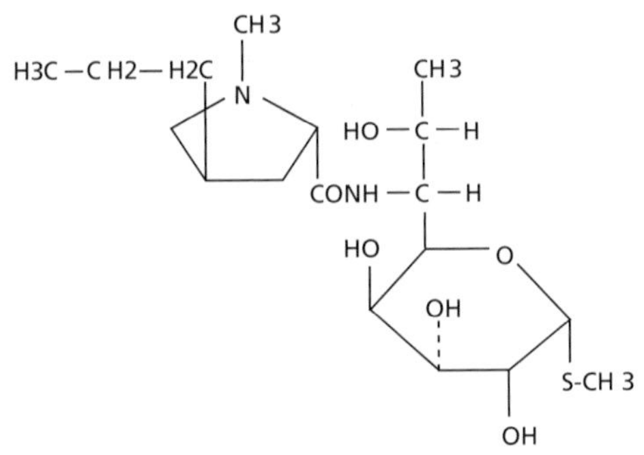

Clindamycin

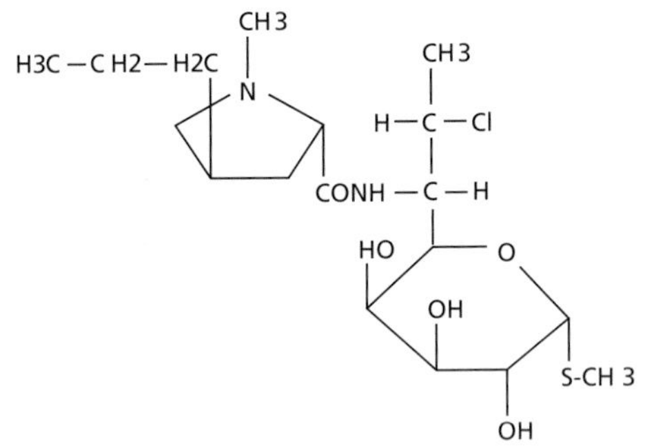

Figure 18.11 *Structures of the lincosamides, lincomycin and clindamycin.*

the site of action appears to be similar since methylation of an adenine residue in ribosomal RNA confers resistance to both types of antibiotic.

RESISTANCE

As well as resistance caused by ribosomal methylase, described above, resistance may also arise by enzymic modification by nucleotidylation (Russell and Chopra 1996). Resistance emerges readily and is common among methicillin-resistant *S. aureus* (Maple et al. 1989).

PHARMACOLOGICAL PROPERTIES

Like chloramphenicol, clindamycin is poorly soluble and very bitter, so that it has to be administered as inactive prodrugs, clindamycin palmitate (for oral use) and clindamycin phosphate (for injection). It is well absorbed when given orally and after hydrolysis, achieves good concentrations in tissues, but not in the cerebrospinal fluid. It is metabolized in the liver and excreted mainly in bile with a half-life of about 2–3 h. Plasma protein binding is 94 percent.

TOXICITY AND SIDE EFFECTS

Rashes and other occasional adverse events have been overshadowed by the reputation of clindamycin for

inducing diarrhea associated with *C. difficile* toxins, which may progress to a life-threatening pseudomembraneous colitis. Other antibiotics, including β-lactam agents, have been implicated in this side effect, but it seems to be more common with clindamycin (Tedesco et al. 1974; Aronsson et al. 1984).

Fusidic acid

STRUCTURE AND SPECTRUM OF ACTIVITY

Fusidic acid is an antibiotic with a steroid-like structure (Figure 18.12) that does not possess steroid-like activity. It exhibits modest activity against streptococci, gram-positive and gram-negative anaerobes, *Nocardia asteroides*, and *M. tuberculosis* and good activity against *Corynebacterium diphtheriae*, but its place in therapy hinges on its excellent activity against staphylococci (Greenwood 1988; Verbist 1990). Various interactions with penicillins (synergy, antagonism, and indifference) have been described (O'Grady and Greenwood 1973). Activity against the protozoa, *Giardia lamblia* (Farthing and Inge 1986) and *Plasmodium falciparum* (Black et al. 1985), have been reported. Enterobacteria and pseudomonads are resistant.

MODE OF ACTION

Unlike other inhibitors of bacterial protein synthesis, fusidic acid does not bind directly to the ribosome. It forms a stable complex with guanosine triphosphate and 'factor G', an elongation factor involved in the translocation of the growing peptide chain (Cundliffe 1972).

RESISTANCE

Resistance may be plasmid mediated, but is more commonly due to chromosomal mutation resulting in alteration in the factor G target. This type of resistance emerges readily in vitro, but the general prevalence of resistance in staphylococci has remained low (Shanson 1990). This may have been helped by the fact that

Fusidic acid

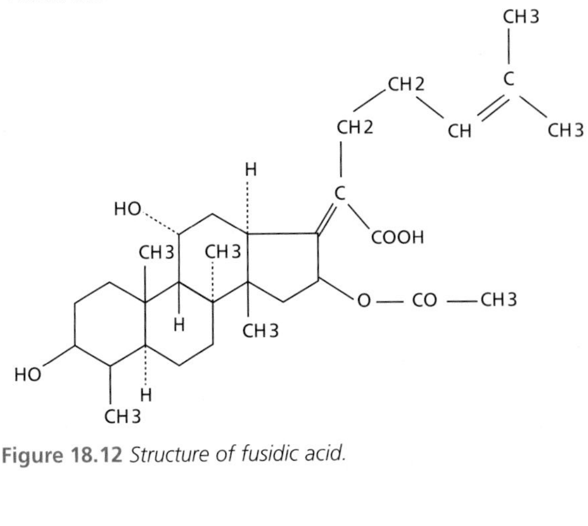

Figure 18.12 *Structure of fusidic acid.*

fusidic acid is usually given in combination (often with a penicillin) to avoid the emergence of resistance during therapy, but topical preparations are normally given alone and there have been recent concerns that increased resistance has been due to such topical use (Mason et al. 2003).

PHARMACOLOGICAL PROPERTIES

Fusidic acid is very well absorbed by the oral route and penetrates well into tissue, including bone. It is excreted via the bile, mostly as inactive metabolites, with a half-life of about 9 h (Reeves 1987; MacGowan et al. 1989). More than 95 percent is bound to plasma protein.

TOXICITY AND SIDE EFFECTS

A small proportion of patients develop transient jaundice by interference with the metabolism and excretion of bilirubin. It is more common in patients receiving intravenous fusidic acid (Humble et al. 1980). Mild gastrointestinal upsets and rashes are occasionally experienced.

Other inhibitors of bacterial protein synthesis

STREPTOGRAMINS

The streptogramins are a family of antibiotics that consist of two synergically interacting macrolactone components: a polyunsaturated peptolide and a hexadepsipeptide. The bactericidal synergy that results from the combined action of the two components of streptogramins is thought to arise through binding to adjacent sites on the bacterial ribosome (Aumercier et al. 1992). The best known members of the group are pristinamycin and virginiamycin, which are used as antistaphylococcal agents in animal husbandry and, in some countries, in human medicine. Lack of aqueous solubility and poor oral absorption have limited their value, but water-soluble derivatives of the two components of pristinamycin (known, respectively, as quinupristin and dalfopristin) have been synthesized to allow parenteral administration and are now available as a combination for parenteral administration in many countries (Barrière et al. 1992; Etienne et al. 1992) (Figure 18.13). Their broad gram-positive spectrum and bactericidal activity ensures their increasing use in difficult to treat cases due to MRSA, PRP, and VRE, but it is only static against *E. faecium* and *E. faecalis* is innately resistant (Eliopoulos 2003). Synergy with glycopeptides and other drugs has been demonstrated (Moyenuddin et al. 2003). Macrolide–lincosamide resistance associated with an adenine methylase also extends to streptogramins (see section on macrolide resistance).

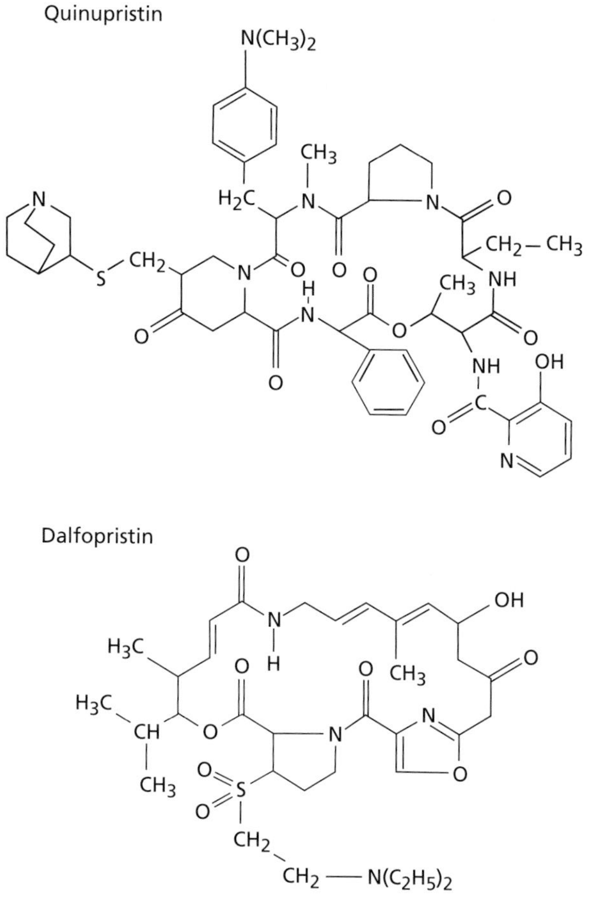

Figure 18.13 *Structure of quinupristin and dalfopristin.*

OXAZOLIDINONES

This group of drugs are being developed in response to clinical need for new agents active against MRSA and other multiresistant gram-positive organisms. Currently only one, linezolid, has been released for general clinical use (Clemett and Markham 2000). These drugs are the first genuinely new class of antibiotics to be successfully developed for many years (Figure 18.14). They act at the messenger RNA level where they are thought to interfere with protein synthesis.

As a result of this there is an interesting association with macrolide resistance. To date, when linezolid resistance develops in MRSA it is associated with a reversion to macrolide susceptibility. Development of resistance is, however, very rare to date and associated with long courses of therapy (Wilson et al. 2003; Boo et al. 2003).

Although linezolid is not bactericidal against MRSA, several preliminary reports of randomized studies show it to be more efficacious than glycopeptide monotherapy for serious MRSA infections such as bacteraemia and ventilatory association pneumonia (Cepeda et al. 2004). It seems well tolerated in general, although long-term use has been associated with marrow suppression and neuropathology. There are also some dietary restrictions to be adhered to when it is used because of its inhibitor

Linezolid

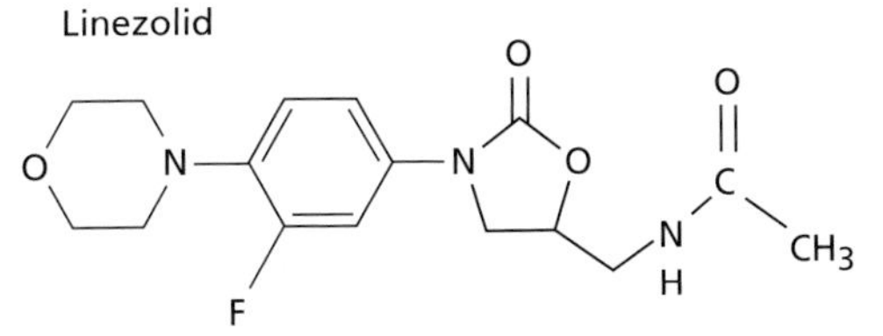

Figure 18.14 *Structure of linezolid.*

of monoamine oxidase. Currently linezolid should be reserved for serious MRSA infections where conventional therapy has failed. Caution should be exerted in its long-term use, although its excellent bioavailibility makes it a good candidate for oral switch. In vitro studies suggest it is often additive or synergistic in combination with other anti-MRSA drugs and consideration should be given to combinations, e.g. with rifampin in high inoculum or recalcitrant infections (Grohs et al. 2003; Cédrick et al. 2003).

MUPIROCIN

Mupirocin is a naturally occurring antibiotic product of *Pseudomonas fluorescens*. It was formerly known as pseudomonic acid. The structure (Figure 18.15) is unrelated to other antibiotics and consists of 'monic acid', the distal portion of which is a structural analog of isoleucine and a short fatty acid (nonanoic acid). The useful activity is restricted to gram-positive cocci, in which it presumably halts protein synthesis by binding to isoleucyl transfer RNA synthetase as it does in *E. coli* (Hughes and Mellows 1980).

Although it displays low toxicity, systemic use of mupirocin is precluded by the fact that it is rapidly inactivated in the body. Consequently, it is used only for topical application in skin infections and to eradicate staphylococci from nasal carriers. For the latter purpose, which has been of particular value in carriers of methicillin-resistant *S. aureus*, a paraffin-based formulation is used, rather than the polyethylene glycol-based ointment applied to skin.

INHIBITORS OF BACTERIAL NUCLEIC ACID SYNTHESIS

A surprising number of antibacterial agents achieve their effect by interacting with bacterial DNA in various ways. Sulfonamides and trimethoprim have an indirect

Mupirocin

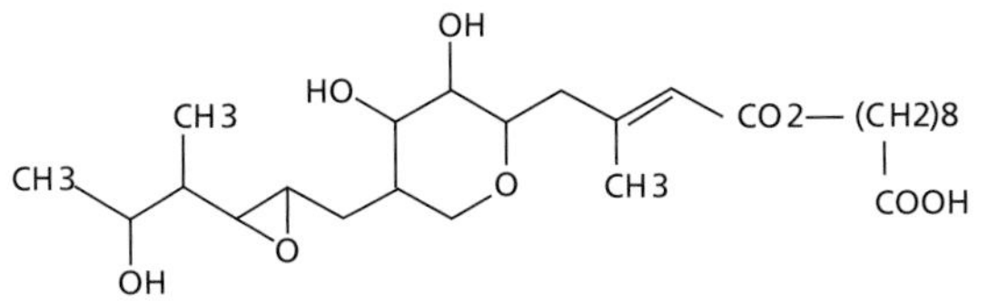

Figure 18.15 *Structure of mupirocin.*

effect on other cellular functions, as well as DNA, through their effect on folic acid synthesis, but for convenience are dealt with under this heading.

Sulfonamides

STRUCTURE AND SPECTRUM OF ACTIVITY

The sulfonamides are a large family of compounds, all of which are derived from the original hydrolysis product of Prontosil red, sulfanilamide. They differ in the nature of the substitution on the amino group of the sulfonamide (SO_2NH_2) moiety (Figure 18.16). The antileprosy drug, dapsone (diaminodiphenylsulphone) and the tuberculostatic *p*-aminosalicylic acid are related substances that are thought to act in a similar way.

They are broad spectrum, predominantly bacteristatic compounds that have a relatively slow effect in halting bacterial growth. Activity in vitro is profoundly affected by the composition of the culture medium, because of the possible presence of interfering substances such as

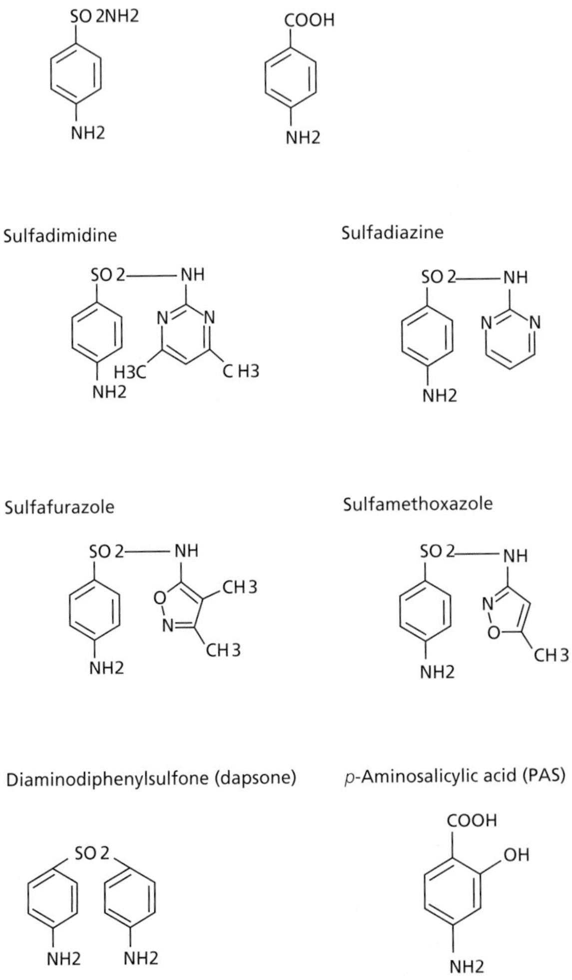

Figure 18.16 *Structures of some sulfonamides and related compounds.*

folic acid, *p*-aminobenzoic acid, and thymidine. Lysed horse blood, which contains the enzyme thymidine phosphorylase, is commonly added to remove thymidine (Waterworth 1978) and media recommended for sensitivity testing have low levels of this and other sulfonamide antagonists (Report 1991).

MODE OF ACTION

Sulfonamides are analogs of *p*-aminobenzoic acid. They inhibit folic acid synthesis by competitive inhibition of dihydropteroic acid synthetase, the enzyme that brings about the condensation of dihydropteridine with *p*-aminobenzoic acid in the early stage of folate production. Since folic acid is conserved in bacterial cells, the inhibitory effects of sulfonamides become apparent only after several generations of growth when the folate pool has been progressively diluted to below a functional level by distribution to the bacterial progeny.

The selective toxicity of sulfonamides arises because bacteria synthesize folic acid de novo, whereas humans absorb the vitamin preformed. Since they block an early stage of the same metabolic pathway as diaminopyrimidines, sulfonamides interact synergically with those compounds.

RESISTANCE

Resistance to sulfonamides occurs readily and was soon apparent in the early life of these agents. There is complete crossresistance among different members of this drug class. Resistance is commonly plasmid mediated and is usually caused by alterations in dihydropteroate synthetase, leading to less efficient binding of sulfonamides, or bypass of the effects of the agents by a duplicate, insensitive version of the enzyme (Huovinen et al. 1995). Chromosomal resistance due to hyperproduction of *p*-aminobenzoic acid is also recognized (Towner 1992a).

PHARMACOLOGICAL PROPERTIES

Differences in pharmacokinetic behavior are the chief distinguishing characteristics of the various members of the sulfonamide family. In particular, they vary in oral absorption (e.g. phthalylsulfathiazole is very poorly absorbed), protein binding (e.g. sulfadimidine and sulfadoxine are more than 90 percent protein bound) and, above all, plasma half-life, which can vary from about 2.5 h (sulfamethizole) to >100 h (sulfadoxine). The most commonly used sulfonamides, sulfadiazine and sulfamethoxazole (as co-trimoxazole) are well absorbed by the oral route and are excreted into urine, partly as inactive *N*-acetylated metabolites and glucuronide conjugates, with a half-life of about 8–10 h. They are well distributed and penetrate into the cerebrospinal fluid in effective concentrations.

TOXICITY AND SIDE EFFECTS

Crystalluria, with renal blockage, is a problem with some of the less soluble compounds, including sulfadiazine and sulfathiazole, especially if excessive dosage is used. Rashes are common and erythema multiforme (Stevens–Johnson syndrome) is a rare, but potentially life-threatening complication. Serious hematological effects are also seen occasionally. Fears of toxicity have led to major reductions in their use in the UK over the past decade, although this has not been the case in most other countries. Their activity (alone and in combination with trimethoprim) against most epidemic strains of MRSA may well see them being used increasingly in the future.

Diaminopyrimidines

STRUCTURE AND SPECTRUM OF ACTIVITY

This family of synthetic pyrimidine derivatives includes trimethoprim, the most familiar member of the group (Figure 18.17), and the closely related antibacterial agents, tetroxoprim and brodimoprim which are in use in some countries, but offer little, if any, advantage over trimethoprim. Also related are the antimalarial agents pyrimethamine and cycloguanil (the in vivo metabolite of proguanil), the antipneumocystis agent, trimetrexate and the anticancer drug, methotrexate.

Trimethoprim and its congeners are active against many gram-positive and gram-negative bacteria, but *P. aeruginosa*, *B. fragilis*, chlamydiae, rickettsiae, mycoplasmas, and mycobacteria are outside the spectrum. Although they exhibit excellent activity in their own right, they are often formulated with sulfonamides, with which they interact synergically, at least in vitro. Among such combination products are: co-trimoxazole (trimethoprim + sulfamethoxazole); co-trimazine (trimethoprim + sulfadiazine); co-trifamole (trimethoprim + sulfamoxole), and co-tetroxazine, (tetroxoprim + sulfadiazine).

Trimethoprim

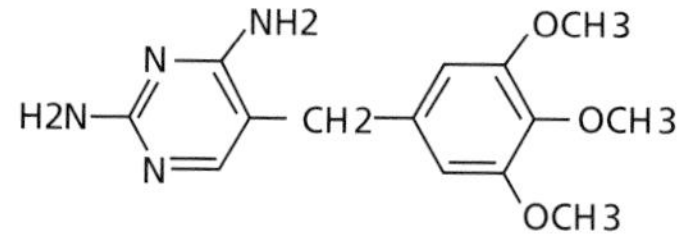

Tetroxoprim

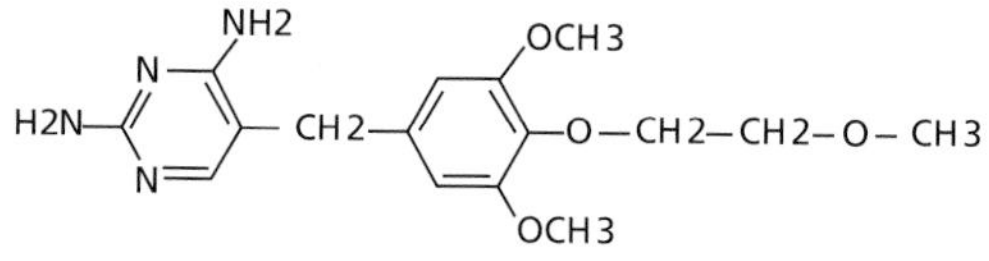

Figure 18.17 *Structures of trimethoprim and tetroxoprim.*

MODE OF ACTION

Diaminopyrimidines act on the same metabolic pathway as sulfonamides, but at a later stage. They inhibit dihydrofolate reductase, the enzyme that converts the precursor form of folic acid, dihydrofolate, to the active cofactor, tetrahydrofolic acid. The affinity of trimethoprim for bacterial dihydrofolate reductase is several thousand times greater than for the corresponding human enzyme. Tetrahydrofolic acid is an essential carrier molecule in many single-carbon transactions within cells and is usually regenerated unchanged. However, in one such transfer, the production of thymidylic acid from deoxyuridylic acid, diaminopyrimidines act as hydrogen donors, as well as methyl-group carriers and emerge from the reaction in the oxidized form, dihydrofolate. Trimethoprim and its relatives prevent regeneration of tetrahydrofolic acid and thus trap the vitamin in the unusable precursor form. Consequently, these drugs influence bacterial growth more quickly than sulfonamides, which rely on dilution of the folate pool to achieve their bacteristatic effect. In the presence of sufficient diaminopyrimidine to completely halt folate activity, the sulfonamides in combined formulations have no opportunity to interfere with bacterial growth, although they still contribute substantially to the overall toxicity of the mixtures.

RESISTANCE

The prevalence of resistance to trimethoprim has steadily increased since its introduction in 1969. Several mechanisms are recognized, the most common of which is attributable to mutations that lead to the production of altered dihydrofolate reductases (Towner 1992a; Huovinen et al. 1995).

PHARMACOLOGICAL PROPERTIES

Trimethoprim can be given by the oral and parenteral routes. It is widely distributed in tissues, including bronchial secretions. Concentrations achieved in cerebrospinal fluid are about 30–50 percent of the corresponding plasma level. Excretion is almost entirely renal, partly as metabolites, some of which retain antibacterial activity. The plasma half-life is about 10 h and it is less than 50 percent protein bound.

TOXICITY AND SIDE EFFECTS

Trimethoprim is well tolerated and most of the side effects of co-trimoxazole are usually attributable to the sulfonamide component. Nevertheless, trimethoprim itself may give rise to idiosyncratic reactions. The potential for exacerbating folate deficiency can normally be countered with folate supplements, but the drug is not recommended in pregnancy.

Quinolones

STRUCTURE AND SPECTRUM OF ACTIVITY

The quinolones are a large family of compounds (Table 18.6), the molecular similarity of which is based on the quinolone nucleus, or related naphthyridine, cinnoline or pyridopyrimidine structures (Smith and Lewin 1988) (Figure 18.18). The first member of the group to be used in therapy, nalidixic acid, is a naphthyridine derivative with a narrow spectrum of activity directed almost exclusively against enterobacteria. With the addition of a fluorine atom at C-6 and a piperazine substituent at C-7, the intrinsic activity and spectrum were substantially altered. Norfloxacin is about 50 times more active than nalidixic acid against sensitive enterobacteria and also exhibits some activity against gram-positive cocci and *P. aeruginosa*. Other derivatives with the 6-fluoro-7-piperazinyl substitutions (now collectively called fluoroquinolones) followed. Ciprofloxacin and ofloxacin offer further improved activity, and recent derivatives such as levofloxacin, moxifloxacin, gatifloxacin, and gemifloxacin have been developed with improved antipneumococcal activity, in response to the world-wide epidemic of PRP. Unfortunately, many other agents in development have dropped by the wayside because of potential neuro, cardiac, and photo-associated toxicity. Certain closely related compounds, including enrofloxacin and sarafloxacin are used in veterinary practice.

Some of the more active quinolones, including ciprofloxacin and ofloxacin, exhibit activity against chlamydiae (Oriel 1989) and mycobacteria (Garcia-Rodriguez and Gomez Garcia 1993), but this does not always translate into therapeutic success (Oriel 1989; Young 1993).

The activity is variably affected by pH (the activity usually being reduced at acid pH values) and by the presence of magnesium and other cations (Smith and Lewin 1988).

MODE OF ACTION

The primary site of action of quinolones is DNA gyrase (topoisomerase II), the remarkable enzyme that engineers the breaking and rejoining of supercoiled DNA. The action mainly involves the DNA gyrase A subunit, although the B subunit can also be affected by some quinolones (Hooper and Wolfson 1989).

RESISTANCE

Resistance to earlier quinolones of the nalidixic acid type occurs readily by chromosomal mutation and restricts the value of these compounds in the treatment of complicated urinary tract infection (Greenwood and O'Grady 1977). Such resistance also affects newer fluoroquinolones, but the reduction of activity is such that variants resistant to nalidixic acid are still inhibited by concentrations achievable therapeutically. Early

Table 18.6 *Proposed mechanisms of fluroquinolone resistance, by organism*

Class of organisms, organism	Drug-target mutations in DNA gyrase or topoisomerase IV				
	Primary target		Secondary target enzyme		
	Altered enzyme subunit	Quinolone affected[a]	Altered enzyme subunit	Quinolone(s) affected[a]	Active efflux[b]
Gram-positive					
Streptococcus pneumoniae	ParC	Cpfx, Lvfx, Nrfx, Tvfx	Par C	Spfx	Yes
	GyrA	Spfx, Gtfx	—	—	—
	ParE	Cpfx	GyrA, GyrB	Cpfx	_[c]
Staphylococcus aureus	GyrA	All	GyrA, GyrB	All	Yes
Enterococcus faecalis	GyrA	All	—	—	Yes
Gram-negative					
Escherichia coli	GyrA	All	GyrB, ParC, ParE	All	Yes
Salmonella	GyrA	All	GyrB	All	_[c]
Klebsiella	GyrA	All	ParC	All	Yes
Pseudomonas aeruginosa	GyrA	All	GyrB	All	Yes
Neisseria gonorrhoeae	GyrA	All	ParC	All	_[c]
Campylobacter	GyrA	All	—	—	Yes
Helicobacter pylori	GyrA	All	—	—	—
Mycobacterium sp.	GyrA	All	GyrB	All	Yes

Cpfx, ciprofloxacin; Gtfx, gatifloxacin; Lvfx, levofloxacin; Nrfx, norfloxacin; Spfx, sparfloxacin; Tvfx, trovafloxacin.

a) Antimicrobials to which the amino acid alterations in the enzyme confer resistance. For many bacteria, the basic resistance mechanism applies for all quinolone agents. However, in some bacteria, such as *S. pneumoniae*, the resistance mechanism will vary depending on the type of quinolone, as indicated here for Cpfx and Spfx.

b) Several different types of active efflux mechanisms exist; these data indicate only whether an active efflux mechanism of resistance has been found in a bacterial organism.

c) No or unknown (after Goldstein and Garbaedian-Ruffalo 2002).

optimism that this, together with the presence of only two convincing reports of plasmid-mediated resistance (Jacoby et al. 2003; Martínez-Martínez et al. 2003) would reduce the likelihood of resistance to fluoroquinolones becoming widespread, have not been realized. As might be expected, the emergence of resistance is particularly common in bacteria, such as *P. aeruginosa*, *Campylobacter jejuni*, and *S. aureus*, for which the minimum inhibitory concentration of fluoroquinolones does not greatly exceed therapeutically achievable concentrations (Wolfson and Hooper 1989; Peterson 1994; Wiedemann and Heisig 1994). Resistance is usually due to sequential mutations in genes for DNA gyrase, but may also follow from alterations in drug accumulation, sometimes associated with alterations in outer-membrane proteins of gram-negative bacilli (Wolfson and Hooper 1989). Early quinolones, which are innately less active, select well for first-step mutations, as these lead to resistance. Although new quinolones remain active, their MICs are raised, making second-step mutations more likely to occur, giving resistance to these new agents. Consequently, there have been calls to stop using these older agents, but this is probably not realistic because of cost. Also, it suggests a narrow view of the effects of antibiotics to make such calls. The targeted pathogen will not be the only bacteria exposed to the new agents – of necessity the whole bacterial flora will be exposed with unknown consequences in terms of resistance. An outbreak of a nalidixic acid resistant *S. dysenteria* with intermediate susceptibility to cipro-floxacin has recently been reported (Sarkar et al. 2003).

PHARMACOLOGICAL PROPERTIES

Quinolones are generally well absorbed when given orally, but some fluoroquinolones, including cipro-floxacin and levofloxacin, are also available in parenteral formulations. Nalidixic acid and the early congeners are extensively metabolized in vivo and sufficient unchanged drug to achieve an antibacterial effect is found only in urine. Fluoroquinolones are, in general, less susceptible to metabolic changes and since they are well distributed in the body including bile, they are of value in many systemic infections. Plasma elimination half-lives of the fluoroquinolones are prolonged, allowing once or twice daily dosing. Protein binding is low.

TOXICITY AND SIDE EFFECTS

They are generally well tolerated, although rashes, gastrointestinal upsets, and photosensitivity may occur and there are persistent reports of occasional neurotoxic side effects (Hooper and Wolfson 1991; Midtvedt and Greenwood 1994). The observation that

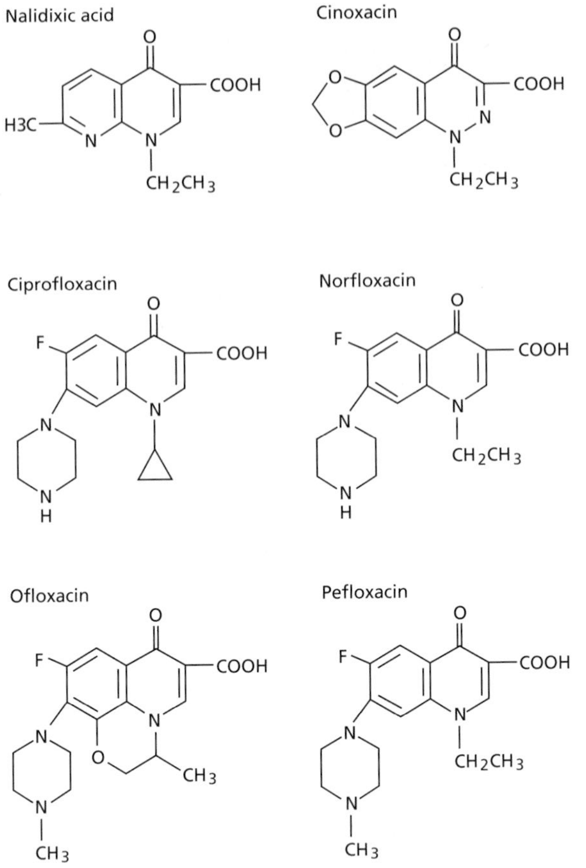

Nalidixic acid

Cinoxacin

Ciprofloxacin

Norfloxacin

Ofloxacin

Pefloxacin

Figure 18.18 *Structures of some quinolones.*

Rifampicin

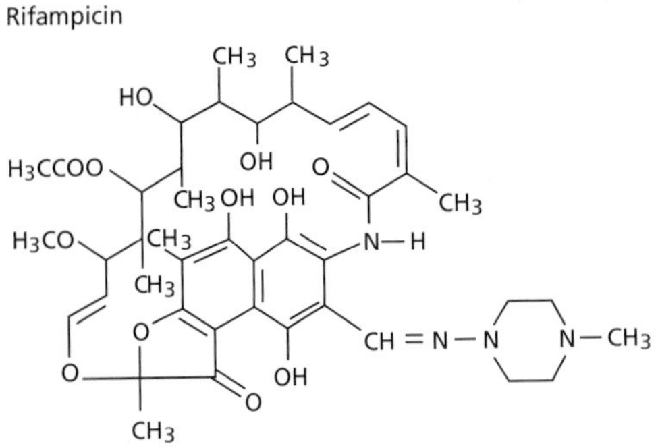

Figure 18.19 *Structure of rifampin.*

MODE OF ACTION

Rifampin and other rifamycins bind to the subunit of DNA-dependent RNA polymerase and prevent initiation of RNA synthesis (Wehrli and Staehelin 1971).

RESISTANCE

Resistance caused by mutational alterations in the target enzyme arises readily and may be a cause of treatment failure if rifampin is used alone. In tuberculosis and leprosy, in which rifampin is normally used in combination with other drugs, resistance is presently uncommon, although multiply-resistant strains of *M. tuberculosis* are causing concern in some places (Warburton et al. 1993; Young 1993; Pablos-Mendez et al. 1998).

PHARMACOLOGICAL PROPERTIES

Rifampin is very well absorbed when given by mouth, but the bioavailability of oral rifabutin is variable (Skinner et al. 1989). Both drugs are well distributed and excreted primarily by the hepatobiliary route, but also renally. They impart a red color to the urine and to tears, so that they may discolor soft contact lenses. They are potent inducers of liver enzymes, which promote self-metabolism of the antibiotics, as well as that of other drugs, including oral contraceptives and warfarin. There is significant binding to plasma proteins (approximately 70 percent). The terminal half-life of rifabutin (approximately 36 h) is much longer than that of rifampin (approximately 3.5 h).

TOXICITY AND SIDE EFFECTS

Rifampin may cause sensitization when used intermittently, as in some recommended regimens for the treatment of tuberculosis. Gastrointestinal upsets, jaundice, and (particularly with rifabutin) hematological effects are recognized. Induction of microsomal liver enzymes may lead to important interactions with other drugs, including failure of oral contraception.

fluoroquinolones can cause arthropathy in young experimental animals has led to a recommendation that these drugs should be avoided in young children and in women of child-bearing age. Cases of Achilles tendinitis, sometimes with rupture of the tendon, have been reported (Ribard et al. 1992).

Rifamycins

STRUCTURE AND SPECTRUM OF ACTIVITY

The rifamycins that are most widely used clinically, rifampin (Figure 18.19) and rifabutin (also known as ansamycin), are semisynthetic derivatives of the naturally occurring antibiotic, rifamycin B. Rifampin exhibits good activity against gram-positive and gram-negative cocci, particularly staphylococci, but is more noted for its antimycobacterial activity. Gram-negative bacilli are much less susceptible. Rifabutin has been introduced specifically because of its activity against organisms of the *M. avium* complex. Rifampin is seeing increased use in combination therapy for MRSA, particularly where medical treatment of prosthetic implant infection is being attempted, as it is thought to retain bactericidal activity in this setting (Widmer 2001).

Nitroimidazoles

STRUCTURE AND SPECTRUM OF ACTIVITY

Azole derivatives of various kinds have wide-ranging antimicrobial activities against fungi, protozoa, and helminths, as well as bacteria. Some may have a role as radiosensitizing agents in cancer therapy. Those that exhibit useful antibacterial activity are 5-nitroimidazoles. Only two, metronidazole and tinidazole (Figure 18.20), are used in human medicine in the UK, but others, such as nimorazole, ornidazole, and secnidazole, are available elsewhere and yet more are used in veterinary practice. They are primarily antiprotozoal agents, but also exhibit potent activity against anaerobic bacteria. Metronidazole offers a useful alternative to vancomycin in the treatment of *C. difficile*-associated colitis.

Microaerophiles and oxygen-tolerant species, such as *Actinomyces* spp. and *Propionibacterium* spp., are mostly insensitive (Greenwood et al. 1991), though metronidazole is used successfully in infections with *H. pylori* and *Gardnerella vaginalis*.

MODE OF ACTION

The narrow spectrum of activity of 5-nitroimidazoles arises because the antibacterial effect is dependent on reduction of the nitro group under anaerobic conditions. The compounds capture electrons from reduced ferredoxin generated in the course of the decarboxylation of pyruvate by the pyruvate–ferredoxin oxidoreductase complex. The short-lived reduction product kills the cell, probably by inducing breaks in the DNA strands (Edwards 1993a). The anomalous susceptibility of certain microaerophiles remains unexplained, although there is evidence that it is related to unusual metabolic pathways in these organisms (Smith and Edwards 1995).

RESISTANCE

Resistance to metronidazole and other 5-nitroimidazoles remains very uncommon despite the widespread use of these compounds. Some reports have subsequently been shown to have been erroneous because sensitivity tests were carried out under inadequately anaerobic conditions. There are, nevertheless, well documented accounts of resistance in clinical isolates of *B. fragilis* and other anaerobes, usually associated with a decreased ability to reduce the drug (Edwards 1993b).

PHARMACOLOGICAL PROPERTIES

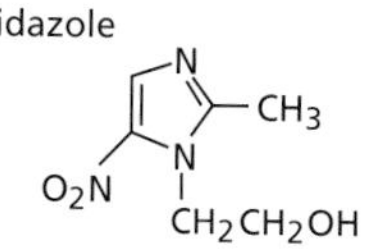

Metronidazole is usually administered by mouth, though intravenous, suppository, and topical preparations are also available for various purposes. Tinidazole is available in the UK only in tablet form. They are virtually completely absorbed by the oral route and are widely distributed, including into cerebrospinal fluid. Plasma protein binding is negligible. Tinidazole has a somewhat longer half-life than metronidazole (12–14 h versus 8–10 h). Both are metabolized and excreted chiefly into urine, partly as glucuronide conjugates.

TOXICITY AND SIDE EFFECTS

Nausea and abdominal cramps occur quite frequently; various other side effects have been reported occasionally. Patients often complain of dryness in the mouth and a metallic taste. Alcohol should be avoided because of a disulfiram-like reaction. Fears that the effect of metronidazole on DNA might lead to mutagenic or teratogenic effects in humans have not been borne out in practice (Morgan 1978; Beard et al. 1979). It is, nonetheless, considered prudent to avoid using these drugs during pregnancy, particularly during the first trimester.

Nitrofurans

STRUCTURE AND SPECTRUM OF ACTIVITY

Various nitrofuran derivatives are in use around the world as antibacterial agents, including furazolidone and nifuratel (which are said to possess antiprotozoal, as well as antibacterial activity) and nitrofurazone. However, the most widely used agent of this type, and the only one available in the UK, is nitrofurantoin (Figure 18.21), which is used as a urinary antiseptic. Nitrofurantoin is bactericidal to most urinary pathogens at concentrations achievable in urine, although activity against *Proteus* spp. is unreliable, partly because the drug is less active in the alkaline conditions produced by urea-splitting organisms. Many strains of MRSA are susceptible.

MODE OF ACTION

Like the nitroimidazoles, the nitrofurans are susceptible to nitroreductases, though in this case reduction takes place in an aerobic environment. The most likely explanation for the bactericidal action of these drugs is, therefore, that a reduced intermediate causes DNA strand breakage in a manner analogous to that of nitroimidazoles. This suggestion has the attraction of accounting for the known mutagenic effects of these compounds in vitro (McCalla 1977). However, more recently it has been suggested that reactive nitrofurantoin metabolites interfere not with DNA but with protein synthesis

Metronidazole

Tinidazole

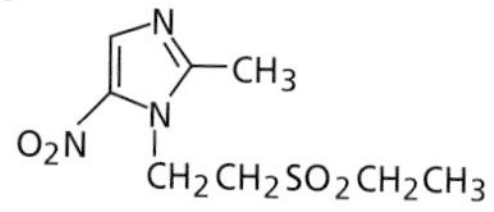

Figure 18.20 *Structures of metronidazole and tinidazole.*

Nitrofurantoin

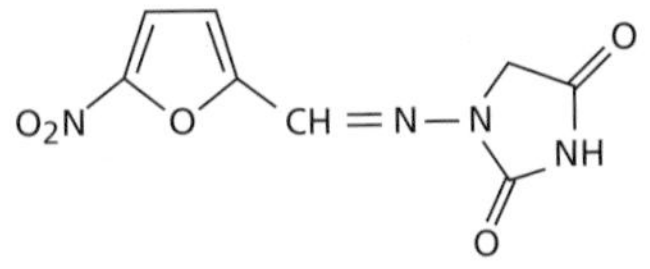

Figure 18.21 *Structure of nitrofurantoin.*

(McOsker and Fitzpatrick 1994). These authors also suggest an alternative mechanism of action that does not depend on bacterial nitroreductases.

RESISTANCE

Acquired resistance in susceptible bacterial species is uncommon and, even when it does occur, is rarely plasmid mediated. For this reason multiresistant strains of enterobacteria usually remain susceptible to nitrofurantoin.

PHARMACOLOGICAL PROPERTIES

Nitrofurantoin is administered by mouth and is rapidly and almost completely absorbed. It is excreted extremely rapidly into urine (half-life 20–60 min) and such drug, as finds its way into body tissues, is metabolized to the inactive derivative, aminofurantoin.

TOXICITY AND SIDE EFFECTS

Nausea is the most frequent complaint of patients receiving nitrofurantoin, but this is less common with a macrocrystalline formulation. Among less common side effects, pulmonary complications are prominent, but even these are rarely seen. The mutagenic potential of nitrofurans has not prevented their widespread and evidently safe use, even in pregnancy.

Novobiocin

Novobiocin is a naturally occurring antibiotic related to the coumarin anticoagulants. It acts on the B subunit of bacterial DNA gyrase. It is quite active against staphylococci and streptococci, but not Enterobacteriaceae or pseudomonads. It was formerly used principally as an antistaphylococcal agent, but toxicity has limited its value. Multiresistant staphylococci may remain susceptible to novobiocin, but mutations to resistance occur readily.

AGENTS THAT ACT ON THE BACTERIAL CELL MEMBRANE

In contrast to antifungal agents, in which the cell membrane is the most common target, few antibacterial agents act at this level, and those that do are quite toxic. Among membrane-active agents used in human medicine, only the polymyxins have been regularly used systemically.

Polymyxins

STRUCTURE AND SPECTRUM OF ACTIVITY

The polymyxins are a family of antibiotics produced by species of *Bacillus*. They are made up of a polypeptide portion, much of which is arranged in a cyclic fashion, with a hydrophobic octanoic acid tail (Figure 18.22). Two members of the family are in therapeutic use: polymyxin B and colistin (polymyxin E). Derivatives in which the diaminobutyric acid residues are sulfomethylated are also available. These sulfomethyl polymyxins exhibit reduced antibacterial activity, but they spontaneously break down to the more active parent compounds. Most gram-negative bacilli (with the exception of *Proteus* spp.) are sensitive to polymyxins, but their chief attraction is their activity against *P. aeruginosa*. Gram-positive organisms are much less susceptible.

MODE OF ACTION

The polymyxins act like cationic detergents to destabilize the cytoplasmic membrane. They also act on the outer membrane of gram-negative bacilli by binding to lipopolysaccharide. Polymyxin B nonapeptide, a derivative in which the fatty acid tail has been removed, binds to lipid A of lipopolysaccharide (but not to the cytoplasmic membrane) and has attracted some attention because of its anti-endotoxin properties (Danner et al. 1989).

RESISTANCE

Primary resistance in sensitive species is uncommon. However, adaptation to resistance readily occurs when dense bacterial populations are exposed to the drug (Greenwood 1975). This type of resistance is readily reversible and is apparently due to phenotypic changes in the membrane structure (Gilleland et al. 1984).

PHARMACOLOGICAL PROPERTIES

Polymyxins are nowadays more widely used as topical than as systemic agents. Polymyxin B, in particular, features as an ingredient of a number of topical preparations. They are not absorbed when given by mouth, but oral suspensions of colistin are used in several selective digestive tract decontamination regimens in neutropenic patients (Donnelly 1993). After injection of parenteral preparations, the polymyxins bind to tissue cells, but sulfomethyl derivatives are less readily bound and are excreted more rapidly. They do not penetrate into cerebrospinal fluid. They are excreted renally with a half-life of about 6 h, but tissue binding ensures that much of the dose is retained for much longer periods. Protein binding is low.

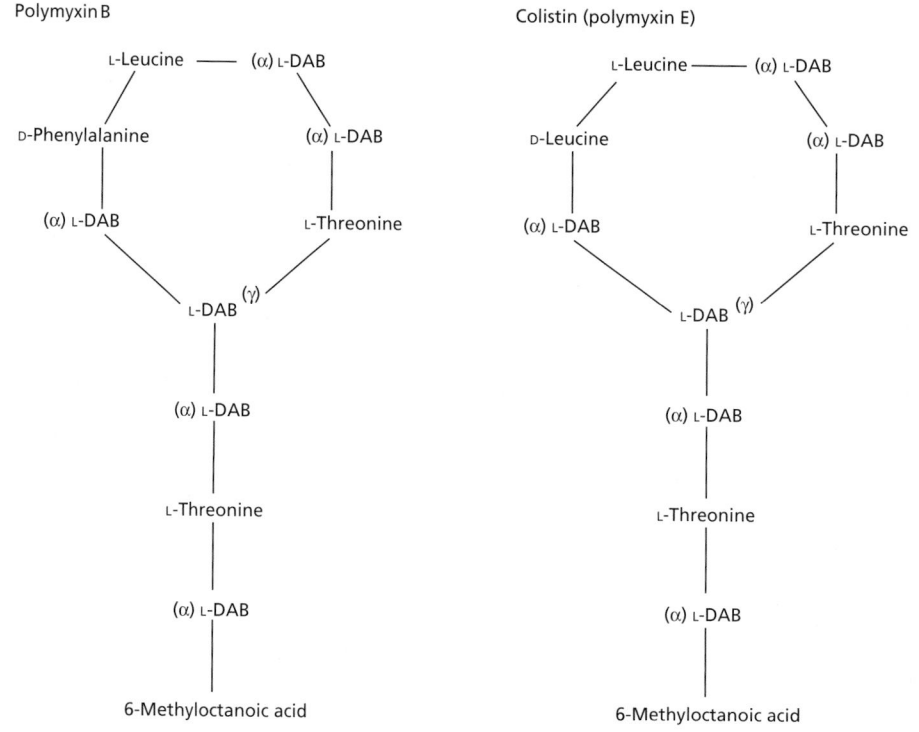

Figure 18.22 *Structures of polymyxin B and colistin (polymyxin E). In the sulfomethyl derivatives each of the β-linked diaminobutyric acid (DAB) residues may be sulfomethylated.*

TOXICITY AND SIDE EFFECTS

The major toxicity problems of the polymyxins relate to their affinity for cell membranes, including those of mammalian cells. Nephrotoxicty and neurotoxicity occur, but are usually reversible. Sulfomethyl polymyxins are somewhat less toxic than the sulfates. Topical preparations are generally well tolerated.

Other agents acting on bacterial membranes

Several antibiotics that are found in topical formulations, including gramicidin and tyrocidine, interfere with the integrity of bacterial cell membranes. Similarly, a number of agents used in animal husbandry, including valinomycin and monensin, achieve their effect at the membrane level. These agents act as ionophores: compounds that form transmembrane channels with consequent leakage of cellular potassium and other cations (Pressman 1973).

In addition, many antiseptics and disinfectants disrupt bacterial membranes. Naturally occurring antimicrobial peptides that are widespread in nature, such as the magainins, cecropins, defensins, and the lanthionine-containing lantibiotics, also act selectively on cell membranes (Boman et al. 1994). There has been much interest in these and related synthetic oligopeptides, but any possible future for these compounds as therapeutic agents is still some way off.

LABORATORY CONTROL OF ANTIMICROBIAL CHEMOTHERAPY

The performance of antimicrobial sensitivity tests and antibiotic assays are among the most important functions of medical microbiology service laboratories. Similarly, the provision of advice to clinicians and to hospital therapeutics committees on the use of antimicrobial agents forms a crucial part of the role of the clinical microbiologist. Indeed, it is essential that this role is expanded if rational antibiotic use is to be encouraged and the threat of multidrug resistance is to be kept at bay. In addition, the microbiology laboratory provides an important service in monitoring local trends of antibiotic drug resistance and in disseminating this information to prescribers (Greenwood 1993).

Sensitivity test methods

ANTIBIOTIC TITRATIONS

The intrinsic activity of antimicrobial drugs in vitro is usually expressed in terms of the minimum inhibitory concentration (MIC); less commonly by the minimum bactericidal concentration (MBC). These values are derived by titration of the drug against a standard inoculum of a pure culture of an isolated bacterial species in broth or on agar plates. The MIC is the lowest concentration that prevents the development of visible growth after a period of incubation that is usually

16–20 h for most common pathogens, but may be longer for slow-growing organisms. The MBC is assessed from broth titrations by subculture of those dilutions of antibiotic that are above the MIC. The lowest concentration of antibiotic that achieves a 1 000-fold or greater reduction in the original bacterial inoculum (i.e. a fall in viable count of at least 3 $\log_{10}$ colony-forming units) is generally taken as the MBC. The rate of killing may be more important than the overnight MBC value. In this case it is necessary to count the number of viable organisms at regular intervals after exposure to appropriate concentrations of the antimicrobial agent. In assessing viability it is good practice to specify colony-forming units rather than bacterial numbers, since bacteria may grow in clumps or chains that give rise to single colonies on solid media.

Alternatively, the E-test, in which a linear gradient of antibiotic is carried on a special strip applied to a plate inoculum of the test organism, can be used. This method is very simple and appears to correlate well with more traditional methods (Baker et al. 1991; Brown and Brown 1991).

Although MIC titrations provide figures that are widely used in comparing the activity of different antimicrobial agents, the results may be extremely variable depending on the test conditions and on the bacterial inoculum (Amsterdam 1991). Moreover, the factors determining the end point of MIC titrations vary from drug to drug and from organism to organism (Greenwood 1981). Selected MIC values shown in Table 18.7 should be interpreted in the light of these provisos. Nevertheless, the MIC is the mainstay of modern pharmacodynamic principles on the best antibiotic dosing schedules and is certainly the single most important measure of an antibiotic's activity.

MIC titrations are generally too time-consuming for routine purposes, although prepared microtitration trays containing suitable dilutions of lyophilized antibiotic can be used to simplify the procedure.

DISK DIFFUSION METHODS

Most medical microbiology laboratories use one of the disk diffusion techniques in which antibiotics incorporated into absorbent disks are applied to the surface of culture plates seeded with the test organism (Acar and Goldstein 1991; Brown 1994). Antibiotic diffuses rapidly from the disk into the surrounding medium, setting up a concentration gradient of the drug. The edge of a zone of inhibition is formed where the concentration is able to prevent visible growth of the microorganism (Barry 1991).

In the USA and world-wide, a highly standardized disk method previously known as the Kirby–Bauer test and now known as the NCCLS method (NCCLS 2000, 2002) is widely used. In the UK it is more usual to use Stokes' comparative disk method, in which the test and control organism are both exposed to the gradient of antibiotic under identical conditions on the same plate (Stokes et al. 1993). Stokes' method is particularly useful in conditions such as exist in many developing countries in which refrigeration facilities are unreliable, so that disks are kept in unsatisfactory, often humid conditions. More standardized methods, such as NCCLS, and the recently described BSAC method are increasingly adopted (BSAC Working Party Report 2001). This is because of the increasing need for comparable surveillance data in the battle against antibiotic resistance. If zone sizes are accurately measured then subtle changes that may become significant over a period can be more easily spotted. Steps can then be taken to reverse such trends before they become a clinical problem.

By comparing disk zone sizes and MIC values for many different organisms, a relationship between the two is evident and this can be formalized by regression analysis. However, since the factors that determine the end point of MIC titrations differ considerably from those that operate in the conditions of the disk diffusion test, the relationship is less than perfect. This is particularly true for those organisms that are of intermediate susceptibility, for which there may be wide discrepancies between the two methods.

BREAKPOINT METHOD

Many laboratories with large workloads find it convenient to use the breakpoint method of sensitivity testing, since this allows many isolates of bacteria to be tested together. In this method, antibiotics are incorporated into agar plates at predetermined 'breakpoint' concentrations based on therapeutically meaningful levels. Plates are spot-inoculated with a multipoint inoculation device and incubated overnight. If no growth occurs at the breakpoint concentration, the organism is scored as sensitive; if growth occurs, the organism is regarded as resistant. Sometimes a second, higher breakpoint is also used in order to establish a category of 'reduced susceptibility'. By use of a plate scanner to read the result, the whole method can be semi-automated.

OTHER METHODS OF SENSITIVITY TESTING

Many different mechanized or semi-automated methods have been devised for determining MICs or otherwise estimating the sensitivity status of bacteria (Tilton 1991). Some of these machines produce results much more rapidly than traditional methods, but the correlation is often less than perfect, partly because of the requirement for reading the end point turbidimetrically (Greenwood 1985).

DNA probes that are able to detect the genetic potential for resistance have also received much attention in recent years. These methods need to be refined if they are to be of use as routine methods, since most probes are too specific to detect the wide range of resistance

Antibacterial agent	Minimum inhibitory concentration (mg/l)									
	Staphylococcus aureus	Streptococcus pyogenes	Streptococcus pneumoniae	Neisseria meningitidis	Haemophilus influenzae	Escherichia coli	Klebsiella pneumoniae	Pseudomonas aeruginosa	Bacteroides fragilis	Chlamydia trachomatis
Benzylpenicillin	0.1	0.01	0.02	0.02	1	64	R	R	16	R
Amoxicillin	0.1	0.01	0.03	0.06	0.25	2	R	R	16	R
Cephalexin	2	1	2	16	8	8	R	R	R	
Cefuroxime	1	0.02	0.02	0.06	0.5	2	4	R	R	
Cefotaxime	1	0.02	0.02	<0.01	0.01	0.06	0.03	R	R	
Ceftazidime	4	0.1	0.25	<0.01	0.1	0.1	0.1	2	R	
Cefepime	2	<0.5	<0.5	...	0.1	0.1	0.1	8	R	
Imipenem	0.03	0.01	0.01	0.03	0.25	0.25	1	2	0.1	R
Ertapenem	0.12	0.03	0.03	<0.03	0.25	<0.03	4	R	0.15	...
Vancomycin	1	0.12	0.12	R	R	R	R	R	R	R
Gentamicin	0.25	8	2	2	0.5	0.5	1	2	R	...
Amikacin	1	R	16	R	0.5	0.5	2	2	R	R
Chloramphenicol	4	4	R	1	0.5	2	R	R	2	R
Doxycycline	0.5	0.25	0.25	0.25	1	1	8	2	0.25	0.25
Tigecycline	0.25	0.06	0.03	1	0.5	0.12	0.5	8	0.25	0.02
Erythromycin	0.25	0.06	0.06	1	1	R	R	R	R	2
Telithromycin	0.12	0.01	0.004	0.03	0.5	R	R	R	R	0.01[a]
Fusidic acid	0.06	8	8	0.25	R	R	R	2	2	...
Linezolid	0.06	0.5	0.12	R	R	R	R	R	1	R
Quinupristin/ dalfopristin	0.03	0.06	0.12	<0.12	1	R	R	R		
Sulfamethoxazole	8	2	16	0.25	2	4	8	R	R	R
Trimethoprim	0.25	0.25	8	1	0.25	0.25	0.5	R	4	R
Nalidixic acid	64	R	R	0.5	1	4	4	R	R	R
Ciprofloxacin	0.5	0.5	1	<0.01	0.01	0.06	0.06	0.25	8	0.5
Moxifloxacin	0.06	0.5	0.25	0.016	0.06	0.06	0.5	8	2	0.5
Rifampin	0.03	0.06	0.06	0.01	0.5	8	16	32	0.25	<0.01
Metronidazole	R	R	R	R	R	R	R	R	0.5	R
Nitrofurantoin	4	4	R	...	...	4	32	R	8	R
Polymyxin B	R	R	R	...	0.03	0.03	0.03	0.1	R	R

These minimum inhibitory concentrations refer to typical values for fully susceptible strains without acquired resistance traits. In using this table it should be borne in mind that inhibitory concentrations may vary considerably depending on the conditions of the test and the bacterial inoculum used. R, intrinsically resistant (agent has no useful activity).

a) C. pneumoniae.

mechanisms that may be present. Moreover, the presence of a resistance gene does not guarantee that it will necessarily be expressed (Towner 1992b). Nonetheless, DNA probes offer an attractive solution to the problem of detecting specific resistance mechanisms, particularly in organisms that are difficult to grow and are already in clinical use for rapid detection of MRSA and rifampin resistance in *M. tuberculosis* (Fluit et al. 2001; Holliman and Johnson 1998).

CLINICAL RELEVANCE OF ANTIBIOTIC SENSITIVITY TESTS

A laboratory report indicating in vitro susceptibility of a microorganism to a particular agent by no means ensures that a therapeutic response will be achieved if that drug is used in treatment. Much will depend on the type of infection, the condition of the patient, and the pharmacological properties of the drug. A statement about in vitro resistance is more likely to indicate that treatment with that agent will not influence the course of infection, although the patient may recover for other reasons: for example, through the influence of normal host defense mechanisms, by the surgical drainage of a collection of pus, or by the removal of a catheter.

In vitro susceptibility tests by their very nature offer only a crude approximation to the complex situation that exists in the infected patient. Laboratory reports merely provide information that needs to be interpreted thoughtfully in the light of the limitations of the tests and the condition of the individual patient (Slack 1995a).

Antibiotic assays

INDICATIONS FOR ANTIBIOTIC ASSAY

Aside from the need to establish the pharmacokinetic behavior of antimicrobial drugs in the early stages of their development, indications for the assay of antimicrobial agents in clinical practice are few. The most common grounds for antibiotic assay are provided by those agents, pre-eminently the aminoglycosides, in which the therapeutic range (the difference between an effective concentration and a toxic concentration) is relatively narrow. With such agents, dosage is often initiated by use of a nomogram that takes account of the sex-related creatinine concentration (a measure of renal function), body weight, and age of the patient (Mawer et al. 1974). Assays are subsequently performed 1 h after the dose, to establish that an adequate peak concentration of the drug has been achieved, and just before the next dose ('trough' concentration) to check that drug accumulation has not occurred (Humphreys and Reeves 1991). In once-daily dosing of aminoglycosides a level between 6 and 14 h post dose is recommended. Reference is then made to a monogram to calculate the dosing schedule (Nicolau et al. 1995).

Other antibiotics that may require laboratory assay during drug therapy include the glycopeptides (although the need for this has been disputed: Ackerman 1994; Cantú et al. 1994) and chloramphenicol in young infants who may be in danger of 'gray baby' syndrome through an immature capacity to metabolize the drug.

ASSAY METHODS

Traditional plate diffusion methods of antibiotic assay have largely been superseded by automated immunoassay procedures that are more specific, more accurate, and much more rapid. High pressure liquid chromatography methods are also available for most antimicrobial agents. They are useful for certain purposes, since metabolites can often be identified and quantified by this means, but they are too cumbersome for routine use. The various assay methods that are in use are described in detail by Chapin-Robertson and Edberg (1991) and Reeves et al. (1999).

SERUM BACTERICIDAL TEST

A type of antibiotic assay commonly called back-titration, in which the bactericidal activity of the patient's serum is titrated against the organism responsible for the infection, is sometimes used in conditions in which it is important to achieve a bactericidal effect (usually because of poor host immune response). The test is commonly used in the management of infective endocarditis and is also sometimes recommended in other types of infection, including osteomyelitis, cystic fibrosis, and sepsis in immunocompromised patients. Peak (usually 1 h after administration of the dose) and trough (immediately before the next dose) concentrations are assayed. The bactericidal end point is measured by subculture of dilutions of patient's serum that inhibit growth of the organism during overnight incubation. The result may reflect the activity of more than one agent if combination therapy is being used.

The value of the serum bactericidal test has been questioned, not least because of the many technical variables that may affect the result (Peterson and Shanholtzer 1992). Moreover, clinical correlates with successful therapy have been hard to come by, although a peak serum bactericidal titer >32, and a trough <16 are usually associated with a bacteriological cure. Performance of these tests is no longer recommended in the treatment of infective endocarditis in the UK (Shanson 1998).

CONCLUSIONS

The twentieth century has seen unprecedented advances in the treatment of infection. More than 250 individual compounds are now on the world market for the systemic treatment of bacterial infection. However, the process of drug discovery and development is now so costly and the return so uncertain that it is unlikely that this abundance will be matched in the next century (Billstein 1994). Already pharmaceutical companies are turning their attention from antibacterial agents to the potentially more profitable areas of antiviral and antifungal compounds. At a recent meeting it was apparent that six major pharmaceutical companies had recently decided to cease their antibiotic programmes and another two to downgrade or sell them off (Shlaes, 2003).

The ability of the drug companies to come up with new and ever more potent agents has, until now, blunted the impact of bacterial resistance and weakened our resolve to preserve the value of antibiotics by circumscribing their use. Such wastefulness can no longer be afforded. The second half of the twentieth century was truly a golden age of antibiotics. The future is largely in the hands of the prescriber.

REFERENCES

Acar, J.F. and Goldstein, F.W. 1991. Disk susceptibility test. In: Lorian, V. (ed.), *Antibiotics in laboratory medicine*, 3rd edn. Baltimore: Williams and Wilkins, 17–52.

Ackerman, B.H. 1994. Clinical value of monitoring serum vancomycin concentrations. *Clin Infect Dis*, **19**, 1180–1.

Alós, J.L., Aracil, B., et al. 2003. Significant increase in the prevalence of erythromycin resistant, C lindamycin- and miocamycin-susceptible (M phenotype) *Streptococcus pyogenes* in Spain. *J Antimicrob Chemother*, **51**, 333–7.

Ambler, R.P. 1980. The structure of β-lactamases. *Philos Trans R Soc Lond [Biol]*, **289**, 321–31.

Amsterdam, D. 1991. Susceptibility testing of antimicrobials in liquid media. In: Lorian, V. (ed.), *Antibiotics in laboratory medicine*, 3rd edn. Baltimore: Williams and Wilkins, 53–105.

Anne, S. and Resman, R.E. 1995. Risk of administering cephalosporin antibiotics to patients with histories of penicillin allergy. *Ann Allergy Asthma Immunol*, **74**, 167–70.

Arnold, S. 2004. Interventions to improve antibiotic prescribing in the community. In *Antibiotic policies: Theory and practice*. Kluwer Academic/Plenum Publishers.

Aronsson, B., Möllby, R. and Nord, C.E. 1984. Diagnosis and epidemiology of *Clostridium difficile* enterocolitis in Sweden. *J Antimicrob Chemother*, **14**, Suppl. D, 85–95.

Aumercier, M., Bouhallab, S., et al. 1992. RP 59500: a proposed mechanism for its bactericidal activity. *J Antimicrob Chemother*, **30**, Suppl. A, 9–14.

Austin, D.J., Kristinsson, K.G. and Anderson, R.M. 1999. The relationship between the volume of antimicrobial consumption in human communities and the frequency of resistance. *Proc Natl Acad Sci USA*, **96**, 1152–6.

Bailey, R.L., Arullendran, P., et al. 1993. Randomised controlled trial of single-dose azithromycin in treatment of trachoma. *Lancet*, **342**, 453–6.

Baker, C.N., Stocker, S.A., et al. 1991. Comparison of the E test to agar dilution, broth microdilution, and agar diffusion susceptibility test techniques by using a special challenge set of bacteria. *J Clin Microbiol*, **29**, 533–8.

Baquero, F., Baquero-Artigao, G., et al. 2002. Antibiotic consumption and resistance selection in *Streptococcus pneumoniae*. *J Antimicrob Chemother*, **50**, Suppl. S2, 27–37.

Barrière, J.C., Bouanchaud, D.H., et al. 1992. Antimicrobial activity against *Staphylococcus aureus* of semisynthetic injectable streptogramins: RP 59500 and related compounds. *J Antimicrob Chemother*, **30**, Suppl. A, 1–8.

Barry, A.L. 1991. Procedures and theoretical considerations for testing antimicrobial agents in agar media. In: Lorian, V. (ed.), *Antibiotics in laboratory medicine*, 3rd edn. Baltimore: Williams and Wilkins, 1–16.

Bartlett, R.C., Quintiliani, R.D., et al. 1991. Effect of including recommendations of antimicrobial therapy in microbiology laboratory reports. *Diagn Microbiol Infect Dis*, **14**, 157–66.

Beard, C., Noller, K.L., et al. 1979. Lack of evidence for cancer due to use of metronidazole. *N Engl J Med*, **301**, 519–22.

Billstein, S.A. 1994. How the pharmaceutical industry brings an antibiotic drug to market in the United States. *Antimicrob Agents Chemother*, **38**, 2679–82.

Black, F.T., Wildfang, I.L. and Borgbjerg, K. 1985. Activity of fusidic acid against *Plasmodium falciparum* in vitro. *Lancet*, **1**, 578–9.

Blondeau, J., Zhao, X., et al. 2001. Mutant prevention concentrations of fluroquinolones for clinical isolates of *Streptococcus pneumoniae*. *Antimicrob Agents Chemother*, **45**, 433–8.

Boman, H.G., Marsh, J. and Goode, J.A. 1994. Antimicrobial peptides. *Ciba Found Symp*, **186**, Chichester: John Wiley & Sons.

Bonhoeffer, S., Lipsitch, M. and Leven, B.R. 1997. Evaluating treatment protocols to prevent resistance. *Proc Natl Acad Sci USA*, **94**, 12106–11.

Boo, T.W., Hone, R., et al. 2003. Isolation of linezolid-resistant *Enterococcus faecalis*. *J Hosp Infect*, **53**, 312–14.

Bowler, L.D., Zhang, Q.Y., et al. 1994. Interspecies recombination between the *penA* genes of *Neisseria meningitidis* and commensal *Neisseria* species during the emergence of penicillin resistance in *N. meningitidis*: natural events and laboratory simulation. *J Bacteriol*, **176**, 333–7.

Brown, D.F.J. 1994. Developments in antimicrobial susceptibility testing. *Rev Med Microbiol*, **5**, 65–75.

Brown, D.F.J. and Brown, L. 1991. Evaluation of the E test, a novel method for quantifying antimicrobial activity. *J Antimicrob Chemother*, **27**, 185–90.

Brown, E.M. 2004. Interventions to optimise prescribing in hospitals: The UK approach. In *Antibiotic policies: Theory and practice*. Kluwer Academic/Plenum Publishers.

BSAC Working Party Report. 2001. Antimicrobial susceptibility testing. *J Antimicrob Chemother*, **48** (Suppl. S1).

Burke, J.P. and Pestotnik, S.L. 1996. Breaking the chain of antibiotic resistance. *Curr Opin Infect Dis*, **9**, 253–5.

Bush, K. 1989. Characterization of β-lactamases. *Antimicrob Agents Chemother*, **33**, 259–76.

Bush, K., Jacoby, G.A. and Medeiros, A.A. 1995. A functional classification scheme for β-lactamases and its correlation with molecular structure. *Antimicrob Agents Chemother*, **39**, 1211–33.

Cantú, T.G., Yamanaka-Yuen, N.A. and Lietman, P.S. 1994. Serum vancomycin concentrations: reappraisal of their clinical value. *Clin Infect Dis*, **18**, 533–43.

Cars, O., Molstad, S. and Melander, A. 2001. Variation in antibiotic use in the European Union. *Lancet*, **357**, 1851–3.

Cédrick, J., Caillon, J., et al. 2003. In vitro activity of linezolid alone and in combination with gentamicin, vancomycin or rifampicin against methicillin-resistant *Staphylococcus aureus* by time-kill curve methods. *J Antimicrob Chemother*, **51**, 857–64.

Cepeda, J.A., Whitehouse, T., et al. 2004. Linezolid versus teicoplanin in the treatment of Gram-positive infections in the critically ill: a randomized, double-blind, multicentre study. *J Antimicrob Chemother*, **53**, 345–55.

Chaibi, E.B., Sirot, D., et al. 1999. Inhibitor-resistant TEM β-lactamases: phenotypic, genetic and biochemical characteristics. *J Antimicrob Chemother*, **43**, 447–58.

Chang, S., Sievert, D.M. and Hageman, J.C. 2003. Infection with vancomycin-resistant *Staphylococcus aureus* containing the *vanA* resistance gene. *N Engl J Med*, **348**, 1342–7.

Chapin-Robertson, K. and Edberg, S.C. 1991. Measurement of antibiotics in human body fluids: techniques and significance. In:

Lorian, V. (ed.), *Antibiotics in laboratory medicine*. Baltimore: Williams and Wilkins, 295–366.

Chopra, I., Hawkey, P.M. and Hinton, M. 1992. Tetracyclines, molecular and clinical aspects. *J Antimicrob Chemother*, **29**, 245–77.

Clemett, D. and Markham, A. 2000. Linezolid. *Drugs*, **59**, 815–27.

Cundliffe, E. 1972. The mode of action of fusidic acid. *Biochem Biophys Res Commun*, **46**, 1794–801.

Danner, R.L., Joiner, K.A., et al. 1989. Purification, toxicity, and antiendotoxin activity of polymyxin B nonapeptide. *Antimicrob Agents Chemother*, **33**, 1428–34.

Davis, B.D. 1988. The lethal action of aminoglycosides. *J Antimicrob Chemother*, **22**, 1–3.

Department of Health. 2000. *UK Antimicrobial resistance strategy and action plan*. June.

Docquier, J., Lamotte-Brasseur, J., et al. 2003. On functional and structural heterogeneity of VIM-type metallo-β-lactamases. *J Antimicrob Chemother*, **51**, 257–66.

Dollery, C. 1991. *Therapeutic drugs*. Edinburgh: Churchill Livingstone, two volumes and supplements.

Donnelly, J.P. 1993. Selective decontamination of the digestive tract and its role in antimicrobial prophylaxis. *J Antimicrob Chemother*, **31**, 813–29.

Doona, M. and Walsh, J.B. 1995. Use of chloramphenicol as topical eye medication: time to cry halt? *Br Med J*, **310**, 1217–18.

Drusano, G.L. 2003. Pharmacodynamics of anti-infectives: target delineation and target attainment. In: Finch, R., Greenwood, D., et al. (eds), *Antibiotics and chemotherapy*. London: Churchill Livingstone, 48–58.

Edwards, D.I. 1993a. Nitroimidazole drugs – action and resistance mechanisms. I. Mechanisms of action. *J Antimicrob Chemother*, **31**, 9–20.

Edwards, D.I. 1993b. Nitroimidazole drugs – action and resistance mechanisms. II. Mechanisms of resistance. *J Antimicrob Chemother*, **31**, 201–10.

Eliopoulos, G.M. 2003. Quinupristin-dalfopristin and linezolid evidence and opinion. *Clin Infect Dis*, **36**, 473–81.

Etienne, S.D., Montay, G., et al. 1992. A phase I, double-blind, placebo-controlled study of the tolerance and pharmacokinetic behaviour of RP 59500. *J Antimicrob Chemother*, **30**, Suppl. A, 123–131.

European Union Conference. 1998. The microbial threat: The Copenhagen recommendations. Rosdahl V.K., Pedersen K.B., (eds), Danish Ministry of Health and Ministry of Food, Agriculture and Fisheries, September, 1–52, www.sum.dk/publika/micro98/.

Farthing, M.J.G. and Inge, P.M.G. 1986. Antigiardial activity of the bile salt-like antibiotic sodium fusidate. *J Antimicrob Chemother*, **17**, 165–71.

Finch, R.G. 1995. General principles of the treatment of infection. In: Greenwood, D. (ed.), *Antimicrobial chemotherapy*, 3rd edn. Oxford: Oxford University Press, 179–87.

Fluckiger, U., Zimmerli, W., et al. 2000. Clinical impact of an infectious disease service on the management of bloodstream infection. *Eur J Clin Microbiol Infect Dis*, **19**, 493–500.

Fluit, A.C., Visser, M.R. and Schmitz, F.J. 2001. Molecular detection of antimicrobial resistance. *Clin Microbiol Rev*, **14**, 836–71.

Fogarty, C.M., Kohno, S., et al. 2003. Community-acquired respiratory tract infections caused by resistant pneumococci: clinical and bacteriological efficacy of the ketolide telithromycin. *J Antimicrob Chemother*, **51**, 947–55.

Garcia-Rodriguez, J.A. and Gomez Garcia, A.C. 1993. In-vitro activities of quinolones against mycobacteria. *J Antimicrob Chemother*, **32**, 797–808.

Giakkoupi, P., Petrikkos, G., et al. 2003. Spread of integron-associated VIM-Type metallo-β-lactamase genes among imipenem-nonsusceptible *Pseudomonas aeruginosa* strains in Greek hospitals. *J Clin Microbiol*, **41**, 822–5.

Gilleland, H.E., Champlin, F.R. and Conrad, R.S. 1984. Chemical alterations in cell envelopes of *Pseudomonas aeruginosa* upon exposure to polymyxin: a possible mechanism to explain adaptive resistance. *Can J Microbiol*, **20**, 869–73.

Goldstein, E.J. and Garbaedian-Ruffalo, S.M. 2002. Widespread use of fluroquinolones versus emerging resistance in pneumococci. *Clin Infect Dis*, **35**, 1505–11.

Gould, I.M. 1998. Determinants of response of antibiotic therapy. *J Chemother*, **10**, 347–53.

Gould, I.M. 1999. A review of the role of antibiotic policies in the control of antibiotic resistance. *J Antimicrob Chemother*, **43**, 459–65.

Gould, I.M. 2001. Minimum antibiotic stewardship measures. *Clin Microbiol Infect*, **7**, Suppl. 6, 22–6.

Gould, I.M. 2002a. Antibiotic policies and control of resistance. *Curr Opin Infect Dis*, **15**, 395–400.

Gould, I.M. 2002b. Antibiotic rotation to control resistance. In: Galley, H.F. (ed.), *Critical care focus*. London: BMJ Books, 41–54.

Gould, I.M. and MacKenzie, F.M. 1997. The response of Enterobacteriaceae to β-lactam antibiotics – 'round forms, filaments and the root of all evil'. *J Antimicrob Chemother*, **40**, 495–9.

Gould, I.M. and MacKenzie, F.M. 2002. Antibiotic exposure as a risk factor for emergence of resistance: the influence of concentration. *J Appl Microbiol*, **92**, 78S–84S.

Gould, I.M., Hampson, J., et al. 1994. Hospital antibiotic control measures in the UK. *J Antimicrob Chemother*, **34**, 21–42.

Gould, I.M., MacKenzie, F.M., et al. 2000. Towards a European strategy for controlling antibiotic resistance. *Clin Microbiol Infect*, **6**, 670–4.

Greenwood, D. 1975. The activity of polymyxins against dense populations of *Escherichia coli*. *J Gen Microbiol*, **91**, 110–18.

Greenwood, D. 1981. In vitro veritas? Antimicrobial susceptibility tests and their clinical relevance. *J Infect Dis*, **144**, 380–5.

Greenwood, D. 1985. The alteration of microbial growth curves by antibiotics. In: Habermehl, K.-O. (ed.), *Rapid methods and automation in microbiology and immunology*. Berlin: Springer-Verlag, 497–503.

Greenwood, D. 1988. Fusidic acid. In: Peterson, P.K. and Verhoef, J. (eds), *The antimicrobial agents annual 3*. Amsterdam: Elsevier, 106–112.

Greenwood, D. 1993. Antimicrobial susceptibility testing: are we wasting our time? *Br J Biomed Sci*, **50**, 31–4.

Greenwood, D. 1995. *Antimicrobial chemotherapy*, 3rd edn. Oxford: Oxford University Press.

Greenwood, D. and O'Grady, F. 1972. The effect of osmolality on the response of *Escherichia coli* and *Proteus mirabilis* to penicillins. *Br J Exp Pathol*, **53**, 457–64.

Greenwood, D. and O'Grady, F. 1973a. Comparison of the responses of *Escherichia coli* and *Proteus mirabilis* to seven β-lactam antibiotics. *J Infect Dis*, **128**, 211–22.

Greenwood, D. and O'Grady, F. 1973b. The two sites of penicillin action in *Escherichia coli*. *J Infect Dis*, **128**, 791–4.

Greenwood, D. and O'Grady, F. 1977. Factors governing the emergence of resistance to nalidixic acid in the treatment of urinary tract infection. *Antimicrob Agents Chemother*, **12**, 678–81.

Greenwood, D., Palfreyman, J., et al. 1988. Activity of teicoplanin against Gram-negative anaerobes. *J Antimicrob Chemother*, **21**, 500–1.

Greenwood, D., Watt, B. and Duerden, B.I. 1991. Antibiotics and anerobes. In: Duerden, B.I. and Drasar, B.S. (eds), *Anaerobes in human disease*. London: Edward Arnold, 415–29.

Grohs, P., Kitzis, M. and Gutmann, L. 2003. In vitro bactericidal activities of linezolid in combination with vancomycin, gentamicin, ciprofloxacin, fusidic acid and rifampicin against *Staphylococcus aureus*. *Antimicrob Agent Chemother*, **47**, 418–20.

Gruteke, P., Goessens, W., et al. 2003. Patterns of resistance associated with integrons, the extended-spectrum β-lactamase SHV-5 gene, and a multidrug efflux pump of *Klebsiella pneumoniae* causing a nosocomial outbreak. *J Clin Microbiol*, **41**, 1161–6.

Guillemot, D., Carbon, C., et al. 1998. Low dosage and long treatment duration of beta-lactam: risk factors for carriage of penicillin-resistant *Streptococcus pneumoniae*. *J Am Med Assoc*, **279**, 365–70.

Hancock, R.E.W. and Bellido, F. 1992. Antibiotic uptake: unusual results for unusual molecules. *J Antimicrob Chemother*, **29**, 235–43.

Handwerger, S. and Tomasz, A. 1985. Antibiotic tolerance among clinical isolates of bacteria. *Rev Infect Dis*, **7**, 368–86.

Harbarth, S., Albrich, W. and Brun-Buisson, C. 2002. Outpatient antibiotic use and prevalence of antibiotic resistant pneumococci in France and Germany: A sociocultural perspective. *Emerg Infect Dis*, **8**, 1460–7.

Harrison, P.F. and Lederberg, J. 1998. *Antimicrobial resistance: issues and options*. Washington DC: National Academy Press.

Healy, S.P. and Zervos, M.J. 1995. Mechanisms of resistance of enterococci to antimicrobial agents. *Rev Med Microbiol*, **6**, 70–6.

Hiramatsu, K. 2004. Has vancomycin-resistant *Staphylococcus aureus* started going it alone? *Lancet*, **364**, 565–6.

Holliman, R.E. and Johnson, J.D. 1998. Bacteriology molecular techniques: sample preparation and application. In: Crocker, J. and Burnett, D. (eds), *The science of laboratory diagnosis*. Oxford: Isis Medical Media Ltd, 163–70.

Honeybourne, D. and Baldwin, D.R. 1992. The site concentrations of antimicrobial agents in the lung. *J Antimicrob Chemother*, **30**, 249–60.

Hooper, D.C. and Wolfson, J.S. 1989. Mode of action of the quinolone antimicrobial agents: review of recent information. *Rev Infect Dis*, **11**, Suppl. 5, S902–11.

Hooper, D.C. and Wolfson, J.S. 1991. Fluoroquinolone antimicrobial agents. *N Engl J Med*, **324**, 384–94.

Hughes, J. and Mellows, G. 1980. Interaction of pseudomonic acid A with *Escherichia coli* B isoleucyl-tRNA synthetase. *Biochem J*, **191**, 209–19.

Humble, M.W., Eykyn, S. and Phillips, I. 1980. Staphylococcal bacteraemia, fusidic acid, and jaundice. *Br Med J*, **280**, 1495–8.

Humphreys, H. and Reeves, D. 1991. Aminoglycoside assays. *Rev Med Microbiol*, **2**, 13–21.

Huovinen, P., Sundström, L., et al. 1995. Trimethoprim and sulphonamide resistance. *Antimicrob Agents Chemother*, **39**, 279–89.

Jacoby, G.A. and Archer, G.L. 1991. New mechanisms of bacterial resistance to antimicrobial agents. *N Engl J Med*, **324**, 601–12.

Jacoby, G.A., Chow, N. and Waites, K.B. 2003. Prevalence of plasmid-mediated quinolone resistance. *Antimicrob Agent Chemother*, **47**, 559–62.

Jeong, S.H., Lee, K., et al. 2003. Characterization of a new integron containing VIM-2, a metallo-β-lactamase gene cassette, in a clinical isolate of *Enterobacter cloacae*. *J Antimicrob Chemother*, **51**, 397–400.

Johnson, A.P., Uttley, A.H.C., et al. 1990. Resistance to vancomycin and teicoplanin: an emerging clinical problem. *Clin Microbiol Rev*, **3**, 280–91.

Kacia, M. 2004. Vancomycin-resistant *Staphylococcus aureus* – New York, 2004. *MMWR*, **53**, 322–3.

Kahan, F.M. and Kahan, J.S. 1979. Thienamycin, a new beta-lactam antibiotic: I. Discovery, taxonomy, isolation and physical properties. *J Antibiot*, **32**, 1–12.

Kahan, F.M., Kropp, H., et al. 1983. Thienamycin: development of imipenem-cilastatin. *J Antimicrob Chemother*, **12**, 1–35.

Kern, W.V., Andriof, E., et al. 1994. Emergence of fluroquinolone-resistance *Escherichia coli* at a cancer center. *Antimicrob Agent Chemother*, **38**, 681–7.

Keuleyan, E. and Gould, I.M. European Study Group on Antibiotic Policy (ESGAP), Subgoup III, 2001. Key issues in developing antibiotic policies: from an institutional level to Europe-wide. *Clin Microbiol Infect*, **7**, Suppl. 6, 16–21.

Kim, J.H. and Gallis, H.A. 1989. Observations on spiraling empiricism: its causes, allure and perils with particular reference to antibiotic therapy. *Am J Med*, **87**, 201–6.

Klugman, K.P. 1990. Pneumococcal resistance to antibiotics. *Clin Microbiol Rev*, **3**, 171–96.

Knudsen, E.T., Rolinson, G.N. and Sutherland, R. 1967. Carbenicillin: a new semisynthetic penicillin active against *Pseudomonas aeruginosa*. *Br Med J*, **3**, 75–8.

Krishnan, P., Thachil, R. and Gillego, V. 2002. Diffuse panbronchiolitis: a treatable sinobronchial disease in need of recognition in the United States. *Chest*, **121**, 659–61.

Kucers, A. and Bennett, N.McK. 1987. *The use of antibiotics*, 4th edn. London: Heinemann.

Kumarasamy, Y., Cadwgan, T., et al. 2003. Optimizing antibiotic therapy – the Aberdeen experience. *Clin Microbiol Infect*, **9**, 406–11.

Lambert, H.P. and O'Grady, F.W. 1997. General principles of chemotherapy. In: O'Grady, F., Lambert, H.P., et al. (eds), *Antibiotics and chemotherapy*, 7th edn. Edinburgh: Churchill Livingstone, 131–5.

Lawrence, D.E., Levin, S., et al. 1973. Ordering patterns and utilization of bacteriological reports. *Arch Intern Med*, **132**, 672–82.

Leverstein-van Hall, M.A., Blok, H.E., et al. 2003. Multidrug resistance among enterobacteriaceae is strongly associated with the presence of integrons and is independent of species or isolate origin. *J Infect Dis*, **187**, 251–9.

Levison, M.E. 1992. New dosing regimens for aminoglycoside antibiotics. *Ann Intern Med*, **117**, 693–4.

Levy, S. 2002. *The antibiotic paradox: how the misuse of antibiotics destroys their curative powers*. Cambridge MA: Perseus Publishing.

Lipsky, J.J. 1988. Antibiotic-associated hypoprothrombinaemia. *J Antimicrob Chemother*, **21**, 281–300.

Lonks, J.R., Garau, J. and Medeiros, A.A. 2002. Implications of antimicrobial resistance in the empirical treatment of community-acquired respiratory tract infections: the case of macrolides. *J Antimicrob Chemother*, **50**, Suppl. S2, 87–91.

Low, D.E. 2002. The era of antimicrobial resistance – implications for the clinical laboratory. *Clin Microbiol Infect*, **8**, Suppl. 3, 9–20.

McCaig, L.F., Besser, R.E. and Huhes, J.M. 2003. Antimicrobial drug prescriptions in ambulatory care settings, United States, 1992–2000. *Emerg Infect Dis*, **9**, 432–7.

McCalla, D.R. 1977. Biological effects of nitrofurans. *J Antimicrob Chemother*, **3**, 517–20.

MacGowan, A.P., Greig, M.A., et al. 1989. Pharmacokinetics and tolerance of a new film-coated tablet of sodium fusidate administered as a single oral dose to healthy volunteers. *J Antimicrob Chemother*, **23**, 409–15.

MacKenzie, F.M. and Gould, I.M. 1998. Extended spectrum β-lactamases. *J Infect*, **36**, 255–8.

MacKenzie, F.M., Reid, J.P., Gould, I.M. 2003. The use of overnight diagnostic tests to guide antibiotic prescription in general practice. Abstract no 0206 13th ECCMID Glasgow.

McOsker, C.C. and Fitzpatrick, P.M. 1994. Nitrofurantoin: mechanism of action and implications for resistance development in common uropathogens. *J Antimicrob Chemother*, **33**, Suppl. A, 23–30.

Manuel, R.J., Tuck, A., et al. 2002. Detection of teicoplanin resistance in UK EMRSA-17 strains. *J Antimicrob Chemother*, **50**, 1089–93.

Maple, P.A.C., Hamilton-Miller, J.M.T. and Brumfitt, W. 1989. World-wide antibiotic resistance in methicillin-resistant *Staphylococcus aureus*. *Lancet*, **1**, 537–40.

Martin, D.H., Mroczkowski, T.F., et al. 1992. A controlled trial of a single dose of azithromycin for the treatment of chlamydial urethritis and cervicitis. *N Engl J Med*, **327**, 921–5.

Martínez-Martínez, L., Pascual, A. and García, I. 2003. Interaction of plasmid and host quinolone resistance. *J Antimicrob Chemother*, **51**, 1037–52.

Mason, B.E., Howard, A.J. and Magee, J.T. 2003. Fusidic acid resistance in community isolates of methicillin-susceptible *Staphylococcus aureus* and fusidic acid prescribing. *J Antimicrob Chemother*, **51**, 1033–6.

Mawer, G.E., Ahmad, R., et al. 1974. Prescribing aids for gentamicin. *Br J Clin Pharmacol*, **1**, 45–50.

Mazzei, T., Mini, E., et al. 1993. Chemistry and mode of action of macrolides. *J Antimicrob Chemother*, **31**, Suppl. C, 1–9.

Midtvedt, T. and Greenwood, D. 1994. Miscellaneous antibacterial drugs. In: Aronson, J.K. and van Boxtel, C.J. (eds), *Side effects of drugs, annual 17*. Amsterdam: Elsevier Science BV, 303–18.

Milatovic, D., Schmitz, F.J., et al. 2003. Activities of the glycylcycline tigecycline (GAR-936) against 1,924 recent European clinical bacterial isolates. *Antimicrob Agent Chemother*, **47**, 400–4.

Mingeot-Leclecq, M.P. and Tulkens, P.M. 1999. Aminoglycosides: nephrotoxicity. *Antimicrob Agent Chemother*, **43**, 1003–12.

Mitchison, D.A. 1992. Understanding the chemotherapy of tuberculosis – current problems. *J Antimicrob Chemother*, **29**, 477–93.

Monnet, D.L. 1998. Methicillin-resistant *Staphylococcus aureus* and its relationship to antimicrobial use: possible implications for control. *Infect Control Hosp Epidemiol*, **19**, 552–9.

Monnet, D.L., MacKenzie, F.M., et al. 2004. Antimicrobial drug use and methicillin-resistant Staphylococcus aureus, Aberdeen, 1996–2000. *Emerg Infect Dis*, **10**, 1432–40.

Morgan, I. 1978. Metronidazole treatment in pregnancy. *Int J Gynaecol Obstet*, **15**, 501–2.

Moyenuddin, M., Ohl, J.C. and Peacock, S.E. 2003. Disseminated oxacillin-resistant *Staphylococcus aureus* infections responsive to vancomycin and quinupristin-dalfopristin combination therapy. *J Antimicrob Chemother*, **51**, 202–3.

Mulla, R.J., Barnes, E., et al. 1995. Is it time to stop using chloramphenicol on the eye. *Br Med J*, **311**, 450–1.

NCCLS. 2000. Performance standards for antimicrobial disk susceptibility tests; approved standard, 7th edn. NCCLS document M2-A7 (ISBN 1-56238-393-0) NCCLS 940 West Valley Road, Suite 1400, Wayne, Pennysylvania 19087-1989 USA.

NCCLS. 2002. Performance standards for antimicrobial susceptibility testing; Twelfth International Supplement. NCCLS document M 100-812 (ISBN 1-56238-454-6) NCCLS 940 West Valley Road, Suite 1400, Wayne, Pennysylvania 19087-1989 USA.

Nicolau, D.P., Freeman, C.D., et al. 1995. Experience with a once-daily aminoglycoside program administered to 2,184 adult patients. *Antimicrob Agent Chemother*, **39**, 650–5.

Nikaido, H. 1985. Role of permeability barriers in resistance to β-lactam antibiotics. *Pharmacol Ther*, **27**, 197–231.

Nikaido, H. and Vaara, M. 1985. Molecular basis of bacterial outer membrane permeability. *Microbiol Rev*, **49**, 1–32.

Noble, W.C., Virani, Z. and Cree, R.G.A. 1992. Co-transfer of vancomycin and other resistance genes from *Enterococcus faecalis* NCTC 12201 to *Staphylococcus aureus*. *FEMS Microbiol Lett*, **93**, 195–8.

O'Grady, F. and Greenwood, D. 1973. Interactions between fusidic acid and penicillins. *J Med Microbiol*, **6**, 441–50.

Oleske, J.M. and Phillips, I. 1983. Clindamycin: bacterial virulence and host defence. *J Antimicrob Chemother*, **12**, Suppl. C, 1–124.

Oriel, J.D. 1989. Use of quinolones in chlamydial infection. *Rev Infect Dis*, **11**, Suppl. 5, S1273–6.

Pablos-Mendez, A., Raviglione, M.C., et al. 1998. Global surveillance of antituberculosis-drug resistance 1994–97. World Health Organisation-International Union against Tuberculosis and Lung Disease Working Group on Anti Tuberculosis Drug Resistance Surveillance. *N Engl J Med*, **338**, 1641–9.

Payne, D.J. 1993. Metallo-β-lactamases – a new therapeutic challenge. *J Med Microbiol*, **39**, 93–9.

Pestka, S. 1971. Inhibitors of ribosome functions. *Annu Rev Microbiol*, **25**, 487–562.

Peterson, L.R. 1994. Quinolone resistance in gram-positive bacteria. *Infect Dis Clin Pract*, **3**, Suppl. 3, S127–37.

Peterson, L.R. and Shanholtzer, C.J. 1992. Tests for bactericidal effects of antimicrobial agents: technical performance and clinical relevance. *Clin Microbiol Rev*, **5**, 420–32.

Pressman, B.C. 1973. Properties of ionophores with broad range cation selectivity. *Fed Proc*, **32**, 1698–703.

Public Health Dispatch. 2002. Vancomycin-resistant *Staphyloccus aureus*. *MMWR* **51**(40) 902.

Reeves, D.S. 1987. The pharmacokinetics of fusidic acid. *J Antimicrob Chemother*, **20**, 467–76.

Reeves, D.S. 1994. Fosfomycin trometamol. *J Antimicrob Chemother*, **34**, 853–8.

Reeves, D.S., Wise, R., et al. 1999. *Clinical antimicrobial assays*. Oxford: Oxford University Press.

Report. 1991. A guide to sensitivity testing. *J Antimicrob Chemother*, **27**, Suppl. D, 1–50.

Reusser, F. 1975. Effect of lincomycin and clindamycin on peptide chain initiation. *Antimicrob Agents Chemother*, **7**, 32–7.

Reynolds, P.E. 1989. Structure, biochemistry and mechanism of action of glycopeptide antibiotics. *Eur J Clin Microbiol Infect Dis*, **8**, 943–50.

Ribard, P., Audisio, F., et al. 1992. Seven achilles tendinitis including three complicated by rupture during fluoroquinolone therapy. *J Rheumatol*, **19**, 1479–81.

Richmond, M.H. and Sykes, R.B. 1973. The β-lactamases of gram-negative bacteria and their possible physiological role. *Adv Microb Physiol*, **9**, 31–88.

Rolinson, G.N. and Stevens, S. 1961. Microbiological studies on a new broad-spectrum penicillin, Penbritin. *Br Med J*, **2**, 191–6.

Ruoff, K.L., Kuritzkes, D.R., et al. 1988. Vancomycin-resistant gram-positive bacteria isolated from human sources. *J Clin Microbiol*, **26**, 2064–8.

Russell, A.D. and Chopra, I. 1996. *Understanding antibacterial action and resistance*, 2nd edn. London: Ellis Horwood.

Sabath, L.D., Wheeler, N., et al. 1977. A new type of penicillin resistance of *Staphylococcus aureus*. *Lancet*, **1**, 443–7.

Sahai, J., Healy, D.P., et al. 1990. Comparison of vancomycin- and teicoplanin-induced histamine release and 'red man syndrome'. *Antimicrob Agents Chemother*, **34**, 765–9.

Saizy-Callaert, S., Causse, R., et al. 2003. Impact of a multidisciplinary approach to the control of antibiotic prescription in a general hospital. *J Hosp Infect*, **53**, 177–82.

Sanyal, D., Johnson, A.P., et al. 1993. In vitro characteristics of glycopeptide resistant strains of *Staphylococcus epidermidis* isolated from patients on CAPD. *J Antimicrob Chemother*, **32**, 267–78.

Sarkar, K., Ghosh, S., et al. 2003. Shigella dysenteriae type 1 with reduced susceptibility to fluroquinolones. *Lancet*, **361**, 785.

Scheld, W.M. 2003. Maintaining fluroquinolone class efficacy: review of influencing factors. *Emerg Infect Dis*, **9**, 1–9.

Shackley, P., Cairns, J. and Gould, I.M. 1997. Accelerated bacteriological evaluation in the management of lower respiratory tract infection in general practice. *J Antimicrob Chemother*, **39**, 663–666.

Shanson, D.C. 1990. Clinical relevance of resistance to fusidic acid in *Staphylococcus aureus*. *J Antimicrob Chemother*, **25**, Suppl. B, 15–21.

Shanson, D.C. 1998. New guidelines for the antibiotic treatment of streptococcal, enterococcal and staphylococcal endocarditis. *J Antimicrob Chemother*, **42**, 292–6.

Shlaes, D.M., Gerding, D.N., et al. 1997. Society for Healthcare Epidemiology of America and Infectious Diseases Society of America Joint Committee on the Prevention of Antimicrobial Resistance: Guidelines for the prevention of antimicrobial resistance in hospitals. *Clin Infect Dis*, **25**, 584–99.

Shlaes, D.N., Bouvet, A., et al. 1989. Inducible, transferable resistance to vancomycin in *Enterococcus faecalis* A256. *Antimicrob Agents Chemother*, **33**, 198–203.

Skinner, M.H., Hsieh, M., et al. 1989. Pharmacokinetics of rifabutin. *Antimicrob Agents Chemother*, **33**, 1237–41.

Slack, R.C.B. 1995a. Use of the laboratory. In: Greenwood, D. (ed.), *Antimicrobial chemotherapy*, 3rd edn. Oxford: Oxford University Press, 128–36.

Slack, R.C.B. 1995b. Chemoprophylaxis. In: Greenwood, D. (ed.), *Antimicrobial chemotherapy*, 3rd edn. Oxford: Oxford University Press, 219–30.

Slater, G.J. and Greenwood, D. 1983. Detection of penicillin tolerance in streptococci. *J Clin Pathol*, **36**, 1353–6.

Smith, M.A. and Edwards, D.I. 1995. Redox potential and oxygen concentration as factors in the susceptibility of *Helicobacter pylori* to nitroheterocyclic drugs. *J Antimicrob Chemother*, **35**, 751–64.

Smith, J.T. and Lewin, C.S. 1988. Chemistry and mechanisms of action of the quinolone antibacterials. In: Andriole, V.T. (ed.), *The quinolones*. London: Academic Press, 23–82.

Solomon, D.H., Van Houten, L., et al. 2001. Academic detailing to improve use of broad spectrum antibiotics at an academic medical centre. *Arch Intern Med*, **161**, 13–27.

Speer, B.S., Shoemaker, N.B. and Salyers, A.A. 1992. Bacterial resistance to tetracycline: mechanisms, transfer, and clinical significance. *Clin Microbiol Rev*, **5**, 387–99.

Spratt, B.G. 1975. Distinct penicillin-binding proteins involved in the division, elongation and shape of *Escherichia coli* K12. *Proc Natl Acad Sci USA*, **72**, 2999–3003.

Stamm, W., Hicks, C.B., et al. 1995. Azithromycin for empirical treatment of the nongonococcal urethritis syndrome in men. *JAMA*, **274**, 545–9.

Steinke, D.T. and Davey, P. 2001. Association between antibiotic resistance and community prescribing: a critical review of bias and confounding in published studies. *Clin Infect Dis*, **33**, S193–205.

Steinman, M.A., Landefeld, S.C. and Gonzales, R. 2003a. Predictors of broad-spectrum antibiotic prescribing for acute respiratory tract infections in adult primary care. *JAMA*, **289**, 719–25.

Steinman, M.A., Gonzales, R., et al. 2003b. Changing use of antibiotics in community-based outpatient practice. *Ann Intern Med*, **138**, 525–66.

Stokes, E.J., Ridgway, G.L. and Wren, M.W.D. 1993. *Clinical microbiology*, 7th edn. London: Edward Arnold.

Stratton, C.W. 2003. Dead bugs don't mutate: susceptibility issues in the emergence of bacterial resistance. *Emerg Infect Dis*, **9**, 10–16.

Suárez, J.E. and Mendoza, M.C. 1991. Plasmid-encoded fosfomycin resistance. *Antimicrob Agents Chemother*, **35**, 791–5.

Tally, F.P. and Malamy, M.H. 1984. Antimicrobial resistance and resistance transfer in anaerobic bacteria. A review. *Scand J Gastroenterol*, **19**, Suppl. 91, 21–30.

Tally, F.T., Ellestad, G.A. and Testa, R.T. 1995. Glycylcyclines: a new generation of tetracyclines. *J Antimicrob Chemother*, **35**, 449–52.

Tedesco, F.J., Barton, R.W. and Alpers, D.H. 1974. Clindamycin-associated colitis. A prospective study. *Ann Intern Med*, **81**, 429–433.

Thomas, J.K., Forrest, A., et al. 1998. Pharmacodynamic evaluation of factors associated with the development of bacterial resistance in acutely ill patients during therapy. *Antimicrob Agent Chemother*, **42**, 521–7.

Tilton, R.C. 1991. Automation and mechanization in antimicrobial susceptibility testing. In: Lorian, V. (ed.), *Antibiotics in laboratory medicine*, 3rd edn. Baltimore: Williams and Wilkins, 106–19.

Tomasz, A. 1979. The mechanism of the irreversible antimicrobial effects of penicillins: how the beta-lactam antibiotics kill and lyse bacteria. *Annu Rev Microbiol*, **33**, 113–37.

Towner, K.J. 1992a. Resistance to antifolate antibacterial agents. *J Med Microbiol*, **36**, 4–6.

Towner, K.J. 1992b. Detection of antibiotic resistance genes with DNA probes. *J Antimicrob Chemother*, **30**, 1–2.

Verbist, L. 1990. The antimicrobial activity of fusidic acid. *J Antimicrob Chemother*, **25**, Suppl. B, 1–5.

Wagenlehner, F.M.E., MacKenzie, F.M., et al. 2002. Molecular epidemiology and antibiotic resistance of *Enterobacter* spp. from three distinct populations in Grampian, UK. *Int J Antimicrob Agent*, **20**, 419–25.

Walsh, C.T. 1993. Vancomycin resistance: decoding the molecular logic. *Science*, **261**, 308–9.

Warburton, A.R., Jenkins, P.A., et al. 1993. Drug resistance in initial isolates of *Mycobacterium tuberculosis* in England and Wales, 1982–1991. *Commun Dis Rev*, **3**, R175–9.

Waterworth, P.M. 1978. Sulphonamides and trimethoprim. In: Reeve, D.S., Philips, I., et al. (eds), *Laboratory methods in antimicrobial chemotherapy*. Edinburgh: Churchill Livingstone, 82–4.

Wehrli, W. and Staehelin, M. 1971. Actions of the rifamycins. *Bacteriol Rev*, **35**, 290–309.

Weisblum, B. 1995a. Erythromycin resistance by ribosome modification. *Antimicrob Agents Chemother*, **39**, 577–85.

Weisblum, B. 1995b. Insights into erythromycin action from studies of its activity as inducer of resistance. *Antimicrob Agents Chemother*, **39**, 797–805.

Weisblum, B. and Davies, J. 1968. Antibiotic inhibitors of the bacterial ribosome. *Bacteriol Rev*, **32**, 493–528.

Wiedemann, B. and Heisig, P. 1994. Quinolone resistance in gram-negative bacteria. *Infect Dis Clin Pract*, **3**, Suppl. 3, S115–26.

Widmer, A.F. 2001. New developments in diagnosis and treatment of infection in orthopaedic implants. *Clin Infect Dis*, **33**, S94–S106.

Williams, J.D., Naber, K.G., et al. 2001. Classification of oral cephalosporins. A matter for debate. *Int J Antimicrob Agent*, **17**, 443–50.

Wilson, P., Andrews, J.A., et al. 2003. Linezolid resistance in clinical isolates of *Staphylococcus aureus*. *J Antimicrob Chemother*, **51**, 186–8.

Wolfson, J.S. and Hooper, D.C. 1989. Bacterial resistance to quinolones: mechanisms and clinical importance. *Rev Infect Dis*, **11**, Suppl. 5, S960–8.

World Health Organization. 1995. The use of essential drugs. Sixth report of the WHO Expert Committee. *WHO Tech Rep Ser*, **850**, WHO, Geneva.

Young, L.S. 1993. Mycobacterial diseases in the 1990s. *J Antimicrob Chemother*, **32**, 179–94.

Zhao, Z. and Drlica, K. 2001. Restricting the selection of antibiotic-resistant mutants: a general strategy derived from fluroquinolone studies. *Clin Infect Dis*, **33**, S146–57.

PART IV

ORGAN AND SYSTEM INFECTIONS

19

Bloodstream infection and endocarditis

HARALD SEIFERT AND HILMAR WISPLINGHOFF

BLOODSTREAM INFECTION

INTRODUCTION

Bloodstream infections (BSI) are a major cause of morbidity and mortality. Based on data from death certificates, these infections represent the tenth leading cause of death in the United States (NNIS 2003), and the age-adjusted death rate has risen by 78 percent over the past two decades (NNIS 2000). The true incidence may even be higher. Assuming that 5 percent of hospitalized patients develop an infection, of which 10 percent are bloodstream infections, with an attributable mortality rate of 15 percent, nosocomial bloodstream infections would represent the eighth leading cause of death in the United States (Wenzel and Edmond 2001). Approximately 250 000 cases of nosocomial bloodstream infection occur annually in the United States (Pittet et al. 1997). Recent studies reported the incidence of BSI to be between 1 percent in intensive care unit (ICU) patients (Warren et al. 2001) and 36 percent in bone marrow transplant patients (Collin et al. 2001). A recent survey reported an overall incidence of six cases of BSI per 1 000 hospital admissions, with pathogen specific attack rates between 0.1 and 1.6 per 1 000 admissions

(Wisplinghoff et al. 2004). In ICU patients, incidences between 32 and 54 BSI per 1 000 ICU admissions have been reported (Daschner et al. 1982; Pittet et al. 1996; Darby et al. 1997; Edgeworth et al. 1999). The crude mortality associated with BSI ranges from 12 percent in general hospital populations to 80 percent in ICU patients (Weinstein et al. 1983a; Pittet et al. 1993; Pittet et al. 1994; Digiovine et al. 1999; Warren et al. 2001). The mortality directly attributable to BSI in these populations has been estimated to be between 16 and 40 percent (Pittet et al. 1994; Digiovine et al. 1999). In ICU patients with BSI, the length of stay is prolonged by 7.5 to 25 days and the total hospitalization time by 4.5 to 32 days (Pittet et al. 1994; Digiovine et al. 1999; Warren et al. 2001). Inappropriate empirical antimicrobial therapy is an important predictor of death in patients with BSI (Bryan et al. 1983; Pittet et al. 1994; Kollef and Fraser 2001).

Throughout the 1960s and 1970s, gram-negative organisms were most frequently isolated from patients with BSI. Since then, gram-positive organisms have become increasingly frequent (Edmond et al. 1999; NNIS 2000; Diekema et al. 2002; Wisplinghoff et al. 2004).

Usually bacteremia or BSI originates from a localized infection at a distant site. Such a secondary BSI can usually only be cured if the site of origin is identified and treated which often requires surgical intervention. Therefore, it is highly important for both clinicians and laboratory personnel to consider and evaluate all possible sources of BSI.

Clinical presentation ranges from benign transient bacteremia with little or no symptoms to fulminant septic shock with high mortality. Transient bacteremia may follow manipulation of or surgery in infected or colonized areas. Intermittent bacteremia is usually seen secondary to abscesses, mostly abdominal or pelvic. Continuous bacteremia is associated with endocarditis and other intravascular infections, but may also occur in the first weeks of typhoid fever or brucellosis. In patients with intermittent or continuous BSI, fever is the most common symptom, followed by other unspecific signs of systemic disease, such as elevated heart and respiratory rate, fatigue, malaise, or failure to thrive in neonates. A substantial proportion of patients with infection may also be euthermic or hypothermic and may present with nondescript clinical conditions. Additional symptoms may develop due to metastatic foci, such as seizures or altered personality or level of consciousness.

In addition to the significant health and economic impact of true BSI, vast resources are also consumed by false-positive blood culture (BC) results. Overall, BCs have a relatively low yield and the proportion of false-positive results is high (Shafazand and Weinacker 2002). Studies in unselected patient populations found that between 7.5 and 12 percent of all BC are positive, and among these about 5 percent are considered contaminated (Weinstein et al. 1983b; Roberts et al. 1991; Darby et al. 1997). Recent studies found about 8.6 percent of blood cultures positive in the ICU, with 3.6 percent being due to contaminants (Schwenzer et al. 1994) while in unselected patient populations 14 percent of all blood cultures were positive with 38 percent of those being due to contaminants (Moser et al. 1998). False positive BCs may increase diagnostic testing, antibiotic use, and prolong hospitalization (Shafazand and Weinacker 2002). Bates et al. (1991) showed that false-positive blood cultures increased the length of stay by a median of 4.5 days, and generated additional hospital, pharmacy, and laboratory charges. Some studies have developed rules to assess the probability of BSI in patients prior to obtaining BCs. The clinical value of these rules remains questionable, since some are retrospective or evaluate subsets of patients only (Shafazand and Weinacker 2002).

DEFINITIONS

Even though definitions for most terms used in the context of BSI are exact, there is still confusion due to imprecise and interchanging usage of these terms. Bacteremia or fungemia denote the presence of viable bacteria or fungi in the blood, with or without clinical symptoms. In contrast, systemic inflammatory response syndrome (SIRS) is a clinical term, defined as the presence of at least two of the following: temperature >38°C or <36°C, respiratory rate >20/min or $PaCO_2$ <32 Torr, heart rate >90/min, and leukocytes >12.0 × 10^9/l or <4.0 × 10^9/l or >10 percent bands. Of note, SIRS is independent of the presence of pathogens in the bloodstream. The combination of SIRS with bacteremia or fungemia is called sepsis, severe sepsis, depending on the additional presence of signs of organ dysfunction, such as hypotension or hypoperfusion, i.e. metabolic acidosis, altered mental status, oliguria, or acute respiratory distress syndrome. For detailed definitions, see Table 19.1. Septic shock is defined as severe sepsis with hypotension that is unresponsive to fluid resuscitation. To assess the extent of organ dysfunction, other scores such as the sequential organ failure assessment (SOFA) score (Vincent et al. 1998) may also be used. Recently, the Sepsis Definition Conference (Society of Critical Care Medicine, European Society of Intensive Care Medicine, American College of Chest Physicians and the Surgical Infection Societies) proposed the PIRO concept to classify sepsis, which is based on predisposing factors, infection, response, and organ dysfunction (Vincent 2002; Levy et al. 2003a). Bloodstream infection is frequently used interchangeably with sepsis, but is not as exactly defined, denoting bacteremia with clinical symptoms. Septicemia is an outdated term that was previously used to describe the systemic symptoms caused by bacteria, fungi or their toxins in the blood.

A bacteremia or BSI is called primary if the point of entry or focus cannot be determined, or if it originates from an intravascular catheter. The latter is also referred to as catheter-related BSI. Recent studies have recommended differentiating catheter-related bacteremia from both primary and secondary bacteremia (Renaud and Brun-Buisson 2001). If a distant site other than an IV catheter can be established as the point of origin, BSI is referred to as secondary. A distant site is usually confirmed to be the focus, if the same pathogen (with the same resistance pattern) is isolated from the blood culture and the primary site of infection.

Another way to classify BSI is by the time or setting of acquisition, i.e. community-acquired or nosocomial BSI. Usually community-acquired BSI is defined as a BSI that is detected (onset of symptoms) within the first 48 h after admission, while BSI is referred to as nosocomial if detected more than 48 h after admission. In recent years the number of people with frequent healthcare contacts, including those receiving healthcare including IV therapy at home or at a nursing home, has considerably increased. Therefore it may be more accurate to differentiate between nosocomial BSI (detected >48 h after admission), healthcare-associated BSI (detected <48 h after admission or during an outpatient visit in a patient with frequent healthcare contacts) and

Table 19.1 *Definitions*

Term	Definition
Bacteremia (fungemia)	Presence of viable bacteria or fungi in the blood, irrespective of clinical symptoms
SIRS	Systemic inflammatory response syndrome, defined as presence of at least two of the following: Temperature >38°C or <36°C Respiratory rate >20/min or $PaCO_2$ <32 Torr Heart rate >90/min Leukocytes >12 000/µl or <4 000/µl or >10% bands
Sepsis	SIRS with (proven or suspected) bacteremia (fungemia)
Severe sepsis	SIRS or sepsis with one or more signs of organ dysfunction, hypotension (systolic arterial pressure <90 mmHg) or hypoperfusion such as lactic acidosis (plasma lactate >2.0 mmol/l, altered mental status (GCS of 11 or below), renal failure (acute increase in serum creatinine >180 µmol/l; doubling in admission creatinine level in patients with chronic renal failure; requirement for acute dialysis or ultrafiltration), oliguria (urinary output <0.5 ml/kg per hour for at least 1 h or <30 ml for 2 h), disseminated intravascular coagulation (decrease in platelet count of 25% or more with any increase in prothrombin time), or adult respiratory distress syndrome (PaO_2/FiO_2 <175 with bilateral infiltrates consistent with pulmonary edema in the absence of heart failure; pulmonary capillary wedge pressure <18 mmHg if measured) (Rangel-Frausto et al. 1995)
Septic shock	Severe sepsis with hypotension that is unresponsive to fluid resuscitation and requires the administration of vasopressors
MODS	Multiple organ dysfunction syndrome; dysfunction of more than one organ
Septicemia	Outdated term, previously used for systemic symptoms caused by bacteria, fungi or their toxins in the blood

true community-acquired (detected <48 h after admission or during an outpatient visit in a patient with no or minor healthcare contacts). A recent study proposed to classify BSI as follows: true community-acquired BSI (detected within the first 48 h after admission, in patients who have not been hospitalized within the past 30 days and did not have a recent history of invasive procedures), BSI in recently discharged patients (2–30 days prior to the recent episode of BSI), BSI in patients with a recent history of invasive procedures (including insertion of Foley catheters, placement of intravascular catheters, long-term devices, central venous catheters, or dialysis), nursing home-acquired bacteremia, and true nosocomial BSI (detected >48 h after admission) (Siegman-Igra et al. 2002).

TECHNICAL ISSUES

Indications

Blood cultures are a valuable diagnostic tool and indicate whenever there is clinical evidence of sepsis or an unknown systemic infection (Aronson and Bor 1987). Signs of sepsis are shown in Table 19.2. Also, clinical suspicion of any infection that may be associated with bacteremia or fungemia such as meningitis, cholangitis, endocarditis, and osteomyelitis warrant the cultural examination of the patient's blood. In addition, blood cultures should be obtained in the following clinical conditions: pneumonia, pyelonephritis, septic arthritis, epiglottitis, and omphalitis (infants and neonates), as well as abscesses, and deep-seated or extended skin and soft tissue infections. Typhoid fever, brucellosis or leptospirosis can also be diagnosed by blood cultures (see below). Since neurological or psychiatric symptoms, such as seizures, acutely altered mental status, or TIA-like symptoms can be due to metastatic foci of BSI or secondary emboli following endocarditis, blood cultures are also highly recommended in these clinical situations. Blood cultures are also imperative in the diagnostic evaluation of fever of unknown origin since this may be caused by a chronic inflammatory process, such as an abscess associated with intermittent BSI. In critically ill patients, blood cultures should be performed in all patients with a new fever, even if the clinical findings do

Table 19.2 *Signs of sepsis (Vincent and Jacobs 2003)*

Category	Sign
General	Chills, fever or hypothermia, tachypnea or respiratory alkalosis, positive fluid balance
Inflammation	Leukocytosis or leukopenia, elevated CRP, IL-6, procalcitonin
Hemodynamic	Hypotension, tachycardia, centralization (cold extremities), oliguria, hyperlactatemia
Organ dysfunction	Hypoxemia, altered mental status, hyperglycemia, thrombocytopenia, altered hepatic function, altered renal function, nutritional imbalance

not strongly suggest an infectious cause (O'Grady et al. 1998). Furthermore, a substantial proportion of patients with infection are euthermic or hypothermic and may present with nondescript clinical symptoms. These include elderly patients or neonates, patients with open abdominal wounds or extensive burns, patients receiving extracorporal membrane oxygenation, or taking anti-inflammatory or antipyretic drugs (Vincent and Jacobs 2003). In these patients blood cultures can also be a valuable tool for detecting potentially life-threatening infection.

Blood cultures should be considered in an even broader range of situations in compromised patients with underlying conditions that predispose for BSI. This group of patients includes neonates (especially with low birth weight) and elderly patients, but also patients with neutropenia and other immunosuppressive conditions, patients with (artificial) grafts, especially vascular grafts or prosthetic valves, or intravascular access devices (central venous catheters or permanently implanted devices, such as ports or Hickman catheters). Patients in the setting of the ICU and patients hospitalized secondary to severe trauma or severe burns should also be considered in this context.

SPECIMEN COLLECTION AND TRANSPORT

The term 'blood culture' denotes a specific volume of blood which is aseptically drawn by a single veni-puncture, inoculated into culture medium or media, and investigated for the presence of microorganisms. Usually an aerobic and an anaerobic vial are used, thus one blood culture – often also referred to as one blood culture set – would be equivalent to two bottles. Each blood culture should be obtained by a separate veni-puncture (Smith-Elekes and Weinstein 1993). Obtaining blood cultures through indwelling intravascular catheters is generally not recommended, unless the involvement of the catheter as a source of BSI is to be investigated.

Disinfection

Since the rapid and reliable recovery of even small numbers of bacteria from the blood is the primary goal of blood cultures, the media also provide excellent growth conditions for contaminants and contamination poses a serious problem. False-positive blood culture results may lead to unnecessary, wrong, and potentially harmful antimicrobial therapy and increase length of stay and costs (Bates et al. 1991). In addition, an increasing number of true bacteremia and BSI in immunocompromised patients is caused by bacteria that are part of the common skin flora, such as coagulase-negative staphylococci (CoNS), corynebacteria, and *Bacillus* species. Correct interpretation of such findings largely depends

on the assumption that proper aseptic techniques have been used for specimen collection and transport.

Blood must always be collected with a sterile needle and syringe or with a proprietary sterile system of needle and evacuated tube or bottle of blood culture medium. Thorough disinfection of the patient's skin is mandatory to reduce the contamination rate (Strand et al. 1993). For scrubbing of the venipuncture site, 70 percent alcohol or iodine solutions such as 10 percent povidone iodine or a 1–2 percent tincture of iodine or a combination of both has been recommended (Weinstein et al. 1983b; Strand et al. 1993; Dunne et al. 1997; O'Grady et al. 1998). Since none of the disinfectant solutions used to date works instantly, it is generally recommended to let the solution dry for at least 30 s before proceeding. Gloves should be worn by the operator after disinfection of the hands; some institutions even recommend sterile surgical gloves.

Time of collection

To obtain the maximum yield of organisms, blood should be collected immediately before the onset of chills. Since prediction of this event is rarely possible, it is generally recommended to draw blood for culture when the patient's temperature rises, as soon as possible after the onset of fever or chills or in connection with other symptoms that indicate sepsis (see Definitions). There is no evidence to support basing the time of collection on specific temperature thresholds (e.g. 39°C). It has also been recommended to draw blood cultures at arbitrary intervals (Strand and Shulman 1988), but others did not find an increase in sensitivity if blood cultures were obtained simultaneously, instead of spaced over time (Li et al. 1994). Intervals should be based on the patient's clinical condition, and should not result in delaying therapy in critically ill patients (O'Grady et al. 1998). Whenever possible, blood cultures should be obtained before antimicrobial chemotherapy is initiated, because ongoing antimicrobial chemotherapy significantly reduces the yield. If a patient is already receiving antimicrobials, blood cultures should be drawn towards the end of a dosing interval. If possible, antimicrobial therapy should be stopped and cultures should be obtained 24–48 h after the last dose was administered.

Collection site

Usually blood for culture is drawn from a peripheral vein, e.g. vena cubitalis. Contamination is more likely if the femoral vein or a vein in close proximity to an inflammation or skin infection site is used (Washington 1975). In neonates, special blood culture systems can be used to obtain blood from better accessible sites. Arterial blood does not offer any advantage with regard to detection of pathogens, not even for endocarditis or fungemia (Tenney et al. 1982; Vaisanen et al. 1985). The reliability and diagnostic impact of blood cultures drawn through an intravascular catheter is controversial (Tonnesen et al. 1976; Bryant and Strand 1987). Contamination rates are usually higher and catheters may be colonized with organisms that do not necessarily invade the bloodstream. However, blood from this source, even via different lumens of the catheter, may be used in conjunction with peripheral venous samples to assess the presence of a catheter-related BSI. In this case quantitative blood cultures (lysis centrifugation method) or differential time to positivity when using a continuously monitoring blood culture system (CMBCS) can be used (Blot et al. 1999; Mermel et al. 2001; Rijnders et al. 2001; Gaur et al. 2003; Seifert et al. 2003a).

Volume

In adults, bacteremia is usually characterized by <1 bacteria per ml of blood (i.e. colony forming units (cfu)/ml) (Ilstrup and Washington 1983; Plorde et al. 1985; O'Hara et al. 2003). The volume of blood obtained for culture is therefore one of the most important variables influencing the diagnostic yield (Reller et al. 1982; Wilson et al. 1994). The diagnostic yield of a blood culture is directly related to the volume of blood obtained for culture (Hall et al. 1976; Tenney et al. 1982; Ilstrup and Washington 1983; Plorde et al. 1985) and increases by approximately 3–5 percent per ml blood taken (Hall et al. 1976; Tenney et al. 1982; Mermel and Maki 1993). In clinical practice, usually 10–20 ml of blood per culture is obtained and distributed evenly into an aerobic and an anaerobic culture bottle, leading to a final volume of 5–10 ml blood per bottle (Tenney et al. 1982). According to guidelines, 10 ml per culture are reasonable with an optimal volume being 20–30 ml per culture, depending on the number and type of bottle used (Washington and Ilstrup 1986; Dunne et al. 1997). The ratio between blood and culture media (blood to broth) should be 1:5–1:10, in order to neutralize the bactericidal activity of human serum (see below). For details on the optimal volume of blood per bottle, the instructions of the respective manufacturers should be consulted. Even though the diagnostic impact of the anaerobic vial is controversial (Sharp et al. 1991; Murray et al. 1992), the use of two vials ensures the culture of an adequate blood volume increasing the probability of recovery.

In pediatric patients a smaller volume (1–5 ml) is usually sufficient, because the concentration of microorganisms in these patients is higher (>100 cfu/ml blood) (Dietzmann et al. 1974; Szymczak et al. 1979; La et al. 1981; Reller et al. 1982). Even though less than 1 ml of blood may allow detection of microorganisms in infants if the concentration is high enough (Dietzmann et al. 1974; Neal et al. 1986), the diagnostic yield in these

patients is also higher if a bigger blood volume is obtained (Szymczak et al. 1979; Isaacman et al. 1996; Kellogg et al. 2000). Recent studies showed that the detection rate in pediatric patients doubled if 6 ml of blood were obtained instead of 2 ml (Isaacman et al. 1996). Showing that low-level BSI does occur in children, some authors recommend obtaining 4–4.5 percent of a child's blood volume for culture (Kellogg et al. 2000). Specific 'pediatric' blood culture bottles contain a decreased volume of broth, and are supplemented with X and V factors and sodium-polyanetholesulfonate (SPS) to enhance detection of microorganisms. In these bottles, a decreased volume is necessary to maintain an optimal broth:blood ratio. However, data proving the necessity of using these bottles remain rare (O'Hara et al. 2003).

The protective covers of the vials should be removed close to inoculation. The membrane should be disinfected using sterile alcohol swabs. As with any disinfection, contact times should be observed and the membranes must be completely dry before inoculation. Based on the data of a meta-analysis, a new needle should be used to inoculate the vials in order to reduce contamination (Spitalnic et al. 1995). However, others have found no significant increase in contamination rates when a single-needle technique was used and have argued that the benefit of the reduced risk of needle-stick injuries of the single-needle technique outweighs the increase in risk of contamination (Mylotte and Tayara 2000b). The atmosphere of aerobic vials used for automated blood culture processing in CMBCS is enriched with oxygen, which eliminates the need for transient venting after inoculation (Rohner et al. 1996).

Number of blood cultures

The detection rate increases with the number of blood cultures obtained. In patients with endocarditis, studies found a sensitivity of 80 percent if one blood culture was used, which increased to 88 and 99 percent if two or three independent cultures, respectively, were obtained over a 24 h period (Washington 1975; Weinstein et al. 1983b). Other studies in patients with BSI found 99 percent sensitivity if only two cultures were drawn (Weinstein et al. 1983b). Therefore, more than three blood cultures do not increase the diagnostic yield, but do increase costs and may cause iatrogenic anemia (Neu 1986). Also, the culture of only a single blood sample should be highly discouraged (Aronson and Bor 1987). There is currently no recommendation for an optimal time difference between the first and the second culture. In patients with suspected endocarditis, two to three blood cultures should be obtained within 24 h. In severely ill patients, it may be necessary to obtain two cultures (with separate venipunctures) in rapid succession in order to start antimicrobial therapy as soon as possible (Smith-Elekes and Weinstein 1993). A recent study failed to show any benefit in yield of obtaining multiple samples, and recommended obtaining a total volume of 35–42 ml blood for six bottles at the same time (Lamy et al. 2002). However, for interpretation of results more than one separate culture (venipuncture) is needed. In neonates, drawing of a second culture is rarely possible. However, in infants or older children, two vials are mandatory, but the inoculation volume may be as low as 1 ml per vial (see Volume).

Media

Depending on the indication, the patient, and the suspected pathogen, the use of special culture media or a different number of bottles per blood culture may be useful. Commercial media that facilitate growth of almost any pathogen are readily available. Nutrient-rich broth in a variety of formulations can be used to culture microorganisms from the blood. These include brain–heart infusion (BHI) broth, tryptic-soy-broth (TSB), soybean casein digest broth, peptone broth, and Columbia media, which may be supplemented with hemin or NADH. Generally, one vial contains 50–100 ml of media and a 10 percent CO_2 atmosphere in the headspace. Therefore, vials can be used to culture aerobic, facultative, and anaerobic bacteria. An aerobic and an anaerobic bottle are usually obtained, i.e. one containing ambient atmosphere in the headspace, the other various amounts of CO_2. Due to the decrease of BSI caused by anaerobic bacteria over the past decades (Dorsher et al. 1991; Lombardi and Engleberg 1992; Murray et al. 1992; Dunne et al. 1997; Weinstein et al. 1997; Wisplinghoff et al. 2004), some authors have suggested that routine use of anaerobic bottles is not necessary, favoring two aerobic bottles to increase the volume of blood cultured (Murray et al. 1992; Morris et al. 1993; Sharp et al. 1993; Zaidi et al. 1995). However, recent surveys found anaerobes accounting for about 4 percent of BSI (Diekema et al. 2003) and anaerobic blood cultures enhance the isolation of other species, such as streptococci including *Streptococcus pneumoniae*, enterococci, and *Listeria monocytogenes* (Cockerill et al. 1997).

Inhibition of antimicrobial agents

Microbial growth in human blood is inhibited by a number of substances including antibodies, leukocytes, complement, lysozyme, and antimicrobial substances, which about one-third of patients already receive, when blood cultures are obtained (Weinstein et al. 1997). Blood clots also reduce the diagnostic yield. Virtually all commercially available blood culture bottles and media contain SPS (Wilson et al. 1994), an anticoagulant that inhibits lysozomal activity, complement, and phagocytosis, in addition to reducing blood clots (Traub

and Kleber 1977; Washington and Ilstrup 1986; Chandrasekar and Brown 1994). Also, the 1:5 to 1:10 dilution of the blood enhances detection of bacteria by neutralizing the endogenous antibacterial activity of the human serum and diluting antimicrobial agents to subinhibitory concentrations. Aminoglycosides and polymyxins are also inhibited by SPS. Side effects of SPS include some inhibition of the growth of *Neisseria gonorrhoeae*, *Neisseria meningitidis*, *Gardnerella vaginalis*, *Peptostreptococcus anaerobius*, and *Moraxella catarrhalis* and a growth enhancement for gram-positive cocci while gram-negative rods are generally inhibited (Staneck and Vincent 1981; Edberg and Edberg 1983; Reimer and Reller 1985; Reimer et al. 1987). Others found that increasing concentrations of SPS increase the recovery of gram-negative bacteria (Chandrasekar and Brown 1994; Wilson et al. 1994). Despite these effects, no other anticoagulant to date is superior. Antimicrobial agents can also be neutralized by adding unspecific adsorbants such as antimicrobial removal device (ARD) (Peterson et al. 1983), antibiotic binding resins on glass beads (BACTEC) (BD, Sparks, MD), or activated charcoal and Fuller's earth (Ecosorb, BacT/ALERT, bioMérieux Inc., Durham, NC). Supplemented media have shown increased detection rates (Weinstein et al. 1995; Doern et al. 1997), but also increased the number of detected contaminants (Weinstein et al. 1995). Studies of the effects of the antimicrobial removal device have shown conflicting results (Appleman et al. 1982; Peterson et al. 1983; Doern 1994) and routine use is generally not recommended (Shafazand and Weinacker 2002).

Transport

Rapid transport of blood cultures to the microbiology laboratory is mandatory. Blood cultures used in CMBCS can be stored at room temperature if direct transport to the laboratory is not possible. Refrigeration of blood cultures should be avoided. In the laboratory, all blood cultures should be processed immediately. Incubation of blood cultures before transport to the microbiology laboratory is not needed with bottles used in CMBCS. If cultures have been incubated prior to arrival at the microbiology laboratory, Gram stains and blind subcultures of all bottles need to be performed before the cultures are processed in the CMBCS.

SAMPLE PROCESSING

In patients with BSI, the appropriateness of antimicrobial chemotherapy is correlated with survival (Ibrahim et al. 2000). The first notification of positive blood cultures has a high impact on the antimicrobial therapy (Schonheyder and Hojbjerg 1995). Samples should therefore be processed as fast as possible. Proces-

Table 19.3 *Comparison of blood culture methods*

Method	Aerobic bacteria	Anaerobic bacteria	Yeast	Molds	Mycobacteria
Conventional	+	+	−/+	−	−
CMBCS	++	+/++	+/++	−	++[a]
Lysis centrifugaton	+/++	−	++	++	++

Modified from (O'Grady et al. 1998).
CMBCS, continuous monitoring blood culture systems; −, not recommended; +, acceptable; ++, good.
a) With use of special mycobacterial culture bottles only.

sing of blood cultures generally includes incubation, Gram staining, and subcultures. Manual processing has continuously decreased over the last decade and most laboratories now use automated blood culture systems. Several authors have compared different processing methods and systems. Table 19.3 compares commonly used blood culture methods.

Manual processing

All manual processing where aerosols could be generated, should be done under a protective hood, both to avoid contamination and to protect laboratory personnel. All blood cultures should be incubated for at least 6–18 h at $36 \pm 1°C$ before processing. Agitation during incubation increases sensitivity and speed of detection (Hawkins et al. 1986; Weinstein et al. 1989; Prag et al. 1991). Afterwards, blind subcultures and microscopic examination of all aerobic cultures should be performed, using approximately 0.2 ml of broth, which should be removed under sterile precautions after gentle shaking. For microscopic examination, Gram or acridinorange stains can be performed. Gram stains are easier to perform, but may lack sensitivity. While acridinorange stains detected 10^4 cfu/ml, Gram stains only detected 10^5 cfu/ml (Seifert et al. 1997). If bacteria are seen, the respective anaerobic (or aerobic) bottle should also be subcultured. Blood-agar, chocolate-agar, anaerobic media, and some differential media (e.g. MacConkey) should be used for subculture of blood cultures and subsequently incubated at $36 \pm 1°C$ in an aerobic atmosphere with 5–10 percent CO_2. Also if signs of turbidity, change of color, or gas production are detected, the respective culture bottle must be processed (i.e. subculture and Gram stain).

Blood cultures do not need to be routinely incubated for more than 7 days. Incubation periods are commonly extended if slow growing organisms or infective endocarditis are suspected, even though some authors noted that in the latter this practice rarely increases sensitivity (Washington 1994). If the cultures remain negative, Gram stains and blind subcultures of the aerobic bottle should be performed again at the end of the incubation period. Blind subcultures of anaerobic bottles do not increase diagnostic yield (Dunne et al. 1997). For special processing requirements, see also individual pathogens.

Automated processing

The first automated blood culture system, BACTEC 460 (Becton Dickinson), was introduced in the 1970s. The use of automated blood culture systems rapidly increased after the introduction of continuous monitoring blood culture systems in the early 1990s. As the earliest systems, most CMBCSs are based on the detection of CO_2, which is produced as microorganisms grow and utilize carbohydrate substrates. While the BACTEC 460 used ^{14}C for detection, the newer instruments of the BACTEC 9000 series use a fluorescence-sensing mechanism. Detection criteria are a linear increase in fluorescence intensity and a change in the increase rate of fluorescence intensity. The BacT/ALERT 3D (bioMérieux) which is the updated version of the BacT/ALERT uses a colorimetric CO_2 sensor system. With emitting and sensing diodes in the incubator units, growth is detected if reflection exceeds an arbitrary threshold, a linear increase or a change in the increase rate are detected. In contrast, the ESP blood culture system (Trek Diagnostic Systems, Inc.) monitors the pressure changes within the bottles' headspace, which changes with the consumption or production of gases (oxygen, hydrogen, nitrogen, CO_2) by growing microorganisms. In all systems cultures are monitored at intervals between 10 and 24 min, 24 h per day, shortening time to detection by 1–1.5 days (Wilson et al. 1992; Morello et al. 1994; Reimer et al. 1997). A 5-day incubation in a CMBCS is usually sufficient to detect most pathogens (Masterson and McGowan 1988; Hardy et al. 1992; Wilson et al. 1993) and even periods of 4 (Dorsher et al. 1991; Johnson et al. 2000) or 3 (Han and Truant 1999; Bourbeau and Pohlman 2001) days have been suggested.

As in manual processing, Gram stains and subcultures must be performed if growth of microorganisms is detected in a culture. Of note, some systems can only detect an increase in CO_2 production, and may not detect a positive culture if the bacterial growth has already entered the static phase. Therefore, if the blood cultures have been incubated previously or obtained more than 24 h prior to processing, it is recommended that Gram stain and blind subculture of both bottles be performed before the culture vials are placed into the system. Blind subcultures of aerobic or anaerobic bottles at the end of the incubation period do not increase the diagnostic yield (Dunne et al. 1997).

Quantitative blood cultures

The lysis centrifugation system (ISOLATOR, Du Pont) is the only commercial quantitative blood culture system. Processing of cultures is intensive in both time and labor and usually not used for routine diagnostic purposes. Vials contain saponin, EDTA, and SPS and must be subcultured on aerobic and anaerobic media after centrifugation. Samples should be processed within 8–10 h after collection. For detailed procedures the reader is referred to the instructions of the respective manufacturer. However, some advantages of the lysis centrifugation method over qualitative blood cultures have been reported: diagnosis is usually faster than with conventional blood cultures. Also diagnostic yield of the ISOLATOR system is higher in some pathogens such as fungi (see Table 19.3).

INTERPRETATION OF RESULTS

Interpretation of the clinical significance of positive and negative blood culture results can be problematic. Misinterpretation can lead to antimicrobial therapy which is expensive and may lead to adverse events in the patient. Recovery of *Staphylococcus aureus*, *S. pneumoniae*, *Escherichia coli* and other Enterobacteriaceae, *Pseudomonas aeruginosa* and *Candida albicans* almost always represents true infection. Other microorganisms including *Corynebacterium* species, *Bacillus* species, and *Propionibacterium* species only rarely cause true BSI (Weinstein et al. 1997). The majority of coagulase-negative staphylococcal isolates also represent contamination rather than true infection. To aid interpretation a number of approaches have been suggested. In addition to the identity of the microorganism, the presence of the same organism in more than one independently obtained culture, or the isolation of the same organism from another usually sterile body site may provide valuable aid for interpretation. Correlation of laboratory findings with the clinical status of the patient is mandatory. Generally, clinically significant BSI is likely if two or more separate blood cultures grow the same organism (Weinstein 1996; Kim et al. 2000; McDonald et al. 2001). Interpretation of CoNS isolated from blood cultures is particularly difficult, since 12–15 percent of CoNS bloodstream isolates are the causative pathogens of clinical significant BSI (Weinstein et al. 1997). While in this instance the number of positive cultures in a set is not a reliable criterion (Peacock et al. 1995; McDonald et al. 2001), the number of positive cultures relative to the number of obtained cultures appears to be useful (Peacock et al. 1995; Weinstein et al. 1983b; Weinstein et al. 1997).

PATHOGENS

Almost any pathogen can be isolated from the bloodstream as a causative agent of clinically significant bacteremia, fungemia or BSI, especially in immunocompromised patients. In addition, in patients with a corresponding history (e.g. travel), parasitic and tropical diseases, which may require a different diagnostic approach should also be considered.

The spectrum of pathogens has changed considerably over the last decades (Figure 19.1). In recent studies, about 13 percent of all episodes were polymicrobial, 60–65 percent were caused by gram-positive and 25–31 percent by gram-negative organisms. Fungi, mainly *Candida* species, were isolated from 7–10 percent of episodes. Anaerobic bacteria accounted for about 1 percent of BSI (Edmond et al. 1999; Diekema et al. 2003; Wisplinghoff et al. 2004). The most common pathogens isolated from BSI are listed in Table 19.4. Proportions of individual organisms vary with age, gender, underlying conditions, and clinical service.

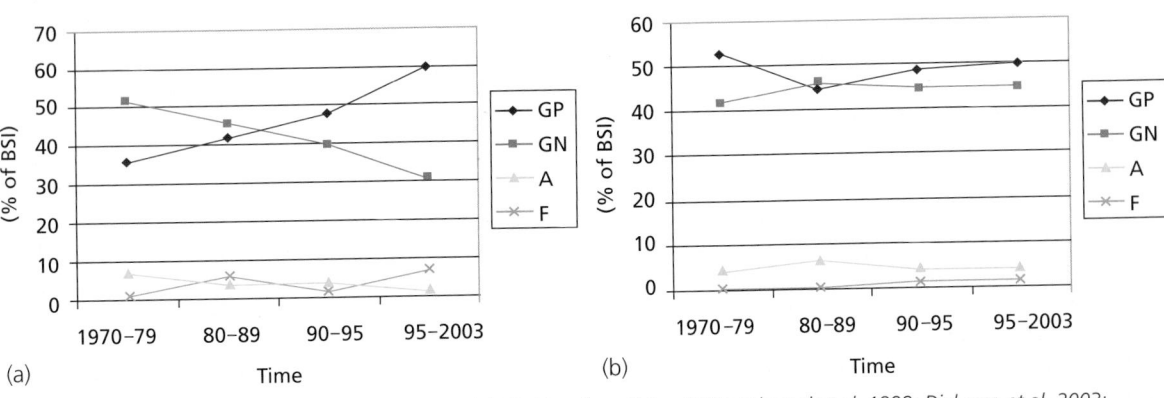

Figure 19.1 (a) *Trends in pathogens causing nosocomial BSI (data from Eykyn 1998; Edmond et al. 1999; Diekema et al. 2003; Wisplinghoff et al. 2004).* **(b)** *Trends in pathogens causing community-acquired BSI. A, anaerobic bacteria; F, fungi; GN, gram-negative pathogens; GP, gram-positive pathogens.*

Table 19.4 *Incidence rates and distribution of pathogens most commonly isolated from monomicrobial BSI and associated crude mortalities for all patients, ICU patients, and patients in the non-ICU setting (Wisplinghoff et al. 2004)*

Pathogen	BSI per 10 000 admissions	Percent of BSI [rank]			Crude mortaility (%)		
		Total (*n* = 20 978)	ICU (*n* = 10 515)	Ward (*n* = 10 442)	Total	ICU	Ward
CoNS	15.8	31.3 [1]	35.9 [1]a	26.6 [1]	20.7	25.7	13.8
S. aureus **	10.3	20.2 [2]	16.8 [2]a	23.7 [2]	25.4	34.4	18.9
Enterococci*	4.8	9.4 [3]	9.8 [4]	9.0 [3]	33.9	43.0	24.0
Candida spp.*	4.6	9.0 [4]	10.1 [3]a	7.9 [4]	39.2	47.1	29.0
E. coli	2.8	5.6 [5]	3.7 [8]a	7.6 [5]	22.4	33.9	16.9
Klebsiella spp.	2.4	4.8 [6]	4.0 [7]a	5.5 [6]	27.6	37.4	20.3
P. aeruginosa	2.1	4.3 [7]	4.7 [5]	3.8 [7]	38.7	47.9	27.6
Enterobacter spp.	1.9	3.9 [8]	4.7 [6]a	3.1 [8]	26.7	32.5	18.0
Serratia spp.**	0.9	1.7 [9]	2.1 [9]a	1.3 [10]	27.4	33.9	17.1
A. baumannii	0.6	1.3 [10]	1.6 [10]a	0.9 [11]	34.0	43.4	16.3

Ward Rank 9: *Bacteroides* spp. (*n* = 150; 1.4%).
BSI, bloodstream infection; CoNS, coagulase-negative staphylococci.
a) $p < 0.05$ for ICU vs. non-ICU patients; more frequent in patients with neutropenia, *, or without neutropenia, **.

GRAM-POSITIVE ORGANISMS

In the 1980s, gram-positive organisms re-emerged as the leading pathogenic causes of nosocomial bloodstream infections (Figure 19.1). Today, gram-positive organisms account for about 65 percent of BSI and the three leading pathogens are gram-positive (Table 19.4). Most gram-positive bacteria causing BSI belong to the genus *Staphylococcus* or *Streptococcus/Enterococcus* (Edmond et al. 1999; Diekema et al. 2003; Wisplinghoff et al. 2004).

Staphylococcus aureus

S. aureus is a major cause of morbidity and mortality. Localized infections have a tendency to spread to the surrounding tissue or frequently via the bloodstream to distant sites. Commonly isolated from patients with both community-acquired and nosocomial BSI, *S. aureus* is the second most common cause of nosocomial BSI, accounting for 20 percent of these infections with an incidence of about 1 per 1 000 hospital admissions (Wisplinghoff et al. 2004). Incidence varies with age ranging from about 10 percent in patients under the age of 1 year to about 24 percent in patients over the age of 65 years (Diekema et al. 2002; Wisplinghoff et al. 2004). *S. aureus* BSI usually originates from a localized infection, an intravascular catheter or a contaminated syringe (e.g. in IV drug users). For example, a high correlation has been reported between *S. aureus* BSI and mediastinitis in patients with coronary artery bypass graft (CABG) (Fowler et al. 2003) and isolation of *S. aureus* even from a single blood culture should prompt the search for a primary focus of infection. Clinical manifestations range from benign transient bacteremia with a single episode of fever and chills to fulminant septic shock. Crude mortality associated with nosocomial *S. aureus* BSI ranges from about 18 to 34 percent; in patients with community-acquired *S. aureus* BSI, mortality rates ranging between 12 percent in unselected patient populations and 58 percent in critically ill patients have been reported (Diekema et al. 2003; Valles et al. 2003; Wisplinghoff et al. 2004).

Methicillin resistance, first detected in the 1960s (Barber 1961), emerged as a major clinical and epidemiological problem in hospitals in the 1980s. In hospitals in the United States, the proportion of isolates of methicillin-resistant *S. aureus* (MRSA) have increased from 2.4 percent in 1975 to up to 55 percent in recent years (NNIS 2003). Whether resistance to methicillin is related to transmission or virulence is controversial. Several clinical studies could not identify significant differences in the mortality rate for patients with infections due to methicillin-resistant strains and that for patients with methicillin-susceptible *S. aureus* (MSSA) strains, after correcting for various confounding variables (Hershow et al. 1992; Mylotte et al. 1996; Pujol et al. 1996; Harbarth et al. 1998; Wisplinghoff et al. 2001). In contrast, other studies have suggested that outcome of methicillin-resistant *S. aureus* BSI is significantly worse and that methicillin resistance is an independent predictor of adverse outcome of patients with *S. aureus* infections (Rello et al. 1994; Romero-Vivas et al. 1995; Blot et al. 1998; Conterno et al. 1998; Blot et al. 2002b; Cosgrove et al. 2003). Even though MRSA has been a nosocomial pathogen, there are an increasing number of reports of MRSA BSI in patients with little or no association to the healthcare system (Frank et al. 1999; Fergie and Purcell 2001; Groom et al. 2001; Naimi et al. 2001; Wu et al. 2002; Aires and De Lencastre 2003; Baggett et al. 2003; CDC 2003; Johnson et al. 2003; Said-Salim et al. 2003), suggesting the emergence of true community-acquired MRSA. In recent years there have been a number of reports of glycopeptide or vancomycin-intermediate *S. aureus* isolates (Rotun et al. 1999; Smith et al. 1999; Wong et al. 2000; Fridkin 2001; Hageman et al. 2001; Fridkin et al. 2003; Naimi et al. 2003) and recently, a *S. aureus* isolate with high-level vancomycin-resistance conferred by the *vanA* gene has been recovered from a catheter tip culture (Johnson et al. 2003).

Small colony variants (SCV), a naturally occurring subpopulation of *S. aureus* have also been identified as a cause of BSI (Seifert et al. 2003b). *S. aureus* SCVs have been implicated in persistent and recurrent infections that give a poor clinical and bacteriological response to standard antimicrobial therapy in patients with BSI, chronic osteomyelitis, cystic fibrosis, and AIDS particularly after prolonged exposure to antibiotics (von Eiff et al. 1997; Kahl et al. 1998; Proctor et al. 1998; Seifert et al. 1999, 2003b). These phenotypic variants are characterized by their fastidious growth and atypical colony morphology on routine media, making recovery as well as correct identification difficult for microbiological laboratories.

Delay of therapy can have deleterious effects on clinical outcomes in patients with *S. aureus* BSI, and appropriate therapy should be initiated without delay (Lodise et al. 2003).

Coagulase-negative staphylococci

BSI caused by Coagulase-negative staphylococci (CoNS) is mostly nosocomial, but CoNS have also been isolated from about 8 percent of patients with community-acquired BSI (Diekema et al. 2003). According to several surveys, CoNS are the most prevalent pathogen isolated from the blood accounting for 15–30 percent of blood culture isolates (Moser et al. 1998; Edmond et al. 1999; NNIS 2000; Diekema et al. 2003; Wisplinghoff et al. 2004). This difference is mostly due to the definitions used in different studies, with lower incidences being reported if only cases were included where CoNS were isolated from two or more separate blood cultures. The fact that CoNS also constitute a major part of the normal microflora makes interpretation of positive blood culture results difficult and if possible detection should be confirmed by more than one blood culture. Although several approaches have been suggested to aid interpretation, this issue remains controversial (see Interpretation of results). *S. epidermidis*, the most frequently isolated species among CoNS has been implicated in a variety of infections, including BSI (Rupp and Archer 1994), mostly associated with intravascular catheters and other prosthetic devices. *S. haemolyticus*, the second most commonly isolated species among CoNS has also been associated with BSI and other infections (Kloos and Bannerman 1994; Rupp and Archer 1994). A variety of other CoNS have also been implicated in BSI, including *S. capitis*, *S. hominis*, *S. lugdunensis*, *S. saprophyticus*, *S. schleiferi*, *S. simulans*, and *S. warneri*. The proportion of CoNS among blood culture isolates seems to decrease with increasing age (Diekema et al. 2002; Wisplinghoff et al. 2004). Crude mortality in patients with CoNS BSI ranges from 14 to 31 percent (Diekema et al. 2003; Wisplinghoff et al. 2004).

Enterococcus species

Enterococcus species are the third most common cause of BSI in the United States (Murray 1990; Schaberg et al. 1991; Moellering 1992; Wisplinghoff et al. 2004), accounting for about 9 percent of nosocomial BSI (Edmond et al. 1999; Wisplinghoff et al. 2004). Enterococcal BSI is mainly nosocomial, and community-acquired BSI due to enterococci should always raise the suspicion of endocarditis. *Enterococcus faecalis* is the most prevalent species, accounting for 80–90 percent of clinical isolates, followed by *E. faecium* (5–10 percent of isolates) (Facklam and Collins 1989; Gordon et al. 1992; Buschelman et al. 1993; Stern et al. 1994). Other enterococci that are rarely isolated from blood include *E. casseliflavus*, *E. gallinarum*, and *E. raffinosus*. As with other gram-positive pathogens, the incidence of nosocomial BSI due to enterococci has been increasing over the past decades (Graninger and Ragette 1992; Maki and Agger 1988; Edmond et al. 1999; Wisplinghoff et al. 2004) and some studies found *Enterococcus* species particularly involved in polymicrobial BSI (Maki and Agger 1988). Primary BSIs are generally seen in highly compromised or immunosuppressed patients and may be due to inapparent bacterial translocation from a gastrointestinal source (Shlaes et al. 1981b; Graninger and Ragette 1992; Linden et al. 1996) or intravascular catheters. Secondary BSI is more frequently seen than primary BSI and usually derives from the urinary or gastrointestinal tract, or from wound infections (Shlaes et al. 1981b; Maki and Agger 1988; Graninger and Ragette 1992). Crude mortality of enterococcal BSI ranges from 34 percent in unselected patients populations to 43 percent in ICU patients (Wisplinghoff et al. 2004). Complications such as septic shock, metastatic infection (except endocarditis, see below) or adverse outcome are rare in monomicrobial enterococcal BSI (Maki and Agger 1988). *E. faecalis* isolates are mostly susceptible to ampicillin and vancomycin, while *E. faecium* isolates are usually susceptible to vancomycin only. Vancomycin-resistant strains emerged in the late 1980s in Europe and have become an increasing nosocomial problem in the United States. Several authors have reported that vancomycin-resistant enterococcal BSI contributes significantly to excess mortality and economic loss (Bhavnani et al. 2000; Lodise et al. 2003; Song et al. 2003) and that early, effective antimicrobial therapy is associated with a significant improvement in survival (Vergis et al. 2001).

Streptococcus pneumoniae

In *S. pneumoniae*, BSI is usually community-acquired and secondary to pneumonia, meningitis or otitis media (Raz et al. 1997). In a recent study about 10 percent of pneumococcal BSI were nosocomial (Canet et al. 2002). The overall incidence of invasive pneumococcal disease is about 15 per 100 000 person-years (Fedson et al. 1998), even though it may be up to ten-fold higher in certain risk populations (Davidson et al. 1994; Torzillo et al. 1995). Neonates and children under the age of 2 years are more likely to develop pneumococcal BSI, as are elderly people over the age of 65 years (Burman et al. 1985; Breiman et al. 1990) with incidences of 160 and 70 per 100 000 person-years, respectively compared to 5 per 100 000 in young adults (Breiman et al. 1990). Incidence of pneumococcal BSI is seasonal with a peak in the cold season (Kim et al. 1996). Meningitis generally occurs in a relatively small number of cases of community-acquired pneumococcal BSI (Gruer et al. 1984), but has been reported in up to 17 percent of patients with BSI by Kuikka et al. (1992). A variety of factors have been identified that predispose to invasive pneumococcal disease, such as immunosuppresion, alcoholism, and chronic lung disease (Burman et al. 1985; Fang et al. 1990; Rahav et al. 1997; Watanakunakorn and Bailey 1997). Severe cases of BSI due to *S. pneumoniae*, often associated with a fatal outcome, occur in children under the age of 2 years or who are physically or functionally asplenic, such as children with sickle cell disease in whom a 100-fold higher incidence of pneumococcal BSI has been reported (Wara 1981; Wong et al. 1992).

While pneumonia and meningitis are the most frequent entities associated with pneumococcal BSI, other manifestations include spontaneous bacterial peritonitis (Wilcox and Dismukes 1987; Gorensek et al. 1988; Shaked and Samra 1988), endo- or pericarditis (Powderly et al. 1986), epidural and brain abscesses, and soft tissue infections (Peters et al. 1989; DiNubile et al. 1991; Grigoriadis and Gold 1997).

Rates of penicillin resistance among peneumococci isolated from the blood increased over the past decade. Recent studies reported a percentage of penicillin-resistant isolates of 25–60 percent with a high degree of variation between individual countries, cities, or hospitals (Doern et al. 1998; Yu et al. 2003). Recent studies in patients with pneumococcal BSI found no correlation between penicillin resistance and outcome (Maugein et al. 2003; Yu et al. 2003).

Other gram-positive organisms

In otherwise healthy adults other gram-positive organisms rarely cause BSI. Among *Streptococcus* species other than *S. pneumoniae*, BSI can also be due to viridans group streptococci as well as serogroups A (e.g. *S. pyogenes*), B (e.g. *S. agalactiae*), D (e.g. enterococci), and G (Facklam 2002). Viridans group streptococci, part of the normal human commensal flora, have been associated with subacute endocarditis (see below), especially *S. gordonii*, *S. mitis*, *S. oralis*, and *S. sanguis* (Douglas et al. 1993; Bouvet et al. 1994). These species have also emerged as important pathogens causing bacteremia and sepsis in neutropenic cancer patients (Awada et al. 1992; Bochud et al. 1994a; Carratala et al. 1995; Richard et al. 1995; Gonzalez-Barca et al. 1996; Bilgrami et al. 1998). Clinical manifestations of viridans group streptococcal bacteremia in these patients may include fever, hypotension, shock, pneumonia, and adult respiratory distress syndrome, with a mortality rate ranging from 6 to 30 percent (Bochud et al. 1997). Serious complications of viridans group streptococcal BSI associated with a high mortality rate occur mainly in patients receiving high-dose antineoplasmatic chemotherapy who develop severe oral mucositis (Marron et al. 2000). The increased frequency of these infections has been attributed to several factors, including high-dose chemotherapy with cytosine arabinoside, oropharyngeal mucositis, and prophylactic therapy with cotrimoxazole or a fluoroquinolone (Elting et al. 1992; Bochud et al. 1994a; Richard et al. 1995). Oral and gastrointestinal mucosal lesions and intravascular catheters have been suggested as the most frequent portals of entry (Elting et al. 1992; Bochud et al. 1994b; Richard et al. 1995; Gonzalez-Barca et al. 1996). In the past, viridans group streptococci were considered to be uniformly susceptible to beta-lactam antimicrobial agents, macrolides, and tetracyclines. However, the emergence of strains resistant to beta-lactams and other antibiotics is a cause of concern and could compromise currently used prophylactic and therapeutic antibiotic regimens (Carratala et al. 1995; Doern et al. 1996; Pfaller et al. 1997; Wisplinghoff et al. 1999). Viridans group streptococcal BSI in both immunocompetent and immunocompromised patients is usually considered to be of endogenous origin.

Streptococci of the *S. anginosus* group (i.e. *S. anginosus*, *S. constellatus*, and *S. intermedius*; previously known as the *S. milleri* group) have been associated with a variety of invasive pyogenic infections. Most episodes of BSI are associated with an identifiable focus (Shlaes et al. 1981a; Kambal 1987; Gossling 1988; Jacobs et al. 1995; Casariego et al. 1996; Salavert et al. 1996; Bancescu et al. 1997), or have been reported in patients with neutropenia or underlying malignancies (Cohen et al. 1983; Awada et al. 1992; Pfaller et al. 1997), but these organisms were also described in neonates and pediatric patients (Hamoudi et al. 1990; Weisman et al. 1990; Raymond et al. 1995). Other clinical features include pulmonary infections, septic arthritis, osteomyelitis, and brain abscesses (Houston et al. 1980; Shlaes et al. 1981a; Brook et al. 1988; Molina et al. 1991; Wong et al. 1995; Jerng et al. 1997).

BSI is relatively uncommon (Weinstein et al. 1983b) in *S. pyogenes* infections. During the 1990s, an increase in frequency and severity of invasive infections due to group A streptococci has been reported (Martin and Hoiby 1990; Demers et al. 1993; Hoge et al. 1993). Patients were predominantly otherwise healthy adults and there has been an association with IV drug abuse (Barg et al. 1985; Craven et al. 1986; Lentnek et al. 1990). Other risk factors include immunosuppression and underlying malignancies, diabetes mellitus, and vascular disease. The clinical course is often fulminant and mortality ranges from 27 to 38 percent with an association of M types 1 and 3 with a more severe clinical course (Francis and Warren 1988; Ispahani et al. 1988; Stevens et al. 1989; Dan et al. 1990; Bucher et al. 1992; Johnson et al. 1992; Demers et al. 1993; Hoge et al. 1993; Martin and Single 1993).

Clinical manifestations of *S. agalactiae* BSI include early and late onset infection in neonates and primary and secondary BSI in adults. BSI is the most common clinical feature of early onset disease (occurring within the first 6 days) which is associated with maternal obstetric complications and infants born at less than 37 weeks. Symptoms are mainly nonspecific and do not allow distinction between infants with and without accompanying meningitis (Baker 1976; Franciosi et al. 1973; Chin and Fitzhardinge 1985). Mortality ranges from 5 to 10 percent and is inversely correlated to birth weight (Yagupsky et al. 1991; Schuchat 1998; Hyde et al. 2002). Occult BSI and meningitis are the most common manifestations of late onset infection (between 7 days and 3 months of age) and mortality is low (Schuchat 1998; Schuchat et al. 1990; Yagupsky et al. 1991; Hyde et al. 2002). In adults, the incidence of group B streptococcal BSI is 0.2 per 1 000 hospital admissions (Opal et al. 1988). About 68 percent of cases are unrelated to pregnancy and there is a predominance of the elderly (Lerner et al. 1977; Gallagher and Watanakunakorn 1985; Verghese et al. 1986; Opal et al. 1988; Schwartz et al. 1991). Predisposing conditions include diabetes, liver disease, neurologic impairment, underlying malignancy, renal failure, cardiovascular disease, and heart failure (Lerner et al. 1977; Gallagher and Watanakunakorn 1985; Verghese et al. 1986; Opal et al. 1988; Schwartz et al. 1991; Farley et al. 1993). Primary BSI accounts for 20–40 percent of cases and is associated with fatal outcome in about 50 percent of cases (Lerner et al. 1977; Opal et al. 1988; Colford et al., 1995; Munoz et al. 1997).

Listeria species, short gram-positive rods, are widely distributed in the environment. *Listeria monocytogenes* is usually seen as a cause of community-acquired BSI and meningitis. Immunocompromised patients including elderly patients are particularly susceptible (Nieman and Lorber 1980; Schuchat et al. 1991). *L. monocytogenes* can cross the placenta, therefore in pregnant woman *L. monocytogenes* BSI can be complicated by placentitis or amnionitis leading to infection of the fetus, abortion or premature birth (Bille et al. 2003). Other clinical manifestations are rare and include cutaneous listeriosis, endocarditis, osteomyelitis, and intraabdominal abscesses (McLauchlin and Low 1994). Early diagnosis is based on maternal blood cultures growing *L. monocytogenes*. Immunoassays based on monoclonal antibodies are also available for rapid detection of *Listeria* species from the blood. Also, polymerase chain reaction (PCR) assays are highly sensitive and specific and can be particularly useful if the patient was receiving antimicrobial therapy prior to culture (Bille et al. 2003).

GRAM-NEGATIVE ORGANISMS

During the 1960s and 1970s, gram-negative organisms were the most frequent blood culture isolates. Today they are isolated from about 30 percent of BSI.

Escherichia coli and other Enterobacteriaceae

Enterobacteriaceae are important causes of both community-acquired and nosocomial infections including BSI. The Enterobacteriaceae combined account for about 17 percent of all BSI (Diekema et al. 1999; Edmond et al. 1999; Wisplinghoff et al. 2004), making them the second most common cause of BSI after CoNS. Besides their involvement in almost any kind of infection, Enterobacteriaceae are an important part of the normal intestinal flora. *E. coli*, the most commonly isolated pathogen in community-acquired BSI (Diekema et al. 2003), accounts for about 7–11 percent of nosocomial BSI to date (Pittet and Wenzel 1995; Edmond et al. 1999; Diekema et al. 2003; Wisplinghoff et al. 2004). Community-acquired BSI due to *E. coli* is usually secondary to urinary tract infection and, less frequently, infection of the gastrointestinal tract (Eisenstein and Zaleznik 2000). *E. coli* strains causing BSI are phenotypically similar to those isolated from urinary tract infections (Johnson et al. 1997; Johnson et al. 2002). In patients with nosocomial BSI, primary sources of infection include IV catheters, the gastrointestinal, respiratory, and urinary tract. A portal of entry is usually absent in patients undergoing chemotherapy, where impairment of the intestinal mucosa may lead to bacterial translocation as a cause of BSI. Metastatic manifestations associated with BSI include septic arthritis, osteomyelitis, septic thrombophlebitis, abscesses in brain and liver, among others (Goldenberg et al. 1974; Saksouk and Salti 1977; Faraawi and Fong 1988; Farstad et al. 2003; Levy et al. 2003b).

Klebsiella species are a frequent cause of gram-negative nosocomial BSI, accounting for about 10 percent of community-acquired BSI and about 6 percent of nosocomial BSI (Diekema et al. 1999; Edmond et al. 1999; Diekema et al. 2003; Wisplinghoff et al. 2004). Similar to *E. coli*, BSI usually originates from a distant site, most commonly from the urinary tract, followed by the lower respiratory tract, bile duct, and wound infections (Eisenstein and Zaleznik 2000). In patients with nosocomial BSI, crude mortality rates of 25 percent have been reported (Diekema et al. 1999; Edmond et al. 1999; Diekema et al. 2003; Wisplinghoff et al. 2004). A recent survey found no difference in mortality between ICU patients with BSI caused by *Klebsiella* spp. and matched controls after adjusting for severity of underlying disease and acute illness (Blot et al. 2002a).

Enterobacter and *Serratia* species are less frequent causes of nosocomial BSI, accounting for 3 and 4 percent of BSI in recent studies, respectively (Edmond et al. 1999; Diekema et al. 2003; Wisplinghoff et al. 2004). Both are opportunistic pathogens and infections occur in patients with numerous predisposing conditions including immunosuppression, intravascular, intraperitoneal, or urinary catheters or following invasive procedures (Rudnick et al. 1996; Fok et al. 1998; Verghese et al. 1998; Eisenstein and Zaleznik 2000; Choi et al. 2002; Ostrowsky et al. 2002). *Serratia* species have also been associated with bacteremic infections in IV drug users, particularly endocarditis (see Infective endocarditis) (Wyler et al. 1975; Ashby 1976; Mills and Drew 1976). A recent study found that after adjustment for severity of underlying disease and acute illness and in the presence of fast and appropriate antibiotic therapy, *Enterobacter* BSI does not seem to adversely affect outcome in ICU patients (Blot et al. 2003b).

Non-typhoid *Salmonella* species usually cause intestinal infections and have been described as etiological agents of community-acquired BSI in both immunocompromised and previously healthy patients. Overall, BSI occurs in 1–4 percent of patients with gastroenteritis and can be caused by any *Salmonella* serotype. Frequencies are higher in patients with severe underlying conditions, immunosuppression, as well as in neonates and in elderly people (Han et al. 1967; Neves et al. 1971; Wolfe et al. 1971; Celum et al. 1987; Sperber and Schleupner 1987; Wheat et al. 1987; Levine et al. 1991). These patients also have a higher risk for severe invasive disease and extraintestinal manifestations including osteomyelitis, septic arthritis, meningitis, and endocarditis (Warren 1970; Wilson and Feldman 1981; Alvarez-Elcoro et al. 1984; Cohen et al. 1987; Stein et al. 1993; Huang 1996; Huang and Chuang 1997). In children, prolonged fever and infection with specific *Salmonella* serotypes is associated with BSI (Yang et al. 2002). Patients with BSI due to non-typhoid *Salmonella* species frequently have persistent bacteremia at follow-up regardless of the initial antibiotic treatment (Zaidi et al. 1999). *Salmonella typhi* is the etiologic agent of typhoid fever, a serious systemic infection, which is mostly seen in Third World countries. Patients usually suffer from continuous high fever and headache, without diarrhea. Typhoid fever has a generally more benign course in pediatric patients (Mermin et al. 1998). Typhoid fever is now rarely seen in the United States and Europe, and >70 percent of cases are travel-related (Mermin et al. 1998; Mead et al. 1999; Ackers et al. 2000).

Pseudomonas aeruginosa

P. aeruginosa is associated with a wide variety of infections, contributing significantly to morbidity and mortality of hospitalized patients. BSI is mainly nosocomial and usually affects patients with underlying clinical conditions such as malignancies, cardiac, or pulmonary diseases, renal failure or diabetes (El Amari et al. 2001). *P. aeruginosa* accounts for between 4 and 7 percent of nosocomial BSI, but for up to 31 percent of BSI in patients with underlying malignancies (Maschmeyer and Braveny 2000; Diekema et al. 2003; Wisplinghoff et al. 2004). Among gram-negatives, *P. aeruginosa* was the third most commonly isolated pathogen in recent surveys in North America (Weinstein et al. 1997; Diekema et al. 1999; Edmond et al. 1999; Wisplinghoff et al. 2004). The crude mortality of monomicrobial nosocomial *P. aeruginosa* BSI was between 28 and 48 percent in recent studies (Edmond et al. 1999; Diekema et al. 2003; Wisplinghoff et al. 2004). In patients with underlying malignancies crude mortality associated with *P. aeruginosa* BSI varied from 5 to 50 percent. *P. aeruginosa* BSI in IV drug users is relatively common and usually associated with endocarditis (see Infective endocarditis) (Komshian et al. 1990). Nosocomial pneumonia due to *P. aeruginosa* is often associated with BSI in neutropenic patients receiving chemotherapy as well as in adult or pediatric patients with AIDS (Iannini et al. 1974; Kielhofner et al. 1992; Flores et al. 1993). The prognosis of *P. aeruginosa* BSI secondary to nosocomial pneumonia is poor, and a fulminant progression to septic shock and death occurring within 3–4 days after the onset of pulmonary symptoms (Kang et al. 2003) is possible. Resistance of *P. aeruginosa* isolated from the blood is usually high, and therapy should be always based on susceptibility testing.

Acinetobacter baumannii

Acinetobacter species are a heterogenous group of organisms that have emerged as significant nosocomial pathogens mainly affecting patients with impaired host defenses in the intensive care unit setting. Members of the genus *Acinetobacter*, particularly *A. baumannii*, are implicated in a wide spectrum of infections, including nosocomial BSI, nosocomial pneumonia, secondary meningitis, skin and soft tissue infections, and urinary tract infections (Bergogne-Berezin and Towner 1996). The characteristics of *Acinetobacter* BSI have been described by researchers from various parts of the world (Tilley and Roberts 1994; Seifert et al. 1995; Cisneros et al. 1996; Ng et al. 1996; Siau et al. 1999; Wisplinghoff et al. 2000). Since *Acinetobacter* species are ubiquitous organisms and most species represent contaminants rather than true pathogens, species identification of all BSI isolates should be attempted. While BSI rates as high as 8.4 percent have been reported for *Acinetobacter* species (Seifert et al. 1995), *A. baumannii* accounted for about 1–2 percent of BSI in recent studies (Edmond et al. 1999; Wisplinghoff et al. 2000, 2004), still being among the ten most prevalent causes of BSI. The crude mortality of *A. baumannii* BSI may be as high as 52 percent (Beck-Sague et al. 1990; Seifert et al. 1995; Cisneros et al. 1996; Cisneros and Rodriguez-Bano 2002), but was 34 percent in nosocomial BSI in unselected patient populations (Wisplinghoff et al. 2004). Others reported that *A. baumannii* BSI is not associated with a significantly increased mortality rate in critically ill patients (Blot et al. 2003a). Nosocomial BSI is often catheter-related or secondary to respiratory or urinary tract infections, as well as wound infections, particularly in burn patients. *Acinetobacter* species other than *A. baumannii* may also be associated with catheter-related BSI, but the clinical course in these cases tends to be less severe (Seifert et al. 1994).

Other gram-negative pathogens

Since the implementation of vaccination the number of invasive *Haemophilus influenzae* infections in the United States and Europe has decreased to below 100 cases per annum (Adams et al. 1993; CDC 1999). Systemic infection due to *H. influenzae* is usually community acquired. Virtually all invasive diseases – usually meningitis and BSI associated with acute epiglottitis – are caused by *H. influenzae* serotype b, which remains highly prevalent in unvaccinated children. Invasive disease due to other serotypes is rare, but has been reported for serotype f, mainly in immunocompromised patients (Gatti et al. 2004). All commercial blood culture media have a high recovery rate of *Haemophilus* species. For primary recovery, X- and V-factor enrichment of culture media is required.

Neisseria species

BSI due to *Neisseria* species is usually community acquired. Clinical manifestation of infections due to *N. meningitidis* may range from benign transient bacteremia to fulminant disease with high mortality (Apicella 2000). Transient, intermittent or persistent oro- or naso-pharyngeal carriage of *N. meningitidis* has been reported in between 8 and 20 percent of healthy individuals (Ala'Aldeen et al. 2000). BSI due to *N. meningitidis* may be associated with meningitis, but transient bacteremia and severe sepsis have also been described without evidence of meningitis. Clinical presentation includes a maculopapular rash that becomes petechial and may be seen in 50 to 60 percent of patients. In severe sepsis, systemic coagulopathy may lead to ecchymotic, hemorrhagic, or necrotic areas. Purpura fulminans with extensive tissue damage is seen in 10 percent of patients. Septic shock can be the most prevalent symptom in patients with acute meningococcal BSI, but also in patients with meningitis (Apicella 2000; van Deuren et al. 2000). Mortality of meningococcal BSI associated with meningitis is usually between 7 and 10 percent, but can be as high as 30 percent (Dominguez et al. 2004). *N. meningitidis* has also been associated with chronic BSI, usually without meningitis (Apicella 2000). The clinical presentation includes fever, headache, leukocytosis, and arthritis.

Gonorrhea is still prevalent with highest incidences in sexually active teenagers between the ages of 15 and 24 (Sparling and Hansfield 2000). BSI due to *N. gonorrhoeae*, also referred to as disseminated gonococcal infection (DGI) occurs in 0.5–3 percent of cases (Hook and Hansfield 1999; Sparling and Hansfield 2000). Risk factors include female gender (especially during menstruation or the third trimester of pregnancy) and certain complement deficiencies (C7, C8, C9) (Ellison et al. 1987). Complications include gonococcal arthritis (in 30–40 percent of DGI) and rarely meningitis or endocarditis (Hook and Hansfield 1999; Sparling and Hansfield 2000).

FUNGI

Blood cultures are an important tool in the diagnosis of disseminated fungal infections (Weinstein et al. 1983b; Ellis 1991; Geha and Roberts 1994). All CMBCSs provide adequate detection of BSI due to yeasts (Mattia 1993; Prevost-Smith and Hutton 1994; McDonald et al. 2001). However, the lysis centrifugation system appears to be the most sensitive method for the detection of filamentous fungi as well as *Histoplasma capsulatum*, and other dimorphic fungi (Guerra-Romero et al. 1987; Graybill et al. 1990; Lyon and Woods 1995). When fungal BSI or disseminated fungal infection is suspected, maximum blood volumes should be collected multiple times and the use of special media should be considered.

Candida species, particularly *C. albicans*, are frequently isolated from the bloodstream, accounting for up to 10 percent of all nosocomial blood culture isolates. Data on the incidence of *Candida* BSI is controversial. A recent NNIS study (Trick et al. 2002) showed significant decreases in BSI due to *Candida* species and *C. albicans*, and reported a significant increase in BSI due to *C. glabrata* in ICUs between 1989 and 1999. Others, studying about 24 000 cases of nosocomial BSI from the United States, found a significant increase in the proportion of *Candida* species isolated from blood cultures between 1995 and 2002. In the same study, there was an increase in *C. albicans* and *C. parapsilosis* BSI, whereas

BSI caused by *C. tropicalis* and *C. glabrata* decreased (Wisplinghoff et al. 2004). Candida BSI is usually associated with long-term TPN and is most frequently encountered in patients with underlying conditions such as malignancies, neutropenia, or HIV/AIDS. However, blood cultures are frequently (50 percent) negative in patients with disseminated *Candida* infection (or *Candida* species may not be detected in blood cultures from patients that are also bacteremic) (Hockey et al. 1982). Therefore repeated cultures should be obtained if clinical suspicion of disseminated *Candida* infection is present. In patients with blood cultures positive for *Candida* a possible source of infection should be established, and intravascular catheters should be removed and reinserted at a different site (Rex et al. 1994; Rex 1996). Of note, candiduria may occur secondary to BSI and blood cultures should be obtained if candidemia cannot be ruled out clinically. Crude mortality in patients with nosocomial BSI due to *Candida* species is about 48 percent (Diekema et al. 2003; Wisplinghoff et al. 2004).

Although other fungi, particularly *Aspergillus* species, may cause disseminated infections, blood cultures often remain negative.

ANAEROBIC BACTERIA

Anaerobic bacteria account for about 4 percent of community-acquired and 2 percent of nosocomial BSI (Diekema et al. 2003). Gram-negative anaerobic rods are the most commonly encountered anaerobes in clinical specimens. *Bacteroides fragilis* is the predominant cause of anaerobic BSI (Felner and Dowell 1971; Wilson et al. 1972; Weinstein et al. 1983b; Finegold et al. 1985; Brook 1989; Lombardi and Engleberg 1992). Other gram-negative anaerobes commonly isolated from the blood include *B. thetaiotaomicron* and *Fusobacterium* species. BSI commonly originates from intraabdominal infections, infections of the female genial tract, soft tissue infections, the oropharynx, and the lower respiratory tract (Galpin et al. 1976; Finegold et al. 1985; Brook 1989). Anaerobic gram-negative bacteria isolated from blood cultures usually represent true pathogens (Lombardi and Engleberg 1992), and if isolated from the blood without a clinically apparent source, intraabdominal infection should be considered. Crude mortality rates as high as 60 percent have been reported (Chow et al. 1974) in BSI caused by *Bacteroides* spp. The attributable mortality in *B. fragilis* BSI has been reported as 19 percent (Redondo et al. 1995).

Clostridium species accounted for up to 4.5 percent of BSI in neutropenic patients (Mathur et al. 2002). Other studies reported that about 1 percent of all clinically significant blood culture isolates were identified as *Clostridium* species, most commonly *C. perfringens*, but less frequently also *C. septicum*, *C. difficile*, *C. novii*, and *C. tertium* (Brook 1989; Lark et al. 2001). Predisposing conditions include chronic alcoholism, abdominal surgery, bowel necrosis, septic abortions, malignancies, and decubital ulcers (Brook 1989; Gorbach et al. 1998). *C. perfringens* is isolated from a variety of infections, including BSI in various clinical settings and fulminant septic shock associated with post-abortion infections (Gorbach et al. 1998). In patients with clostridial myonecrosis, most commonly caused by *C. perfringens*, blood cultures are positive in 15 percent of patients (Summanen et al. 2002). However, *C. perfringens* BSI is not always associated with a serious underlying condition (Gorbach et al. 1998).

C. septicum is rarely isolated from anaerobic blood cultures and BSI is usually seen in patients with underlying malignancies, especially leukemia and lymphoma, or carcinoma of the large bowel. Underlying malignancies are present in 70–85 percent of patients with *C. septicum* BSI (Bodey et al. 1991; Kirchner 1991). Other associations include neutropenia and enterocolitis (King et al. 1984; Kudsk 1992). However *C. septicum* has also been isolated from BSI in patients with diabetes mellitus, severe arthritis or gas gangrene (Koransky et al. 1979).

BSI due to anaerobic gram-positive cocci, e.g. *Peptostreptococcus* species, is usually associated with obstetric or gynecological infections (Moll et al. 1996). The clinical significance of *Peptostreptococcus* species recovered from blood cultures is not clear and their isolation should be interpreted with caution.

MICROORGANISMS WITH SPECIAL REQUIREMENTS

Brucellosis

Isolation of *Brucella* species from blood, bone marrow, or tissue cultures in the only irrefutable proof of brucellosis (Young 1995; Solera et al. 1999). Organisms were historically cultured in biphasic bottles (Castaneda) containing a liquid and a solid agar phase. Today, commercial biphasic bottles, lysis centrifugation tubes, and CMBCSs, such as Bact/ALERT and BACTEC9000, are used to isolate *Brucella* species from the blood (Arnow et al. 1984; Solomon and Jackson 1992; Bannatyne et al. 1997; Ozturk et al. 2002). While lysis centrifugation is superior to the Castenada bottles, modern CMBCSs, such as the BACTEC9000, achieve detection times that are similar to those of the lysis centrifugation method (Yagupsky et al. 1997). Details on the performance of automated blood culture systems can be found in a recent review (Yagupsky 1999). Biosafety levels 2 and 3 are recommended for handling of specimens potentially containing *Brucella* species and processing of cultures of *Brucella* species, respectively. In addition to blood cultures, serological testing is mandatory.

Leptospirosis

Since leptospiremia only occurs during the acute phase of the illness (usually 4–7 days), cultures drawn beyond day 10 are likely to be of low yield. Dark-field or phase-contrast microscopy may be used to directly examine the blood. Of note, the detection level is 10^4 organisms/ml (Turner 1970). For culture, media with added rabbit serum or albumin and fatty acids are recommended with the addition of SPS. Tubes are incubated for 4–6 weeks in air at 28–29°C and examined weekly by removing a drop of culture from 1–3 cm below the surface of the medium (Dunne et al. 1997). In addition to blood cultures, serological testing is mandatory.

Bartonellosis

Bartonella henselae, the most common cause of cat scratch disease and *B. quintana*, the etiologic agent of trench fever, have both been associated with bacillary angiomatosis and bacteremia. Both species, as well as *B. elisabethae*, have been isolated from the blood, using conventional, lysis centrifugation or CMBC systems (Dunne et al. 1997). Nutrient-rich media, such as chocolate agar, should be used for subcultures, and prolonged incubation for at least 14 days is required (Anderson and Neuman 1997; Maurin and Raoult 1996).

Mycobacteria

While *M. tuberculosis* is rarely isolated from the bloodstream, *Mycobacterium avium* complex (MAC) is responsible for the majority of disseminated mycobacterial infections and can always be regarded as significant when isolated from the blood (Havlik et al. 1992). Rapidly growing mycobacteria can be cultured from the blood using all commercially available CMBCSs (Waite and Woods 1998; Archibald et al. 2000; Pintado et al. 2001). Survival times of MAC are longer if lysis centrifugation (ISOLATOR) tubes are used (Havlik et al. 1993). Today a variety of systems can be used to process mycobacterial blood cultures, including the ISOLATOR system, the radiometric BACTEC 460TB (BACTEC 13A; detection of $^{14}CO_2$), and the nonradiometric BACTEC 9000MB (BACTEC Myco/F lytic), ESP Culture System II (detection of pressure changes; ESP Myco, Trek Diagnostic Systems) or MB/BacT ALERT 3D (colorimetric CO_2 detection; bioMérieux). All systems use modified Middlebrook 7H9 broth or a similar broth supplemented with a variety of growth factors and antimicrobial agents (Pfyffer et al. 2003). The diagnostic yield of the CMBCSs is similar to that of the BACTEC 460TB and superior to conventional solid media (Pfyffer et al. 1997; Woods et al. 1997; Hanna et al. 1999; Whyte et al. 2000; Piersimoni et al. 2001). The average time to positivity is 5 to 14 days for *M. tuberculosis* and below 7 days for other mycobacteria, when using the BACTEC 460TB and similar detection times for the CMBCSs have been reported (Pfyffer et al. 2003). It is not recommended to transport blood before inoculation of blood culture media; however, if necessary SPS, heparin, EDTA or citrate may be used as anticoagulants (Pfyffer et al. 2003) depending on the culture medium used.

MANAGEMENT OF BLOODSTREAM INFECTION

Progression from bacteremia to sepsis, septic shock, and death can be rapid depending on underlying medical conditions of the host, virulence of the pathogen, and treatment parameters. In a recent study, mortality was double when inappropriate empirical antibiotics were administered to ICU patients with BSI (Ibrahim et al. 2000). Therefore, successful management must include early and appropriate antimicrobial therapy in addition to supportive care and the elimination of a removable focus, if present.

Resistance rates to antimicrobial agents have been rising for the last two decades in all predominant organisms, including *S. aureus* (Edmond et al. 1999; Smith et al. 1999; Trick et al. 2002), CoNS (Diekema et al. 2001), enterococci (Doern et al. 1998), and gram-negative pathogens (Diekema et al. 1999, 2002; Heinemann et al. 2000). In the face of emerging multiresistant organisms, antimicrobial prophylaxis and treatment has become increasingly difficult, and timely and accurate epidemiological information is needed for guiding appropriate empirical therapy. Analyses of *Candida* BSIs have shown trends to selection of non-*albicans* species, some of which are difficult to treat with first generation azoles (Trick et al. 2002). In this context, surveillance programs have become important in defining the species distribution and resistance patterns of pathogens causing BSI, thus providing the basis for appropriate empirical therapy.

INFECTIVE ENDOCARDITIS

INTRODUCTION

Infective endocarditis is a microbial infection of the endocardium, generally of a heart valve or valves, but sometimes of the cordae tendineae or at the site of a septal defect or other congenital abnormalities. With the advent of intracardiac valvular prostheses it was soon realized that these, too, or the tissues into which they are sewn, can also become infected. The term *infective endocarditis* covers the various clinical subcategories of the disease and has replaced the old term *bacterial endocarditis* since it became clear that chlamydiae, rickettsiae, fungi, and probably viruses can also be responsible for the disease. However, endocarditis caused by these latter agents is not within the focus of this chapter.

Historically, infective endocarditis has been classified as *acute*, *subacute*, and *chronic* endocarditis; this classification was related to the usual course of the untreated disease. The *acute* form is a fulminating infection, characterized by high fever, severe malaise, leukocytosis, and rapid destruction of heart valves and is usually caused by a virulent organism, such as *Staphylococcus aureus*, *Streptococcus pneumoniae*, *S. pyogenes*, or *Neisseria gonorrhoeae*. Without treatment, the disease was uniformly fatal resulting in the death of the patient in less than 6 weeks, but sometimes in days. *Subacute* endocarditis referred to infection with unspecific clinical findings and an indolent course with death from the disease occurring after 6 weeks to 3 months, and in the *chronic* form after 3 months and up to 2 years. The acute presentation of the disease usually affected patients without a previous cardiac abnormality, whereas the slowly progressing form of the disease was usually caused by relatively avirulent organisms such as the 'viridans' streptococci and occurred in patients with damaged heart valves.

Although this concept is still very useful mainly for educational purposes, it is less helpful nowadays for the individual patient. Currently, the diagnosis of infective endocarditis based on modern imaging techniques is reached earlier in the course of the disease. In addition, there is considerable overlap in clinical manifestation that is influenced by host factors, virulence determinants, and antimicrobial susceptibility of the offending pathogen, and the timely institution of appropriate antimicrobial therapy, as well as surgical intervention. A classification based on the nature of the causative pathogen may be more appropriate particularly with regard to the different treatment regimens recommended for specific pathogens and clinical prognosis. Mortality associated with infective endocarditis has considerably decreased in the past two decades. With modern diagnostic imaging techniques, the use of standardized antimicrobial therapy, and timely surgical valve replacement if necessary, cure rates approach 90–95 percent for streptococcal endocarditis, whereas infective endocarditis caused by enterococci and in particular *S. aureus* still carries a much higher mortality (Fowler et al. 1999; Cabell et al. 2002). Overall mortality of patients with infective endocarditis, however, has decreased to less than 20 percent in recent years (Cabell et al. 2002; Hoen et al. 2002; Wallace et al. 2002).

EPIDEMIOLOGY

Infective endocarditis is an uncommon disease. In a review of 10 large American series, a frequency of about one case per 1 000 hospital admissions with a range of 0.16–5.4 cases per 1 000 admissions was reported (Bayer and Scheld 2000). Several factors have led to dramatic changes in the epidemiologic features and clinical manifestations of infective endocarditis over the past few decades (Finland and Barnes 1970; Watanakunakorn and Burkert 1993). Among these are the increased life expectancy, a shift in the at-risk population with their predisposing valvular lesions illustrated by a decline of rheumatic carditis and an increase in degenerative valvular disease, an increase in nosocomial cases of infective endocarditis, refined diagnostic imaging techniques, and major advances in cardiothoracic valvular surgery. Most cases of infective endocarditis are acquired in the community, but there are now an increasing number of cases of hospital-acquired (nosocomial) endocarditis and this can affect both native and prosthetic valves (Lamas and Eykyn 1998; Gouello et al. 2000). The incidence of infective endocarditis is difficult to determine due to different patient populations surveyed and since the diagnostic criteria may vary in different series. In the United States and Western Europe, the incidence of infective endocarditis has been estimated to range between two and six cases per 100 000 person-years (van der Meer et al. 1992; Berlin et al. 1995; Hogevik et al. 1995; Hoen et al. 2002) with incidences differing between community-acquired (4.5 cases/100 000 patients-years) versus nosocomial (0.94 cases) endocarditis, as well as native-valve versus prosthetic-valve (0.94 cases) endocarditis (Berlin et al. 1995). The increased longevity has resulted in an increased median age of patients with infective endocarditis, a growing number of patients with cardiovascular disease and prosthetic heart valves, and an increased cumulative exposure to nosocomial bacteremia. Among the population with infective endocarditis associated with intravenous drug abuse, in contrast, there is a trend towards younger patients. In the United States, the incidence of infective endocarditis in this group of patients is estimated at 150–2 000 per 100 000 person-years (Frontera and Gradon 2000); it is much lower in most European countries.

There is a well recognized increased incidence of infective endocarditis with age, and over all the infection is more common in males than females (Hoen et al. 2002). Fifty years ago native valve infective endocarditis was mainly a disease of young people: in Cates and Christie's study (1951) of 442 cases of 'subacute bacterial endocarditis' seen between 1945 and 1948, 62 percent of patients were aged between 15 and 35 years. Since then native valve infective endocarditis has become a disease of the middle-aged and elderly (Gagliardi et al. 1998; Dhawan 2003; Gregoratos 2003).

In earlier series, 60–80 percent of patients with native-valve endocarditis had an identifiable predisposing cardiac lesion (Weinstein and Schlesinger 1974; Watanakunakorn 1977). In a recent population-based survey from France, however, 47 percent among 390 patients with infective endocarditis had no underlying valvular disease (Hoen et al. 2002). Almost any type of structural heart disease that leads to turbulence of blood flow resulting in endocardial damage may predispose to infective endocarditis, including rheumatic heart disease, congenital heart disease, and arteriosclerotic cardiovascular disease. The proportion of cases of infective endocarditis related to rheumatic heart disease, the primary underlying condition in the past has now declined to less than 25 percent (Kaye 1985), however rheumatic heart disease is still the most common underlying cardiac condition in developing countries (Jalal et al. 1998). Congenital heart disease predisposing to infective endocarditis includes bicuspid aortic valve (Lamas and Eykyn 2000), ventricular septal defect, and tetralogy of Fallot. These conditions were seen previously in 6–24 percent of cases, but in the most recent series reported by Hoen et al. (2002) the frequency of underlying congenital heart disease was only 1 percent. The majority of mainly elderly patients with infective endocarditis and without a history of underlying valvular disease have arteriosclerotic cardiovascular disease with calcified nodular endocardial lesions or an intraventricular thrombus following myocardial infarction. Other conditions associated with infective endocarditis include idiopathic hypertophic subaortic stenosis (IHSS) (Morgan-Hughes and Motwani 2002) and mitral-valve prolapse (Baddour and Bisno 1986), in particular with the presence of mitral valve regurgitation. Hemodialysis shunts (Doulton et al. 2003), intracardiac pacemaker wires (Arber et al. 1994), and central venous catheters (Miele et al. 2001) may also predispose to endocarditis.

Nosocomial infective endocarditis is a rare complication of nosocomial bacteremia. Whilst many aspects of nosocomial infective endocarditis are similar to community-acquired infective endocarditis, both entities differ in their predisposing factors, microbial etiology and prognosis. The increasing use of intravascular access devices for hemodynamic monitoring, administration of fluids, and in particular those used for long-term hyperalimentation and dialysis, has brought a parallel increase in associated nosocomial bacteremia and has contributed largely to the increasing number of hospital-acquired cases of infective endocarditis (Gilleece and Fenelon 2000; Gouello et al. 2000). Other predisposing conditions include advanced age, surgical wounds, genitourinary instrumentation, major abdominal surgery, the implantation of prosthetic valves and cardiac devices, and major burn injury (Cartotto et al. 1998; Fernandez-Guerrero et al. 2002; Giamarellou et al. 2002). Nosocomial endocarditis can affect both previously normal and damaged heart valves, as well as prosthetic valves. S. aureus in native valve endocarditis and S. epidermidis in the presence of foreign bodies are the main implicated pathogens. Fungal endocarditis can also result from infected intravascular catheters with Candida spp. as the most usual pathogens. In their review of 21 cases of endocarditis associated with central venous catheterization, Tsao and Katz (1984) found that two-thirds of cases had right-sided infection, suggesting a parallel with the animal model in which catheter-induced traumatic sterile vegetations become infected after bacteremia. In fact, although catheter-related bacteremia may result in right-sided endocarditis, the infection is not invariably right-sided. In the series reported by Gouello et al. (2000), the overall mortality in patients with nosocomial endocarditis was 68 percent (n = 15) and mortality was directly associated with endocarditis in 36 percent of cases (n = 8).

Intravenous drug users (IVDU) represent a specific risk group for the acquisition of right-sided but also left-sided infective endocarditis usually without predisposing cardiac disease (Brown and Levine 2002; Moss and Munt 2003). Endocarditis was one of the earliest recognized medical complications of intravenous drug abuse and it is the most common cause of bacteremia in IVDUs. Overall, the predominant pathogen is S. aureus (including MRSA), but as with endocarditis not associated with IVDU, a wide spectrum of organisms has been reported. These include pyogenic streptococci, pseudomonads, other gram-negative bacilli and fungi, as well as polymicrobial infections. Most cases involve a previously normal tricuspid valve and result from repeated intravenous injection of foreign material and organisms. Left-sided endocarditis is uncommon in IVDUs unless they have previous heart disease, but appears to be increasing (Seghatol and Grinberg 2002). Tricuspid endocarditis presents with the signs of systemic infection and respiratory symptoms from septic pulmonary emboli and not with the 'classical signs' of endocarditis. Such symptoms (and the chest X-ray findings) may be misdiagnosed as pneumonia by clinicians unfamiliar with right-sided endocarditis. The disease is more frequent in HIV-positive or than in HIV-negative IVDUs, but the presentation and clinical manifestations are similar (Wilson et al. 2002). The causative organisms are usually acquired from the patient's own flora although the drug 'works' have been implicated in cases of P. aeruginosa infection. Most IVDUs with tricuspid endocarditis have a good prognosis, but the mortality rate is markedly increased in those with very large vegetations and left-sided involvement.

Prosthetic-valve endocarditis accounts for 10–25 percent of cases of infective endocarditis in developed countries (Hogevik et al. 1995; Fefer et al. 2002; Hoen et al. 2002). It has been estimated that between 3 and 6 percent of patients develop prosthetic-valve endocarditis within the first 5 years following surgery (Calderwood et al. 1985; Agnihotri et al. 1995).

PATHOGENESIS

The development of infective endocarditis usually follows a multifactorial course. Alterations of the valve surface and less frequently the epithelial lining of the heart chamber (mural endocarditis) represent the first step in this series of events. These lesions, induced by high velocity turbulent blood flow across the valve results from a variety of local or systemic stresses, such as a small ventricular defect or a valvular stenosis. The location of valvular jet lesions is determined by the direction of the accelerated bloodstream, i.e. they are usually located on the low-pressure side of the cardiac valves, such as the atrial surface of an insufficient mitral or tricuspid valve and the ventricular surface of an incompetent aortic valve. Any endothelial lesion results in exposure of the underlying extracellular matrix (ECM) proteins, the production of tissue factor, the deposition of platelets and fibrin, and the formation of so-called *sterile vegetation*. These deposits are characteristic of nonbacterial thrombotic endocarditis (NBTE) (Blanchard et al. 1992), a condition also seen to be associated with chronic wasting diseases, particularly malignancy (marantic endocarditis), connective tissue diseases, and intracardiac catheters.

Infective endocarditis occurs when microorganisms are deposited on to the sterile vegetation during an episode of transient bacteremia. Transient bacteremia results from mechanical manipulation of mucosal surfaces colonized with bacteria, such as with dental extractions or other dental procedures (Coulter et al. 1990; Bayliss et al. 1983), endoscopy and/or biopsy of the gastrointestinal (Rigilano et al. 1984; Schlaeffer et al. 1996) or the urogenital tract (Zimhony et al. 1996), and surgical procedures leading to injury of these organs. Such a history, however, is only obtained in a minority of patients with infective endocarditis. Studies have also demonstrated spontaneous transient bacteremia in patients with poor dentition even in the absence of dental procedures whilst chewing or brushing their teeth (Bhanji et al. 2002). These procedures may also induce bacteremia in people with good oral hygiene. Bacteremia is low grade (<10 organisms per ml) and transient, i.e. the bloodstream is usually cleared from the invading bacteria within 15–30 min.

Infective endocarditis can also occur in the absence of a known or identifiable pre-existing valve lesion. This is particularly true for virulent, invasive pathogens such as *S. aureus*. This organism may bind directly to endothelial cells, a property that explains the ability of *S. aureus* to initiate endocarditis on 'normal' cardiac valves, in particular those of the elderly with some degree of degenerative valve sclerosis and IVDUs in whom valvular damage may be induced by repeated injection of impure materials (Clifford et al. 1994).

The adherence of a microorganism to platelets and fibrin is the next critical step in the development of endocarditis. Numerous studies have demonstrated a correlation between the ability of bacteria to adhere in vitro to platelet-fibrin matrices mimicking NTBE and their propensity to induce infective endocarditis in humans and in animal models (Scheld et al. 1978; Ramirez-Ronda 1978). Certain organisms such as

viridans group streptococci, enterococci, and staphylococci – the predominant pathogens involved in infective endocarditis – adhere more avidly to normal canine endocardial surfaces in vitro than enteric bacilli, such as *Escherichia coli* and *Klebsiella pneumoniae* (Gould et al. 1975). *S. aureus* and *S. pneumoniae* appear to have a particular propensity to adhere to fibrin-platelet deposits and thus may produce infective endocarditis with a lower inoculum. Adherence is mediated by bacterial surface molecules interacting with host ECM molecules. These are termed microbial surface components recognizing adhesive matrix molecules (MSCRAMM). Streptococci and staphylococci carry multiple MSCRAMMs that mediate binding to NTBE and directly to endothelial cells.

The adherence of oral streptococci to NBTE lesions may be promoted by their ability to produce dextran or glucan, a complex extracellular surface polysaccharide, which in the case of *S. mutans* is also involved in the pathogenesis of dental caries (Gibbons and Nygaard 1968). Glucan production may thus be one of several virulence factors required for infection. Since not all species of streptococci that cause infective endocarditis are glucan producers (including most strains of *S. oralis*, one of the most common of the oral streptococci to cause infective endocarditis), other factors must also be involved in the pathogenesis of infective endocarditis caused by viridans streptococci and enterococci. One example is FimA, a surface adhesin located at the tips of the fimbriae that has been shown to mediate the attachment of *S. parasanguis* to fibrin-platelet matrices (Burnette-Curley et al. 1995). Inactivation of FimA greatly decreased the organisms' ability to induce endocarditis in rats. FimA-like proteins were found in a number of streptococcal species. Another factor involved in the pathogenesis of streptococcal endocarditis is the ability to bind to fibronectin which probably acts as the host receptor within the NBTE (Lowrance et al. 1990).

In contrast to less virulent organisms, such as the streptococci that usually implant only on cardiac sites with pre-existing endothelial lesions (Sullam et al. 1987; Kitada et al. 1997a; McCormick et al. 2000), *S. aureus* can either colonize damaged endothelium or invade physically intact endothelial cells (Yao et al. 1995). The major adhesins involved in the binding of *S. aureus* to NBTE lesions are its fibrinogen-binding proteins such as clumping factor and coagulase (Devitt et al. 1994; Moreillon et al. 1995). The expression of these factors is regulated by the global regulators: accessory gene regulator (*agr*), staphylococcal accessory regulator (*sar*) and probably other determinants (Cheung and Projan 1994). *S. aureus* may also bind directly to and invade endothelial cells, a property mediated by fibronectin-binding proteins FnBPA and FnBPA (Vercellotti et al. 1984; Kuypers and Proctor 1989; Flock et al. 1996; Menzies 2003).

Local bacterial survival and persistence is the next step involved in the development of infective endocarditis. Following bacterial adherence to and colonization of NBTE lesions, the organisms are rapidly covered with more fibrin and platelets which form a protective sheath that shelters the bacteria from host phagocytes and soluble host defenses, as well as from antimicrobials and permits further bacterial proliferation and vegetation growth. Local production of tissue factor (TF) by host monocytes and endothelial cells and aggregation of platelets are mainly involved in this process (Sullam et al. 1996). Additional factors produced by staphylococci, such as hemolysins, may disrupt the endothelial envelope of adjacent cells and allow further tissue invasion and enlargement of the infectious process (Moreillon et al. 2002). These conditions allow for unimpaired bacterial multiplication leading to extremely high colony counts of 10^9 to 10^{11} bacteria per gram of tissue. Bacteria residing deeply within the vegetation can adopt a reduced metabolic state that enables them to resist the attack of beta-lactam antibiotics which exert their action mainly on proliferating bacteria. These fresh vegetations are thus composed of bacterial masses, fibrin, platelet aggregates, rarely red blood cells, and inflammatory cells. The morphology of the vegetations can range from small, flat, granular lesions a few millimeters in size, to large, soft, pedunculated, friable masses several centimeters in size. They may or may not progress to further valvular, perivalvular, or extracardiac complications, resulting from embolic phenomena such as embolization into the coronary arteries or the systemic arterial tree leading to ischemia, infarcts, bleeding, and abscess formation. The same factors involved in both vegetation growth and tissue destruction are likely to be responsible for bacterial dissemination to remote sites where colonization of target tissue probably involves the same infection cycle starting with tissue adherence. During the course of healing, infiltration by polymorphonuclear leukocytes and fibroblasts results in fibrosis and sometimes calcification of the lesion which finally will become covered by endothelium.

Infective endocarditis, like other infectious diseases, is associated with circulating immune complexes that are found in high titers in virtually all patients (Bayer et al. 1976; Bayer and Theofilopoulos 1990). Some of the peripheral manifestations of infective endocarditis, such as Osler nodes and Roth spots as well as glomerulonephritis, may result from deposition of circulating immune complexes.

MICROBIOLOGY

A multitude of microorganisms has been implicated in infective endocarditis, and heart valves, in particular prosthetic valves, can become infected with just about any microbe or fungus. The majority of cases are caused by a fairly limited number of pathogens, i.e. streptococci

Table 19.5 *Distribution of etiologic agents in patients with infective endocarditis*

Microorganism	% of cases	
	Mean	Range
Streptococci	50	35–53
Viridans streptococci	33	17–48
Other streptococci	17	5–33
Enterococci	8	6–10
Staphylococci	30	29–38
Staphylococcus aureus	23	22–31
Coagulase-negative staphylococci	7	6-8
Gram-negative aerobic bacilli	3	1–3
Fungi	1	0–1
Other bacteria	2	1–4
Polymicrobial infections	3	3–4
Culture-negative	8	6–12

Data are compiled from three population-based studies comprising 927 patients with infective endocarditis (van der Meer et al. 1991; Hogevik et al. 1995; Hoen et al. 2002).

and staphylococci. The most common etiologic agents are listed in Table 19.5.

Streptococci and *enterococci*

Nearly two-thirds of cases of community-acquired native valve infective endocarditis are caused by streptococci and enterococci: 214 of 326 (65.6 percent) in a Dutch series and 226 of 390 (58 percent) in a recent population-based survey in France (van der Meer et al. 1991; Hoen et al. 2002). However, streptococcal and enterococcal infective endocarditis cannot be considered as a homogeneous entity since both the pathogenesis of the infection and the disease itself differ according to the species responsible.

The largest group are the oral streptococci which are usually referred to as the 'viridans' group streptococci because most produce α-hemolysis (greening) when cultured on blood agar. While clearly predominating in the past, oral streptococci were responsible for less than 20 percent of the cases in more recent series and declined significantly from 7.8 cases per million person-years in 1991 to 5.1 cases per million person-years in 1999 (Hoen et al. 2002). For many years the oral streptococci were known as '*Streptococcus viridans*', but this is inaccurate because this term suggests a species designation. In fact, the oral streptococci constitute a heterogeneous group of different species within the genus *Streptococcus* which are usually nontypable by the Lancefield system. The taxonomy of the streptococci is confusing and has undergone major taxonomic changes. These taxonomic changes are well reviewed by Facklam (2002), but it seems that further reclassification of some species is inevitable. The oral streptococci include *S. gordonii*, *S. mitis*, *S. mutans*, *S. oralis*, *S. salivarius*,

S. sanguinis (previously *S. sanguis*), and the 'nutritionally variant streptococci' (now classified as *Abiotrophia* and *Granulicatella* spp.), as well as some others. *S. morbillorum* is now classified as *Gemella morbillorum*. In Parker and Ball's study (1976) of streptococci associated with systemic disease, the oral streptococci so defined accounted for about two-thirds of the 317 streptococci isolated from the blood in patients with infective endocarditis. Among the 225 streptococci reported in the French series (Hoen et al. 2002), 68 (30 percent) were identified as oral streptococci, while 98 (44 percent) were group D streptococci (including *S. bovis*, *S. gallolyticus*, and *S. infantarius*) and 22 (10 percent) belonged to the pyogenic streptococci. Many of the oral streptococci seldom cause infection other than infective endocarditis. Thus, the presence of these organisms in blood cultures, particularly in more than one set, should raise the suspicion of infective endocarditis even in the absence of any other clinical sign and symptom and prompt appropriate diagnostics such as transesophageal echocardiography (TEE) (see below). In fact, in such cases the microbiologist may often be able to provide the clinician with the necessary clue as to the etiology of the patient's illness and thus initiate an appropriate diagnostic approach and antimicrobial therapy.

The different species of oral streptococci differ in their habitat and in their propensity to cause infective endocarditis. A recent report of the identification of 47 strains of oral streptococci from cases of infective endocarditis found that 31.9 percent were *S. sanguis*, 29.8 percent *S. oralis*, and 12.7 percent *S. gordonii*. Other species were much less common (Douglas et al. 1993).

Cates and Christie (1951) in their study of 'subacute bacterial endocarditis' 40 years ago found that only 12.3 percent of their 187 cases of 'viridans' streptococcal endocarditis had had a tooth extraction less than 3 months before the onset of symptoms, whereas 34.8 percent had dental sepsis or caries. More than 30 years later, Bayliss et al. (1983) reported a similar incidence of 13.7 percent of cases occurring within 3 months of a dental procedure. Even if there is no overt dental focus it can still be assumed that the oral streptococci have originated in the mouth. Transient bacteremia with oral streptococci may also follow procedures such as tonsillectomy or adenoidectomy and (rarely) bronchoscopy or gastroscopy, but there is no evidence that these procedures result in infective endocarditis.

There are currently three recognized species that are traditionally referred to as the *S. milleri* group, *S. anginosus*, *S. constellatus*, and *S. intermedius*. Unlike the viridans group streptococci, organisms of the *S. milleri* group seldom cause native valve endocarditis. These organisms, in particular *S. anginosus*, are commonly isolated from septic lesions at a variety of sites, including abscesses of internal organs, such as the liver and brain, as well as from cholangitis, peritonitis, and empyema, either in pure culture or more commonly in polymicrobial infections with anaerobes (Casariego et al. 1996; Bert et al. 1998). However, *S. milleri* group infective endocarditis does not occur in association with such septic conditions; it seems to be an entirely separate entity (Kitada et al. 1997b; Marinella 1997; Lefort et al. 2002). A Dutch study (van der Meer et al. 1991) is the only published series with data on species distribution within this group. In a total of 326 cases of native valve infective endocarditis they reported 12 cases of *S. intermedius* and three of *S. constellatus* (interestingly no *S. anginosus*), giving an overall incidence for the *S. milleri* group of 4.6 percent. This is similar to the 2–5 percent incidence in the five reports quoted by Douglas et al. (1993), although they included prosthetic infections as well as native. Infective endocarditis caused by *S. anginosus* tends to be a more acute and severe infection associated with intracardiac complications that are more commonly seen with *S. aureus* (Levandowski 1985; Hurle et al. 1996).

Nutritionally variant streptococci (NVS), recently classified as *Granulicatella (Abiotrophia) adiacens* and *A. defectiva*, do not grow on subculture unless specific growth factors are added, therefore isolation and correct identification of these species is often difficult. These organisms accounted for 2 percent of cases of infective endocarditis in the French series (Hoen et al. 2002). The disease usually runs an indolent course and is associated with pre-existing valvular heart disease and illicit drug use (Leonard et al. 2001; Chang et al. 2002). However, complications such as systemic embolization frequently occur and are responsible for the relatively high mortality rate (Christensen et al. 1999).

S. bovis is a normal inhabitant of the gastrointestinal tract and is only seldom found in the mouth. In the past, this organism was often misidentified as *Enterococcus* (previously *Streptococcus*) *faecalis* since the two species share the Lancefield group D antigen and are biochemically quite similar. Other members of the group D streptococci with pathogenic significance include *S. gallolyticus* and *S. infantarius*, previously classified as different *S. bovis* biotypes. With the advent of commercial streptococcal identifiation kits such as the API 20 Strep and Rapid ID 32 Strep (bioMérieux), which are commonly used in many laboratories, correct identification of *S. bovis*, which is important since it involves a different therapeutic approach, has become much easier. *S. bovis* was as common as *S. sanguis* and *S. oralis* (*S. mitior*) as a cause of streptococcal infective endocarditis in Parker and Ball's (1976) series in which it was responsible for 15.8 percent of 317 cases and for nearly a quarter of the cases in patients over 55 years. Comparing the etiologic agents of infective endocarditis between two surveys performed in 1991 and 1999, Hoen et al. (2002) noted an increase of Group D streptococci from 5.3 in 1991 to 6.2 cases per million person-years in 1999. The clinical manifestations of *S. bovis* infective endocarditis are similar to those caused by 'viridans' streptococci (Ballet et al. 1995; Duval et al. 2001; Pergola et al. 2001). In 1977, Klein et al. reported two patients with *S. bovis* endocarditis and adenocarcinoma of the colon. Since then others have noted an association between *S. bovis* bacteremia and not only colonic cancer but also polyposis and other gastrointestinal disorders (Leport et al. 1987; Waisberg et al. 2002). In their study of 90 cases of *S. bovis* bacteremia during the period 1951–1980, Honberg and Gutschik (1987) identified 15 patients (17 percent) with gastrointestinal cancer and concluded that there was a high frequency of neoplasms in patients with a history of *S. bovis* bacteremia, particularly in those with endocarditis, even several years after the episode of bacteremia. Leport et al. (1987), in a study of 77 cases of group D streptococcal endocarditis, found a significantly higher frequency of colonic polyps and neoplasms in patients with *S. bovis* endocarditis than in those with enterococcal endocarditis and they and others emphasized the need for colonoscopy in any patient with *S. bovis* bacteremia (Klein et al. 1977; Beeching et al. 1985).

In the pre-antibiotic era, *S. pneumoniae* accounted for about 10 percent of cases of infective endocarditis (Powderly et al. 1986), but it is now rarely seen. In a nationwide survey conducted in France between 1991 and 1998, 30 cases of native valve pneumococcal endocarditis were observed; the majority of patients (66.7 percent) had no predisposing cardiac condition (Lefort et al. 2000). The primary focus of infection in one-third of patients was pneumonia. Pneumococcal endocarditis causes an acute fulminating infection, generally of a previously normal valve, most often the aortic valve, and complications such as large arterial emboli, perivalvular abscess formation, and rapid valve destruction are frequent (Aronin et al. 1998; Lindberg et al. 1998; Lefort et al. 2000). The infection tends to occur in alcoholics and these patients usually have concomitant pneumonia and meningitis, a triad called Austrian's syndrome (Munoz et al. 1999). The overall mortality remains high at approximately 25–50 percent, valve replacement is frequently necessary, and penicillin resistance is a cause of concern (Lefort et al. 2000; Martinez et al. 2002).

Alpha-hemolytic pyogenic streptococci of Lancefield groups A, B, C, and G only occasionally cause endocarditis, usually attacking normal valves. The patterns of infective endocarditis caused by these organisms were recently reviewed by Lefort et al. (2002). The group B streptococci (*S. agalactiae*) are by far the most common and their incidence has recently increased (Gallagher and Watanakunakorn 1986; Scully et al. 1987; Baddour 1998; Hoen et al. 2002; Sambola et al. 2002). In the Spanish series (Sambola et al. 2002) *S. agalactiae* accounted for 2.6 percent of native valve endocarditis and a similar figure of 2.8 percent was found in a Dutch series (van der Meer et al. 1991). Although group B streptococci have a well recognized though unexplained predilection for diabetics, most cases of infective endo-

carditis do not occur in diabetic patients. Other risk factors for group B streptococcal bacteremia and infective endocarditis include carcinoma, alcoholism, hepatic failure, and intravenous drug use (Farley 2001). They are aggressive pathogens and, as in staphylococcal endocarditis, other sites of infection are often found; these include septic arthritis, vertebral osteomyelitis, meningitis, and endophthalmitis. The overall mortality has decreased in the Spanish series from 61 percent in 1975–1992 to 8 percent in 1993–1998 probably due to earlier cardiac surgery. Similar clinical manifestations are seen with group G streptococcal endocarditis with left-sided involvement, frequent complications, and considerable mortality (Venezio et al. 1986; Smyth et al. 1988; Liu et al. 1995). Group A streptococci (*S. pyogenes*) remain a very rare cause of infective endocarditis (Burkert and Watanakunakorn 1991; Liu et al. 1992; Mohan et al. 2000).

In a nationwide prospective survey in Sweden (Olaison and Schadewitz 2002), enterococci accounted for 11 percent of cases of infective endocarditis; in the Dutch series they accounted for 7.4 percent of 326 cases (van der Meer et al. 1991), and a nearly identical figure of 8 percent was reported in the most recent survey from France (Hoen et al. 2002). Enterococcal endocarditis is usually considered a community-acquired disease; however the incidence of nosocomial acquisition appears to be increasing (Fernandez-Guerrero et al. 2002). Enterococci are normal inhabitants of the gastrointestinal tract and occasionally the anterior urethra and they may cause urinary tract infection, particularly after instrumentation. Enterococcal bacteremia has been reported after colonoscopy, proctoscopy, sigmoidoscopy and barium enema, prostate biopsy, and abdominal tract surgery (Caballero-Granado et al. 2001). Most enterococcal bacteremias are nosocomial in origin, associated with invasive procedures, and often polymicrobial. Factors suggesting infective endocarditis in patients with enterococcal bacteremia include community acquisition, underlying valvular heart disease, the absence of another likely portal of entry, and monomicrobial bacteremia. Most cases of infective endocarditis are caused by *Enterococcus faecalis*, but there are occasional cases of other species such as *E. faecium*, *E. avium*, *E. durans*, and *E. gallinarum* (Dargere et al. 2002). Enterococcal endocarditis is most commonly seen in elderly men. Enterococcal bacteruria may be a helpful diagnostic clue. The disease usually runs a subacute course and classic peripheral manifestations are often absent. In cases of enterococcal infective endocarditis, both normal and previously damaged valves, as well as prosthetic valves can be involved (Rice et al. 1991). Enterococcal endocarditis is more difficult to treat than streptococcal endocarditis because enterococci are intrinsically more resistant to antibiotics than streptococci. However, with standard treatment and the appropriate use of valve replacement, a cure rate of approximately 85 percent can be expected (Megran 1992).

Staphylococci

An ever increasing number of cases of infective endocarditis, both community-acquired and of nosocomial origin, are caused by staphylococci (Cabell et al. 2002). These organisms accounted for nearly 30 percent of infective endocarditis in the large population-based survey in France (Hoen et al. 2002) and their incidence increased from 4.9 cases per million person-years in 1991 to 5.7 cases per million person-years in 1999. Most of these staphylococci are *S. aureus*, but an increasing proportion are now coagulase-negative staphylococci (Hogevik et al. 1995; Cabell et al. 2002; Hoen et al. 2002).

Some 80–90 percent of the staphylococci isolated from cases of community-acquired native valve infective endocarditis are *S. aureus*. This organism can attack 'normal' heart valves (i.e. in patients without detectable cardiac disease) in one-third of the patients or damaged valves in patients with an underlying cardiac condition, producing severe fulminating infection that results in death in approximately 40 percent of patients (Petti and Fowler 2003; Hasbun et al. 2003). The infection arises from a bacteremia which itself stems from a trivial septic lesion such as wound infection, from an episode of catheter-related bacteremia occurring during a previous hospitalization, and often with no detectable focus (Mylotte et al. 1987; Eykyn 1988; Fowler et al., 1998; Mylotte and Tayara 2000a). Infective endocarditis caused by *S. aureus* has been traditionally regarded as a largely community-acquired disease. However, a recent study from Duke University revealed that among 59 definite cases of *S. aureus* infective endocarditis nearly half of the cases were hospital-acquired and were frequently caused by methicillin-resistant *S. aureus* (Fowler et al. 1999). The authors underscore the association of intravascular-device related *S. aureus* bacteremia and the development of infective endocarditis and recommend complete diagnostic evaluation including TEE (see below) for underlying infective endocarditis in any case of nosocomial bloodstream infection caused by *S. aureus*.

The disease often runs a rapidly progressive, *acute* course (Thompson 1982; Roder et al. 1999). Heart murmurs are seldom present initially unless there was a previous cardiac abnormality. The physician is thus confronted with a toxic ill patient with *S. aureus* recovered from a blood culture and sometimes without localizing signs. In about a quarter of cases the urine will contain both pus cells and staphylococci, the result of microabscesses in the kidney, and this can lead to further diagnostic confusion. Embolic phenomena involving the brain, spleen, or kidney, as well as the classic peripheral stigmata of infective endocarditis are more common in staphylococcal endocarditis than in endocarditis caused by other microorganisms. Intracardiac complications such as valve ring or myocardial abscess formation, and purulent pericarditis are frequently observed. Valvular destruction leading to congestive heart failure and death may occur within days (Watanakunakorn et al. 1973; Bayer 1982; Sanabria et al. 1990). This, the classical presentation of staphylococcal endocarditis, was accurately described over a century ago by William Osler in his Gulstonian lectures on 'malignant endocarditis' (1885), but it still seems largely unknown to many clinicians. Any patient who is unwell with a *S. aureus* bacteremia but no obvious site of infection should be assumed to have infective endocarditis and treated accordingly. Native valve infective endocarditis with *S. aureus* may occur at any age, but is more common in the middle-aged and elderly where mortality may exceed 50 percent (Julander 1985; Sanabria et al. 1990); it occurs in the previously healthy, as well as in those with underlying diseases such as diabetes mellitus and end-stage renal disease (Marr et al. 1998; Maraj et al. 2002; Doulton et al. 2003).

S. aureus is also the most common cause of infective endocarditis in intravenous drug users (IVDU) and is often associated with HIV infection (Mathew et al. 1995; Miro et al. 2003). However, since right-sided involvement mainly affecting the tricuspid valve predominates in these patients, the disease tends to be less severe than in non-addicts and usually responds to antimicrobial therapy alone with mortality rates below 10 percent.

Although still regarded by many as pathogens involved in prosthetic rather than in native valve endocarditis, coagulase-negative staphylococci also cause native valve infection and this has become more common, or certainly more commonly recognized, since the early 1980s (Caputo et al. 1987; Etienne and Eykyn 1990; Whitener et al. 1993; Miele et al. 2001). In the French series, coagulase-negative staphylococci accounted for 6.4 percent of cases of infective endocarditis (Hoen et al. 2002), and the Dutch series (van der Meer et al. 1992) reported a similar incidence (4.9 percent). In both studies it was not specified whether the infections were community- or hospital-acquired. Coagulase-negative staphylococci are normal inhabitants of the skin and mucous membranes and different species vary in their distribution throughout the body. Thus, it seems likely that native valve coagulase-negative staphylococcal infective endocarditis is an endogenous infection, even in cases of nosocomial acquisition associated with the use of intravascular catheters. In community-acquired cases a predisposing skin lesion is rarely detected and the oropharynx can be considered a likely portal of entry similar to the case with viridans streptococcal endocarditis. Approximately two-thirds of patients have pre-existing cardiac valvular disease

(Baddour et al. 1986). The clinical course is usually indolent (*subacute*) in onset and complications such as embolic disease and congestive heart failure are rare. Separating infective endocarditis from uncomplicated bacteremia due to coagulase-negative staphylococci or even from contamination may be difficult. Continuously positive blood cultures – in nosocomial cases in particular after intravascular catheters have been removed – may be an important clinical clue to the diagnosis.

The infecting species is most often *S. epidermidis* (Etienne and Eykyn 1990). However, in many reports the designation *S. epidermidis* tends to be used for any unspeciated coagulase-negative staphylococcus. With the advent of commercial identification systems, species identification of coagulase-negative staphylococci has become much easier. Thus, many other species have been reported to cause rare cases of native valve infective endocarditis, including *S. capitis*, *S. caprae*, *S. hominis*, *S. lugdunensis*, *S. saprophyticus*, *S. sciuri*, *S. schleiferi*, *S. simulans*, and *S. warneri* (Singh and Raad 1990; Leung et al. 1999; Sandoe et al. 1999; Hoen et al. 2002). Of these, *S. lugdunensis* appears to cause a more fulminate disease than that due to other coagulase-negative staphylococci (Vandenesch et al. 1993; Burgert et al. 1999; Farrag et al. 2001; Seenivasan and Yu 2003). These organisms are frequently misidentified as *S. aureus* owing to their morphological appearance when cultured on blood agar with yellow pigmentation and complete β-hemolysis.

Gram-negative bacilli

A small proportion of cases – 6.1 percent in the Dutch series (van der Meer et al. 1992), and 1.5 percent in the French series (Hoen et al. 2002) – of community-acquired native valve endocarditis are caused by gram-negative bacilli, the vast majority of these are fastidious, nonenteric organisms that tend to grow slowly on primary isolation from blood cultures or that cannot readily be isolated on conventional culture media. They include species that have a predilection for the heart valve and rarely cause infections other than infective endocarditis, as, for example, those known as the HACEK group (*Haemophilus aphrophilus*, *H. paraphrophilus*, *H. parainfluenzae*, *Actinobacillus actinomycetemcomitans*, *Cardiobacterium hominis*, *Eikenella corrodens*, and *Kingella kingae*), whose isolation from blood cultures is virtually diagnostic of infective endocarditis (Das et al. 1997; Feder et al. 2003). The disease usually runs a subacute course and occurs in the setting of pre-existing valvular disease. Large friable vegetations are common, as are frequent emboli, and often the need for valve replacement. However, with early diagnosis and institution of appropriate antimicrobial therapy, prognosis is usually favorable. The slow growth of these organisms in blood cultures necessitates incubation of blood cultures from all patients with presumptive infective endocarditis for at least 3 weeks.

Other fastidious gram-negative rods that have been implicated in infective endocarditis include *Acinetobacter* spp. (Gradon et al. 1992), *Brucella* spp. (Jacobs et al. 1990), *Campylobacter fetus* (Farrugia et al. 1994), *Francisella tularensis* (Tancik and Dillaha 2000), *Haemophilus influenzae* (Geraci et al. 1977), and *Streptobacillus moniliformis* (Rupp 1992). Valve replacement is usually necessary for cure of *Brucella* endocarditis (Jacobs et al. 1990; Keles et al. 2001).

Gram-negative enteric bacilli, although frequently isolated from blood cultures, are only rarely a cause of infective endocarditis (Cohen et al. 1980). Drug addicts, patients with prosthetic heart valves, and patients with liver disease are at increased risk for the development of enterobacterial endocarditis. Sustained gram-negative bacillary bacteremia despite adequate antimicrobial therapy should herald possible infective endocarditis. A fulminate clinical course resulting in congestive heart failure is common and the prognosis usually poor (Carruthers 1977; Geraci and Wilson 1982). *Salmonella* spp., in particular *S. cholerasuis*, *S. typhimurium*, and *S. enteritidis* were most frequently encountered in earlier reports (Schneider et al. 1967), with valvular destruction, intracardiac complications, and rapidly progressive congestive heart failure, commonly observed. *Salmonella* endocarditis has been found in both native and prosthetic valve endocarditis (Goerre et al. 1998) and is also associated with HIV infection (Fernandez Guerrero et al. 1996). High overall mortality was also seen with infective endocarditis caused by *Serratia marcescens* (Mills and Drew 1976), as well as *Pseudomonas aeruginosa* (Wieland et al. 1986; Komshian et al. 1990). Both organisms have been mainly recovered from patients with infective endocarditis who have abused intravenous drugs. *Pseudomonas aeruginosa* frequently affects normal valves, and complications include embolic phenomena, splenic abscesses, valve ring abscesses, and congestive heart failure. Due to the poor prognosis with medical therapy alone, early surgical intervention has been advocated if left-sided cardiac involvement is present (Wieland et al. 1986; Komshian et al. 1990).

Other bacteria

The nondiphtheriae corynebacteria are major components of the normal flora of human skin and mucous membranes. As such, they are frequently dismissed as contaminants if isolated from clinical specimens. However, these organisms are also uncommon, but increasingly recognized as agents of endocarditis in patients with underlying structural heart disease or prosthetic valves (Ross et al. 2001; Knox and Holmes 2002). Among the species reported are *Corynebacterium amycolatum*, *C. jekeium*, *C. striatum*, and *C. xerosis*.

However, methods that reliably differentiate related species are not available in most routine diagnostic laboratories. Other rare, gram-positive rods implicated in infective endocarditis include *Erysipelothrix rusiopathiae* (Venditti et al. 1990), *Lactobacillus* spp. (Husni et al. 1997), and *Listeria* spp. (Spyrou et al. 1997).

Anaerobes are found as normal inhabitants in the human mouth with or without dental caries and they represent the predominant flora of the bowel. However, these organisms have rarely been reported to cause endocarditis. According to Felner and Dowell (1970), non-streptococcal anaerobic bacteria accounted for 1.3 percent of cases of infective endocarditis and some other authors have reported a higher incidence, especially in IVDUs and polymicrobial infections. *Bacteroides fragilis* was the predominant pathogen in a review of 67 cases published by Nastro and Finegold (1973). Other anaerobes recovered from patients with infective endocarditis include other *Bacteroides* spp., *Clostridium* spp., *Fusobacterium* spp., and *Propionibacterium acnes* (Lortholary et al. 1995; Mohsen et al. 2001; Brook 2002).

Endocarditis due to rare and fastidious bacteria has recently been reviewed by Berbari et al. (1997) and by Brouqui and Raoult (2001). Some 350 cases of endocarditis caused by *Coxiella burnetii*, the etiologic agent of Q fever, have been described in the medical literature. Endocarditis is a recognized, though rare (and very late) sequel of acute *C. burnetii* infection. Most infections occur in middle-aged men usually (88.5 percent) with pre-existing heart disease (Marrie and Raoult 2002). The reservoir of the organism is sheep and cattle, but the source and mode of transmission of this anthropozoonosis to humans often remains unclear. Occupational exposure was reported in 12.5 percent of endocarditis patients. In contrast to most other cases of infective endocarditis, the diagnosis is usually made serologically although *C. burnetii* can be recovered on culture with special techniques (Houpikian and Raoult 2003). In view of this it is likely that *C. burnetii* endocarditis is underdiagnosed and some cases falsely labeled 'culture-negative' endocarditis.

Other organisms occasionally causing infective endocarditis that cannot be isolated in routine laboratories include *Bartonella* spp. (Spach et al. 1995; Raoult et al. 2003), *Chlamydia* spp. such as *C. psittaci*, the agent of psittacosis, and *C. pneumoniae* (Marrie et al. 1990; Brouqui and Raoult 2001), *Mycoplasma* spp. (Popat et al. 1980; Cohen et al. 1989), and *Tropheryma whipplei*, the agent of Whipple's disease (Fenollar et al. 2001; Richardson et al. 2003).

'CULTURE-NEGATIVE' ENDOCARDITIS

Most reported series of infective endocarditis include a variable number of cases that are designated *culture-negative*, that is they have negative blood cultures while other criteria of infective endocarditis such as echocardiographic demonstration of vegetations or histological evidence at surgery (see Clinical manifestations and complications) are met (Pesanti and Smith 1979; Tunkel and Kaye 1992; Hoen et al. 1995; Kupferwasser and Bayer 2000; Lamas and Eykyn 2003; Werner et al. 2003). The reported incidences varied ranging from 12 to 24 percent. Blood cultures remained negative in 9 percent of the cases recently reported by Hoen et al. (2002). Some earlier studies used poorly defined diagnostic criteria and possibly inadequate culture techniques. There are various explanations for the phenomenon of culture-negative endocarditis.

Previous administration of antibiotics will reduce the incidence of positive blood cultures, particularly in streptococcal infective endocarditis. The diagnosis of infective endocarditis is often initially missed in patients who do not present with classical 'textbook' signs, and these will be treated with antibiotics empirically before this diagnosis is being considered and blood cultures are taken. In the series reported by Werner et al. (2003), 45 percent of patients with culture-negative endocarditis had received prior antimicrobial therapy. The duration of previous antibiotic therapy is also an important factor. After only 2–3 days of antibiotic treatment, blood cultures that were initially negative after prolonged incubation often then become positive, but after longer courses they may remain persistently negative. In such cases, the antibiotics should be stopped and repeated blood cultures performed for a period up to 5 days.

Fastidious slow growing organisms are uncommon causes of infective endocarditis, but may take much longer to grow in blood cultures than the more usual 'viridans' streptococci or staphylococci. Examples include the HACEK group, the nutritionally variant streptococci such as *Granulicatella (Abiotrophia) adiacens* and *A. defectiva*, *Legionella* spp. (Chen et al. 1996) as well as mycobacteria (Cope et al. 1990) and some anaerobes. In such cases infective endocarditis only initially appears *culture-negative*, but cultures eventually become positive if they are held for the recommended 3 weeks and/or are subcultured on to appropriate media, respectively.

Organisms requiring special isolation techniques or serological diagnosis include *Bartonella* spp., *Chlamydia* spp., *C. burnetii*, *Mycoplasma* spp., and *T. whipplei* (Raoult et al. 1996; Brouqui and Raoult 2001; Fenollar et al. 2003; Houpikian and Raoult 2003). Although specialized culture techniques are available for some of these obligate intracellular pathogens in certain reference laboratories, most laboratories will still rely on serological diagnosis. Blood cultures also remain negative in over 50 percent of cases of fungal endocarditis such as that caused by *Aspergillus* spp. (Rubinstein et al. 1975).

An incorrect diagnosis of infective endocarditis based on a false interpretation of clinical or echocardiographic findings may also be the reason for an apparently *culture-negative* endocarditis. This is probably quite common since physicians are rightly concerned not to miss a case of infective endocarditis.

In conclusion, in all cases of defined or suspected endocarditis with negative blood cultures in the absence of previous antimicrobial therapy serum should be analyzed for *Bartonella*, *Coxiella*, and *Chlamydia* species antibodies, and the excised valve should be analyzed by microscopy, culture, and histology. In addition, specialized culture techniques and relevant PCR techniques might be attempted at a reference laboratory (see below).

Clinical manifestations and complications

The clinical presentation of infective endocarditis is determined by the local infection of the valve and its complications, the occurrence of septic embolization of valvular vegetations to distant organs, constant bacteremia that may result in distant foci of infection, and the consequences of circulating immune complexes. The symptoms and signs of infective endocarditis are therefore extremely variable (Table 19.6), and the differential diagnosis is broad.

Fever is the most common sign but may be absent in severely compromised patients, patients with chronic renal or hepatic failure, infective endocarditis with less virulent organisms and in particular with previous antimicrobial therapy. Fever usually abates within 3–7 days of appropriate antimicrobial treatment, but prolonged fever may be associated with major intracardiac or embolic complications, such as myocardial or splenic abscess, and should therefore prompt further diagnostic

evaluation. Nonspecific findings seen especially in subacute cases of infective endocarditis include weight loss, weakness, anorexia, malaise, nausea, and night sweats.

Most patients with infective endocarditis have a heart murmur, that may rarely present as the classic *changing murmur* or *new regurgitant murmur*, but in most cases is a pre-existing murmur. Other classical peripheral findings include splinter hemorrhages (Figure 19.2), i.e. red to brown streaks below the finger nails or toe nails, petechiae on the skin, the conjunctivae, and oral mucosa that both result from local vasculitis or emboli, and the only rarely seen Osler nodes, small, painful subcutaneous nodular lesions found in the pulp of the digits or in the thenar eminence. Janeway lesions are nontender, hemorrhagic or pustular lesions, often found on the palms and soles, also resulting from peripheral septic emboli (Figure 19.3). Janeway lesions consist of bacteria, neutrophils, necrosis, and subcutaneous hemorrhage and are

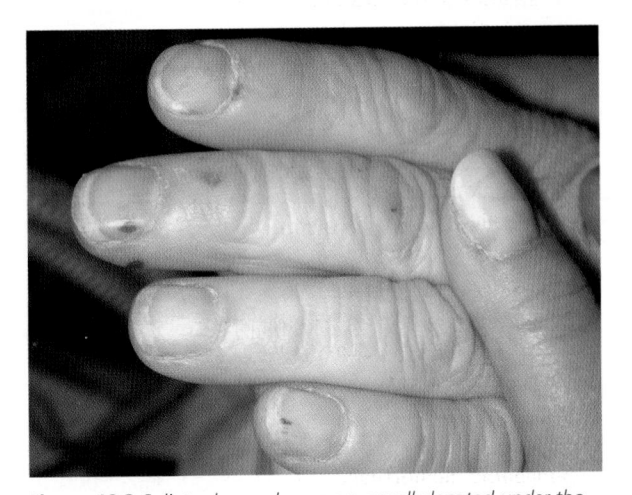

Figure 19.2 *Splinter hemorrhages are usually located under the finger or toe nails.*

Table 19.6 *Signs and symptoms of infective endocarditis*

Sign/symptom	% of patients affected
Fever	80
Malaise	25
Chills	40
Weakness	40
Sweats	25
Anorexia/weight loss	25
Myalgia/arthralgia	15
Heart murmur	85
Changing or new murmur	3–10
Skin manifestations	15–30
Janeway lesions	<10
Osler nodes	1–23
Petechiae	20–40
Splinter hemorrhages	15
Roth spots	2–10
Splenomegaly	20–57
Embolic phenomena	>50

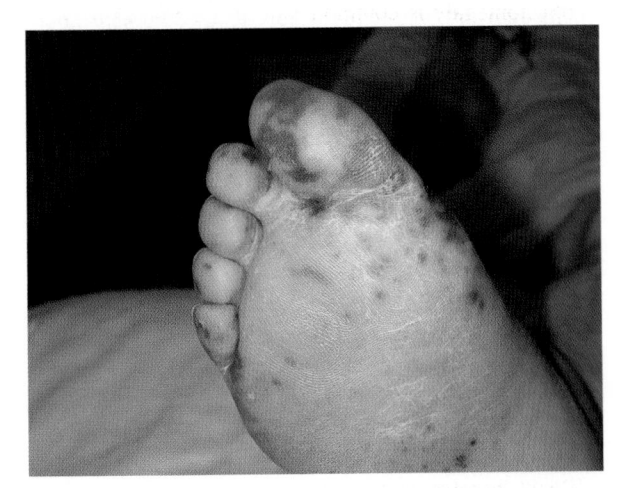

Figure 19.3 *Janeway lesions are nontender, erythematous, hemorrhagic, or pustular lesions, often located on the palm or soles.*

often seen in staphylococcal endocarditis. The offending microorganisms may be readily cultured from these lesions. Roth spots are oval retinal lesions surrounded by hemorrhage occurring only in a minority of cases.

The complications of infective endocarditis have been reviewed recently by Mansur et al. (1992). The leading cardiac complication is congestive heart failure caused by a progressive infection-induced valvular destruction such as perforation of the valve leaflet, the cordae tendineae, or papillary muscle. Severe congestive heart failure was observed in 34 percent of patients with infective endocarditis in the French series (Hoen et al. 2002). Aortic-valve involvement is more frequently associated with congestive heart failure than is mitral-valve involvement in infective endocarditis. Other complications include rupture of the intraventricular septum, myocarditis or myocardial infarction resulting from coronary artery embolization or mycotic aneurysm, and myocardial and valve ring abscesses. Extension of the infection into the septum may cause arrhythmias and conduction abnormalities. Other distant complications mainly result from embolic events and metastatic infections (Millaire et al. 1997). Overall, vascular complications were seen in 44 percent of patients with infective endocarditis reported by Hoen et al. (2002), the vast majority of which were systemic arterial emboli.

Neurologic manifestations usually resulting from cerebral emboli occur in 20–40 percent of cases and often represent the dominating clinical syndrome, especially in staphylococcal endocarditis (Roder et al. 1997; Heiro et al. 2000). A stroke syndrome in a febrile patient with underlying valvular disease suggests the possibility of infective endocarditis. Cerebral manifestations include cerebral infarction, abscesses, intracerebral mycotic aneurysms, cerebritis, meningitis, and intracerebral or subarachnoid hemorrhage. Resulting symptoms encompass hemiplegia, aphasia, sensory loss, headache, seizures, and altered mental status, ranging from slight personality changes to frank psychosis.

Splenomegaly is common among patients with infective endocarditis and mostly seen in patients with prolonged disease. Splenic abscess may develop from bacteremic seeding of a previously infarcted area or directly from an infected embolus. Splenic abscess was among the rarer complications reported by Mansur et al. (1992) with clinical signs or symptoms often absent; this complication was observed in 10 percent of patients with infective endocarditis in the French series (Hoen et al. 2002). However, splenic septic emboli and infarction were reported in 44 percent of patients at autopsy (Weinstein and Schlesinger 1974). Prolonged fever and sepsis may be a clinical clue to this complication and should prompt ultrasound investigation, abdominal CT scan or MRI (Yilmaz et al. 2003).

Renal involvement includes infarction and abscess formation leading to bacteruria, as well as immune complex glomerulonephritis. Renal failure is mainly seen in long-standing disease, but is usually reversible with antimicrobial treatment (Conlon et al. 1998).

Musculoskeletal symptoms in infective endocarditis include diffuse myalgias, low back pain, arthalgia, and septic arthritis.

Mycotic aneurysms may result from direct bacterial invasion of the arterial wall with subsequent abscess formation and rupture, from embolic occlusion of the vasa vasorum, or immune complex deposition into the arterial wall. They are found most commonly at bifurcation points primarily of the cerebral vessels and are often clinically asymptomatic until perforation or rupture occurs. Major vessel emboli are rare in bacterial endocarditis and are more frequently encountered in fungal endocarditis (Pierrotti and Baddour 2002).

Right-sided endocarditis is not associated with peripheral emboli and other peripheral vascular phenomena. Instead, pulmonary findings are usually present, such as pulmonary embolism with and without infarction, pneumonia, pleuritic chest pain, pleural effusion, and empyema (Miro et al. 2003; Moss and Munt 2003). Community-acquired *S. aureus* bloodstream infection in the presence of multiple pulmonary infiltrates should always suggest right-sided infective endocarditis, and in the at-risk person intravenous drug abuse as well. Heart murmurs are frequently absent in right-sided endocarditis.

Diagnosis

DEFINITION OF DIAGNOSTIC CRITERIA

The diagnosis of infective endocarditis requires consideration of clinical, laboratory, and echocardiographic findings. To help the clinician to establish the diagnosis of infective endocarditis and to allow for comparison of published reports, attempts have been made to define hard diagnostic criteria. Von Reyn et al. (1981) introduced the categories of *definite*, *probable*, and *possible* IE but their 'definite' category relied on direct evidence from histology or microbiology at surgery or autopsy. The so-called Beth Israel criteria proved very specific, but less sensitive. Their *definite* category was too restrictive. Over a decade later with the advent and more common usage of echocardiography, Durack et al. (1994) proposed new criteria of *definite* and *possible* infective endocarditis based on a much more comprehensive assessment that included clinical and echocardiographic findings, and these have gained widespread acceptance. The so-called Duke criteria and their suggested modifications (Fournier et al. 1996; Lamas and Eykyn 1997; Li et al. 2000) are built on two major criteria, i.e. microbiologic criteria and clinical or echocardiographic evidence of endocardial involvement.

The microbiologic criteria include a positive blood culture for infective endocarditis as defined by the recovery of (1) a typical microorganism from two separate blood cultures in the absence of a primary focus (i.e. 'viridans' streptococci, *S. bovis*, HACEK group, or community-acquired *S. aureus* or enterococci); (2) a persistently positive blood culture defined as the recovery of a microorganism consistent with endocarditis from either blood cultures drawn more than 12 h apart or all three or a majority of four or more separate blood cultures with the first and last drawn at least 1 h apart; or (3) serology for Q fever by IFA showing phase 1 IgG antibodies at >1:800.

Evidence of endocardial involvement includes a positive echocardiogram for infective endocarditis, i.e. (1) an oscillating intracardiac vegetation on valve or supporting structures, in the path of regurgitant jets, or on implanted material; (2) a periannular abscess; or (3) a new dehiscence of a prosthetic valve; or new valvular regurgitation on auscultation (increase or change in pre-existing murmur is not sufficient).

Minor diagnostic criteria include (1) a predisposing cardiac condition – such as previous endocarditis, aortic-valve disease, rheumatic heart disease, prosthetic heart valve, complex congenital heart disease, and many others – or injection-drug use; (2) fever ≥38°C; (3) vascular phenomena such as major arterial emboli, septic pulmonary infarcts, mycotic aneurysm, intracranial hemorrhage, conjunctival hemorrhages, or Janeway lesions; (4) immunologic phenomena such as glomerulonephritis, Osler's nodes, Roth spots, or rheumatoid factor; and (5) microbiologic evidence such as a positive blood culture not meeting the major criteria or serologic evidence of an active infection with organism consistent with infective endocarditis, such as *Brucella* spp., *Chlamydia* spp., *Legionella* spp., and *Bartonella* spp.; and (6) echocardiographic findings consistent with infective endocarditis, but not meeting major criteria.

According to the Duke Endocarditis Service, the diagnosis of infective endocarditis is definite (1) when a microorganism is demonstrated by culture or histologic testing in a vegetation; (2) when active endocarditis is histologically confirmed at surgery or at autopsy; or (3) in the presence of two major criteria, one major and three minor criteria, or five minor criteria as outlined above (Durack et al. 1994).

The usefulness of the Duke criteria has been validated and compared to the von Reyn criteria in several subsequent studies (Bayer 1996; Martos-Perez et al. 1996; Heiro et al. 1998; Habib et al. 1999; Perez-Vazquez et al. 2000). Overall, the specificity of the proposed criteria was high (99 percent) and the negative predictive value greater than 92 percent.

MEDICAL HISTORY

The patient's medical history might be an important clue for the etiologic diagnosis of infective endocarditis. A history of contact with cattle or sheep may suggest infection caused by *C. burnetii* (Q fever) or *Brucella* species. Contact with a cat, in particular a history of scratches or bites, suggests *Bartonella henselae* or *Pasteurella* spp. Exposure to birds has been related to infection with *C. psittaci*. Further clues as to patients' medical history related to endocarditis can be found in the review by Brouqui and Raoult (2001).

LABORATORY FINDINGS

Unspecific laboratory findings are usually present in infective endocarditis, but none is diagnostic. Hematological parameters include anemia in 70–90 percent of cases and less often leukocytosis which is present in only 20 to 30 percent of cases of subacute endocarditis, but is more frequent in the acute form. The erythrocyte sedimentation rate (ESR) is elevated in 90–100 percent of cases, so that a normal ESR usually rules out the diagnosis of infective endocarditis. Similarly, the C-reactive protein (CRP) concentration is virtually always elevated. Serial determination of both parameters, ESR and CRP, are useful for monitoring antimicrobial therapy and may also be a diagnostic clue for the detection of clinically silent embolic complications.

ECHOCARDIOGRAPHY

Cardiac imaging, specifically echocardiography, has greatly enhanced the ability of clinicians to effectively diagnose and manage infective endocarditis. Echocardiograms should generally be obtained in all patients suspected of having infective endocarditis, both to establish the diagnosis and to identify complicated cardiac involvement that may warrant surgical intervention (Bayer et al. 1998). Manifestations of endocardial involvement include vegetations, abscesses, aneurysms, fistulae, leaflet perforations, and valvular dehiscence. Transthoracic echocardiography (TTE) is a noninvasive and rapid method with excellent specificity. The sensitivity however, drops to less than 60 percent in patients with obesity, chronic obstructive lung disease, and chest wall deformities. It has therefore been recommended to perform transesophageal echocardiography if TTE is inconclusive. Transesophageal echocardiography (TEE) is more invasive and also more expensive than TTE, but also more sensitive than the transthoracic approach (Shively et al. 1991; Chamis et al. 1999; Roe et al. 2000; Humpl et al. 2003) and currently represents the optimal approach to echocardiographic imaging. Initial use of TEE is more cost-effective and diagnostically efficient in all patients with an intermediate or high pretest probability of infective endocarditis, particularly in the setting of prosthetic valves (Heidenreich et al. 1999). In a prospective study of patients with *S. aureus* bacteremia, Fowler and colleagues (1997) reported that sensitivity of TTE for detecting infective endocarditis was 32 percent, and the specificity was 100 percent. The addition of TEE increased the sensitivity to 100 percent, but resulted in one false-positive result (specificity 99 percent). Echocardiographic imaging has also been used to guide therapeutic management of patients with infective endocarditis (Harris et al. 2003; Kim et al. 2003). TEE is especially useful for detecting perivalvular extension of infective endocarditis and the presence of an intramyocardial abscess, as well as for evaluation of prosthetic-valve endocarditis.

CULTURE METHODS

The blood culture is the single most important method applied in the diagnostic work-up for infective endocarditis. The bacteremia in patients with culture-positive endocarditis is usually continuous and low-grade. The application of quantitative blood culture techniques has demonstrated that between one and 10 bacteria per ml are present in the blood of patients with infective endocarditis and this bacterial density remains constant during the entire course of the untreated disease (Werner et al. 1967). The details of taking blood cultures have been presented elsewhere in this chapter. It has been recommended that at least 10 ml of blood be obtained for each culture and that at least three blood cultures sets (no more than two bottles per venipuncture) are obtained in a patient with suspected endocarditis (Washington 1982, 1987). If clinical conditions permit, blood cultures should be drawn at least 1 h apart. In cases where the institution of antimicrobial therapy can be postponed for 24 h, another three blood cultures should be obtained the following day. More blood cultures may be necessary if the patient has received antimicrobials in the preceding 2 weeks. Blood cultures should be incubated for at least 3 weeks. An evidence-based approach for sequential testing in infective endocarditis has recently been proposed by Watkin et al. (2003).

If no previous antimicrobial therapy was administered and blood cultures remain negative in a patient with probable or definite endocarditis on clinical or echocardiographic grounds, i.e. culture-negative endocarditis by definition, fastidious organism should be suspected, and special cultivation methods be considered. The simplest approach to isolate *Abiotrophia* spp. is to inoculate the blood culture broth on to blood agar that is cross-streaked with *S. aureus* where these species grow as satellite colonies. Most members of the HACEK group and other fastidious organisms can be recovered after prolonged incubation of blood cultures for 3 to 4 weeks both with conventional, as well as with automated blood culture systems. Prolonged incubation is therefore recommended as a routine procedure in the diagnostic work-up of blood cultures from patients with suspected endocarditis. The isolation of rare pathogens such as *Legionella* spp., mycobacteria, and *Bartonella* spp. requires special techniques recently reviewed by Brouqui and Raoult (2001). These authors also have developed tissue cell cultures for the isolation of obligate intracellular pathogens such as *Bartonella* spp., *Chlamydia* spp., and *C. burnetii*.

Pathogens can also be cultured from resected valves or biopsy specimens by inoculation on to agar or into tissue culture. Histologic examination of resected valve material is mandatory, particularly in the case of culture-negative endocarditis. A discussion of the appropriate staining techniques and immunohistological methods is beyond the scope of this chapter.

SEROLOGY

The most common serologic methods applied to the microbiological diagnosis of infective endocarditis include indirect immunofluorescence for *B. henselae*, the tube agglutination test for *Brucella melitensis* infections, ELISA or indirect immunofluorescence for *Chlamydia* spp., indirect immunofluorescence for *C. burnetii* and *Legionella* spp., and ELISA for *Mycoplasma pneumoniae*. Systematic serologic testing for these organisms is particularly warranted in cases of clinically suspected culture-negative infective endocarditis (Houpikian and Raoult 2003). Among 19 cases of culture-negative endocarditis in the French series, eight microorganisms were identified by serological methods: *Bartonella* spp. ($n = 3$), *C. burnetii* ($n = 1$), *Legionella* spp. ($n = 1$), *Chlamydia* spp. ($n = 2$), and *M. pneumoniae* ($n = 1$).

MOLECULAR METHODS

The advent of molecular amplification methods has enlarged the spectrum of nonculture laboratory methods for the diagnosis of infective endocarditis (Fournier and Raoult 1999; Lisby et al. 2002). Such techniques have been applied to most types of clinical materials including blood and valvular specimens obtained at surgery (Grijalva et al. 2003) or autopsy. Broad-range PCR that incorporates broad-spectrum primers targeting bacterial 16S rRNA allows for the detection of almost all bacteria in a single reaction. This method combined with sequence analysis of the amplified DNA has proved to be a promising tool for patients with culture-negative endocarditis (including instances of prior antimicrobial treatment) and allows the detection of rare, noncultivable organisms (Bosshard et al. 2003). Such novel approaches may lead to more comprehensive patient evaluations and the discovery of new etiologic agents of infective endocarditis. It has therefore been suggested to include positive identification of an organism by molecular biology methods as an additional major criterion in the revised Duke criteria of infective endocarditis (Millar et al. 2001; Bosshard et al. 2003).

In contrast to its use directly on clinical samples, PCR with subsequent sequencing of the amplification product can also be used for identification of bacteria isolated by culture methods in cases when species identification proves difficult by conventional biochemical methods. Such techniques clearly have increased the identification of rare and fastidious microorganisms recovered from patients with infective endocarditis in recent years.

PROSTHETIC VALVE ENDOCARDITIS

Prosthetic valve endocarditis (PVE) is an uncommon but potentially disastrous complication of cardiac valve replacement. It accounts for 7 to 25 percent of cases of infective endocarditis in most developed countries (Vlessis et al. 1997; Mylonakis and Calderwood 2001; Hoen et al. 2002). The risk of an individual patient developing PVE at any particular time is difficult to determine from reported studies of PVE. As Rutledge et al. (1985) pointed out, to cite a frequency of PVE of 4 percent as many studies do is not very valuable, since the risk of developing PVE is not uniform over time; it is higher during the initial 6 to 12 months after surgery and subsequently decreases. It has been estimated that the cumulative risk developing PVE ranges between 3 and 6 percent by 5 years after surgery (Ivert et al. 1984; Calderwood et al. 1985; Rutledge et al. 1985; Agnihotri et al. 1995). The rate of PVE was higher for recipients of a mechanical valve during the initial 3 months after surgery compared to patients with a bioprosthetic valve, but after 12 months the risk of infection was greater for the latter group (Calderwood et al. 1985). It appears that patients with a prosthetic heart valve carry a life-long risk of acquiring PVE.

PVE has conventionally been considered as *early* and *late*, with 60 days regarded as the time limit for early cases (Dismukes et al. 1973) because of differences between the microbiology and pathogenesis of infection in the two time periods. Early PVE results from perioperative contamination, as well as from bacteremia arising during the early postoperative period, and is thus usually a hospital-acquired infection (Gordon et al. 2000). Early after valve replacement the intracardiac structures involved in the operation are not endothelialized and therefore especially vulnerable to the deposition of fibrin platelet aggregates and subsequent bacterial colonization. Not all these 'early' infections present within 60 days of operation and this has led some investigators to suggest that the time limit for early disease should be extended to 6 months or even a year. PVE occurring later than 12 months after valve replacement is more likely to result from transient community-acquired bacteremia, such as is the case with community-acquired native valve endocarditis. Whatever classification is used, it is clear that the causative microbes of PVE acquired in hospital and the severity of disease differ from PVE acquired in the community, often many years after valve replacement.

The incidence of early PVE has fallen since the advent of valve replacement surgery over 40 years ago, but the most common infecting organisms are still likely to be staphylococci, both coagulase-negative staphylococci (mostly *S. epidermidis*) and *S. aureus*, followed by gram-negative bacilli, enterococci, corynebacteria, and fungi (Wolff et al. 1995). Coagulase-negative staphylococci account for approximately 30 percent of cases. These are skin organisms and may arise both from the patient's own commensal flora, as well as from health care personnel. Interestingly, the majority of these strains are methicillin-resistant (Calderwood et al. 1985).

The pathogenesis of late PVE differs from that of early PVE and is similar to that of native valve endocarditis with bacteria from the bloodstream localizing on the prosthesis or damaged endocardium. As with community-acquired native valve endocarditis, the most common pathogens in late PVE are streptococci, enterococci, and staphylococci, but a larger percentage of cases than in native valve disease are caused by coagulase-negative staphylococci.

Typical features of PVE are large vegetations, embolic complications, and invasive infection extending into the annulus and adjacent myocardium resulting in paravalvular abscess formation and valve dehiscence (Graupner et al. 2002). Invasive infection has been associated with clinical findings of valvular dysfunction, persistent fever despite appropriate antimicrobial therapy, and echocardiographic evidence of abscess formation. Invasive infection is more commonly observed during the first 12 months after surgery.

The Duke criteria, primarily established for the diagnosis of native valve infective endocarditis can also be applied for the diagnosis of PVE (Nettles et al. 1997; Perez-Vazquez et al. 2000). Diagnosis mainly relies on blood cultures and echocardiographic findings. Blood cultures must be obtained frequently and in the absence of antimicrobial therapy, in particular in the febrile postoperative patient. This is of particular importance related to the fact that the organisms most frequently involved in early PVE, i.e. coagulase-negative staphylococci and diphtheroids, are also the most frequently encountered contaminants of blood cultures. The superiority of TEE in comparison to TTE relates to the increased diagnostic sensitivity in detecting periprosthetic leaks, abscesses, and fistulas (Lengyel 1997; Bach 2000; Greaves et al. 2003).

The mortality of PVE is high, particularly in patients with onset of infection within 2 months of surgery and if infection is caused by *S. aureus* (Wolff et al. 1995). Other predictors of mortality are severe congestive heart failure and signs of invasive disease such as persisting fever, despite appropriate antimicrobial therapy. Mortality rates as high as 77 percent have been described in earlier series (Dismukes et al. 1973). Mortality rates ranging from 10 to 30 percent have been observed in more recent surveys, where earlier surgery may have contributed to a more favorable outcome (Akowuah et al. 2003). Overall, 40 to 65 percent of patients with PVE may be candidates for surgical intervention (Lytle et al. 1996; Fowler et al. 1998). Patients that may be treated by medical therapy alone are those with late-onset infection, with infection caused by less virulent organisms, and those without evidence of invasive disease. The mortality of late PVE is less than that of early infection and not much different than for native valve infection. Patients with late PVE are likely to present to doctors earlier than many of those with native valve infection and hence to be referred to specialist departments earlier. Both these factors are relevant to outcome.

TREATMENT OF INFECTIVE ENDOCARDITIS

In the pre-antibiotic era, infective endocarditis was almost uniformly fatal and no-one can deny that antibiotics have dramatically reduced the mortality of the disease. Antimicrobial therapy should consider some specific conditions in infective endocarditis that are unique among bacterial infections, i.e. the infection involves an area of impaired host defense – the vegetation – with bacteria embedded in a fibrin meshwork where they are sheltered from phagocytic cells, where the bacteria reach enormous population densities, and where they may exist in a state of reduced metabolic activity. Therefore certain general treatment principles should be followed: (1) high-dose parenteral antimicrobial therapy is recommended to reach sustained antibacterial activity; (2) prolonged administration of antimicrobial therapy is necessary to prevent relapse; (3) bactericidal antimicrobial agents are generally preferred over bacteriostatic drugs; (4) antibiotic combination therapy is recommended to produce a rapid bactericidal effect (Le and Bayer 2003). Unless the patient is very unwell, toxic, and septic, antibiotics should preferably be withheld until the blood culture results are available. Many patients with infective endocarditis will have been ill for several weeks or even months, and thus a delay of 1–2 days in starting antibiotics cannot make any difference to the course of the disease.

Wilson et al. (1995) have recently reviewed the treatment of infective endocarditis due to streptococci, enterococci, staphylococci, and fastidious organisms of the HACEK group. Detailed guidelines are also given in the report from the Working Party in Endocarditis of the British Society for Antimicrobial Chemotherapy (BSAC) which has recently been updated (1998).

The recommended approach to 'penicillin-sensitive' streptococcal native valve endocarditis (i.e. oral streptococci and group D streptococci including S. bovis with a minimum inhibitory concentration (MIC) for penicillin ≤0.1 mg/l) consists of administration of aqueous penicillin G, 10–30 million units IV daily, or ceftriaxon, 2 g once daily IV or IM, for 4 weeks. The addition of gentamicin to penicillin exerts a synergistic killing effect on viridans streptococci in vitro and results in more rapid sterilization of cardiac vegetations in an animal model of endocarditis, but has not reached clinical cure rates superior to those obtained with penicillin alone in clinical studies. Short-course therapy of penicillin combined with gentamicin (3 mg/kg IV divided into three daily doses) for 2 weeks has been advocated in uncomplicated cases with comparable cure rates as high as 98 percent. This approach is not recommended for patients with myocardial abscess, embolism, or extracardiac metastatic foci of infection, and impaired renal function. In patients with prosthetic valve endocarditis a 6-week regimen of penicillin together with gentamicin for the first 2 weeks is recommended. Vancomycin therapy, 1 g IV twice daily for 4 weeks is indicated for patients with confirmed immediate hypersensitivity reactions to betalactam antibiotics.

Endocarditis due to viridans group streptococci with an MIC for penicillin >0.1 to 0.5 mg/l usually requires a 4-week course of penicillin together with gentamicin for the first 2 weeks. For patients with endocarditis with viridans streptococci with an MIC for penicillin >0.5 mg/l and nutritionally variant streptococci treatment with the regimen recommended for enterococcal endocarditis is advised (see below).

Enterococci are less susceptible to betalactam antimicrobial agents than streptococci (median MIC for penicillin: 2 mg/l), and in general these agents are bacteriostatic against the enterococci; combination regimens are therefore always recommended for treating enterococcal endocarditis. Cell wall-active antibiotics, such as beta-lactams or glycopeptides plus an aminoglycoside, are synergistic and exert a bactericidal effect in vitro against most enterococci unless high-level resistance to gentamicin (MIC >500 mg/l) or streptomycin (MIC >2 000 mg/l) is present. In these cases, penicillin- or vancomycin-aminoglycoside synergy and bactericidal activity is not apparent. The standard antimicrobial recommendation for enterococcal endocarditis is penicillin (20–30 million units daily) or ampicillin (12 g daily) combined with gentamicin (3 mg/kg per day) administered IV for 4–6 weeks. The 6-week regimen is recommended for patients with duration of symptomatic illness exceeding 3 months, as well as for complicated and prosthetic valve endocarditis. In patients allergic to beta-lactams or infected with penicillin-resistant strains such as E. faecium, vancomycin is administered in combination with gentamicin. If endocarditis is caused by a strain exhibiting high-level aminoglycoside resistance, the addition of an aminoglycoside is no longer beneficial, and therapy should be prolonged to 8–12 weeks instead. However, clinical cure is often not obtained without valve replacement. Endocarditis due to vancomycin-resistant enterococci (VRE) is a cause of concern, and the antimicrobial regimen of choice for these cases remains unsettled.

The great majority of staphylococci, acquired either in the hospital or in the community, produce β-lactamase, and they are highly resistant to penicillin. The current recommended regimen for treatment of native valve staphylococcal endocarditis includes a penicillinase-resistant penicillin, nafcillin or oxacillin, 12 g daily IV, or cefazolin (or another first generation cephalosporin), 6–8 g daily IV, given for 4–6 weeks. The addition of gentamicin induced a synergistic killing effect both in vitro and in an animal model of endocarditis and has also been shown to result in more rapid clearing of bacteremia from the bloodstream in patients. Many authorities therefore recommend the addition of gentamicin for the first 3–5 days of therapy, although the combination did not improve the survival rate in a clinical study involving a small group of patients. In the patient allergic to penicillin or when the isolate is methicillin-resistant (MRSA or MRSE in the case of *S. epidermidis*), vancomycin, 1 g twice daily IV for 4–6 weeks is the agent of choice. However, vancomycin is less rapidly bactericidal than nafcillin in vitro against *S. aureus*, and prolonged bacteremia and increased failure rates have been associated with the use of this agent in infective endocarditis caused by MRSA. Accordingly, caution is advised when considering treatment with vancomycin only for reasons of convenience related to pharmacokinetics such as in hemodialysis patients where once weekly administration of vancomycin is possible. Although routine use of rifampin in native valve endocarditis is not recommended, it has been used, however, as supplemental therapy for patients who did not respond to conventional antimicrobial therapy or those with suppurative complications, such as valve ring abscess, intracerebral abscess formation, and meningitis. However, the adjunctive role of rifampin remains controversial.

Staphylococcal endocarditis in the presence of intracardiac prosthetic material is most often caused by coagulase-negative staphylococci that are usually resistant to methicillin. In these cases, vancomycin and rifampicin are administered for 6 weeks combined with gentamicin for the first 2 weeks of therapy. If the organism is susceptible to methicillin, nafcillin or oxacillin in combination with rifampin and gentamicin is advocated. Based on the experimental finding that rifampin results in the complete sterilization of foreign bodies infected with *S. aureus*, a similar combination therapy of nafcillin or oxacillin in combination with rifampin and gentamicin is also considered the regimen of choice for prosthetic valve endocarditis due to methicillin-susceptible *S. aureus*. If strains are resistant to methicillin or the patient is allergic to penicillin, vancomycin is substituted for the penicillinase-resistant penicillin. If the organism is resistant to gentamicin, another third agent should be chosen based on in vitro susceptibility testing since staphylococci may become resistant to rifampin during combination therapy of prosthetic valve endocarditis.

Treatment options for endocarditis caused by fastidious gram-negative bacilli of the HACEK group include the administration of ampicillin (12 g daily) combined with gentamicin (3 g/kg per day) administered IV for 4 weeks, or preferably ceftriaxone to cover β-lactamase producing strains, 2 g daily for 4 weeks.

Therapy should be extended to 6 weeks in the case of prosthetic valve endocarditis.

Treatment regimens for endocarditis caused by less common pathogens such as gram-negative enteric bacilli, *Brucella* and *Bartonella* spp., *C. burnetii*, *Legionella* spp., obligate intracellular pathogens, and fungi is still not adequately defined, and the readers are referred to the most recent review articles for details (Bayer et al. 1998; Brouqui and Raoult 2001; Mylonakis and Calderwood 2001).

The choice of antibiotics for the treatment of infective endocarditis is governed by the susceptibility of the causative organism. Determination of the minimum inhibitory concentration is recommended to define optimal treatment, while standard disk sensitivity testing is unreliable. The determination of the minimum bactericidal concentration (MBC), as well as performance of the serum bactericidal titer (SBT) test is not routinely advocated, the latter correlating poorly with clinical outcome. Since adverse events are rather frequent during prolonged antimicrobial therapy, it is often necessary to revise therapy. For additional testing, the causative organism should therefore be retained in the laboratory until cure has been achieved.

Several studies suggest that combined medical and surgical therapy for infective endocarditis can improve outcome. A major reduction in mortality followed the introduction of valve replacement surgery some 40 years ago, and even a further decline in mortality observed in some of the most recent surveys has been associated with earlier valve replacement and more refined surgical techniques (Cabell et al. 2002; Hoen et al. 2002). Currently, surgical intervention during treatment is required in 25–30 percent of patients with infective endocarditis. The generally accepted indications for surgical intervention during active infective endocarditis include (1) refractory congestive heart failure; (2) more than one serious embolic event; (3) uncontrolled infection or sepsis despite maximal antibiotic therapy; (4) significant valve dysfunction, perforation or rupture as demonstrated by echocardiography; (5) ineffective antimicrobial therapy (e.g. as in Q fever endocarditis, fungal endocarditis, and enterococcal endocarditis caused by high-level gentamicin- or vancomycin-resistant enterococci); (6) local suppurative complications including periannular extension of the infection, perivalvular or myocardial abscess; and (7) most cases of prosthetic valve endocarditis caused by antibiotic-resistant pathogens such as MRSA, gram-negative enteric bacilli, and *P. aeruginosa*. The optimal time for surgical intervention is before severe hemodynamic compromise or extension of the infection to adjacent tissue has occurred. Medical and surgical management decisions can be guided by echocardiographic detection of abscesses, prosthetic valve dehiscence, valvular dysfunction, and obstructive vegetations. If there is evidence of active endocarditis at the time of valve replacement surgery based on a culture-positive vegetation or histo-logic evidence of significant polymorphonuclear inflammation, antibiotic therapy should be continued for at least several weeks. Excellent reviews on the indications for surgery during therapy for infective endocarditis are available (Moon et al. 1997; Bayer et al. 1998; Alexiou et al. 2000; Delay et al. 2000; Guerra et al. 2001; Olaison and Pettersson 2003).

PREVENTION OF ENDOCARDITIS

Since infective endocarditis is a disease with significant morbidity and mortality, its prevention is a laudable objective, but how best to achieve this, or indeed whether it can be achieved, is another matter. A variety of preventive measures could theoretically be used to interrupt the sequence of events that finally results in endocarditis. These include elimination or treatment of predisposing conditions, elimination of portals of entry for organisms, e.g. by maintaining good oral hygiene, immunization against bacteria that cause endocarditis, and administration of antibiotics during high risk procedures. Of these, administration of antibiotics has received the most attention. There are numerous national guidelines for prophylaxis and these vary both in their antibiotic regimens and in the range of procedures for which prophylaxis is required. Over the past decade, however, there has been some uniformity of recommendations both within Europe and elsewhere. The diversity of the recommendations serves to emphasize the lack of critical data on which to base them. A detailed discussion of antimicrobial prophylactic strategies is beyond the scope of this chapter, and readers are referred to the above-mentioned guidelines (Simmons 1993; Leport et al. 1995; Dajani et al. 1997).

The guidelines for the prevention of late PVE are essentially the same as those outlined for endocarditis on native valves though some differentiate between the regimens for the two groups (Hyde et al. 1998; Segreti 1999). The prevention of early PVE concerns prophylactic antibiotics at the time of valve replacement. This has been a controversial issue for many years and has been plagued by many small poorly conducted trials. The available evidence suggests that antibiotic prophylaxis reduces the postoperative wound infection in cardiothoracic surgery in general (Kreter and Woods 1992) and, by inference, in valve replacement surgery in particular. It is assumed, though never proved, that a reduction in wound infection will equate with a reduction in PVE and this seems a reasonable assumption. Various antibiotics are used and no single regimen has been shown to be superior. The increasing prevalence of methicillin-resistant staphylococci in hospitals, both *S. aureus* (MRSA) and coagulase-negative staphylococci, has led many to use vancomycin rather than cefazolin and cefuroxime which are the most cost-effective agents

for perioperative prophylaxis. However, this issue remains controversial.

REFERENCES

Ackers, M.L., Puhr, N.D., et al. 2000. Laboratory-based surveillance of *Salmonella* serotype Typhi infections in the United States: antimicrobial resistance on the rise. *JAMA*, **283**, 2668–73.

Adams, W.G., Deaver, K.A., et al. 1993. Decline of childhood *Haemophilus influenzae* type b (Hib) disease in the Hib vaccine era. *JAMA*, **269**, 221–6.

Agnihotri, A.K., McGiffin, D.C., et al. 1995. The prevalence of infective endocarditis after aortic valve replacement. *J Thorac Cardiovasc Surg*, **110**, 1708–20.

Aires, D.S. and De Lencastre, H. 2003. Evolution of sporadic isolates of methicillin-resistant *Staphylococcus aureus* (MRSA) in hospitals and their similarities to isolates of community-acquired MRSA. *J Clin Microbiol*, **41**, 3806–15.

Akowuah, E.F., Davies, W., et al. 2003. Prosthetic valve endocarditis: early and late outcome following medical or surgical treatment. *Heart*, **89**, 269–72.

Ala'Aldeen, D.A., Neal, K.R., et al. 2000. Dynamics of meningococcal long-term carriage among university students and their implications for mass vaccination. *J Clin Microbiol*, **38**, 2311–16.

Alexiou, C., Langley, S.M., et al. 2000. Surgery for active culture-positive endocarditis: determinants of early and late outcome. *Ann Thorac Surg*, **69**, 1448–54.

Alvarez-Elcoro, S., Soto-Ramirez, L. and Mateos-Mora, M. 1984. Salmonella bacteremia in patients with prosthetic heart valves. *Am J Med*, **77**, 61–3.

Anderson, B.E. and Neuman, M.A. 1997. *Bartonella* spp. as emerging human pathogens. *Clin Microbiol Rev*, **10**, 203–19.

Apicella, M.A. 2000. *Neisseria meningitidis*. In: Mandell, G.L., Bennet, J.E. and Dolin, R. (eds), *Mandell, Dougals, and Bennett's Principles and practice of infectious diseases*. Philadelphia: Churchill Livingstone, 2228–41.

Appleman, M.D., Swinney, R.S. and Heseltine, P.N. 1982. Evaluation of the antibiotic removal device. *J Clin Microbiol*, **15**, 278–81.

Arber, N., Pras, E., et al. 1994. Pacemaker endocarditis. Report of cases and review of the literature. *Medicine (Balt)*, **44**, 299–305.

Archibald, L.K., McDonald, L.C., et al. 2000. Comparison of BACTEC MYCO/F LYTIC and WAMPOLE ISOLATOR 10 (lysis-centrifugation) systems for detection of bacteremia, mycobacteremia, and fungemia in a developing country. *J Clin Microbiol*, **38**, 2994–7.

Arnow, P.M., Smaron, M. and Ormiste, V. 1984. Brucellosis in a group of travelers to Spain. *JAMA*, **251**, 505–7.

Aronin, S.I., Mukherjee, S.K., et al. 1998. Review of pneumococcal endocarditis in adults in the penicillin era. *Clin Infect Dis*, **26**, 165–71.

Aronson, M.D. and Bor, D.H. 1987. Blood cultures. *Ann Intern Med*, **106**, 246–53.

Ashby, M.E. 1976. *Serratia* osteomyelitis in heroin users. A report of two cases. *J Bone Joint Surg Am*, **58**, 132–4.

Awada, A., van der Auwera, A.P., et al. 1992. Streptococcal and enterococcal bacteremia in patients with cancer. *Clin Infect Dis*, **15**, 33–48.

Bach, D.S. 2000. Transesophageal echocardiographic (TEE) evaluation of prosthetic valves. *Cardiol Clin*, **18**, 751–71.

Baddour, L.M. 1998. Infective endocarditis caused by beta-hemolytic streptococci. The Infectious Diseases Society of America's Emerging Infections Network. *Clin Infect Dis*, **26**, 66–71.

Baddour, L.M. and Bisno, A.L. 1986. Infective endocarditis complicating mitral valve prolapse: epidemiologic, clinical, and microbiologic aspects. *Rev Infect Dis*, **8**, 117–37.

Baddour, L.M., Phillips, T.N. and Bisno, A.L. 1986. Coagulase-negative staphylococcal endocarditis. Occurrence in patients with mitral valve prolapse. *Arch Intern Med*, **146**, 119–21.

Baggett, H.C., Hennessy, T.W., et al. 2003. An outbreak of community-onset methicillin-resistant *Staphylococcus aureus* skin infections in southwestern Alaska. *Infect Control Hosp Epidemiol*, **24**, 397–402.

Baker, C.J. 1976. Group B streptococcal infections: is prevention possible. *South Med J*, **69**, 1527–9.

Ballet, M., Gevigney, G., et al. 1995. Infective endocarditis due to *Streptococcus bovis*. A report of 53 cases. *Eur Heart J*, **16**, 1975–80.

Bancescu, G., Lofthus, B., et al. 1997. Isolation and characterization of 'Streptococcus milleri' group strains from oral and maxillofacial infections. *Adv Exp Med Biol*, **418**, 165–7.

Bannatyne, R.M., Jackson, M.C. and Memish, Z. 1997. Rapid diagnosis of *Brucella* bacteremia by using the BACTEC 9240 system. *J Clin Microbiol*, **35**, 2673–4.

Barber, M. 1961. Methicillin-resistant staphylococci. *J Clin Pathol*, **14**, 385–93.

Barg, N.L., Kish, M.A., et al. 1985. Group A streptococcal bacteremia in intravenous drug abusers. *Am J Med*, **78**, 569–74.

Bates, D.W., Goldman, L. and Lee, T.H. 1991. Contaminant blood cultures and resource utilization. The true consequences of false-positive results. *JAMA*, **265**, 365–9.

Bayer, A.S. 1982. Staphylococcal bacteremia and endocarditis. State of the art. *Arch Intern Med*, **142**, 1169–77.

Bayer, A.S. 1996. Diagnostic criteria for identifying cases of endocarditis-revisiting the Duke criteria two years later. *Clin Infect Dis*, **23**, 303–4.

Bayer, A.S. and Scheld, W.M. 2000. Endocarditis and intravascular infections. In: Mandell, G.L., Bennett, J.E. and Dolin, R. (eds), *Mandell, Dougals, and Bennett's Principles and practice of infectious diseases*, 5th edn. New York: Churchill Livingstone, 857–902.

Bayer, A.S. and Theofilopoulos, A.N. 1990. Immunopathogenetic aspects of infective endocarditis. *Chest*, **97**, 204–12.

Bayer, A.S., Theofilopoulos, A.N., et al. 1976. Circulating immune complexes in infective endocarditis. *N Engl J Med*, **295**, 1500–5.

Bayer, A.S., Bolger, A.F., et al. 1998. Diagnosis and management of infective endocarditis and its complications. *Circulation*, **98**, 2936–48.

Bayliss, R., Clarke, C., et al. 1983. The teeth and infective endocarditis. *Br Heart J*, **50**, 506–12.

Beck-Sague, C.M., Jarvis, W.R., et al. 1990. Epidemic bacteremia due to *Acinetobacter baumannii* in five intensive care units. *Am J Epidemiol*, **132**, 723–33.

Beeching, N.J., Christmas, T.I., et al. 1985. *Streptococcus bovis* bacteraemia requires rigorous exclusion of colonic neoplasia and endocarditis. *Q J Med*, **56**, 439–50.

Berbari, E.F., Cockerill 3rd, F.R. and Steckelberg, J.M. 1997. Infective endocarditis due to unusual or fastidious microorganisms. *Mayo Clin Proc*, **72**, 532–42.

Bergogne-Berezin, E. and Towner, K.J. 1996. *Acinetobacter* spp. as nosocomial pathogens: microbiological, clinical, and epidemiological features. *Clin Microbiol Rev*, **9**, 148–65.

Berlin, J.A., Abrutyn, E., et al. 1995. Incidence of infective endocarditis in the Delaware Valley, 1988–1990. *Am J Cardiol*, **76**, 933–6.

Bert, F., Bariou-Lancelin, M. and Lambert-Zechovsky, N. 1998. Clinical significance of bacteremia involving the 'Streptococcus milleri' group: 51 cases and review. *Clin Infect Dis*, **27**, 385–7.

Bhanji, S., Williams, B., et al. 2002. Transient bacteremia induced by toothbrushing. A comparison of the Sonicare toothbrush with a conventional toothbrush. *Pediatr Dent*, **24**, 295–9.

Bhavnani, S.M., Drake, J.A., et al. 2000. A nationwide, multicenter, case-control study comparing risk factors, treatment, and outcome for vancomycin-resistant and -susceptible enterococcal bacteremia. *Diagn Microbiol Infect Dis*, **36**, 145–58.

Bilgrami, S., Feingold, J.M., et al. 1998. Streptococcus viridans bacteremia following autologous peripheral blood stem cell transplantation. *Bone Marrow Transplant*, **21**, 591–5.

Bille, J., Rocourt, J. and Swaminathan, B. 2003. *Listeria* and *Erysipelothrix*. In: Murray, P.R., Baron, E.J., et al. (eds), *Manual of clinical microbiology*. Washington DC: ASM Press, 461–71.

Blanchard, D.G., Ross, R.S. and Dittrich, H.C. 1992. Nonbacterial thrombotic endocarditis. Assessment by transesophageal echocardiography. *Chest*, **102**, 954–6.

Blot, S., Vandewoude, K. and Colardyn, F. 1998. *Staphylococcus aureus* infections. *N Engl J Med*, **339**, 2025–6.

Blot, F., Nitenberg, G., et al. 1999. Diagnosis of catheter-related bacteraemia: a prospective comparison of the time to positivity of hub-blood versus peripheral-blood cultures. *Lancet*, **354**, 1071–7.

Blot, S.I., Vandewoude, K.H., et al. 2002a. Effects of nosocomial candidemia on outcomes of critically ill patients. *Am J Med*, **113**, 480–5.

Blot, S.I., Vandewoude, K.H., et al. 2002b. Outcome and attributable mortality in critically ill patients with bacteremia involving methicillin-susceptible and methicillin-resistant *Staphylococcus aureus*. *Arch Intern Med*, **162**, 2229–35.

Blot, S., Vandewoude, K. and Colardyn, F. 2003a. Nosocomial bacteremia involving *Acinetobacter baumannii* in critically ill patients: a matched cohort study. *Intensive Care Med*, **29**, 471–5.

Blot, S.I., Vandewoude, K.H. and Colardyn, F.A. 2003b. Evaluation of outcome in critically ill patients with nosocomial *Enterobacter* bacteremia: results of a matched cohort study. *Chest*, **123**, 1208–13.

Bochud, P.Y., Calandra, T. and Francioli, P. 1994a. Bacteremia due to viridans streptococci in neutropenic patients: a review. *Am J Med*, **97**, 256–64.

Bochud, P.Y., Eggiman, P., et al. 1994b. Bacteremia due to viridans streptococcus in neutropenic patients with cancer: clinical spectrum and risk factors. *Clin Infect Dis*, **18**, 25–31.

Bochud, P.Y., Cometta, A. and Francioli, P. 1997. Virulent infections caused by alpha-haemolytic streptococci in cancer patients and their management. *Curr Opin Infect Dis*, **10**, 422–30.

Bodey, G.P., Rodriguez, S., et al. 1991. Clostridial bacteremia in cancer patients. A 12-year experience. *Cancer*, **67**, 1928–42.

Bosshard, P.P., Kronenberg, A., et al. 2003. Etiologic diagnosis of infective endocarditis by broad-range polymerase chain reaction: a 3-year experience. *Clin Infect Dis*, **37**, 167–72.

Bourbeau, P.P. and Pohlman, J.K. 2001. Three days of incubation may be sufficient for routine blood cultures with BacT/Alert FAN blood culture bottles. *J Clin Microbiol*, **39**, 2079–82.

Bouvet, A., Durand, A., Group d'Enquete sur l'endocardite en France 1990–1991, et al. 1994. In vitro susceptibility to antibiotics of 200 strains of streptococci and enterococci isolated during infective endocarditis. In: Totalian, A. (ed.), *Pathogenic streptococci: present and future*. St. Petersburg, Russia: Lancer Publications, 72–3.

Breiman, R.F., Spika, J.S., et al. 1990. Pneumococcal bacteremia in Charleston County, South Carolina. A decade later. *Arch Intern Med*, **150**, 1401–5.

Brook, I. 1989. Anaerobic bacterial bacteremia: 12-year experience in two military hospitals. *J Infect Dis*, **160**, 1071–5.

Brook, I. 2002. Endocarditis due to anaerobic bacteria. *Cardiology*, **98**, 1–5.

Brook, M.G., Lucas, R.E. and Pain, A.K. 1988. Clinical features and management of two cases of *Streptococcus milleri* chest infection. *Scand J Infect Dis*, **20**, 345–6.

Brouqui, P. and Raoult, D. 2001. Endocarditis due to rare and fastidious bacteria. *Clin Microbiol Rev*, **14**, 177–207.

Brown, P.D. and Levine, D.P. 2002. Infective endocarditis in the injection drug user. *Infect Dis Clin North Am*, **16**, 645–65.

Bryan, C.S., Reynolds, K.L. and Brenner, E.R. 1983. Analysis of 1,186 episodes of gram-negative bacteremia in non-university hospitals: the effects of antimicrobial therapy. *Rev Infect Dis*, **5**, 629–38.

Bryant, J.K. and Strand, C.L. 1987. Reliability of blood cultures collected from intravascular catheter versus venipuncture. *Am J Clin Pathol*, **88**, 113–16.

Bucher, A., Martin, P.R., et al. 1992. Spectrum of disease in bacteraemic patients during a *Streptococcus pyogenes* serotype M-1 epidemic in Norway in 1988. *Eur J Clin Microbiol Infect Dis*, **11**, 416–26.

Burgert, S.J., LaRocco, M.T. and Wilansky, S. 1999. Destructive native valve endocarditis caused by *Staphylococcus lugdunensis*. *South Med J*, **92**, 812–14.

Burkert, T. and Watanakunakorn, C. 1991. Group A streptococcus endocarditis: report of five cases and review of literature. *J Infect*, **23**, 307–16.

Burman, L.A., Norrby, R. and Trollfors, B. 1985. Invasive pneumococcal infections: incidence, predisposing factors, and prognosis. *Rev Infect Dis*, **7**, 133–42.

Burnette-Curley, D., Wells, V., et al. 1995. FimA, a major virulence determinant associated with *Streptococus parasanguis* endocarditis. *Infect Immun*, **63**, 4669–74.

Buschelman, B.J., Bale, M.J. and Jones, R.N. 1993. Species identification and determination of high-level aminoglycoside resistance among enterococci. Comparison study of sterile body fluid isolates, 1985–1991. *Diagn Microbiol Infect Dis*, **16**, 119–22.

Caballero-Granado, F.J., Becerril, B., et al. 2001. Case-control study of risk factors for the development of enterococcal bacteremia. *Eur J Clin Microbiol Infect Dis*, **20**, 83–90.

Cabell, C.H., Jollis, J.G., et al. 2002. Changing patient characteristics and the effect on mortality in endocarditis. *Arch Intern Med*, **162**, 90–4.

Calderwood, S.B., Swinski, L.A., et al. 1985. Risk factors for the development of prosthetic valve endocarditis. *Circulation*, **72**, 31–7.

Canet, J.J., Juan, N., et al. 2002. Hospital-acquired pneumococcal bacteremia. *Clin Infect Dis*, **35**, 697–702.

Caputo, G.M., Archer, G.L., et al. 1987. Native valve endocarditis due to coagulase-negative staphylococci. Clinical and microbiologic features. *Am J Med*, **83**, 619–25.

Carratala, J., Alcaide, F., et al. 1995. Bacteremia due to viridans streptococci that are highly resistant to penicillin: increase among neutropenic patients with cancer. *Clin Infect Dis*, **20**, 1169–73.

Carruthers, M. 1977. Endocarditis due to enteric bacilli other than salmonellae: Case reports and literature review. *Am J Med Sci*, **273**, 203–11.

Cartotto, R.C., Macdonald, D.B. and Wasan, S.M. 1998. Acute bacterial endocarditis following burns: case report and review. *Burns*, **24**, 369–73.

Casariego, E., Rodriguez, A., et al. 1996. Prospective study of *Streptococcus milleri* bacteremia. *Eur J Clin Microbiol Infect Dis*, **15**, 194–200.

Cates, J.E. and Christie, R.V. 1951. Subacute bacterial endocarditis: a review of 442 patients treated in 14 centres appointed by the Penicillin Trials Committee of the Medical Research Council. *Q J Med (NS)*, **20**, 93–130.

CDC. 1999. From the Centers for Disease Control and Prevention. Impact of vaccines universally recommended for children – United States, 1900–1998. *JAMA*, **281**, 1482–1483.

CDC. 2003. From the Centers for Disease Control and Prevention. Public health dispatch: outbreaks of community-associated methicillin-resistant *Staphylococcus aureus* skin infections – Los Angeles County, California, 2002–2003. *JAMA*, **289**, 1377.

Celum, C.L., Chaisson, R.E., et al. 1987. Incidence of salmonellosis in patients with AIDS. *J Infect Dis*, **156**, 998–1002.

Chamis, A.L., Gesty-Palmer, D., et al. 1999. Echocardiography for the diagnosis of *Staphylococcus aureus* infective endocarditis. *Curr Infect Dis Rep*, **1**, 129–35.

Chandrasekar, P.H. and Brown, W.J. 1994. Clinical issues of blood cultures. *Arch Intern Med*, **154**, 841–9.

Chang, H.H., Lu, C.Y., et al. 2002. Endocarditis caused by *Abiotrophia defectiva* in children. *Pediatr Infect Dis J*, **21**, 697–700.

Chen, T.T., Schapiro, J.M. and Loutit, J. 1996. Prosthetic valve endocarditis due to *Legionella pneumophila*. *J Cardiovasc Surg (Torino)*, **37**, 631–3.

Cheung, A.L. and Projan, S.J. 1994. Cloning and sequencing of sarA of *Staphylococcus aureus*, a gene required for the expression of agr. *J Bacteriol*, **176**, 4168–72.

Chin, K.C. and Fitzhardinge, P.M. 1985. Sequelae of early-onset group B hemolytic streptococcal neonatal meningitis. *J Pediatr*, **106**, 819–22.

Choi, S.H., Kim, Y.S., et al. 2002. Serratia bacteremia in a large university hospital: trends in antibiotic resistance during 10 years and implications for antibiotic use. *Infect Control Hosp Epidemiol*, **23**, 740–7.

Chow, A.W., Montgomerie, J.Z. and Guze, L.B. 1974. Parenteral clindamycin therapy for severe anaerobic infections. *Arch Intern Med*, **134**, 78–82.

Christensen, J.J., Gruhn, N. and Facklam, R.R. 1999. Endocarditis caused by *Abiotrophia* species. *Scand J Infect Dis*, **31**, 210–12.

Cisneros, J.M., Reyes, M.J., et al. 1996. Bacteremia due to *Acinetobacter baumannii*: epidemiology, clinical findings, and prognostic features. *Clin Infect Dis*, **22**, 1026–32.

Cisneros, J.M. and Rodriguez-Bano, J. 2002. Nosocomial bacteremia due to *Acinetobacter baumannii*: epidemiology, clinical features and treatment. *Clin Microbiol Infect*, **8**, 687–93.

Clifford, C.P., Eykyn, S.J. and Oakley, C.M. 1994. Staphylococcal tricuspid valve endocarditis in patients with structurally normal hearts and no evidence of narcotic abuse. *QJM*, **87**, 755–7.

Cockerill III, F.R., Hughes, J.G., et al. 1997. Analysis of 281,797 consecutive blood cultures performed over an eight-year period: trends in microorganisms isolated and the value of anaerobic culture of blood. *Clin Infect Dis*, **24**, 403–18.

Cohen, P.S., Maquire, J.H. and Weinstein, L. 1980. Infective endocarditis caused by gram-negative bacteria: A review of the literature, 1945–1977. *Prog Cardiovasc Dis*, **22**, 205–42.

Cohen, J., Donnelly, J.P., et al. 1983. Septicaemia caused by viridans streptococci in neutropenic patients with leukaemia. *Lancet*, **2**, 1452–4.

Cohen, J.I., Bartlett, J.A. and Corey, G.R. 1987. Extra-intestinal manifestations of salmonella infections. *Medicine (Balt)*, **66**, 349–88.

Cohen, J.I., Sloss, L.J., et al. 1989. Prosthetic valve endocarditis caused by *Mycoplasma hominis*. *Am J Med*, **86**, 819–21.

Colford, J.M. Jr., Mohle-Boetani, J., et al. 1995. Group B streptococcal bacteremia in adults. Five years' experience and a review of the literature. *Medicine (Balt)*, **74**, 176–90.

Collin, B.A., Leather, H.L., et al. 2001. Evolution, incidence, and susceptibility of bacterial bloodstream isolates from 519 bone marrow transplant patients. *Clin Infect Dis*, **33**, 947–53.

Conlon, P.J., Jefferies, F., et al. 1998. Predictors of prognosis and risk of acute renal failure in bacterial endocarditis. *Clin Nephrol*, **49**, 96–101.

Conterno, L.O., Wey, S.B. and Castelo, A. 1998. Risk factors for mortality in *Staphylococcus aureus* bacteremia. *Infect Control Hosp Epidemiol*, **19**, 32–7.

Cope, A.P., Heber, M. and Wilkins, E.G. 1990. Valvular tuberculous endocarditis: a case report and review of the literature. *J Infect*, **21**, 293–6.

Cosgrove, S.E., Sakoulas, G., et al. 2003. Comparison of mortality associated with methicillin-resistant and methicillin-susceptible *Staphylococcus aureus* bacteremia: a meta-analysis. *Clin Infect Dis*, **36**, 53–9.

Coulter, W.A., Coffey, A., et al. 1990. Bacteremia in children following dental extraction. *J Dent Res*, **69**, 1691–5.

Craven, D.E., Rixinger, A.I., et al. 1986. Bacteremia caused by group G streptococci in parenteral drug abusers: epidemiological and clinical aspects. *J Infect Dis*, **153**, 988–92.

Dajani, A.S., Taubert, K.A., et al. 1997. Prevention of bacterial endocarditis. Recommendations by the American Heart Association. *JAMA*, **277**, 1794–801.

Dan, M., Maximova, S., et al. 1990. Varied presentations of sporadic group A streptococcal bacteremia: clinical experience and attempt at classification. *Rev Infect Dis*, **12**, 537–42.

Darby, J.M., Linden, P., et al. 1997. Utilization and diagnostic yield of blood cultures in a surgical intensive care unit. *Crit Care Med*, **25**, 989–94.

Dargere, S., Vergnaud, M., et al. 2002. *Enterococcus gallinarum* endocarditis occurring on native heart valves. *J Clin Microbiol*, **40**, 2308–10.

Das, M., Badley, A.D., et al. 1997. Infective endocarditis caused by HACEK microorganisms. *Annu Rev Med*, **48**, 25–33.

Daschner, F.D., Frey, P., et al. 1982. Nosocomial infections in intensive care wards: a multicenter prospective study. *Intensive Care Med*, **8**, 5–9.

Davidson, M., Parkinson, A.J., et al. 1994. The epidemiology of invasive pneumococcal disease in Alaska, 1986–1990 – ethnic differences and opportunities for prevention. *J Infect Dis*, **170**, 368–76.

Delay, D., Pellerin, M., et al. 2000. Immediate and long-term results of valve replacement for native and prosthetic valve endocarditis. *Ann Thorac Surg*, **70**, 1219–23.

Demers, B., Simor, A.E., et al. 1993. Severe invasive group A streptococcal infections in Ontario, Canada: 1987–1991. *Clin Infect Dis*, **16**, 792–800.

Devitt, D., Francois, P., et al. 1994. Molecular characterization of the clumping factor (fibrinogen receptor) of *Staphylococcus aureus*. *Mol Microbiol*, **11**, 237–48.

Dhawan, V.K. 2003. Infective endocarditis in elderly patients. *Curr Infect Dis Rep*, **5**, 285–92.

Diekema, D.J., Pfaller, M.A., et al. 1999. Survey of bloodstream infections due to gram-negative bacilli: frequency of occurrence and antimicrobial susceptibility of isolates collected in the United States, Canada and Latin America for the SENTRY Antimicrobial Surveillance Program, 1997. *Clin Infect Dis*, **29**, 595–607.

Diekema, D.J., Pfaller, M.A., et al. 2001. Survey of infections due to *Staphylococcus* species: frequency of occurrence and antimicrobial susceptibility of isolates collected in the United States, Canada, Latin America, Europe, and the Western Pacific region for the SENTRY Antimicrobial Surveillance Program, 1997–1999. *Clin Infect Dis*, **32**, Suppl. 2 , S114–32.

Diekema, D.J., Pfaller, M.A. and Jones, R.N. 2002. Age-related trends in pathogen frequency and antimicrobial susceptibility of bloodstream isolates in North America: SENTRY Antimicrobial Surveillance Program, 1997–2000. *Int J Antimicrob Agents*, **20**, 412–18.

Diekema, D.J., Beekmann, S.E., et al. 2003. Epidemiology and outcome of nosocomial and community-onset bloodstream infection. *J Clin Microbiol*, **41**, 3655–60.

Dietzmann, D.E., Fischer, G.W. and Schoenknecht, F.D. 1974. Neonatal *Escherichia coli* septicemia: bacterial counts in the blood. *J Pediatr*, **85**, 128–30.

Digiovine, B., Chenoweth, C., et al. 1999. The attributable mortality and costs of primary nosocomial bloodstream infections in the intensive care unit. *Am J Respir Crit Care Med*, **160**, 976–81.

DiNubile, M.J., Albornoz, M.A., et al. 1991. Pneumococcal soft-tissue infections: possible association with connective tissue diseases. *J Infect Dis*, **163**, 897–900.

Dismukes, W.E., Karchmer, A.W., et al. 1973. Prosthetic valve endocarditis. Analysis of 38 cases. *Circulation*, **48**, 365–77.

Doern, G.V. 1994. Manual blood culture systems and the antimicrobial removal device. *Clin Lab Med*, **14**, 133–47.

Doern, G.V., Ferraro, M.J., et al. 1996. Emergence of high rates of antimicrobial resistance among viridans group streptococci in the United States. *Antimicrob Agents Chemother*, **40**, 891–4.

Doern, G.V., Brueggemann, A.B., et al. 1997. Four-day incubation period for blood culture bottles processed with the Difco ESP blood culture system. *J Clin Microbiol*, **35**, 1290–2.

Doern, G.V., Pfaller, M.A., et al. 1998. Prevalence of antimicrobial resistance among respiratory tract isolates of *Streptococcus pneumoniae* in North America: 1997 results from the SENTRY antimicrobial surveillance program. *Clin Infect Dis*, **27**, 764–70.

Dominguez, A., Cardenosa, N., et al. 2000. The case-fatality rate of meningococcal disease in Catalonia, 1990–1997. *Scand J Infect Dis*, **36**, 274–9.

Dorsher, C.W., Rosenblatt, J.E., et al. 1991. Anaerobic bacteremia: decreasing rate over a 15-year period. *Rev Infect Dis*, **13**, 633–6.

Douglas, C.W., Heath, J., et al. 1993. Identity of viridans streptococci isolated from cases of infective endocarditis. *J Med Microbiol*, **39**, 179–82.

Doulton, T., Sabharwal, N., et al. 2003. Infective endocarditis in dialysis patients: new challenges and old. *Kidney Int*, **64**, 720–7.

Dunne, M., Nolte, F. and Wilson, M.L. 1997. Cumitech 1B. In: Hindler, J.A. (ed.), *Blood cultures III*. Washington DC: ASM Press, 1–21.

Durack, D.T., Lukes, A.S., et al. 1994. New criteria for diagnosis of infective endocarditis. *Am J Med*, **96**, 200–9.

Duval, X., Papastamopoulos, V., et al. 2001. Definite *Streptococcus bovis* endocarditis: characteristics in 20 patients. *Clin Microbiol Infect*, **7**, 3–10.

Edberg, S.C. and Edberg, M.K. 1983. Inactivation of the polyanionic detergent sodium polyanetholsulfonate by hemoglobin. *J Clin Microbiol*, **18**, 1047–50.

Edgeworth, J.D., Treacher, D.F. and Eykyn, S.J. 1999. A 25-year study of nosocomial bacteremia in an adult intensive care unit. *Crit Care Med*, **27**, 1421–8.

Edmond, M.B., Wallace, S.E., et al. 1999. Nosocomial bloodstream infections in United States hospitals: a three-year analysis. *Clin Infect Dis*, **29**, 239–44.

Eisenstein, B.I. and Zaleznik, D.F. 2000. Enterobacteriaceae. In: Mandell, G.L., Bennet, J.E. and Dolin, R. (eds), *Mandell, Dougals, and Bennett's Principles and practice of infectious diseases*. Philadelphia: Churchill Livingstone, 2294–309.

El Amari, E.B., Chamot, E., et al. 2001. Influence of previous exposure to antibiotic therapy on the susceptibility pattern of *Pseudomonas aeruginosa* bacteremic isolates. *Clin Infect Dis*, **33**, 1859–64.

Ellis, C.J. 1991. The use and abuse of blood cultures. *Infect Dis Newsl*, **20**, 27–30.

Ellison III, R.T., Curd, J.G., et al. 1987. Underlying complement deficiency in patients with disseminated gonococcal infection. *Sex Transm Dis*, **14**, 201–4.

Elting, L.S., Bodey, G.P. and Keefe, B.H. 1992. Septicemia and shock syndrome due to viridans streptococci: a case-control study of predisposing factors. *Clin Infect Dis*, **14**, 1201–7.

Etienne, J. and Eykyn, S.J. 1990. Increase in native valve endocarditis caused by coagulase negative staphylococci: an Anglo-French clinical and microbiological study. *Br Heart J*, **64**, 381–4.

Eykyn, S.J. 1988. Staphylococcal sepsis. The changing pattern of disease and therapy. *Lancet*, **1**, 100–4.

Eykyn, S.J. 1998. Bacteraemia, septicaemia and endocarditis. In: Collier, L., Balows, A. and Sussman, M. (eds), *Topley and Wilson's Microbiology and microbial infections*. London, Sydney, Auckland: Arnold.

Facklam, R. 2002. What happened to the streptococci: overview of taxonomic and nomenclature changes. *Clin Microbiol Rev*, **15**, 613–30.

Facklam, R.R. and Collins, M.D. 1989. Identification of *Enterococcus* species isolated from human infections by a conventional test scheme. *J Clin Microbiol*, **27**, 731–4.

Fang, G.D., Fine, M., et al. 1990. New and emerging etiologies for community-acquired pneumonia with implications for therapy. A prospective multicenter study of 359 cases. *Medicine (Balt)*, **69**, 307–16.

Faraawi, R. and Fong, I.W. 1988. *Escherichia coli* emphysematous endophthalmitis and pyelonephritis. Case report and review of the literature. *Am J Med*, **84**, 636–9.

Farley, M.M. 2001. Group B streptococcal disease in nonpregnant adults. *Clin Infect Dis*, **33**, 55661.

Farley, M.M., Harvey, R.C., et al. 1993. A population-based assessment of invasive disease due to group B Streptococcus in nonpregnant adults. *N Engl J Med*, **328**, 1807–11.

Farrag, N., Lee, P., et al. 2001. *Staphylococcus lugdunensis* endocarditis. *Postgrad Med J*, **77**, 259–60.

Farrugia, D.C., Eykyn, S.J. and Smyth, E.G. 1994. *Campylobacter fetus* endocarditis: two case reports and review. *Clin Infect Dis*, **18**, 443–6.

Farstad, H., Gaustad, P., et al. 2003. Cerebral venous thrombosis and *Escherichia coli* infection in neonates. *Acta Paediatr*, **92**, 254–7.

Feder, H.M. Jr., Roberts, J.C., et al. 2003. HACEK endocarditis in infants and children: two cases and a literature review. *Pediatr Infect Dis J*, **22**, 557–62.

Fedson, D.S., Musher, D.M. and Eskola, J. 1998. Pneumococcal vaccine. In: Plotkin, S.A. and Orenstein, W.A. (eds), *Vaccines*. Philadelphia: WB Saunders.

Fefer, P., Raveh, D., et al. 2002. Changing epidemiology of infective endocarditis: a retrospective survey of 108 cases, 1990–1999. *Eur J Clin Microbiol Infect Dis*, **21**, 432–7.

Felner, J.M. and Dowell, V.R. Jr. 1970. Anaerobic bacterial endocarditis. *N Engl J Med*, **283**, 1188–92.

Felner, J.M. and Dowell, V.R. Jr. 1971. 'Bacteroides' bacteremia. *Am J Med*, **50**, 787–96.

Fenollar, F., Lepidi, H. and Raoult, D. 2001. Whipple's endocarditis: review of the literature and comparisons with Q fever, Bartonella infection, and blood culture-positive endocarditis. *Clin Infect Dis*, **33**, 1309–16.

Fenollar, F., Birg, M.L., et al. 2003. Culture of *Tropheryma whipplei* from human samples: a 3-year experience (1999 to 2002). *J Clin Microbiol*, **41**, 3816–22.

Fergie, J.E. and Purcell, K. 2001. Community-acquired methicillin-resistant *Staphylococcus aureus* infections in south Texas children. *Pediatr Infect Dis J*, **20**, 860–3.

Fernandez Guerrero, M.L., Torres Perea, R., et al. 1996. Infectious endocarditis due to non-typhi Salmonella in patients infected with human immunodeficiency virus: report of two cases and review. *Clin Infect Dis*, **22**, 853–5.

Fernandez-Guerrero, M.L., Herrero, L., et al. 2002. Nosocomial enterococcal endocarditis: a serious hazard for hospitalized patients with enterococcal bacteraemia. *J Intern Med*, **252**, 510–15.

Finegold, S.M., George, W.L. and Mulligan, M.E. 1985. Anaerobic infections. Part I. *Dis Mon*, **31**, 1–77.

Finland, M. and Barnes, M.W. 1970. Changing etiology of bacterial endocarditis in the antibacterial era. Experiences at Boston City Hospital 1933–1965. *Ann Intern Med*, **72**, 341–8.

Flock, J.I., Hienz, S.A., et al. 1996. Reconsideration of the role of fibronectin binding in endocarditis caused by *Staphylococcus aureus*. *Infect Immun*, **64**, 1876–8.

Flores, G., Stavola, J.J. and Noel, G.J. 1993. Bacteremia due to *Pseudomonas aeruginosa* in children with AIDS. *Clin Infect Dis*, **16**, 706–8.

Fok, T.F., Lee, C.H., et al. 1998. Risk factors for *Enterobacter* septicemia in a neonatal unit: case-control study. *Clin Infect Dis*, **27**, 1204–9.

Fournier, P.E., Casalta, J.P., et al. 1996. Modification of the diagnostic criteria proposed by the Duke Endocarditis Service to permit improved diagnosis of Q fever endocarditis. *Am J Med*, **100**, 629–33.

Fournier, P.E. and Raoult, D. 1999. Nonculture laboratory methods for the diagnosis of infectious endocarditis. *Curr Infect Dis Rep*, **1**, 136–41.

Fowler, V.G., Li, J., et al. 1997. Role of echocardiography in evaluation of patients with *Staphylococcus aureus* bacteremia: experience in 103 patients. *J Am Coll Cardiol*, **30**, 1072–8.

Fowler, V.G., Sanders, L.L., et al. 1998. Outcome of *Staphylococcus aureus* bacteremia according to compliance with recommendations of infectious diseases specialists: experience with 244 patients. *Clin Infect Dis*, **27**, 478–86.

Fowler, V.G., Sanders, L.L., et al. 1999. Infective endocarditis due to *Staphylococcus aureus*: 59 prospectively identified cases with follow-up. *Clin Infect Dis*, **28**, 106–14.

Fowler, V.G. Jr., Kaye, K.S., et al. 2003. *Staphylococcus aureus* bacteremia after median sternotomy: clinical utility of blood culture results in the identification of postoperative mediastinitis. *Circulation*, **108**, 73–8.

Franciosi, R.A., Knostman, J.D. and Zimmerman, R.A. 1973. Group B streptococcal neonatal and infant infections. *J Pediatr*, **82**, 707–18.

Francis, J. and Warren, R.E. 1988. *Streptococcus pyogenes* bacteraemia in Cambridge – a review of 67 episodes. *Q J Med*, **68**, 603–13.

Frank, A.L., Marcinak, J.F., et al. 1999. Community-acquired and clindamycin-susceptible methicillin-resistant *Staphylococcus aureus* in children. *Pediatr Infect Dis J*, **18**, 993–1000.

Fridkin, S.K. 2001. Vancomycin-intermediate and -resistant *Staphylococcus aureus*: what the infectious disease specialist needs to know. *Clin Infect Dis*, **32**, 108–15.

Fridkin, S.K., Hageman, J., et al. 2003. Epidemiological and microbiological characterization of infections caused by

Staphylococcus aureus with reduced susceptibility to vancomycin, United States, 1997–2001. *Clin Infect Dis*, **36**, 429–39.

Frontera, J.A. and Gradon, J.D. 2000. Right-side endocarditis in injection drug users: review of proposed mechanisms of pathogenesis. *Clin Infect Dis*, **30**, 374–9.

Gagliardi, J.P., Nettles, R.E., et al. 1998. Native valve infective endocarditis in elderly and younger adult patients: comparison of clinical features and outcomes with use of the Duke criteria and the Duke Endocarditis Database. *Clin Infect Dis*, **26**, 1165–8.

Gallagher, P.G. and Watanakunakorn, C. 1985. Group B streptococcal bacteremia in a community teaching hospital. *Am J Med*, **78**, 795–800.

Gallagher, P.G. and Watanakunakorn, C. 1986. Group B streptococcal endocarditis: Report of seven cases and review of the literature, 1962–1985. *Rev Infect Dis*, **8**, 175–88.

Galpin, J.E., Chow, A.W., et al. 1976. Sepsis associated with decubitus ulcers. *Am J Med*, **61**, 346–50.

Gatti, B.M., Ramirez Gronda, G.A., et al. 2004. Isolation of *Haemophilus influenzae* serotypes from deep sites in sick children. *Rev Argent Microbiol*, **36**, 20–3.

Gaur, A.H., Flynn, P.M., et al. 2003. Difference in time to detection: a simple method to differentiate catheter-related from non-catheter-related bloodstream infection in immunocompromised pediatric patients. *Clin Infect Dis*, **37**, 469–75.

Geha, D.J. and Roberts, G.D. 1994. Laboratory detection of fungemia. *Clin Lab Med*, **14**, 83–97.

Geraci, J.E. and Wilson, W.R. 1982. Symposium on infective endocarditis. III. Endocarditis due to gram-negative bacteria. Report of 56 cases. *Mayo Clin Proc*, **57**, 145–8.

Geraci, J.E., Wilkowske, C.J., et al. 1977. *Haemophilus* endocarditis. Report of 14 cases. *Mayo Clin Proc*, **52**, 209–15.

Giamarellou, H. 2002. Nosocomial cardiac infections. *J Hosp Infect*, **50**, 91–105.

Gibbons, R.J. and Nygaard, M. 1968. Synthesis of insoluble dextran and its significance in the formation of gelatinous deposits by plaque-forming streptococci. *Arch Oral Biol*, **13**, 1249–62.

Gilleece, A. and Fenelon, L. 2000. Nosocomial infective endocarditis. *J Hosp Infect*, **46**, 83–8.

Goerre, S., Malinverni, R. and Aeschbacher, B.C. 1998. Successful conservative treatment of nontyphoid salmonella endocarditis involving a bioprosthetic valve. *Clin Cardiol*, **21**, 368–70.

Goldenberg, D.L., Brandt, K.D., et al. 1974. Acute arthritis caused by gram-negative bacilli: a clinical characterization. *Medicine (Balt)*, **53**, 197–208.

Gonzalez-Barca, E., Fernandez-Sevilla, A., et al. 1996. Prospective study of 288 episodes of bacteremia in neutropenic cancer patients in a single institution. *Eur J Clin Microbiol Infect Dis*, **15**, 291–6.

Gorbach, P.M., Hoa, D.T., et al. 1998. Reproduction, risk and reality: family planning and reproductive health in northern Vietnam. *J Biosoc Sci*, **30**, 393–409.

Gordon, S., Swenson, J.M., Enterococcal Study Group, et al. 1992. Antimicrobial susceptibility patterns of common and unusual species of enterococci causing infections in the United States. *J Clin Microbiol*, **30**, 2373–8.

Gordon, S.M., Serkey, J.M., et al. 2000. Early onset prosthetic valve endocarditis: the Cleveland Clinic experience 1992–1997. *Ann Thorac Surg*, **69**, 1388–92.

Gorensek, M.J., Lebel, M.H. and Nelson, J.D. 1988. Peritonitis in children with nephrotic syndrome. *Pediatrics*, **81**, 849–56.

Gossling, J. 1988. Occurrence and pathogenicity of the *Streptococcus milleri* group. *Rev Infect Dis*, **10**, 257–85.

Gouello, J.P., Asfar, P., et al. 2000. Nosocomial endocarditis in the intensive care unit: an analysis of 22 cases. *Crit Care Med*, **28**, 377–82.

Gould, K., Ramirez-Ronda, C.H., et al. 1975. Adherence of bacteria to heart valves in vitro. *J Clin Invest*, **56**, 1364–70.

Gradon, J.D., Chapnick, E.K. and Lutwick, L.I. 1992. Infective endocarditis of a native valve due to *Acinetobacter*: case report and review. *Clin Infect Dis*, **14**, 1145–8.

Graninger, W. and Ragette, R. 1992. Nosocomial bacteremia due to *Enterococcus faecalis* without endocarditis. *Clin Infect Dis*, **15**, 49–57.

Graupner, C., Vilacosta, I., et al. 2002. Periannular extension of infective endocarditis. *J Am Coll Cardiol*, **39**, 1204–11.

Graybill, J.R., Sharkey, P.K., et al. 1990. The major endemic mycosis in the setting of AIDS: clinical manifestations. In: Bossche, H.V. (ed.), *Mycoses in AIDS patients*. New York, NY: Plenum Press.

Greaves, K., Mou, D., et al. 2003. Clinical criteria and the appropriate use of transthoracic echocardiography for the exclusion of infective endocarditis. *Heart*, **89**, 273–5.

Gregoratos, G. 2003. Infective endocarditis in the elderly: diagnosis and management. *Am J Geriatr Cardiol*, **12**, 183–9.

Grigoriadis, E. and Gold, W.L. 1997. Pyogenic brain abscess caused by *Streptococcus pneumoniae*: case report and review. *Clin Infect Dis*, **25**, 1108–12.

Grijalva, M., Horvath, R., et al. 2003. Molecular diagnosis of culture negative infective endocarditis: clinical validation in a group of surgically treated patients. *Heart*, **89**, 263–8.

Groom, A.V., Wolsey, D.H., et al. 2001. Community-acquired methicillin-resistant *Staphylococcus aureus* in a rural American Indian community. *JAMA*, **286**, 1201–5.

Gruer, L.D., McKendrick, M.W. and Geddes, A.M. 1984. Pneumococcal bacteraemia – a continuing challenge. *Q J Med*, **53**, 259–70.

Guerra, J.M., Tornos, M.P., et al. 2001. Long term results of mechanical prostheses for treatment of active infective endocarditis. *Heart*, **86**, 63–8.

Guerra-Romero, L., Edson, R.S., et al. 1987. Comparison of Du Pont Isolator and Roche Septi-Chek for detection of fungemia. *J Clin Microbiol*, **25**, 1623–5.

Habib, G., Derumeaux, G., et al. 1999. Value and limitations of the Duke criteria for the diagnosis of infective endocarditis. *J Am Coll Cardiol*, **33**, 2023–9.

Hageman, J.C., Pegues, D.A., et al. 2001. Vancomycin-intermediate *Staphylococcus aureus* in a home health-care patient. *Emerg Infect Dis*, **7**, 1023–5.

Hall, M.M., Ilstrup, D.M. and Washington, J.A. 1976. Effect of volume of blood cultured on detection of bacteremia. *J Clin Microbiol*, **3**, 643–5.

Hamoudi, A.C., Hribar, M.M., et al. 1990. Clinical relevance of viridans and nonhemolytic streptococci isolated from blood and cerebrospinal fluid in a pediatric population. *Am J Clin Pathol*, **93**, 270–2.

Han, T., Sokal, J.E. and Neter, E. 1967. Salmonellosis in disseminated malignant diseases. A seven-year review (1959–1965). *N Engl J Med*, **276**, 1045–52.

Han, X.Y. and Truant, A.L. 1999. The detection of positive blood cultures by the AccuMed ESP-384 system: the clinical significance of three-day testing. *Diagn Microbiol Infect Dis*, **33**, 1–6.

Hanna, B.A., Ebrahimzadeh, A., et al. 1999. Multicenter evaluation of the BACTEC MGIT 960 system for recovery of mycobacteria. *J Clin Microbiol*, **37**, 748–52.

Harbarth, S., Rutschmann, O., et al. 1998. Impact of methicillin resistance on the outcome of patients with bacteremia caused by *Staphylococcus aureus*. *Arch Intern Med*, **158**, 182–9.

Hardy, D.J., Hulbert, B.B. and Migneault, P.C. 1992. Time to detection of positive BacT/Alert blood cultures and lack of need for routine subculture of 5- to 7-day negative cultures. *J Clin Microbiol*, **30**, 2743–5.

Harris, K.M., Li, D.Y., et al. 2003. The prospective role of transesophageal echocardiography in the diagnosis and management of patients with suspected infective endocarditis. *Echocardiography*, **20**, 57–62.

Hasbun, R., Vikram, H.R., et al. 2003. Complicated left-sided native valve endocarditis in adults: risk classification for mortality. *JAMA*, **289**, 1933–40.

Havlik, J.A. Jr., Horsburgh, C.R. Jr., et al. 1992. Disseminated *Mycobacterium avium* complex infection: clinical identification and epidemiologic trends. *J Infect Dis*, **165**, 577–80.

Havlik, J.A. Jr., Metchock, B., et al. 1993. A prospective evaluation of *Mycobacterium avium* complex colonization of the respiratory and gastrointestinal tracts of persons with human immunodeficiency virus infection. *J Infect Dis*, **168**, 1045–8.

Hawkins, B.L., Peterson, E.M. and de la Maza, L.M. 1986. Improvement of positive blood culture detection by agitation. *Diagn Microbiol Infect Dis*, **5**, 207–13.

Heidenreich, P.A., Masoudi, F.A., et al. 1999. Echocardiography in patients with suspected endocarditis: a cost-effectiveness analysis. *Am J Med*, **107**, 198–208.

Heinemann, B., Wisplinghoff, H., et al. 2000. Comparative activities of ciprofloxacin, clinafloxacin, gatifloxacin, gemifloxacin, levofloxacin, moxifloxacin, and trovafloxacin against epidemiologically defined *Acinetobacter baumannii* strains. *Antimicrob Agents Chemother*, **44**, 2211–13.

Heiro, M., Nikoskelainen, J., et al. 1998. Diagnosis of infective endocarditis. Sensitivity of the Duke vs von Reyn criteria. *Arch Intern Med*, **158**, 18–24.

Heiro, M., Nikoskelainen, J., et al. 2000. Neurologic manifestations of infective endocarditis: a 17-year experience in a teaching hospital in Finland. *Arch Intern Med*, **160**, 2781–7.

Hershow, R.C., Khayr, W.F. and Smith, N.L. 1992. A comparison of clinical virulence of nosocomially acquired methicillin-resistant and methicillin-sensitive *Staphylococcus aureus* infections in a university hospital. *Infect Control Hosp Epidemiol*, **13**, 587–93.

Hockey, L.J., Fujita, N.K., et al. 1982. Detection of fungemia obscured by concomitant bacteremia: in vitro and in vivo studies. *J Clin Microbiol*, **16**, 1080–5.

Hoen, B., Selton-Suty, C., et al. 1995. Infective endocarditis in patients with negative blood cultures – analysis of 88 cases from a one-year nationwide survey in France. *Clin Infect Dis*, **20**, 501–6.

Hoen, B., Alla, F., Association pour l'Etude et la Prevention de l'Endocardite Infectieuse (AEPEI) Study Group, et al. 2002. Changing profile of infective endocarditis: results of a-year survey in France. *JAMA*, **288**, 75–81.

Hoge, C.W., Schwartz, B., et al. 1993. The changing epidemiology of invasive group A streptococcal infections and the emergence of streptococcal toxic shock-like syndrome. A retrospective population-based study. *JAMA*, **269**, 384–9.

Hogevik, H., Olaison, L., et al. 1995. Epidemiologic aspects of infective endocarditis in an urban population. A 5-year prospective study. *Medicine (Balt)*, **74**, 324–39.

Honberg, P.Z. and Gutschik, E. 1987. *Streptococcus bovis* bacteraemia and its association with alimentary-tract neoplasm. *Lancet*, **1**, 8525, 163–4.

Hook, E.W. and Hansfield, H.H. 1999. Gonococcal infections in the audit. In: Holmes, K.K., Mardh, P.A., et al. (eds), *Sexually transmitted diseases*. New York, NY: McGraw-Hill Book Co, 451–66.

Houpikian, P. and Raoult, D. 2003. Diagnostic methods. Current best practices and guidelines for identification of difficult-to-culture pathogens in infective endocarditis. *Cardiol Clin*, **21**, 207–17.

Houston, B.D., Crouch, M.E. and Finch, R.G. 1980. Streptococcus MG-intermedius (*Streptococcus milleri*) septic arthritis in a patient with rheumatoid arthritis. *J Rheumatol*, **7**, 89–92.

Huang, L.T. 1996. *Salmonella* meningitis complicated by brain infarctions. *Clin Infect Dis*, **22**, 194–5.

Huang, C.B. and Chuang, J.H. 1997. Acute scrotal inflammation caused by *Salmonella* in young infants. *Pediatr Infect Dis J*, **16**, 1091–2.

Humpl, T., McCrindle, B.W. and Smallhorn, J.F. 2003. The relative roles of transthoracic compared with transesophageal echocardiography in children with suspected infective endocarditis. *J Am Coll Cardiol*, **41**, 2068–71.

Hurle, A., Nistal, J.F., et al. 1996. Isolated apical intracavitary left ventricular abscess in a normal heart: a rare complication of *Streptococcus milleri* endocarditis. *Cardiovasc Surg*, **4**, 61–3.

Husni, R.N., Gordon, S.M., et al. 1997. *Lactobacillus* bacteremia and endocarditis: review of 45 cases. *Clin Infect Dis*, **25**, 1048–55.

Hyde, J.A., Darouiche, R.O. and Costerton, J.W. 1998. Strategies for prophylaxis against prosthetic valve endocarditis: a review article. *J Heart Valve Dis*, **7**, 316–26.

Hyde, T.B., Hilger, T.M., et al. 2002. Trends in incidence and antimicrobial resistance of early-onset sepsis: population-based surveillance in San Francisco and Atlanta. *Pediatrics*, **110**, 690–5.

Iannini, P.B., Claffey, T., et al. 1974. Bacteremic *Pseudomonas* pneumonia. *JAMA*, **230**, 558–61.

Ibrahim, E.H., Sherman, G., et al. 2000. The influence of inadequate antimicrobial treatment of bloodstream infections on patient outcomes in the ICU setting. *Chest*, **118**, 146–55.

Ilstrup, D.M. and Washington, J.A. 1983. The importance of volume of blood cultured in the detection of bacteremia and fungemia. *Diagn Microbiol Infect Dis*, **1**, 107–10.

Isaacman, D.J., Karasic, R.B., et al. 1996. Effect of number of blood cultures and volume of blood on detection of bacteremia in children. *J Pediatr*, **128**, 190–5.

Ispahani, P., Donald, F.E. and Aveline, A.J. 1988. *Streptococcus pyogenes* bacteraemia: an old enemy subdued, but not defeated. *J Infect*, **16**, 37–46.

Ivert, T.S., Dismukes, W.E., et al. 1984. Prosthetic valve endocarditis. *Circulation*, **69**, 223–32.

Jacobs, F., Abramowicz, D., et al. 1990. *Brucella* endocarditis. The role of combined medical and surgical treatment. *Rev Infect Dis*, **12**, 740–3.

Jacobs, J.A., Pietersen, H.G., et al. 1995. *Streptococcus anginosus*, *Streptococcus constellatus* and *Streptococcus intermedius*. Clinical relevance, hemolytic and serologic characteristics. *Am J Clin Pathol*, **104**, 547–53.

Jalal, S., Khan, K.A., et al. 1998. Clinical spectrum of infective endocarditis: 15 years experience. *Indian Heart J*, **50**, 516–19.

Jerng, J.S., Hsueh, P.R., et al. 1997. Empyema thoracis and lung abscess caused by viridans streptococci. *Am J Respir Crit Care Med*, **156**, 1508–14.

Johnson, A.S., Touchie, C., et al. 2000. Four-day incubation for detection of bacteremia using the BACTEC 9240. *Diagn Microbiol Infect Dis*, **38**, 195–9.

Johnson, D.R., Stevens, D.L. and Kaplan, E.L. 1992. Epidemiologic analysis of group A streptococcal serotypes associated with severe systemic infections, rheumatic fever, or uncomplicated pharyngitis. *J Infect Dis*, **166**, 374–82.

Johnson, J.R., Russo, T.A., et al. 1997. Discovery of disseminated J96-like strains of uropathogenic *Escherichia coli* O4:H5 containing genes for both PapG(J96) (class I) and PrsG(J96) (class III) Gal(alpha1-4)Gal-binding adhesins. *J Infect Dis*, **175**, 983–8.

Johnson, J.R., Oswald, E., et al. 2002. Phylogenetic distribution of virulence-associated genes among *Escherichia coli* isolates associated with neonatal bacterial meningitis in the Netherlands. *J Infect Dis*, **185**, 774–84.

Johnson, L.B., Bhan, A., et al. 2003. Changing epidemiology of community-onset methicillin-resistant *Staphylococcus aureus* bacteremia. *Infect Control Hosp Epidemiol*, **24**, 431–5.

Julander, I. 1985. Unfavourable prognostic factors in *Staphylococcus aureus* septicemia and endocarditis. *Scand J Infect Dis*, **17**, 179–87.

Kahl, B., Herrmann, M., et al. 1998. Persistent infection with small colony variant strains of *Staphylococcus aureus* in patients with cystic fibrosis. *J Infect Dis*, **177**, 1023–9.

Kambal, A.M. 1987. Isolation of *Streptococcus milleri* from clinical specimens. *J Infect*, **14**, 217–23.

Kang, C.I., Kim, S.H., et al. 2003. *Pseudomonas aeruginosa* bacteremia: risk factors for mortality and influence of delayed receipt of effective antimicrobial therapy on clinical outcome. *Clin Infect Dis*, **37**, 745–51.

Kaye, D. 1985. Changing pattern of infective endocarditis. *Am J Med*, **78**, Suppl. 6B, 157–62.

Keles, C., Bozbuga, N., et al. 2001. Surgical treatment of *Brucella* endocarditis. *Ann Thorac Surg*, **71**, 1160–3.

Kellogg, J.A., Manzella, J.P. and Bankert, D.A. 2000. Frequency of low-level bacteremia in children from birth to fifteen years of age. *J Clin Microbiol*, **38**, 2181–5.

Kielhofner, M., Atmar, R.L., et al. 1992. Life-threatening *Pseudomonas aeruginosa* infections in patients with human immunodeficiency virus infection. *Clin Infect Dis*, **14**, 403–11.

Kim A.I., Adal K.A., Schmitt S.K. 2003. *Staphylococcus aureus* bacteremia: using echocardiography to guide length of therapy. *Cleve Clin J Med* **70**, 517, 520–1, 525–6.

Kim, P.E., Musher, D.M., et al. 1996. Association of invasive pneumococcal disease with season, atmospheric conditions, air pollution, and the isolation of respiratory viruses. *Clin Infect Dis*, **22**, 100–6.

Kim, S.D., McDonald, L.C., et al. 2000. Determining the significance of coagulase-negative staphylococci isolated from blood cultures at a community hospital: a role for species and strain identification. *Infect Control Hosp Epidemiol*, **21**, 213–17.

King, A., Rampling, A., et al. 1984. Neutropenic enterocolitis due to *Clostridium septicum* infection. *J Clin Pathol*, **37**, 335–43.

Kirchner, J.T. 1991. *Clostridium septicum* infection. Beware of associated cancer. *Postgrad Med*, **90**, 157–60.

Kitada, K., Inoue, M. and Kitano, M. 1997a. Experimental endocarditis induction and platelet aggregation by *Streptococcus anginosus*, *Streptococcus constellatus* and *Streptococcus intermedius*. *FEMS Immunol Med Microbiol*, **19**, 25–32.

Kitada, K., Inoue, M. and Kitano, M. 1997b. Infective endocarditis-inducing abilities of 'Streptococcus milleri' group. *Adv Exp Med Biol*, **418**, 161–3.

Klein, R.S., Recco, R.A., et al. 1977. Association of *Streptococcus bovis* with carcinoma of the colon. *N Engl J Med*, **297**, 800–2.

Kloos, W.E. and Bannerman, T.L. 1994. Update on clinical significance of coagulase-negative staphylococci. *Clin Microbiol Rev*, **7**, 117–40.

Knox, K.L. and Holmes, A.H. 2002. Nosocomial endocarditis caused by *Corynebacterium amycolatum* and other nondiphtheriae corynebacteria. *Emerg Infect Dis*, **8**, 97–9.

Kollef, M.H. and Fraser, V.J. 2001. Antibiotic resistance in the intensive care unit. *Ann Intern Med*, **134**, 298–314.

Komshian, S.V., Tablan, O.C., et al. 1990. Characteristics of left-sided endocarditis due to *Pseudomonas aeruginosa* in the Detroit Medical Center. *Rev Infect Dis*, **12**, 693–702.

Koransky, J.R., Stargel, M.D. and Dowell, V.R. Jr. 1979. *Clostridium septicum* bacteremia. Its clinical significance. *Am J Med*, **66**, 63–6.

Kreter, B. and Woods, M. 1992. Antibiotic prophylaxis for cardiothoracic operations. Meta-analysis of thirty years of clinical trials. *J Thorac Cardiovasc Surg*, **104**, 590–9.

Kudsk, K.A. 1992. Occult gastrointestinal malignancies producing metastatic *Clostridium septicum* infections in diabetic patients. *Surgery*, **112**, 765–70.

Kuikka, A., Syrjanen, J., et al. 1992. Pneumococcal bacteraemia during a recent decade. *J Infect*, **24**, 157–68.

Kupferwasser, L.I. and Bayer, A.S. 2000. Update on culture-negative endocarditis. *Curr Clin Top Infect Dis*, **20**, 113–33.

Kuypers, J.M. and Proctor, R.A. 1989. Reduced adherence to traumatized rat heart valves by a low-fibronectin-binding mutant of *Staphylococcus aureus*. *Infect Immun*, **57**, 2306–12.

La, S.L. Jr., Dryja, D., et al. 1981. Diagnosis of bacteremia in children by quantitative direct plating and a radiometric procedure. *J Clin Microbiol*, **13**, 478–82.

Lamas, C.C. and Eykyn, S.J. 1997. Suggested modifications to the Duke criteria for the clinical diagnosis of native valve and prosthetic valve endocarditis: analysis of 118 pathologically proven cases. *Clin Infect Dis*, **25**, 713–19.

Lamas, C.C. and Eykyn, S.J. 1998. Hospital acquired native valve endocarditis: analysis of 22 cases presenting over 11 years. *Heart*, **79**, 442–7.

Lamas, C.C. and Eykyn, S.J. 2000. Bicuspid aortic valve – A silent danger: analysis of 50 cases of infective endocarditis. *Clin Infect Dis*, **30**, 336–41.

Lamas, C.C. and Eykyn, S.J. 2003. Blood culture negative endocarditis: analysis of 63 cases presenting over 25 years. *Heart*, **89**, 258–62.

Lamy, B., Roy, P., et al. 2002. What is the relevance of obtaining multiple blood samples for culture? A comprehensive model to optimize the strategy for diagnosing bacteremia. *Clin Infect Dis*, **35**, 842–50.

Lark, R.L., McNeil, S.A., et al. 2001. Risk factors for anaerobic bloodstream infections in bone marrow transplant recipients. *Clin Infect Dis*, **33**, 338–43.

Le, T. and Bayer, A.S. 2003. Combination antibiotic therapy for infective endocarditis. *Clin Infect Dis*, **36**, 615–21.

Lefort, A., Mainardi, J.L., The Pneumococcal Endocarditis Study Group, et al. 2000. *Streptococcus pneumoniae* endocarditis in adults. A multicenter study in France in the era of penicillin resistance (1991–1998). *Medicine (Balt)*, **79**, 327–37.

Lefort, A., Lortholary, O., Beta-Hemolytic Streptococci Infective Endocarditis Study Group, et al. 2002. Comparison between adult endocarditis due to beta-hemolytic streptococci (serogroups A, B, C and G) and *Streptococcus milleri*: a multicenter study in France. *Arch Intern Med*, **162**, 2450–6.

Lengyel, M. 1997. The impact of transesophageal echocardiography on the management of prosthetic valve endocarditis: experience of 31 cases and review of the literature. *J Heart Valve Dis*, **6**, 204–11.

Lentnek, A.L., Giger, O. and O'Rourke, E. 1990. Group A beta-hemolytic streptococcal bacteremia and intravenous substance abuse. A growing clinical problem? *Arch Intern Med*, **150**, 89–93.

Leonard, M.K., Pox, C.P. and Stephens, D.S. 2001. *Abiotrophia* species bacteremia and a mycotic aneurysm in an intravenous drug abuser. *N Engl J Med*, **344**, 233–4.

Leport, C., Bure, A., et al. 1987. Incidence of colonic lesions in *Streptococcus bovis* and enterococcal endocarditis. *Lancet*, **1**, 8535, 748.

Leport, C., Horstkotte, D. and Burckhardt, D. Group of Experts of the International Society for Chemotherapy, 1995. Antibiotic prophylaxis for infective endocarditis from an international group of experts towards a European consensus. *Eur Heart J*, **16**, Suppl. B, 126–31.

Lerner, P.I., Gopalakrishna, K.V., et al. 1977. Group B streptococcus (*S. agalactiae*) bacteremia in adults: analysis of 32 cases and review of the literature. *Medicine (Balt)*, **56**, 457–73.

Leung, M.J., Nuttall, N., et al. 1999. Case of *Staphylococcus schleiferi* endocarditis and a simple scheme to identify clumping factor-positive staphylococci. *J Clin Microbiol*, **37**, 3353–6.

Levandowski, R.A. 1985. *Streptococcus milleri* endocarditis complicated by myocardial abscess. *South Med J*, **78**, 892–3.

Levine, W.C., Buehler, J.W., et al. 1991. Epidemiology of nontyphoidal *Salmonella* bacteremia during the human immunodeficiency virus epidemic. *J Infect Dis*, **164**, 81–7.

Levy, M.M., Fink, M.P., et al. 2003a. SCCM/ESICM/ACCP/ATS/SIS International Sepsis Definitions Conference. *Intensive Care Med*, **29**, 530–8.

Levy, V., Reed, C., et al. 2003b. *Escherichia coli* myonecrosis in alcoholic cirrhosis. *J Clin Gastroenterol*, **36**, 443–5.

Li, J., Plorde, J.J. and Carlson, L.G. 1994. Effects of volume and periodicity on blood cultures. *J Clin Microbiol*, **32**, 2829–31.

Li, J.S., Sexton, D.J., et al. 2000. Proposed modifications to the Duke criteria for the diagnosis of infective endocarditis. *Clin Infect Dis*, **30**, 633–8.

Lindberg, J., Prag, J. and Schonheyder, H.C. 1998. Pneumococcal endocarditis is not just a disease of the past: an analysis of 16 cases diagnosed in Denmark 1986–1997. *Scand J Infect Dis*, **30**, 469–72.

Linden, P.K., Pasculle, A.W., et al. 1996. Differences in outcomes for patients with bacteremia due to vancomycin-resistant *Enterococcus faecium* or vancomycin-susceptible *E. faecium*. *Clin Infect Dis*, **22**, 663–70.

Lisby, G., Gutschik, E. and Durack, D.T. 2002. Molecular methods for diagnosis of infective endocarditis. *Infect Dis Clin North Am*, **16**, 393–412.

Liu, C.E., Jang, T.N., et al. 1995. Invasive group G streptococcal infections: a review of 37 cases. *Zhonghua Yi Xue Za Zhi (Taipei)*, **56**, 173–8.

Liu, V.C., Stevenson, J.G. and Smith, A.L. 1992. Group A Streptococcus mural endocarditis. *Pediatr Infect Dis J*, **11**, 1060–2.

Lodise, T.P., McKinnon, P.S., et al. 2003. Outcome analysis of delayed antibiotic treatment for hospital-acquired *Staphylococcus aureus* bacteremia. *Clin Infect Dis*, **36**, 1418–23.

Lombardi, D.P. and Engleberg, N.C. 1992. Anaerobic bacteremia: incidence, patient characteristics, and clinical significance. *Am J Med*, **92**, 53–60.

Lortholary, O., Buu-Hoi, A., et al. 1995. Endocarditis caused by multiply resistant *Bacteroides fragilis*: case report and review. *Clin Microbiol Infect*, **1**, 44–7.

Lowrance, J.H., Baddour, L.M. and Simpson, W.A. 1990. The role of fibronectin binding in the rat model of experimental endocarditis caused by *Streptococcus sanguis*. *J Clin Invest*, **86**, 7.

Lyon, R. and Woods, G. 1995. Comparison of the BacT/Alert and Isolator blood culture systems for recovery of fungi. *Am J Clin Pathol*, **103**, 660–2.

Lytle, B.W., Priest, B.P., et al. 1996. Surgical treatment of prosthetic valve endocarditis. *J Thorac Cardiovasc Surg*, **111**, 198–207.

Maki, D.G. and Agger, W.A. 1988. Enterococcal bacteremia: clinical features, the risk of endocarditis, and management. *Medicine (Balt)*, **67**, 248–69.

Mansur, A.J., Grinberg, M., et al. 1992. The complications of infective endocarditis. A reappraisal in the 1980s. *Arch Intern Med*, **152**, 2428–32.

Maraj, S., Jacobs, L.E., et al. 2002. Epidemiology and outcome of infective endocarditis in hemodialysis patients. *Am J Med Sci*, **324**, 254–60.

Marinella, M.A. 1997. *Streptococcus constellatus* endocarditis presenting as acute embolic stroke. *Clin Infect Dis*, **24**, 1271–2.

Marr, K.A., Kong, L., et al. 1998. Incidence and outcome of *Staphylococcus aureus* bacteremia in hemodialysis patients. *Kidney Int*, **54**, 1684–9.

Marrie, T.J., Harczy, M., et al. 1990. Culture-negative endocarditis probably due to *Chlamydia pneumoniae*. *J Infect Dis*, **161**, 127–9.

Marrie, T.J. and Raoult, D. 2002. Update on Q fever, including Q fever endocarditis. *Curr Clin Top Infect Dis*, **22**, 97–124.

Marron, A., Carratala, J., et al. 2000. Serious complications of bacteremia caused by viridans streptococci in neutropenic patients with cancer. *Clin Infect Dis*, **31**, 1126–30.

Martin, D.R. and Single, L.A. 1993. Molecular epidemiology of group A streptococcus M type 1 infections. *J Infect Dis*, **167**, 1112–17.

Martin, P.R. and Hoiby, E.A. 1990. Streptococcal serogroup A epidemic in Norway 1987–1988. *Scand J Infect Dis*, **22**, 421–9.

Martinez, E., Miro, J.M., Spanish Pneumococcal Endocarditis Study Group, et al. 2002. Effect of penicillin resistance of *Streptococcus pneumoniae* on the presentation, prognosis, and treatment of pneumococcal endocarditis in adults. *Clin Infect Dis*, **35**, 130–9.

Martos-Perez, F., Reguera, J.M. and Colmenero, J.D. 1996. Comparable sensitivity of the Duke criteria and the modified Beth Israel criteria for diagnosing infective endocarditis. *Clin Infect Dis*, **23**, 410–11.

Maschmeyer, G. and Braveny, I. 2000. Review of the incidence and prognosis of *Pseudomonas aeruginosa* infections in cancer patients in the 1990s. *Eur J Clin Microbiol Infect Dis*, **19**, 915–25.

Masterson, K.C. and McGowan, J.E. Jr. 1988. Detection of positive blood cultures by the Bactec NR660. The clinical importance of five versus seven days of testing. *Am J Clin Pathol*, **90**, 91–4.

Mathew, J., Addai, T., et al. 1995. Clinical features, site of involvement, bacteriologic findings, and outcome of infective endocarditis in intravenous drug users. *Arch Intern Med*, **155**, 1641–8.

Mathur, P., Chaudhry, R., et al. 2002. A study of bacteremia in febrile neutropenic patients at a tertiary-care hospital with special reference to anaerobes. *Med Oncol*, **19**, 267–72.

Mattia, A.R. 1993. FDA review criteria for blood culture systems. *Clin Microbiol Newsl*, **15**, 132–6.

Maugein, J., Guillemot, D., et al. 2003. Clinical and microbiological epidemiology of *Streptococcus pneumoniae* bacteremia in eight French counties. *Clin Microbiol Infect*, **9**, 280–8.

Maurin, M. and Raoult, D. 1996. *Bartonella (Rochalimea) quintana* infections. *Clin Microbiol Rev*, **9**, 273–92.

McCormick, J.K., Hirt, H., et al. 2000. Pathogenic mechanisms of enterococcal endocarditis. *Curr Infect Dis Rep*, **2**, 315–21.

McDonald, L.C., Weinstein, M.P., et al. 2001. Controlled comparison of BacT/ALERT FAN aerobic medium and BATEC fungal blood culture medium for detection of fungemia. *J Clin Microbiol*, **39**, 622–4.

McLauchlin, J. and Low, J.C. 1994. Primary cutaneous listeriosis in adults: an occupational disease of veterinarians and farmers. *Vet Rec*, **135**, 615–17.

Mead, P.S., Slutsker, L., et al. 1999. Food-related illness and death in the United States. *Emerg Infect Dis*, **5**, 607–25.

Megran, D.W. 1992. Enterococcal endocarditis. *Clin Infect Dis*, **15**, 63–71.

Menzies, B.E. 2003. The role of fibronectin binding proteins in the pathogenesis of *Staphylococcus aureus* infections. *Curr Opin Infect Dis*, **16**, 225–9.

Mermel, L.A. and Maki, D.G. 1993. Detection of bacteremia in adults: consequences of culturing an inadequate volume of blood. *Ann Intern Med*, **119**, 270–2.

Mermel, L.A., Farr, B.M., et al. 2001. Guidelines for the management of intravascular catheter-related infections. *Clin Infect Dis*, **32**, 1249–72.

Mermin, J.H., Townes, J.M., et al. 1998. Typhoid fever in the United States, 1985–1994: changing risks of international travel and increasing antimicrobial resistance. *Arch Intern Med*, **158**, 633–8.

Miele, P.S., Kogulan, P.K., et al. 2001. Seven cases of surgical native valve endocarditis caused by coagulase-negative staphylococci: An underappreciated disease. *Am Heart J*, **142**, 571–6.

Millaire, A., Leroy, O., et al. 1997. Incidence and prognosis of embolic events and metastatic infections in infective endocarditis. *Eur Heart J*, **18**, 677–84.

Millar, B., Moore, J., et al. 2001. Molecular diagnosis of infective endocarditis – a new Duke's criterion. *Scand J Infect Dis*, **33**, 673–80.

Mills, J. and Drew, D. 1976. *Serratia marcescens* endocarditis: a regional illness associated with intravenous drug abuse. *Ann Intern Med*, **84**, 29–35.

Miro, J.M., Moreno, A. and Mestres, C.A. 2003. Infective endocarditis in intravenous drug abusers. *Curr Infect Dis Rep*, **5**, 307–16.

Moellering, R.C. Jr. 1992. Emergence of *Enterococcus* as a significant pathogen. *Clin Infect Dis*, **14**, 1173–6.

Mohan, U.R., Walters, S. and Kroll, J.S. 2000. Endocarditis due to group A beta-hemolytic *Streptococcus* in children with potentially lethal sequelae: 2 cases and review. *Clin Infect Dis*, **30**, 624–5.

Mohsen, A.H., Price, A., et al. 2001. *Propionibacterium acnes* endocarditis in a native valve complicated by intraventricular abscess: a case report and review. *Scand J Infect Dis*, **33**, 379–80.

Molina, J.M., Leport, C., et al. 1991. Clinical and bacterial features of infections caused by *Streptococcus milleri*. *Scand J Infect Dis*, **23**, 659–66.

Moll, W.M., Ungerechts, J., et al. 1996. Comparison of BBL Crystal ANR ID Kit and API rapid ID 32 A for identification of anaerobic bacteria. *Zentralbl Bakteriol*, **284**, 329–47.

Moon, M.R., Stinson, E.B. and Miller, D.C. 1997. Surgical treatment of endocarditis. *Prog Cardiovasc Dis*, **40**, 239–64.

Moreillon, P., Entenza, J.M., et al. 1995. Role of *Staphylococcus aureus* coagulase and clumping factor in pathogenesis of experimental endocarditis. *Infect Immun*, **63**, 4738–43.

Moreillon, P., Que, Y.A. and Bayer, A.S. 2002. Pathogenesis of streptococcal and staphylococcal endocarditis. *Infect Dis Clin North Am*, **16**, 297–318.

Morello, J.A., Leitch, C., et al. 1994. Detection of bacteremia by Difco ESP blood culture system. *J Clin Microbiol*, **32**, 811–18.

Morgan-Hughes, G. and Motwani, J. 2002. Mitral valve endocarditis in hypertrophic cardiomyopathy: case report and literature review. *Heart*, **87**, e8.

Morris, A.J., Wilson, M.L., et al. 1993. Rationale for selective use of anaerobic blood cultures. *J Clin Microbiol*, **31**, 2110–13.

Moser S., Fätkenheuer G. et al. 1998. *Die Klinische Signifikanz positiver Blutkulturen an einem Universitätsklinikum: eine prospektive Untersuchung zu Mikrobiologie, Epidemiologie und Prognose*. 1998. Annual Meeting of the DGHM. Abstract 252.

Moss, R. and Munt, B. 2003. Injection drug use and right sided endocarditis. *Heart*, **89**, 577–81.

Munoz, P., Llancaqueo, A., et al. 1997. Group B streptococcus bacteremia in nonpregnant adults. *Arch Intern Med*, **157**, 213–16.

Munoz, P., Sainz, J., et al. 1999. Austrian syndrome caused by highly penicillin-resistant *Streptococcus pneumoniae*. *Clin Infect Dis*, **29**, 1591–2.

Murray, B.E. 1990. The life and times of the *Enterococcus*. *Clin Microbiol Rev*, **3**, 46–65.

Murray, P.R., Traynor, P. and Hopson, D. 1992. Critical assessment of blood culture techniques: analysis of recovery of obligate and facultative anaerobes, strict aerobic bacteria, and fungi in aerobic and anaerobic blood culture bottles. *J Clin Microbiol*, **30**, 1462–8.

Mylonakis, E. and Calderwood, S.B. 2001. Infective endocarditis in adults. *N Engl J Med*, **345**, 1318–30.

Mylotte, J.M. and Tayara, A. 2000a. *Staphylococcus aureus* bacteremia: predictors of 30-day mortality in a large cohort. *Clin Infect Dis*, **31**, 1170–4.

Mylotte, J.M. and Tayara, A. 2000b. Blood cultures: clinical aspects and controversies. *Eur J Clin Microbiol Infect Dis*, **19**, 157–63.

Mylotte, J.M., McDermott, C. and Spooner, J.A. 1987. Prospective study of 114 consecutive episodes of *Staphylococcus aureus* bacteremia. *Rev Infect Dis*, **9**, 891–908.

Mylotte, J.M., Aeschlimann, J.R. and Rotella, D.L. 1996. *Staphylococcus aureus* bacteremia: factors predicting hospital mortality. *Infect Control Hosp Epidemiol*, **17**, 165–8.

Naimi, T.S., LeDell, K.H., et al. 2001. Epidemiology and clonality of community-acquired methicillin-resistant *Staphylococcus aureus* in Minnesota, 1996–1998. *Clin Infect Dis*, **33**, 990–6.

Naimi, T.S., Anderson, D., et al. 2003. Vancomycin-intermediate *Staphylococcus aureus* with phenotypic susceptibility to methicillin in a patient with recurrent bacteremia. *Clin Infect Dis*, **36**, 1609–12.

Nastro, L.J. and Finegold, S.M. 1973. Endocarditis due to anaerobic gram-negative bacilli. *Am J Med*, **54**, 482–96.

Neal, P.R., Kleiman, M.B., et al. 1986. Volume of blood submitted for culture from neonates. *J Clin Microbiol*, **24**, 353–6.

Nettles, R.E., McCarty, D.E., et al. 1997. An evaluation of the Duke criteria in 25 pathologically confirmed cases of prosthetic valve endocarditis. *Clin Infect Dis*, **25**, 1401–3.

Neu, H.C. 1986. Cost effective blood cultures – is it possible or impossible to modify behavior? *Infect Control*, **7**, 32–3.

Neves, J., Raso, P. and Marinho, R.P. 1971. Prolonged septicaemic salmonellosis intercurrent with *Schistosomiasis mansoni* (intestinal polyposis, hepatic and cardiopulmonary forms) Chagas' disease, cerebral cysticercosis, taeniasis, shigellosis, ancylostomiasis, ascariasis and chronic malnutrition. Clinopathologic discussion. *J Trop Med Hyg*, **74**, 9–18.

Ng, T.K., Ling, J.M., et al. 1996. A retrospective study of clinical characteristics of *Acinetobacter* bacteremia. *Scand J Infect Dis Suppl*, **101**, 26–32.

Nieman, R.E. and Lorber, B. 1980. Listeriosis in adults: a changing pattern. Report of eight cases and review of the literature, 1968–1978. *Rev Infect Dis*, **2**, 207–27.

NNIS. 2000. National Nosocomial Infections Surveillance (NNIS) system report, data summary from January 1992–April 2000, issued June 2000. *Am J Infect Control* **28**, 429–48.

NNIS. 2003. National vital statistics reports, Vol. 49, No. 12. Available at: http://www.cdc.gov/ nchs/data/nvsr/nvsr49/nvsr49_12.pdf. Accessed April 10, 2003.

O'Grady, N.P., Barie, P.S., et al. 1998. Practice guidelines for evaluating new fever in critically ill adult patients. Task Force of the Society of Critical Care Medicine and the Infectious Diseases Society of America. *Clin Infect Dis*, **26**, 1042–59.

O'Hara, C.M., Weinstein, M.P. and Miller, J.M. 2003. Manual and automated systems for detection and identification of microorganisms. In: Murray, P.R., Baron, E.J., et al. (eds), *Manual of clinical microbiology*. Washington DC: ASM Press, 185–207.

Olaison, L. and Schadewitz, K. Swedish Society of Infectious Diseases Quality Assurance Study Group for Endocarditis, 2002. Enterococcal endocarditis in Sweden, 1995–1999: can shorter therapy with aminoglycosides be used? *Clin Infect Dis*, **34**, 159–66.

Olaison, L. and Pettersson, G. 2003. Current best practices and guidelines. Indications for surgical intervention in infective endocarditis. *Cardiol Clin*, **21**, 235–51.

Opal, S.M., Cross, A., et al. 1988. Group B streptococcal sepsis in adults and infants. Contrasts and comparisons. *Arch Intern Med*, **148**, 641–5.

Osler, W. 1885. The Gulstonian lectures on malignant endocarditis. *Br Med J* **1**, 467–70, 522–26 and 577–79.

Ostrowsky, B.E., Whitener, C., et al. 2002. *Serratia marcescens* bacteremia traced to an infused narcotic. *N Engl J Med*, **346**, 1529–37.

Ozturk, R., Mert, A., et al. 2002. The diagnosis of brucellosis by use of BACTEC 9240 blood culture system. *Diagn Microbiol Infect Dis*, **44**, 133–5.

Parker, M.T. and Ball, L.C. 1976. Streptococci and aerococci associated with systemic infection in man. *J Med Microbiol*, **9**, 275–302.

Peacock, S.J., Bowler, I.C. and Crook, D.W. 1995. Positive predictive value of blood cultures growing coagulase-negative staphylococci. *Lancet*, **346**, 191–2.

Perez-Vazquez, A., Farinas, M.C., et al. 2000. Evaluation of the Duke criteria in 93 episodes of prosthetic valve endocarditis: could sensitivity be improved? *Arch Intern Med*, **160**, 1185–91.

Pergola, V., Di Salvo, G., et al. 2001. Comparison of clinical and echocardiographic characteristics of *Streptococcus bovis* endocarditis with that caused by other pathogens. *Am J Cardiol*, **88**, 871–5.

Pesanti, E.L. and Smith, I.M. 1979. Infective endocarditis with negative blood cultures. An analysis of 52 cases. *Am J Med*, **66**, 43–50.

Peters, N.S., Eykyn, S.J. and Rudd, A.G. 1989. Pneumococcal cellulitis: a rare manifestation of pneumococcaemia in adults. *J Infect*, **19**, 57–9.

Peterson, L.R., Shanholtzer, C.J., et al. 1983. Improved recovery of microorganisms from patients receiving antibiotics with the antimicrobial removal device. *Am J Clin Pathol*, **80**, 692–6.

Petti, C.A. and Fowler, V.G. Jr. 2003. *Staphylococcus aureus* bacteremia and endocarditis. *Cardiol Clin*, **21**, 219–33.

Pfaller, M.A., Jones, R.N., The SCOPE Hospital Study Group, et al. 1997. Nosocomial streptococcal blood stream infections in the SCOPE Program: species occurrence and antimicrobial resistance. *Diagn Microbiol Infect Dis*, **29**, 259–63.

Pfyffer, G.E., Cieslak, C., et al. 1997. Rapid detection of mycobacteria in clinical specimens by using the automated BACTEC 9000 MB system and comparison with radiometric and solid-culture systems. *J Clin Microbiol*, **35**, 2229–34.

Pfyffer, G.E., Brown-Elliot, B.A., et al. 2003. *Mycobacterium*: general characteristics, isolation, and staining procedures. In: Murray, P.R., Baron, E.J., et al. (eds), *Manual of clinical microbiology*. Washington DC: ASM Press, 532–59.

Pierrotti, L.C. and Baddour, L.M. 2002. Fungal endocarditis, 1995–2000. *Chest*, **122**, 302–10.

Piersimoni, C., Scarparo, C., et al. 2001. Comparison of MB/Bact alert 3D system with radiometric BACTEC system and Lowenstein–Jensen medium for recovery and identification of mycobacteria from clinical specimens: a multicenter study. *J Clin Microbiol*, **39**, 651–7.

Pintado, V., Fortun, J., et al. 2001. *Mycobacterium kansasii* pericarditis as a presentation of AIDS. *Infection*, **29**, 48–50.

Pittet, D., Li, N. and Wenzel, R.P. 1993. Association of secondary and polymicrobial nosocomial bloodstream infections with higher mortality. *Eur J Clin Microbiol Infect Dis*, **12**, 813–19.

Pittet, D., Tarara, D. and Wenzel, R.P. 1994. Nosocomial bloodstream infection in critically ill patients. Excess length of stay, extra costs, and attributable mortality. *JAMA*, **271**, 1598–601.

Pittet, D. and Wenzel, R.P. 1995. Nosocomial bloodstream infections. Secular trends in rates, mortality, and contribution to total hospital deaths. *Arch Intern Med*, **155**, 1177–84.

Pittet, D., Thievent, B., et al. 1996. Bedside prediction of mortality from bacteremic sepsis. A dynamic analysis of ICU patients. *Am J Respir Crit Care Med*, **153**, 684–93.

Pittet, D., Li, N., et al. 1997. Microbiological factors influencing the outcome of nosocomial bloodstream infections: a 6-year validated, population-based model. *Clin Infect Dis*, **24**, 1068–78.

Plorde, J.J., Tenover, F.C. and Carlson, L.G. 1985. Specimen volume versus yield in the BACTEC blood culture system. *J Clin Microbiol*, **22**, 292–5.

Popat, K., Barnardo, D. and Webb-Peploe, M. 1980. *Mycoplasma pneumoniae* endocarditis. *Br Heart J*, **44**, 111–12.

Powderly, W.G., Stanley, S.L. Jr. and Medoff, G. 1986. Pneumococcal endocarditis: report of a series and review of the literature. *Rev Infect Dis*, **8**, 786–91.

Prag, J., Nir, M., et al. 1991. Should aerobic blood cultures be shaken intermittently or continuously? *APMIS*, **99**, 1078–82.

Prevost-Smith, E. and Hutton, N. 1994. Improved detection of *Cryptococcus neoformans* in the BACTEC NR 660 blood culture system. *Am J Clin Pathol*, **102**, 741–5.

Proctor, R.A., Kahl, B., et al. 1998. Staphylococcal small colony variants have novel mechanisms for antibiotic resistance. *Clin Infect Dis*, **27**, Suppl. 1, S68–74.

Pujol, M., Pena, C., et al. 1996. Nosocomial *Staphylococcus aureus* bacteremia among nasal carriers of methicillin-resistant and methicillin-susceptible strains. *Am J Med*, **100**, 509–16.

Rahav, G., Toledano, Y., et al. 1997. Invasive pneumococcal infections. A comparison between adults and children. *Medicine (Balt)*, **76**, 295–303.

Ramirez-Ronda, C.H. 1978. Adherence of glucan-positive and glucan-negative streptococcal strains to normal and damaged heart valves. *J Clin Invest*, **62**, 805–14.

Rangel-Frausto, S.M., Pittet, D., et al. 1995. The natural history of the systemic inflammatory response syndrome (SIRS) – A prospective study. *JAMA*, **273**, 117–23.

Raoult, D., Fournier, P.E., et al. 1996. Diagnosis of 22 new cases of *Bartonella endocarditis*. *Ann Intern Med*, **125**, 646–52.

Raoult, D., Fournier, P.E., et al. 2003. Outcome and treatment of *Bartonella endocarditis*. *Arch Intern Med*, **163**, 226–30.

Raymond, J., Bergeret, M., et al. 1995. Neonatal infection with *Streptococcus milleri*. *Eur J Clin Microbiol Infect Dis*, **14**, 799–801.

Raz, R., Elhanan, G., Israeli Adult Pneumococcal Bacteremia Group, et al. 1997. Pneumococcal bacteremia in hospitalized Israeli adults: epidemiology and resistance to penicillin. *Clin Infect Dis*, **24**, 1164–8.

Redondo, M.C., Arbo, M.D., et al. 1995. Attributable mortality of bacteremia associated with the *Bacteroides fragilis* group. *Clin Infect Dis*, **20**, 1492–6.

Reimer, L.G. and Reller, L.B. 1985. Effect of sodium polyanetholesulfonate and gelatin on the recovery of *Gardnerella vaginalis* from blood culture media. *J Clin Microbiol*, **21**, 686–8.

Reimer, L.G., Reller, L.B., et al. 1987. Controlled evaluation of trypticase soy broth with and without gelatin and yeast extract in the detection of bacteremia and fungemia. *Diagn Microbiol Infect Dis*, **8**, 19–24.

Reimer, L.G., Wilson, M.L. and Weinstein, M.P. 1997. Update on detection of bacteremia and fungemia. *Clin Microbiol Rev*, **10**, 444–65.

Reller, L.B., Murray, P. and MacLowery, J. 1982. Cumitech 1A. In: Washington II, J. (ed.), *Blood cultures II*. Washington DC: American Society of Microbiology, 1–11.

Rello, J., Torres, A., et al. 1994. Ventilator-associated pneumonia by *Staphylococcus aureus*. Comparison of methicillin-resistant and methicillin-sensitive episodes. *Am J Respir Crit Care Med*, **150**, 1545–9.

Renaud, B. and Brun-Buisson, C. 2001. Outcomes of primary and catheter-related bacteremia. A cohort and case-control study in critically ill patients. *Am J Respir Crit Care Med*, **163**, 1584–90.

Rex, J.H. 1996. Editorial response: catheters and candidemia. *Clin Infect Dis*, **22**, 467–70.

Rex, J.H., Bennett, J.E., Candidemia Study Group and the National Institute, et al. 1994. A randomized trial comparing fluconazole with amphotericin B for the treatment of candidemia in patients without neutropenia. *N Engl J Med*, **331**, 1325–30.

Rice, L.B., Calderwood, S.B., et al. 1991. Enterococcal endocarditis: a comparison of prosthetic and native valve disease. *Rev Infect Dis*, **13**, 1–7.

Richard, P., Amador, D.V., et al. 1995. Viridans streptococcal bacteraemia in patients with neutropenia. *Lancet*, **345**, 1607–9.

Richardson, D.C., Burrows, L.L., et al. 2003. *Tropheryma whippelii* as a cause of culture-negative endocarditis: the evolving spectrum of Whipple's disease. *J Infect*, **47**, 170–3.

Rigilano, J., Mahapatra, R., et al. 1984. Enterococcal endocarditis following sigmoidoscopy and mitral valve prolapse. *Arch Intern Med*, **144**, 850–1.

Rijnders, B.J., Verwaest, C., et al. 2001. Difference in time to positivity of hub-blood versus nonhub-blood cultures is not useful for the diagnosis of catheter-related bloodstream infection in critically ill patients. *Crit Care Med*, **29**, 1399–403.

Roberts, F.J., Geere, I.W. and Coldman, A. 1991. A three-year study of positive blood cultures, with emphasis on prognosis. *Rev Infect Dis*, **13**, 34–46.

Roder, B.L., Wandall, D.A., et al. 1997. Neurologic manifestations in *Staphylococcus aureus* endocarditis: a review of 260 bacteremic cases in nondrug addicts. *Am J Med*, **102**, 379–86.

Roder, B.L., Wandall, D.A., et al. 1999. Clinical features of *Staphylococcus aureus* endocarditis: a 10-year experience in Denmark. *Arch Intern Med*, **159**, 462–9.

Roe, M.T., Abramson, M.A., et al. 2000. Clinical information determines the impact of transesophageal echocardiography on the diagnosis of infective endocarditis by the Duke criteria. *Am Heart J*, **139**, 945–51.

Rohner, P., Pepey, B. and Auckenthaler, R. 1996. Comparative evaluation of BACTEC aerobic Plus/F and Septi-Chek Release blood culture media. *J Clin Microbiol*, **34**, 126–9.

Romero-Vivas, J., Rubio, M., et al. 1995. Mortality associated with nosocomial bacteremia due to methicillin-resistant *Staphylococcus aureus*. *Clin Infect Dis*, **21**, 1417–23.

Ross, M.J., Sakoulas, G., et al. 2001. *Corynebacterium jeikeium* native valve endocarditis following femoral access for coronary angiography. *Clin Infect Dis*, **32**, E120–1.

Rotun, S.S., McMath, V., et al. 1999. *Staphylococcus aureus* with reduced susceptibility to vancomycin isolated from a patient with fatal bacteremia. *Emerg Infect Dis*, **5**, 147–9.

Rubinstein, E., Noriega, E.R., et al. 1975. Fungal endocarditis: Analysis of 24 cases and review of the literature. *Medicine (Balt)*, **54**, 331–4.

Rudnick, J.R., Beck-Sague, C.M., et al. 1996. Gram-negative bacteremia in open-heart-surgery patients traced to probable tap-water contamination of pressure-monitoring equipment. *Infect Control Hosp Epidemiol*, **17**, 281–5.

Rupp, M.E. 1992. *Streptobacillus moniliformis* endocarditis: Case report and review. *Clin Infect Dis*, **14**, 769–72.

Rupp, M.E. and Archer, G.L. 1994. Coagulase-negative staphylococci: pathogens associated with medical progress. *Clin Infect Dis*, **19**, 231–43.

Rutledge, R., Kim, B.J. and Applebaum, R.E. 1985. Actuarial analysis of the risk of prosthetic valve endocarditis in 1,598 patients with mechanical and bioprosthetic valves. *Arch Surg*, **120**, 469–72.

Said-Salim, B., Mathema, B. and Kreiswirth, B.N. 2003. Community-acquired methicillin-resistant *Staphylococcus aureus*: an emerging pathogen. *Infect Control Hosp Epidemiol*, **24**, 451–5.

Saksouk, F. and Salti, I.S. 1977. Acute suppurative thyroiditis caused by *Escherichia coli*. *Br Med J*, **2**, 23–4.

Salavert, M., Gomez, L., et al. 1996. Seven-year review of bacteremia caused by *Streptococcus milleri* and other viridans streptococci. *Eur J Clin Microbiol Infect Dis*, **15**, 365–71.

Sambola, A., Miro, J.M., et al. 2002. *Streptococcus agalactiae* infective endocarditis: analysis of 30 cases and review of the literature, 1962–1998. *Clin Infect Dis*, **34**, 1576–84.

Sanabria, T.J., Alpert, J.S., et al. 1990. Increasing frequency of staphylococcal infective endocarditis. Experience at a university hospital, 1981 through 1988. *Arch Intern Med*, **150**, 1305–9.

Sandoe, J.A., Kerr, K.G., et al. 1999. *Staphylococcus capitis* endocarditis: two cases and review of the literature. *Heart*, **82**, e1.

Schaberg, D.R., Culver, D.H. and Gaynes, R.P. 1991. Major trends in the microbial etiology of nosocomial infection. *Am J Med*, **91**, 72S–5S.

Scheld, W.M., Valone, J.A. and Sande, M.A. 1978. Bacterial adherence in the pathogenesis of endocarditis. Interaction of bacterial dextran, platelets, and fibrin. *J Clin Invest*, **61**, 1394–404.

Schlaeffer, F., Riesenberg, K., et al. 1996. Serious bacterial infections after endoscopic procedures. *Arch Intern Med*, **156**, 572–4.

Schneider, P.J., Nernoff, J. and Gold, J.A. 1967. Acute salmonella endocarditis. Report of a case and review. *Arch Intern Med*, **120**, 478–86.

Schonheyder, H.C. and Hojbjerg, T. 1995. The impact of the first notification of positive blood cultures on antibiotic therapy. A one-year survey. *APMIS*, **103**, 37–44.

Schuchat, A. 1998. Epidemiology of group B streptococcal disease in the United States: shifting paradigms. *Clin Microbiol Rev*, **11**, 497–513.

Schuchat, A., Oxtoby, M., et al. 1990. Population-based risk factors for neonatal group B streptococcal disease: results of a cohort study in metropolitan Atlanta. *J Infect Dis*, **162**, 672–7.

Schuchat, A., Swaminathan, B. and Broome, C.V. 1991. Epidemiology of human listeriosis. *Clin Microbiol Rev*, **4**, 169–83.

Schwartz, B., Schuchat, A., et al. 1991. Invasive group B streptococcal disease in adults. A population-based study in metropolitan Atlanta. *JAMA*, **266**, 1112–14.

Schwenzer, K.J., Gist, A. and Durbin, C.G. 1994. Can bacteremia be predicted in surgical intensive care unit patients? *Intensive Care Med*, **20**, 425–30.

Scully, B.E., Spriggs, D. and Neu, H.C. 1987. *Streptococcus agalactiae* (group B) endocarditis – a description of twelve cases and review of the literature. *Infection*, **15**, 169–76.

Seenivasan, M.H. and Yu, V.L. 2003. *Staphylococcus lugdunensis* endocarditis – the hidden peril of coagulase-negative Staphylococcus in blood cultures. *Eur J Clin Microbiol Infect Dis*, **22**, 489–91.

Seghatol, F. and Grinberg, I. 2002. Left-sided endocarditis in intravenous drug users: a case report and review of the literature. *Echocardiography*, **19**, 509–11.

Segreti, J. 1999. Is antibiotic prophylaxis necessary for preventing prosthetic device infection? *Infect Dis Clin North Am*, **13**, 871–7.

Seifert, H., Strate, A., et al. 1994. Bacteremia due to *Acinetobacter* species other than *Acinetobacter baumannii*. *Infection*, **22**, 379–85.

Seifert, H., Strate, A. and Pulverer, G. 1995. Nosocomial bacteremia due to *Acinetobacter baumannii*. Clinical features, epidemiology, and predictors of mortality. *Medicine (Balt)*, **74**, 340–9.

Seifert, H., Shah, P., et al. 1997. MiQ 3: Sepsis – Blutkulturdiagnostik. In: Mauch, H., Lüttiken, R. and Gatermann, S. (eds), *MiQ Qualitätsstandards in der mikrobiologisch-infektiologischen Diagnostik*. Stuttgart, Jena, Lübeck, Ulm: Gustav Fischer Verlag.

Seifert, H., von Eiff, C. and Fatkenheuer, G. 1999. Fatal case due to methicillin-resistant *Staphylococcus aureus* small colony variants in an AIDS patient. *Emerg Infect Dis*, **5**, 450–3.

Seifert, H., Cornely, O., et al. 2003a. Bloodstream infection in neutropenic cancer patients related to short-term nontunnelled catheters determined by quantitative blood cultures, differential time to positivity, and molecular epidemiological typing with pulsed-field gel electrophoresis. *J Clin Microbiol*, **41**, 118–23.

Seifert, H., Wisplinghoff, H., et al. 2003b. Small colony variants (SCVs) of *Staphylococcus aureus* as a cause of pacemaker-related infection. *Emerg Infect Dis*, **9**, 10, 1316–18.

Shafazand, S. and Weinacker, A.B. 2002. Blood cultures in the critical care unit: improving utilization and yield. *Chest*, **122**, 1727–36.

Shaked, Y. and Samra, Y. 1988. Primary pneumococcal peritonitis in patients with cardiac ascites: report of 2 cases. *Cardiology*, **75**, 372–4.

Sharp, S.E., Goodman, J.M. and Poppiti, R.J. Jr. 1991. Comparison of blood-culture results after five and seven days of incubation using the BACTEC NR660. *Diagn Microbiol Infect Dis*, **14**, 177–9.

Sharp, S.E., McLaughlin, J.C., et al. 1993. Clinical assessment of anaerobic isolates from blood cultures. *Diagn Microbiol Infect Dis*, **17**, 19–22.

Shively, B.K., Gurule, F.T., et al. 1991. Diagnostic value of transesophageal compared with transthoracic echocardiography in infective endocarditis. *J Am Coll Cardiol*, **18**, 391–7.

Shlaes, D.M., Lerner, P.I., et al. 1981a. Infections due to Lancefield group F and related Streptococci (*S. milleri*, *S. anginosus*). *Medicine (Balt)*, **60**, 197–207.

Shlaes, D.M., Levy, J. and Wolinsky, E. 1981b. Enterococcal bacteremia without endocarditis. *Arch Intern Med*, **141**, 578–81.

Siau, H., Yuen, K.Y., et al. 1999. *Acinetobacter* bacteremia in Hong Kong: prospective study and review. *Clin Infect Dis*, **28**, 26–30.

Siegman-Igra, Y., Fourer, B., et al. 2002. Reappraisal of community-acquired bacteremia: a proposal of a new classification for the spectrum of acquisition of bacteremia. *Clin Infect Dis*, **34**, 1431–9.

Simmons, N.A. The Endocarditis Working Party for Antimicrobial Chemotherapy, 1993. Recommendations for endocarditis prophylaxis. *J Antimicrob Chemother*, **31**, 437–8.

Singh, V.R. and Raad, I. 1990. Fatal *Staphylococcus saprophyticus* native valve endocarditis in an intravenous drug addict. *J Infect Dis*, **162**, 783–4.

Smith, T.L., Pearson, M.L., Glycopeptide-Intermediate *Staphylococcus aureus* Working Group, et al. 1999. Emergence of vancomycin resistance in *Staphylococcus aureus*. *N Engl J Med*, **340**, 493–501.

Smith-Elekes, S. and Weinstein, M.P. 1993. Blood cultures. *Infect Dis Clin North Am*, **7**, 221–34.

Smyth, E.G., Pallett, A.P. and Davidson, R.N. 1988. Group G streptococcal endocarditis: Two case reports, a review of the literature and recommendations for treatment. *J Infect*, **16**, 169–76.

Solera, J., Lozano, E., et al. 1999. *Brucellar* spondylitis: review of 35 cases and literature survey. *Clin Infect Dis*, **29**, 1440–9.

Solomon, H.M. and Jackson, D. 1992. Rapid diagnosis of *Brucella melitensis* in blood: some operational characteristics of the BACT/ALERT. *J Clin Microbiol*, **30**, 222–4.

Song, X., Srinivasan, A., et al. 2003. Effect of nosocomial vancomycin-resistant enterococcal bacteremia on mortality, length of stay, and costs. *Infect Control Hosp Epidemiol*, **24**, 251–6.

Spach, D.H., Kanter, A.S., et al. 1995. *Bartonella* (*Rochalimaea*) species as a cause of apparent 'culture-negative' endocarditis. *Clin Infect Dis*, **20**, 1044–7.

Sparling, P.F. and Hansfield, H.H. 2000. *Neisseria gonorrhoeae*. In: Mandell, G.L., Bennet, J.E. and Dolin, R. (eds), *Mandell, Dougals, and Bennett's Principles and practice of infectious diseases*. Philadelphia: Churchill Livingstone, 2242–58.

Sperber, S.J. and Schleupner, C.J. 1987. Salmonellosis during infection with human immunodeficiency virus. *Rev Infect Dis*, **9**, 925–34.

Spitalnic, S.J., Woolard, R.H. and Mermel, L.A. 1995. The significance of changing needles when inoculating blood cultures: a meta-analysis. *Clin Infect Dis*, **21**, 1103–6.

Spyrou, N., Anderson, M. and Foale, R. 1997. *Listeria* endocarditis: current management and patient outcome – world literature review. *Heart*, **77**, 380–3.

Staneck, J.L. and Vincent, S. 1981. Inhibition of *Neisseria gonorrhoeae* by sodium polyanetholsulfonate. *J Clin Microbiol*, **13**, 463–7.

Stein, M., Houston, S., et al. 1993. HIV infection and *Salmonella* septic arthritis. *Clin Exp Rheumatol*, **11**, 187–9.

Stern, C.S., Carvalho, M.G. and Teixeira, L.M. 1994. Characterization of enterococci isolated from human and nonhuman sources in Brazil. *Diagn Microbiol Infect Dis*, **20**, 61–7.

Stevens, D.L., Tanner, M.H., et al. 1989. Severe group A streptococcal infections associated with a toxic shock-like syndrome and scarlet fever toxin A. *N Engl J Med*, **321**, 1–7.

Strand, C.L. and Shulman, J.A. 1988. Collection of blood culture specimens. In *Bloodstream infections: Laboratory detection and clinical considerations*. Chicago, IL: American Society of Clinical Pathology, 15–21.

Strand, C.L., Wajsbort, R.R. and Sturmann, K. 1993. Effect of iodophor vs iodine tincture skin preparation on blood culture contamination rate. *JAMA*, **269**, 1004–6.

Sullam, P.M., Valone, F.H. and Mills, J. 1987. Mechanisms of platelet aggregation by viridans group streptococci. *Infect Immun*, **55**, 1743–50.

Sullam, P.M., Bayer, A.S., et al. 1996. Diminished platelet binding in vitro by *Staphylococcus aureus* is associated with reduced virulence in a rabbit model of infective endocarditis. *Infect Immun*, **64**, 4915–21.

Summanen, P., Baron, E.J., et al. 2002. *Wadsworth anaerobic bacteriology manual*, 6th edn. Belmont, CA: Star Publishing.

Szymczak, E.G., Barr, J.T., et al. 1979. Evaluation of blood culture procedures in a pediatric hospital. *J Clin Microbiol*, **9**, 88–92.

Tancik, C.A. and Dillaha, J.A. 2000. *Francisella tularensis* endocarditis. *Clin Infect Dis*, **30**, 399–400.

Tenney, J.H., Reller, L.B., et al. 1982. Controlled evaluation of the volume of blood cultured in detection of bacteremia and fungemia. *J Clin Microbiol*, **15**, 558–61.

Thompson, R.L. 1982. Staphylococcal infective endocarditis. *Mayo Clin Proc*, **57**, 106–14.

Tilley, P.A. and Roberts, F.J. 1994. Bacteremia with *Acinetobacter* species: risk factors and prognosis in different clinical settings. *Clin Infect Dis*, **18**, 896–900.

Tonnesen, A., Peuler, M. and Lockwood, W.R. 1976. Cultures of blood drawn by catheters vs venipuncture. *JAMA*, **235**, 1877.

Torzillo, P.J., Hanna, J.N., et al. 1995. Invasive pneumococcal disease in central Australia. *Med J Aust*, **162**, 182–6.

Traub, W.H. and Kleber, I. 1977. Inactivation of classical and alternative pathway-activated bactericidal activity of human serum by sodium polyanetholsulfonate. *J Clin Microbiol*, **5**, 278–84.

Trick, W.E., Fridkin, S.K., et al. 2002. Secular trend of hospital-acquired candidemia among intensive care unit patients in the United States during 1989–1999. *Clin Infect Dis*, **35**, 627–30.

Tsao, M.M. and Katz, D. 1984. Central venous catheter-induced endocarditis: human correlate of the animal experimental model of endocarditis. *Rev Infect Dis*, **6**, 783–90.

Tunkel, B.R. and Kaye, D. 1992. Endocarditis with negative blood cultures. *N Engl J Med*, **326**, 1215–17.

Turner, L.H. 1970. Leptospirosis. 3. Maintenance, isolation and demonstration of leptospires. *Trans R Soc Trop Med Hyg*, **64**, 623–46.

Vaisanen, I.T., Michelsen, T., et al. 1985. Comparison of arterial and venous blood samples for the diagnosis of bacteremia in critically ill patients. *Crit Care Med*, **13**, 664–7.

Valles, J., Rello, J., et al. 2003. Community-acquired bloodstream infection in critically ill adult patients: impact of shock and inappropriate antibiotic therapy on survival. *Chest*, **123**, 1615–24.

van der Meer, J.T., van Vianen, W., et al. 1991. Distribution, antibiotic susceptibility and tolerance of bacterial isolates in culture-positive cases of endocarditis in The Netherlands. *Eur J Clin Microbiol Infect Dis*, **10**, 728–34.

van der Meer, J.T., Thompson, J., et al. 1992. Epidemiology of bacterial endocarditis in The Netherlands. I. Patient characteristics. *Arch Intern Med*, **152**, 1863–8.

van Deuren, M., Brandtzaeg, P. and van der Meer, J.W. 2000. Update on meningococcal disease with emphasis on pathogenesis and clinical management. *Clin Microbiol Rev*, **13**, 144–66, table.

Vandenesch, F., Etienne, J., et al. 1993. Endocarditis due to *Staphylococcus lugdunensis*: Report of 11 cases and review. *Clin Infect Dis*, **17**, 871–6.

Venditti, M., Gelfusa, V., et al. 1990. *Erysipelothrix rhusiopathiae* endocarditis. *Eur J Clin Microbiol Infect Dis*, **9**, 50–2.

Venezio, F.R., Gullberg, R.M., et al. 1986. Group G streptococcal endocarditis and bacteremia. *Am J Med*, **81**, 29–34.

Vercellotti, G.M., Lussenhop, D., et al. 1984. Bacterial adherence to fibronectin and endothelial cells: A possible mechanism for bacterial tissue tropism. *J Lab Clin Med*, **103**, 34–43.

Verghese, A., Mireault, K. and Arbeit, R.D. 1986. Group B streptococcal bacteremia in men. *Rev Infect Dis*, **8**, 912–17.

Verghese, S.L., Padmaja, P. and Koshi, G. 1998. Central venous catheter related infections in a tertiary care hospital. *J Assoc Phys India*, **46**, 445–7.

Vergis, E.N., Hayden, M.K., et al. 2001. Determinants of vancomycin resistance and mortality rates in enterococcal bacteremia. A prospective multicenter study. *Ann Intern Med*, **135**, 484–92.

Vincent, J.L., de Mendonca, A., Working group on 'sepsis-related problems' of the European Society of Intensive Care Medicine, et al. 1998. Use of the SOFA score to assess the incidence of organ dysfunction/failure in intensive care units: results of a multicenter, prospective study. *Crit Care Med*, **26**, 1793–800.

Vincent, J.L. 2002. Sepsis definitions. *Lancet Infect Dis*, **2**, 135.

Vincent, J.L. and Jacobs, F. 2003. Infection in critically ill patients: clinical impact and management. *Curr Opin Infect Dis*, **16**, 309–13.

Vlessis, A.A., Khaki, A., et al. 1997. Risk, diagnosis and management of prosthetic valve endocarditis: a review. *J Heart Valve Dis*, **6**, 443–65.

von Eiff, C., Bettin, D., et al. 1997. Recovery of small colony variants of *Staphylococcus aureus* following gentamicin bead placement for osteomyelitis. *Clin Infect Dis*, **25**, 1250–1.

Von Reyn, C.F., Levy, B.S., et al. 1981. Infective endocarditis: An analysis based on strict case definitions. *Ann Intern Med*, **94**, 505–18.

Waisberg, J., de Matheus, C.O. and Pimenta, J. 2002. Infectious endocarditis from *Streptococcus bovis* associated with colonic carcinoma: case report and literature review. *Arq Gastroenterol*, **39**, 177–80.

Waite, R.T. and Woods, G.L. 1998. Evaluation of BACTEC MYCO/F lytic medium for recovery of mycobacteria and fungi from blood. *J Clin Microbiol*, **36**, 1176–9.

Wallace, S.M., Walton, B.I., et al. 2002. Mortality from infective endocarditis: clinical predictors of outcome. *Heart*, **88**, 53–60.

Wara, D.W. 1981. Host defense against *Streptococcus pneumoniae*: the role of the spleen. *Rev Infect Dis*, **3**, 299–309.

Warren, C.P. 1970. Arthritis associated with *Salmonella* infections. *Ann Rheum Dis*, **29**, 483–7.

Warren, D.K., Zack, J.E., et al. 2001. Nosocomial primary bloodstream infections in intensive care unit patients in a nonteaching community medical center: a 21-month prospective study. *Clin Infect Dis*, **33**, 1329–35.

Washington, J.A. 1975. Blood cultures: principles and techniques. *Mayo Clin Proc*, **50**, 91–8.

Washington, J.A. 1982. The role of the microbiology laboratory in the diagnosis and antimicrobial treatment of infective endocarditis. *Mayo Clin Proc*, **57**, 22–32.

Washington, J.A. 1987. The microbiological diagnosis of infective endocarditis. *J Antimicrob Chemother*, **20**, Suppl. A, 29–39.

Washington, J.A. 1994. Collection, transport, and processing of blood cultures. *Clin Lab Med*, **14**, 59–68.

Washington, J.A. and Ilstrup, D.M. 1986. Blood cultures: issues and controversies. *Rev Infect Dis*, **8**, 792–802.

Watanakunakorn, C. 1977. Changing epidemiology and newer aspects of infective endocarditis. *Adv Intern Med*, **22**, 21–47.

Watanakunakorn, C. and Burkert, T. 1993. Infective endocarditis at a large community teaching hospital, 1980–1990. *Medicine*, **72**, 90–102.

Watanakunakorn, C. and Bailey, T.A. 1997. Adult bacteremic pneumococcal pneumonia in a community teaching hospital, 1992–1996. A detailed analysis of 108 cases. *Arch Intern Med*, **157**, 1965–71.

Watanakunakorn, C., Tan, J.S. and Phair, J.P. 1973. Some salient features of *Staphylococcus aureus* endocarditis. *Am J Med*, **54**, 473–81.

Watkin, R.W., Lang, S., et al. 2003. The microbial diagnosis of infective endocarditis. *J Infect*, **47**, 1–11.

Weinstein, L. and Schlesinger, J.J. 1974. Pathoanatomic, pathophysiologic and clinical correlations in endocarditis (first of two parts). *N Engl J Med*, **291**, 832–7.

Weinstein, M.P. 1996. Current blood culture methods and systems: clinical concepts, technology, and interpretation of results. *Clin Infect Dis*, **23**, 40–6.

Weinstein, M.P., Murphy, J.R., et al. 1983a. The clinical significance of positive blood cultures: a comprehensive analysis of 500 episodes of bacteremia and fungemia in adults. II. Clinical observations, with special reference to factors influencing prognosis. *Rev Infect Dis*, **5**, 54–70.

Weinstein, M.P., Reller, L.B., et al. 1983b. The clinical significance of positive blood cultures: a comprehensive analysis of 500 episodes of bacteremia and fungemia in adults. I. Laboratory and epidemiologic observations. *Rev Infect Dis*, **5**, 35–53.

Weinstein, M.P., Mirrett, S., et al. 1989. Effect of agitation and terminal subcultures on yield and speed of detection of the Oxoid Signal blood culture system versus the BACTEC radiometric system. *J Clin Microbiol*, **27**, 427–30.

Weinstein, M.P., Mirrett, S., et al. 1995. Controlled evaluation of BacT/Alert standard aerobic and FAN aerobic blood culture bottles for detection of bacteremia and fungemia. *J Clin Microbiol*, **33**, 978–81.

Weinstein, M.P., Towns, M.L., et al. 1997. The clinical significance of positive blood cultures in the 1990s: a prospective comprehensive evaluation of the microbiology, epidemiology, and outcome of bacteremia and fungemia in adults. *Clin Infect Dis*, **24**, 584–602.

Weisman, S.J., Scoopo, F.J., et al. 1990. Septicemia in pediatric oncology patients: the significance of viridans streptococcal infections. *J Clin Oncol*, **8**, 453–9.

Wenzel, R.P. and Edmond, M.B. 2001. The impact of hospital-acquired bloodstream infections. *Emerg Infect Dis*, **7**, 174–7.

Werner, A.S., Cobbs, C.G., et al. 1967. Studies on the bacteremia of bacterial endocarditis. *JAMA*, **202**, 199–203.

Werner, M., Andersson, R., et al. 2003. A clinical study of culture-negative endocarditis. *Medicine (Balt)*, **82**, 263–73.

Wheat, L.J., Rubin, R.H., et al. 1987. Systemic salmonellosis in patients with disseminated histoplasmosis. Case for 'macrophage blockade' caused by *Histoplasma capsulatum*. *Arch Intern Med*, **147**, 561–4.

Whitener, C., Caputo, G.M., et al. 1993. Endocarditis due to coagulase-negative staphylococci. Microbiologic, epidemiologic, and clinical considerations. *Infect Dis Clin North Am*, **7**, 81–96.

Whyte, T., Hanahoe, B., et al. 2000. Evaluation of the BACTEC MGIT 960 and MB BAC/T systems for routine detection of *Mycobacterium tuberculosis*. *J Clin Microbiol*, **38**, 3131–2.

Wieland, M., Lederman, M.M., et al. 1986. Left-sided endocarditis due to *Pseudomonas aeruginosa*. A report of 10 cases and review of the literature. *Medicine (Balt)*, **65**, 180–9.

Wilcox, C.M. and Dismukes, W.E. 1987. Spontaneous bacterial peritonitis. A review of pathogenesis, diagnosis, and treatment. *Medicine (Balt)*, **66**, 447–56.

Wilson, L.E., Thomas, D.L., et al. 2002. Prospective study of infective endocarditis among injection drug users. *J Infect Dis*, **185**, 1761–6.

Wilson, M.L., Weinstein, M.P., et al. 1992. Controlled comparison of the BacT/Alert and BACTEC 660/730 nonradiometric blood culture systems. *J Clin Microbiol*, **30**, 323–9.

Wilson, M.L., Mirrett, S., et al. 1993. Recovery of clinically important microorganisms from the BacT/Alert blood culture system does not require testing for seven days. *Diagn Microbiol Infect Dis*, **16**, 31–4.

Wilson, M.L., Weinstein, M.P. and Reller, L.B. 1994. Automated blood culture systems. *Clin Lab Med*, **14**, 149–69.

Wilson, R. and Feldman, R.A. From the Centers for Disease Control, 1981. Reported isolates of *Salmonella* from cerebrospinal fluid in the United States, 1968–1979. *J Infect Dis*, **143**, 504–6.

Wilson, W.R., Martin, W.J., et al. 1972. Anaerobic bacteremia. *Mayo Clin Proc*, **47**, 639–46.

Wilson, W.R., Karchmer, A.W., et al. 1995. Antibiotic treatment of adults with infective endocarditis due to streptococci, enterococci, staphylococci, and HACEK microorganisms. American Heart Association. *JAMA*, **274**, 1706–13.

Wisplinghoff, H., Reinert, R.R., et al. 1999. Molecular relationships and antimicrobial susceptibilities of viridans group streptococci isolated from blood of neutropenic cancer patients. *J Clin Microbiol*, **37**, 1876–80.

Wisplinghoff, H., Edmond, M.B., et al. 2000. Nosocomial bloodstream infections caused by *Acinetobacter* species in United States hospitals: clinical features, molecular epidemiology, and antimicrobial susceptibility. *Clin Infect Dis*, **31**, 690–7.

Wisplinghoff, H., Seifert, H., et al. 2001. Systemic inflammatory response syndrome in adult patients with nosocomial bloodstream infection due to *Staphylococcus aureus*. *Clin Infect Dis*, **33**, 733–6.

Wisplinghoff, H., Bischoff, T., et al. 2004. Nosocomial bloodstream infections in United States hospitals: analysis of 24,000 cases from a prospective nationwide surveillance study. *Clin Infect Dis*, **39**, 309–17.

Wolfe, M.S., Louria, D.B., et al. 1971. Salmonellosis in patients with neoplastic disease. A review of 100 episodes at Memorial Cancer Center over a 13-year period. *Arch Intern Med*, **128**, 546–54.

Wolff, M., Witchitz, S., et al. 1995. Prosthetic valve endocarditis in the ICU. Prognostic factors of overall survival in a series of 122 cases and consequences for treatment decision. *Chest*, **108**, 688–94.

Wong, C.A., Donald, F. and Macfarlane, J.T. 1995. *Streptococcus milleri* pulmonary disease: a review and clinical description of 25 patients. *Thorax*, **50**, 1093–6.

Wong, S.S., Ng, T.K., et al. 2000. Bacteremia due to *Staphylococcus aureus* with reduced susceptibility to vancomycin. *Diagn Microbiol Infect Dis*, **36**, 261–8.

Wong, W.Y., Overturf, G.D. and Powars, D.R. 1992. Infection caused by *Streptococcus pneumoniae* in children with sickle cell disease: epidemiology, immunologic mechanisms, prophylaxis, and vaccination. *Clin Infect Dis*, **14**, 1124–36.

Woods, G.L., Fish, G., et al. 1997. Clinical evaluation of Difco ESP culture system II for growth and detection of mycobacteria. *J Clin Microbiol*, **35**, 121–4.

Working Party of the British Society for Antimicrobial Chemotherapy. 1998. Antibiotic treatment of streptococcal, enterococcal, and staphylococcal endocarditis. *Heart* **79**, 207–10.

Wu, K.C., Chiu, H.H., et al. 2002. Characteristics of community-acquired methicillin-resistant *Staphylococcus aureus* in infants and children without known risk factors. *J Microbiol Immunol Infect*, **35**, 53–6.

Wyler, D.J., Glickman, M.G. and Brewin, A. 1975. Persistent *Serratia* bacteremia associated with drug abuse. *West J Med*, **122**, 70–3.

Yagupsky, P. 1999. Detection of brucellae in blood cultures. *J Clin Microbiol*, **37**, 3437–42.

Yagupsky, P., Menegus, M.A. and Powell, K.R. 1991. The changing spectrum of group B streptococcal disease in infants: an eleven-year experience in a tertiary care hospital. *Pediatr Infect Dis J*, **10**, 801–8.

Yagupsky, P., Peled, N., et al. 1997. Comparison of BACTEC 9240 Peds Plus medium and isolator 1.5 microbial tube for detection of Brucella melitensis from blood cultures. *J Clin Microbiol*, **35**, 1382–4.

Yang, Y.J., Huang, M.C., et al. 2002. Analysis of risk factors for bacteremia in children with nontyphoidal *Salmonella* gastroenteritis. *Eur J Clin Microbiol Infect Dis*, **21**, 290–3.

Yao, L., Bengualid, V., et al. 1995. Internalization of *Staphylococcus aureus* by endothelial cells induces cytokine gene expression. *Infect Immun*, **63**, 1835–9.

Yilmaz, M.B., Kisacik, H.L. and Korkmaz, S. 2003. Persisting fever in a patient with brucella endocarditis: occult splenic abscess. *Heart*, **89**, e20.

Young, E.J. 1995. An overview of human brucellosis. *Clin Infect Dis*, **21**, 283–9.

Yu, V.L., Chiou, C.C., et al. 2003. An international prospective study of pneumococcal bacteremia: correlation with in vitro resistance, antibiotics administered, and clinical outcome. *Clin Infect Dis*, **37**, 230–7.

Zaidi, A.K., Knaut, A.L., et al. 1995. Value of routine anaerobic blood cultures for pediatric patients. *J Pediatr*, **127**, 263–8.

Zaidi, E., Bachur, R. and Harper, M. 1999. Non-typhi *Salmonella* bacteremia in children. *Pediatr Infect Dis J*, **18**, 1073–7.

Zimhony, O., Goland, S., et al. 1996. Enterococcal endocarditis after extracorporeal shock wave lithotripsy for nephrolithiasis. *Postgrad Med J*, **72**, 51–2.

20

Bacterial meningitis

KEITH A.V. CARTWRIGHT

INTRODUCTION

More than 2 000 cases of bacterial meningitis, with at least 150 deaths, are notified each year in the UK and similar incidence data are reported by many other countries. The emphasis in this chapter will be on the epidemiological, pathophysiological, and diagnostic aspects of the disease. Fuller accounts of the key organisms can be found in Chapters 33, *Streptococcus* and *Lactobacillus*; 37, *Listeria*; 48, *Neisseria*; and 65, *Haemophilus*. The various clinical syndromes are well described in Lambert (1991)).

In meningitis, the membranes that surround the brain and spinal cord become inflamed. Most meningitis is viral, usually with mild illness, but viral and bacterial meningitis may be confused (Maxson and Jacobs 1993).

The commonest cause of bacterial meningitis worldwide is *Neisseria meningitidis*, the meningococcus. In unimmunized populations, capsulate strains of *Haemophilus influenzae* type b (Hib) may cause as many, or more cases, especially in young children. *Streptococcus pneumoniae* and *Mycobacterium tuberculosis* rank next in order of frequency. Many other microbes and organisms, including bacteria, fungi, protozoa, and worms, may also cause meningitis or meningeal inflammation.

In neonatal meningitis, group B streptococci and *Escherichia coli* predominate, whereas *Listeria monocytogenes* causes meningitis in neonates, during pregnancy, and in old age. Fungal meningitis is particularly associated with severe immunosuppression as in the late stages of human immunodeficiency virus (HIV) infection. Many other bacteria, including staphylococci (Jensen et al. 1993), coliforms, and environmental gram-negative bacilli such as *Pseudomonas aeruginosa* and *Acinetobacter* spp. (Siegman-Ingra et al. 1993), occasionally cause meningitis, often in associations with inpatient hospital care, severe immunosuppression, neurosurgery, cranial trauma, infection of cerebrospinal fluid (CSF) reservoirs or shunts.

In economically advanced countries, the mortality from bacterial meningitis is less than 10 percent but it may be 30 percent or more in developing countries (Greenwood 1987; Bryan et al. 1990; Bijlmer 1991). The epidemiology is changing swiftly following the development of conjugated polysaccharide vaccines for invasive Hib disease, for meningococcal meningitis caused by serogroups A, C, Y, and W-135, and for pneumococcal meningitis caused by a range of common serotypes. In the UK, following the control of serogroup C meningococcal disease through vaccination, there are now good prospects for further reductions in the incidence of bacterial meningitis in the near future, particularly if serogroup B meningococcal vaccines can be developed, but also if the incidence of pneumococcal meningitis can be reduced through vaccination (Figure 20.1).

PATHOPHYSIOLOGY

The evolution of bacterial meningitis follows a sequence of exposure to the pathogen, acquisition, and then invasion. Invasion can be subdivided into three processes: mucosal penetration is followed by invasion of the

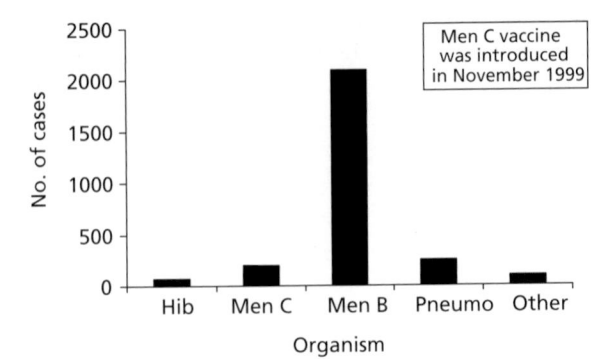

Figure 20.1 *Estimated incidence of bacterial meningitis and meningococcal disease in the UK, end 2002.*

bloodstream and then invasion of the meninges (Tunkel and Scheld 1993). The human nasopharynx is the natural habitat of the meningococcus, Hib, and the pneumococcus. Exposure is more frequent than acquisition, which is more frequent than bloodstream invasion. Understanding the process of establishment of bacterial meningitis has been greatly assisted by the development of animal models of infection (O'Donoghue et al. 1974; Moxon et al. 1977; Tauber and Zwahlen 1994).

Exposure

Exposure depends on the frequency and intimacy of interactions between colonized and susceptible individuals, the capacity of colonized individuals to disseminate bacteria from the nasopharynx, and the capacity of others around them to become colonized. Surprisingly little is known about these aspects, but dissemination may be influenced by the intensity of colonization, intercurrent viral infections (Gwaltney et al. 1975), and perhaps the ambient temperature and humidity.

Acquisition

Carriage rates reflect both acquisition rates and the duration of carriage. The duality is important; much evidence suggests that if bacterial invasion is to occur, it follows very shortly after acquisition. Individuals colonized for weeks or months are unlikely to develop invasive disease, though anecdotes do exist to show that invasion may sometimes follow weeks after acquisition and colonization (Neal et al. 1999).

With increasing age, the nasopharynx becomes progressively less susceptible to colonization by meningococci, Hib, and pneumococci. This may be due to the development of mucosal immunity.

Mucosal invasion

Bacterial strains vary greatly in their capacity to invade. Meningococci that do not express a capsule are essentially avirulent, while strains of serogroups A, B, and C

are more invasive than strains of serogroups X, Y, and W-135 and other groups of meningococci (see Chapter 48, *Neisseria*). Similarly, acapsulate strains of *H. influenzae* have little invasive potential, and though there are six capsular serotypes of *H. influenzae*, type b strains account for more than 95 percent of all invasive hemophilus disease in unimmunized populations.

The interaction of bacteria with the nasopharyngeal mucosa and other tissues has been investigated with tissue explants (Stephens and Farley 1991; Read et al. 1995), cultured epithelial and endothelial cells (Virji et al. 1991; Pujol et al. 1997), and the use of bacterial transformants and microarray technology (Grandi 2001). This has allowed comparisons between the behavior of fully virulent bacteria and strains that differ from them by the deletion of single genes or operons.

Capsulate meningococci bind to epithelial cells by adhesins, including fimbriae and the Opc outer-membrane protein (Virji et al. 1991, 1992). They cross the mucosal barrier by direct invasion of epithelial cells (endocytosis), whereas Hib strains migrate between epithelial cells by breaking down intercellular tight junctions (Stephens and Farley 1991). Once in the submucosa, the invading bacteria cross the capillary vessel basement membrane and the endothelium to gain access to the bloodstream.

The three major bacterial meningitis pathogens produce an IgA1 protease, whereas closely related nasopharyngeal 'commensals,' such as *Haemophilus parainfluenzae*, *Neisseria lactamica*, and viridans streptococci, do not (Mulks and Plaut 1978; Mulks 1985). Invasion potential and production of IgA1 protease (and the converse) are so closely associated that a causal relationship has long been sought. One hypothesis is that circulating IgA may block binding sites that would otherwise be accessible to lethal, bacteriolytic complement-fixing IgG or IgM antibodies (Griffiss 1982). Bacteria that invade may be further protected by cleavage of the blocking immunoglobulin molecules by IgA protease.

Respiratory viral infections may play a part in precipitating bacterial meningitis (Rolleston 1919; Krasinski et al. 1987; Moore et al. 1990; Scholten et al. 1999), in particular influenza A, which increases both the risk (Cartwright et al. 1991) and the severity (Hubert et al. 1992) of meningococcal disease. The mechanism in this latter situation is probably postviral immune suppression.

Meningeal invasion

Normal CSF contains few phagocytic cells and very low levels of complement and antibody; the normal CSF:blood IgG ratio is about 1:800. The integrity of the normal blood–brain barrier, and thus of the subarachnoid space, is maintained by a combination of tight intercellular junctions between capillary endothelial cells of the cerebral microvasculature and by the comparative rarity of pinocytosis by these cells.

The cerebral capillary endothelium of the choroid plexus can bind some of the bacteria that cause meningitis. Bound bacteria may then undergo fimbrial phase variation in order to cross the blood–brain barrier (Saukkonnen et al. 1988). Bacteria may also gain access to the subarachnoid space within host phagocytic cells (Tunkel and Scheld 1993; McNeil et al. 1994).

Once bacteria are established within the subarachnoid space they are transiently sheltered from the host defenses. They can release substances capable of disrupting the blood–brain barrier. In the CSF, *H. influenzae* type b lipopolysaccharide (an endotoxin) rapidly induces the production of inflammatory cytokines, including interleukin-1 and tumor necrosis factor. Pneumococcal cell walls, which contain peptidoglycan and teichoic acid, are more potent than capsular polysaccharide in inducing subarachnoid inflammation (Tuomanen et al. 1985). High CSF cytokine concentrations are reached within 1–2 h of the arrival of bacteria in the CSF (Tunkel and Scheld 1993; Spellerberg and Tuomanen 1994).

Disruption of the blood–brain barrier occurs as a result of breakdown of the intercellular tight junctions of the cerebral microvascular epithelium, permitting an influx of larger molecules from the bloodstream and migration of phagocytic cells into the CSF.

Factors predisposing to bacterial meningitis

Many factors predispose to bacterial meningitis, including age, male gender, winter season, exposure to smokers and smoking, low socioeconomic status, and (probably) stress (Stanwell-Smith et al. 1994). Contact with a case confers a substantially increased risk of meningococcal disease (de Wals et al. 1981; Cooke et al. 1989), though less so for *H. influenzae* meningitis; in contrast, pneumococcal meningitis is not normally regarded as a transmissible infection and clusters of cases of any form of invasive pneumococcal disease (including meningitis) are rare.

Host genetic factors that confer increased susceptibility to bacterial meningitis or to a poor outcome are increasingly the subject of investigation in the context of new investigative methods now becoming available (Hill 2001). This is an area where much information will accrue over the next few years. Specific host factors conferring susceptibility to meningococcal disease are addressed in the relevant section of this chapter.

CLINICAL FEATURES

The presenting features of meningitis are determined largely by the age of the patient and, to a lesser extent,

by the causative organism and the route by which it has reached the meninges. In the great majority of cases, the early signs and symptoms are nonspecific. Fever, lassitude, malaise, muscular aches and pains, nausea, vomiting, and headache occur in both viral and bacterial meningitis but are more commonly due to the former. Identification of the symptoms that may give warning of impending serious illness can be very difficult. In temperate countries, the difficulty is compounded by the fact that influenza, viral respiratory infections, and bacterial meningitis all peak in the winter months. Specific symptoms that should immediately raise the suspicion of bacterial meningitis include photophobia, neck stiffness, drowsiness, and fitting. Fever accompanied by pallor, inconsolable crying, a fit, or any evidence of decreased consciousness in a febrile infant or young child are highly suspicious and should warrant immediate careful investigation. The anterior fontanelle may bulge as meningitis progresses in infants.

As bacterial meningitis develops, symptoms become more severe; most patients with bacterial meningitis are much iller than those with viral infections. The highest attack rates of bacterial meningitis are in infants and very young children who cannot give a verbal account of their symptoms. Familiarity with the symptoms, repeated examinations, and a high index of suspicion are the best safeguards for parents and family doctors. The combination of fever and a vasculitic rash (petechiae or purpura), which may herald meningococcal disease, constitutes a medical emergency. Meningococcal septicemia may occasionally progress so rapidly that patients can become moribund within 4–6 h of the first symptoms.

In contrast, tuberculous and fungal meningitis more commonly develop insidiously over days, weeks, or even months.

DIAGNOSIS OF MENINGITIS

Cerebrospinal fluid

LUMBAR PUNCTURE

Examination of CSF has long been considered as routine in the investigation of suspected bacterial meningitis (Gray and Fedorko 1992) but is not without its dangers and drawbacks, especially 'coning' (Rennick et al. 1993). If intracranial pressure is raised, lumbar puncture may occasionally be followed by herniation of the tentorium through the *falx cerebri* or of the brain stem through the *foramen magnum* (coning); either may be a fatal event or may lead to catastrophic neurological morbidity. Papilloedema is an insensitive indicator of raised intracranial pressure and clinicians will normally be wary of lumbar puncture in patients with evidence of diminished consciousness or fluctuating neurological status. A computed tomography (CT) scan is helpful in determining whether there is raised intracranial pressure but does not provide absolute assurance of safety to proceed with lumbar puncture. An increasing proportion of pediatric patients with clinical evidence of meningitis are now managed without lumbar puncture (Cartwright and Kroll 1997; Wylie et al. 1997; Nadel 2001), though the procedure is still recommended in suspected meningitis in adults (Begg et al. 1999) because of the wider range of potential pathogens.

Despite the hazards, examination of CSF still offers the best chance of observing, isolating, and identifying the causative organism in bacterial meningitis (Kaplan et al. 1986b; British Society for the Study of Infection Research Committee 1995). Lumbar puncture is particularly valuable if a dose of a parenteral antibiotic has been given either before or on hospital admission, but if indicated, must be done swiftly, since cultures, even of CSF, rapidly become negative; CSF sterilization times of 2 h for meningococci and 4 h for pneumococci have been suggested (Kanegaye et al. 2001). Even if bacteria cannot be recovered on culture, microscopy is likely to confirm the diagnosis of bacterial meningitis and may indicate the likely causative organism. Lumbar puncture should never be used as a reason to defer commencing antibiotic treatment and other resuscitation measures. The 'door-to-needle' time for patients hospitalized for suspected bacterial meningitis should be less than 1 h.

Polymerase chain reaction (PCR) tests of CSF for the common bacterial meningitis pathogens remain positive for several days after cultures become negative. Thus, if a patient is considered too sick to undergo lumbar puncture on admission, it may be preferable to initiate empirical antibiotic treatment and to reconsider the need for lumbar puncture after 24–48 h, when the patient's condition has stabilized.

A portion of the CSF sample should be examined promptly to determine the number and type of leukocytes present. Bacterial meningitis is usually accompanied by the presence of neutrophils, but about 5 percent of CSF samples subsequently found to be culture-positive contain only a few lymphocytes or no cells at all when first examined. The absence of leukocytes should never be used to justify withholding antibiotic treatment if there is reasonable clinical suspicion of bacterial meningitis. The cell population may consist of a mixture of polymorphs and lymphocytes; this is more likely if there has been a long prodrome or if the patient has been part-treated with antibiotics. A mixture of cells is sometimes seen in viral meningitis. The CSF leukocyte count may be raised in patients who have had a cerebral hemorrhage a few days earlier and in patients who have had recent fits. In these cases, the cells are normally predominantly lymphocytes. Other conditions that give rise to a lymphocytic pleocytosis are listed in Table 20.1.

The CSF leukocyte concentration falls much more slowly than the concentration of viable bacteria. It is not uncommon for lumbar puncture 1–2 weeks after the onset of bacterial meningitis to show persistence of leukocytes (Chartrand and Cho 1976; Connolly 1979), especially when meningitis has been caused by *H. influenzae* or by a pneumococcus. Persistence of pleocytosis is less likely in meningococcal meningitis.

BIOCHEMICAL ANALYSIS

Estimation of CSF glucose and protein can be helpful. The CSF:blood glucose ratio (normally about 70 percent) is often, though not invariably, reduced in bacterial meningitis, commonly to about 25 percent (Kaplan et al. 1986b), but a normal CSF:blood glucose ratio is common in tuberculous meningitis. The CSF

Table 20.1 *Conditions causing cerebrospinal fluid lymphocytic pleocytosis*

Infections	Drugs and vaccines	Inflammatory processes	Malignancy
Viral meningitis	Trimethoprim, isoniazid	Postictal, postmyelogram	Meningeal carcinomatosis
Partly treated bacterial meningitis	MMR and polio vaccine	Post-CNS bleed	CNS leukemia and lymphoma
Syphilis	Ibuprofen, NSAIDs	Sarcoidosis, Wegener's granulomatosis	
Parameningeal infection (e.g. abscess, subdural empyema)	Intravenous immunoglobulin	Malignant hypertension, migraine	
Protozoa (e.g. toxoplasmosis)		Kawasaki disease	
Acute and chronic bacterial meningitis		Post-neurosurgery, intrathecal drugs	
Tuberculous meningitis		Multiple sclerosis	
Lyme borreliosis		SLE, Behçet's syndrome	
Fungal meningitis		Benign intracranial hypertension	
Worm infection		Mollaret's meningitis	

CNS, central nervous system; NSAID, nonsteroidal anti-inflammatory drug; SLE, systemic lupus erythemastosus.

protein is often raised, initially reflecting leakage of blood protein into the CSF; later it may reflect local antibody production. CSF lactate concentration may help to differentiate between bacterial and viral meningitis.

Urine reagent strips have been used for the direct examination of CSF (Moosa et al. 1995) to discriminate accurately between viral and bacterial meningitis. This cheap and simple method may have an important role where laboratory facilities are limited or distant.

MICROSCOPY

Direct microscopy of uncentrifuged or centrifuged CSF may reveal the presence of bacteria or fungi and can provide immediate confirmation of the diagnosis. Staining with acridine orange is more sensitive than the Gram stain (Kleinman et al. 1984). The organisms seen on microscopy in antibiotic-treated patients may fail to grow on culture, but the morphology and Gram reaction of the organism, the age of the patient, and the clinical features often permit an educated guess at the identity of the causative organism.

DETECTION OF BACTERIAL ANTIGEN

Tests for bacterial antigen in CSF can provide a quick diagnosis, but they are less sensitive than the Gram stain and do not often alter clinical management (Maxson et al. 1994). Many tests are based on the agglutination of antibody-coated latex particles and work quite well for pneumococci, for meningococci of serogroups A or C, and for group B streptococci. They are less successful for the detection of serogroup B meningococci. The *Limulus* lysate test is a sensitive test for endotoxin from gram-negative organisms but has not found wide acceptance to date.

CULTURE

CSF should be inoculated on to good-quality culture media, always including at least Columbia blood agar and a heated blood agar. Plates should be incubated in five percent CO_2 for a minimum of 48 h. If a ruptured cerebral abscess is suspected, or if meningitis has followed neurosurgery or a history of previous meningeal trauma, a second blood agar plate should be incubated anaerobically for 5–7 days.

The possibility of parameningeal infection, especially subdural empyema and brain abscess, must be considered and pursued actively if CSF shows an inflammatory response, if bacteria are not seen, cultured, or detected by PCR tests, and the patient's condition is not improving. CT scans may sometimes fail to detect intracranial collections of pus. Subdural empyema (or brain abscess) should be suspected if 'meningitis' is diagnosed in a child or young adult in whom there is a recent history of sinusitis or middle ear infection. The importance of not missing these conditions lies in the need for urgent neurosurgical and ear, nose, and throat (ENT) assessment as part of the management protocol. Most cases of subdural empyema are initially misdiagnosed as bacterial meningitis.

POLYMERASE CHAIN REACTION

PCR tests for the detection of meningococcal and pneumococcal DNA in CSF are now used routinely in the UK. Meningococcal PCR is specific and sensitive for the diagnosis of meningococcal meningitis (Borrow et al. 1997) and provides serogroup information in the majority of cases. Amplification of sections of 16S ribosomal RNA, common to most species of pathogenic bacteria, may also prove to be of value (Greisen et al. 1994).

Peripheral blood

BLOOD CULTURE

Blood cultures are positive in only about half of patients with meningococcal disease who have not previously received parenteral antibiotic therapy, and they are almost invariably sterile in those who have. In contrast, in Hib and pneumococcal meningitis the causative organisms are commonly isolated from peripheral blood of untreated patients (British Society for the Study of Infection Research Committee 1995).

MICROSCOPY

When bacteremia is heavy, meningococci and other bacteria may be detected within polymorphs in Giemsa-stained films of peripheral blood (Bush and Bailey 1944) and they can be seen in and grown from tissue fluid obtained from the hemorrhagic skin rash (van Deuren et al. 1993) and from other normally sterile sites such as joints and pericardial fluid.

PCR OF PERIPHERAL BLOOD

PCR tests for the detection of meningococcal DNA (Borrow et al. 1997) are of particular value when the patient has received parenteral antibiotic treatment before hospital admission, an increasingly frequent event. They are now offered routinely in the UK (Kaczmarski et al. 1998), with high sensitivity and specificity (Hackett et al. 2002a). About half of all cases of laboratory-confirmed meningococcal disease in England and Wales are now diagnosed by PCR methodology. This is related to the increasing use of preadmission parenteral antibiotic therapy and to a decline in the use of lumbar puncture, especially in pediatric patients (Wylie et al. 1997). Quantitative meningococcal blood PCR has recently confirmed historical quantitative culture findings, indicating that meningococcal bacterial load at presentation in hospital correlates with disease severity (Hackett et al. 2002b).

Though pneumococcal and Hib meningitis are very frequently accompanied by bacteremia, blood PCR tests for these bacteria are not used widely. Blood PCR tests for pneumococci are only moderately sensitive, and the specificity is low in children (better in adults), either because they fail to differentiate between carriage and invasive infection or, more likely, 'occult' bacteremia (or spillover of bacteria from the nasopharynx) is a frequent occurrence in colonized infants and children.

Serology

Serological tests for meningococcal disease were long hampered by the failure of serogroup B capsular polysaccharide to provoke a good immune response. Second-generation serological tests now available in the UK can reliably detect the presence of acute-phase (IgM) meningococcal antibodies in a single serum specimen obtained after the first week of the illness; alternatively, a rising titer of antibody to meningococcal outer-membrane proteins may be sought (Jones and Kaczmarski 1994).

Serology can be used to confirm invasive Hib infection but antibody responses in pneumococcal infection are highly unreliable because they fail to differentiate between carriage and invasive illness.

Nasopharyngeal swabs

Since the main bacterial meningitis pathogens are nasopharyngeal commensals, their isolation from the nasopharynx in cases of clinical meningitis does not provide evidence of causation. However, when lumbar puncture is not undertaken and blood cultures are negative, a throat swab may give the only chance of obtaining an isolate. Meningococci are rarely recovered from healthy infants and the isolation of a capsulate meningococcus of serogroup B or C from the posterior pharynx of a young child with symptoms of meningitis with or without a hemorrhagic rash is very strong evidence of a meningococcal etiology. Similarly, the isolation of Hib from a child with meningitis aged 12 months or less is unlikely to be coincidental. However, isolation of a pneumococcus from the nasopharynx of a child or adult is sufficiently common that it does not assist in management.

Swabs of the posterior pharynx are ideally taken through the mouth (Olcén et al. 1979). Patients with clinical meningitis are often too ill to cooperate with peroral swabbing, but pernasal swabs are simple to obtain, though an assistant is required to steady the patient's head. A single negative nasopharyngeal swab is an insensitive predictor of freedom from carriage of any specific bacterial meningitis pathogen, and decisions on the prophylaxis of contacts of meningococcal or Hib disease should never be based on the results of nasopharyngeal swabbing of the contacts.

MENINGOCOCCAL DISEASE

Historical aspects

Vieusseux (1805) described an outbreak of meningococcal disease with 33 deaths in the small community of Eaux Vives, near Lake Geneva in Switzerland, in the spring of 1805. Before this, meningococcal disease may have been confused with other spotted fevers, including typhus, which also sometimes occurred in clusters and outbreaks, especially in the military, and was often characterized by a hemorrhagic skin rash. It seems most likely that meningococcal disease has afflicted man for centuries.

August Hirsch (1886) documented many outbreaks of infectious diseases including cerebrospinal fever in Europe and the New World up to 1882. The Italian pathologists Marchiafava and Celli (1884) are credited with the first description of intracellular oval micrococci in a sample of CSF. Three years later Weichselbaum, in Vienna, reported the isolation of an organism he described as *Diplococcus intracellularis meningitidis* from six of eight cases of primary sporadic community-acquired meningitis (Weichselbaum 1887). A period of confusion followed when Jaeger isolated a gram-positive chain-forming coccus from meningitis cases in an outbreak in Stuttgart. Subsequent reports showed that Weichselbaum was correct: Jaeger's gram-positive organism was probably a contaminant.

Eight years later came the first account of lumbar puncture in a living patient (Quincke 1893), and in 1896 meningococci were isolated for the first time from the CSF of patients with meningitis (Heubner 1896). In the same year, meningococci were also isolated from human throat cultures (Kiefer 1896), offering an explanation for the spread of the bacteria in human populations.

Before World War I, German and American bacteriologists were developing serum therapy for meningococcal meningitis in an effort to reduce its extremely high mortality (Kolle and Wasserman 1906; Flexner and Jobling 1908). They were able to reduce mortality from about 80 to 25 percent by repeated intrathecal instillation of immune horse serum (Flexner and Jobling 1908). Not surprisingly, a number of successfully treated patients developed 'serum disease' (serum sickness), with fever and arthritis, and a few developed secondary bacterial meningitis.

In Britain, nationwide notification of a range of infectious diseases including cerebrospinal fever was introduced in 1912. Two years later, during the first winter of World War I, there were several outbreaks of cerebrospinal fever amongst military recruits. Glover (1920) carried out a detailed investigation at the Guards Depot, Caterham, in south London. His report is a masterpiece of early meningococcal epidemiology; he concluded that the disease flared up at times of extreme overcrowding and was preceded by very high nasopharyngeal carriage rates. Effective interventions, including increasing the amount of sleeping accommodation and the space between beds, fixing open windows in huts, and shortening of parades, brought the outbreaks to a halt, even during the 1918–19 winter in which there was a major influenza epidemic. Later, Dudley and Brennan (1934) and others documented very high nasopharyngeal carriage rates without the occurrence of cases of meningococcal disease. All the major combatant countries in World War I experienced clusters of cases and outbreaks of meningococcal disease in recruit camps, but there was little disease in seasoned troops (Rolleston 1919). At the end of the war, French microbiologists published a serogrouping scheme, which forms the basis for the current classification system (Nicolle et al. 1918).

In the 1930s, Rake and colleagues showed that freshly isolated meningococci were 'smooth' (capsulate), that they produced a polysaccharide capsule and that such strains produced more potent antisera when used to immunize horses (Rake 1931; Rake and Scherp 1933). Rake also carried out studies on the duration of nasopharyngeal meningococcal carriage and showed that, though very variable, carriage could persist for years.

Rake's work, which was leading towards the development of vaccines, was halted by the discovery of sulfonamides. After their first use in the treatment of human meningococcal infections (Schwentker 1937), sulfonamides were used to treat very large numbers of cases of meningococcal disease at the beginning of World War II and, subsequently, also for chemoprophylaxis. In Britain in 1941, there were 11 000 notified cases and more than 2 000 cases in each subsequent year up to 1945 (Figure 20.2).

Germany, France, and the USA also experienced outbreaks of unprecedented size, in each case shortly after becoming a combatant country. These epidemics occurred in association with mobilization (in 1940–41 in the UK, France, and Germany, and in 1942–43 in the USA) and usually began among recruits in training, later spilling over into civilian populations. At the same time, most noncombatant countries, such as Sweden, did not experience any great change in the incidence of meningococcal disease.

Sulfonamides transformed the treatment of meningococcal disease – mortality rates fell to about 10–20 percent. When penicillin became available later in the war it gradually superseded sulfonamides for treatment of invasive disease, but sulfonamides continued to be used for the treatment of carriers.

Epidemiology

Meningococcal disease remains a worldwide problem, occurring sporadically as clusters of cases and as epidemics. Annual rates of disease in many western countries and in the North American continent are about one to two per 100 000 of the total population, though there are wide fluctuations from year to year and from country to country. Even within countries, the disease rate can fluctuate remarkably, with clustering of cases in small communities or regions, giving high attack rates (5–25 per 100 000) for months or years. Norway experienced high rates of disease, mainly due to strains of serogroup B (ET-5 clone), for almost 20 years from the 1970s to the 1990s (Lystad and Aasen 1991), whereas the disease incidence in Sweden remained low throughout the same period. At the start of a period of increased disease incidence, the average age of cases rises (Peltola et al. 1982) and the severity of disease is greater. These phenomena may reflect the accumulation of susceptible people in the population. New Zealand has experienced hyperendemic serogroup B

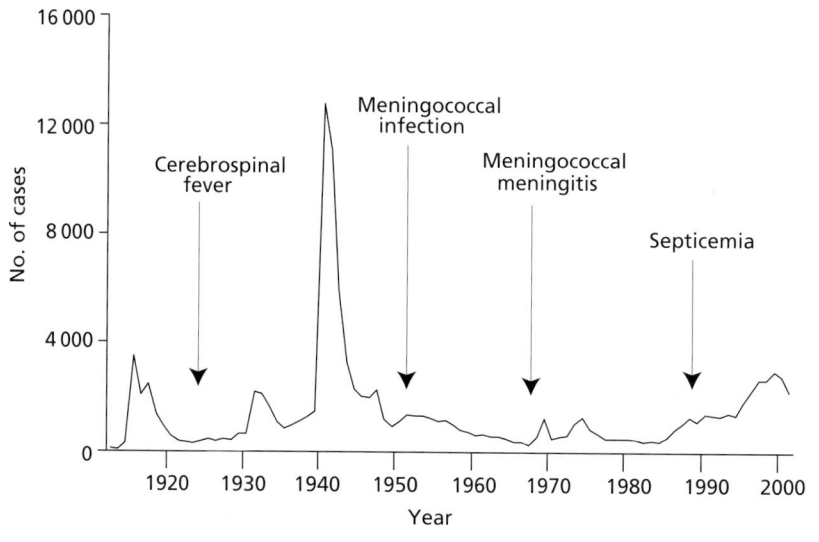

Figure 20.2 *Notifications of meningococcal disease, England and Wales 1912–2001.*

meningococcal disease since the early 1990s, with particularly high attack rates amongst the Pacific Islander and Maori populations (Martin et al. 1998).

THE 'MENINGITIS BELT'

The pattern of disease is quite different in some tropical countries, with large epidemics of serogroup A disease occurring at irregular intervals, separated by periods of much lower disease activity (Peltola 1983; Moore 1992). Lapeysonnie (1963) drew attention to an area of Africa south of the Sahara and extending almost from coast to coast, where epidemics of meningococcal disease occur every 5–10 years – the 'meningitis belt' (Figure 20.3). Epidemic waves may last 2 or 3 years.

The boundaries of the meningitis belt are probably defined by climatic factors; continuous high absolute humidity may reduce meningococcal transmission (Cheesbrough et al. 1995). This leads to the situation in sub-Saharan Africa, where countries such as Ghana and Nigeria may experience epidemics of meningococcal disease in their northern regions but have low attack rates in their southern coastal areas.

The disease is highly seasonal in the meningitis belt, with an upsurge during the period of dry, dusty winds ('harmattan'), with abrupt cessation of the epidemic with the start of the rainy season. The numbers of cases are staggering by developed country standards – attack rates of 500 per 100 000 or more have been recorded. The highest attack rate is in the age group 5–15 years; the sexes are affected equally. Multilocus electrophoretic typing (Caugant et al. 1987) and multilocus sequence typing (Maiden et al. 1998), which permit the identification of meningococcal clones, demonstrate that these epidemics are usually due to the spread of new clones, almost always of serogroup A (Olyhoek et al. 1987; Achtman 1990, 1995).

EPIDEMIOLOGY IN TEMPERATE CLIMATES

In temperate and colder climates, the disease is strongly seasonal, with peak rates in the winter months. In Britain, disease rates are highest in December and January, falling to a low point in the autumn months. In western countries, the age-specific incidence shows two peaks, with the first, and larger, in infants and young children. A second smaller, but important, peak occurs in late teenage. There is no obvious relationship between the age-specific prevalence of carriage and the incidence of disease (Figure 20.4). In most western countries, affected males outnumber females by up to 1.5:1.

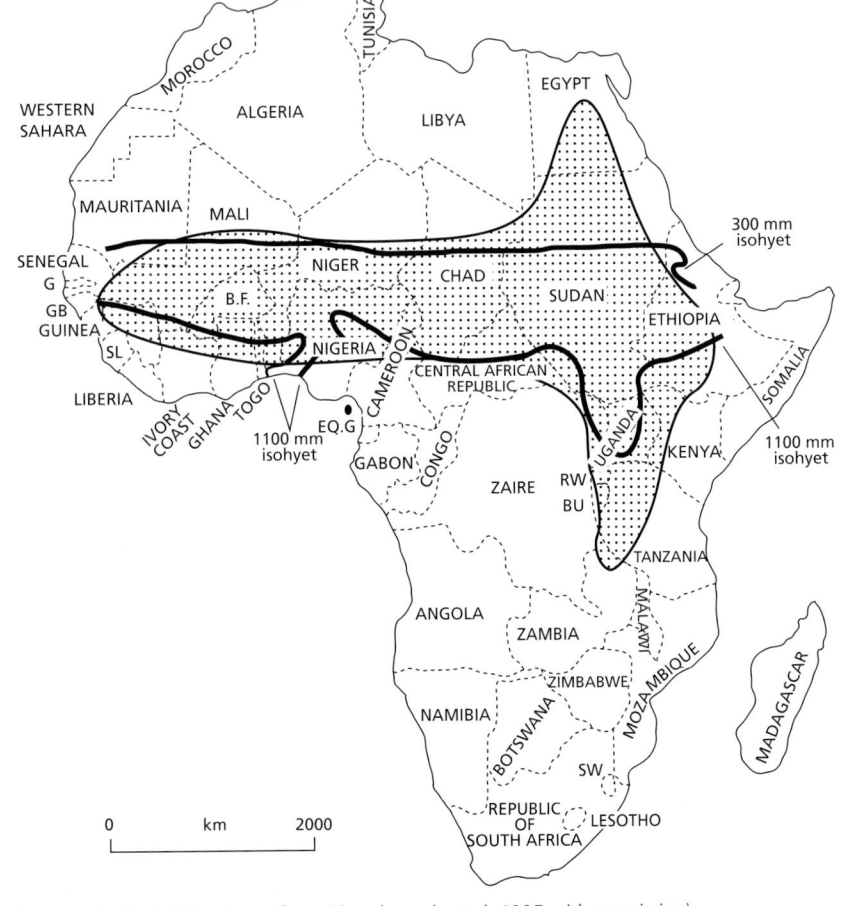

Figure 20.3 *Africa: the 'Meningitis Belt' (redrawn from Cheesbrough et al. 1995 with permission).*

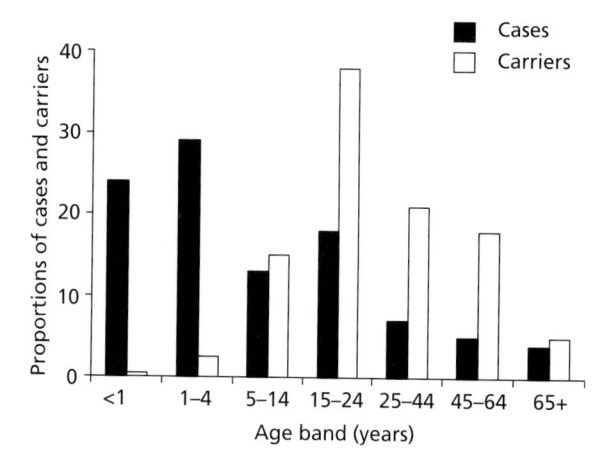

Figure 20.4 *Proportions of meningococcal cases and carriers within age groups. (Carriage data from Cartwright et al. 1987, with permission; age incidence from Jones and Mallard 1993, with permission.)*

In the UK, prior to the introduction of conjugated meningococcal serogroup C vaccines in 1999, about 70 percent of invasive disease was caused by serogroup B meningococci, a further 25 percent by serogroup C strains, and the remainder by a mixture of other serogroups, including W-135, Y, and X (Jones and Kaczmarski 1994). The large epidemics that occurred during World War I and World War II were caused by strains of serogroup A, with a major change in the epidemiology occurring in the 1950s to the currently observed pattern. During the late 1990s in the UK, the proportion of cases attributable to serogroup C strains has been as high as 40 percent. This was accompanied by a higher incidence of cases in older children and adults and an increase in the number of outbreaks in schools and universities. The introduction into the national childhood immunization schedule in November 1999 of highly effective conjugated serogroup C vaccines has dramatically reduced serogroup C disease in the UK (see below). It has not been accompanied by replacement of serogroup C disease by strains of other serogroups, e.g. serogroup B, either amongst disease-causing strains or amongst carried meningococci (Maiden et al. 2002).

In the UK, the USA, and a number of European countries, there was a notable upsurge in meningococcal disease caused by strains of W-135 amongst Muslim pilgrims making the Haj pilgrimage to Mecca and Medina in 2000 (Taha et al. 2000) and 2001. A requirement by the Saudi Arabian health authorities later in 2001 that all pilgrims should receive quadrivalent A, C, Y, W-135 vaccine as a condition of entry to the country brought these outbreaks of disease to a close. The clone causing this upsurge of W-135 disease was the same as that responsible for most serogroup C meningococcal disease in the UK over the preceding few years (ET-37). This clone is highly virulent and case-fatality rates were high.

Carriage

This subject has been dealt with in two in-depth reviews (Broome 1986; Cartwright 1995). The human nasopharynx is the natural habitat of the meningococcus. Meningococci are not carried by animals; nor are they recovered from the environment. Humans may be selectively colonized because meningococci are only able to acquire iron, an essential nutrient (van Putten 1990), from human (and primate) transferrin and lactoferrin (Schryvers and Gonzalez 1990). Fimbriae facilitate binding to human epithelial cells, but capsulation, essential for invasion, reduces adherence. Colonization by a single strain may last for months or even years. Capsule expression may be downregulated during prolonged carriage.

AGE-SPECIFIC INCIDENCE OF CARRIAGE

Carriage rates are low in infants and young children but rise with age to peak in late teenage and early adult life. They then decline slowly over the next 20–30 years; carriage is rare after the age of 65 years (Greenfield et al. 1971; Cartwright et al. 1987; Caugant et al. 1994). Most nasopharyngeal meningococci have very low invasive potential. Many express little or no capsular polysaccharide and are nongroupable or belong to serogroups of low virulence such as X, Y, or W-135. Exposure to such strains probably boosts levels of antibody against noncapsular surface antigens.

FACTORS AFFECTING CARRIAGE RATES

Throat-swabbing is an insensitive procedure and swabs from individuals and groups underestimate carriage frequency. Swabbing the posterior pharyngeal wall is much more reliable for the detection of meningococcal carriage than swabbing the tonsils or sampling the front of the mouth (J.M. Stuart, unpublished), but a single negative throat swab, even if collected by an experienced sampler, cannot be relied upon as proof of freedom from colonization. Somewhat surprisingly, carriage appears to be unaffected by season (Blakebrough et al. 1982; de Wals et al. 1983) or by intercurrent viral infection. Vaccination with purified capsular polysaccharides also appears to have little effect, but conjugated serogroup C vaccines have reduced dramatically the prevalence of carriage of homologous meningococcal strains (Maiden et al. 2002).

Smoking has a strong dose-related effect on carriage (Stuart et al. 1989), but contact with a case of meningococcal disease is the most important risk factor for carriage. Rates of carriage are high in family members and close contacts of cases of meningococcal disease (Greenfield and Feldman 1967; Munford et al. 1974; Kristiansen et al. 1998). Most (but not all) such isolates are indistinguishable from the index case strain (Olcén et al. 1981). Carriage rates rise rapidly when teenagers and young adults are brought into close contact and close social interaction, such as that facilitated by entry into a military recruit camp (Riordan et al. 1998) or commencing study at university (Neal et al. 2000), explaining the high rates of disease seen in these groups. For students, high carriage rates and high disease rates are most marked in the freshman (first) year.

Immunity

The best serological correlate of protection from meningococcal disease is the presence of bactericidal antibodies (Goldschneider et al. 1969a) (Figure 20.5). Opsonizing antibodies are also important (Halstensen et al. 1989). For serogroup A and C meningococci, most bactericidal antibody is IgG directed against the capsular polysaccharide and is capable of causing bacterial lysis through the activation of complement (Gotschlich et al. 1969a). Serogroup B capsular polysaccharide is a very weak antigen for humans, probably because it shares epitopes with host cell antigens (Finne et al. 1983; Azmi et al. 1995). Antibodies present after serogroup B meningococcal disease are directed against a variety of surface-expressed antigens, including outer-membrane proteins and lipopolysaccharide (Kasper et al. 1973; Griffiss et al. 1984).

THE ROLE OF *NEISSERIA LACTAMICA*

Bactericidal antibodies are acquired progressively during childhood (Goldschneider et al. 1969a), beginning at a time when exposure to meningococci is slight. Thus, protection is probably acquired first through exposure to other bacteria that express cross-reacting surface antigens (Glode et al. 1977; Guirguis et al. 1985; Devi et al. 1991).

The carriage rate of *Neisseria lactamica* is high when bactericidal meningococcal antibodies are being acquired. Colonization with *N. lactamica*, a noncapsulate species, generates antibody that is bactericidal against a range of serogroups and serotypes of meningococci (Gold et al. 1978), indicating the presence of cross-protective subcapsular antigenic determinants. Later in

life, exposure to avirulent meningococci induces antibodies against a wide range of meningococcal serogroups (Reller et al. 1973), providing further confirmation that some bactericidal antibody is directed against noncapsular antigens. A number of oral and enteric bacteria also express cross-reactive antigens and probably play a part in generating and maintaining immunity.

BLOCKING IGA ANTIBODIES

Complement-dependent immune lysis is not initiated by IgA. However, secretory or circulating IgA antibodies may block meningococcal surface antigens that would otherwise be accessible to IgG or IgM antibodies that activate complement. Blocking IgA antibodies generated in response to cross-reacting enteric bacteria may be important in some cases of meningococcal disease in older children and adults (Griffiss 1995). The timing of exposure to the meningococcus and the cross-reacting enteric bacterium may be critical (Kilian and Reinholdt 1987; Kilian et al. 1988).

HOST SUSCEPTIBILITY

Various congenital or acquired immune deficiency syndromes are associated with an increased risk of meningococcal disease (Figueroa and Densen 1991). Deficiency of any of the terminal complement components (C5 to C9) is associated with an increased risk of meningococcal disease, especially of serotypes not normally regarded as invasive, including X, Y, and W-135 (Fijen et al. 1989). The risk of second and subsequent attacks is also increased but disease tends to be milder (Ross and Densen 1984). Half the individuals with the very rare deficiency of properdin (factor P) develop meningococcal disease, but this tends to occur in older children or adults and is often overwhelming, with high fatality rates (Densen et al. 1987). There are at least three genetic mechanisms leading to properdin deficiency; all three identified to date are inherited in an X-linked manner (Linton and Morgan 1999).

Recent evidence suggests that mannose-binding lectin, part of a phylogenetically older innate immune system,

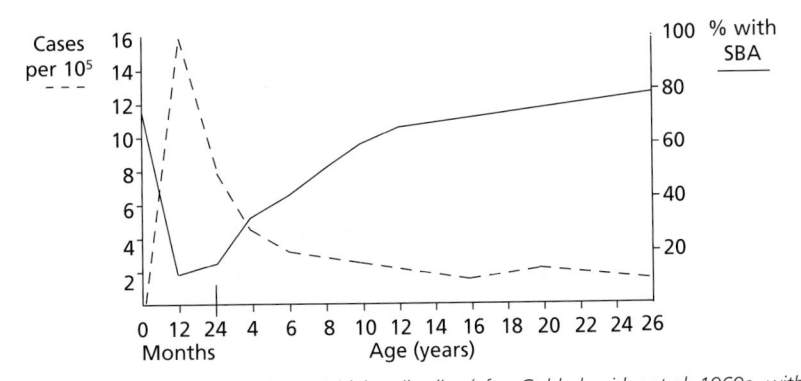

Figure 20.5 *Meningococcal disease: relationship to bactericidal antibodies (after Goldschneider et al. 1969a, with permission).*

plays an important role in early defense from meningo-coccal (and other invasive bacterial) infections (Summerfield et al. 1997), and mutations in the mannose-binding lectin gene may be associated with an increased susceptibility to meningococcal infection. There is also evidence that particular patterns of inher-ited cytokine profile may be associated with a higher risk of a fatal outcome in meningococcal disease (Westendorp et al. 1997). More recently, polymorphisms of genes encoding plasminogen activator inhibitor 1 and IL-1β have been shown to be associated with the devel-opment of septic shock or with increased mortality in meningococcal disease (Westendorp et al. 1999; Read et al. 2000), and FcγRIIa (CD32) polymorphisms appear to be associated with susceptibility to, and severity of, meningococcal disease (Bredius et al. 1994; Platonov et al. 1998).

SUSCEPTIBILITY AT DIFFERENT AGES

Meningococcal infection is infrequent in the first 2 months of life, indicating the likely key role of passively transferred maternal antibodies in initial protection. Subsequently, meningococcal disease in infants and young children probably results from exposure of an immunologically virgin child to a virulent meningo-coccus. Carriage studies suggest that exposure to viru-lent meningococci is probably uncommon in infancy. This seems less likely in teenagers and young adults in whom there is also a peak of infection (see Figure 20.4), but at a time when meningococcal carriage is most fre-quent. Most, if not all, individuals aged 10 years or more will probably have encountered *N. lactamica*, poorly capsulate meningococci, and/or other immunizing bacte-ria on numerous occasions. Therefore, meningococcal disease in this age group may result from inherited inc-reased susceptibility, a high level of exposure to menin-gococci, subversion of pre-existing defense mechanisms by environmental factors such as antecedent or inter-current viral infection, or a combination of these factors.

Meningitis and septicemia

Meningococcal infection most commonly presents as meningitis (75 percent) or septicemia (20 percent), though the two syndromes show considerable overlap; clinical differentiation may be difficult or impossible, and UK clinicians vary in their practice as to what disease they choose to notify; recently, proportionately more cases have been notified as septicemic (Figure 20.6).

A few patients have other pyogenic infections such as primary septic arthritis, conjunctivitis, and pericarditis. Primary meningococcal pneumonia is usually caused by strains of serogroup Y. Increasing numbers of cases have been reported from the USA in recent years, and this condition may be underdiagnosed in the UK.

The outlook in meningococcal meningitis is very good. Mortality in the UK is less than 5 percent, and the inci-dence of long-term neurological morbidity, mainly deaf-ness, is low (3–5 percent), though a small percentage of survivors suffer severe or even catastrophic brain damage. Most meningococcal meningitis occurs in infants and young children in whom the symptoms and signs are similar to those of other types of bacterial meningitis occurring at the same age. A common, though not invariable, feature of meningococcal menin-gitis is the presence of petechiae or purpura in the skin. Though an ecchymotic skin rash may also occur occa-sionally in Hib or pneumococcal meningitis (and in a wide range of noninfectious conditions), it is seen in more than 50 percent of cases of meningococcal menin-gitis and in all cases of meningococcal septicemia. The pathology of the skin lesions resembles a series of loca-lized Shwartzman reactions (DeVoe 1982). Nondescript, macular, flat or raised nonvasculitic, evanescent skin rashes are very common in the early stages of develop-ment of meningococcal infection in children (Marzouk et al. 1991). They may be mistaken for the rash of viral infection and may cause delay in diagnosis. Such rashes

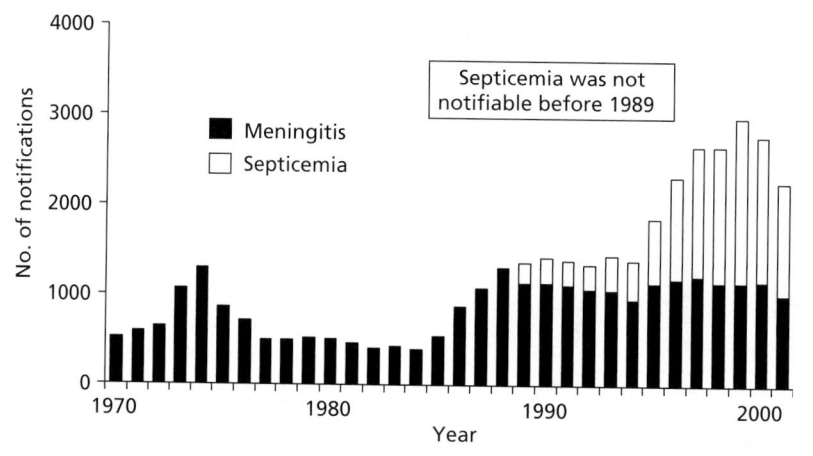

Figure 20.6 *Meningococcal disease, England and Wales: annual numbers of notifications, 1970–2001*

disappear with the onset of the more characteristic vasculitic rash.

Meningococcal meningitis in older patients usually presents with symptoms of malaise, muscular aches and pains, diarrhea together with vomiting, fever, and headache, progressing if unrecognized to confusion, coma, shock, and circulatory collapse. In the later stages, an ecchymotic skin rash is almost inevitable.

Septicemia in children or adults is a far more serious condition, with a mortality greater than 20–25 percent. It can occur with or without evidence of meningitis. Fever, shock, and an ecchymotic skin rash are the hallmarks of this condition. Onset is frequently abrupt, over a few hours. Multisystem failure is common. In addition to antibiotics, oxygen, aggressive fluid replacement, correction of electrolyte imbalances, inotropes, mechanical ventilation, control of disseminated intravascular coagulation, and hemodialysis may all be required for considerable periods. Survivors of severe septicemia may need amputations of necrotic fingers, toes, or sometimes limbs; a few suffer severe brain damage, probably associated with prolonged periods of hypotension and hypoxia.

Management

ANTIBIOTIC TREATMENT

Benzylpenicillin has traditionally been regarded as the drug of choice in meningococcal disease. The England Chief Medical Officer recommends that general practitioners and casualty doctors should administer a dose of benzylpenicillin to any patient with suspected meningococcal disease, without awaiting the results of laboratory investigations. This course of action may halve mortality in the more seriously ill and is strongly endorsed, though it makes confirmation of the diagnosis somewhat more difficult (Cartwright et al. 1992) (Figure 20.7). Fortunately, the availability of molecular diagnostic tests on blood and CSF have largely obviated this problem. The combination of fever and a vasculitic (nonblanching) skin rash must be taken to indicate the possibility of

meningococcal disease until an alternative diagnosis is established. This combination of symptoms constitutes a medical emergency. As well as administering benzylpenicillin, general practitioners are urged to arrange immediate ('blue-light') transfer of the patient to hospital.

Since the etiology will probably not be known for certain at the start of inpatient treatment, it is wise to use a broad-spectrum antibiotic for the first 24–48 h, especially in young children. Cefotaxime, ceftriaxone, or, if cost is a factor, chloramphenicol are suitable alternatives that give high rates of cure; these agents can also be used in patients who are allergic to penicillin. Cefuroxime is inferior. Treatment may be switched to benzylpenicillin when a meningococcal etiology has been confirmed. Wall has argued for the use of cefotaxime or ceftriaxone as agents of first choice for treatment of meningococcal meningitis throughout on grounds of high activity and high penetration into the subarachnoid space (Wall 2001). In view of the very high success rates achieved with benzylpenicillin in this condition, prospective randomized trials would be very difficult or impossible.

'Penicillin-resistant' meningococci

So-called 'penicillin-resistant' meningococci have been reported in a number of countries and regions, including Spain (Sáez-Nieto et al. 1992), South Africa, the UK, and the North American continent. These strains are better described as showing reduced sensitivity to penicillin, and it is likely that such reduced sensitivity results from acquisition of genetic material from related neisserial species including *Neisseria flavescens*, an oropharyngeal commensal, leading to production of an altered penicillin-binding protein 2 (Spratt et al. 1989). Illness caused by these strains responds almost invariably to treatment with adequate doses of benzylpenicillin.

Production of β-lactamase by meningococci isolated from invasive infections has been reported from South Africa and Spain (Botha 1988; Fontanals et al. 1989). Neither strain has survived for detailed investigation. In view of the known transformability of meningococci, it is more than a little surprising (albeit very welcome) that β-lactamase-producing meningococci have not been encountered more often.

Duration of antibiotic treatment

Antibiotic treatment for 7 days usually suffices for children with mild illness (McCracken et al. 1987; O'Neill 1993). Longer periods of treatment, of at least 10 days, are often used in more seriously affected patients, including those with septicemia, even though all meningococci are likely to have been killed soon after the commencement of treatment. Very occasionally, meningococci may survive within the subarachnoid

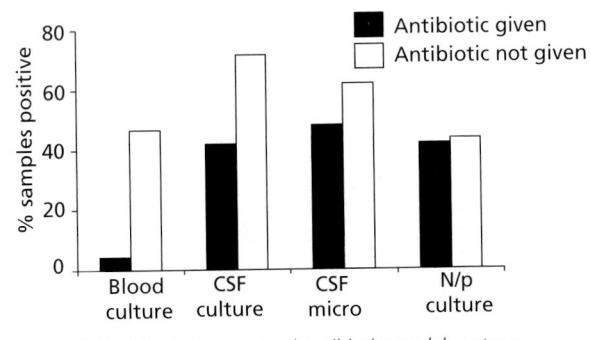

Figure 20.7 *Effect of parenteral antibiotics on laboratory investigations in meningococcal disease (from Cartwright et al. 1992, with permission).*

space in a loculated space after a short period of antibiotic treatment, subsequently to re-emerge and to cause relapse. Normally, they are killed and cleared rapidly from the subarachnoid space (Feldman 1977; Kanegaye et al. 2001).

STEROIDS AND IMMUNOMODULATORS

The role of steroids in meningococcal disease remains controversial. A meta-analysis of trials undertaken in the 1990s showed a reduction in the incidence of neurological sequelae, mainly deafness, in patients with bacterial meningitis when dexamethasone was given before or with the first dose of antibiotics (Feigin et al. 1992; Schaad et al. 1993; Jafari and McCracken 1994). Most patients had Hib meningitis. Since the pathology of neurological damage is probably similar for the three main bacterial pathogens, it seems logical to endorse this addition to the treatment of meningococcal and pneumococcal meningitis. Steroids are ineffective in bacterial meningitis if given more than about an hour after the first parenteral dose of antibiotic. A 2-day course is probably as effective as treatment for 4 days (Syrogiannopoulos et al. 1994). After 48 h the cytokine inflammatory cascade in the CSF should have stabilized.

There is no evidence to support the use of dexamethasone in meningococcal septicemia (Nadel et al. 1995). Various novel therapies for severe meningococcal disease are under consideration or evaluation, including antiendotoxins, anticytokine response agents, leukocyte activation antagonists, cardiovascular support agents, and many others (Nadel et al. 1995). Some encouraging results have been achieved in small uncontrolled studies (Giroir et al. 1997; Goldman et al. 1997; Smith et al. 1997).

Recent evidence indicates that case-fatality rates in meningococcal infection have fallen significantly in centers of excellence (Thorburn et al. 2001; Booy et al. 2001). These analyses indicate that this is almost certainly due to improved clinical management, since the trend to lower mortality was based on children with comparable disease severity scores. The marked trend to lower case-fatality rates that has been observed nationally across the UK could reflect relatively more efficient ascertainment of surviving cases than fatal cases but is also likely to include a number of other factors, including earlier recognition of possible meningococcal infection by parents of young children, the greater use of pre-admission parenteral penicillin, and improved management in hospital. The UK national case fatality rate is now below 7 percent and maintaining a falling trend.

SECONDARY CASES AND CLUSTERS

Close contacts of cases of meningococcal disease run an increased risk of developing disease in the days and weeks after exposure to an index case, especially a first-degree relative living in the same household. The increased risk may be due to exposure to a known pathogenic strain, an inherited susceptibility to meningococcal infection, transient exposure to a common environmental factor (e.g. influenza virus) (Harrison et al. 1991), or a combination of these. An attempt has been made to estimate the proportion of risk attributable to host factors by calculating the sibling familial risk ratio (Haramboulos et al. 2003); these authors estimated that host genetic factors contributed approximately one third of the risk.

The nasopharyngeal carriage rate of meningococci is high in close family contacts of cases (Kristiansen et al. 1998). Rifampicin, ciprofloxacin, and ceftriaxone are all more than 90 percent successful in eliminating nasopharyngeal colonization in contacts (Cuevas and Hart 1993). They also reduce the incidence of secondary cases of disease in the 30–60 days after the illness of the index case. Though there have been no large prospective trials to establish the benefit of chemoprophylaxis, and though there is no evidence to show that the risk of secondary disease is reduced in the longer term, it is accepted policy to offer chemoprophylaxis to defined close contacts (Meningococcal Infections Working Group 1989). If the index case strain is of serogroup A or C, close contacts should also be offered A + C vaccine. Regardless of the causative organism, all contacts should be informed about their increased risk of meningococcal disease, which may persist for many months, especially when chemoprophylaxis has been given.

Clusters of cases sometimes occur in playgroups, schools, universities, military establishments, and other closed and semiclosed communities. Serogroup C strains are more commonly isolated in such outbreaks. Consideration should be given to the value of chemoprophylaxis, with or without vaccination, but each situation must be judged on its merits. The level of public alarm is often at variance with the actual risk (Hastings et al. 1997; Stuart et al. 1997). Cases of meningococcal disease occur occasionally in laboratory workers who have been handling the organism. Safety cabinets must always be used when handling liquid suspensions of virulent meningococci (Boutet et al. 2001).

Multilocus sequence typing (MLST) (Maiden et al. 1998) and other molecular characterization methods will play an increasingly important role in elucidating meningococcal microepidemiology and facilitating rational management of clusters and community outbreaks. MLST can also be applied to other genetically diverse bacterial species such as pneumococci, *Haemophilus influenzae*, etc.

Vaccines

The first vaccines for the prevention of meningococcal disease were developed in 1912 (Sophian and Black

1912). Uncontrolled field trials of a whole-cell meningo-coccal vaccine took place in England during World War I (Greenwood 1916). In the 1930s, Scherp and Rake (1935) identified capsular material from a meningo-coccus as a polysaccharide, but work on vaccine development, driven by outbreaks of disease amongst military recruits, waned when sulfonamides and then penicillin became available. With the emergence of sulfonamide resistance in the 1960s, interest in prevention by vaccination was reawakened, stimulated by the inability to control clusters of cases in recruit camps by sulfonamide prophylaxis. In a series of classical papers, Gotschlich, Goldschneider, and their colleagues at the Walter Reed Army Institute of Research, Washington, USA, demonstrated the fundamental importance of bactericidal antibodies in determining protection from invasive meningococcal disease. They went on to develop the first serogroup C polysaccharide vaccine to be properly evaluated in humans (Goldschneider et al. 1969a, b; Gotschlich et al. 1969a, b; Artenstein et al. 1970). Soon after this, a large clinical trial of a serogroup A poly-saccharide vaccine was reported from Finland (Peltola et al. 1977), and further serogroup A vaccine studies in Africa followed.

SEROGROUP A AND C VACCINES

Vaccines containing serogroup A and C polysaccharides have been used quite widely in the UK, and quad-rivalent A, C, Y, and W-135 vaccine is also available. Their use has declined with the availability of conjugated serogroup C vaccines (see below). Purified poly-saccharide (i.e. nonconjugated) vaccines are not immunogenic in infants and give only short-term (3–4 years) protection to older children and adults. They are not suitable for universal use and are employed in a more limited set of circumstances (see Table 20.2).

A conjugated serogroup A and C vaccine has undergone preliminary evaluation in the Gambia (Twumasi et al. 1995). In the UK, conjugated serogroup C vaccines were evaluated in the latter half of the 1990s, were found to be safe and effective, and were then offered to all infants and older children commencing in November 1999 in the face of a high incidence of serogroup C disease (Miller et al. 2002). Their impact has been immediate and dramatic in the targeted age groups, with a reduction in serogroup C disease incidence now

Table 20.2 Uses of meningococcal serogroup A + C nonconjugated vaccine

Uses
Protection of contacts
Protection of travelers to endemic areas
Control of epidemics
Outbreak control in defined communities
Outbreak control in open communities

exceeding 80 percent (Figure 20.8). These conjugated vaccines also reduce the carriage of serogroup C meningo-cocci (Maiden et al. 2002), and an impact on those older people who were not initially offered the vaccine is now becoming apparent, driven by reduced exposure, i.e. herd immunity. The exploitation of herd immunity as a deliberate strategy in the vaccine control of meningococcal disease has been advocated by Stephens (1999).

Prior to the introduction of conjugated serogroup C vaccines in the UK, most serogroup C disease was caused by a single clone (ET-37). Since it was known that naturally occurring strains of this clone were capable of expressing a serogroup B capsule (as opposed to the much commoner expression of a serogroup C capsule), fears were expressed that though serogroup C disease might be reduced, this could be followed by an expansion of the population of ET-37 strains expressing the serogroup B capsule, with a consequent increase in the incidence of serogroup B disease. Despite very careful ongoing monitoring of both invasive and carried meningococcal strains, to date there is no evidence either of an increase in serogroup B disease overall or of any expansion in the numbers of cases of ET-37 serogroup B disease.

Conjugated meningococcal serogroup A vaccines are now being developed and subjected to clinical trials in the sub-Saharan meningitis belt in a program enjoying substantial funding from a US charitable foundation. The ultimate aim of the program is to offer serogroup A (or possibly A + C) conjugated vaccines to at-risk individuals across the whole meningitis belt, with a view to controlling epidemic meningococcal disease in this region.

Conjugated A, C, Y, and W-135 vaccines are also being developed and could have applications in the USA, where serogroup Y strains contribute significantly to the total burden of meningococcal disease, and also in travelers.

SEROGROUP B VACCINES

Serogroup B meningococcal polysaccharide is a weak antigen in humans and prospects for effective serogroup B vaccines are more distant. Occasionally, serogroup B disease can become hyperendemic, and a single strain variant can become responsible for the great majority of cases. In such circumstances it is possible to develop outer-membrane vesicle vaccines against the causative strain of meningococcus. Such vaccines have been developed in Norway (Bjune et al. 1991) and in Cuba (de Moraes et al. 1992); however, neither was effective in infants and there is some evidence that the duration of protection may be limited. In most countries (including the UK), serogroup B disease is caused by a hetero-geneous range of serogroup B strains, and the approach of using a single serogroup B variant as the basis for a vaccine is inapplicable.

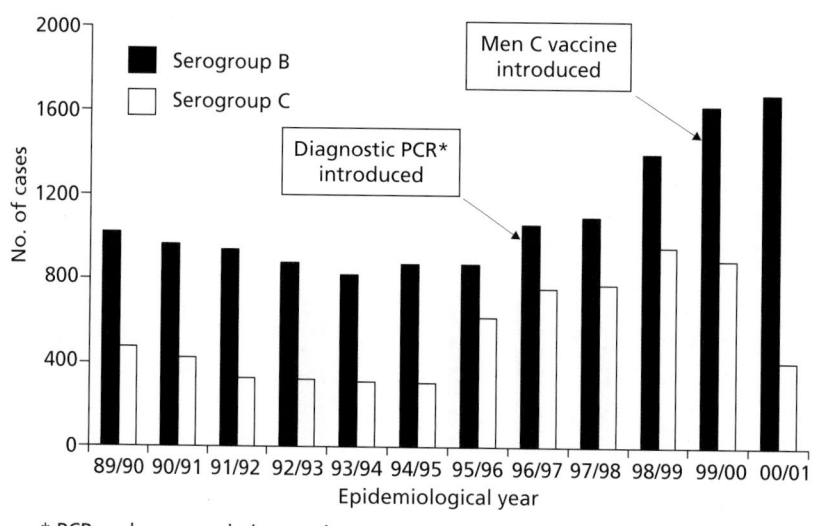

* PCR: polymerase chain reaction

Figure 20.8 *Laboratory-confirmed cases of meningococcal disease by epidemiological year (July–June), England and Wales, 1989–90 to 2000–01.*

Since bactericidal antibody to serogroup B meningococci is directed against surface components other than the capsular polysaccharide, much effort is being invested in the search for stable, surface-expressed serogroup B antigens or epitopes that will evoke the development of human bactericidal antibodies (Poolman et al. 1995). However, meningococci are very well adapted human colonizers and many outer-membrane proteins have hypervariable regions, presumably to evade the host immune response. A further problem hampering development of serogroup B vaccines is the lack of correlation between the immune responses to candidate antigens in laboratory animals and in man.

Among the antigens currently under investigation are the class 1 outer-membrane protein (van der Ley and Poolman 1992; van der Ley et al. 1993) and iron-binding and iron-regulated proteins (Frasch 1995). Vaccines based on conjugated, chemically modified B polysaccharide are also being explored (Fusco et al. 1997). The recent publication of the entire genetic sequence of a well-characterized serogroup B meningococcus, MC58 (Tettelin et al. 2000) has facilitated the search for genes encoding surface-expressed, well-conserved antigens and may result in the identification of a number of novel targets relevant for vaccine development (Pizza et al. 2000). Progress in the development of serogroup B (and other meningococcal) vaccines has been reviewed by Morley and Pollard (2002) and by Jódar et al. (2002)

PNEUMOCOCCAL MENINGITIS

Unlike Hib meningitis, the incidence of pneumococcal meningitis in the UK has not declined in recent years. In countries that have introduced conjugated Hib vaccines,

pneumococci are now the second most common cause of bacterial meningitis. Mortality remains high. Antibiotic resistance, not only to penicillin but also to chloramphenicol and third-generation cephalosporins, is becoming a serious therapeutic problem in many parts of the world (Jacobs and Appelbaum 1995; Tomasz 1995).

Rapid progress in the development of conjugated pneumococcal polysaccharide vaccines offers good prospects for reducing the overall incidence of invasive pneumococcal disease and more modest but still interesting possibilities for reducing the incidence of pneumococcal meningitis. For all these reasons, pneumococcal meningitis is an extremely important condition and will remain so for the foreseeable future.

Epidemiology

Streptococcus pneumoniae causes bacteremia and meningitis at all ages, but attack rates are highest in infants, falling to low levels in children and young adults. Bacteremia rates then rise again to very high levels in the elderly, though rates of meningitis do not (Figure 20.9).

Meningitis occurs in about one-sixth of all invasive pneumococcal infections. The highest meningitis attack rate occurs in the first week of life, when the invading organisms are probably acquired from the maternal birth canal. Males are affected more often than females and infection is more common in the winter months. In the UK, pneumococcal meningitis attack rates have remained fairly constant over the period 1991–2000 at around 0.5–0.6 per 100 000 of the total population per annum (Aszkenasy et al. 1995; PHLS unpublished data).

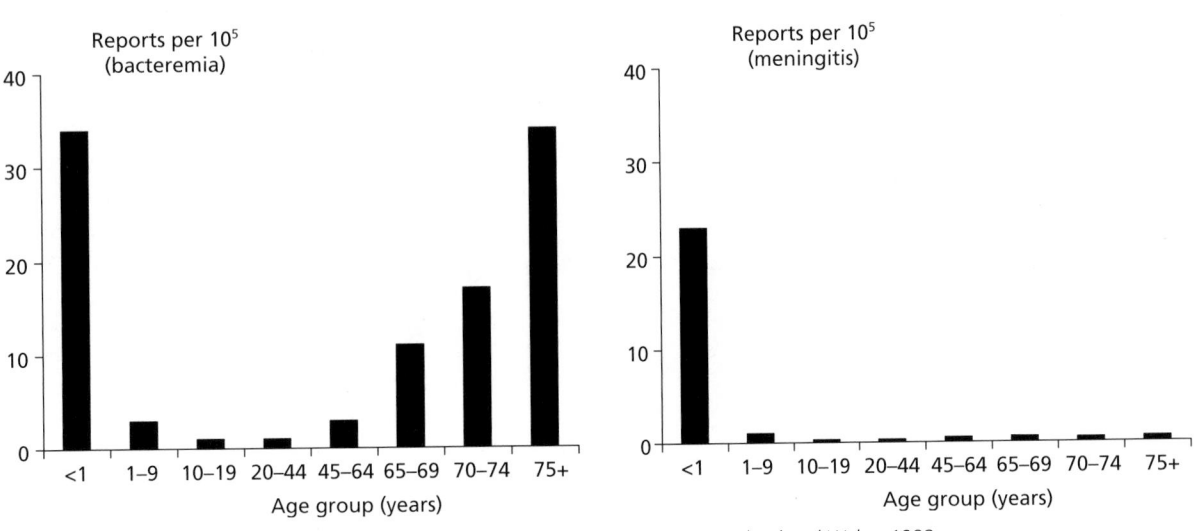

Figure 20.9 *Age incidence of* (**a**) *pneumococcal bacteremia and* (**b**) *meningitis: England and Wales, 1992.*

RISK FACTORS

Though age is the most important determinant of pneumococcal meningitis, patients with pneumococcal meningitis often have a pre-existing medical condition; recognized associations include splenectomy, thalassemia, sickle cell disease, alcoholism, myeloma, hypogammaglobulinemia, and complement deficiency. In the USA, rates of pneumococcal meningitis are much higher in nonwhites, though whether this is associated in part with socioeconomic factors is not clear.

Though pneumococci usually gain access to the subarachnoid space via the bloodstream, they may also spread directly from adjacent tissues, such as an infected paranasal sinus, the middle ear, or from a more distant focus of infection such as the lung. Recently it has been recognized that the risk of pneumococcal meningitis may be raised in recipients of some types of cochlear implants. Pneumococcal meningitis may follow any breach of the integrity of the subarachnoid space; this occurs most frequently following trauma (road accidents, etc.) resulting in dural tear. Pneumococcal meningitis following extrameningeal pneumococcal infection elsewhere in the body (and particularly pneumonia or endocarditis) carries a poor prognosis (Lepper and Dowling 1951; Bohr et al. 1985).

Only a small number of the 84 pneumococcal serotypes (1–9, 12, 14, 18, 19, 22, 23) are responsible for most invasive pneumococcal disease. Particular serotypes may be associated with central nervous system infection, some with a higher case fatality rate.

Clinical presentations and outcome

Clinical presentations depend on age but are similar to those of other types of acute pyogenic meningitis. Typically the presentation is abrupt, but symptoms may develop over several days. At presentation, patients with pneumococcal meningitis tend to be iller than those with Hib or meningococcal meningitis; fits are commoner (Pedersen and Henrichsen 1983) and there may be focal neurological signs, especially cranial nerve palsies. Comatose or semicomatose patients are more likely to have pneumococcal than meningococcal meningitis (Carpenter and Petersdorf 1962). Petechial rashes, though much less frequently encountered than in meningococcal meningitis, may also occur, especially in asplenic patients. In the neonatal period, clinical presentations are similar to those seen in other types of neonatal bacterial meningitis.

Elderly patients, patients with impaired consciousness or with convulsions on admission, and patients with a distant focus of pneumococcal infection are all more likely to die. Deafness and other permanent neurological sequelae are encountered frequently in survivors of pneumococcal meningitis.

Morbidity and mortality in pneumococcal meningitis are much worse than in Hib or meningococcal meningitis. Sangster and colleagues reported a mortality rate of 25 percent in the UK in the 1970s (Sangster et al. 1982) and mortality rates of 40–50 percent are commonplace in pneumococcal meningitis in tropical Africa (Greenwood 1987). Case fatality rates are still probably in excess of 20 percent in the UK and higher in the elderly. A similar proportion of survivors are left with moderate or severe neurological disability. This may be related in part to the much higher CSF bacterial load in pneumococcal meningitis than in meningococcal meningitis (Hassan-King et al. 1984) and to the long persistence of live bacteria within the subarachnoid space after the commencement of antibiotic treatment (Greenwood et al. 1986), the latter possibly associated with the particular capacity of pneumococci to induce a viscid, gelatinous exudate into which antibiotic penetration is poor.

Treatment

ANTIBIOTICS

Until quite recently, pneumococci were sensitive to penicillin in almost all parts of the world. When the causative strain is fully sensitive (minimum inhibitory concentration (MIC) 0.1 μg/ml), benzylpenicillin in large doses remains the antibiotic treatment of first choice. Schmidt and Sesler (1943) were the first to report penicillin resistance. Multiple antibiotic-resistant strains are now encountered frequently in the USA, Australia, New Guinea, and South Africa (Lister 1995; Schwartz and Tunkel 1995). The prevalence of moderately penicillin-resistant strains (MIC 0.1–1.0 μg/ml) varies widely from country to country; in England and Wales it was 2 percent in 1995 (Aszkenasy et al. 1995) rising to 7 percent in 2000 (PHLS unpublished data), 25 percent in the USA, and 40–45 percent in Spain and South Africa (Friedland and Klugman 1992a; Tomasz 1995); in almost all countries the prevalence of moderately resistant and of highly resistant strains (MIC 2.0 μg ml/1 or greater) is increasing (Breiman et al. 1994). Penicillin resistance may have arisen through transfer of genetic material from strains of *Streptococcus mitis* (Coffey et al. 1995).

Pneumococci resistant to third-generation cephalosporins have been documented in the USA and in South Africa (Friedland and Klugman 1992a; Tenover et al. 1992). Detection of resistance to β-lactam antibiotics (Tenover et al. 1992) is not merely of interest in the laboratory: it is associated with treatment failure (Bradley and Connor 1991; Sloas et al. 1992). In countries with a high prevalence of low-level pneumococcal penicillin resistance, third-generation cephalosporins such as cefotaxime and ceftriaxone are widely used as empirical agents of first choice in bacterial meningitis. However, as the level of resistance to penicillin rises, cephalosporins and chloramphenicol lose their efficacy (Friedland and Klugman 1992b; Lister 1995). Antibiotic treatment options for penicillin and cephalosporin-resistant pneumococcal meningitis have been reviewed (Jacobs and Appelbaum 1995; Lister 1995; Schwartz and Tunkel 1995). Chloramphenicol should not be used, even though strains may appear sensitive in the laboratory. Vancomycin, with or without the addition of rifampicin, becomes the antibiotic of choice, even though it penetrates poorly into the CSF. McCracken's group advocate the use of vancomycin accompanied by cefotaxime or ceftriaxone, and with the addition of dexamethasone for the empirical treatment of suspected pneumococcal meningitis (Paris et al. 1995). To cover meningococci, benzylpenicillin or a cephalosporin should always be incorporated into any empirical treatment regimen for bacterial meningitis when the causative organism has not been identified. Though data are scanty, most experts treat pneumococcal meningitis with parenteral antibiotics for at least 10–14 days (McCracken et al. 1987; O'Neill 1993).

STEROIDS

Dexamethasone reduces morbidity in pneumococcal meningitis. A recent pan-European study has demonstrated significantly improved outcome when dexamethasone is given before or with the first dose of parenteral antibiotic in pneumococcal meningitis in adults (de Gans and de Beek 2002; Tunkel and Scheld 2002). If a number of practical issues can be addressed, this may be advocated more widely as part of management protocols if cases of suspected pneumococcal meningitis in adults can be identified reasonably accurately.

MANAGEMENT OF CONTACTS

Contact with a case of pneumococcal meningitis does not usually confer an increased risk of disease. Neither antibiotic prophylaxis nor vaccination is normally recommended to protect intimate contacts of patients with pneumococcal meningitis.

RECURRENT PNEUMOCOCCAL MENINGITIS

The occurrence of a second episode of pneumococcal meningitis in a patient should raise the strong suspicion of a dural tear and there is usually, but not always, a history of cranial trauma. Such patients should be referred to a neurosurgeon for evaluation. A few patients with a history of recurrent pneumococcal meningitis but without a dural defect respond poorly to pneumococcal capsular polysaccharides; usually they respond to conjugated pneumococcal vaccines.

Prevention

PNEUMOCOCCAL VACCINES

Pneumococcal capsular polysaccharides are major virulence determinants and are immunogenic; vaccines based on capsular polysaccharides have been used for more than 20 years. Multivalent vaccines are used because there are more than 80 serotypes and protection is serotype specific. Vaccines containing 6–14 common serotypes were used in the 1970s but were superseded by 23-valent vaccines in the 1980s. Though the frequency with which the different serotypes cause invasive disease varies from country to country and across different age groups (Scott et al. 1996), the currently available 23-valent polysaccharide vaccines offer a high degree of protection in most, if not all, countries.

Purified pneumococcal polysaccharides do not make ideal vaccines (Mitchell and Andrew 1995). In common with other bacterial polysaccharides, they are high-molecular-weight compounds with repeating structures that stimulate B lymphocytes directly without

Table 20.3 *Characteristics of T-cell-independent (TI) and T-cell-dependent (TD) antibody responses*

Characteristics	TI	TD
T cells required	–	+
Ontogeny of response	Late	Early
Induction of memory	–	+
Isotype restriction	+	–
Affinity maturation	–	+

Purified polysaccharides are converted from TI to TD antigens by conjugation to carrier proteins, e.g. tetanus toxoid.

stimulating T cells. T-cell-independent antigens have a number of defects as vaccine candidates (Table 20.3).

In older children and adults, the antibody induced by plain polysaccharide vaccines does not persist, leading to waning of protection over time. The capacity of the 23-valent vaccine to induce good immune responses declines progressively after the age of 60–65 years, particularly in patient groups at increased risk of invasive disease, including those with asplenia, hyposplenism, and a variety of immunodeficiency conditions, including HIV infection. Patients with AIDS and CD4 counts greater than $500/mm^3$ respond well, those with lower counts less satisfactorily (Rodriguez-Barradas et al. 1992). Thus, the groups at highest risk of invasive pneumococcal disease are the most difficult to protect with purified polysaccharide vaccines. Nevertheless, the 23-valent polysaccharide vaccine has been recommended for a wide variety of groups at increased risk of invasive or severe disease. These include patients aged over 2 years with functional or anatomical asplenia or hyposplenism (including sickle cell disease), patients with AIDS (regardless of the stage of illness) and other immunodeficiency states, malignancies, chronic respiratory, renal, and liver diseases, and some metabolic diseases such as diabetes (Department of Health 1996). Plain polysaccharide vaccines offer healthy elderly people protection against invasive pneumococcal disease (and may thereby protect against meningitis) but may not protect against nonbacteremic pneumonia.

Vaccination does not protect against recurrent pneumococcal meningitis associated with dural tears; repair of the defect is the only effective treatment known. Revaccination with 23-valent pneumococcal polysaccharide vaccine within 18 months of a previous dose is associated with a higher incidence of side effects (Borgono et al. 1978), but there is no evidence that revaccination after 4–6 years causes increased side effects (Kaplan et al. 1986a; Mufson et al. 1991).

Conjugated pneumococcal polysaccharide vaccines have now been subjected to clinical trials. A study of a seven-valent conjugated pneumococcal vaccine in infants in California showed excellent protection against invasive disease (Black et al. 2000). Similar conjugated vaccines containing a greater number of pneumococcal serotypes are expected to become available shortly.

They should give rise to long-lasting, though serotype-specific, protection at almost all ages. Conjugated pneumococcal vaccines also reduce nasopharyngeal carriage of homologous pneumococcal serotypes (Dagan et al. 1996) and thus a herd immunity effect may occur if such vaccines were to be widely used in children. Pneumococcal conjugated vaccines will be expensive and may need to be tailored to the serotypes prevalent in different countries.

Attempts to produce pneumococcal vaccines based on surface-exposed proteins expressed by all pneumococci, regardless of serotype, have so far been unsuccessful. Research is continuing on a number of candidate proteins, including pneumolysin, neuraminidase, pneumococcal surface protein A, and a 37-kDa outer-membrane protein. Such outer-membrane proteins may become constituents of pneumococcal vaccines in the future, but are unlikely stand-alone vaccine candidates (Mitchell and Andrew 1995).

Pneumococcal meningitis in the first month of life is probably not preventable by infant vaccination, even at birth. Alternative strategies, such as maternal immunization during or before pregnancy, or passive protection of newborn infants with immunoglobulin followed by active immunization at a later date, may need to be considered.

ANTIBIOTIC PROPHYLAXIS

Long-term oral penicillin or amoxycillin prophylaxis has been used in an attempt to reduce the risk of invasive pneumococcal disease in patients at increased risk, but there is no evidence that prophylactic antibiotics can reduce the risk of primary or recurrent pneumococcal meningitis.

Haemophilus influenzae type b meningitis

EPIDEMIOLOGY

Before the widespread introduction of effective vaccines, Hib was the commonest cause of bacterial meningitis in children under the age of 5 years in almost all countries; in countries such as the USA, Australia, Brazil, and Zaire, it was the commonest cause of bacterial meningitis in the population as a whole. Within countries, attack rates of invasive Hib disease vary widely. In the UK, attack rates of 24–36 per 10^5 children under 5 years of age were observed in six regions in 1990–92 (Anderson et al. 1995), with almost 90 percent of invasive *H. influenzae* infections occurring in children under 5 years of age.

In contrast, attack rates of Hib disease in 'native American' and Australian aboriginal populations of 150–450 per 10^5 children aged under 5 years, i.e. rates ten-fold higher than in Europeans, have been recorded (Bijlmer 1991). In the USA, higher disease rates in blacks than whites may have been due to socioeconomic rather than racial factors (Cochi et al. 1986). Though Hib meningitis has a relatively low case fatality rate in economically advanced countries (3–5 percent), high fatality rates (20–30 percent) are common in tropical Africa (Bijlmer 1991).

Before vaccination was introduced, most UK cases of Hib meningitis occurred in infants, with a peak attack rate in the 6–11 months age group (Anderson et al. 1995); in population groups with very high disease attack rates, peak attack rates occurred in children under 6 months old (Bijlmer 1991). A peak of disease at such an early age presented problems in the deployment of vaccines.

The increasing incidence of invasive Hib disease in the USA and the UK in the 1970s and 1980s was probably due to social changes as well as more accurate diagnosis. Attendance at daycare centers or preschool nurseries may have conferred an increased risk of disease; conversely, breast feeding was protective (Redmond and Pichichero 1984; Cochi et al. 1986).

Bacteriology and immunology

Acapsulate strains of *H. influenzae* are part of the human nasopharyngeal commensal flora. They are of low pathogenicity and cause mainly localized infections in the respiratory tract in young children and in patients with pre-existing disease.

A small minority of *H. influenzae* strains possess a polysaccharide capsule, of which structural variants define six serotypes designated a–f. The type b polysaccharide is a polymer of ribose and ribitol phosphate (polyribosephosphate (PRP)). Before vaccination was introduced, type b strains (Hib) were responsible for more than 95 percent of invasive hemophilus disease. Meningitis was the commonest clinical manifestation, followed by epiglottitis, bacteremia without localizing features, cellulitis, pneumonia, and septic arthritis (Anderson et al. 1995). Most, if not all, invasive Hib disease is accompanied by a bacteremic phase.

Hib is carried mainly by children, but the overall carriage rate in unimmunized populations is only 1 percent in children under 6 years of age; the peak of carriage occurs in children aged 24–36 months (Howard et al. 1988).

The early classical investigations of Fothergill and Wright (1933) showed that blood samples from children aged between 3 months and 3 years lacked bactericidal activity against Hib, in contrast to samples from neonates, older children, and adults. Subsequently Alexander et al. (1942) showed that administration of Hib antiserum caused a substantial increase in the phagocytosis of Hib bacteria in the subarachnoid space, suggesting the importance of opsonizing type-specific antibodies.

The carriage rate of Hib strains in young children is so low that many who acquire antibodies to PRP must do so without exposure to Hib strains. Their protective antibodies probably result from exposure to heterologous bacteria that produce polysaccharides with antigenic determinants identical to those on PRP. Amongst several other bacterial species, *Escherichia coli* K100 expresses an almost identical polysaccharide and, if fed to adult volunteers or laboratory animals, induces antibodies that show bactericidal and opsonic activity against Hib (Schneerson and Robbins 1975).

Bacteremia precedes Hib meningitis, epiglottitis, and the other invasive Hib syndromes. Why the bacteria localize in the meninges, or other organs, is not well understood (Moxon 1992). Experiments in an infant rat model indicate that the intensity of bacteremia may be an important factor in the development of meningitis (Moxon and Ostrow 1977).

Clinical features

The disease may develop insidiously over 24–48 h or it may present more abruptly. Most cases are in infants, who often start by being 'snuffly.' Fever and vomiting are common, early, nonspecific symptoms (Haggerty and Ziai 1964) that may be accompanied by failure to feed, pallor, irritability, and persistent crying. As the infection progresses, shock may supervene and the level of consciousness may become clearly depressed with listlessness and inability to maintain eye contact; infants may be extremely ill by the time the diagnosis is made. Seizures occur in a minority of cases (Feigin 1987), more commonly in Hib and pneumococcal meningitis than in meningococcal meningitis.

In older children, the more traditional symptoms of meningitis – headache, neck stiffness, and photophobia – may be encountered.

Treatment

Intravenous chloramphenicol replaced ampicillin when resistance of Hib to the latter due to β-lactamase production began to increase; in 2002 it was 15 percent in the UK but is higher elsewhere. Chloramphenicol fell into disuse partly because of a small but increasing percentage of resistant strains (1–2 percent), but also because of fears of irreversible bone-marrow suppression. It has been replaced by third-generation cephalosporins. Cefotaxime and ceftriaxone are superior to cefuroxime; the latter takes longer to sterilize the CSF and is associated with a higher incidence of hearing loss (Schaad et al. 1990). Ceftriaxone may be given as a single daily dose and, in carefully selected patients with milder illness, it is suitable for outpatient completion of a course of parenteral antibiotic treatment for bacterial meningitis. Antimicrobial treatment of Hib meningitis should normally be continued for 7–10 days (McCracken et al. 1987; O'Neill 1993).

Administration of dexamethasone to children with Hib meningitis before, or at the same time as, the first dose of antibiotic is given reduces the incidence of neurological sequelae, principally deafness (Lebel et al. 1988; Feigin et al. 1992; Schaad et al. 1993; McIntyre et al. 1997). Steroids probably act by blocking the inflammatory cascade induced by the release of endotoxin into the subarachnoid compartment. If given after the first dose of antibiotic, they are ineffective.

Susceptible contacts of cases of invasive Hib disease run a considerably increased risk of developing Hib disease in the weeks after the contact. The degree of increased risk depends on the intimacy of the contact; household contacts are at greater risk than daycare-center or preschool playgroup contacts. Unvaccinated close contacts aged up to 4 years should be offered rifampicin prophylaxis and an immediate course of Hib vaccine: three doses if aged under 1 year or one dose if aged 13–48 months. Adult contacts do not need Hib vaccine and require chemoprophylaxis only if an unimmunized contact aged under 4 years is in the household (Cartwright et al. 1994).

Vaccines

The first Hib vaccines consisted of purified type b capsular polysaccharide (PRP) and suffered from all the defects of T-cell-independent antigens (see Table 20.3). In spite of their disadvantages, PRP vaccines were widely used in the USA in the 1980s to immunize children aged over 18 months and they did reduce the incidence of invasive Hib disease albeit modestly.

Conjugated polysaccharides are T-cell-dependent antigens and their introduction as vaccines changed the situation dramatically, since such vaccines are immunogenic even in infants under 3 months of age. They induce high-avidity IgG antibodies and long-term immunity.

The impact of Hib vaccination is almost immediate, because almost all invasive Hib disease occurs in infants and very young children. In England and Wales, Hib vaccines were introduced in October 1992 and by 1994 there was a two-thirds reduction in invasive Hib disease and a continuing falling trend. Invasive Hib disease in unimmunized older people also fell, due to a reduction in the prevalence of Hib nasopharyngeal carriage and the beneficial effect of herd immunity

Hib vaccine is offered to all UK infants aged 2, 3, and 4 months and to selected at-risk groups such as asplenic adults. For financial and logistic reasons, these highly effective vaccines have hardly been used in poorer countries, where the need is greatest (Booy 1998).

Over the past 3 years there has been some evidence in the UK of a slight resurgence of invasive Hib disease (Figure 20.10) and it is now planned to offer a fourth,

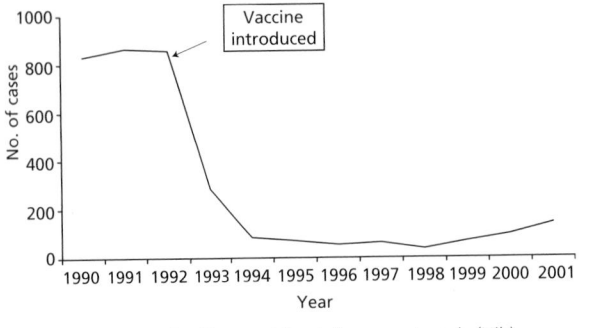

Figure 20.10 *Invasive* Haemophilus influenzae *type b (Hib) infections, England and Wales: Combined Health Protection Agency (HPA) Haemophilus Reference Unit & HPA Communicable Disease Surveillance Center data.*

reinforcing dose of Hib vaccine to all children. This will be given at the same time as the preschool doses of diphtheria, tetanus, and polio vaccines.

NEONATAL MENINGITIS

Neonatal bacterial meningitis (bacterial meningitis in the first month of life) is rare but serious, with a mortality up to 30–40 percent and permanent sequelae in up to 30 percent of survivors. The causative organisms are different from those of bacterial meningitis at other ages, since most bacterial meningitis in this age group is due to organisms derived from ascending infection in utero or from the birth canal during delivery (de Louvois 1994). Occasionally, outbreaks occur in hospital nurseries. Neonatal bacterial meningitis is almost always preceded by bacteremia. In the UK, the principal causative organisms are gram-negative enteric bacilli and group B streptococci. The latter can cause early-onset or late-onset disease as long as 3–4 months after birth. Other bacteria, including pneumococci, *Listeria monocytogenes*, meningococci, other streptococci, and *Staphylococcus aureus*, can also cause neonatal meningitis.

The overall rate of infection in the UK is about 0.25 per 1 000 live births (de Louvois 1994), whereas in the USA it may be two to three times higher (Riley 1972). Twenty years ago, gram-negative enteric bacteria were the most frequent cause of neonatal meningitis in the USA, but more recently, group B streptococci have predominated. The neonatal sepsis rate with group B streptococci is much higher in the USA than in the UK. In the USA, screening late in pregnancy and treatment of carriers has been advocated, but such a policy presents considerable practical problems (Towers 1995).

Prematurity, low birth weight, prolonged rupture of membranes, and prolonged labor all increase the risk of neonatal meningitis. Babies born to mothers who are pyrexial during labor are also at risk, and some infections are acquired in utero.

It is unclear why particular babies develop meningitis. Almost all strains of *E. coli* that cause neonatal meningitis express the K1 capsular antigen (Mulder and Zanen 1984). Group B streptococci of type III are more commonly isolated from cases of neonatal meningitis than other group B streptococci, and both neonatal and maternal antibody levels to these organisms are lower in affected than in control babies.

Neonates frequently show little specific evidence of meningeal irritation. They may be 'jittery' or may present with listlessness, pallor, crying that may be high-pitched, failure to feed, vomiting, and sometimes jaundice; respiration, heart rate, and temperature control may become unstable. These nonspecific symptoms may progress to more obvious depression of consciousness. Raised intracranial pressure, as evidenced by a bulging fontanelle, occurs late, and only in a minority of cases.

Blood cultures are part of the standard investigation of any sick baby. If meningitis is suspected, lumbar puncture should be considered at an early stage, though failure to detect an abnormality on initial examination should never lead to the withholding of antibiotic treatment. Infection at another site such as the urinary tract should be excluded.

Once meningitis is confirmed, the choice of antibiotics depends on the causative organism. A combination of benzylpenicillin and gentamicin is often used while the results of blood and CSF culture are awaited and can be continued if group B streptococci are isolated. If a gram-negative bacillus is isolated, a combination of a cephalosporin and an aminoglycoside may be appropriate, but treatment should be guided by the antibiotic sensitivity of the particular organism.

TUBERCULOUS MENINGITIS

Tuberculous meningitis is rare. In the UK there are around 100 cases each year, the majority in people from the Indian subcontinent. After a steady decline over many years in the incidence of tuberculous disease, there has been a rising trend in notifications over the period 1988–2000 (9.4–12.0 cases per 10^5 per annum). In the USA, tuberculous meningitis is a well-recognized late complication of AIDS (Bishburg et al. 1986); atypical mycobacteria are sometimes responsible. So far, there is little evidence of an overlap between tuberculosis and AIDS in the UK.

Tuberculous meningitis probably always arises as a result of dissemination of tubercle bacilli from a distant focus of infection, often the lung. The disease is caused by liberation of tubercle bacilli from subpial tuberculomata into the subarachnoid space (Rich and McCordock 1933). Postmortem examination confirms the gelatinous nature of the exudate, particularly around the base of the brain, consistent with the high incidence of cranial nerve palsies seen in tuberculous meningitis.

The disease almost always presents insidiously with persistent fever and increasing headache over a period of weeks or months, but, rarely, the presentation may be acute. The cellular exudate in the CSF is most frequently lymphocytic, but it may be mixed, and the protein concentration is often very high (2–3 g/l). The CSF:blood glucose ratio is usually, but not always, reduced. In the majority of cases tubercle bacilli are not seen on direct microscopy. The culture of CSF for tubercle bacilli should be extended because about 10 percent of cultures will become positive only after 6 weeks. A mycobacterial PCR test is now available from the Regional Centres for Mycology in the UK for the investigation of suspected tuberculous meningitis.

Plain chest and abdominal radiographs may reveal a primary tuberculous focus, so increasing suspicions of tuberculous meningitis. Computed tomography and nuclear magnetic resonance imaging of the cranium are useful investigative techniques (Berger 1994). Since direct microscopy is insensitive, empirical anti-tuberculous treatment may have to be started if the clinical picture is strongly suggestive. Treatment is usually based on a combination of rifampicin, isoniazid, and pyrazinamide, but the use of steroids is controversial. The prevalence of isoniazid and multidrug resistance has remained stable between 1994 and 1999 (Djuretic et al. 2002). In the UK, tuberculous meningitis is usually managed by chest physicians because of their familiarity with antituberculous therapy.

OTHER TYPES OF BACTERIAL MENINGITIS

In this brief review it is not possible to list all the more unusual types of bacterial meningitis. *Listeria monocytogenes* continues to cause a small number of cases of meningitis, mainly in the elderly and in dependent patients (Calder 1997). Fulminant listeria meningitis is accompanied by a high case fatality rate. Three nonviral causes of aseptic (lymphocytic) meningitis should be briefly mentioned – Lyme borreliosis (Pachner 1995), syphilis (Lukehart et al. 1988), and primary amebic meningoencephalitis (Carter 1968). These may all present acutely, but all are very rare in the UK. The history, such as that of a tick bite in Lyme meningitis, or of swimming in warm water before amebic meningoencephalitis, may provide diagnostic clues.

For cryptococcal and fungal meningitis see the Medical Mycology Volume, Chapter 32, Cryptococcosis.

REFERENCES

Achtman, M. 1990. Molecular epidemiology of epidemic bacterial meningitis. *Rev Med Microbiol*, **1**, 29–38.

Achtman, M. 1995. Global epidemiology of meningococcal disease. In: Cartwright, K. (ed.), *Meningococcal disease*. Chichester: J. Wiley & Sons, 159–75.

Alexander, H.E., Ellis, C., et al. 1942. Treatment of type-specific *Haemophilus influenzae* infections in infancy and childhood. *J Pediatr*, **20**, 673–98.

Anderson, E.C., Begg, N.T., et al. 1995. Epidemiology of invasive *Haemophilus influenzae* infections in England and Wales in the pre-vaccination era (1990–2). *Epidemiol Infect*, **115**, 89–100.

Artenstein, M.S., Gold, R., et al. 1970. Prevention of meningococcal disease by serogroup C polysaccharide vaccine. *N Engl J Med*, **121**, 372–7.

Aszkenasy, O.M., George, R.C. and Begg, N.T. 1995. Pneumococcal bacteraemia and meningitis in England and Wales 1982 to 1992. *Commun Dis Rep Rev*, **5**, R45–50.

Azmi, F.H., Lucas, A.H., et al. 1995. Human immunoglobulin M paraproteins cross-reactive with *Neisseria meningitidis* group B polysaccharide and fetal brain. *Infect Immun*, **63**, 1906–13.

Begg, N., Cartwright, K.A.V., et al. 1999. Consensus statement on diagnosis, investigation, treatment and prevention of acute bacterial meningitis in immunocompetent adults. *J Infect*, **39**, 1–15.

Berger, J.R.R. 1994. Tuberculous meningitis. *Curr Opin Neurol*, **7**, 191–200.

Bijlmer, H.A. 1991. Worldwide epidemiology of *Haemophilus influenzae* meningitis. *Vaccine*, **9**, Suppl., S5–9.

Bishburg, E., Sunderam, G., et al. 1986. Central nervous system tuberculosis in the acquired immunodeficiency syndrome and its related complex. *Ann Intern Med*, **105**, 210–13.

Bjune, G., Høiby, E.A., et al. 1991. Effect of outer membrane vesicle vaccine against serogroup B meningococcal disease in Norway. *Lancet*, **338**, 1093–6.

Black, S., Shinefeld, H., et al. 2000. Efficacy, safety and immunogenicity of heptavalent pneumococcal conjugate vaccine in children. Northern California Kaiser Permanente Vaccine Study Center Group. *Pediatr Infect Dis J*, **19**, 187–95.

Blakebrough, I.S., Greenwood, B.M., et al. 1982. The epidemiology of infections due to *Neisseria meningitidis* and *Neisseria lactamica* in a northern Nigerian community. *J Infect Dis*, **146**, 626–37.

Bohr, V., Rasmussen, N., et al. 1985. Pneumococcal meningitis: an evaluation of prognostic factors in 164 cases based on mortality and on a study of lasting sequelae. *J Infect*, **10**, 143–57.

Booy, R. 1998. Getting Hib vaccine to those who need it. *Lancet*, **351**, 1446–7.

Booy, R., Habibi, P., et al. 2001. Reduction in case fatality rate from meningococcal disease associated with improved healthcare delivery. *Arch Dis Child*, **85**, 386–90.

Borgono, J.M., McLean, A.A., et al. 1978. Vaccination and revaccination with polyvalent pneumococcal polysaccharide vaccines in adults and infants. *Proc Soc Exp Biol Med*, **157**, 148–54.

Borrow, R., Claus, H., et al. 1997. Non-culture diagnosis and serogroup determination of meningococcal B and C infection by a sialyltransferase (*siaD*) PCR ELISA. *Epidemiol Infect*, **118**, 111–17.

Botha, P. 1988. Penicillin-resistant *Neisseria meningitidis* in southern Africa. *Lancet*, **1**, 54.

Boutet, R., Stuart, J.M., et al. 2001. Risk of laboratory-acquired meningococcal disease. *J Hosp Infect*, **49**, 282–4.

Bradley, J.S. and Connor, J.D. 1991. Ceftriaxone failure in meningitis caused by *Streptococcus pneumoniae* with reduced susceptibility to beta-lactam antibiotics. *Pediatr Infect Dis J*, **10**, 871–3.

Bredius, R.G.M., Derkx, B.H.F., et al. 1994. FCγ receptor IIa (CD32) polymorphism in fulminant meningococcal septic shock in children. *J Infect Dis*, **170**, 848–53.

Breiman, R.F., Butler, J.C., et al. 1994. Emergence of drug-resistant pneumococcal infections in the United States. *J Am Med Assoc*, **271**, 1831–5.

British Society for the Study of Infection Research Committee, 1995. Bacterial meningitis: causes for concern. *J Infect*, **30**, 89–94.

Broome, C.V. 1986. The carrier state: *Neisseria meningitidis*. *J Antimicrob Chemother*, **18**, Suppl. A, 25–34.

Bryan, J.P., de Silva, H.R., et al. 1990. Etiology and mortality of bacterial meningitis in northeastern Brazil. *Rev Infect Dis*, **12**, 128–35.

Bush, F.W. and Bailey, F.R. 1944. The treatment of meningococcus infections with especial reference to the Waterhouse–Friderichsen syndrome. *Ann Intern Med*, **20**, 619–31.

Calder, J.A.M. 1997. Listeria meningitis in adults. *Lancet*, **350**, 307–8.

Carpenter, R.R. and Petersdorf, R.G. 1962. The clinical spectrum of bacterial meningitis. *Am J Med*, **33**, 262–75.

Carter, R.F. 1968. Primary amoebic meningoencephalitis: clinical, pathological and epidemiological features of six fatal cases. *J Pathol Bacteriol*, **96**, 1–25.

Cartwright, K. 1995. Meningococcal carriage and disease. In: Cartwright, K. (ed.), *Meningococcal disease*. Chichester: J. Wiley & Sons, 115–46.

Cartwright, K. and Kroll, S. 1997. Optimising the investigation of meningococcal disease. *Br Med J*, **315**, 757–8.

Cartwright, K.A.V., Stuart, J.M., et al. 1987. The Stonehouse survey: nasopharyngeal carriage of meningococci and *Neisseria lactamica*. *Epidemiol Infect*, **99**, 591–601.

Cartwright, K.A., Jones, D.M., et al. 1991. Influenza A and meningococcal disease. *Lancet*, **338**, 554–7.

Cartwright, K., Reilly, S., et al. 1992. Early treatment with parenteral penicillin in meningococcal disease. *Br Med J*, **305**, 143–7.

Cartwright, K.A.V., Begg, N.T. and Rudd, P. 1994. Use of vaccines and antibiotic prophylaxis in contacts and cases of Haemophilus influenzae type b (Hib) disease. *Commun Dis Rep Rev*, **4**, R16–17.

Caugant, D.A., Mocca, L.F., et al. 1987. Genetic structure of *Neisseria meningitidis* populations in relation to serogroup, serotype, and outer membrane protein pattern. *J Bacteriol*, **169**, 2781–92.

Caugant, D.A., Høiby, E.A., et al. 1994. Asymptomatic carriage of *Neisseria meningitidis* in a randomly sampled population. *J Clin Microbiol*, **32**, 323–30.

Chartrand, S.A. and Cho, C.T. 1976. Persistent pleocytosis in bacterial meningitis. *J Pediatr*, **88**, 424–6.

Cheesbrough, J.S., Morse, A.P., et al. 1995. Meningococcal meningitis and carriage in western Zaire: a hypoendemic zone related to climate? *Epidemiol Infect*, **114**, 75–92.

Cochi, S.L., Fleming, D.W., et al. 1986. Primary invasive *Haemophilus influenzae* type b disease: a population-based assessment of risk factors. *J Pediatr*, **108**, 887–96.

Coffey, T.J., Dowson, C.G., et al. 1995. Genetics and molecular biology of β-lactam-resistant pneumococci. *Microb Drug Resist*, **1**, 29–34.

Connolly, K.D. 1979. Lumbar punctures, meningitis and persisting pleocytosis. *Arch Dis Child*, **54**, 792–3.

Cooke, R.P.D., Riordan, T., et al. 1989. Secondary cases of meningococcal infection among close family and household contacts in England and Wales, 1984–7. *Br Med J*, **298**, 555–8.

Cuevas, L.E. and Hart, C.A. 1993. Chemoprophylaxis of bacterial meningitis. *J Antimicrob Chemother*, **31**, Suppl. B, 79–91.

Dagan, R., Melamed, R., et al. 1996. Reduction of nasopharyngeal carriage of pneumococci during the second year of life by a heptavalent conjugate pneumococcal vaccine. *J Infect Dis*, **174**, 1271–8.

De Gans, J. and van de Beek, D. 2002. Dexamethasone in adults with bacterial meningitis. *N Engl J Med*, **347**, 1549–56.

De Louvois, J. 1994. Acute bacterial meningitis in the newborn. *J Antimicrob Chemother*, **34**, Suppl. A, 61–73.

De Moraes, J.C., Perkins, B.A., et al. 1992. Protective efficacy of a serogroup B meningococcal vaccine in São Paolo, Brazil. *Lancet*, **340**, 1074–8.

Densen, P., Weiler, J.M., et al. 1987. Familial properdin deficiency and fatal meningococcemia. Correction of the bactericidal defect by vaccination. *N Engl J Med*, **316**, 922–6.

Department of Health, 1996. *Immunisation against infectious disease*. London: HMSO, 168–9.

Devi, S.J.N., Schneerson, R., et al. 1991. Identity between polysaccharide antigens of Moraxella nonliquefaciens, group B *Neisseria meningitidis* and *Escherichia coli* K1 (non-O acetylated). *Infect Immun*, **59**, 732–6.

DeVoe, I.W. 1982. The meningococcus and mechanisms of pathogenicity. *Microbiol Rev*, **46**, 162–90.

De Wals, P., Hertoghe, L., et al. 1981. Meningococcal disease in Belgium. Secondary attack rate among household, day-care nursery and pre-elementary school contacts. *J Infect*, **3**, Suppl. 1, S53–61.

De Wals, P., Gilquin, C., et al. 1983. Longitudinal study of asymptomatic meningococcal carriage in two Belgian populations of schoolchildren. *J Infect*, **6**, 147–56.

Djuretic, T., Herbert, J., et al. 2002. Antibiotic resistant tuberculosis in the United Kingdom: 1993–1999. *Thorax*, **57**, 477–82.

Dudley, S.F. and Brennan, J.R. 1934. High and persistent carrier rates of *Neisseria meningitidis* unaccompanied by cases of meningitis. *J Hyg Camb*, **34**, 525–41.

Feigin, R.D. 1987. Bacterial meningitis beyond the neonatal period. In: Feigin, R.D. and Cherry, J.D. (eds), *Textbook of pediatric infections*, 2nd edn. Philadelphia: Saunders, 439–65.

Feigin, R.D., McCracken, G.H. and Klein, J.O. 1992. Diagnosis and management of meningitis. *Pediatr Infect Dis J*, **11**, 785–814.

Feldman, W.E. 1977. Relation of concentrations of bacteria and bacterial antigen in cerebrospinal fluid to prognosis in patients with bacterial meningitis. *N Engl J Med*, **296**, 433–5.

Figueroa, J.E. and Densen, P. 1991. Infectious diseases associated with complement deficiencies. *Clin Microbiol Rev*, **4**, 359–95.

Fijen, C.A.P., Kuijper, E.J., et al. 1989. Complement deficiencies in patients over ten years old with meningococcal disease due to uncommon serogroups. *Lancet*, **2**, 585–8.

Finne, J., Leionen, M. and Mäkelä, H. 1983. Antigenic similarities between brain components and bacteria causing meningitis. *Lancet*, **2**, 355–7.

Flexner, S. and Jobling, J.W. 1908. An analysis of four hundred cases of epidemic meningitis treated with the anti-meningitis serum. *J Exp Med*, **10**, 690–733.

Fontanals, D., Pineda, V., et al. 1989. Penicillin resistant beta-lactamase producing *Neisseria meningitidis* in Spain. *Eur J Clin Microbiol Infect Dis*, **8**, 90–1.

Fothergill, L.D. and Wright, J. 1933. Influenzal meningitis: the relation of age incidence to the bactericidal power of blood against causal organisms. *J Immunol*, **24**, 273–84.

Frasch, C.E. 1995. Meningococcal vaccines: past, present and future. In: Cartwright, K. (ed.), *Meningococcal disease*. Chichester: J. Wiley & Sons, 21–34.

Friedland, I.R. and Klugman, K.P. 1992a. Antibiotic-resistant pneumococcal disease in South African children. *Am J Dis Child*, **146**, 920–3.

Friedland, I.R. and Klugman, K.P. 1992b. Failure of chloramphenicol therapy in penicillin-resistant pneumococcal meningitis. *Lancet*, **339**, 405–8.

Fusco, P.C., Michon, F., et al. 1997. Preclinical evaluation of a novel group B meningococcal conjugate vaccine that elicits bactericidal activity in both mice and nonhuman primates. *J Infect Dis*, **175**, 364–72.

Giroir, B.P., Quint, P.A., et al. 1997. Preliminary evaluation of recombinant amino-terminal fragment of human bactericidal/permeability-increasing protein in children with severe meningococcal sepsis. *Lancet*, **350**, 1439–43.

Glode, M.P., Robbins, J.B., et al. 1977. Cross-antigenicity and immunogenicity between capsular polysaccharides of group C *Neisseria meningitidis* and of *Escherichia coli* K92. *J Infect Dis*, **135**, 94–102.

Glover, J.A. 1920. Observations of the meningococcus carrier rate and their application to the prevention of cerebro-spinal fever. *Special Report Series of the Medical Research Council (London)*, **50**, 133–65.

Gold, R., Goldschneider, I., et al. 1978. Carriage of *Neisseria meningitidis* and *Neisseria lactamica* in infants and children. *J Infect Dis*, **137**, 112–21.

Goldman, A.P., Kerr, S.J., et al. 1997. Extracorporeal support for interactable cardiorespiratory failure due to meningococcal disease. *Lancet*, **349**, 466–9.

Goldschneider, I., Gotschlich, E.C. and Artenstein, M.S. 1969a. Human immunity to the meningococcus. I. The role of humoral antibodies. *J Exp Med*, **129**, 1307–26.

Goldschneider, I., Gotschlich, E.C. and Artenstein, M.S. 1969b. Human immunity to the meningococcus. II. Development of natural immunity. *J Exp Med*, **129**, 1327–48.

Gotschlich, E.C., Goldschneider, I. and Artenstein, M.S. 1969a. Human immunity to the meningococcus. IV. Immunogenicity of group A and group C meningococcal polysaccharides in human volunteers. *J Exp Med*, **129**, 1367–84.

Gotschlich, E.C., Liu Teh, Yung and Artenstein, M.S. 1969b. Human immunity to the meningococcus. III. Preparation and immunochemical properties of the group A, group B and group C meningococcal polysaccharides. *J Exp Med*, **129**, 1349–65.

Grandi, G. 2001. Antibacterial vaccine design using genomics and proteomics. *Trends Biotechnol*, **19**, 181–8.

Gray, L.D. and Fedorko, D.P. 1992. Laboratory diagnosis of bacterial meningitis. *Clin Microbiol Rev*, **5**, 130–45.

Greenfield, S. and Feldman, H.A. 1967. Familial carriers and meningococcal meningitis. *N Engl J Med*, **277**, 498–502.

Greenfield, S., Sheehe, P.R. and Feldman, H.A. 1971. Meningococcal carriage in a population of 'normal' families. *J Infect Dis*, **123**, 67–73.

Greenwood, M. 1916. The outbreak of cerebrospinal fever at Salisbury in 1914–15. *Proc R Soc Med*, **10**, 2, 44–60.

Greenwood, B.M. 1987. The epidemiology of acute bacterial meningitis in tropical Africa. In: Williams, J.D. and Burnie, J. (eds), *Bacterial meningitis*. London: Academic Press, 61–91.

Greenwood, B.M., Hassan-King, M., et al. 1986. Sequential bacteriological findings in the cerebrospinal fluid of Nigerian patients with pneumococcal meningitis. *J Infect*, **12**, 49–56.

Greisen, K., Loeffelholz, M., et al. 1994. PCR primers and probes for the 16S rRNA gene of most species of pathogenic bacteria, including bacteria found in cerebrospinal fluid. *J Clin Microbiol*, **32**, 335–51.

Griffiss, J.M. 1982. Epidemic meningococcal disease: synthesis of a hypothetical immunoepidemiologic model. *Rev Infect Dis*, **4**, 159–72.

Griffiss, J.M. 1995. Mechanisms of host immunity. In: Cartwright, K. (ed.), *Meningococcal disease*. Chichester: J. Wiley & Sons, 35–70.

Griffiss, J.M., Brandt, B.L., et al. 1984. Immune response of infants and children to disseminated infections with *Neisseria meningitidis*. *J Infect Dis*, **150**, 71–9.

Guirguis, N., Schneerson, R., et al. 1985. *Escherichia coli* K51 and K93 capsular polysaccharides are cross-reactive with the group A capsular polysaccharide of *Neisseria meningitidis*. *J Exp Med*, **162**, 1837–51.

Gwaltney, J.M., Sande, M.A., et al. 1975. Spread of *Streptococcus pneumoniae* in families. II. Relation of transfer of *S. pneumoniae* to incidence of colds and serum antibody. *J Infect Dis*, **132**, 62–8.

Hackett, S.J., Carrol, E.D., et al. 2002a. Improved case confirmation in meningococcal disease with whole blood Taqman PCR. *Arch Dis Child*, **86**, 449–52.

Hackett, S.J., Guiver, M., et al. 2002b. Meningococcal bacterial DNA load at presentation correlates with disease severity. *Arch Dis Child*, **86**, 44–6.

Haggerty, R.J. and Ziai, M. 1964. Acute bacterial meningitis. *Adv Pediatr*, **13**, 129–81.

Halstensen, A., Sjursen, H. and Vollset, S.E. 1989. Serum opsonins to serogroup B meningococci in meningococcal disease. *Scand J Infect Dis*, **21**, 267–76.

Haramboulos, E., Weiss, H.A., et al. 2003. Sibling familial risk ratio of meningococcal disease in UK caucasians. *Epidemiol Infect*, **130**, 413–18.

Harrison, L.H., Armstrong, C.W., et al. 1991. A cluster of meningococcal disease on a school bus following epidemic influenza. *Arch Intern Med*, **151**, 1005–9.

Hassan-King, M., Whittle, H.C. and Greenwood, B.M. 1984. Inhibitory effect of cerebrospinal fluid on the growth of meningococci and pneumococci. *J Clin Pathol*, **37**, 428–32.

Hastings, L., Stuart, J., et al. 1997. A retrospective survey of clusters of meningococcal disease in England and Wales, 1993 to 1995: estimated risks of further cases in household and educational settings. *Commun Dis Rep Rev*, **7**, R195–200.

Heubner, J.O.L. 1896. Beobachtungen und versuche über den Meningokokkus intracellularis (Weichselbaum-Jaeger). *Jb Kinderheilk*, **43**, 1–22.

Hill, A.V.S. 2001. The genomics and genetics of human infectious disease susceptibility. *Ann Rev Genomics Hum Genet*, **2**, 373–400.

Hirsch, A. 1886. Epidemic cerebro-spinal meningitis, *Handbook of geographical and historical pathology. Vol. III – Diseases of organs and parts*, translated from the German by Creighton, C. London: New Sydenham Society, 547–94.

Howard, A.J., Dunkin, K.T. and Millar, G.W. 1988. Nasopharyngeal carriage and antibiotic resistance of *Haemophilus influenzae* in healthy children. *Epidemiol Infect*, **100**, 193–203.

Hubert, B., Watier, L., et al. 1992. Meningococcal disease and influenza-like syndrome: a new approach to an old question. *J Infect Dis*, **166**, 542–5.

Jacobs, M.R. and Appelbaum, P.C. 1995. Antibiotic-resistant pneumococci. *Rev Med Microbiol*, **6**, 77–93.

Jafari, H.S. and McCracken, G.H. 1994. Dexamethasone therapy in bacterial meningitis. *Pediatr Ann*, **23**, 83–8.

Jensen, A.G., Espersen, F., et al. 1993. *Staphylococcus aureus* meningitis. A review of 104 nationwide, consecutive cases. *Arch Intern Med*, **153**, 1902–8.

Jódar, L., Feavers, I.M., et al. 2002. Development of vaccines against meningococcal disease. *Lancet*, **359**, 1499–508.

Jones, D.M. and Mallard, R.H. 1993. Age incidence of meningococcal infection England and Wales, 1984–1991. *J Infect*, **27**, 83–8.

Jones, D.M. and Kaczmarski, E.B. 1994. Meningococcal infections in England and Wales: 1993. *Commun Dis Rep Rev*, **4**, R97–100.

Kaczmarski, E.B., Ragunathan, P.L., et al. 1998. Creating a national service for the diagnosis of meningococcal disease by polymerase chain reaction. *Commun Dis Public Health*, **1**, 54–6.

Kanegaye, J.T., Soliemanzadeh, P. and Bradley, J.S. 2001. Lumbar puncture in pediatric bacterial meningitis: defining the time interval for recovery of cerebrospinal fluid pathogens after parenteral antibiotic pretreatment. *Pediatrics*, **108**, 1169–74.

Kaplan, J., Sarnaik, S. and Schiffman, G. 1986a. Revaccination with polyvalent pneumococcal vaccine in children with sickle cell anaemia. *Am J Pediatr Hematol Oncol*, **8**, 80–2.

Kaplan, S.L., Smith, E.O.'B. et al. 1986b. Association between preadmission oral antibiotic therapy and cerebrospinal fluid findings and sequelae caused by *Haemophilus influenzae* type b meningitis. *Pediatr Infect Dis*, **5**, 626–32.

Kasper, D.L., Winkelhake, J.L., et al. 1973. Antigenic specificity of bactericidal antibodies in antisera to *Neisseria meningitidis*. *J Infect Dis*, **127**, 378–87.

Kiefer, F. 1896. Zur Differentialdiagnose des Erregers der epidemischen Cerebrospinal-meningitis und der Gonorrhoe. *Berl Klin Wochenschr*, **33**, 628–30.

Kilian, M. and Reinholdt, J. 1987. A hypothetical model for the development of invasive infection due to IgA1 protease-producing bacteria. *Adv Exp Med Biol*, **216B**, 1261–9.

Kilian, M., Mestecky, J. and Russell, M.W. 1988. Defense mechanisms involving Fc-dependent functions of immunoglobulin A and their subversion by bacterial immunoglobulin A proteases. *Microbiol Rev*, **52**, 296–303.

Kleinman, M.B., Reynolds, J.K., et al. 1984. Superiority of acridine orange versus Gram stain in partially treated bacterial meningitis. *J Pediatr*, **104**, 401–4.

Kolle, W. and Wasserman, A. 1906. Versuche zur Gewinnung und Wertbestimmung eines Meningokokkenserums. *Dtsch Med Wochenschr*, **32**, 609–12.

Krasinski, K., Nelson, J.D., et al. 1987. Possible association of mycoplasma and viral respiratory infections with bacterial meningitis. *Am J Epidemiol*, **125**, 499–509.

Kristiansen, B.-E., Tveten, Y. and Jenkins, A. 1998. Which contacts of patients with meningococcal disease carry the pathogenic strain of *Neisseria meningitidis*? A population based study. *Br Med J*, **317**, 621–5.

Lambert, H.P. In: Lambert, H.P. (ed.), *Infections of the central nervous system*. London: Arnold.

Lapeysonnie, L. 1963. La méningite cérébrospinale en Afrique. *Bull WHO*, **28**, Suppl., 53–114.

Lebel, M.H., Freij, B.J., et al. 1988. Dexamethasone therapy for bacterial meningitis: results of two double-blind, placebo-controlled trials. *N Engl J Med*, **319**, 964–71.

Lepper, M.H. and Dowling, F.H. 1951. Treatment of pneumococci meningitis with penicillin compared with penicillin and aureomycin. *Arch Intern Med*, **88**, 489–94.

Linton, S.M. and Morgan, B.P. 1999. Properdin deficiency and meningococcal disease – identifying those most at risk. *Clin Exp Immunol*, **118**, 189–91.

Lister, P.D. 1995. Multiply-resistant pneumococcus: therapeutic problems in the management of serious infections. *Eur J Clin Microbiol Infect Dis*, **14**, Suppl. 1, 18–25.

Lukehart, S., Hook, E.W., et al. 1988. Invasion of the central nervous system by *Treponema pallidum*. Implications for diagnosis and therapy. *Ann Intern Med*, **109**, 855–62.

Lystad, A. and Aasen, S. 1991. The epidemiology of meningococcal disease in Norway 1975–91. *NIPH Ann*, **14**, 57–65.

Martin, D.R., Walker, S.J., et al. 1998. New Zealand epidemic of meningococcal disease identified by a strain with phenotype B:4:P1.4. *J Infect Dis*, **177**, 497–500.

Marzouk, O., Thomson, A.P.J., et al. 1991. Features and outcome in meningococcal disease presenting with maculopapular rash. *Arch Dis Child*, **66**, 485–7.

Maiden, M.C.J., Bygraves, J.A., et al. 1998. Multilocus sequence typing: A portable approach to the identification of clones within populations of pathogenic microorganisms. *Proc Natl Acad Sci USA*, **95**, 3140–5.

Maiden, M.C.J., Stuart, J.M., et al. 2002. Carriage of serogroup C meningococci year after meningococcal C conjugate polysaccharide vaccination. *Lancet, 359,* **1**, 1829–30.

Marchiafava, E. and Celli, A. 1884. Spra i micrococchi della meningite cerebrospinale epidemica. *Gazzdegli Ospedali*, **5**, 59.

Maxson, S. and Jacobs, R.F. 1993. Viral meningitis. Tips to rapidly diagnose treatable causes. *Postgrad Med*, **93**, 153–66.

Maxson, S., Lewno, M.J. and Schutze, G.E. 1994. Clinical usefulness of cerebrospinal fluid bacterial antigen studies. *J Pediatr*, **125**, 235–8.

McCracken, G.H., Nelson, J.D., et al. 1987. Consensus report: antimicrobial therapy for bacterial meningitis in infants and children. *Pediatr Infect Dis J*, **6**, 501–5.

McIntyre, P.B., Berkey, C.S., et al. 1997. Dexamethasone as adjunctive therapy in bacterial meningitis: a meta-analysis of randomized clinical trials since 1988. *J Am Med Assoc*, **278**, 925–31.

McNeil, G., Virji, M. and Moxon, E.R. 1994. Interactions of *Neisseria meningitidis* with human monocytes. *Microb Pathogen*, **16**, 153–63.

Meningococcal Infections Working Group, 1989. The epidemiology and control of meningococcal disease. Internal publication of the Public Health Laboratory Service, London. *Commun Dis Rep*, **8**, 3–6.

Miller, E., Salisbury, D. and Ramsay, M. 2002. Planning, registration, and implementation of an immunisation campaign against meningococcal serogroup C disease in the UK: a success story. *Vaccine*, **20**, S58–67.

Mitchell, T.J. and Andrew, P.W. 1995. Vaccines against *Streptococcus pneumoniae*. In: Ala'Aldeen, D.A.A. and Hormaeche, C.E. (eds), *Molecular and clinical aspects of bacterial vaccine development*. Chichester: J. Wiley & Sons, 93–117.

Moore, P.S. 1992. Meningococcal disease in sub-Saharan Africa: a model for the epidemic process. *Clin Infect Dis*, **14**, 515–25.

Moore, P.S., Hierholzer, J., et al. 1990. Respiratory viruses and mycoplasma as cofactors for epidemic group A meningococcal meningitis. *J Am Med Assoc*, **264**, 1271–5.

Moosa, A.A., Quortum, H.A. and Ibrahim, M.D. 1995. Rapid diagnosis of bacterial meningitis with reagent strips. *Lancet*, **345**, 1290–1.

Morley, S.L. and Pollard, A.J. 2002. Vaccine prevention of meningococcal disease, coming soon? *Vaccine*, **20**, 666–87.

Moxon, E.R. 1992. Molecular basis of invasive *Haemophilus influenzae* type b disease. *J Infect Dis*, **165**, Suppl. 1, S77–81.

Moxon, E.R. and Ostrow, P.T. 1977. *Haemophilus influenzae* meningitis in infant rats: role of bacteremia in pathogenesis of age-dependent inflammatory responses in cerebrospinal fluid. *J Infect Dis*, **135**, 303–7.

Moxon, E.R., Glode, M.P., et al. 1977. The infant rat as a model of bacterial meningitis. *J Infect Dis*, **136**, Suppl., S186–92.

Mufson, M.A., Hughey, D.F., et al. 1991. Revaccination with pneumococcal vaccine of elderly persons 6 years after primary vaccination. *Vaccine*, **9**, 403–7.

Mulder, C.J.J. and Zanen, H.C. 1984. Neonatal meningitis caused by *Escherichia coli* in the Netherlands. *J Infect Dis*, **150**, 935–9.

Mulks, M.H. 1985. Microbial IgA proteases. In: Holder, I.A. (ed.), *Bacterial enzymes and virulence*. Boca Raton, FL: CRC Press, 81–104.

Mulks, M.H. and Plaut, A.G. 1978. IgA protease production as a characteristic distinguishing pathogenic from harmless Neisseriaceae. *N Engl J Med*, **299**, 973–6.

Munford, R.S., Taunay, A.E., et al. 1974. Spread of meningococcal infection within households. *Lancet*, **1**, 1275–8.

Nadel, S. 2001. Lumbar puncture should not be performed in meningococcal disease. *Arch Dis Child*, **84**, 373.

Nadel, S., Levin, M. and Habibi, P. 1995. Treatment of meningococcal disease in childhood. In: Cartwright, K. (ed.), *Meningococcal disease*. Chichester: J. Wiley & Sons, 207–43.

Neal, K.R., Nguyen-Van-Tam, J.S., et al. 1999. Seven-week interval between acquisition of a meningococcus and the onset of invasive disease. A case report. *Epidemiol Infect*, **123**, 507–9.

Neal, K.R., Nguyen-Van-Tam, J.S., et al. 2000. Changing carriage rate of *Neisseria meningitidis* among university students during the first week of term: a cross-sectional study. *Br Med J*, **320**, 846–9.

Nicolle, M., Debains, E. and Jouan, C. 1918. Etudes sur les méningococciques et les serums anti-méningococciques. *Ann Inst Pasteur*, **32**, 150–69.

O'Donoghue, J.M., Schweid, A.I. and Beaty, H.N. 1974. Experimental pneumococcal meningitis. I. A rabbit model. *Proc Soc Exp Biol Med*, **146**, 571–6.

Olcén, P., Kjellander, J., et al. 1979. Culture diagnosis of meningococcal carriers: yield from different sites and influence of storage in transport medium. *J Clin Pathol*, **32**, 1222–5.

Olcén, P., Kjellander, J., et al. 1981. Epidemiology of *Neisseria meningitis*: prevalence and symptoms from the upper respiratory tract in family members to patients with meningococcal disease. *Scand J Infect Dis*, **13**, 105–9.

Olyhoek, T., Crowe, B.A. and Achtman, M. 1987. Clonal population structure of *Neisseria meningitidis* serogroup A isolated from epidemics and pandemics between 1915 and 1983. *Rev Infect Dis*, **9**, 665–92.

O'Neill, P. 1993. How long to treat bacterial meningitis? *Lancet*, **341**, 530.

Pachner, A.R. 1995. Early disseminated Lyme disease: Lyme meningitis. *Am J Med*, **98**, 30S–7S.

Paris, M.M., Ramilo, O. and McCracken, G.H. 1995. Management of meningitis caused by penicillin-resistant *Streptococcus pneumoniae*. *Antimicrob Agents Chemother*, **39**, 2171–5.

Pedersen, F.K. and Henrichsen, J. 1983. Pneumococcal meningitis and bacteraemia in Danish children 1969–78. *Acta Pathol Microbiol Immunol Scand*, **91**, 129–34.

Peltola, H. 1983. Meningococcal disease: still with us. *Rev Infect Dis*, **5**, 71–91.

Peltola, H., Kataja, J.M. and Mäkelä, P.H. 1982. Shift in the age distribution of meningococcal disease as predictor of an epidemic. *Lancet*, **2**, 595–7.

Peltola, H., Mäkelä, P.H., et al. 1977. Clinical efficacy of meningococcus serogroup A capsular polysaccharide vaccine in children three months to five years of age. *N Engl J Med*, **297**, 686–91.

Pizza, M., Scarlato, V., et al. 2000. Identification of vaccine candidates against serogroup B meningococcus by whole-genome sequencing. *Science*, **287**, 1816–20.

Platonov, A.E., Shipulin, G.A., et al. 1998. Association of human FCγRIIa (CD32) polymorphism with susceptibility to and severity of meningococcal disease. *Clin Infect Dis*, **27**, 746–50.

Poolman, J.T., van der Ley, P.A. and Tommassen, J. 1995. Surface structures and secreted products of meningococci. In: Cartwright, K. (ed.), *Meningococcal disease*. Chichester: J. Wiley & Sons, 21–34.

Pujol, C., Eugène, E., et al. 1997. Interaction of *Neisseria meningitidis* with a polarized monolayer of epithelial cells. *Infect Immun*, **65**, 4836–42.

Quincke, H.I. 1893. Ueber meningitis serosa. *Samml Klin Vort (Leipzig)*, **67**, 655–94.

Rake, G. 1931. Biological properties of 'fresh' and 'stock' strains of the meningococcus. *Proc Soc Exp Biol Med*, **29**, 287–9.

Rake, G. and Scherp, H.W. 1933. Studies on meningococcus infection. III. The antigenic complex of the meningococcus – a type-specific substance. *J Exp Med*, **58**, 341–60.

Read, R.C., Fox, A., et al. 1995. Experimental infection of human nasal mucosal explants with *Neisseria meningitidis*. *J Med Microbiol*, **42**, 353–61.

Read, R.C., Camp, N.J., et al. 2000. An interleukin-1 genotype is associated with fatal outcome of meningococcal disease. *J Infect Dis*, **182**, 1557–60.

Redmond, S.R. and Pichichero, M.E. 1984. *Haemophilus influenzae* type b disease, an epidemiologic study with special reference to day-care centers. *JAMA*, **252**, 2581–4.

Reller, L.B., MacGregor, R.R. and Beaty, H.N. 1973. Bactericidal antibody after colonization with *Neisseria meningitidis*. *J Infect Dis*, **127**, 56–62.

Rennick, G., Shann, F. and de Campo, J. 1993. Cerebral herniation during bacterial meningitis in children. *Br Med J*, **306**, 953–5.

Rich, A.R. and McCordock, H.A. 1933. The pathogenesis of tuberculous meningitis. *Bull Johns Hopkins Hosp*, **52**, 5–37.

Riley, H.D. 1972. Neonatal meningitis. *J Infect Dis*, **125**, 420–5.

Riordan, T., Cartwright, K., et al. 1998. Acquisition and carriage of meningococci in marine commando recruits. *Epidemiol Infect*, **121**, 495–505.

Rodriguez-Barradas, M.C., Musher, D.M., et al. 1992. Antibody to capsular polysaccharides of *Streptococcuspneumoniae* after vaccination of human immunodeficiency virus infected subjects with 23-valent pneumococcal vaccine. *J Infect Dis*, **165**, 553–6.

Rolleston, H. 1919. Lumleian lectures on cerebro-spinal fever. Lecture 1. *Lancet*, **1**, 541–9.

Ross, S.C. and Densen, P. 1984. Complement deficiency states and infection: epidemiology, pathogenesis and consequences of neisserial and other infections in an immune deficiency. *Medicine (Baltimore)*, **63**, 243–73.

Sáez-Nieto, J.A., Lujan, R., et al. 1992. Epidemiology and molecular basis of penicillin-resistant *Neisseria meningitidis* in Spain: a 5-year history (1985–1989). *Clin Infect Dis*, **14**, 394–402.

Sangster, G., Murdoch, J.M.C.C. and Gray, J.A. 1982. Bacterial meningitis 1940–79. *J Infect*, **5**, 245–56.

Saukkonnen, K.M., Nowicki, B. and Leinonen, M. 1988. Role of type 1 and S fimbriae in the pathogenesis of *Escherichia coli* O18:K1 bacteremia and meningitis in the infant rat. *Infect Immun*, **56**, 892–7.

Schaad, U.B., Suter, S., et al. 1990. A comparison of ceftriaxone and cefuroxime for the treatment of bacterial meningitis in children. *N Engl J Med*, **322**, 141–7.

Schaad, U.B., Lips, U., et al. 1993. Dexamethasone therapy for bacterial meningitis in children. *Lancet*, **342**, 457–61.

Scherp, H. and Rake, G. 1935. Studies on meningococcus infection. VIII. The type I specific substance. *J Exp Med*, **61**, 753–69.

Schmidt, L.H. and Sesler, C.L. 1943. Development of resistance to penicillin by pneumococci. *Proc Soc Exp Biol Med*, **53**, 353–7.

Schneerson, R. and Robbins, J.B. 1975. Induction of serum *Haemophilus influenzae* type b capsular antibodies in adult volunteers fed cross-reacting *Escherichia coli* 075.K100:H5. *N Engl J Med*, **29**, 1093–6.

Scholten, R.J.P.M., Bijlmer, H.A., et al. 1999. Upper respiratory tract infection, heterologous immunisation and meningococcal disease. *J Med Microbiol*, **48**, 943–6.

Schryvers, A.B. and Gonzalez, G.C. 1990. Receptors for transferrin in pathogenic bacteria are specific for the host's protein. *Can J Microbiol*, **36**, 145–7.

Schwartz, M.T. and Tunkel, A.R. 1995. Therapy of penicillin-resistant pneumococcal meningitis. *Int J Med Microbiol Virol Parasitol Infect Dis*, **282**, 7–12.

Schwentker, F.F. 1937. Treatment of meningococci meningitis with sulfanilamide. *J Pediatr*, **11**, 874–80.

Scott, J.A.G., Hall, A.J., et al. 1996. Serogroup-specific epidemiology of *Streptococcus pneumoniae*: associations with age, sex and geography in 7,000 episodes of invasive disease. *Clin Infect Dis*, **22**, 973–81.

Siegman-Ingra, Y., Bar-Yosef, S., et al. 1993. Nosocomial acinetobacter meningitis secondary to invasive procedures: report of 25 cases and review. *Clin Infect Dis*, **17**, 843–9.

Sloas, M.M., Barrett, F.F., et al. 1992. Cephalosporin treatment failure in penicillin- and cephalosporin-resistant *Streptococcus pneumoniae* meningitis. *Pediatr Infect Dis J*, **11**, 662–6.

Smith, O.P., White, B., et al. 1997. Use of protein-C concentrate, heparin, and haemodiafiltration in meningococcus-induced purpura fulminans. *Lancet*, **350**, 1590–3.

Sophian, A. and Black, J. 1912. Prophylactic vaccination against epidemic meningitis. *J Am Med Assoc*, **59**, 527–32.

Spellerberg, B. and Tuomanen, E. 1994. The pathophysiology of pneumococcal meningitis. *Ann Med*, **26**, 411–18.

Spratt, B.G., Zhang, Q.-Y., et al. 1989. Recruitment of a penicillin-binding protein gene from *Neisseria flavescens* during the emergence of penicillin resistance in *Neisseria meningitidis*. *Proc Natl Acad Sci USA*, **86**, 8988–92.

Stanwell-Smith, R.E., Stuart, J.M., et al. 1994. Smoking, the environment and meningococcal disease: a case-control study. *Epidemiol Infect*, **112**, 315–28.

Stephens, D.S. 1999. Uncloaking the meningococcus: dynamics of carriage and disease. *Lancet*, **353**, 941–2.

Stephens, D.S. and Farley, M.M. 1991. Pathogenic events during infection of the human nasopharynx with *Neisseria meningitidis* and *Haemophilus influenzae*. *Rev Infect Dis*, **13**, 22–33.

Stuart, J.M., Cartwright, K.A.V., et al. 1989. Effect of smoking on meningococcal carriage. *Lancet*, **2**, 723–5.

Stuart, J.M., Monk, P.N., et al. 1997. Management of clusters of meningococcal disease. *Commun Dis Rep Rev*, **7**, R3–5.

Summerfield, J.A., Sumiya, M., et al. 1997. Association of mutations in mannose binding protein gene with childhood infection in consecutive hospital series. *Br Med J*, **314**, 1229–32.

Syrogiannopoulos, G.A., Lourida, A.N., et al. 1994. Dexamethasone therapy for bacterial meningitis in children: 2- versus 4-day regimen. *J Infect Dis*, **169**, 853–8.

Taha, M.-K., Achtman, M., et al. 2000. Serogroup W135 meningococcal disease in Hajj pilgrims. *Lancet*, **356**, 2159.

Tauber, M.G. and Zwahlen, A. 1994. Animal models for meningitis. *Methods Enzymol*, **235**, 93–106.

Tenover, F.C., Swenson, J.M. and McDougal, L.K. 1992. Screening for extended spectrum cephalosporin resistance in pneumococci. *Lancet*, **340**, 1420.

Tettelin, H., Saunders, N.J., et al. 2000. Complete genome sequence of *Neisseria meningitidis* serogroup B strain MC58. *Science*, **287**, 1809–15.

Thorburn, K., Baines, P., et al. 2001. Mortality in severe meningococcal disease. *Arch Dis Child*, **85**, 382–5.

Tomasz, A. 1995. The pneumococcus at the gates. *N Engl J Med*, **333**, 514–15.

Towers, C.V. 1995. Group B streptococcus: the US controversy. *Lancet*, **346**, 197–8.

Tunkel, A.R. and Scheld, W.M. 1993. Pathogenesis and pathophysiology of bacterial meningitis. *Clin Microbiol Rev*, **6**, 118–36.

Tunkel, A.R. and Scheld, W.M. 2002. Corticosteroids for everyone with meningitis? *N Engl J Med*, **347**, 1613–14.

Tuomanen, E., Tomasz, A., et al. 1985. The relative role of bacterial cell wall and capsule in the induction of inflammation in pneumococcal meningitis. *J Infect Dis*, **151**, 535–40.

Twumasi, P.A., Kumah, S., et al. 1995. A trial of a group A plus group C meningococcal polysaccharide-protein conjugate vaccine in African infants. *J Infect Dis*, **171**, 632–8.

Van der Ley, P. and Poolman, J.T. 1992. Construction of a multivalent meningococcal vaccine strain based on the class 1 outer membrane protein. *Infect Immun*, **60**, 3156–61.

Van der Ley, P., van der Biezen, J., et al. 1993. Use of transformation to construct antigenic hybrids of the class 1 outer membrane protein in *Neisseria meningitidis*. *Infect Immun*, **61**, 4217–24.

Van Deuren, M., van Dijke, BJ, et al. 1993. Rapid diagnosis of acute meningococcal infections by needle aspiration of biopsy of skin lesions. *Br Med J*, **306**, 1229–32.

Van Putten, J.P.M. 1990. Iron acquisition and the pathogenesis of meningococcal and gonococcal disease. *Med Microbiol Immunol*, **179**, 289–95.

Vieusseux, G. 1805. Mémoire sur la maladie qui a régné à Genève au printemps de 1805. *J Méd Chir Pharm*, **xi**, 163–82.

Virji, M. and Kayhty, H. 1991. The role of pili in the interactions of pathogenic neisseria with cultured human endothelial cells. *Mol Microbiol*, **5**, 1831–41.

Virji, M., Makepeace, K., et al. 1992. Expression of the Opc protein correlates with invasion of epithelial and endothelial cells by *Neisseria meningitidis*. *Mol Microbiol*, **6**, 2785–95.

Wall, R.A. 2001. Meningococcal disease – some issues in treatment. *J Infect*, **42**, 87–99.

Weichselbaum, A. 1887. Ueber die Aetiologie der akuten Meningitis cerebro-spinalis, *Fortschr Med*, **5**, 573–83, 620–6.

Westendorp, R.G.J., Langermans, J.A.M., et al. 1997. Genetic influence on cytokine production and fatal meningococcal disease. *Lancet*, **349**, 170–3.

Westendorp, R.G.J., Hottenga, J.J. and Slagboom, P.E. 1999. Variation in plasminogen-activator-inhibitor-1 gene and risk of meningococcal septic shock. *Lancet*, **354**, 561–3.

Wylie, P.A.L., Stevens, D., et al. 1997. Epidemiology and clinical management of meningococcal disease in Gloucestershire: retrospective population-based study. *Br Med J*, **315**, 774–9.

21

Other CNS bacterial infections

FIONA E. DONALD

INTRODUCTION

Focal pyogenic infections of the central nervous system (CNS) are life-threatening conditions that require treatment both surgically and with antibiotics. The most common three conditions are brain abscess, subdural empyema, and spinal abscess; however, these serious infections are uncommon with a cumulative lifetime incidence of approximately 1 percent for brain abscess.

Until the last quarter of the nineteenth century successful treatment for a brain abscess, commonly traumatic in origin, was rare. The first surgeon said to operate successfully on an otitic abscess was a Frenchman, Morand, in 1752 (Canale 1996). Further occasional successes were reported over the next hundred years (Weeds 1872). In 1893, William MacEwan, a surgeon from Glasgow, published a monograph that included details of his cases successfully treated by surgery (MacEwan 1893). Of 19 patients with brain abscess, 18 survived surgery, and this was a remarkable achievement at the time. By the beginning of the twentieth century surgical drainage of brain abscess was accepted as the treatment of choice.

Antibiotics began to play a part in treatment of abscesses as they were discovered throughout the first part of the twentieth century (Canale 1996). Mortality still remained high, however, until antibiotics such as penicillin and chloramphenicol were more widely used. The advent of better imaging techniques such as CT scans, and improved surgical techniques have been responsible for improving the outcome of these serious infections. In the field of microbiology an important advance was the upsurge of interest in obligate anaerobes and their role in human disease (Heineman and Braude 1963). The importance of anaerobic bacteria in CNS infections was emphasized and allowed antibiotic regimens to be tailored accordingly (De Louvois et al. 1977a; Ingham et al. 1977). Further advances which have been made in the last 10 years include the wider availability of MRI scanning which is able to pick up abscesses in the early stages, minimally invasive surgical techniques for abscess drainage, and new antimicrobial agents with better activity and CNS penetration.

INTRACRANIAL ABSCESS (BRAIN ABSCESS)

Epidemiology

Brain abscess accounts for about one in 10 000 general hospital admissions in the UK and USA each year and this incidence has remained relatively stable in the antibiotic era. Approximately four to ten cases are seen annually in neurosurgical units of developed countries each year (Donald et al. 1990; O'Donoghue et al. 1992; Richards et al. 1990). Some series have noted a slight increase in cases in recent years (Neilsen et al. 1982; Chun et al. 1986) that may be accounted for by better diagnostic techniques. Conversely other series have noted a decline in incidence (Nicolosi et al. 1991).

Mortality rates have ranged from 30 to 60 percent until the late 1970s (Garfield 1969) when the overall mortality dropped, and is now 0–24 percent (Seydoux and Francioli 1992; Donald et al. 1990; Ibrahim et al. 1990). This improvement has been attributed to better radiological techniques to diagnose and localize the abscesses, better surgical techniques, and new more effective antimicrobial agents (Mampalam and Rosenblum 1988).

A male predominance exists in most series, with brain abscesses being two to three times more common in men than in women in most recent series (O'Donoghue et al. 1992; Seydoux and Francioli 1992). The median age is said to be 30–40 years; however, this will vary with predisposing factor and site of abscess. Incidence among other countries varies greatly and is related to differences in predisposing factors and underlying medical conditions (Ibrahim et al. 1990; Lu et al. 2002); for example, in China, 65 percent of brain abscesses are secondary to otitis media, while only 0.5 percent are associated with sinusitis (Yang 1981). In European countries, 20–40 percent occur in association with otitic infections and 15–25 percent with sinusitis (Bradley and Shaw 1983; Van Alphen and Driessen 1976).

Overall about 25 percent of brain abscess occur in children mostly in the 4–7 year age group (Neilsen et al. 1982; Kaplan 1985). Neonatal brain abscess is rare and is usually associated with gram-negative bacillary meningitis (Renier et al. 1988).

Changes in the epidemiology and causative organisms have occurred in recent years with the advent of HIV-related diseases and a larger group of immunosuppressed patients for example, following bone marrow transplant. In a series of bone marrow transplants, brain abscess was found to occur in one of every 49 patients (Hagensee et al. 1994). Solid organ transplant recipients are at increased risk of developing CNS infections – approximately 2–12 percent of solid organ recipients will develop CNS infections including brain abscess (Selby et al. 1997; Conti and Rubin 1988).

Predisposing conditions

The most common predisposing conditions are those which form a contiguous focus of infection, in other words infection in the ear, sinuses, or teeth (Table 21.1). Ear infections, such as acute or chronic otitis media, or

mastoiditis have been most commonly reported in series with numbers ranging from 14 up to 60 percent (Lu et al. 2002; Alderson et al. 1981; Chun et al. 1986; O'Donoghue et al. 1992; Yang 1981; Ariza et al. 1986). The numbers vary with country of origin; overall, ear (otogenic) infections are the most common factor associated with contiguous spread.

The age incidence of otogenic brain abscesses shows a bimodal distribution with cases peaking in the pediatric age group then again in the over-40s (Small and Dale 1984). Otogenic brain abscess may be decreasing in relative frequency, as shown by a series from Sweden where otogenic abscesses accounted for only 7.4 percent (Schliamser et al. 1988). Areas where otitis media is neglected or therapy is delayed still see the serious complications associated with this condition. A series from Saudia Arabia found that chronic otitis media and sinusitis accounted for 57 percent of cases (Ibrahim et al. 1990).

The majority of otogenic brain abscesses (35–75 percent) are located in the temporal lobe (Figure 21.1). The cerebellum is the next most common site (20–30 percent) and otogenic infections account for 85–99 percent of cerebellar abscesses. Most of these are solitary lesions (Wispelwey et al. 1997).

Sinus disease is generally the next most common predisposing factor and is reported to account for 10–36 percent of cases in recent series (Alderson et al. 1981; Chun et al. 1986; Lu et al. 2002), but these figures may be showing a decrease as the early use of antibiotics may modify the disease process. Sinus disease is a major predisposing factor for subdural empyema. The age group most commonly affected is 10–30 years (Bradley et al. 1984). Generally cases of frontoethmoidal sinusitis will localize to the frontal lobe, and the rarely seen condition of sphenoidal sinusitis will produce abscesses in the frontal or temporal lobes (Table 21.1).

Dental infections as a source for intracranial abscess (odontogenic) are much less common but may lead to cavernous sinus thrombosis, subdural empyema, and meningitis. Rarely a brain abscess may complicate cavernous sinus thrombosis, but it is not clear whether the abscess develops as a coexisting event rather than as a consequence of the thrombosis (DiNubile 1988).

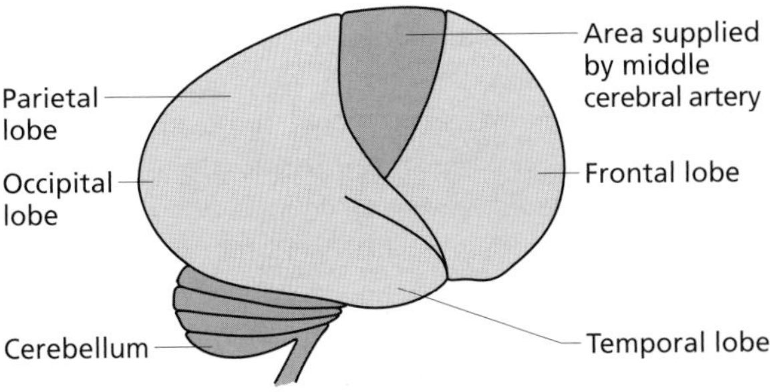

Table 21.1 *Brain abscess: predisposing conditions and site of abscess*

Predisposing condition	Site of abscess
Chronic ear infection, mastoiditis	Temporal lobe, cerebellum
Paranasal sinusitis	Frontal lobe
frontoethmoidal	frontal
sphenoidal	temporal
Dental infection	Frontal lobe
Trauma or post surgical	Related to wound/operation site
Hematogenous spread	Middle cerebral artery distribution

Figure 21.1 *Diagram of the brain showing the anatomical areas where abscesses occur.*

Odontogenic abscesses were found in six of 45 patients in one series (Chun et al. 1986), and in only five of 81 in another (Small and Dale 1984). Infection of the molar teeth is more likely to lead to intracranial complications as the bacteria can spread between the muscle planes to the base of the skull. Unrecognized dental infection may account for some abscesses where no other clear predisposing factor is evident and the bacteria cultured resemble those of dental origin (Ingham et al. 1978). Odontogenic abscesses are generally found in the frontal lobe. An odontogenic source should be considered in patients presenting with septic cavernous sinus thrombosis and any abscess of uncertain origin.

Hematogenous spread from a distant focus of infection usually causes multiple abscesses in the region of the middle cerebral artery distribution and at the junction of the gray matter and white matter (Figure 21.1). Hematogenous abscesses account for around 25 percent of cases in most series (Chun et al. 1986; Yang 1981). Endocarditis and pyogenic pulmonary infections such as lung abscess, bronchiectasis, empyema, and cystic fibrosis are common predispositions. Chronic suppurative lung disease has in the past accounted for one half of all hematogenous cases, but now may be becoming less common as a predisposing factor, and a study shows the number of cases dropping from 18 percent in the early 1970s to 2 percent in the 1980s (Mampalam and Rosenblum 1988). Congenital cyanotic heart disease, commonly seen as a predisposing factor in children, and pulmonary arterio-venous (AV) malformations increase the risk of hematogenous spread to brain and may account for 3–13 percent of cases (Park and Neches 1993). A report of 50 cases from India had 18 percent caused by congenital heart disease and 12 percent by pulmonary disease (Lakshmi et al. 1993). Other remote foci of infection which may lead to brain abscess are septic foci in abdomen, pelvis, urinary tract, or bones. Rarely cases have been reported following injection of bleeding varices (Cohen et al. 1985) and following tongue piercing (Martinello and Cooney 2003).

Endocarditis is an important infection that may be complicated by brain abscess; 1–5 percent of cases may go on to develop brain abscess (Lerner 1985). The lesions are mostly multiple. Endocarditis and subsequent central nervous system infection is increasingly being seen in injecting drug users (Tunkel and Pradhan 2002).

Hereditary hemorrhagic telangiectasia is a condition that carries a significant risk for developing brain abscess; this complication may develop in 1–6 percent of patients (Press and Ramsey 1984) most often in the third to fifth decade of life. AV malformations may occur in the brain as a result of this condition, providing a nidus for bacteria.

Trauma, either accidental injury to the brain or post neurosurgical procedure is an important risk factor, but less common than those listed above. Clean neurosurgical procedures, for example, craniotomy not associated with trauma, is complicated by infection in 0.6–1.7 percent of cases (Tenney 1986). Most of these cases are meningitis, about 10 percent are brain abscess.

Cases ascribed to trauma in published series vary greatly and will depend on epidemiological factors and availability of specialist neurosurgical units. Published figures for the incidence of cases related to trauma range from 12 to 37 percent (Schliamser et al. 1988; Lu et al. 2002; Ariza et al. 1986; O'Donoghue et al. 1992) and may be showing an increase as abscesses secondary to chronic lung disease and sinusitis decline. Accidental trauma, such as head injury with skull fractures and penetrating wounds with sharp objects, for example darts, are all reported to cause brain abscess. Abscesses may develop up to 113 days after the injury (Patir et al. 1995).

Bacterial meningitis as a predisposing factor for intracranial abscess is quite unusual, but may predispose to subdural empyema. The exception is in the neonatal age group when gram-negative bacilli may cause focal intracranial sepsis after meningitis (Renier et al. 1988).

Over the last decade there has been an increase in intracranial sepsis seen in immunosuppressed patients. This is an enlarging group of patients who may be immunosuppressed by disease such as HIV-related disease, by hematological malignancies and their treatment, and by immunosuppression following solid organ transplant. A varied range of microorganisms may be seen in these patients to include bacteria, fungi, and protozoa (Calfee and Wispelwey 2000).

In most cases of intracranial sepsis the primary source can be found, but in approximately 15–20 percent of cases no predisposing factor is identified. These are referred to as cryptogenic abscesses.

Pathogenesis and pathophysiology

The factors necessary for establishment of a brain abscess are a source of virulent microorganisms and the presence of ischemic or devitalized brain tissue. There may be a variety of primary foci and there is a close relationship between the primary focus and the bacterial flora isolated. Experimental evidence suggests that infection is difficult to establish in normal brain tissue (Molinari et al. 1973).

There are differences in the ability of microorganisms to produce brain abscess. Animal models of abscess formation studied *Escherichia coli* and *Staphylococcus aureus* injected into skin and brain tissue and found that the brain is much more susceptible to bacterial challenge than the skin with as few as 10^2 colony-forming units (cfu) of either organism resulting in abscess formation (Mendes et al. 1980). Capsular strains of *E. coli* were more virulent (Costello et al. 1983), the capsule of anaerobes such as *Bacteroides fragilis* may play a part in pathogenicity. Bacteria more commonly found in brain abscess such as *Bacteroides fragilis* or *Streptococcus intermedius* failed to produce infection and abscess formation in an animal model (Onderonk et al. 1979). In most abscesses mixtures of organisms are found and the synergistic interactions undoubtedly play an important role in the pathogenesis of these mixed infections. The aerobic and facultative organisms may aid the anaerobes by using the available oxygen.

The role of other virulence factors in brain abscess formation has not been evaluated. Bacterial lipopolysaccharide has been studied in detail with regard to pathogenesis of bacterial meningitis, but similar studies looking at its effect on pathogenesis of brain abscess have not been done.

The development of a brain abscess from the earliest stages to the development of a capsule has been carefully studied (Britt et al. 1981) using an animal model and CT scanning. These are the stages that were described:

1 Early cerebritis days 1–3: There is a marked inflammatory infiltrate of polymorphonuclear cells, lymphocytes, and plasma cells from capillaries. Bacteria can be seen on Gram stain and there is marked edema of the white matter surrounding the lesion.
2 Late cerebritis days 4–9: A well-formed necrotic center has developed and enlarged and there is new vessel formation around the developing abscess.
3 Early capsule formation 10–13 days: The necrotic center has decreased slightly in size and there is development of a layer of fibroblasts around the center that will eventually form the capsule. The edema is diminishing.
4 Late capsule stage 14 days and later: The process continues with the development of the capsule.

Bacteriology of brain abscess

The range of bacteria isolated from intracerebral abscesses has undergone a marked change during the last 50 years. Several factors may account for this. There have been improvements in microbiology techniques especially anaerobic culture; increased use of antibiotics early in treating predisposing conditions, such as otitis media, may change the flora of the ear or render cultures negative; changes in patient susceptibility, for example, more immunocompromised patients; predisposing factors such as access to neurosurgery; and availability of laboratory facilities and anaerobic culture methods.

In the preantibiotic era the common isolates from brain abscess were *Staphylococcus aureus*, reportedly found in 25–30 percent of cases, streptococci in 30 percent, and coliforms in 12 percent: 50 percent of pus samples were culture negative (De Louvois et al. 1977a; De Louvois 1978). With improvements in anaerobic techniques the role of anaerobes has become apparent

(Heineman and Braude 1963). Series from the UK have stressed the importance of anaerobes particularly in otogenic infections (O'Donoghue et al. 1992; Ingham et al. 1977; Donald et al. 1990; Richards et al. 1990). The site of the abscess and the predisposing conditions are important predictors of the likely organisms (Table 21.2). There seems to be a predominance of aerobic organisms reported but this varies with the center and anaerobes have been reported as being present in 33–60 percent of cases (Seydoux and Francioli 1992; Sofianou et al. 1996; Ariza et al. 1986; Chaudhry et al. 1998), higher numbers

Table 21.2 *Brain abscess: predisposing condition and likely bacterial isolates*

Predisposing condition	Likely bacterial isolates
Ear disease	Streptococci aerobic and anaerobic
	Bacteroides fragilis and other *Bacteroides* spp.
	Enterobacteriaceae especially *Proteus* spp.
	Pseudomonas aeruginosa
	Actinomyces spp.
Sinusitis	Microaerophilic and anaerobic streptococci
	Bacteroides spp.
	Fusobacterium spp.
	Haemophilus spp.
Dental infection	*Streptococci*
	Fusobacterium spp.
	Bacteroides spp.
	Actinomyces spp.
Hematogenous spread Congenital heart disease	Viridans streptococci
	Anaerobic, microaerophilic streptococci
	Haemophilus spp.
Lung disease	*Actinomyces* spp.
	Fusobacterium spp.
Endocarditis	Viridans streptococci
	Staphylococcus aureus
Immunosuppression HIV disease	*Listeria monocytogenes*
Neutropenia	Enterobacteriaceae
	Pseudomonas aeruginosa
	Streptococcus pneumoniae
	Nocardia spp.
	Mycobacteria
Trauma surgical	*Staphylococcus aureus*
	Propionibacterium
Trauma by penetrating injury	*Pseudomonas* spp.
	Clostridium spp.
	Enterobacteriaceae
Neonate	*Citrobacter diversus*
	Pseudomonas aeruginosa

being reported in later studies. This may reflect the increasing availability of anaerobic culture facilities and better diagnostic laboratory support. One recent study identified nine anaerobes in 18 patients, these included *Prevotella melaninogenicus*, *Bacteroides preacutus*, *Fusobacterium nucleatum*, and *Peptostreptococcus* spp. (Chaudhry et al. 1998). The samples were received in the laboratory within 30 min, inoculated into anaerobic broth and enriched thioglycollate broth which undoubtedly improved the isolation rate. Four out of six of their patients had chronic suppurative otitis media as the predisposing condition.

The bacteriology of otogenic abscesses can be complex with multiple isolates of bacteria being recorded (Alderson et al. 1981) and represent the microbial flora of a chronically infected ear, but usually with the exception of *Pseudomonas* spp. Isolates include gram-negative aerobic bacilli, such as *Escherichia coli* and *Proteus* spp., aerobic and anaerobic streptococci, and other anaerobes such as *Bacteroides fragilis*, other *Bacteroides* spp., *Eubacterium*, and *Clostridium* spp. It is unusual to find organisms normally associated with upper respiratory tract infection, such as *Haemophilus* spp. and pneumococci. The isolation rate of anaerobes does vary and Seydoux reported almost equal numbers of aerobes and anaerobes in his paper (Seydoux and Francioli 1992). *Fusobacterium* spp. were the most common anaerobes recovered, whilst viridans streptococci and *Streptococcus milleri* were the most common aerobic isolates.

The presence of anaerobes may be detected by noncultural methods such as gas-liquid chromatography (Donald et al. 1990); this investigation showed that 13 of 18 specimens of pus were found to be positive by this method.

Gram stain is an essential part of the investigation of brain abscess pus, particularly if antibiotics have been given. The morphology of bacteria seen on Gram stain will be helpful in choosing antibiotics especially if the cultures are subsequently negative (Lakshmi et al. 1993).

Actinomyces spp. are reported in mixed culture with other aerobic and anaerobic bacteria (Donald et al. 1990; Smego and Foglia 1998) and represent part of the normal flora of the mouth or respiratory tract. Actinomycosis is of course a recognized disease entity and may present with abscess formation in the central nervous system. Usually, in this condition *Actinomyces* spp. would be present either as a single isolate or in conjunction with associate anaerobic organisms.

A series from Taiwan (Lu et al. 2002) looked at isolates over two time periods. During the second time period there was an increase in aerobic gram-negative bacilli. Viridans streptococci and *Klebsiella pneumoniae* were the most common isolates from hematogenous spread. Nosocomial infections and traumatic infections showed an increase in the second time period.

In sinugenic brain abscess the bacteriology resembles that of sinugenic subdural empyema. *S. milleri* is the most common isolate and is often found in pure culture. Mixed cultures are less commonly reported and may be associated with chronic sinusitis.

In general, streptococci are common isolates in brain abscess and are reported in most series. A detailed study of streptococci associated with man reported *Streptococcus milleri* in 13 of 16 streptococcal brain abscesses (Parker and Ball 1976). Streptococci are commonly reported in more than 31 percent of brain abscesses – the actual figure may be higher than this as microaerophilic streptococci may be classified as anaerobes in some series. In some cases the reported figure is as high as 70 percent (De Louvois et al. 1977a; Ariza et al. 1986; Mathisen et al. 1985). The most frequently isolated are those of the *Streptococcus milleri* group (*anginosus*, *constellatus*, *intermedius*), which are well recognized to cause focal suppuration. An evaluation of infections caused by *Streptococcus milleri* found that 10 percent of them were brain abscesses (Molina et al. 1991).

Streptococcus pneumoniae is infrequently found in brain abscess despite being a common cause of meningitis. A review of 23 pneumococcal brain abscesses found that 50 percent of them had a contiguous focus such as sinusitis, otitis media or meningitis (Grigoriadis and Gold 1997).

Abscesses that follow trauma will show a range of bacteria related to the type of trauma. *Staphylococcus aureus* will be a common isolate after surgery, and is isolated in 10–15 percent of cases of postoperative brain abscess. Some of these will be methicillin resistant *Staphylococcus aureus* (MRSA) strains and are most likely to be in pure culture. *S. aureus* is also found in pure culture in intracerebral abscesses as a complication of endocarditis, and has recently been described in injecting drug users (Tunkel and Pradhan 2002).

Aerobic gram-negative bacilli are usually reported in mixed culture. Abscesses associated with chronic ear infections are most likely to yield these organisms, particularly *Proteus* spp. *Escherichia coli*, *Klebsiella* spp., and *Enterobacter* spp. have also been recorded. *Klebsiella pneumoniae* has been associated with brain abscess in diabetics and produces a gas-forming appearance on X-ray imaging (Lu et al. 2002). Gram-negative bacilli have also been reported following neurosurgical procedures, accounting for 30 percent of cases in one report (Rau et al. 2002). *Pseudomonas aeruginosa* is rarely reported from otogenic abscesses despite its presence in the external ear canal of patients with chronic ear disease. It has been reported as a cause of meningitis and brain abscess in a premature baby (Shah et al. 1999). *Citrobacter freundii* has also been associated with neonatal brain abscess following meningitis (Curless 1980).

Listeria monocytogenes is a recognized cause of meningitis and brain stem encephalitis in the immunosuppressed patient. Brain abscesses account for approximately 10 percent of cases (Eckburg et al. 2001) and are usually associated with concomitant meningitis and bacteremia in immunosuppressed patients.

Propionibacterium acnes is a true opportunist pathogen and is increasingly seen causing indolent infections in patients with prosthetic material. It is reported to cause post-operative brain abscesses which may be insidious in onset (Berenson and Bia 1989).

Focal intracranial infections with *Salmonella* spp. are rare; a review of the world literature identified only 43 cases, 11 of which were brain abscess mostly in adults (Rodriguez et al. 1986). Patients with HIV disease seem to be more at risk (Aliaga et al. 1997). *Salmonella typhi, typhimurium*, and *enteritidis* were the most common isolates. *S. typhi* has been reported to cause a brain abscess in a 2 month old (Hanel et al. 2000).

Nocardia spp. cause single or multiple abscesses and are associated with pulmonary disease or immunosuppression, or indeed as a primary infection in otherwise healthy individuals (Beaman and Beaman 1994). An increase in cases has been seen over the last decade in association with HIV disease. The causative organism is mostly *Nocardia asteroides* (Mamelak et al. 1994).

Rhodococcus equi is an unusual isolate from a brain abscess but was reported first in patients with HIV and recently in an immunocompetent patient (Corne et al. 2002).

Rare causes which have been reported in case reports are *Vibrio cholerae* non-0:1 (Ismail et al. 2001); *Mycoplasma hominis* and *Ureaplasma* (Rao et al. 2002); *Eikenella corrodens* (Asensi et al. 2002); *Capnocytophaga* (Engelhardt et al. 2002); and *Bacillus cereus* (Sakai et al. 2001).

Mycobacterium tuberculosis is a recognized cause of intracerebral lesions (tuberculomas) and the incidence is geographically determined. More cases are seen in Asia and the Far East than in Europe and North America. Acid-fast bacilli can be demonstrated in the pus from these lesions and should be actively looked for if cultures are negative and there are other risk factors for mycobacterial disease, such as immunosuppression and AIDS (Mathisen and Johnson 1997).

Clinical presentation

The clinical manifestations are caused by the presence of a space-occupying lesion in the brain, and the presentation may vary from indolent to fulminant. Symptoms may be nonspecific which can lead to delays in diagnosis. The most common symptom is headache, either hemicranial or generalized, and this may occur in up to 70 percent of patients (Table 21.3). Sudden worsening of the headache with associated meningism may indicate rupture of the abscess into the ventricle which carries a high mortality rate.

Table 21.3 *Common signs and symptoms of brain abscess*

Signs or symptoms	Frequency (%)
Headache	70–90
Focal neurology	>60
Fever	45–50
Triad of fever, headache, and focal neurology	<50
Seizures	25–35
Nausea and vomiting	25–50
Neck stiffness	25
Papilledema	25

Other common symptoms are fever, focal neurological signs, nausea, vomiting, and seizures which are usually generalized. If the abscess is near the meninges, neck stiffness may be present. The classic triad of fever, headache, and focal neurology is only seen in <50 percent of cases.

Other symptoms and signs vary with the location, size of the abscess, virulence of the infecting organisms, and the patient's immune status. Fever occurs in <50 percent of patients and the absence of fever should not exclude the diagnosis.

Patients with frontal lobe abscess are more likely to have mental status changes, speech disturbances or hemiparesis with unilateral motor signs. Those with a temporal lobe abscess may have headache, aphasia, and visual field defects. Cerebellar abscesses cause vomiting, ataxia, nystagmus, and dysmetria. Brain stem abscesses cause vomiting, hemiparesis, dysphagia, and facial weakness.

The symptoms caused by the contiguous focus of infection may predominate (for example, discharging ear). The duration of symptoms varies from a few days to as long as 6 weeks. About 75 percent of cases will have symptoms for 2 weeks (Wispelwey et al. 1997). In the immunocompromised patient the presentation may be occult because of reduced inflammatory response (Conti and Rubin 1988).

Diagnostic methods

The diagnosis of intracranial sepsis has been revolutionized by the introduction of CT and MRI scanning. These new techniques have rendered older more invasive diagnostic tests such as angiography and ventriculography virtually obsolete. Scanning methods also allow examination of the paranasal sinuses, mastoids, and middle ear.

The characteristic appearance of a brain abscess on a CT scan is that of a hypodense center with peripheral uniform ring enhancement following injection of contrast material, surrounded by a variable hypodense area of edema (Figure 21.2). MRI scanning is more sensitive particularly in the early stages and will detect

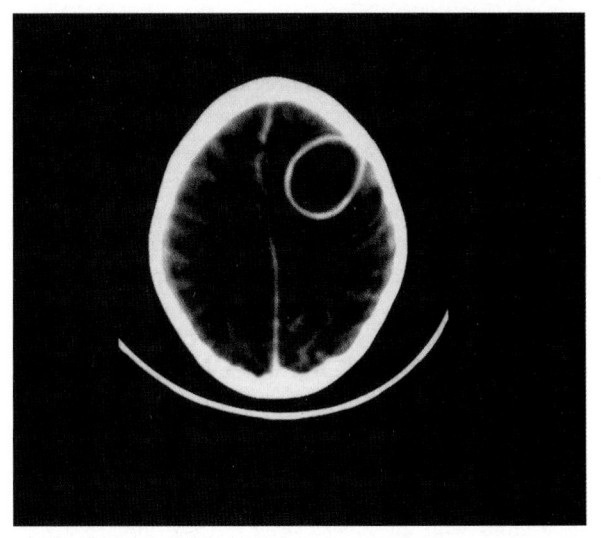

Figure 21.2 *Contrast enhanced CT scan showing ring enhancing frontal lobe lesion.*

abscesses in the early cerebritis stage (Figure 21.3). It is also better at defining brain stem abscesses. These forms of imaging are now the main diagnostic tool.

Diagnostic samples may be sent to microbiology; blood cultures will usually be negative. Lumbar puncture to obtain a sample of cerebrospinal fluid (CSF) is contraindicated as there is a significant risk of brain stem herniation (Chun et al. 1986), which outweighs any possible benefit of examining the CSF, which will show a negative or nonspecific result. The CSF may be positive if the abscess has ruptured into the ventricles or subarachnoid space.

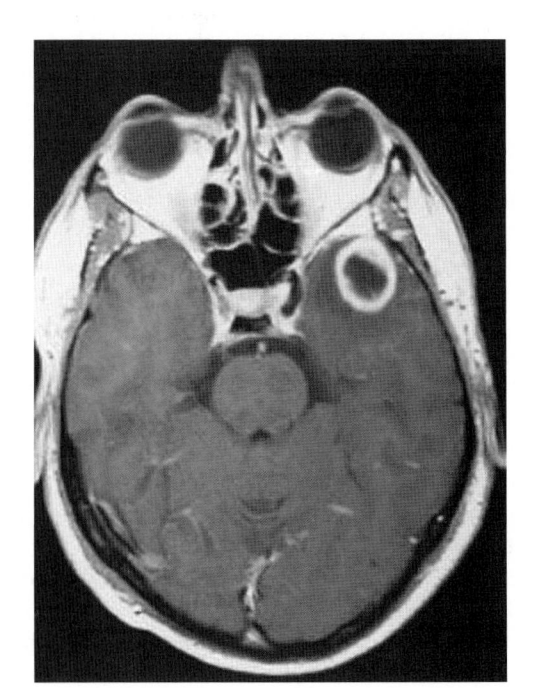

Figure 21.3 *Gadolinium enhanced MRI scan showing temporal lobe abscess.*

There may be a peripheral leukocytosis, but 40 percent of patients will have a normal cell count. Other inflammatory indicators such as C-reactive protein (CRP) and erythrocyte sedimentation rate (ESR) may show elevation. The CRP may be more often raised in intracranial sepsis than in patients with neoplasm and may be useful in differentiating the diagnosis (Hirschberg and Bosnes 1987).

Samples of pus are readily sent for microbiology from surgery and are essential to define the infecting microorganisms, particularly in immunocompromised patients where a variety of bacteria, fungi, or parasites may be causing the infection.

Treatment

ANTIBIOTICS

The availability of antibiotics has undoubtedly contributed to the reduction in mortality seen with brain abscess. Prior to the discovery of antibiotics treatment was only surgical and usually led to death. In the early days of antibiotic use, large amounts of penicillin and streptomycin were instilled into abscess cavities and showed some promising results (Garfield 1969). The availability of more active agents in recent years has shown more improvements and now regimens can be tailored to cover the polymicrobial flora of brain abscesses.

When choosing an antibiotic to treat brain abscess, consideration has to be given to the penetration of the antibiotic into brain tissue. The cerebral capillary endothelial cells constitute the blood–brain barrier, which limits the penetration of numerous substances including antibiotics, into the brain. Inflammation as in meningitis disrupts the tight junctions and allows easier penetration of antibiotics, such as penicillin, into the brain. Other factors regulating antibiotic penetration into the brain include molecular size and structure, lipid solubility, degree of ionization at physiological pH, protein binding, and active transport mechanisms (Everett and Strausbaugh 1980). Antibiotic penetration into CSF has been extensively studied, but the blood–brain and the blood–CSF barriers are not identical and results may not be extrapolated (Gortvai et al. 1987). Therapeutic concerns about abscesses would be whether the results of CSF drug penetration would apply directly to brain abscess therapy, what would be the effect of the abscess environment on activity of antibiotics, and the possibility of negative drug interactions when multiple agents are administered. Certain antibiotics, for example penicillin and chloramphenicol, may be inactivated by abscess pus (Gortvai et al. 1987) and aminoglycosides are ineffective in the acid environment of an abscess. Chloramphenicol may antagonize the effects of penicillin and gentamicin as noted in experimental meningitis models (Tunkel and Scheld 1989). These are of course theoretical concerns and few studies have addressed these issues in clinical practice.

The key question to ask is what is the concentration of antibiotic in the pus, and overall there have been only a few studies looking directly at this. The earliest study looked at brain tissue, CSF, and serum from 27 patients having a prefrontal lobotomy who had been given a dose of antibiotic preoperatively (Wellman et al. 1954). Tetracycline was detectable in the CSF and brain, but penicillin was not detected in either, perhaps reflected by the low dose that was given. Some other antibiotics were studied (Kramer et al. 1969) in patients having an intracranial neoplasm excised, they were given a bolus antibiotic dose of 2 g, and achieved the following brain:-blood levels: chloramphenicol 9:1, cephalothin 1:10, penicillin 1:23, ampicillin 1:56. On the basis of this study chloramphenicol has been used in the treatment of brain abscess for its excellent penetration into brain tissue; however, it lacks bactericidal activity and has the potential for serious side effects. Black and colleagues (Black et al. 1973) analyzed antibiotic concentrations in brain abscess pus from six patients. Chloramphenicol, methicillin, and penicillin were detectable after standard doses, but nafcillin was not. All the patients deteriorated clinically until surgical drainage was carried out; the cultures were still positive at the time of surgery. This highlighted the importance of surgery as part of the management. Another group analyzed pus from 32 patients (De Louvois et al. 1977b) for various antibiotics. Penicillin G was found to be present, but only if the dose exceeded 24 mega units per day. Incubating the pus for 1 h inactivated the penicillin by 90 percent in four of 22 samples (De Louvois and Hurley 1977). Fusidic acid was found to enter readily but cephaloridine, cloxacillin, and gentamicin showed only low levels. Metronidazole achieves high levels in pus (Ingham et al. 1977) and is widely used in combination to ensure bactericidal anaerobe activity. A single report looked at vancomycin levels in one case and found excellent concentrations (Levy et al. 1986). The first generation cephalosporins have not been recommended for treatment as they have poor CNS penetration, but the second and third generation agents have been shown to have very good CSF penetration and are widely used for treating meningitis. Of course the activity in brain pus cannot be predicted from that. Cefotaxime has been evaluated in brain pus (Sjolin et al. 1991) and it was found that cefotaxime and its active metabolite, desacetylcefotaxime, were present in the pus in concentrations above the minimum inhibitory concentrations (MIC) for the likely infecting organisms. Cefotaxime was evaluated in combination with metronidazole in 15 patients who also had surgery, and the outcomes were good (Sjolin et al. 1993). Ceftazidime, ceftriaxone, and ceftizoxime have all been used successfully in the treatment of small numbers of patients (Green et al. 1989; Donald and Ispahani 1988, 1990). Moxalactam has been shown to reach adequate

concentrations in brain abscess pus and has been successful in treatment, but the toxicity of this agent precludes its use in treatment regimens (Yamamoto et al. 1993). Ampicillin/sulbactam was used in 11 patients with brain abscess with success and concentrations in the pus were variable, but adequate (Akova et al. 1993).

Imipenem-cilastatin, a carbapenem antibiotic, has been used successfully in the treatment of brain abscess (Asensi et al. 1996) and also in cerebral nocardiosis (Krone et al. 1989). The drug has a side effect of neurotoxicity so should be reserved for cases of resistant organisms. Meropenem, which is related to imipenem, does not have the same neurotoxic effects. Although limited evidence has been published for its efficacy in brain abscess, it has however been used successfully in treating meningitis. There is one report of successful use in a child with an *Enterobacter cloacae* abscess (Meis et al. 1995).

Cerebral nocardia has also been successfully treated with trimethoprim-sulphamethoxazole combined with surgery (Maderazo and Quintiliani 1974) and this combination may be used for Enterobacteriaceae depending on sensitivities.

Newer antibiotics, such as the quinolones, have not been studied in detail. There is evidence that they have good CSF penetration, and are highly active against the Enterobacteriaceae and Pseudomonads. There are inadequate data to support their use in brain abscess, but anecdotal evidence and case reports provide evidence of their clinical efficacy (Wessalowski et al. 1993). They should be used with some caution as they can lower seizure thresholds.

Undoubtedly some of the newer antimicrobial agents will be useful in the future as they offer additional options for treatment and possibly increased efficacy. There is a growing trend for nonoperative treatment of brain abscess and the new agents may have a part to play.

The antimicrobial regimens recommended for the treatment of intracerebral abscess are largely empiric and are based on the aforementioned studies (Table 21.4). No controlled trials have been undertaken; the type of surgery performed also influences the outcomes. The predisposing condition and the site of the abscess is a guiding factor. The initial Gram stain on a pus sample is essential information for guiding empiric antibiotic therapy, which can be modified once culture results are available. Antibiotics should be started as soon as the diagnosis is made: there is no reason to wait for a pus sample before starting therapy, as cultures may still be positive despite antibiotic treatment.

Penicillin remains a mainstay of therapy because of its excellent activity against streptococci; the *Streptococcus milleri* group are commonly found in abscesses. It has a low incidence of side effects, and for many years was combined successfully with chloramphenicol. However, chloramphenicol is associated with side effects and is not

Table 21.4 *Suggested empiric antimicrobial therapy for brain abscess*

Predisposing condition	Antimicrobial regimen
Chronic ear disease	Third generation cephalosporin plus metronidazole
Sinusitis	Penicillin or third generation cephalosporin plus metronidazole
Dental infection	Penicillin or third generation cephalosporin plus metronidazole
Hematogenous spread	Penicillin or third generation cephalosporin plus metronidazole. Flucloxacillin or vancomycin if *S. aureus* suspected
Trauma or post surgery	Third generation cephalosporin plus flucloxacillin or vancomycin. Consider ceftazidime if *P. aeruginosa* suspected
Immunosuppression	Third generation cephalosporin or meropenem. Ampicillin or meropenem if *Listeria* is suspected
	Meropenem or trimethoprim/sulfamethoxazole if *Nocardia* is suspected

Based on Calfee and Wispelwey (2000) and Tunkel et al. (2000).

bactericidal so should not now be used as a first line agent. Metronidazole is added to give anaerobic cover, as it should be assumed that anaerobes would be present in most cases. The third generation cephalosporins are now accepted as empiric therapy, either cefotaxime or ceftriaxone, and in combination with metronidazole provide an adequate empiric therapy for most sinugenic, otogenic, and odontogenic abscesses (Donald 1990). Adequate coverage for enterobacteria found in otogenic abscesses is provided by the cephalosporins. For abscesses where *Staphylococcus aureus* is suspected, for example, postoperative, then an antistaphylococcal penicillin such as flucloxacillin should be used, in combination with fusidic acid or rifampicin. If MRSA is likely then vancomycin should be used, again in combination with rifampicin depending on local sensitivities.

Local instillation of antibiotics into the brain abscess cavity is not generally recommended as local irritation can occur which will increase the incidence of seizures. Rarely, the local route may be indicated in a difficult, poorly resolving abscess which remains culture positive despite appropriate antibiotics.

SURGERY

Surgical drainage is an important part of the management of these abscesses and is recommended for all abscesses of 2.5 cm or greater. As well as draining pus and relieving the effects of intracranial pressure, a pus sample will provide valuable microbiological information to guide therapy. The procedures commonly used are aspiration by burr hole or open drainage and excision by craniotomy. The technique used will depend on the location of the abscess, whether multiple abscesses are present, and on the clinical state of the patient. A recent technique of closed aspiration using CT guidance is increasingly being used instead of open craniotomy and stereotactic guidance can be employed for greater accuracy (Wispelwey et al. 1997). Follow up CT scans should be done to monitor the size of the abscess.

Sometimes antibiotic treatment alone is used without surgery (Boom and Tuazon 1985). This would be in certain categories of patients, which includes those at high risk for surgery, those with concomitant meningitis or ependymitis, those with hydrocephalus requiring a shunt, and those with inaccessible or multiple lesions (Rosenblum et al. 1980).

Guidance as to the duration of therapy is variable; no controlled trials have been done. Past recommendations have been to give parenteral treatment for 6–8 weeks provided that the infecting organisms are susceptible and that adequate surgical drainage is achieved. Recent studies suggest that 3–4 weeks of parenteral antibiotics may be adequate for those patients whose abscesses have been excised, with 4–6 weeks recommended for those whose abscesses have been aspirated. Those given antibiotics alone should be treated for a minimum of 4 weeks with parenteral antibiotics (Report 2000). The benefit of using oral antibiotics for follow-on treatment is unproven: in practice oral antibiotics based on sensitivity testing may be given for a further 4–6 weeks (Skoutelis et al. 2000). The serum C-reactive protein may be used to monitor resolution of the inflammation (Brown et al. 1994).

Outpatient parenteral antibiotic treatment (OPAT) is used increasingly for treating patients outside hospitals, thus reducing the costs and morbidity associated with in-patient stay. It has been reported to be successful in a carefully selected group of patients with central nervous system infections including 19 with brain abscesses (Tice et al. 1999).

SUBDURAL EMPYEMA AND INTRACRANIAL EPIDURAL ABSCESS

These two conditions will be considered together as they very often occur together in the patient. Subdural empyema is the second most common type of intracranial bacterial infection, followed by intracranial epidural abscess – it is difficult to establish the true incidence. The subdural space lies between the two outer layers of the meninges: the dura mater and the arachnoid mater (Figure 21.4). There is a large potential space for pus to

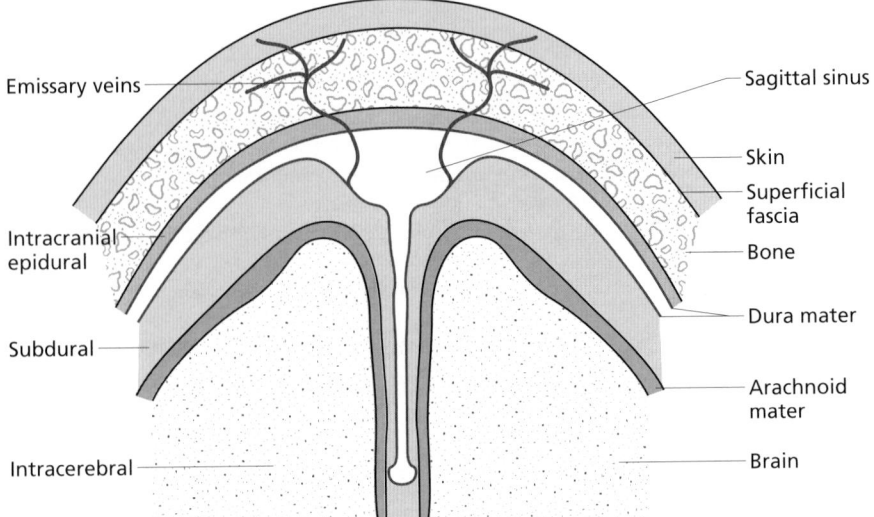

Figure 21.4 *Diagram of a cross-section through the skull showing the anatomy of the subdural space.*

spread through with only a few anatomical restrictions. Thus subdural empyemas behave as rapidly expanding space-occupying lesions. Overall, subdural empyemas account for 15–25 percent of localized bacterial infections of the CNS (Helfgott et al. 1997).

An epidural abscess represents infection between the outer layer of the dura mater and the overlying skull bone. There is little space there for pus to spread, so they are always very localized and usually associated with osteomyelitis. Subdural empyema is usually found in conjunction as the infection readily crosses the emissary veins. Virtually all cases follow sinusitis, craniotomy, or mastoiditis (Gallagher et al. 1998).

Epidemiology

Before the introduction of effective antimicrobial therapy the mortality from subdural empyema was uniformly 100 percent. After the introduction of penicillin the rate began to drop, and as with other forms of focal intracranial suppuration advances in diagnosis, surgical technique, and antibiotic therapy have brought the mortality rate down to 10–40 percent (Helfgott et al. 1997). Mortality is higher if the patient presents late or is comatose on admission, and in the elderly. Seizures may develop as a late complication (Dill et al. 1995).

Subdural empyema may occur at any age, but is most commonly seen in the second and third decades of life. Men are four times more commonly affected than women (Nathoo et al. 1999); it is seen in children (Smith and Hendrick 1983; Brook 1995). The most common predisposing factor, accounting for up to 60 percent of cases, is sinusitis, either frontal or ethmoidal. The rate may be higher in children. Ear infections account for a smaller number than with brain abscess, about 10–20 percent, but may be a more common predisposing factor

in developing countries (Pathak et al. 1990). Hematogenous spread accounts for a small number of cases, around 5 percent, and this route may infect pre-existing subdural hematomas. Other cases have been reported following trauma, burns, and dental procedures. A series of 41 patients with postoperative subdural empyema showed that sinusitis accounted for only 29 percent of cases, 66 percent were postcraniotomy (Hlavin et al. 1994). Meningitis is an important predisposing factor in infants and subdural empyema may occur in 2 percent of infants with bacterial meningitis.

Uncomplicated intracranial epidural abscesses are, however, rare in children and have been reported from a wide age range with a median in the sixth decade. Recent case series suggest that a common predisposition for epidural abscess is surgery, up to 2 percent of craniotomies may be complicated by this infection (Hlavin et al. 1994).

Pathogenesis

Infection may spread directly or indirectly into the subdural space. Direct extension from the sinuses occurs by erosion of the posterior bony wall of the frontal sinus by infection, then further erosion of the dura mater to spread into the subdural space. It is thought that the most likely route is that of indirect spread by infection leading to thrombophlebitis of the emissary veins which connect the facial and dural venous systems. An epidural abscess always arises from infection of contiguous structures, such as sinuses and ear, and osteomyelitis is usually present. Infection from the ears occurs by the same mechanisms, either direct through the bone of the mastoid air cells, or indirect spread via the venous system (Bleck and Greenlee 2000a).

Bacteriology

Many different types of aerobic and anaerobic bacteria have been reported from epidural abscesses (Table 21.5); however, up to 30 percent may be negative on culture. Many of these patients will have had antibiotics prior to a sample of pus being taken. The types of bacteria grown will reflect the predisposing source. Streptococci are the most common isolates, being found in around 35 percent of cases, and are the most common isolate in children (Smith and Hendrick 1983). Most of these will belong to the *Streptococcus milleri* group. Sinugenic subdural empyemas most often grow streptococci, aerobic or anaerobic, often in pure culture. Staphylococci, in particular *Staphylococcus aureus*, are less commonly reported and account for 12–17 percent of cases. Most of these will be after surgery, trauma or related to hematogenous spread. *Staphylococcus aureus* is a frequent isolate from intracranial epidural abscess. This is due to the higher incidence of postoperative cases.

Skin flora such as *Propionibacterium acnes* have been reported in postoperative cases of subdural empyema (Critchley and Strachan 1996).

Aerobic gram-negative bacilli are less commonly reported, and usually in association with trauma and surgery (Hlavin et al. 1994). *Salmonella* spp. have been reported in neonates and infants (Mahapatra et al. 2002). Other unusual isolates are *Serratia marcescens* (Safani and Ehrensaft 1982) and *Pasteurella multocida* (Stern et al. 1981).

Subdural empyema following meningitis will reflect the causative organism of the meningitis, i.e. *Haemophilus influenzae*, *Streptococcus pneumoniae* or *Neisseria meningitidis* (Jacobson and Farmer 1981).

Polymicrobial infections are much less commonly reported compared to intracranial abscesses, but may be associated with surgery or trauma. As with other intracranial abscesses anaerobes play an important part and may be present either in pure culture or as part of a polymicrobial infection (Yoshikawa et al. 1975).

Mycobacterium tuberculosis is not often associated with subdural empyema despite it being a common cause of other CNS infections (Van Dellen et al. 1998).

Clinical features

In subdural empyema, fever and focal headache are prominent early features, the headache may then become generalized (Table 21.6). Focal neurological deficits may develop within 24–48 h with vomiting and signs of meningeal irritation. The symptoms can progress rapidly and focal fits can develop in 50 percent of patients. The presentation of localized epidural abscess may be more insidious and the initial presentation will be of the predisposing infection such as sinusitis.

Diagnostic methods

It is important to consider the diagnosis of subdural empyema in patients with sinusitis who go on to develop signs of meningeal irritation, for example, neck stiffness. The investigation of choice is gadolinium enhanced MRI scan which is more sensitive than contrast enhanced CT scan for this condition (Figure 21.5). Lumbar puncture should not be performed as it is dangerous and will not help in the diagnosis. Blood cultures may be positive in around 10 percent of cases,

Table 21.5 *Microbial cause and initial antibiotic therapy for cranial subdural empyema and epidural abscess*

Site of primary infection	Probable organism	Suggested initial therapy
Paranasal sinuses	Aerobic, microaerophilic, and anaerobic streptococci (especially *S. milleri* group); *Bacteroides fragilis* and other anaerobes; *Staphylococcus aureus*; *Haemophilus* spp.	Third generation cephalosporin plus metronidazole
Otitis media or mastoiditis	Aerobic, microaerophilic, and anaerobic streptococci (especially *S. milleri* group); *Bacteroides fragilis* and other anaerobes; *S. aureus*; Enterobacteriaceae	Third generation cephalosporin plus metronidazole
Cranial surgery or trauma	*S. aureus*; coagulase negative staphylococci, *Clostridium* spp.; Enterobacteriaceae; *Propionibacterium acnes*	Flucloxacillin or vancomycin plus third generation cephalosporin plus metronidazole
Dental sepsis	*Fusobacterium* sp.; *Actinomyces* sp.; aerobic and anaerobic streptococci	Metronidazole plus penicillin or third generation cephalosporin
Meningitis in infants	*Haemophilus influenzae*; *S. pneumoniae*; *Neisseria meningitidis*	Third generation cephalosporin
Meningitis in neonates	*Salmonella* spp; group B streptococci, Enterobacteriaceae	Third generation cephalosporin

Adapted from Bleck and Greenlee (2000b).

Table 21.6 *Clinical signs of subdural empyema*

Signs and symptoms	% patients
Fever	77
Headache	74
Hemiparesis	71
Altered consciousness	69
Neck stiffness	63
Fits	48
Other focal neurology	45

Adapted from Helfgott et al. (1997).

particularly in children, and may be helpful in identifying the causative organism.

Treatment

The principles of antibiotic treatment have already been outlined, and it is necessary to choose an antibiotic which will penetrate the blood–brain barrier. The antibiotic choices for subdural and epidural empyema are similar to those for intracranial abscess (Table 21.5).

Early surgical treatment is essential to remove the pus and relieve the pressure on the brain (Nathoo et al. 1999).

The clinical symptoms, range of bacteria isolated, and management of intracranial epidural abscess mirrors that of subdural empyema. The mortality for uncomplicated intracranial epidural abscess is, however, lower than that for subdural empyema and a recent series showed a mortality rate of 18 percent (Hlavin et al. 1994). A good

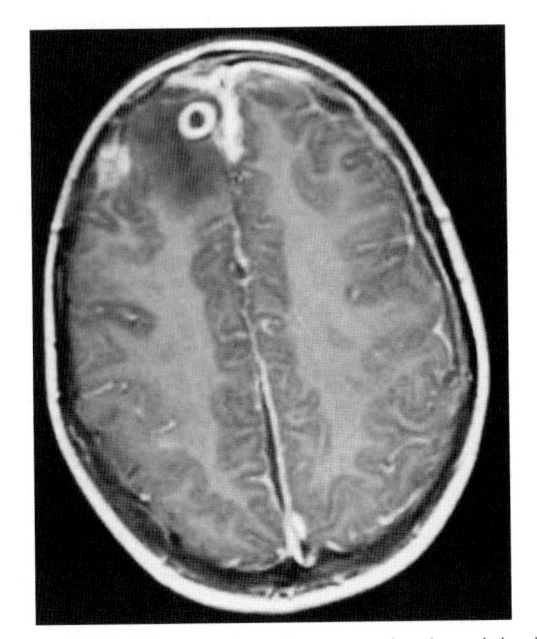

Figure 21.5 *Gadolinium enhanced MRI scan showing subdural empyema and frontal lobe abscess.*

outcome depends on early diagnosis before irreversible neurological deficits are present.

SPINAL EPIDURAL ABSCESS

A spinal epidural abscess is a localized collection of pus in the paraspinal epidural space. This surrounds the spinal cord between the dura and the bone (Figure 21.6). This condition is a medical emergency because irreversible damage can be caused to the spinal cord by pressure of the abscess. In the preantibiotic era, these conditions were progressive and universally fatal, but with the advent of antibiotics and better diagnostic and surgical techniques these infections have a better outcome with mortality rates as low as 5 percent (Darouiche et al. 1992); although in some reports the mortality rate remains high at 18–31 percent (Hlavin et al. 1990). The higher mortality rates are thought to be in cases where the diagnosis was delayed; delays in diagnosis also lead to a worse outcome in terms of neurological sequelae.

Epidemiology

Spinal epidural abscess is an uncommon condition. Estimated incidence rates from case series range from 0.2 cases per 10 000 population in the mid 1970s (Baker et al. 1975) to 2.8 cases per 10 000 population more recently as observed in large tertiary care centers (Hlavin et al. 1990). The incidence is reported to be increasing; factors which may account for this are an increase in cases in injecting drug users; more patients with cancer who receive epidural analgesia, and more patients having spinal surgery or invasive diagnostic procedures.

There is a broad age range reported from 14-day-old infants (Rubin et al. 1993) to 78 year olds (Del Curling et al. 1990). It is most common in adults and the average age is now over 50 years of age, reflecting the increasing age of the at risk population. It is a condition which is increasingly seen in injecting drug users who tend to have an average age of 35 years (Koppel et al. 1988).

The male to female ratio has stayed consistent with a male predominance (Reihsaus et al. 2000).

Predisposing conditions

Spinal epidural abscess follows infection elsewhere in the body. Most cases reach the epidural space by the hematogenous route, either by direct seeding of the epidural space or by causing vertebral osteomyelitis which then extends into the epidural space. Infection may also spread from contiguous foci of infection, such as decubitus ulcers, retropharyngeal abscesses, and retroperitoneal infection. Increasingly, cases are seen following direct inoculation from surgery, from spinal procedures, such as epidural catheters for analgesia or

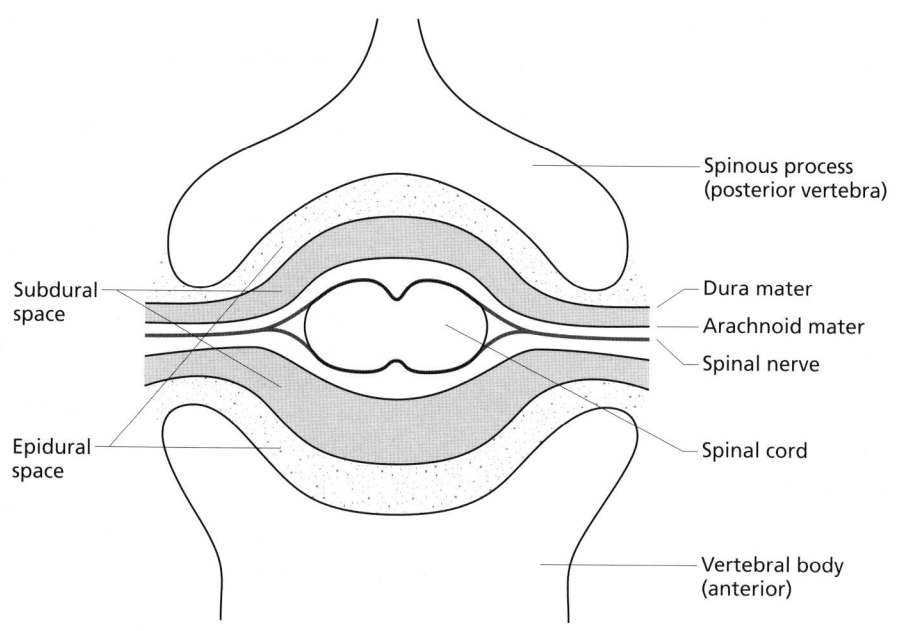

Figure 21.6 *Diagram showing the anatomy of the spine and the relationships of the epidural space.*

after spinal biopsy procedures (Bleck and Greenlee 2000a). Sometimes there is an underlying medical condition such as diabetes, degenerative joint disease, cirrhosis of the liver, or malignancy. Hematogenous seeding of the epidural space may occur in patients with endocarditis. In children, vertebral osteomyelitis is rare and most cases are associated with direct hematogenous spread.

The primary source is not apparent in 20–40 percent of cases (Gellin et al. 1997).

Pathogenesis

Spinal abscesses may be acute or chronic; chronic abscesses are more often associated with tuberculous osteomyelitis and epidural abscess.

The most common site for an abscess is posterior to the spinal cord; this occurs in 70 percent of cases. Anterior abscesses are more common in tuberculous infections and occur in the thoracic and lumbar regions. The abscess usually occupies about three to six vertebral segments, but rarely may extend more widely up and down the spinal column. Children are less likely to have adjacent vertebral osteomyelitis and usually have abscesses in the cervical or lumbar regions. The thoracic area is most common in adults.

The damage caused by the infection is often out of proportion to the size of the abscess. This may be due to several factors including thrombosis and thrombophlebitis of veins draining the spinal cord leading to edema and venous infarction; focal areas of vasculitis induced by the inflammatory mass; or bacterial exotoxin production particularly that due to *Staphylococcus aureus* (Gellin et al. 1997).

Bacteriology

A microbiological diagnosis is made from blood cultures, which may be positive in up to 90 percent of cases particularly if *S. aureus* is present, or by examination of pus drained at operation. Infections are rarely polymicrobial, only about 10 percent of cases having multiple isolates (Darouiche et al. 1992). *Staphylococcus aureus* is the most common isolate in all series accounting for up to 90 percent of cases; MRSA is also reported (Danner and Hartman 1987; Tang et al. 2002). Aerobic streptococci including *Streptococcus pneumoniae* (Turner et al. 1999) and *Streptococcus milleri* (Gelfand et al. 1991) are reported in up to 18 percent of series. The occurrence of aerobic gram-negative bacilli, including *Pseudomonas aeruginosa*, is on the increase both in postoperative cases and in injecting drug users; up to 37 percent of isolates may be gram-negative bacilli (Gellin et al. 1997). Coagulase-negative staphylococci are associated increasingly with postoperative cases. *Mycobacterium tuberculosis* is a well recognized cause of vertebral osteomyelitis and epidural abscess and is reported in most series.

The increase in infections seen in injecting drug users is represented in most series (Tunkel and Pradhan 2002) and may account for up to 40 percent of cases (Nussbaum et al. 1992).

Clinical features

The initial features may be subtle. There are four characteristic stages described (Gellin et al. 1997). The first symptom, with fever and malaise, is focal back pain. This is followed by nerve root pain; the next stage is

Table 21.7 *Bacterial isolates and suggested antibiotic therapy for spinal epidural abscess*

Bacteria	Frequency	Suggested antibiotics
S. aureus including MRSA	60–90%	Flucloxacillin or vancomycin plus rifampicin
Aerobic and anaerobic streptococci	18%	Benzylpenicillin plus metronidazole
Aerobic gram-negative bacilli	13%	Third generation cephalosporin or ceftazidime or quinolone
Polymicrobial	10%	Third generation cephalosporin plus metronidazole

spinal cord dysfunction, such as incontinence or sensory loss. The final stage is cord compression leading to paralysis. All this may happen over a short timescale; the progression of symptoms to nerve root pain can be over 3 to 4 days, then signs of cord compression over the next 4 to 5 days. At this stage the neurological effects are usually reversible but urgent surgical intervention is still needed as paralysis can develop rapidly. Acute infections, usually due to hematogenous spread, will progress rapidly in under 2 weeks with prominent signs of local infection, whereas abscesses arising from vertebral osteomyelitis will have a longer duration of back pain lasting more than 2 weeks before neurological symptoms appear. Back pain is a very common symptom, so epidural infection is often not considered as a diagnosis until neurological symptoms appear.

Diagnosis

Routine tests will show that the inflammatory markers, such as ESR and CRP, will be raised, as will the peripheral blood leukocyte count. Cerebrospinal fluid examination is sometimes performed and will show evidence of parameningeal inflammation, in other words raised white cell count and protein. The CSF cultures should be negative unless the abscess was directly sampled in which case the white cell count would be very high and the cultures will yield the causative organism.

Radiological diagnosis is made by MRI scanning using gadolinium enhanced scans which are very specific for this type of abscess.

Management

Surgical drainage to relieve pressure on the spinal cord is essential and is an emergency procedure. Antibiotics should be started as soon as possible to cover the common pathogen, *S. aureus*, with antipseudomonal cover added in postoperative cases and injecting drug users (Table 21.7). The spinal epidural space is outside the blood–brain barrier which will make antibiotic choice easier. Recommended duration of therapy is intravenous antibiotics for 3 to 4 weeks, then oral agents for a further 2 to 3 months if vertebral osteomyelitis is present. Some studies have recommended a shorter duration of therapy (Del Curling et al. 1990).

Some reports suggest that nonsurgical management may be successful (Wheeler et al. 1992), but the patients were carefully selected and the true success of this is difficult to ascertain. Some patients in another series showed progression of symptoms while on antibiotics alone (Hlavin et al. 1990).

It is well reported that a delay in diagnosis leads to a worse outcome. The preoperative status of the patient is a good predictor of outcome; those that are paralyzed before surgery do worse (Mackenzie et al. 1998).

REFERENCES

Akova, M., Akalin, H.E., et al. 1993. Treatment of intracranial abscesses: experience with sulbactam/ampicillin. *J Chemother*, **5**, 181–5.

Alderson, D., Strong, A.J., et al. 1981. Fifteen year review of the mortality of brain abscess. *Neurosurgery*, **8**, 1–6.

Aliaga, L., Mediaville, J.D., et al. 1997. Nontyphoidal salmonella intracranial infections in HIV infected patients. *Clin Infect Dis*, **25**, 1118–20.

Ariza, J., Casanova, A., et al. 1986. Etiologic agent and primary source of infection in 42 cases of focal intracranial suppuration. *J Clin Microbiol*, **24**, 899–902.

Asensi, V., Carton, J.A., et al. 1996. Imipenem therapy of brain abscesses. *Eur J Clin Microbiol Infect Dis*, **15**, 653–7.

Asensi, V., Alvarez, M., et al. 2002. *Eikenella corrodens* brain abscess after repeated periodontal manipulations cured with imipenem and neurosurgery. *Infection*, **30**, 240–2.

Baker, A.S., Ojemann, R.G., et al. 1975. Spinal epidural abscess. *N Engl J Med*, **293**, 463–8.

Beaman, B.L. and Beaman, L.V. 1994. Nocardia species: host–parasite relationships. *Clin Micro Rev*, **7**, 213–64.

Berenson, C.S. and Bia, F.J. 1989. *Propionibacterium acnes* causes postoperative brain abscesses unassociated with foreign bodies: case reports. *Neurosurgery*, **25**, 130–4.

Black, P., Graybill, J.R. and Charache, P. 1973. Penetration of brain abscess by systemically administered antibiotics. *J Neurosurg*, **38**, 705–9.

Bleck, T.P. and Greenlee, J.E. 2000a. Epidural abscess. In: Mandell, G.L., Bennett, J.E. and Dolin, R. (eds), *Principles and practice of infectious diseases*, 5th edn. London: Churchill Livingstone, 1031–4.

Bleck, T.P. and Greenlee, J.E. 2000b. Subdural empyema. In: Mandell, G.L., Bennett, J.E. and Dolin, R. (eds), *Principles and practice of infectious diseases*, 5th edn. London: Churchill Livingstone, 1028–31.

Boom, W.H. and Tuazon, C.U. 1985. Successful treatment of multiple brain abscesses with antibiotics alone. *Rev Infect Dis*, **7**, 189–99.

Bradley, P.J. and Shaw, M.D.M. 1983. Three decades of brain abscess in Merseyside. *J R Coll Surg Edinb*, **28**, 223–8.

Bradley, P.J., Manning, K.P. and Shaw, M.D.M. 1984. Brain abscess secondary to paranasal sinusitis. *J Laryngol Otol*, **98**, 719–25.

Britt, R., Enzmann, D. and Yeager, A. 1981. Neuropathological and computed tomographic findings in experimental brain abscess. *J Neurosurg*, **55**, 590–603.

Brook, I. 1995. Brain abscess in children: microbiology and management. *J Child Neurol*, **10**, 283–8.

Brown, E.M., Strangelis, G., et al. 1994. Short-course antimicrobial therapy for brain abscess and subdural empyema. Proceedings of the 123rd meeting of the Society of British Neurological Surgeons. *J Neurol Neurosurg Psychiat*, **57**, 390–1.

Canale, D.J. 1996. William Macewan and the treament of brain abscesses: Revisited after one hundred years. *J Neurosurg*, **84**, 133–42.

Calfee, D.P. and Wispelwey, B. 2000. Brain abscess. *Semin Neurol*, **20**, 353–60.

Chaudhry, R., Dhawan, B., et al. 1998. The microbial spectrum of brain abscess with special reference to anaerobic bacteria. *Br J Neurosurg*, **12**, 127–30.

Chun, C.H., Johnson, J.D., et al. 1986. Brain abscess. Study of 45 consecutive cases. *Medicine*, **65**, 415–31.

Cohen, F.L., Koerner, R.S. and Taub, S.J. 1985. Solitary brain abscess following endoscopic injection of oesophageal varices. *Gastro Endosc*, **31**, 331–3.

Conti, D.J. and Rubin, R.H. 1988. Infection of the central nervous system in organ transplant recipients. *Neurol Clin*, **6**, 241–60.

Corne, P., Rajeebally, I. and Jonquet, O. 2002. *Rhodococcus equi* brain abscess in an immunocompetent patient. *Scand J Infect Dis*, **34**, 300–2.

Costello, G.T., Heppe, R., et al. 1983. Susceptibility of brain to aerobic, anaerobic and fungal organisms. *Infect Immun*, **41**, 535–9.

Critchley, G. and Strachan, R. 1996. Postoperative subdural empyema caused by *Propionibacterium acnes* – a report of two cases. *Br J Neurosurg*, **10**, 321–3.

Curless, R.G. 1980. Neonatal intracranial abscess: two cases caused by *Citrobacter* and a literature review. *Ann Neurol*, **8**, 269–72.

Danner, R.L. and Hartman, B.J. 1987. Update of spinal epidural abscess: 35 cases and review of the literature. *Rev Infect Dis*, **9**, 265–74.

Darouiche, R.O., Hamill, R.J., et al. 1992. Bacterial spinal epidural abscess. Review of 43 cases and literature survey. *Medicine*, **71**, 369–85.

De Louvois, J. 1978. The bacteriology and chemotherapy of brain abscess. *J Antimicrob Chemother*, **4**, 395–413.

De Louvois, J. and Hurley, R. 1977. Inactivation of penicillin by purulent exudates. *Br Med J*, **1**, 998–1000.

De Louvois, J., Gortvai, P. and Hurley, R. 1977a. Bacteriology of abscesses of the central nervous system: a multicentre prospective study. *Br Med J*, **2**, 981–4.

De Louvois, J., Gortvai, P. and Hurley, R. 1977b. Antibiotic treatment of abscesses of the central nervous system. *Br Med J*, **2**, 985–7.

Del Curling, O. Jr., Gower, D.J. and McWhorter, J.M. 1990. Changing concepts in spinal epidural abscess: a report of 29 cases. *Neurosurgery*, **27**, 185–92.

Dill, S.R., Cobbs, C.G. and McDonald, C.K. 1995. Subdural empyema: analysis of 32 cases and review. *Clin Infect Dis*, **20**, 372–86.

DiNubile, M.J. 1988. Cavernous sinus thrombosis – a review. *Arch Neurol*, **45**, 567–72.

Donald, F.E. 1990. Treatment of brain abscess. *J Antimicrob Chemother*, **25**, 310–12.

Donald, F.E. and Ispahani, P. 1988. Use of cefotaxime in brain abscess (letter). *Br J Neurosurg*, **2**, 539–40.

Donald, F.E. and Ispahani, P. 1990. Penetration of ceftazidime into intracranial abscess (letter). *J Antimicrob Chemother*, **25**, 297–303.

Donald, F.E., Firth, J.L., et al. 1990. Brain abscess in the 1980s. *Br J Neurosurg*, **4**, 265–72.

Eckburg, P.B., Montoya, J.G. and Vosti, K.L. 2001. Brain abscess due to *Listeria monocytogenes:* five cases and a review of the literature. *Medicine*, **80**, 223–35.

Engelhardt, K., Kampfl, A., et al. 2002. Brain abscess due to *Capnocytophaga* species, *Actinomyces* species and *Streptococcus intermedius* in a patient with cyanotic congenital heart disease. *Eur J Clin Micro Infect Dis*, **21**, 236–7.

Everett, E.D. and Strausbaugh, L.J. 1980. Antimicrobial agents and the central nervous system. *Neurosurgery*, **6**, 691–714.

Gallagher, R.M., Gross, C.W. and Phillips, C.D. 1998. Suppurative intracranial complications of sinusitis. *Laryngoscope*, **108**, 1635–42.

Garfield, J. 1969. Management of supratentorial intracranial abscess: a review of 200 cases. *Br Med J*, **2**, 7–11.

Gelfand, M.S., Bakhtian, B.J. and Simmons, B.P. 1991. Spinal sepsis due to *Streptococcus milleri:* two cases and review. *Rev Infect Dis*, **13**, 559–63.

Gellin, B.G., Weingarten, K., et al. 1997. Epidural abscess. In: Scheld, W.M., Whitley, R.J. and Durack, D.T. (eds), *Infections of the central nervous system*, 2nd edition. Philadelphia: Lippincott Raven, 507–22.

Gortvai, P., De Louvois, J. and Hurley, R. 1987. The bacteriology and chemotherapy of acute pyogenic brain abscess. *Br J Neurosurg*, **1**, 189–203.

Green, H.T., O'Donoghue, M.A.T., et al. 1989. Penetration of ceftazidime into intracranial abscess. *J Antimicrob Chemother*, **24**, 431–6.

Grigoriadis, E. and Gold, W.L. 1997. Pyogenic brain abscess caused by *Streptococcus pneumoniae*: case report and review. *Clin Infect Dis*, **25**, 1108–12.

Hagensee, M.E., Bauwens, J.E., et al. 1994. Brain abscess following marrow transplantation: experience at the Fred Hutchinson Cancer Research Center 1984–1992. *Clin Infect Dis*, **19**, 402–8.

Hanel, R.A., Araujo, J.C., et al. 2000. Multiple brain abscesses caused by *Salmonella typhi*: case report. *Surg Neurol*, **53**, 86–90.

Heineman, H.S. and Braude, A.I. 1963. Anaerobic infection of the brain. *Am J Med*, **35**, 682–97.

Helfgott, D.C., Weingarten, K. and Hartman, B.J. 1997. Subdural empyema. In: Scheld, W.M., Whitley, R.J. and Durack, D.T. (eds), *Infections of the central nervous system*, 2nd edn. Philadelphia: Lipincott Raven, 495–505.

Hirschberg, H. and Bosnes, V. 1987. C-reactive protein levels in the differential diagnosis of brain abscesses. *J Neurosurg*, **67**, 358–60.

Hlavin, M.L., Kaminski, H.J., et al. 1990. Spinal epidural abscess: a ten year perspective. *Neurosurgery*, **27**, 177–84.

Hlavin, M.L., Kaminski, H.J., et al. 1994. Intracranial suppuration: A modern decade of postoperative subdural empyema and epidural abscess. *Neurosurgery*, **34**, 974–81.

Ibrahim, A.W., al-Rajeh, S.M., et al. 1990. Brain abscess in Saudi Arabia. *Neurosurg Rev*, **13**, 103–7.

Ingham, H.R., Selkon, J.B. and Roxby, C.M. 1977. Bacteriological study of otogenic cerebral abscesses: chemotherapeutic role of metronidazole. *Br Med J*, **2**, 991–3.

Ingham, H.R., High, A.S., et al. 1978. Abscess of the frontal lobe secondary to covert dental sepsis. *Lancet*, **2**, 497–9.

Ismail, E.A., Shafik, M.H. and Al-Mutavi, G. 2001. A case of non-01 *Vibrio cholerae* septicaemia with meningitis, cerebral abscess and unilateral hydrocephalus in a preterm baby. *Eur J Clin Micro Infect Dis*, **20**, 598–600.

Jacobson, P.L. and Farmer, T.W. 1981. Subdural empyema complicating meningitis in infants: improved prognosis. *Neurology*, **31**, 190-19, 3.

Kaplan, K. 1985. Brain abscess. *Med Clin N Am*, **69**, 345–60.

Koppel, B.S., Tuchman, A.J., et al. 1988. Epidural spinal infection in intravenous drug users. *Arch Neurol*, **45**, 1331–7.

Kramer, P.W., Griffith, R.S. and Campbell, R.I. 1969. Antibiotic penetration of the brain: a comparative study. *J Neurosurg*, **31**, 295–302.

Krone, A., Schaal, K.P., et al. 1989. Nocardial cerebral abscess cured with imipenem/amikacin and enucleation. *Neurosurg Rev*, **12**, 333–40.

Lakshmi, V., Rao, R.R. and Dinaker, I. 1993. Bacteriology of brain abscess – observations on 50 cases. *J Med Micro*, **38**, 187–90.

Lerner, P.J. 1985. Neurologic complications of infective endocarditis. *Med Clin North Am*, **69**, 385–99.

Levy, R.M., Gutin, P.H., et al. 1986. Vancomycin penetration of a brain abscess: case report and review of the literature. *Neurosurgery*, **18**, 632–6.

Lu, C.H., Chang, W.N., et al. 2002. Bacterial brain abscess: microbiological features, epidemiological trends and therapeutic outcomes. *QJ Med*, **95**, 501–9.

MacEwan, W. 1893. *Pyogenic infective diseases of the brain and spinal cord. Meningitis, abscess of the brain, infective sinus thrombosis.* Glasgow: James Maclehose and Sons.

Mackenzie, A.R., Laing, R.B., et al. 1998. Spinal epidural abscess: The importance of early diagnosis and treatment. *J Neurol Neurosurg Psych*, **65**, 209–12.

Maderazo, E.G. and Quintiliani, R. 1974. Treatment of nocardial infection with trimethoprim and sulphamethoxazole. *Am J Med*, **57**, 671–5.

Mahapatra, A.K., Pawar, S.J., et al. 2002. Intracranial *Salmonella* infections: meningitis, subdural collections and brain abscess. A series of six surgically managed cases with follow-up results. *Pediatr Neurosurg*, **36**, 8–13.

Mamelak, A.N., Obana, W.G., et al. 1994. Nocardial brain abscess: treatment strategies and factors influencing outcome. *Neurosurgery*, **35**, 622–31.

Mampalam, T.J. and Rosenblum, M.L. 1988. Trends in the management of bacterial brain abscesses: a review of 102 cases over 17 years. *Neurosurgery*, **23**, 451–8.

Martinello, R.A. and Cooney, E.L. 2003. Cerebellar brain abscess associated with tongue piercing. *Clin Infect Dis*, **36**, 32–4.

Mathisen, G.E. and Johnson, J.P. 1997. Brain abscess. *Clin Infect Dis*, **25**, 763–81.

Mathisen, G.E., Meyer, R.D., et al. 1985. Brain abscess and cerebritis. *Rev Infect Dis*, **6**, Suppl. 1, S101–6.

Meis, J.F.G.M., Groot-Loonen, J. and Hoogkamp-Korstanje, J.A.A. 1995. A brain abscess due to multiply-resistant *Enterobacter cloacae* successfully treated with meropenem (letter). *Clin Infect Dis*, **20**, 1567.

Mendes, M., Moore, P., et al. 1980. Susceptibility of brain and skin to bacterial challenge. *J Neurosurg*, **52**, 772–5.

Molina, J.M., Leport, C., et al. 1991. Clinical and bacterial features of infection caused by *Streptococcus milleri*. *Scand J Infect Dis*, **23**, 659–66.

Molinari, G.F., Smith, L., et al. 1973. Brain abscess from septic cerebral embolism: an experimental model. *Neurology*, **23**, 1205–10.

Nathoo, N., Nadvi, S.S., et al. 1999. Intracranial subdural empyemas in the era of computed tomography: a review of 699 cases. *Neurosurgery*, **44**, 529–35.

Neilsen, H., Gyldensted, C. and Harmsen, A. 1982. Cerebral abscess: aetiology and pathogenesis, symptoms, diagnosis and treatment. *Acta Neurol Scand*, **65**, 609–22.

Nicolosi, A., Hauser, W.A., et al. 1991. Incidence and prognosis of brain abscess in a defined population: Olmsted county, Minnesota. *Neuroepidemiology*, **10**, 122–31.

Nussbaum, E.S., Rigamonti, D., et al. 1992. Spinal epidural abscess: a report of 40 cases and review. *Surg Neurol*, **38**, 225–31.

O'Donoghue, M.A.T., Green, H.T. and Shaw, M.D.M. 1992. Cerebral abscess on Merseyside 1980–1988. *J Infect*, **25**, 163–72.

Onderonk, A.B., Kasper, D.L., et al. 1979. Experimental animal models for anaerobic infections. *Rev Infect Dis*, **1**, 291–301.

Park, S.C. and Neches, W.H. 1993. The neurologic complications of congenital heart disease. *Neurol Clin*, **11**, 441–62.

Parker, M.T. and Ball, L.C. 1976. Streptococci and aerococci associated with systemic infection in man. *J Med Micro*, **9**, 275–302.

Pathak, A., Sharma, B.S., et al. 1990. Controversies in the management of subdural empyema. A study of 41 cases with review of literature. *Acta Neurochir*, **102**, 25–32.

Patir, R., Sood, S. and Bhatia, R. 1995. Post traumatic brain abscess: experience of 36 patients. *Br J Neurosurg*, **9**, 29–35.

Press, O.W. and Ramsey, P.G. 1984. Central nervous system infections associated with hereditary haemorrhagic telangiectasia. *Am J Med*, **77**, 86–92.

Rao, R.P., Ghanayem, N.S., et al. 2002. *Mycoplasma hominis* and *Ureaplasma* species brain abscess in a neonate. *Pediatr Infect Dis J*, **21**, 1083–5.

Rau, C.S., Chang, W.N., et al. 2002. Brain abscess caused by aerobic Gram-negative bacilli: clinical features and therapeutic outcomes. *Clin Neurol Neurosurg*, **105**, 60–5.

Reihsaus, E., Waldbaur, H., et al. 2000. Spinal epidural abscess: a metanalysis of 915 patients. *Neurosurg Rev*, **23**, 175–204.

Renier, D., Flaudin, C., et al. 1988. Brain abscesses in neonates – a study of 30 cases. *J Neurosurg*, **69**, 877–82.

Report by the 'Infection in Neurosurgery' Working Party of the British Society for Antimicrobial Chemotherapy. 2000. Review article. The rational use of antibiotics in the treatment of brain abscess. *Br J Neurosurg*, **14**, 525–30.

Richards, J., Sisson, P.R., et al. 1990. Microbiology, chemotherapy and mortality of brain abscess in Newcastle upon Tyne between 1979 and 1988. *Scand J Infect Dis*, **22**, 511–18.

Rodriguez, R.E., Valero, V. and Watanakunakorn, C. 1986. *Salmonella* focal intracranial infections: review of the world literature (1884–1984) and report of an unusual case. *Rev Infect Dis*, **8**, 31–41.

Rosenblum, M.L., Hoff, J.T., et al. 1980. Nonoperative treament of brain abscesses in selected high-risk patients. *J Neurosurg*, **52**, 217–25.

Rubin, G., Michowiz, S.D., et al. 1993. Spinal epidural abscess in the pediatric age group: case report and review of the literature. *Pediatr Infect Dis J*, **12**, 1007–11.

Safani, M.M. and Ehrensaft, D.V. 1982. Successful treatment of subdural empyema due to *Serratia marcescens*. *Drug Intell Clin Pharm*, **16**, 777–9.

Sakai, C., Iuchi, T., et al. 2001. *Bacillus cereus* brain abscesses occurring in a severely neutropenic patient: successful treatment with antimicrobial agents, granulocyte colony-stimulating factor and surgical drainage. *Intern Med*, **40**, 654–7.

Schliamser, S.E., Backman, K. and Norrby, S.R. 1988. Intracranial abscesses in adults: an analysis of 54 consecutive cases. *Scand J Infect Dis*, **20**, 1–9.

Selby, R., Ramirez, C.B., et al. 1997. Brain abscess in solid organ transplant recipients receiving cyclosporine-based immunosuppression. *Arch Surg*, **132**, 304–10.

Seydoux, C. and Francioli, P. 1992. Bacterial brain abscesses: factors influencing mortality and sequelae. *Clin Infect Dis*, **15**, 394–401.

Shah, S.S., Gloor, P. and Gallagher, P.G. 1999. Bacteremia, meningitis, and brain abscesses in a hospitalised infant: complications of *Pseudomonas aeruginosa* conjunctivitis. *J Perinatol*, **19**, 462–5.

Sjolin, J., Eriksson, N., et al. 1991. Penetration of cefotaxime and desacetylcefotaxime into brain abscesses in humans. *Antimicrob Agent Chemother*, **35**, 2606–10.

Sjolin, J., Lilja, A., et al. 1993. Treatment of brain abscess with cefotaxime and metronidazole: prospective study on 15 consecutive patients. *Clin Infect Dis*, **17**, 857–63.

Skoutelis, A.T., Gogos, C.A., et al. 2000. Management of brain abscesses with sequential intravenous/oral antibiotic therapy. *Eur J Clin Micro Infect Dis*, **19**, 332–5.

Small, M. and Dale, B.A.B. 1984. Intracranial suppuration 1968–1982: a fifteen year review. *Clin Otolaryngol*, **9**, 315–21.

Smego, R.A. Jr. and Foglia, G. 1998. Actinomycosis. *Clin Infect Dis*, **26**, 1255–63.

Smith, H.P. and Hendrick, E.B. 1983. Subdural empyema and epidural abscess in children. *J Neurosurg*, **58**, 392–7.

Sofianou, D., Selviarides, P., et al. 1996. Etiological agents and predisposing factors of intracranial abscesses in a Greek university hospital. *Infection*, **24**, 144–6.

Stern, J., Bernstein, C.A., et al. 1981. *Pasteurella multocida* subdural empyema. *J Neurosurg*, **54**, 550–2.

Tang, H.J., Lin, H.J., et al. 2002. Spinal epidural abscess – experience with 46 patients and evaluation of prognostic factors. *J Infect*, **45**, 76–81.

Tenney, J.H. 1986. Bacterial infections of the central nervous system in neurosurgery. *Neurol Clin*, **4**, 91–114.

Tice, A.D., Strait, K., et al. 1999. Outpatient parenteral antimicrobial therapy for central nervous system infections. *Clin Infect Dis*, **29**, 1394–9.

Tunkel, A.R. and Pradhan, S.K. 2002. Central nervous system infections in injection drug users. *Infect Dis Clin North Am*, **16**, 589–605.

Tunkel, A.R. and Scheld, W.M. 1989. Therapy of bacterial meningitis: principles and practice. *Infect Control Hosp Epidemiol*, **10**, 565–71.

Tunkel, A.R., Wispelwey, B. and Scheld, W.M. 2000. Brain abscess. In: Mandell, G.L., Bennett, J.E. and Dolin, R. (eds), *Principles and practice of infectious diseases*.
London: Churchill Livingstone, 1016–28.

Turner, D.P.J., Weston, V.C. and Ispahani, P. 1999. *Streptococcus pneumoniae spinal infection* in Nottingham, United Kingdom: not a rare event. *Clin Infect Dis*, **28**, 873–81.

Van Alphen, H.A.M. and Driessen, J.J.R. 1976. Brain abscess and subdural empyema. *J Neurol Neurosurg Psych*, **39**, 481–90.

Van Dellen, A., Nadvi, S.S., et al. 1998. Intracranial tuberculous subdural empyema: case report. *Neurosurgery*, **43**, 370–3.

Weeds, J.F. 1872. Case of cerebral abscess. *Nashville J Med Surg*, **9**, 156–71.

Wellman, W.E., Dodge, H.W., et al. 1954. Concentration of antibiotics in the brain. *J Lab Clin Med*, **43**, 275–9.

Wessalowski, R., Thomas, L., et al. 1993. Multiple brain abscesses caused by *Salmonella enteritidis* in a neonate: successful treatment with ciprofloxacin. *Ped Infect Dis J*, **12**, 683–8.

Wheeler, D., Keiser, P., et al. 1992. Medical management of spinal epidural abscesses: case report and review. *Clin Infect Dis*, **15**, 22–7.

Wispelwey, B., Dacey, R.G. Jr. and Scheld, W.M. 1997. Brain abscess. In: Scheld, W.M., Whitley, R.J. and Durack, D.T. (eds), *Infections of the central nervous system*, 2nd edn. Philadelphia: Lippincott Raven, 463–93.

Yamamoto, M., Jimbo, M., et al. 1993. Penetration of intravenous antibiotics into brain abscesses. *Neurosurgery*, **33**, 44–9.

Yang, S.Y. 1981. Brain abscess: a review of 400 cases. *J Neurosurg*, **55**, 794–9.

Yoshikawa, T.T., Chow, A.W. and Guze, L.B. 1975. Role of anaerobic bacteria in subdural empyema. Report of four cases and review of 327 cases from the English literature. *Am J Med*, **58**, 99–104.

22

Bacterial infections of the eye

SUSAN E. SHARP AND MICHAEL R. DRIKS

INTRODUCTION

This chapter will focus on bacterial infections of the eye, specifically those associated with conjunctivitis, keratitis, and endophthalmitis. Within each of these infections, the areas of predisposing factors, clinical manifestations, infecting agents, diagnosis, and treatment will be reviewed.

Infections of the eye can be exogenously acquired, contiguous (from infections of the face, eyelids, etc.), or from hematogenous spread (O'Brien 2000). One or more of these routes can be associated with each of the above conditions.

NORMAL FLORA AND COMMON PATHOGENIC BACTERIA OF THE EYE

The bacterial organisms most commonly associated with 'normal flora' and the 'common pathogenic' bacteria of the eye are indicated in Table 22.1 (Sharp 1999; O'Brien 2000).

BACTERIAL CONJUNCTIVITIS

Predisposing factors

Conjunctivitis is defined as inflammation of the conjunctiva, usually caused by infections, allergens, or irritative substances. Predisposition to bacterial conjunctivitis primarily involves contact exposure to airborne fomites, upper respiratory infections, skin flora on hands, or genital secretions (O'Brien 2000). A patient's age is an important determinant of the likely infecting organism. In children, bacterial conjunctivitis is more common than viral, whereas in adults, the opposite is true (Wishart et al. 1984; Fitch et al. 1989; Woodland et al. 1992). The clinical setting can also determine the microbiology of bacterial conjunctivitis. Institutional outbreaks have been described, presumably due to contact transmission between residents, patients, and medical personnel (King et al. 1988; Schwartz et al. 1988). Community outbreaks of *Haemophilus influenzae* biogroup *aegyptius* occur in children.

Clinical manifestations

All forms of conjunctivitis may be characterized by irritation, itching, foreign body sensation, tearing, and discharge. Bacterial conjunctivitis can be distinguished from other forms by the presence of mucopurulent drainage and a papillary reaction, characterized by many small bumps on the tarsal conjunctiva, creating a velvety appearance (Chung and Cohen 2000). In contrast to viral causes, bacterial conjunctivitis is typically unilateral. In severe cases, there is often significant lid edema with copious purulent drainage, marked conjunctival erythema, chemosis, membrane formation, and associated keratitis (O'Brien 2000). *Neisseria* conjunctival infections cause especially severe inflammation (Valenton and Abendanior 1973; Barquet et al. 1990). Streptococci and *Corynebacterium diphtheriae* are notable for causing membrane formation (Boralkar 1989). Nonpyogenic bacterial organisms, including

Table 22.1 *Bacterial organisms most commonly associated with 'normal flora' and the 'common pathogenic' bacteria of the eye*

Normal flora	Most common pathogens
Aerobic bacteria	Aerobic bacteria
Corynebacterium	*Streptococcus pneumoniae*
Staphylococcus species	*Staphylococcus aureus*
Micrococcus	*Staphylococcus*, coagulase-negative
Nonpathogenic *Neisseria* species	*Streptococcus* species
Aerococcus	*Pseudomonas aeruginosa* and other gram-negative bacilli
	Haemophilus influenzae
	Moraxella catarrhalis
	Neisseria gonorrhoeae
	Neisseria cinerea
	Bacillus species
	Chlamydia trachomatis
Anaerobic bacteria	Anaerobic bacteria
Propionibacterium	*Propionibacterium*
Clostridium	*Peptostreptococcus*
Peptostreptococcus	

Mycobacterium tuberculosis, *Francisella tularensis*, *Treponema pallidum*, and *Bartonella henselae* can cause an atypical conjunctivitis with unilateral nodules that ulcerate and ipsilateral preauricular lymphadenopathy.

Chlamydial conjunctivitis deserves special mention. Trachoma refers to conjunctival and limbic scarring leading to trichiasis and corneal opacification in recurrent or chronic *Chlamydia trachomatis* infection (Bobo et al. 1997). Trachoma is among the leading causes of blindness in the world. Inclusion conjunctivitis is an acute infection in adults, most often transmitted by contact with genital secretions. In a neonate infected during the birthing process, the disease is grouped together with gonococcal conjunctivitis and referred to as opthalmia neonatorum. As of the 1970s, the incidence of chlamydial conjunctivitis in the newborn was increasing, involving 2.8 percent of all births in one clinic and 35 percent of infants born to mothers with known chlamydial cervicitis (Schachter et al. 1979). The usual incubation period is 5–12 days from birth (Chandler et al. 1977). In infants born to infected mothers, 22–44 percent develop neonatal conjunctivitis (Harrison and Alexander 1990).

Infecting agents

In the healthy and nonimmunocompromised population, *Staphylococcus aureus*, *Streptococcus pneumoniae*, and *H. influenzae* are the most common causes of acute bacterial conjunctivitis. However, differences in causative organisms are seen based on the age of the patient.

In the newborn, acute conjunctivitis can be caused by *C. trachomatis*, *Neisseria gonorrhoeae*, *Neiserria cinerea*, or *Pseudomonas aeruginosa*, although *C. trachomatis* and *N. gonorrhoeae* are the most frequently encountered organisms in this population (de Toledo and Chandler 1992; Ratelle et al. 1997). Studies have shown *H. influenzae*, *S. pneumoniae*, and *Moraxella catarrhalis* to be the bacterial agents most often associated with acute conjunctivitis in children (Weiss et al. 1993; Gigliotti et al. 1981; Wald et al. 2001). *H. influenzae* has been reported to be the most prevalent isolated pathogen, accounting for between 54 and 74 percent of infections in children (Gigliotti et al. 1981; Wald 1997; Wald et al. 2001). Ocular *H. influenzae* isolates also appear to be mostly serologically nontypable strains, probably as a result of the widespread use of the *H. influenzae* b vaccine (Alrawi et al. 2002; Robinson et al. 2001). Conjunctivitis in young children may also be caused by *N. cinerea*, *S. aureus*, or *P. aeruginosa* (Trottier et al. 1991; Weiss et al. 1993; Iroha et al. 1998; Shah and Gallagher 1998). In addition, the *Streptococcus* viridans group have been shown to be associated with conjunctivitis in patients aged less than 1 year (Seal et al. 1982).

In adults, the most common causes of bacterial conjunctivitis are *S. pneumoniae*, *S. aureus*, and coagulase-negative staphylococci, while in sexually active teenagers and adults, *C. trachomatis* and *N. gonorrhoeae* can also be found (Seal et al. 1982; O'Brien 2000). Additional bacteria such as those listed at the beginning of this chapter can also be found to cause conjunctivitis in a variety of community and nosocomial settings. For example, hospital- and institutionally acquired cases of conjunctivitis have been associated with infections caused by *Serratia marcescens*, *S. pneumoniae*, *P. aeruginosa*, and *Moraxella* spp. (van Ogtrop et al. 1997; King et al. 1988; Schwartz et al. 1988). Mycobacterial species, especially *M. tuberculosis*, have been shown to be an etiologic agent of atypical conjunctivitis and may be associated with ulcerations of the eye.

Organisms thought to be nonpathogenic can occasionally cause conjunctivitis. However, those bacteria normally considered skin flora may be cultured from diseased eyes without causing infection. For example, 40 percent of specimens in one study were associated with recovery of *S. epidermidis* or mixed skin flora, which also occurred in cultures of normal eyes (Seal et al. 1982). In another study of children, staphylococci, corynebacteria, and alpha-hemolytic streptococci were the predominant organisms recovered from the eyelids of uninfected control subjects (Weiss et al. 1993). In addition, a study by Gigliotti and colleagues found that *S. aureus* was as commonly isolated from normal eyes as from diseased eyes in children (Gigliotti et al. 1981). Thus, care must be taken when interpreting the results of bacterial cultures.

Diagnosis

The value of laboratory diagnostic procedures in mild conjunctivitis is not well established and is therefore not done routinely. Cases of mild conjunctivitis are often treated without the knowledge of bacterial culture information. In severe conjunctivitis, culture of the superior and inferior tarsal conjunctiva is recommended. In all cases of ophthalmia neonatorum, smears and cultures for bacteria, including *Chlamydia*, should be performed.

If culture diagnosis of bacterial conjunctivitis is desired, it can be attempted by performing conjunctiva cultures. Samples of both eyes should be taken with separate swabs that have been premoistened with sterile saline. For routine bacteriology collections, swabs that are used for normal bacteriology specimen collection should be used. If cultures for *Chlamydia* are being collected, cotton swabs should not be used, but dacron or calcium alginate swabs would be acceptable. Topical ointments and analgesics should be avoided prior to collection. Samples are collected by rolling the swabs over each conjunctiva. They should be directly inoculated on to blood agar and chocolate agar plates (warmed to room temperature prior to inoculation) or added to transport media for travel to, and inoculation in, the laboratory. After inoculation, the swabs are rolled across two separate slides for staining purposes. Plates inoculated at the bedside should be transported at room temperature to the laboratory as soon as possible, preferably within 15 min, or stored for up to 24 h at room temperature prior to delivery. If swabs are collected and not directly inoculated at the bedside, they should be placed back into the swab device sleeve, transported at room temperature, and delivered to the laboratory immediately and no longer than 2 h after collection. If samples are being collected for *Chlamydia* culture, appropriate *Chlamydia* transport media should be utilized for delivery of the specimens to the laboratory, again within 2 h of collection. In addition, molecular testing for the presence of *Chlamydia* can be performed if the sample is collected in specimen collection/transfer devices specific for this purpose. Swab samples are normally not appropriate for mycobacterial stain or culture. If *M. tuberculosis* conjunctivitis is suspected, ulcer/lesion scrapings can be collected for appropriate stains and cultures.

As stated earlier, if possible both conjunctiva should be sampled, even if one seems to be uninfected. This will allow the determination of the normal flora of the uninfected eye. This can be used for comparison with the organisms cultured from the infected eye and assist in the determination of possible pathogens. Anaerobic cultures are not performed in cases of conjunctivitis, as the causative bacterial pathogens are aerobic in nature. An exception might exist with *Peptostreptococcus*, an anaerobic gram-positive cocci; however, this is not a frequent cause of conjunctivitis.

The Gram stain can assist in the determination of the causative agent(s) of conjunctivitis, especially if only the infected conjunctiva is sampled. In acute bacterial conjunctivitis, specimens show large numbers of neutrophils, and bacteria may be seen inside or outside of these cells. With chronic conjunctivitis there are more lymphocytes and large mononuclear cells seen in the Gram stain as compared with neutrophils. Organisms seen on the Gram stain are more likely to be involved in the infective process and can be used to help interpret the bacterial culture information. Gram stains of conjunctival scrapings have been shown to provide a rapid means of predicting the causative agent in 90 percent of cases of bacterial conjunctivitis (Weiss et al. 1993).

Treatment

Mild bacterial conjunctivitis is often self-limiting, requiring no treatment. Multiple topical antibiotics, including aminoglycosides for Gram-negative organisms, eythromycin, bacitracin, polymyxin/trimethoprim, or neomycin/polymyxin for Gram-positive organisms, are available in ophthalmic preparations, given every 2–4 h for 7–10 days. Empiric therapy for children should focus on *H. influenzae*, *S. pneumoniae*, and *S. aureus*. Fluoroquinolone solutions offer broad-spectrum activity but have been no more effective than aminogycosides. Gonococcal conjunctivitis requires parenteral treatment, usually with a single 125-mg dose of intramuscular ceftriaxone. Adult inclusion conjunctivitis and trachoma are treated systemically with doxycycline or macrolide antibiotics. Neonatal chlamydial conjunctivitis is treated with erythromycin. Prophylaxis of both gonococcal and chlamydial opthalmia neonatorum is best achieved with topical erythromycin (O'Brien 2000).

BACTERIAL KERATITIS

Predisposing factors

Keratitis is defined as inflammation of the cornea, produced by infectious and noninfectious causes. Microbial keratitis is a common, potentially sight-threatening infection. Bacterial keratitis accounts for 65–90 percent of all corneal infections (Jones 1981; Asbell and Stenson 1982; Liesegang and Forster 1980). Direct penetration of intact corneal epithelium occurs only by organisms with particular virulence factors. Most organisms gain access via corneal epithelial defects created by mechanical, thermal, or chemical injury or by epithelial erosion secondary to underlying ocular conditions (O'Brien 2000).

Various forms of external trauma, including obvious or subtle injury, contact lenses, trichiasis, entropion, and abnormal lid margins, predispose to infection. Radial keratotomy, now infrequently performed, is associated with infection in 0.25–0.7 percent of cases (Jain and Azar 1996). Infectious complications of photoreactive keratectomy using the excimer laser are rare (Lin and Maloney 1999).

Corneal surface defects may also result from severe dry eyes, exposure, bullous keratopathy, or neurogenic corneal anesthesia. Comatose patients with chemosis that prevents lid occlusion are prone to bacterial keratitis (Parkin et al. 1997). Underlying leprosy or trachoma with lagophthalmos or trichiasis are also at risk.

Clinical manifestations

There are no clinical symptoms or signs that reliably distinguish microbial keratitis from noninfectious inflammatory conditions. Due to the rich nervous innervation of the cornea, the most common complaint is pain. Unlike conjunctivitis, keratitis is usually accompanied by a variable decrease in vision. Examination of the patient is greatly facilitated by use of topical anesthesia. The cornea is assessed for epithelial ulceration and the presence and location of stromal suppuration. Again, unlike conjunctivitis, discharge is uncommon in keratitis. Reflex tearing, photophobia, and blepharospasm are common. Loss of corneal transparency and epithelial defects are best observed with a cobalt blue light after instillation of fluorescein. Neovascularization of the cornea may be seen in severe cases. Slit-lamp examination may reveal associated intraocular inflammation with flare and cells in the anterior chamber and hypopyon formation (O'Brien 2000).

Infecting agents

Both gram-positive and gram-negative bacteria can be associated with infective keratitis, and many organisms can cause keratitis when introduced by trauma or ocular abnormality. The most prevalent causative organisms include *Staphylococcus* spp., *Streptococcus* spp., *Pseudomonas* spp., and members of the Enterobacteriaceae (O'Brien 2000). Several studies have shown *P. aeruginosa* to be the dominant bacterial pathogen causing bacterial keratitis (Lam et al. 2002; Boonpasart et al. 2002; Martins et al. 2002; Sun et al. 2002), whereas others have found the gram-positive bacteria (staphylococci, *S. pneumoniae*, and other streptococci) to be the most frequent agents of bacterial keratitis (Schaefer et al. 2001; Sun et al. 2002; O'Brien 2000). Other infectious agents found to be associated with bacterial keratitis include *Moraxella* spp., *Serratia marcescens*, *Citrobacter* spp., *Enterobacter* spp., *Proteus* spp., *Morganella morganii*, *Bacillus cereus* (and other *Bacillus* spp.), *Corynebacterium* spp., *N. gonorrhoeae*, *Alcaligenes xylosoxidans*, *Acinetobacter* spp., *H. influenzae*, and *C. trachomatis* (Schaefer et al. 2001; Pinna et al. 2001; O'Brien 2000).

P. aeruginosa is a frequent cause of bacterial keratitis in contact-lens wearers and can cause a quick suppurative course due to the production of proteolytic enzymes that damage the eye (O'Brien 2000). *Streptococcus pyogenes*, also a common cause of bacterial keratitis in adults, may induce a severe inflammatory response due to the production of exotoxins (Bachman and Gabriel 2002). *B. cereus* is recognized as a primary pathogen of ocular infections and is commonly seen in infections associated with post-tramatic events (Tauzon 2000). Extremely serious and virulent ocular infections with this organism can be seen following a foreign body injury (O'Brien 2000).

Staphylococci that normally colonize other parts of the body can occasionally reach the eye and cause corneal infection. *N. gonorrhoeae* may cause keratitis resulting from an inappropriately treated conjunctivitis. Polymicrobic infections can be seen in association with 'bandage' contact lenses in diseased corneas, often associated with a mixture of bacterial and fungal organisms (O'Brien 2000). Several mycobacterial species have also been associated with chronic ulcerative keratitis, including, most commonly, *M. fortuitum* and *M. chelonae* and, much less commonly, *M. gordonae*, *M. avium complex*, and *M. tuberculosis*.

Diagnosis

Superficial keratectomy or corneal biopsy specimens may be necessary for difficult to diagnose nonsuppurative cases. For bacterial culture it is recommended that samples of both the conjunctiva and cornea be collected. Swabs of the conjunctiva should be collected prior to anesthetic applications, whereas corneal scrapings can be obtained afterward. Conjunctiva swab specimens should be collected as described earlier for conjunctivitis. For routine bacteriology collections, swabs that are used for normal bacteriology specimen collection should be used. If cultures for *Chlamydia* are being collected, cotton swabs should not be used, but dacron or calcium alginate swabs would be acceptable. Again, if possible, both conjunctiva should be sampled, even if one seems to be uninfected to allow the determination of normal flora in the uninfected eye and possible pathogens in the affected eye. After collection of the conjunctiva swabs, instillation of two drops of local anesthetic should be administered. Using a sterile spatula, ulcers or lesions from the cornea are scraped and some of the scrapings are inoculated directly on to culture media (multiple samples from areas of suppurations should be obtained for each stain and medium). Direct culture inoculation should be performed on to brain–heart infusion agar supplemented with 10 percent sheep blood and chocolate agar. All media should be warmed to room temperature prior to inoculation. Anaerobic media is not routinely necessary as the majority of organisms associated with keratitis are aerobic in nature. The remaining material should be applied to two clean glass slides for staining. Inoculated specimens should be held at room temperature and transported to the laboratory within 15 min. If this is not possible, the inoculated specimens should be held at room temperature for not more than 24 h prior to transport to the laboratory. The majority of bacteria that cause keratitis will grow on standard media within 12–48 h.

The sensitivity and specificity of the Gram stain for the diagnosis of keratitis was determined in a study that examined corneal scrapings (Sharma et al. 2002). In this study, the sensitivity of the Gram stain in the detection of bacteria was 36.0 percent in early keratitis and 40.9 percent in advanced keratitis cases. However, the specificity was higher in both groups (84.9 and 87.1 percent, respectively), while predictive values were poor. This study determined that the Gram stain had limited value in determining therapeutic options for bacterial keratitis. However, it is still held that the diagnostic laboratory evaluation of corneal scrapings (including Gram stain and culture) is imperative if there is suspicion of microbial keratitis (O'Brien 2000).

Special methods are necessary for the detection of *C. trachomatis* or the mycobacterial species. Specimens can be inoculated into cell culture or tested by molecular detection techniques for *Chlamydia* detection. If samples are being collected for *Chlamydia* culture, appropriate *Chlamydia* transport media should be utilized for delivery of the specimens to the laboratory, within 2 h of collection. In addition, molecular testing for the presence of *Chlamydia* can be performed if the sample is collected in specimen collection/transfer devices specific for this purpose. For the detection of mycobacterial infections, routine acid-fast bacilli growth media/systems should be utilized.

Treatment

Bacterial keratitis is potentially sight-threatening and demands prompt attention. Corneal perforation with resulting endophthalmitis and eye loss can occur rapidly. Topical antibiotics remain the mainstay of treatment, but the potential for rapid deterioration and the need for high-frequency dosing often requires hospitalization with trained nursing personnel. Several routes of topical administration include drops, ointments, continuous lavage, antibiotic-soaked corneal shields, therapeutic soft contact lenses, and iontophoresis. Subconjunctival injection and parenteral routes are also utilized. Empiric antibiotics are usually begun pending return of diagnostic information. The antimicrobial agent selected should have bactericidal activity against common gram-positive and gram-negative corneal pathogens. A common initial choice includes a topical first-generation cephalosporin (e.g. cefazolin) and an aminoglycoside (e.g. gentamicin, tobramycin, or amikacin). Fluoroquinolone ophthalmic solutions are available for broad-spectrum coverage. Various penicillin, cephalosporin, vancomycin, aminoglycoside, and fluoroquinolone preparations are commercially available or can be formulated and fortified by the pharmacist for directed therapy once a microbiologic etiology is established (Forster 1998). The utility of susceptibility testing is unclear, due to the high local antibiotic concentrations afforded by topical therapy. Response to treatment is based on serial slit-lamp examinations. Decreasing frequency of antibiotic administration and conversion from fortified to commercial strength drops can usually occur within a few days.

BACTERIAL ENDOPHTHALMITIS

Predisposing factors

Endophthalmitis is an inflammatory process involving the deep structures of the eye. Bacteria are the most common cause of infectious endophthalmitis (O'Brien 2000). By far the most common predisposing factor for bacterial endophthalmitis is antecedent eye surgery. The incidence of postoperative endophthalmitis is low (<0.9 percent), but the high frequency of cataract extraction, glaucoma surgery, vitrectomy, and keratoplasty make it an important problem (Aaberg et al. 1998). As for keratitis, radial keratotomy and laser keratectomy are uncommonly complicated by endophthalmitis. Penetrating nonsurgical trauma is another significant presdisposing event, with endophthalmitis developing in 4.2 percent of cases (Duch-Samper et al. 1997). The presence of a distant focus of infection may predispose to endogenous bacterial endophthalmitis, secondary to hematogenous seeding. Cases associated with infection at virtually any site have been described, prominently including meningitis, respiratory tract infection, abdominal infection, skin and soft-tissue infection, and endocarditis (Barnard et al. 1997). Occasionally, presumed hematogenous endophthalmitis occurs in a previously healthy patient with no apparent focus (Piczenik et al. 1997).

Clinical manifestations

Bacterial endophthalmitis typically develops suddenly and progresses rapidly. Symptoms and signs of ocular inflammation usually manifest within one to several days after surgical or nonsurgical trauma. Patients complain of pain and progressive blurring of vision. Examination reveals intense conjunctival hyperemia, chemosis, lid and corneal edema, hypopyon, and decreased visual acuity. Ophthalmoscopic or slit-lamp examination reveals anterior chamber and vitreous inflammatory cell reaction. Sudden onset of endophthalmitis in an unoperated, nontraumatized eye suggests hematogenous spread. The clinical picture of endogenous endophthalmitis is similar to postoperative cases, except for more prominent involvement of the posterior segment of the eye. These cases normally occur in systemically ill and/or immunocompromised patients (O'Brien 2000). Delayed-onset endophthalmitis can follow cataract extraction with intraocular lens implantation. These patients have a chronic course with low-grade inflammation (Fox et al. 1991; Sawusch et al. 1989).

Diagnosis

Once bacterial endophthalmitis is clinically suspected, prompt diagnostic aspirates should be obtained for thorough microbiologic investigation. Vitreous fluid has the highest yield for microbial pathogens and should be collected for the diagnosis of endophthalmitis along with aqueous aspiration. Microbiological studies should include a Gram stain and appropriate culture media for aerobic and anaerobic organisms. In one study, 78 of 140 specimens (56 percent) collected by vitreous and aqueous paracentesis were positive for a bacterial agent, with the vitreous having the higher yield as compared with the aqueous culture (O'Brien 2000). In another study that cultured vitreous fluid from over 400 patients with endophthalmitis, the overall yields of bacterial organisms recovered from solid media and broth culture were similar. This study also found that a positive Gram stain was highly predictive of a positive culture from the eye; however, a negative Gram stain had little predictive value for the culture result (Braza et al. 1997). Although culture results have little relevance on initial therapy choices, modifications of therapy can be directed by the results of bacterial culture and antimicrobial susceptibility testing. As stated previously, media for both aerobic and anaerobic bacteria should be included in the culture of these specimens; however, the use of broth cultures is not routinely necessary (Braza et al. 1997). The usual pathogens causing endophthalmitis should grow in 24–48 h, but cultures for anaerobic organisms may need to be held for up to 5–7 days. For suspected cases of mycobacterial infection, acid-fast stains and appropriate cultures should be performed.

Polymerase chain reaction (PCR) technology has been studied by several investigators for the diagnosis of bacterial endophthalmitis. Vitreous fluid and aqueous humor have both been shown to be reliable samples for use in PCR followed by sequencing procedures to detect the presence of bacterial organisms, especially in smear- and culture-negative specimens (Therese et al. 1998; Lohmann et al. 1998; Lohmann et al. 1997; Okhravi et al. 2000; Knox et al. 1999). A case of culture-proven *Listeria monocytogenes* endophthalmitis was identified more than 2 days sooner using PCR followed by direct sequencing protocols as compared with conventional culture (Lohmann et al. 1998). PCR was shown to be helpful in identifying *Propionibacterium acnes* in nine of 17 eubacterial genome-positive specimens that were smear- and culture-negative for bacterial organisms (Therese et al. 1998). The study by Lohmann and colleagues tested aqueous humor or vitreous in cases of endophthalmitis (21 eyes) and identified the causative organisms as *P. acnes*, *Staphylococcus epidermidis*, or *Actinomyces israelli* by PCR and comparing the determined sequences with all GenBank entries (Lohmann et al. 2000). The study by Okhravi and colleagues identified *Escherichia coli*, coagulase-negative staphylococci, *B. cereus*, *Pseudomonas* spp., *Comomonas* spp., *Proteus* spp., *Methylobacterium* spp., *Aeromonas* spp., and *Streptococcus faecalis* through detecting bacterial DNA by PCR and identifying sequences by restriction fragment length polymorphism and/or sequencing techniques (Okhravi et al. 2000). In addition, PCR techniques are a useful adjunct to diagnostic techniques, as unlike conventional culture, results are not affected by previously administered antibiotics (Lohmann et al. 2000). Caution must be taken when using PCR or other nucleic amplification techniques to detect 'causative' organisms, especially from nonsterile specimens such as eye collections. Identification of a certain genomic sequence proves that the specimen harbors this piece of nucleic acid (if not contaminated); however, it does not prove that the organism caused the patient's disease or that the organism is viable. Organisms determined to be present in clinical samples can be consistent with infection, colonization, or contamination, thus the results of molecular diagnostic techniques, as well as conventional bacterial culture, must be interpreted with caution. To date, these techniques are expensive and not part of the routine diagnostic microbiology laboratory. It is unclear at this time whether resources will be utilized for the development of PCR and sequencing commercial diagnostic kits for ophthalmic pathogens (Van Gelder 2001).

Treatment

Acute bacterial endophthalmitis is an ophthalmologic emergency, requiring early recognition and immediate and aggressive therapy. Traditional treatment options have included mechanical debridement via vitrectomy, intravitreal, subconjunctival, topical, and/or systemic antibiotics, and steroids by various routes. The Endophthalmitis Vitrectomy Study (Endophthalmitis Vitrectomy Study Group 1995) demonstrated that systemic antibiotics, such as ceftazidime and amikacin, did not improve visual outcome and that early vitrectomy benefited patients with light-perception vision or less on presentation. Patients with better than light perception (hand motions or better) fared equally well with vitreous tap and intravitreal antibiotics, with 50 percent achieving 20/40 or better vision, 75 percent with 20/100 or better, and only 12 percent with severe vision loss (5/200 or worse). Empiric intravitreal antibiotics are aimed at both gram-positive pathogens, including methicillin-resistant staphylococci, and gram-negative pathogens, including *Pseudomonas* spp. Vancomycin and an aminoglycoside is a common therapeutic combination. Amikacin is often chosen due to its possibly decreased retinal toxicity as compared with the other aminoglycosides. Similarly, concern for retinal toxicity has led to a trend toward ceftazidime use as a replacement for the aminoglycosides. Clindamycin is often included in nonsurgical post-traumatic cases because of the risk for *B. cereus* endophthalmitis. Adjunctive subconjunctival and topical antibiotics are often utilized, but their contribution beyond intravitreal antibiotics is unknown. Steroids are commonly added to hypothetically reduce destruction secondary to intraocular inflammation, but their efficacy is not well established. In delayed-onset chronic *P. acnes* endophthalmitis, surgical removal of intraocular lens implants may be necessary to remove sequestered organisms.

Prophylactic antibiotics and antiseptics before, during, and after surgery are widely utilized, but their efficacy is unproven. The low incidence of postoperative bacterial endophthalmitis makes demonstration of efficacy difficult. Povidone–iodine appears to be effective in reducing rates of these infections (Speaker and Menikoff 1991).

CONCLUSIONS

The eye is an anatomically and functionally complex structure, with variable and unique infectious manifestations. Bacterial infections of the eye are common and may be associated with significant morbidity. Predisposing factors, clinical presentation, microbiology, diagnosis, and treatment all depend on the ocular structure involved. Whether dealing with a relatively benign conjunctivitis, or a more threatening keratitis or endophthalmitis, accurate diagnosis and treatment can limit discomfort and disfigurement, and preserve vision.

REFERENCES

Aaberg, T.M., Flynn, H.W., et al. 1998. Nosocomial acute-onset postoperative endophthalmitis survey: a 10-year review of incidence and outcomes. *Opthalmology*, **105**, 1004–10.

Alrawi, A.M., Chern, K.C., et al. 2002. Biotypes and serotypes of *Haemophilus influenzae* ocular isolates. *Br J Ophthalmol*, **86**, 3, 276–7.

Asbell, P. and Stenson, S. 1982. Ulcerative keratitis. Survey of 30 years laboratory experience. *Arch Opthalmol*, **100**, 77.

Bachman, J.A. and Gabriel, H. 2002. A 10-year case report and current clinical review of chronic beta-hemolytic streptococcal keratoconjunctivitis. *Optometry*, **73**, 5, 303–10.

Barnard, T., Das, A. and Hickey, S. 1997. Bilateral endophthalmitis as an initial presentation in meningococcal meningitis. *Arch Ophthalmol*, **115**, 1472–3.

Barquet, N., Gasser, I., et al. 1990. Primary meningococcal conjunctivitis: Report of 21 patients and review. *Rev Infect Dis*, **12**, 838–47.

Bobo, L.D., Novak, N., et al. 1997. Severe disease in children with trachoma is associated with persistent *Chlamydia trachomatis* infection. *J Infect Dis*, **176**, 1524–30.

Boonpasart, S., Kasetsuwan, N., et al. 2002. Infectious keratitis at King Chulalongkorn Memorial Hospital: a 12-year retrospective study of 391 cases. *J Med Assoc Thai*, **85**, Suppl 1, S217–30.

Boralkar, A.W. 1989. Diptheritic conjunctivitis: a rare case report in Indian literature. *Indian J Ophthalmol*, **37**, 49–50.

Braza, M., Pavan, P.R., et al. 1997. Evaluation of microbiological diagnostic techniques in prospective endophthalmitis in the Endophthalmitis Vitrectomy Study. *Arch Ophthalmol*, **115**, 1142–50.

Chandler, J.W., Alexander, E.R., et al. 1977. *Ophthalmia neonatorum* associated with maternal chlamydial infections. *Trans Am Acad Ophthalmol Otolaryngol*, **83**, 302–8.

Chung, C.W. and Cohen, E.J. 2000. Eye disorders: Conjunctivitis. *West J Med*, **173**, 202–5.

De Toledo, A.R. and Chandler, J.W. 1992. Conjunctivitis of the newborn. *Infect Dis Clin North Am*, **6**, 4, 807–13.

Duch-Samper, A.M., Menezo, J.L. and Hurtado-Sarrio, M. 1997. Endophthalmitis following penetrating eye injuries. *Acta Ophthalmol Scand*, **75**, 104–6.

Endophalmitis Vitrectomy Study Group. 1995. Results of the endophthalmitis vitrectomy study: a randomized trial of immediate vitrectomy and of postoperative bacterial endophthalmitis. *Arch Opthalmol* **13**, 1479–86.

Fitch, C.P., Rapoza, P.A., et al. 1989. Epidemiology and diagnosis of acute conjunctivitis at an inner-city hospital. *Ophthamology*, **96**, 1215–20.

Forster, R.K. 1998. Conrad Beren's lecture. The management of infectious keratitis as we approach the 21st century. *Clao J*, **24**, 175–80.

Fox, G.M., Joondeph, B.C., et al. 1991. Delayed-onset pseudophakic endophthalmitis. *Am J Opthalmol*, **111**, 163–73.

Gigliotti, F., Williams, W.T., et al. 1981. Etiology of acute conjunctivitis in children. *J Pediatr*, **98**, 4, 531–6.

Harrison, H.R. and Alexander, E.R. 1990. Chlamydial infections in infants and children. In: Sparling, P.F., Weisner, P.J., et al. (eds), *Sexually transmitted diseases*, 2nd edn. New York: McGraw-Hill, 811–20.

Iroha, E.O., Kesah, C.N., et al. 1998. Bacterial eye infection in neonates, a prospective study in a neonatal unit. *West Afr J Med*, **17**, 3, 168–72.

Jain, S. and Azar, D.T. 1996. Eye infections after refractive keratotomy. *J Refract Surg*, **12**, 148–55.

Jones, D.B. 1981. Polymicrobial keratitis. *Trans Am Opthalmol Soc*, **79**, 153.

King, S., Devi, S.P., et al. 1988. Nosocomial *Pseudomonas aeruginosa* conjunctivitis in a pediatric hospital. *Infect Control Hosp Epidemiol*, **9**, 77–80.

Knox, C.M., Cevallos, V., et al. 1999. Identification of bacterial pathogens in patients with endophthalmitis by 16S ribosomal DNA typing. *Am J Ophthalmol*, **128**, 511–12.

Lam, D.S., Houang, E., The Hong Kong Microbial Keratitis Study Group, et al. 2002. Incidence and risk factors for microbial keratitis in Hong Kong: Comparison with Europe and North America. *Eye*, **16**, 5, 608–18.

Liesegang, T.J. and Forster, R.K. 1980. Spectrum of microbial keratitis in south Florida. *Am J Opthalmol*, **90**, 38.

Lin, R.T. and Maloney, R.K. 1999. Flap complications associated with lamellar refractive surgery. *Am J Ophthalmol*, **127**, 129–36.

Lohmann, C.P., Heeb, M., et al. 1998. Diagnosis of infectious endophthalmitis after cataract surgery by polymerase chain reaction. *J Cataract Refract Surg*, **24**, 821–6.

Lohmann, C.P., Linde, H.J. and Reischl, U. 1997. Rapid diagnosis of infectious endophthalmitis using polymerase chain reaction (PCR): a supplement to conventional microbiologic diagnostic methods. *Klin Monatsbl Augenheilkd*, **211**, 22–7.

Lohmann, C.P., Linde, H.J. and Reischl, U. 2000. Improved detection of microorganisms by polymerase chain reaction in delayed endophthalmitis after cataract surgery. *Ophthalmology*, **107**, 1047–51.

Martins, E.N., Farah, M.E., et al. 2002. Infectious keratitis: correlation between corneal and contact lens cultures. *CLAO J*, **28**, 3, 146–8.

O'Brien, T.P. 2000. Eye infections. In: Mandell, G.L., Bennett, J.E. and Dolin, R. (eds), *Principles and practice of infectious diseases*, 5th edn. New York: Churchill Livingstone, 1251–72.

Okhravi, N., Adamson, P. and Lightman, S. 2000. Use of PCR in endophthalmitis. *Ocul Immunol Inflamm*, **8**, 189–200.

Parkin, B., Turner, A., et al. 1997. Bacterial keratitis in the critically ill. *Br J Opthalmol*, **81**, 1060–3.

Pinna, A., Sechi, L.A., et al. 2001. *Bacillus cereus* keratitis associated with contact lens wear. *Ophthalmology*, **108**, 10, 1830–4.

Piczenik, Y., Kjer, B. and Fledelius, H.C. 1997. Metastatic bacterial endophthalmitis: A report of four cases all leading to blindness. *Acta Opthalmol Scand*, **75**, 466–9.

Ratelle, S., Keno, D., et al. 1997. Neonatal chlamydial infections in Massachusetts, 1992–1993. *Am J Prev Med*, **13**, 3, 221–4.

Robinson, N., Miller, M. and Levett, P.N. 2001. Antimicrobial sensitivity of *Haemophilus influenzae* isolates from bacterial conjunctivitis. *West Indian Med J*, **50**, 2, 137–9.

Sawusch, M.R., Michels, R.G., et al. 1989. Endophthalmitis due to *Propionibacterium acnes* sequestered between IOL optic and posterior capsule. *Ophthal Surg*, **20**, 90–2.

Schachter, J., Holt, J. and Goodner, E. 1979. Prospective study of chlamydial infection in neonates. *Lancet*, **2**, 377.

Schaefer, F., Bruttin, O., et al. 2001. Bacterial keratitis: A prospective clinical and microbiological study. *Br J Ophthalmol*, **85**, 7, 842–7.

Schwartz, B., Harrison, L.H., et al. 1988. Investigation of an outbreak of *Moraxella* conjunctivitis at a Navajo boarding school. *Am J Ophthalmol*, **107**, 341–7.

Seal, D.V., Barrett, S.P. and McGill, J.I. 1982. Aetiology and treatment of acute bacterial infection of the external eye. *Br J Ophthalmol*, **66**, 6, 357–60.

Sharma, S., Kunimoto, D.Y., et al. 2002. Evaluation of corneal scraping smear examination methods in the diagnosis of bacterial and fungal keratitis: a survey of eight years of laboratory experience. *Cornea*, **21**, 7, 643–7.

Shah, S.S. and Gallagher, P.G. 1998. Complications of conjunctivitis caused by *Pseudomonas aeruginosa* in a newborn intensive care unit. *Pediatr Infect Dis J*, **17**, 2, 97–102.

Sharp, S.E. 1999. Commensal and pathogenic microorganisms of humans. In: Murray, P.R., Baron, E.J., et al. (eds), *Manual of clinical microbiology*. Washington, DC: ASM Press, 23–32.

Speaker, M.G. and Menikoff, J.A. 1991. Prophylaxis of endophthalmitis with topical povidone-iodine. *Ophthalmology*, **98**, 1769–75.

Sun, X., Wang, Z., et al. 2002. Distribution and shifting trends of the pathogens for bacterial keratitis. *Chung Hua Yen Ko Tsa Chih*, **38**, 5, 292–4.

Tauzon, C. 2000. Other *Bacillus* species. In: Mandell, G.L., Bennett, J.E. and Dolin, R. (eds), *Principles and practice of infectious diseases*, 5th edn. New York: Churchill Livingstone, 2220–6.

Therese, K.L., Anand, A.R. and Madhavan, H.N. 1998. Polymerase chain reaction in the diagnosis of bacterial endophthalmitis. *Br J Ophthalmol*, **82**, 1078–82.

Trottier, S., Stenberg, K., et al. 1991. *Haemophilus influenzae* causing conjunctivitis in day-care children. *Pediatr Infect Dis J*, **10**, 8, 578–84.

Valenton, M.J. and Abendanior, R. 1973. Gonorrhea conjunctivitis. *Can J Ophthalmol*, **8**, 421.

Van Gelder, R.N. 2001. Applications of polymerase chain reaction to diagnosis of ophthalmic disease. *Surv Ophthalmol*, **46**, 248–58.

Van Ogtrop, M.L., van Zoeren-Grobben, D., et al. 1997. *Serratia marcescens* infections in neonatal departments: description of an outbreak and review of the literature. *J Hosp Infect*, **36**, 2, 95–103.

Wald, E.R., Greenberg, D. and Hoberman, A. 2001. Short term oral cefixime therapy for treatment of bacterial conjunctivitis. *Pediatr Infect Dis J*, **20**, 11, 1039–42.

Wald, E.R. 1997. Conjunctivitis in infants and children. *Pediatr Infect Dis J*, **16**, 2, S17–20.

Weiss, A., Brinser, J.H. and Nazar-Stewart, V. 1993. Acute conjunctivitis in childhood. *J Pediatr*, **122**, 1, 10–14.

Wishart, P.K., James, C., et al. 1984. Prevalence of acute conjunctivitis caused by *Chlamydia*, Adenovirus and Herpes Simplex virus in an ophthalmic casualty department. *Br J Ophthalmol*, **68**, 653–5.

Woodland, R.M., Darougar, S., et al. 1992. Causes of conjunctivitis and keratoconjunctivitis in Karachi, Pakistan. *Trans R Soc Trop Med Hyg*, **86**, 317–20.

23

Bacterial infections of the upper respiratory tract

RICHARD B. THOMSON Jr

INTRODUCTION

Bacterial infections of the upper respiratory tract are many and varied, ranging from benign pharyngitis to life-threatening epiglottis (Chow 2000; Berman 1995; Peterson and Thomson 1999). In spite of these differences, many concepts pertaining to host defense, pathogenesis, and bacterial virulence mechanisms are shared among diseases. In addition, anatomic features of the upper respiratory tract and normal commensal flora of epithelial surfaces are similar factors in all infections. Finally, infections of the upper respiratory tract may disseminate to distant foci, underscoring the potential seriousness of seemingly simple clinical conditions.

Infections of the upper respiratory tract can be divided into those that occur in the nasal airways and sinuses, middle ear and associated tissues, oral cavity, and in spaces adjacent to the oral cavity as abscesses (Table 23.1). Descriptions of specific infections will include the etiologies, epidemiology, unique defense mechanisms and pathogenesis, normal flora, a brief summary of clinical features, recommended antimicrobial therapy, and means of prevention. The goal of this chapter is to provide a complete understanding of factors contributing to the development and resolution of upper respiratory tract infection, the accurate and rapid diagnosis of infection, and treatment and prevention of subsequent infection.

ANATOMY

The upper respiratory tract is the area proximal to and including the epiglottis. It includes the middle ear, paranasal sinuses, nasopharynx, and oral cavity. Understanding the anatomy of the upper respiratory tract is necessary in order to understand the pathogenesis and prevention of disease at these sites (Thadepalli and Mandal 1984, 1988). The upper respiratory tract is composed of two distinct types of epithelial surfaces (Thadepalli and Mandal 1984; Thadepalli and Mandal 1988; Chow 2000). A stratified squamous epithelium lines the oropharynx and nasopharynx, and a respiratory epithelium, composed of ciliated columnar cells, goblet cells, and mucous glands, lines the paranasal sinuses and the middle ear. Stratified squamous epithelium is found in areas filled with normal microbial flora, representing sites easy to sample for bacteriologic culture. Epithelium of the sinuses and middle ear is usually sterile and is difficult to access if culture is needed.

The nasal airways and paranasal sinuses include the nasal septum, which divides the nose at the midline, turbinates, and four sinus complexes (Gwaltney 1996;

Table 23.1 *Upper respiratory tract diseases and etiologies*

Anatomic site/disease	Bacterial etiologies
Oral cavity	
Pharyngitis	*Streptococcus pyogenes*
	Occasionally other beta-hemolytic streptococci,
	Arcanobacterium haemolyticum, Mycoplasma pneumoniae
Diphtheria	*Corynebacterium diphtheriae*
Epiglottitis	*Haemophilus influenzae*
	Occasionally *Streptococcus pneumoniae, Neisseria meningitidis,*
	Staphylococcus aureus, beta-hemolytic streptococci
Gingivitis	Mixed anaerobes, including spirochetes, *Actinobacillus*
	actinomycetemcomitans, Eikenella corrodens
Abscess	Mixed anaerobic, and aerobic bacteria, especially *Fusobacterium* spp. and
	viridans streptococci
Stomatitis	Anaerobaic bacteria, especially *Treponema vincentii, Prevotella*
	melaninogenica, and *Fusobacterium nucleatum*
Ludwig's angina	Streptococci, mouth flora
	Occasionally *H. influenzae,* staphylococci and facultative
	Gram-negative bacilli
Paranasal sinuses	
Sinusitis	*S. pneumoniae, H. influenzae, Moraxella catarrhalis, S. pyogenes,* other
	streptococci, *S. aureus,* anaerobic bacteria
Rhinoscleroma	*Klebsiella pneumoniae* subsp. *rhinoscleromatis*
Ear	
Otitis media and acute mastoiditis	*S. pneumoniae, H. influenzae, M. catarrhalis, S. pyogenes, S. aureus,* anaerobes
Malignant external otitis	*Pseudomonas aeruginosa*
Abscesses adjacent to the oral cavity	
Peritonsillar, pharyngeal, retropharyngeal, and parapharyngeal	Mixed aerobic and anaerobic bacteria especially, *S. pyogenes, S. aureus,* *S. anginosus* (*milleri*) group
Lemierre's syndrome	*F. necrophorum,* other anaerobes
Suppurative paratitis	*S. aureus,* viridans streptococci, anaerobic bacteria

Wald 1996; Thadepalli and Mandal 1988; Thadepalli and Mandal 1984). The inferior, middle, and superior turbinates are three shelf-like projections that extend from the septum. Beneath the middle turbinate is a meatus that drains the maxillary, anterior ethmoids, and frontal sinuses. Beneath the superior turbinate is a meatus that drains the posterior ethmoids and sphenoid sinuses. The lacrimal duct drains to the inferior meatus. The maxillary sinus has a volume of approximately 15 ml and drains through an ostium high on the medial wall of the sinus. This positioning impedes normal gravitational drainage and, most likely, predisposes to bacterial infection. The tubular passage that connects the ostium to the meatus is called the infundibulum. The ethmoid sinus is composed of multiple cells, each of which drains by a separate ostium into the middle meatus. The frontal sinus is not fully developed until late adolescence. The sphenoid sinuses are located just behind the posterior ethmoids. The sphenoid and frontal sinuses can be a source for the spread of infection to the central nervous system.

The relevant structures of the ear include the external auditory canal, the tympanic membrane (eardrum), middle ear, mastoid, and Eustachian tube (Hendley 2002). The oral cavity includes the nasopharynx, pharynx, epiglottis, tongue, teeth, periodontium, and adjacent salivary and parotid glands (Roscoe and Chow 1988).

DEFENSE MECHANISMS

Protective mechanisms found in the upper respiratory tract include the primary defenses that are constantly active and secondary mechanisms that are activated or recruited at the time of infectious insult (Roscoe and Chow 1988). Primary defense mechanisms include the anatomy of the upper airways, epithelial surfaces, mucociliary clearance, and soluble factors in airway secretions, such as degratory enzymes, complement, and immunoglobulins (Roscoe and Chow 1988). Secondary mechanisms include recruitment of polymorphonuclear leukocytes and lymphocytes and the secretion of chemical mediators of inflammation by these cells (Chow 2000).

During breathing, the turbulence of air flowing through the nasal turbinates, nasopharynx, and oropharynx results in the sedimentation of bacteria within the upper respiratory tract. Tight junctions between epithelial cells provide an efficient barrier to bacterial penetration. The epithelium is continually replenished if damaged. Throughout the nasal sinuses, middle ear, and Eustachian tubes, mucus-secreting cells secrete a blanket of mucus that is propelled along toward the oropharynx by ciliated cells, where it is coughed out or swallowed. Mucous contains soluble factors that augment the clearance of bacteria from normally sterile areas and prevents invasion past the epithelial layer (Feldman 2001). Other nonspecific defense mechanisms include the continuous cell shedding and turnover of the mucosal epithelium and the constant flow of saliva-containing lysozyme, lactoferrin, beta-lysin, lactoperoxidease, and other antimicrobial constituents (Chow 2000; Roscoe and Chow 1988). Immunoglobulin A (secretory IgA) is the predominant immunoglobulin in the upper respiratory tract. In addition to complement-independent neutralization of respiratory tract viruses, secretory IgA interacts with lysozyme and complement to improve bactericidal activity, augment bacterial adherence to mucus, and inhibit attachment to epithelial surfaces (Bernstein 1992; Roscoe and Chow 1988).

If primary defense mechanisms are unable to eradicate a bacterial pathogen, polymorphonuclear leukocytes and lymphocytes, constituting secondary defense mechanisms, are recruited to bolster the host's response (Roscoe and Chow 1988). Neutrophils migrate from the vascular compartment following the release of chemotactic factors by degraded bacteria. Bacterial antigen is transported to local lymph nodes either within phagocytic cells or free in afferent lymphatic fluid. Antigen processed in lymph nodes results in the expansion and differentiation of specific B and T cell clones. One to two weeks after infection, effector T and B cells migrate from lymph nodes to provide specific antibody locally in the respiratory tract.

PATHOGENESIS

Disease occurs when the balance between host and bacteria tips in favor of the microbes. This results from a defect in host defenses, an overwhelming inoculum of bacteria, or the presence of a novel bacterial virulence mechanism. Host defects range from increased colonization of the respiratory epithelium and failure of epithelial barriers to immunoglobulin deficiency and reduced phagocytic activity. In addition, environmental influences, such as tobacco smoke, may modify host defenses (Obeid and Bercy 2000).

Colonization of the upper respiratory tract is the first step in the pathogenesis of most bacterial infections. Individuals with poor nutrition, particularly vitamin A deficiency, and those with local areas of inflammation, such as gingivitis, are more likely to have increased bacterial colonization densities and a higher concentration of pathogenic species (Enwonwu et al. 2002). Antimicrobial therapy may change the relative numbers of bacterial commensals providing an advantage for some

bacteria. It is also known that epithelial cells from patients with severe or chronic illness are more receptive to bacteria, especially gram-negative microorganisms (Niederman 1990).

Compromise of the respiratory epithelium results from trauma, radiation therapy, damage from chemotherapeutic agents, and environmental factors (Niederman 1990). Nasal intubation traumatizes the nasal airway, obstructs normal drainage, and provides a conduit for bacteria to gain access and to persist in deep respiratory spaces (Westergren et al. 1998a, b). Chemotherapy may damage or destroy rapidly dividing epithelial cells resulting in oral ulcers and breaches in barrier defense. Tobacco smoking, secondary tobacco smoke, and air pollution restrict ciliary clearance by direct toxicity, increase mucus production, and attract neutrophils (Obeid and Bercy 2000). Nasal discharge is more frequent in heavily polluted urban areas and more common upper respiratory tract infection is associated with increased pollution (Ng and Tan 1994).

BACTERIAL VIRULENCE MECHANISMS

Most bacteria that are pathogens in the upper respiratory tract are capable of adherence to and colonization of a mucus membrane. Factors that govern colonization are many and varied, including adherence characteristics of bacteria and epithelial cells, oxygen tension and oxidation-reduction potential, pH, age of patient, diet, oral hygiene, smoking habits, antimicrobial therapy, pregnancy, and genetic factors (Chow 2000). Colonization represents a 'period of waiting' for an opportunity to invade deeper tissues following trauma, or to be 'sequestered' in a closed space, such as a paranasal sinus or middle ear, as a result of inflammation caused by viral infection, allergy, or other insult (Bluestone 1996; Saez-Llorens 1994). Once positioned for disease, bacterial characteristics such as avoidance of complement deposition and phagocytosis, toxin production, synthesis of IgA proteases, and avoidance of mucociliary clearance augment disease severity (Kilian 1981).

Most bacteria are pathogenic throughout the upper respiratory tract, rather than one specific site (Table 23.1). *Streptococcus pneumoniae* resides in the pharynx and nasopharynx and persists in presumably sterile sites such as the paranasal sinuses and middle ear primarily because of an antiphagocytic capsule. *H. influenzae* possess fimbriae that bind to respiratory epithelium and mucin and reduce ciliary beating, which minimizes mucociliary clearance (Rodriguez et al. 2003). Group A streptococcal M protein binds to sialic acid moieties on mucin, followed by local invasion and toxin production that destroy local tissues (Ryan et al. 2001). Anaerobes provide 'colonization resistance,' i.e. prevent introduction or persistence of new bacterial populations, yet they are pathogens if they gain access to subepithelial spaces. The hallmark of anaerobic infection is the abscess. As a rule, multiple anaerobic species, rather than a single species, are involved in upper respiratory infection. *Corynebacterium diphtheriae* is a specialized respiratory tract pathogen, capable of causing local disease in the pharynx or nasopharynx, yet exerting its most serious damage at distant sites as a result of toxin production (Hadfield et al. 2000). Viridans streptococci are included as part of a mixed etiology in abscesses, yet *Streptococcus mutans*, specifically, is the sole etiology of dental caries (Shaw 1987). Finally, *Staphylococcus aureus* is an abscess-forming, toxin-producing pathogen occasionally involved in infection of paranasal sinus, middle ear, and abscesses of the oropharyngeal spaces.

NORMAL FLORA

The anterior nares are colonized by staphylococci, micrococci, and miscellaneous *Corynebacterium* spp. (Roscoe and Chow 1988; Schuster 1999). The paranasal sinuses are normally sterile. In the mouth, anaerobes colonize periodontal areas along with a variety of other bacteria, including viridans group streptococci, *Haemophilus* spp., *Neisseria* spp., and diphtheroids (Schuster 1999; Roscoe and Chow 1988). The flora of the nasopharynx is much more complex. It may include *Streptococcus* spp., *Neisseria* spp., *Enterobacteriaceae*, *Bacteroides* spp., and *Fusobacterium* spp. (Sutter 1984). In addition, all individuals are colonized intermittently by one or more potential pathogens, such as group A streptococci, *S. pneumoniae*, *H. influenzae*, *N. meningitidis*, and *Moraxella catarrhalis* (Neto et al. 2003). Point prevalence surveys in communities demonstrate the ever-present nature of these pathogens.

Quantitative studies of the oral cavity indicate that certain species predominate in specific sites (Table 23.2) (Roscoe and Chow 1988; Schuster 1999). Obligate anaerobes constitute a large and important part of the oral flora. As an example, in the gingival crevice of healthy adults, microscopic counts average 2.7×10^{11} bacteria/gram wet weight, with anaerobes outnumbering facultative bacteria by a 8:1 margin (Chow 2000). Identification of specific genera shows *Streptococcus* spp., *Peptostreptococcus* spp., *Veillonella* spp., *Lactobacillus* spp., *Corynebacterium* spp., and *Actinomyces* spp. accounting for more than 80 percent of the cultivatable oral flora (Chow 2000). Other unique niches include *Streptococcus sanguis*, *Streptococcus mutans*, *Streptococcus mitis*, and *A. viscosus*, which preferentially colonize the surfaces of teeth (Chow 2000). *Streptococcus salivarius* and *Veillonella* spp. have a predilection for the tongue and buccal mucosa. *Fusobacterium* spp., *Porphyromonas* spp., *Prevotella* spp., and anaerobic spirochetes concentrate in the gingival crevice (Chow 2000). Facultative gram-negative bacilli are uncommon in the oral cavity of healthy adults; however, this group becomes a more prominent part of the flora in seriously ill, hospitalized, and elderly patients (Lipchik and Kuzo 1996). A number of exogenous factors may disturb the ecology of the mouth and nasopharynx. Antimicrobials can suppress normal flora, allowing yeasts and resistant bacteria to proliferate.

GENERAL CONSIDERATIONS: COLLECTION, TRANSPORT, AND PROCESSING OF SPECIMENS FOR BACTERIOLOGIC EXAMINATION

All specimens for bacteriologic culture should be collected from the specific site of inflammation, as free of contaminating flora as possible (Carroll and Reimer 1996; Peterson and Thomson 1999; Thomson and Miller 2003). In practice, most specimens from the upper

Table 23.2 *Oral flora*

Site	Predominant bacteria	Bacteria present but not predominant
Gingiva	Viridans group streptococci	Anaerobic gram-negative bacilli
	Lactobacilli	Other aerobic and anaerobic bacteria
	Anaerobic gram-positive rods	
	Anaerobic cocci	
Tongue	Viridans group cocci	Anaerobic gram-negative bacilli
	Lactobacilli	Anaerobic gram-positive cocci
	Anaerobic gram-negative cocci	Anaerobic gram-positive rods
		Other aerobic and anaerobic bacteria
Saliva	Viridans group cocci	Lactobacilli
	Anaerobic cocci	Anaerobic gram-positive bacilli
		Anaerobic gram-negative bacilli
		Other aerobic and anaerobic bacteria
Dental plaque	Viridans group streptococci	Anaerobic cocci
	Lactobacilli	Anaerobic gram-negative bacilli
	Anaerobic gram-positive bacilli	Other aerobic and anaerobic bacteria

Data from Chow (2000), Roscoe and Chow (1988), and Schuster (1999).

respiratory tract will contain at least low numbers of contaminating bacteria. Purulence should be aspirated or gently washed and aspirated. Sterile, nonbacteriostatic saline is an acceptable wash/lavage fluid. Specimens for anaerobic culture are collected by aspiration (Thomson and Miller 2003). Oxygen-free transport vials should be used to hold the aspirate during transport to the laboratory regardless of how quickly this will occur. Specimens for the detection of aerobic and facultative bacteria may be collected with a swab only when absolutely necessary, e.g. pharyngeal swab.

Specimens must be transported to the laboratory as quickly as possible. Receipt within 1 h is optimal (Thomson and Miller 2003; Carroll and Reimer 1996). Longer transport requires a holding medium. Specimens should not be allowed to dry or be frozen. Refrigerated or 'on-ice' temperature (approximately 40°C) is used unless anaerobes or *H. influenzae* are expected. Pathogenic *Neisseria* spp., *H. influenzae*, *S. pneumoniae*, and anaerobes are especially sensitive to extended holding and transport (Thomson and Miller 2003).

Once in the laboratory, processing includes a gram-stained smear, whenever purulence is submitted from an abscess or closed space, and a culture using appropriate media for the pathogens suspected. Gram stain results are used to determine inflammation and the presence of morphotypes suggesting a potential pathogen. Gram stain results are important when selecting empiric therapy, previous antimicrobial therapy inhibits growth, or anaerobes are present but have not survived exposure to oxygen during collection or transport.

Hospital microbiology laboratories are equipped to perform usual aerobic and anaerobic cultures. Staphylococci, streptococci, and nonfastidious gram-negative bacilli as predominant bacterial populations are readily detected. Less common pathogens, such as *Arcanobacterium haemolyticum*, *Neisseria gonorrhoea*, and *Corynebacterium diphtheriae*, can be detected by culture, stain, or molecular methods but only when the laboratory is notified in advance that a specific microorganism is being sought. Multiple anaerobes in mixed culture are not readily isolated and identified with usual methods. Newer selective media and techniques have been introduced to identify specific anaerobic pathogens in oral infections (Moore and Moore 1994; Chow 2000). These are available in a limited number of research, university or commercial settings. A recent, comprehensive review of collection, transport, and processing for bacterial specimens is available.

INFECTIONS OF THE ORAL CAVITY

Streptococcal pharyngitis

Pharyngitis is one of the most common reasons for physician visits among pediatric patients, with the group A streptococcus (*S. pyogenes*) the most likely treatable etiology (Gerber 1998). However, considering all patient groups and etiologies, including viruses, group A streptococci may account for as few as 5 percent of cases of pharyngitis (Carroll and Reimer 1996). Group A streptococcal pharyngitis is important not only because of the morbidity associated with acute disease but also because it may be followed by nonsuppurative sequelae (acute rheumatic fever or acute glomerulonephritis). Accurate detection and treatment of group A streptococcal pharyngitis is necessary to shorten the duration of symptoms, prevent the spread of the organism to other patients, prevent nonsuppurative sequelae, and avoid antimicrobial therapy in patients with viral infections (Gerber 1998). In spite of the frequency of disease and the potential for morbidity, a surprising number of practical questions dealing with the disease and its diagnosis remain unanswered.

Streptococcal pharyngitis is caused by *S. pyogenes*, a Lancefield group A Streptococcus based on cell-wall carbohydrate analysis. Group A carbohydrate is composed of N-acetylglucosamine linked to a rhamnose polymer backbone (Cunningham 2000). Disease is spread by person-to-person contact with infectious nasal or oral secretions, and is more common in situations of crowding. Disease occurs primarily in children between 5 and 15 years of age, mostly during winter and spring months in temperate climates (Gerber 1998). Further identification of group A streptococci can be performed by typing on the presence of M protein in the cell wall or by sequencing the *emm* gene that encodes the M protein (Bisno 1991; Bisno 2001). Currently, more than 80 M protein serotypes have been identified. Typing demonstrates that multiple serotypes circulate in a community at any one time (Haukness et al. 2002). During a 3-month winter period, typing of 63 group A streptococcal strains from three large pediatric practices in the Chicago area identified 16 different pulsed-field gel electrophoresis patterns. The resurgence of invasive, life-threatening streptococcal necrotozing fasciitis, and toxic shock over the past 15 years has stimulated a study of community pharyngeal strains to detect the location and prevalence of potentially invasive strains. Invasive strains constitute 25 percent of pharyngeal pathogens circulating in the community (Haukness et al. 2002).

M protein is the chief virulence factor of group A streptococci (Bisno 1991; Bisno 2001). Strains rich in M protein resist phagocytosis by polymorphonuclear leukocytes. M protein exerts its antiphagocytic effect by diminishing complement activation by the alternative complement pathway, thus limiting the deposition of complement on the bacterial surface (Bisno 1991; Bisno 2001). Immunity to infection is associated with the development of opsonic antibodies to antiphagocytic epitopes of M protein and, with few exceptions, such immunity is type-specific. Immunity lasts for many years, possibly for the life of the patient. In addition, lipoteichoic acid is expressed on the surface of group A streptococci and is the adhesin responsible for binding of the organism to fibronectin present on the surface of oral epithelial cells (Bisno 1991). The streptococcal cell wall is surrounded by a hyaluronic acid capsule that is also antiphagocytic, serving as an additional virulence factor. Strains with large capsules are mucoid colonies on blood agar plates. Group A streptococcal strains that are both rich in M protein and heavily encapsulated are readily transmitted from person to person and tend to produce severe infections (Bisno 2001).

Patients with group A streptococcal pharyngitis present with sore throat (usually of sudden onset), pain on swallowing, and fever (Gerber 1998). Headache, nausea, vomiting, and abdominal pain (especially in children) also may be present. Clinical examination reveals tonsillopharyngeal erythema, with or without exudate, a red, swollen uvula, petechiae on the palate, and tender, enlarged anterior cervical lymph nodes. Tonsils, if present, are enlarged and erythematous. When a fine, diffuse red rash accompanies streptococcal pharyngitis, the syndrome is called scarlet fever (Gerber 1998).

The gold standard laboratory diagnostic test for group A streptococcal infection is the throat culture (Gerber 1998). If performed correctly, it has a sensitivity of 90–95 percent. Unfortunately, the culture takes 24–48 h to complete. If the patient leaves the physician or healthcare facility office with a prescription for antibiotics, the treatment would then begin before the microbial entity is known. In the age of emerging drug resistance, unnecessary antimicrobial use must be avoided. As a result, rapid streptococcal detection methods were developed to give nearly immediate results allowing for antimicrobial decision making while the patient is in the office or clinic (Peterson and Thomson 1999). Many rapid direct tests for group A streptococci are commercially available, including enzyme immunoassay (EIA), optical immunoassay, and nucleic acid-based probe assays (Thomson and Miller 2003). Although specificities are >95 percent, the reported sensitivities of EIAs and an optical immunoassay vary between 60 and 95 percent but can be as low as 31 percent. The nucleic acid-based assay has a sensitivity greater than 90 percent. When a rapid test is requested, two throat swabs should be collected. If only one swab is received, the culture plate should be inoculated first. Material remaining on the swab is used for the direct test. If the rapid test is positive, the second swab can be discarded, but if the rapid test is negative, the second swab must be used for culture to confirm the negative direct test (Gerber 1998; Thomson and Miller 2003). The nucleic acid-based probe test is considered sensitive and specific enough by many to obviate the need for confirmatory culture (Thomson and Miller 2003). A position paper by representatives of the American Academy of Family Physicians, the American College of Physicians–American Society of Internal Medicine, and the Centers for Disease Control and Prevention, states that rapid tests do not require confirmatory culture when used with specimens from adult patients (Cooper et al. 2001). This recommendation does not hold for specimens from children.

To culture group A streptococci, either sheep blood agar or selective blood agar may be used (Thomson and Miller 2003). Selective agar makes the organism easier to visualize by inhibiting the accompanying flora but may delay the appearance of colonies of *S. pyogenes*. Cultures should be incubated for 48 h at 35°C in an environment of reduced oxygen achieved by incubating anaerobically, in 5 percent CO_2, or in air with multiple 'stabs' through the agar surface. Stabbing the agar surface with the inoculating loop pushes inoculum-containing streptococci below the surface, where the oxygen concentration is reduced compared with ambient.

Group A streptococci are uniformly susceptible to penicillin in vitro and the treatment of choice for acute group A streptococcal pharyngitis remains penicillin, as recommended by the American Heart Association, the Committee on Infectious Diseases of the American Academy of Pediatrics, and the World Health Organization (Gerber 1998). Cephalosporins and macrolides, such as erythromycin, are recommended in patients unable to take penicillin. Certain antimicrobials are not recommended for the treatment of group A streptococcal pharyngitis. Tetracyclines and trimethoprim-sulfamethoxazole have been shown to be ineffective in the treatment of this infection. Although in vitro testing is not necessary for penicillin, worldwide surveys show that up to 30 percent of stains are resistant to erythromycin and other macrolides (Alos et al. 2003; Hoban et al. 2003).

During the winter and spring in temperate climates, up to 20 percent of asymptomatic school-aged children may be streptococcal carriers (McEwan et al. 2003; Kaplan et al. 1981; Gerber 1998). These individuals have group A *Streptococcus* in their pharynx but have no evidence of immunologic response to this organism. They may be colonized for several months, and during that period they may experience episodes of intercurrent viral pharyngitis. When tested, these patients have group A streptococci in their pharynx and would seem to be suffering from acute streptococcal pharyngitis. Streptococcal carriers are unlikely to spread the organisms to their close contacts and are at very low risk, if any, of developing suppurative or nonsuppurative complications (Gerber 1998). Streptococcal carriers do not ordinarily require antimicrobial therapy. When eradication of streptococcal carriage is desirable, a short course of rifampin in conjunction with penicillin or a 10-day course of oral clindamycin have been shown to be effective (Gerber 1998).

Non-group A streptococcal pharyngitis caused by beta-hemolytic streptococci groups B, C, G, and F have been reported (Peterson and Thomson 1999). Added to this list of uncommon causes of bacterial pharyngitis are mixed anaerobes (gangrenous pharyngitis), *N. gonorrhoeae*, *C. diphtheriae*, *Arcanobacterium haemolyticum*, *Yersinia pestis*, *Francisella tularensis*, *Mycoplasma pneumoniae*, *Chlamydophila psittaci*, and *C. pneumoniae* (Peterson and Thomson 1999). The clinical laboratory does not routinely look for any of these other agents in throat swabs, except when specifically requested by the clinician.

Diphtheria

Although much less common than streptococcal pharyngitis, *C. diphtheriae* can still be isolated from patients with a sore throat as well as those with a more serious systemic disease. Epidemic diphtheria re-emerged in the former Soviet Union beginning in 1990, resulting in approximately 125 000 cases and 4 000 deaths between 1990 and 1995 (Markina et al. 2000). The number of new cases began to decrease in 1996 following a World Health Organization strategy to vaccinate 90 percent of people aged 3 years or older (Markina et al. 2000). The introduction of new strains into countries with no endemic disease from exogenous reservoirs can cause outbreaks in developing nations, homeless populations, or in specific communities where vaccination is not accepted (Galazka 2000; Harnisch et al. 1989). The changing epidemiology of diphtheria includes the high proportion of infected adults, a progressive spread of disease from urban centers to rural areas, and transition from well-distinguished outbreaks to a more generalized epidemic (Galazka 2000).

The upper respiratory tract mucosa is the most common site of infection with *C. diphtheriae* in children but was a rare site for localization in adults patients in the epidemic that occurred in the former Soviet Union (Hadfield et al. 2000). In adults with oral mucosal lesions, buccal mucosa, upper and lower lips, hard and soft palate, and the tongue were sites of infection. *C. diphtheriae* localizes in the upper respiratory tract, ulcerates the mucosa, and induces the formation of an inflammatory pseudomembrane. The pseudomembrane is initially white, becoming a dirty gray color over time.

Late in the course of infection the membrane may have patches of green or black necrosis. The growth of the organism remains localized, but exotoxin is absorbed and is disseminated through the blood to evoke severe systemic pathology. The exotoxin inhibits cellular protein synthesis by stimulating adenosine diphosphate ribosylation, thus inactivating protein synthesis elongation factor 2 (EF-2) (Harnisch et al. 1989).

The microbiologic diagnosis of diphtheria is designed to provide early and accurate diagnosis, identification of contacts and carriers, and appropriate clinical management (Efstratiou et al. 2000). The recommended microbiologic procedure includes obtaining a clinical specimen from the throat and nasopharynx. If present, membranous material should be sampled, including material from beneath the membrane. Blood agar and selective tellurite medium (Hoyle's tellurite medium is recommended) should be used (Efstratiou et al. 2000). Most normal flora will be inhibited on the tellurite medium allowing only *C. diphtheriae* and some other corynebacteria, staphylococci, and yeasts to be identified by the production of black colonies. The diagnosis of diphtheria on the basis of direct microscopy of a smear is unreliable because both false-positive and false-negative results may occur (Efstratiou et al. 2000). Rapid screening tests recommended for the differentiation of potentially toxigenic species of corynebacteria (*C. diphtheriae*, *C. ulcerans*, and *C. psuedotuberculosis*) are the presence of cystinase and the absence of pyrazinamidase activity (Efstratiou et al. 2000). Although there are four biotypes of *C. diphtheriae* (*C. diphtheriae* var. *gravis*, var. *mitis*, var. *belfanti*, and var. *intermedius*), biotyping is of limited use in epidemiologic investigations because discrimination is poor (Efstratiou et al. 2000). The minimum laboratory information required to report a specimen as positive for *C. diphtheriae* is as follows: Catalase-positive (most biotypes), urea-negative, nitrate-positive (except biotype *belfanti*), pyrazinamidase-negative, and cystinase-positive (Efstratiou et al. 2000). They will also ferment glucose, maltose, and starch (Table 23.3).

Testing for toxigenicity, which detects the phage-encoded exotoxin, is the most important test and should be carried out without delay on any suspect isolate that

Table 23.3 *Laboratory identification of* Corynebacterium diphtheriae *and related species*

Corynebacterium species	Nitrate	Urea	Gelatin Liquefaction	Glucose	Maltose	Sucrose	Trehalose
C. diphtheriae var. *gravis*	+	−	−	+	+	−	−
C. diphtheriae var. *mitis*	+	−	−	+	+	−	−
C. diphtheriae var. *intermedius*	+	−	−	+	+	−	−
C. diphtheriae var. *belfanti*	−	−	−	+	+	−	+
C. ulcerans	−	+	+	+	+	−	+
C. pseudotuberculosis	−	+	−	+	+	−	

Data from Efstratiou et al. (2000).

is found by routine screening or while investigating a possible case. The potentially toxigenic species (*C. diphtheriae*, *C. ulcerans*, and *C. pseudotuberculosis*) acquire this characteristic when infected by bacteriophage. Although rapid polymerase chain reaction (PCR) assays have been used to detect biologically active fragment A of the toxin, positive results must be confirmed by phenotypic testing with the modified Elek immunoprecipitation test, since strains may possess toxin genes but do not express biologically active toxin (Efstratiou et al. 2000).

Epiglottitis

Acute bacterial epiglottitis, defined as inflammation and edema of the epiglottis, can result in life-threatening airway obstruction (MayoSmith et al. 1986). Inflammation can also occur in the arytenoid cartilage (paired jug or pitcher-shaped structures that move vocal cords apart) or pharyngeal wall, resulting in acute supraglottitis (Donnelly and Crausman 1997). Acute epiglottitis in children is caused by *H. influenzae* type b in most cases. In bacteremic cases an even higher percentage of cases is caused by *H. influenzae* (Bamberger and Jackson 2001). In adults, only 25 percent of cases are associated with *H. influenzae*. Disease caused by *S. pneumoniae*, bete-hemolytic streptococci (groups A, B and C), *S. aureus*, *H. parainfluenzae*, and *N. meningitidis* have been described (Warner and Finlay 1985; Sack and Brock 2002). Unusual pathogens such as *Klebsiella pneumoniae* and *Pasteurella multocida* have been reported in immunocompromised patients (Wine et al. 1997).

Epiglottitis occurs in people of all ages, with most cases found in children 2–7 years of age before widespread use of the *Haemophilus influenzae* type b (Hib) vaccine (Bamberger and Jackson 2001). Annual attack rates of 2–13 per 100 000 children were reported. With the introduction of the vaccine in the mid 1980s, the incidence of acute epiglottis in the pediatric population has decreased to 0.3–0.6 per 100 000. The incidence in adults is reported to be two to three cases per 100 000 and possibly increasing due to greater awareness and recognition of disease in adults (MayoSmith et al. 1986; Bamberger and Jackson 2001). For adults, the mean age at the time of presentation of disease ranges from 42 to 50 years. Cigarette smoking in diseased patients is more frequent than in the general population. Although cases occur in all months, spring and winter seasonal peaks have been reported (Bamberger and Jackson 2001).

Haemophilus species are exclusively adapted to the human mucosal membranes (Kilian 1981). *H. influenzae* is the species most frequently associated with disease in humans. The portal of entry for *H. influenzae* is the nasopharynx following person-to-person spread via respiratory droplets. Upper respiratory tract colonization rates for all species in humans may be as high as 50

percent. Nonencapsulated *H. influenzae* and *H. parainfluenzae* are the species most commonly found as commensals. Colonization by serotype b encapsulated *H. influenzae* strains is uncommon in healthy children and adults (Bamberger and Jackson 2001). The occurrence of disease after acquisition of an *H. influenzae* serotype b strain depends on the ability to colonize the nasopharynx and the ability to infect target tissue. The presence of the polysaccharide capsule and the production of IgA proteases are postulated to play a role in the pathogenesis of serotype b strains.

Epiglottitis in a previously well child is characterized by the sudden onset of high fever, severe sore throat with dysphagia, and drooling (MayoSmith et al. 1986; Ward 2002). In adults, epiglottitis manifests as a sore throat with odynophagia. Inflammation in adult disease is not confined to the epiglottis, as is the case with children, but rather can also affect other laryngeal and oropharyngeal structures. Consequently, a more accurate term than epiglottitis for adults is supraglottitis (Donnelly and Crausman 1997). If the diagnosis of epiglottitis is in question (pediatric or adult patient), radiography may be used but only if personnel experienced in intubation and resuscitation equipment are available, since airway compromise may occur (MayoSmith et al. 1986). Intensive care unit (ICU)-level care for the initial 24–48 h is prudent. In addition, oropharyngeal examination should not be performed, attempts should not be made to lay the patient down, and venipuncture should be avoided until the patient's airway is secured. In one review of acute epiglottitis in adults, two of 44 patients receiving medial treatment with no artificial airway died of sudden respiratory failure (MayoSmith et al. 1986). Epiglottitis must be differentiated from viral laryngotracheitis, bacterial tracheitis, diphtheria, and a foreign body lodged in the larynx. Penetrating pharyngeal injury and parapharyngeal or paravertebral abscess may mimic epiglottitis (Chung 2002).

The likelihood of beta-lactamase-producing *H. influenzae* and the inclusion of multiple etiologies in adults has led to the practice of empiric treatment with a beta-lactamase-resistant, third-generation cephalosporin (Bamberger and Jackson 2001). Most disease in children can be prevented by routine use of Hib conjugate vaccine (McEwan et al. 2003).

Gingivitis/Vincent's disease

Inflammation of the gingiva begins with irritation and microbial invasion of the gums (Fenesy 1998; Chow 2000). Subgingival plaque is always present. Simple gingivitis is characterized by swelling and thickening of the gingiva with a tendency for the gums to bleed after eating or toothbrushing. There is usually no pain. Patients with acute necrotozing gingivitis, also known as Vincent's disease or trench mouth, experience pain in the gingiva that interferes with normal chewing. Necrosis of the gingiva between the teeth, a grayish pseudomembrane, and a characteristic halitosis with altered taste are present. Fever, malaise, and regional lymphadenopathy may occur (Fenesy 1998).

Vincent described the disease in the late nineteenth century as a fusospirochetal infection due to the prominence of these microorganisms in smears of material removed from the lesion (Loesche and Grossman 2001). Mixed aerobic and anaerobic flora, especially spirochetes, with some that have never been cultured, *Fusobacterium* spp., *Prevotella* spp. and *Porphyromonas* spp., *Actinobacillus acintomycetemcomitans*, and *Eikenella corrodens* are frequently implicated as prominent etiologies of acute necrotozing ulcerative gingivitis and other periodontal diseases (Loesche and Grossman 2001; Darby and Curtis 2001).

Patient defense mechanisms against gingivitis include the highly vascular gingival tissue that presents an oxidative barrier to the penetration of anaerobic flora from the dental plaque (Loesche and Grossman 2001). While certain bacteria, such as *A. actinomycetemcomitans*, can be detected within the tissue, they are rarely able to cause tissue necrosis without the association of other anaerobic bacteria. Conditions that cause vasoconstriction of peripheral arterioles, such as smoking and stress, are risk factors for periodontal disease, most likely because of reduced blood flow allowing invading anaerobes to survive (Loesche and Grossman 2001; Fenesy 1998).

Treatment includes local debridement and lavage with oxidizing agents, which usually brings relief from pain within 24 h. Antimicrobial therapy with penicillin or metronidazole is highly effective (Chow 2000; Duckworth et al. 1966; Loesche and Grossman 2001).

Periodontal abscess

Periodontal disease is a general term that refers to all diseases involving the periodontum (supporting structures of the teeth) (Suzuki 1988). Subgingival plaque and gingivitis precede infection of the underlying supporting tissue, ultimately leading to destruction of the periodontium and permanent loss of teeth. Chronic inflammation of the periodontium is the major cause of tooth loss in adults. Periodontal infections generally drain freely, resulting in little or no discomfort to the patient. The microbial specificity in different periodontal infections has only recently been appreciated (Chow 2000; Moore and Moore 1994). In the healthy periodontum, normal flora is sparse and consists mainly of gram-positive bacteria such as *S. oralis*, *S. sanguis*, and *Actinomyces* spp. In patients with gingivitis, the predominant subgingival flora includes anaerobic gram-negative bacilli, *Capnocytophaga* spp., and *Peptostreptococcus* spp. With progression to

periodontitis, the flora increases further in complexity, with the addition of spirochetes (*Treponema denticola*), *Porphyromonas gingivalis*, and *A. actinomycetemcomitans*. With periodontal abscess, the polymicrobial flora also includes *Fusobacterium* spp. and viridans streptococci. Facultative gram-negative bacilli and staphylococci are uncommon in all periodontal infections. This microbial specificity has been defined by improved sampling and anaerobic culture techniques (Moore and Moore 1994; Maiden et al. 1997). The progression from supragingival dental plaque to subgingival dental plaque represents saccharolytic bacteria that can adhere to the tooth versus asaccharolytic bacteria that do not adhere to teeth, respectively. Periodontitis can be treated with systemic tetracycline or metronidazole combined with local treatment involving root debridement and surgical resection of inflamed tissues. Loculated abscesses must be drained (Loesche and Grossman 2001).

Stomatitis

Stomatitis refers to the inflammation of the oral mucosa including the buccal and labial mucosa, palate, tongue, floor of the mouth, and the gingivae. Severe forms occur in those with underlying malnutrition, debility, or immunocompromising disease such as human immunodeficiency virus (HIV) (Wexler et al. 1997). A small gingival ulcer progresses to a necrotic ulcer and then to a painful cellulitis of the lips and cheeks. Sloughing of necrotic soft tissues occurs, exposing underlying bone and teeth. *Borrelia vincentii*, *Prevotella melaninogenica*, and *Fusobacterium nucleatum* are common in the lesions (Chow 2000). Other bacteria are commonly seen in biopsy specimens but can not be cultured and identified. At this time, a comprehensive understanding of the pathogens involved does not exist. Treatment involves debridement, irrigation with a povidone–iodine solution, and systemic antibiotics with anaerobic coverage such as metronidazole, amoxicillin with clavulanic acid, or clindamycin (Chow 2000, Patton and van der Horst 1999).

Ludwig's angina

Ludwig's angina is a cellulitis of the submandibular, sublingual, and submental regions (Hartmann 1999; Spitalnic and Sucov 1995). In most patients, the infection originates from a dental focus. Clinical features include fever, toxicity, and a rapidly progressive edema in the floor of the mouth and the anterior neck. Elevation of the tongue impedes swallowing, and airway obstruction may be fatal. Streptococci and mouth flora are the most common etiologic agents, but *H. influenzae*, staphylococci, and gram-negative bacilli have also been implicated. Broad-spectrum antimicrobial therapy is needed and tracheostomy may be necessary to preserve the airway (Barakate et al. 2001).

PARANASAL SINUSES AND ASSOCIATED AIRWAYS

Sinusitis

Sinusitis is a common disease with significant morbidity. It is estimated that 20 million cases of acute community-acquired bacterial sinusitis occur each year in the USA (Gwaltney 1996). In addition, bacterial nosocomial sinusitis is commonly recognized. Although bacterial infections of the sinuses are common, they are greatly overdiagnosed by patients and physicians. An understanding of the etiology, pathogenesis, and methods of diagnosis are important factors in preventing overdiagnosis.

Predominant pathogens in adult acute maxillary sinusitis are *S. pneumoniae* and *H. influenzae*, accounting for more than 50 percent of cases. *M. catarrhalis*, other streptococci such as *S. pyogenes* and viridans streptococci, *S. aureus*, and anaerobic bacteria account for the remainder of cases (Table 23.4). *Moraxella catarrhalis* is more common in pediatric patients, while anaerobes are less common. In spite of comprehensive bacteriologic analysis, cultures are negative from a high percentage of patients with the clinical diagnosis of sinusitis (Gwaltney 1996; Gwaltney et al. 1992). Some of these patients may have had viral infection rather than bacterial, and many may have had ostial occlusion preventing infected inflammatory material from draining and being accessible for culture. Others have suggested that *Chlamydophila pneumoniae* and *Mycoplasma pneumoniae* represent likely etiologies when bacterial cultures are negative (Gwaltney 1996). These bacteria have not been confirmed as causes of sinusitis by careful aspiration of sinus material and are not established as true pathogens. Although the relative importance in sinus infection of

different bacteria has not changed in decades, there have been important changes in resistance to antimicrobials. The emergence of beta-lactamase production by *H. influenzae* and *M. catarrhalis* and the emergence of multiply resistant pneumococci have focused significant attention on the management of sinus infections (Brook 2002).

The bacteriology of chronic sinusitis is less clear because many of the patients have had prolonged and varied courses of antimicrobial therapy and culture material has not been collected by endoscopic or surgical drainage that avoids contamination with airway flora. Nevertheless, anaerobic pathogens are common in some studies of adults and uncommon in others (Bamberger and Jackson 2001; Nadel et al. 1998). Bacteria more frequently detected in chronic sinusitis include *S. aureus*, facultative gram-negative bacilli (*Enterobacteriaceae*) and *Pseudomonas aeruginosa*, and nonpneumococcal streptococci (Nadel et al. 1998).

Sphenoid, or frontal sinusitis, differs from acute maxillary sinusitis in that *S. aureus* is a frequent pathogen, in addition to pneumococci, *H. influenzae*, and anaerobic streptococci. Chronic bacterial sphenoid sinusitis is caused most commonly by gram-negative pathogens (Lew et al. 1983).

Bacteria associated with nosocomial sinusitis include usual nosocomial pathogens such as *P. aeruginosa*, *Acinetobacter* spp., enteric gram-negative bacilli, and *S. aureus*. Streptococci and anaerobes are less common. (Westergren et al. 1998a, b). Sinusitis is reported to be present in as many as 68 percent of HIV infected patients, especially those with CD4 cell counts below 200 per mm^3. A variety of bacterial and nonbacterial pathogens can be detected, with *P. aeruginosa* a common bacterial etiology (Rubin and Honigberg 1990). *P. aeruginosa* and *S. aureus* are common in patients with cystic fibrosis or syndromes that compromise ciliary motility (Mak and Henig 2001; Bertrand et al. 2000).

The pathogenesis of sinus infection is dominated by mucosal swelling and mechanical obstruction that occur following upper respiratory tract viral infection (Gwaltney 1996). Seventy-seven percent of patients have occlusion of the infundibulum and 87 percent have sinus cavity disease accompanying the common cold (Gwaltney 1996; Gwaltney et al. 1994). Also important is the fact that rhinovirus infection induces goblet cells to secrete an increased amount of thick mucus that is inadequately cleared by ciliary function. Thus, a major part of the disease process appears to be a function of mucocilliary clearance as a result of increased amounts of viscous material, which occurs in addition to infundibular and ostiomeatal obstruction (Gwaltney 1996). An estimated 0.5–2.0 percent of colds are complicated by bacterial infection of the sinuses arising from normal flora of the nasopharynx and nasal passages. Although the paranasal sinuses are normally sterile, transient inoculation with commensals from the nasal passages

Table 23.4 *Etiology of adult and pediatric acute sinusitis*

Etiology	Approximate percentage of cases	
	Adult patients	**Pediatric patients**
Streptococcus pneumoniae	30	36
Haemophilus influenzae	20	23
Moraxella catarrhalis	5	20
Staphylococcus aureus	4	0
Streptococcus pyogenes	2	<5
Anaerobic bacteria	6	0
Other bacteria	5–10	<5
Viruses	25	3–5

Data from Bamberger and Jackson (2001) and Gwaltney (1996).

can be effectively cleared by ciliary action within the sinus. Secretions and inflammation following rhinovirus infection, and presumably other viruses causing the common cold, impede effective clearing resulting in trapped bacteria in a 'closed' space (Puhakka et al. 1998; Diaz and Bamberger 1995; Monto et al. 2001).

Other less common causes of bacterial sinusitis include extension of dental infections into the maxillary sinus, ostial obstruction due to allergic rhinitis, anatomic abnormalities including as deviated nasal septum or nasal polyps, and barotrauma due to flying or deep-water diving (Bamberger and Jackson 2001). Immunodeficiency diseases also predispose patients to sinusitis, including HIV infection, diabetes, chronic granulomatous disease, and hypogammaglobulinemias. Risk factors for nosocomial sinusitis include nasogastric tubes and nasoendotracheal tubes (Rubin and Honigberg 1990; Westergren et al. 1998a).

Clinical features of sinusitis represent combined features of viral upper respiratory tract infection and bacterial sinusitis (Andre et al. 2002). Sneezing, rhinorrhea, nasal obstruction, facial pressure, and headache are common sinonasal complaints. In addition, those with confirmed acute community-acquired bacterial sinusitis are described to have purulent nasal discharge, a temperature of $\geqslant 38°C$, facial pain or erythema, cough, and a decreased sense of smell (Gwaltney 1996).

Diagnosis of acute bacterial sinusitis is difficult despite modern technology (Gwaltney 1996; Andre et al. 2002). Differentiating allergy from bacterial infection and viral from bacterial etiology, and determining what specific bacterial etiology is involved, can be challenging. Sinus puncture provides definitive identification of etiology in cases serious enough to require invasive diagnostic techniques. Purulent sinus secretions carefully smeared and gram-stained have been shown to improve culture specificity (Jousimies-Somer et al. 1988). Morphotypes seen in the Gram stain, in association with polymorphonuclear leukocytes, predicted pathogens that eventually grew in concentrations above 1 000 CFU/ml on culture media. This suggests that smear, staining, and culture of sinus secretions may help differentiate bacterial from viral acute maxillary sinusitis. In addition, studies have shown that patients with sinus infection who have secretions containing *H. influenzae*, *M. catarrhalis*, or *S. pneumoniae* do benefit from antimicrobial therapy (Kaiser et al. 1996).

Appropriate antimicrobials given according to accepted regimens has been shown to reduce bacterial concentrations in the sinus cavity compared with incorrect antimicrobials or regimens (Gwaltney 1996). Because of the possibility of progressive damage to ciliated epithelium in the sinus, early therapy and eradication of a bacterial etiology are beneficial (Gwaltney et al. 1992; Hinni et al. 1992). Although multiply resistant pneumococci are an ever-increasing problem, therapy with amoxicillin-clavulante, an oral cephalosporin with

maximal pneumococcal activity, or a newer fluoroquinolone are recommended antibacterial regimens (Pfaller et al. 2001; Poole 1997).

Prevention of acute sinusitis requires preventing upper respiratory tract viral infection and entry of nasal flora into the sinuses (Puhakka et al. 1998). To some extent, colds can be prevented by hygienic measures such as handwashing and covering one's mouth when sneezing. There is no proven measure for preventing secondary bacterial infection of the sinuses.

Rhinoscleroma

Rhinoscleroma is a chronic, progressive, disfiguring, and debilitating granulomatous infectious disease that has affected man for over 1 000 years (Hart and Rao 2000). It predominately affects the upper airways. It is caused by a gram-negative bacterium, *Klebsiella pneumoniae* subsp. *rhinoscleromatis*, most often referred to as *K. rhinoscleromatis*. Although test kits may not reliably identify this subspecies of *Klebsiella*, it can be differentiated from *K. pneumoniae* sensu stricto by a positive methyl red test and negative Voges-Proskauer, urease, citrate, and lysine decarboxylase tests (Toohill 1993).

More than 16 000 cases of rhinoscleroma have been reported since 1960 from most parts of the world (Hart and Rao 2000). It is found predominantly in rural areas and is more common in regions where socioeconomic conditions are poor. In developed countries in temperate zones, cases are detected but are generally imported. The disease has been widely reported from many Middle Eastern countries, tropical Africa, India, Southeast Asia, and Central and South America (Muzyka and Gubina 1971). Females are more commonly affected than males (ratio 13:1) and disease most commonly presents in the second and third decades of life (Hart and Rao 2000).

Disease usually occurs at the junction between epithelial surfaces, such as the interface between the squamous epithelium of the anterior nares and the deeper columnar ciliated epithelium (Hart and Rao 2000). Other explanations of disease prevalence include the fact that iron deficiency leads to squamous metaplasia, suggesting that poor nutrition and pregnancy may play a roll in the preponderance of disease occurring in women of childbearing age (Hart and Rao 2000) (Akhnoukh and Saad 1987).

Rhinoscleroma affects most areas of the respiratory tract (Hart and Rao 2000). The nose is involved in 95–100 percent of cases and the pharynx in up to half of cases. Occasionally, other areas of the respiratory tract are involved, such as the Eustachian tubes, sinuses, mouth, orbit, larynx, trachea, and bronchi. Although it may be considered an opportunistic infection in patients with acquired immunodeficiency syndrome (AIDS), there are no reports of dissemination despite low CD4-lymphocyte (Andraca et al. 1993; Paul et al. 1993). Rhinoscleroma is divided into three stages: catarrhal-atrophic (referred to as ozaena), granulomatous, and sclerotic (Hart and Rao 2000). The first stage is characterized by a foul-smelling purulent nasal discharge. Most cases are diagnosed in the granulomatous stage as the patient complains of epistaxis and nasal deformity with lesions appearing as bluish-red, rubbery granulomas. Histologic examination during the granulomatous stage reveals Mikulicz cells, which are large macrophages containing vacuoles filled with bacteria.

Treatment involves surgical debridement and prolonged antimicrobial therapy. In vitro, *K. rhinoscleromatis* is susceptible to many antimicrobials. With most therapeutic regimens, relapse is common. Prolonged therapy with fluoroquinolones has produced the best results with low relapse rates (Hart and Rao 2000; Borgstein et al. 1993).

EAR

Otitis media and acute mastoiditis

Otitis media is a common reason for very young children to seek medical treatment. The peak age of disease is between 6 and 15 months of age. By the age of 7 years, 90 percent of children have had at least one episode of otitis media (Klein 1994). Otitis media is categorized as acute, serous, or chronic (Hendley 2002; Klein 1994; Shurin et al. 1983). Acute disease is characterized simply by suppuration in the middle ear, fever, ear pain, and a bulging tympanic membrane. Serous otitis media is defined as the presence of middle ear effusion behind an intact tympanic membrane without acute signs or symptoms. Chronic otitis media occurs when long-standing discharge through a perforated tympanic membrane is noted. It is important to understand that, in spite of the widespread use of antimicrobial therapy for otitis, acute mastoiditis remains the most common suppurative complication of acute otitis media (Klein 1994).

Streptococcus pneumoniae and *H. influenzae* are the most common etiologies detected in all patient groups. Many pneumococcal serotypes are involved and nontypeable strains of *H. influenzae* predominate, similar to strains found as normal upper respiratory tract flora in children and adults. *Moraxella catarrhalis* may be recovered in up to 10 percent of cases (Shurin et al. 1983). Other etiologies occasionally encountered include *S. pyogenes*, *S. aureus*, facultative gram-negative bacilli (Enterobacteriaceae), and anaerobes (Bluestone 1998). *Pseudomonas aeruginosa* is a common etiology in chronic suppurative otitis media (Chan and Hadley 2001). In addition, *Mycobacterium tuberculosis* is an occasional cause of chronic suppurative otitis media (Jeang and Fletcher 1983).

The pathogenesis of otitis media is multifactoral (Bluestone 1996). Prior or concurrent viral infection, host differences, anatomic factors, and environmental exposures all impact the pathogenesis of middle ear infection. Viral infection predisposes the patient to development of acute otitis media by causing Eustachian tube dysfunction and enhancing nasopharyngeal colonization with middle ear pathogens (Saez-Llorens 1994; Faden et al. 1998). Otitis media is more common at a young age, in patients with immature or impaired immunologic status, in those with upper respiratory allergy, in those with a familial predisposition or other siblings in the household, in males, in Native Americans and Australian Aborigines, and in babies who are bottle-fed (Bamberger and Jackson 2001). Otitis media is also more common in children who attend daycare and in those exposed to passive cigarette smoke. The pathogenesis of acute otitis media may be summarized as follows: The patient has an upper respiratory tract viral infection that results in congestion of the respiratory mucosa, including the nasopharynx and Eustachian tube, resulting in obstruction of the tube. Negative pressure in the middle ear develops, and, when prolonged, results in 'aspiration' of potential bacterial and viral pathogens into the middle ear. Because of the obstruction, clearance of infected middle ear fluid is impaired. Fluid accumulates in the middle ear, followed by bacterial growth and resulting inflammation. Suppurative and symptomatic otitis media result. A complete discussion of the pathogenesis of otitis media can be found in a review by Bluestone (1996).

Acute mastoiditis can be a natural extension and part of the pathological process of acute otitis media. In fact, the mastoid air cells are involved in many children who have acute otitis media (Bluestone 1998). Infection of the mastoid air cells usually resolves spontaneously or after effective antimicrobial treatment. The effusion within the mastoid air cells drains into the middle ear and, in turn, into the Eustachian tube. The communication between the middle ear and the mastoid air cells is narrow and is called the aditus-ad-antrum (Bluestone 1996, 1998). Functionally, the mastoid air cell system is a reservoir of gas for the middle ear and the narrow passage makes it difficult for middle ear infection to enter the mastoid. The aditus-ad-antrum can become obstructed as a result of swelling and granulation tissue. This swelling is implicated in the pathogenesis of acute mastoiditis. If the obstruction persists, mastoiditis with periosteitis (infection of the periosteum covering the mastoid process) can develop. This condition can lead to mastoid osteitis and further develop into infection of the temporal bone and intracranial cavity (Bluestone 1998).

Overall, 85 percent of cases of acute otitis media in children will resolve without antimicrobial therapy. Importantly, spontaneous resolution occurs in 90, 70, and 20 percent of cases caused by *M. catarrhalis*, nontypeable *H. influenzae*, and *S. pneumoniae*, respectively

(Faden et al. 1998). It is clear that the pneumococcus is the most important pathogen. It is also clear that pneumococci collected worldwide can be resistant to multiple antimicrobial agents (Zhanel et al. 2003; Muhlemann et al. 2003). Since the widespread use of antimicrobial agents to treat acute otitis media, the incidence of mastoiditis and other complications has decreased from 40 percent early in the twentieth century to less than 0.5 percent at present (Faden et al. 1998). Antimicrobial treatment of acute otitis media is recommended to prevent suppurative sequelae, shorten the time spent with middle ear effusion, and prevent recurrent disease. Amoxycillin remains an appropriate choice for first-line therapy for acute otitis media (Hendley 2002; Klein 1994; Berman 1995; Hoberman et al. 2002). For patients in whom amoxicillin is unsuccessful, second-line therapy should have demonstrated activity against penicillin-resistant *S. pneumoniae* as well as beta-lactamase-producing pathogens. Appropriate options include high-dose amoxycillin-clavulanate and ceftriaxone. Tympanocentesis is useful for identifying persistent and resistant pathogens (Hoberman et al. 2002). The pneumococcal conjugate vaccine was approved recently for use in children and should be administered to all children less than 2 years old. Introduction of this vaccine is likely to have some beneficial effect on the proportion of resistant *S. pneumoniae* isolates by decreasing carriage of vaccine-related serotypes (Pelton 2002). In addition, the possible use of fluoroquinolones in pediatric patients and the introduction of ketolides may impact therapeutic recommendations for acute otitis media (Pelton 2002).

Malignant external otitis

Malignant external otitis is a serious infection caused by *P. aeruginosa* that occurs primarily in elderly diabetic subjects. Infection can progress to a skullbase osteomyelitis with resultant cranial neuropathies. In spite of long-term systemic antimicrobial therapy, the disease is recurrent and may be associated with significant mortality (Rubin and Yu 1988).

Since the natural reservoir of *P. aeruginosa* is water, the accumulation of moisture in the external ear canal is the likely inciting event (*P. aeruginosa* is also the most common etiologic agent of chronic external otitis). Spread from the external canal occurs as infection crosses the cartilaginous-osseus junction into the temporal bone, bypassing the tympanic membrane and middle ear. Infection then passes to the mastoid. Because of its proximity to the initial site of infection, the facial nerve is the first, and often only, cranial nerve to be involved. Small-vessel disease that results in cutaneous hypoperfusion, diminished local host resistance, and increased susceptibility to infection, is a likely reason that most patients have malignant external otitis. The pathogenesis of malignant external otitis in nondia-

betic patients appears related to immune dysfunction or a combination of impairments resulting from underlying disease or malnutrition (Rubin and Yu 1988).

Therapy should include surgery for local debridement and excision of foci of infection (Rubin and Yu 1988; Raines and Schindler 1980). *P. aeruginosa* can be cultured from specimens representing material from the external auditory canal and removed during surgical debridement. However, the mainstay of therapy is systemic antimicrobial therapy with two anti-pseudomonal drugs (such as antipseudomonal beta-lactam and aminoglycoside) for up to 8 weeks or more if extensive disease is present.

ABSCESSES ADJACENT TO AND ORIGINATING FROM THE ORAL CAVITY

Peritonsillar, pharyngeal, retropharyngeal, and parapharyngeal abscesses

Infections of the deep spaces of the neck are potentially serious because of the likelihood of spread to critical structures, such as major blood vessels and the mediastinum, leading to thrombosis, mediastinitis, purulent pericarditis, and pleural empyema (Peterson and Thomson 1999; Chow 1992; Parhiscar and Har-El 2001). Peritonsillar abscess is caused by mixed aerobic and anaerobic bacteria, with *S. pyogenes*, *S. aureus*, and *S. anginosus* group streptococci playing a prominent role (Fujiyoshi et al. 2001). Pharyngeal, retropharyngeal, and parapharyngeal abscesses involve structures extending into the neck that may be life-threatening. These infectious sites are associated with anaerobes, which outnumber aerobes three to one, with an average of more than seven bacterial species in the abscess material. As expected, oral flora cause these infections, primarily represented by *Peptostreptococcus* spp., *Prevotella* spp., *Porphyromonas* spp., *Fusobacterium* spp., and *Actinomyces* spp. Streptococci, primarily of the viridans group, are also important. When these abscesses develop in the hospitalized patient, *S. aureus* and various aerobic gram-negative bacilli are more likely to be involved (Peterson and Thomson 1999). Immediate surgical drainage and antimicrobial therapy directed at oral anaerobic flora and streptococci are indicated since complications can be life threatening (Parhiscar and Har-El 2001; Matsuda et al. 2002).

Lemierre's syndrome (jugular vein thrombophlebitis)

Lemierre's syndrome is a specialized oropharyngeal infection characterized by secondary septic thrombophlebitis of the internal jugular vein and frequent metastatic spread of infection (Chirinos et al. 2002).

Fusobacterium necrophorum, a normal inhabitant of the oral cavity, is the usual etiologic agent, along with other oral anaerobes that are occasionally detected. The disease progresses most commonly from pharyngitis, to local invasion of the lateral pharyngeal space, to internal jugular vein thrombophlebitis and subsequent metastatic complications involving the lungs, joints, liver, and mediastinal muscle. A swollen and/or tender neck and noncavitating pulmonary infiltrates in patients with recent or current pharyngitis are the most common findings. Blood collected by venipuncture and specimen representing material debrided during surgery, when performed, should be transported and cultured for aerobic and anaerobic bacteria. In the preantibiotic era the outcome was fatal with rapidly progressing septicemia. Mortality with antimicrobials and surgery (if necessary) is approximately 6 percent (Chirinos et al. 2002). Antimicrobial therapy with activity against anaerobic gram-negative bacilli and gram-positive cocci is recommended, such as metronidazole or clindamycin (Chirinos et al. 2002; Williams et al. 1998).

Suppurative parotitis

Acute suppurative parotitis (sialadenitis) is seen in very ill patients, especially those that are dehydrated, malnourished, elderly, or recovering from surgery (Thomson and Miller 2003; Raad et al. 1990). It is characterized by painful, tender swelling of the parotid gland. Purulent drainage may be evident at the opening of the duct of the gland in the mouth. *S. aureus* is the major pathogen, but on occasion viridans streptococci and oral anaerobes may play a role (Brook 2003). A chronic bacterial parotitis has been described that also is caused by *S. aureus*. Less often, other salivary glands may be involved, usually because of ductal obstruction. *Mycobacterium tuberculosis* may involve the parotid gland in conjunction with pulmonary tuberculosis (Mert et al. 2000).

REFERENCES

Akhnoukh, S. and Saad, E.F. 1987. Iron-deficiency in atrophic rhinitis and scleroma. *Indian J Med Res*, **85**, 576–9.

Alos, J.I., Aracil, B., et al. 2003. Significant increase in the prevalence of erythromycin-resistant, clindamycin- and miocamycin-susceptible (M phenotype) Streptococcus pyogenes in Spain. *J Antimicrob Chemother*, **51**, 333–7.

Andraca, R., Edson, R.S. and Kern, E.B. 1993. Rhinoscleroma: a growing concern in the United States? Mayo Clinic experience. *Mayo Clin Proc*, **68**, 1151–7.

Andre, M., Odenholt, I., et al. 2002. Upper respiratory tract infections in general practice: diagnosis, antibiotic prescribing, duration of symptoms and use of diagnostic tests. *Scand J Infect Dis*, **34**, 880–6.

Bamberger, D.M. and Jackson, M.A. 2001. Upper respiratory tract infections: pharyngitis, sinusitis, otitis media, and epiglottitis. In: Niederman, M., Sarosi, G.A. and Glassroth, J. (eds), *Respiratory infections*. Philadelphia: Lippincott Williams & Wilkins, 125–39.

Barakate, M.S., Jensen, M.J., et al. 2001. Ludwig's angina: report of a case and review of management issues. *Ann Otol Rhinol Laryngol*, **110**, 453–6.

Berman, S. 1995. Otitis media in children. *N Engl J Med*, **332**, 1560–5.

Bernstein, J.M. 1992. Mucosal immunology of the upper respiratory tract. *Respiration*, **59**, Suppl 3, 3–13.

Bertrand, B., Collet, S., et al. 2000. Secondary ciliary dyskinesia in upper respiratory tract. *Acta Otorhinolaryngol Belg*, **54**, 309–16.

Bisno, A.L. 1991. Group A streptococcal infections and acute rheumatic fever. *N Engl J Med*, **325**, 783–93.

Bisno, A.L. 2001. Acute pharyngitis. *N Engl J Med*, **344**, 205–11.

Bluestone, C.D. 1996. Pathogenesis of otitis media: role of eustachian tube. *Pediatr Infect Dis J*, **15**, 281–91.

Bluestone, C.D. 1998. Acute and chronic mastoiditis and chronic suppurative otitis media. *Semin Ped Infect Dis*, **9**, 12–26.

Borgstein, J., Sada, E. and Cortes, R. 1993. Ciprofloxacin for rhinoscleroma and ozena. *Lancet*, **342**, 122.

Brook, I. 2002. Antibiotic resistance of oral anaerobic bacteria and their effect on the management of upper respiratory tract and head and neck infections. *Semin Respir Infect*, **17**, 195–203.

Brook, I. 2003. Acute bacterial suppurative parotitis: microbiology and management. *J Craniofac Surg*, **14**, 37–40.

Carroll, K. and Reimer, L. 1996. Microbiology and laboratory diagnosis of upper respiratory tract infections. *Clin Infect Dis*, **23**, 442–8.

Chan, J. and Hadley, J. 2001. The microbiology of chronic rhinosinusitis: results of a community surveillance study. *Ear Nose Throat J*, **80**, 143–5.

Chirinos, J.A., Lichtstein, D.M., et al. 2002. The evolution of Lemierre syndrome: report of 2 cases and review of the literature. *Medicine (Baltimore)*, **81**, 458–65.

Chow, A. 2000. Infections of the oral cavity, neck, and head. In: Mandell, G.B.J. and Dolin, R. (eds), *Principles and practices of infectious diseases*. Vol. 1. Philadelphia: Churchill Livingstone, 690.

Chow, A.W. 1992. Life-threatening infections of the head and neck. *Clin Infect Dis*, **14**, 991–1002.

Chung, C.H. 2002. Adult acute epiglottitis and foreign body in the throat – chicken or egg? *Eur J Emerg Med*, **9**, 167–9.

Cooper, R.J., Hoffman, J.R., et al. 2001. Principles of appropriate antibiotic use for acute pharyngitis in adults: background. *Ann Emerg Med*, **37**, 711–19.

Cunningham, M.W. 2000. Pathogenesis of group A streptococcal infections. *Clin Microbiol Rev*, **13**, 470–511.

Darby, I. and Curtis, M. 2001. Microbiology of periodontal disease in children and young adults. *Periodontol*, **2000**, 26, 33–53.

Diaz, I. and Bamberger, D.M. 1995. Acute sinusitis. *Semin Respir Infect*, **10**, 14–20.

Donnelly, T.J. and Crausman, R.S. 1997. Acute supraglottitis: when a sore throat becomes severe. *Geriatrics*, **52**, 65-6, 69.

Duckworth, R., Waterhouse, J.P., et al. 1966. Acute ulcerative gingivitis. A double-blind controlled clinical trial of metronidazole. *Br Dent J*, **120**, 599–602.

Efstratiou, A., Engler, K.H., et al. 2000. Current approaches to the laboratory diagnosis of diphtheria. *J Infect Dis*, **181**, Suppl 1, S138–45.

Enwonwu, C. O., Phillips, R. S. and Falkler, W. A., Jr. 2002. Nutrition and oral infectious diseases: state of the science. *Compend Contin Educ Dent*, **23**, 431–4, 436, 438 passim; quiz 448.

Faden, H., Duffy, L. and Boeve, M. 1998. Otitis media: back to basics. *Pediatr Infect Dis J*, **17**, 1105–12, quiz 1112-3.

Feldman, C. 2001. Nonspecific host defenses: mucociliary clearance and cough. In: Niederman, M.S., Sarosi, G.A. and Glassroth, J. (eds), *Respiratory infections*. Philadelphia: Lippincott Williams & Wilkins, 13–26.

Fenesy, K.E. 1998. Periodontal disease: an overview for physicians. *Mt Sinai J Med*, **65**, 362–9.

Fujiyoshi, T., Inaba, T., et al. 2001. Clinical significance of the Streptococcus milleri group in peritonsillar abscesses. *Nippon Jibiinkoka Gakkai Kaiho*, **104**, 866–71.

Galazka, A. 2000. The changing epidemiology of diphtheria in the vaccine era. *J Infect Dis*, **181**, Suppl 1, S2–9.

Gerber, M.A. 1998. Diagnosis of group A streptococcal pharyngitis. *Pediatr Ann*, **27**, 269–73.

Gwaltney, J.M. Jr. 1996. Acute community-acquired sinusitis. *Clin Infect Dis*, **23**, 1209–23, quiz 1224-5.

Gwaltney, J.M. Jr., Phillips, C.D., et al. 1994. Computed tomographic study of the common cold. *N Engl J Med*, **330**, 25–30.

Gwaltney, J.M. Jr., Scheld, W.M., et al. 1992. The microbial etiology and antimicrobial therapy of adults with acute community-acquired sinusitis: a fifteen-year experience at the University of Virginia and review of other selected studies. *J Allergy Clin Immunol*, **90**, 457–61, discussion 462.

Hadfield, T.L., McEvoy, P., et al. 2000. The pathology of diphtheria. *J Infect Dis*, **181**, Suppl 1, S116–20.

Harnisch, J.P., Tronca, E., et al. 1989. Diphtheria among alcoholic urban adults. A decade of experience in Seattle. *Ann Intern Med*, **111**, 71–82.

Hart, C.A. and Rao, S.K. 2000. Rhinoscleroma. *J Med Microbiol*, **49**, 395–6.

Hartmann, R.W. Jr. 1999. Ludwig's angina in children. *Am Fam Physician*, **60**, 109–12.

Haukness, H.A., Tanz, R.R., et al. 2002. The heterogeneity of endemic community pediatric group a streptococcal pharyngeal isolates and their relationship to invasive isolates. *J Infect Dis*, **185**, 915–20.

Hendley, J.O. 2002. Clinical practice. Otitis media. *N Engl J Med*, **347**, 1169–74.

Hinni, M.L., McCaffrey, T.V. and Kasperbauer, J.L. 1992. Early mucosal changes in experimental sinusitis. *Otolaryngol Head Neck Surg*, **107**, 537–48.

Hoban, D., Waites, K. and Felmingham, D. 2003. Antimicrobial susceptibility of community-acquired respiratory tract pathogens in North America in 1999–2000: findings of the PROTEKT surveillance study. *Diagn Microbiol Infect Dis*, **45**, 251–9.

Hoberman, A., Marchant, C.D., et al. 2002. Treatment of acute otitis media consensus recommendations. *Clin Pediatr (Phila)*, **41**, 373–90.

Jeang, M.K. and Fletcher, E.C. 1983. Tuberculous otitis media. *J Am Med Assoc*, **249**, 2231–2.

Jousimies-Somer, H.R., Savolainen, S. and Ylikoski, J.S. 1988. Macroscopic purulence, leukocyte counts, and bacterial morphotypes in relation to culture findings for sinus secretions in acute maxillary sinusitis. *J Clin Microbiol*, **26**, 1926–33.

Kaiser, L., Lew, D., et al. 1996. Effects of antibiotic treatment in the subset of common-cold patients who have bacteria in nasopharyngeal secretions. *Lancet*, **347**, 1507–10.

Kaplan, E.L., Gastanaduy, A.S. and Huwe, B.B. 1981. The role of the carrier in treatment failures after antibiotic for group A streptococci in the upper respiratory tract. *J Lab Clin Med*, **98**, 326–35.

Kilian, M. 1981. Degradation of immunoglobulins A2, A2 and G by suspected principal periodontal pathogens. *Infect Immun*, **34**, 757–65.

Klein, J.O. 1994. Otitis media. *Clin Infect Dis*, **19**, 823–33.

Lew, D., Southwick, F.S., et al. 1983. Sphenoid sinusitis. A review of 30 cases. *N Engl J Med*, **309**, 1149–54.

Lipchik, R.J. and Kuzo, R.S. 1996. Nosocomial pneumonia. *Radiol Clin North Am*, **34**, 47–58.

Loesche, W.J. and Grossman, N.S. 2001. Periodontal disease as a specific, albeit chronic, infection: diagnosis and treatment. *Clin Microbiol Rev*, **14**, 727–52, table of contents.

Maiden, M.F., Macuch, P.J., et al. 1997. "Checkerboard" DNA-probe analysis and anaerobic culture of initial periodontal lesions. *Clin Infect Dis*, **25**, Suppl 2, S230–2.

Mak, G.K. and Henig, N.R. 2001. Sinus disease in cystic fibrosis. *Clin Rev Allergy Immunol*, **21**, 51–63.

Markina, S.S., Maksimova, N.M., et al. 2000. Diphtheria in the Russian Federation in the 1990s. *J Infect Dis*, **181**, Suppl 1, S27–34.

Matsuda, A., Tanaka, H., et al. 2002. Peritonsillar abscess: a study of 724 cases in Japan. *Ear Nose Throat J*, **81**, 384–9.

MayoSmith, M.F., Hirsch, P.J., et al. 1986. Acute epiglottitis in adults. An eight-year experience in the state of Rhode Island. *N Engl J Med*, **314**, 1133–9.

McEwan, J., Giridharan, W., et al. 2003. Paediatric acute epiglottitis: not a disappearing entity. *Int J Pediatr Otorhinolaryngol*, **67**, 317–21.

Mert, A., Ozaras, R., et al. 2000. Primary tuberculosis of the parotid gland. *Int J Infect Dis*, **4**, 229–30.

Monto, A.S., Fendrick, A.M. and Sarnes, M.W. 2001. Respiratory illness caused by picornavirus infection: a review of clinical outcomes. *Clin Ther*, **23**, 1615–27.

Moore, W.E. and Moore, L.V. 1994. The bacteria of periodontal diseases. *Periodontol*, **2000**, 5, 66–77.

Muhlemann, K., Matter, H.C., et al. 2003. Nationwide surveillance of nasopharyngeal Streptococcus pneumoniae isolates from children with respiratory infection, Switzerland, 1998–1999. *J Infect Dis*, **187**, 589–96.

Muzyka, M.M. and Gubina, K.M. 1971. Problems of the epidemiology of scleroma. I. Geographical distribution of scleroma. *J Hyg Epidemiol Microbiol Immunol*, **15**, 233–42.

Nadel, D.M., Lanza, D.C. and Kennedy, D.W. 1998. Endoscopically guided cultures in chronic sinusitis. *Am J Rhinol*, **12**, 233–41.

Neto, A.S., Lavado, P., et al. 2003. Risk factors for the nasopharyngeal carriage of respiratory pathogens by Portuguese children: phenotype and antimicrobial susceptibility of Haemophilus influenzae and Streptococcus pneumoniae. *Microb Drug Resist*, **9**, 99–108.

Ng, T.P. and Tan, W.C. 1994. Epidemiology of allergic rhinitis and its associated risk factors in Singapore. *Int J Epidemiol*, **23**, 553–8.

Niederman, M.S. 1990. Gram-negative colonization of the respiratory tract: pathogenesis and clinical consequences. *Semin Respir Infect*, **5**, 173–84.

Obeid, P. and Bercy, P. 2000. Effects of smoking on periodontal health: a review. *Adv Ther*, **17**, 230–7.

Parhiscar, A. and Har-El, G. 2001. Deep neck abscess: a retrospective review of 210 cases. *Ann Otol Rhinol Laryngol*, **110**, 1051–4.

Patton, L.L. and van der Horst, C. 1999. Oral infections and other manifestations of HIV disease. *Infect Dis Clin North Am*, **13**, 879–900.

Paul, C., Pialoux, G., et al. 1993. Infection due to Klebsiella rhinoscleromatis in two patients infected with human immunodeficiency virus. *Clin Infect Dis*, **16**, 441–2.

Pelton, S.I. 2002. Acute otitis media in an era of increasing antimicrobial resistance and universal administration of pneumococcal conjugate vaccine. *Pediatr Infect Dis J*, **21**, 599–604, discussion 613-4.

Peterson, L.R. and Thomson, R.B. Jr. 1999. Use of the clinical microbiology laboratory for the diagnosis and management of infectious diseases related to the oral cavity. *Infect Dis Clin North Am*, **13**, 775–95.

Pfaller, M.A., Ehrhardt, A.F. and Jones, R.N. 2001. Frequency of pathogen occurrence and antimicrobial susceptibility among community-acquired respiratory tract infections in the respiratory surveillance program study: microbiology from the medical office practice environment. *Am J Med,9A*, **111**, Suppl 9A, 4S–12S, discussion 36S-38S.

Poole, M.D. 1997. Antimicrobial therapy for sinusitis. *Otolaryngol Clin North Am*, **30**, 331–9.

Puhakka, T., Makela, M.J., et al. 1998. Sinusitis in the common cold. *J Allergy Clin Immunol*, **102**, 403–8.

Raad, I.I., Sabbagh, M.F. and Caranasos, G.J. 1990. Acute bacterial sialadenitis: a study of 29 cases and review. *Rev Infect Dis*, **12**, 591–601.

Raines, J.M. and Schindler, R.A. 1980. The surgical management of recalcitrant malignant external otitis. *Laryngoscope*, **90**, 369–78.

Rodriguez, C.A., Avadhanula, V., et al. 2003. Prevalence and distribution of adhesins in invasive non-type b encapsulated Haemophilus influenzae. *Infect Immun*, **71**, 1635–42.

Roscoe, D.L. and Chow, A.W. 1988. Normal flora and mucosal immunity of the head and neck. *Infect Dis Clin North Am*, **2**, 1–19.

Rubin, J. and Yu, V.L. 1988. Malignant external otitis: insights into pathogenesis, clinical manifestations, diagnosis, and therapy. *Am J Med*, **85**, 391–8.

Rubin, J.S. and Honigberg, R. 1990. Sinusitis in patients with the acquired immunodeficiency syndrome. *Ear Nose Throat J*, **69**, 460–3.

Ryan, P.A., Pancholi, V. and Fischetti, V.A. 2001. Group A streptococci bind to mucin and human pharyngeal cells through sialic acid-containing receptors. *Infect Immun*, **69**, 7402–12.

Sack, J.L. and Brock, C.D. 2002. Identifying acute epiglottitis in adults. High degree of awareness, close monitoring are key. *Postgrad Med*, **112**, 81-2, 85–6.

Saez-Llorens, X. 1994. Pathogenesis of acute otitis media. *Pediatr Infect Dis J*, **13**, 1035–8.

Schuster, G.S. 1999. Oral flora and pathogenic organisms. *Infect Dis Clin North Am*, **13**, 757–74.

Shaw, J.H. 1987. Causes and control of dental caries. *N Engl J Med*, **317**, 996–1004.

Shurin, P.A., Marchant, C.D., et al. 1983. Emergence of beta-lactamase-producing strains of Branhamella catarrhalis as important agents of acute otitis media. *Pediatr Infect Dis*, **2**, 34–8.

Spitalnic, S.J. and Sucov, A. 1995. Ludwig's angina: case report and review. *J Emerg Med*, **13**, 499–503.

Sutter, V.L. 1984. Anaerobes as normal oral flora. *Rev Infect Dis*, **6**, Suppl 1, S62–6.

Suzuki, J.B. 1988. Diagnosis and classification of the periodontal diseases. *Dent Clin North Am*, **32**, 195–216.

Thadepalli, H. and Mandal, A.K. 1988. Anatomic basis of head and neck infections. *Infect Dis Clin North Am*, **2**, 21–34.

Thadepalli, H. and Mandal, A. 1984. Head and neck. In: Thadepalli, H. (ed.), *Anatomical basis of infectious disease*. Springfield, IL: Charles Thomas, 1–46.

Thomson, R.B. Jr. and Miller, M. 2003. Specimen collection, transport, and processing: bacteriology. In: Murray, P.R.M. (ed.), *Manual of clinical microbiology*. Vol. 1. Washington, DC: ASM Press, 318.

Toohill, R.J. 1993. Rhinoscleroma in perspective. *Mayo Clin Proc*, **68**, 1219.

Wald, E.R. 1996. Diagnosis and management of sinusitis in children. *Adv Pediatr Infect Dis*, **12**, 1–20.

Ward, M.A. 2002. Emergency department management of acute respiratory infections. *Semin Respir Infect*, **17**, 65–71.

Warner, J.A. and Finlay, W.E. 1985. Fulminating epiglottitis in adults. Report of three cases and review of the literature. *Anaesthesia*, **40**, 348–52.

Westergren, V., Lundblad, L. and Forsum, U. 1998a. Ventilator-associated sinusitis: antroscopic findings and bacteriology when excluding contaminants. *Acta Otolaryngol*, **118**, 574–80.

Westergren, V., Lundblad, L., et al. 1998b. Ventilator-associated sinusitis: a review. *Clin Infect Dis*, **27**, 851–64.

Wexler, H.M., Molitoris, E. and Molitoris, D. 1997. Susceptibility testing of anaerobes: old problems, new options? *Clin Infect Dis*, **25**, Suppl 2, S275–8.

Williams, A., Nagy, M., et al. 1998. Lemierre syndrome: a complication of acute pharyngitis. *Int J Pediatr Otorhinolaryngol*, **45**, 51–7.

Wine, N., Lim, Y. and Fierer, J. 1997. Pasteurella multocida epiglottitis. *Arch Otolaryngol Head Neck Surg*, **123**, 759–61.

Zhanel, G.G., Palatnick, L., et al. 2003. Antimicrobial resistance in respiratory tract Streptococcus pneumoniae isolates: results of the Canadian Respiratory Organism Susceptibility Study, 1997 to 2002. *Antimicrob Agents Chemother*, **47**, 1867–74.

24

Bacterial infections of the lower respiratory tract

ROBERT C. READ

INTRODUCTION

Despite the very large surface area of the lung and airways, few microorganisms colonize this tissue and cause disease. Thousands of microorganisms are inhaled with the air and aspirated with pharyngeal secretions during sleep (Kikuchi et al. 1994); for each of these, there is at least one of four possible outcomes. First, the microorganisms may be rapidly cleared from the lung. Second, they may lodge in the upper respiratory tract and asymptomatically colonize the site. Third, they may persist in regions that are normally sterile, such as below the larynx, as is observed in chronic obstructive pulmonary disease. Finally, they may penetrate epithelium and initiate parenchymal disease, as in pneumonia, or they may invade the bloodstream to cause systemic infection. Microbial load and host defenses are the subject of a constant balancing act. Successful microbes have adapted to avoid rapid clearance, gain nutrients, and survive in the respiratory tract sufficiently to grow and disseminate to other hosts. The aim of host defenses is to maintain the sterility of the gas-exchange areas. Disease results when local defenses are overcome and inflammatory cells are recruited to eradicate the infection.

PULMONARY DEFENSE MECHANISMS

Pulmonary defense mechanisms are of two types (Reynolds 1994): resident defenses, such as airway architecture and mucociliary clearance, which are constantly operative, and recruited defenses, which augment the resident defenses when the host detects danger and result in inflammation with resulting disease.

Resident defenses

AIRWAY ARCHITECTURE

Movements of the glottis and epiglottis during swallowing and the cough reflex are important barriers to bacterial invasion of the airway, especially by organisms aspirated from the alimentary tract.

During nose breathing, particles larger than 5 μm are efficiently removed by impaction on the walls of the nose and nasopharynx as a result of airflow rendered turbulent by the nasal turbinates and the configuration of the nasopharynx and oropharynx. Sedimentation is the main mechanism of deposition of particles, such as bacteria in the size range 5–0.6 μm, and takes place in parts of the lung in which airflow is slow, i.e. between the fifth bronchial division and the terminal lung units. Invasion of the lung by sedimented particles is prevented by the combination of an efficient epithelial barrier, mucociliary clearance, and phagocytic mechanisms.

THE EPITHELIAL BARRIER

The respiratory tract, from the nose to the respiratory bronchiole, is lined by ciliated, pseudostratified, columnar cells interspersed with occasional mucus-secreting cells (see Figure 24.1). The remainder is lined by nonciliated epithelium including (1) a small zone of stratified epithelium below the pharyngeal fornix and (2) nonkeratinizing squamous epithelium in the oropharynx, the anterior surface of the epiglottis and the upper half of its posterior surface, the upper half of the aryepiglottic folds, and the vocal cords. Tight junctions between epithelial cells provide an efficient barrier to bacterial penetration. This epithelium is continually replenished; if damaged, fully differentiated ciliated cells are replaced within 2–6 weeks, though basal cells continue to protect the basement membrane.

MUCOCILIARY CLEARANCE

Throughout the airway mucus-secreting cells secrete a blanket of mucus that is propelled along by ciliary action towards the larynx where it is coughed up or swallowed. Inhaled bacteria adhere to mucus and are conveyed along this mucociliary escalator (Figure 24.2). Efficient mucociliary transport of inhaled foreign bodies depends

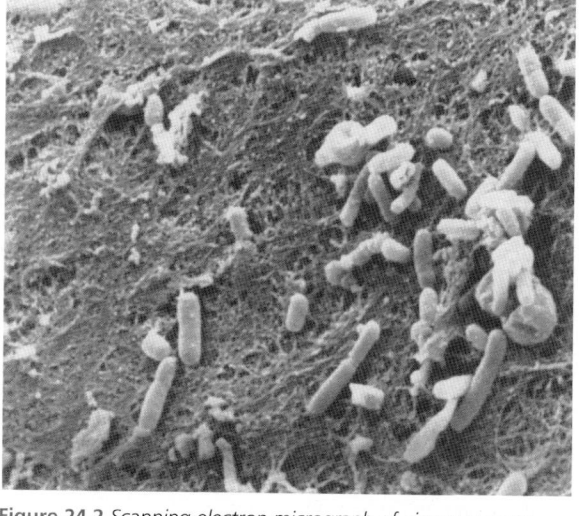

Figure 24.2 *Scanning electron micrograph of airway mucosa showing* Haemophilus influenzae *attached to mucus. (A. Brain and R.C. Read.)*

on coordinated ciliary beating, the depth and constituents of the periciliary fluid that lies beneath the blanket of mucus, and the rheological qualities of the mucus. Viscid mucus, such as that in cystic fibrosis, is difficult for the cilia to clear. Cilia beat at 12–17 Hz, but the frequency is slightly lower in more peripheral airways. In normal airways, clearance times of inhaled aerosols are 30 min from the lobar bronchi, 4–6 h from 1–5-mm airways, and 1 day to several months or more, depending on the substance inhaled, from airways distal to the terminal bronchioles. Clearly, disruption of mucociliary clearance leads to stasis of infected mucus and potentially to disease.

SOLUBLE FACTORS IN AIRWAY SECRETIONS

Mucus contains many factors that enhance the clearance of microorganisms from the airway; most of these are plasma proteins and include α_1-antitrypsin, a low-molecular-weight chymotrypsin inhibitor. Neutral protease and elastase are secreted by polymorphonuclear leukocytes and alveolar macrophages. α_1-Antitrypsin deficiency is associated with chronic pulmonary disease, probably because of failure to inhibit the neutrophil products during recurrent inflammatory responses to lung infection.

Lysozyme is secreted by neutrophils and found in most mucosal secretions, where it is directly active against some bacteria. Lactoferrin is secreted by mucosal epithelial cells and neutrophils. Like transferrin, it has a high affinity for iron and inhibits bacterial replication within the airway by restricting iron availability.

Surfactant is secreted by pneumocytes within alveoli and has the vital function of reducing surface tension in gas exchange areas and keeping these patent. Surfactant proteins A and D are significant defense molecules present in alveolar fluid that agglutinate and opsonize a

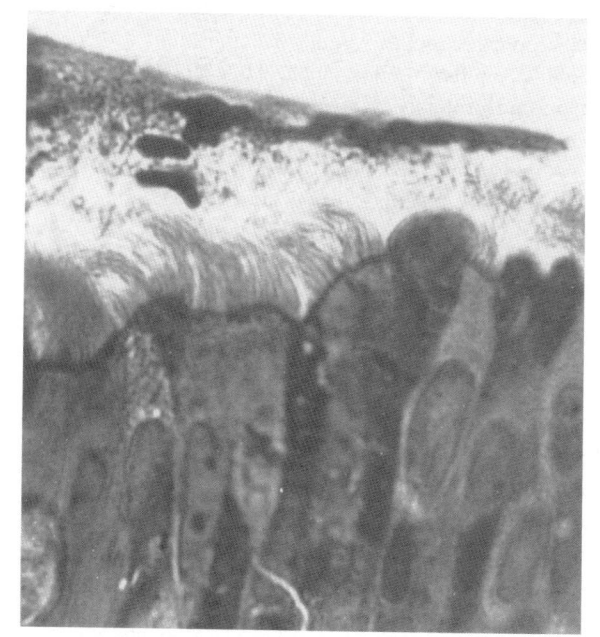

Figure 24.1 *Micrograph of typical human pseudostratified columnar epithelium within the respiratory tract, showing ciliated cells interspersed with mucus-secreting (goblet) cells. (Courtesy of A. Rutman.)*

wide range of bacteria and fungi (Lawson and Reid 2000).

Complement

Many of the components of the complement system are present at very low levels in normal bronchial secretions. They are probably derived directly from serum, but alveolar macrophages also secrete a number of the components. In terms of defense against bacteria, complement has two broad functions. Deposition of C3b on bacteria ultimately results in their recognition by macrophage CR1, CR3, and CR4 receptors, and then phagocytosis. Additionally, there is probably sufficient C5 to initiate bacterial lysis by the terminal membrane attack complex. Complement concentrations rise dramatically during periods of lung damage, e.g. during serious infection of the lung, when airways are relatively leaky for serum proteins.

Immunoglobulins

Immunoglobulins IgA, IgG, IgM, and IgE, including specific antibody, can be recovered from respiratory tract secretions (Lipscomb et al. 1995). IgM is found in lung washings in only very low concentrations.

Secretory IgA (sIgA) is synthesized by lymphoid cells beneath the basement membrane, predominantly in the upper and middle portions of the respiratory tract. Whereas IgA is the predominant immunoglobulin in the upper airways, lower-airway secretions more resemble serum in that the IgG to IgA ratio is much higher. The major function of sIgA is complement-independent neutralization of respiratory viruses. It also opsonizes organisms for phagocytosis by macrophages, but it does not contribute to polymorph opsonization; nor does it activate complement. Secretory IgA interacts with lysozyme and complement to augment their bactericidal activity, mediates bacterial adherence to mucus, and inhibits bacterial adherence to epithelial surfaces. Once challenged by a given agent, the host produces sIgA in response to homologous challenge. However, the period of immune protection is relatively short in contrast to serum IgG and booster responses are variable after later challenge. Certain bacteria, notably *Streptococcus pneumoniae*, *Haemophilus influenzae*, and *Neisseria meningitidis*, secrete IgA proteases capable of inactivating mucosal IgA and other activities.

IgG gains access to the airways by transudation from the circulation, but it can be produced locally in the lung. It efficiently agglutinates bacteria, mediates bacterial opsonization for macrophages and neutrophils, neutralizes bacterial exotoxins, and activates complement. After a primary challenge, specific antibody is released into blood from lung-associated lymph nodes and transudes into the airway. Subsequently, memory B cells traffic to the lung and produce specific IgG in response to further specific challenge (Bice et al. 1991).

ALVEOLAR MACROPHAGES

Alveolar macrophages are responsible for noninflammatory clearance of microorganisms, including those that sediment in the most distal airways beyond mucociliary clearance (Gordon and Read 2002). They comprise about 80–90 percent of the cell population of fluid obtained by bronchoalveolar lavage. Four populations of pulmonary macrophages can be distinguished:

1 alveolar macrophages
2 interstitial macrophages
3 dendritic macrophages and
4 intravascular macrophages.

Pulmonary macrophages are derived from peripheral blood monocytes; they differentiate after chemotaxis to pulmonary tissue and can replicate in the lung. Alveolar macrophages reside within the airspace, whereas interstitial macrophages are located in the lung connective tissues. Although alveolar and interstitial macrophages both undertake Fc receptor-dependent phagocytosis, they differ in their other functions. Alveolar macrophages are capable of Fc receptor-independent phagocytosis and the production of cytokines, such as tumor necrosis factor α (TNF-α) and interferons α and β, which are released in order to recruit an inflammatory response. Production of oxygen radicals is greatest in alveolar macrophages. Interstitial macrophages are adapted for antigen presentation (Lohmann-Mathes et al. 1994), whereas dendritic cells are predominantly nonphagocytic and specialized for antigen presentation. Macrophages in the vascular compartment are located on the capillary endothelium. They are highly phagocytic and probably remove foreign material that enters the lung via the bloodstream.

Alveolar macrophages possess three classes of surface receptor that facilitate phagocytosis of bacteria. The Fc receptor, of which there are four subclasses with differing affinities, are responsible for the recognition of antibody-coated bacteria. They recognize the Fc region of IgG and IgA on antibody-coated bacteria. The second are complement receptors CR1, CR3, and CR4, which recognize bacteria coated with C3b. The third are lectin-binding receptors such as the mannose-6-phosphate receptor. After phagocytosis, bacteria are enclosed in phagosomes, which ultimately acquire the characteristics of lysosomes (Gordon et al. 2000a). Bacteria are killed by a combination of products of the oxidative burst and bactericidal enzymes and other proteins.

The bactericidal activity of alveolar macrophages depends crucially on the inoculum size. Up to about 10^5 bacteria are eliminated by alveolar macrophage phagocytosis alone, whereas the killing of an inoculum of 10^6 requires a modest influx of neutrophils into the alveoli.

Inocula greater than 10^8 organisms, however, necessitate activation of local T and B lymphocytes in addition to phagocytosis by macrophages and neutrophils. Apart from the number of microorganisms, the species of the bacterium can affect the efficiency of killing by macrophages. *Staphylococcus aureus* can be killed adequately by macrophage phagocytosis alone, but *Pseudomonas aeruginosa* and *Klebsiella pneumoniae* additionally require neutrophils to aid intrapulmonary killing. Other bacteria, such as mycobacterial species and *Legionella pneumophila*, are readily phagocytosed by macrophages but cannot be killed. Killing occurs once the macrophage has been activated by T lymphocyte-derived macrophage-activating cytokines, e.g. interleukin-1.

Alveolar macrophages, in common with other cellular components of innate immune defense, recognize pathogens via a family of Toll-like receptors and cofactors and coreceptors such as MD-2 and CD14. Following activation of these receptors by microbial ligands (e.g. lipoteichoic acid of *S. pneumoniae* or lipopolysaccharide of *H. influenzae*), there is transcriptional activation of a number of important host response genes, including those that encode cytokines and chemokines directly relevant to eradication of bacteria (Read and Wyllie 2001).

Recruited defenses

If the combined effect of mucociliary clearance, complement, and macrophage phagocytosis fails to clear an inoculum of bacteria, or if the infecting organism is particularly virulent, for example some strains of pneumococcus, polymorphonuclear neutrophils and lymphocytes are recruited to augment the host response. Neutrophils migrate from the vascular compartment into the alveolus by chemotaxis. The stimulus for chemotaxis originates within the alveolus and is due to direct generation of chemotactic factors by microorganisms entering the alveoli, and the release of chemotactic factors from alveolar macrophages after phagocytosis. Leukotriene B4 is an important chemotactic factor that also alters pulmonary capillary permeability. This, together with the secretion of TNF-α by macrophages, promotes the accumulation of neutrophils and fluid and other humoral substances in alveoli. Neutrophils kill ingested bacteria very much faster than macrophages with a combination of oxidative metabolites. This process, together with injury due to proteolytic enzymes, e.g. neutrophil elastase, contributes to consolidation of the lung.

Bronchoalveolar lavage fluid from the normal resting human lung contains predominantly macrophages but also lymphocytes (20 percent) including CD4 (helper), CD8 (suppressor), and a few B lymphocytes (Reynolds 1994). There is a complex interplay between lymphocytes, macrophages, and neutrophils in the management and termination of inflammation. Lymphocytes can

regulate the activation of macrophages and subsequently coordinate the inflammatory response in a given infection. The resolution of inflammation is accompanied by neutrophil apoptosis, with subsequent consumption by macrophages. This process is conducted by inflammatory macrophages, which employ novel phagocytic recognition mechanisms that fail to provoke a macrophage proinflammatory response (Haslett 1999). Additionally, macrophages may themselves undergo apoptosis following successful phagocytic killing of some organisms, e.g. *S. pneumoniae* (Dockrell et al. 2001).

A host infected with organisms capable of intracellular survival, such as *Mycobacterium tuberculosis*, *Pneumocystis carinii*, and *Legionella pneumophila*, requires cell-mediated immunity to eradicate the infections (Sim 1998).

At the time of primary challenge, antigen is transported to lung-associated lymph nodes either contained within alveolar and dendritic macrophages or recruited neutrophils, or free in the afferent lymphatic fluid (Lipscomb et al. 1995). In the presence of lung inflammation, this process is accelerated. Within lymph nodes, antigen is reprocessed by antigen-presenting cells. Specific T and B cell clones expand and differentiate, and effector T cells and B lymphoblasts migrate out (maximally by 10–14 days), the latter to provide specific antibody locally in lung tissue. Subsequent responses to the same infecting organism are produced by immune memory B cells lodged within the lung parenchyma.

PATHOGENESIS OF BACTERIAL RESPIRATORY INFECTIONS

Disease results when the equilibrium between organisms that enter the respiratory tract and resident host defenses is disturbed. This may be due to a defect in host defenses, an overwhelming microbial challenge, or a combination of the two. A normal host may not be able to deal with the microbial load, due to the large number of organisms or because the organisms possess potent virulence determinants.

Host defense defects

HOST FACTORS THAT PERMIT INCREASED COLONIZATION OF THE UPPER RESPIRATORY TRACT

Colonization of the upper respiratory tract is probably the first step in the pathogenesis of most bacterial pulmonary infections and adherence of bacteria to epithelial cells is a key event in colonization (Woods 1994). Individuals with poor nutrition, particularly vitamin A deficiency, are more likely to have upper respiratory tract colonization. Epithelial cells from patients with severe or chronic illness appear to be particularly receptive to bacteria, especially gram-negative organisms. This may be due to a reduction in local IgA, local fibronectin production, or a change in the glycoprotein and glycosphingolipid composition of epithelial cell surfaces.

FAILURE OF EPITHELIAL BARRIERS AND REFLEXES

Bypassing the upper respiratory tract with an endotracheal tube or by tracheostomy vastly increases the host dependence on resident host defenses of the lower respiratory tract. The distal airways in intubated patients rapidly become colonized by gram-negative bacteria including *P. aeruginosa* (Johanson et al. 1980). Failure of the swallowing reflex in patients with neurological defects permits aspiration of the commensal oropharyngeal flora and alimentary tract contents.

FAILURE OF MUCOCILIARY CLEARANCE

Ciliary dyskinesia may be due to intrinsic structural abnormalities, including defective dynein arms, defective radial spokes, microtubular transposition, and random orientation of the central tubule (Rutland and de Longh 1990). Such abnormalities can result in complications that range from relatively mild recurrent chest infections and sinusitis to severe clinical problems such as Kartagener's syndrome (sinusitis, bronchiectasis, and malrotation of viscera – *situs inversus*). Respiratory viral infection, particularly influenza, can also result in marked depletion of ciliated epithelial cells, reduced mucociliary clearance, and secondary bacterial infection, for example with *S. aureus* (Shanley 1995).

Cystic fibrosis is due to abnormalities of chloride-ion transport by glandular and ciliated epithelial cells in various tissues, resulting in viscid external secretions. In the airways, this results in inadequate mucociliary clearance, colonization by pathogens with subsequent chronic inflammation, and eventual bronchiectasis. These processes lead to virtually continuous inflammation and infection of the respiratory tract of cystic fibrosis patients from infancy onwards.

IMMUNOGLOBULIN DEFICIENCY

Patients with agammaglobulinemia and dysgammaglobulinemia produce decreased amounts of some or all of the serum immunoglobulins, notably IgG. They are susceptible to respiratory infections, including those due to encapsulated organisms, such as *S. pneumoniae* and *H. influenzae*, and others such as *Mycoplasma pneumoniae* (Buckley 1992). Chronic respiratory infection due to common variable immunodeficiency can lead to bronchiectasis. Selective IgA deficiency, particularly when associated with deficiency of IgG2, is also associated with recurrent respiratory tract infections. The IgG2 includes antibodies against capsulate organisms such as pneumococcus and *Haemophilus* spp., and also to lipoteichoic

acid of *Streptococcus* spp. IgG4 deficiency is also associated with recurrent respiratory tract infection.

REDUCED PHAGOCYTIC ACTIVITY

The most common cause of failure of the phagocytic response results from iatrogenic immunosuppression. Neutropenia results in a reduced inflammatory response in the lung and an increased incidence of gram-negative bacillary and fungal infection. Macrophages can also be depleted as a result of immunosuppression, which results in pneumonia due to intracellular microorganisms, including *Legionella* spp.

DISTURBED CELLULAR IMMUNITY

Disorders of cell-mediated immunity can result from hematological malignancies, e.g. leukemia, human immunodeficiency virus (HIV) infection, or from drug therapy, e.g. steroids. The consequence for the lung is that an inflammatory response cannot be adequately recruited or controlled, and macrophages cannot be activated to kill intracellular pathogens. The reduction in CD4 (T helper cells) in acquired immune deficiency syndrome (AIDS) is associated with a variety of respiratory infections by intracellular pathogens, notably *P. carinii* and *M. tuberculosis*.

ENVIRONMENTAL INFLUENCES ON THE HOST

Tobacco smoking is the major cause of airway morbidity that leads to recurrent respiratory tract infection. Smoke reduces ciliary clearance by direct toxicity, oxidizes α_1-antitrypsin, increases mucus production, and is a chemoattractant for neutrophils and macrophages in the peripheral airways. These lead to squamous metaplasia of ciliated epithelium, goblet cell hyperplasia, and intraepithelial inflammation. The consequences are frequent cough, sputum production and intercurrent infection, particularly with *H. influenzae* and *Moraxella catarrhalis* (Floreani et al. 1994).

The effects of air pollution are far more subtle than those of smoking. Upper respiratory tract infections (coryza and nasal discharge) and lower respiratory tract infection (pneumonia and purulent sputum production) are observed more frequently in polluted urban areas (Lunn et al. 1967). High ambient pollutant levels correlate with upper respiratory infections (Ponka 1990). Pollutants, including sulfur dioxide, nitrogen dioxide, acid aerosols, and particulates, reduce mucociliary clearance, damage epithelial cells, and impair phagocytosis in vitro (Schlesinger 1990).

Bacterial virulence mechanisms

Respiratory pathogens produce a variety of virulence factors that promote infection by mechanisms that include:

- increased adherence to mucosal epithelial cells
- avoidance of mucociliary clearance

- increased nutrient acquisition
- avoidance of complement deposition and phagocytosis and
- perturbation of neutrophil and macrophage apoptosis.

Some pathogens produce factors that are directly toxic to the host, such as streptococcal hyaluronidase and *Pseudomonas* elastase, which allow expansion of the ecological niche and damage host connective tissue.

STREPTOCOCCUS PNEUMONIAE

S. pneumoniae produces a protein adhesin that binds to the *N*-acetylgalactose–galactose component of respiratory tract epithelial cell-membrane glycolipids (Krivan et al. 1988). It also produces IgA protease, which inactivates respiratory mucosal IgA. Pneumolysin, a 52.8-kDa protein, is a thiol-activated cytotoxin that shares amino acid homology with bacterial thiol-activated toxins and is released on autolysis. In experimental animals it can induce inflammation in the lungs independently of intact pneumococci (Feldman et al. 1991). Virulent pneumococci have an antiphagocytic capsule, which permits evasion of recruited neutrophils, and some capsular serotypes, e.g. serotype 3, are more virulent than others. This may be due to the relative composition of capsules, particularly their choline content (Bruyn et al. 1992). Pneumococci produce a range of other products important in pathogenesis, including pneumococcal surface protein A, autolysin, choline binding protein A (CbpA), pneumococcal surface antigen A, and neuraminidase (Jedrzejas 2001). Toxic products of *S. pneumoniae* also include hydrogen peroxide (Duane et al. 1993). The cell-wall lipoteichoic acid of *S. pneumoniae* is potently pro-inflammatory in the lung by activation of the alternative complement pathway and, like lipopolysaccharide (LPS), it elicits production of interleukin-1 (IL-1) and TNF-α (Tuomanen et al. 1995). The quantity of strain cell-wall lipoteichoic acid and capsular polysaccharide has been correlated with virulence (Weiser et al. 1996). *S. pneumoniae* expresses a C3-binding protein that elicits interleukin-8 secretion by pulmonary epithelial cells (Madsen et al. 2000), and CbpA also induces chemokine release from these cells (Murdoch et al. 2002). The complete genome sequence of *S. pneumoniae* has been published and contains 2236 predicted coding regions. Approximately 5 percent of the genome is composed of insertion sequences and there are approximately 70 predicted surface proteins (Tettelin et al. 2001).

HAEMOPHILUS INFLUENZAE

H. influenzae possesses fimbriae that are structurally and serologically related to the P and mannose-sensitive fimbriae of *Escherichia coli* and promote adherence to respiratory epithelium and mucin. *H. influenzae* also appears to bind specifically to GalNAcβ1-4Gal sequences of human lung glycolipids. The organism may also secrete a soluble toxin that reduces ciliary beating

and so promotes avoidance of mucociliary clearance (Wilson 1988). Lipo-oligosaccharide of *H. influenzae* is toxic to human respiratory tract epithelium. This organism also secretes IgA protease, which contributes to virulence (Vitovski et al. 2002).

BORDETELLA PERTUSSIS

B. pertussis produces several adhesins including a filamentous hemagglutinin (220 kDa), fimbriae, and an array of toxins, including pertussis toxin (105 kDa), which binds to respiratory tract ciliated cells and macrophages and is clearly important in airway colonization. Release of filamentous hemagglutinin from the surface of individual bacteria within adherent microcolonies facilitates dispersal of the organism through the respiratory tract (Coutte et al. 2003). Pertussis toxin also causes epithelial cell cytotoxicity by increasing host cell cyclic adenosine monophosphate (cAMP) levels (Masure 1992), an action broadly similar to cholera toxin. Tracheal cytotoxin is a peptidoglycan fragment toxic for ciliated cells that directly induces inflammation in animal models (Wilson et al. 1991). *B. pertussis* can survive in human phagocytes by inhibiting phagosome–lysosome fusion, but it does not inhibit the respiratory burst (Steed et al. 1992).

PSEUDOMONAS AERUGINOSA

P. aeruginosa produces fimbrial and nonfimbrial adhesins that mediate binding to epithelial cell gangliosides after its sialic acid residues have been removed by neuraminidase. It also produces a number of toxins, including exotoxin A, which damages respiratory epithelial tissue and inhibits phagocytosis (Coburn 1992), and a number of elastases that act together to damage epithelial cells and blood vessel walls (Galloway 1991). *P. aeruginosa* directly activates NF-kappaB, which in turn activates MUC2 mucin transcription in epithelial cells (Li et al. 1998). A membrane glycolipid of *P. aeruginosa*, rhamnolipid, can reduce ciliary beating and mucociliary clearance (Read et al. 1992), and secreted pigments, including pyocyanin and 1-hydroxyphenazine, also reduce ciliary beating (Wilson 1988). Alginate, which forms a viscous gel around bacteria and gives them a mucoid appearance, functions as an adhesin and prevents phagocytosis. Alginate also appears to contribute to biofilm formation on epithelial surfaces (Nivens et al. 2001), a phenomenon that appears to be regulated by quorum-sensing genes (Davies et al. 1998).

LEGIONELLA PNEUMOPHILA

L. pneumophila appears to survive macrophage phagocytosis and kill human macrophages. It has a 24-kDa outer-membrane macrophage invasion protein that allows invasion of macrophages in the absence of opsonization, but this does not appear to improve intracellular survival once the organism is endocytosed. The bacterium enters macrophages by coiling phagocytosis (a form of phagocytosis in which the cell envelops the organism by coiling its pseudopodia around it) and then resides within a unique phagosome, which does not become highly acidic. Therein, legionella express stationary-phase proteins that encourage intracellular survival and transmission to new cells (Swanson and Hammer 2000). Several bacterial enzymes, including acid phosphatase, phospholipase C, protein kinases, and superoxide dismutase, potentially enhance intracellular survival. *L. pneumophila* proteases cause lung damage and secretes a metalloproteinase similar to *P. aeruginosa* elastase (Dowling et al., 1992). Adaption and replication of *L. pneumophila* in vivo are enhanced by phase variation of expression of LPS (Luneberg et al., 1998). A 22-kb DNA locus contains at least 16 genes involved in macrophage killing (Segal et al. 1998).

RESPIRATORY TRACT COMMENSALS

The anterior nares are colonized by staphylococci, micrococci, and miscellaneous *Corynebacterium* spp., but the paranasal sinuses are normally sterile.

In the mouth, anaerobes colonize periodontal areas and a variety of other organisms including 'viridans' streptococci (e.g. *Streptococcus mitis*, *Streptococcus sanguis*), *Haemophilus* spp., *Neisseria* spp., and diphtheroids are present in saliva. The practical consequence is that sputum can be heavily contaminated and make laboratory interpretation very difficult.

The flora of the oronasopharynx is much more complex; it may be colonized by *Streptococcus* spp., *Neisseria* spp., coliform bacteria, *Bacteroides* spp., fusobacteria, actinomycetes, and yeasts. From time to time, however, individuals may be colonized by one or more pathogens, such as β-hemolytic streptococci, *S. pneumoniae*, *H. influenzae*, *N. meningitidis*, and *M. catarrhalis*. Periods of asymptomatic colonization may be very short or may last for several months, and although the colonization is asymptomatic, it is generally accepted that stable residence at this site is a prerequisite for invasive disease by such organisms as *H. influenzae*, *S. pneumoniae*, and *N. meningitidis*. Point-prevalence studies in communities indicate the relative prevalence of these pathogens in the nasopharynx of colonized individuals (Table 24.1).

A number of exogenous factors may disturb the ecology of the mouth and nasopharynx. Antibiotics can suppress oral bacteria and allow yeasts to proliferate and thrush to occur. Smoking and passive smoking increase the prevalence of colonization with pathogens, including *N. meningitidis* (Caugant et al. 1994).

The upper respiratory tract is colonized by an abundant resident microbial flora that varies as a result of changing endogenous and exogenous conditions. Below the larynx, the lower respiratory tract is sterile in normal individuals because inhaled organisms are rapidly removed by clearance mechanisms.

INFECTIONS OF THE UPPER RESPIRATORY TRACT

Infections of the oronasopharynx

SINUSITIS

The paranasal sinuses are normally sterile. Bacteria recovered by sinus puncture before treatment of acute community-acquired sinusitis include *S. pneumoniae*, *H. influenzae*, anaerobes, *Streptococcus* spp., *M. catarrhalis*, and *S. aureus*. In chronic sinusitis, bacteriological cultures are more exotic and polymicrobial than in acute sinusitis and may include anaerobes and *Pseudomonas* spp. (van Cauwenberge et al. 1993). Conservative management of acute sinusitis involves use of appropriate antibiotics such as ampicillin, amoxycillin, trimethoprim–sulfamethoxazole, cefaclor, cefuroxime axetil, amoxycillin–clavulanate, and loracarbef (Gwaltney et al. 1992). An alternative strategy is immediate sinus puncture, which may permit more rapid resolution and provide material for the determination of antimicrobial sensitivities. Chronic disease should prompt allergic and immunological investigation and prolonged therapy, including antibiotics for anaerobic infection. Ultimately, chronic disease may require surgery to establish free drainage.

OTITIS MEDIA

Upper respiratory tract infections, often viral, are most commonly responsible for the events that lead to otitis media. Such infections impair the function of the eustachian tube, creating a negative pressure and transudation into the middle ear. Bacterial contamination results by reflux from the oropharynx, and this leads to further accumulation of fluid and pus.

Table 24.1 *Point prevalence of nasopharyngeal carriage of pathogens*

	Prevalence (%)
Nontypable *Haemophilus influenzae*	25–48 (Turk 1984)
Haemophilus influenzae type b (prevaccine era)	2–4 (Moxon 1986)
Streptococcus pneumoniae	60–100 (Austrian 1986)
Neisseria meningitidis	5–10 (Broome 1986)

The four stages of otitis media are:

1 Myringitis – inflammation of the tympanic membrane
2 Acute suppurative otitis media, which denotes a middle ear infection behind the reddened tympanic membrane
3 Secretory (serous) otitis media, which refers to chronic middle ear effusion behind an intact tympanic membrane with acute signs and symptoms and
4 Chronic suppurative otitis media – a chronic discharge from the middle ear through a perforation of the tympanic membrane. Chronic disease can lead to loss of aeration of the middle ear cleft, loss of hearing, and speech retardation during infancy.

Otitis media primarily affects children but can lead to lifelong sequelae. In acute otitis media the major organisms are *H. influenzae*, *S. pneumoniae*, and *M. catarrhalis*. These organisms can also be recovered in chronic otitis media but in addition, *S. aureus*, *E. coli*, *K. pneumoniae*, *P. aeruginosa*, and anaerobic bacteria can all be present (Brook and Van de Heyning 1994). Empirical therapy may be given in most cases, including the aminopenicillins and second-generation cephalosporins and aminopenicillins in combination with β-lactamase inhibitors where β-lactamase producing organisms are present. About half the patients recover within 10 days with appropriate therapy, whereas most of the remainder have a residual exudate that resolves over 3 months. About 5 percent develop secretory otitis media. A meta-analysis has suggested that 5 days of short-acting antibiotic use is effective treatment for uncomplicated otitis media in children (Kozyrskyj et al. 1998).

Chronic otitis media requires surgical correction, drainage, and treatment for anaerobic bacteria with agents such as aminopenicillins plus β-lactamase inhibitors. In such cases, the microbiological flora should always be determined to rule out the presence of anaerobes or *Pseudomonas* spp.

RHINOSCLEROMA

This is a chronic granulomatous disease of the nose found in eastern Europe and parts of the tropics. It leads to nasal obstruction and marked secretion with postnasal drip. It is due to *K. pneumoniae* subsp. *rhinoscleromatis* (see Chapter 57, Citrobacter, Klebsiella, Enterobacter, Serratia and other enterobacteriaceae).

Infections of the perioral and peripharyngeal spaces

Infections of oral soft tissues are uncommon and caused by endogenous oral bacteria including oral anaerobes. Such infections can lead to pain and swelling of the sublingual and submandibular space (Vincent's angina) and also collections within the lateral pharyngeal space. An extreme form of this is the rare Lemierre's syndrome, which consists of oropharyngeal infection and anaerobic bacteremia and septic thrombophlebitis of the jugular vein, with embolization to the lungs and other areas. *Fusobacterium necrophorum* is usually the cause and prolonged treatment with parenteral benzylpenicillin and metronidazole is necessary.

Peritonsillar abscess can occur in patients with recurrent tonsillitis or inadequately treated pharyngotonsillitis. Bacteria that are recovered include *Streptococcus pyogenes*, *Streptococcus milleri*, *H. influenzae*, and viridans streptococci. Anaerobes, including *F. necrophorum*, *Prevotella* spp., and *Peptostreptococcus* spp., can also be isolated from peritonsillar abscesses. The clinical features are high fever, intense pain in the throat, voice change, and occasionally airway obstruction (Jousimies-Somer et al. 1993). Treatment with needle aspiration and oral penicillin is sufficient for most patients with peritonsillar abscesses (Maharaj et al. 1991). When incipient airway obstruction is present, this is best treated by incision and drainage.

Laryngotracheal infections

Most laryngitis is due to virus infection, but *M. catarrhalis* and *H. influenzae* may be recovered from such patients and some symptoms of laryngitis can be improved with oral erythromycin (Schalen et al. 1992). Tracheitis is usually due to viruses, including respiratory syncytial virus. Bacterial pathogens associated with tracheitis include *H. influenzae*, *S. aureus*, and *Streptococcus* spp., including *S. pneumoniae*.

PNEUMONIA

This is a general term for disease that includes consolidation of the lung parenchyma. There is acute inflammation in the gas-exchanging areas of the lung (pneumonitis), with a polymorphonuclear leukocyte exudate in and around the alveoli and terminal and respiratory bronchioles.

The most useful clinical classification of pneumonia divides cases into community-acquired and nosocomial. Other considerations include whether the pneumonia results from aspiration or follows from an acute viral infection, is acquired in specific geographical settings, or occurs in the context of immunosuppression, including AIDS.

Community-acquired pneumonia

The true incidence of community-acquired pneumonia is not easy to determine, but approximately one per 1 000 of the population is admitted to hospital with pneumonia annually in the UK (Woodhead et al. 1987). In the USA it is estimated that there are 2–3 million episodes of pneumonia per annum (an attack rate of 12 per 1 000 persons per year). Each year the bacterial pneumonias account for over half a million hospital admissions of patients aged 15 years or older in the USA (Pennington 1994a) and 45 000 deaths (Pinner et al. 1996; Marston et al. 1997). Pneumonia is substantially more common in the winter and affects more males than females (ratio 2–3:1). It is more common amongst older persons; the annual incidence of pneumonia that requires hospitalization of those older than 75 years is 11.6 cases per 1 000; for those aged 35–44 years it is 0.54 cases per 1 000 persons (Marrie 1994).

The symptoms include cough, sputum, dyspnea, and pleuritic pain. Classically the sputum is rusty-colored in pneumococcal pneumonia but is usually mucoid, scanty, or absent, especially at an early stage, and in mycoplasma and legionella infections it may be absent. Extrapulmonary symptoms occasionally predominate and patients may present with severe headache, confusion, and myalgia. Until recently, cases in which these symptoms have predominated have been described as 'atypical pneumonia' and therefore pathognomonic of mycoplasma, coxiella, and legionella infection. It is now accepted that atypical presentations may be seen with pneumococcal pneumonia and the distinction is regarded as unhelpful.

Mortality due to community-acquired pneumonia was markedly decreased by the introduction of antibiotics. In the antibiotic era, mortality from ambulatory pneumonia has been of the order of 1 percent (Woodhead et al. 1987). Mortality in hospitalized patients is about 13–15 percent (Austrian and Gold 1964; Macfarlane et al., 1982). In patients who need intensive therapy, mortality ranges from 22 to 54 percent (Woodhead et al. 1985; Torres et al., 1991). The outcome for patients admitted to hospital with pneumonia can be greatly modified by prompt antibiotic therapy. In a large multicenter study, none of the patients who died of pneumococcal, staphylococcal, or *M. pneumoniae* pneumonia had received appropriate antibiotics before hospital admission (British Thoracic Society 1987). A number of clinical and laboratory features of community-acquired pneumonia are associated with increased mortality (Table 24.2). Three independent studies have shown a

21-fold increase in the risk of death or the need for intensive therapy when two or more of the following factors are present:

- respiratory rate greater than 30/min
- diastolic blood pressure below 60 mmHg
- serum urea above 7 mmol/l
  (British Thoracic Society 1987; Farr et al. 1991; Karalus et al. 1991).

Fine and colleagues have designed a scoring system that enables physicians to identify patients with a low risk of death within 30 days (Fine et al. 1997) and to stratify patients by severity of illness.

Community-treated pneumonia

Accurate epidemiological data for community-treated pneumonia are difficult to obtain. Results of three large studies in which pathogens were identified in 55–67 percent of hospitalized patients are summarized in Table 24.3. The predominant organisms are *S. pneumoniae*, which is by far the most common bacterial cause, *M. pneumoniae*, and viruses. These series record a preponderance of bacterial causes in hospitalized patients, but it has been estimated that at least half of nonhospitalized cases are due to viruses or mycoplasma (Pennington 1994a).

STREPTOCOCCUS PNEUMONIAE
Epidemiology

S. pneumoniae is frequently isolated from the nasopharynx of asymptomatic individuals (point prevalence 60–100 percent). The American serotyping system assigns serotype numbers in sequence. There are now in excess of 90 serotypes (see Chapter 33, Streptococci and lactobacilli). The lower-numbered serotypes are most frequently implicated in pneumococcal diseases. In Papua New Guinea, which has a high rate of pneumococcal disease, serotypes 2, 3, 5, 8, and 14 infrequently colonize the nasopharynx but are likely to cause invasive disease. Other serotypes are commonly carried by the population and cause disease in children, but not as frequently in adults (serotypes 6, 19F, and 23F). The remaining serotypes are carried relatively rarely in the nasopharynx and do not cause disease (Montgomery et al. 1990).

A high incidence of pneumococcal bacteremia occurs in infants under 2 years of age. The incidence is low in teenage children and young adults, but increases again in males and females in their seventies. The incidence of pneumococcal disease in immunocompetent young adults is five per 100 000; the incidence of pneumococcal bacteremia is 1 000 per 100 000 patients with AIDS (Redd et al. 1990). Certain occupational groups, including military recruits, have higher annual rates of infection but the highest recorded incidence is in South African gold miners (Farr and Mandell 1994). Cigarette

Table 24.2 *Clinical and laboratory features of community-acquired pneumonia associated with increased risk of death*

Clinical features	Laboratory features
Respiratory rate >30[a]	Urea >7 mmol/l[a]
Diastolic BP <60[a] mmHg	Serum albumin <35 g/l
Age >60 years	Hypoxemia Po_2 <8 kpa
Underlying disease	Leukopenia WBC <4 $\times$ 10^9/l
Confusion	Leukocytosis WBC >20 $\times$ 10^9/l
Atrial fibrillation	Bacteremia
Multilobar involvement	

a) Twenty-one-fold increase in risk of death or requirement of admission to an intensive therapy unit if two or three of these factors are present. Data from Farr et al. (1991); Karalus et al. (1991).

Table 24.3 *Microbial causes of community-acquired pneumonia*

No. of patients	Location	Year of study	Microbial causes (%)	Ref.
236	Nottingham, UK	1984–85	*S. pneumoniae* (36), Viral (13), *H. influenzae* (10), GNB (1.5), *S. aureus* (1), Unknown (45)	Woodhead et al. 1987
359	Pittsburgh, USA	1986–87	*S. pneumoniae* (15), *H. influenzae* (11), *Legionella* spp. (7), *Chlamydia* spp. (6), GNB (6), Unknown (33)	Fang et al. 1990
331	Leiden, the Netherlands	1991–93	*S. pneumoniae* (27), Viral (8), *H. influenzae* (8), *Mycoplasma* spp. (6), *Chlamydia* spp. (3), Unknown (45)	Bohte et al., 1995

After Brown and Lerner (1998).
GNB, gram-negative bacilli.

smoking is a strong independent risk factor for invasive pneumococcal disease in immunocompetent non-elderly adults (Nuorti et al. 2000).

Pathogenesis

The first step in the pathogenesis of pneumococcal disease is believed to be nasopharyngeal colonization. The average duration of carriage is about 6 weeks in adults, but some individuals may carry a strain for more than a year. Most infections occur during the first week of carriage, while sensitization and production of IgG proceed (Musher et al. 1990). People who have antibody against pneumococcal surface protein A appear to be protected against experimental nasopharyngeal carriage (McCool et al. 2002). In normal individuals, with intact mucociliary clearance, there is a much lower risk of pneumococcal disease once colonization is established in comparison with those who have pre-existing pulmonary disease or immunosuppression. However, viral respiratory infections, particularly due to influenza virus, predispose even more individuals to pneumonia. Malnourished individuals, e.g. alcoholics and residents of nursing homes, and those with chronic liver or kidney disease, cancer, or diabetes mellitus, are also vulnerable to invasive disease. Splenectomy renders individuals vulnerable to catastrophic pneumococcal bacteremia.

Highly specific enzyme-linked immunosorbent assay (ELISA) techniques have shown that most healthy young adults lack antibody to the majority of serotype-specific capsular polysaccharides (Musher 1992). Invasive pneumococcal infection probably occurs only in individuals who are deficient of serotype-specific antibody to the capsular polysaccharide of their colonizing serotype.

Clinical features

Pneumococcal pneumonia usually develops over several days with cough and sputum production, dyspnea, pleuritic chest pain, weakness, malaise, and often myalgia. Occasionally a dramatic rigor may be the first symptom; this hyperacute presentation is more common in otherwise healthy young adults. In older patients the presentation is typically more insidious with minimal cough and absence of fever; confusion and hypothermia are often presenting features in this group. Physical examination may reveal evidence of consolidation and chest radiograph commonly reveals an area of infiltration of less than a whole segment (Figure 24.3), and several areas may be involved (Ort et al. 1983).

Laboratory findings

Most patients have a polymorphonuclear leukocytosis; leukopenia ($<4 \times 10^9$/l) is a poor prognostic indicator. There may be hyperbilirubinemia and elevated liver enzymes.

Identification by Gram stain or culture of pneumococci in appropriate specimens remains the principal microbiological investigation. Examination of sputum requires specimens with at least 15–25 white blood cells

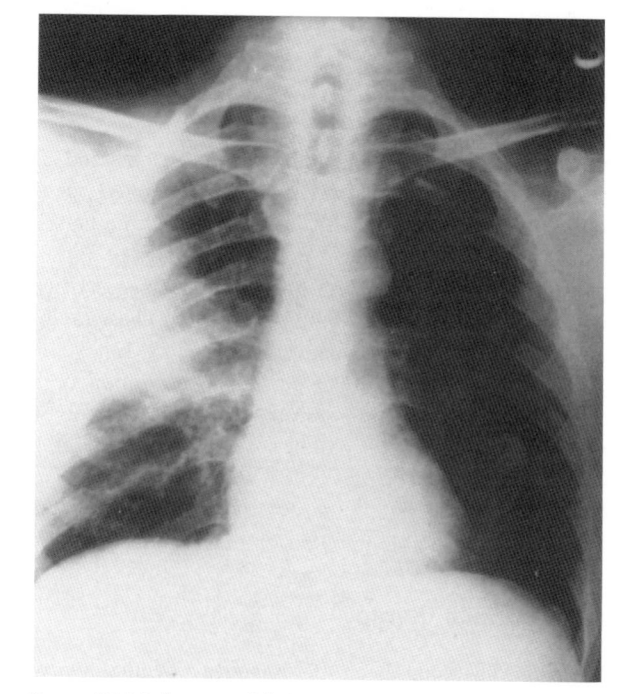

Figure 24.3 *Lobar consolidation in a male, age 25 years, with pneumococcal pneumonia*

and fewer than ten epithelial cells in a standard microscopic field. If lanceolate-shaped gram-positive cocci (i.e. pneumococci) are seen with more than ten cocci per oil immersion field (×100), the likelihood of culture of pneumococci in the sputum is about 60 percent and the specificity about 85 percent (Rein et al. 1978). The presence of pneumococci can be confirmed by the capsule swelling (quellung) reaction with polyvalent pneumococcal antiserum, but in practice this is rarely used. The success of culture depends on sputum quality – if saliva is present, viridans streptococci may outnumber the pneumococci. Transtracheal aspiration and transcutaneous aspiration are more sensitive but are of value only as research tools.

Pneumococcal polysaccharides can be identified rapidly in sputum, blood, or urine, but these investigations have poor sensitivity and specificity. Countercurrent immunoelectrophoresis (CIE) is positive in approximately 50 percent of patients with non-bacteremic pneumococcal pneumonia when performed on urine and serum, and in 50 percent of bacteremic patients when performed only on serum. Sputum CIE is positive in approximately 75 percent of pneumonia patients with positive pneumococcal culture and remains positive after antibiotic therapy. Latex agglutination, coagglutination and ELISA are as reliable as CIE. A pneumolysin-based ploymerase chain reaction (PCR) assay of whole blood, buffy coat, or plasma has comparable sensitivity and specificity to other antigen tests and remains positive after antimicrobial therapy (Michelow et al. 2002). Clonal origin of pneumococci can be identified by multilocus sequence typing, which is a useful tool in molecular epidemiology (Meats et al. 2003).

Complications

Empyema is the most common complication of pneumococcal pneumonia. A reactive effusion can occur but is trivial; empyema is potentially more serious and is presumably due to bacteria reaching the pleural space via the lymphatics. Clinically, it is signaled by the persistence of fever and leukocytosis after 4–5 days of appropriate antibiotic therapy. Empyema is also suggested by large amounts of pleural fluid seen on the chest radiograph and can be confirmed by ultrasound. Empyema should be drained by repeated needle aspiration or a chest tube; thoracotomy is rarely necessary. Fibrinolytic agents to irrigate the plural space may be helpful. Complications of pneumococcal bacteremia such as endocarditis, pericarditis, peritonitis, and brain abscess are very uncommon in immunocompetent individuals.

Antimicrobial therapy

Penicillin has long been the mainstay of therapy of pneumococcal disease but *S. pneumoniae* has become more resistant to penicillin over recent years (Markie-wicz and Tomasz 1989). Strains with intermediate resistance are now prevalent throughout the world and highly resistant isolates are common in certain geographical areas, e.g. Southern Europe, South Africa, and the Far East, where up to 70 percent of isolates are resistant (Felmingham and Gruneberg 2000). For this reason an oxacillin disk is used routinely to screen for penicillin-resistant pneumococci. Such strains are often resistant to other antibiotics, such as chloramphenicol, erythromycin, and clindamycin (Whitney et al. 2000). In practice, pneumococcal pneumonia almost always responds to penicillin or cephalosporins in high doses (Friedland and McCracken, 1994; Pallares et al., 1995), in contrast to meningitis. Organisms highly resistant to penicillin are susceptible to vancomycin. New fluoroquinolones such as moxifloxacin are active against penicillin-resistant pneumococci, but resistance to quinolone-class antibiotics is rising in some parts of the world (Chen et al. 1999).

Prevention

The 23-valent pneumococcal vaccine contains antigens from 23 serotypes of pneumococci: 1, 2, 3, 4, 5, 6B, 7F, 8, 9N, 9V, 10A, 11A, 12F, 14, 15B, 17F, 18C, 19A, 19F, 20, 22F, 23F, and 33F. These types account for the majority of cases of bacteremic pneumococcal infection and the vaccine produces antibody responses in approximately 85 percent of those vaccinated. It is effective in splenectomized individuals though lower antibody levels are achieved (Giebink et al. 1980) but not in children under 2 years of age. Its use is, therefore, recommended in adults with splenic dysfunction or splenectomy and other chronic diseases, including diabetes mellitus, chronic cardiopulmonary disease, renal failure, nephrotic syndrome, liver disease, and AIDS, but its use in the immunocompromised is controversial (Hirschmann and Lipsky 1994). A large case–control field study revealed a 56 percent efficacy in preventing bacteremic pneumococcal infection; efficacy in immunocompetent patients (61 percent) was higher than in immunosuppressed patients (21 percent) (Shapiro et al. 1991). In contrast, a large randomized control trial among nonimmunocompromised middle-aged and elderly people in Sweden did not demonstrate any effect on prevention of pneumonia overall or pneumococcal pneumonia in middle-aged and elderly individuals (Ortqvist et al. 1998). The vaccines are given by intramuscular or subcutaneous injection. Repeat vaccination is recommended after 3–6 years in certain individuals, e.g. splenectomized adults or children under 10 with nephrotic syndrome or hyposplenism, but local intolerance at the injection site can be a problem. Pneumococcal disease is a particular problem in the African population with AIDS (Gordon et al. 2000b), and in this population the 23-valent vaccine was shown to be ineffective in preventing pneumococcal events (French et al. 2000).

Conjugated polysaccharide vaccines containing seven, nine, and eleven separate polysaccharides possess T-cell-dependent properties and exhibit enhanced induction of B-cell memory and high levels of immunoglobulin G even in very young children (Eskola 2000). Serogroups in the 7-valent formulation (4, 6, 9, 14, 18, 19, and 23) cause 70–88 percent of invasive disease in young children in the USA and Canada, Oceania, Africa, and Europe, but less than 65 percent in Latin America and Asia (Hausdorff et al. 2000). The 7-valent pneumococcal vaccine prevents invasive pneumococcal events in children (Black et al. 2000) and is now routinely used in children from the age of 2 months in the USA (MMWR 2000).

MYCOPLASMA PNEUMONIAE

M. pneumoniae causes tracheobronchopneumonia with severe constitutional symptoms. The clinical syndrome of primary atypical pneumonia was first described by Reimann (1938). *M. pneumoniae* causes a wide spectrum of clinical disease involving the respiratory tract, skin, central nervous system, and blood-forming elements.

Epidemiology

M. pneumoniae spreads with ease in institutions such as schools, universities, and military installations. The highest incidence is during the first two decades of life; along with *C. pneumoniae*, it is a highly significant cause of community-acquired lower respiratory tract infection in children (Principi et al. 2001). The infection can occur during any season but there is generally a 4-year cycle in which 1 year of high prevalence is followed by 2 or 3 years of low incidence (Figure 24.4). Explosive community outbreaks of pneumonia due to *M. pneumoniae* can occur (MMWR 2001).

Pathogenesis

The organism is carried in the nasopharynx and infection is transmitted by droplets. *M. pneumoniae* binds to epithelial cells via neuraminic acid receptors and subsequently causes damage to cells locally by hydrogen peroxide production. Once the bronchial mucosa is penetrated, a local inflammatory response is initiated. Many of the disease features are a result of a complex immune response to the organism (Tuazon and Murray 1994). Repeated infections result in the generation of sensitized T lymphocytes and autoantibodies, which initiate the pneumonitis. Immune complexes may also contribute to injury of the lung, brain, and synovia and elsewhere. The host-mediated facets of the disease explain why some patients develop only upper respiratory tract symptoms, whereas others may develop pneumonitis with or without severe extrapulmonary manifestations.

Clinical features

The clinical features are at first similar to those of influenza, except that the onset of *M. pneumoniae* infection tends to be gradual over several days or a week (Clyde 1993). Constitutional symptoms initially include headache, malaise, myalgia, and pharyngitis, but rigors are uncommon. The cough, which is initially dry and dominates the disease, is the last symptom to clear. Despite antibiotics clinical relapse can occur a week after the initial response. The physical signs are often minimal and include sinus tenderness, pharyngitis without exudate, and occasionally myringitis. The chest is initially clear, but wheezing and rhonchi may develop in the second week. Although the pneumonia is usually self-limited, occasionally adult respiratory distress syndrome associated with *M. pneumoniae* may develop. Other

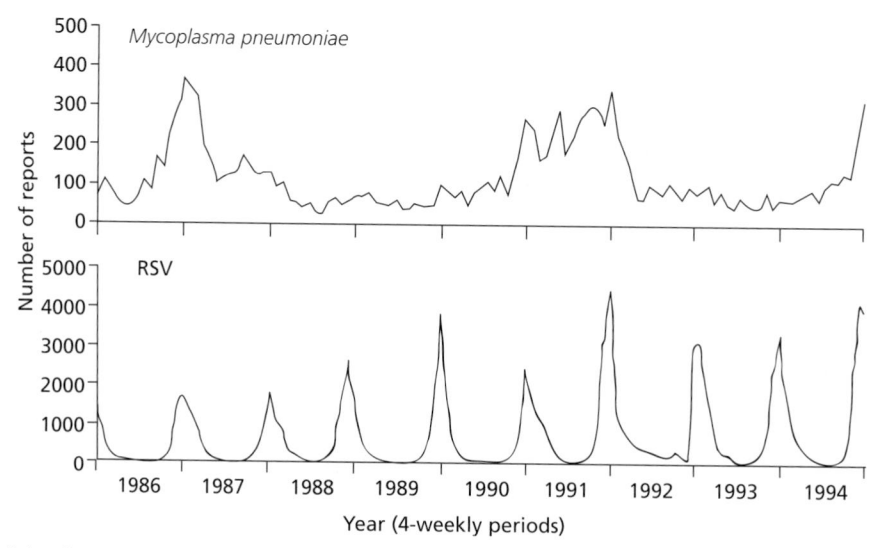

Figure 24.4 *Periodicity of* Mycoplasma pneumoniae *infection, in comparison with respiratory syncytial virus (RSV). (Reproduced with permission from CDSC 1995.)*

pulmonary complications are unusual (Tuazon and Murray 1994).

Extrapulmonary manifestations

Pneumonia with rash is most commonly due to *M. pneumoniae*. Stevens–Johnson syndrome in association with *M. pneumoniae* is potentially life-threatening and is probably due to the cell-mediated immune response to *M. pneumoniae*, but since the rash often develops after drug therapy, antibiotics, particularly erythromycin, may be responsible (Cherry 1993).

Hemolytic anemia characteristically occurs 2–3 weeks after the onset of illness, is usually self-limiting, and coincides with high cold agglutinin titers. It is complement-dependent and involves IgM antibody to the I antigen of red blood cells (Cherry 1993). Although hemolytic anemia is uncommon, approximately three-quarters of patients with *M. pneumoniae* pneumonia develop high titers of cold agglutinins.

Central nervous system symptoms are present in up to 70 percent of patients with *M. pneumoniae* infection treated in hospital (Koskiniemi 1993). Encephalitis is the most frequent CNS manifestation, but meningitis, myelitis, and polyradiculitis and many other symptoms, including coma, ataxia, and stroke due to cerebral infarction, have been reported. *M. pneumoniae* has not been isolated from brain tissue but has been recovered from the cerebrospinal fluid (CSF). In aseptic meningitis the CSF usually reveals a modest pleocytosis.

Laboratory findings

The white blood cell count is usually greater than 10×10^9 per liter and neutrophilia, lymphocytosis, and monocytosis can all occur, but leukopenia is rare. Routine sputum culture is negative, but *M. pneumoniae* can be grown in broth media in long-term cultures.

A useful bedside test detects the presence of cold agglutinins. A few drops of blood are placed into a citrate tube and placed in a refrigerator for a few minutes. If cold agglutinins are present, then agglutination of red cells will be seen. The IgM cold agglutinins first appear 7–9 days after infection with a peak after 4–6 weeks.

The mainstay of diagnosis is serological. A four-fold increase in complement fixation titer is diagnostic. Unfortunately serology is frequently negative during the acute phase; even laboratory assay for cold hemagglutinins, the first acute-phase marker, is positive in only about 50 percent of acutely ill patients. Some centers offer PCR for *M. pneumoniae*, a technique that is improving in value (Loen et al. 2002). The combination of PCR detection within nasopharyngeal samples together with ELISA-based IgM detection greatly enhances diagnostic yield in children (Ferwerda et al. 2001).

Antimicrobial therapy

The mainstays of treatment for *M. pneumoniae* infection are erythromycin and tetracycline. Recovery with erythromycin is quite slow. Newer macrolides such as azithromycin and clarithromycin are better tolerated and have equivalent activity in vitro. Newer fluoroquinolones are also active against *M. pneumoniae* (Bébéar et al. 1993).

Prevention

Killed and inactivated vaccines do not evoke effective serum or local respiratory tract immune responses (Tuazon and Murray 1994). Future vaccines may be based on recently discovered peptide adhesins.

LEGIONELLA PNEUMOPHILA

This organism is named after an epidemic of pneumonia in legionnaires who attended the American Legion Convention in Philadelphia in the late summer of 1976 (Fraser et al. 1977). Among the 4500 persons attending, 192 cases were observed and 29 proved fatal. The syndrome was characterized by nonproductive cough, pulse temperature dissociation, abnormalities of liver function, diarrhea, hyponatremia, hypophosphatemia, myalgia, confusion, and multiple rigors. The condition became known as legionnaires' disease. A previously unknown pleomorphic gram-negative bacillus was isolated from four of the original patients and named *L. pneumophila*. Some 3 years later an acute, short-lived, febrile illness consisting of myalgia but not pneumonia that affected staff and visitors at the County Health Department at Pontiac, Michigan, was shown to be due to *L. pneumophila*. It had a shorter incubation period and there were no deaths (Broome and Fraser 1979). Subsequently devised serological techniques implicated *Legionella* spp. in pneumonia not only in outbreaks associated with the water facilities of large buildings such as hospitals and hotels but also with sporadic community-acquired infections. Legionella are aerobic, fastidious, gram-negative bacilli, of which 34 species have been characterized (see Chapter 68, Legionella). Rarely, infections may be caused by *L. micdadei*, *L. longbeachae*, *L. dumoffii*, *L. bozemanii*, and other species (Edelstein and Meyer 1994).

Pathogenesis

Infection results from inhalation of *Legionella*-contaminated aerosols, e.g., disseminated by cooling towers, humidifiers, respiratory therapy equipment, showerheads, and cooling sprays. *L. pneumophila* is phagocytosed by macrophages but they fail to kill the organism. In macrophages infected in vitro, legionella can multiply between 100 and 1 000 times in 48–72 h (Yamamoto et al. 1994). Defense against the infection, therefore, crucially depends on cell-mediated immunity, as illustrated by the susceptibility to legionellosis of patients in whom such immunity is deficient; those with neutropenia do not appear to be susceptible.

Epidemiology

Legionella spp. survive in water for prolonged periods, as intracellular parasites of free-living protozoa (Fields et al. 2002). Although human infection occurs sporadically in the community (Macfarlane et al. 1982), outbreaks are dramatic and associated with water facilities. Of the 129 cases of legionnaires' disease reported in England and Wales in 1993, 51 percent were associated with travel, mainly to Spain, but also to Greece, Turkey, the USA, and within the UK; six were nosocomial and the remainder were sporadic community cases (Joseph et al. 1994a). Outbreaks have occurred in the vicinity of cooling towers (Broome and Fraser 1979), a power station (Morton et al. 1986), and, notably, Broadcasting House, London, in 1988. The role of air-conditioning equipment was demonstrated in Memphis in 1978 (Dondero et al. 1980) and by a large hospital outbreak in Staffordshire, UK (Editorial 1986). Showerheads and hot-water taps spread aerosols contaminated with *Legionella* (Bolling et al. 1985). This has caused outbreaks in hospitals and is the probable cause of disease in holidaymakers returning from Mediterranean resorts. Contaminated showerheads and tap water led to a prolonged and disastrous outbreak in the VA Wadsworth Medical Center, Los Angeles, with a high rate of nosocomial *Legionella* pneumonia, particularly in immunosuppressed patients. The outbreak was terminated by changing showerheads and by hyperchlorination of potable water (Shands et al., 1985). Between 1980 and 1992, 218 cases were reported in the UK, 68 of whom died. Approximately half were sporadic, but three hospital outbreaks contributed to the majority of the remaining cases. Hospital domestic water systems were the source of infection in 19 outbreaks. The cooling tower was thought to be the source in one hospital and probably contributed to two other outbreaks in which the domestic water system was suspected to be the main source (Joseph et al. 1994b). It is considered that contamination of water only presents a risk at a threshold concentration of 10^4–10^5 cfu per liter, though low concentrations can be underestimated because of insensitive culture techniques (Leoni and Legnani 2001).

Clinical features

The majority of patients with *L. pneumophila* infection have a pneumonia virtually indistinguishable from pneumococcal pneumonia (Edelstein 1993; Stout and Yu 1997). Clinical features range from asymptomatic infection and nonpneumonic myalgia (Pontiac fever) to legionnaires' disease with severe pneumonia and extrapulmonary manifestations. After an incubation period of 2–10 days, there is a gradual onset of malaise, lethargy, fever, headache, myalgia, and weakness, with evolution of a dry nonproductive cough after 1–2 days. There may be purulent sputum, which can be bloody (Edelstein 1993). In about 25 percent of patients there may be disorientation, agitation, confusion, hallucinations, ataxia, and seizure. The systemic manifestations are probably due to several exotoxins and, possibly, endotoxins (Edelstein and Meyer 1994). Extrapulmonary manifestations include pancreatitis, peritonitis, cellulitis, myositis, and focal abscesses, and prosthetic valve endocarditis is well recognized. Pneumonia due to species other than *L. pneumophila* is clinically indistinguishable. *Legionella micdadei* infection occurs in immunosuppressed patients and causes nodular pulmonary infiltrates (Pittsburgh pneumonia) (Myerowitz et al. 1979).

Diagnosis

Nonspecific laboratory abnormalities include elevated white blood count (usually with neutrophilia), hyponatremia, elevation of alkaline phosphatase and liver transaminases, and proteinuria. The gold standard for the diagnosis of legionnaires' disease is culture of *Legionella* spp. from sputum samples on charcoal yeast extract medium (BCYE), but this is only possible in patients with productive cough. Antigen can be detected in sputum, respiratory secretions, bronchoalveolar lavage, or lung biopsy using immunofluorescence with genus-specific monoclonal antibodies, even after several days of chemotherapy (Fields et al. 2002). Techniques to detect nucleic acid, for example PCR, are of relatively low sensitivity and are practiced in very few laboratories. Urinary antigen detection has an overall sensitivity of 80 percent (Helbig et al. 2003) but detects only serogroup 1 *L. pneumophila*. Serological diagnosis can be conducted with a variety of techniques, including complement fixation and ELISA, and relies on a fourfold rise between acute and convalescent titres (which can take up to 2 months) or a single titer of $\geqslant$1:256. In culture confirmed cases, the sensitivity of serology is 70–80 percent.

Antimicrobial chemotherapy

In the legionnaires' disease outbreak in Philadelphia in 1976, the case fatality rate was highest for patients treated with cephalosporins and lowest with erythromycin. Erythromycin should be given at a dose of 2–4 g per day, initially intravenously in severely ill patients, for at least 3 weeks. Rifampicin and a fluoroquinalone should be added for patients who are critically ill. Patients usually respond to erythromycin therapy within 2–3 days.

Antimicrobial sensitivity testing is not applicable to *L. pneumophila*. Most of the β-lactam agents are active in vitro against *L. pneumophila*. However, animal models have shown that these agents are ineffective because they do not enter macrophages sufficiently well. Therefore, the potential efficacy of new agents must be based upon their success in intracellular infection and experimental animal models. Azithromycin, clarithromycin, fluoroquinolone antibiotics, trimethoprim–sulfamethoxazole, tetracyclines including doxycycline, and rifampicin are effective in animal models (Edelstein 1993). In patients with *Legionella* who are severely ill or immunocompromised, a fluoroquinolone must be included in the chemotherapeutic regimen (Edelstein 1995).

Prevention

Person-to-person transmission of legionnaires' disease has not been observed. The most important preventive measure is to render water sources safe. Thirty to 80 percent of hospital cooling towers and water supplies are to some extent contaminated with *Legionella*, and at total viable concentrations >10^8 cfu/l the hazard of infection is increased (Edelstein and Meyer 1994). Chlorination, pasteurization, and heating can be used to reduce the *Legionella* burden (Muraca and Stout 1988).

The report of a single case of legionnaires' disease requires investigation to determine whether it is part of an outbreak, which involves tracing the patient's activities during the incubation period. If hospital-acquired infection is suspected, then the hospital water maintenance programme must be reviewed and a search made for associated cases, beginning with the immunosuppressed population, who are most vulnerable to infection (Saunders et al. 1994).

COXIELLA BURNETII

C. burnetii (see Chapter 79, Rickettsia and Orienta) is the cause of Q fever, which was first described in Australia in the 1930s in abattoir workers. The organism spreads between farm animals and man by air-borne transmission, by infected milk, feces, and urine, and is probably also transmitted by tick bite. It is highly resistant to drying and in agricultural facilities contaminates dust, which can cause infection by inhalation. After an incubation period of 2–4 weeks there is abrupt onset of high fever, malaise, headache, and neck stiffness. There is dry cough and pleuritic chest pain and on examination there is usually high fever, hepatosplenomegaly, but few chest signs. Some patients develop a transient maculopapular rash, but the exanthem typical of other rickettsial infections is absent. Most illness resolves within 2 weeks but fever can persist for up to 3 months. Q fever can be complicated by extrapulmonary lesions, including endocarditis. There may be hepatitis, which is usually mild but can be severe or even fulminant. *C. burnetii* may cause blood culture-negative endocarditis many months after the primary infection.

Complement fixation and indirect fluorescent antibody tests reach a peak between 1 and 3 months after original infection. Phase 2 antibodies are raised in acute Q fever; phase 1 antibodies are prominent in patients with chronic Q fever, e.g. endocarditis.

Tetracycline or doxycycline is used to treat acutely ill patients and those with persistent symptoms. Long-term treatment with tetracycline is desirable for patients with Q fever endocarditis, but valve replacement may become necessary.

STAPHYLOCOCCUS AUREUS

Although *S. aureus* pneumonia is usually nosocomial, it accounts for a small proportion of community-acquired pneumonia, particularly during influenza epidemics and in certain patient groups, e.g. diabetics. Although 15–30 percent of adults are nasal carriers of *S. aureus*, resulting pneumonia is rare and presumably requires an underlying host defect, e.g. diabetes. Staphylococcal pneumonia is also seen in patients with right-sided endocarditis (e.g. in intravenous drug abusers) and in patients with septic thrombophlebitis or an infected vascular prosthesis (Farr et al. 1989). Patients with staphylococcal pneumonia more commonly have pleural effusions and cavitation on chest radiograph and 25 percent develop abscesses (Waldvogel 1990). Staphylococcal pneumonia secondary to septic embolization from right-sided endocarditis causes multiple small round cavitary lesions.

The appearance of clusters of staphylococci on Gram stain of sputum is diagnostic of staphylococci pneumonia. If this diagnosis is suspected after clinical assessment of a patient, sputum microscopy can rapidly confirm the diagnosis and a penicillinase-resistant penicillin can be added to the empiric antibiotic regimen. Vancomycin should be used only when pneumonia due to methicillin-resistant *S. aureus* is suspected.

HAEMOPHILUS INFLUENZAE

H. influenzae causes community-acquired pneumonia in patients with chronic bronchitis and those with impaired host defenses, such as smokers, alcoholics, or HIV infection. In the pre-Hib vaccine era, it was also a relatively common cause of pneumonia in children, particularly those aged 4 years and younger. Nontypable *H. influenzae* tends to cause mucosal infections, such as those affecting the middle ear and the airway of bronchitics. Capsulate organisms, particularly of type b, cause a significant proportion of invasive parenchymal disease, i.e. pneumonia (Smith 1994). In adults and children the illness tends to be preceded by coryza, and sudden onset of pleuritic chest pain is the predominant chest symptom (Crowe and Lavitz 1987). Microbiological diagnosis is notoriously difficult because of sputum contamination by commensal *Haemophilus* spp. and blood culture is usually negative. β-Lactamase production by infecting strains may be very frequent in some areas (Felmingham and Gruneberg 2000). If such strains may be involved, second-generation cephalosporins should be used to treat moderately severe pneumonia.

CHLAMYDIA SPECIES

Pneumonia due to chlamydia is common. See Chapter 78, Chlamydia.

GRAM-NEGATIVE BACILLARY PNEUMONIAS

Although gram-negative bacteria are a prominent cause of nosocomial pneumonia, they may also cause disease in the community, particularly in alcoholics, diabetics, residents of nursing homes, and patients with underlying diseases, such as malignancy and cardiac or renal failure. The upper respiratory tract becomes colonized followed by aspiration and subsequent pneumonia. The most common gram-negative causes of community pneumonia are *K. pneumoniae*, *E. coli*, *Serratia marcesens*, *Enterobacter* spp., and *Pseudomonas* spp.

K. pneumoniae classically causes pneumonia in debilitated middle-aged males, with sudden fever and rigors, dyspnea, and large volumes of bloody sputum. Often there is a necrotizing process with multiple abscess formation and dramatic lobar consolidation. The sputum contains large numbers of capsulate gram-negative bacilli and blood cultures are often positive. The infection is destructive, often is severe, and has a high mortality (Feldman et al. 1989).

E. coli is the cause of up to 3 percent of community pneumonias in American series (Eisenstadt and Crane 1994). It may result from bacteremic spread from the genitourinary and gastrointestinal tracts and is often bilateral. Other causes of community-acquired gram-negative pneumonia include *Proteus* spp. (*Morganella* and *Providencia* spp.) and *Acinetobacter baumannii*.

Pseudomonas spp. can, rarely, cause dramatic community-acquired pneumonia. *P. aeruginosa* causes pneumonia in patients admitted from nursing homes or individuals with malignancy. The disease is often necrotizing and there is diffuse bilateral consolidation on chest radiograph with occasional multiple abscesses. *Pseudomonas stutzeri* may cause disease in patients with malignancy (Noble and Overman 1994). For a discussion of pneumonia due to *Burkholderia pseudomallei*, the reader is referred to Chapter 63, Burkholderia spp., and related genera.

STREPTOCOCCUS SPECIES

Both group A and group B streptococci may cause pneumonia, the latter particularly in elderly debilitated patients with diabetes mellitus, stroke, dementia, and malignancies, and is often associated with other organisms, particularly *S. aureus* and *S. pneumoniae*. Enterococci can cause pneumonia in elderly debilitated patients with multiple chronic problems, often with

nasogastric tubes. Treatment is with high-dose penicillin or ampicillin and an aminoglycoside.

BACILLUS ANTHRACIS

Most (95 percent) natural infection by *B. anthracis* results in the cutaneous form of anthrax, but inhalation of spores can result in inhalational anthrax within 60 days of exposure (Friedlander et al. 1993). Inhalation anthrax usually begins with mild fever and malaise, myalgia, and nonproductive cough. Within 2 or 3 days there is rapid evolution of fever, acute dyspnea, cyanosis, and shock. Mediastinal and cervical lymphadenopathy may be sufficiently severe to cause stridor. The chest radiograph reveals mediastinal widening and occasionally parenchymal infiltrates. Microbiological diagnosis of inhalational anthrax is usually made by blood culture of a heavily encapsulated gram-positive bacillus or by PCR detection in tissue samples (Levine et al. 2002). Confirmation can be done by lysis by gamma phage or direct fluorescent antibody staining of cell-wall polysaccharide antigen. Treatment of cases is with ciprofloxacin 400 mg twice daily intravenously (Swartz 2001). Inhalation exposure of individuals without disease can be investigated by nasopharyngeal swabbing and culture or PCR detection. Post-exposure prophylaxis with ciprofloxacin 500 mg or doxycycline 100 mg both orally and twice daily for 60 days is recommended for persons exposed to spores (Inglesby et al. 2002).

FRANCISELLA TULARENSIS

Exposure to as few as one to ten aerosolized organisms may cause tularemia, which may present as pneumonia or in an ulcerglandular or typhoidal form. The incubation period is 1–21 days, with initial symptoms of fever, malaise, and nonproductive cough. The chest radiograph usually reveals pneumonia with or without mediastinal lymphadenopathy. The organism is difficult to culture from blood but serology is positive in 50–70 percent of cases by the second week of illness. Standard therapy is with gentamicin, streptomycin, or doxycycline.

OTHER RARE CAUSES OF BACTERIAL PNEUMONIA

N. meningitidis can cause a typical pneumonia in the absence of the classic or cutaneous rash and without evidence of meningitis or shock. The disease occurs as outbreaks in military and school dormitories (Koppes et al. 1977). The lungs may be involved in up to 20 percent of cases of brucellosis (Lulu et al. 1988) (See Chapter 66, Brucella). *Pasteurella multocida*, which causes cellulitis after dog bites, can cause a necrotizing pneumonia, usually in patients with pre-existing respiratory disease and with history of exposure to domestic animals. *Yersinia pestis* can cause pneumonia as part of a septicemic process (see Chapter 56, Yersinia).

THE APPROACH TO PATIENTS WITH COMMUNITY-ACQUIRED PNEUMONIA

In most patients presenting with symptoms and signs of pneumonia the likely cause is *S. pneumoniae*. Routine laboratory investigation should include sputum examination and culture, blood culture, acute and convalescent serology to detect antibodies to viruses, *Mycoplasma* spp., *Chlamydia* spp., *Legionella* spp., and *Coxiella burnetii*, urinary antigen detection, and examination of pleural fluid, if available (Carroll 2002), Yersina.

ANTIMICROBIAL THERAPY

The British, European, and American thoracic societies, and also the Infectious Disease Society of America, have each issued guidelines for the management of adult community-acquired pneumonia, and these are summarized in Table 24.4. Each of these guidelines is opinion-based, as the evidence available to discriminate between therapies is limited (Read 1999). All modern guidelines stratify patients by demographic and epidemiological features and include suggested therapies tailored to the likely pathogens that are seen in more complex clinical settings. The British Thoracic Society guidelines recommend the use of an aminopenicillin (e.g. amoxycillin) in patients with simple community-treated pneumonia who are ambulatory, with erythromycin recommended as an alternative choice. In view of the fact that the greatest cause of mortality is due to *S. pneumoniae*, aminopenicillins are probably adequate in the treatment of the majority of patients with mild community-acquired pneumonia treated in the UK. A problem in southern and eastern Europe, the USA, the Asia-Pacific region, and South Africa has been the emergence of penicillin-resistant pneumococci. Aminopenicillins have not been widely employed in the USA for the management of community-acquired pneumonia, largely because of the perceived frequency of atypical pathogens as microbial etiologies in this condition, and the difficulty in distinguishing, on clinical grounds alone, those patients who are infected with such pathogens. The emergence of penicillin-resistant pneumococci, especially in countries other than the UK, has led to the inclusion of fluoroquinolones as choices in first-line management, even for ambulatory patients. The treatment guidelines issued by the Infectious Diseases Society of America (IDSA) (see Table 24.4) recommend the use of either doxycycline, a macrolide drug, or a fluoroquinolone for first-line therapy of patients with mild ambulant (outpatient) pneumonia. While such recommendations would probably not be germane in, for example, the UK, in certain parts of the world they are certainly appropriate – the frequency of penicillin resistance in Spain varies from 50 to 70 percent. In such regions, the frequency of macrolide resistance is also high, and clinical failures with this class have been reported.

As can be seen in Table 24.4, the American Thoracic Society (ATS) has produced a sophisticated set of guidelines in which the likelihood of involvement of penicillin-resistant pneumococci or relatively unusual pathogens in certain patient groups is acknowledged. The ATS recommends the use of either macrolides or tetracyclines for patients who are managed as outpatients and have no pulmonary comorbidity. This approach was validated in the early 1990s (Gleason et al. 1997) in a study of 864 outpatients with community-acquired pneumonia (55.4 percent of whom yielded a positive microbiological diagnosis); 547 younger patients without comorbidity were treated; of these younger patients, 62 percent were prescribed therapy recommended by the ATS – predominantly erythromycin. There were no significant differences in medical outcome between these patients and outpatients who had been erroneously prescribed more sophisticated therapy, but their treatment was associated with a three- to four-fold lower treatment cost.

For outpatients who have pulmonary comorbidities or other risk factors for more exotic flora, including penicillin-resistant pneumococci (e.g. older age), the current ATS guidelines recommend the use of more sophisticated treatments, including oral cephalosporins or a β-lactam/β-lactamase inhibitor combination in addition to a macrolide or doxycycline. An alternative choice for this group is a fluoroquinolone with antipneumococcal activity. The ATS specifies available fluoroquinolones with such activity and ranks them in descending order of antipneumococcal potency as moxifloxacin, gatifloxacin, sparfloxacin, then levofloxacin.

This approach is also followed in the recent guidelines from the IDSA, in which a fluoroquinolone such as moxifloxacin, gatifloxacin, or levofloxacin is recommended in elderly patients or those with underlying disease (see Table 24.4).

HOSPITALIZED PATIENTS

Patients who require hospitalization are often in groups at risk of unusual pathogens and/or more severe disease. All guidelines recommend broader-spectrum (often combination) therapy for this group (see Table 24.4), but the ATS guidelines stratify inpatients further into those with or without cardiopulmonary comorbidities or other modifying factors, including nursing-home residency. It is recommended that these inpatients should receive aggressive therapy with an IV β-lactam/macrolide combination or an IV antipneumococcal fluoroquinolone given as monotherapy.

SEVERE COMMUNITY-ACQUIRED PNEUMONIA

All patients with severe community-acquired pneumonia should be managed with adequate ventilatory support and fluid resuscitation, in addition to antibiotics. This group has a high mortality and the pathogens most frequently implicated as the microbial cause include *S. pneumoniae*, *Legionella* spp., *H. influenzae*, enteric

Table 24.4 *Guidelines for first-line empirical management of community-acquired pneumonia*

Society	Mild ambulant pneumonia (outpatient)	Ambulant pneumonia with risk factors (outpatient)	Inpatient pneumonia (medical ward)	Severe pneumonia (intensive care unit)
European Respiratory Society (1998) (ERS Task Force Report 1998)	Aminopenicillin, or new tetracycline or newer FQ, oral streptogramins, or macrolide	Inpatients with COPD: β-lactam/β-lactamase inhibitor	IV second- or third-generation cephalosporin or IV β-lactam/β-lactamase or IV aminopenicillin and IV macrolide	Either a second- or third-generation cephalosporin and a newer FQ, or macrolide ± rifampicin
Infectious Diseases Society of America (2000) (Bartlett et al. 2000)	Doxycycline or a macrolide, or a newer FQ	Suspected drug-resistant *Streptococcus pneumoniae*, or older patients, or patients with an underlying disease: a newer FQ	A macrolide plus either an extended-spectrum cephalosporin or a β-lactam/β-lactamase inhibitor, plus either a newer FQ or a macrolide	Either an extended-spectrum cephalosporin or β-lactam/β-lactamase inhibitor, plus either a newer FQ or a macrolide
American Thoracic Society (2001)	Macrolide (azithromycin or clarithromycin) or doxycycline	With cardiopulmonary disease: β-lactam (oral cefpodoxime, cefuroxime, amoxil, co-amoxiclav) or parenteral ceftriaxone followed by oral cefpodoxime plus macrolide or doxycycline or antipneumococcal FQ monotherapy	Cardiopulmonary disease or nursing home: IV β-lactam (cefotaxime, ceftriaxone, ampicillin/sulbactam, high-dose ampicillin plus IV/PO macrolide or doxycycline or IV antipneumococcal FQ monotherapy; no modifying factors: IV azithromycin or doxycycline plus β-lactam or IV antipneumococcal FQ	No risk of *P. aeruginosa*: IV β-lactam plus IV azithromycin or fluoroquinolone. Risk of *P. aeruginosa*: IV antipseudomonal β-lactam (cefepime, imipenem, pip/tazo) plus IV ciprofloxacin or IV antipseudomonal β-lactam plus IV aminoglycoside plus IV azithromycin or IV FQ
British Thoracic Society (2001)	Amoxycillin or erythromycin	Not specified	Oral amoxycillin plus erythromycin or IV ampicillin plus IV clarithromycin or IV FQ plus IV penicillin	Co-amoxiclav or 2/3 cephalosporin plus clarithromycin ± rifampicin or FQ plus IV penicillin

COPD, chronic obstructive pulmonary disease; FQ, fluoroquinolones (include moxifloxacin, gatifloxacin, levofloxacin); IV, intravenous; PO, orally, 2/3 2nd or 3rd generation.

gram-negative bacilli, *Mycoplasma* spp., and *S. aureus*. There is a paucity of clinical trial data to support individual choices of empirical antimicrobial therapy but, in general, patients should be managed with broad-spectrum therapy: usually an IV cephalosporin with the addition of an IV macrolide. In patients for whom there is clinical or laboratory evidence leading to suspicion of *S. aureus*, IV flucloxacillin (nafcillin) or vancomycin should be added.

Nosocomial pneumonia

Pneumonia is the second most common nosocomial infection after urinary tract infection (Horan et al. 1986), causes considerable mortality and morbidity, and increases hospital stay and its cost. The US Centers for Disease Control have published the following definition of nosocomial pneumonia:

Onset of pneumonia more than 72 h after hospital admission with lung consolidation or an infiltrate on the chest radiograph plus at least one of the following:

- Infected sputum
- Isolation of a pathogen from the blood, transtracheal aspirate, biopsy, or bronchial lavage specimen
- Isolation of a virus in respiratory secretions
- Diagnostic antibody titers or
- Histopathological evidence of pneumonia (Centers for Disease Control 1988).

According to the US National Nosocomial Infection Surveillance (NNIS) system, nosocomial pneumonia occurs at a frequency of 0.6–1.0 episodes per 100 hospitalizations and in 18 percent of postoperative patients (Horan et al. 1986; Craven et al., 1991). Intubated patients may have rates of pneumonia seven- to 21-fold higher than patients without a respiratory therapy device; the incidence of ventilator-associated pneumonia (VAP) varies from ten to 25/1 000 ventilator days (Chastre and Fagon 2002). Infection rates are twice as high in large teaching hospitals compared with smaller institutions. Factors that lead to nosocomial pneumonia include older age, chronic lung disease, depressed consciousness, mechanical ventilation, the use of H2 antagonists, frequent changes of ventilator circuits, and the winter season (Celis et al. 1988). Mortality in patients with nosocomial pneumonia is high (20–50 percent), but only about one-third of fatalities are due directly to pneumonia (Pennington 1994b).

PATHOGENESIS

Most nosocomial pneumonia results from aspiration of pathogens from the upper airways. Patients admitted to intensive care units become colonized with aerobic gram-negative bacilli within 1 week of entering hospital, and of these about 25 percent develop nosocomial pneumonia; there is a correlation between severity of illness and the likelihood of subsequent colonization with gram-negative organisms (Johanson et al. 1969). Enterobacteriaceaea probably reach the upper respiratory tract by fecal–oral transmission. However, *P. aeruginosa* and *S. aureus* probably colonize the patient from environmental sources and contamination by attending staff, fomites, or supporting equipment (Pennington 1994b). Any factors that reduce gastric acidity may lead to increased oropharyngeal colonization: these include advanced age, achlorhydria, malnutrition, and various pharmacological agents.

MICROBIAL ETIOLOGY

The majority of hospital-acquired pneumonias are caused by aerobic gram-negative bacilli; Enterobacteriaceae cause 40 percent of these pneumonias, *S. aureus* 25 percent, and *P. aeruginosa* 15 percent; *S. pneumoniae*, *H. influenzae*, fungi, viruses, and anaerobic bacteria rarely cause nosocomial pneumonia. The results of three large studies of nosocomial pneumonia are shown in Table 24.5. Polymicrobial infections are common. Nosocomial pneumonia that occurs during the first 4 days in hospital is usually due to one of *S. pneumoniae* and

Table 24.5 *Bacteriology from three nosocomial pneumonia studies*

| Culture material | Sputum (%) | TTA (%) | Blood (%) |
Study	Hughes et al. (1983)	Bartlett et al. (1986)	Bryan and Reynolds (1984)
Pseudmonas aeruginosa	17	9	15
Staphylococcus aureus	13	26	27
Enterobacteriaceae[a]	37	48	42
Fungi	6	ND	ND
Anaerobic bacteria	2	35	2
Streptococcus pneumoniae	<3	31	12
Haemophilus influenzae	<3	17	<4
Viruses	<3	ND	ND

ND, not done; TTA, transtracheal aspirate.
Adapted from Dal Nogare (1994).
a) Includes *Klebsiella pneumoniae*, *Enterobacter* spp., *Escherichia coli*, *Serratia*, and *Proteus*.

H. influenzae. Nosocomial pneumonia occurring later than 4 days after admission is more commonly caused by hospital-acquired, aerobic gram-negative bacilli, *S. aureus*, *Legionella* spp., or, rarely, fungi. A number of host risk factors can predispose to infection with particular pathogens (reviewed by Niederman 1994). Staphylococcal pneumonia is associated with diabetes, renal failure, a recent history of influenza, and recent head injury and trauma, whereas *P. aeruginosa* is seen in malnourished patients and those on steroids or who have chronic structural lung disease or are subject to prolonged mechanical ventilation or tracheostomy. *L. pneumophila* may cause nosocomial pneumonia in hospitals with contaminated water supplies, particularly in patients immunosuppressed by treatment with steroids. Anaerobes are seen particularly in patients with gross aspiration of acid gastric contents and recent thoraco-abdominal surgery. *Acinetobacter* spp. are a significant cause of nosocomial pneumonia on surgical intensive care units and are a particular problem because of multiple antibiotic resistance (Fagon et al. 1989). Viruses may contribute to nosocomial pneumonia, particularly influenza A, parainfluenza, and adenovirus (Dal Nogare 1994).

Methods for the diagnosis of nosocomial pneumonia in severely ill patients include transtracheal aspiration, bronchoscopic sampling with or without protective brush, and transcutaneous aspiration of lung material, but in many studies these have not proved superior to clinical evaluation (Pennington 1994b). In patients who are conscious, sputum production may yield gram-positive (*S. aureus*) or gram-negative bacteria. In intubated patients, aspirated respiratory secretions and blood cultures may provide a bacteriological diagnosis without the need for bronchoalveolar lavage.

TREATMENT AND PREVENTION

Appropriate therapy should include antibiotics active against *S. aureus* and gram-negative bacteria, including *P. aeruginosa*; common regimens include a third-generation cephalosporin or fluoroquinolone plus an aminoglycoside. In patients infected with *Acinetobacter* spp., imipenem and tetracycline are among the most active drugs (Vila et al. 1993). A schema for empiric treatment based on severity of disease has been suggested by the ATS (American Thoracic Society 1996) (see Table 24.6).

Prevention of nosocomial pneumonia depends on prevention of colonization and aspiration and spread of pathogens between patients and staff. Hand washing by staff members should be encouraged. If swallowing is impaired, oral intake should be curtailed and unnecessary nasogastric and endotracheal tubes should be removed. Elevation of the patient's head reduces risk of aspiration (Dal Nogare 1994). Gastric acidity should be maintained by avoidance of H2 antagonists in seriously ill patients.

Aspiration pneumonia

Most aspiration pneumonia is polymicrobial and usually includes anaerobes (Finegold 1994). The latter thrive in lungs injured by chemical pneumonitis. Where there is periodontal disease and gingivitis, anaerobic colonization of the mouth is increased. In hospitalized patients and in nursing homes there is increased oropharyngeal colonization with gram-negative bacilli. A full list of potential organisms that may be involved in aspiration pneumonitis is shown in Table 24.7.

Pneumonia in patients with a predisposition to aspiration is insidious and mainly affects dependent zones of the lung. The most common site is the posterior segment of the right upper lobe, but the apical segments of the lower lobes can be affected. Lung abscess or empyema with high swinging fever may occur, often with foul-smelling sputum and hemoptysis. Microbiological diagnosis is confounded by the normal anaerobic flora of the mouth and the Gram stain of the sputum is difficult to interpret. Gram stain of empyema fluid or transtracheal aspirate may reveal the characteristic appearance of *Prevotella* spp., *Porphyromonas* spp., and *Fusobacterium* spp. Bacteremia is uncommon in aspiration pneumonia.

Treatment of early uncomplicated aspiration or pneumonia in previously normal individuals is relatively straightforward because the majority of oral anaerobes in this situation are sensitive to benzylpenicillin. In patients with underlying disease or who are hospitalized the frequency of infections with nonpenicillin-sensitive anaerobes is increased, including the *Bacteroides fragilis* group and gram-negative organisms including *Pseudomonas* spp. The possibility of infection with *S. aureus* should also be covered by administration of clindamycin or an aminoglycoside. Alternatively, combination therapy with a broad-spectrum penicillin, such as ticarcillin or piperacillin, plus an aminoglycoside can be given. Newer antibiotics such as imipenem, or a combination of a broad-spectrum antibiotic with a β-lactamase inhibitor, are also effective.

Complications of pneumonia

PLEURAL EFFUSION

Pleural effusions occur very commonly in pneumococcal pneumonia. Usually the effusion is sterile and is absorbed spontaneously within two weeks. Occasionally the effusion is large and requires aspiration. Empyema should be suspected if an effusion persists beyond two weeks, especially if there is continuing fever and pain.

EMPYEMA

Pus in the pleural space leads to symptoms of chest pain, fever, and general malaise. Most empyemas occur as a complication of pneumonia or lung abscess, but 15–30

Table 24.6 *Summary of clinical setting, organisms, and empiric treatment of nosocomial pneumonia based on severity of disease (adapted in part from the American Thoracic Society 1996)*

Mild or moderate nosocomial pneumonia		
No risk factors	*Core organisms*[a]	*Initial antibiotics*
No unusual risk factors; early onset	S. pneumoniae	Second- or third -generation cephalosporin or new
Early onset	H. influenzae	fluoroquinolone or β-lactam + β-lactamase inhibitor
	S. aureus	
	E. coli, Klebsiella	
	Enterobacter spp.	
	Serratia spp.	
With risk factors	*Pathogens*	*Initial antibiotics*
Recent abdominal surgery	Core organisms[a] + anaerobes	Second- or third-generation cephalosporin +
Witnessed aspiration		clindamycin or new fluoroquinolone alone or
		β-lactam + β-lactamase inhibitor
High-dose steroids	L. pneumophila	Use macrolide or quinolone
Prior hospitalization	P. aeruginosa and MRSA	Treat as severe nosocomial pneumonia
ICU stay		
Steroids		
Prior antibiotics		
Chronic lung disease		
Severe nosocomial pneumonia		
Clinical setting		*Initial antibiotics*
ICU	Core organisms + P. aeruginosa +	Third- or fourth-generation antipseudomonal
Nursing home	Legionella pneumophila + MRSA	cephalosporin + quinolone ± vancomycin or
Prior antibiotics		third- or fourth-generation antipseudomonal
Respiratory failure		cephalosporin + macrolide ±
Progressive pneumonia		vancomycin ± aminoglycoside or
Sepsis, shock		carbapenem + quinolone ± vancomycin
Organ failure		

ICU, intensive care unit; MRSA, methicillin-resistant *S. aureus*.
a) Core organisms are those pathogens commonly implicated in nosocomial pneumonia with no unusual risk factors.

Table 24.7 *Potential microbial etiology of aspiration pneumonitis*

Anaerobes	Aerobes
Gram-negative bacilli	Gram-positive bacilli
Pigmented *Prevotella* and *Porphyromonas*	*Staphylococcus aureus*
Prevotella aureus, P. buccae	*Streptococcus pyogenes*
Prevotella oralis group	'viridans' streptococci
Bacteroides ureolyticus group (especially *B. gracilis*)	Gram-negative bacilli
Bacteroides fragilis group	*Klebsiella pneumoniae*
Fusobacterium nucleatum	*Enterobacter* spp.
Fusobacterium necrophorum, F. naviforme, F. gonidiaformans	*Serratia* spp.
Gram-positive cocci	*Pseudomonas aeruginosa*
Peptostreptococcus (especially *P. magnus, P. asaccharolyticus,*	*Escherichia coli*
P. prevotii, P. anaerobius, and *P. micros*)	*Proteus* spp.
Microaerophilic streptococci (*Streptococcus intermedius*)	
Gram-positive nonsporing bacilli	
Actinomyces spp.	
Propionibacterium propionicum	
Bifidobacterium dentium	
Gram-positive spore-forming bacilli	
Clostridium (especially *C. perfringens, C. ramosum*)	

Modified after Finegold (1994).

percent occur after thoracic surgery and 10 percent occur in association with intraabdominal infection. Empyema following pneumonia is mostly polymicrobial and anaerobes are present in 75 percent of cases. Empyema after thoracic surgery is more likely to be monomicrobial and due to common nosocomial pathogens such as *S. aureus* and aerobic gram-negative bacilli. Progress of pneumonia to empyema is related to delay in appropriate antimicrobial therapy. Once infection is established the empyema rapidly becomes fibrinopurulent with eventual locule formation in the pleural space and fibrous adhesions between the visceral and parietal pleura. Diagnosis is based on demonstration of purulent pleural fluid. Microbiological diagnosis should include Gram and acid-fast stains, wet mount for fungi, and culture for aerobic and anaerobic bacteria, *Mycobacterium* spp., and fungi. Once an empyema has developed, the aim should be to sterilize the space with antibiotics and establish early and adequate pleural space drainage. Diffusion of antibiotics into the pleural space is good, but aminoglycosides and some β-lactams may be inactivated in the presence of pus, a low pH, and β-lactamase enzymes (Hughes and van Scoy 1991). Empirical treatment of empyema should include an effective agent against anaerobic bacteria and aerobic gram-negative bacteria. Metronidazole may not be reduced to its active metabolite in a partially oxygenated environment and so clindamycin in combination with a fluoroquinolone for 4–6 weeks is a suitable combination. Alternatively, imipenem–cilastatin can be used. Drainage of pus is the mainstay of treatment and in the early exudative phase, closed drainage may suffice. If the empyema is loculated, operation can sometimes be avoided by the judicious use of urokinase to break down adhesions (Robinson and Moulton 1994).

NECROTIZING PNEUMONITIS AND LUNG ABSCESS

Pus-filled cavities in the lung can evolve when there is suppurative lung infection together with destruction of lung parenchyma. Necrotizing pneumonitis is arbitrarily defined as multiple cavities of less than 2 cm in diameter; lung abscesses are larger. Aspiration is the most common predisposing factor and in such cases the microbial etiology reflects that seen in aspiration pneumonitis (see Table 24.5). Tooth decay and gingivitis also predispose to necrotizing infections. Abscess may also occur in patients with necrotic neoplasms, in patients with bronchiectasis, during septic embolization from another source (e.g. endocarditis), as a complication of pulmonary embolus, and as direct spread from the abdomen (e.g. amoebic abscess).

Pathogenesis

Necrotizing pneumonia after aspiration is common because of the destructive nature of gastric contents and the large microbial load. In the unconscious patient, bacteria can replicate in static fluid in dependent lung segments. Some microorganisms produce necrotizing infections by virtue of their virulence factors, including *F. necrophorum, K. pneumoniae,* and *S. aureus. P. aeruginosa* secretes toxins that create a local vasculitis in addition to pneumonitis. Bronchiectasis can lead to lung abscess by virtue of markedly impaired local airway clearance.

Microbial etiology

In a study of lung abscess caused by transtracheal aspiration, the majority of patients had multiple isolates but in every case one or more anaerobes was recovered (Bartlett et al. 1974). The most common were gram-negative bacilli, including *Bacteroides* spp. and *Fusobacterium* spp., and also gram-positive cocci, including *Peptostreptococcus* spp. and microaerophilic streptococci. The aerobic organisms that can cause necrotizing pneumonitis are *S. aureus*, *S. pyogenes*, *K. pneumoniae*, *P. aeruginosa*, *Proteus* spp., and *E. coli*. Occasionally the more common causes of community-acquired pneumonia such as *S. pneumoniae* and *L. pneumophila* can be involved in a necrotizing process. *Actinomyces* spp., *Arachnia* spp., and *Nocardia* spp. can cause necrotizing pneumonitis as part of a chronic infection. A number of pathogens cause necrotizing pneumonitis in the context of immunosuppression, including *Burkholderia cepacia*, in patients receiving nebulized therapy (Yamagishi et al. 1993). *Legionella* spp. and *Yersinia enterocolitica* can cause necrotizing pneumonias in such patients. Necrotizing pneumonitis is a prominent feature of *B. pseudomallei* infection (melioidosis) and as part of infection with *B. anthracis* in anthrax and *Y. pestis* in pneumonic plague.

Where multiple abscesses are present in the lung, hematogenous spread from an extrapulmonary focus is the probable cause. *S. aureus* abscesses can complicate right-sided endocarditis, particularly in intravenous drug abusers. Enterobacteriaceae can produce multiple abscesses in association with urinary tract or bowel surgery and anaerobes can do so in patients with pelvic infections.

Diagnosis

Microbiological diagnosis can be achieved by transtracheal aspiration, bronchoscopic alveolar lavage, or with a protected brush specimen. Material should be rendered anaerobic as quickly as possible.

Therapy

For patients who are not severely ill, penicillin plus metronidazole or clindamycin is the initial therapy. Treatment must be continued for up to 3 months to achieve cure. In severely ill patients, empirical antibiotic therapy should aim to reduce the burden of β-lactamase producing anaerobes and also *S. aureus* and gram-negative bacilli, including *P. aeruginosa*. Some anaerobic bacteria, including *Actinomyces* spp. and *Propionibacterium* spp., are resistant to metronidazole. Some strains of the *B. fragilis* group are also resistant to the broad-spectrum penicillins, including piperacillin (Finegold 1994). Severely ill patients with necrotizing pneumonia can be given imipenem or a broad-spectrum β-lactam–β-lactamase inhibitor combination.

Chronic pneumonia

Apart from mycobacteria (see Chapters 46, Mycobacterium tuberculosis complex, Mycobacterium leprae and other slow growing mycobacteria and 50, Rapidly growing mycobacteria), *Actinomyces* spp., notably *A. israelii* and *Propionibacterium propionicum*, and *Nocardia asteroides* may cause chronic pneumonia (see Chapter 41, Actinomyces and related genera, and 45, Nocardia and other aerobic actinomycetes.

Pneumonia in immunocompromised patients

Most pneumonias in non-HIV immunocompromised patients occur in the context either of neutropenia or T-cell defects. These occur in malignancies, particularly hematological malignancies, with and without chemotherapy; they also occur in patients undergoing organ transplantation and attendant immunosuppressive therapy and in other patients receiving steroids. Neutropenic patients are susceptible to pneumonia due to gram-negative organisms, particularly *E. coli*, *Klebsiella* spp., *Enterobacter* spp., *Serratia* spp., *Proteus* spp., and *Pseudomonas*. They are also susceptible to gram-positive infections, including *S. aureus* and *Staphylococcus epidermidis* and viridans streptococci (Verhoef 1993). Patients with cellular immune defects such as T-cell defects due to cyclosporin, purine antagonists, corticosteroids, or hematological malignancy are vulnerable to pneumonia mainly due to intracellular pathogens such as *Mycobacterium* spp. and *Legionella* spp., but more importantly to viruses, including cytomegalovirus (CMV), commonly derived from the transplanted organ, fungi (*Candida*, *Cryptococcus*) and *P. carinii* (Hughes 1993). It is difficult to distinguish on clinical grounds alone between pneumonia of bacterial or nonbacterial origin, but in general bacterial infections produce a rapidly advancing clinical course with focal infiltrates on chest radiograph (Fanta and Pennington 1994). *P. carinii* pneumonia in the non-HIV setting can also cause a rapidly advancing pneumonia. Bronchoscopy, with bronchoalveolar lavage, is necessary to identify the agents responsible because several may be involved. A number of unusual bacteria cause pneumonia in the immunocompromised host, including *Nocardia* spp., *Bordetella* spp., and *B. cepacia*. Nosocomial outbreaks of pneumonia in immunocompromised patients can be due to *L. pneumophila* (Carratala et al. 1994) and tuberculosis in renal units (Hall et al. 1994). Fanta and Pennington (1994) have proposed a scheme for initiating empirical therapy in immunocompromised patients. Patients with a focal infiltrate should be given antibacterial treatment while results of lung sampling are awaited. A combination of antibiotics to cover staphylococci and resistant gram-negative bacilli, including *Pseudomonas*, should be given, for example, a third-generation cephalosporin and an aminoglycoside, or a combination of broad-spectrum penicillin and an aminoglycoside. If *Legionella* is suspected, intravenous erythromycin should be added; if *Aspergillus* is suspected, amphotericin B should also be given. The challenging problem of aspergillosis is discussed in the Medical Mycology volume, Chapter 34, Aspergillosis.

Bacterial pneumonia in HIV infection

In late-stage unmodified HIV infection, particularly in patients with CD4 counts $<0.2 \times 10^9$ per liter, lung disease is dominated by infection with *P. carinii*, *M. tuberculosis*, *Toxoplasma gondii*, and *Cryptococcus neoformans*. Patients are also susceptible to conventional bacterial infections. HIV-seropositive individuals have increased susceptibility to bacterial pneumonia, particularly caused by *S. pneumoniae*, *H. influenzae*, *M. catarrhalis*, *K. pneumoniae*, *N. meningitidis*, *Rhodococcus* spp., and *S. aureus* (Mitchell and Miller 1995). This susceptibility is particularly so in patients with a history of intravenous drug abuse (Caiaffa et al. 1994). Smoking of illicit drugs, such as marijuana and cocaine, is also a risk factor. Patients with AIDS in Africa appear particularly susceptible to pneumococcal disease (Gordon et al. 2002). In patients with relatively normal CD4 lymphocyte counts and no other HIV-related symptoms there is an appropriate response to antibiotic therapy. More advanced immunosuppression increases the risk of bacterial pneumonia (Rosen 1994). In the era of highly active antiretroviral therapy (HAART), there has been a declining morbidity and mortality of patients with AIDS (Palella et al. 1998).

Rhodococcus equi emerged as an important pathogen as a result of the HIV pandemic but is rarely seen in other patients. It is an anaerobic, nonmotile gram-positive organism that causes typical pneumonia which can be necrotizing and associated with pleural effusion. It is sensitive to vancomycin, erythromycin, chloramphenicol, and aminoglycosides. Treatment requires prolonged administration of vancomycin and erythromycin (Guerra et al. 1994). Infection with *H. influenzae* in HIV infection is somewhat unusual in that there is a high frequency of capsulate type b organisms in this group (Casadevall et al. 1992).

HIV-positive individuals have an increased incidence of sinusitis. The organisms involved are similar to those in non-HIV infected individuals, but *P. aeruginosa* is more frequent in the late stages of HIV disease. Occasionally nonbacterial pathogens such as *Cryptococcus*, *Aspergillus*, *Alternaria*, and CMV are responsible.

Chronic sinusitis is particularly troublesome in patients with low CD4 counts. Treatment follows general principles.

Patients with HIV infection appear to be more susceptible to chronic bronchitis and bronchiectasis. The most common pathogens responsible for bacterial bronchitis include *H. influenzae* and *S. pneumoniae*, though *Pseudomonas* spp. may also be involved (Verghese et al. 1994). Exacerbations respond promptly to appropriate antibiotic therapy but may recur. Bronchiectasis tends to occur as discrete lesions in single lobes.

PERTUSSIS

Pertussis (whooping cough) is caused by *Bordetella pertussis*. It causes prolonged paroxysms of cough and in the very young can lead to prolonged apnea and death; long-term sequelae include bronchiectasis. There is no effective antibiotic treatment for the paroxysmal stage of the disease and prevention is the lynchpin of management.

Epidemiology

B. pertussis is transmitted by droplets and is highly infectious, with attack rates in susceptible individuals between 50 and 100 percent. Worldwide, there are approximately 50 million cases and 600 000 deaths annually (Crowcroft et al. 2003). Epidemics occur in cycles of 3–5 years, reflecting the appearance of susceptible individuals in a given population. Since 1950 the incidence of pertussis has been modified dramatically by the introduction of the whole-cell vaccine. Vaccination is given to children older than 1 month and provides immunity for approximately 12 years. Before vaccination pertussis was seen mainly in children 1–5 years of age, because of passive protection by maternal antibody in those aged less than 1 year. Adults, including mothers, are generally immune because of natural infection in childhood. In Sweden, where the pertussis programme was terminated in 1979, the incidence of pertussis is 60 cases per 100 000 population; age-specific attack rates per 100 000 have been 630 infants >1 year, 670 in children 1–4 years, and 258 in children 5–9 years (Romanus and Jonsell 1987). In contrast, in Massachusetts, USA, which has an active pertussis programme, the incidence of bacteriologically confirmed pertussis is 0.92 cases per 100 000 population, with a peak incidence in infants aged 1 month (104 per 100 000 population), which then declines to five per 100 000 (Marchant et al. 1994). Vaccination limits disease in children but the peak incidence is now in infants aged 1 month probably because of the absence of passive immunity from mothers in whom vaccine-induced immunity has declined. Even in countries with high coverage, adult pertussis does occur (Gilberg et al. 2002) and there is evidence of transmission of pertussis from adults to susceptible infants (Mokotoff et al. 1995). During the mid-1970s the general public was alerted by media publicity to the possibility of vaccine-induced brain damage, and this led to a decline in vaccine uptake. Subsequently, there were three large epidemics in the UK at characteristic 4-year intervals in 1977, 1981, and 1985 (Preston 1994). In both Massachusetts (Marchant et al. 1994) and Cincinnati (Christie et al. 1994), there has been a resurgence of pertussis in the past 15 years in the face of vaccination programmes. These are due either to over-reporting or, more worrying, failure of vaccine efficacy.

Pathogenesis

B. pertussis produces a number of virulence factors (see Chapter 69, Bordetella). The first step in B. pertussis infection is circumvention of mucociliary clearance by binding to ciliated cells and multiplying on their surfaces. This is achieved by two adhesins: filamentous hemagglutinin and pertussis toxin. Filamentous hemagglutinin release facilitates dispersal through the respiratory tract (Coutte et al. 2003). In addition, B. pertussis produces serotype 2 and 3 fimbriae, which also facilitate adherence. Colonization of ciliated cells results in their death because of toxin production by the bacterium. This damage to the airway epithelium and direct activity against local neurons by pertussis toxin is responsible for the paroxysmal cough (Salyers and Whitt 1994). B. pertussis survives for long periods in phagocytes. Pertussis toxin inhibits phagocyte killing in vitro and the migration of monocytes. The organism can survive in human polymorphs by inhibiting phagosome–lysosome fusion but it does not inhibit the respiratory burst (Steed et al. 1992). Similarly, B. pertussis survives for 3–4 days in human monocyte-derived macrophages (Friedman et al. 1992).

Clinical features

After an incubation period of 1–3 weeks a nonspecific catarrhal illness occurs with symptoms of malaise, lethargy, low-grade fever, dry cough, and poor feeding. A week later paroxysms of coughing occur, which increase in pitch before terminating in a deep inspiratory movement accompanied by the typical whoop. Chest examination is usually normal unless pneumonia or lobar collapse has occurred, as it occasionally does in infants. In infants, mortality is due to a combination of apnea and aspiration of vomitus. In adults, the disease tends to be less severe, though typical whooping cough may be seen. Children may be susceptible to bronchial infections for up to a year after pertussis, as the result of reduced host defenses of the airway. Bronchiectasis in some adults has been attributed to severe pertussis in childhood.

Diagnosis

In the presence of the typical whoop, diagnosis presents no difficulty, but isolation of B. pertussis by culture is diagnostic. This is most likely during the late catarrhal and early paroxysmal phases, but later becomes progressively less likely. Pernasal pharyngeal swabs should be cultured on Bordet–Gengou agar, supplemented with methicillin or cephalexin to inhibit the normal flora, or patients may be asked to cough directly on to the medium. Cultures must be examined daily for 5–7 days to identify the tiny, slow-growing, hemolytic colonies. In nasopharyngeal smears, B. pertussis can be detected directly with fluorescent antibody. Serological tests are not used routinely for the diagnosis of pertussis, but agglutinins and complement-fixing antibodies can be detected in serum and antipertussis secretory IgA can be detected in nasopharyngeal smears. PCR-based detection is also available but is not more sensitive than culture (Lingappa et al. 2002). Laboratory confirmation of pertussis can be difficult, but a cough lasting 2 weeks during community outbreaks is a sensitive and specific marker. Other organisms, including coxsackie viruses, echoviruses, adenoviruses serotypes 1–3, 5, and 7, and M. pneumoniae, can cause illness clinically indistinguishable from pertussis.

Management

Treatment during the early stage of the disease can reduce its course and restrict transmission. Erythromycin is recommended for cases and their household contacts, regardless of age or vaccination status. It is important to protect children aged less than 6 months who are too young to vaccinate. Chemoprophylaxis should be given to contacts as early as possible because more than 21 days after first contact it is of limited value (Mokotoff et al. 1995).

Prevention

Whole-cell vaccine can be combined with diphtheria and tetanus toxoids and aluminum-containing adjuvants (DTP vaccine), but an acellular vaccine has been introduced in several parts of the world. The pertussis component of this contains the three major agglutinogens, 1, 2, and 3. It is administered as three primary doses at intervals of 2 months beginning at 6–8 weeks of age. Booster doses are given 6–12 months later, and the third dose is given at 4–6 years of age. The vaccine is believed to have an effectiveness greater than 80 percent in preventing pertussis, but there has been a decline in efficacy in recent years (reviewed by Preston 1994). The duration of protection is limited and wanes by approximately 12 years after immunization. Vaccine failures have been attributed to omission of one of the three agglutinogens or omission of the adjuvant.

A major problem with whole-cell vaccine has been public reaction to reports of adverse effects. Pertussis incidence has been 10–100 times lower in countries that have maintained higher coverage (Gangarosa et al. 1998). The most common of these are local reactions at the injection site and high fever; convulsions and brain damage are rare. The risk of permanent neurological damage in normal children given a full course of DTP vaccination is approximately one in 100 000, which is less than that of complications from the disease. Vaccination should, however, be avoided in children with a history of convulsions or brain damage in the neonatal period or other neurological disorder and during a febrile illness. A history of idiopathic epilepsy or developmental delay due to neurological disease in first-degree relatives is a relative contraindication that should be individually assessed. Those who have had a serious local or systemic reaction to a dose of vaccine should not be given further doses.

These considerations led to an effort to develop acellular vaccines. Pertussis toxin (PT), filamentous hemagglutinin (FHA), 68-kDa outer-membrane protein (OMP), and fimbrial agglutinogens are the most important antigens that elicit protective immunity against pertussis. FHA and OMP are, together with chemically detoxified PT, the components of some new acellular pertussis vaccines. During a pertussis outbreak in previously immunized children, those with higher serum levels of anti-FHA antibodies remained healthy, whereas those with lower antibody levels developed the disease (He et al. 1994). Formulations of acellular vaccine containing FHA, PT, and agglutinogens have been effective since 1981 in Japan in children 24 months of age and older. Pertussis toxoid alone or in combination with FHA has been tested in Swedish children from 6 months of age and gave partial protection against infection and good protection against severe disease (Ad hoc Group for the Study of Vaccines 1988). Five- and three-component acellular vaccines have been shown to be as effective as the whole-cell vaccine against culture-confirmed pertussis with at least 21 days of paroxysmal cough, though the five-component vaccine was superior to the three-component vaccine in preventing mild disease (Olin et al. 1997). Acellular vaccines can be effectively used as boosters. They are expensive and do not give rise to type-specific agglutinins. However, multicomponent acellular pertussis vaccines do appear to be effective and show fewer adverse effects than whole-cell pertussis vaccines. In areas where whooping cough is more likely to be fatal, it is possible that the higher toxicity of some whole-cell vaccines may be offset by their increased effectiveness (Tinnion and Hanlon 2000).

Infections with *Bordetella* species other than *B. pertussis*

Bordetella parapertussis has been incriminated as a minor cause of whooping cough, particularly a milder form than that caused by *B. pertussis*. Some evidence suggests that *B. parapertussis* may not be a separate species; this includes the isolation of both *B. pertussis* and *B. parapertussis* from whooping cough patients and the apparent conversion of *B. pertussis* to *B. parapertussis* by loss of a prophage. Doubt has been cast on *Bordetella bronchiseptica* as a possible cause of whooping cough (Woolfrey and Moody 1991).

ACUTE BRONCHITIS

Acute bronchitis is manifest as productive cough in a patient without a history of chronic chest disease. The microbial etiology of acute bronchitis includes viruses, particularly adenovirus, influenza virus, respiratory syncytial virus, and parainfluenza virus, the latter two being particularly common in children. *M. pneumoniae* and the pneumococcus are the prominent bacterial causes of acute bronchitis (Pennington 1994a). On balance, antibiotic therapy is regarded as unnecessary (reviewed by Gonzales and Sande, 1995). Randomized double-blind studies have failed to show a major clinical role for antibiotics in uncomplicated acute bronchitis. Some patients with acute bronchitis experience a prolonged episode of bronchial hyperreactivity. Antibiotics do not alter this syndrome.

CHRONIC BRONCHITIS AND CHRONIC OBSTRUCTIVE PULMONARY DISEASE

Chronic bronchitis is defined as the daily production of sputum for at least three consecutive months in two consecutive years and its essential feature is chronic bronchial mucin hypersecretion. This may be accompanied by bacterial infection, which causes an increase in the purulence of sputum, and also generalized small-airway obstruction. As a result, patients may experience one of three entities of increasing severity. These are simple chronic bronchitis with cough productive of non-purulent phlegm, chronic or recurrent mucopurulent bronchitis with cough, and chronic obstructive bronchitis, in which sputum is mostly purulent and there is airways obstruction (Medical Research Council 1965). Each of these is subject to acute exacerbation with a decline in the clinical state of the patient. The term chronic obstructive pulmonary disease (COPD) describes a range of chronic entities associated with airflow obstruction, including chronic bronchitis, emphysema, chronic asthma, and bronchiectasis. The disease COPD is characterized by episodic exacerbations in which there is fever plus any combination of increased dyspnea, increased sputum purulence, or increased sputum volume. It is unclear whether exacerbations are associated with accelerated loss of lung function during the natural history of chronic bronchitis (Bates 1973), but following an exacerbation most patients experience at least a temporary decrease in functional status and quality of life (Seemungal et al. 1998).

Pathogenesis

Smoking plays a part in the generation of chronic bronchitis (Floreani et al. 1994). Gradually there is increased mucus hypersecretion, depressed ciliary function, and epithelial cell injury. Bacteria that are not normally present in the airway below the larynx are then able to colonize the diseased epithelium (Read et al. 1991) (Figure 24.5) and this leads to further damage.

Bacterial infection and exacerbations of chronic bronchitis

Nontypable *H. influenzae* is most commonly present in acute, purulent exacerbations of chronic bronchitis, but *S. pneumoniae* and *M. catarrhalis* can also be isolated. *H. influenzae* adheres to mucus via fimbriae (Barsum et al. 1995) and galactoside sequences on epithelial cells (Krivan et al. 1988). Soluble products of *H. influenzae* reduce ciliary beat frequency (Wilson et al. 1985) and mucociliary clearance. Nontypable *H. influenzae* are present in the lower respiratory tract of most patients with chronic bronchitis and in all patients at some time, even between exacerbations. Molecular typing of strains of *H. influenzae* isolated from sputum of patients with COPD has demonstrated that a single strain may persist for many months within the lower respiratory tract

mucosa and remains present even if there have been acute exacerbations of COPD that have been treated with antibiotics (Groenveld et al. 1990). Multiple strains of *H. influenzae* can be present simultaneously in the sputum of an individual with COPD, and these distinct isolates may even vary in their antimicrobial sensitivity (Murphy et al. 1999). Their presence in the sputum during periods of increased symptoms does not necessarily imply a pathogenic role, but such a role for *H. influenzae* has been suggested by Musher et al. (1983), who showed that the serum of patients with acute febrile bronchitis contains opsonizing antibody for their own sputum isolates. There is evidence that acquisition of strain variants of *H. influenzae*, *M. catarrhalis*, or *S. pneumoniae* is associated with exacerbated disease, based upon molecular typing of sputum isolates from patients with COPD (Sethi et al. 2002). Virus and mycoplasma infections are responsible for up to a third of the acute exacerbations of chronic obstructive airways disease; the viruses more commonly isolated include influenza A, parainfluenza virus, coronavirus, rhinovirus, and herpes simplex virus (reviewed by Floreani et al. 1994). This suggests that bacteria, such as *H. influenzae*, flourish in the airways when viral infection further damages already compromised host defenses. The role of *H. influenzae* and *S. pneumoniae* in bacterial infection during chronic obstructive pulmonary disease has been reviewed by Murphy and Sethi (1992).

M. catarrhalis is present in pure culture in some transtracheal aspirates and is a recognized pathogen in chronic bronchitis; improvement is seen with specific treatment for *M. catarrhalis* infection. The organism has also been isolated from the blood and pleural fluid of patients with chronic obstructive airways disease (Murphy and Sethi 1992). In patients with more severe chronic bronchitis, a wider range of bacteria, including gram-negative enteric bacilli, *Pseudomonas* spp., and *Stenotrophomonas* spp., can be isolated by protective specimen brush (Soler et al. 1998).

Culture of sputum is of doubtful value because it is likely to be contaminated with the potential pathogens that are also present in oropharyngeal secretions. If such examination is undertaken, care should be taken to ensure that only purulent sputum is examined. This should be homogenized and serial dilutions examined. Examination of mucoid sputum will not yield useful information.

Therapy of exacerbations of chronic bronchitis

The use of antibiotics in all but the most severely ill patients is controversial (Saint et al. 1995). Small studies have either shown a positive effect of antibiotics (Berry et al. 1960; Pines et al. 1968), or no effect (Elmes et al. 1965; Nicotra et al. 1982). In a large placebo-controlled study,

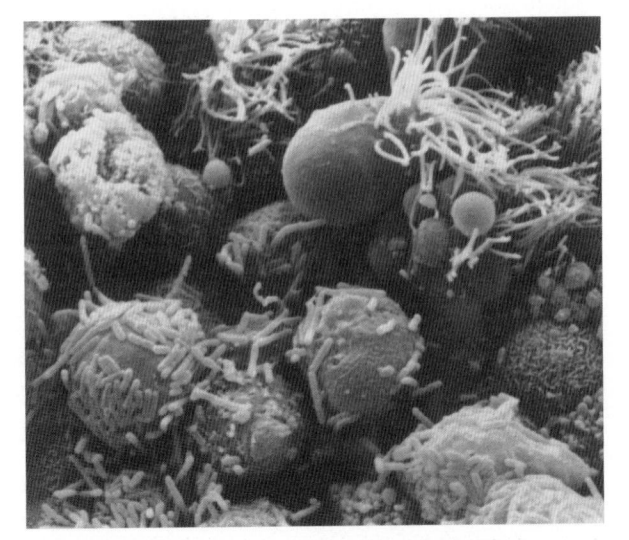

Figure 24.5 *Scanning electron micrograph of nontypable* Haemophilus influenzae *adhering to damaged areas of airway mucosa. (A. Brain and R.C. Read.)*

Anthonisen et al. (1987) found significant benefit from antibiotics in severe but not in mild exacerbations. Airway function also recovered more rapidly with antibiotic treatment. The use of prophylactic antibiotics is controversial; their use can reduce the frequency and length of exacerbations to a very small degree but is complicated by side-effects of the antibiotics (Black et al. 2002).

It is appropriate to treat severe exacerbations of COPD with antibiotics capable of covering *H. influenzae* and *S. pneumoniae* (Bach et al. 2001). Rates of resistance to aminopenicillins and macrolides of sputum isolates of *H. influenzae* and *S. pneumoniae* are rising throughout the world (Felmingham and Gruneberg 2000). Although antibiotic therapy in patients with mild disease is controversial, appropriate antibiotics include amoxycillin (provided local β-lactamase activity is at low level), co-amoxiclav, or the oral cephalosporins and the macrolides, but erythromycin has relatively poor activity against *H. influenzae*. Newer macrolides such as clarithromycin and azithromycin are better orally tolerated and have greater activity against *H. influenzae* in vitro. New fluoroquinolones are highly active against *H. influenzae* and *S. pneumoniae* and penetrate tissue and sputum well.

BRONCHIECTASIS

Bronchiectasis is permanent abnormal dilation of bronchi accompanied by suppurative inflammation that results in intermittent production of large quantities of purulent sputum. The most common causes are severe parenchymal infections of the lung, including pertussis and measles in childhood and also – most importantly in the developing world – tuberculosis. Foreign bodies, carcinoma, AIDS, allergic bronchopulmonary aspergillosis, and certain congenital abnormalities, including immunoglobulin deficiencies and ciliary abnormalities, can also give rise to bronchiectasis. Cystic fibrosis is associated with bronchiectasis (see Cystic fibrosis). It may also be seen in patients with the α_1-antiprotease deficiencies and is associated with rheumatoid arthritis and inflammatory bowel disease. Once bronchiectasis is present, repeated infections serve to maintain a vicious circle of inflammation and damage to the already compromised airway, which leads to further infection (Cole and Wilson 1989).

The sputum flora of such patients often yields a mixed growth, but *H. influenzae* is probably the most important in sustaining continued infection and inflammation. *S. aureus* and *S. pneumoniae* are also prominent. In established disease, *P. aeruginosa* is associated with a poor prognosis (Angrill et al. 2002). Diagnosis of bronchiectasis is made on the basis of chronic daily production of mucopurulent sputum and confirmed by high-resolution computed tomography (CT), of the thorax (Barker 2002) (Figure 24.6).

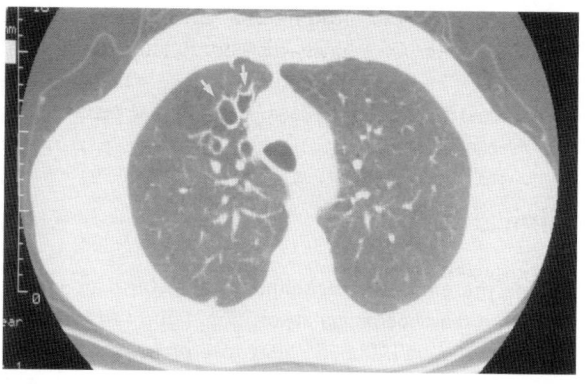

Figure 24.6 *Computed tomography (CT) scan demonstrating dilated airways and bronchial wall thickening in a patient with bronchiectasis. (Courtesy of A. Nakielny.)*

Management

The mainstay of treatment of bronchiectasis is drainage, which reduces the frequency of exacerbations, and antibiotic therapy. β-Lactam antibiotics do not penetrate diseased tissue at conventional dosages, but high doses of amoxycillin are effective in terminating exacerbations (Cole and Wilson 1989). High doses of oral cephalosporins or quinolones are effective when β-lactamase-producing strains are isolated. Long-term oral antibiotics are well tolerated and lead to symptomatic improvement but ciprofloxacin-resistant *P. aeruginosa* may be troublesome (Rayner et al. 1995). The use of prophylactic aminoglycosides given by nebulizer can reduce bacterial load but does not alter pulmonary function (Barker et al. 2000). In very severe exacerbations the use of intravenous antibiotics effective against *H. influenzae* and *P. aeruginosa* is indicated. Mucolytics have been suggested as adjunctive therapy for bronchiectasis but there is not currently enough evidence to evaluate this. High doses of bromhexine coupled with antibiotics may help with sputum production and clearance (Crockett et al. 2000).

CYSTIC FIBROSIS

Cystic fibrosis (CF) is the most common lethal genetic disease in Caucasians. It is autosomal recessive and caused by mutations in the cystic fibrosis transmembrane conductance regulator (*CFTR*) gene located on chromosome 7. Classic CF is characterized by chronic bacterial infection of the airways and sinuses, fat malabsorbtion due to pancreatic exocrine insufficiency, infertility in males due to obstructive azoospermia, and elevated concentrations of chloride in sweat. Patients with partial expression of *CFTR* usually have no overt signs of pancreatic insufficiency and later onset of bacterial infections of the airways (Knowles and Durie 2002). The incidence is approximately one in every 3 300 live Caucasian births. In the airways thick tenacious sputum causes obstruction, leading to recurrent infection. Patients with CF have airway inflammation that commences at birth (Cantin 1995). The carrier rate is about 5 percent in Caucasians; about 6 000 people have the disease in the UK.

Clinical features

Infants present with meconium ileus at birth or a little later with diarrhea or malabsorption, failure to thrive, or progressive cough. Patients may also present with recurrent pneumonia. Acute exacerbations occur with increased sputum production, cyanosis, dyspnea, fever, and weight loss. Patients also suffer with nasal polyposis and chronic sinusitis.

Role of microbial infection

The first pathogen to colonize children with CF is *S. aureus*. This produces α- and δ-toxins that induce bronchial wall injury and abscess formation. The intense inflammatory response to *S. aureus* leads to further tissue destruction.

Staphylococcal airway injury permits colonization by *P. aeruginosa*. This is the major pathogen implicated in lung function decline and ultimate mortality in CF patients (Lyczak et al. 2002). Patients are at first colonized with nonmucoid strains, but later mucoid variants emerge. The extracellular alginate that they produce increases their adherence to ciliated epithelium (Marcus and Baker 1985). Despite the high antibody levels present against such strains and intensive antibiotic therapy, they cannot be eradicated completely. They reappear after treatment but it is not clear whether this is due to recolonization or acquisition of a new strain (Ojeniyi 1994). Alginate also interferes with antibody coating and inhibits phagocytosis of *P. aeruginosa* (Baltimore and Mitchell, 1980). Bacterial proteases produced by *P. aeruginosa* cause significant damage to the airways. Pseudomonas elastase and alkaline protease are secreted in vivo over prolonged periods in the airways (Suter 1994). *Pseudomonas* elastase increases the permeability of epithelial cells, destroys tight junctions, and induces shedding of epithelial cell-surface heparan sulfate proteoglycans, each of which encourages persistent colonization (Park et al. 2001).

The *Pseudomonas* phenazine pigments pyocyanin and 1-hydroxyphenazine interfere with mucociliary function (Wilson et al. 1987); the membrane glycolipid, the hemolysin rhamnolipid, also reduces mucociliary transport in vivo (Read et al. 1992). Exotoxin A produced by most colonizing strains of *P. aeruginosa* causes tissue injury and is highly immunogenic. Patients become colonized for years with *P. aeruginosa*, often with the same strain, and isolates from sputum relentlessly exhibit multiple antibiotic resistance. It has been demonstrated that the lungs of patients with CF provide an environment ideal for hypermutable strains of *P. aeruginosa* (Oliver et al. 2000).

Noncapsulate *H. influenzae* is most frequently isolated from infants more than a year old. *Burkholderia cepacia* also colonizes CF patients and is transmissible between close contacts. It is occasionally responsible for a fulminating septicemia. This organism produces a number of potential virulence factors, including adhesins, extracellular polysaccharide, lipopolysaccharide, and extracellular enzymes, including proteases and lipases. Some *B. cepacia* genomovars worsen prognosis of patients, whereas others do not (Jones et al. 2001). It is intrinsically highly resistant to antibiotics, including those effective against *P. aeruginosa* (Wilkinson and Pitt 1995), but it is usually sensitive to trimethoprim–sulfamethoxazole and chloramphenicol. *Stenotrophomonas* (*Xanthomonas*) *maltophilia* also colonizes patients with CF and can result in disease, though it does not adversely affect overall prognosis. A range of other microorganisms, including the Enterobacteriaceae, *Candida albicans*, *Aspergillus* spp., and environmental mycobacteria, may also be isolated from these patients, but their significance is uncertain.

Although CF patients produce markedly elevated titers of local and systemic antipseudomonal antibodies, the elastase produced by *P. aeruginosa* interferes with their activity by cleaving off the Fab and Fc fragments of immunoglobulins so that they are unable to interact with receptor sites on pulmonary macrophages or neutrophils (Thick et al. 1985).

Antimicrobial therapy

Antimicrobial therapy of exacerbations should always be guided by sensitivity data from cultured specimens. There is continuing controversy over the use of antibiotics as prophylaxis in CF. Once disease is established, some physicians attempt to suppress sputum colonization with long-term administration of antibiotics. Such antibiotics include cephalexin, oral chloramphenicol, trimethoprim–sulfamethoxazole, and co-amoxiclav, but controlled evaluations of this approach are lacking because of obvious methodological problems with such studies (Chartrand and Marks 1994). The use of oral cephalexin as prophylaxis suppresses *S. aureus* colonization but is associated with increased likelihood of *P. aeruginosa* colonization (Stutman et al. 2002). Most centers treat only acute exacerbations of pulmonary infection, with inclusion of empirical antipseudomonas and *S. aureus* antibiotics. This requires combined antibiotic therapy with a β-lactam plus an aminoglycoside. With the genesis of monotherapy capable of activity against both these pathogens, outpatient administration has flourished. Such antibiotics include ceftazidime, the penems, monobactams, and some β-lactam–β-lactamase inhibitors. Unfortunately monotherapy can result in the spread of resistant organisms, e.g. *P. aeruginosa*, and this has been demonstrated following the use of ceftazidime (Pedersen et al. 1986). Emergence of *P. aeruginosa* resistance is also seen after monotherapy with imipenem and aztreonam (Chartrand and Marks 1994). The use of oral quinolones has permitted more flexible management of exacerbations. They have been used in children with CF (Schaad et al. 1991) without side effects and their use in moderately severe CF exacerbations can be as effective as traditional intravenous treatment with β-lactams plus aminoglycosides. In vitro susceptibility of isolates returns when the quinolones are withheld for 3 months. The use of aerosolized antibiotics has been advocated but controlled data to support this are limited. Aerosolized *tobramycin* given by nebulizer has been shown to eradicate *P. aeruginosa* (Ramsey et al. 1999) and improve pulmonary function tests (Smith et al. 1989), but emergence of resistance is a problem.

Other forms of therapy

Inhaled corticosteroids are being increasingly used in adults and children with CF to reduce lung damage arising from inflammation, but there is no clear evidence for their efficacy (Dezateux et al. 2000). There is potential to ameliorate the pulmonary disease in CF with somatic gene therapy. The optimal mode of delivery of the normal gene into airway epithelium is under investigation. Adenovirus vectors, DNA–liposome complexes, adeno-associated viral vectors, and DNA–ligand complexes have been used effectively in vitro and have been tested in animals. Adenovirus vectors and DNA–liposome complexes are currently in phase 1 clinical trial. There has been transient correction of electrophysiological defects in human CF nasal epithelium (O'Neal and Beaudet 1994). At present, CF patients face the prospect of lung transplantation as their only salvation once extreme disease ensues. Single-lung, double-lung, and heart–lung transplantation have all been used with success in such patients, and because the transplanted lung retains the physiological characteristics of the donor, there is no tendency for colonization by *P. aeruginosa* over and above that seen in transplanted lungs in non-CF patients.

REFERENCES

Ad hoc Group for the Study of Pertussis Vaccines. 1988. Placebo-controlled trial of two acellular pertussis vaccines in Sweden – protective efficacy and adverse effects. *Lancet*, **1**, 955-7.

American Thoracic Society. 1996. Hospital-acquired pneumonia in adults: diagnosis, assessment of severity, initial antibiotic therapy and prevention strategies: a consensus statement. *Am J Respir Crit Care Med*, **153**, 1711-25.

American Thoracic Society. 2001. Guidelines for the management of adults with community-acquired pneumonia: diagnosis, assessment of severity, antimicrobial therapy and prevention. *Am J Respir Crit Care Med*, **163**, 1730-54.

Angrill, J., Agusti, C., et al. 2002. Bacterial colonisation in patients with bronchiectasis: Microbiological pattern and risk factors. *Thorax*, **57**, 15–19.

Anthonisen, N.R., Manfreda, J., et al. 1987. Antibiotic therapy in exacerbations of chronic obstructive pulmonary disease. *Ann Intern Med*, **106**, 196–204.

Austrian, R. 1986. Some aspects of the pneumococcal carrier state. *J Antimicrob Chemother*, **18**, Suppl. A, 25–34.

Austrian, R. and Gold, J. 1964. Pneumococcal bacteraemia with special reference to bacteraemic pneumococcal pneumonia. *Ann Intern Med*, **60**, 759–76.

Bach, P.B., Brown, C., et al. 2001. Management of acute exacerbations of chronic obstructive pulmonary disease: a summary and appraisal of published evidence. *Ann Intern Med*, **134**, 600–20.

Baltimore, R.S. and Mitchell, M. 1980. Immunologic investigations of mucoid strains of *Pseudomonas aeruginosa*: comparison of susceptibility to opsonic antibody in mucoid and non-mucoid strains. *J Infect Dis*, **141**, 238–47.

Barker, A.F., Couch, L., et al. 2000. Tobramycin solution for inhalation reduces sputum *Pseudomonas aeruginosa* density in bronchiectasis. *Am J Resp Crit Care Med*, **162**, 481–5.

Barker, A.F. 2002. Bronchiectasis. *New Engl J Med*, **346**, 1383–93.

Barsum, W., Wilson, R., et al. 1995. Interaction of fimbriate and non-fimbriate strains of *H. influenzae* with human airway *mucin in vitro*. *Eur Respir J*, **8**, 709–14.

Bartlett, J.G., Gorbach, S.L., et al. 1974. Bacteriology and treatment of primary lung abscess. *Am Rev Respir Dis*, **109**, 510–18.

Bartlett, J.G., O'Keefe, P., et al. 1986. Bacteriology of hospital acquired pneumonia. *Arch Intern Med*, **146**, 868–71.

Bartlett, J.G., Dowell, S.F., et al. 2000. Practice guidelines for the management of community-acquired pneumonia in adults. *Clin Infect Dis*, **31**, 347–82.

Bates, D.V. 1973. The fate of the chronic bronchitic: a report of the 10 year follow-up in the Canadian Department of Veterans' Affairs co-ordinated study of chronic bronchitis. *Am Rev Respir Dis*, **108**, 1043–65.

Bébéar, C., Dupon, M., et al. 1993. Potential improvements in therapeutic options for mycoplasma respiratory infections. *Clin Infect Dis*, **1**, Suppl. 1, S202–7.

Berry, D.G., Fry, J., et al. 1960. Exacerbation of chronic bronchitis treatment with oxytetracycline. *Lancet*, **1**, 137–9.

Bice, D.E., Weissman, D.N. and Muggenburg, B.A. 1991. Long term maintenance of localised antibody responses in the lung. *Immunology*, **74**, 215–22.

Black, S., Shinefield, H., et al. 2000. Efficacy, safety and immunogenicity of heptavalent pneumococcal conjugate vaccine in children. *Pediatr Infect Dis J*, **19**, 187–95.

Black, P., Staykova, T., et al. 2002. Prophylactic antibiotic therapy for chronic bronchitis (Cochrane Review). *Cochrane Database Syst Rev*, **1**, CD004105.

Bohte, R., van Furth, R., et al. 1995. Aetiology of community-acquired pneumonia: a prospective study among adults requiring admission to hospital. *Thorax*, **50**, 543–7.

Bolling, G.E., Plouffe, J.F., et al. 1985. Aerosols containing *Legionella pneumophila* generated by showerheads and hot water fawcetts. *Appl Environ Microbiol*, **50**, 1128–31.

British Thoracic Society. 1987. The hospital management of community acquired pneumonia. *JR Coll Physicians Lond*, **21**, 267-9.

British Thoracic Society. 2001. Guidelines for the management of community-acquired pneumonia in adults. *Thorax*, **56**, Suppl. IV.

Brook, I. and Van de Heyning, P.H. 1994. Microbiology and management of otitis media. *Scand J Infect Dis Suppl*, **93**, 20–32.

Broome, C.V. 1986. The carrier state: *Neisseria meningitidis*. *J Antimicrob Chemother*, **18**, Suppl. A, 25–34.

Broome, C.V. and Fraser, D.W. 1979. Epidemiologic aspects of legionellosis. *Epidemiol Rev*, **1**, 1–16.

Brown, P.D. and Lerner, S.A. 1998. Community acquired pneumonia. *Lancet*, **352**, 1295–302.

Buckley, H. 1992. Immunodefiency diseases. *J Am Med Assoc*, **268**, 2797–806.

Bruyn, G.A.W., Zegers, B.J.M., et al. 1992. Mechanisms of host defence against infection with *Streptococcus pneumoniae*. *Clin Infect Dis*, **14**, 251–62.

Bryan, C.S. and Reynolds, K.L. 1984. Bacteremic nosocomial pneumonia. *Am Rev Resp Dis*, **129**, 668–71.

Caiaffa, W.T., Vlahov, D., et al. 1994. Bacterial pneumonia among HIV-seropositive injection drug users. *Am J Respir Crit Care Med*, **150**, 1493–8.

Cantin, A. 1995. Cystic fibrosis lung inflammation: early, sustained and severe. *Am J Respir Crit Care Med*, **151**, 939–41.

Carratala, J., Gudiol, F., et al. 1994. Risk factors for nosocomial *Legionella pneumophila* pneumonia. *Am J Respir Crit Care Med*, **149**, 625–9.

Carroll, K.C. 2002. Laboratory diagnosis of lower respiratory tract infections: controversy and conundrums. *J Clin Microbiol*, **40**, 3115–20.

Casadevall, A., Dobroszycki, J., et al. 1992. *Haemophilus influenzae* type B bacteraemia in adults with AIDS and at risk for AIDS. *Am J Med*, **92**, 587–90.

Caugant, D.A., Hoiby, E.A., et al. 1994. Asymptomatic carriage of *Neisseria meningitidis* in a randomly sampled population. *J Clin Microbiol*, **32**, 323–50.

CDSC. 1995. Current respiratory infections. *Commun Dis Rep*, **5**, 21.

Celis, R., Torres, A., et al. 1988. Nosocomial pneumonia: a multivariate analysis of risk and prognosis. *Chest*, **93**, 318–24.

Centers for Disease Control. 1988. CDC definitions for nosocomial infections 1988. *Am Rev Respir Dis*, **139**, 1058-9.

Chartrand, S.A. and Marks, M.I. 1994. Pulmonary infections in cystic fibrosis: pathogenesis and therapy. In: Pennington, J.E. (ed.), *Respiratory infections: diagnosis and management*, 3rd edn. New York: Raven Press, 323–48.

Chastre, J. and Fagon, J.Y. 2002. Ventilator-associated pneumonia. *Am J Respir Crit Care Med*, **165**, 867–903.

Chen, D.K., McGeer, A., et al. 1999. Decreased susceptibility of *Streptococcus pneumoniae* to fluoroquinolones in Canada. *New Engl J Med*, **341**, 233–9.

Cherry, J.D. 1993. Anaemia and mucocutaneous lesions due to *Mycoplasma pneumoniae* infections. *Clin Infect Dis*, **17**, Suppl. 1, S47–51.

Christie, C.D.C., Marx, M.L., et al. 1994. The 1993 epidemic of pertussis in Cincinnati: resurgence of disease in a highly immunized population of children. *N Engl J Med*, **331**, 16–21.

Clyde, W.A. 1993. Clinical overview of typical *Mycoplasma pneumoniae* infections. *Clin Infect Dis*, **17**, Suppl. 1, S32–6.

Coburn, J. 1992. Pseudomonas aeruginosa exoenzymes. *Curr Top Microbiol Immunol*, **175**, 133–43.

Cole, P.J. and Wilson, R. 1989. Host-microbial interrelationships in respiratory infection. *Chest*, **95**, 2175–83.

Coutte, L., Alonso, S., et al. 2003. Role of adhesin release for mucosal colonization for a bacterial pathogen. *J Exp Med*, **197**, 735–42.

Craven, D.E., Steager, K.A. and Barber, T.W. 1991. Preventing nosocomial pneumonia: state of the art and perspectives for the 1990s. *Am J Med*, **91**, Suppl. 3b, 44s–53s.

Crockett, A.J., Cranston, J.M., et al. 2000. Mucolytics for bronchiectasis. *Cochrane Database Syst Rev*, **2**, CD001289.

Crowcroft, N.S., Stein, C., et al. 2003. How best to estimate the global burden of pertussis. *Lancet Infect Dis*, **3**, 413–18.

Crowe, H.L. and Lavitz, R.E. 1987. Invasive *Haemophilus influenzae* disease in adults. *Arch Intern Med*, **147**, 241–4.

Dal Nogare, A.R. 1994. Nosocomial pneumonia in the medical and surgical patient. *Med Clin North Am*, **78**, 1081–90.

Davies, D.G., Parsek, M.R., et al. 1998. The involvement of cell to cell signals in the development of a bacterial biofilm. *Science*, **280**, 295–8.

Dezateux, C., Walters, S. and Balfour-Lynn, I. 2000. Inhaled corticosteroids for cystic fibrosis. *Cochrane Database Syst Rev*, **2**, CD001915.

Dockrell, D.H., Lee, M., et al. 2001. Immune mediated phagocytosis and killing of *Streptococcus pneumoniae* are associated with directed and bystander macrophage phagocytosis. *J Infect Dis*, **184**, 713–22.

Dondero, T.J., Rendtorff, R.C., et al. 1980. An outbreak of legionnaires' disease associated with a contaminated air conditioning cooling tower. *N Engl J Med*, **302**, 365–70.

Dowling, J.N., Saha, A.K. and Glew, R.H. 1992. Virulence factors of the family Legionellaceae. *Microbiol Rev*, **56**, 32–60.

Duane, P.G., Rubins, J.B., et al. 1993. Identification of hydrogen peroxide as a *Streptococcus pneumoniae* toxin for rat alveolar epithelial cells. *Infect Immun*, **61**, 4392–7.

Edelstein, P.H. 1993. Legionnaires' disease. *Clin Infect Dis*, **16**, 741–9.

Edelstein, P.H. 1995. Antimicrobial chemotherapy for Legionnaires' disease: a review. *Clin Infect Dis*, **21**, Suppl3, 5265–76.

Edelstein, P.H. and Meyer, R.D. 1994. *Legionella* pneumonias. In: Pennington, J.E. (ed.), *Respiratory infections: diagnosis and management*, 3rd edn. New York: Raven Press, 455–84.

Editorial. 1986. Lessons from Stafford. *Lancet*, **327**, 1363-4.

Eisenstadt, J. and Crane, L.R. 1994. Gram-negative bacillary pneumonias. In: Pennington, J.E. (ed.), *Respiratory infections: diagnosis and management*, 3rd edn. New York: Raven Press.

Elmes, P.C., King, T.K.C., et al. 1965. Value of ampicillin in the hospital treatment of exacerbations of chronic bronchitis. *Br Med J*, **2**, 904–8.

ERS Task Force Report. 1998. Guidelines for management of acult community-acquired lower respiratory tract infections, *Eur Respir J*, **11**, 986-91.

Eskola, J. 2000. Immunogenicity of pneumococcal conjugate vaccines. *Pediatr Infect Dis J*, **19**, 388–93.

Fagon, J.Y., Chastre, J., et al. 1989. Nosocomial pneumonia in patients receiving continuous mechanical ventilation. Prospective analysis of 52 episodes with the use of a protected specimen brush and quantitative culture techniques. *Am Rev Respir Dis*, **1398**, 77–84.

Fang, F., Fine, M., et al. 1990. New and emerging aetiologies for community acquired pneumonia with implications for therapy: a prospective multicenter study of 359 cases. *Medicine*, **69**, 307–16.

Fanta, C.H. and Pennington, J.E. 1994. Pneumonia in the immunocompromised host. In: Pennington, J.E. (ed.), *Respiratory infections: diagnosis and management*, 3rd edn. New York: Raven Press, 275–94.

Farr, B.M. and Mandell, G.L. 1994. Gram-positive pneumonia. In: Pennington, J.E. (ed.), *Respiratory infections: diagnosis and management*, 3rd edn. New York: Raven Press, 349–67.

Farr, B.M., Sloman, A.J. and Frisch, M.J. 1991. Predicting death in patients hospitalized for community-acquired pneumonia. *Ann Intern Med*, **115**, 428–36.

Farr, B.M., Kaiser, D.L., et al. 1989. Prediction of microbial aetiology at admission to hospital for pneumonia from the presenting clinical features. *Thorax*, **44**, 1031–5.

Feldman, C., Kallenbach, J.M., et al. 1989. Community-acquired pneumonia of diverse etiology: prognostic features in patients admitted to an intensive care unit and 'severity of illness' score. *Intensive Care Med*, **15**, 302–7.

Feldman, C., Munro, N., et al. 1991. Pneumolysin induces the salient features of pneumococcal infection in the rat lung *in vivo*. *Am J Respir Cell Mol Biol*, **5**, 416–23.

Felmingham, D. and Gruneberg, R.N. 2000. The Alexander Project 1996-1997: latest susceptibility data from this international study of bacterial pathogens from lower respiratory tract infections. *J Antimicrob Chemother,*, **45**, 191–203.

Ferwerda, A., Moll, H.A., et al. 2001. Respiratory tract infection by *Mycoplasma pneumoniae* in children; a review of diagnostic and therapeutic measures. *Eur J Paediatrics*, **160**, 483–91.

Fields, B.S., Benson, R.F., et al. 2002. Legionella and Legionnaires' disease: 25 years of investigation. *Clin Microbiol Rev*, **15**, 506–26.

Fine, M.J., Auble, T.E., et al. 1997. A prediction rule to identify low risk patients with community acquired pneumonia. *N Engl J Med*, **336**, 243–50.

Finegold, S.M. 1994. Aspiration pneumonia, lung abscess and empyema. In: Pennington, J.E. (ed.), *Respiratory infections: diagnosis and management*, 3rd edn. New York: Raven Press, 311–22.

Floreani, A.A., Buchalter, S.E., et al. 1994. Chronic bronchitis. In: Pennington, J.E. (ed.), *Respiratory infection: diagnosis and management*, 3rd edn. New York: Raven Press, 149–92.

Fraser, D.W., Tsai, T.R., et al. 1977. Legionnaires' disease: description of an epidemic of pneumonia. *N Engl J Med*, **297**, 1189–97.

French, N., Nakiyingi, J., et al. 2000. 23-valent pneumococcal vaccine in HIV-infected Ugandan adults: double blind randomized and placebo-controlled trial. *Lancet*, **355**, 2106–11.

Friedland, I.R. and McCracken, G.H. 1994. Management of infections caused by antibiotic-resistant *Streptococcus pneumoniae*. *N Engl J Med*, **331**, 377–82.

Friedlander, A.M., Welkos, S.L., et al. 1993. Post exposure prophylaxis against experimental inhalation anthrax. *J Infect Dis*, **167**, 1239–43.

Friedman, R.L., Nordensson, K., et al. 1992. Uptake and intracellular survival of *Bordetella pertussis* in human macrophages. *Infect Immun*, **60**, 4578–85.

Galloway, D. 1991. Pseudomonas aeruginosa elastase and elastolysis: recent developments. *Mol Microbiol*, **5**, 2315–21.

Gangarosa, E.J., Galazka, A.M., et al. 1998. Impact of anti-vaccine movements on pertussis control: the untold story. *Lancet*, **351**, 356–61.

Giebink, G.S., Foker, J.E., et al. 1980. Serum antibody and opsonic responses to vaccination with pneumococcal capsular polysaccharide in normal and splenectomized children. *J Infect Dis*, **141**, 404–12.

Gilberg, S., Njamkepo, E., et al. 2002. Evidence of *Bordetella pertussis* infection in adults presenting with persistent cough in a French area with very high whole-cell vaccine coverage. *J Infect Dis*, **186**, 415–18.

Gleason, P.P., Kapoor, W.M., et al. 1997. Medical outcomes and antimicrobial costs for the use of the American Thoracic Society Guidelines for outpatients with community-acquired pneumonia. *J Am Med Assoc*, **278**, 32–9.

Gonzales, R. and Sande, M. 1995. What will it take to stop physicians prescribing antibiotics in acute bronchitis? *Lancet*, **345**, 665–6.

Gordon, S.B. and Read, R.C. 2002. Macrophage defences against respiratory tract infections. *Br Med Bull*, **61**, 45–61.

Gordon, S.B., Irving, G.R.B., et al. 2000a. Intracellular trafficking and killing of *Streptococcus pneumoniae* by human alveolar macrophages is influenced by opsonins. *Infect Immun*, **68**, 2286–93.

Gordon, S.B., Walsh, A.L., et al. 2000b. Bacterial meningitis in Malawian adults. Pneumococcal disease is common, severe and seasonal. *Clin Infect Dis*, **31**, 53–7.

Gordon, S.B., Chaponda, M., et al. 2002. Pneumococcal disease in HIV-infected Malawian adults, acute mortality and long-term survival. *AIDS*, **16**, 409–17.

Groenveld, K., van Alphen, L., et al. 1990. Endogenous reinfections by *Haemophilus influenzae* in patients with chronic obstructive pulmonary disease: the effect of antibiotic treatment on persistence. *J Infect Dis*, **161**, 512–17.

Guerra, L.G., Ho, H. and Verghese, A. 1994. New pathogens in pneumonia. *Med Clin North Am*, **78**, 967–85.

Gwaltney, J.M. Jr, Scheld, W.M., et al. 1992. The microbial etiology and antimicrobial therapy of adults with acute community-acquired sinusitis: a fifteen-year experience at the University of Virginia and review of other selected studies. *J Allergy Clin Immunol*, **90**, 457–61.

Hall, C.M., Willcox, P.A., et al. 1994. Mycobacterial infection in renal transplant recipients. *Chest*, **106**, 435–9.

Haslett, C. 1999. Granulocyte apoptosis and its role in the resolution and control of lung inflammation. *Am J Respir Crit Care Med*, **16**, S5–S11.

Hausdorff, W.P., Bryant, J., et al. 2000. Which pneumococcal serogroups cause the most invasive disease: implications for conjugate vaccine formulation and use. Part 1. *Clin Infect Dis*, **30**, 100–21.

He, Q., Viljanen, M.K., et al. 1994. Antibodies to filamentous haemagglutinin of *Bordetella pertussis* and protection against whooping cough in school children. *J Infect Dis*, **170**, 705–8.

Helbig, J.H., Uldum, S.A., et al. 2003. Clinical utility of urinary antigen detection for diagnosis of community acquired, travel-associated, and nosocomial Legionnaires' disease. *J Clin Microbiol*, **41**, 838–40.

Hirschmann, J.V. and Lipsky, B.A. 1994. The pneumococcal vaccine after 15 years. *Arch Intern Med*, **154**, 373–7.

Horan, T.C., White, J.W., et al. 1986. Nosocomial infection surveillance 1984. *MMWR CD Surveill Summ*, **35**, SS17–29.

Hughes, C.E. and van Scoy, R.E. 1991. Antibiotic therapy of pleural empyema. *Semin Respir Infect*, **6**, 94–102.

Hughes, W.T. 1993. Prevention of infection in patients with T cell defects. *Clin Infect Dis*, **17**, Suppl. 2, S367–71.

Hughes, J.M., Cullver, D.H., et al. 1983. Nosocomial infection surveillance 1980–1982. *Mort Morb Wkly Rep*, **32**, 1–15.

Inglesby, T.V., O'Toole, T., et al. 2002. Anthrax as a biological weapon 2002: updated recommendations for management. *J Am Med Assoc*, **287**, 2236–52.

Jedrzejas, M.J. 2001. Pneumococcal virulence factors: structure and function. *Microbiol Mol Biol Rev*, **65**, 187–207.

Johanson, W.G., Pierce, A.K. and Sanford, J.P. 1969. Changing pharyngeal bacterial flora of hospitalized patients. Emergence of gram-negative bacilli. *N Engl J Med*, **281**, 1137–40.

Johanson, W.G., Higuchi, J.H., et al. 1980. Bacterial adherence to epithelial cells in bacillary colonization of the respiratory tract. *Am Rev Respir Dis*, **121**, 55–63.

Jones, A.M., Dodd, M.E., et al. 2001. *Burkholderia cepacia*, current clinical issues, environmental controversies and ethical dilemmas. *Eur Resp J*, **17**, 295–301.

Joseph, C.A., Dedman, D., et al. 1994a. Legionnaires' disease surveillance: England and Wales, 1993. *Commun Dis Rep CDR Rev*, **4**, 10, R109–11.

Joseph, C.A., Watson, J.M., et al. 1994b. Nosocomial legionnaires' diseases in England and Wales 1980-1992. *Epidemiol Infect*, **112**, 329–45.

Jousimies-Somer, H., Savolainen, S., et al. 1993. Bacteriologic findings in peritonsillar abscesses in young adults. *Clin Infect Dis*, **6**, Suppl. 4, S292–8.

Karalus, N.C., Cursons, R.T., et al. 1991. Community-acquired pneumonia: aetiology and prognostic index evaluation. *Thorax*, **46**, 413–18.

Kikuchi, R., Watabe, N., et al. 1994. High incidence of silent aspiration in elderly patients with community acquired pneumonia. *Am J Respir Crit Care Med*, **150**, 251–3.

Knowles, M.R. and Durie, P.R. 2002. What is cystic fibrosis? *New Engl J Med*, **347**, 439–42.

Koppes, G.M., Ellenbogen, C., et al. 1977. Group Y meningococcal disease in United States Airforce recruits. *Am J Med*, **62**, 661–6.

Koskiniemi, M. 1993. CNS manifestations associated with *Mycoplasma pneumoniae* infections: summary of cases at the University of Helsinki. A review. *Clin Infect Dis*, **17**, Suppl. 1, S52–7.

Kozyrskyj, A.L., Hildes-Ripstein, G.E., et al. 1998. Therapy of acute otitis media with a shortened course of antibiotics: a meta-analysis. *J Am Med Assoc*, **279**, 1736–42.

Krivan, H.C., Roberts, D.D. and Ginsburg, V. 1988. Many pulmonary pathogenic bacteria bind specifically to the carbohydrate sequence GalNAcβ-1-4 Gal found in some glycolipids. *Proc Natl Acad Sci USA*, **85**, 6157–61.

Lawson, P.R. and Reid, K.B. 2000. The roles of surfactant proteins A and D in innate immunity. *Immunol Rev*, **173**, 66–78.

Leoni, E. and Legnani, P.P. 2001. Comparison of selective procedures for isolation and enumeration of *Legionella* species from hot water systems. *J Appl Microbiol*, **90**, 27–33.

Levine, S.M., Perez-Perez, G., et al. 2002. PCR-based detection of *Bacillus anthracis* in formalin-fixed tissue from a patient receiving ciprofloxacin. *J Clin Microbiol*, **40**, 4360–2.

Li, J.D., Feng, W., et al. 1998. Activiation of NF-kappaB via a Src-dependent Ras-MPAK-pp90vsk pathway is required for *Pseudomonas aeruginosa*-induced mucin overproduction in epithelial cells. *Proc Natl Acad Sc USA*, **95**, 5718–23.

Lingappa, J.R., Lawrence, W., et al. 2002. Diagnosis of community-acquired pertussis infection: comparison of both culture and fluorescent-antibody assays with PCR detection using electrophoresis or dot blot hybridisation. *J Clin Microbiol*, **40**, 2908–12.

Lipscomb, M.F., Bice, D.E., et al. 1995. The regulation of pulmonary immunity. *Adv Immunol*, **59**, 369–455.

Loen, S.K., Ursi, D., et al. 2002. Detection of *M. pneumoniae* in spiked clinical samples by nucleic acid sequence-based amplification. *J Clin Microbiol*, **40**, 1339–45.

Lohmann-Mathes, M.L., Steinmüller, C., et al. 1994. Pulmonary macrophages. *Eur Respir J*, **7**, 1678–89.

Lulu, A.R., Araj, G.F., et al. 1988. Human brucellosis in Kuwait: a prospective study of 400 cases. *Q J Med*, **66**, 39–54.

Luneberg, E., Zahringer, U., et al. 1998. Phase-variable expression of lipopolysaccharide contributes to the virulence of *Legionella pneumophila*. *J Exp Med*, **188**, 49–60.

Lunn, J.E., Knowlden, J. and Handyside, A.J. 1967. Patterns of respiratory illness in Sheffield infant school children. *Br J Prev Soc Med*, **21**, 7–16.

Lyczak, J.B., Cannon, C.L., et al. 2002. Lung infections associated with cystic fibrosis. *Clin Microbiol Rev*, **15**, 194–222.

Macfarlane, J.T., Finch, R.G., et al. 1982. Study of adult community-acquired pneumonia. *Lancet*, **2**, 255–8.

Madsen, M., Lebenthal, Y., et al. 2000. A pneumococcal protein that elicits interleukin-8 from pulmonary epithelial cells. *J Infect Dis*, **181**, 1330–6.

Maharaj, D., Raja, V. and Hemsley, S. 1991. Management of peritonsillar abscess. *J Laryngol Otol*, **105**, 743–5.

Marchant, C.D., Loghlin, A.N., et al. 1994. Pertussis in Massachusetts 1981-1991: incidence, serologic diagnosis and vaccine effectiveness. *J Infect Dis*, **169**, 1297–305.

Marcus, H. and Baker, N.R. 1985. Quantitation of adherence of mucoid and non-mucoid *Pseudomonas aeruginosa* to hamster tracheal epithelium. *Infect Immun*, **47**, 723–9.

Markiewicz, A.Z. and Tomasz, A. 1989. Variation in penicillin-binding protein patterns of penicillin-resistant clinical isolates of pneumococci. *J Clin Microbiol*, **27**, 405–10.

Marrie, T.J. 1994. Community acquired pneumonia. *Clin Infect Dis*, **18**, 501–15.

Marston, B.J., Plouffe, J.F., et al. 1997. Incidence of community acquired pneumonia requiring hospitalisation: results of a population-based active surveillance study in Ohio. *Arch Intern Med*, **157**, 1509–18.

Masure, H.R. 1992. Modulation of adenylate cyclase toxin production as *Bordetella pertussis* enters human macrophages. *Proc Natl Acad Sci USA*, **89**, 6521–5.

McCool, T.L., Cate, T.R., et al. 2002. The immune response to pneumococcal proteins during experimental human carriage. *J Exp Med*, **195**, 359–65.

Meats, E., Brueggemann, A.B., et al. 2003. Stability of serotyping during nasopharyngeal carriage of *Streptococcus pneumoniae*. *J Clin Microbiol*, **41**, 386–92.

Medical Research Council. 1965. Definition and classification of chronic bronchitis and clinical epidemiological purposes. *Lancet*, **1**, 775-9.

Michelow, I.C., Lozano, J., et al. 2002. Diagnosis of *Streptococcus pneumoniae* lower respiratory infection in hospitalized children by culture, polymerase chain reaction, serological testing, and urinary antigen detection. *Clin Infect Dis*, **34**, 1–11.

Mitchell, D.M. and Miller, R.F. 1995. New developments in the pulmonary diseases affecting HIV-infected individuals. *Thorax*, **50**, 294–302.

Mokotoff, E.D., Dunn, R.A., et al. 1995. Transmission of pertussis from adult to infant – Michigan 1993. *Morb Mort Wkly Rep*, **44**, 74–6.

Montgomery, J.M., Lehmann, D., et al. 1990. Bacterial colonization of the upper respiratory tract and its association with acute lower respiratory tract infections in highland children of Papua New Guinea. *Rev Infect Dis*, **12**, Suppl., S1006–16.

Morbidity and Mortality Weekly Report. 2000. Preventing pneumococcal disease among infants and young children. *Morb Mort Wkly Rep*, **49**, RR-9.

Morbidity and Mortality Weekly Report. 2001. An outbreak of *Mycoplasma pneumoniae* in Colorado. *Morb Mort Wkly Rep*, **50**, 227-230.

Morton, S., Bartlett, C.L., et al. 1986. Outbreak of legionnaires' disease from a cooling water system in a power station. *Br J Ind Med*, **43**, 630–5.

Moxon, E.R. 1986. The carrier state: *Haemophilus influenzae*. *J Antimicrob Chemother*, **18**, Suppl. A, 124.

Muraca, P.W. and Stout, J.E. 1988. Legionnaires' disease in the working environment. Implications for environmental health. *Am Ind Hyg Assoc J*, **49**, 584–91.

Murdoch, C., Read, R.C., et al. 2002. Choline binding protein A of *Streptococcus pneumoniae* elicits chemokine production and expression of intercellular adhesion molecule 1 (CD54) by human alveolar epithelial cells. *J Infect Dis*, **186**, 1253–60.

Murphy, T.F. and Sethi, S. 1992. Bacterial infection in chronic obstructive pulmonary disease. *Am Rev Respir Dis*, **146**, 1067–83.

Murphy, T.F., Sethi, S., et al. 1999. Simultaneous respiratory tract colonization by multiple strains of non-typable *H. influenzae* in chronic obstructive pulmonary disease. *J Infec Dis*, **180**, 404–9.

Musher, D.M. 1992. Infections caused by *Streptococcus pneumoniae*: clinical spectrum, pathogenesis, immunity and treatment. *Clin Infect Dis*, **14**, 801–9.

Musher, D.M., Kubeitschek, K.R., et al. 1983. Pneumonia and acute febrile tracheal bronchitis due to *Haemophilus influenzae*. *Ann Intern Med*, **99**, 444–50.

Musher, D.M., Watson, D.A. and Baughn, R.E. 1990. Does naturally acquired IgG antibody to cell wall polysaccharide protect human subjects against pneumococcal infection? *J Infect Dis*, **161**, 736–40.

Myerowitz, R.L., Pasculle, A.W., et al. 1979. Opportunistic lung infection due to Pittsburgh pneumonia agent. *N Engl J Med*, **301**, 953–8.

Nicotra, M.B., Rivera, M., et al. 1982. Antibiotic therapy of acute exacerbations of chronic bronchitis: a controlled study using tetracycline. *Ann Intern Med*, **97**, 18–21.

Niederman, M.S. 1994. An approach to empiric therapy of nosocomial pneumonia. *Med Clin North Am*, **78**, 1123–41.

Nivens, D.E., Ohman, D.E., et al. 2001. Role of alginate and its O acetylation in the formation of *Pseudomonas aeruginosa* microcolonies and biofilms. *J Bacteriol*, **183**, 1047–57.

Noble, R.C. and Overman, S.B. 1994. Pseudomonas stutzeri infection. A review of hospital isolates and a review of the literature. *Diagn Microbiol Infect Dis*, **19**, 51–64.

Nuorti, S.P., Butler, J.C., et al. 2000. Cigarette smoking and invasive pneumococcal disease. *New Engl J Med*, **342**, 732–4.

Ojeniyi, B. 1994. Polyagglutinable *Pseudomonas aeruginosa* from cystic fibrosis patients. A survey. *APMIS Suppl*, **46**, 1–44.

Olin, P., Rasmussen, F., et al. 1997. Randomised control trial of 2-component, 3-component and 5-component acellular pertussis vaccines compared with whole-cell pertussis vaccine. *Lancet*, **350**, 1569–77.

Oliver, A., Canton, R., et al. 2000. High frequency of hypermutable *Pseudomonas aeruginosa* in cystic fibrosis lung infection. *Science*, **288**, 1251–4.

O'Neal, W.K. and Beaudet, A.L. 1994. Somatogene therapy for cystic fibrosis. *Hum Mol Genet*, **3**, 1497–502.

Ort, S., Ryan, J.L., et al. 1983. Pneumococcal pneumonia in hospitalized patients: clinical and radiological presentations. *J Am Med Assoc*, **249**, 214–18.

Ortqvist, A., Hedlund, J., et al. 1998. Randomised trial of 23-valent pneumococcal capsular polysaccharide vaccine in prevention of pneumonia in middle-aged and elderly people. *Lancet*, **351**, 399–403.

Palella, F.J., Delaney, K.M., et al. 1998. Declining morbidity and mortality amongst patients with advanced HIV infection. *N Engl J Med*, **338**, 853–60.

Pallares, R., Linares, J., et al. 1995. Resistance to penicillin and cephalosporin and mortality from severe pneumococcal pneumonia in Barcelona, Spain. *N Engl J Med*, **333**, 474–80.

Park, P.W., Pier, G.B., et al. 2001. Exploitation of syndecan-1 shedding by *Pseudomonas aeruginosa* enhances virulence. *Nature*, **411**, 98–102.

Pedersen, S.S., Koch, C., et al. 1986. An epidemic spread of multi-resistant *Pseudomonas aeruginosa* in a cystic fibrosis centre. *J Antimicrob Chemother*, **17**, 505–16.

Pennington, J.E. 1994a. Community-acquired pneumonia. In Pennington, J.E. (ed.), *Respiratory infections: diagnosis and management*, 3rd edn. New York: Raven Press, 193–206.

Pennington, J.E. 1994b. Hospital-acquired pneumonia. In Pennington, J.E. (ed.), *Respiratory infections: diagnosis and management*, 3rd edn. New York: Raven Press, 207–27.

Pines, A., Raafat, H., et al. 1968. Antibiotic regimes in severe and acute purulent exacerbations of chronic bronchitis. *Br Med J*, **2**, 735–8.

Pinner, R.W., Teutsch, S.M., et al. 1996. Trends in infectious disease mortality in the United States. *J Am Med Assoc*, **275**, 189–93.

Ponka, A. 1990. Absenteeism and respiratory disease among children and adults in relation to low level air pollution and temperature. *Environ Res*, **52**, 34–46.

Preston, N.W. 1994. Pertussis vaccination: neither panic nor complacency. *Lancet*, **344**, 491–2.

Principi, N., Esposito, S., et al. 2001. Role of *Mycoplasma pneumoniae* and *Chlamydia pneumoniae* in children with community acquired lower respiratory tract infection. *Clin Infect Dis*, **32**, 281–9.

Ramsey, B.W., Pepe, M.S., et al. 1999. Intermittent administration of inhaled tobramycin in patients with cystic fibrosis. *New Engl J Med*, **340**, 23–30.

Rayner, C.F., Tillotson, G., et al. 1995. Efficacy and safety of long-term ciprofloxacin the management of severe bronchiectasis. *J Antimicrob Chemother*, **34**, 149–56.

Read, R.C. 1999. Evidence based medicine: Empiric antibiotic therapy in community acquired pneumonia. *J Infect*, **39**, 171–8.

Read, R.C. and Wyllie, D.H. 2001. Toll receptors and sepsis. *Curr Opin Crit Care*, **7**, 371–5.

Read, R.C., Wilson, R., et al. 1991. Interaction of non-typable *Haemophilus influenzae* with human respiratory mucosa *in vitro*. *J Infect Dis*, **163**, 549–58.

Read, R.C., Roberts, P., et al. 1992. Effect of *Pseudomonas aeruginosa* rhamnolipids on mucociliary transport and ciliary beating. *J Appl Physiol*, **72**, 2271–7.

Redd, S.C., Rutherford III, G.W., et al. 1990. The role of human immunodeficiency virus infection in pneumococcal bacteremia in San Francisco residents. *J Infect Dis*, **162**, 1012–17.

Reimann, H.A. 1938. An acute infection of the respiratory tract with atypical pneumonia. A disease entity probably caused by a filterable virus. *J Am Med Assoc*, **111**, 2377–84.

Rein, M.F., Gwaltney, J.M., et al. 1978. Accuracy of Gram's stain in identifying pneumococci in sputum. *J Am Med Assoc*, **239**, 2671–3.

Reynolds, H.Y. 1994. Normal and defective respiratory host defenses. In: Pennington, J.E. (ed.), *Respiratory infections: diagnosis and management*, 3rd edn. New York: Raven Press, 1–34.

Robinson, L.A. and Moulton, A.L. 1994. Intrapleural fibrinolytic treatment of multi-loculated thoracic empyemas. *Ann Thorac Surg*, **57**, 803–13.

Romanus, V. and Jonsell, R. 1987. Pertussis in Sweden after cessation of general immunization in 1979. *Pediatr Infect Dis J*, **6**, 364–71.

Rosen, M.J. 1994. Pneumonia in patients with HIV infection. *Med Clin North Am*, **78**, 1067–79.

Rutland, J. and de Longh, R.U. 1990. Random ciliary orientation. A cause of respiratory tract disease. *N Engl J Med*, **323**, 1681–4.

Saint, S., Bent, S., et al. 1995. Antibiotics in chronic obstructive pulmonary disease exacerbations: a meta-analysis. *J Am Med Assoc*, **273**, 957–60.

Salyers, A.A. and Whitt, D.D. 1994. *Bacterial pathogenesis: a molecular approach*. Washington, DC: ASM Press.

Saunders, C.J.P., Joseph, C.A. and Watson, J.M. 1994. Investigating a single case of legionnaire's disease: guidance for consultants in communicable disease control. *Commun Dis Rep CDR Rev*, **4**, 10, R112–14.

Schaad, V.B., Stoupis, C., et al. 1991. Clinical, radiological and magnetic resonance monitoring for skeletal toxicity in pediatric patients with cystic fibsosis receiving a three month course of ciprofloxacin. *Pediatr Infect Dis J*, **10**, 723–9.

Schalen, L., Eliasson, I., et al. 1992. Acute laryngitis in adults: results of erythromycin treatment. *Acta Otolaryngol Suppl*, **492**, 55–7.

Schlesinger, R.B. 1990. The interaction of inhaled toxicants with respiratory tract clearance mechanisms. *Rev Toxicol*, **20**, 257–68.

Seemungal, T.A., Donaldson, G.C., et al. 1998. Effect of exacerbation on quality of life in patients with chronic obstructive pulmonary disease. *Am J Respir Crit Care Med*, **157**, 1418–22.

Segal, G., Purcell, M. and Shuman, H.A. 1998. Host cell killing and bacterial conjugation require overlapping sets of genes within a 22 kb region of the *Legionella pneumophila* genome. *Proc Natl Acad Sci USA*, **95**, 1669–74.

Sethi, S., Evans, N., et al. 2002. New strains of bacteria and exacerbations of chronic obstructive pulmonary disease. *N Engl J Med*, **347**, 465–71.

Shands, K.N., Ho, J.L., et al. 1985. Potable water as a source of legionnaires' disease. *J Am Med Assoc*, **253**, 1412–16.

Shanley, J.D. 1995. Mechanisms of injury by virus infections of the lower respiratory tract. *Rev Med Virol*, **5**, 41–50.

Shapiro, E.D., Berg, A.T., et al. 1991. The protective efficacy of polyvalent pneumococcal polysaccharide vaccine. *N Engl J Med*, **325**, 1453–60.

Sim, R.B. 1998. *Complement*. In: Hausler, W.J. Jr and Sussman, M. (eds), *Topley & Wilson's microbiology and microbial infections*, vol. 3. Bacterial infections, 9th edn. London: Edward-Arnold, 37–46.

Smith, A.L. 1994. *Haemophilus influenzae*. In: Pennington, J.E. (ed.), *Respiratory infections: diagnosis and management*, 3rd edn. New York: Raven Press, 435–54.

Smith, A.L., Ramsey, B.W., et al. 1989. Safety of aerosol tobramycin administration for three months to patients with cystic fibrosis. *Pediatr Pulmonol*, **7**, 265–71.

Soler, N., Torres, A., et al. 1998. Bronchial microbial patterns in the severe exacerbation of chronic obstructive pulmonary disease requiring mechanical ventilation. *Am J Respir Crit Care Med*, **157**, 1598–605.

Steed, L.L., Akporiaye, E.T., et al. 1992. Bordetella pertussis induces respiratory burst activity in human polymorphonuclear leucocytes. *Infect Immun*, **60**, 2101–5.

Stout, J.E. and Yu, V.L. 1997. Legionellosis. *N Engl J Med*, **337**, 682–7.

Stutman, H.R., Lieberman, M.J., et al. 2002. Antibiotic prophylaxis in infants and young children with cystic fibrosis: a randomized controlled trial. *J Pediatrics*, **140**, 299–305.

Suter, S. 1994. The role of bacterial proteases in the pathogenesis of cystic fibrosis. *Am J Respir Crit Care Med*, **150**, S118–22.

Swanson, M.S. and Hammer, B.K. 2000. Legionella pneumophila pathogenesis: a fateful journey from amoebae to macrophages. *Annu Rev Microbiol*, **54**, 567–613.

Swartz, M.N. 2001. Recognition and management of anthrax - an update. *New Engl J Med*, **345**, 1621–6.

Tettelin, H., Nelson, K.E., et al. 2001. Complete genome sequence of a virulent isolate of *Streptococcus pneumoniae. Science*, **293**, 498–505.

Thick, R.B., Baltimore, R.S., et al. 1985. IgG proteolytic activity of *Pseudomonas aeruginosa* in cystic fibrosis. *J Infect Dis*, **151**, 589–98.

Tinnion, O.N. and Hanlon, M. 2000. Acellular vaccines for preventing whooping cough in children. *Cochrane Database Syst Rev*, **2**, CD001478.

Torres, A., Serra-Batilles, J., et al. 1991. Severe community acquired pneumonia: epidemiology and prognostic factors. *Am Rev Respir Dis*, **144**, 312–18.

Tuazon, C.U. and Murray, H.W. 1994. Atypical pneumonias. In: Pennington, J.E. (ed.), *Respiratory infections: diagnosis and management*, 3rd edn. New York: Raven Press, 407–34.

Tuomanen, E.I., Austrian, R., et al. 1995. Pathogenesis of pneumococcal infection. *N Engl J Med*, **332**, 1280–4.

Turk, D.C. 1984. The pathogenicity of *Haemophilus influenzae. J Med Microbiol*, **18**, 1–16.

Van Cauwenberge, P.B., Vander Mijnsbrugge, A.M. and Ingels, K.J. 1993. The microbiology of acute and chronic sinusitis and otitis media: A review. *Eur Arch Otorhinolaryngol*, **250**, Suppl. 1, 3–6.

Verghese, A., Al-Samman, M., et al. 1994. Chronic bronchitis and bronchiectasis in HIV infection. *Arch Intern Med*, **154**, 2086–91.

Verhoef, J. 1993. Prevention of infection in the neutropenic patient. *Clin Infect Dis*, **17**, Suppl. 2, S359–67.

Vila, J., Markos, A., et al. 1993. *In vitro* antimicrobial production of β-lactamases, aminoglycoside-modifying enzymes, and chloramphenicol acetyl transferase by and susceptibility of clinical isolates of *Acinetobacter baumannii. Antimicrob Agents Chemother*, **37**, 138–41.

Vitovski, S., Dunkin, K.T., et al. 2002. Non-typable *Haemophilus influenzae* in carriage and disease: a difference in IgA1 protease levels. *J Am Med Assoc*, **287**, 1699–705.

Waldvogel, F.A. 1990. *Staphylococcal aureus* (including toxic shock syndrome). In: Mandell, G.L., Douglas, R.G. and Bennett, J. (eds), *Principles and practice of infectious diseases*, 3rd edn. New York: John Wiley & Sons, 1489–510.

Weiser, J.N., Markiewicz, Z., et al. 1996. Relationship between phase variation in colony morphology, intrastrain variation in cell wall physiology, and nasopharyngeal colonization by *Streptococcus pneumoniae. Infect Immun*, **64**, 2240–5.

Whitney, C.G., Farley, M.M., et al. 2000. Increasing prevalence of multidrug resistant *Streptococcus pneumoniae* in the United States. *N Engl J Med*, **343**, 1917–24.

Wilkinson, S.G. and Pitt, T.L. 1995. *Burkholderia (Pseudomonas) cepacia* pathogenicity and resistance. *Rev Med Microbiol*, **6**, 10–17.

Wilson, R. 1988. Secondary ciliary dysfunction. *Clin Sci*, **75**, 113–20.

Wilson, R., Roberts, D.E., et al. 1985. The effect of bacterial products on human ciliary function *in vitro. Thorax*, **40**, 125–31.

Wilson, R., Pitt, T., et al. 1987. Pyocyanin and 1-hydroxyphenazine produced by *Pseudomonas aeruginosa* inhibit the beating of human respiratory cilia *in vitro. J Clin Invest*, **79**, 221–9.

Wilson, R., Read, R.C., et al. 1991. Effects of *Bordetella pertussis* infection on human respiratory epithelium *in vivo* and *in vitro. Infect Immun*, **59**, 337–45.

Woodhead, M.A., MacFarlane, J.T., et al. 1985. Aetiology and outcome of severe community-acquired pneumonia. *J Infect*, **10**, 204–10.

Woodhead, M.A., MacFarlane, J.T., et al. 1987. Prospective study of the aetiology and outcome of pneumonia in the community. *Lancet*, **1**, 671–4.

Woods, D.E. 1994. Bacterial colonization of the respiratory tract: clinical significance. In: Pennington, J.E. (ed.), *Respiratory infections: diagnosis and management*, 3rd edn. New York: Raven Press, 35–41.

Woolfrey, B.T. and Moody, J.A. 1991. Human infections associated with *Bordetella bronchiseptica. Clin Microbiol Rev*, **4**, 243–55.

Yamagishi, Y., Fujita, J., et al. 1993. Epidemic of nosocomial *Pseudomonas cepacia* in immunocompromised patients. *Chest*, **103**, 1706–9.

Yamamoto, Y., Cline, T.W., et al. 1994. *Legionella* and macrophages. In: Eisenstein, T.K. and Zwilling, B. (eds), *Macrophages and infection*. New York: Marcel Dekker, 329–48.

25

Bacterial infections of the genital tract

CATHERINE ISON

The lower genital tract in women has its own microbial flora and in men is often colonized, although this may be a transient flora following sexual intercourse. In contrast, the upper genital tract in men and women is believed to be sterile. It was Doderlein in 1894 who first described lactobacilli as part of the 'normal flora' of the vagina. He believed that the lactic acid produced by lactobacilli from glucose in the vaginal fluid kept the vagina at an acidic pH, which discouraged colonization by other organisms. Glucose is produced from glycogen by the vaginal epithelium and this is utilized by the lactobacilli producing lactic acid. Glycogen is found on vaginal epithelium after puberty and is under estrogen control, hence is seldom present in premenarchal girls or postmenopausal women, unless on hormone replacement therapy. Lactobacilli are therefore found mostly in the vaginal fluid of women of child-bearing years and are considered 'normal' flora. In latter years it has been evident that the production of hydrogen peroxide (H_2O_2) by some lactobacilli is toxic to other bacteria and that this, in addition to lactic acid, may exert a protective effect in the vagina (Hillier 1998). The prevalence of H_2O_2-producing lactobacilli has been shown to be between 42–74 percent in pregnant women (Hillier et al. 1992, 1993; Puapermpoonsiri et al. 1996) and 65 percent in nonpregnant women (Hawes et al. 1996). In premenarchal girls the prevalence is low, 11 percent (Hill et al. 1995), although postmenopausal women have intermediate levels, 38 percent (Hillier and Lau 1997). Hydrogen peroxide has been shown to be toxic to

bacteria associated with bacterial vaginosis, *Gardnerella vaginalis* and *Bacteroides bivius* (Klebanoff et al. 1991) and to the human immunodeficiency virus (Klebanoff and Coombs 1991) supporting the hypothesis that lactobacilli play a major role in protecting the vagina from specific pathogens.

Lactobacillus acidophilus and *Lactobacillus fermentum* have been demonstrated in early studies to be the predominant species in the vagina (Ragosa and Sharpe 1960; Giorgi et al. 1987). However, this may be inaccurate as the use of molecular methods identified *Lactobacillus crispatus* and *Lactobacillus jensenii* as the predominant species (Johnson et al. 1980). It appears that the capacity for the organism to produce H_2O_2 is more important than the individual speciation, being more persistent in the vagina than H_2O_2-negative strains.

GENITAL TRACT INFECTIONS

Disruption of the normal ecology of the genital tract can result by acquisition of a sexually transmitted pathogen such as *Neisseria gonorrhoeae*, *Chlamydia trachomatis*, *Treponema pallidum* (Table 25.1) or by increased numbers of bacteria that comprise part of the normal flora, generally in small numbers, such as *G. vaginalis* and group B streptococci (Table 25.1). Infections due to chlamydia are dealt with in Chapter 78, Chlamydia. In many instances the change in ecology will result in symptoms, encouraging the individual to seek care, but

Table 25.1 *Clinical presentation of bacterial infections of the genital tract*

Organism	Main clinical presentation
Primary pathogens	
Neisseria gonorrhoeae	Uncomplicated infection (*cervicitis*, urethritis, pharyngeal, rectal)
	Complicated infection (salpingitis, PID, prostatitis, epididymitis)
Treponema pallidum	Genital ulceration
	Systemic infection
Haemophilus ducreyi	Genital ulceration
Klebsiella granulomatis	Genital ulceration
Secondary pathogens	
Gardnerella vaginalis	Bacterial vaginosis
Prevotella spp.	
Peptostreptococci	
Mobliuncus spp.	
Streptococcus agalactiae	Aerobic vaginitis
	Asymptomatic carriage
Staphylococcus aureus	Toxic shock syndrome
Clostridium perfringens	Post-partum infection
	Septic abortion
Actinomyces spp.	Discharge associated with IUCD

in some instances the bacteria remain silent causing asymptomatic infection which may remain undetected. The majority of infections present as mucosal infections of the lower genital tract where the challenge is to diagnose the infecting agent from amongst the resident microflora. However, many of these bacteria can ascend to the upper genital tract, which is normally sterile, to cause more complicated infections such as pelvic inflammatory disease with the possible sequelae of infertility, or, on occasions, can invade the blood causing systemic infections such as toxic shock syndrome. This chapter will first consider those bacteria that are primary pathogens, where their isolation always indicates the need for antimicrobial therapy, and then the bacteria that are secondary or opportunistic pathogens, where isolation needs interpretation together with clinical information before therapy is initiated.

NEISSERIA GONORRHOEAE

Neisseria gonorrhoeae and *Chlamydia trachomatis* (see Chapter 48, Neisseria and Chapter 78, Chlamydia) are the major causes of bacterial sexually transmitted infection in the industrialized world. *N. gonorrhoeae* is primarily a mucosal pathogen that causes infection of the lower genital tract in men and women, occasionally ascends to the upper genital tract to cause complicated infection, and rarely invades the blood to cause disseminated disease. Uncomplicated gonococcal infection (UGI) is predominantly a mucosal infection of the lower genital tract. In men, the primary site of infection is the urethra causing a discharge and/or dysuria and is symptomatic in most of the patients. Rectal infections are common among men who have sex with men and pharyngeal infection can occur in men or women who have oral sexual intercourse and is generally asymptomatic. In women, the primary site of colonization is the cervix, where the organism attaches preferentially to the columnar epithelium rather than the squamous epithelium of the vagina. In contrast to gonorrhea in men, infection in women is often asymptomatic in at least 50 percent of patients. Colonization of the urethra and rectum in women does occur but it is often unclear whether this is a true colonization or is leakage of cervical secretions from the vagina.

N. gonorrhoeae presents as a lower genital tract infection in >90 percent of patients in most parts of the world but can ascend to cause complicated gonococcal infection (CGI) of the upper genital tract. It presents as salpingitis or pelvic inflammatory disease in women and epididymitis or prostatitis in men. Complicated infection occurs more commonly in women than in men, largely because of undiagnosed asymptomatic infection. Disseminated gonococcal infection, where the organism invades into the blood to cause systemic infection, is a separate entity and is rare. It presents primarily as a rash, septicemia, and arthritis and is associated with certain strains of gonococci and is more prevalent in individuals with deficiency in the terminal complement components.

N. gonorrhoeae is an obligate human pathogen with no other natural host, which can cause repeated infection in the same host with no apparent immunity. This is believed to result from its ability for phase and antigenic variation, allowing the organism to appear novel at each infection. *N. gonorrhoeae* is genetically diverse and highly competent for genetic exchange and recombination at all stages of its life cycle, resulting in a nonclonal or panmictic population. In vivo, this probably occurs during mixed infections which are more likely to occur in sexually active individuals with high rates of partner change.

N. gonorrhoeae possess pili, small appendages, which enable the organism to overcome the hydrophobic interactions between the organism and the epithelial cell and hence come into close proximity, allowing specific receptors to attach to the host cell surface. Attachment is then enhanced by a family of outer membrane proteins, Opa proteins. *N. gonorrhoeae* are then believed to adhere to the epithelial cell, mediated by Opa proteins which interact with two receptors on the epithelial cells, CD66 and the heparan sulfate proteoglycans. The bacteria–epithelial cell interaction also induces a proinflammatory cytokine response. Invasion then occurs and the major porin, Por, is believed to play an important role in this and also induces apototic cell death of both epithelial cells and polymorphs (Naumann et al. 1999).

Diagnosis

PRESUMPTIVE DIAGNOSIS

Microscopy is used for the presumptive diagnosis of gonorrhea in most parts of the world and is simple to perform and gives a result quickly. The presence of intracellular gram-negative diplococci in smears is considered diagnostic and in urethral smears from symptomatic men has a sensitivity of >95 percent. However, in women or asymptomatic men, the sensitivity is between 30–50 percent (Ison 1990; Sherrard and Barlow 1996; Jephcott 1997). This difference is probably due to higher numbers of bacteria being present in symptomatic individuals, which are easily detected by microscopy, compared to considerably lower numbers present in asymptomatic individuals which are below the detection limit. The sensitivity is also lower for rectal smears due to the large numbers of other bacteria present, including other gram-negative cocci, and is really only reliable when specimens have been taken from a lesion through a proctoscope, rather than perianal or blind rectal swabs. Microscopy is not commonly used for examining pharyngeal specimens. The specificity of the Gram stained smear for the presumptive diagnosis of gonorrhea is usually high when performed by experienced personnel (Hook and Handsfield 1999; Knapp and Koumans 1999). Comparison of numbers of positives obtained by Gram stained smears and culture can give an indication of either problems with reading smears or of isolation techniques, the correlation between smear and culture for urethral specimens from symptomatic men should be more than 95 percent.

CONFIRMATORY DIAGNOSIS

Isolation of the causative organism, *N. gonorrhoeae*, is still considered necessary for the confirmation of the diagnosis of gonorrhea, whenever the resources are available. The sensitivity of culture is thought to be high, 80–100 percent, but is dependent on a good specimen and isolation procedure. A nutritious agar base is

required for the growth of *N. gonorrhoeae*, which is a fastidious organism, supplements containing an iron source and selective agents. The mostly commonly used agar base is GC agar base, which contains a mixture of peptones, starch, and agar, alternatively Columbia agar can be used. Blood, either lysed, heated (chocolatized) or pure hemoglobin, is often used as a supplement providing both amino acids and iron. Nonblood based supplements include IsoVitaleX or Vitox, which are an equivalent alternative. The use of selective agents to suppress the normal flora is recommended, particularly for specimens taken from the lower genital tract in women or from the rectum. Vancomycin or lincomycin is added to suppress gram-positive cocci, colistin and/or trimethoprim to inhibit other gram-negative organisms, and nystatin or amphotericin to suppress yeasts. Strains susceptible to the vancomycin, *env* mutants, and trimethoprim have been reported and can cause false negative results. In an ideal situation, both nonselective and selective isolation media would be used but there are seldom resources available for both and if a single medium is to be used it should contain selective agents to prevent overgrowth of the *N. gonorrhoeae* by the resident normal flora. Incubation at 36–37°C in a high humidity (>90 percent) and increased carbon dioxide concentration (5–10 percent) should be for a minimum of 48 h before the isolation medium should be discarded as negative.

Preliminary identification of any resulting colonies as oxidase positive gram-negative cocci is a presumptive identification of *Neisseria* spp. In many countries where resources are limited, no further identification is performed but it is still standard practice, where possible, to confirm the identity as *N. gonorrhoeae*. Historically, *N. gonorrhoeae* was identified by its ability to produce acid from glucose but not from other carbohydrates such as maltose, sucrose, and lactose. Carbohydrate utilization has now been combined with other preformed enzymes, such as the aminopeptidases, in identification kits, which give a result within 4 h. These tests will fully speciate the organism under test and are particularly useful for laboratories that encounter only occasional isolates. Biochemical reactions of the clinically important *Neisseria* spp. are shown in Table 25.2.

Table 25.2 *Identification of* Neisseria gonorrhoeae *from the clinically important* Neisseria *spp. and* Branhamella catarrhalis

Organism	Acid production from				Aminopeptidase		DNAase	Butyrate esterase
	Glucose	Maltose	Sucrose	Lactose	γ-glutamyl	Propyl		
N. gonorrhoeae	+	−	−	−	−	+	−	−
N. meningitidis	+	+	−	−	+	V	−	−
N. lactamica	+	+	−	+	−	−	−	−
N. cinerea	−	−	−	−	NK	NK	−	−
B. catarrhalis	−	−	−	−	−	−	+	+

NK, not known; V, variable result.

Alternatively, there are immunological reagents available that can identify *N. gonorrhoeae* and eliminate *N. meningitidis* and *N. lactamica* but do not give a full speciation. There are two reagents commonly used which both contain monoclonal antibodies to the major porin, Por. The antibodies are either linked to staphylococcal protein A and used in a coagglutination test (Phadebact Monoclonal GC Test, distributed by Boule in the UK) or linked to fluorescein and used in an immunofluoresence test (Syva Microtrak *Neisseria gonorrhoeae* Culture Confirmation Test, distributed by Launch Diagnostics in the UK). Both these reagents contain mixtures of antibodies to specific epitopes rather than a single antibody to a conserved epitope, hence strains giving false negative results do occasionally occur. Fortunately, the mixtures in each of these reagents differ and so they complement each other, rarely both giving a negative result. The coagglutination test has the advantage that it will also divide gonococcal isolates into two groups, PIA (WI) or PIB (WII/III), allowing a preliminary typing of strains.

MOLECULAR DETECTION

Detection of *N. gonorrhoeae* directly in the specimen can be achieved using probes or by amplification of specific sequences of DNA. The advantage of these tests are that they are highly sensitive and can detect small amounts of DNA and that they can detect nonviable organisms. Unlike previous attempts at antigen detection using immunological approaches, molecular detection methods are also highly specific. This is particularly important for screening low prevalence populations. The disadvantage of using molecular detection for *N. gonorrhoeae* is that susceptibility testing still requires a viable organism.

Molecular detection for gonorrhea has not been as widely used as for chlaymdial infection, largely because existing diagnostic testing has performed satisfactorily and has been less expensive. However, nucleic acid amplification tests (NAAT) has been shown to have a higher sensitivity than culture in some studies, particularly in situations where the resources or facilities for culture are not ideal. They have the added advantage that they can be used with noninvasive samples such as urine or self-taken swabs or tampons. The widespread use of NAATs for chlamydial infection has encouraged the use for gonorrhea as most of the commercial kits can test for both organisms on the same sample.

The nucleic acid hybridization test (Gene Pace 2) has a high sensitivity and specificity when used with endocervical or urethral specimens (Koumans et al. 1998). However because of the lack of amplification, they are likely to be less sensitive than the NAATs. The sensitivity of NAATs using amplification by polymerase chain reaction (Amplicor, Roche) or strand displacement amplification (BD Probtec, Becton Dickinsen) on endocervical specimens (van Dyck et al. 2001) or self-taken swabs or tampons (Knox et al. 2002) is high and often higher than culture. However, they are less satisfactory for detection of *N. gonorrhoeae* in urine from women (Van Doornum et al. 2001). Sensitivity is high when either urethral swabs or urine are used in men (Van Doornum et al. 2001; Palladino et al. 1999; Akduman et al. 2002). The specificty is high with all these tests but confirmation of positive results should be considered, particularly when low prevalence populations are screened.

SUSCEPTIBILITY TESTING

The purpose of susceptibility testing is to predict the outcome of therapy. In many instances for gonorrhea the patient will have been treated before the result of the susceptibility testing is known. The choice of a first line therapy should be based on good surveillance data and should give a high chance of therapeutic success, >95 percent. Surveillance programs mostly test a sample of gonococcal isolates each year using standard methods which detect emergence of resistance and drifts in susceptibility. There are a number of well established programs that have produced longitudinal data (Members of the Australian Gonococcal Surveillance Programme 1984; Schwarcz et al. 1990; Dillon 1992; Van de Laar et al. 1997) that inform therapy. In many parts of the world, penicillin is no longer suitable therapy and in recent years ciprofloxacin resistance has become widespread (WHO Western and Pacific Region Gonococcal Antimicrobial Surveillance Programme 2001).

However, most laboratories continue to test individual isolates on a daily basis in order to categorize strains as susceptible, exhibiting reduced susceptibility or resistance, particularly to the antibiotic used for therapy. Commonly used treatments include ciprofloxacin, ceftriaxone, and in some areas penicillin (or ampicillin). Routine susceptibility testing is largely performed using disc diffusion using a recommended method such as BSAC (British Society of Antimicrobial Chemotherapy) (King 2001) or NCCLS (National Committee for Control of Laboratory Standards 2003). The methods differ in the base medium and the disc content of antibiotic used, but if used according to the standard protocols and with appropriate controls, both give acceptable results. Problems can occur with susceptibility testing for *N. gonorrhoeae* because different strains vary in their nutritional requirements and growth rates, making comparison with control strains difficult. A quick and simple alternative is to use agar dilution breakpoints where growth or no growth is scored on agar plates incorporating one or two concentrations of antibiotic to categorize the isolates. It is a useful method for detecting high level resistance, for example, resistance to ciprofloxacin can be detected using a single concentration of 0.5 mg/l (minimum inhibitory concentration (MIC) $\geqslant 1$ mg/l).

Antibiotic resistance

PENICILLIN

Penicillin or ampicillin was for many years the treatment of choice for gonorrhea until the emergence of chromosomal resistance due to continual usage of the antibiotic and the acquisition of plasmids from *Haemophilus* spp. Chromosomal resistance was the first to appear but was low level and initially responded to the use of increased dosage for treatment. Therapeutic failure due to chromosomally-mediated resistant *N. gonorrhoeae* (CMRNG) was reported in 1985 (Faruki et al. 1985) and became an increasing problem. Chromosomal resistance in *N. gonorrhoeae* is the result of additive effect of mutations at a number of loci including *penA*, *mtr*, *penB*, *penC*, and *ponA1* (Ropp et al. 2002), *penA* affects the affinity of penicillin for penicillin binding protein (PBP)2, *mtr* involves over expression of the efflux pump, *penB* is a mutation in the *por* gene, *ponA1* affects PBP1 and the function of *penC* is unknown. Together, mutations in these loci decrease the permeability of the cell wall, acting as a barrier to the penicillin.

In contrast, plasmid-mediated resistance was not detected until 1976, is enzyme-mediated and high level with a greater chance of therapeutic failure. In many countries the greatest prevalence was also in the 1980s but the prevalence remains high in some areas such as the Western Pacific, and penicillin cannot be used for therapy. Penicillinase-producing *N. gonorrhoeae* (PPNG) were first detected in infections acquired in Africa and Asia and were shown to be carrying plasmids of 3.2 and 4.4 MDa, respectively (Roberts 1989). These plasmids require the presence of a conjugative plasmid to transfer between gonococci and have now become disseminated worldwide. PPNG carrying plasmids of differing molecular weights have been isolated but have not spread so widely. All penicillinase plasmids in *N. gonorrhoeae* encode for the TEM-1 β-lactamase and differences in size reflect deletions in nonfunctional parts of the gene (Pagotto et al. 2000).

CIPROFLOXACIN

Ciprofloxacin, a fluoroquinolone, is an attractive alternative for the treatment of gonorrhea because it is given orally as a single dose, and hence compliance is good and can be observed. When ciprofloxacin was first introduced, *N. gonorrhoeae* were exquisitely sensitive and resistance was unknown. However, probably due to the misuse and overuse, particularly of the earlier generation of the quinolones, resistance due to mutations in the DNA gyrase gene, *gyrA*, and the topoisomerase IV gene, *parC*, has emerged (Knapp et al. 1997). High-level resistance (minimum inhibitory concentration (MIC) $\geqslant 1$ mg/l) is due to multiple mutations in both *gyrA* and/or *parC* and may be enhanced by over expression of efflux systems and changes in the porin. Gonorrhea can be treated with other quinolones such as ofloxacin but *N. gonorrhoeae* exhibit cross-resistance between the quinolones and hence these are inactive against resistant strains. The newer quinolones such as moxifloxacin act primarily against *parC* and so are unlikely to be useful alternatives. Currently resistance to quinolones in *N. gonorrhoeae* is solely chromosomally mediated but reports of plasmid-mediated resistance in other organisms is worrying (Martinez-Martinez et al. 1998). Quinolone-resistant *N. gonorrhoeae* (QRNG) are most prevalent in the Western Pacific Region where the prevalence in some countries is above 50 percent. However, the prevalence of QRNG is increasing rapidly in countries in the western world such that ciprofloxacin is unlikely to remain an effective therapy for gonorrhea for very many more years.

CEPHALOSPORINS

Ceftriaxone, a third generation cephalosporin, is a recommended treatment for gonorrhea. It has proved highly active against *N. gonorrhoeae* including PPNG, and hence the original recommended dosage of 250 mg has been reduced to 125 mg in some guidelines (Centers for Disease Control and Prevention 2002a). Therapeutic resistance has not been documented although reduced susceptibility has been detected using in vitro susceptibility testing (Schwebke et al. 1995). This is of concern because cross-resistance has already been noted between cefuroxime, an earlier cephalosporin, and penicillin (Ison et al. 1990), particularly if lower dosages become common and usage increases because of resistance to other agents such as ciprofloxacin. Ceftriaxone is given by injection and some patients may prefer an oral third generation cephalosporin. However, data on efficacy and availability are still limited.

ALTERNATIVE TREATMENTS

Spectinomycin, 2 g intramuscularly, is still recommended as an alternative therapy for gonorrhea. Resistance has been documented and is single-step and high-level but it has occurred sporadically (Easmon et al. 1982; Boslego et al. 1987; Zenilman et al. 1987) and has not spread widely suggesting that it may be a suitable alternative. However, it has not been extensively used in recent years and has become difficult to obtain.

Azithromycin is used more commonly for chlamydial infection but has been suggested as an alternative for gonorrhea particularly when resistance to penicillin and ciprofloxacin is high. The recommended dosage is 2 g, but patients often find this difficult to tolerate and 1 g has been used (Young et al. 1997; Xia et al. 2000). However, resistance has been noted with both dosages and increased usage is likely to increase the selective pressure.

Tetracycline is also more commonly used as treatment for chlamydial infection but has been used for gonorrhea where resources are limited. Resistance can be both chromosomal and low level (Johnson and Morse 1988) or plasmid-mediated and high-level (Morse et al. 1986). Plasmid-mediated resistance is due to the insertion of the *tetM* determinant, probably from streptococci, into the gonococcal conjugative plasmid, hence the plasmid can mobilize itself between gonococci and other genera. Since its first description in 1985 (Morse et al. 1986), tetracycline-resistant *N. gonorrhoeae* (TRNG) have spread widely, reducing its efficacy as an inexpensive alternative therapy.

Aminoglycosides such as gentamicin and kanamycin have been used in resource poor settings but information on resistance and relationship to in vitro susceptibility testing is lacking (Daly et al. 1997; Lkhamsuren et al. 2001). Co-trimoxazole is also used but the true prevalence of resistance is difficult to determine because of the difficulties in susceptibility testing (West et al. 1995).

Typing

Techniques used to discriminate between strains of *N. gonorrhoeae* from different sources have been used to monitor temporal changes, detect clusters of antibiotic resistant strains, and confirm or dispute contact, largely in cases of child or sexual abuse. Phenotypic methods of auxotyping, differentiation by nutritional requirement (Catlin 1973) and serotyping, reactivity with a panel of monoclonal antibodies directed at specific epitopes on Por, the major outer membrane protein (Knapp et al. 1984), have been used extensively for these purposes. However, these tests lacked discriminatory power and even when used in combination, a few auxotype/serovar classes predominate in most populations. Both auxotyping and serotyping had subjective endpoints presenting some problems of reproducibility, particularly in longitudinal studies. In recent years a range of molecular techniques have been described, all of which have enhanced discriminatory ability compared to the phenotypic methods (Ison 1998). However, many of the techniques examine variation in genes that are exposed to the immune response and hence are subject to greater diversity. While this approach does not give valuable information for studies over long time periods, where change is likely to occur through genetic exchange in mixed infections, it is more useful for identifying sexual contacts, short chains of transmission, and for studying sexual networks (Ward et al. 2000). There is increasing interest in using a molecular approach to identifying clusters of isolates from linked patients to aid public health interventions. Combined studies using epidemiological and molecular data suggest that molecular typing will identify clusters that are larger than clusters identified by epidemiological information alone, which may give useful data to clinicians if used prospectively.

TREPONEMA PALLIDUM

Treponema pallidum subsp. *pallidum* is the causative agent of syphilis, a mucocutaneous infection, which is highly infectious in the early stages and has serious sequelae if left untreated. The primary stage of syphilis usually presents as a painless ulcer, which occurs 9–90 days after exposure, and may be accompanied by a regional lymphadenopathy. The secondary stage has many symptoms including localized or diffuse ulcers, generalized lymphadenopathy, rash, and condylomata lata and presents 6 weeks to 6 months after exposure. Latent syphilis, by definition, has no clinical signs but gives a positive serological reaction and is present either less than (early) or more than (late) 2 years after exposure. Tertiary syphilis is rare when good antibiotic therapy is available but can include gummatous syphilis, neurosyphilis, and cardovascular syphilis.

T. pallidum is a spirochete of approximately 0.2 μm in diameter and 5–15 μm in length. It is an obligate human pathogen and has never been grown successfully on artificial media, the only successful cultivation being in rabbits. The cell envelope structure is unique and unlike other gram-negative organisms, it has an outer membrane, but lacks lipopolysaccharide and porins have not been detected. The organism possesses flagella, for motility, which are located in the periplasmic space instead of externally as is normally found in motile bacteria. It has a very restricted temperature range for growth, 30–35°C, and in rabbits the generation time has been demonstrated as 35 h. The complete genome sequence is known (Fraser et al. 1998) and the genome is determined to be 1.14 Mb, approximately 25 percent of the size of the *E. coli* genome, and encodes for 1040 genes or open reading frames.

Epidemiology

Syphilis is a worldwide public health problem (Gerbase et al. 1998). It is a major cause of genital ulcer disease in the developing world and the prevalence is high in the newly independent states of the former Soviet Union and is a concern because of the strong association with the acquisition of HIV infection. After a peak in incidence following World War II there was a decline in many countries, particularly in the industrialized world, and syphilis became largely a disease of men who have sex with men. Following the advent of AIDS in 1984 and subsequent changes in sexual behaviour, syphilis declined. In the United States there were efforts to eliminate syphilis. In recent years there have been an increasing numbers of outbreaks of syphilis reported in the USA and many countries in Europe (Doherty et al. 2002). Many of the outbreaks are among men who have sex with men, possibly reflecting an increase in risk behaviour following a perception that HIV infection is now a treatable infection. Penicillin is the mainstay of treatment for all stages of syphilis infection (Centers for Disease Control and Prevention 2002a)

Diagnosis

MICROSCOPY

Examination of lesion exudate or lymph nodes by dark-ground microscopy is a useful presumptive diagnostic tool for infectious (primary) syphilis. However, it should be performed by experienced workers because of the interference with commensal spirochetes that are found as part of the normal flora of the genital and gastro-intestinal tract. It is not considered as suitable for use with rectal or nongenital lesions and is not recommended for oral lesions.

Exudate from the syphilitic ulcer should be collected after cleansing of the ulcer with sterile saline using a gauze swab. The ulcer should then be squeezed to allow sufficient serous fluid to be produced and then this should be collected by a loop and placed onto a microscope slide. It is not necessary to add any saline to the microscope slide. After placing a coverslip onto the fluid it should be examined by dark ground microscopy within 10 min in order to identify the characteristic morphology and motility of *T. pallidum*. If the initial examination is negative, this test can be repeated daily for at least 3 days as long as antibiotics are not given.

SEROLOGICAL DIAGNOSIS

The inability to cultivate *T. pallidum* has led to the use of the detection of the host's immune response as an indicator of exposure to infection. Serological tests for syphilis can be divided into screening tests and confirmatory tests. Historically screening tests employed nontreponemal antigen, which detects antilipoidal antibodies that are produced in response to the tissue damage caused by treponemal diseases including syphilis but also other unrelated diseases, whereas confirmatory tests use a specific treponemal antigen. However, the development of enzyme immunoassays using specific antigen that are automated and can be used to screen large numbers has in some instances replaced the nontreponemal tests for screening.

NONTREPONEMAL TESTS

The nontreponemal tests using the Venereal Disease Research Laboratory (VDRL) antigen, whose main components are cardiolipin, cholesterol, and lecithin, have been widely used for screening for syphilis. Plasma is mixed with the VDRL antigen and if antibody is present flocculation results which can be read microscopically (VDRL test) or macroscopically (Rapid Plasma Reagin (RPR) test), which is applicable for use in resource poor settings with little or no laboratory facilities. The tests differ in the particle size used and the pigment used to enhance visualization of the result. These tests are mostly used qualitatively but if used quantitatively can give a measure of the response to therapy. False positive results do occur, for instance during pregnancy or with autoimmune disease, such that all positive results should be confirmed using a specific treponemal test.

SPECIFIC TREPONEMAL TESTS

Enzyme immunoassays (EIA) using specific treponemal antigen, which has been cloned or sonicated and adsorbed onto a solid phase, are a relatively new approach to detection of antibody in syphilis. The EIAs can detect both IgG and/or IgM but those that detect IgG and IgM together are recommended because they tend to be more sensitive in primary infection (Schmidt et al. 2000). The *T. pallidum* particle assay (TPPA) and the *T. pallidum* haemagglutination assay (TPHA) both utilize a specific antigen which is attached to gelatin beads (TPPA) or erythrocytes (TPHA). The fluorescent-treponemal antibody-absorbed (FTA-ABS) test is an indirect immunofluorescence test which detects antibody that attaches to treponemes coating a microscope slide and the resulting antigen–antibody complexes are detected by a secondary antibody linked to a fluorochrome. The TPHA and TPPA have been widely used as screening and confirmatory tests, although the TPPA has been shown to be more sensitive than the TPHA in primary syphilis. The EIA is attractive because it has the capacity to screen large numbers using a specific antigen in an automated system. The TPHA/TPPA can be used to confirm a positive EIA result and vice versa. The FTA-ABS test is more time consuming and used mostly to confirm discrepant results between other tests. An additional test often used in research laboratories is based on immunoblotting to detect antibody to recombinant antigens. This test has a sensitivity similar to the FTA-ABS test. The choice of screening and confirmatory tests is dependent on workload and facilities available and will differ considerably dependent on the resources available.

HAEMOPHILUS DUCREYI

Haemophilus ducreyi is a fastidious organism which is the causative agent of the sexually transmitted disease, chancroid. It presents primarily as a genital ulcer but can also present as a regional lymphadenopathy with bubo formation. *H. ducreyi* is thought to establish itself through small abrasions, which may occur in the epidermis during sexual intercourse. Once colonization has been achieved, a small papule develops after a short incubation period of 5–10 days, which may progress to form an ulcer. The ulcers are painful and have ragged and undermined edges, and can persist for many months in the absence of antibiotic therapy. Lesions usually occur on the prepuce and frenulum of men and the vulva, cervix, and perianal region in women. Extra-genital chancroid is rare but lesions have been reported on the inner thighs, breasts, and fingers (Lewis 2000). Unlike other sexually transmitted infections, chancroid has few sequelae and has largely been considered a trea-table infection. However, the strong association of chan-croid, together with syphilis, with the acquisition of the human immunodeficiency virus (HIV) has renewed interest both in providing good diagnostic tests and research studies. This has become more important with the studies which have suggested that aggressive treat-ment of bacterial sexually transmitted infections (STI), such as chancorid, can reduce the acquisition of HIV (Grosskurth et al. 1995).

Chancroid is one of the most common STIs but is found predominantly in the Tropics and found only occasionally in North America and Europe (Trees and Morse 1995). Chancroid is more common in men than women and there is considerable debate over whether asymptomatic carriage occurs in women, which is largely unresolved. The high prevalence in resource-poor settings raises challenges for diagnosis as good labora-tory facilities are often unavailable. Isolation of *H. ducreyi* is still regarded as the gold standard, if available, because microscopy is nonspecific and molecular methods are not commercially available. However, in resource poor settings diagnosis is usually made on the clinical symptoms and signs and treated syndromically.

Isolation of *H. ducreyi*

It is possible to cultivate *H. ducreyi* but it is a fastidious organism that requires a highly nutritious medium. It also appears that different strains of *H. ducreyi* appear to grow preferentially on some culture media, particu-larly on primary isolation, and so the highest isolation rates (approximately 80 percent) have used multiple media. Two culture media, which include serum, that have been used are GC agar base supplemented with 1 percent hemoglobin, 5 percent fetal calf serum, and 1 percent IsoVitaleX and Mueller–Hinton agar supple-mented with 5 percent chocolatized horse blood and 1 percent IsoVitaleX (Dangor et al. 1992). A serum free medium which is less expensive to prepare, which has been used successfully, contains GC agar supplemented with 1 percent hemoglobin, 0.2 percent charcoal, and 1 percent IsoVitaleX (Lockett et al. 1991). Selective agents to inhibit the growth of skin organisms, particu-larly gram-positive cocci, and allow the growth of *H. ducreyi* have been difficult to choose, but 3 mg/l of vancomycin is usually added.

Isolation of *H. ducreyi* appears most successful if the ulcer is cleaned with a cotton swab to remove superficial pus and the material taken from the undermined edge of the ulcer and inoculated directly from the patient onto the culture medium. Primary isolation media should be incubated at 33°C for optimal growth condition in high humidity with 5 percent carbon dioxide for a minimum of 48–72 h before being discarded as negative. There is no transport medium that has been widely tested although some success has been demonstrated using a thioglycollate-hemin-based medium (Dangor et al. 1993).

After incubation, gray translucent colonies will appear which are gram-negative cocco-bacilli and weakly oxidase positive. The most characteristic sign is friable colonies that can easily be pushed around the plate. These three criteria are highly predictive of *H. ducreyi*, and in most instances no further identification is performed.

Microscopy

Microscopy is often used in clinical settings for the presumptive diagnosis of an STI but has proved poorly specific for the diagnosis of chancroid. The gram-negative cocco-bacilli are described as appearing in sheets, which resemble 'schools of fish' or 'railroad tracks' but this can be very difficult to detect in clinical specimens direct from the patient and is not recom-mended for the routine diagnosis (Albritton 1989).

Molecular detection

Commercial tests for the molecular detection of *H. ducreyi* are not widely available. However, a multiplex polymerase chain reaction (PCR) assay has been eval-uated which gives the simultaneous amplification of DNA from *H. ducreyi*, *T. pallidum*, and herpes simplex virus types 1 and 2 which is more sensitive than existing methods (Orle et al. 1996). Comparison between multi-plex-PCR and culture has shown the sensitivity of culture for *H. ducreyi* to be around 75 percent but this is dependent on good isolation procedures and has been less in some studies (Orle et al. 1996; Morse et al. 1997).

Antimicrobial resistance and therapy

H. ducreyi are innately sensitive to most antibiotics but, in a similar manner to *N. gonorrhoeae*, have become resistant to many antimicrobial agents through the acquisition of plasmids or selection of chromosomal mutations. Plasmid-mediated resistance has been reported to tetracycline, chloramphenicol, sulphona-mides, aminoglycosides, and β-lactam antibiotics. Chro-mosomal resistance is less well documented but decreased susceptibilities to trimethoprim, penicillin, ciprofloxacin, and ofloxacin have been reported (Trees and Morse 1995; Lewis 2000).

Despite the emergence of resistance over the last decade, chancroid is still treatable. Azithromycin, ciprofloxacin, erythromycin, and ceftriaxone form part of the recommendations for treatment from the Centers for Disease Control and Prevention and from the World Health Organization, although the dosage and regimens differ slightly.

Susceptibility testing

Testing the susceptibility of *H. ducreyi* can be difficult because of the fastidious nature of the organism and its tendency to adhere to the agar. Most studies have used agar dilution rather than disc diffusion and have been performed in specialist centers. A prerequisite for susceptibility testing is a viable organism and only a limited number of centers have routine culture established and hence the antibiogram of many strains is not tested. A standardized method has been recommended for determination of the MIC which includes, standardization of the inoculum, use of enriched Mueller–Hinton agar containing a range of antibiotic concentrations, incubation at 35°C in a humid atmosphere containing 5 percent carbon dioxide, and examination of the result after 48 h of incubation (Dangor et al. 1990).

KLEBSIELLA (CALYMMATOBACTERIUM) GRANULOMATIS

K. granulomatosis is a gram-negative bacillus which is the causative agent of donovanosis, which usually presents as genital ulceration but may also be oral or found at extra-genital sites. Genital ulceration is often associated with inguinal lymphadenopathy and if left untreated considerable tissue destruction can occur. Donovanosis is largely considered a tropical disease having not been reported in Europe and the USA although its true prevalence is unknown.

K. granulomatis has been included in several genera and was for a long time called *Calymmatobacterium granulomatis*. However, its similarity with *Klebsiella* species, originally detected using serology (Goldberg 1962) and ultrastructural studies (Kuberski et al. 1980) has been confirmed by molecular studies and the name change to *K. granulomatis* (Carter et al. 1999) proposed, although it remains controversial (Kharsany et al. 1999).

Culture of the organism in vitro is difficult and was originally performed in the yolk sacs of chick embryos but more recently in human peripheral blood mononuclear cells (Kharsany et al. 1997) and in Hep-2 cells (Carter et al. 1997). However, diagnosis is rarely made by culture but usually by the examination of smears stained using a Giemsa or related stain for the charactersitic donovan bodies or more recently by a newly described nucleic amplification test (Carter and Kemp 2000). The lack of a culture has prevented studies on susceptibility to antibiotics although a range have been used for treatment including ampicillin, chloramphenicol, and ceftraixone.

BACTERIAL VAGINOSIS

Gardnerella vaginalis, *Prevotella* spp., *Peptostreptococci*, the anaerobic curved rods, *Mobiluncus* spp. and *Mycoplasma hominis* are closely associated with the syndrome known as bacterial vaginosis (BV) which is characterized by a change in vaginal ecology. These organisms can be found in the 'normal' vagina but in small numbers, which in BV multiply and increase 100–1000 fold to become the predominant flora. These large numbers of bacteria adhere to squamous epithelial cells giving them a studded appearance that led to them being described as 'clue cells', as an indicator of the condition. The pH of the vaginal fluid also increases from below 4.5 to 5.0 or greater. This is almost certainly maintained by the large numbers of bacteria associated with BV that produce amines, but it is unclear which occurs first, the rise in pH or the increase in bacterial numbers. BV was previously known as 'non-specific vaginitis' or *Haemophilus vaginalis* vaginitis (Gardner and Dukes 1955) but in 1983 the term bacterial vaginosis became popular as it reflected the strong association with overgrowth of certain bacteria and the absence of an inflammatory response. Other terms which are commonly used are anaerobic vaginosis and Gardnerella vaginosis or vaginitis.

BV is the commonest cause of vaginal discharge of women of child-bearing years in the western world. BV is a clinical syndrome which presents as an increased vaginal discharge which has a distinctive fishy smell which is often reported as worse after sexual intercourse. The true prevalence is probably unknown but has been reported as being 10–20 percent of sexually active women (Hay et al. 1992). Predisposing factors for BV have been difficult to identify but of the many investigated, the association with sexual activity is the strongest. This is supported by a higher prevalence in sexually active women including those attending specialized clinics for sexually transmitted diseases or for termination of pregnancy (Blackwell et al. 1993). It has also been found to be more common among women using intra-uterine contraceptive devices (Amsel et al. 1983) and in a number of studies among lesbian women (Berger et al. 1995; Skinner et al. 1996). It is generally considered to be a sexually associated condition rather than sexually transmitted. This is supported by its presence in some virgin girls (Bump and Buesching 1988), its high prevalence in lesbian women who have a low incidence of sexually transmitted infections, and the failure to prevent recurrent episodes after treatment of sexual partners (Colli et al. 1997).

G. vaginalis was the first bacterium to be associated with the syndrome of BV, but it became clear in the 1970s, with the improvement of cultural techniques, that a more complex bacterial flora existed including anaerobes and *M. hominis*. It was not until 1984 that *Mobiluncus*, anaerobic curved rods, were described (Spiegel and Roberts 1984) and consists of two species, *Mobiluncus curtisii* and *Mobiluncus mulieris*. *M. curtisii* are small curved rods which are gram-variable and are divided into two subspecies, *M. curtisii* subsp. *curtisii* and *M. curtisii* subsp *holmesii*. *M. mulieris* is a longer curved rod which is gram-negative. Anaerobic curved rods show a strong association with BV with few women without vaginosis apparently being colonized. However, *Mobiluncus* spp. are very difficult to cultivate and more sensitive nucleic acid amplification methods have shown that although a strong association with BV exists, more women without BV are colonized than previously found (Schwebke and Lawing 2001). The strong association between colonization with *Mobiluncus* spp. and BV has led to the hypothesis that they may represent a separate syndrome, but no evidence exists for this.

Diagnosis

For many years a variety of methods were used for the diagnosis of BV, many of which arose from the original description of the condition by Gardner and Dukes (1955). Recognition of the clinical signs and symptoms, presence of 'clue cells' in a wet mount of vaginal discharge, and isolation of *G. vaginalis* were among the methods used. These methods were unsatisfactory because reporting of signs and symptoms is subjective and varies between patients and while 'clue cells' and *G. vaginalis* are good markers for BV they can also be present in small numbers in women without BV.

Since 1983, the accepted standard for the diagnosis of BV has been the use of four composite criteria (Amsel et al. 1983); a thin homogeneous vaginal discharge, elevated vaginal pH, presence of amines detected by a fishy smell on the addition of potassium hydroxide to the discharge, and the presence of 'clue cells'. If three or more of the four criteria are present then the patient is diagnosed with BV, if less than three are detected the patient is considered not to have BV or to be 'normal'. These criteria have formed the basis of the diagnosis of BV for nearly two decades because they are simple to perform, particularly in a clinic or consultation room, and require minimal material with the exception of a microscope. However, the composite or Amsel's criteria do have some disadvantages as they require a vaginal examination and recognition of the vaginal discharge and 'fishy' smell have a subjective endpoint. There has also been a tendency to use only one or at least not all four criteria, hence invalidating the method. Each of the individual criteria are predictive of BV but do not have

the same sensitivity and specificity as when they are used in combination.

An alternative approach for diagnosis is the grading or scoring of the Gram stained vaginal smears to detect the change in vaginal ecology and the strong bacterial associations. The method first described divided smears into those with a predominantly *Lactobacillus* morphotype, which was considered normal, and those with a mixed microbial flora and no lactobacilli, which were graded as BV (Spiegel et al. 1983). This initial method was modified to include an intermediate category, where the smears showed a predominantly mixed microbial flora but had significant numbers of lactobacilli (Nugent et al. 1991). These methods quantify the numbers of different morphotypes of bacteria, lactobacilli, *Gardnerella/Bacteroides*, and *Mobiluncus*, assign a score for each which when combined gives a number which defines the grading. These methods have been used extensively for research studies and have given valuable data but they require considerable time and skill, which is not always available in clinical or diagnostic situations. Simpler methods have been described where the microbial flora is assessed qualitatively and graded into normal, intermediate or BV (Thomason et al. 1992; Hay et al. 1994). Both the quantitative and qualitative methods have been validated in comparison with the composite criteria and found to have a high sensitivity, specificity, and predictive values for the definition of 'normal' and 'BV' (Schwebke et al. 1996; Thomason et al. 1992; Ison and Hay 2002). However, the inclusion of the intermediate category with BV reduces the specificity and is more frequent in women defined as normal by the composite criteria (Ison and Hay 2002; Taylor-Robinson et al. 2003). This suggests that the vaginal flora associated with the intermediate grade is distinct from that of BV and does not equate with the full clinical syndrome of BV and hence cannot be diagnosed by clinical criteria. A recommended grading system is shown in Table 25.3 and Figure 25.1.

Treatment

In the absence of a clear causative agent, the treatment for BV is primarily aimed at the eradication of the bacteria associated with the clinical condition in an attempt to re-establish the normal 'lactobacillary' flora. The inadequacy of antimicrobial agents to effect a cure is highlighted by the high relapse rates but alternative approaches using replacement flora (bacteriotherapy) are not yet widely available.

Metronidazole is the agent of choice by many clinicians and is given either as a 5 day course of 400 mg twice a day or as a single dose, 2 g. The 5 day course may give a slightly higher cure rate but metronidazole causes nausea and the single dose has a higher compliance rate. Topical gels have been used increasingly and

Table 25.3 *Grading scheme for Gram-stained vaginal smears (Ison and Hay 2002)*

Grade of flora	Definition of flora	Description
0	Normal[a]	Epithelial cells only – no bacteria seen
I	Normal	Lactobacillus morphotype only
II	Intermediate	Reduced lactobacillus morphotype with mixed bacterial flora
III	Bacterial vaginosis	Mixed bacterial flora with few or absent lactobacillus morphotype
IV	Normal in asymptomatic women[b]	Epithelial cells covered with gram-positive cocci with few or absent lactobacillus morphotype

a) Normally found during or immediately post antibiotic therapy.
b) May be a cause of aerobic vaginitis or be of importance in pregnant women.

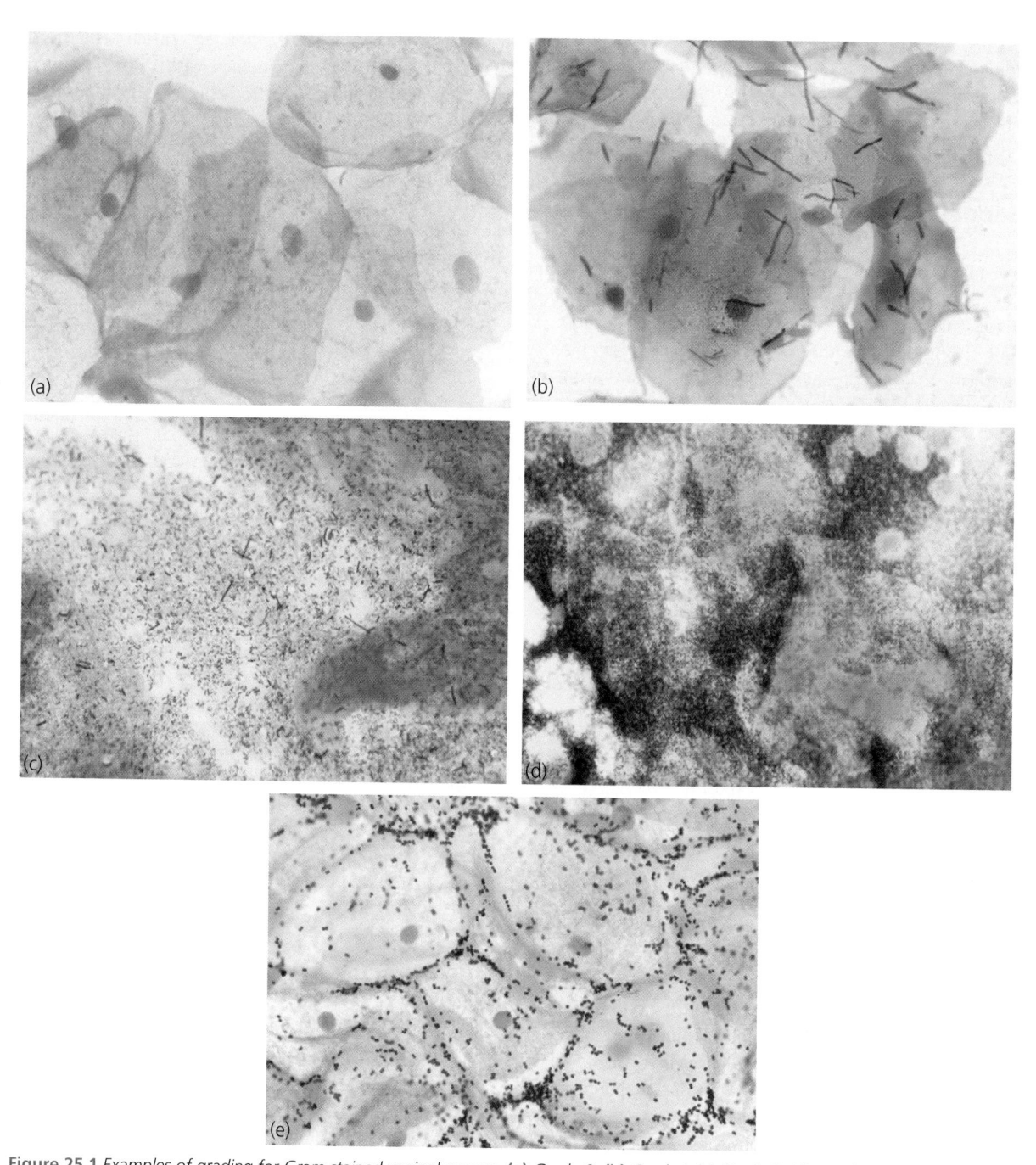

Figure 25.1 *Examples of grading for Gram stained vaginal smears,* **(a)** *Grade 0,* **(b)** *Grade I,* **(c)** *Grade II,* **(d)** *Grade III, and* **(e)** *Grade IV.*

offer a more acceptable alternative particularly for women with recurrent episodes or for pregnant women. Two agents are widely available, clindamycin 2 percent cream, which is applied twice a day for 7 days or a metronidazole gel, 0.75 percent. Clindamycin is active against lactobacilli and hence it takes a few days for them to re-establish in the vagina. Cure rates with topical treatment are similar to the oral treatment (Hay 1998). An attractive approach is the use of lactobacilli as a probiotic, and has been tried with H_2O_2 producing lactobacilli, but has not been proven to be successful (Hallen et al. 1992) and is not widely available. Treatment of male partners does not appear to improve relapse rates even in double-blind, placebo-controlled trials. Unfortunately, for many women, antimicrobial therapy only offers a short term resolution and recurrence is common with all treatments and probably reflects our ignorance at the underlying cause of BV.

Sequelae

For many years BV was considered a mild if unpleasant condition which had no sequelae and was often not taken seriously as a clinical entity. However, there is now strong evidence that BV is associated with complications in both pregnant and nonpregnant women. Women with BV who are pregnant have an increased risk of late miscarriage and preterm birth (Hay et al. 1994; Hillier et al. 1995), with women at greatest risk being those who have had a previous preterm birth. Complications, such as endometritis, are also more common among women with BV (Watts et al. 1990; Soper 1993; Persson et al. 1996).

STREPTOCOCCUS AGALACTIAE

S. agalactiae (group B streptococci) are known to colonize the vagina of some women, although the prevalence varies between 10–40 percent in different studies (Honig et al. 2002; Bayo et al. 2002). It has generally been regarded as part of the normal flora but its role in vaginitis is unclear. Group B streptococci have been isolated from women with symptoms and it has been suggested as a cause of aerobic vaginitis (Maniatis et al. 1996; Donder et al. 2002) with characteristics distinct from BV (Donder et al. 2002). Its importance lies in its ability to cause neonatal infection, particularly meningitis, which can be acquired from the mother during passage through the birth canal. Of neonates that are colonized, 1–3 percent develop disease caused by group B streptococci, mostly early-onset infection within 24 h of birth. Intervention of vaginal colonization during pregnancy has been found to be ineffective but introduction of guidelines for screening of vaginal and rectal swabs in 1996 at 35–37 weeks gestation followed by treatment of colonized women has reduced incidence (Schrag et al. 2002).

Isolation and identification

Group B streptococci grow easily on enriched culture medium such as blood agar, producing reasonably large colonies and a diffuse β-hemolysis. In women who are colonized, group B stretococci are often found in large numbers, but for screening purposes and for detection of small numbers in mixed cultures, selective media can be useful, blood agar with neomycin (30 mg/l) and nalidixic acid (15 mg/l), neomycin nalidixic acid (NNA) agar or colistin (10 mg/l) and nalidixic acid (10 mg/l), colistin nalixidic acid (CNA) agar. NNA has been reported to be more sensitive than CNA (Dunne 1999). Some studies favor prior incubation in enrichment broth, Todd–Hewitt broth supplemented with 15 mg/l nalidixic acid and 8 mg/l gentamicin, to achieve maximum identification of carriers (Dunne and Holland-Staley 1998; Votava et al. 2001). Detection of group B streptococci using rapid immunological methods such as latex agglutination have only proved useful for specimens from heavily colonized women. However, recent development of real-time PCR has proved a possible alternative to culture (Ke and Bergeron 2001).

Commercial kits, that have good quality control and rarely give false positives, are available for serogrouping β-hemolytic colonies. Other tests for identifying group B streptococci include inoculation of media containing starch and horse serum, which produce a distinctive orange pigment after incubation (Islam 1977), and the Christie–Atkins–Munch-Petersen (CAMP) test, where group B streptococci produce a diffusible extracellular protein (CAMP factor) that acts synergistically with staphylococcal beta-lysin to enhance lysis of red cells, and hydrolysis of sodium hippurate.

Treatment

Treatment of nonpregnant women colonized with group B streptococci who do not have symptoms is not considered necessary. Group B streptococci are not known to be resistant to penicillin and this remains the treatment of choice, if required. Prophylaxis with penicillin is recommended in women colonized when screened at 35–37 weeks gestation. Ampicillin can be used as an alternative (Centers for Disease Control and Prevention 2002b).

STAPHYLOCOCCUS AUREUS

Colonization of the vagina by *S. aureus* is relatively unusual but in menstruating young women, often associated with tampon use, has been linked to toxic shock syndrome. The syndrome presents as an acute febrile illness resulting in shock and organ failure and requires rapid diagnosis for successful treatment. It is caused by *S. aureus* that produce a number of toxins including toxic-shock syndrome toxin-1 (TSST-1) which can be detected using radio-immunoassays, immunoassays or reverse passive latex agglutination tests.

OTHER BACTERIAL CAUSES OF GENITAL TRACT INFECTIONS

There are a few bacteria that cause infections of the genital tract that are encountered relatively rarely. *Listeria monocytogenes*, which can cause infection in pregnant women and lead to serious infection such as septicemia, may affect the fetus, resulting in stillbirth or neonatal meningitis. It is not routine practice to screen vaginal swabs for *L. monocytogenes* but blood cultures should be taken if clinically indicated. *Actinomyces* spp. are occasionally encountered associated with infections related to the presence of an intrauterine device and *Clostridium perfringens* has been a common cause of infection following delivery of gynecological surgery, which has largely been controlled by good surgical practice.

Mycoplasma spp. are commonly found in the genital tract, particularly *M. hominis*, and their role in infection has been debated. However, *M. hominis* is considered a cause of pelvic inflammatory disease and *M. genitalium* is associated with urethritis in men (see Chapter 77, Mycoplasma).

REFERENCES

Akduman, D., Ehret, J.M., et al. 2002. Evaluation of a strand displacement amplification (BD Probtec-SDA) for detection of *Neisseria gonorrhoeae* in urine specimens. *J Clin Microbiol*, **40**, 281–3.

Albritton, W.L. 1989. Biology of *Haemophilus ducreyi*. *Microbiol Rev*, **53**, 377–89.

Amsel, R., Totten, P.A., et al. 1983. Nonspecific vaginitis. Diagnostic criteria and microbial and epidemiologic associations. *Am J Med*, **74**, 14–22.

Bayo, M., Berlanga, M. and Agut, M. 2002. Vaginal microbiota in healthy pregnant women and prenatal screening of group B streptococci (GBS). *Int Microbiol*, **5**, 87–90.

Berger, B.J., Kolton, S., et al. 1995. Bacterial vaginosis in lesbians: a sexually transmitted disease. *Clin Infect Dis*, **21**, 1402–5.

Blackwell, A.L., Thomas, P.D., et al. 1993. Health gains from screening for infection of the lower genital tract in women attending for termination of pregnancy. *Lancet*, **342**, 206–10.

Boslego, J.W., Tramont, E.C., et al. 1987. Effect of spectinomycin use on the prevalence of spectinomycin resistant and of penicillinase-producing *Neisseria gonorrhoeae*. *N Engl J Med*, **317**, 272–8.

Bump, R.C. and Buesching, W.J. 1988. Bacterial vaginosis in virginal and sexually active adolescent females: evidence against exclusive sexual transmission. *Am J Obstet Gynecol*, **158**, 935–9.

Carter, J.S. and Kemp, D.J. 2000. A colorimetric detection system for *Calymmatobacterium granulomatis*. *Sex Transm Infect*, **76**, 134–6.

Carter, J., Hutton, S., et al. 1997. Culture of the causative organism of donovanosis (*Calymmatobacterium granulomatis*) in Hep-2 cells. *J Clin Microbiol*, **35**, 2915–17.

Carter, J.S., Bowden, F.J., et al. 1999. Phylogenetic evidence for reclassification of *Calymmatobacterium granulomatis* as *Klebsiella granulomatis* com. nov. *Int J Syst Bacteriol*, **49**, 1695–700.

Catlin, B.W. 1973. Nutritional profiles of *Neisseria gonorrhoeae*, *Neisseria meningitidis*, and *Neisseria lactamica* in chemically defined

media and the use of growth requirements for gonococcal typing. *J Infect Dis*, **128**, 178–93.

Centers for Disease Control and Prevention, 2002a. Guidelines for the treatment of sexually transmitted diseases. *MMWR*, **51**, RR6.

Centers for Disease Control and Prevention, 2002b. Prevention of perinatal group B streptococcal disease. *MMWR*, **51**, 1–28.

Colli, E., Landoni, M. and Parazzini, F. 1997. Treatment of male partners and recurrence of bacterial vaginosis. *Genitourin Med*, **73**, 267–70.

Daly, C.C., Hoffman, I., et al. 1997. Development of an antimicrobial susceptibility surveillance system for *Neisseria gonorrhoeae* in Malawi: comparison of methods. *J Clin Microbiol*, **35**, 2985–8.

Dangor, Y., Ballard, R.C., et al. 1990. Antimicrobial susceptibility of *Haemophilus ducreyi*. *Antimicrob Agents Chemother*, **34**, 1303–7.

Dangor, Y., Miller, D., et al. 1992. A simple medium for the primary isolation of *Haemophilus ducreyi*. *Eur J Clin Microbiol Infect Dis*, **11**, 930–4.

Dangor, Y., Radebe, F. and Ballard, R.C. 1993. Transport media for *Haemophilus ducreyi*. *Sex Transm Dis*, **20**, 5–9.

Dillon, J.R. 1992. National microbiological surveillance of the susceptibility of gonococcal isolates to antimicrobial agents. *Can J Infect Dis*, **3**, 202–6.

Doderlein, A. 1894. Die Scheidensekretunterschugen. *Zentralbl Gynakol*, **18**, 10–14.

Doherty, L., Fenton, K.A., et al. 2002. Syphilis: an old problem, new strategy. *Br Med J*, **325**, 153–6.

Donder, G.G., Vereecken, A., et al. 2002. Definition of a type of abnormal vaginal flora that is distinct from bacterial vaginosis: aerobic vaginitis. *Br J Obstet Gynaecol*, **109**, 34–43.

Dunne, W.M. Jr 1999. Comparison of selective broth medium plus neomycin-nalidixic acid agar and selective broth medium plus Columbia colistin-nalidixic acid agar for detection of group B colonization in women. *J Clin Microbiol*, **37**, 3705–6.

Dunne, W.M. Jr and Holland-Staley, C.A. 1998. Comparison of NNA agar culture and selective broth culture for detection of group B streptococcal colonization in women. *J Clin Microbiol*, **36**, 2298–300.

Easmon, C.S.F., Ison, C.A., et al. 1982. Emergence of resistance after spectinomycin treatment for gonorrhoea due to β-lactamase-producing strain of *Neisseria gonorrhoeae*. *Br Med J*, **284**, 1604–5.

Faruki, H., Kohmescher, R.N., et al. 1985. A community-based outbreak of infection with penicillin-resistant *Neisseria gonorrhoeae* not producing penicillinase (chromosomally mediated resistance). *New Engl J Med*, **313**, 607–11.

Fraser, C.M., Norris, S.J., et al. 1998. Complete genome sequence of *Treponema pallidum*, the syphilis spirochete. *Science*, **281**, 375–88.

Gardner, H.L. and Dukes, C.D. 1955. *Haemophilus vaginalis* vaginitis. A newly defined specific infection previously classified as 'nonspecific' vaginitis. *Am J Obstet Gynecol*, **69**, 962–76.

Gerbase, A.C., Rowley, J.T. and Mertens, T.E. 1998. Global epidemiology of sexually transmitted diseases. *Lancet*, **351**, Suppl III, 2–4.

Giorgi, A., Torriani, S., et al. 1987. Identification of vaginal lactobacilli from asymptomatic women. *Microbiologica*, **10**, 377–84.

Goldberg, J. 1962. Studies on granuloma inguinale V. Isolation of the bacterium resembling *Donvania granulomatis* from the faeces of a patient with granuloma inguinale. *Br J Vener Dis*, **38**, 99–102.

Grosskurth, H., Mosha, F., et al. 1995. Impact of improved treatment of sexually transmitted diseases on HIV infection in rural Tanzania: randomized controlled trial. *Lancet*, **346**, 530–6.

Hallen, A., Jarstrand, C. and Pahlson, C. 1992. Treatment of bacterial vaginosis with lactobacilli. *Sex Transm Dis*, **19**, 146–8.

Hay, P.E. 1998. Therapy of bacterial vaginosis. *J Antimicrob Chemother*, **41**, 6–9.

Hay, P.E., Taylor-Robinson, D. and Lamont, R.F. 1992. Diagnosis of bacterial vaginosis in a gynaecology clinic. *Br J Obstet Gynaecol*, **99**, 63–6.

Hay, P.E., Lamont, R.F., et al. 1994. Abnormal bacterial colonisation of the lower genital tract and subsequent preterm delivery and late miscarriage. *Br Med J*, **308**, 295–8.

Hawes, S.E., Hillier, S.L., et al. 1996. H$_2$O$_2$-producing lactobacilli and acquisition of vaginal infections. *J Infect Dis*, **174**, 1058–63.

Hill, G.B., St. Claire, K.K. and Gutman, L. 1995. Anaerobes predominate among the vaginal flora of prepubertal girls. *Clin Infect Dis*, **20**, Suppl 2, S269–270.

Hillier, S.L. 1998. The vaginal microbial ecosystem and resistance to HIV. *AIDS Res Hum Retroviruses*, **14**, Suppl 1, S17–21.

Hillier, S.L. and Lau, R.J. 1997. Vaginal microflora in postmenopausal women who have not received estrogen replacement therapy. *Clin Infect Dis*, **25**, Suppl 2, S123–126.

Hillier, S.L., Krohn, M.A., et al. 1992. The relationship of hydrogen peroxide-producing lactobacilli to bacterial vaginosis and genital microflora in pregnant women. *Obstet Gynecol*, **79**, 369–73.

Hillier, S.L., Krohn, M.A., et al. 1993. The normal vaginal flora H$_2$O$_2$-producing lactobacilli, and bacterial vaginosis in pregnant women. *Clin Infect Dis*, **16**, Suppl 4, S273–281.

Hillier, S.L., Nugent, R.P., et al. 1995. Association between bacterial vaginosis and preterm delivery of a low-birth-weight infant. *New Engl J Med*, **333**, 1737–42.

Honig, E., Mouton, J.W. and van der Meijden, W.M. 2002. The epidemiology of vaginal colonisation with group B streptococci in a sexually transmitted disease clinic. *Eur J Obstet Gynecol Reprod Biol*, **105**, 177–80.

Hook III, E.W. and Handsfield, H.H. 1999. Gonococcal infections in the adult. In: Holmes, K.K., et al. (eds), *Sexually transmitted diseases*, 3rd edn. New York, NY: McGraw Hill, 451–66.

Islam, A.K.M.S. 1977. Rapid recognition of group B streptococci. *Lancet*, **1**, 256–7.

Ison, C.A. 1990. Laboratory methods in genitourinary medicine: methods of diagnosing gonorrhoea. *Genitourin Med*, **66**, 453–9.

Ison, C.A. 1998. Genotyping of *Neisseria gonorrhoeae*. *Curr Opin Infect Dis*, **11**, 43–6.

Ison, C.A. and Hay, P.E. 2002. Validation of a simplified grading of Gram stained vaginal smears for use in genitourinary medicine clinics. *Sex Transm Infect*, **78**, 413–15.

Ison, C.A., Bindayna, K.M., et al. 1990. Penicillin and cephalosporin resistance in gonococci. *Genitourin Med*, **66**, 351–6.

Jephcott, A.E. 1997. Microbiological diagnosis of gonorrhoea. *Genitourin Med*, **73**, 245–52.

Johnson, J.L., Phelps, C.F., et al. 1980. Taxonomy of the *Lactobacillus acidophilus* group. *Int J Syst Bacteriol*, **30**, 53–68.

Johnson, S.R. and Morse, S.A. 1988. Antibiotic resistance in *Neisseria gonorrhoeae*, genetics and mechanisms of resistance. *Sex Transm Dis*, **15**, 217–24.

Ke, D. and Bergeron, M.G. 2001. Molecular methods for rapid detection of group B streptococci. *Expert Rev Mol Diagn*, **1**, 175–81.

Kharsany, A.B., Hoosen, A.A., et al. 1997. Growth and cultural characteristics of *Calymmatobacterium granulomatis* – the aetiological agent of granuloma inguinale (Donovanosis). *J Med Microbiol*, **46**, 579–85.

Kharsany, A.B., Hoosen, A.A., et al. 1999. Phylogentic analysis of *Calymmatobacterium granulomatis* based on 16S rRNA gene sequences. *J Med Microbiol*, **48**, 841–7.

King, A. 2001. Recommendations for susceptibility tests on fastidious organisms and those requiring special handling. *J Antimicrob Chemother*, **48**, Suppl. 1, 77–80.

Klebanoff, S.J. and Coombs, R.W. 1991. Virucidal effect of *Lactobacillis acidophilus* on human immunodeficiency virus type-1: Possible role in heterosexual transmission. *J Exp Med*, **174**, 289–92.

Klebanoff, S.J., Hillier, S.L., et al. 1991. Control of microbial flora of the vagina by H$_2$O$_2$-generating lactobacilli. *J Infect Dis*, **164**, 94–100.

Knapp, J.S. and Koumans, E.H. 1999. *Neisseria* and *Branhamella*. In: Murray, P.R., et al. (eds), *Manual of clinical microbiology*. Washington DC: ASM Press, 586–603.

Knapp, J.S., Tam, M.R., et al. 1984. Serological classification of *Neisseria gonorrhoeae* with use of monoclonal antibodies to gonococcal outer membrane protein I. *J Infect Dis*, **150**, 44–8.

Knapp, J.S., Fox, K.K., et al. 1997. Fluoroquinolone resistance in *Neisseria gonorrhoeae*. *Emerg Infect Dis*, **3**, 33–8.

Knox, J., Tabrizi, S.N., et al. 2002. Evaluation of self-collected samples in contrast to practitioner-collected samples for the detection of *Chlamydia trachomatis*, *Neisseria gonorrhoeae*, and *Trichomonas vaginalis* by polymerase chain reaction among women living in remote areas. *Sex Transm Dis*, **29**, 697–54.

Koumans, E.H., Johnson, R.E., et al. 1998. Laboratory testing for *Neisseria gonorrhoeae* by recently introduced nonculture tests: a performance review with clinical and public health considerations. *Clin Infect Dis*, **27**, 1171–80.

Kuberski, T., Papadimitriou, J.M. and Phillips, P. 1980. Ultrastructure of *Calymmatobacterium granulomatis* in lesions of granuloma inguinale. *J Infect Dis*, **142**, 744–9.

Lewis, D.A. 2000. Chancroid: From clinical practice to basic science. *AIDS Patient Care and STDs*, **14**, 1936.

Lkhamsuren, E., Shultz, T.R., et al. 2001. The antibiotic susceptibility of *Neisseria gonorrhoeae* isolated in Ulaanbaatar, Mongolia. *Sex Transm Infect*, **77**, 218–19.

Lockett, A.E., Dance, D.A.B., et al. 1991. Serum-free media for the isolation of *Haemophilus ducreyi*. *Lancet*, **338**, 326.

Maniatis, A.N., Palermos, J., et al. 1996. *Streptococcus agalactiae*: a vaginal pathogen? *J Med Microbiol*, **44**, 199–202.

Martinez-Martinez, L., Pascual, A. and Jacoby, G.A. 1998. Quinolone resistance from a transferable plasmid. *Lancet*, **351**, 797–9.

Members of the Australian Gonococcal Surveillance Programme, 1984. Penicillin sensitivity of gonococci in Australia: development of Australian gonococcal surveillance programme. *Br J Vener Dis*, **60**, 226–30.

Morse, S.A., Johnson, S.R., et al. 1986. High-level tetracycline resistance in *Neisseria gonorrhoeae* is result of acquisition of streptococcal *tetM* determinant. *Antimicrob Agents Chemother*, **30**, 664–70.

Morse, S.A., Trees, D.L., et al. 1997. Comparison of clinical diagnosis and standard laboratory and molecular methods for the diagnosis of genital ulcer disease in Lesotho: association with human immunodeficiency virus infection. *J Infect Dis*, **175**, 583–9.

National Committee for Clinical Laboratory Standards, 2003 Approved Standard: Performance Standards for Antimicrobial Disk Susceptibility Tests, 8th ed. Document M2-A8. National Committee for Clinical Laboratory Standards, Villanova, Pa.

Naumann, M., Rudel, T. and Meyer, T. 1999. Host cell interactions and signalling with *Neisseria gonorrhoeae*. *Curr Opin Microbiol*, **2**, 62–70.

Nugent, R.P., Krohn, M.A., et al. 1991. Reliability of diagnosing bacterial vaginosis is improved by a standardized method of Gram stain. *J Clin Microbiol*, **29**, 297–301.

Orle, K.A., Gates, C.A., et al. 1996. Simultaneous PCR detection of *Haemophilus ducreyi*, *Treponema pallidum* and herpes simplex virus types 1 and 2 from genital ulcers. *J Clin Microbiol*, **34**, 49–54.

Palladino, S., Pearman, J.W., et al. 1999. Diagnosis of *Chlamydia trachomatis* and *Neisseria gonorrhoeae*, genitourinary infections in males by the Amplicor PCR assay of urines. *Diagn Microbiol Infect Dis*, **33**, 141–6.

Pagotto, F., Aman, A.T., et al. 2000. Sequence analysis of the family of penicillinase-producing plasmids of *Neisseria gonorrhoeae*. *Plasmid*, **43**, 24–34.

Persson, E., Bergstrom, M., et al. 1996. Infections after hysterectomy. A prospective nation-wide Swedish study. *Acta Obstet Gynecol Scand*, **75**, 757–61.

Puapermpoonsiri, S., Kato, N., et al. 1996. Vaginal microflora associated with bacterial vaginosis in Japanese and Thai pregnant women. *Clin Infect Dis*, **23**, 748–52.

Ragosa, M. and Sharpe, M. 1960. Species differentiation of human vaginal lactobacilli. *J Gen Microbiol*, **23**, 197–201.

Roberts, M.C. 1989. Plasmids of *Neisseria gonorrhoeae* and other *Neisseria* species. *Clin Microbiol Rev*, **2**, Suppl, S18–23.

Ropp, P.A., Hu, M., et al. 2002. Mutations in *ponA*, the gene encoding penicillin-binding protein 1, and a novel locus, *penC*, are required for high-level chromosomally mediated penicillin resistance. *Antimicrob Agents Chemother*, **46**, 769–77.

Schmidt, B.L., Edjlalipour, M. and Luger, A. 2000. Comparative evaluation of nine different enzyme-linked immunosorbent assays for determination of antibodies against *Treponema pallidum* in patients with primary syphilis. *J Clin Microbiol*, **38**, 1279–82.

Schrag, S.J., Zell, E.R., et al. 2002. A population-based comparison of strategies to prevent early-onset group B streptococcal disease in neonates. *New Engl J Med*, **347**, 233–9.

Schwarcz, S.K., Zenilman, J.M., et al. 1990. National surveillance of antimicrobial resistance in *Neisseria gonorrhoeae*. The Gonococcal Isolate Surveillance Project. *JAMA*, **264**, 1413–17.

Schwebke, J.R. and Lawing, L.F. 2001. Prevalence of *Mobiluncus* spp. among women with and without bacterial vaginosis as detected by polymerase chain reaction. *Sex Transm Dis*, **28**, 195–9.

Schwebke, J.R., Whittington, W., et al. 1995. Trends in susceptibility of *Neisseria gonorrhoeae* to ceftriaxone from 1985 through 1991. *Antimicrob Agents Chemother*, **39**, 917–20.

Schwebke, J.R., Hillier, S.L., et al. 1996. Validity of the vaginal Gram stain for the diagnosis of bacterial vaginosis. *Obstet Gynecol*, **88**, 573–6.

Sherrard, J. and Barlow, D. 1996. Gonorrhoea in men: clinical and diagnostic aspects. *Genitourin Med*, **72**, 422–6.

Skinner, C.J., Stokes, J., et al. 1996. A case-controlled study of the sexual health needs of lesbians. *Genitourin Med*, **72**, 277–80.

Soper, D.E. 1993. Bacterial vaginosis and postoperative infections. *Am J Obstet Gynecol*, **169**, 467–9.

Spiegel, C.A. and Roberts, M. 1984. *Mobiluncus* gen. nov., *Mobiluncus curtisii* subsp. curtisii sp. nov., *Mobiluncus curtisii* subsp. *holmesii* sp. nov. and *Mobiluncus mulieris* sp. nov., curved rods from the vagina. *Int J Syst Bacteriol*, **34**, 177–84.

Spiegel, C.A., Amsel, R. and Holmes, K.K. 1983. Diagnosis of bacterial vaginosis by direct Gram stain of vaginal fluid. *J Clin Microbiol*, **18**, 170–7.

Taylor-Robinson, D., Morgan, D.J., et al. 2003. Relation between Gram stain and clinical criteria for diagnosing bacterial vaginosis with special reference to Gram grade II evaluation. *Int J STD AIDS*, **14**, 6–10.

Thomason, J.L., Anderson, R.J., et al. 1992. Simplified Gram stain interpretative method for the diagnosis of bacterial vaginosis. *Am J Obstet Gynecol*, **167**, 16–19.

Trees, D.L. and Morse, S.A. 1995. Chancroid and *Haemophilus ducreyi*: an update. *Clin Microbiol Rev*, **8**, 357–75.

Van de Laar, M.J.W., van Duynhoven, Y.T.P.H., et al. 1997. Surveillance of antibiotic resistance in *Neisseria gonorrhoeae* in the Netherlands, 1977–95. *Genitourin Med*, **73**, 510–17.

Van Doornum, G.J., Schouls, L.M., et al. 2001. Comparison between the LCx Probe system and the COBAS AMPLICOR system for the detection of *Chlamydia trachomatis* and *Neisseria gonorrhoeae* infections in patients attending a clinic for sexually transmitted diseases in Amsterdam, The Netherlands. *J Clin Microbiol*, **39**, 829–35.

Van Dyck, E., Ieven, M., et al. 2001. Detection of *Chlamydia trachomatis* and *Neisseria gonorrhoeae* by enzyme immunoassay, culture and three nucleic acid amplification tests. *J Clin Microbiol*, **39**, 1751–6.

Votava, M., Tejlaova, M., et al. 2001. Use of GBS media for rapid detection of group B streptococci in vaginal and rectal swabs from women in labor. *Eur J Clin Microbiol Infect Dis*, **20**, 120–2.

Ward, H., Ison, C.A., et al. 2000. A prospective social and molecular investigation of gonococcal transmission. *Lancet*, **356**, 1812–17.

Watts, D.H., Krohn, M.A., et al. 1990. Bacterial vaginosis as a risk factor for post-cesarean endometritis. *Obstet Gynecol*, **75**, 52–8.

West, B., Changalucha, J., et al. 1995. Antimicrobial susceptibility, auxotype and plasmid content of *Neisseria gonorrhoeae* in northern Tanzania: emergence of high-level plasmid mediated tetracycline resistance. *Genitourin Med*, **71**, 9–12.

WHO Western and Pacific Region Gonococcal Antimicrobial Surveillance Programme, 2001. Surveillance of antibiotic resistance in Neisseria gonorrhoeae in the WHO Pacific Region, 2000. *Commun Dis Intell*, **25**, 274–7.

Young, H., Moyes, A. and McMillan, A. 1997. Azithromycin and erythromycin resistant *Neisseria gonorrhoeae* following treatment with azithromycin. *Int J STD AIDS*, **8**, 299-30, 2.

Xia, M., Whittington, W.L., et al. 2000. Gonorrhoea among men who have sex with men: an outbreak caused by a single genotype of erythromycin-resistant *Neisseria gonorrhoeae* with a single-base pair deletion in the *mtrR* promotor region. *J Infect Dis*, **181**, 2080–2.

Zenilman, J.M., Nims, L.J., et al. 1987. Spectinomycin-resistant gonococcal infections in the United States, 1985–86. *J Infect Dis*, **156**, 1002–4.

26

Bacterial infections of the urinary tract

SÖREN G. GATERMANN

Urinary tract infections are amongst the most common infections encountered in clinical practice. About 50 percent of all women experience at least one episode of urinary tract infection during their lifetime. In women, the incidence of symptomatic infections increases with age. In men, urinary tract infections tend to occur more often in young boys or elderly persons and homosexuals. Symptomatic urinary tract infections are characterized by symptoms like dysuria, urgency, painful and frequent voiding, accompanied by the presence of bacteria in the normally sterile urine. Apart from symptomatic infections, there is a condition called asymptomatic bacteriuria that is characterized by the presence of bacteria in the urine but the lack of clinical symptoms. This condition may be a predisposition for symptomatic infection.

It is of utmost importance to distinguish between so-called uncomplicated urinary tract infections, which are characterized by the absence of functional or anatomical obstacles to urinary flow, and complicated urinary tract infections in which such obstacles are present, because patients with complicated urinary tract infections more often develop urosepsis and renal damage and bear a higher mortality. In addition, in complicated urinary tract infections the infecting organisms tend to be less susceptible to antimicrobial drugs.

Urinary tract infections encompass a wide array of clinical entities such as urethritis, cystitis, pyelonephritis, epididymitis, prostatitis, and perinephritic abscess. Most commonly, infections affect the bladder, urethra, and kidney.

ANATOMICAL CONSIDERATIONS

The epithelia of the urinary tract are bathed in a constant flow of urine originating in the kidneys, flowing via both ureters into the bladder where the fluid accumulates until it is voided at intervals via the urethra. Normally, only a minute residual volume of urine remains within the urinary tract after voiding. Anatomical or functional obstacles to urinary flow may cause an increased volume of urine to be retained within the urinary tract. If such residual volume does contain bacteria, the organisms may multiply, attain high bacterial densities, and ultimately, damage the epithelial surface causing an immune response and symptoms. The pressure in the bladder is increased during voiding and reflux into the ureters is normally avoided by the oblique insertion of the ureter into the bladder wall, providing an efficient valve mechanism. Reflux into the kidneys may cause renal scarring and malfunction. A number of conditions may lead to an increased residual urinary volume. These can be pathological conditions that impair voiding such as an enlarged prostate in men, urinary stones, or tumors, as well as neurological or functional conditions such as the neurogenic bladder present in spinal cord injury, multiple sclerosis or cerebrovascular accidents, or the atonic bladder of diabetics.

HOST DEFENSE MECHANISMS

The constant urinary flow and the low residual volume are major factors that counteract bacterial colonization.

In addition, the urine contains protective factors such as immunoglobulins that may interfere with bacterial colonization. Also high concentrations of urea are inhibitory to bacterial growth.

The kidneys produce and secrete a protein, the Tamm–Horsefall protein, which acts as a receptor for bacterial adhesins thus trapping the invading organisms before they reach their cellular targets (Leeker et al. 1997; Pak et al. 2001). It has also been shown that bacterial adhesion to uroepithelial cells causes cells to exfoliate and thus effectively expel the more invasive organisms (Kunin et al. 2002). In addition, adhesion to and invasion of urothelial cells triggers secretion of IL8 which is a signal for transmigration of leukocytes into the urinary tract (Frendéus et al. 2000).

It is known that the distal portion of the female urethra is colonized by bacteria. Colonization becomes less dense in the proximal portions. Several studies have shown that the slope of the curve representing the bacterial counts in different portions of the urethra is steeper in women who do not suffer from recurrent urinary tract infection (UTI) compared to those who do (reviewed in Kunin et al. 2002), suggesting that the urethra has an important function in preventing UTI. The epithelium of the urethra responds to hormonal influences like the vaginal epithelium.

An important defense mechanism is the secretion of blood group antigens that may function as receptors for bacterial attachment. It has been shown that women suffering from recurrent urinary tract infections more frequently do not secrete Lewis blood group antigens than controls (Sheinfeld et al. 1989; Raz et al. 2000).

CLINICAL SYNDROMES

Acute cystitis

Cystitis is the most common presentation of urinary tract infection. It is characterized by dysuria, increased frequency and urgency of voiding, suprapubic discomfort, and a cloudy urine. The condition is far more common in women than in men. In some women, cystitis may be associated with sexual activity (Remis et al. 1987), the use of contraceptive diaphragms (Latham et al. 1985), or spermicides. About one-third of women with cystitis experience a recurrence of the infection.

Urethral syndrome

A striking proportion of women (30–50 percent) who present with symptoms suggestive of a urinary tract infection do not have significant bacteriuria (O'Grady et al. 1970). In some patients the bladder urine obtained by suprapubic aspiration shows that some of these women have cystitis with a low bacterial count, but in the remainder the urine is sterile (Stamm et al. 1980).

Various microbial and nonmicrobial etiologies have been suggested to be responsible for the syndrome in these patients. Since the first episodes of urethral syndrome and acute cystitis are virtually indistinguishable in clinical practice, antimicrobial therapy is recommended for both conditions (Baerheim et al. 1999).

Acute pyelonephritis

Patients with acute pyelonephritis often present with loin pain, fever, and a positive urine culture. The microbial spectrum is similar to that of acute cystitis. Acute pyelonephritis is a common complication in pregnant women with previously untreated asymptomatic bacteriuria. In patients with diabetes, upper urinary tract infection is common (Patterson and Andriole 1997) and severe complications such as necrotizing papillitis may occur.

Other infections

Renal abscess is a hematogenous infection most often caused by *S. aureus*. The lesion is localized in the renal cortex and urine culture may be negative even in the presence of pyuria. *Enterobacteria* only rarely cause renal infections via the hematogenous route.

Perinephric abscess does not involve the renal parenchyma, but it may complicate acute pyelonephritis, in which case the causative pathogen derives from the urinary tract.

Acute prostatitis is most often caused by the same spectrum of microorganisms that is responsible for acute cystitis. In chronic prostatitis, bacteria often cannot be isolated by conventional techniques, but the symptoms and the presence of leukocytes suggests infection. In a molecular study, Krieger et al. have found the DNA of fastidious and rarely isolated species in the prostatic fluid of such patients (Krieger et al. 1996). As other authors (Lee et al. 2003) did not find differences of bacterial colonization between men with and without prostatitis, it is unclear at present how the results of Krieger et al. should be interpreted.

Urinary tract infections in special hosts

Diabetics are reported to have a somewhat higher prevalence of asymptomatic bacteriuria. They may also suffer from a nerve dysfunction leading to a lower voiding frequency and an increased residual urinary volume (Patterson and Andriole 1997; Zhanel et al. 1991).

Urinary tract infections in solid organ transplant patients tend to occur more often during the first few months after transplantation, especially after renal transplantation (Muñoz 2001; Takai et al. 1998).

MICROBIOLOGY OF URINARY TRACT INFECTIONS

The vast majority of urinary tract infections is caused by a single organism from a very limited spectrum of bacteria. In uncomplicated cystitis, the organism is almost always *Escherichia coli* or *Staphylococcus saprophyticus*, whereas in complicated cases, the spectrum is more diverse, encompassing several Enterobacteriaceae and gram-positive microorganisms such as enterococci (Table 26.1). Polymicrobial infections occur virtually exclusively in complicated cases, especially in patients with chronic indwelling catheters. Pyelonephritis is most often caused by *E. coli*, but *Proteus mirabilis* may also be encountered.

Many other species are infrequently or only rarely found (Bottone et al. 1998; Bézian et al. 1990; Cappelli et al. 1999; Fraimow et al. 1994; Funke et al. 1998; Janda et al. 1993; Morgan and Hamilton-Miller 1990; Salahuddin et al. 1996; Snehalatha et al. 1992; Stewart et al. 1993; Sturm 1989; Wong and Yuen 1995). If an infection with an unusual organism is suspected, etiology can be confirmed by repeated culture and/or suprapubic aspiration. However, as most microbiologists only expect the usual bacteria, fastidious or difficult-to-identify bacteria may be overlooked.

In nosocomial infections, gram-positive cocci, especially enterococci are the second most commonly isolated species (Table 26.1). It is not always clear whether an isolate represents a true pathogen or is merely contamination with an organism selected by chemotherapy. However, urinary tract infections with enterococci are a predisposition for the development of enterococcal bacteremia (Maki and Agger 1988).

PATHOGENESIS

Most often the infecting microorganisms ascend via the urethra into the bladder. Of course, their invasion may be facilitated by the presence of devices such as urinary catheters. Before the bacteria ascend they usually colonize the periurethral area and the vaginal introitus (Bollgren and Winberg 1976b). In addition, rectal colonization is often present during infection episodes. Bacteria that cause infections in otherwise healthy women express distinct virulence properties enabling them to establish an infection, whereas bacterial strains that cause infections in patients with complicating factors do not need to possess the same armamentarium of virulence factors (Dalet et al. 1991; Hull et al. 2002; Sandberg et al. 1988). Although the presence of special virulence factors is often accepted, this theory is not unchallenged. Selection of the infecting strain may just be a consequence of the heavy colonization by the pathogen. However, in recurrent asymptomatic bacteriuria the isolated strains appear to be a random sample of the fecal flora, whereas infecting strains do not appear to be such random samples (Lidin-Janson et al. 1977).

Adhesion

It is commonly accepted that the adherence of *E. coli* to uroepithelium is crucial to counteract the natural clearance of the urinary tract and to allow its colonization (Reid and Sobel 1987; Dalet et al. 1991; Tullus et al. 1991). Two classes of adhesins can be found on the surface of *E. coli*: fimbrial and nonfimbrial adhesins. While fimbrial adhesins carry ligands on their tips that bind to receptors present on cells, the structure and binding domains of nonfimbrial adhesins are less clear. Traditionally, *E. coli* adhesins are grouped according to the specificity of the hemagglutination they cause. Type 1 fimbriae, which are present on many Enterobacteriaceae, are inhibited by D-mannose and various mannosides. Hemagglutination by mannose-resistant fimbriae is not inhibited by D-mannose or mannosides and their receptors contain various carbohydrates.

The best studied of these are P fimbriae that bind to neutral globoseries glycolipids, including globote-

Table 26.1 *Percentage of bacterial species in urinary samples obtained in different clinical situations*

Species	Community acquired[a]	Nosocomial[b]	Catheter associated[c]	Own data nosocomial[d]
E. coli	69–80	36	30–35	45
Other Enterobacteria[e]	6–26	15	26–30	16
S. saprophyticus	2.3–15			
Enterococci[f]		15–23	12–22	21
Staphylococci[g]		11	8–12	6
Pseudomonas[h]		7–12	11	6

a) Kahlmeter, 2000; Ronald, 2002; uncomplicated urinary tract infections in women.
b) Wagenlehner et al. 2002; Bouza et al. 2002.
c) Wazait et al. 2003.
d) Data from inpatients, 2002 (author's observation).
e) Mainly *Proteus mirabilis*, *Klebsiella* spp., *Enterobacter* spp.
f) Mainly *E. faecalis*.
g) Not *S. saprophyticus*.
h) Mainly *P. aeruginosa*.

traosylceramide and trihexosylceramide, which are antigens of the P blood group system (Leffler and Svanborg-Eden 1981). The minimal receptor to which these fimbriae bind is α-D-Gal-(1→4)-β-D-Gal (Gal-Gal) (Bock et al. 1985; Källenius et al. 1981).

P fimbriae, like many other fimbriae, are synthesized from a polycistronic operon that codes for the structural proteins as well as for chaperone/usher proteins required for assembly of the final pilus. Figure 26.1 shows a model of the pilus structure. The pilus is anchored into the outer membrane by a protein that functions as an usher during pilus synthesis. The helical shaft is composed of the major structural subunit. The end of the pilus is fibrillar shaped and carries the adhesin (for detailed information see Mühldorfer et al. 2002; Saulino et al. 2000; Soto et al. 1998).

Donnenberg and Welch (1996)) used pooled data from many studies to show that P fimbriae are more often expressed by *E. coli* isolated from pyelonephritis (80.8 percent) than by strains from the fecal flora (relative risk of expression of P fimbriae in pyelonephritis, 6.1). In strains isolated from pyelonephritis complicated by bacteremia, the proportion of P frimbriated strains is even higher (88.9 percent). The prevalence of P-fimbriate *E. coli* in lower urinary tract infections is lower

(31.1 percent), but these isolates still produce P fimbriae more often than fecal isolates (relative risk 1.98).

Immunuofluorescence with antibody to P fimbriae was used to show expression of these appendages by *E. coli* in vivo (Pere et al. 1987). In addition, patients with pyelonephritis caused by P-fimbriate *E. coli* mount an antibody response toward these antigens, pointing in the same direction.

Experimental evidence for the importance of P fimbriae comes from in vitro studies showing binding of fimbriated *E. coli* to tissue from human kidney, but not to the bladder epithelium (Donnenberg and Welch 1996) and from experimental infections using suspensions of P-fimbriate bacteria suspended in a solution of a receptor analogue, globotriosylceramide, to show reduced recovery of the infecting organism. In contrast, studies employing isogenic mutants did not always give unequivocal results. Mobley et al. (1993) constructed a mutant that lacked fimbrial components crucial to binding but could not show reduced pathogenicity in experimental infections of mice. Another group used a mutant that produced a truncated adhesin to infect monkeys. The parent and the mutant both established infections but the strain expressing P fimbriae persisted significantly longer (Winberg et al. 1995). These conflicting results may be explained by the presence of additional virulence factors in the strain used by Mobley's group.

Recently, Wullt et al. (2001) used a nonfimbriate clinical isolate and a transformant of the same strain expressing P fimbriae to infect volunteers. Bacteria expressing P fimbriae required lower inocula to reach significant densities within the urine and persisted longer than the wild-type strain. This experiment can be taken as a fulfillment of the molecular Koch's postulates (Falkow 1989) for P fimbriae.

The F1C fimbriae are one additional type of fimbria correlated with uropathogenicity. They interact with two minor glycosphingolipids, galactosylceramide and globotriaosylceramide, isolated from the rat, canine, and human urinary tract (Bäckhed et al. 2002). The authors also showed that binding of F1C-fimbriated *Escherichia coli* to renal cells induced interleukin-8 production, thus suggesting a role for F1C-mediated attachment in mucosal defense against bacterial infections. Strains producing F1C fimbriae are also more often isolated from patients with urinary tract infections than from fecal samples (22.3 percent in pyelonephritis, 13.9 percent in cystitis, and 6.2 percent in fecal samples). This indirect evidence of a role in pathogenesis is corroborated by an experimental infection showing that a mutant devoid of F1C fimbriae colonized the kidneys of rats in lower numbers than the mutant; however, reintroduction of a functional gene did not fully restore virulence (Marre et al. 1986).

Several other fimbrial and nonfimbrial adhesins (e.g. Dr, X, M, G) that confer mannose-resistant hemaggluti-

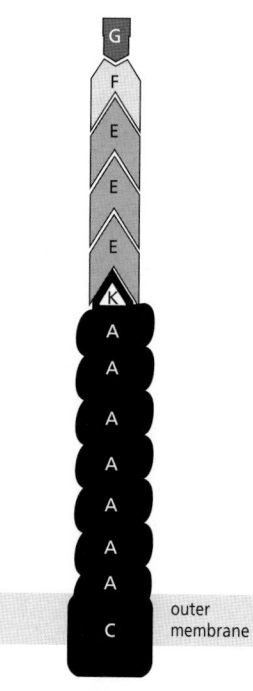

Figure 26.1 *General structure of* E. coli *fimbriae. The P pilus of* E. coli *consists of a helical shaft that is made up of PapA and a fibrillar structure projecting outwards that carries the adhesin, PapG, at its end. The diverse subunits are assembled by the major usher PapC that also anchors the pilus in the outer membrane. Between the shaft and the fibrillum and between the fibrillum and the adhesin there are adaptors, PapK and PapF. Synthesis of the shaft is terminated by PapH, which is buried in the channel formed by PapC (Soto et al. 1998). Assembly of type 1 pili is similar (Saulino et al. 2000).*

nation have been described in pathogenic *E. coli*. Association of these adhesins with disease or pathogenicity has not been shown conclusively although adhesins of the Dr series cause internalization of *E. coli* and thus may contribute to persistence of the organism (Goluszko et al. 1997).

Type I fimbriae cause mannose-sensitive hemagglutination. Although they are ubiquitous and can be found in many enterobacterial species and are present in most strains of *E. coli*, they are more often (odds ratio, 1.22) expressed by isolates from cystitis (64.1 percent) than by control isolates. In contrast, there is no such predilection in pyelonephritis (Donnenberg and Welch 1996). Type I fimbriae are known to undergo phase variation that is due to an invertible promoter (Eisenstein 1981; Abraham et al. 1985). Production of fimbriae is selected for by culture in liquid media as opposed to agar. When bacteria were grown under conditions favoring expression of type I fimbriae, the strains colonized the bladder of infected animals in higher numbers than phase variants not expressing fimbriae (Hultgren et al. 1985). Organisms colonizing the bladder were usually fimbriated while those recovered from the kidneys lost their pilation. When a mixture of mutants that expressed either phase constitutively was used to infect animals, the fimbriated strains were recovered in higher numbers than their nonfimbriated counterparts, especially during the early phase of the infection (Gunther et al. 2002). Type I fimbriae bind to uroplakins, major glycoproteins of the urothelial surface (Wu et al. 1996; Zhou et al. 2002). More recently, a study showed that *E. coli* expressing type I fimbriae persist longer and cause more severe symptoms than their nonfimbriated counterparts (Connell et al. 1996). These experiments included epidemiological data, as well as experimental infections. In addition, signature-tagged mutagenesis identified type I fimbriae amongst the most important virulence factors for urinary tract infections (Bahrani-Mougeot et al. 2002). In keeping with these results are observations that type I fimbriae mediate invasion of bladder epithelial cells by *E. coli* facilitating persistence and recurrent infections (Mulvey et al. 2000, 2001; Mulvey 2002). Once within the urothelial cell the bacteria may form intracellular biofilm-like pods in which they are embedded in a polysaccharide (Anderson et al. 2003). Accordingly, recurrent urinary tract infections in infancy seem to be caused by the same persisting strain (Jantunen et al. 2002).

In other uropathogenic bacteria, surface factors necessary for adherence have also been identified. *Proteus mirabilis* produces several fimbriae, some of which have a role in experimental urinary tract infections (Mobley 1996). For *Staphylococcus saprophyticus* a hemagglutinin that recognizes a protein receptor on the erythrocyte has been identified (Gatermann et al. 1992; Meyer et al. 1997). This adhesin has multiple functions as it is also the major autolysin of the species (Hell et al. 1998). It is also responsible for the attachment to uroepithelial cells (Meyer et al. 1996). Enterococci express an aggregation substance that contains an RGDS and an RGDV motif and thus binds to cells expressing appropriate integrins, which are receptors for these motifs (for a review see Gatermann 1996).

Hemolysin

It is a long standing observation that *E. coli* isolated from urinary tract infections more often show hemolysis on blood agar than strains from other infections. Analysis of pooled data from several epidemiological studies showed a higher prevalence of hemolysin-producing strains in pyelonephritis or cystitis than in feces. In addition, two studies used genetically manipulated strains to show that the hemolysin contributes to pathogenicity (Marre et al. 1986; O'Hanley et al. 1991). Recently, it was shown that expression of the hemolysin may contribute to renal failure (Kreft and Pagel 2000).

Cytotoxic necrotizing factor 1

A proportion of 40 percent of urinary isolates of *E. coli* produce this apoptosis-inducing cytotoxin (Caprioli et al. 1987). In experiments by Rippere-Lampe et al. (2001a, b), a isogenic mutant lacking expression of cytotoxic necrotizing factor 1 (CNF1) was less pathogenic in a mouse model than the parent strain. These findings were in contrast to experiments by Johnson et al. (2000). Thus, the function of this toxin remains controversial.

Iron siderophores

Iron is needed by almost all bacteria and free iron is present in exceedingly low concentrations in human serum and tissue. Aerobactin, a siderophore, is more often produced by *E. coli* causing infections, including urinary tract infections, than in fecal isolates (Carbonetti et al. 1986). Recently, Torres et al. (2001) showed that siderophores and TonB, a protein that facilitates transport of iron through the cytoplasmic membrane, are required in experimental infections.

Urease

As early as 1960 it was suggested that bacteria producing urease cause more severe urinary tract infections than urease negative bacteria (Braude and Siemienski 1960). Most studies used urease negative mutants generated by chemical mutagenesis or urease inhibitors to show a contribution of the enzyme to virulence (Griffith and Musher 1975; Griffith et al. 1976; Musher et al. 1975). In 1989, Mobley's group (Jones et al. 1990) as well as that of Gatermann (Gatermann and Marre 1989) used genetically manipulated strains of *Proteus mirabilis* and *Staphylococcus saprophyticus*, respectively, to show that urease positive strains caused more severe damage to the urinary tract than their urease-negative derivatives. Urease is thought to act by increasing the urinary pH and the concentration of ammonium ions which may lead to precipitation of struvite ($MgNH_4PO_4$) or carbonate-apatite ($Ca_{10}(PO_4)_6 \cdot CO_3$) (Griffith et al. 1976; Johnson et al. 1993). In addition, urease may increase adhesion of bacteria to the urothelial cells (Parsons et al. 1984).

Organization of virulence factors

It has been shown in uropathogenic *E. coli* that virulence factors such as fimbriae and hemolysin are clustered in large segments of DNA, termed pathogenicity islands (Knapp et al. 1986). A pathogen may contain one or more of these islands and a crosstalk between genes encoded by the pathogenicity islands and chromosomal loci has been described (Morschhäuser et al. 1994).

THE PHYSIOLOGICAL FLORA OF THE URINARY TRACT

Normally, the bladder urine is sterile. In adults the distal urethra is colonized by a variety of microorganisms, including coagulase-negative staphylococci (excluding *S. saprophyticus*), viridans streptococci, diphtheroids, nonpathogenic *Neisseria* spp., anaerobic cocci, and lactobacilli. Commensal mycobacteria and mycoplasmas may also be present. In the postpartal period the flora consists predominantly of gram-negative bacilli, esp. *E. coli* (Bollgren and Winberg 1976a). Thereafter, in girls, the flora is replaced by anaerobic bacilli (Bollgren and Winberg 1976b; Bollgren et al. 1979).

Studies have shown that periurethral colonization is a prerequisite for the development of urinary tract infections (Bollgren and Winberg 1976b).

DIAGNOSIS

Sample collection

SUPRAPUBIC ASPIRATION

This technique is considered the diagnostic gold standard. Any species in any number is considered significant in urine obtained by suprapubic aspiration.

MIDSTREAM CLEAN-VOIDED SPECIMEN

As the distal urethra, the periurethral area, and the skin may be colonized by potentially pathogenic bacteria it is of paramount importance to avoid any contamination of the specimen by these bacteria. The most commonly used technique, the midstream urine, requires that the first 10–30 ml of the voided urine be discarded and the second, midstream portion be sampled. The first portion washes away nonadherent colonizing bacteria and the second portion mostly reflects the bacterial colonization of the bladder urine.

In the female patient, adequate periurethral cleansing is necessary to reduce the probability of contamination. For cleaning, water or a soap solution without antibacterial activity should be used as residues of antibacterial soap or disinfectants may impair culture results.

In men, the foreskin should be retracted and the external meatus should be cleansed. There has been some debate whether cleansing is really necessary (Lipsky et al. 1984); however, as it is easily done and decreases the number of contaminated specimens, some authorities recommend this procedure (Schaeffer 1994).

STRAIGHT CATHETERIZATION

Although this procedure yields specimens of a quality similar to that of suprapubic aspiration, it bears the risk of introduction of bacteria into the bladder and bacteriuria after straight catheterization is not uncommon (Eisenstadt and Washington 1996). The technique may be considered if correct clean-voided specimens cannot be obtained and suprapubic aspiration is contraindicated or not feasible.

Specimen transport

Urine should be cultured within 2–4 h. If longer delays cannot be avoided the specimen should be refrigerated to preserve the bacterial density. Transport devices containing boric acid as a preservative have been developed to allow unrefrigerated transport of specimes. However, culture results after prolonged storage of specimens may be altered significantly (Gillespie et al. 1999).

Culture

Bacteriuria is defined as the presence of bacteria within the urinary tract. The central problem in interpreting bacteriuria by the examination of voided urine is to distinguish between contamination originating from the physiological flora and bacteriuria that originates in the normally sterile parts of the urinary tract. Kass (1957) observed that some healthy women have bacteriuria, but the number of colony-forming units (cfu) in their urine was considerably lower than that in urine collected from women with pyelonephritis. In pyelonephritis the urine of 95 percent of patients contained $>10^5$ bacteria per ml, whereas this density was only rarely reached in asymptomatic women (6 percent). The number of 100 000 organisms per ml in a carefully collected sample of clean-voided or midstream urine has since then become the threshold for 'significant' bacteriuria.

Later studies have shown, however, that this threshold may not be applicable to cases of uncomplicated cystitis (Stamm 1988), or posterior urethritis (Stamm et al. 1980; Latham et al. 1985). In these infections colony counts as low as 10^2/ml have been found. Therefore colony counts of $<10^5$/ml may be significant when the clinical diagnosis of uncomplicated cystitis has been established, in cases of complicated urinary tract infections colony counts are usually $>10^5$/ml (Naber et al. 2002). However, cases of uncomplicated cystitis are usually treated empirically, without taking urinary cultures (Warren et al. 1999; Le and Miller 2001).

Consequently, many guidelines for microbiological diagnostic workup of clean-voided urine require a species (or isolate) specific enumeration to distinguish contamination from colonization (Clarridge et al. 1987; Gatermann et al. 1997; Kouri et al. 2000; Naber et al. 2002). The diagnostic interpretation is made relative to the clinical diagnosis or situation. In complicated urinary tract infections the number of 10^5 colony-forming units per ml is considered significant, whereas in clearly uncomplicated cases the threshold is 10^2. Specimens from indwelling catheters may present a special situation since colonization with multiple organisms even in high densities is common (Warren et al. 1982).

The diagnosis of urinary tract infections thus requires a method that is capable of simultaneously determining the colony-forming units of all viable bacterial species present in the sample. Traditionally, this has been done by the pour plate method, where a sample of the urine is mixed with a fixed volume of nutritient agar and plates are poured. The number of bacteria per volume can be determined by multiplying the number of colonies with the dilution factor of the urine. Although this method is most precise, it is also cumbersome and time-consuming. Many laboratories, therefore, use the surface streak technique where a calibrated loop is dipped into the urine and the volume held by the loop is applied to the surface of the appropriate agar media. Commercial single-use loops hold 1 or 10 µl of urine, thus the number of cfu per ml is calculated after incubation by multiplying the number of colonies by 1 000 if a 1-µl loop was used or by 100 if a 10-µl loop was used. There is some evidence that 1-µl loops lack precision and underestimate the true number of bacteria in the sample (Frimodt-Møller and Espersen 2001). Most often, media that permit growth of gram-positive bacteria, including β-hemolytic streptococci, e.g. sheep blood agar, and selective media for Enterobacteriaceae, e.g. MacConkey agar, are inoculated. Cystine lactose electrolyte-deficient agar may also be used, although some gram-positive species grow less well than on blood agar.

Simple, semiquantitative culture methods, such as the filter paper method (Kunin and Buesching 2000) and dipslide cultures have also been developed. These techniques are useful as screening techniques but lack sensitivity for fastidious organisms (Eisenstadt and Washington 1996). In addition, technical problems, such as the detachment of the agar from the paddle or dried out agar in dipslide cultures, can occur.

Due to the limited spectrum of bacteria that cause urinary tract infections identification of causative species can usually be done with a reduced set of tests (Baron 2001). However, care must be exercised not to overlook the unusual or rarely encountered species (Damron et al. 1986). Corynebacteria and some gram-positive cocci may require prolonged incubation and other species such as *Haemophilius* or *Gardnerella* or even a fastidious enterobacterial strain may not grow on the routinely used media. It is therefore advisable to be suspicious, especially in cases where UTI is clinically suspected, pyuria is present, but routine cultures are negative. In such cases an infection with mycobacteria, mycoplasmas, or chlamydia should be ruled out and the unusual should be sought.

Some authors recommend testing of urine samples for the presence of antimicrobial substances (Ansorg et al. 1975; Abu Shaqra 2001) which may mask the presence of bacteria in the urine. To this end, a small volume of urine is placed on an agar plate inoculated with an indicator bacterium (e.g. *Bacillus subtilis* or *Staphylococcus aureus*) and any inhibition of bacterial growth is taken as evidence for the presence of antibacterial substances.

Extrapulmonary tuberculosis may affect the kidneys, prostate, epidymis, or testes. Mycobacterial infections of the kidneys often present with pyuria, but fast-growing bacteria cannot be isolated from the urine. Clinical symptoms suggestive of cystitis may be present and the intravenous pyelogram is usually abnormal (El Khader et al. 1997). Diagnosis is based on cultural detection of *Mycobacterium tuberculosis* in urine, if renal tuberculosis or in ejaculate, if epididymitis is suspected.

Infections with mycoplasmas and chlamydia usually affect the urethra but symptoms may suggest cystitis. As expected, bacterial densities are highest in the first portion of a voided urine and suprapubic aspiration is usually negative. Sensitive and specific tests, molecular genetic amplication techniques, have been described that detect chlamydia in genital specimens (Van Der Pol et al. 2001; Vincelette et al. 1999). Mycoplasmas may be grown from urethral samples although a swab offers higher sensitivities than urine cultures. In addition, *M. hominis* is part of the normal flora of the urethra.

In many countries, standardized protocols for the culture and interpretation of urine samples have been developed (Clarridge et al. 1987; Gatermann et al. 1997; Kouri et al. 2000).

Deprecated procedures and unacceptable specimens

Midstream urine or urine obtained via a catheter should not be cultured anaerobically and should not be inoculated into enrichment broths. Urine should not be centrifuged to culture the sediment.

Foley catheter tips, urine inoculated into broth media, urine sediment, native urine that has been stored unrefrigerated for more than 2 h, or dried dip slides are unacceptable for culture.

Interpretation of urine cultures

For a meaningful interpretation of culture results the clinical situation (complicated, uncomplicated UTI, asymptomatic patient), the patient's sex and the type of the specimen (clean-voided, catheter, suprapubic aspiration) must be available. Also, the number of different bacterial species and their respective number of colony-forming units must be known. Interpretation guidelines aggregate these pieces of information into reasonable laboratory interpretations of culture results and give suggestions for further diagnostic workup, e.g. identification and susceptibility testing. Generally, these guidelines consider pure cultures of typical urinary pathogens to be significant if present in $\geq 10^4$ cfu per ml. In acute, uncomplicated cystitis the thresholds are lower, e.g. 10^3 per ml, whereas in complicated infections they may be higher (10^5 per ml) (Clarridge et al. 1987; Gatermann et al. 1997; Kouri et al. 2000; Naber et al. 2002). Repeated isolation of the same microorganism increases the specificity of the result and is absolutely necessary if asymptomatic bacteriuria is to be detected (Pattaragarn and Alon 2003).

In patients with chronic indwelling catheters, cultures often yield more than one potential pathogen regardless of patients' symptoms (Warren et al. 1982). These specimens therefore reflect colonization rather than infection. It is rarely necessary to perform urinary cultures in asymptomatic patients with indwelling catheters because high colony counts may reflect infection or colonization and low bacterial densities usually increase to high ($>10^5$) levels within a few days (Stark and Maki 1984). In symptomatic patients with flank pain or fever, a diagnostic workup of all organisms may be indicated (Gatermann et al. 1997).

Candida spp. is often isolated from the urine of catheterized patients who receive antibiotic treatment and antibiotic therapy is the most important risk factor for candiduria (Lundstrom and Sobel 2001; Weinberger et al. 2003). However, since candiduria may also be a symptom of deep-seated candidal infection, patients should be carefully evaluated to exclude such condition (Fisher et al. 1995).

MANAGEMENT OF URINARY TRACT INFECTIONS

A plethora of literature exists concerning the treatment of urinary tract infections. Most current guidelines recommend either trimethoprim/sulfmethoxazol or a fluoroquinolone for empiric therapy of uncomplicated cystitis. These antibiotics are usually given for 3 days. β-lactam antibiotics are less satisfactory, because recurrences are more frequent (Johnson and Stamm 1989) and longer therapies may be needed. During pregnancy trimethoprim/sulfmethoxazol or quinolones cannot be used and a susceptible β-lactam antibiotic is the therapy of choice (Naber et al. 2002). In uncomplicated cystitis in nonpregnant women, the narrow bacterial spectrum permits empiric therapy without the aid of a urine culture. In pregnant women a urine culture may be needed to select the most appropriate agent (Naber et al. 2002).

In patients with acute uncomplicated pyelonephritis, oral therapy can be an alternative to the classical intravenous regimens (Warren et al. 1999; Naber et al. 2002). As with uncomplicated cystitis, a fluoroquinolone, or, if susceptibility has been established or is likely, trimethoprim/sulfmethoxazole is usually appropriate. However, a urine culture should be obtained to guide therapy in poorly responding individuals.

Although most standards recommend the use of trimethoprim/sulfamethoxazole or fluoroquinolones as a first-line treatment (Johnson and Stamm 1989; Warren et al. 1999), the most appropriate therapy is contingent on the local situation, because susceptibility towards antimicrobials may vary widely between geographic regions. For trimethoprim/sulfamethoxazole resistance rates for *E. coli* as high as 45 percent have been described in some regions (Mazzulli 2002; Farrell et al. 2003; Gales et al. 2003) and it has been recommended that fluoroquinolones be used if resistance rates to trimethoprim/sulfamethoxazole exceed 20 percent (Le and Miller 2001). However, resistance to fluoroquinolones is also emerging and resistance rates of 10 percent in *E. coli* are common (Mazzulli 2002; Farrell et al. 2003), and up to 20 percent have been described (Gales et al. 2003).

In complicated urinary tract infections treatment should be based on the results of a urinary culture, because the spectrum of causative organisms is much larger and resistance to antimicrobial agents is frequent (Naber et al. 2002).

Asymptomatic bacteriuria should be sought for and treated in pregnant women (Smaill 2002), because it is a risk factor for pyelonephritis, low birth weight, and early delivery. In contrast, screening for and treatment of asymptomatic bacteriuria in diabetic women (Harding et al. 2002) and patients with renal transplants is probably not indicated (Muñoz 2001).

Prophylaxis

Given the fact that most patients experience only one UTI in their lifetimes, prophylactic measures are needed only for women suffering frequent recurrences. If it has been established in an individual that episodes of acute UTI frequently occur after sexual intercourse, a prophylactic dose of an antibiotic may be (Stapleton and Stamm 1997) indicated. In some women, for periods of clustering of UTIs a regular dose of an antibiotic may suppress infections. Currently there is no evidence that cranberry juice reduces the frequency of urinary tract infections (Jepson et al. 2002).

There have been many attempts at developing vaccines against UTI with *E. coli* and *P. mirabilis*. Whole cell killed vaccines or vaccines based on single virulence factors have been used in animal models to show that the animals were protected against challanges with strains expressing the respective virulence factor (Langermann et al. 2000; Li and Mobley 2002; Uehling et al. 1994a, b, 1997). Since many different virulence factors contribute to pathogenicity it is unlikely that one single factor will be sufficient to offer protection.

REFERENCES

Abraham, J.M., Freitag, C.S., et al. 1985. An invertible element of DNA controls phase variation of type 1 fimbriae of *Escherichia coli*. *Proc Natl Acad Sci USA*, **82**, 5724–7.

Abu Shaqra, Q.M. 2001. Antimicrobial activity in urine: effect on leukocyte count and bacterial culture results. *New Microbiol*, **24**, 137–42.

Anderson, G.G., Palermo, J.J., et al. 2003. Intracellular bacterial biofilm-like pods in urinary tract infections. *Science*, **301**, 105–7.

Ansorg, R., Zappel, H. and Thomssen, R. 1975. Bedeutung des Nachweises antibakterieller Stoffe im Urin fur die bakteriologische Diagnostik und die Kontrolle der Chemotherapie von Harnweginfektionen. *Zentralbl Bakteriol [Orig A]*, **230**, 492–507.

Bäckhed, F., Alsén, B., et al. 2002. Identification of target tissue glycosphingolipid receptors for uropathogenic, F1C-fimbriated *Escherichia coli* and its role in mucosal inflammation. *J Biol Chem*, **277**, 18198–205.

Baerheim, A., Digranes, A. and Hunskaar, S. 1999. Equal symptomatic outcome after antibacterial treatment of acute lower urinary tract infection and the acute urethral syndrome in adult women. *Scand J Prim Health Care*, **17**, 170–3.

Bahrani-Mougeot, F.K., Buckles, E.L., et al. 2002. Type 1 fimbriae and extracellular polysaccharides are preeminent uropathogenic *Escherichia coli* virulence determinants in the murine urinary tract. *Mol Microbiol*, **45**, 1079–93.

Baron, E.J. 2001. Rapid identification of bacteria and yeast: summary of a National Committee for Clinical Laboratory Standards proposed guideline. *Clin Infect Dis*, **33**, 220–5.

Bézian, M.C., Ribou, G., et al. 1990. Isolation of a urease positive thermophilic variant of *Campylobacter lari* from a patient with urinary tract infection. *Eur J Clin Microbiol Infect Dis*, **9**, 895–7.

Bock, K., Breimer, M.E. and Brignole, A. 1985. Specificity of binding of a strain of uropathogenic *Escherichia coli* to Gal alpha 1–4Gal-containing glycosphingolipids. *J Biol Chem*, **260**, 8545–51.

Bollgren, I. and Winberg, J. 1976a. The periurethral aerobic bacterial flora in healthy boys and girls. *Acta Paediatr Scand*, **65**, 74–80.

Bollgren, I. and Winberg, J. 1976b. The periurethral aerobic flora in girls highly susceptible to urinary infections. *Acta Paediatr Scand*, **65**, 81–7.

Bollgren, I., Källenius, G., et al. 1979. Periurethral anaerobic microflora of healthy girls. *J Clin Microbiol*, **10**, 419–24.

Bottone, E.J., Patel, L., et al. 1998. Mucoid encapsulated *Enterococcus faecalis*: an emerging morphotype isolated from patients with urinary tract infections. *Diagn Microbiol Infect Dis*, **31**, 429–30.

Bouza, E., San Juan, R., et al. 2002. A European perspective on nosocomial urinary tract infections I. Report on the microbiology workload, etiology and antimicrobial susceptibility (ESGNI-003 study). European Study Group on Nosocomial Infections. *Clin Microbiol Infect*, **7**, 523–31.

Braude, A.I. and Siemienski, J. 1960. Role of bacterial urease in experimental pyelonephritis. *J Bacteriol*, **80**, 171–9.

Cappelli, E.A., Barros, R.R., et al. 1999. *Leuconostoc pseudomesenteroides* as a cause of nosocomial urinary tract infections. *J Clin Microbiol*, **37**, 4124–6.

Caprioli, A., Falbo, V., et al. 1987. Cytotoxic necrotizing factor production by hemolytic strains of *Escherichia coli* causing extraintestinal infections. *J Clin Microbiol*, **25**, 146–9.

Carbonetti, N.H., Boonchai, S., et al. 1986. Aerobactin-mediated iron uptake by *Escherichia coli* isolates from human extraintestinal infections. *Infect Immun*, **51**, 966–8.

Clarridge, J.E., Pezzlo, M.T., et al. 1987. *Laboratory diagnosis of urinary tract infections*. Washington DC: American Society of Microbiology.

Connell, I., Agace, W., et al. 1996. Type 1 fimbrial expression enhances *Escherichia coli* virulence for the urinary tract. *Proc Natl Acad Sci USA*, **93**, 9827–32.

Dalet, F., Segovia, T. and Del Río, G. 1991. Frequency and distribution of uropathogenic *Escherichia coli* adhesins: a clinical correlation over 2000 cases. *Eur Urol*, **19**, 295–303.

Damron, D.J., Warren, J.W., et al. 1986. Do clinical microbiology laboratories report complete bacteriology in urine from patients with long-term urinary catheters? *J Clin Microbiol*, **24**, 400–4.

Donnenberg, M.S. and Welch, R.A. 1996. Virulence determinants of uropathogenic *E. coli*. In: Mobley, H.L.T. and Warren, J.W. (eds), *Urinary tract infections: molecular pathogenesis and clinical management*. Washington DC: ASM Press, 135–74.

Eisenstein, B.I. 1981. Phase variation of type 1 fimbriae in *Escherichia coli* is under transcriptional control. *Science*, **214**, 337–9.

Eisenstadt, P.O. and Washington, J.A. 1996. Diagnostic microbiology for bacteria and yeasts causing urinary tract infections. In: Mobley, H.L.T. and Warren, J.W. (eds), *Urinary tract infections: molecular pathogenesis and clinical management*. Washington DC: ASM Press, 29–66.

El Khader, K., El Fassi, J., et al. 1997. Urogenital tuberculosis. Apropros of 40 cases. *Ann Urol*, **31**, 339–43.

Falkow, S. 1989. Molecular Koch's postulates applied to microbial pathogenicity. *Rev Infect Dis*, **10**, Suppl. 2, 274–6.

Farrell, D.J., Morrissey, I., et al. 2003. A UK multicentre study of the antimicrobial susceptibility of bacterial pathogens causing urinary tract infection. *Infection*, **46**, 94–100.

Fisher, J.F., Newman, C.L. and Sobel, J.D. 1995. Yeast in the urine: solutions for a budding problem. *Clin Infect Dis*, **20**, 183–9.

Fraimow, H.S., Jungkind, D.L., et al. 1994. Urinary tract infection with an *Enterococcus faecalis* isolate that requires vancomycin for growth. *Ann Intern Med*, **121**, 22–6.

Frendéus, B., Godaly, G., et al. 2000. Interleukin 8 receptor deficiency confers susceptibility to acute experimental pyelonephritis and may have a human counterpart. *J Exp Med*, **192**, 881–90.

Frimodt-Møller, N. and Espersen, F. 2001. Evaluation of calibrated 1 and 10 microl loops and dipslide as compared to pipettes for detection of low count bacteriuria in vitro. *APMIS*, **108**, 525–30.

Funke, G., Lawson, P.A. and Collins, M.D. 1998. *Corynebacterium riegelii* sp. nov., an unusual species isolated from female patients with urinary tract infections. *J Clin Microbiol*, **36**, 624–7.

Gales, A.C., Sader, H.S., et al. 2003. Urinary tract infection trends in Latin American hospitals: report from the SENTRY antimicrobial surveillance program (1997–2000). *Diagn Microbiol Infect Dis*, **44**, 289–99.

Gatermann, S. 1996. Virulence factors of *Staphylococcus saprophyticus*, *Staphylococcus epidermidis* and enterococci. In: Mobley, H.L.T. and Warren, J.W. (eds), *Urinary tract infections: molecular pathogenesis and clinical management*. Washington DC: ASM Press, 313–40.

Gatermann, S. and Marre, R. 1989. Cloning and expression of *Staphylococcus saprophyticus* urease gene sequences in *Staphylococcus carnosus* and contribution of the enzyme to virulence. *Infect Immun*, **57**, 2998–3002.

Gatermann, S., Meyer, H.G. and Wanner, G. 1992. *Staphylococcus saprophyticus* hemagglutinin is a 160-kilodalton surface polypeptide. *Infect Immun*, **60**, 4127–32.

Gatermann, S., Podschun, R., et al. 1997. Harnwegsinfektionen. In: Mauch, H., Lütticken, R. and Gatermann, S. (eds), *MiQ2 Qualitätsstandards in der mikrobiologisch-infektiologischen Diagnostik*. Jena und Stuttgart: Gustav Fischer.

Gillespie, T., Fewster, J. and Masterton, R.G. 1999. The effect of specimen processing delay on borate urine preservation. *J Clin Pathol*, **52**, 95–8.

Goluszko, P., Popov, V., et al. 1997. Dr fimbriae operon of uropathogenic *Escherichia coli* mediate microtubule-dependent invasion to the HeLa epithelial cell line. *J Infect Dis*, **176**, 158–67.

Griffith, D.P. and Musher, D.M. 1975. Acetohydroxamic acid. Potential use in urinary infection caused by urea-splitting bacteria. *Urology*, **5**, 299–302.

Griffith, D.P., Musher, D.M. and Itin, C. 1976. Urease. The primary cause of infection-induced urinary stones. *Invest Urol*, **13**, 346–50.

Gunther, N.W., Snyder, J.A., et al. 2002. Assessment of virulence of uropathogenic *Escherichia coli* type 1 fimbrial mutants in which the invertible element is phase-locked on or off. *Infect Immun*, **70**, 3344–54.

Harding, G.K., Zhanel, G.G., et al. 2002. Antimicrobial treatment in diabetic women with asymptomatic bacteriuria. *N Engl J Med*, **347**, 1576–83.

Hell, W., Meyer, H.G. and Gatermann, S.G. 1998. Cloning of *aas*, a gene encoding a *Staphylococcus saprophyticus* surface protein with adhesive and autolytic properties. *Mol Microbiol*, **29**, 871–81.

Hull, R.A., Donovan, W.H., et al. 2002. Role of type 1 fimbria- and P fimbria-specific adherence in colonization of the neurogenic human bladder by *Escherichia coli*. *Infect Immun*, **70**, 6481–4.

Hultgren, S.J., Porter, T.N., et al. 1985. Role of type 1 pili and effects of phase variation on lower urinary tract infections produced by *Escherichia coli*. *Infect Immun*, **50**, 370–7.

Janda, W.M., Senseng, C., et al. 1993. Asymptomatic *Neisseria subflava* biovar *perflava* bacteriuria in a child with obstructive uropathy. *Eur J Clin Microbiol Infect Dis*, **12**, 540–5.

Jantunen, M.E., Saxén, H., et al. 2002. Recurrent urinary tract infections in infancy: relapses or reinfections? *J Infect Dis*, **185**, 375–9.

Jepson, R.G., Mihaljevic, L. and Craig, J. 2002. Cranberries for preventing urinary tract infections. *Cochrane Database Syst Rev*, **30**, CD001321.

Johnson, D.E., Drachenberg, C., et al. 2000. The role of cytotoxic necrotizing factor-1 in colonization and tissue injury in a murine model of urinary tract infection. *FEMS Immunol Med Microbiol*, **28**, 37–41.

Johnson, J.R. and Stamm, W.E. 1989. Urinary tract infections in women: diagnosis and treatment. *Ann Intern Med*, **111**, 906–17.

Johnson, D.E., Russell, R.G., et al. 1993. Contribution of *Proteus mirabilis* urease to persistence, urolithiasis, and acute pyelonephritis in a mouse model of ascending urinary tract infection. *Infect Immun*, **61**, 2748–54.

Jones, B.D., Lockatell, C.V., et al. 1990. Construction of a urease-negative mutant of *Proteus mirabilis*: analysis of virulence in a mouse model of ascending urinary tract infection. *Infect Immun*, **58**, 1120–3.

Kahlmeter, G. 2000. The ECO.SENS Project: a prospective, multinational, multicentre epidemiological survey of the prevalence and antimicrobial susceptibility of urinary tract pathogens – interim report. *J Antimicrob Chemother*, **46**, Suppl. 1, 15–22.

Kass, E.H. 1957. Bacteriuria and the diagnosis of infections of the urinary tract. *Arch Intern Med*, **100**, 708–14.

Knapp, S., Hacker, J., et al. 1986. Large, unstable inserts in the chromosome affect virulence properties of uropathogenic *Escherichia coli* O6 strain 536. *J Bacteriol*, **168**, 22–30.

Kouri, T., Fogazzi, G., et al. 2000. European urinalysis guidelines. *Scand J Clin Lab Invest Suppl*, **146**, 1–86.

Kreft, B. and Pagel, H. 2000. Virulence factors of *Escherichia coli* contribute to acute renal failure. *Exp Nephrol*, **8**, 244–51.

Krieger, J.N., Riley, D.E., et al. 1996. Prokaryotic DNA sequences in patients with chronic idiopathic prostatitis. *J Clin Microbiol*, **34**, 3120–8.

Kunin, C.M. and Buesching, W.J. 2000. Novel screening method for urine cultures using a filter paper dilution system. *J Clin Microbiol*, **38**, 1187–90.

Kunin, C.M., Evans, C., et al. 2002. The antimicrobial defense mechanism of the female urethra: a reassessment. *J Urol*, **168**, 413–19.

Källenius, G., Svenson, S., et al. 1981. Structure of carbohydrate part of receptor on human uroepithelial cells for pyelonephritogenic *Escherichia coli*. *Lancet*, **2**, 604–6.

Langermann, S., Möllby, R., et al. 2000. Vaccination with FimH adhesin protects cynomolgus monkeys from colonization and infection by uropathogenic *Escherichia coli*. *J Infect Dis*, **181**, 774–8.

Latham, R.H., Wong, E.S. and Larson, A. 1985. Laboratory diagnosis of urinary tract infection in ambulatory women. *JAMA*, **254**, 3333–6.

Le, T.P. and Miller, L.G. 2001. Empirical therapy for uncomplicated urinary tract infections in an era of increasing antimicrobial resistance: a decision and cost analysis. *Clin Infect Dis*, **33**, 615–21.

Lee, J.C., Muller, C.H., et al. 2003. Prostate biopsy culture findings of men with chronic pelvic pain syndrome do not differ from those of healthy controls. *J Urol*, **169**, 584–7.

Leeker, A., Kreft, B., et al. 1997. Tamm-Horsfall protein inhibits binding of S- and P-fimbriated *Escherichia coli* to human renal tubular epithelial cells. *Exp Nephrol*, **5**, 38–46.

Leffler, H. and Svanborg-Eden, C. 1981. Glycolipid receptors for uropathogenic *Escherichia coli* on human erythrocytes and uroepithelial cells. *Infect Immun*, **34**, 3, 920–9.

Li, X. and Mobley, H.L. 2002. Vaccines for *Proteus mirabilis* in urinary tract infection. *Int J Antimicrob Agents*, **19**, 461–5.

Lidin-Janson, G., Hanson, L.A., et al. 1977. Comparison of *Escherichia coli* from bacteriuric patients with those from feces of healthy schoolchildren. *J Infect Dis*, **136**, 346–53.

Lipsky, B.A., Inui, T.S., et al. 1984. Is the clean-catch midstream void procedure necessary for obtaining urine culture specimens from men? *Am J Med*, **76**, 257–62.

Lundstrom, T. and Sobel, J. 2001. Nosocomial candiduria: a review. *Clin Infect Dis*, **32**, 1602–7.

Maki, D.G. and Agger, W.A. 1988. Enterococcal bacteremia: clinical features, the risk of endocarditis, and management. *Medicine (Balt)*, **67**, 248–69.

Marre, R., Hacker, J., et al. 1986. Contribution of cloned virulence factors from uropathogenic *Escherichia coli* strains to nephropathogenicity in an experimental rat pyelonephritis model. *Infect Immun*, **54**, 761–7.

Mazzulli, T. 2002. Resistance trends in urinary tract pathogens and impact on management. *J Urol*, **168**, 1720–2.

Meyer, H.G., Muthing, J. and Gatermann, S.G. 1997. The hemagglutinin of *Staphylococcus saprophyticus* binds to a protein receptor on sheep erythrocytes. *Med Microbiol Immunol (Berl)*, **186**, 37–43.

Meyer, H.G., Wengler-Becker, U. and Gatermann, S.G. 1996. The hemagglutinin of *Staphylococcus saprophyticus* is a major adhesin for uroepithelial cells. *Infect Immun*, **64**, 3893–6.

Mobley, H.L.T. 1996. Virulence of *Proteus mirabilis*. In: Mobley, H.L.T. and Warren, J.W. (eds), *Urinary tract infections: molecular pathogenesis and clinical management*. Washington DC: ASM Press, 245–70.

Mobley, H.L., Jarvis, K.G., et al. 1993. Isogenic P-fimbrial deletion mutants of pyelonephritogenic *Escherichia coli*: the role of alpha Gal(1-4) beta Gal binding in virulence of a wild-type strain. *Mol Microbiol*, **10**, 143–55.

Morgan, M.G. and Hamilton-Miller, J.M. 1990. *Haemophilus influenzae* and *H. parainfluenzae* as urinary pathogens. *J Infect*, **20**, 143–5.

Morschhäuser, J., Vetter, V., et al. 1994. Adhesin regulatory genes within large, unstable DNA regions of pathogenic *Escherichia coli*: cross-talk between different adhesin gene clusters. *Mol Microbiol*, **11**, 555–66.

Mühldorfer, I., Ziebuhr, W. and Hacker, J. 2002. *Escherichia coli* in urinary tract infections. In *Molecular medical microbiology*. London: Academic Press, 1515–40.

Mulvey, M.A. 2002. Adhesion and entry of uropathogenic *Escherichia coli*. *Cell Microbiol*, **4**, 257–71.

Mulvey, M.A., Schilling, J.D., et al. 2000. Bad bugs and beleaguered bladders: interplay between uropathogenic *Escherichia coli* and innate host defenses. *Proc Natl Acad Sci USA*, **97**, 8829–35.

Mulvey, M.A., Schilling, J.D. and Hultgren, S.J. 2001. Establishment of a persistent *Escherichia coli* reservoir during the acute phase of a bladder infection. *Infect Immun*, **69**, 4572–9.

Musher, D.M., Griffith, D.P., et al. 1975. Role of urease in pyelonephritis resulting from urinary tract infection with *Proteus*. *J Infect Dis*, **131**, 177–81.

Muñoz, P. 2001. Management of urinary tract infections and lymphocele in renal transplant recipients. *Clin Infect Dis*, **33**, Suppl. 1 , S53–7.

Naber, K.G., Bergman, B., Urinary Tract Infection (UTI) Working Group of the Health Care Office (HCO) of the European Association of Urology (EAU), et al. 2002. EAU guidelines for the management of urinary and male genital tract infections. *Eur Urol*, **40**, 576–88.

O'Grady, F.W., Mcherry, M.A., et al. 1970. Introital enterobacteria, urinary infection, and the urethral syndrome. *Lancet*, **2**, 1208–10.

O'Hanley, P., Lalonde, G. and Ji, G. 1991. Alpha-hemolysin contributes to the pathogenicity of piliated digalactoside-binding *Escherichia coli* in the kidney: efficacy of an alpha-hemolysin vaccine in preventing renal injury in the BALB/c mouse model of pyelonephritis. *Infect Immun*, **59**, 1153–61.

Pak, J., Pu, Y., et al. 2001. Tamm-Horsfall protein binds to type 1 fimbriated *Escherichia coli* and prevents *E. coli* from binding to uroplakin Ia and Ib receptors. *J Biol Chem*, **276**, 9924–30.

Parsons, C.L., Stauffer, C., et al. 1984. Effect of ammonium on bacterial adherence to bladder transitional epithelium. *J Urol*, **132**, 365–6.

Pattaragarn, A. and Alon, U.S. 2003. Urinary tract infection in childhood. Review of guidelines and recommendations. *Minerva Pediatr*, **54**, 401–13.

Patterson, J.E. and Andriole, V.T. 1997. Bacterial urinary tract infections in diabetes. *Infect Dis Clin North Am*, **11**, 735–50.

Pere, A., Nowicki, B., et al. 1987. Expression of P, type-1, and type-1C fimbriae of *Escherichia coli* in the urine of patients with acute urinary tract infection. *J Infect Dis*, **156**, 567–74.

Raz, R., Gennesin, Y. and Wasser, J. 2000. Recurrent urinary tract infections in postmenopausal women. *Clin Infect Dis*, **30**, 152–6.

Reid, G. and Sobel, J.D. 1987. Bacterial adherence in the pathogenesis of urinary tract infection: a review. *Rev Infect Dis*, **9**, 470–87.

Remis, R.S., Gurwith, M.J., et al. 1987. Risk factors for urinary tract infection. *Am J Epidemiol*, **126**, 685–94.

Rippere-Lampe, K.E., O'Brien, A.D., et al. 2001a. Mutation of the gene encoding cytotoxic necrotizing factor type 1 (cnf(1)) attenuates the virulence of uropathogenic *Escherichia coli*. *Infect Immun*, **69**, 3954–64.

Rippere-Lampe, K.E., Lang, M., et al. 2001b. Cytotoxic necrotizing factor type 1-positive *Escherichia coli* causes increased inflammation and tissue damage to the prostate in a rat prostatitis model. *Infect Immun*, **69**, 6515–19.

Ronald, A. 2002. The etiology of urinary tract infection: traditional and emerging pathogens. *Am J Med*, **113**, Suppl. 1A, 14S–9S.

Salahuddin, F., Sen, P. and Chechko, S. 1996. Urinary tract infection with an unusual pathogen (*Nocardia asteroides*). *J Urol*, **155**, 654–5.

Sandberg, T., Kaijser, B., et al. 1988. Virulence of *Escherichia coli* in relation to host factors in women with symptomatic urinary tract infection. *J Clin Microbiol*, **26**, 1471–6.

Saulino, E.T., Bullitt, E. and Hultgren, S.J. 2000. Snapshots of usher-mediated protein secretion and ordered pilus assembly. *Proc Natl Acad Sci USA*, **97**, 9240–5.

Schaeffer, A.J. 1994. Urinary tract infection in men – state of the art. *Infection*, **22**, Suppl. 1 , S19–21.

Sheinfeld, J., Schaeffer, A.J., et al. 1989. Association of the Lewis blood-group phenotype with recurrent urinary tract infections in women. *N Engl J Med*, **320**, 773–7.

Smaill, F. 2002. Antibiotics for asymptomatic bacteriuria in pregnancy. *Cochrane Database Syst Rev*, **30**, CD000490.

Soto, G.E., Dodson, K.W. and Ogg, D. 1998. Periplasmic chaperone recognition motif of subunits mediates quaternary interactions in the pilus. *EMBO J*, **17**, 6155–67.

Snehalatha, S., Mathai, E., et al. 1992. *Salmonella choleraesuis* subsp. indica serovar bornheim causing urinary tract infection. *J Clin Microbiol*, **30**, 2504–5.

Stamm, W.E. 1988. Protocol for diagnosis of urinary tract infection: reconsidering the criterion for significant bacteriuria. *Urology*, **32**, 6–12.

Stamm, W.E., Wagner, K.F., et al. 1980. Causes of the acute urethral syndrome in women. *N Engl J Med*, **303**, 409–15.

Stapleton, A. and Stamm, W.E. 1997. Prevention of urinary tract infection. *Infect Dis Clin North Am*, **11**, 719–33.

Stark, R.P. and Maki, D.G. 1984. Bacteriuria in the catheterized patient. What quantitative level of bacteriuria is relevant? *N Engl J Med*, **311**, 560–4.

Stewart, R.G., Nowbath, V., et al. 1993. *Corynebacterium* group D2 urinary tract infection. *S Afr Med J*, **83**, 95–6.

Sturm, A.W. 1989. *Gardnerella vaginalis* in infections of the urinary tract. *J Infect*, **18**, 45–9.

Takai, K., Aoki, A., et al. 1998. Urinary tract infections following renal transplantation. *Transplant Proc*, **30**, 3140–1.

Torres, A.G., Redford, P., et al. 2001. TonB-dependent systems of uropathogenic *Escherichia coli*: aerobactin and heme transport and TonB are required for virulence in the mouse. *Infect Immun*, **69**, 6179–85.

Tullus, K., Jacobson, S.H., et al. 1991. Relative importance of eight virulence characteristics of pyelonephritogenic *Escherichia coli* strains assessed by multivariate statistical analysis. *J Urol*, **146**, 1153–5.

Uehling, D.T., Hopkins, W.J., et al. 1994a. Phase I clinical trial of vaginal mucosal immunization for recurrent urinary tract infection. *J Urol*, **152**, 2308–11.

Uehling, D.T., Hopkins, W.J., et al. 1994b. Vaginal immunization of monkeys against urinary tract infection with a multi-strain vaccine. *J Urol*, **151**, 214–16.

Uehling, D.T., Hopkins, W.J., et al. 1997. Vaginal mucosal immunization for recurrent urinary tract infection: phase II clinical trial. *J Urol*, **157**, 2049–52.

Van Der Pol, B., Ferrero, D.V., et al. 2001. Multicenter evaluation of the BDProbeTec ET System for detection of *Chlamydia trachomatis* and *Neisseria gonorrhoeae* in urine specimens, female endocervical swabs, and male urethral swabs. *J Clin Microbiol*, **39**, 1008–16.

Vincelette, J., Schirm, J., et al. 1999. Multicenter evaluation of the fully automated COBAS AMPLICOR PCR test for detection of *Chlamydia trachomatis* in urogenital specimens. *J Clin Microbiol*, **37**, 74–80.

Wagenlehner, F.M., Niemetz, A., et al. 2002. Spectrum and antibiotic resistance of uropathogens from hospitalized patients with urinary tract infections: 1994–2000. *Int J Antimicrob Agents*, **19**, 557–64.

Wazait, H.D., Patel, H.R., et al. 2003. Catheter-associated urinary tract infections: prevalence of uropathogens and pattern of antimicrobial resistance in a UK hospital (1996–2001). *BJU Int*, **91**, 9, 806–9.

Warren, J.W., Tenney, J.H., et al. 1982. A prospective microbiologic study of bacteriuria in patients with chronic indwelling urethral catheters. *J Infect Dis*, **146**, 719–23.

Warren, J.W., Abrutyn, E., et al. 1999. Guidelines for antimicrobial treatment of uncomplicated acute bacterial cystitis and acute pyelonephritis in women. Infectious Diseases Society of America (IDSA). *Clin Infect Dis*, **29**, 745–58.

Weinberger, M., Sweet, S. and Leibovici, L. 2003. Correlation between candiduria and departmental antibiotic use. *J Hosp Infect*, **53**, 183–6.

Winberg, J., Mollby, R., et al. 1995. PapG-adhesin at the tip of P-fimbriae provides *Escherichia coli* with a competitive edge in experimental bladder infections of cynomolgus monkeys. *J Exp Med*, **182**, 6, 1695–702.

Wong, S.S. and Yuen, K.Y. 1995. Acute pyelonephritis caused by *Mycoplasma hominis*. *Pathology*, **27**, 61–3.

Wu, X.R., Sun, T.T. and Medina, J.J. 1996. In vitro binding of type 1-fimbriated *Escherichia coli* to uroplakins Ia and Ib: relation to urinary tract infections. *Proc Natl Acad Sci USA*, **93**, 9630–5.

Wullt, B., Bergsten, G., et al. 2001. P fimbriae enhance the early establishment of *Escherichia coli* in the human urinary tract. *Mol Microbiol*, **38**, 456–64.

Zhanel, G.G., Harding, G.K. and Nicolle, L.E. 1991. Asymptomatic bacteriuria in patients with diabetes mellitus. *Rev Infect Dis*, **13**, 150–4.

Zhou, G., Mo, W.J., et al. 2002. Uroplakin Ia is the urothelial receptor for uropathogenic *Escherichia coli*: evidence from in vitro FimH binding. *J Cell Sci*, **114**, 4095–103.

27

Bacterial infections of bones and joints

ANTHONY R. BERENDT

INTRODUCTION

Bone and joint infections pose a number of challenges to the academic and the clinical microbiologist. From the academic standpoint the pathogenesis of musculoskeletal infection, and the pathophysiology of the osteomyelitic lesion or the infected prosthetic joint, remain incompletely understood (Mandal et al. 2002). So too do the molecular mechanisms of infection-related bone destruction (Josefsson and Tarkowski 1999; Nair et al. 2000; Tarkowski et al. 2002) or bacterial persistence in the face of host defenses and antibiotic use.

From the clinical microbiology perspective, there can be significant difficulties to be overcome in the interpretation of microbial cultures, notably because of the role of skin commensals in device-related infection. Additional problems are generated by those clinical scenarios where bone, infected as a consequence of soft tissue loss, is then sampled through the resulting soft tissue defect.

Having isolated bacteria and formed a view about their relevance, the clinical microbiologist is frequently challenged when feeding that information into a management plan. This is partly because of a relative lack of a strong evidence base for recommendations about durations of treatment or the mode of delivery of antimicrobials. It is also because of the strongly surgical nature of the specialty. The microbiologist must exercise skills in multidisciplinary working and to be credible, must demonstrate an understanding not only of his own field, but also of the major orthopedic imperatives and drivers.

Given the very significant morbidity caused by bone and joint infection, and the substantial health gains for patients who are treated successfully, there are major benefits for patients and clinicians (of all disciplines) when the clinical microbiologist can participate in, and if need be lead, effective multidisciplinary working for the management of these problems (Ziran et al. 2003).

CLINICAL MANIFESTATIONS

Musculoskeletal infection, affecting bone and joint, may be acute or chronic. Any part of the skeleton may be affected, and bacteria may reach the tissues of the skeleton through hematogenous or contiguous spread. The range of manifestations most commonly seen is outlined in Table 27.1.

EPIDEMIOLOGY

The epidemiology of musculoskeletal infection is a reflection of the prevalence, in the population, of underlying risk factors for bacteremia, trauma, and instrumentation on the one hand, and the geographical distribution of pathogens on the other.

In industrialized societies, in which antibiotics are relatively available, child health is good, and orthopedic

Table 27.1 Commonly seen range of manifestations

Manifestation	Site of infection	Presentation	Soft tissue component	Causative pathogen(s)
Septic arthritis	Native	Acute	−	MSSA, BHS, HIB, GC
	Prosthetic		+	MRSA, MSSA, CoNS GNR
	Native	Chronic	−	MTB, Lyme, MSSA
	Prosthetic		++	CoNS, SA, BHS, EC, GNR
Osteomyelitis	Hematogenous	Acute	−	MSSA, BHS, HIB, *Salmonella*
	Contiguous focus		+	MSSA, MRSA, GNR
	Hematogenous	Chronic	++	MSSA, BHS, GNR
	Contiguous focus		+++	MSSA, MRSA, CoNS, BHS, GNR, EC
Vertebral osteomyelitis and discitis	Hematogenous	Chronic	−	MSSA, GNR, MRSA
	Contiguous focus (post surgical)	Acute	−	MRSA, MSSA, CoNS
Fracture-fixation infection	Contiguous focus	Acute	−/+	MSSA, MRSA, GNR
		Chronic	++	MSSA, MRSA, CoNS, GNR
Osteomyelitis 2° to pressure sore including diabetic foot ulcer		Chronic	+++	MSSA, MRSA, CoNS, BHS, GNR, AnO2, EC

ANO2, anaerobic rods; BHS, β-hemolytic streptococci; CONS, coagulase-negative staphylococci; EC, *E. coli*; GC, gonococci; GNR, gram-negative rods; HiB, *H. influenzae* type b; Lyme, Lyme borreliosis; MRSA, methicillin-resistant *S. aureus*; MSSA, methicillin-susceptible *S. aureus*; MTB, *Mycobacterium tuberculosis*; SA, *S. aureus*.

management includes energetic trauma management and joint replacement, there has been a decline in 'classical' presentations. Advances in fracture management mean that infections of fracture fixations have replaced infections of open fractures (Gustilo and Anderson 1976); childhood hematogenous bone and joint infection is in decline and more frequently presents in a subacute manner (Gonzalez-Lopez et al. 2001; Blyth et al. 2001; Rasool 2001); hematogenous infections of the spine in the elderly are increasing in incidence (Jensen et al. 1997; Chelsom and Solberg 1998; Krogsgaard et al. 1998), as is osteomyelitis secondary to pressure sores (Sugarman 1987). Infections of prosthetic joints are now the most formidable manifestation of septic arthritis (Malchau et al. 1993), a role occupied by tuberculosis some 50 years ago, and by *Staphylococcus aureus* in the pre-antibiotic era.

Much of this remains a contrast to less developed areas of the world, where higher incidences of soft tissue infection, and less universal access to healthcare, perhaps explain the higher recorded incidences of conditions such as acute septic arthritis and osteomyelitis (Gillespie 1990). Limited healthcare resources may restrict access to joint replacement or complex orthopedic instrumentation, but correspondingly make all the more serious the impact of infection on traumatic wounds. Combat and landmine injuries from military conflict are important in this regard, as are injuries from increasing numbers of industrial and road traffic accidents, though systematic data on this are hard to find. As a result, chronic osteomyelitis remains a considerable problem (Museru and Mcharo 2001).

Certain manifestations appear to have a truly worldwide importance. The global epidemic of Type 2 (non-insulin dependent) diabetes is responsible for a dramatic increase in problems due to neuropathic foot ulceration, including osteomyelitis (Lipsky et al. 1990). Because patients with this disease survive for relatively long periods without expensive treatments, developing countries share with the industrialized nations a rising prevalence of diabetics with neuropathy and secondary foot ulceration, frequently leading to osteomyelitis (Abbas et al. 2002).

From a microbiological perspective, although certain pathogens such as *Staphylococcus aureus* are universally important, others are restricted by geographical or occasionally socioeconomic factors. These include the predominant distribution of methicillin-resistant *Staphylococcus aureus* (MRSA) within that group of society in frequent contact with healthcare institutions and nursing homes (Blanc et al. 2002); and the greater prevalence of infections due to *Mycobacterium tuberculosis*, *Mycobacterium leprae*, and *Brucellosis* spp. in tropical countries (McGill 2003).

Table 27.2 summarizes available epidemiological data on incidences, prevalence, and risk factors of the various manifestations of musculoskeletal infection.

PATHOGENESIS

The pathogenesis of bone and joint infection, as understood to date, is a multistep process (Mandal et al. 2002). Organisms gain access to musculoskeletal structures and proliferate. This may occur in both planktonic

Table 27.2 *Epidemiology of musculoskeletal infection*

Manifestation	Incidence	Prevalence	Risk factors
Acute septic arthritis	10–100/100 000/yr	N/A	In Australasia, indigenous race (Gillespie 1990)
Chronic septic arthritis	N/A	Unknown	Delayed diagnosis or inadequate treatment of acute septic arthritis
Acute prosthetic joint infection	1–2% of joint replacements	N/A	Prior joint surgery; adverse wound score, ASA status or prolonged surgery; history of cancer; superficial wound infection (Berbari et al. 1998a)
Chronic prosthetic joint infection	N/A	Unknown	As for acute infection
Acute osteomyelitis		N/A	None proven. Soft tissue infection and bacteremia presumed
Chronic osteomyelitis: hematogenous	N/A	Unknown	Delayed diagnosis or inadequate treatment of acute osteomyelitis
Acute fracture fixation infection	Unknown percentage of high energy tibial fractures without free flap soft tissue cover	N/A	Open fracture; high energy injury; Inadequate debridement; failure to establish early soft tissue cover (Gustilo and Anderson 1976)
Chronic fracture fixation infection	N/A	Unknown	Inadequate management or delayed diagnosis of acute fracture fixation infection
Chronic osteomyelitis due to diabetic foot ulceration	Unknown	Unknown; 4–10% of diabetics have ulcers; 10–40% of ulcers have osteomyelitis	Increasing size, depth and chronicity of ulcer; failure to heal despite correction of ischemia and adequate offloading
Chronic osteomyelitis: other pressure sores	Unknown		Increasing size, depth, and chronicity
Vertebral osteomyelitis		Unknown	Increased age; intravascular catheterization; intravenous drug use

N/A, not applicable.

and sessile states. In each situation, intercellular signaling within the bacterial population orchestrates the expression of genes involved in host tissue damage (Abdelnour et al. 1993) and in the formation of biofilms (Chopp et al. 2003; Kjelleberg and Molin 2002), believed relevant to persistence of infection. Host inflammatory responses have both protective components (in terms of the prevention of septicemia and the long-term preservation of function) and harmful ones (joint damage and bone death) (Josefsson and Tarkowski 1999). Reparative and resorptive responses are seen in bone, the mechanisms of which are beginning to be elucidated.

Access of organisms to bone or joint

Microorganisms can gain access to bone, joint or muscle via hematogenous spread or from a contiguous source of infection or contamination.

Direct inoculation of pathogens via the latter mechanisms, though apparently self-evident, needs a little more consideration. It is noteworthy that it appears to be relatively difficult to infect healthy bone. In many experimental models, it is necessary to injure bone in some way (Norden 1988; Rissing 1990); one of the

classic models involves injecting the sclerosing agent sodium morrhuate into the bone marrow prior to a bacterial challenge. Other models have involved surgically creating a cavity, a devitalized segment of a long bone, or a fracture (which can then be plated or pinned), or even resecting a joint and inserting foreign material to mimic joint replacement (Mader 1985; Belmatoug et al. 1996; Smeltzer et al. 1997b; Kaarsemaker et al. 1997; Monzon et al. 2002). In the majority of cases the pathogens are then directly inoculated, though hematogenous challenge can also be employed (Bremell et al. 1991).

These observations appear to find a direct parallel in human disease, in which the majority of contiguous source infections are in contexts where bone is indeed injured. This includes both elective orthopedic and emergency trauma surgery, and infections related to chronic soft tissue defects. Elective surgery usually involves cutting, drilling or instrumenting bone to achieve the desired end, whether that be joint replacement, realignment of the axis of a long bone, fusion of one or more joints, or the removal of a painful bony pressure point or joint. In trauma, fractures may involve one or more bones with variable severity depending on the energy of the injury, and may extend into the joint line, forming a potential route for the spread of infection. Bone may be directly devitalized by dissipation of the high fracture energies through it, and acutely devascularized by stripping of the soft tissue attachments normally providing the periosteal circulation. This vascular compromise may be compounded by intramedullary nailing (affecting the endosteal circulation) or by further soft tissue stripping performed to gain access to the fracture for its reduction.

Soft tissue loss is also the key initiating element in the pathogenesis of osteomyelitis secondary to decubitus and diabetic foot ulceration. In the former case, immobility, poor prior skin condition, incontinence, and reductions in sensation, all play roles in allowing excess chronic pressure to develop and lead to ischemic soft tissue necrosis. In the case of the diabetic foot, peripheral neuropathy is the key driver (Boulton 1996). Motor neuropathy causes an imbalance of intrinsic and extrinsic musculature, leading to a high arched foot with subluxation at the metatarsophalangeal joints and clawing of the toes. There are also alterations in the compliance of skin and deeper tissues as a result of long-term hyperglycemia and autonomic neuropathy. These changes in the biomechanics of the foot cause excessive pressure on metatarsal heads, heel, tips of clawed toes, and dorsal aspects of interphalangeal joints (on the inner upper surface of footwear) (Murray et al. 1996). Sensory neuropathy causes a lack of protective sensation (without necessarily total numbness) and repetitive loading of the new pressure points leads to ulceration.

Once ulceration develops, it may worsen either through continuing loading, or infection, or both. Once the tissues directly overlying and attaching to bone become involved, the periosteal circulation is compromised and the most superficial cortical layer of bone is at risk. The high numbers of bacteria located in the ulcer base are then situated directly adjacent to injured bone for the initiation of infection.

It is also relevant, in considering infections relating to direct inoculation, to recall the important potentiating role played by devitalized tissue or foreign inanimate matter if present. Since the elegant experiments of Elek and Conen (1957) it has been clear that foreign materials greatly lower the inoculum required to initiate infection, and increase its severity. In elective orthopedic surgery, the use of metal ware and bone cement provides extensive surfaces for colonization with microbes if a breach in sterile technique occurs. In trauma and its management, bony injury also includes variable soft tissue loss and damage, and environmental contamination with dirt, soil, vegetable matter or debris from road surfaces or projectiles. All these factors potentiate infection just as in animal models. They also provide an explanation for the microbiology of this group of conditions, since pathogens are derived from the contaminating external environment or from the skin, and include low-virulence organisms when metal ware is involved (Christensen et al. 1994).

Hematogenous spread is clearly of importance in a number of manifestations. A range of bacteria can, under certain conditions, translocate into a joint space, or set up a focus of infection within bone, usually at the metaphysis just proximal to the growth plate. Although the in vivo mechanisms of this are poorly understood, *Staphylococcus aureus* does display the ability to adhere to and invade cultured human endothelial cells in vitro (Peacock et al. 1999). This is mediated by interactions between the staphylococcal cell wall-anchored fibronectin-binding protein and cell surface-associated fibronectin (Peacock et al. 1999), an interaction involving specific modules within each of these macromolecules (Massey et al. 2001a). Internalization is dependent upon endothelial cell membrane integrins ($\alpha5\beta1$), which appear to generate signals that drive active bacterial uptake on engagement of the fibronectin (Fowler et al. 2000; Massey et al. 2001a). Once within the endothelial cell, bacteria can escape into the cytosol (Bayles et al. 1998), and can persist for prolonged periods of time (several days at least) (Ellington et al. 2003).

The fate of endothelial cells following bacterial uptake varies according to toxin elaboration by the bacteria. If α-toxin is produced, endothelial cell death follows rapidly (Buerke et al. 2002). If cells do not die, they upregulate adhesion molecules and become pro-adhesive for white cells (Beekhuizen et al. 1997). In vivo, this would lead to adhesion of white cells, platelets and the formation of microthrombus; significantly, *Staphylococcus aureus* also has receptors for components of thrombus such as fibrinogen and thrombospondin. If there is no toxin production, persistence or translocation are possible outcomes.

Adhesion is considered to be the next crucial step once bacteria have reached bone or joint (Gristina et al. 1985), the importance of which appears to be borne out by the possession, in the case of *Staphylococcus aureus*, of numerous cell wall-associated molecules that act as proteinaceous receptors for a range of cell matrix or cell-surface associated proteins (Smeltzer et al. 1997a). These include receptors for collagen, fibronectin, bone matrix sialoprotein, and osteopontin (Herrmann et al. 1988; Buxton et al. 1990; Park et al. 1991, 1996; Patti et al. 1992; Switalski et al. 1993; Yacoub et al. 1994). As with direct inoculation of bacteria onto wounded bone, there is a plausible relationship between these adhesive properties and the observation that a proportion of cases of hematogenous infection of bone occur following blunt trauma. In healthy bone the components of the osteoid matrix are not exposed, being covered in mineral, but on injury, the underlying matrix becomes accessible to external agents including bacteria. For *Staphylococcus aureus*, some data support the importance of the collagen binding protein, Cna, in the pathogenesis of bone and joint infection (Patti et al. 1994; Elasri et al. 2002), though other carefully conducted studies of clinical isolates have not confirmed that this specific protein is either necessary or sufficient (Thomas et al. 1999). The gene *cna*, encoding the collagen binding protein, may however be a more general marker of virulence. It was identified, along with the fibronectin binding protein gene *fnb*, as one of seven genes that appeared to be independently and cumulatively associated with the potential to cause invasive staphylococcal disease (Peacock et al. 2002); in the study concerned, some of the cases certainly had bone and joint manifestations of infection. It is possible that microfracture following blunt trauma produces areas of microscopic thrombus, or exposed bone matrix proteins, that provide a site for bacterial adhesion via Cna or Fnbp.

Adhesion is also of importance when considering infection of orthopedic metal ware of any kind. It has been shown that foreign materials are rapidly 'conditioned', after insertion into the body, by the deposition of plasma proteins such as fibronectin (Delmi et al. 1994). These proteins provide an ideal substrate for bacterial adhesion, though some important device-related pathogens, such as coagulase-negative staphylococci, have relatively weak adhesion mechanisms for fibronectin. They are, however, able to adhere relatively well to unconditioned metal and plastic, via a number of macromolecules, and subsequently establish very strong associations through the production of biofilm (see below) (Christensen et al. 1994).

Quorum sensing

A well-recognized clinical pattern in bone and joint infection is a prodromal illness that may include features suggestive of transient bacteremia, resolving completely, and then followed by acute infection of bone or joint. In such cases there is a high prevalence of bacteremia at the time of presentation, but without clinical evidence to suggest that this has persisted since the initial seeding of the joint. This is compatible with current understanding of the importance of quorum sensing, and global regulation of gene expression, in many pathogens. In *Staphylococcus aureus*, a cyclic octapeptide that is the product of the *agr* locus acts as a density-dependent signal to the whole bacterial population (Ji et al. 1995), ultimately resulting in the transcription of a regulatory RNA molecule, RNA III (Novick et al. 1993). This affects the expression of a wide variety of genes, many responsible for adhesin or toxin production (Novick 2003). Expressed during logarithmic phase, and triggering RNA III expression as the population moves towards stationary phase, *agr* upregulates toxin production and downregulates adhesin expression. Teleologically, as their growth rate slows through nutrient depletion, organisms become less adhesive and more destructive to surrounding cells and tissues, and so more able to disseminate to new sites. This fits with the observed sequence of a clinically unapparent bacteremia that leads to the onset of symptoms within a joint, followed by a secondary bacteremia detected at presentation. The *agr* gene is present in coagulase-negative staphylococci (Van Wamel et al. 1998) where it may be able to interfere with the *agr* expression of *Staphylococcus aureus* (Tegmark et al. 1998). It has been speculated that in this way, coagulase-negative staphylococci could act as negative regulators of the natural virulence of *Staphylococcus aureus*.

Biofilm formation and other mechanisms of persistence

Quorum sensing is also likely to be important within biofilms and during biofilm formation (Davies et al. 1998; Vuong et al. 2003). A biofilm is an adherent consortium of microorganisms embedded in an exocellular polysaccharide (glycocalyx) (Costerton et al. 1999; Gotz 2002). This material protects bacteria from complement and from phagocytes and, within the biofilm, organisms are present that are able to act as 'persisters', with growth states that confer relative resistance to antibiotics and other microbicides. The initial events in the formation of these structures occurs surprisingly rapidly; in *Pseudomonas aeruginosa*, studies have demonstrated how immediately after adhesion to a substrate, bacterial cells begin to express genes involved in alginate synthesis. The genetics of this process are well understood for *Pseudomonas aeruginosa*, and are increasingly so for *Staphylococcus epidermidis* (Heilmann et al. 1996), in which the *ica* gene cluster orchestrates the production of biofilms (reviewed in Gotz 2002). The *ica* operon is also represented in *Staphylococcus aureus*, and is another of the seven virulence factors previously mentioned that are associated with invasive characteristics (Peacock et al. 2002).

Other mechanisms of bacterial persistence have been suggested as potentially relevant. The clinical picture of osteomyelitis includes chronic active sepsis, but also periods of relative or complete remission, with subsequent relapse. Recurrent infection can occur many years after the initial episode, prompting speculation as to the mechanism of such long-term survival. One hypothesis is that small colony variants (SCV), which under certain conditions have been shown to emerge rapidly in populations of staphylococci, might survive for prolonged periods inside long lived cells such as fibroblasts. Certainly SCVs can be demonstrated as a surviving intracellular bacterial population in cultures of endothelial cells when host cell death is prevented through external antibiotic use. *Staphylococcus aureus* can invade a wide range of host cells including fibroblasts, and SCVs are known to have high rates not only of induction, but also of reversion to wild type characteristics. Furthermore, SCVs have been shown capable of causing osteoarticular infection in vivo in animal models

(Jonsson et al. 2003). Finally, a careful study has shown that SCVs can be isolated, in some cases in almost pure growth, from individual cases of relapsed chronic osteomyelitis (Proctor et al. 1995). The general relevance of this observation remains to be demonstrated, but it is an appealing hypothesis that SCVs, located within cells or indeed deep within a biofilm, provide a mechanism for remission and relapse. They may even, through rapid switching to and from wild type, allow populations of bacteria to manifest antibiotic resistance (a feature of SCVs) (Chuard et al. 1997) without the fitness costs usually associated with true antibiotic resistance (Massey et al. 2001b).

Inflammatory responses

BONE RESPONSES

Acute infection in bone and joint leads to acute inflammation. An influx of neutrophils, accompanied by increased vascular permeability and edema, is seen both in infected synovium and in cancellous bone. These responses are important in protection of the host against the secondary bacteremia commonly seen in acute infections, but they also contribute to the immediate damage to the infected part of the skeleton (Josefsson and Tarkowski 1999). The inflammatory response within the joint leads to an outpouring of synovial fluid and a tense effusion. In joints such as the hip and shoulder, where the head of the femur and humerus, respectively, are supplied with blood via recurrent vessels running up the neck of the bone, the high intra-articular pressure may be enough to compromise the circulation in those vessels. Within cancellous bone at any site, edema also leads to high pressures and potential vascular compromise. The procoagulant effects of inflammatory cytokines on vascular endothelium exacerbate this. The consequence is that the inflammatory response frequently leads to areas of bone death, through microvascular thrombosis around the infected focus. If the area of infection is not contained, progressive bone death can result.

The tracking of infection through areas of cancellous or cortical bone leads to other key features of bone infection. Infection originating in the metaphysis of a long bone, the common site for hematogenous infection, can track via the Haversian canals through the cortex and into the soft tissue. Where the periosteum is loosely connected, the accumulation of pus results in a subperiosteal abscess, stripping off the periosteum from the bone. As with traumatic periosteal stripping, this results in further vascular compromise to the bone, already affected internally by the inflammatory focus. A whole segment of bone death can result from the combination of endosteal infection and a circumferentially developing sub-periosteal abscess.

Bone is a highly dynamic structure adapted to respond to biomechanical forces and to preserve function. The response to periosteal stripping is to form new bone beneath the elevated tissue. In a fracture, this would be the beginning of fracture callus, which would then involve the fracture hematoma and so lead to healing. In infection, the periosteal reaction leads ultimately to the formation of new bone that externally bridges the involved segment. Where an area of dead bone would compromise the mechanics of the bone, this is therefore compensated for by the production of the new bony involucrum. Although this may mean that there is dead bone (now relatively trapped within the new bony shell), it does also mean that function, the basic role of the skeleton, is preserved.

Developing in parallel with the process of new bone formation is bony resorption in the areas contiguous with the dead bone. This is a direct consequence of the inflammatory response. There is evidence that inflammation both enhances the activity of professional bone resorbing cells (osteoclasts) and triggers bone resorption by 'semi-professional' cells such as recruited polymorphs. Bone resorption in the area of infection and inflammation acts to separate the dead bone from living bone. The resulting fragment(s) of dead bone, called a sequestrum, lying within the living bone, may now act as a site for permanent infection from a resident population of adherent bacteria, but this phenomenon also opens up the possibility of self-cure. If the sequestrum can be fragmented or is of sufficiently small size (since not all infections cause large areas of bone death), it can potentially be extruded, leaving behind living bone capable of healing. Spontaneous extrusion of sequestra is seen to occur in both chronic hematogenous and contiguous focus osteomyelitis. Bone is discharged either through the soft tissue defect that caused the infection, or through sinus tracts, the formation of which is another characteristic of chronic osteomyelitis. Sinuses occur because of the destructive nature of pus under pressure, which will track along tissue planes (e.g. beneath the periosteum), but will frequently then breach fascial barriers and eventually find a route to the outside world.

Bone resorption is also of key importance in the pathogenesis of bone infections related to orthopedic metal ware. Here, the impact of the infection is not merely how it affects the host systemically and through local damage. It is also that bone resorption, occurring adjacent to the infected orthopedic device, results in mechanical loosening of that device. In infected prosthetic joints, loosening causes additional pain and generates 'wear particles' of metal and acrylic cement, which themselves cause more inflammation (Panday et al. 1996). In infections of fracture-fixations and fusions, loosening leads to instability and increased risk of nonunion. In both situations, loosening also appears to further reduce resistance of the immediate environment to infection, making control of infection more difficult.

MOLECULAR BASIS OF BONE RESORPTION

There have been considerable advances in our understanding of the process by which bone is resorbed in health and disease. Central in this is the osteoclast, which differentiates from monocytic precursor cells in response to a range of endocrine and paracrine signals. Osteoclasts form tight associations with underlying bone, and acidify a physical compartment beneath them, leading to bone resorption (Panday et al. 1996). In vitro, this results in a visible pit on the surface of the bone, and the release of calcium into the medium, either of which can be measured. Using models of this type, the response of osteoclasts to bacterial products has been studied. Cell-associated materials eluted from a number of different bacteria trigger osteoclast activity. In the case of *Mycobacterium tuberculosis*, the chaperonin Hsp 60 is a potent stimulant. Less well-characterized constituents associated with the cell walls of *Staphylococcus aureus* and *Staphylococcus epidermidis* also stimulate resorption, though intriguingly, by different mechanisms as demonstrated by the effects of prostaglandin inhibitors (Meghji et al. 1997).

CELL AND MOLECULAR BASIS OF JOINT DESTRUCTION

Improved understanding in this area owes much to the systematic use of a rodent model of hematogenous septic arthritis induced by tail vein challenge with *Staphylococcus aureus*. This model has proved tractable to manipulation using a range of knockout mice, cytokine antagonists, and varying triggers of articular inflammation (Tarkowski et al. 2002). Clear themes emerging from this large body of work are that the inflammatory response protects the host from septicemic death, but is the major mechanism of articular cartilage destruction. T cells play a major role in this, and their actions are orchestrated by inflammatory cytokines such as TNF. T cell responses are triggered by CpG motifs in bacterial DNA (Deng and Tarkowski 2000) and by peptidoglycan cell wall components (Liu et al. 2001). Downmodulation of inflammatory response protects the infected joint, but only if antibiotics are used to keep the whole animal alive. In addition to demonstrating this in complex situations such as mice defective in adhesion molecules critical for immune function, such as intercellular adhesion molecule-1 (ICAM-1) (Verdrengh et al. 1996), it has also been shown that steroid therapy, if combined with antibiotics, reduces joint destruction, a finding that may be of clinical relevance (Odio et al. 2003).

MICROBIOLOGY

Despite the wide range of bacteria that are known to infect bone, *Staphylococcus aureus* stands as the pre-eminent cause of these infections (Bouza and Munoz 1999). It is the dominant cause of hematogenous and contiguous focus bone and joint infections. The bulk of hematogenous infections are monomicrobial. In addition to *Staphylococcus aureus*, aerobic gram-negative rods and group B β-hemolytic streptococci play important roles in neonatal infections; in children group A β-hemolytic streptococci, *Kingella kingae* (Birgisson et al. 1997; Dodman et al. 2000) (and in the unimmunized child, *Haemophilus influenzae* type B (HIB) though vaccine use has led to dramatic falls in the incidence of this (Bowerman et al. 1997)); *Salmonella* spp. in those with sickle cell anemia. Chronic infections complicating orthopedic devices have a preponderance of coagulase-negative staphylococci as causes (Christensen et al. 1994), but *Staphylococcus aureus*, including MRSA, remains important in acute device-related infection. In contiguous focus infections secondary to chronic soft tissue loss, infections are commonly polymicrobial, with aerobic gram-negative rods (mainly the Enterobacteriaciae and *Pseudomonas* spp.), anaerobes and enterococci represented alongside *Staphylococcus aureus*. This includes not only osteomyelitis complicating decubitus and diabetic foot ulcers, but also chronic osteomyelitis and prosthetic joint infection when draining sinuses have been present for prolonged periods of time. Gram-negative aerobes also play an important role in vertebral osteomyelitis in the elderly, presumably because of the increased rates of bacteremia secondary to more frequent urinary tract and intra-abdominal infection.

Clearly other pathogens may be of importance. *Mycobacterium tuberculosis* is the commonest worldwide cause of chronic septic arthritis and is of importance elsewhere in the skeleton, notably in the spine (Potts disease). Where endemic, *Brucella* spp. are also relatively common causes of infective spondylitis and septic arthritis (Sankaran-Kutty et al. 1991; Geyik et al. 2002). Brucellosis, Lyme disease (Franz and Krause 2003), and gonococcal infection (Bardin 2003) should always be considered in culture negative cases of septic arthritis. Appropriate epidemiological risk factors should be enquired after in culture-negative cases, or when the pathogens are distinctively unusual. Reports of osteomyelitis caused by *Salmonella arizonae*, linked to pet reptiles, are good examples of how individual risk factors can be identified and used in subsequent public health awareness (Nowinski and Albert 2000; Nowinski 2001).

In addition to the common, there is a long list of rare causes of bone and joint infection, including *Listeria* spp. (Mereghetti et al. 1998), pneumococci (Ispahani et al. 1999; Ross et al. 2003), *Pasteurella* spp. (Shapero and Fox 1999), *Fusobacterium* (Koornstra et al. 1998), and *Tropheryma whipplei* (Lange and Teichmann 2003). This fact is important only because it emphasizes the need for a rational approach to microbiological diagnosis and empiric treatment. Essentially any organism capable of causing a bacteremia or a soft tissue infection can occasionally involve skeletal structures, including atypical mycobacteria (Clark et al. 1997; Ekerot et al. 1998), *Nocardia* spp. (Wilkerson et al. 1985; Ostrum 1993) and *Actinomyces* spp. The compromise to local host responses caused by an implanted device or by debris means that a wide range of environmental or low virulence pathogens are, from time to time, implicated in infections related to prosthetic joints, fracture fixations, and puncture wounds. Although coagulase-negative staphylococci are the prime movers in this regard (Christensen et al. 1994), many other organisms including *Clostridium difficile, Pasteurella* (Takwale et al. 1997) *Listeria, Brucella* (Malizos et al. 1997; Orti et al. 1997), corynebacteria (von Graevenitz et al. 1998), *Mycobacterium tuberculosis* (Berbari et al. 1998b) and other mycobacteria (Guerra et al. 1998) have been identified. Probably the most common atypical pathogen in this context, and a paradigm for the interaction between environmental contamination and host compromise, would be the high incidence of osteochondritis, progressing to osteomyelitis, caused by *Pseudomonas aeruginosa* in puncture wounds of the foot associated with the wearing of training shoes (sneakers) (Fisher et al. 1985; Niall et al. 1997). High densities of *Pseudomonas aeruginosa* living in the insole allow for inoculation into the weight bearing area of the foot if a nail penetrates the foot through the shoe. There is frequently also some debris present in addition to the bacterial contamination (Chang et al. 2001), and the result is that a pathogen normally only associated with significant host compromise is able to cause a potentially serious infection in an otherwise healthy individual (usually a child).

PREVENTION

Given that no vaccines yet exist against most of the causes of skeletal infection, prevention is geared at restricting the access of organisms to exposed bone or orthopedic implants (Norden 1985). This means taking measures to prevent the breakdown and infection of soft tissue in the at-risk patient; preventing pressure ulcers through risk assessment and risk reduction in hospitalized, bed-bound and other immobile persons, organizing foot care programs for diabetics, and taking measures to prevent surgical wound and deep infection after orthopedic surgery or trauma (Gillespie 1997). The latter include meticulous sterile theater technique, the use of ultraclean air for implant work, prophylactic antibiotics (chosen with regard to local sensitivity patterns among key pathogens) (Glenny and Song 1999), and careful handling of tissues to minimize necrosis and hematoma formation. Preventive measures of this kind involve considerable effort, and regular feedback of results to clinical teams is essential.

Post-operative hematomas need to be drained (Saleh et al. 2002), while leaking and clinically infected wounds should be explored (to prevent the development of deeper infection, or to diagnose it early and minimize the risk of progression to more severe or chronic disease). Traumatic wounds need careful exploration and debridement of devitalized tissue and debris, with priority given to early soft tissue cover of fractures with the involvement of plastic surgeons if necessary (Gustilo and Anderson 1976; Khatod et al. 2003). The relationship of some cases of hematogenous septic arthritis and osteomyelitis to episodes of clinically apparent bacteremia also implies that general measures to reduce staphylococcal bacteremia associated with intravascular catheterization (through appropriate infection control measures) would be valuable, though proof of efficacy would be hard to establish.

There is considerable interest at present in the cell wall-associated adhesins as potential vaccine candidates, hoping that specific antibody might be able to play a role in preventing localization of infection to deeper structures (Flock 1999). Progress is encouraging (Nilsson et al. 1998), but has not yet translated into clinically applied products.

Finally, one specific area for prevention concerns the implantation of prosthetic joints in patients with histories of prior septic arthritis or osteomyelitis, including osteoarticular tuberculosis (Lee et al. 2002; Kim et al. 2003). There is some evidence based on case series that joint replacement can reactivate tuberculosis or pyogenic infection (Wang 1997; Su et al. 1996). Active infection should ideally be treated before joint replacement is undertaken (Hardinge et al. 1979; Jupiter et al. 1981). Where this is impossible, or where the history is recent (raising the suspicion that cultures will in fact be positive), it may be necessary to stage the joint replacement, resecting the infected area, treating according to the results of cultures, and implanting the joint replacement after an interval (Nazarian et al. 2003). However, some centers have described acceptable results using a single stage approach, antibiotic loaded cement, and systemic therapy.

CLINICAL FEATURES

Skeletal infections, as seen above, cause local and sometimes systemic features, together with destruction of bone and joint (irrespective of attempts at preservation of function), and have the potential rapidly to establish chronic active infection. Clinical features are in keeping with these characteristics, notably pain, loss of function, acute inflammatory changes in overlying soft tissue, systemic illness due either to bacteremia or to chronic suppuration, sinus tract drainage, and progressive soft tissue and osteoarticular destruction. There may be other local complications, such as spinal cord or nerve root compromise from vertebral osteomyelitis when there is complicating epidural abscess (Baker et al. 1975), retropulsion of infected disk material, or mechanical instability. Pathological fracture, growth plate disturbance, or secondary septic arthritis may complicate osteomyelitis; soft tissue abscesses frequently develop during the evolution of undiagnosed or untreated acute osteomyelitis and are also commonly found in the context of infected metal ware.

When joints are infected, the features are of pain, swelling, and considerable reluctance to undergo active or passive movement. Joint effusion is followed by the development of synovitis, which may persist for weeks after successful treatment of the infection. If the diagnosis is missed or delayed, or treatment unsuccessful, progressive joint destruction ensues, with increased pain and disability.

The evolution of acute to chronic osteomyelitis has been detailed above. Acute osteomyelitis classically presents as a prostrating febrile illness with pain in a long bone. In the young, this may manifest simply as refusal to use the affected limb. There may be little in the way of soft tissue change, though edema and eventually abscess formation will develop once infection tracks from the interior of the bone out through the cortex. Some 50 percent of cases of acute osteomyelitis have an accompanying bacteremia. Chronic osteomyelitis much more commonly has suspicious features visible in the soft tissues, whether this is a chronic wound or ulcer overlying bone, a sinus tract or simply old scars that indicate previous surgery or episodes of drainage. Chronic osteomyelitis presents with fever rarely, unless a collection of pus has been trapped through premature sinus closure, or has not yet discharged since a recurrence began. Bone may be visibly exposed in chronic osteomyelitis, be unexpectedly palpated with a gloved finger on inspection of a pressure sore, or be identified by palpation with a sterile metal probe, which in diabetic foot ulcers is moderately predictive of underlying osteomyelitis (Grayson et al. 1995a).

DIAGNOSIS

General principles underlying diagnosis, including the desirability of understanding the performance and limitations of specific tests, are taken as understood and are dealt with elsewhere in this text. These considerations are of particular importance in relation to imaging studies and the interpretation of cultures, where there are significant problems with sensitivity and specificity to consider. One important point that must be made, however, is that there is a relative shortage of high-quality diagnostic studies, despite a considerable number of publications. Part of the problem is a lack of understanding of the requirements of studies aiming to assess a diagnostic test rigorously. But part too is a lack of consensus definitions of osteomyelitis or prosthetic joint infection, such that although expert clinicians would recognize most of these infections when they saw them, they would not necessarily agree on the formal requirements for making the diagnosis. Imprecision in

this matter makes it hard to interpret the majority of studies once the goal is to compare one study with another. This problem besets all the diagnostic modalities that will be discussed below, and must constantly be borne in mind.

Clinical features

Clinical suspicion is of paramount importance in arriving at a diagnosis. Certain features, such as sinus tract formation, are pathognomonic of infection. More commonly, clinical features are nonspecific (pain, loss of function, soft tissue swelling, even fever) and are sometimes insensitive, for example in chronic prosthetic joint or diabetic foot infection (Lavery et al. 1996). It follows that a high level of clinical suspicion for infection must be maintained whenever patients present with acute changes in musculoskeletal function, especially if accompanied by fever; with unexplained pain, stiffness or soft tissue swelling; with destructive bony lesions especially if accompanied by reparative responses; or whenever orthopedic implants fail rapidly with no obvious technical explanation. Night pain is indicative of very significant inflammation and is a warning symptom; pain on first using a prosthetic joint that eases as the joint is used, only to return after each period of rest (so-called 'start-up' pain) is highly suggestive of loosening; and some patients with chronic osteomyelitis will complain that the severity of bone pain is linked to changes in the weather. But none of these features is universal or pathognomonic for infection, and so the diagnosis must rest on recognizing the possibility, and taking the necessary steps, to confirm or refute infection within an appropriate timescale depending on the nature of the diagnosis being entertained. For acute presentations in which early intervention is key in preventing damage and chronicity, diagnosis must be expedited and may even need to follow on from the initiation of treatment. For chronic conditions there is rarely urgency to make the diagnosis or commence treatment unless the differential diagnosis of a tumor is a significant possibility.

Laboratory

Most laboratory studies provide only supportive information alongside the clinical impression. With the exception of serological testing for minority causes of infection such as *Brucella* spp., tests are nonspecific. Measurement of full blood count may reveal anemia of chronic disease. Acute phase responses (elevations of white cell count and C-reactive protein (CRP)) are generally elevated in acute septic arthritis and acute osteomyelitis or if a deep abscess containing a high-grade pathogen has formed. Elevations in the erythrocyte sedimentation rate are often seen, and if no other cause is found, support a diagnosis of osteomyelitis.

Generally, more chronic infections are associated with less perturbation of the inflammatory markers, a feature also commonly seen in diabetic foot infections, where fever and elevations in white cell count are indicative of more severe disease, with many infections not eliciting a strong systemic response (Edelson et al. 1996).

There are no other generally accepted serological diagnostic tests for pathogens prevalent in bone and joint infection. Serodiagnosis of prosthetic joint infection using antibody responses to glyocalyx has been put forward, but has yet to be widely validated and adopted. The majority of pathogens diagnosable by current serological techniques are exceedingly rare causes of bone and joint infection, and so the use of these tests cannot be justified in routine practice, though specific clinical indications may make them relevant.

Imaging

The aims of diagnostic imaging are both to diagnose infection and ultimately to delineate its anatomical extent (Santiago Restrepo et al. 2003). Few single modalities provide all the required information, and all have flaws in their performance depending on the circumstances of the case.

Plain XR

Plain radiological abnormalities take time to develop; in the case of bone this is several days for sufficient bone to be resorbed or to form to the extent that it becomes visible. The first signs of acute osteomyelitis are soft tissue swelling, followed at 7–10 days by the development of lucency. When periosteal reaction produces new bone, it is not visible for 10–14 days. For this reason, a single normal radiograph does not rule out an acute osteomyelitis. A chronic osteomyelitis is generally identifiable from the mixed picture of lucency and periosteal reaction (Figure 27.1). Serial films are useful, because osteomyelitis is an aggressive process that generally shows rapid evolution. An evolving mixed destructive and reparative picture is highly suspicious of active infection.

The location of the lesion depends on the chronicity, the age of presentation and the specific manifestation. Acute childhood osteomyelitis is a metaphyseal disease, but the skeletal abnormality will become increasingly diaphyseal as the bone grows, with the growth plate moving away from the old focus of infection. Ongoing activity or episodes of recurrence will maintain a metaphyseal component to the lesion. More active disease is characterized by 'fluffiness' of the periosteal reaction and of the margins of bony lucencies. Healed infection has much 'crisper' margins, with the periosteal reaction being fully organized into involucrum, and lucent areas either being filled in again or becoming more demarcated. Some subacute forms of infection present with

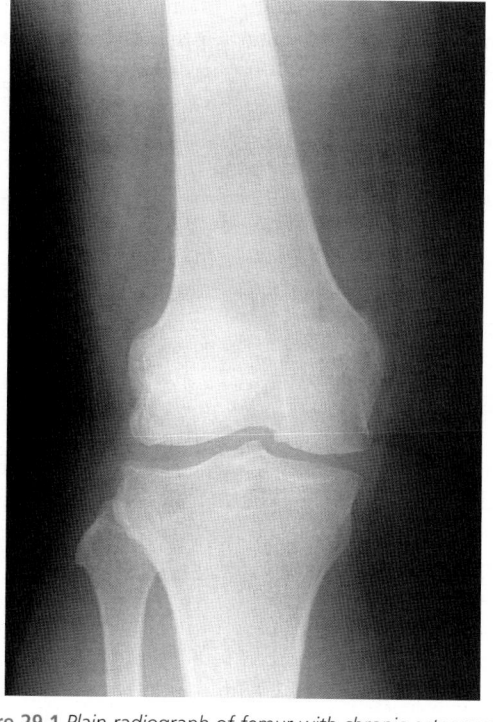

Figure 29.1 *Plain radiograph of femur with chronic osteomyelitis present. A minor periosteal reaction is seen, but no other changes.*

obvious cavities within bone, classically surrounded by a sclerotic rim (the whole known as a Brodie's abscess). Even more indolent presentations are found in the form of chronic multifocal osteomyelitis, a poorly understood condition in which cultures are negative, and multiple sites may be involved over a period of time, generally without the kind of progressive destruction usually seen in osteomyelitis, but with intense sclerotic changes in bone and expansion of the bone due to chronic periosteal reaction (Kozlowski et al. 1983; Golla et al. 2002; Huber et al. 2002).

Bony changes in septic arthritis consist of accelerated arthritis with loss of joint space and secondary changes in the underlying bone. In hip or shoulder, there may be avascular necrosis of the femoral or humeral heads, with collapse.

In prosthetic joint infection, the cardinal features are the development of lucency at the bone–cement interface. This reflects loss of bone due to chronic inflammation, and its replacement, at the interface, with connective tissue (so-called 'membrane') that is edematous, hyperemic, and inflamed. As loosening progresses, mechanical failure of one or both components can occur. In hip replacements, one may see evidence of rotation of the acetabular component, or cortical erosion at the tip of the femoral component ultimately leading to femoral fracture. In knee replacement, the tibial component may collapse into valgus or varus.

Although the changes outlined here are consistent with infection, they are rarely diagnostic. In long bones, the differential diagnosis of tumor must always be borne in mind, while in prosthetic joints it is the accelerated timescale that is particularly suggestive of infection; the changes of loosening are nonspecific and may occur in the absence of infection. In diabetic foot disease, neuropathy can lead to changes in bone and joint; healed fractures, subluxation with secondary degenerative changes, and sometimes of a destructive nature including acute Charcot (neuropathic) arthropathy. The appearances of this diabetic neuro-osteoarthropathy can easily be mistaken for infection.

Ultrasound

Appropriately powered ultrasound can obtain relatively good images of many external parts of the skeleton. Sinus tracts can be identified and followed down to likely points of origin on bone or prosthesis. Collections can be imaged, and periosteal elevation can frequently be seen. Cellulitis or edema within deeper structures (muscle) can be visualized and joint effusions distinguished from synovitis. Ultrasound, if available, can therefore play an important role in the management of acute osteomyelitis, identifying a subperiosteal abscess, the early periosteal edema consistent with the evolution of one, or an abscess that has already tracked out into the soft tissues (Kaiser and Rosenborg 1994; Mah et al. 1994; Cardinal et al. 2001). Diagnostic aspiration under guidance may be possible, and this application is also of great value in attempting to obtain a microbiological diagnosis in suspected or proven prosthetic joint infection, and sometimes in chronic osteomyelitis.

Isotope scanning

The history of nuclear medicine in osteoarticular infection encompasses several isotope scans that have failed to live up to early enthusiasm. The fundamental problem is the relatively nonspecific nature of musculoskeletal responses as currently detected. Noninfective processes such as diabetic neuro-osteoarthropathy, fracture nonunion, and prosthesis loosening, also generate abnormal scans (Turpin and Lambert 2001). The need for high levels of sensitivity and specificity has not yet been met reliably, despite studies with triple-phase Technetium bone scanning (sensitive but very nonspecific) (Schauwecker 1992), Indium-labeled white scan (more specific but less sensitive, and so sometimes combined with conventional bone scan) (Johnson et al. 1996), marrow colloid scan (designed to exclude marrow white cell packing as a confounding cause of false positive white cell scans), labeled immunoglobulins and anti-white cell monoclonal antibody scans (Palestro et al. 2003), and Technetium-labeled white cell scans (Devillers et al. 1995) (designed, in different ways, also to localize the white cells present in inflammatory lesions). Of

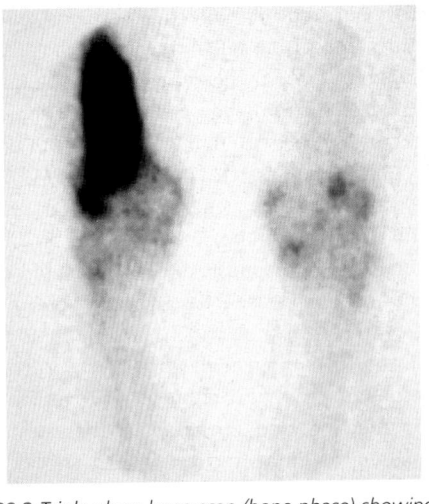

Figure 29.2 *Triple phase bone scan (bone phase) showing intense uptake*

all currently available isotope scanning protocols, the greatest specificity can be obtained with the combination of indium white cell scanning and nanocolloid scan; alternative labeling techniques for white cells may offer promise. White cell labeling scans are not trivial matters, however, given the need to remove, purify, label, and reinfuse white cells, which opens up obvious possibilities of infection and cross-infection. For prosthetic joint infection, no scan is recommended; for acute osteomyelitis in the otherwise normal skeleton (e.g. in childhood) triple phase bone scan (Figure 27.2) may be an acceptable confirmatory test that will follow a clinical decision to treat; for chronic osteomyelitis, a labeled white cell scan combined with nanocolloid scanning is acceptable if magnetic resonance imaging (MRI) is unavailable.

MRI

MRI is able to produce images in axial, sagittal, and coronal planes, and gives information that is physiologically derived (based on fat and water content of tissues) but anatomically portrayed. With few exceptions, it is the imaging modality of first choice in bone and joint infection. It will show soft tissue edema and fluid collections; cortical breaches in bone; edema (Figures 27.3 and 27.4), abscess and gas within bone; and through gadolinium enhancement, will distinguish inflammatory from noninflammatory high signal. By demonstrating intraosseous edema, it can diagnose osteomyelitis before plain radiological changes are evident, and is also of great value in assessing relapse in cases of chronic osteomyelitis, where plain films are already abnormal. In some areas, such as the spine, MRI has transformed the diagnosis of infection, with superior simultaneous resolution of disk space, paravertebral structures, nerve root foraminae, epidural space (important in identifying epidural abscess) and the anterior and posterior bony columns.

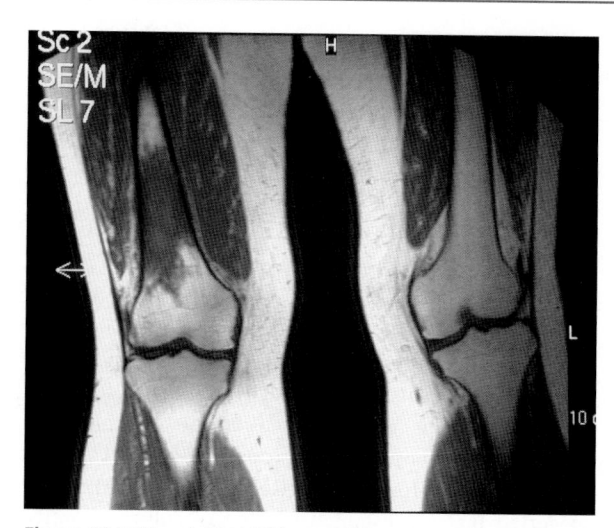

Figure 29.3 *T1 weighted MRI scan of the same region as in Figure 27.1, showing intramedullary edema as low signal*

MRI does suffer some pitfalls, notably when there has been previous surgery, when follow-up scans are performed after treatment of infection (images take many months to return to normal after either) or when metal is or has been present. Titanium implants produce minimal signal void, but more standard materials can produce dramatic metal artefact, even when the metal ware has long been removed (small amounts of metal-losis in the tissues being enough to disrupt the image).

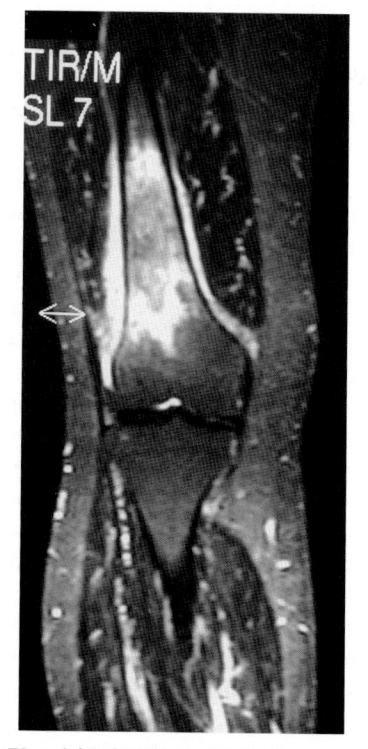

Figure 29.4 *T2 weighted MRI scan (STIR) of the same region as in Figure 27.1, showing intramedullary edema as high signal. Additionally, high signal periosteal reaction is seen.*

There are also problems in the interpretation of diabetic foot osteomyelitis, where underlying neuro-osteoarthro-pathy can mislead the unwary; some stress-related bone marrow changes and bony fragmentation are to be expected, and do not necessarily indicate infection with as much reliability as in the non-neuropathic patient. Overall, however, when available, MRI is of great value in suspected acute osteomyelitis (as a confirmatory test after the onset of treatment) and in chronic osteomye-litis (assessment of activity and surgical planning). Its role in prosthetic joint infection is minimal because the primary interest is at the bone–cement interface, which is usually affected by signal void, though occasionally distant collections or osteomyelitis can be identified.

Culture

This is clearly the 'gold' or criterion standard in the diagnosis of almost all infections, excepting only those rare cases caused by nonculturable organisms. The major issues that beset bacterial culture are as follows.

LACK OF SENSITIVITY OF STANDARD METHODS OF SAMPLING AND OF CULTURE

Some 15 percent of prosthetic joint infections, and a similar proportion of cases of chronic osteomyelitis, are culture-negative, yet show histological signs of active infection (Atkins et al. 1998). One potential explanation is the very low densities of organisms in some manifesta-tions of infection. Experimental studies in animal models suggest that the density of bacteria in peripros-thetic tissues is considerably lower than the density in an acute abscess, and explains why a swab of fluid near the implant or infected bone will not always yield a useful result. To increase sensitivity, a number of pieces of the periprosthetic tissues should be obtained for culture, and the culture routine should include cooked meat broth or an alternative enrichment medium that is supportive to damaged and fastidious organisms (Atkins et al. 1998). To increase specificity, each sample should be taken with a separate set of instruments to avoid cross-contam-ination between sampling. A carefully conducted prospective study of hip and knee replacements under-going revision, suggested high levels of both sensitivity and specificity can be achieved through this rigorous multiple sampling approach (Atkins et al. 1998). Other approaches have focused on increasing the yield from cultures by using ultrasonication to dislodge organisms from the surface of the explanted prosthesis. This has suggested that the presence of bacteria is greatly under-estimated in loose prostheses, but the work currently awaits independent confirmation.

INTERPRETATION OF POSITIVE AND MIXED CULTURE RESULTS

Many infections are active in the context of chronic wounds that overlie the involved bone or prosthesis. The wound will acquire a dynamic flora of its own, and irrespective of the original pathogens, secondary ones may establish themselves in the bone after prolonged access via the wound. In our experience, a transition from gram-positive to gram-negative pathogen, or from methicillin-sensitive gram-positive, to methicillin- (and occasionally vancomycin-) resistant gram-positive pathogen is frequently seen. Antibiotic selection is an important cause of this dynamic evolution.

How to interpret microbiology results in this situation is problematic. Studies suggest that superficial cultures are misleading and poorly predictive of the deep flora (Mackowiak et al. 1978). The microbiologist is therefore dependent upon sensible sampling behavior and accurate specimen labeling by the surgeon. Wounds should be thoroughly debrided before cultures are sent, or cultures obtained through a quite separate, uninfected field, as with a percutaneous biopsy. The surgeon must understand the crucial importance of cross-contamination. Tissue samples should be encouraged over swabs, which do not facilitate the isolation of anaerobes. Material must be sent to the laboratory without delay. Results should be fed back to surgeons in person as often as is possible, to allow discussion of therapeutic options and to reinforce joint working with the specialist in infection. Where the pathogens are not distinct from the likely skin or ulcer flora, and the biopsies have had to come through this field, a good rule of thumb is to demand the isolation of an indistinguishable organism (by biochemical identity and antibiogram) in at least two independent samples. Confirmatory histological examination of tissue is also invaluable.

Histology

The majority of pyogenic infections cause an acute inflammatory response. This is in contrast, in the skeleton, with conditions that give more chronic inflammation, where the inflammatory response shows remarkably few neutrophils. It is possible to distinguish infected from uninfected periprosthetic tissue with considerable accuracy, mainly on the basis of the numbers of infiltrating polymorphs (Mirra et al. 1982; Panday et al. 2000). This forms the basis of being able to diagnose infection in a prosthetic joint in real time, using frozen section, during joint revision surgery (Fehring and McAlister 1994; Athanasou et al. 1995; Lonner et al. 1996). It also provides an essential second diagnostic modality to match up with cultures in all contexts (White et al. 1995). The importance of this should be stressed, as it is not unknown for lesions that do not respond to appropriate treatment for infection to turn out to be tumors. In some situations, either because of sampling error, prior antibiotic use or presumed fastidious organisms, histology suggestive of infection may be the only evidence available, because the cultures are

negative (White et al. 1995). Histology is also of great importance in the diagnosis of mycobacterial disease, given the slowness of culture results in most mycobacterial infections.

Polymerase chain reaction

The use of polymerase chain reaction (PCR) for diagnosis is appealing given the limitations in sensitivity of conventional culture. There are surprisingly few published studies in this area, suggesting that researchers are encountering significant technical difficulties in achieving a robust and reproducible test. Using a universal primer for the 16S rRNA gene, an early study suggested that a substantial number of culture-negative explanted prostheses had evidence of bacterial DNA on PCR, but with an incomplete correlation between a positive conventional culture and a positive PCR (Mariani et al. 1996; Hoeffel et al. 1999). Other studies have shown detection of bacterial DNA, but have not evaluated this as a formal diagnostic test (Tunney et al. 1999). Although an absolute indicator of infection would be useful to answer certain clinical questions, an identity and preferably an antibiogram are both of importance in definitive treatment planning, so more sophisticated techniques may ultimately have to be employed.

TREATMENT

The fundamental principles of treatment follow on from the special features of pathogenesis outlined earlier. Treatment of acute presentations must be prompt, effective, and predicated on appropriate empiric antibiotics based on clinical suspicion. The aim in acute cases is to prevent the progression to chronicity, via the death of bone, with its correspondingly worse prognosis. Provided treatment is effective and timely, it often does not need to be protracted. The use of clear guidelines and protocols can help ensure consistent treatment approaches to acute infections (Peltola et al. 1997; Levy et al. 2000; Kocher et al. 2003).

Treatment of chronic presentations is based on the principles of establishing a clear microbiological diagnosis, using histology in parallel; excising dead and foreign infected tissue; establishing soft tissue cover and skeletal stability; managing surgical dead space; preserving functional implants where possible while they are necessary; using appropriate antibiotics for often extended periods of time, according to the biology of the infection and the host; and carrying out further reconstructive surgery, usually when infection is controlled. For all forms of chronic bone and joint infection, it is important to set realistic goals and expectations, which must include the possibility of recurrence and plans for the management of such an eventuality.

In treating infected orthopedic implants, the principles are a fusion of the two. For acute infections, urgent appropriate diagnosis and treatment is needed, aimed at preventing chronic infection, often with retention of the implant if still functional. For chronic infections, a more measured approach is required in which realistic goals are set with the patient, reconstructive options considered and a treatment plan delivered as with other chronic skeletal infections.

Surgical treatment

This is a key element of successful treatment in most situations, especially in chronic cases. In acute osteomyelitis and in vertebral osteomyelitis, provided there are no collections or abscesses, surgery can be deferred and may not be needed. Indeed, in acute osteomyelitis, gratuitous windowing or drilling of cortical bone, in the absence of other evidence of intramedullary abscess, is no longer in favor among most surgeons. In all other situations of chronic osteomyelitis, with the possible exception of the diabetic foot with its predominant involvement of small bones, surgical removal of dead infected bone is crucial.

Debridement and excision

Careful exploration of infected tissues allows determination of the extent of infection, scarring, abnormal soft tissue, and bone, and hence permits decisions about the degree of resection needed. The aim should be that at the end of surgery, bone at the resection margins is viable as indicated by bleeding, provided this leaves a limb that is either still functional or is reconstructable. Recurrent infection is often linked to failures of resection. In chronic prosthetic joint infection, the same thinking applies. The prosthesis and all cement, infected soft tissue, and dead bone should be removed, and abscesses drained. Rarely, it is appropriate to attempt salvage of a chronically infected prosthetic joint, if solidly fixed, pain free, and in situations where the expected morbidity of revision surgery exceeds that of ongoing infection and long-term antibiotics (see below). This is a more standard approach in an acute infection with a well-fixed prosthesis, where the damage caused by removal of the implant may be significant, and where studies show success rates of 50–70 percent if the debridement is thorough and timely (Brandt et al. 1997; Mont et al. 1997; Crockarell et al. 1998; Tattevin et al. 1999; Krasin et al. 2001).

Soft tissue cover

Some failures of osteomyelitis management are linked to persisting soft tissue defects or areas of extensive (and by definition relatively avascular) scar tissue. A tissue defect takes a prolonged period of time to heal and in some cases cannot (for example, when it overlies extensive dead bone or metal ware), while scar tissue limits the access of inflammatory mediators and antimicrobials to the infected area. Fortunately, modern plastic surgical techniques allow for the transposition of skeletal muscle with the corresponding vascular pedicle, using microvascular anastomosis, into tissue defects (Kuokkanen et al. 2002). Muscle flaps provide healthy, well-vascularized soft tissue cover and the physical characteristics of muscle make it ideal for laying into a bony defect, to the walls of which it rapidly adheres. Alternatively, muscle can be moved through lateral release or rotation upon the vascular pedicle, in this context most commonly in the case of the gastrocnemius, one head of which can be tunneled anteriorly to provide cover for the knee (McPherson et al. 1997; Casanova et al. 2001). This is of crucial value when wound breakdown compromises the soft tissue over a knee replacement, either acutely to aid implant salvage, or as part of a revision strategy.

Dead space management and reconstruction

Surgical 'dead spaces', namely cavities in tissue that are left to fill with blood, exudate or serous fluid, are risk factors for renewed or recurrent infection. Elimination of these dead spaces is an additional priority of surgical technique. Muscle flaps are excellent means of dead space management, but other means are sometimes employed. These include antibiotic loaded acrylic bone cement, which combines antibiotic delivery with dead space management. This can be combined with subsequent bone grafting if required, either for structural indications or to eliminate larger cavities definitively (McNally et al. 1993; Alonge et al. 2002).

Reconstructive options depend upon the specific situation. Fracture non-unions require stabilization; external fixation is generally preferred, to avoid instrumenting an already infected field, but the key priority is effective stabilization, which is essential for comfort, for union to proceed, and it seems, for control of infection.

For prosthetic joints that have been removed, reconstructive options rest with reimplanting a new joint, fusing the joint, or leaving an excision arthroplasty. In the hip, the latter results in a Girdlestone's pseudarthrosis, which is compatible with function in most patients, albeit with walking aids; for the other major joints, excision is a relatively poor option. For reimplantation, there is significant uncertainty over the choice between one- and two-stage revision surgery (Buchholz et al. 1981; McDonald et al. 1989; Raut et al., 1995; Goldman et al. 1996; Younger et al. 1997; Haddad et al., 2000). The one-stage technique is favored when highly sensitive organisms infect relatively healthy hosts, with two-stage revision favored for less fit hosts and more resistant bacteria (Hanssen and Osmon 2000; Fisman et al. 2001). Frustratingly, it is precisely in the less fit patient that one stage revision is most inherently attractive, to avoid the considerable morbidities of spending weeks without the function afforded by a functioning prosthesis, and of a second general anesthetic. For many, this limits the usefulness of the one-stage concept (Jackson and Schmalzried 2000), especially in view of difficulties in defining the infecting pathogen before definitive surgery. Some go so far as to take additional samples at the end of the first stage and accompanying antibiotic treatment, to determine whether to carry out additional debridements prior to reimplantation (Mont et al. 1997).

One important advance has been in the management of segmental defects of long bones using the Ilizarov method. This involves external stabilization of the limb using fine wires traversing the bone, attached to circular rings. A number of such rings, arrayed axially along the limb, are linked by longitudinal rods, and the resulting frame stabilizes the limb. The power of this method is that it is possible not only to remove an entire segment of bone, but also to grow new bone. By creating a corticotomy through bone adjacent to the defect, a viable segment can be produced that is itself stabilized by wires and a ring as part of the frame. By altering the relative positions of the rings, this 'transport segment' can be moved through the defect, a few millimeters a day, until it is 'docked' at the far end of the bone defect. The traction on the intact periosteum at the corticotomy site results in new bone formation. Using the method, large segmental defects can be regenerated with healthy bone of normal structure.

Antimicrobial treatment

Although bone infection can be managed without antibiotics, relapse rates were high in the pre-antibiotic era. Furthermore, the prolonged nature of resection and reconstruction surgery would lead to a high postoperative infection rate, highly likely to involve the surgically injured bone. Despite the rise in antibiotic resistance, there are still several good drugs available for use in osteomyelitis. The historical standard of care is prolonged intravenous therapy, but there is growing confidence in the use of shorter treatment regimens and highly bioavailable oral therapies in some situations (Kim et al. 2000; Jaberi et al. 2002; Le Saux et al. 2002; Vinod et al. 2002), alongside technologies that permit early discharge from hospital despite a decision to continue parenteral treatment.

It is notable that relatively few antibiotics have specific licensed indications for either osteomyelitis or prosthetic joint infection. This is more a reflection of the high costs and risks to pharmaceutical companies (in seeking indications for treatment in conditions that require prolonged treatment and may still have appreciable failure rates) than an indication that the drugs inherently will not work. Although the literature frequently raises the issue of 'bone penetration', there are uncertainties about the meaning and relevance of this term. Measuring antibiotic levels in cortical and cancellous bone, accounting for the levels within marrow stroma, and understanding whether the key targets of antibiotics are in the interstitial fluid, the bony matrix, or even the cellular elements within bone are all topics fraught with difficulty.

It has been hard to escape the view that in the antibiotic era, it has not been the development of drugs with bone penetration, but the renewed understanding of the critical nature of adequate surgical resection, and improvements in reconstructive techniques, that have led to improved outcomes for many with chronic infection. Nonetheless, antibiotics are still rightly regarded as a key element of treatment, even though a meta-analysis of published studies concluded that there is insufficient evidence to favor one drug, route of administration or duration over another (Stengel et al. 2001). They do play a pivotal role in treating those manifestations of infection not amenable to surgical excision; acute infections where bone and joint destruction have not yet occurred, and where a decision is made to undertake long-term suppression of the infection, with implant retention (Segreti et al. 1998; Tattevin et al. 1999). This strategy has been useful in acute prosthetic joint infection, and seems particularly likely to succeed when there is a short interval (48 hours or less) between onset of symptoms and the surgical debridement (Brandt et al. 1997; Mont et al. 1997; Crockarell et al. 1998; Krasin et al. 2001).

Systemic: intravenous

The lack of oral bioavailability of many β-lactam antibiotics, aminoglycosides, and glycopeptides, make them good choices for intravenous therapy against many gram-positive organisms. In many situations, such as prosthetic joint infection and some cases of osteomyelitis, courses of intravenous therapy of 4–6 weeks duration are commonplace. This requires the placement of central venous access, most commonly with a tunneled subclavian (Hickman or Broviac) line, but increasingly with peripherally inserted central catheters (PICC) (Parker and Gaines 1995) or shorter lines terminating in the axillary vein, so-called 'mid'-lines. The dependable nature of such vascular access means that they can be used to underpin programs for administration of intravenous antibiotics in the home or outpatient setting (Grayson et al. 1995b). Ceftriaxone and teicoplanin are useful drugs in this regard because of their once-daily dosing regimens, but more frequent dosing is possible, especially if patients can be taught to self-administer, and various technologies permit the use of drugs that must be delivered by continuous or intermittent infusion if need be. Recent analysis indicates a high success rate in cases of osteomyelitis treated with prolonged courses of ceftriaxone (Tice et al. 2003). Outpatient parenteral antibiotic therapy (OPAT), outpatient and home parenteral antibiotic therapy (OHPAT) and a number of other acronyms describe this mode of treatment, which is in established use in many centers around the world (Bernard et al. 2001). Osteomyelitis and prosthetic joint infection patients are major users of such programs.

Systemic: oral

Although the availability of OPAT transforms the prospects for giving prolonged intravenous treatment (in an era when prolonged hospital stays are, in the main, neither culturally nor financially acceptable), there are obvious issues about the costs and risks of intravenous therapy. There is now a range of highly bioavailable oral therapies, many of which are used in the treatment of osteomyelitis or prosthetic joint infection (Shuford and Steckelberg 2003). With the exception of first-generation cephalosporins, the orally available drugs do not act on the bacterial cell wall but on DNA or protein metabolism. The nature of these targets is, in the main, associated with easier acquisition of resistance, and it is important to bear this in mind in selecting treatment, especially if there is a high bioburden of infection at the start (for example, when debridement surgery is not carried out). Useful drugs include high-dose trimethoprim-sulfamethoxazole, doxycycline, rifampin, and fusidic acid for staphylococci, especially MRSA and MRSE. The published role of linezolid remains anecdotal (Bassetti et al. 2001).

In addition to these drugs, there is a substantial body of evidence supporting the use of ciprofloxacin. This has proved extremely useful for the treatment of gram-negative osteomyelitis (Lew and Waldvogel 1995). It has also assumed a role, in combination with rifampin, in the treatment of gram-positive infections where biofilms are believed to be an important element of the pathophysiology. The use dates from animal models of device-related infection in which cure of staphylococcal infection was observed with fluroquinolone-rifampin combinations. This led to observational studies in orthopedic device-related infection (Widmer et al. 1992; Drancourt et al. 1993), culminating in a randomized controlled trial of ciprofloxacin-rifampin therapy compared to ciprofloxacin monotherapy in patients with staphylococcal infection and a mechanically stable implant, who underwent prompt surgical debridement with retention of the implant (Zimmerli et al. 1998). The results clearly showed the ability of the combined regimen to cure infection with 3–6 months of therapy in selected cases. All the failures were seen in the monotherapy group,

with the evolution of resistance in the same bacterial strain, as demonstrated by pulse-field gel analysis of the initial and relapsing isolates. It remains to be seen in larger series of orthopedic device-related infection how generalizable these findings will prove. Applying the same reasoning to chronic osteomyelitis, quinolone–rifampin combinations have also been used in this condition, particularly in diabetic foot osteomyelitis (Senneville et al. 2001). In this condition, several separate studies have shown that antibiotic therapy alone frequently suffices to treat osteomyelitis, though the lack of standardized definitions does cause some difficulty (Venkatesan et al. 1997; Pittet et al. 1999)

Topical

The need for surgical dead space management has already been discussed. A logical extension of this requirement, and arising from a view that if high doses of antibiotics are required these are best provided locally, is the incorporation of antibiotics into a suitable carrier that can be implanted at the site of infection. An ideal carrier would release high levels of active drug for a prolonged period, without generating high systemic levels of the drug. It would not in its own right be systemically or locally toxic, and would be absorbable and support or stimulate the production of new bone. The carrier would be compatible either with one or two agents that singly, or in combination, produce a very broad-spectrum regimen, or with a range of drugs that could be mixed in at the time of application, based on prior microbiological information (Wininger and Fass 1996). In practice, no carriers yet fulfill all these requirements. The most widely used system is polymethylmethacrylate (PMMA) bone cement, which can be loaded with a range of antibiotics (Figure 27.5). These include cefuroxime, gentamicin, amikacin, tobramycin, clindamycin, and vancomycin. Of these, gentamicin was first developed commercially (in Europe) and can be obtained as a mix for reconstitution, or precast into beads of varying sizes threaded into chains on wire (Wininger and Fass 1996; Walenkamp et al. 1998; Klemm 2001). Antibiotic loaded cements are not available pre-mixed in the USA for regulatory reasons, so surgeons who make use of them there have to hand mix antibiotic powder into standard cement, a procedure that can obviously lead to variations in the properties of the cement and the elution of the drug.

There are extensive data on the elution properties of gentamicin and to a lesser extent other antibiotics from PMMA cement, which indicate that very high local levels of drug are maintained in the immediate environment of the cement and in wound drainage fluid. These are maintained for upwards of two weeks before levels start to fall. Levels are so high that bacteria that are antibiotic resistant may nonetheless be killed. Elution of

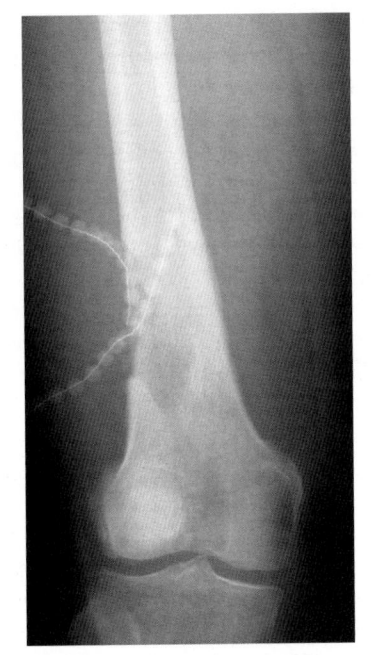

Figure 27.5 *Post-operative radiograph, showing bone loss subsequent to debridement via a cortical bone window, and antibiotic loaded PMMA beads to manage the dead space*

antibiotic is from the exposed surface of the cement, with little drug diffusing out from the depths of the bead or block. This has implications for removal of spacers, which may potentially release bactericidal levels of drug if cracked or otherwise damaged (when new surfaces are exposed), affecting culture results.

In addition to its temporary use in spacers and beads, antibiotic loaded cement has also been used in revision of prosthetic joints. This may be in one- or two-stage exchange, or in rigid or articulating spacers. The latter are designed to preserve some joint movement, in the belief that morbidity is reduced, technical difficulties at reimplantation ameliorated and ultimate range of movement improved (Fehring et al. 2000; Emerson et al. 2002). There are even kits available with molds and internal metal scaffolds around which antibiotic loaded cement can be cast to give a relatively functioning, cement-surfaced implant, the PROSTALAC (prosthesis of antibiotic loaded acrylic cement) (Masri et al. 1994; Haddad et al. 2000; Meek et al. 2003). These may prove invaluable in managing major proximal femoral loss, by preserving function, muscle bulk, and an anatomic space for eventual implantation of a large definitive prosthesis. There are no controlled data on the use of functional spacers compared to static alternatives.

Alternatives to the use of PMMA cement as an antibiotic carrier are available. Gentamicin can also be incorporated into collagen fleeces, which can be packed into cavities and which should, over time, be absorbed (Kollenberg 1998). The kinetics of release differ from antibiotic-loaded PMMA cement, with high levels of antibiotic eluting over a much shorter period of time.

Hydroxyapatite can be used (Yamashita et al. 1998; Shirtliff et al. 2002), and 'low technology' versions also exist, in the form of calcium phosphate (plaster of Paris) (Alonge et al. 2002; Santschi and McGarvey 2003), which can be reconstituted, with antibiotics incorporated, into sterile pellets for packing into bone defects (Benoit et al. 1997). This may be particularly valuable in parts of the world where resources are scarce (Alonge et al. 2002; Heybeli et al. 2003). Pellets have also been studied in the treatment of bony defects produced after debridement of diabetic foot ostomyelitis (Roeder et al. 2000).

The evidence base for the use of antibiotic-impregnated PMMA cement is relatively poor, but surprisingly, little different from that supporting the use of systemic therapies. An attempt at a randomized trial of gentamicin-loaded PMMA beads compared to systemic therapy was unsuccessful due to substantial protocol violations and concerns about the randomization (Blaha et al. 1993). Another study was abandoned when nursing staff refused to continue the traditional suction drainage approach in favor of the much simpler management of antibiotic-loaded PMMA beads (Walenkamp et al. 1998). Although systemic and local therapies have therefore never been adequately compared, nor have intravenous and oral therapies except in small studies.

One final concern in the use of antibiotic-loaded cement is the potential for persistence of the infection around the new foreign body. This certainly can occur if the pathogen has absolute resistance to the antibiotic in use (even at the very high doses released) but there is also the potential that small colony variants could be induced by the antibiotic and then persist. Clearly, for this mechanism to be plausible, residual infection would have to be present at the point the beads were used, which would be a risk factor for recurrence anyway. Nonetheless, small colony variants have been recovered from relapsed osteomyelitis in which antibiotic-loaded beads had previously been used (von Eiff et al. 1997).

Duration of treatment

The duration of treatment required varies according to the severity of the disease at the beginning of treatment, a concept that for chronic osteomyelitis at least, includes some sense of the local and systemic host factors that contribute to outcome. Taking this into account, the overall trend of treatment has, in general, been to reserve very prolonged treatment for situations where tissue that is dead or otherwise inanimate is retained in the infected field, and/or where the host has major systemic compromise. Where surgical resection produces remaining tissues that are minimally contaminated and viable, antibiotic treatment durations are reducing. In acute osteomyelitis, overall durations of therapy for uncomplicated infection are also falling (Le Saux et al.

2002; Vinod et al. 2002), in some series to as little as two weeks. Suggestions are given in Table 27.3, but it must be stressed that these have not been determined by a large series of clinical trials.

Adjunctive treatment

A range of adjunctive treatments has been proposed, the most common one being hyperbaric oxygen (Mader et al. 1990; Stone and Cianci 1997; Bakker 2000; Wunderlich et al. 2000; Wang 1997). This has its strong advocates, and there is laboratory evidence to suggest enhanced bacterial killing with hyperbaric oxygen. It is also believed to support increased function of phagocytes. Despite the existence of these data, the role of hyperbaric oxygen in the patient who has had debridement surgery and antibiotics is unclear. Randomized controlled trials will be crucial in allowing an informed view of the role of this treatment to emerge.

Other commonly advocated adjunctive treatments include Manuka honey, which is believed to have broad antimicrobial activity, and Tea Tree Oil. There is little available information on the value of these agents.

A final modality worthy of consideration is the use of maggots (larvae of the greenbottle fly, *Lucilia serricata*). The larvae of this species eat only dead tissue, and can be reared from eggs on chemically defined, sterile media; furthermore, the eggs themselves are highly resistant to external agents and so can be sterilized of microbiological contamination (Jukema et al. 2002). Larvae reared in this way are thus effectively sterile at the point of application to a wound, and are capable of carrying out efficient debridement. Some believe they also stimulate the formation of healthy granulation tissue. The use of maggots of this sort for the treatment of osteomyelitis was first put forward in the pre-antibiotic era, as an adjunct to sequestrectomy. Their role without surgery, and the extent to which they are necessary in the era of antibiotics and modern plastic surgical methods, is not defined; they certainly have proponents in the diabetic foot world and may be of value in situations where conventional surgery is deemed to carry too high a risk. Microbiologically, maggots develop a bowel flora that includes *Proteus* spp., and it must be assumed that the wound will be contaminated with this. Certainly the maggot secretions and excreta can be highly irritant to skin surrounding the wound, which itself becomes very exudative, and measures must be taken not only to contain the maggots (for obvious therapeutic and aesthetic reasons) but also to manage these wound secretions. In addition, maggots can carry a range of serious pathogenic *Clostridium* spp. in their gut; there were historical accounts of tetanus, botulism, and gas gangrene secondary to maggot use before the need for absolute sterility from egg onwards was identified. Much more recently, outbreaks of septicemia have been associated with the use of other species of maggot, stressing the need for this form of treatment to be subject to careful design and quality control. As with so much else in the management of osteomyelitis, randomized controlled studies are lacking. Even if maggots are effective in the management of diffusely necrotic soft tissue, their role in removing dead bone is open to question; this function is likely to require human intervention in most cases.

PROGNOSIS

The prognosis for acute infections is excellent provided the diagnosis is made promptly and the treatment is appropriate. Cure rates are over 95 percent in modern series of acute childhood hematogenous osteomyelitis. Function is generally good on treatment of infection, but in childhood osteomyelitis, annual follow up is required to look for growth plate disturbance. Acute septic arthritis also carries a good prognosis, with poor outcomes highly associated with delay in diagnosis and treatment (Christiansen et al. 1999). Acute prosthetic joint infection similarly requires rapid intervention for optimal outcomes, with salvage rates from debridement and retention procedures that can reach 70 percent (Karchmer 1998). Certain organisms, for example *Staphylococcus aureus*, confer worse prognosis; this is also seen in prognosis following one-stage revision of prosthetic joints. In relation to cure of infection, suppression (retention of chronically infected joints with long-term antibiotics) fares worse than debridement and retention of acute infections (Deirmengian et al. 2003), which are themselves outperformed by one-, and finally, two-stage revision. These achieve cure rates of 80–95 percent when antibiotic loaded cement is used.

Chronic osteomyelitis has high recurrence rates unless radical excision of dead bone is combined with antibiotic treatment. Recurrence rates in expert series are as low as 10 percent. Diabetic foot osteomyelitis has a poorer outlook, and responds to treatment in only 60–70 percent of cases (Grayson et al. 1994; Venkatesan et al. 1997; Pittet et al. 1999). This is due to the very much less favorable host factors operating, notably degrees of ischemia, and the relatively muted responses well recognized in diabetics (Armstrong et al. 1996). As a result, treatment frequently culminates in minor or major amputation; this is in part due to failure to eliminate the bone infection, but in a substantial number of cases, amputation is precipitated by the soft tissue loss that led to the osteomyelitis. If not fully surgically resected, osteomyelitis in many locations is well known to be able to recur decades after the initial event (Donati et al. 1999). There is no good evidence to suggest that the likelihood of this phenomenon is dictated by the initial treatment; the mechanisms of such prolonged dormancy remain unclear.

Table 27.3 *Indications for surgery and suggested duration of treatment*

Condition	Indications for surgery	Host biology after any surgery	Duration
Acute septic arthritis	Hip, shoulder, and when antibiotics alone fail. Arthroscopic washout now common	Normal	2 weeks for streptococci, *Neisseria* spp., *Haemophilus* spp.; 3 weeks for *Staphylococcus aureus* and gram-negative rods
		Arthritic joint, exposed bone or devitalized fibro-cartilage	4–6 weeks
Chronic septic arthritis	May need resection and fusion/arthroplasty if joint destroyed	Generally abnormal joint surface, exposed bone, and debris	As chronic osteomyelitis for pyogenic bacteria; *Brucella* spp. 3 months; *Mycobacterium* spp. as for osteomyelitis
Acute prosthetic joint infection	Mandatory, to establish diagnosis, microbiology, debride infected tissue, provide soft tissue cover. Remove implant if loose	Implant removed	6 weeks
		Implant retained	6 months–2 years (varying practice, lifelong treatment sometimes given)
Chronic prosthetic joint infection	Loose implant; pain; sinus or abscess formation	Implant removed definitively	6 weeks
		Implant revised in one stage	3–6 months
		Implant revised in two stages	6–12 weeks
		Implant retained	Indefinite
Acute osteomyelitis	Abscess formation	Infected but viable bone, no abscess	2–4 weeks
		Complicated (surgery required, dead bone)	4–12 weeks
Chronic osteomyelitis: long bone and foot	Dead bone; loss of soft tissue cover; abscess; pain; multiple acute flares	The whole of the involved bone removed (ablative treatment, e.g. amputation), with normal remaining soft tissue	1–3 days
		The whole of the involved bone removed, but with residual soft tissue infection	1–2 weeks
		All dead bone removed, but viable portion of infected bone remains	4–6 weeks
		Residual dead bone currently or previously involved in infected field	12 weeks or more
Chronic osteomyelitis: spine	Abscess; spinal instability; pain; progressive deformity; spinal cord compression	No metal ware	6–12 weeks
		Retained metal in infected field	3–12 months

FUTURE DIRECTIONS

The area of musculoskeletal infection still needs definitive answers to questions such as how best to diagnose and treat infection in the most cost-effective way. The necessary duration, and the favored route of administration (if one exists) are yet to be determined, as is the difference in outcome and morbidity between one- and two-stage revision for prosthetic joint infection. Prediction of successful cure with medical therapy alone remains difficult, and there is yet to be a significant translation into clinical practice of recent advances in understanding the mechanisms of bone loss and bacterial pathogenesis.

CONCLUSIONS

Bacterial infections of bone and joint continue to pose major challenges and questions, which understandings of pathogenesis, diagnosis, and treatment options go only some way to answering. It is to be hoped that improved multidisciplinary working will increase clinical confidence in the management of these problems, and pave the way for the definitive studies that are required to establish a strong evidence base available to all.

REFERENCES

Abbas, Z.G., Gill, G.V. and Archibald, L.K. 2002. The epidemiology of diabetic limb sepsis: an African perspective. *Diabet Med*, **19**, 11, 895–9.

Abdelnour, A., Arvidson, S., et al. 1993. The accessory gene regulator (*agr*) controls *Staphylococcus aureus* virulence in a murine arthritis model. *Infect Immun*, **61**, 9, 3879–85.

Alonge, T.O., Ogunlade, S.O., et al. 2002. Management of chronic osteomyelitis in a developing country using ceftriaxone-PMMA beads: an initial study. *Int J Clin Pract*, **56**, 3, 181–3.

Armstrong, D.G., Perales, T.A., et al. 1996. Value of white blood cell count with differential in the acute diabetic foot infection. *J Am Podiatr Med Assoc*, **86**, 224–7.

Athanasou, N.A., Pandey, P., et al. 1998. Diagnosis of infection by frozen section surgery. *J Bone Joint Surg*, **77**, 28–33.

Atkins, B.L., Athanasou, N. and Deeks, J.J. 1998. Prospective evaluation of criteria for microbiological diagnosis of prosthetic-joint infection at revision arthroplasty. *J Clin Microbiology*, **36**, 2932–9.

Baker, A.S., Ojemann, R.G., et al. 1975. Spinal epidural abscess. *N Eng J Med*, **293**, 463–8.

Bakker, D.J. 2000. Hyperbaric oxygen therapy and the diabetic foot. *Diab Metab Res Rev*, **16**, Suppl 1, S55–58.

Bardin, T. 2003. Gonococcal arthritis. *Best Pract Res Clin Rheumatol*, **17**, 2, 201–8.

Bassetti, M., Di Biagio, A., et al. 2001. Linezolid treatment of prosthetic hip infections due to methicillin-resistant *Staphylococcus aureus* (MRSA). *J Infect*, **43**, 2, 148–9.

Bayles, K.W., Wesson, C.A., et al. 1998. Intracellular *Staphylococcus aureus* escapes the endosome and induces apoptosis in epithelial cells. *Infect Immun*, **66**, 1, 336–42.

Beekhuizen, H., van de Gevel, J.S., et al. 1997. Infection of human vascular endothelial cells with *Staphylococcus aureus* induces hyper-adhesiveness for human monocytes and granulocytes. *J Immunol*, **158**, 2, 774–82.

Belmatoug, N., Cremieux, A.C., et al. 1996. A new model of experimental prosthetic joint infection due to methicillin-resistant *Staphylococcus aureus*: a microbiologic, histopathologic, and magnetic resonance imaging characterization. *J Infect Dis*, **174**, 2, 414–17.

Benoit, M.A., Mousset, B., et al. 1997. Antibiotic-loaded plaster of Paris implants coated with polylactide-co-glycolide as a controlled release delivery system for the treatment of bone infections. *Int Orthop*, **21**, 6, 403–8.

Berbari, E.F., Hanssen, A.D., et al. 1998a. Risk factors for prosthetic joint infection: case control study. *Clin Infect Dis*, **27**, 1247–54.

Berbari, E.F., Hanssen, A.D., et al. 1998b. Prosthetic joint infection due to *Mycobacterium tuberculosis*: a case series and review of the literature. *Am J Orthop*, **27**, 3, 219–27.

Bernard, L., El-Hajj, et al. 2001. Outpatient parenteral antimicrobial therapy (OPAT) for the treatment of osteomyelitis: evaluation of efficacy, tolerance and cost. *J Clin Pharm Ther*, **26**, 445–51.

Birgisson, H., Steingrimsson, O. and Gudnason, T. 1997. *Kingella kingae* infections in paediatric patients: 5 cases of septic arthritis, osteomyelitis and bacteraemia. *Scand J Infect Dis*, **29**, 5, 495–498.

Blaha, J.D., Calhoun, J.H., et al. 1993. Comparison of the clinical efficacy and tolerance of Gentamicin PMMA beads on surgical wire versus combined and systemic treatment for osteomyelitis. *Clin Orthop*, **295**, 8–12.

Blanc, D.S., Pittet, D., et al. 2002. Epidemiology of methicillin-resistant *Staphylococcus aureus*: results of a nation-wide survey in Switzerland. *Swiss Med Wkly*, **132**, 17-18, 223–9.

Blyth, M.J., Kincaid, R., et al. 2001. The changing epidemiology of acute and subacute haematogenous osteomyelitis in children. *J Bone Joint Surg Br*, **83**, 99–102.

Boulton, A.J.M. 1996. The pathogenesis of diabetic foot problems: an overview. *Diabetic Med*, **13**, Suppl. 1, S12–16.

Bouza, E. and Munoz, P. 1999. Micro-organisms responsible for osteo-articular infections. *Baillieres Best Pract Res Clin Rheumatol*, **13**, 1, 21–35.

Bowerman, S.G., Green, N.E. and Mencio, G.A. 1997. Decline of bone and joint infections attributable to *Haemophilus influenzae* type b. *Clin Orthop*, Aug, **341**, 128–33.

Brandt, C.M., Sistrunk, W.W., et al. 1997. *Staphylococcus aureus* prosthetic joint infection treated with debridement and prosthesis retention. *Clin Infect Dis*, **24**, 914–19.

Bremell, T., Lange, S., et al. 1991. Experimental *Staphylococcus aureus* arthritis in mice. *Infect Immun*, **59**, 8, 2615–23.

Buchholz, H.W., Elson, R.A., et al. 1981. Management of deep infection of total hip replacement. *J Bone Joint Surg [Br]*, **63B**, 342–53.

Buerke, M., Sibelius, U., et al. 2002. *Staphylococcus aureus* alpha toxin mediates polymorphonuclear leukocyte-induced vasocontraction and endothelial dysfunction. *Shock*, **17**, 30–5.

Buxton, T.B., Rissing, J.P., et al. 1990. Binding of a *Staphylococcus aureus* bone pathogen to type I collagen. *Microb Pathog*, **8**, 441–8.

Cardinal, E., Bureau, N.J., et al. 2001. Role of ultrasound in musculoskeletal infections. *Radiol Clin North Am*, **39**, 191–201.

Casanova, D., Hulard, O., et al. 2001. Management of wounds of exposed or infected knee prostheses. *Scand J Plast Reconstr Surg Hand Surg*, **35**, 1, 71–7.

Chang, H.C., Verhoeven, W. and Chay, W.M. 2001. Rubber foreign bodies in puncture wounds of the foot in patients wearing rubber-soled shoes. *Foot Ankle Int*, **22**, 5, 409–14.

Chelsom, J. and Solberg, C.O. 1998. Vertebral osteomyelitis at a Norwegian university hospital 1987–97: clinical features, laboratory findings and outcome. *Scand J Infect Dis*, **30**, 2, 147–51.

Christiansen, P., Frederiksen, B., et al. 1999. Epidemiologic, bacteriologic, and long-term follow-up data of children with acute hematogenous osteomyelitis and septic arthritis: a ten-year review. *J Pediatr Orthop B*, **8**, 4, 302–5.

Christensen, G.D., Baldassarri, L. and Simpson, W.A. 1994. Colonization of medical devices by coagulase-negative staphylocci. In: Bisno, A.I. and Waldwogel, F.A. (eds), *Infections associated with indwelling medical devices*. Washington, DC: ASM Press, 45–78.

Chopp, D.L., Kirisits, M.J., et al. 2003. The dependence of quorum sensing on the depth of a growing biofilm. *Bull Math Biol*, **65**, 6, 1053–79.

Chuard, C., Vaudaux, P.E., et al. 1997. Decreased susceptibility to antibiotic killing of a stable small colony variant of *Staphylococcus aureus* in fluid phase and on fibronectin-coated surfaces. *J Antimicrob Chemother*, **39**, 5, 603–8.

Clark, J.E., Abinun, M., et al. 1997. *Mycobacterium xenopi* osteomyelitis. *Pediatr Infect Dis J*, **16**, 10, 1011.

Costerton, J.W., Stewart, P.S. and Greenberg, E.P. 1999. Bacterial biofilms: a common cause of persistent infections. *Science*, **284**, 1318–22.

Crockarell, J.R., Hanssen, A.D., et al. 1998. Treatment of infection with debridement and retention of the components following hip arthroplasty. *J Bone Joint Surg Am*, **80**, 1306–13.

Davies, D.G., Parsek, M.R., et al. 1998. The involvement of cell-to-cell signals in the development of a bacterial biofilm. *Science*, **280**, 295–8.

Deirmengian, C., Greenbaum, J., et al. 2003. Limited success with open debridement and retention of components in the treatment of acute *Staphylococcus aureus* infections after total knee arthroplasty. *J Arthroplasty*, **18**, 22–6.

Delmi, M., Vaudaux, P., et al. 1994. Role of fibronectin in staphylococcal adhesion to metallic surfaces used as models of orthopaedic devices. *J Orthop Res*, **12**, 3, 432–8.

Deng, G.M. and Tarkowski, A. 2000. The features of arthritis induced by CpG motifs in bacterial DNA. *Arthritis Rheum*, **43**, 2, 356–64.

Devillers, A., Moisan, A., et al. 1995. Technetium-99m hexamethylpropylene amine oxime leukocyte scintigraphy for the diagnosis of bone and joint infections: a retrospective study in 116 patients. *Eur J Nucl Med*, **22**, 4, 302–7.

Dodman, T., Robson, J. and Pincus, D. 2000. *Kingella kingae* infections in children. *J Paediatr Child Health*, **36**, 87–90.

Donati, L., Quadri, P. and Reiner, M. 1999. Reactivation of osteomyelitis caused by *Staphylococcus aureus* after 50 years. *J Am Geriatr Soc*, **47**, 1035–7.

Drancourt, M., Stein, A., et al. 1993. Oral rifampin plus ofloxacin for treatment of *Staphylococcus aureus*-infected orthopaedic implants. *Antimicrob Agents Chemother*, **37**, 1214–18.

Edelson, G.W., Armstrong, D.G., et al. 1996. The acutely infected diabetic foot is not adequately evaluated in an inpatient setting. *Arch Intern Med*, **156**, 2373–8.

Ekerot, L., Jacobsson, L. and Forsgren, A. 1998. *Mycobacterium marinum* wrist arthritis: local and systematic dissemination caused by concomitant immunosuppressive therapy. *Scand J Infect Dis*, **30**, 1, 84–7.

Elasri, M.O., Thomas, J.R., et al. 2002. *Staphylococcus aureus* collagen adhesin contributes to the pathogenesis of osteomyelitis. *Bone*, **30**, 1, 275–80.

Elek, S.D. and Conen, P.E. 1957. The virulence of *Staphylococcus pyogenes* for man: a study of the problems of wound infection. *Br J Exp Pathol*, **38**, 573–86.

Ellington, J.K., Harris, M., et al. 2003. Intracellular *Staphylococcus aureus*. A mechanism for the indolence of osteomyelitis. *J Bone Joint Surg Br*, **85**, 918–21.

Emerson, R.H. Jr, Muncie, M., et al. 2002. Comparison of a static with a mobile spacer in total knee infection. *Clin Orthop*, Nov, **404**, 132–8.

Fehring, T.K. and McAlister, J.A. 1994. Frozen histologic section as a guide to sepsis in revision joint arthroplasty. *Clin Orthop*, **304**, 229–37.

Fehring, T.K., Odum, S., et al. 2000. Articulating versus static spacers in revision total knee arthroplasty for sepsis. The Ranawat Award. *Clin Orthop*, **380**, 9–16.

Fisher, M.C., Goldsmith, J.F. and Gilligan, P.H. 1985. Sneakers as a source of *Pseudomonas aeruginosa* in children with osteomyelitis following puncture wounds. *J Pediatr*, **106**, 607–9.

Fisman, D.N., Reilly, D.T., et al. 2001. Clinical effectiveness and cost-effectiveness of 2 management strategies for infected total hip arthroplasty in the elderly. *Clin Infect Dis*, **32**, 3, 419–30.

Flock, J.I. 1999. Extracellular-matrix-binding proteins as targets for the prevention of *Staphylococcus aureus* infections. *Mol Med Today*, **5**, 12, 532–7.

Fowler, T., Wann, E.R., et al. 2000. Cellular invasion by *Staphylococcus aureus* involves a fibronectin bridge between the bacterial fibronectin-binding MSCRAMMs and host cell beta1 integrins. *Eur J Cell Biol*, **79**, 10, 672–9.

Franz, J.K. and Krause, A. 2003. Lyme disease (Lyme borreliosis). *Best Pract Res Clin Rheumatol*, **17**, 2, 241–64.

Geyik, M.F., Gur, A., et al. 2002. Musculoskeletal involvement of brucellosis in different age groups: a study of 195 cases. *Swiss Med Wkly*, **132**, 7-8, 98–105.

Gillespie, W.J. 1990. Epidemiology in bone and joint infection. *Infect Dis Clin North Am*, **4**, 361–76.

Gillespie, W.J. 1997. Prevention and management of infection after total joint replacement. *Clin Infect Dis*, **25**, 1310–17.

Glenny, A. and Song, F. 1999. Antimicrobial prophylaxis in total hip replacement; a systematic review. *Health Technol Assess*, **3**, 1–57.

Goldman, R.T., Scuderi, G.R. and Insall, J.N. 1996. Two-stage reimplantation for infected total knee replacement. *Clin Orthop*, **331**, 118–24.

Golla, A., Jansson, A., et al. 2002. Chronic recurrent multifocal osteomyelitis (CRMO): evidence for a susceptibility gene located on chromosome 18q21.3-18q22. *Eur J Hum Genet*, **10**, 217–21.

Gonzalez-Lopez, J.L., Soleto-Martin, F.J., et al. 2001. Subacute osteomyelitis in children. *J Pediatr Orthop B*, **10**, 2, 101–4.

Gotz, F. 2002. *Staphylococcus* and biofilms. *Mol Microbiol*, **43**, 1367–1378.

Grayson, M.L., Gibbons, G.W., et al. 1994. Use of ampicillin/sulbactam versus imipenem/cilastatin in the treatment of limb-threatening foot infections in diabetic patients. *Clin Infect Dis*, **18**, 683–93, (Erratum in *Clin Infect Dis* 1994, **19**, 820).

Grayson, M.L., Gibbons, G.W., et al. 1995a. Probing to bone in infected pedal ulcers. A clinical sign of underlying osteomyelitis in diabetic patients. *JAMA*, **273**, 721–3.

Grayson, M.L., Silvers, J. and Turnidge, J. 1995b. Home intravenous antibiotic therapy. A safe and effective alternative to inpatient care. *Med J Aust*, **162**, 5, 249–53.

Gristina, A.G., Oga, M., et al. 1985. Adherent bacterial colonization in the pathogenesis of osteomyelitis. *Science*, **228**, 990–3.

Guerra, C.E., Betts, R.F., et al. 1998. *Mycobacterium bovis* osteomyelitis involving a hip arthroplasty after intravesicular bacille Calmette–Guerin for bladder cancer. *Clin Infect Dis*, **27**, 3, 639–40.

Gustilo, R.B. and Anderson, J.T. 1976. Prevention of infection in the treatment of one thousand and twenty five open fractures of long bones: Retrospective and prospective analyses. *J Bone Joint Surg*, **58**, 453–8.

Haddad, F.S., Masri, B.A., et al. 2000. The PROSTALAC functional spacer in two-stage revision for infected knee replacements. Prosthesis of antibiotic-loaded acrylic cement. *J Bone Joint Surg Br*, **82**, 807–12.

Hanssen, A.D. and Osmon, D.R. 2000. Assessment of patient selection criteria for treatment of the infected hip arthroplasty. *Clin Orthop*, **381**, 91–100.

Hardinge, K., Cleary, J. and Charnley, J. 1979. Low-friction arthroplasty for healed septic and tuberculous arthritis. *J Bone Joint Surg Br*, **61-B**, 144–7.

Heilmann, C., Schweitzer, O., et al. 1996. Molecular basis of intercellular adhesion in the biofilm-forming *Staphylococcus epidermidis*. *Mol Microbiol*, **20**, 1083–91.

Herrmann, M., Vaudaux, P.E., et al. 1988. Fibronectin, fibrinogen and laminin act as mediators of adherence of clinical staphylococcal isolates to foreign material. *J Infect Dis*, **158**, 693–701.

Heybeli, N., Oktar, F.N., et al. 2003. Low-cost antibiotic loaded systems for developing countries. *Technol Health Care*, **11**, 3, 207–16.

Hoeffel, D.P., Hinrichs, S.H. and Garvin, K.L. 1999. Molecular diagnostics for the detection of musculoskeletal infection. *Clin Orthop*, **360**, 37–46.

Huber, A.M., Lam, P.Y., et al. 2002. Chronic recurrent multifocal osteomyelitis: clinical outcomes after more than five years of follow-up. *J Paediatr*, **141**, 198–203.

Ispahani, P., Weston, V.C., et al. 1999. Septic arthritis due to *Streptococcus pneumoniae* in Nottingham, United Kingdom, 1985–1998. *Clin Infect Dis*, **29**, 6, 1450–4.

Jaberi, F.M., Shahcheraghi, G.H. and Ahadzadeh, M. 2002. Short-term intravenous antibiotic treatment of acute hematogenous bone and joint infection in children: a prospective randomized trial. *J Pediatr Orthop*, **22**, 3, 317–20.

Jackson, W.O. and Schmalzried, T.P. 2000. Limited role of direct exchange arthroplasty in the treatment of infected total hip replacements. *Clin Orthop*, **381**, 101–5.

Jensen, A.G., Espersen, F., et al. 1997. Increasing frequency of vertebral osteomyelitis following *Staphylococcus aureus* bacteraemia in Denmark 1980–1990. *J Infect*, **34**, 113–18.

Ji, G., Beavis, R.C. and Novick, R.P. 1995. Cell density control of staphylococcal virulence mediated by an octapeptide pheromone. *Proc Natl Acad Sci USA*, **92**, 26, 12055–9.

Johnson, J.E., Kennedy, E.J., et al. 1996. Prospective study of bone, indium-111-labeled white blood cell, and gallium-67 scanning for the evaluation of osteomyelitis in the diabetic foot. *Foot Ankle Int*, **17**, 10–16.

Jonsson, I.M., von Eiff, C., et al. 2003. Virulence of a *hemB* mutant displaying the phenotype of a *Staphylococcus aureus* small colony variant in a murine model of septic arthritis. *Microb Pathog*, **34**, 73–79.

Josefsson, E. and Tarkowski, A. 1999. *Staphylococcus aureus*-induced inflammation and bone destruction in experimental models of septic arthritis. *J Periodontal Res*, **34**, 7, 387–92.

Jukema, G.N., Menon, A.G., et al. 2002. Amputation-sparing treatment by nature: 'surgical' maggots revisited. *Clin Infect Dis*, **35**, 12, 1566–1571.

Jupiter, J.B., Karchmer, A.W., et al. 1981. Total hip arthroplasty in the treatment of adult hips with current or quiescent sepsis. *J Bone Joint Surg Am*, **63**, 194–200.

Kaarsemaker, S., Walenkamp, G.H. and van de Bogaard, A.E. 1997. New model for chronic osteomyelitis with *Staphylococcus aureus* in sheep. *Clin Orthop*, **339**, 246–52.

Kaiser, S. and Rosenborg, M. 1994. Early detection of subperiosteal abscesses by ultrasonography. *Pediatr Radiol*, **24**, 336–9.

Karchmer, A.W. 1998. Salvage of infected orthopedic devices. *Clin Infect Dis*, **27**, 4, 714–16.

Khatod, M., Botte, M.J., et al. 2003. Outcomes in open tibia fractures: relationship between delay in treatment and infection. *J Trauma*, **55**, 5, 949–54.

Kim, H.K., Alman, B. and Cole, W.G. 2000. A shortened course of parenteral antibiotic therapy in the management of acute septic arthritis of the hip. *J Pediatr Orthop*, **20**, 1, 44–7.

Kim, Y.H., Oh, S.H. and Kim, J.S. 2003. Total hip arthroplasty in adult patients who had childhood infection of the hip. *J Bone Joint Surg Am*, **85-A**, 2, 198–204.

Kjelleberg, S. and Molin, S. 2002. Is there a role for quorum sensing signals in bacterial biofilms? *Curr Opin Microbiol*, **5**, 3, 254–8.

Klemm, K. 2001. The use of antibiotic-containing bead chains in the treatment of chronic bone infections. *Clin Microbiol Infect*, **7**, 1, 28–31.

Kocher, M.S., Mandiga, R., et al. 2003. A clinical practice guideline for treatment of septic arthritis in children: efficacy in improving process of care and effect on outcome of septic arthritis of the hip. *J Bone Joint Surg Am*, **85-A**, 994–9.

Kollenberg, L.O. 1998. A new topical antibiotic delivery system. *World Wide Wounds*, **1**, 1–19, (http://www.worldwidewounds.com/1998/july/Topical-Antibiotic-Delivery-System/topical).

Koornstra, J.J., Veenendaal, D., et al. 1998. Septic arthritis due to *Fusobacterium nucleatum*. *Br J Rheumatol*, **37**, 11, 1249.

Kozlowski, K., Masel, J., et al. 1983. Multifocal chronic osteomyelitis of unknown etiology. *Pediatr Radiol*, **13**, 130–6.

Krasin, E., Goldwirth, M., et al. 2001. Could irrigation, debridement and antibiotic therapy cure an infection of a total hip arthroplasty? *J Hosp Infect*, **47**, 3, 235–8.

Krogsgaard, M.R., Wagn, P. and Bengtsson, J. 1998. Epidemiology of acute vertebral osteomyelitis in Denmark: 137 cases in Denmark 1978–1982, compared to cases reported to the National Patient Register 1991–1993. *Acta Orthop Scand*, **69**, 513–17.

Kuokkanen, H.O., Tukiainen, E.J. and Asko-Seljavaara, S. 2002. Radical excision and reconstruction of chronic tibial osteomyelitis with microvascular muscle flaps. *Orthopedics*, **25**, 137–40.

Lange, U. and Teichmann, J. 2003. Whipple arthritis: diagnosis by molecular analysis of synovial fluid – current status of diagnosis and therapy. *Rheumatology (Oxford)*, **42**, 473–80.

Lavery, L.A., Armstrong, D.G., et al. 1996. Puncture wounds: normal laboratory values in the face of severe infection in diabetics and non-diabetics. *Am J Med*, **101**, 5, 521–5.

Lee, G.C., Pagnano, M.W. and Hanssen, A.D. 2002. Total knee arthroplasty after prior bone or joint sepsis about the knee. *Clin Orthop*, **404**, 226–31.

Le Saux, N., Howard, A., et al. 2002. Shorter courses of parenteral antibiotic therapy do not appear to influence response rates for children with acute hematogenous osteomyelitis: a systematic review. *BMC Infect Dis*, **2**, 16.

Levy, J., Peetermans, W.E., et al. 2000. Treatment of bone and joint infections: recommendations of a Belgian panel. *Acta Orthop Belg*, **66**, 127–36.

Lew, D. and Waldvogel, F.A. 1995. Quinolones and osteomyelitis: state-of-the-art. *Drugs*, **49**, 100–11.

Lipsky, B.A., Pecoraro, R.E. and Wheat, L.J. 1990. The diabetic foot. Soft tissue and bone infection. *Infect Dis Clin North Am*, **4**, 409–32.

Liu, Z.Q., Deng, G.M., et al. 2001. Staphylococcal peptidoglycans induce arthritis. *Arthritis Res*, **3**, 375–80.

Lonner, J.H., Desai, P., et al. 1996. The reliability of analysis of intraoperative frozen sections for identifying active infection during revision hip or knee arthroplasty. *J Bone Joint Surg Am*, **78**, 1553–8.

Mackowiak, P.A., Jones, S.R. and Smith, J.W. 1978. Diagnostic valve of sinus-tract cultures in chronic osteomyelitis. *JAMA*, **239**, 2772–5.

Mader, J.T. 1985. Animal models of osteomyelitis. *Am J Med*, **78**, 213–17.

Mader, J.T., Adams, K.R., et al. 1990. Hyperbaric oxygen as adjunctive therapy for osteomyelitis. *Infect Dis Clin North Am*, **4**, 433–40.

Mah, E.T., Lequesne, G.W., et al. 1994. Ultrasonic features of acute osteomyelitis in children. *J Bone Joint Surg [Br]*, **76B**, 969–74.

Malchau, H., Herberts, P. and Ahnfelt, L. 1993. Prognosis of total hip replacement in Sweden. Follow-up of 92,675 operations performed between 1978–1990. *Acta Orthop Scand*, **64**, 497–506.

Malizos, K.N., Makris, C.A. and Soucacos, P.N. 1997. Total knee arthroplasties infected by *Brucella melitensis*: a case report. *Am J Orthop*, **26**, 4, 283–5.

Mandal, S., Berendt, A.R. and Peacock, S.J. 2002. *Staphylococcus aureus* bone and joint infection. *J Infect*, **44**, 143–51.

Mariani, B.D., Martin, D.S., et al. 1996. The Coventry Award. Polymerase chain reaction detection of bacterial infection in total knee arthroplasty. *Clin Orthop*, **331**, 11–22.

Masri, B.A., Kendall, R.W., et al. 1994. Two-stage exchange arthroplasty using a functional antibiotic-loaded spacer in the treatment of the infected knee replacement: the Vancouver experience. *Semin Arthroplasty*, **5**, 126–36.

Massey, R.C., Kantzanou, M.N., et al. 2001a. Fibronectin-binding protein A of *Staphylococcus aureus* has multiple, substituting, binding regions that mediate adherence to fibronectin and invasion of endothelial cells. *Cell Microbiol*, **3**, 839–51.

Massey, R.C., Buckling, A. and Peacock, S.J. 2001b. Phenotypic switching of antibiotic resistance circumvents permanent costs in *Staphylococcus aureus*. *Curr Biol*, **11**, 1810–14.

McDonald, D.J., Fitzgerald, R.H. Jr and Ilstrup, D.M. 1989. Two-stage reconstruction of a total hip arthroplasty because of infection. *J Bone Joint Surg Am*, **71**, 828–34.

McGill, P.E. 2003. Geographically specific infections and arthritis, including rheumatic syndromes associated with certain fungi and parasites, *Brucella* species and *Mycobacterium leprae*. *Best Pract Res Clin Rheumatol*, **17**, 289–307.

McNally, M.A., Small, J.O., et al. 1993. Two-stage management of chronic osteomyelitis of the long bones. The Belfast technique. *J Bone Joint Surg Br*, **75**, 375–80.

McPherson, E.J., Patzakis, M.J., et al. 1997. Infected total knee arthroplasty. Two-stage reimplantation with a gastrocnemius rotational flap. *Clin Orthop*, **341**, 73–81.

Meek, R.M., Masri, B.A., et al. 2003. Patient satisfaction and functional status after treatment of infection at the site of a total knee arthroplasty with use of the PROSTALAC articulating spacer. *J Bone Joint Surg Am*, **85-A**, 1888–92.

Meghji, S., Crean, S.J., et al. 1997. *Staphylococcus epidermidis* produces a cell-associated proteinaceous fraction which causes bone resorption by a prostanoid-independent mechanism: relevance to the treatment of infected orthopaedic implants. *Br J Rheumatol*, **36**, 957–63.

Mereghetti, L., Marquet-Van Der Mee, N., et al. 1998. *Listeria monocytogenes* septic arthritis in a natural joint: report of a case and review. *Clin Microbiol Infect*, **4**, 3, 165–8.

Mirra, J.M., Marder, R.A. and Amstutz, H.C. 1982. The pathology of failed joint arthroplasty. *Clin Orthop*, **170**, 504–46.

Mont, M.A., Waldman, B., et al. 1997. Multiple irrigation, debridement, and retention of components in infected total knee arthroplasty. *J Arthroplasty*, **12**, 426–33.

Monzon, M., Garcia-Alvarez, F., et al. 2002. Evaluation of four experimental osteomyelitis infection models by using precolonized implants and bacterial suspensions. *Acta Orthop Scand*, **73**, 11–19.

Murray, H.J., Young, M.J., et al. 1996. The association between callus formation, high pressure and neuropathy in diabetic foot ulceration. *Diabetic Med*, **13**, 979–82.

Museru, L.M. and Mcharo, C.N. 2001. Chronic osteomyelitis: a continuing orthopaedic challenge in developing countries. *Int Orthop*, **25**, 127–31.

Nair, S.P., Williams, R.J. and Henderson, B. 2000. Advances in our understanding of the bone and joint pathology caused by *Staphylococcus aureus* infection. *Rheumatology*, **39**, 821–34.

Nazarian, D.G., De Jesus, D., et al. 2003. A two-stage approach to primary knee arthroplasty in the infected arthritic knee. *J Arthroplasty*, **18**, Suppl 1, 16–21.

Niall, D.M., Murphy, P.G., et al. 1997. Puncture wound related pseudomonas infections of the foot in children. *Ir J Med Sci*, **166**, 98–101.

Nilsson, I.M., Patti, J.M., et al. 1998. Vaccination with a recombinant fragment of collagen adhesin provides protection against *Staphylococcus aureus*-mediated septic death. *J Clin Invest*, **101**, 12, 2640–9.

Norden, C. 1985. Prevention of bone and jont infections. *Am J Med*, **78**, 229–31.

Norden, C.W. 1988. Lessons learned from animal models of osteomyelitis. *Rev Infect Dis*, **10**, 103–10.

Novick, R.P. 2003. Autoinduction and signal transduction in the regulation of staphylococcal virulence. *Mol Microbiol*, **48**, 6, 1429–1449.

Novick, R.P., Ross, H.F., et al. 1993. Synthesis of staphylococcal virulence factors is controlled by a regulatory RNA molecule. *EMBO J*, **12**, 3967–75.

Nowinski, R.J. 2001. Iguana-transmitted *Salmonella* osteomyelitis. *Orthopedics*, **24**, 7, 694.

Nowinski, R.J. and Albert, M.C. 2000. *Salmonella* osteomyelitis secondary to iguana exposure. *Clin Orthop*, **37**, 2, 250–3.

Odio, C.M., Ramirez, T., et al. 2003. Double blind, randomized, placebo-controlled study of dexamethasone therapy for hematogenous septic arthritis in children. *Pediatr Infect Dis J*, **22**, 883–8.

Orti, A., Roig, P., et al. 1997. Brucellar prosthetic arthritis in a total knee replacement. *Eur J Clin Microbiol Infect Dis*, **16**, 11, 843–845.

Ostrum, R.F. 1993. *Nocardia* septic arthritis of the hip with associated avascular necrosis. A case report. *Clin Orthop*, **288**, 282–6.

Palestro, C.J., Caprioli, R., et al. 2003. Rapid diagnosis of pedal osteomyelitis in diabetics with a technetium-99m-labeled monoclonal antigranulocyte antibody. *J Foot Ankle Surg*, **42**, 2–8.

Panday, R., Quinn, J., et al. 1996. Arthroplasty implant biomaterial particle associated macrophages differentiate into lacunar bone resorbing cells. *Ann Rheum Dis*, **55**, 388–95.

Panday, R., Berendt, A.R. and Athanasou, N.A. 2000. Histological and microbiological findings in non-infected and infected revision arthroplasty tissues. The OSIRIS Collaborative Study Group. Oxford Skeletal Infection Research and Intervention Service. *Arch Orthop Trauma Surg*, **120**, 570–4.

Park, P.W., Roberts, D.D., et al. 1991. Binding of elastin to *Staphylococcus aureus*. *J Biol Chem*, **266**, 34, 23399–406.

Park, P.W., Rosenbloom, J., et al. 1996. Molecular cloning and expression of the gene for elastin-binding protein (*ebpS*) in *Staphylococcus aureus*. *J Biol Chem*, **271**, 26, 15803–9.

Parker, J.W. and Gaines, R.W. Jr. 1995. Long-term intravenous therapy with use of peripherally inserted silicone-elastomer catheters in orthopaedic patients. *J Bone Joint Surg Am*, **77**, 572–7.

Patti, J.M., Jonsson, H., et al. 1992. Molecular characterization and expression of a gene encoding a *Staphylococcus aureus* collagen adhesin. *J Biol Chem*, **267**, 4766–72, (erratum in: *J Biol Chem* 1994: **269**, 11672).

Patti, J.M., Bremell, T., et al. 1994. The *Staphylococcus aureus* collagen adhesin is a virulence determinant in experimental septic arthritis. *Infect Immun*, **62**, 1, 152–61.

Peacock, S.J., Foster, T.J., et al. 1999. Bacterial fibronectin-binding proteins and endothelial cell surface fibronectin mediate adherence of *Staphylococcus aureus* to resting human endothelial cells. *Microbiology*, **145**, 3477–86.

Peacock, S.J., Moore, C.E., et al. 2002. Virulent combinations of adhesin and toxin genes in natural populations of *Staphylococcus aureus*. *Infect Immun*, **70**, 4987–96.

Peltola, H., Unkila-Kallio, L. and Kallio, M.J. 1997. Simplified treatment of acute staphylococcal osteomyelitis of childhood. The Finnish Study Group. *Pediatrics*, **99**, 6, 846–50.

Pittet, D., Wyssa, B., et al. 1999. Outcome of diabetic foot infections treated conservatively. A retrospective cohort study with long-term follow-up. *Arch Intern Med*, **159**, 851–6.

Proctor, R.A., van Langevelde, P., et al. 1995. Persistent and relapsing infections associated with small-colony variants of *Staphylococcus aureus*. *Clin Infect Dis*, **20**, 95–102.

Rasool, M.N. 2001. Primary subacute haematogenous osteomyelitis in children. *J Bone Joint Surg Br*, **83**, 93–8.

Raut, V.V., Siney, P.D. and Wroblewski, B.M. 1995. One-stage revision of total hip arthroplasty for deep infection. *Clin Orthop*, **321**, 202–7.

Rissing, J.P. 1990. Animal models of osteomyelitis. Knowledge, hypothesis, and speculation. *Infect Dis Clin North Am*, **4**, 3, 377–390.

Roeder, B., Van Gils, C.C. and Maling, S. 2000. Antibiotic beads in the treatment of diabetic pedal osteomyelitis. *J Foot Ankle Surg*, **39**, 124–30.

Ross, J.J., Saltzman, C.L., et al. 2003. Pneumococcal septic arthritis: review of 190 cases. *Clin Infect Dis*, **36**, 3, 318–27.

Saleh, K., Olson, M., et al. 2002. Predictors of wound infection in hip and knee joint replacement: results from a 20 year surveillance program. *J Orthop Res*, **20**, 506–15.

Sankaran-Kutty, M., Marwah, S. and Kutty, M.K. 1991. The skeletal manifestations of brucellosis. *Int Orthop*, **15**, 1, 17–19.

Santiago Restrepo, C., Gimenez, C.R. and McCarthy, K. 2003. Imaging of osteomyelitis and musculoskeletal soft tissue infections: current concepts. *Rheum Dis Clin North Am*, **29**, 89–109.

Santschi, E.M. and McGarvey, L. 2003. In vitro elution of gentamicin from Plaster of Paris beads. *Vet Surg*, **32**, 2, 128–33.

Schauwecker, D.J. 1992. The scintigraphic diagnosis of osteomyelitis. *Am J Roentgenol*, **158**, 9–18.

Segreti, J., Nelson, J.A. and Trenholme, G.M. 1998. Prolonged suppressive antibiotic therapy for infected orthopedic prostheses. *Clin Infect Dis*, **27**, 711–13.

Senneville, E., Yazdanpanah, Y., et al. 2001. Rifampicin-ofloxacin oral regimen for the treatment of mild to moderate diabetic foot osteomyelitis. *J Antimicrob Chemother*, **48**, 927–30.

Shapero, C. and Fox, I.M. 1999. *Pasteurella multocida* and gout in the first metatarsophalangeal joint. *J Am Podiatr Med Assoc*, **89**, 318–20.

Shirtliff, M.E., Calhoun, J.H. and Mader, J.T. 2002. Experimental osteomyelitis treatment with antibiotic-impregnated hydroxyapatite. *Clin Orthop*, **401**, 239–47.

Shuford, J.A. and Steckelberg, J.M. 2003. Role of oral antimicrobial therapy in the management of osteomyelitis. *Curr Opin Infect Dis*, **16**, 515–19.

Smeltzer, M.S., Gillaspy, A.F., et al. 1997a. Prevalence and chromosomal map location of *Staphylococcus aureus* adhesin genes. *Gene*, **196**, 1-2, 249–59.

Smeltzer, M.S., Thomas, J.R., et al. 1997b. Characterization of a rabbit model of staphylococcal osteomyelitis. *J Orthop Res*, **15**, 414–21.

Stengel, D., Bauwens, K., et al. 2001. Systematic review and meta-analysis of antibiotic therapy for bone and joint infections. *Lancet Infect Dis*, **1**, 175–88.

Stone, J.A. and Cianci, P. 1997. The adjunctive role of hyperbaric oxygen in the treatment of lower extremity wounds in patients with diabetes. *Diab Spectrum*, **10**, 118–23.

Su, J.Y., Huang, T.L. and Lin, S.Y. 1996. Total knee arthroplasty in tuberculous arthritis. *Clin Orthop*, **323**, 181–7.

Sugarman, B. 1987. Pressure sores and underlying bone infection. *Arch Intern Med*, **147**, 553–5.

Switalski, L.M., Patti, J.M., et al. 1993. A collagen receptor on *Staphylococcus aureus* strains isolated from patients with septic arthritis mediates adhesion to cartilage. *Mol Microbiol*, **7**, 99–107.

Takwale, V.J., Wright, E.D., et al. 1997. *Pasteurella multocida* infection of a total hip arthroplasty following cat scratch. *J Infect*, **34**, 3, 263–264.

Tarkowski, A., Bokarewa, M., et al. 2002. Current status of pathogenetic mechanisms in staphylococcal arthritis. *FEMS Microbiol Lett*, **217**, 125–32.

Tattevin, P., Cremieux, A.-C., et al. 1999. Prosthetic joint infection: when can prosthesis salvage be considered? *Clin Infect Dis*, **29**, 292–5.

Tegmark, K., Morfeldt, E. and Arvidson, S. 1998. Regulation of *agr*-dependent virulence genes in *Staphylococcus aureus* by RNAIII from coagulase-negative staphylococci. *J Bacteriol*, **180**, 12, 3181–6.

Thomas, M.G., Peacock, S., et al. 1999. Adhesion of *Staphylococcus aureus* to collagen is not a major virulence determinant for septic arthritis, osteomyelitis, or endocarditis. *J Infect Dis*, **179**, 291–3.

Tice, A.D., Hoaglund, P.A. and Shoultz, D.A. 2003. Outcomes of osteomyelitis among patients treated with outpatient parenteral antimicrobial therapy. *Am J Med*, **114**, 723–8.

Tunney, M.M., Patrick, S., et al. 1999. Detection of prosthetic hip infection at revision arthroplasty by immunofluorescence microscopy and PCR amplification of the bacterial 16S rRNA gene. *J Clin Micro*, **37**, 3281–90.

Turpin, S. and Lambert, R. 2001. Role of scintigraphy in musculoskeletal and spinal infections. *Radiol Clin North Am*, **39**, 169–89.

Van Wamel, W.J., van Rossum, G., et al. 1998. Cloning and characterization of an accessory gene regulator (*agr*)-like locus from *Staphylococcus epidermidis*. *FEMS Microbiol Lett*, **163**, 1–9.

Venkatesan, P., Lawn, S., et al. 1997. Conservative management of osteomyelitis in the feet of diabetic patients. *Diabetic Med*, **14**, 487–490.

Verdrengh, M., Springer, T.A., et al. 1996. Role of intercellular adhesion molecule 1 in pathogenesis of staphylococcal arthritis and in host defense against staphylococcal bacteremia. *Infect Immun*, **64**, 2804–7.

Vinod, M.B., Matussek, J., et al. 2002. Duration of antibiotics in children with osteomyelitis and septic arthritis. *J Paediatr Child Health*, **38**, 363–7.

von Eiff, C., Bettin, D., et al. 1997. Recovery of small colony variants of *Staphylococcus aureus* following gentamicin bead placement for osteomyelitis. *Clin Infect Dis*, **25**, 1250–1.

von Graevenitz, A., Frommelt, L., et al. 1998. Diversity of coryneforms found in infections following prosthetic joint insertion and open fractures. *Infection*, **26**, 1, 36–8.

Vuong, C., Gerke, C., et al. 2003. Quorum-sensing control of biofilm factors in *Staphylococcus epidermidis*. *J Infect Dis*, **188**, 706–18.

Walenkamp, G.H., Kleijn, L.L. and de Leeuw, M. 1998. Osteomyelitis treated with gentamicin-PMMA beads: 100 patients followed for 1–12 years. *Acta Orthop Scand*, **69**, 518–22.

Wang, J.W. 1997. Uncemented total arthroplasty in old quiescent infection of the hip. *J Formos Med Assoc*, **96**, 634–40.

White, L.M., Schweitzer, M.E., et al. 1995. Study of osteomyelitis: utility of combined histologic and microbiologic evaluation of percutaneous biopsy samples. *Radiology*, **197**, 840–2.

Widmer, A.F., Gaechter, A., et al. 1992. Antimicrobial treatment of orthopaedic implant-related infections with rifampin combinations. *Clin Infect Dis*, **14**, 1251–3.

Wilkerson, R.D., Taylor, D.C., Opal, S.M. and Curl, W.W. 1985. *Nocardia asteroides* sepsis of the knee. *Clin Orthop*, **197**, 206–208.

Wininger, D.A. and Fass, R.J. 1996. Antibiotic-impregnated cement and beads for orthopedic infections. *Antimicrob Agents Chemother*, **40**, 2675–9.

Wunderlich, R.P., Peters, E.J.G. and Lavery, L. 2000. Systemic hyperbaric oxygen therapy. Lower-extremity wound healing and the diabetic foot. *Diabetes Care*, **23**, 1551–5.

Yacoub, A., Lindahl, P., et al. 1994. Purification of a bone sialoprotein-binding protein from *Staphylococcus aureus*. *Eur J Biochem*, **222**, 3, 919–25.

Yamashita, Y., Uchida, A., et al. 1998. Treatment of chronic osteomyelitis using calcium hydroxyapatite ceramic implants impregnated with antibiotic. *Int Orthop*, **22**, 247–51.

Younger, A.S., Duncan, C.P., et al. 1997. The outcome of two-stage arthroplasty using a custom-made interval spacer to treat the infected hip. *J Arthroplasty*, **12**, 615–23.

Zimmerli, W., Widmer, A.F., et al. 1998. Role of rifampin for treatment of orthopedic implant-related staphylococcal infections: a randomised controlled trial. Foreign-Body infection (FBI) study group. *JAMA*, **279**, 1537–41.

Ziran, B.H., Rao, N. and Hall, R.A. 2003. A dedicated team approach enhances outcomes of osteomyelitis treatment. *Clin Orthop*, **41**, 4, 31–6.

PART V

LABORATORY ASPECTS

28

Conventional laboratory diagnosis of infection

JOSEPH D.C. YAO

INTRODUCTION

Diagnosis of a microbial infection usually begins with an assessment of the clinical and epidemiologic features, leading to formation of a diagnostic hypothesis. Physical and radiological findings provide clues in determining the anatomic site of infection. This clinical diagnosis suggests a number of possible etiologic agents causing the infection, based on knowledge of infectious syndromes and their usual clinical courses. By applying various diagnostic microbiologic tests on clinical specimens, a specific microbial etiology can be established. Therefore, clear communication between the clinician and the clinical microbiology laboratory is essential to obtain optimal test results for arriving at an accurate microbiologic diagnosis. The clinician must select the appropriate tests and specimens to be processed and suggest the suspected microbial pathogens to the laboratory. The clinical microbiology laboratory must utilize a series of tests that would demonstrate the probable etiologic agents and be prepared to explore other possible agents suggested by the clinical features or findings of the laboratory examinations.

This chapter provides an overview of the appropriate use of the clinical microbiology laboratory and the various common conventional laboratory methods used for establishing the microbial causes of infection. Molecular laboratory methods are discussed separately in Chapter 29, Molecular laboratory diagnosis of infection, while additional specialized tests for identifying specific pathogens are described in the chapters devoted to specific microorganisms. Detailed descriptions of the

principles and techniques of these diagnostic laboratory methods are available in several standard textbooks on this subject (Forbes et al. 2002; Murray et al. 2002).

CLINICAL SPECIMENS

Because the objective of most microbiologic laboratory tests is to isolate viable organisms, proper specimens must be collected and delivered promptly in a suitable transport system for processing in the laboratory. The results of these laboratory tests are limited by the quality of specimens and their condition on arrival in the laboratory. Errors in specimen collection and transportation are the most common reasons for failure to ascertain etiologic diagnoses of infections. Therefore, proper collection and transport of clinical specimens from the patient's bedside to the laboratory are the critical first steps in establishing a microbiologic diagnosis.

Collection

Specimens should be obtained from the patient with care to minimize the possibility of introducing extraneous or contaminating microorganisms that are not involved in the infectious process. This precaution is needed especially to distinguish resident (commensal) or 'normal' organisms from those causing the infection. The presence of commensal or normal flora at the anatomic site of infection will mask the true etiological agent, making its isolation more difficult. Of particular

Table 28.1 *Sites of infection and common sources of contamination*

Site of infection	Source of contamination
Middle ear	External ear canal
Nasal sinus	Nasopharynx
Lower respiratory tract	Oropharynx
Endometrium	Vagina
Superficial wounds/ subcutaneous infections	Skin and mucous membranes
Abdominal fistulae	Gastrointestinal tract
Bladder	Urethra and external genitalia

difficulty are specimens from sites listed in Table 28.1, which are adjacent to body sites having more resident flora than others. For example, expectorated sputum specimens are frequently contaminated during collection by organisms that are part of the indigenous or normal flora of the oropharynx. Knowledge of the potential contaminating organisms and the probable pathogens to be sought in these 'indirect' specimens is essential to avoid misinterpretation of results. Guidelines to assess the quality of such 'indirect' specimens have been developed to determine the suitability of a specimen for processing. Such an example is the use of microscopic examination of direct Gram-stained smear of expectorated sputum to indicate the high probability of contaminating flora when >10 squamous epithelial cells are present per low-power field (10× magnification).

Use of special techniques or procedures that bypass anatomic areas containing normal flora whenever feasible (e.g. protected brush bronchoscopy for diagnosis of ventilator-associated pneumonia) can prevent false-positive results. Careful skin antisepsis before collection procedures, such as blood cultures and lumbar punctures, will decrease the risk of false-positive findings from contamination by organisms normally present on the skin. Whenever possible, specimens should be collected directly from normally sterile tissues (e.g. lung, liver, bone marrow) and body fluids (e.g. blood, CSF, synovial fluid) by methods ranging from needle biopsy or aspiration to surgical biopsy. Positive findings obtained from these specimens are almost always diagnostic.

Though convenient and most commonly used for specimen collection, the sterile swab absorbs a small volume of specimen material and provides the poorest conditions for microbial survival. If available, infected tissue or needle aspirates of infected fluid should be submitted for microbiologic cultures. An adequate volume of specimen is important for optimal test results because infecting organisms present in small quantities may not be detected in a small sample. Specimens should be collected during the early (acute) phase of an infection and before initiation of antimicrobial therapy, whenever possible.

Transport and handling

Ideally, specimens should be transported to the laboratory as soon as possible, preferably within 30 min of collection. Some microorganisms can only survive briefly outside the body (e.g. *N. gonorrhoeae*) and are susceptible to environmental conditions, such as presence of oxygen (anaerobic bacteria), changes in temperature (*N. meningitidis*), or changes in pH (*Shigella*). To minimize the effects of delay between specimen collection and laboratory processing, various transport media have been developed for specimens that could not reach the laboratory in less than 30 min. These media are usually buffered fluids or semisolid media that contain minimal nutrients and special preservatives to prevent drying, maintain a neutral pH, and minimize bacterial growth during transport. Specialized media providing oxygen-free atmosphere are needed for transporting obligate anaerobes. Detailed comprehensive lists of optimal specimen selection and transport requirements are available in several authoritative references (Miller et al. 2003; Murray 1998).

When received in the laboratory, clinical specimens are processed with priority given to those that are most critical, such as blood, cerebrospinal fluid (CSF), sterile body fluids, and tissue. After macroscopic examination of each specimen to ensure appropriateness and adequate volume for the tests requested, some specimens may need to undergo initial processing by procedures before inoculation on to culture media. These procedures include homogenization (grinding) of tissue, concentration by centrifugation or filtration of large volumes of sterile fluids (ascites, pleural fluid), or decontamination to remove commensal flora that may interfere with recovery of fastidious pathogenic microbes (e.g. mycobacteria, *Legionella*).

DIAGNOSTIC METHODS

The general approaches to laboratory diagnosis of infectious diseases vary with different infective pathogens and infections. The types of conventional diagnostic methods employed for testing clinical specimens are usually a combination of direct examination, culture, microbial antigen detection, and detection of patient's antibody response (serology). In recent years, molecular laboratory methods (see Chapter 29, Molecular laboratory diagnosis of infection) capable of directly detecting microbial genomic components have gained increasing importance and popularity for the sensitive and rapid detection of fastidious pathogens that are otherwise difficult to detect by conventional approaches.

Direct examination

Other than some of the parasites, none of the microbial pathogens present in clinical specimens are large enough

Table 28.2 *Utility of microscopic methods for diagnostic microbiology*

Microscopic methods	Bacteria	Fungi	Viruses	Parasites
Bright-field (light) microscopy	+	+	−	+
Dark-field microscopy	±	−	−	−
Phase-contrast microscopy	−	±	−	±
Fluorescent microscopy	+	+	+	+
Electron microscopy	−	−	±	±

+, commonly used; ±, limited use; −, rarely used.
Adapted from Forbes et al. (2002).

to be visible with the naked eye. Therefore, microscopic examination of clinical specimens is necessary for the initial detection and preliminary or definitive identification of bacterial cells in infected cells. In diagnostic microbiology, five microscopic methods are used with varying utility for each of the four major microbial pathogen groups (Table 28.2). To better visualize and differentiate these microbes in smears of clinical specimens and histologic sections of tissues, various stains are used commonly in the diagnostic laboratory. As shown in Table 28.3, these staining techniques are adapted for

specific microscopic methods in detecting and identifying certain microorganisms.

LIGHT MICROSCOPY

Direct examination of unstained or stained preparation of clinical specimens by bright-field (light) microscopy is used commonly for detection of bacteria, fungi, and parasites. Due to the minute sizes of bacteria (usually 0.3–0.5 µm wide), oil immersion magnification (500× to 1 000×) is necessary to visualize these organisms. Unstained bacteria are too transparent to be seen directly, but two important staining methods, Gram stain and acid-fast stains, are frequently employed to stain and classify bacteria by light microscopy.

Gram stain

As the principal stain used for microscopic examination of bacteria, the Gram stain divides most bacteria into two groups: gram-positive (purple-colored) when the alkaline dye, crystal violet, is retained by the organism, and gram-negative (pink-colored) when the crystal violet dye is lost with uptake of counterstaining safranin dye. Due to abundant peptidoglycan and teichoic acid cross-linkages in their cell wall, gram-positive bacteria are able to retain the crystal violet dye which is chemically bonded to the Gram's iodine reagent, while resisting alcohol or acetone decolorization in the subsequent

Table 28.3 *Microscopic preparations and staining methods commonly used for examination of clinical specimens in diagnostic microbiology*

Staining method	Detectable microorganisms
Direct examination	
Wet mount	Parasites and ova (examined by bright-field, dark-field, or phase-contrast microscopy)
10% KOH	Fungal hyphae
Lugol's iodine	*Entamoeba*, other amoeba
Differential stains	
Gram stain	Bacteria, yeast cells
Iron hematoxylin stain	Fecal protozoa
Toluidine blue O stain	*Pneumocystis jiroveci*
Trichrome stain	Fecal protozoa
Wright–Giemsa stain	Blood parasites (filaria, *Plasmodium*, *Babesia*), *Borrelia*, *Rickettsia*, *Ehrlichia*, *Anaplasma*, *Toxoplasma*, *Leishmania*, *Pneumocystis*, viral and chlamydial inclusion bodies
Acid-fast stains	
Ziehl–Neelsen stain	*Mycobacterium* spp.
Kinyoun stain	*Mycobacterium* spp.
Modified acid-fast stain	*Nocardia*, *Rhodococcus*, *Gordonia*, *Tsukamurella*, *Cryptosporidium*, *Isospora*, *Sarcocystis*, *Cyclospora*
Fluorescent stains	
Acridine orange stain	Bacteria cells, fungi elements
Auramine-rhodamine stain	*Mycobacterium* spp.
Calcofluor white stain	Fungal elements, *Pneumocystis*
Direct fluorescent antibody stain	*Chlamydia*, *Bordetella*, *Francisella*, *Legionella*, *Neisseria*, *S. pyogenes*, *Cryptosporidium*, *Giardia*, RSV, influenza virus, HSV, VZV

Adapted from Murray et al. (2002).

staining procedure. Gram-negative bacteria, having a thinner cell wall layer of peptidoglycan, lose the crystal violet dye easily during the decolorization process and appear pink in color from uptake of the safranin counterstain dye. Gram-positive bacteria without an intact cell wall because of cell damage, old age, or antibiotic therapy, may fail to retain the crystal violet dye during decolorization and appear 'gram-variable,' with a mixture of purple- and pink-staining cells.

Acid-fast stain

The acid-fast stain is used specifically for direct light microscopic detection of mycobacteria and related bacteria in clinical specimens. The presence of mycolic acids in their cell wall renders this group of bacteria 'acid-fast' (resistant to acid alcohol decolorization) and difficult to stain with dyes, including Gram stain. However, they can be stained by prolonged application of highly concentrated dyes, with or without the use of heat treatment to facilitate the staining process. Once stained, acid-fast bacteria resist acid alcohol decolorization which removes the same dyes from other bacteria. Ziehl–Neelsen stain is the classic acid-fast stain with heated carbol-fuchsin (red) applied as the primary stain, followed by decolorization with 3 percent HCl–alcohol and counterstaining. A modified method, also known as the Kinyoun acid-fast stain, involves using carbol-fuchsin containing a higher concentration of phenol to facilitate penetration of the primary dye without use of heat.

DARK-FIELD MICROSCOPY

Using special optical features of dark-field microscopy, bacteria that are usually too thin to be visualized by light microscopy can be seen as objects surrounded by a bright halo amid a dark background. This method is used to detect spirochetes, such as *Treponema pallidum* (syphilis) and *Borrelia* spp. in clinical specimens.

PHASE-CONTRAST MICROSCOPY

Phase-contrast is an optical technique that yields visible contrast among microbial cells and cell structures in a given specimen. Objects with different thickness or densities have different refractive indices that produce different light intensities (phases) in an image, allowing better visualization of cell structures in unstained specimens. Whereas microscopic examination of stained specimens permits observation of dead organisms, phase-contrast microscopy offers the advantage of observing viable organisms of interest. This method is used mainly for the identification of fungi and parasites.

FLUORESCENT MICROSCOPY

Using similar optical features as dark-field microscopy, fluorescent microscopy is used to detect microorganisms that are labeled by a fluorescent dye (e.g. acridine orange, calcofluor white, auramine-rhodamine) or by microbial antigen-specific antibody conjugated with a fluorescent dye (immunofluorescence). Against a dark background, the organism of interest appears with a halo emitting color of the fluorescent compound. This method is used to detect organisms that are difficult or slow to grow or to identify those already grown in culture.

ELECTRON MICROSCOPY

With its high magnification of $\geqslant 100\,000\times$, electron microscopy is used mainly for the detection and identification of viruses in clinical specimens and viral cultures. It is also a useful method for the confirmation of microsporidia (e.g. *Enterocytozoon*, *Encephalitozoon*) in clinical specimens. However, the need for specialized equipment limits the availability of this method for routine use in diagnostic microbiology.

Isolation and identification

The most definitive and specific means of diagnosis of an infection is the isolation and identification of the infecting organism in vitro from clinical specimens. To identify and characterize the organism, it must be grown and isolated in the clinical laboratory. Although most bacteria can be cultivated by various artificial media, strict intracellular organisms such as *Chlamydia*, *Rickettsia*, and *Ehrlichia* can be isolated only from cultures of living eukaryotic cells (cell cultures).

Cultivation of bacteria in the laboratory involves inoculating clinical specimens obtained from the infection site into culture media and growing the microorganisms of interest in vitro. Once they are grown in culture, the organisms are present in sufficient amounts to allow further diagnostic laboratory testing to be carried out. Their survival depends on the availability of essential nutrients and appropriate environmental conditions in vitro. The nutritional needs include different gases, water, ions, sources of carbon and nitrogen, and energy. For certain bacteria, the needs are complex, so that exceptional nutrient components must be added in the culture media to support their growth. Pathogens with such growth requirements are said to be fastidious. To date, the growth requirements of certain clinically relevant bacteria, such as *T. pallidum* and *M. leprae*, remain poorly understood such that these pathogens cannot be grown in vitro.

BACTERIOLOGIC CULTURE

Various artificial culture media are employed in the diagnostic laboratory to facilitate the growth of medically important bacteria to be detectable by the naked eye. These media are essentially recipes prepared from digests of animal or plant protein supplemented with nutrients such as glucose, yeast extract, serum, or blood, to meet the metabolic requirements of the organisms (Atlas 1997; Chapin and Lauderdale 2003; Chapin and

Murray 2003). Bacteriologic media can be prepared in solid (agar), semisolid, and liquid (broth) forms to which bacteria or clinical specimens may be added directly. On solid agar plates, bacteria grow as discrete sediment (colonies) or as a film on the agar surface, while their growth is not visible in semisolid and broth media until there are 10^6 to 10^7 bacteria per milliliter of media to produce turbidity or macroscopic clumps.

Compared to broth cultures, solid media offer advantages for diagnostic work, since discrete colonies of bacteria in pure culture grow well separated from one another on the agar surface and arise from a single organism or an organism cluster (colony-forming unit). Bacterial colonies differ significantly among different genera or species, varying greatly in size, shape, color, texture, and other characteristics. For example, colonies of *S. pneumoniae* possessing large polysaccharide capsules are usually mucoid in appearance, while those of organisms that fail to separate after division are frequently granular. These features are usually consistent for colonies derived from the same strain of organism. Therefore, differences in colonial morphology are very useful as clues to bacterial identity and for separating them in mixtures. Broth media offer the advantage of providing a favorable homogeneous environment for enhancing bacterial growth in recovering organisms that are present in minute quantities in clinical specimens. However, broth cultures of clinical specimens frequently yield mixtures of bacteria that are not readily separated for identification.

Bacteriology media may be classified as nutrient, enrichment, selective, or differential media according to their main function in the laboratory. Nutrient media, also known as supportive media, are designed to provide the growth requirements of most nonfastidious bacteria without giving growth advantage to any particular organism. Mueller–Hinton medium and brain–heart infusion in agar or broth are such examples. Enrichment media contain specific nutrients (e.g. trace metals, coenzymes, vitamins) necessary for the growth of certain bacteria. They are used to enhance the growth of a particular bacterial pathogen from a mixture of organisms on the basis of nutrient specificity. Typical examples are thioglycolate broth, chocolate agar (*Neisseria* and *Haemophilus* species), and buffered charcoal-yeast extract agar (*Legionella* spp.). Selective media contain dyes, bile salts, acids, alcohols, or antimicrobial agents that are inhibitory to all organisms except those being sought. Selecting for the growth of certain bacteria to the disadvantage of others, these media are useful for isolating specific pathogenic organisms from clinical specimens containing extensive contaminating flora. *Campylobacter*–blood agar, mannitol salt agar or broth (staphylococci), and phenylethyl alcohol agar (gram-positive cocci) are commonly used selective media. Differential media contain substances that allow colonies of specific pathogens or organism groups to exhibit

certain biochemical or other culture characteristics distinguishable from other organisms growing in the same media. Such culture features may be based on changes in pH as a result of carbohydrate metabolism, enhancement of pigment production, or hemolysis of red blood cells present in the media. Examples of such media are bile esculin agar (enterococci), MacConkey sorbitol agar (*E. coli* O157:H7), and sheep blood agar (streptococci). Some media have multiple functions, such as enriched and selective (Thayer–Martin agar for *N. gonorrhoeae* and *N. meningitidis*, Columbia colistin-nalidixic acid with 5 percent blood for gram-positive cocci), enriched and differential (sheep blood agar for hemolytic streptococci), selective and differential (eosin methylene blue agar and MacConkey agar for gram-negative bacilli, Hektoen enteric agar for *Salmonella* and *Shigella* spp., thiosulfate citrate-bile salts agar for *Vibrios* spp.).

BACTERIAL IDENTIFICATION

In addition to Gram-stained smears, conventional methods of bacterial identification involve examination of the cultural characteristics, as well as biochemical and physiologic testing on pure cultures obtained from single colonies. The exact tests and the sequence of testing depend on the suspected group of bacteria and the taxonomic level of identification (e.g. genus, species, subspecies) needed for diagnosis and treatment of the infection. Important properties of bacterial cultures that aid in identification include unique nutritional requirements, pigment production, and the ability to grow in the presence of certain substances (e.g. bile, sodium chloride) or on certain selective or differential media (e.g. MacConkey agar, Hektoen enteric agar), growth at particular temperature, and presence and type of hemolysis (alpha, beta, gamma) on blood agar.

The ability to react with certain chemical substrates and the production of particular metabolites form the basis of biochemical tests for bacterial identification. Such tests include utilization of carbohydrates (e.g. sucrose, glucose, lactose) and citrate, ability to metabolize certain amino acids (e.g. lysine, ornithine, arginine), oxidase reaction, and production of catalase, coagulase, proteinase, urease, indole, and hydrogen sulfide (Chapin and Lauderdale 2003). The results of these tests are analyzed by reference to published algorithms for identification of many bacteria (Ruoff 2003; Funke 2003; Schreckenberger and Wong 2003; Citron 2003; Versalovic 2003).

Bacteria may also be identified by the presence of their antigenic structures, such as capsular polysaccharides, flagellar proteins, and cell wall components. In addition, in vitro detection of specific bacterial toxins, with neutralization of the toxic effect by specific antitoxin, may be necessary to confirm the identity of pathogenic bacteria (e.g. Elek immunodiffusion test for

Corynebacterium diphtheriae toxin) or a clinical diagnosis (e.g. *Clostridium difficile* toxin). Serologic techniques using bacteria-specific antibodies can detect these antigens present on whole bacteria or free in bacterial lysates (soluble antigens). Further details of these methods are discussed in a later section.

Blood cultures

The primary means for establishing diagnosis of bloodstream infections due to bacteria and fungi is by blood culture, which is based on the same microbiologic principles of any culture. A sample of patient's blood is obtained by aseptic venipuncture and inoculated into an enriched broth or agar plate. Once growth is detected by manual or automated methods, the organisms are isolated, identified, and tested for antimicrobial susceptibility. Careful attention must be paid to details of obtaining blood from the patient in order to maximize the yield of blood culture and properly interpret the results.

Proper antiseptic techniques should be used to prepare the skin over the vein for phlebotomy, so that contamination of the blood culture by skin microbial flora is minimized. Blood should not be drawn through indwelling venous or arterial catheters unless it cannot be obtained by venipuncture, or if it is for detection of intravascular catheter-associated sepsis. Many commercially available blood culture media contain an anticoagulant, sodium polyanethol sulfonate (SPS), which is free of antimicrobial properties and has anticomplementary and antiphagocytic activities. SPS also interferes with the activity of some antibiotics, notably aminoglycosides, that are present in the blood of patients receiving antibiotic therapy. Blood cultures obtained from patients after initiation of antimicrobial therapy are not useful unless none was collected before starting therapy or the clinical course of the infection suggests superinfection with a drug-resistant or other microorganism.

Since the number of organisms present in blood is often low (<10 organisms/ml), the yield of blood cultures in a given patient is also dependent on the amount of blood and the number of blood cultures obtained. The diagnostic yield increases as the blood volume drawn is increased. Collection of a minimum 5–10 and 20 ml of blood per culture is strongly recommended for pediatric and adult patients, respectively. Because microorganisms causing infection may enter the blood continuously, intermittently, or transiently, blood infections may not be detectable by a single blood culture. If the blood volume drawn per culture is adequate, two or three blood cultures collected at brief intervals (30–60 min apart) are usually sufficient to yield positive results.

Most diagnostic laboratories currently utilize automated blood culture systems, such as BacT/ALERT® 3D System (bioMerieux, Durham, NC, USA), Bactec™

9000 Systems (Becton Dickinson Diagnostic Systems, Sparks, MD, USA), and ESP® Culture System II (TREK Diagnostic Systems, Cleveland, OH, USA), for consistent quality control and optimal work efficiency. However, some pathogens, especially intracellular microorganisms such as *Histoplasma*, *Legionella*, and *Mycobacterium* spp., are not easily recovered by these automated culture systems. The lysis centrifugation blood culture method (e.g. ISOLATOR™ Microbial Tubes; Wampole Laboratories, Cranbury, NJ, USA) would be helpful for detection of bloodstream infections due to these pathogens. The ISOLATOR™ tubes contain SPS and saponin which causes lysis of blood cells, releasing intracellular pathogens into the blood. After centrifugation, the supernatant is discarded and the sediment containing the pathogens is inoculated on to solid agar plates capable of supporting the growth of the pathogens of interest. This manual blood culture method provides the following benefits: more rapid and greater recovery of the pathogens, presence of colonies for direct identification and susceptibility testing after initial incubation, and ability to quantify the number of colony-forming units present in the blood.

Results of blood cultures yielding growth of coagulase-negative staphylococci, *Corynebacterium* spp., *Propionibacterium acnes*, *Bacillus* spp. should be interpreted with caution because these organisms are part of the normal skin flora and are frequent contaminants in blood cultures. Growth of multiple species from only one of several cultures would also suggest probable contaminants, since polymicrobial bloodstream infections are uncommon. When the same isolate is recovered from multiple blood cultures obtained at different times or from different anatomic sites, bloodstream infection from this organism is probably real.

Blood cultures obtained for the diagnosis of infections due to fastidious pathogens require special processing in the laboratory, so that special procedures or culture media are used to support the growth of these organisms. Routine automated blood cultures are unlikely to recover such pathogens as *Brucella*, *Borrelia*, *Leptospira*, *Bartonella*, *Mycoplasma hominis*, *Legionella*, *Mycobacterium* spp., and the HACEK group of fastidious gram-negative bacilli. The laboratory should be alerted to process the blood cultures accordingly when one of these pathogens is suspected to cause bloodstream infection in a given patient.

Antimicrobial susceptibility testing

In vitro susceptibility testing of microbial pathogens to antimicrobial agents is an important function of the clinical microbiology laboratory. Testing guidelines, standardized procedures, and interpretative criteria are regularly updated and available from the National Committee for Clinical Laboratory Standards (NCCLS) in the USA (NCCLS, 2000, 2001, 2002a, b, 2003a, b, c, d, e) and from its counterparts in other countries (EUCAST 2000, 2003). The goal of susceptibility testing is to provide data on antimicrobial resistance in conjunction with other diagnostic information to optimize therapy. Of the methods available to detect and evaluate antimicrobial resistance in bacteria, the conventional methods of directly measuring the activity of the antibiotics against a given infecting isolate are the most widely employed by laboratories. These conventional testing methods include broth dilution, agar dilution, and disk diffusion.

The broth dilution susceptibility test consists of inhibiting the growth of the isolate of interest by a series of doubling broth dilutions (e.g. 0.25, 0.5, 1, 2, 4, 8, 16 µg/ml) of a given antibiotic. The concentration range tested for a particular drug depends on the following criteria: (1) the drug concentration that is safely achievable in a patient's serum (i.e. pharmacokinetic properties of the drug); and (2) the microorganism and its associated antimicrobial resistance features that the test is attempting to detect. The lowest antibiotic concentration that completely inhibits visible pathogen growth in the broth solution is known as the minimal inhibitory concentration (MIC). Broth dilution testing can be performed in two approaches with the same principles: macrodilution and microdilution. The macrodilution method involves testing the pathogen in test tubes containing antibiotic dilution in a broth volume of $\geqslant 1$ ml per tube. This approach is cumbersome and labor intensive, and it is rarely used in clinical laboratories. For microdilution testing, the total broth volume is in the range of 0.05 to 0.1 ml for each antibiotic dilution. This method is usually adapted in a convenient, single microtiter tray format with multiple sample wells for several serial dilutions of various antibiotics. Several commercially available manual and automated susceptibility testing systems in use at many laboratories are developed according to this format: MicroScan® WalkAway System (Dade International, Sacramento, CA, USA), BD PHOENIX™ Automated Microbiology System (BD Diagnostic Systems, Sparks, MD, USA), Vitek® 2 System (bioMérieux, Inc., Durham, NC, USA), Sensitive® Microbiology Systems (TREK Diagnostic Systems Inc., Cleveland, OH, USA), and PASCO MIC Panels (PASCO Laboratories, Wheat Ridge, CO, USA).

In agar dilution susceptibility testing, each dilution of a given antibiotic is incorporated into a single agar plate, on which multiple isolates can be inoculated each at a standardized inoculum of 10^4 colony-forming units (cfu). A series of six dilutions of an antibiotic would require using six antibiotic-impregnated agar plates and one positive growth control plate without antibiotic, with $\geqslant 1$ isolates tested per plate. After incubation, the plates are examined for growth, with the MIC being the lowest concentration of an antibiotic in agar that completely inhibits visible growth. The main advantage of this

method is that MICs can be determined for fastidious organisms, such as *N. gonorrhoeae* and *E. corrodens*, that do not grow well in the broth media used in the broth dilution method. However, preparation of agar dilution plates is labor intensive, so this testing method is limited to use by reference laboratories testing large numbers of isolates daily.

The disk diffusion method is a practical and convenient approach of antimicrobial susceptibility testing, in which antibiotic-impregnated filter paper disks ('antibiotic disks') are placed on the surface of an agar plate that has been inoculated with a lawn of bacteria. Each antibiotic disk contains a known concentration of antibiotic, and the drug diffuses out of the disk into the surrounding agar, creating a concentration gradient around the disk. During incubation, the bacteria grow on the surface of the agar plate, except where the antibiotic concentration gradient is sufficiently high to inhibit growth. Findings of in vitro studies are used to correlate the size (in millimeters) of the resulting zone of inhibition around each disk with the MIC of the drug obtained by broth or agar dilution method.

A commercial susceptibility testing method that combines the convenience of disk diffusion with the ability to determine MIC data is the Etest® (AB Biodisk, Solna, Sweden). It is a unique patented antimicrobial gradient diffusion technique in which a predefined gradient of 15 dilutions of an antibiotic is impregnated on to a single plastic reagent strip. The strip is then placed on to the surface of an inoculated agar plate, allowing a gradient of antibiotic to diffuse from the strip into the surrounding agar. After incubation, the resulting elliptical zone of inhibition intersects the scale of antibiotic dilution markings at the indicated MIC on the strip. Reagent strips of various antibiotic agents are available for testing both fastidious and nonfastidious bacteria and *Candida* spp. The main disadvantage of this method is the relatively high cost of the reagent strips when several antibiotic strips are needed to test a given pathogen isolate.

To interpret test results reliably, antibacterial susceptibility testing methods are standardized and carefully controlled according to the appropriate inoculum size of the organism, conditions of growth medium (pH, calcium and magnesium cation concentrations, blood and serum supplements, thymidine content), incubation atmosphere, temperature and duration, and antibiotic concentrations tested (NCCLS 2003d). Procedural standards and interpretive guidelines have been published for the broth and agar dilution methods (NCCLS 2001, 2002a, b, 2003b) and the disk diffusion testing method (NCCLS 2003a, c). Use of the interpretive guidelines enables laboratories to report susceptibility test results on a pathogen as 'susceptible,' 'intermediate,' or 'resistant' to a given antibiotic. However, in some instances, the MIC value is more valuable in directing therapy than a simple qualitative result because the clinician can determine the potential success or failure of antibiotic therapy when comparing the MIC to the antibiotic level achievable at the anatomic site of infection. Because no one method can provide adequate testing of all clinically important bacteria, most diagnostic laboratories usually employ a combination of the susceptibility testing methods described above.

Detection of microbial antigens

Immunologic techniques are frequently used to detect, identify, and quantitate antigens of microbes directly in clinical specimens or from microbiologic cultures. These techniques are based on binding interactions between antibody and specific antigen, forming antibody–antigen complexes that can be detected directly by precipitation methods or by labeling of the antibody with a radio-active, fluorescent, or enzyme probe. The complexes can also be detected indirectly by measurement of an anti-body-directed reaction, such as complement fixation.

PRECIPITATION AND IMMUNODIFFUSION METHODS

The principle of these tests is based on the formation of precipitate from interaction between soluble antigen and antibody. The precipitate is visible only because the antigen–antibody complexes formed are too large to remain soluble and therefore precipitate. At antigen-to-antibody concentration ratios above and below an equivalence concentration, the complexes are soluble. Further details are described in the section on Serologic methods below.

Immunodiffusion (ID) assays are adaptations of the precipitation technique to determine the identity of an antigen or the presence of antibody. In single radial immunodiffusion tests, antigen is placed into a well and allowed to diffuse into surrounding agar containing a specific antibody. The higher the concentration of antigen, the farther it diffuses in the agar to reach equivalence with the antibody and forms a precipitate ring around the well.

AGGLUTINATION TESTS

When relatively large particles coated with antigen or antibody are allowed to react with the specimen of interest (e.g. serum, CSF, urine), the resulting antigen–antibody reactions are detectable visually by agglutina-tion. Soluble antigen or antibody can be fixed on to the surface of red blood cells or microscopic latex particles, and the assays are known as passive hemagglutination and latex agglutination, respectively. Bacterial antigens present on surfaces of whole bacteria that are large enough to behave as particles may also agglutinate with bacteria-specific antibody (bacterial agglutination). These assays are rapid and technically simple tests for identification of bacteria isolated in pure cultures (e.g.

coagglutination test for staphylococcal surface protein A) and for detecting microbial antigens directly in clin-ical specimens.

IMMUNOFLUORESCENCE ASSAYS

Antigens present on the cell surface or within the cell can be detected by direct or indirect immuno-fluorescence (IF). In direct IF, antibody to the specific antigen is bonded directly with a fluorescent molecule, usually fluorescein isothiocyanate (FITC), whereas in indirect IF, a second antibody specific for the primary antibody is labeled and used to detect the primary anti-microbial antibody and locate the antigen. Indirect IF assays are usually more sensitive than direct IF tests, but the latter are faster, involving only one incubation step. IF tests are used to identify *Bordetella pertussis* in culture and to detect *Legionella pneumophila* in clinical specimens.

ENZYME IMMUNOASSAYS

Enzyme immunoassays (EIA), also known as enzyme-linked immunosorbent assay (ELISA), utilize antibodies conjugated to enzymes (e.g. horseradish peroxidase, alkaline phosphatase) capable of catalyzing a reaction yielding a visible end product, while the antibody-binding sites remain free to react with their specific antigen. The advantages of such assays are: ability to amplify the antigen–antibody reaction and enhance the sensitivity of detection, relatively long-term stability of the enzyme-conjugated antibodies during storage, and formation of a colored end product that allows both visible detection or automated spectrophotometric measurement. The use of monoclonal antibodies also helps to increase the specificity of these assays. Exam-ples of commercial EIA developed for the direct detec-tion of bacteria in clinical specimens include *Strepto-coccus pyogenes*, *Streptococcus agalactiae*, *Helicobacter pylori*, and *Chlamydia trachomatis*.

Serologic methods

Immunocompetent humans produce both IgM and IgG antibodies against pathogens during infection. In general, IgM is produced during the first exposure to a given pathogen and is no longer detectable after this relatively short period of time. Over time, production of IgG replaces that of IgM in the immunologic response of the exposed patients. This humoral immune response in the infected hosts is the basis of serologic diagnosis of infection. Serologic testing is commonly used to identify infections due to bacteria that are difficult to be detected by other conventional methods, to evaluate the course of an infection, and to determine the nature of the infection (primary infection versus reinfection, acute versus chronic infection). Serologic results are usually expressed as a titer, which is the inverse of the greatest dilution, or lowest concentration (e.g. dilution of 1:16 = titer of 16) of a patient's serum that retains the specific antibody–antigen reactivity. Patient's specific IgM or IgG titers can be determined separately through the use of labeled antihuman antibody specific for the antibody isotype in the particular serologic assays.

Serologic diagnosis of infection is confirmed usually by occurrence of seroconversion during a primary infection. Seroconversion is defined as a minimum four-fold increase in antibody titer between serum during the acute phase of infection and that during the convalescent phase (≥2 weeks later). Reinfection or recurrent infection later in life causes an anamnestic (secondary or booster) immune response which can be detected as increase in antibody titer.

Most of the immunologic methods for detecting and identifying microbial antigens are also applied to serologic diagnosis by reversing the detection system: using a known antigen to detect the presence of the specific antibody. These serologic methods used in diagnostic laboratories include precipitation, immunodiffusion, agglutination, hemagglutination inhibition, EIA, indirect IF, complement fixation, radioimmunoassay, and Western blot. Some of these methods have been discussed in detail in the above sections.

PRECIPITIN TESTS

As described above, these tests are based on formation of a visible precipitate when antigen and antibody are combined in proper proportions. Two variations of the precipitin tests, flocculation and counterimmunoelectrophoresis (CIE), are available for serologic studies. In flocculation tests, the precipitin forms macroscopically and microscopically visible clumps. An example of such a test is the Venereal Disease Research Laboratory test (VDRL), in which the test antigen (cardiolipin–lecithin-coated cholesterol particles) binds to an antibody-like protein (reagin) in sera of patients with syphilis, causing flocculation of the particles. However, since reagin is not an antibody specific for *T. pallidum* antigens, the test is not highly specific, but is a good screening serologic test. A comparable qualitative serologic test, the rapid plasma reagin (RPR) test, is more widely used than VDRL in laboratories, because of increased specificity and ease of performing the assay.

CIE takes advantage of the net electric charge of the antigen and antibody being tested in a particular test buffer solution, when the two components migrate towards each other in a semisolid matrix under the influence of an electrical current. When the antigen and antibody meet in optimal proportions, a line of precipitation appears. The optimal performance of CIE tests depends on many variables, such as buffer pH, type of agarose gel, amounts of antigen and antibody placed, and amount of electrical current. These tests are difficult to develop and perform and are not commonly used.

COMPLEMENT FIXATION TEST

Complement fixation (CF) assays utilize two properties of complement: complement is inactivated in the presence of antigen–antibody complexes, and complement causes hemolysis of sheep red blood cells (RBC) coated with anti-sheep RBC antibodies (sensitized RBCs). In this assay system, laboratory-derived antigen and patient's serum containing the antibody of interest are allowed first to react in the presence of excess complement, followed by addition of sensitized RBCs to detect residual complement. Hemolysis indicates that complement was present and therefore antigen–antibody complexes were not formed in the reaction. Useful for detecting and quantitating antibody, CF is a test method used in reference laboratories for detecting infections due to *Mycoplasma pneumoniae*, *Chlamydia* and *Chlamydophila* spp., and *Coxiella burnetii*. The CF assay is a technically difficult serologic test and it has been replaced by the use of other simpler methods in most laboratories.

IMMUNOFLUORESCENCE ASSAYS

Due to increased specificity, indirect IF is the preferred technique of IF for detection of specific antibodies in patient serum. In these tests, specific microbial antigens are affixed to the surface of the microscope slide, on which serially dilutions of patient serum are added to form antigen–antibody complexes. After removal of unbound antibodies from the slide surface, fluorescein-conjugated antihuman globulin (directed against human IgG or IgM antibody) is added, and the slide is visualized under fluorescent microscopy. Commercially available indirect IF tests are widely used for detection of antibodies against *Legionella* spp., *B. burgdorferi*, *C. trachomatis*, *M. pneumoniae*, *Rickettsia* spp., and *Ehrlichia* spp.

ENZYME IMMUNOASSAYS

Using specific antigen immobilized on the surface of plastic wells, bead, or some other solid matrix, the antibody of interest in patient serum can be separated from other nonspecific antibodies, captured and quantified according to the intensity of the color produced from the enzymatic reaction. Various EIA are available commercially for detecting specific antibodies to *B. burgdorferi*, *H. pylori*, *M. pneumoniae*, chlamydiae, and other bacteria.

RADIOIMMUNOASSAY

Radioimmunoassay (RIA) is performed as a capture assay, such as that for ELISA, or as a competitive assay. Radioactively labeled antigen or antibody is used to quantitate the amount of antigen–antibody complexes formed. In the competitive assay, the antibody of interest in patient's serum is measured by its ability to compete with laboratory-prepared radiolabeled antibody for the antigen. The antigen–antibody complexes are precipitated and separated from free antibody, and radioactivity is measured for both fractions. The amount of serum antibody is determined from standard calibration curves obtained from testing known quantities of competing antibody. RIA has been largely replaced by other equally sensitive methods, such as ELISA and indirect IF test, that are not associated with the disposal problems of radioactive reagents and the health hazards of radioactive exposure.

WESTERN BLOT IMMUNOASSAY

This assay involves separating microbial proteins (antigens) by electrophoresis according to molecular weight or electrical charge in a two-dimensional agarose gel matrix. These protein 'bands' are then transferred (blotted) from the gel on to a membrane (e.g. filter paper, nitrocellulose, nylon) to immobilize the proteins. When allowed to react with patient's serum, antibodies will bind to the immobilized proteins and are visualized as bands on the membrane with an enzyme-conjugated antihuman antibody. Western blot assays are mainly used for the confirmation of microbial antigen-specific antibodies in sera of patients who usually produce numerous cross-reacting antibodies in response to certain infections, such as Lyme disease.

REFERENCES

Atlas, R.M. 1997. *Handbook of microbiological media*, 2nd edn. Boca Raton, FL: CRC Press.

Chapin, K.C. and Lauderdale, T.L. 2003. Reagents, stains, and media: bacteriology. In: Murray, P.R., Baron, E.J., et al. (eds), *Manual of clinical microbiology*, 8th edn. Washington, DC: American Society for Microbiology Press, 354–83.

Chapin, K.C. and Murray, P.R. 2003. Principles of stains and media. In: Murray, P.R., Baron, E.J., et al. (eds), *Manual of clinical microbiology*, 8th edn. Washington, DC: American Society for Microbiology Press, 257–66.

Citron, D.M. 2003. Algorithm for identification of aerobic gram-positive cocci. In: Murray, P.R., Baron, E.J., et al. (eds), *Manual of clinical microbiology*, 8th edn. Washington, DC: American Society for Microbiology Press, 343–4.

European Committee on Antimicrobial Susceptibility Testing. 2000. Determination of minimum inhibitory concentrations (MICs) of antibacterial agents by agar dilution. EUCAST definitive document E. Def 3.1. *Clin Microbiol Infect*, **6**, 509–15.

European Committee on Antimicrobial Susceptibility Testing. 2003. Determination of minimum inhibitory concentrations (MICs) of antibacterial agents by broth dilution. EUCAST discussion document E.Dis 5.1. *Clin Microbiol Infect*, **9**, 1–7.

Forbes, B.A., Sahm, D.F. and Weissfeld, A. (eds) 2002. *Bailey & Scott's Diagnostic microbiology*, 11th edn. St Louis, MO: Mosby Inc.

Funke, G. 2003. Algorithm for identification of aerobic gram-positive rods. In: Murray, P.R., Baron, E.J., et al. (eds), *Manual of clinical microbiology*, 8th edn. Washington, DC: American Society for Microbiology Press, 334–6.

Miller, J.M., Holmes, H.T. and Krisher, K. 2003. General principles of specimen collection and handling. In: Murray, P.R., Baron, E.J., et al. (eds), *Manual of clinical microbiology*, 8th edn. Washington, DC: American Society for Microbiology Press, 55–66.

Murray, P.R. 1998. *Pocket guide to clinical microbiology*, 2nd edn. Washington, DC: American Society for Microbiology Press.

Murray, P.R., Rosenthal, K.S., et al. 2002. *Medical microbiology*, 4th edn. St Louis, MO: Mosby, Inc.

National Committee for Clinical Laboratory Standards (NCCLS). 2000. *Antiviral susceptibility testing; proposed standard*. NCCLS document M33-P. Wayne, PA: National Committee for Clinical Laboratory Standards.

NCCLS. 2001. *Methods for antimicrobial susceptibility testing of anaerobic bacteria; approved standard*, 5th edn. NCCLS document M11-A5. Wayne, PA: National Committee for Clinical Laboratory Standards.

NCCLS. 2002a. *Reference method for broth dilution antifungal susceptibility testing of filamentous fungi; approved standard*. NCCLS document M38-A. Wayne, PA: National Committee for Clinical Laboratory Standards.

NCCLS. 2002b. *Reference method for broth dilution antifungal susceptibility testing of yeasts; approved standard*, 2nd edn. NCCLS document M27-A2. Wayne, PA: National Committee for Clinical Laboratory Standards.

NCCLS. 2003a. *Method for antifungal disk diffusion susceptibility testing of yeasts; proposed guideline*. NCCLS document M44-P. Wayne, PA: National Committee for Clinical Laboratory Standards.

NCCLS. 2003b. *Methods for dilution antimicrobial susceptibility tests for bacteria that grow aerobically; approved standard*, 6th edn. NCCLS document M7-A6. Wayne, PA: National Committee for Clinical Laboratory Standards.

NCCLS. 2003c. *Performance standards for antimicrobial disk susceptibility tests; approved standard*, 8th edn. NCCLS document M2-A8. Wayne, PA: National Committee for Clinical Laboratory Standards.

NCCLS. 2003d. *Performance standards for antimicrobial susceptibility testing; 13th informational supplement*. NCCLS document M100-S14. Wayne, PA: National Committee for Clinical Laboratory Standards.

NCCLS. 2003e. *Susceptibility testing of mycobacteria, nocardia, and other aerobic actinomycetes; approved standard*. NCCLS document M24-A. Wayne, PA: National Committee for Clinical Laboratory Standards.

Ruoff, K.L. 2003. Algorithm for identification of aerobic gram-positive cocci. In: Murray, P.R., Baron, E.J., et al. (eds), *Manual of clinical microbiology*, 8th edn. Washington, DC: American Society for Microbiology Press, 331–3.

Schreckenberger, P.C. and Wong, J.D. 2003. Algorithm for identification of aerobic gram-negative bacteria. In: Murray, P.R., Baron, E.J., et al. (eds), *Manual of clinical microbiology*, 8th edn. Washington, DC: American Society for Microbiology Press, 337–42.

Versalovic, J. 2003. Algorithm for identification of curved and spiral-shaped gram-negative rods. In: Murray, P.R., Baron, E.J., et al. (eds), *Manual of clinical microbiology*, 8th edn. Washington, DC: American Society for Microbiology Press, 345–8.

29

Molecular laboratory diagnosis of infection

ROBIN PATEL

BACKGROUND

For a variety of reasons, molecular diagnostic tests are being increasingly used for the diagnosis of infectious diseases. Advances in testing formats, including automated specimen preparation instrumentation and rapid amplification and detection technologies, have enabled the introduction of molecular microbiologic testing into even the most routine diagnostic laboratories. For some infections (e.g. herpes simplex viral encephalitis), molecular testing is *the* diagnostic approach of choice. For others, such as hepatitis C virus (HCV) infection, aside from serologic testing (which detects the host's immune response to the infectious agent), molecular diagnostics are routinely used. Even for infectious agents where conventional tests exist (e.g. culture for *Bordetella pertussis*), molecular diagnostic tests are increasingly favored because of improved sensitivity and speed as compared with conventional approaches (Sloan et al. 2002). Besides qualitative detection of microbes in clinical specimens, nucleic acid technologies are also used to quantify microorganisms (e.g. human immunodeficiency

virus-1 (HIV-1)) in clinical specimens, to classify microorganisms isolated from clinical specimens, to detect and characterize antimicrobial resistance, and for epidemiology studies.

Nucleic acid amplification technologies have enabled substantial advances in the diagnosis of infectious disease agents because of their ability to detect small amounts of nucleic acid in a wide variety of clinical specimens. Molecular assays are independent of the presence of viable (e.g. able to be cultured) microorganisms. Quantification of microbial nucleic acid in patient specimens enables myriad potential innovations in clinical practice including assessment of disease severity, progression and response to therapy. Quantification of RNA levels is valuable both for measuring organisms with genomic RNA (e.g. HIV-1), and, potentially, for studies of microbial and host gene expression.

The polymerase chain reaction (PCR) was the first nucleic acid amplification technique to be broadly applied to the molecular detection of microorganisms and is a standard method used in clinical microbiology laboratories today. Since the original description of

PCR, numerous technological advances have facilitated its application to routine clinical molecular microbiology diagnostics. Additionally, a number of non-PCR amplification molecular methods are commonly used in clinical microbiology laboratories.

Although molecular tests substantially enhance diagnostic capabilities, as with any laboratory test, results should be interpreted within the clinical context and on the basis of individual assay performance characteristics. Clinical research and strict adherence to guidelines for method validation are needed to compare molecular diagnostic techniques with conventional methods, to validate molecular approaches where comparable conventional techniques are unavailable, and to determine the clinical usefulness of molecular diagnostic testing.

Although molecular diagnostic tests have classically been considered more expensive (i.e. 'boutique' tests) than conventional diagnostic techniques, the ease with which molecular tests can now be performed and the rapid results generated enable timely diagnosis, resulting in overall savings. Cost savings may be realized as rapid diagnoses prevent invasive diagnostic procedures, limit unnecessary or potentially toxic treatment, shorten hospital stays, replace labor-intensive conventional (e.g. cell culture) methods, and/or shorten hands-on laboratory technologist time. Early detection of contagious diseases, such as pulmonary tuberculosis, using molecular techniques would also limit the spread of infection.

ASSAY PERFORMANCE

Molecular techniques developed in research settings have been applied as 'in-house' developed assays in many clinical microbiology laboratories. The reason for this is that for many microorganisms, commercial assays are unavailable. Furthermore, it is relatively easy for laboratories with research and development expertise to develop and validate their own molecular assays based on nucleic acid amplification technology. The downside is that results derived from such assays may not be comparable from laboratory to laboratory, while no universal standards exist for these assays. The National Committee for Clinical Laboratory Standards (NCCLS) published guidelines for molecular diagnostic methods for infectious diseases in 1995 (Enns et al. 1995) and for quantitative molecular methods for infectious diseases in 2001 (Madej et al., 2001). Strategies to provide quality and to standardize molecular diagnostic tests include laboratory requirements for certification, proficiency surveys, and checklists, such as those published by the American College of Pathologists (www.cap.org/html/ftpdirectory/checklistftp.html) (College of American Pathologists 2003). The availability of state-of-the-art Food and Drug Administration (FDA)-approved

commercial molecular diagnostic assays substantially contributes to interlaboratory standardization. A limited number of nucleic acid amplification assays for detection of microbes in human specimens are currently available (Table 29.1).

In order to understand and interpret molecular microbiology diagnostic tests, certain universal assay parameters must be defined (see also Wolk et al. 2001). Analytical sensitivity, also referred to as the lower limit of detection, refers to the lowest number of microorganisms that can be reproducibly detected by the assay. Clinical sensitivity is the proportion of specimens that yield positive test results from patients who have a specified clinical entity. High analytical sensitivity may exist with inadequate clinical sensitivity if false-negative results occur because the target nucleic acid copy number in the clinical specimen is low. Analytical specificity refers to the ability of the test to detect only the microorganism it purports to measure. Clinical specificity is the proportion of specimens that yield negative results from patients who do not have a specified clinical entity.

For quantitative molecular microbiologic diagnostic testing, additional parameters bear consideration. The linear range refers to the quantitative spectrum over which the test provides results detecting a direct relationship between input target concentration and output signal. The upper and lower limits of quantification reflect the upper and lower ends of this linear range. The lower limit of quantification of a quantitative assay may be higher than the analytical sensitivity of a qualitative assay, confusing the interpretation of results. This is exemplified by molecular testing for HCV. The lower limit of quantification of the VERSANT™ HCV RNA 3.0 assay (bDNA) (Bayer Healthcare LLC, Tarrytown, NY, USA), a quantitative HCV RNA assay, is higher than the lower limit of detection of the COBAS AMPLICOR™ HCV Test, version 2.0 (Roche Diagnostic Corporation, Indianapolis, IN, USA), a qualitative HCV RNA assay (Germer et al. 2002).

Precision refers to the agreement between replicate measurements of the same material. Accuracy refers to the ability of a method to reliably determine the true value of the particular target. The tolerance limit is the difference between two results that can be considered to be significantly different and is the sum of the biological variation in quantification and intra-assay variability. For quantitative assays of HIV-1 RNA, for example, increases or decreases in the RNA titer of at least three-fold typically reflect biologically relevant changes in the level of viral replication (Wolk et al. 2001).

In an effort to standardize quantitative testing, inter-laboratory collaborations with the World Health Organization have established the World Health Organization International Standards, which are standard reference materials with concentrations expressed as international units (IU) per milliliter, which can be used to calibrate,

Table 29.1 *Commercial amplification assays for molecular detection of microbes in human specimens (United States, 2003)*[a]

Test	Manufacturer	Detection	Application (specimen)
Amplicor™ and Amplicor Monitor™	Roche Diagnostic Corporation www.roche-diagnostics.com/ba_rmd/products.html	Qualitative or quantitative (varies with target) polymerase chain reaction can be automated on the COBAS instrument	*Mycobacterium tuberculosis* (sputum, bronchial specimen)[b] *Chlamydia trachomatis* (urine, cervical, male urethral specimen)[b] *Neisseria gonorrhoeae* (urine, cervical, male urethral specimen)[b] *Mycobacterium intracellulare* (sputum, bronchial specimen) *Mycobacterium avium* (sputum, bronchial specimen) Human immunodeficiency virus-1 (HIV-1) (plasma)[b] Hepatitis B virus (serum, plasma) Hepatitis C virus (serum, plasma)[b] Cytomegalovirus (plasma)
LightCycler®	Roche Diagnostic Corporation www.lightcycler-online.com	Rapid cycle real-time qualitative or quantitative (varies with target) polymerase chain reaction	Hepatitis A virus (serum, plasma) Parvovirus B19 (serum, plasma) *Bacillus anthracis* (multiple) Herpes simplex virus type 1 and type 2 (multiple) Epstein–Barr virus (multiple) *Streptococcus pyogenes* *Staphylococcus aureus* and coagulase-negative *Staphylococcus* spp. Methicillin-resistant *Staphylococcus aureus* (*mecA*) *Enterococcus faecalis* and *Enterococcus faecium* Vancomycin-resistant enterococci (*vanA/B*) *Pseudomonas aeruginosa* (multiple) *Candida albicans* (multiple)
VERSANT® TMA™	Distributed by Bayer Healthcare LLC under license from Gen-Probe www.bayerdiag.com/products/index.html	Qualitative transcription-mediated amplification	Hepatitis C virus (serum, plasma)[b]
VERSANT® bDNA	Bayer Healthcare LLC www.bayerdiag.com/products/index.html	Quantitative branched DNA assay	HIV-1 (plasma)[b] Hepatitis B virus (serum, plasma) Hepatitis C virus (serum, plasma)[b] HIV-1 (plasma)
NucliSens®	BioMérieux www.biomerieux-usa.com/clinical/nucleicacid/index.htm	Qualitative or quantitative (varies with target) nucleic acid sequence-based analysis	Cytomegalovirus pp67 mRNA (whole blood)[b]
Amplified™ MTD, Amplified CT	Gen-Probe www.gen-probe.com	Qualitative transcription-mediated amplification	*M. tuberculosis* (sputum, bronchial specimen, tracheal aspirate)[b] *C. trachomatis* (urine, cervical, male urethral specimen)[b]
Group A Streptococcus Direct (GASD)	Gen-Probe www.gen-probe.com	Qualitative hybridization protection assay	*Streptococcus pyogenes* (throat swab)
BDProbe Tec™	BD Biosciences www.bd.com/biosciences	Qualitative strand-displacement amplification	*C. trachomatis* and *N. gonorrhoeae* (urine, cervical, male urethral specimen)[b]

(Continued over)

Table 29.1 *Commercial amplification assays for molecular detection of microbes in human specimens (United States, 2003)*[a] *(Continued)*

Test	Manufacturer	Detection	Application (specimen)
Hybrid Capture®	Digene www.digene.com/lab.html	Quantitative DNA–RNA hybrid capture assay	Human papilloma virus (cervical specimen)[b]
			Cytomegalovirus (peripheral white blood cells)[b]
			Hepatitis B virus (serum)
			C. trachomatis and *N. gonorrhoeae* (cervical specimen)
HCV ASR	Abbott Laboratories www.abbott diagnostics.com	Real-time quantitative polymerase chain reaction with TaqMan™ probe	Hepatitis C virus

Modified with permission from Wolk et al. 2001.
a) Culture confirmation, genotyping and mutation detection assays are not listed.
b) FDA-approved assay as of February, 2003.

validate, and compare different quantitative molecular assays measuring very similar targets in the same pathogen. Currently, standards exist for HIV-1, hepatitis A virus, hepatitis B virus (HBV), HCV, and parvovirus B19. Further information is available from the National Institute for Biological Standards and Controls (www.nibsc.ac.uk).

SPECIMEN SUITABILITY

For each molecular diagnostic assay, criteria for the appropriate specimen, including the optimal specimen source, specimen volume, collection method, transport and storage conditions, specimen stability, and nucleic acid preparation method must be considered. The sensitivity and specificity of an assay may vary considerably if any of these conditions are altered. The choice of specimen plays a key role in the performance and interpretation of test results. Typically, molecular microbiologic assays are clinically validated using one or more specimen sources (e.g. spinal fluid, blood) that may contain a target organism. For blood specimens, sensitivity and specificity may vary depending on the specific fraction (i.e. plasma, serum, whole blood, various leukocyte fractions) analyzed. The viral load, for example, will vary depending on the specific specimen tested; plasma or serum is typically used for detection of HBV DNA, HCV RNA, or HIV-1 RNA. The volume of specimen from which nucleic acid is extracted can also affect the sensitivity and specificity of the assay. Traditional molecular diagnostic tests typically analyze relatively small specimen volumes (e.g. 200 µl or less). For low-copy-number targets, increasing the amount of specimen analyzed may increase sensitivity (Germer et al. 2003).

Inhibitory substances may interfere with various steps in molecular diagnostic assays yielding false-negative results (Wolk et al. 2001). Inhibitors present in the patient specimen (e.g. hemoglobin, lactoferrin, immuno-globulin G, bile salts, polysaccharides, proteinases, urea, leukocyte DNA) or introduced during specimen collection or processing (e.g. ethylene diamine tetraacetic acid (EDTA), heparin, guanidinium HCl, sodium dode-cylsulfate) may interact with nucleic acid or critical enzymes (e.g. DNA polymerase) to prevent target amplification or remove reaction components (e.g. metals) affecting enzymatic substrates (Al-Soud et al. 2000, 2001). The presence of inhibitors, and therefore the negative predictive value of the assay, can be assessed in several ways. The patient's specimen can be spiked with known target nucleic acid and the spiked specimen tested along with the native patient specimen. Alternatively, internal amplification controls can be added to the patient specimen and amplified concurrently with the target nucleic acid. Detection of housekeeping genes (e.g. human β-globin, interleukin-2) has been incorporated into some assays to confirm the presence of human cellular DNA in the patient specimen. Although this approach can be used to assess the presence of inhibitors, since the amount of human DNA generally exceeds the amount of microbial DNA, this approach to inhibitor detection may be less sensitive than the use of spiked specimens or internal amplification controls.

SPECIMEN PREPARATION

Specimen preparation techniques are a critical component of clinical molecular microbiologic diagnostics (Wolk et al. 2001). Optimal specimen preparation efficiently releases microbial nucleic acid while preserving its integrity, removes inhibitors, sterilizes the specimen of viable organisms, concentrates the target nucleic acid into a small volume (if appropriate), and places the target into an environment suitable for amplification. For many assays, specimen preparation may be more labor-intensive and time-consuming than actual amplification and detection. Many specimen-preparation methods exist; some require considerable technical skill, as they are labor-intensive, nonstandardized, and prone to manipulation steps where template cross-contamination may occur. A detailed discussion of available nucleic acid preparation methods for molecular microbiology diagnostic purposes is beyond the scope of this chapter. In general, specimens are treated with alkaline pH, chaotropic agents, sonication, heat, detergents, freeze–thaw cycles, and/or proteolytic enzymes to disrupt the cell membranes and cell walls and to release nucleic acid. Thereafter, nucleic acid may or may not be extracted with organic solvents, adsorbed to silica in the presence of chaotropic agents such as guanidinium salts, and/or precipitated in the presence of salts and alcohol. Nucleic acid capture technologies such as automated silica-based membrane capture technologies or magnetic bead nucleic acid capture can also be used. For specimens that represent a potential biohazard for laboratory personnel (e.g. specimens submitted for testing for agents of bioterrorism), autoclaving the specimen prior to molecular testing (121°C, 15 min, 20 psi) enables inactivation of potentially viable hazardous infectious agents without destruction of nucleic acids (Espy et al. 2002).

Automation of specimen preparation provides rapid, cost-effective, and consistent results. Several automated extraction instruments are available. The ability to interface with high-throughput amplification systems is an important advance of these new automated extraction systems. For example, the MagNA Pure LC (Roche Diagnostic Corporation) integrates with the Light-Cycler® PCR instrument (Roche Diagnostic Corporation) and can be used to automate or partially automate both extraction and PCR setup processes (Espy et al. 2001).

PROBE HYBRIDIZATION

Nucleic acid probes are short segments of single-stranded DNA or RNA that can be 'labeled' (i.e. with antigen, chemiluminescent substrate, radioisotope, enzyme) and can bind with high specificity to complementary (target) nucleic acid sequences. In situ hybridization uses intact cells (e.g. in formalin-fixed, paraffin-embedded tissue) containing specific DNA or RNA as a target for probe hybridization. The technique allows visualization of infected cells within tissue and thereby association of hybridization results with other morphologic changes. Because in situ hybridization is not an amplification-based technique, sensitivity may be limited. Probe-based commercial kits, such as the PACE® System from Gen-Probe, Inc. (San Diego, CA, USA), which directly detects *Chlamydia trachomatis* and *Neisseria gonorrhoeae* in clinical samples, produce results that are generally equivalent to culture techniques but in much less time. Nucleic acid probes are also

an integral component of the amplification techniques described below.

POLYMERASE CHAIN REACTION

PCR is the original and most widely used nucleic acid target amplification technology. A buffered reaction mixture containing the target DNA sequence (template), oligonucleotide primers, thermostable DNA polymerase, deoxynucleotide triphosphates (dNTPs), and magnesium or manganese ions with or without other additives, is placed into a thermal cycler, which heats and cools the components, exposing them to consecutive cycles of alternating temperatures (Wolk et al. 2001). In each cycle, three processes occur:

1 Denaturation: heating to high temperatures to separate double-stranded DNA into single strands.
2 Primer annealing: lowering the temperature to allow for synthetic oligonucleotide primers to anneal to the single-stranded DNA and create a partial double strand.
3 Primer extension: addition of dNTPs to the 3′ ends of the bound primers by DNA polymerase, thereby creating a new synthetic piece of double-stranded DNA, which is complementary to the original template strand.

With the exception of the first cycle, the amount of target DNA theoretically doubles with each cycle, resulting in an exponential increase in the quantity of amplified DNA until the reaction reaches a plateau phase due to depletion of reaction components. The amount of product that accumulates is $N_{total} = N_{initial}(1 + Y)^X$, where Y is the efficiency per cycle. Examples of some specific types of PCR are shown in Table 29.2.

Following amplification, amplified DNA is detected. Many of the classic methods for detection of amplified DNA are manual, requiring manipulation of amplified DNA and adding substantial risk of contamination to subsequent PCR reactions, rendering them suboptimal for use in clinical laboratories. Agarose or polyacrylamide gel electrophoresis use electric current to separate DNA fragments according to molecular size; DNA fragments are visualized by staining with a DNA binding dye (e.g. ethidium bromide). Generally, agarose or polyacrylamide gel electrophoresis (alone) does not provide ideal sensitivity or specificity for clinical microbiology diagnostic testing. To increase specificity, the amplified DNA can be digested with a restriction enzyme prior to electrophoresis; the resultant restriction digest yields a unique pattern of DNA fragments by gel electrophoresis. To increase sensitivity as well as specificity, PCR products may be subjected to Southern blotting followed by probe hybridization. Briefly, DNA from a gel is transferred to a synthetic membrane, on which labeled oligonucleotide probes

Table 29.2 *Modifications of the polymerase chain reaction (PCR)*

Modification	Principle
Broad-range bacterial PCR	Use of PCR primers targeted to highly conserved regions of bacterial DNA such that, depending on the primer sets used, the target can be amplified from most bacteria
In situ PCR	Amplification of DNA (or complementary DNA) inside an intact cell (e.g. in a paraffin-embedded tissue section) to determine the disposition of a microbe within tissue
Multiplex PCR	Use of two or more primer sets in the same reaction mix to identify multiple microorganisms or multiple nucleic acid targets within one microorganism in a single assay
Nested PCR	Use of one set of primers followed by reamplification with a second set of primers, internal to the first
Quantitative PCR	A technique used to assess the amount of a microbe present in a patient specimen
Semiquantitative or limiting dilution PCR	Titer determination of target template by endpoint specimen dilution prior to PCR
Quantitative competitive PCR	Use of an internal quantitative standard (or calibrator) in the amplification reaction along with the target sequences. The standard competes with target sequences in the PCR reaction. The quantity of internal quantitative standard amplified correlates inversely with the quantity of target
Real-time quantitative PCR	Quantitation of amplified target determined by cycle number at which the amplification product is first detected (using fluorescence detection) in comparison with a quantitative standard
Randomly amplified polymorphic DNA (RAPD) analysis (arbitrarily primed PCR)	Random amplification of segments of target DNA using arbitrary primers that do not have any known homology to the target sequence. Amplified DNA is visualized on a gel following electrophoresis and patterns are compared to one another
Real-time PCR	Simultaneous amplification and amplification product detection using fluorescence detection (Figure 29.1)
Repetitive element sequence-based PCR	Uses consensus sequence-based PCR primers to amplify DNA sequences located between successive repetitive elements
Reverse transcriptase-PCR (RT-PCR)	Uses reverse transcriptase to create complementary DNA from an RNA target. The complementary DNA is then amplified by PCR
	Used to detect RNA (either because an infectious agent's nucleic acid is naturally present as RNA, or to detect transcribed microbial DNA as a marker of active infection)

complementary to an internal region of the amplified sequence are subsequently hybridized to the DNA. Hybridized probes are then detected by one of a variety of methods (e.g. autoradiography, chemiluminescence). Another option for detection is incorporation of various chemicals (e.g. digoxigenen-11-dUTP) into the DNA during amplification. Amplified DNA can then be detected on the membrane by a second chemical reaction (e.g. anti-digoxigenin antibody:alkaline phosphatase conjugate and a chemiluminescent substrate).

Besides membrane-based amplified DNA detection methods, a number of other approaches can be used (Wolk et al. 2001). Probes can be bound to microtiter plate wells or beads to capture amplified DNA. A specific 'signal' probe can then be added to react with the captured target nucleic acid. In the PCR-enzyme-linked immunosorbent assay, amplified DNA containing digoxigenen-11-dUTP is hybridized to oligonucleotide capture probes labeled with biotin. Biotinylated DNA hybrids are then bound to a streptavidin-coated micro-titer plate so those specific DNA complexes can be detected in a microtiter plate format.

Solution- (liquid-) phase hybridization is another option. An example is the hybridization protection assay wherein chemiluminescent probes hybridize to original target nucleic acid or to amplified DNA, which have been produced by a previous amplification method. Following chemical destruction of the nonhybridized probes, the chemiluminescent label, protected within the hybrid, reacts under specific conditions to produce light, which can be measured by a luminometer.

Other methods that can be used to separate, quantify, or identify amplified DNA include high-performance liquid chromatography, mass spectrometry, and DNA sequencing methods.

Homogeneous or real-time PCR assays incorporate amplification and detection within one closed system. Such assays are being used increasingly in clinical microbiology laboratories because of their rapid turnaround time and the lack of postamplification manipulation and

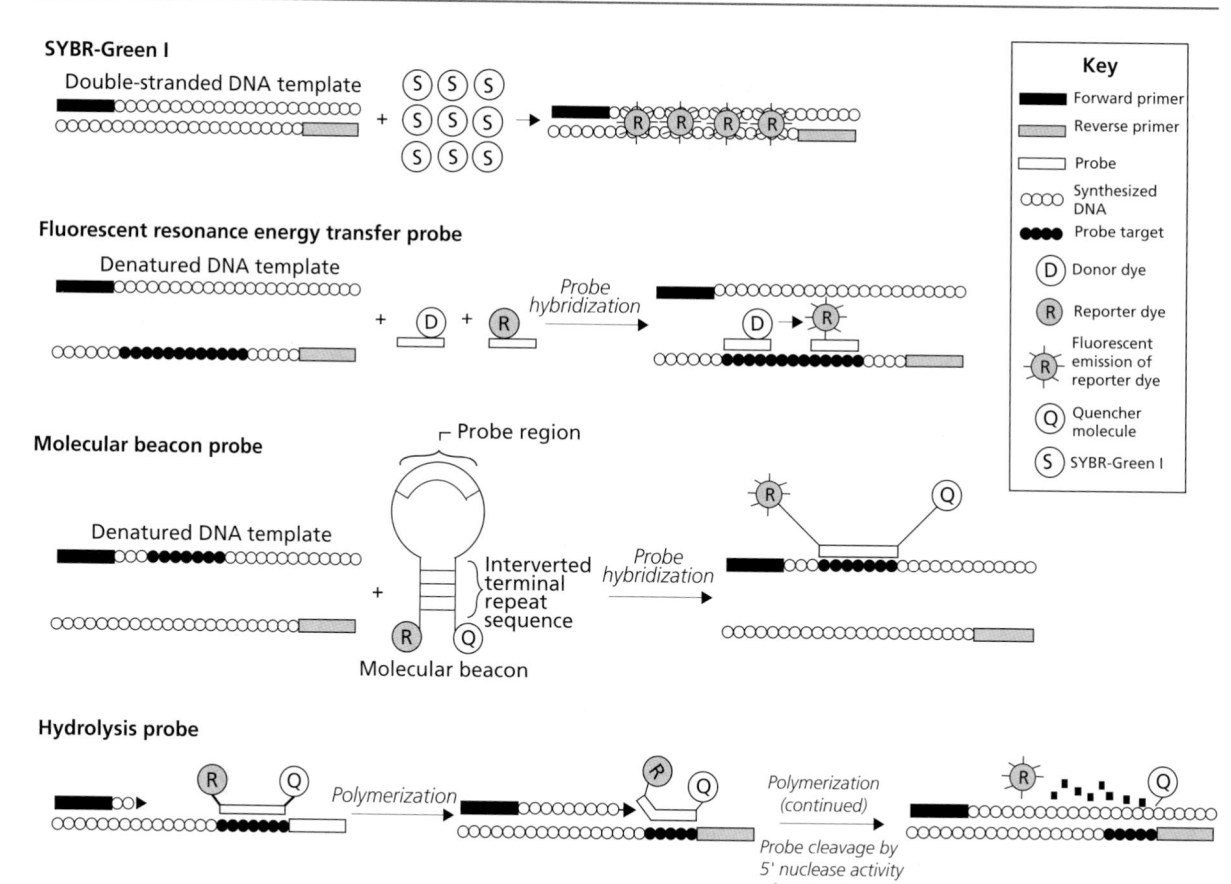

SYBR-Green I

Double-stranded DNA template

Fluorescent resonance energy transfer probe

Denatured DNA template

Probe hybridization

Molecular beacon probe

Probe region

Denatured DNA template

Interverted terminal repeat sequence

Probe hybridization

Molecular beacon

Hydrolysis probe

Polymerization

Polymerization (continued)

Probe cleavage by 5' nuclease activity of DNA polymerase

Key

◼ Forward primer
▨ Reverse primer
▢ Probe
〇〇〇〇 Synthesized DNA
●●●● Probe target
Ⓓ Donor dye
Ⓡ Reporter dye
Ⓡ Fluorescent emission of reporter dye
Ⓠ Quencher molecule
Ⓢ SYBR-Green I

Figure 29.1 *Real-time polymerase chain reaction detection (see text for details).*

CONTAMINATION AND CONTAMINATION CONTROL

The strength of nucleic acid amplification is high sensitivity (Wolk et al. 2001). Because of this sensitivity and the large amount of amplified DNA generated, prevention of nucleic acid contamination of other molecular assays performed in the same laboratory is essential. Sources of contamination include crossover contamination from specimens containing large numbers of target molecules or, most importantly, from manipulation of amplified nucleic acid. Amplified nucleic acid may contaminate reagents, laboratory surfaces, as well as the laboratory personnel themselves. The setup of the laboratory should be such that the movement of patient specimens and personnel proceeds from pre-amplification to postamplification areas. Separate areas (preferably rooms) for reagent preparation, sample processing, and amplification and detection are ideal. Besides these physical approaches to contamination control, enzymatic (e.g. uracil-*N*-glycosylase (UNG)) or photochemical (e.g. isopsoralen) methods of inactivating amplified DNA may be applied (Wolk et al. 2001). For

enzymatic inactivation utilizing UNG, dUTP in lieu of dTTP is incorporated into the amplified DNA such that amplified target will contain dUTP (whereas native target contains dTTP). Any dUTP-containing amplification products (i.e. contaminants from prior PCR reactions) are cleaved by UNG prior to amplification in subsequent reactions. For photochemical inactivation, amplified DNA is generated in the presence of isopsoralen. Inactivation occurs by postamplification irradiation of the reaction vessel with ultraviolet light prior to opening the reaction vessel; irradiation causes the isopsoralen to modify the amplified DNA such that the DNA is unable to serve as a template for further amplification but is still capable of hybridization in various detection formats. As discussed above, closed-system PCR reactions can also be used to prevent contamination.

All PCR testing should include a sufficient number of negative controls processed along with patient specimens to provide a measure of assurance that assay contamination is not occurring. For each assay, a positive control in low copy number should also be included.

RAPID REAL-TIME PCR

A significant advance in PCR technology is the development of rapid PCR with simultaneous probe detection, so-called 'real-time PCR' (Wolk et al. 2001). In real-time PCR assays, amplification and product detection occur simultaneously in the same sealed reaction vessel, tube, or well. A fluorescent signal generated by specific binding of chemicals or probes to amplified DNA is used for real-time monitoring of production of amplified DNA.

Because amplification and product detection are performed in the same vessel, these systems can be closed. Closed PCR systems lessen the risk of contamination by avoiding postamplification manipulation and lessen the needs for extensively engineered molecular laboratory space. Nevertheless, careful attention to unidirectional workflow, amplification product control, and amplification product-inactivation protocols remain important. Template carryover contamination can still occur, and liberation of amplified DNA through broken reaction vessels can lead to contamination events unless DNA-inactivating methods are included in the assay.

Many real-time PCR systems have incorporated technology that enables a dramatic reduction in the time required for assay performance. For example, realtime PCR performed using the LightCycler® instrument (Roche Diagnostic Corporation) can be completed in 30 to 40 min (versus several hours to days for conventional PCR and amplification product detection).

Several fluorescent chemistry approaches can be utilized to detect amplified DNA in real time (Figure 29.1) (Wolk et al. 2001). Real-time monitoring of nonspecific fluorescent dye (e.g. SYBR-Green I) incorporation can be used (Figure 29.1). Nonspecific dyes bind to double-stranded DNA generated during PCR and emit fluorescence, which can be monitored by various detection systems. Since the dyes bind to all nonspecific amplification products produced in the reaction, as well as to primer-dimers, dye incorporation alone does not ensure detection of a specific PCR product. To improve specificity, melting-curve analysis can be performed. The amplified product will have a characteristic melting temperature (Tm), the sequence-dependent temperature at which half of the nucleic acid strands become separated in solution, differentiating it from nonspecific amplification products. To further ensure more specific hybridization and product identification, protocols that incorporate specific internal probes are used. Several internal probe formats exist. They include dual hybridization (fluorescence resonance energy transfer (FRET)) probes, molecular beacon probes, and hydrolysis (e.g. TaqMan™) probes (Figure 29.1).

Dual hybridization uses two probes (one coupled to a fluorescent donor dye and the other to a fluorescent reporter dye) designed to recognize single strands of amplified DNA internal to the primer binding sites (Figure 29.1). As DNA is amplified, probes hybridize to amplified DNA in close proximity to one another (usually one to five nucleotides apart) during the annealing step. An external light source then stimulates the fluorescent donor dye, and energy emitted from the donor dye excites the fluorescent reporter dye to emit a specific wavelength of light.

Molecular beacons are single-stranded oligonucleotide probes that are created using a specific hybridization probe sequence, which is flanked by two inverted terminal repeat sequences such that a stem-loop (or hairpin) structure is formed (Figure 29.1). As the molecular beacon probe opens and binds to a denatured DNA template, a quencher molecule (located on one arm) is removed from proximity of a fluorescent reporter dye (located on the other arm), allowing the reporter dye to fluoresce upon stimulation from an external light source. In contrast, while the probes are in a hairpin formation, the reporter dye and quencher molecules are proximal and fluorescence is quenched.

Using FRET-based probe pairs or molecular beacons, melting-curve analyses can be performed by monitoring the Tm at which the probe separates from the amplified DNA. This analysis permits specific identification of the amplified DNA. After PCR is performed, the amplified DNA is heated to its denaturation temperature and diminished fluorescent reporter signal is observed. Because mismatches, such as point mutations or polymorphisms, combined with other amplified DNA characteristics, such as fragment size and percentage GC content, can affect the Tm of the probe:DNA hybrid, Tm curves can be used to distinguish different PCR products including those with as few as 1 bp difference.

TaqMan™ chemistry combines DNA polymerase with 5'-exonuclease activity and a linear oligonucleotide hybridization probe carrying both a fluorescent reporter dye and a quencher molecule (Figure 29.1). The hybridization probe binds to the denatured amplified target DNA strands. Fluorescence of the reporter dye is prevented (quenched) by its proximity to the quencher molecule. After each annealing cycle, DNA polymerase, with 5'-exonuclease activity, extends the specific target strand and cleaves the hybridization probe, which is specifically bound to the strand. The exonuclease activity fragments the probe, separating the quencher molecule from the reporter dye. The cleaved reporter dye fluoresces when stimulated from an external light source. If specific amplification of target DNA has occurred along with cleavage of the hybridization probe, reporter dye fluorescence increases and accumulates with each PCR cycle. Melting-curve analysis cannot be performed with TaqMan™ probes.

While SYBR-Green I, TaqMan™, FRET, and molecular beacon technology can be used to monitor real-time endpoint (qualitative) PCR, these techniques can also be used to quantify target nucleic acid. The amount of microbe-specific nucleic acid in a given patient specimen is determined by comparing target amplification signal to internal or external quantitative standards and, through mathematical algorithms, determining the exact number of amplification cycles at which amplified DNA is detected above a baseline or threshold value. Real-time quantitative PCR assays offer advantages over nonreal-time semiquantitative PCR approaches (Table 29.2), which analyze endpoint amplification products at a point when net synthesis is significantly reduced, inhibitory effects accumulate, and differences in the initial starting template concentrations are masked.

Several automated real-time PCR instruments that use the aforementioned technologies are currently available: LightCycler® (Roche Diagnostic Corporation), SmartCycler® (Cepheid, Sunnyvale, CA, USA), ABI PRISM® Sequence Detection Systems (SDS) (Applied Biosystems, Foster City, CA, USA), iCycler (Bio-Rad Laboratories, Hercules, CA, USA), Mx4000™ Multi-plex Quantitative PCR System (Stratagene, La Jolla, CA, USA), and Rotor-Gene™ (Corbett Research, Sydney, Australia). Real-time PCR assays are run on highly specialized automated instruments that include optical systems to excite fluorescent dyes and detect fluorescent emissions at a variety of wavelengths specific to the dyes used. These systems have the potential to provide rapid results and to facilitate multiplex PCR through the use of different fluorophores.

NUCLEIC ACID SEQUENCE-BASED AMPLIFICATION AND TRANSCRIPTION-MEDIATED AMPLIFICATION

Nucleic acid sequence-based amplification (NASBA) and transcription-mediated amplification (TMA) are similar isothermal reactions that use a RNA target as the substrate for reverse transcriptase, creating a complementary DNA (cDNA) strand from the RNA template (Figure 29.2) (Wolk et al. 2001). RNA from the resulting RNA:DNA hybrid is then degraded by RNAse H, and cDNA is used as a template for RNA polymerase, producing multiple RNA copies from the cDNA. Results can be qualitative or quantitative. Modifications of TMA also exist for use with DNA targets.

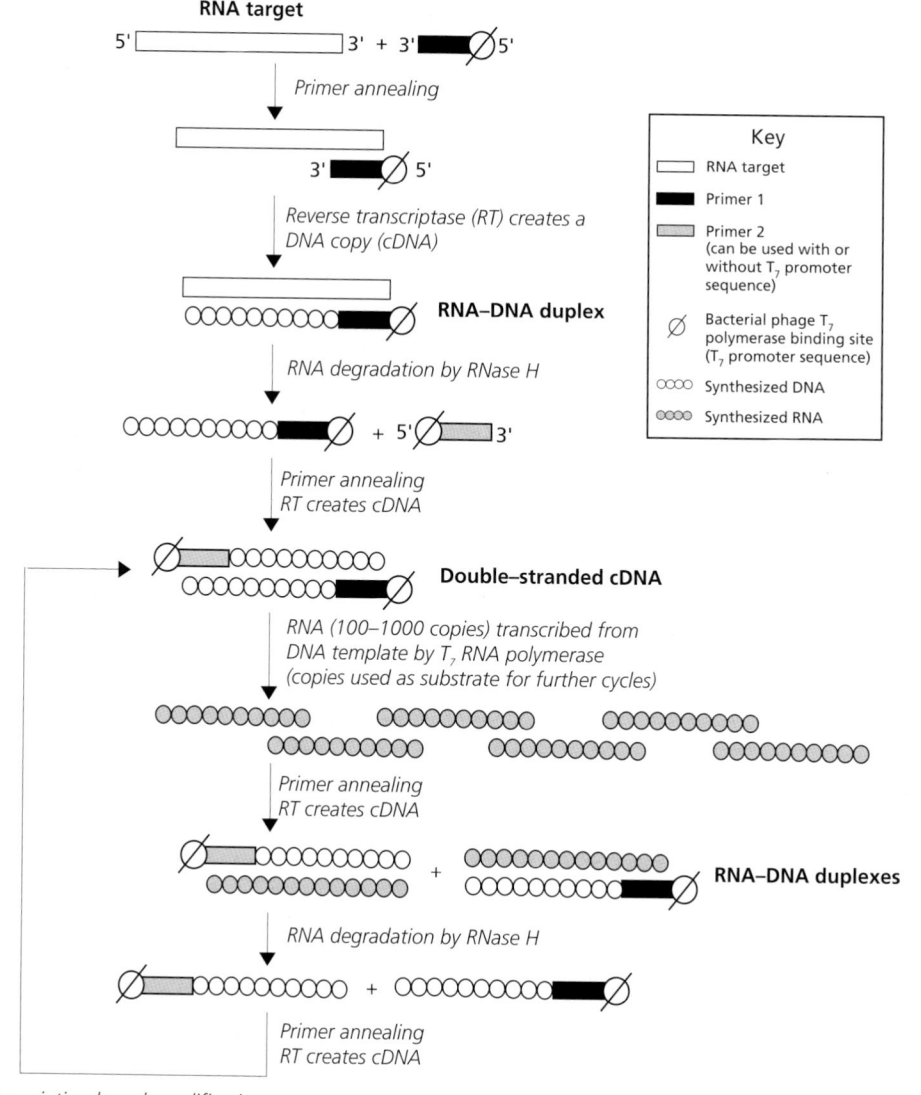

Figure 29.2 *Transcription-based amplification systems: nucleic acid sequence-based amplification (NASBA) and transcription-mediated amplification (TMA) (see text for details). Reprinted with permission from Wolk et al. 2001.*

NASBA uses separate reverse transcriptase and RNase H enzymes, whereas TMA uses a reverse transcriptase enzyme with endogenous RNase H activity. Gen-Probe, Inc. currently offers commercial TMA assays for detection of *Mycobacterium tuberculosis* and *Chlamydia trachomatis*. A TMA assay for qualitative detection of HCV RNA is available from Bayer Healthcare LLC. BioMérieux (Boxtel, The Netherlands and Durham, NC, USA) offers NASBA testing for quantitative detection of HIV-1 RNA and qualitative detection of CMV pp67 mRNA.

STRAND-DISPLACEMENT AMPLIFICATION

Strand-displacement amplification (SDA) is an isothermal amplification technique in which primers, containing both a target-specific region and a hemimodified restriction enzyme site, combine with a DNA polymerase to create a synthetic double-stranded DNA that contains the restriction enzyme site (Figure 29.3) (Wolk et al. 2001). A restriction enzyme generates a site-specific 'nick' in one of the two strands of double-stranded DNA. Once the strand is nicked, DNA polymerase initiates synthesis at the 3' end of the nick site to create a new DNA strand. Because the DNA polymerase used is deficient in 5' to 3' exonuclease activity, it does not destroy the existing strand as it polymerizes the new strand; it simply displaces the downstream nontemplate strand. Each displaced strand is then available to anneal with more primers and the process continues with repeated nicking, extension, and displacement of newly formed DNA strands. The entire process results in exponential amplification of the original DNA target.

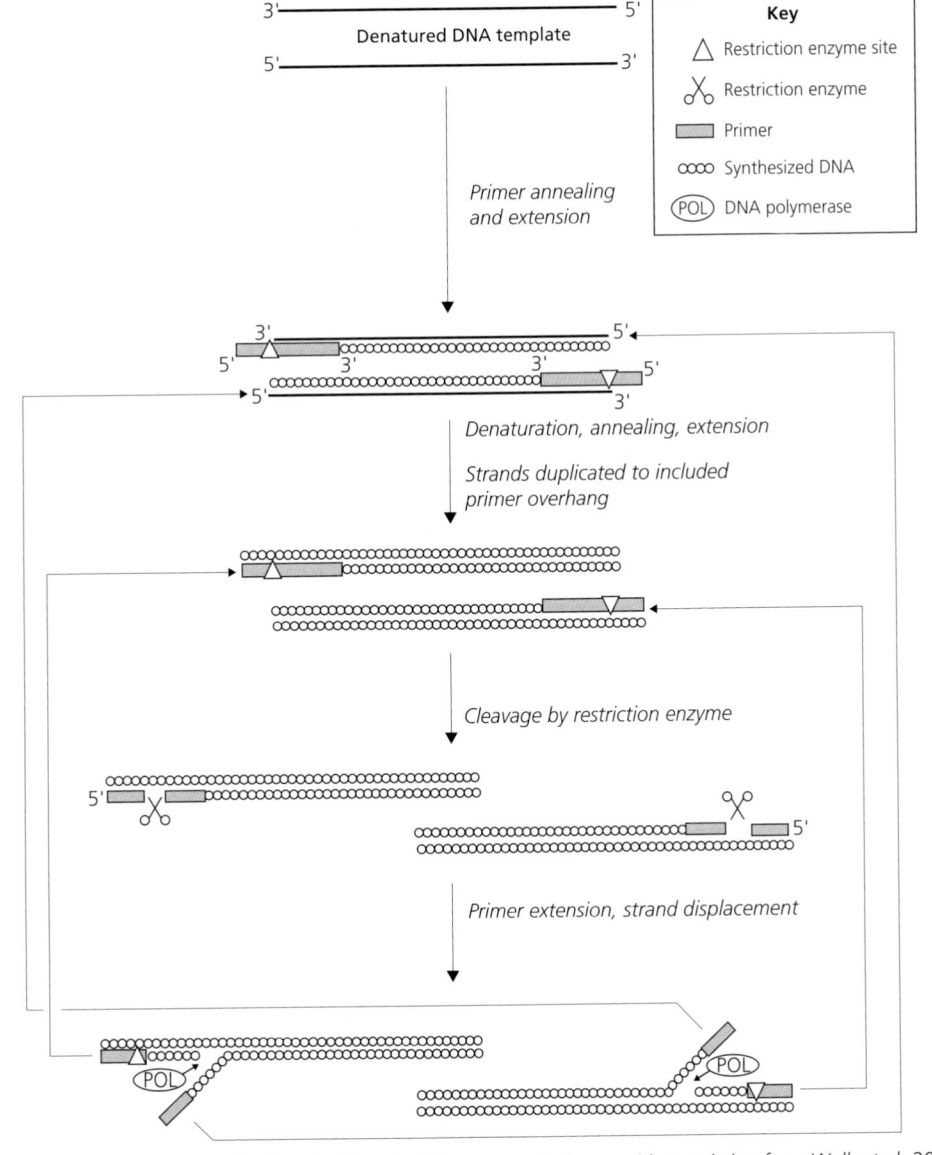

Figure 29.3 *Strand-displacement amplification (SDA) (see text for details). Redrawn with permission from Wolk et al. 2001.*

HYBRID CAPTURE ASSAY

In the hybrid capture assay, an RNA or DNA probe hybridizes to a DNA or RNA target, respectively (Wolk et al. 2001) (Figure 29.4). Hybrids are captured on a solid phase coated with capture antibodies specific for the RNA:DNA hybrids (duplexes). Captured hybrids are detected using chemiluminescence. Digene Corporation (Silver Spring, MD, USA) is a manufacturer for hybrid capture assays for human papilloma virus (HPV), CMV, HBV, *C. trachomatis*, and *N. gonorrhoeae*.

LIGASE CHAIN REACTION

In the ligase chain reaction (LCR), DNA template is denatured and subsequently hybridized to probes that are located next to each other on the target strand (Figure 29.5) (Wolk et al. 2001). The hybridized probes are ligated (joined) by DNA ligase to form a ligation

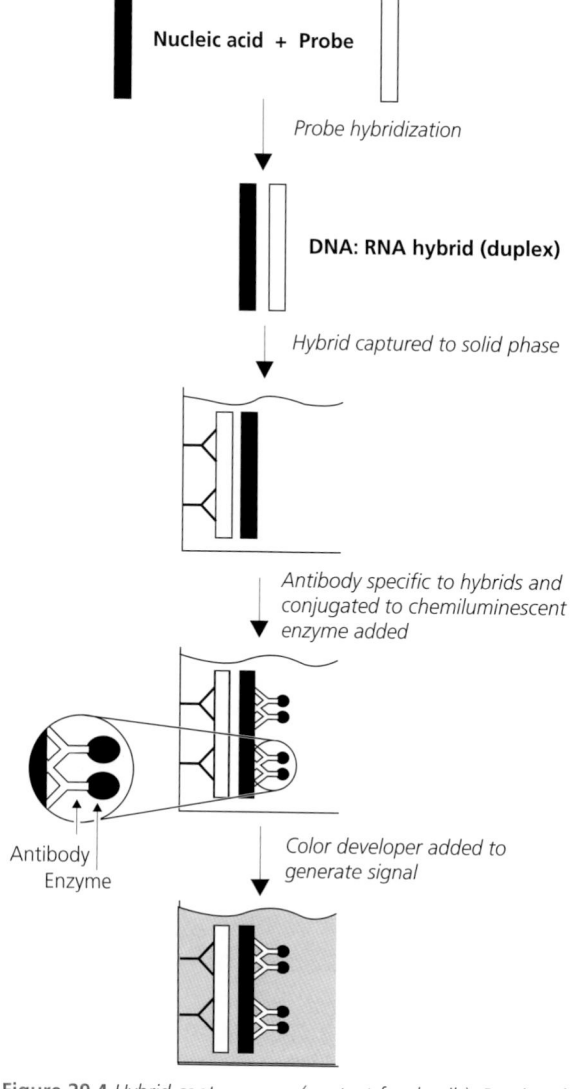

Figure 29.4 *Hybrid capture assay (see text for details). Reprinted with permission from Wolk et al. 2001.*

product that mimics one strand of original target sequence and can serve as a template for ligation of more probes. Successful ligation of the probes depends on the proximal positioning and perfect base pairing of the 3′ end of one probe with the 5′ end of the other to complete a complementary DNA strand. DNA polymerase and dNTPs are used to extend the probe and fill the gap region prior to ligation with DNA ligase. Accumulated ligation products are detected by incorporation of labeled probes.

BRANCHED DNA ASSAY

Quantitative branched DNA (bDNA) assays are commercially available from Bayer for HBV, HCV, and HIV-1 (Table 29.1). With bDNA technology, organisms are disrupted, releasing nucleic acid to be captured by multiple capture probes that are bound to a solid surface (Figure 29.6) (Wolk et al. 2001). Target probes hybridize to both microbial nucleic acid and signal amplification multimer (branched DNA or bDNA). Finally, enzyme-labeled probes hybridize with bDNA structures. Detection occurs via a chemiluminescent process. The amount of signal generated is directly related to the amount of target nucleic acid present in the specimen.

CYCLING PROBE TECHNOLOGY

Cycling Probe Technology (CPT) (Velogene™; ID Biochemical, Vancouver, British Columbia, Canada) is an isothermal process that can be used to detect target DNA (Wolk et al. 2001). Commercial applications of CPT are available as Velogene test kits for methicillin-resistant *Staphylococcus aureus* (MRSA) and vancomycin-resistant enterococci (VRE). Briefly, the technology utilizes a synthetic probe, a chimera of DNA-RNA-DNA, which is labeled at the 5′ end with fluorescein dye and at the 3′ end with a biotin molecule (Figure 29.7). The fluorescein/biotin-labeled probes are designed to anneal with a specific DNA template. Once the probes anneal, RNase H is added to cleave the RNA portion of the bound probe. The shorter, cleaved probe fragments will then dissociate from the target, releasing the target for use in subsequent cycles. In the presence of a specific target molecule, cleaved probes accumulate with each cycle. The test is designed so that a colorimetric reaction will occur in the presence of uncleaved probe; no color will occur in the presence of cleaved probes (specific target). To detect the presence or absence of cleaved probes, an antifluorescein antibody coupled to horseradish peroxidase (Ab/HRP) is added and the reaction is transferred to microtiter wells coated with streptavidin molecules. An alternative test using a nitrocellulose strip has been manufactured for detection of uncleaved probes. All probes will bind to streptavidin, but only the uncleaved probes will bind to the Ab/HRP. A substrate for horseradish peroxidase is added, and

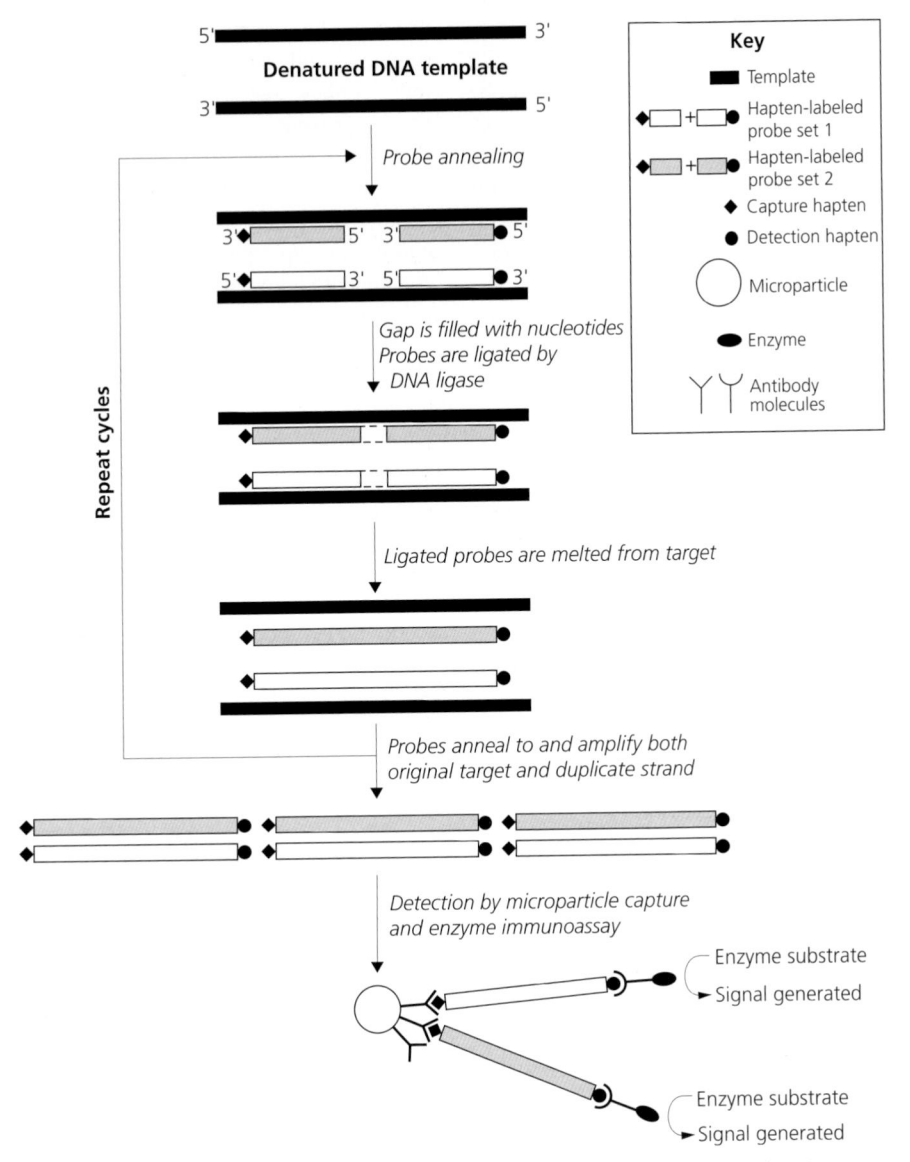

Figure 29.5 *Ligase chain reaction (LCR) (see text for details). Reprinted with permission from Wolk et al. 2001.*

color development is measured to determine the amount of uncleaved probe present in the well (or in the region of the strip). The color development is inversely proportional to the original amount of target DNA present.

NUCLEIC ACID SEQUENCING

Nucleic acid sequencing has several applications in the clinical microbiology laboratory (Wolk et al. 2001). For example, the technology can be used to determine HBV, HCV, HIV-1, and HPV genotypes, to determine the DNA sequences of bacteria and fungi for microbial classification postculture, and to identify mutations associated with resistance to antimicrobics.

Most DNA sequencing methods are based on a modification of Sanger dideoxynucleotide triphosphate (ddNTP) chain termination chemistry in which synthetic ddNTP molecules that lack 3′ hydroxyl groups normally

present in the natural molecules are added to a DNA synthesis reaction (Figure 29.8) (Wolk et al. 2001). Whenever a ddNTP molecule is incorporated into the growing DNA strand (in lieu of the usual dNTPs), the synthesis reaction is terminated. For dye-terminator cycle sequencing, four different fluorescent dyes are incorporated into each of the four ddNTPs (ddATP, ddCTP, ddGTP, and ddTTP), and the end result is a collection of DNA fragments of different lengths labeled with different dyes depending on the terminal nucleotide. Reaction end products are separated by electrophoresis, excited and detected by a laser. The resulting digital data are processed into electronic DNA sequence data. Sequence data are analyzed by special computer software and are displayed either as linear strings of one-letter nucleotide codes or as a graphic electropherogram.

The generation of large numbers of DNA sequences has created new needs for archiving and updating

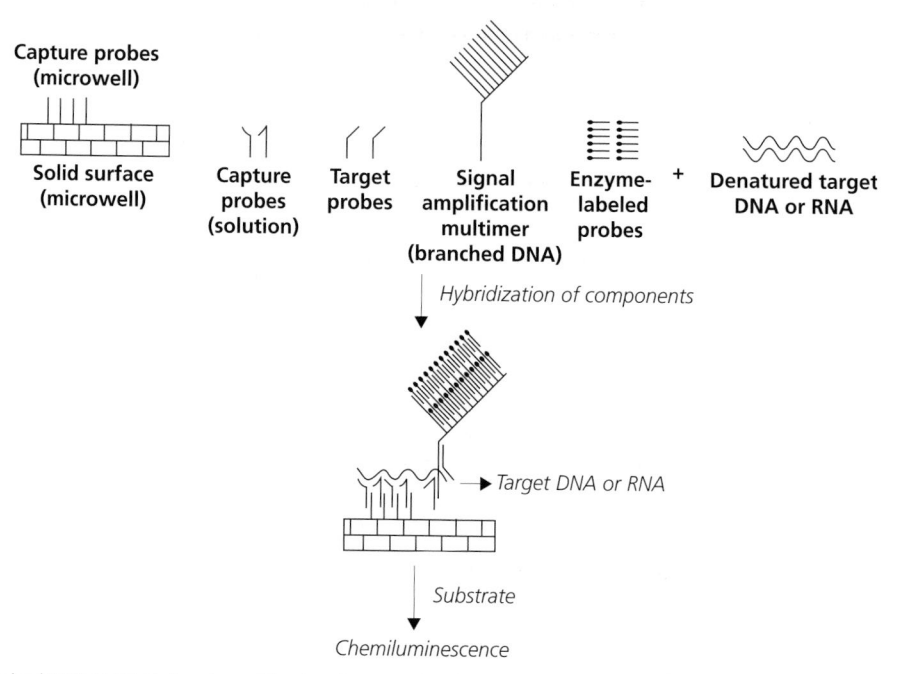

Figure 29.6 *Branched DNA (bDNA) signal amplification (see text for details). Redrawn with permission from Wolk et al. 2001.*

bioinformatics data and validated sequence information. Although several microbes have been sequenced, the vast majority of microbes, even clinically important microbes, have not yet been sequenced. Furthermore, although microbial genomic databases exist, there are limitations to these databases. The web-based genomic database, GenBank, established in 1988, is available as a national resource of molecular biology information from the National Center for Biotechnology Information (www.ncbi.nlm.nih.gov). Related databases such as the rRNA WWW server (www.oberon.rug.ac.be:8080/rRNA), the Ribosomal Database Project II (www.rdp.cme.msu.edu/html), the Institute for Genomic Research (www.rdp.cme.msu.edu/html), and the Ribosomal Differentiation of Medical Microorganisms (www.ridom.de) provide nonproprietary sources of genomic information (Wolk et al. 2001).

While public databases contain a wealth of information, certain caveats must be noted if the data are to be used by clinical microbiology laboratories (Wolk et al. 2001). The sequence data and the descriptive data included with each entry may not be peer-reviewed, so any use of the information contained in these databases must be made with caution. The quality of the sequence data can vary and some sequences may contain errors. Furthermore, although these are large databases, one cannot be assured that they contain information for every gene for every organism. For this reason, well-validated and comprehensive databases are needed for use in clinical microbiology laboratories. The commercially available MicroSeq® 16S rDNA Bacterial Sequencing kit (Applied Biosystems) combines a PCR and cycle sequencing kit that can be used with the commercially available MicroSeq® software and the commercially available MicroSeq® 16S rDNA database for organism identification. Although this approach is useful to aid in characterization of many organisms, its use is limited by the relatively small size of the MicroSeq® database.

Several commercial systems have been designed for nucleic acid sequencing and sequence analysis. The TRUGENE™ assays (Bayer Healthcare LLC) use dye primer chemistry (as compared with dye terminator chemistry) and a two-dye system in a proprietary (CLIP™) sequencing system. These assays identify mutations in the HIV-1 protease and reverse transcriptase genes, mutations in the HBV surface antigen and polymerase genes, and genotypes of HCV (based on the 5' NC region) and HBV. Sequence data are obtained and compared with a proprietary sequence. Sequence data are regularly updated and interpreted using proprietary interpretation codes and software.

The line probe assay (VERSANT™ genotype LiPA assay, Bayer Healthcare LLC) is a modification of Southern blot technology in which reverse hybridization occurs with multiple specific nucleic acid probes arranged in a multiline format on a nitrocellulose strip (Mitchell et al. 2002). After PCR is performed using a biotinylated primer, the amplified DNA is hybridized to a strip containing immobilized oligonucleotide probes. Subsequently, streptavidin-alkaline phosphatase is used to detect the hybridized biotinylated amplified DNA. Alkaline phosphatase acts on a chromogenic substrate added to the system to enable color development at locations on the strip where specific DNA is bound. The result is a fingerprint of the hybridization

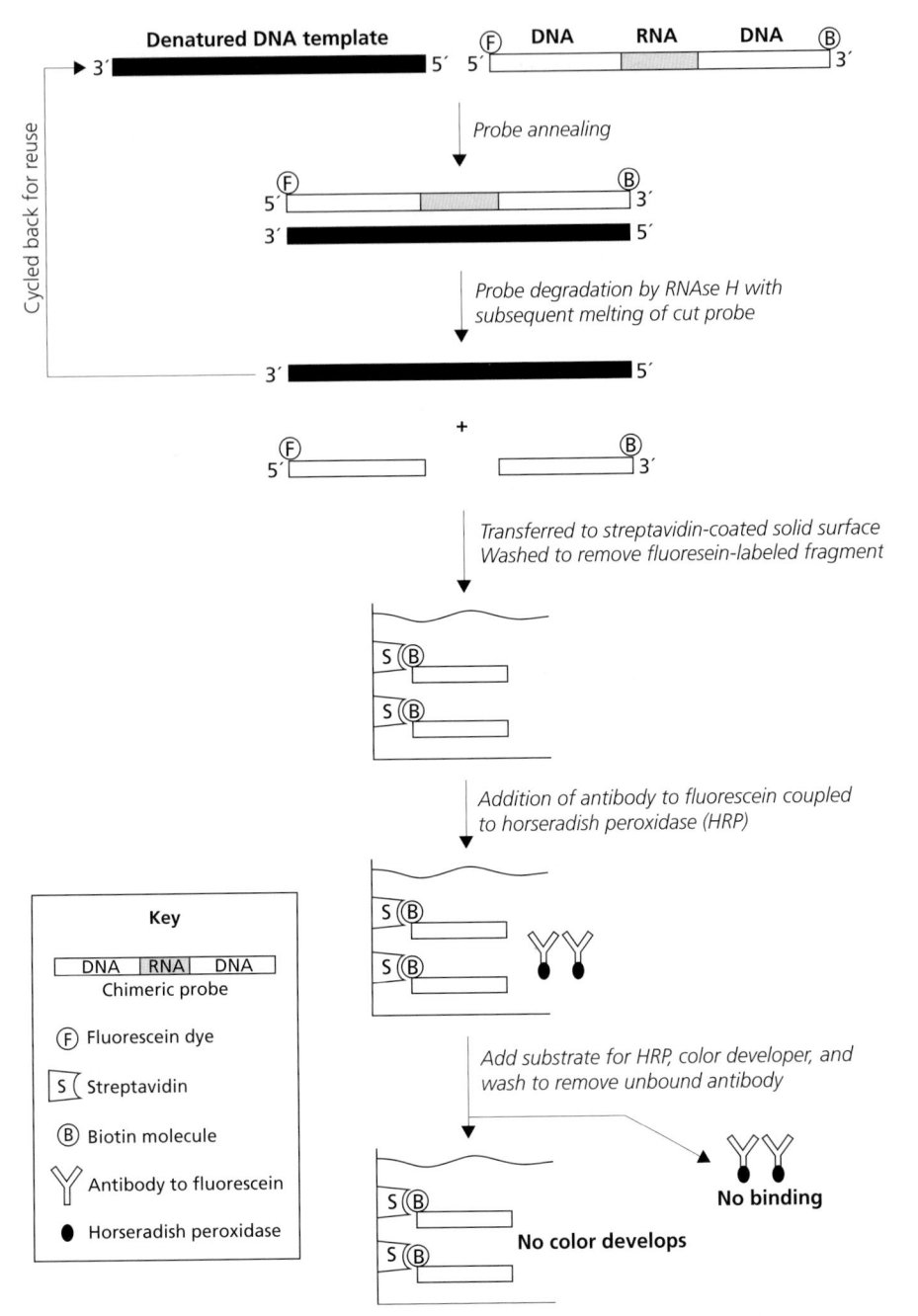

Figure 29.7 *Cycling probe technology (see text for details). Reprinted with permission from Wolk et al. 2001.*

pattern from the PCR reaction product. Patterns are compared with standard patterns. Kits are currently available for HCV genotyping (based on the 5′ NC region), HIV protease and reverse transcriptase mutation analysis, HBV precore promotor and precore region and polymerase mutation analysis and genotyping, and HPV genotyping.

Pyrosequencing™ (Pyrosequencing, Inc., Westborough, MA and Uppsala, Sweden) is a process by which short fragments of DNA sequence (i.e. 20–30 bases) are determined (Figure 29.9) (Wolk et al. 2001). With this technique, a sequencing primer is hybridized to a DNA

template and incubated with DNA polymerase, ATP sulfurylase, luciferase, apyrase, adenosine 5′-phosphosulfate, and luciferin. Complementary dNTPs (dATP shown in Figure 29.9) are serially added to the reaction mixture. If a dNTP is incorporated into the strand, pyrophosphate is released in a quantity equal to the amount of incorporated dNTP. ATP is generated and drives a luciferase reaction so that light is generated, monitored by a charged-coupled device camera, and converted to graphic data of peaks called a pyrogram. The enzyme apyrase modifies residual dNTPs prior to incorporation of dNTPs in subsequent reactions.

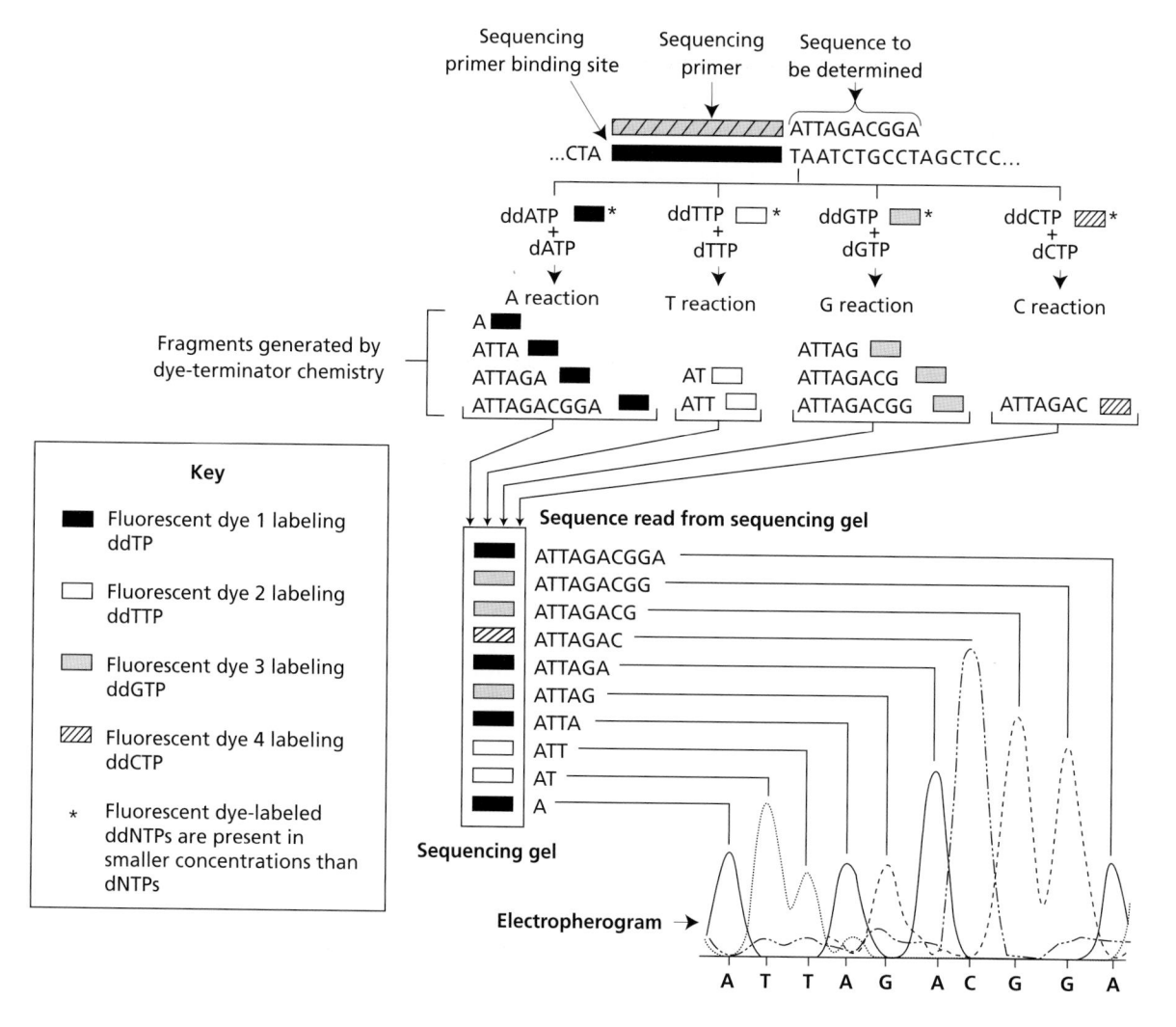

Figure 29.8 *Dye-terminated cycle sequencing of DNA (see text for details). Redrawn with permission from Wolk et al. 2001.*

MOLECULAR EPIDEMIOLOGY

A variety of molecular techniques for epidemiological strain typing are currently available. Genomic DNA restriction site analysis involves comparison of the number and sizes of DNA fragments produced by digestion of DNA with a restriction enzyme. Because of the specificity of restriction enzymes, complete digestion of an organism's genomic DNA by a specific restriction enzyme provides a reproducible array of fragments that can be separated by agarose gel electrophoresis and visualized by staining with ethidium bromide. Variations in the array of fragments generated by a specific restriction enzyme are called restriction fragment length polymorphisms (RFLP). Changes to RFLP patterns can result from gene sequence rearrangements, insertions or deletions of DNA, or base substitutions within the restriction enzyme cleavage sites. The disadvantage of this technique is that the genomic restriction fragments are usually too numerous and too closely spaced for easy analysis.

To simplify genomic DNA RFLP analysis, several techniques are available to limit the number of fragments for comparison. One approach is to probe the genomic DNA restriction fragments with labeled probes targeting multicopy nucleic acid sequences such as insertion sequences, bacteriophage DNA, or rDNA (ribotyping). The RiboPrinter® System (DuPont Qualicon, Wilmington, DE, USA) is a commercially available ribotyping system that can generate, analyze, compare, and store genetic 'fingerprint' information.

Another approach is to generate large fragments of chromosomal DNA by using 'rare cutting' restriction enzymes that recognize infrequent restriction sites and allow separation of fragments by special electrophoretic procedures. Due to shearing, intact DNA required for the generation of large DNA fragments cannot be isolated using conventional methods. Furthermore, large DNA fragments cannot be efficiently resolved by conventional agarose gel electrophoresis. Pulsed-field gel electrophoresis (PFGE) solves both of these problems. Organisms are embedded in agarose and lysed in situ,

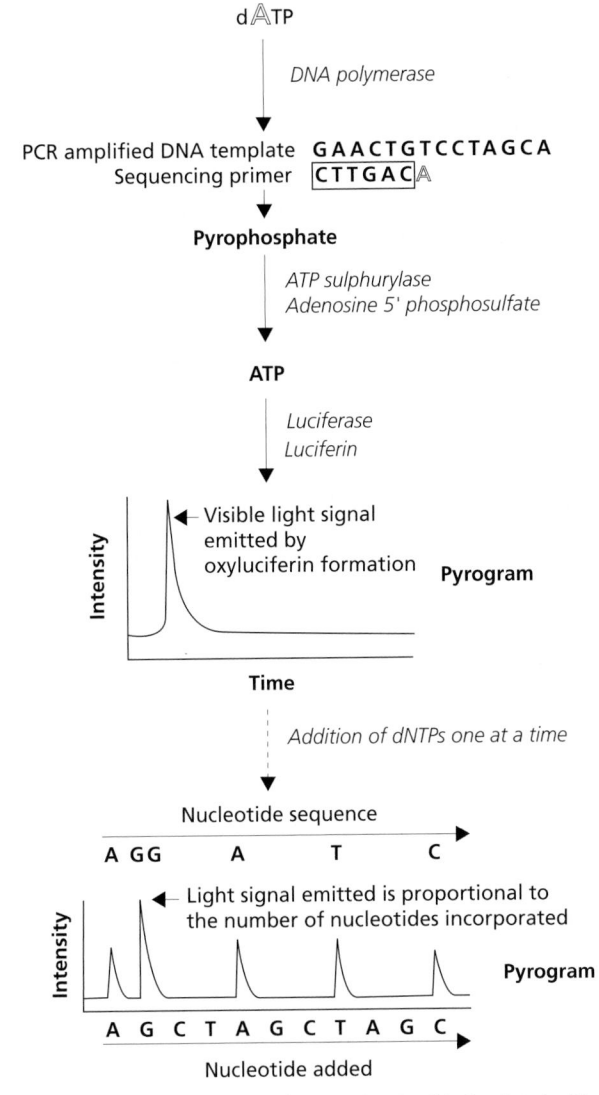

dATP

DNA polymerase

PCR amplified DNA template   G A A C T G T C C T A G C A
Sequencing primer   C T T G A C A

Pyrophosphate

ATP sulphurylase
Adenosine 5' phosphosulfate

ATP

Luciferase
Luciferin

← Visible light signal
emitted by
oxyluciferin formation   **Pyrogram**

Intensity

Time

Addition of dNTPs one at a time

Nucleotide sequence

A GG   A   T   C

← Light signal emitted is proportional to
the number of nucleotides incorporated

Intensity   **Pyrogram**

A G C T A G C T A G C

Nucleotide added

Figure 29.9 *Pyrosequencing (see text for details). Reprinted with permission from Wolk et al. 2001.*

and the chromosomal DNA is digested with restriction enzymes that cleave infrequently, thereby generating large (macrorestriction) restriction fragments (Wolk et al. 2001). Slices of agarose containing chromosomal DNA fragments are inserted into the wells of an agarose gel. Restriction fragments are resolved into a pattern of discrete bands on the gel by an apparatus that switches the direction of current in a predetermined 'pulsed-field' pattern, resulting in the smaller fragments migrating faster through the gel than the larger ones. The DNA restriction patterns of the isolates are then compared with one another to determine their relatedness. PFGE analysis provides a global chromosomal overview, scanning most of the chromosome, but it has only moderate sensitivity since minor genetic changes may go undetected. Interpretative guidelines for PFGE have been published (Tenover et al. 1995).

AFLP® Microbial Fingerprinting (Applied Bioscience) produces a distinctive DNA fingerprint by selective PCR amplification of restriction fragments of the entire microbial genome. The AFLP® procedure includes the preparation of an AFLP template where genomic DNA is digested with two restriction enzymes ('rare cutter' and 'frequent cutter'), which produce cohesive fragment ends and cut DNA with different frequencies. Following digestion, genomic restriction fragments are modified by ligation (joining) of synthetic, double-stranded oligonucleotide adapters with ends complementary to those of the restriction fragments. Thus, after the ligation step, genomic restriction fragments have termini of known sequences. Such an AFLP template is submitted to highly stringent PCR amplification with primers complementary to the adapter/restriction site sequences. Amplified fragments are separated by electrophoresis and visualized either directly by gel electrophoresis or by the laser detection system of an automated sequencing instrument yielding specific patterns (fingerprints). The different patterns are compared with one another to determine the relatedness of strains.

'Sequence-based' typing involves determining the sequence of portions of several microbial housekeeping genes, a procedure termed multilocus sequence typing (MLST). This technique can generate reproducible isolate profiles that can be compared against standardized databases.

MICROARRAYS

DNA microarray technology has the potential to assess and visualize thousands of nucleic acid templates (targets) simultaneously (Wolk et al. 2001). The underlining principle is nucleotide hybridization, which takes place on a 'chip,' a solid platform (e.g. glass, silica, nylon) containing genetic material (e.g. cDNA, oligonucleotides) arrayed in a predetermined fashion. When the chip is probed with cDNAs that are made from an unknown sample of RNA from a tissue or cell, complementary sequences will hybridize with various sequences already positioned on the chip. By monitoring the hybridization pattern of the unknown cDNA, the identity of the hybridized nucleic acids can be determined. The patterns can indicate which RNAs were in the tissue and therefore which genes are being expressed. Microarray technology can also be used to identify gene sequences and mutations associated with HIV-1 drug resistance (e.g. GeneChip® HIV PRT Plus Probe Array Design for analysis of subtype B protease and reverse transcriptase genes; Affymetrix, Santa Clara, CA, USA), but there is only limited support for its clinical use. Currently, a limitation of microarray technology is the high cost for instrumentation and disposable supplies. In addition, there is an inability to easily analyze the enormous amount of information potentially available through this technology.

CONCLUSIONS

Molecular diagnostic testing is revolutionizing clinical microbiology enabling testing that is faster and more sensitive than conventional methods. The field is rapidly changing as new technology evolves for specimen preparation, nucleic acid amplification, and nucleic acid sequence analysis and interpretation.

ACKNOWLEDGMENTS

We acknowledge Jeffrey J. Germer BS, P. Shawn Mitchell MS, and Joseph D. Yao MD for their insightful review of this manuscript. Text based on Wolk et al. (2001).

REFERENCES

Al-Soud, W.A. and Radstrom, P. 2001. Purification and characterization of PCR-inhibitory components in blood cells. *J Clin Microbiol*, **39**, 2, 485–93.

Al-Soud, W.A., Jonsson, L.J. and Radstrom, P. 2000. Identification and characterization of immunoglobulin G in blood as a major inhibitor of diagnostic PCR. *J Clin Microbiol*, **38**, 1, 345–50.

College of American Pathologists. 2001. Molecular pathology checklist, http://www.cap.org/apps/docs/laboratory_accreditation/checklists/checklistftp.html.

Enns, R.K., Bromley, S.E., et al. 1995. *Molecular diagnostic methods for infectious diseases; approved guideline NCCLS document MM3-A*. Wayne, PA: NCCLS.

Espy, M.J., Rys, P.N., et al. 2001. Detection of herpes simplex virus DNA in genital and dermal specimens by LightCycler PCR after extraction using the IsoQuick, MagNA Pure, and BioRobot 9604 methods. *J Clin Microbiol*, **39**, 6, 2233–6.

Espy, M.J., Uhl, J.R., et al. 2002. Detection of vaccinia virus, herpes simplex virus, varicella-zoster virus, and *Bacillus anthracis* DNA by LightCycler polymerase chain reaction after autoclaving: Implications for biosafety of bioterrorism agents. *Mayo Clin Proc*, **77**, 7, 624–8.

Germer, J.J., Heimgartner, P.J., et al. 2002. Comparative evaluation of the VERSANT HCV RNA 3.0, QUANTIPLEX HCV RNA 2.0, and COBAS AMPLICOR HCV MONITOR version 2.0 Assays for quantification of hepatitis C virus RNA in serum. *J Clin Microbiol*, **40**, 2, 495–500.

Germer, J.J., Lins, M.M., et al. In press. Evaluation of the MagNA Pure LC instrument for extraction of hepatitis C virus RNA for the COBAS AMPLICOR™ Hepatitis C Virus Test, version 2.0. *J Clin Microbiol*, **41**, 3503–8.

Madej, R.M., Caliendo, A.M., et al. 2001. *Quantitative molecular methods for infectious diseases; proposed guideline MM6-P*. Wayne, PA: NCCLS.

Mitchell, P.S., Sloan, L.M., et al. 2002. Comparison of line probe assay and DNA sequencing of 5' untranslated region for genotyping hepatitis C virus: description of novel line probe patterns. *Diagn Microbiol Infect Dis*, **42**, 3, 175–9.

Sloan, L.M., Hopkins, M.K., et al. 2002. Multiplex LightCycler PCR assay for detection and differentiation of *Bordetella pertussis* and *Bordetella parapertussis* in nasopharyngeal specimens. *J Clin Microbiol*, **40**, 1, 96–100.

Tenover, F.C., Arbeit, R.D., et al. 1995. Interpreting chromosomal DNA restriction patterns produced by pulsed-field gel electrophoresis: Criteria for bacterial strain typing. *J Clin Microbiol*, **33**, 2233–9.

Wolk, D., Mitchell, S. and Patel, R. 2001. Principles of molecular microbiology testing methods. *Infect Dis Clin N Am*, **15**, 4, 1157–204.

30

Bacterial immunoserology

ROGER FREEMAN

INTRODUCTION AND OVERVIEW

Bacterial immunoserology has long been an important component of the diagnosis of infectious diseases. As is so often the case in medicine, the subject is not precisely defined or confined by its title, nor is it always or necessarily underpinned by a thorough scientific understanding of the mechanisms on which it is based. Indeed, rather than immunology being the science out of which bacterial immunoserology emerged, the reverse is often more true. Empirical observations made by early bacteriologists in the pursuit of clinical diagnosis were often the platforms for subsequent scientific immunological exploration.

Bacterial immunoserology is also part of the long tradition in diagnostic microbiology, whereby parallel developments in other medical sciences have been adopted into the subject to satisfy the understandable demands of clinicians for ever more rapid results.

Bacterial immunoserology also illustrates well the need to revisit subject areas in medicine regularly in the light of technological advances. The most recent is the advent of molecular biology. To give two examples: First, the advent of recombinant antigen techniques now allows the use of antigens in amounts well beyond those naturally available, thus unlocking the ability to devise tests which were theoretically, even obviously, attractive

in the past but simply impossible to deploy on any useful scale. Obvious examples include antigens not expressed extracellularly (e.g. pneumolysin in *Streptococcus pneumoniae*) or antigens from organisms never or hardly ever cultured successfully (e.g. *Treponema pallidum*). Second, the developments in proteomics, transcriptomics, and microarrays will permit a more logical and precise definition of which antigens (and in which form) might be useful for further study.

All this being so, bacterial immunoserology currently comprises a whole range of methods, often empirical, occasionally paradoxical, but above all practical, deployed to aid in the diagnosis of infectious diseases.

The subject can be divided into two clear areas. First, aiding diagnosis by measuring the response of the patient (or patients), for instance by finding antibodies to a particular organism or its products. Second, by using immunological reagents (usually produced by animals or from another nonhuman source) to detect organisms or their products. These two approaches will be considered separately.

First, however, it is necessary, briefly, to consider the various responses that bacterial invasion evokes. Detailed and comprehensive accounts of the many components of the immune response are found elsewhere in these volumes. What follows is a greatly simplified overview, suitable only for the present purpose.

THE MULTILAYERED RESPONSE TO INFECTION

The immunology of infection is frequently portrayed as a battle in which the aggressors (microbes) gain footholds in the body by overcoming the standing defenses, such as skin, mucous membranes, etc., and then begin to multiply, spreading further, creating more local damage. At the same time, a wide range of microbial products (enzymes, toxins, etc.) are released, which may enter the circulation and cause damage at sites distant from the portal of entry. The body responds in diverse ways, each designed to limit the damage, localize the extent of invasion, and, where necessary, neutralize the toxicity of elaborated bacterial products. Outcomes range from resolution (implying a return to total normality, as though no infection had ever occurred, as for instance in pneumococcal pneumonia) through varying degrees of residual damage (with consequent morbidity) to death.

There is an order and a chronology to these events, which allows bacterial immunoserology to be applied.

THE ACUTE-PHASE RESPONSE

The acute-phase response (APR) is the body's generic response to any significant injury, but its magnitude is greatly enhanced if infection is involved or supervenes. There are three pathological components: the development of local hyperemia as blood vessels dilate, the local exudation of certain plasma proteins, and the attraction and diapedesis of white blood cells (predominantly neutrophils) into the tissues at the site of damage. This is not a passive process. Indeed, it is the consequential responses of the body in augmenting the white cell response and the plasma protein response that make the APR a valuable phenomenon in the diagnosis of bacterial infections.

As neutrophils are attracted to the site and consumed, the bone-marrow neutrophil reserves are immediately released into the circulation and the bone-marrow stem cells respond by producing more neutrophils. This results in a peripheral leukocytosis with a pronounced neutrophilia, containing predominantly juvenile neutrophils, the so-called 'shift to the left.' In inflammation due to infection, these events are considerably exaggerated, probably due to the ability of most bacteria to degrade complement (a prominent acute-phase protein), the degraded or 'split' products of which are an extremely powerful chemotactic stimulus for neutrophils.

Within 2–4 h of injury/infection, the liver begins to synthesize and release large amounts of a range of plasma proteins (acute-phase proteins), some of which are normally found in the plasma in small amounts (e.g. complement) and some of which only appear under these circumstances, e.g. C reactive protein (CRP). The magnitude of this response (i.e. the levels of these easily measured proteins) gives an index of the degree and extent of the inflammatory damage being undergone by the body, from which inferences can be drawn. Thus, an apparently inappropriately large APR response (for instance, a very high white blood cell count, a very pronounced neutrophilia, and a very high CRP level) would imply that either a noninfective injury was much larger or more severe than originally suspected, for instance having gone on to necrosis, or, commonly, that infection was present.

The ease and speed with which both the leukocytosis and CRP can be measured and their equally rapid response to successful control of the APR, for instance by correct chemotherapy of an infection, make them useful and versatile tools in clinical microbiology. They are especially useful when the patient has not responded as well as expected to antibiotic therapy, say by continuing to be febrile. There being many other reasons for the fever, a normal or rapidly declining leukocytosis and CRP might eliminate lack of control of the infection as the reason. Conversely, if both remained high, it might be wise to alter therapy, take further samples, or even to contemplate superinfection with a different organism.

A major disadvantage of the APR is that it is nonspecific, i.e. the organism is not identified, or not, at any rate, by this means. However, the APR can be very specific indeed in a negative sense. Thus, if an organism is isolated from a blood culture or another pathologically very significant site, such as cerebrospinal fluid, it will be very unusual indeed if all parameters of the APR, and particularly the peripheral white blood cell count (WBC) and the CRP, are entirely normal.

This very combination of events occurred many times several years ago, when the organism *Stenotrophomonas maltophilia* (then called *Pseudomonas maltophilia*) was isolated from blood cultures in apparently well, or at least apparently nonsepticemic, patients. The clearly negative APR parameters led to a search for another explanation and it emerged that blood culture samples were being cross-contaminated with *Ps. maltophilia* from nonsterile biochemistry sample tubes, filled at the same venepuncture. The condition dubbed 'pseudobacteremia' was thus identified (Jumaa and Chattopadhay 1994). Similar intelligent uses of the APR have undoubtedly saved many patients from unnecessary antibiotic therapy.

Problems in interpretation of the APR

Problems may be encountered with interpreting the APR under several circumstances.

First, if either of the major contributing organ systems is grossly deranged, then production of the APR components may be inappropriately low or even absent. Thus, the WBC response may be abnormally low in marrow failure. Acute-phase proteins may not be produced in normal amounts in patients with liver disease, whether acute or chronic. Rare conditions in which complement or its components are not manufactured may greatly dampen the amplitude of the APR.

Second, there are rare instances in which organisms, especially *Staphylococcus aureus*, can survive often for many years, in a 'privileged site,' immured from the APR. A good example is the so-called Brodie's abscess (Stephens and McAuley 1988) in which the infection survives as an indolent localized pocket of infection in a sclerosed area of bone, often the residue of a childhood bacteremia that did not progress to acute osteomyelitis. After a period the APR declines, although the CRP is often not entirely normal. However, the patient very occasionally has a febrile attack or even a rigor. If blood cultures are taken, they may be positive. Brodie's abscesses are easily seen on X-rays. The erythrocyte sedimentation rate (ESR) is often raised, but this is as a result of the dysglobulinemia and secondary anemia (often called the 'anemia of inflammation') due to longstanding production of humoral antibody (see below), rather than the presence of any acute-phase protein.

Third, in severe infection due to gram-negative bacilli, such as *Escherichia coli*, the WBC may fall precipitately rather than rise. This effect is thought to be due to the direct effect of endotoxin (lipopolysaccharide (LPS)) on the neutrophils, causing their degranulation and resulting in some of the features of disseminated intravascular coagulation (DIC). Prompt and effective therapy of the presumptive gram-negative septicemia will often result in a rapid and gratifying resolution of the leukopenia.

Recent developments in measuring the acute-phase response

There has been a longstanding wish to find a component within the APR that is specific to infection and, therefore, will allow the distinction to be made between noninfective inflammation and sepsis without recourse to the subjective estimates of magnitude alone. In any case, it is frequently the case that infective and noninfective values overlap. This desire is particularly strong in those caring for infants and children (Nuntnarumit et al. 2002) and those dealing with intensive care patients of any age, in all of whom physical signs are notoriously unreliable. There have been two lines of investigation.

First, several candidate acute-phase proteins have been put forward as being disproportionately raised or even raised at all in bacterial infection. The most recent candidate is procalcitonin (PCT). The evaluation of PCT has parallelled that of many other candidates over the years, in that it seems that PCT does indeed rise higher in bacterial inflammation than in noninfective syndromes (or viral infection), but still not exclusively so (Balc et al. 2003; Schuttrumpf et al. 2003).

Second, it is known that one component of the APR does behave very differently in infection. This is the neutrophil. When neutrophils phagocytose a live bacterial cell, they must first kill the bacterium prior to being able to digest it. The intracellular killing is achieved by a respiratory energy burst fuelling an oxidative microbicidal system, which results in the organism being killed by halides (probably by iodination). This efficient system will kill *S. aureus* within approximately 20 min (Klebanoff 1980).

It is possible microscopically to identify such 'activated neutrophils' by supravital staining methods wherein tetrazolium (for instance, nitroblue tetrazolium (NBT)) salts are reduced to insoluble formazan, which is visible as a deposit in the neutrohil phagosome. It was, therefore, hoped that estimation of the percentage of peripheral blood neutrophils reducing NBT would be much higher in patients with severe infection. In general, this is so, but extensive evaluation concluded that, once again, the test was not specific enough for the purpose, although it retains an important place in the testing of neutrophil function (see below). Attempts to find an 'infection-specific' component of the APR will doubtless continue. The prize is immense.

Final comments on the APR

Clinicians and microbiologists evaluate the APR in most patients daily without formally recognizing the exercise. The mundane nature of the APR is, in fact, one of its most important aspects. The ability rapidly to detect inflammatory damage, including that due to, or complicated by, bacterial invasion, rapidly to deploy an effective systemic response to the site that can begin to impede the progress of the damage and moreover deliver effector cells that can kill pathogenic bacteria and scavenge debris, is a triumph of evolution. This mechanism deals, mostly successfully, with the vast majority of infective assaults on the human body. Its manifestations range from small pustules on the skin to large areas of suppuration, such as peritonitis. Studies of neutrophil function and of other aspects of the APR suggest that low-level ongoing activity of this system, outside the occasions when it is recognizably evoked, is probably essential to the maintenance of the integrity of mucosal surfaces and the skin. The development of 'mucositis' in patients in whom neutrophil function or numbers is depleted also suggests this.

If the APR is successful, invading organisms do not, largely, survive to present an ongoing threat and, hence, do not require further response. There are, however, organisms that survive the APR.

THE GRANULOMATOUS RESPONSE

In an immunocompetent host, the granulomatous response is the exception rather than the rule and is induced by organisms that can survive intracellularly. Thus, *Mycobacterium tuberculosis*, *Listeria monocytogenes*, and *Brucella* spp. will be the organisms predominantly involved in temperate countries. Other species will be more common in other climates and environments.

The unity of the above hypothesis is strikingly supported by the occasional example of what happens when immune defects in, for instance, the neutrophil bactericidal mechanisms, detected by a demonstrable inability to reduce NBT, even in the presence of live bacteria, force the body to react to unkilled common organisms, such as staphylococci, as though they were intracellular bacteria, such as mycobacteria. This results in granulomatous reactions to these mundane bacteria and produces a clinical disease called chronic granulomatous disease (CGD) (Gallin and Fauci 1983). CGD was for some time thought to be an atypical form of tuberculosis, but the discovery of the intracellular killing system and of defects within it showed that affected children were forced to deal with all bacteria in a way in which their healthy peers reserved for rarities. Other examples of deficits, such as myeloperoxidase deficiency, and so on, serve to emphasize the pivotal importance of the neutrophil bactericidal pathway in determining how the body reacts to bacteria.

There are two consequences of the granulomatous response relevant to bacterial immunoserology.

First, the surviving organisms will enter the active specific immunity system, antigen will be processed, and both cellular immunity and humoral immunity will result. Thus, it will be possible to devise tests based on (1) humoral antibody detection and/or (2) evocation of specific cellular immune responses to antigens to diagnose infection with these organisms.

Second, it is highly possible that the bacteria involved will remain alive within the body, albeit localized and contained by the granulomatous reaction. Frequently, it seems, these contained organisms adopt a different physiological and metabolic state, and this may even extend in some species to a viable but noncultivable state. In terms of the infection or disease associated with the organism, it is often described as a latent phase. Latent infections may and do reactivate, often when an intercurrent illness, medical procedure, acquired immunoincompetence, or simply advancing age depresses the immune system below the level required for continued containment of the organism. Thus, it will be possible not only to devise tests based upon (1) humoral antibody detection and/or (2) evocation of specific cellular immune responses to diagnose infection with these organisms, but also, under certain circumstances, to identify latently infected but seemingly healthy individuals who will be at risk of subsequent symptomatic disease. Pre-emptive therapy may therefore be possible.

Humoral antibody production

Figure 30.1 is a diagrammatic and idealized depiction of humoral antibody production after the introduction of pathogenic bacteria into the body, subsequent bacterial infection, and development of active acquired immunity in a patient never before exposed to this particular organism.

All specimens for serological studies must be seen as individual static 'snapshots' in a dynamic process, the natural history of the disease. This is easier to appreciate in virology since viral diseases have, until recently, run

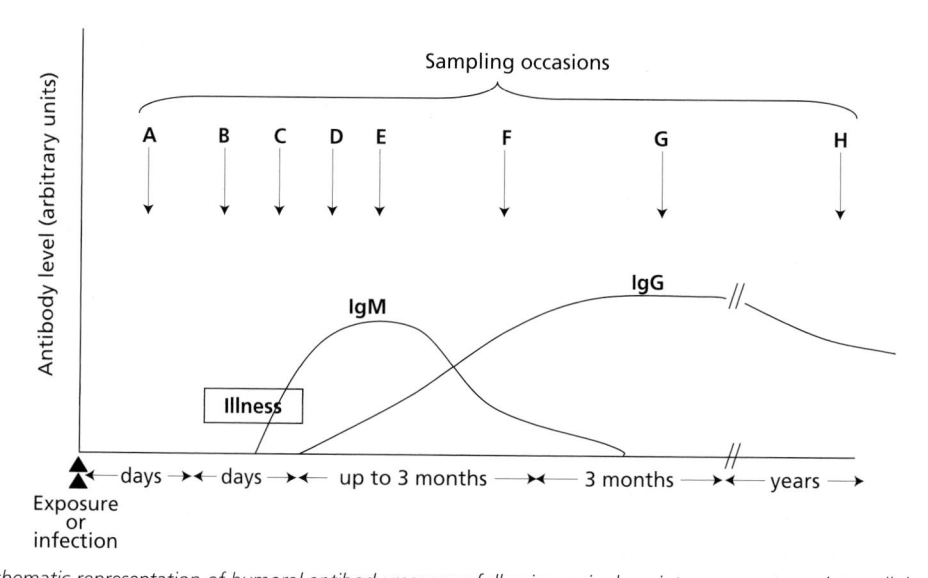

Figure 30.1 *Schematic representation of humoral antibody response following a single point exposure to an intracellular infectious agent.*

their course, there being no possibility of intervention. Nonetheless, the same view of many nonpyogenic, granulomatous bacterial diseases is equally important.

Before considering the differing significance of results from specimens taken at the sampling occasions given in Figure 30.1, it is important to clarify two matters of simple practical importance.

First, this account will shortly be referring to antibody 'titers,' a measure of the amount of antibody/unit volume determined by serial dilution (titration) of a sample to the endpoint of reactivity. These will be referred to as, for example, 'a titer of 1 in 32 (or 1/32).' This is a scientifically incorrect statement that has become correct by common usage. Strictly, titers are the *reciprocal* of serum dilutions, so the preceding should be 'a titer of 32,' the titer being a measure of the strength of the reaction studied, in this case 32 units/volume. It is a way of comparing the amounts of antibody in two or more different samples. This is an essential concept in the quest to study and compare serial samples, but it can be confusing. Despite many laudable efforts to correct this problem, it persists. To summarize, a sample with a titer of 1/32 has a stronger reaction (i.e. contains more antibody/unit volume) than one with a titer of 1/8.

Second, the technical reproducibility of doubling dilution is one dilution either side of the intended value. Hence, a serum with a titer of 1/32 may, some of the time give a titer of 1/64 and some of the time 1/16. In order, therefore, to be certain that two titers are significantly different, it is insisted that they differ by at least two dilutions. This leads to the stipulation that a significant rise in antibody titer between two samples must be at least four-fold, say 1/8 to 1/32, or 1/10 to 1/40.

Figure 30.1 shows that after infection, there is often a short period in which the patient remains ostensibly unaffected and well (the incubation period). Next follows the clinical illness, during which it is often possible to directly detect the causative agent and so make the diagnosis without recourse to bacterial immunoserology. However, if it is assumed that the organism cannot be cultured, or that inadvertent antibiotic therapy has rendered culture useless, or for any other reason diagnosis cannot be made by direct detection, bacterial immunoserology may be useful.

It is, therefore, important to note that IgM is the first antibody produced by the host, often whilst the clinical illness is still ongoing. IgM is the classical first response and, because of its large molecular size, remains confined to the bloodstream even after its production declines. It can usually be detected for at least 3 months after its first appearance. Unfortunately, however, IgM is a notoriously difficult immunoglobulin with which to contend in assays, mainly because it is so avid or 'sticky' that it frequently reacts with antigens other than that which evoked it. Nonetheless, its appearance (and, therefore, detection) strongly suggests both a first infection with the appropriate organism and its recency.

Shortly thereafter, specific IgG becomes detectable. IgG is a much more easily assayed immunoglobulin and assays based upon it are usually very specific. However, its later appearance and the fact that it is the 'memory' immunoglobulin (i.e. the antibody that persists for years, perhaps for life) make its detection difficult to interpret in a single specimen. To settle a judgment of de novo production (and, therefore, recency) on IgG levels requires evidence of increasing levels over a short period of time. This is the classical immunoserological concept of the rising titer.

Some recent work has drawn attention to the differences in avidity for the same antigen between the IgG produced for the first time and that produced later or after rechallenge with the same antigen. These differences can be demonstrated in technically simple ways. It may, therefore, be possible to devise IgG assays that will also indicate recency of infection. However, although this interesting approach has been pursued in a small number of instances (Prince and Leber 2002), it has not yet gained wide support.

Turning, therefore, to the details in Figure 30.1, it is now possible to discuss the results of assays for both IgM and IgG at the different sampling occasions and to interpret the significance of the results. It will already be evident that the same assay result (especially for IgG) may have an entirely different significance, depending on when it was taken in relation to the disease.

SAMPLE A

No antibody of any sort is detected. Absence of antibody in the incubation period is to be expected, implying that if known to have been exposed, the patient is likely to become ill soon. This is not often a useful observation in bacterial infection, but it is of immense use in the childhood viral exanthema, allowing isolation.

SAMPLE B

Early in the illness and no antibody is yet detectable. Although such samples will hardly ever themselves make a diagnosis, such so-called 'baseline' or 'acute' samples are vital for subsequent comparative tests (see below). Many experienced investigators simply store such samples until later samples are obtained.

SAMPLE C

In later illness, specific IgM is detected. Provided that the specificity of the assay is beyond question, the diagnosis can be made, the presence of specific IgM being evidence of both a novel antibody and of a recently produced antibody. However, further samples should be requested to confirm subsequent IgG production so that specificity is confirmed (see below).

SAMPLE D

In late illness or early convalescence, both IgM and IgG are detected, the specificity of the IgG complementing and confirming the implications of the IgM result. This is a common format for commercial test kits. However, the IgG levels found at this stage may border on the limits of detectability.

SAMPLE E

In established convalescence, IgM can still be detected and IgG is plentiful. It is also now possible to explore an alternative and highly specific way to make the diagnosis. If it can be shown between successive samples that the levels of IgG are significantly increasing with time, this is good evidence that the infection is ongoing or recent. Thus, if the IgG antibody level is greater in E than in D, this will be so and will secure the diagnosis. It will be even more certain if E is compared with B, since B contains no antibody. Hence, the prudence in taking samples early in the illness, though they alone cannot make the diagnosis (see above).

It might be argued that samples taken prior to A might also serve. For instance, serum may be available from an unrelated event in previous years. However, all that can then be said is that the infection has occurred some time between the sample dates of 'pre-A' and E. If this is a long time (months or even years), the significance might be lost in terms of any one specific episode of illness if several have occurred in the interval.

SAMPLES F AND G

IgM is declining to low levels and its index of recency is becoming eroded. Low levels of IgM antibodies (and IgM-like antibodies) can be found in normal states. However, the relatively high levels of specific IgG make recent infection likely, but unless earlier samples are available for comparison, the results on a single sample taken at F might be misleading. Since IgG production is now nearly maximal, comparison of the titer at F with that at G may not help. It might be observed that the need to make the diagnosis is now long past and, therefore, the exercise is pointless. However, it is occasionally very important to establish the nature of an illness undergone several months before, perhaps in a distant location, to ascertain the need to test contacts and to gauge prognosis and the need for further, perhaps invasive, investigations. In earlier days, a common such example was enteric fever.

The serological picture has now gone full circle because similar questions can now be asked as were also asked at point A, although the answers will be quite the opposite. If antibody is detectable during the incubation period, the patient is immune, assuming the antibody detected correlates with protection. It is remarkable how often a previous serum specimen, often submitted for a totally unrelated reason, such as,

for instance, antenatal screening, can be recalled and tested to solve a current problem. Consequently, storage of sera is good practice.

SEROEPIDEMIOLOGY

A very important aspect of humoral antibody testing is the use of the results from individual patients being viewed collectively as those of a population. This is the basis of the very important subject of seroepidemiology, whereby antibody testing of carefully structured populations is designed to address specific questions in relation to the incidence and prevalence of a given illness within the population. Good evidence can be produced to assess the extent of disease, those groups most likely to contract it, and, therefore, those groups most likely to benefit from interventions such as vaccination. This is yet another reason for sera to be stored over a longer period. If storage of all sera is impracticable, the preservation of a collection representative of the wider population and extending through a period of several years is a useful surrogate and will greatly inform epidemiology studies.

Three other points require to be addressed before progressing to look at the application of these principles in real examples.

First, in disease the antigen persists. The situation is not precisely analogous to the single antigen injection given under experimental conditions on which Figure 30.1 is based. Patients may receive both primary and further 'doses' of antigen as the illness continues. Therefore, the speed and magnitude of antibody production will be greater than in experimental observations.

Second, in clinical disease the host is required to respond to a variety of antigens of differing immunogenicity, both polysaccharides and proteins. A spectrum of antibodies results, each directed against a different antigen. These different antibodies may exhibit markedly different degrees of persistence, some being detectable for years, others for only a few months, perhaps reflecting the different immunogenicities of the antigens which elicited them. Whatever the explanation, these phenomena can be exploited to good practical effect in bacterial immunoserology, an excellent example being the contrasts between the O and H antibodies in enteric fever (see below). A side issue of this point is the realization that antibodies evoked by weak (poor) or incomplete immunogens may not only be transient in their appearance in the circulation but also may escape immunological memory and be re-evoked in subsequent infective episodes.

Third, there is little true knowledge of the functions of the many antibodies which can be detected. The point re-emerges that bacterial immunoserology is an empirical, albeit useful, practice. What follows are several examples of the practical application of humoral antibody testing in diagnostic bacteriology.

Enteric fever (infection with *Salmonella* Typhi) is here used as the paradigm. Infections with *Ss. paratyphi A, B*, and *C* behave similarly but are omitted for clarity.

S. Typhi exhibits three well-studied antigens. H, or flagellar, antigens, are protein in nature and unique to *S.* Typhi; O, or somatic, antigens, are polysaccharides and more widely distributed within the *Salmonella* species than in *S.* Typhi alone. The Vi, or virulence, antigen is thought to be capsular in nature and is found only in recently isolated strains.

It is relatively straightforward to produce suspensions of *S.* Typhi that exhibit either only H or only O antigens at a time (the details are unnecessary). Since these suspensions will be particulate, antibody can be detected by demonstrating the ability of the serum to agglutinate the suspension (see below). The test to be described (and many variations on this theme) is often called the Widal reaction.

Typhoid fever follows the ingestion of the organism and has an asymptomatic incubation period of 7–10 days, during which the organisms multiply within the gut lymphoid tissues (Peyer's patches) and subsequently invade the bloodstream to give rise to a particularly serious form of gram-negative septicemia, often characterized by a leukopenia. Before ingestion of *S.* Typhi and during the incubation period, no antibodies will be detectable. As the fever breaks, diarrhea ensues and at this time, or shortly afterwards, both O and H antibodies may be detectable. As the disease continues and, hopefully, thereafter subsides, both antibodies can be detected, although typically O antibody may be more plentiful than H. In late convalescence the H titer continues to rise, but the O titer plateaus. A year later, the H antibody will remain easily detected, but O antibody may have become undetectable once more.

The preceding commentary serves to explain these results. O antibodies rise first but, more importantly, decline quickly after the illness, perhaps because of their nonprotein nature. Though less specific than H antibodies they serve as a good index of recency. Their specificity, however, must be confirmed by the copresence of the highly specific H antibodies, which also serve as the memory antibody in future years, when O antibody is gone. Four-fold or greater rises in antibody titers in either antibody indicate ongoing production. Thus, it is possible to distinguish between a nontyphoidal enteritis, early typhoid fever, recent typhoid fever, and past typhoid fever. No IgM detection was typically employed in the Widal reaction, although it might well have been in the same way as that for the brucella test (see below).

Vaccination against typhoid (and paratyphoid) fever has been commonly practiced since the days of Almroth Wright. Most such vaccines were intended to evoke H antibody, which is at least partially protective. However, such antibodies can confuse the interpretation of the Widal reaction, but in cases of doubt antibodies to the

Vi antigen can be sought. This antigen was traditionally excluded from the vaccine but is present in great amounts in a natural and ongoing infection.

The Widal reaction has fallen from favor for three reasons. Modern cultural and noncultural detection methods make it increasingly outmoded, vaccine changes make it increasingly difficult to interpret, and, most of all, the development of antibiotics capable of being used widely and effectively (especially fluoroquinolones) limits the necessity to make serological and retrospective diagnoses.

BRUCELLOSIS

Brucellosis is a zoonosis. Humans at risk include farm workers, vets, slaughter men, and those in other occupations concerned with livestock and farms. The organism can also be transmitted via the ingestion of unpasteurized milk.

Brucellosis is a disease with discernible clinical stages. It begins with an acute bacteremic phase in which the patient has chills and rigors, is usually severely debilitated, and sometimes has a rash. This phase may then remit and then recur. Sometimes this cycle is repeated several times, the characteristic temperature pattern earning the disease its alternative name of undulant fever. Brucella organisms can usually be isolated from the blood at this stage but successful culture of this very fastidious bacterium is very difficult and cultures are often pronounced negative. Additionally, culture is often discouraged because of a high risk of laboratory-acquired infection. Untreated, the acute phase eventually peters out and the organisms localize into the reticuloendothelial system, principally into the liver and the spleen, together with involvement of bones and joints, especially of the spine. Unsurprisingly, as the infection becomes chronic, the patient becomes debilitated with anemia, low-grade fever, weight loss, and increasing problems with arthritis, especially of the spine. If still unsuspected the disease continues, eventually shortening the life span and occasionally ending in brucella endocarditis.

The difficulties in culturing the organisms make the diagnosis of brucellosis an ideal opportunity for bacterial immunoserology, and it is a particularly instructive example. The main technique used is that of agglutination of whole brucella organisms. Serum taken in the acute phase will contain high levels of specific agglutinins, and this result will of itself strongly suggest the diagnosis of acute brucellosis, the antibody being unusual in the general population. Since at this stage the organism is predominantly in the blood, it also follows that the agglutinins will be largely of IgM. IgM is a pentameric immunoglobulin rich in disulfide bonds, and exposure to 2-mercaptoethanol (2ME) destroys its activity. If, therefore the agglutination test is performed in parallel, first using normal saline as diluent and second using saline plus 2ME, a substantial drop in the agglutination titer in the second test strongly suggests that the agglutinins are IgM in nature and confirms that the patient has acute brucellosis. This is doubly important since the disease responds better and more completely to antibiotic treatment in the acute phase.

As the disease (if undiagnosed and untreated) enters the chronic phase, the titer of agglutinins will fall, but not usually to an undetectable level. However, faced with a patient with a suggestive clinical picture but only very modest levels of antibody in the agglutination test, the diagnosis may be difficult to sustain by this means alone. Brucella organisms (or, perhaps more accurately, processed brucella antigens in the tissues) evoke IgG antibodies in high titers, which react in a complement-fixation test (CFT). Thus, the detection of low-titer agglutinins, the activity of which is not abolished by mercaptoethanol, and a strong reaction to high titer in the brucella CFT is the classical immunoserology of chronic brucellosis, a condition requiring very prolonged antibiotic therapy with drugs that will penetrate granulomata and one that is difficult to declare cured even after many years. Thus, bacterial immunoserology can not only diagnose but also stage the disease and, thereby, give useful guidance to therapy and cure.

There is one caveat. As will be seen below, it is a characteristic of intracellular infection that both cellular immunity and hypersensitivity will be evoked, especially in the chronic phase. Repeated exposure to brucella antigens can result in such hypersensitivity without the necessity of preceding infection with the organism. This can be manifest as a positive 'Brucellin' skin test, which is performed and read in a way analogous to the Mantoux test in tuberculosis (see below). Low levels of agglutinins can accompany this, although the typically negative brucella CFT will suggest an absence of genuine chronic infection. The hypersensitivity nature of the phenomenon can be positively shown by the detection of specific IgA in the serum.

Thankfully, acute brucellosis is now a rare human disease in most advanced countries due to the eradication of the infection in animals by a slaughter and restocking approach. However, patients with chronic brucellosis will continue to be discovered for some time to come. In many laboratories, a single screening test using an enzyme immunoassay technique (see below) is now used.

SYPHILIS

The history of the serological diagnosis of this medically and socially very important disease is almost the history of bacterial immunoserology in microcosm. The protean clinical manifestations of syphilis, resulting in its nickname of the 'Great Imitator,' made the need for an objective and impartial laboratory test essential for differential diagnosis in illness and for the detection of a latent phase of indeterminate duration (up to decades).

The first tests were based on the empirical observation that a liver extract from syphilitic fetal liver appeared to fix complement in the presence of sera from patients with syphilis. The obvious, but erroneous, assumption was that antigens from the then (and still) uncultivable *Treponema pallidum* were the basis of the test, subsequently universally dubbed the Wassermann reaction (WR). In fact, even some of the early observations hinted at two anomalies with the WR. These were the finding of false-positive results (reactors who clearly did not have syphilis) and occasional claims that the test worked almost as well if nonsyphilitic tissue was used as the antigen. Finally, it emerged that the relevant antigen (called cardiolipin) could be found in other normal tissues and was eventually able to be synthesized. It also emerged that the substance reacting with cardiolipin was not, in fact, a true antibody but a serum component evoked, possibly due to tissue damage, in syphilis and some other states and called reagin. (Note: this substance must be clearly distinguished from the reaginic antibodies encountered in allergic states.)

Reagin is nowadays detected by a slide or tube flocculation technique (this latter being a virtually unique serological appearance in which 'agglutination' of a much coarser than normal nature is produced) using a cardiolipin preparation produced to a standard defined by the Venereal Disease Research Laboratories (VDRL) several years ago. This test is inevitably referred to as the 'VDRL test' or, simply, the VDRL.

As experience grew with cardiolipin tests, it became obvious that specificity was a major problem and that false-positives were common. These so-called biological false-positives (BFP) were of two sorts. Many were transient and, interestingly, often associated with diseases or states in which liver function was temporarily deranged. Common examples included viral hepatitis, glandular fever, and pregnancy. When the associated condition resolved, these acute BFPs became negative. A timeframe of 3 months usually sufficed for this. Conversely, chronic BFPs by definition of more than 3 months' duration were frequently associated with much more serious and chronic diseases, typically of the autoimmune variety, such as systemic lupus erythematosus (SLE).

Nonetheless, the VDRL was also still positive in almost all cases of syphilis and retained and still retains its place in the diagnosis and management of the disease because the VDRL titer is an extremely useful index of the activity of syphilis. It has been justifiably called 'the ESR of syphilis.' Its retention, however, made mandatory the devising of other much more specific tests to set the VDRL result in a proper context.

First efforts centered on attempts to grow *T. pallidum* and thereby prepare a highly specific antigen for (at that time) the inevitable CFT. An early partial success was the isolation of what came to be called Reiter's strain. This is not *T. pallidum* but at least the extracted antigen (Reiter protein) was more specific than cardiolipin and conferred a treponemal group status to any antibody detected. This was the basis of the Reiter protein CFT (RPCFT).

Although *T. pallidum* was, and yet remains, uncultivable, the organism can be made to infect and multiply within rabbit testes and on this basis artificial serial passage can be accomplished. The resultant suspensions of live *T. pallidum* so produced were used in a form of neutralization test, whereby the characteristic motility of these live spirochetes was shown under dark ground microscopy to be inhibited by the presence of immune serum. This very laborious and expensive assay was the *Treponema pallidum* immobilization (TPI) test. Immunological technical developments have since allowed the *T. pallidum* suspension to be used as the basis of an indirect immunofluorescent test (fluorescent treponemal antibody (FTA) test). An even more specific version of this approach was then devised, whereby treponemal group antibodies were removed from the sample, prior to testing, by adsorption with the Reiter protein antigen so that any remaining antibodies must be to antigens found on *T. pallidum* only. This is the principle of the FTA (abs) test. *T. pallidum* material was also made to adhere to red cells, allowing the development of a passive hemagglutination test (the TPHA).

In terms of *T. pallidum*-specific tests, it is now possible, using genomics, proteomics, and recombinant antigen technology, to produce a range of tests, utilizing a range of different formats (most commonly enzyme immunoassay assay (EIA)), which detect antibodies specific to only a few epitopes of *T. pallidum*. As will be appreciated, many of these kit-based tests will also be capable of being made specific for IgM antibodies, if needed. Thus, over 100 years of immunoserological development has furnished the tools to decipher the diagnosis of syphilis, but it is testimony to the practicality of the bacterial immunoserology approach that the oldest, least specific test and one that in retrospect never detected an antibody at all, the VDRL, is still an essential component of the syphilis testing portfolio.

Tests of hypersensitivity and cellular immunity

In intracellular infections, a range of immunological phenomena occurs in addition to the production of humoral antibodies. The granulomatous nature of the histological response has been mentioned already. Two further characteristics can also be capitalized upon to devise diagnostic tests. These are the behavior of lymphocytes and other cells in response to an appropriate antigen and the development of hypersensitivity phenomena.

Other chapters will discuss these matters in detail. As always in bacterial immunoserology, this section will concentrate on the practical applications. This account will deal almost exclusively with tests in tuberculosis, which is currently by far the major disease in which these tests are used. Hypersensitivity tests were the first to be developed.

HYPERSENSITIVITY

The fundamentally different immunological processes involved in tuberculosis, the archetypal model of cell-mediated immunity, were recognized from the outset. Koch and other early workers were able to show that primary infection with *M. tuberculosis* could be shown to have occurred by the alteration in the body's response to intradermal injection of a protein extract of the organism (originally, Koch's old tuberculin; later purified protein derivative (PPD)). Before infection, no reaction occurred, whereas after infection, an area of redness and induration within the skin appeared 72 h after injection. It was also noticed that reinfection with *M. tuberculosis* in animals with a positive skin test provoked a brisk, even necrotic, reaction at the site of reinfection (Koch's phenomenon), again indicating that the body's reaction to the organisms was altered forever. The intradermal test was later standardized and done by differing techniques (Mantoux test, Heaf test), all of which can variously be described as tuberculin skin tests (TST).

It is important to note that this 'tuberculin conversion,' i.e. the acquisition of hypersensitivity to tuberculoproteins, occurs whether or not the primary TB infection progresses to manifest clinical illness (progressive primary tuberculosis, including tuberculous meningitis and miliary tuberculosis). If conversion occurs and the subject remains healthy, including nowadays no radiological evidence of disease, the supposition is that the subject has latent TB infection (LTBI). The clinical significance of these different outcomes and the results of reactivation of LTBI to produce 'adult-type' tuberculosis in later life are found elsewhere in these volumes. From the bacterial immunoserology viewpoint, the significance is that TSTs can act as a quasi-serological marker that infection has occurred.

It is remarkable that a test devised in the nineteenth century is still in daily use in the twenty-first century. It is also testimony to the dearth of any better methods for the diagnosis of primary tuberculosis. It should be remembered that over 90 percent of cases of primary tuberculosis are asymptomatic and that in symptomatic primary disease the bacillary load in the body is so small that even the most sensitive of modern cultural techniques are usually unhelpful.

TSTs were the only method until very recently, but there are considerable problems with them. First, they are cumbersome to perform, require an injection (particularly unpopular with children, the most tested group), require a second visit for the 'reading,' and are subjective. Second, infection due to nontuberculous mycobacteria, such as *M. avium-intracellulare*, *M. malmoense*, and other environmental species, especially in childhood, can also elicit positive TSTs. Third, BCG vaccination, which is, after all, an artificial primary mycobacterial infection, will also give positive TSTs, rendering the TST approach virtually useless for the detection of primary tuberculosis in a universally vaccinated population after the age of screening. Fourth, since the TST itself introduces the test antigen to the immune system, regular and repeated TSTs can produce false-positive results, the so-called 'booster effect.' Lastly, whereas it was at one time thought and, indeed, observed that once TST conversion has occurred (including by BCG vaccination) the TST would be positive for life, it is now thought that in low-incidence countries, the lack of repetitive contact with the disease allows hypersensitivity to fade. Hypersensitivity ceases, thereby, to be a reliable surrogate for cell-mediated immunity, itself a marker of past primary infection. Nonetheless, despite these many drawbacks, TSTs have proved very helpful in investigating, diagnosing and controlling tuberculosis.

EVOKED GAMMA INTERFERON ASSAYS

It is against this background that a new approach to the 'serodiagnosis' of primary infection with *M. tuberculosis* has been adopted. The new method makes use of the fact that the 'sensitized' or primed T cells, especially those of certain subsets, will produce and release gamma interferon in the presence of the appropriate antigen. The production of gamma interferon can then be assayed or otherwise detected and thus confirm or deny the previous exposure to the antigen.

Unfortunately, the composition and comparative composition of mycobacterial cell walls is ill understood. Consequently, although these evoked gamma interferon assays (EGIA) can more precisely, reliably, and objectively assess responses to tuberculin and, moreover, do so without injection and at a single patient visit and be endlessly repeated without affecting the test result, differentiation between TB and infections due to nontuberculous mycobacterial species is difficult. A common approach is to expose the cells to a range of 'tuberculins' from different mycobacteria and allocate species specificity to that eliciting the highest response.

The most desirable goal is to differentiate between responses due to BCG and other members of the *M. tuberculosis* complex. Fortunately, it is known that the molecular hallmark of BCG is the lack of the RD1 region in its genome. It is likely, therefore, that if proteins coded for by this region (for instance, early secretory antigen type 6 (ESAT6)) elicit a gamma interferon response, then the primary exposure was to *M. tuberculosis* or *M. bovis* and not to BCG (Pollock et al. 2001).

Extensive studies have shown excellent correlation between TSTs and at least one EGIA (Mazurek et al. 2001) and assessments of the later *M. tuberculosis*-specific EGIAs are under way. Assays based on this technology will benefit many patients with LTBI and others. They may, in fact, be the key to the eradication of tuberculosis rather than its control (Institute of Medicine 2002).

As will be clear from the earlier section on the granulomatous response, tuberculosis may well not be the only disease in which these assays can be applied with benefit.

PRIVILEGED SITE INFECTIONS

Both the APR and the granulomatous reaction require access to the site of infection for cells. A blood supply to the infected site is, therefore, essential for these reactions.

Very rarely, examples occur whereby organisms gain access to the body, typically as a result of bacteremia, and lodge in sites that either physically or functionally are beyond the normal immune system. The nidus of *S. aureus* infection encased within sclerotic bone that persists for years, even decades, known as Brodie's abscess has already been mentioned, but the best naturally occurring example is subacute bacterial endocarditis (SBE).

In SBE, classically the organism lodges on a previously damaged area of a heart valve. The organisms involved are typically of very low pathogenicity and would normally be extinguished rapidly by the APR/ pyogenic response, but the almost avascular nature of the distal third of a heart valve cusp, exacerbated by fibrosis due to previous scarring, means that the site is 'privileged,' i.e. beyond the reach of the blood-delivered components of the pyogenic response, and, because cells of all kinds are denied access, even of the granulomatous response. Organisms multiply unhampered and a lesion comprising fibrin, platelets, and bacteria grows at the site – literally, a vegetation. Occasionally a fragment of the vegetation embolizes into the tissues, where the indolent organisms in the embolus are killed almost immediately. This paradox – a septic embolus resulting in a sterile infarct – illustrates well the unusual but typical nature of a privileged site infection. Accounts elsewhere in these volumes discuss the details of SBE. For the present purposes, it is important to note that the release of soluble antigens from the vegetation (bacterial cell-wall material, toxins, and other extracellular products) will stimulate humoral antibody production and this will typically go on over many weeks, if not months, before the true diagnosis is made by blood culture. This hyperglobulinemia is often unaccompanied by a significant APR (there is not the usual stimulus) and, for instance, the CRP is not very high (but it is raised) in SBE. Similarly, there may not be a leukocytosis of any note, although the dysglobulinemia usually induces a so-called 'secondary' or sideropenic anemia. Prolonged IgG production may eventually result in damaged or abnormal IgG being circulated, and this may provoke an IgM response, leading to the classic setting for production of rheumatoid factor (RF). A positive RF is found in approximately 60 percent of patients with SBE.

Since much of the IgG produced in SBE is nonspecific and also because the antigenic analysis of streptococcal cell walls is little studied, the humoral antibody response in SBE is hardly ever used for etiological diagnosis. Fortunately, blood culture is a very reliable test under these circumstances and little is therefore lost. However, serology can occasionally be useful, typically when it is either difficult or dangerous to isolate the organism from the blood. The best illustration is the diagnosis of Q fever endocarditis, since this obligate intracellular Rickettsia-like organism is both difficult to grow and very highly infectious to laboratory workers. However, the serological diagnosis of Q fever has an additional merit that, once again, illustrates the practical utility of bacterial immunoserology.

It has been found that *Coxiella burneti*, the causative agent of Q fever, can exist in two quite distinct antigenic phases. Serial passage of the organism in guinea pigs, results in a preparation said to be in phase I, whereas culture in fertile hens' eggs yields an organism in phase II. In uncomplicated Q fever infection, which usually presents as an atypical pneumonia, only antibodies to phase II antigens can be detected by CFT. However, in either of the two recognized chronic complications of Q fever infection, endocarditis or granulomatous hepatitis, antibodies to phase I antigens also appear (Raoult et al. 2000). Thus, bacterial immunoserology, in this case, will confirm Q fever infection, say in a respiratory infection serology screen, but will also alert the investigator to the possibility of complications.

From the above account, it should now be clear that much bacterial immunoserology depends on the detection, characterization, and quantitation of humoral antibody. It is now necessary to discuss in outline how this is accomplished.

THE TECHNOLOGY OF ANTIBODY DETECTION

Whenever tests are said to 'detect antibody,' they are in fact techniques that allow us to visualize or to infer, often indirectly, that an antigen–antibody reaction has taken place. With a few exceptions (antibody-mediated lysis and serum bactericidal tests, for instance, neither of which are routinely used in bacterial immunoserology), the test-tube event does not replicate an in vivo event.

It is common practice for all serological tests to be conducted on an initial dilution, even if only for screening. This is because neat human serum contains low titers of a wide range of nonspecific antibodies of no diagnostic significance, and these are best disposed of by a simple initial dilution step. Thus, serological tests are often reported as 'titer <1/8,' rather than 'negative,' but the meaning is the same. There are, however, exceptions, the most common being the VDRL, which is reported as 'positive/negative in neat serum or at a specific titer.' However, it should be remembered that reagin is not an antibody in any case.

The following brief account gives a range of examples of techniques commonly used, but it does not give other than outline details.

Direct visualization of an antigen–antibody reaction

Agglutination is one of the oldest techniques and depends on the simple observation that whole bacterial cells will form aggregates in the presence of an antibody specific to a cell-surface antigen, the lattice formed by the action of the divalent antibody and the bacterial particles becoming a visible aggregate. Classically, antibodies to flagellar (H) antigens produce a delicate floccular appearance, those to somatic (O) antigens a granular appearance, and those to capsules a disk-like appearance within the fluid. The appearance of the supernatant fluid (after settling) is said to be 'water clear' in contrast to the milky and smooth appearance of the original bacterial suspension. The Widal test best exhibits these different agglutination phenomena. Agglutination can be demonstrated in test tubes (special tubes such as Dreyer's tubes were devised for this) but can also be shown on slides. However, for the latter, the antibody levels in the serum must be very high and the bacterial suspension must be very dense. It will still be necessary to do a tube test to assess the titer, but slide agglutination is often used as a rapid preliminary screening technique.

Variations on the agglutination theme include passive (heme)agglutination in which an antigenic extract of the organism is coated (by various chemically assisted processes) on to the surface of inert carrier particles, most commonly red cells, which will then (heme)agglutinate in the presence of the appropriate antibody. Tests can be made very precise indeed if a recombinant antigen is used. Other commonly used inert particles include latex and gelatin.

A new, very exciting, and highly sophisticated development of these relatively simple principles is the binding, severally, of a range of different antigens to the surfaces of a matched range of differently colored inert particles, the chromophores used being detectable by fluorescence. Antibody-induced or antibody-dependent changes in any one or more of the subpopulations of particles of a specific 'color' can be rapidly, accurately, and objectively detected using multiplex differential flow analysis (Opalka et al. 2003). In this way, antibody to one or more antigens within a panel of a much greater number can be measured in one reaction. There are now several commercial systems available. The system can be reversed, so that the particles are coated with diverse antibodies, thus allowing the detection of more than one antigen in clinical material within a single screening test.

Lastly, the agglutination tests described thus far all detect divalent antibodies (IgG and IgM) in which the antibody links two (or more) particles to thus build up the lattice. Agglutination tests can also be used to detect monovalent antibody (IgA), which will ordinarily only coat the antigen-bearing particles, whether bacterial cells or carrier particles, but not cross-link them and, therefore, not result in visible agglutination. If, however, following washing of the IgA-coated suspension, anti-human (e.g. mouse, rabbit, goat, and so on) divalent IgG is added, a lattice will form. The resultant agglutination will indicate the presence of specific IgA.

Precipitation results when a soluble antigen meets the appropriate antibody. It can be demonstrated simply by allowing two liquids to meet at an interface whereupon a precipitate will form.

If both liquids are allowed passively to diffuse towards each other in a solid but transparent matrix, for instance clear agar or agarose gel, a visible precipitin line will form where they meet. This passive (immuno)diffusion, or double diffusion, has two additional properties. First, it will reveal the presence of more than one antigen–antibody system within the same reaction since they will form separate lines. Second, double diffusion will confirm the identity/nonidentity of any two systems if they meet. If they differ, the lines will not merge, but if they are identical, the lines will smoothly merge (the so-called 'reaction of identity'). This is because the precipitate will always redissolve in an excess of either reactant (antigen or antibody). These two reactions and the third possible outcome indicating partial identity are all well illustrated in the classical double diffusion method for the detection of diphtheria toxin production, the Elek plate, although it is a test for antigen and not antibody.

Precipitation methods can be speeded up considerably by applying an electric field across the gel, so driving the antigen and presumed antibody towards one another (immunoelectroosmophoresis (IEOP) or countercurrent immunoelectrophoresis (CIE)).

Indirect visualization of an antigen–antibody reaction

The general principle used here is that a secondary visible system, itself an antigen–antibody reaction, is used to infer that the sought for and invisible reaction has occurred.

CFT is the oldest and still the most widely used example. In this method the patient's serum and the antigen are mixed and allowed to react, but in the presence of complement. If the (invisible) reaction occurs, complement will be consumed ('fixed'). The fact that the complement has or has not been consumed can be demonstrated by the addition of a second system comprising animal red cells coated ('sensitized') with an antibody to the red cells raised in another (animal) species. If complement remains, the red cells will undergo complement-mediated immune lysis; if it has been consumed, no lysis will follow and, commonly, the red cells sediment to a button. No lysis therefore implies that the antibody sought has been detected and lysis implies that it is absent.

CFTs are inexpensive and popular because the reagents are relatively cheap and easily accessible. However, CFTs can be very difficult to standardize and there are many pitfalls in their use. Nonetheless, the CFT approach was the first semiautomated and generic antibody detection and screening system in microbiology.

Indirect immunofluorescent antibody tests (IFAT) utilize a similar principle to CFTs but confirm that immunoglobulin has been bound to the antigen by detecting it using an antihuman immunoglobulin tagged with fluorescein. The antigen is almost always solid, being cells from a culture or even a sample of tissue within which it is known that the whole organism can be seen. It is important to ensure that all unbound traces of serum are washed from the preparation prior to applying the detection system. Antihuman antibodies can be raised that are specifically of either the IgM or IgG class, thus enhancing the usefulness of the technique.

Obvious disadvantages of this method are the fading of the fluorescence after some time, making standardization and archiving very difficult. Quantitation of the antibody detected, even by titration to a poorly defined endpoint of fluorescence extinction, is subjective. The method, however, has the advantage of detecting exactly where in the tissues the antibody can be found.

EIAs take the indirect detection principle in a direction that solves many of the problems outlined above. Although there are many variations on the theme, the essence of EIAs is the production of a sandwich comprising (1) the antigen (whether a crude organism preparation, an extract of the organism or, increasingly, a recombinant antigen); (2) the patient's serum as a putative source of the antibody; and (3) an antihuman globulin tagged with an enzyme. As with IFATs, if the antigen–antibody reaction occurs, the enzyme-tagged antibody will become attached. The enzyme is then made to convert an appropriate substrate to a visible product. Enzymes commonly used include alkaline phosphatase and peroxidase. In both cases, the color intensity can be read on a colorimeter and the amount of antibody (a value equivalent to the 'titer') can be determined and compared with that of other samples.

EIAs have several distinct advantages and have become one of the workhorses of bacterial immunoserology. First, the 'antiantibody' nature of the detector system makes it possible to produce EIAs specific for IgM, IgG, or IgA antibody detection. Second, the stability, reproducibility, and standardizability of the endproduct make EIAs excellent assays for automation, in turn making them quicker, cheaper, and less subjective.

However, EIAs remain individual reactions, capable of detecting only one antibody at a time. The multiplex differential flow analysis technology outlined above is, therefore, likely to be a strong competitor, able to detect several antigen–antibody reactions in one assay.

Other antibody assay techniques

Other, rarer, methods have been devised to detect antibody. Many are historical curiosities and can be dismissed, having been replaced by one or more of the above techniques. Nonetheless, they demonstrate the basis of bacterial immunoserology as a practical response to a clinical need and so will be briefly mentioned.

Toxin neutralization tests allow the detection of an antibody, in this case literally an antitoxin, by showing that the patient's serum will abolish the toxicity of the antigen. Thus, a cytopathic effect (CPE) of a toxin on a tissue culture monolayer will be prevented by the addition of immune patients' sera. Other examples include the inhibition of hemolysis due to a wide range of hemolytic extracellular toxins (such as streptolysin O, staphylococcal α-lysin), intradermal tests in animals, and even tests designed to show that immune serum will prevent a specific lethal or morbid reaction in an animal (such as the paralytic effects of tetanus or botulinum toxins, or the lethality of diphtheria toxin).

Thankfully, almost all of these animal assays have been successfully replaced by modern assays in which animals are not used, although equivalent sensitivity has not always or easily been achieved. In those bioassays that remain, it should be obvious that control reactions are absolutely vital, more so than in any other assays, because of the immense biological variation that there must inevitably be in the assay components.

Immobilization assays make use of the fact that many microbes have the property of locomotion (motility) and an immune serum will immobilize them. The TPI has been discussed above. Immobilization assays are prone to the particular problem that other inimical substances in the patient's serum, most commonly antibiotics, can produce an identical result to that of antibody.

Metabolic assays are a derivative of toxin and mobility assays since they capitalize on the fact that immune serum will either kill the organism or, as in this case, prevent it expressing a particular and easily measured characteristic. Thus, antibody to

Mycoplasma pneumoniae can be detected by showing that the patient's serum will prevent the mycoplasma growing and metabolizing glucose, this property being manifest by a color change in the reaction tubes. Clearly, antibiotics will produce an identical effect.

COMMON APPLICATIONS OF ANTIBODY ASSAYS IN ORGANISM GROUPS

Staphylococci

In general, because of the pyogenic nature of staphylococcal infection, immunoserology is not needed. There are two exceptions. First, in suspected long-standing and deep-seated infection (see Brodie's abscess, above) in which samples for culture are difficult to obtain, recourse is made to the antistaphylolysin test, a hemolysin inhibition test, utilizing the α-hemolysin of *S. aureus* as antigen. When antibody is detected, the test is very predictive of chronic infection with *S. aureus*. Unfortunately, a significant minority of *S. aureus* strains do not produce the α-hemolysin, so negative results do not exclude infection. EIAs have also been devised testing for antibodies to the universal *S. aureus* antigen, ribitoltechoic acid (Ayyagari and Pal 1991), and seem promising.

Second, the propensity of coagulase-negative staphylococci to cause chronic infections of prosthetic material (a 'privileged site infection'), especially in replacement joints, but also in cerebrospinal fluid (CSF) shunts and indwelling intravascular devices, has led to the development of EIAs for antibodies to the glycocalyx ('slime') (Elliott et al. 2002) that such strains produce in abundance. These assays have the extra merit of confirming or denying the likely significance of isolates of these otherwise ubiquitous contaminants in tissue and blood cultures.

Streptococci

The need to relate recent *Streptococcus pyogenes* infection to postinfective sequelae, such as rheumatic fever, acute type 1 nephritis, and rarer events such as erythema nodosum and Henoch–Schoenlein purpura, has provoked the development of a range of assays, all originally based on the principle of neutralizing the effect of an extracellular product of *S. pyogenes*.

The most common by far is the antistreptolysin O titer (ASOT). Since, however, *S. pyogenes* infections of the throat and middle ear are common and the number of different (M protein) types is so large, some background level of ASOT is almost inevitable in any population. Thus, a 'positive' ASOT is usually defined as a value in excess of the norm for the population from which the individual to be tested comes. Alternative tests such as antiDNAse B and antihyaluronidase are more rarely performed but remain necessary since *S. pyogenes* infections of skin (streptococcal pyoderma), which can be followed by nephritis but not rheumatic fever, rarely give rise to a raised ASOT, whereas the other antibodies will rise (Ferrieri et al. 1970). *S. pyogenes* throat infections, however, reliably provoke ASOT rises. Unrelated infections and, rarely, infections due to streptococci of other Lancefield groups can also occasionally provoke a raised ASOT, giving another useful role for the second line tests. Nowadays, especially with the availability of recombinant antigen technology, these antibodies are more commonly detected by passive particle agglutination tests or EIAs rather than the classical neutralization approach.

Much recent interest has been targeted on devising a reliable antibody test for infection due to *S. pneumoniae* (the pneumococcus). New vaccines offer the prospect of broad and enduring protection but will need accurate evaluation. The diversity (>90 types) of the capsular polysaccharide antigens makes simple tests unlikely, although the multiplex differential flow analysis technology may greatly help, and attention has centered on the universal pneumococcal product, pneumolysin, which is of the same family of thiol-activated toxins as Streptolysin O. Progress has been hampered until recently because it is found only intracellularly and cannot, therefore, be produced in large amounts. However, recombinant protein technology has solved this problem and assays are being developed (Lankinen et al. 1999). Other targets are also under study.

Corynebacteria

The principal immunoserological need is for the detection of diphtheria antitoxin prior to, and subsequent to, immunization. Although EIAs have been developed, experience suggests that (in contrast to the detection of the toxin itself; see below) they are insufficiently sensitive. The standard assay presently consists of a toxin neutralization assay using the CPE in Vero cells. This assay has replaced the subjective and complex skin test previously used, the Schick test.

Clostridia

The highly virulent exotoxins of *C. tetani* and *C. botulinum* produce the characteristic diseases with which they are associated at concentrations below those necessary to immunize – a conundrum that should lead all doctors to remember that their last duty to a patient recovering from clinical tetanus is to immunize them with tetanus toxoid prior to discharge. Therefore, antitoxin assays are not used in diagnosis. However, assays to detect high titers of *C. tetani* antibodies following repeated immunization with toxoid have been devised to select donors for the production of human tetanus immunoglobulin for treatment purposes.

Neisseria

Antibodies to *Neisseria meningitidis* are not commonly assayed for diagnostic purposes, although they have occasionally helped in retrospective diagnosis of the serogroup involved in an outbreak. The most common circumstance requiring a noncultural approach, the failure to culture *N. meningitidis* from blood or CSF, is now addressed through polymerase chain reaction (PCR) of these specimens. PCR is quicker, more revealing, and more sensitive than immunoserology.

However, antibody tests for *N. meningitidis* have been invaluable in assessing the efficacy of vaccines, particularly those for *N. meningitidis* serogroup C. Although many different assays for antibodies to many different antigens have been devised, the classical bactericidal antibody test remains the specified method for vaccine licensing purposes.

There are no currently useful assays for antibodies to *N. gonorrhoeae*. The gonococcal complement fixation test (GCFT) continues to be requested and even promoted in some textbooks. However, many investigations have shown that it is very rarely helpful and often is misleading in an area of diagnosis that has considerable medicolegal potential.

Enterobacteria (here taken to encompass most nonfastidious aerobic gram-negative bacilli)

The Widal reaction for enteric fevers (typhoid and paratyphoid) has already been discussed (see above). It should be emphasized that these methods cannot and should not be applied to nonenteric fever *Salmonella* species. The sharing of antigens (especially O antigens) between species makes the results impossible to interpret. No useful routine antibody assays are in current use for infections with shigellae.

The recent recognition of serious infection with *Escherichia coli* O157 has stimulated the development of an EIA to detect antibody to the lipopolysaccharide antigen of its cell wall, incorporating the O antigen. Judicious use of this Reference Laboratory test has given invaluable help in identifying unrecognized cases prior to an outbreak and in rapidly determining the true extent of an outbreak by testing apparently unaffected subjects (Chart 1999).

Although numerous antibody tests have been developed from time to time for various research studies, there are few other established assays used for diagnosis of infections with the remaining organisms within this group. However, an assay for antibodies to *Campylobacter jejuni* is frequently requested for a tangential purpose. It has emerged that the most common preceding event to the onset of Guillian–Barre syndrome is *C. jejuni* infection. The relationship is thought to be based upon cross-specificity of *C. jejuni* antigens and certain epitopes in neural tissue, a situation dubbed molecular mimicry (Schwerer 2002).

Parvobacteria (fastidious gram-negative cocco-bacilli)

Serology is still used for the diagnosis (using paired sera) of infection with *Bordetella pertussis*, but as with *N. meningitidis* (see above), the most common reason for the test is for confirmation of clinically suspicious but culture-negative cases, especially in adults, older children, and 'partially vaccinated' children. As with *N. meningitidis*, this need is increasingly being met by PCR-based testing. The demand for serology, which is complex and the results of which are difficult to interpret, is likely to diminish.

Apart from research settings and studies for the evaluation of vaccines, serology is little used in the diagnosis of infections with *Haemophilus* spp. The immunoserology of *Brucella* was discussed earlier. There is an occasional rare need for antibody assays in melioidosis, and a passive hemagglutination test using *Yersinia pestis* Fraction 1 antigen is used in the serodiagnosis of plague. In both diseases, isolation of the causative organism is much the most common diagnostic method.

In tularemia, however, diagnosis is commonly made by demonstration of a rising antibody titer, since the clinical appearances are often confusing. Cross-reactions with *Proteus* spp. (see below) and *Brucella* spp. antigens can occur. Such is the complexity and diversity of *Francisella tularensis* strains and their antigens that interpretation is an expert matter. Nonetheless, serodiagnosis is an attractive option given the highly infectious nature of *F. tularensis* and the danger of laboratory-acquired disease.

Legionella

Antibodies to *Legionella pneumophila* were originally detected by IFAT using a formalized yolk-sac culture of the organism on a slide as the antigen. High and, preferably, rising titers are diagnostic. It was noted from the start that antibody might take up to 3 weeks to appear after the onset of the illness.

EIAs have now largely replaced the IFAT, and PCR of tracheal secretions or bronchoalveolar lavage fluid and urine antigen assays have supplemented serology. However, all these tests are specific for *L. pneumophila* infections, which account for nearly all cases of legionnaires' disease. Investigating infections with one or more of the nearly 40 'other legionella species' (sometimes

called fluorobacters) in highly immunosuppressed patients is a task for an expert.

Low and unchanging levels of antibodies to *L. pneumophila* can be found in Pontiac fever, an extrinsic alveolitis (but not a pneumonia) triggered by inhalation of *L. pneumophila* antigens from colonized or contaminated air conditioning systems and, as such, one of many forms of 'sick building syndrome' (Friedman et al. 1987).

Mycoplasma pneumoniae

Interest in the serology of this common organism is three-fold. The technology is rather mundane, most cases being diagnosed by a rising antibody titer using a CFT or by detection of specific IgM using an EIA.

The first point of note is the association between serological evidence of infection with *M. pneumoniae* and various neurological conditions, most notably Guillian–Barre. As with the previously mentioned association with *C. jejuni* infection, there is good evidence that cross-reacting antibodies are the basis of this problem (molecular mimicry) (Kusunoki and Kanazawa 2001). Second, there is a similar association with acute pancreatitis.

The third point of interest harks back to the early empirical days of bacterial immunoserology. It was noticed many years ago that patients with 'atypical pneumonia,' a condition not then known to be due to *M. pneumoniae*, often exhibited cold agglutinin antibodies (CAA) in the early stage of the disease, the titer of which appeared to correlate with the severity of the illness. In fact, the CAAs are IgM class antibodies produced in the acute stage of the infection. Their specificity and evocation are still uncertain. They do not, for instance, react with the antigen used in the CFT. They are, however, relatively specific, being found only in pneumonias and nearly always only in those due to *M. pneumoniae*. Their very early appearance in the illness can be capitalized upon in a very simple bedside test (Davidsohn and Wells 1962).

Freshly drawn blood is mixed four parts to one with citrate in a small glass tube, the anticoagulated blood sample then being placed in a refrigerator or on ice for a few minutes. If CAAs are present, large aggregates will form and can be seen if the blood is run over the glass surface. These readily disappear on hand-warming of the tube and can be redeveloped by its return to the cold. Samples with this result can be taken to contain CAA titers of at least one in 80 and a presumptive diagnosis can be made. Other causes of cold agglutinins must also, obviously, be considered in certain settings.

Rickettsia

Diagnosis of rickettsial disease is usually by IFAT, but this may not distinguish between the different forms of typhus (epidemic louse-borne, murine, scrub, etc.). Washed and adsorbed antigens can be used to achieve these distinctions, which are of both diagnostic and epidemiological importance. However, recourse can still be made to yet another empirical bacterial immunoserology test, the Weil–Felix reaction. This relies on the observation that the serum of patients with these differing rickettsial infections will agglutinate different suspensions of correspondingly different strains of *Proteus vulgaris*. Thus, sera from patients with epidemic louse-borne typhus agglutinate *Proteus* OX-19 strains, whereas sera from patients with scrub or mite-borne typhus agglutinate *Proteus* OXK strains. These easily maintained antigens were invaluable before rickettsial culture was possible and still retain their usefulness on occasion today.

The usefulness of bacterial immunoserology in the diagnosis and management of Q fever infection has already been discussed above.

Spirochetes

Serology is a common method for the diagnosis of leptospiral infection, since the best opportunity to see leptospira in clinical material, such as urine, is usually gone by the time the diagnosis is considered and microscopy (and culture, if attempted) is notoriously difficult.

At a group identity level, CFT can be used but it is increasingly overtaken by EIA, which can offer IgM-specific testing. However, microscopic agglutination tests with specific leptospiral antigens are necessary to identify the particular species involved, for instance *L. canicola*, *L. interrogans*, *L. icterohaemorrhagicae*, and so on. Borrelioses such as relapsing fever (*B. recurrentis*) are not usually diagnosed by serology.

The complex serological approach to syphilis (*T. pallidum*) has already been discussed in some detail. It should be added here that the same tests are used to investigate suspected infection with *T. pertenue* and *T. carateum*, the agents of, respectively, yaws and pinta, although serology cannot discriminate between them.

In recent years, Lyme disease has been added to the list of major spirichetal diseases. Lyme disease can be diagnosed by serology using EIAs with *Borrelia burgdoferi* antigens. Like syphilis, Lyme disease evolves through recognizable acute, latent, and chronic stages. Whilst humoral antibody assays are invaluable in diagnosis and following the disease, they may be unpredictable in both the time and magnitude of their appearance. EIAs are particularly troublesome in chronic Lyme disease and/or treated cases. There is a need to develop an equivalent to the VDRL to act as an index of activity. The advent of the evoked gamma interferon assays measuring cell-mediated immunity may be helpful in this.

Chlamydiae

Until recently, it was held that only two groups of *Chlamydiae* existed. These were group A (now known as *Chlamydia trachomatis*) and group B (*C. psittaci*). Since *C. trachomatis* largely caused infections of mucosal surfaces (trachoma in the eye, inclusion conjunctivitis, and nongonococcal urethritis), the need for serology was limited. Occasional invasive infection was recognized in the forms of salpingitis and infant pneumonitis and a specific subgroup of *C. trachomatis* was held responsible for the granulomatous genital infection known as lymphogranuloma venereum (LGV).

C. psittaci was accepted as an uncommon but well-recognized cause of atypical pneumonia (psittacosis or ornithosis), and almost all such cases were diagnosed serologically. It was convenient, therefore, to use a common CFT antigen, a mixture of an LGV strain and a *C. psittaci* strain, to detect all antibodies, the likely specificity being decided by the markedly different clinical circumstances. The recent discovery of a third species, *C. pneumoniae*, which causes pneumonia, appears to be of purely human origin, and antibodies to which may not always react with the old antigen (Blasi et al. 1998), has spurred the development of species-specific antibody tests, which are usually of a microIFAT basis.

Other comments

SALIVA TESTING

It is important to note that along with all the many advances in assay technology, there has also been an advance in specimen technology. This account thus far has assumed that all tests are carried out on serum (excluding those concerned with cytokine evocation from cells). Recently, there has been extensive investment in detecting antibodies of all immunoglobulin classes in 'saliva' or, more precisely, crevicular fluid (Akingbade et al. 2003). Special oral devices allow adequate volumes of sample to be obtained. Salivary testing is convenient, noninvasive, and particularly suitable for children and others with aversions to venepuncture. It is being taken up with increasing frequency.

NEAR-PATIENT TESTING

Unsurprisingly, the development of sophisticated systems for the direct or indirect visualization of antigen–antibody reactions, the ever-increasing precision and robustness of these technologies, and the benefits of modern mass-production techniques have all combined to promote near-patient testing kits. In addition to simply refining the methods outlined above, commercial houses have introduced some truly innovative approaches. For instance, it is possible to tag detection monoclonal antibodies with gold particles, so that when the antibody localizes (having reacted with an immobilized antigen), its presence becomes macroscopically visible. All the necessary reagents can be offered in a cartridge format, including control reagents, and the addition of the sample under tightly specified circumstances can give a result within minutes, with no laboratory involvement.

Even more sophisticated approaches include embedding antigens or antibodies within membranes, the optical properties of which will alter if a reaction occurs, thus giving a visual signal of the event.

The potential variations on this theme are myriad and need not be listed here. However, whatever the technology employed, it is important to address three areas of concern:

1 Safety. Many of the near-patient kits require venepuncture, needle-stick, or other means of drawing blood. Since the instant diagnosis of diseases such as hepatitis B or HIV in the field is one of the attractive settings for such kits, it is important to attend to waste disposal of the kits (often catered for by the manufacturer) and the attendant materials (not always catered for).

2 Controls. Although all reputable kits come with control reagents and the results obtained will be reproducible if the instructions are carefully followed, all serologists and laboratory diagnosticians will be aware that these alone are an inadequate control framework. Diagnostic tests need to be continually challenged with archived and current 'wild' samples, thus allowing for possible events such as the effects of target mutation, epitope loss, and so on. These checks are part of laboratory systems but are unlikely to be available to individual clinicians. Arranging for patients to be cohorted or even to attend a clinic for testing destroys the essence of near-patient testing.

3 Inappropriate use in inappropriate contexts. The devolution of bacterial immunoserology testing to the bedside implies that a test will be performed by interested and well-motivated nonlaboratory personnel, the experience of whom will likely be confined to one medical specialty. Occasionally, therefore, a near-patient test system may be used outside the testing scenario for which it was designed. Under these conditions, very misleading results can occur, the consequences of which can be both unfortunate and embarrassing.

With careful planning and forethought and, most importantly, the ready cooperation of the conventional laboratory, all these problems can be resolved. Done well, near-patient testing will be an increasing boon to clinical care (Borriello 1999).

Lastly, it must also be remembered that the need to avoid sending specimens some distance to a laboratory can also be avoided by careful clinical assessment and the development of algorithms, which eventually may result in no specimens being required at all. An excellent example is seen in the long story of the attempts to optimize treatment of the 'strep throat' (Needham et al. 1998; Bisno et al. 2002; McGinn et al. 2003).

METHODS FOR THE DETECTION OF ORGANISMS AND THEIR PRODUCTS AND FOR THE CHARACTERIZATION OF ORGANISMS BY IMMUNOSEROLOGICAL MEANS

In this aspect of bacterial immunoserology, the principles, and largely, the technologies remain as before, but the purpose of the assays is to detect the antigen using the antibody as the tool. First, it is important to recognize the settings within which this noncultural approach is useful and then to understand the limitations and the alternative, nonimmunoserological means by which it can also be attempted.

Noncultural methods of diagnosis

Broadly, noncultural methods are used because of their speed or specificity or because they give useful information in circumstances in which culture is not possible. In this latter instance, this might be because the organism has never been successfully cultured or, more commonly, because prior antibiotic therapy has rendered culture useless.

Until recently, the major approach for noncultural diagnosis was by detecting a component of the organism by immunological means. Since antibodies were the essential reagent, it followed that the target had to be an antigen and this almost invariably dictated the choice of a surface component such as a capsule, flagellum, somatic antigen, and so on, or a secreted antigen, such as an exotoxin. Medical microbiologists continue to use a very large number of such assays. The technology used to detect the antigen–antibody reaction comprises the whole range described above (agglutination, precipitation, EIA, etc.), but there are some notable additional aspects.

Advances in antibody production and application

MONOCLONAL ANTIBODIES

In many cases a test used will have begun as a method using a polyclonal antibody raised in animals. The polyclonal antibody is now almost always replaced by a monoclonal antibody. This change has several important advantages but one important limitation. The advantages include the 'immortalization' of the antibody, or at least the hybridoma, the standardization of the antibody, thus removing batch variation problems at one stroke, and the potential to generate antibodies of such potency that the sensitivity of the test can be exquisite and well beyond that of its polyclonal antibody-based predecessor.

However, since the monoclonal antibody is, by definition, directed towards only one epitope of the target antigen, the whole assay can be vitiated if this epitope is not universally present or, if, as occasionally happens, the organism mutates and loses it (Smith et al. 1996). The considerable and well-deserved reputation for the accuracy of monoclonal antibody assays can then be a hindrance to good practice. It is, therefore, often recommended that very important results, for instance identifications for medicolegal purposes or in the diagnosis of life-threatening conditions, be confirmed by alternative pathway tests or by a second monoclonal assay directed towards a different epitope.

COAGGLUTINATION

Protein A on the surface of well-characterized strains of *S. aureus* binds to the Fc component of IgG molecules, making it possible to coat suspensions of the inactivated organism with monoclonal antibody. The resultant preparation becomes a particulate suspension in which each particle exhibits on its surface the free Fab component of the IgG used. Thus, in the presence of the appropriate antigen (in whatever physical form), the staphylococcal suspension will agglutinate, a lattice having been formed by the antigen and Fab receptors, which also brings down the suspension. This assay format, known as coagglutination, is widely used. Suspensions can be stained in different colors for different Fab specificities and then mixed so that several antigens can be tested for at one time. In some assays, protein A is itself coated on to inert or passive particles (for instance latex or gelatin).

PHAGE ANTIBODY DISPLAY (PINI AND BRACCI 2000)

Phage antibody display (PAD) works on a simple principle, which is that it is possible to insert recombinant DNA sequences coding for antigen-binding fragments (Fab) of IgG into bacteriophages, which will then express the appropriate Fabs on their protein coats. In this way, a 'library' of such modified phages can be produced, which will contain an almost limitless variety of Fab specificities. Selecting one individual phage and cultivating it in a bacterial host becomes akin to growing a monoclonal antibody, but without the rest of the immunoglobulin in the final product.

The Fab-bearing phage can then be tagged with fluorescein to give the equivalent of a fluorescein-tagged monoclonal antibody for microscopy assays. Alternatively, immobilization of the phage on a surface, such as in the wells of an assay plate, can form the basis of an EIA. The details of the process whereby the appropriate phage is selected from within the original library are complex but are analogous to the old serological method of making a polyclonal antiserum more specific by serial adsorption of antibodies to closely related but ultimately undesirable antigens (technically described as 'panning'). This offers the prospect of producing equivalent reagents to monoclonal antibodies without the use of animals and more quickly.

However, PAD offers one further, potentially revolutionary advantage. The epitopes and 'antigens' to which PAD 'antibodies' can be 'raised' need not be constrained by their acceptability, or not, to any human or animal immune system. Thus, antigenicity and immunogenicity are irrelevant. As importantly, self- or auto-antigens are not excluded as is the case in conventional raising of antibodies. The only real prerequisite is that the chemical group can be recognized by a Fab grouping on one of the phages.

Consequently, as yet unrealized differences in surface structures between two populations of bacterial cells can be explored as the basis of their reliable distinction, and reagents can be produced to do this reliably, quickly, and relatively cheaply. In the future, we may refer to nonimmunoserology tests.

APTAMERS (JAYASENA 1999; OSBORNE ET AL. 1997)

Aptamers are nucleic acid molecules that bind to specific ligands. Consequently, they can be used to link oligonucleotides to specific sites. If a pair of aptamers is targeted on to a protein (for instance, an epitope or an antigen), the PCR can be used to amplify a stretch of DNA or RNA corresponding to the nucleotide sequence that bridges the 'gap' between the aptamer-linked oligonucleotides. By this means, assays detecting zeptomolar (10^{-21} mol) amounts of the target can be devised.

Since the aptamers themselves can be synthesized, this system can rival antibody-based assays, yet, as with PAD, animals are not necessary and diversity is not necessarily confined by antigenicity or immunogenicity of the target.

COMPETITION WITH MOLECULAR ASSAYS

The major new change in recent years has been the influence of molecular biology, especially PCR. The ability to detect small amounts of DNA and the development of assays in which the PCR detects specific sequences indicative of certain organisms or their characteristic products has challenged the previous primacy of the antigen-detection immunoassays. PCR assays are, of course, also very quick, sensitive, and helpful if the organism cannot be cultured, is dead, or its growth is inhibited by antibiotics. However, it should be remembered that PCR assays also have problems. The reaction is very vulnerable to the presence within clinical material of substances inimical to the Taq polymerase enzyme ('inhibitors'). These are relatively common (heparin, IgG, and so on), whereas immunoassays do not suffer these problems.

Furthermore, the genomic assay can detect the gene that confers on the organism the ability to produce a given product, e.g. a toxin, but the gene may not be expressed. In diseases such as diphtheria, which is defined almost totally by the actions of the toxin, mere detection of the gene is insufficient to make the diagnosis. It is the phenotype that infects and/or causes disease.

It seems likely that the immunoassays will be needed for some time to come and that the two approaches should be complementary rather than competitive.

COMMON APPLICATIONS OF IMMUNODETECTION ASSAYS IN ORGANISM GROUPS

Staphylococci

The never-ending search for a test with the simplicity and speed of the slide coagulase reaction and the accuracy of definition of the tube coagulase reaction in this most ubiquitous of human pathogens has led to the development of a whole range of assays. Almost all of the recognized formats have been deployed. Since speed was an important objective, reverse passive hemagglutination, passive particle agglutination, and coagglutination, have been especially popular. These assays have variously targeted coagulase itself (here being detected as an antigen and not an enzyme), capsular antigens, and fibronectin receptors, amongst others. Almost all such tests are very nearly as accurate as the definitive tube coagulase test (Summers et al. 1998).

A widely used innovation has been the identification of methicillin resistance in *S. aureus* by EIA or passive particle agglutination immunodetection of penicillin binding protein (PBP) 2A on its surface by monoclonal antibodies, thus identifying methicillin-resistant *S. aureus* (MRSA) in a few minutes (Naratomi and Sugiyama 1998).

Streptococci

One of the classical applications of the precipitin reaction was the Lancefield grouping of β-hemolytic streptococci in which the extracted 'C substance' (a carbohydrate) was identified by layering a series of group-specific antisera on to aliquots of the extract in capillary tubes and observing a precipitation reaction at the appropriate interface. Nowadays, expectedly, this has been replaced by a passive particle assay, usually by slide agglutination, using particles coated with a series of group-specific monoclonal antibodies.

S. pneumoniae is commonly identified by a series of rapid tests, one of which is a slide agglutination test targeting a C substance common to all pneumococci.

Pneumococci also provide a good example of an unusual form of the precipitin reaction. They can be typed by the capsule swelling or Quellung reaction. This is based on the observation that if pneumococci in broth culture are microscopically observed before and after the addition of the homologous capsular antiserum, the capsule appears to swell dramatically in size. In fact, the capsule is at least 90 percent water and continually dissolving into the surrounding medium. It is a soluble antigen that precipitates when in contact with the appropriate antibody. This increases its refractivity and thus it appears to swell in the presence of the antiserum.

Pneumococcal infections illustrate another use of immunodetection. During invasive pneumococcal disease, it is often possible directly to detect capsular antigen in CSF, blood and urine, by passive particle agglutination (usually latex) or by EIA. The diversity of the capsular antigens (>90 serotypes) and their polysaccharide nature mitigate against a monoclonal antibody approach so that antigen detection is not as sensitive as culture. It has the merit of being suitable when culture is likely to be unrewarding. Students of the potential pitfalls of immunoserology will realize that antigen will be detected in the urine of the recently vaccinated.

Corynebacteria

Mention has already been made of the Elek double diffusion plate method for the detection of diphtheria toxin production in isolates of *Corynebacterium diphtheriae* or, more rarely, in some strains of *C. ulcerans*.

More recently, toxin production has been assayed by EIA. In either instance, it is important to understand that these tests demonstrate toxin synthesis, whilst the PCR often used to screen strains merely demonstrates the presence of the gene (see above).

Clostridia

The many exotoxins of clostridia are highly antigenic and are a potentially rich field for immunoserology. However, many of those of human interest, classically tetanus toxin and botulinum toxins, are still assayed when required by animal tests, although EIAs are being increasingly applied. These are Reference Laboratory tests. However, routine bacterial immunoserology includes two clostridial assays.

The first is the Nagler reaction. In this unusual format, the production of a lecithinase by an organism thought to be *Clostridium perfringens* is demonstrated by streaking the organism on to an egg-yolk medium. The further demonstration that the lecithinase produced is the α toxin of *C. perfringens* (there are other lecithinases) is shown by the specific neutralization of the effect of the enzyme on the egg yolk on one half of the medium that was pretreated with specific antitoxin. The organism will grow on both halves of the plate. The Nagler reaction thus permits a secure identification of *C. perfringens* overnight. It should be noted that mere demonstration of lecithinase activity without neutralization is not the Nagler reaction.

Second, detection of the toxin(s) in the liquid stool of an affected patient is an essential prerequisite for the diagnosis of *Clostridium difficile*-associated disease (CDAD). Originally, this was accomplished by detecting a CPE in cell cultures (several were used, the most common being Vero cells) and then confirming that the CPE was neutralized by appropriate antitoxin. Although *C. difficile* antitoxin can be produced, it is common practice to use *C. sordellii* antitoxin, which was first used. There is a very high degree of homology between the two toxins. It is also important to note, contrary to widespread belief, that the neutralized cytotoxin assay so produced detects both *C. difficile* toxins, A and B.

Although the neutralized cytotoxin assay remains the gold standard, its possible subjectivity and relatively long turnround time have stimulated the search for more standardized and quicker tests. These are usually in EIA format. An adequate replacement assay for the cytotoxin test must detect both toxins A and B by an immunological pathway.

Neisseria

There are several monoclonal antibody-based tests using a variety of formats, which can confirm the identity of a suspect isolate as *N. gonorrhoeae*. Over-reliance on any one such assay, the accuracy of which is, after all, vulnerable to a change in only one epitope, can be problematic in practice. Gonococcal identity should, therefore, be investigated in parallel by an alternative pathway test, for instance sugar fermentation tests. A second monoclonal test (directed to a different epitope) might also be necessary on occasion.

Rapid serogrouping of isolates of *N. meningitidis* is very important since it influences case management, especially the decision to vaccinate close contacts. It is usually performed with assays similar to those used in streptococcal grouping or the identification of pneumococci (see above).

Enterobacteria

The use of antibodies to analyze the antigenic make-up of Enterobacteria is very widely used. A classic application is the identification of salmonellas by analysis of their O and H antigens (the Kaufmann–White scheme). Shigellas can be identified in an analogous way,

although since these nonmotile organisms do not display H antigens, it is necessary to use adsorbed O antisera and some supplementary biochemical tests.

Within the species *Escherichia coli*, there are several identification schemes based on serological analysis. Thus, the identification of enteropathogenic *E. coli* is based on O antigen analysis. Serological analysis plays a part in identifying enteroinvasive and other pathogenic strains.

It is in identifying *E. coli* O157 isolates, however, that serology is most commonly used. Colonies of putative *E. coli* O157 growing on a selective medium can be screened for further investigation by slide agglutination with an O157 antiserum. However, before being accepted as the cause of disease, the organism must next formally be confirmed to be *E. coli* and, secondly, to be elaborating one or both of the verocytoxins (VT1 and/or VT2). Final proof is obtained by toxin gene PCRs and an EIA for VT1 and VT2. This process re-emphasizes the important principle that O antigens are shared within and even between species of Enterobacteria. Serological analysis must, therefore, always be assessed within a context.

Parvobacteria

Serology can be used to identify and type *Brucella*, *Yersinia*, *Francisella*, and other Parvobacteria, but in common practice it is *Haemophilus influenzae* that attracts the most attention. *H. influenzae* is divisible into six capsular serotypes, A–F, often called Pittman types. These can be identified in capsule swelling (Quellung) reactions using specific antisera. *H. influenzae* capsule type B, often abbreviated to HiB, is responsible for the overwhelming majority of severe human infection, which is predominantly in childhood. HiB meningitis, septicemia, pneumonia, osteomyelitis, and epiglottitis are all life-threatening diseases. Recently, HiB vaccination has successfully controlled these diseases, but it is necessary to maintain vigilance. All isolates of *H. influenzae* from frankly pathogenic circumstances merit examination for the B antigen. Serotyping can now be performed in simpler formats than capsule swelling.

It should be noted that 'noncapsulate' *H. influenzae* is very commonly isolated from the sputum of adult patients with exacerbations of chronic bronchitis and bronchiectasis, but this is not the setting for HiB.

Meningitis screening

It is common practice to examine samples of CSF, especially those samples with a pleocytosis or other markers of bacterial meningitis, for capsular antigens of *N. meningitidis*, *S. pneumoniae*, and HiB. Whilst these tests are occasionally valuable and the effort is laudable, it should be remembered that these tests are designed primarily to identify organisms, i.e. under circumstances in which antigen is in plentiful supply.

In contrast, in CSF for instance, it has been estimated that up to 10^6 organisms/ml may be needed to detect antigens reliably (West and Nasrallah 1988). Therefore, negative results do not exclude the disease. It is in such circumstances that the new approach of PCR is often preferable, being more sensitive and, when necessary, also capable of pointing to other properties of the causative organism, such as antibiotic sensitivity or resistance (Kearns et al. 2002).

Legionella

The slow growth of these organisms in culture and the unusually slow appearance of detectable antibody in the patient's blood have been stimuli to develop alternative diagnostic means. The detection by EIA of a cryptic antigen of *L. pneumophila* in the urine of infected patients has proven to be a very successful approach. These tests (several are available) have greatly enhanced patient management in individual cases. Their simplicity and rapidity have also assisted outbreak investigation.

Two caveats must be applied. First, the tests are effectively restricted to diagnosing infections with *L. pneumophila*. These tests cannot, therefore, assist in the diagnosis of infections, usually of highly immunosuppressed patients, with other *Legionella* species, such as *L. bozemanii* (Humphreys et al. 1992). Second, the antigen may continue to be excreted for extraordinarily long periods after first becoming detected, even if the clinical course has been benign and response to treatment has been good.

Monoclonal antibodies are used in Reference Laboratories as an important tool in identifying and typing legionella species.

CONCLUDING REMARKS

The above account is not exhaustive, but it gives some idea of the extent, diversity, and utility of current bacterial immunoserology practice. Although, inevitably, future technical advances will continue to be incorporated into this area of diagnostic microbiology, the principles remain the same. It is likely that bacterial immunoserology will continue to be an important element in medical practice.

REFERENCES

Akingbade, D., Cohen, B.J. and Brown, D.W. 2003. Detection of low-avidity immunoglobulin g in oral fluid samples: a new approach to rubella diagnosis and surveillance. *Clin Diagn Lab Immunol*, **10**, 189–90.

Ayyagari, A. and Pal, N. 1991. Antiribitol-teichoic acid antibody (ARTA) in diagnosis of deep-seated *Staphylococcus aureus* infections. *Indian J Pathol Microbiol*, **34**, 176–80.

Balc, I.C., Sungurtekin, H., et al. 2003. Usefulness of procalcitonin for diagnosis of sepsis in the intensive care unit. *Crit Care*, **7**, 85–90.

Bisno, A.L., Peter, G.S. and Kaplan, E.L. 2002. Diagnosis of strep throat in adults: are clinical criteria really good enough? *Clin Infect Dis*, **36**, 126–9.

Blasi, F., Tarsia, P., et al. 1998. Epidemiology of *Chlamydia pneumoniae*. *Clin Microbiol Infect*, **4**, 1–6.

Borriello, S.P. 1999. Near patient microbiological tests. *Br Med J*, **319**, 298–301.

Chart, H. 1999. Evaluation of a latex agglutination kit for the detection of human antibodies to the lipopolysaccharide of *Escherichia coli* O157, following infection with verocytotoxin-producing *E. coli* O157. *Lett Appl Microbiol*, **29**, 434–6.

Davidsohn, I. and Wells, B.B. (eds) 1962. *Clinical diagnosis by laboratory methods*. Philadelphia: W.B. Saunders Co.

Elliott, T., Worthington, T. and Lambert, P. 2002. Antibody response to staphylococcal slime and lipoteichoic acid. *Lancet*, **360**, 1977.

Ferrieri, P., Dajani, A.S., et al. 1970. Appearance of nephritis associated with type 57 streptococcal impetigo in North America. *N Engl J Med*, **283**, 832–6.

Friedman, S., Spitalney, K., et al. 1987. Pontiac fever outbreak associated with a cooling tower. *Am J Public Health*, **77**, 568–72.

Gallin, J.I. and Fauci, A.S. (eds) 1983. *Chronic granulomatous disease*. New York: Raven Press.

Humphreys, H., Marshall, R.J., et al. 1992. Pneumonia due to *Legionella bozemanii* and *Chlamydia psittaci*/TWAR following renal transplantation. *J Infect*, **25**, 67–71.

Institute of Medicine. 2002. *Ending neglect: the elimination of tuberculosis in the United States*. Washington, DC: National Academy Press.

Jayasena, S.D. 1999. Aptamers: an emerging class of molecules that rival antibodies in diagnostics. *Clin Chem*, **45**, 1628–50.

Jumaa, P.A. and Chattopadhay, B. 1994. Pseudobacteraemia. *J Hosp Infect*, **27**, 167–77.

Kearns, A.M., Graham, C., et al. 2002. Rapid real-time PCR for determination of penicillin susceptibility in pneumococcal meningitis, including culture-negative cases. *J Clin Microbiol*, **40**, 682–4.

Klebanoff, S.J. 1980. Oxygen metabolism and the toxic properties of phagocytes. *Ann Intern Med*, **93**, 480–9.

Kusunoki, S. and Kanazawa, S.M. 2001. Anti-Gal-C antibodies in Guillain–Barre syndrome subsequent to Mycoplasma infection: Evidence of molecular mimicry. *Neurology*, **57**, 736–8.

Lankinen, K.S., Ruutu, P., et al. 1999. Pneumococcal pneumonia diagnosis by demonstration of pneumolysin antibodies in precipitated immune complexes: A study in 350 Philippine children with lower respiratory tract infections. *Scand J Infect Dis*, **31**, 155–61.

McGinn, T.G., Deluca, J., et al. 2003. Validation and modification of streptococcal pharyngitis clinical prediction rules. *Mayo Clin Proc*, **78**, 289–93.

Mazurek, G.H., LoBue, P.A., et al. 2001. Comparison of a whole-blood interferon gamma assay with tuberculin skin testing for latent *Mycobacterium tuberculosis* infection. *J Am Med Assoc*, **286**, 1740–7.

Naratomi, Y. and Sugiyama, J. 1998. A rapid latex agglutination assay for the detection of penicillin-binding protein 2A. *Microbiol Immunol*, **42**, 739–43.

Needham, C.A., McPherson, K.A. and Webb, K.H. 1998. Streptococcal pharyngitis: Impact of a high-sensitivity antigen test on physician outcome. *J Clin Microbiol*, **36**, 3468–73.

Nuntnarumit, P., Pinkaew, O. and Kitiwanwanich, S. 2002. Predictive values of serial C-reactive protein in neonatal sepsis. *J Med Assoc Thai*, **85**, S1151–8.

Opalka, D., Lachman, C.E., et al. 2003. Simultaneous quantitation of antibodies to neutralizing epitopes on virus-like particles for human papillomaviruses types 6, 11, 16 and 18, by a multiplexed Luminex assay. *Clin Diagn Lab Immunol*, **10**, 108–15.

Osborne, S.E., Matsumura, I. and Ellington, A.D. 1997. Aptamers as therapeutic and diagnostic reagents: Problems and prospects. *Curr Opin Chem Biol*, **1**, 5–9.

Pini, A. and Bracci, L. 2000. Phage display of antibody fragments. *Curr Protein Pept Sci*, **1**, 155–69.

Pollock, J.M., Buddle, B.M. and Andersen, P. 2001. Towards more accurate diagnosis of bovine tuberculosis using defined antigens. *Tuberculosis (Edinb)*, **81**, 65–9.

Prince, H.E. and Leber, A.L. 2002. Validation of an in-house assay for cytomegalovirus immunoglobulin G (CMV IgG) avidity and relationship to CMV IgM levels. *Clin Diagn Lab Immunol*, **9**, 824–7.

Raoult, D., Tissot-Dupont, H., et al. 2000. Q fever 1985–1998. Clinical and epidemiologic features of 1,383 infections. *Medicine (Balt)*, **79**, 109–23.

Schuttrumpf, S., Binder, L., et al. 2003. Procalcitonin: a useful discriminator between febrile conditions of different origin in haemato-oncological patients? *Ann Haematol*, **82**, 98–103.

Schwerer, B. 2002. Antibodies against gangliosides: a link between preceding infection and immunopathogenesis of Guillain–Barre syndrome. *Microbes Infect*, **4**, 373–84.

Smith, K.R., Fisher, H.C. and Hook, E.W. 1996. Prevalence of fluorescent monoclonal antibody non-reactive *Neisseria gonorrhoeae* in five North American sexually transmitted diseases clinics. *J Clin Microbiol*, **34**, 1551–2.

Stephens, M.M. and McAuley, P. 1988. Brodie's abscess. A long-term review. *Clin Orthop*, **234**, 211–16.

Summers, W.C., Brookings, E.S. and Waites, K.B. 1998. Identification of oxacillin-susceptible and oxacillin-resistant *Staphylococcus aureus* using commercial latex agglutination kits. *Diagn Microbiol Infect Dis*, **30**, 131–4.

West, P.W. and Nasrallah, A.Y. 1988. Evaluation of a latex kit reagent for the detection and identification of pneumococci. *APMIS*, **3**, Suppl, 13–16.

31

Biological safety for the clinical laboratory

MICHAEL NOBLE

INTRODUCTION

Laboratory biosafety is the active, assertive process based on evidence-based principles to ensure safety from microbial contamination or infection, or toxic reaction for workers, the public, and the environment as a result of the active manipulations of live microorganisms or their products while pursuing academic, research, industrial, and clinical investigations. The goals are to prevent laboratory-acquired infections in workers and to prevent accidental releases of live agents that can potentially endanger and have severe negative impact on humans, animals, and plants. Laboratory safety involves all aspects of the laboratory cycle, starting from before microorganisms arrive in the facility, through to the training of personnel, and the establishment and monitoring of safe working practices, through the proper use of reagents, materials, and equipment, through the safe storage and transport of agents, and ultimately to the terminal sterilization and destruction of microorganisms.

EPIDEMIOLOGY OF LABORATORY-ACQUIRED INFECTIONS

Laboratory-acquired infections are generally defined as infections that likely occur while handling offending organisms during the analytic phase of laboratory testing. Usually, this is an epidemiological conclusion, based on the presumption of low likelihood of a concurrent community exposure, although with molecular or other identifying techniques the likelihood of another source can be reduced or eliminated. It is important to appreciate that the testing cycle begins well before the sample actually reaches the laboratory (the preanalytic phase of laboratory testing). In one of the original studies of laboratory-acquired infections, infections acquired during the collection of some samples were included if it could be ascertained that the collection was solely for the purpose of a laboratory investigation. Infections experienced by phlebotomists as a result of needlestick injuries are now routinely included as laboratory acquired infections. Arbitrarily, phlebotomist infections with chicken pox acquired while collecting samples in patients rooms are often not included (Collins and Kennedy 1999).

Although difficult to date precisely, the first microbiology laboratories were active by 1840–60, with the laboratories run by pioneers including Pasteur and Koch. The first report of a laboratory-acquired infection was in 1899, with the reporting of Mediterranean fever.

The study of laboratory-acquired infections was not carried out in a systematic fashion until 1951 with the most extensive and historic survey of 5 000 American laboratories by Sulkin and Pike (Collins 1983). This work was extended by three additional series (Sulkin

1964; Pike et al. 1965; Pike 1976). Through these reports, the authors cited 3 921 laboratory infections dating between 1930 and 1974, with a mortality of 164 (4.1 percent).

Of note, 2 307 (58.8 percent) of the infections were reported from research facilities, 677 (17.3 percent) were from diagnostic facilities, 134 (3.4 percent) were from the generation of biological products (industry), and 106 (2.7) were from teaching facilities. The remaining 697 (17.8 percent) were unspecified by source.

From four series performed in the UK between 1971 and 1991, it was noted that within clinical facilities, the majority of infections were reported from workers in the microbiology laboratory, with the second most common department being morbid anatomy. Of note, over the 20-year period, using the format of single-year surveys, the number of infections reported dropped over 80 percent, from 104 to only 17.

All studies are hampered, however, by having access only to those laboratory-acquired infections that are officially reported, either in literature or through other database systems. In no jurisdiction is there open access to required laboratory-acquired infections reporting. It is assumed that even complete listings reflect an immeasurable minority of infections that indeed occur (Sewell 1995).

Microorganisms are transmitted by a limited number of routes, which includes generation of aerosols of respirable size affecting the lungs or nonrespirable size resulting in contamination of external surfaces. Contamination of surfaces also result from spills, leaks, and untidy work practices. Ingestion results from contamination of edible or nonedible oral substances that reach the mouth as a consequence of contaminated hands, contaminated foods, or unsafe practices such as mouth pipetting. Laboratory accidents, especially involving the use and abuse of sharp instruments or breakable glass, result in puncture wounds of potentially fatal outcome. Inadvertent exposures resulting from inadequately identified or improperly stored reagents and materials, can result in serious, preventable infection. A laboratory risk assessment recognizes the potential routes and limits their ability to function.

While laboratory factors including aerosols, splashes, and accidents are described as major causes of laboratory-acquired infection, so too are human factors. In four series between 1952 and 1976 (Pike and Sulkin 1952; Reid 1957; Pike 1976; Harrington and Shannon 1976), it was repeatedly demonstrated that most workers who developed infection had had prior laboratory training. While at first blush this would appear to be problematic, it is also this group that would be performing the vast majority of hands-on activity. Disturbingly, it has also been pointed out that three of four surveys have indicated that 7–8 percent of laboratory infections occur in clerical and maintenance personnel (Collins 1983).

Briggs Phillips used a matched case–control study of 33 laboratory workers who experienced a laboratory associated injury over a 2-year period (Briggs Phillips, 1986). No differences were noted related to age, length of employment, years of formal education, wearing of glasses, use of prescription medications, off-the-job accidents, or driving record. On the other hand, accident-involved people were significantly more likely to have had a laboratory accident or laboratory infection prior to the 2-year study period and were significantly more likely to have a low opinion of laboratory safety programs. When the conditions surrounding the accidents were examined, 36 percent occurred when the employee was working too quickly, either just before lunch or at the end of the day. In 30 percent of accidents, the employee acknowledged a breech in safety regulations. In summary, in this study, attitudes and work habits were important contributing factors to laboratory accidents. Of note, to date there have been no published studies that confirm or refute these findings.

Quality assurance materials

Clinical laboratory workers are aware of the risks and hazards associated with handling clinical samples; however, laboratory-acquired infections in clinical laboratories can occur from improper handling not only of clinical samples but also of quality control materials, especially proficiency testing samples. In 1980, Blaser and Feldman (1980) reported 24 cases of infection with *Salmonella typhi* resulting from national proficiency testing program samples. A similar experience was repeated by Chin (1988) with an infection in a worker with *Corynebacterium diphtheriae*. In 2002, Richardson et al. (2002) reported two workers in separate laboratories infected with *Shigella flexneri* from external quality assessment samples that clinically simulated fecal samples. All materials within a laboratory, regardless of source, must be handled with all required safety and diligence.

BIOSAFETY QUALITY MANAGEMENT

Risk assessment

Laboratory biosafety programs may be driven by the recognition of good practices, in some jurisdictions may be directed by the application of established international standards, and in others may be driven by regulatory requirements. All laboratorians should regularly survey the facilities for potential hazards and implement corrective plans. Assessments should be formal, systematic, and planned, and cover all laboratory elements, including safety manual policies and procedures, training requirements and performance, supervision, maintenance of hazardous materials and substances, first aid services and equipment, personnel safety, and health reports (ISO 2002).

Laboratory safety officer

All laboratories should designate a knowledgeable, trained, and experienced individual with the responsibility of ensuring laboratory safety. The position may require only part-time activity or may be a full-time responsibility. The responsibilities and authorities of the safety officer vis-à-vis the laboratory workers and management, and their role with respect to reporting and ability to initiate action, should be clearly defined. The laboratory safety officer should execute their activities on behalf of the laboratory management and at the same time have both the authority and responsibility to act independently and effectively. The roles of a laboratory safety officer may vary depending on the size and complexity of the facility; however, as a minimum, they should include (WHO 2003a):

- establishing and monitoring a laboratory safety program to a process of regular audit;
- performance of regular risk assessment, and development of recommendations for laboratory safety improvement;
- participating in investigation of accidental exposures, in order to prevent their occurrence;
- identifying the training needs of laboratory workers, and assisting with training implementation;
- ensuring that laboratory procedures for decontamination and disinfection are followed.

CLASSIFICATION SYSTEMS TO CLARIFY RISK

Three commonly recognized classification schemes are integral to defining the degree of risk associated with the handling of infectious materials and agents. The first is the classification of organisms, the second is the classification of environments, and the third is the classification of behaviors.

Classification of organisms

Given the variety of experiences and possible exposures, it is accepted that some microorganisms are more pathogenic than others. This may in part be due to the broad range of virulence factors that affect attachment, invasions, tissue tropisms, and production of exotoxins. While laboratorians need to have cautionary respect for everything they handle, it is accepted that some are inherently more dangerous and need to be addressed with an increased level of safety procedures and concern.

In order to designate increasing risk, national bodies including the Health Canada Bureau of Biosafety, Centers for Disease Control, National Institutes of Health, European Union, and Australia (ABSA 1998), have developed and harmonized a classification system for microorganisms that is subclassified across the spectrum of virulence. Classifications of microorganisms have traditionally been based on four stages:

Category 1: Microbes not known to consistently cause disease in healthy adults. Examples of category 1 organisms include *Bacillus subtilis* and *Naeglaria gruberi*. Microorganisms assigned to category 1 can be safely addressed in all categories of laboratories.

Category 2: Microbes that may commonly cause disease as a consequence of injury or ingestion. Therapy is readily available. Organisms tend not to spread or represent a significant community concern. Examples of category 2 organisms include *Staphylococcus* spp, most *Enterobacteriaceae*, pseudomonads, and hepatitis B virus.

Category 3: Microbes that have the potential for aerosol transmission. Therapy may be less readily available. May be indigenous or exotic agents. Examples of category 3 organisms include *Mycobacterium tuberculosis* and *Brucella* spp.

Category 4: Microbes that are exotic and dangerous, and that pose a high risk of life-threatening disease. May be transmitted by aerosol spread. Examples of category 4 agents include tick-borne encephalitis virus, Marburg virus, and Ebola fever virus.

While the classification of most agents is relatively straightforward, some organisms, such as rabies virus, dengue virus, and *Bacillus anthracis*, may be classified differently by different agencies. In these situations, appropriate handling of these agents depends on the circumstances, the availability of settings, and, above all, the practice behaviors. Note that complete listings of organisms can be obtained through examination of the tables drafted for the original documents

Classification of laboratory environments

Having established a microbial risk-classification system, it is appropriate to develop parallel and coordinated systems for addressing the handling of microorganisms safely. All laboratory workers require a safe working environment; however, it is critical that the intensity of the safety is seen to be appropriate to the inherent level of risk to avoid on the one hand undue potential of exposure to dangerous organisms and on the other hand unreasonable interference in work practices when handling commonplace pathogens. Classification of laboratory environments is relatively straightforward in research and reference laboratories where the workload and pathogen awareness are predictable. Specific requirements are more problematic in the clinical laboratory setting because of the potential for a broad range of samples and microorganisms; that being said, even in this setting, there exists a high probability that most microorganisms processed are of category 2. While infrequent, clinical laboratories do receive risk group 3 pathogens, including *Mycobacterium tuberculosis* and *Francesella tulerensis*, and need to be prepared for the possibility.

Clinical and research laboratories can also be distinguished based on the types of organisms likely to be handled or processed. Biosafety level (BSL) 1 requirements are the most basic and general and are required for all laboratories. Each subsequent BSL builds upon the level below. BSL 4 provides the highest level of sophistication and safety.

BSL 1 facilities should be separated from public access by a lockable door. A biohazard sign should be posted at the entrance. There should be handwashing sinks readily available. Work surfaces should be constructed of materials that allow for proper disinfection. There should be equipment available for the safe decontamination of wastes prior to disposal. Special containment equipment need not be available; however, mechanical pipetting devices are mandatory for all pipetting procedures. Personal protective equipment, including laboratory gowns, coats, gloves, and eyewear, are required to be available for appropriate use. Laboratories designated as BSL 1 would be appropriate for undergraduate or secondary education and training but would be inappropriate for work involving samples of human blood or other tissues or body fluids.

BSL 2 facilities need to adhere to the requirements as above. In addition, door signage should include the name and contact information of the primary responsible person. Biological safety cabinets and other containment equipments are recommended to be available if procedures involving large inocula are to be performed, and emergency eyewash equipment should be available. While regional regulations would vary, most clinical microbiology laboratories would be required to meet the requirements of BSL 2 facilities.

BSL 3 facilities are designed to a higher level of security. A higher degree of door security is required, often under key, key card, or electronic access. BSL 3 laboratories are maintained under negative pressure, and access to the laboratories is through a vestibule or anteroom in order to protect the integrity of the pressure gradient. Exhaust air pressure handling systems is monitored and controlled. All interior surfaces, including walls, floors, and ceilings, need to be impact-resistant and without separations that would allow microorganisms to reside. Office areas are not allowed within BSL 3 facilities.

BSL 4 facilities are designed for the handling of dangerous and exotic organisms. Often, BSL 4 facilities are in separate buildings built to the required specification. In addition to the BSL 3 requirements, access to the laboratory must be through an anteroom with interlocking airtight doors. The anteroom area provides space for changing containment suits and for both chemical and water showers. Air supply into the laboratory is filtered with high-efficiency particulate air (HEPA). Outbound air passes through a two-stage HEPA filter process. Autoclaves need to be of a double-door design with interlocking doors to prevent accidental release of waste. All plumbing drains must pass through an effluent sterilization system.

Classification of laboratory practices

Just as handling of organisms and materials with increasing levels of risk requires increased levels of laboratory security, the practices of laboratorians can be scaled to risk.

General practices for all laboratory practice requires that all personnel working within a facility are trained and knowledgeable of the potential hazards to which they may be exposed, at least to the level of their competence and expertise. Eating and drinking and similar practices are not allowed within the laboratory workspace. The laboratory environment should be maintained tidily. Oral pipetting is prohibited. Long hair should be tied back or otherwise restrained. Protective garb and equipment should be available and be worn and used properly. Use of sharps should be strictly limited. All injuries should be reported without delay. All equipment, and especially autoclaves, should be maintained in good working order in order to prevent leaks, spills, and release of viable organisms. There should be an up-to-date laboratory safety manual that is available to all. The manual should include information on hazards within the specific laboratory, required safety information including material safety data sheets, and policies and procedures for all safety practices. Handwashing should be a mandatory practice when hands are either known or suspected to be contaminated and when gloves are removed.

In addition, level 2 practices would ensure that entry to the laboratory is restricted to only those on official business and that hazard warning signs are clearly posted at all entry locations. A health and medical surveillance program should be provided for all those working in the facility. An emergency procedure for spills and accidents should be written and accessible.

Level 3 practices would ensure that, in addition, there is a designated biosafety officer with the appropriate authority to oversee practices, and that all persons have successfully completed training courses in the specific requirements for added precaution. All employees should not only follow required procedures but also certify in writing that they understand the protocols. Personnel need to understand that when performing level 3 practices, the laboratory should be locked and they must stay within the laboratory in order to maintain traffic at a minimum. Personal items are not brought into the laboratory. Personal clothing and jewelry should not be worn while performing level 3 practices. Laboratory-specific protective equipment and footwear should not be worn outside the laboratory.

In most situations and most facilities, practices and facilities would be in sync with the most common agents that pass through the facility. On some occasions, where a heightened level of concern exists, the level of recommended practices would raise. A case in point are the recommendations for laboratories likely to receive and process samples from patients with severe acute respiratory syndrome (SARS) containing coronavirus SARS-CoV. This agent, first recognized in the spring of 2003, was noted to transmit by both aerosols and contact and was often associated with a severe atypical pneumonia. While routine handling of diagnostic samples for serology or bacteriology and mycology agents was acceptable, it was deemed appropriate to recommend that workers performing procedures likely to result in aerosols follow practices more consistent with level 3, which included wearing disposable gloves, a mask or respirator, and eye protection, or a full-face shield (WHO 2003b). Despite this, there have been laboratory-acquired infections recorded while working with live strains of SARS-CoV in level 3 facilities.

At the highest level of security, level 4 practices entail all the aforementioned, but also require establishment of protocols to cover emergencies to cover possible damage to positive pressure suits. Staff should be highly trained and knowledgeable about all entry and exit procedures. They should be knowledgeable and experienced in working with all small and large animals, as may be required. They must keep a detailed log book of all agents handled, including date and time and procedures performed. A daily log should be maintained for all workers to monitor and report any and all illnesses (Health Canada 1996).

PREVENTION OF INFECTIONS

Biosafety measures to prevent laboratory-acquired infections are a combination of ready awareness, application of knowledge and education, proper use of personal protective equipment, devices, and procedures, and application of specific public-health measures. Breakdown in any domain can result in significant injury and illness.

Respiratory protective devices

Arguably, the most dangerously formed structure in the laboratory is the microbially contaminated airborne aerosol of respirable size. Microbially contaminated droplets larger than 5 μm settle rapidly, and while they tend not to be inhaled, they may be associated with contamination of the environment. Droplets smaller than 1–5 μm remain suspend in air longer and are more likely associated with inhalation (Hatch 1961).

Individual procedures tend to be associated with aerosols of certain size. Protection from droplets may be addressed by a variety of equipments. Large liquid splatters may be contained by splash guards or face masks. Smaller aerosols are far better contained by a properly functioning biological safety cabinet. The decision with respect to the most appropriate equipment would need to take into consideration the procedure being performed, potential agents, possible concentration, and availability of equipment.

Masks that are properly worn, covering the nose and mouth, provide the most direct respiratory protection. Disposable masks are constructed of a variety of materials and styles, either molded or pleated. Additional changes may include the application of hydrophobic coating to make the mask more water-resistant or the application of a face mask to provide eye protection. Application of masks may be by ear-loops, tie-on straps, or snug elastic bands.

Masks can be quantitatively measured to determine their ability to block the penetration of particles, including bacteria. N95 masks are measured to be at least 95 percent efficient. Many surgical masks, dependant upon their material and composition, may meet the filtering level of 95 percent efficiency; however, in order to be effective, masks must be snug to the face in order to prevent unfiltered air penetrating around the periphery. Beards, hair, or a thin face may result in gaps that impede snugness. Simulation studies based upon head models with external spray and upstream microbial and particle capture units clearly indicate the efficiency of form-fitted masks. Masks can be fit-tested to ensure the most effecting fitting. In many settings, disposable masks that provide nontested fitness can be considered as providing a sufficient level of safety, although in situations of higher security concerns, fitted masks are more appropriate. Unfortunately, observation studies indicate that many procedures are performed in the absence of any respiratory protection. There is ample real case evidence to indicate that the absence of wearing any personal protective equipment can result in the acquisition of infectious disease. Evidence that demonstrates the clinical inferiority of surgical masks in comparison to NIOSH approved N95 masks is more difficult to cite. In the absence of approved N95 masks, it is appropriate to use surgical masks as an alternative.

Biological safety cabinets (BSC) are units to protect the operator and the environment from aerosol contamination. BSCs prevent contamination through the presence of a combination of a physical glass containment barrier, an air curtain of 75 linear feet per minute face velocity or greater on the front access point, negative air flow, and a process of exhaust filtration.

Table 31.1 *Classification of biological safety cabinets (BSCs)*

Class	Front	Air curtain (linear feet per minute)	Supply air through HEPA filter	Exhaust air through HEPA filter	Ducted exhaust
I	Open	75	No	Yes	No
IIa	Open	75–100	Yes	Yes	No
IIb	Open	100	Yes	Yes	Yes
IIb[a]	Open	100	Yes	Yes	Yes
III	Closed	Not required	Yes	Double or followed by incineration	Yes

HEPA, high-efficiency particulate air.
a) IIb BSCs are subclassified as IIB-1, IIB-2, and IIB-3 based upon the complexity of air supply.

BSCs are classified by their degree to protection. Class 1 and class 2A units are designed for laboratories working with low or moderate risk. As class 2B and class III units are exhausted through ducts to outside the laboratory, they can also be used with volatile chemicals. Class III units are the most secure and are the only acceptable BSCs for use with risk group 3 and risk group 4 agents (see Table 31.1).

In many laboratories, small, inexpensive equipment can be readily available and significantly reduce the exposure of workers to biological hazards. Splashguards of plastic or glass can provide an effective physical barrier to block workers eyes from aerosols when opening blood tubes. This same function can be served well by full-face protectors or goggles or other eye protectors. Mechanical or electrical pipetting aids must always be used for all pipetting procedures. Mouth pipetting, once common practice, is completely unacceptable regardless of reagent or chemical or body fluid. Mouth pipetting not only creates the opportunity to inhale or swallow dangerous materials while applying suction, but it also puts others at risk by inhalation or contamination as large amounts of aerosols are generated with each clearing blow. Homgenizers, vortex stirrers, blenders, shakers, and centrifuges are all important laboratory equipment, but all create great risk for generation of aerosols unless they are used with sealed vessels. For some equipment, it is appropriate that it be housed only within a biological safety cabinet. For others, such as large centrifuges, it is critical that all tubes and vials are used with sealed protective units that are opened under a biological safety cabinet. In the microbiology laboratory, sterilization of platinum loops by thrusting the tip into a Bunsen burner flame can result in aerosol splatter. This activity can be avoided in many laboratories by the use of disposable loops or the use of microincinerators.

PERSONAL PROTECTIVE CLOTHING AND EQUIPMENT

Exposing personal street clothing to splashes and accidents can result in damage or contamination that could endanger the user or others that have contact with

soiled laundry. Workers should be in the practice of wearing protective garb while performing all activities within the laboratory. Protective garb should be, as a minimum, long enough to cover the lap while sitting, be fluid-resistant, be closed at the neck, and have long sleeves, preferably with cuffs. Workers should not leave the laboratory wearing their gowns. Gowns can be used throughout the day, provided there have been no spills or obvious contamination, and should be hung in a fashion to prevent contamination when not being used. Gowns should be laundered regularly, in many facilities at the end of each day's use. It is inappropriate to take home soiled gowns for personally laundering.

Glove use is a valuable technique to protect hands from contact exposure to microbial agents, provided that the gloves are selected properly and used appropriately. Readily available gloves are made of a variety of materials including vinyl, latex, and nitrile.

The quality of gloves can be affected by a wide range of variables, including composition, manufacturer, design, degree of manipulation, chemical exposure, and type of glove analysis performed.

For most tasks within the laboratory, the most appropriate gloves to wear are disposable, made of light-weight materials that allow the greatest degree of flexibility and dexterity, and yet provide a safe protective barrier from microorganisms.

While there was little evidence to suggest that individuals wearing vinyl gloves had developed seroconversion to human immunodeficiency virus (HIV), early studies demonstrated that vinyl gloves were significantly more likely to leak than latex gloves, especially after being exposed to isopropyl alcohol. The impact of these studies contributed to a rush on latex gloves and the recognition of glove use contributing to latex allergy. Thermoplastic elastomer materials were implemented for glove use as an alternative to both vinyl and latex. Studies based upon the penetration by bacteriophage phi X174 have suggested that thermoplastic elastomer gloves provide a superior mechanical barrier to either vinyl or latex.

While gloves can provide a barrier for the user, the exterior surface does become contaminated. While some will tolerate exposure to isopropyl alcohol, most will not, and glove disinfection is not considered an appropriate practice. Glove use should be task-specific, and they should be removed and discarded before performing tasks that can lead to environmental contamination of keyboards, telephones, and other equipment.

For many tasks, protection from injury due to either sharps or chemical irritation requires more durable gloves. Fabric undergloves can be worn along with a disposable glove to provide improved grip compared with double-gloving with disposable gloves and some protection from disposable-glove tears. Metal-mesh gloves provide substantial protection when using cutting tools and other sharps, although they compromise dexterity. Household rubber and rubber-blend gloves can provide added protection against a variety of chemical irritants.

Regardless of the glove materials, or the reason for which they are worn, removal of gloves should always be followed by adequate handwashing.

HANDWASHING

Handwashing is commonly recognized as the single most important procedure to prevent the acquisition of infection. One might conclude that handwashing is sufficiently commonplace; however, it has been the focus of study and guidelines for many years.

The single most important procedure for the prevention of infection in oneself and in others is to wash hands. The wisdom of handwashing was first demonstrated and reported in 1847 by Semelweis (Lechevalier and Solotorovsky 1974), and yet occult observation studies within both the hospital and community consistently show poor handwashing practices by all professional groups (Muto et al. 2000; Eckmanns et al. 2001). The reasons for inappropriate lack of handwashing have a degree of justification: excessive work load, lack of time, poor access to handwashing sinks and materials, concerns for skin irritation, and other problems associated with frequent handwashing (Pittet 2001).

Laboratories tend to rely primarily on a brief list of agents for handwashing, including plain soap, chlorhexidine- or triclosan-containing compounds, and more recently liquid or gel forms of alcohol, either ethanol or isopropanol.

Plain soap with running water is an effective technique to remove soil and organic substances but is not effective for the removal of bacterial pathogens.

Chlorhexidine is a chlorine-containing cationic organic compound with broad-spectrum antimicrobial activity. It is inhibited by anionic soaps but is insensitive to interference by organic materials.

Triclosan is an organic phenyl ether also with broad antimicrobial activity, although its activity against gram-negative bacilli is reduced. It is also insensitive to organic matter.

Replacing plain soap with antimicrobial soaps that contain antiseptics such as chlorhexidine, chloroxylenol, hexachlorophene, quaternary ammonium compounds, and triclosan, are more effective for the removal of pathogens, although individual products may have their own concerns. Chlorhexidine, triclosan, and quaternary ammonium compounds can be contaminated with gram-negative bacteria. Skin irritation is recognized with higher concentrations of chlorhexidine and free iodine concentrations with iodophors. Hexachlorophene can manifest neurotoxicity when bathing neonates.

On some occasions, especially in older facilities, ready access to handwashing sinks is not immediately available. Waterless alcohol-based antiseptics are handwashing agents. Alcohols used in these products include ethanol, isopropanol, n-propanol, or combinations, all usually between 60 and 90 percent alcohol. Alcohol preparations are available in both gel and liquid formats.

Alcohols are germicidal against a broad range of microorganisms, including gram-positive and gram-negative bacteria and lipid-enveloped viruses. The activity is rapidly germicidal through the denaturation of proteins. Bacillus species tend to be resistant to alcohols. Alcohols do not have persistent activity, although bacterial regrowth following alcohol wash tends to be slow. Activity is dependent upon alcohol concentration, volume, and contact time. Frequent use can lead to skin dryness if glycerol is not incorporated as a conditioning agent. It is important to remember at all times that all alcohols are flammable and the potential for flash fires may exist.

Ultimately, the agent that is the most successful in reducing carriage and transmission of microbial pathogens may not necessarily be the most antimicrobially active agent but the one that is used most effectively by the greatest proportion of individuals. Indeed, there is little information that clearly links a specific mean log-reduction in bacterial count associated with any specific product and a correlate reduction in nosocomial or occupationally related infection. In-use comparison of 2 percent chlorhexidine versus 61 percent ethanol has demonstrated no significant differences in bacterial counts but less skin irritation with the alcohol and a significant reduction in required wash time with the alcohol (Larson et al. 2001).

In vitro studies of handwashing have to address many issues, including gender, age, underlying condition of skin, presence of jewelry, test organisms, inoculum, method and duration of application, and retrieval and enumeration methods. Campaigns to encourage improved handwashing unfortunately tend to have short-lived efficacy and require persistent energy and activity (Muto et al. 2000).

In many institutions, the number and placement of handwashing sinks has been found to be less convenient than desirable. As an alternative to expensive plumbing alterations, ready access to alcohol hand rubs has been associated with increased compliance with hand hygiene (Bissett 2002).

Immunization

A regular and maintained program of protective immunization is an essential part of a biosafety program. As a minimum, laboratory workers should be protected from diphtheria, hepatitis B, measles, mumps, poliomyelitis, rubella, tetanus, and typhoid fever. Many workers will have protection from either childhood infection or an immunization program. For the remainder, an adult immunization program is appropriate. While laboratory-acquired influenza does not occur more frequently than community-acquired infection, annual influenza immunization provides protection from loss of work activity.

All workers in laboratories that may handle specimens or other samples containing *Mycobacterium tuberculosis* should participate in a regular skin-test surveillance program. Chest X-rays need not be preformed other than at the time of skin-test conversion or at the onset of symptoms suggestive of respiratory disease. While BCG immunization is available and used in many countries, others find that BCG obscures the interpretation of skin tests without providing long-term protection and as such is often no longer provided.

As a minimum, all laboratory workers required to work with animals, especially including large and small mammals, should receive pre-exposure human diploid cell rabies vaccine. Rabies vaccine is not required for workers handling fish, birds, amphibians, and small rodents.

Laboratories of special interest in other specific organisms including, but not limited to, *Neisseria meningitidis*, *Haemophilus influenzae*, *Coxiella burnettii*, *Francisella tularensis*, and *Yersinia pestis*, may consider providing additional specific immunization. Vaccinia vaccine can be used postexposure to prevent clinical manifestation of human monkey pox as a result of handling prairie dogs and exotic rodents (Wilson 2003).

Sharps injuries and blood-borne infection

Blood-borne infection is a frequent cause of occupation acquired infection in healthcare workers. It is estimated that between 800 and 1 000 healthcare workers acquire hepatitis B on an annual basis, along with an additional 500–1 000 who acquire hepatitis C. Worldwide since 1978, fewer than 60 healthcare workers have acquired HIV from occupational exposure (Sewell et al. 2001). This disproportionate distribution is related to the concentration of the virus found in blood samples and

the established risk of infection following a known single occupational percutaneous exposure (see Table 31.2).

Blood-borne agents are most often transmitted through sharps injuries; however, mucus-membrane exposure and contact with apparently intact skin are also associated. While blood splatter from accidents and spills may lead to skin or mucus-membrane exposure, there are no descriptions of occupationally acquired hepatitis or HIV associated with aerosols. Hepatitis B can be transferred from common environmental surfaces, although studies have clearly demonstrated the absence of transmission associated with cardiopulmonary resuscitation training manikins (Glaser and Nadler 1985)

Laboratory workers, including phlebotomists, are second only to nurses as the most common professional group to have occupationally acquired HIV infection.

While antiviral therapy may play a role in the clinical suppression and treatment of blood-borne disease, clearly avoidance and prevention are superior alternatives. Diligence while handling all samples, and assuming that all samples are contaminated, provides a mental framework for preventing infection in oneself and others.

It is instructive to note that the first paper citing laboratory-acquired infections, published in the *Lancet* in 1899, described two laboratory workers who acquired Mediterranean fever (brucellosis) by needlestick injury (Birt and Lamb 1899). As much as possible, sharps instruments should be avoided when handling blood and body fluids. Recapping of sharps is almost always inappropriate. Where recapping is not avoidable, glove-use procedures to recap single handedly make procedures safer.

BIOTERRORISM AND LABORATORY SAFETY

The intentional use of infectious agents long predates the actual knowledge and history of microbiology or laboratories. The deliberate use of biological agents used for war has been documented through the twelfth century (Marty 2001). The use of laboratories for the deliberate development of biological agents for warfare has been documented in many countries.

Table 31.2 *Relative risk factors with human immunodeficiency virus (HIV), hepatitis B virus (HBV), and hepatitis virus C (HCV)*

	HBV	HBC	HIV
Concentration in blood particles per ml	10^8–10^9	10^2–10^3	10^0–10^4
Risk per percutaneous exposure (%)	18	1.8	0.3
Healthcare cases per year	800–1 000	500–1 000	Fewer than 2

In the latter part of the twentieth century, several key events were associated with biowarfare agents. From 1941 to 1969, the USA, the UK, Canada, and others formed national programs of biological warfare research. In 1979, at least 66 Soviet citizens living downwind of a top-secret biological warfare research center in Sverdlovsk died as a result of an accidental release of *Bacillus anthracis*. In 1984, Iraqi forces were accused of deploying biological weapons against Iran.

The poignancy of biological risk was again heightened with the spread of *B. anthracis* through the postal service in the USA in 2001. Events at this time have resulted in a new and active level of laboratory awareness with respect to safety and security, and with the detection and proper handling of samples with specific identified biological warfare agents.

While virtually any microorganism deployed creatively could be used for bioterror or biowarfare, the predominant microorganisms of concern include *B. anthacis*, *Escherichia coli* O157:H7, *Francisella tulerensis*, *Yersinia pestis*, and smallpox.

The Laboratory Response Network (LRN) within the USA asserted that hospital-based clinical microbiology laboratories should be prepared to respond to events through the collection, preservation, transport, and testing of human specimens that potentially contain agents of biowarfare potential. A four-tier system was developed and categorized from front-line facilities (level A), through local public health laboratories (level B), and state health laboratories with advanced capability (level C), to federal high-level security facilities (level D). Level A laboratories are recommended to have developed bioterror response plans with standard operating procedures. Level A laboratories would be expected to have the technical capability to rule out or refer suspicious isolates to the next level. It is expected that all hospital-associated laboratories would be able to meet the requirements of BSL 2 and also have access to a properly functioning biological safety cabinet. The LRN system is predicated upon the workers in a level A facility having access to guidance and assistance from a level B facility.

REFERENCES

American Biological Safety Association, 1998. Risk group classification for infectious agents. www.ABSA.org

Birt, C. and Lamb, G. 1899. Mediterranean or Malta fever. *Lancet*, **2**, 701–10.

Bissett, L. 2002. Can alcohol handrubs increase compliance with hand hygiene? *Br J Nurs*, **11**, 1072–7.

Blaser, M.J. and Feldman, R.A. 1980. Acquisition of typhoid fever form proficiency testing specimens. *N Engl J Med*, **303**, 1481–3.

Briggs Phillips, G. 1986. Human factors in microbiological laboratory accidents. In: Miller, B.M., Gröschel, D.H.M., et al. (eds), *Laboratory safety: principles and practices*. Washington, DC: American Society for Microbiology.

Chin, J. 1988. Throat infection with toxigenic *Coynebacterium diphtheriae*. *Comm Dis Reports*, **8**, 7.

Collins, C.H. 1983. *Laboratory acquired infections: history, incidence causes and preventions*. London: Butterworth.

Collins, C.H. and Kennedy, D.A. 1999. *Laboratory acquired infections: history, incidence, causes and preventions*, 4th edn. Oxford: Butterworth Heinemann.

Eckmanns, T., Rath, A., et al. 2001. Compliance with hand hygiene in intensive care units. *Dtsch Med Wochenschr*, **126**, 745–9.

Glaser, M. and Nadler, J.P. 1985. Hepatitis B virus in a cardiopulmonary resuscitation training course. Risk of transmission from a surface antigen-positive participant. *Arch Intern Med*, **145**, 1653–5.

Harrington, J.M. and Shannon, H.S. 1976. Incidence of tuberculosis, hepatitis, brucellosis and shigellosis in British medical laboratory workers. *Br Med J*, **1**, 759–62.

Hatch, T.F. 1961. Distribution and dispostion of inhaled particles in respiratory tract. *Bacteriol Rev*, **25**, 237–40.

Health Canada, In: Kennedy, M.E. (ed.), *Health biosafety guidelines*, 2nd edn. Ottawa: Health Canada.

International Organization for Standardization, ISO/DIS 15190, 2002, *Clinical Laboratory Medicine – Safety in Medical Laboratories*, Geneva: International Organization for Standardization.

Larson, E.L., Aiello, A.E., et al. 2001. Assessment of two hand hygiene regimens for intensive care unit personnel. *Crit Care Med*, **29**, 944–51.

Lechevalier, H.A. and Solotorovsky, M. 1974. *Three centuries of microbiology*. New York: Dover Publications.

Marty, A.M. 2001. History of the development and use for biological weapons. *Clin Lab Med*, **21**, 421–34.

Muto, C.A., Sistrom, M.G. and Farr, B.M. 2000. Hand hygiene rates unaffected by installation of dispensers of a rapidly acting hand antiseptic. *Am J Infect Control*, **28**, 273–6.

Pike, R.M. and Sulkin, S.E. 1952. Occupational hazards in microbiology. *Science Monthly*, **75**, 222–8.

Pike, R.M., Sulkin, S.E. and Schulze, M.L. 1965. Continuing importance of laboratory-acquired infections. *Am J Pub Health*, **55**, 190–9.

Pike, R.M. 1976. Laboratory-acquired infections, summary and analysis of 3921 cases. *Lab Health Sci*, **3**, 105–14.

Pittet, D. 2001. Compliance with hand disinfection and its impact on hospital acquired infections. *J Hosp Infect*, **48**, S40–6.

Reid, D.D. 1957. The incidence of tuberculosis among workers in medical laboratories. *Brit Med J*, **2**, 10–14.

Richardson, H., Fleming, C.A., et al. 2002. Laboratory acquired *Shigella flexneri* (Abstract). 70th Conjoint Meeting on Infectious Diseases. Halifax, Nova Scotia. November.

Sulkin, S.E. 1964. Laboratory acquired infections. *Bacteriol Rev*, **25**, 203–11.

Sewell, D.L. 1995. Laboratory-associated infections and biosafety. *Clin Microbiol Rev*, **8**, 389–405.

Sewell, D.L., Callihan, D.R. and Denys, G.A. 2001. *M29-A2, Protection of laboratory workers from occupationally acquired infections; approved guideline*, 2nd edn. Wayne PA: NCCLS.

Wilson, J.M. 2003. Wisconsin: suspected case of human-to-human transmission of monkeypox, www.promedmail.org, Archive number 20030612.1450.

World Health Organization. 2003a. Biosafety Guidelines 25 April 2003. www.who.ch.

World Health Organization. 2003b. *Laboratory biosafety manual*, 2nd edn (revised). Geneva: World Health Organization.

Index

Complete table of contents for *Topley & Wilson's Microbiology and Microbial Infections*

VIROLOGY, VOLUMES 1 AND 2

BACTERIOLOGY, VOLUMES 1 AND 2

MEDICAL MYCOLOGY

PARASITOLOGY

IMMUNOLOGY